SHOCK AND VIBRATION HANDBOOK

Other McGraw-Hill Reference Books of Interest

Handbooks

Avallone & Baumeister · Marks' Standard Handbook for Mechanical Engineers
Bovay · Handbook of Mechanical and Electrical Systems for Buildings
Brady and Clauser · Materials Handbook
Brater and King · Handbook of Hydraulics
Dudley · Handbook of Practical Gear Design
Fink and Beaty · Standard Handbook for Electrical Engineers
Hicks · Standard Handbook of Engineering Calculations
Juran · Quality Control Handbook
Karassik, Krutzsch, Fraser, and Messina · Pump Handbook
Kurtz · Handbook of Engineering Economics
Maynard · Industrial Engineering Handbook
Parmley · Standard Handbook of Fastening and Joining
Perry and Green · Perry's Chemical Engineers' Handbook
Rohsenow, Hartnett and Ganic · Handbook of Heat Transfer Fundamentals
Schwartz · Metals Joining Manual
Seidman and Mahrous · Handbook of Electric Power Calculations
Shand and McLellan · Glass Engineering Handbook
Shigley and Mischke · Standard Handbook of Machine Design
Smeaton · Motor Application and Control Handbook
Smeaton · Switchgear and Control Handbook
Transamerica Delaval, Inc · Transamerica Delaval Engineering Handbook
Tuma · Engineering Mathematics Handbook
Tuma · Technology Mathematics Handbook
Tuma · Handbook of Physical Calculations

Encyclopedias

Concise Encyclopedia of Science and Technology
Encyclopedia of Engineering
Encyclopedia of Physics

Dictionaries

Dictionary of Scientific and Technical Terms
Dictionary of Mechanical and Design Engineering

SHOCK AND VIBRATION HANDBOOK

THIRD EDITION

Edited by

CYRIL M. HARRIS

*Charles Batchelor Professor of Electrical Engineering
and Professor of Architecture
Columbia University*

*First Edition of the Shock and Vibration Handbook (1961)
Edited by
Cyril M. Harris and Charles E. Crede*

McGRAW-HILL BOOK COMPANY
New York   St. Louis   San Francisco   Auckland   Bogotá
Hamburg   London   Madrid   Mexico   Milan
Montreal   New Delhi   Panama
Paris   São Paulo   Singapore
Sydney   Tokyo   Toronto

Library of Congress Cataloging-in-Publication Data

Shock and vibration handbook.

Includes bibliographical references and index.
1. Vibration—Handbooks, manuals, etc.   2. Shock
(Mechanics)—Handbooks, manuals, etc.   I. Harris,
Cyril Manton.
TA355.S5164   1987   620.3   86-27707
ISBN 0-07-026801-0

1234567890   DOC/DOC   893210987

ISBN 0-07-026801-0

*The editors for this book were Harold B. Crawford and David E. Fogarty,
and the production supervisor was Richard A. Ausburn.
It was set in Century Expanded by University Graphics, Inc.*

It was printed and bound by R. R. Donnelley & Sons Company.

CONTENTS

Preface   xi

1. INTRODUCTION TO THE HANDBOOK   1–1

> Cyril M. Harris, *Charles Batchelor Professor of Electrical Engineering; Columbia University, New York, N.Y.*
> AND
> Charles E. Crede, *Late Professor of Mechanical Engineering and Applied Mechanics; California Institute of Technology, Pasadena, Calif.*

2. BASIC VIBRATION THEORY   2–1

> Ralph E. Blake, *Consultant; Technical Center of Silicon Valley, San Jose, Calif.*

3. VIBRATION OF A RESILIENTLY SUPPORTED RIGID BODY   3–1

> Harry Himelblau, Jr., *Member of the Technical Staff, Space Transportation Systems Division, Rockwell International Corp., Downey, Calif.*
> AND
> Sheldon Rubin, *Senior Project Engineer; Aerospace Corp., Los Angeles, Calif.*

4. NONLINEAR VIBRATION   4–1

> H. Norman Abramson, *Executive Vice President; Southwest Research Institute, San Antonio, Texas*

5. SELF-EXCITED VIBRATION   5–1

> Frederick F. Ehrich, *Staff Engineer; Aircraft Engine Group, General Electric Co., Lynn, Mass.*

6. DYNAMIC VIBRATION ABSORBERS AND AUXILIARY MASS DAMPERS   6–1

> F. Everett Reed, *President; Littleton Research and Engineering Corp., Littleton, Mass.*

7. VIBRATION OF SYSTEMS HAVING DISTRIBUTED MASS AND ELASTICITY   7–1

> William F. Stokey, *Professor of Mechanical Engineering Emeritus; Carnegie-Mellon University, Pittsburgh, Penn.*

8. TRANSIENT RESPONSE TO STEP AND PULSE FUNCTIONS   8-1

   Robert S. Ayre, *Professor Emeritus of Civil Engineering; University of Colorado, Boulder, Colo.*

9. EFFECTS OF IMPACT ON STRUCTURES   9-1

   William H. Hoppmann II, *Professor Emeritus of Engineering; University of South Carolina, Columbia, S.C.*

10. MECHANICAL IMPEDANCE   10-1

    Elmer L. Hixson, *Professor of Electrical Engineering; The University of Texas at Austin, Austin, Texas*

11. STATISTICAL CONCEPTS IN VIBRATION   11-1

    John W. Miles, *Professor of Applied Mathematics and Geophysics; University of California, San Diego, La Jolla, Calif.*
    AND
    William T. Thomson, *Professor Emeritus of Mechanical Engineering; University of California, Santa Barbara, Calif.*

12. PIEZOELECTRIC AND PIEZORESISTIVE TRANSDUCERS   12-1

    Eldon E. Eller, *Senior Project Engineer; Endevco Corp., San Juan Capistrano, Calif.*
    AND
    Robert M. Whittier, *Research and Development Manager, Silicon Products; Endevco Corp., San Juan Capistrano, Calif.*

13. VIBRATION MEASUREMENT EQUIPMENT AND SIGNAL
    ANALYZERS   13-1

    Robert B. Randall, *Vibration Applications Engineer; Brüel & Kjaer Pty. Ltd., Concord, N.S.W., Australia.*

14. SPECIAL-PURPOSE TRANSDUCERS   14-1

    John E. Judd, *President; Vibra-Metrics, Inc., Hamden, Conn.*
    AND
    John D. Ramboz, *Consultant*

15. MEASUREMENT TECHNIQUES   15-1

    Cyril M. Harris, *Charles Batchelor Professor of Electrical Engineering; Columbia University, New York, N.Y.*

16. CONDITION MONITORING OF MACHINERY   16-1

    Joëlle Courrech, *Senior Applications Engineer; Brüel & Kjaer, Naerum, Denmark*

17. STRAIN-GAGE INSTRUMENTATION   17-1

   Earl J. Wilson, *National Aeronautics and Space Administration, Flight Research Center, Edwards, Calif.*

18. CALIBRATION OF PICKUPS   18-1

   John D. Ramboz, *Consultant*
   M. R. Serbyn, *Research Engineer; National Bureau of Standards, Gaithersburg, Md.*
   AND
   Robert W. Lally, *Consultant; PCB Piezotronics, Inc., Depew, N.Y.*

19. VIBRATION STANDARDS   19-1

   Paul H. Maedel, Jr., *Westinghouse Electric Corp., Philadelphia, Penn.*
   AND
   Ronald L. Eshleman, *President; The Vibration Institute, Clarendon Hills, Ill.*

20. INTRODUCTION TO DATA REDUCTION, TESTING, AND
   SPECIFICATIONS   20-1

   Charles T. Morrow, *Consultant, Encinitas, Calif.*

21. EXPERIMENTAL MODAL ANALYSIS   21-1

   Randall J. Allemang, *Associate Professor of Mechanical Engineering; Structural Dynamics Research Laboratory, University of Cincinnati, Cincinnati, Ohio*
   AND
   David L. Brown, *Professor of Mechanical Engineering; Structural Dynamics Laboratory, University of Cincinnati, Cincinnati, Ohio*

22. CONCEPTS IN VIBRATION DATA ANALYSIS   22-1

   Allen J. Curtis, *Chief Scientist; Hughes Aircraft Company, El Segundo, Calif.*

23. CONCEPTS IN SHOCK DATA ANALYSIS   23-1

   Sheldon Rubin, *Senior Project Engineer; The Aerospace Corporation, El Segundo, Calif.*

24. PART I: VIBRATION OF STRUCTURES INDUCED BY GROUND
   MOTION   24-1

   W. J. Hall, *Professor of Civil Engineering; University of Illinois, Urbana, Ill.*

PART II: SEISMIC QUALIFICATION OF EQUIPMENT   24–21

George D. Shipway, *Technical Director, Western Test and Engineering;
Wyle Laboratories, Norco, Calif.*
AND
Daniel D. Kana, *Institute Engineer; Southwest Research Institute, San
Antonio, Texas*

25. VIBRATION TESTING MACHINES   25–1

Karl Unholtz, *Unholtz-Dickie Corporation, Hamden, Conn.*

26. SHOCK TESTING MACHINES   26–1

Edward W. Clements, *Head; Equipment Integrity Section, Shock and
Vibration Branch, U.S. Naval Research Laboratory, Washington, D.C.*
Jerry R. Sullivan, *Head; Ship Combat Survivability Section; U.S. Naval
Ship Engineering Center, Washington, D.C.*
AND
Irwin Vigness, *Late Head, Shock and Vibration Branch; U.S. Naval
Research Laboratory, Washington, D.C.*

27. APPLICATIONS OF DIGITAL COMPUTERS   27–1

Allen J. Curtis, *Chief Scientist; Hughes Aircraft Company, Elsegundon,
Calif.*

28. MATRIX METHODS OF ANALYSIS   28–1

Stephen H. Crandall, *Ford Professor of Engineering; Massachusetts
Institute of Technology, Cambridge, Mass.*
AND
Rober B. McCalley, Jr., *Senior Technical Consultant; General Electric
Co., Schenectady, N.Y.*

29. PART I: VIBRATION OF STRUCTURES INDUCED BY FLUID FLOW   29–1

Robert D. Blevins, *Consultant; San Diego, Calif.*

PART II: VIBRATION OF STRUCTURES INDUCED BY WIND   29–17

A. G. Davenport, *Director, Boundary Layer Wind Tunnel Laboratory,
and Chairman of the Department of Civil Engineering; University of
Western Ontario, London, Ont.*
AND
Milos Novak, *Professor of Civil Engineering; University of Western
Ontario, London, Ont.*

30. THEORY OF VIBRATION ISOLATION   30–1

Charles E. Crede, *Late Professor of Mechanical Engineering and Applied
Mechanics; California Institute of Technology, Pasadena, Calif.*
AND
Jerome E. Ruzicka, *Barry Controls Division, Barry Wright Corp.,
Watertown, Mass.*

31. THEORY OF SHOCK ISOLATION 31-1

R. E. Newton, *Professor of Mechanical Engineering; United States Naval Postgraduate School, Monterey, Calif.*

32. TYPES AND CHARACTERISTICS OF VIBRATION ISOLATORS 32-1

Romulus H. Racca, *Principal Engineer; Barry Controls Division, Barry Wright Corp., Watertown, Mass.*

33. AIR SUSPENSION AND ACTIVE VIBRATION-ISOLATION SYSTEMS 33-1

Rodney W. Horning, *Senior Project Engineer; Barry Wright Corp., Watertown, Mass.*
AND
Dale W. Schubert, *Senior Development Engineer, Barry Wright Corp., Watertown, Mass.*

34. APPLICATION OF ISOLATORS 34-1

Harry L. Hain, *Manager; Avionics/Marine Products Design, Lord Corp., Erie, Penn.*
John J. Heintzel, *Senior Engineering Specialist, Lord Corp., Erie, Penn.*
AND
Charles J. Leingang, *Manager; Products Design/Industrial Products Engineering, Lord Corp., Erie, Penn.*

35. PROPERTIES OF RUBBER 35-1

William A. Frye, *Manager; Quality & Reliability System, Inland Division, General Motors Corporation, Dayton, Ohio*

36. MATERIAL DAMPING AND SLIP DAMPING 36-1

Lawrence E. Goodman, *Professor of Mechanics; University of Minnesota, Minneapolis, Minn.*

37. APPLICATION OF DAMPING TREATMENTS 37-1

David I. G. Jones, *Materials Research Engineer; Materials Laboratory, Wright-Patterson Air Force Base, Ohio*

38. TORSIONAL VIBRATION IN RECIPROCATING AND ROTATING MACHINES 38-1

Ronald L. Eshleman, *President; Vibration Institute, Clarendon Hills, Ill.*
AND
Frank M. Lewis, *Late Professor Emeritus of Marine Engineering, Massachusetts Institute of Technology, Cambridge, Mass.*

39. BALANCE OF ROTATING MACHINERY 39-1

Douglas G. Stadelbauer, *Vice President; Schenck Trebel Corp., Deer Park, N.Y.*
AND
Douglas Muster, *Brown and Root Professor of Mechanical Engineering; University of Houston, Houston, Texas*

40. MACHINE-TOOL VIBRATION 40-1

John G. Bollinger, *Bascom Professor of Mechanical Engineering; University of Wisconsin, Madison, Wisc.*

41. PACKAGING DESIGN 41-1

Masaji T. Hatae, *Manager, Transportability Engineering; Advanced Systems Division, Northrop Corp., Pico Rivera, Calif.*

42. THEORY OF EQUIPMENT DESIGN 42-1

Edward G. Fischer, *Consulting Engineer; Westinghouse Electrical Corp., Pittsburg, Penn.*

43. PRACTICE OF EQUIPMENT DESIGN 43-1

Edward G. Fischer, *Consulting Engineer; Westinghouse Electric Corp., Pittsburg, Penn.*

44. EFFECTS OF SHOCK AND VIBRATION ON MAN 44-1

Henning E. von Gierke, *Director; Biodynamics and Bioengineering Division, Harry G. Armstrong Aerospace Medical Research Laboratories, Wright-Patterson Air Force Base, Ohio*
AND
David E. Goldman, *Professor Emeritus of Physiology; Medical College of Pennsylvania and Hospital, Philadelphia, Penn.*

INDEX—follows Chapter 44

PREFACE

The First Edition of the "Shock and Vibration Handbook," published in 1961, recognized for the first time the full scope of the field of shock and vibration and brought together under one title classical vibration theory combined with modern applications of the theory to current engineering practice, including the topics of mechanical shock and the instrumentation of shock and vibration. Each chapter covered the particular subject matter in the most comprehensive manner possible. Many chapters contained more material than had been found collectively in all books previously published on that topic; others contained considerable material that had not been summarized in the literature. As a result, the Handbook became a standard reference work throughout the world. My co-editor for the First Edition, Charles E. Crede, a remarkable engineer and a special friend, died three years after its publication.

The Second Edition was published in 1976. Rising production costs dictated that the total number of pages of the original three-volume edition of the Handbook be reduced by one-third and that the new edition be bound in a single volume. This was done by eliminating archival material and highly specialized information of interest to relatively few engineers; at the same time, the coverage of problems of current general interest was increased. Since the Second Edition, technical developments have continued at a rapid pace. For example, modal analysis (the subject of Chap. 21 of this volume) was then relatively unknown; in recent years, conferences on this topic have drawn large numbers of attendees and contributed papers. Other topics also have become important, warranting inclusion in the Handbook as new chapters, for example, vibration induced by fluid flow, seismic qualifications of equipment, condition monitoring of machinery, and measurement instrumentation. In addition, some chapters required major revisions because of new technological developments. These include piezoelectric and piezoresistive transducers, special-purpose transducers, measurement techniques, transducer calibration, vibration standards, introduction to testing, data analysis and environmental specification, vibration of structures induced by ground motion, shock testing machines, types and characteristics of isolators, application of isolators, air springs and servo-controlled isolation systems, application of damping treatments, and torsional vibration in reciprocating and rotating machines. These and other developments in many specialized areas of shock and vibration have brought about the need for a Third Edition of the Handbook to reflect current engineering practice. Comprehensive coverage has continued to be a primary objective.

The Third Edition, a unified treatment of the subject of shock and vibration, is the equivalent in content of at least five textbooks of the usual size. The 44 chapters were written by 60 authorities from industry, government laboratories, and universities. Each chapter covers a particular topic. Chapters dealing with related topics are grouped together. The first group of chapters provides a theoretical basis for shock and vibration. The second group considers instrumentation and measurements. This is followed by a chapter on vibration standards. The next group of chapters deals with analysis and testing, concepts in the treatment of data obtained from measurements, and procedures for analyzing and testing systems subjected to vibration and shock, including the use of digital computers. Vibration which is induced by ground motion and by fluid flow is considered next. Then, the important subject of methods of controlling shock and vibration is discussed in a group of chapters dealing with isolation, damping, and balancing. This is followed by chapters devoted to equipment design, packaging, and the effects of shock and vibration on humans. Duplication of material between chapters is avoided insofar as this is desirable, cross-references to other chapters being used frequently. There are extensive references to available technical literature.

The new edition of the Handbook reflects changes in terminology which have taken place during the past decade. Thus in the new chapters and those which have been completely rewritten, the term *hertz* (abbreviated *Hz*) has replaced *cycles per second* (abbreviated *cps*). In the other chapters, replacement of *cps* by *Hz* in the text and illustrations would have resulted in a significant increase in the cost of the Third Edition. Making the Handbook less affordable for students for the sake of such uniformity did not seem warranted at this time, since clarity or quality of presentation is not affected.

Although this Handbook is not intended primarily as a textbook, many teachers will find the classical and rigorous treatment suitable for classroom use; in particular, the extensive discussion of practical examples will be of value as a supplement to the usual classroom theory. The control of shock and vibration is of practical importance in many aspects of engineering. The Handbook is particularly intended to be used as a working reference by engineers and scientists in the mechanical, civil, acoustical, aeronautical, electrical, air-conditioning, transportation, and chemical fields. Engineers in manufacturing, including plant maintenance, measurement and control, environmental testing, and packing and shipping, will find much of value, as will those engaged in development and design work.

In the preparation of the Third Edition, many persons and organizations made useful suggestions and contributions; these are far too numerous to acknowledge individually, but I am grateful to them all. However one individual deserves special thanks for his suggestions for the content of the new edition—Rudolph H. Volin of the Shock and Vibration Information Center at the Naval Research Laboratory. The contributors worked diligently with me toward the objective of making each chapter the definitive treatment in its field, and I appreciate their commitment to excellence. Finally I wish to express my appreciation to the government agencies and industrial organizations with whom some of our contributors are associated for clearing the material presented in chapters written by these contributors.

Cyril M. Harris

1

INTRODUCTION TO THE HANDBOOK

Cyril M. Harris
Columbia University

Charles E. Crede
California Institute of Technology

CONCEPTS OF SHOCK AND VIBRATION

Vibration is a term that describes oscillation in a mechanical system. It is defined by the frequency (or frequencies) and amplitude. Either the motion of a physical object or structure or, alternatively, an oscillating force applied to a mechanical system is vibration in a generic sense. Conceptually, the time-history of vibration may be considered to be sinusoidal or simple harmonic in form. The frequency is defined in terms of cycles per unit time, and the magnitude in terms of amplitude (the maximum value of a sinusoidal quantity). The vibration encountered in practice often does not have this regular pattern. It may be a combination of several sinusoidal quantities, each having a different frequency and amplitude. If each frequency component is an integral multiple of the lowest frequency, the vibration repeats itself after a determined interval of time and is called *periodic*. If there is no integral relation among the frequency components, there is no periodicity and the vibration is defined as *complex*.

Vibration may be described as *deterministic* or *random*. If it is deterministic, it follows an established pattern so that the value of the vibration at any designated future time is completely predictable from the past history. If the vibration is random, its future value is unpredictable except on the basis of probability. Random vibration is defined in statistical terms wherein the probability of occurrence of designated magnitudes and frequencies can be indicated. The analysis of random vibration involves certain physical concepts that are different from those applied to the analysis of deterministic vibration.

Vibration of a physical structure often is thought of in terms of a model consisting of a mass and a spring. The vibration of such a model, or system, may be "free" or "forced." In *free vibration*, there is no energy added to the system but rather the vibration is the continuing result of an initial disturbance. An *ideal system* may be considered undamped for mathematical purposes; in such a system the free vibration is assumed to continue indefinitely. In any *real system*, damping (i.e., energy dissipation) causes the amplitude of free vibration to decay continuously to a negligible value. Such free vibration sometimes is referred to as *transient vibration*. *Forced vibration*, in contrast to free vibration, continues under "steady-state" conditions because energy is supplied to the system continuously to compensate for that dissipated by damping in the system. In general, the frequency at which energy is supplied (i.e., the forcing frequency) appears in the vibration of the system. Forced vibration may be either deterministic or random. In either instance, the vibration of the system depends upon the relation of the excitation or forcing function to the properties of the system. This relationship is a prominent feature of the analytical aspects of vibration.

Shock is a somewhat loosely defined aspect of vibration wherein the excitation is non-periodic, e.g., in the form of a pulse, a step, or transient vibration. The word "shock" implies a degree of suddenness and severity. These terms are relative rather than absolute measures of the characteristic; they are related to a popular notion of the characteristics of shock and are not necessary in a fundamental analysis of the applicable principles. From the analytical viewpoint, the important characteristic of shock is that the motion of the system upon which the shock acts includes both the frequency of the shock excitation and the natural frequency of the system. If the excitation is brief, the continuing motion of the system is free vibration at its own natural frequency.

The technology of shock and vibration embodies both theoretical and experimental facets prominently. Thus, methods of analysis and instruments for the measurement of shock and vibration are of primary significance. The results of analysis and measurement are used to evaluate shock and vibration environments, to devise testing procedures and testing machines, and to design and operate equipment and machinery. Shock and/or vibration may be either wanted or unwanted, depending upon circumstances. For example, vibration is involved in the primary mode of operation of such equipment as conveying and screening machines; the setting of rivets depends upon the application of impact or shock. More frequently, however, shock and vibration are unwanted. Then the objective is to eliminate or reduce their severity or, alternatively, to design equipment to withstand their influences. These procedures are embodied in the control of shock and vibration. Methods of control are emphasized throughout this Handbook.

CONTROL OF SHOCK AND VIBRATION

Methods of shock and vibration control may be grouped into three broad categories:

1. Reduction at the Source
   a. *Balancing of Moving Masses.* Where the vibration originates in rotating or reciprocating members, the magnitude of a vibratory force frequently can be reduced or possibly eliminated by balancing or counterbalancing. For example, during the manufacture of fans and blowers, it is common practice to rotate each rotor and to add or subtract material as necessary to achieve balance.
   b. *Balancing of Magnetic Forces.* Vibratory forces arising in magnetic effects of electrical machinery sometimes can be reduced by modification of the magnetic path. For example, the vibration originating in an electric motor can be reduced by skewing the slots in the armature laminations.
   c. *Control of Clearances.* Vibration and shock frequently result from impacts involved in operation of machinery. In some instances, the impacts result from inferior design or manufacture, such as excessive clearances in bearings, and can be reduced by closer attention to dimensions. In other instances, such as the movable armature of a relay, the shock can be decreased by employing a rubber bumper to cushion motion of the plunger at the limit of travel.

2. Isolation
   a. *Isolation of Source.* Where a machine creates significant shock or vibration during its normal operation, it may be supported upon isolators to protect other machinery and personnel from shock and vibration. For example, a forging hammer tends to create shock of a magnitude great enough to interfere with the operation of delicate apparatus in the vicinity of the hammer. This condition may be alleviated by mounting the forging hammer upon isolators.
   b. *Isolation of Sensitive Equipment.* Equipment often is required to operate in an environment characterized by severe shock or vibration. The equipment may be protected from these environmental influences by mounting it upon isolators. For example, equipment mounted in ships of the navy is subjected to shock of great severity during naval warfare and may be protected from damage by mounting it upon isolators.

3. Reduction of the Response

a. *Alteration of Natural Frequency.* If the natural frequency of the structure of an equipment coincides with the frequency of the applied vibration, the vibration condition may be made much worse as a result of resonance. Under such circumstances, if the frequency of the excitation is substantially constant, it often is possible to alleviate the vibration by changing the natural frequency of such structure. For example, the vibration of a fan blade was reduced substantially by modifying a stiffener on the blade, thereby changing its natural frequency and avoiding resonance with the frequency of rotation of the blade. Similar results are attainable by modifying the mass rather than the stiffness.

b. *Energy Dissipation.* If the vibration frequency is not constant or if the vibration involves a large number of frequencies, the desired reduction of vibration may not be attainable by altering the natural frequency of the responding system. It may be possible to achieve equivalent results by the dissipation of energy to eliminate the severe effects of resonance. For example, the housing of a washing machine may be made less susceptible to vibration by applying a coating of damping material on the inner face of the housing.

c. *Auxiliary Mass.* Another method of reducing the vibration of the responding system is to attach an auxiliary mass to the system by a spring; with proper tuning the mass vibrates and reduces the vibration of the system to which it is attached. For example, the vibration of a textile-mill building subjected to the influence of several hundred looms was reduced by attaching large masses to a wall of the building by means of springs; then the masses vibrated with a relatively large motion and the vibration of the wall was reduced. The incorporation of damping in this auxiliary mass system may further increase its effectiveness.

CONTENT OF HANDBOOK

The chapters of this Handbook each deal with a discrete phase of the subject of shock and vibration. Frequent references are made from one chapter to another, to refer to basic theory in other chapters, to call attention to supplementary information, and to give illustrations and examples. Therefore, each chapter when read with other referenced chapters presents one complete facet of the subject of shock and vibration.

Chapters dealing with similar subject matter are grouped together. The first eleven chapters following this introductory chapter deal with fundamental concepts of shock and vibration. Chapter 2 discusses the free and forced vibration of linear systems that can be defined by lumped parameters with similar types of coordinates. The properties of rigid bodies are discussed in Chap. 3, together with the vibration of resiliently supported rigid bodies wherein several modes of vibration are coupled. Nonlinear vibration is discussed in Chap. 4, and self-excited vibration in Chap. 5. Chapter 6 discusses two degree-of-freedom systems in detail—including both the basic theory and the application of such theory to dynamic absorbers and auxiliary mass dampers. The vibration of systems defined by distributed parameters, notably beams and plates, is discussed in Chap. 7. Chapters 8 and 9 relate to shock; Chap. 8 discusses the response of lumped parameter systems to step- and pulse-type excitation, and Chap. 9 discusses the effects of impact on structures. Chapter 10, entitled "Mechanical Impedance," discusses the concepts whereby the characteristics of a composite system may be determined from the characteristics of the component parts. Random vibration is discussed in Chap. 11 under the title of "Statistical Concepts in Vibration."

The second group of chapters is concerned with instrumentation for the measurement of shock and vibration, and Chap. 12 describes piezoelectric and piezoresistive transducers. Other transducers of a more specialized nature, are considered in Chaps. 14 and 17. The instrumentation to which such transducers are connected (including various types of amplifiers, signal conditioners, analyzers, and recorders) is described in detail in Chap. 13. Techniques for the use of such equipment in making measurements are given in Chap. 15. Within the past decade there has been an increase in the use of vibration measurement equipment to monitor the mechanical condition of machinery, as an aid in preventive maintenance. This is the subject of a new chapter, Chap. 16. The calibration of transducers, Chap. 18, is followed by a chapter that considers national and international standards related to shock and vibration.

The subjects of testing, data reduction, and environmental specification are introduced in Chap. 20, followed by a comprehensive chapter on modal analysis and testing. Chapters 22 and 23 discuss the applicable concepts of data analysis. The first of these two chapters is concerned with the analysis of data defining vibration conditions, and discusses the transformation of a time-history of a vibration measurement into more compact forms of data; the second of these two chapters is an analogous presentation—treating the transformation of time-histories defining conditions of shock. Vibration that is induced as a result of ground motion is described in Chap. 24, and vibration that is induced as a result of wind and liquid flow is presented in Chap. 29. Chapter 24 is in two parts: the first describes how structures respond to seismic activity, and the second describes the process by which equipment is qualified to meet government regulations designed to ensure that such equipment will function when subject to seismic activity. Vibration testing machines, also called vibration exciters or *shakers*, are presented in Chap. 25, followed in the next chapter by an analogous presentation describing shock testing machines.

The next two chapters deal with computational methods. Chapter 27 is concerned with the use of computers, presenting information which is useful in both analytical and experimental work. This is followed by a chapter describing modern matrix methods of analysis, dealing largely with the formulation of matrices for use with digital computers and other numerical calculation methods.

The theory of vibration isolation is discussed in detail in Chap. 30 and an analogous presentation for the isolation of mechanical shock is given in Chap. 31. Commercially available isolators are described in Chap. 32; the selection of such isolators is presented in Chap. 34. Chapter 33 considers special types of isolation systems (i.e., those using air springs and servocontrol). The properties of rubber used in isolators are described in detail in Chap. 35.

An important method of controlling shock and vibration involves the addition of damping or energy-dissipating means to structures that are susceptible to vibration. Chapter 36 discusses the general concepts of damping together with the application of such concepts to hysteresis and slip damping. The application of damping materials to structures is discussed in Chap. 37.

The latter chapters of the Handbook deal with the specific application of fundamentals of analysis, methods of measurement, and control techniques—where these are developed sufficiently to form a separate and discrete subject. Torsional vibration is discussed in Chap. 38, with particular application to internal-combustion engines. The balancing of rotating equipment is discussed in Chap. 39, and balancing machines are described. Chapter 40 describes the special vibration problems associated with the design and operation of machine tools. Among the most prominent occurrences of shock and vibration are those that arise during handling and shipping of merchandise. Packaging of equipment to protect against such shock and vibration is discussed in Chap. 41. Chapters 42 and 43 describe procedures for the design of equipment to withstand shock and vibration—the former considering primarily the theory of design and the latter considering practical aspects. A comprehensive discussion of the human aspect of shock and vibration is considered in Chap. 44 which describes the effect of shock and vibration on people.

SYMBOLS

This section includes a list of symbols with their usual English units as used generally in the Handbook; metric units are given as alternates in the text. In special circumstances, some of the following symbols have different meanings in certain chapters but are defined in those chapters. Other symbols of special or limited application are defined in the respective chapters as they are used.

a	radius	in.
B	bandwidth	cps (Hz)
B	magnetic flux density	gauss
c	damping coefficient	lb-sec/in.
c	velocity of sound	in./sec
c_c	critical damping coefficient	lb-sec/in.
C	capacitance	farads
D	diameter	in.
e	electrical voltage	volts
e	eccentricity	in.
E	energy	in.-lb
E	modulus of elasticity in tension and compression (Young's modulus)	lb/in.^{2}
f	frequency	cps (Hz)
f_n	undamped natural frequency	cps (Hz)
f_i	undamped natural frequencies in a multiple degree-of-freedom system, where $i = 1, 2, \ldots$	cps (Hz)
f_d	damped natural frequency	cps (Hz)
f_r	resonance frequency	cps (Hz)
F	force	lb
f_f	Coulomb friction force	lb
g	acceleration of gravity	in./sec^2
G	modulus of elasticity in shear	lb/in.^{2}
h	height, depth	in.
H	magnetic field strength	oersteds
i	electric current	amperes
I_i	area or mass moment of inertia (subscript indicates axis)	in.^{4}
I_p	polar moment of inertia	in.^{4}
I_{ij}	area or mass product of inertia (subscripts indicate axes)	in.^{4}
$\mathscr{I}$	imaginary part of	
j	$\sqrt{-1}$	
J	inertia constant (weight moment of inertia)	lb-in.^{2}
J	impulse	lb-sec
k	linear stiffness	lb/in.
k_t	rotational (torsional) stiffness	lb-in./rad
l	length	in.
L	inductance	henrys
m	mass	lb-sec^2/in.

m_u	unbalanced mass	lb-sec^2/in.
M	torque	lb-in.
M	mutual inductance	henrys
$\mathfrak{M}$	mobility	in./lb-sec
n	number of coils, supports, etc.	
p	alternating pressure	lb/in.^{2}
p	probability density	
P	probability distribution	
P	static pressure	lb/in.^{2}
q	electric charge	coulombs
Q	resonance factor (also ratio of reactance to resistance)	
r	electrical resistance	ohms
R	radius	in.
$\mathfrak{R}$	real part of	
s	arc length	in.
S	area of diaphragm, tube, etc.	in.^{2}
t	thickness	in.
t	time	sec
T	transmissibility	
T	kinetic energy	in.-lb
v	linear velocity	in./sec
V	potential energy	in.-lb
w	width	in.
W	weight	lb
W	power	in.-lb/sec
W_e	spectral density of the excitation	(rms)^2/unit freq.
W_r	spectral density of the response	(rms)^2/unit freq.
x	linear displacement in direction of X axis	in.
y	linear displacement in direction of Y axis	in.
z	linear displacement in direction of Z axis	in.
Z	impedance	lb-sec/in.
α	rotational displacement about X axis	radians
β	rotational displacement about Y axis	radians
γ	rotational displacement about Z axis	radians
γ	shear strain	
γ	weight density	lb/in.^{3}
δ	deflection	in.
δ_{st}	static deflection	in.
Δ	logarithmic decrement	
ϵ	tension or compression strain	
ζ	fraction of critical damping	
η	stiffness ratio, loss factor	
θ	phase angle	radians
λ	wavelength	in.
μ	coefficient of friction	
μ	mass density	lb-sec^2/in.^{4}
ν	Poisson's ratio	
ρ	mass density	lb-sec^2/in.^{4}
ρ_i	radius of gyration (subscript indicates axis)	in.
σ	Poisson's ratio	
σ	normal stress	lb/in.^{2}
σ	root-mean-square (rms) value	
τ	period	sec
τ	shear stress	lb/in.^{2}
ϕ	magnetic flux	maxwell
ψ	phase angle	radians
ω	forcing frequency—angular	rad/sec
ω_n	undamped natural frequency—angular	rad/sec

ω_i undamped natural frequencies—angular—in a multiple degree-of-freedom system, where $i = 1, 2, \ldots$ rad/sec

ω_d damped natural frequency—angular....................   rad/sec

ω_r resonance frequency—angular ....   rad/sec

Ω rotational speed......................................   rad/sec

$\simeq$ approximately equal to

CHARACTERISTICS OF HARMONIC MOTION

Harmonic functions are employed frequently in the analysis of shock and vibration. A body that experiences simple harmonic motion follows a displacement pattern defined by

$$x = x_0 \sin (2\pi ft) = x_0 \sin \omega t \qquad (1.1)$$

where f is the *frequency* of the simple harmonic motion, $\omega = 2\pi f$ is the corresponding *angular frequency*, and x_0 is the *amplitude* of the displacement.

The velocity $\dot{x}$ and acceleration $\ddot{x}$ of the body are found by differentiating the displacement once and twice, respectively:

$$\dot{x} = x_0(2\pi f) \cos 2\pi ft = x_0\omega \cos \omega t \qquad (1.2)$$

$$\ddot{x} = -x_0(2\pi f)^2 \sin 2\pi ft = -x_0\omega^2 \sin \omega t \qquad (1.3)$$

The maximum absolute values of the displacement, velocity, and acceleration of a body undergoing harmonic motion occur when the trigonometric functions in Eqs. (1.1) to (1.3) are numerically equal to unity. These values are known, respectively, as displacement, velocity, and acceleration amplitudes; they are defined mathematically as follows:

$$x_0 = x_0 \qquad \dot{x}_0 = (2\pi f)x_0 \qquad \ddot{x}_0 = (2\pi f)^2 x_0 \qquad (1.4)$$

It is common to express the displacement amplitude x_0 in inches when the English system of units is used and in centimeters or millimeters when the metric system is used. Accordingly, the velocity amplitude x_0 is expressed in inches per second in the English system (centimeters per second or millimeters per second in the metric system). The acceleration amplitude x_0 usually is expressed as a dimensionless multiple of the gravitational acceleration g, where $g = 386$ in./sec^2 or 980 cm/sec^2. For example, an acceleration of $3,860$ in./sec^2 is written $10g$.

Table 1.1. Conversion Factors for Translational Velocity and Acceleration

Multiply Value in → or → By ↘ To obtain value in ↓	g-sec, g	ft/sec ft/sec^2	in./sec in./sec^2	cm/sec cm/sec^2	m/sec m/sec^2
g-sec, g	1	0.0311	0.00259	0.00102	0.102
ft/sec ft/sec^2	32.16	1	0.0833	0.0328	3.28
in./sec in./sec^2	386	12.0	1	0.3937	39.37
cm/sec cm/sec^2	980	30.48	2.540	1	100
m/sec m/sec^2	9.80	0.3048	0.0254	0.010	1

Factors for converting values of rectilinear velocity and acceleration to different units are given in Table 1.1; similar factors for angular velocity and acceleration are given in Table 1.2. Displacement, velocity, and acceleration amplitudes (expressed in the English system of units) as a function of frequency are shown graphically in Figs. 1.1A and 1.1B and by the nomograph of Fig. 1.2.

For certain purposes in analysis, it is convenient to express the amplitude in terms of the average value of the harmonic function, the root-mean-square (rms) value, or two times the amplitude (i.e., peak-to-peak value). These terms are defined mathematically in Chap. 22; numerical conversion factors are set forth in Table 1.3 for ready reference.

Table 1.2. Conversion Factors for Rotational Velocity and Acceleration

Multiply Value in→ or→ By↘ To obtain value in ↓	rad/sec rad/sec^2	degree/sec degree/sec^2	rev/sec rev/sec^2	rev/min rev/min/sec
rad/sec rad/sec^2	1	0.01745	6.283	0.1047
degree/sec degree/sec^2	57.30	1	360	6.00
rev/sec rev/sec^2	0.1592	0.00278	1	0.0167
rev/min rev/min/sec	9.549	0.1667	60	1

Table 1.3. Conversion Factors for Simple Harmonic Motion

Multiply numerical value in terms of→ By↘ To obtain value in terms of ↓	Amplitude	Average value	Root-mean-square value (rms)	Peak-to-peak value
Amplitude	1	1.571	1.414	0.500
Average value	0.637	1	0.900	0.318
Root-mean-square value (rms)	0.707	1.111	1	0.354
Peak-to-peak value	2.000	3.142	2.828	1

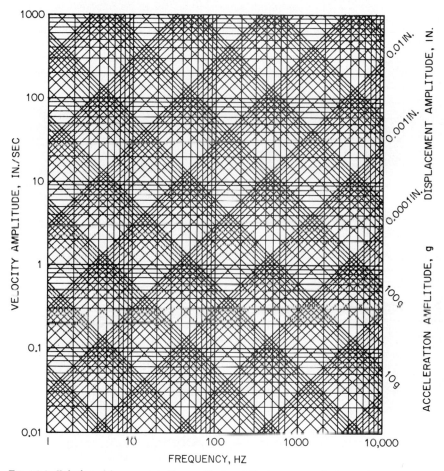

FIG. 1.1A. Relation of frequency to the amplitudes of displacement, velocity, and acceleration in harmonic motion (English system of units).

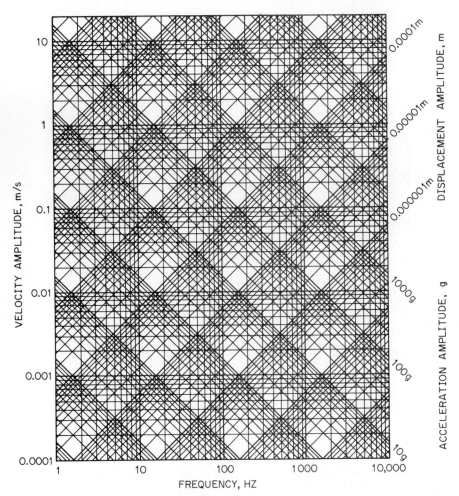

FIG. 1.1*B*. Relation of frequency to the amplitudes of displacement, velocity, and acceleration in harmonic motion (Standard International units).

Fig. 1.2. Nomogram for harmonic motion showing relation of frequency and acceleration amplitude to displacement amplitude (extreme left scale) and relation of frequency and velocity amplitude to displacement amplitude (middle scale on left). Use of the nomogram is indicated by the broken lines. For example, harmonic motion with a displacement amplitude of 0.03 in. and a frequency of 50 Hz has an acceleration amplitude of 7.7g; harmonic motion with a displacement amplitude of 0.10 in. and a frequency of 40 Hz has a velocity amplitude of 25.1 in./sec.

APPENDIX 1.1

NATURAL FREQUENCIES OF COMMONLY USED SYSTEMS

The most important aspect of vibration analysis often is the calculation or measurement of the natural frequencies of mechanical systems. Natural frequencies are discussed prominently in many chapters of the Handbook. Appendix 1.1 includes in tabular form, convenient for ready reference, a compilation of frequently used expressions for the natural frequencies of common mechanical systems. The data for beams and plates are abstracted from Chap. 7.

MASS – SPRING SYSTEMS IN TRANSLATION
(RIGID MASS AND MASSLESS SPRING)

k = SPRING STIFFNESS, LB/IN.

m = MASS, LB-SEC^2/IN.

ω_n = ANGULAR NATURAL FREQUENCY, RAD/SEC

$$\omega_n = \sqrt{\frac{k}{m}}$$

$$\omega_n = \sqrt{\frac{k(m_1+m_2)}{m_1 m_2}}$$

$$\omega_n = \frac{a}{b}\sqrt{\frac{k}{m}}$$

$$\omega_n = \sqrt{\frac{1}{2}\left[\frac{k_1}{m_1}+\frac{k_2}{m_2}\left(1+\frac{m_2}{m_1}\right)\pm\sqrt{\left[\frac{k_1}{m_1}+\frac{k_2}{m_2}\left(1+\frac{m_2}{m_1}\right)\right]^2-\frac{4k_1 k_2}{m_1 m_2}}\right]}$$

$$\omega_n = \sqrt{\frac{2k}{m}}$$

SPRINGS IN COMBINATION

k_r = RESULTANT STIFFNESS OF COMBINATION

$$k_r = k_1 + k_2$$

$$k_r = \frac{k_1 k_2}{k_1 + k_2}$$

$$k_r = \frac{(k_1 + k_2)k_3}{k_1 + k_2 + k_3}$$

HELICAL SPRINGS

d = WIRE DIAMETER, IN

D = MEAN COIL DIAMETER, IN.

n = NUMBER OF ACTIVE COILS

E = YOUNG'S MODULUS, LB/IN.^{2}

G = MODULUS OF ELASTICITY IN SHEAR, LB/IN.^{2}

AXIAL (TENSION OR COMPRESSION)	LATERAL	TORSION	BENDING
$k = \dfrac{Gd^4}{8nD^3}$ LB/IN.	SEE EQS (34.10) AND (34.11), AND FIG 34.22	$k_t = \dfrac{Ed^4}{64nD}$ $\dfrac{\text{LB-IN.}}{\text{RAD}}$	$k_t = \dfrac{Ed^4}{32nD}\left(\dfrac{1}{1+E/2G}\right)$ $\dfrac{\text{LB-IN.}}{\text{RAD}}$

ROTOR-SHAFT SYSTEMS
(RIGID ROTOR AND MASSLESS SHAFT)

k_t = TORSIONAL STIFFNESS OF SHAFT, LB-IN./RAD

I = MASS MOMENT OF INERTIA OF ROTOR, LB-IN.-SEC^2

ω_n = ANGULAR NATURAL FREQUENCY, RAD/SEC

$$\omega_n = \sqrt{\frac{k_t}{I}}$$

$$\omega_n = \sqrt{\frac{1}{2}\left[\frac{k_{t1}}{I_1} + \frac{k_{t2}}{I_2}\left(1+\frac{I_2}{I_1}\right) \pm \sqrt{\left[\frac{k_{t1}}{I_1} + \frac{k_{t2}}{I_2}\left(1+\frac{I_2}{I_1}\right)\right]^2 - \frac{4 k_{t1} k_{t2}}{I_1 I_2}}\,\right]}$$

$$\omega_n = \sqrt{\frac{k_t (I_1 + I_2)}{I_1 I_2}}$$

$$\omega_n = \sqrt{\frac{1}{2}\left[B \pm \sqrt{B^2 - \frac{4 k_{t1} k_{t2}}{I_1 I_2 I_3}(I_1 + I_2 + I_3)}\,\right]}$$

WHERE $B = \dfrac{k_{t1}}{I_1} + \dfrac{k_{t2}}{I_3} + \dfrac{k_{t1} + k_{t2}}{I_2}$

(MOMENT OF INERTIA OF SHAFT = I_s)

$$\omega_n = \sqrt{\frac{k_t}{I + \frac{1}{3} I_s}}$$

GEARED SYSTEM WITH MASSLESS GEARS (SPEED OF ROTOR 2 IS n TIMES SPEED OF ROTOR 1)

$$\omega_n = \sqrt{\frac{k_1 k_2 (I_1 + n^2 I_2)}{I_1 I_2 (n^2 k_2 + k_1)}}$$

STIFFNESS OF SHAFTS IN TORSION

G = MODULUS OF ELASTICITY IN SHEAR, LB/IN.^{2}

l = LENGTH OF SHAFT, IN.

I_p = POLAR MOMENT OF INERTIA OF SHAFT CROSS-SECTION, IN.^{4}

SOLID CIRCULAR

$$k_t = \frac{\pi G D^4}{32 l}$$

HOLLOW CIRCULAR

$$k_t = \frac{\pi G (D_0^4 - D_i^4)}{32 l}$$

RECTANGULAR

$$k_t = \frac{G w t^3}{3 l}$$

ANY SOLID SECTION

S = AREA

$$k_t = \frac{G S^4}{4 \pi^2 l I_p}$$

<u>MASSLESS BEAMS WITH CONCENTRATED MASS LOADS</u>

m = MASS OF LOAD, LB-SEC^2/IN.

ℓ = LENGTH OF BEAM, IN.

I = AREA MOMENT OF INERTIA OF BEAM CROSS SECTION, IN.^{4}

E = YOUNG'S MODULUS, LB/IN.^{2}

ω_n = ANGULAR NATURAL FREQUENCY, RAD/SEC

FIXED-FREE END LOAD	HINGED-HINGED CENTER LOAD	HINGED-HINGED OFF-CENTER LOAD	FIXED-FIXED CENTER LOAD	FIXED-FIXED OFF-CENTER LOAD
$\omega_n = \sqrt{\dfrac{3EI}{m\ell^3}}$	$\omega_n = 4\sqrt{\dfrac{3EI}{m\ell^3}}$	$\omega_n = \dfrac{1}{ab}\sqrt{\dfrac{3EI\ell}{m}}$	$\omega_n = 8\sqrt{\dfrac{3EI}{m\ell^3}}$	$\omega_n = \dfrac{1}{ab}\sqrt{\dfrac{3EI\ell^3}{mab}}$

<u>MASSIVE SPRINGS (BEAMS) WITH CONCENTRATED MASS LOADS</u>

m = MASS OF LOAD, LB-SEC^2/IN.

$m_s(m_b)$ = MASS OF SPRING (BEAM), LB-SEC^2/IN.

k = STIFFNESS OF SPRING LB/IN.

ℓ = LENGTH OF BEAM, IN.

I = AREA MOMENT OF INERTIA OF BEAM CROSS SECTION, IN.^{4}

E = YOUNG'S MODULUS, LB/IN.^{2}

ω_n = ANGULAR NATURAL FREQUENCY, RAD/SEC

MASS-HELICAL SPRING	FIXED-FREE END LOAD	HINGED-HINGED CENTER LOAD	FIXED-FIXED CENTER LOAD
$\omega_n = \sqrt{\dfrac{k}{\left(m + \dfrac{m_s}{3}\right)}}$	$\omega_n = \sqrt{\dfrac{3EI}{\ell^3(m + 0.23 m_b)}}$	$\omega_n = \sqrt{\dfrac{48EI}{\ell^3(m + 0.5 m_b)}}$	$\omega_n = 14\sqrt{\dfrac{EI}{\ell^3(m + 0.375 m_b)}}$

<u>AREA MOMENT OF INERTIA OF BEAM SECTIONS</u>
(WITH RESPECT TO AXIS a-a)

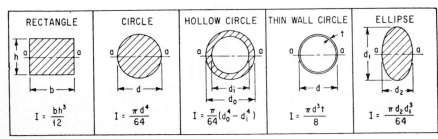

RECTANGLE	CIRCLE	HOLLOW CIRCLE	THIN WALL CIRCLE	ELLIPSE
$I = \dfrac{bh^3}{12}$	$I = \dfrac{\pi d^4}{64}$	$I = \dfrac{\pi}{64}(d_0^4 - d_i^4)$	$I = \dfrac{\pi d^3 t}{8}$	$I = \dfrac{\pi d_2 d_1^3}{64}$

BEAMS OF UNIFORM SECTION AND UNIFORMLY DISTRIBUTED LOAD

ANGULAR NATURAL FREQUENCY $\omega_n = A\sqrt{\dfrac{EI}{\mu \ell^4}}$ RAD/SEC

WHERE E = YOUNG'S MODULUS, LB/IN²
 I = AREA MOMENT OF INERTIA OF BEAM CROSS SECTION, IN.
 ℓ = LENGTH OF BEAM, IN.
 μ = MASS PER UNIT LENGTH OF BEAM, LB-SEC²/IN²
 A = COEFFICIENT FROM TABLE BELOW

NODES ARE INDICATED IN TABLE BELOW AS A PROPORTION OF LENGTH ℓ MEASURED FROM LEFT END

	Mode 1	Mode 2	Mode 3	Mode 4	Mode 5
FIXED–FREE (CANTILEVER)	A = 3.52	0.774 A = 22.0	0.500 0.868 A = 61.7	0.356 0.644 0.906 A = 121.0	0.279 0.500 0.723 0.926 A = 200.0
HINGED–HINGED (SIMPLE)	A = 9.87	0.500 A = 39.5	0.333 0.667 A = 88.9	0.25 0.50 0.75 A = 158	0.20 0.40 0.60 0.80 A = 247
FIXED–FIXED (BUILT-IN)	A = 22.4	0.500 A = 61.7	0.359 0.641 A = 121	0.278 0.500 0.722 A = 200	0.227 0.409 0.591 0.773 A = 298
FREE–FREE	0.224 0.776 A = 22.4	0.132 0.500 0.868 A = 61.7	0.094 0.356 0.644 0.906 A = 121	0.073 0.277 0.500 0.723 0.927 A = 200	0.060 0.227 0.409 0.591 0.773 0.940 A = 298
FIXED–HINGED	A = 15.4	0.560 A = 50.0	0.384 0.692 A = 104	0.294 0.529 0.765 A = 178	0.238 0.429 0.619 0.810 A = 272
HINGED–FREE	0.736 A = 15.4	0.446 0.853 A = 50.0	0.308 0.616 0.898 A = 104	0.235 0.471 0.707 0.922 A = 178	0.190 0.381 0.581 0.763 0.937 A = 272

NATURAL FREQUENCIES OF THIN FLAT PLATES OF UNIFORM THICKNESS

$$\omega_n = B \sqrt{\frac{E t^2}{\rho a^4 (1-\nu^2)}} \quad \text{RAD/SEC}$$

E = YOUNG'S MODULUS, LB /IN.^{2}
t = THICKNESS OF PLATE, IN.
ρ = MASS DENSITY, LB-SEC^2/IN.^{4}
a = DIAMETER OF CIRCULAR PLATE OR SIDE OF SQUARE PLATE, IN.
ν = POISSON'S RATIO

SHAPE OF PLATE	DIAGRAM	EDGE CONDITIONS	VALUE OF B FOR MODE:							
			1	2	3	4	5	6	7	8
CIRCULAR		CLAMPED AT EDGE	11.84	24.61	40.41	46.14	103.12			
CIRCULAR		FREE	6.09	10.53	14.19	23.80	40.88	44.68	61.38	69.44
CIRCULAR		CLAMPED AT CENTER	4.35	24.26	70.39	138.85				
CIRCULAR		SIMPLY SUPPORTED AT EDGE	5.90							
SQUARE		ONE EDGE CLAMPED–THREE EDGES FREE	1.01	2.47	6.20	7.94	9.01			
SQUARE		ALL EDGES CLAMPED	10.40	21.21	31.29	38.04	38.22	47.73		
SQUARE		TWO EDGES CLAMPED–TWO EDGES FREE	2.01	6.96	7.74	13.89	18.25			
SQUARE		ALL EDGES FREE	4.07	5.94	6.91	10.39	17.80	18.85		
SQUARE		ONE EDGE CLAMPED–THREE EDGES SIMPLY SUPPORTED	6.83	14.94	16.95	24.89	28.99	32.71		
SQUARE		TWO EDGES CLAMPED–TWO EDGES SIMPLY SUPPORTED	8.37	15.82	20.03	27.34	29.54	37.31		
SQUARE		ALL EDGES SIMPLY SUPPORTED	5.70	14.26	22.82	28.52	37.08	48.49		

MASSLESS CIRCULAR PLATE WITH CONCENTRATED CENTER MASS

CLAMPED EDGES $\omega_n = 4.09 \sqrt{\dfrac{E h^3}{m a^2 (1-\nu^2)}}$

SIMPLY SUPPORTED EDGES $\omega_n = 4.09 \sqrt{\dfrac{E h^3}{m a^2 (1-\nu)(3+\nu)}}$

NATURAL FREQUENCIES OF MISCELLANEOUS SYSTEMS
(ω_n = ANGULAR NATURAL FREQUENCY, RAD/SEC)

SIMPLE PENDULUM

$$\omega_n = \sqrt{\frac{g}{\ell}}$$

g = ACCELERATION OF GRAVITY
(g AND ℓ IN CONSISTENT UNITS)

COMPOUND PENDULUM

AXIS OF SUPPORT
CG

$$\omega_n = \sqrt{\frac{ag}{\rho_0^2}} = \sqrt{\frac{aW}{I_0}}$$

ρ_0 = RADIUS OF GYRATION ABOUT AXIS OF SUPPORT
g = ACCELERATION OF GRAVITY
W = WEIGHT OF PENDULUM
I_0 = MOMENT OF INERTIA ABOUT AXIS OF SUPPORT

SPHERE IN CYLINDRICAL TRACK

SPHERE
R_s

$$\omega_n = \sqrt{\frac{5g}{7(R_t - R_s)}}$$

g = ACCELERATION OF GRAVITY
R_t = RADIUS OF TRACK
R_s = RADIUS OF SPHERE

PNEUMATIC SYSTEM

m

$$\omega_n = \sqrt{\frac{2p\,S^2}{m\,V_0}}$$

p = PRESSURE AT EACH END OF CYLINDER, LB/IN.^{2}
S = AREA OF PISTON, IN.^{2}
m = MASS OF PISTON, LB-SEC^2/IN.
V_0 = VOLUME OF EACH END OF CYLINDER, IN.^{3}

U–TUBE WITH LIQUID

S

$$\omega_n = \sqrt{\frac{2g}{S}}$$

g = ACCELERATION OF GRAVITY

PLANK ON ROTATING DRUMS

ω W ω
a

$$\omega_n = \sqrt{\frac{2\mu g}{a}}$$

g = ACCELERATION OF GRAVITY
μ = COEFFICIENT OF FRICTION BETWEEN PLANK AND DRUM

TANKS WITH CONNECTING CONDUIT

TANK 1   TANK 2
ℓ h

$$\omega_n = \sqrt{\frac{g(1 + S_1/S_2)}{h(1 + S_1/S_2) + \ell(S_1/S_0)}}$$

g = ACCELERATION OF GRAVITY
S_1 = AREA OF TANK 1
S_2 = AREA OF TANK 2
S_0 = AREA OF CONDUIT

CYLINDER IN CYLINDRICAL TRACK

R_t CYLINDER R_c

$$\omega_n = \sqrt{\frac{2g}{3(R_t - R_c)}}$$

g = ACCELERATION OF GRAVITY
R_t = RADIUS OF TRACK
R_c = RADIUS OF CYLINDER

1-17

APPENDIX 1.2

TERMINOLOGY

For convenience, definitions of terms which are used frequently in the field of shock and vibration are assembled here. Many of the definitions in this appendix are identical with those contained in a document prepared by ISO, TC-108, Mechanical vibration and shock.† For terms which are not listed below, the reader is referred to the Index.

Acceleration.   Acceleration is a vector quantity that specifies the time rate of change of velocity.

Amplitude.   Amplitude is the maximum value of a sinusoidal quantity.

Analog.   If a first quantity or structural element is analogous to a second quantity or structural element belonging in another field of knowledge, the second quantity is called the analog of the first, and vice versa.

Analogy.   An analogy is a recognized relationship of consistent mutual similarity between the equations and structures appearing within two or more fields of knowledge, and an identification and association of the quantities and structural elements that play mutually similar roles in these equations and structures, for the purpose of facilitating transfer of knowledge of mathematical procedures of analysis and behavior of the structures between these fields.

Angular Frequency (Circular Frequency).   The angular frequency of a periodic quantity, in radians per unit time, is the frequency multiplied by 2π.

Angular Mechanical Impedance (Rotational Mechanical Impedance).   Angular mechanical impedance is the impedance involving the ratio of torque to angular velocity.   (See *Impedance*.)

Antinode (Loop).   An antinode is a point, line, or surface in a standing wave where some characteristic of the wave field has maximum amplitude.

Antiresonance.   For a system in forced oscillation, antiresonance exists at a point when any change, however small, in the frequency of excitation causes an increase in the response at this point.

Audio Frequency.   An audio frequency is any frequency corresponding to a normally audible sound wave.

Autocorrelation Coefficient.   The autocorrelation coefficient of a signal is the ratio of the autocorrelation function to the mean-square value of the signal:

$$R(\tau) = \overline{x(t)x(t + \tau)}/\overline{[x(t)]^2}$$

Autocorrelation Function.   The autocorrelation function of a signal is the average of the product of the value of the signal at time t with the value at time $t + \tau$:

$$R(\tau) = \overline{x(t)x(t + \tau)}$$

For a stationary random signal of infinite duration, the power spectral density (except for a constant factor) is the cosine Fourier transform of the autocorrelation function.

Auxiliary Mass Damper (Damped Vibration Absorber).   An auxiliary mass damper is a system consisting of a mass, spring, and damper which tends to reduce vibration by the dissipation of energy in the damper as a result of relative motion between the mass and the structure to which the damper is attached.

Background Noise.   Background noise is the total of all sources of interference in a system used for the production, detection, measurement, or recording of a signal, independent of the presence of the signal.

Balancing.   Balancing is a procedure for adjusting the mass distribution of a rotor so that vibration of the journals, or the forces on the bearings at once-per-revolution, are reduced or controlled.   (See *Chap. 39* for a complete list of definitions related to *balancing*.)

Band-pass Filter.   A band-pass filter is a wave filter that has a single transmission band extending from a lower cutoff frequency greater than zero to a finite upper cutoff frequency.

Bandwidth, Effective.   (See *Effective Bandwidth*.)

Beats.   Beats are periodic variations that result from the superposition of two simple harmonic quantities of different frequencies f_1 and f_2. They involve the periodic increase and decrease of amplitude at the beat frequency $(f_1 - f_2)$.

† Copies of this document, as well as others in the field of mechanical shock and vibration, are available from the American National Standards Institute, Inc., 1430 Broadway, New York 10018.

Broad-band Random Vibration. Broad-band random vibration is random vibration having its frequency components distributed over a broad frequency band. (See *Random Vibration*.)

Circular Frequency. (See *Angular Frequency*.)

Complex Angular Frequency. As applied to a function $a = Ae^{\sigma t} \sin(\omega t - \phi)$, where σ, ω, and ϕ are constant, the quantity $\omega_c = \sigma + j\omega$ is the complex angular frequency where j is an operator with rules of addition, multiplication, and division as suggested by the symbol $\sqrt{-1}$. If the signal decreases with time, σ must be negative.

Complex Function. A complex function is a function having real and imaginary parts.

Complex Vibration. Complex vibration is vibration whose components are sinusoids not harmonically related to one another. (See *Harmonic*.)

Compliance. Compliance is the reciprocal of stiffness.

Continuous System (Distributed System). A continuous system is one that is considered to have an infinite number of possible independent displacements. Its configuration is specified by a function of a continuous spatial variable or variables in contrast to a discrete or lumped parameter system which requires only a finite number of coordinates to specify its configuration.

Correlation Coefficient. The correlation coefficient of two variables is the ratio of the correlation function to the product of the averages of the variables:

$$\overline{x_1(t) \cdot x_2(t)} / \overline{x_1(t)} \cdot \overline{x_2(t)}$$

Correlation Function. The correlation function of two variables is the average value of their product:

$$\overline{x_1(t) \cdot x_2(t)}$$

Coulomb Damping (Dry Friction Damping). Coulomb damping is the dissipation of energy that occurs when a particle in a vibrating system is resisted by a force whose magnitude is a constant independent of displacement and velocity, and whose direction is opposite to the direction of the velocity of the particle.

Coupled Modes. Coupled modes are modes of vibration that are not independent but which influence one another because of energy transfer from one mode to the other. (See *Mode of Vibration*.)

Coupling Factor, Electromechanical. The electromechanical coupling factor is a factor used to characterize the extent to which the electrical characteristics of a transducer are modified by a coupled mechanical system, and vice versa.

Critical Damping. Critical damping is the minimum viscous damping that will allow a displaced system to return to its initial position without oscillation.

Critical Speed. Critical speed is a speed of a rotating system that corresponds to a resonant frequency of the system.

Cycle. A cycle is the complete sequence of values of a periodic quantity that occur during a period.

Damped Natural Frequency. The damped natural frequency is the frequency of free vibration of a damped linear system. The free vibration of a damped system may be considered periodic in the limited sense that the time interval between zero crossings in the same direction is constant, even though successive amplitudes decrease progressively. The frequency of the vibration is the reciprocal of this time interval.

Damping. Damping is the dissipation of energy with time or distance.

Damping Ratio. (See *Fraction of Critical Damping*.)

Decibel (db). The decibel is a unit which denotes the magnitude of a quantity with respect to an arbitrarily established reference value of the quantity, in terms of the logarithm (to the base 10) of the ratio of the quantities. For example, in electrical transmission circuits a value of power may be expressed in terms of a power level in decibels; the power level is given by 10 times the logarithm (to the base 10) of the ratio of the actual power to a reference power (which corresponds to 0 db).

Degrees-of-freedom. The number of degrees-of-freedom of a mechanical system is equal to the minimum number of independent coordinates required to define completely the positions of all parts of the system at any instant of time. In general, it is equal to the number of independent displacements that are possible.

Deterministic Function. A deterministic function is one whose value at any time can be predicted from its value at any other time.

Displacement. Displacement is a vector quantity that specifies the change of position of a body or particle and is usually measured from the mean position or position of rest. In general, it can be represented as a rotation vector or a translation vector, or both.

Distortion. Distortion is an undesired change in waveform. Noise and certain desired changes in waveform, such as those resulting from modulation or detection, are not usually classed as distortion.

Distributed System. (See *Continuous System.*)

Driving Point Impedance. Driving point impedance is the impedance involving the ratio of force to velocity when both the force and velocity are measured at the same point and in the same direction. (See *Impedance.*)

Dry Friction Damping. (See *Coulomb Damping.*)

Duration of Shock Pulse. The duration of a shock pulse is the time required for the acceleration of the pulse to rise from some stated fraction of the maximum amplitude and to decay to this value. (See *Shock Pulse.*)

Dynamic Vibration Absorber (Tuned Damper). A dynamic vibration absorber is an auxiliary mass-spring system which tends to neutralize vibration of a structure to which it is attached. The basic principle of operation is vibration out-of-phase with the vibration of such structure, thereby applying a counteracting force.

Effective Bandwidth. The effective bandwidth of a specified transmission system is the bandwidth of an ideal system which (1) has uniform transmission in its pass band equal to the maximum transmission of the specified system and (2) transmits the same power as the specified system when the two systems are receiving equal input signals having a uniform distribution of energy at all frequencies.

Electromechanical Coupling Factor. (See *Coupling Factor, Electromechanical.*)

Electrostriction. Electrostriction is the phenomenon wherein some dielectric materials experience an elastic strain when subjected to an electric field, this strain being independent of the polarity of the field.

Ensemble. A collection of signals. (Also see *Process.*)

Environment. (See *Natural Environments* and *Induced Environment.*)

Equivalent System. An equivalent system is one that may be substituted for another system for the purpose of analysis. Many types of equivalence are common in vibration and shock technology: (1) equivalent stiffness; (2) equivalent damping; (3) torsional system equivalent to a translational system; (4) electrical or acoustical system equivalent to a mechanical system; etc.

Equivalent Viscous Damping. Equivalent viscous damping is a value of viscous damping assumed for the purpose of analysis of a vibratory motion, such that the dissipation of energy per cycle at resonance is the same for either the assumed or actual damping force.

Ergodic Process. An ergodic process is a random process that is stationary and of such a nature that all possible time averages performed on one signal are independent of the signal chosen and hence are representative of the time averages of each of the other signals of the entire random process.

Excitation (Stimulus). Excitation is an external force (or other input) applied to a system that causes the system to respond in some way.

Filter. A filter is a device for separating waves on the basis of their frequency. It introduces relatively small insertion loss to waves in one or more frequency bands and relatively large insertion loss to waves of other frequencies. (See *Insertion Loss.*)

Force Factor. The force factor of an electromechanical transducer is: (a) the complex quotient of the force required to block the mechanical system divided by the corresponding current in the electric system; (b) the complex quotient of the resulting open-circuit voltage in the electric system divided by the velocity in the mechanical system. Force factors (a) and (b) have the same magnitude when consistent units are used and the transducer satisfies the principle of reciprocity. It is sometimes convenient in an electrostatic or piezoelectric transducer to use the ratios between force and charge or electric displacement, or between voltage and mechanical displacement.

Forced Vibration (Forced Oscillation). The oscillation of a system is forced if the response is imposed by the excitation. If the excitation is periodic and continuing, the oscillation is steady-state.

Foundation (Support). A foundation is a structure that supports the gravity load of a mechanical system. It may be fixed in space, or it may undergo a motion that provides excitation for the supported system.

Fraction of Critical Damping. The fraction of critical damping (damping ratio) for a system with viscous damping is the ratio of actual damping coefficient c to the critical damping coefficient c_c.

Free Vibration. Free vibration of a system is vibration that occurs in the absence of forced vibration.

Frequency. The frequency of a function periodic in time is the reciprocal of the period. The unit is the cycle per unit time and must be specified; the unit *cycle per second* is called *Hertz* (Hz).

Frequency, Angular. (See *Angular Frequency.*)

Fundamental Frequency. (1) The fundamental frequency of a periodic quantity is the frequency of a sinusoidal quantity which has the same period as the periodic quantity. (2) The fundamental frequency of an oscillating system is the lowest natural frequency. The normal mode of vibration associated with this frequency is known as the fundamental mode.

Fundamental Mode of Vibration. The fundamental mode of vibration of a system is the mode having the lowest natural frequency.

g. The quantity g is the acceleration produced by the force of gravity, which varies with the latitude and elevation of the point of observation. By international agreement, the value 980.665 cm/sec^2 = 386.087 in./sec^2 = 32.1739 ft/sec^2 has been chosen as the standard acceleration due to gravity.

Harmonic. A harmonic is a sinusoidal quantity having a frequency that is an integral multiple of the frequency of a periodic quantity to which it is related.

Harmonic Motion. (See *Simple Harmonic Motion.*)

Harmonic Response. Harmonic response is the periodic response of a vibrating system exhibiting the characteristics of resonance at a frequency that is a multiple of the excitation frequency.

High-pass Filter. A high-pass filter is a wave filter having a single transmission band extending from some critical or cutoff frequency, not zero, up to infinite frequency.

Image Impedances. The image impedances of a structure or device are the impedances that will simultaneously terminate all of its inputs and outputs in such a way that at each of its inputs and outputs the impedances in both directions are equal.

Impact. An impact is a single collision of one mass in motion with a second mass which may be either in motion or at rest.

Impedance. Mechanical impedance is the ratio of a force-like quantity to a velocity-like quantity when the arguments of the real (or imaginary) parts of the quantities increase linearly with time. Examples of force-like quantities are: force, sound pressure, voltage, temperature. Examples of velocity-like quantities are: velocity, volume velocity, current, heat flow. *Impedance* is the reciprocal of *mobility*. (Also see *Angular Mechanical Impedance, Linear Mechanical Impedance, Driving Point Impedance,* and *Transfer Impedance.*)

Impulse. Impulse is the product of a force and the time during which the force is applied; more specifically, the impulse is $\int_{t_1}^{t_2} F dt$ where the force F is time dependent and equal to zero before time t_1 and after time t_2.

Induced Environments. Induced environments are those conditions generated as a result of the operation of a structure or equipment.

Insertion Loss. The insertion loss, in decibels, resulting from insertion of an element in a transmission system is 10 times the logarithm to the base 10 of the ratio of the power delivered to that part of the system that will follow the element, before the insertion of the element, to the power delivered to that same part of the system after insertion of the element.

Isolation. Isolation is a reduction in the capacity of a system to respond to an excitation, attained by the use of a resilient support. In steady-state forced vibration, isolation is expressed quantitatively as the complement of transmissibility.

Jerk. Jerk is a vector that specifies the time rate of change of acceleration; jerk is the third derivative of displacement with respect to time.

Line Spectrum. A line spectrum is a spectrum whose components occur at a number of discrete frequencies.

Linear Mechanical Impedance. Linear mechanical impedance is the impedance involving the ratio of force to linear velocity. (See *Impedance.*)

Linear System. A system is linear if for every element in the system the response is proportional to the excitation. This definition implies that the dynamic properties of each element in the system can be represented by a set of linear differential equations with constant coefficients, and that for the system as a whole superposition holds.

Logarithmic Decrement. The logarithmic decrement is the natural logarithm of the ratio of any two successive amplitudes of like sign, in the decay of a single-frequency oscillation.

Longitudinal Wave. A longitudinal wave in a medium is a wave in which the direction of displacement at each point of the medium is normal to the wave front.

Low-pass Filter. A low-pass filter is a wave filter having a single transmission band extending from zero frequency up to some critical or cutoff frequency which is not infinite.

Magnetic Recorder. A magnetic recorder is equipment incorporating an electromagnetic transducer and means for moving a ferromagnetic recording medium relative to the transducer for recording electric signals as magnetic variations in the medium.

Magnetostriction. Magnetostriction is the phenomenon wherein ferromagnetic materials experience an elastic strain when subjected to an external magnetic field. Also, magnetostriction is the converse phenomenon in which mechanical stresses cause a change in the magnetic induction of a ferromagnetic material.

Mechanical Admittance. (See *Mobility.*)

Mechanical Impedance. (See *Impedance.*)

Mechanical Shock. Mechanical shock is a nonperiodic excitation (e.g., a motion of the foundation or an applied force) of a mechanical system that is characterized by suddenness and severity, and usually causes significant relative displacements in the system.

Mechanical System. A mechanical system is an aggregate of matter comprising a defined configuration of mass, stiffness, and damping.

Mobility (Mechanical Admittance). Mobility is the ratio of a velocity-like quantity to a force-like quantity when the arguments of the real (or imaginary) parts of the quantities increase linearly with time. *Mobility* is the reciprocal of *impedance.* The terms *Angular Mobility*, *Linear Mobility*, *Driving-point Mobility*, and *Transfer Mobility* are used in the same sense as corresponding impedances.

Modal Numbers. When the normal modes of a system are related by a set of ordered integers, these integers are called modal numbers.

Mode of Vibration. In a system undergoing vibration, a mode of vibration is a characteristic pattern assumed by the system in which the motion of every particle is simple harmonic with the same frequency. Two or more modes may exist concurrently in a multiple degree-of-freedom system.

Modulation. Modulation is the variation in the value of some parameter which characterizes a periodic oscillation. Thus, amplitude modulation of a sinusoidal oscillation is a variation in the amplitude of the sinusoidal oscillation.

Multiple Degree-of-freedom System. A multiple degree-of-freedom system is one for which two or more coordinates are required to define completely the position of the system at any instant.

Narrow-band Random Vibration (Random Sine Wave). Narrow-band random vibration is random vibration having frequency components only within a narrow band. It has the appearance of a sine wave whose amplitude varies in an unpredictable manner. (See *Random Vibration.*)

Natural Environments. Natural environments are those conditions generated by the forces of nature and whose effects are experienced when the equipment or structure is at rest as well as when it is in operation.

Natural Frequency. Natural frequency is the frequency of free vibration of a system. For a multiple degree-of-freedom system, the natural frequencies are the frequencies of the normal modes of vibration.

Noise.* Noise is any undesired signal. By extension, noise is any unwanted disturbance within a useful frequency band, such as undesired electric waves in a transmission channel or device.

Nominal Bandwidth. The nominal bandwidth of a filter is the difference between the nominal upper and lower cutoff frequencies. The difference may be expressed (1) in cycles per second; (2) as a percentage of the pass-band center frequency; or (3) in octaves.

Nominal Pass-band Center Frequency. The nominal passband center frequency is the geometric mean of the nominal cutoff frequencies.

Nominal Upper and Lower Cutoff Frequencies. The nominal upper and lower cutoff frequencies of a filter pass-band are those frequencies above and below the frequency of maximum response of a filter at which the response to a sinusoidal signal is 3 db below the maximum response.

Nonlinear Damping. Nonlinear damping is damping due to a damping force that is not proportional to velocity.

Normal Mode of Vibration. A normal mode of vibration is a mode of vibration that is uncoupled from (i.e., can exist independently of) other modes of vibration of a system. When vibration of the system is defined as an eigenvalue problem, the normal modes are the eigenvectors and the normal mode frequencies are the eigenvalues. The term "classical normal mode" is sometimes applied to the normal modes of a vibrating system characterized by vibration of each element of the system at the same frequency and phase. In general, classical normal modes exist only in systems having no damping or having particular types of damping.

Oscillation. Oscillation is the variation, usually with time, of the magnitude of a quantity with respect to a specified reference when the magnitude is alternately greater and smaller than the reference.

Partial Node. A partial node is the point, line, or surface in a standing-wave system where some characteristic of the wave field has a minimum amplitude differing from zero. The appropriate modifier should be used with the words "partial node" to signify the type that is intended; e.g., displacement partial node, velocity partial node, pressure partial node.

Peak-to-peak Value. The peak-to-peak value of a vibrating quantity is the algebraic difference between the extremes of the quantity.

Period. The period of a periodic quantity is the smallest increment of the independent variable for which the function repeats itself.

Periodic Quantity. A periodic quantity is an oscillating quantity whose values recur for certain increments of the independent variable.

Phase of a Periodic Quantity. The phase of a periodic quantity, for a particular value of the independent variable, is the fractional part of a period through which the independent variable has advanced, measured from an arbitrary reference.

Pickup. (See *Transducer*.)

Piezoelectric (Crystal) (Ceramic) Transducer. A piezoelectric transducer is a transducer that depends for its operation on the interaction between the electric charge and the deformation of certain asymmetric crystals having piezoelectric properties.

Piezoelectricity. Piezoelectricity is the property exhibited by some asymmetrical crystalline materials which when subjected to strain in suitable directions develop electric polarization proportional to the strain. Inverse piezoelectricity is the effect in which mechanical strain is produced in certain asymmetrical crystalline materials when subjected to an external electric field; the strain is proportional to the electric field.

Power Spectral Density. Power spectral density is the limiting mean-square value (e.g., of acceleration, velocity, displacement, stress or other random variable) per unit bandwidth, i.e., the limit of the mean-square value in a given rectangular bandwidth divided by the bandwidth, as the bandwidth approaches zero.

Power Spectral Density Level. The spectrum level of a specified signal at a particular frequency is the level in decibels of that part of the signal contained within a band 1 cycle per second wide, centered at the particular frequency. Ordinarily this has significance only for a signal having a continuous distribution of components within the frequency range under consideration.

Process. A process is a collection of signals. The word *process* rather than the word *ensemble* ordinarily is used when it is desired to emphasize the properties the signals have or do not have as a group. Thus, one speaks of a stationary process rather than a stationary ensemble.

Pulse Rise Time. The pulse rise time is the interval of time required for the leading edge of a pulse to rise from some specified small fraction to some specified larger fraction of the maximum value.

Q (Quality Factor). The quantity Q is a measure of the sharpness of resonance or frequency selectivity of a resonant vibratory system having a single degree of freedom, either mechanical or electrical. In a mechanical system, this quantity is equal to one-half the reciprocal of the damping ratio. It is commonly used only with reference to a lightly damped system, and is then approximately equal to the following:

(1) Transmissibility at resonance,
(2) π/logarithmic decrement,
(3) $2\pi W/\Delta W$ where W is the stored energy and ΔW the energy dissipation per cycle, and
(4) $f_r/\Delta f$ where f_r is the resonant frequency and Δf is the bandwidth between the half-power points.

Quasi-ergodic Process. A quasi-ergodic process is a random process which is not necessarily stationary but of such a nature that some time averages performed on a signal are independent of the signal chosen.

Quasi-periodic Signal. A signal consisting only of quasi-sinusoids.

Quasi-sinusoid. A function of the form $a = A \sin(2\pi ft - \phi)$ where either A or f, or both, is not a constant but may be expressed readily as a function of time. Ordinarily ϕ is considered constant.

Random Sine Wave. (See *Narrow-band Random Vibration.*)

Random Vibration. Random vibration is vibration whose instantaneous magnitude is not specified for any given instant of time. The instantaneous magnitudes of a random vibration are specified only by probability distribution functions giving the probable fraction of the total time that the magnitude (or some sequence of magnitudes) lies within a specified range. Random vibration contains no periodic or quasi-periodic constituents. If random vibration has instantaneous magnitudes that occur according to the Gaussian distribution, it is called "Gaussian random vibration."

Ratio of Critical Damping. (See *Fraction of Critical Damping.*)

Rayleigh Wave. A Rayleigh wave is a surface wave associated with the free boundary of a solid, such that a surface particle describes an ellipse whose major axis is normal to the surface, and whose center is at the undisturbed surface. At maximum particle displacement away from the solid surface the motion of the particle is opposite to that of the wave.

Recording Channel. The term "recording channel" refers to one of a number of independent recorders in a recording system or to independent recording tracks on a recording medium.

Recording System. A recording system is a combination of transducing devices and associated equipment suitable for storing signals in a form capable of subsequent reproduction.

Relaxation Time. Relaxation time is the time taken by an exponentially decaying quantity to decrease in amplitude by a factor of $1/e = 0.3679$.

Re-recording. Re-recording is the process of making a recording by reproducing a recorded signal source and recording this reproduction.

Resonance. Resonance of a system in forced vibration exists when any change, however small, in the frequency of excitation causes a decrease in the response of the system.

Resonance Frequency (Resonant Frequency). A frequency at which resonance exists.

Response. The response of a device or system is the motion (or other output) resulting from an excitation (stimulus) under specified conditions.

Response Spectrum. (See *Shock Spectrum.*)

Rotational Mechanical Impedance. (See *Angular Mechanical Impedance.*)

Self-induced (Self-excited) Vibration.* The vibration of a mechanical system is self-induced if it results from conversion, within the system, of nonoscillatory excitation to oscillatory excitation.

Shake Table. (See *Vibration Machine.*)

Shear Wave (Rotational Wave). A shear wave is a wave in an elastic medium which causes an element of the medium to change its shape without a change of volume.

Shock. (See *Mechanical Shock.*)

Shock Absorber. A shock absorber is a device which dissipates energy to modify the response of a mechanical system to applied shock.

Shock Isolator (Shock Mount). A shock isolator is a resilient support that tends to isolate a system from a shock motion.

Shock Machine. A shock machine is a device for subjecting a system to controlled and reproducible mechanical shock.

Shock Motion. Shock motion is an excitation involving motion of a foundation. (See *Foundation* and *Mechanical Shock.*)

Shock Mount. (See *Shock Isolator.*)

Shock Pulse. A shock pulse is a substantial disturbance characterized by a rise of acceleration from a constant value and decay of acceleration to the constant value in a short period of time. Shock pulses are normally displayed graphically as curves of acceleration as functions of time.

Shock-pulse Duration. (See *Duration of Shock Pulse.*)

Shock Spectrum (Response Spectrum). A shock spectrum is a plot of the maximum response experienced by a single degree-of-freedom system, as a function of its own natural frequency, in response to an applied shock. The response may be expressed in terms of acceleration, velocity or displacement.

Signal. A signal is (1) a disturbance used to convey information; (2) the information to be conveyed over a communication system.

Simple Harmonic Motion. A simple harmonic motion is a motion such that the displacement is a sinusoidal function of time; sometimes it is designated merely by the term *harmonic motion*.

Single Degree-of-freedom System. A single degree-of-freedom system is one for which only one coordinate is required to define completely the configuration of the system at any instant.

Sinusoidal Motion. (See *Simple Harmonic Motion.*)

Snubber. A snubber is a device used to increase the stiffness of an elastic system (usually by a large factor) whenever the displacement becomes larger than a specified value.

Spectrum. A spectrum is a definition of the magnitude of the frequency components that constitute a quantity.

Spectrum Density. (See *Power Spectral Density.*)

Standard Deviation. Standard deviation is the square root of the variance; i.e., the square root of the mean of the squares of the deviations from the mean value of a vibrating quantity.

Standing Wave. A standing wave is a periodic wave having a fixed distribution in space which is the result of interference of progressive waves of the same frequency and kind. Such waves are characterized by the existence of nodes or partial nodes and antinodes that are fixed in space.

Stationary Process. A stationary process is an ensemble of signals such that an average of values over the ensemble at any given time is independent of time.

Stationary Signal. A stationary signal is a random signal of such nature that averages over samples of finite time intervals are independent of the time at which the sample occurs.

Steady-state Vibration. Steady-state vibration exists in a system if the velocity of each particle is a continuing periodic quantity.

Stiffness. Stiffness is the ratio of change of force (or torque) to the corresponding change in translational (or rotational) deflection of an elastic element.

Subharmonic. A subharmonic is a sinusoidal quantity having a frequency that is an integral submultiple of the fundamental frequency of a periodic quantity to which it is related.

Subharmonic Response. Subharmonic response is the periodic response of a mechanical system exhibiting the characteristic of resonance at a frequency that is a submultiple of the frequency of the periodic excitation.

Superharmonic Response. Superharmonic response is a term sometimes used to denote a particular type of harmonic response which dominates the total response of the system; it frequently occurs when the excitation frequency is a submultiple of the frequency of the fundamental resonance.

Transducer (Pickup). A transducer is a device which converts shock or vibratory motion into an optical, a mechanical, or most commonly to an electrical signal that is proportional to a parameter of the experienced motion.

Transfer Impedance. Transfer impedance between two points is the impedance involving the ratio of force to velocity when force is measured at one point and velocity at the other point. The term *transfer impedance* also is used to denote the ratio of force to velocity measured at the same point but in different directions. (See *Impedance.*)

Transient Vibration. Transient vibration is temporarily sustained vibration of a mechanical system. It may consist of forced or free vibration or both.

Transmissibility. Transmissibility is the nondimensional ratio of the response amplitude of a system in steady-state forced vibration to the excitation amplitude. The ratio may be one of forces, displacements, velocities, or accelerations.

Transmission Loss. Transmission loss is the reduction in the magnitude of some characteristic of a signal, between two stated points in a transmission system.

Transverse Wave. A transverse wave is a wave in which the direction of displacement at each point of the medium is parallel to the wave front.

Tuned Damper. (See *Dynamic Vibration Absorber.*)

Uncorrelated. Two signals or variables $a_1(t)$ and $a_2(t)$ are said to be uncorrelated if the average value of their product is zero: $\overline{a_1(t) \cdot a_2(t)} = 0$. If the correlation coefficient is equal to unity, the variables are said to be completely correlated. If the coefficient is less than unity but larger than zero, they are said to be partially correlated. (See *Correlation Coefficient.*)

Uncoupled Mode. An uncoupled mode of vibration is a mode that can exist in a system concurrently with and independently of other modes.

Undamped Natural Frequency. The undamped natural frequency of a mechanical system is the frequency of free vibration resulting from only elastic and inertial forces of the system.

Variance. Variance is the mean of the squares of the deviations from the mean value of a vibrating quantity.

Velocity. Velocity is a vector quantity that specifies the time rate of change of displacement with respect to a reference frame. If the reference frame is not inertial, the velocity is often designated "relative velocity."

Velocity Shock. Velocity shock is a particular type of shock motion characterized by a sudden velocity change of the foundation. (See *Foundation* and *Mechanical Shock.*)

Vibration. Vibration is an oscillation wherein the quantity is a parameter that defines the motion of a mechanical system. (See *Oscillation.*)

Vibration Isolator. A vibration isolator is a resilient support that tends to isolate a system from steady-state excitation.

Vibration Machine. A vibration machine is a device for subjecting a mechanical system to controlled and reproducible mechanical vibration.

Vibration Meter. A vibration meter is an apparatus for the measurement of displacement, velocity, or acceleration of a vibrating body.

Vibration Mount. (See *Vibration Isolator.*)

Vibration Pickup. (See *Transducer.*)

Viscous Damping. Viscous damping is the dissipation of energy that occurs when a particle in a vibrating system is resisted by a force that has a magnitude proportional to the magnitude of the velocity of the particle and direction opposite to the direction of the particle.

Viscous Damping, Equivalent. (See *Equivalent Viscous Damping.*)

Wave. A wave is a disturbance which is propagated in a medium in such a manner that at any point in the medium the quantity serving as measure of disturbance is a function of the time, while at any instant the displacement at a point is a function of the position of the point. Any physical quantity that has the same relationship to some independent variable (usually time) that a propagated disturbance has, at a particular instant, with respect to space, may be called a wave.

Wave Interference. Wave interference is the phenomenon which results when waves of the same or nearly the same frequency are superposed; it is characterized by a spatial or temporal distribution of amplitude of some specified characteristic differing from that of the individual superposed waves.

Wavelength.   The wavelength of a periodic wave in an isotropic medium is the perpendicular distance between two wave fronts in which the displacements have a difference in phase of one complete period.

White Noise.   White noise is a noise whose power spectral density is substantially independent of frequency over a specified range.

2

BASIC VIBRATION THEORY

Ralph E. Blake
Technical Center of Silicon Valley

INTRODUCTION

This chapter presents the theory of free and forced steady-state vibration of single degree-of-freedom systems. Undamped systems and systems having viscous damping and structural damping are included. Multiple degree-of-freedom systems are discussed, including the normal-mode theory of linear elastic structures and Lagrange's equations.

ELEMENTARY PARTS OF VIBRATORY SYSTEMS

Vibratory systems are comprised of means for storing potential energy (spring), means for storing kinetic energy (mass or inertia), and means by which the energy is gradually lost (damper). The vibration of a system involves the alternating transfer of energy between its potential and kinetic forms. In a damped system, some energy is dissipated at each cycle of vibration and must be replaced from an external source if a steady vibration is to be maintained. Although a single physical structure may store both kinetic and potential energy, and may dissipate energy, this chapter considers only *lumped parameter systems* comprised of ideal springs, masses, and dampers wherein each element has only a single function. In translational motion, displacements are defined as linear distances; in rotational motion, displacements are defined as angular motions.

FIG. 2.1. Linear spring.

TRANSLATIONAL MOTION

SPRING. In the linear spring shown in Fig. 2.1, the change in the length of the spring is proportional to the force acting along its length:

$$F = k(x - u) \qquad (2.1)$$

The ideal spring is considered to have no mass; thus, the force acting on one end is equal and opposite to the force acting on the other end. The constant of proportionality k is the *spring constant or stiffness*.

FIG. 2.2. Rigid mass.

MASS. A mass is a rigid body (Fig. 2.2) whose acceleration $\ddot{x}$ according to Newton's second law is proportional to the resultant F of all forces acting on the mass: †

$$F = m\ddot{x} \qquad (2.2)$$

† It is common to use the word "mass" in a general sense to designate a rigid body. Mathematically, the mass of the rigid body is defined by m in Eq. (2.2).

FIG. 2.3. Viscous damper.

DAMPER. In the viscous damper shown in Fig. 2.3, the applied force is proportional to the relative velocity of its connection points:

$$F = c(\dot{x} - \dot{u}) \qquad (2.3)$$

The constant c is the *damping coefficient*, the characteristic parameter of the damper. The ideal damper is considered to have no mass; thus the force at one end is equal and opposite to the force at the other end. "Structural damping" is considered below and several other types of damping are considered in Chap. 30.

ROTATIONAL MOTION

The elements of a mechanical system which moves with pure rotation of the parts are wholly analogous to the elements of a system that moves with pure translation. The property of a rotational system which stores kinetic energy is inertia; stiffness and damping coefficients are defined with reference to angular displacement and angular velocity, respectively. The analogous quantities and equations are listed in Table 2.1.

Table 2.1. Analogous Quantities in Translational and Rotational Vibrating Systems

Translational quantity	*Rotational quantity*
Linear displacement x	Angular displacement α
Force F	Torque M
Spring constant k	Spring constant k_r
Damping constant c	Damping constant c_r
Mass m	Moment of inertia I
Spring law $F = k(x_1 - x_2)$	Spring law $M = k_r(\alpha_1 - \alpha_2)$
Damping law $F = c(\dot{x}_1 - \dot{x}_2)$	Damping law $M = c_r(\dot{\alpha}_1 - \dot{\alpha}_2)$
Inertia law $F = m\ddot{x}$	Inertia law $M = I\ddot{\alpha}$

Inasmuch as the mathematical equations for a rotational system can be written by analogy from the equations for a translational system, only the latter are discussed in detail. Whenever translational systems are discussed, it is understood that corresponding equations apply to the analogous rotational system, as indicated in Table 2.1.

SINGLE DEGREE-OF-FREEDOM SYSTEM

The simplest possible vibratory system is shown in Fig. 2.4; it consists of a mass m attached by means of a spring k to an immovable support. The mass is constrained to translational motion in the direction of the X axis so that its change of position from an initial reference is described fully by the value of a single quantity x. For this reason it is called a *single degree-of-freedom system*. If the mass m is displaced from its equilibrium position and then allowed to vibrate free from further external forces, it is said to have *free vibration*. The vibration also may be forced; i.e., a continuing force acts upon the mass or the foundation experiences a continuing motion. Free and forced vibration are discussed below.

FIG. 2.4. Undamped single degree-of-freedom system.

FREE VIBRATION WITHOUT DAMPING [1,2,3]

Considering first the free vibration of the undamped system of Fig. 2.4, Newton's equation is written for the mass m. The force $m\ddot{x}$ exerted by the mass on the spring is equal and opposite to the force kx applied by the spring on the mass:

$$m\ddot{x} + kx = 0 \qquad (2.4)$$

where $x = 0$ defines the equilibrium position of the mass.

The solution of Eq. (2.4) is

$$x = A \sin \sqrt{\frac{k}{m}}\, t + B \cos \sqrt{\frac{k}{m}}\, t \tag{2.5}$$

where the term $\sqrt{k/m}$ is the *angular natural frequency* defined by

$$\omega_n = \sqrt{\frac{k}{m}} \qquad \text{rad/sec} \tag{2.6}$$

The sinusoidal oscillation of the mass repeats continuously, and the time interval to complete one cycle is the *period:*

$$\tau = \frac{2\pi}{\omega_n} \tag{2.7}$$

The reciprocal of the period is the *natural frequency:*

$$f_n = \frac{1}{\tau} = \frac{\omega_n}{2\pi} = \frac{1}{2\pi}\sqrt{\frac{k}{m}} = \frac{1}{2\pi}\sqrt{\frac{kg}{W}} \tag{2.8}$$

where $W = mg$ is the weight of the rigid body forming the mass of the system shown in Fig. 2.4. The relations of Eq. (2.8) are shown by the solid lines in Fig. 2.5.

INITIAL CONDITIONS. In Eq. (2.5), B is the value of x at time $t = 0$, and the value of A is equal to $\dot{x}/\omega_n$ at time $t = 0$. Thus, the conditions of displacement and velocity which exist at zero time determine the subsequent oscillation completely.

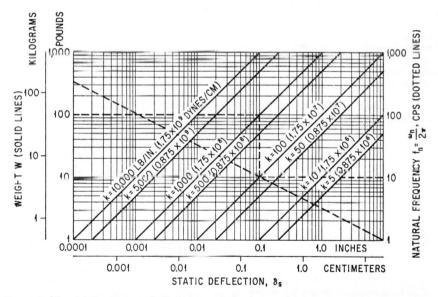

FIG. 2.5. Natural frequency relations for a single degree-of-freedom system. Relation of natural frequency to weight of supported body and stiffness of spring [Eq. (2.8)] is shown by solid lines. Relation of natural frequency to static deflection [Eq. (2.10)] is shown by diagonal-dashed line. Example: To find natural frequency of system with $W = 100$ lb and $k = 1,000$ lb/in., enter at $W = 100$ on left ordinate scale; follow the dashed line horizontally to solid line $k = 1,000$, then vertically down to diagonal-dashed line, and finally horizontally to read $f_n = 10$ cps from right ordinate scale.

PHASE ANGLE. Equation (2.5) for the displacement in oscillatory motion can be written, introducing the frequency relation of Eq. (2.6),

$$x = A \sin \omega_n t + B \cos \omega_n t = C \sin (\omega_n t + \theta) \qquad (2.9)$$

where $C = (A^2 + B^2)^{1/2}$ and $\theta = \tan^{-1}(B/A)$. The angle θ is called the *phase angle*.

STATIC DEFLECTION. The static deflection of a simple mass-spring system is the deflection of spring k as a result of the gravity force of the mass, $\delta_{st} = mg/k$. (For example, the system of Fig. 2.4 would be oriented with the mass m vertically above the spring k.) Substituting this relation in Eq. (2.8),

$$f_n = \frac{1}{2\pi} \sqrt{\frac{g}{\delta_{st}}} \qquad (2.10)$$

The relation of Eq. (2.10) is shown by the diagonal-dashed line in Fig. 2.5. This relation applies only when the system under consideration is both linear and elastic. For example, rubber springs tend to be nonlinear or exhibit a dynamic stiffness which differs from the static stiffness; hence, Eq. (2.10) is not applicable.

FIG. 2.6. Single degree-of-freedom system with viscous damper.

FREE VIBRATION WITH VISCOUS DAMPING [1,2,3]

Figure 2.6 shows a single degree-of-freedom system with a viscous damper. The differential equation of motion of mass m, corresponding to Eq. (2.4) for the undamped system, is

$$m\ddot{x} + c\dot{x} + kx = 0 \qquad (2.11)$$

The form of the solution of this equation depends upon whether the damping coefficient is equal to, greater than, or less than the *critical damping coefficient* c_c:

$$c_c = 2\sqrt{km} = 2m\omega_n \qquad (2.12)$$

The ratio $\zeta = c/c_c$ is defined as the *fraction of critical damping*.

LESS-THAN-CRITICAL-DAMPING. If the damping of the system is less than critical, $\zeta < 1$; then the solution of Eq. (2.11) is

$$x = e^{-ct/2m}(A \sin \omega_d t + B \cos \omega_d t)$$

$$= Ce^{-ct/2m} \sin (\omega_d t + \theta) \qquad (2.13)$$

where C and θ are defined with reference to Eq. (2.9). The *damped natural frequency* is related to the undamped natural frequency of Eq. (2.6) by the equation

$$\omega_d = \omega_n(1 - \zeta^2)^{1/2} \qquad \text{rad/sec} \quad (2.14)$$

Equation (2.14), relating the damped and undamped natural frequencies, is plotted in Fig. 2.7.

CRITICAL DAMPING. When $c = c_c$, there is no oscillation and the solution of Eq. (2.11) is

$$x = (A + Bt)e^{-ct/2m} \qquad (2.15)$$

FIG. 2.7. Damped natural frequency as a function of undamped natural frequency and fraction of critical damping.

GREATER-THAN-CRITICAL-DAMPING. When $\zeta > 1$, the solution of Eq. (2.11) is

$$x = e^{-ct/2m}(Ae^{\omega_n\sqrt{\zeta^2-1}\ t} + Be^{-\omega_n\sqrt{\zeta^2-1}\ t}) \qquad (2.16)$$

This is a nonoscillatory motion; if the system is displaced from its equilibrium position, it tends to return gradually.

LOGARITHMIC DECREMENT. The degree of damping in a system having $\zeta < 1$ may be defined in terms of successive peak values in a record of a free oscillation. Substituting the expression for critical damping from Eq. (2.12), the expression for free vibration of a damped system, Eq. (2.13), becomes

$$x = Ce^{-\zeta\omega_n t}\sin(\omega_d t + \theta) \qquad (2.17)$$

Consider any two maxima (i.e., value of x when $dx/dt = 0$) separated by n cycles of oscillation, as shown in Fig. 2.8. Then the ratio of these maxima is

$$\frac{x_n}{x_0} = e^{-2\pi n\zeta/(1-\zeta^2)^{\frac{1}{2}}} \qquad (2.18)$$

Values of x_n/x_0 are plotted in Fig. 2.9 for several values of n over the range of ζ from 0.001 to 0.10.

The *logarithmic decrement* Δ is the natural logarithm of the ratio of the amplitudes of two successive cycles of the damped free vibration:

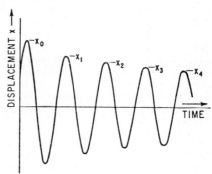

FIG. 2.8. Trace of damped free vibration showing amplitudes of displacement maxima.

$$\Delta = \ln\frac{x_1}{x_2} \quad \text{or} \quad \frac{x_2}{x_1} = e^{-\Delta} \qquad (2.19)$$

[Also see Eq. (37.6).] A comparison of this relation with Eq. (2.18) when $n = 1$ gives the following expression for Δ:

$$\Delta = \frac{2\pi\zeta}{(1 - \zeta^2)^{\frac{1}{2}}} \qquad (2.20)$$

The logarithmic decrement can be expressed in terms of the difference of successive amplitudes by writing Eq. (2.19) as follows:

$$\frac{x_1 - x_2}{x_1} = 1 - \frac{x_2}{x_1} = 1 - e^{-\Delta}$$

Writing $e^{-\Delta}$ in terms of its infinite series, the following expression is obtained which gives a good approximation for $\Delta < 0.2$:

$$\frac{x_1 - x_2}{x_1} = \Delta \qquad (2.21)$$

For small values of ζ (less than about 0.10), an approximate relation between

FIG. 2.9. Effect of damping upon the ratio of displacement maxima of a damped free vibration.

the fraction of critical damping and the logarithmic decrement, from Eq. (2.20), is

$$\Delta \simeq 2\pi\zeta \qquad (2.22)$$

FORCED VIBRATION

Forced vibration in this chapter refers to the motion of the system which occurs in response to a continuing excitation whose magnitude varies sinusoidally with time. (See Chaps. 8 and 23 for a treatment of the response of a simple system to step, pulse, and transient vibration excitations.) The excitation may be, alternatively, force applied to the system (generally, the force is applied to the mass of a single degree-of-freedom system) or motion of the foundation that supports the system. The resulting response of the system can be expressed in different ways, depending upon the nature of the excitation and the use to be made of the result.

1. If the excitation is a force applied to the mass of the system shown in Fig. 2.4, the result may be expressed in terms of (*a*) the amplitude of the resulting motion of the mass or (*b*) the fraction of the applied force amplitude that is transmitted through the system to the support. The former is termed the *motion response* and the latter is termed the *force transmissibility*.

2. If the excitation is a motion of the foundation, the resulting response usually is expressed in terms of the amplitude of the motion of the mass relative to the amplitude of the motion of the foundation. This is termed the *motion transmissibility* for the system.

In general, the response and transmissibility relations are functions of the forcing frequency, and vary with different types and degrees of damping. Results are presented in this chapter for undamped systems, and for systems with either viscous or structural damping. Corresponding results are given in Chap. 30 for systems with Coulomb damping, and for systems with either viscous or Coulomb damping in series with a linear spring.

FORCED VIBRATION WITHOUT DAMPING

FORCE APPLIED TO MASS. When the sinusoidal force $F = F_0 \sin \omega t$ is applied to the mass of the undamped single degree-of-freedom system shown in Fig. 2.10, the differential equation of motion is

$$m\ddot{x} + kx = F_0 \sin \omega t \qquad (2.23)$$

The solution of this equation is

$$x = A \sin \omega_n t + B \cos \omega_n t + \frac{F_0/k}{1 - \omega^2/\omega_n^{\cdot\cdot}} \sin \omega t \qquad (2.24)$$

FIG. 2.10. Undamped single degree-of-freedom system excited in forced vibration by force acting on mass.

where $\omega_n = \sqrt{k/m}$. The first two terms represent an oscillation at the undamped natural frequency ω_n. The coefficient B is the value of x at time $t = 0$, and the coefficient A may be found from the velocity at time $t = 0$. Differentiating Eq. (2.24) and setting $t = 0$,

$$\dot{x}(0) = A\omega_n + \frac{\omega F_0/k}{1 - \omega^2/\omega_n^2} \qquad (2.25)$$

The value of A is found from Eq. (2.25).

The oscillation at the natural frequency ω_n gradually decays to zero in physical systems because of damping. The steady-state oscillation at forcing frequency ω is

$$x = \frac{F_0/k}{1 - \omega^2/\omega_n^2} \sin \omega t \qquad (2.26)$$

This oscillation exists after a condition of equilibrium has been established by decay of the oscillation at the natural frequency ω_n and persists as long as the force F is applied.

The force transmitted to the foundation is directly proportional to the spring deflection: $F_t = kx$. Substituting x from Eq. (2.26) and defining transmissibility $T = F_t/F$,

$$T = \frac{1}{1 - \omega^2/\omega_n^2} \qquad (2.27)$$

If the mass is initially at rest in the equilibrium position of the system (i.e., $x = 0$ and $\dot{x} = 0$) at time $t = 0$, the ensuing motion at time $t > 0$ is

$$x = \frac{F_0/k}{1 - \omega^2/\omega_n^2}\left(\sin \omega t - \frac{\omega}{\omega_n}\sin \omega_n t\right) \qquad (2.28)$$

FIG. 2.11. Undamped single degree-of-freedom system excited in forced vibration by motion of foundation.

For large values of time, the second term disappears because of the damping inherent in any physical system, and Eq. (2.28) becomes identical to Eq. (2.26).

When the forcing frequency coincides with the natural frequency, $\omega = \omega_n$ and a condition of resonance exists. Then Eq. (2.28) is indeterminate and the expression for x may be written as

$$x = -\frac{F_0\omega}{2k}\,t\cos\omega t + \frac{F_0}{2k}\sin\omega t \qquad (2.29)$$

According to Eq. (2.29), the amplitude x increases continuously with time, reaching an infinitely great value only after an infinitely great time.

MOTION OF FOUNDATION. The differential equation of motion for the system of Fig. 2.11 excited by a continuing motion $u = u_0 \sin \omega t$ of the foundation is

$$m\ddot{x} = -k(x - u_0 \sin \omega t)$$

The solution of this equation is

$$x = A_1 \sin \omega_n t + B_2 \cos \omega_n t + \frac{u_0}{1 - \omega^2/\omega_n^2}\sin \omega t$$

where $\omega_n = \sqrt{k/m}$ and the coefficients A_1, B_1 are determined by the velocity and displacement of the mass, respectively, at time $t = 0$. The terms representing oscillation at the natural frequency are damped out ultimately, and the ratio of amplitudes is defined in terms of transmissibility T:

$$\frac{x_0}{u_0} = T = \frac{1}{1 - \omega^2/\omega_n^2} \qquad (2.30)$$

where $x = x_0 \sin \omega t$. Thus, in the forced vibration of an undamped single degree-of-freedom system, the motion response, the force transmissibility, and the motion transmissibility are numerically equal.

FIG. 2.12. Single degree-of-freedom system with viscous damping, excited in forced vibration by force acting on mass.

FORCED VIBRATION WITH VISCOUS DAMPING

FORCE APPLIED TO MASS. The differential equation of motion for the single degree-of-freedom system with viscous damping shown in Fig. 2.12, when the excitation is a force $F = F_0 \sin \omega t$ applied to the mass, is

$$m\ddot{x} + c\dot{x} + kx = F_0 \sin \omega t \qquad (2.31)$$

Equation (2.31) corresponds to Eq. (2.23) for forced vibration of an undamped system; its solution would correspond to Eq. (2.24) in that it includes terms representing oscillation at the natural frequency. In a damped system, however, these terms are damped out rapidly and only the steady-state solution usually is considered. The resulting motion occurs at the forcing frequency ω; when the damping coefficient c is greater than zero, the phase between the force and resulting motion is different than zero. Thus, the

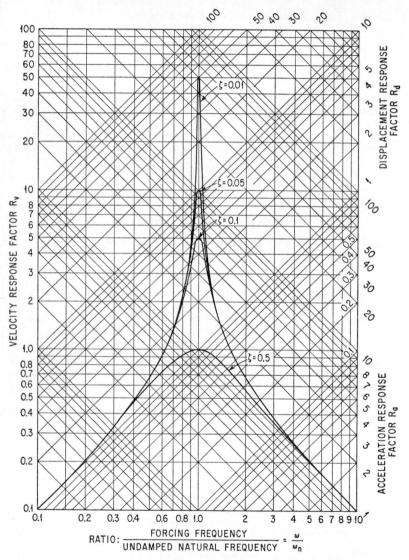

FIG. 2.13. Response factors for a viscous-damped single degree-of-freedom system excited in forced vibration by a force acting on the mass. The velocity response factor shown by horizontal lines is defined by Eq. (2.36); the displacement response factor shown by diagonal lines of positive slope is defined by Eq. (2.33); and the acceleration response factor shown by diagonal lines of negative slope is defined by Eq. (2.37).

response may be written

$$x = R \sin(\omega t - \theta) = A_1 \sin \omega t + B_1 \cos \omega t \tag{2.32}$$

Substituting this relation in Eq. (2.31), the following result is obtained:

$$\frac{x}{F_0/k} = \frac{\sin(\omega t - \theta)}{\sqrt{(1 - \omega^2/\omega_n^2)^2 + (2\zeta\omega/\omega_n)^2}} = R_d \sin(\omega t - \theta) \tag{2.33}$$

where

$$\theta = \tan^{-1}\left(\frac{2\zeta\omega/\omega_n}{1 - \omega^2/\omega_n^2}\right)$$

and R_d is a *dimensionless response factor* giving the ratio of the amplitude of the vibratory displacement to the spring displacement that would occur if the force F were applied statically. At very low frequencies R_d is approximately equal to 1; it rises to a peak near ω_n and approaches zero as ω becomes very large. The displacement response is defined at these frequency conditions as follows:

$$x \simeq \left(\frac{F_0}{k}\right) \sin \omega t \qquad [\omega \ll \omega_n]$$

$$x = \frac{F_0}{2k\zeta} \sin\left(\omega_n t + \frac{\pi}{2}\right) = -\frac{F_0 \cos \omega_n t}{c\omega_n} \qquad [\omega = \omega_n] \tag{2.34}$$

$$x \simeq \frac{\omega_n^2 F_0}{\omega^2 k} \sin(\omega t + \pi) = \frac{F_0}{m\omega^2} \sin \omega t \qquad [\omega \gg \omega_n]$$

For the above three frequency conditions, the vibrating system is sometimes described as *spring-controlled*, *damper-controlled*, and *mass-controlled*, respectively, depending on which element is primarily responsible for the system behavior.

Curves showing the dimensionless response factor R_d as a function of the frequency ratio ω/ω_n are plotted in Fig. 2.13 on the coordinate lines having a positive 45° slope. Curves of the phase angle θ are plotted in Fig. 2.14. A phase angle between 180 and 360°

FIG. 2.14. Phase angle between the response displacement and the excitation force for a single degree-of-freedom system with viscous damping, excited by a force acting on the mass of the system.

cannot exist in this case since this would mean that the damper is furnishing energy to the system rather than dissipating it.

An alternative form of Eqs. (2.33) and (2.34) is

$$\frac{x}{F_0/k} = \frac{(1 - \omega^2/\omega_n^2)\sin\omega t - 2\zeta(\omega/\omega_n)\cos\omega t}{(1 - \omega^2/\omega_n^2)^2 + (2\zeta\omega/\omega_n)^2} \tag{2.35}$$

$$= (R_d)_x \sin\omega t + (R_d)_R \cos\omega t$$

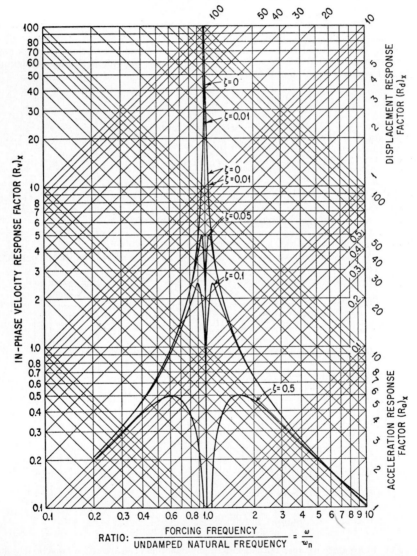

FIG. 2.15. In-phase component of response factor of a viscous-damped system in forced vibration. All values of the response factor for $\omega/\omega_n > 1$ are negative but are plotted without regard for sign. The fraction of critical damping is denoted by ζ.

This shows the components of the response which are in phase $[(R_d)_x \sin \omega t]$ and $90°$ out of phase $[(R_d)_R \cos \omega t]$ with the force. Curves of $(R_d)_x$ and $(R_d)_R$ are plotted as a function of the frequency ratio ω/ω_n in Figs. 2.15 and 2.16.

Velocity and Acceleration Response. The shape of the response curves changes distinctly if velocity $\dot{x}$ or acceleration $\ddot{x}$ is plotted instead of displacement x. Differentiating Eq. (2.33),

$$\frac{\dot{x}}{F_0/\sqrt{km}} = \frac{\omega}{\omega_n} R_d \cos(\omega t - \theta) = R_v \cos(\omega t - \theta) \qquad (2.36)$$

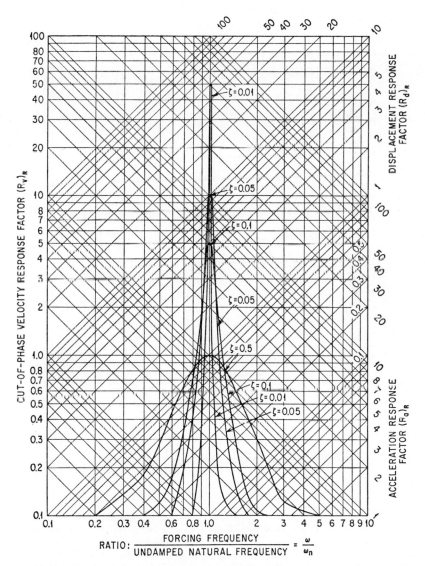

Fig. 2.16. Out-of-phase component of response factor of a viscous-damped system in forced vibration. The fraction of critical damping is denoted by ζ.

The acceleration response is obtained by differentiating Eq. (2.36):

$$\frac{\ddot{x}}{F_0/m} = -\frac{\omega^2}{\omega_n{}^2} R_d \sin(\omega t - \theta) = -R_a \sin(\omega t - \theta) \tag{2.37}$$

The velocity and acceleration response factors defined by Eqs. (2.36) and (2.37) are shown graphically in Fig. 2.13, the former to the horizontal coordinates and the latter to the coordinates having a negative 45° slope. Note that the velocity response factor approaches zero as $\omega \to 0$ and $\omega \to \infty$, whereas the acceleration response factor approaches 0 as $\omega \to 0$ and approaches unity as $\omega \to \infty$.

FIG. 2.17. Transmissibility of a viscous-damped system. Force transmissibility and motion transmissibility are identical numerically. The fraction of critical damping is denoted by ζ.

Force Transmission. The force transmitted to the foundation of the system is

$$F_T = c\dot{x} + kx \tag{2.38}$$

Since the forces $c\dot{x}$ and kx are 90° out of phase, the magnitude of the transmitted force is

$$|F_T| = \sqrt{c^2\dot{x}^2 + k^2x^2} \tag{2.39}$$

The ratio of the transmitted force F_T to the applied force F_0 can be expressed in terms of transmissibility T:

$$\frac{F_T}{F_0} = T \sin(\omega t - \psi) \tag{2.40}$$

where

$$T = \sqrt{\frac{1 + (2\zeta\omega/\omega_n)^2}{(1 - \omega^2/\omega_n^2)^2 + (2\zeta\omega/\omega_n)^2}} \tag{2.41}$$

and

$$\psi = \tan^{-1}\frac{2\zeta(\omega/\omega_n)^3}{1 - \omega^2/\omega_n^2 + 4\zeta^2\omega^2/\omega_n^2}$$

The transmissibility T and phase angle ψ are shown in Figs. 2.17 and 2.18, respectively, as a function of the frequency ratio ω/ω_n and for several values of the fraction of critical damping ζ.

Hysteresis. When the viscous damped, single degree-of-freedom system shown in Fig. 2.12 undergoes vibration defined by

$$x = x_0 \sin \omega t \tag{2.42}$$

the net force exerted on the mass by the spring and damper is

$$F = kx_0 \sin \omega t + c\omega x_0 \cos \omega t \tag{2.43}$$

Equations (2.42) and (2.43) define the relation between F and x; this relation is the ellipse

Fig. 2.18. Phase angle of force transmission (or motion transmission) of a viscous-damped system excited (1) by force acting on mass and (2) by motion of foundation. The fraction of critical damping is denoted by ζ.

FIG. 2.19. Hysteresis curve for a spring and viscous damper in parallel.

shown in Fig. 2.19. The energy dissipated in one cycle of oscillation is

$$W = \int_{T}^{T+2\pi/\omega} F \frac{dx}{dt}\, dt = \pi c \omega x_0^2 \quad (2.44)$$

MOTION OF FOUNDATION. The excitation for the elastic system shown in Fig. 2.20 may be a motion $u(t)$ of the foundation. The differential equation of motion for the system is

$$m\ddot{x} + c(\dot{x} - \dot{u}) + k(x - u) = 0 \quad (2.45)$$

Consider the motion of the foundation to be a displacement that varies sinusoidally with time, $u = u_0 \sin \omega t$. A steady-state condition exists after the oscillations at the natural frequency ω_n are damped out, defined by the displacement x of mass m:

$$x = T u_0 \sin(\omega t - \psi) \quad (2.46)$$

where T and ψ are defined in connection with Eq. (2.40) and are shown graphically in Figs. 2.17 and 2.18, respectively. Thus, the motion transmissibility T in Eq. (2.46) is identical numerically to the force transmissibility T in Eq. (2.40). The motion of the foundation and of the mass m may be expressed in any consistent units, such as displacement, velocity, or acceleration, and the same expression for T applies in each case.

FIG. 2.20. Single degree-of-freedom system with viscous damper, excited in forced vibration by foundation motion.

FIG. 2.21. Single degree-of-freedom system with viscous damper, excited in forced vibration by rotating eccentric weight.

VIBRATION DUE TO A ROTATING ECCENTRIC WEIGHT. In the mass-spring-damper system shown in Fig. 2.21, a mass m_u is mounted by a shaft and bearings to the mass m. The mass m_u follows a circular path of radius e with respect to the bearings. The component of displacement in the X direction of m_u relative to m is

$$x_3 - x_1 = e \sin \omega t \quad (2.47)$$

where x_3 and x_1 are the absolute displacements of m_u and m, respectively, in the X direction; e is the length of the arm supporting the mass m_u; and ω is the angular velocity of the arm in radians per second. The differential equation of motion for the system is

$$m\ddot{x}_1 + m_u\ddot{x}_3 + c\dot{x}_1 + kx_1 = 0 \quad (2.48)$$

Differentiating Eq. (2.47) with respect to time, solving for $\ddot{x}_3$, and substituting in Eq. (2.48):

$$(m + m_u)\ddot{x}_1 + c\dot{x}_1 + kx_1 = m_u e \omega^2 \sin \omega t \quad (2.49)$$

Equation (2.49) is of the same form as Eq. (2.31); thus, the response relations of Eqs. (2.33), (2.36), and (2.37) apply by substituting $(m + m_u)$ for m and $m_u e \omega^2$ for F_0. The resulting displacement, velocity, and acceleration responses are

$$\frac{x_1}{m_u e \omega^2} = R_d \sin (\omega t - \theta) \qquad \frac{\dot{x}_1 \sqrt{km}}{m_u e \omega^2} = R_v \cos (\omega t - \theta)$$

(2.50)

$$\frac{\ddot{x}_1 m}{m_u e \omega^2} = -R_a \sin (\omega t - \theta)$$

RESONANT FREQUENCIES. The peak values of the displacement, velocity, and acceleration response of a system undergoing forced, steady-state vibration occur at slightly different forcing frequencies. Since a *resonant frequency* is defined as the frequency for which the response is a maximum, a simple system has three resonant frequencies if defined only generally. The natural frequency is different from any of the resonant frequencies. The relations among the several resonant frequencies, the damped natural frequency, and the undamped natural frequency ω_n are:

Displacement resonant frequency: $\omega_n (1 - 2\zeta^2)^{1/2}$
Velocity resonant frequency: ω_n
Acceleration resonant frequency: $\omega_n / (1 - 2\zeta^2)^{1/2}$
Damped natural frequency: $\omega_n (1 - \zeta^2)^{1/2}$

For the degree of damping usually embodied in physical systems, the difference among the three resonant frequencies is negligible.

RESONANCE, BANDWIDTH, AND THE QUALITY FACTOR Q. Damping in a system can be determined by noting the maximum response, i.e., the response at the resonant frequency as indicated by the maximum value of R_v in Eq. (2.36). This is defined by the factor Q sometimes used in electrical engineering terminology and defined with respect to mechanical vibration as

$$Q = (R_v)_{\text{max}} = 1/2\zeta$$

The maximum acceleration and displacement responses are slightly larger, being

$$(R_d)_{\text{max}} = (R_a)_{\text{max}} = \frac{(R_v)_{\text{max}}}{(1 - \zeta^2)^{1/2}}$$

The damping in a system is also indicated by the sharpness or width of the response curve in the vicinity of a resonant frequency ω_n. Designating the width as a frequency increment $\Delta\omega$ measured at the "half-power point" (i.e., at a value of R equal to $R_{\text{max}}/\sqrt{2}$), as illustrated in Fig. 2.22, the damping of the system is defined to a good approximation by

$$\frac{\Delta\omega}{\omega_n} = \frac{1}{Q} = 2\zeta \qquad (2.51)$$

for values of ζ less than 0.1. The quantity $\Delta\omega$, known as the *bandwidth*, is commonly represented by the letter B.

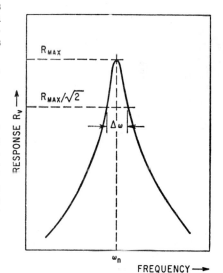

FIG. 2.22. Response curve showing bandwidth at "half-power point."

STRUCTURAL DAMPING. The energy dissipated by the damper is known as *hysteresis loss*; as indicated by Eq. (2.44), it is proportional to the forcing frequency ω. However, the hysteresis loss of many engineering structures has been found to be independent of frequency. To provide a better model for defining the *structural damping* experienced during vibration, an arbitrary damping term $k\mathfrak{g} = c\omega$ is introduced. In effect, this defines the damping force as being equal to the viscous damping force at some frequency, depending upon the value of $\mathfrak{g}$, but being invariant with frequency. The relation of the damping force F to the displacement x is defined by an ellipse similar to Fig. 2.19, and the displacement response of the system is described by an expression corresponding to Eq. (2.33) as follows:

$$\frac{x}{F_0/k} = R_g \sin (\omega t - \theta) = \frac{\sin (\omega t - \theta)}{\sqrt{(1 - \omega^2/\omega_n^2)^2 + \mathfrak{g}^2}} \tag{2.52}$$

where $\mathfrak{g} = 2\zeta\omega/\omega_n$. The resonant frequency is ω_n, and the value of R_g at resonance is $1/\mathfrak{g} = Q$.

The equations for the hysteresis ellipse for structural damping are

$$F = kx_0 (\sin \omega t + \mathfrak{g} \cos \omega t)$$
$$x = x_0 \sin \omega t \tag{2.53}$$

UNDAMPED MULTIPLE DEGREE-OF-FREEDOM SYSTEMS

An elastic system sometimes cannot be described adequately by a model having only one mass but rather must be represented by a system of two or more masses considered to be *point masses* or *particles* having no rotational inertia. If a group of particles is bound together by essentially rigid connections, it behaves as a rigid body having both mass (significant for translational motion) and moment of inertia (significant for rotational motion). There is no limit to the number of masses that may be used to represent a system. For example, each mass in a model representing a beam may be an infinitely thin slice representing a cross section of the beam; a differential equation is required to treat this continuous distribution of mass.

DEGREES-OF-FREEDOM

The number of independent parameters required to define the distance of all the masses from their reference positions is called the number of *degrees-of-freedom N*. For example, if there are N masses in a system constrained to move only in translation in the X and Y directions, the system has $2N$ degrees-of-freedom. A continuous system such as a beam has an infinitely large number of degrees-of-freedom.

For each degree-of-freedom (each coordinate of motion of each mass) a differential equation can be written in one of the following alternative forms:

$$m_j\ddot{x}_j = F_{xj} \qquad I_k\ddot{\alpha}_k = M_{\alpha k} \tag{2.54}$$

where F_{xj} is the component in the X direction of all external, spring, and damper forces acting on the mass having the jth degree-of-freedom, and $M_{\alpha k}$ is the component about the α axis of all torques acting on the body having the kth degree-of-freedom. The moment of inertia of the mass about the α axis is designated by I_k. (This is assumed for the present analysis to be a principal axis of inertia, and product of inertia terms are neglected. See Chap. 3 for a more detailed discussion.) Equations (2.54) are identical in form and can be represented by

$$m_j\ddot{x}_j = F_j \tag{2.55}$$

where F_j is the resultant of all forces (or torques) acting on the system in the jth degree-of-freedom, $\ddot{x}_j$ is the acceleration (translational or rotational) of the system in the jth degree-of-freedom, and m_j is the mass (or moment of inertia) in the jth degree-of-freedom. Thus, the terms defining the motion of the system (displacement, velocity, and acceleration) and the deflections of structures may be either translational or rotational, depending upon the type of coordinate. Similarly, the "force" acting on a system may be either a force or a torque, depending upon the type of coordinate. For example, if a system has n bodies each free to move in three translational modes and three rotational modes, there would be $6n$ equations of the form of Eq. (2.55), one for each degree-of-freedom.

DEFINING A SYSTEM AND ITS EXCITATION

The first step in analyzing any physical structure is to represent it by a mathematical model which will have essentially the same dynamic behavior. A suitable number and distribution of masses, springs, and dampers must be chosen, and the input forces or foundation motions must be defined. The model should have sufficient degrees-of-freedom to determine the modes which will have significant response to the exciting force or motion.

The properties of a system that must be known are the natural frequencies ω_n, the normal mode shapes D_{jn}, the damping of the respective modes, and the mass distribution m_j. The detailed distributions of stiffness and damping of a system are not used directly but rather appear indirectly as the properties of the respective modes. The characteristic properties of the modes may be determined experimentally as well as analytically.

STIFFNESS COEFFICIENTS

The spring system of a structure of N degrees-of-freedom can be defined completely by a set of N^2 *stiffness coefficients*.[5] A stiffness coefficient K_{jk} is the change in spring force acting on the jth degree-of-freedom when only the kth degree-of-freedom is slowly displaced a unit amount in the negative direction. This definition is a generalization of the linear, elastic spring defined by Eq. (2.1). Stiffness coefficients have the characteristic of reciprocity, i.e., $K_{jk} = K_{kj}$. The number of independent stiffness coefficients is $(N^2 + N)/2$.

The total elastic force acting on the jth degree-of-freedom is the sum of the effects of the displacements in all of the degrees-of-freedom:

$$F_{el} = -\sum_{k=1}^{N} K_{jk}x_k \tag{2.56}$$

Inserting the spring force F_{el} from Eq. (2.56) in Eq. (2.55) together with the external forces F_j results in the n equations:

$$m_j\ddot{x}_j = F_j - \sum_k K_{jk}x_k \tag{2.56a}$$

FREE VIBRATION

When the external forces are zero, the preceding equations become

$$m_j\ddot{x}_j + \sum_k K_{jk}x_k = 0 \tag{2.57}$$

Solutions of Eq. (2.57) have the form

$$x_j = D_j \sin(\omega t + \theta) \tag{2.58}$$

Substituting Eq. (2.58) in Eq. (2.57),

$$m_j\omega^2 D_j = \sum_k K_{jk}D_k \tag{2.59}$$

This is a set of n linear algebraic equations with n unknown values of D. A solution of these equations for values of D other than zero can be obtained only if the determinant of the coefficients of the D's is zero:

$$\begin{vmatrix} (m_1\omega^2 - K_{11}) & -K_{12} & \cdot & \cdot & -K_{in} \\ -K_{21} & (m_2\omega^2 - K_{22}) & \cdot & \cdot & \cdot \\ \cdot & \cdot & \cdot & \cdot & \cdot \\ \cdot & \cdot & \cdot & \cdot & \cdot \\ -K_{ni} & \cdot & \cdot & \cdot & (m_n\omega^2 - K_{nn}) \end{vmatrix} = 0 \tag{2.60}$$

Equation (2.60) is an algebraic equation of the nth degree in ω^2; it is called the *frequency equation* since it defines n values of ω which satisfy Eq. (2.57). The roots are all real; some may be equal, and others may be zero. These values of frequency determined from Eq. (2.60) are the frequencies at which the system can oscillate in the absence of external forces. These frequencies are the *natural frequencies* ω_n of the system. Depending upon the initial conditions under which vibration of the system is initiated, the oscillations may occur at any or all of the natural frequencies and at any amplitude.

Example 2.1. Consider the three degree-of-freedom system shown in Fig. 2.23; it consists of three equal masses m and a foundation connected in series by three equal springs k. The absolute displacements of the masses are x_1, x_2, and x_3. The stiffness coefficients (see section entitled *Stiffness Coefficients*) are thus $K_{11} = 2k$, $K_{22} = 2k$, $K_{33} = k$, $K_{12} = K_{21} = -k$, $K_{23} =$

FIG. 2.23. Undamped three degree-of-freedom system on foundation.

$K_{32} = -k$, and $K_{13} = K_{31} = 0$. The frequency equation is given by the determinant, Eq. (2.60),

$$\begin{vmatrix} (m\omega^2 - 2k) & k & 0 \\ k & (m\omega^2 - 2k) & k \\ 0 & k & (m\omega^2 - k) \end{vmatrix} = 0$$

The determinant expands to the following polynomial:

$$\left(\frac{m\omega^2}{k}\right)^3 - 5\left(\frac{m\omega^2}{k}\right)^2 + 6\left(\frac{m\omega^2}{k}\right) - 1 = 0$$

Solving for ω,

$$\omega = 0.445\sqrt{\frac{k}{m}}, \quad 1.25\sqrt{\frac{k}{m}}, \quad 1.80\sqrt{\frac{k}{m}}$$

NORMAL MODES OF VIBRATION. A structure vibrating at only one of its natural frequencies ω_n does so with a characteristic pattern of amplitude distribution called a *normal mode of vibration*.[3,4,5] A normal mode is defined by a set of values of D_{jn} [see Eq. (2.58)] which satisfy Eq. (2.59) when $\omega = \omega_n$:

$$\omega_n{}^2 m_j D_{jn} = \sum_k K_{jn} D_{kn} \qquad (2.61)$$

A set of values of D_{jn} which form a normal mode is independent of the absolute values of D_{jn} but depends only on their relative values. To define a mode shape by a unique set of numbers, any arbitrary *normalizing condition* which is desired can be used. A condition often used is to set $D_{1n} = 1$ but $\sum_j m_j D_{jn}{}^2 = 1$ and $\sum_j m_j D_{jn}{}^2 = \sum_j m_j$ also may be found convenient.

ORTHOGONALITY OF NORMAL MODES. The usefulness of normal modes in dealing with multiple degree-of-freedom systems is due largely to the orthogonality of the normal modes. It can be shown [3,4,5] that the set of inertia forces $\omega_n{}^2 m_j D_{jn}$ for one mode does no work on the set of deflections D_{jm} of another mode of the structure:

$$\sum_j m_j D_{jm} D_{jn} = 0 \qquad [m \neq n] \qquad (2.62)$$

This is the *orthogonality condition*.

NORMAL MODES AND GENERALIZED COORDINATES. Any set of N deflections x_j can be expressed as the sum of normal mode amplitudes:

$$x_j = \sum_{n=1}^{N} q_n D_{jn} \tag{2.63}$$

The numerical values of the D_{jn}'s are fixed by some normalizing condition, and a set of values of the N variables q_n can be found to match any set of x_j's. The N values of q_n constitute a set of *generalized coordinates* which can be used to define the position coordinates x_j of all parts of the structure. The q's are also known as the amplitudes of the normal modes, and are functions of time. Equation (2.63) may be differentiated to obtain

$$\ddot{x}_j = \sum_{n=1}^{N} \ddot{q}_n D_{jn} \tag{2.64}$$

Any quantity which is distributed over the j coordinates can be represented by a linear transformation similar to Eq. (2.63). It is convenient now to introduce the parameter γ_n relating D_{jn} and F_j/m_j as follows:

$$\frac{F_j}{m_j} = \sum_{n} \gamma_n D_{jn} \tag{2.65}$$

where F_j may be zero for certain values of n.

FORCED MOTION

Substituting the expressions in generalized coordinates, Eqs. (2.63) to (2.65), in the basic equation of motion, Eq. (2.56a),

$$m_j \sum_{n} \ddot{q}_n D_{jn} + \sum_{k} k_{jk} \sum_{n} q_n D_{kn} - m_j \sum_{n} \gamma_n D_{jn} = 0 \tag{2.66}$$

The center term in Eq. (2.66) may be simplified by applying Eq. (2.61) and the equation rewritten as follows:

$$\sum_{n} (\ddot{q}_n + \omega_n^2 q_n - \gamma_n) m_j D_{jn} = 0 \tag{2.67}$$

Multiplying Eqs. (2.67) by D_{jm} and taking the sum over j (i.e., adding all the equations together),

$$\sum_{n} (\ddot{q}_n + \omega_n^2 q_n - \gamma_n) \sum_{j} m_j D_{jn} D_{jm} = 0$$

All terms of the sum over n are zero, except for the term for which $m = n$, according to the orthogonality condition of Eq. (2.62). Then since $\sum_{j} m_j D_{jn}^2$ is not zero, it follows that

$$\ddot{q}_n + \omega_n^2 q_n - \gamma_n = 0$$

for every value of n from 1 to N.

An expression for γ_n may be found by using the orthogonality condition again. Multiplying Eq. (2.65) by $m_j D_{jm}$ and taking the sum taken over j,

$$\sum_{j} F_j D_{jm} = \sum_{n} \gamma_n \sum_{j} m_j D_{jn} D_{jm} \tag{2.68}$$

All the terms of the sum over n are zero except when $n = m$, according to Eq. (2.62), and Eq. (2.68) reduces to

$$\gamma_n = \frac{\sum_{j} F_j D_{jn}}{\sum_{j} m_j D_{jn}^2} \tag{2.69}$$

Then the differential equation for the response of any generalized coordinate to the externally applied forces F_j is

$$\ddot{q}_n + \omega_n{}^2 q_n = \gamma_n = \frac{\sum_j F_j D_{jn}}{\sum_j m_j D_{jn}{}^2} \qquad (2.70)$$

where $\Sigma F_j D_{jn}$ is the generalized force, i.e., the total work done by all external forces during a small displacement δq_n divided by δq_n, and $\Sigma m_j D_{jn}{}^2$ is the generalized mass.

Thus the amplitude q_n of each normal mode is governed by its own equation, independent of the other normal modes, and responds as a simple mass-spring system. Equation (2.70) is a generalized form of Eq. (2.23).

The forces F_j may be any functions of time. Any equation for the response of an undamped mass-spring system applies to each mode of a complex structure by substituting:

The *generalized coordinate* q_n for x

The *generalized force* $\sum_j F_j D_{jn}$ for F $\qquad\qquad$ (2.71)

The *generalized mass* $\sum_j m_j D_{jn}$ for m

The *mode natural frequency* ω_n for ω_n

RESPONSE TO SINUSOIDAL FORCES. If a system is subjected to one or more sinusoidal forces $F_j = F_{0j} \sin \omega t$, the response is found from Eq. (2.26) by noting that $k = m\omega_n{}^2$ [Eq. (2.6)] and then substituting from Eq. (2.71):

$$q_n = \frac{\sum_j F_{0j} D_{jn}}{\omega_n{}^2 \sum_j m_j D_{jn}{}^2} \frac{\sin \omega t}{(1 - \omega^2/\omega_n{}^2)} \qquad (2.72)$$

Then the displacement of the kth degree-of-freedom, from Eq. (2.63), is

$$x_k = \sum_{n=1}^{N} \frac{D_{kn} \sum_j F_{0j} D_{jn} \sin \omega t}{\omega_n{}^2 \sum_j m_j D_{jn}{}^2 (1 - \omega^2/\omega_n{}^2)} \qquad (2.73)$$

This is the general equation for the response to sinusoidal forces of an undamped system of N degrees-of-freedom. The application of the equation to systems free in space or attached to immovable foundations is discussed below.

Example 2.2. Consider the system shown in Fig. 2.24; it consists of 3 equal masses m connected in series by 2 equal springs k. The system is free in space and a force $F \sin \omega t$ acts on the first mass. Absolute displacements of the masses are x_1, x_2, and x_3. Determine the ac-

FIG. 2.24. Undamped three degree-of-freedom system acted on by sinusoidal force.

celeration $\ddot{x}_3$. The stiffness coefficients (see section entitled *Stiffness Coefficients*) are $K_{11} = K_{33} = k$, $K_{22} = 2k$, $K_{12} = K_{21} = -k$, $K_{13} = K_{31} = 0$, and $K_{23} = K_{32} = -k$. Substituting in Eq. (2.60), the frequency equation is

$$\begin{vmatrix} (m\omega^2 - k) & k & 0 \\ k & (m\omega^2 - 2k) & k \\ 0 & k & (m\omega^2 - k) \end{vmatrix} = 0$$

The roots are $\omega_1 = 0$, $\omega_2 = \sqrt{k/m}$, and $\omega_3 = \sqrt{3k/m}$. The zero value for one of the natural frequencies indicates that the entire system translates without deflection of the springs. The mode shapes are now determined by substituting from Eq. (2.58) in Eq. (2.57), noting that $\ddot{x} = -D\omega^2$, and writing Eq. (2.59) for each of the three masses in each of the oscillatory modes 2 and 3:

$$mD_{21}\left(\frac{k}{m}\right) = K_{11}D_{21} + K_{21}D_{22} + K_{31}D_{23}$$

$$mD_{22}\left(\frac{k}{m}\right) = K_{12}D_{21} + K_{22}D_{22} + K_{32}D_{23}$$

$$mD_{23}\left(\frac{k}{m}\right) = K_{13}D_{21} + K_{23}D_{22} + K_{33}D_{23}$$

$$mD_{31}\left(\frac{3k}{m}\right) = K_{11}D_{31} + K_{21}D_{32} + K_{31}D_{33}$$

$$mD_{32}\left(\frac{3k}{m}\right) = K_{12}D_{31} + K_{22}D_{32} + K_{32}D_{33}$$

$$mD_{33}\left(\frac{3k}{m}\right) = K_{13}D_{31} + K_{23}D_{32} + K_{33}D_{33}$$

where the first subscript on the D's indicates the mode number (according to ω_1 and ω_2 above) and the second subscript indicates the displacement amplitude of the particular mass. The values of the stiffness coefficients K are calculated above. The mode shapes are defined by the relative displacements of the masses. Thus, assigning values of unit displacement to the first mass (i.e., $D_{21} = D_{31} = 1$), the above equations may be solved simultaneously for the D's:

$$D_{21} = 1 \qquad D_{22} = 0 \qquad D_{23} = -1$$

$$D_{31} = 1 \qquad D_{32} = -2 \qquad D_{33} = 1$$

Substituting these values of D in Eq. (2.71), the generalized masses are determined: $M_2 = 2m$, $M_3 = 6m$.

Equation (2.73) then can be used to write the expression for acceleration $\ddot{x}_3$:

$$\ddot{x}_3 = \left[\frac{1}{3m} + \frac{(\omega^2/\omega_2{}^2)(-1)(+1)}{2m(1 - \omega^2/\omega_2{}^2)} + \frac{(\omega^2/\omega_3{}^2)(+1)(+1)}{6m(1 - \omega^2/\omega_3{}^2)}\right] F_1 \sin \omega t$$

FREE AND FIXED SYSTEMS. For a structure which is free in space, there are six "normal modes" corresponding to $\omega_n = 0$. These represent motion of the structure without relative motion of its parts; this is rigid body motion with six degrees-of-freedom.

The rigid body modes all may be described by equations of the form

$$D_{jm} = a_{jm}D_m \qquad [m = 1, 2, \ldots, 6]$$

where D_m is a motion of the rigid body in the m coordinate and a is the displacement of the jth degree-of-freedom when D_m is moved a unit amount. The geometry of the structure determines the nature of a_{jm}. For example, if D_m is a rotation about the Z axis, $a_{jm} = 0$ for all modes of motion in which j represents rotation about the X or Y axis and $a_{jm} = 0$ if j represents translation parallel to the Z axis. If D_{jm} is a translational mode of motion parallel to X or Y, it is necessary that a_{jm} be proportional to the distance r_j of m_j from the Z axis and to the sine of the angle between r_j and the jth direction. The above relations may be applied to an elastic body. Such a body moves as a rigid body in the gross sense in that all particles of the body move together generally but may experience relative vibratory motion. The orthogonality condition applied to the relation between any rigid body mode D_{jm} and any oscillatory mode D_{jn} yields

$$\sum_j m_j D_{jn} D_{jm} = \sum_j m_j a_{jm} D_{jn} = 0 \qquad \begin{bmatrix} m \le 6 \\ n > 6 \end{bmatrix} \tag{2.74}$$

These relations are used in computations of oscillatory modes, and show that normal modes of vibration involve no net translation or rotation of a body.

A system attached to a fixed foundation may be considered as a system free in space in which one or more "foundation" masses or moments of inertia are infinite. Motion of the system as a rigid body is determined entirely by the motion of the foundation. The amplitude of an oscillatory mode representing motion of the foundation is zero; i.e., $M_j D_{jn}{}^2 = 0$ for the infinite mass. However, Eq. (2.73) applies equally well regardless of the size of the masses.

FOUNDATION MOTION. If a system is small relative to its foundation, it may be assumed to have no effect on the motion of the foundation. Consider a foundation of large but unknown mass m_0 having a motion $x_0 \sin \omega t$, the consequence of some unknown force

$$F_0 \sin \omega t = -m_0 x_0 \omega^2 \sin \omega t \tag{2.75}$$

acting on m_0 in the x_0 direction. Equation (2.73) is applicable to this case upon substituting

$$-m_0 x_0 \omega^2 D_{0n} = \sum_j F_{0j} D_{jn} \tag{2.76}$$

where D_{0n} is the amplitude of the foundation (the 0 degree-of-freedom) in the nth mode.
The oscillatory modes of the system are subject to Eqs. (2.74):

$$\sum_{j=0} m_j a_{jm} D_{jn} = 0$$

Separating the 0th degree-of-freedom from the other degrees-of-freedom:

$$\sum_{j=0} m_j a_{jm} D_{jn} = m_0 a_{0m} D_{0n} + \sum_{j=1} m_j a_{jm} D_{jn}$$

If m_0 approaches infinity as a limit, D_{0n} approaches zero and motion of the system as a rigid body is identical with the motion of the foundation. Thus, a_{0m} approaches unity for motion in which $m = 0$, and approaches zero for motion in which $m \neq 0$. In the limit:

$$\lim_{m_0 \to \infty} m_0 D_{0n} = -\sum_j m_j a_{j0} D_{jn} \tag{2.77}$$

Substituting this result in Eq. (2.76),

$$\lim_{m_0 \to \infty} \sum_j F_{0j} D_{jn} = x_0 \omega^2 \sum_j m_j a_{j0} D_{jn} \tag{2.78}$$

The generalized mass in Eq. (2.73) includes the term $m_0 D_{0n}{}^2$, but this becomes zero as m_0 becomes infinite.

The equation for response of a system to motion of its foundation is obtained by substituting Eq. (2.78) in Eq. (2.73):

$$x_k = \sum_{n=1}^{N} \frac{\omega^2}{\omega_n{}^2} D_{kn} \frac{\sum_j m_j a_{j0} D_{jn} x_0 \sin \omega t}{\sum_j m_j D_{jn}{}^2 (1 - \omega^2/\omega_n{}^2)} + x_0 \sin \omega t \tag{2.79}$$

DAMPED MULTIPLE DEGREE-OF-FREEDOM SYSTEMS

Consider a set of masses interconnected by a network of springs and acted upon by external forces, with a network of dampers acting in parallel with the springs. The viscous dampers produce forces on the masses which are determined in a manner analogous to that used to determine spring forces and summarized by Eq. (2.56). The damping force acting on the jth degree-of-freedom is

$$(F_d)_j = -\sum_k C_{jk} \dot{x}_k \tag{2.80}$$

where C_{jk} is the resultant force on the jth degree-of-freedom due to a unit velocity of the kth degree-of-freedom.

In general, the distribution of damper sizes in a system need not be related to the spring or mass sizes. Thus, the dampers may couple the normal modes together, allowing motion of one mode to affect that of another. Then the equations of response are not easily separable [6] into independent normal mode equations. However, there are two types of damping distribution which do not couple the normal modes.[6,7,8] These are known as *uniform viscous damping* and *uniform mass damping*.

UNIFORM VISCOUS DAMPING

Uniform damping is an appropriate model for systems in which the damping effect is an inherent property of the spring material. Each spring is considered to have a damper acting in parallel with it, and the ratio of damping coefficient to stiffness coefficient is the same for each spring of the system. Thus, for all values of j and k,

$$\frac{C_{jk}}{k_{jk}} = 2\mathcal{G} \tag{2.81}$$

where $\mathcal{G}$ is a constant.

Substituting from Eq. (2.81) in Eq. (2.80),

$$-(F_d)_j = \sum_k C_{jk}\dot{x}_k = 2\mathcal{G}\sum_k k_{jk}\dot{x}_k \tag{2.82}$$

Since the damping forces are "external" forces with respect to the mass-spring system, the forces $(F_d)_j$ can be added to the external forces in Eq. (2.70) to form the equation of motion:

$$\ddot{q}_n + \omega_n{}^2 q_n = \frac{\sum_j (F_d)_j D_{jn} + \sum_j F_j D_{jn}}{\sum_j m_j D_{jn}{}^2} \tag{2.83}$$

Combining Eqs. (2.61), (2.63), and (2.82), the summation involving $(F_d)_j$ in Eq. (2.83) may be written as follows:

$$\sum_j (F_d)_j D_{jn} = -2\mathcal{G}\omega_n{}^2 \dot{q}_n \sum_j m_j D_{jn}{}^2 \tag{2.84}$$

Substituting Eq. (2.84) in Eq. (2.83),

$$\ddot{q}_n + 2\mathcal{G}\omega_n{}^2 \dot{q}_n + \omega_n{}^2 q_n = \gamma_n \tag{2.85}$$

Comparison of Eq. (2.85) with Eq. (2.31) shows that each mode of the system responds as a simple damped oscillator.

The damping term $2\mathcal{G}\omega_n{}^2$ in Eq. (2.85) corresponds to $2\zeta\omega_n$ in Eq. (2.31) for a simple system. Thus, $\mathcal{G}\omega_n$ may be considered the critical damping ratio of each mode. Note that the effective damping for a particular mode varies directly as the natural frequency of the mode.

FREE VIBRATION. If a system with uniform viscous damping is disturbed from its equilibrium position and released at time $t = 0$ to vibrate freely, the applicable equation of motion is obtained from Eq. (2.85) by substituting $2\zeta\omega$ for $2\mathcal{G}\omega_n{}^2$ and letting $\gamma_n = 0$:

$$\ddot{q}_n + 2\zeta\omega_n\dot{q}_n + \omega_n{}^2 q_n = 0 \tag{2.86}$$

The solution of Eq. (2.86) for less than critical damping is

$$x_j(t) = \sum_n D_{jn}e^{-\zeta\omega_n t}(A_n \sin \omega_d t + B_n \cos \omega_d t) \tag{2.87}$$

where $\omega_d = \omega_n(1 - \zeta^2)^{\frac{1}{2}}$.

The values of A and B are determined by the displacement $x_j(0)$ and velocity $\dot{x}_j(0)$ at time $t = 0$:

$$x_j(0) = \sum_n B_n D_{jn}$$

$$\dot{x}_j(0) = \sum_n (A_n \omega_{dn} - B_n \zeta \omega_n) D_{jn}$$

Applying the orthogonality relation of Eq. (2.62) in the manner used to derive Eq. (2.69),

$$B_n = \frac{\sum_j x_j(0) m_j D_{jn}}{\sum_j m_j D_{jn}^2}$$

$$(2.88)$$

$$A_n \omega_{dn} - B_n \zeta \omega_{dn} = \frac{\sum_j \dot{x}_j(0) m_j D_{jn}}{\sum_j m_j D_{jn}^2}$$

Thus each mode undergoes a decaying oscillation at the damped natural frequency for the particular mode, and the amplitude of each mode decays from its initial value, which is determined by the initial displacements and velocities.

UNIFORM STRUCTURAL DAMPING

To avoid the dependence of viscous damping upon frequency, as indicated by Eq. (2.85), the uniform viscous damping factor $\mathfrak{g}$ is replaced by $\mathfrak{g}/\omega$ for uniform structural damping. This corresponds to the structural damping parameter $\mathfrak{g}$ in Eqs. (2.52) and (2.53) for sinusoidal vibration of a simple system. Thus, Eq. (2.85) for the response of a mode to a sinusoidal force of frequency ω is

$$\ddot{q}_n + \frac{2\mathfrak{g}}{\omega} \omega_n^2 \dot{q}_n + \omega_n^2 q_n = \gamma_n \qquad (2.89)$$

The amplification factor at resonance ($Q = 1/\mathfrak{g}$) has the same value in all modes.

UNIFORM MASS DAMPING

If the damping force on each mass is proportional to the magnitude of the mass,

$$(F_d)_j = -B m_j \dot{x}_j \qquad (2.90)$$

where B is a constant. For example, Eq. (2.90) would apply to a uniform beam immersed in a viscous fluid.

Substituting as $\dot{x}_j$ in Eq. (2.90) the derivative of Eq. (2.63),

$$\Sigma(F_d)_j D_{jn} = -B \sum_j m_j D_{jn} \sum_m \dot{q}_m D_{jm} \qquad (2.91)$$

Because of the orthogonality condition, Eq. (2.62):

$$\Sigma(F_d)_j D_{jn} = -B \dot{q}_n \sum_j m_j D_{jn}^2$$

Substituting from Eq. (2.91) in Eq. (2.83), the differential equation for the system is

$$\ddot{q}_n + B \dot{q}_n + \omega_n^2 q_n = \gamma_n \qquad (2.92)$$

where the damping term B corresponds to $2\zeta\omega$ for a simple oscillator, Eq. (2.31). Then $B/2\omega_n$ represents the fraction of critical damping for each mode, a quantity which diminishes with increasing frequency.

GENERAL EQUATION FOR FORCED VIBRATION

All the equations for response of a linear system to a sinusoidal excitation may be regarded as special cases of the following general equation:

$$x_k = \sum_{n=1}^{N} \frac{D_{kn}}{\omega_n^2} \frac{F_n}{m_n} R_n \sin(\omega t - \theta_n) \tag{2.93}$$

where x_k = displacement of structure in kth degree-of-freedom
N = number of degrees-of-freedom, including those of the foundation
D_{kn} = amplitude of kth degree-of-freedom in nth normal mode
F_n = generalized force for nth mode
m_n = generalized mass for nth mode
R_n = response factor, a function of the frequency ratio ω/ω_n (Fig. 2.13)
θ_n = phase angle (Fig. 2.14)

Equation (2.93) is of sufficient generality to cover a wide variety of cases, including excitation by external forces or foundation motion, viscous or structural damping, rotational and translational degrees-of-freedom, and from one to an infinite number of degrees-of-freedom.

LAGRANGIAN EQUATIONS

The differential equations of motion for a vibrating system sometimes are derived more conveniently in terms of kinetic and potential energies of the system than by the application of Newton's laws of motion in a form requiring the determination of the forces acting on each mass of the system. The formulation of the equations in terms of the energies, known as Lagrangian equations,[3,4,b] is expressed as follows:

$$\frac{d}{dt} \frac{\partial T}{\partial \dot{q}_n} - \frac{\partial T}{\partial q_n} + \frac{\partial V}{\partial q_n} = F_n \tag{2.94}$$

where T = total kinetic energy of system
V = total potential energy of system
q_n = generalized coordinate—a displacement
$\dot{q}_n$ = velocity at generalized coordinate q_n
F_n = generalized force, the portion of the total forces not related to the potential energy of the system (gravity and spring forces appear in the potential energy expressions and are not included here)

The method of applying Eq. (2.94) is to select a number of independent coordinates (generalized coordinates) equal to the number of degrees-of-freedom, and to write expressions for total kinetic energy T and total potential energy V. Differentiation of these expressions successively with respect to each of the chosen coordinates leads to a number of equations similar to Eq. (2.94), one for each coordinate (degree-of-freedom). These are the applicable differential equations and may be solved by any suitable method.

Example 2.3. Consider free vibration of the three degree-of-freedom system shown in Fig. 2.23; it consists of three equal masses m connected in tandem by equal springs k. Take as coordinates the three absolute displacements x_1, x_2, and x_3. The kinetic energy of the system is

$$T = \tfrac{1}{2}m(\dot{x}_1^2 + \dot{x}_2^2 + \dot{x}_3^2)$$

The potential energy of the system is

$$V = \frac{k}{2}[x_1^2 + (x_1 - x_2)^2 + (x_2 - x_3)^2] = \frac{k}{2}(2x_1^2 + 2x_2^2 + x_3^2 - 2x_1x_2 - 2x_2x_3)$$

Differentiating the expression for the kinetic energy successively with respect to the velocities,

$$\frac{\partial T}{\partial \dot{x}_1} = m\dot{x}_1 \qquad \frac{\partial T}{\partial \dot{x}_2} = m\dot{x}_2 \qquad \frac{\partial T}{\partial \dot{x}_3} = m\dot{x}_3$$

The kinetic energy is not a function of displacement; therefore, the second term in Eq. (2.94) is zero. The partial derivatives with respect to the displacement coordinates are

$$\frac{\partial V}{\partial x_1} = 2kx_1 - kx_2 \qquad \frac{\partial V}{\partial x_2} = 2kx_2 - kx_1 - kx_3 \qquad \frac{\partial V}{\partial x_3} = kx_3 - kx_2$$

In free vibration, the generalized force term in Eq. (2.93) is zero. Then, substituting the derivatives of the kinetic and potential energies from above into Eq. (2.94),

$$m\ddot{x}_1 + 2kx_1 - kx_2 = 0$$
$$m\ddot{x}_2 + 2kx_2 - kx_1 - kx_3 = 0$$
$$m\ddot{x}_3 + kx_3 - kx_2 = 0$$

The natural frequencies of the system may be determined by placing the preceding set of simultaneous equations in determinant form, in accordance with Eq. (2.60):

$$\begin{vmatrix} (m\omega^2 - 2k) & k & 0 \\ k & (m\omega^2 - 2k) & k \\ 0 & k & (m\omega^2 - k) \end{vmatrix} = 0$$

The natural frequencies are equal to the values of ω that satisfy the preceding determinant equation.

Example 2.4. Consider the compound pendulum of mass m shown in Fig. 2.25, having its center-of-gravity located a distance l from the axis of rotation. The moment of inertia is I

Fig. 2.25. Forces and motions of a compound pendulum.

about an axis through the center-of-gravity. The position of the mass is defined by three coordinates, x and y to define the location of the center-of-gravity, and θ to define the angle of rotation.

The *equations* of *constraint* are $y = l \cos \theta$; $x = l \sin \theta$. Each equation of constraint reduces the number of degrees-of-freedom by one; thus the pendulum is a one degree-of-freedom system whose position is defined uniquely by θ alone.

The kinetic energy of the pendulum is

$$T = \tfrac{1}{2}(I + ml^2)\dot{\theta}^2$$

The potential energy is

$$V = mgl(1 - \cos \theta)$$

Then

$$\frac{\partial T}{\partial \theta} = (I + ml^2)\dot{\theta} \qquad \frac{d}{dt}\left(\frac{\partial T}{\partial \dot{\theta}}\right) = (I + ml^2)\ddot{\theta}$$

$$\frac{\partial T}{\partial \theta} = 0 \qquad \frac{\partial V}{\partial \theta} = mgl \sin \theta$$

Substituting these expressions in Eq. (2.94), the differential equation for the pendulum is

$$(I + ml^2)\ddot{\theta} + mgl \sin \theta = 0$$

Example 2.5. Consider oscillation of the water in the U tube shown in Fig. 2.26. If the displacements of the water levels in the arms of a uniform-diameter U tube are h_1 and h_2, then conservation of matter requires that $h_1 = -h_2$. The kinetic energy of the water flowing in the tube with velocity $\dot{h}_1$ is

$$T = \tfrac{1}{2}\rho S l \dot{h}_1{}^2$$

where ρ is the water density, S is the cross-section area of the tube, and l is the developed length of the water column. The potential energy (difference in potential energy between arms of tube) is

$$V = S\rho g h_1{}^2$$

Taking h_1 as the generalized coordinate, differentiating the expressions for energy, and substituting in Eq. (2.94),

$$S\rho l \ddot{h}_1 + 2\rho g S h_1 = 0$$

Dividing through by $\rho S l$,

$$\ddot{h}_1 + \frac{2g}{l} h_1 = 0$$

This is the differential equation for a simple oscillating system of natural frequency ω_n, where

$$\omega_n = \sqrt{\frac{2g}{l}}$$

Fig. 2.26. Water column in a U tube.

REFERENCES

1. Timoshenko, S.: "Vibration Problems in Engineering," D. Van Nostrand Company, Inc., Princeton, N.J., 1937.
2. Hansen, H. M., and P. F. Chenea: "Mechanics of Vibration," John Wiley & Sons, Inc., New York, 1952.
3. Kármán, T. V., and M. A. Biot: "Mathematical Methods in Engineering," McGraw-Hill Book Company, Inc., New York, 1940.
4. Slater, J. C., and N. H. Frank: "Mechanics," McGraw-Hill Book Company, Inc., New York, 1947.
5. Pipes, L. A.: "Applied Mathematics for Engineers and Physicists," McGraw-Hill Book Company, Inc., New York, 1958.
6. Foss, K. A.: "Coordinates Which Uncouple the Equations of Motion of Damped Linear Dynamic Systems," *ASME Applied Mechanics Paper* 57-A-86, December, 1957.
7. Rayleigh: "The Theory of Sound," vol. I, Macmillan & Co., Ltd., London, 1894.
8. Crumb, S. F.: "A Study of the Effects of Damping on Normal Modes of Electrical and Mechanical Systems," California Institute of Technology, Pasadena, Calif., 1955.

3

VIBRATION OF A RESILIENTLY SUPPORTED RIGID BODY

Harry Himelblau, Jr.
Rockwell International Corporation

Sheldon Rubin
The Aerospace Corporation

INTRODUCTION

SCOPE

This chapter discusses the vibration of a rigid body on resilient supporting elements, including (1) methods of determining the inertial properties of a rigid body, (2) discussion of the dynamic properties of resilient elements, and (3) motion of a single rigid body on resilient supporting elements for various dynamic excitations and degrees of symmetry.

The general equations of motion for a rigid body on linear massless resilient supports are given; these equations are general in that they include any configuration of the rigid body and any configuration and location of the supports. They involve six simultaneous equations with numerous terms, for which a general solution is impracticable without the use of high-speed automatic computing equipment. Various degrees of simplification are introduced by assuming certain symmetry, and results useful for engineering purposes are presented. Several topics are considered: (1) determination of undamped natural frequencies and discussion of coupling of modes of vibration; (2) forced vibration where the excitation is a vibratory motion of the foundation; (3) forced vibration where the excitation is a vibratory force or moment generated within the body; and (4) free vibration caused by an instantaneous change in velocity of the system (velocity shock). Results are presented mathematically and, where feasible, graphically.

SYSTEM OF COORDINATES

The motion of the rigid body is referred to a fixed "inertial" frame of reference. The inertial frame is represented by a system of cartesian coordinates $\overline{X}, \overline{Y}, \overline{Z}$. A similar system of coordinates X, Y, Z fixed in the body has its origin at the center-of-mass. The two sets of coordinates are coincident when the body is in equilibrium under the action of gravity alone. The motions of the body are described by giving the displacement of the body axes relative to the inertial axes. The translational displacements of the center-of-mass of the body are x_c, y_c, z_c in the $\overline{X}, \overline{Y}, \overline{Z}$ directions, respectively. The rotational displacements of the body are characterized by the angles of rotation α, β, γ of

CENTER OF MASS
OF RIGID BODY

FIG. 3.1. System of coordinates for the motion of a rigid body consisting of a fixed inertial set of reference axes $(\bar{X}, \bar{Y}, \bar{Z})$, and a set of axes (X, Y, Z) fixed in the moving body with its origin at the center-of-mass. The axes $\bar{X}, \bar{Y}, \bar{Z}$ and X, Y, Z are coincident when the body is in equilibrium under the action of gravity alone. The displacement of the center-of-mass is given by the translational displacements x_c, y_c, z_c and the rotational displacements α, β, γ as shown. A positive rotation about an axis is one which advances a right-handed screw in the positive direction of the axis.

the body axes about the $\bar{X}$, $\bar{Y}$, $\bar{Z}$ axes, respectively. These displacements are shown graphically in Fig. 3.1.

Only small translations and rotations are considered. Hence, the rotations are commutative (i.e., the resulting position is independent of the order of the component rotations) and the angles of rotation about the body axes are equal to those about the inertial axes. Therefore, the displacements of a point b in the body (with the coordinates b_x, b_y, b_z in the X, Y, Z directions, respectively) are the sums of the components of the center-of-mass displacement in the directions of the $\bar{X}$, $\bar{Y}$, $\bar{Z}$ axes plus the tangential components of the rotational displacement of the body:

$$x_b = x_c + b_z\beta - b_y\gamma$$

$$y_b = y_c - b_z\alpha + b_x\gamma \qquad (3.1)$$

$$z_b = z_c - b_x\beta + b_y\alpha$$

EQUATIONS OF SMALL MOTION OF A RIGID BODY

The equations of motion for the translation of a rigid body are

$$m\ddot{x}_c = \mathbf{F}_x \qquad m\ddot{y}_c = \mathbf{F}_y \qquad m\ddot{z}_c = \mathbf{F}_z \quad (3.2)$$

where m is the mass of the body, $\mathbf{F}_x$, $\mathbf{F}_y$, $\mathbf{F}_z$ are the summation of all forces acting on the body, and $\ddot{x}_c$, $\ddot{y}_c$, $\ddot{z}_c$ are the accelerations of the center-of-mass of the body in the $\bar{X}$, $\bar{Y}$, $\bar{Z}$ directions, respectively. The motion of the center-of-mass of a rigid body is the same as the motion of a particle having a mass equal to the total mass of the body and acted upon by the resultant external force.

The equations of motion for the rotation of a rigid body are

$$I_{xx}\ddot{\alpha} - I_{xy}\ddot{\beta} - I_{xz}\ddot{\gamma} = \mathbf{M}_x$$

$$-I_{xy}\ddot{\alpha} + I_{yy}\ddot{\beta} - I_{yz}\ddot{\gamma} = \mathbf{M}_y \qquad (3.3)$$

$$-I_{xz}\ddot{\alpha} - I_{yz}\ddot{\beta} + I_{zz}\ddot{\gamma} = \mathbf{M}_z$$

where $\ddot{\alpha}$, $\ddot{\beta}$, $\ddot{\gamma}$ are the rotational accelerations about the X, Y, Z axes, as shown in Fig. 3.1; $\mathbf{M}_x$, $\mathbf{M}_y$, $\mathbf{M}_z$ are the summation of torques acting on the rigid body about the axes X, Y, Z, respectively; and $I_{xx} \ldots, I_{xy} \ldots$ are the moments and products of inertia of the rigid body as defined below.

INERTIAL PROPERTIES OF A RIGID BODY

The properties of a rigid body that are significant in dynamics and vibration are the mass, the position of the center-of-mass (or center-of-gravity), the moments of inertia, the products of inertia, and the directions of the principal inertial axes. This section discusses the properties of a rigid body, together with computational and experimental methods for determining the properties.

MASS

COMPUTATION OF MASS. The mass of a body is computed by integrating the product of mass density $\rho(V)$ and elemental volume dV over the body:

$$m = \int_V \rho(V)\, dV \tag{3.4}$$

If the body is made up of a number of elements, each having constant or an average density, the mass is

$$m = \rho_1 V_1 + \rho_2 V_2 + \cdots + \rho_n V_n \tag{3.5}$$

where ρ_1 is the density of the element V_1, etc. Densities of various materials may be found in handbooks containing properties of materials.[1]

If a rigid body has a common geometrical shape, or if it is an assembly of subbodies having common geometrical shapes, the volume may be found from compilations of formulas. Typical formulas are included in Tables 3.1 and 3.2. Tables of areas of plane sections as well as volumes of solid bodies are useful.

If the volume of an element of the body is not given in such a table, the integration of Eq. (3.4) may be carried out analytically, graphically, or numerically. A graphical approach may be used if the shape is so complicated that the analytical expression for its boundaries is not available or is not readily integrable. This is accomplished by graphically dividing the body into smaller parts, each of whose boundaries may be altered slightly (without change to the area) in such a manner that the volume is readily calculable or measurable.

The weight W of a body of mass m is a function of the acceleration of gravity g at the particular location of the body in space:

$$W = mg \tag{3.6}$$

Unless otherwise stated, it is understood that the weight of a body is given for an average value of the acceleration of gravity on the surface of the earth. For engineering purposes, $g = 32.2$ ft/sec^2 or 386 in./sec^2 is usually used.

EXPERIMENTAL DETERMINATION OF MASS. Although Newton's second law of motion, $F = m\ddot{x}$, may be used to measure mass, this usually is not convenient. The mass of a body is most easily measured by performing a static measurement of the weight of the body and converting the result to mass. This is done by use of the value of the acceleration of gravity at the measurement location [Eq. (3.6)].

CENTER-OF-MASS

COMPUTATION OF CENTER-OF-MASS. The center-of-mass (or center-of-gravity) is that point located by the vector

$$\mathbf{r}_c = \frac{1}{m} \int_m \mathbf{r}(m)\, dm \tag{3.7}$$

where $\mathbf{r}(m)$ is the radius vector of the element of mass dm. The center-of-mass of a body in a cartesian coordinate system X, Y, Z is located at

$$X_c = \frac{1}{m} \int_V X(V)\rho(V)\, dV$$

$$Y_c = \frac{1}{m} \int_V Y(V)\rho(V)\, dV \tag{3.8}$$

$$Z_c = \frac{1}{m} \int_V Z(V)\rho(V)\, dV$$

where $X(V)$, $Y(V)$, $Z(V)$ are the X, Y, Z coordinates of the element of volume dV and m is the mass of the body.

If the body can be divided into elements whose centers-of-mass are known, the center-of-mass of the entire body having a mass m is located by equations of the following type:

$$X_c = \frac{1}{m}\,(X_{c1}m_1 + X_{c2}m_2 + \cdots + X_{cn}m_n),\ \text{etc.} \qquad (3.9)$$

where X_{c1} is the X coordinate of the center-of-mass of element m_1. Tables (see Tables 3.1 and 3.2) which specify the location of centers of area and volume (called centroids) for simple sections and solid bodies often are an aid in dividing the body into the submasses indicated in the above equation. The centroid and center-of-mass of an element are coincident when the density of the material is uniform throughout the element.

EXPERIMENTAL DETERMINATION OF CENTER-OF-MASS. The location of the center-of-mass is normally measured indirectly by locating the center-of-gravity of the body, and may be found in various ways. Theoretically, if the body is suspended by a flexible wire attached successively at different points on the body, all lines represented by the wire in its various positions when extended inwardly into the body intersect at the center-of-gravity. Two such lines determine the center-of-gravity, but more may be used as a check. There are practical limitations to this method in that the point of intersection often is difficult to designate.

FIG. 3.2. Three-scale method of locating the center-of-gravity of a body. The vertical forces F_1, F_2, F_3 at the scales result from the weight of the body. The vertical line located by the distances a_o and b_o [see Eqs. (3.10)] passes through the center-of-gravity of the body.

Other techniques are based on the balancing of the body on point or line supports. A point support locates the center-of-gravity along a vertical line through the point; a line support locates it in a vertical plane through the line. The intersection of such lines or planes determined with the body in various positions locates the center-of-gravity. The greatest difficulty with this technique is the maintenance of the stability of the body while it is balanced, particularly where the height of the body is great relative to a horizontal dimension. If a perfect point or edge support is used, the equilibrium position is inherently unstable. It is only if the support has width that some degree of stability can be achieved, but then a resulting error in the location of the line or plane containing the center-of-gravity can be expected.

Another method of locating the center-of-gravity is to place the body in a stable position on three scales. From static moments the vector weight of the body is the resultant of the measured forces at the scales as shown in Fig. 3.2. The vertical line through the center-of-gravity is located by the distances a_0 and b_0:

$$a_0 = \frac{F_2}{F_1 + F_2 + F_3}\,a_1$$

$$b_0 = \frac{F_3}{F_1 + F_2 + F_3}\,b_1 \qquad (3.10)$$

This method cannot be used with more than three scales.

MOMENT AND PRODUCT OF INERTIA

COMPUTATION OF MOMENT AND PRODUCT OF INERTIA.[2] The moments of inertia of a rigid body with respect to the orthogonal axes X, Y, Z fixed in the body are

$$I_{xx} = \int_m (Y^2 + Z^2)\, dm \qquad I_{yy} = \int_m (X^2 + Z^2)\, dm \qquad I_{zz} = \int_m (X^2 + Y^2)\, dm \quad (3.11)$$

where dm is the infinitesimal element of mass located at the coordinate distances X, Y, Z; and the integration is taken over the mass of the body. Similarly, the products of inertia are

$$I_{xy} = \int_m XY\, dm \qquad I_{xz} = \int_m XZ\, dm \qquad I_{yz} = \int_m YZ\, dm \qquad (3.12)$$

It is conventional in rigid body mechanics to take the center of coordinates at the center-of-mass of the body. Unless otherwise specified, this location is assumed and the moments of inertia and products of inertia refer to axes through the center-of-mass of the body. For a unique set of axes, the products of inertia vanish. These axes are called the principal inertial axes of the body. The moments of inertia about these axes are called the principal moments of inertia.

The moments of inertia of a rigid body can be defined in terms of radii of gyration as follows:

$$I_{xx} = m\rho_x{}^2 \qquad I_{yy} = m\rho_y{}^2 \qquad I_{zz} = m\rho_z{}^2$$

$$(3.13)$$

where I_{xx}, ... are the moments of inertia of the body as defined by Eqs. (3.11), m is the mass of the body, and ρ_x, ... are the radii of gyration. The radius of gyration has the dimension of length, and often leads to convenient expressions in dynamics of rigid bodies when distances are normalized to an appropriate radius of gyration. Solid bodies of various shapes have characteristic radii of gyration which sometimes are useful intuitively in evaluating dynamic conditions.

FIG. 3.3. Axes required for moment and product of inertia transformations. Moments and products of inertia with respect to the axes X'', Y'', Z'' are transferred to the mutually parallel axes X', Y', Z' by Eqs. (3.14) and (3.15), and then to the inclined axes X, Y, Z by Eqs. (3.16) and (3.17).

Unless the body has a very simple shape, it is laborious to evaluate the integrals of Eqs. (3.11) and (3.12). The problem is made easier by subdividing the body into parts for which simplified calculations are possible. The moments and products of inertia of the body are found by first determining the moments and products of inertia for the individual parts with respect to appropriate reference axes chosen in the parts, and then summing the contributions of the parts. This is done by selecting axes through the centers-of-mass of the parts, and then determining the moments and products of inertia of the parts relative to these axes. Then the moments and products of inertia are transferred to the axes chosen through the center-of-mass of the whole body, and the transferred quantities summed. In general, the transfer involves two sets of non-parallel coordinates whose centers are displaced. Two transformations are required as follows.

Transformation to Parallel Axes. Referring to Fig. 3.3, suppose that X, Y, Z is a convenient set of axes for the moment of inertia of the whole body with its origin at the center-of-mass. The moments and products of inertia for a part of the body are $I_{x''x''}$,

$I_{y''y''}$, $I_{z''z''}$, $I_{x''y''}$, $I_{x''z''}$, and $I_{y''z''}$, taken with respect to a set of axes X'', Y'', Z'' fixed in the part and having their center at the center-of-mass of the part. The axes X', Y', Z' are chosen parallel to X'', Y'', Z'' with their origin at the center-of-mass of the body. The perpendicular distance between the X'' and X' axes is a_x; between Y'' and Y' is a_y; between Z'' and Z' is a_z. The moments and products of inertia of the part of mass m_n with respect to the X', Y', Z' axes are

$$I_{x'x'} = I_{x''x''} + m_n a_z{}^2$$

$$I_{y'y'} = I_{y''y''} + m_n a_y{}^2 \tag{3.14}$$

$$I_{z'z'} = I_{z''z''} + m_n a_z{}^2$$

The corresponding products of inertia are

$$I_{x'y'} = I_{x''y''} + m_n a_x a_y$$

$$I_{x'z'} = I_{x''z''} + m_n a_x a_z \tag{3.15}$$

$$I_{y'z'} = I_{y''z''} + m_n a_y a_z$$

If X'', Y'', Z'' are the principal axes of the part, the product of inertia terms on the right-hand side of Eqs. (3.15) are zero.

Transformation to Inclined Axes. The desired moments and products of inertia with respect to axes X, Y, Z are now obtained by a transformation theorem relating the properties of bodies with respect to inclined sets of axes whose centers coincide. This theorem makes use of the direction cosines λ for the respective sets of axes. For example, $\lambda_{xx'}$ is the cosine of the angle between the X and X' axes. The expressions for the moments of inertia are

$$
\begin{aligned}
I_{xx} = {}& \lambda_{xx'}{}^2 I_{x'x'} + \lambda_{xy'}{}^2 I_{y'y'} + \lambda_{xz'}{}^2 I_{z'z'} \\
& - 2\lambda_{xx'}\lambda_{xy'} I_{x'y'} - 2\lambda_{xx'}\lambda_{xz'} I_{x'z'} - 2\lambda_{xy'}\lambda_{xz'} I_{y'z'}
\end{aligned}
$$

$$
\begin{aligned}
I_{yy} = {}& \lambda_{yx'}{}^2 I_{x'x'} + \lambda_{yy'}{}^2 I_{y'y'} + \lambda_{yz'}{}^2 I_{z'z'} \\
& - 2\lambda_{yx'}\lambda_{yy'} I_{x'y'} - 2\lambda_{yx'}\lambda_{yz'} I_{x'z'} - 2\lambda_{yy'}\lambda_{yz'} I_{y'z'}
\end{aligned} \tag{3.16}
$$

$$
\begin{aligned}
I_{zz} = {}& \lambda_{zx'}{}^2 I_{x'x'} + \lambda_{zy'}{}^2 I_{y'y'} + \lambda_{zz'}{}^2 I_{z'z'} \\
& - 2\lambda_{zx'}\lambda_{zy'} I_{x'y'} - 2\lambda_{zx'}\lambda_{zz'} I_{x'z'} - 2\lambda_{zy'}\lambda_{zz'} I_{y'z'}
\end{aligned}
$$

The corresponding products of inertia are

$$
\begin{aligned}
-I_{xy} = {}& \lambda_{xx'}\lambda_{yx'} I_{x'x'} + \lambda_{xy'}\lambda_{yy'} I_{y'y'} + \lambda_{xz'}\lambda_{yz'} I_{z'z'} \\
& - (\lambda_{xx'}\lambda_{yy'} + \lambda_{xy'}\lambda_{yx'}) I_{x'y'} - (\lambda_{xy'}\lambda_{yz'} + \lambda_{xz'}\lambda_{yy'}) I_{y'z'} \\
& - (\lambda_{xz'}\lambda_{yx'} + \lambda_{xx'}\lambda_{yz'}) I_{x'z'}
\end{aligned}
$$

$$
\begin{aligned}
-I_{xz} = {}& \lambda_{xx'}\lambda_{zx'} I_{x'x'} + \lambda_{xy'}\lambda_{zy'} I_{y'y'} + \lambda_{xz'}\lambda_{zz'} I_{z'z'} \\
& - (\lambda_{xx'}\lambda_{zy'} + \lambda_{xy'}\lambda_{zx'}) I_{x'y'} - (\lambda_{xy'}\lambda_{zz'} + \lambda_{xz'}\lambda_{zy'}) I_{y'z'} \\
& - (\lambda_{xx'}\lambda_{zz'} + \lambda_{xz'}\lambda_{zx'}) I_{x'z'}
\end{aligned} \tag{3.17}
$$

$$
\begin{aligned}
-I_{yz} = {}& \lambda_{yx'}\lambda_{zx'} I_{x'x'} + \lambda_{yy'}\lambda_{zy'} I_{y'y'} + \lambda_{yz'}\lambda_{zz'} I_{z'z'} \\
& - (\lambda_{yx'}\lambda_{zy'} + \lambda_{yy'}\lambda_{zx'}) I_{x'y'} - (\lambda_{yy'}\lambda_{zz'} + \lambda_{yz'}\lambda_{zy'}) I_{y'z'} \\
& - (\lambda_{yz'}\lambda_{zx'} + \lambda_{yx'}\lambda_{zz'}) I_{x'z'}
\end{aligned}
$$

EXPERIMENTAL DETERMINATION OF MOMENTS OF INERTIA. The moment of inertia of a body about a given axis may be found experimentally by suspending the body as a pendulum so that rotational oscillations about that axis can occur. The period of free oscillation is then measured, and is used with the geometry of the pendulum to calculate the moment of inertia.

Two types of pendulums are useful: the compound pendulum and the torsional pendulum. When using the compound pendulum, the body is supported from two overhead points by wires, illustrated in Fig. 3.4. The distance l is measured between the axis of support O–O and a parallel axis C–C through the center-of-gravity of the body. The moment of inertia about C–C is given by

$$I_{cc} = ml^2 \left[\left(\frac{\tau_0}{2\pi} \right)^2 \left(\frac{g}{l} \right) - 1 \right] \quad (3.18)$$

where τ_0 is the period of oscillation in seconds, l is the pendulum length in inches, g is the gravitational acceleration in in./sec^2, and m is the mass in lb-sec^2/in., yielding a moment of inertia in lb-in.-sec^2.

The accuracy of the above method is dependent upon the accuracy with which the distance l is known. Since the center-of-gravity often is an inaccessible point, a direct measurement of l may not be practicable. However, a change in l can be measured quite readily. If the experiment is repeated with a different support axis O'–O', the length l becomes $l + \Delta l$ and the period of oscillation becomes τ_0'. Then, the distance l can be written in terms of Δl, and the two periods τ_0, τ_0':

FIG. 3.4. Compound pendulum method of determining moment of inertia. The period of oscillation of the test body about the horizontal axis O–O and the perpendicular distance l between the axis O–O and the parallel axis C–C through the center-of-gravity of the test body give I_{cc} by Eq. (3.18).

$$l = \Delta l \left[\frac{(\tau_0'^2/4\pi^2)(g/\Delta l) - 1}{[(\tau_0^2 - \tau_0'^2)/4\pi^2][g/\Delta l] - 1} \right] \quad (3.19)$$

This value of l can be substituted into Eq. (3.18) to compute I_{cc}.

Note that accuracy is not achieved if l is much larger than the radius of gyration ρ_c of the body about the axis C–C ($I_{cc} = m\rho_c^2$). If l is large, then $(\tau_0/2\pi)^2 \simeq l/g$ and the expression in brackets in Eq. (3.18) is very small; thus, it is sensitive to small errors in the measurement of both τ_0 and l. Consequently, it is highly desirable that the distance l be chosen as small as convenient, preferably with the axis O–O passing through the body.

A torsional pendulum may be constructed with the test body suspended by a single torsional spring (in practice, a rod or wire) of known stiffness, or by three flexible wires. A solid body supported by a single torsional spring is shown in Fig. 3.5. From the known torsional stiffness k_t and the measured period of torsional oscillation τ, the moment of inertia of the body about the vertical torsional axis is

$$I_{cc} = \frac{k_t \tau^2}{4\pi^2} \quad (3.20)$$

FIG. 3.5. Torsional pendulum method of determining moment of inertia. The period of torsional oscillation of the test body about the vertical axis C–C passing through the center-of-gravity and the torsional spring constant k_t give I_{cc} by Eq. (3.20).

A platform may be constructed below the torsional spring to carry the bodies to be measured, as shown in Fig. 3.6. By repeating the experiment with two different bodies placed on the platform, it becomes unnecessary to measure the torsional stiffness k_t. If a body with a *known* moment of inertia I_1 is placed on the platform and an oscillation period τ_1 results, the moment of inertia I_2 of a body which produces a period τ_2 is given by

$$I_2 = I_1 \left[\frac{(\tau_2/\tau_0)^2 - 1}{(\tau_1/\tau_0)^2 - 1} \right] \quad (3.21)$$

where τ_0 is the period of the pendulum comprised of platform alone.

Fig. 3.6. A variation of the torsional pendulum method shown in Fig. 3.5 wherein a light platform is used to carry the test body. The moment of inertia I_{cc} is given by Eq. (3.20).

A body suspended by three flexible wires, called a trifilar pendulum, as shown in Fig. 3.7, offers some utilitarian advantages. Designating the perpendicular distances of the wires to the vertical axis C–C through the center-of-gravity of the body by R_1, R_2, R_3, the angles between wires by ϕ_1, ϕ_2, ϕ_3, and the length of each wire by l, the moment of inertia about axis C–C is

$$I_{cc} = \frac{mgR_1R_2R_3\tau^2}{4\pi^2 l} \frac{R_1 \sin \phi_1 + R_2 \sin \phi_2 + R_3 \sin \phi_3}{R_2R_3 \sin \phi_1 + R_1R_3 \sin \phi_2 + R_1R_2 \sin \phi_3} \quad (3.22)$$

Apparatus that is more convenient for repeated use embodies a light platform supported by three equally spaced wires. The body whose moment of inertia is to be measured is placed on the platform with its center-of-gravity equidistant from the wires. Thus $R_1 = R_2 = R_3 = R$ and $\phi_1 = \phi_2 = \phi_3 = 120°$. Substituting these relations in Eq. (3.22), the moment of inertia about the vertical axis C–C is

$$I_{cc} = \frac{mgR^2\tau^2}{4\pi^2 l} \quad (3.23)$$

where the mass m is the sum of the masses of the test body and the platform. The moment of inertia of the platform is subtracted from the test result to obtain the moment of inertia of the body being measured. It becomes unnecessary to know the distances R and l in Eq. (3.23) if the period of oscillation is measured with the platform empty, with the body being measured on the platform, and with a second body of known mass m_1 and known moment of inertia I_1 on the platform. Then the desired moment of inertia I_2 is

$$I_2 = I_1 \left[\frac{[1 + (m_2/m_0)][\tau_2/\tau_0]^2 - 1}{[1 + (m_1/m_0)][\tau_1/\tau_0]^2 - 1} \right] \quad (3.24)$$

where m_0 is the mass of the unloaded platform, m_2 is the mass of the body being measured, τ_0 is the period of oscillation with the platform unloaded, τ_1 is the period when loaded with known body of mass m_1, and τ_2 is the period when loaded with the unknown body of mass m_2.

EXPERIMENTAL DETERMINATION OF PRODUCT OF INERTIA. The experimental determination of a product of inertia usually requires the measurement of moments of inertia. (An exception is the balancing machine technique described later.) If possible, symmetry of the body is used to locate directions of principal inertial axes, thereby simplifying the relationship between the moments of inertia as known and the products of inertia to be found. Several alternative procedures are described below, depending on the number of principal inertia axes whose directions are known. Knowledge of two principal axes implies a knowledge of all three since they are mutually perpendicular.

If the directions of all three principal axes (X', Y', Z') are known and it is desirable to use another set of axes (X, Y, Z), Eqs. (3.16) and (3.17) may be simplified because the products of inertia with respect to the principal directions are zero. First, the three principal moments of inertia $(I_{x'x'}, I_{y'y'}, I_{z'z'})$ are measured by one of the above techniques; then

Fig. 3.7. Trifilar pendulum method of determining moment of inertia. The period of torsional oscillation of the test body about the vertical axis C–C passing through the center-of-gravity and the geometry of the pendulum give I_{cc} by Eq. (3.22); with a simpler geometry, I_{co} is given by Eq. (3.23).

the moments of inertia with respect to the X, Y, Z axes are

$$I_{xx} = \lambda_{xx'}{}^2 I_{x'x'} + \lambda_{xy'}{}^2 I_{y'y'} + \lambda_{xz'}{}^2 I_{z'z'}$$

$$I_{yy} = \lambda_{yx'}{}^2 I_{x'x'} + \lambda_{yy'}{}^2 I_{y'y'} + \lambda_{yz'}{}^2 I_{z'z'} \tag{3.25}$$

$$I_{zz} = \lambda_{zx'}{}^2 I_{x'x'} + \lambda_{zy'}{}^2 I_{y'y'} + \lambda_{zz'}{}^2 I_{z'z'}$$

The products of inertia with respect to the X, Y, Z axes are

$$-I_{xy} = \lambda_{xx'}\lambda_{yx'} I_{x'x'} + \lambda_{xy'}\lambda_{yy'} I_{y'y'} + \lambda_{xz'}\lambda_{yz'} I_{z'z'}$$

$$-I_{xz} = \lambda_{xx'}\lambda_{zx'} I_{x'x'} + \lambda_{xy'}\lambda_{zy'} I_{y'y'} + \lambda_{xz'}\lambda_{zz'} I_{z'z'} \tag{3.26}$$

$$-I_{yz} = \lambda_{yx'}\lambda_{zx'} I_{x'x'} + \lambda_{yy'}\lambda_{zy'} I_{y'y'} + \lambda_{yz'}\lambda_{zz'} I_{z'z'}$$

The direction of one principal axis Z may be known from symmetry. The axis through the center-of-gravity perpendicular to the plane of symmetry is a principal axis. The product of inertia with respect to X and Y axes, located in the plane of symmetry, is determined by first establishing another axis X' at a counterclockwise angle θ from X, as shown in Fig. 3.8. If the three moments of inertia I_{xx}, $I_{x'x'}$, and I_{yy} are measured by any applicable means, the product of inertia I_{xy} is

$$I_{xy} = \frac{I_{xx} \cos^2 \theta + I_{yy} \sin^2 \theta - I_{x'x'}}{\sin 2\theta} \tag{3.27}$$

where $0 < \theta < \pi$. For optimum accuracy, θ should be approximately $\pi/4$ or $3\pi/4$. Since the third axis Z is a principal axis, I_{zz} and I_{yz} are zero.

Another method is illustrated in Fig. 3.9.[3, 4] The plane of the X and Z axes is a plane of symmetry, or the Y axis is otherwise known to be a principal axis of inertia. For determining I_{xz}, the body is suspended by a cable so that the Y axis is horizontal and the Z axis is vertical. Torsional stiffness about the Z axis is provided by four springs acting in the Y direction at the points shown. The body is oscillated about the Z axis with various positions of the springs so that the angle θ can be varied. The spring stiffnesses and locations must be such

Fig. 3.8. Axes required for determining the product of inertia with respect to the axes X and Y when Z is a principal axis of inertia. The moments of inertia about the axes X, Y, and X', where X' is in the plane of X and Y at a counterclockwise angle θ from X, give I_{xy} by Eq. (3.27).

that there is no net force in the Y direction due to a rotation about the Z axis. In general, there is coupling between rotations about the X and Z axes, with the result that oscillations about both axes occur as a result of an initial rotational displacement about the Z axis. At some particular value of $\theta = \theta_0$, the two rotations are uncoupled; i.e., oscillation about the Z axis does not cause oscillation about the X axis. Then

SIDE VIEW   END VIEW

Fig. 3.9. Method of determining the product of inertia with respect to the axes X and Z when Y is a principal axis of inertia. The test body is oscillated about the vertical Z axis with torsional stiffness provided by the four springs acting in the Y direction at the points shown. There should be no net force on the test body in the Y direction due to a rotation about the Z axis. The angle θ is varied until, at some value of $\theta = \theta_0$, oscillations about X and Z are uncoupled. The angle θ_0 and the moment of inertia about the Z axis give I_{xz} by Eq. (3.28).

$$I_{xz} = I_{zz} \tan \theta_0 \qquad (3.28)$$

The moment of inertia I_{zz} can be determined by one of the methods described under *Experimental Determination of Moments of Inertia*.

When the moments and product of inertia with respect to a pair of axes X and Z in a principal plane of inertia XZ are known. the orientation of a principal axis P is given by

$$\theta_p = \tfrac{1}{2} \tan^{-1} \left(\frac{2I_{xz}}{I_{zz} - I_{xx}} \right) \qquad (3.29)$$

where θ_p is the counterclockwise angle from the X axis to the P axis. The second principal axis in this plane is at $\theta_p + 90°$.

Consider the determination of products of inertia when the directions of all principal axes of inertia are unknown. In one method, the moments of inertia about two independent sets of three mutually perpendicular axes are measured, and the direction cosines between these sets of axes are known from the positions of the axes. The values for the six moments of inertia and the nine direction cosines are then substituted into Eqs. (3.16) and (3.17). The result is six linear equations in the six unknown products of inertia, from which the values of the desired products of inertia may be found by simultaneous solution of the equations. This method leads to experimental errors of relatively large magnitude because each product of inertia is, in general, a function of all six moments of inertia, each of which contains an experimental error.

Fig. 3.10. Balancing machine technique for determining products of inertia. The test body is rotated about the Z axis with angular velocity Ω. The dynamic reactions F_x and F_y measured at the bearings, which are a distance l apart, give I_{xz} and I_{yz} by Eq. (3.30).

An alternative method is based upon the knowledge that one of the principal moments of inertia of a body is the largest and another is the smallest that can be obtained for any axis through the center-of-gravity. A trial-and-error procedure can be used to locate the orientation of the axis through the center-of-gravity having the maximum and/or minimum moment of inertia. After one or both are located, the moments and products of inertia for any set of axes are found by the techniques previously discussed.

The products of inertia of a body also may be determined by rotating the body at a constant angular velocity Ω about an axis passing through the center of gravity, as illustrated in Fig. 3.10. This method is similar to the balancing machine technique used to balance a body dynamically (see Chap. 39). If the bearings are a distance l apart and the dynamic reactions F_x and F_y are measured, the products of inertia are

$$I_{xz} = -\frac{F_x l}{\Omega^2} \qquad I_{yz} = -\frac{F_y l}{\Omega^2} \qquad (3.30)$$

Limitations to this method are (1) the size of the body that can be accommodated by the balancing machine and (2) the angular velocity that the body can withstand without damage from centrifugal forces. If the angle between the Z axis and a principal axis of inertia is small, high rotational speeds may be necessary to measure the reaction forces accurately.

PROPERTIES OF RESILIENT SUPPORTS

A resilient support is considered to be a three-dimensional element having two terminals or end connections. When the end connections are moved one relative to the other in any direction, the element resists such motion. In this chapter, the element is considered to be massless; the force that resists relative motion across the element is considered to consist of a spring force that is directly proportional to the relative displacement (deflection across the element) and a damping force that is directly proportional to the relative velocity (velocity across the element). Such an element is defined as a *linear resilient support*. Nonlinear elements are discussed in Chap. 4; elements with mass are discussed in Chap. 30; and nonlinear damping is discussed in Chaps. 2 and 30.

In a single degree-of-freedom system or in a system having constraints on the paths of motion of elements of the system (Chap. 2), the resilient element is constrained to deflect in a given direction and the properties of the element are defined with respect to the force opposing motion in this direction. In the absence of such constraints, the application of a force to a resilient element generally causes a motion in a different direction. The *principal elastic axes* of a resilient element are those axes for which the element, when unconstrained, experiences a deflection colineal with the direction of the applied force.[5] Any axis of symmetry is a principal elastic axis.

In rigid body dynamics, the rigid body sometimes vibrates in modes that are coupled by the properties of the resilient elements as well as by their location. For example, if the body experiences a static displacement x in the direction of the X axis only, a resilient element opposes this motion by exerting a force $k_{xx}x$ on the body in the direction of the X axis where one subscript on the spring constant k indicates the direction of the force exerted by the element and the other subscript indicates the direction of the deflection. If the X direction is not a principal elastic direction of the element and the body experiences a static displacement x in the X direction, the body is acted upon by a force $k_{yx}x$ in the Y direction if no displacement y is permitted. The stiffnesses have reciprocal properties; i.e., $k_{xy} = k_{yx}$. In general, the stiffnesses in the directions of the coordinate axes can be expressed in terms of (1) principal stiffnesses and (2) the angles between the coordinate axes and the principal elastic axes of the element. (See Chap. 30 for a detailed discussion of a biaxial stiffness element.) Therefore, the stiffness of a resilient element can be represented pictorially by the combination of three mutually perpendicular, idealized springs oriented along the principal elastic directions of the resilient element. Each spring has a stiffness equal to the principal stiffness represented.

FIG. 3.11. Pictorial representation of the properties of an undamped resilient element in the XZ plane including a torsional spring k_t. An analysis of the motion of the supported body in the XZ plane shows that the torsional spring can be neglected if $k_t \ll a_z{}^2 k_x$.

A resilient element is assumed to have damping properties such that each spring representing a value of principal stiffness is paralleled by an idealized viscous damper, each damper representing a value of principal damping. Hence, coupling through damping exists in a manner similar to coupling through stiffness. Consequently, the viscous damping coefficient c is analogous to the spring coefficient k; i.e., the force exerted by the damping of the resilient element in response to a velocity $\dot{x}$ is $c_{xx}\dot{x}$ in the direction of the X axis and $c_{yx}\dot{x}$ in the direction of the Y axis if $\dot{y}$ is zero. Reciprocity exists; i.e., $c_{xy} = c_{yx}$.

The point of intersection of the principal elastic axes of a resilient element is designated as the *elastic center of the resilient element*. The elastic center is important since it defines the theoretical point location of the resilient element for use in the equations of motion of a resiliently supported rigid body. For example, the torque on the rigid body about the Y axis due to a force $k_{xx}x$ transmitted by a resilient element in the X direction is $k_{xx}a_z x$, where a_z is the Z coordinate of the elastic center of the resilient element.

In general, it is assumed that a resilient element is attached to the rigid body by means of "ball joints"; i.e., the resilient element is incapable of applying a couple to the body. If this assumption is not made, a resilient element would be represented not only by translational springs and dampers along the principal elastic axes but also by torsional springs and dampers resisting rotation about the principal elastic directions.

Figure 3.11 shows that the torsional elements usually can be neglected. The torque which acts on the rigid body due to a rotation β of the body and a rotation $\boldsymbol{\beta}$ of the support is $(k_t + a_z^2 k_x)(\beta - \boldsymbol{\beta})$, where k_t is the torsional spring constant in the β direction. The torsional stiffness k_t usually is much smaller than $a_z^2 k_x$ and can be neglected. Treatment of the general case indicates that if the torsional stiffnesses of the resilient element are small compared with the product of the translational stiffnesses times the square of distances from the elastic center of the resilient element to the center-of-gravity of the rigid body, the torsional stiffnesses have a negligible effect on the vibrational behavior of the body. The treatment of torsional dampers is completely analogous.

EQUATIONS OF MOTION FOR A RESILIENTLY SUPPORTED RIGID BODY

The differential equations of motion for the rigid body are given by Eqs. (3.2) and (3.3) where the $\mathbf{F}$'s and $\mathbf{M}$'s represent the forces and moments acting on the body, either directly or through the resilient supporting elements. Figure 3.12 shows a view of a rigid body at rest with an inertial set of axes $\bar{X}$, $\bar{Y}$, $\bar{Z}$ and a coincident set of axes X, Y, Z fixed in the rigid body, both sets of axes passing through the center-of-mass. A typical resilient element (2) is represented by parallel spring and viscous damper combinations arranged respectively parallel with the $\bar{X}$, $\bar{Y}$, $\bar{Z}$ axes. Another resilient element (1) is shown with its principal axes not parallel with $\bar{X}$, $\bar{Y}$, $\bar{Z}$.

The displacement of the center-of-gravity of the body in the $\bar{X}$, $\bar{Y}$, $\bar{Z}$ directions is in Fig. 3.1 indicated by x_c, y_c, z_c, respectively; and rotation of the rigid body about these axes is indicated by α, β, γ, respectively. In Fig. 3.12, each resilient element is represented by three mutually perpendicular spring-damper combinations. One end of each such combination is attached to the rigid body; the other end is considered to be attached to a foundation whose corresponding translational displacement is defined by u, v, w in the $\bar{X}$, $\bar{Y}$, $\bar{Z}$ directions, respectively, and whose rotational displacement about these axes is defined by $\boldsymbol{\alpha}$, $\boldsymbol{\beta}$, $\boldsymbol{\gamma}$, respectively. The point of attachment of each of the idealized resilient elements is located at the coordinate distances a_x, a_y, a_z of the elastic center of the resilient element.

Consider the rigid body to experience a translational displacement x_c of its center-of-gravity and no other displacement, and neglect the effects of the viscous dampers. The force developed by a resilient element has the effect of a force $-k_{xx}(x_c - u)$ in the X direction, a moment $k_{xx}(x_c - u)a_y$ in the γ coordinate (about the Z axis) and a moment $-k_{xx}(x_c - u)a_z$ in the β coordinate (about the Y axis). Furthermore, the coupling stiffness causes a force $-k_{xy}(x_c - u)$ in the Y direction and a force $-k_{xz}(x_c - u)$ in the Z direction. These forces have the moments: $k_{xy}(x_c - u)a_z$ in the α coordinate; $-k_{xy}(x_c - u)a_x$ in the γ coordinate; $k_{xz}(x_c - u)a_x$ in the β coordinate; and $-k_{xz}(x_c - u)a_y$ in the α coordinate. By considering in a similar manner the forces and moments developed by a resilient element for successive displacements of the rigid body in the three translational and three rotational coordinates, and summing over the number of resilient elements, the equations of motion are written as follows:[6, 7, 8]

$$m\ddot{x}_c + \Sigma k_{xx}(x_c - u) + \Sigma k_{xy}(y_c - v) + \Sigma k_{xz}(z_c - w)$$
$$+ \Sigma(k_{xz}a_y - k_{xy}a_z)(\alpha - \boldsymbol{\alpha}) + \Sigma(k_{xx}a_z - k_{xz}a_x)(\beta - \boldsymbol{\beta})$$
$$+ \Sigma(k_{xy}a_x - k_{xx}a_y)(\gamma - \boldsymbol{\gamma}) = F_x \tag{3.31a}$$

$$I_{xx}\ddot{\alpha} - I_{xy}\ddot{\beta} - I_{xz}\ddot{\gamma} + \Sigma(k_{xz}a_y - k_{xy}a_z)(x_c - u)$$
$$+ \Sigma(k_{yz}a_y - k_{yy}a_z)(y_c - v) + \Sigma(k_{zz}a_y - k_{yz}a_z)(z_c - w)$$
$$+ \Sigma(k_{yy}a_z^2 + k_{zz}a_y^2 - 2k_{yz}a_ya_z)(\alpha - \boldsymbol{\alpha})$$
$$+ \Sigma(k_{xz}a_ya_z + k_{yz}a_xa_z - k_{zz}a_xa_y - k_{xy}a_z^2)(\beta - \boldsymbol{\beta})$$
$$+ \Sigma(k_{xy}a_ya_z + k_{yz}a_xa_y - k_{yy}a_xa_z - k_{xz}a_y^2)(\gamma - \boldsymbol{\gamma}) = M_x \tag{3.31b}$$

$$m\ddot{y}_c + \Sigma k_{xy}(x_c - u) + \Sigma k_{yy}(y_c - v) + \Sigma k_{yz}(z_c - w)$$
$$+ \Sigma(k_{yz}a_y - k_{yy}a_z)(\alpha - \mathbf{a}) + \Sigma(k_{xy}a_z - k_{yz}a_x)(\beta - \boldsymbol{\beta})$$
$$+ \Sigma(k_{yy}a_x - k_{xy}a_y)(\gamma - \boldsymbol{\gamma}) = F_y \qquad (3.31c)$$

$$I_{yy}\ddot{\beta} - I_{xy}\ddot{\alpha} - I_{yz}\ddot{\gamma} + \Sigma(k_{xx}a_z - k_{xz}a_x)(x_c - u)$$
$$+ \Sigma(k_{xy}a_z - k_{yz}a_x)(y_c - v) + \Sigma(k_{xz}a_z - k_{zz}a_x)(z_c - w)$$
$$+ \Sigma(k_{zz}a_ya_z + k_{yz}a_xa_z - k_{zz}a_xa_y - k_{xy}a_z^2)(\alpha - \mathbf{a})$$
$$+ \Sigma(k_{xx}a_z^2 + k_{zz}a_x^2 - 2k_{xz}a_xa_z)(\beta - \boldsymbol{\beta})$$
$$+ \Sigma(k_{xy}a_za_z + k_{xz}a_xa_y - k_{xx}a_ya_z - k_{yz}a_x^2)(\gamma - \boldsymbol{\gamma}) = M_y \qquad (3.31d)$$

$$m\ddot{z}_c + \Sigma k_{xz}(x_c - u) + \Sigma k_{yz}(y_c - v) + \Sigma k_{zz}(z_c - w)$$
$$+ \Sigma(k_{zz}a_y - k_{yz}a_z)(\alpha - \mathbf{a}) + \Sigma(k_{xz}a_z - k_{zz}a_x)(\beta - \boldsymbol{\beta})$$
$$+ \Sigma(k_{yz}a_x - k_{xz}a_y)(\gamma - \boldsymbol{\gamma}) = F_z \qquad (3.31e)$$

$$I_{zz}\ddot{\gamma} - I_{xz}\ddot{\alpha} - I_{yz}\ddot{\beta} + \Sigma(k_{xy}a_x - k_{xx}a_y)(x_c - u)$$
$$+ \Sigma(k_{yy}a_x - k_{xy}a_y)(y_c - v) + \Sigma(k_{yz}a_x - k_{xz}a_y)(z_c - w)$$
$$+ \Sigma(k_{xy}a_ya_z + k_{yz}a_xa_y - k_{yy}a_xa_z - k_{xz}a_y^2)(\alpha - \mathbf{a})$$
$$+ \Sigma(k_{xy}a_xa_z + k_{xz}a_xa_y - k_{xx}a_ya_z - k_{yz}a_x^2)(\beta - \boldsymbol{\beta})$$
$$+ \Sigma(k_{xx}a_y^2 + k_{yy}a_x^2 - 2k_{xy}a_xa_y)(\gamma - \boldsymbol{\gamma}) = M_z \qquad (3.31f)$$

where the moments and products of inertia are defined by Eqs. (3.11) and (3.12) and the stiffness coefficients are defined as follows:

$$k_{xx} = k_p\lambda_{xp}^2 + k_q\lambda_{xq}^2 + k_r\lambda_{xr}^2$$
$$k_{yy} = k_p\lambda_{yp}^2 + k_q\lambda_{yq}^2 + k_r\lambda_{yr}^2$$
$$k_{zz} = k_p\lambda_{zp}^2 + k_q\lambda_{zq}^2 + k_r\lambda_{zr}^2$$
$$k_{xy} = k_p\lambda_{xp}\lambda_{yp} + k_q\lambda_{xq}\lambda_{yq} + k_r\lambda_{xr}\lambda_{yr} \qquad (3.32)$$
$$k_{xz} = k_p\lambda_{xp}\lambda_{zp} + k_q\lambda_{xq}\lambda_{zq} + k_r\lambda_{xr}\lambda_{zr}$$
$$k_{yz} = k_p\lambda_{yp}\lambda_{zp} + k_q\lambda_{yq}\lambda_{zq} + k_r\lambda_{yr}\lambda_{zr}$$

where the λ's are the cosines of the angles between the principal elastic axes of the resilient supporting elements and the coordinate axes. For example, λ_{xp} is the cosine of the angle between the X axis and the P axis of principal stiffness.

The equations of motion, Eqs. (3.31), do not include forces applied to the rigid body by damping forces from the resilient elements. To include damping, appropriate damping terms analogous to the corresponding stiffness terms are added to each equation. For example, Eq. (3.31a) would become

$$m\ddot{x}_c + \Sigma c_{xx}(\dot{x}_c - \dot{u}) + \Sigma k_{xx}(x_c - u) + \cdots$$
$$+ \Sigma(c_{xz}a_y - c_{xy}a_z)(\dot{\alpha} - \dot{\mathbf{a}}) + \Sigma(k_{xz}a_y - k_{xy}a_z)(\alpha - \mathbf{a}) + \cdots = F_x \qquad (3.31a')$$

where
$$c_{xx} = c_p\lambda_{xp}^2 + c_q\lambda_{xq}^2 + c_r\lambda_{xr}^2$$
$$c_{xy} = c_p\lambda_{xp}\lambda_{yp} + c_q\lambda_{xq}\lambda_{yq} + c_r\lambda_{xr}\lambda_{yr}$$

The number of degrees-of-freedom of a vibrational system is the minimum number of coordinates necessary to define completely the positions of the mass elements of the system in space. The system of Fig. 3.12 requires a minimum of six coordinates $(x_c, y_c, z_c, \alpha, \beta, \gamma)$ to define the position of the rigid body in space; thus, the system is said to vibrate in six degrees-of-freedom. Equations (3.31) may be solved simultaneously for the three components x_c, y_c, z_c of the center-of-gravity displacement and the three components α, β, γ of the rotational displacement of the rigid body. In most practical instances, the equations are simplified considerably by one or more of the following simplifying conditions:

1. The reference axes X, Y, Z are selected to coincide with the principal inertial axes of the body; then

$$I_{xy} = I_{xz} = I_{yz} = 0 \qquad (3.33)$$

2. The resilient supporting elements are so arranged that one or more planes of symmetry exist; i.e., motion parallel to the plane of symmetry has no tendency to excite motion perpendicular to it, or rotation about an axis lying in the plane does not

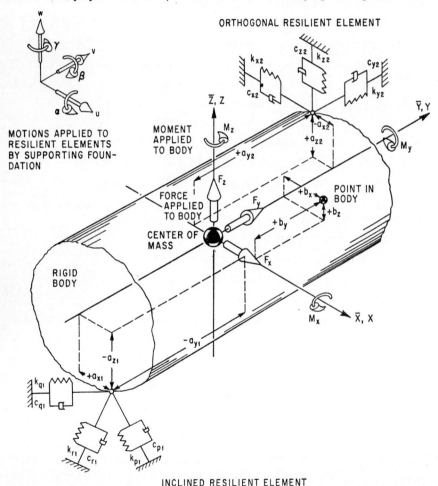

Fig. 3.12. Rigid body at rest supported by resilient elements, with inertial axes $\overline{X}$, $\overline{Y}$, $\overline{Z}$ and coincident reference axes X, Y, Z passing through the center-of-mass. The forces F_x, F_y, F_z and the moments M_x, M_y, M_z are applied directly to the body; the translations u, v, w and rotations α, β, γ in and about the X, Y, Z axes, respectively, are applied to the resilient elements located at the coordinates a_x, a_y, a_z. The principal directions of resilient element (2) are parallel to the $\overline{X}$, $\overline{Y}$, $\overline{Z}$ axes (orthogonal) and those of resilient element (1) are not parallel to the $\overline{X}$, $\overline{Y}$, $\overline{Z}$ axes (inclined).

excite motion parallel to the plane. For example, in Eq. (3.31a), motion in the XY plane does not tend to excite motion in the XZ or YZ plane if Σk_{zz}, $\Sigma(k_{zz}a_y - k_{xy}a_z)$, and $\Sigma(k_{xx}a_z - k_{xz}a_x)$ are zero.

3. The principal elastic axes P, Q, R of all resilient supporting elements are orthogonal with the reference axes X, Y, Z of the body, respectively. Then, in Eqs. (3.32),

$$k_{xx} = k_p = k_x \qquad k_{yy} = k_q = k_y \qquad k_{zz} = k_r = k_z$$

$$k_{xy} = k_{xz} = k_{yz} = 0 \tag{3.34}$$

where k_x, k_y, k_z are defined for use when orthogonality exists. The supports are then called *orthogonal supports*.

4. The forces F_x, F_y, F_z and moments M_x, M_y, M_z are applied directly to the body and there are no motions ($u = v = w = \alpha = \beta = \gamma = 0$) of the foundation; or alternatively, the forces and moments are zero and excitation results from motion of the foundation.

In general, the effect of these simplifications is to reduce the numbers of terms in the equations and, in some instances, to reduce the number of equations that must be solved simultaneously. Simultaneous equations indicate coupled modes; i.e., motion cannot exist in one coupled mode independently of motion in other modes which are coupled to it.

MODAL COUPLING AND NATURAL FREQUENCIES

Several conditions of symmetry resulting from zero values for the product of inertia terms in Eq. (3.33) are discussed in the following sections.

ONE PLANE OF SYMMETRY WITH ORTHOGONAL RESILIENT SUPPORTS

When the YZ plane of the rigid body system in Fig. 3.12 is a plane of symmetry, the following terms in the equations of motion are zero:

$$\Sigma k_{yy}a_x = \Sigma k_{zz}a_x = \Sigma k_{yy}a_x a_z = \Sigma k_{zz}a_x a_y = 0 \tag{3.35}$$

Introducing the further simplification that the principal elastic axes of the resilient elements are parallel with the reference axes, Eqs. (3.34) apply. Then the motions in the three coordinates y_c, z_c, α are coupled but are independent of motion in any of the other coordinates; furthermore, the other three coordinates x_c, β, γ also are coupled. For example, Fig. 3.13 illustrates a resiliently supported rigid body, wherein

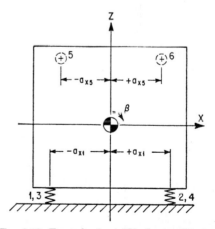

Fig. 3.13. Example of a rigid body on orthogonal resilient supporting elements with one plane of symmetry. The YZ plane is a plane of symmetry since each resilient element has properties identical to those of its mirror image in the YZ plane; i.e., $k_{x1} = k_{x2}$, $k_{x3} = k_{x4}$, $k_{x5} = k_{x6}$, etc. The conditions satisfied are Eqs. (3.33) to (3.35).

FIG. 3.14A. Graphical method of determining solutions of the cubic Eq. (3.36). Calculate A, B, C for the appropriate set of coupled coordinates, enter the abscissa at B/A^2 (values less than 0.2 on Fig. 3.14A, values greater than 0.2 on Fig. 3.14B), and read three values of $(f_n/f_z)/\sqrt{A}$ from the curve having the appropriate value of C/A^3.

FIG. 3.14B. Using the above nomograph with values of $(f_n/f_z)/\sqrt{A}$ (see Fig. 3.14A), a diagonal line is drawn from each value of $(f_n/f_z)/\sqrt{A}$ on the left scale of the nomograph to the value of A on the right scale, as indicated by the dotted lines. The three roots f_n/f_z of Eq. (3.36) are given by the intercept of these dotted lines with the center scale of the nomograph. (*After F. F. Vane.*[9])

the YZ plane is a plane of symmetry that meets the requirements of Eq. (3.35). The three natural frequencies for the y_c, z_c, α coupled directions are found by solving Eqs. (3.31b), (3.31c), and (3.31e) [or Eqs. (3.31a), (3.31d), and (3.31f) for the x_c, β, γ coupled directions] simultaneously.[9, 10]

$$\left(\frac{f_n}{f_z}\right)^6 - A\left(\frac{f_n}{f_z}\right)^4 + B\left(\frac{f_n}{f_z}\right)^2 - C = 0 \tag{3.36}$$

where

$$f_z = \frac{1}{2\pi}\sqrt{\frac{\Sigma k_z}{m}} \tag{3.37}$$

is a quantity having mathematical rather than physical significance if translational motion in the direction of the Z axis is coupled to other modes of motion. (Such coupling exists for the system of Fig. 3.13.) The roots f_n represent the natural frequencies of the system in the coupled modes. The coefficients A, B, C for the coupled modes in the y_c, z_c, α coordinates are

$$A_{yz\alpha} = 1 + \frac{\Sigma k_y}{\Sigma k_z} + D_{zx}$$

$$B_{yz\alpha} = D_{zx} + \frac{\Sigma k_y}{\Sigma k_z}(1 + D_{zx}) - \frac{(\Sigma k_y a_z)^2 + (\Sigma k_z a_y)^2}{\rho_x{}^2(\Sigma k_z)^2}$$

$$C_{yz\alpha} = \frac{\Sigma k_y}{\Sigma k_z}\left(D_{zx} - \frac{(\Sigma k_z a_y)^2}{\rho_x{}^2(\Sigma k_z)^2}\right) - \frac{(\Sigma k_y a_z)^2}{\rho_x{}^2(\Sigma k_z)^2}$$

where

$$D_{zx} = \frac{\Sigma k_y a_z{}^2 + \Sigma k_z a_y{}^2}{\rho_x{}^2 \Sigma k_z}$$

and ρ_x is the radius of gyration of the rigid body with respect to the X axis.

The corresponding coefficients for the coupled modes in the x_c, β, γ coordinates are

$$A_{x\beta\gamma} = \frac{\Sigma k_x}{\Sigma k_z} + D_{zy} + D_{zz}$$

$$B_{x\beta\gamma} = \frac{\Sigma k_x}{\Sigma k_z}(D_{zy} + D_{zz}) + D_{zy}D_{zz}$$

$$- \frac{(\Sigma k_x a_z)^2}{\rho_y{}^2(\Sigma k_z)^2} - \frac{(\Sigma k_x a_y)^2}{\rho_z{}^2(\Sigma k_z)^2} - \frac{(\Sigma k_x a_y a_z)^2}{\rho_y{}^2\rho_z{}^2(\Sigma k_z)^2}$$

$$C_{x\beta\gamma} = \frac{\Sigma k_x}{\Sigma k_z}\left[D_{zy}D_{zz} - \frac{(\Sigma k_x a_y a_z)^2}{\rho_y{}^2\rho_z{}^2(\Sigma k_z)^2}\right] - \frac{(\Sigma k_x a_y)^2}{\rho_z{}^2(\Sigma k_z)^2}D_{zy}$$

$$- \frac{(\Sigma k_x a_z)^2}{\rho_y{}^2(\Sigma k_z)^2}D_{zz} + 2\frac{(\Sigma k_x a_y)(\Sigma k_x a_z)(\Sigma k_x a_y a_z)}{\rho_y{}^2\rho_z{}^2(\Sigma k_z)^3}$$

where

$$D_{zy} = \frac{\Sigma k_x a_z{}^2 + \Sigma k_z a_x{}^2}{\rho_y{}^2 \Sigma k_z} \qquad D_{zz} = \frac{\Sigma k_x a_y{}^2 + \Sigma k_y a_x{}^2}{\rho_z{}^2 \Sigma k_z}$$

and ρ_y, ρ_z are the radii of gyration of the rigid body with respect to the Y, Z axes.

The roots of the cubic equation Eq. (3.36) may be found graphically from Fig. 3.14.[9] The coefficients A, B, C are first calculated from the above relations for the appropriate set of coupled coordinates. Figure 3.14 is entered on the abscissa scale at the appropriate value for the quotient B/A^2. Small values of B/A^2 are in Fig. 3.14A, and large values in Fig. 3.14B. The quotient C/A^3 is the parameter for the family of curves. Upon selecting the appropriate curve, three values of $(f_n/f_z)/\sqrt{A}$ are read from the ordinate and transferred to the left scale of the nomograph in Fig. 3.14B. Diagonal lines are drawn for each root to the value of A on the right scale, as indicated by dotted lines, and the roots f_n/f_z of the equation are indicated by the intercept of these dotted lines with the center scale of the nomograph.

The coefficients A, B, C can be simplified if all resilient elements have equal stiffness in the same direction. The stiffness coefficients always appear to equal powers in numerator and denominator, and lead to dimensionless ratios of stiffness. For n resilient elements, typical terms reduce as follows:

$$\frac{\Sigma k_y}{\Sigma k_z} = \frac{k_y}{k_z} \qquad \frac{\Sigma k_z a_y{}^2}{\rho_x{}^2 \Sigma k_z} = \frac{\Sigma a_y{}^2}{n\rho_x{}^2}$$

$$\frac{(\Sigma k_x a_y a_z)^2}{\rho_y{}^2 \rho_z{}^2 (\Sigma k_z)^2} = \left(\frac{k_x}{n k_z}\frac{\Sigma a_y a_z}{\rho_y \rho_z}\right)^2, \text{ etc.}$$

TWO PLANES OF SYMMETRY WITH ORTHOGONAL RESILIENT SUPPORTS

Two planes of symmetry may be achieved if, in addition to the conditions of Eqs. (3.33) to (3.35), the following terms of Eqs. (3.31) are zero:

$$\Sigma k_{xx} a_y = \Sigma k_{zz} a_y = \Sigma k_{xx} a_y a_z = 0 \quad (3.38)$$

Under these conditions, Eqs. (3.31) separate into two independent equations, Eqs. (3.31e) and (3.31f), and two sets each consisting of two coupled equations [Eqs. (3.31a) and (3.31d); Eqs. (3.31b) and (3.31c)]. The planes of symmetry are the XZ and YZ planes. For example, a common system is illustrated in Fig. 3.15, where four identical resilient supporting elements are located symmetrically about the Z axis in a plane not containing the center-of-gravity.[6, 11, 12] Coupling exists between translation in the X direction and rotation about the Y axis (x_c,β), as well as between translation in the Y direction and rotation about the X axis (y_c,α). Translation in the Z direction (z_c) and rotation about the Z axis (γ) are each independent of all other modes.

The natural frequency in the Z direction is found by solving Eq. (3.31e) to obtain Eq. (3.37), where $\Sigma k_{zz} = 4k_z$. The rotational natural frequency f_γ about the Z axis is found by solving Eq. (3.31f); it can be expressed with respect to the natural frequency in the direction of the Z axis:

$$\frac{f_\gamma}{f_z} = \sqrt{\frac{k_x}{k_z}\left(\frac{a_y}{\rho_z}\right)^2 + \frac{k_y}{k_z}\left(\frac{a_x}{\rho_z}\right)^2} \quad (3.39)$$

where ρ_z is the radius of gyration with respect to the Z axis.

FIG. 3.15. Example of a rigid body on orthogonal resilient supporting elements with two planes of symmetry. The XZ and YZ planes are planes of symmetry since the four resilient supporting elements are identical and are located symmetrically about the Z axis. The conditions satisfied are Eqs. (3.33), (3.34), (3.35), and (3.38). At any single frequency, coupled vibration in the x_c, β direction due to X vibration of the foundation is equivalent to a pure rotation of the rigid body with respect to an axis of rotation as shown. Points 1, 2, and 3 refer to the example of Fig. 3.26.

FIG. 3.16. Curves showing the ratio of each of the two coupled natural frequencies $f_{x\beta}$ to the decoupled natural frequency f_z, for motion in the XZ plane of symmetry for the system in Fig. 3.15 [see Eq. (3.40)]. Calculate the abscissa $(\rho_y/a_x)\sqrt{k_x/k_z}$ and the parameter a_z/ρ_y, where a_x, a_z are indicated in Fig. 3.15; k_x, k_z are the stiffnesses of the resilient supporting elements in the X, Z directions, respectively; and ρ_y is the radius of gyration of the body about the Y axis. The two values read from the ordinate when divided by ρ_y/a_x give the natural frequency ratios $f_{x\beta}/f_z$. (After C. E. Crede.[12])

The natural frequencies in the coupled x_c, β modes are found by solving Eqs. (3.31a) and (3.31d) simultaneously; the roots yield the following expression for natural frequency:

$$\frac{f_{x\beta}{}^2}{f_z{}^2} = \frac{1}{2}\left\{\frac{k_x}{k_z}\left(1 + \frac{a_x{}^2}{\rho_y{}^2}\right) + \frac{a_x{}^2}{\rho_y{}^2} \pm \sqrt{\left[\frac{k_x}{k_z}\left(1 + \frac{a_z{}^2}{\rho_y{}^2}\right) + \frac{a_x{}^2}{\rho_y{}^2}\right]^2 - 4\,\frac{k_x}{k_z}\,\frac{a_x{}^2}{\rho_y{}^2}}\right\} \quad (3.40)$$

Figure 3.16 provides a convenient graphical method for determining the two coupled natural frequencies $f_{x\beta}$. An expression similar to Eq. (3.40) is obtained for $f_{y\alpha}{}^2/f_z{}^2$ by

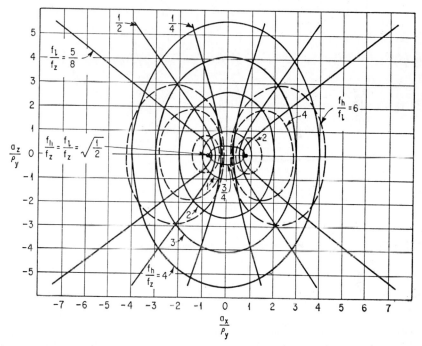

Fig. 3.17. Space-plot for the system in Fig. 3.15 when the stiffness ratio $k_x/k_z = 0.5$, obtained from Eqs. (3.40a) to (3.40c). With all dimensions divided by the radius of gyration ρ_y about the Y axis, superimpose the outline of the rigid body in the XZ plane on the plot; the center-of-gravity of the body is located at the coordinate center of the plot. The elastic centers of the resilient supporting elements give the natural frequency ratios f_l/f_z, f_h/f_z and f_h/f_l for x_c, β coupled motion, each ratio being read from one of the three families of curves as indicated on the plot. Replacing k_z, ρ_y, a_x with k_y, ρ_x, a_y, respectively, allows the plot to be applied to motions in the YZ plane.

solving Eqs. (3.31b) and (3.31d) simultaneously. By replacing ρ_y, a_x, k_x, $f_{x\beta}$ with ρ_x, a_y, k_y, $f_{y\alpha}$, respectively, Fig. 3.16 also can be used to determine the two values of $f_{y\alpha}$.

It may be desirable to select resilient element locations a_x, a_y, a_z which will produce coupled natural frequencies in specified frequency ranges, with resilient elements having specified stiffness ratios k_x/k_z, k_y/k_z. For this purpose it is convenient to plot solutions of Eq. (3.40) in the form shown in Figs. 3.17 to 3.19. These plots are termed *space-plots* and their use is illustrated in Example 3.1.[13, 14]

The space-plots are derived as follows: In general, the two roots of Eq. (3.40) are numerically different, one usually being greater than unity and the other less than unity. Designating the root associated with the positive sign before the radical (higher value)

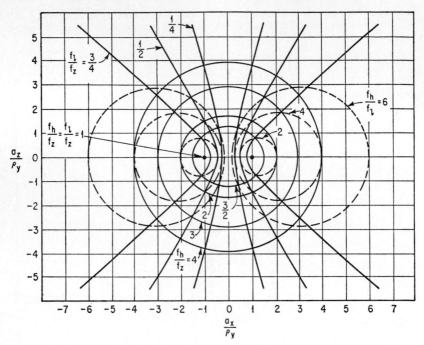

FIG. 3.18. Space-plot for the system in Fig. 3.15 when the stiffness ratio $k_x/k_z = 1$. See caption for Fig. 3.17.

as f_h/f_z, Eq. (3.40) may be written in the following form:

$$\frac{(a_x/\rho_y)^2}{(f_h/f_z)^2} + \frac{(a_z/\rho_y)^2}{(k_z/k_x)(f_h/f_z)^2 - 1} = 1 \tag{3.40a}$$

Equation (3.40a) is shown graphically by the large ellipses about the center of Figs. 3.17 to 3.19, for stiffness ratios k_x/k_z of $\frac{1}{2}$, 1, and 2, respectively. A particular type of resilient element tends to have a constant stiffness ratio k_x/k_z; thus, Figs. 3.17 to 3.19 may be used by cut-and-try methods to find the coordinates a_x, a_z of such elements to attain a desired value of f_h.

Designating the root of Eq. (3.40) associated with the negative sign (lower value) by f_l, Eq. (3.40) may be written as follows:

$$\frac{(a_x/\rho_y)^2}{(f_l/f_z)^2} - \frac{(a_z/\rho_y)^2}{1 - (k_z/k_x)(f_l/f_z)^2} = 1 \tag{3.40b}$$

Equation (3.40b) is shown graphically by the family of hyperbolas on each side of the center in Figs. 3.17 to 3.19, for values of the stiffness ratio k_x/k_z of $\frac{1}{2}$, 1, and 2.

The two roots f_h/f_z and f_l/f_z of Eq. (3.40) may be expressed as the ratio of one to the other. This relationship is given parametrically as follows:

$$\left[\frac{2\dfrac{a_x}{\rho_y} \pm \sqrt{\dfrac{k_x}{k_z}}\left(\dfrac{f_h}{f_l} + \dfrac{f_l}{f_h}\right)}{\sqrt{\dfrac{k_x}{k_z}}\left(\dfrac{f_h}{f_l} - \dfrac{f_l}{f_h}\right)}\right]^2 + \left[\frac{2\dfrac{a_z}{\rho_y}}{\dfrac{f_h}{f_l} - \dfrac{f_l}{f_h}}\right]^2 = 1 \tag{3.40c}$$

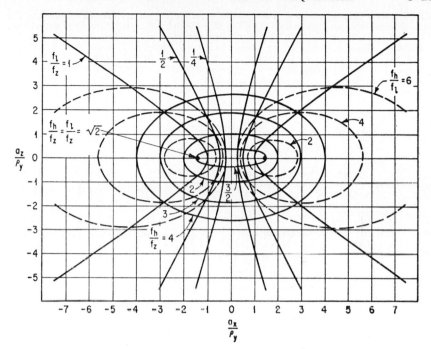

FIG. 3.19. Space-plot for the system in Fig. 3.15 when the stiffness ratio $k_x/k_z = 2$. See caption for Fig. 3.17.

Equation (3.40c) is shown graphically by the smaller ellipses (shown dotted) displaced from the vertical center line in Figs. 3.17 to 3.19.

Example 3.1. A rigid body is symmetrical with respect to the XZ plane; its width in the X direction is 13 in. and its height in the Z direction is 12 in. The center-of-gravity is $5\frac{1}{2}$ in. from the lower side and $6\frac{3}{4}$ in. from the right side. The radius of gyration about the Y axis through the center-of-gravity is 5.10 in. Use a space-plot to evaluate the effects of the location for attachment of resilient supporting elements having the characteristic stiffness ratio $k_x/k_z = \frac{1}{2}$.

Superimpose the outline of the body on the space-plot of Fig. 3.20, with its center-of-gravity at the coordinate center of the plot. (Figure 3.20 is an enlargement of the central portion of Fig. 3.17.) All dimensions are divided by the radius of gyration ρ_y. Thus, the four corners of the body are located at coordinate distances as follows:

Upper right corner:
$$\frac{a_z}{\rho_y} = \frac{+6.50}{5.10} = +1.28 \qquad \frac{a_x}{\rho_y} = \frac{+6.75}{5.10} = +1.32$$

Upper left corner:
$$\frac{a_z}{\rho_y} = \frac{+6.50}{5.10} = +1.28 \qquad \frac{a_x}{\rho_y} = \frac{-6.25}{5.10} = -1.23$$

Lower right corner:
$$\frac{a_z}{\rho_y} = \frac{-5.50}{5.10} = -1.08 \qquad \frac{a_x}{\rho_y} = \frac{+6.75}{5.10} = +1.32$$

Lower left corner:
$$\frac{a_z}{\rho_y} = \frac{-5.50}{5.10} = -1.08 \qquad \frac{a_x}{\rho_y} = \frac{-6.25}{5.10} = -1.23$$

The resilient supports are shown in heavy outline at A in Fig. 3.20, with their elastic centers indicated by the solid dots. The horizontal coordinates of the resilient supports are $a_x/\rho_y =$

± 0.59, or $a_x = \pm 0.59 \times 5.10 = \pm 3$ in. from the vertical coordinate axis. The corresponding natural frequencies are $f_h/f_z = 1.25$ (from the ellipses) and $f_l/f_z = 0.33$ (from the hyperbolas). An alternative position is indicated by the hollow circles B. The natural frequencies for this position are $f_h/f_z = 1.43$ and $f_l/f_z = 0.50$. The natural frequency f_z in vertical translation is found from the mass of the equipment and the summation of stiffnesses in the Z direction, using Eq. (3.37). This example shows how space-plots make it possible to determine the locations of the resilient elements required to achieve given values of the coupled natural frequencies with respect to f_z.

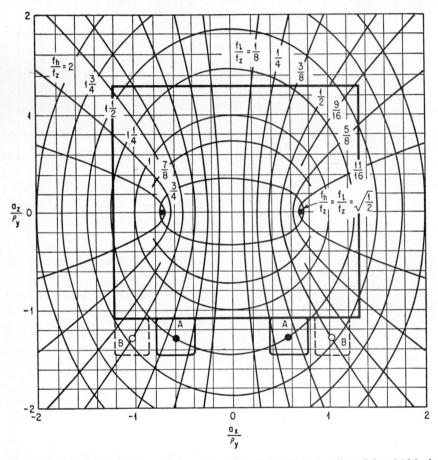

FIG. 3.20. Enlargement of the central portion of Fig. 3.17 with the outline of the rigid body discussed in Example 3.1.

THREE PLANES OF SYMMETRY WITH ORTHOGONAL RESILIENT SUPPORTS

A system with three planes of symmetry is defined by six independent equations of motion. A system having this property is sometimes called a *center-of-gravity system*. The equations are derived from Eqs. (3.31) by substituting, in addition to the conditions of Eqs. (3.33), (3.34), (3.35), and (3.38), the following condition:

$$\Sigma k_{xx}a_z = \Sigma k_{yy}a_z = 0 \tag{3.41}$$

The resulting six independent equations define six uncoupled modes of vibration, three in translation and three in rotation. The natural frequencies are:

Translation along X axis:

$$f_x = \frac{1}{2\pi}\sqrt{\frac{\Sigma k_x}{m}}$$

Translation along Y axis:

$$f_y = \frac{1}{2\pi}\sqrt{\frac{\Sigma k_y}{m}}$$

Translation along Z axis:

$$f_z = \frac{1}{2\pi}\sqrt{\frac{\Sigma k_z}{m}}$$

Rotation about X axis:

$$f_\alpha = \frac{1}{2\pi}\sqrt{\frac{\Sigma(k_y a_z{}^2 + k_z a_y{}^2)}{m\rho_x{}^2}} \qquad (3.42)$$

Rotation about Y axis:

$$f_\beta = \frac{1}{2\pi}\sqrt{\frac{\Sigma(k_x a_z{}^2 + k_z a_x{}^2)}{m\rho_y{}^2}}$$

Rotation about Z axis:

$$f_\gamma = \frac{1}{2\pi}\sqrt{\frac{\Sigma(k_x a_y{}^2 + k_y a_x{}^2)}{m\rho_z{}^2}}$$

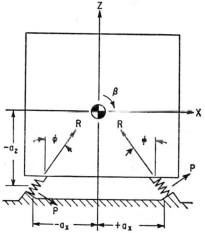

FIG. 3.21. Example of a rigid body on resilient supporting elements inclined toward the YZ plane. The resilient supporting elements are identical and are located symmetrically about the Z axis, making XZ and YZ planes of symmetry. The principal stiffnesses in the XZ plane are k_p and k_r. The conditions satisfied are Eqs. (3.33), (3.35), and (3.38).

TWO PLANES OF SYMMETRY WITH RESILIENT SUPPORTS INCLINED IN ONE PLANE ONLY

When the principal elastic axes of the resilient supporting elements are inclined with respect to the X, Y, Z axes, the stiffness coefficients k_{xy}, k_{xz}, k_{yz} are nonzero. This introduces elastic coupling which must be considered in evaluating the equations of motion. Two planes of symmetry may be achieved by meeting the conditions of Eqs. (3.33), (3.35), and (3.38). For example, consider the rigid body supported by four identical resilient supporting elements located symmetrically about the Z axis, as shown in Fig. 3.21. The XZ and the YZ planes are planes of symmetry and the resilient elements are inclined toward the YZ plane so that one of their principal elastic axes R is inclined at the angle ϕ with the Z direction as shown; hence $k_{yy} = k_q$, and $k_{xy} = k_{yz} = 0$.

Because of symmetry, translational motion z_c in the Z direction and rotation γ about the Z axis are each decoupled from the other modes. The pairs of translational and rotational modes in the x_c, β and y_c, α coordinates are coupled. The natural frequency in the Z direction is

$$\frac{f_z}{f_r} = \sqrt{\frac{k_p}{k_r}\sin^2\phi + \cos^2\phi} \qquad (3.43)$$

where f_r is a fictitious natural frequency used for convenience only; it is related to Eq (3.37) wherein $4k_r$ is substituted for Σk_z:

$$f_r = \frac{1}{2\pi}\sqrt{\frac{4k_r}{m}}$$

Equation (3.43) is plotted in Fig. 3.22 where the angle ϕ is indicated by the upper of the abscissa scales.

FIG. 3.22. Curves showing the ratio of the decoupled natural frequency f_z of translation z_c to the fictitious natural frequency f_r for the system shown in Fig. 3.21 [see Eq. (3.43)] when the resilient supporting elements are inclined at the angle ϕ. The curves also indicate the ratio of the decoupled natural frequency f_x of translation x_c to f_r when ϕ has a value ϕ' (use lower abscissa scale) which decouples x_c, β motions [see Eqs. (3.47) and (3.48)]. (*After C. E. Crede.*[15])

The rotational natural frequency about the Z axis is obtained from

$$\frac{f_\gamma}{f_r} = \sqrt{\left(\frac{k_p}{k_r}\cos^2\phi + \sin^2\phi\right)\left(\frac{a_y}{\rho_z}\right)^2 + \frac{k_q}{k_r}\left(\frac{a_x}{\rho_z}\right)^2} \tag{3.44}$$

For the x_c, β coupled mode, the two natural frequencies are

$$\frac{f_{x\beta}}{f_r} = \frac{1}{2}\left[A \pm \sqrt{A^2 - 4\frac{k_p}{k_r}\left(\frac{a_x}{a_y}\right)^2}\right] \tag{3.45}$$

where $\quad A = \left(\frac{k_p}{k_r}\cos^2\phi + \sin^2\phi\right)\left[1 + \left(\frac{a_z}{\rho_y}\right)^2\right] + \left(\frac{k_p}{k_r}\sin^2\phi + \cos^2\phi\right)\left(\frac{a_x}{\rho_y}\right)^2$

$$+ 2\left(1 - \frac{k_p}{k_r}\right)\left|\frac{a_x}{\rho_y}\right|\sin\phi\cos\phi$$

For the y_c, α coupled mode, the natural frequencies are

$$\frac{f_{y\alpha}}{f_r} = \frac{1}{2}\left[B \pm \sqrt{B^2 - 4\frac{k_q}{k_r}\left(\frac{k_p}{k_r}\sin^2\phi + \cos^2\phi\right)\left(\frac{a_y}{\rho_x}\right)^2} \right] \qquad (3.46)$$

where

$$B = \frac{k_q}{k_r}\left[1 + \left(\frac{a_z}{\rho_x}\right)^2\right] + \left(\frac{k_p}{k_r}\sin^2\phi + \cos^2\phi\right)\left(\frac{a_y}{\rho_x}\right)^2$$

DECOUPLING OF MODES IN A PLANE USING INCLINED RESILIENT SUPPORTS

The angle ϕ of inclination of principal elastic axes (see Fig. 3.21) can be varied to produce changes in the amount of coupling between the x_c and β coordinates. Decoupling

FIG. 3.23. Curves showing the angle of inclination ϕ' of the resilient elements which achieves decoupling of the x_c, β motions in Fig. 3.21 [see Eq. (3.47)]. Calculate the ordinate $|a_z/a_x|$ and with the stiffness ratio k_p/k_r determine two values of ϕ' for which decoupling is possible. Decoupling is not possible for a particular value of k_p/k_r if $|a_z/a_y|$ has a value greater than the maximum ordinate of the k_p/k_r curve.

of the x_c and β coordinates is effected if[15]

$$\left| \frac{a_z}{a_x} \right| = \frac{[1 - (k_p/k_r)] \cot \phi'}{1 + (k_p/k_r) \cot^2 \phi'} \tag{3.47}$$

where ϕ' is the value of the angle of inclination ϕ required to achieve decoupling. When Eq. (3.47) is satisfied, the configuration is sometimes called an "equivalent center-of-gravity system" in the YZ plane since all modes of motion in that plane are decoupled. Figure 3.23 is a graphical presentation of Eq. (3.47). There may be two values of ϕ' that decouple the x_c and β modes for any combination of stiffness and location for the resilient supporting elements.

The decoupled natural frequency for translation in the X direction is obtained from

$$\frac{f_x}{f_r} = \sqrt{\frac{k_p}{k_r} \cos^2 \phi' + \sin^2 \phi'} \tag{3.48}$$

The relation of Eq. (3.48) is shown graphically in Fig. 3.22 where the angle ϕ' is indicated by the lower of the abscissa scales. The natural frequency in the β mode is obtained from

$$\frac{f_\beta}{f_r} = \frac{a_x}{\rho_y} \sqrt{\frac{1}{(k_r/k_p) \sin^2 \phi' + \cos^2 \phi'}} \tag{3.49}$$

COMPLETE DECOUPLING OF MODES USING RADIALLY INCLINED RESILIENT SUPPORTS

In general, the analysis of rigid body motion with the resilient supporting elements inclined in more than one plane is quite involved. A particular case where sufficient symmetry exists to provide relatively simple yet useful results is the configuration illustrated in Fig. 3.24. From symmetry about the Z axis, $I_{xx} = I_{yy}$. Any number n of resilient supporting elements greater than 3 may be used. For clarity of illustration, the rigid body is shown as a right circular cylinder with $n = 3$.

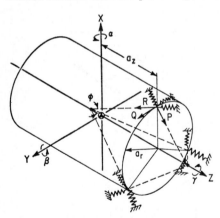

The resilient supporting elements are arranged symmetrically about the Z axis; they are attached to one end face of the cylinder at a distance a_r from the Z axis and a distance a_z from the XY reference plane. The resilient elements are inclined so that their principal elastic axes R intersect at a common point on the Z axis; thus, the angle between the Z axis and the R axis for each element is ϕ. The principal elastic axes P also intersect at a common point on the Z axis, the angle between the Z axis and the P axis for each element being $90° - \phi$. Consequently, the Q principal elastic axes are each tangent to the circle of radius a_r which bounds the end face of the cylinder.

Fig. 3.24. Example of a rigid cylindrical body on radially inclined resilient supports. The resilient supports are attached symmetrically about the Z axis at one end face of the cylinder at a distance a_r from the Z axis and a distance a_z from the XY plane. The resilient elements are inclined so that their principal elastic axes R and P intersect the Z axis at common points. The angle between the R axes and the Z axis is ϕ; and the angle between the P axis and Z axis is $90° - \phi$. The Q principal elastic axes are each tangent to the circle of radius a_r.

The use of such a configuration permits decoupling of all six modes of vibration of the rigid body. This complete decoupling is achieved if the angle of inclination ϕ has the value ϕ' which satisfies the following equation:

$$\left|\frac{a_z}{a_r}\right| = \frac{(\frac{1}{2})[1 - (k_p/k_r)]\sin 2\phi'}{(k_q/k_r) + (k_p/k_r) + [1 - (k_p/k_r)]\sin^2 \phi'} \tag{3.50}$$

Since complete decoupling is effected, the system may be termed an "equivalent center-of-gravity system."[16, 17] The natural frequencies of the six decoupled modes are

$$\frac{f_x}{f_r} = \frac{f_y}{f_r} = \sqrt{\frac{1}{2}\left(\frac{k_p}{k_r}\cos^2 \phi' + \sin^2 \phi' + \frac{k_q}{k_r}\right)} \tag{3.51}$$

$$\frac{f_\alpha}{f_r} = \frac{f_\beta}{f_r} = \left\{\frac{a_r}{2\rho_x}\left[\frac{k_p}{k_r}\sin \phi'\left(\frac{a_r}{\rho_x}\sin \phi' + \frac{a_z}{\rho_x}\cos \phi'\right) + \cos \phi'\left(\frac{a_r}{\rho_x}\cos \phi' - \frac{a_z}{\rho_x}\sin \phi'\right)\right]\right\}^{\frac{1}{2}} \tag{3.52}$$

$$\frac{f_\gamma}{f_r} = \sqrt{\frac{k_q}{k_r}\frac{a_r}{\rho_z}} \tag{3.53}$$

The frequency ratio f_z/f_r is given by Eq. (3.43) or Fig. 3.22. The fictitious natural frequency f_r is given by

$$f_r = (1/2\pi)\sqrt{nk_r/m}$$

Similar solutions are also available for the configuration of four resilient supports located in a rectangular array and inclined to achieve complete decoupling.[18]

FORCED VIBRATION

Forced vibration results from a continuing excitation that varies sinusoidally with time. The excitation may be a vibratory displacement of the foundation for the resiliently supported rigid body (*foundation-induced vibration*), or a force or moment applied to or generated within the rigid body (*body-induced vibration*). These two forms of excitation are considered separately.

FOUNDATION-INDUCED SINUSOIDAL VIBRATION

This section includes an analysis of foundation-induced vibration for two different systems, each having two planes of symmetry. In one system, the principal elastic axes of the resilient elements are parallel to the X, Y, Z axes; in the other system, the principal elastic axes are inclined with respect to two of the axes but in a plane parallel to one of the reference planes. The excitation is translational movement of the foundation in its own plane, without rotation. No forces or moments are applied directly to the rigid body; i.e., in the equations of motion [Eqs. (3.31)], the following terms are equal to zero:

$$F_x = F_y = F_z = M_x = M_y = M_z = \alpha = \beta = \gamma = 0 \tag{3.54}$$

TWO PLANES OF SYMMETRY WITH ORTHOGONAL RESILIENT SUPPORTS
The system is shown in Fig. 3.15. The excitation is a motion of the foundation in the direction of the X axis defined by $u = u_0 \sin \omega t$. (Alternatively, the excitation may be the displacement $v = v_0 \sin \omega t$ in the direction of the Y axis, and analogous results are obtained.) The resulting motion of the resiliently supported rigid body involves translation x_c and rotation β simultaneously. The conditions of symmetry are defined by Eqs. (3.33), (3.34), (3.35), and (3.38); these conditions decouple Eqs. (3.31) so that only Eqs. (3.31a) and (3.31d), and Eqs. (3.31b) and (3.31c), remain coupled. Upon substituting $u = u_0 \sin \omega t$ as the excitation, the response in the coupled modes is of a form

$x_c = x_{c0} \sin \omega t$, $\beta = \beta_0 \sin \omega t$ where x_{c0} and β_0 are related to u_0 as follows:[19]

$$\frac{x_{c0}}{u_0} = \frac{\dfrac{k_x}{k_z}\left[\left(\dfrac{a_x}{\rho_y}\right)^2 - \left(\dfrac{f}{f_z}\right)^2\right]}{\left(\dfrac{f}{f_z}\right)^4 - \left[\dfrac{k_x}{k_z} + \dfrac{k_x}{k_z}\left(\dfrac{a_z}{\rho_y}\right)^2 + \left(\dfrac{a_x}{\rho_y}\right)^2\right]\left(\dfrac{f}{f_z}\right)^2 + \dfrac{k_x}{k_z}\left(\dfrac{a_x}{\rho_y}\right)^2} \tag{3.55}$$

$$\frac{\beta_0}{u_0/\rho_y} = \frac{-\dfrac{k_x}{k_z}\dfrac{a_z}{\rho_y}\left(\dfrac{f}{f_z}\right)^2}{\left(\dfrac{f}{f_z}\right)^4 - \left[\dfrac{k_x}{k_z} + \dfrac{k_x}{k_z}\left(\dfrac{a_z}{\rho_y}\right)^2 + \left(\dfrac{a_x}{\rho_y}\right)^2\right]\left(\dfrac{f}{f_z}\right)^2 + \dfrac{k_x}{k_z}\left(\dfrac{a_x}{\rho_y}\right)^2} \tag{3.56}$$

where $f_z = \dfrac{1}{2\pi}\sqrt{4k_z/m}$ in accordance with Eq. (3.37). A similar set of equations apply for vibration in the coupled y_c, α coordinates. There is no response of the system in the z_c or γ modes since there is no net excitation in these directions; that is, $\mathbf{F}_z$ and $\mathbf{M}_z$ are zero.

FIG. 3.25. Curve showing the position of the axis of pure rotation of the rigid body in Fig. 3.15 as a function of the frequency ratio f/f_z when the excitation is sinusoidal motion of the foundation in the X direction [see Eq. (3.57)]. The axis of rotation is parallel to the Y axis and in the XZ plane, and its coordinate along the Z axis is designated by b_z'.

As indicated by Eqs. (3.1), the displacement at any point in a rigid body is the sum of the displacement at the center-of-gravity and the displacements resulting from motion of the body in rotation about axes through the center-of-gravity. Equations (3.55) and (3.56) together with analogous equations for y_{c0}, α_0 provide the basis for calculating these displacements. Care should be taken with phase angles, particularly if two or more excitations u, v, w exist concurrently.

At any single frequency, coupled vibration in the x_c, β modes is equivalent to a pure rotation of the rigid body with respect to an axis parallel to the Y axis, in the YZ plane and displaced from the center-of-gravity of the body (see Fig. 3.15). As a result, the rigid body has zero displacement x in the horizontal plane containing this axis. Therefore, the Z coordinate of this axis b_z' satisfies $x_{c0} + b_z'\beta_0 = 0$, which is obtained from the first of Eqs. (3.1) by setting $x_b = 0$ (γ_0 motion is not considered). Substituting Eqs. (3.55) and (3.56) for x_{c0} and β_0, respectively, the axis of rotation is located at

$$\frac{b_z'}{\rho_y} = \frac{(a_x/\rho_y)^2 - (f/f_z)^2}{(a_z/\rho_y)(f/f_z)^2} \tag{3.57}$$

Figure 3.25 shows the relation of Eq. (3.57) graphically. At high values of frequency f/f_z, the axis does not change position significantly with frequency; b_z'/ρ_y approaches a positive value as f/f_z becomes large since a_z is negative (see Fig. 3.15).

When the resilient supporting elements have damping as well as elastic properties, the solution of the equations of motion [see Eq. (3.31a)] becomes too laborious for general use. Responses of systems with damping have been obtained for several typical cases using a digital computer.[20] Figures 3.26A, B, and C show the response at three points in the body of the system shown in Fig. 3.15, with the excitation $u = u_0 \sin \omega t$. The weight of the body is 45 lb; each of the four resilient supporting elements has a stiffness $k_z = 1,050$ lb/in. and stiffness ratios $k_x/k_z = k_y/k_z = \frac{1}{2}$. The critical damping coefficients in the X, Y, Z directions are taken as $c_{cx} = 2\sqrt{4k_x m}$, $c_{cy} = 2\sqrt{4k_y m}$, $c_{cz} = 2\sqrt{4k_z m}$, respectively, where the expression for c_{cz} follows from the single degree-of-freedom case defined by Eq. (2.12). The fractions of critical damping are $c_x/c_{cx} = c_y/c_{cy} = c_z/c_{cz} = c/c_c$, the parameter of the curves in Figs.

3.26A, B, and C. Coordinates locating the resilient elements are $a_z = \pm 5.25$ in., $a_y = \pm 3.50$ in., and $a_z = -6.50$ in. The radii of gyration with respect to the X, Y, Z axes are $\rho_x = 4.40$ in., $\rho_y = 5.10$ in., and $\rho_z = 4.60$ in.

Natural frequencies calculated from Eqs. (3.37) and (3.40) are $f_z = 30.0$ cps; $f_{x\beta} = 43.7$ cps, 15.0 cps; and $f_{y\alpha} = 43.2$ cps, 11.7 cps. The fraction of critical damping c/c_c varies between 0 and 0.25. Certain characteristic features of the response curves in Figs. 3.26A, B, and C are:

1. The relatively small response at the frequency of 24.2 cps in Fig. 3.26C occurs because point 3 lies near the axis of rotation of the rigid body at that frequency. Point 2 lies near the axis of rotation at higher frequencies and the response becomes correspondingly low, as shown in Fig. 3.26B. The position of the axis of rotation changes rapidly for small changes of fre-

Fig. 3.26A. Response curves for point 1 with damping in the resilient supports in the system shown in Fig. 3.15. The response is the ratio of the amplitude at point 1 of the rigid body in the X direction to the amplitude of the foundation in the X direction (x_0/u_0). The fraction of critical damping c/c_c is the same in the X, Y, Z directions.

quency in the low- and intermediate-frequency range (indicated by the sharp dip in the curves for small damping in Fig. 3.26C) and varies asymptotically toward a final position as the forcing frequency increases (see Fig. 3.25).

2. The effect of damping on the magnitude of the response at the higher and lower natural frequencies in coupled modes is illustrated. When the fraction of critical damping is between 0.01 and 0.10, the response at the lower of the coupled natural frequencies is approximately 10 times as great as the response at the higher of the coupled natural frequencies. With greater damping ($c/c_c \geq 0.15$), the effect of resonance in the vicinity of the higher coupled natural frequency becomes so slight as to be hardly discernible.

TWO PLANES OF SYMMETRY WITH RESILIENT SUPPORTS INCLINED IN ONE PLANE ONLY. The system is shown in Fig. 3.21, and the excitation is $u = u_0 \sin \omega t$. The conditions of symmetry are defined by Eqs. (3.33), (3.35), and (3.38).

The response is entirely in the x_c, β coupled mode with the following amplitudes:

$$\frac{x_{c0}}{u_0} = \frac{\frac{k_p}{k_r}\left(\frac{a_x}{\rho_y}\right)^2 - \left(\frac{k_p}{k_r}\cos^2\phi + \sin^2\phi\right)\left(\frac{f}{f_r}\right)^2}{\left(\frac{f}{f_r}\right)^4 - A\left(\frac{f}{f_r}\right)^2 + \frac{k_p}{k_r}\left(\frac{a_x}{\rho_y}\right)^2}$$

$$\frac{\beta_0}{u_0/\rho_y} = \frac{-\left[\left(\frac{k_p}{k_r}\cos^2\phi + \sin^2\phi\right)\left(\frac{a_z}{\rho_y}\right) + \left(1 - \frac{k_p}{k_r}\right)\left|\frac{a_x}{\rho_y}\right|\sin\phi\cos\phi\right]\left(\frac{f}{f_r}\right)^2}{\left(\frac{f}{f_r}\right)^4 - A\left(\frac{f}{f_r}\right)^2 + \frac{k_p}{k_r}\left(\frac{a_x}{\rho_y}\right)^2}$$

$$(3.58)$$

where A is defined after Eq. (3.45). A similar set of expressions may be written for the response in the y_c, α coupled mode when the excitation is the motion $v = v_0 \sin \omega t$ of the foundation:

$$\frac{y_{c0}}{v_0} = \frac{\frac{k_q}{k_r}\left(\frac{k_p}{k_r}\sin^2\phi + \cos^2\phi\right)\left(\frac{a_y}{\rho_x}\right)^2 - \frac{k_q}{k_r}\left(\frac{f}{f_r}\right)^2}{\left(\frac{f}{f_r}\right)^4 - B\left(\frac{f}{f_r}\right)^2 + \frac{k_q}{k_r}\left(\frac{k_p}{k_r}\sin^2\phi + \cos^2\phi\right)\left(\frac{a_y}{\rho_x}\right)}$$

$$(3.59)$$

$$\frac{\alpha_0}{v_0/\rho_x} = \frac{\frac{k_q}{k_r}\frac{a_z}{\rho_x}\left(\frac{f}{f_r}\right)^2}{\left(\frac{f}{f_r}\right)^4 - B\left(\frac{f}{f_r}\right)^2 + \frac{k_q}{k_r}\left(\frac{k_p}{k_r}\sin^2\phi + \cos^2\phi\right)\left(\frac{a_y}{\rho_x}\right)}$$

Fig. 3.26B. Response curves at point 2 in the system shown in Fig. 3.15. See caption for Fig. 3.26A.

where B is defined after Eq. (3.46). No motion occurs in the z_c or γ mode since the quantities F_z and M_z are zero in Eqs. (3.31e) and (3.31f).

Response curves for the system shown in Fig. 3.21 when damping is included are qualitatively similar to those shown in Figs. 3.26.[21] The significant advantage in the use of inclined resilient supports is the additional versatility gained from the ability to vary the angle of inclination ϕ which directly affects the degree of coupling in the x_c, β coupled mode. For example, a change in the angle ϕ produces a change in the position of the axis of pure rotation of the rigid body. In a manner similar to that used to derive Eq. (3.57), Eqs. (3.58) yield the following expression defining the location of the axis of rotation:

$$
\frac{b_z'}{\rho_y} = \frac{\dfrac{k_p}{k_r}\left(\dfrac{a_x}{\rho_y}\right)^2 - \left(\dfrac{k_p}{k_r}\cos^2\phi + \sin^2\phi\right)\left(\dfrac{f}{f_r}\right)^2}{\left[\left(\dfrac{k_p}{k_r}\cos^2\phi + \sin^2\phi\right)\dfrac{a_z}{\rho_y} + \left(1 - \dfrac{k_p}{k_r}\right)\left|\dfrac{a_x}{\rho_y}\right|\sin\phi\cos\phi\right]\left(\dfrac{f}{f_r}\right)^2} \tag{3.60}
$$

BODY-INDUCED SINUSOIDAL VIBRATION

This section includes the analysis of a resiliently supported rigid body wherein the excitation consists of forces and moments applied directly to the rigid body (or originating within the body). The system has two planes of symmetry with orthogonal resilient supports; the modal coupling and natural frequencies for such a system are considered above. Two types of excitation are considered: (1) a force rotating about an axis parallel to one of the principal inertial axes and (2) an oscillatory moment acting about one of the

Fig. 3.26C. Response curves at point 3 in the system shown in Fig. 3.15. See caption for Fig. 3.26A.

principal inertial axes.[22, 23]   There is no motion of the foundation that supports the resilient elements; thus, the following terms in Eqs. (3.31) are equal to zero:

$$u = v = w = \alpha = \beta = \gamma = 0 \qquad (3.61)$$

TWO PLANES OF SYMMETRY WITH ORTHOGONAL RESILIENT ELEMENTS EXCITED BY A ROTATING FORCE.   The system excited by the rotating force is illustrated in Fig. 3.27.   The force F_0 rotates at frequency ω about an axis parallel to the Y

FIG. 3.27. Example of a rigid body on orthogonal resilient supports with two planes of symmetry, excited by body-induced sinusoidal excitation.   Alternative excitations are (1) the force F_0 in the XZ plane rotating with angular velocity ωt about an axis parallel to the Y axis and (2) the oscillatory moment $M_0 \sin \omega t$ acting about the Y axis.   There is no motion of the foundation that supports the resilient elements.

axis but spaced therefrom by the coordinate distances d_x, d_z; the force is in the XZ plane. The forces and moments applied to the body by the rotating force F_0 are

$$F_x = F_0 \cos \omega t \qquad\qquad M_x = 0$$

$$F_y = 0 \qquad\qquad M_y = F_0(d_z \cos \omega t - d_x \sin \omega t) \qquad (3.62)$$

$$F_z = F_0 \sin \omega t \qquad\qquad M_z = 0$$

The conditions of symmetry are defined by Eqs. (3.33), (3.34), (3.35), and (3.38); and the excitation is defined by Eqs. (3.61) and (3.62).   Substituting these conditions into the equations of motion, Eqs. (3.31) show that vibration response is not excited in the coupled y_c, α mode or in the γ mode.   In the Z direction, the motion z_{c0} of the body and the force F_{tz} transmitted through the resilient elements can be found from Eq. (2.30) and Fig. 2.17 since single degree-of-freedom behavior is involved.   The horizontal displacement amplitude x_{c0} of the center-of-gravity in the X direction and the rotational displacement amplitude β_0 about the Y axis are given by

$$\frac{x_{c0}}{F_0/4k_x} = \frac{k_x}{k_z} \frac{\sqrt{\left[\frac{k_x}{k_z}\frac{a_z}{\rho_y}\left(\frac{a_z}{\rho_y} - \frac{d_z}{\rho_y}\right) + \left(\frac{a_x}{\rho_y}\right)^2 - \left(\frac{f}{f_z}\right)^2\right]^2 + \left[\frac{k_x}{k_z}\frac{d_x}{\rho_y}\frac{a_z}{\rho_y}\right]^2}}{\left(\frac{f}{f_z}\right)^4 - \left[\frac{k_x}{k_z} + \frac{k_x}{k_z}\left(\frac{a_z}{\rho_y}\right)^2 + \left(\frac{a_x}{\rho_y}\right)^2\right]\left(\frac{f}{f_z}\right)^2 + \frac{k_x}{k_z}\left(\frac{a_x}{\rho_y}\right)^2}$$

$$(3.63)$$

$$\frac{\beta_0}{F_0/4k_x\rho_y} = \frac{k_x}{k_z} \frac{\sqrt{\left[\frac{k_x}{k_z}\left(\frac{a_z}{\rho_y} - \frac{d_z}{\rho_y}\right) + \frac{d_z}{\rho_y}\left(\frac{f}{f_z}\right)^2\right]^2 + \left[\frac{d_x}{\rho_y}\left(\frac{k_x}{k_z} - \frac{f^2}{f_z^2}\right)\right]^2}}{\left(\frac{f}{f_z}\right)^4 - \left[\frac{k_x}{k_z} + \frac{k_x}{k_z}\left(\frac{a_z}{\rho_y}\right)^2 + \left(\frac{a_x}{\rho_y}\right)^2\right]\left(\frac{f}{f_z}\right)^2 + \frac{k_x}{k_z}\left(\frac{a_x}{\rho_y}\right)^2}$$

where a_x, a_z are location coordinates of the resilient supports, and

$$f_z = \frac{1}{2\pi}\sqrt{\frac{4k_z}{m}} \qquad (3.64)$$

The amplitude of the oscillating force F_{tx} in the X direction and the amplitude of the oscillating moment M_{ty} about the Y axis which are transmitted to the foundation by the combination of resilient elements are

$$F_{tx} = 4k_x\sqrt{x_{c0}^2 - 2a_zx_{c0}\beta_0 \cos(\phi_x - \phi_\beta) + a_z^2\beta_0^2}$$

$$M_{ty} = 4k_za_x^2\beta_0$$

$$(3.65)$$

where F_{tx} is the sum of the forces transmitted by the individual resilient elements and M_{ty} is a moment formed by forces in the Z direction of opposite sign at opposite resilient supports. The angles ϕ_x and ϕ_β are defined by

$$\tan \phi_x = \frac{\frac{k_x}{k_z}\frac{a_z}{\rho_y}\left(\frac{a_z}{\rho_y} - \frac{d_z}{\rho_y}\right) + \left(\frac{a_x}{\rho_y}\right)^2 - \left(\frac{f}{f_z}\right)^2}{\frac{k_x}{k_z}\frac{a_z}{\rho_y}\frac{d_x}{\rho_y}} \qquad [0° \leq \phi_x \leq 360°]$$

$$\tan \phi_\beta = \frac{\frac{k_x}{k_z}\left(\frac{a_z}{\rho_y} - \frac{d_z}{\rho_y}\right) + \frac{d_z}{\rho_y}\left(\frac{f}{f_z}\right)^2}{\frac{d_z}{\rho_y}\left[\frac{k_x}{k_z} - \left(\frac{f}{f_z}\right)^2\right]} \qquad [0° \leq \phi_\beta \leq 360°]$$

To obtain the correct value of $(\phi_x - \phi_\beta)$ in Eq. (3.65), the signs of the numerator and denominator in each tangent term must be inspected to determine the proper quadrant for ϕ_x and ϕ_β.

Example 3.2. Consider an electric motor which has an unbalanced rotor, creating a centrifugal force. The motor weighs 3,750 lb, and has a radius of gyration $\rho_y = 9.10$ in. The distances $d_x = d_y = d_z = 0$, that is, the axis of rotation is the Y principal axis and the center-of-gravity of the rotor is in the XZ plane. The resilient supports each have a stiffness ratio of $k_x/k_z = 1.16$, and are located at $a_z = -14.75$ in., $a_x = \pm 12.00$ in. The resulting displacement amplitudes of the center-of-gravity, expressed dimensionlessly, are shown in Fig. 3.28; the force and moment amplitudes transmitted to the foundation, expressed dimensionlessly, are shown in Fig. 3.29. The displacements of the center-of-gravity of the body are dimensionalized with respect to the displacements at zero frequency:

$$z_{c0}(0) = \frac{F_0}{4k_z}$$

$$x_{c0}(0) = \frac{F_0}{4k_x}\left[1 + \frac{k_x}{k_z}\left(\frac{a_z}{a_x}\right)^2\right] \tag{3.66}$$

$$\beta_0(0) = \frac{F_0}{4k_z a_z}\left(\frac{a_z}{a_x}\right)^2$$

At excitation frequencies greater than the higher natural frequency of the x_c,β coupled motion, the displacements, forces, and moment all continuously decrease as the frequency increases.

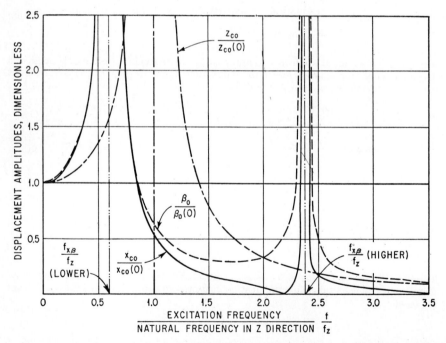

FIG. 3.28. Response curves for the system shown in Fig. 3.27 when excited by a rotating force F_0 acting about the Y axis. The parameters of the system are $k_x/k_z = 1.16$, $a_x/\rho_y = \pm 1.32$, $a_z/\rho_y = -1.62$. Only x_c, z_c, β displacements of the body are excited [see Eqs. (3.63)]. The displacements are expressed dimensionlessly by employing the displacements at zero frequency [see Eqs. (3.66)].

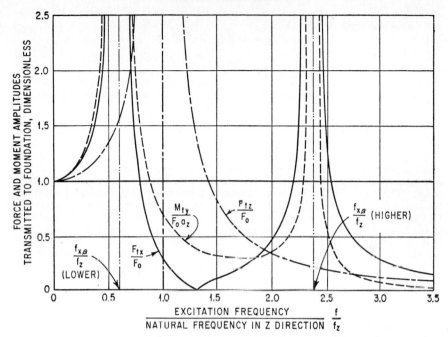

FIG. 3.29. Force and moment amplitudes transmitted to the foundation for the system shown in Fig. 3.27 when excited by a rotating force F_0 acting about the Y axis. The parameters of the system are $k_x/k_z = 1.16$, $a_z/\rho_y = \pm 1.32$, $a_z/\rho_y = -1.62$. The amplitudes of the oscillating forces in the X and Z directions transmitted to the foundation are F_{tx} and F_{tz}, respectively. The amplitude of the total oscillating moment about the Y axis transmitted to the foundation is M_{ty}.

TWO PLANES OF SYMMETRY WITH ORTHOGONAL RESILIENT ELEMENTS EXCITED BY AN OSCILLATING MOMENT. Consider the oscillatory moment M_0 acting about the Y axis with forcing frequency ω. The resulting applied forces and moments acting on the body are

$$M_y = M_0 \sin \omega t$$

$$F_x = F_y = F_z = M_x = M_z = 0 \tag{3.67}$$

Substituting conditions of symmetry defined by Eqs. (3.33), (3.34), (3.35), and (3.38), and the excitation defined by Eqs. (3.61) and (3.67), the equations of motion [Eqs. (3.31)] show that oscillations are excited only in the x_c,β coupled mode. Solving for the resulting displacements,

$$\frac{x_{c0}}{M_0/4k_x\rho_y} = \frac{\left(\dfrac{k_x}{k_z}\right)^2 \dfrac{a_z}{\rho_y}}{\left(\dfrac{f}{f_z}\right)^4 - \left[\dfrac{k_x}{k_z} + \dfrac{k_x}{k_z}\left(\dfrac{a_z}{\rho_y}\right)^2 + \left(\dfrac{a_x}{\rho_y}\right)^2\right]\left(\dfrac{f}{f_z}\right)^2 + \dfrac{k_x}{k_z}\left(\dfrac{a_x}{\rho_y}\right)^2}$$

$$\frac{\beta_0}{M_0/4k_x\rho_y{}^2} = \frac{\dfrac{k_x}{k_z}\left[\dfrac{k_x}{k_z} - \left(\dfrac{f}{f_z}\right)^2\right]}{\left(\dfrac{f}{f_z}\right)^4 - \left[\dfrac{k_x}{k_z} + \dfrac{k_x}{k_z}\left(\dfrac{a_z}{\rho_y}\right)^2 + \left(\dfrac{a_x}{\rho_y}\right)^2\right]\left(\dfrac{f}{f_z}\right)^2 + \dfrac{k_x}{k_z}\left(\dfrac{a_x}{\rho_y}\right)^2} \tag{3.68}$$

The amplitude of the oscillating force F_{tx} in the X direction and the amplitude of the oscillating moment M_{ty} about the Y axis transmitted to the foundation by the combination of resilient supports are

$$F_{tx} = 4k_x(x_{c0} - a_z\beta_0)$$

$$M_{ty} = 4k_za_x{}^2\beta_0$$

$$(3.69)$$

where F_{tx} and M_{ty} have the same meaning as in Eqs. (3.65). Low vibration transmission of force and moment to the foundation is decreased at the higher frequencies in a manner similar to that shown in Fig. 3.29.

FOUNDATION-INDUCED VELOCITY SHOCK

A velocity shock is an instantaneous change in the velocity of one portion of a system relative to another portion. In this section, the system is a rigid body supported by orthogonal resilient elements within a rigid container; the container experiences a velocity shock. The system has one plane of symmetry; modal coupling and natural frequencies for such a system are considered above. Two types of velocity shock are analyzed: (1) a sudden change in the translational velocity of the container and (2) a sudden change in the rotational velocity of the container. In both instances the change in velocity is from an initial velocity to zero. No forces or moments are applied directly to the resiliently supported body; i.e., only the forces transmitted by the resilient supports act. Thus, in the equations of motion, Eqs. (3.31):

$$F_x = F_y = F_z = M_x = M_y = M_z = 0 \qquad (3.70)$$

The modal coupling and natural frequencies for this system have been determined when the YZ plane is a plane of symmetry and the conditions of symmetry of Eqs. (3.33) to (3.35) apply. It is assumed that the velocity components of the body ($\dot{x}_c$, $\dot{y}_c$, $\dot{z}_c$, $\dot{\alpha}$, $\dot{\beta}$, $\dot{\gamma}$) and the velocity components of the supporting container ($\dot{u}$, $\dot{v}$, $\dot{w}$, $\dot{\alpha}$, $\dot{\beta}$, $\dot{\gamma}$) are respectively equal at time $t < 0$. At $t = 0$, all velocity components of the supporting container are brought to zero instantaneously. To determine the subsequent motion of the resiliently supported body, the natural frequencies f_n in the coupled modes of response are first calculated using Eq. (3.36). Then the response motion of the resiliently supported body to the two types of velocity shock can be found by the analyses which follow.[24]

ONE PLANE OF SYMMETRY WITH ORTHOGONAL RESILIENT SUPPORTS EXCITED BY A TRANSLATIONAL VELOCITY SHOCK. Figure 3.30 shows a rigid body supported within a rigid container by resilient supports in such a manner that the YZ plane is a plane of symmetry. The entire system moves with constant velocity $\dot{v}_0$ and without relative motion. At time $t = 0$, the container impacts inelastically against the rigid wall shown at the right. The following initial conditions of displacement and velocity apply at the instant of impact ($t = 0$):

$$\dot{y}_c(0) = \dot{v}_0$$

$$x_c(0) = y_c(0) = z_c(0) = \alpha(0) = \beta(0) = \gamma(0) = 0 \qquad (3.71)$$

$$\dot{x}_c(0) = \dot{z}_c(0) = \dot{\alpha}(0) = \dot{\beta}(0) = \dot{\gamma}(0) = 0$$

As a result of the impact, the velocity of the supported rigid body tends to continue and is responsible for excitation of the system in the coupled mode of the y_c, z_c, α motions. The maximum displacements of the center-of-gravity of the supported body are

$$\frac{y_{cm}}{2\pi\dot{v}_0/f_z} = \frac{1}{B}\sum_{n=1}^{3}\left(\frac{|A_n|}{f_n/f_z}\right)$$

$$\frac{z_{cm}}{2\pi\dot{v}_0/f_z} = \frac{1}{B}\sum_{n=1}^{3}\left(\frac{|M_nA_n|}{f_n/f_z}\right) \qquad (3.72)$$

$$\frac{\alpha_m}{2\pi\dot{v}_0/\rho_xf_z} = \frac{1}{B}\sum_{n=1}^{3}\left(\frac{|N_nA_n|}{f_n/f_z}\right)$$

The maximum accelerations of the center-of-gravity of the supported body are

$$\frac{\ddot{y}_{cm}}{2\pi f_z \dot{v}_0} = \frac{1}{B} \sum_{n=1}^{3} \left(|A_n| \frac{f_n}{f_z} \right)$$

$$\frac{\ddot{z}_{cm}}{2\pi f_z \dot{v}_0} = \frac{1}{B} \sum_{n=1}^{3} \left(|M_n A_n| \frac{f_n}{f_z} \right) \qquad (3.73)$$

$$\frac{\ddot{\alpha}_m}{2\pi f_z \dot{v}_0/\rho_x} = \frac{1}{B} \sum_{n=1}^{3} \left(|N_n A_n| \frac{f_n}{f_z} \right)$$

where the subscript m denotes maximum value and

$$M_n = \frac{1}{1 - (f_n/f_z)^2} \left[\frac{\Sigma k_y}{\Sigma k_z} - \left(\frac{f_n}{f_z} \right)^2 \right] \frac{\Sigma k_z a_y}{\Sigma k_y a_z}$$

$$N_n = \left[\frac{\Sigma k_y}{\Sigma k_z} - \left(\frac{f_n}{f_z} \right)^2 \right] \frac{\rho_x \Sigma k_z}{\Sigma k_y a_z} \qquad (3.74)$$

$$A_n = M_{n+1} N_{n+2} - M_{n+2} N_{n+1}$$

$$B = \left| \sum_{n=1}^{3} M_n (N_{n+1} - N_{n+2}) \right|$$

The fictitious natural frequency f_z is defined for mathematical purposes by Eq. (3.37). The numerical values of the subscript numbers n, $n + 1$, $n + 2$ denote the three natural frequencies in the coupled mode of the y_c, z_c, α motions determined from Eq. (3.36). These natural frequencies are arbitrarily assigned the values $n = 1, 2, 3$. When $n + 1$

FIG. 3.30. Example of a rigid body supported within a rigid container by resilient elements with YZ a plane of symmetry. Excitation is by a translational velocity shock in the Y direction. Prior to impact the entire system moves with constant velocity $\dot{v}_0$ and without relative motion. The rigid container impacts inelastically against the wall shown at the right, and y_c, z_c, α motions of the internally supported body result, as described mathematically by Eqs. (3.72) and (3.73).

or $n + 2$ equals 4, use 1 instead; when $n + 2$ equals 5, use 2 instead. Maximum displacements and accelerations may be calculated for other points in the supported rigid body by using Eqs. (3.1) except that each of the terms must be made numerically additive. For example, the maximum value of the y displacement at the point b having the Z coordinate b_z is

$$y_{bm} = y_{cm} + |b_z|\alpha_m \qquad (3.75)$$

since $\gamma = 0$.

Since the system is assumed undamped, the response of the suspended body in terms of displacement or acceleration consists of a superposition of three sinusoidal components at the three natural frequencies in the coupled y_c, z_c, α mode. The absolute values of terms appear in Eq. (3.75) because the maximum response is the sum of the amplitudes

Fig. 3.31. System shown in Fig. 3.30 excited by a rotational velocity shock about the X axis. The shock is induced by lifting and dropping one end of the rigid container to make inelastic impact with the foundation. If the height of drop is h, the rotational velocity of the system about the corner A at the instant of impact is given by Eq. (3.79). The response of the resiliently supported body is described mathematically by Eqs. (3.77) and (3.78).

of the three component vibrations which comprise the over-all response. In general, the maximum response occurs when the three component vibrations reach their maximum positive or negative values at the same instant. Thus, the maximum values of response apply both in positive and negative directions.

ONE PLANE OF SYMMETRY WITH ORTHOGONAL RESILIENT SUPPORTS EXCITED BY A ROTATIONAL VELOCITY SHOCK. Alternative to the type of impact illustrated in Fig. 3.30, the system may be excited by imparting a rotational velocity shock (e.g., by lifting and dropping one end of the container), as illustrated in Fig. 3.31. It is assumed that the container impacts inelastically. The system has the same form of symmetry as that shown in Fig. 3.30, and only the y_c, z_c, α modes are excited. The initial conditions at the instant of impact ($t = 0$), based upon the angular velocity $\dot{\alpha}_0$ of the rigid container about point A in Fig. 3.31, are

$$\dot{y}_c(0) = -d_z\dot{\alpha}_0 \qquad \dot{z}_c(0) = d_y\dot{\alpha}_0 \qquad \dot{\alpha}(0) = \dot{\alpha}_0$$

$$x_c(0) = y_c(0) = z_c(0) = \alpha(0) = \beta(0) = \gamma(0) = 0 \qquad (3.76)$$

$$\dot{x}_c(0) = \dot{\beta}(0) = \dot{\gamma}(0) = 0$$

Note that d_y and d_z are negative quantities. The initial conditions in Eqs. (3.76) are based on the assumption that motion of the rigid body relative to the container during the fall is negligible compared to that which occurs after the impact. The maximum displacements of the center-of-gravity of the supported body are

$$\frac{y_{cm}}{2\pi\rho_x\dot{\alpha}_0/f_z} = \frac{1}{B}\sum_{n=1}^{3}\left[\left|\frac{d_z}{\rho_x}A_n + \frac{d_y}{\rho_x}(N_{n+1} - N_{n+2}) + (M_{n+2} - M_{n+1})\right|\frac{f_z}{f_n}\right]$$

$$\frac{z_{cm}}{2\pi\rho_x\dot{\alpha}_0/f_z} = \frac{1}{B}\sum_{n=1}^{3}\left[\left|M_n\left(\frac{d_z}{\rho_x}A_n + \frac{d_y}{\rho_x}(N_{n+1} - N_{n+2}) + (M_{n+2} - M_{n+1})\right)\right|\frac{f_z}{f_n}\right] \quad (3.77)$$

$$\frac{\alpha_m}{2\pi\dot{\alpha}_0/f_z} = \frac{1}{B}\sum_{n=1}^{3}\left[\left|N_n\left(\frac{d_z}{\rho_x}A_n + \frac{d_y}{\rho_x}(N_{n+1} - N_{n+2}) + (M_{n+2} - M_{n+1})\right)\right|\frac{f_z}{f_n}\right]$$

The maximum accelerations of the center-of-gravity of the supported body are

$$\frac{\ddot{y}_{cm}}{2\pi\rho_x f_z\dot{\alpha}_0} = \frac{1}{B}\sum_{n=1}^{3}\left[\left|\frac{d_z}{\rho_x}A_n + \frac{d_y}{\rho_x}(N_{n+1} - N_{n+2}) + (M_{n+2} - M_{n+1})\right|\frac{f_n}{f_z}\right]$$

$$\frac{\ddot{z}_{cm}}{2\pi\rho_x f_z\dot{\alpha}_0} = \frac{1}{B}\sum_{n=1}^{3}\left[\left|M_n\left(\frac{d_z}{\rho_x}A_n + \frac{d_y}{\rho_x}(N_{n+1} - N_{n+2}) + (M_{n+2} - M_{n+1})\right)\right|\frac{f_n}{f_z}\right] \quad (3.78)$$

$$\frac{\ddot{\alpha}_m}{2\pi f_z\dot{\alpha}_0} = \frac{1}{B}\sum_{n=1}^{3}\left[\left|N_n\left(\frac{d_z}{\rho_x}A_n + \frac{d_y}{\rho_x}(N_{n+1} - N_{n+2}) + (M_{n+2} - M_{n+1})\right)\right|\frac{f_n}{f_z}\right]$$

where d_z and d_y are the Z and Y coordinates, respectively, of the edges of the container, as shown in Fig. 3.31, and the other quantities are the same as those appearing in Eqs. (3.72) and (3.74). The maximum response at any point in the suspended body can be found in the manner of Eq. (3.75).

The rotational velocity $\dot{\alpha}_0$ of the container about the corner A in Fig. 3.31 may be induced by lifting the opposite end to a height h and dropping it. The resulting velocity $\dot{\alpha}_0$ is

$$\dot{\alpha}_0 = \left\{\frac{2g}{\rho_A^2}\left[\frac{h}{l}d_y' + d_z'\sqrt{1 - \left(\frac{h}{l}\right)^2} - d_z'\right]\right\}^{1/2} \quad (3.79)$$

where g is the acceleration of gravity, ρ_A is the radius of gyration of the rigid body plus container about the corner A, h is the initial elevation of the raised end of the container, l is the length of the container, and d_y' and d_z' are the Y and Z coordinates, respectively, of the edges of the container with respect to the center-of-gravity of the assembly of rigid body plus container (see Fig. 3.31).

Example 3.3. The rigid body shown in Fig. 3.31 weighs 1,500 lb and has a radius of gyration $\rho_x = 42$ in. with respect to the X axis. The resilient supporting elements apply forces parallel to their longitudinal axes *only*. Each element with its longitudinal axis in the X or Y direction has a stiffness of $k_x = k_y = 500$ lb/in. Each element whose longitudinal axis extends in the Z direction has a stiffness $k_z = 1,000$ lb/in. The resilient elements are positioned as shown in Fig. 3.30, and $l = 168$ in., $d_y = d_y' = -84$ in., $d_z = d_z' = -21$ in., $\rho_A = 308$ in. The rotational velocity shock results from a height of drop $h = 36$ in.

The fictitious natural frequency f_z is obtained from Eq. (3.37), yielding $f_z = 7.22$ cps. From Eq. (3.36) or Fig. 3.14, the natural frequencies in the y_c, z_c, α mode are $f_1 = 3.58$ cps, $f_2 = 6.02$ cps, and $f_3 = 9.75$ cps. From Eqs. (3.74), it is determined that $M_1 \simeq 0$, $M_2 = 11.7$, $M_3 = -15.3$, $N_1 = -0.1$, $N_2 = 7.1$, $N_3 = 25.1$, $A_1 = 402$, $A_2 = 2$, $A_3 = 1$, $B = 405$. Sample calculations for M_1 and A_1 are

$$M_1 = \frac{1}{1 - (3.58/7.22)^2}\left[\frac{4(500)}{8(1,000)} - \left(\frac{3.58}{7.22}\right)^2\right]\frac{4(1,000)(68 - 26)}{4(500)(-10.5)} = -0.04$$

$$A_1 = M_2N_3 - M_3N_2 = (11.7)(25.1) - (-15.3)(7.1) = 402$$

From Eq. (3.79), $\dot{\alpha}_0 = 0.38$ rad/sec. Then Eqs. (3.78) give the maximum acceleration of the center-of-gravity in the Y direction of the supported body as follows:

$$\ddot{y}_{cm} = \frac{2\pi\rho_x f_z \dot{\alpha}_0}{B} \left[\begin{array}{l} \left| \dfrac{d_z}{\rho_x} A_1 + \dfrac{d_y}{\rho_x}(N_2 - N_3) + (M_3 - M_2) \right| \dfrac{f_1}{f_z} \\[2mm] + \left| \dfrac{d_z}{\rho_x} A_2 + \dfrac{d_y}{\rho_x}(N_3 - N_1) + (M_1 - M_3) \right| \dfrac{f_2}{f_z} \\[2mm] + \left| \dfrac{d_z}{\rho_x} A_3 + \dfrac{d_y}{\rho_x}(N_1 - N_2) + (M_2 - M_1) \right| \dfrac{f_3}{f_z} \end{array} \right]$$

$$= \frac{724 \text{ in./sec}^2}{405} \left[\begin{array}{l} \left| \dfrac{-21}{42}(402) + \dfrac{-84}{42}(7.1 - 25.1) + (-15.3 - 11.7) \right| \dfrac{3.58}{7.22} \\[2mm] + \left| \dfrac{-21}{42}(2) + \dfrac{-84}{42}(25.1 + 0.1) + (0 + 15.3) \right| \dfrac{6.02}{7.22} \\[2mm] + \left| \dfrac{-21}{42}(1) + \dfrac{-84}{42}(-0.1 - 7.1) + (11.7 - 0) \right| \dfrac{9.75}{7.22} \end{array} \right]$$

$$= 286 \text{ in./sec}^2 = 0.74g$$

In a similar manner:

$$z_{cm} = 1{,}580 \text{ in./sec}^2 = 4.09g$$
$$\ddot{\alpha}_m = 45.9 \text{ rad/sec}^2$$

REFERENCES

1. Marks, L. S., and T. Baumeister: "Mechanical Engineer's Handbook," 6th ed., p. **6**-6, McGraw-Hill Book Company, Inc., New York, 1958.
2. Housner, G. W., and D. E. Hudson: "Applied Mechanics-Dynamics," 2d ed., chap. 7, D. Van Nostrand Company, Inc., Princeton, N. J., 1959.
3. Boucher, R. W., D. A. Rich, H. L. Crane, and C. E. Matheny: *NACA Tech. Note* 3084, 1954.
4. Woodward, C. R.: "Handbook of Instructions for Experimentally Determining the Moments and Products of Inertia of Aircraft by the Spring Oscillation Method," *WADC Tech. Rept.* 55–415, June 1955.
5. Rubin, S.: *SAE Preprint* 197, 1957.
6. Macduff, J. N.: *Product Eng.*, **17**:106, 154 (1946).
7. Vane, F. F.: "A Guide for the Selection and Application of Resilient Mountings to Shipboard Equipment—Revised," *David W. Taylor Model Basin Rept.* 880, February, 1958, p. 98.
8. Smollen, L. E.: *J. Acoust. Soc. Amer.*, **40**:195 (1966).
9. Ref. 7, p. 50.
10. Crede, C. E.: "Vibration and Shock Isolation," p. 68, John Wiley & Sons, Inc., New York, 1951.
11. Ref. 7, pp. 37–49.
12. Ref. 10, pp. 53–58.
13. Lewis, R. C., and K. Unholtz: *Trans. ASME*, **69**:813 (1947).
14. Klein, E., R. S. Ayre, and I. Vigness: "Fundamentals of Guided Missile Packaging," *Dept. Defense (U.S.) Rept.* RD 219/3, July, 1955, appendix 8, pp. 49–52.
15. Ref. 10, p. 73.
16. Taylor, E. S., and K. A. Browne: *J. Aeronaut. Sci.*, **6**:43 (1938).
17. Browne, K. A.: *Trans. SAE*, **44**:185 (1939).
18. Derby, T. F.: "Decoupling the Three Translational Modes from the Three Rotational Modes of a Rigid Body Supported by Four Corner-Located Isolators," *Shock and Vibration Bull.* 43, pt. 4, June 1973, pp. 91–108.
19. Ref. 10, p. 50.
20. Himelblau, H.: "A Reliable Approach to Protecting Fragile Equipment from Aircraft Vibration," *North American Aviation, Inc., Rept.* NA-56-1030, 1957, pp. 16, 86.
21. Ref. 20, pp. 22, 95.
22. Ref. 10, pp. 43, 61.
23. Himelblau, H.: *Product Eng.*, **23**:151 (1952).
24. Ref. 14, chap. 11, November, 1955.
25. Ref. 2, appendix IV.

Table 3.1. Properties of Plane Sections

(After G. W. Housner and D. E. Hudson.[25])

The dimensions X_c, Y_c are the X, Y coordinates of the centroid, A is the area, $I_z \ldots$ is the area moment of inertia with respect to the $X \ldots$ axis, $\rho_z \ldots$ is the radius of gyration with respect to the $X \ldots$ axis; uniform solid cylindrical bodies of length l in the Z direction having the various plane sections as their cross sections have mass moment and product of inertia values about the Z axis equal to ρl times the values given in the table, where ρ is the mass density of the body; the radii of gyration are unchanged.

Plane section	Area and centroid	Area moment of inertia	Square of radius of gyration	Area product of inertia
1 — Right triangle	$A = \tfrac{1}{2}bh$ $X_c = \tfrac{2}{3}b$ $Y_c = \tfrac{1}{3}h$	$I_{x_c} = \dfrac{bh^3}{36}$ $I_{y_c} = \dfrac{b^3h}{36}$	$\rho_{x_c}^2 = \tfrac{1}{18}h^2$ $\rho_{y_c}^2 = \tfrac{1}{18}b^2$	$I_{x_c y_c} = \dfrac{A}{36}\,hb = \dfrac{h^2 b^2}{72}$
2	$A = \tfrac{1}{2}bh$ $X_c = \tfrac{1}{3}b$ $Y_c = \tfrac{1}{3}h$	$I_{x_c} = \dfrac{bh^3}{36}$ $I_{y_c} = \dfrac{b^3h}{36}$	$\rho_{x_c}^2 = \tfrac{1}{18}h^2$ $\rho_{y_c}^2 = \tfrac{1}{18}b^2$	$I_{x_c y_c} = -\dfrac{A}{36}\,hb = -\dfrac{h^2 b^2}{72}$
3 — Triangle	$A = \tfrac{1}{2}bh$ $X_c = \tfrac{1}{3}(a + b)$ $Y_c = \tfrac{1}{3}h$	$I_{x_c} = \dfrac{bh^3}{36}$ $I_{y_c} = \dfrac{bh}{36}(b^2 - ab + a^2)$	$\rho_{x_c}^2 = \tfrac{1}{18}h^2$ $\rho_{y_c}^2 = \tfrac{1}{18}(b^2 - ab + a^2)$	$I_{x_c y_c} = \dfrac{Ah}{36}(2a - b) = \dfrac{bh^2}{72}(2a - b)$

Table 3.1 Properties of Plane Sections (Continued)

Plane Section	Area and centroid	Area moment of inertia	Square of radius of gyration	Area product of inertia
4 Square	$A = a^2$ $X_c = \frac{1}{2}a$ $Y_c = \frac{1}{2}a$	$I_{z_c} = I_{y_c} = \frac{a^4}{12}$	$\rho_{z_c}^2 = \rho_{y_c}^2 = \frac{1}{12}a^2$	$I_{z_c y_c} = 0$
5 Rectangle	$A = bh$ $X_c = \frac{1}{2}b$ $Y_c = \frac{1}{2}h$	$I_{z_c} = \frac{bh^3}{12}$ $I_{y_c} = \frac{b^3 h}{12}$	$\rho_{z_c}^2 = \frac{1}{12}h^2$ $\rho_{y_c}^2 = \frac{1}{12}h^2$	$I_{z_c y_c} = 0$
6 Parallelogram	$A = ab\sin\theta$ $X_c = \frac{1}{2}(b + a\cos\theta)$ $Y_c = \frac{1}{2}(a\sin\theta)$	$I_{z_c} = \frac{a^3 b}{12}\sin^3\theta$ $I_{y_c} = \frac{ab}{12}\sin\theta\,(b^2 + a^2\cos^2\theta)$	$\rho_{z_c}^2 = \frac{1}{12}(a\sin\theta)^2$ $\rho_{y_c}^2 = \frac{1}{12}(b^2 + a^2\cos^2\theta)$	$I_{z_c y_c} = \frac{a^3 b}{12}\sin^2\theta\cos\theta$
7 Trapezoid	$A = \frac{1}{2}h(a + b)$ $Y_c = \frac{1}{3}h\left(\frac{2a + b}{a + b}\right)$	$I_{z_c} = \frac{h^3(a^2 + 4ab + b^2)}{36(a + b)}$	$\rho_{z_c}^2 = \frac{h^2(a^2 + 4ab + b^2)}{18(a + b)^2}$	

8 Circle	$A = \pi a^2$ $X_s = a$ $Y_c = a$	$I_{z_c} = I_{y_c} = \frac{1}{4}\pi a^4$	$\rho_{z_c}^2 = \rho_{y_c}^2 = \frac{1}{4}a^2$	$I_{z_c y_c} = 0$
9 Annulus	$A = \pi(a^2 - b^2)$ $X_c = a$ $Y_c = a$	$I_{z_c} = I_{y_c} = \frac{\pi}{4}(a^4 - b^4)$	$\rho_{z_c}^2 = \rho_{y_c}^2 = \frac{1}{4}(a^2 + b^2)$	$I_{z_c y_c} = 0$
10 Semicircle	$A = \frac{1}{2}\pi a^2$ $X_c = a$ $Y_c = \frac{4a}{3\pi}$	$I_{z_c} = \frac{a^4(9\pi^2 - 64)}{72\pi}$ $I_{y_c} = \frac{1}{8}\pi a^4$	$\rho_{z_c}^2 = \frac{a^2(9\pi^2 - 64)}{36\pi^2}$ $\rho_{y_c}^2 = \frac{1}{4}a^2$	$I_{z_c y_c} = 0$
11 Circular sector	$A = a^2\theta$ $X_c = \frac{2a}{3}\frac{\sin\theta}{\theta}$ $Y_c = 0$	$I_x = \frac{1}{4}a^4(\theta - \sin\theta\cos\theta)$ $I_y = \frac{1}{4}a^4(\theta + \sin\theta\cos\theta)$	$\rho_x^2 = \frac{1}{4}a^2\left(\frac{\theta - \sin\theta\cos\theta}{\theta}\right)$ $\rho_y^2 = \frac{1}{4}a^2\left(\frac{\theta + \sin\theta\cos\theta}{\theta}\right)$	$I_{z_c y_c} = 0$ $I_{xy} = 0$

Table 3.1. Properties of Plane Sections (Continued)

Plane section	Area and centroid	Area moment of inertia	Square of radius of gyration	Area product of inertia
12 Circular segment	$A = a^2(\theta - \tfrac{1}{2}\sin 2\theta)$ $X_c = \dfrac{2a}{3}\left(\dfrac{\sin^3\theta}{\theta - \sin\theta\cos\theta}\right)$ $Y_c = 0$	$I_x = \dfrac{Aa^2}{4}\left[1 - \dfrac{2\sin^3\theta\cos\theta}{3(\theta - \sin\theta\cos\theta)}\right]$ $I_y = \dfrac{Aa^2}{4}\left[1 + \dfrac{2\sin^3\theta\cos\theta}{\theta - \sin\theta\cos\theta}\right]$	$\rho_x^2 = \dfrac{a^2}{4}\left[1 - \dfrac{2\sin^3\theta\cos\theta}{3(\theta - \sin\theta\cos\theta)}\right]$ $\rho_y^2 = \dfrac{a^2}{4}\left[1 + \dfrac{2\sin^3\theta\cos\theta}{\theta - \sin\theta\cos\theta}\right]$	$I_{x_cy_c} = 0$ $I_{xy} = 0$
13 Ellipse	$A = \pi ab$ $X_c = a$ $Y_c = b$	$I_{x_c} = \dfrac{\pi}{4}ab^3$ $I_{y_c} = \dfrac{\pi}{4}a^3b$	$\rho_c^2 = \tfrac{1}{4}b^2$ $\rho_{y_c}^2 = \tfrac{1}{4}a^2$	$I_{x_cy_c} = 0$
14 Semiellipse	$A = \tfrac{1}{2}\pi ab$ $X_c = a$ $Y_c = \dfrac{4b}{3\pi}$	$I_{x_c} = \dfrac{ab^3}{72\pi}(9\pi^2 - 64)$ $I_{y_c} = \dfrac{\pi}{8}a^3b$	$\rho_c^2 = \dfrac{b^2}{36\pi^2}(9\pi^2 - 64)$ $\rho_{y_c}^2 = \tfrac{1}{4}a^2$	$I_{x_cy_c} = 0$
15 Parabola	$A = \tfrac{4}{3}ab$ $X_c = \tfrac{3}{5}a$ $Y_c = 0$	$I_{x_c} = \tfrac{4}{15}ab^3$ $I_{y_c} = \tfrac{16}{175}a^3b$	$\rho_{x_c}^2 = \tfrac{1}{5}b^2$ $\rho_{y_c}^2 = \tfrac{12}{175}a^2$	$I_{x_cy_c} = 0$
16 Semiparabola	$A = \tfrac{2}{3}ab$ $X_c = \tfrac{3}{8}a$ $Y_c = \tfrac{3}{5}b$	$I_x = \tfrac{2}{7}ab^3$ $I_y = \tfrac{2}{15}a^3b$	$\rho_x^2 = \tfrac{3}{7}b^2$ $\rho_y^2 = \tfrac{1}{5}a^2$	$I_{xy} = \dfrac{A}{4}ab = \tfrac{1}{6}a^2b^2$

| 17 |
$Y = \dfrac{h}{b^n} X^n$
nth-degree parabola | $A = \dfrac{bh}{n+1}$

$X_c = \dfrac{n+1}{n+2}\, b$

$Y_c = \dfrac{h}{2}\left(\dfrac{n+1}{2n+1}\right)$ | $I_x = \dfrac{bh^3}{3(3n+1)}$

$I_y = \dfrac{hb^3}{n+3}$ | $\rho_x^2 = \dfrac{h^2(n+1)}{3(3n+1)}$

$\rho_y^2 = \dfrac{n+1}{n+3}\, b^2$ |
| 18 |
$Y = \dfrac{h}{b^{\frac{1}{n}}} X^{\frac{1}{n}}$
nth-degree parabola | $A = \dfrac{n}{n+1}\, bh$

$X_c = \dfrac{n+1}{2n+1}\, b$

$Y_c = \dfrac{n+1}{2(n+2)}\, h$ | $I_x = \dfrac{n}{3(n+3)}\, bh^3$

$I_y = \dfrac{n}{3n+1}\, hb^3$ | $\rho_x^2 = \dfrac{n+1}{3(n+3)}\, h^2$

$\rho_y^2 = \dfrac{n+1}{3n+1}\, b^2$ |

Table 3.2. Properties of Homogeneous Solid Bodies

(After G. W. Housner and D. E. Hudson.[25])

The dimensions X_c, Y_c, Z_c are the X, Y, Z coordinates of the centroid, S is the cross-sectional area of the thin rod or hoop in cases 1 to 3, V is the volume, I_x... is the mass moment of inertia with respect to the X... axis, ρ_x... is the radius of gyration with respect to the X... axis, ρ is the mass density of the body.

	Solid body	Volume and centroid	Mass moment of inertia	Radius of gyration squared	Mass product of inertia
1	Thin rod	$V = Sl$ $X_c = \frac{1}{2}l$ $Y_c = 0$ $Z_c = 0$	$I_{z_c} = 0$ $I_{y_c} = I_{x_c} = \frac{\rho V}{12}l^2$	$\rho_{z_c}^2 = 0$ $\rho_{y_c}^2 = \rho_{x_c}^2 = \frac{1}{12}l^2$	$I_{z_c y_c}$, etc. $= 0$
2	Thin circular rod	$V = 2SR\theta$ $X_c = \dfrac{R \sin\theta}{\theta}$ $Y_c = 0$ $Z_c = 0$	$I_z = I_{z_c}$ $= \dfrac{\rho VR^2(\theta - \sin\theta\cos\theta)}{2\theta}$ $I_y = \dfrac{\rho VR^2(\theta + \sin\theta\cos\theta)}{2\theta}$ $I_z = \rho VR^2$	$\rho_z^2 = \rho_{z_c}^2 = \dfrac{R^2(\theta - \sin\theta\cos\theta)}{2\theta}$ $r_y^2 = \dfrac{R^2(\theta + \sin\theta\cos\theta)}{2\theta}$ $\rho_z^2 = R^2$	$I_{z_c y_c}$, etc. $= 0$ I_{xy}, etc. $= 0$
3		$V = 2\pi SR$ $X_c = R$ $Y_c = R$ $Z_c = 0$	$I_{z_c} = I_{y_c} = \frac{\rho V}{2}R^2$ $I_{z_c} = \rho VR^2$	$\rho_{x_c}^2 = \rho_{y_c}^2 = \frac{1}{2}R^2$ $\rho_{z_c}^2 = R^2$	$I_{z_c y_c}$, etc. $= 0$

 Cube	$V = a^3$ $X_c = \frac{1}{2}a$ $Y_c = \frac{1}{2}a$ $Z_c = \frac{1}{2}a$	$I_{x_c} = I_{y_c} = I_{z_c} = \frac{1}{6}\rho V a^2$	$\rho_{x_c}^2 = \rho_{y_c}^2 = \rho_{z_c}^2 = \frac{1}{6}a^2$	$I_{z_c y_c},$ etc. $= 0$
 Rectangular prism	$V = abc$ $X_c = \frac{1}{2}a$ $Y_c = \frac{1}{2}b$ $Z_c = \frac{1}{2}c$	$I_{z_c} = \frac{1}{12}\rho V(b^2 + c^2)$	$\rho_{z_c}^2 = \frac{1}{12}(b^2 + c^2)$	$I_{z_c y_c},$ etc. $= 0$
 Right rectangular pyramid	$V = \frac{1}{3}abh$ $X_c = 0$ $Y_c = \frac{1}{4}h$ $Z_c = 0$	$I_{z_c} = \frac{1}{80}\rho V(4b^2 + 3h^2)$ $I_{y_c} = \frac{1}{20}\rho V(a^2 + b^2)$	$\rho_{z_c}^2 = \frac{1}{80}(4b^2 + 3h^2)$ $\rho_{y_c}^2 = \frac{1}{20}(a^2 + b^2)$	$I_{z_c y_c},$ etc. $= 0$

Table 3.2. Properties of Homogeneous Solid Bodies (Continued)

Solid body	Volume and centroid	Mass moment of inertia	Radius of gyration squared	Mass product of inertia
7 Right circular cone	$V = \frac{1}{3}\pi R^2 h$ $X_c = 0$ $Y_c = \frac{3}{4}h$ $Z_c = 0$	$I_{z_c} = I_{x_c} = \dfrac{3\rho V}{80}(4R^2 + h^2)$ $I_{y_c} = \frac{3}{10}\rho V R^2$	$\rho_{z_c}^2 = \rho_{x_c}^2 = \frac{3}{80}(4R^2 + h^2)$ $\rho_{y_c}^2 = \frac{3}{10}R^2$	$I_{z_c y_c}$, etc. $= 0$
8 Right circular cylinder	$V = \pi R^2 h$ $X_c = 0$ $Y_c = \frac{1}{2}h$ $Z_c = 0$	$I_{z_c} = I_{x_c}$ $= \frac{1}{12}\rho V(3R^2 + h^2)$ $I_{y_c} = \frac{1}{2}\rho V R^2$	$\rho_{z_c}^2 = \rho_{x_c}^2 = \frac{1}{12}(3R^2 + h^2)$ $\rho_{y_c}^2 = \frac{1}{2}R^2$	$I_{z_c y_c}$, etc. $= 0$
9 Hollow right circular cylinder	$V = \pi h(R_1^2 - R_2^2)$ $X_c = 0$ $Y_c = \frac{1}{2}h$ $Z_c = 0$	$I_{z_c} = I_{x_c}$ $= \frac{1}{12}\rho V(3R_1^2 + 3R_2^2 + h^2)$ $I_{y_c} = \frac{1}{2}\rho V(R_1^2 + R_2^2)$	$\rho_{z_c}^2 = \rho_{x_c}^2 = \frac{1}{12}(3R_1^2 + 3R_2^2 + h^2)$ $\rho_{y_c}^2 = \frac{1}{2}(R_1^2 + R_2^2)$	$I_{z_c y_c}$, etc. $= 0$

No.	Figure	Properties	Moments of inertia	Radii of gyration	Products of inertia
10	Sphere	$V = \tfrac{4}{3}\pi R^3$ $X_c = 0$ $Y_c = 0$ $Z_c = 0$	$I_{z_c} = \tfrac{2}{5} VR^2$ $I_{y_c} = \tfrac{2}{5} VR^2$ $I_{t_c} = \tfrac{2}{5} VR^2$	$\rho_{z_c}^2 = \tfrac{2}{5} R^2$ $\rho_{y_c}^2 = \tfrac{2}{5} R^2$ $\rho_{t_c}^2 = \tfrac{2}{5} R^2$	$I_{z_c y_c},\ \text{etc.} = 0$
11	Hollow sphere	$V = \tfrac{4}{3}\pi (R_1^3 - R_2^3)$ $X_c = 0$ $Y_c = 0$ $Z_c = 0$	$I_z = I_y = I_s$ $= \dfrac{2}{5} \rho V \dfrac{R_1^5 - R_2^5}{R_1^3 - R_2^3}$	$\rho_x^2 = \rho_y^2 = \rho_z^2$ $= \dfrac{2}{5} \dfrac{R_1^5 - R_2^5}{R_1^3 - R_2^3}$	$I_{xy},\ \text{etc.} = 0$
12	Hemisphere	$V = \tfrac{2}{3}\pi R^3$ $X_c = 0$ $Y_c = \tfrac{3}{8} R$ $Z_c = 0$	$I_x = I_y = I_s = \tfrac{2}{5}\rho VR^2$	$\rho_x^2 = \rho_y^2 = \rho_z^2 = \tfrac{2}{5} R^2$	$I_{z_c y_c},\ \text{etc.} = 0$ $I_{xy},\ \text{etc.} = 0$

Table 3.2. Properties of Homogeneous Solid Bodies (Continued)

	Solid body	Volume and centroid	Mass moment of inertia	Radius of gyration squared	Mass product of inertia
13	Ellipsoid	$V = \frac{4}{3}\pi abc$ $X_c = 0$ $Y_c = 0$ $Z_c = 0$	$I_x = \frac{1}{5}\rho V(b^2 + c^2)$ $I_y = \frac{1}{5}\rho V(a^2 + c^2)$ $I_z = \frac{1}{5}\rho V(a^2 + b^2)$	$\rho_x^2 = \frac{1}{5}(b^2 + c^2)$ $\rho_y^2 = \frac{1}{5}(a^2 + c^2)$ $\rho_z^2 = \frac{1}{5}(a^2 + b^2)$	I_{xy}, etc. $= 0$
14	Paraboloid of revolution	$V = \frac{1}{2}\pi R^2 h$ $X_c = \frac{2}{3}h$ $Y_c = 0$ $Z_c = 0$	$I_{x_c} = \frac{1}{3}\rho V R^2$ $I_{y_c} = I_{z_c} = \frac{1}{18}\rho V(3R^2 + h^2)$	$\rho_{x_c}^2 = \frac{1}{3}R^2$ $\rho_{y_c}^2 = \rho_{z_c}^2 = \frac{1}{18}(3R^2 + h^2)$	$I_{z_c y_c}$, etc. $= 0$
15	Elliptic paraboloid	$V = \frac{1}{2}\pi abc$ $X_c = \frac{2}{3}a$ $Y_c = 0$ $Z_c = 0$	$I_{x_c} = \frac{1}{6}\rho V(b^2 + c^2)$ $I_{y_c} = \frac{1}{18}\rho V(3c^2 + a^2)$ $I_{z_c} = \frac{1}{18}\rho V(3b^2 + a^2)$	$\rho_{x_c}^2 = \frac{1}{6}(b^2 + c^2)$ $\rho_{y_c}^2 = \frac{1}{18}(3c^2 + a^2)$ $\rho_{z_c}^2 = \frac{1}{18}(3b^2 + a^2)$	$I_{z_c y_c}$, etc. $= 0$

4

NONLINEAR VIBRATION

H. Norman Abramson
Southwest Research Institute

INTRODUCTION

A vast body of scientific knowledge has been developed over a long period of time devoted to a description of natural phenomena. In the field of mechanics, rapid progress in the past two centuries has occurred, due in large measure to the ability of investigators to represent physical laws in terms of rather simple equations. In many cases the governing equations were not so simple; therefore, certain assumptions, more or less consistent with the physical situation, were employed to reduce the equations to types more easily soluble. Thus, the process of linearization has become an intrinsic part of the rational analysis of physical problems. An analysis based on linearized equations, then, may be thought of as an analysis of a corresponding but idealized problem.

In many instances the linear analysis is insufficient to describe the behavior of the physical system adequately. In fact, one of the most fascinating features of a study of nonlinear problems is the occurrence of new and totally unsuspected phenomena; i.e., new in the sense that the phenomena are not predicted, or even hinted at, by the linear theory. On the other hand, certain phenomena observed physically are unexplainable except by giving due consideration to nonlinearities present in the system.

The branch of mechanics that has been subjected to the most intensive attack from the nonlinear viewpoint is the theory of vibration of mechanical and electrical systems. Other branches of mechanics, such as incompressible and compressible fluid flow, elasticity, plasticity, wave propagation, etc., also have been studied as nonlinear problems, but the greatest progress has been made in treating vibration of nonlinear systems. The systems treated in this chapter are systems with a finite number of degrees-of-freedom which can be defined by a finite number of simultaneous ordinary differential equations; on the other hand, the mechanics of continua involves partial differential equations. Nonlinear ordinary differential equations are easier to handle than nonlinear partial differential equations. Interesting surveys of the entire realm of nonlinear mechanics are given in Refs. 1 and 2.

This chapter provides information concerning features of nonlinear vibration theory likely to be encountered in practice, and methods of nonlinear vibration analysis which find ready application.

EXAMPLES OF SYSTEMS POSSESSING NONLINEAR CHARACTERISTICS

SIMPLE PENDULUM

As a first example of a system possessing nonlinear characteristics, consider a simple pendulum of length l having a bob of mass m, as shown in Fig. 4.1. The well-known differential equation governing free vibration is

$$ml^2\ddot{\theta} + mgl\theta = 0 \tag{4.1}$$

This equation holds only for small oscillations about the position of equilibrium since the actual restoring moment is characterized by the quantity $\sin\theta$. Equation (4.1) thus employs the assumption $\sin\theta \simeq \theta$. The exact, but nonlinear, equation of motion is

$$ml^2\ddot{\theta} + mgl\sin\theta = 0 \tag{4.2}$$

FIG. 4.1. Simple pendulum.   FIG. 4.2. Simple spring-mass system.

SIMPLE SPRING-MASS SYSTEM

A simple spring-mass system, as shown in Fig. 4.2, is characterized by the equation

$$m\ddot{x} + kx = 0$$

This equation is based on the assumption that the elastic spring obeys Hooke's law; i.e., the characteristic curve of restoring force versus displacement is a straight line. However, many materials do not exhibit such a linear characteristic. Further, in the case of a simple coil spring, a deviation from linearity occurs at large compression as the coils begin to close up, or conversely, when the extension becomes so great that the coils begin to lose their individual identity. In either case, the spring exhibits a characteristic such that the restoring force increases more rapidly than the displacement. Such a characteristic is called *hardening*. In a similar manner, certain systems (e.g., a simple pendulum) exhibit a *softening* characteristic. Both types of characteristic are shown in Fig. 4.3. A simple system with either softening or hardening restoring force may be described approximately by an

FIG. 4.3. Restoring force characteristic curves for linear, hardening, and softening vibration systems.

equation of the form

$$m\ddot{x} + k(x \pm \mu^2 x^3) = 0$$

where the upper sign refers to the hardening characteristic and the lower to the softening characteristic.

It is possible for a system with only linear components to exhibit nonlinear characteristics, by snubber action for example, as shown in Fig. 4.4. A system undergoing vibration of small amplitude also may exhibit nonlinear characteristics; for example, in the pendulum shown in Fig. 4.5, the length depends on the amplitude.

FIG. 4.4. Nonlinear mechanical system with snubber action showing piecewise linear restoring force characteristic curve.

FIG. 4.5. Pendulum with non-linear characteristic resulting from dependence of length on vibration amplitude.

STRETCHED STRING WITH CONCENTRATED MASS

The large amplitude vibration of a stretched string with a concentrated mass, as shown in Fig. 4.6, offers another example of a nonlinear system. The governing nonlinear differential equation is, approximately,

$$m\ddot{w} + F_0 \left(\frac{l}{ab}\right) w + (SE - F_0) \left(\frac{a^3 + b^3}{2a^3 b^3}\right) w^3 = 0$$

where F_0 is the initial tension, S is the cross-sectional area, and E is the elastic modulus of the string. Consider now the special case of $a = b$ and denote the unstretched length

FIG. 4.6. Vibration of a weighted string as an example of a nonlinear system.

of the half string by l_0. Then the initial tension and the restoring force become

$$F_0 = SE \left(\frac{a - l_0}{l_0}\right)$$

$$F_r \simeq SE \left[2 \left(\frac{a}{l_0} - 1\right)\left(\frac{w}{a}\right) + \left(2 - \frac{a}{l_0}\right)\left(\frac{w}{a}\right)^3\right]$$

FIG. 4.7. Restoring force characteristics for the weighted string shown in Fig. 4.6.

An interesting feature of this system is that it exhibits a wide variety of either hardening or softening characteristics, depending upon the value of a/l_0, as shown in Fig. 4.7.

SYSTEM WITH VISCOUS DAMPING

The foregoing examples all involve nonlinearities in the elastic components, either as a result of appreciable amplitudes of vibration or as a result of peculiarities of the elastic element. Consider a simple spring-mass system which also includes a dashpot. The usual assumptions pertaining to this system are that the spring is linear and that the motion is sufficiently slow that the viscous resistance provided by the dashpot is proportional to the velocity; therefore, the governing equation of motion is linear. Frequently, the dashpot resistance is more correctly expressed by a term proportional to the square of the velocity. Further, the resistance is always such as to oppose the motion; therefore, the nonlinear equation of motion may be written

$$m\ddot{x} + c|\dot{x}|\dot{x} + kx = 0$$

BELT FRICTION SYSTEM

The system shown in Fig. 4.8 involves a nonlinearity depending upon the dry friction between the mass and the moving belt. The belt has a constant speed v_0, and the applicable equation of motion is

$$m\ddot{x} + F(\dot{x}) + kx = 0$$

where the friction force $F(\dot{x})$ is shown in Fig. 4.9. For large values of displacement, the damping term is positive, has positive slope, and removes energy from the system; for small values of displacement, the damping term is negative, has negative slope, and actually puts energy into the system. Even though there is no external stimulus, the system can have an oscillatory solution, and thus corresponds to a nonlinear *self-excited* system.

FIG. 4.8. Belt friction system which exhibits self-excited vibration.

FIG. 4.9. Damping force characteristic curve for the belt friction system shown in Fig. 4.8.

Many other examples of nonlinear systems are given in the references of this chapter, particularly Refs. 3 to 5.

DESCRIPTION OF NONLINEAR PHENOMENA

This section describes briefly, largely in nonmathematical terms, certain of the more important features of nonlinear vibration. Further details and methods of analysis are given later.

FREE VIBRATION

In so far as the free vibration of a system is concerned, one distinguishing feature between linear and nonlinear behavior is the dependence of the period of the motion in nonlinear vibration on the amplitude. For example, the simple pendulum of Fig. 4.1 may be analyzed on the basis of the linearized equation of motion, Eq. (4.1), from which it is found that the period of the vibration is given by the constant value $\tau_0 = 2\pi/\omega_n$. An analysis on the basis of the nonlinear equation of motion, Eq. (4.2), leads to an expression for the period of the form

$$\frac{\tau}{\tau_0} = 1 + \tfrac{1}{4}(U)^2 + \tfrac{9}{64}(U)^4 + \tfrac{25}{256}(U)^6 + \cdots \qquad (4.3)$$

where U is related to the amplitude of the vibration Θ by the relation $U = \sin(\Theta/2)$. The linear solution thus corresponds to the first term of Eq. (4.3). The dependence of

FIG. 4.10. Period of free vibration of a simple pendulum according to Eq. (4.3) and showing the effect of nonlinear terms.

FIG. 4.11. Deflection time-history for free damped vibration of the nonlinear system described by Duffing's equation [Eq. (4.16)].

the period of vibration on amplitude is shown in Fig. 4.10. Systems in which the period of vibration is independent of the amplitude are called *isochronous*, while those in which the period τ is dependent on the amplitude are called *nonisochronous*.

The dependence of period on amplitude also may be seen from the vibration trace shown in Fig. 4.11, which corresponds to a solution of the equation

$$m\ddot{x} + c\dot{x} + k(x + \mu^2 x^3) = 0$$

CHARACTER OF THE RESPONSE CURVES FOR FORCED VIBRATION

Representations of vibration behavior in the form of curves of response amplitude versus exciting frequency are called *response curves*. The response curves for an undamped linear system acted on by a harmonic exciting force of amplitude p and frequency ω may be derived from the equation of motion

$$\ddot{x} + \omega_n^2 x = \frac{p}{m} \cos \omega t \qquad (4.4)$$

The solution has the form shown in Fig. 4.12. The vertical line at $\omega = \omega_n$ corresponds not only to resonance, but also to free vibration ($p = 0$); the amplitude in this instance is determined by the initial conditions of the motion. In a nonlinear system the character of the motion is dependent upon the amplitude. This requires that the natural frequency likewise be amplitude-dependent; hence, it follows that the free vibration curve $p = 0$ for nonlinear systems cannot be a straight line. Figure 4.13 shows free vibration

FIG. 4.12. Family of response curves for the undamped linear system defined by Eq. (4.4).

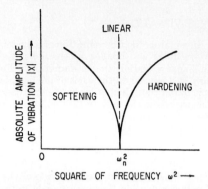

FIG. 4.13. Free vibration curves (natural frequency as a function of amplitude) in the response diagram for linear, hardening, and softening vibration systems [see Eq. (4.49)].

(A) SOFTENING SYSTEM

(B) HARDENING SYSTEM

FIG. 4.14. Response curves for undamped nonlinear systems with hardening and softening restoring force characteristics [see Eq. (4.50)].

(A) SOFTENING SYSTEM

(B) HARDENING SYSTEM

FIG. 4.15. Response curves for damped nonlinear systems with hardening and softening restoring force characteristics [see Eq. (4.52)].

curves (i.e., natural frequency as a function of amplitude) for hardening and softening systems.

Figures 4.12 and 4.13 suggest that the forced vibration response curves for systems with nonlinear restoring forces have the general form of those of a linear system, but are "swept over" to the right or left, depending on whether the system is hardening or softening. These are shown in Fig. 4.14. The principal effect of damping in forced vibration of a nonlinear system is to limit the amplitude at resonance, as shown in Fig. 4.15.

STABILITY

Consider a hardening system whose response curve is shown in Fig. 4.15B. Suppose that the exciting frequency starts at a low value, and increases continuously at a slow rate. The amplitude of the vibration also increases, but only up to a point. In particular, at

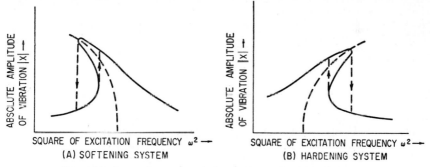

FIG. 4.16. Jump phenomenon in hardening and softening systems.

the point of vertical tangency of the response curve, a slight increase in frequency requires that the system perform in an unusual manner; i.e., that it "jump" down in amplitude to the lower branch of the response curve. This experiment may be repeated by starting with a large value of exciting frequency, but requiring that the forcing frequency be continuously reduced. A similar situation again is encountered; the system must jump up in amplitude in order to meet the conditions of the experiment. This *jump phenomenon* is shown in Fig. 4.16 for both the hardening and softening systems. The jump is not instantaneous in time, but requires a few cycles of vibration to establish a steady-state vibration at the new amplitude.

There is a portion of the response curve which is "unattainable"; it is not possible to obtain that particular amplitude by a suitable choice of forcing frequency. Thus, for certain values of ω there appear to be three possible amplitudes of vibration but only the upper and lower can actually exist. If by some means it were possible to initiate a steady-state vibration with just the proper amplitude and frequency to correspond to the middle branch, the condition would be unstable; at the slightest disturbance the motion would jump to either of the other two states of motion. The direction of the jump depends on the direction of the disturbance. Thus, of the three possible states of motion, one in phase and two out-of-phase with the exciting force, the one having the larger amplitude of the two out-of-phase motions is

FIG. 4.17. Instability region defined by the loci of vertical tangents of the damped response curves for the hardening system.

unstable. This region of instability in the response diagram is defined by the loci of vertical tangents to the response curves, and is shown for a hardening system in Fig. 4.17.

SUBHARMONIC AND SUPERHARMONIC VIBRATIONS

In the preceding discussion, only the harmonic solutions of nonlinear differential equations were considered; i.e., the solutions described were those whose frequency is the same as that of the exciting force $F \cos \omega t$. However, permanent oscillations whose frequency is a fraction ($\frac{1}{2}, \frac{1}{3}, \ldots, 1/n$) of that of the exciting force also can occur in nonlinear systems. The term *subharmonic vibration* usually is applied to this phenomenon, although the term *frequency demultiplication* is also used. In a similar manner, *superharmonic vibration* of frequency equal to some multiple ($2, 3, \ldots, n$) of the exciting frequency also can occur.

Although subharmonic and superharmonic oscillations occur in nonlinear systems, it is not a simple matter to give a plausible physical explanation for their occurrence. In a linear system, if the frequency of the free vibration is ω_n, then a periodic external force of frequency ω can excite free oscillation in addition to steady-state forced vibration of frequency ω. But since every physical system possesses some damping, the free vibration eventually disappears, leaving only the steady-state forced vibration. In a nonlinear system, the free vibration contains many higher harmonics; hence, it is possible that an exciting force with a frequency corresponding to one of these higher harmonics might be able to excite and sustain that particular harmonic component of the free vibration, in addition to the normal forced vibration.

Two essential differences between subharmonic and superharmonic oscillations may be noted: (1) subharmonics may occur under very special conditions while superharmonics always occur; (2) damping only diminishes the amplitudes of superharmonic vibrations, but may completely prohibit the existence of subharmonic vibrations if it is greater than a certain amount. Subharmonic and superharmonic vibrations are discussed analytically later in this chapter. Jump phenomena also can exist in both superharmonic and subharmonic vibrations.

OTHER PHENOMENA

SELF-EXCITED VIBRATION.* Consider the nonlinear equation of motion

$$m\ddot{x} + c(x^2 - 1)\dot{x} + kx = 0$$

This is known as Van der Pol's equation and may be written alternatively

$$\ddot{x} - \epsilon(1 - x^2)\dot{x} + \kappa^2 x = 0 \tag{4.5}$$

The principal feature of this self-excited system resides in the damping term; for small displacements the damping is negative, and for large displacements the damping is positive. Thus, even an infinitesimal disturbance causes the system to oscillate; however, when the displacement becomes sufficiently large, the damping becomes positive and limits further increase in amplitude. This is shown in Fig. 4.18. Such systems, which start in a spontaneous manner, often are called "soft" systems in contrast with "hard" systems which exhibit sustained oscillations only if a shock in excess of a certain level is applied. Note that stability questions arise here (which are different from those discussed earlier in connection with jump phenomena) concerning the existence of one or more limiting amplitudes, such as the one noted above in the Van der Pol oscillator.

FIG. 4.18. Displacement time-histories for Van der Pol's equation [Eq. (4.5)] for various values of the damping.

* A general treatment of self-excited vibration is given in Chap. 5.

RELAXATION OSCILLATIONS. As shown in Fig. 4.18, the motion of the Van der Pol oscillator is very nearly harmonic for $c/m = 0.1$ while the motion is made up of relatively sudden transitions between deflections of opposite sign for $c/m = 10.0$. The period of the harmonic motion for $c/m = 0.1$ is determined essentially by the linear spring stiffness k and the mass m; the period of the motion corresponding to $c/m = 10.0$ is very much larger and depends also on c. Thus, it is possible to obtain an undamped periodic oscillation in a damped system as a result of the particular behavior of the damping term. Such oscillations are often called *relaxation oscillations.*

ASYNCHRONOUS EXCITATION AND QUENCHING. In linear systems, the principle of superposition is valid, and there is no interaction between different oscillations. Moreover, the mathematical existence of a periodic solution always indicates the existence of a periodic phenomenon. In nonlinear systems, there is an interaction between oscillations; the mathematical existence of a periodic solution is only a necessary condition for the existence of corresponding physical phenomena. When supplemented by the condition of stability, the conditions become both necessary and sufficient for the appearance of the physical oscillation. Therefore, it is conceivable that under these conditions the appearance of one oscillation may either create or destroy the stability condition for another oscillation. In the first case, the other oscillation appears (*asynchronous excitation*), and in the second case, disappears (*asynchronous quenching*). The term asynchronous is used to indicate that there is no relation between the frequencies of these two oscillations.

ENTRAINMENT OF FREQUENCY. According to linear theory, if two frequencies ω_1 and ω_2 are caused to beat in a system, the period of beating increases indefinitely as ω_2 approaches ω_1. In nonlinear systems, the beats disappear as ω_2 reaches certain values. Thus, the frequency ω_1 falls in synchronism with, or is entrained by, the frequency ω_2 within a certain range of values. This is called *entrainment of frequency*, and the band of frequencies in which entrainment occurs is called the zone of entrainment or the interval of synchronization. In this region, the frequencies ω_1 and ω_2 combine and only vibration at a single frequency ensues.

EXACT SOLUTIONS

It is possible to obtain exact solutions for only a relatively few second-order nonlinear differential equations. In this section, some of the more important of these exact solutions are listed. They are exact in the sense that the solution is given either in closed form or in an expression that can be evaluated numerically to any desired degree of accuracy. The examples given here are fairly general in nature; more specialized examples are given in Ref. 4.

FREE VIBRATION

Consider the free vibration of an undamped system with a general restoring force $f(x)$ as governed by the differential equation

$$\ddot{x} + \kappa^2 f(x) = 0$$

This can be rewritten as

$$\frac{d(\dot{x}^2)}{dx} + 2\kappa^2 f(x) = 0 \tag{4.6}$$

and integrated to yield

$$\dot{x}^2 = 2\kappa^2 \int_x^X f(\xi)\, d\xi$$

where ξ is an integration variable and X is the value of the displacement when $\dot{x} = 0$. Thus

$$|\dot{x}| = \kappa \sqrt{2} \sqrt{\int_x^X f(\xi)\, d\xi}$$

This may be integrated again to yield

$$t - t_0 = \frac{1}{\kappa\sqrt{2}} \int_0^x \frac{d\zeta}{\sqrt{\int_\zeta^X f(\xi)\,d\xi}} \tag{4.7}$$

where ζ is an integration variable and t_0 corresponds to the time when $x = 0$. The displacement-time relation may be obtained by inverting this result. Considering the restoring force term to be an odd function, i.e.,

$$f(-x) = -f(x)$$

and considering Eq. (4.7) to apply to the time from zero displacement to maximum displacement, the period τ of the vibration is

$$\tau = \frac{4}{\kappa\sqrt{2}} \int_0^X \frac{d\zeta}{\sqrt{\int_\zeta^X f(\xi)\,d\xi}} \tag{4.8}$$

Exact solutions can be obtained in all cases where the integrals in Eq. (4.8) can be expressed explicitly in terms of X.

CASE 1. PURE POWERS OF DISPLACEMENT. Consider the restoring force function

$$f(x) = x^n$$

Equation (4.8) then becomes

$$\tau = \frac{4}{\kappa} \sqrt{\frac{n+1}{2}} \int_0^X \frac{d\zeta}{\sqrt{X^{n+1} - \zeta^{n+1}}}$$

Setting $u = \zeta/X$,

$$\tau = \frac{4}{\kappa\sqrt{X^{n-1}}} \left(\sqrt{\frac{n+1}{2}} \int_0^1 \frac{du}{\sqrt{1 - u^{n+1}}} \right)$$

The expression within the parentheses depends only on the parameter n and is denoted by $\psi(n)$. Thus

$$\tau = \frac{4}{\kappa\sqrt{X^{n-1}}}\, \psi(n) \tag{4.9}$$

The factor $\psi(n)$ may be evaluated numerically to any desired degree of accuracy, and is tabulated in Table 4.1.

CASE 2. POLYNOMIALS OF DISPLACEMENT. Consider the binomial restoring force

$$f(x) = x^n + \mu x^m \qquad [m > n \geq 0]$$

Introducing this expression into Eq. (4.8) and performing the integrations: [5]

$$\tau = \frac{4}{\kappa\sqrt{X^{n+1}}} \left(\sqrt{\frac{n+1}{2}} \int_0^1 \frac{du}{\sqrt{(1 + \bar{\mu}) - (u^{n+1} + \bar{\mu} u^{m+1})}} \right) \tag{4.10}$$

where

$$\bar{\mu} = \mu X^{m-n} \left(\frac{n+1}{m+1} \right) \tag{4.11}$$

For particular values of n, m, and $\bar{\mu}$, the expression within the parentheses can be evaluated to any desired degree of accuracy by numerical methods. The extension of this method to higher-order polynomials can be made quite readily.

CASE 3. **HARMONIC FUNCTION OF DISPLACEMENT.** Consider now the problem of the simple pendulum which has a restoring force of the form

$$f(x) = \sin x$$

Introducing this relation into Eq. (4.7):

$$t - t_0 = \frac{1}{2\kappa} \int_0^x \frac{d\zeta}{\sqrt{\sin^2 \frac{X}{2} - \sin^2 \frac{\zeta}{2}}}$$

If $x = X$ and $t_0 = 0$, this integral can be reduced to the standard form of the complete elliptic integral of the first kind:

$$\hat{K}(\alpha) = \int_0^{\pi/2} \frac{dv}{\sqrt{1 - \sin^2 \alpha \sin^2 v}} \tag{4.12}$$

Thus, the period of vibration is

$$\tau = \frac{1}{\kappa} \hat{K} \left(\frac{X}{2} \right) \tag{4.13}$$

The displacement-time function can be obtained by inversion and leads to the inverse elliptic functions. Replacing $\sin \alpha$ by U in Eq. (4.12), expanding by the binomial theorem, and then integrating yields Eq. (4.3).

CASE 4. **VELOCITY SQUARED DAMPING.** As indicated by Eq. (4.6), the introduction of any other function of $\dot{x}^2$ does not complicate the problem. Thus, the differential equation *

$$\ddot{x} \pm \frac{\delta}{2} \dot{x}^2 + \kappa^2 f(x) = 0$$

can be reduced to

$$\frac{d(\dot{x}^2)}{dx} \pm \delta \dot{x}^2 = 2\kappa^2 f(x)$$

Integrating the above equation,

$$\dot{x}^2 = 2\kappa^2 e^{\mp \delta x} \int_x^X e^{\pm \delta \xi} f(\xi) \, d\xi$$

Integrating again,

$$l = \int_{x_0}^x \frac{d\eta}{\dot{x}(\eta)}$$

where η is an integration variable.

FORCED VIBRATION

Exact solutions for forced vibration of nonlinear systems are virtually nonexistent, except as the system can be represented in a stepwise linear manner. For example, consider a system with a stepwise linear symmetrical restoring force characteristic, as shown in Fig. 4.4. Denote the lower of the two stiffnesses by k_1, the upper by k_2, and the dis-

---

* The $\pm$ sign is employed here, and elsewhere in this chapter, to account for the proper direction of the resisting force. Consequently, reference frequently is made to upper or lower sign rather than to plus or minus.

placement at which the change in stiffness occurs by x_1. Thus, the problem reduces to the solution of two linear differential equations:

$$m\ddot{x}' + k_1 x' = \pm P \sin \omega t \qquad [x_1 \geq x' \geq 0] \qquad (4.14a)$$

$$m\ddot{x}'' + (k_1 - k_2)x_1 + k_2 x'' = \pm P \sin \omega t \qquad [x'' \geq x_1] \qquad (4.14b)$$

where the upper sign refers to in-phase exciting force and the lower sign to out-of-phase exciting force. The appropriate boundary conditions are

$$x'(t = 0) = 0$$

$$x'(t = t_1) = x''(t = t_1) = x_1$$

$$\dot{x}'(t = t_1) = \dot{x}''(t = t_1)$$

$$\dot{x}''\left(t = \frac{\pi}{2\omega}\right) = 0$$

$$(4.15)$$

The solutions of Eqs. (4.14) are

$$x' = \frac{\pm P/k_1}{1 - \omega^2/\omega_1{}^2} \sin \omega t + A_1 \cos \omega_1 t + B_1 \sin \omega_1 t$$

$$x'' = \frac{\pm P/k_2}{1 - \omega^2/\omega_2{}^2} \sin \omega t + A_2 \cos \omega_2 t + B_2 \sin \omega_2 t + \left(1 - \frac{k_1}{k_2}\right) x_1$$

where $\omega_1{}^2 = k_1/m$, $\omega_2{}^2 = k_2/m$, and the constants A_1, A_2, B_1, B_2 may be evaluated from the boundary conditions, Eq. (4.15).

This analysis also applies to the case of free vibration by setting $P = 0$. By assigning various values to k_1 and k_2, a wide variety of specific problems may be treated; a collection of such solutions is given in Refs. 7 and 8. It is not necessary to restrict the restoring forces to odd functions.

APPROXIMATE METHODS OF NONLINEAR VIBRATION ANALYSIS

Any second-order ordinary differential equation can be integrated numerically in a stepwise manner to yield a time-history of the motion. Known methods of stepwise integration [9] may be employed for this purpose and result in a solution of any desired degree of accuracy, depending on the size of the steps. Numerical methods apply only to specific equations, and are not useful for general studies of the behavior of nonlinear vibrating systems.

A large number of approximate analytical methods of nonlinear vibration analysis exist, each of which may or may not possess advantages for certain classes of problems. Some of these are restricted techniques which may work well with some types of equations, but not with others. The methods which are outlined below are among the better known and possess certain advantages as to ranges of applicability.[4-21]

DUFFING'S METHOD

Consider the nonlinear differential equation

$$\ddot{x} + \kappa^2(x \pm \mu^2 x^3) = p \cos \omega t \qquad (4.16)$$

where the $\pm$ sign indicates either a hardening or softening system.* As a first approximation to a harmonic solution, assume that

$$x_1 = A \cos \omega t \qquad (4.17)$$

* Equation (4.16) is known as Duffing's equation.

and rewrite Eq. (4.16) to obtain an equation for the second approximation:

$$\ddot{x}_2 = -(\kappa^2 A \pm \tfrac{3}{4}\kappa^2\mu^2 A^3 - p)\cos \omega t - \tfrac{1}{4}\kappa^2\mu^2 A^3 \cos \omega t$$

This equation may now be integrated to yield

$$x_2 = \frac{1}{\omega^2}(\kappa^2 A \pm \tfrac{3}{4}\kappa^2\mu^2 A^3 - p)\cos \omega t + \tfrac{1}{36}\kappa^2\mu^2 A^3 \cos 3\omega t \tag{4.18}$$

where the constants of integration have been taken as zero to insure periodicity of the solution.

This may be regarded as an iteration procedure by reinserting each successive approximation into Eq. (4.16) and obtaining a new approximation. For this iteration procedure to be convergent, the nonlinearity must be small; i.e., κ^2, μ^2, A, and p must be small quantities. This restricts the study to motions in the neighborhood of linear vibration (but not near $\omega = \kappa$, since A would then be large); thus, Eq. (4.17) must represent a reasonable first approximation. It follows that the coefficient of the cos ωt term in Eq. (4.18) must be a good second approximation, and should not be far different from the first approximation.[22] Since this procedure furnishes the exact result in the linear case, it might be expected to yield good results for the "slightly nonlinear" case. Thus, a relation between frequency and amplitude is found by equating the coefficients of the first and second approximations:

$$\omega^2 = \kappa^2(1 \pm \tfrac{3}{4}\mu^2 A^2) - \frac{p}{A} \tag{4.19}$$

This relation describes the response curves, as shown in Fig. 4.14.

The above method applies equally well when linear velocity damping is included. Further details concerning this method, and an analysis of its applicability when only μ^2 and p are considered small, may be found in Ref. 10.

RAUSCHER'S METHOD [23]

Duffing's method considered above is based on the idea of starting the iteration procedure from the linear vibration. More rapid convergence might be expected if the approximations were to begin with free nonlinear vibration; Rauscher's method [23] is based on this idea.

Consider a system with general restoring force described by the differential equation

$$\omega^2 x'' + \kappa^2 f(x) = p \cos \omega t \tag{4.20}$$

where primes denote differentiation with respect to ωt, and $f(x)$ is an odd function. Assume that the conditions at time $t = 0$ are $x(0) = A$, $x'(0) = 0$. Start with the free nonlinear vibration as a first approximation, i.e., with the solution of the equation

$$\omega_0^2 x'' + \kappa^2 f(x) = 0 \tag{4.21}$$

such that $x = x_0(\phi)$ (where $\omega t = \phi$) has the period 2_π and $x_0(0) = A$, $x_0'(0) = 0$. Equation (4.21) may be solved exactly in the form of quadratures according to Eq. (4.7):

$$\phi = \phi_0(x) = \frac{\omega_0}{\kappa\sqrt{2}} \int_A^x \frac{d\zeta}{\sqrt{\int_\zeta^A f(\xi)\, d\xi}} \tag{4.22}$$

Since $f(x)$ is an odd function and noting that ωt varies from 0 to $\pi/2$ as x varies from 0 to A,

$$\frac{1}{\omega_0} = \frac{\sqrt{2}}{\kappa\pi} \int_0^A \frac{d\zeta}{\sqrt{\int_\zeta^A f(\xi)\, d\xi}} \tag{4.23}$$

With ω_0 and ϕ_0 determined by Eqs. (4.23) and (4.22), respectively, the next approximation may be found from the equation

$$\omega_1^2 x'' + \kappa^2 f(x) - p \cos \phi_0 = 0 \tag{4.24}$$

In the original differential equation, Eq. (4.20), ωt is replaced by its first approximation ϕ_0 and ω_0 (now known) is replaced by its second approximation ω_1, thus giving Eq. (4.24). This equation is again of a type which may be integrated explicitly; therefore, the next approximation ω_1 and ϕ_1 may be determined. In those cases where $f(x)$ is a complicated function, the integrals may be evaluated graphically.[23]

This method involves reducing nonautonomous systems to autonomous ones * by an iteration procedure in which the solution of the free vibration problem is used to replace the time function in the original equation which is then solved again for $t(x)$. The method is accurate and frequently two iterations will suffice.

THE PERTURBATION METHOD

In one of the most common methods of nonlinear vibration analysis, the desired quantities are developed in powers of some parameter which is considered small; then the coefficients of the resulting power series are determined in a stepwise manner. The method is straightforward, although it becomes cumbersome for actual computations if many terms in the perturbation series are required to achieve a desired degree of accuracy.

Consider Duffing's equation, Eq. (4.16), in the form

$$\omega^2 x'' + \kappa^2 (x + \mu^2 x^3) - p \cos \phi = 0 \tag{4.25}$$

where $\phi = \omega t$ and primes denote differentiation with respect to ϕ. The conditions at time $t = 0$ are $x(0) = A$ and $x'(0) = 0$, corresponding to harmonic solutions of period $2\pi/\omega$. Assume that μ^2 and p are small quantities, and define $\kappa^2 \mu^2 \equiv \epsilon$, $p \equiv \epsilon p_0$. The displacement $x(\phi)$ and the frequency ω may now be expanded in terms of the small quantity ϵ:

$$x(\phi) = x_0(\phi) + \epsilon x_1(\phi) + \epsilon^2 x_2(\phi) + \cdots$$
$$\omega = \omega_0 + \epsilon \omega_1 + \epsilon^2 \omega_2 + \cdots \tag{4.26}$$

The initial conditions are taken as $x_i(0) = x_i'(0) = 0$ $[i = 1, 2, \ldots]$.

Introducing Eq. (4.26) into Eq. (4.25) and collecting terms of zero order in ϵ gives the linear differential equation

$$\omega_0^2 x_0'' + \kappa^2 x_0 = 0$$

Introducing the initial conditions into the solution of this linear equation gives $x_0 = A \cos \omega t$ and $\omega_0 = \kappa$. Collecting terms of the first order in ϵ,

$$\omega_0^2 x_1'' + \kappa^2 x_1 - (2\omega_0 \omega_1 A - \tfrac{3}{4} A^3 + p_0) \cos \phi + \tfrac{1}{3} A^3 \cos 3\phi = 0 \tag{4.27}$$

The solution of this differential equation has a nonharmonic term of the form $\phi \cos \phi$, but since only harmonic solutions are desired, the coefficient of this term is made to vanish so that

$$\omega_1 = \frac{1}{2\kappa} \left(\tfrac{3}{4} A^2 - \frac{p_0}{A} \right)$$

Using this result and the appropriate initial conditions, the solution of Eq. (4.27) is

$$x_1 = \frac{A^3}{32\kappa^2} (\cos 3\phi - \cos \phi)$$

---

* An autonomous system is one in which the time *does not* appear explicitly, while a nonautonomous system is one in which the time *does* appear explicitly.

To the first order in ϵ, the solution of Duffing's equation, Eq. (4.25), is

$$x = A \cos \omega t + \epsilon \frac{A^3}{32\kappa^2} (\cos 3\omega t - \cos \omega t)$$

$$\omega = \kappa + \frac{\epsilon}{2\kappa} \left(\frac{3}{4} A^2 - \frac{p_0}{A} \right)$$

This agrees with the results obtained previously [Eqs. (4.18) and (4.19)]. The analysis may be carried beyond this point, if desired, by application of the same general procedures.

As a further example of the perturbation method, consider the self-excited system described by Van der Pol's equation

$$\ddot{x} - \epsilon(1 - x^2)\dot{x} + \kappa^2 x = 0 \tag{4.5}$$

where the initial conditions are $x(0) = 0$, $\dot{x}(0) = A\kappa_0$. Assume that

$$x = x_0 + \epsilon x_1 + \epsilon^2 x_2 + \cdots$$

$$\kappa^2 = \kappa_0^2 + \epsilon \kappa_1^2 + \epsilon^2 \kappa_2^2 + \cdots$$

Inserting these series into Eq. (4.5) and equating coefficients of like terms, the result to the order ϵ^2 is

$$x = \left(2 - \frac{29\epsilon^2}{96\kappa_0^2} \right) \sin \kappa_0 t + \frac{\epsilon}{4\kappa_0} \cos \kappa_0 t + \frac{\epsilon}{4\kappa_0} \left(\frac{3\epsilon}{4\kappa_0} \sin 3\kappa_0 t - \cos 3\kappa_0 t \right) - \frac{5\epsilon^3}{124\kappa_0^2} \sin 5\kappa_0 t \tag{4.28}$$

An application of the perturbation method which employs operational calculus to solve the resulting linear differential equations is given in Ref. 21, while other applications of the method are given in Refs. 24 and 25. An application to the problem of subharmonic response is outlined in Ref. 1.

THE METHOD OF KRYLOFF AND BOGOLIUBOFF [11]

Consider the general autonomous differential equation

$$\ddot{x} + F(x, \dot{x}) = 0$$

which can be rewritten in the form

$$\ddot{x} + \kappa^2 x + \epsilon f(x, \dot{x}) = 0 \qquad [\epsilon \ll 1] \tag{4.29}$$

For the corresponding linear problem ($\epsilon = 0$), the solution is

$$x = A \sin (\kappa t + \theta) \tag{4.30}$$

where A and θ are constants.

The procedure employed often is used in the theory of ordinary linear differential equations, and known variously as the method of variation of parameters or the method of Lagrange. In the application of this procedure to a nonlinear equation of the form of Eq. (4.29), assume the solution to be of the form of Eq. (4.30) but with A and θ as time-dependent functions rather than constants. This procedure, however, introduces an excessive variability into the solution; consequently, an additional restriction may be introduced. The assumed solution, of the form of Eq. (4.30), is differentiated once considering A and θ as time-dependent functions; this is made equal to the corresponding relation from the linear theory (A and θ constant) so that the additional restriction

$$\dot{A}(t) \sin [\kappa t + \theta(t)] + \dot{\theta}(t) A(t) \cos [\kappa t + \theta(t)] = 0 \tag{4.31}$$

is placed on the solution. The second derivative of the assumed solution is now formed and these relations are introduced into the differential equation, Eq. (4.29). Combining this result with Eq. (4.31),

$$\dot{A}(t) = -\left(\frac{\epsilon}{\kappa}\right) f[A(t) \sin \Phi, A(t)\kappa \cos \Phi] \cos \Phi$$

$$\dot{\theta}(t) = \frac{\epsilon}{\kappa A(t)} f[A(t) \sin \Phi, A(t)\kappa \cos \Phi] \sin \Phi$$

where
$$\Phi = \kappa t + \theta(t)$$

Thus, the second-order differential equation, Eq. (4.29), has been transformed into two first-order differential equations for $A(t)$ and $\theta(t)$.

The expressions for $\dot{A}(t)$ and $\dot{\theta}(t)$ may now be expanded in Fourier series:

$$\dot{A}(t) = -\left(\frac{\epsilon}{\kappa}\right) \left\{ K_0(A) + \sum_{n=1}^{r} [K_n(A) \cos n\Phi + L_n(A) \sin n\Phi] \right\}$$

$$\dot{\theta}(t) = \frac{\epsilon}{\kappa A} \left\{ P_0(A) + \sum_{n=1}^{r} [P_n(A) \cos n\Phi + Q_n(A) \sin n\Phi] \right\}$$

(4.32)

where

$$K_0(A) = \frac{1}{2\pi} \int_0^{2\pi} f[A \sin \Phi, A\kappa \cos \Phi] \cos \Phi \, d\Phi$$

$$P_0(A) = \frac{1}{2\pi} \int_0^{2\pi} f[A \sin \Phi, A\kappa \cos \Phi] \sin \Phi \, d\Phi$$

It is apparent that A and θ are periodic functions of time of period $2\pi/\kappa$; therefore, during one cycle, the variation of $\dot{A}$ and $\dot{\theta}$ is small because of the presence of the small parameter ϵ in Eqs. (4.32). Hence, the average values of $\dot{A}$ and $\dot{\theta}$ are considered. Since the motion is over a single cycle, and since the terms under the summation signs are of the same period and consequently vanish, then approximately:

$$\dot{A} \simeq -\left(\frac{\epsilon}{\kappa}\right) K_0(A)$$

$$\dot{\theta} \simeq \frac{\epsilon}{\kappa A} P_0(A)$$

$$\dot{\Phi} \simeq \kappa + \frac{\epsilon}{\kappa A} P_0(A)$$

For example, consider Rayleigh's equation

$$\ddot{x} - (\alpha - \beta \dot{x}^2)\dot{x} + \kappa^2 x = 0$$

(4.33)

By application of the above procedures:

$$\dot{A} = -\left(\frac{1}{\kappa}\right) K_0(A) = -\frac{1}{\kappa}\left[\frac{1}{2\pi} \int_0^{2\pi} (-\alpha + \beta A^2 \kappa^2 \cos^2 \Phi) A\kappa \cos^2 \Phi \, d\Phi\right]$$

$$= \frac{A}{2}(\alpha - \tfrac{3}{4}\beta A^2 \kappa^2)$$

(4.34)

Equation (4.34) may be integrated directly:

$$t = 2 \int_{A_0}^{A} \frac{dA}{A(\alpha - \gamma A^2)} = \frac{1}{\alpha} \ln \frac{A^2}{\alpha - \gamma A^2}$$

Solving for A,

$$A = \frac{\alpha}{\gamma}\left[\frac{1}{1 + \left(\frac{\alpha}{\gamma A_0^2} - 1\right)e^{-\alpha t}}\right]^{\frac{1}{2}} \tag{4.35}$$

where

$$\gamma = \tfrac{3}{4}\beta^2\kappa^2$$

The application of the method to Van der Pol's equation, Eq. (4.5), is easily accomplished and leads to a solution in the first approximation of the form

$$x = 2\cos o + \theta) \tag{4.36}$$

This may be compared with the perturbation solution given by Eq. (4.28).
The method may be applied equally well to nonautonomous systems.[5,19,20]

THE RITZ METHOD

In addition to methods of nonlinear vibration analysis stemming from the idea of small nonlinearities and from extensions of methods applicable to linear equations, other methods are based on such ideas as satisfying the equation at certain points of the motion [13] or satisfying the equation in the average. The Ritz method is an example of the latter method, and is quite powerful for general studies.

One method of determining such "average" solutions is to multiply the differential equation by some "weight function" $\psi_n(t)$ and then integrate the product over a period of the motion. If the differential equation is denoted by E, this procedure leads to

$$\int_0^{2\pi} E \cdot \psi_n(t)\, dt = 0 \tag{4.37}$$

A second method of obtaining such "average" solutions can be derived from the calculus of variations by seeking functions that minimize a certain integral:

$$I = \int_{t_0}^{t_1} F(\dot{x}, x, t)\, dt = \text{minimum}$$

Consider a function of the form

$$\bar{x}(t) = a_1\psi_1(t) + a_2\psi_2(t) + \cdots + a_n\psi_n(t)$$

where the $\psi_k(t)$ are prescribed functions. If $\bar{x}$ is now introduced for x, then

$$I = I(a_1, a_2, \ldots, a_n)$$

and a necessary condition for I to be a minimum is

$$\frac{\partial I}{\partial a_1} = 0, \qquad \frac{\partial I}{\partial a_2} = 0, \ldots, \qquad \frac{\partial I}{\partial a_n} = 0 \tag{4.38}$$

This gives n equations of the form

$$\frac{\partial I}{\partial a_k} = \int_{t_0}^{t_1}\left(\frac{\partial F}{\partial \bar{x}}\psi_k + \frac{\partial F}{\partial \dot{\bar{x}}}\dot{\psi}_k\right)dt = 0 \tag{4.39}$$

for determining the n unknown coefficients. Integrating Eq. (4.39),

$$\frac{\partial I}{\partial a_k} = \left[\frac{\partial F}{\partial \dot{\bar{x}}}\psi_k\right]_{t_0}^{t_1} + \int_{t_0}^{t_1}\left[\frac{\partial F}{\partial \bar{x}} - \frac{d}{dt}\left(\frac{\partial F}{\partial \dot{\bar{x}}}\right)\right]\psi_k\, dt = 0$$

The first term is zero because ψ_k must satisfy the boundary conditions; the expression in brackets under the integral in the second term is Euler's equation. The conditions given in Eqs. (4.38) then reduce to

$$\int_{t_0}^{t} E(\bar{x})\psi_k \, dt = 0 \qquad [k = 1, 2, \ldots, n] \tag{4.40}$$

This is the same as Eq. (4.37); thus, it is not necessary to "know" the variational problem, but only the differential equation. The conditions given in Eqs. (4.40) then yield average solutions based on variational concepts.

EXAMPLES. As a first example of the application of the Ritz method, consider the equation

$$\ddot{x} + \kappa^2 x^n = 0$$

for which an exact solution was given earlier in this chapter [Eq. (4.9)]. Assume a single-term solution of the form

$$\bar{x} = A \cos \omega t$$

The Ritz procedure, defined by Eq. (4.40), gives

$$\int_0^{2\pi} (-\omega^2 A \cos^2 \omega t + \kappa^2 A^n \cos^{n+1} \omega t) \, d(\omega t) = 0$$

from which

$$\frac{\omega^2}{\kappa^2} = \frac{4}{\pi} A^{n-1} \int_0^{\pi/2} \cos^{n+1} \omega t \, d(\omega t) = A^{n-1}\varphi(n) \tag{4.41}$$

The comparable exact solution obtained previously by introducing in Eq. (4.9) the quantity $2\pi/\omega$ for the period τ is

$$\frac{\omega^2}{\kappa^2} = \left[\frac{\pi^2/4}{\psi^2(n)}\right] X^{n-1} = \Phi(n)X^{n-1} \tag{4.42}$$

Values of $\varphi(n)$ from the approximate analysis and $\Phi(n)$ from the exact analysis are compared directly in Table 4.1, affording an appraisal of the accuracy of the method.

Table 4.1. Values of the Functions $\psi(n)$, $\Phi(n)$, $\varphi(n)$ *

n	$\psi(n)$	$\Phi(n)$	$\varphi(n)$
0	1.4142	1.2337	1.2732
1	1.5708	1.0000	1.0000
2	1.7157	0.8373	0.8488
3	1.8541	0.7185	0.7500
4	1.9818	0.6282	0.6791
5	2.1035	0.5577	0.6250
6	2.2186	0.5013	0.5820
7	2.3282	0.4552	0.5469

* The mathematical expressions for $\psi(n)$, $\Phi(n)$, and $\varphi(n)$ and the equations to which they refer are:

$$\psi(n) = \sqrt{\frac{n+1}{2}} \int_0^1 \frac{du}{\sqrt{1 - u^{n+1}}} \qquad [\text{Eq. (4.9)}]$$

$$\Phi(n) = \frac{\pi^2/4}{\psi^2(n)} \qquad [\text{Eq. (4.42)}]$$

$$\varphi(n) = \frac{4}{\pi} \int_0^{\pi/2} \cos^{n+1} \sigma \, d\sigma \qquad [\text{Eq. (4.41)}]$$

Consider now the nonautonomous system described by Duffing's equation

$$E \equiv \ddot{x} + \kappa^2(x + \mu^2 x^3) - p \cos \omega t = 0$$

Assuming

$$\ddot{x} = A \cos \phi, \qquad \phi = \omega t$$

the Ritz condition, Eq. (4.40), leads to

$$\int_0^{2\pi} \{[(1 - \eta^2)A - s] \cos \phi + \mu^2 A^3 \cos^3 \phi\} \cos \phi \, d\phi$$

from which the amplitude-frequency relation is

$$(1 - \eta^2)A + \tfrac{3}{4}\mu^2 A^3 = \pm s \tag{4.43}$$

where

$$s = \frac{p}{\kappa^2}, \qquad \eta^2 = \frac{\omega^2}{\kappa^2} \tag{4.44}$$

The upper sign indicates vibration in phase with the exciting force. Equation (4.43) describes the response curves shown in Fig. 4.14*A*, and corresponds to Eq. (4.19) obtained by Duffing's method.

Application of the Ritz method to Van der Pol's equation, Eq. (4.5), leads to the identical result given by Eq. (4.36).

GENERAL EQUATIONS FOR RESPONSE CURVES

The Ritz method has been applied extensively in studies of nonlinear differential equations.[6,8] Some of the general equations for response curves thereby obtained are given here, both as a further example of the application of the method and as a collection of useful relations.

SYSTEM WITH LINEAR DAMPING AND GENERAL RESTORING FORCES

Consider a system with general elastic restoring force (an odd function) and described by the equation of motion

$$a\ddot{x} + b\dot{x} + cf(x) - P \cos \omega t = 0$$

A solution may be assumed in the form

$$\ddot{x} = A \cos (\omega t - \theta) = B \cos \phi + C \sin \phi \tag{4.45}$$

where $\phi = \omega t$, $B = A \cos \theta$, $C = A \sin \theta$. Introducing Eq. (4.45) according to the Ritz conditions, and recalling that $f(x)$ is to be an odd function,

$$-a\omega^2 A \cos \theta + b\omega A \sin \theta + cAF(A) \cos \theta = P$$

$$-a\omega^2 A \sin \theta - b\omega A \cos \theta + cAF(A) \sin \theta = 0 \tag{4.46}$$

where

$$F(A) = \frac{1}{\pi A} \int_0^{2\pi} f(A \cos \sigma) \cos \sigma \, d\sigma$$

and σ is simply an integration variable.

Some algebraic manipulations with Eqs. (4.46) give independent equations for the two unknowns A and θ:

$$[F(A) - \eta^2]^2 + 4D^2\eta^2 = \left(\frac{s}{A}\right)^2 \tag{4.47}$$

$$\tan \theta = \frac{2D\eta}{F(A) - \eta^2} \tag{4.48}$$

where η^2 and s are defined according to Eq. (4.44) and

$$\kappa^2 = \frac{c}{a} \qquad p = \frac{P}{a} \qquad D = \frac{b}{2\sqrt{ac}}$$

Equation (4.47) describes response curves of the form shown in Fig. 4.15, and Eq. (4.48) gives the corresponding phase angle relationships. These two equations also yield other special relations which describe various curves in the response diagram:

Undamped free vibration curve (Fig. 4.13),

$$\eta^2 = F(A) \tag{4.49}$$

Undamped response curves (Fig. 4.14),

$$\eta^2 = F(A) \mp \frac{s}{A} \tag{4.50}$$

Locus of vertical tangents of undamped response curves (Fig. 4.17),

$$\eta^2 = F(A) + A\frac{\partial F(A)}{\partial A} \tag{4.51}$$

Damped response curves (Fig. 4.15),

$$\eta^2 = [F(A) - 2D^2] \mp \sqrt{\left(\frac{s}{A}\right)^2 - 4D^2[F(A) - D^2]} \tag{4.52}$$

Locus of vertical tangents of damped response curves (Fig. 4.17),

$$[F(A) - \eta^2]\left[F(A) + A\frac{\partial F(A)}{\partial A} - \eta^2\right] = -4D^2\eta^2 \tag{4.53}$$

The maximum amplitude of vibration is of interest. The amplitude at the point at which a response curve crosses the free vibration curve is termed the *resonance amplitude,* and is determined in the nonlinear case by solving Eqs. (4.49) and (4.52) simultaneously. This leads to [27]

$$2D\eta = \frac{s}{A} \qquad \theta = \frac{\pi}{2} \tag{4.54}$$

The first of these two equations defines a hyperbola in the response diagram, describing the locus of crossing points, as shown in Fig. 4.19; hence, the intersection of this curve with the free vibration curve gives the resonance amplitude. The phase angle at resonance has the value $\pi/2$, as in the linear case. This result is of great help in computing response curves since the effect of damping (except for very large values) is negligible except in the neighborhood of resonance. Therefore, one may compute only the undamped curves (which is not difficult) and the hyperbola (which does not contain the nonlinearity); then, the effect of damping may be sketched in from knowledge of the crossing point.

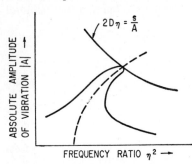

FIG. 4.19. Determination of the resonant amplitude in accordance with Eq. (4.54).

SYSTEM WITH GENERAL DAMPING AND GENERAL RESTORING FORCES

The preceding analysis may be extended to include the more general differential equation

$$E \equiv \ddot{x} + 2D\kappa g(\dot{x}) + \kappa^2 f(x) - p\cos\omega t = 0$$

By procedures similar to those employed above:

$$[F(A) - \eta^2]^2 + 4D^2S^2(A) = \left(\frac{s}{A}\right)^2 \tag{4.55}$$

$$\tan \theta = \frac{2DS(A)}{F(A) - \eta^2} \tag{4.56}$$

where

$$S(A) = \frac{1}{\pi \kappa A} \int_0^{2\pi} g(\omega A \sin \sigma) \sin \sigma \, d\sigma$$

In the case of linear velocity damping, $S(A) = \eta$, and Eqs. (4.55) and (4.56) reduce to Eqs. (4.47) and (4.48). The results for various types of damping forces are:

Coulomb damping: $\qquad g(\dot{x}) = \pm \nu_0 \qquad\qquad S(A) = \dfrac{4}{\pi} \dfrac{\nu_0}{\kappa A}$

Linear velocity damping: $\qquad g(\dot{x}) = \nu_1 \dot{x} \qquad\qquad S(A) = \nu_1 \eta$

Velocity squared damping: $\qquad g(\dot{x}) = \nu_2 \dot{x} |\dot{x}| \qquad\quad S(A) = \dfrac{8}{3\pi} \nu_2 \eta (A\omega)$

nth-power velocity damping: $\quad g(\dot{x}) = \nu_n \dot{x} |\dot{x}|^{n-1} \quad S(A) = \nu_n \eta (A\omega)^{n-1} \varphi(n)$

where $\varphi(n)$ is defined in Eq. (4.41) and values are given in Table 4.1. The locus of resonance amplitudes or crossing points is now given by [27]

$$2DS(A) = \frac{s}{A} \qquad\qquad \theta = \frac{\pi}{2}$$

GRAPHICAL METHODS OF INTEGRATION

INTRODUCTION

Graphical methods may be employed in the analysis of nonlinear vibration and often prove to be of great value both for general studies of the behavior of a particular system and for actual integration of the equation of motion.

A single degree-of-freedom system requires two parameters to describe completely the state of the motion. When these two parameters are used as coordinate axes, the graphical representation of the motion is called a *phase-plane* representation. In dealing with ordinary dynamical problems, these parameters frequently are taken as the displacement and velocity. First consider an undamped linear system having the equation of motion

$$\ddot{x} + \omega_n^2 x = 0 \tag{4.57}$$

and the solution

$$x = A \cos \omega_n t + B \sin \omega_n t$$

$$y = \frac{\dot{x}}{\omega_n} = -A \sin \omega_n t + B \cos \omega_n t \tag{4.58}$$

Eliminating time as a variable between Eqs. (4.58):

$$x^2 + y^2 = c^2$$

Thus, the phase-plane representation is a family of concentric circles with centers at the origin. Such curves are called *trajectories*. The necessary and sufficient conditions in the phase-plane for periodic motions are (1) closed trajectories and (2) paths described in finite time.

Now, suppose that the solution of Eq. (4.57) is not known. By introducing $y = \dot{x}/\omega_n$,

$$\frac{dx}{dt} = \omega_n y \qquad \frac{dy}{dt} = -\omega_n x$$

Therefore

$$\frac{dy}{dx} = -\frac{x}{y} \tag{4.59}$$

Thus, the path in the phase-plane is described by a simple first-order differential equation. This process of eliminating the time always can be done in principle, but frequently the problem is too difficult. Since the time is to be eliminated, *only autonomous systems can be treated by phase-plane methods.* When an equation of the type of Eq. (4.59) can

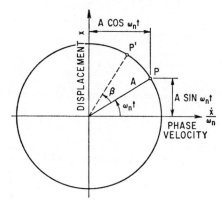

Fig. 4.20. Phase-plane using oblique co-ordinates which results in a logarithmic spiral trajectory for a linear system with viscous damping [Eq. (4.60)].

Fig. 4.21. Phase-plane solution for a linear undamped vibrating system.

be found, a direct solution of the problem follows since slopes of the trajectories can be sketched in the phase-plane and the trajectories determined by connecting the tangents; this is known as the method of *isoclines.* It sometimes happens that $dx = 0$, $dy = 0$ simultaneously so that there is no knowledge of the direction of the motion; such points in the phase-plane are called *singular* points. In the present example, the origin constitutes a singular point.

Consider now a damped linear system having the equation of motion

$$\ddot{x} + 2\zeta\omega_n\dot{x} + \omega_n^2 x = 0$$

and the solution

$$x = Ce^{-\delta t}\cos\Phi$$

$$y = \frac{\dot{x}}{\omega_n} = Ce^{-\delta t}\cos(\Phi + \sigma) \tag{4.60}$$

where

$$\delta = \zeta\omega_n = -\omega_n\cos\sigma \qquad \left[\sigma > \frac{\pi}{2}\right]$$

$$\Phi = \nu t + \theta$$

$$\nu = \omega_n\sqrt{1 - \zeta^2} = \omega_n\sin\sigma$$

Equations (4.60) indicate that the trajectories in the phase-plane are some form of spiral (one of the simplest known of which is the logarithmic spiral). By referring to the oblique coordinate system shown in Fig. 4.20, and recalling that $\sin\sigma$ is a constant and

$$r^2 = x^2 + y^2 - 2xy\cos\sigma$$

Eqs. (4.60) reduce to

$$r = Ce^{-\delta t} \sin \sigma$$

This is a form of a logarithmic spiral.

The trajectories also could be found in a rectangular coordinate system, by the method of isoclines, without knowledge of the solution [Eq. (4.60)]. The governing differential equation is

$$\frac{dy}{dx} = -\frac{2\zeta y + x}{y} \tag{4.61}$$

The resulting trajectories can be sketched in the phase-plane. On the other hand, Eq. (4.61) also can be integrated analytically by use of the substitution $z = y/x$ and separation of the variables:

$$y^2 + 2\zeta xy + x^2 = C \exp\left(\frac{2\zeta}{\sqrt{1-\zeta^2}} \tan^{-1} \frac{x\zeta + y}{x\sqrt{1-\zeta^2}}\right)$$

This is a spiral of the form of Eq. (4.60).

The method of isoclines is extremely useful in studying the behavior of solutions in the neighborhood of singular points and for the related questions of stability of solutions. In this sense, phase-plane methods may be thought of as topological methods.[5, 10, 12, 19]

However, it is desirable also to study the over-all solutions, rather than solutions in the neighborhood of special points, and preferably by some straightforward method of graphical integration. Such integration methods are given in the following sections of this chapter, while topological studies, with special reference to questions of stability, are treated in Chap. 5.

PHASE-PLANE INTEGRATION OF STEPWISE LINEAR SYSTEMS

Consider the undamped linear system described by Eq. (4.57). The known solution $x = A \sin \omega_n t$, $\dot{x} = A\omega_n \cos \omega_n t$ may be shown graphically in the phase-plane representation of Fig. 4.21. The point P moves with constant angular velocity ω_n, and the deflection increases to P' in the time β/ω_n.

If the system has a nonlinear restoring force composed of straight lines (as in Fig. 4.4), the motion within the region represented by any one linear segment can be described as above. For example, consider a system with the force-deflection characteristic shown at the top of Fig. 4.22. If the motion starts with initial velocity q_6 and zero initial deflection, the motion is described by a circular arc with center at 0 and angular velocity $\omega_{n_1} = \sqrt{\dfrac{1}{m} \tan \alpha_1}$, from q_6 to q_5. At the point q_5, it is seen that $\dot{x}_A/\omega_{n_1} = \overline{x_A q_5}$ and $\dot{x}_A/\omega_{n_2} = \overline{x_A q_4}$. Therefore

$$\frac{\overline{x_A q_4}}{\overline{x_A q_5}} = \sqrt{\frac{\tan \alpha_2}{\tan \alpha_1}}$$

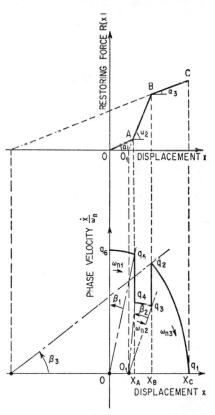

FIG. 4.22. Phase-plane solution for the stepwise linear restoring force characteristic curve shown at the top. The motion starts with zero displacement but finite velocity.

In this example, $\tan \alpha_1 < \tan \alpha_2$ so that $\overline{x_A q_4} < \overline{x_A q_5}$. The circular arc from q_4 to q_3 corresponds to the segment AB of the restoring force characteristic with center at the intersection 0_1 of the segment (extended) with the X axis. The radius of this circle is $\overline{0_1 q_4}$, where

$$q_4 = \frac{\dot{x}_B}{\omega_{n_2}} \qquad \text{and} \qquad \omega_{n_2} = \sqrt{\frac{1}{m} \tan \alpha_2}$$

The total time required to go from q_6 to q_1 is

$$t = \frac{\beta_1}{\omega_{n_1}} + \frac{\beta_2}{\omega_{n_2}} + \frac{\beta_3}{\omega_{n_3}}$$

For a symmetrical system this is one quarter of the period.

If the force-deflection characteristic of a nonlinear system is a smooth curve, it may be approximated by straight line segments and treated as above. It should be noted that the time required to complete one cycle is strongly influenced by the nature of the curve in regions where the velocity is low; therefore, linear approximations near the equilibrium position do not greatly affect the period.

The time-history of the motion (i.e., the x,t representation) may be obtained quite readily by projecting values from the X axis to an x,t plane.

Inasmuch as phase-plane methods are restricted to autonomous systems, only free vibration is discussed above. However, if a constant force were to act on the system, the nature of the vibration would be unaffected, except for a displacement of the equilibrium position in the direction of the force and equal to the static deflection produced by that force. Thus, the trajectory would remain a circular arc but with its center displaced from the origin. Therefore, *nonautonomous systems may be treated by phase-plane methods, if the time function is replaced by a series of stepwise constant values.* The degree of accuracy attained in such a procedure depends only on the number of steps assumed to represent the time function.

A system having a bilinear restoring force and acted upon by an external stepwise function of time, treated by the method described above, is shown in Fig. 4.23. Phase-plane methods therefore offer the possibility of treating transient as well as free vibrations.

Phase-plane methods have been widely used for the analysis of control mechanisms.[29] A comprehensive analysis of discontinuous-type systems possessing various types of nonlinearities is given in Ref. 30.

FIG. 4.23. Phase-plane solution for transient motion. The bilinear restoring force characteristic curve is shown at the left, and the exciting force $F(t)$ and the resulting motion of the system $X(t)$ are shown at the right. (*After Evaldson et al.*[28])

PHASE-PLANE INTEGRATION OF AUTONOMOUS SYSTEMS WITH NONLINEAR DAMPING

Consider the differential equation

$$\ddot{x} + g(\dot{x}) + \kappa^2 x = 0$$

Introducing $y = \dot{x}/\kappa$, the following isoclinic equation is obtained:

$$\frac{dy}{dx} = -\frac{g(y) + x}{y} \qquad (4.62)$$

For points of zero slope in the phase-plane, the numerator of Eq. (4.62) must vanish; therefore, the condition for zero slope is

$$x_0 = -g(y)$$

Points of infinite slope correspond to the X axis. Singular points occur where the x_0 curve intersects the X axis.

To construct the trajectory, the slope at any point P_i must be determined first. This is done as illustrated in Fig. 4.24: A line is drawn parallel to the X axis through P_i. The intersection of this line with the x_0 curve determines a point S_i on the X axis. With S_i as the center, a circular arc of short length is drawn through P_i; the tangent to this arc is the required slope. The termination of this short arc may be taken as the point P_{i+1}, etc. The accuracy of the construction is dependent on the lengths of the arcs. This construction is known as Liénard's method.[5,10]

As an example of Liénard's method, consider Rayleigh's equation, Eq. (4.33), in the form

$$\ddot{x} + \epsilon\left(\frac{\dot{x}^3}{3} - \dot{x}\right) + x = 0$$

The corresponding isoclinic equation is

$$\frac{dy}{dx} = \frac{\epsilon(y - y^3/3) - x}{y}$$

The x_0 curve is given by

$$x_0 = \epsilon\left(y - \frac{y^3}{3}\right) \qquad (4.63)$$

This is illustrated in Fig. 4.25.

A little experimentation shows that if a point P_1 is taken near the origin, the slope is such as to take the trajectory away from the origin (as compared with the undamped

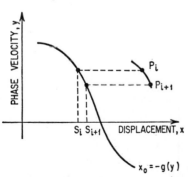

FIG. 4.24. Liénard's construction for phase-plane integration of autonomous systems with nonlinear damping.

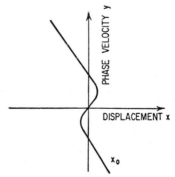

FIG. 4.25. Curve of x_0 for Rayleigh's equation [Eq. (4.33)] as given by Eq. (4.63).

vibration); by the same reasoning, a point P_2 far from the origin tends to take the trajectory toward the origin (again as compared with the undamped vibration). Therefore, there is some neutral curve, describing a periodic motion, toward which the trajectories tend; this neutral curve is called a *limit cycle* and is illustrated in Fig. 4.26. Such a limit cycle is obtained when x_0 has a different sign for different parts of the Y axis.

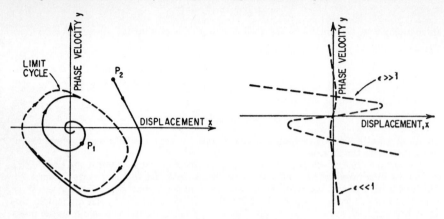

FIG. 4.26. Limit cycle for Rayleigh's equation [Eq. (4.33)].

FIG. 4.27. Curves of x_0 for extreme values of ϵ in Rayleigh's equation [Eq. (4.33)]. See Fig. 4.25 for a solution with a moderate value of ϵ.

For extreme values of ϵ, the x_0 curves would appear as shown in Fig. 4.27. For $\epsilon \gg 1$, introduce the notation $\xi = x/\epsilon$; then

$$\frac{dy}{d\xi} = \frac{\epsilon}{y}\left(y - \frac{y^3}{3} - \xi\right)$$

This leads to a trajectory as shown in Fig. 4.28. This type of motion is known as a *relaxation oscillation*. Note from Fig. 4.28 that for this case of large ϵ the slope changes quickly from horizontal to vertical. Hence, for a motion starting at some point P_i, a vertical

FIG. 4.28. Relaxation oscillations of Rayleigh's equation [Eq. (4.33)].

trajectory is followed until it intersects the ξ_0 curve; then, the trajectory turns and follows the ξ_0 curve until it enters the vertical field at the lower knee in the curve. The trajectory then moves straight up until it intersects ξ_0 again after which it swings right and down again. A few circuits bring the trajectory into the limit cycle.

There is a possibility that more than one limit cycle may exist. If the x_0 curve crosses the X axis more than three times, it can be shown that at least two limit cycles may exist. Further discussion of limit cycles is given in Chap. 5.

GENERALIZED PHASE-PLANE ANALYSIS

The following method of integrating second-order differential equations by phase-plane techniques has general application.[31, 32] Consider the general equation

$$\ddot{x} + F(x,\dot{x},t) = 0 \tag{4.64}$$

Equation (4.64) can be converted to the form

$$\ddot{x} + \kappa^2 x = g(x,\dot{x},t)$$

by adding $\kappa^2 x$ to both sides where

$$\kappa^2 x - F(x,\dot{x},t) = g(x,\dot{x},t)$$

Let

$$g(x_0,\dot{x}_0,t_0) = -\kappa^2 \Delta_0$$

where κ is chosen arbitrarily. At some point P_0 on the trajectory,

$$\ddot{x} + \kappa^2(x + \Delta_0) = 0$$

and

$$\frac{dy}{dx} = -\frac{x + \Delta_0}{y}$$

Referring to Fig. 4.29,

$$dt = \frac{1}{\kappa} \frac{dx}{y} = \frac{1}{\kappa} d\theta$$

Therefore, the time may be obtained by integration of the **angular displacements.** Thus, at a nearby point P_1 on the trajectory:

$$x_1 = x_0 + dx$$
$$y_1 = y_0 + dy$$
$$t_1 = t_0 + dt$$

Now, compute Δ_1 for the new center, and repeat the process.

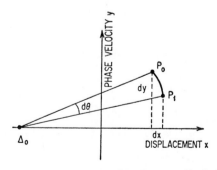

FIG. 4.29. Method of construction employed in the generalized phase-plane analysis.

This method has been applied to a very wide variety of linear and nonlinear equations.[31, 33] For example, Fig. 4.30 shows the solution of Bessel's equation

$$\ddot{x} + \frac{1}{t}\dot{x} + \left(p^2 - \frac{n^2}{t^2}\right)x = 0$$

of order zero. The angle (or time) projection of x yields $J_0(pt)$, while the $\dot{x}/p$ projection yields $J_1(pt)$; that is, the Bessel functions of the zeroth and first order of the first kind. Bessel functions of the second kind also can be obtained.

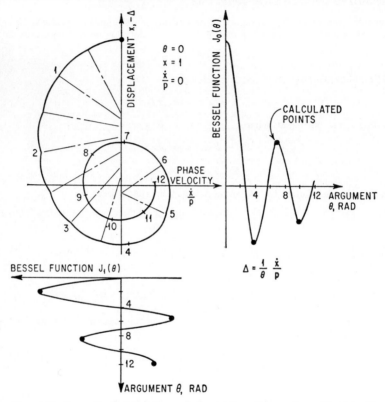

FIG. 4.30. Generalized phase-plane solution of Bessel's equation. (*Jacobsen.*[31])

STABILITY OF PERIODIC NONLINEAR VIBRATION

Certain systems having nonlinear restoring forces and undergoing forced vibration exhibit unstable characteristics for certain combinations of amplitude and exciting frequency. The existence of such an instability leads to the "jump phenomenon" shown in Fig. 4.16. To investigate the stability characteristics of the response curves, consider Duffing's equation

$$\ddot{x} + \kappa^2(x + \mu^2 x^3) = p \cos \omega t \tag{4.65}$$

Assume that two solutions of this equation exist and have slightly different initial conditions:

$$x_1 = x_0$$

$$x_2 = x_0 + \delta \qquad [\delta \ll x_0]$$

Introducing the second of these into Eq. (4.65) and employing the condition that x_0 is also a solution,

$$\ddot{\delta} + \kappa^2(1 + 3\mu^2 x_0^2)\delta = 0 \tag{4.66}$$

Now an expression for x_0 must be obtained; assuming a one-term approximation of the form $x_0 = A \cos \omega t$, Eq. (4.66) becomes

$$\frac{d^2\delta}{d\varphi^2} + (\lambda + \gamma \cos \varphi)\delta = 0 \tag{4.67}$$

where

$$\kappa^2(1 + \tfrac{3}{2}\mu^2 A^2) = 4\omega^2\lambda \tag{4.68}$$

and

$$\tfrac{3}{2}\kappa^2\mu^2 A^2 = 4\omega^2\gamma \qquad 2\omega t = \varphi$$

Equation (4.67) is known as Mathieu's equation.

Mathieu's equation has appeared in this analysis as a variational equation characterizing small deviations from the given periodic motion whose stability is to be investigated; thus, the stability of the solutions of Mathieu's equation must be studied. A given periodic motion is stable if *all* solutions of the variational equation associated with it tend toward zero for all positive time and unstable if there is at least one solution which does not tend toward zero. The stability characteristics of Eq. (4.67) often are represented in a chart as shown in Fig. 4.31.[10]

From the response diagram of Duffing's equation, the out-of-phase motion having the larger amplitude appears to be unstable. This portion of the response diagram (Fig. 4.17) corresponds to unstable motion in the Mathieu stability chart (Fig. 4.31), and the locus of vertical tangents of the response curves (considering undamped vibration for simplicity) corresponds exactly to the boundaries between stable and unstable regions in the stability chart. Thus, the region of interest in the response diagram is described by the free vibration

$$\omega^2 = \kappa^2(1 + \tfrac{3}{4}\mu^2 A^2) \tag{4.69}$$

and the locus of vertical tangents

$$\tfrac{3}{2}\kappa^2\mu^2 A^2 + \frac{p}{A} = 0 \tag{4.70}$$

The corresponding curves in the stability chart are taken as those for small positive values of γ and λ which have the approximate equations

$$\gamma = \tfrac{1}{2} - 2\lambda \tag{4.71}$$

$$\gamma = -\tfrac{1}{2} + 2\lambda \tag{4.72}$$

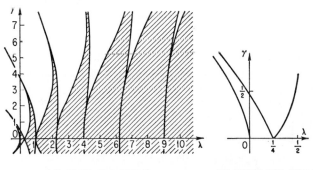

FIG. 4.31. Stability chart for Mathieu's equation [Eq. (4.67)].

Now, if Eq. (4.69) is introduced into Eqs. (4.68), the resulting equations expanded by the binomial theorem (assuming μ^2 small), and Eq. (4.72) introduced, the result is an identity. Therefore, the free vibration-response curve maps onto the curve of positive slope in the stability chart. The locus of vertical tangents to the response curves maps into the curve of negative slope in the stability chart; this may be seen from the identity obtained by introducing the equations obtained above by binomial expansion into Eq. (4.71) and then employing Eq. (4.70).

In any given case, it can be determined whether a motion is stable or unstable on the basis of the values of γ and λ, according to the location of the point in the stability chart.

The question of stability of response also can be resolved by means of a "stability criterion" developed from the Kryloff-Bogoliuboff procedures.[34] The differential equation of motion is considered in the form

$$\ddot{x} + \kappa^2 x + f(x,\dot{x}) = p \cos \omega t$$

Proceeding in the manner of the Kryloff-Bogoliuboff procedure described earlier,

$$\dot{A} = \frac{1}{\kappa} f(x,\dot{x}) \sin (\kappa t + \theta) - \frac{p}{\kappa} \cos \omega t \sin (\kappa t + \theta)$$

$$\dot{\theta} = \frac{1}{\kappa} f(x,\dot{x}) \cos (\kappa t + \theta) - \frac{p}{A\kappa} \cos \omega t \cos (\kappa t + \theta)$$

Expanding the last terms of these equations, the result contains motions of frequency κ, $\kappa + \omega$, and $\kappa - \omega$. The motion over a long interval of time is of interest, and the motions of frequencies $\kappa + \omega$ and $\kappa - \omega$ may be averaged out; this is accomplished by integrating over the period $2\pi/\omega$:

$$\dot{A} = S(A) - \frac{p}{2\kappa} \sin (\Phi - \omega t)$$

$$\dot{\theta} = \frac{C(A)}{A} - \frac{p}{2\kappa A} \cos (\Phi - \omega t)$$

where

$$S(A) = \frac{1}{2\pi\kappa} \int_0^{2\pi} f(A \cos \Phi, -A\kappa \sin \Phi) \sin \Phi \, d\Phi$$

$$C(A) = \frac{1}{2\pi\kappa} \int_0^{2\pi} f(A \cos \Phi, -A\kappa \sin \Phi) \cos \Phi \, d\Phi$$

The steady-state solution may be determined by employing the conditions $A = A_0$, $\psi = \Phi - \omega t = \psi_0$:

$$\frac{p^2}{4\kappa^2} = S^2(A_0) + [C(A_0) + A_0(\kappa - \omega)]^2$$

$$\tan \psi_0 = \frac{S(A_0)}{C(A_0) + A_0(\kappa - \omega)}$$

This steady-state solution will now be perturbed and the stability of the ensuing motion investigated. Let

$$A(t) = A_0 + \xi(t) \qquad [\xi \ll A_0]$$

$$\psi(t) = \psi_0 + \eta(t) \qquad [\eta \ll \psi_0]$$

By Taylor's series expansion:

$$\dot{\xi} = \xi S'(A_0) - \frac{p}{2\kappa} \eta \cos \psi_0$$

$$\dot{\eta} = \frac{\xi}{A_0} [(\kappa - \omega) + C'(A_0)] + \frac{p}{2\kappa A_0} \eta \sin \psi_0$$

where primes indicate differentiation with respect to A. These two differential equations are satisfied by the solutions

$$\xi = \mathcal{Q}e^{zt} \qquad \eta = \mathcal{B}e^{zt}$$

where $\mathcal{Q}$ and $\mathcal{B}$ are arbitrary constants and

$$z = \frac{1}{2A_0}\left\{[S(A_0) + A_0 S'(A_0)] \pm \sqrt{[S(A_0) + A_0 S'(A_0)]^2 - 4A_0\bar{p}\frac{d\bar{p}}{dA_0}}\right\}$$

and $\bar{p} = p/2\kappa$.

For stability, the real parts of z must be negative; hence, the following criteria can be established:[34]

$$[S(A_0) + A_0 S'(A_0)] < 0,\ \frac{d\bar{p}}{dA_0} > 0,\ \text{ensures stability}$$

$$[S(A_0) + A_0 S'(A_0)] < 0,\ \frac{d\bar{p}}{dA_0} < 0,\ \text{ensures instability}$$

$$[S(A_0) + A_0 S'(A_0)] > 0,\ \frac{d\bar{p}}{dA_0} \gtrless 0,\ \text{ensures instability}$$

$$[S(A_0) + A_0 S'(A_0)] = 0,\ \frac{d\bar{p}}{dA_0} > 0,\ \text{ensures stability}$$

These criteria can be interpreted in terms of response curves by reference to Fig. 4.14. For systems of this type, $[S(A_0) + A_0 S'(A_0)] < 0$; when $\frac{d\bar{p}}{dA_0} > 0$, $\bar{p}$ increases as A_0 also increases. This does not hold for the middle branch of the response curves, thus confirming the earlier results. Other analyses of stability are found in Refs. 35 to 38.

SUPERHARMONIC AND SUBHARMONIC VIBRATIONS

INTRODUCTION

Steady-state response of a nonlinear system also may occur at some multiple or submultiple of the forcing frequency. The response depends on both the frequency and the amplitude of the excitation, and may vary with time as well as with different initial conditions. Such vibration can be analyzed by several of the methods discussed above: The Kryloff and Bogoliuboff method,[39] the Ritz method,[40] a method employing Fourier series with either a perturbation or an iteration procedure used to determine the coefficients,[10] etc. Other studies are given in Refs. 41 to 43, as well as in many of the other references of this chapter.

SUPERHARMONIC RESPONSE

Consider forced vibration of a system defined by the differential equation

$$\ddot{x} + 2\delta\kappa\dot{x} + \kappa^2(x + \mu^2 x^3) = p\cos\omega t \tag{4.73}$$

The solution of Eq. (4.73) may be of the form

$$x = A_1\cos\omega t + A_3\cos 3\omega t \tag{4.74}$$

Typical values for the coefficients A_1 and A_3 are illustrated graphically in Fig. 4.32[2] as a function of the frequency ratio ω/ω_n, where ω_n is the frequency of free vibration obtained from Eq. (4.73) by letting $\mu^2 = 0$.

If the term with frequency 3ω (i.e., the third harmonic) in Eq. (4.74) is neglected, the response can be described by plotting the coefficient A_1 against frequency, as shown by the broken lines a in Fig. 4.32. Such response is designated *harmonic response*; the resonance condition (at $\omega/\omega_n = 1$) is designated *harmonic resonance* and the peak is bent to the right as a consequence of the assumed hardening type of nonlinearity. When the third harmonic is included, the response is described by the two coefficients A_1 and A_3,

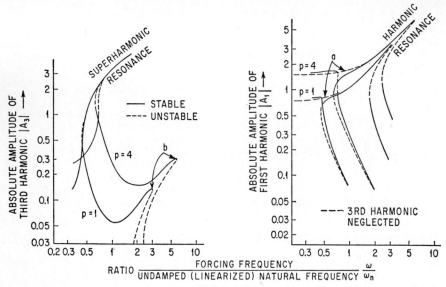

FIG. 4.32. Response curves for system defined by Eq. (4.73) with harmonic response. (*After Clauser*.[2])

FIG. 4.33. Response curves for system with harmonic response in the third and fifth harmonics, where $\bar{s} = \dfrac{p}{\kappa^2}\sqrt{\dfrac{3}{4}}\,\mu^2$ and $\delta = 0.01$. (*After Atkinson*.[43])

plotted separately in Fig. 4.32. The latter is characterized by the following features:

1. A peak in the region $\omega/\omega_n = 1$; this represents a vibration at a frequency of approximately $3\omega_n$ superimposed upon the resonance of the harmonic response whose frequency is ω_n. This peak (b in Fig. 4.32) is designated the *third-order component of the harmonic resonance*.

2. A peak in the region $\omega/\omega_n = \frac{1}{3}$; this represents vibration at a frequency of approximately ω_n when the forcing frequency is $\omega_n/3$. This region of the response is designated *superharmonic resonance*.* It is accompanied by a decrease in A_1, as indicated in Fig. 4.32; thus, the value of A_3 at the frequency of the superharmonic resonance may exceed the value of A_1.

If terms of frequency 5ω, 7ω ... with corresponding coefficients A_5, A_7, ... had been included in Fig. 4.32, the same general pattern would have been repeated at the frequency ratios $\omega/\omega_n = \frac{1}{5}$, $\frac{1}{7}$ .... The vibration amplitude of the response may then be expressed as $A = A_1 + A_3 + \cdots$ and is illustrated graphically as a sequence of superharmonic resonances superposed on the approximate harmonic response, as shown in Fig. 4.33.[43] The jump phenomenon described with reference to Fig. 4.16 also may occur with superharmonic resonance. If the damping is very small, the superharmonic resonances may be of greater magnitude than the first-order component of the harmonic resonance because the nonlinearity provides a mechanism whereby energy from the excitation may be transferred to the higher harmonics.†

SUBHARMONIC RESPONSE

The subharmonic response of the system described by Eq. (4.73) may be studied by assuming a solution of the form

$$x = A_{\frac{1}{3}} \cos \tfrac{1}{3}\omega t + A_1 \cos \omega t \tag{4.75}$$

Typical values for the coefficients $A_{\frac{1}{3}}$ and A_1 are illustrated graphically in Fig. 4.34 as a function of the frequency ratio ω/ω_n, where ω_n is again the frequency of the free vibration obtained from Eq. (4.73) by letting $\mu^2 = 0$.

If the term with frequency $(\frac{1}{3})\omega$ (i.e., the one-third harmonic) in Eq. (4.75) is neglected, the response can be described by the coefficient A_1 plotted against frequency, as shown by the lines c in Fig. 4.34.[2] Such response is designated harmonic response, and the resonance condition is designated harmonic resonance. When the one-third harmonic is included, the response is defined by the coefficients $A_{\frac{1}{3}}$ and A_1, plotted separately in Fig. 4.34. The response is characterized by a significant vibration $A_{\frac{1}{3}} \sin (\frac{1}{3})\omega t$ in the region of $\omega/\omega_n = 3$; physically, this means that the system vibrates at the frequency $\omega = \omega_n$ when the forcing frequency is $3\omega_n$. This is the *subharmonic resonance*. The harmonic response of the system is modified by the subharmonic resonance and is characterized by the solid (stable condition) and dotted (unstable condition) lines D in Fig. 4.34. The subharmonic response offers an alternate mode of vibration rather than a modification of the fundamental response. Since the subharmonic response is unstable at the junction with the fundamental response, the subharmonic cannot simply arise but must be induced by a disturbance of sufficient magnitude. Again, jump phenomena and subharmonics of higher orders are possible.

Consider the undamped system defined by

$$\omega^2 x'' + \kappa^2(x + \mu^2 x^3) = p \cos \omega t \tag{4.76}$$

Introducing an assumed solution of the form of Eq. (4.75):

$$\left(\kappa^2 - \frac{\omega^2}{9}\right) A_{\frac{1}{3}} + \tfrac{3}{4}\kappa^2\mu^2(A_{\frac{1}{3}}^3 + A_{\frac{1}{3}}^2 A_1 + 2A_{\frac{1}{3}}A_1^2) = 0$$

$$(\kappa^2 - \omega^2)A_1 + \tfrac{1}{4}\kappa^2\mu^2(A_{\frac{1}{3}}^3 + 6A_{\frac{1}{3}}^2 A_1 + 3A_1^3) = p \tag{4.77}$$

---

* The term *ultraharmonic resonance* sometimes is used in place of the term superharmonic resonance.

† Even as well as odd harmonics may appear in the superharmonic response.[40, 43]

These equations can be studied by an iteration procedure starting from the linear forced vibration.[10] Since $A_{1/3}$ is presumed to exist, Eqs. (4.77) can be rewritten

$$\omega^2 = 9\kappa^2 + 2\tfrac{7}{4}\kappa^2\mu^2(A_{1/3}^2 + A_{1/3}A_1 + 2A_1^2)$$

$$-8\kappa^2 A_1 = p + (\omega^2 - 9\kappa^2)A_1 - \tfrac{1}{4}\kappa^2\mu^2(A_{1/3}^3 + 6A_{1/3}^2 A_1 + 3A_1^3)$$

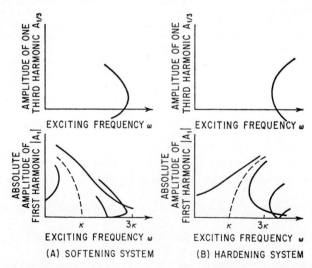

Fig. 4.34. Response curves for system defined by Eq. (4.74) with subharmonic response. (*After Clauser.*[2])

Fig. 4.35. Response curves for hardening and softening systems with subharmonic response, defined by Eq. (4.76). (*After Stoker.*[10])

Beginning the iteration with $\mu = 0$ (linear forced vibration) yields

$$\omega = 3\kappa \qquad A_1 = -\frac{p}{8\kappa}$$

The next step becomes

$$\omega^2 = 9\kappa + {}^{27}\!/_4\kappa^2\mu^2\left(A_{1/3}^2 - \frac{p}{8\kappa}A_{1/3} + \frac{p^2}{32\kappa^2}\right)$$

$$A_1 = -\frac{p}{8\kappa} + \frac{1}{3}{}_2\mu^2\left(A_{1/3}^3 + \frac{21}{8}\frac{p}{\kappa}A_{1/3}^2 - \frac{27}{64}\frac{p^2}{\kappa^2}A_{1/3} + \frac{51}{512}\frac{p^3}{\kappa^3}\right)$$

This solution applies to both the hardening system $(x + \mu^2 x^3)$ and the softening system $(x - \mu^2 x^3)$.

In the response diagram showing $A_{1/3}$ as a function of ω, ω^2 has a minimum for $+\mu^2$ and a maximum for $-\mu^2$ corresponding to

$$A_{1/3} = \frac{p}{16\kappa} \qquad \omega^2 = 9[\kappa^2 + ({}^{21}\!/_{1024})\mu^2 p^2]$$

Thus, the subharmonic vibration can exist only for

$$\omega < 9[\kappa^2 + ({}^{21}\!/_{1024})\mu^2 p^2] \qquad \text{(hardening system)}$$

$$\omega > 9[\kappa^2 + ({}^{21}\!/_{1024})\mu^2 p^2] \qquad \text{(softening system)}$$

Response curves for both systems are shown in Fig. 4.35.

This analysis can be extended to include linear damping [10]; then the subharmonic of order $\frac{1}{3}$ cannot occur unless the damping is very small.

SYSTEMS OF MORE THAN A SINGLE DEGREE-OF-FREEDOM

Interest in systems of more than one degree-of-freedom arises from the problem of the dynamic vibration absorber. The earliest studies of nonlinear two degree-of-freedom systems were those of vibration absorbers having nonlinear elements.[44-46]

The analysis of multiple degree-of-freedom systems can be carried out by various of the methods described earlier in this chapter; thus, a stepwise linear system is treated in Ref. 47, and more general systems are treated in Refs. 47 to 51. The extension of phase-plane methods to such systems is given in Refs. 29, 30, and 52.

All the essential features of nonlinear vibration of single degree-of-freedom systems described earlier occur in the multiple degree-of-freedom systems as well. An analysis which considers subharmonic vibration in such systems by an iteration procedure is given in Ref. 53 and is completely analogous to that given here for the single degree-of-freedom system, with analogous results.

REFERENCES

Additional citations to the literature are given in Refs. 4, 29, 32, 54, and 55, and in past and current issues of the journal *Applied Mechanics Reviews*.

1. von Kármán, T.: *Bull. Am. Math. Soc.*, **46**: 615 (1940).
2. Clauser, F. H.: *J. Aeronaut. Sci.*, **23**:411 (1956).
3. Davis, S. A.: *Product Eng.*, **25**:181 (1954).
4. McLachlan, N. W.: "Ordinary Nonlinear Differential Equations," Oxford University Press, New York, 1956.
5. Minorsky, N.: "Introduction to Nonlinear Mechanics," Edwards Bros., Ann Arbor, Mich., 1947.
6. Klotter, K.: *Proc. 1st U.S. Natl. Congr. Appl. Mechs.*, 1951.
7. Klotter, K.: *Stanford Univ., Rept. 17*, Contract N6 ONR-251-II, 1951.
8. Klotter, K.: *Proc. Symposium on Nonlinear Circuit Analysis*, 1953.

9. Levy, H., and E. A. Baggott: "Numerical Solutions of Differential Equations," Dover Publications, New York, 1950.
10. Stoker, J. J.: "Nonlinear Vibrations," Interscience Publishers, Inc., New York, 1950.
11. Kryloff, N., and N. Bogoliuboff: "Introduction to Nonlinear Mechanics," Princeton University Press, Princeton, N.J., 1943.
12. Andronow, A. A., and C. E. Chaikin: "Theory of Oscillations," Princeton University Press, Princeton, N.J., 1949.
13. Rudenberg, R.: *ZAMM*, **3**:454 (1923).
14. Brock, J. E.: *J. Appl. Mechanics*, **18** (1951).
15. Roberson, R. E.: *J. Appl. Mechanics*, **20**:237 (1953).
16. Wylie, C. R.: *J. Franklin Inst.*, **236**:273 (1943).
17. Young, D.: *Proc. 1st Midwest. Conf. Solid Mechanics*, 1953.
18. Fifer, S.: *J. Appl. Phys.*, **22**:1421 (1951).
19. Minorsky, N.: "Advances in Applied Mechanics," Vol. I, Academic Press, Inc., New York, 1948.
20. Bellin, A. I.: "Advances in Applied Mechanics," Vol. III, Academic Press, Inc., New York, 1953.
21. Pipes, L. A.: *J. Appl. Phys.*, **13**:117 (1942).
22. Duffing, G.: "Erzwungene Schwingungen bei veranderlicher Eigenfrequenz," F. Vieweg u Sohn, Brunswick, 1918.
23. Rauscher, M.: *J. Appl. Mechanics*, **5**:169 (1938).
24. Minorsky, N.: *J. Franklin Inst.*, **248**:205 (1949).
25. Carrier, G. F.: *Quart. Appl. Math.*, **3**:157 (1945).
26. Schwesinger, G.: *J. Appl. Mechanics*, **17**:202 (1950).
27. Abramson, H. N.: *Product Eng.*, **25**:179 (1954).
28. Evaldson, R. L., R. S. Ayre, and L. S. Jacobsen: *J. Franklin Inst.*, **248**:473 (1949).
29. Ku, Y. H.: "Analysis and Control of Nonlinear Systems," The Ronald Press Company, New York, 1958.
30. Flügge-Lotz, I.: "Discontinuous Automatic Control," Princeton University Press, Princeton, N.J., 1953.
31. Jacobsen, L. S.: *J. Appl. Mechanics*, **19**:543 (1952).
32. Jacobsen, L. S., and R. S. Ayre: "Engineering Vibrations," McGraw-Hill Book Company, Inc., New York, 1958.
33. Bishop, R. E. D.: *Proc. Inst. Mech. Engrs.*, **168**:299 (1954).
34. Klotter, K., and E. Pinney: *J. Appl. Mechanics*, **20**:9 (1953).
35. John, F.: "Studies in Nonlinear Vibration Theory," New York University, 1946.
36. Rosenberg, R. M.: *Proc. 2nd Natl. Congr. Appl. Mechanics*, 1954.
37. Young, D., and P. N. Hess: *Proc. 2nd Natl. Congr. Appl. Mechanics*, 1954.
38. Hayashi, C.: "Forced Oscillations in Nonlinear Systems," Nippon Pub. Co., Osaka, Japan, 1953.
39. Caughey, T. K.: *J. Appl. Mechanics*, **21**:327 (1954).
40. Burgess, J. C.: *Stanford Univ., Rept.* 27, Contract N6 ONR-251-II, 1954.
41. Wu, M. H. L.: *Proc. 1st Natl. Congr. Appl. Mechanics*, 1951.
42. Rosenberg, R. M.: *Proc. 2nd Midwest. Conf. Solid Mechanics*, 1955.
43. Atkinson, C. P.: *J. Appl. Mechanics*, **24**:520 (1957).
44. Roberson, R. E.: *J. Franklin Inst.*, **254**:205 (1952).
45. Pipes, L. A.: *J. Appl. Mechanics*, **20**:515 (1953).
46. Arnold, F. R.: *J. Appl. Mechanics*, **22**:487 (1955).
47. Soroka, W. W.: *J. Appl. Mechanics*, **17**:185 (1950).
48. Sethna, P. R.: *Proc. 2nd Natl. Congr. Appl. Mechanics*, 1954.
49. Huang, T. C.: *J. Appl. Mechanics*, **22**:107 (1955).
50. Klotter, K.: *Trans. IRE on Circuit Theory*, **CT-1** (4):13 (1954).
51. Stoker, J. J.: *Proc. 2nd Natl. Congr. Appl. Mechanics*, 1954.
52. Ku, Y. H.: *J. Franklin Inst.*, **259**:115 (1955).
53. Huang, T. C.: *Proc. 2nd Natl. Congr. Appl. Mechanics*, 1954.
54. Minorsky, N.: *Appl. Mechanics Revs.*, **4**:266 (1951).
55. Klotter, K.: *Appl. Mechanics Revs.*, **10**:495 (1957).

ADDITIONAL REFERENCES

Minorsky, N.: "Nonlinear Oscillations," D. Van Nostrand Company, Inc., Princeton, N.J., 1962.
Hayashi, C.: "Nonlinear Oscillations in Physical Systems," McGraw-Hill Book Co., New York, 1964.
Bogoliubov, N. N. and Y. A. Mitropolsky: "Asymptotic Methods in the Theory of Non-linear Oscillations" (translation), Hindustan Publishing Corp., Delhi, 1961.

5

SELF-EXCITED VIBRATION

F. F. Ehrich
General Electric Co.

INTRODUCTION

GENERAL NATURE

Self-excited systems begin to vibrate of their own accord spontaneously, the amplitude increasing until some nonlinear effect limits any further increase.   The energy supplying these vibrations is obtained from a uniform source of power associated with the system which, due to some mechanism inherent in the system, gives rise to oscillating forces. The nature of self-excited vibration compared to forced vibration is:[1]

In self-excited vibration the alternating force that sustains the motion is created or controlled by the motion itself; when the motion stops, the alternating force disappears.

In a forced vibration the sustaining alternating force exists independent of the motion and persists when the vibratory motion is stopped.

The occurrence of self-excited vibration in a physical system is intimately associated with the stability of equilibrium positions of the system.   If the system is disturbed from a position of equilibrium, forces generally appear which cause the system to move either toward the equilibrium position or away from it.   In the latter case the equilibrium position is said to be unstable; then the system may either oscillate with increasing amplitude or monotonically recede from the equilibrium position until nonlinear or limiting restraints appear.   The equilibrium position is said to be stable if the disturbed system approaches the equilibrium position either in a damped oscillatory fashion or asymptotically.

The forces which appear as the system is displaced from its equilibrium position may depend on the displacement or the velocity, or both.   If displacement-dependent forces appear and cause the system to move away from the equilibrium position, the system is said to be statically unstable.   For example, an inverted pendulum is statically unstable. Velocity-dependent forces which cause the system to recede from a statically stable equilibrium position lead to dynamic instability.

Self-excited vibrations are characterized by the presence of a mechanism whereby a system will vibrate at its own natural or critical frequency, essentially *independent* of the *frequency* of any external stimulus.   In mathematical terms, the motion is described by the unstable *homogeneous* solution to the homogeneous equations of motion.   In contradistinction, in the case of "forced," or "resonant," vibrations, the *frequency* of the oscillation is *dependent* on (equal to, or a whole number ratio of) the frequency of a forcing function external to the vibrating system (e.g., shaft rotational speed in the case of rotating shafts).   In mathematical terms, the forced vibration is the *particular* solution to the *nonhomogeneous* equations of motion.

Self-excited vibrations pervade all areas of design and operations of physical systems where motion or time-variant parameters are involved—aeromechanical systems (flutter, aircraft flight dynamics); aerodynamics (separation, stall, musical wind instruments, diffuser and inlet chugging); aerothermodynamics (flame instability, combustor screech); mechanical systems (machine-tool chatter); and feedback networks (pneumatic, hydraulic, and electromechanical servomechanisms).

ROTATING MACHINERY

One of the more important manifestations of self-excited vibrations, and the one that is the principal concern in this chapter, is that of rotating machinery, specifically, the self-excitation of lateral, or flexural, vibration of rotating shafts (as distinct from torsional, or longitudinal, vibration).

In addition to the description of a large number of such phenomena in standard vibrations textbooks (most typically and prominently, Ref. 1), the field has been subject to several generalized surveys.[2, 3, 4] The mechanisms of self-excitation which have been identified can be categorized as follows:

WHIRLING OR WHIPPING
Hysteretic whirl
Fluid trapped in the rotor
Dry friction whip
Fluid bearing whip
Seal and blade-tip-clearance effect in turbomachinery
Propeller and turbomachinery whirl

PARAMETRIC INSTABILITY
Asymmetric shafting
Pulsating torque
Pulsating longitudinal loading

STICK-SLIP RUBS AND CHATTER

INSTABILITIES IN FORCED VIBRATIONS
Bistable vibration
Unstable imbalance

In each instance, the physical mechanism is described and aspects of its prevention or its diagnosis and correction are given. Some exposition of its mathematical analytic modeling is also included.

WHIRLING OR WHIPPING

ANALYTIC MODELING

In the most important subcategory of instabilities (generally termed whirling or whipping), the unifying generality is the generation of a tangential force, normal to an arbitrary radial deflection of a rotating shaft, whose magnitude is proportional to (or varies monotonically with) that deflection. At some "onset" rotational speed, such a force system will overcome the stabilizing external damping forces which are generally present and induce a whirling motion of ever-increasing amplitude, limited only by nonlinearities which ultimately limit deflections.

A close mathematical analogy to this class of phenomena is the concept of "negative damping" in linear systems with constant coefficients, subject to *plane* vibration.

A simple mathematical representation of a self-excited vibration may be found in the concept of negative damping. Consider the differential equation for a damped, free vibration:

$$m\ddot{x} + c\dot{x} + kx = 0 \tag{5.1}$$

This is generally solved by assuming a solution of the form

$$x = Ce^{st}$$

Substitution of this solution into Eq. (5.1) yields the characteristic (algebraic) equation

$$s^2 + \frac{c}{m} s + \frac{k}{m} = 0 \qquad (5.2)$$

If $c < 2\sqrt{mk}$, the roots are complex:

$$s_{1,2} = -\frac{c}{2m} \pm iq$$

where

$$q = \sqrt{\frac{k}{m} - \left(\frac{c}{2m}\right)^2}$$

The solution takes the form

$$x = e^{-ct/2m}(A \cos qt + B \sin qt) \qquad (5.3)$$

This represents a decaying oscillation because the exponential factor is negative, as illustrated in Fig. 5.1A. If $c < 0$, the exponential factor has a positive exponent and the vibration appears as shown in Fig. 5.1B. The system, initially at rest, begins to oscillate spontaneously with ever-increasing amplitude. Then, in any physical system, some nonlinear effect enters and Eq. (5.1) fails to represent the system realistically. Equation (5.4) defines a nonlinear system with negative damping at small amplitudes but with large positive damping at larger amplitudes, thereby limiting the amplitude to finite values:

$$m\ddot{x} + (-c + ax^2)\dot{x} + kx = 0 \qquad (5.4)$$

Thus, the fundamental criterion of stability in linear systems is that the roots of the characteristic equation have negative real parts, thereby producing decaying amplitudes.

In the case of a whirling or whipping shaft, the equations of motion (for an idealized shaft with a single lumped mass m) are more appropriately written in polar coordinates for the radial force balance,

$$-m\omega^2 r + m\ddot{r} + c\dot{r} + kr = 0 \qquad (5.5)$$

and for the tangential force balance,

$$2m\omega\dot{r} + c\omega r - F_n = 0 \qquad (5.6)$$

Fig. 5.1. (A) Illustration showing a decaying vibration (stable) corresponding to negative real parts of the complex roots. (B) Increasing vibration corresponding to positive real parts of the complex roots (unstable).

where we presume a constant rate of whirl ω.

In general, the whirling is predicated on the existence of some physical phenomenon which will induce a force F_n that is normal to the radial deflection r and is in the direction of the whirling motion — i.e., in opposition to the damping force, which tends to inhibit the whirling motion. Very often, this normal force can be characterized or approximated as being proportional to the radial deflection:

$$F_n = f_n r \qquad (5.7)$$

The solution then takes the form

$$r = r_0 e^{at} \qquad (5.8)$$

For the system to be stable, the coefficient of the exponent

$$a = \frac{f_n - c\omega}{2m\omega} \qquad (5.9)$$

must be negative, giving the requirement for stable operation as

$$f_n < \omega c \tag{5.10}$$

As a rotating machine increases its rotational speed, the left-hand side of this inequality (which is generally also a function of shaft rotation speed) may exceed the right-hand side, indicative of the onset of instability. At this onset condition,

$$a = 0+ \tag{5.11}$$

so that whirl speed at onset is found to be

$$\omega = \left(\frac{k}{m}\right)^{1/2} \tag{5.12}$$

That is, the whirling speed at onset of instability is the shaft's natural or critical frequency, irrespective of the shaft's rotational speed (rpm). The direction of whirl may

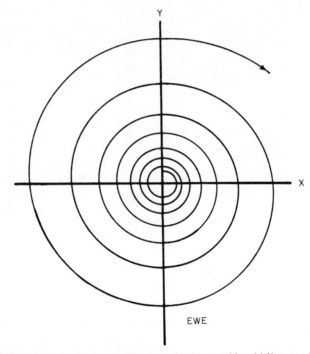

FIG. 5.2. Trajectory of rotor center of gravity in unstable whirling or whipping.

be in the same rotational direction as the shaft rotation (*forward* whirl) or opposite to the direction of shaft rotation (*backward* whirl) depending on the direction of the de-stabilizing force F_n.

When the system is unstable, the solution for the trajectory of the shaft's mass is, from Eq. (5.8), an exponential spiral as in Fig. 5.2. Any planar component of this two-dimensional trajectory takes the same form as the unstable planar vibration shown in Fig. 5.1*B*.

GENERAL DESCRIPTION

The most important examples of whirling and whipping instabilities are
  Hysteretic whirl
  Fluid trapped in the rotor
  Dry friction whip
  Fluid bearing whip
  Seal and blade-tip-clearance effect in turbomachinery
  Propeller and turbomachinery whirl

All these self-excitation systems involve friction or fluid energy mechanisms to generate the destabilizing force.

These phenomena are rarer than forced vibration due to unbalance or shaft misalignment, and they are difficult to anticipate before the fact or diagnose after the fact

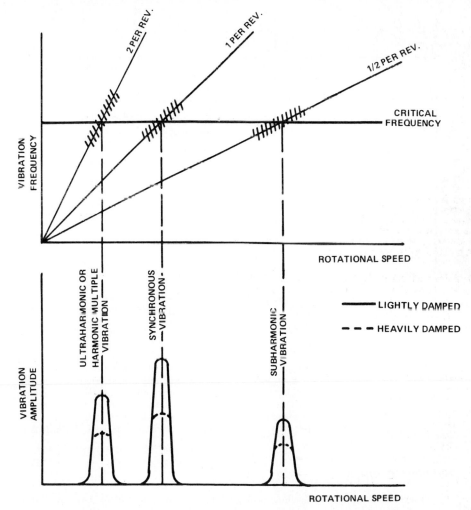

Fig. 5.3. (A) Attributes of forced vibration or resonance in rotating machinery.

FIG. 5.3. (*B*) Attributes of whirling or whipping in rotating machinery.

because of their subtlety. Also, self-excited vibrations are potentially more destructive, since the asynchronous whirling of self-excited vibration induces alternating stresses in the rotor and can lead to fatigue failures of rotating components. Synchronous forced vibration typical of unbalance does not involve alternating stresses in the rotor and will rarely involve rotating element failure. The general attributes of these instabilities, insofar as they differ from forced excitations, are summarized in Table 5.1 and Fig. 5.3*A* and *B*.

HYSTERETIC WHIRL

The mechanism of hysteretic whirl, as observed experimentally,[5] defined analytically,[6] or described in standard texts,[7] may be understood from the schematic representation

Table 5.1. Characterization of Two Categories of Vibration of Rotating Shafts

	Forced or resonant vibration	Whirling or whipping
Vibration frequency–rpm relationship	Frequency is equal to (i.e., synchronous with) rpm or a whole number or rational fraction of rpm, as in Fig. 5.3A.	Frequency is nearly constant and relatively independent of rotor rotational speed or any external stimulus and is at or near one of the shaft critical or natural frequencies, as in Fig. 5.3B.
Vibration amplitude–rpm relationship	Amplitude will peak in a narrow band of rpm wherein the rotor's critical frequency is equal to the rpm or to a whole-number multiple or a rational fraction of the rpm or an external stimulus, as in Fig. 5.3A.	Amplitude will suddenly increase at an onset rpm and continue at high or increasing levels as rpm is increased, as in Fig. 5.3B.
Influence of damping	Addition of damping may reduce peak amplitude but not materially affect rpm at which peak amplitude occurs, as in Fig. 5.3A.	Addition of damping may defer onset to a higher rpm but not materially affect amplitude after onset, as in Fig. 5.3B.
System geometry	Excitation level and hence amplitude are dependent on some lack of axial symmetry in the rotor mass distribution or geometry, or external forces applied to the rotor. Amplitudes may be reduced by refining the system to make it more perfectly axisymmetric.	Amplitudes are independent of system axial symmetry. Given an infinitesimal deflection to an otherwise symmetric system, the amplitude will self-propagate.
Rotor fiber stress	For synchronous vibration, the rotor vibrates in frozen, deflected state, without oscillatory fiber stress.	Rotor fibers are subject to oscillatory stress at a frequency equal to the difference between rotor rpm and whirling speed.
Avoidance or elimination	1. Tune the system's critical frequencies to be out of the rpm operating range. 2. Eliminate all deviations from axial symmetry in the system as built or as induced during operation (e.g., balancing). 3. Introduce damping to limit peak amplitudes at critical speeds which must be traversed.	1. Restrict operating rpm to below instability onset rpm. 2. Defeat or eliminate the instability mechanism. 3. Introduce damping to raise the instability onset speed to above the operating speed range.

of Fig. 5.4. With some nominal radial deflection of the shaft, the flexure of the shaft would induce a neutral strain axis normal to the deflection direction. From first-order considerations of elastic-beam theory, the neutral axis of stress would be coincident with the neutral axis of strain. The net elastic restoring force would then be perpendicular to the neutral stress axis, i.e., parallel to and opposing the deflection. In actual fact, hysteresis, or internal friction, in the rotating shaft will cause a phase shift in the development of stress as the shaft fibers rotate around through peak strain to the neutral strain axis. The net effect is that the neutral stress axis is displaced in angle orientation from the neutral strain axis, and the resultant force is not parallel to the deflection. In particular, the resultant force has a tangential component *normal* to deflection which is the fundamental precondition for whirl. This tangential force component is in the direction of rotation and induces a *forward* whirling motion which in-

Fig. 5.4. Hysteretic whirl.

creases centrifugal force on the deflected rotor, thereby increasing its deflection. As a consequence, induced stresses are increased, thereby increasing the whirl-inducing force component.

Several surveys and contributions to the understanding of the phenomenon have been published in Refs. 8, 9, 10, and 11. It has generally been recognized that hysteretic whirl can occur only at rotational speeds above the first-shaft critical speed (the lower the hysteretic effect, the higher the attainable whirl-free operating rpm). It has been shown[12] that once whirl has started, the critical whirl speed that will be induced (from among the spectrum of criticals of any given shaft) will have a frequency approximately half the onset rpm.

A straightforward method for hysteretic whirl avoidance is that of limiting shafts to subcritical operation, but this is unnecessarily and undesirably restrictive. A more effective avoidance measure is to limit the hysteretic characteristic of the rotor. Most investigators (e.g., Ref. 5) have suggested that the essential hysteretic effect is caused by working at the interfaces of joints in a rotor rather than within the material of that rotor's components. Success has been achieved in avoiding hysteretic whirl by minimizing the number of separate elements, restricting the span of concentric rabbets and shrunk fitted parts, and providing secure lockup of assembled elements held together by tie bolts and other compression elements. Bearing-foundation characteristics also play a role in suppression of hysteretic whirl.[8]

WHIRL DUE TO FLUID TRAPPED IN ROTOR

There has always been a general awareness that high-speed centrifuges are subject to a special form of instability. It is now appreciated that the same self-excitation may be experienced more generally in high-speed rotating machinery where liquids (e.g., oil from bearing sumps, or steam condensate, etc.) may be inadvertently trapped in the internal cavity of hollow rotors.   The mechanism of instability is shown schematically in Fig. 5.5.   For some nominal deflection of the rotor, the fluid is flung out radially in

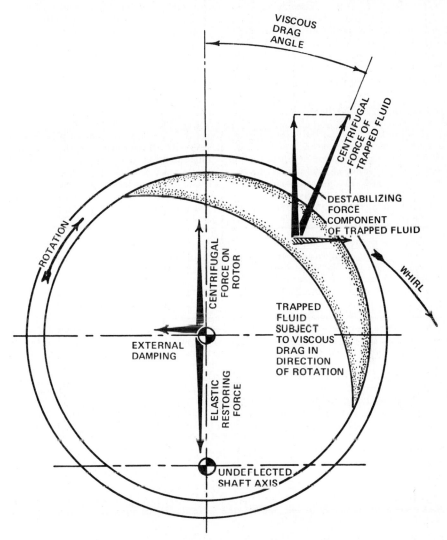

FIG. 5.5. Whirl due to fluid trapped in rotor.

the direction of deflection.   But the fluid does not remain in simple radial orientation. The spinning surface of the cavity drags the fluid (which has some finite viscosity) in the direction of rotation.   This angle of advance results in the centrifugal force on the fluid having a component in the tangential direction in the direction of rotation.   This force then is the basis of instability, since it induces forward whirl which increases centrifugal force on the fluid and thereby increases the whirl-inducing force.

Contributions to the understanding of the phenomenon as well as a complete history of the phenomenon's study are available in Ref. 13. It has been shown[14] that onset speed for instability is always above critical rpm and below twice-critical rpm. Since the whirl is at the shaft's critical frequency, the ratio of whirl frequency to rpm will be in the range of 0.5 to 1.0.

Avoidance of this self-excitation can be accomplished by running shafting subcritically, although this is generally undesirable in centrifuge-type applications when further consideration is made of the role of trapped fluids as unbalance in forced vibration of rotating shafts (as described in Ref. 14). Where the trapped fluid is not fundamental to the machine's function, the appropriate avoidance measure, if the particular application permits, is to provide drain holes at the outermost radius of all hollow cavities where fluid might be trapped.

DRY FRICTION WHIP

As described in standard vibration texts (e.g., Ref. 15), dry friction whip is experienced when the surface of a rotating shaft comes in contact with an unlubricated stationary guide or shroud or stator system. This can occur in an unlubricated journal bearing or with loss of clearance in a hydrodynamic bearing or inadvertent closure and contact in the radial clearance of labyrinth seals or turbomachinery blading or power screws.[16]

The phenomenon may be understood with reference to Fig. 5.6. When radial contact is made between the surface of the rotating shaft and a static part, Coulomb friction will induce a tangential force on the rotor. Since the friction force is approximately proportional to the radial component of the contact force, we have the preconditions for instability. The tangential force induces a whirling motion which induces larger centrifugal force on the rotor, which in turn induces a large radial contact and hence larger whirl-inducing friction force.

It is interesting to note that this whirl system is *counter* to the shaft rotation direction (i.e., *backward* whirl). One may envision the whirling system as the rolling (accompanied by appreciable slipping) of the shaft in the stator system.

The same situation can be produced by a thrust bearing where angular deflection is combined with lateral deflection.[17] If contact occurs on the same side of the disc as the virtual pivot point of the deflected disc, then backward whirl will result. Conversely, if contact occurs on the side of the disc opposite to the side where the virtual pivot point of the disc is located, then forward whirl will result.

It has been suggested (but not concluded)[18] that the whirling frequency is generally less than the critical speed.

The vibration is subject to various types of control. If contact between rotor and stator can be avoided or the contact area can be kept well lubricated, no whipping will occur. Where contact must be accommodated, and lubrication is not feasible, whipping may be avoided by providing abradability of the rotor or stator element to allow disengagement before whirl. When dry friction is considered in the context of the dynamics of the stator system in combination with that of the rotor system,[19] it is found that whirl can be inhibited if the independent natural frequencies of the rotor and stator are kept dissimilar, that is, a very stiff rotor should be designed with a very soft mounted stator element that may be subject to rubs. No first-order interdependence of whirl speed with rotational speed has been established.

FLUID BEARING WHIP

As described in experimental and analytic literature,[20] and in standard texts (e.g., Ref. 21), fluid bearing whip can be understood by referring to Fig. 5.7. Consider some nominal radial deflection of a shaft rotating in a fluid (gas- or liquid-) filled clearance. The entrained, viscous fluid will circulate with an average velocity of about half the shaft's surface speed. The bearing pressures developed in the fluid will not be symmetric about the radial deflection line. Because of viscous losses of the bearing fluid

Fig. 5.6. Dry friction whip.

circulating through the close clearance, the pressure on the upstream side of the close clearance will be higher than that on the downstream side. Thus, the resultant bearing force will include a tangential force component in the direction of rotation which tends to induce *forward* whirl in the rotor. The tendency to instability is evident when this tangential force exceeds inherent stabilizing damping forces. When this happens, any induced whirl results in increased centrifugal forces; this, in turn, closes the clearance further and results in ever-increasing destabilizing tangential force. Detailed reviews of the phenomenon are available in Refs. 22 and 23.

These and other investigators have shown that to be unstable, shafting must rotate at an rpm equal to or greater than approximately twice the critical speed, so that one would expect the ratio of frequency to rpm to be equal to less than approximately 0.5.

The most obvious measure for avoiding fluid bearing whip is to restrict rotor maximum rpm to less than twice its lowest critical speed. Detailed geometric variations in the bearing runner design, such as grooving and tilt-pad configurations, have also been found effective in inhibiting instability. In extreme cases, use of rolling contact bearings instead of fluid film bearings may be advisable.

Various investigators (e.g., Ref. 24) have noted that fluid seals as well as fluid bearings are subject to this type of instability.

FIG. 5.7. Fluid bearing whip.

SEAL AND BLADE-TIP-CLEARANCE EFFECT IN TURBOMACHINERY

Axial-flow turbomachinery may be subject to an additional whirl-inducing effect by virtue of the influence of tip clearance on turbopump or compressor or turbine efficiency.[25] As shown schematically in Fig. 5.8, some nominal radial deflection will close the radial clearance on one side of the turbomachinery component and open the clearance 180° away on the opposite side. We would expect the closer clearance zone to operate more efficiently than the open clearance zone. For a compressor, the more efficient zone requires a lesser blade force to achieve the average pressure rise. For a turbine, a greater work extraction and blade force level is achieved in the more efficient region for a given average pressure drop. In either case, a resultant net tangential force may be generated to induce whirl in the direction of rotor rotation (i.e., forward whirl). In the case of radial-flow turbomachinery, it has been suggested[26] that destabilizing forces are exerted on an eccentric (i.e., dynamically deflected) impeller due to variations of loading of the diffuser vanes.

One text[27] describes several manifestations of this class of instability—in the thrust balance piston of a steam turbine; in the radial labyrinth seal of a radial-flow Ljungstrom counterrotating steam turbine; in the Kingsbury thrust bearing of a vertical-shaft hydraulic turbogenerator; and in the tip seals of a radial-inflow hydraulic Francis turbine.

A survey paper[3] includes a bibliography of several German papers on the subject from 1958 to 1969.

FIG. 5.8. Turbomachinery tip clearance effect's contribution to whirl.

An analysis is available[28] dealing with the possibility of stimulating flexural vibrations in the seals themselves, although it is not clear if the solutions pertain to gross deflections of the entire rotor.

It is reasonable to expect that such destabilizing forces may at least contribute to instabilities experienced on high-powered turbomachines. If this mechanism were indeed a key contributor to instability, one would conjecture that very small or very large initial tip clearances would minimize the influence of tip clearance on the unit's performance and, hence, minimize the contribution to destabilizing forces.

PROPELLER AND TURBOMACHINERY WHIRL

Propeller whirl has been identified both analytically[29] and experimentally.[30] In this instance of shaft whirling, a small deflection of the shaft will generally be accompanied by incremental velocities of the propeller blades. Where the plane of the propeller rotates as a result of the slope deflection of the shaft centerline, incremental axial velocities are induced, as shown in Fig. 5.9.   The incremental velocities result in changes in relative velocities between the blades and the airstream which change the aerodynamic forces on the blades.   At high airspeeds, the forces are destabilizing and can overcome damping forces to cause destructive whirling of the entire system.   The whirling is generally found to be counter to the shaft-rotation direction.   It has been suggested[31] that equivalent stimulation is possible in turbomachinery in the process of diagnosing and correcting whirl in an aircraft-engine compressor.   An attempt has been made[32] to generalize the analysis for axial-flow turbomachinery.   Although it has been shown that this analysis is in error, the general deduction seems appropriate that forward whirl may also be possible if the virtual pivot point of the deflected rotor is forward of the rotor (i.e., on the side of the approaching fluid).

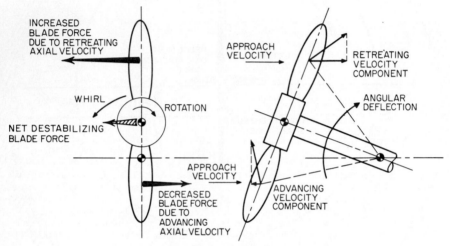

FIG. 5.9. Propeller whirl.

Instability is found to be load-sensitive in the sense of being a function of the velocity and density of the impinging flow.   It is not thought to be sensitive to the torque level of the turbomachine since, for example, experimental work[30] was on an unloaded windmilling rotor.   Corrective action is generally recognized to be stiffening the entire system and manipulating the effective pivot center of the whirling mode to inhibit angular motion of the propeller (or turbomachinery) disc as well as system damping.

PARAMETRIC INSTABILITY

ANALYTIC MODELING

There are systems in engineering and physics which are described by linear differential equations having periodic coefficients,

$$\frac{d^2y}{dz^2} + p(z)\frac{dy}{dz} + q(z)y = 0 \tag{5.13}$$

where $p(z)$ and $q(z)$ are periodic in z.   These systems also may exhibit self-excited vibrations, but the stability of the system cannot be evaluated by finding the roots of a characteristic equation.   A specialized form of this equation, which is representative of a variety of real physical problems in rotating machinery, is Mathieu's equation:

$$\frac{d^2f}{dz^2} + (a - 2q\cos 2z)f = 0 \tag{5.14}$$

Mathematical treatment and applications of Mathieu's equation are given in Refs. 33 and 34.

This general subcategory of self-excited vibrations is termed "parametric instability," since instability is induced by the effective periodic variation of the system's parameters (stiffness, inertia, natural frequency, etc.). Three particular instances of interest in the field of rotating machinery are

Lateral instability due to asymmetric shafting and/or bearing characteristics
Lateral instability due to pulsating torque
Lateral instabilities due to pulsating longitudinal compression

LATERAL INSTABILITY DUE TO ASYMMETRIC SHAFTING

If a rotor or its stator contains sufficient levels of asymmetry in the flexibility associated with its two principal axes of flexure as illustrated in Fig. 5.10, self-excited vibration may take place. This phenomenon is completely independent of any unbalance, and independent of the forced vibrations associated with twice-per-revolution excitation of such shafting mounted horizontally in a gravitational field.

FIG. 5.10. Shaft system possessing unequal rigidities, leading to a pair of coupled inhomogeneous Mathieu equations.

As a described in standard vibration texts,[35] we find that presupposing a nominal whirl amplitude of the shaft at some whirl frequency, the rotation of the asymmetric shaft at an rpm different from the whirling speed will appear as periodic change in flexibility in the plane of the whirling shaft's radial deflection. This will result in an instability in certain specific ranges of rpm as a function of the degree of asymmetry. In general, instability is experienced when the rpm is approximately one-third and one-half the critical rpm and approximately equal to the critical rpm (where the critical rpm is defined with the average value of shaft stiffness), as in Fig. 5.11. The ratios of whirl frequency to rotational speed will then be approximately 3.0, 2.0, and 1.0. But with gross asymmetries, and with the additional complication of asymmetrical inertias with principal axes in arbitrary orientation to the shaft's principal axes' flexibility, no simple generalization is possible.

There is a considerable literature dealing with many aspects of the problem and substantial bibliographies.[36, 37, 38]

Stability is accomplished by minimizing shaft asymmetries and avoiding rpm ranges of instability.

LATERAL INSTABILITY DUE TO PULSATING TORQUE

Experimental confirmation[39] has been achieved that establishes the possibility of inducing first-order lateral instability in a rotor-disc system by the application of a proper combination of constant and pulsating torque. The application of torque to a shaft affects its natural frequency in lateral vibration so that the instability may also be characterized as "parametric." Analytic formulation and description of the phenomenon are available in Ref. 40 and in the bibliography of Ref. 3. The experimental work (Ref. 39) explored regions of shaft speed where the disc always whirled at the first critical speed of the rotor-disc system, regardless of the torsional forcing frequency or the rotor speed within the unstable region.

It therefore appears that combinations of ranges of steady and pulsating torque, which have been identified[40] as being sufficient to cause instability, should be avoided in the narrow-speed bands where instability is possible in the vicinity of twice the critical speed and lesser instabilities at 2/2, 2/3, 2/4, 2/5, . . . times the critical frequency, as in Fig. 5.12, implying frequency/speed ratios of approximately 0.5, 1.0, 1.5, 2.0, 2.5, ...

FIG. 5.11. Instability regimes of rotor system induced by asymmetric stiffness (Ref. 37).

LATERAL INSTABILITY DUE TO PULSATING LONGITUDINAL LOADS

Longitudinal loads on a shaft which are of an order of magnitude of the buckling will tend to reduce the natural frequency of that lateral, flexural vibration of the shaft. Indeed, when the compressive buckling load is reached, the natural frequency goes to zero. Therefore pulsating longitudinal loads effectively cause a periodic variation in stiffness, and they are capable of inducing "parametric instability" in rotating as well as stationary shafts,[41] as noted above in Fig. 5.13.

STICK-SLIP RUBS AND CHATTER

GENERAL DESCRIPTION

Mention is appropriate of another family of instability phenomena—stick-slip or chatter. Though the instability mechanism is associated with the dry friction contact force at the point of rubbing between a rotating shaft and a stationary element, it must not be confused with dry friction whip, previously discussed. In the case of stick-slip, as is described in standard texts (e.g., Ref. 42), the instability is caused by the irregular nature of the friction force developed at very low rubbing speeds.

At high velocities, the friction force is essentially independent of contact speed. But at very low contact speeds we encounter the phenomenon of "stiction," or breakaway friction, where higher levels of friction force are encountered, as in Fig. 5.14. Any periodic motion of the rotor's point of contact, superimposed on the basic relative contact velocity, will be self-excited. In effect, there is negative damping (as illustrated in Fig. 5.1B) since motion of the rotor's contact point in the direction of rotation will increase

FIG. 5.12. Instability regimes of rotor system induced by pulsating torque (Ref. 40).

relative contact velocity and reduce stiction and the net force resisting motion. Rotor motion counter to the contact velocity will reduce relative velocity and increase friction force, again reinforcing the periodic motion. The ratio of vibration frequency to rotation speed will be much larger than unity.

While the vibration associated with stick-slip or chatter is often reported to be torsional, planar lateral vibrations can also occur. Surveys of the phenomenon are included in Refs. 43 and 44.

Measures for avoidance are similar to those prescribed for dry friction whip: avoid contact where feasible and lubricate the contact point where contact is essential to the function of the apparatus.

FIG. 5.13. Long column with pinned ends. A periodic force is superimposed upon a constant axial pull. (*After McLachlan.*[41])

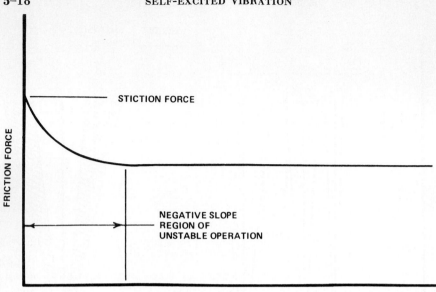

FIG. 5.14. Dry friction characteristic giving rise to stick-slip rubs or chatter.

INSTABILITIES IN FORCED VIBRATIONS

GENERAL DESCRIPTION

In a middle ground between the generic categories of force vibrations and self-excited vibrations is the category of *instabilities in force vibrations*. These instabilities are characterized by forced vibration at a frequency equal to rotor rotation (generally induced by unbalance), but with the amplitude of that vibration being unsteady or unstable. Such unsteadiness or instability is induced by the interaction of the forced vibration on the mechanics of the system's response, or on the unbalance itself. Two manifestations of such instabilities and unsteadiness have been identified in the literature—bistable vibration and unstable imbalance.

BISTABLE VIBRATION

A classical model of one type of unstable motion is the "relaxation oscillator," or *"multivibrator."* A system subject to relaxation oscillation has *two* fairly stable states, separated by a zone where stable operation is impossible.[45] Furthermore, in each of the stable states, a mechanism exists which will induce the system to drift toward the unstable state. The system will develop a periodic motion of the general form shown in Fig. 5.15.

An idealized formulation of this class of vibration with nonlinear damping is[46]

$$m\ddot{x} + c(x^2 - 1)\dot{x} + kx = 0 \tag{5.15}$$

When the deflection amplitude x is greater than $+1$ or less than -1, as in *A-B* and *C-D*, the damping coefficient is positive, and the system is stable, although presence of a

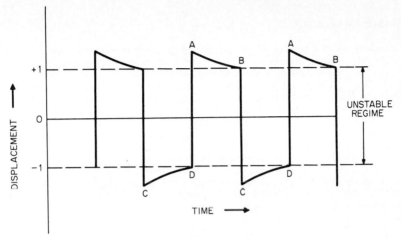

FIG. 5.15. General form of relaxation oscillations.

spring system k will always tend to drag the mass to a smaller absolute deflection amplitude. When the deflection amplitude lies between -1 and $+1$, as in B-C or D-A, the damping coefficient is negative and the system will move violently until it stabilizes in one of the damped stable zones.

While such systems are common in electronic circuitry, they are rather rare in the field of rotating machinery. One instance has been observed[47] in a rotor system supported by rolling element bearings with finite internal clearance. In this situation, the effective stiffness of the rotor is small for small deflections (within the clearance) but large for large deflections (when full contact is made between the rollers and the rotor

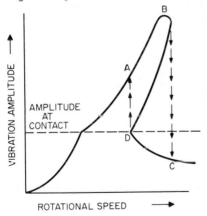

FIG. 5.16. Response of a rotor, in bearings with (constant) internal clearance, to unbalance excitation in the vicinity of its critical speed.

and stator). Such a nonlinearity in stiffness causes a "rightward leaning" peak in the response curve when the rotor is operating in the vicinity of its critical speed and being stimulated by unbalance. In this region, two stable modes of operation are possible, as in Fig. 5.16. In region A-B, the rotor and stator are in solid contact through the rollers. In region C-D, the rotor is whirling within the clearance, out of contact. A jump in amplitude is experienced when operating from B to C or D to A. When operating at constant speed, either of the nominally stable states can drift toward instability by virtue of thermal effects on the rollers. When the rollers are unloaded, they will skid and heat up, thereby reducing the clearance. When the rollers are loaded, they will be cooled by lubrication and will tend to contract and increase clearance. In combination, these mechanisms are sufficient to cause a relaxation oscillation in the amplitude of the forced vibration.

The remedy for this type of self-excited vibration is to eliminate the precondition of skidding rollers by reducing bearing geometric clearance, by preloading the bearing, or by increasing the temperature of any recirculating lubricant.

UNSTABLE IMBALANCE

A standard text[48] describes the occurrence of unstable vibration of steam turbines where the rotor "would vibrate with the frequency of its rotation, obviously caused by unbalance, but the intensity of the vibration would vary periodically and extremely slowly." The instability in the vibration amplitude is attributable to thermal bowing of the shaft, which is caused by the heat input associated with rubbing at the rotor's deflected "high spot," or by the mass of accumulated steam condensate in the inside of a hollow rotor at the rotor's deflected high spot. In either case, there is basis for continuous variation of amplitude, since unbalance gives rise to deflection and the deflection is, in turn, a function of that imbalance.

The phenomenon is sometimes referred to as the Newkirk effect in reference to its early recorded experimental observation.[49] A manifestation of the phenomenon in a steam turbine has been diagnosed and reported in Ref. 50 and a bibliography is available in Ref. 51. An analytic study[52] shows the possibility of both spiraling, oscillating, and constant modes of amplitude variability.

Table 5.2. Diagnostic Table of Rotating Machinery Self-excited Vibrations

	R, characteristic ratio: whirl frequency/rpm	Whirl direction
Whirling or whipping:		
Hysteretic whirl	$R \approx 0.5$	Forward
Fluid trapped in rotor	$0.5 < R < 1.0$	Forward
Dry friction whip	No functional relationship; whirl frequency a function of coupled rotor-stator system; onset rpm is a function of rpm at contact	Backward—axial contact on disc side nearest virtual pivot; Forward—axial contact on disc side opposite to virtual pivot; Backward—radial contact
Fluid bearing whip	$R < 0.5$	Forward
Seal and blade-tip-clearance effect in turbomachinery	Load-dependent	Forward—blade tip clearance; Unspecified—for seal clearance
Propeller and turbomachinery whirl	Load-dependent	Backward—virtual pivot aft of rotor; Forward—virtual pivot front of rotor (where front is source of impinging flow)
Parametric instability:		
Asymmetric shafting	$R \approx 1.0, 2.0, 3.0, \ldots$	Unspecified
Pulsating torque	$R \approx 0.5, 1.0, 1.5, 2.0, \ldots$	Unspecified
Pulsating longitudinal load	A function of pulsating load frequency rather than rpm	Unspecified
Stick-slip rubs and chatter	$R \ll 1$	Essentially planar rather than whirl motion
Instabilities in forced vibrations:		
Bistable vibration	$R = 1$ with periodic square wave fluctuations in amplitude of frequency much lower than rotation rate	Forward
Unstable imbalance	$R = 1$ with slow variation in amplitude	Forward

IDENTIFICATION OF SELF-EXCITED VIBRATION

Even with the best of design practice and application of the most effective methods of avoidance, the conditions and mechanisms of self-excited vibrations in rotating machinery are so subtle and pervasive that incidents continue to occur, and the major task for the vibrations engineer is diagnosis and correction.

Figure 5.3B suggests the forms for display of experimental data to perceive the patterns characteristic of whirling or whipping, so as to distinguish it from forced vibration, Fig. 5.3A. Table 5.2 summarizes particular quantitative measurements that can be made to distinguish between the various types of whirling and whipping, and other types of self-excited vibrations. The table includes the characteristic ratio of whirl speed to rotation speed at onset of vibration, and the direction of whirl with respect to the rotor rotation. The latter parameter can generally be sensed by noting the phase relation between two stationary vibration pickups mounted at 90° to one another at similar radial locations in a plane normal to the rotor's axis of rotation. Table 5.1 and specific prescriptions in the foregoing text and references suggest corrective action based on these diagnoses. Reference 53 gives additional description of corrective actions.

REFERENCES

1. Den Hartog, J. P.: "Mechanical Vibrations," 4th ed., p. 346, McGraw-Hill Book Company, Inc., New York, 1956.
2. Ehrich, F. F.: "Identification and Avoidance of Instabilities and Self-Excited Vibrations in Rotating Machinery," *ASME Paper* 72-DE-21, May 1972.
3. Kramer, E.: "Instabilities of Rotating Shafts," *Proc. Conf. Vib. Rotating Systems, Inst. Mech. Eng., London*, February 1972.
4. Vance, J. M.: "High Speed Rotor Dynamics—Assessment of Current Technology for Small Turboshaft Engines," USAAMRDL-TR-74-66, Ft. Eustis, Va., July 1974.
5. Newkirk, B. L.: *Gen. Elec. Rev.*, **27**:169 (1924).
6. Kimball, A. L.: *Gen. Elec. Rev.*, **17**:244 (1924).
7. Ref. 1, pp. 205–206.
8. Gunter, E. J.: "Dynamic Stability of Rotor-Bearing Systems," NASA SP-113, chap. 4, 1966.
9. Bolatin, V. V.: "Non-Conservative Problems of the Theory of Elastic Stability," Pergamon Press, New York, 1964.
10. Bentley, D. E.: "The Re-Excitation of Balance Resonance Regions by Internal Friction," *ASME Paper* 72-PET-49, September 1972.
11. Vance, J. M., and J. Lee: "Stability of High Speed Rotors with Internal Friction," *ASME Paper* 73-DET-127, September 1973.
12. Ehrich, F. F.: *J. Appl. Mech.*, (E) **31**(2):279 (1964).
13. Wolf, J. A.: "Whirl Dynamics of a Rotor Partially Filled with Liquids," *ASME Paper* 68-WA/APM-25, December 1968.
14. Ehrich, F. F.: *J. Eng. Ind.*, (B) **89**(4):806 (1967).
15. Ref. 1, pp. 292–293.
16. Sapetta, L. P., and R. J. Harker: "Whirl of Power Screws Excited by Boundary Lubrication at the Interface," *ASME Paper* 67-Vibr-37, March 1967.
17. Ref. 1, pp. 293–295.
18. Begg, I. C.: "Friction Induced Rotor Whirl—A Study in Stability," *ASME Paper* 73-DET-10, September 1973.
19. Ehrich, F. F.: "The Dynamic Stability of Rotor/Stator Radial Rubs in Rotating Machinery," *ASME Paper* 69-Vibr-56, April 1969.
20. Newkirk, B. L., and H. D. Taylor: *Gen. Elec. Rev.*, **28**:559–568 (1925).
21. Ref. 1, pp. 297–298.
22. Ref. 8, chap. 5.
23. Pinkus, O., and B. Sternlight: "Theory of Hydrodynamic Lubrication," chap. 8, McGraw-Hill Book Company, Inc., New York, 1961.
24. Black, H. F., and D. N. Jenssen: "Effects of High Pressure Seal Rings on Pump Rotor Vibrations," ASME Paper 71-WA/FE-38, December 1971.
25. Alford, J. S.: *J. Eng. Power*, **87**(4):333 October 1965.

26. Black, H. F.: "Calculation of Forced Whirling and Stability of Centrifugal Pump Rotor Systems," *ASME Paper* 73-DET-131, September 1973.
27. Ref. 1, pp. 317–321.
28. Ehrich, F. F.: *Trans. ASME*, (A) **90**(4):369 (1968).
29. Taylor, E. S., and K. A. Browne: *J. Aeronaut. Sci.*, **6**(2):43–49 (1938).
30. Houbolt, J. C., and W. H. Reed: "Propeller Nacelle Whirl Flutter," *I.A.S. Paper* 61-34, January 1961.
31. Trent, R., and W. R. Lull: "Design for Control of Dynamic Behavior of Rotating Machinery," *ASME Paper* 72-DE-39, May 1972.
32. Ehrich, F. F.: "An Aeroelastic Whirl Phenomenon in Turbomachinery Rotors," *ASME Paper* 73-DET-97, September 1973.
33. Floquet, G.: *Ann. l'école normale supérieure*, **12**:47 (1883).
34. McLachlan, N. W.: "Theory and Applications of Mathieu Functions," p. 40, Oxford University Press, New York, 1947.
35. Ref. 1, pp. 336–339.
36. Brosens, P. J., and Crandall, H. S.: *J. Appl. Mech.*, **83**(4):567 (1961).
37. Messal, E. E., and Bronthon, R. J.: "Subharmonic Rotor Instability Due to Elastic Asymmetry," *ASME Paper* 71-Vibr-57, September 1971.
38. Arnold, R. C., and E. E. Haft: "Stability of an Unsymmetrical Rotating Cantilever Shaft Carrying an Unsymmetrical Rotor," *ASME Paper* 71-Vibr-57, September 1971.
39. Eshleman, R. L., and R. A. Eubanks: "Effects of Axial Torque on Rotor Response: An Experimental Investigation," *ASME Paper* 70-WA/DE-14, December 1970.
40. Wehrli, V. C.: "Uber Kritische Drehzahlen unter Pulsierender Torsion," *Ing. Arch.*, **33**:73–84 (1963).
41. Ref. 34, p. 292.
42. Ref. 1, p. 290.
43. Conn, H.: *Tool Eng.*, **45**:61–65 (1960).
44. Sadowy, M.: *Tool Eng.*, **43**:99–103 (1959).
45. Ref. 1, pp. 365–368.
46. Van der Pol, B.: *Phil. Mag.*, **2**:978 (1926).
47. Ehrich, F. F.: "Bi-Stable Vibration of Rotors in Bearing Clearance," *ASME Paper* 65-WA/MD-1, November 1965.
48. Ref. 1, pp. 245–246.
49. Newkirk, B. L.: "Shaft Rubbing," *Mech. Eng.*, **48**:830 (1926).
50. Kroon, R. P., and W. A. Williams: "Spiral Vibration of Rotating Machinery," *5th Int. Congr. Appl. Mech.*, John Wiley & Sons, Inc., New York, 1939, p. 712.
51. Dimarogonas, A. D., and Sander, G. N.: *Wear*, **14**(3):153 (1969).
52. Dimarogonas, A. D.: "Newkirk Effect: Thermally Induced Dynamic Instability of High-Speed Rotors," *ASME Paper* 73-GT-26, April 1973.
53. Ehrich, F. and Childs, D.: *Mech. Eng.*, **106**(5):66 (1984).

6

DYNAMIC VIBRATION ABSORBERS AND AUXILIARY MASS DAMPERS

F. Everett Reed

Littleton Research and
Engineering Corp.

INTRODUCTION

Auxiliary masses are frequently attached to vibrating systems by springs and damping devices to assist in controlling the amplitude of vibration of the system. Depending upon the application, these auxiliary mass systems fall into two distinct classes.

1. If the primary system is excited by a force or displacement that has a constant frequency, or in some cases by an exciting force that is a constant multiple of a rotational speed, then it is possible to modify the vibration pattern and to reduce its amplitude significantly by the use of an auxiliary mass on a spring tuned to the frequency of the excitation. When the auxiliary mass system has as little damping as possible, it is called a *dynamic absorber*.

2. If it is impossible to incorporate damping into a structure that vibrates excessively, it may be possible to provide the damping in an auxiliary system attached to the structure. When used in this manner, the auxiliary mass system is one form of a damper. (Other forms may be incorporated as an integral part of the system.) The names *damped absorber* or *auxiliary mass damper* are given to this type of system.

It is sometimes useful to analyze the auxiliary mass system in terms of its electrical analog.

FORMS OF DYNAMIC ABSORBERS AND AUXILIARY MASS DAMPERS

In its simplest form, as applied to a single degree-of-freedom system, the character of the auxiliary mass system is the same as that of the primary system. Thus a torsional system has a torsionally connected auxiliary mass; a linear system has a linear-spring connected mass; and a pendulum has an auxiliary pendulum. Examples of undamped auxiliary mass systems attached to single degree-of-freedom systems are shown in Figs. 6.1 and 6.2; examples of damped auxiliary mass systems are shown in Figs. 6.3 and 6.4. With multiple degree-of-freedom systems the attachment of the auxiliary masses is not as conventional as with the single degree-of-freedom system. For example, consider the two degree-of-freedom system shown in Fig. 6.5*A* consist-

FIG. 6.1. Dynamic vibration absorbers in pendulum form (*A*) and linear form (*B*).

PRINCIPAL SYSTEM   AUXILIARY SYSTEM

(A)

PRINCIPAL SYSTEM

AUXILIARY
SYSTEMS

(B)

FIG. 6.2. Typical dynamic vibration absorbers. The principal and auxiliary systems vibrate in torsion in the arrangement at (A); the auxiliary system is in the form of masses and beams at (B).

ing of two masses m_1 and m_2 on a rigid, massless bar. A dynamic absorber of the type shown in Fig. 6.5B is effective for the vertical translational motion; however, if the auxiliary masses are on cantilever beams mounted on the rigid bar, as shown in Fig. 6.5C, the absorber can be made effective for both vertical translational motion and rotational motion about an axis normal to the page.

WAYS OF EXPRESSING THE EFFECTS OF AUXILIARY MASS SYSTEMS

Suppose a linear auxiliary mass system, consisting of one or more masses, springs, and dampers, is attached to a vibrating primary system. The reaction back on the primary system is proportional to the amplitude of motion at the point of attachment. It is a function of the frequency of excitation and of the masses, spring stiffnesses, and damping constants of the auxiliary mass system. If there is no damping in the auxiliary mass system, the reaction forces are either in phase or 180° out of phase with the displacement and the acceleration at the point of attachment. However, where there is damping in the auxiliary system, the reaction has a component that is 90° out of phase with the acceleration and the displacement.

Since the reaction is proportional to the amplitude of motion, it is possible to express the properties of the auxiliary mass system in terms of the motion at the point of attachment. This can be done in three ways: (1) the ratio of the reaction force to the displacement at the point of attachment, (2) the ratio of the reaction force to the velocity at the point of attachment, or (3) the ratio of the reaction force to the acceleration at the point of attachment. The first ratio can be considered equivalent to a spring whose stiffness changes with frequency. The second ratio can be considered equivalent to a damper; at any frequency it is equal in magnitude to the force-displacement ratio divided by the angular frequency. The phase angle between the force and the velocity is 90° from the phase angle between the force and the displacement.

(A)   (B)

FIG. 6.3. Damped auxiliary mass systems corresponding to the undamped vibration absorbers shown in Fig. 6.1.

This force-velocity ratio is called the *mechanical impedance* Z of the auxiliary system. The third ratio corresponds to a mass and is designated *equivalent mass* m_{eq}. The equivalent mass of a system is $-1/\omega^2$ that of the equivalent spring k_{eq} of the system.

Because of the phase relations between the force and the displacement, velocity, and acceleration at the point of connection, it is customary to represent the ratios as complex quantities. Thus $Z = k_{eq}/j\omega = j\omega m_{eq}$. Most dynamic analyses of mechanical systems are made on purely reactive systems, i.e., systems having masses and stiffnesses only, and no damping. The effects of auxiliary mass systems are most easily understood if the effect of the auxiliary system is represented as a reactive subsystem. For this reason, and because the hypothetical addition of a mass to a system is often more easily comprehended than the addition of a spring, the effects of auxiliary mass systems are treated in terms of the equivalent masses in this chapter, i.e., in terms of the ratio of the force exerted by the auxiliary system upon the primary system to the acceleration at the point of attachment of the auxiliary system.

FIG. 6.4. Typical damped auxiliary mass systems. In the torsional system at (A), damping is provided by relative motion of the flywheels J, J_a. In the antiroll tanks for ships shown at (B), water flows from one tank to the other and damping is provided by a constriction in the connecting pipe.

FIG. 6.5. Application of a dynamic absorber to reduce the vibration of the spring-mounted bar at (A) in both vertical translational and rotational modes. The linear mass-spring system at (B) is effective for only translational motion, whereas the cantilever beams at (C) are effective for rotational as well as translational motion.

THE INFLUENCE OF A SIMPLE AUXILIARY MASS SYSTEM UPON A VIBRATING SYSTEM

The magnitude of the equivalent mass of a simple auxiliary mass system, consisting of a mass m_a, spring k_a, and viscous damper c_a, can be determined readily by evaluating the forces exerted by such a system upon a foundation vibrating at a frequency $f = \omega/2\pi$. The system with its assumed constants and displacements is shown in Fig. 6.6A. The spring and damping forces acting on m are shown in Fig. 6.6B, and the equation of motion is

$$(-k_a x_r - c_a j\omega x_r)e^{j\omega t} = -m_a(x_0 + x_r)\omega^2 e^{j\omega t}$$

Solving for x_r,

$$x_r = \frac{m_a\omega^2 x_0}{-m_a\omega^2 + jc_a\omega + k_a} \quad (6.1)$$

The force acting on the foundation is

$$Fe^{j\omega t} = (k_a + jc_a\omega)x_r e^{j\omega t}$$

FIG. 6.6. Auxiliary mass damper. The arrangement of the damper is shown at (A), and the forces acting on the mass are indicated at (B).

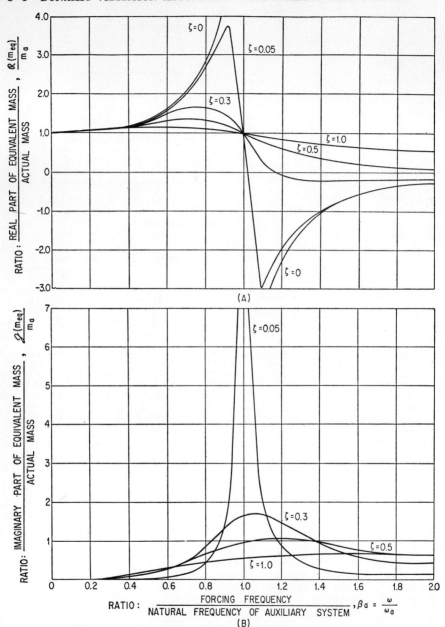

FIG. 6.7. Equivalent mass m_{eq} of the auxiliary-mass system shown in Fig. 6.6. The real part oft he equivalent mass is shown at (A) and the imaginary part at (B).

Eliminating x_r from the preceding equations,

$$F = \frac{(k_a + jc_a\omega)m_a\omega^2}{-m_a\omega^2 + jc_a\omega + k_a} x_0 \tag{6.2}$$

Since the force exerted by an equivalent mass m_{eq} rigidly attached to the moving foundation is $F = m_{eq}\omega^2 x_0$:

$$m_{eq} = \frac{(k_a + jc_a\omega)}{k_a + jc_a\omega - m_a\omega^2} m_a \tag{6.3}$$

Equation (6.3) can be written in terms of nondimensional quantities:

$$m_{eq} = \frac{1 + 2\zeta\beta_a j}{(1 - \beta_a^2) + 2\zeta\beta_a j} m_a \tag{6.4}$$

where $\beta_a = \dfrac{\omega}{\omega_a}$, a tuning parameter

$\omega_a{}^2 = \dfrac{k_a}{m_a}$, the natural frequency of the auxiliary system

$\zeta = \dfrac{c_a}{c_{ca}}$, a damping parameter

$c_{ca} = 2\sqrt{k_a m_a}$, critical damping of the auxiliary system

Equation (6.4) can be divided into the following real and imaginary components:

$$m_{eq} = \frac{(1 - \beta_a^2) + (2\zeta\beta_a)^2}{(1 - \beta_a^2)^2 + (2\zeta\beta_a)^2} m_a - \frac{2\zeta\beta_a^3}{(1 - \beta_a^2)^2 + (2\zeta\beta_a)^2} jm_a \tag{6.5}$$

The real and imaginary parts of m_{eq} are shown in Fig. 6.7A and Fig. 6.7B, respectively. If there is no damping, $\zeta = 0$ and

$$m_{eq} = \frac{1}{1 - \beta_a^2} m_a \tag{6.6}$$

If $\beta_a = 1$ in Eq. (6.6), m_{eq} becomes infinite and a finite force produces no displacement. Thus, the auxiliary mass enforces a point of no motion (i.e., a node) at its point of attachment.

This concept can be applied to reduce the amplitude of the forced vibration of a single degree-of-freedom system by attaching a damped absorber.[1,2] A sketch of the system with a damped auxiliary mass system is shown in Fig. 6.8A. In the equivalent system shown in Fig. 6.8B, there is no force acting on the mass m but instead the support is given a motion $ue^{j\omega t}$. The equations for the system of Fig. 6.8B are similar to those for the system of Fig. 6.8A with the value ku substituted for F. The amplitude of forced vibration of a single degree-of-freedom system, Eq. (2.24), is

$$x_0 = \frac{F/k}{1 - m\omega^2/k}$$

The effect of the auxiliary mass system is to increase the mass m of the primary system by the equivalent mass of the auxiliary system as

FIG. 6.8. Schematic diagram of auxiliary mass m_a coupled by a spring k_a and viscous damper c_a to a primary system k, m. The primary system is excited by the force F at (A), or alternatively by the foundation motion u at (B).

given by Eq. (6.4):

$$x_0 = \frac{F/k}{1 - \dfrac{\omega^2}{k}\left[m + m_a \dfrac{(1 + 2\zeta\beta_a j)}{(1 - \beta_a^2) + 2\zeta\beta_a j}\right]}$$

Substituting $\mu = m_a/m$, the mass ratio, $\delta_{st} = F/k$, the static deflection of the spring of the primary system, and $\beta = \sqrt{m\omega^2/k}$, the ratio of the forcing frequency to the natural frequency of the primary system, and writing in dimensionless form,

$$\frac{x_0}{\delta_{st}} = \frac{(1 - \beta_a^2) + 2\zeta\beta_a j}{(1 - \beta_a^2) + 2\zeta\beta_a j - \beta^2[(1 - \beta_a^2) + 2\zeta\beta_a j + \mu(1 + 2\zeta\beta_a j)]}$$

The amplitude of motion of the primary mass, without regard to phase, is

$$\frac{x_0}{\delta_{st}} = \left\{\frac{(1 - \beta_a^2)^2 + (2\zeta\beta_a)^2}{[(1 - \beta_a^2)(1 - \beta^2) - \beta^2\mu]^2 + (2\zeta\beta_a)^2[1 - \beta^2 - \beta^2\mu]^2}\right\}^{\frac{1}{2}} \qquad (6.7)$$

If $\zeta = 0$ (no damping), then

$$\frac{x_0}{\delta_{st}} = \frac{1 - \beta_a^2}{(1 - \beta_a^2)(1 - \beta^2) - \beta^2\mu} \qquad (6.8)$$

If $\beta_a = 1$, $x_0 = 0$; that is, the vibration of the primary system is eliminated entirely when the auxiliary system is undamped and is tuned to the forcing frequency.

THE DYNAMIC ABSORBER

If the auxiliary mass system has no damping and is tuned to the forcing frequency, it acts as a dynamic absorber and enforces a node at its point of attachment. The auxiliary mass must be sufficiently large so that it will not have an excessive amplitude.[3] For a dynamic absorber attached to the primary system at the point where the excitation is introduced, the required mass of the auxiliary body is easily determined. Since the primary mass is motionless, the force exerted by the absorber, when the amplitude of motion of the auxiliary mass is u_0, is equal and of opposite sign to the exciting force F. Hence

$$F = m_a\omega^2 u_0 \qquad (6.9)$$

Since the frequency is known, the mass and amplitude of motion necessary to neutralize a given excitation force are determined by Eq. (6.9). The spring stiffness in the auxiliary system is determined by the requirement that the auxiliary system be tuned to the frequency of the exciting force:

$$k_a = m_a\omega^2 \qquad (6.10)$$

Although the concept of tuning a dynamic absorber appears simple, practical considerations make it difficult to tune any system exactly. When the auxiliary mass is small relative to the mass of the primary system, its effectiveness depends upon accurate tuning. If the tuning is incorrect, the addition of the auxiliary mass may bring the composite system (primary and auxiliary systems) into resonance with the exciting force.

Consider the natural frequencies of the composite system. The natural frequency of the primary system is $\omega_0 = \sqrt{k/m}$. With this relation, Eq. (6.8) in which the damping is zero ($\zeta = 0$) becomes

$$\frac{x_0}{\delta_{st}} = \frac{1 - \omega^2/\omega_a^2}{(1 - \omega^2/\omega_a^2)(1 - \omega^2/\omega_0^2) - (\omega^2/\omega_0^2)\mu}$$

At resonance the denominator is zero and ω is designated ω_n:

$$(\omega_n^2 - \omega_a^2)(\omega_n^2 - \omega_0^2) - \omega_n^2\omega_a^2\mu = 0 \qquad (6.11)$$

The natural frequencies are found from the roots ω_n^2 of Eq. (6.11):

$$\omega_n^2 = \frac{\omega_a^2(1 + \mu) + \omega_0^2}{2}$$

$$\pm \sqrt{\left[\frac{\omega_a^2(1 + \mu) - \omega_0^2}{2}\right]^2 + \omega_a^2\omega_0^2\mu} \quad (6.12)$$

This last relation may be represented by Mohr's circle, Fig. 6.9.

Since the absorber is nominally tuned to the frequency of the excitation, the root ω_{n2}^2 that is closer to the forcing frequency is of interest. The ratio ω_{n2}/ω_a is a measure of the sensitivity of the tuning required to avoid resonance. This is given as a function of μ for various ratios of ω_0/ω_a in Fig. 6.10. Dynamic absorbers are most generally used when the primary system without the absorber is nearly in resonance with the excitation. If the natural frequency of the primary system is less than the forcing frequency, it is preferable to tune the dynamic absorber to a frequency slightly lower than the forcing frequency to avoid the resonance that lies above the natural frequency of the primary system. Likewise if the natural frequency of the primary system is above the forcing frequency, it is well to tune the damper to a frequency slightly greater than the forcing frequency. Figure 6.10

Fig. 6.9. Representation of the natural frequencies ω_n of the composite system by Mohr's circle. The circle is constructed on the diameter located by the natural frequencies ω_0, ω_a of the primary and auxiliary systems, respectively. The natural frequencies of the composite system are indicated by the intercept of the circle with the horizontal axis.

shows that the tuning for a primary system with high natural frequency is more sensitive than that for a primary system with low natural frequency. Mohr's circle of Fig. 6.9 provides a useful graphical representation.

Where the natural frequency of the composite system is nearly equal to the tuned frequency of the absorber, the amplitude of motion of the primary mass at resonance is much smaller than that of the absorber. Consequently, the motion of the primary mass does not become large even at resonance; but the motion of the absorber, unless limited by damping, may become so large that failure occurs.

The use of the dynamic absorber is not restricted to single degree-of-freedom systems or to locations in simple systems where the exciting forces act. However, dynamic absorbers are most effective if located where the excitation force acts. For example, consider a dynamic absorber that is attached to the spring in the simple system shown in Fig. 6.11. When the absorber is tuned so that $\sqrt{k_a/m_a} = \omega$, the equivalent mass is infinite at its point of attachment and enforces a node at point A. If the stiffness of the spring between A and the mass m is k_1, then the force F' exerted by the absorber to

Fig. 6.10. Curves showing effect of mass ratio m_a/m on the natural frequencies ω_n of the composite system, for several ratios of the natural frequency ω_a of the auxiliary system to the natural frequency ω_0 of the primary system.

RATIO: $\dfrac{\text{AUXILIARY MASS}}{\text{MASS OF PRIMARY SYSTEM}}$, $\mu = \dfrac{m_a}{m}$

RATIO: NATURAL FREQUENCY OF COMPOSITE SYSTEM / NATURAL FREQUENCY OF AUXILIARY SYSTEM $\dfrac{\omega_n}{\omega_a}$

$\omega_0 = \omega_a$

$\omega_0 = \frac{1}{2}\omega_a$

$\omega_0 = 2\omega_a$

$\omega_0 = \omega_n$

FIG. 6.11. Dynamic absorber attached to the spring of the primary system. The analysis shows that this is not as effective as if it were attached to the rigid body on which the force acts.

enforce the node is equal to that exerted by a system composed of the mass m and the spring k_1 attached to a fixed foundation at A and acted upon by the force $Fe^{j\omega t}$. The force F' is

$$F' = \frac{F}{1 - (m\omega^2/k_1)}$$

Thus the amplitude of motion of the auxiliary mass is

$$u_0 = \frac{F}{1 - (m\omega^2/k_1)} \times \frac{1}{m_a\omega^2} \qquad (6.13)$$

The amplitude of motion of the primary mass is

$$x = \frac{F}{k_1}\left(1 - \frac{m\omega^2}{k_1}\right)^{-1} \qquad (6.14)$$

Hence, an absorber attached to the spring is not as effective as one attached to the body where the force is acting. It is possible for the primary system to come into resonance about the new node at A.

AUXILIARY MASS DAMPERS

In general, the dynamic absorber is effective only for a system that is subjected to a constant frequency excitation. In the special case of a pendulum absorber (discussed later in this chapter), it is effective for an excitation that is a constant multiple of a rotating shaft speed. When excited at frequencies other than the frequency to which it is tuned, the absorber acts as an attached mass of positive value at frequencies below the tuned frequency and of negative value at frequencies above the tuned frequency. It introduces an additional degree-of-freedom and an additional natural frequency into the primary system.

In a multiple degree-of-freedom system, the introduction of an auxiliary mass system tends to lower those original natural frequencies of the primary system that are below the tuned frequency of the auxiliary system. This is because the auxiliary mass system adds a positive equivalent mass at frequencies below the tuned frequency. The original natural frequencies of the primary system that are higher than the tuned frequency of the auxiliary system are raised by adding the auxiliary mass system, because the equivalent mass of the auxiliary system is negative. A new natural mode of vibration corresponding to the vibration of the auxiliary mass system against the primary system is injected between the displaced initial natural frequencies of the primary system. Because the equivalent mass of the auxiliary mass system is large only at frequencies near the tuned frequency, those frequencies of the primary system that are closest to the tuned frequency are most strongly influenced by the auxiliary mass system. The addition of damping in the auxiliary mass system can be effective in reducing the amplitudes of motion of the primary system at the natural frequencies. For this reason auxiliary mass dampers are used quite commonly to reduce over-all vibration stresses and amplitudes.

Studies of the effects of a damped auxiliary mass system upon the amplitude of motion of an undamped, single degree-of-freedom system [1–5] have been applied to a multimass system.[6,7] In analyzing dampers utilizing auxiliary masses, it is desirable to consider a composite system in which the characteristics of both the primary and auxiliary systems are fixed. This composite system is excited by a harmonic force of varying frequency. It is desirable to express the tuned frequency of the auxiliary mass system in terms of the natural frequency of the primary system rather than the ratio β_a of the excitation frequency ω to the tuned frequency ω_a of the auxiliary system. Defining a new ratio α,

$$\alpha = \frac{\omega_a}{\omega_0} = \frac{\beta}{\beta_a}$$

Then Eq. (6.7) becomes

$$\frac{x_0}{\delta_{st}} = \left\{ \frac{(\alpha^2 - \beta^2)^2 + (2\zeta\alpha\beta)^2}{[(\alpha^2 - \beta^2)(1 - \beta^2) - \alpha^2\beta^2\mu]^2 + (2\zeta\alpha\beta)^2(1 - \beta^2 - \beta^2\mu)^2} \right\}^{\frac{1}{2}} \tag{6.15}$$

This equation is plotted in Fig. 6.12. Note that all curves pass through two points A, B on the graph, independent of the damping parameter ζ. These points are known as *fixed points*. Their locations are independent of the value of ζ if the ratio of the coeffi-

FIG. 6.12. Curves for auxiliary mass damper showing amplitude of vibration of mass of primary system, as given by Eq. (6.15), as a function of the ratio of forcing frequency ω to natural frequency of primary system $\omega = \sqrt{k/m}$. The mass ratio $m_a/m = 0.05$, and the natural frequency ω_a of the auxiliary mass system is equal to the natural frequency ω_0 of the primary system. Curves are included for several values of damping in the auxiliary system.

cient of ζ^2 to the term independent of ζ is the same in both numerator and denominator of Eq. (6.15):

$$\frac{(2\alpha\beta)^2}{(\alpha^2 - \beta^2)^2} = \frac{2\alpha\beta(1 - \beta^2 - \beta^2\mu)^2}{[(\alpha^2 - \beta^2)(1 - \beta^2) - \alpha^2\beta^2\mu]^2} \tag{6.16}$$

This equation is satisfied if

$$(2\alpha\beta)^2 = 0$$

$$\frac{1}{\alpha^2 - \beta^2} + \frac{(1 - \beta^2 - \beta^2\mu)}{(\alpha^2 - \beta^2)(1 - \beta^2) - \alpha^2\beta^2\mu} = 0$$

$$\frac{1}{\alpha^2 - \beta^2} - \frac{(1 - \beta^2 - \beta^2\mu)}{(\alpha^2 - \beta^2)(1 - \beta^2) - \alpha^2\beta^2\mu} = 0$$

The first two solutions are trivial. The third yields the equation

$$\beta^4 \left(1 + \frac{\mu}{2}\right) - \beta^2(1 + \alpha^2 + \alpha^2\mu) + \alpha^2 = 0 \tag{6.17}$$

The solution of this equation gives two values of β, designated β_c, one corresponding to each fixed point.

The amplitude of motion at each fixed point may be found by substituting each value of β_c given by Eq. (6.17) into Eq. (6.15). Since the amplitude is independent of ζ, the value that gives the simplest calculation (namely, $\zeta = \infty$) can be used for the calculation:

$$\frac{x_0}{\delta_{st}}\bigg|_c = \left[\frac{1}{(1 - \beta_c^2 - \beta_c^2\mu)^2}\right]^{\frac{1}{2}} \tag{6.18}$$

For the auxiliary mass damper to be most effective in limiting the value of x_0/δ_{st} over a full range of excitation frequencies, it is necessary to select the spring and damping constants of the system as given by the parameters α and ζ, respectively, so that the amplitude x_0 of the primary mass is a minimum. First consider the influence of the ratio α. As α is varied, the values of β_c computed from Eq. (6.17) are substituted in Eq. (6.18) to obtain values of x_0/δ_{st} for the fixed points A and B. The optimum value of α is that for which the amplitude x_0 at A is equal to that at B.

Let the two roots of Eq. (6.17) be β_1^2 and β_2^2, where β_1^2 is less than one and β_2^2 is greater than one. When x_0/δ_{st} has the same value for both β_1 and β_2 in Eq. (6.18),

$$\beta_1^2 + \beta_2^2 = \frac{2}{1 + \mu}$$

In an equation having unity for the coefficient of its highest power, the sum of the roots is equal to the coefficient of the second term with its sign changed:

$$\beta_1^2 + \beta_2^2 = \frac{1 + \alpha^2 + \alpha^2\mu}{1 + \mu/2}$$

From the two preceding equations, the optimum tuning (i.e., that required to give the same amplitude of motion at both fixed points) is obtained:

$$\alpha_{opt} = \frac{1}{1 + \mu} \tag{6.19}$$

where α is defined by the equation preceding Eq. (6.15).

If the effect of the damping is considered, it is possible to choose a value of the damping parameter ζ that will make the fixed points nearly the points of greatest amplitude of the motion. Consider Fig. 6.13, which represents the curves defining the motion of a single degree-of-freedom system to which an ideally tuned damped vibration absorber is attached (Fig. 6.8). The solid curves (1) represent the response of a system fitted with an undamped absorber. Curve 2 represents infinite damping of the auxiliary system. Curves 3 have horizontal tangents at the fixed points A and B, respectively. Since it is difficult to determine the required damping from maxima at the fixed points, the assumption is made that an optimum damping gives the same value of x_0/δ_{st} at a convenient point between A and B as at these fixed points. First find the values of β at A and B. This is done by solving Eq. (6.17) with the values of α as determined by Eq. (6.19) substituted:

$$\beta^4 - \frac{2\beta^2}{1 + \mu} + \frac{2}{(2 + \mu)(1 + \mu)^2} = 0$$

Solving for β to obtain the abscissas at the fixed points,

$$\beta^2 = \frac{1}{1 + \mu}\left(1 \pm \sqrt{\frac{\mu}{2 + \mu}}\right) \tag{6.20}$$

A convenient value for β lying between the two fixed points A and B is defined by

$$\beta_l^2 = \frac{1}{1 + \mu} \tag{6.21}$$

The frequency corresponding to this frequency ratio β_l is the natural frequency of the composite system when the damping is infinite; it is called the locked frequency.[7] The value of x_0/δ_{st} at the fixed points is found by substituting Eq. (6.20) into Eq. (6.18):

$$\frac{x_0}{\delta_{st}} \text{ at fixed point } = \sqrt{1 + \frac{2}{\mu}} \tag{6.22}$$

An approximate value for the maximum damping is obtained by solving for the value of ζ in Eq. (6.15) that gives a value of $x_0/\delta_{st} = \sqrt{1 + 2/\mu}$ when β_l^2 (the locked frequency)

FIG. 6.13. Curves similar to Fig. 6.12 but with optimum tuning. Curves 1 apply to an undamped absorber, curve 2 represents infinite damping in the auxiliary system, and curves 3 have horizontal tangents at the fixed points A and B.

is given by Eq. (6.21) and α has the optimum value given by Eq. (6.19). This gives the following value for the optimum damping parameter:

$$\zeta_{opt} = \sqrt{\frac{\mu}{2(1 + \mu)}} \tag{6.23}$$

It is possible to find the value of ζ^2 that makes the fixed point A a maximum on the x_0/δ_{st} vs. β plot, Fig. 6.13, and also to find the value of ζ^2 that makes the point B a maximum. The average of the two values so obtained indicates optimum damping:[4]

$$\zeta_{opt} = \sqrt{\frac{3\mu}{8(1 + \mu)^3}} \tag{6.24}$$

OPTIMUM DAMPING FOR AN AUXILIARY MASS ABSORBER CONNECTED TO THE PRIMARY SYSTEM WITH DAMPING ONLY. In general, the most effective damping is obtained where the auxiliary mass damping system includes a spring in its connection to the primary system. However, such a design requires a calculation of the optimum stiffness of the spring. Sometimes it is more expedient to add an oversize mass, coupled only by damping to the primary system, than it is to compute the optimum system. However, if use is made of such a simplified damper by taking it from a list of standard dampers and applying it with a minimum of calculations, the stock dampers should be as efficient as the application will permit.

In computing the optimum damping characteristic for an auxiliary mass absorber, attached to a single degree-of-freedom system by damping only, from the relations that have been developed, note in Eq. (6.4) that $\zeta = \infty$ and $\beta_a = \infty$ when $k = 0$. Then $\alpha = \beta/\beta_a = 0$. However, the product $\zeta\alpha = \zeta\beta/\beta_a$ is finite; thus, substituting $\alpha = 0$ but retaining the product $\zeta\alpha$ in Eq. (6.15),

$$\frac{x_0}{\delta_{\text{st}}} = \sqrt{\frac{\beta^2 + 4(\zeta\alpha)^2}{\beta^2(1 - \beta^2)^2 + 4(\zeta\alpha)^2[1 - \beta^2(1 + \mu)]^2}} \tag{6.25}$$

The value of x_0/δ_{st} is independent of $\zeta\alpha$ where the ratio of the coefficient of $\zeta\alpha$ to the term independent of $\zeta\alpha$ in the numerator is the same as the corresponding ratio in the denominator:

$$\frac{4}{\beta^2} = \frac{4[1 - \beta^2(1 + \mu)]^2}{\beta^2(1 - \beta^2)^2}$$

The solution of this equation for β gives the fixed points

$$\beta^2 = 0 \quad \text{and} \quad \beta^2 = \frac{2}{2 + \mu} \tag{6.26}$$

The amplitude of motion of the primary mass where $\beta^2 = 2/(2 + \mu)$ is

$$\frac{x_0}{\delta_{\text{st}}} = \frac{2 + \mu}{\mu} \tag{6.27}$$

Curves showing the motion of the mass of a primary system fitted with an auxiliary mass system connected by damping only are given in Fig. 6.14. The optimum damping is that

FIG. 6.14. Curves similar to Fig. 6.12 for system having auxiliary mass coupled by damping only. Several values of damping are included.

FIG. 6.15. Displacement amplitude of the primary mass as a function of the size of the auxiliary mass: (a) auxiliary system coupled only by Coulomb friction ($\alpha = 0$) with optimum damping; (b) auxiliary system coupled only by viscous damping ($\alpha = 0$) of optimum value; (c) auxiliary system coupled by spring and damper tuned to frequency of primary system ($\alpha = 1$) with optimum damping; (d) auxiliary system coupled by spring and damper with optimum tuning [$\alpha = 1/(1 + \mu)$] and optimum damping.

FIG. 6.16. Relative displacement amplitude between the primary mass and the auxiliary mass as a function of the size of the auxiliary mass: (a) auxiliary system coupled by spring and damper with optimum tuning [$\alpha = 1/(1 + \mu)$] and optimum damping; (b) auxiliary system coupled only by viscous damping ($\alpha = 0$) of optimum value; (c) auxiliary system coupled by spring and damper tuned to frequency of primary system ($\alpha = 1$) with optimum damping.

which makes the maximum amplitude occur at the fixed point B. By finding the value of $\zeta\alpha$ that makes the slope of x_0/δ_{st} versus β equal to zero at $\beta^2 = 2/(2 + \mu)$, the optimum damping is defined by

$$(\zeta\alpha)_{\text{opt}} = \sqrt{\frac{1}{2(2 + \mu)(1 + \mu)}} \qquad (6.28)$$

The values for the amplitude of vibration of the primary mass, the relative amplitude between the primary and auxiliary masses, and the optimum damping constants are given in Figs. 6.15 to 6.17 as functions of the mass ratio $\mu = m_a/m$.

FIG. 6.17. Curves showing damping required in auxiliary mass systems to minimize vibration amplitude of primary system: (a) auxiliary mass coupled by viscous damping only ($\alpha = 0$); (b) auxiliary system coupled by spring and damper tuned to frequency of primary system ($\alpha = 1$); (c) auxiliary system coupled by spring and damper with optimum tuning [$\alpha = 1/(1 + \mu)$]. The ordinate of the curves is $\zeta\alpha$, where ζ is the fraction of critical damping in the auxiliary system [Eq. (6.4)] and α is the tuning parameter [Eq. (6.15)].

THE USE OF AUXILIARY MASS ABSORBERS FOR VIBRATION ENERGY DISSIPATION. When a complicated mass-spring system is analyzed for possible vibration troubles, it is customary to compute the natural frequencies of the several modes of vibration of the system. The vibration amplitudes and stresses are estimated by making an energy balance between the energy input from the various exciting forces and the energy dissipated in the system and external reactions. From this point of view, it is desirable to know how much energy is dissipated in auxiliary mass systems and what value the damping constant should have in an auxiliary mass system of limited size to give maximum energy absorption. This is not the best criterion for determining the optimum damping because it neglects the effects of damping upon the mode shapes and the frequencies of the system, but it is generally adequate when compared with the other uncertainties of the calculations. Methods of designing dampers for torsional systems are given in Chap. 38.

Optimum Viscous Damping to Give Large Energy Absorption in an Auxiliary Mass Absorber.[8] Suppose the amplitude of motion of the primary system is unaffected by the auxiliary mass system which is attached to it. All energy absorption occurs in the damping element of the auxiliary mass system and is obtained by integrating the differential work done in the damper over a vibration cycle. The force exerted by damping is $c\dot{x}_r$ (the subscripts a are dropped), where x_r is the relative motion and the increment of work is $c\dot{x}_r\,dx_r = c\dot{x}_r^2\,dt$. If $x_r = x_{r0}\cos\omega t$, the work done over a cycle is

$$V = \oint c\omega^2 x_{r0}^2 \sin^2 \omega t\,dt = \pi c x_{r0}^2 \omega \tag{6.29}$$

For a damper attached to a support moving in harmonic motion of amplitude x_0, the relative motion x_r is given by Eq. (6.1). The amplitude of relative motion is

$$x_{r0} = \frac{m\omega^2 x_0}{\sqrt{(k - m\omega^2)^2 + c^2\omega^2}} = \frac{\beta_a^2 x_0}{\sqrt{(1 - \beta_a^2)^2 + (2\zeta\beta_a)^2}}$$

Substituting the above value of x_{r0} in Eq. (6.29) and integrating,

$$V = \frac{\pi c\omega x_0^2 m^2 \omega^4}{(k - m\omega^2)^2 + c^2\omega^2} = \frac{\pi x_0^2 m\omega^2 (2\zeta\beta_a)\beta_a^2}{(1 - \beta_a^2)^2 + (2\zeta\beta_a)^2} \tag{6.30}$$

Equation (6.30) can be used to find the tuning and the damping that gives the maximum energy dissipation when the amplitude of the forcing motion remains constant. Placing $\partial V/\partial\beta_a = 0$, the optimum value of β_a for given values of ζ is found from

$$(\beta_a)_{\text{opt}}^2 = (2\zeta^2 - 1) \pm 2\sqrt{1 - \zeta^2 + \zeta^4} \tag{6.31}$$

Placing $\partial V/\partial\zeta = 0$, the optimum value of ζ for a given value of β_a is

$$\zeta_{\text{opt}} = \frac{1 - \beta_a^2}{2\beta_a} \tag{6.32}$$

Where $k = 0$, the optimum damping is determined most conveniently by setting $\partial V/\partial c = 0$, using the dimensional form of Eq. (6.30), and determining c for maximum energy absorption:

$$c_{\text{opt}} = m\omega \tag{6.33}$$

AUXILIARY MASS DAMPER USING COULOMB FRICTION DAMPING.[9]

Dampers relying on Coulomb friction (i.e., friction whose force is constant) have been widely used. A damper relying on dry friction and connected to its primary system with a spring is too complicated to be analyzed or to be adjusted by experiment. For this reason, a damper with Coulomb friction has been used with only friction damping connecting the seismic mass (usually in a torsional application) to the primary system.[1,2,9] Because the motion is irregular, it is necessary to use energy methods of analysis. The analysis given here applies to the case of linear vibration. By analogy, the application to torsional or other vibration can be made easily (see Table 2.1 for analogous parameters).

FIG. 6.18. Schematic diagram of auxiliary mass absorber with Coulomb friction damping.

Consider the system shown in Fig. 6.18. It consists of a mass resting on wheels that provide no resistance to motion and connected through a friction damper to a wall that is moving sinusoidally. The friction damper consists of two friction facings that are held on opposite sides of a plate by a spring that can be adjusted to give a desired clamping force. The maximum force that can be transmitted through each interface of the damper is the product of the normal force and the coefficient of friction; the maximum total force for the damper is the summation over the number of interfaces.

Consider the velocity diagrams shown in Fig. 6.19A, B, and C. In these diagrams the velocity of the moving wall, $\dot{x} = x_0\omega\sin\omega t$, is shown by curve 1; the velocity $\dot{u}$ of the mass is shown by curve 2. The force exerted by the damper when slipping occurs is F_s.

When $F_s \geq m\ddot{u}$, the mass moves sinusoidally with the wall. When $F_s < m\ddot{u}$, slipping occurs in the damper and the mass is accelerated at a constant rate. Since a constant acceleration produces a uniform change in velocity, the velocity of the mass when the damper is slipping is shown by straight lines. The relative velocity between the wall and the mass is shown by the vertical shading.

Figure 6.19A applies to a damper with a low friction force. The damper slips continuously. In Fig. 6.19B the velocities resulting from a larger friction force are shown. Slipping disappears for certain portions of the cycle. Where the wall and the mass have the same velocity, their accelerations also are equal. Slipping occurs when the force

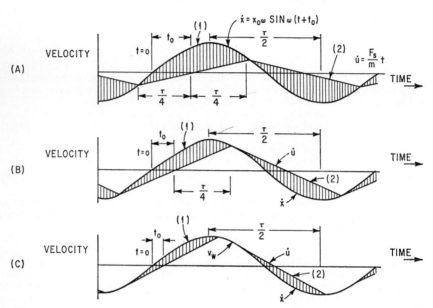

FIG. 6.19. Velocity-time diagrams for motion of wall (curve 1) and mass (curve 2) of Fig. 6.18. The conditions for a small damping force are shown at (A), for an intermediate damping force at (B), and for a large damping force at (C). The relative velocity between the wall and the mass is indicated by vertical shading.

transmitted by the damper is not large enough to keep the mass accelerating with the wall. Since at the breakaway point the accelerations of the wall and mass are equal, their velocity-time curves have the same slope; i.e., the curves are tangent at this point. In Fig. 6.19C, the damping force is so large that the mass follows the wall for a considerable portion of the cycle, and slips only where its acceleration becomes greater than the value of F_s/m. A slight increase in the clamping force or in the coefficient of friction locks the mass to the wall; then there is no relative motion and no damping.

Because of the nature of the damping force, the damping provided by the friction damper can be computed most practically in terms of energy. If the friction force exerted through the damper is F_s, the energy dissipated by the damper is the product of the friction force and the total relative motion between the mass and the moving wall. The time reference is taken at the moment when the auxiliary mass m has a zero velocity and is being accelerated to a positive velocity, Fig. 6.19A. Let the period of the vibratory motion of the wall be $\tau = 2\pi/\omega$, where ω is the angular frequency of the wall motion. By symmetry, the points of no slippage in the damper occur at times $-\tau/4$, $\tau/4$, and $3\tau/4$. Let the time when the velocity of the wall is zero be $-t_0$; then the velocity of the wall $\dot{x}$ is

$$\dot{x} = + x_0\omega \sin \omega(t + t_0)$$

The velocity $\dot{u}$ of the mass for $-\tau/4 < t < \tau/4$ is

$$\dot{u} = \ddot{u}t = \frac{F_s}{m}t$$

The velocities of the wall and the mass are equal at time $t = \tau/4$:

$$x_0\omega \sin \omega \left(\frac{\tau}{4} + t_0\right) = \frac{F_s}{m}\frac{\tau}{4}$$

Since $\omega\tau/4 = \pi/2$, $\sin \omega(\tau/4 + t_0) = \cos \omega t_0$.   Therefore

$$\cos \omega t_0 = \frac{F_s}{m}\frac{\pi}{2x_0\omega^2}$$

The relative velocity between the moving wall and the mass is $\dot{x} - \dot{u}$, and the total relative motion is the integral of the relative velocity over a cycle.   Note that the area between the two curves for the second half of the cycle is the same as for the first.   Hence, the work V per cycle is

$$V = 2\int_{-\tau/4}^{\tau/4} F_s(\dot{x} - \dot{u})\, dt = 4F_s x_0 \sqrt{1 - \left(\frac{F_s\pi}{2mx_0\omega^2}\right)^2} \tag{6.34}$$

Optimum damping occurs when the work per cycle is a maximum.   It can be determined by setting the derivative of V with respect to F_s in Eq. (6.34) equal to zero and solving for F_s:

$$(F_s)_{\text{opt}} = \frac{\sqrt{2}}{\pi}m\omega^2 x_0 \tag{6.35}$$

Energy absorption per cycle with optimum damping is, from Eq. (6.34),

$$V_{\text{opt}} = \frac{4}{\pi}m\omega^2 x_0^2 \tag{6.36}$$

A comparison of the effectiveness of the Coulomb friction damper with other types is given in Fig. 6.15.

EFFECT OF NONLINEARITY IN THE SPRING OF AN AUXILIARY MASS DAMPER

It is possible to extend the range of frequency over which a dynamic absorber is effective by using a nonlinear spring.[10-12]   When a nonlinear spring is used, the natural frequency of the absorber is a function of the amplitude of vibration; it increases or decreases, depending upon whether the spring stiffness increases or decreases with deflection.   Figure 6.20A shows a typical response curve for a system with increasing spring stiffness; Fig. 6.20B illustrates types of systems having increasing spring stiffness and shows typical force-deflection curves.   Figure 6.21A shows a typical response curve for a system of decreasing spring stiffness; Fig. 6.21B illustrates types of systems having decreasing stiffnesses and shows typical force-deflection curves.

To compute the equivalent mass at a given frequency when a nonlinear spring is used, it is necessary to use a trial-and-error procedure.   By the methods given in Chap. 4, compute the natural frequency of the auxiliary mass system, assuming the point of attachment fixed, as a function of the amplitude of motion of the auxiliary mass.   This will result in a curve similar to the dotted curves in Figs. 6.20A and 6.21A. At the given frequency, compute β_a in Eq. (6.4) in terms of the tuned frequency of the absorber at zero amplitude. (The tuned frequency will change with amplitude because the spring constant changes.) With this value of β_a compute the equivalent mass from Eq. (6.6). With this mass in the system, compute the amplitude of motion x_0 of the primary mass to which the auxiliary system is attached [Eq. (6.7)] and the amplitude of the relative motion $x_{r0} = \nu^2(1 - \nu^2)x_0$. Using this value of x_{r0}, ascertain the corresponding value of

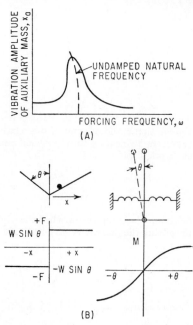

FIG. 6.20. Auxiliary mass damper with non-linear spring having stiffness that increases as deflection increases. The response to forced vibration and the natural frequency are shown at (A). Several arrangements of non-linear systems with the corresponding force-deflection curves are shown at (B).

FIG. 6.21. Auxiliary mass damper with nonlinear spring having stiffness that decreases as deflection increases. The response to forced vibration and the natural frequency are shown at (A). Two arrangements of nonlinear systems with the corresponding force-deflection curves are shown at (B).

resonant frequency of the system from the computed curve, and compute the new value of β_a. Repeat the process until the value of β_a remains unchanged upon repeated calculation.

A dynamic absorber having a nonlinear characteristic can be used to introduce nonlinearity into a resonant system. This can be useful in the case where a machine passes through a resonance rapidly as the speed is increased but slowly as the speed is decreased. In bringing this machine up to speed, there is a natural frequency that comes into strong resonance, giving a critical speed. A strongly nonlinear dynamic absorber tuned at low amplitudes to the optimum frequency for the damped absorber can be used to reduce the effects of the critical speed. Two resonant peaks will be introduced, as shown on curve 1 of Fig. 6.13. By making the dynamic absorber nonlinear, so that the stiffness becomes greater as the amplitude of vibration is increased, the peaks are bent over to provide the response curve shown in Fig. 6.22. In starting, the machine is accelerated through the two critical speeds so fast that a resonance is unable to build up. In coasting to a stop, there would be ample time for significant amplitudes to build up if the nonlinearity did not exist. Because of the nonlinearity, the amplitude of vibration as a function of speed (since β is proportional to speed) follows the path A, B, C, D, E, F, G and never reaches the extreme amplitudes H_1 and H_2.

FIG. 6.22. Motion of the primary mass, as a function of forcing frequency, in a system having a nonlinear dynamic absorber whose natural frequency increases with amplitude. The mass of the absorber is 0.25 times the mass of the primary system ($\mu = 0.25$).

MULTIMASS ABSORBERS

In general, only one mass is used in a dynamic absorber. However, it is possible to provide a dynamic absorber that is effective for two or more frequencies by attaching an auxiliary mass system that resonates at the frequencies that are objectionable. The principle that would make such a dynamic absorber effective is utilized in the design of the elastic system of a ship's propulsion plant driven by independent high-pressure and low-pressure turbines. By making the frequencies of the two branches about the reduction gear identical, the gear becomes a node for one of the resonant modes. Then it is impossible to excite the mode of vibration where one turbine branch vibrates against the other as a result of excitation transmitted by the propeller shaft to that node.

DISTRIBUTED MASS ABSORBERS

It is possible to use distributed masses as vibration dampers. Consider an undamped rod of distributed mass and elasticity attached to a foundation that vibrates the rod axially, as shown in Fig. 6.23. The differential equation for the motion of this rod is derived in Chap. 7. The values of the constants are set by the boundary conditions:

$$\text{Stress} = E\,\frac{\partial u}{\partial x} = 0 \qquad \text{where } x = l$$

$$u = u_0 \cos t \qquad \text{where } x = 0 \tag{6.37}$$

The solution of the equation of motion is

$$u = u_0 \cos \omega t \left(\cos \sqrt{\frac{\gamma \omega^2}{Eg}}\, x + \tan \sqrt{\frac{\gamma \omega^2}{Eg}}\, l \sin \sqrt{\frac{\gamma \omega^2}{Eg}}\, x \right)$$

$$= u_0 \cos \omega t\, \frac{\cos \sqrt{(\gamma \omega^2/Eg)}\,(1 - x)}{\cos \sqrt{(\gamma \omega^2/Eg)}\, l} \tag{6.38}$$

MOVING FOUNDATION

$u_0 \cos \omega t$

X

l

DISTRIBUTED ELASTIC MASS

FIG. 6.23. Elastic body with distributed mass used as auxiliary mass damper.

where E is the modulus of elasticity and γ is the weight density of the material. When $x = 0$, the force F on the foundation is

$$F = SE \frac{\partial u}{\partial x}\Big|_0 = SEu_0 \sqrt{\frac{\gamma\omega^2}{Eg}}\left(\tan \sqrt{\frac{\gamma\omega^2}{Eg}}\, l\right)$$
(6.39)

where S is the cross-sectional area of the bar. It is apparent that as the argument of the tangent has successive values of $\pi/2$, $3\pi/2$, $5\pi/2$, ..., the force exerted on the foundation becomes infinite. The distributed mass acts as a dynamic absorber enforcing a node at its point of attachment. By tuning the mass so that

$$\sqrt{\frac{\gamma\omega^2}{Eg}}\, l = \frac{n\pi}{2} \quad \text{or} \quad l = \frac{n\pi}{2\omega}\sqrt{\frac{Eg}{\gamma}} \quad (6.40)$$

the distributed mass acts as a dynamic absorber for not only the fundamental frequency $\omega/2\pi$, but also for the third, fifth, seventh, ... harmonics of the fundamental.

The above solution neglects damping. It is possible to consider the effect of damping by including a damping term in the differential equation. The stress in an element is assumed to be the sum of a deformation stress and a stress related to the velocity of strain:

$$\sigma = E\epsilon + \mu \frac{d\epsilon}{dt}$$
(6.41)

where $\epsilon = \partial u/\partial x$ is the strain. Then the differential equation becomes

$$E \frac{\partial^2 u}{\partial x^2} + \mu \frac{\partial^3 u}{\partial x^2 \partial t} = \frac{\gamma}{g}\frac{\partial^2 u}{\partial t^2}$$
(6.42)

Since the absorber is excited by a foundation moving with a frequency $f = \omega/2\pi$, u may be expressed as $\Re u_1 e^{j\omega t}$ and the partial differential equation can be written as the ordinary linear differential equation

$$E \frac{d^2 u_1}{dx^2} + j\omega\mu \frac{d^2 u_1}{dx^2} + \frac{\gamma\omega^2 u_1}{g} = 0$$

This equation may be written

$$\left(1 + \frac{\mu\omega j}{E}\right)\frac{d^2 u_1}{dx^2} + \frac{\gamma\omega^2}{Eg} u_1 = 0$$
(6.43)

Since Eq. (6.43) is a second-order linear differential equation, the solution may be written

$$u = A_1 e^{\beta_1 x} + A_2 e^{\beta_2 x}$$
(6.44)

where β_1 and β_2 are the two roots of the equation

$$\beta^2 = \frac{-(\gamma/Eg)\omega^2}{1 + (j\mu\omega/E)}$$
(6.45)

For small values of μ, by a binomial expansion of the denominator,

$$\pm \beta = \frac{1}{2}\sqrt{\frac{\gamma}{Eg}}\frac{\mu\omega^2}{E} + j\sqrt{\frac{\gamma}{Eg}}\omega \qquad (6.46)$$

where μ is defined by Eq. (6.41).

The boundary conditions to be met by the damper are:

At $x = 0$: $u = u_0$ therefore, $A_1 + A_2 = u_0$

$$(6.47)$$

At $x = l$: $\sigma = (E + j\omega\mu)\dfrac{\partial u}{\partial x} = 0$ therefore, $A_1 e^{\beta l} - A_2 e^{-\beta l} = 0$

Solving Eqs. (6.47) for A_1 and A_2 and substituting the result in (6.44),

$$u = u_0 \frac{\cosh \beta(l - x)}{\cosh \beta l} \qquad (6.48)$$

The force exerted on the foundation by the damper is

$$F_{(x=0)} = S\sigma_{(x=0)} = -Su_0(E + j\omega\mu)\frac{\beta \sinh \beta l}{\cosh \beta l} \qquad (6.49)$$

FIG. 6.24 Real and imaginary components of the force applied to a vibrating body by the distributed mass damper shown in Fig. 6.23. These relations are given mathematically by Eq. (6.50), and the terms are defined in connection with Eq. (6.38). The curves are for a value of the damping coefficient $\mu\omega/E = 0.1$, where μ is defined by Eq. (6.41).

where S is the cross-section area of the bar. When the complex value of β as given in Eq. (6.46) is substituted in Eq. (6.49), the following value for the dynamic force exerted on the foundation is obtained:

$$
\frac{F_{(x=0)}}{SE\sqrt{\dfrac{\gamma}{Eg}}\,\omega u_0} = \frac{\left(1+\dfrac{\mu^2\omega^2}{2E^2}\right)\sin 2\left(\sqrt{\dfrac{\gamma}{Eg}}\,\omega l\right) + \dfrac{\mu\omega}{2E}\sinh\left(\dfrac{\mu\omega}{E}\sqrt{\dfrac{\gamma}{Eg}}\,\omega l\right)}{\cos 2\left(\sqrt{\dfrac{\gamma}{Eg}}\,\omega l\right) + \cosh\left(\dfrac{\mu\omega}{E}\sqrt{\dfrac{\gamma}{Eg}}\,\omega l\right)}
$$

$$
+ j\,\frac{\dfrac{\mu\omega}{2E}\sin 2\left(\sqrt{\dfrac{\gamma}{Eg}}\,\omega l\right) - \left(1+\dfrac{\mu^2\omega^2}{2E^2}\right)\sinh\left(\dfrac{\mu\omega}{E}\sqrt{\dfrac{\gamma}{Eg}}\right)}{\cos 2\left(\sqrt{\dfrac{\gamma}{Eg}}\,\omega l\right) + \cosh\left(\dfrac{\mu\omega}{E}\sqrt{\dfrac{\gamma}{Eg}}\,\omega l\right)} \qquad (6.50)
$$

A plot of the real and imaginary values of $F_{(x=0)}/SE\sqrt{\dfrac{\gamma}{Eg}}\,\omega u_0$ is given in Fig. 6.24 for zero damping and for a damping coefficient $\mu\omega/E = 0.1$ as a function of a tuning parameter $\sqrt{\gamma/Eg}\,(\omega l)$. Damping decreases the effectiveness of the distributed mass damper substantially, particularly for the higher modes.

PRACTICAL APPLICATIONS OF AUXILIARY MASS DAMPERS AND ABSORBERS TO SINGLE DEGREE-OF-FREEDOM SYSTEMS

THE DYNAMIC ABSORBER

The dynamic absorber, because of its tuning, can be used to eliminate vibration only where the frequency of the vibration is constant. Many pieces of equipment to which it is applied are operated by alternating current. So that it can be used for time keeping, the frequency of this a-c is held remarkably constant. For this reason, most applications of dynamic absorbers are made to mechanisms that operate in synchronism from an a-c power supply.

An application of a dynamic absorber to the pedestal of an a-c generator having considerable vibration is shown in Fig. 6.25, where the relative sizes of absorber and pedestal are shown approximately to scale. In this case, the application is made to a complicated structure and the mass of the absorber is much less than that of the primary system; however, since the frequency of the excitation is constant, the dynamic absorber reduces the vibration. When the mass ratio is small, it is important that the absorber be accurately tuned and that the damping be small. In this case, the excitation was the unbalance in the turbine rotor which was elastically connected to the pedestal through the flexibility of the shaft. If the absorber were ideally effective, there would be no forces at the frequency of the shaft speed; therefore, there would be no displacements from the pedestal where the force is neutralized through the remainder of the structure.

The dynamic absorber has been applied to the electric clipper shown in Fig. 6.26. The structure consisting of the cutter blade and its driving mechanism is actuated by the magnetic field at a frequency of 120 cps, as a

FIG. 6.25. Application of a dynamic absorber to the bearing pedestal of an a-c generator.

result of the 60-cps a-c power supply. The forces and torques required to move the blade are balanced by reactions on the housing, causing it to vibrate. The dynamic absorber tuned to a frequency of 120 cps enforces a node at the location of its mass. Since this is approximately the center-of-gravity of the assembly of the cutter and its driving mechanism, the absorber effectively neutralizes the unbalanced force. The

FIG. 6.26. Application of a dynamic absorber to a hair clipper.

moment caused by the rotation of the moving parts is still unbalanced. A second very small dynamic absorber placed in the handle of the clipper could enforce a node at the handle and substantially eliminate all vibration. The design of these absorbers is simple after the unbalanced forces and torques generated by the cutter mechanism are computed. The sum of the inertia forces generated by the two absorbers, $m_1x_1\omega^2 + m_2x_2\omega^2$ (where m_1 and x_1 are the mass and amplitude of motion of the first absorber, m_2 and x_2 are the corresponding values for the second absorber, and $\omega = 240\pi$), must equal the unbalanced force generated by the clipper mechanism. The torque generated by the two absorbers must balance the torque of the mechanism. Since the value of ω^2 is known, the values of m_1x_1 and m_2x_2 can be determined. Weights that fit into the available space with adequate room to move are chosen, and a spring is designed of such stiffness that the natural frequency is 120 cps.

Because of the desirable balancing properties of the simple dynamic absorber and the constancy of frequency of a-c power, it might be expected that devices operating at a frequency of 120 cps would be used more widely. However, their application is limited because the frequency of vibration is too high to allow large amplitudes of motion.

REDUCTION OF ROLL OF SHIPS BY AUXILIARY TANKS

An interesting application of auxiliary mass absorbers is found in the auxiliary tanks used to reduce the rolling of ships,[1, 13] as shown in Fig. 6.27. When a ship is heeled, the restoring moment $k_r\phi$ acting on it is proportional to the angle of heel (or roll). This restoring moment acts to return the ship (and the water that moves with it) to its equilibrium position. If I_s represents the polar moment of inertia of the ship and its entrained water, the differential equation for the rolling motion of the ship is

FIG. 6.27. Cross section of ship equipped with antiroll tanks. The flow of water from one tank to the other tends to counteract rolling of the ship.

$$I_s\ddot{\phi} + k_r\phi = M_s \qquad (6.51)$$

where M_s represents the rolling moments exerted on the ship, usually by waves.

To reduce rolling of the ship, auxiliary wing tanks connected by pipes are used. The water flowing from one tank to another has a natural frequency that is determined by the length and cross-sectional area of the tube connecting the tanks. The damping is controlled by restricting the flow of water, either with a valve S in the line that allows air to flow between the tanks (Fig. 6.27) or with a valve V in the water line. Since the tanks occupy valuable space, the mass ratio of the water in the tanks to the ship is small. Fortunately, the excitation from waves generally is not large relative to the restoring moments, and roll becomes objectionable only because the normal damping of a ship in rolling motion is not very large. The use of antirolling tanks in the German luxury liners *Bremen* and *Europa* reduced the maximum roll from 15 to 5°.

REDUCTION OF ROLL OF SHIPS BY GYROSCOPES

A large gyroscope may be used to reduce roll in ships, as shown in Fig. 6.28.[1,14]   In response to the velocity of roll of a ship, the gyroscope precesses in the plane of symmetry of the ship.   By braking this precession, energy can be dissipated and the roll reduced.   The torque exerted by the gyroscope is proportional to the rate of change of the angular momentum about an axis perpendicular to the torque.   Letting I represent the polar moment of inertia of the gyroscope about its spin axis and θ the angular velocity of precession of the gyroscope, then the equation of motion of the ship is

FIG. 6.28. Application of a gyroscope to a ship to reduce roll.

$$I_s\ddot{\phi} + k_r\phi + I\Omega\dot{\theta} = M_s \qquad (6.52)$$

Assume that the gyroscope has (1) a moment of inertia about the precession axis of I_g, (2) a weight of W, and (3) that its center-of-gravity is below the gimbal axis (as it must be for the gyro to come to equilibrium in a working position) a distance a, as shown in Fig. 6.28.   Then the equation of motion of the gyroscope is

$$I_g\ddot{\theta} + W_a\theta + c\dot{\theta} - I\Omega\phi = 0 \qquad (6.53)$$

where Ω is the spin velocity of the gyroscope.   From Eq. (6.53), for a roll frequency of ω, the angle of precession of the gyroscope is

$$\theta = \frac{jI\Omega\omega\phi}{-I_g\omega^2 + Wa - jc\omega} \qquad (6.54)$$

The torque exerted on the ship is

$$I\Omega\dot{\theta} = \frac{-(I\Omega)^2\omega^2\phi}{-I_g\omega^2 + Wa + jc\omega} \qquad (6.55)$$

The equivalent moment of inertia of the gyroscope system in its reaction on the ship is

$$\frac{I\Omega^2}{-I_g\omega^2 + Wa + cj\omega} \qquad (6.56)$$

By analogy with the steps of Eqs. (6.2) through (6.7), it follows that

$$\frac{\phi}{\phi_{st}} = \sqrt{\frac{(1 - \beta_g^2)^2 + (2\zeta\beta_g)^2}{[(1 - \beta_g^2)(1 - \beta^2) - \beta^2\mu]^2 + (2\zeta\beta_g)^2(1 - \beta^2)^2}} \qquad (6.57)$$

where the parameters are defined in terms of ship and gyro constants as follows:

$$\beta_g = \frac{\omega}{\sqrt{Wa/I_g}} \qquad \beta = \frac{\omega}{\sqrt{k_r/I_s}} \qquad \zeta = \frac{c}{2\sqrt{WaI_g}} \qquad \mu = \frac{(I\Omega)^2}{WaI_s} \qquad \psi_{st} = \frac{M_s}{k_r}$$

Because $I\Omega$ can be made large by using a large gyro rotor and spinning it at a high speed, and Wa can be made small by choice of a design, the value of μ can be made quite large even though I_s is large.   In one experimental ship, $\mu = 20$ was obtained.   Even with this large value of μ, the precession angle of the gyroscope would become very large for optimum damping.   Therefore it is necessary to use much more damping than optimum.

Gyro stabilizers were used on the Italian ship *Conte di Savoia;* they are sometimes installed on yachts.

Both antirolling tanks and gyro stabilizers are more effective if they are active rather than passive. Activated dampers are considered below.

AUXILIARY MASS DAMPERS APPLIED TO ROTATING MACHINERY

An important industrial use of auxiliary mass systems is to neutralize the unbalance of centrifugal machinery. A common application is the balance ring in the spin dryer of home washing machines. The operation of such a balancer is dependent upon the basket

of the washer rotating at a speed greater than the natural frequency of its support. The balance ring is attached to the washing machine basket concentric with its axis of rotation, as shown in Fig. 6.29.

Consider the washing machine basket shown in Fig. 6.29. When its center-of-gravity does not coincide with its axis of rotation and it is rotating at a speed lower than its critical speed (corresponding to the natural frequency in rocking motion about the spherical seat), the centrifugal force tends to pull the rotational axis in the direction of the unbalance. This effect increases with an increase in rotational speed until the critical speed is reached. At this speed the amplitude would become infinite if it were not for the damping in the system. Above the critical speed, the phase position of the axis of rotation relative to the

Fig. 6.29. Schematic diagram showing location of balance ring on basket of a spin dryer.

center-of-gravity shifts so that the basket tends to rotate about its center-of-gravity with the flexibly supported bearing moving in a circle about an axis through the center-of-gravity. The relative positions of the bearing center and the center-of-gravity are shown in Fig. 6.30A and B.

Since the balance ring is circular with a smooth inner surface, any weights or fluids contained in the ring can be acted upon only by

(A)

(B)

Fig. 6.30. Diagram in plane normal to axis of rotation of spin dryer in Fig. 6.29. Relative positions of axes when rotating speed is less than natural frequency are shown at (A); corresponding diagram for rotation speed greater than natural frequency is shown at (B).

forces directed radially. When the ring is rotated about a vertical axis, the weights or fluids will move within the ring in such a manner as to be concentrated on the side farthest from the axis of rotation. If this concentration occurs below the natural frequency (Fig. 6.30A), the weights tend to move further from the axis and the resultant center-of-gravity is displaced so as to give a greater eccentricity. The points A and G rotate about the axis O at the frequency ω. The initial eccentricity of the center-of-gravity of the washer basket and its load from the axis of rotation is represented by e, and ρ is the elastic displacement of this center of rotation due to the centrifugal force. Where the off-center rotating weight is W, the unbalanced force is $(W/g)(\rho + e)\omega^2$ [where $\rho = e/(1 - \beta^2)$ and $\beta^2 = \omega^2/\omega_n^2 < 1$] and acts in the direction from A to G.

If the displacement of the weights or fluids in the balance ring occurs above the natural frequency, the center-of-gravity tends to

(A)

(B)

(C)

FIG. 6.31. Examples of balancing means for rotating machinery: (*A*) spheres (or cylinders) in a race; (*B*) annular discs rotating on shaft; (*C*) damping fluid in torus.

move closer to the dynamic location of the axis. The action in this case is shown in Fig. 6.30*B*. Then the points *A* and *G* rotate about *O* at the frequency ω. The unbalanced force is $(W/g)(\rho + e)\omega^2$ [where $\rho = e/(1 - \beta^2)$ and $\beta^2 = \omega^2/\omega_n^2 > 1$]. This gives a negative force that acts in a direction from *G* to *A*. Thus the eccentricity is brought toward zero and the rotor is automatically balanced. Because it is necessary to pass through the critical speed in bringing the rotor up to speed and in stopping it, it is desirable to heavily damp the balancing elements, either fluid or weights.

In practical applications, the balancing elements can take several forms. The earliest form consisted of two or more spheres or cylinders free to move in a race concentric with the axis of the rotor, as shown in Fig. 6.31*A*. A later modification consists of three annular discs that rotate about an enlarged shaft concentric with the axis, as indicated in Fig. 6.31*B*. These are contained in a sealed compartment with oil for lubrication and damping. A fluid type of damper is shown in Fig. 6.31*C*, the fluid usually being a high-density viscous material. With proper damping, mercury would be excellent, but it is too expensive. Therefore a more viscous, high-density halogenated fluid is used.

The balancers must be of sufficient weight and operate at such a radius that the product of their weight and the maximum eccentricity they can attain is equivalent to the unbalanced moment of the load. This requirement makes the use of the spheres or cylinders difficult because they cannot be made large; it makes the annular plates large because they are limited in the amount of eccentricity that can be obtained.

In a cylindrical volume 24 in. (61 cm) in diameter and 2 in. (5 cm) thick, seven spheres 2 in. (5 cm) in diameter can neutralize 98.6 lb-in. (114 kg-cm) of unbalance; three cylinders 4 in. (10 cm) in diameter by 2 in. (5 cm) thick can neutralize 255 lb-in. (295 kg-cm); three annular discs, each $\frac{5}{8}$ in. (1.6 cm) thick with an outside diameter of 19.55 in. (50 cm) and an inside diameter of 10.45 in. (26.5 cm) [the optimum for a center post 6 in. (15.2 cm) in diameter], can neutralize 250 lb-in. (290 kg-cm); and half of a 2-in. (5-cm) diameter torus filled with fluid of density 0.2 lb/in³ (5.5 gm/cm³) can neutralize 609 lb-in. (700 kg-cm). Only the fluid-filled torus would be initially balanced.

AUXILIARY MASS DAMPERS APPLIED TO TORSIONAL VIBRATION

Dampers and absorbers are used widely for the control of torsional vibration of internal-combustion engines. The most common absorber is the viscous-damped, untuned auxiliary mass unit shown in Fig. 6.32. The device is comprised of a cylindrical housing carrying an inertia mass that is free to rotate. There is a preset clearance between the housing and the inertia mass that is filled with a silicone oil of proper viscosity. Silicone oil is used because of its high viscosity index; i.e., its viscosity changes relatively little with temperature. With the inertia mass and the damping medium contained, the housing is seal-welded to provide a leakproof and simple absorber. However, the silicone oil has poor boundary lubricating properties and if decomposed by a local hot spot (such as might be caused by a reduced clearance at some particular spot), the decomposed damping fluid is abrasive.

Because of the simplicity of this untuned damper, it is commonly used in preference to the more effective tuned absorber. However, it is possible to use the same construction methods for a tuned damper, as shown in Fig. 6.33. It is also possible to mount the standard damper with the housing for the unsprung inertia mass attached to the main

FIG. 6.32. Untuned auxiliary mass damper with viscous damping. The application to a torsional system is shown at (A), and the linear analog at (B).

FIG. 6.33. Tuned auxiliary mass damper with viscous damping. The application to a torsional system is shown at (A), and the linear analog at (B).

mass by a spring, as shown in Fig. 6.34. If the viscosity of the oil and the dimensions of the masses and the clearance spaces are known, the damping effects of the dampers shown in Figs. 6.32 and 6.34 can be computed directly in terms of the equations previously developed. The damper in Fig. 6.34 can be analyzed by treating the spring and housing as additional elements in the main system and the untuned mass as a viscous damped auxiliary mass. If the inertia of the housing is negligible, the inertia mass is

FIG. 6.34. Auxiliary mass damper with viscous damping and spring-mounted housing. The application to a torsional system is shown at (A), and the linear analog at (B).

FIG. 6.35. Schematic cross section through Lanchester damper.

effectively connected to the main mass through a spring and a dashpot in series. The two elements in series can be represented by a complex spring constant equal to

$$\frac{1}{(1/jc\omega) + (1/k)} = \frac{kcj\omega}{k + cj\omega}$$

Where there is no damping in parallel with the spring, Eq. (6.3) becomes

$$m_{eq} = km/(k - m\omega^2)$$

Substituting the complex value of the spring constant, the effective mass is

$$m_{eq} = \frac{ckj\omega}{k + cj\omega} \left[\frac{m}{-m\omega^2 + cjk\omega/(k + cj\omega)} \right] \tag{6.58}$$

In terms of the nondimensional parameters defined in Eq. (6.4):

$$m_{eq} = \frac{(2\zeta\beta_a)^2(1 - \beta_a{}^2)}{\beta_a{}^4 - (2\zeta\beta_a)^2(1 - \beta_a{}^2)} m + \frac{-2\zeta\beta_a{}^3 m}{\beta_a{}^4 - (2\zeta\beta_a)^2(1 - \beta_a{}^2)} j \tag{6.59}$$

Before the advent of silicone oil with its chemical stability and relatively constant viscosity over service temperature conditions, the damper most commonly used for absorbing torsional vibration energy was the dry friction or Lanchester damper shown in Fig. 6.35. The damping is determined by the spring tension and the coefficient of friction at the sliding interfaces. Its optimum value is determined by the equation for a torsional system analogous to Eq. (6.35) for a linear system:

$$(T_s)_{opt} = \frac{\sqrt{2}}{\pi} I\omega^2\theta_0 \tag{6.60}$$

where T_s is the slipping torque, I is the moment of inertia of the flywheels, and θ_0 is the amplitude of angular motion of the primary system. The dry-friction-based Lanchester damper requires frequent adjustment, as the braking material wears, to maintain a constant braking force.

It is possible to use torque-transmitting couplings that can absorb vibration energy, as the spring elements for tuned dampers. The Bibby coupling (Fig. 6.36) is used in this manner. Since the stiffness of this coupling is nonlinear, the optimum tuning of such an absorber is secured for only one amplitude of motion.

A discussion of dampers and of their application to engine systems is given in Chap. 38.

HUB

TEETH

HUB

UNDULATING STRIP SPRING

FIG. 6.36. Coupling used as elastic and damping element in auxiliary mass damper for torsional vibration. The torque is transmitted by an undulating strip of thin steel interposed between the teeth on opposite hubs. The stiffness of the strip is nonlinear, increasing as torque increases. Oil pumped between the strip and teeth dissipates energy.

DYNAMIC ABSORBERS TUNED TO ORDERS OF VIBRATION RATHER THAN CONSTANT FREQUENCIES

In the torsional vibration of rotating machinery, it is generally found that exciting torques and forces occur at the same frequency as the rotational speed, or at multiples of this frequency. The ratio of the frequency of vibration to the rotational speed is called the *order of the vibration* q. Thus a power plant driving a four-bladed propeller may have a torsional vibration whose frequency is four times the rotational speed of the drive shaft; sometimes it may have a second torsional vibration whose frequency is eight times the rotational speed. These are called the fourth-order and eighth-order torsional vibrations.

If a dynamic absorber in the form of a pendulum acting in a centrifugal field is used, then its natural frequency increases linearly with speed. Therefore it can be used to neutralize an order of vibration.[15-19]

Consider a pendulum of length l and of mass m attached at a distance R from the center of a rotating shaft, as shown in Fig. 6.37. Since the pendulum is excited by torsional vibration in the shaft, let the radius R be rotating at a constant speed Ω with a superposed vibration $\theta = \theta_0 \cos q\Omega t$, where q represents the order of the vibration. Then the angle of R with respect to any desired reference is $\Omega t + \theta_0 \cos q\Omega t$. The angle of the pendulum with respect to the radius R is defined as $\psi = \psi_0 \cos q\Omega t$, as shown by Fig. 6.37.

The acceleration acting on the mass m at

(A)

(B)

(C)

Fig. 6.37. Schematic diagram of pendulum absorber.

Fig. 6.38. Velocity vectors for the pendulum absorber: (A) velocities at time t; (B) velocities at time $t + \Delta t$; (C) change in velocities during time increment Δt.

position B is most easily ascertained by considering the change in velocity during a short increment of time Δt. The components of velocity of the mass m at time t are shown graphically in Fig. 6.38A; at time $t + \Delta t$, the corresponding velocities are shown in Fig. 6.38B. The change in velocity during the time interval Δt is shown in Fig. 6.38C. Since the acceleration is the change in velocity per unit of time, the accelerations along and perpendicular to l are:

Acceleration along l:

$$\frac{-l(\Omega + \dot\theta + \dot\psi^2)\,\Delta t - R(\Omega + \dot\theta)^2\,\Delta t \cos\psi + R\ddot\theta\,\Delta t \sin\psi}{\Delta t} \tag{6.61}$$

Acceleration perpendicular to l:

$$\frac{l(\ddot\theta + \ddot\psi)\,\Delta t + R(\Omega + \dot\theta)^2\,\Delta t \sin\psi + R\ddot\theta\,\Delta t \cos\psi}{\Delta t} \tag{6.62}$$

Only the force $-F$, directed along the pendulum, acts on the mass m. Therefore the equations of motion are

$$-F = -ml(\Omega + \dot\theta + \dot\psi)^2 - mR(\Omega + \dot\theta)^2 \cos\psi + R\ddot\theta \sin\psi$$
$$0 = ml(\ddot\theta + \ddot\psi) + mR(\Omega + \dot\theta)^2 \sin\psi + mR\ddot\theta \cos\psi \tag{6.63}$$

Assuming that ψ and θ are small, Eqs. (6.63) simplify to

$$Ft = m(R + l)\Omega^2$$
$$l(\ddot\theta + \ddot\psi) + R\Omega^2\psi + R\ddot\theta = 0 \tag{6.64}$$

The second of Eqs. (6.64) upon substitution of $\theta = \theta_0 \cos q\Omega t$ and $\psi = \psi_0 \cos q\Omega t$ yields

$$\frac{\psi_0}{\theta_0} = \frac{(q\Omega)^2(l + R)}{-(q\Omega)^2 l + \Omega^2 R} = \frac{q^2(l + R)}{R - q^2 l} \tag{6.65}$$

The torque M exerted at point 0 by the force F is

$$M = RF \sin \psi = RF\psi \qquad \text{when } \psi \text{ is small}$$

From Eqs. (6.64) and (6.65), when ψ is small,

$$M = \frac{mq^2 R(R + l)^2 \Omega^2}{R - q^2 l} \tag{6.66}$$

If a flywheel having a moment of inertia I is accelerated by a shaft having an amplitude of angular vibratory motion θ_0 and a frequency $q\Omega$, the torque amplitude exerted on the shaft is $I(q\Omega)^2\theta_0$. Therefore, the equivalent moment of inertia I_{eq} of the pendulum is

$$I_{eq} = \frac{mR(R + l)^2}{R - q^2 l} = \frac{m(R + l)^2}{1 - q^2 l/R} \tag{6.67}$$

When

$$\frac{R}{l} = q^2 \tag{6.68}$$

the equivalent inertia is infinite and the pendulum acts as a dynamic absorber by enforcing a node at its point of attachment.

Where the pendulum is damped, the equivalent moment of inertia is given by an equation analogous to Eqs. (6.4) and (6.5):

$$I_{eq} = \frac{1 + 2\zeta\nu j}{(1 - \nu^2) + 2\zeta\nu j} \, m(R + l)^2$$

$$= m(R + l)^2 \left[\frac{1 - \nu^2 + (2\zeta\nu)^2}{(1 - \nu^2)^2 + (2\zeta\nu)^2} - \frac{2\zeta\nu^3 j}{(1 - \nu^2)^2 + (2\zeta\nu)^2} \right] \tag{6.69}$$

where $\nu^2 = q^2 l/R$ and $\zeta = (c/2m\Omega)\sqrt{l/R}$.

When the pendulum is attached to a single degree-of-freedom system as is shown in Fig. 6.39, the amplitude of motion θ_a of the flywheel of inertia I is given, by analogy to Eq. (6.7), as

$$\frac{\theta_a}{\theta_{st}} = \sqrt{\frac{(1 - \nu^2)^2 + (2\zeta\nu)^2}{[(1 - \nu^2)(1 - \beta_p{}^2) - \beta_p{}^2 \mu]^2 + (2\zeta\nu)^2[1 - \beta_p{}^2 - \beta_p{}^2 \mu_p]^2}} \tag{6.70}$$

where

$$2\zeta\nu = \frac{cql}{mR}$$

$$\mu_p = \frac{m(R + l)^2}{I}$$

$$\beta_p = \frac{q}{k_r I}$$

$$\theta_{st} = \frac{m_0}{k_r}$$

The pendulum tends to detune when the amplitude of motion of the pendulum is large, thereby introducing harmonics of the torque that it neutralizes.[17] Suppose the shaft rotates at a constant speed Ω, i.e., $\theta_0 = 0$, and consider the torque exerted on the shaft as m moves through a large amplitude ψ_0 about its equilibrium position. Equations (6.63) become

$$F = ml(\Omega + \dot{\psi})^2 + mR\Omega^2 \cos \psi$$

$$l\ddot{\psi} + R\Omega^2 \sin \psi = 0$$

(6.71)

A solution for the second of Eqs. (6.71) is

$$\dot{\psi} = \sqrt{\frac{2\Omega^2 R}{l}} \sqrt{\cos \psi - \cos \psi_0}$$

(6.72)

The solution of Eq. (6.72) involves elliptic integrals and is given approximately by

$$\psi = \psi_0 \sin \omega t$$

where

$$\omega = \sqrt{\frac{R}{l}} \frac{\pi/2}{F(\psi_0/2, \pi/2)} \Omega$$

and $F(\psi_0/2, \pi/2)$ is an elliptic function of the first kind whose value may be obtained from tables.

Since $\omega/\Omega = q$ (the order of the disturbance), the tuning of the damper will be changed for large angles and becomes

$$q^2 = \frac{R}{l} \left(\frac{\pi/2}{F(\psi_0/2, \pi/2)} \right)^2$$

(6.73)

The value of $q^2 l/R = \nu^2$ used in Eqs. (6.69) and (6.70) is given in Fig. 6.40 as a function of the amplitude of the pendulum.

FIG. 6.39. Application of pendulum absorber to a rotational single degree-of-freedom system.

FIG. 6.40. Tuning function for a pendulum absorber used in Eqs. (6.69) and (6.70).

Since the force exerted by the mass m is directed along the rod connecting it to the pivot A (Fig. 6.37), the reactive torque on the shaft is

$$M = FR \sin \psi$$

$$= mR^2\Omega^2 \left[\frac{l}{R} \left(1 + \frac{\psi}{\Omega} \right)^2 \sin \psi + \sin \psi \cos \psi \right]$$

$$= mR^2\Omega^2(A_1 \sin q\Omega t + A_2 \sin 2q\Omega t + A_3 \sin 3q\Omega t + \cdots) \qquad (6.74)$$

The values of the fundamental torque corresponding to the tuned frequency and to the second and third harmonics of this tuned frequency are given in Fig. 6.41 as a function of the angle of swing of the pendulum, for a typical installation. In this case, the pendulum is tuned to the $4\frac{1}{2}$ order of vibration. (The $4\frac{1}{2}$ order of vibration is one whose frequency is $4\frac{1}{2}$ times the rotational frequency and 9 times the fundamental frequency. The latter is called the half order and occurs at half of the rotational frequency. This is common in four-cycle engines.)

Two types of pendulum absorber are used. The one most commonly used is shown in Fig. 6.42. The counterweight, which also is used to balance rotating forces in the engine, is suspended from a hub carried by the crankshaft by pins that act through holes with clearance, Fig. 6.42A. By suspending the pendulum from two pins, the pendulum when oscillating does not rotate but rather moves as shown in Fig. 6.42B. Since it is not subjected to angular acceleration, it may be treated as a particle located at its center-

FIG. 6.41. Harmonic components of torque generated by a pendulum absorber as a function of its angle of swing. The torque is expressed by the parameters used in Eq. (6.74).

FIG. 6.42. Bifilar type of pendulum absorber. The mechanical arrangement is shown at (A), and a schematic diagram at (B).

of-gravity. Referring to Fig. 6.42A and B, the expression for acceleration [Eqs. (6.61) and (6.62)] and the equations of motion [Eqs. (6.63)] apply if

$$R = H_1 + H_2$$
$$l = \frac{D_c + D_p}{2} - D_b \tag{6.75}$$

where H_1 = distance from center of rotation to center of holes in crank hub

H_2 = distance from center of holes in pendulum to center-of-gravity of pendulum

D_c = diameter of hole in crank hub

D_p = diameter of hole in pendulum

D_b = diameter of pin

In practice, difficulty arises from the wear of the holes and the pin. Moreover, the motion on the pins generally is small and the loads due to centrifugal forces are large so that fretting is a problem. Because the radius of motion of the pendulum is short, only a small amount of wear can be tolerated. Hardened pins and bushings are used to reduce the wear.

The pendulum is most easily designed if it is recognized that the inertia torques generated by the pendulum must neutralize the forcing torques. Thus

$$m\omega^2 l\psi_0 R = M \tag{6.76}$$

Fig. 6.43. Roller type of pendulum absorber.

The radii l and R are set by the design of the crank and the order of vibration to be neutralized. The original motion ψ_0 is generally limited to a small angle, approximately 20°. It is probable that the most stringent condition is at the lowest operating speed, although the absorber may be required only to avoid difficulty at some particular critical speed. Knowing the excitation M, it is possible to compute the required mass of the pendulum weight.

A second type of pendulum absorber is a cylinder that rolls in a hole in a counterweight, as shown in Fig. 6.43. In this type, the radius of the pendulum corresponds to the difference in the radii of the hole and of the cylinder. It is found, by observing tests and checking the tuning of actual systems using cylindrical pendulums, that the weight rotates with a uniform angular velocity. Therefore the tuning is independent of the moments of inertia of the cylinder. It is common to allow a larger amplitude of motion with the absorber of Fig. 6.43 than with the absorber of Fig. 6.40.

Applications of pendulum absorbers to torsional-vibration problems are given in Chap. 38.

PENDULUM ABSORBER FOR LINEAR VIBRATION

Fig. 6.44. Application of pendulum absorbers to counteract linear vibration.

The principle of the pendulum absorber can be applied to linear vibration as well as to torsional vibration. To neutralize linear vibration, pendulums are rotated about an axis parallel to the direction of vibration, as shown in Fig. 6.44. This can be accomplished with an absorber mounted on the moving body. Two or more pendulums are used so that centrifugal forces are balanced. Free rotational movement of each pendulum in the plane of the axis allows the axial forces to be neutralized. The pendulum assembly must rotate about the axis at some submultiple of the frequency of vibration. The size of the absorber is determined by the condition that the components of the inertia forces of the weights in the axial direction $[\Sigma m\omega^2 r\theta]$ must balance the exciting forces. This device can be applied where the vibration is generated by the action of rotating members, but the magnitude of the vibratory forces is uncertain. A discussion of this absorber, including the influence of moments of inertia and damping of the pendulum, together with some applications to the elimination of vibration in special locations on a ship, is given in Ref. 20.

APPLICATIONS OF DAMPERS TO MULTIPLE DEGREE-OF-FREEDOM SYSTEMS

Auxiliary mass dampers as applied to systems of several degrees-of-freedom can be represented most effectively by equivalent masses or moments of inertia, as determined by Eq. (6.5) or Eq. (6.6). The choice of proper damping constants is more difficult. For the case of torsional vibration, the practical problems of designing dampers and selecting the proper damping are considered in Chap. 38.

There are many applications of dampers to vibrating structures that illustrate the use of different types of auxiliary mass damper. One such application has been to ships.[21] These absorbers had low damping and were designed to be filled with water so that they could be tuned to the objectionable frequencies. In one case, the absorber was located near the propeller (the source of excitation) and when properly tuned was found to be effective in reducing the resonant vibration of the ship. In another case, the absorber was located on an upper deck but was not as effective. It enforced a node at its point of attachment but, because of the flexibility between the upper deck and the bottom of the ship, there was appreciable motion in the vicinity of the propeller and vibratory energy was fed to the ship's structure. To operate properly, the absorbers must be closely tuned and the propeller speed closely maintained. Because the natural frequencies of the ship vary with the types of loading, it is not sufficient to install a fixed frequency absorber that is effective at only one natural frequency of the hull, corresponding to a particular loading condition.

An auxiliary mass absorber has been applied to the reduction of vibration in a heavy building that vibrated at a low frequency under the excitation of a number of looms.[22] The frequency of the looms was substantially constant. However, the magnitude of the excitation was variable as the looms came into and out of phase. The dynamic absorber, consisting of a heavy weight hung as a pendulum, was tuned to the frequency of excitation. Because the frequency was low and the forces large, the absorber was quite large. However, it was effective in reducing the amplitude of vibration in the building and was relatively simple to construct.

ACTIVATED VIBRATION ABSORBERS

The cost and space that can be allotted to ship antirolling devices are limited. Therefore it is desirable to activate the absorbers so that their full capacity is used for small amplitudes as well as large. Activated dampers can be made to deliver as large restoring forces for small amplitudes of motion of the primary body as they would be required to deliver if the motions were large. For example, the gyrostabilizer that is used in the ship is precessed by a motor through its full effective range, in the case of small angles as well as large. Thus, it introduces a restoring torque that is much larger than would be introduced by the normal damped precession.[14] In the same manner the water in antiroll tanks is always pumped to the tank where it will introduce the maximum torque to counteract the roll. By pumping, much larger quantities of water can be transferred and larger damping moments obtained than can be obtained by controlled gravity flows.

Devices for damping the roll are desirable for ships. It has been common practice to install bilge keels (long fins which extend into the water) in steel ships. Some ships are now fitted with activated, retractable hydrofoils located at the bilge at the maximum beam. Both these devices are effective only when the ship is in motion and add to the resistance of the ship.

Activated vibration absorbers are essentially servomechanisms designed to maintain some desired steady state. Steam and gas turbine speed governors, wicket gate controls for frequency regulation in water turbines, and temperature control equipment can be considered as special forms of activated vibration absorbers.[23]

THE USE OF AUXILIARY MASS DEVICES TO REDUCE TRANSIENT AND SELF-EXCITED VIBRATIONS

Where the vibration is self-excited or caused by repeated impact, it is necessary to have sufficient damping to prevent a serious build-up of vibration amplitude. This damping, which need not always be large, may be provided by a loosely coupled auxiliary mass. A simple application of this type is the ring fitted to the interior of a gear, as shown in Fig. 6.45. By fitting this ring with the proper small clearance so that relative motion occurs between it and the gear, it is possible to obtain enough energy dissipation to damp the high-frequency, low-energy vibration that causes the gear to ring. The rubbery coatings applied to large, thin-metal panels such as automobile doors to give them a solid rather than a "tinny" sound depend for their effectiveness on a proper balance of mass, elasticity, and damping (see Chap. 37).

SOUND DEADENING RING

FIG. 6.45. Application of auxiliary mass damper to deaden noise in gear.

Another application where auxiliary mass dampers are useful is in the prevention of fatigue failures in turbines. At the high-pressure end of an impulse turbine, steam or hot gas is admitted through only a few nozzles. Consequently, as the blade passes the nozzle it is given an impulse by the steam and set into vibration at its natural frequency. It is a characteristic of alloy steels that they have very little internal damping at high operating temperature. For this reason the free vibration persists with only a slightly diminished amplitude until the blade again is subjected to the steam impulse. Some of these second impulses will be out of phase with the motion of the blade and will reduce its amplitude; however, successive impulses may increase the amplitude on subsequent passes until failure occurs. Damping can be increased by placing a number of loose wires in a cylindrical hole cut in the blade in a radial direction. The damping of a number of these wires has been computed in terms of the geometry of the application (number of wires, density of wires, size of the hole, radius of the blade, rotational speed, etc.) and the amplitude of vibration.[24] These computations show reasonable agreement with experimental results.

An auxiliary mass has been used to damp the cutting tool chatter set up in a boring bar.[25] Because of the characteristics of the metal-cutting process or of some coupling between motions of the tool parallel and perpendicular to the work face, it is sometimes found that a self-excited vibration is initiated at the natural frequency of the cutter system. Since the self-excitation energy is low, the vibration usually is initiated only if the damping is small. Chatter of the tool is most common in long, poorly supported tools, such as boring bars (see Chap. 40). To eliminate this chatter, a loose auxiliary mass is incor-

FIG. 6.46. Application of auxiliary mass damper to reduce chatter in boring bar.

porated in the boring bar, as shown in Fig. 6.46. This may be air-damped or fluid-damped. Since the excitation is at the natural frequency of the tool, the damping should be such that the tool vibrates with a minimum amplitude at this frequency. The damping requirement can be estimated by substituting $\beta = 1$ in Eq. (6.25),

$$\frac{x_0}{\delta_{st}} = \sqrt{\frac{1 + 4(\zeta\alpha)^2}{4(\zeta\alpha)^2\mu^2}} \tag{6.77}$$

The optimum value of the parameter $(\zeta\alpha)$ is infinity. Thus when the frequency of excitation is constant, a greater reduction in amplitude can be obtained by a shift in natural

FIG. 6.47. Schematic diagram of damper shown in Fig. 6.46. The arrangement is shown at (A), and the forces acting on the boring bar and auxiliary mass are shown at (B).

frequency than by damping. However, such a shift cannot be attained because the frequency of the excitation always coincides with the natural frequency of the complete system. Instead, a better technique is to determine the damping that gives the maximum decrement of the free vibration.

Let the boring bar and damper be represented by a single degree-of-freedom system with a damper mass coupled to the main mass by viscous damping, as shown in Fig. 6.47A. The forces acting on the masses are shown in Fig. 6.47B. The equations of motion are

$$-kx_1 - c\dot{x}_1 + c\dot{x}_2 = m_1\ddot{x}_1$$
$$c\dot{x}_1 - c\dot{x}_2 = m_2\ddot{x}_2 \tag{6.78}$$

Substituting $x = Ae^{st}$, the resulting frequency equation is

$$s^3 + \frac{c(m_1 + m_2)}{m_1 m_2}s^2 + \frac{k}{m_1}s + \frac{kc}{m_1 m_2} = 0 \tag{6.79}$$

Where chatter occurs, this equation has three roots, one real and two complex. The complex roots correspond to decaying free vibrations. Let the roots be as follows:

$$\alpha_1, \quad \alpha_2 + j\beta, \quad \alpha_2 - j\beta$$

The value of β determines the frequency of the free vibration, and the value of α_2 determines the decrement (rate of decrease of amplitude) of the free vibration. The decrement α_2 is of primary interest; it is most easily found from the conditions that when the coefficient of s^3 is unity, (1) the sum of the roots is equal to the negative of the coefficient of s^2, (2) the sum of the products of the roots taken two at a time is the negative of the coefficient of s, and (3) the product of the roots is the negative of the constant term. The equations thus obtained are

$$\alpha_1 + 2\alpha_2 = -\frac{c(1 + \mu)}{\mu m_1} \tag{6.80}$$

$$2\alpha_1\alpha_2 + \alpha_2{}^2 + \beta^2 = -\omega_n{}^2 \tag{6.81}$$

$$\alpha_1(\alpha_2{}^2 + \beta^2) = -\omega_n{}^2 \frac{c}{m_1\mu} \tag{6.82}$$

where $\omega_n{}^2 = k/m_1$ and $\mu = m_2/m_1$. It is not practical to find the optimum damping by solving these equations for α_2 and then setting the derivative of α_2 with respect to c equal to zero. However, it is possible to find the optimum damping by the following process. Eliminate $(\alpha_2{}^2 + \beta^2)$ between Eqs. (6.81) and (6.82) to obtain

$$2\alpha_1{}^2\alpha_2 = \omega_n{}^2 \left(\frac{c}{\mu m_1} - \alpha_1 \right) \tag{6.83}$$

Substituting the value of α_1 from Eq. (6.80) in Eq. (6.83),

$$2\alpha_2 \left[2\alpha_2 + \frac{c(1 + \mu)}{\mu m_1} \right]^2 = \frac{c\omega_n{}^2}{\mu m_1} + \omega_n{}^2 \left[2\alpha_2 + \frac{c(1 + \mu)}{\mu m_1} \right] \tag{6.84}$$

To find the damping that gives the maximum decrement, differentiate with respect to c and set $d\alpha_2/dc = 0$:

$$2\alpha_2 \left[2\alpha_2 + \frac{c(1 + \mu)}{\mu m_1} \right] = \tfrac{1}{2}\omega_n{}^2 \frac{2 + \mu}{1 + \mu} \tag{6.85}$$

Solving Eqs. (6.84) and (6.85) simultaneously,

$$c_{\text{opt}} = \frac{\mu^2 m_1 \omega_n}{2(1 + \mu)^{3/2}} \tag{6.86}$$

$$(\alpha_2)_{\text{opt}} = -\frac{(2 + \mu)\omega_n}{4(1 + \mu)^{1/2}} \tag{6.87}$$

These values may be obtained by proper choice of clearance between the auxiliary mass and the hole in which it is located. Air damping is preferable to oil because it

PRIMARY MASS
AUXILIARY MASS
AUXILIARY SPRING
FLUID DASHPOT
STEADY TABLE
BASE

FIG. 6.48. Application of auxiliary mass to spring-mounted table to reduce vibration of table. (*Macinante.*[26])

requires less clearance. Therefore the plug is not immobilized by the centrifugal forces that, with the rotating boring bar, become larger as the clearance is increased.

In precision measurements, it is necessary to isolate the instruments from effects of shock and vibration in the earth and to damp any oscillations that might be generated in the measuring instruments. A heavy spring-mounted table fitted with a heavy auxiliary mass that is attached to the table by a spring and submerged in an oil bath (Fig. 6.48) has proved to be effective.[26] In this example the table has a top surface of 13½ in. (34 cm) by 13½ in. (34 cm) and a height of 6 in. (15 cm). Each auxiliary mass weighs about 70 lb (32 kg). The springs for both the primary table and the auxiliary system are designed to give a natural frequency between 2 and 4 cps in both the horizontal and vertical directions. By trying different fluids in the bath, suitable damping may be obtained experimentally.

REFERENCES

1. Timoshenko, S.: "Vibration Problems in Engineering," p. 240, D. Van Nostrand Company, Inc., Princeton, N.J., 1937.
2. Den Hartog, J. P.: "Mechanical Vibrations," chap. III, McGraw-Hill Book Company, Inc., New York, 1956.
3. Ormondroyd, J., and J. P. Den Hartog: *Trans. ASME*, **50**:A9 (1928).
4. Brock, J. E.: *J. Appl. Mechanics*, **13**(4):A-284 (1946).
5. Brock, J. E.: *J. Appl. Mechanics*, **16**(1):86 (1949).
6. Saver, F. M., and C. F. Garland: *J. Appl. Mechanics*, **16**(2):109 (1949).
7. Lewis, F. M.: *J. Appl. Mechanics*, **22**(3):377 (1955).
8. Georgian, J. C.: *Trans. ASME*, **16**:389 (1949).
9. Den Hartog, J. P., and J. Ormondroyd: *Trans. ASME*, **52**:133 (1930).
10. Roberson, R. E.: *J. Franklin Inst.*, **254**:205 (1952).
11. Pipes, L. A.: *J. Appl. Mechanics*, **20**:515 (1953).
12. Arnold, F. R.: *J. Appl. Mechanics*, **22**:487 (1955).
13. Hort, W.: "Technische Schwingungslehre," 2d ed., Springer-Verlag, Berlin, 1922.
14. Sperry, E. E.: *Trans. SNAME*, **30**:201 (1912).
15. Solomon, B.: *Proc. 4th Intern. Congr. Appl. Mechanics, Cambridge, England*, 1934.
16. Taylor, E. S.: *Trans. SAE*, **44**:81 (1936).
17. Den Hartog, J. P.: "Stephen Timoshenko 60th Anniversary Volume," The Macmillan Company, New York, 1939.
18. Porter, F. P.: "Evaluation of Effects of Torsional Vibration," SAE War Engineering Board, SAE, New York, 1945, p. 269.
19. Crossley, F. R. E.: *J. Appl. Mechanics*, **20**(1):41 (1953).
20. Reed, F. E.: *J. Appl. Mechanics*, **16**:190 (1949).
21. Constanti, M.: *Trans. Inst. of Naval Arch.*, **80**:181 (1938).
22. Crede, C. E.: *Trans. ASME*, **69**:937 (1947).
23. Brown, G. S., and D. P. Campbell: "Principles of Servomechanisms," John Wiley & Sons, Inc., New York, 1948.
24. DiTaranto, R. A.: *J. Appl. Mechanics*, **25**(1):21 (1958)
25. Hahn, R. S.: *Trans. ASME*, **73**:331 (1951).
26. Macinante, J. A.: *J. Sci. Instr.*, **35**:224 (1958)

7

VIBRATION OF SYSTEMS HAVING
DISTRIBUTED MASS AND ELASTICITY

William F. Stokey
Carnegie-Mellon University

INTRODUCTION

Preceding chapters consider the vibration of lumped parameter systems; i.e., systems that are idealized as rigid masses joined by massless springs and dampers. Many engineering problems are solved by analyses based on ideal models of an actual system, giving answers that are useful though approximate. In general, more accurate results are obtained by increasing the number of masses, springs, and dampers; i.e., by increasing the number of degrees-of-freedom. As the number of degrees-of-freedom is increased without limit, the concept of the system with distributed mass and elasticity is formed. This chapter discusses the free and forced vibration of such systems. Types of systems include rods vibrating in torsional modes and in tension-compression modes, and beams and plates vibrating in flexural modes. Particular attention is given to the calculation of the natural frequencies of such systems for further use in other analyses. Numerous charts and tables are included to define in readily available form the natural frequencies of systems commonly encountered in engineering practice.

FREE VIBRATION

DEGREES-OF-FREEDOM. Systems for which the mass and elastic parts are lumped are characterized by a finite number of degrees-of-freedom. In physical systems, all elastic members have mass, and all masses have some elasticity; thus, all real systems have distributed parameters. In making an analysis, it is often assumed that real systems have their parameters lumped. For example, in the analysis of a system consisting of a mass and a spring, it is commonly assumed that the mass of the spring is negligible so that its only effect is to exert a force between the mass and the support to which the spring is attached, and that the mass is perfectly rigid so that it does not deform and exert any elastic force. The effect of the mass of the spring on the motion of the system may be considered in an approximate way, while still maintaining the assumption of one degree-of-freedom, by assuming that the spring moves so that the deflection of each of its elements can be described by a single parameter. A commonly used assumption is that the deflection of each section of the spring is proportional to its distance from the support, so that if the deflection of the mass is given, the deflection of any part of the spring is defined. For the exact solution of the problem, even though the mass is considered to be perfectly rigid, it is necessary to consider that the deformation of the spring can occur in any manner consistent with the requirements of physical continuity.

Systems with distributed parameters are characterized by having an infinite number of degrees-of-freedom. For example, if an initially straight beam deflects laterally, it may be necessary to give the deflection of each section along the beam in order to define completely the configuration. For vibrating systems, the coordinates usually are defined in such a way that the deflections of the various parts of the system from the equilibrium position are given.

NATURAL FREQUENCIES AND NORMAL MODES OF VIBRATION. The number of natural frequencies of vibration of any system is equal to the number of degrees-of-freedom; thus, any system having distributed parameters has an infinite number of natural frequencies. At a given time, such a system usually vibrates with appreciable amplitude at only a limited number of frequencies, often at only one. With each natural frequency is associated a shape, called the normal or natural mode, which is assumed by the system during free vibration at the frequency. For example, when a uniform beam with simply supported or hinged ends vibrates laterally at its lowest or fundamental natural frequency, it assumes the shape of a half sine wave; this is a normal mode of vibration. When vibrating in this manner, the beam behaves as a system with a single degree-of-freedom, since its configuration at any time can be defined by giving the deflection of the center of the beam. When any linear system, i.e., one in which the elastic restoring force is proportional to the deflection, executes free vibration in a single natural mode, each element of the system except those at the supports and nodes executes simple harmonic motion about its equilibrium position. All possible free vibration of any linear system is made up of superposed vibrations in the normal modes at the corresponding natural frequencies. The total motion at any point of the system is the sum of the motions resulting from the vibration in the respective modes.

There are always nodal points, lines, or surfaces, i.e., points which do not move, in each of the normal modes of vibration of any system. For the fundamental mode, which corresponds to the lowest natural frequency, the supported or fixed points of the system usually are the only nodal points; for other modes, there are additional nodes. In the modes of vibration corresponding to the higher natural frequencies of some systems, the nodes often assume complicated patterns. In certain problems involving forced vibrations, it may be necessary to know what the nodal patterns are, since a particular mode usually will not be excited by a force acting at a nodal point. Nodal lines are shown in some of the tables.

METHODS OF SOLUTION. The complete solution of the problem of free vibration of any system would require the determination of all the natural frequencies and of the mode shape associated with each. In practice, it often is necessary to know only a few of the natural frequencies, and sometimes only one. Usually the lowest frequencies are the most important. The exact mode shape is of secondary importance in many problems. This is fortunate, since some procedures for finding natural frequencies involve assuming a mode shape from which an approximation to the natural frequency can be found.

Classical Method. The fundamental method of solving any vibration problem is to set up one or more equations of motion by the application of Newton's second law of motion. For a system having a finite number of degrees-of-freedom, this procedure gives one or more ordinary differential equations. For systems having distributed parameters partial differential equations are obtained. Exact solutions of the equations are possible for only a relatively few configurations. For most problems other means of solution must be employed.

Rayleigh's and Ritz's Methods. For many elastic bodies, Rayleigh's method is useful in finding an approximation to the fundamental natural frequency. While it is possible to use the method to estimate some of the higher natural frequencies, the accuracy often is poor; thus, the method is most useful for finding the fundamental frequency. When any elastic system without damping vibrates in its fundamental normal mode, each part of the system executes simple harmonic motion about its equilibrium position. For example, in lateral vibration of a beam the motion can be expressed as $y = X(x) \sin \omega_n t$ where X is a function only of the distance along the length of the beam. For lateral vibration of a plate, the motion can be expressed as $w = W(x,y) \sin \omega_n t$ where x and y are the coordinates in the plane of the plate. The equations show that when the deflection from equilibrium is a maximum, all parts of the body are motionless. At that time all the energy associated with the vibration is in the form of elastic strain energy. When the body is passing through its equilibrium position, none of the vibrational energy is in the form of strain energy so that all of it is in the form of kinetic energy. For conservation of energy, the strain energy in the position of maximum deflection must equal the kinetic energy when passing through the equilibrium position. Rayleigh's method of finding the natural frequency is to compute these maximum energies, equate them, and solve for the frequency. When the kinetic-energy term is evaluated, the frequency always appears as a factor. Formulas for finding the strain and kinetic energies of rods, beams, and plates are given in Table 7.1.

If the deflection of the body during vibration is known exactly, Rayleigh's method gives the true natural frequency. Usually the exact deflection is not known, since its determination involves the solution of the vibration problem by the classical method. If the classical solution is available, the natural frequency is included in it, and nothing is gained by applying Rayleigh's method. In many problems for which the classical solution is not available, a good approximation to the deflection can be assumed on the basis of physical reasoning. If the strain and kinetic energies are computed using such an assumed shape, an approximate value for the natural frequency is found. The correctness of the approximate frequency depends on how well the assumed shape approximates the true shape.

In selecting a function to represent the shape of a beam or a plate, it is desirable to satisfy as many of the boundary conditions as possible. For a beam or plate supported at a boundary, the assumed function must be zero at that boundary; if the boundary is built in, the first derivative of the function must be zero. For a free boundary, if the conditions associated with bending moment and shear can be satisfied, better accuracy

Table 7.1. Strain and Kinetic Energies of Uniform Rods, Beams, and Plates

Member	Strain energy V	Kinetic energy T	
		General	Maximum *
Rod in tension or compression	$\dfrac{SE}{2}\displaystyle\int_0^l \left(\dfrac{\partial u}{\partial x}\right)^2 dx$	$\dfrac{S\gamma}{2g}\displaystyle\int_0^l \left(\dfrac{\partial u}{\partial t}\right)^2 dx$	$\dfrac{S\gamma\omega_n^2}{2g}\displaystyle\int_0^l V^2 \, dx$
Rod in torsion	$\dfrac{GI_p}{2}\displaystyle\int_0^l \left(\dfrac{\partial \phi}{\partial x}\right)^2 dx$	$\dfrac{I_p\gamma}{2g}\displaystyle\int_0^l \left(\dfrac{\partial \phi}{\partial t}\right)^2 dx$	$\dfrac{I_p\gamma\omega_n^2}{2g}\displaystyle\int_0^l \Phi^2 \, dx$
Beam in bending	$\dfrac{EI}{2}\displaystyle\int_0^l \left(\dfrac{\partial^2 y}{\partial x^2}\right)^2 dx$	$\dfrac{S\gamma}{2g}\displaystyle\int_0^l \left(\dfrac{\partial y}{\partial t}\right)^2 dx$	$\dfrac{S\gamma\omega_n^2}{2g}\displaystyle\int_0^l Y^2 \, dx$
Rectangular plate in bending [1]	$\dfrac{D}{2}\displaystyle\int_S\!\!\int \left\{ \left(\dfrac{d^2 w}{dx^2} + \dfrac{d^2 w}{dy^2}\right)^2 - 2(1-\mu)\left[\dfrac{\partial^2 w}{\partial x^2}\dfrac{\partial^2 w}{\partial y^2} - \left(\dfrac{\partial^2 w}{\partial x\,\partial y}\right)^2\right] \right\} dx\,dy$	$\dfrac{\gamma h}{2g}\displaystyle\int_S\!\!\int \left(\dfrac{\partial w}{\partial t}\right)^2 dx\,dy$	$\dfrac{\gamma h\omega_n^2}{2g}\displaystyle\int_S\!\!\int W^2 \, dx\,dy$
Circular plate (deflection symmetrical about center) [1]	$\pi D\displaystyle\int_0^a \left\{ \left(\dfrac{\partial^2 w}{\partial r^2} + \dfrac{1}{r}\dfrac{\partial w}{\partial r}\right)^2 - 2(1-\mu)\dfrac{\partial^2 w}{\partial r^2}\dfrac{1}{r}\dfrac{\partial w}{\partial r} \right\} r\,dr$	$\dfrac{\pi\gamma h}{g}\displaystyle\int_0^a \left(\dfrac{\partial w}{\partial t}\right)^2 r\,dr$	$\dfrac{\pi\gamma h\omega_n^2}{g}\displaystyle\int_0^a W^2 r\,dr$

u = longitudinal deflection of cross section of rod
ϕ = angle of twist of cross section of rod
y = lateral deflection of beam
w = lateral deflection of plate
    Capitals denote values at extreme deflection
    for simple harmonic motion.
l = length of rod or beam
a = radius of circular plate
h = thickness of beam or plate

S = area of cross section
I_p = polar moment of inertia
I = moment of inertia of beam
γ = weight density
E = modulus of elasticity
G = modulus of rigidity
μ = Poisson's ratio
$D = \dfrac{Eh^3}{12(1-\mu^2)}$

* This is the maximum kinetic energy in simple harmonic motion.

usually results. It can be shown [2] that the frequency that is found by using any shape except the correct shape always is higher than the actual frequency. Therefore, if more than one calculation is made, using different assumed shapes, the lowest computed frequency is closest to the actual frequency of the system.

In many problems for which a classical solution would be possible, the work involved is excessive. Often a satisfactory answer to such a problem can be obtained by the application of Rayleigh's method. In this chapter several examples are worked using both the classical method and Rayleigh's method. In all, Rayleigh's method gives a good approximation to the correct result with relatively little work. Many other examples of solutions to problems by Rayleigh's method are in the literature.[3, 4, 5]

Ritz's method is a refinement of Rayleigh's method. A better approximation to the fundamental natural frequency can be obtained by its use, and approximations to higher natural frequencies can be found. In using Ritz's method, the deflections which are assumed in computing the energies are expressed as functions with one or more undetermined parameters; these parameters are adjusted to make the computed frequency a minimum. Ritz's method has been used extensively for the determination of the natural frequencies of plates of various shapes, and is discussed in the section on the lateral vibrations of plates.

Lumped Parameters. A procedure that is useful in many problems for finding approximations to both the natural frequencies and the mode shapes is to reduce the system with distributed parameters to one having a finite number of degrees-of-freedom. This is done by lumping the parameters for each small region into an equivalent mass and elastic element. Several formalized procedures for doing this and for analyzing the re-

Table 7.2. Approximate Formulas for Natural Frequencies of Systems Having Both Concentrated and Distributed Mass

TYPE OF SYSTEM	NATURAL FREQUENCY $\omega_n = 2\pi f_n$	STIFFNESS
k, m M (spring with mass attached) SPRING WITH MASS ATTACHED	$\sqrt{\dfrac{k}{M + m/3}}$	$k = \dfrac{Gd^4}{8nD^3}$ D = COIL DIA d = WIRE DIA n = NUMBER OF TURNS
k_r, I_s I (circular rod with disc) CIRCULAR ROD, WITH DISC ATTACHED, IN TORSION	$\sqrt{\dfrac{k_r}{I + I_s/3}}$	$k_r = \dfrac{G\pi D^4}{32\,l}$ D = ROD DIAMETER l = ROD LENGTH
m/2 M m/2 UNIFORM SIMPLY SUPPORTED BEAM WITH MASS IN CENTER	$\sqrt{\dfrac{k}{M + m/2}}$	$k = \dfrac{48EI}{l^3}$ l = BEAM LENGTH I = MOMENT OF INERTIA
m M UNIFORM CANTILEVER BEAM WITH MASS ON END	$\sqrt{\dfrac{k}{M + 0.23m}}$	$k = \dfrac{3EI}{l^3}$ l = BEAM LENGTH I = MOMENT OF INERTIA

sulting systems are described in Chap. 28. If a system consists of a rigid mass supported by a single flexible member whose mass is not negligible, the elastic part of the system sometimes can be treated as an equivalent spring; i.e., some of its mass is lumped with the rigid mass. Formulas for several systems of this kind are given in Table 7.2.

ORTHOGONALITY. It is shown in Chap. 2 that the normal modes of vibration of a system having a finite number of degrees-of-freedom are orthogonal to each other. For a system of masses and springs having n degrees-of-freedom, if the coordinate system is selected in such a way that X_1 represents the amplitude of motion of the first mass, X_2 that of the second mass, etc., the orthogonality relations are expressed by $(n-1)$ equations as follows:

$$m_1 X_1{}^a X_1{}^b + m_2 X_2{}^a X_2{}^b + \cdots = \sum_{i=1}^{n} m_i X_i{}^a X_i{}^b = 0 \qquad [a \neq b]$$

where $X_1{}^a$ represents the amplitude of the first mass when vibrating only in the ath mode, $X_1{}^b$ the amplitude of the first mass when vibrating only in the bth mode, etc.

For a body such as a uniform beam whose parameters are distributed only lengthwise; i.e., in the X direction, the orthogonality between two normal modes is expressed by

$$\int_0^l \rho \phi_a(x) \phi_b(x) \, dx = 0 \qquad [a \neq b] \tag{7.1}$$

where $\phi_a(x)$ represents the deflection in the ath normal mode, $\phi_b(x)$ the deflection in the bth normal mode, and ρ the density.

For a system, such as a uniform plate, in which the parameters are distributed in two dimensions, the orthogonality condition is

$$\int_A \int \rho \phi_a(x,y) \phi_b(x,y) \, dx \, dy = 0 \qquad [a \neq b] \tag{7.2}$$

LONGITUDINAL AND TORSIONAL VIBRATIONS OF UNIFORM CIRCULAR RODS

Equations of Motion. A circular rod having a uniform cross section can execute longitudinal, torsional, or lateral vibrations, either individually or in any combination. The equations of motion for longitudinal and torsional vibrations are similar in form, and the solutions are discussed together. The lateral vibration of a beam having a uniform cross section is considered separately.

In analyzing the longitudinal vibration of a rod, only the motion of the rod in the longitudinal direction is considered. There is some lateral motion because longitudinal stresses induce lateral strains; however, if the rod is fairly long compared to its diameter, this motion has a minor effect.

Consider a uniform circular rod, Fig. 7.1A. The element of length dx, which is formed by passing two parallel planes A–A and B–B normal to the axis of the rod, is shown in Fig. 7.1B. When the rod executes only longitudinal vibration, the force acting on the

(A)   (B)   (C)

FIG. 7.1. (A) Rod executing longitudinal or torsional vibration. (B) Forces acting on element during longitudinal vibration. (C) Moments acting on element during torsional vibration.

face A–A is F, and that on face B–B is $F + (\partial F/\partial x)\, dx$. The net force acting to the right must equal the product of the mass of the element $(\gamma/g)S\, dx$ and its acceleration $\partial^2 u/\partial t^2$, where γ is the weight density, S the area of the cross section, and u the longitudinal displacement of the element during the vibration:

$$\left(F + \frac{\partial F}{\partial x}\, dx\right) - F = \frac{\partial F}{\partial x}\, dx = \left(\frac{\gamma}{g}\right) S\, dx\, \frac{\partial^2 u}{\partial t^2} \quad \text{or} \quad \frac{\partial F}{\partial x} = \frac{\gamma S}{g}\frac{\partial^2 u}{\partial t^2} \tag{7.3}$$

This equation is solved by expressing the force F in terms of the displacement. The elastic strain at any section is $\partial u/\partial x$, and the stress is $E\partial u/\partial x$. The force F is the product of the stress and the area, or $F = ES\, \partial u/\partial x$, and $\partial F/\partial x = ES\, \partial^2 u/\partial x^2$. Equation (7.3) becomes $Eu'' = \gamma/g\ddot{u}$, where $u'' = \partial^2 u/\partial x^2$ and $\ddot{u} = \partial^2 u/\partial t^2$. Substituting $a^2 = Eg/\gamma$,

$$a^2 u'' = \ddot{u} \tag{7.4}$$

The equation governing the torsional vibration of the circular rod is derived by equating the net torque acting on the element, Fig. 7.1C, to the product of the moment of inertia J and the angular acceleration $\ddot{\phi}$, ϕ being the angular displacement of the section. The torque on the section A–A is M and that on section B–B is $M + (\partial M/\partial x)\, dx$. By an analysis similar to that for the longitudinal vibration, letting $b^2 = Gg/\gamma$,

$$b^2 \phi'' = \ddot{\phi} \tag{7.5}$$

Solution of Equations of Motion. Since Eqs. (7.4) and (7.5) are of the same form, the solutions are the same except for the meaning of a and b. The solution of Eq. (7.5) is of the form $\phi = X(x)T(t)$ in which X is a function of x only and T is a function of t only. Substituting this in Eq. (7.5) gives $b^2 X'' T = X\ddot{T}$. By separating the variables,[6]

$$T = A\cos(\omega_n t + \theta)$$

$$X = C\sin\frac{\omega_n x}{b} + D\cos\frac{\omega_n x}{b} \tag{7.6}$$

The natural frequency ω_n can have infinitely many values, so that the complete solution of Eq. (7.5) is, combining the constants,

$$\phi = \sum_{n=1}^{n=\infty}\left(C_n\sin\frac{\omega_n x}{b} + D_n\cos\frac{\omega_n x}{b}\right)\cos(\omega_n t + \theta_n) \tag{7.7}$$

The constants C_n and D_n are determined by the end conditions of the rod and by the initial conditions of the vibration. For a built-in or clamped end of a rod in torsion, $\phi = 0$ and $X = 0$ because the angular deflection must be zero. The torque at any section of the shaft is given by $M = (GI_p)\phi'$, where GI_p is the torsional rigidity of the shaft; thus, for a free end, $\phi' = 0$ and $X' = 0$. For the longitudinal vibration of a rod, the boundary conditions are essentially the same; i.e., for a built-in end the displacement is zero $(u = 0)$ and for a free end the stress is zero $(u' = 0)$.

Example 7.1. The natural frequencies of the torsional vibration of a circular steel rod of 2-in. diameter and 24-in. length, having the left end built in and the right end free, are to be determined.

Solution. The built-in end at the left gives the condition $X = 0$ at $x = 0$ so that $D = 0$ in Eq. (7.6). The free end at the right gives the condition $X' = 0$ at $x = l$. For each mode of vibration, Eq. (7.6) is $\cos\dfrac{\omega_n l}{b} = 0$ from which $\dfrac{\omega_n l}{b} = \dfrac{\pi}{2}, \dfrac{3\pi}{2}, \dfrac{5\pi}{2}, \ldots$ Since $b^2 = \dfrac{Gg}{\gamma}$, the natural frequencies for the torsional vibration are

$$\omega_n = \frac{\pi}{2l}\sqrt{\frac{Gg}{\gamma}}, \frac{3\pi}{2l}\sqrt{\frac{Gg}{\gamma}}, \frac{5\pi}{2l}\sqrt{\frac{Gg}{\gamma}}, \cdots \quad \text{rad/sec}$$

For steel, $G = 11.5 \times 10^6$ lb/in.^{2} and $\gamma = 0.28$ lb/in.^{3} The fundamental natural frequency is

$$\omega_n = \frac{\pi}{2(24)} \sqrt{\frac{(11.5 \times 10^6)(386)}{0.28}} = 8,240 \text{ rad/sec} = 1,311 \text{ cps}$$

The remaining frequencies are 3, 5, 7, etc., times ω_n.

Since Eq. (7.4), which governs longitudinal vibration of the bar, is of the same form as Eq. (7.5), which governs torsional vibration, the solution for longitudinal vibration is the same as Eq. (7.7) with u substituted for ϕ and $a = \sqrt{Eg/\gamma}$ substituted for b. The natural frequencies of a uniform rod having one end built in and one end free are obtained by substituting a for b in the frequency equations found above in Example 7.1:

$$\omega_n = \frac{\pi}{2l} \sqrt{\frac{Eg}{\gamma}}, \ \frac{3\pi}{2l} \sqrt{\frac{Eg}{\gamma}}, \ \frac{5\pi}{2l} \sqrt{\frac{Eg}{\gamma}}, \ \cdots$$

The frequencies of the longitudinal vibration are independent of the lateral dimensions of the bar, so that these results apply to uniform noncircular bars. Equation (7.5) for torsional vibration is valid only for circular cross sections.

TORSIONAL VIBRATIONS OF CIRCULAR RODS WITH DISCS ATTACHED.

An important type of system is that in which a rod which may twist has mounted on it one or more rigid discs or members that can be considered as the equivalents of discs. Many systems can be approximated by such configurations. If the moment of inertia of the rod is small compared to the moments of inertia of the discs, the mass of the rod may be neglected and the system considered to have a finite number of degrees-of-freedom. Then the methods described in Chaps. 2 and 38 are applicable. Even if the moment of inertia of the rod is not negligible, it usually may be lumped with the moment of inertia of the disc. For a shaft having a single disc attached, the formula in Table 7.2 gives a close approximation to the true frequency.

The exact solution of the problem requires that the effect of the distributed mass of the rod be considered. Usually it can be assumed that the discs are rigid enough that their elasticity can be neglected; only such systems are considered. Equation (7.5) and its solution, Eq. (7.7), apply to the shaft where the constants are determined by the end conditions. If there are more than two discs, the section of shaft between each pair of discs must be considered separately; there are two constants for each section. The constants are determined from the following conditions:

1. For a disc at an end of the shaft, the torque of the shaft at the disc is equal to the product of the moment of inertia of the disc and its angular acceleration.

2. Where a disc is between two sections of shaft, the angular deflection at the end of each section adjoining the disc is the same; the difference between the torques in the two sections is equal to the product of the moment of inertia of the disc and its angular acceleration.

Example 7.2. The fundamental frequency of vibration of the system shown in Fig. 7.2 is to be calculated and the result compared with the frequency obtained by considering that each half of the system is a simple shaft-disc system with the end of the shaft fixed. The system consists of a steel shaft 24 in. long and 4 in. in diameter having attached to it at each end a rigid steel disc 12 in. in diameter and 2 in. thick. For the approximation, add one-third of the moment of inertia of half the shaft to that of the disc (Table 7.2). (Because of symmetry, the center of the shaft is a nodal point; i.e., it does not move. Thus, each half of the system can be considered as a rod-disc system.)

Fig. 7.2. Rod with disc attached at each end.

Exact Solution. The boundary conditions are: at $x = 0$, $M = GI_p\phi' = I_1\ddot{\phi}$; at $x = l$, $M = GI\phi' = -I_2\ddot{\phi}$, where I_1 and I_2 are the moments of inertia of the discs. The signs are opposite for the two boundary conditions because, if the shaft is twisted in a certain direction, it will tend to accelerate the disc at the left end in one direction and the disc at the right end in the other. In the present example, $I_1 = I_2$; however, the solution is carried out in general terms.

Using Eq. (7.7), the following is obtained for each value of n:

$$\phi' = \frac{\omega_n}{b} \left(C \cos \frac{\omega_n x}{b} - D \sin \frac{\omega_n x}{b} \right) \cos (\omega_n t + \theta)$$

$$\ddot{\phi} = \omega_n^2 \left(C \sin \frac{\omega_n x}{b} + D \cos \frac{\omega_n x}{b} \right) [- \cos (\omega_n t + \theta)]$$

The boundary conditions give the following:

$$GI_p \frac{\omega_n}{b} C = -\omega_n^2 D I_1 \qquad \text{or} \qquad C = -\frac{b\omega_n I_1}{GI_p} D$$

$$\frac{\omega_n}{b} GI_p \left(C \cos \frac{\omega_n l}{b} - D \sin \frac{\omega_n l}{b} \right) = \omega_n^2 I_2 \left(C \sin \frac{\omega_n l}{b} + D \cos \frac{\omega_n l}{b} \right)$$

These two equations can be combined to give

$$-\frac{\omega_n}{b} GI_p \left(\frac{b\omega_n I_1}{GI_p} \cos \frac{\omega_n l}{b} + \sin \frac{\omega_n l}{b} \right) = \omega_n^2 I_2 \left(-\frac{b\omega_n I_1}{GI_p} \sin \frac{\omega_n l}{b} + \cos \frac{\omega_n l}{b} \right)$$

The preceding equation can be reduced to

$$\tan \alpha_n = \frac{(c + d)\alpha_n}{cd\alpha_n^2 - 1} \tag{7.8}$$

where $\alpha_n = (\omega_n l)/b$, $c = I_1/I_s$, $d = I_2/I_s$, and I_s is the polar moment of inertia of the shaft as a rigid body. There is a value for X in Eq. (7.6) corresponding to each root of Eq. (7.8) so that Eq. (7.7) becomes

$$\theta = \sum_{n=1}^{n=\infty} A_n \left(\cos \frac{\omega_n x}{b} - c\alpha_n \sin \frac{\omega_n x}{b} \right) \cos (\omega_n t + \theta_n)$$

For a circular disc or shaft, $I = \frac{1}{2}mr^2$ where m is the total mass; thus $c = d = \frac{D^4}{d^4} \frac{h}{l} = 6.75$. Equation (7.8) becomes $(45.56\alpha_n^2 - 1) \tan \alpha_n = 13.5\alpha_n$, the lowest root of which is $\alpha_n = 0.538$. The natural frequency is $\omega_n = 0.538\sqrt{Gg/\gamma l^2}$ rad/sec.

Approximate Solution. From Table 7.2, the approximate formula is

$$\omega_n = \left(\frac{k_r}{I + I_s/3} \right)^{\frac{1}{2}} \qquad \text{where } k_r = \frac{\pi d^4}{32} \frac{G}{l}$$

For the present problem where the center of the shaft is a node, the values of moment of inertia I_s and torsional spring constant for half the shaft must be used:

$$\frac{1}{2}I_s = \frac{\pi d^4}{32} \frac{\gamma}{g} \frac{l}{2} \qquad \text{and} \qquad k_r = 2 \left[\frac{\pi d^4}{32} \frac{G}{l} \right]$$

From the previous solution:

$$I_1 = 6.75 I_s \qquad I_1 + \frac{1}{2} \left(\frac{I_s}{3} \right) = \frac{\pi d^4}{32} \frac{\gamma}{g} \frac{l}{2} [2(6.75) + 0.333]$$

Substituting these values into the frequency equation and simplifying gives

$$\omega_n = 0.538 \sqrt{\frac{Gg}{\gamma l^2}}$$

In this example, the approximate solution is correct to at least three significant figures. For larger values of I_s/I, poorer accuracy can be expected.

For steel, $G = 11.5 \times 10^6$ lb/in.^{2} and $\gamma = 0.28$ lb/in.^{3}; thus

$$\omega_n = 0.538 \sqrt{\frac{(11.5 \times 10^6)(386)}{(0.28)(24)^2}} = 0.538 \times 5,245 = 2,822 \text{ rad/sec} = 449 \text{ cps}$$

LONGITUDINAL VIBRATION OF A ROD WITH MASS ATTACHED. The natural frequencies of the longitudinal vibration of a uniform rod having rigid masses attached to it can be solved in a manner similar to that used for a rod in torsion with discs attached. Equation (7.4) applies to this system; its solution is the same as Eq. (7.7) with a substituted for b. For each value of n,

$$u = \left(C_n \sin \frac{\omega_n x}{a} + D_n \cos \frac{\omega_n x}{a} \right) \cos (\omega_n t + \theta)$$

In Fig. 7.3, the rod of length l is fixed at $x = 0$ and has a mass m_2 attached at $x = l$. The boundary conditions are: at $x = 0$, $u = 0$ and at $x = l$, $SEu' = -m_2\ddot{u}$. The latter expresses the condition that the force in the bar equals the product of the mass and its acceleration at the end with the mass attached. The sign is negative because the force is tensile or positive when the acceleration of the mass is negative. From the first boundary condition, $D_n = 0$. The second boundary condition gives

$$\frac{\omega_n SE}{a} C_n \cos \frac{\omega_n l}{a} = m_2 \omega_n{}^2 C_n \sin \frac{\omega_n l}{a}$$

from which

$$\frac{SEl}{m_2 a^2} = \frac{\omega_n l}{a} \tan \frac{\omega_n l}{a}$$

Since $a^2 = Eg/\gamma$, this can be written

$$\frac{m_1}{m_2} = \frac{\omega_n l}{a} \tan \frac{\omega_n l}{a}$$

FIG. 7.3. Rod, with mass attached to end, executing longitudinal vibration.

where m_1 is the mass of the rod. This equation can be applied to a simple mass-spring system by using the relation that the constant k of a spring is equivalent to SE/l for the rod, so that $l/a = (m_1/k)^{\frac{1}{2}}$, where m_1 is the mass of the spring:

$$\frac{m_1}{m_2} = \omega_n \sqrt{\frac{m_1}{k}} \tan \omega_n \sqrt{\frac{m_1}{k}} \tag{7.9}$$

Rayleigh's Method. An accurate approximation to the fundamental natural frequency of this system can be found by using Rayleigh's method. The motion of the mass can be expressed as $u_m = u_0 \sin \omega t$. If it is assumed that the deflection u at each section of the rod is proportional to its distance from the fixed end, $u = u_0(x/l) \sin \omega_n t$. Using this relation in the appropriate equation from Table 7.1, the strain energy V of the rod at maximum deflection is

$$V = \frac{SE}{2} \int_0^l \left(\frac{\partial u}{\partial x} \right)^2 dx = \frac{SE}{2} \int_0^l \left(\frac{u_0}{l} \right)^2 dx = \frac{SEu_0{}^2}{2l}$$

The maximum kinetic energy T of the rod is

$$T = \frac{S\gamma}{2g} \int_0^l V^2{}_{\max} dx = \frac{S\gamma}{2g} \int_0^l \left(\omega_n u_0 \frac{x}{l} \right)^2 dx = \frac{S\gamma}{2g} \omega_n{}^2 u_0{}^2 \frac{l}{3}$$

The maximum kinetic energy of the mass is $T_m = m_2 \omega_n{}^2 u_0{}^2/2$. Equating the total maximum kinetic energy $T + T_m$ to the maximum strain energy V gives

$$\omega_n = \left(\frac{SE}{l(m_2 + m_1/3)} \right)^{\frac{1}{2}}$$

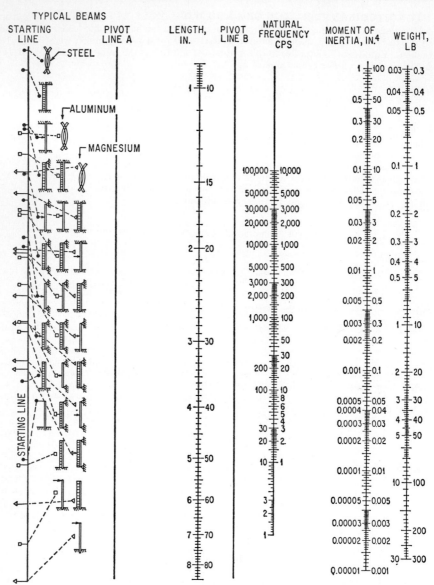

FIG. 7.4. Nomograph for determining fundamental natural frequencies of beams. From the point on the starting line which corresponds to the loading and support conditions for the beam, a straight line is drawn to the proper point on the length line. (If the length appears on the left side of this line, subsequent readings on all lines are made to the left; and if the length appears to the right, subsequent readings are made to the right.) From the intersection of this line with pivot line A, a straight line is drawn to the moment of inertia line; from the intersection of this line with pivot line B, a straight line is drawn to the weight line. (For concentrated loads, the weight is that of the load; for uniformly distributed loads, the weight is the total load on the beam, including the weight of the beam.) The natural frequency is read where the last line crosses the natural frequency line. (*J. J. Kerley.*[7])

where $m_1 = S\gamma l/g$ is the mass of the rod. Letting $SE/l = k$,

$$\omega_n = \sqrt{\frac{k}{M + m/3}} \tag{7.10}$$

This formula is included in Table 7.2. The other formulas in that table are also based on analyses by the Rayleigh method.

Example 7.3. The natural frequency of a simple mass-spring system for which the weight of the spring is equal to the weight of the mass is to be calculated and compared to the result obtained by using Eq. (7.10).

Solution. For $m_1/m_2 = l$, the lowest root of Eq. (7.9) is $\omega_n\sqrt{m/k} = 0.860$. When $m_2 = m_1$,

$$\omega_n = 0.860 \sqrt{\frac{k}{m_2}}$$

Using the approximate equation,

$$\omega_n = \sqrt{\frac{k}{m_2(1 + \frac{1}{3})}} = 0.866 \sqrt{\frac{k}{m_2}}$$

LATERAL VIBRATION OF STRAIGHT BEAMS

Natural Frequencies from Nomograph. For many practical purposes the natural frequencies of uniform beams of steel, aluminum, and magnesium can be determined with sufficient accuracy by the use of the nomograph, Fig. 7.4. This nomograph applies to many conditions of support and several types of load. Figure 7.4A indicates the procedure for using the nomograph.

GIVEN:
1. E = 30 × 10⁶ (STEEL) PSI
2. BEAM: DOUBLE END FIXITY
3. I = 10 IN.⁴
4. l = 80 IN.
5. W = 100 LB

GIVEN:
1. E = 10.5 × 10⁶ (ALUM) PSI
2. I = 0.04 IN.⁴
3. l = 8 IN.
4. W = 20 LB
5. DOUBLE END FIXITY

Fig. 7.4A. Example of use of Fig. 7.4. The natural frequency of the steel beam is 105 cps and that of the aluminum beam is 280 cps. (*J. J. Kerley.*[7])

Classical Solution.   In the derivation of the necessary equation, use is made of the relation

$$EI \frac{d^2y}{dx^2} = M \tag{7.11}$$

This equation relates the curvature of the beam to the bending moment at each section of the beam. This equation is based upon the assumptions that the material is homogeneous, isotropic and obeys Hooke's law, and that the beam is straight and of uniform cross section. The equation is valid for small deflections only, and for beams that are long compared to cross-sectional dimensions since the effects of shear deflection are neg-

(A)   (B)

FIG. 7.5. (A) Beam executing lateral vibration. (B) Element of beam showing shear forces and bending moments.

lected. The effects of shear deflection and rotation of the cross sections are considered later.

The equation of motion for lateral vibration of the beam shown in Fig. 7.5A is found by considering the forces acting on the element, Fig. 7.5B, which is formed by passing two parallel planes A–A and B–B through the beam normal to the longitudinal axis. The vertical elastic shear force acting on section A–A is V, and that on section B–B is V + (∂V/∂x) dx. Shear forces acting as shown are considered to be positive. The total vertical elastic shear force at each section of the beam is composed of two parts: that caused by the static load including the weight of the beam and that caused by the vibration. The part of the shear force caused by the static load exactly balances the load, so that these forces need not be considered in deriving the equation for the vibration if all deflections are measured from the position of equilibrium of the beam under the static load. The sum of the remaining vertical forces acting on the element must equal the product of the mass of the element $S\gamma/g\,dx$ and the acceleration $\partial^2y/\partial t^2$ in the lateral direction: $V + (\partial V/\partial x)\,dx - V = (\partial V/\partial x)\,dx = -(S\gamma/g)(\partial^2y/\partial t^2)\,dx$, or

$$\frac{\partial V}{\partial x} = -\frac{\gamma S}{g}\frac{\partial^2 y}{\partial t^2} \tag{7.12}$$

If moments are taken about point 0 of the element in Fig. 7.5B, $V\,dx = (\partial M/\partial x)\,dx$, and $V = \partial M/\partial x$. Other terms contain differentials of higher order, and can be neglected. Substituting this in Eq. (7.12) gives $-\partial^2 M/\partial x^2 = (S\gamma/g)(\partial^2y/\partial t^2)$. Substituting Eq. (7.11) gives

$$-\frac{\partial^2}{\partial x^2}\left(EI\frac{\partial^2 y}{\partial x^2}\right) = \frac{\gamma S}{g}\frac{\partial^2 y}{\partial t^2} \tag{7.13}$$

Equation (7.13) is the basic equation for the lateral vibration of beams. The solution of this equation, if EI is constant, is of the form $y = X(x)\,[\cos(\omega_n t + \theta)]$, in which X is a function of x only. Substituting

$$\kappa^4 = \frac{\omega_n{}^2 \gamma S}{EIg} \tag{7.14}$$

and dividing Eq. (7.13) by $\cos(\omega_n t + \theta)$:

$$\frac{d^4 X}{dx^4} = \kappa^4 X \tag{7.15}$$

where X is any function whose fourth derivative is equal to a constant multiplied by the function itself. The following functions satisfy the required conditions and represent the solution of the equation:

$$X = A_1 \sin \kappa x + A_2 \cos \kappa x + A_3 \sinh \kappa x + A_4 \cosh \kappa x$$

The solution can also be expressed in terms of exponential functions, but the trigonometric and hyperbolic functions usually are more convenient to use.

For beams having various support conditions, the constants A_1, A_2, A_3, and A_4 are found from the end conditions. In finding the solutions, it is convenient to write the equation in the following form in which two of the constants are zero for each of the usual boundary conditions:

$$X = A \, (\cos \kappa x + \cosh \kappa x) + B(\cos \kappa x - \cosh \kappa x)$$
$$+ \, C(\sin \kappa x + \sinh \kappa x) + D(\sin \kappa x - \sinh \kappa x) \quad (7.16)$$

In applying the end conditions, the following relations are used where primes indicate successive derivatives with respect to x:

The deflection is proportional to X and is zero at any rigid support.
The slope is proportional to X' and is zero at any built-in end.
The moment is proportional to X'' and is zero at any free or hinged end.
The shear is proportional to X''' and is zero at any free end.

The required derivatives are:

$$X' = \kappa[A(- \sin \kappa x + \sinh \kappa x) + B(- \sin \kappa x - \sinh \kappa x)$$
$$+ \, C(\cos \kappa x + \cosh \kappa x) + D(\cos \kappa x - \cosh \kappa x)]$$

$$X'' = \kappa^2[A(- \cos \kappa x + \cosh \kappa x) + B(- \cos \kappa x - \cosh \kappa x)$$
$$+ \, C(- \sin \kappa x + \sinh \kappa x) + D(- \sin \kappa x - \sinh \kappa x)]$$

$$X''' = \kappa^3[A(\sin \kappa x + \sinh \kappa x) + B(\sin \kappa x - \sinh \kappa x)$$
$$+ \, C(- \cos \kappa x + \cosh \kappa x) + D(- \cos \kappa x - \cosh \kappa x)]$$

For the usual end conditions, two of the constants are zero, and there remain two equations containing two constants. These can be combined to give an equation which contains only the frequency as an unknown. Using the frequency, one of the unknown constants can be found in terms of the other. There always is one undetermined constant, which can be evaluated only if the amplitude of the vibration is known.

Example 7.4. The natural frequencies and modes of vibration of the rectangular steel beam shown in Fig. 7.6 are to be determined and the fundamental frequency compared with that obtained from Fig. 7.4. The beam is 24 in. long, 2 in. wide, and ¼ in. thick, with the left end built in and the right end free.

FIG. 7.6. First mode of vibration of beam with left end clamped and right end free.

Solution. The boundary conditions are: at $x = 0$, $X = 0$ and $X' = 0$; at $x = l$, $X'' = 0$ and $X''' = 0$. The first condition requires that $A = 0$ since the other constants are multiplied by zero at $x = 0$. The second condition requires that $C = 0$. From the third and fourth conditions, the following equations are obtained:

$$0 = B(- \cos \kappa l - \cosh \kappa l) + D(- \sin \kappa l - \sinh \kappa l)$$

$$0 = B(\sin \kappa l - \sinh \kappa l) + D(- \cos \kappa l - \cosh \kappa l)$$

Solving each of these for the ratio D/B and equating, or making use of the mathematical condition that for a solution the determinant of the two equations must vanish, the following equation results:

$$\frac{D}{B} = -\frac{\cos \kappa l + \cosh \kappa l}{\sin \kappa l + \sinh \kappa l} = \frac{\sin \kappa l - \sinh \kappa l}{\cos \kappa l + \cosh \kappa l} \quad (7.17)$$

Table 7.3. Natural Frequencies and Normal Modes of Uniform Beams

SUPPORTS	MODE n	(A) SHAPE AND NODES (NUMBERS GIVE LOCATION OF NODES IN FRACTION OF LENGTH FROM LEFT END)	(B) BOUNDARY CONDITIONS EQ (7.16)	(C) FREQUENCY EQUATION	(D) CONSTANTS EQ (7.16)	(E) kl EQ (7.14) $\omega_n = \kappa^2 \sqrt{\dfrac{EIg}{A\gamma}}$	(F) R RATIO OF NON-ZERO CONSTANTS COLUMN (D)
HINGED-HINGED	1	0.50	$x=0 \begin{cases} X=0 \\ X''=0 \end{cases}$	$\text{SIN } kl = 0$	$A=0$	3.1416	1.0000
	2	0.333 0.667			$B=0$	6.283	1.0000
	3	0.25 0.50 0.75			$\dfrac{C}{D}=1$	9.425	1.0000
	4		$x=l \begin{cases} X=0 \\ X''=0 \end{cases}$			12.566	1.0000
	n>4					$\approx n\pi$	1.0000
CLAMPED-CLAMPED	1	0.50	$x=0 \begin{cases} X=0 \\ X'=0 \end{cases}$	$(\text{COS } kl)$ $(\text{COSH } kl)$ $=1$	$A=0$	4.730	-0.9825
	2	0.359 0.641			$C=0$	7.853	-1.0008
	3	0.278 0.50 0.722			$\dfrac{D}{B}=R$	10.996	-1.0000-
	4		$x=l \begin{cases} X=0 \\ X'=0 \end{cases}$			14.137	-1.0000+
	n>4					$\approx \dfrac{(2n+1)\pi}{2}$	-1.0000-
CLAMPED-HINGED	1	0.558	$x=0 \begin{cases} X=0 \\ X'=0 \end{cases}$	$\text{TAN } kl =$ $\text{TANH } kl$	$A=0$	3.927	-1.0008
	2	0.386 0.692			$C=0$	7.069	-1.0000+
	3	0.294 0.529 0.765			$\dfrac{D}{B}=R$	10.210	-1.0000
	4		$x=l \begin{cases} X=0 \\ X''=0 \end{cases}$			13.352	-1.0000
	n>4					$\approx \dfrac{(4n+1)\pi}{4}$	-1.0000
CLAMPED-FREE	1	0.783	$x=0 \begin{cases} X=0 \\ X'=0 \end{cases}$	$(\text{COS } kl)$ $(\text{COSH } kl)$ $=-1$	$A=0$	1.875	-0.7341
	2	0.504 0.868			$C=0$	4.694	-1.0185
	3	0.358 0.644 0.906			$\dfrac{D}{B}=R$	7.855	-0.9992
	4		$x=l \begin{cases} X''=0 \\ X'''=0 \end{cases}$			10.996	-1.0000+
	n>4					$\approx \dfrac{(2n-1)\pi}{2}$	-1.0000-
FREE-FREE	1	0.224 0.776	$x=0 \begin{cases} X''=0 \\ X'''=0 \end{cases}$	$(\text{COS } kl)$ $(\text{COSH } kl)$ $=1$	$B=0$	0 (REPRESENTS TRANSLATION)	
	2	0.132 0.50 0.868			$D=0$	4.730	-0.9825
	3	0.094 0.356 0.644 0.906			$\dfrac{C}{A}=R$	7.853	-1.0008
	4	0.0734 0.277 0.50 0.723 0.927	$x=l \begin{cases} X''=0 \\ X'''=0 \end{cases}$			10.996	-1.0000-
	5					14.137	-1.0000+
	n>5					$\approx \dfrac{(2n-1)\pi}{2}$	-1.0000-

Equation (7.17) reduces to $\cos \kappa l \cosh \kappa l = -1$. The values of κl which satisfy this equation can be found by consulting tables of hyperbolic and trigonometric functions. The first five are: $\kappa_1 l = 1.875$, $\kappa_2 l = 4.694$, $\kappa_3 l = 7.855$, $\kappa_4 l = 10.996$, $\kappa_5 l = 14.137$. The corresponding frequencies of vibration are found by substituting the length of the beam to find each κ, and then solving Eq. (7.14) for ω_n:

$$\omega_n = \kappa_n{}^2 \sqrt{\frac{EIg}{\gamma S}}$$

For the rectangular section, $I = bh^3/12 = 1/384$ in.[4] and $S = bh = 0.5$ in.[2] For steel, $E = 30 \times 10^6$ lb/in.[2] and $\gamma = 0.28$ lb/in.[3] Using these values,

$$\omega_1 = \frac{(1.875)^2}{(24)^2} \sqrt{\frac{(30 \times 10^6)(386)}{(0.28)(384)(0.5)}} = 89.6 \text{ rad/sec} = 14.26 \text{ cps}$$

The remaining frequencies can be found by using the other values of κ. Using Fig. 7.4, the fundamental frequency is found to be about 12 cps.

To find the mode shapes, the ratio D/B is found by substituting the appropriate values of κl in Eq. (7.17). For the first mode:

$$\cosh 1.875 = 3.33710 \qquad \sinh 1.875 = 3.18373$$

$$\cos 1.875 = -0.29953 \qquad \sin 1.875 = 0.95409$$

Therefore, $D/B = -0.73410$. The equation for the first mode of vibration becomes

$$y = B_1[(\cos \kappa x - \cosh \kappa x) - 0.73410(\sin \kappa x - \sinh \kappa x)] \cos (\omega_1 t + \theta_1)$$

in which B_1 is determined by the amplitude of vibration in the first mode. A similar equation can be obtained for each of the modes of vibration; all possible free vibration of the beam can be expressed by taking the sum of these equations.

Frequencies and Shapes of Beams. Table 7.3 gives the information necessary for finding the natural frequencies and normal modes of vibration of uniform beams having various boundary conditions. The various constants in the table were determined by computations similar to those used in Example 7.4. The table includes (1) diagrams showing the modal shapes including node locations; (2) the boundary conditions; (3) the frequency equation that results from using the boundary conditions in Eq. (7.16); (4) the constants that become zero in Eq. (7.16); (5) the values of κl from which the natural frequencies can be computed by using Eq. (7.14); and (6) the ratio of the nonzero constants in Eq. (7.16). By the use of the constants in this table, the equation of motion for any normal mode can be written. There always is a constant which is determined by the amplitude of vibration.

Values of characteristic functions representing the deflections of beams, at fifty equal intervals, for the first five modes of vibration have been tabulated.[8] Functions are given for beams having various boundary conditions, and the first three derivatives of the functions are also tabulated.

Rayleigh's Method. This method is useful for finding approximate values of the fundamental natural frequencies of beams. In applying Rayleigh's method, a suitable function is assumed for the deflection, and the maximum strain and kinetic energies are calculated, using the equations in Table 7.1. These energies are equated and solved for the frequency. The function used to represent the shape must satisfy the boundary conditions associated with deflection and slope at the supports. Best accuracy is obtained if other boundary conditions are also satisfied. The equation for the static deflection of the beam under a uniform load is a suitable function, although a simpler function often gives satisfactory results with less numerical work.

Example 7.5. The fundamental natural frequency of the cantilever beam in Example 7.4 is to be calculated using Rayleigh's method.

Solution. The assumed deflection $Y = (a/3l^4)[x^4 - 4x^3l + 6x^2l^2]$ is the static deflection of a cantilever beam under uniform load and having the deflection $Y = a$ at $x = l$. This deflection satisfies the conditions that the deflection Y and the slope Y' be zero at $x = 0$. Also, at $x = l$, Y'' which is proportional to the moment and Y''' which is proportional to the shear are zero. The second derivative of the function is $Y'' = (4a/l^4)[x^2 - 2xl + l^2]$. Using this in the expression from Table 7.1, the maximum strain energy is

$$V = \frac{EI}{2} \int_0^l \left(\frac{d^2Y}{dx^2}\right)^2 dx = \frac{8}{5} \frac{EIa^2}{l^3}$$

The maximum kinetic energy is

$$T = \frac{\omega_n^2 \gamma S}{2g} \int_0^l Y^2 \, dx = \frac{52}{405} \frac{\omega_n^2 \gamma S l a^2}{g}$$

Equating the two energies and solving for the frequency,

$$\omega_n = \sqrt{\frac{162}{13} \times \frac{EIg}{\gamma Sl^4}} = \frac{3.530}{l^2} \sqrt{\frac{EIg}{\gamma S}}$$

The exact frequency as found in Example 4 is $(3.516/l^2)\sqrt{EIg/\gamma S}$; thus, Rayleigh's method gives good accuracy in this example.

If the deflection is assumed to be $Y = a[1 - \cos(\pi x/2l)]$, the calculated frequency is $(3.66/l^2)\sqrt{EIg/\gamma S}$. This is less accurate, but the calculations are considerably shorter. With this function, the same boundary conditions at $x = 0$ are satisfied; however, at $x = l$, $Y'' = 0$ but Y''' does not equal zero. Thus, the condition of zero shear at the free end is not satisfied. The trigonometric function would not be expected to give as good accuracy as the static deflection relation used in the example, although for most practical purposes the result would be satisfactory.

EFFECTS OF ROTARY MOTION AND SHEARING FORCE. In the preceding analysis of the lateral vibration of beams it has been assumed that each element of the beam moves only in the lateral direction. If each plane section that is initially normal to the axis of the beam remains plane and normal to the axis, as assumed in simple beam theory, then each section rotates slightly in addition to its lateral motion when the beam deflects.[9] When a beam vibrates, there must be forces to cause this rotation, and for a complete analysis these forces must be considered. The effect of this rotation is small except when the curvature of the beam is large relative to its thickness; this is true either for a beam that is short relative to its thickness or for a long beam vibrating in a higher mode so that the nodal points are close together.

Another factor that affects the lateral vibration of a beam is the lateral shear force. In Eq. (7.11) only the deflection associated with the bending stress in the beam is included. In any beam except one subject only to pure bending, a deflection due to the shear stress in the beam occurs. The exact solution of the beam vibration problem requires that this deflection be considered. The analysis of beam vibration including both the effects of rotation of the cross section and the shear deflection is called the "Timoshenko beam theory." The following equation governs such vibration: [10]

$$a^2 \frac{\partial^4 y}{\partial x^4} + \frac{\partial^2 y}{\partial t^2} - \rho^2\left(1 + \frac{E}{\varkappa G}\right)\frac{\partial^4 y}{\partial x^2 \partial t^2} + \rho^2 \frac{\gamma}{g\varkappa G}\frac{\partial^4 y}{\partial t^4} = 0 \qquad (7.18)$$

where $a^2 = EIg/S\gamma$, E = modulus of elasticity, G = modulus of rigidity, and $\rho = \sqrt{I/S}$, the radius of gyration; $\varkappa = F_s/GS\beta$, F_s being the total lateral shear force at any section and β the angle which a cross section makes with the axis of the beam because of shear deformation. Under the assumptions made in the usual elementary beam theory, $\varkappa$ is $\frac{2}{3}$ for a beam with a rectangular cross section and $\frac{3}{4}$ for a circular beam. More refined analysis shows [11] that, for the present purposes, $\varkappa = \frac{5}{6}$ and $\frac{9}{10}$ are more accurate values for rectangular and circular cross sections, respectively. Using a solution of the form $y = C \sin(n\pi x/l) \cos \omega_n t$, which satisfies the necessary end conditions, the following frequency equation is obtained for beams with both ends simply supported:

$$a^2 \frac{n^4\pi^4}{l^4} - \omega_n^2 - \omega_n^2 \frac{n^2\pi^2\rho^2}{l^2} - \omega_n^2 \frac{n^2\pi^2\rho^2}{l^2}\frac{E}{\varkappa G} + \frac{\rho^2\gamma}{g\varkappa G}\omega_n^4 = 0 \qquad (7.18a)$$

If it is assumed that $nr/l \ll 1$, Eq. (7.18a) reduces to

$$\omega_n = \frac{a\pi^2}{(l/n)^2}\left[1 - \frac{\pi^2 n^2}{2}\left(\frac{\rho}{l}\right)^2\left(1 + \frac{E}{\varkappa G}\right)\right] \qquad (7.18b)$$

When $nr/l < 0.08$, the approximate equation gives less than 5 per cent error in the frequency.[11]

Values of the ratio of ω_n to the natural frequency uncorrected for the effects of rotation and shear have been plotted,[11] using Eq. (7.18a) for three values of $E/\varkappa G$, and are shown in Fig. 7.7.

FIG. 7.7. Influence of shear force and rotary motion on natural frequencies of simply supported beams. The curves relate the corrected frequency to that given by Eq. (7.14). (*J. G. Sutherland and L. E. Goodman.*[11])

FIG. 7.8. Influence of shear force and rotary motion on natural frequencies of uniform cantilever beams ($E/\kappa G = 3.20$). The curves relate the corrected frequency to that given by Eq. (7.14). (*J. G. Sutherland and L. E. Goodman.*[11])

For a cantilever beam the frequency equation is quite complicated. For $E/\varkappa G = 3.20$, corresponding approximately to the value for rectangular steel or aluminum beams, the curves in Fig. 7.8 show the effects of rotation and shear on the natural frequencies of the first six modes of vibration.

Example 7.6. The first two natural frequencies of a rectangular steel beam 40 in. long, 2 in. wide, and 6 in. thick, having simply supported ends, are to be computed with and without including the effects of rotation of the cross sections and shear deflection.

Solution. For steel $E = 30 \times 10^6$ lb/in.^{2}, $G = 11.5 \times 10^6$ lb/in.^{2}, and for a rectangular cross section $\varkappa = 5/6$; thus $E/\varkappa G = 3.13$. For a rectangular beam $\rho = h/\sqrt{12}$ where h is the thickness; thus $\rho/l = 6/(40\sqrt{12}) = 0.0433$. The approximate frequency equation, Eq. (7.18b), becomes

$$\omega_n = \frac{a\pi^2}{(l/n)^2}\left[1 - \frac{\pi^2}{2}(0.0433n)^2(1 + 3.13)\right]$$

$$= \frac{a\pi^2}{(l/n)^2}(1 - 0.038n^2)$$

Letting $\omega_0 = a\pi^2/(l/n)^2$ be the uncorrected frequency obtained by neglecting the effect of n in Eq. (7.18b):

For $n = 1$:
$$\frac{\omega_n}{\omega_0} = 1 - 0.038 = 0.962$$

For $n = 2$:
$$\frac{\omega_n}{\omega_0} = 1 - 0.152 = 0.848$$

Comparing these results with Fig. 7.7, using the curve for $E/\varkappa G = 3.00$, the calculated frequency for the first mode agrees with the curve as closely as the curve can be read. For the second mode, the curve gives $\omega_n/\omega_0 = 0.91$; therefore the approximate equation for the second mode is not very accurate. The uncorrected frequencies are, since $I/S = \rho^2 = h^2/12$:

For $n = 1$: $\omega_0 = \dfrac{\pi^2}{l^2}\sqrt{\dfrac{EIg}{S\gamma}} = \dfrac{\pi^2}{(40)^2}\sqrt{\dfrac{(30 \times 10^6)(36)386}{(12)(0.28)}} = 2{,}170$ rad/sec $= 345$ cps

For $n = 2$: $\omega_0 = 345 \times 4 = 1{,}380$ cps

The frequencies corrected for rotation and shear, using the value from Fig. 7.7 for correction of the second mode, are:

For $n = 1$: $f_n = 345 \times 0.962 = 332$ cps

For $n = 2$: $f_n = 1{,}380 \times 0.91 = 1{,}256$ cps

EFFECT OF AXIAL LOADS. When an axial tensile or compressive load acts on a beam, the natural frequencies are different from those for the same beam without such load. The natural frequencies for a beam with hinged ends, as determined by an energy analysis, assuming that the axial force F remains constant, are [12]

$$\omega_n = \frac{\pi^2 n^2}{l^2}\sqrt{\frac{EIg}{S\gamma}}\sqrt{1 \pm \frac{\alpha^2}{n^2}} = \omega_0\sqrt{1 \pm \frac{\alpha^2}{n^2}}$$

where $\alpha^2 = Fl^2/EI\pi^2$, n is the mode number, ω_0 is the natural frequency of the beam with no axial force applied, and the other symbols are defined in Table 7.1. The plus sign is for a tensile force and the minus sign for a compressive force.

For a cantilever beam with a constant axial force F applied at the free end, the natural frequency is found by an energy analysis [13] to be $[1 + \frac{5}{14}(Fl^2/EI)]^{1/2}$ times the natural frequency of the beam without the force applied. If a uniform axial force is applied along the beam, the effect is the same as if about seven-twentieths of the total force were applied at the free end of the beam.

If the amplitude of vibration is large, an axial force may be induced in the beam by the supports. For example, if both ends of a beam are hinged but the supports are rigid enough so that they cannot move axially, a tensile force is induced as the beam deflects. The force is not proportional to the deflection; therefore, the vibration is of the type characteristic of nonlinear systems in which the natural frequency depends on the amplitude of vibration. The natural frequency of a beam having immovable hinged ends is given in the following table where the axial force is zero at zero deflection of the beam [14] and where x_0 is the amplitude of vibration, I the moment of inertia, and S the area of the cross section; ω_0 is the natural frequency of the unrestrained bar.

$\dfrac{x_0}{\sqrt{I/S}}$	0	0.1	0.2	0.4	0.6	0.8
$\dfrac{\omega_n}{\omega_0}$	1	1.0008	1.0038	1.015	1.038	1.058
$\dfrac{x_0}{\sqrt{I/S}}$	1.0	1.5	2	3	4	5
$\dfrac{\omega_n}{\omega_0}$	1.089	1.190	1.316	1.626	1.976	2.35

BEAMS HAVING VARIABLE CROSS SECTIONS. The natural frequencies for beams of several shapes having cross sections that can be expressed as functions of the distance along the beam have been calculated.[15] The results are shown in Table 7.4. In the analysis, Eq. (7.13) was used, with EI considered to be variable.

Rayleigh's method or Ritz's method can be used to find approximate values for the frequencies of such beams. The frequency equation becomes, using the equations in Table 7.1, and letting $Y(x)$ be the assumed deflection,

$$\omega_n{}^2 = \frac{Eg}{\gamma}\frac{\displaystyle\int_0^l I\left(\frac{d^2Y}{dx^2}\right)^2 dx}{\displaystyle\int_0^l SY^2\,dx}$$

where $I = I(x)$ is the moment of inertia of the cross section and $S = S(x)$ is the area of the cross section. Examples of the calculations are in the literature.[18] If the values of $I(x)$ and $S(x)$ cannot be defined analytically, the beam may be divided into two or more sections, for each of which I and S can be approximated by an equation. The strain and kinetic energies of each section may be computed separately, using an appropriate function for the deflection, and the total energies for the beam found by adding the values for the individual sections.

CONTINUOUS BEAMS ON MULTIPLE SUPPORTS. In finding the natural frequencies of a beam on multiple supports, the section between each pair of supports is considered as a separate beam with its origin at the left support of the section. Equation (7.16) applies to each section. Since the deflection is zero at the origin of each section, $A = 0$ and the equation reduces to

$$X = B(\cos \kappa x - \cosh \kappa x) + C(\sin \kappa x + \sinh \kappa x) + D(\sin \kappa x - \sinh \kappa x)$$

There is one such equation for each section, and the necessary end conditions are as follows:

1. At each end of the beam the usual boundary conditions are applicable, depending on the type of support.

2. At each intermediate support the deflection is zero. Since the beam is continuous, the slope and the moment just to the left and to the right of the support are the same.

General equations can be developed for finding the frequency for any number of spans.[19,20] Table 7.5 gives constants for finding the natural frequencies of uniform continuous beams on uniformly spaced supports for several combinations of end supports.

Table 7.4. Natural Frequencies of Variable-section Steel Beams
(J. N. Macduff and R. P. Felgar.[16,17])

BEAM STRUCTURE	$\dfrac{b}{b_0}$	$\dfrac{h}{h_0}$	$(f_n l^2/\rho)/10^4$		
			$n = 1$	$n = 2$	$n = 3$
	1	$\dfrac{x}{l}$	17.09	48.89	96.57
	$\dfrac{x}{l}$	$\dfrac{x}{l}$	26.08	68.08	123.64
	$\left(\dfrac{x}{l}\right)^{\frac{1}{2}}$	$\dfrac{x}{l}$	22.30	58.18	109.90
	$e^{x/l}$	1	15.23	77.78	206.07
	1	$\dfrac{x}{l}$	21.21* 35.05†	56.97	
	$\dfrac{x}{l}$	$\dfrac{x}{l}$	32.73* 49.50†	76.57	
	$\left(\dfrac{x}{l}\right)^{\frac{1}{2}}$	$\dfrac{x}{l}$	25.66* 42.02†	66.06	

* SYMMETRIC   † ANTISYMMETRIC

f_n = natural frequency, cps
$\rho = \sqrt{I/S}$ = radius of gyration, in.
h = depth of beam, in.

l = beam length, in.
n = mode number
b = width of beam, in.

For materials other than steel: $f_n = f_{n_s} \sqrt{\dfrac{E\gamma_s}{E_s\gamma}}$

E = modulus of elasticity, lb/in.^{2}
γ = density, lb/in.^{3}
Terms with subscripts refer to steel
Terms without subscripts refer to other material

Table 7.5. Natural Frequencies of Continuous Uniform Steel * Beams (J. N. Macduff and R. P. Felgar.[16,17])

Beam structure	$(f_n l^2/\rho)/10^4$					
	N	$n = 1$	$n = 2$	$n = 3$	$n = 4$	$n = 5$
Extreme Ends Simply Supported						
	1	31.73	126.94	285.61	507.76	793.37
	2	31.73	49.59	126.94	160.66	285.61
	3	31.73	40.52	59.56	126.94	143.98
	4	31.73	37.02	49.59	63.99	126.94
	5	31.73	34.99	44.19	55.29	66.72
	6	31.73	34.32	40.52	49.59	59.56
	7	31.73	33.67	38.40	45.70	53.63
	8	31.73	33.02	37.02	42.70	49.59
	9	31.73	33.02	35.66	40.52	46.46
	10	31.73	33.02	34.99	39.10	44.19
	11	31.73	32.37	34.32	37.70	41.97
	12	31.73	32.37	34.32	37.02	40.52
Extreme Ends Clamped						
	1	72.36	198.34	388.75	642.63	959.98
	2	49.59	72.36	160.66	198.34	335.20
	3	40.52	59.56	72.36	143.98	178.25
	4	37.02	49.59	63.99	72.36	137.30
	5	34.99	44.19	55.29	66.72	72.36
	6	34.32	40.52	49.59	59.56	67.65
	7	33.07	38.40	45.70	53.63	62.20
	8	33.02	37.02	42.70	49.59	56.98
	9	33.02	35.66	40.52	46.46	52.81
	10	33.02	34.99	39.10	44.19	49.59
	11	32.37	34.32	37.70	41.97	47.23
	12	32.37	34.32	37.02	40.52	44.94
Extreme Ends Clamped-Supported						
	1	49.59	160.66	335.2	573.21	874.69
	2	37.02	63.99	137.30	185.85	301.05
	3	34.32	49.59	67.65	132.07	160.66
	4	33.02	42.70	56.98	69.51	129.49
	5	33.02	39.10	49.59	61.31	70.45
	6	32.37	37.02	44.94	54.46	63.99
	7	32.37	35.66	41.97	49.59	57.84
	8	32.37	34.99	39.81	45.70	53.63
	9	31.73	34.32	38.40	43.44	49.59
	10	31.73	33.67	37.02	41.24	46.46
	11	31.73	33.67	36.33	39.81	44.19
	12	31.73	33.02	35.66	39.10	42.70

* For materials other than steel, use equation at bottom of Table 7.4.

f_n = natural frequency, cps
$\rho = \sqrt{I/S}$ = radius of gyration, in.
l = span length, in.

n = mode number
N = number of spans

BEAMS WITH PARTLY CLAMPED ENDS. For a beam in which the slope at each end is proportional to the moment, the following empirical equation gives the natural frequency: [21]

$$f_n = f_0 \left[n + \frac{1}{2} \left(\frac{\beta_L}{5n + \beta_L} \right) \right] \left[n + \frac{1}{2} \left(\frac{\beta_R}{5n + \beta_R} \right) \right]$$

where f_0 is the frequency of the same beam with simply supported ends and n is the mode number. The parameters $\beta_L = k_L l / EI$ and $\beta_R = k_R l / EI$ are coefficients in which k_L and k_R are stiffnesses of the supports as given by $k_L = M_L / \theta_L$, where M_L is the moment and θ_L the angle at the left end, and $k_R = M_R / \theta_R$, where M_R is the moment and θ_R the angle at the right end. The error is less than 2 per cent except for bars having one end completely or nearly clamped ($\beta > 10$) and the other end completely or nearly hinged ($\beta < 0.9$).

LATERAL VIBRATION OF BEAMS WITH MASSES ATTACHED

The use of Fig. 7.4 is a convenient method of estimating the natural frequencies of beams with added loads.

Exact Solution. If the masses attached to the beam are considered to be rigid so that they exert no elastic forces, and if it is assumed that the attachment is such that the bending of the beam is not restrained, Eqs. (7.13) and (7.16) apply. The section of the beam between each two masses, and between each support and the adjacent mass, must be considered individually. The constants in Eq. (7.16) are different for each section. There are $4N$ constants, N being the number of sections into which the beam is divided. Each support supplies two boundary conditions. Additional conditions are provided by:

1. The deflection at the location of each mass is the same for both sections adjacent to the mass.

2. The slope at each mass is the same for each section adjacent thereto.

3. The change in the lateral elastic shear force in the beam, at the location of each mass, is equal to the product of the mass and its acceleration $\ddot{y}$.

4. The change of moment in the beam, at each mass, is equal to the product of the moment of inertia of the mass and its angular acceleration $(\partial^2 / \partial t^2)(\partial y / \partial x)$.

Setting up the necessary equations is not difficult, but their solution is a lengthy process for all but the simplest configurations. Even the solution of the problem of a beam with hinged ends supporting a mass with negligible moment of inertia located anywhere except at the center of the beam is fairly long. If the mass is at the center of the beam, the solution is relatively simple because of symmetry, and is illustrated to show how the result compares with that obtained by Rayleigh's method.

Rayleigh's Method. Rayleigh's method offers a practical method of obtaining a fairly accurate solution of the problem, even when more than one mass is added. In carrying out the solution, the kinetic energy of the masses is added to that of the beam. The strain and kinetic energies of a uniform beam are given in Table 7.1. The kinetic energy of the ith mass is $(m_i/2)\omega_n^2 Y^2(x_i)$, where $Y(x_i)$ is the value of the amplitude at the location of mass. Equating the maximum strain energy to the total maximum kinetic energy of the beam and masses, the frequency equation becomes

$$\omega_n{}^2 = \frac{EI \int_0^l (Y'')^2 \, dx}{\dfrac{\gamma S}{g} \int_0^l Y^2 \, dx + \sum_{i=1}^{n} m_i Y^2(x_i)} \tag{7.19}$$

where $Y(x)$ is the maximum deflection. If $Y(x)$ were known exactly, this equation would give the correct frequency; however, since Y is not known, a shape must be assumed. This may be either the mode shape of the unloaded beam or a polynomial that satisfies the necessary boundary conditions, such as the equation for the static deflection under a load.

Beam as Spring. A method for obtaining the natural frequency of a beam with a single mass mounted on it is to consider the beam to act as a spring, the stiffness of which is found by using simple beam theory. The equation $\omega_n = \sqrt{k/m}$ is used. Best accuracy is obtained by considering m to be made up of the attached mass plus some portion of the mass of the beam. The fraction of the beam mass to be used depends on the type of beam. The equations for simply supported and cantilevered beams with masses attached are given in Table 7.2.

(A)

(B)

FIG. 7.9. (A) Beam having simply supported ends with mass attached at center. (B) Forces exerted on mass, at extreme deflection, by shear stresses in beam.

Example 7.7. The fundamental natural frequencies of a beam with hinged ends 24 in. long, 2 in. wide, and ¼ in. thick having a mass m attached at the center (Fig. 7.9) are to be calculated by each of the three methods, and the results compared for ratios of mass to beam mass of 1, 5, and 25. The result is to be compared with the frequency from Fig. 7.4.

Exact Solution. Because of symmetry, only the section of the beam to the left of the mass has to be considered in carrying out the exact solution. The boundary conditions for the left end are: at $x = 0$, $X = 0$, and $X'' = 0$. The shear force just to the left of the mass is negative at maximum deflection (Fig. 7.9B) and is $F_s = -EIX'''$; to the right of the mass, because of symmetry, the shear force has the same magnitude with opposite sign. The difference between the shear forces on the two sides of the mass must equal the product of the mass and its acceleration. For the condition of maximum deflection,

$$2EIX''' = m\ddot{y}_{max} \tag{7.20}$$

where X''' and $\ddot{y}_{max}$ must be evaluated at $x = l/2$. Because of symmetry the slope at the center is zero. Using the solution $y = X \cos \omega_n t$ and $\ddot{y}_{max} = -\omega_n^2 X$, Eq. (7.20) becomes

$$2EIX''' = -m\omega_n^2 X \tag{7.21}$$

The first boundary condition makes $A = 0$ in Eq. (7.16) and the second condition makes $B = 0$. For simplicity, the part of the equation that remains is written

$$X = C \sin \kappa x + D \sinh \kappa x \tag{7.22}$$

Using this in Eq. (7.20) gives

$$2EI\left(-C\kappa^3 \cos\frac{\kappa l}{2} + D\kappa^3 \cosh\frac{\kappa l}{2}\right) = -m\omega_n^2\left(C \sin\frac{\kappa l}{2} + D \sinh\frac{\kappa l}{2}\right) \tag{7.23}$$

The slope at the center is zero. Differentiating Eq. (7.22) and substituting $x = l/2$,

$$\kappa\left(C \cos\frac{\kappa l}{2} + D \cosh\frac{\kappa l}{2}\right) = 0 \tag{7.24}$$

Solving Eqs. (7.23) and (7.24) for the ratio C/D and equating, the following frequency equation is obtained:

$$2\frac{m_b}{m} = \frac{\kappa l}{2}\left(\tan\frac{\kappa l}{2} - \tanh\frac{\kappa l}{2}\right)$$

where $m_b = \gamma S l/g$ is the total mass of the beam. The lowest roots for the specified ratios m/m_b are as follows:

m/m_b	1	5	25
$\kappa l/2$	1.1916	0.8599	0.5857

The corresponding natural frequencies are found from Eq. (7.14) and are tabulated, with the results obtained by the other methods, at the end of the example.

Solution by Rayleigh's Method. For the solution by Rayleigh's method it is assumed that $Y = B \sin (\pi x/l)$. This is the fundamental mode for the unloaded beam (Table 7.3). The terms in Eq. (7.19) become

$$\int_0^l (Y'')^2\, dx = B^2 \left(\frac{\pi}{l}\right)^4 \int_0^l \sin^2 \frac{\pi x}{l}\, dx = B^2 \frac{l}{2}\left(\frac{\pi}{l}\right)^4$$

$$\int_0^l Y^2\, dx = B^2 \int_0^l \sin^2 \frac{\pi x}{l}\, dx = B^2 \frac{l}{2}$$

$$Y^2(x_1) = B^2$$

Substituting these terms, Eq. (7.19) becomes

$$\omega_n = \sqrt{\frac{EIB^2 \frac{l}{2}\left(\frac{\pi}{l}\right)^4}{\frac{S\gamma B^2 l}{2g} + mB^2}} = \frac{\pi^2}{\sqrt{1 + 2m/m_b}} \sqrt{\frac{EIg}{S\gamma l^4}}$$

The frequencies for the specified values of m/m_b are tabulated at the end of the example. Note that if $m = 0$, the frequency is exactly correct, as can be seen from Table 7.3. This is to be expected since, if no mass is added, the assumed shape is the true shape.

Lumped Parameter Solution. Using the appropriate equation from Table 7.2, the natural frequency is

$$\omega_n = \sqrt{\frac{48EI}{l^3(m + 0.5m_b)}}$$

Since $m_b = \gamma Sl/g$, this becomes

$$\omega_n = \sqrt{\frac{48}{(m/m_b) + 0.5}} \sqrt{\frac{EIg}{S\gamma l^4}}$$

Comparison of Results. The results for each method can be expressed as a coefficient α multiplied by $\sqrt{EIg/S\gamma l^4}$. The values of α for the specified values by m/m_b for the three methods of solution are:

m/m_b	1	5	25
Exact.....	5.680	2.957	1.372
Rayleigh ..	5.698	2.976	1.382
Spring	5.657	2.954	1.372

The results obtained by all the methods agree closely. For large values of m/m_b the third method gives very accurate results.

Numerical Calculations. For steel, $E = 30 \times 10^6$ lb/in.^{2}, $\gamma = 0.28$ lb/in.^{3}; for a rectangular beam, $I = bh^3/12 = 1/384$ in.^{4} and $S = bh = \frac{1}{2}$ in.^{2} The fundamental frequency using the value of α for the exact solution when $m/m_b = 1$ is

$$\omega_1 = \frac{\alpha}{l^2} \sqrt{\frac{EIg}{S\gamma}} = \frac{5.680}{576} \sqrt{\frac{(30 \times 10^6)(386)}{(0.5)(384)(0.28)}} = 145 \text{ rad/sec} = 23 \text{ cps}$$

Other frequencies can be found by using the other values of α. Nearly the same result is obtained by using Fig. 7.4, if half the mass of the beam is added to the additional mass.

LATERAL VIBRATION OF PLATES

General Theory of Bending of Rectangular Plates. For small deflections of an initially flat plate of uniform thickness (Fig. 7.10) made of homogeneous isotropic material and subjected to normal and shear forces in the plane of the plate, the following equation relates the lateral deflection w to the lateral loading:[22]

$$D\nabla^4 w = D\left(\frac{\partial^4 w}{\partial x^4} + 2\frac{\partial^4 w}{\partial x^2\, \partial y^2} + \frac{\partial^4 w}{\partial y^4}\right) = P + N_x \frac{\partial^2 w}{\partial x^2} + 2N_{xy}\frac{\partial^2 w}{\partial x\, \partial y} + N_y \frac{\partial^2 w}{\partial y^2} \quad (7.25)$$

where $D = Eh^3/12(1 - \mu^2)$ is the plate stiffness, h being the plate thickness and μ Poisson's ratio. The parameter P is the loading intensity, N_x the normal loading in the X direction per unit of length, N_y the normal loading in the Y direction, and N_{xy} the shear load parallel to the plate surface in the X and Y directions.

FIG. 7.10. Element of plate showing bending moments, normal forces, and shear forces.

The bending moments and shearing forces are related to the deflection w by the following equations: [23]

$$M_{1x} = -D\left(\frac{\partial^2 w}{\partial x^2} + \mu \frac{\partial^2 w}{\partial y^2}\right) \qquad M_{1y} = -D\left(\frac{\partial^2 w}{\partial y^2} + \mu \frac{\partial^2 w}{\partial x^2}\right)$$

$$T_{1xy} = D(1 - \mu)\frac{\partial^2 w}{\partial x\, \partial y} \tag{7.26}$$

$$S_{1x} = -D\left(\frac{\partial^3 w}{\partial x^3} + \frac{\partial^3 w}{\partial x\, \partial y^2}\right) \qquad S_{1y} = -D\left(\frac{\partial^3 w}{\partial y^3} + \frac{\partial^3 w}{\partial x^2\, \partial y}\right)$$

As shown in Fig. 7.10, M_{1x} and M_{1y} are the bending moments per unit of length on the faces normal to the X and Y directions, respectively, T_{1xy} is the twisting or warping moment on these faces, and S_{1x}, S_{1y} are the shearing forces per unit of length normal to the plate surface.

The boundary conditions that must be satisfied by an edge parallel to the X axis, for example, are as follows:

Built-in edge:

$$w = 0 \qquad \frac{\partial w}{\partial y} = 0$$

Simply supported edge:

$$w = 0 \qquad M_{1y} = -D\left(\frac{\partial^2 w}{\partial y^2} + \mu \frac{\partial^2 w}{\partial x^2}\right) = 0$$

Free edge:

$$M_{1y} = -D\left(\frac{\partial^2 w}{\partial y^2} + \mu \frac{\partial^2 w}{\partial x^2}\right) = 0 \qquad T_{1xy} = 0 \qquad S_{1y} = 0$$

which together give

$$\frac{\partial}{\partial y}\left[\frac{\partial^2 w}{\partial y^2} + (2 - \mu)\frac{\partial^2 w}{\partial x^2}\right] = 0$$

Similar equations can be written for other edges. The strains caused by the bending of the plate are

$$\epsilon_x = -z\frac{\partial^2 w}{\partial x^2} \qquad \epsilon_y = -z\frac{\partial^2 w}{\partial y^2} \qquad \gamma_{xy} = 2z\frac{\partial^2 w}{\partial x\, \partial y} \tag{7.27}$$

where z is the distance from the center plane of the plate.

Hooke's law may be expressed by the following equations:

$$\epsilon_x = \frac{1}{E}(\sigma_x - \mu\sigma_y) \qquad \sigma_x = \frac{E}{1 - \mu^2}(\epsilon_x + \mu\epsilon_y)$$

$$\epsilon_y = \frac{1}{E}(\sigma_y - \mu\sigma_x) \qquad \sigma_y = \frac{E}{1 - \mu^2}(\epsilon_y + \mu\epsilon_x) \qquad (7.28)$$

$$\gamma_{xy} = \frac{\tau_{xy}}{G} \qquad \tau_{xy} = G\gamma_{xy}$$

Substituting the expressions giving the strains in terms of the deflections, the following equations are obtained for the bending stresses in terms of the lateral deflection:

$$\sigma_x = -\frac{Ez}{1 - \mu^2}\left(\frac{\partial^2 w}{\partial x^2} + \mu\frac{\partial^2 w}{\partial y^2}\right) = \frac{12M_{1x}}{h^3}z$$

$$\sigma_y = -\frac{Ez}{1 - \mu^2}\left(\frac{\partial^2 w}{\partial y^2} + \mu\frac{\partial^2 w}{\partial x^2}\right) = \frac{12M_{1y}}{h^3}z \qquad (7.29)$$

$$\tau_{xy} = 2G\frac{\partial^2 w}{\partial x\,\partial y}z = \frac{12T_{1xy}}{h^3}z$$

Table 7.6 gives values of maximum deflection and bending moment at several points in plates which have various shapes and conditions of support, and which are subjected to uniform lateral pressure. The results are all based on the assumption that the deflections are small and that there are no loads in the plane of the plate. The bending stresses are found by the use of Eqs. (7.29). Bending moments and deflections for many other types of load are in the literature.[22]

The stresses caused by loads in the plane of the plate are found by assuming that the stress is uniform through the plate thickness. The total stress at any point in the plate is the sum of the stresses caused by bending and by the loading in the plane of the plate.

For plates in which the lateral deflection is large compared to the plate thickness but small compared to the other dimensions, Eq. (7.25) is valid. However, additional equations must be introduced because the forces N_x, N_y, and N_{xy} depend not only on the initial loading of the plate but also upon the stretching of the plate due to the bending. The equations of equilibrium for the X and Y directions in the plane of the plate are

$$\frac{\partial N_x}{\partial x} + \frac{\partial N_{xy}}{\partial y} = 0 \qquad \frac{\partial N_{xy}}{\partial x} + \frac{\partial N_y}{\partial y} = 0 \qquad (7.30)$$

It can be shown [27] that the strain components are given by

$$\epsilon_x = \frac{\partial u}{\partial x} + \frac{1}{2}\left(\frac{\partial w}{\partial x}\right)^2 \qquad \epsilon_y = \frac{\partial v}{\partial y} + \frac{1}{2}\left(\frac{\partial w}{\partial y}\right)^2$$

$$\gamma_{xy} = \frac{\partial u}{\partial y} + \frac{\partial v}{\partial x} + \frac{\partial w}{\partial x}\frac{\partial w}{\partial y} \qquad (7.31)$$

where u is the displacement in the X direction and v is the displacement in the Y direction. By differentiating and combining these expressions, the following relation is obtained:

$$\frac{\partial^2 \epsilon_x}{\partial y^2} + \frac{\partial^2 \epsilon_y}{\partial x^2} - \frac{\partial^2 \gamma_{xy}}{\partial x\,\partial y} = \left(\frac{\partial^2 w}{\partial x\,\partial y}\right)^2 - \frac{\partial^2 w}{\partial x^2}\frac{\partial^2 w}{\partial y^2} \qquad (7.32)$$

If it is assumed that the stresses caused by the forces in the plane of the plate are uni-

Table 7.6. Maximum Deflection and Bending Moments in Uniformly Loaded Plates under Static Conditions

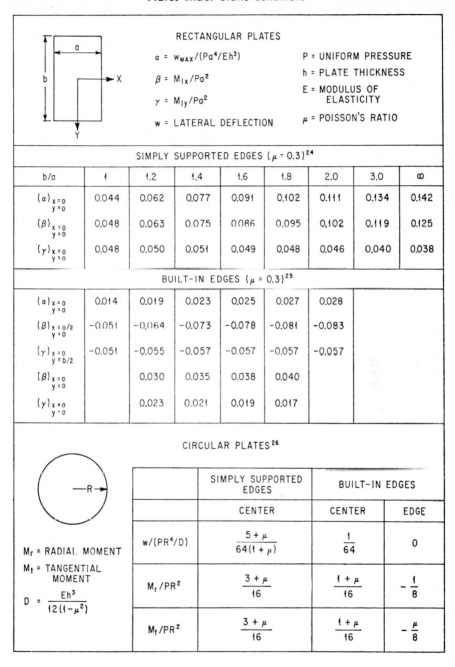

RECTANGULAR PLATES

$\alpha = w_{MAX}/(Pa^4/Eh^3)$

$\beta = M_{lx}/Pa^2$

$\gamma = M_{ly}/Pa^2$

w = LATERAL DEFLECTION

P = UNIFORM PRESSURE

h = PLATE THICKNESS

E = MODULUS OF ELASTICITY

μ = POISSON'S RATIO

SIMPLY SUPPORTED EDGES $(\mu = 0.3)$ [24]

b/a	1	1.2	1.4	1.6	1.8	2.0	3.0	∞
$(\alpha)_{x=0, y=0}$	0.044	0.062	0.077	0.091	0.102	0.111	0.134	0.142
$(\beta)_{x=0, y=0}$	0.048	0.063	0.075	0.086	0.095	0.102	0.119	0.125
$(\gamma)_{x=0, y=0}$	0.048	0.050	0.051	0.049	0.048	0.046	0.040	0.038

BUILT-IN EDGES $(\mu = 0.3)$ [25]

$(\alpha)_{x=0, y=0}$	0.014	0.019	0.023	0.025	0.027	0.028		
$(\beta)_{x=a/2, y=0}$	−0.051	−0.064	−0.073	−0.078	−0.081	−0.083		
$(\gamma)_{x=0, y=b/2}$	−0.051	−0.055	−0.057	−0.057	−0.057	−0.057		
$(\beta)_{x=0, y=0}$		0.030	0.035	0.038	0.040			
$(\gamma)_{x=0, y=0}$		0.023	0.021	0.019	0.017			

CIRCULAR PLATES [26]

M_r = RADIAL MOMENT

M_t = TANGENTIAL MOMENT

$D = \dfrac{Eh^3}{12(1-\mu^2)}$

	SIMPLY SUPPORTED EDGES	BUILT-IN EDGES	
	CENTER	CENTER	EDGE
$w/(PR^4/D)$	$\dfrac{5+\mu}{64(1+\mu)}$	$\dfrac{1}{64}$	0
M_r/PR^2	$\dfrac{3+\mu}{16}$	$\dfrac{1+\mu}{16}$	$-\dfrac{1}{8}$
M_t/PR^2	$\dfrac{3+\mu}{16}$	$\dfrac{1+\mu}{16}$	$-\dfrac{\mu}{8}$

formly distributed through the thickness, Hooke's law, Eqs. (7.28), can be expressed:

$$\epsilon_x = \frac{1}{hE}(N_x - \mu N_y) \qquad \epsilon_y = \frac{1}{hE}(N_y - \mu N_x) \qquad \gamma_{xy} = \frac{1}{hG} N_{xy} \qquad (7.33)$$

The equilibrium equations are satisfied by a stress function ϕ which is defined as follows:

$$N_x = h\frac{\partial^2 \phi}{\partial y^2} \qquad N_y = h\frac{\partial^2 \phi}{\partial x^2} \qquad N_{xy} = -h\frac{\partial^2 \phi}{\partial x\,\partial y} \qquad (7.34)$$

If these are substituted into Eqs. (7.33) and the resulting expressions substituted into Eq. (7.32), the following equation is obtained:

$$\frac{\partial^4 \phi}{\partial x^4} + 2\frac{\partial^4 \phi}{\partial x^2\,\partial y^2} + \frac{\partial^4 \phi}{\partial y^4} = E\left[\left(\frac{\partial^2 w}{\partial x\,\partial y}\right)^2 - \frac{\partial^2 w}{\partial x^2}\frac{\partial^2 w}{\partial y^2}\right] \qquad (7.35)$$

A second equation is obtained by substituting Eqs. (7.34) in Eq. (7.25):

$$D\nabla^4 w = P + h\left(\frac{\partial^2 \phi}{\partial y^2}\frac{\partial^2 w}{\partial x^2} - 2\frac{\partial^2 \phi}{\partial x\,\partial y}\frac{\partial^2 w}{\partial x\,\partial y} + \frac{\partial^2 \phi}{\partial x^2}\frac{\partial^2 w}{\partial y^2}\right) \qquad (7.36)$$

Equations (7.35) and (7.36), with the boundary conditions, determine ϕ and w, from which the stresses can be computed. General solutions to this set of equations are not known, but some approximate solutions can be found in the literature.[28]

Free Lateral Vibrations of Rectangular Plates. In Eq. (7.25), the terms on the left are equal to the sum of the rates of change of the forces per unit of length in the X and Y directions where such forces are exerted by shear stresses caused by bending normal to the plane of the plate. For a rectangular element with dimensions dx and dy, the net force exerted normal to the plane of the plate by these stresses is $D\nabla^4 w\,dx\,dy$. The last three terms on the right-hand side of Eq. (7.25) give the net force normal to the plane of the plate, per unit of length, which is caused by the forces acting in the plane of the plate. The net force caused by these forces on an element with dimensions dx and dy is $(N_x\,\partial^2 w/\partial x^2 + 2N_{xy}\,\partial^2 w/\partial x\,\partial y + N_y\,\partial^2 w/\partial y^2)\,dx\,dy$. As in the corresponding beam problem, the forces in a vibrating plate consist of two parts: (1) that which balances the static load P including the weight of the plate and (2) that which is induced by the vibration. The first part is always in equilibrium with the load and together with the load can be omitted from the equation of motion if the deflection is taken from the position of static equilibrium. The force exerted normal to the plane of the plate by the bending stresses must equal the sum of the force exerted normal to the plate by the loads acting in the plane of the plate; i.e., the product of the mass of the element $(\gamma h/g)\,dx\,dy$ and its acceleration $\ddot{w}$. The term involving the acceleration of the element is negative, because when the bending force is positive the acceleration is in the negative direction. The equation of motion is

$$D\nabla^4 w = -\frac{\gamma}{g}h\ddot{w} + \left(N_x\frac{\partial^2 w}{\partial x^2} + 2N_{xy}\frac{\partial^2 w}{\partial x\,\partial y} + N_y\frac{\partial^2 w}{\partial y^2}\right) \qquad (7.37)$$

This equation is valid only if the magnitudes of the forces in the plane of the plate are constant during the vibration. For many problems these forces are negligible and the term in parentheses can be omitted.

When a system vibrates in a natural mode, all parts execute simple harmonic motion about the equilibrium position; therefore, the solution of Eq. (7.37) can be written as $w = AW(x,y)\cos(\omega_n t + \theta)$ in which W is a function of x and y only. Substituting this in Eq. (7.37) and dividing through by $A\cos(\omega_n t + \theta)$ gives

$$D\nabla^4 W = \frac{\gamma h\omega_n^2}{g}W + \left(N_x\frac{\partial^2 W}{\partial x^2} + 2N_{xy}\frac{\partial^2 W}{\partial x\,\partial y} + N_y\frac{\partial^2 W}{\partial y^2}\right) \qquad (7.38)$$

The function W must satisfy Eq. (7.38) as well as the necessary boundary conditions.

The solution of the problem of the lateral vibration of a rectangular plate with all edges simply supported is relatively simple; in general, other combinations of edge conditions require the use of other methods of solution. These are discussed later.

Example 7.8. The natural frequencies and normal modes of small vibration of a rectangular plate of length a, width b, and thickness h are to be calculated. All edges are hinged and subjected to unchanging normal forces N_x and N_y.

Solution. The following equation, in which m and n may be any integers, satisfies the necessary boundary conditions:

$$W = A \sin \frac{m\pi x}{a} \sin \frac{n\pi y}{b} \tag{7.39}$$

Substituting the necessary derivatives into Eq. (7.38),

$$D\left[\left(\frac{m}{a}\right)^4 + 2\left(\frac{m}{a}\right)^2\left(\frac{n}{b}\right)^2 + \left(\frac{n}{b}\right)^4\right]\pi^4 \sin\frac{m\pi x}{a}\sin\frac{n\pi y}{b}$$

$$= \frac{\gamma h \omega_n^2}{g}\sin\frac{m\pi x}{a}\sin\frac{n\pi y}{b} - \pi^2\left[N_x\left(\frac{m}{a}\right)^2 + N_y\left(\frac{n}{b}\right)^2\right]\sin\frac{m\pi x}{a}\sin\frac{n\pi y}{b}$$

Solving for ω_n^2,

$$\omega_n^2 = \frac{g}{\gamma h}\left\{\pi^4 D\left[\left(\frac{m}{a}\right)^2 + \left(\frac{n}{b}\right)^2\right]^2 + \pi^2\left[N_x\left(\frac{m}{a}\right)^2 + N_y\left(\frac{n}{b}\right)^2\right]\right\} \tag{7.40}$$

By using integral values of m and n, the various frequencies are obtained from Eq. (7.40) and the corresponding normal modes from Eq. (7.39). For each mode, m and n represent the number of half sine waves in the X and Y directions, respectively. In each mode there are $m - 1$ evenly spaced nodal lines parallel to the Y axis, and $n - 1$ parallel to the X axis.

Rayleigh's and Ritz's Methods. The modes of vibration of a rectangular plate with all edges simply supported are such that the deflection of each section of the plate parallel to an edge is of the same form as the deflection of a beam with both ends simply supported. In general, this does not hold true for other combinations of edge conditions. For example, the vibration of a rectangular plate with all edges built in does not occur in such a way that each section parallel to an edge has the same shape as does a beam with both ends built in. A function that is made up using the mode shapes of beams with built-in ends obviously satisfies the conditions of zero deflection and slope at all edges, but it cannot be made to satisfy Eq. (7.38).

The mode shapes of beams give logical functions with which to formulate shapes for determining the natural frequencies, for plates having various edge conditions, by the Rayleigh or Ritz methods. By using a single mode function in Rayleigh's method an approximate frequency can be determined. This can be improved by using more than one of the modal shapes and using Ritz's method as discussed below.

The strain energy of bending and the kinetic energy for plates are given in Table 7.1. Finding the maximum values of the energies, equating them, and solving for ω_n^2 gives the following frequency equation:

$$\omega_n^2 = \frac{V_{\max}}{\dfrac{\gamma h}{2g}\displaystyle\int_A\int W^2\, dx\, dy} \tag{7.41}$$

where V is the strain energy.

In applying the Rayleigh method, a function W is assumed that satisfies the necessary boundary conditions of the plate. An example of the calculations is given in the section on circular plates. If the shape assumed is exactly the correct one, Eq. (7.41) gives the exact frequency. In general, the correct shape is not known and a frequency greater than the natural frequency is obtained. The Ritz method involves assuming W to be of the form $W = a_1 W_1(x,y) + a_2 W_2(x,y) + \ldots$ in which W_1, W_2, $\ldots$ all satisfy the boundary conditions, and a_1, a_2, $\ldots$ are adjusted to give a minimum frequency. Reference 28a is an extensive compilation, with references to sources, of calculated and experimental results for plates of many shapes. Some examples are cited in the following sections.

Table 7.7. Natural Frequencies and Nodal Lines of Square Plates with Various Edge Conditions (After D. Young.[29])

	1ST MODE	2ND MODE	3RD MODE	4TH MODE	5TH MODE	6TH MODE
$\omega_n/\sqrt{Dg/\gamma h a^4}$	3.494	8.547	21.44	27.46	31.17	
NODAL LINES						
$\omega_n/\sqrt{Dg/\gamma h a^4}$	35.99	73.41	108.27	131.64	132.25	165.15
NODAL LINES						
$\omega_n/\sqrt{Dg/\gamma h a^4}$	6.958	24.08	26.80	48.05	63.14	
NODAL LINES						

$\omega_n = 2\pi f_n$ h = PLATE THICKNESS

$D = Eh^3/12(1-\mu^2)$ a = PLATE LENGTH

γ = WEIGHT DENSITY

SQUARE, RECTANGULAR, AND SKEW RECTANGULAR PLATES. Tables of the functions necessary for the determination of the natural frequencies of rectangular plates by the use of the Ritz method are available,[29] these having been derived by using the modal shapes of beams having end conditions corresponding to the edge conditions of the plates. Information is included from which the complete shapes of the vibrational modes can be determined. Frequencies and nodal patterns for several modes of vibration of square plates having three sets of boundary conditions are shown in Table 7.7. By the use of functions which represent the natural modes of beams, the frequencies and nodal patterns for rectangular and skew cantilever plates have been determined [30] and are shown in Table 7.8. Comparison of calculated frequencies with experimentally determined values shows good agreement. Natural frequencies of rectangular plates having other boundary conditions are given in Table 7.9.

TRIANGULAR AND TRAPEZOIDAL PLATES. Nodal patterns and natural frequencies for triangular plates have been determined [32] by the use of functions derived from the mode shapes of beams, and are shown in Table 7.10. Certain of these have been compared with experimental values and the agreement is excellent. Natural frequencies and nodal patterns have been determined experimentally for six modes of vibration of a number of cantilevered triangular plates [33] and for the first six modes of cantilevered trapezoidal plates derived by trimming the tips of triangular plates parallel to the clamped edge.[34] These triangular and trapezoidal shapes approximate the shapes of various delta wings for aircraft and of fins for missiles.

CIRCULAR PLATES. The solution of the problem of small lateral vibration of circular plates is obtained by transforming Eq. (7.38) to polar coordinates and finding the solution that satisfies the necessary boundary conditions of the resulting equation.

Table 7.8. Natural Frequencies and Nodal Lines of Cantilevered Rectangular and Skew Rectangular Plates ($\mu = 0.3$)* (M. V. Barton.[30])

a/b MODE	1/2	1	2	5
FIRST	3.508	3.494	3.472	3.450
SECOND	5.372	8.547	14.93	34.73
THIRD	21.96	21.44	21.61	21.52
FOURTH	10.26	27.46	94.49	563.9
FIFTH	24.85	31.17	48.71	105.9

MODE	FIRST	SECOND	FIRST	SECOND	FIRST	SECOND
$\omega_n \sqrt{Dg/\gamma ha^4}$	3.601	8.872	3.961	10.190	4.824	13.75
NODAL LINES						

* For terminology, see Table 7.7.

Table 7.9. Natural Frequencies of Rectangular Plates (R. F. S. Hearmon.[31])

	b/a	1.0	1.5	2.0	2.5	3.0	∞
s, s, s (plate)	$\omega_n/\sqrt{Dg/\gamma h a^4}$	19.74	14.26	12.34	11.45	10.97	9.87

	b/a	1.0	1.5	2.0	2.5	3.0	∞
c, s, s (plate)	$\omega_n/\sqrt{Dg/\gamma h a^4}$	23.65	18.90	17.33	16.63	16.26	15.43
	a/b	1.0	1.5	2.0	2.5	3.0	∞
	$\omega_n/\sqrt{Dg/\gamma h b^4}$	23.65	15.57	12.92	11.75	11.14	9.87

	b/a	1.0	1.5	2.0	2.5	3.0	∞
c, s, c (plate)	$\omega_n/\sqrt{Dg/\gamma h a^4}$	28.95	25.05	23.82	23.27	22.99	22.37
	a/b	1.0	1.5	2.0	2.5	3.0	∞
	$\omega_n/\sqrt{Dg/\gamma h b^4}$	28.95	17.37	13.69	12.13	11.36	9.87

	b/a	1.0	1.5	2.0	2.5	3.0	∞
c, c, c (plate)	$\omega_n/\sqrt{Dg/\gamma h a^4}$	35.98	27.00	24.57	23.77	23.19	22.37

s DENOTES SIMPLY SUPPORTED EDGE

c DENOTES BUILT-IN OR CLAMPED EDGE

a = LENGTH OF PLATE

b = WIDTH OF PLATE

FOR OTHER TERMINOLOGY SEE TABLE 7.7

Omitting the terms involving forces in the plane of the plate,[35]

$$\left(\frac{\partial^2}{\partial r^2} + \frac{1}{r}\frac{\partial}{\partial r} + \frac{1}{r}\frac{\partial^2}{\partial \theta^2}\right)\left(\frac{\partial^2 W}{\partial r^2} + \frac{1}{r}\frac{\partial W}{\partial r} + \frac{1}{r}\frac{\partial^2 W}{\partial \theta^2}\right) = \varkappa^4 W \tag{7.42}$$

where

$$\varkappa^4 = \frac{\gamma h \omega_n^2}{gD}$$

The solution of Eq. (7.42) is [35]

$$W = A \cos (n\theta - \beta)[J_n(\varkappa r) + \lambda J_n(i\varkappa r)] \tag{7.43}$$

where J_n is a Bessel function of the first kind. When $\cos (n\theta - \beta) = 0$, a mode having a nodal system of n diameters, symmetrically distributed, is obtained. The term in

Table 7.10. Natural Frequencies and Nodal Lines of Triangular Plates (B. W. Anderson.[32])

MODE k	2	4	8	14
FIRST (a/k, a)	7.194	7.122	7.080	7.068
SECOND	30.803	30.718	30.654	30.638
THIRD	61.131	90.105	157.70	265.98
FOURTH	148.8	259.4	493.4	853.6

k	2	4	7	
FIRST	5.887	6.617	6.897	
SECOND (a/k, a)	25.40	28.80	30.28	

a - LENGTH OF TRIANGLE
k - RATIO OF LENGTH TO WIDTH OF TRIANGLE
FOR OTHER TERMINOLOGY SEE TABLE 7.7

brackets represents modes having concentric nodal circles. The values of $\varkappa$ and λ are determined by the boundary conditions, which are, for radially symmetrical vibration:

Simply supported edge:

$$W = 0 \qquad M_{1r} = D\left(\frac{d^2W}{dr^2} + \frac{\mu}{a}\frac{dW}{dr}\right) = 0$$

Fixed edge:

$$W = 0 \qquad \frac{dW}{dr} = 0$$

Free edge:

$$M_{1r} = D\left(\frac{d^2W}{dr^2} + \frac{\mu}{a}\frac{dW}{dr}\right) = 0 \qquad \frac{d}{dr}\left(\frac{d^2W}{dr^2} + \frac{1}{r}\frac{dW}{dr}\right) = 0$$

Example 7.9. The steel diaphragm of a radio earphone has an unsupported diameter of 2.0 in. and is 0.008 in. thick. Assuming that the edge is fixed, the lowest three frequencies for the free vibration in which only nodal circles occur are to be calculated, using the exact method and the Rayleigh and Ritz methods.

Exact Solution. In this example $n = 0$, which makes $\cos(n\theta - \beta) = 1$; thus, Eq. (7.43) becomes

$$W = A[J_0(\varkappa r) + \lambda I_0(\varkappa r)]$$

where $J_0(i\varkappa r) = I_0(\varkappa r)$ and I_0 is a modified Bessel function of the first kind.

At the boundary where $r = a$,

$$\frac{\partial W}{\partial r} = A\varkappa[-J_1(\varkappa a) + \lambda I_1(\varkappa a)] = 0 \qquad -J_1(\varkappa a) + \lambda I_1(\varkappa a) = 0$$

The deflection at $r = a$ is also zero:

$$J_0(\varkappa a) + \lambda I_0(\varkappa a) = 0$$

The frequency equation becomes

$$\lambda = \frac{J_1(\varkappa a)}{I_1(\varkappa a)} = -\frac{J_0(\varkappa a)}{I_0(\varkappa a)}$$

The first three roots of the frequency equation are: $\varkappa a = 3.196, 6.306, 9.44$. The corresponding natural frequencies are, from Eq. (7.42),

$$\omega_n = \frac{10.21}{a^2}\sqrt{\frac{Dg}{\gamma h}}, \quad \frac{39.77}{a^2}\sqrt{\frac{Dg}{\gamma h}}, \quad \frac{88.9}{a^2}\sqrt{\frac{Dg}{\gamma h}}$$

For steel, $E = 30 \times 10^6$ lb/in.^{2}, $\gamma = 0.28$ lb/in.^{3}, and $\mu = 0.28$. Hence

$$D = \frac{Eh^3}{12(1 - \mu^2)} = \frac{30 \times 10^6 (0.008)^3}{12(1 - 0.078)} = 1.38 \text{ lb-in.}$$

Thus, the lowest natural frequency is

$$\omega_1 = 10.21\sqrt{\frac{(1.38)(386)}{(0.28)(0.008)}} = 4,960 \text{ rad/sec} = 790 \text{ cps}$$

The second frequency is 3,070 cps, and the third is 6,880 cps.

Solution by Rayleigh's Method. The equations for strain and kinetic energies are given in Table 7.1. The strain energy for a plate with clamped edges becomes

$$V = \pi D \int_0^a \left(\frac{\partial^2 W}{\partial r^2} + \frac{1}{r}\frac{\partial W}{\partial r}\right)^2 r \, dr$$

The maximum kinetic energy is

$$T = \frac{\omega_n^2 \pi \gamma h}{g} \int_0^a W^2 r \, dr$$

An expression of the form $W = a_1[1 - (r/a)^2]^2$, which satisfies the conditions of zero deflection and slope at the boundary, is used. The first two derivatives are $\partial W/\partial r = a_1(-4r/a^2 + 4r^3/a^4)$ and $\partial^2 W/\partial r^2 = a_1(-4/a^2 + 12r^2/a^4)$. Using these values in the equations for strain and kinetic energy, $V = 32\pi D a_1^2/3a^2$ and $T = \omega_n^2 \pi \gamma h a^2 a_1^2/10g$. Equating these values and solving for the frequency,

$$\omega_n = \sqrt{\frac{320 \, Dg}{3a^4 \, \gamma h}} = \frac{10.33}{a^2}\sqrt{\frac{Dg}{\gamma h}}$$

This is somewhat higher than the exact frequency.

Solution by Ritz's Method. Using an expression for the deflection of the form

$$W = a_1[1 - (r/a)^2]^2 + a_2[1 - (r/a)^2]^3$$

and applying the Ritz method, the following values are obtained for the first two frequencies:

$$\omega_1 = \frac{10.21}{a^2}\sqrt{\frac{Dg}{\gamma h}} \qquad \omega_2 = \frac{43.04}{a^2}\sqrt{\frac{Dg}{\gamma h}}$$

The details of the calculations giving this result are in the literature.[36] The first frequency agrees with the exact answer to four significant figures, while the second frequency is somewhat high. A closer approximation to the second frequency and approximations of the higher frequencies could be obtained by using additional terms in the deflection equation.

The frequencies of modes having n nodal diameters are:[36]

$n = 1$:
$$\omega_1 = \frac{21.22}{a^2}\sqrt{\frac{Dg}{\gamma h}}$$

$n = 2$:
$$\omega_2 = \frac{34.84}{a^2}\sqrt{\frac{Dg}{\gamma h}}$$

For a plate with its center fixed and edge free, and having m nodal circles, the frequencies are: [37]

m	0	1	2	3
$\omega_n a^2 \Big/ \sqrt{\dfrac{Dg}{\gamma h}}$	3.75	20.91	60.68	119.7

STRETCHING OF MIDDLE PLANE. In the usual analysis of plates, it is assumed that the deflection of the plate is so small that there is no stretching of the middle plane. If such stretching occurs, it affects the natural frequency. Whether it occurs depends

FIG. 7.11. Influence of amplitude on period of vibration of uniform rectangular plates with immovable hinged edges. The aspect ratio r is the ratio of width to length of the plate. (*H. Chu and G. Herrmann* [38])

FIG. 7.12. Influence of amplitude on maximum total stress in rectangular plates with immovable hinged edges. The aspect ratio r is the ratio of width to length of the plate. (*H. Chu and G. Herrmann.* [38])

on the conditions of support of the plate, the amplitude of vibration, and possibly other conditions. In a plate with its edges built in, a relatively small deflection causes a significant stretching. The effect of stretching is not proportional to the deflection; thus, the elastic restoring force is not a linear function of deflection. The natural frequency is not independent of amplitude, but becomes higher with increasing amplitudes. If a plate is subjected to a pressure on one side, so that the vibration occurs about a deflected position, the effect of stretching may be appreciable. The effect of stretching in rectangular plates with immovable hinged supports has been discussed. [38] The effect of the amplitude on the natural frequency is shown in Fig. 7.11; the effect on the total stress in the plate is shown in Fig. 7.12. The natural frequency increases rapidly as the amplitude of vibration increases.

ROTATIONAL MOTION AND SHEARING FORCES. In the foregoing analysis, only the motion of each element of the plate in the direction normal to the plane of the plate is considered. There is also rotation of each element, and there is a deflection associated with the lateral shearing forces in the plate. The effects of these factors becomes

Table 7.11. Natural Frequencies of Complete Circular Rings Whose Thickness in Radial Direction Is Small Compared to Radius

TYPE OF VIBRATION	SHAPE OF LOWEST MODE	RECTANGULAR CROSS SECTION ω_n	CIRCULAR CROSS SECTION ω_n
FLEXURAL IN PLANE OF RING WITH n COMPLETE WAVELENGTH IN CIRCUMFERENCE	$n=2$	$\sqrt{\dfrac{Eg}{\gamma}\dfrac{I}{AR^4}\dfrac{n^2(n^2-1)^2}{n^2+1}}$ n ANY INTEGER >1	$\sqrt{\dfrac{E\pi r^4}{4mR^4}\dfrac{n^2(n^2-1)^2}{n^2+1}}$ n ANY INTEGER >1
FLEXURAL NORMAL TO PLANE OF RING	$n=2$		$\sqrt{\dfrac{E\pi r^4}{4mR^4}\dfrac{n^2(n^2-1)^2}{n^2+1+\mu}}$ n ANY INTEGER >1
TORSIONAL		FIRST MODE $\sqrt{\dfrac{Eg}{\gamma R^2}\dfrac{I_x}{I_p}}$	$\sqrt{\dfrac{G\pi r^2}{mR^2}}\,(n^2+1+\mu)$ $n=0$, OR ANY INTEGER
EXTENSIONAL		$\sqrt{\dfrac{Eg}{\gamma R^2}}$	$\sqrt{\dfrac{E\pi r^2}{mR^2}(1+n^2)}$ n $=0$, OR ANY INTEGER

E = MODULUS OF ELASTICITY
G = MODULUS OF RIGIDITY
γ = WEIGHT DENSITY
n : DEFINED FOR EACH TYPE OF VIBRATION
R = RADIUS OF RING
μ = POISSON'S RATIO

PROPERTIES OF CROSS SECTIONS
I = MOMENT OF INERTIA WITH RESPECT TO AXIS OF SECTION
I_x = MOMENT OF INERTIA WITH RESPECT TO RADIAL LINE
I_p = POLAR MOMENT OF INERTIA
A = AREA
r = RADIUS
m = MASS PER UNIT OF LENGTH

significant if the curvature of the plate is large relative to its thickness, i.e., for a plate in which the thickness is large compared to the lateral dimensions or when the plate is vibrating in a mode for which the nodal lines are close together. These effects have been analyzed for rectangular plates [39] and for circular plates.[40]

COMPLETE CIRCULAR RINGS. Equations have been derived [41, 42] for the natural frequencies of complete circular rings for which the radius is large compared to the thickness of the ring in the radial direction. Such rings can execute several types of free vibration, which are shown in Table 7.11 with the formulas for the natural frequencies.

TRANSFER MATRIX METHOD

In some assemblies which consist of various types of elements, e.g., beam segments, the solution for each element may be known. The transfer matrix method[43, 44] is a procedure by means of which the solution for such elements can be combined to yield a frequency equation for the assembly. The associated mode shapes can then be determined. The method is an extension to distributed systems of the Holzer method, described in Chap. 38, in which torsional problems are solved by dividing an assembly into lumped masses and elastic elements, and of the Myklestad method,[45] in which a similar procedure is applied to beam problems. The method has been used[46] to find the natural frequencies and mode shapes of the internals of a nuclear reactor by modeling the various elements of the system as beam segments.

The method will be illustrated by setting up the frequency equation for a cantilever beam, Fig. 7.13, composed of three segments, each of which has uniform section properties. Only the effects of bending will be considered, but the method can be extended to include other effects, such as shear deformation and rotary motion of the cross section.[44] Application to other geometries is described in Ref. 44.

Depending on the type of element being considered, the values of appropriate parameters must be expressed at certain sections of the piece in terms of their values at other sections. In the beam problem, the deflection and its first three derivatives must be used.

Transfer Matrices. Two types of transfer matrix are used. One, which for the beam problem is called the **R** matrix (after Lord Rayleigh[43]), yields the values of the parameters at the right end

FIG. 7.13. Cantilever beam made up of three segments having different section properties.

of a uniform segment of the beam in terms of their values at the left end of the segment. The other type of transfer matrix is the point matrix, which yields the values of the parameters just to the right of a joint between segments in terms of their values just to the left of the joint.

As can be seen by looking at the successive derivatives, the coefficients in Eq. (7.16) are equal to the following, where the subscript 0 indicates the value of the indicated parameter at the left end of the beam:

$$A = \frac{X_0}{2} \qquad C = \frac{X_0{}'}{2\kappa} \qquad B = \frac{-X_0{}''}{2\kappa^2} \qquad D = \frac{-X_0{}'''}{2\kappa^3}$$

Using the following notation, X and its derivatives at the right end of a beam segment can be expressed, by the matrix equation, in terms of the values at the left end of the segment. The subscript n refers to the number of the segment being considered, the subscript l to the left end of the segment and the subscript r to the right end.

$$C_{0n} = \frac{\cos \kappa_n l_n + \cosh \kappa_n l_n}{2}$$

$$S_{1n} = \frac{\sin \kappa_n l_n + \sinh \kappa_n l_n}{2\kappa_n}$$

$$C_{2n} = \frac{-(\cos \kappa_n l_n - \cosh \kappa_n l_n)}{2\kappa_n{}^2}$$

$$S_{3n} = \frac{-(\sin \kappa_n l_n - \sinh \kappa_n l_n)}{2\kappa_n{}^3}$$

where κ_n takes the value shown in Eq. (7.14) with the appropriate values of the parameters for the segment and l_n is the length of the segment.

$$\begin{bmatrix} X \\ X' \\ X'' \\ X''' \end{bmatrix}_{rn} = \begin{bmatrix} C_{0n} & S_{1n} & C_{2n} & S_{3n} \\ \kappa_n{}^4 S_{3n} & C_{0n} & S_{1n} & C_{2n} \\ \kappa_n{}^4 C_{2n} & \kappa_n{}^4 S_{3n} & C_{0n} & S_{1n} \\ \kappa_n{}^4 S_{1n} & \kappa_n{}^4 C_{2n} & \kappa_n{}^4 S_{3n} & C_{0n} \end{bmatrix} \begin{bmatrix} X \\ X' \\ X'' \\ X''' \end{bmatrix}_{ln}$$

or $x_{rn} = R_n x_{ln}$, where the boldface capital letter denotes a square matrix and the boldface lowercase letters denote column matrices.   Matrix operations are discussed in Chap. 28.

At a section where two segments of a beam are joined, the deflection, the slope, the bending moment, and the shear must be the same on the two sides of the joint.   Since $M = EI \cdot X''$ and $V = EI \cdot X'''$, the point transfer matrix for such a joint is as follows, where the subscript jn refers to the joint to the right of the nth segment of the beam.

$$
\begin{bmatrix} X \\ X' \\ X'' \\ X''' \end{bmatrix}_{rjn}
=
\begin{bmatrix}
1 & 0 & 0 & 0 \\
0 & 1 & 0 & 0 \\
0 & 0 & (EI)_l/(EI)_r & 0 \\
0 & 0 & 0 & (EI)_l/(EI)_r
\end{bmatrix}
\begin{bmatrix} X \\ X' \\ X'' \\ X''' \end{bmatrix}_{ljn}
$$

or $x_{rjn} = J_n x_{ljn}$.

The Frequency Equation.   For the cantilever beam shown in Fig. 7.13, the coefficients relating the values of X and its derivatives at the right end of the beam to their values at the left end are found by successively multiplying the appropriate R and J matrices, as follows:

$$ x_{r3} = R_3 J_2 R_2 J_1 R_1 x_{l1} $$

Carrying out the multiplication of the square R and J matrices, and calling the resulting matrix P yields

$$
\begin{bmatrix} X \\ X' \\ X'' \\ X''' \end{bmatrix}_{r3}
=
\begin{bmatrix}
P_{11} & P_{12} & P_{13} & P_{14} \\
P_{21} & P_{22} & P_{23} & P_{24} \\
P_{31} & P_{32} & P_{33} & P_{34} \\
P_{41} & P_{42} & P_{43} & P_{44}
\end{bmatrix}
\begin{bmatrix} X \\ X' \\ X'' \\ X''' \end{bmatrix}_{l1}
$$

The boundary conditions at the fixed left end of the cantilever beam are $X = X' = 0$. Using these, and performing the multiplication of P by x_{l1} yields the following:

$$
\begin{aligned}
X_{r3} &= P_{13} X_{l1}'' + P_{14} X_{l1}''' \\
X_{r3}' &= P_{23} X_{l1}'' + P_{24} X_{l1}''' \\
X_{r3}'' &= P_{33} X_{l1}'' + P_{34} X_{l1}''' \\
X_{r3}''' &= P_{43} X_{l1}'' + P_{44} X_{l1}'''
\end{aligned}
\tag{7.44}
$$

The boundary conditions for the free right end of the beam are $X'' = X''' = 0$.   Using these in the last two equations results in two simultaneous homogeneous equations, so that the following determinant, which is the frequency equation, results.

$$
\begin{vmatrix} P_{33} & P_{34} \\ P_{43} & P_{44} \end{vmatrix} = 0
$$

It can be seen that for a beam consisting of only one segment, this determinant yields a result which is equivalent to Eq. (7.17).

While in theory it would be possible to multiply the successive R and J matrices and obtain the P matrix in literal form, so that the transcendental frequency equation could be written, the process, in all but the simplest problems, would be long and time-consuming.   A more practicable procedure is to perform the necessary multiplications with numbers, using a digital computer, and finding the roots by trial and error.

Mode Shapes.   Either of the last two equations of Eq. (7.44) may be used to find the ratio X_{l1}''/X_{l1}'''.   These are used in Eq. (7.16), with $\kappa = \kappa_1$ to find the shape of the first segment.   By the use of the R and J matrices the values of the coefficients in Eq. (7.16) are found for each of the other segments.

With intermediate rigid supports or pinned connections, numerical difficulties occur in the solution of the frequency equation.   These difficulties are eliminated by the use of delta matrices, the elements of which are combinations of the elements of the R matrix. These delta matrices, for various cases, are tabulated in Refs. 43 and 44.

In Ref. 46 transfer matrices are developed and used for structures which consist, in part, of beams that are parallel to each other.

FORCED VIBRATION

CLASSICAL SOLUTION. The classical method of analyzing the forced vibration that results when an elastic system is subjected to a fluctuating load is to set up the equation of motion by the application of Newton's second law. During the vibration, each element of the system is subjected to elastic forces corresponding to those experienced during free vibration; in addition, some of the elements are subjected to the disturbing force. The equation which governs the forced vibration of a system can be obtained by adding the disturbing force to the equation for free vibration. For example, in Eq. (7.13) for the free vibration of a uniform beam, the term on the left is due to the elastic forces in the beam. If a force $F(x,t)$ is applied to the beam, the equation of motion is obtained by adding this force to Eq. (7.13), which becomes, after rearranging terms,

$$EI \frac{\partial^4 y}{\partial x^4} + \frac{\gamma S}{g} \frac{\partial^2 y}{\partial t^2} = F(x,t)$$

where EI is a constant. The solution of this equation gives the motion that results from the force F. For example, consider the motion of a beam with hinged ends subjected to a sinusoidally varying force acting at its center. The solution is obtained by representing the concentrated force at the center by its Fourier series:

$$EI y'''' + \frac{\gamma S}{g} \ddot{y} = \frac{2F}{l} \left[\sin \frac{\pi x}{l} - \sin \frac{3\pi x}{l} + \sin \frac{5\pi x}{l} \cdots \right] \sin \omega t$$

$$= \frac{2F}{l} \sum_{n=1}^{n=\infty} \left(\sin \frac{n\pi}{2} \sin \frac{n\pi x}{l} \right) \sin \omega t \qquad (7.45)$$

where $\sin(n\pi/2)$, which appears in each term of the series, makes the nth term positive, negative, or zero. The solution of Eq. (7.45) is

$$y = \sum_{n=1}^{n=\infty} \left[A_n \sin \frac{n\pi x}{l} \sin \omega_n t + B_n \sin \frac{n\pi x}{l} \cos \omega_n t \right.$$

$$\left. + \sin \frac{n\pi}{2} \frac{2Fg/S\gamma l}{(n\pi/l)^4(EIg/S\gamma) - \omega^2} \sin \frac{n\pi x}{l} \sin \omega t \right] \qquad (7.46)$$

The first two terms of Eq. (7.46) are the values of y which make the left side of Eq. (7.45) equal to zero. They are obtained in exactly the same way as in the solution of the free-vibration problem and represent the free vibration of the beam. The constants are determined by the initial conditions; in any real beam, damping causes the free vibration to die out. The third term of Eq. (7.46) is the value of y which makes the left-hand side of Eq. (7.45) equal the right-hand side; this can be verified by substitution. The third term represents the forced vibration. From Table 7.3, $\kappa_n l = n\pi$ for a beam with hinged ends; then from Eq. (7.14), $\omega_n^2 = n^4\pi^4 EIg/S\gamma l^4$. The term representing the forced vibration in Eq. (7.46) can be written, after rearranging terms,

$$y = \frac{2Fg}{S\gamma l} \sum_{n=1}^{n=\infty} \frac{\sin(n\pi/2)}{\omega_n^2[1 - (\omega/\omega_n)^2]} \sin \frac{n\pi x}{l} \sin \omega t \qquad (7.47)$$

From Table 7.3 and Eq. (7.16), it is evident that this deflection curve has the same shape as the nth normal mode of vibration of the beam since, for free vibration of a beam with hinged ends, $X_n = 2C \sin \kappa x = \sin(n\pi x/l)$.

The equation for the deflection of a beam under a distributed static load $F(x)$ can be obtained by replacing $-(\gamma S/g)\ddot{y}$ with F in Eq. (7.12); then Eq. (7.13) becomes

$$y_s'''' = \frac{F(x)}{EI} \qquad (7.48)$$

where EI is a constant.   For a static loading $F(x) = 2F/l \sin \dfrac{n\pi}{2} \sin \dfrac{n\pi x}{l}$ corresponding to the nth term of the Fourier series in Eq. (7.45), Eq. (7.48) becomes $y_{sn}'''' = 2F/EIl$ $\sin \dfrac{n\pi}{2} \sin \dfrac{n\pi x}{l}$.   The solution of this equation is

$$y_{sn} = \frac{2F}{EIl} \left(\frac{l}{n\pi}\right)^4 \sin \frac{n\pi}{2} \sin \frac{n\pi x}{l}$$

Using the relation $\omega_n{}^2 = n^4\pi^4 EIg/S\gamma l^4$, this can be written

$$y_{sn} = \frac{2Fg}{\omega_n{}^2 S\gamma l} \sin \frac{n\pi x}{l} \sin \frac{n\pi}{2}$$

Thus, the nth term of Eq. (7.47) can be written

$$y_n = y_{sn} \frac{1}{1 - (\omega/\omega_n)^2} \sin \omega t$$

Thus, the amplitude of the forced vibration is equal to the static deflection under the Fourier component of the load multiplied by the "amplification factor" $1/[1 - (\omega/\omega_n)^2]$. This is the same as the relation that exists, for a system having a single degree-of-freedom, between the static deflection under a load F and the amplitude under a fluctuating load $F \sin \omega t$.   Therefore, in so far as each mode alone is concerned, the beam behaves as a system having a single degree-of-freedom.   If the beam is subjected to a force fluctuating at a single frequency, the amplification factor is small except when the frequency of the forcing force is near the natural frequency of a mode.   For all even values of n, $\sin (n\pi/2)$ $= 0$; thus, the even-numbered modes are not excited by a force acting at the center, which is a node for those modes.   The distribution of the static load that causes the same pattern of deflection as the beam assumes during each mode of vibration has the same form as the deflection of the beam.   This result applies to other beams since a comparison of Eqs. (7.15) and (7.48) shows that if a static load $F = (\omega_n{}^2 \gamma S/g)y$ is applied to any beam it will cause the same deflection as occurs during the free vibration in the nth mode.

The results for the simply supported beam are typical of those which are obtained for all systems having distributed mass and elasticity.   Vibration of such a system at resonance is excited by a force which fluctuates at the natural frequency of a mode, since nearly any such force has a component of the shape necessary to excite the vibration. Even if the force acts at a nodal point of the mode, vibration may be excited because of coupling between the modes.

METHOD OF WORK.   Another method of analyzing forced vibration is by the use of the theorem of virtual work and D'Alembert's principle.   The theorem of virtual work states that when any elastic body is in equilibrium, the total work done by all external forces during any virtual displacement equals the increase in the elastic energy stored in the body.   A virtual displacement is an arbitrary small displacement that is compatible with the geometry of the body, and which satisfies the boundary conditions.

In applying the principle of work to forced vibration of elastic bodies, the problem is made into one of equilibrium by the application of D'Alembert's principle.   This permits a problem in dynamics to be considered as one of statics by adding to the equation of static equilibrium an "inertia force" which, for each part of the body, is equal to the product of the mass and the acceleration.   Using this principle, the theorem of virtual work can be expressed in the following equation:

$$\Delta V = \Delta(F_I + F_E) \tag{7.49}$$

in which V is the elastic strain energy in the body, F_I is the inertia force, F_E is the external disturbing force, and Δ indicates the change of the quantity when the body undergoes a virtual displacement.   The various quantities can be found separately.

For example, consider the motion of a uniform beam having hinged ends with a sinusoidally varying force acting at the center, and compare the result with the solution obtained by the classical method. All possible motions of any beam can be represented by a series of the form

$$y = q_1 X_1 + q_2 X_2 + q_3 X_3 + \cdots = \sum_{n=1}^{n=\infty} q_n X_n \tag{7.50}$$

in which the X's are functions representing displacements in the normal modes of vibration and the q's are coefficients which are functions of time. The determination of the values of q_n is the problem to be solved. For a beam having hinged ends, Eq. (7.50) becomes

$$y = \sum_{n=1}^{n=\infty} q_n \sin \frac{n\pi x}{l} \tag{7.51}$$

This is evident by using the values of $\kappa_n l$ from Table 7.3 in Eq. (7.16). A virtual displacement, being any arbitrary small displacement, can be assumed to be

$$\Delta y = \Delta q_m X_m = \Delta q_m \sin \frac{m\pi x}{l}$$

The elastic strain energy of bending of the beam is

$$V = \frac{EI}{2} \int_0^l \left(\frac{\partial^2 y}{\partial x^2} \right)^2 dx = \frac{EI}{2} \sum_{n=1}^{n=\infty} q_n{}^2 \int_0^l \left[\frac{\partial^2}{\partial x^2} \left(\sin \frac{n\pi x}{l} \right) \right]^2 dx$$

$$= \frac{EI}{2} \sum_{n=1}^{n=\infty} q_n{}^2 \left(\frac{n\pi}{l} \right)^4 \int_0^l \left(\sin \frac{n\pi x}{l} \right)^2 dx = \frac{EI}{2} \sum_{n=1}^{n=\infty} q_n{}^2 \left(\frac{n\pi}{l} \right)^4 \frac{l}{2}$$

For the virtual displacement, the change of elastic energy is

$$\Delta V = \frac{\partial V}{\partial q_m} \Delta q_m = \frac{EI}{2l^3} (n\pi)^4 q_m \Delta q_m = \frac{EI}{2l^3} (\kappa_n l)^4 q_m \Delta q_m$$

The value of the "inertia force" at each section is

$$F_I = -\frac{\gamma S}{g} \ddot{y} = -\frac{\gamma S}{g} \sum_{n=1}^{n=\infty} \frac{d^2 q_n}{dt^2} \sin \frac{n\pi x}{l}$$

The work done by this force during the virtual displacement Δy is

$$\Delta F_I = F_I \Delta y = -\frac{\gamma S}{g} \sum_{n=1}^{n=\infty} \frac{d^2 q_n}{dt^2} \Delta q_m \int_0^l \sin \frac{n\pi x}{l} \sin \frac{m\pi x}{l} dx$$

$$= -\frac{\gamma S l}{2g} \frac{d^2 q_m}{dt^2} \Delta q_m$$

The orthogonality relation of Eq. (7.1) is used here, making the integral vanish when $n = m$. For a disturbing force F_E, the work done during the virtual displacement is

$$\Delta F_E = F_E \Delta y = F(X_m)_{x=c} \Delta q_m$$

in which $(X_m)_{x=c}$ is the value of X_m at the point of application of the load. Substituting the terms into Eq. (7.49),

$$\frac{\gamma S l}{2g} \ddot{q}_m + \frac{EI}{2l^3} (\kappa_m l)^4 q_m = F(X_m)_{x=c}$$

Rearranging terms and letting $EI/S\gamma = a^2$,

$$\ddot{q}_m + \kappa_m{}^4 a^2 q_m = \frac{2g}{\gamma S l} F(X_m)_{x=c} \tag{7.52}$$

If F_E is a force which varies sinusoidally with time at any point $x = c$,

$$F(X_m)_{x=c} = \bar{F} \sin \frac{m\pi c}{l} \sin \omega t$$

and Eq. (7.52) becomes

$$\ddot{q}_m + \kappa_m{}^4 a^2 q_m = \frac{2g\bar{F}}{\gamma Sl} \sin \frac{m\pi c}{l} \sin \omega t$$

The solution of this equation is

$$q_m = A_m \sin \kappa_m{}^2 at + B_m \cos \kappa_m{}^2 at + \frac{2\bar{F}g}{\gamma Sl} \frac{\sin (m\pi c/l)}{\kappa_m{}^4 a^2 - \omega^2} \sin \omega t$$

Since $\kappa_m{}^2 a = \omega_m$,

$$q_m = A_m \sin \omega_m t + B_m \cos \omega_m t + \frac{2\bar{F}g}{\gamma Sl} \frac{\sin (m\pi c/l)}{\omega_m{}^2 - \omega^2} \sin \omega t$$

when the force acts at the center $c/l = \frac{1}{2}$. Substituting the corresponding values of q in Eq. (7.51), the solution is identical to Eq. (7.46), which was obtained by the classical method.

VIBRATION RESULTING FROM MOTION OF SUPPORT. When the supports of an elastic body are vibrated by some external force, forced vibration may be induced in the body.[47]   For example, consider the motion that results in a uniform beam, Fig. 7.14,

FIG. 7.14. Simply supported beam undergoing sinusoidal motion induced by sinusoidal motion of the supports.

when the supports are moved through a sinusoidally varying displacement $(y)_{x=0,l} = Y_0 \sin \omega t$. Although Eq. (7.13) was developed for the free vibration of beams, it is applicable to the present problem because there is no force acting on any section of the beam except the elastic force associated with the bending of the beam.   If a solution of the form $y = X(x) \sin \omega t$ is assumed and substituted into Eq. (7.13):

$$X'''' = \frac{\omega^2 \gamma S}{EIg} X \tag{7.53}$$

This equation is the same as Eq. (7.15) except that the natural frequency $\omega_n{}^2$ is replaced by the forcing frequency ω^2.   The solution of Eq. (7.53) is the same except that κ is replaced by $\kappa' = (\omega^2 \gamma S/EIg)^{1/4}$:

$$X = A_1 \sin \kappa' x + A_2 \cos \kappa' x + A_3 \sinh \kappa' x + A_4 \cosh \kappa' x \tag{7.54}$$

The solution of the problem is completed by finding the constants, which are determined by the boundary conditions.   Certain boundary conditions are associated with the supports of the beam and are the same as occur in the solution of the problem of free vibration.   Additional conditions are supplied by the displacement through which the supports are forced.   For example, if the supports of a beam having hinged ends are moved sinusoidally, the boundary conditions are: at $x = 0$ and $x = l$, $X'' = 0$, since the moment exerted by a hinged end is zero, and $X = Y_0$, since the amplitude of vibration is prescribed at each end.   By the use of these boundary conditions, Eq. (7.54) becomes

$$X = \frac{Y_0}{2} \left[\tan \frac{\kappa' l}{2} \sin \kappa' x + \cos \kappa' x - \tanh \frac{\kappa' l}{2} \sinh \kappa' x + \cosh \kappa' x \right] \tag{7.55}$$

The motion is defined by $y = X \sin \omega t$. For all values of κ', each of the coefficients except the first in Eq. (7.55) is finite. The tangent term becomes infinite if $\kappa'l = n\pi$, for odd values of n. The condition for the amplitude to become infinite is $\omega = \omega_n$ because $\kappa'/\kappa = \omega^2/\omega_n^2$ and, for natural vibration of a beam with hinged ends, $\kappa_n l = n\pi$. Thus, if the supports of an elastic body are vibrated at a frequency close to a natural frequency of the system, vibration at resonance occurs.

DAMPING. The effect of damping on forced vibration can be discussed only qualitatively. Damping usually decreases the amplitude of vibration, as it does in systems having a single degree-of-freedom. In some systems, it may cause coupling between modes, so that motion in a mode of vibration that normally would not be excited by a certain disturbing force may be induced.

REFERENCES

1. Timoshenko, S.: "Vibration Problems in Engineering," 3d ed., pp. 442, 448, D. Van Nostrand Company, Inc., Princeton, N.J., 1955.
2. Den Hartog, J. P.: "Mechanical Vibrations," 4th ed., p. 161, McGraw-Hill Book Company, Inc., New York, 1956.
3. Ref. 2, p. 152.
4. Hansen, H. M., and P. F. Chenea: "Mechanics of Vibration," p. 274, John Wiley & Sons, Inc., New York, 1952.
5. Jacobsen, L. S., and R. S. Ayre: "Engineering Vibrations," p. 73, McGraw-Hill Book Company, Inc., New York, 1958.
6. Ref. 4, p. 256.
7. Kerley, J. J.: Prod. Eng., Design Digest Issue, Mid-October, 1957, p. F34.
8. Young, D., and R. P. Felgar: "Tables of Characteristic Functions Representing Normal Modes of Vibration of a Beam," Univ. Texas Bur. Eng. Research Bull. 44, July 1, 1949.
9. Rayleigh, Lord: "The Theory of Sound," 2d rev. ed., vol. 1, p. 293; reprinted by Dover Publications, New York, 1945.
10. Timoshenko, S.: Phil. Mag. (ser. 6), 41:744 (1921); 43:125 (1922).
11. Sutherland, J. G., and L. E. Goodman: "Vibrations of Prismatic Bars Including Rotatory Inertia and Shear Corrections," Department of Civil Engineering, University of Illinois, Urbana, Ill., April 15, 1951.
12. Ref. 1, p. 374.
13. Ref. 1, 2d ed., p. 366.
14. Woinowsky-Krieger, S.: J. Appl. Mechanics, 17:35 (1950).
15. Cranch, E. T., and A. A. Adler: J. Appl. Mechanics, 23:103 (1956).
16. Macduff, J. N., and R. P. Felgar: Trans. ASME, 79:1459 (1957).
17. Macduff, J. N., and R. P. Felgar: Machine Design, 29(3):109 (1957).
18. Ref. 1, p. 386.
19. Darnley, E. R.: Phil. Mag., 41:81 (1921).
20. Smith, D. M.: Engineering, 120:808 (1925).
21. Newmark, N. M., and A. S. Veletsos. J. Appl. Mechanics, 19:503 (1952).
22. Timoshenko, S.: "Theory of Plates and Shells," p. 301, McGraw-Hill Book Company, Inc., New York, 1940.
23. Ref. 22, p. 88.
24. Ref. 22, p. 133.
25. Evans, T. H.: J. Appl. Mechanics, 6:A-7 (1939).
26. Ref. 22, p. 58.
27. Ref. 22, p. 304.
28. Ref. 22, p. 344.
28a. Leissa, A. W.: "Vibration of Plates," NASA SP-160, 1969.
29. Young, D.: J. Appl. Mechanics, 17:448 (1950).
30. Barton, M. V.: J. Appl. Mechanics, 18:129 (1951).
31. Hearmon, R. F. S.: J. Appl. Mechanics, 19:402 (1952).
32. Anderson, B. W.: J. Appl. Mechanics, 21:365 (1954).
33. Gustafson, P. N., W. F. Stokey, and C. F. Zorowski: J. Aeronaut. Sci., 20:331 (1953).
34. Gustafson, P. N., W. F. Stokey, and C. F. Zorowski: J. Aeronaut. Sci., 21:621 (1954).
35. Ref. 9, p. 359.
36. Ref. 1, p. 449.
37. Southwell, R. V.: Proc. Roy. Soc. (London), A101:133 (1922).
38. Chu, Hu-Nan, and G. Herrmann: J. Appl. Mechanics, 23:532 (1956).

39. Mindlin, R. D., A. Schacknow, and H. Deresiewicz: *J. Appl. Mechanics*, **23**:430 (1956).
40. Deresiewicz, H., and R. D. Mindlin: *J. Appl. Mechanics*, **22**:86 (1955).
41. Love, A. E. H.: "A Treatise on the Mathematical Theory of Elasticity," 4th ed., p. 451, reprinted by Dover Publications, New York, 1944.
42. Ref. 1, p. 425.
43. Marguerre, K.: *J. Math. Phys.*, **35**:28 (1956).
44. Pestel, E. C., and F. A. Leckie: "Matrix Methods in Elastomechanics," McGraw-Hill Book Company, Inc., New York, 1963.
45. Myklestad, N. O.: *J. Aeronaut. Sci.*, **11**:153 (1944).
46. Bohm, G. J.: *Nucl. Sci. Eng.*, **22**:143 (1965).
47. Mindlin, R. D., and L. E. Goodman: *J. Appl. Mech.*, **17**:377 (1950).

8

TRANSIENT RESPONSE TO
STEP AND PULSE FUNCTIONS *

Robert S. Ayre
University of Colorado

INTRODUCTION

In analyses involving shock and transient vibration, it is essential in most instances to begin with the time-history of a quantity that describes a motion, usually displacement, velocity, or acceleration. The method of reducing the time-history depends upon the purpose for which the reduced data will be used. When the purpose is to compare shock motions, to design equipment to withstand shock, or to formulate a laboratory test as means to simulate an environmental condition, the *response spectrum* is found to be a useful concept. This concept in data reduction is discussed in Chap. 23, and its application to the simulation of environmental conditions is discussed in Chap. 24.

This chapter deals briefly with methods of analysis for obtaining the response spectrum from the time-history, and includes in graphical form certain significant spectra for various regular step- and pulse-type excitations. The usual concept of the response spectrum is based upon the single degree-of-freedom system, usually considered linear and undamped, although useful information sometimes can be obtained by introducing nonlinearity or damping. The single degree-of-freedom system is considered to be subjected to the shock or transient vibration, and its response determined.

The *response spectrum* is a graphical presentation of a selected quantity in the response taken with reference to a quantity in the excitation. It is plotted as a function of a dimensionless parameter that includes the natural period of the responding system and a significant period of the excitation. The excitation may be defined in terms of various physical quantities, and the response spectrum likewise may depict various characteristics of the response.

LINEAR, UNDAMPED, SINGLE DEGREE-OF-FREEDOM SYSTEMS

DIFFERENTIAL EQUATION OF MOTION

It is assumed that the system is linear and undamped. The excitation, which is a known function of time alone, may be a force function $F(t)$ acting directly on the mass of the system (Fig. 8.1A) or it may be a ground motion, i.e., foundation or base motion, acting on the spring anchorage. The ground motion may be expressed as a ground dis-

---

* Chapter 8 is based on Chaps. 3 and 4 of "Engineering Vibrations," by L. S. Jacobsen and R. S. Ayre, McGraw-Hill Book Company, Inc., 1958, and on the report, "A Comparative Study of Pulse and Step-type Loads on a Simple Vibratory System," by L. S. Jacobsen and R. S. Ayre, U.S. Navy Contract N6-ori-154, T. O. 1, Stanford University, 1952.

placement function $u(t)$ (Fig. 8.1B). In many cases, however, it is more useful to express it as a ground acceleration function $\ddot{u}(t)$ (Fig. 8.1C).

The differential equation of motion, written in terms of each of the types of excitation, is given in Eqs. (8.1a), (8.1b), and (8.1c).

$$m\ddot{x} = -kx + F(t) \qquad \text{or} \qquad \frac{m\ddot{x}}{k} + x = \frac{F(t)}{k} \tag{8.1a}$$

$$m\ddot{x} = -k[x - u(t)] \qquad \text{or} \qquad \frac{m\ddot{x}}{k} + x = u(t) \tag{8.1b}$$

$$m[\ddot{\delta}_x + \ddot{u}(t)] = -k\delta_x \qquad \text{or} \qquad \frac{m\ddot{\delta}_x}{k} + \delta_x = -\frac{m\ddot{u}(t)}{k} \tag{8.1c}$$

where x is the displacement (absolute displacement) of the mass relative to a *fixed reference* and δ_x is the displacement relative to a *moving anchorage* or ground. These displacements are related to the ground displacement by $x = u + \delta_x$. Similarly, the accelerations are related by $\ddot{x} = \ddot{u} + \ddot{\delta}_x$.

Furthermore, if Eq. (8.1b) is differentiated twice with respect to time, a differential equation is obtained in which ground acceleration $\ddot{u}(t)$ is the excitation and the absolute acceleration $\ddot{x}$ of the mass m is the variable. The equation is

$$\frac{m}{k}\frac{d^2\ddot{x}}{dt^2} + \ddot{x} = \ddot{u}(t) \tag{8.1d}$$

If Eq. (8.1d) is treated as a second-order equation in $\ddot{x}$ as the dependent variable, it is of the same general form as Eqs. (8.1a), (8.1b) and (8.1c).

Occasionally, the excitation is known in terms of ground velocity $\dot{u}(t)$. Differentiating Eq. (8.1b) once with respect to time, the following second-order equation in $\dot{x}$ is obtained:

$$\frac{m}{k}\frac{d^2\dot{x}}{dt^2} + \dot{x} = \dot{u}(t) \tag{8.1e}$$

The analogy represented by Eqs. (8.1b), (8.1d), and (8.1e) may be extended further since it is generally possible to differentiate Eq. (8.1b) any number of times n:

$$\frac{m}{k}\frac{d^2}{dt^2}\left(\frac{d^nx}{dt^n}\right) + \left(\frac{d^nx}{dt^n}\right) = \left(\frac{d^nu}{dt^n}\right)(t) \tag{8.1f}$$

Fig. 8.1. Simple oscillator acted upon by known excitation functions of time: (A) force $F(t)$, (B) ground displacement $u(t)$, (C) ground acceleration $\ddot{u}(t)$.

This is of the same general form as the preceding equations if it is considered to be a second-order equation in (d^nx/dt^n) as the response variable, with $(d^nu/dt^n)(t)$, a known function of time, as the excitation.

ALTERNATE FORMS OF THE EXCITATION AND OF THE RESPONSE

The foregoing equations are alike, mathematically, and a solution in terms of one of them may be applied to any of the others by making simple substitutions. Therefore, the equations may be expressed in the single general form:

$$\frac{m}{k}\ddot{\nu} + \nu = \xi(t) \tag{8.2}$$

where ν and ξ are the *response* and the *excitation*, respectively, at time t.

A general notation (ν and ξ) is desirable in the presentation of response functions and response spectra for general use. However, in the discussion of examples of solution, it sometimes is preferable to use more specific notations. Both types of notation are used in this chapter. For ready reference, the alternate forms of the excitation and the response are given in Table 8.1 where $\omega_n{}^2 = k/m$.

Table 8.1.   Alternate Forms of Excitation and Response in Eq. (8.2)

Excitation $\xi(t)$		Response ν	
Force	$\dfrac{F(t)}{k}$	Absolute displacement	x
Ground displacement	$u(t)$	Absolute displacement	x
Ground acceleration	$\dfrac{-\ddot{u}(t)}{\omega_n{}^2}$	Relative displacement	δ_x
Ground acceleration	$\ddot{u}(t)$	Absolute acceleration	$\ddot{x}$
Ground velocity	$\dot{u}(t)$	Absolute velocity	$\dot{x}$
nth derivative of ground displacement	$\dfrac{d^n u}{dt^n}(t)$	nth derivative of absolute displacement	$\dfrac{d^n x}{dt^n}$

METHODS OF SOLUTION OF THE DIFFERENTIAL EQUATION

A brief review of four methods of solution is given in the following sections.

CLASSICAL SOLUTION. The complete solution of the linear differential equation of motion consists of the sum of the *particular integral* x_1 and the *complementary function* x_2, that is, $x = x_1 + x_2$. Since the differential equation is of second order, two constants of integration are involved. They appear in the complementary function and are evaluated from a knowledge of the initial conditions.

Example 8.1. Versed-sine Force Pulse. In this case the differential equation of motion, applicable for the duration of the pulse, is

$$\frac{m\ddot{x}}{k} + x = \frac{F_p}{k}\frac{1}{2}\left(1 - \cos\frac{2\pi t}{\tau}\right) \qquad [0 \leq t \leq \tau] \tag{8.3a}$$

where, in terms of the general notation, the excitation function $\xi(t)$ is

$$\xi(t) \equiv \frac{F(t)}{k} = \frac{F_p}{k}\frac{1}{2}\left(1 - \cos\frac{2\pi t}{\tau}\right)$$

and the response ν is displacement x. The maximum value of the pulse excitation force is F_p.

The particular integral (particular solution) for Eq. (8.3a) is of the form

$$x_1 = M + N\cos\frac{2\pi t}{\tau} \tag{8.3b}$$

By substitution of the particular solution into the differential equation, the required values of the coefficients M and N are found.

The complementary function is

$$x_2 = A\cos\omega_n t + B\sin\omega_n t \tag{8.3c}$$

where A and B are the constants of integration. Combining x_2 and the explicit form of x_1 gives the complete solution:

$$x = x_1 + x_2 = \frac{F_p/2k}{1 - \tau^2/T^2}\left(1 - \frac{\tau^2}{T^2} + \frac{\tau^2}{T^2}\cos\frac{2\pi t}{\tau}\right) + A\cos\omega_n t + B\sin\omega_n t \quad (8.3d)$$

If it is assumed that the system is initially at rest, $x = 0$ and $\dot{x} = 0$ at $t = 0$, and the constants of integration are

$$A = -\frac{F_p/2k}{1 - \tau^2/T^2} \quad \text{and} \quad B = 0 \quad (8.3e)$$

The complete solution takes the following form:

$$\nu \equiv x = \frac{F_p/2k}{1 - \tau^2/T^2}\left(1 - \frac{\tau^2}{T^2} + \frac{\tau^2}{T^2}\cos\frac{2\pi t}{\tau} - \cos\omega_n t\right) \quad (8.3f)$$

If other starting conditions had been assumed, A and B would have been different from the values given by Eqs. (8.3e). It may be shown that if the starting conditions are general, namely, $x = x_0$ and $\dot{x} = \dot{x}_0$ at $t = 0$, it is necessary to superimpose on the complete solution already found, Eq. (8.3f), only the following additional terms:

$$x_0\cos\omega_n t + \frac{\dot{x}_0}{\omega_n}\sin\omega_n t \quad (8.3g)$$

For values of time equal to or greater than τ, the differential equation is

$$m\ddot{x} + kx = 0 \quad [\tau \leq t] \quad (8.4a)$$

and the complete solution is given by the complementary function alone. However, the constants of integration must be redetermined from the known conditions of the system at time $t = \tau$. The solution is

$$\nu \equiv x = \frac{F_p}{k}\frac{\sin(\pi\tau/T)}{1 - \tau^2/T^2}\sin\omega_n\left(t - \frac{\tau}{2}\right) \quad [\tau \leq t] \quad (8.4b)$$

The additional terms given by expressions (8.3g) may be superimposed on this solution if the conditions at time $t = 0$ are general.

DUHAMEL'S INTEGRAL. The use of Duhamel's integral (convolution integral or superposition integral) is a well-known approach to the solution of transient vibration problems in linear systems. Its development [7,33] is based on the *superposition* of the responses of the system to a sequence of impulses.

A general excitation function is shown in Fig. 8.2, where $F(t)$ is a known force function of time, the variable of integration is t_v between the limits of integration 0 and t and the elemental impulse is $F(t_v)\,dt_v$. It may be shown that the complete solution of the differential equation is

$$x = \left(x_0 - \frac{1}{m\omega_n}\int_0^t F(t_v)\sin\omega_n t_v\,dt_v\right)\cos\omega_n t + \left(\frac{\dot{x}_0}{\omega_n} + \frac{1}{m\omega_n}\int_0^t F(t_v)\cos\omega_n t_v\,dt_v\right)\sin\omega_n t \quad (8.5)$$

where x_0 and $\dot{x}_0$ are the initial conditions of the system at zero time.

FIG. 8.2. General excitation and the elemental impulse.

Example 8.2. Half-cycle Sine, Ground Displacement Pulse. Consider the following excitation:

$$\xi(t) \equiv u(t) = \begin{cases} u_p \sin \dfrac{\pi t}{\tau} & [0 \le t \le \tau] \\ 0 & [\tau \le t] \end{cases}$$

The maximum value of the excitation displacement is u_p. Assume that the system is initially at rest, so that $x_0 = \dot{x}_0 = 0$. Expressing the excitation function in terms of the variable of integration t_v, Eq. (8.5) may be rewritten for this particular case in the following form:

$$x = \frac{ku_p}{m\omega_n} \left(-\cos \omega_n t \int_0^t \sin \frac{\pi t_v}{\tau} \sin \omega_n t_v \, dt_v + \sin \omega_n t \int_0^t \sin \frac{\pi t_v}{\tau} \cos \omega_n t_v \, dt_v \right) \qquad (8.6a)$$

Equation (8.6a) may be reduced, by evaluation of the integrals, to

$$\nu \equiv x = \frac{u_p}{1 - T^2/4\tau^2} \left(\sin \frac{\pi t}{\tau} - \frac{T}{2\tau} \sin \omega_n t \right) \qquad [0 \le t \le \tau] \qquad (8.6b)$$

where $T = 2\pi/\omega_n$ is the natural period of the responding system.

For the second era of time, where $\tau \le t$, it is convenient to choose a new time variable $t' = t - \tau$. Noting that $u(t) = 0$ for $\tau \le t$, and that for continuity in the system response the initial conditions for the second era must equal the closing conditions for the first era, it is found from Eq. (8.5) that the response for the second era is

$$x = x_\tau \cos \omega_n t' + \frac{\dot{x}_\tau}{\omega_n} \sin \omega_n t' \qquad (8.7a)$$

where x_τ and $\dot{x}_\tau$ are the displacement and velocity of the system at time $t = \tau$ and hence at $t' = 0$. Equation (8.7a) may be rewritten in the following form:

$$\nu \equiv x = u_p \frac{(T/\tau) \cos (\pi\tau/T)}{(T^2/4\tau^2) - 1} \sin \omega_n \left(t - \frac{\tau}{2} \right) \qquad [\tau \le t] \qquad (8.7b)$$

PHASE-PLANE GRAPHICAL METHOD. Several numerical and graphical methods,[18, 23, 27, 37] all related in general but differing considerably in the details of procedure, are available for the solution of linear transient vibration problems. Of these methods, the phase-plane graphical method [28] is one of the most useful. The procedure is basically

FIG. 8.3. General excitation approximated by a sequence of finite rectangular steps.

very simple, it gives a clear physical picture of the response of the system, and it may be applied readily to some classes of *nonlinear* systems.[3, 5, 6, 8, 13, 15, 21, 22]

In Fig. 8.3 a general excitation in terms of ground displacement is represented, approximately, by a sequence of finite steps. The ith step has the total height u_i, where u_i is constant for the duration of the step. The differential equation of motion and its complete solution, applying for the duration of the step, are

$$\frac{m\ddot{x}}{k} + x = u_i \qquad [t_{i-1} \le t \le t_i] \qquad (8.8a)$$

$$x - u_i = (x_{i-1} - u_i) \cos \omega_n(t - t_{i-1}) + \frac{\dot{x}_{i-1}}{\omega_n} \sin \omega_n(t - t_{i-1}) \qquad (8.8b)$$

where x_{i-1} and $\dot{x}_{i-1}$ are the displacement and velocity of the system at time t_{i-1}; consequently, they are the initial conditions for the ith step. The system velocity (divided by ω_n) during the ith step is

$$\frac{\dot{x}}{\omega_n} = - (x_{i-1} - u_i) \sin \omega_n(t - t_{i-1}) + \frac{\dot{x}_{i-1}}{\omega_n} \cos \omega_n(t - t_{i-1}) \qquad (8.8c)$$

Squaring Eqs. (8.8b) and (8.8c) and adding them,

$$\left(\frac{\dot{x}}{\omega_n}\right)^2 + (x - u_i)^2 = \left(\frac{\dot{x}_{i-1}}{\omega_n}\right)^2 + (x_{i-1} - u_i)^2 \qquad (8.8d)$$

This is the equation of a circle in a rectangular system of coordinates $\dot{x}/\omega_n$, x. The center is at 0, u_i; and the radius is

$$R_i = \left[\left(\frac{\dot{x}_{i-1}}{\omega_n}\right)^2 + (x_{i-1} - u_i)^2\right]^{1/2} \qquad (8.8e)$$

The solution for Eq. (8.8a) for the ith step may be shown, as in Fig. 8.4, to be the arc of the circle of radius R_i and center 0, u_i, subtended by the angle $\omega_n(t_i - t_{i-1})$ and starting at the point $\dot{x}_{i-1}/\omega_n$, x_{i-1}. Time is positive in the counterclockwise direction.

Fig. 8.4. Graphical representation in the phase-plane of the solution for the ith step.

Example 8.3. Application to a General Pulse Excitation. Figure 8.5 shows an application of the method for the general excitation $u(t)$ represented by seven steps in the time-displacement plane. Upon choice of the step heights u_i and durations $(t_i - t_{i-1})$, the arc-center locations can be projected onto the X axis in the phase-plane and the arc angles $\omega_n(t_i - t_{i-1})$ can be computed. The graphical construction of the sequence of circular arcs, the *phase trajectory*, is then carried out, using the system conditions at zero time (in this example, 0,0) as the starting point.

Projection of the system displacements from the phase-plane into the time-displacement plane at once determines the time-displacement response curve. The time-velocity response can also be determined by projection as shown. The velocities and displacements at particular instants of time can be found directly from the phase trajectory coordinates without the necessity for drawing the time-response curves. Furthermore, the times of occurrence and the magnitudes of all the maxima also can be obtained directly from the phase trajectory.

Good accuracy is obtainable by using reasonable care in the graphical construction and in the choice of the steps representing the excitation. Usually, the time intervals should not be longer than about one-fourth the natural period of the system.[22]

THE LAPLACE TRANSFORMATION. The Laplace transformation provides a powerful tool for the solution of linear differential equations. The following discussion of the technique of its application is limited to the differential equation of the type applying to the undamped linear oscillator. Application to the linear oscillator with viscous damping is illustrated in a later part of this chapter.

Definitions. The Laplace transform $F(s)$ of a known function $f(t)$, where $t > 0$, is defined by

$$F(s) = \int_0^\infty e^{-st} f(t)\, dt \qquad (8.9a)$$

where s is a complex variable. The transformation is abbreviated as

$$F(s) = \mathcal{L}[f(t)] \qquad (8.9b)$$

The limitations on the function $f(t)$ are not discussed here. For the conditions for existence of $\mathcal{L}[f(t)]$, for complete accounts of the technique of application, and for extensive tables of function-transform pairs, the references should be consulted.[16, 17, 34, 36]

FIG. 8.5. Example of phase-plane graphical solution.[2]

General Steps in Solution of the Differential Equation. In the solution of a differential equation by Laplace transformation, the first step is to transform the differential equation, in the variable t, into an algebraic equation in the complex variable s. Then, the algebraic equation is solved, and the solution of the differential equation is determined by an *inverse* transformation of the solution of the algebraic equation. The process of inverse Laplace transformation is symbolized by

$$\mathcal{L}^{-1}[F(s)] = f(t) \qquad (8.10)$$

Tables of Function-transform Pairs. The processes symbolized by Eqs. (8.9b) and (8.10) are facilitated by the use of tables of function-transform pairs. Table 8.2 is a brief

Table 8.2. Pairs of Functions $f(t)$ and Laplace Transforms $F(s)$

	$f(t)$	$F(s)$
	Operation Transforms	
1	Definition, $f(t)$	$F(s) = \int_0^\infty e^{-st} f(t)\, dt$
2	First derivative, $f'(t)$	$sF(s) - f(0)$
3	nth derivative, $f^{(n)}(t)$	$s^n F(s) - s^{n-1} f(0) - s^{n-2} f'(0) - \cdots$ $\qquad\qquad - sf^{(n-2)}(0) - f^{(n-1)}(0)$ †
4	Superposition, $C_1 f_1(t) + C_2 f_2(t) + \cdots$ $\qquad\qquad + C_n f_n(t)$	$C_1 F_1(s) + C_2 F_2(s) + \cdots + C_n F_n(s)$
5	Shifting in s plane, $e^{at} f(t)$	$\int_0^\infty e^{-st} e^{at} f(t)\, dt = \int_0^\infty e^{-(s-a)t} f(t)\, dt$ $\qquad\qquad = F(s - a)$
6	Shifting in t plane $\begin{cases} f(t-b) \text{ when } t > b, \\ 0 \text{ when } t < b \end{cases}$	$e^{-bs} F(s)$
	Function Transforms	
7	1	$\dfrac{1}{s}$
8	$\dfrac{t^{n-1}}{(n-1)!}$	$\dfrac{1}{s^n}$, for $n = 1, 2, \ldots$
9	e^{-at}	$\dfrac{1}{s+a}$
10	$\dfrac{1}{a}(1 - e^{-at})$	$\dfrac{1}{s(s+a)}$
11	te^{-at}	$\dfrac{1}{(s+a)^2}$
12	$\dfrac{1}{a}\sin at$	$\dfrac{1}{s^2 + a^2}$
13	$\dfrac{1}{a^2}(1 - \cos at)$	$\dfrac{1}{s(s^2 + a^2)}$
14	$\dfrac{1}{a^3}(at - \sin at)$	$\dfrac{1}{s^2(s^2 + a^2)}$
15	$\dfrac{1}{(b-a)}(e^{-at} - e^{-bt})$	$\dfrac{1}{(s+a)(s+b)}$
16	$\dfrac{1}{ab} + \dfrac{be^{-at} - ae^{-bt}}{ab(a-b)}$	$\dfrac{1}{s(s+a)(s+b)}$
17	$\dfrac{a\sin bt - b\sin at}{ab(a^2 - b^2)}$	$\dfrac{1}{(s^2 + a^2)(s^2 + b^2)}$
18	$e^{-at}(1 - at)$	$\dfrac{s}{(s+a)^2}$
19	$\cos at$	$\dfrac{s}{s^2 + a^2}$
20	Rectangular pulse	$\dfrac{1 - e^{-s\tau}}{s}$
21	Sine pulse	$\dfrac{\pi/\tau}{s^2 + \pi^2/\tau^2}(1 + e^{-s\tau})$

† $f(t)$ and its derivatives through $f^{(n-1)}(t)$ must be continuous.

example. Transforms for general operations, such as differentiation, are included as well as transforms of explicit functions.

In general, the transforms of the explicit functions can be obtained by carrying out the integration indicated by the definition of the Laplace transformation. For example:

For $f(t) = 1$:

$$F(s) = \int_0^\infty e^{-st}\, dt = -\frac{1}{s} e^{-st} \Big]_0^\infty = \frac{1}{s}$$

Transformation of the Differential Equation. The differential equation for the undamped linear oscillator is given in general form by

$$\frac{1}{\omega_n^2}\ddot{v} + v = \xi(t) \tag{8.11}$$

Applying the operational transforms (items 1 and 3, Table 8.2), Eq. (8.11) is transformed to

$$\frac{1}{\omega_n^2} s^2 F_r(s) - \frac{1}{\omega_n^2} sf(0) - \frac{1}{\omega_n^2} f'(0) + F_r(s) = F_e(s) \tag{8.12a}$$

where $F_r(s)$ = the transform of the unknown response $v(t)$, sometimes called the *response transform*

$s^2 F_r(s) - sf(0) - f'(0)$ = the transform of the second derivative of $v(t)$

$f(0)$ and $f'(0)$ = the known *initial values* of v and $\dot{v}$, i.e., v_0 and $\dot{v}_0$

$F_e(s)$ = the transform of the known excitation function $\xi(t)$, written $F_e(s) = \mathcal{L}[\xi(t)]$, sometimes called the *driving transform*

It should be noted that the initial conditions of the system are explicit in Eq. (8.12a).

The Subsidiary Equation. Solving Eq. (8.12a) for $F_r(s)$,

$$F_r(s) = \frac{sf(0) + f'(0) + \omega_n^2 F_e(s)}{s^2 + \omega_n^2} \tag{8.12b}$$

This is known as the *subsidiary equation* of the differential equation. The first two terms of the transform derive from the initial conditions of the system, and the third term derives from the excitation.

Inverse Transformation. In order to determine the response function $v(t)$, which is the solution of the differential equation, an inverse transformation is performed on the subsidiary equation. The entire operation, applied explicitly to the solution of Eq. (8.11), may be abbreviated as follows:

$$v(t) = \mathcal{L}^{-1}[F_r(s)] = \mathcal{L}^{-1}\left[\frac{sv_0 + \dot{v}_0 + \omega_n^2 \mathcal{L}[\xi(t)]}{s^2 + \omega_n^2}\right] \tag{8.13}$$

Example 8.4. Rectangular Step Excitation. In this case $\xi(t) = \xi_c$ for $0 \le t$ (Fig. 8.6A). The Laplace transform $F_e(s)$ of the excitation is, from item 7 of Table 8.2,

$$\mathcal{L}[\xi_c] = \xi_c \mathcal{L}[1] = \xi_c \frac{1}{s}$$

Assume that the starting conditions are general, that is, $v = v_0$ and $\dot{v} = \dot{v}_0$ at $t = 0$. Substituting the transform and the starting conditions into Eq. (8.13), the following is obtained:

$$v(t) = \mathcal{L}^{-1}\left[\frac{sv_0 + \dot{v}_0 + \omega_n^2 \xi_c(1/s)}{s^2 + \omega_n^2}\right] \tag{8.14a}$$

The foregoing may be rewritten as three separate inverse transforms:

$$v(t) = v_0 \mathcal{L}^{-1}\left[\frac{s}{s^2 + \omega_n^2}\right] + \dot{v}_0 \mathcal{L}^{-1}\left[\frac{1}{s^2 + \omega_n^2}\right] + \xi_c \omega_n^2 \mathcal{L}^{-1}\left[\frac{1}{s(s^2 + \omega_n^2)}\right] \tag{8.14b}$$

The inverse transforms in Eq. (8.14b) are evaluated by use of items 19, 12 and 13, respectively, in Table 8.2. Thus, the time-response function is given explicitly by

$$\nu(t) = \nu_0 \cos \omega_n t + \frac{\dot{\nu}_0}{\omega_n} \sin \omega_n t + \xi_c(1 - \cos \omega_n t) \tag{8.14c}$$

The first two terms are the same as the starting condition response terms given by expressions (8.20a). The third term agrees with the response function shown by Eq. (8.22), derived for the case of a start from rest.

Example 8.5. Rectangular Pulse Excitation. The excitation function, Fig. 8.6B, is given by

$$\xi(t) = \begin{cases} \xi_p & \text{for } 0 \le t \le \tau \\ 0 & \text{for } \tau \le t \end{cases}$$

For simplicity, assume a start from rest, i.e., $\nu_0 = 0$ and $\dot{\nu}_0 = 0$ when $t = 0$.

During the first time interval, $0 \le t \le \tau$, the response function is of the same form as Eq. (8.14c) except that, with the assumed start from rest, the first two terms are zero.

During the second interval, $\tau \le t$, the transform of the excitation is obtained by applying the delayed-function transform (item 6, Table 8.2) and the transform for the rectangular step function (item 7) with the following result:

$$F_e(s) = \mathcal{L}[\xi(t)] = \xi_p \left(\frac{1}{s} - \frac{e^{-s\tau}}{s} \right)$$

This is the transform of an excitation consisting of a rectangular step of height $-\xi_p$ starting at time $t = \tau$, superimposed on the rectangular step of height $+\xi_p$ starting at time $t = 0$.

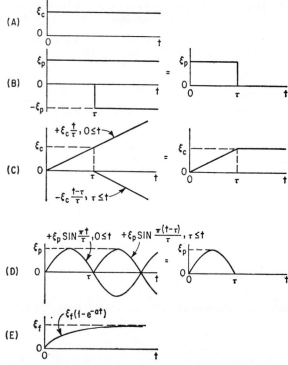

FIG. 8.6. Excitation functions in examples of use of the Laplace transform: (A) rectangular step, (B) rectangular pulse, (C) step with constant-slope front, (D) sine pulse, and (E) step with exponential asymptotic rise.

Substituting for $\mathcal{L}[\xi(t)]$ in Eq. (8.13),

$$\nu(t) = \xi_p\omega_n{}^2 \left\{ \mathcal{L}^{-1}\left[\frac{1}{s(s^2 + \omega_n{}^2)}\right] - \mathcal{L}^{-1}\left[\frac{e^{-s\tau}}{s(s^2 + \omega_n{}^2)}\right]\right\} \tag{8.15a}$$

The first inverse transform in Eq. (8.15a) is the same as the third one in Eq. (8.14b) and is evaluated by use of item 13 in Table 8.2. However, the second inverse transform requires the use of items 6 and 13. The function-transform pair given by item 6 indicates that when $t < b$ the inverse transform in question is zero, and when $t > b$ the inverse transform is evaluated by replacing t by $t - b$ (in this particular case, by $t - \tau$). The result is as follows:

$$\nu(t) = \xi_p\omega_n{}^2 \left\{\frac{1}{\omega_n{}^2}(1 - \cos\omega_n t) - \frac{1}{\omega_n{}^2}[1 - \cos\omega_n(t - \tau)]\right\}$$

$$= 2\xi_p \sin\frac{\pi\tau}{T}\sin\omega_n\left(t - \frac{\tau}{2}\right) \qquad [\tau \leq t] \tag{8.15b}$$

Theorem on the Transform of Functions Shifted in the Original (t) Plane. In Example 8.5, use is made of the theorem on the transform of functions shifted in the original plane. The theorem (item 6 in Table 8.2) is known variously as the second shifting theorem, the theorem on the transform of delayed functions, and the time-displacement theorem. In determining the transform of the excitation, the theorem provides for shifting, i.e., displacing the excitation or a component of the excitation in the positive direction along the time axis. This suggests the term *delayed function.* Examples of the shifting of component parts of the excitation appear in Fig. 8.6B, 8.6C, and 8.6D. Use of the theorem also is necessary in determining, by means of inverse transformation, the response following the delay in the excitation. Further illustration of the use of the theorem is shown by the next two examples.

Example 8.6. Step Function with Constant-slope Front. The excitation function (Fig. 8.6C) is expressed as follows:

$$\xi(t) = \begin{cases} \xi_c\dfrac{t}{\tau} & [0 \leq t \leq \tau] \\ \xi_c & [\tau \leq t] \end{cases}$$

Assume that $\nu_0 = 0$ and $\dot{\nu}_0 = 0$.

The driving transforms for the first and second time intervals are

$$\mathcal{L}[\xi(t)] = \begin{cases} \xi_c\dfrac{1}{\tau}\dfrac{1}{s^2} & [0 \leq t \leq \tau] \\ \xi_c\dfrac{1}{\tau}\left(\dfrac{1}{s^2} - \dfrac{e^{-s\tau}}{s^2}\right) & [\tau \leq t] \end{cases}$$

The transform for the second interval is the transform of a *negative* constant slope excitation, $-\xi_c(t - \tau)/\tau$, starting at $t = \tau$, superimposed on the transform for the positive constant slope excitation, $+\xi_c t/\tau$, starting at $t = 0$.

Substituting the transforms and starting conditions into Eq. (8.13), the responses for the two time eras, in terms of the transformations, are

$$\nu(t) = \begin{cases} \xi_c\dfrac{\omega_n{}^2}{\tau}\mathcal{L}^{-1}\left[\dfrac{1}{s^2(s^2 + \omega_n{}^2)}\right] & [0 \leq t \leq \tau] \\ \xi_c\dfrac{\omega_n{}^2}{\tau}\left\{\mathcal{L}^{-1}\left[\dfrac{1}{s^2(s^2 + \omega_n{}^2)}\right] - \mathcal{L}^{-1}\left[\dfrac{e^{-s\tau}}{s^2(s^2 + \omega_n{}^2)}\right]\right\} & [\tau \leq t] \end{cases} \tag{8.16a}$$

Evaluation of the inverse transforms by reference to Table 8.2 [item 14 for the first of Eqs. (8.16a), items 6 and 14 for the second] leads to the following:

$$\nu(t) = \begin{cases} \xi_c\dfrac{\omega_n{}^2}{\tau}\dfrac{1}{\omega_n{}^3}(\omega_n t - \sin\omega_n t) & [0 \leq t \leq \tau] \\ \xi_c\dfrac{\omega_n{}^2}{\tau}\left\{\dfrac{1}{\omega_n{}^3}(\omega_n t - \sin\omega_n t) - \dfrac{1}{\omega_n{}^3}[\omega_n(t - \tau) - \sin\omega_n(t - \tau)]\right\} & [\tau \leq t] \end{cases}$$

Simplifying,

$$
\nu(t) = \begin{cases} \xi_c \dfrac{1}{\omega_n \tau}(\omega_n t - \sin \omega_n t) & [0 \le t \le \tau] \\[2mm] \xi_c \left[1 + \dfrac{2}{\omega_n \tau} \sin \dfrac{\omega_n \tau}{2} \cos \omega_n \left(t - \dfrac{\tau}{2}\right)\right] & [\tau \le t] \end{cases}
\tag{8.16b}
$$

Example 8.7. Half-cycle Sine Pulse. The excitation function (Fig. 8.6D) is

$$
\xi(t) = \begin{cases} \xi_p \sin \dfrac{\pi t}{\tau} & [0 \le t \le \tau] \\[2mm] 0 & [\tau \le t] \end{cases}
$$

Let the system start from rest. The driving transforms are

$$
\mathcal{L}[\xi(t)] = \begin{cases} \xi_p \dfrac{\pi}{\tau} \dfrac{1}{s^2 + \pi^2/\tau^2} & [0 \le t \le \tau] \\[2mm] \xi_p \dfrac{\pi}{\tau}\left(\dfrac{1}{s^2 + \pi^2/\tau^2} + \dfrac{e^{-s\tau}}{s^2 + \pi^2/\tau^2}\right) & [\tau \le t] \end{cases}
$$

The driving transform for the second interval is the transform of a sine wave of *positive* amplitude ξ_p and frequency π/τ starting at time $t = \tau$, superimposed on the transform of a sine wave of the same amplitude and frequency starting at time $t = 0$.

By substitution of the driving transforms and the starting conditions into Eq. (8.13), the following are found:

$$
\nu(t) = \begin{cases} \xi_p \dfrac{\pi}{\tau} \omega_n^2 \mathcal{L}^{-1}\left[\dfrac{1}{s^2 + \pi^2/\tau^2} \cdot \dfrac{1}{s^2 + \omega_n^2}\right] & [0 \le t \le \tau] \\[2mm] \xi_p \dfrac{\pi}{\tau} \omega_n^2 \left\{\mathcal{L}^{-1}\left[\dfrac{1}{s^2 + \pi^2/\tau^2} \cdot \dfrac{1}{s^2 + \omega_n^2}\right] + \mathcal{L}^{-1}\left[\dfrac{e^{-s\tau}}{s^2 + \pi^2/\tau^2} \cdot \dfrac{1}{s^2 + \omega_n^2}\right]\right\} & [\tau \le t] \end{cases}
\tag{8.17a}
$$

Determining the inverse transforms from Table 8.2 [item 17 for the first of Eqs. (8.17a), items 6 and 17 for the second]:

$$
\nu(t) = \begin{cases} \xi_p \dfrac{\pi}{\tau} \omega_n^2 \dfrac{\omega_n \sin (\pi t/\tau) - (\pi/\tau) \sin \omega_n t}{(\pi \omega_n/\tau)(\omega_n^2 - \pi^2/\tau^2)} & [0 \le t \le \tau] \\[2mm] \xi_p \dfrac{\pi}{\tau} \omega_n^2 \left[\dfrac{\omega_n \sin (\pi t/\tau) - (\pi/\tau) \sin \omega_n t}{(\pi \omega_n/\tau)(\omega_n^2 - \pi^2/\tau^2)}\right. \\[2mm] \qquad \left. + \dfrac{\omega_n \sin [\pi(t - \tau)/\tau] - (\pi/\tau) \sin \omega_n (t - \tau)}{(\pi \omega_n/\tau)(\omega_n^2 - \pi^2/\tau^2)}\right] & [\tau \le t] \end{cases}
$$

Simplifying,

$$
\nu(t) = \begin{cases} \xi_p \dfrac{1}{1 - T^2/4\tau^2}\left(\sin \dfrac{\pi t}{\tau} - \dfrac{T}{2\tau} \sin \omega_n t\right) & [0 \le t \le \tau] \\[2mm] \xi_p \dfrac{(T/\tau) \cos (\pi \tau/T)}{(T^2/4\tau^2) - 1} \sin \omega_n \left(t - \dfrac{\tau}{2}\right) & [\tau \le t] \end{cases}
\tag{8.17b}
$$

where $T = 2\pi/\omega_n$ is the natural period of the responding system. Equations (8.17b) are equivalent to Eqs. (8.6b) and (8.7b) derived previously by the use of Duhamel's integral.

Example 8.8. Exponential Asymptotic Step. The excitation function (Fig. 8.6E) is

$$
\xi(t) = \xi_f(1 - e^{-at}) \qquad\qquad [0 \le t]
$$

Assume that the system starts from rest. The driving transform is

$$
\mathcal{L}[\xi(t)] = \xi_f\left(\dfrac{1}{s} - \dfrac{1}{s + a}\right) = \xi_f a \dfrac{1}{s(s + a)}
$$

It is found by Eq. (8.13) that

$$\nu(t) = \xi_f a \omega_n^2 \mathcal{L}^{-1} \left[\frac{1}{s(s+a)(s^2+\omega_n^2)} \right] \qquad [0 \le t] \qquad (8.18a)$$

It frequently happens that the inverse transform is not readily found in an available table of transforms. Using the above case as an example, the function of s in Eq. (8.18a) is first *expanded in partial fractions*; then the inverse transforms are sought, thus:

$$\frac{1}{s(s+a)(s^2+\omega_n^2)} = \frac{\kappa_1}{s} + \frac{\kappa_2}{s+a} + \frac{\kappa_3}{s+j\omega_n} + \frac{\kappa_4}{s-j\omega_n} \qquad (8.18b)$$

where $j = \sqrt{-1}$

$$\kappa_1 = \left[\frac{1}{(s+a)(s+j\omega_n)(s-j\omega_n)} \right]_{s=0} = \frac{1}{a\omega_n^2}$$

$$\kappa_2 = \left[\frac{1}{s(s+j\omega_n)(s-j\omega_n)} \right]_{s=-a} = \frac{1}{-a(a^2+\omega_n^2)}$$

$$\kappa_3 = \left[\frac{1}{s(s+a)(s-j\omega_n)} \right]_{s=-j\omega_n} = \frac{1}{-2\omega_n^2(a-j\omega_n)}$$

$$\kappa_4 = \left[\frac{1}{s(s+a)(s+j\omega_n)} \right]_{s=+j\omega_n} = \frac{1}{-2\omega_n^2(a+j\omega_n)}$$

Consequently, Eq. (8.18a) may be rewritten in the following expanded form:

$$\nu(t) = \xi_f \left\{ \mathcal{L}^{-1}\left[\frac{1}{s}\right] - \frac{\omega_n^2}{a^2+\omega_n^2} \mathcal{L}^{-1}\left[\frac{1}{s+a}\right] - \frac{a}{2(a-j\omega_n)} \cdot \right.$$

$$\left. \mathcal{L}^{-1}\left[\frac{1}{s+j\omega_n}\right] - \frac{a}{2(a+j\omega_n)} \mathcal{L}^{-1}\left[\frac{1}{s-j\omega_n}\right] \right\} \qquad (8.18c)$$

The inverse transforms may now be found readily (items 7 and 9, Table 8.2):

$$\nu(t) = \xi_f \left[1 - \frac{\omega_n^2}{a^2+\omega_n^2} e^{-at} - \frac{a}{2(a-j\omega_n)} e^{-j\omega_n t} - \frac{a}{2(a+j\omega_n)} e^{j\omega_n t} \right]$$

Rewriting,

$$\nu(t) = \xi_f \left[1 - \frac{\omega_n^2 e^{-at} + a^2 \tfrac{1}{2}(e^{j\omega_n t} + e^{-j\omega_n t}) - aj\omega_n \tfrac{1}{2}(e^{j\omega_n t} - e^{-j\omega_n t})}{a^2+\omega_n^2} \right]$$

Making use of the relations, $\cos z = (\tfrac{1}{2})(e^{jz} + e^{-jz})$ and $\sin z = -j(\tfrac{1}{2})(e^{jz} - e^{-jz})$, the equation for $\nu(t)$ may be expressed as follows:

$$\nu(t) = \xi_f \left[1 - \frac{(a/\omega_n)[\sin \omega_n t + (a/\omega_n) \cos \omega_n t] + e^{-at}}{1 + a^2/\omega_n^2} \right] \qquad (8.18d)$$

Partial Fraction Expansion of $F(s)$. The partial fraction expansion of $F_r(s)$, illustrated for a particular case in Eq. (8.18b), is a necessary part of the technique of solution. In general $F_r(s)$, expressed by the subsidiary equation (8.12b) and involved in the inverse transformation, Eqs. (8.10) and (8.13), is a quotient of two polynomials in s, thus

$$F_r(s) = \frac{A(s)}{B(s)} \qquad (8.19)$$

The purpose of the expansion of $F_r(s)$ is to divide it into simple parts, the inverse transforms of which may be determined readily. The general procedure of the expansion is to factor $B(s)$ and then to rewrite $F_r(s)$ in partial fractions.[16, 17, 34, 36]

INITIAL CONDITIONS OF THE SYSTEM

In all the solutions for response presented in this chapter, unless otherwise stated, it is assumed that the initial conditions (v_0 and $\dot{v}_0$) of the system are both zero. Other starting conditions may be accounted for merely by superimposing on the time-response functions given the additional terms

$$v_0 \cos \omega_n t + \frac{\dot{v}_0}{\omega_n} \sin \omega_n t \qquad (8.20a)$$

These terms are the complete solution of the homogeneous differential equation, $m\ddot{v}/k + v = 0$. They represent the free vibration resulting from the initial conditions.

The two terms in Eq. (8.20a) may be expressed by either one of the following combined forms:

$$\sqrt{v_0{}^2 + \left(\frac{\dot{v}_0}{\omega_n}\right)^2} \; \sin(\omega_n t + \theta_1) \qquad \text{where } \tan \theta_1 = \frac{v_0 \omega_n}{\dot{v}_0} \qquad (8.20b)$$

$$\sqrt{v_0{}^2 + \left(\frac{\dot{v}_0}{\omega_n}\right)^2} \; \cos(\omega_n t - \theta_2) \qquad \text{where } \tan \theta_2 = \frac{\dot{v}_0}{v_0 \omega_n} \qquad (8.20c)$$

where $\sqrt{v_0{}^2 + \left(\frac{\dot{v}_0}{\omega_n}\right)^2}$ is the resultant amplitude and θ_1 or θ_2 is the phase angle of the *initial-condition free vibration*.

PRINCIPLE OF SUPERPOSITION

When the system is linear, the *principle of superposition* may be employed. Any number of component excitation functions may be superimposed to obtain a prescribed total excitation function, and the corresponding component response functions may be superimposed to arrive at the total response function. However, the superposition must be carried out on a time basis and with complete regard for algebraic sign. The superposition of maximum component responses, disregarding time, may lead to completely erroneous results. For example, the response functions given by Eqs. (8.31) to (8.34) are defined completely with regard to time and algebraic sign, and may be superimposed for any combination of the excitation functions from which they have been derived.

COMPILATION OF RESPONSE FUNCTIONS AND RESPONSE SPECTRA; SINGLE DEGREE-OF-FREEDOM, LINEAR UNDAMPED SYSTEMS

STEP-TYPE EXCITATION FUNCTIONS

CONSTANT-FORCE EXCITATION (SIMPLE STEP IN FORCE). The excitation is a constant force applied to the mass at zero time, $\xi(t) \equiv F(t)/k = F_c/k$. Substituting this excitation for $F(t)/k$ in Eq. (8.1a) and solving for the absolute displacement x,

$$x = \frac{F_c}{k}(1 - \cos \omega_n t) \qquad (8.21a)$$

CONSTANT-DISPLACEMENT EXCITATION (SIMPLE STEP IN DISPLACEMENT). The excitation is a constant displacement of the ground which occurs at zero time, $\xi(t) \equiv u(t) = u_c$. Substituting for $u(t)$ in Eq. (8.1b) and solving for the absolute displacement x,

$$x = u_c(1 - \cos \omega_n t) \tag{8.21b}$$

CONSTANT-ACCELERATION EXCITATION (SIMPLE STEP IN ACCELERATION). The excitation is an instantaneous change in the ground acceleration at zero time, from zero to a constant value $\ddot{u}(t) = \ddot{u}_c$. The excitation is thus

$$\xi(t) \equiv -m\ddot{u}_c/k = -\ddot{u}_c/\omega_n{}^2.$$

Substituting in Eq. (8.1c) and solving for the *relative* displacement δ_x,

$$\delta_x = \frac{-\ddot{u}_c}{\omega_n}(1 - \cos \omega_n t) \tag{8.21c}$$

When the excitation is defined by a function of acceleration $\ddot{u}(t)$, it is often convenient to express the response in terms of the absolute acceleration $\ddot{x}$ of the system. The force acting on the mass in Fig. 8.1C is $-k\,\delta_x$; the acceleration $\ddot{x}$ is thus $-k\,\delta_x/m$ or $-\delta_x\omega_n{}^2$. Substituting $\delta_x = -\ddot{x}/\omega_n{}^2$ in Eq. (8.21c),

$$\ddot{x} = \ddot{u}_c(1 - \cos \omega_n t) \tag{8.21d}$$

The same result is obtained by letting $\xi(t) \equiv \ddot{u}(t) = \ddot{u}_c$ in Eq. (8.1d) and solving for $\ddot{x}$. Equation (8.21d) is similar to Eq. (8.21b) with acceleration instead of displacement on both sides of the equation. This analogy generally applies in step- and pulse-type excitations.

The absolute displacement of the mass can be obtained by integrating Eq. (8.21d) twice with respect to time, taking as initial conditions $x = \dot{x} = 0$ when $t = 0$,

$$x = \frac{\ddot{u}_c}{\omega_n{}^2}\left[\frac{\omega_n{}^2 t^2}{2} - (1 - \cos \omega_n t)\right] \tag{8.21e}$$

Equation (8.21e) also may be obtained from the relation $x = u + \delta_x$, noting that in this case $u(t) = \ddot{u}_c t^2/2$.

CONSTANT-VELOCITY EXCITATION (SIMPLE STEP IN VELOCITY). This excitation, when expressed in terms of ground or spring anchorage motion, is equivalent to prescribing, at zero time, an instantaneous change in the ground velocity from zero to a constant value $\dot{u}_c$. The excitation is $\xi(t) \equiv u(t) = \dot{u}_c t$, and the solution for the differential equation of Eq. (8.1b) is

$$x = \frac{\dot{u}_c}{\omega_n}(\omega_n t - \sin \omega_n t) \tag{8.21f}$$

For the velocity of the mass,

$$\dot{x} = \dot{u}_c(1 - \cos \omega_n t) \tag{8.21g}$$

The result of Eq. (8.21g) could have been obtained directly by letting $\xi(t) \equiv \dot{u}(t) = \dot{u}_c$ in Eq. (8.1e) and solving for the *velocity* response $\dot{x}$.

GENERAL STEP EXCITATION. A comparison of Eqs. (8.21a), (8.21b), (8.21c), (8.21d), and (8.21g) with Table 8.1 reveals that the response ν and the excitation ξ are related in a common manner. This may be expressed as follows:

$$\nu = \xi_c(1 - \cos \omega_n t) \tag{8.22}$$

where ξ_c indicates a constant value of the excitation. The excitation and response of the system are shown in Fig. 8.7.

ABSOLUTE DISPLACEMENT RESPONSE TO VELOCITY-STEP AND ACCELERATION-STEP EXCITATIONS. The absolute displacement responses to the velocity-step and the acceleration-step excitations are given by Eqs. (8.21*f*) and (8.21*e*) and are shown in Figs. 8.8 and 8.9, respectively. The comparative effects of displacement-

FIG. 8.7. Time response to a simple step excitation (general notation).

FIG. 8.8. Time-displacement response to a constant-velocity excitation (simple step in velocity).

step, velocity-step, and acceleration-step excitations, in terms of *absolute displacement* response, may be seen by comparing Figs. 8.7 to 8.9.

In the case of the velocity-step excitation, the *velocity* of the system is always positive, except at $t = 0, T, 2T, \ldots$, when it is zero. Similarly, an acceleration-step excitation results in system *acceleration* that is always positive, except at $t = 0, T, 2T, \ldots$, when it is zero. The natural period of the responding system is $T = 2\pi/\omega_n$.

RESPONSE MAXIMA. In the response of a system to step or pulse excitation, the maximum value of the response often is of considerable physical significance. Several kinds of maxima are important. One of these is the *residual response amplitude,* which is the amplitude of the free vibration about the final position of the excitation as a base. This is designated ν_R, and for the response given by Eq. (8.22):

$$\nu_R = \pm \xi_c \qquad (8.22a)$$

FIG. 8.9. Time-displacement response to a constant-acceleration excitation (simple step in acceleration).

Another maximum is the *maximax response*, which is the greatest of the maxima of ν attained at *any time* during the response.

In general, it is of the same sign as the excitation. For the response given by Eq. (8.22), the maximax response ν_M is

$$\nu_M = 2\xi_c \qquad (8.22b)$$

ASYMPTOTIC STEP. In the exponential function $\xi(t) = \xi_f(1 - e^{-at})$, the maximum value ξ_f of the excitation is approached asymptotically. This excitation may be defined alternatively by $\xi(t) = (F_f/k)(1 - e^{-at})$; $u_f(1 - e^{-at})$; $(-\ddot{u}_f/\omega_n^2)(1 - e^{-at})$; etc. (see Table 8.1). Substituting the excitation $\xi(t) = \xi_f(1 - e^{-at})$ in Eq. (8.2), the response ν is

$$\nu = \xi_f \left[1 - \frac{(a/\omega_n)[\sin \omega_n t + (a/\omega_n) \cos \omega_n t] + e^{-at}}{1 + a^2/\omega_n^2} \right] \qquad (8.23a)$$

The excitation and the response of the system are shown in Fig. 8.10. For large values of the exponent at, the motion is nearly simple harmonic. The residual amplitude, relative to the final position of equilibrium, approaches the following value asymptotically.

$$\nu_R \rightarrow \xi_f \frac{1}{\sqrt{1 + \omega_n^2/a^2}} \tag{8.23b}$$

The maximax response $\nu_M = \nu_R + \xi_f$ is plotted against ω_n/a to give the response spectrum in Fig. 8.11.

FIG. 8.10. Time response to an exponentially asymptotic step for the particular case $\omega_n/a = 2$.

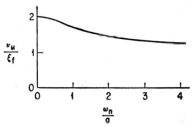

FIG. 8.11. Spectrum for maximax response resulting from exponentially asymptotic step excitation.

STEP-TYPE FUNCTIONS HAVING FINITE RISE TIME. Many step-type excitation functions rise to the constant maximum value ξ_c of the excitation in a finite length of time τ, called the *rise time*. Three such functions and their first three time derivatives are shown in Fig. 8.12. The step having a *cycloidal* front is the only one of the three that does not include an infinite third derivative; i.e., if the step is a ground displacement, it does not have an infinite rate of change of ground acceleration (infinite "jerk").

The excitation functions and the expressions for maximax response are given by the following equations:

Constant-slope front:

$$\xi(t) = \begin{cases} \xi_c \dfrac{t}{\tau} & [0 \leq t \leq \tau] \\[2mm] \xi_c & [\tau \leq t] \end{cases} \tag{8.24a}$$

$$\frac{\nu_M}{\xi_c} = 1 + \left| \frac{T}{\pi\tau} \sin \frac{\pi\tau}{T} \right| \tag{8.24b}$$

Versed-sine front:

$$\xi(t) = \begin{cases} \dfrac{\xi_c}{2}\left(1 - \cos \dfrac{\pi t}{\tau}\right) & [0 \leq t \leq \tau] \\[2mm] \xi_c & [\tau \leq t] \end{cases} \tag{8.25a}$$

$$\frac{\nu_M}{\xi_c} = 1 + \left| \frac{1}{(4\tau^2/T^2) - 1} \cos \frac{\pi\tau}{T} \right| \tag{8.25b}$$

Cycloidal front:

$$\xi(t) = \begin{cases} \dfrac{\xi_c}{2\pi}\left(\dfrac{2\pi t}{\tau} - \sin\dfrac{2\pi t}{\tau}\right) & [0 \le t \le \tau] \\ \xi_c & [\tau \le t] \end{cases} \tag{8.26a}$$

$$\frac{\nu_M}{\xi_c} = 1 + \left| \frac{T}{\pi\tau(1 - \tau^2/T^2)} \sin\frac{\pi\tau}{T} \right| \tag{8.26b}$$

where $T = 2\pi/\omega_n$ is the natural period of the responding system.

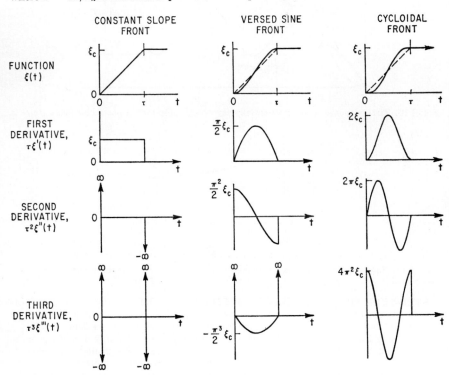

FIG. 8.12. Three step-type excitation functions and their first three time-derivatives.   (*Jacobsen and Ayre.*[22])

In the case of step-type excitations, the maximax response occurs after the excitation has reached its constant maximum value ξ_c and is related to the residual response amplitude by

$$\nu_M = \nu_R + \xi_c \tag{8.27}$$

Figure 8.13 shows the spectra of maximax response versus step rise time τ expressed relative to the natural period T of the responding system. In Fig. 8.13A the comparison is based on *equal rise times*, and in Fig. 8.13B it relates to *equal maximum slopes of the step fronts*. The residual response amplitude has values of zero ($\nu_M/\xi_c = 1$) in all three cases; for example, the step excitation having a constant-slope front results in zero residual amplitude at $\tau/T = 1, 2, 3, \ldots$.

Figure 8.12 and the other illustrations crediting Ref. 22 are reprinted by permission from "Engineering Vibrations," by L. S. Jacobsen and R. S. Ayre. Copyright, 1958. McGraw-Hill Book Company, Inc., New York.

Fig. 8.13. Spectra of maximax response resulting from the step excitation functions of Fig. 8.12. (A) For step functions having equal rise time τ. (B) For step functions having equal maximum slope ξ_c/τ_a. (*Jacobsen and Ayre.*[22])

A FAMILY OF EXPONENTIAL STEP FUNCTIONS HAVING FINITE RISE TIME.

The inset diagram in Fig. 8.14 shows and Eqs. (8.28a) define a family of step functions having fronts which rise exponentially to the constant maximum ξ_c in the rise time τ. Two limiting cases of vertically fronted steps are included in the family: When $a \to -\infty$, the vertical front occurs at $t = 0$; when $a \to +\infty$, the vertical front occurs at $t = \tau$. An intermediate case has a constant-slope front ($a = 0$). The maximax responses are given

FIG. 8.14. Spectra of maximax response for a family of step functions having exponential fronts, including the vertical fronts $a \to \pm \infty$, and the constant-slope front $a = 0$, as special cases. (*Jacobsen and Ayre.*[22])

by Eq. (8.28b) and by the response spectra in Fig. 8.14. The values of the maximax response are independent of the sign of the parameter a.

$$\xi(t) = \begin{cases} \xi_c \dfrac{1 - e^{at/\tau}}{1 - e^a} & [0 \leq t \leq \tau] \\ \xi_c & [\tau \leq t] \end{cases} \qquad (8.28a)$$

$$\frac{\nu_M}{\xi_c} = 1 + \left| \frac{a}{1 - e^a} \left[\frac{1 - 2e^a \cos(2\pi\tau/T) + e^{2a}}{a^2 + 4\pi^2\tau^2/T^2} \right]^{1/2} \right| \qquad (8.28b)$$

where T is the natural period of the responding system.

There are zeroes of residual response amplitude ($\nu_M/\xi_c = 1$) at finite values of τ/T only for the constant-slope front ($a = 0$). Each of the step functions represented in Fig. 8.13 results in zeroes of residual response amplitude, and each function has antisymmetry with respect to the half-rise time $\tau/2$. This is of interest in the selection of cam and control-function shapes, where one of the criteria of choice may be *minimum residual amplitude of vibration* of the driven system.

PULSE-TYPE EXCITATION FUNCTIONS

THE SIMPLE IMPULSE. If the duration τ of the pulse is short relative to the natural period T of the system, the response of the system may be determined by equating the impulse J, i.e., the force-time integral, to the momentum $m\dot{x}_J$:

$$J = \int_0^\tau F(t)\, dt = m\dot{x}_J \qquad (8.29a)$$

Thus, it is found that the impulsive velocity $\dot{x}_J$ is equal to J/m. Consequently, the velocity-time response is given by $\dot{x} = \dot{x}_J \cos \omega_n t = (J/m) \cos \omega_n t$. The displacement-time response is obtained by integration, assuming a start from rest,

$$x = x_J \sin \omega_n t$$

where

$$x_J = \frac{J}{m\omega_n} = \omega_n \int_0^\tau \frac{F(t)\, dt}{k} \tag{8.29b}$$

The impulse concept, used for determining the response to a short-duration force pulse, may be generalized in terms of ν and ξ by referring to Table 8.1. The *generalized impulsive response* is

$$\nu = \nu_J \sin \omega_n t \tag{8.30a}$$

where the amplitude is

$$\nu_J = \omega_n \int_0^\tau \xi(t)\, dt \tag{8.30b}$$

The impulsive response amplitude ν_J and the generalized impulse $k \int_0^\tau \xi(t)\, dt$ are used in comparing the effects of various pulse shapes when the pulse durations are short.

SYMMETRICAL PULSES. In the following discussion a comparison is made of the responses caused by single symmetrical pulses of rectangular, half-cycle sine, versed-sine, and triangular shapes. The excitation functions and the time-response equations are given by Eqs. (8.31) to (8.34). Note that the residual response amplitude factors are set in brackets and are identified by the time interval $\tau \leq t$.

Rectangular:

$$\left.\begin{aligned}\xi(t) &= \xi_p \\ \nu &= \xi_p(1 - \cos \omega_n t)\end{aligned}\right\} \qquad [0 \leq t \leq \tau] \tag{8.31a}$$

$$\left.\begin{aligned}\xi(t) &= 0 \\ \nu &= \xi_p \left[2 \sin \frac{\pi \tau}{T} \right] \sin \omega_n \left(t - \frac{\tau}{2} \right)\end{aligned}\right\} \qquad [\tau \leq t] \tag{8.31b}$$

Half-cycle sine:

$$\left.\begin{aligned}\xi(t) &= \xi_p \sin \frac{\pi t}{\tau} \\ \nu &= \frac{\xi_\mu}{1 - T^2/4\tau^2} \left(\sin \frac{\pi t}{\tau} - \frac{T}{2\tau} \sin \omega_n t \right)\end{aligned}\right\} \qquad [0 \leq t \leq \tau] \tag{8.32a}$$

$$\left.\begin{aligned}\xi(t) &= 0 \\ \nu &= \xi_p \left[\frac{(T/\tau) \cos (\pi \tau/T)}{(T^2/4\tau^2) - 1} \right] \sin \omega_n \left(t - \frac{\tau}{2} \right)\end{aligned}\right\} \qquad [\tau \leq t] \tag{8.32b}$$

Versed-sine:

$$\left.\begin{aligned}\xi(t) &= \frac{\xi_p}{2} \left(1 - \cos \frac{2\pi t}{\tau} \right) \\ \nu &= \frac{\xi_p/2}{1 - \tau^2/T^2} \left(1 - \frac{\tau^2}{T^2} + \frac{\tau^2}{T^2} \cos \frac{2\pi t}{\tau} - \cos \omega_n t \right)\end{aligned}\right\} \qquad [0 \leq t \leq \tau] \tag{8.33a}$$

$$\left.\begin{aligned}\xi(t) &= 0 \\ \nu &= \xi_p \left[\frac{\sin \pi \tau/T}{1 - \tau^2/T^2} \right] \sin \omega_n \left(t - \frac{\tau}{2} \right)\end{aligned}\right\} \qquad [\tau \leq t] \tag{8.33b}$$

Triangular:

$$\xi(t) = 2\xi_p \frac{t}{\tau}$$

$$\nu = 2\xi_p \left(\frac{t}{\tau} - \frac{T}{\tau} \frac{\sin \omega_n t}{2\pi} \right) \qquad \left[0 \leq t \leq \frac{\tau}{2} \right] \quad (8.34a)$$

$$\xi(t) = 2\xi_p \left(1 - \frac{t}{\tau} \right)$$

$$\nu = 2\xi_p \left(1 - \frac{t}{\tau} - \frac{T}{\tau} \frac{\sin \omega_n t}{2\pi} + \frac{T}{\tau} \frac{\sin \omega_n (t - \tau/2)}{\pi} \right) \qquad \left[\frac{\tau}{2} \leq t \leq \tau \right] \quad (8.34b)$$

$$\xi(t) = 0$$

$$\nu = \xi_p \left[2 \frac{\sin^2 (\pi\tau/2T)}{\pi\tau/2T} \right] \sin \omega_n (t - \tau/2) \qquad [\tau \leq t] \quad (8.34c)$$

where T is the natural period of the responding system.

Equal Maximum Height of Pulse as Basis of Comparison. Examples of time response, for six different values of τ/T, are shown separately for the rectangular, half-cycle sine, and versed-sine pulses in Fig. 8.15, and for the triangular pulse in Fig. 8.22B. The basis of comparison is equal maximum height of excitation pulse ξ_p.

Residual Response Amplitude and Maximax Response. The spectra of maximax response ν_M and residual response amplitude ν_R are given in Fig. 8.16 by (A) for the rectangular pulse, by (B) for the sine pulse, and by (C) for the versed-sine pulse. The maximax response may occur either within the duration of the pulse or after the pulse function has dropped to zero. In the latter case the maximax response is equal to the residual response amplitude. In general, the maximax response is given by the residual response amplitude only in the case of short-duration pulses; for example, see the case $\tau/T = \frac{1}{4}$ in Fig. 8.15 where T is the natural period of the responding system. The response spectra for the triangular pulse appear in Fig. 8.24.

Maximax Relative Displacement When the Excitation Is Ground Displacement. When the excitation $\xi(t)$ is given as *ground displacement* $u(t)$, the response ν is the absolute displacement x of the mass (Table 8.1). It is of practical importance in the investigation of the maximax *distortion* or *stress* in the elastic element to know the maximax value of the relative displacement. In this case the relative displacement is a *derived quantity* obtained by taking the difference between the response and the excitation, that is, $x - u$ or, in terms of the general notation, $\nu - \xi$.

If the excitation is given as ground acceleration, the response is determined directly as relative displacement and is designated δ_x (Table 8.1). To avoid confusion, relative displacement determined as a *derived quantity*, as described in the first case above, is designated by $x - u$; relative displacement determined directly as the *response variable* (second case above) is designated by δ_x. The distinction is made readily in the general notation by use of the symbols $\nu - \xi$ and ν, respectively, for relative response and for response. The maximax values are designated $(\nu - \xi)_M$ and ν_M, respectively.

The maximax relative response may occur *either* within the duration of the pulse or during the residual vibration era ($\tau \leq t$). In the latter case the maximax relative response is equal to the residual response amplitude. This explains the discontinuities which occur in the spectra of maximax relative response shown in Fig. 8.16 and elsewhere.

The meaning of the relative response $\nu - \xi$ may be clarified further by a study of the time-response and time-excitation curves shown in Fig. 8.15.

Equal Area of Pulse as Basis of Comparison. In the preceding section on the comparison of responses resulting from pulse excitation, the pulses are assumed of equal maximum height. Under some conditions, particularly if the pulse duration is short relative to the natural period of the system, it may be more useful to make the comparison on the basis of equal pulse area; i.e., equal impulse (equal time integral).

The *areas* for the pulses of maximum height ξ_p and duration τ are as follows: rectangle, $\xi_p\tau$; half-cycle sine, $(2/\pi)\xi_p\tau$; versed-sine $(\frac{1}{2})\xi_p\tau$; triangle, $(\frac{1}{2})\xi_p\tau$. Using the area of the

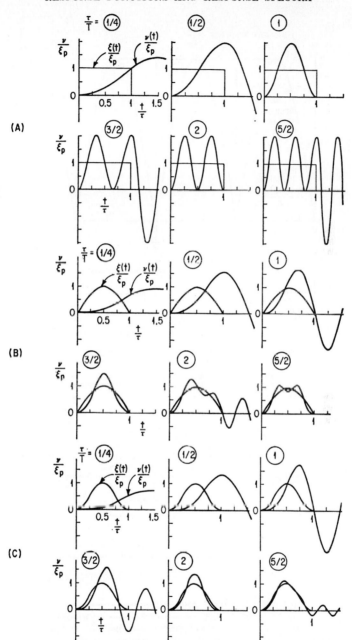

FIG. 8.15. Time response curves resulting from single pulses of (A) rectangular, (B) half-cycle sine, and (C) versed-sine shapes.[19]

FIG. 8.16. Spectra of maximax response, residual response amplitude, and maximax relative response resulting from single pulses of (A) rectangular, (B) half-cycle sine, and (C) versed-sine shapes.[19] The spectra are shown on another basis in Fig. 8.18.

triangular pulse as the basis of comparison, and requiring that the areas of the other pulses be equal to it, it is found that the pulse *heights*, in terms of the height ξ_{po} of the *reference triangular pulse*, must be as follows: rectangle, $(\tfrac{1}{2})\xi_{po}$; half-cycle sine, $(\pi/4)\xi_{po}$; versed-sine, ξ_{po}.

Figure 8.17 shows the time responses, for four values of τ/T, redrawn on the basis of *equal pulse area* as the criterion for comparison. Note that the response reference is the constant ξ_{po}, which is the height of the triangular pulse. To show a direct comparison, the response curves for the various pulses are superimposed on each other. For the shortest duration shown, $\tau/T = \tfrac{1}{4}$, the response curves are nearly alike. Note that the responses to two different rectangular pulses are shown, one of duration τ and height $\xi_{po}/2$, the other of duration $\tau/2$ and height ξ_{po}, both of area $\xi_{po}\tau/2$.

The response spectra, plotted on the basis of equal pulse area, appear in Fig. 8.18. The residual response spectra are shown altogether in (A), the maximax response spectra in (B), and the spectra of maximax relative response in (C).

Since the pulse area is $\xi_{po}\tau/2$, the generalized impulse is $k\xi_{po}\tau/2$, and the amplitude of vibration of the system computed on the basis of the generalized impulse theory, Eq. (8.30b), is given by

$$\nu_J = \omega_n \xi_{po} \frac{\tau}{2} = \pi \frac{\tau}{T} \xi_{po} \tag{8.35}$$

A comparison of this straight-line function with the response spectra in Fig. 8.18B shows that *for values of τ/T less than one-fourth the shape of the symmetrical pulse is of little concern.*

Family of Exponential, Symmetrical Pulses. A continuous variation in shape of pulse may be investigated by means of the family of pulses represented by Eqs. (8.36a) and shown in the inset diagram in Fig. 8.19A:

$$\zeta(t) = \begin{cases} \xi_p \dfrac{1 - e^{2at/\tau}}{1 - e^a} & \left[0 \le t \le \dfrac{\tau}{2}\right] \\[3ex] \xi_p \dfrac{1 - e^{2a(1-t/\tau)}}{1 - e^a} & \left[\dfrac{\tau}{2} \le t \le \tau\right] \\[3ex] 0 & [\tau \le t] \end{cases} \tag{8.36a}$$

The family includes the following special cases:

$a \to -\infty$: rectangle of height ξ_p and duration τ;
$a = 0$: triangle of height ξ_p and duration τ;
$a \to +\infty$: spike of height ξ_p and having zero area.

The residual response amplitude of vibration of the system is

$$\frac{\nu_R}{\xi_p} = \frac{2aT}{\pi\tau} \left(\frac{e^a - \cos(\pi\tau/T) - (aT/\pi\tau)\sin(\pi\tau/T)}{(1 - e^a)(1 + a^2 T^2/\pi^2\tau^2)} \right) \tag{8.36b}$$

where T is the natural period of the responding system. Figure 8.19A shows the spectra for residual response amplitude for seven values of the parameter a, compared on the basis of *equal pulse height*. The zero-area spike ($a \to +\infty$) results in zero response.

The area of the general pulse of height ξ_p is

$$A_p = \xi_p \frac{\tau}{a} \left(\frac{1 - e^a + a}{1 - e^a} \right) \tag{8.36c}$$

If a comparison is to be drawn on the basis of *equal pulse area* using the area $\xi_{po}\tau/2$ of the

FIG. 8.17. Time response to various symmetrical pulses having equal pulse area, for four different values of τ/T. *(Jacobsen and Ayre.*[22]*)*

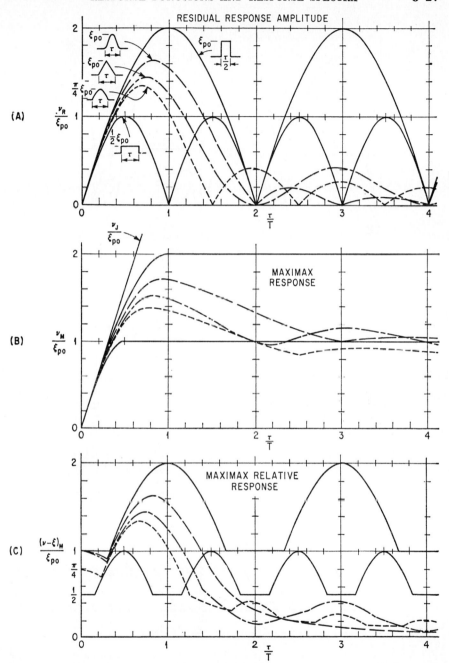

FIG. 8.18. Response spectra for various symmetrical pulses having equal pulse area: (A) residual response amplitude, (B) maximax response, and (C) maximax relative response. (*Jacobsen and Ayre.*[22])

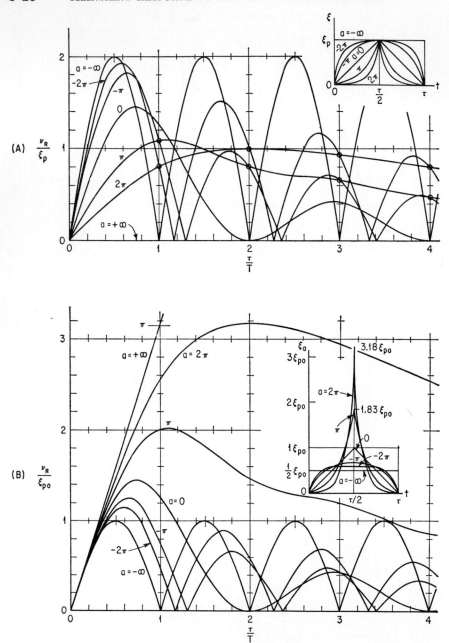

Fig. 8.19. Spectra for residual response amplitude for a family of exponential, symmetrical pulses: (A) pulses having equal height; (B) pulses having equal area.   (*Jacobsen and Ayre.*[22])

triangular pulse as the reference, the height ξ_{pa} of the general pulse is

$$\xi_{pa} = \xi_{po}\frac{a}{2}\left(\frac{1 - e^a}{1 - e^a + a}\right) \tag{8.36d}$$

The residual response amplitude spectra, based on the equal-pulse-area criterion, are shown in Fig. 8.19B. The case $a \rightarrow +\infty$ is equivalent to a generalized impulse of value $k\xi_{po}\tau/2$ and results in the straight-line spectrum given by Eq. (8.35).

Symmetrical Pulses Having a Rest Period of Constant Height. In the inset diagrams of Fig. 8.20 each pulse consists of a rise, a central rest period or "dwell" having constant height, and a decay. The expressions for the pulse *rise* functions may be obtained from Eqs. (8.24a), (8.25a), and (8.26a) by substituting $\tau/2$ for τ. The pulse *decay* functions are available from symmetry.

If the *rest* period is long enough for the maximax displacement of the system to be reached during the duration τ_r of the pulse rest, the maximax may be obtained from Eqs. (8.24b), (8.25b), and (8.26b) and, consequently, from Fig. 8.13. The substitution of $\tau/2$ for τ is necessary.

Equations (8.37) to (8.39) give the residual response amplitudes. The spectra computed from these equations are shown in Fig. 8.20.

Constant-slope rise and decay:

$$\frac{\nu_R}{\xi_p} = \frac{2T}{\pi\tau}\left[1 - \cos\frac{\pi\tau}{T} + \frac{1}{2}\cos\frac{2\pi\tau_r}{T} - \cos\frac{\pi(\tau + 2\tau_r)}{T} + \frac{1}{2}\cos\frac{2\pi(\tau + \tau_r)}{T}\right]^{\frac{1}{2}} \tag{8.37}$$

Versed-sine rise and decay:

$$\frac{\nu_R}{\xi_p} = \frac{1}{1 - \tau^2/T^2}\left[1 + \cos\frac{\pi\tau}{T} - \frac{1}{2}\cos\frac{2\pi\tau_r}{T} - \cos\frac{\pi(\tau + 2\tau_r)}{T} - \frac{1}{2}\cos\frac{2\pi(\tau + \tau_r)}{T}\right]^{\frac{1}{2}} \tag{8.38}$$

Cycloidal rise and decay:

$$\frac{\nu_R}{\xi_p} = \frac{2T/\pi\tau}{1 - \tau^2/4T^2}\left[\cos\frac{\pi\tau_r}{T} - \cos\frac{\pi(\tau + \tau_r)}{T}\right] \tag{8.39}$$

Note that τ in the abscissa is the sum of the rise time and the decay time and is *not* the total duration of the pulse. Attached to each spectrum is a set of values of τ_r/T where T is the natural period of the responding system.

When $\tau_r/T = 1, 2, 3, \ldots$, the residual response amplitude is equal to that for the case $\tau_r = 0$, and the spectrum starts at the origin. If $\tau_r/T = \frac{1}{2}, \frac{3}{2}, \frac{5}{2}, \ldots$, the spectrum has the maximum value 2.00 at $\tau/T = 0$. The *envelopes* of the spectra are of the same forms as the *residual-response-amplitude* spectra for the related *step functions*; see the spectra for $[(\nu_M/\xi_c) - 1]$ in Fig. 8.13A. In certain cases, for example, at $\tau/T = 2, 4, 6, \ldots$, in Fig. 8.20A, $\nu_R/\xi_p = 0$ for all values of τ_r/T.

UNSYMMETRICAL PULSES. Pulses having only slight asymmetry may often be represented adequately by symmetrical forms. However, if there is considerable asymmetry, resulting in appreciable steepening of either the rise or the decay, it is necessary to introduce a parameter which defines the *skewing* of the pulse.

The ratio of the rise time to the pulse period is called the *skewing constant*, $\sigma = t_1/\tau$. There are three special cases:

$\sigma = 0$: the pulse has an *instantaneous* (vertical) *rise*, followed by a decay having the duration τ. This case may be used as an elementary representation of a *blast pulse*.

$\sigma = \frac{1}{2}$: the pulse may be *symmetrical*.

$\sigma = 1$: the pulse has an *instantaneous decay*, preceded by a rise having the duration τ.

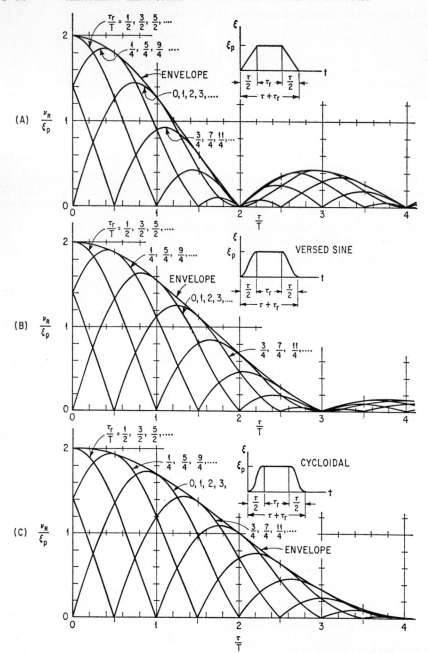

FIG. 8.20. Residual response amplitude spectra for three families of symmetrical pulses having a central rest period of constant height and of duration τ_r. Note that the abscissa is τ/T, where τ is the sum of the rise time and the decay time. (A) Constant-slope rise and decay. (B) Versed-sine rise and decay. (C) Cycloidal rise and decay. (*Jacobsen and Ayre.*[22])

Triangular Pulse Family. The effect of asymmetry in pulse shape is shown readily by means of the family of triangular pulses (Fig. 8.21). Equations (8.40) give the excitation and the time response.

Rise era: $0 \leq t \leq t_1$

$$\xi(t) = \xi_p \frac{t}{t_1}$$

$$\nu = \xi_p \left(\frac{t}{t_1} - \frac{T}{2\pi t_1} \sin \omega_n t \right)$$

(8.40a)

Decay era: $0 \leq t' \leq t_2$, where $t' = t - t_1$

$$\xi(t) = \xi_p \left(1 - \frac{t'}{t_2} \right)$$

$$\nu = \xi_p \left[1 - \frac{t'}{t_2} + \frac{T}{2\pi t_2} \left(1 + 4 \frac{t_2}{t_1} \frac{\tau}{t_1} \sin^2 \frac{\pi t_1}{T} \right)^{\frac{1}{2}} \sin (\omega_n t' + \theta') \right]$$

(8.40b)

where

$$\tan \theta' = \frac{\sin (2\pi t_1/T)}{\cos (2\pi t_1/T) - \tau/t_2}$$

Residual-vibration era: $0 \leq t''$, where $t'' = t - \tau = t - t_1 - t_2$

$$\xi(t) = 0$$

$$\nu = \xi_p \frac{1}{\pi} \left[\frac{T}{t_1} \frac{T}{t_2} \left(\frac{\tau}{t_1} \sin^2 \frac{\pi t_1}{T} + \frac{\tau}{t_2} \sin^2 \frac{\pi t_2}{T} - \sin^2 \frac{\pi \tau}{T} \right) \right]^{\frac{1}{2}} \sin (\omega_n t'' + \theta_R)$$ (8.40c)

where

$$\tan \theta_R = \frac{(\tau/t_2) \sin (2\pi t_2/T) - \sin (2\pi \tau/T)}{(\tau/t_2) \cos (2\pi t_2/T) - \cos (2\pi \tau/T) - t_1/t_2}$$

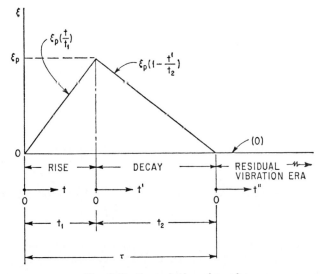

FIG. 8.21. General triangular pulse.

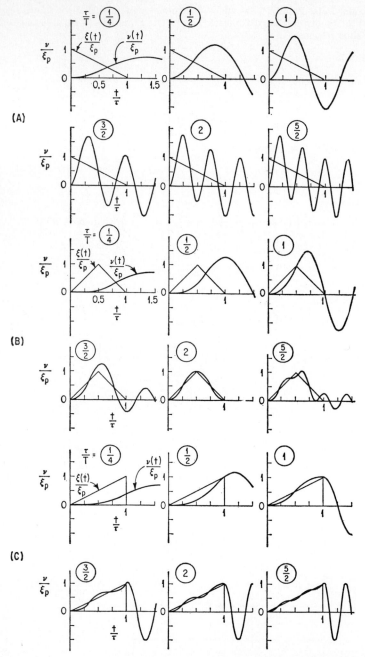

Fig. 8.22. Time response curves resulting from single pulses of three different triangular shapes: (A) vertical rise (elementary blast pulse), (B) symmetrical, and (C) vertical decay.[15]

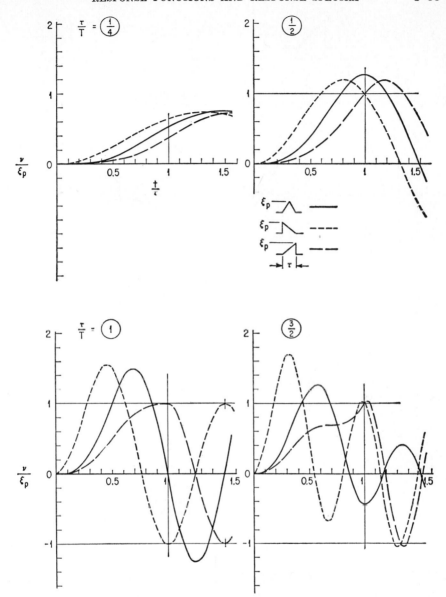

Fig. 8.23. Time response curves of Fig. 8.22 superposed, for four values of τ/T. (*Jacobsen and Ayre.*[22])

For the special cases $\sigma = 0$, ½ and 1, the time responses for six values of τ/T are shown in Fig. 8.22, where T is the natural period of the responding system. Some of the curves are superposed in Fig. 8.23 for easier comparison. The response spectra appear in Fig. 8.24. The straight-line spectrum v_J/ξ_p for the amplitude of response based on the impulse theory also is shown in Fig. 8.24A. In the two cases of extreme skewing, $\sigma = 0$ and

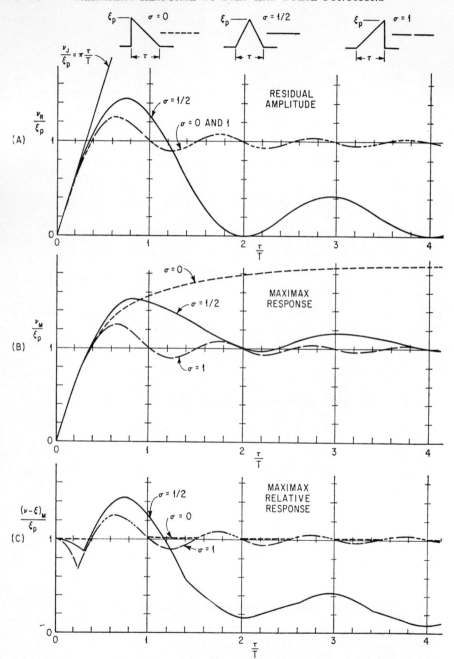

FIG. 8.24. Response spectra for three types of triangular pulse: (A) Residual response amplitude. (B) Maximax response. (C) Maximax relative response. (*Jacobsen and Ayre.*[22])

$\sigma = 1$, the residual amplitudes are *equal* and are given by Eq. (8.41a). For the symmetrical case, $\sigma = \frac{1}{2}$, ν_R is given by Eq. (8.41b).

$$\sigma = 0 \text{ and } 1: \qquad \frac{\nu_R}{\xi_p} = \left[1 - \frac{T}{\pi\tau} \sin \frac{2\pi\tau}{T} + \left(\frac{T}{\pi\tau}\right)^2 \sin^2 \frac{\pi\tau}{T} \right]^{\frac{1}{2}} \qquad (8.41a)$$

$$\sigma = \frac{1}{2}: \qquad \frac{\nu_R}{\xi_p} = 2 \frac{\sin^2 (\pi\tau/2T)}{\pi\tau/2T} \qquad (8.41b)$$

The residual response amplitudes for other cases of skewness may be determined from the amplitude term in Eqs. (8.40c); they are shown by the response spectra in Fig. 8.25. The residual response amplitudes resulting from single pulses that are mirror images of

FIG. 8.25. Spectra for residual response amplitude for a family of triangular pulses of varying skewness. (*Jacobsen and Ayre.*[22])

each other in time are equal. In general, the phase angles for the residual vibrations are unequal.

Note that in the cases $\sigma = 0$ and $\sigma = 1$ for *vertical rise* and *vertical decay*, respectively, there are *no zeroes of residual amplitude*, except for the trivial case, $\tau/T = 0$.

The family of triangular pulses is particularly advantageous for investigating the effect of varying the skewness, because both criteria of comparison, equal pulse height and equal pulse area, are satisfied simultaneously.

Various Pulses Having Vertical Rise or Vertical Decay. Figure 8.26 shows the spectra of residual response amplitude plotted on the basis of *equal pulse area*. The rectangular pulse is included for comparison. The expressions for residual response amplitude for the rectangular and the triangular pulses are given by Eqs. (8.31b) and (8.41a), and for the quarter-cycle sine and the half-cycle versed-sine pulses by Eqs. (8.42) and (8.43).

Quarter-cycle "sine":

$$\xi(t) = \xi_p \begin{cases} \sin \dfrac{\pi t}{2\tau} & \text{for vertical decay} \\[2mm] \text{or} \\[1mm] \cos \dfrac{\pi t}{2\tau} & \text{for vertical rise} \end{cases} \qquad [0 \leq t \leq \tau]$$

$$\xi(t) = 0 \qquad [\tau \leq t]$$

$$\frac{\nu_R}{\xi_p} = \frac{4\tau/T}{(16\tau^2/T^2) - 1} \left(1 + \frac{16\tau^2}{T^2} - \frac{8\tau}{T} \sin \frac{2\pi\tau}{T} \right)^{\frac{1}{2}} \qquad (8.42)$$

Half-cycle "versed-sine":

$$\xi(t) = \xi_p \begin{cases} \dfrac{1}{2}\left(1 - \cos\dfrac{\pi t}{\tau}\right) & \text{for vertical decay} \\[2mm] \text{or} \\[2mm] \dfrac{1}{2}\left(1 + \cos\dfrac{\pi t}{\tau}\right) & \text{for vertical rise} \end{cases} \qquad [0 \le t \le \tau]$$

$$\xi(t) = 0 \qquad [\tau \le t]$$

$$\frac{v_R}{\xi_p} = \frac{1/2}{(4\tau^2/T^2) - 1}\left[1 + \left(1 - \frac{8\tau^2}{T^2}\right)^2 - 2\left(1 - \frac{8\tau^2}{T^2}\right)\cdot\cos\frac{2\pi\tau}{T}\right]^{\frac{1}{2}} \qquad (8.43)$$

where T is the natural period of the responding system.

Note again that the residual response amplitudes, caused by single pulses that are mirror images in time, are equal. Furthermore, it is seen that the unsymmetrical pulses,

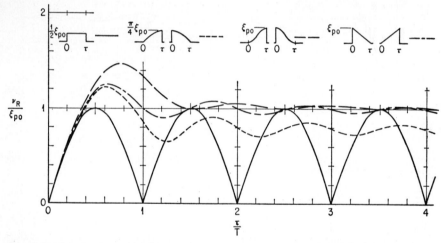

FIG. 8.26. Spectra for residual response amplitude for various unsymmetrical pulses having either vertical rise or vertical decay. Comparison on the basis of equal pulse area.[19]

having either vertical rise or vertical decay, result in no zeroes of residual response amplitude, except in the trivial case $\tau/T = 0$.

Exponential Pulses of Finite Duration, Having Vertical Rise or Vertical Decay. Families of exponential pulses having either a vertical rise or a vertical decay, as shown in the inset diagrams in Fig. 8.27, can be formed by Eqs. (8.44a) and (8.44b).

Vertical rise with exponential decay:

$$\xi(t) = \begin{cases} \xi_p\left(\dfrac{1 - e^{a(1-t/\tau)}}{1 - e^a}\right) & [0 \le t \le \tau] \\[2mm] 0 & [\tau \le t] \end{cases} \qquad (8.44a)$$

Exponential rise with vertical decay:

$$\xi(t) = \begin{cases} \xi_p\left(\dfrac{1 - e^{at/\tau}}{1 - e^a}\right) & [0 \le t \le \tau] \\[2mm] 0 & [\tau \le t] \end{cases} \qquad (8.44b)$$

FIG. 8.27. Spectra for residual response amplitude for unsymmetrical exponential pulses having either vertical rise or vertical decay. Comparison on the basis of equal pulse height.[19]

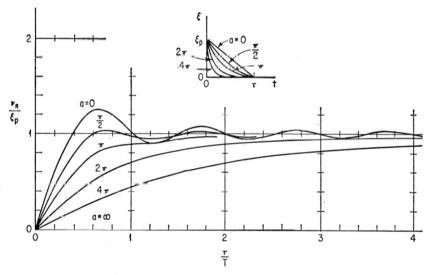

FIG. 8.28. Spectra for residual response amplitude for a family of simple blast pulses, the same family shown in Fig. 8.27 but limited to positive values of the exponential decay parameter a. Comparison on the basis of equal pulse height. (These spectra also apply to mirror-image pulses having vertical decay.) (*Jacobsen and Ayre.*[22])

Residual response amplitude for *either form of pulse:*

$$\frac{v_R}{\xi_p} = \frac{a}{1 - e^a} \left\{ \frac{[(2\pi\tau/T)(1 - e^a)/a + \sin(2\pi\tau/T)]^2 + [1 - \cos(2\pi\tau/T)]^2}{a^2 + 4\pi^2\tau^2/T^2} \right\}^{1/2} \quad (8.44c)$$

When $a = 0$, the pulses are triangular with vertical rise or vertical decay. If $a \to +\infty$ or $-\infty$, the pulses approach the shape of a zero-area spike or of a rectangle, respectively. The spectra for residual response amplitude, plotted on the basis of *equal pulse height*, are shown in Fig. 8.27.

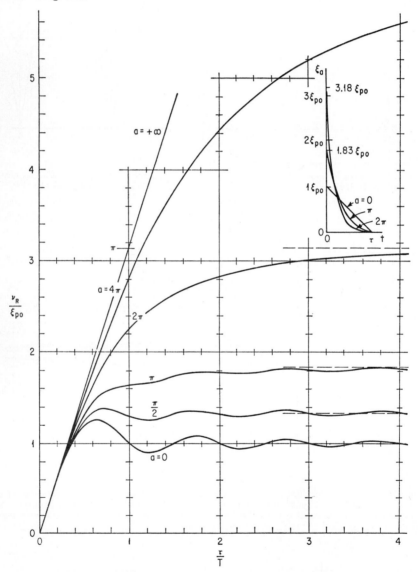

FIG. 8.29. Spectra for residual response amplitude for the family of simple blast pulses shown in Fig. 8.28, compared on the basis of equal pulse area. (*Jacobsen and Ayre.*[22])

Figure 8.28 shows the spectra of residual response amplitude in greater detail for the range in which the parameter a is limited to positive values. This group of pulses is of interest in studying the effects of a simple form of blast pulse, in which the peak height and the duration are constant but the rate of decay is varied.

The areas of the pulses of equal height ξ_p, and the heights of the pulses of equal area $\xi_{po}\tau/2$ are the same as for the symmetrical exponential pulses [see Eqs. (8.36c) and (8.36d)]. If the spectra in Fig. 8.28 are redrawn, using *equal pulse area* as the criterion for comparison, they appear as in Fig. 8.29. The limiting pulse case $a \to +\infty$ repre-

Fig. 8.30. Pulse consisting of vertical rise followed by exponential decay of infinite duration.

Fig. 8.31. Time response to the pulse having a vertical rise and an exponential decay of infinite duration (Fig. 8.30), for the particular case $\omega_n/a = 2$.

sents a generalized impulse of value $k\xi_{po}\tau/2$. The asymptotic values of the spectra are equal to the peak heights of the equal area pulses and are given by

$$\frac{\nu_R}{\xi_{po}} \to \frac{a(1 - e^a)}{2(1 - e^a + a)} \quad \text{as} \quad \frac{\tau}{T} \to \infty \qquad (8.44d)$$

Exponential Pulses of Infinite Duration. Five different cases are included as follows:
1. The excitation function, consisting of a vertical rise followed by an exponential decay, is

$$\xi(t) = \xi_p e^{-at} \qquad [0 \le t] \qquad (8.45a)$$

It is shown in Fig. 8.30. The response time equation for the system is

$$\nu = \xi_p \frac{(a/\omega_n)\sin\omega_n t - \cos\omega_n t + e^{-at}}{1 + a^2/\omega_n^2} \qquad (8.45b)$$

and the asymptotic value of the residual amplitude is given by

$$\frac{\nu_R}{\xi_p} \to \frac{1}{\sqrt{1 + a^2/\omega_n^2}} \qquad (8.45c)$$

The maximax response is the first maximum of ν. The time response, for the particular case $\omega_n/a = 2$, and the response spectra are shown in Figs. 8.31 and 8.32, respectively.

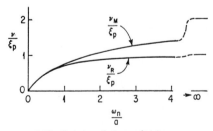

Fig. 8.32. Spectra for maximax response and for asymptotic residual response amplitude, for the pulse shown in Fig. 8.30.

Fig. 8.33. Pulse formed by taking the difference of two exponentially decaying functions.

2. The *difference of two exponential functions,* of the type of Eq. (8.45a), results in the pulse given by Eq. (8.46a):

$$\xi(t) = \xi_0(e^{-bt} - e^{-at}) \tag{8.46a}$$

$$a > b \qquad [0 \leq t]$$

The shape of the pulse is shown in Fig. 8.33. Note that ξ_0 is the ordinate of each of the exponential functions at $t = 0$; it is *not* the pulse maximum. The asymptotic residual response amplitude is

$$\nu_R \to \xi_0 \frac{(b/\omega_n) - (a/\omega_n)}{[(1 + a^2/\omega_n{}^2)(1 + b^2/\omega_n{}^2)]^{1/2}} \tag{8.46b}$$

3. The *product of the exponential function* e^{-at} *by time* results in the excitation given by Eq. (8.47a) and shown in Fig. 8.34.

$$\xi(t) = C_0 t e^{-at} \tag{8.47a}$$

where C_0 is a constant. The peak height of the pulse ξ_p is equal to C_0/ae, and occurs at the time $t_1 = 1/a$. Equations (8.47b) and (8.47c) give the time response and the asymptotic residual response amplitude:

$$\nu = \xi_p \frac{ae/\omega_n}{(1 + a^2/\omega_n{}^2)^2} \left\{ \left[\frac{2a}{\omega_n} + \left(1 + \frac{a^2}{\omega_n{}^2}\right) \omega_n t \right] e^{-at} - \frac{2a}{\omega_n} \cos \omega_n t - \left(1 - \frac{a^2}{\omega_n{}^2}\right) \sin \omega_n t \right\} \tag{8.47b}$$

$$\frac{\nu_R}{\xi_p} \to \frac{e}{(a/\omega_n) + (\omega_n/a)} \tag{8.47c}$$

The *maximum* value of ν_R occurs in the case $a/\omega_n = 1$, and is given by

$$(\nu_R)_{\max} = \xi_p e/2 = 1.36\xi_p$$

Both of the excitation functions described by Eqs. (8.46a) and (8.47a) include finite times of rise to the pulse peak. These rise times are dependent on the exponential decay constants.

4. The rise time may be made independent of the decay by inserting a separate rise function before the decay function, as in Fig. 8.35, where a *straight-line rise precedes the exponential decay.* The response-time equations are as follows:

Pulse rise era:

$$\nu = \xi_p \frac{\omega_n t - \sin \omega_n t}{\omega_n t_1} \qquad [0 \leq t \leq t_1] \tag{8.48a}$$

FIG. 8.34. Pulse formed by taking the product of an exponentially decaying function by time.

FIG. 8.35. Pulse formed by a straight-line rise followed by an exponential decay asymptotic to the time axis.

Pulse decay era:

$$\nu = \xi_p \left[\frac{e^{-at'}}{1 + a^2/\omega_n^2} + \left(\frac{a^2/\omega_n^2}{1 + a^2/\omega_n^2} - \frac{\sin \omega_n t_1}{\omega_n t_1} \right) \cos \omega_n t' \right.$$

$$\left. + \left(\frac{a/\omega_n}{1 + a^2/\omega_n^2} + \frac{1 - \cos \omega_n t_1}{\omega_n t_1} \right) \sin \omega_n t' \right] \quad (8.48b)$$

where $t' = t - t_1$ and $0 \leq t'$.

5. Another form of pulse, which is a more complete representation of a blast pulse since it includes the possibility of a negative phase of pressure,[14] is shown in Fig. 8.36. It consists of a straight-line rise, followed by an exponential decay through the positive phase, into the negative phase, finally becoming asymptotic to the time axis. The rise time is t_1 and the duration of the positive phase is $t_1 + t_2$.

Unsymmetrical Exponential Pulses with Central Peak. An interesting family of unsymmetrical pulses may be formed by using Eqs. (8.36a) and changing the sign of the exponent of e in both the numerator and the denominator of the second of the equations. The resulting family consists of pulses whose maxima occur at the mid-period time and which satisfy simultaneously both criteria for comparison (equal pulse height and equal pulse area).

FIG. 8.36. Pulse formed by a straight-line rise followed by a continuous exponential decay through positive and negative phases. (*Frankland.*[14])

Figure 8.37 shows the spectra of residual response amplitude and, in the inset diagrams, the pulse shapes. The limiting cases are the symmetrical triangle of duration τ and height ξ_p, and the rectangles of duration $\tau/2$ and height ξ_p. All pulses in the family have the area $\xi_p \tau/2$. Zeroes of residual response amplitude occur for all values of a, at

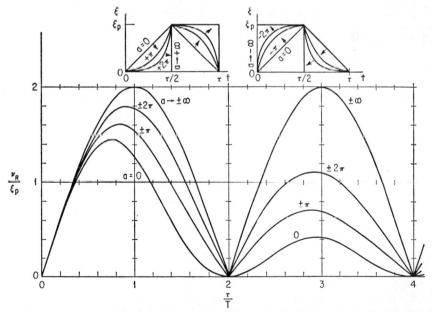

FIG. 8.37. Spectra of residual response amplitude for a family of unsymmetrical exponential pulses of equal area and equal maximum height, having the pulse peak at the mid-period time. (*Jacobsen and Ayre.*[22])

even integer values of τ/T. The residual response amplitude is

$$\frac{\nu_R}{\xi_p} = \frac{aT/\pi\tau}{1 - \cosh a} \cdot$$

$$\left[\frac{\cosh 2a - \cosh a - (1 - \cosh a)\cos(2\pi\tau/T) + (1 - \cosh 2a)\cos(\pi\tau/T)}{1 + a^2 T^2/\pi^2\tau^2}\right]^{\frac{1}{2}} \quad (8.49)$$

Pulses which are mirror images of each other in time result in equal residual amplitudes.

Skewed Versed-sine Pulse. By taking the product of a decaying exponential and the versed-sine function, a family of pulses with varying skewness is obtained.[13, 22] The

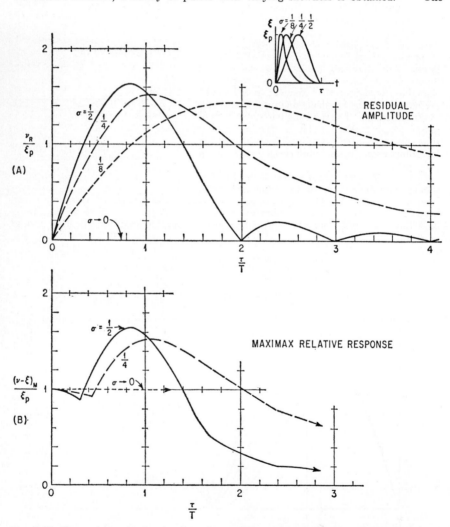

FIG. 8.38. Response spectra for the skewed versed-sine pulse, compared on the basis of equal pulse height: (A) Residual response amplitude. (B) Maximax relative response. (*Jacobsen and Ayre.*[22])

family is described by the following equation:

$$\xi(t) = \begin{cases} \xi_p \dfrac{e^{2\pi(\sigma - t/\tau)\cot \pi\sigma}}{1 - \cos 2\pi\sigma}(1 - \cos 2\pi t/\tau) & [0 \le t \le \tau] \\[2mm] 0 & [\tau \le t] \end{cases} \tag{8.50}$$

These pulses are of particular interest when the excitation is a ground displacement function because they have continuity in both velocity and displacement; thus, they do not involve theoretically infinite accelerations of the ground. When the skewing constant σ equals one-half, the pulse is the symmetrical versed sine. When $\sigma \to 0$, the front of the pulse approaches a straight line with infinite slope, and the pulse area approaches zero.

The spectra of residual response amplitude and of maximax relative response, plotted on the basis of equal pulse height, are shown in Fig. 8.38 for several values of σ. The residual response amplitude spectra are reasonably good approximations to the spectra of maximax relative response except at the lower values of τ/T.

Figure 8.39 compares the residual response amplitude spectra on the basis of equal pulse area. The required pulse heights, for a constant pulse area of $\xi_{po}\tau/2$, are shown in the inset diagram. On this basis, the pulse for $\sigma \to 0$ represents a generalized impulse of value $k\xi_{po}\tau/2$.

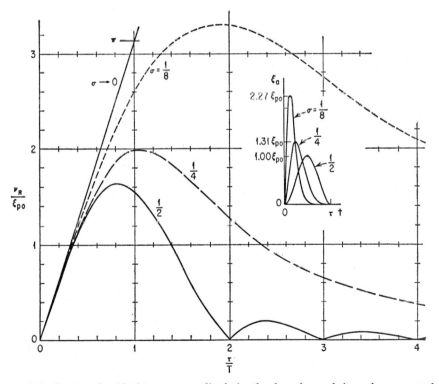

Fig. 8.39. Spectra of residual response amplitude for the skewed versed-sine pulse, compared on the basis of equal pulse area.[19]

FULL-CYCLE PULSES (FORCE-TIME INTEGRAL = 0). The residual response amplitude spectra for three groups of *full-cycle pulses* are shown as follows: in Fig. 8.40 for the rectangular, the sinusoidal, and the symmetrical triangular pulses; in Fig. 8.41 for three types of pulse involving sine and cosine functions; and in Fig. 8.42 for three forms of triangular pulse. The pulse shapes are shown in the inset diagrams. Expressions for the residual response amplitudes are given in Eqs. (8.51) to (8.53).

Full-cycle rectangular pulse:

$$\frac{\nu_R}{\xi_p} = 2 \sin \frac{\pi \tau}{T} \left[2 \sin \frac{\pi \tau}{T} \right] \tag{8.51}$$

Full-cycle "sinusoidal" pulses:
Symmetrical half cycles

$$\frac{\nu_R}{\xi_p} = 2 \sin \frac{\pi \tau}{T} \left[\frac{T/\tau}{(T^2/4\tau^2) - 1} \cos \frac{\pi \tau}{T} \right] \tag{8.52a}$$

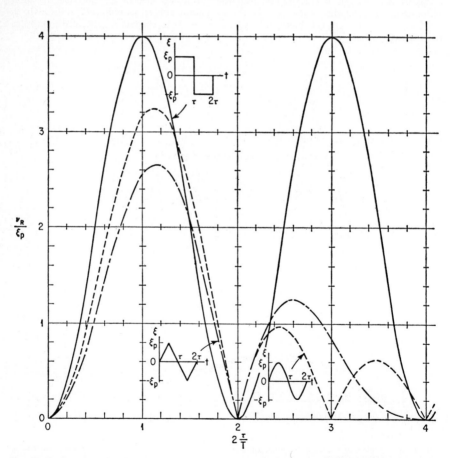

FIG. 8.40. Spectra of residual response amplitude for three types of full-cycle pulses. Each half cycle is symmetrical.[19]

Vertical front and vertical ending

$$\frac{\nu_R}{\xi_p} = \frac{2}{1 - T^2/16\tau^2} \cos \frac{2\pi\tau}{T} \tag{8.52b}$$

Vertical jump at mid-cycle

$$\frac{\nu_R}{\xi_p} = \frac{2}{1 - T^2/16\tau^2} \left(1 - \frac{T}{4\tau} \sin \frac{2\pi\tau}{T} \right) \tag{8.52c}$$

Full-cycle triangular pulses:
Symmetrical half cycles

$$\frac{\nu_R}{\xi_p} = 2 \sin \frac{\pi\tau}{T} \left[\frac{4T}{\pi\tau} \sin^2 \frac{\pi\tau}{2T} \right] \tag{8.53a}$$

Vertical front and vertical ending

$$\frac{\nu_R}{\xi_p} = 2 \left(\frac{T}{2\pi\tau} \sin \frac{2\pi\tau}{T} - \cos \frac{2\pi\tau}{T} \right) \tag{8.53b}$$

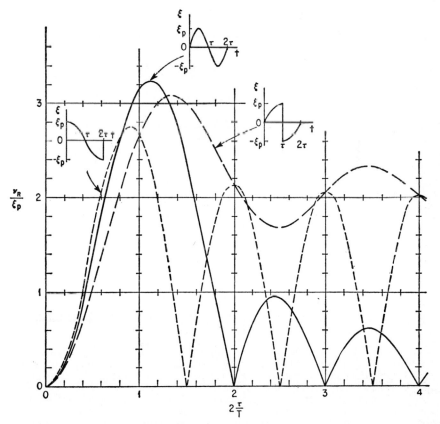

Fig. 8.41. Spectra of residual response amplitude for three types of full-cycle "sinusoidal" pulses.[19]

Vertical jump at mid-cycle

$$\frac{\nu_R}{\xi_p} = 2 \left(1 - \frac{T}{2\pi\tau} \sin \frac{2\pi\tau}{T} \right) \tag{8.53c}$$

In the case of full-cycle pulses having symmetrical half cycles, note that the residual response amplitude equals the residual response amplitude of the symmetrical one-half-cycle pulse of the same shape, multiplied by the dimensionless residual response amplitude function $2 \sin (\pi\tau/T)$ for the single rectangular pulse. Compare the bracketed functions in Eqs. (8.51), (8.52a), and (8.53a) with the bracketed functions in Eqs. (8.31b), (8.32b), and (8.34c), respectively.

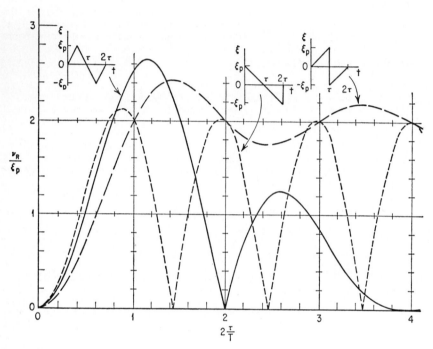

Fig. 8.42. Spectra of residual response amplitude for three types of full-cycle triangular pulses.[19]

SUMMARY OF TRANSIENT RESPONSE SPECTRA FOR THE SINGLE DEGREE-OF-FREEDOM, LINEAR, UNDAMPED SYSTEM

INITIAL CONDITIONS. The following conclusions are based on the assumption that the system is initially at rest.

STEP-TYPE EXCITATIONS:

1. The maximax response ν_M occurs *after* the step has risen (monotonically) to full value ($\tau \leq t$, where τ is the step rise time). It is equal to the residual response amplitude plus the constant step height ($\nu_M = \nu_R + \xi_c$).

2. The extreme values of the ratio of maximax response to step height ν_M/ξ_c are 1 and 2. When the ratio of step rise time to system natural period τ/T approaches zero, the step approaches the simple rectangular step in shape and ν_M/ξ_c approaches the upper extreme of 2. If τ/T approaches infinity, the step loses the character of a dynamic ex-

citation; consequently, the inertia forces of the system approach zero and ν_M/ξ_c approaches the lower extreme of 1.

3. For some particular shapes of step rise, ν_M/ξ_c is equal to 1 at certain finite values of τ/T. For example, for the step having a constant-slope rise, $\nu_M/\xi_c = 1$ when $\tau/T = 1, 2, 3, \ldots$. The lowest values of $\tau/T = (\tau/T)_{\min}$, for which $\nu_M/\xi_c = 1$, are, for three shapes of step rise: constant-slope, 1.0; versed-sine, 1.5; cycloidal, 2.0. The lowest possible value of $(\tau/T)_{\min}$ is 1.

4. In the case of step-type excitations, when $\nu_M/\xi_c = 1$ the residual response amplitude ν_R is zero. Sometimes it is of practical importance in the design of cams and dynamic control functions to achieve the smallest possible residual response.

SINGLE-PULSE EXCITATIONS:

1. When the ratio τ/T of pulse duration to system natural period is less than $\frac{1}{2}$, the time shapes of certain types of equal area pulses are of secondary significance in determining the maxima of system response [maximax response ν_M, maximax relative response $(\nu - \xi)_M$, and residual response amplitude ν_R]. If τ/T is less than $\frac{1}{4}$, the pulse shape is of little consequence in almost all cases and the system response can be determined to a fair approximation by use of the simple impulse theory. If τ/T is larger than $\frac{1}{2}$, the pulse shape may be of great significance.

2. The maximum value of maximax response for a given shape of pulse, $(\nu_M)_{\max}$, usually occurs at a value of the period ratio τ/T between $\frac{1}{2}$ and 1. The maximum value of the ratio of maximax response to the reference excitation, $(\nu_M)_{\max}/\xi_p$, is usually between 1.5 and 1.8.

3. If the pulse has a *vertical rise*, ν_M is the first maximum occurring, and $(\nu_M)_{\max}$ is an asymptotic value approaching $2\xi_p$ as τ/T approaches infinity. In the special case of the rectangular pulse, $(\nu_M)_{\max}$ is equal to $2\xi_p$ and occurs at values of τ/T equal to or greater than $\frac{1}{2}$.

4. If the pulse has a *vertical decay*, $(\nu_M)_{\max}$ is equal to the maximum value $(\nu_R)_{\max}$ of the residual response amplitude.

5. The maximum value $(\nu_R)_{\max}$ of the residual response amplitude, for a given shape of pulse, often is a reasonably good approximation to $(\nu_M)_{\max}$, except if the pulse has a steep rise followed by a decay. A few examples are shown in Table 8.3. Furthermore, if $(\nu_M)_{\max}$ and $(\nu_R)_{\max}$ for a given pulse shape are approximately equal in magnitude, they occur at values of τ/T not greatly different from each other.

6. Pulse shapes that are mirror images of each other in time result in equal values of residual response amplitude.

7. The residual response amplitude ν_R generally has zero values for certain finite values of τ/T. However, if the pulse has either a vertical rise or a vertical decay, but not both, there are no zero values except the trivial one at $\tau/T = 0$. In the case of the rectangular

Table 8.3. Comparison of Greatest Values of Maximax Response and Residual Response Amplitude

Pulse shape	$(\nu_M)_{\max}/(\nu_R)_{\max}$
Symmetrical:	
Rectangular	1.00
Sine	1.04
Versed sine	1.05
Triangular	1.06
Vertical-decay pulses	1.00
Vertical-rise pulses:	
Rectangular	1.00
Triangular	1.60
Asymptotic exponential	
decay	2.00

pulse, $\nu_R = 0$ when $\tau/T = 1, 2, 3, \ldots$. For several shapes of pulse the values of $(\tau/T)_{\min}$ (lowest values of τ/T for which $\nu_R = 0$) are as follows: rectangular, 1.0; sine, 1.5; versed-sine, 2.0; symmetrical triangle, 2.0. The lowest possible value of $(\tau/T)_{\min}$ is 1.

8. In the formulation of pulse as well as of step-type excitations, it may be of practical consequence for the residual response to be as small as possible; hence, attention is devoted to the case, $\nu_R = 0$.

SINGLE DEGREE-OF-FREEDOM LINEAR SYSTEM WITH DAMPING

The calculation of the effects of damping on transient response may be laborious. If the investigation is an extensive one, use should be made of an analog computer.

DAMPING FORCES PROPORTIONAL TO VELOCITY (VISCOUS DAMPING)

In the case of steady forced vibration, even very small values of the viscous damping coefficient have great effect in limiting the system response at or near resonance. If the excitation is of the single step- or pulse-type, however, the effect of damping on the maximax response may be of relatively less importance, unless the system is highly damped.

For example, in a system under steady sinusoidal excitation at resonance, a tenfold increase in the fraction of critical damping c/c_c from 0.01 to 0.1 results in a theoretical tenfold decrease in the magnification factor from 50 to 5. In the case of the same system, initially at rest and acted upon by a half-cycle sine pulse of "resonant duration" $\tau = T/2$, the same increase in the damping coefficient results in a decrease in the maximax response of only about 9 per cent.

HALF-CYCLE SINE PULSE EXCITATION. Figure 8.43 shows the spectra of maximax response for a viscously damped system excited by a half-cycle sine pulse.[12,29] The

FIG. 8.43. Spectra of maximax response for a viscously damped single degree-of-freedom system acted upon by a half-cycle sine pulse. (*R. D. Mindlin, F. W. Stubner, and H. L. Cooper.*[29])

system is initially at rest. The results apply to the cases indicated by the following differential equations of motion:

$$\frac{m\ddot{x}}{k} + \frac{c\dot{x}}{k} + x = \frac{F_p}{k} \sin \frac{\pi t}{\tau} \tag{8.54a}$$

$$\frac{m\ddot{x}}{k} + \frac{c\dot{x}}{k} + x = u_p \sin \frac{\pi t}{\tau} \tag{8.54b}$$

$$\frac{m\ddot{\delta}_x}{k} + \frac{c\dot{\delta}_x}{k} + \delta_x = \frac{-m\ddot{u}_p}{k}\sin\frac{\pi t}{\tau} \tag{8.54c}$$

and in general

$$\frac{m\ddot{\nu}}{k} + \frac{c\dot{\nu}}{k} + \nu = \xi_p \sin\frac{\pi t}{\tau} \tag{8.54d}$$

where $0 \leq t \leq \tau$.

For values of t greater than τ, the excitation is zero. The distinctions among these cases may be determined by referring to Table 8.1. The fraction of critical damping c/c_c in Fig. 8.43 is the ratio of the damping coefficient c to the critical damping coefficient $c_c = 2\sqrt{mk}$. The damping coefficient must be defined in terms of the velocity $(\dot{x}, \dot{\delta}_x, \dot{\nu})$ appropriate to each case. For $c/c_c = 0$, the response spectrum is the same as the spectrum for maximax response shown for the undamped system in Fig. 8.16B.

OTHER FORMS OF EXCITATION; METHODS. Qualitative estimates of the effects of viscous damping in the case of other forms of step or pulse excitation may be made by the use of Fig. 8.43 and of the appropriate spectrum for the undamped response to the excitation in question.

Quantitative calculations may be effected by extending the methods described for the undamped system. If the excitation is of general form, given either numerically or graphically, the *phase-plane-delta* [21, 22] method described in a later section of this chapter may be used to advantage. Of the analytical methods, the *Laplace transformation* is probably the most useful. A brief discussion of its application to the viscously damped system follows.

LAPLACE TRANSFORMATION. The differential equation to be solved is

$$\frac{m\ddot{\nu}}{k} + \frac{c\dot{\nu}}{k} + \nu = \xi(t) \tag{8.55a}$$

Rewriting Eq. (8.55a),

$$\frac{\ddot{\nu}}{\omega_n{}^2} + \frac{2\zeta\dot{\nu}}{\omega_n} + \nu = \xi(t) \tag{8.55b}$$

where $\zeta = c/c_c$ and $\omega_n{}^2 = k/m$.

Applying the operation transforms of Table 8.2 to Eq. (8.55b), the following algebraic equation is obtained:

$$\frac{1}{\omega_n{}^2}[s^2 F_r(s) - sf(0) - f'(0)] + \frac{2\zeta}{\omega_n}[sF_r(s) - f(0)] + F_r(s) = F_e(s) \tag{8.56a}$$

The *subsidiary* equation is

$$F_r(s) = \frac{(s + 2\zeta\omega_n)f(0) + f'(0) + \omega_n{}^2 F_e(s)}{s^2 + 2\zeta\omega_n s + \omega_n{}^2} \tag{8.56b}$$

where the initial conditions $f(0)$ and $f'(0)$ are to be expressed as ν_0 and $\dot{\nu}_0$, respectively.

By performing an *inverse* transformation of Eq. (8.56b), the response is determined in the following operational form:

$$\nu(t) = \mathcal{L}^{-1}[F_r(s)]$$

$$= \mathcal{L}^{-1}\left[\frac{(s + 2\zeta\omega_n)\nu_0 + \dot{\nu}_0 + \omega_n{}^2 F_e(s)}{s^2 + 2\zeta\omega_n s + \omega_n{}^2}\right] \tag{8.57}$$

Example 8.9. Rectangular Step Excitation. Assume that the damping is less than critical ($\zeta < 1$), that the system starts from rest ($\nu_0 = \dot{\nu}_0 = 0$), and that the system is acted upon by the rectangular step excitation: $\xi(t) = \xi_c$ for $0 \leq t$. The transform of the excitation is given by

$$F_e(s) = \mathcal{L}[\xi(t)] = \mathcal{L}[\xi_c] = \xi_c \frac{1}{s}$$

Substituting for ν_0, $\dot{\nu}_0$ and $F_e(s)$ in Eq. (8.57), the following equation is obtained:

$$\nu(t) = \mathcal{L}^{-1}[F_r(s)] = \xi_c \omega_n{}^2 \mathcal{L}^{-1}\left[\frac{1}{s(s^2 + 2\zeta\omega_n s + \omega_n{}^2)}\right] \tag{8.58a}$$

Rewriting,

$$\nu(t) = \xi_c \omega_n{}^2 \mathcal{L}^{-1}\left[\frac{1}{s[s + \omega_n(\zeta - j\sqrt{1 - \zeta^2})][s + \omega_n(\zeta + j\sqrt{1 - \zeta^2})]}\right] \tag{8.58b}$$

where $j = \sqrt{-1}$.

To determine the inverse transform $\mathcal{L}^{-1}[F_r(s)]$, it may be necessary to expand $F_r(s)$ in partial fractions as explained previously. However, in this particular example the transform pair is available in Table 8.2 (see item 16). Thus, it is found readily that $\nu(t)$ is given by the following:

$$\nu(t) = \xi_c \omega_n{}^2 \left[\frac{1}{ab} + \frac{be^{-at} - ae^{-bt}}{ab(a - b)}\right] \tag{8.59a}$$

where $a = \omega_n(\zeta - j\sqrt{1 - \zeta^2})$ and $b = \omega_n(\zeta + j\sqrt{1 - \zeta^2})$. By using the relations, $\cos z = (\tfrac{1}{2})(e^{jz} + e^{-jz})$ and $\sin z = -(\tfrac{1}{2})j(e^{jz} - e^{-jz})$, Eq. (8.59a) may be expressed in terms of *cosine* and *sine* functions:

$$\nu(t) = \xi_c \left[1 - e^{-\zeta\omega_n t}\left(\cos \omega_d t + \frac{\zeta}{\sqrt{1 - \zeta^2}} \sin \omega_d t\right)\right] \qquad [\zeta < 1] \tag{8.59b}$$

where the *damped* natural frequency $\omega_d = \omega_n\sqrt{1 - \zeta^2}$.

If the damping is negligible, $\zeta \to 0$ and Eq. (8.59b) reduces to the form of Eq. (8.22) previously derived for the case of zero damping:

$$\nu(t) = \xi_c(1 - \cos \omega_n t) \qquad [\zeta = 0] \tag{8.22}$$

CONSTANT (COULOMB) DAMPING FORCES; PHASE-PLANE METHOD

The phase-plane method is particularly well suited to the solving of transient response problems involving Coulomb damping forces.[21, 22] The problem is truly a stepwise linear one, provided the usual assumptions regarding Coulomb friction are valid. For example, the differential equation of motion for the case of ground displacement excitation is

$$m\ddot{x} \pm F_f + kx = ku(t) \tag{8.60a}$$

where F_f is the Coulomb friction force. In Eq. (8.60b) the friction force has been moved to the right side of the equation and the equation has been divided by the spring constant k:

$$\frac{m\ddot{x}}{k} + x = u(t) \mp \frac{F_f}{k} \tag{8.60b}$$

The effect of friction can be taken into account readily in the construction of the phase trajectory by modifying the ordinates of the stepwise excitation by amounts equal to $\mp F_f/k$. The quantity F_f/k is the Coulomb friction "displacement," and is equal to one-fourth the decay in amplitude in each cycle of a *free* vibration under the influence of Coulomb friction. The algebraic sign of the friction term changes when the velocity changes sign. When the friction term is placed on the right-hand side of the differential equation, it must have a *negative* sign when the velocity is positive.

Example 8.10. Free Vibration. Figure 8.44 shows an example of free vibration with the initial conditions $x = x_0$ and $\dot{x} = 0$. The locations of the arc centers of the phase trajectory alternate each half cycle from $+F_f/k$ to $-F_f/k$.

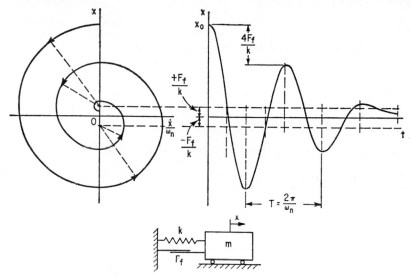

FIG. 8.44. Example of phase-plane solution of free vibration with Coulomb friction;[2] the natural frequency is $\omega_n = \sqrt{k/m}$.

Example 8.11. General Transient Excitation. A general stepwise excitation $u(t)$ and the response x of a system under the influence of a friction force F_f are shown in Fig. 8.45. The case of zero friction is also shown. The initial conditions are $x = 0$, $\dot{x} = 0$. The arc centers are located at ordinates of $u(t) \mp F_f/k$. During the third step in the excitation, the velocity of the system changes sign from positive to negative (at $t = t_2'$); consequently, the friction displacement must also change sign, but from negative to positive.

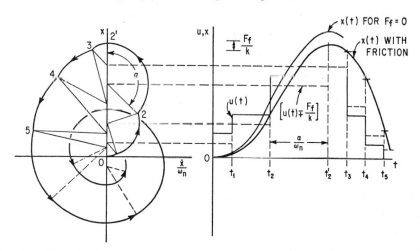

FIG. 8.45. Example of phase-plane solution for a general transient excitation with Coulomb friction in the system.[2]

SINGLE DEGREE-OF-FREEDOM NONLINEAR SYSTEMS

PHASE-PLANE-DELTA METHOD

The transient response of damped linear systems and of nonlinear systems of considerable complexity can be determined by the *phase-plane-delta* method.[21, 22] Assume that the differential equation of motion of the system is

$$m\ddot{x} = G(x,\dot{x},t) \tag{8.61a}$$

where $G(x,\dot{x},t)$ is a general function of x, $\dot{x}$, and t to any powers. The coefficient of $\ddot{x}$ is constant, either inherently or by a suitable division.

In Eq. (8.61a) the general function may be replaced by another general function minus a linear, constant-coefficient, restoring force term:

$$G(x,\dot{x},t) = g(x,\dot{x},t) - kx$$

By moving the linear term kx to the left side of the differential equation, dividing through by m, and letting $k/m = \omega_n^2$, the following equation is obtained:

$$\ddot{x} + \omega_n^2 x = \omega_n^2 \delta \tag{8.61b}$$

where the *operative displacement* δ is given by

$$\delta = \frac{1}{k} g(x,\dot{x},t) \tag{8.61c}$$

The separation of the kx term from the G function does not require that the kx term exist physically. Such a term can be separated by first adding to the G function the *fictitious* terms, $+kx - kx$.

With the differential equation of motion in the δ form, Eq. (8.61b), the response problem can now be solved readily by stepwise linearization. The left side of the equation represents a simple, undamped, linear oscillator. Implicit in the δ function on the right side of the equation are the nonlinear restoration terms, the linear or nonlinear dissipation terms, and the excitation function.

If the δ function is held constant at a value δ for an interval of time Δt, the response of the linear oscillator in the phase-plane is an arc of a circle, with its center on the X axis at δ and subtended by an angle equal to $\omega_n \, \Delta t$. The graphical construction may be similar in general appearance to the examples already shown for linear systems in Figs. 8.5, 8.44, and 8.45. Since in the general case the δ function involves the dependent variables, it is necessary to estimate, before constructing each step, appropriate average values of the system displacement and/or velocity to be expected during the step. In some cases, more than one trial may be required before suitable accuracy is obtained.

Many examples of solution for various types of systems are available in the literature.[3, 5, 6, 8, 13, 15, 20, 21, 22, 25]

MULTIPLE DEGREE-OF-FREEDOM, LINEAR, UNDAMPED SYSTEMS

Some of the transient response analyses, presented for the single degree-of-freedom system, are in complete enough form that they can be employed in determining the responses of linear, undamped, multiple degree-of-freedom systems. This can be done by the use of *normal (principal) coordinates*. A system of normal coordinates is a system of generalized coordinates chosen in such a way that vibration in each normal mode involves only one coordinate, a normal coordinate. The differential equations of motion, when written in normal coordinates, are all independent of each other. Each differential equation is related to a particular normal mode and involves only one coordinate. The differential equations are of the same general form as the differential equation of motion for the single degree-of-freedom system. The response of the system in terms of the physical coordinates, for example, displacement or stress at various locations in the system, is determined by superposition of the normal coordinate responses. Normal coordinates are discussed in Chaps. 2 and 7, and in Refs. 37 and 38.

Example 8.12. Sine Force Pulse Acting on a Simple Beam. Consider the flexural vibration of a prismatic bar with simply supported ends, Fig. 8.46. A sine-pulse concentrated force F_p $\sin(\pi t/\tau)$ is applied to the beam at a distance c from the left end (origin of coordinates). Assume that the beam is initially at rest. The displacement response of the beam, during the time of action of the pulse, is given by the following series:

$$y = \frac{2F_p l^3}{\pi^4 EI} \sum_{i=1}^{i=\infty} \frac{1}{i^4} \sin \frac{i\pi c}{l} \sin \frac{i\pi x}{l} \left[\frac{1}{1 - T_i^2/4\tau^2} \left(\sin \frac{\pi t}{\tau} - \frac{T_i}{2\tau} \sin \omega_i t \right) \right] \qquad [0 \le t \le \tau] \quad (8.62a)$$

where

$$i = 1, 2, 3, \ldots; \quad T_i = \frac{2\pi}{\omega_i} = \frac{2l^2}{i^2\pi} \sqrt{\frac{A\gamma}{EIg}} = \frac{T_1}{i^2}, \text{ sec}$$

A comparison of Eqs. (8.62a) and (8.32a) shows that the time function $[\sin(\pi t/\tau) - (T_i/2\tau) \sin \omega_i t]$ for the ith term in the beam-response series is of exactly the same form as the time function $[\sin(\pi t/\tau) - (T/2\tau) \sin \omega_n t]$ in the response of the single degree-of-freedom system. Furthermore, the magnification factors $1/(1 - T_i^2/4\tau^2)$ and $1/(1 - T^2/4\tau^2)$ in the two equations have identical forms.

Following the end of the pulse, beginning at $t = \tau$, the vibration of the beam is expressed by

$$y = \frac{2F_p l^3}{\pi^4 EI} \sum_{i=1}^{i=\infty} \frac{1}{i^4} \sin \frac{i\pi c}{l} \sin \frac{i\pi x}{l} \left[\frac{(T_i/\tau) \cos (\pi\tau/T_i)}{(T_i^2/4\tau^2) - 1} \sin \omega_i \left(t - \frac{\tau}{2} \right) \right] \qquad [\tau \le t] \quad (8.62b)$$

A comparison of Eqs. (8.62b) and (8.32b) leads to the same conclusion as found above for the time era $0 \le t \le \tau$.

Excitation and Displacement at Mid-span. As a specific case, consider the displacement at mid-span when the excitation is applied at mid-span ($c = x = l/2$). The even-numbered terms of the series now are all zero and the series take the following forms:

$$y_{l/2} = \frac{2F_p l^3}{\pi^4 EI} \sum_{i=1,3,5,\ldots}^{\infty} \frac{1}{i^4} \left[\frac{1}{1 - T_i^2/4\tau^2} \left(\sin \frac{\pi t}{\tau} - \frac{T_i}{2\tau} \sin \omega_i t \right) \right] \qquad [0 \le \le \tau] \quad (8.63a)$$

$$y_{l/2} = \frac{2F_p l^3}{\pi^4 EI} \sum_{i=1,3,5,\ldots}^{\infty} \frac{1}{i^4} \left[\frac{(T_i/\tau) \cos (\pi\tau/T_i)}{(T_i^2/4\tau^2) - 1} \sin \omega_i \left(t - \frac{\tau}{2} \right) \right] \qquad [\tau \le t] \quad (8.63b)$$

Assume, for example, that the pulse period τ equals two-tenths of the fundamental natural period of the beam ($\tau/T_1 = 0.2$). It is found from Fig. 8.16B, by using an abscissa value of 0.2, that the maximax response in the *fundamental* mode ($i = 1$) occurs in the residual vibration era ($\tau \le t$). The value of the corresponding ordinate is 0.75. Consequently, the maximax response for $i = 1$ is 0.75 $(2F_p l^3/\pi^4 EI)$.

In order to determine the maximax for the *third* mode ($i = 3$), an abscissa value of $\tau/T_i = i^4\tau/T_1 - 3^2 \times 0.2 = 1.8$, is used. It is found that the maximax is greater than the residual amplitude and consequently that it occurs during the time era $0 \le t \le \tau$. The value of the corresponding ordinate is 1.36; however, this must be multiplied by $1/3^4$, as indicated by the series. The maximax for $i = 3$ is thus 0.017 $(2F_p l^3/\pi^4 EI)$.

The maximax for $i = 5$ also occurs in the time era $0 \le t \le \tau$ and the ordinate may be estimated to be about 1.1. Multiplying by $1/5^4$, it is found that the maximax for $i = 5$ is approximately 0.002 $(2F_p l^3/\pi^4 EI)$, a negligible quantity when compared with the maximax value for $i = 1$.

To find the maximax total response to a reasonable approximation, it is necessary to sum on a time basis several terms of the series. In the particular example above, the maximax total response occurs in the residual vibration era and a reasonably accurate value can be obtained by considering only the first term ($i = 1$) in the series, Eq. (8.63b).

FIG. 8.46. Simply supported beam loaded by a concentrated force sine pulse of half-cycle duration.

GENERAL INVESTIGATION OF TRANSIENTS

An extensive (and efficient) investigation of transient response in multiple degree-of-freedom systems requires the use of an automatic computer.   In some of the simpler cases, however, it is feasible to employ numerical or graphical methods.   For example, the phase-plane method may be applied to multiple degree-of-freedom linear systems [1,2] through the use of normal coordinates.   This involves independent phase-planes having the coordinates q_i and $\dot{q}_i/\omega_i$, where q_i is the ith normal coordinate.

REFERENCES

1. Ayre, R. S.: *J. Franklin Inst.*, **253**:153 (1952).
2. Ayre, R. S.: *Proc. World Conf. Earthquake Eng.*, 1956, p. 13–1.
3. Ayre, R. S., and J. I. Abrams: *Proc. ASCE*, EM 2, Paper 1580, 1958.
4. Biot, M. A.: *Trans. ASCE*, **108**:365 (1943).
5. Bishop, R. E. D.: *Proc. Inst. Mech. Engrs. (London)*, **168**:299 (1954).
6. Braun, E.: *Ing.-Arch.*, **8**:198 (1937).
7. Bronwell, A.: "Advanced Mathematics in Physics and Engineering," McGraw-Hill Book Company, Inc., New York, 1953.
8. Bruce, V. G.: *Bull. Seismol. Soc. Amer.*, **41**:101 (1951).
9. Cherry, C.: "Pulses and Transients in Communication Circuits," Dover Publications, New York, 1950.
10. Crede, C. E.: "Vibration and Shock Isolation," John Wiley & Sons, Inc., New York, 1951.
11. Crede, C. E.: *Trans. ASME*, **77**:957 (1955).
12. Criner, H. E., G. D. McCann, and C. E. Warren: *J. Appl. Mechanics*, **12**:135 (1945).
13. Evaldson, R. L., R. S. Ayre, and L. S. Jacobsen: *J. Franklin Inst.*, **248**:473 (1949).
14. Frankland, J. M.: *Proc. Soc. Exptl. Stress Anal.*, **6**:2, 7 (1948).
15. Fuchs, H. O.: *Product Eng.*, August, 1936, p. 294.
16. Gardner, M. F., and J. L. Barnes: "Transients in Linear Systems," vol. I, John Wiley & Sons, Inc., New York, 1942.
17. Hartman, J. B.: "Dynamics of Machinery," McGraw-Hill Book Company, Inc., New York, 1956.
18. Hudson, G. E.: *Proc. Soc. Exptl. Stress Anal.*, **6**:2, 28 (1948).
19. Jacobsen, L. S., and R. S. Ayre: "A Comparative Study of Pulse and Step-type Loads on a Simple Vibratory System," *Tech. Rept.* N16, under contract N6-ori-154, T. O. 1, U.S. Navy, Stanford University, 1952.
20. Jacobsen, L. S.: *Proc. Symposium on Earthquake and Blast Effects on Structures*, 1952, p. 94.
21. Jacobsen, L. S.: *J. Appl. Mechanics*, **19**:543 (1952).
22. Jacobsen, L. S., and R. S. Ayre: "Engineering Vibrations," McGraw-Hill Book Company, Inc., New York, 1958.
23. Kelvin, Lord: *Phil. Mag.*, **34**:443 (1892).
24. Kornhauser, M.: *J. Appl. Mechanics*, **21**:371 (1954).
25. Lamoen, J.: *Rev. universelle mines*, ser. 8, **11**:7, 3 (1935).
26. McCann, G. D., and J. M. Kopper: *J. Appl. Mechanics*, **14**:A127 (1947).
27. Meissner, E.: *Schweiz. Bauzt.*, **99**:27, 41 (1932).
28. Mindlin, R. D.: *Bell System Tech. J.*, **24**:353 (1945).
29. Mindlin, R. D., F. W. Stubner, and H. L. Cooper: *Proc. Soc. Exptl. Stress Anal.*, **5**:2, 69 (1948).
30. Morrow, C. T.: *J. Acoust. Soc. Amer.*, **29**:596 (1957).
31. Muller, J. T.: *Bell System Tech. J.*, **27**:657 (1948).
32. Rothbart, H. A.: "Cams—Design, Dynamics and Accuracy," John Wiley & Sons, Inc., New York, 1956.
33. Salvadori, M. G., and R. J. Schwarz: "Differential Equations in Engineering Problems," Prentice-Hall, Inc., Englewood Cliffs, N.J., 1954.
34. Scott, E. J.: "Transform Calculus," Harper & Brothers, New York, 1955.
35. Shapiro, H., and D. E. Hudson: *J. Appl. Mechanics*, **20**:422 (1953).
36. Thomson, W. T.: "Laplace Transformation," Prentice-Hall, Inc., Englewood Cliffs, N.J., 1950.
37. Timoshenko, S. P., and D. H. Young: "Advanced Dynamics," McGraw-Hill Book Company, Inc., New York, 1948.
38. Timoshenko, S. P., and D. H. Young: "Vibration Problems in Engineering," 3d ed., D. Van Nostrand Company, Inc., Princeton, N.J., 1955.
39. Walsh, J. P., and R. E. Blake: *Proc. Soc. Exptl. Stress Anal.*, **6**:2, 151 (1948).
40. Williams, H. A.: *Trans. ASCE*, **102**:838 (1937).

9

EFFECTS OF IMPACT ON STRUCTURES

W. H. Hoppmann II
University of South Carolina

INTRODUCTION

This chapter discusses a particular phenomenon in the general field of shock and vibration usually referred to as impact.[1] An impact occurs when two or more bodies collide. An important characteristic of an impact is the generation of relatively large forces at points of contact for relatively short periods of time. Such forces sometimes are referred to as *impulse-type* forces.

Three general classes of impact are considered in this chapter: (1) impact between spheres or other rigid bodies, where a body is considered to be rigid if its dimensions are large relative to the wavelengths of the elastic stress waves in the body; (2) impact of a rigid body against a beam or plate that remains substantially elastic during the impact; and (3) impact involving yielding of structures.

DIRECT CENTRAL IMPACT OF TWO SPHERES

The elementary analysis of the central impact of two bodies is based upon an experimental observation of Newton.[2] According to that observation, the relative velocity of two bodies after impact is in constant ratio to their relative velocity before impact and is in the opposite direction. This constant ratio is the *coefficient of restitution;* usually it is designated by e.[3]

Let $\dot{u}$ and $\dot{x}$ be the components of velocity along a common line of motion of the two bodies before impact, and $\dot{u}'$ and $\dot{x}'$ the component velocities of the bodies in the same direction after impact. Then, by the observation of Newton,

$$\dot{u}' - \dot{x}' = -e(\dot{u} - \dot{x}) \qquad (9.1)$$

Now suppose that a smooth sphere of mass m_u and velocity $\dot{u}$ collides with another smooth sphere having the mass m_x and velocity $\dot{x}$ moving in the same direction. Let the coefficient of restitution be e, and let $\dot{u}'$ and $\dot{x}'$ be the velocities of the two spheres, respectively, after impact. Figure 9.1 shows the

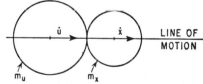

FIG. 9.1. Positions of two solid spheres at instant of central impact.

condition of the two spheres just before collision. The only force acting on the spheres during impact is the force at the point of contact, acting along the line through the centers of the spheres.

According to the law of conservation of linear momentum:

$$m_u\dot{u}' + m_x\dot{x}' = m_u\dot{u} + m_x\dot{x} \qquad (9.2)$$

Solving Eqs. (9.1) and (9.2) for the two unknowns, the velocities $\dot{u}'$ and $\dot{x}'$ after impact,

$$\dot{u}' = \frac{(m_u\dot{u} + m_x\dot{x}) - em_x(\dot{u} - \dot{x})}{m_u + m_x}$$

$$\dot{x}' = \frac{(m_u\dot{u} + m_x\dot{x}) + em_u(\dot{u} - \dot{x})}{m_u + m_x}$$

(9.3)

This analysis yields the resultant velocities for the two spheres on the basis of an experimental law and the principle of the conservation of momentum, without any specific reference to the force of contact F. A similar result is obtained for a ballistic pendulum used to measure the muzzle velocity of a bullet. A bullet of mass m_u and velocity $\dot{u}$ is fired into a block of wood of mass m_x which is at rest initially and finally assumes a velocity $\dot{x}'$ after the impact. Using only the principle of the conservation of momentum,

$$\dot{u} = \frac{(m_u + m_x)\dot{x}'}{m_u}$$

(9.4)

No knowledge of the complicated pattern of force acting on the bullet and the pendulum during the embedding process is required.

These simple facts are introductory to the more complicated problem involving the vibration of at least one of the colliding bodies, as discussed in a later section.

HERTZ THEORY OF IMPACT OF TWO SOLID SPHERES

The theory of two solid elastic spheres which collide with one another is based upon the results of an investigation of two elastic bodies pressed against one another under purely statical conditions.[4] For these static conditions, the relations between the sum of the displacements at the point of contact in the direction of the common line of motion and the resultant total pressure have been derived. The sum of these displacements is equal to the relative approach of the centers of the spheres, assuming that the spheres act as rigid bodies except for elastic compression at the point of contact. The relative approach varies as the two-thirds power of the total pressure; a formula is given for the time of duration of the contact.[4] The theory is valid only if the duration of contact is long in comparison with the period of the fundamental mode of vibration of either sphere.

The range of validity of the Hertz theory is related to the possibility of exciting vibration in the spheres.[5] The dimensionless ratio of the maximum kinetic energy of vibration to the sum of the kinetic energies of the two spheres just before collision is approximately

$$R = \frac{1}{50}\frac{\dot{u} - \dot{x}}{\sqrt{E/\rho}}$$

(9.5)

where $\dot{u} - \dot{x}$ = relative velocity of approach, in./sec

E = Young's modulus of elasticity, assumed to be the same for each sphere, lb/in.²

ρ = density of each sphere, lb-sec²/in.⁴

$\sqrt{E/\rho}$ = approximate velocity of propagation of dilatational waves, in./sec

The ratio R usually is a very small quantity; thus, the theory of impact set forth by Eq. (9.5) has wide application because vibration is not generated in the spheres to an appreciable degree under ordinary conditions. The energy of the colliding spheres remains translational, and the velocities after impact are deducible from the principles of energy and of momentum. The important point of plastic deformation at the point of contact is discussed in a later section.

Formulas for force between the spheres, the radius of the circular area of contact, and the relative approach of the centers of the spheres, all as functions of time, can be determined for any two given spheres.[6]

IMPACT OF A SOLID SPHERE ON AN ELASTIC PLATE

An extension of the Hertz theory of impact to include the effect of vibration of one of the colliding bodies involves a study of the transverse impact of a solid sphere upon an infinitely extended plate.[7] The plate has the role of the vibrating body. The coefficient of restitution is an important element in any analysis of the motion ensuing after the collision of two bodies.

The analysis is based on the assumption that the principal clastic waves of importance are flexural waves of half-period equal to the duration of impact. Let $2h$ and $2D$ be the thickness of plate and diameter of sphere, respectively, ρ_1, ρ_2 their densities, E_1, E_2 their Young's moduli, ν_1, ν_2 their values of Poisson's ratio, and τ_H the duration of impact. The velocity c of long flexural waves of wavelength λ in the plate is given by

$$c^2 = \frac{4\pi^2}{3}\frac{h^2}{\lambda^2}\frac{E_1}{\rho_1(1-\nu_1^2)} \tag{0.6}$$

The radius a of the circle on the plate over which the disturbance has spread at the termination of impact is given by

$$a = c\tau_H = \frac{\lambda}{2} \tag{9.7}$$

Combining Eqs. (9.6) and (9.7),

$$a^2 = \pi\tau_H h \sqrt{\frac{E_1}{3\rho_1(1-\nu_1^2)}} \tag{9.8}$$

The next step is to find the kinetic and potential energies of the wave motion of the plate. The kinetic energy may be determined from the transverse velocity of the plate at each point over the circle of radius a covered by the wave. Figure 9.2 shows an approximate distribution of velocity over the circle of radius a at the end of impact.[8] The direction of the impact also is shown. The kinetic energy in the wave at the end of impact is

$$T = \int_0^a \tfrac{1}{2}\cdot 2h\cdot\rho_1\cdot 2\pi R\cdot \dot{w}^2\, dR \tag{9.9}$$

where $\dot{w}$ is the transverse velocity at distance R from the origin. As an approxima- tion it is assumed that the sum of the potential energy and the kinetic energy in the wave is $2T$. With considerable effort these energies can be calculated in terms of the motion of the plate, although the cal- culation may be laborious.

FIG. 9.2. Distribution of transverse veloci- ties in plate as a result of impact by a moving body. (*After Lamb.*)

The impulse in the plate produced by the colliding body is

$$J = \int_0^a \tfrac{1}{2}\cdot 2h\cdot\rho_1\cdot 2\pi R\cdot \dot{w}\, dr \tag{9.10}$$

The integration should be carried out with due regard to the sign of velocity. If m_u is the mass of colliding body, $\dot{u}$ its velocity before impact, and e the coefficient of restitu- tion, the following relations are obtained on the assumption that the energy is conserved:

$$\tfrac{1}{2}m_u\dot{u}^2(1-e^2) = 2T \tag{9.11}$$

$$m_u\dot{u}(1+e) = J \tag{9.12}$$

Equation (9.11) represents the energy lost to the moving sphere as a result of impact and Eq. (9.12) represents the change in momentum of the sphere.

The coefficient of restitution e is determined by evaluating the integrals for T and J and substituting their values in Eq. (9.12). The necessary integrations can be performed by taking the function for transverse velocity in Fig. 9.2 as arcs of sine curves. The resultant expression for e is

$$e = \frac{h\rho_1 a^2 - 0.56 m_u}{h\rho_1 a^2 + 0.56 m_u} \tag{9.13}$$

where a, the radius of the deformed region, is given by Eq. (9.8) and τ_H, the time of contact between sphere and plate, is given by Hertz's theory of impact to a first approximation.[4] The mass of the sphere is m_u; the mass of the plate is assumed to be infinite. Large discrepancies between theory and experiment occur when the diameter of the sphere is large compared with the thickness of the plate. The duration of impact τ_H is

$$\tau_H = 2.94 \frac{\alpha}{u} \tag{9.14}$$

where
$$\alpha = \left[\frac{15}{16} v_1^2 \left(\frac{1 - v_1^2}{E_1} + \frac{1 - v_2^2}{E_2} \right) m_u \right]^{2/5} R_s^{-1/5}$$

The radius of the striking sphere is R_s and its velocity before impact is $\dot{u}$. Subscripts 1 and 2 represent the properties of the sphere and plate, respectively. The value of τ_H may be substituted in Eq. (9.8) above.

Experimental results verify the theory when the limitations of the theory are not violated. The velocity of impact must be sufficiently small to avoid plastic deformation. When the collision involves steel on steel, the velocity usually must be less than 1 ft/sec. However, useful engineering results can be obtained with this approach even though plastic deformation does occur locally.[9, 10]

TRANSVERSE IMPACT OF A MASS ON A BEAM

If $F(t)$ is the force acting between the sphere and the beam during contact, the distance traveled by the sphere in time t after collision is [11]

$$\dot{u}t - \frac{1}{m_u} \int_0^t F(t_v)(t - t_v) \, dt_v \tag{9.15}$$

where $\dot{u}$ = velocity of sphere before collision (beam assumed to be at rest initially)

m_u = mass of solid sphere

The beam is assumed to be at rest initially.

For example, the deflection of a simply supported beam under force $F(t_v)$ at its center is

$$\sum_{1,3,5\ldots}^{\infty} \frac{1}{m_b} \int_0^t F(t_v) \frac{\sin \omega_n(t - t_v)}{\omega_n} \, dt_v \tag{9.16}$$

where m_b = one-half of mass of beam

ω_n = angular frequency of the nth mode of vibration

Equation (9.16) represents the transverse vibration of a beam. While the present case is only for direct central impact, the cases for noncentral impact depend only on the corresponding solution for transverse vibration. Oblique impact also is treated readily.

The expression for the relative approach of the sphere and beam; i.e., penetration of beam by sphere, is [11]

$$\alpha = \kappa_1 F(t)^{2/3} \tag{9.17}$$

where κ_1 is a constant depending on the elastic and geometrical properties of the sphere and the beam at the point of contact, and α is given by Eq. (9.14). Consequently, the equation that defines the problem is

$$\alpha = K_1 F^{3/2} = \dot{u}t - \frac{1}{m_u} \int_0^t F(t_v)(t - t_v) \, dt_v - \sum_{1,3,5}^{\infty} \frac{1}{m_b} \int_0^t F(t_v) \frac{\sin \omega_n(t - t_v)}{\omega_n} \, dt_v \tag{9.18}$$

Equation (9.18) has been solved numerically for two specific problems by subdividing the time interval 0 to t into small elements and calculating, step by step, the displacements of the sphere.[11]   The results are not general but rather apply only to the cases of beam and sphere.

For the impact of a mass on a beam, the sum of the kinetic and the potential energies may be expressed in terms of the unknown contact force.[12]   Also, the impulse integral J in terms of the contact force may be expressed as

$$J = \int_0^t F(t)\, dt = m_u \dot{u}(1 + e) \tag{9.19}$$

A satisfactory approximation to $F(t)$ is defined in terms of a normalized force $\bar{F}$:

$$F(t) = m_u \dot{u}(1 + e)\bar{F}(t) \tag{9.20}$$

Thus, from Eqs. (9.19) and (9.20),

$$\int_0^t \bar{F}\, dt = 1 \tag{9.21}$$

The value of this integral is independent of the shape of $F(t)$. The normalized force is defined such that its maximum value equals the maximum value of the corresponding normalized Hertz force.[12] To perform the necessary integrations, a suitable function for defining $F(t)$ is chosen as follows:

$$\bar{F}(t) = \frac{\pi}{2\tau_L} \sin \frac{\pi}{\tau_L} t \quad [0 < t < \tau_L]$$

$$\bar{F}(t) = 0 \qquad [|t| > \tau_L] \tag{9.22}$$

Results for particular problems solved in this manner agree well with those obtained for the same problems by the numerical solution of the exact integral equation.[12]

To apply these results to a specific beam impact problem, it is necessary to express the deflection equation for the beam in terms of known quantities. One of these quantities is the coefficient of restitution; a formula must be provided for its determination in terms of known functions. This is given by Eq. (9.31).

IMPACT OF A RIGID BODY ON A DAMPED ELASTICALLY SUPPORTED BEAM

For the more general case of impact of a rigid body on a damped, elastically supported beam, it is assumed that there is external damping, damping determined by the Stokes' law of stress-strain, and an elastic support attached to the beam along its length in such a manner that resistance is proportional to deflection.[13] The differential equation for the deflection of the beam is

$$EI \frac{\partial^4 w}{\partial x^4} + c_1 I \frac{\partial^5 w}{\partial x^4\, \partial t} + c_2 \frac{\partial w}{\partial t} + kw + \rho S \frac{\partial^2 w}{\partial t^2} = F(x,t) \tag{9.23}$$

where w = deflection, in.
E = Young's modulus, lb/in.^{2}
I = moment of inertia for cross section (constant), in.^{4}
c_1 = internal damping coefficient, lb/in.^{2}-sec (Stokes' law)
c_2 = external damping coefficient, lb/in.^{2}-sec
k = foundation modulus, lb/in.^{2}
ρ = density, lb-sec^2/in.^{4}
S = area of cross section (constant), in.^{2}
$\dfrac{\partial^2 w}{\partial t^2}$ = acceleration, in./sec^2
t = time, sec
$F(x,t)$ = driving force per unit length of beam, lb/in.

For example, to illustrate the application of specific boundary conditions, consider a simply supported beam of length l. The moments and deflections must vanish at the ends. The beam is assumed undeflected and at rest just before impact, and central impact is assumed although with some additional computation this restriction may be dropped. The solution may be written as follows:

$$w(x,t) = \sum_{}^{\infty} \sin \frac{n\pi x}{l} \sin \frac{n\pi}{2} \frac{1}{m} \frac{1}{\sqrt{\omega_n{}^2 - \delta_n{}^2}}$$

$$\times \int_0^t e^{-\delta_n(t-\tau)} \sin \left[\sqrt{\omega_n{}^2 - \delta_n{}^2} \cdot (t - \tau) \right] F_1(\tau) \, d\tau \quad (9.24)$$

where e = base of natural logarithms

δ_n = damping numbers = $\dfrac{1}{2} \left(r_i \dfrac{n^4 \pi^4}{l^4} + r_e \right)$

$r_i = \dfrac{c_1 I}{\rho S}$

$r_e = \dfrac{c_2}{\rho S}$

ω_n = angular frequencies

$m = \frac{1}{2}\rho A l$

A satisfactory analytical expression for the contact force $F_1(t)$, a particularization of $F(x,t)$ in Eq. (9.23), must be developed. Although $F_1(t)$ is assumed to act at the center of the beam, the methods apply with only minor alterations if the impact occurs at any other point of the beam.

One of the conditions which the contact force must satisfy is that its time integral for the duration of impact equal the change in momentum of the striking body. The change of momentum is

$$m\dot{z} - m\dot{z}' = m\dot{z} \left(1 - \frac{\dot{z}'}{\dot{z}} \right) \quad (9.25)$$

where m = mass of rigid body, lb-sec^2/in.

$\dot{z}$ = velocity of rigid body just before collision, in./sec

$\dot{z}'$ = velocity of rigid body just after collision, in./sec

When the velocity of the beam is zero, Eq. (9.1) may be written

$$e = -\frac{\dot{z}'}{\dot{z}} \quad (9.26)$$

Equation (9.26) may be written

$$m\dot{z} \left(1 - \frac{\dot{z}'}{\dot{z}} \right) = m\dot{z}(1 + e) \quad (9.27)$$

From the equivalence of impulse and momentum:

$$\int_0^{\tau_0} F_1(t) \, dt = m\dot{z}(1 + e) \quad (9.28)$$

where τ_0 is the time of contact.

It can then be shown [13] that the impact force may be written

$$F_1(t) = m\dot{z}(1 + e) \frac{\pi}{2\tau_L} \sin \frac{n\pi t}{\tau_L} \quad [0 < t < \tau_L]$$

$$F_1 = 0 \quad\quad\quad [t > \tau_L] \quad (9.29)$$

It can be shown further [13] that

$$\tau_L = 3.28 \left[\frac{m^2}{\dot{z}R} \cdot \frac{(1 - \nu^2)}{E^2} \right]^{1/5}$$

(9.30)

where R = radius of sphere, in.

ν = Poisson's ratio

The time interval τ_L is a special value of the time of contact T_0. It agrees well with experimental results.

The coefficient of restitution e is [13]

$$e = \frac{1 - \dfrac{m}{m_b} \sum_1^\infty \Phi_n - \dfrac{m}{m_b} \sum_1^\infty \Psi_n}{1 + \dfrac{m}{m_b} \sum_1^\infty \Phi_n + \dfrac{m}{m_b} \sum_1^\infty \Psi_n}$$

(9.31)

where m = mass of sphere

m_b = half mass of beam

The functions Φ_n and Ψ_n are given in the form of curves in Figs. 9.3 and 9.4; the symbol $\beta_n = \delta_n/\omega_n$ represents fractional damping and $Q_n = \omega_n \tau_L/2\pi$ is a dimensionless frequency where ω_n = angular frequency of nth mode of vibration of undamped vibration

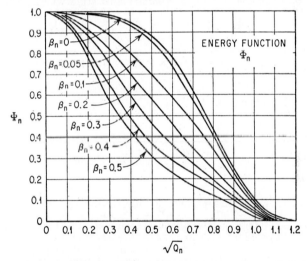

Fig. 9.3. Energy functions ϕ_n used with Eq. (9.31) to determine the coefficient of restitution from the impact of a rigid body on a damped elastically supported beam.

Fig. 9.4. Dissipative (damping) functions ψ_n used with Eq. (9.31) to determine the coefficient of restitution from the impact of a rigid body on a damped elastically supported beam.

of beam, rad/sec, and τ_L = length of time the sinusoidal pulse is assumed to act on beam [see Eq. (9.30)]. If damping is neglected, the functions Ψ_n vanish from Eq. (9.31).

The above theory may be generalized to apply to the response of plates to impact. The deflection equation of a plate subjected to a force applied at a point is required. The various energy distributions at the end of impact are arrived at in a manner analogous to that for the beam.

The theory has been applied to columns and continuous beams [14, 15] and also could be applied to transverse impact on a ring. Measurement of the force of impact illustrates the large number of modes of vibration that can be excited by an impact.[16, 17, 22]

Principal qualitative results of the foregoing analysis are:

1. Impacts by bodies of relatively small mass moving with low velocities develop significant bending strains in beams.

2. External damping of the type assumed above has a rapidly decreasing effect on reducing deflection and strain as the number of the mode increases.

3. Internal damping of the viscous type here assumed reduces deflection and strain appreciably in the higher modes. For a sufficiently high mode number, the vibration becomes aperiodic.

4. Increasing the modulus for an elastic foundation reduces the energy absorbed by the structure from the colliding body.

5. Impacts from collision produce sharp initial rises in strain which are little influenced by damping.

6. Because of result 5, the fatigue problem for machines and structures, in which the impact conditions are repeated many times, can be serious. Ordinary damping affords little protection.

7. The structure seldom can be treated as a single degree-of-freedom system with any degree of reliability in predicting strain.[18, 19]

LONGITUDINAL AND TORSIONAL IMPACT ON BARS

If a mass strikes the end of a long bar, the response may be investigated by means of the Hertz contact theory.[11] The normal modes of vibration must be known so the displacement at each part of the bar can be calculated in terms of a contact force. In a similar manner, the torsional vibration of a long bar can be studied, using the normal modes of torsional vibration.

PLASTIC DEFORMATION RESULTING FROM IMPACT

Many problems of interest involve plastic deformation rather than elastic deformation as considered in the preceding analyses. Using the concept of the plastic hinge, the large plastic deformation of beams under transverse impact [23] and the plastic deformation of free rings under concentrated dynamic loads [24] have been studied. In such analyses, the elastic portion of the vibration usually is neglected. To make further progress in analyses of large deformations as a result of impact, a realistic theory of material behavior in the plastic phase is required.

FIG. 9.5. Longitudinal impact of moving body on end of rod.

An attempt to solve the problem for the longitudinal impact on bars has been made using the static engineering-type stress-strain curve as a part of the analysis.[25] An extension of the work to transverse impact also was attempted.[26]

Figure 9.5 illustrates the impact of a large body m colliding axially with a long rod. The body m has an initial velocity $\dot{u}$ and is sufficiently large that the end of the rod may be assumed to move with constant velocity $\dot{u}$. At any time t a stress wave will have moved into the bar a definite distance; by the condition of continuity (no break in the material), the struck end of the bar will have moved a distance equal to the total elongation of the end portion of the bar:

$$\dot{u}t = \epsilon \cdot l \tag{9.32}$$

The velocity c of a stress wave is $c = l/t$, and Eq. (9.32) becomes

$$\epsilon = \frac{\dot{u}}{c} \tag{9.33}$$

The stress and strain in an elastic material are related by Young's modulus. Substituting for strain from Eq. (9.33),

$$\sigma = \epsilon \cdot E = E\frac{\dot{u}}{c} \tag{9.34}$$

where $\dot{u}$ = velocity of end of rod, in./sec
     l = distance stress wave travels in time t, in.
     t = time, sec
     σ = stress, lb/in.^{2}
     ϵ = strain (uniform), in./in.
     E = Young's modulus, lb/in.^{2}
     c = velocity of stress wave (dilatational), in./sec

When the yield point of the material is exceeded, Eq. (9.34) is inapplicable. Extensions of the analysis, however, lead to some results in the case of plastic deformation.[25] The differential equation for the elastic case is

$$E\frac{\partial^2 u}{\partial x^2} = \rho\frac{\partial^2 u}{\partial t^2} \tag{9.35}$$

where u = displacement, in.
     x = coordinate along rod, in.
     t = time, sec
     E = Young's modulus, lb/in.^{2}
     ρ = mass density, lb-sec^2/in.^{4}

The velocity of the elastic dilatational wave obtained from Eq. (9.35) is

$$c = \sqrt{\frac{E}{\rho}}$$

The modulus E is the slope of the stress-strain curve in the initial linear elastic region. Replacing E by $\partial\sigma/\partial\epsilon$ for the case in which plastic deformation occurs, the slope of the static stress-stress curve can be determined at any value of the strain ϵ.[25] Equation (9.35) then becomes

$$\frac{\partial\sigma}{\partial\epsilon}\frac{\partial^2 u}{\partial x^2} = \rho\frac{\partial^2 u}{\partial t^2} \tag{9.36}$$

Equation (9.36) is nonlinear; its general solution never has been obtained. For the simple type of loading discussed above and an infinitely long bar, the theory predicts a so-called critical velocity of impact because the velocities of the plastic waves are much smaller than those for the elastic waves and approach zero as the strain is indefinitely increased.[25] Since the impact velocity $\dot{u}$ is an independent quantity, it can be made larger and larger while the wave velocities are less than the velocity for elastic waves. Hence a point must be reached at which the continuity of the material is violated. Experimental data illustrate this point.[27]

ENERGY METHOD

Many problems in the design of machines and structures require knowledge of the deformation of material in the plastic condition. In statical problems the method of limit design [28] may be used. In dynamics, the most useful corresponding concept is less theoretical and may be termed the energy method; it is based upon the impact test used for the investigation of brittleness in metals. Originally, the only purpose of this test was to break a standard specimen as an index of brittleness or ductility. The general method, using a tension specimen, may be used in studying the dynamic resistance of materials.[27] An axial force is applied along the length of the specimen and causes the material to rupture ultimately. The energy of absorption is the total amount of energy taken out of the loading system and transferred to the specimen to cause the plastic deformation. The elastic energy and the specific mode of build-up of stress to the final plastic state are ignored. Such an approach has value only to the extent that the material has ductility. For example, in a long tension-type specimen of medium steel, the energy absorbed before neck-down and rupture is of the order of 500 ft-lb per cubic inch of material. Thus, if the moving body in Fig. 9.5 weighs 200 lb and has an initial velocity of 80 ft/sec, it represents 20,000 ft-lb of kinetic energy. If the tension bar subjected to the impact is 10 in. long and 0.5 in. in diameter, it will absorb approximately 1,000 ft-lb of energy. Under these circumstances it will rupture. On the other hand, if the moving body m weighs only 50 lb and has an initial velocity of 30 ft/sec, its kinetic energy is approximately 700 ft-lb and the bar will not rupture.

If the tension specimen were severely notched at some point along its length, it would no longer absorb 500 ft-lb per cubic inch to rupture. The material in the immediate neighborhood of the notch would deform plastically; a break would occur at the notch with the bulk of the material in the specimen stressed below the yield stress for the material. A practical structural situation related to this problem occurs when a butt weld is located at some point along an unnotched specimen. If the weld is of good quality, the full energy absorption of the entire bar develops before rupture; with a poor weld, the rupture occurs at the weld and practically no energy is absorbed by the remainder of the material. This is an important consideration in applying the energy method to design problems.

REFERENCES

1. Love, A. E. H.: "The Mathematical Theory of Elasticity," p. 25, Cambridge University Press, New York, 1934.
2. Timoshenko, S., and D. H. Young: "Engineering Mechanics," p. 334, McGraw-Hill Book Company, Inc., New York, 1940.
3. Loney, S. L.: "A Treatise on Elementary Dynamics," Cambridge University Press, p. 199, New York, 1900.
4. Hertz, H.: *J. Math. (Crelle)*, pp. 92, 155, 1881.
5. Rayleigh, Lord: *Phil. Mag.* (ser. 6), **11**:283 (1906).
6. Timoshenko, S.: "Theory of Elasticity," p. 350, McGraw-Hill Book Company, Inc., New York, 1934.
7. Raman, C. V.: *Phys. Rev.*, **15**, 277 (1920).
8. Lamb, H.: *Proc. London Math. Soc.*, **35**, 141 (1902).
9. Hoppmann, II, W. H.: *Proc. SESA*, **9**:2, 21 (1952).
10. Hoppmann, II, W. H.: *Proc. SESA*, **10**:1, 157 (1952).
11. Timoshenko, S.: "Vibration Problems in Engineering," 3d ed., p. 413, D. Van Nostrand Company, Inc., Princeton, N.J., 1955.
12. Zener, C., and H. Feshbach: *Trans. ASME*, **61**:a-67 (1939).
13. Hoppmann, II, W. H.: *J. Appl. Mechanics*, **15**:125 (1948).
14. Hoppmann, II, W. H.: *J. Appl. Mechanics*, **16**:370 (1949).
15. Hoppmann, II, W. H.: *J. Appl. Mechanics*, **17**:409 (1950).
16. Goldsmith, W., and D. M. Cunningham: *Proc. SESA*, **14**:1, 179 (1956).
17. Barnhart, Jr., K. E., and Werner Goldsmith: *J. Appl. Mechanics*, **24**:440 (1957).
18. Emschermann, H. H., and K. Ruhl: *VDI-Forschungsheft* 443, Ausgabe B, Band 20, 1954.
19. Hoppmann, II, W. H.: *J. Appl. Mechanics*, **19** (1952).
20. Wenk, E., Jr.: Dissertation, The Johns Hopkins University, 1950, and *David W. Taylor Model Basin Rept.* 704, July, 1950.

21. Compendium, "Underwater Explosion," O. N. R., Department of the Navy, 1950.
22. Prager, W.: James Clayton Lecture, *The Institution of Mechanical Engineers, London*, 1955.
23. Lee, E. H., and P. S. Symonds: *J. Appl. Mechanics*, **19**:308 (1952).
24. Owens, R. H., and P. S. Symonds: *J. Appl. Mechanics*, **22** (1955).
25. Von Kármán, T.: *NDRC Rept.* A-29, 1943.
26. Duwez, P. E., D. S. Clark, and H. F. Bohnenblust: *J. Appl. Mechanics*, **17**, 27 (1950).
27. Hoppmann, II, W. H.: *Proc. ASTM*, **47**:533 (1947).
28. Symposium on the Plastic Theory of Structures, Cambridge University, September, 1956, *British Welding J.*, **3**(8) (1956); **4**(1) (1957).

10

MECHANICAL IMPEDANCE

Elmer L. Hixson
The University of Texas at Austin

INTRODUCTION

The differential equations of motion of a linear mechanical system can be expressed in terms of the driving force and the acceleration, velocity, or displacement of the elements of the system. With a sinusoidal force driving a linear system, the steady-state response exhibits sinusoidal acceleration, velocity, and displacement at the driving frequency with a fixed phase relation to the driving force. The relationship between driving force and the motion resulting at various points of the system can be expressed by algebraic equations involving complex numbers. The force and motion characteristics of the system are defined as described in the following paragraphs.

Force. A sinusoidal force can be represented as a rotating phasor (to differentiate from a vector having magnitude and direction) on the complex plane, as shown in Fig. 10.1. The phasor has a length or magnitude F_0; it rotates counterclockwise with an angular velocity ω, and at some time t subtends an angle ωt with the $(+)$ real axis. The instantaneous force is the projection of the phasor on the real axis, $F_0 \cos \omega t$. The complex representation of the force in terms of real and imaginary components is

FIG. 10.1. Complex-plane representation for sinusoidal force and velocity.

$$F = F_0 \cos \omega t + j F_0 \sin \omega t \tag{10.1}$$

The force also can be expressed in terms of magnitude and phase angle:

$$F = F_0 e^{j\omega t} \tag{10.2}$$

Velocity. A sinusoidal velocity also can be represented as a phasor, as shown in Fig. 10.1. When v_0 is rotating at the same angular velocity ω as F_0, it can be represented as

$$v = v_0[\cos(\omega t + \phi) + j\sin(\omega t + \phi)] = v_0 e^{j(\omega t + \phi)} \tag{10.3}$$

where ϕ is the phase angle between F and v. The phase reference in Eq. (10.3) is the $(+)$

10–1

real axis. When F is taken as the phase reference, v may be written

$$v = v_0 e^{j\phi} \tag{10.4}$$

Displacement. Displacement in sinusoidal motion is readily determined from velocity by expressing velocity in the form of Eq. (10.2), $v = v_0 e^{j\omega t}$, and noting that $x = \int v \, dt$:

$$x = \frac{v}{j\omega} \tag{10.5}$$

Acceleration. Acceleration is derived from velocity by using the relation $\ddot{x} = dv/dt$:

$$\ddot{x} = j\omega v \tag{10.6}$$

Mechanical Driving-point Impedance. The ratio of the driving force acting on a system to the resulting velocity of the system is *mechanical impedance.** If the velocity is measured at the point of application of the driving force, the ratio of force to velocity is *mechanical driving-point impedance* Z defined as follows: [1,2,3]

$$Z = \frac{F}{v} \tag{10.7}$$

where F is the applied force and v is the resultant velocity in the direction of the force. The quantities F and v are represented by complex numbers, as indicated by Eqs. (10.1) to (10.3). When F is taken as the phase reference, $\phi = \theta$ where θ is defined as the *impedance angle.*

Mechanical Transfer Impedance. The ratio of the driving force to the velocity at another point in the system (or in a direction different from the direction of the applied force) may be expressed as an impedance:

$$Z_{12} = \frac{F_1}{v_2} \tag{10.8}$$

where F_1 is the sinusoidal force applied at one point in a system and v_2 is the resulting sinusoidal velocity at another point in the system, F_1 and v_2 being complex quantities.

Mechanical Driving-point Mobility. Mobility, the inverse of impedance, is determined as follows: [4,5,6]

$$\mathfrak{M} = \frac{v}{F} \tag{10.9}$$

where F and v are as defined by Eqs. (10.1) and (10.3). The concept of mobility is useful in representing electromechanical systems. Mobility sometimes is called *mechanical admittance.*

Dynamic Modulus. Dynamic modulus is defined as follows: [7,8]

$$D = \frac{F}{x} \tag{10.10}$$

where F is the applied sinusoidal force and x is the resulting displacement in the direction of the force, both expressed as complex quantities.

Receptance. Receptance is the reciprocal of dynamic modulus; it is defined as follows: [9,10]

$$R = \frac{x}{F} \tag{10.11}$$

where F and x are defined as above.

---

* The ratio of driving force to resulting displacement sometimes is termed "mechanical impedance." This is not recommended usage.

Table 10.1. Units of Parameters Used in Analysis by Mechanical Impedance and Mobility Methods

Quantity	Units		
	cgs	mks	English gravitational
Velocity, v	cm/sec	meters/sec	in./sec
Force, F	dynes	newtons	lb
Mass, m	grams	kilograms	lb-sec^2/in.
Spring stiffness, k	dynes/cm	newtons/meter	lb/in.
Damping coefficient, c	dynes/cm/sec	newtons/meter/sec	lb/in./sec
Impedance, Z	dyne-sec/cm	newton-sec/meter	lb-sec/in.
	grams/sec	kilograms/sec	
Mobility, $\mathfrak{M}$	cm/dyne/sec	meters/newton/sec	in./lb-sec
	sec/gram	sec/kilogram	
Power, P	dyne-cm/sec	newton-meters/sec	in.-lb/sec
	ergs/sec	joules/sec	

In the ensuing discussion, only impedance and mobility are used; the discussion is limited to linear systems in which the motion is translational in the direction of the driving force. Other systems that are analogous to translational systems may be analyzed by the same methods using analogous parameters. The several quantities used in impedance methods may be expressed in any consistent system of units. Three such systems are given in Table 10.1.

Impedance methods allow a complete analysis of the motions and forces acting on each part of a physical system; however, the differential equations of Newtonian mechanics become algebraic equations. In addition, a "black box" concept is introduced in which the forces and motions at one or two points of major interest in a system are determined without a complete analysis of the entire system. These methods are adaptable to the use of results of physical measurements on systems that are so complicated that a complete analysis becomes impractical. Typical curves illustrating the quantitative aspect of impedance as measured on a wide range of structure types are presented.

IMPEDANCE OF MECHANICAL ELEMENTS

Three idealized mechanical system elements with lumped constants are conventionally assembled to form linear physical systems. These elements are resistance (or damping), spring, and mass.

MECHANICAL RESISTANCE (DAMPER). A mechanical resistance is a device in which the relative velocity between its end points is proportional to the force applied to the end points. Such a device can be represented by the dashpot of Fig. 10.2 in which the force resisting the extension (or compression) of the dashpot is the result of viscous friction. An ideal resistance is assumed to be made of massless, infinitely rigid elements.

In Fig. 10.2 the velocity of point A with respect to point B is

$$v = (v_1 - v_2) = \frac{F_a}{c} \qquad (10.12)$$

where c is the constant of proportionality called the *mechanical resistance* or *damping constant*. For there to be a relative velocity v as a result of force at A, there must be an equal reaction force at B. Thus, it may be considered that a transmitted force F_b is

FIG. 10.2. Two alternate schematic representations of an ideal mechanical resistance element. The upper one is that recommended by the American National Standards Institute, ANSI Y-32.18 (1972).

equal to F_a. The velocities v_1 and v_2 are measured with respect to the stationary reference G; their difference is the relative velocity between the end points of the resistance.

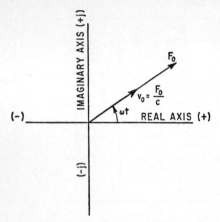

With the sinusoidal force of Eq. (10.2) applied to point A of Fig. 10.2 and point B attached to a fixed (immovable) point, the velocity v_1 is obtained from Eq. (10.12):

$$v_1 = \frac{F_0 e^{j\omega t}}{c} = v_0 e^{j\omega t} \qquad (10.13)$$

The force and velocity phasors thus are rotating at the same angular frequency and are said to be "in phase," as shown in Fig. 10.3.

The mechanical impedance of the resistance is obtained by substituting from Eqs. (10.2) and (10.13) in Eq. (10.7):

$$Z_c = \frac{F}{v} = \frac{F_0 e^{j\omega t}}{\dfrac{F_0}{c} e^{j\omega t}} = c \qquad (10.14)$$

FIG. 10.3. Force and velocity relationships at the connections of an ideal resistance element.

The mechanical impedance of a resistance is the value of its damping constant c; i.e., when a resistance is placed in a mechanical system, the ratio of the transmitted force to the relative velocity of its end points is its impedance c.

SPRING. A linear spring is a device for which the relative displacement between its end points is proportional to the force applied. It is illustrated in Fig. 10.4 and can be represented mathematically as follows:

$$\delta x = x_1 - x_2 = \frac{F_a}{k} \qquad (10.15)$$

FIG. 10.4. Two alternate schematic representations of an ideal spring. The upper one is that recommended by the American National Standards Institute, ANSI Y-32.18 (1972).

where k is the *spring stiffness* and x_1, x_2 are displacements relative to the reference point G. The stiffness k can be expressed alternately in terms of a *compliance* $C = 1/k$. The spring transmits the applied force so that $F_b = F_a$.

With the force of Eq. (10.2) applied to point A and with point B fixed, the displacement of point A is given by Eq. (10.15):

$$x_1 = \frac{F_0 e^{j\omega t}}{k} = x_0 e^{j\omega t} \qquad (10.16)$$

The displacement is thus sinusoidal and in phase with the force. The relative velocity of the end connections is required for impedance calculations and is given by the differentiation of x with respect to time:

$$\dot{x} = v = \frac{j\omega F_0 e^{j\omega t}}{k} = j v_0 e^{j\omega t} \qquad (10.17)$$

From Eqs. (10.2) and (10.17) the impedance of the spring is

$$Z_k = \frac{F_0 e^{j\omega t}}{j\omega F_0 e^{j\omega t}/k} = \frac{k}{j\omega} = -\frac{jk}{\omega} \qquad (10.18)$$

The ratio of force to velocity for a spring thus depends on the stiffness and frequency, and is an imaginary number.

To find the phase relation between v and F for a spring, Eq. (10.17) may be written in the form of Eq. (10.1):

$$v = \frac{F_0}{k}(-\omega \sin \omega t + j\omega \cos \omega t) = \frac{\omega}{k} F_0 e^{j(\omega t + 90°)}$$

The velocity is always 90° ahead of, or leading, the applied force, as shown in Fig. **10.5**. This is indicated by the factor j in Eq. (10.17).

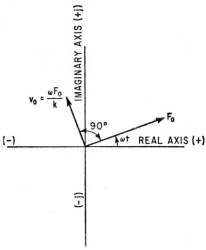

Fig. 10.5. Force and velocity relationships at the connections of an ideal spring.

MASS. In the ideal mass illustrated in Fig. 10.6A, the acceleration $\ddot{x}$ of the rigid body is proportional to the applied force F:

$$\ddot{x}_1 = \frac{F_a}{m} \tag{10.19}$$

where m is the mass of the body and $\ddot{x}_1 = \ddot{x}_2$ since the element is rigid. By Eq. (10.19), the force F_a is required to give the mass the acceleration $\ddot{x}_1$ and the transmitted force F_b is zero. The representations of Fig. 10.6B and C are used commonly to indicate that one

Fɪɢ. 10.6. Schematic representations of an ideal mass. The one shown in C is that recommended by the American National Standards Institute, ANSI Y-32.18 (1972).

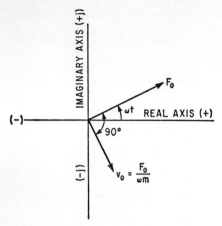

FIG. 10.7. Force and velocity relationships for an ideal mass.

end of a mass is connected rigidly to an inertial reference point, leaving it with only one end which is free to move. When a sinusoidal force is applied, Eq. (10.19) becomes

$$\ddot{x}_1 = \frac{F_0 e^{j\omega t}}{m} \qquad (10.20)$$

The acceleration is sinusoidal and in phase with the applied force.

Integrating Eq. (10.20) to find velocity

$$\dot{x} = v = \frac{F_0 e^{j\omega t}}{j\omega m} = -jv_0 e^{j\omega t} \quad (10.21)$$

The mechanical impedance of the mass is the ratio of F to v:

$$Z_m = \frac{F_0 e^{j\omega t}}{F_0 e^{j\omega t}/j\omega m} = j\omega m \quad (10.22)$$

Thus, the impedance of a mass is an imaginary quantity that depends on the magnitude of the mass and on the frequency.

To find the phase relation between F and v, Eq. (10.21) may be written:

$$v = \frac{F_0}{\omega m}(\sin \omega t - j \cos \omega t) = \frac{F_0}{\omega m} e^{j(\omega t - 90°)}$$

Thus the velocity of the mass lags behind the applied force by 90° as shown in Fig. 10.7, and indicated by the $-j$ in Eq. (10.21).

The impedances of a resistance, a spring, and a mass at any particular frequency may be represented on the complex plane as shown in Fig. 10.8. Their variation with frequency is indicated in Fig. 10.9. The impedances of elements and several combinations of elements are given in Table 10.2.

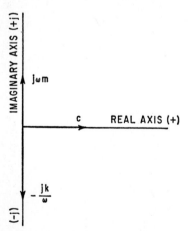

FIG. 10.8. A complex-plane representation for the impedance of ideal resistance, spring, and mass at a fixed frequency.

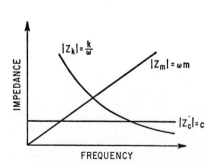

FIG. 10.9. The variation with frequency of the magnitude of the impedance of ideal resistance, spring, and mass.

Table 10.2. Driving-point Impedance and Mobility of Ideal Mechanical Elements and Lumped Parameter Systems

When the system of elements shown includes a mass, the impedance or mobility given is the relationship between force and velocity at the one available connection, the other connection being attached to the inertial reference plane. When no mass is included, the impedance or mobility describes the relation between the force applied to the two system connections and the resulting relative velocity between these connections. Graphs of magnitude of impedance vs. ω and magnitude of mobility vs. ω are plotted on a log-log scale.

DIAGRAM OF SYSTEM	MATHEMATIC FORMULAS: IMPEDANCE Z – EQ.(10.7) MOBILITY $\mathcal{M}$ – EQ.(10.9)	IMPEDANCE IN THE COMPLEX PLANE	MOBILITY IN THE COMPLEX PLANE	MAGNITUDE OF IMPEDANCE	MAGNITUDE OF MOBILITY	IMPEDANCE ANGLE θ FIG.10.34
1.	$Z = c$ $\mathcal{M} = 1/c$					
2.	$Z = \dfrac{k}{j\omega}$ $\mathcal{M} = \dfrac{j\omega}{k}$					
3.	$Z = j\omega m$ $\mathcal{M} = \dfrac{1}{j\omega m}$					

10–7

Table 10.2. (Continued)

DIAGRAM OF SYSTEM	MATHEMATIC FORMULAS: IMPEDANCE Z – EQ. (10.7) MOBILITY $\mathcal{M}$ – EQ. (10.9)	IMPEDANCE IN THE COMPLEX PLANE	MOBILITY IN THE COMPLEX PLANE	MAGNITUDE OF IMPEDANCE	MAGNITUDE OF MOBILITY	IMPEDANCE ANGLE θ FIG. 10.34
4.	$Z = c + \dfrac{k}{j\omega}$ $\mathcal{M} = \dfrac{c - \dfrac{k}{j\omega}}{c^2 + (k/\omega)^2}$ $\omega_1 = \dfrac{k}{c}$					
5.	$Z = \dfrac{1/c - j\omega/k}{(1/c)^2 + (\omega/k)^2}$ $\mathcal{M} = \dfrac{1}{c} + \dfrac{j\omega}{k}$ $\omega_1 = \dfrac{k}{c}$					
6.	$Z = c + j\omega m$ $\mathcal{M} = \dfrac{c - j\omega m}{c^2 + \omega^2 m^2}$ $\omega_1 = \dfrac{c}{m}$					

	7.	8.	9.
DIAGRAM OF SYSTEM	c—m (damper and mass)	k, m (spring and mass in parallel)	k—m (spring and mass)
MATHEMATIC FORMULAS: IMPEDANCE Z - EQ.(10.7) MOBILITY M - EQ.(10.9)	$Z = \dfrac{1/c + j\omega m}{(1/c)^2 + (1/\omega m)^2}$ $M = 1/c + 1/j\omega m$ $\omega_1 = \dfrac{c}{m}$	$Z = j\omega m + \dfrac{k}{j\omega}$ $M = \dfrac{-j}{\omega m - k/\omega}$ $\omega_0 = \sqrt{\dfrac{k}{m}}$	$Z = \dfrac{-j}{\omega/k - 1/\omega m}$ $M = j\omega/k + 1/j\omega m$ $\omega_0 = \sqrt{\dfrac{k}{m}}$
IMPEDANCE IN THE COMPLEX PLANE			
MOBILITY IN THE COMPLEX PLANE			
MAGNITUDE OF IMPEDANCE $|Z|$			
MAGNITUDE OF MOBILITY $|M|$			
IMPEDANCE ANGLE θ FIG.10.34			

Table 10.2. (Continued)

DIAGRAM OF SYSTEM	MATHEMATIC FORMULAS: IMPEDANCE Z - EQ.(10.7) MOBILITY $\mathcal{M}$ - EQ.(10.9)	IMPEDANCE IN THE COMPLEX PLANE	MOBILITY IN THE COMPLEX PLANE	MAGNITUDE OF IMPEDANCE	MAGNITUDE OF MOBILITY	IMPEDANCE ANGLE θ FIG.10.34
10.	$Z = c + j(\omega m - k/\omega)$ $\mathcal{M} = \dfrac{c - j(\omega m - k/m)}{c^2 + (\omega m - k/m)^2}$ $\omega_0 = \sqrt{\dfrac{k}{m}}$	$(Z)\mathcal{S}$	$(\mathcal{M})\mathcal{S}$	$|Z|$	$|\mathcal{M}|$	θ
11.	$Z = \dfrac{1/c - j(\omega/k - 1/\omega m)}{(1/c)^2 + (\omega/k - 1/\omega m)^2}$ $\mathcal{M} = 1/c + j(\omega/k - 1/\omega m)$ $\omega_0 = \sqrt{\dfrac{k}{m}}$	$(Z)\mathcal{S}$	$(\mathcal{M})\mathcal{S}$	$|Z|$	$|\mathcal{M}|$	θ
12.	$Z = \dfrac{c_1 + c_2 + \dfrac{c_2}{\omega^2 m^2} + \dfrac{j}{\omega m}}{(1/c_1)^2 + (1/\omega m)^2}$ $\mathcal{M} = \dfrac{\dfrac{c_1+c_2}{c_1^2} + \dfrac{c_2}{\omega^2 m^2} - \dfrac{j}{\omega m}}{\left(\dfrac{c_1+c_2}{c_1}\right)^2 + \left(\dfrac{c_2}{\omega m}\right)^2}$	$(Z)\mathcal{S}$	$(\mathcal{M})\mathcal{S}$	$|Z|$	$|\mathcal{M}|$	θ

Table 10.34 — Impedance and mobility characteristics of mechanical systems (continued)

	System 13	System 14	System 15
IMPEDANCE ANGLE θ (FIG. 10.34)	Plot of θ vs ω; axes $+90°$, 0, $-90°$	Plot of θ vs ω; axes $+90°$, 0, $-90°$	Plot of θ vs ω; axes $+90°$, 0, $-90°$; marks ω_0
MAGNITUDE OF MOBILITY	$|W|$ vs ω; asymptotes $-\frac{1}{c_1}+\frac{c_2}{\cdots}$, $-\frac{1}{c_2}$	$|W|$ vs ω; asymptotes $\frac{1}{c}(m_1+m_2)$, $\frac{m_1^2}{c(m_1+m_2)}$, $\frac{1}{j\omega m_2}$, $\frac{1}{j\omega(m_1+m_2)}$	$|W|$ vs ω; asymptotes $1/c$, $\frac{1}{j\omega m}$, $\frac{\omega}{k}$; marks ω_0
MAGNITUDE OF IMPEDANCE	$|Z|$ vs ω; marks c_2, $\frac{c_1 c_2}{c_1+c_2}$	$|Z|$ vs ω; $j\omega(m_1+m_2)$, $j\omega m_2$, c, ω_1	$|Z|$ vs ω; $j\omega m$, $\frac{k}{\omega}$, c; marks ω_0
MOBILITY IN THE COMPLEX PLANE	$\mathcal{R}(W)$, $\mathcal{J}(W)$; $\omega=C$, $\omega=\infty$; $-\frac{1}{c_2}$, $\frac{1}{c_1}+\frac{1}{c_2}$	$\mathcal{R}(M)$, $\mathcal{J}(W)$; $\omega=\infty$; $\frac{1}{c}\left(\frac{m_1}{m_1+m_2}\right)^2$	$\mathcal{R}(M)$, $\mathcal{J}(W)$; $\omega=0$, $\omega=\infty$; $1/c$, ω_0
IMPEDANCE IN THE COMPLEX PLANE	$\mathcal{R}(Z)$, $\mathcal{J}(Z)$; $\omega=0$, $\omega=\infty$; c_2, $\frac{c_1 c_2}{c_1+c_2}$	$\mathcal{R}(Z)$, $\mathcal{J}(Z)$; $\omega=0$; c	$\mathcal{R}(Z)$, $\mathcal{J}(Z)$; $\omega=\infty$, ω_0; c
MATHEMATIC FORMULAS: IMPEDANCE Z – EQ.(10.7); MOBILITY M – EQ.(10.9)	$Z = c_1\left(\dfrac{c_1+c_2}{c_2}\right) + \dfrac{\omega^2 m^2}{c_2} + j\omega m$ $M = \dfrac{c_1\left(\dfrac{c_1+c_2}{c_2}\right) + \dfrac{\omega^2 m^2}{c_2} - j\omega m}{\left(c_1\dfrac{c_1+c_2}{c_2}\right)^2 + \left(\dfrac{\omega m}{c_2}\right)^2}$ $= \dfrac{c_1\left(\dfrac{c_1+c_2}{c_2}\right) + \dfrac{\omega^2 m^2}{c_2} - j\omega m}{c_1^2 + \omega^2 m^2}$	$Z = \dfrac{\dfrac{1}{c} + j\left[\dfrac{\omega m_2}{c^2} + \dfrac{1}{\omega m_1}\left(\dfrac{m+m_2}{m_1}\right)\right]}{(1/c)^2 + (1/\omega m_1)^2}$ $M = \dfrac{1}{c} - j\left[\dfrac{\omega m_2}{c^2} + \left(\dfrac{m_1+m_2}{m_1}\right)\dfrac{1}{\omega m_1}\right]^2 ... \left(\dfrac{m_1+m_2}{m_1}\right)$ $\omega_1 = \dfrac{c}{m_2}$ $\omega_2 = \dfrac{c}{m_1}$	$Z = \dfrac{\dfrac{1}{c} + \dfrac{j}{\omega}\left[\dfrac{1}{m} - \left(\dfrac{1}{c^2} + \dfrac{1}{\omega^2 - m^2}\right)\right]}{(1/c)^2 + (1/\omega m)^2}$ $M = \dfrac{\omega^2 m^2 + j\left(\dfrac{\omega^2 m k}{c} + \dfrac{\gamma}{c} - m\right)}{(mk/c)^2 + (\omega m - k/\omega)^2}$ $\omega_0 = \sqrt{\dfrac{k}{m - \dfrac{km^2}{c^2}}}$
DIAGRAM OF SYSTEM	13. spring c_2 in series with parallel m and c_1, to ground	14. spring c in series with parallel m_1 and m_2, to ground	15. parallel spring c, mass m, and spring k, to ground

Table 10.2. (Continued)

DIAGRAM OF SYSTEM	MATHEMATIC FORMULAS: IMPEDANCE Z - EQ.(10.7) MOBILITY $\mathcal{M}$ - EQ.(10.9)	IMPEDANCE IN THE COMPLEX PLANE	MOBILITY IN THE COMPLEX PLANE	MAGNITUDE OF IMPEDANCE	MAGNITUDE OF MOBILITY	IMPEDANCE ANGLE θ FIG.10.34
16.	$Z = \dfrac{\frac{ck^2}{\omega^2} - jkm\left[(\omega m - k/\omega) + \frac{c^2 k}{\omega m}\right]}{c^2 + (\omega m - k/\omega)^2}$ $\mathcal{M} = \dfrac{c + j\omega\left(\frac{c^2 + \omega^2 m^2}{k} - m\right)}{c^2 + \omega^2 m^2}$ $\omega_0 = \sqrt{\dfrac{k}{m} - \dfrac{c^2}{m^2}}$					
17.	$Z = \dfrac{\frac{c_1 + c_2}{c_1^2} + \frac{c_2\omega^2}{k^2} - j\frac{\omega}{k}}{(1/c_1)^2 + (\omega/k)^2}$ $\mathcal{M} = \dfrac{\frac{c_1+c_2}{c_1^2} + \frac{c_2\omega^2}{k^2} + j\frac{\omega}{k}}{\left(\frac{c_1+c_2}{c_1}\right)^2 + \left(\frac{c_2\omega}{k}\right)^2}$					
18.	$Z = \dfrac{c_1\left(\frac{c_1+c_2}{c_2}\right) + \frac{k^2}{c_2\omega^2} - j\frac{k}{\omega}}{\left(\frac{c_1+c_2}{c_2}\right)^2 + \left(\frac{k}{\omega c_2}\right)^2}$ $\mathcal{M} = \dfrac{c_1\left(\frac{c_1+c_2}{c_2}\right) + \frac{k^2}{c_2\omega^2} + j\frac{k}{\omega}}{c_1^2 + (k/\omega)^2}$					

	19.	20.	21.
IMPEDANCE ANGLE θ FIG. 10.34	$+90°$, 0, $-90°$; θ vs ω	$+90°$, 0, $-90°$; θ vs ω	$+90°$, 0, $-90°$; θ vs ω, ω_0
MAGNITUDE OF MOBILITY	$\frac{j\omega}{k_2}$, $\frac{j\omega}{k_1+k_2}$; $\lvert \mathcal{M}\rvert$ vs ω	$j\omega\left(\frac{1}{k_1}+\frac{1}{k_2}\right)$, $\frac{j\omega}{k_2}$; $\lvert \mathcal{M}\rvert$ vs ω	$\frac{1}{j\omega m}$, $\frac{j\omega}{k}$, $\frac{1}{c}$; $\lvert \mathcal{M}\rvert$ vs ω, ω_0
MAGNITUDE OF IMPEDANCE	$\frac{k_1+k_2}{j\omega}$, $\frac{k_2}{j\omega}$, c; $\lvert Z\rvert$ vs ω	$\frac{1}{j\omega}\left(\frac{k_1 k_2}{k_1+k_2}\right)$, $\frac{k_2}{j\omega}$; $\lvert Z\rvert$ vs ω	$j\omega m$, $\frac{k}{j\omega}$, c; $\lvert Z\rvert$ vs ω, ω_0
MOBILITY IN THE COMPLEX PLANE	$\mathcal{R}(\mathcal{M})$, $\omega=0$, $\frac{1}{c}\left(\frac{k_1}{k_1+k_2}\right)$; $\mathfrak{J}(\mathcal{M})$	$\mathcal{R}(\mathcal{M})$, $1/c$, $\omega=0$; $\mathfrak{J}(\mathcal{M})$	$\mathcal{R}(\mathcal{M})$, ω_0, $\omega=0$, $\omega=\infty$, $1/c$; $\mathfrak{J}(\mathcal{M})$
IMPEDANCE IN THE COMPLEX PLANE	$\mathcal{R}(Z)$, $\omega=\infty$ c; $\mathfrak{J}(Z)$	$\mathcal{R}(Z)$, $\omega=\infty$, $c\left(\frac{k_2}{k_1+k_2}\right)^2$; $\mathfrak{J}(Z)$	$\mathcal{R}(Z)$, ω_0 c, $\omega=0$; $\mathfrak{J}(Z)$
MATHEMATIC FORMULAS: IMPEDANCE Z – EQ. (10.7) MOBILITY $\mathcal{M}$ – EQ. (10.9)	$Z=\dfrac{\frac{1}{c}-j\left[\frac{k_2}{\omega c^2}+\frac{\omega}{k_1}\left(\frac{k_1+k_2}{k_1}\right)\right]}{(1/c)^2+(\omega/k_1)^2}$ $\mathcal{M}=\dfrac{\frac{1}{c}+j\left[\frac{k_2}{\omega c^2}+\frac{\omega}{k_1}\left(\frac{k_1+k_2}{k_1}\right)\right]}{\left(\frac{k_1+k_2}{k_1}\right)^2+\left(\frac{k_2}{\omega c}\right)^2}$	$Z=\dfrac{c-j\left[\frac{\omega c^2}{k_2}+\frac{k_1}{\omega}\left(\frac{k_1+k_2}{k_2}\right)\right]}{\left(\frac{k_1+k_2}{k_2}\right)^2+\left(\frac{\omega c}{k_2}\right)^2}$ $\mathcal{M}=\dfrac{c+j\left[\frac{\omega c^2}{k_2}+\frac{k_1}{\omega}\left(\frac{k_1+k_2}{k_2}\right)\right]}{c^2+(k_1/\omega)^2}$	$Z=\dfrac{\frac{1}{c}+j\omega\left[m\left(\frac{1}{c^2}+\frac{\omega^2}{k^2}\right)-\frac{1}{k}\right]}{(1/c)^2+(\omega/k)^2}$ $\mathcal{M}=\dfrac{\frac{1}{c}-j\omega\left[m\left(\frac{1}{c^2}+\frac{\omega^2}{k^2}\right)-\frac{1}{k}\right]}{\left(\frac{\omega m}{c}\right)^2+\left(1-\frac{\omega^2 m}{k}\right)^2}$ $\omega_0=\sqrt{\frac{k}{m}-\frac{k^2}{c^2}}$
DIAGRAM OF SYSTEM	c, k_1, k_2	k_2, k_1, c	k, c, m

Table 10.2. (Continued)

DIAGRAM OF SYSTEM	MATHEMATIC FORMULAS: IMPEDANCE Z - EQ.(10.7) MOBILITY $\mathcal{M}$ - EQ.(10.9)	IMPEDANCE IN THE COMPLEX PLANE	MOBILITY IN THE COMPLEX PLANE	MAGNITUDE OF IMPEDANCE	MAGNITUDE OF MOBILITY	IMPEDANCE ANGLE θ FIG.10.34
22.	$Z = c + \dfrac{j\omega mk}{k - \omega^2 m}$ $\mathcal{M} = \dfrac{\dfrac{j\omega mk}{k-\omega^2 m}}{c^2 + \left(\dfrac{\omega mk}{k-\omega^2 m}\right)^2}$ $\omega_0 = \sqrt{\dfrac{k}{m}}$					
23.	$Z = \dfrac{\dfrac{1}{c} - \dfrac{j\omega}{k-\omega^2 m}}{\left(\dfrac{1}{c}\right)^2 + \left(\dfrac{\omega}{k-\omega^2 m}\right)^2}$ $\mathcal{M} = \dfrac{1}{c} + \dfrac{j\omega}{k-\omega^2 m}$ $\omega_0 = \sqrt{\dfrac{k}{m}}$					
24.	$Z = \dfrac{-j\left(\dfrac{k_1+k_2}{k_1} - \dfrac{k_2}{\omega^2 m}\right)}{\omega/k_1 - 1/\omega m}$ $\mathcal{M} = \dfrac{+j\left(\omega/k_1 - 1/\omega m\right)}{\left(\dfrac{k_1+k_2}{k_1}\right) - \dfrac{k_2}{\omega^2 m}}$ $\omega_1 = \sqrt{\dfrac{1}{m}\left(\dfrac{k_1 k_2}{k_1+k_2}\right)}$ $\omega_2 = \sqrt{\dfrac{k_1}{m}}$					

DIAGRAM OF SYSTEM	MATHEMATIC FORMULAS: IMPEDANCE Z – EQ.(10.7) MOBILITY $\mathcal{M}$ – EQ.(10.9)	IMPEDANCE IN THE COMPLEX PLANE	MOBILITY IN THE COMPLEX PLANE	MAGNITUDE OF IMPEDANCE	MAGNITUDE OF MOBILITY	IMPEDANCE ANGLE θ FIG.10.34
25.	$Z = j\,\dfrac{\omega m - k_1/\omega}{1 - \dfrac{\omega}{k_2}\left(\omega m - \dfrac{k_1}{\omega}\right)}$ $\mathcal{M} = j\,\dfrac{\dfrac{\omega}{k_2}\left(\omega m - \dfrac{k_1}{\omega}\right) - 1}{\omega m - k_1/\omega}$ $\omega_1 = \sqrt{\dfrac{k_1}{m}}\qquad \omega_2 = \sqrt{\dfrac{k_1 + k_2}{m}}$					
26.	$Z = +j\left(\omega m_2 + 1/\omega m_1 - \omega/k\right)$ $\mathcal{M} = +j\,\dfrac{\omega/k - 1/\omega m_1}{\dfrac{m_1 + m_2}{m_2} - \dfrac{\omega^2 m_2}{k}}$ $\omega_1 = \sqrt{\dfrac{k}{m_1}}\quad \omega_2 = \sqrt{k\left(\dfrac{1}{m_1} + \dfrac{1}{m_2}\right)}$					
27.	$Z = \dfrac{\omega m\left\{\omega m c + j\left[c^2 - \dfrac{k}{\omega}\left(\omega m - \dfrac{k}{\omega}\right)\right]\right\}}{c + \dfrac{1}{j\omega}\left[c^2 - \dfrac{k}{\omega}\left(\omega m - \dfrac{k}{\omega}\right)\right]}$ $\mathcal{M} = \dfrac{c + (k/\omega)^2}{k}$ $\omega_0 = \sqrt{\dfrac{k}{m - c^2/k}}$					

10–15

MOBILITY OF MECHANICAL ELEMENTS

Mobility as defined by Eq. (10.9) is the reciprocal of impedance. Expressions for the mobility of the three basic elements can be written directly from the respective expressions for impedance:

Resistance: $$\mathfrak{M}_c = \frac{1}{c} \tag{10.23}$$

Spring: $$\mathfrak{M}_k = \frac{j\omega}{k} \tag{10.24}$$

Mass: $$\mathfrak{M}_m = \frac{1}{j\omega m} = \frac{-j}{\omega m} \tag{10.25}$$

These mobilities are plotted on the complex plane in Fig. 10.10A, and represented as functions of frequency in Fig. 10.10B. Mobility of various elements and combinations of elements are summarized in Table 10.2.

(A)

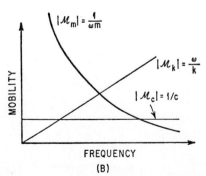

(B)

Fig. 10.10. The complex-plane representation of the mobility of an ideal resistance, spring, and mass at a fixed frequency is shown in (A); the variation of the mobility magnitude with frequency is shown in (B).

MECHANICAL SYSTEMS

The properties of a mechanical system that are used in analyses involving mechanical impedance and mobility are points of force application, paths for transmitting forces, and points of common velocities. The velocity at each connection point and the forces exerted on each element may be calculated by the proper application of Eq. (10.7) or Eq. (10.9) if the impedance of each element in the system and the forces or velocities imposed on points in the system are known. In many applications, only the motion at a few points is of primary interest. Then it is advantageous to combine groups of elements into single impedances. Methods for calculating the impedance of such combined elements are given in this section.

PARALLEL ELEMENTS

To illustrate the concept of parallel elements only ideal elements, multiply connected across two points, will be considered. Two ideal elements connected in parallel are shown in Fig. 10.11. Both are constrained to have the same relative velocities between their connections. The force required to give the resistance the velocity v is found from Eqs. (10.7) and (10.14):

Fig. 10.11. Schematic representation of a parallel spring-resistance combination.

$$F_c = vZ_c = vc$$

The force required to give the spring this same velocity is, from Eq. (10.18),

$$F_k = vZ_k = \frac{vk}{j\omega}$$

The total force F is

$$F = F_c + F_k = v\left(c + \frac{k}{j\omega}\right)$$

Since $Z = F/v$,

$$Z = c + \frac{k}{j\omega}$$

Thus, the over-all impedance is the sum of the impedances of the two elements.

By extending this concept to any number of parallel elements, the driving force F equals the sum of the resisting forces:

$$F = \sum_{i=1}^{n} vZ_i = v\sum_{i=1}^{n} Z_i \quad \text{and} \quad Z_p = \sum_{i=1}^{n} Z_i \tag{10.26}$$

where Z_p is the total mechanical impedance of the parallel combination of the individual elements Z_i.

When the properties of the parallel elements are expressed as mobilities, the total mobility of the combination follows from Eqs. (10.9) and (10.26):

$$\frac{1}{\mathfrak{M}_p} = \sum_{i=1}^{n} \frac{1}{\mathfrak{M}_i} \tag{10.27}$$

SERIES ELEMENTS

In determining the impedance presented by the end of a number of series-connected elements, consider the arrangement shown in Fig. 10.12. Elements Z_1 and Z_2 must have no mass since a mass always has one end connected to a stationary inertial reference. However, the impedance Z_3 may be a mass. The relative velocities between the end connections of each element are indicated by v_a, v_b, and v_c; the velocities of the connections with respect to the stationary reference point G are indicated by v_1, v_2, and v_3:

$$v_3 = v_c \qquad v_2 = v_3 + (v_2 - v_3) = v_c + v_b$$

$$v_1 = v_2 + (v_1 - v_2) = v_a + v_b + v_c$$

The impedance at point 1 is F/v_1, and the force F is transmitted to all three elements. The relative velocities are

$$v_a = \frac{F}{Z_1} \qquad v_b = \frac{F}{Z_2} \qquad v_c = \frac{F}{Z_3}$$

Thus, the total impedance is defined by

$$\frac{1}{Z} = \frac{F/Z_1 + F/Z_2 + F/Z_3}{F} = \frac{1}{Z_1} + \frac{1}{Z_2} + \frac{1}{Z_3}$$

FIG. 10.12. Generalized three-element system of series-connected mechanical impedances.

Extending this principle to any number of massless series elements,

$$\frac{1}{Z_s} = \sum_{i=1}^{n} \frac{1}{Z_i} \qquad (10.28)$$

where Z_s is the total mechanical impedance of the elements Z_i connected in series.

Using Eq. (10.9), the total mobility of series connected elements (expressed as mobilities) is

$$\mathfrak{M}_s = \sum_{i=1}^{n} \mathfrak{M}_i \qquad (10.29)$$

SYSTEM SIMPLIFICATION

When one point of attachment of a multiple element system is of interest, the above methods for calculating the impedance of combined elements may be used to give one expression for the impedance at that point. When a force is applied to the point of attachment, the resulting velocity can be calculated from Eq. (10.7). In general, such an expression involves real and imaginary terms. When an *impedance* expression is obtained, the two terms can be considered to represent a purely resistive element connected in *parallel* with a purely imaginary or reactive element. A *mobility* expression may be considered to represent a resistive element in *series* with a reactive element. Either of these two representations has the characteristics of the original multiple element system and is called an *equivalent system.*

The calculation of the impedance of a multielement system using Eqs. (10.26) through (10.29) requires considerable manipulation of complex quantities.

FIG. 10.13. System of several ideal elements analyzed in Example 10.1.

Example 10.1. For the system of Fig. 10.13, calculate the impedance at point A so that the velocity v_1 as a result of the application of force F_A may be determined. The procedure to be followed is basically as follows:

1. When series elements such as k_4 and c_3 are paralleled by other elements (such as m), add the mobilities of the series elements and convert the resulting expression to an impedance which may be added directly to the impedance of the parallel elements.

2. When parallel connected elements such as k_4, c_3, and m are connected in series with other elements (such as k_3 and c_2), convert the impedance of the parallel elements to a mobility and add to the mobility of the series elements.

In accordance with (1), add the mobilities of k_4 and c_3 to obtain the mobility of the branch at B:

$$\mathfrak{M}_{B1} = \frac{1}{c_3} + \frac{j\omega}{k_4}$$

The impedance of this branch is then

$$Z_{B1} = \frac{1}{\mathfrak{M}_{B1}} = \frac{1/c_3 - j\omega/k_4}{(1/c_3)^2 + (\omega/k_4)^2}$$

$$= \frac{1/c_3}{(1/c_3)^2 + (\omega/k_4)^2} - \frac{j\omega/k_4}{(1/c_3)^2 + (\omega/k_4)^2}$$

The system at B is then the parallel combination shown in Fig. 10.14A where $\mathfrak{R}(Z_{B1})$ represents the real part of Z_{B1}, and $\mathfrak{g}(Z_{B1})$ is the imaginary part of Z_{B1}. The parallel impedance at B including the impedance of mass m is

$$Z_B = \frac{1/c_3}{(1/c_3)^2 + (\omega/k_4)^2} + j\left[\omega m - \frac{\omega/k_4}{(1/c_3)^2 + (\omega/k_4)^2}\right]$$

The corresponding mobility is

$$\mathfrak{M}_B = \frac{\mathfrak{R}(Z_B)}{|Z_B|^2} - \frac{j\mathfrak{g}(Z_B)}{|Z_B|^2}$$

where $|Z_B|$ is the magnitude of Z_B which is represented by Fig. 10.14B. Adding the mobilities of c_2 and k_3, the mobility presented at A by this branch is

$$\mathfrak{M}_{A2} = \mathfrak{R}(\mathfrak{M}_B) + \frac{1}{c_2} + j\left(\mathfrak{g}(\mathfrak{M}_B) + \frac{\omega}{k_3}\right)$$

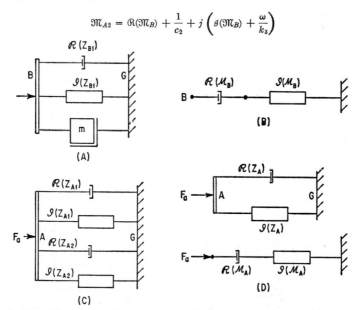

Fɪɢ. 10.14. Equivalent representations of parts of the system of Fig. 10.13 used in reducing that system to a representation equivalent to a single impedance or mobility.

To add the effect of this branch to that of the parallel branch consisting of k_1, k_2, and c_1, the impedances of both branches must be found:

$$Z_{A2} = \frac{1}{\mathfrak{M}_{A2}} = \frac{\Re(\mathfrak{M}_{A2})}{|\mathfrak{M}_{A2}|^2} - j\frac{\mathcal{J}(\mathfrak{M}_{A2})}{|\mathfrak{M}_{A2}|^2}$$

$$\mathfrak{M}_{A1} = \frac{1}{c_1} + \frac{k_1 + k_2}{j\omega}$$

$$Z_{A1} = \frac{1/c_1}{(1/c_1)^2 + [(k_1 + k_2)/\omega]^2} + j\frac{(k_1 + k_2)/\omega}{(1/c_1)^2 + [(k_1 + k_2)/\omega]^2}$$

The impedances Z_{A1} and Z_{A2} then can be represented by the four parallel elements of Fig. 10.14C. The total impedance at A is

$$Z_A = Z_{A1} + Z_{A2} = \Re(Z_{A1}) + \Re(Z_{A2}) + j[\mathcal{J}(Z_{A1}) + \mathcal{J}(Z_{A2})]$$

and can be represented by the two parallel elements of Fig. 10.14D. The mobility expression at A is $1/Z_A$; it can be represented by the series elements of Fig. 10.14D.

As an aid to system simplification problems, the impedances and mobilities of several combinations of elements are given in Table 10.2.

EQUIVALENT SYSTEMS

The ratio of an applied force to a resulting velocity at the point A of the eight-element system in Fig. 10.13 is finally expressed analytically by an impedance having real and imaginary terms. The impedance characteristics at A are represented by the two-element systems of Fig. 10.14D. The two elements are not necessarily the ideal resistance, spring, or mass. The resistance term may vary with frequency and the reactance term may appear springlike, masslike, or be zero—depending on frequency. Even the impedance Z_{B1} of the two series elements c_3 and k_4 in Example 10.1 exhibits some of these characteristics.

The advantage of the two-element equivalent representation is that the response of complex systems to single frequency excitation may be calculated easily and the system may be represented by an ideal mass or spring and a resistance. In many cases there are ranges of frequencies over which complex systems can be represented by one or two ideal elements. Thus in Example 10.1, when $1/c_3 \gg \omega/k_4$ (which occurs when $\omega \ll k_4/c_3$), $Z_{B1} = c_3$, and the two elements can be represented by the resistance c_3. At higher frequencies, $\omega \gg k_4/c_3$, $Z_{B1} = k_4/j\omega$, and the impedance is springlike.

A further advantage of the equivalent-system representation lies in considering the effect of other elements to be attached to the point of interest. Qualitative considerations become apparent immediately. For example, when a mass is attached to a structure that appears springlike at the excitation frequency, the possibility of system resonance exists. This may or may not be desirable, but in either case the size of the mass is very important. The quantitive determination of the response can be determined by combining the added element with the equivalent system.

PHYSICAL ELEMENTS

In general, the characteristics of real masses, springs, and mechanical resistance elements differ from those of ideal elements in three respects:

1. A spring may have a nonlinear force-deflection characteristic; a mass may suffer plastic deformation with motion; and the force presented by a resistance may not be exactly proportional to velocity.

2. All materials have some mass; thus, a perfect spring or resistance cannot be made. Some compliance or spring effect is inherent in all elements. Energy can be dissipated in a system in several ways: friction, acoustic radiation, hysteresis, etc. Such a loss can be represented as a resistive component of the element impedance.

3. Elements can differ from the ideal elements at high frequencies or when long connecting elements are used because the element length becomes comparable to a wavelength of stress waves in the material. Then a lumped system analysis does not apply, and analysis on the basis of wave motion in a transmitting medium must be made.

LUMPED ELEMENTS

A piece of high-density metal of spherical or cubical shape provides a nearly ideal mass. Wave motion effects do not occur until very high frequencies are reached, and a rigid attachment is feasible. However, supporting means for the mass may involve friction, and the attaching elements may have significant compliance. A parallel representation of the impedance of a physical mass at low frequencies is

$$Z_m = c + j\omega m + \frac{k}{j\omega}$$

If the mass is freely suspended by fairly rigid supports, the $j\omega m$ term predominates over a wide frequency range and Eqs. (10.22) and (10.25) may be used.

A helical wound spring can be made to have a nearly linear force-displacement characteristic, thereby attaining a constant value of k. But, under dynamic conditions the mass of the spring must be considered. When one end of a spring is rigidly fixed and the other end attached to a vibrating point, one-third of the mass of the spring may be considered to be lumped at the vibrating end of the spring. When both ends are vibrating, from one-third to one-half the mass of the spring may be considered lumped at each end. At low frequencies the mass reactance may be small compared to the compliant effect so that either approximation for the mass of the spring may be adequate.

Hysteresis effects in the spring metal can cause energy loss resulting in a resistive component in the spring impedance. The impedance of a spring can be represented by a parallel connection of the three ideal elements, but in most cases the resistive term is negligible. At low frequencies, the impedance of a spring when one end is fixed is approximately

$$Z_k = \frac{k}{j\omega} + j\omega \frac{m_k}{3} \tag{10.30}$$

In many cases a mechanical resistance represents damping which is inherent in physical systems. In the case of sliding friction the reaction force may not be proportional to velocity but may be a constant, independent of velocity. Such a characteristic can be represented by a constant-force generator, 180° out-of-phase with the relative velocity across the sliding elements.

Resistive effects also may arise from the losses in rubber, felt, cork, or other vibration isolation materials. With such materials the damping coefficient may vary with the amplitude of motion.

One type of resistive element can be constructed by an arrangement of parts so arranged that the applied force pushes a body through a liquid or subjects it to other viscous forces. Because of the mass and compliance of the parts, such a device will not constitute an ideal resistance. Elements can be made sufficiently rigid so that the spring term is negligible; however, the effects of the mass of the elements may not be negligible.

Although ideal elements do not actually exist, it is common in the use of impedance methods to represent physical members by one or more ideal elements. Figure 10.15A shows a system of five physical elements as represented, for analysis at low frequency, by ideal elements. Many of the masses are essentially in parallel and can be added together as follows to produce the system of Fig. 10.15B:

$$m' = \frac{m_{k1}}{2} + \frac{m_{c1}}{2} \qquad m'' = m + \frac{m_{c2}}{2} + \frac{m_{k2}}{3} + \frac{m_{k1}}{2} + \frac{m_{c1}}{2}$$

Since one end of k_2 is rigidly fixed, one-third of its mass is lumped on the moving end. The masses on the fixed ends of c_2 and k_2 are not included in Fig. 10.15B because they do not enter into the dynamic problem.

(B)

FIG. 10.15. The schematic representation applicable at low frequencies of a system of physical elements is shown in (A); the redrawn system suitable for impedance method analysis is shown in (B).

ELEMENTS WITH DISTRIBUTED MASS

All mechanical elements have distributed mass since mass cannot be concentrated at a point. If the vibration frequency is low enough that the time of travel of stress waves in an element of length l is very small compared to a period (i.e., if $l/\lambda \ll 1/2\pi$), the mass can be considered as a lumped element.[11]

The velocity of compressional stress waves in an elastic bar is $c_s = \sqrt{E/\rho}$, where E is Young's modulus and ρ is mass density. In long thin rods the velocity is c_s if the diameter is small compared to a wavelength, but when the diameter is as great as 2λ, the velocity drops to about $0.6c_s$.[12]

The force transmitted by and the velocities at the ends of a distributed mass element may be determined by transmission line methods.[13, 14, 15] For example, consider a uniform bar, as shown in Fig. 10.16A. (Other element shapes may be approximated by sections of uniform bars.) The bar has a length l, a cross-sectional area S, a density ρ, and a stress-wave velocity c_s. The force and velocity at the driven end (vibrational source attachment) are represented by F_i and v_i; F_r and v_r are the force and velocity at the load end. The material

FIG. 10.16. The schematic representation of an element having distributed mass and its representation when it is connected to a source of vibration energy.

is assumed to be lossless, and the bar is characterized by a *propagation constant* $\gamma = j\beta = j\omega/c_s$ (where $\beta = \omega/c_s$) and a *characteristic impedance* (the impedance seen at the driven end for an infinitely long bar) $Z_0 = \rho c_s S$. Analysis by transmission line methods with the load Z_r attached gives the following relationships:

Load velocity in terms of the driven end velocity:

$$v_r = \frac{v_i}{\cos \beta \ell + j(Z_r/Z_0) \sin \beta \ell} \tag{10.31}$$

Load force in terms of the driving end force:

$$F_r = \frac{F_i}{\cos \beta \ell + j(Z_0/Z_r) \sin \beta \ell} \tag{10.32}$$

Driving end impedance:

$$Z_i = Z_0 \frac{Z_r \cos \beta \ell + jZ_0 \sin \beta \ell}{Z_0 \cos \beta \ell + jZ_r \sin \beta \ell} \tag{10.33}$$

When the bar is driven by a force generator with a paralleled impedance Z_g the schematic representation is that of Fig. 10.16B. The load element force and velocity, and the driven end force and velocity, may be determined in terms of the known force F by the following expressions:

$$F_i = \frac{F}{1 + Z_i/Z_g} \qquad V_i = \frac{F}{Z_i(1 + Z_g/Z_i)}$$

$$F_r = \frac{F}{(1 + Z_g/Z_i)[\cos \beta \ell + j(Z_0/Z_r) \sin \beta \ell]} \tag{10.34}$$

$$v_r = \frac{F}{(Z_g + Z_i)[\cos \beta \ell + j(Z_r/Z_0) \sin \beta \ell]}$$

where Z_g is the impedance of the generator.

When the bar is driven from a velocity generator with a series mobility $\mathfrak{M}_g$, as in Fig. 10.16C, the quantities of Eq. (10.34) are given in terms of the known velocity v and the mobility:

$$\mathfrak{M}_0 = \frac{1}{Z_0} \qquad \mathfrak{M}_r = \frac{1}{Z_r} \qquad \mathfrak{M}_g = \frac{1}{Z_g} \qquad \text{and} \qquad \mathfrak{M}_i = \frac{1}{Z_i}$$

$$v_i = \frac{v}{1 + \mathfrak{M}_g/\mathfrak{M}_i} \qquad F_i = \frac{v}{\mathfrak{M}_g + \mathfrak{M}_i}$$

$$v_r = \frac{v}{(1 + \mathfrak{M}_g/\mathfrak{M}_i)[\cos \beta \ell + j(\mathfrak{M}_0/\mathfrak{M}_r) \sin \beta \ell]} \tag{10.35}$$

$$F_r = \frac{v}{(\mathfrak{M}_g + \mathfrak{M}_i)[\cos \beta \ell + j(\mathfrak{M}_r/\mathfrak{M}_0) \sin \beta \ell]}$$

Equation (10.33) can be applied to a mass at frequencies for which the lumped approximation is no longer valid. In this case $Z_r = 0$ and

$$Z_m = jZ_0 \tan \beta \ell \tag{10.36}$$

where $Z_0 = \rho c_s S$. This expression may be compared with Eq. (10.22), the impedance of an ideal mass, $Z_m = j\omega m$.

THEOREMS

The following theorems are statements of basic principles (or combination of them) which apply to elements of a mechanical system; they are in a form convenient to analysis by the impedance method. In all but Kirchhoff's laws, these theorems apply only to systems made up of linear, bilateral elements. Linearity implies that the system elements can be represented by combinations of ideal elements in which c, k, and m are constants regardless of motion amplitude. A bilateral element is one in which forces are transmitted equally well in either direction across its connections.

Kirchhoff's Laws. 1. *The sum of all the forces acting at a point (common connection of several elements) is zero:*

$$\sum_i^n F_i = 0 \qquad \text{(at a point)} \tag{10.37}$$

This follows directly from the consideration leading to Eq. (10.26).

2. *The sum of the relative velocities across the connections of series mechanical elements taken around a closed loop is zero:*

$$\sum_i^n v_i = 0 \qquad \text{(around a closed loop)} \tag{10.38}$$

This follows from the considerations leading to Eq. (10.28).

Kirchhoff's laws apply to any system even when the elements are not linear or bilateral.

Superposition Theorem. *If a mechanical system of linear bilateral elements includes more than one vibration source, the force or velocity response at a point in the system can be determined by adding the response to each source, taken one at a time (the other sources supplying no energy but replaced by their internal impedances).*
The internal impedance of a vibrational generator is that impedance presented at its connection point when the generator is supplying no energy. This theorem finds useful application in systems having several sources. A very important application arises when the applied force is nonsinusoidal but can be represented by a Fourier series. Each term in the series can be considered a separate sinusoidal generator. The response at any point in the system can be calculated for each generator by using the impedance values at that frequency. Each response term becomes a term in the Fourier series representation of the total response function. The over-all response as a function of time then can be synthesized from the series.

Reciprocity Theorem. *If a force generator operating at a particular frequency at some point (1) in a system of linear bilateral elements produces a velocity at another point (2), the generator can be removed from (1) and placed at (2); then the former velocity at (2) will exist at (1), provided the impedances at all points in the system are unchanged.* This theorem also can be stated in terms of a vibration generator that produces a certain velocity at its point of attachment (1), regardless of force required, and the force resulting on some element at (2).

Reciprocity is an important characteristic of linear bilateral elements. It indicates that a system of such elements can transmit energy equally well in both directions. It further simplifies calculation on two-way energy transmission systems since the characteristics need be calculated for only one direction.

Thévenin's Equivalent System. *If a mechanical system of linear bilateral elements contains vibration sources and produces an output to a load at some point at any particular frequency, the whole system can be represented at that frequency by a single constant-force generator F_c in parallel with a single impedance Z_i connected to the load.* Thévenin's equivalent-system representation for a physical system may be determined by the following experimental procedure: Denote by F_c the force which is transmitted by the attachment point of the system to an infinitely rigid fixed point; this is called the *clamped force*. When the load connection is disconnected and perfectly free to move, a free velocity v_f is measured. Then the parallel impedance Z_i is F_c/v_f. The impedance Z_i also can be determined by measuring the internal impedance of the system when no source is supplying motional energy.

If the values of all the system elements in terms of ideal elements are known, F_c and Z_i may be determined analytically. A great advantage is derived from this representation in that attention is focused on the characteristics of a system at its output point and not on the details of the elements of the system. This allows an easy prediction of the response when different loads are attached to the output connection. After a final load condition has been determined, the system may be analyzed in detail for strength considerations.

Norton's Equivalent System. *A mechanical system of linear bilateral elements having vibration sources and an output connection may be represented at any particular frequency by a single constant-velocity generator v_f in series with an internal impedance Z_i.*

This is the series system counterpart of Thévenin's equivalent system where v_f is the free velocity and Z_i is the impedance as defined above. The same advantages in analysis exist as with Thévenin's parallel representation. The most advantageous one to use depends upon the type of structure to be analyzed. In the experimental determination of an equivalent system, it is usually easier to measure the free velocity than the clamped force on large heavy structures, while the converse is true for light structures. In any case, one representation is easily derived from the other. When v_f and Z_i are determined, $F_c = v_f Z_i$.

T and π Equivalent System. *Any mechanical system having an input and output connection and which is composed of linear bilateral elements may be represented by a three-element system of the T or π configuration.* The required equivalent system impedances

Fig. 10.17. Generalized two-connection mechanical system.

Fig. 10.18. A π equivalent system representation for a passive two-connection system.

determined below may not be obtainable physically, but the mathematical model represents the original system accurately.

π System. A generalized system is shown in Fig. 10.17. If a force F_1 is applied to the input connection, a velocity v_1 results. The ratio of F_1 to v_1 is the *input impedance*. If a force F_2 is applied to the output connection, a velocity v_2 results; the ratio of F_2 to v_2 is the *output impedance*. When the force F_1 is applied to the input, a velocity v_2 at the output results; the ratio of F_1 to v_2 is the *reverse transfer impedance*. If the force F_2 is applied to the output and a velocity v_1 results, the ratio of F_2 to v_1 is the *forward transfer impedance*.

These definitions can be used to find the impedances of an equivalent π system, as shown in Fig. 10.18, that are required to represent any general system, such as Fig. 10.17. First, find the input, output, and transfer impedances of the π system under the conditions specified below:

$$Z_{11} = \frac{F_1}{v_1}\bigg|_{v_2=0} \tag{10.39}$$

where Z_{11} is the input impedance with the output connected to a rigid point so that

$$v_2 = 0$$

i.e., output clamped.

$$Z_{22} = \frac{F_2}{v_2}\bigg|_{v_1=0} \tag{10.40}$$

where Z_{22} is the output impedance with the input clamped.

$$Z_{12} = \frac{F_1}{v_2}\bigg|_{v_1=0} \tag{10.41}$$

where Z_{12} is the reverse transfer impedance with the input clamped and F_1 is the force required to maintain the input velocity $v_1 = 0$.

$$Z_{21} = \frac{F_2}{v_1}\bigg|_{v_2=0} \tag{10.42}$$

where Z_{21} is the forward transfer impedance with the output clamped and F_2 is the force required to clamp the output.

Applying Eq. (10.39) to the system of Fig. 10.18, the input impedance when the output is clamped consists of Z_a and Z_c in parallel:

$$Z_{11} = Z_a + Z_c \tag{10.43}$$

Applying Eq. (10.40),

$$Z_{22} = Z_b + Z_c \tag{10.44}$$

When the input is clamped and a force F_2 is applied to the output, the velocity of the output v_2 results. The force F_1 then is transmitted by Z_c to the clamping point, and v_2 is the relative velocity of the connections of Z_c. By Eq. (10.7),

$$\frac{F_1}{v_2} = Z_c = Z_{12} \tag{10.45}$$

By the same procedure, $Z_{21} = Z_c$. Solving Eqs. (10.43) to (10.45) for Z_a, Z_b, and Z_c,

$$Z_a = Z_{11} - Z_{12} \qquad Z_b = Z_{22} - Z_{21} \qquad Z_c = Z_{12} = Z_{21} \tag{10.46}$$

If the impedances defined by Eqs. (10.39) to (10.42) can be determined for a system by either analytical or experimental means, the three impedances of an equivalent π system can be determined by Eqs. (10.46).

T System. An equivalent T system is shown in Fig. 10.19. The system values are most conveniently found in terms of mobilities. The necessary mobilities for a generalized system such as Fig. 10.17 are as follows:
The input mobility with the output "free" (no restraining force):

$$\mathfrak{M}_{11} = \frac{v_1}{F_1}\bigg|_{F_2=0} \tag{10.47}$$

The output mobility with the input "free":

$$\mathfrak{M}_{22} = \frac{v_2}{F_2}\bigg|_{F_1=0} \tag{10.48}$$

The reverse transfer mobility with the input "free":

$$\mathfrak{M}_{12} = \frac{v_1}{F_2}\bigg|_{F_1=0} \tag{10.49}$$

where v_1 is the velocity of the "free" input point when the force F_2 is applied to the output. The forward transfer mobility with the output "free":

$$\mathfrak{M}_{21} = \frac{v_2}{F_1}\bigg|_{F_2=0} \tag{10.50}$$

Finding these quantities for the T system of Fig. 10.19,

$$\mathfrak{M}_{11} = \mathfrak{M}_x + \mathfrak{M}_z \qquad \mathfrak{M}_{22} = \mathfrak{M}_y + \mathfrak{M}_z \qquad \mathfrak{M}_{12} = \mathfrak{M}_{21} = \mathfrak{M}_z \tag{10.51}$$

Solving Eqs. (10.51) for $\mathfrak{M}_x$, $\mathfrak{M}_y$, and $\mathfrak{M}_z$,

$$\mathfrak{M}_x = \mathfrak{M}_{11} - \mathfrak{M}_{12} \qquad \mathfrak{M}_y = \mathfrak{M}_{22} - \mathfrak{M}_{21} \qquad \mathfrak{M}_z = \mathfrak{M}_{12} = \mathfrak{M}_{21} \tag{10.52}$$

If the mobilities required by Eqs. (10.47) to (10.50) can be determined analytically or experimentally for a system, the three mobilities of an equivalent T system can be obtained from Eqs. (10.52).

The forward and reverse transfer impedances and mobilities are equal for both π and T systems. This is a characteristic of even very complicated systems when the system elements are linear and bilateral; it is not true if the system contains sources of vibration energy. However, systems with vibration sources can be represented by π and T equivalent systems of a slightly more complex form.

A system of mechanical elements with one connection point can be represented by a single impedance. Thévenin's or Norton's equivalent systems represent systems having vibration sources and an output connection by a source with a constant output and a single impedance element. The π or T equivalent systems can represent connecting elements between a vibration source and some load system. Combining the three representations gives a system that is simple to analyze; it focuses attention on the force and motion at the load connection and at the output connection of the source. The effects of changes in the source, load, or connecting elements then can be readily calculated.

Another advantage of the π and T representations is that the equivalent system can be obtained from direct measurements on complicated systems that are almost impossible to represent by a collection of idealized mechanical elements. An analytical expression for the system thus is obtained that allows an analysis by the methods discussed in the next section. After the values of the equivalent elements are determined at the operating frequencies, it may become evident that the original system may be replaced by a much simpler system of physical elements, based on the equivalent representation.

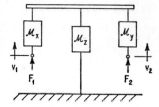

Fig. 10.19. A T equivalent system representation for a passive two-connection system.

Despite the many advantages of equivalent system representations, the complete schematic representation of a system often is desired. This representation is necessary for a complete analysis to determine the stresses on all of the parts of the system. To draw such a representation, a detailed study is required that can lead to a better understanding of the operation of the system.

In all but Kirchhoff's laws, the above theorems require linear, bilateral elements. This does not prevent their use in nonlinear, nonbilateral systems. Equivalent system representations with linear bilateral elements can be determined in many cases that represent accurately the nonlinear system over a limited range of operation.

ANALYSIS METHODS

The analysis of a system by impedance methods leads to the determination of forces acting on elements and the velocities of their connection points. From these quantities stresses may be determined, and accelerations and displacements may be calculated from Eqs. (10.5) and (10.6). When the motion is nonsinusoidal, the forces and velocities can be determined with the impedance approach, using the methods of analysis presented in this section. These methods are not the only ones available, but they are basic and adequate for many problems. Other methods may be found in the literature on electric circuit theory.

COMPLETE ANALYSIS

Kirchhoff's laws provide a sufficient number of simultaneous equations to determine all forces and velocities in a mechanical system. Equation (10.37) allows a force equation to be written for each independent connection point (force node). These equations usually are written in terms of impedances and velocities since a connection point ensures that one end of each element connected there has the same velocity when referred to a stationary reference. Equation (10.38) allows a velocity equation to be written for each independent loop, usually in terms of mobilities and forces. Velocity equations, best suited to series systems, do not find much usage in physical systems. Because the masses associated with physical elements have one connection effectively attached to a stationary reference (inertial reference), such a system seldom includes ideal series connected elements.

Example 10.2. Find the velocity of all the connection points and the forces acting on the elements of the system shown in Fig. 10.20. The system contains two velocity generators v_1 and v_6. Their magnitudes are known, their frequencies are the same, and they are 180° out-of-phase.

A. Using Eq. (10.37), write a force equation for each connection point except a and e. At point b: $F_1 - F_2 - F_3 = 0$. In terms of velocities and impedances:

$$(v_1 - v_2)Z_1 - (v_2 - v_3)Z_2 - (v_2 - v_4)Z_4 = 0 \tag{a}$$

At c, the two series elements have the same force acting: $F_2 - F_2 = 0$. In terms of velocities and impedances:

$$(v_2 - v_3)Z_2 - (v_3 - v_4)Z_3 = 0 \tag{b}$$

At d: $F_2 + F_3 - F_4 - F_5 = 0$. In terms of velocities and impedances:

$$(v_3 - v_4)Z_3 + (v_2 - v_4)Z_4 - (v_4 + v_6)Z_5 - (v_4 - v_5)Z_6 = 0 \tag{c}$$

Note that v_6 is (+) because of the 180° phase relation to v_1.
At f: $F_5 - F_5 = 0$. In terms of velocities and impedances:

$$(v_4 - v_5)Z_6 - v_5Z_7 = 0 \tag{d}$$

Since v_1 and v_6 are known, the four unknown velocities v_2, v_3, v_4, and v_5 may be determined by solving the four simultaneous equations above. After the velocities are obtained, the forces may be determined from the following:

$$F_1 = (v_1 - v_2)Z_1 \qquad\qquad F_2 = (v_2 - v_3)Z_2 = (v_3 - v_4)Z_3$$

$$F_3 = (v_2 - v_4)Z_4 \qquad\qquad F_4 = (v_4 + v_6)Z_5$$

$$F_5 = (v_4 - v_5)Z_6 = v_5Z_7$$

B. The method of *node forces.* Equations (a) through (d) above can be rewritten as follows:

$$v_1Z_1 = (Z_1 + Z_2 + Z_3)v_2 - Z_2v_3 - Z_4v_4 \tag{a'}$$

$$0 = -Z_2v_2 + (Z_2 + Z_3)v_3 - Z_3v_4 \tag{b'}$$

$$0 = -Z_4v_2 - Z_3v_3 + (Z_3 + Z_4 + Z_5 + Z_6)v_4 - Z_6v_5 \tag{c'}$$

$$-v_6Z_5 = -Z_6v_4 + (Z_6 + Z_7)v_5 \tag{d'}$$

These equations can be written by inspection of the schematic diagram by the following rule: *At each point with a common velocity (force node), equate the force generators to the sum of the impedances attached to the node multiplied by the velocity of the node, minus the impedances multiplied by the velocities of their other connection points.*

When the equations are written so that the unknown velocities form columns, the equations are in the proper form for a determinant solution for any of the unknowns. Note that the

Fig. 10.20. System of mechanical elements and vibration sources analyzed in Example 10.2 to find the velocity of each connection and the force acting on each element.

determinant of the Z's is symmetrical about the main diagonal. This condition always exists and provides a check for the correctness of the equations.

C. Using Eq. (10.38), write a velocity equation in terms of force and mobility around enough closed loops to include each element at least once. In Fig. 10.20, note that

$$F_3 = F_1 - F_2 \quad \text{and} \quad F_5 = F_1 - F_4$$

Around loop (1):

$$F_2(\mathfrak{M}_2 + \mathfrak{M}_3) - (F_1 - F_2)\mathfrak{M}_4 = 0 \tag{e}$$

The minus sign preceding the second term results from going across the element 4 in a direction opposite to the assumed force acting on it.

Around loop (2):

$$F_4\mathfrak{M}_5 - v_6 - (F_1 - F_4)(\mathfrak{M}_6 + \mathfrak{M}_7) = 0 \tag{f}$$

A summation of velocities from A to G along the upper path forms the following closed loop:

$$v_1 + F_1\mathfrak{M}_1 + F_2(\mathfrak{M}_2 + \mathfrak{M}_3) + F_4\mathfrak{M}_5 - v_6 = 0 \tag{g}$$

Equations (e), (f), and (g) then may be solved for the unknown forces F_1, F_2, and F_4. The other forces are $F_3 = F_1 - F_2$ and $F_5 = F_1 - F_4$. The velocities are:

$$v_2 = v_1 - F_1\mathfrak{M}_1 \qquad v_3 = v_2 - F_2\mathfrak{M}_2 \qquad v_4 = v_2 - F_3\mathfrak{M}_4 \qquad v_5 = F_5\mathfrak{M}_7$$

When a system includes more than one source of vibration energy, a Kirchhoff's law analysis with impedance methods can be made only if all the sources are operating at the same frequency. This is the case because sinusoidal forces and velocities can add as phasors only when their frequencies are identical. However, they may differ in magnitude and phase. Kirchhoff's laws still hold for instantaneous values and can be used to write the differential equations of motion for any system.

SUPERPOSITION

The superposition method of system analysis is advantageous when there are several vibration sources in a system. This method is illustrated by an example involving two force generators.

Example 10.3. In the system of Fig. 10.21A, the force generators F_1 and F_2 operate at the same frequency and their magnitudes are known. Determine the velocity of the common connection point c by the superposition theorem. First, determine the velocity v_c by Kirchhoff's laws for comparison with the results obtained by superposition. Writing the force-node equations at a, b, and c,

At a: $F_1 = Z_1v_a - Z_1v_c$

At c: $0 = -Z_1v_a + (Z_1 + Z_2 + Z_3)v_c - Z_2v_b$

At b: $F_2 = -Z_2v_c + Z_3v_b$

The impedance Z_1 is c_1 and k_1 in parallel, Z_2 is c_2 and k_2 in parallel, and Z_3 is c_3, k_3, and m_3 in parallel. Solving the equations for forces at a, b, and c to obtain v_c,

$$v_c = \frac{F_1 + F_2}{Z_3}$$

Fig. 10.21. System of mechanical elements including two force generators used to illustrate the principle of superposition in Example 10.3.

Using the superposition theorem, let the force F_2 be zero, and redraw the system as in Fig. 10.21B. Then let the velocity at point c be v_c' when only F_1 is applied. Since the force F_1 is transmitted by Z_1 to Z_3, v_c' becomes

$$v_c' = \frac{F_1}{Z_3}$$

Now letting $F_1 = 0$, applying F_2 as in Fig. 10.21C, and denoting the velocity of point c by v_c'',

$$v_c'' = \frac{F_2}{Z_3}$$

By the superposition theorem,

$$v_c = v_c' + v_c'' = \frac{F_1 + F_2}{Z_3}$$

If the force generators in Example 10.3 operate at different frequencies, the velocities are $v_c' = F_1/Z_3'$ and $v_c'' = F_2/Z_3''$, where Z_3' is the value of Z_3 at the frequency of force generator 1 and Z_3'' is the value of Z_3 at the frequency of generator 2. The instantaneous value of v_c, when phase angles are neglected, is

$$v_c = |v_c'| \sin \omega_1 t + |v_c''| \sin \omega_2 t$$

The mean square value is

$$v_c{}^2 = |v_c'|^2 + |v_c''|^2$$

Since the solution for a velocity or force by the superposition method involves only one source at a time, Kirchhoff's laws and impedance methods can be used in finding the responses to each source.

SYSTEMS WITH ONE CONNECTION

In one-connection systems considered here, one connection is attached to a rigid immovable point or inertial reference and another connection is available for external connection. These systems differ from multiple-connection systems in that they do not transmit a force or velocity from one system to another. One-connection systems can be divided into two types:

1. Passive systems, in which no sources of vibration energy are included. These systems can be represented by a single impedance or mobility function as shown in the section on *system simplification*. Such systems often are connected to, and considered as loads on, active systems as defined below.

2. Active systems, in which one or more sources of vibration energy is included. A complete schematic representation of such a system and its external connections may be drawn and analyzed by Kirchhoff's laws and superposition. The simplified representations of Thévenin's and Norton's equivalent systems are considered here.

THÉVENIN'S EQUIVALENT SYSTEM. Example 10.4. The system of Fig. 10.22A has one force generator, a number of elements, and a connection point b. Find Thévenin's equivalent system for this system. First combine the parallel elements c_1, k_1, and m_1 into one impedance Z_1, combine c_2 and k_2 into Z_2, and express m_3 in terms of Z_3. The schematic representation is shown in Fig. 10.22B. Determine the value of the force transmitted to a rigid immovable structure when point b is clamped to it. Point b is shown clamped in Fig. 10.22C, and Z_3 is omitted because it has no motion. Impedances Z_1 and Z_2 then are in parallel, and the force F divides between them to give the force transmitted to Z_2:

$$F_c = \frac{FZ_2}{Z_1 + Z_2} \qquad (a)$$

The impedance required for the Thévenin equivalent system can be determined from the system as rearranged in Fig. 10.22D. The force generator is removed and the impedance at b is calculated:

$$Z_i = Z_3 + \frac{1}{1/Z_1 + 1/Z_2} = \frac{Z_1 Z_2 + Z_1 Z_3 + Z_2 Z_3}{Z_1 + Z_2} \qquad (b)$$

FIG. 10.22. System of mechanical elements including a force generator for which a Thévenin equivalent system is determined in Example 10.4 and a Norton equivalent system is determined in Example 10.5.

The clamped force is given by Eq. (a) above, and the internal impedance is given by Eq. (b). The Thévenin equivalent system is shown in Fig. 10.22E.

When a load Z_4 is connected to point b, the equivalent representation is shown in Fig. 10.23. Under these conditions the velocity v_b of point b and the force applied to Z_4 can be calculated. Writing the force-node equation at b,

$$F_c = (Z_i + Z_4)v_b$$

Solving for v_b,

$$v_b = \frac{F_c}{Z_i + Z_4} \qquad (c)$$

The force at b is then $F_b = v_b Z_4$, or

$$F_b = \frac{F_c Z_4}{Z_i + Z_4} \qquad (d)$$

For any load impedance Z_4, the force and velocity at b can be calculated from Eqs. (c) and (d).

FIG. 10.23. A Thévenin equivalent system connected to a load impedance.

NORTON'S EQUIVALENT SYSTEM. Example 10.5. Find Norton's equivalent representation for the system of Fig. 10.22A, and find the force and velocity of point b when a load is attached to that point.

For this representation the free velocity at point b is required. This velocity v_f can be calculated from the system as drawn in Fig. 10.22B. Write the force-node equations at a and b, and let v_b be called v_f:

$$F = (Z_1 + Z_2)v_a - Z_2 v_f$$
$$0 = -Z_2 v_a + (Z_2 + Z_3)v_f$$

Solving for v_f,

$$v_f = \frac{FZ_2}{Z_1 Z_2 + Z_1 Z_3 + Z_2 Z_3} \qquad (a)$$

The internal impedance is given by Eq. (b) of Example 10.4, the free velocity is given by Eq. (a) above, and the Norton equivalent representation is shown in Fig. 10.22F.

When the load Z_4 is connected to b, the equivalent system is shown in Fig. 10.24.

NORTON
EQUIVALENT
SOURCE

FIG. 10.24. A Norton equivalent system
connected to a load mobility.

Since the elements are in series, they all
transmit the same force F_b:

$$F_b = \frac{v_f}{\mathfrak{M}_i + \mathfrak{M}_4} = \frac{v_f}{1/Z_i + 1/Z_4}$$

$$= \frac{v_f Z_i Z_4}{Z_i + Z_4} \qquad (b)$$

The velocity at b is

$$v_b = \frac{F_b}{Z_4} = \frac{v_f Z_i}{Z_i + Z_4} \qquad (c)$$

The force and velocity of b can be determined for any load Z_4 in terms of the constant velocity
source and internal impedance of the equivalent system representation.

SYSTEMS WITH TWO CONNECTIONS

Only two-connection systems used to transmit vibration energy are considered here.
Active and passive systems of this type are possible. Systems that include vibration
sources that are unrelated to the force or velocity to be transmitted are not considered
active in the usual sense. They are best analyzed using superposition. This involves
determining the responses at the two connections due to the internal sources by Kirchhoff's laws, finding the transmitted motion and force by the methods that follow, and
adding the results.

An active two-connection system by the usual definition is one in which an input function applied to one connection controls an energy source so as to produce an output
function of the same shape as the input, but
with an increased amplitude. Such a device
is called an *amplifier*. The hydraulically
powered vibration generator that has its
control valve driven by a small electrodynamic vibration exciter is an example of
this type of system. Many amplifying systems produce mechanical output functions,
but they involve other forms of energy and
dynamic media. In such cases, analysis is
possible by impedance methods when compatible dynamical analogies and suitable
coupling factors between dynamic media are
used.

**EQUIVALENT π SYSTEMS. Example
10.6.** When a passive two-connection system
is represented by an equivalent π system as
shown in Fig. 10.18, its operation as a device
to transmit vibration energy may be analyzed
by adding the vibration source and the load
impedance to the schematic representation.
With the equivalent π system, it is convenient
to use a Thévenin equivalent system for the
source and a single impedance Z_c to represent

FIG. 10.25. A π equivalent system representation for a two-connection element
used to transmit vibration energy, as analyzed in Example 10.6.

the load. The complete system then is represented as shown in Fig. 10.25A. Since Z_i and
Z_a as well as Z_b and Z_L are in parallel, they may be combined as in Fig. 10.25B where

$$Z_a' = Z_i + Z_a \quad \text{and} \quad Z_b' = Z_L + Z_b$$

Writing force node equations in terms of v_1 and v_L,

$$F_c = (Z_a' + Z_c)v_1 - Z_c v_L$$

$$0 = -Z_c v_1 + (Z_b' + Z_c)v_L$$

Solving for v_1 and v_L,

$$v_1 = \frac{F_c(Z_b' + Z_c)}{Z_a'Z_b' + Z_a'Z_c + Z_b'Z_c}$$

$$v_L = \frac{F_cZ_c}{Z_a'Z_b' + Z_a'Z_c + Z_b'Z_c}$$

The force applied to the load is $F_L = v_LZ_L$.

EQUIVALENT T SYSTEM. Example 10.7. When the T equivalent system representation is used, it is convenient to use the Norton equivalent representation for the vibration source and a mobility representation for the load. The complete representation is shown in Fig. 10.26A. Mobilities $\mathfrak{M}_i$ and $\mathfrak{M}_x$ as well as $\mathfrak{M}_z$ and $\mathfrak{M}_y$ are in series and can be combined as in Fig. 10.26B in which $\mathfrak{M}_x' = \mathfrak{M}_x + \mathfrak{M}_i$ and $\mathfrak{M}_y' = \mathfrak{M}_y + \mathfrak{M}_L$. A determination of the velocity v of the common connection allows a simple determination of F_1 and F_L. Since $\mathfrak{M}_y$ and $\mathfrak{M}_L$ are in series, the force F_L is the force applied to the load. The force F_1 is supplied by the vibration source. Proceeding to find v, write a force equation at the common connection by Eq. (10.37):

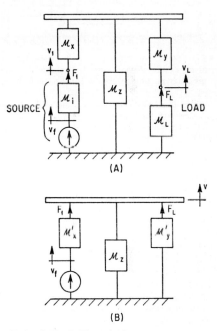

$$\frac{v_f - v}{\mathfrak{M}_x'} = \frac{v}{\mathfrak{M}_y'} + \frac{v}{\mathfrak{M}_z}$$

$$v = \frac{v_f}{1 + \mathfrak{M}_x'/\mathfrak{M}_y' + \mathfrak{M}'_x/\mathfrak{M}_z}$$

The force acting on the load is then $F_L = v/\mathfrak{M}_y'$, and the load velocity is

$$v_L = F_L\mathfrak{M}_L = v_L\mathfrak{M}_L/\mathfrak{M}_y'$$

The force produced by the source which acts at the input connection is

$$F_1 = (v_f - v)/\mathfrak{M}_x'$$

and the velocity of the input connection v_1 is

$$v_1 = v + F_1\mathfrak{M}_i = \frac{v\mathfrak{M}_i + v_f\mathfrak{M}_x}{\mathfrak{M}_x'}$$

Fig. 10.26. A T equivalent system representation for the two-connection system analyzed in Example 10.7.

IMPEDANCE AND MOBILITY PARAMETERS. In Examples 10.6 and 10.7 the actual system is represented by an equivalent system made up of fictitious impedance or mobility elements. Another approach is to represent a two-connection passive system by a "black box," attached to an inertial reference, that has input and output connections. Such a "black box" is illustrated in Fig. 10.27. When the elements in the system are linear and bilateral, and force generators are attached to the connections, the relationship between the velocities and forces may be expressed:

$$F_1 = Z_{11}v_1 + Z_{12}v_2$$
$$F_2 = Z_{21}v_1 + Z_{22}v_2$$

(10.53)

where the Z's are defined by Eqs. (10.39) to (10.42) and are called the *impedance parameters*. When the impedance parameters and two forces are known, the two velocities may be obtained by solving Eqs. (10.53).

The forces and velocities of a system considered as a "black box" also may be expressed as

$$v_1 = \mathfrak{M}_{11}F_1 + \mathfrak{M}_{12}F_2$$
$$v_2 = \mathfrak{M}_{21}F_1 + \mathfrak{M}_{22}F_2 \tag{10.54}$$

Here the $\mathfrak{M}$'s are defined by Eqs. (10.47) to (10.50) and are called the *mobility parameters*. This representation is useful when the velocities at connections (1) and (2) are known.

When a "black box" system is used to transmit vibration energy in only one direction, there is a source attached to one connection and a load connected to the other. Such a case is shown in Fig. 10.27.

Example 10.8. Find the force and velocity at connections 1 and 2 for the system of Fig. 10.27. A Norton equivalent system is used to represent the vibration source and the load is represented by a mobility. The mobility parameters are thus the most advantageous to use.

Fig. 10.27. A "black box" representation of a two-connection system analyzed by the mobility parameter method in Example 10.8.

The load and internal mobility of the source may be included with the "black box" system by measuring or calculating the mobility parameters with $\mathfrak{M}_L$ and $\mathfrak{M}_i$ in place. Equations (10.54) then become

$$v_f = \mathfrak{M}_{11}'F_1 + \mathfrak{M}_{12}F_2$$
$$0 = \mathfrak{M}_{21}F_1 + \mathfrak{M}_{22}'F_2$$

Forces and velocities are considered at points 1' and 2. The velocity v_1 becomes v_f, and v_2 is zero since no external force is applied. The mobility $\mathfrak{M}_{11}'$ is $\mathfrak{M}_{11}$ of Eq. (10.47) determined at point 1', and $\mathfrak{M}_{22}'$ is $\mathfrak{M}_{22}$ of Eq. (10.48) with $\mathfrak{M}_L$ in place. Mobilities $\mathfrak{M}_{12}$ and $\mathfrak{M}_{21}$ are not dependent on the external connections. Solving for F_1 and F_2:

$$F_1 = \frac{v_f \mathfrak{M}_{22}'}{\mathfrak{M}_{11}'\mathfrak{M}_{22}' - \mathfrak{M}_{12}\mathfrak{M}_{21}}$$

$$F_2 = \frac{-v_f \mathfrak{M}_{21}}{\mathfrak{M}_{11}'\mathfrak{M}_{22}' - \mathfrak{M}_{12}\mathfrak{M}_{21}}$$

The force applied to the load is thus $F_L = F_2$, and the load velocity is $v_L = F_2\mathfrak{M}_L$. The force at connection 1 is F_1, since $\mathfrak{M}_i$ transmits this force and the input connection velocity is

$$v_1 = v_f - F_1\mathfrak{M}_i$$

FOUR-POLE PARAMETERS. The relationship between input and output forces and velocities can be written in still another form:

$$F_1 = \alpha_{11}F_2 + \alpha_{12}v_2$$
$$v_1 = \alpha_{21}F_2 + \alpha_{22}v_2 \tag{10.55}$$

The α's are called *four-pole parameters* and are defined as follows:

$$\alpha_{11} = \frac{F_1}{F_2}\bigg|_{v_2=0} \qquad \alpha_{12} = \frac{F_1}{v_2}\bigg|_{F_2=0}$$

$$\alpha_{21} = \frac{v_1}{F_2}\bigg|_{v_2=0} \qquad \alpha_{22} = \frac{v_1}{v_2}\bigg|_{F_2=0}$$

The notation $v_2 = 0$ indicates that the output connection (2) is *clamped* and $F_2 = 0$ indicates the output is *free*. The quantities α_{11} and α_{22} are the force and velocity transfer functions, while α_{12} is an impedance and α_{21} is a mobility. "Black box" systems may be analyzed by the use of these parameters and Eqs. (10.55). In addition, the system may be analyzed by the method that uses the four-pole parameters of the ideal elements and rules for combining their parameters when they are series or parallel connected.[16]

To establish the four-pole parameters for the ideal elements, the relationships between applied and transmitted forces and the velocities of the connections are written in the form of Eqs. (10.55); then the α's are noted as the coefficients of the F_2 and v_2 terms.

Example 10.9. Determine the four-pole parameters for a mass, spring, and resistance.

Mass. Since a mass may be considered as a rigid body, the velocities of its connections are equal. The force required to give the mass a velocity v_1 is $j\omega m v_1$ (or $j\omega m v_2$). If a force F_2 is transmitted, this must be added to $j\omega m v_2$ to determine F_1. Writing these relationships in the form of Eqs. (10.55),

$$F_1 = F_2 + j\omega m v_2$$

$$v_1 = v_2$$

The α's are $\alpha_{11} = 1$, $\alpha_{12} = j\omega m = Z_m$, $\alpha_{21} = 0$, and $\alpha_{22} = 1$.

Spring. The ideal spring transmits an applied force so that $F_1 = F_2$. The relative velocity is $(v_1 - v_2) = j\omega F_2/k$. Expressing these relationships in the form of Eqs. (10.55),

$$F_1 = F_2$$

$$v_1 = \frac{j\omega}{k} F_2 + v_2$$

The α's are $\alpha_{11} = 1$, $\alpha_{12} = 0$, $\alpha_{21} = j\omega/k = \mathfrak{M}_k$, and $\alpha_{22} = 1$.

Resistance. For the resistance, $F_1 = F_2$ and the relative velocity is $(v_1 - v_2) = F_2/c$. In the form of Eqs. (10.55):

$$F_1 = F_2$$

$$v_1 = \frac{F_2}{c} + v_2$$

The α's are $\alpha_{11} = 1$, $\alpha_{12} = 0$, $\alpha_{21} = 1/c = \mathfrak{M}_c$, and $\alpha_{22} = 1$.

Note that the determinant of the α's in each case in Example 10.9 is equal to unity. This is a characteristic of all systems made up of linear bilateral elements and is another indication of reciprocity. The unity value of the α determinant greatly simplifies the solution of Eqs. (10.55) by the determinant or matrix method.

Matrix methods are convenient in system analysis by the four-pole parameter methods. The parameters for a series-connected system, expressed in terms of the individual element parameters, are determined as follows:

SERIES-CONNECTED SYSTEMS. Two such elements are shown in Fig. 10.28A. The four-pole parameter equations for both elements in matrix form are

$$\begin{bmatrix} F_1 \\ v_1 \end{bmatrix} = \begin{bmatrix} \alpha_{11} & \alpha_{12} \\ \alpha_{21} & \alpha_{22} \end{bmatrix} \begin{bmatrix} F_2 \\ v_2 \end{bmatrix}$$

$$\begin{bmatrix} F_2 \\ v_2 \end{bmatrix} = \begin{bmatrix} \alpha_{11}' & \alpha_{12}' \\ \alpha_{21}' & \alpha_{22}' \end{bmatrix} \begin{bmatrix} F_3 \\ v_3 \end{bmatrix}$$

Combining these to find the output at (3) in terms of the input at (1):

$$\begin{bmatrix} F_1 \\ v_1 \end{bmatrix} = \begin{bmatrix} \alpha_{11} & \alpha_{12} \\ \alpha_{21} & \alpha_{22} \end{bmatrix} \begin{bmatrix} \alpha_{11}' & \alpha_{12}' \\ \alpha_{21}' & \alpha_{22}' \end{bmatrix} \begin{bmatrix} F_3 \\ v_3 \end{bmatrix}$$

$$\begin{bmatrix} F_1 \\ v_1 \end{bmatrix} = \begin{bmatrix} \alpha_{11}'' & \alpha_{12}'' \\ \alpha_{21}'' & \alpha_{22}'' \end{bmatrix} \begin{bmatrix} F_3 \\ v_3 \end{bmatrix}$$

(B)

Fig. 10.28. Series-connected mechanical elements considered in the four-pole parameter method.

The single system of Fig. 10.28B is represented by the α''''s and replaces the two series elements. The final four-pole parameters in terms of the original element parameters are

$$\alpha_{11}'' = \alpha_{11}\alpha_{11}' + \alpha_{12}\alpha_{21}' \qquad \alpha_{12}'' = \alpha_{11}\alpha_{12}' + \alpha_{12}\alpha_{22}'$$
$$\alpha_{21}'' = \alpha_{21}\alpha_{11}' + \alpha_{22}\alpha_{21}' \qquad \alpha_{22}'' = \alpha_{22}\alpha_{22}' + \alpha_{21}\alpha_{12}'$$

$$(10.56)$$

This process can be continued to combine any number of elements in a series string. Note that this method allows for a series mass and any number of elements to be considered series-connected.

PARALLEL-CONNECTED ELEMENTS. The canonical equations for parallel connected elements, as shown in Fig. 10.29A, can be written in the form of Eqs. (10.55)

FIG. 10.29. Parallel-connected mechanical elements considered in the four-pole parameter method.

where the α's for the single resultant element are defined in terms of the individual elements in the system:

$$\alpha_{11} = \frac{A}{B} \qquad \alpha_{12} = \frac{AC}{B} - B \qquad \alpha_{21} = \frac{1}{B} \qquad \alpha_{22} = \frac{C}{B}$$

$$(10.57)$$

where $\qquad A = \sum_{i=1}^{n} \frac{\alpha_{11}^{(i)}}{\alpha_{21}^{(i)}} \qquad B = \sum_{i=1}^{n} \frac{1}{\alpha_{21}^{(i)}} \qquad C = \sum_{i=1}^{n} \frac{\alpha_{22}^{(i)}}{\alpha_{21}^{(i)}}$

The factors A, B, and C depend on the parameters of the individual elements as indicated by $i = 1, 2, \ldots, n$. The parallel combination then is represented by a single element as in Fig. 10.29B.

VIBRATION SOURCES. Thévenin's and Norton's equivalent system representations are convenient in this method. Consider a Thévenin representation such as Fig. 10.23. The force F_1 applied to a load and the velocity of the load input v_1 are related to the clamped force F_c and internal impedance Z_i as follows:

$$F_1 = F_c - Z_i v_1 \qquad (10.58)$$

For the Norton representation in Fig. 10.24, the force F_1 is related to the v_1 as follows:

$$F_1 = Z_i v_c - Z_i v_1 \qquad (10.59)$$

Example 11.10. By the four-pole parameter method, determine the velocity of a mass driven by a vibration source. The mass is shown driven by a Thévenin equivalent source in

Fig. 10.30. When there is no force transmitted by the mass, the four-pole parameter equations for the mass are

$$\begin{bmatrix} F_1 \\ v_1 \end{bmatrix} = \begin{bmatrix} 1 & j\omega m \\ 0 & 1 \end{bmatrix} \begin{bmatrix} 0 \\ v_2 \end{bmatrix}$$

In conventional form, $F_1 = j\omega m v$ and $v_1 = v_2$. The source equation $F_1 = F_c - Z_i v_1$ is combined with the mass equation to give

$$F_c - Z_i v_1 = j\omega m v_2 = j\omega m v_1$$

Solving for v_1,

$$v_1 = \frac{F_c}{Z_i + j\omega m}$$

When the mass is driven by a Norton equivalent system with free velocity v_f, the combination of equations yields

$$Z_i v_f - Z_i v_1 = j\omega m v_1$$

Solving for v_1,

$$v_1 = \frac{Z_i v_f}{Z_i + j\omega m}$$

The "black box" concept is evident in each of the above methods of analysis for two-connection systems. This approach focuses attention on two points in a system, usually the more important points. In the case of vibration isolators, the performance is measured in terms of the velocities or forces at the two connections when the devices are used to connect two systems. The above methods allow a calculation of the response of such a device in any application when the impedances of the systems to be connected are known, and one of the sets of the above parameters is known. All the parameters lend themselves to direct measurement. Vibration response characteristics of very complex systems under many different conditions may be determined with a minimum of analysis time when one of the sets of parameters is known. All the sets of parameters are defined in terms of sinusoidal forces and velocities, and they must be measured with this type of driving function.

Fig. 10.30. A mass driven by a Thévenin equivalent source, as analyzed by the four-pole parameter method in Example 10.10.

NONSINUSOIDAL FUNCTIONS

The forces generated by the sources of vibration energy encountered in the normal operation of many mechanical systems are not sinusoidal in form. The responses of systems to nonsinusoidal driving functions can be determined by the impedance method. The Fourier analysis methods and the principle of superposition make this possible.

Excitations of three types frequently encountered are:

1. Continuous, periodic but nonsinusoidal force or velocity functions. Such functions can be represented by a Fourier series and the response to each term found by the impedance method.

2. Nonrepeated transient excitation, usually called shock excitation. The Fourier integral transform of the function yields a continuous frequency spectrum. When impedance expressions are considered as frequency functions, the frequency spectrum of the response may be determined. By the inverse Fourier integral transform the response as an instantaneous time function may be obtained.

3. Continuous, nonperiodic random excitation. The Fourier integral transform applies, and the methods are similar to those for (2) above. Random vibration is discussed in Chap. 11.

PRESENTATION AND EVALUATION OF IMPEDANCE DATA

At a particular frequency the mechanical impedance (or mobility) of a system is a complex number which can be expressed as the sum of a real and an imaginary component or as a magnitude and impedance or mobility angle.

$$Z = \mathcal{R}(Z) + j\mathcal{I}(Z) = |Z|e^{j\theta}$$

The magnitude $|Z|$ and angle θ are

$$|Z| = \sqrt{[\mathcal{R}(Z)]^2 + [\mathcal{I}(Z)]^2} \qquad \theta = \tan^{-1}\frac{\mathcal{I}(Z)}{\mathcal{R}(Z)}$$

These relations also may be expressed

$$\mathcal{R}(Z) = |Z|\cos\theta \qquad \mathcal{I}(Z) = |Z|\sin\theta$$

Impedance at a particular frequency may be represented as a directed line on the complex plane, as shown in Fig. 10.31. Thus, two quantities are required to specify completely an impedance or mobility.

The impedances of the three ideal elements (mass, spring, and resistance) are functions of frequency as indicated by Eqs. (10.14), (10.18), and (10.22). It is convenient to plot

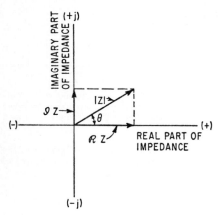

FIG. 10.31. The complex-plane representation of magnitude and angle of impedance, by real and imaginary parts.

FIG. 10.32. The magnitude of the impedance as a function of frequency of ideal resistance, spring, and mass plotted to a log-log scale.

impedance data on graph paper with logarithmic impedance and frequency scales (usually referred to as log-log paper). The impedance curves of Fig. 10.9 are replotted on log-log paper in Fig. 10.32. Curves for impedance of masses have a slope of $(+1)$; they cross the line $f = 1$ cps at $Z_m = 2\pi m$. Thus $m = Z_m/2\pi$. The value of m for an unknown mass thus can be obtained from the impedance curve. The curves for impedance of ideal springs have slopes of (-1) and they intercept the line $f = 1$ cps at $Z_k = k/2\pi$; thus, $k = 2\pi Z_k$. Resistance is a constant with frequency; thus, the impedance Z_c is represented by a horizontal line.

The impedances of the ideal elements can be represented by single curves because they are entirely real or imaginary; however, for combinations of elements it often is necessary to keep the real and imaginary parts separate. For example, consider the parallel

Fig. 10.33. Impedance characteristics of a parallel resistance-spring combination.

spring-dashpot arrangement of Fig. 10.33A. The mechanical impedance of the system is

$$Z = c - \frac{jk}{2\pi f}$$

Expressed as magnitude and angle:

$$|Z| = \sqrt{c^2 + \left(\frac{k}{2\pi f}\right)^2} \qquad \theta = \tan^{-1}\left(-\frac{k}{2\pi fc}\right)$$

where the real part $\mathcal{R}(Z) = c$ and the imaginary part $\mathcal{I}(Z) = -k/2\pi f$ are plotted in **Fig. 10.33B** to linear coordinates. The magnitude $|Z|$ is plotted on the same set of coordinates. The imaginary part $\mathcal{I}(Z)$ is shown dotted in the positive impedance region to indicate that $|Z|$ approaches the imaginary term at low frequencies and the real term at high frequencies. The angle of the impedance is plotted in Fig. 10.33C. The curves

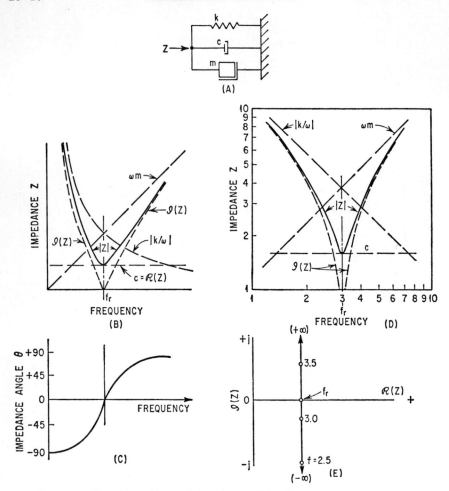

Fig. 10.34. Impedance characteristics of a parallel resistance-spring-mass system.

shown in Fig. 10.33B to linear coordinates are shown in Fig. 10.33D to logarithmic coordinates.

An alternate method for presenting impedance data is shown in Fig. 10.33E. All four factors required to specify impedance are presented on one curve. At each frequency the real and imaginary parts of impedance are plotted as a point on the complex plane, corresponding to Fig. 10.31. These points are connected to form the impedance curve. As shown in Fig. 10.33E for a frequency of 4 cps, the coordinates of the point are the real and imaginary parts of Z. The length of the radial line from the origin to the point is then $|Z|$, and the angle that this line makes with the $(+)$ real axis is the angle of impedance. Frequency appears as a parameter in this presentation.

As shown in Fig. 10.33E, the impedance becomes real at high frequencies since $|Z|$ approaches c and θ approaches $360°$ $(0°)$. At low frequencies $|Z|$ becomes large, θ approaches $270°$ $(-90°)$, and the negative imaginary impedance indicates springlike action. The asymptotes of $|Z|$ indicate the characteristics of the system at frequency extremes.

The addition of a mass to the parallel system of Fig. 10.33A gives the system of Fig.

10.34A. The impedance at the driving point is

$$Z = c + j\left(2\pi f m - \frac{k}{2\pi f}\right)$$

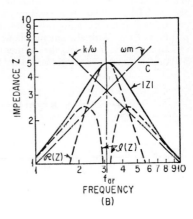

(A)

The magnitude and angle are

$$|Z| = \sqrt{c^2 + \left(2\pi f m - \frac{k}{2\pi f}\right)^2}$$

$$\theta = \tan^{-1}\frac{2\pi f m - k/2\pi f}{c}$$

The impedance of each element is plotted without regard for sign on the linear coordinates of Fig. 10.34B and the logarithmic coordinates of Fig. 10.34D. At the frequency of

$$f_r = \frac{\sqrt{k/m}}{2\pi}$$

the imaginary term is zero. This frequency is the *resonant frequency;* the impedance Z is real and equal to c. At very low frequencies, $|Z|$ approaches the spring impedance and the system is said to be "stiffness controlled." At high frequencies, $|Z|$ approaches the mass impedance and the system is "mass controlled." The impedance angle is plotted on a linear plot in Fig. 10.34C. The asymptotes of this curve are $\theta = -90°$ at low frequencies, and $\theta = +90°$ at high frequencies. At $f = f_r$, $\theta = 0$.

A plot in the complex plane is given in Fig. 10.34E. The curve, parallel to the imaginary axis, indicates a constant real term. The imaginary term approaches infinity at both extremes and becomes zero at f_r.

In Figs. 10.33 and 10.34, the real part of impedance is constant with frequency. This is not the case for the series elements of Fig. 10.35A. The mobility of this system is

$$\mathfrak{M} = \frac{1}{c} + j\left(\frac{2\pi f}{k} - \frac{1}{2\pi f m}\right)$$

The impedance is the reciprocal of the mobility:

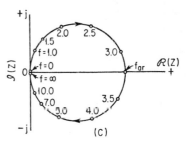

(C)

Fig. 10.35. Impedance characteristics of a series resistance-spring-mass system.

$$Z = \frac{1}{\mathfrak{M}} = \frac{1/c}{1/c^2 + (2\pi f/k - 1/2\pi f m)^2} - j\frac{(2\pi f/k - 1/2\pi f m)}{1/c^2 + (2\pi f/k - 1/2\pi f m)^2}$$

The impedances of the individual elements are plotted to logarithmic coordinates in Fig. 10.35B, together with the magnitude of the impedance $|Z|$ and its real and imaginary parts. The impedance $|Z|$ reaches a maximum at the *antiresonant frequency*

$$f_{ar} = \frac{\sqrt{k/m}}{2\pi}$$

When the system is driven by a sinusoidal excitation of constant-force amplitude, the velocity amplitude at the driving point is at a minimum at the antiresonant frequency.

The complex plane representation is shown in Fig. 10.35C. At low frequencies the impedance is small, the imaginary term is positive, and the angle of the impedance is nearly $+90°$. Masslike action is indicated. At high frequencies, the impedance is small but the imaginary term is negative and the angle is $-90°$. The system is then "stiffness controlled." At intermediate frequencies the locus of the impedance graph is a circle having a diameter c with its center on the $+$ real axis. At f_{ar} the impedance is real and equal to c.

Note that the mobility of the series-connected elements of Fig. 10.35A is given by an expression of the same form as that for the impedance of the parallel-connected elements of Fig. 10.34A. Thus, when mobility is plotted in the same way as impedance, the shapes of the resulting curves for the series elements (Fig. 10.36) are the same as for the impedance of the parallel elements shown in Fig. 10.34.

The mobility of the parallel elements of Fig. 10.34A takes the same form as the impedance of the series elements of Fig. 10.35A. Conversely, the mobility curves for the

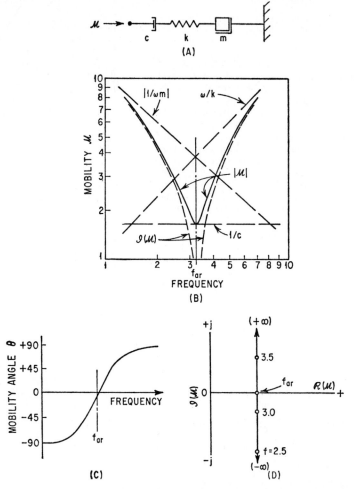

Fig. 10.36. Mobility characteristics of a series resistance-spring-mass system.

parallel-connected elements have the same form as the impedance curves for the series-connected elements of Fig. 10.35.

Complicated mechanical structures may have many resonances at different frequencies. If the impedance is measured and plotted on log-log paper, results similar to those presented in Figs. 10.34 and 10.36 are useful in analyzing the system qualitatively. Positive peaks on the impedance curve indicate the resonance of series elements, and negative peaks indicate the resonance of parallel elements. Frequency regions where the impedance curve has a constant positive slope of $+1$ indicate masslike action with a positive impedance angle. Similarly, regions of the curve with a slope of -1 indicate springlike action with a negative impedance angle.

On the complex plane, each crossing of the real axis indicates a resonance since the reactance $\mathscr{I}(Z)$ is zero at that frequency. In general, circles such as Fig. 10.35C on the impedance curve indicate series element resonances even if they do not cross the real axis. Fig. 10.37A shows a system in which there are parallel resonant elements but one parallel branch has a series resonance at a frequency at which the imaginary part of Z is masslike. The arrows in Fig. 10.37B indicate the direction the locus takes as frequency is increased. The parallel resonance is indicated by the crossing of the real axis, and the circle indicates the resonance of the series branch.

The characteristics of the mobility plotted on the complex plane take on a form that is essentially opposite to the form of the impedance plot. Since mobility and impedance are related by the transformation $Z = 1/\mathfrak{M}$, straight lines on the Z plot transform to circles on the $\mathfrak{M}$ plot, and vice versa. Thus, real axis crossings on the $\mathfrak{M}$ plot indicate series resonances and circles indicate parallel resonances.

FIG. 10.37. Impedance characteristics of a system having series and parallel resonances.

The graphical representation of impedance and mobility is useful in predicting the response of physical systems. If a vibration generator with a constant velocity with respect to frequency is applied to a system at the point at which the impedance is known, the force required to produce this velocity is directly proportional to the impedance. Thus, the impedance plot is a force plot when multiplied by velocity. The mobility plot serves a similar purpose in becoming a velocity plot when a constant force is applied to the system.

The graphical representation of measured impedance data is useful in the study of complicated systems. Dips in the curves indicate parallel resonances. At these frequencies large motions can be obtained with small forces. Peaks in the impedance curves indicate antiresonances. At these frequencies large forces are required to produce significant motion. In some cases the curve shape as determined by measurement may be approximated by the impedance curve of a simple two- or three-element equivalent system. Such an equivalent system of ideal elements can represent a complicated system quite accurately if the frequency range is small. Once the equivalent system is determined, an analytical expression for impedance can be obtained from known theory and the response predicted when the original system is combined with another system of known impedance.

MEASUREMENT OF MECHANICAL IMPEDANCE

In the measurement of mechanical impedance a sinusoidal force is applied to the point at which the impedance is desired. Then a measurement is made of the applied force and resulting motion at the point of application. From this information, the value of the impedance is calculated.[17] Every point of attachment in a system has six possible directions of motion with respect to any set of reference coordinates: three translational and three rotational. A measured impedance in each of these directions is necessary for a complete description of a system. The process is complex and time-consuming and the technology for rotational impedance measurements is undeveloped, so that such a complete description is seldom attempted. The description here is therefore limited to measurement in one dimension. In many cases this is adequate and provides a powerful tool for dynamic analysis and design. Experience is a great aid in making accurate mechanical impedance measurements. Good experience may be obtained by making measurements on simple structures having known characteristics.

FORCE OR MOTION GENERATION

Since impedance is defined in terms of sinusoidal variables, the force or motion applied should have, as nearly as possible, a perfectly sinusoidal waveshape. Its frequency should be easily controllable over the range of interest and its amplitude conveniently variable. The amplitude of force and motion should be kept within the linear range. The most widely used means of excitation is an electromagnetic shaker driven by an electronic oscillator and power amplifier. In general, large forces are not required;[18] a peak force of 10 lb is usually sufficient. The mechanical exciter should be mounted in such a way that the application of the exciting force or motion is restricted to the point at which the impedance is desired.

MEASUREMENT OF FORCE AND MOTION

Force transducers and accelerometers are available commercially for use in impedance measurements.* Usually they are employed in an arrangement such as that shown in

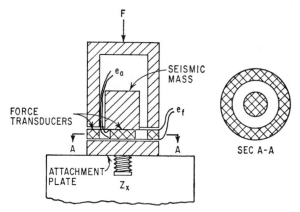

Fig. 10.38. Device for the measurement of mechanical impedance in which force and acceleration are measured. (*Plunkett.*[19])

Fig. 10.38. Here the force is applied through a force transducer and the resulting motion of the point to which it is attached is measured by an accelerometer consisting of a force transducer and seismic mass. The electrical outputs of the force transducer and the accelerometer may be filtered electrically by very narrow bandpass filters, centered at the excitation frequency, to ensure that only the signals at the driving frequency are measured and to reduce interfering noise. One disadvantage of the arrangement of Fig. 10.38 is that the force required to move the accelerometer and the attachment plate is also measured by the force transducer. Thus, the measured force may be in error.

* For example, from B & K Instruments, Inc., Cleveland, Ohio 44142; Wilcoxon Research, Bethesda, Md. 20014.

However, such an error can be compensated to some extent, as indicated in the next section.

IMPEDANCE CALCULATIONS

Mechanical impedance is defined as the ratio of force to velocity [see Eq. (10.7)]. Since the impedance transducer of Fig. 10.38 measures force and acceleration, the velocity must be determined from the acceleration measurement. For sinusoidal motion, the velocity v is related to acceleration a as follows:

$$v = \frac{a}{j\omega}$$

so that Eq. (10.7) becomes

$$Z = \frac{j\omega F}{a} \tag{10.60}$$

For the device of Fig. 10.38, the voltage produced by the force transducer is

$$e_f = K_f F$$

where K_f is the force-transducer calibration constant in volts per unit force.* The voltage produced by the acceleration transducer is

$$e_a = K_a a$$

where K_a is the accelerometer calibration constant in volts per unit acceleration. The impedance then is obtained from Eq. (10.60) by substituting the values of force and acceleration obtained from these voltage measurements, taking phase angle into account; the transducer constants must be known.

If the accelerometer and attachment plate in Fig. 10.38 had no mass, the unknown impedance Z_x would be given by

$$Z_x - \frac{\omega |e_f|/K_f}{|e_a|/K_a} \exp j \left(\theta + \frac{\pi}{2} \right) \tag{10.61}$$

Here θ is the phase angle between the voltage produced by the force and acceleration transducers; the angle $\pi/2$ accounts for the 90° phase shift between acceleration and velocity.

Suppose the mass of the accelerometer and attachments in Fig. 10.38 is represented by m_0. This additional mass may then be taken into account in Eq. (10.61) by subtracting the impedance of the additional mass. Then the unknown impedance becomes

$$Z_x - \omega \frac{|e_f|/K_f}{|e_a|/K_a} \exp j \left(\theta + \frac{\pi}{2} \right) - j\omega m_0 \tag{10.62}$$

The correction term, $j\omega m_0$, in Eq. (10.62) can be accounted for electronically by subtracting a voltage equal to $m_0(K_f/K_a)e_a$ from the voltage produced by the force transducer; this subtraction is known as *mass cancellation*.† Then a new force voltage is obtained:

$$e_f{}' = e_f - m_0 \frac{K_f}{K_a} e_a \tag{10.63}$$

---

* When K_f is expressed as volts/newton, and K_a is volts/meter/sec² the calculated impedance is in units of newton-sec/meter. For K_f in volts/lb and K_a in volts/in./sec², the units of Z are lb-sec/in.

† Mass cancellation is performed electrically by subtracting a variable fraction of the voltage produced by the acceleration transducer from the voltage produced by the force transducer *without* the unknown impedance attached; this fraction is adjusted until $e_f{}'$ is zero. An exact null is generally not possible over a large range of frequencies because of differential phase shifts in the two electrical circuits and because the accelerometer plus its attachments do not behave as an ideal mass.

When this corrected value of force voltage is substituted in Eq. (10.62), the corrected value of Z_x (taking the effects of m_0 into account) is

$$Z_x = \frac{\omega|e_f'|/K_f}{|e_a|/K_a} \exp j \left(\theta + \frac{\pi}{2} \right) \tag{10.64}$$

CALIBRATION

In order to determine the unknown impedance by means of Eq. (10.64), the transducer calibration constants K_f and K_a must be known. These constants can be determined from measurements made on a known impedance, such as on a spherical or cubical block of stiff, dense material, (steel or tungsten is nearly ideal). The mass m_s of such a block is given by W/g, where W is the weight of the block of material and g is the gravitational constant. Well below resonance of the mass, its impedance is given by $j\omega m_s$. Then Eq. (10.64) becomes

$$j\omega m_s = j\omega \frac{e_{f0}K_a}{e_{a0}K_f}$$

so that

$$\frac{K_a}{K_f} = m_s \frac{e_{a0}}{e_{f0}}$$

where e_{f0} and e_{a0} are the measured low-frequency values of e_f and e_a with the calibrated mass attached.

MEASUREMENTS SYSTEMS

Complete systems are commercially available which provide the automatic computation and plotting of mechanical impedance. (Such systems generally provide a sinusoidal excitation voltage which can be swept over the frequency range of interest.) Firms which manufacture such systems include: Bafco, Inc., Warminster, Pa. 18971; B & K Instruments, Inc., Cleveland, Ohio 44142; Hewlett-Packard, Inc., Palo Alto, Calif. 94306; Nicolet Scientific Corp., Northvale, N.J. 07647; Spectral Dynamics Corp., San Diego, Calif. 92112.

REFERENCES

1. Thomson, W. T.: "Mechanical Vibrations," p. 70, Prentice-Hall, Inc., Englewood Cliffs, N.J., 1948.
2. Rocard, Y.: "Dynamique générale des vibrations," p. 62, Masson et Cie, Paris, 1949.
3. McLachlan, N. W.: "Theory of Vibrations," p. 27, Dover Publications, New York, 1941.
4. Firestone, F. A.: *J. Appl. Phys.*, **9**:373 (1938).
5. Freeberg, C. R., and E. N. Kemler: "Elements of Mechanical Vibrations," p. 179, John Wiley & Sons, Inc., New York, 1949.
6. Firestone, F. A.: *J. Acoust. Soc. Amer.*, **28**:1117 (1956).
7. Biot, M. A.: *J. Aeronaut. Sci.*, **7**:376 (1940).
8. Von Kármán, T., and M. A. Biot: "Mathematical Methods in Engineering," McGraw-Hill Book Company, Inc., New York, 1940.
9. Duncan, W. J.: *British Aeronautical Research Committee R & M 2000*, 1947.
10. Bishop, R. E. D.: *J. Roy. Aeronaut. Soc.*, **58**:703 (1954).
11. Harrison, M., A. O. Sykes, and P. G. Marcotti: *J. Acoust. Soc. Amer.*, **24**:384 (1952).
12. Bancroft, D.: *Phys. Revs.*, **59**:588 (1941).
13. Sykes, A. O.: "The Effects of Machine and Foundation Resilience and of Wave Propagation on the Isolation Provided by Vibration Mounts," SAE National Aeronautic Meeting, Los Angeles, Calif., Oct. 2, 1957.
14. Wright, D. V.: *Colloq. Mechanical Impedance Methods Mechanical Vibrations, ASME,* Dec. 2, 1958, pp. 19–42.
15. Chenea, P. F.: *J. Appl. Mechanics*, **20**:233 (1953).
16. Molloy, C. T.: *J. Acoust. Soc. Amer.*, **29**:842 (1957).
17. "Nomenclature and Symbols for Specifying the Mechanical Impedance of Structures," American National Standards Institute, S2.6–1963 (R 1971):
18. Muster, D. F.: *Colloq. Mechanical Impedance Methods Mechanical Vibrations, ASME,* Dec. 2, 1958, p. 87.
19. Plunkett, R.: ASME Paper 53-A-45, *J. Appl. Mechanics*, **21**:250 (1954).

11

STATISTICAL CONCEPTS IN VIBRATION

John W. Miles
University of California, San Diego

W. T. Thomson
University of California, Santa Barbara

INTRODUCTION

A random vibration is one whose instantaneous value is not predictable. Such vibration is generated, for example, by the rocket engines of guided missiles and by aerodynamic turbulence and gusts. Figure 11.1A is an acceleration-time record of a vibration measured in the vicinity of a rocket engine. Clearly there are no well-defined periodicities, and the value of the acceleration at any instant of time is unrelated to that at any other instant and cannot be predicted.

Although the time variation of a random vibration cannot be predicted, the probability of a given value being within a certain range is predictable on a statistical basis. For example, consider a short sample of the above record. The acceleration $\ddot{x}$, at any instant of time t, varies in a random manner about a mean value $\bar{\ddot{x}}$. Suppose the vertical scale is divided into small divisions $\Delta \ddot{x}$. Now determine the statistical probability that the value will be within $\Delta \ddot{x}$. This can be done by measuring the time Δt_n during which the signal has amplitude values between $\ddot{x}$ and $\ddot{x} + \Delta \ddot{x}$ relative to the total length of time T over which the phenomenon is studied. Thus *amplitude probability* P in this example is given by

$$P(\ddot{x}, \ddot{x} + \Delta \ddot{x}) = \frac{\sum_s \Delta t_n}{T}$$

and the *amplitude density* p is given by

$$p(\ddot{x}) = \lim_{\Delta \ddot{x} \to 0} \frac{P(\ddot{x}, \ddot{x} + \Delta \ddot{x})}{\Delta \ddot{x}}$$

The probability of the instantaneous value being infinite is zero; the probability of its peak value being zero or infinite is zero; and the probability of any other value may be distributed as indicated in Fig. 11.1B.

This chapter deals with the problem of a linear, vibrating system subjected to a random excitation, including (1) a statistical description of the random input (*excitation*), (2) a description of the linear system in terms that permit the calculation of its response over the frequency spectrum of the input, and (3) a statistical description of the random output (*response*). Mathematical concepts are discussed first, followed by illustrative applications to mechanical systems characterized by one or more resonances and applica-

FIG. 11.1. Examples of random vibration.   (A) Acceleration vs. time record of vibration measured in the vicinity of a rocket engine.   (B) Example of narrow-band random vibration; the curves along the vertical axis give the probability distribution of the instantaneous (dashed curve) and peak (solid curve) values.

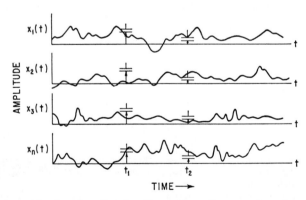

FIG. 11.2. Assembly of random time functions representing a random phenomenon $x(t)$.

tions to the problem of fatigue. A study of the statistical properties of the spectra of a vibrating system often is useful in leading to an understanding of the physical mechanism producing the vibration. Furthermore, it provides information which is useful in the simulation of vibration and useful for design purposes.

STATISTICAL CONCEPTS

FIRST PROBABILITY DISTRIBUTION

The statistical properties of a random time function, such as the acceleration-time record shown in Fig. 11.1A, are related to its so-called "probability distribution." For the determination of the statistical properties of a random function, it is generally necessary to consider a large assembly of random time records, as shown in Fig. 11.2. In this illustration, x is a function that may represent displacement, velocity, or acceleration. Let n represent the number of the members of this family of functions (called an *assembly* or *ensemble*). This chapter is restricted to a consideration of "stationary random processes."

A random process (an ensemble of time functions) is said to be *stationary* if any translation of the time origin leaves its statistical properties unaffected. The *ergodic hypothesis* states that assembly averaging in stationary random processes can be replaced by time averaging over a single record of very long duration. Time averaging of random processes greatly simplifies the task of data analysis.

At any time t_1, the value of $x(t_1)$ will differ in the various records; however, it is possible to count the fraction ν/n of the total number of records for which $x(t_1)$ will be less than some specified value x and plot it as shown in Fig. 11.3A. Such a curve is known as the *first probability distribution*, $P_1(x,t_1)$, at time t_1, and it will tend to a stationary value as $n \to \infty$. The first probability distribution $P_1(x,t_1)$ is a nondecreasing function of x with the following bounds (see Fig. 11.3):

The fraction having a value less than ∞ at time t_1 is 1, i.e., the probability distribution is given by

$$P_1(\infty,t_1) = 1$$

The fraction having a value less than $-\infty$ at time t_1 is 0:

$$P_1(-\infty,t_1) = 0$$

The fraction having a value less than some intermediate value x at time t_1 is given by

$$0 \leq P_1(x,t_1) \leq 1$$

If the derivative of $P_1(x,t_1)$ exists, the density of the *first probability distribution* $p_1(x,t_1)$ is given by

$$p_1(x,t_1) = \frac{d}{dx} P_1(x,t_1) \tag{11.1}$$

as shown in Fig. 11.3B.

The probability of $x(t_1)$ having a value between x_1 and $x_1 + dx$, as shown in Fig. 11.3, is

$$dP_1 = P_1(x_1 + dx, t_1) - P_1(x_1,t_1) = p_1(x_1,t_1)\, dx \tag{11.2}$$

FIG. 11.3. (A) First probability distribution of the assembly of time functions shown in Fig. 11.2 at time t. (B) The probability density function corresponding to (A).

It follows that

$$P_1(x,t_1) = \int_{-\infty}^{x} p_1(x,t_1)\, dx \tag{11.3}$$

$$P_1(\infty,t_1) = \int_{-\infty}^{\infty} p_1(\xi,t_1)\, d\xi = 1 \tag{11.4}$$

SECOND PROBABILITY DISTRIBUTION

The *second probability distribution* P_2 determines the likelihood of pairs of values occurring a specified time interval apart. Suppose this time interval to be τ sec, so that $t_2 = t_1 + \tau$. Then a count of the number of records in which $x(t)$ lies between x_1 and $(x_1 + dx_1)$ at t_1 and between x_2 and $(x_2 + dx_2)$ at t_2, as shown in Fig. 11.2, defines the probability of a pair of values in any record being in the specified ranges at the specified times as

$$dP_2 = p_2(x_1,t_1,\, x_2,t_2)\, dx_1\, dx_2 \tag{11.5}$$

In a similar manner, higher-order probability functions can be determined.

AVERAGE VALUE

Given the first probability density $p_1(x)$, the average value $\bar{x}$ of the random function $x(t)$ can be determined from

$$\bar{x} = \int_{-\infty}^{\infty} x p_1(x,t_1)\, dx \tag{11.6}$$

The nth moment $\overline{x^n}$ is given by

$$\overline{x^n} = \int_{-\infty}^{\infty} x^n p_1(x,t_1)\, dx \tag{11.7}$$

For stationary random processes, Eqs. (11.6) and (11.7) reduce to the following time averages by letting $n = 2$ in Eq. (11.7):
Average value:

$$\bar{x} = \lim_{T \to \infty} \frac{1}{2T} \int_{-T}^{T} x(t)\, dt \tag{11.8}$$

Mean-square value:

$$\overline{x^2} = \lim_{T \to \infty} \frac{1}{2T} \int_{-T}^{T} x^2(t)\, dt \tag{11.9}$$

GAUSSIAN (NORMAL) DISTRIBUTION

The probability distribution that is most frequently encountered (at least as an acceptable approximation) in random vibration problems and the only such distribution that is amenable to extensive analysis is the *Gaussian or normal distribution*. Suppose the amplitude-density curve is normalized:

$$\int_{-\infty}^{\infty} p(x)\, dx = 1$$

where $p(x)$ is the amplitude density. Then for a Gaussian distribution the amplitude density at time t_1 is given by

$$p_1(x,t_1) = \frac{1}{\sigma\sqrt{2\pi}}\, e^{-\frac{(x-\bar{x})^2}{2\sigma^2}} \tag{11.10}$$

where $\bar{x}$ is the mean value of x and σ, the *standard deviation* (i.e., the rms value), is defined by

$$\sigma^2 = \int_{-\infty}^{\infty} (x - \bar{x})^2 p_1(x,t_1) \, dx = \overline{x^2} - (\bar{x})^2 \tag{11.11}$$

The quantity σ^2 is a measure of the dispersion of the random function and is called the *variance*.

The Gaussian distribution defined by Eq. (11.10) is plotted in Fig. 11.4. It is the only distribution (for instantaneous values) considered in this chapter. The assumption that a distribution is Gaussian automatically excludes such properties as skewness (asymmetry of the distribution). Conversely, the existence of some degree of skewness in the plot of probability density vs. $(x - \bar{x})$ is a priori evidence of departure from a Gaussian distribution (see Figs. 11.5 and 22.9).

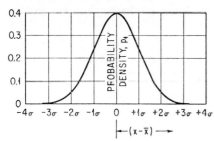

FIG. 11.4. Gaussian (also called normal) distribution. This represents the probability that the magnitude of the instantaneous value of x exceeds its rms value. For example, the probability of x exceeding 2σ is 4.6 per cent.

FIG. 11.5. Example of skewed distribution compared with normal distribution.

POWER SPECTRAL DENSITY

Power, which is the rate of doing work, is proportional to the square of the amplitude of a harmonic vibration. If two frequencies are present in a vibration, the power is proportional to the sum of the squares of the individual amplitudes associated with the two frequencies.

A random vibration can be considered to be a sum of a large number (tending to infinity) of harmonic vibrations of appropriate amplitudes and phase. The total power is again the sum of the power of the component harmonic vibrations, and it is of interest to know how this power is distributed as a function of frequency. Therefore *power spectral density* is defined as the power per unit frequency interval. A plot of this quantity indicates the frequency distribution of power [see Eq. (22.17)].

DETERMINATION OF POWER SPECTRAL DENSITY

The analysis of random vibration always involves a record of finite length. Assume that such a record has been recorded on magnetic tape and is to be analyzed by an electronic analyzer as a stationary process; the tape can be formed into a loop of arbitrary length and made to repeat indefinitely on the analyzer. By such a procedure, an arbitrary fundamental period corresponding to the length of the loop is established, and the contents of the record may be defined in terms of the Fourier coefficients of integral multiple harmonics of the loop frequency ω_0.

The function $x(t)$ then can be represented by the real part of the Fourier series,

$$x(t) = \Re \sum_{n=-\infty}^{\infty} C_n e^{jn\omega_0 t} = C_0 + 2\sum_{n=1}^{\infty} |C_n| \cos(n\omega_0 t - \alpha_n) \tag{11.12}$$

(A) FREQUENCY →

(B) FREQUENCY →

FIG. 11.6. Spectrum analysis of random vibration, as determined from a magnetic-tape loop. In (A) the period is T. In (B) the period is $2T$. Note that doubling loop period doubles number of spectral lines and reduces $|C_n|^2$ to one-half the former value.

where the complex amplitude C_n is defined by the equation

$$C_n = \frac{1}{2T} \int_{-T}^{T} x(\xi) e^{-jn\omega_0 \xi} \, d\xi \quad (11.13)$$

and $2T$ is the loop period.

The mean-square value $\overline{x^2}$ is a stationary property of $x(t)$ that is determined by Eq. (11.9) as

$$\overline{x^2} = \frac{1}{2T} \int_{-T}^{T} x^2(t) \, dt = C_0^2 + 2 \sum_{n=1}^{\infty} |C_n|^2 \quad (11.14)$$

This equation indicates that the contribution to the mean-square value in any frequency interval is the summation of all the components within that frequency band.

If now the length of the loop and the loop period are doubled, the number of spectral lines will be doubled. Since the mean-square value of the entire spectrum must remain essentially constant, the values of the new coefficients $|C_n'|^2$ must decrease to approximately one-half their former values, as illustrated in Fig. 11.6. The sum of the lengths of the $|C_n|^2$ lines, up to any frequency, is the same in each diagram of Fig. 11.6, and the increment in $\overline{x^2}$ divided by the increment in frequency is essentially the same in each case,

$$\frac{\Delta \overline{x^2}}{\Delta \omega} = \frac{2|C_n|^2}{\omega_0} = \frac{2|C_n'|^2}{\frac{1}{2}\omega_0} = W(\omega_n) \quad (11.15)$$

The quantity $W(\omega_n)$ is the *power spectral density* at $\omega = \omega_n$. The dimensions are the square of the parameter represented by $\Delta \overline{x^2}$ in Eq. (11.15) per unit of frequency.

As the length of the loop is increased, more and more spectral lines are introduced; the magnitude of $|C_n|^2$ decreases correspondingly, but $W(\omega_n)$ remains finite. In the limiting case, as $T \to \infty$ or $\Delta \omega \to 0$, the spectrum becomes continuous, and the discrete values of $W(\omega_n)$ approach a smooth function $W(\omega)$:

$$W(\omega) = \lim_{\Delta \omega \to 0} \frac{\Delta \overline{x^2}}{\Delta \omega} = \frac{d\overline{x^2}}{d\omega} \quad (11.16)$$

The mean-square value can be expressed by the integral

$$\overline{x^2} = \int_0^\infty W(\omega) \, d\omega \quad (11.17)$$

RANDOM
INPUT $x(t)$

$\Delta \overline{x^2} = W(\omega)\Delta\omega$

(A)

(B)

FIG. 11.7. The experimental determination of the power spectral density $W(\omega)$ of a random function $x(t)$. The function $x(t)$ is recorded on a magnetic tape and fed into a spectrum analyzer that transmits only those frequencies within the passband $\overline{\omega} \pm \frac{1}{2}\Delta\omega$. The output in the passband, namely, $\Delta \overline{x^2}$, is recorded on a mean-square meter, and the mean power spectral density over the passband is determined by dividing the meter reading by $\Delta\omega$. Thus, as shown in (B), the meter reading, $\Delta \overline{x^2}$, appears in the power spectral density vs. frequency plane as an increment of area of width $\Delta\omega$ and mean ordinate $f(\overline{\omega})$.

The power spectral density of a random time function may be determined by (see Fig. 11.7) feeding the function $x(t)$ into a spectrum analyzer that transmits only those frequency components within the passband of the analyzer, $\overline{\omega} \pm \frac{1}{2}\Delta\omega$. The output in the passband, namely, $\Delta \overline{x^2}$ is then indicated by a "mean-square" meter. The mean power spectral density $W(\overline{\omega})$ over the passband, at $\overline{\omega}$, is determined by dividing the meter reading by $\Delta\omega$. This division may be incorporated directly in the calibration of the meter.

The complete distribution of $W(\omega)$ then is determined by changing the frequency $\bar{\omega}$ in increments of $\Delta\omega$.

TYPICAL POWER SPECTRAL DENSITY

Typical power spectral densities may be characterized *broadly* by two parameters: the mean-square value, given by Eq. (11.17), and some characteristic frequency—say ω_1. For example, the *white noise* * distribution with a cutoff frequency ω_1 is described by the equation

$$W(\omega) = \begin{cases} \overline{x^2}/\omega_1 & [\omega < \omega_1] \\ 0 & [\omega > \omega_1] \end{cases} \tag{11.18}$$

If $W(\omega)$ is (at least approximately) a monotonically decreasing function of ω, it is convenient to choose $\omega_1 = \omega_{1/2}$, where $\omega_{1/2}$ defines (at least roughly) the *half-power point* according to

$$\int_0^{\omega_{1/2}} W(\omega)\, d\omega \doteq \int_{\omega_{1/2}}^\infty W(\omega)\, d\omega \tag{11.19}$$

If $W(\omega)$ exhibits a definite peak at some frequency appreciably greater than zero, say ω_m, it is convenient to choose $\omega_1 = \omega_m$.

Examples of these two possible types of power spectral density are given by

Monotonic:

$$W_A(\omega) = \left(\frac{2\overline{x^2}}{\pi^{1/2}\omega_1}\right) e^{-(\omega/\omega_1)^2} \tag{11.20}$$

Peaked:

$$W_B(\omega) = \left(\frac{4\overline{x^2}}{\pi^{1/2}\omega_1}\right)\left(\frac{\omega}{\omega_1}\right)^2 e^{-(\omega/\omega_1)^2} \tag{11.21}$$

and are plotted in Fig. 11.8. The actual half-power point for $W_A(\omega)$ occurs at $\omega = 0.48\omega_1$, while the peak of $W_B(\omega)$ occurs at $\omega = \omega_1$. It should be emphasized that these examples are hypothetical; although they are representative of the more important types, most spectra are so complicated that they can be defined only graphically.

Whether or not either of the (closely related) methods discussed in this section is

FIG. 11.8. Typical power spectral densities: *(A)* as given by Eq. (11.20) and *(B)* by Eq. (11.21). Note that $\overline{x^2} = \int_0^\infty W(\omega)\, d\omega$.

the most expedient for the actual determination of the power spectral density from a given random function depends on the equipment available, the accuracy, and the ultimate application of the data. Another procedure for obtaining $W(\omega)$ is based on the autocorrelation of $x(t)$, making use of Eq. (11.60).

INPUT-OUTPUT RELATIONSHIP

The relationship between the input and output of a linear system for any time function $x(t)$, whether random or not, may be expressed in terms of the integral

$$y(t) = \int_{-\infty}^t x(\tau)h(t-\tau)\, d\tau = \int_0^\infty x(t-\xi)h(\xi)\, d\xi \tag{11.22}$$

where $h(t)$ is the response of the system to a unit impulse $\delta(t)$ and $x(t)$ is the input, which may have existed from $t = -\infty$. In the usual form of the above integral, the input is

* The power spectral density of true white noise is a constant for all ω.

assumed to be zero for $t < 0$, in which case the range of integration becomes 0 to t in both integrals.

The spectral aspects of the system are defined by a response function which is directly significant only for harmonic excitation. Letting $x(t) = e^{j\omega t}$ represent such a harmonic excitation, and noting that there is no provision for starting or stopping such a function, its substitution into Eq. (11.22) results in the following:

$$y(t) = \int_0^\infty e^{j\omega(t-\tau)} h(\tau)\, d\tau = e^{j\omega t} \int_0^\infty e^{-j\omega\tau} h(\tau)\, d\tau$$

$$= A(j\omega)e^{j\omega t} \qquad (11.23)$$

where
$$A(j\omega) = \int_0^\infty e^{-j\omega\tau} h(\tau)\, d\tau \qquad (11.24)$$

is defined as the response function of the system. Thus the response to a harmonic excitation is the excitation multiplied by the response function $A(j\omega)$ which then serves to define the spectral characteristics of the system. However, a sinusoidal input is given by either the real or the imaginary part of $e^{j\omega t}$, and either the real or the imaginary part of the response $y(t)$ represents the solution to such an excitation. Since $h(\tau)$ is zero for $\tau < 0$, the lower limit of Eq. (11.24) can be extended to $-\infty$ without altering the integral. Thus $A(j\omega)$ is the Fourier transform of the impulsive response function $h(\tau)$.

Suppose that a random time function (over a finite time) is expressible in terms of a Fourier series. Determine the spectral form of the output after the function has passed through a linear system having the response function $A(j\omega)$. Let Eq. (11.12) represent the input, and substitute into the integral of Eq. (11.22) for the response:

$$y(t) = \int_0^\infty h(\tau) \sum_{n=-\infty}^\infty C_n e^{jn\omega_0(t-\tau)}\, d\tau$$

$$= \sum_{n=-\infty}^\infty C_n e^{jn\omega_0 t} \int_0^\infty h(\tau) e^{-jn\omega_0\tau}\, d\tau$$

$$= \sum_{n=-\infty}^\infty C_n e^{jn\omega_0 t} A(jn\omega_0) \qquad (11.25)$$

Note that Eq. (11.25) is merely a superposition of harmonic components, one of which is given by Eq. (11.23). For a real input, the output must be real; then Eq. (11.25) becomes

$$y(t) = C_0 A(0) + 2 \sum_{n=1}^\infty |C_n|\, |A(jn\omega_0)|\, \cos(n\omega_0 t + \beta_n) \qquad (11.26)$$

From Eq. (11.9), the mean-square value of the response is

$$\overline{y^2} = \frac{1}{2T} \int_{-T}^T y^2(t)\, dt = C_0^2 A^2(0) + 2 \sum_{n=1}^\infty |C_n|^2 |A(jn\omega_0)|^2 \qquad (11.27)$$

The increment of the mean-square value of the response at $\omega = n\omega_0$ is

$$\Delta\overline{y^2} = 2|C_n|^2 |A(jn\omega_0)|^2 \qquad (11.28)$$

and the power spectral density of the output for the discrete spectrum, with the frequency increment of $\omega_0 = \pi/T$, is

$$W_r(\omega_n) = \frac{2|C_n|^2 |A(jn\omega_0)|^2}{\omega_0} \qquad (11.29)$$

Since the power spectral density of the input $W_e(\omega_n)$ is given by Eq. (11.15), the input and output spectra are related by the equation

$$W_r(\omega_n) = W_e(\omega_n) |A(jn\omega_0)|^2 \qquad (11.30)$$

If the period $2T$ is increased without limit, the discrete spectrum becomes continuous; then $n\omega_0 = \omega_n = \omega$, so that Eq. (11.30) may be written as

$$W_r(\omega) = W_e(\omega)|A(j\omega)|^2 \tag{11.31}$$

The mean-square output is given by the integral

$$\overline{y^2} = \int_0^\infty W_r(\omega)\, d\omega = \int_0^\infty W_e(\omega)|A(j\omega)|^2\, d\omega \tag{11.32}$$

Thus the mean-square output (which is a statistical property of the output) is determined by the integral, over frequency, of the power spectral density of the input as modified by the response function of the system. When the time variation of the input is random and unknown, the time variation of the output is also random and unknown, and the relationship between the output and input is the statistical one given by Eqs. (11.31) and (11.32).

RESPONSE OF A SINGLE DEGREE-OF-FREEDOM SYSTEM

According to Eq. (2.27), the differential equation for a mechanical system of one degree-of-freedom (referred to as a "simple oscillator") is

$$m\ddot{y} + c\dot{y} + ky = F_0 \sin \omega t \tag{11.33}$$

Let

$$\omega_n = \sqrt{\frac{k}{m}} = \text{natural angular frequency of the oscillator}$$

$$c_c = 2\sqrt{km} = \text{critical damping}$$

$$\zeta = \frac{c}{c_c} = \text{fraction of critical damping}$$

If the excitation is due to the motion $u(t)$ of the support instead of the force $F_0 \sin \omega t$, according to Eq. (2.47), the equation of motion of the mass has the same general form:

$$m\ddot{z} + c\dot{z} + kz = -m\ddot{u} \tag{11.34}$$

where $z = (y - u)$ is the relative displacement.

The response function of the simple oscillator can be determined from its Fourier transform, or more simply from the relation

$$z(t) = \ddot{u}_0 e^{j\omega t} A_1(j\omega) \tag{11.35}$$

which expresses its steady-state response to the harmonic input $\ddot{u}_0 e^{j\omega t}$. The square of the absolute value of the response function for Eqs. (11.33) and (11.34) is

$$|A_1(j\omega)|^2 = \frac{1}{\omega_n^4 \left\{ \left[1 - \left(\dfrac{\omega}{\omega_n} \right)^2 \right]^2 + \left(2\zeta \dfrac{\omega}{\omega_n} \right)^2 \right\}} \tag{11.36}$$

The power spectral density of the response then may be obtained from Eq. (11.32). In general this integration can be performed graphically, but it sometimes may be possible to effect analytical approximations, as in the following paragraph.

RESPONSE OF A NARROW-BAND OSCILLATOR

An important case, illustrated in Fig. 11.9, arises when the bandwidth of the oscillator, $\zeta\omega_n$, is small compared with the frequency interval over which the power spectral density is substantial and when $W_e(\omega_n)$, the power spectral density of the input at the resonant frequency of the oscillator, is of the same order of magnitude as $W_e(\omega_1)$. The response

FIG. 11.9. Power spectral density of the output response of a simple oscillator for an input having the power spectral density shown by the dashed curve [see Eq. (11.37)].

of the oscillator then is essentially resonant, and it receives energy only in the neighborhood of its resonant frequency. Then its power spectral density of relative displacement may be approximated by

$$W_r(\omega) = W_e(\omega_n) |A_1(j\omega)|^2 \qquad (11.37)$$

where $W_r(\omega)$ is the mean-square displacement density of relative motion z; $W_e(\omega)$ is the mean-square acceleration density of the input u; and $A_1(j\omega)$ is the response function defined by Eq. (11.36). The corresponding mean-square response then is given by Eq. (11.32) as

$$\overline{z^2} = W_e(\omega_n) \int_0^\infty |A_1(j\omega)|^2 \, d\omega = \frac{\pi}{4\zeta} \frac{W_e(\omega_n)}{\omega_n^3} \qquad (11.38)$$

where the integral has been evaluated by substituting $|A(j\omega)|^2$ from Eq. (11.36) and assuming $\zeta \ll 1$ *after* carrying out the integration (which is given in integral tables). It may be assumed, without loss of generality, that the mean response $\bar{z}$ is zero (since, if $\bar{z} \neq 0$, it is necessary only to measure z from $\bar{z}$, i.e., replace z in the following analysis by $z - \bar{z}$); it then follows that $\sigma^2 = \overline{z^2}$.

The resonant response of Eq. (11.38) is simple in that, whatever the probability distribution of the input $x(t)$, the probability distribution of the response $z(t)$ tends to the Gaussian distribution of Eq. (11.10) as $\zeta\omega_n/\omega_1$ tends to zero, i.e., as the bandwidth of the oscillator becomes small compared with that of the input.[*] (Note that this last qualification excludes a periodic input.) The statistical properties of this essentially resonant response then are *completely* characterized by its resonant frequency ω_n and its mean-square value (compare the *broad* characterization of the input by its characteristic frequency ω_1 and mean-square value $\overline{x^2}$).

Note that if a random input is applied to a lightly damped oscillator, it acts as a narrow bandpass filter—passing only those frequencies in the neighborhood $\omega_n \pm \zeta\omega_n$ [cf. Eq. (11.37) and Fig. 11.9]. Because of the phenomenon of *beats* (wherein the addition of two harmonic waves of approximately the same frequency gives rise to a wave oscillating with the mean of the two frequencies and having an amplitude envelope that fluctuates at a rate equal to the difference frequency), the addition of a large number (tending to infinity) of harmonic waves having frequencies in the approximate band $\omega_n \pm \zeta\omega_n$ and randomly distributed phases will give rise to a wave oscillating with frequency ω_n and having an amplitude envelope that exhibits a random fluctuation, the rapidity of which must be of the order of $2\zeta\omega_n$. Such a response (illustrated in Fig. 11.1B) is called a *random sine wave*.

The probability distribution of this envelope (as opposed to the distribution of the instantaneous response) may be determined by synthesizing it from the randomly oriented vectors representing the individual harmonic waves.[†]

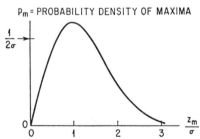

FIG. 11.10. Probability density distribution of an idealized random amplitude vibration of narrow bandwidth (nearly single frequency). This curve, known as a "Rayleigh distribution," gives the probability density of the maxima for essentially resonant loading as defined by Eq. (11.39).

---

[*] It is generally argued that this result follows directly from the "central-limit theorem."
[†] This problem, known as the two-dimensional *random walk* problem, was first solved by Lord Rayleigh.

RAYLEIGH DISTRIBUTION

The probability that the envelope $z_m(t)$ lies between z and $z + dz$ is

$$P\{z < z_m(t) < z + dz\} = p_m\, dz = \sigma^{-2} e^{-\frac{1}{2}(z/\sigma)^2} z\, dz \tag{11.39}$$

where σ, the standard deviation or rms value, is defined by Eq. (11.11). The quantity p_m, known as a *Rayleigh distribution* density, is plotted in Fig. 11.10. Also see Fig. 22.12.

The Rayleigh distribution of Eq. (11.39) describes the variation in amplitudes of the envelope of the random sine wave; the probability distribution of the instantaneous value is given by the Gaussian distribution of Eq. (11.10). Thus, the probability that $|z|$ will exceed a specified value z_1 is determined by

$$P\{|z| > z_1\} = \frac{2}{\sqrt{2\pi}} \int_{z_1}^{\infty} e^{-\frac{1}{2}(z/\sigma)^2} d\left(\frac{z}{\sigma}\right) = erfc\, \frac{z_1}{\sigma\sqrt{2}} \tag{11.40a}$$

$$P\{|z| > z_1\} \simeq \sqrt{\frac{2}{\pi}} \left(\frac{\sigma}{z_1}\right) e^{-\frac{1}{2}(z_1/\sigma)^2} \left[1 - \left(\frac{\sigma}{z_1}\right)^2 + \cdots\right] \tag{11.40b}$$

where *erfc* denotes the complementary error function and Eq. (11.40b) gives its asymptotic approximation for $z_1 \gg \sigma$. Equations (11.40a) and (11.40b) may be used to determine the probability of exceeding the ultimate stress, whereas the Rayleigh distribution P_m is employed in studies of fatigue failure.

RESPONSE OF MULTIPLE DEGREE-OF-FREEDOM AND CONTINUOUS SYSTEMS

Every elastic system has associated with it certain normal-mode properties. Because of the orthogonal character of normal modes, such a system can be treated in terms of simple oscillators.[2]

The displacement response of a linear structure can be expressed in terms of its normal modes $\varphi_n(x,y,z)$: *

$$u(x,y,z,t) = \sum_n q_n(t)\varphi_n(x,y,z) \tag{11.41}$$

Assuming structural damping (damping in phase with the velocity and proportional to the generalized displacement of each mode), the Lagrange equations of the system have the form

$$\ddot{q}_n + (1 + j\gamma_n)\omega_n^2 q_n = \frac{Q_n(t)}{M_n} \tag{11.42}$$

where γ_n = coefficient of structural damping

ω_n = nth normal angular frequency

$$M_n = \int \varphi_n^2(x,y,z)\, dm = \text{generalized mass} \tag{11.43}$$

$$Q_n(t) = \int f(x,y,z,t)\varphi_n(x,y,z)\, dv = \text{generalized force} \tag{11.44}$$

$$dv = dx\, dy\, dz$$

The generalized mass and generalized force are obtained by integration over the structure.

---

* The concept of normal modes is discussed in detail under *Normal Modes of Vibration* in Chaps. 2 and 7. In particular, see Eq. (2.52) for a discussion of structural damping and Eq. (2.71) for a discussion of generalized mass and generalized force.

To establish the response function, a harmonic force can be applied. In this case the generalized force becomes

$$Q_n(t) = e^{j\omega t} \int F(x,y,z)\varphi_n(x,y,z)\, dv = e^{j\omega t} w_0 w_n \tag{11.45}$$

where $\qquad w_0 = \int F(x,y,z)\, dv = $ amplitude of the total applied force $\qquad$ (11.46)

and $\qquad w_n = \dfrac{1}{w_0} \int F(x,y,z)\varphi_n(x,y,z)\, dv = $ mode participation factor of load $\qquad$ (11.47)

The steady-state solution for q_n is

$$q_n = \frac{w_0 w_n e^{j(\omega t + \theta_n)}}{M_n \omega_n^{\,2}\left\{\left[1 - \left(\dfrac{\omega}{\omega_n}\right)^2\right]^2 + \gamma_n^{\,2}\right\}^{\frac{1}{2}}} \tag{11.48}$$

which results in the response function

$$A(j\omega) = \sum_n \frac{w_n \varphi_n(x,y,z) e^{j\theta_n}}{M_n \omega_n^{\,2}\left\{\left[1 - \left(\dfrac{\omega}{\omega_n}\right)^2\right]^2 + \gamma_n^{\,2}\right\}} \tag{11.49}$$

The square of the response function, required for the evaluation of the mean-square response, is then

$$|A(j\omega)|^2 = A(j\omega)A(-j\omega)$$

$$= \sum_n \frac{w_n^{\,2}\varphi_n^{\,2}(x,y,z)}{M_n^{\,2}\omega_n^{\,4}\left\{\left[1 - \left(\dfrac{\omega}{\omega_n}\right)^2\right]^2 + \gamma_n^{\,2}\right\}} \tag{11.50}$$

Utilizing again the approximations allowed for small damping, the mean-square response for the structure becomes

$$\overline{u^2} = \frac{\pi}{2}\sum_n \frac{w_n^{\,2}\varphi_n^{\,2}(x,y,z)W(\omega_n)}{\gamma_n M_n^{\,2}\omega_n^{\,3}} \tag{11.51}$$

Equation (11.51) is expressed in terms of the normal modes of the structure and the power spectral density of the excitation, which allows a simple calculation for any linear structure.

CORRELATION FUNCTIONS

In statistical studies, the correlation function indicates the correlation between two sets of numbers. The autocorrelation, $R(t_1,\tau)$, of a random function $x(t)$ is a measure of the correlation between two values of $x(t)$ spaced τ apart. Crosscorrelation indicates the correlation between sets of numbers in two quantities $x(t)$ and $y(t)$.

AUTOCORRELATION FUNCTIONS *

Autocorrelation is found by averaging the product of the two values $x(t_1)$ and $x(t_1 + \tau)$ of each record over the assembly of records. The autocorrelation function $R(t,\tau)$ is related to the second probability density as follows:

$$R(t_1,\tau) = \overline{x_i(t_1)x_i(t_1 + \tau)}$$

$$= \int_{-\infty}^{\infty}\int_{-\infty}^{\infty} x_1 x_2 p_2(x_1,t_1,x_2,t_2)\, dx_1\, dx_2 \tag{11.52}$$

* Also see Chap. 22.

For a stationary function, the autocorrelation function $R(\tau)$ is determined by time averaging.

$$R(\tau) = \lim_{T \to \infty} \frac{1}{2T} \int_{-T}^{T} x(t)x(t + \tau)\, dt \qquad (11.53)$$

From Eq. (11.12) the two terms to be multiplied and averaged are

$$x(t) = C_0 + 2 \sum_{n=1}^{\infty} |C_n|\, \cos (n\omega_0 t - \alpha_n) \qquad (11.54)$$

$$x(t + \tau) = C_0 + 2 \sum_{n=1}^{\infty} |C_n|\, \{\cos (n\omega_0 t - \alpha_n) \cos n\omega_0 \tau - \sin (n\omega_0 t - \alpha_n) \sin n\omega_0 \tau\} \quad (11.55)$$

and the correlation function $R(\tau_n)$ for a finite period τ_n becomes

$$R(\tau_n) = \frac{1}{2T} \int_{-T}^{T} x(t)x(t + \tau)\, dt = C_0^2 + 2 \sum_{n=1}^{\infty} |C_n|^2 \cos n\omega_0 \tau \qquad (11.56)$$

The increment in $R(\tau_n)$ divided by the increment in frequency is then

$$\frac{\Delta R(\tau_n)}{\Delta \omega} = \frac{2|C_n|^2}{\omega_0} \cos n\omega_0 \tau = W(\omega_n) \cos n\omega_0 \tau \qquad (11.57)$$

As $T \to \infty$, $n\omega_0 = \omega$, $W(\omega_n) \to W(\omega)$, $R(\tau_n) \to R(\tau)$, and

$$\frac{dR(\tau)}{d\omega} = W(\omega) \cos \omega \tau \qquad (11.58)$$

Thus, by integration, the relationship between the power spectral density $W(\omega)$ and the autocorrelation function $R(\tau)$ is

$$R(\tau) = \int_{0}^{\infty} W(\omega) \cos \omega \tau\, d\omega \qquad (11.59)$$

It is evident from this expression that $R(\tau)$ is an even function of τ. In any practical mechanical system $W(\omega)$ approaches zero as $\omega \to \infty$, and $R(\infty) \to 0$. Also if $\tau \to 0$, the value of the function at each time increment is multiplied by itself; then Eq. (11.59) and Eq. (11.17) are identical, indicating that $R(0)$ is equal to the mean-square value $\overline{x^2}$. For $0 < \tau < \infty$, $R(\tau)$ may experience a change in sign and the curve for $R(\tau)$ may appear as in Fig. 11.11.

FIG. 11.11. Typical autocorrelation, an even function of τ [see Eq. (11.59)].

Equation (11.59) may be identified as a Fourier cosine transform, whence the power spectral density may be determined from the autocorrelation function by taking the inverse transform to obtain

$$W(\omega) = \frac{2}{\pi} \int_0^\infty R(\tau) \cos \omega\tau \, d\tau \tag{11.60}$$

It is often desirable to obtain $W(\omega)$ for a given record by first obtaining $R(\tau_n)$ according to Eq. (11.56), with T finite, and then evaluating $W(\omega)$ from Eq. (11.60). These operations are mathematically equivalent to those discussed under *power spectral density*, and the ultimate choice depends on the equipment available, the required accuracy, and the intended application of the results as indicated in Chap. 22.

REFERENCES

1. Rice, S. O.: *Bell System Tech. J.*, **23**:282 (1944) and **24**:44 (1945); reprinted in "Noise and Stochastic Processes," N. Wax (ed.), Dover Publications, New York, 1954.
2. Thomson, W. T., and M. V. Barton: *J. Appl. Mechanics*, **24**:248 (1957).

12

PIEZOELECTRIC AND PIEZORESISTIVE TRANSDUCERS

Eldon E. Eller
Endevco Corp.

Robert M. Whittier
Endevco Corp.

INTRODUCTION

This chapter is the first in a group of seven chapters on the measurement of shock and vibration. The following chapters describe in detail various types of instruments and measurement systems, and set forth primary considerations in the calibration and field use of such instruments and systems. This chapter defines the terms and describes the general principles of piezoelectric and piezoresistive transducers; it also sets forth the mathematical basis for the use of shock and vibration transducers and includes a brief description of piezoelectric accelerometers, piezoresistive accelerometers, and piezoelectric force and impedance gages, along with a review of their performance and characteristics.

Certain solid-state materials are electrically responsive to mechanical force; they often are used as the mechanical-to-electrical transduction elements in shock and vibration transducers. Generally exhibiting high elastic stiffness, these materials can be divided into two categories: the *self-generating type*, in which electric charge is generated as a direct result of applied force, and the *passive-circuit type*, in which applied force causes a change in the electrical characteristics of the material.

A *piezoelectric* material is one which produces an electric charge proportional to the stress applied to it, within its linear elastic range. Piezoelectric materials are of the self-generating type. A *piezoresistive* material is one whose electrical resistance depends upon applied force. Piezoresistive materials are of the passive-circuit type.

A *transducer* (sometimes called a *pickup* or *sensor*) is a device which converts shock or vibratory motion into an optical, a mechanical, or, most commonly, an electrical signal that is proportional to a parameter of the experienced motion.

A *transducing element* is the part of the transducer that accomplishes the conversion of motion into the signal.

A *measuring instrument* or *measuring system* converts shock and vibratory motion into an observable form that is directly proportional to a parameter of the experienced motion. It may consist of a transducer with transducing element, signal-conditioning equipment, and device for displaying the signal. An instrument contains all of these elements in one package, while a system utilizes separate packages.

An *accelerometer* is a transducer whose output is proportional to the acceleration input. The output of a force gage is proportional to the force input; an impedance gage contains both an accelerometer and a force gage.

CLASSIFICATION OF MOTION TRANSDUCERS

In principle, shock and vibration motions are measured with reference to a point fixed in space by either of two fundamentally different types of transducers:

1. *Fixed-reference transducer.* One terminal of the transducer is attached to a point that is fixed in space; the other terminal is attached (e.g., mechanically, electrically, optically) to the point whose motion is to be measured.

2. *Mass-spring transducer (seismic transducer).* The only terminal is the base of a mass-spring system; this base is attached at the point where the shock or vibration is to be measured. The motion at the point is inferred from the motion of the mass relative to the base.

MASS-SPRING TRANSDUCERS (SEISMIC TRANSDUCERS)

In many applications, such as moving vehicles or missiles, it is impossible to establish a fixed reference for shock and vibration measurements. Therefore, many transducers use the response of a mass-spring system to measure shock and vibration. A mass-spring transducer is shown schematically in Fig. 12.1; it consists of a mass m suspended from the transducer case a by a spring of stiffness k. The motion of the mass within the case may be damped by a viscous fluid or electric current symbolized by a dashpot with damping coefficient c. It is desired to measure the motion of the moving part whose displacement with respect to fixed space is indicated by u. When the transducer case is attached to the moving part, the transducer may be used to measure displacement, velocity, or acceleration, depending on the portion of the frequency range which is utilized and whether the relative displacement or relative velocity $d\delta/dt$ is sensed by the transducing element. The typical response of the mass-spring system is analyzed in the following paragraphs and applied to the interpretation of transducer output.

Fɪɢ. 12.1. Mass-spring type of vibration-measuring instrument consisting of a mass m supported by spring k and viscous damper c. The case a of the instrument is attached to the moving part whose vibratory motion u is to be measured. The motion u is inferred from the relative motion δ between the mass m and the case a.[1]

Consider a transducer whose case experiences a displacement motion u, and let the relative displacement between the mass and the case be δ. Then the motion of the mass with respect to a reference fixed in space is $\delta + u$, and the force causing its acceleration is $m[d^2(\delta + u)/dt^2]$. Thus, the force applied by the mass to the spring and dashpot assembly is $-m[d^2(\delta + u)/dt^2]$. The force applied by the spring is $-k\delta$, and the force applied by the damper is $-c(d\delta/dt)$, where c is the damping coefficient. Adding all force terms and equating the sum to zero,

$$-m\frac{d^2(\delta + u)}{dt^2} - c\frac{d\delta}{dt} - k\delta = 0 \tag{12.1}$$

Equation (12.1) may be rearranged:

$$m\frac{d^2\delta}{dt^2} + c\frac{d\delta}{dt} + k\delta = -m\frac{d^2u}{dt^2} \tag{12.2}$$

Assume that the motion u is sinusoidal, $u = u_0 \cos \omega t$, where $\omega = 2\pi f$ is the angular frequency in radians per second and f is expressed in cycles per second. Neglecting transient terms, the response of the instrument is defined by $\delta = \delta_0 \cos(\omega t - \theta)$; then the solution of Eq. (12.2) is

$$\frac{\delta_0}{u_0} = \frac{\omega^2}{\sqrt{\left(\frac{k}{m} - \omega^2\right)^2 + \left(\omega\frac{c}{m}\right)^2}} \tag{12.3}$$

$$\theta = \tan^{-1}\frac{\omega\frac{c}{m}}{\frac{k}{m} - \omega^2} \tag{12.4}$$

The undamped natural frequency f_n of the instrument is the frequency at which

$$\frac{\delta_0}{u_0} = \infty$$

when the damping is zero ($c = 0$), or the frequency at which $\theta = 90°$. From Eqs. (12.3) and (12.4), this occurs when the denominators are zero:

$$\omega_n = 2\pi f_n = \sqrt{\frac{k}{m}} \quad \text{rad/sec} \tag{12.5}$$

Thus, a stiff spring and/or light mass produces an instrument with a high natural frequency. A heavy mass and/or compliant spring produces an instrument with a low natural frequency.

The damping in a transducer is specified as a *fraction of critical damping*. Critical damping c_c is the minimum level of damping that prevents a mass-spring transducer from oscillating when excited by a step function or other transient. It is defined by

$$c_c = 2\sqrt{km} \tag{12.6}$$

Thus, the fraction of critical damping ζ is

$$\zeta = \frac{c}{c_c} = \frac{c}{2\sqrt{km}} \tag{12.7}$$

It is convenient to define the excitation frequency ω for a transducer in terms of the undamped natural frequency ω_n by using the dimensionless frequency ratio ω/ω_n. Substituting this ratio and the relation defined by Eq. (12.7), Eqs. (12.3) and (12.4) may be written:

$$\frac{\delta_0}{u_0} = \frac{\left(\frac{\omega}{\omega_n}\right)^2}{\sqrt{\left[1 - \left(\frac{\omega}{\omega_n}\right)^2\right]^2 + \left(2\zeta\frac{\omega}{\omega_n}\right)^2}} \tag{12.8}$$

$$\theta = \tan^{-1}\frac{2\zeta\frac{\omega}{\omega_n}}{1 - \left(\frac{\omega}{\omega_n}\right)^2} \tag{12.9}$$

The response of the mass-spring transducer given by Eq. (12.8) may be expressed in terms of the acceleration $\ddot{u}$ of the moving part by substituting $\ddot{u}_0 = -u_0\omega^2$. Then the ratio of the relative displacement amplitude δ_0 between the mass m and transducer case a to the impressed acceleration amplitude $\ddot{u}_0$ is

$$\frac{\delta_0}{u_0} = -\frac{1}{\omega_n^2}\left[\frac{1}{\sqrt{\left[1-\left(\dfrac{\omega}{\omega_n}\right)^2\right]^2 + \left(2\zeta\dfrac{\omega}{\omega_n}\right)^2}}\right] \tag{12.10}$$

The relation between δ_0/u_0 and the frequency ratio ω/ω_n is shown graphically in Fig. 12.2 for several values of the fraction of critical damping ζ. Corresponding curves for $\delta_0/\ddot{u}_0$ are shown in Fig. 12.3. The phase angle θ defined by Eq. (12.9) is shown graphically in Fig. 12.4, using the scale at the left side of the figure. Corresponding phase angles between the relative displacement δ, and the velocity $\dot{u}$ and acceleration $\ddot{u}$, are indicated by the scales at the right side of the figure.

FIG. 12.2. Displacement response δ_0/u_0 of a mass-spring system subjected to a sinusoidal displacement $\ddot{u} = u_0 \sin \omega t$. The fraction of critical damping ζ is indicated for each curve.

ACCELERATION-MEASURING TRANSDUCERS

As indicated in Fig. 12.3, the relative displacement amplitude δ_0 is directly proportional to the acceleration amplitude $\ddot{u}_0 = -u_0\omega^2$ of the sinusoidal vibration being measured, at small values of the frequency ratio ω/ω_n. Thus, when the natural frequency ω_n of the transducer is high, the transducer is an accelerometer. If the transducer is undamped, the response curve of Fig. 12.3 is substantially flat when $\omega/\omega_n < 0.2$, approximately. Consequently, an undamped accelerometer can be used for the measurement of acceleration when the vibration frequency does not exceed approximately 20 per cent of the natural frequency of the accelerometer. The range of measurable frequency increases as the damping of the accelerometer is increased, up to an optimum value of damping. When the fraction of critical damping is approximately 0.65, an accelerometer gives accurate

FIG. 12.3. Relationship between the relative displacement amplitude δ_0 of a mass-spring system and the acceleration amplitude $\ddot{u}_0$ of the case. The fraction of critical damping ζ is indicated for each response curve.

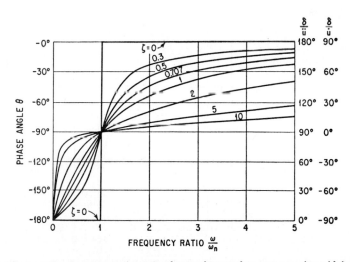

FIG. 12.4. Phase angle of a mass-spring transducer when used to measure sinusoidal vibration. The phase angle θ on the left-hand scale relates the relative displacement δ to the impressed displacement, as defined by Eq. (12.9). The right-hand scales relate the relative displacement δ to the impressed velocity and acceleration.

results in the measurement of vibration at frequencies as great as approximately 60 per cent of the natural frequency of the accelerometer.

As indicated in Fig. 12.3, the useful frequency range of an accelerometer increases as its natural frequency ω_n increases. However, the deflection of the spring in an accelerometer is inversely proportional to the square of the natural frequency; i.e., for a given value of $\ddot{u}_0$, the relative displacement is directly proportional to $1/\omega_n^2$ [see Eq. (12.10)]. As a consequence, the electrical signal from the transducing element may be very small, thereby requiring a large amplification to increase the signal to a level at which recording is feasible. For this reason, a compromise usually is made between high sensitivity and the highest attainable natural frequency, depending upon the desired application.

ACCELEROMETER REQUIREMENTS FOR SHOCK

HIGH-FREQUENCY RESPONSE. The capability of an accelerometer to measure shock may be evaluated by observing the response of the accelerometer to acceleration pulses. Ideally, the response of the accelerometer (i.e., the output of the transducing element) should correspond identically with the pulse. In general, this result may be approached but not attained exactly. Three typical pulses and the corresponding responses of accelerometers are shown in Fig. 12.5 to 12.7. The pulses are shown in dashed lines. A sinusoidal pulse is shown in Fig. 12.5, a triangular pulse in Fig. 12.6, and a rectangular pulse in Fig. 12.7. Curves of the response of the accelerometer are shown in solid lines. For each of the three pulse shapes, the response is given for ratios τ_n/τ of 1.014, 0.338, and 0.203, where τ is the pulse duration and $\tau_n = 1/f_n$ is the natural period of the accelerometer. These response curves, computed for the fraction of critical damping $\zeta = 0$, 0.4, 0.7, and 1.0, indicate the following general relationships:

1. The response of the accelerometer follows the pulse most faithfully when the natural period of the accelerometer is smallest relative to the period of the pulse. For example, the responses at A in Figs. 12.5 to 12.7 show considerable deviation between the pulse and the response; this occurs when τ_n is approximately equal to τ. However, when τ_n is small

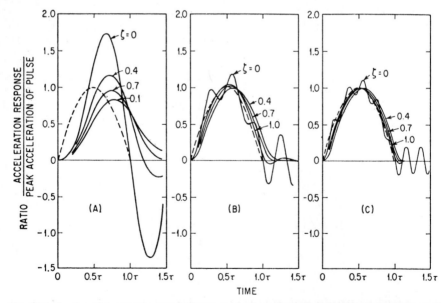

FIG. 12.5. Acceleration response to a half-sine pulse of acceleration of duration τ (dashed curve) of a mass-spring transducer whose natural period τ_n is equal to: (A) 1.014 times the duration of the pulse, (B) 0.338 times the duration of the pulse, and (C) 0.203 times the duration of the pulse. The fraction of critical damping ζ is indicated for each response curve. (*Levy and Kroll.*[1])

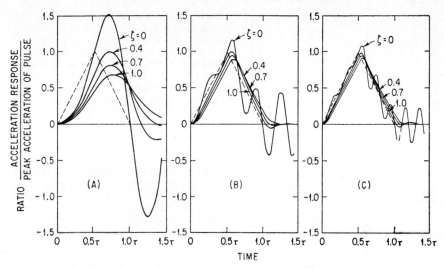

Fig. 12.6. Acceleration response to a triangular pulse of acceleration of duration τ (dashed curve) of a mass-spring transducer whose natural period is equal to: (A) 1.014 times the duration of the pulse, (B) 0.338 times the duration of the pulse, and (C) 0.203 times the duration of the pulse. The fraction of critical damping ζ is indicated for each response curve. (*Levy and Kroll.*[1])

relative to τ (Figs. 12.5C to 12.7C), the deviation between the pulse and the response is much smaller.

2. Damping in the transducer reduces the response of the transducer at its own natural frequency; i.e., it reduces the transient vibration superimposed upon the pulse. Damping also reduces the maximum value of the response to a value lower than the actual pulse in the case of large damping. For example, in some cases a fraction of critical damping $\zeta = 0.7$ provides an instrument response that does not reach the peak value of the acceleration pulse.

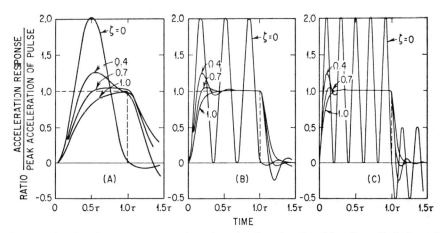

Fig. 12.7. Acceleration response to a rectangular pulse of acceleration of duration τ (dashed curve) of a mass-spring transducer whose natural period τ_n is equal to: (A) 1.014 times the duration of the pulse, (B) 0.338 times the duration of the pulse, and (C) 0.203 times the duration of the pulse. The fraction of critical damping ζ is indicated for each response curve. (*Levy and Kroll.*[1])

FIG. 12.8. Response of a typical instrument for measuring shock, illustrating drop-off at low frequencies. The low-frequency cutoff point f_c is the frequency at which the response is down 3 dB or 30 per cent below the flat portion of the response curve.

LOW-FREQUENCY RESPONSE. The measurement of shock requires that the accelerometer and its associated equipment have good response at low frequencies because pulses and other types of shock motions characteristically include low-frequency components. Such pulses can be measured accurately only with an instrumentation system whose response is flat down to the lowest frequency of the spectrum; in general, this lowest frequency is zero for pulses.

The response of an instrumentation system is defined by a plot of output voltage vs. excitation frequency; the response of a typical system is shown in Fig. 12.8. For purposes of shock measurement, the decrease in response at low frequencies is significant. The decrease is defined quantitatively by the frequency f_c at which the response is down 3 dB or approximately 30 per cent below the flat response which exists at the higher frequencies. The distortion which occurs in the measurement of a pulse is related to the frequency f_c as illustrated in Fig. 12.9.

IMPORTANT CHARACTERISTICS OF ACCELEROMETERS

SENSITIVITY

The *sensitivity* of a shock- and vibration-measuring instrument is the ratio of its electrical output to its mechanical input. The output usually is expressed in terms of voltage per unit of displacement, velocity, or acceleration. This specification of sensitivity is sufficient for instruments which generate their own voltage independent of an external voltage power source. However, the sensitivity of an instrument requiring an external voltage usually is specified in terms of output voltage per unit of voltage supplied to the instrument per unit of displacement, velocity, or acceleration, e.g., millivolts per volt per g of acceleration. It is important to note the terms in which the respective parameters are expressed, e.g., average, rms, or peak. The relation between these terms is shown in Fig. 12.10. Also see Table 1.3.

FIG. 12.9. Response of an accelerometer to a half-sine acceleration pulse for RC time constants equal to τ, 5τ, 10τ, 50τ, and ∞, where τ is equal to the duration of the half-sine pulse.[1]

RESOLUTION

The *resolution* of a transducer is the smallest change in mechanical input (e.g., acceleration) for which a change in the electrical output is discernible. The resolution of an accelerometer is a function of the transducing element and the mechanical design.

Recording equipment, indicating equipment, and other auxiliary equipment used with accelerometers often establish the resolution of the overall measurement system. If the electrical output of an instrument is indicated by a meter, the resolution may be established by the smallest increment that can be read from the meter. Resolution can be limited by noise levels in the instrument or in the system. In general, any signal change smaller than the noise level will be obscured by the noise, thus determining the resolution of the system.

FIG. 12.10. Relationships between average, rms, peak, and peak-to-peak values for a simple sine wave. These values are used in specifying sensitivities of shock and vibration transducers (e.g., peak millivolts per peak g, or rms millivolts per peak-to-peak displacement). These relationships do not hold true for other than simple sine waves.

TRANSVERSE SENSITIVITY

If a transducer is subjected to vibration of unit amplitude along its axis of maximum sensitivity, the amplitude of the voltage output e_{max} is the sensitivity. The sensitivity e_θ along the X axis, inclined at an angle θ to the axis of e_{max}, is $e_\theta = e_{max} \cos \theta$, as illustrated in Fig. 12.11. Similarly, the sensitivity along the Y axis is $e_t = e_{max} \sin \theta$. In general, the sensitive axis of a transducer is designated. Ideally, the X axis would be designated the sensitive axis, and the angle θ would be zero. Practically, θ can be made only to approach zero because of manufacturing tolerances and/or unpredictable variations in the characteristics of the transducing element. Then the transverse sensitivity is expressed as the tangent of the angle, i.e., the ratio of e_t to e_θ:

$$\frac{e_t}{e_\theta} = \tan \theta \tag{12.11}$$

In practice, $\tan \theta$ is between 0.01 and 0.05 and is expressed as a percentage. For example, if $\tan \theta = 0.05$, the transducer is said to have a transverse sensitivity of 5 per cent. Figure 12.12 is a typical polar plot of transverse sensitivity.

AMPLITUDE LINEARITY AND LIMITS

When the ratio of the electrical output of a transducer to the mechanical input (i.e., the sensitivity) remains constant within specified limits, the transducer is said to be "linear" within those limits, as illustrated in Fig. 12.13. A transducer is linear only over a certain range of amplitude values. The lower end of this range is determined by the electrical noise of the measurement system.

The upper limit of linearity may be imposed by the electrical characteristics of the transducing element and by the size or the fragility of the instrument. Generally, the greater the sensitivity of a transducer, the more nonlinear it will be. Similarly, for very large acceleration values, the large forces produced by the spring of the mass-spring system may exceed the yield strength of a part of the instrument, causing nonlinear behavior or complete failure.

FREQUENCY RANGE

The operating frequency range is the range over which the sensitivity of the transducer does not vary more than a stated percentage from the rated sensitivity. This range may be limited by the electrical or mechanical characteristics of the transducer or by its associated auxiliary equipment. These limits can be added to amplitude linearity limits to define completely the operating ranges of the instrument, as illustrated in Fig. 12.14.

LOW-FREQUENCY LIMIT. The mechanical response of a mass-spring transducer does not impose a low-frequency limit for an acceleration transducer because the transducer responds to vibration with frequencies less than the natural frequency of the transducer.

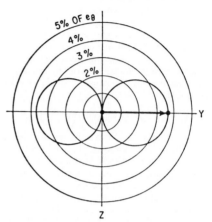

FIG. 12.11. The designated sensitivity e_θ and cross-axis sensitivity e_t that result when the axis of maximum sensitivity $e_{\max}$ is not aligned with the axis of e_θ.

FIG. 12.12. Plot of transducer sensitivity in all axes normal to the designated axis e_θ plotted according to axes shown in Fig. 12.11. Cross-axis sensitivity reaches a maximum e_t along the Y axis and a minimum value along the Z axis.

In evaluating the low-frequency limit, it is necessary to consider the electrical characteristics of both the transducer and the associated equipment. In general, a transducing element that utilizes external power or a carrier voltage does not have a lower frequency limit, whereas a self-generating transducing element is not operative at zero frequency. The frequency response of amplifiers and other circuit components may limit the lowest usable frequency of an instrumentation system.

HIGH-FREQUENCY LIMIT. An acceleration transducer (accelerometer) has an upper usable frequency limit because it responds to vibration whose frequency is less than the natural frequency of the transducer. The limit is a function of (1) the natural frequency and (2) the damping of the transducer, as discussed with reference to Fig. 12.3. An attempt to use such a transducer beyond this frequency limit may result in distortion of the signal, as illustrated in Fig. 12.15.

The upper frequency limit for slightly damped vibration-measuring instruments is important because these instruments exaggerate the small amounts of harmonic con-

FIG. 12.13. Typical plot of sensitivity as a function of amplitude for a shock and vibration transducer. The *linear range* is established by the intersection of the sensitivity curve and the specified limits (dashed lines).

tent that may be contained in the motion, even when the operating frequency is well within the operating range of the instrument. The result of exciting an undamped instrument at its natural frequency may either damage the instrument or obscure the desired measurement. Figure 12.15 shows how a small amount of harmonic distortion in the vibratory motion may be exaggerated by an undamped transducer.

PHASE SHIFT. Phase shift is the time delay between the mechanical input and the electrical output signal of the instrumentation system. Unless the phase-shift characteristics of an instrumentation system meet certain requirements, a distortion may be introduced that consists of the superposition of vibration at several different frequencies. Consider first an accelerometer, for which the phase angle θ_1 is given by Fig. 12.4. If the accelerometer is undamped, $\theta_1 = 0$ for values of ω/ω_n less than 1.0; thus, the phase of the relative displacement δ is equal to that of the acceleration being measured, for all values of frequency within the useful range of the accelerometer. Therefore, an undamped accelerometer measures acceleration without distortion of phase. If the fraction of critical damping ζ for the accelerometer is 0.65, the phase angle θ_1 increases approximately linearly with the frequency ratio ω/ω_n within the useful frequency range of the accelerometer. Then the expression for the relative displacement may be written

$$\delta = \delta_0 \cos{(\omega t - \theta)} = \delta_0 \cos{(\omega t - a\omega)} = \delta_0 \cos{\omega\,(t - a)} \qquad (12.12)$$

where a is a constant. Thus, the relative motion δ of the instrument is displaced in phase relative to the acceleration $\ddot{u}$ being measured; however, the increment along the time axis is a constant independent of frequency. Consequently, the waveform of the accelerom-

eter output is undistorted but is delayed with respect to the waveform of the vibration being measured. As indicated by Fig. 12.4, any value of damping in an accelerometer other than $\zeta = 0$ or $\zeta = 0.65$ (approximately) results in a nonlinear shift of phase with frequency and a consequent distortion of the waveform.

ENVIRONMENTAL EFFECTS

TEMPERATURE. The sensitivity, natural frequency, and damping of a transducer may be affected by temperature. The specific effects produced depend on the type of transducer and the details of its design. The sensitivity may increase or decrease with temperature, or remain relatively constant. Figure 12.16 shows the variation of damping with temperature for several different damping media. Either of two methods may be employed to compensate for temperature effects.

1. The temperature of the pickup may be held constant by local heating or cooling.

2. The pickup characteristics may be measured as a function of temperature; if necessary, the appropriate corrections can then be applied to the measured data.

Fig. 12.14. Linear operating range of a transducer. Amplitude linearity limits are shown as a combination of displacement and acceleration values. The lower amplitude limits usually are expressed in acceleration values as shown.

HUMIDITY. Humidity may affect the characteristics of certain types of vibration instruments. In general, a transducer which operates at a high electrical impedance is affected by humidity more than a transducer which operates at a low electrical impedance. It usually is impractical to correct the measured data for humidity effects. However, instruments that might otherwise be adversely affected by humidity often are sealed hermetically to protect them from the effects of moisture.

ACOUSTIC NOISE.   Sound waves of high intensity often accompany vibration. If the structure of the transducer or auxiliary equipment can be set into vibration by acoustic excitation, large error signals may result.   Sometimes the vibration to be measured is the source of the acoustic energy; in other cases, the acoustic energy induces the vibration to be measured. Only in the latter case is the ratio of the acoustic energy to the vibration energy high enough to produce significant errors.   In general, an accelerometer subjected to acoustic noise does not produce an electrical output equivalent to the voltage produced by an acceleration of $1g$ until the sound pressure level of the acoustic noise exceeds 150 dB.[3] With such a high sound pressure level it is likely that the vibration level is large and the error introduced by the acoustic noise is not apt to be important.   Where it is necessary to measure vibration of small magnitude in a high-intensity acoustic field, the response of the accelerometer to the acoustic field alone should be measured to determine the effect of the noise on the measurement of vibration.   In performing such a test, the transducer must be mounted so that its mounting will not be set into vibration by the acoustic field.

Fig. 12.15. Distorted response (solid line) of a lightly damped ($\zeta < 0.1$) mass-spring accelerometer to vibration (dashed line) containing a small harmonic content of the small frequency as the natural frequency of the accelerometer.

PHYSICAL PROPERTIES

Size and weight of the transducer are very important considerations in many vibration and shock measurements.   A large instrument may require a mounting structure that will change the local vibration characteristics of the structure whose vibration is being measured.   Similarly, the added mass of the transducer may also produce substantial changes in the vibratory response of such a structure.   Generally, the natural frequency of a structure is lowered by the addition of mass; specifically, for a simple spring-mass structure:

$$\frac{f_n - \Delta f_n}{f_n} = \sqrt{\frac{m}{m + \Delta m}} \qquad (12.13)$$

where f_n = natural frequency of structure
Δf_n = change in natural frequency
m = mass of structure
Δm = increase in mass resulting from addition of transducer

In general, for a given type of transducing element, the sensitivity increases approximately in proportion to the mass of the transducer.   In most applications, it is more important that the transducer be small in size than that it have high sensitivity because amplification of the signal increases the output to a usable level.

Mass-spring-type transducers for the measurement of displacement usually are larger and heavier than similar transducers for the measurement of acceleration.   In the former, the mass must remain substantially stationary in space while the instrument case moves about it; this requirement does not exist with the latter.

For the measurement of shock and vibration in aircraft or missiles, the size and weight of not only the transducer but also the auxiliary equipment are important.   In these applications, self-generating instruments that require no external power may have a significant advantage.

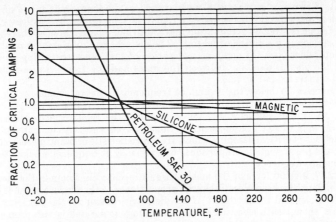

FIG. 12.16. Variation of damping with temperature for different damping means. The ordinate indicates the fraction of critical damping ζ at various temperatures assuming $\zeta = 1$ at 70°F (21°C).

PIEZOELECTRIC ACCELEROMETERS

PRINCIPLE OF OPERATION

An accelerometer of the type shown in Fig. 12.17A is a linear seismic transducer utilizing a piezoelectric element in such a way that an electric charge is produced which is proportional to the applied acceleration. This "ideal" seismic piezoelectric transducer can be represented (over most of its frequency range) by the elements shown in Fig. 12.17B. A mass is supported on a linear spring which is fastened to the frame of the

FIG. 12.17. (A) Schematic diagram of a linear seismic piezoelectric accelerometer. (B) A simplified representation of the accelerometer shown in (A) which applies over most of the useful frequency range. A mass m rests on the piezoelectric element which acts as a spring having a spring constant k. The damping in the system, represented by the dashpot, has a damping coefficient c. (C) The frame is accelerated upward, producing a displacement u of the frame, moving the mass from its initial position by an amount x, and compressing the spring by an amount δ.

instrument. The piezoelectric crystal which produces the charge acts as the spring. Viscous damping between the mass and the frame is represented by dashpot c. In Fig. 12.17C the frame is given an acceleration upward to a displacement of u, thereby producing a compression in the spring equal to δ. The displacement of the mass relative to the frame is dependent upon the applied acceleration of the frame, the spring stiffness, the mass, and the viscous damping between the mass and the frame, as indicated in Eq. (12.10) and illustrated in Fig. 12.3.

For frequencies far below the resonance frequency of the mass and spring, this displacement is directly proportional to the acceleration of the frame and is independent of frequency. At low frequencies, the phase angle of the relative displacement δ, with respect to the applied acceleration, is proportional to frequency. As indicated in Fig. 12.4, for low fractions of critical damping which are characteristic of many piezoelectric accelerometers, the phase angle is proportional to frequency at frequencies below 30 per cent of the resonance frequency.

In Fig. 12.17, inertial force of the mass causes a mechanical strain in the piezoelectric element which produces an electric charge proportional to the stress and, hence, proportional to the strain and acceleration.[4] If the dielectric constant of the piezoelectric material does not change with electric charge, the voltage generated is also proportional to acceleration. Metallic electrodes are applied to the piezoelectric element, and electrical leads are connected to the electrodes for measurement of the electrical output of the piezoelectric element.

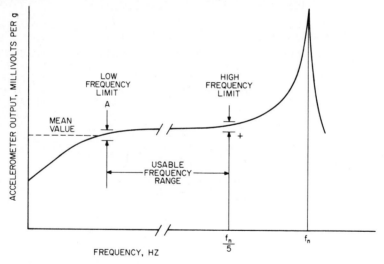

FIG. 12.18. Typical response curve for a piezoelectric accelerometer. The resonance frequency is denoted by f_n. The useful range depends on the acceptable deviation from the mean value of the response over the "flat" portion of the response curve.

In the ideal seismic system shown in Fig. 12.17, the mass and the frame have infinite stiffness, the spring has zero mass, and viscous damping exists only between the mass and the frame. In practical piezoelectric accelerometers, these assumptions cannot be fulfilled. For example, the mass may have as much compliance as the piezoelectric element. In some seismic elements, the mass and spring are inherently a single structure. Furthermore, in many practical designs where the frame is used to hold the mass and piezoelectric element, distortion of the frame may produce mechanical forces upon the seismic element.[5] All these factors may change the performance of the seismic system from those calculated using equations based on an ideal system. In particular, the resonance frequency of the piezoelectric combination may be substantially lower than that indicated by theory. Nevertheless, the equations for an ideal system are useful both in design and application of piezoelectric accelerometers.

Figure 12.18 shows a typical frequency response curve for a piezoelectric accelerometer. In this illustration, the electrical output in millivolts per g acceleration is plotted as a function of frequency. The resonance frequency is denoted by f_n. If the accelerometer is properly mounted on the device being tested, then the upper frequency limit of the useful frequency range usually is taken to be $f_n/3$ for a deviation of 12 per cent (1 dB) from the mean value of the response. For a deviation of 6 per cent (0.5 dB) from the mean value, the upper frequency limit usually is taken to be $f_n/5$. As indicated in Fig. 12.1, the type of mounting can have a significant effect on the value of f_n.

The decrease in response at low frequencies (i.e., the "roll-off") depends primarily on the characteristics of the preamplifier that follows the accelerometer. The low-frequency limit also is usually expressed in terms of the deviation from the mean value of the response over the flat portion of the response curve, being the frequency at which the response is either 12 per cent (1 dB) or 6 per cent (0.5 dB) below the mean value.

TYPICAL ACCELEROMETER CONSTRUCTIONS

Piezoelectric accelerometers utilize a variety of seismic element configurations. Most are constructed of polycrystalline ceramic piezoelectric materials because of their ease of manufacture, high piezoelectric sensitivity, and excellent time and temperature stability. These seismic devices may be classified in two modes of operation: compression- or shear-type accelerometers.

COMPRESSION-TYPE ACCELEROMETER. The compression-type seismic accelerometer, in its simplest form, consists of a piezoelectric disc and a mass placed on a frame as shown in Fig. 12.17. Motion in the direction indicated causes compressive (or tensile) forces to act on the piezoelectric element, producing an electrical output proportional to acceleration. In this example, the mass is cemented with a conductive material to the piezoelectric element which, in turn, is cemented to the frame. The components must be cemented firmly so as to avoid being separated from each other by the applied acceleration.

In the unit shown in Fig. 12.19, the mass is held in place by means of a stud extending from the frame through the ceramic. Accelerometers of this design often use quartz, tourmaline, or ferroelectric ceramics as the sensing material. Typical characteristics are a sensitivity of 2 to 100 picocoulombs/g, a resonance frequency of 30,000 Hz, and a total weight of 30 gm.

Typical commercial accelerometers are shown in Fig. 12.20. The unit shown in Fig. 12.20B consists of a housing, two piezoelectric ceramic elements, a seismic mass, and a preloading spring. The resonance frequency of the transducer is always lower than that of the mass-ceramic combination alone. This type of accelerometer must be attached to the structure with care in order to minimize distortion of the housing and base which can cause an electrical output.

FIG. 12.19. NRL-type piezoelectric accelerometer. This pickup has a sensitivity of 100 millivolts/ g, a resonance frequency of 14,000 Hz, and a useful acceleration up to 5,000g. *(Courtesy of the U.S. Naval Research Laboratory.)*

The temperature characteristics of compression-type accelerometers have been improved greatly in recent years; it is now possible to measure acceleration over a temperature range of -425 to $+1400°F$ (-254 to $+760°C$). This wider range has been primarily a result of the use of two piezoelectric materials: tourmaline and lithium niobate.

SHEAR-TYPE ACCELEROMETERS. One shear-type accelerometer, illustrated in Fig. 12.21, employs a cylindrically shaped piezoelectric element fitted around a middle mounting post; a loading ring (or mass) is cemented to the outer diameter of the piezoelectric element. The cylinder is made of ceramic and is polarized along its length; the output voltage of the accelerometer is taken from its inner and outer walls. This type of design is generally known as an axially poled shear-mode annular accelerometer.

Another shear-type accelerometer utilizes flat-plate shear-sensing elements. Manufacturers preload these against a flattened post element in several ways. One method is shown in Fig. 12.20A. Accelerometers of this style have low cross-axis response, excellent temperature characteristics, and negligible output from strain sensitivity or base bending. The temperature range of the bolted shear design can be from -425 to $+1400°F$ (-254 to $+760°C$). The following are typical specifications: sensitivity, 10 to 500 picocoulombs/g; acceleration range, 1 to 500g; resonance frequency, 25,000 Hz; useful frequency range, 3 to 5,000 Hz; temperature range, -425 to $+1400°F$ (-254 to 760°C); transverse response, 3 per cent.

FIG. 12.20. Piezoelectric accelerometers of the (A) delta-shear type and (B) compression type. *(Courtesy of Bruel & Kjaer.)*

FIG. 12.21. An annular shear accelerometer. The piezoelectric element is cemented to the post and mass. Electrical connections (not shown) are made to the inner and outer diameters of the piezoelectric element. *(Courtesy of Endevco Corp.)*

PHYSICAL CHARACTERISTICS OF PIEZOELECTRIC ACCELEROMETERS

SHAPE, SIZE, AND WEIGHT. Commercially available piezoelectric accelerometers usually are cylindrical in shape. They are available with both attached and detachable mounting studs at the bottom of the cylinder. A coaxial cable connector is provided at either the top or side of the housing.

Most commercially available piezoelectric accelerometers are relatively light in weight, ranging from approximately 0.03 to 2 oz (1 to 60 gm). Usually, the larger the accelerometer, the higher its sensitivity and the lower its resonance frequency. The smallest units have a diameter of less than about 0.2 in. (5 mm); the larger units have a diameter of about 1 in. (25.4 mm) and a height of about 1 in. (25.4 mm).

RESONANCE FREQUENCY. The highest fundamental resonance frequency of an accelerometer may be above 100,000 Hz. The higher the resonance frequency, the lower will be the sensitivity and the more difficult it will be to provide mechanical damping.

DAMPING. The *amplification ratio* of an accelerometer is defined as the ratio of the sensitivity at its resonance frequency to the sensitivity in the frequency band in which sensitivity is independent of frequency. This ratio depends on the amount of damping in the seismic system; it decreases with increasing damping. Most piezoelectric accelerometers are essentially undamped, having amplification ratios between 5 and 50, or a fraction of critical damping less than 0.1. The frequency response of a typical piezoelectric accelerometer is shown in Fig. 12.22.

FIG. 12.22. The frequency response of a typical accelerometer is shown with damping (solid curve) and without damping (dashed curve). Damping with a fluid having a kinematic viscosity of 100,000 centistokes greatly reduces the amplification ratio and shifts the undamped resonance frequency to a slightly lower value.

ELECTRICAL CHARACTERISTICS OF PIEZOELECTRIC ACCELEROMETERS

DEPENDENCE OF VOLTAGE SENSITIVITY ON SHUNT CAPACITANCE. The *sensitivity* of an accelerometer is defined as the electrical output per unit of applied acceleration. The sensitivity of a piezoelectric accelerometer can be expressed either as a *charge sensitivity* or *voltage sensitivity*. Charge sensitivity usually is expressed in units of coulombs generated per g of applied acceleration; voltage sensitivity usually is expressed in volts per g (where g is the acceleration of gravity). Voltage sensitivity often is expressed as open-circuit voltage sensitivity, i.e., in terms of the voltage produced across the electrical terminals per unit acceleration when the electrical load impedance is infinitely high. Open-circuit voltage sensitivity may be given either with or without the connecting cable.

An electrical capacitance often is placed across the output terminals of a piezoelectric transducer. This added capacitance (called *shunt capacitance*) may result from the connection of an electrical cable between the pickup and other electrical equipment (all electrical cables exhibit interlead capacitance). The effect of shunt capacitance in reducing the sensitivity of a pickup is shown in Fig. 12.23.

The charge equivalent circuits, with shunt capacitance C_S, are shown in Fig. 12.23A. The charge sensitivity is not changed by addition of shunt capacitance. The total capacitance C_T of the pickup including shunt is given by

$$C_T - C_E + C_S \tag{12.14}$$

where C_E is the capacitance of the transducer without shunt capacitance.

The voltage equivalent circuits are shown in Fig. 12.23B. With the shunt capacitance C_S, the total capacitance is given by Eq. (12.14) and the open-circuit voltage sensitivity is given by

$$\frac{e_s}{\ddot{x}} = \frac{q_s}{\ddot{x}} \frac{1}{C_E + C_S} \tag{12.15}$$

where $q_s/\ddot{x}$ is the charge sensitivity. The voltage sensitivity without shunt capacitance is given by

$$\frac{e}{\ddot{x}} = \frac{q_s}{\ddot{x}} \frac{1}{C_E} \tag{12.16}$$

Therefore, the effect of the shunt capacitance is to reduce the voltage sensitivity by a factor

$$\frac{e_s/\ddot{x}}{e/\ddot{x}} = \frac{C_E}{C_E + C_S} \tag{12.17}$$

Piezoelectric accelerometers are used with both voltage-sensing and charge-sensing signal conditioners, although charge sensing is by far the most common because the sensitivity does not change with external capacitance (up to a limit). These factors are discussed in Chap. 13. In addition, electronic circuitry can be placed within the case of the accelerometer, as discussed below.

LOW-IMPEDANCE PIEZOELECTRIC ACCELEROMETERS CONTAINING INTERNAL ELECTRONICS

Piezoelectric accelerometers are available with simple electronic circuits internal to their cases to provide signal amplification and low-impedance output. Some designs operate from low-current dc voltage supplies and are designed to be intrinsically safe when coupled by appropriate barrier circuits. Other designs have common power and signal lines and use coaxial cables.

The principal advantages of piezoelectric accelerometers with integral electronics are that they are relatively immune to cable-induced noise and spurious response, they can be used with lower-cost cable, and they have a lower signal conditioning cost. In the simplest case the power supply might consist of a battery, a resistor, and a capacitor. Chapter 14 describes such accelerometers in which the integral electronics provide a velocity or displacement output. These advantages do not come without compromise. Because the impedance-matching circuitry is built into the transducer, gain cannot be adjusted to utilize the wide dynamic range of the basic transducer. Ambient temperature is limited to that which the circuit will withstand, and this is considerably lower than that of the piezoelectric sensor itself. In order to retain the advantages of

Fɪɢ. 12.23. Equivalent circuits which include shunt capacitance across a piezoelectric pickup. (*A*) Charge equivalent circuit. (*B*) Voltage equivalent circuit.

small size, the integral electronics must be kept relatively simple. This precludes the use of multiple filtering and dynamic overload protection and thus limits their application. But when conditions are relatively benign, these accelerometers can economically provide excellent noise immunity and signal fidelity.

ACCELERATION-AMPLITUDE CHARACTERISTICS

AMPLITUDE RANGE. Piezoelectric accelerometers are generally useful for the measurement of acceleration of magnitudes of from $10^{-4}g$ to more than $10^4 g$. The lowest value of acceleration which can be measured is approximately that which will produce an output voltage equivalent to the electrical input noise of the coupling amplifier connected to the accelerometer when the pickup is at rest. Over its useful operating range, the output of a piezoelectric accelerometer is directly and continuously proportional to the input acceleration. A single accelerometer often can be used to provide measurements over a dynamic amplitude range of 10,000 to 1 (80 dB) or more, which is substantially greater than the dynamic range of most of the associated transmission, recording, and analysis equipment. Commerical accelerometers generally exhibit excellent linearity of electrical output vs. input acceleration under normal usage.

At very high values of acceleration (depending upon the design characteristics of the particular transducer), nonlinearity or damage may occur. For example, large dynamic forces may produce electrical outputs sufficient to reduce permanently the sensitivity of the piezoelectric material. Further, if the dynamic forces exceed the biasing or clamping forces, the seismic element may "chatter" or fracture, although such a fracture might not be observed in subsequent low-level acceleration calibrations. High dynamic accelerations also may cause a slight physical shift in position of the piezoelectric element in the accelerometer—sometimes sufficient to cause a change in sensitivity. The upper limit of acceleration measurements depends upon the specific design and construction details of the pickup and may vary considerably from one accelerometer to another, even though the design is the same. It is not always possible to calculate the upper acceleration limit of a pickup. Therefore one cannot assume linearity of acceleration levels for which calibration data cannot be obtained.

EFFECTS OF TEMPERATURE

TEMPERATURE RANGE. Piezoelectric accelerometers are available which may be used in the temperature range from $-425°F$ ($-254°C$) to above $+1400°F$ ($+760°C$) without the aid of external cooling. The voltage sensitivity, charge sensitivity, capacitance, and frequency response depend upon the ambient temperature of the transducer. This temperature dependence is due primarily to variations in the characteristics of the piezoelectric material, but it also may be due to variations in the insulation resistance of cables and connectors—especially at high temperatures.

EFFECTS OF TEMPERATURE ON CHARGE SENSITIVITY. The charge sensitivity of a piezoelectric accelerometer is directly proportional to the d piezoelectric constant of the material used in the piezoelectric element. The d constants of most piezoelectric materials vary with temperature.

EFFECTS OF TEMPERATURE ON VOLTAGE SENSITIVITY. The open-circuit voltage sensitivity of an accelerometer is the ratio of its charge sensitivity to its total capacitance ($C_S + C_E$). Hence, the temperature variation in voltage sensitivity depends on the temperature dependence of both charge sensitivity and capacitance. The voltage sensitivity of most piezoelectric accelerometers decreases with temperature.

EFFECTS OF TRANSIENT TEMPERATURE. The voltage sensitivity of a piezoelectric transducer is not affected by the rate of change of the temperature of the instrument. However, electrical voltages (as large as 10 volts) due to a temperature fluctuation may be produced at the output terminals of the transducer in the following ways.

1. Differential expansion or compression of the mechanical and piezoelectric elements of the transducer may produce slowly varying forces on the piezoelectric element, and hence produce an output.

2. Many piezoelectric materials exhibit a pyroelectric effect, which is the generation of electric charge when the temperature of the piezoelectric material is changed. In general, the charge produced is proportional to the temperature change.

The voltage produced by temperature fluctuations occurs at a very slow rate and follows the temperature changes. Thermal insulation of the seismic element often is used to increase the time lag. Such pyroelectrically generated voltages may be filtered out electrically by means of a high-pass filter. Such a filter (having a cutoff frequency of the order of 5 Hz) may be incorporated in the coupling amplifier. However, if the input voltage which results from this source is sufficiently high, the operation of the amplifier may be affected even though such a filter is used.

PIEZORESISTIVE ACCELEROMETERS

PRINCIPLE OF OPERATION

A piezoresistive accelerometer differs from the piezoelectric type in that it is not self-generating. In this type of transducer a semiconductor material, usually silicon, is used as the strain-sensing element. Such a material changes its resistivity in proportion to an applied stress or strain. The equivalent electric circuit of a piezoresistive transducing element is a variable resistor. Piezoresistive elements are almost always arranged in pairs; a given acceleration places one element in tension and the other in compression. This causes the resistance of one element to increase while the resistance of the other decreases. Often two pairs are used and the four elements are connected electrically in a Wheatstone-bridge circuit, as shown in Fig. 12.24B. When only one pair is used, it forms half of a Wheatstone bridge, the other half being made up of fixed value resistors, either in the transducer or in the signal conditioning equipment. The use of transducing elements by pairs not only increases the sensitivity, but also cancels most of the errors due to temperature changes.

At one time, wire or foil strain gages were used as the transducing elements in resistive accelerometers. Now silicon elements are usually used because of their higher sensitivity. (Metallic gages made of foil or wire change their resistance with strain because the dimensions change. The resistance of a piezoresistive material changes because the material's electrical nature changes.) Sensitivity is given by the gage factor; the *gage factor* is the ratio of the fractional change in resistance to the fractional change in length that produced it. The gage factor of a typical wire or foil strain gage is approximately 2.5; the gage factor of silicon is approximately 100.

A major advantage of piezoresistive accelerometers is that they have good frequency response down to dc (0 Hz) along with a relatively good high-frequency response.

FIG. 12.24. (A) Schematic drawing of a piezoresistive accelerometer of the cantilever-beam type. (B) Four piezoresistive elements are used—two cemented to each side of the uniformly stressed beam. These elements are connected in the above bridge circuit.

DESIGN PARAMETERS

Many different configurations are possible for an accelerometer of this type. For purposes of illustration, the design parameters are considered for a piezoresistive accelerometer which has a cantilever arrangement as shown in Fig. 12.24A. This uniformly stressed cantilever beam is loaded at its end with mass m. In this arrangement, four identical piezoresistive elements are used—two cemented to each side of the beam whose length is L in. These elements, whose resistance is R, form the active arms of the balanced bridge shown in Fig. 12.24B. A change of length L of the beam produces a change in resistance R in each element. The gage factor K for each of the elements [defined by Eq. (17.1)] is

$$K = \frac{\Delta R/R}{\Delta L/L} = \frac{\Delta R/R}{\epsilon} \tag{12.18}$$

where ϵ is the strain induced in the beam, expressed in inches/inch, at the surface where the elements are cemented. If the resistances in the four arms of the bridge are equal, then the ratio of the output voltage E_o of the bridge circuit to the input voltage is E_i

$$\frac{E_o}{E_i} = \frac{\Delta R}{R} = \epsilon K \tag{12.19}$$

TYPICAL ACCELEROMETER CONSTRUCTIONS

Figure 12.25 shows three basic piezoresistive accelerometer designs which illustrate several of the many types available for various applications.

FIG. 12.25. Three basic types of piezoresistive accelerometers. (*A*) Bending-beam type. The strain elements are bonded to the beam. (*B*) Stress-concentrated type. The thin section on the neutral axis acts as a hinge; the strain energy is concentrated in the piezoresistive elements. (*C*) Micromachined type. The entire mechanism is etched from a single crystal of silicon. The thin section on the neutral axis acts as a hinge. The pedestal serves as a mounting base.

BENDING-BEAM TYPE. This design approach is described by Fig. 12.25A. The advantages of this type are simplicity and ruggedness. The disadvantage is relatively low sensitivity for a given resonance frequency. The relatively lower sensitivity results from the fact that much of the strain energy goes into the beam rather than the strain gages bonded to it.

STRESS-CONCENTRATED STOPPED AND DAMPED TYPE. To provide higher sensitivities and resonance frequencies than are possible with the bending-beam type, designs are provided which place most of the strain energy in the piezoresistive elements. This is described by Fig. 12.25B. This approach is used to provide sensitivities more suitable for the measurement of acceleration below 100g. To provide environmental shock resistance overload, stops are added. To provide wide frequency response, damping is added by surrounding the mechanism with silicone oil. The advantages of these designs are high sensitivity, broad frequency response for the sensitivity, and over-range protection. The disadvantages are complexity and limited temperature range. The high sensitivity results from the relatively large mass with the strain energy mostly coupled into the strain gages. (The thin section on the neutral axis acts as a hinge; it contributes very little stiffness.) The broad frequency response results from the relatively high damping (0.7 times critical damping), which allows the accelerometer to be used to frequencies nearer the resonance frequency without excessive increase in sensitivity. The over-range protection is provided by stops which are designed to stop the motion of the mass before it overstresses the gages. (Stops are omitted from Fig. 12.25B in the interest of clarity.) Over-range protection is almost mandatory in sensitive piezoresistive accelerometers; without it they would not survive ordinary shipping and handling. The viscosity of the damping fluid does change with temperature; as a result, the damping coefficient changes significantly with temperature. The damping is at 0.7 times critical only near room temperature.

MICROMACHINED TYPE. The entire working mechanism (mass, spring, and support) of a micromachined-type accelerometer is etched from a single crystal of silicon, a process known as *micromachining*. This produces a very tiny and rugged device, shown in Fig. 12.25C. The advantages of the micromachined type are very small size, very high resonance frequency, ruggedness, and high range. The disadvantage is low sensitivity. Accelerometers of this design are used to measure high-amplitude shock accelerations of from about 1,000 to 200,000g (1.97 Mm/s^2). The sensitivity is low and the resonance frequency is high because of the small mass. The ruggedness follows from this and from the fact that no adhesives are required to bond the strain gages to the structure.

ELECTRICAL CHARACTERISTICS OF PIEZORESISTIVE ACCELEROMETERS

EXCITATION. Piezoresistive transducers require an external power supply to provide the necessary current or voltage excitation in order to operate. These energy sources must be well regulated and stable since they may introduce sensitivity errors and secondary effects at the transducer which will result in error signals at the output.

Fig. 12.26. Loading effects on piezo-resistive accelerometers.

Traditionally, the excitation has been provided by a battery or a constant voltage supply. Other sources of excitation, such as constant current supplies or ac excitation generators, may be used. The sensitivity and temperature response of a piezoresistive transducer may depend on the kind of excitation applied. Therefore, it should be operated in a system which provides the same source of excitation as used during temperature compensation and calibration of the transducer. The most common excitation source is 10 volts dc.

SENSITIVITY. The *sensitivity* of an accelerometer is defined as the ratio of its electrical output to its mechanical input. Specifically, in the case of piezoresistive accelerometers, it is expressed as voltage per unit of acceleration at the rated excitation (i.e., mV/g or peak mV/peak g at 10 volts dc excitation).

LOADING EFFECTS. An equivalent circuit of a piezoresistive accelerometer, for use when considering loading effects, is shown in Fig. 12.26. Using the equivalent circuit and the measured output resistance of the transducer, the effect of loading may be directly calculated:

$$E_{oL} = E_o \frac{R_L}{R_o + R_L} \qquad (12.20)$$

where R_o = output resistance of accelerometer, including cable resistance
E_o = sensitivity into an infinite load
E_oL = loaded output sensitivity
R_L = load resistance

Because the resistance of the strain-gage elements varies with temperature, output resistance should be measured at the operating temperature.

Effect of Cable on Sensitivity. Long cables may result in the following effects:

1. A reduction in sensitivity because of resistance in the input wires. The fractional reduction in sensitivity is equal to

$$\frac{R_i}{R_i + 2R_{ci}} \qquad (12.21)$$

where R_i is the input resistance of the transducer and R_{ci} is the resistance of one input (excitation) wire. This effect may be overcome by using remote sensing leads.

2. Signal attenuation resulting from resistance in the output wires. This fractional reduction in signal is given by

$$\frac{R_L}{R_o + R_L + 2R_{co}} \qquad (12.22)$$

where R_{co} is the resistance of one output wire between transducer and load.

3. Attenuation of the high-frequency components in the data signal as a result of R-C filtering in the shielded instrument leads. The stray and distributed capacitance present in the transducer and a short cable are such that any filtering effect is negligible to frequencies well beyond the usable range of the accelerometer. However, when long leads are connected between transducer and readout equipment, the frequency response at higher frequencies may be affected significantly.

WARMUP TIME. The excitation voltage across the piezoresistive elements causes a current to flow through each element. The I^2R heating results in an increase in temperature of the elements above ambient which slightly increases the resistance of the elements. Differentials in this effect may cause the output voltage to vary slightly with time until the temperature is stabilized. Therefore, resistance measurements and shock and vibration data should not be taken until stabilization is reached.

INPUT AND OUTPUT RESISTANCE. For an equal-arm Wheatstone bridge, the input and output resistances are equal. However, temperature compensating and zero balance resistors may be internally connected in series with the input leads or in series with the sensing elements. These additional resistors will usually result in unequal input and output resistance.

ZERO BALANCE. Although the resistance elements in the bridge of a piezoresistive accelerometer may be closely matched during manufacture, slight differences in resistance will exist. These differences result in a small offset or residual dc voltage at the output of the bridge. Circuitry within associated signal conditioning instruments may provide compensation or adjustment of the electrical zero.

INSULATION. The case of the accelerometer acts as a mechanical and electrical shield for the sensing elements. It is normally electrically insulated from the elements but connected to the shield of the cable. If the case is grounded at the structure, the shield of the connecting cable may be left floating. When connecting the cable shield at the end away from the transducer, care must be taken to prevent ground loops.

STRAIN SENSITIVITY. The strain sensitivity of piezoresistive accelerometers is very small and may be disregarded in almost all applications. The magnitude of the output due to base or case strain is usually less than the normal noise level of the transducer and associated electronics.

THERMAL SENSITIVITY SHIFT. The sensitivity of a piezoresistive accelerometer varies as a function of temperature. This change in the sensitivity is caused by changes in the gage factor and resistance and is determined by the temperature characteristics of the modulus of elasticity and piezoresistive coefficient of the sensing elements. The sensitivity deviations are minimized by installing compensating resistors in the bridge circuit within the accelerometer.

Figure 12.27 shows the typical temperature response of a piezoresistive accelerometer. Note that over a relatively wide range of temperature, the change of sensitivity with temperature is small.

THERMAL ZERO SHIFT. Because of small differences in resistance change of the sensing elements as a function of temperature, the bridge may become slightly unbalanced when subjected to temperature changes. This unbalance produces small changes in the dc voltage output of the bridge. Transducers are usually compensated during manufacture to minimize the change in dc voltage output (zero balance) of the accelerometer with temperature. Adjustment of external balancing circuitry should not be necessary in most applications.

Fig. 12.27. Effect of temperature on the sensitivity of a piezoresistive accelerometer. The sensitivity deviation is referred to the sensitivity at 70°F (24°C).

DAMPING. The frequency-response characteristics of piezoresistive accelerometers having damping near zero are similar to those obtained with piezoelectric accelerometers. Viscous damping is provided in accelerometers having relatively low resonance frequencies to increase the useful high-frequency range of the accelerometer and to reduce the output at resonance. At room temperature this damping is usually 0.7 of critical damping or less. With damping, the sensitivity of the accelerometer is "flat" to greater than one-fifth of its resonance frequency.

The piezoresistive accelerometer using viscous damping is intended for use in a limited temperature range, usually $+20$ to $+200°F$ (-7 to $+94°C$). At the high temperatures the viscosity of the oil decreases, resulting in low damping; and at low temperatures the viscosity increases, which causes high damping. Accordingly, the frequency-response characteristics change as a function of temperature.

FILTERING OUT THERMAL ZERO DRIFT. Many piezoresistive transducers have enough signal to be used at one-tenth or less of rated range. The limiting factor is the zero drift with temperature which becomes significant when operating at a small fraction of full range. This temperature shift is, in most cases, essentially a dc phenomenon. Most applications do not require dc information but do require frequency response down to ½ or ¹⁄₁₀ Hz. The solution suggested is the use of a high-pass filter in the electronic system so that the dc thermal phenomenon is rejected but the low-frequency vibration signal is preserved. The method used is a simple high-pass *R-C* filter consisting of a series capacitor followed by a shunt resistor. The cutoff frequency (3 dB down) is

$$f_c = \frac{1}{2\pi RC} \tag{12.23}$$

where R is the shunt resistance and C is the series capacitor.

When an amplifier cannot tolerate a high resistance at its input, the filter can be connected to the output of the amplifier using a suitable value of shunt resistance. The input resistance of the electronic equipment following the amplifier must be included in the calculation of the resistance.

FORCE GAGES AND IMPEDANCE HEADS

MECHANICAL IMPEDANCE MEASUREMENT

Mechanical impedance measurements are made to relate the force applied to a structure to the motion of a point on the structure. If the motion and force are measured at the same point, the relationship is called the *driving-point impedance;* otherwise it is called the *transfer impedance.* Any given point on a structure has six degrees-of-freedom: translations along three orthogonal axes and rotations around the axes, as explained in Chap. 2. A complete impedance measurement requires measurement of all six excitation forces and response motions. In practice, rotational forces and motions are rarely measured, and translational forces and motions are measured in a single direction, usually normal to the surface of the structure under test.

Mechanical impedance is the ratio of input force to resulting output velocity. *Mobility* is the ratio of output velocity to input force, the reciprocal of mechanical impedance. *Dynamic stiffness* is the ratio of input force to output displacement. *Receptance*, or *admittance*, is the ratio of output displacement to input force, the reciprocal of dynamic stiffness. *Dynamic mass*, or *apparent mass*, is the ratio of input force to output acceleration. All of these quantities are complex and functions of frequency. All are often loosely referred to as impedance measurements. They all require the measurement of input force obtained with a force gage (an instrument which produces an output proportional to the force applied through it). They also require the measurement of output motion. This is usually accomplished with an accelerometer; if velocity or displacement is the desired measure of motion, either can be determined from the acceleration.

Impedance measurements usually are made for one of these reasons:

1. To determine the natural frequencies and mode shapes of a structure
2. To measure a specific property, such as stiffness or damping, of a material or structure
3. To measure the dynamic properties of a structure in order to develop an analytical model of it

Chapter 10 discusses mechanical impedance and mobility in detail, and Chap. 21 discusses modal analysis and testing.

FORCE EXCITATION

The input force (excitation) applied to a structure under test should be capable of exciting the structure over the frequency range of interest. This excitation may be either a vibratory force or a transient impulse force (shock). If vibration excitation is used, the frequency is swept over the range of interest while the output motion (response) is measured. If shock excitation is used, the transient input excitation and resulting transient output response are measured. The frequency spectra of the input and output are then calculated by Fourier analysis.

PRINCIPLES OF FORCE-GAGE OPERATION

A force gage measures the force which is being applied to a structural point. Force gages used for impedance measurements invariably utilize piezoelectric transducing elements. A piezoelectric force gage is, in principle, a very simple device. The transducing element generates an output charge or voltage proportional to the applied force. Piezoelectric transducing elements are discussed in detail earlier in this chapter.

TYPICAL FORCE-GAGE AND IMPEDANCE-HEAD CONSTRUCTIONS

FORCE GAGES FOR USE WITH VIBRATION EXCITATION.   Force gages for use with vibration excitation are designed with provision for attaching one end to the structure and the other end to a force driver (vibration exciter). A thin film of oil or grease is often used between the gage and structure to improve the coupling at high frequencies.

FORCE GAGES FOR USE WITH SHOCK EXCITATION.   Force gages for use with shock excitation are usually built into the head of a hammer. Excitation is provided by striking the structure with the hammer. The hammer is often available with interchangeable faces of various materials to control the waveform of the shock pulse generated. Hard materials produce a short-duration, high-amplitude shock with fast rise and fall times; soft materials produce longer, lower-amplitude shocks with slower rise and fall times. Short-duration shocks have a broad frequency spectrum extending to high frequencies. Long-duration shocks have a narrower spectrum with energy concentrated at lower frequencies.

Shock excitation by a hammer with a built-in force gage requires less equipment than sinusoidal excitation and requires no special preparation of the structure.

IMPEDANCE HEADS.   Impedance heads combine a force gage and an accelerometer in a single instrument. They are convenient for measuring driving-point impedance because only a single instrument is required and the force gage and accelerometer are mounted as nearly as possible at a single point.

FORCE-GAGE CHARACTERISTICS

AMPLITUDE RESPONSE, SIGNAL CONDITIONING, AND ENVIRONMENTAL EFFECTS.   The amplitude response, signal conditioning requirements, and environmental effects associated with force gages are the same as those associated with piezoelectric accelerometers. They are described in detail earlier in this chapter. The sensitivity is expressed as charge or voltage per unit of force, e.g., picocoulomb/newton or millivolt/lb.

Near a resonance, usually a point of particular interest, the input force may be quite low; it is important that the force-gage sensitivity be high enough to provide accurate readings, unobscured by noise.

FREQUENCY RESPONSE.   A force gage, unlike an accelerometer, does not have an inertial mass attached to the transducing element. Nevertheless, the transducing element is loaded by the mass of the output end of the force gage. This is called the *end dynamic mass*. Therefore, it has a frequency response that is very similar to that of an accelerometer, as described earlier in this chapter.

EFFECT OF MASS LOADING.   The dynamic mass of a transducer (force gage, accelerometer, or impedance head) affects the motion of the structure to which the transducer is attached. Neglecting the effects of rotary inertia, the motion of the structure with the transducer attached is given by

$$A = A_o \frac{m_s}{m_s + m_t} \tag{12.24}$$

where a = amplitude of motion with transducer attached
A_o = amplitude of motion without transducer attached

m_s = dynamic mass of structure at point of transducer attachment in direction of sensitive axis of transducer

m_t = dynamic mass of the transducer in its sensitive direction

These are all complex quantities and functions of frequency. Near a resonance the dynamic mass of the structure becomes very small; therefore, the mass of the transducer should be as small as possible. The American National Standards Institute recommends that the dynamic mass of the transducer be less than 10 times the dynamic mass of the structure at resonance.

REFERENCES

1. Levy, S., and W. D. Kroll: *Research Paper* 2138, *J. Research Natl. Bur. Standards.*, **45**:4 (1950).
2. Lawrence, A. F.: "Crystal Accelerometer Response to Mechanical Shock Impulses," *Shock and Vibration Bull.* 24, Office of the Secretary of Defense, February, 1957, p. 298.
3. Bradley, W., Jr.: "Effects of High Intensity Acoustic Fields on Crystal Vibration Pickups," *Proc. 26th Shock and Vibration Symposium*, Office of Secretary of Defense, 1958.
4. Guttwein, G. K., and A. I. Dranetz: *Electronics*, **24**:120 (1951).
5. Orlacchio, A. W., and G. Hieber: *Electronic Inds.*, **11**:75 (1957).

13

VIBRATION MEASUREMENT EQUIPMENT AND SIGNAL ANALYZERS

Robert B. Randall
Brüel & Kjaer Australia Pty. Ltd.

INTRODUCTION

This chapter describes the principles of operation of typical instrumentation used in the measurement of shock and vibration. The first section deals with the measurement of parameters which characterize the total (broad-band) signal; the second section deals with signal analysis—primarily frequency analysis. Considerable reference is made to Chaps. 22 and 23, which give the mathematical background for various signal descriptors and for techniques such as Fourier analysis, for vibration and shock analysis, respectively. Some reference is also made to the digital techniques of Chap. 27. Many of the techniques introduced here are applied in Chap. 16.

VIBRATION MEASUREMENT EQUIPMENT

Figure 13.1 shows a typical measurement system consisting of a preamplifier, a signal conditioner, a detector, and an indicating meter. Most or all of these elements often are combined into a single unit called a *vibration meter*, which is described in a following section.

The preamplifier is required to convert the very weak signal at high impedance from a typical piezoelectric transducer into a voltage signal at low impedance, which is less prone to the influence of external effects such as electromagnetic noise pickup. The signal conditioner is used to limit the frequency range of the signal (possibly to integrate it from acceleration to velocity and/or displacement) and to provide extra amplification. The detector is used to extract from the signal, parameters which characterize it, such as rms value, peak values, and crest factor. The so-called "dc" or slowly varying signal from the detector can be viewed on a meter, graphically recorded, or digitized and stored in a digital memory.

ACCELEROMETER PREAMPLIFIERS

Types of accelerometer preamplifiers include: *voltage preamplifiers, charge preamplifiers,* and *line-drive preamplifiers.* Voltage preamplifiers now are little used because, as indicated in Chap. 12, the voltage sensitivity of an accelerometer plus a cable is very dependent on the cable length. The sensitivity of the other two types is virtually independent of cable length, and this is of considerable practical importance.

Figure 13.2 shows the equivalent circuit of a charge preamplifier with an accelerometer and cable. The charge preamplifier consists of an operational amplifier having an amplification A, back-coupled across a condenser C_f; the input voltage to the amplifier is e_i. The output voltage e_o of this circuit can be expressed as

$$e_o = e_i A = \frac{q_a A}{C_a + C_c + C_i - C_f(A-1)} \tag{13.1}$$

which is proportional to the charge q_a generated by the accelerometer. If A is very large, then the capacitances C_a, C_c, and C_i become negligible in comparison with A. C_f and the expression can be simplified to

$$e_o \approx -\frac{q_a}{C_f} \tag{13.2}$$

which is independent of the cable capacitance.

Although with a charge preamplifier the sensitivity is independent of cable length, the noise pickup in the high-impedance circuit increases with cable length, so it is an advantage to have the preamplifier mounted as close to the transducer as is practicable. The line-drive amplifier represents an excellent solution to this problem, made possible by the development of miniaturized thick-film circuits. The amplifier can thus be attached to or even included internally in the transducer. In principle the initial amplifier can either be of charge- or voltage-type, but it can be advantageous to have the option of separating the amplifier from the transducer by a short length of cable, in which case the amplifier

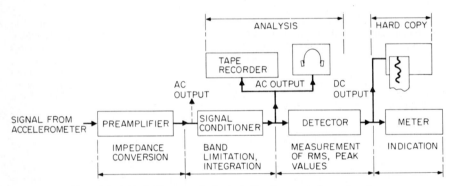

FIG. 13.1. A block diagram of a typical vibration measurement system.

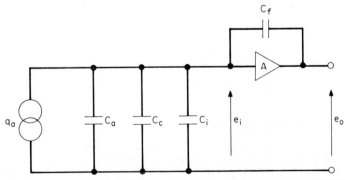

FIG. 13.2. Diagram of a charge amplifier with accelerometer and cable. A = amplification of operational amplifier; C_f = shunt capacitance across amplifier; C_a = accelerometer capacitance; C_c = cable capacitance; C_i = preamplifier input capacitance; q_a = charge generated by accelerometer; e_i = amplifier input voltage; e_o = amplifier output voltage.

should be of the charge-type. If the output signal from the initial amplifier is used to modulate the current or voltage of the power supply, then a single cable can be used both to power the amplifier and to carry the signal; the modulation is converted to a voltage signal in the power supply at the other end of this cable, which can be very long, e.g., up to a kilometer.

The output cable from a line-drive preamplifier is less subject to electromagnetic noise pickup than the cable connecting the transducer to a charge preamplifier. On the other hand line-drive preamplifiers typically have some restriction of dynamic range and frequency range in comparison with a high-quality general-purpose charge preamplifier, and so reference should be made to the manufacturer's specifications when this choice is being made.

SIGNAL CONDITIONERS. A signal-conditioning section is often required to band-limit the signal, possibly to integrate it (to velocity and/or displacement) and to adjust the gain. High- and low-pass filters normally are required to remove extraneous low- and high-frequency signals and to restrict the measurement to within the frequency range of interest. For broad-band measurements the frequency range is often specified, while for tape-recording and/or subsequent analysis the main reason for the restriction in frequency range is to remove extraneous components which may dominate and restrict the available dynamic range of the useful part of the signal.

Examples of extraneous low-frequency signals (see Chap. 12) are thermal transient effects, triboelectric effects, and accelerometer base strain. There may also be some low-frequency vibrations transmitted through the foundations from external sources. At the high-frequency end, the accelerometer resonance at least must be filtered out by an appropriate low-pass filter. This high- and low-pass filtering does not affect the signal in the input amplifier, which must be able to cope with the full dynamic range of the signal from the transducer. It is thus possible for a preamplifier to overload even when the output signal is relatively small. Consequently, it is important that the preamplifier indicates overload when it does occur.

INTEGRATION. Although an accelerometer, in general, is the best transducer to use, it is often preferable to evaluate vibration in terms of velocity or displacement. Most criteria for evaluating machine housing vibration (Chap. 16) are effectively constant velocity criteria, as are many criteria for evaluating the effects of vibration on buildings and on man, at least within certain frequency ranges (Chaps. 24 and 44). Some vibration criteria (e.g., for aircraft engines) are expressed in terms of displacement. For rotating machines, it is sometimes desired to add the absolute displacement of the bearing housing to the relative displacement of the shaft in its bearing (measured with proximity probes) to determine the absolute motion of the shaft in space.

Acceleration signals can be integrated electronically to obtain velocity and/or displacement signals; an accelerometer plus integrator can produce a velocity signal which is valid over a range of three decades (1000:1) in frequency—a capability which generally is not possessed by velocity transducers. Moreover, simply by switching the lower limiting frequency (for valid integration) on the preamplifier, the three decades can be moved by a further decade, without changing the transducer.

A typical sinusoidal vibration component may be represented by the phasor $Ae^{j\omega t}$. Integrating this once gives $\dfrac{1}{j\omega} Ae^{j\omega t}$, and thus integration corresponds in the frequency

Fɪɢ. 13.3. Electrical integration network of the simple R-C type.

domain to a division by $j\omega$. This is the same as a phase shift of $-\pi/2$ and an amplitude weighting inversely proportional to frequency, and thus electronic integrating circuits must have this property.

One of the simplest integrating circuits is a simple $R\text{-}C$ circuit, as illustrated in Fig. 13.3. If e_i represents the input voltage, then the output voltage e_o is given by

$$e_o = e_i \frac{1}{1 + j\omega RC} \tag{13.3}$$

which for high frequencies ($\omega RC \gg 1$) becomes

$$e_o \approx \frac{e_i}{j\omega RC} \tag{13.4}$$

which represents an integration, apart from the scaling constant $1/RC$.

The characteristic of Eq. (13.3) is shown in Fig. 13.4; it is that of a low-pass filter with a slope of -20 dB/decade and a cutoff frequency $f_n = 1/(2\pi RC)$ (corresponding to $\omega RC = 1$).

The limits f_L (below which no integration takes place) and f_T (above which the signal is integrated) can be taken as roughly a factor of 3 on either side of f_n, for normal measurements where amplitude accuracy is most important. Where phase accuracy is important (e.g., to measure true peak values), the factor should be somewhat greater.

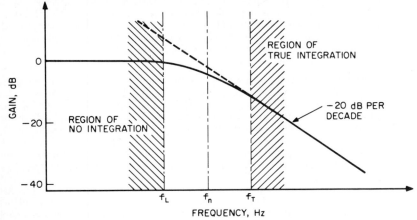

FIG. 13.4. Frequency characteristic of the circuit shown in Fig. 13.3. f_T = lower frequency limit for true integration; f_L = upper frequency limit for no integration.

Modern integrators tend to use active filters with a more localized transition between the region of no integration and the region of integration.

One situation where the choice of the low frequency limit is important is in the integration of impulsive signals, for example, in the determination of peak velocity and displacement from an input acceleration pulse. Figure 13.5 shows the effect of single and double integration on a 10-millisecond single-period sine burst, with both 1- and 10-Hz cutoff frequencies, in comparison with the true results. The deviations depend to some extent on the actual amplitude and phase characteristics of the integrator, but the following values can be used as a rough guide to select the integrator cutoff frequency f_T:

For single integration (acceleration to velocity)

$$f_T < \frac{1}{30t_p} \tag{13.5}$$

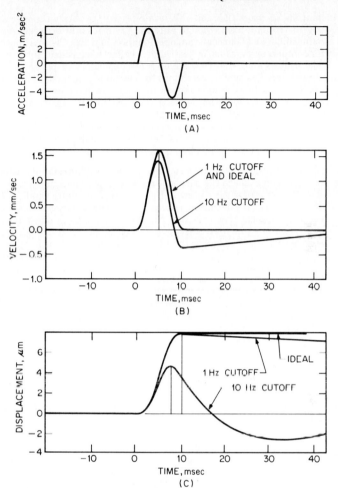

Fɪɢ. 13.5. Integration and double integration of a 10-millisecond acceleration pulse using lower frequency limits of 1 and 10 Hz, respectively. (*A*) Input acceleration signal. (*B*) Velocity signal resulting from a single integration with different cutoff frequencies compared with ideal integration. (*C*) Displacement signal resulting from a double integration compared with ideal integration.

For double integration (acceleration to displacement)

$$f_T < \frac{1}{50t_p} \tag{13.6}$$

where t_p is the time from the start of the pulse to the measured peak. For the case shown in Fig. 13.5 these values of f_T are <6.7 Hz and <2 Hz, respectively.

DETECTORS

Detectors are used to extract parameters which characterize a signal, such as arithmetic average, mean-square, and root-mean-square (rms) values, as defined in Chap. 22. The arithmetic average value is the simplest to measure, using a full-wave rectifier to obtain the instantaneous magnitude and a smoothing circuit to obtain the average. However, even though there is a fixed (though different) relationship between average and rms values for sinusoidal and Gaussian random signals (Chap. 22), the relationship varies considerably for complex signals and, in particular, is affected considerably by phase relationships. Since mean-square and rms values are independent of phase relationships, they are usually preferred as signal descriptors for stationary signals; where an average detector is used, it is usually as an approximation of an rms detector.

Mean-square values have the advantage that they are directly additive when two signals are added together (in particular different frequency bands or components), while rms values have the advantage that they have the same dimensions and units as the original signal. Thus, a "true rms" detector must include a squaring section and averager to obtain the mean-square value, followed by a square-root extractor.

SQUARING. One of the earliest true rms detectors, the Wahrman detector,[1] used a piecewise linear approximation to the parabola representing true squaring. For moderate signal values, the errors for the piecewise linear circuit are quite small, but past the last breakpoint on the curve the deviation becomes progressively larger. The breakpoints are dimensioned for a typical rms level, and the errors are thus greatest for relatively large instantaneous values which are characteristic of signals with a high *crest factor* (ratio of peak-to-rms value). The higher the crest factor, the larger the number of breakpoints required. As an example, four breakpoints give an accuracy within ½ dB for crest factors up to 5. Thus, for accurate results, this type of detector must be specified with respect to both dynamic range and crest factor.

Later designs of analog squaring circuits, so-called "log-mean-square" or "lms" detectors, make use of the logarithmic characteristic of certain diodes to achieve squaring by doubling the logarithmic value of the rectified signal. This type generally has no limitation on crest factor other than that given by the dynamic range. In a similar manner, digital instruments achieve true squaring and are only limited by the dynamic range of the detector.

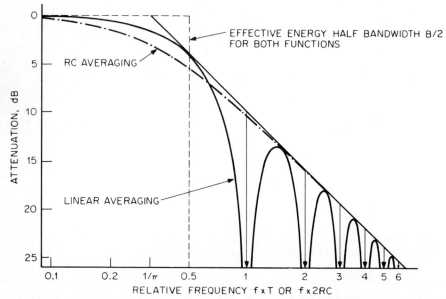

FIG. 13.6. Comparison of linear averaging (over time T) with exponential averaging (time constant RC) in the frequency domain for the case where $T = 2RC$. The low-pass filter characteristics have the same asymptotic curves and the same bandwidth $B = 1/T = 1/2RC$.

AVERAGING. The definition of mean-square value given in Eq. (22.6) assumes a uniform weighting for the whole of the averaging time T. In practice, for measurements on continuous signals, it is often desired to have a running average, giving at any time the average value over the previous T seconds. It is extremely difficult to achieve a linearly weighted running average, and so recourse is usually made to the two alternatives:

1. Exponentially weighted running average. This is achieved by an R-C smoothing circuit in most analog instruments, and also by exponential averaging in digital instruments such as FFT analyzers.

2. Linearly weighted average over a fixed time period of length T. The result is only available at the end of each period and is usually held until processed further, so new incoming data may be lost.

The averaging process acts as a low-pass filter to remove high-frequency ripple components and leave the slowly varying dc or average value. Figure 13.6 compares the low-pass filter characteristics of exponential and linear averaging and demonstrates that they are equivalent for the case where $T = 2RC$ (where RC is the time constant of the exponential decay). This low-pass filtration in the frequency domain corresponds to a convolution in the time domain with the impulse response of the averaging circuit. The two impulse responses (reversed in time because of the convolution) are compared in Fig. 13.7

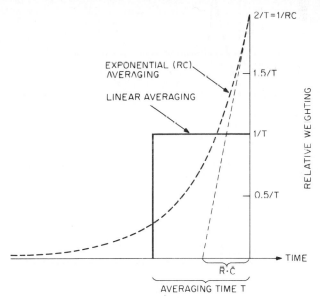

FIG. 13.7. Comparison of linear averaging (over time T) with exponential averaging (time constant RC) in the time domain for the case where $T = 2RC$. The weighting curves represent the impulse-response functions reversed in time.

for the same case where $T = 2RC$. When scaled to give the same result on stationary signals (same area under the curve), the peak output for exponential averaging is twice that for linear averaging. Account msut be taken of this in the analysis of impulses.

A method of checking the effective averaging time of an exponential averager is to remove the excitation and measure the rate of decay of the output. This will be 4.34 dB per RC time constant, or 8.7 dB per averaging time T. This does not apply to FFT analyzers operating above their real-time frequency, in the same way that the effective linear averaging time is then less than the time required to obtain the result.

PEAK DETECTORS. In some cases it is desired to measure the true peak values of the original signal (for example, to avoid overloading a tape recorder). Peak detectors are available which capture the highest value encountered and either hold it until reset or have it decay slowly enough that the eye can read the peak value from a meter. Care should be taken to distinguish between maximum positive peak, maximum negative peak, maximum peak (positive or negative), and peak-to-peak values (Fig. 13.8). Care should also be taken to distinguish between true peak values and what is roughly referred to as peak-to-peak shaft vibration, which is often assumed to be sinusoidal and is measured with an average detector.

CREST FACTOR.   The *crest factor* is the ratio of peak-to-rms value. The maximum peak (positive or negative) should be used. It is only meaningful where peak values are reasonably uniform and repeatable from one signal sample to another. The crest factor yields a measure of the spikiness of a signal and is often used to characterize signals containing repetitive impulses in addition to a lower-level continuous signal. Examples of such vibration signals are those from reciprocating machines and those produced by localized faults in gears and rolling element bearings.

FIG. 13.8. Illustration of various peak values.

KURTOSIS.   *Kurtosis*, a statistical parameter akin to the mean and mean-square values defined in Chap. 22, is defined as[2]

$$\beta_2 = \frac{\int_{-\infty}^{\infty} (\xi - \bar{\xi})^4 p(\xi) \, d\xi}{\sigma^4} \tag{13.7}$$

using the terminology of Chap. 22.

For signals with zero mean value $\bar{\xi}$, a practical estimator for this can be expressed as

$$\frac{\dfrac{1}{T} \int_0^T \xi^4(t) \, dt}{\left[\dfrac{1}{T} \int_0^T \xi^2(t) \, dt \right]^2} \tag{13.8}$$

Because of the fourth power, considerable weight is given to large amplitude values, and the kurtosis thus is a good indicator of the spikiness of the signal. Because it takes the whole signal into account, rather than just isolated peaks, it generally gives a more stable value than the crest factor, but this must be weighed against the more complicated measurement procedure. Kurtosis is normally calculated digitally, but can be measured by a more complex version of the log-mean-square detector described above.

ENVELOPE DETECTORS.   Many machine vibration signals of interest contain repetitive high-frequency bursts, as a result of exciting high-frequency resonances at regular intervals. Direct frequency analysis of the signal does not always give much information on the repetition frequencies, in particular when the resonances excited are at very high frequencies. These repetition frequencies are, however, easily measurable in the envelope signal illustrated in Fig. 13.9. Quite often, the signal is first bandpass filtered in a frequency region dominated by the repetitive bursts (i.e., one of the regions containing resonances which are excited and where the extraneous background signal is low).

The true envelope signal can be obtained using a peak detector with a decay time constant set sufficiently short so that it is able to follow the relatively slow variations in the (rectified) signal envelope. If the signal is first passed through a bandpass filter, it will have a roughly sinusoidal form with slowly varying amplitude and there will be a fixed ratio of peak-to-(short-term)-rms or average value, in which case an rms or average detector can be used instead of the peak detector. Moreover, where a frequency analysis of the signal is to be obtained with an FFT analyzer, it is not necessary to apply a smoothing circuit, as the antialiasing filters (described below) will automatically remove high-frequency ripple components in the rectified signal. Thus, a tunable bandpass filter of, say, one-third–octave bandwidth, followed by a full-wave rectifier, can be used as an enve-

FIG. 13.9. Illustration of the envelope signal for an impulsive signal containing repetitive high-frequency bursts.

lope detector in cases where it is primarily the burst repetition frequencies which are of interest. It is shown below that envelope signals can also be calculated by Hilbert transform techniques in an FFT analyzer.

VIBRATION METERS

Vibration meters are instruments which receive a signal from a vibration transducer and process it so as to give an indication of relevant vibration parameters. They are sometimes made specifically to meet certain standards, for example, ISC 2372 on "Vibration Severity of Rotating Machines" or ISO 2631 on "Human Vibration." In these cases, the requirements are specified in the relevant standard; the discussion here is aimed at more general-purpose vibration meters.

For measurements on most rotating machines, a frequency range of 10 Hz to 10 kHz is desirable. The lower limit includes the shaft speed for all machines operating over 600 rpm and any subharmonic components such as oil whirl for higher-speed plain bearing machines where such effects are most prevalent. The upper frequency of 10 kHz includes tooth-meshing frequencies and their harmonics in gearboxes, bladepassing frequencies in most bladed machines, and resonance frequencies typically excited by rolling element bearing faults.

It can be advantageous to be able to choose a number of upper and lower limiting frequencies within the over-all range. For example, restriction of the upper frequency to 1 kHz allows measurements in accordance with ISO2372 previously cited. For special purposes, it may be necessary to go to frequencies lower than 10 Hz, for example, in measurements on slow-speed machines and on bridges and other structures. It is possible to cover a total range of 1 Hz to 10 kHz with one accelerometer; if the meter is able to accept a range of transducers, its own frequency range can be even wider.

If restriction is to be made to one vibration parameter, then velocity usually is the best choice, as most machine vibration signals have a roughly uniform velocity spectrum so that an increase at any frequency has a roughly equal chance of influencing over-all vibration levels.

It is also desirable to be able to measure acceleration and displacement; changes at low frequency reflect themselves primarily in the displacement value, while changes at high frequency have the most effect on the acceleration value.

In addition to the measurement of rms levels in each of the vibration parameters it is of advantage to be able to measure some parameter indicating the spikiness of the signal, such as peak values (and hence crest factor), kurtosis, spike energy, or shock-pulse value. Finally, it is useful if the meter has an ac output, to allow the signal to be fed to a tape recorder or headphones. In the absence of frequency analysis, the human ear can discern a great deal about the characteristics of a signal, and this setup provides an excellent stethoscope. The ac signal should preferably be of selected parameter (acceleration, velocity, or displacement); the frequency range should be restricted as little as possible.

TAPE RECORDERS

The two most widely used techniques for instrumentation tape recorders are direct recording and frequency modulation (FM). In direct recording, the signal amplitude reflects itself directly in the local degree of magnetization of the tape, while in FM recording the amplitude information is contained in the deviation of the frequency of a carrier tone from its nominal value. Thus, the degree of magnetization of the tape is less critical for FM recording, and the recorded blips are normally saturated. Hence, one of the advantages of FM recording is that the recorded signals are less susceptible to change due to poor storage conditions (heat, light, and stray magnetic fields). On the other hand, since the carrier frequency is typically three to five times higher than the maximum signal frequency in FM recording, tape speeds (and hence tape quantities used) must be three to five times greater for a given frequency range.

The major difference between the two techniques is in their ability to record low-frequency signals. Since on playback of direct recordings it is the rate of change of tape magnetization which is detected, this technique cannot record down to dc; a typical lower frequency limit is 25 Hz. In contrast, FM recording can record down to dc; a dc signal is simply represented by a constant deviation of the carrier frequency.

Since on playback of direct recordings it is necessary to integrate the detected signal and compensate for other effects such as tape magnetic properties, this is usually done by equalization networks designed primarily to provide amplitude linearity; phase linearity is poor. Thus, the actual form of signals is likely to be modified by direct recording; peak values cannot be relied upon. The phase linearity of FM recording is excellent for all except the highest part of the frequency range, where the effects of the required low-pass filter become significant.

One of the most important characteristics of a tape recorder is its dynamic range, since the tape recorder is likely to be the element in the measurement chain whose dynamic range is restricted the most. The dynamic range usually is expressed in terms of a signal-to-noise ratio, which is typically 40 dB for FM recording and up to 50 dB for direct recording. These figures can be somewhat misleading, however, as the noise referred to is a total figure over the entire frequency range and has less influence in a narrow-band analysis. After narrow-band analysis, the noise level for FM recording typically is more than 60 dB below full scale, as compared with 70 to 80 dB for the digitization noise in a modern frequency analyzer.

Table 13.1 summarizes the most important features of FM and direct recording. Some recorders are able to record using both techniques, in which case the heads normally are optimized for FM and the signal-to-noise ratio for direct recording is reduced somewhat. The most important addition provided by direct recording is the possibility of recording considerably higher frequencies, typically 50 to 100 kHz.

Both techniques are limited by the accuracy of the tape transport system, and small variations in tape orientation and speed give rise to so-called "wow and flutter."

DIGITAL RECORDERS. In digital recorders, the signal is passed through an analog-to-digital (A/D) converter and stored in a digital storage medium as a sequence of digital values. These are later played back through a digital-to-analog converter to regenerate the analog signal. Alternatively, they could be passed directly to a digital signal analyzer for further processing, without signal deterioration. To avoid problems with *aliasing* (discussed in a later section) the sampling rate must be such as to give at least two samples per period of the highest frequency in the signal. In practice the figure is usually 2.5 to 3, and even this higher figure requires that the signal is first low-pass filtered with very steep *antialiasing filters* to ensure that signal components above half the sampling frequency are negligible. In theory, digital recording solves the problem of dynamic range limitation inherent in the other two techniques; it is purely a question of the number of bits used to quantize the signal values. As a rough guide, approximately 6 dB of dynamic range is obtained for each bit in the digital sample; A/D converters giving 16 bits (approximately 96 dB) are currently in use. The problems of wow and flutter are also largely solved by digital recording, as the sampling frequency on playback is not directly dependent on the tape transport mechanism and can be made extremely accurate. The problem associated with digital recording is the vast amount of data which have to be stored; for example, a frequency bandwidth of 20 kHz requires 50,000 samples per second, each sample of 16 bits. However, the technology is now available, both on video cassettes and on (compact) laser discs, to store gigabytes of information, thus making digital recording feasible.

Table 13.1. Comparison of Direct and FM Recording Techniques

	Direct	FM
Dynamic range (typical, narrow-band)	70 dB	60 dB
Lower frequency limit (typical)	25 Hz	DC
Upper frequency limit (typical)	50 kHz	10 kHz
Amplitude stability	Acceptable	Excellent
Phase linearity	Poor	Good
Preservation of recorded information	Acceptable	Good

SIGNAL ANALYZERS AND THEIR USE

This section deals primarily with frequency analysis, but a number of related analysis techniques, namely, synchronous averaging, cepstrum analysis, and Hilbert transform techniques are also considered briefly.

As discussed in Chaps. 22 and 23, frequency analysis may be carried out by two different processes: (1) Passage of the signal through a number of filters with different center frequencies, and measurement of the power transmitted at each frequency by an appropriate detector. (2) Direct calculation of the Fourier transform of blocks of data taken from the signal. Both approaches are considered here.

FILTERS

An ideal bandpass filter is a circuit which transmits that part of the input signal within its passband and completely attenuates components at all other frequencies. Practical electronic filters differ slightly from the ideal as discussed below. An analysis may be performed over a frequency range either by using a single filter with a tunable center frequency which is swept over the entire frequency range or by using banks of fixed filters having contiguous (or overlapping) passbands.

For general vibration analysis it is common to use tunable filters whose center frequency is either tuned by hand or synchronized with the X position of the pen on a graphic recorder so that the spectrum is plotted automatically by sweeping the center frequency over the desired frequency range. Alternatively, the center frequency may be synchronized with an external signal, e.g., a trigger pulse once per revolution of a shaft, in which case the filter becomes a tracking filter which can be used to filter out the component corresponding to a designated harmonic, or multiple, of the synchronizing signal. A tracking adaptor, or frequency multiplier/divider, is normally required to generate the tuning frequency from the original synchronizing signal, if other than the fundamental or first harmonic is required.

In the past, banks of filters with fixed center frequencies, each with its own detector, were widely used for parallel analysis of all frequency bands in real time. This arrangement is costly, however, and has largely been superseded by digital filter analyzers (described in the following section). If real-time analysis is not required, a less-expensive alternative is to switch the output of each filter in turn to a single detector and record the outputs sequentially on paper. The individual filters can then also share many components, which are selected in appropriate combinations by the switching process. Sequentially stepped fixed filters are typically used for relatively broad-band analysis and are rarely used with less than one-third-octave bandwidth. This type of analysis finds most application in acoustics and in studies of the effects of vibration on humans (Chap. 44).

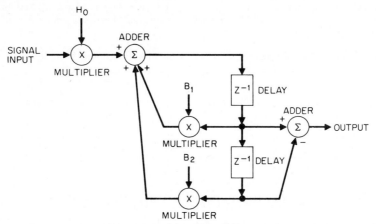

FIG. 13.10. Block diagram of a typical two-pole digital filter section, consisting of multipliers, adders, and delay units. H_0, B_1, and B_2 are constants by which the appropriate signal sample is multiplied. Z^{-1} indicates a delay of one sample interval before the following operation.

DIGITAL FILTERS. Digital filters (in particular, recursive digital filters) are devices which process a continuous digitized signal and provide another signal as an output which is filtered in some way with respect to the original. The relationship between the output and input samples can be expressed as a difference equation (in general, involving previous output and input values) with properties similar to those of a differential equation which might describe an analog filter. Figure 13.10 shows a typical two-pole section used in a one-third-octave digital filter analyzer (three of these are cascaded to give six-pole filtration).

Two ways of changing the properties of a given digital filter circuit such as that shown in Fig. 13.10 are:

1. For a given sampling frequency, the characteristics can be changed by changing the coefficients of the difference equation. (In the circuit of Fig. 13.10 there are three, effectively defining the resonance frequency, damping, and scaling.)

2. For given coefficients, the filter characteristic is only defined with respect to the sampling frequency. Thus, halving the sampling frequency will halve the cutoff frequencies, center frequencies, and bandwidths; consequently, the constant-percentage characteristics are maintained one octave lower in frequency. For this reason, digital filters are

well-adapted to constant-percentage bandwidth analysis on a logarithmic (i.e., octave-based) frequency scale.

Thus, the 3 one-third-octave characteristics within each octave are generated by changing coefficients, while the various octaves are covered by repetitively halving the sampling frequency. Every time the sampling frequency is halved, it means that only half the number of samples must be processed in a given time; the total number of samples for all octaves lower than the highest is ($\frac{1}{2} + \frac{1}{4} + \frac{1}{8} + \cdots$) which in the limit is the same as the number in the highest octave. By being able to calculate twice as fast as is necessary for the upper octave alone, it is possible to cover any number of lower octaves in real time. This is the other reason why digital filters are so well-adapted to real-time constant-percentage bandwidth analysis over a wide frequency range.

FILTER PROPERTIES. Figure 13.11 illustrates what is meant by the 3-dB bandwidth and the effective noise bandwidth, the first being most relevant when separating discrete frequencies, and the second when dealing with random signals. For filters having good selectivity (i.e., having steep filter flanks) there is not a great difference between the two values, so in the following discussion no distinction is made between them.

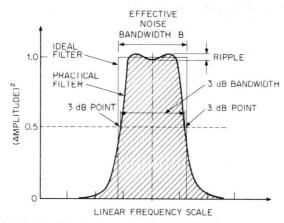

FIG. 13.11. Bandwidth definitions for a practical filter characteristic. The *3-dB bandwidth* is the width at the 3-dB (half power) points. The *effective noise bandwidth* is the width of ideal filter with the same area as the (hatched) area under the practical filter characteristic on an amplitude squared (power) scale.

The response time T_R of a filter of bandwidth B is on the order of $1/B$ as illustrated in Fig. 13.12, and thus the delay introduced by the filter is also on this order. This relationship can be expressed in the form

$$BT_R \approx 1 \tag{13.9}$$

which is most applicable to constant-bandwidth filters, or in the form

$$bn_r \approx 1 \tag{13.10}$$

where $b = B/f_0$ = relative bandwidth
$n_r = f_0 T_R$ = number of periods of frequency f_0 in time T_R
f_0 = center frequency of filter

This form is more applicable to constant-percentage bandwidth filters. Thus, the response time of a 10-Hz bandwidth filter is approximately 100 milliseconds, while the response time of a 1 per cent bandwidth filter is approximately 100 periods. Figure 13.12 also illustrates that the effective length of the impulse T_E is also approximately $1/B$, while to integrate all of the energy contained in the filter impulse response it is necessary to integrate over at least $3T_R$.

CHOICE OF BANDWIDTH AND FREQUENCY SCALE.   In general it is found that analysis time is governed by expressions of the type $BT \geq K$, where K is a constant [see, for example, Eq. (13.9)] and T is the time required for each measurement with bandwidth B. Thus, it is important to choose the maximum bandwidth which is consistent with obtaining an adequate resolution, because not only is the analysis time per bandwidth proportional to $1/B$ but so is the number of bandwidths required to cover a given frequency range—a squared effect.

It is difficult to give precise rules for the selection of filter bandwidth, but the following discussion provides some general guidelines: For stationary deterministic and, in particular, periodic signals containing equally spaced discrete frequency components, the aim is to separate adjacent components; this can best be done using a constant bandwidth on a linear frequency scale. The bandwidth should, for example, be chosen as one-fifth to one-third of the minimum expected spacing (e.g., the lowest shaft speed, or its half-order if this is to be expected) (see Fig. 13.13A). For stationary random or transient signals,

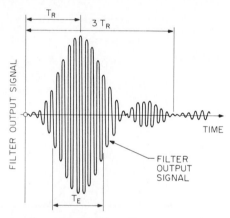

Fig. 13.12. Typical filter impulse response. T_R = filter-response time ($\approx 1/B$); T_E = effective duration of the impulse ($\approx 1/B$); B = bandwidth.

the shape of the spectrum will most likely be determined by resonances in the transmission path between the source and the pickup, and the bandwidth B should be chosen so that it is about one-third of the bandwidth B_r of the narrowest resonance peak (Fig. 13.13B). For constant damping these tend to have a constant Q or constant-percentage bandwidth character, and thus constant-percentage bandwidth on a logarithmic frequency scale often is most appropriate.

A linear frequency scale is normally used together with a constant bandwidth, while a logarithmic frequency scale is normally used together with a constant-percentage bandwidth, as each combination gives uniform resolution along the scale. A logarithmic scale may be selected in order to cover a wide frequency range, and then constant-percentage bandwidth is virtually obligatory. A logarithmic frequency scale may, however, occasionally be chosen in conjunction with constant bandwidth (though over a limited frequency range) in order to demonstrate a relationship which is linear on log-log scales (e.g., conversions between acceleration, velocity, and displacement).

CHOICE OF AMPLITUDE SCALE.   Externally measured vibrations, on a machine casing for example, are almost always the result of internal forces acting on a structure whose frequency-response function modifies the result. Because the structural response functions vary over a very wide dynamic range, it is almost always an advantage to depict the vibration spectra on a logarithmic amplitude axis. This applies particularly when the vibration measurements are used as an indicator of machine condition (and thus, internal forces and stresses) since the largest vibration components by no means necessarily represent the largest stresses. Even where the vibration is of direct interest itself, in vibration measurements on humans the amplitude axis should be logarithmic because this is the way the body perceives the vibration level.

It is a matter of personal choice (though sometimes dictated by standards) whether the logarithmic axes are scaled directly in linear units or in logarithmic units expressed in decibels (dB) relative to a reference value. Another aspect to be considered is dynamic range. The signal from an accelerometer (plus preamplifier) can very easily have a valid dynamic range of 120 dB (and more than 60 dB over three frequency decades when integrated to velocity). The only way to utilize this wide range of information is on a logarithmic amplitude axis. Figure 13.14 illustrates both these considerations; it shows spectra measured at two different points on the same gearbox (and representing the same internal condition), on both logarithmic and linear amplitude axes. The logarithmic representations of the two spectra are quite similar, while the linear representations are not only different but hide a number of components which could be important.

An exception where a linear amplitude scale usually is preferable to a logarithmic scale is in the analysis of relative displacement signals, measured using proximity probes, for

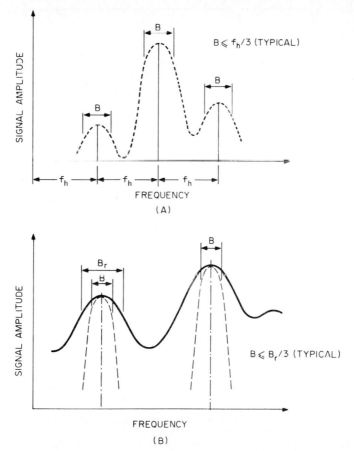

FIG. 13.13. Choice of filter bandwidth B for different types of signals. (A) Discrete frequency signals—harmonic spacing f_h. (B) Stationary random and impulsive signals.

Fɪɢ. 13.14. Comparison of rms logarithmic and rms linear amplitude scales for the depiction of vibration velocity spectra from two measurement points [(A) and (B)] on the same gearbox (thus representing the same internal condition). The logarithmic representations in terms of velocity level are similar and show all components of interest. The linear spectra in terms of velocity amplitude are quite different, and both hide many components which could be important.

the following reasons: (1) The parameter being measured is directly of interest for comparison with the results of rotor dynamic and bearing hydrodynamic calculations. (2) The dynamic range achievable with relative shaft vibration measurements (as limited by mechanical and electrical runout) does not justify or necessitate depiction on a logarithmic axis.

ANALYSIS SPEED. There are three basic elements in a filter analyzer which can give rise to significant delays and thus influence the speed of analysis.

The *filter* introduces a delay on the same order as its response time T_R (see Fig. 13.12). This is most likely to dominate in the analysis of stationary deterministic signals, where the filter only contains one discrete frequency component at a time and only a short averaging time is required. It also dominates the analysis of transients, in the case where a narrow bandwidth is used such that the filter impulse response is longer than the input transient itself. Details are given in the appropriate section on the choice of analysis parameters.

The *detector* introduces a delay on the same order as the averaging time T_A. The choice of averaging time depends on the type of signal being analyzed, namely, stationary deterministic (discrete frequency), stationary random, or transient, and is also discussed in detail in later sections.

In a swept-frequency analysis, the delays resulting from both filter response time and averaging time mean that the recorded spectrum departs from the true spectrum as illustrated in Fig. 13.15. The rate at which the pen traverses the frequency scale is called the

FIG. 13.15. Differences between recorded and actual spectra in a swept frequency analysis with definition of the *delay* and *error* in recorded peaks and valleys.

paper speed, denoted by P, even in situations where the pen carriage moves with respect to stationary paper, as with an X-Y recorder. The following considerations are important in selecting an appropriate paper speed, which determines the analysis time:

1. Errors in the level of recorded peaks and valleys must be limited. Generally, valleys are less important because they are often determined by artifacts, such as the filter bandwidth and characteristic.

2. The delay, or frequency offset, of recorded peaks and valleys must also be limited to, say, one-fourth of the bandwidth.

3. The maximum rate of decay of the detector output signal is 8.7 dB per averaging time; this limits the maximum slope which can be traced on the trailing edge of peaks in the spectrum. In the limiting case, the maximum slope which is to be recorded corresponds to the filter characteristic, and the paper speed should be limited to allow this.

4. The *graphic level recorder* may have a limited pen speed (referred to herein as *writing speed* (denoted by W); this speed should be selected to be sufficiently greater than the paper speed so that the maximum slope of the filter characteristic can be reproduced. In some graphic level recorders, the writing speed is used indirectly to set the averaging time. In this case it is probably the pen speed as such, and not the averaging time, which limits the paper speed.

Further details concerning the choice of the parameters which limit analysis speed are given in following sections which deal with the analysis of the various types of signals.

FFT ANALYZERS

FFT analyzers make use of the FFT (fast Fourier transform) algorithm to calculate the spectra of blocks of data. The FFT algorithm is an efficient way of calculating the discrete Fourier transform (DFT). As described in Chap. 27, this is a finite, discrete approximation of the Fourier integral transform. The equations given there for the DFT [Eqs. (27.3) and (27.4)] assume real-valued time signals. The FFT algorithm makes use of the following versions, which apply equally to real or complex time series:

$$S_x(m) = \Delta t \sum_{n=0}^{N-1} x(n\,\Delta t) \exp(-j2\pi m\,\Delta f\,n\,\Delta t) \tag{13.11}$$

$$x(n) = \Delta f \sum_{m=0}^{N-1} S_x(m\,\Delta f) \exp(j2\pi m\,\Delta f\,n\,\Delta t) \tag{13.12}$$

These equations give the spectrum values $S_x(m)$ at the N discrete frequencies $m\,\Delta f$ and give the time series $x(n)$ at the N discrete time points $n\,\Delta t$.

Whereas the Fourier transform equations are infinite integrals of continuous functions, the DFT equations are finite sums but otherwise have similar properties. The function being transformed is multiplied by a rotating unit vector $\exp(\pm j2\pi m\,\Delta f\,n\,\Delta t)$, which rotates (in discrete jumps for each increment of the time parameter n) at a speed proportional to the frequency parameter m. The direct calculation of each frequency component from Eq. (13.11) requires N complex multiplications and additions, and so to calculate the whole spectrum requires N^2 complex multiplications and additions.

The FFT algorithm factors the equation in such a way that the same result is achieved in roughly $N \log_2 N$ operations.[3] This represents a speedup by a factor of more than 100 for the typical case where $N = 1024 = 2^{10}$. However, the properties of the FFT result are the same as those of the DFT.

INHERENT PROPERTIES OF THE DFT.   Figure 13.16 graphically illustrates the differences between the DFT and the Fourier integral transform.

Because the spectrum is only available at discrete frequencies $m\,\Delta f$ (where m is an integer), the time function is implicitly periodic (as for the Fourier series). The periodic time

$$T = N\,\Delta t = 1/\Delta f \tag{13.13}$$

where N = number of samples in time function and frequency spectrum
T = corresponding record length of time function
Δt = time sample spacing
Δf = line spacing = $1/T$

In an analogous manner, the discrete sampling of the time signal means that the spectrum is implicitly periodic, with a period equal to the sampling frequency f_s, where

$$f_s = N\,\Delta f = 1/\Delta t \tag{13.14}$$

Note from Fig. 13.16 that because of the periodicity of the spectrum, the latter half ($m = N/2$ to N) actually represents the negative frequency components ($m = -N/2$ to 0). For real-valued time samples (the usual case) the negative frequency components are determined in relation to the positive frequency components by the equation

$$S_x(-m) = S_x^*(m) \tag{13.15}$$

and the spectrum is said to be *conjugate even*.

In the usual case where the $x(n)$ are real, it is only necessary to calculate the spectrum from $m = 0$ to $N/2$ and the transform size may be halved by one of the following two procedures:

1. The N real samples are transformed as though representing $N/2$ complex values, and that result is then manipulated to give the correct result.[4]

2. A zoom analysis (discussed in a later section) is performed which is centered on the middle of the base-band range to achieve the same result.

Thus, most FFT analyzers produce a (complex) spectrum with a number of spectral lines equal to half the number of (real) time samples tranformed. To avoid the effects of aliasing (see next section) not all the spectrum values calculated are valid, and it is usual to display, say, 400 lines for a 1024-point transform or 800 lines for a 2048-point transform.

Aliasing. *Aliasing* is an effect introduced by the sampling of the time signal, whereby high frequencies after sampling appear as lower ones (as with a stroboscope). The DFT algorithm of Eq. (13.11) cannot distinguish between a component which rotates, say, seven-eights of a revolution between samples and one which rotates a negative one-eighth of a revolution. Aliasing is normally prevented by low-pass filtering the time signal before sampling to exclude all frequencies above half the sampling frequency (i.e., $-N/2 < m < N/2$). From Fig. 13.16 it will be seen that this removes the ambiguity. In order to utilize up to 80 per cent of the calculated spectrum components (e.g., 400 lines from 512 calculated) it is necessary to use very steep antialiasing filters with a slope of about 120 dB/octave.

Normally, the user does not have to be concerned with aliasing because suitable antialiasing filters automatically are applied by the analyzer. One situation where it does have to be allowed for, however, is in tracking analysis (discussed in a following section) where, for example, the sampling frequency varies in synchronism with machine speed.

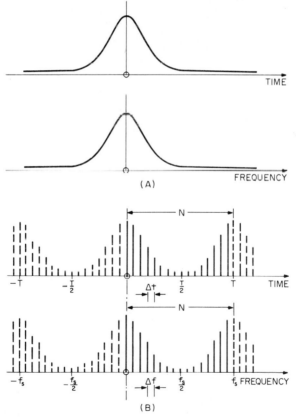

Fig. 13.16. Graphical comparison of (*A*) the Fourier transform with (*B*) the discrete Fourier transform (DFT) (see text).

Leakage.   *Leakage* is an effect whereby the power in a single frequency component appears to leak into adjacent bands. It is caused by the finite length of the record transformed (N samples) whenever the original signal is longer than this; the DFT implicitly assumes that the data record transformed is one period of a periodic signal, and the leakage depends on what is actually captured within the time window, or data window.

Figure 13.17 illustrates this for three different sinusoidal signals. In (A) the data window corresponds to an exact integer number of periods, and a periodic repetition of this produces an infinitely long sinusoid with only one frequency. For (B) and (C) (which have a slightly higher frequency) there is an extra half-period in the data record, which gives a discontinuity where the ends are effectively joined into a loop, and considerable leakage is apparent. The leakage would be somewhat less for intermediate frequencies. The difference between the cases of Fig. 13.17B and C lies in the phase of the signal, and other phases give an intermediate result.

Fig. 13.17. Time-window effects when analyzing a sinusoidal signal in an FFT analyzer using rectangular weighting. (A) Integer number of periods, no discontinuity. (B) and (C) Half integer number of periods but with different phase relationships, giving a different discontinuity when the ends are joined into a loop.

When analyzing a long signal using the DFT, it can be considered to be multiplied by a (rectangular) time window of length T, and its spectrum consequently is convolved with the Fourier spectrum of the rectangular time window,[5] which thus acts like a filter characteristic. The actual filter characteristic depends on how the resulting spectrum is sampled in the frequency domain, as illustrated in Fig. 13.18.

In practice, leakage may be counteracted:

1. By forcing the signal in the data window to correspond to an integer number of periods of all important frequency components. This can be done in tracking analysis (discussed in a later section) and in modal analysis measurements (Chap. 21), for example, where periodic excitation signals can be synchronized with the analyzer cycle.

2. (For long transient signals) By increasing the length of the time window (for example, by zooming) until the entire transient is contained within the data record.

3. By applying a special time window, which has better leakage characteristics than the rectangular window already discussed.

Later sections deal with the choice of data windows for both stationary and transient signals.

Picket Fence Effect. The *picket fence effect* is a term used to describe the effects of discrete sampling of the spectrum in the frequency domain. It has two connotations:

1. It results in a nonuniform frequency weighting corresponding to a set of overlapping filter characteristics, the tops of which have the appearance of a picket fence (Fig. 13.19).

2. It is as though the spectrum is viewed through the slits in a picket fence, and thus peak values are not necessarily observed.

One extreme example is in fact shown in Fig. 13.18 where in (A) the side lobes are completely missed, while in (B) the side lobes are sampled at their maxima and the peak value is missed.

The picket fence effect is not a unique feature of FFT analysis; it occurs whenever discrete fixed filters are used, such as in normal one-third octave analysis. The maximum amplitude error which can occur depends on the overlap of the adjacent filter characteristics, and this is one of the factors taken into account in the following discussion on the choice of data window.

DATA WINDOWS FOR ANALYSIS OF STATIONARY SIGNALS. A *data window* is a weighting function by which the data record is effectively multiplied before trans-

FIG. 13.18. Frequency sampling of the continuous spectrum of a time-limited sinusoid, length T. (A) Integer number of periods, side lobes sampled at zero points (compare with Fig. 13.17A). (B) Half integer number of periods, side lobes sampled at maxima (compare with Fig. 13.17B and C).

FIG. 13.19. Illustration of the picket fence effect. Each analysis line has a filter characteristic associated with it which depends on the weighting function used. If a frequency coincides exactly with a line, it is indicated at its full level. If it falls midway between two lines, it is represented in each at a lower level corresponding to the point where the characteristics cross.

formation. (It is sometimes more efficient to apply it by convolution in the frequency domain). The purpose of a data window is to minimize the effects of the discontinuity which occurs when a section of continuous signal is joined into a loop.

For stationary signals, a good choice is the *Hanning window* (one period of a cosine squared function) which has a zero value and slope at each end and thus gives a gradual transition over the discontinuity. In Fig. 13.20 it is compared with a rectangular window, both in the time and frequency domains. Even though the main lobe (and thus the bandwidth) of the frequency function is wider, the side lobes fall off much more rapidly and the highest is at −32 dB, compared with −13.4 dB for the rectangular. Figure 13.21 shows what this means, in terms of actual filter characteristics, for a case similar to Fig. 13.17*C*.

Other time-window functions may be chosen, usually with a trade-off between the steepness of filter characteristic on the one hand and effective bandwidth on the other.

Fig. 13.20. Comparison of rectangular and Hanning window functions of length T seconds. Full line—rectangular weighting; dotted line—Hanning weighting. The inset shows the weighting functions in the time domain.

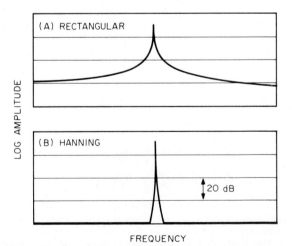

Fig. 13.21. Comparison of worst-case filter characteristics for rectangular and Hanning weighting functions for a 80-dB dynamic range.

Table 13.2 compares the time windows most commonly used for stationary signals. The most highly selective window, giving the best separation of closely spaced components of widely differing levels, is the *Kaiser-Bessell window*. On the other hand, it is usually possible to separate closely spaced components by zooming, at the expense of a slightly increased analysis time.

Another window, the *flattop window*, is designed specifically to minimize the picket fence effect so that the correct level of sinusoidal components will be indicated, independent of where their frequency falls with respect to the analysis lines. This is particularly useful with calibration signals. Nonetheless, by taking account of the distribution of samples around a spectrum peak, it is possible to compensate for picket fence effects with other windows as well. Figure 13.22, which is specifically for the Hanning window, is a nomogram giving both amplitude and frequency corrections, based on the decibel difference (ΔdB) between the two highest samples around a peak. For stable single-frequency components this allows determination of the frequency to an accuracy of an order of magnitude better than the line spacing.

Table 13.2. Properties of Various Data Windows

Window type	Highest side lobe, dB	Side lobe fall-off, dB/decade	Noise bandwidth*	Maximum amplitude error, dB
Rectangular	−13.4	−20	1.00	3.9
Hanning	−32	−60	1.50	1.4
Hamming	−43	−20	1.36	1.8
Kaiser-Bessel	−69	−20	1.80	1.0
Truncated Gaussian	−69	20	1.90	0.9
Flattop	−93	0	3.70	<0.1

* Relative to line spacing.

DATA WINDOWS FOR ANALYSIS OF TRANSIENT SIGNALS. When using impulsive (e.g., hammer) excitation of structures for determining their frequency-response characteristics (e.g., see Chap. 21), it is common to use the following special data windows:

1. A *short rectangular* window may be applied over the very short excitation impulse in order to exclude noise from the remaining portion of the record.

2. An *exponential* window can be applied where the response is very long (i.e., lightly damped structures) to reduce the signal to practically zero at the end of the record, and thus avoid discontinuities. The effect is the same as adding extra damping which is very precisely known and can be subtracted from the measurement results.

ZOOM ANALYSIS.[6] Zoom analysis is the term given to a spectrum analysis having increased resolution over a restricted part of the frequency range. The following are two techniques used to generate zoom analyses.

1. *Real-time zoom* (illustrated in the block diagram of Fig. 13.23) is a zoom process in which the entire signal is first modified to shift its frequency origin to the center of the zoom range. Then it is passed through a low-pass filter (usually a digital filter in real time) which has a passband corresponding to the original zoom-band (Fig. 13.24). Because of the low-pass filtration, the signal then can be resampled at a lower sampling rate without aliasing, and the resampled signal processed by an FFT transform. The original frequency shift is accomplished by multiplying the incoming signal by a unit vector (phasor) rotating at $-f_0$ (thereby subtracting f_0 from all frequencies in it), and the modified time signal is thus complex. This is one situation where the FFT transform of complex time data is used.

Nondestructive zoom is effectively a way of achieving a larger transform size without modification of the original data record. For a typical case, data are first captured in a 10K (i.e., 10240-point) buffer. Ten 1K records obtained by taking every 10th sample are transformed using a 1K (1024-point) FFT transform. Even though this gives only 512 frequency values per transform, the rest are generated by periodic repetition (because of aliasing resulting from the undersampling). After compensating the phase of the results for the small time shift of each of the undersampled records, the entire spectrum of the 10K record can in principle be obtained by addition. In practice, only a part of the whole spectrum is normally obtained at any one time in order to save on memory requirements, but the whole spectrum can be generated by repetitive operation on exactly the same data.

Real-time zoom has the advantage that the zoom factor obtainable is virtually unlimited. A procedure is often employed (as illustrated in Fig. 13.23) whereby the signal samples are repeatedly circulated around a loop containing a low-pass filter which cuts off at one-half the previous maximum frequency, after which the sampling frequency is halved by dropping every second sample. Each circulation doubles the zoom factor and at the same time doubles the length of original signal required to fill the transform buffer. It is this time requirement which places a limit on the zoom factor, as well as on the stability of the signal itself. A zoom factor of 10 in a 400-line spectrum, for example, gives the equivalent of a 4000-line spectrum; a finer resolution is not required to analyze the vibration spectrum of a machine whose speed fluctuates by, say, 0.1 per cent.

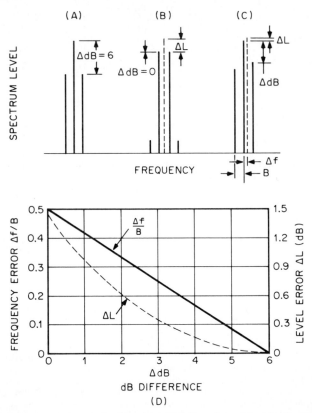

FIG. 13.22. Picket fence corrections for Hanning weighting, where ΔL = level correction, dB; Δf = frequency correction, Hz; B = line spacing, Hz; ΔdB = difference in decibels between the two highest samples around a peak representing a discrete frequency component. Three examples are shown: (A) Actual frequency coincides with center line. (B) Actual frequency midway between two lines. (C) General situation. Note that the frequency correction $\Delta f/B$ is almost linear.

Real-time zoom suffers the disadvantage that the entire signal must be reprocessed to zoom in another band. This has two detrimental consequences:

1. For very narrow bandwidths (long record lengths) the analysis time is very long for each zoom analysis.

2. There is no certainty that exactly the same signal is processed each time.

On the other hand, nondestructive zoom has the advantage that for zoom analysis in different bands exactly the same data record is used. Thus it is known, for example, that there will be an exact integer relationship between the various harmonics of a periodic signal. This can be useful, as a typical example, in separating the various harmonics of shaft speed from those of line frequency, in induction motor vibrations. Furthermore, the long analysis time is required only once (to fill the data buffer); further zoom analyses on the same record are limited only by the calculation speed.

Nondestructive zoom suffers the disadvantage that the zoom factor is limited by the size of the memory buffer in the analyzer. Where the memory buffer is 10 times the normal transform size, for example, the zoom factor is equal to 10.

Thus, both types of zoom are advantageous for different purposes. Nondestructive zoom is probably best for diagnostic analysis of machine vibration signals, whereas real-

Fig. 13.23. Block diagram for real-time zoom with bandwidth B centered on frequency f_o. M is the zoom factor and also the factor by which the sampling frequency is reduced.

Fig. 13.24. Principle of real-time zoom, using a low-pass filter to filter out the portion of the original signal in the zoom-band of width B. Prior to this, the frequency origin is shifted to frequency f_o (the desired center frequency of the zoom-band) by multiplying the (digitized) time signal by $e^{-j2\pi f_o t}$.

time zoom gives more flexibility in frequency-response measurements (system frequency response should not change even where the excitation signals change). Real-time zoom also gives the possibility of very large zoom factors when they are required.

In real-time zoom, it is only the preprocessing of the signal which has to be in real time; the actual FFT analysis of the signal, once it is stored in the transform buffer, does not have to be in real time.

ANALYSIS OF STATIONARY SIGNALS

USING FILTERS. The choice of bandwidth was discussed in connection with Fig. 13.13; it depends largely on whether the signal being analyzed is deterministic (i.e., made up of discrete frequency components) or random (power distributed over a frequency range).

For *stationary deterministic signals*, where only one component is in the filter at a time (e.g., following the recommendations of Fig. 13.13A), the squaring of the signal by the detector doubles its frequency and suitable attenuation of the ripple is achieved if

$$T_A \geq 3/f_m \qquad (13.16)$$

where f_m is the lowest frequency being analyzed. For constant-percentage bandwidth analysis at higher frequencies (and other situations where the filter passband encompasses more than one discrete frequency component), the lowest frequency fluctuations of the filter output are the "beat" variety and their frequency will not be doubled by squaring. It is then preferable to use

$$T_A \geq 5/f_m \qquad (13.17)$$

where f_m is now the minimum spacing between any two components in the passband (and thus the lowest beat frequency). Where the filter contains many discrete frequency components of approximately uniform level, it can be treated in the same way as a random signal.

For *stationary random signals*, T_A should be chosen to limit the standard deviation ϵ of the error in the (rms) result. This can be calculated from the formula

$$\epsilon = \frac{1}{2\sqrt{BT_A}} \qquad (13.18)$$

As an example, to limit ϵ to 12½ per cent, T_A can be calculated from

$$BT_A \geq 16 \qquad (13.19)$$

The *dwell time* T_D is the time taken to sweep over a frequency span equal to one filter bandwidth; the choice of T_D should take into account the delays introduced in the filter and the detector.

With respect to the filter response time T_R, if T_D is made equal to T_R (i.e., $BT_D = 1$), the whole spectrum will be shifted by one bandwidth. It would normally be more acceptable to limit the delay to one-fourth of the bandwidth, in which case T_D can be calculated from the relationship

$$BT_D \geq 4 \qquad (13.20)$$

In this case, errors in the level of recorded peaks will be negligible.

With respect to the averaging time T_A, T_D can be calculated from the relationship

$$T_D \geq 2T_A \qquad (13.21)$$

This should give acceptable errors and delays of recorded peaks and also make it possible for the pen to follow the slope of typical filter characteristics.

The correct choice of T_D must then satisfy both Eqs. (13.20) and (13.21). By comparison of Eqs. (13.19) and (13.21) with Eq. (13.20), it can be seen that the averaging time always dominates for random signals. However, the filter response time may dominate for deterministic signals, in which case T_A should be increased to $T_D/2$.

The analysis sweep speed, df/dt, is then given by

$$df/dt \le B/T_D \qquad (13.22)$$

where df/dt is in Hz/sec for B expressed in Hz. Often it is more practical to calculate df/dt directly as a paper speed by expressing the bandwidth B in terms of its equivalent width on the paper. With this paper speed, a suitably high pen speed should be selected so that the steepest slope of filter characteristic can be reproduced; if this is not possible, the paper speed must be restricted even further.

USING FFT. Equation (13.13) shows that for a single FFT transform, the product (*bandwidth* times *averaging time*) $BT_A = 1$, at least for rectangular weighting where B is equal to the line spacing Δf (Table 13.2). The same applies for any weighting function, the increased bandwidth being exactly compensated by a corresponding decrease in effective record length.[7]

For *stationary deterministic signals*, a single transform having a BT_A product equal to unity is theoretically adequate, although a small number of averages is sometimes performed if the signal is not completely stable. Figure 13.25 illustrates the effect of averaging for a deterministic signal and demonstrates that the sinusoidal components are unaffected; the only effect is to smooth out the (nondeterministic) noise at the base of the spectrum (Fig. 13.25B).

For *stationary random signals*, Eq. (13.18) applies if n, the number of independent spectra averaged, replaces BT_A. Figure 13.26 illustrates (A) an instantaneous spectrum, (B) the average of eight spectra, and (C) the average of 128 spectra. The meaning of the

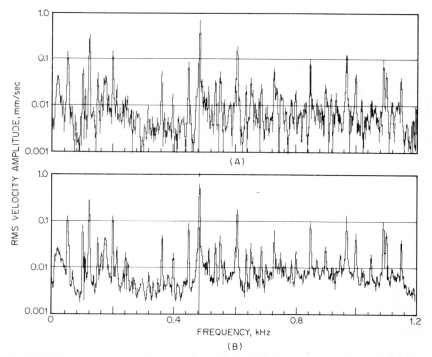

FIG. 13.25. Effect of averaging with a stationary deterministic signal. (*A*) Instantaneous spectrum (average of 1). (*B*) The linear average of eight spectra.

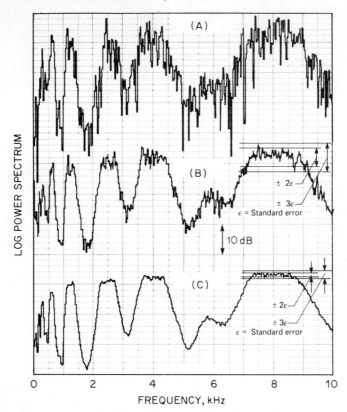

FIG. 13.26. Effect of averaging with a stationary random signal. (A) Instantaneous spectrum. (B) Average of eight spectra. (C) Average of 128 spectra.

standard error ϵ (Eq. 13.18) is illustrated in (B) and (C). Statistically, there is a 68 per cent probability that the actual error will be less than ϵ, a 95.5 per cent probability that it will be less than 2ϵ, and a 99.7 per cent probability that it will be less than 3ϵ.

For rectangular (flat) weighting, independent spectra are those from nonoverlapping time records; when other weighting functions are used, the situation is different. For example, Fig. 13.27A illustrates the over-all (power) weighting obtained when Hanning windows are applied to contiguous records. Note that virtually half of the incoming signal is excluded from the analysis, whereas a 50 per cent overlapping of consecutive records regains most of the lost information. Thus, when using window functions similar to Hanning (as recommended for stationary signals), it is almost always advantageous to average the results from 50 per cent overlapping records. A method for calculating the effective number of averages obtained in this way is given in Ref. 8; for 50 per cent overlapping Hanning windows the error is very small in treating them as independent records.

Real-time Analysis. An FFT analyzer is said to operate in real time when it is able to process all the incoming data, even though presentation of the results is delayed by an amount corresponding to the calculation time. This implies that the time taken to analyze a data record, T_a, is less than the time taken to collect the data transformed, T. It also implies that the analysis process should not interrupt the continuous recording of data, so that recording can continue in one part of the memory at the same time as analysis is being performed in another. T is inversely proportional to the selected frequency range, and the highest frequency range for which T_a is less than T is called the *real-time* *frequency.* This condition will only ensure that all the incoming data are analyzed when rectangular weighting is used. With Hanning weighting, for example, where 50 per cent overlap analysis must be employed to analyze all the data, the true real-time frequency will be halved, since twice as many transforms must be performed for the same length of data record. In yet another sense, the analysis is not truly real-time unless the over-all weighting function is uniform. As illustrated in Fig. 13.27, the minimum overlap of Hanning windows to achieve this is two-thirds, which reduces the true real-time frequency to one-third of the commonly understood definition given above.

In practice, with stationary signals, there is no advantage in more than a 50 percent overlap, since (1) statistical reliability is not significantly improved and (2) all sections of the record are statistically equivalent, so the over-all weighting function is not important. It can be important for nonstationary signals, such as transients, as discussed below. For stationary signals, where any data missed are statistically no different from the data analyzed, the only advantages of real-time analysis are that (1) results with a given accuracy are obtained in the minimum possible time and (2) maximum information is extracted from a record of limited length.

SCALING AND CALIBRATION FOR STATIONARY SIGNALS. *Scaling* is the process of determining the correct units for the Y axis of a frequency analysis, while *calibration* is the process of setting and confirming the numerical values along the axis. In the most general case, spectra can be scaled in terms of mean-square or rms values at each frequency (or, strictly speaking, for each filter-band). For signals dominated by discrete frequency components, with no more than one component per filter-band, this yields the mean-square or rms value of each component. It may be necessary to compensate for picket fence effects, for example, using Fig. 13.22.

For FFT analysis, direct application of Eq. (13.11) only provides the positive frequency

FIG. 13.27. Over-all weighting functions for spectrum averaging with overlapping Hanning windows. (*A*) Zero overlap (step length *T*). (*B*) 50 per cent overlap (step length *T*/2). (*C*) 66.7 per cent overlap (step length *T*/3). (*D*) 75 per cent overlap (step length *T*/4). *T* is the record length for the FFT transform.

contribution. The squared amplitude of the complex frequency component calculated must be doubled to include the negative frequency contribution in the over-all mean-square value as measured using filters. Thus, the rms value is $\sqrt{2}$ times greater than the calculated amplitude.

A spectrum of mean-square values is known as a *power spectrum* since physical power often is related to the mean-square value of parameters such as voltage, current, force, and velocity.

For random signals, the power spectrum values vary with the bandwidth, but can be normalized to a *power spectral density*, $W(f)$, by dividing by the bandwidth [see Eq. (22.17)]. The results then are independent of the analysis bandwidth, provided the latter is narrower than the width of peaks in the spectrum being analyzed (e.g., following Fig. 13.13B). As examples, power spectral density is expressed in g^2 per hertz when the input signal is expressed in gs acceleration, and in volts squared per hertz when the input signal is in volts.

The concept of power spectral density is meaningless in connection with discrete frequency components (with infinitely narrow bandwidth); it can be applied only to the random parts of signals containing mixtures of discrete frequency and random components.

FIG. 13.28. Typical one-third–octave power spectrum (units, g^2) on log-log scales with lines of constant power spectral density (units, g^2/Hz). The power spectral density scaling of the peak between 8 and 10 kHz is not valid since its width is less than three times the analysis bandwidth.

Nevertheless, it is possible to calibrate a power spectral density scale using a discrete frequency calibration signal. For example, when analyzing a $1g$ sinusoidal signal with a 10-Hz analyzer bandwidth, the height of the discrete frequency peak may be labeled: $1^2 g^2/$ 10 Hz = $0.1 g^2$/Hz.

For constant-bandwidth analysis (including FFT analysis), the scaling thus achieved is valid for all frequencies; for constant-percentage bandwidth analysis, the bandwidth and power spectral density scaling vary with frequency. On log-log axes, it is perhaps simplest to draw straight lines representing constant power spectral density, which slope upwards at 10 dB per frequency decade from the calibration point. Figure 13.28 illustrates this for a particular one-third–octave (23 per cent bandwidth) spectrum. The results are not valid for the peak at 8 to 10 kHz, since its width is less than three times the analyzer bandwidth (as recommended in Fig. 13.13B).

ANALYSIS OF SHORT TRANSIENTS

USING FILTERS. When using a single swept filter (see Fig. 23.12), the signal must be repeated for every position of the analyzer filter. First, the signal to be analyzed is recorded on a loop of magnetic tape (or in a digital single-event recorder); then it is played back repeatedly while the filter of the analyzer is changed in frequency (as in Fig. 11.7). The length of the loop should be chosen such that the repetition time T_{rep} is longer than the transient itself; the value of the signal should be zero on either side of the loop junction. In the case of analog tape recording, a splice noise-suppression circuit usually is necessary to avoid mixing the splice noise with the actual signal. This is particularly true in the case of FM recording, where the carrier frequency has a discontinuity at the splice, even when the signal there is zero.

Analog Methods of Transient Analysis. Figure 13.29 illustrates three methods of analysis and indicates conditions for the selection of the analysis bandwidth, B; the averaging time, T_A; and the loop repetition time, T_{rep}. Figure 23.13A is also relevant.

Method A (Fig. 13.29). The looped signal is played back periodically, and the averaging time is made sufficiently long ($T_A \gg T_{rep}$) so that the signal seen by the analysis system is effectively periodic. Nevertheless, the filter bandwidth B is made greater than the spacing between the harmonics ($B > 1/T_{rep}$) in order to obtain a continuous rather than a line spectrum. The analysis can be treated like that for a stationary signal, with sweep speed given by Eqs. (13.21) and (13.22). If the loop repetition time is very long (there may be physical constraints on its length), problems may be encountered with the crest factor/dynamic range limitations of the analyzer.

The spectrum as measured can be expressed in power units (engineering units squared or U^2). The energy (per filter-band) per loop length is obtained by multiplication by T_{rep} giving engineering units squared times seconds. Where the bandwidth, B, is at least three times smaller than the effective bandwidth of the transient, B_{eff}, it is valid to convert the results to energy spectral density by dividing by B. The units will then be engineering units squared times seconds per hertz. Where the condition is not satisfied, the results may still be interpreted as mean energy spectral density in the filter-band.

Method B (Fig. 13.29). In this method a spectral measurement is made for each playback of the looped signal. It is limited to constant-bandwidth analysis with $B \ll B_{eff}$, but is very efficient. In this case, the filter output resembles its impulse response for every application of the signal, with a varying carrier frequency corresponding to the tuned center frequency of the filter and an amplitude corresponding to the desired frequency component. Fig. 13.30 shows such a filter output pulse (compare with Fig. 13.12); the maximum peak value V_{peak} and the maximum short-term rms value V_{rms} are illustrated. The spectral component can be detected either with a peak detector or an rms

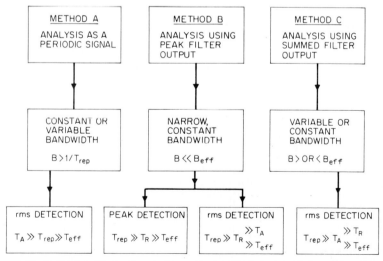

Fig. 13.29. Block diagram illustrating analog methods of transient analysis—choice of analysis method and parameters.

detector with short averaging time T_A. Where a peak detector is used, it must either be reset at the start of each playback or have a decay time constant corresponding to T_A.

The spectrum is thus obtained as a series of peaks, but the conditions indicated for the selection of analysis parameters (Fig. 13.29) ensure that there are at least nine recorded peaks for each true spectrum peak; the actual spectrum can be interpreted as the envelope of the recorded peaks.

As regards scaling, for an ideal filter of bandwidth B, the energy spectral density (ESD) of the result is given by[9]

$$\text{ESD} = \frac{V_{\text{peak}}^2}{2B^2} = \frac{V_{\text{rms}}^2}{B^2} \tag{13.23}$$

where the units of V_{peak} and V_{rms} should be converted from volts to engineering units (U). For an actual filter there may be a slightly different scaling factor; it is advisable to calibrate it using a known signal such as a rectangular pulse (Chap. 23).

As in the analysis of stationary signals, the dwell time T_D is the time taken to sweep the filter center frequency an amount corresponding to the filter bandwidth B. In this case it should be greater than or equal to the repetition time T_{rep}. The sweep speed then can be calculated from Eq. (13.22).

Method C (Fig. 13.29).   As in method B, a spectral measurement is obtained for each playback of the looped signal, but unlike method B, the analyzer bandwidth, B, can be greater than or less than the effective bandwidth, B_{eff}, of the transient; method C can thus be used for constant-percentage bandwidth analysis. The filter output signal does not necessarily resemble its impulse response, but the averaging time is chosen to be sufficiently long so that all the energy in the filter output pulse is summed (independent of the pulse shape). The basic principles and scaling are the same as those described in the next section *Real-time Digital Filter Analysis*, the principal difference being that only one frequency component is obtained at a time.

For both methods B and C it is necessary to ensure that the recorder writing speed is fast enough (e.g., > 50 dB per T_{rep}) so that the recording of peak and maximum rms values is not restricted.

FIG. 13.30. Short-term rms and peak values for a typical filter impulse response.

Real-time Digital Filter Analysis. Suppose a digital filter analyzer has a constant-per-centage bandwidth (e.g., one-third-octave or one-twelfth-octave) and covers a frequency range of three or four decades. As with method C above, the filter output signal varies greatly with the bandwidth. At low frequencies (where $B \ll B_{\text{eff}}$) the filter output resembles its impulse response, with a length dominated by the filter response time T_R. At high frequencies (where $B \gg B_{\text{eff}}$) the filter output signal follows the input more closely and has a length dominated by T_I, the duration of the input impulse.

This is illustrated in Fig. 13.31 which traces the path of a typical impulsive signal (an N-wave) through the complete analysis system of filter, squarer, and averager for both a narrow-band (low-frequency) and a broad-band (high-frequency) filter.

The averaging time T_A must always satisfy

$$T_A \geq T_I + 3T_R \tag{13.24}$$

Thus the averaging time is determined by the lowest frequency to be analyzed. The ideal solution would be running linear integration [with T_A selected using Eq. (13.24)] followed by a maximum-hold circuit (which retains the maximum value experienced). The output of such a running linear averager is shown in Fig. 13.31. Note that during the time the entire filter ouptut is contained within the averaging time T_A, the averager output provides the correct result, which is held by the maximum-hold circuit. However, a running linear average is very difficult to achieve, and normally it is necessary to choose between fixed linear averaging and running exponential averaging.

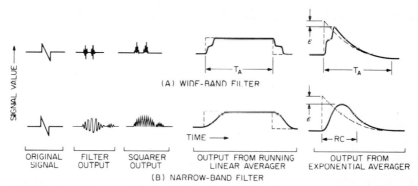

FIG. 13.31. Passage of a transient signal through an analyzer comprising a filter, squarer, and averager (alternatively running linear averaging and exponential averaging). The dotted curves represent the averager impulse responses. RC is the time constant for exponential averaging. ϵ is the error in peak response. (A) With a wide-band filter. (B) With a narrow-band filter.

The problem with fixed linear averaging is that it must be started just before the arrival of the impulse and thus cannot be triggered from the signal itself (unless use is made of a delay line before the analyzer). It is, however, possible to record the signal first and then insert a trigger signal (for example, on another channel of a tape recorder).

In order to extract all the information from a given signal, it may be necessary to make the total analysis in two passes. For example, Fig. 13.32 shows the analysis of a 220-millisecond N-wave (the pressure signal from a sonic boom). For an averaging time T_A = 0.5 seconds, the spectrum is only valid down to about 50 Hz, but it includes frequency components up to 5 kHz. This illustration also shows an analysis of the same signal using T_A = 8 seconds; this is valid down to about 1.6 Hz. However, as a result of this longer averaging time, there is a 12-dB loss of dynamic range, so all the frequency components above 500 Hz are lost. The result (with scaling adjusted by 12 dB) is given as a dotted line in Fig. 13.32; it shows that the two spectra are identical over the mutually valid range.

Where the analysis is carried out in real time on randomly occurring impulses, exponential averaging may be used followed by a maximum-hold circuit, but then there is the

FIG. 13.32. Transient analysis of a sonic boom (length 218 milliseconds) using a one-third-octave digital filter analyzer. T_A = selected averaging time. The dotted curve (T_A = 8 seconds) has been raised 12 dB to compensate for the longer averaging time.

added complication that the averager leaks energy at a (maximum) rate of 8.7 dB per averaging time T_A, and thus the total impulse duration must be short with respect to T_A. The error is less than 0.5 dB if

$$T_A \geq 10(T_I + T_R) \tag{13.25}$$

Note that the peak output of an exponential averager is a factor of 2 (i.e., 3 dB) higher than for the equivalent linear averager (Figs. 13.7 and 13.31); thus the equivalent averaging time to be used in converting from power to energy units is $T_A/2$ for exponential averaging and T_A for linear averaging. Conversion from energy to energy spectral density is only valid for that part of the spectrum where the analyzer bandwidth is appreciably less than the signal bandwidth ($B \ll B_{\text{eff}}$), although outside that range the results may be interpreted as the mean energy spectral density in the band.

USE OF FFT ANALYSIS. As for the case of stationary signals, discussed above, analysis by use of the FFT is much simpler than by analog techniques. This is particularly the case when the entire transient fits into the transform size T without loss of high-frequency information. Figure 13.33 shows such an example where the duration of the transient is less than the analyzer record length of 2048 samples (2K) in a frequency range which does not exclude high-frequency information in the signal. Flat (i.e., rectangular) weighting should be used in such a case, where the signal value is zero at each end, so that no discontinuity arises from making the record into a loop (an inherent property of the FFT process). Exponential weighting sometimes may be used to force the signal down to zero at the end of the record, but the frequency spectrum will then include the effects of the extra damping which this represents.

With rectangular weighting, the analysis bandwidth is equal to the line spacing $1/T$, which is always less than the effective signal bandwidth B_{eff} (because the effective length of the transient $T_{\text{eff}} < T$). Conversion of the results to energy spectral density, therefore, is valid in most practical situations. Some analyzers provide the results in terms of energy spectral density, but if the results are only available in terms of power (U^2), they must be multiplied by the time T corresponding to the record length to convert them to energy and divided by the bandwidth $1/T$ to convert them to energy spectral density, expressed in engineering units squared times seconds per hertz. Altogether, this represents a multiplication by T^2.

Where a transient is longer than the normal transform size T, it can be analyzed in one of the following ways:

1. *Zoom FFT* (see *Zoom Analysis*, above, for background information). A suitable zoom factor is chosen such that the transform length ($1/\Delta f$) is greater than the duration

of the transient. Thus in the case of nondestructive zoom, the entire transient can be recorded in the long memory and the entire narrow-band spectrum can be obtained by repetitive analysis in contiguous zoom-bands. In the case of real-time zoom, analysis in more than one zoom band requires that the transient be recorded in an external medium and played back for each zoom analysis. Rectangular weighting should be used (thus $B = \Delta f$) and energy spectral density (as above) is always valid using a value of T corresponding to the zoom record length ($1/B$). The narrow bandwidth may give a restriction of dynamic range of the result. Figure 13.34 shows a typical energy spectrum, obtained by repetitive nondestructive zoom analysis.

2. *Scan Averaging.* When the entire transient is stored in digital form in a long memory (as for nondestructive zoom), it is possible to obtain its spectrum by scanning a short time window (e.g., a Hanning window) of length T over the entire record; this is done in overlapping steps, and the results are averaged. As already demonstrated for stationary signals (Fig. 13.27), this procedure yields a result with uniform weighting for step lengths $T/3$ and $T/4$. The same applies to step lengths $T/5$, $T/6$, etc., but there is a slight difference with respect to the over-all weighting function for the different step lengths. Figure 13.35 illustrates the over-all time weighting function for different step lengths T/n (where n is an integer greater than 2) and shows the length of the uniform section (within which the entire transient should ideally be located) and the effective length T_{eff} by which power units should be multiplied to convert them to energy. For a conversion to energy spectral density to be valid, the width of spectrum peaks must be somewhat greater than the analysis bandwidth; this can be seen by inspection of the analysis results. For example, for the Hanning window, the bandwidth B is 1.5 times the line spacing Δf (see Table 13.2), so spectrum peaks should have a 3-dB bandwidth of more than five lines.

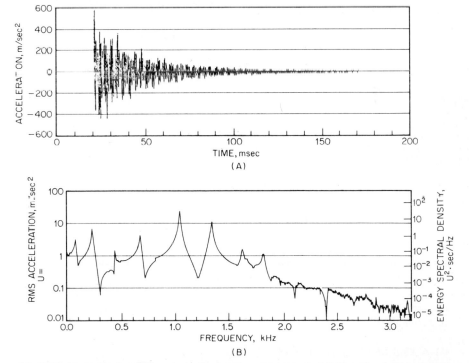

FIG. 13.33. Example of an FFT analysis of a short transient signal. (*A*) Time signal of length 2048 samples (2K) corresponding to 250 milliseconds ($T = 250$ milliseconds). (*B*) 800-line FFT spectrum with bandwidth $B = \Delta f = 4$ Hz (flat weighting). Scaling on left is in rms units. Scaling on right is converted to energy spectral density (ESD) by multiplying mean-square values by T^2.

FIG. 13.34. Analysis of a long transient signal using nondestructive zoom FFT. (*A*) Envelope of time signal of length 10240 samples (10K) corresponding to 2 seconds ($T = 2$ seconds). (*B*) 4000-line composite zoom spectrum with bandwidth $B = \Delta f = 0.5$ Hz (flat weighting). Scaling on right is converted to energy spectral density (ESD).

Even though the broader bandwidth obtained by scan averaging may result in a loss of spectrum detail, it provides considerable improvement in the dynamic range of the result. Figure 13.36 (using scan averaging) illustrates these points for the same signal as Fig. 13.34 (using zoom). The spectrum obtained by scan averaging generally has 12 dB more dynamic range than that obtained by zoom (with factor 10), but the level of peaks does not differ by this amount; this confirms that their resolution is not sufficient to allow scaling in terms of energy spectral density.

To obtain Fig. 13.36, scan averaging with a step length of $T/4$ was used (an overlap of successive records of 75 per cent). Even though a step length of $T/3$ (overlap 66.7 per cent) is theoretically more efficient, $T/4$ is usually more convenient because the number of samples in T generally is a power of 2.

ANALYSIS OF NONSTATIONARY SIGNALS

A typical nonstationary signal results from measurements made during a machine run-up or coast-down (here, the primary cause of the nonstationary signal is a change in shaft speed). The signal can be analyzed by dividing it up into a series of short quasi-station-

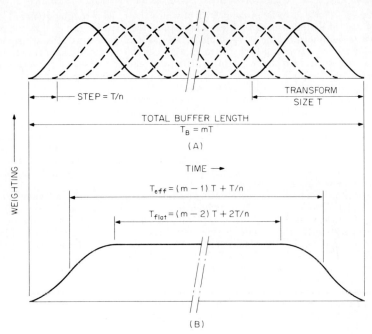

FIG. 13.35. Over-all weighting function for scan averaging of a transient. (A) Overlapping Hanning windows of length T with definition of parameters m and n. (B) Over-all weighting function with indication of T_{eff} and T_{flat} in terms of T, m, and n. T_{eff} is the effective length of the time window for conversion of power to energy units. T_{flat} is the length of the section with uniform weighting within which the transient ideally should be located.

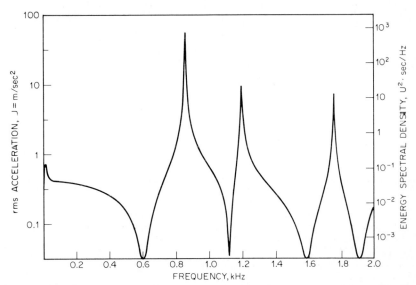

FIG. 13.36. Analysis of a long transient by scan averaging (same signal as Fig. 13.34). The energy spectral density (ESD) scaling on the right can be compared with that in Fig. 13.34, although the peaks are not valid because of insufficient resolution.

ary time periods (often overlapping), in each of which the speed is roughly constant. The length of the time window used to select a portion of the continuous signal may have to be chosen so as to ensure this. The simplest way to analyze a nonstationary signal of this type is to use a tracking filter tuned to a specific harmonic of shaft speed and to record the results vs. rpm of the machine. If a phase meter is inserted between the filtered signal and the tracking signal, it is possible to record phase as well as amplitude against rpm to give what is called a *Bode plot.*[10]

Using an FFT analyzer, the behavior of several harmonics may be studied simultaneously. One way to do this, using an FFT analyzer having a long memory (such as is required for nondestructive zoom), is with a simple scan analysis; a short Hanning window is scanned through the record (as for a *scan average*) and successive instantaneous spectra (from each window position) are viewed on the display screen. The speed of the scan may be changed by varying the step length; this is one situation (in contrast to scan averaging) where very short step lengths may be of advantage, for example, in slowing down the passage through a resonance.

A highly effective method of representing such a scan analysis is by a "waterfall," or "cascade," plot as shown in Fig. 13.37 (which represents a typical machine run-up). As indicated, the third dimension of such a three-dimensional plot can either be time or rpm; for a simple scan analysis it usually is time, but if the spectra are spaced at equal intervals of rpm, a number of advantages result. Harmonically related components (whose bases follow radial lines) then can be separated easily from constant-frequency components (e.g., related to line frequency or resonances) whose bases follow lines parallel with the rpm axis. Such a cascade plot, with rpm as the third axis, is sometimes referred to as a *Campbell diagram*, although strictly speaking a Campbell diagram has a vertical frequency axis, a horizontal rpm axis, and a signal amplitude represented as the diameter of a circle (or square) centered on the appropriate point in the diagram.

Ideally, each of the spectra in a cascade plot such as Fig. 13.37 should be obtained with constant shaft speed at the respective rpm. This is sometimes possible, for example, dur-

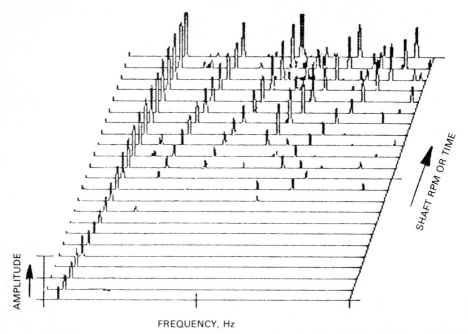

FIG. 13.37. Three-dimensional spectral map or waterfall plot, showing how spectra change with shaft rpm or time.

ing the very slow start-up of a large steam turbine, but usually each spectrum is a windowed section of a continuously varying signal with a small speed change within the window length. Consequently, the peak corresponding to each harmonic is not always localized in one analysis line; in particular, the higher harmonics are likely to be spread over progressively more lines. Thus, the height of each peak cannot be used directly as a measure of the strength of each component; it would be necessary to integrate over the whole of a distributed peak to measure the total power contained in it.

A way of overcoming this problem is to use *tracking analysis*, where the sampling rate of the FFT analyzer is related directly to shaft speed. A frequency multiplier is used to produce a sampling frequency signal (controlling the A/D converter of the analyzer) which is a specified multiple of the shaft speed.

Figure 13.38 illustrates the basic principles. Figure 13.38*B* shows a hypothetical signal produced by a rotating shaft during a run-up (in practice, the amplitude normally also would vary with shaft speed). Figure 13.38*A* shows the samples obtained by sampling the signal value at a constant sampling frequency (as for normal frequency analysis) and the spectrum resulting from FFT analysis of these samples. The spectral peak is seen to spread over a number of lines corresponding to the speed change along the time record.

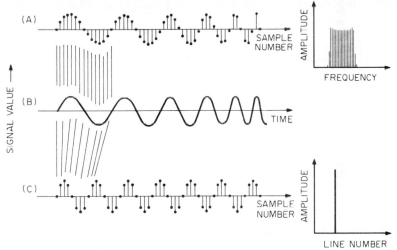

FIG. 13.38. Analysis of a fundamental component which is increasing in frequency. (*A*) Data record resulting from a uniform sampling rate, and its spectrum which spreads over a frequency band corresponding to the speed change. (*B*) The original time signal. (*C*) Data record resulting from sampling eight times per fundamental cycle, and its spectrum which is concentrated in one analysis line.

Figure 13.38*C* shows the samples obtained by sampling the signal a fixed number of times per shaft revolution (in this case, eight). The samples are indistinguishable from those obtained from normal analysis of a constant-frequency component, and thus the frequency spectrum is concentrated in one line.

When the sampling frequency varies with shaft speed, however, special precautions must be taken to avoid problems with aliasing. One possibility is to use a tracking low-pass filter with a cutoff frequency suitably less than half the sampling frequency. Because of the difficulty of obtaining a tracking filter having a very steep roll-off (e.g., 120 dB/octave), it is often simpler to choose one of a series of filters with a fixed cutoff frequency, depending on the current shaft speed. Such a series of filters (in, for example, a 2, 5, 10 sequence) often is available in the analyzer to determine the normal frequency ranges. Taking the case of a 400-line analyzer, for example, all 400 lines in the measured spectrum are valid when the sampling frequency is appropriate to the selected filter (Fig.

13.39*A*). If the sampling frequency is higher than the ideal for a given filter, the upper part of the spectrum is affected by the filter (Fig. 13.39*B*). If it is lower, the upper part of the spectrum may be contaminated by aliasing components (Fig. 13.39*C*). Nevertheless, by arranging for the selection of the optimum filter at all times (either manually or automatically), at least 60 per cent of the measured spectrum (i.e., in this case 240 lines) is always valid. The analysis parameters can be selected so that the desired number of harmonics is contained within this range, based on the fact that the line number in the spectrum of a given component is equal to the number of periods it represents in the data record of length N samples. If, for example, the 30th harmonic is to be located in line no. 240, the fundamental must be in line no. 8; there must be eight periods of the fundamental component along the data record. Where the data record contains 1024 samples (i.e., $N = 1024$), the sampling frequency must then be 128 times the shaft speed; thus a frequency multiplier with a multiplication factor of 128 should be used in this specific case.

For FFT analyzers with zoom, a simpler approach can be used, as illustrated in Fig. 13.40. An analog low-pass filter is applied to the signal with a cutoff frequency corresponding to the highest required harmonic at maximum shaft speed. However, a frequency multiplying factor is chosen so as to make the sampling frequency, say, 10 or 20 times this cutoff frequency (instead of the normal 2.56). The spectrum then is obtained

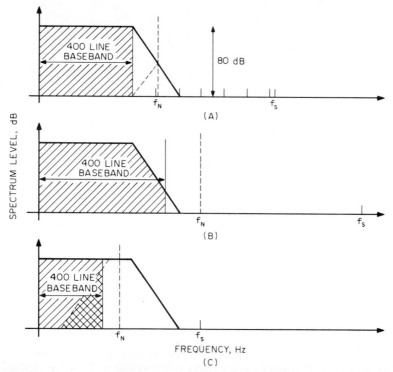

FIG. 13.39. Effect of sampling frequency on the validity of spectral components, assuming an FFT analyzer with 400 lines and 80-dB dynamic range. f_s = sampling frequency. f_N = Nyquist folding frequency = $f_s/2$. (*A*) Normal situation with optimum choice of sampling frequency for the low-pass filter. (*B*) Situation with increased sampling frequency. The upper lines in the spectrum are influenced by the low-pass filter. (*C*) Situation with decreased sampling frequency. The upper lines in the spectrum are influenced by aliasing components folded around f_N (double cross-hatched area).

by zooming in a range corresponding to the highest required harmonic. As shown in Fig. 13.40, the shaft speed (and thus the sampling frequency) can then be varied over a wide range, without aliasing components affecting the measurement results.

Figure 13.41 shows the results of tracking FFT analysis on a large turbogenerator. It was made using nondestructive zoom with zoom factor 10. A frequency multiplying factor of 256 was used, giving 40 periods of the fundamental component in the 10K (10240-point) memory of the FFT analyzer. The fundamental is thus located in line no. 40 of the 400-line zoom spectrum. Because the harmonics coincide exactly with analysis lines, rectangular weighting could have been used in place of the Hanning weighting actually used (all harmonics have exact integer numbers of periods along the record length); Hanning weighting can, however, be advantageous for nonsynchronous components such as constant-frequency components. Such a component at 150 Hz (initially coinciding with the third harmonic of shaft speed) is shown in Fig. 13.41. Constant-frequency components follow a hyperbolic locus in cascade plots employing order tracking.

FIG. 13.40. Use of a fixed low-pass filter to prevent aliasing when tracking with an FFT analyzer employing zoom to analyze in a lower-frequency band. For illustration purposes, the sampling frequency at maximum shaft speed has been made four times greater than that appropriate to the analog LP filter. The shaft speed range could be made proportionally greater by increasing this factor. (A) Situation at maximum shaft speed. All harmonics of interest must be contained in the display range. (B) Situation at one-fourth maximum shaft speed. The analog filter characteristics overlap, but are well-separated from the display range. (C) Situation at three-sixteenths maximum shaft speed. The aliasing range almost intrudes on the display range.

RELATED ANALYSIS TECHNIQUES

Signal analysis techniques other than those described above, which are useful as an adjunct to frequency analysis, include synchronous averaging, cepstrum analysis, and Hilbert transform techniques.

SYNCHRONOUS AVERAGING (SIGNAL ENHANCEMENT). *Synchronous averaging* is an averaging of digitized time records, the start of which is defined by a repetitive trigger signal. One example of such a trigger signal is a once-per-revolution synchronizing pulse from a rotating shaft. This process serves to enhance the repetitive part of the signal (whose period coincides with that of the trigger signal) with respect to nonsynchronous effects. That part of the signal which repeats each time adds directly, in proportion to the number of averages, n. The nonsynchronous components, both random

Fig. 13.41. Tracking FFT analysis of the rundown of a large turbogenerator. The superimposed hyperbolic curve represents a fixed-frequency component at 150 Hz.

noise and periodic signals with a different period, add like noise, with random phase; the amplitude increase is in proportion to $\sqrt{n}$. The over-all improvement in the signal-to-noise ratio is thus $\sqrt{n}$, resulting in an improvement of $10 \log_{10} n$ dB, i.e., 10 dB for 10 averages, 20 dB for 100, 30 dB for 1000.

Figure 13.42 shows the application of synchronous averaging to vibration signals from similar gearboxes in good and faulty condition. Figure 13.42*A* shows the enhanced time signal (120 averages) for the gear on the output shaft. The signal is fairly uniform and gives evidence of periodicity corresponding to the tooth-meshing. Figure 13.42*B* is a similarly enhanced time signal for a faulty gear; a localized defect on the gear is revealed. By way of comparison, Fig. 13.42*C* shows a single time record, without enhancement, for the same signal as in Fig. 13.42*B*; neither the tooth-meshing effect nor the fault is readily seen.

For best results, synchronous averaging should be combined with tracking. Where there is no synchronization between the digital sampling and the (analog) trigger signal, an uncertainty of up to one sample spacing can occur between successive digitized records. This represents a phase change of 360° at the sampling frequency, and approximately 140° at the highest valid frequency component in the signal, even with perfectly stable speed. Where speed varies, an additional phase shift occurs; for example, a speed fluctuation of 0.1 per cent would cause a shift of one sample spacing at the end of a typical 1024-sample record. The use of tracking analysis (generating the sampling frequency from the synchronizing signal) reduces both effects to a minimum.

CEPSTRUM ANALYSIS. Originally the *cepstrum* was defined as the power spectrum of the logarithmic power spectrum.[11] A number of other terms, commonly found in the cepstrum literature (and with an equivalent meaning in the cepstrum domain) are derived in an analogous way, e.g., *cep*strum from *spec*trum, *quef*rency from *freq*uency, *rah*monic from *har*monic. The distinguishing feature of the cepstrum is not just that it is a spectrum of a spectrum, but rather that it is the spectrum of a spectrum on a logarithmic amplitude axis; by comparison, the autocorrelation function (see Eq. 27.14) is the inverse Fourier transform of the power spectrum without logarithmic conversion.

F IG. 13.42. Use of signal enhancement in gear fault diagnosis. (*A*) Enhanced signal (120 averages) for a gear in normal condition. (*B*) Enhanced signal (120 averages) for a similar gear with a local fault. (*C*) Section of raw signal corresponding to (*B*).

Most commonly, the *power cepstrum* is defined as the inverse Fourier transform of the logarithmic power spectrum,[12] which differs primarily from the original definition in that the result of the second Fourier transformation is not modified by obtaining the amplitude squared at each quefrency; it is thus reversible back to the logarithmic spectrum. Another type of cepstrum, the *complex cepstrum*, discussed below, is reversible to a time signal.

Figure 13.43, the analysis of a vibration signal from a faulty bearing, shows the advantage of the power cepstrum over the autocorrelation function. In Fig. 13.43A, the same power spectrum is depicted on both linear and logarithmic amplitude axes; in (B) and (C) the autocorrelation and cepstrum, respectively, are shown. In (C), the logarithmic depiction of the power spectrum reveals a family of harmonics which are concealed in the linear depiction. The presence of the family of harmonics is made evident by a corresponding series of rahmonics in the cepstrum (denoted ①, ②, etc.), but is not detected in the autocorrelation function. The quefrency axis of the cepstrum is a time axis, most

Fig. 13.43. Effect of linear vs. logarithmic amplitude scale in power spectrum. (A) Power spectrum on linear scale (lower curve) and logarithmic scale (upper curve). (B) Autocorrelation function (obtained from linear representation). (C) Cepstrum (obtained from logarithmic representation)—①, ②, etc., are rahmonics corresponding to harmonic series in spectrum (4.85 milliseconds equivalent to 1/206 Hz). The harmonics result from a fault in a bearing.

closely related to the X axis of the autocorrelation function (i.e., time delay or periodic time rather than absolute time). The reciprocal of the quefrency of any component gives the equivalent *frequency spacing* in the spectrum, not the absolute frequency.

Most of the applications of the power cepstrum derive from its ability to detect a periodic structure in the spectrum, for example, families of uniformly spaced harmonics and/or side-bands. The application of the cepstrum to the diagnosis of faults in gears and rolling element bearings is discussed in Chap. 16 and Ref. 13.

To obtain a distinct peak in the cepstrum, a reasonable number of the members of the corresponding harmonic or side-band family must be present (although the fundamental may be absent). These uniformly spaced components must be adequately resolved in the spectrum. As a guide, the spacing of components to be detected should be a minimum of eight lines in the original spectrum. For this reason, it is often advantageous to perform a cepstrum analysis on a spectrum obtained by *zoom FFT*. In this case it is desirable to use a slightly modified definition of the cepstrum corresponding to the amplitude of the *analytic signal*.[13] (See the next section on *Hilbert Transform Techniques*.)

The *complex cepstrum*[12,14] (referred to above) is defined as the inverse Fourier transform of the complex logarithm of the complex spectrum. Despite its name it is a real-valued function of time, differing from the power cepstrum primarily in that it uses phase as well as logarithmic amplitude information at each frequency in the spectrum. It is thus reversible to a time function (from which the complex spectrum is obtained by direct Fourier transformation).

FIG. 13.44. Diesel engine cylinder pressure signal, derived from an externally measured vibration-acceleration signal using cepstrum techniques. (*From R. H. Lyon and A. Ordubadi.*[15])

Measured vibration signals generally represent a combination of source and transmission path effects; for example, internal forces in a machine (the source effect) act on a structure whose properties may be described by a frequency-response function between the point of application and the measurement point (the transmission path effect). As shown in Refs. 12 and 14, the source and transmission path effects are convolved in the time signals, multiplicative in the spectra, and additive in the logarithmic spectra and in the cepstra (both power cepstra and complex cepstra). In the cepstra, they quite often separate into different regions, which in principle allows a separation of source and transmission path effects in an externally measured signal.

Figure 13.44 shows an example of an internal cylinder pressure signal in a diesel engine, derived from an externally measured vibration acceleration signal making use of cepstrum techniques to generate the inverse filter.[15]

Reference 16 gives similar results for the tooth-mesh signal in a gearbox and also shows that a frequency-response function derived by windowing in the cepstrum of an output signal compares favorably with a direct measurement (which requires measurement of both an input and an output signal).

HILBERT TRANSFORM TECHNIQUES. The *Hilbert transform* is the relationship between the real and imaginary parts of the Fourier transform of a one-sided signal.[17] An example is a causal signal such as the impulse response of a vibratory system (a *causal signal* is one whose value is zero for negative time). The real and imaginary parts of the frequency response (the Fourier transform of the impulse response) are related by the Hilbert transform; thus, only one part need be known—the other can be calculated.

Analogously, the time function obtained by an inverse Fourier transformation of a one-sided spectrum (positive frequencies only) is complex, but the imaginary part is the Hilbert transform of the real part. Such a complex time signal is known as an *analytic signal.*

An analytic signal can be thought of as a rotating vector (or phasor) described by the formula $A(t)e^{j\phi(t)}$ whose amplitude $A(t)$ and rotational speed $\omega(t) = d\phi(t)/dt$, in general, vary with time. Analytic signals are useful in vibration studies to describe modulated signals. For example, a *phase-coherent* signal [Eq. (22.3)] can be represented as the real part of an analytic signal, in which case the imaginary part can be obtained by a Hilbert transform. Therefore, from a measured time signal, $a(t)$, it is possible to obtain the amplitude and phase (or frequency) modulation components from the relationship

$$A(t)e^{j\phi(t)} = a(t) + j\tilde{a}(t) \tag{13.26}$$

where $\tilde{a}(t)$ is the Hilbert transform of $a(t)$.

The Hilbert transform may be evaluated directly from the equation

$$\tilde{a}(t) = \frac{1}{\pi} \int_{-\infty}^{\infty} a(\tau) \frac{1}{t - \tau} \, d\tau \tag{13.27}$$

but it can be more readily evaluated by a phase shift in the frequency domain, in particular in an FFT analyzer.[18] An alternative way of generating analytic signals using an FFT analyzer is by an inverse Fourier transformation of the equivalent one-sided spectrum formed from the spectrum of the real part only. The time signals resulting from the real-time zoom process (described above) automatically have the same amplitude function $A(t)$ as the equivalent bandpass-filtered analytic signal, since they are obtained

Fig. 13.45. Analytic signal for simple amplitude modulation. (A) Analytic signal $a(t + j\tilde{a}(t) = A(t)\,e^{j\phi(t)}$. (B) Real part $a(t)$. (C) Imaginary part $\tilde{a}(t)$. (D) Amplitude $A(t)$. (E) Phase $\phi(t)$.

from the positive frequency components only [Fig. (13.24)]. The frequency-shifting operation only affects the phase function $e^{j\phi(t)}$.

The major applications of Hilbert transform techniques in vibration studies involve either amplitude demodulation or phase demodulation.

Amplitude Demodulation. Figure 13.45 shows the analytic signal for the case of single frequency amplitude modulation of a higher frequency carrier component. The imaginary part is the Hilbert transform of the real part; this manifests itself as a 90° phase lag. The amplitude function is the envelope of both the real and imaginary parts and represents the modulating signal plus a dc offset. The phase function is a linear function of time (whose slope represents the speed of rotation, or frequency, of the carrier component); it is, however, shown modulo 2π, as is conventional.

One area of application of amplitude demodulation where it is advantageous to view the signal envelope rather than the time signal itself is in the interpretation of such oscillating time functions as autocorrelation and crosscorrelation functions (see Chap. 22). Figure 13.46[19] shows a typical case where peaks indicating time delays are difficult to

FIG. 13.46. Example of a crosscorrelation function expressed as follows:[19] (A) The real part of an analytic signal, i.e., the normal definition (Eq. 22.27). (B) The amplitude of the analytic signal. The peaks corresponding to time delays are more easily seen in this representation. The signal was obtained by bandpass filtering (using FFT zoom) in the frequency range from 512 to 13,312 Hz.

identify in a crosscorrelation function as defined in Eq. (22.27), because of the oscillating nature of the basic function (Fig. 13.46A). The peaks are much more easily seen in the envelope or magnitude of the analytic signal (Fig. 13.46B). Another advantage of the analytic signal is that its magnitude can be displayed on a logarithmic axis; this allows low-level peaks to be detected and converts exponential decays to straight lines.[19]

Another area of application of amplitude demodulation is in *envelope analysis* (discussed in the earlier section on *Envelope Detectors*). In particular, when the signal is to be bandpass-filtered before forming the envelope, this can be done by real-time zoom in the appropriate passband. Figure 13.47 shows an example from the same vibration source as was analyzed in Fig. 13.43. Figure 13.47A shows a typical envelope signal obtained from zooming in a 1,600-Hz band centered at 3 kHz. The spectrum of Fig. 13.43A shows that this frequency range is dominated by the harmonic family which results from a fault in a bearing. Consequently, the corresponding envelope signal (Fig. 13.47A) indicates a series of bursts with the same period, 4.85 milliseconds (compare with the cepstrum of Fig. 13.43C). Figure 13.47B shows the average spectrum of a number of such envelope signals; this gives a further indication that the dominant periodicity is 206 Hz.

Fɪɢ. 13.47. Envelope analysis using Hilbert transform techniques. (*A*) Typical envelope signal showing bursts with a period of 4.85 milliseconds from a fault in a ball bearing. (*B*) Average spectrum of the envelope signal showing corresponding harmonics of 206 Hz. Signal obtained by bandpass filtering (using FFT zoom) in the frequency range from 2,200 to 3,800 Hz (compare with Fig. 13.43*A* which shows a base-band analysis of this same signal).

Phase Demodulation. For a purely phase-modulated signal, the amplitude function $A(t)$ is constant and the phase function $\phi(t)$ is given by the sum of a carrier component of constant frequency f_c and the modulation signal $\phi_m(t)$. Thus

$$\phi(t) = 2\pi f_c t + \phi_m(t) \tag{13.28}$$

Real-time zoom analysis centered on frequency f_0 subtracts this frequency from all components in the signal; consequently, by zooming at the carrier frequency f_c, only the modulation signal $\phi_m(t)$ remains. In general it is only possible to zoom exactly at the carrier frequency when the latter is made to coincide exactly with an analysis line (for example, by employing order tracking). Otherwise, the small difference in frequency gives a residual slope to the phase signal. Figure 13.48 shows an example of the application of the phase demodulation technique to the determination of the torsional vibration signal of a ship's diesel engine. Figure 13.48*A* shows the phase of the complex time signal obtained by zooming at a frequency as close as possible to the toothpass frequency from an accurately machined toothed wheel mounted on the shaft. The toothpass signal was obtained from a stationary proximity transducer which generated a pulse for the passage of each tooth. For comparison purposes, Fig. 13.48*B* shows the equivalent signal obtained from an instrument specifically designed to extract torsional vibration signals (by frequency demodulation and integration). In this case, tracking was not used; the torsional vibra-

FIG. 13.48. Measurement of torsional vibration from a ship's diesel engine over approximately two cycles (four revolutions) determined from the toothpass signal of a 63-tooth wheel. (*A*) The phase angle of the complex time signal using zoom FFT as close as possible to the toothpass frequency. The small slope corresponds to the difference of the latter from the zoom center frequency. (*B*) Torsional vibration signal determined by frequency demodulation and integration. The scales of the two results are different because of the different analysis techniques.

tion signal in Fig. 13.48*A* thus has a superimposed slope corresponding to the small frequency difference.

REFERENCES

1. Wahrman, C. G.: *Brüel & Kjaer Tech. Rev.*, (3) (1958).
2. Dyer, D., and R. M. Stewart: *J. Mech. Des.*, **100**:229 (1978).
3. Cooley, J. W., and J. W. Tukey: *Math. Computing*, **19**(90):297 (1965).
4. Cooley, J. W., P. A. W. Lewis, and P. D. Welch: *J. Sound Vibration*, **12**(3):315 (1970).
5. Brigham, E. O.: "The Fast Fourier Transform," Prentice-Hall, Inc., Englewood Cliffs, N.J., 1974.
6. Thrane, N.: "Zoom-FFT," *Brüel & Kjaer Tech. Rev.*, (2) (1980).
7. Sloane, E. A.: *IEEE Trans. Audio Electroacoust.* **AU-17**(2):133 (1969).
8. Welch, P. D.: *IEEE Trans. Audio Electroacoust.* **AU-15**(2):70 (1967).
9. Randall, R. B.: "Frequency Analysis," Brüel & Kjaer, Naerum, Denmark, 1987.
10. Mitchell, J. S.: "An Introduction to Machinery Analysis and Monitoring," Penwell Publishing Company, Tulsa, Okla., 1981.
11. Bogert, B. P., M. J. R. Healy, and J. W. Tukey: In M. Rosenblatt (ed.), "Proceedings of the Symposium on Time Series Analysis," John Wiley & Sons, Inc., New York, 1963, pp. 209–243.
12. Childers, D. G., D. P. Skinner, and R. C. Kemerait: *Proc. IEEE*, **65**(10):1428 (1977).
13. Randall, R. B.: *Maintenance Management Int.* **3**:183 (1982/1983).

14. Oppenheim, A. V., R. W. Schafer, and T. G. Stockham Jr.: *Proc. IEEE,* **56**(August):1264 (1968).
15. Lyon, R. H., and A. Ordubadi: *J. Mech. Des.,* **104**(Trans. ASME)(April):303 (1982).
16. DeJong, R. G., and J. E. Manning: "Gear Noise Analysis using Modern Signal Processing and Numerical Modeling Techniques," *SAE Paper* No. 840478, 1984.
17. Papoulis, A.: "The Fourier Integral and its Applications," McGraw-Hill Book Company, Inc., New York, 1962.
18. Thrane, N.: *Brüel & Kjaer Tech. Rev.,* (3) (1984).
19. Herlufsen, H.: *Brüel & Kjaer Tech. Rev.,* (1 and 2) (1984).

14

SPECIAL-PURPOSE TRANSDUCERS

John E. Judd
Vibra-Metrics, Inc.

John D. Ramboz
Consultant

INTRODUCTION

This chapter describes special-purpose and miscellaneous transducers used in the measurement of shock and vibration. These include:

Piezoelectric accelerometers with internal electronics
Optical-electronic transducer systems
Inductive-type transducers
Servo accelerometers
Potentiometer-type transducers
Capacitance-type transducers

PIEZOELECTRIC ACCELEROMETERS WITH INTERNAL ELECTRONICS

A broad range of piezoelectric accelerometers containing microelectronic circuits within their cases is available. These circuits perform some of the functions described under *Accelerometer Preamplifiers* and *Signal Conditioners* in Chap. 13, such as signal amplification, equalization, and integration. They typically provide a low-impedance output as well. The low-impedance output gives freedom from cable-induced noise, permits the use of lower-cost coaxial cable, and usually results in lower signal-conditioning costs. However, the maximum temperature at which they can be operated is limited by the electronic circuitry. This temperature is usually somewhat lower than that for the piezoelectric sensor. Furthermore, the gain cannot be adjusted to utilize the full dynamic range of the sensor. Units of this type are available from most manufacturers.

VELOCITY-OUTPUT DEVICES

Piezoelectric accelerometers are available with internal electronic circuitry which integrates the output signal provided by the accelerometer, thereby yielding a velocity or displacement output. These transducers have several advantages not possessed by ordinary velocity pickups. They are smaller, have a wider frequency response, have better resolution, have no moving parts, and are relatively unaffected by magnetic fields where measurements are made. Some units of this type have provisions for internal calibration *in situ.* Filtering and equalization may be included. For example, Fig. 14.1*A* shows the frequency-response characteristics with and without a low-pass filter having a high-frequency cutoff of 2.5 Hz. Figure 14.1*B* shows the velocity output without the low-pass filter; note the presence of considerable high-frequency energy. Figure 14.1*C* shows the velocity output with the use of the low-pass filter. The use of such filters can be important in a situation where high-frequency, high-level signals from the transducer may overload the input of the amplifier which follows. The application of filtering at this point in the system is sometimes referred to as "prefiltering."

OPTICAL-ELECTRONIC TRANSDUCER SYSTEMS

LASER DOPPLER SYSTEM

The laser Doppler vibration measurement system provides a real-time analog output proportional to the instantaneous velocity of the test surface. The instrument is designed for noncontact vibration measurements of rough surfaces which reflect light diffusively and for velocity measurements of continuously moving liquid and solid surfaces. Measurements of small amplitude (10 nm) to large amplitude (1 m) can be made in the frequency range from 0 Hz to 740 kHz. The dynamic velocity range is from 10^{-6} to 3 m/s. Measurements can be made at distances between 1.2 and 20 m from the vibrating surface.

This measurement technique is based on the detection of the Doppler frequency shift of light from a laser beam which is backscattered from the moving object, as illustrated in Fig. 14.2. The Doppler frequency shift is detected by comparison of the reflected beam with the original beam, using an optical heterodyning technique. The output current of the detector contains the Doppler frequency information as a linear function of the moving target velocity. The Doppler frequency is processed by a frequency tracker, which generates an output voltage proportional to the instantaneous velocity.

DISPLACEMENT MEASUREMENT SYSTEM

The electro-optical displacement measurement system consists of an electro-optical sensor and a servo-control unit designed to track the displacement of the motion of a light-dark target. This target provides a light discontinuity in the intensity of reflected light from an object. If such a light-dark discontinuity is not inherent to the object under study, a light-dark target may be applied on the object. An image of the light-dark target is formed by a lens on the photocathode of an image dissector photomultiplier tube, as shown in Fig. 14.3. The photocathode emits electrons in proportion to the intensity of the light striking the tube, causing an electron image to be generated in real time. The electron image is accelerated through a small aperture that is centrally located within the phototube. The number of electrons that enter the aperture constitute a small electric current that is directly proportional to the amount of light striking the corresponding area on the photocathode. This signal current is then amplified. As the light-dark target moves across the face of the phototube, the output current changes from high (light) to low (dark). When the target is exactly at the center of tube, the output current represents half light and half dark covering the aperture. If the target moves away from this position, the output current changes. This change is detected by the control unit which feeds a compensation current back to the optical tracking head. The current that is needed for this deflection is directly

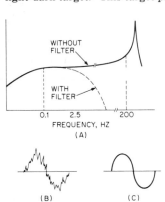

Fig. 14.1. (*A*) Frequency-response characteristic of a piezoelectric accelerometer with internal electronics which provides a velocity output, with and without a low-pass filter. (*B*) Output waveform without filter. (*C*) Same as (*B*) but with filter.

proportional to the distance that the image has moved away from the center. Therefore it is a direct measure of displacement.

The displacement amplitudes that can be measured range from a few micrometers to several meters; the exact value is determined by the lens selected. Systems are available which measure displacements in one, two, or three directions.

FIG. 14.2. Laser Doppler vibration measurement system. *(Courtesy of DISA Electronics.)*

FIG. 14.3. Image dissector tube of an electro-optical displacement measurement system. *(Courtesy of Optron Corp.)*

FIBER-OPTIC REFLECTIVE DISPLACEMENT SENSOR

A fiber-optic reflective displacement sensor measures the amount of light normal to, and vibrating along, the optical axis of the device. The amount of reflected light is related to the distance between the surface and the fiber-optic transmitting/receiving element, as illustrated in Fig. 14.4. The sensor is comprised of two bundles of single optical fibers. One of these bundles transmits light to the reflecting target; the other traps reflected light and transmits it to a detector. The intensity of the detected light depends on how far the reflecting surface is from the fiber-optic probe. Light is transmitted from the bundle of fibers in a solid cone defined by a numerical aperture. Since the angle of reflection is equal to the angle of incidence, the size of the spot that strikes the bundle after reflection is twice the size of the spot that hits the target initially. As the distance from the reflecting surface increases, the spot size increases as well. The amount of reflected light is inversely proportional to the spot size.

FIG. 14.4. Fiber-optic displacement sensor. *(Courtesy of EOTEC Corp.)*

As the probe tip comes closer to the reflecting target, there is a position in which the reflected light rays are not coupled to the receiving fiber bundle. At the onset of this occurrence, a maximum forms which drops to zero as the reflecting surface contacts the probe. The output-current sensitivity can be varied by using various optical configurations.

While sensitivities approaching 1 microinch are possible, such extreme sensitivities limit the corresponding dynamic range. If the sensor is used at a distance from the reflecting target, a lens system is required in conjunction with a fiber-optic probe. With available lenses, the instruments have displacement measurement ranges from 0 to 0.015 in. (0 to 0.38 mm) and 0 to 5.0 in. (0 to 12.7 cm). Resolution typically is better than one one-hundredth of the full-scale range. The sensor is sensitive to rotation of the reflecting target. For rotations of $\pm 3°$ or less, the error is less than ± 3 per cent.

INDUCTIVE-TYPE PICKUPS[1]

ELECTRODYNAMIC PICKUPS

The output voltage of the electrodynamic pickup is proportional to the relative velocity between the coil and the magnetic flux lines being cut by the coil. For this reason it is commonly called a velocity pickup. The principle of operation of the device is illustrated in Fig. 14.5. A magnet has an annular gap in which a coil wound on a hollow cylinder of nonmagnetic material moves. Usually a permanent magnet is used, although an electromagnet may be used. The pickup also can be designed with the coil stationary and the magnet movable. The open-circuit voltage e generated in the coil is[2, 3]

$$e = -Blv(10^{-8}) \qquad \text{volts} \qquad (14.1)$$

Fig. 14.5. Principle of operation of an electrodynamic pickup. The voltage e generated in the coil is proportional to the velocity of the coil relative to the magnet.

where B is the flux density in gausses; l is the total length in centimeters of the conductor in the magnetic field; and v is the relative velocity in centimeters per second between the coil and magnetic field. The magnetic field decreases sharply outside the space between the pole pieces; therefore, the length of coil wire outside the gap generates only a very small portion of the total voltage.

SEISMIC ELECTRODYNAMIC PICKUPS. The electrodynamic principle is used in the velocity-type seismic pickup. There are a number of variations in the design of transducers of this type, two of which are illustrated in Fig. 14.6. The permanent magnet of the pickup shown in Fig. 14.6A is attached to the housing. The coil is suspended by springs from the housing. Above its natural frequency the coil stands still in space when the housing is vibrated. In the design shown in Fig. 14.6B the coil is attached rigidly to the housing, and the permanent magnet is mounted seismically through a spring to the housing of the pickup. Usually the pickup is used only at frequencies above its natural frequency, and it is not very useful at frequencies above several thousand hertz. The sensitivity of most pickups of this type is quite high, particularly at low frequencies where their output voltage is greater than that of many other types of pickups. The coil impedance is low even at relatively high frequencies so that the output voltage can be measured directly with a high-impedance voltmeter.

This type of pickup is designed to measure quite large displacement amplitudes. Its use in some applications is limited by the effect of its relatively large weight on the structural response. For a pickup in which the coil is seismically mounted, the mechanical loading on the structure corresponds closely to the total weight of the pickup. When the permanent magnet is the seismic mass element, the effective mass attached to the vibrating structure is less than the total weight of the pickup. The actual effect on the structure is dependent on the mechanical impedance of the pickup. The pickup must be attached quite rigidly to the structure, particularly for measurements at very high frequencies. The contact resonance, i.e., the resonance between the pickup and the structure, should be considerably higher than the highest frequency of motion to be measured.

The contact resonance of a typical pickup, for example, is about 3,000 Hz when attached to the structure by a screw of approximately 0.4 in. in diameter.[4]

Most of the pickups can be used to measure vibration in either the vertical or horizontal direction. In some cases it is necessary to adjust the springs in the pickup to compensate for the deflection of the mass element due to its own weight when the position of the pickup is changed. Regardless of position, the transverse sensitivity of the pickup is usually near zero. However, for a pickup in which frictional contact can exist on the moving parts inside the pickup, the output may not indicate very accurately the motion along the sensitive axis when the pickup is vibrated in another direction.

EDDY-CURRENT PICKUPS

Consider a nonferrous plate of low resistivity which is moved in a direction perpendicular to the flux lines of a magnet, as indicated in Fig. 14.7. A current is generated in the plate which is proportional to the velocity of the plate; these eddy currents set up a magnetic field in a direction opposing the magnetic field that creates them. Thus, the resulting magnetic field varies as the generated current changes. The output voltage e

(A)

(B)

FIG. 14.6. Rectilinear velocity pickup based on the electrodynamic principle. The pickup is used at frequencies above its natural frequency. *(Courtesy of Consolidated Electrodynamics Corp.)*

of a coil of wire wound on the magnet is proportional to the time rate of change of the eddy currents, and therefore to the acceleration of the motion of the plate relative to the flux field. The above principle forms the basis of the eddy-current pickup. Such a device has an advantage over an electromagnetic pickup in that the voltage is linearly related to the motion since the gap between the plate and electromagnet remains constant.

A significant feature of a pickup based on the eddy-current principle is that the force produced by the flux field from the eddy currents opposes the motion of the plate and is proportional to its velocity. Thus mechanical damping is achieved. Such a pickup has inherent damping in the same plate that is used to cause a voltage to be generated in the coil.

In the pickup which is based on the eddy-current principle, a permanent magnet and coil are rigidly attached to the housing of the pickup. A rectangular plate is seismically mounted on the housing. The eddy currents generated in the plate induce a voltage in the coil and at the same time provide damping in the pickup. Above the natural frequency of the pickup, the generated voltage is proportional to the acceleration applied to the pickup. The frequency response of this type of pickup is nearly ideal for measuring transient motion. The phase lag between the output voltage and acceleration applied to the pickup is close to 0° at frequencies above

FIG. 14.7. Eddy-current principle. Current generated in the plate is proportional to the velocity of the plate as it cuts the flux field in the gap of the magnet. The current reduces the flux field, and the voltage e generated in the coil is proportional to the acceleration of the plate.

its natural frequency, and the percentage increase in the sensitivity is very small near the natural frequency of the pickup. These characteristics reduce the distortion in the output voltage of the pickup.

DIFFERENTIAL-TRANSFORMER PICKUPS

The output of a differential-transformer pickup depends on the mutual inductance between a primary and secondary coil. The basic components are shown in Fig. 14.8. The pickup consists of a core of magnetic material, a primary coil, and two secondary coils. As the core moves, a voltage is induced in the secondary coils. When the core is exactly in the center, each secondary coil contains the same length of core. Therefore, the mutual inductances of both secondary coils are equal in magnitude. However, they are connected in series opposition so that the output voltage is zero. As the core is moved up or down, both the inductance and induced voltage of one secondary coil are increased while those of the other are decreased. The output voltage is the difference between these two induced voltages. In this type of transducer, the output voltage is proportional to the displacement of the core over an appreciable range. In practice, the output voltage at the carrier frequency of the primary current is not exactly zero when the core is centered, and the output near the center position is not exactly linear. When the core is vibrated, the output voltage is a carrier wave, modulated at a frequency and amplitude corresponding to the motion of the core relative to the coils. The cores and coils can be arranged in several ways, two of which are described below.

FIG. 14.8. Differential-transformer principle. The inductance of the coils changes as the core is moved. For constant input current i_p to the primary coil, the output voltage e is the difference of the voltages in the two secondary coils which are wound in series opposition. *(Courtesy of Automatic Timing and Controls, Inc.)*

MOVABLE-CORE   DIFFERENTIAL-TRANSFORMER PICKUPS. These pickups are used for very low frequency measurements. The sensitivity varies with the carrier frequency of the current in the primary coil. The carrier frequency should be at least 10 times the highest frequency of the motion to be measured. Since this range is usually between 0 and 60 Hz, the carrier frequency is usually above 600 Hz.

SEISMIC DIFFERENTIAL-TRANSFORMER PICKUPS. A design of this type is shown in Fig. 14.9. The flat diaphragm spring permits the mass-element iron core to move axially in the differential transformer. A pickup of this type is used in the frequency range from 0 Hz up to about two-thirds of its natural frequency. Pickups with high acceleration ranges have wider usable frequency ranges; however, their sensitivities are correspondingly lower.

The effect of temperature on these pickups is small. If the pickup were used only for sinusoidal motion measurements at frequencies up to one-third of this undamped natural frequency, it would be preferable to use no damping. However, in general-purpose testing and in measuring transient motions when significant components of the motion are present at frequencies near the natural frequency, damping is necessary to avoid excessive voltage outputs and to avoid contact between the mass element and housing of the pickup as a result of excessive displacements. For accurate measurements of transient motion, it is also necessary for the phase angle of the output voltage and input motion to be zero or to be directly proportional to frequency within its operating range.

SERVO ACCELEROMETER

The servo accelerometer (sometimes called a "force-balance accelerometer") contains a seismically suspended mass which has a displacement sensor (e.g., a capacitance-type transducer) attached to it. If the accelerometer is subject to motion, the mass is displaced with respect to the accelerometer case; this displacement generates a servo-loop error signal, establishing a restoring force that is applied to the mass in such a way as to reduce the initial error signal to zero. The acceleration, which is proportional to the restoring force, may be determined from the voltage drop across a resistance in series with a coil which generates the restoring force.

Servo accelerometers can be made very sensitive, some having threshold sensitivities on the order of a few micro-g.[5] Excellent amplitude linearity is attainable, usually on the order of a few hundredths of one per cent with peak acceleration amplitudes up to 40 g. Typical frequency ranges are from 0 to 500 Hz.

POTENTIOMETER-TYPE TRANSDUCER

The potentiometer-type transducer is a variable-voltage divider in which the output voltage is a function (usually linear) of a displacement. If the displacement varies with time, then the output voltage will vary with time in a similar manner.

This type of transducer is used in the study of the relative motion of one part of a structure with respect to another. It is also used as the sensing elements in force gages, pressure gages, etc., and in remote position-indicating systems.

The potentiometer-type transducer may incorporate a seismic system consisting of a spring mounted mass whose motion with respect to the base is measured. In this application, the transducer output voltage is proportional to the vibration acceleration of the base, provided that the vibration frequency is substantially below the natural frequency of the seismic system (see Chap. 12). If the vibration frequency is substantially above the natural frequency, the transducer output voltage is proportional to the vibration velocity.

The potentiometer-type transducer circuit is shown schematically in Fig. 14.10A and its equivalent circuit in Fig. 14.10B. The potentiometer is energized with the dc voltage e. Its total circuit resistance is r_p, which includes the resistance of the dc source, the potentiometer, and the line between the source and potentiometer. The potentiometer senses a displacement by moving a variable contact through a distance which is proportional to the displacement; this movement causes a corresponding change in resistance r_2, or in the resistance ratio α = r_2/r_p in Fig. 14.10A.

Fig. 14.9. Typical differential-transformer acceleration pickup used below its natural frequency. The output voltage of the differential transformer is proportional to the acceleration of the housing in the operating frequency range of the pickup. (*Courtesy of Gulton Industries, Inc.*)

When there is no electrical circuit load on the output terminal of the potentiometer, the output voltage $e_{\text{out}} = e\alpha$ (the equivalent generator voltage of Fig. 14.10B). If an indicating instrument or load having a resistance r_L is connected across the potentiometer output terminals, the loading effect[6–8] introduces an error or nonlinearity in the transducer action as shown by Eq. (14.2). Then the output voltage e_{out} is not a linear function of α, but is given by

$$e_{\text{out}} = \frac{e}{\dfrac{r_p}{r_2} + \dfrac{(r_p - r_2)}{r_L}}$$

(14.2)

The full-scale travel of the movable contact in the potentiometer depends upon the particular transducer. In some accelerometers, a full-scale travel of less than 0.1 in. is used. If the transducer is used in the study of the motion of controls, control surfaces, actuators, etc., a travel up to 10 in. may be required.

The sensitivity in terms of output voltage per unit travel of the movable contact is seldom listed by the manufacturer. The general practice is to give the operating limitations, and to let users calculate the sensitivity according to their own specific operating conditions. Manufacturers recommend either the maximum energizing voltage, the maximum energizing current, or the maximum power input to the potentiometer. Knowing these and the full-scale travel, it is possible to determine the sensitivity.

The resolution of potentiometers used in transducers is usually between 0.1 and 0.5 per cent of the full range. It represents the step change in voltage obtained when the movable contact moves from one turn to the next in the potentiometer resistance winding.

The potentiometer can be linear to within 0.1 per cent full scale in the more accurate transducers, but the average value is about 1.0 per cent full scale.

FIG. 14.10. Potentiometer-type transducer circuits: (A) typical circuit and (B) equivalent circuit.

Typical potentiometers in such transducers have a resistance r_p between 100 and 10,000 ohms. The actual value of r_p which is selected is a compromise between current drain on the battery and the resulting heat dissipation within the potentiometer, and the nonlinearity due to the loading effect of r_L in the indicating instrument.

Friction is a serious drawback in potentiometers when they are used with seismic instruments. The sticking friction determines the threshold sensitivity, while both sticking and rubbing friction result in a hysteresis loop.

CAPACITANCE-TYPE TRANSDUCER

The capacitance-type transducer is basically a displacement-sensitive device. Its output is proportional to the change in capacitance between two plates caused by the change of relative displacement between them as a result of the motion to be measured. Appropriate electronic equipment is used to generate a voltage corresponding to the change in capacitance.

Transducers based upon the variation of capacitance are primarily special-purpose devices. This is principally due to the fact that auxiliary electronic equipment required for their operation is inherently more complicated than that required for the more popular types of transducers. The capacitance-type transducer's main advantages are (1) its simplicity in installation, (2) its negligible effect on the operation of the vibrating system since it is a proximity-type pickup which adds no mass or restraints, (3) its extreme sensitivity, (4) its wide displacement range, due to its low background noise, and (5) its wide frequency range, which is limited only by the electric circuit used.

The capacitance-type transducer often is applied to a conducting surface of a vibrating system by using this surface as the ground plate of the capacitor. In this arrangement, the insulated plate of the capacitor should be supported on a rigid structure close to the vibrating system. Figure 14.11A shows the construction of a typical capacitance pickup; Figs. 14.11B, C, D, and E show a number of possible methods of applying this type of transducer. In each of these, the metallic vibrating system is the ground plate of the capacitor. Where the vibrating system at the point of instrumentation is an electrical

FIG. 14.11. Capacitance-type transducers and their application: (A) construction of typical assembly, (B) gap length or spacing sensitive pickup for transverse vibration, (C) area sensitive pickup for transverse vibration, (D) area sensitive pickup for axial vibration, and (E) area sensitive pickup for torsional vibration.

Table 14.1. Formulas for Calculating Capacitance and Change in Capacitance for Various Geometries of Air-dielectric-type Capacitors

	Capacitance in air, pF		Change in capacitance due to change in position ΔC, pF
	Dimensions, in.	Dimensions, cm	
	$C = 0.225\,\dfrac{ab}{d}$	$C = 0.0885\,\dfrac{ab}{d}$	$\Delta C \simeq C\,\dfrac{\Delta d}{d}$
	$C = 0.225\,\dfrac{ab}{d}$	$C = 0.0885\,\dfrac{ab}{d}$	$\Delta C \simeq C\,\dfrac{\Delta a}{a}$
	$C = 0.225\,\dfrac{\pi R^2}{d}$	$C = 0.0885\,\dfrac{\pi R^2}{d}$	$\Delta C \simeq C\,\dfrac{\Delta d}{d}$
	$C = 0.353\,\dfrac{R_1{}^2 - R_2{}^2}{d}\left(\dfrac{\theta}{\pi}\right)$	$C = 0.139\,\dfrac{R_1{}^2 - R_2{}^2}{d}\left(\dfrac{\theta}{\pi}\right)$	$\Delta C \simeq C\,\dfrac{\Delta\theta}{\theta}$
	$C = \dfrac{0.613l}{\log\dfrac{R_1}{R_2}}$	$C = \dfrac{0.242l}{\log\dfrac{R_1}{R_2}}$	$\Delta C \simeq C\,\dfrac{\Delta l}{l}$

14–9

insulator, the surface can be made slightly conducting and grounded by using a metallic paint or by rubbing the surface with graphite.

The maximum operating temperature of the transducer is limited by the insulation breakdown of the plate supports and leads. Bushings made of alumina are commercially available and provide adequate insulation at temperatures as high as 2000°F (1093°C).

THE CAPACITOR

DISPLACEMENT-TO-CAPACITANCE SENSITIVITY.   The capacitance C between two parallel conducting plates that are insulated from each other is directly proportional to the effective area S of the plates and is inversely proportional to the distance d between the plates, i.e.,

$$C = 0.225 \frac{S(\text{in.}^2)}{d(\text{in.})} = 0.0885 \frac{S(\text{cm}^2)}{d(\text{cm})} \quad \text{pF} \tag{14.3}$$

for an air-dielectric capacitor at standard conditions.   Equation (14.3) shows that the capacitance variation can be obtained either by changing the effective area S or by changing the spacing d.   Table 14.1 shows this equation applied to various geometric shapes of capacitors; it also gives the relationship for the change in capacitance ΔC due to a small change in plate position.   (Fringing effects are neglected.)

CAPACITANCE VERSUS DISPLACEMENT NONLINEARITY.   Variable-capacitance transducers have three principal sources of nonlinearity in the relationship of output voltage to mechanical displacement which should be considered in any application of the transducer: (1) the effect of fringing in the electric field between the capacitor plates, (2) the fact that the capacitance is inversely, and not directly, proportional to the spacing, and (3) the electric circuit associated with the transducer.   The effect of fringing depends to such an extent upon the transducer plate design and its shielding that it is difficult to determine analytically; fringing effects should be checked experimentally in each installation.

The capacitance variation, which is obtained when the spacing between the plates is changed, is not a linear function of the spacing.   This can lead to serious distortion unless the percentage change in the spacing is limited.

Electric circuits are available commercially in which a small change in capacitance will result in a corresponding change in output voltage.   Such devices are used with condenser microphones.

REFERENCES

1. Based on chap. 15, R. R. Bouche, "Shock and Vibration Handbook," 2d ed., C. M. Harris and C. E. Crede (eds.), McGraw-Hill Book Company, Inc., New York, 1976.
2. Mason, W. P.: "Electromechanical Transducers and Wave Filters," p. 187, D. Van Nostrand Company, Inc., Princeton, N.J., 1948.
3. Nelson, H. M.: "Moving Coil Vibration Generators and Pickups," *Instruments Note* 54, Aeronautical Research Laboratories, Australia, 1954.
4. Anon.: "Philips Vibration Measuring and Exciting Instruments," Philips, Eindhoven, Holland.
5. Pope, K. E.: *Control Eng.*, **5**(11):97 (1958).
6. Nettleton, L. A., and F. E. Dole: *Rev. Sci. Instr.*, **11**(5):332 (1947).
7. Kaufman, A. B.: *Radio-Electronic Engr.*, **44**(3):12 (1952).
8. Post, R. E., and J. Zdzieborski: *Electrical Design News*, **4**(6):52 (1959).

15

MEASUREMENT TECHNIQUES

Cyril M. Harris
Columbia University

INTRODUCTION

Earlier chapters describe equipment used in vibration measurements. For example, detailed information concerning transducers, their characteristics, and how these characteristics are influenced by environmental factors is given in Chaps. 12 and 14. The various measurement system components and the characteristics which determine their selection are described in Chap. 13. The use of such measurement systems in vibration problems may involve only one or two engineers as in monitoring the condition of machinery in a factory (Chap. 16), in some problems in modal testing (Chap. 21), in measurements in building structures (Chap. 24), in measuring torsional vibration in reciprocating and rotating engines (Chap. 38), and in the balancing of rotating machinery (Chap. 39). In contrast, in the aerospace industry, some measurement problems are so complex that teams of engineers and several divisions of the company may be involved. Yet all these examples share certain basic measurement procedures. It is these basic procedures (rather than measurement details, which vary from problem to problem) that are considered here. Thus, this chapter includes a general discussion of (1) planning measurements to achieve stated objectives, (2) selecting the type of measurements which should be made to achieve these objectives, (3) selecting transducers, (4) mounting transducers, (5) mounting cable and wiring (including shielding and grounding), and (6) selecting techniques for the field calibration of the overall measurement system.

MEASUREMENT PLANNING

Careful pretest planning (and, in the case of a complex measurement program, detailed documentation) can save much time in making measurements and in ensuring that the most useful information is obtained from the test data. In many cases, as in environmental testing, measurement procedures are contained in test specifications. In other cases (as in balancing rotating machinery), measurement procedures are outlined in detail in national or international standards. In general, the first step in planning is to define the purpose of the test and to define what is to be measured. Planning should start with a clear definition of the test objectives, including the required accuracy and reliability. The second step is to define those nonequipment related factors which influence the selection of measurement equipment and measurement techniques. These include availability of trained personnel; cost considerations; length of time available for measurements; scheduling considerations; and available techniques for data analysis, validation, and presentation.

Next, the various factors listed in Table 15.1 should be considered. For example, it is important to have some estimate of the characteristics of the motion to be measured— e.g., its frequency range, amplitude, dynamic range, duration, and principal direction of

Table 15.1. Factors Which Are Important Considerations in the Selection of Measurement Equipment and Measurement Techniques for Mechanical Shock and Vibration Measurements

Parameter to Be Measured

Acceleration	Strain
Velocity	Force
Displacement	Mechanical impedance

Characteristics of Motion to Be Measured

Frequency range	Direction of motion
Amplitude range	Transient characteristics
Phase	Duration

Environmental Conditions

Temperature (ambient and transient)	Magnetic and radio-frequency fields
Humidity	Corrosive and abrasive media
Ambient pressure	Nuclear radiation
Acoustic noise	Sustained acceleration

Transducer Characteristics (See Chap. 12)

Electrical characteristics (sensitivity, resolution, cross-axis sensitivity, amplitude linearity,
  dynamic range, frequency response, phase response)
Physical characteristics (e.g., size and mass)
Self-generating or auxiliary power required
Electrically grounded to case, or isolated
Self-contained amplifier

Transducer Mountings and Locations of Mountings

Effect of mounting on transducer characteristics
Effect of mounting on vibratory characteristics of item under test
Number of measurement locations
Space availability for measurement locations
Accessibility for installation and maintenance
Ease of installation
Possibility of mounting misalignment with respect to intended direction of measurement

System Components (Preamplifiers, Signal Conditioners, Filters, Analyzers (See Chap. 13)

Electrical characteristics (e.g., input and output impedances)
Power availability
Noise interference (shielding, avoidance of ground loops)
Number of channels required for measurement and recording
Possible requirement for real-time information

Method of Data Transmission

Coaxial cable
Twisted pair of wires
Telemetry (channels assigned)
Optical fiber

Recording Equipment (See Chap. 13)

Recording-time capability
Electrical characteristics (e.g., signal-to-noise ratio)
Portability; power requirements
Correlation between recorded information and physical phenomena
Redundancy to minimize the risk of loss of vital information

Table 15.1.   Factors Which Are Important Considerations in the Selection of Measurement Equipment and Measurement Techniques for Mechanical Shock and Vibration Measurements (*Continued*)

Field Calibration
Transducers
Over-all measurement system

Data Analysis, Presentation, and Validation (See Chap. 20)
Manual or automatic
Type of presentation required

motion.   Such information is needed to provide the basis for the optimum selection of measurement equipment.   Yet often very little is known about the characteristics of the motion to be measured.   Previous experience may provide a guide in estimating signal characteristics.   Where this is not available, preliminary measurements may be carried out to obtain information which serves as a guide for further measurements.   For example, suppose preliminary measurements show a frequency spectrum having considerable content in the region of the lowest frequency measured.   This would indicate that the instrumentation capability should be extended to a somewhat lower frequency in subsequent measurements.   Thus an iterative process often takes place in a shock and vibration measurement program.   To speed this process, it is helpful to employ equipment whose characteristics cover a wide range and which has considerable flexibility.   Failure to take this feedback process into account can sometimes result in the acquisition of meaningless test results.   For example, a measurement program was carried out by one organization over a period of many weeks.   The objective was to correlate building vibration data, measured in the organization's own laboratories, with the acceptability of these laboratories as sites for ultrasensitive galvanometers and other motion-sensitive equipment.   No correlation was found and the entire measurement program was a waste of time, for two reasons:

1. The measurements were made with equipment with a frequency limit which was not sufficiently low, so that important spectral components of building vibration could not be measured.

2. Measurements were made only in the vertical direction, whereas it was the horizontal component which was dominant and which made certain laboratory areas unacceptable for the location of vibration-sensitive equipment.

Various factors that should be considered in planning instrumentation for shock and vibration measurements are listed in Table 15.1   Many of these are discussed in earlier chapters and are cross-referenced, rather than repeated, here.   For example, Chap. 13 describes various components which follow the transducer in a measurement system (such as preamplifiers, signal conditioners, filters, analyzers, and recorders) and describes the selection of the appropriate analyzer bandwidth, frequency scale, amplitude scale, selection of data windows, etc.

SELECTION OF THE PARAMETER TO BE MEASURED

Often, the selection of the parameter to be measured (displacement, velocity, acceleration, or strain) is predetermined by specifications or by standards.   When this is not the case, it is often helpful to apply the considerations given in Table 15.2 or to apply the *flatness rule* described in Chap. 16.   According to this rule, the best motion parameter to use is the one whose spectrum is closest to being uniform (i.e., the one having the flattest spectrum).   This is important for two reasons: If the spectrum is relatively flat, then (1) an increase at any frequency has a roughly even chance of influencing overall vibration levels, and (2) minimum demands are placed on the required dynamic range of the equipment which follows the transducer. For example, Fig. 16.2 shows two spectra obtained under identical conditions—one a velocity spectrum, the other, a displacement spectrum. The spectrum obtained using a velocity transducer is the more uniform of the two; therefore, velocity would be the appropriate motion parameter to select.

SELECTING THE TRANSDUCER

In selecting a transducer best suited for a given measurement, various factors listed in Table 15.1 must be taken into consideration, particularly those under *Parameter to Be Measured, Characteristics of Motion to Be Measured, Environmental Conditions*, and *Transducer Considerations*. Each of these factors (as well as cost and availability) influences the selection process. If consideration of different factors leads to recommendations which are in opposition, then their relative importance must be determined and a decision made on this basis. For example, consider two factors which enter into the selection of a piezoelectric accelerometer—*sensitivity* and *mass*. Sensitivity considerations would suggest that a transducer of large size be selected since transducer sensitivity generally increases with size (and therefore with mass) for an accelerometer of this type.

Table 15.2.
A Guide for the Selection of the Parameter to Be Measured

Acceleration Measurements
Used at high frequencies where acceleration measurements provide the highest signal outputs
Used where forces, loads, and stresses must be analyzed—where force is proportional to acceleration (which is not always the case)
Used where a transducer of small size and small mass is required—since accelerometers usually are somewhat smaller than velocity or displacement pickups

Velocity Measurements
Used where vibration measurements are to be correlated with acoustic measurements since sound pressure is proportional to the velocity of the vibrating surface
Used at intermediate frequencies where displacement measurements yield transducer outputs which may be too small to measure conveniently
Used extensively in measurements on machinery where the velocity spectrum usually is more uniform than either the displacement or acceleration spectra

Displacement Measurements
Used where amplitude of displacement is particularly important—e.g., where vibrating parts must not touch or where displacement beyond a given value results in equipment damage
Used where the magnitude of the displacement may be an indication of stresses to be analyzed
Used at low frequencies—where the output of accelerometers or velocity pickups may be too small for useful measurement
Used to measure relative motion between rotating bodies and structure of a machine

Strain Measurements
Used where a portion of the specimen being tested undergoes an appreciable variation in strain caused by vibration—usually limited to low frequencies

In contrast, mass considerations would suggest that a transducer of small size be selected in order to minimize the mass loading on the test item; a small size is advantageous since, as Eq. (12.13) indicates, the natural frequency of a structure is lowered by the addition of mass. Therefore in this case one should choose the most sensitive transducer (and therefore the largest size) which produces no significant mass loading. In special cases, even the smallest transducer may result in an unacceptable load. Then one of the devices described in Chap. 14 which make no contact with the test surface may be selected.

Consider another example. Suppose a specification requires that vibration displacement be measured. It is reasonable to assume that a displacement transducer (such as the one described in Chap. 14) should be chosen since (depending on the frequency spectrum) such a selection could yield the highest signal-to-noise ratio. On the other hand, in many measurement problems it is more convenient and equally satisfactory to select an accelerometer having a wide dynamic range and to employ an electric circuit which obtains displacement by double integration of the signal from the transducer's output.

TRANSDUCER MOUNTINGS

Various methods of mounting a transducer on a test surface include (1) screwing the transducer to the test surface by means of a threaded stud, (2) cementing the transducer to the test surface, (3) mounting the transducer on the test surface by means of a layer of wax, (4) attaching the transducer to a ferromagnetic surface by means of a permanent magnet, (5) mounting the transducer on a bracket which, in turn, is mounted on the test surface, and (6) holding the transducer against the test surface by hand. Several of these mounting techniques are illustrated in Fig. 15.1, and their frequency-response characteristics are shown in Fig. 15.2. Two types of mechanical brackets are illustrated in Fig. 15.3

The method of mounting affects the resonance frequency and, hence, the useful frequency range of the transducer. Therefore it is important to ensure that the frequency response is adequate before measurements are taken. Each of the above methods of mounting has its advantages and disadvantages. The appropriate choice for a given measurement problem depends on a number of factors, including the following:

Effect of the mounting on the useful frequency range of the transducer
Effect of mass loading of the transducer mounting on the test surface
Maximum level of vibration the mounting can withstand
Maximum operating temperature
Measurement accuracy
Repeatability of measurements (Can the transducer be remounted at exactly the same position with the same orientation?)
Stability of the mounting with time
Requirement that the test surface not be damaged by screw holes
Requirement for electrical insulation of the transducer
Time required to prepare mounting
Time required to remove mounting
Difficulty in cleaning the transducer after removal from test surface
Difficulty in cleaning test surface after transducer removed
Skill required to prepare mounting
Cost of mounting
Environmental problems (dirt, dust, oil, moisture)

For example, the above "requirement for electrical insulation of the transducer" would be a major consideration in the selection of a method of mounting if the insulation, so obtained, would result in the breaking of a ground loop, as explained in a following section.

STUD MOUNTING. Figure 15.1A illustrates a typical stud-mounted transducer; the transducer is fixed to the test surface by means of a threaded metal screw. One method of insulating the stud-mounted transducer from the test surface is shown in Fig. 15.1B. The metal stud is replaced with one which is fabricated of insulating material, and a mica washer is inserted betwen the transducer and the test surface. Other manufacturers employ a threaded, insulated stud with a flange made of the same material; the flange, midway along the length of the stud, serves as the base for the accelerometer.

Where stud mounting is practical, it is the best type to use for the following reasons:

1. It provides the highest resonance frequency (up to 100 kHz) of any of the mounting techniques and, therefore, the widest possible measurement frequency range (up to 50 kHz).

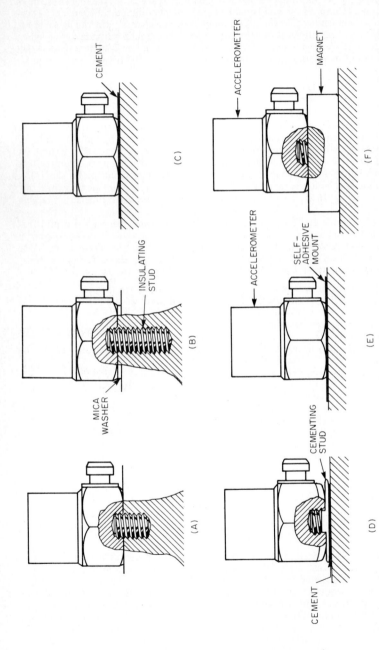

FIG. 15.1. Various methods of mounting a transducer on a test surface: (A) Stud mounting; transducer screws directly to the surface by a threaded stud. (B) Same as (A) but with a transducer insulated from test surface by use of stud fabricated of insulating material and by a mica washer between the surface and transducer. (C) Cement mounting of a transducer; the cement bonds the transducer directly to the surface. (D) Similar to (C) but here cement bonds the surface to a cementing stud screwed into the transducer. (E) Transducer mounted to surface by means of double-sided adhesive tape or disc. (F) Transducer mounted to surface by means of a magnet. (Courtesy of Brüel & Kjaer.)

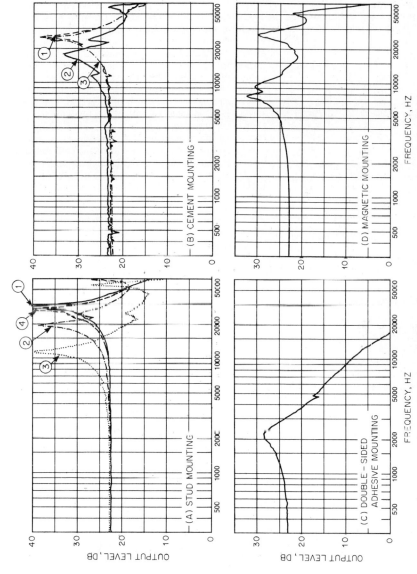

Fig. 15.2. Frequency-response curves for the same piezoelectric accelerometer mounted by the different methods illustrated in Fig. 15.1: (A) stud mounting; (B) cement mounting; (C) double-sided adhesive mounting; (D) magnetic mounting. (Courtesy of Brüel & Kjaer.)

2. It permits measurements at very high vibration levels without the loosening of the transducer from the test surface.

3. It does not reduce the maximum permissible operating temperature at which measurements can be made.

4. It permits accurate and reproducible results since the measurement position can always be duplicated.

In preparing a stud mounting, the test surface must be drilled and tapped. A standard 10-32 thread is widely used. (Also see International Standards Organization Standard ISO 1101.) Distortion of the transducer as mounted may produce strains that affect the transducer's response. Therefore, it is important (1) to ensure that the test surface is very flat (which can be done by grinding or lapping), (2) to prevent the mounting stud from bottoming in the transducer case—this can lead to strain, and (3) to screw the stud into the hole in the test surface, and then the accelerometer onto the stud using the torque recommended by the transducer's manufacturer. The application of a silicone grease (such as Dow-Corning DC-4) or a light machine oil between the transducer and the test surface usually provides better response at high frequencies—say, above 2,000 Hz.

FIG. 15.3. Two types of mounting brackets. In this example, a velocity-type transducer is shown; the arrows indicate the direction of sensed motion.

Figure 15.2*A* shows response curves for a stud-mounted accelerometer for the following conditions: ① spanner tight, which has the highest resonance frequency, ② finger tight, ③ mounted with a mica washer to provide electrical insulation between the transducer and the vibrating surface, and ④ mounted on a somewhat thinner mica washer—which results in a higher resonance frequency than for ③.

CYANOACRYLATE, DENTAL, AND EPOXY CEMENT MOUNTINGS. Where it is not possible to use a stud mounting, a transducer can be bonded to a test surface by means of a thin layer of cement (for example, a cyanoacrylate, dental, or epoxy cement), as shown in Fig. 15.1*C*. If the test surface is not flat and a miniature accelerometer is used, it is not difficult to build up a layer of dental cement around the accelerometer so as to provide firm attachment for the accelerometer.

Another method of mounting is to use a cementing stud which is threaded into the transducer; the flat side of the stud is then cemented to the test surface as shown in Fig. 15.1*D*. This is a useful technique where repeated measurements are required at the same point. The transducer may be removed for measurements elsewhere, but the cementing stud is left in place. This provides assurance that future measurements will be made at precisely the same point.

The cement method of mounting a transducer provides excellent frequency response, as shown in Fig. 15.2*B* for three conditions: ① accelerometer cemented directly to test surface, ② accelerometer cemented with a "soft" adhesive (not recommended), and ③ accelerometer with a cementing stud which is cemented to the surface with a hard cement.

This type of mounting may be used at high levels of vibration if the cementing surfaces are carefully prepared. Cement mounting may or may not provide electrical insulation; if insulation is required, the electrical resistance between the transducer and the test surface should be checked with an ohmmeter. The maximum temperature at which measurements can be made is limited by the physical characteristics of the cement employed—usually about 176°F (80°C). This type of mounting has good stability with time.

Methyl cyanoacrylate cements, such as Eastman Kodak 910MHT 3M, Cyanolite 101, and Permabond 747, dry much more rapidly than epoxy cements and therefore require less time to mount a transducer. They may be removed easily and the surface cleaned with a solvent such as acetone. Removal of epoxy from the test surface and from the transducer may be time-consuming. In fact, the epoxy bond may be so good that the transducer can be damaged in removing it from the test surface. When encased in epoxy, an accelerometer may be subject to considerable strain, which will significantly alter its characteristics. On the other hand, unless the cemented surfaces are very smooth, an epoxy can provide a superior bond since it will fill in a rough surface far better than a cyanoacrylate cement. With either bonding agent, the surfaces must be very clean before application of the cement.

WAX MOUNTING. Beeswax or a petroleum-based petrowax may be used to attach a transducer to a flat test surface. If the bonding layer is thin (say, no greater than 0.2 mm), it is possible to obtain a resonance frequency almost as high as that for the stud mounting, but if the test surface is not smooth, a thicker wax layer is required and the resonance frequency will be reduced. If the mating surfaces are very clean and free from moisture, the transducer can be mounted fairly easily, although some practice may be required. The transducer can be removed rapidly with a naphtha-type solvent. Disadvantages include the possibility of disattachment of the transducer at high vibration levels, a temperature limitation because of the relatively low melting point of wax, and poor long-time stability of the mounting.

ADHESIVE MOUNTING. An adhesive film may be used to mount a small transducer on a flat, clean test surface—usually by means of a double-sided adhesive tape. Double-sided adhesive discs are supplied by some transducer manufacturers. This mounting technique, illustrated in Fig. 15.1E, is rapid and easy to apply. Furthermore, such a mounting has the advantage of providing electrical insulation between the transducer and test surface, and it does not require the drilling of a hole in the test surface; it is particularly applicable for use with a transducer having no tapped hole in its base. Such adhesives can provide secure attachment over a limited temperature range. In preparing an adhesive mounting, it is important to clean both the accelerometer and the test surface so the adhesive will adhere firmly. When this is done, the frequency response can be fairly good, as illustrated in Fig. 15.2C, but not as good as with a wax mounting.

MAGNETIC MOUNTING. With magnetic mounting, illustrated in Fig. 15.1E, a permanent magnet attaches the transducer to the test surface, which must be ferromagnetic, flat, free from dirt particles, and reasonably smooth. Magnetic mounting is useful in measuring low acceleration levels. The transducer can be attached to the test surface easily and moved quickly from one measurement point to another. For example, in a condition-monitoring system (described in Chap. 16) it can be used to determine a suitable measurement location for a transducer to be mounted permanently on a large rotating machine. In a heavy machine of this type, the added mass of the magnet is not important, but in other problems, the additional mass loading on the test surface may make the use of magnetic mounting unacceptable. Furthermore, if the acceleration levels are sufficiently high, as in impact testing, the magnet may become loosened momentarily. This can result in an inaccurate reading and possibly a slight change in position of the transducer which would also change the reading. The frequency response for this type of mounting is fair, as shown in Fig. 15.2D, but not as good as the wax mounting. The magnet, often available from the transducer's manufacturer, usually is attached to the transducer by means of (1) a projecting screw on the magnet, which is threaded into the base of the transducer, or (2) a machine screw, one end of which is threaded into the transducer and the other end into the magnet. Application of a light machine oil or silicone grease usually improves the frequency response above about 2,000 Hz.

BRACKET OR BLOCK MOUNTING. Physical conditions may make it impractical to mount a transducer by any of the above methods. In such cases, a mounting bracket or block that has been especially prepared for use on the test surface may be employed (see Fig. 15.3). Instead of using a triaxial accelerometer, sometimes it is more convenient to mount three transducers on a single block having sensitivities in three orthogonal directions. Any such mounting must couple the transducer to the test surface so that the transducer accurately follows the motion of the surface to which it is attached. This requires that the effective stiffness of the transducer mounting be high so that the mounting does not deflect under the inertial load of the transducer mass. This is not a problem in many transducer installations.

Mounting brackets may have resonance frequencies which are below 2,000 Hz and have little damping. Under such conditions, their use may result in significant measurement error as a result of resonant amplification or because of attenuation of vibration in the mounting. This is illustrated in Fig. 15.4, which shows the frequency response of a transducer mounted on brackets which are identical in geometry but which are fabricated from different materials. Note that a change in material from (*A*) steel to (*B*) a phenolic plastic halves the resonance frequency of the mounting. A change in the method of attachment, from (*B*) screw mounting to (*C*) an epoxy resin adhesive bond, significantly increases the frequency of the mounting resonance. Although these results are not of a general nature, they show that such minor variations in the transducer mounting may produce significant changes in the output characteristics of the transducer.

HAND-HELD TRANSDUCER. A transducer which is held against the test surface by hand provides the poorest performance of any of the techniques described here, but it sometimes can be useful in making a rapid survey of a test surface because the measurement location can be changed more rapidly than with any other method of mounting. Usually, a rod (called a *probe*), which is threaded at one end, is screwed into the transducer; the other end has a tip that is pressed against the test surface. The frequency response is highly restricted—about 20 to 1,000 Hz. Thus, this technique is used when measurement accuracy is not essential, e.g., in finding the nodal points on a vibrating surface.

FIG. 15.4. Relative frequency response of a velocity transducer mounted on three brackets which have identical geometry but are fabricated of different materials: (*A*) steel bracket, screw mounted, (*B*) cloth-reenforced phenolic plastic bracket, screw mounted, and (*C*) same as (*B*) but attached with epoxy resin adhesive.

MASS-LOADING. The effect of the mounting on the accuracy of measurement can be estimated roughly if it is assumed that the combination of the transducer (having a mass m) and the mounting (having a stiffness k) behaves as a simple spring-mass system driven at the spring end of the system. Then the acceleration of the transducer $\ddot{x}$ is given by

$$\ddot{x} = \ddot{u}\,\frac{k}{k + m(2\pi f)^2} \tag{15.1}$$

where $\ddot{u}$ is the acceleration of the test item, and f is its frequency of vibration. If the acceleration of the transducer is to be within 10 per cent of the acceleration of the test item, then from Eq. (15.1), k must have a value at least 10 times greater than the term $m(2\pi f)^2$. Since the undamped natural frequency f_n of the transducer-mounting system is given by $f_n = \frac{1}{2\pi}(k/m)^{1/2}$, the value of the natural frequency of the system must be at least 10 times the frequency of vibration of the test item—especially for the measurement of transients.

FIELD CALIBRATION TECHNIQUES

TRANSDUCERS

Various methods of calibrating transducers are described in Chap. 18. Of these, the following are particularly convenient for use in the field.

COMPARISON METHOD. This is a rapid and convenient method of obtaining the sensitivity of a transducer. It is one of the most commonly used calibration techniques. Calibration is obtained by a direct comparison of the output generated when the transducer is attached to a vibration exciter with the output generated by a secondary standard transducer which is attached to the same vibration exciter and which is subject to precisely the same motion. The two transducers are mounted back to back, as illustrated in Fig. 18.30. Calibration by this method is limited to the frequency and amplitude ranges for which the secondary standard has been calibrated and for which the vibration exciter has adequate rectilinear motion.

FREE-FALL CALIBRATION METHOD. The gravimetric free-fall calibration method (sometimes called a *drop test*) is a simple and rapid method of calibrating motion and force sensors. The transducer under test is allowed to fall freely for an instant of time under the influence of gravity; the peak signal then is measured for an acceleration of gravity having a value of $1g$. This technique is illustrated in Figs. 18.6 to 18.8.

EARTH'S GRAVITATIONAL FIELD METHOD. The following technique (sometimes called the *"inversion method"* of calibration) is a variation of the *tilting-support calibrator calibrator* illustrated in Fig. 18.3. The sensitive axis of the transducer is first aligned vertically in one direction of the earth's gravitational field, as shown in Fig. 15.5A. Then it is inverted so that its sensitive axis is aligned in the opposite direction, as shown in Fig. 15.5B. The transducer output is observed for a $2g$ change in acceleration, as shown in Fig. 15.5C. This method is limited in application to accelerometers having sensitivity down to 0 Hz; it is not recommended for calibration of accelerometers having significant transverse sensitivity.

OVERALL SYSTEM

Calibration of a complete vibration measurement system usually is referred to as *overall calibration* or *end-to-end calibration*. It is good practice to perform such a calibration at periodic intervals—particularly both before and after an extensive series of measurements. In such a calibration, the amplitude characteristics, phase characteristics, and linearity of the overall system are determined when the transducer is subject to a known acceleration, velocity, or displacement. When this is not possible, then the overall characteristics sometimes are determined by combining the characteristics of the individual components of the system, as described below, or the system is calibrated employing a simulated transducer output [see *Voltage Substitution Method of Calibration* below, and *Calibration of Auxiliary Circuits* (Chap. 18)].

There may be a significant electrical signal at the output of a measurement system though no signal is supplied by the transducer to the input; such electrical signals, which represent noise, (1) may result from a coupling between circuits in the measurement system with power circuits, (2) may be generated by vibration sensitive elements (such as cable) other than the transducer, or (3) may be the result of improper selection of system components, or the improper setting of one or more of these components, so that the signal-to-noise ratio that the overall system is capable of attaining is not achieved.

Where a single component of a measurement system is the source of noise, it can sometimes be located by using an oscilloscope which is first connected to the transducer output with no vibration applied. Then the oscilloscope connection is moved, component by component, through the measurement system until the noise is observed. Another approach is to short-circuit the signal path at various points in the system (where this is practical), one at a time, until the system noise disappears. Usually this pinpoints the source as the component next nearest the transducer from the last short circuit.

Spurious mechanical sources and acoustic noise sources must be eliminated or controlled if they result in noise in the measurement system. Spurious resonances in the response of the overall system may result from improper seating of the transducer on the test surface or from resonances in the transducer mounting. It is often very useful to excite the transducer-mounting system by giving it a blow and then to observe the transducer's output—look for resonances other than the resonance frequency of the transducer. The other resonance frequencies which appear may be due to (1) resonances in

INITIAL CONDITION
(A)

FINAL CONDITION
(B)

ΔV

VOLTAGE

TIME

−1g   TRANSITION   +1g

INITIAL
CONDITION

FINAL
CONDITION

(C)

FIG. 15.5. Gravitational field method (inversion test for calibrating an accelerometer having useful sensitivity down to 0 Hz. Inversion of the accelerometer, initially aligned in one direction, as in (A), to the opposite direction, as in (B), produces a change in acceleration of $2g$. The transducer output for this change is measured in (C). *(Courtesy of Quixote Measurement Dynamics, Inc.)*

the test specimen or (2) resonances in the transducer mounting. Loose mountings usually produce "noisy" signals and may produce audible buzzing sounds. Often it is difficult to determine the difference between resonances in the mounting and resonances in the item under test. If serious doubt exists, the test should be repeated with a different mounting or a different measurement location for the transducer. If the resonance frequencies are identical for the new mounting, the resonances are probably due to the test specimen, and the original mounting probably was satisfactory.

COMBINING CALIBRATION CHARACTERISTICS OF A MEASUREMENT SYSTEM'S COMPONENTS. An overall system can be calibrated by combining the measured electrical characteristics of all components in the measurement system from one end to the other. Obtaining a system calibration in this way circumvents the difficulties of precise field calibration, but it requires that each element in the system be calibrated in the laboratory with extreme care and that the effects of the source and load impedances be completely accounted for. Thus, a system calibration is subject to the sum of the experimental errors introduced by the calibration of each element in addition to any errors resulting from improper simulation of, or accounting for, loading effects. In general, the calibration of each element is performed before the system is assembled, and so this method is subject to error resulting from (1) undetected damage to components between calibration and use and/or (2) improper connections, misidentifications, or confusion in polarity.

VOLTAGE SUBSTITUTION METHOD OF CALIBRATION. A suitable simulated transducer for use in field checkout must duplicate the electrical outputs of the actual transducer for the various vibration conditions to be simulated. The simulated transducer must either (1) reproduce the electrical voltage or current-generating characteristics of the actual transducer and have the same output impedance or (2) duplicate the electrical quantity generated by the actual transducer when connected to its load. Failure to meet these conditions will result in a different loading of the actual and simulated transducers and will probably cause calibration errors. It is important that the simulated transducer have the same electrical grounding configuration as the actual transducer; otherwise, electric-circuit noise and cross talk* will not be represented accurately when the simulated transducer is in use.

Typical examples of circuits which simulate transducers are shown in Fig. 15.6. The simulated transducer introduces an electrical signal into the measurement system, thereby simulating the response of the actual transducer.

CABLE AND WIRING CONSIDERATIONS

The method of data transmission between a transducer and the electronic instrumentation which follows it depends on the complexity of the problem. In general, cable is used for most problems, but the aerospace industry often relies on telemetry for data transmission. Many types of cable are available. The choice of a suitable cable depends primarily on the particular application, the transducer, the cable length, whether the transducer is followed by a voltage amplifier or charge amplifier, and on environmental conditions. For example, cable jackets may be made of silicone rubber having a useful temperature range from -100 to $500°F$ (-73 to $260°C$), of polyvinylchloride having a useful range from -65 to $175°F$ (-54 to $79°C$), or of fused Teflon having a useful range from -450 to $500°F$ (-268 to $260°C$). Special-purpose cables are available that can be used at much higher temperatures. In general, cable should be as light and flexible as possible—consistent with other requirements. The effect of the shunt capacitance of the cable following the transducer on the sensitivity of the transducer depends on the type of amplifier connected to the cable. If a voltage amplifier is used, there is a reduction in sensitivity of the transducer, given by Eq. (12.17). In contrast, when a charge amplifier is used, the effect of the shunt capacitance of the cable in reducing the sensitivity of the transducer is negligible, as shown in Eq. (13.2) (although the noise pickup in the high-impedance circuit increases with cable length).

In the audio-frequency range, the series inductance L and the shunt leakage G of short, good-quality cables are negligibly small in comparison with other parameters and may be neglected. Figure 15.7A shows the equivalent low-frequency representation of a cable with distributed constants. For most purposes the simpler lumped-constant configuration of Fig. 15.7B is a sufficiently accurate representation. The quantities R_c and C_c are the total resistance of the conductors and the total capacitance between them, respectively. Values for a typical coaxial cable having a Teflon dielectric are $R_c = 0.01$ ohm/ft (0.03 ohm/m) and $C_c = 29$ mmfd/ft (88 mmfd/m).

The normal characteristic impedance of about 50 ohms for such cable has no significance in most measurement problems where cables usually are relatively short. The open-circuit input impedance of the cable is almost exclusively capacitive. When ter-

---

* *Cross talk* is the output of one measurement channel when a signal is applied to another measurement channel. Cross talk can be distinguished from other electrical disturbances because it is a function of the applied signal in the other measurement channel and disappears when this applied signal is removed.

Fig. 15.6. Electrical schematic diagrams of some common types of transducers and typical circuits used to simulate them during field calibration. Terminals labeled *A* and *B* are the signal lead connections to which either the transducer or simulated-transducer is connected.

minated, it takes on the impedance of the load, modified by the series and shunt parameters.

In general, cables should be treated with the same care given transducers in shock and vibration measurement systems. The following are based on recommendations given in Ref. 1; they represent good engineering practice.

1. Attach a coaxial cable to a transducer by turning the connector nut onto the threads of the transducer (no vice versa) to avoid damage to the pins.

2. Avoid cable whip by tying down the cable at a point near the transducer and at regular intervals to avoid induced cable noise.

3. Screw the cable connection to the tightness specified by the manufacturer.

4. Loop the cable near the connector, in a high-humidity environment, to allow condensation to drip off before reaching the connector.

5. Clean the cable connector before use (with Freon, acetone, or chlorothene) to remove contamination as a result of handling; the contamination can create a low impedance between the signal path and ground.

6. Check electrical continuity of cable conductors and shield if intermittent signals are observed. Then, flex the cable—especially near the connector—and observe if the signal is affected by flexing.

(A)   (B)

FIG. 15.7. Successive approximations in the representation of a short, high-quality transmission line at audio frequencies. (A) Distributed constant configuration neglecting series inductance and shunt leakage. (B) Lumped-constant configuration.

CABLE NOISE GENERATION

When two dissimilar substances are rubbed together, they become oppositely charged—a phenomenon known as *triboelectricity*, illustrated in Fig. 15.8. Thus a charge may be generated when a cable is flexed, bent, struck, squeezed, or otherwise distorted, for then such friction takes place between the dielectric and the outer shield or between the dielectric and the center conductor.[2] A charge is generated across the cable capacitance so that a voltage appears across the termination of the cable.

Another mechanism by which noise may be induced in the cable results from the change in capacitance of the cable when it is flexed. If the transducer produces a charge across the cable, the change in capacitance results in a voltage change across the output of the cable, appearing as noise at the input of a voltage amplifier; it will not produce a similar change if a charge amplifier is used.

Suppose the dielectric surfaces within the cable are coated so that an electrical leakage path is provided along the dielectric surface. Then if the cable shield is separated from the outer surface of the dielectric, the charges flow along the surface to the nearest point of contact of the dielectric and shield; without this leakage path, the charges would flow to the terminating impedance where they would give rise to a noise signal. Such coatings are provided in low-noise cables which are available commercially. Cables of this type are capable of withstanding considerable abuse before becoming noisy. Usually they are tested by the manufacturer continuously along their lengths to assure meeting the low-

FIG. 15.8. A section of cable during distortion, showing how separation of triboelectric charge leads to the production of cable noise across the termination resistance. (*After T. T. Perls.*[2])

noise characteristics. It is important in fitting such a cable with a connector, or in splicing such a cable, that no conducting material be allowed to form a leakage path between the conductors. Carbon tetrachloride and xylene are satisfactory solvents and cleaning agents.

NOISE-SUPPRESSION TECHNIQUES

Under certain conditions of use and environment, spurious signals (noise) may be induced in wiring and cables in a measurement system. Then there will be signals at the termination of the system that were not present in the transducer output.

Electrical noise may be generated by motion of some parts of the wiring because of variation in contact resistance in connectors, because of changes in geometry of the wiring, or because of voltages induced by motion through, or changes in, the electrostatic fields or magnetic fields which may be present. In general, such electrical noise will be reduced if the cable is securely fastened to the structure at frequent intervals and if connectors are provided with mechanical locks and strain-relief loops in their cables. Precautions taken to avoid interference usually include the use of shielding, cables which are only as long as necessary, and proper grounding. In addition, the use of a transducer containing an internal amplifier (described in Chaps. 13 and 14) can provide advantages in noise supression.

SHIELDING. A change in the electric field or a change in the magnetic field around a circuit or cable may induce a voltage within it and thus be a source of electrical noise. Such electrical interference can be avoided by completely surrounding the circuit or cable with a conductive surface which keeps the space within it free of external electrostatic or magnetic fields. This is called *shielding*. Protection against changes in each type of field is different.

ELECTROSTATIC SHIELDS. Electrostatic shields provide a conducting surface for the termination of electrostatic lines of flux. Stranded braid, mesh, and screens of good electrical conductors such as copper or aluminum are good electrostatic shields. Most shielded cables use copper braid as the outer conductor and electrostatic shield. A good magnetic shield is also a good electrostatic shield, but the converse is not true. For installations where cable lengths are especially long, where impedances are high, or where noise interference is highly objectionable, double-shielded cable is sometimes used. In this type of cable, a second shielding braid is woven over the cable jacket, electrically insulating it from the inner shield; the inner braid furnishes additional shielding against electrostatic fields which penetrate the first shield. The shields should be connected to ground at one point only, as explained below under *Ground Loops*.

MAGNETIC SHIELDS. Magnetic shields are effective partly because of the short circuiting of magnetic lines of flux by low-reluctance paths and partly because of the cancellation resulting from opposing fields set up by eddy currents. Accordingly, they are made from high-permeability materials such as Permalloy, are as thick as possible, and contain a minimum of joints, holes, etc.

Magnetic fields associated with current-carrying power lines, electronic equipment, and power transformers are among the most troublesome sources of magnetic interference in instrumentation setups—chiefly at the frequency of the power line and its harmonics. Since these fields attenuate rapidly with distance from the source, the most practical solution for this type of interference usually is to keep the signal cables as far from the power source as possible.

GROUND LOOPS. A circuit is said to be grounded when one terminal of the circuit is connected to the "earth." Grounding removes the potential difference between that side of the circuit and earth, and the variable stray capacitances which tend to induce voltages in "floating" (i.e., ungrounded) systems. Water pipes make good ground connections because of their intimate contact with the earth.

Ground loops are formed when a common connection in a system is grounded at more than one point, as illustrated in Fig. 15.9 where the cable shield is grounded at both ends.

FIG. 15.9. Ground loop in a system as a result of grounding the cable shield at two point. Then, the input signal e_1 is modulated by the potential difference e_{gnd} which develops between these two points.

Since it is unlikely that the two grounds will be at a common potential, their potential difference, e_{gnd}, will be the source of circulating currents in the ground loop. Then a signal produced by the transducer will be modulated by the potential e_{gnd}, thereby introducing noise in the measurement system. Such a condition may occur when one end of a cable is connected to one side of the electrical output of a transducer that has been grounded to the transducer's housing and when the other end of the cable is connected to a voltage amplifier or signal conditioner which is also grounded (usually to the case of the instrument). Then, a ground loop will be formed. *Such a condition must be avoided by grounding the circuit at only one point.* Thus the circuit shown in Fig. 15.10A will result in noise because of the ground loop, but by insulating the transducer as shown in Fig. 15.10B the ground loop has been broken.

FIG. 15.10. (A) A ground loop formed by the "low" side of both the transducer and the amplifier are connected to their respective cases which are grounded. (B) The ground loop shown in (A) is broken by isolating the case of the transducer from ground.

REFERENCES

1. Endevco Instruction Manual for Piezoelectric Accelerometer, No. 1, 12-72, San Juan Capistrano, Calif., 1972.

2. Perls, T. A.: *J. Appl. Phys.*, **23**(6):674 (1952).

16

CONDITION MONITORING OF MACHINERY

Joëlle Courrech

Brüel & Kjaer, Denmark

INTRODUCTION

Condition monitoring of machinery is the measurement of various parameters related to the mechanical condition of the machinery such as vibration, bearing temperature, oil pressure, oil debris, and performance, which make it possible to determine whether the machinery is in good or bad mechanical condition; if the mechanical condition is bad, then condition monitoring makes it possible to determine the cause of the problem.

Condition monitoring is used in conjunction with *on-condition maintenance*, i.e., maintenance of machinery when conditions indicate that a problem is about to occur. In many plants this philosophy now tends to replace run-to-breakdown maintenance and also preventive maintenance in which mechanical parts are replaced periodically at fixed time intervals regardless of the machinery's mechanical condition. On-condition maintenance provides the following advantages:

1. Avoids unexpected catastrophic breakdowns with expensive or dangerous consequences to follow.

2. Reduces the amount of overhauls on machines to a minimum and thereby reduces maintenance costs.

3. Eliminates unnecessary interventions and risks of introducing faults on smoothly operating machines.

4. Allows spare parts to be ordered in time and thus eliminates costly stocks.

5. Reduces the intervention time and thereby minimizes production loss as the fault to be repaired is known in advance and the overhauls can be scheduled when most convenient.

This chapter describes the use of vibration measurements for monitoring the condition of machinery; this is the parameter which can be used to predict the broadest range of faults in machines or machine elements most successfully. This chapter also describes (1) selection of type of monitoring system (permanent or intermittent); (2) establishment of a condition-monitoring program including the selection of transducers, the selection of the appropriate vibration parameter, the selection of measurement location on the machine, and the selection of time interval between measurements; (3) fault detection; (4) spectrum interpretation and fault diagnosis; and (5) trend analysis.

SELECTION OF TYPE OF MONITORING SYSTEM

Condition-monitoring systems are of two types: intermittent and permanent. In an *intermittent monitoring system*, machinery vibration is measured (or recorded, and later analyzed) in the field at selected time intervals; then, an analysis is made either in the field or in the laboratory. Advanced analysis techniques usually are required for fault diagnosis and the analysis of trends in the measured data. Intermittent monitoring provides information at every stage about incipient failure and usually is used (1) where very early warning of faults is required, (2) where advanced diagnostics are required, (3) on complex machines, and (4) where measurements must be made at many locations on a machine.

In a *permanent monitoring system,* machinery vibration is measured continuously at selected points of the machine and constantly compared with permitted levels. The measurement system may be permanent (as in parallel acquisition systems where one transducer and one measurement chain are used for each measurement point) or it may be quasipermanent (as in multiplexed systems where one transducer is used for each measurement point but the rest of the measurement chain is shared between a few points with a multiplexing interval of a few seconds).

In a permanent monitoring system, transducers are mounted permanently at each selected measurement point. For this reason such a system can be very costly, so it is usually only used in critical applications where: (1) no personnel are available to perform measurements (e.g., in remote offshore pumping stations), (2) it is necessary to stop the machine before a breakdown occurs in order to avoid a catastrophic accident, (3) an instantaneous fault may occur that requires machine shutdown, and (4) the environment (explosive, toxic, or high temperature) does not permit intermittent measurements.

Before selecting a permanent monitoring system, make preliminary measurements intermittently over a period of time to become acquainted with the vibration characteristics of the machine. This procedure will make it possible to select the most appropriate vibration measurement parameter, frequency range, and normal alarm and trip levels.

ESTABLISHMENT OF A CONDITION-MONITORING PROGRAM

The first step in establishing a condition-monitoring program for rotating machinery is to determine what shall be measured: shaft vibration or bearing vibration. If it is shaft vibration, then displacement is measured; in general, a displacement transducer usually will be most suitable. If it is bearing vibration, then either a velocity transducer or an accelerometer will be the most suitable.

SELECTION OF MEASUREMENT PARAMETER: DISPLACEMENT, VELOCITY, OR ACCELERATION

When an accelerometer is employed as the sensing device in a condition-monitoring system, the resulting signal can be electronically integrated to obtain velocity or displacement, so any one of these three parameters may be used in measurements. The appropriate parameter may be selected by application of the following simple rule: Use the parameter which provides the flattest spectrum. The flattest spectrum requires the least dynamic range from the instrumentation which follows the transducer. For example, Fig. 16.1 shows a displacement spectrum and a velocity spectrum obtained under identical conditions. The dynamic range required to measure the displacement spectrum is much larger than either the velocity spectrum or the acceleration spectrum and may exceed the

FIG. 16.1. Displacement and velocity spectra obtained under identical conditions. The velocity spectrum requires a smaller dynamic range of the equipment which follows the transducer. Therefore it is preferable.

available dynamic range of the instrumentation. Therefore, either of the latter would be a better choice.

The above rule of using the parameter which provides the flattest spectrum applies only to the frequency range of interest. Therefore the parameter selection, to some extent, depends on the type of machine and the type of faults considered.

SELECTION OF TRANSDUCERS

DISPLACEMENT TRANSDUCERS. Displacement transducers are commonly used to measure shaft vibration, particularly those of the eddy-current type, which have noncontacting probes. Such transducers provide information about the relative motion between shaft and bearing. This information can be related directly to physical values such as mechanical clearance or oil-film thickness; e.g., it can give an indication of incipient rubbing. The use of displacement transducers is essential in machinery having journal bearings. However, noncontacting probes of this type (1) are difficult to calibrate

FIG. 16.2. Spectra of the same vibration source, obtained with different types of transducers. (*A*) A typical displacement spectrum showing a negative slope. Even if a displacement transducer has a large frequency range (typically up to 10 kHz), its limited dynamic range (40 dB) results in a spectrum whose useful frequency range is restricted (typically up to 500 Hz). (*B*) A velocity spectrum. Although velocity transducers have a good dynamic range (60 dB), their frequency range is limited, between 10 and 1,500 Hz (typically). (*C*) An acceleration spectrum. Accelerometers cover the frequency range from 0.1 Hz (or below) to 15 kHz and above (depending on types), with a dynamic range of typically 150 dB. Even after integration of the signal to velocity, it provides a 90-dB dynamic range over three frequency decades.

absolutely, (2) have a limited dynamic range because of the influence of electrical and mechanical runout on the shaft, and (3) have a limited frequency range. The useful upper frequency is typically limited to the frequency of the first few harmonics of the shaft speed as a consequence of the limited dynamic range, as shown in Fig. 16.2*A*.

Shaft vibration provides information about the current condition of the machine and is used principally in permanent monitoring systems which immediately shut the machine down in the event of trouble.

ACCELEROMETERS AND VELOCITY PICKUPS. Accelerometers and velocity pickups are used to make bearing or casing vibration measurements. Accelerometers have a wider dynamic range and a broader frequency range than either displacement or velocity transducers as illustrated in Fig. 16.2. They are lightweight and rugged; they should be selected for detecting faults which occur at high frequencies such as in rolling element bearing deterioration or gearbox wear. The magnitude of the acceleration of the bearing or casing may be directly related to the forces acting inside the machine which are transmitted through the bearings. Accurate measurements of bearing vibration, obtained with an accelerometer and which are adequately processed, can provide very early warning of incipient faults in a machine.

SELECTION OF MEASUREMENT LOCATION

In employing an intermittent monitoring system it is not necessary to reduce the number of points at which measurements are made for reasons of economy except for keeping measurement time to a minimum. A general rule for bearing vibration measurements consists of measuring in the radial direction on each accessible bearing, and in the axial direction on thrust bearings. When measuring bearing vibration, it is not usually necessary to measure in both the horizontal and vertical directions since both measurements give the same information regarding the forces within the machine; this information is merely transmitted through two different transmission paths. When measuring shaft vibrations with proximity transducers, it is convenient to use two probes on each bearing, located at 90° from each other, thereby providing an indication of the orbit of the shaft within the bearing.

In employing a permanent monitoring system the number of measurement points usually is minimized for reasons of economy. A study of the vibration spectra of different bearings must be made in order to find the points where all significant components related to the different expected faults are transmitted at measurable vibration levels.

SELECTION OF TIME INTERVAL BETWEEN MEASUREMENTS

The selection of the time interval between measurements requires knowledge of the specific machine. Some machines develop faults quickly and others run trouble-free for years. A compromise must be found between the safety of the system and the time taken for measurements and analysis. The following rough rule of thumb is useful: Select a time interval between measurements which is one-sixth to one-tenth of the expected period between overhauls. In any case, measurements should be made frequently in the initial stages of a condition-monitoring program to ensure that the vibration levels measured are stable and that no fault is developing. When a significant change is detected, the time interval between measurements should be reduced to follow more closely the development of the fault by the trend in the measurements.

FAULT DETECTION IN ROTATING MACHINERY

It is highly desirable to be able to detect all types of faults likely to occur during the operation of rotating machinery. Such faults range from very low frequencies (subsynchronous components, indicating looseness, oil whirl, faulty belt drive, etc.) to very high frequencies (tooth-meshing frequencies, blade-passing frequencies, structural resonances excited by faulty rolling element bearings, etc.). Such detection should be applicable to the complete range of machines in a plant, which operate from very low to very high speeds. This can only be achieved by selecting equipment and analysis techniques which cover a very broad frequency range.

Measurements of absolute vibration levels of bearings provide no indication of the machine's condition since they are influenced by the transmission path between the force and the measurement point. Bearing vibration levels change from one measurement point to another on a given machine; they also change from machine to machine for measurements made at the same point.[3] Therefore in estimating the condition of a machine, it is essential to follow changes in level from a reference value established when the machine was known to be in good condition. Changes are given as a ratio, or more commonly as the logarithm of a ratio, expressed in decibels.

OVER-ALL MEASUREMENT AND SPECTRUM COMPARISON

The objective of condition monitoring of machinery is to predict when a fault will occur well in advance of failure. A measurement of the over-all level employing a simple vibration meter will not be successful because the highest component within the frequency range will dominate the over-all measurement. This is illustrated in Fig. 16.3 which shows in (A) the vibration velocity spectrum of a gearbox and in (B) a similar spectrum obtained 3 months later. In Fig. 16.3C, the 121-Hz component, an indication of a developing failure, increases rapidly with time. In contrast, the over-all level increases much more slowly. The latter leads to a wrong or far too optimistic prediction. The detection of faults in machinery can only be made successfully by comparing spectra. This section compares the types of spectrum analysis for this purpose.

FIG. 16.3. (A) The velocity spectrum of vibration measured on a gearbox after installation. Note the high amplitude of the 480-Hz component, dominating the reference spectrum. (B) The velocity spectrum 3 months later. Note the dramatic increase in the 121-Hz component which is the output shaft speed of the gearbox. (C) Graph comparing the increase in the 121-Hz component in the velocity spectrum (indicating a developing fault) with the increase in over-all velocity in the band from 10 to 1,000 Hz.

CONSTANT-BANDWIDTH VS. CONSTANT-PERCENTAGE–BANDWIDTH ANALYSIS

A constant-bandwidth analyzer such as a fast Fourier transform (FFT) analyzer provides spectra having constant bandwidth on a linear frequency scale. This yields very fine resolution at high frequencies but very poor resolution at low frequencies. To cover a large frequency range with sufficient resolution using constant-bandwidth analysis, a number of different frequency analyses are required—typically one analysis per frequency decade. This is time-consuming since an enormous amount of data must be stored and compared.

A constant-percentage-bandwidth analyzer typically covers three decades with equal resolution in a single spectrum on a logarithmic frequency range. In this way considerable data reduction is achieved, yet adequate resolution is maintained. A further advantage of the constant-percentage-bandwidth logarithmic frequency scale is the simplicity

FIG. 16.4. Analysis of gearbox vibration. (A) Velocity spectrum obtained with an analyzer having a constant-percentage bandwidth of 4 percent covering a frequency range from 20 Hz to 20 kHz; the frequency scale is logarithmic. (B) Velocity spectrum obtained with an analyzer having a constant bandwidth of 50 Hz, covering a frequency range from 0 to 20 kHz; the frequency scale is linear.

of compensating for speed changes. A speed variation of the machine manifests itself as a global translation of all speed-related components in the vibration spectrum. For the above reasons, the best analysis method for the comparison of spectra and fault detection is the use of constant-percentage bandwidth with a logarithmic frequency scale. Figure 16.4 shows an example which compares measurements on the same gearbox made with two different analysis techniques. Fig. 16.4*A* employs a constant-percentage bandwidth; Fig. 16.4*B* employs constant bandwidth. Note in Fig. 16.4*B* the loss of information in the range from 0 to 2 kHz.

Sufficient frequency resolution must be maintained to separate important speed-related components. As occurred when performing over-all measurements (Fig. 16.3), if two components, equally important but largely different in level, happen to be in the same filter bandwidth, the amplitude of the higher one will conceal changes in the lower one. A bandwidth of between 3 and 6 per cent usually is sufficient to cover small speed fluctuations (on the order of 1 per cent), yet it has sufficient resolution to separate significant components.

HOW SPECTRUM CHANGES ARE RELATED TO THE CONDITION OF A MACHINE

To obtain information about changes in the condition of a machine, vibration spectra should be compared only for similar operating conditions. The influence of the operating condition of the machine (such as machine speed, load, or temperature) on the vibration parameter being measured varies greatly for different types of machines. Speed changes of up to 10 per cent usually can be compensated for, and spectra can be compared. If the speed changes are greater than this value, the operating condition of the machine should be considered to be different and a new reference spectrum used as a basis of comparison. The reference spectrum need not be measured when the machine is new (after allowing for a run-in period). The reference spectrum can be determined at any time during the life of a machine provided the vibrations are stable, since a stable spectrum is a sign of stable operation of the machine. The principal problem is to establish when changes in the spectrum are sufficiently large to warrant stopping the machine.

Most national and international standards for the measurement of bearing vibration do not consider frequency spectra; instead, they give values for vibration changes of the rms value of the velocity amplitude from 10 Hz to 1 kHz (or 10 kHz) for machines in good and bad condition. These ratios have successfully been transposed to characteristic components in the vibration spectrum such as unbalance or gear tooth-meshing frequency. Usually, a change in the bearing vibration (acceleration, velocity, or displacement) amplitude by a factor of 2 to 2.5 (6 to 8 dB in level) in any characteristic component of the spectrum is considered significant and a change by a factor of 8 to 10 (18 to 20 dB in level) is considered critical, unless specified otherwise by the manufacturer. Limits for shaft vibration measurements, giving the relative motion of the shaft inside the bearing, directly relate to physical bearing clearance in the machine. Since the required time interval between measurements varies greatly from one machine to another and depends directly on the expected mean time between failure and the deterioration rate of the expected failures, measurements should be made more frequently as soon as incipient deterioration is noticed.

Successful fault detection in machinery is the first step toward establishment of a successful condition-monitoring program. Early recognition of deterioration is the key to valuable fault diagnosis and efficient trend analysis. Consequently, this phase of condition monitoring should not be neglected although sometimes it may seem unimportant.

SPECTRUM INTERPRETATION AND FAULT DIAGNOSIS

Although constant-percentage-bandwidth analyzers are better for fault detection and spectrum comparison than constant-bandwidth analyzers, the latter are essential for efficient diagnosis. Commercially available FFT analyzers provide a suitable tool for spectrum interpretation. They provide constant bandwidth (on a linear frequency scale) and, by means of zoom, they also provide very high resolution in any frequency range of interest. This permits (1) early recognition and separation of harmonic patterns or side-band patterns and (2) separation of closely spaced individual components. FFT analyzers also provide diagnostic tools such as synchronous time averaging, cepstrum analysis, or Hilbert transform for amplitude and phase demodulation (see Chap. 13).

Table 16.1 classifies different types of faults and indicates at which frequency the faults are displayed in a vibration spectrum. Although such a table is of considerable help in spectrum interpretation, any such simplified presentation must be used with care as illustrated by the examples considered below. The various faults can be classified according to their spectral components, as discussed below.

SUBSYNCHRONOUS COMPONENTS

Subsynchronous components of vibration (at frequencies below the rotational speed of the machine) occur in cases where there is oil whirl, hysteresis whirl, or mechanical looseness; most subsynchronous components usually occur where sleeve bearings are used. Figure 16.5 shows a spectrum measured on the journal bearing of a centrifugal compressor with mechanical looseness. A characteristic pattern of half-order harmonics of the rotational speed can be clearly seen. Figure 16.6 shows a spectrum of the journal bearing of a pump in which a developing oil whirl shows up clearly at 21 Hz (42 per cent of the rotational speed) and its second harmonic.

Both examples clearly indicate how the use of a linear frequency scale facilitates the diagnosis of the fault by providing a clear indication of the different types of harmonic patterns. High resolution is required to separate a half-order harmonic component due

Fig. 16.5. Acceleration spectrum of a journal bearing on a centrifugal compressor: (A) In good condition. Before shutdown the vibration pattern is normal with few harmonics of the compressor's rotation speed and broad-band noise at higher frequencies due to inherent turbulences (B) With looseness. After shutdown higher-order harmonics can be seen with increased amplitude as well as an interharmonic pattern (half-order harmonic components).

FIG. 16.6. Spectrum analysis showing component due to an oil whirl at 42 percent of the rotation speed measured on the journal bearing of a pump.

to looseness (exactly 50 per cent of rotational speed) from a component due to oil whirl (42 to 48 per cent of rotational speed).

LOW HARMONICS OF ROTATIONAL SPEED

Low harmonics of the rotational speed occur in cases such as unbalance, misalignment, bent shaft, eccentricity, and cracks in shafts. These various faults may be difficult to separate since they may be mechanically related. A bad coupling may result in misalignment. A bent shaft may result in unbalance. Even a well-known and well-defined fault such as unbalance may give misleading vibration components. The exciting fault due to eccentric masses is a centrifugal force (thus radial) rotating at the shaft speed and is therefore expected to result in a component in the vibration spectrum at the machine speed. However, dynamic unbalance may also result in a rocking motion and consequently in vibration in both the radial and axial directions. In the same way, if there is a nonlinear transmission path from the point where the force is applied to the point of measurement, a rise in the harmonics of the rotational speed can be observed in the vibration spectrum, due to a distortion of the signal.

HARMONICS OF THE LINE (MAINS) FREQUENCY

Vibrational components which are harmonics of the electrical line frequency are due to electromagnetic forces in electrical machines. Usually there is a strong component at two times the line frequency (often called the *pole passing frequency*). Sidebands at a frequency which is equal to the number of poles times the slip frequency also appear in asynchronous machines; these components represent an indication of rotor eccentricity or magnetic problems.[4]

Figure 16.7 shows a vibration signal measured on the rolling element bearing of an asynchronous electric motor. By zooming in the region of the high-level 100-Hz component (i.e., twice the line frequency in Europe), this component can be diagnosed as the pole passing frequency of 100 Hz and not as the two times rotational speed at 99.6 Hz which could have been an indication of a faulty alignment. This demonstrates the value of being able to zoom around any component of interest with sufficient resolution to separate closely spaced components. The use of zoom, however, is of no help in synchronous

Table 16.1. Example of a Vibration Trouble-Shooting Chart

Nature of fault	Frequency of dominant vibration, Hz = rpm/60	Direction	Remarks
Rotating members out of balance	1 × rpm	Radial	A common cause of excess vibration in machinery
Misalignment and bent shaft	Usually 1 × rpm Often 2 × rpm Sometimes 3 & 4 × rpm	Radial and axial	A common fault
Damaged rolling element bearings (ball, roller, etc.)	Impact rates for the individual bearing component Also vibrations at high frequencies (2 to 60 kHz) often related to radial resonances in bearings	Radial and axial	Uneven vibration levels, often with shocks Impact Rates f (Hz): For Outer Race Defect $$f(\text{Hz}) = \frac{n}{2} f_r \left(1 - \frac{\text{BD}}{\text{PD}} \cos \beta \right)$$ For Innter Race Defect $$f(\text{Hz}) = \frac{n}{2} f_r \left(1 + \frac{\text{BD}}{\text{PD}} \cos \beta \right)$$ For Ball Defect $$f(\text{Hz}) = \frac{\text{PD}}{\text{BD}} f_r \left[1 - \left(\frac{\text{BD}}{\text{PD}} \cos \beta \right)^2 \right]$$ n = number of balls or rollers f_r = relative rps between inner & outer races
Journal bearings loose in housing	Subharmonics of shaft rpm, exactly ½ or ⅓ × rpm	Primarily radial	Looseness may only develop at operating speed and temperature (e.g., turbomachines)

Fault	Frequency	Direction	Remarks
Oil-film whirl or whip in journal bearings	Slightly less than half shaft speed (42 to 48%)	Primarily radial	Applicable to high-speed (e.g., turbo) machines
Hysteresis whirl	Shaft critical speed	Primarily radial	Vibrations excited when passing through critical shaft speed are maintained at higher shaft speeds. Can sometimes be cured by tightening the rotor components
Damaged or worn gears	Tooth-meshing frequencies (shaft rpm × number of teeth) and harmonics	Radial and axial	Side-bands around tooth-meshing frequencies indicate modulation (e.g., eccentricity) at frequency corresponding to side-band spacings. Normally only detectable with very narrow-band analysis and cepstrum analysis
Mechanical looseness	2 × rpm		Also sub- and interharmonics, as for loose journal bearings
Faulty belt drive	1, 2, 3, & 4 × rpm of belt	Radial	The precise problem can usually be identified visually with the help of a stroboscope
Unbalanced reciprocating forces and couples	1 × rpm and/or multiples for higher-order unbalance	Primarily radial	
Increased turbulence	Blade & Vane passing frequencies and harmonics	Radial and axial	An increased level indicates increased turbulence
Electrically induced vibrations	1 × rpm or 1 or 2 times synchronous frequency	Radial and axial	Should disappear when power turned off

F<small>IG</small>. 16.7. Spectrum analysis employing a zoom around the 100-Hz component, measured on the rolling element bearing of an asynchronous electric motor. Note the simultaneous presence of the pole passing frequency (100 Hz) and the second harmonic of rotation speed 99.6 Hz. A lesser resolution would not permit a separation.

machines or generators since the rotational speed and the line (mains) frequency are identical. In such a case, the machine should be permitted to coast to a stop. When the power is cut, electrically induced components of vibration disappear and the harmonics of the rotational speed decrease in frequency.

HIGHER HARMONICS OF THE ROTATIONAL SPEED

Higher harmonics of the rotational speed typically occur in the case of gearboxes, compressors, and turbines where characteristic frequencies are an integral multiple of the rotational speed of the machine, corresponding to the number of teeth, blades, lobes, etc. An increase in components, such as tooth-meshing frequencies or blade-passing frequencies, indicates deterioration acting on all teeth or blades, e.g., as uniform wear or increased turbulences, respectively.

Ghost components sometimes are observed in vibration spectra obtained from measurements on gearboxes; these components appear as tooth-meshing frequencies, but at frequencies where no gear in the gearbox has the corresponding number of teeth. Such components arise from faults on the gear cutting equipment which have been transmitted to the new gear. Since the faults are geometric, they are neither load-sensitive nor do they increase with wear; rather, as the surface wears, the faults tend to decrease with time. The frequency of the components are an integral multiple of the number of teeth on the index wheel and, therefore, appear as harmonics of the rotational speed of the faulty gear.

SIDEBAND PATTERNS DUE TO MODULATION

Modulations, frequently seen in vibration measurements on gearboxes, are caused by eccentricities, varying gear tooth spacing, pitch errors, varying load, etc. Such modulations manifest themselves as families of sidebands around gear tooth-meshing frequencies with a frequency spacing equal to the modulating frequency (e.g., the rotational speed of the faulty gear in the case of an eccentric gear). Figure 16.8*A* shows the distribution of the sidebands for such a condition. Any gear in a gearbox can be a source of modulation. In order to distinguish all possible sidebands the analysis must be carried out with sufficient resolution to detect even the lowest rotational speed inside the gearbox, and, therefore, zoom cannot be spared.

Local faults, such as cracked or broken gear teeth, also appear as a family of sidebands with a spacing equal to the rotational speed of the faulty gear. The sidebands shown in Fig. 16.8*B* are low in level and cover a broad frequency range. Very often the influence of the transmission path modifies the shape of the sideband pattern and does not permit a precise diagnosis.[5]

Local faults in turbines (for example, missing or distorted blades) result in a similar family of harmonics or sidebands.[6, 7]

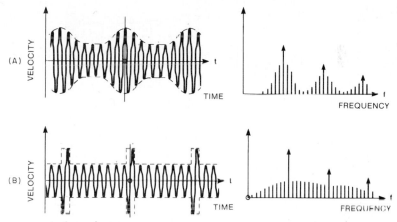

FIG. 16.8. Distribution of sideband patterns for distributed and local faults on a gear. (*A*) In the case of a distributed fault, sidebands have a high level and are grouped around tooth-meshing frequency and harmonics with a spacing equal to the speed of the faulty gear. (*B*) In the case of a local fault such as a cracked or broken tooth, the sidebands have a low level and expand widely over a large frequency range.

HARMONIC PATTERNS NOT HARMONICALLY RELATED TO THE ROTATIONAL SPEED

Harmonic patterns which are not harmonically related to the speed of rotation typically appear where there are local faults in rolling element bearings.[8,9] A local fault produces an impulse having a repetition rate equal to the characteristic frequencies of the bearing: ball-passing frequency for outer race, ball-passing frequency for inner race, ball-spin frequency (see Table 16.1). Such faults appear as a series of harmonics separated by the impact frequency with an amplitude proportional to the spectrum of the single impulse. As illustrated in Fig. 16.9, such an impact tends to excite any structural resonances in the frequency range covered, and the harmonic pattern around these resonances thus are emphasized. This provides two methods of detecting rolling element bearing faults:

1. By finding the fundamental of the impact rate in the low frequency range; this is often difficult as the fundamental is normally rather small and lost among the other components in the background vibration.

2. By finding the harmonic pattern at the impact frequency in the high frequency range; where resonances are excited, this may also be difficult because speed fluctuations tend to smear these components.

FIG. 16.9. Effect of local fault in a rolling element bearing. (A) In the time domain, there is a repeated impact having a period T. (B) In the frequency domain, the repeated impact results in a line spectrum extending very high in frequency (the shorter the impact, the higher cover of frequency range) with a spacing equal to the frequency of impact $1/T$. Any structural resonance in this frequency range is excited as shown by the peak in this picture.

ENVELOPE DETECTION

Envelope detection* is particularly useful for fault diagnosis in machinery. It permits elimination of a signal resulting from background vibration, placing greatest emphasis on the harmonic pattern of the impact frequency. This can be done either by analog or digital means.

Figure 16.10 illustrates the analog process. Figure 16.10A shows the unfiltered (acceleration versus time) signal generated in a machine by small impacts due to a ball-bearing fault. The frequency spectrum of the signal in Fig. 16.10A is shown as the dashed curve in Fig. 16.10B. The solid curve shows the frequency spectrum for a similar machine having good bearings. Note that the faulty bearing causes an increase in spectrum level in a restricted frequency range. If the signal shown in Fig. 16.10A is passed through a bandpass filter covering this restricted frequency range, almost all of the background vibration is eliminated. The filtered signal, which now contains only the ringing of selected resonances excited by repetitive impacts, is shown in Fig. 16.10C. The frequency analysis of Fig. 16.10C is shown in Fig. 16.10D. If the signal of Fig.10C is rectified, the envelope shown in Fig. 16.10E is obtained. Its spectrum in the low-frequency range, illustrated in Fig. 16.10F, shows the impact frequency and harmonics.

The advantages of envelope detection are:

1. The elimination of background noise resulting from other vibration sources (for example, from unbalance or gear vibration) by the use of bandpass filters. All that remains is the repetition rate of the impacts exciting the structural resonance, possibly amplitude-modulated.

2. High-frequency analysis is not required since only the envelope of the signal is of importance, not the signal itself, which can extend upward to hundreds of kilohertz.

3. Diagnosis is possible since the impact frequencies are determined and can be related

* Envelope detectors are discussed in Chap. 13.

to a specific source (ball-passing frequency for outer race, ball-passing frequency for inner race, ball-spin frequency, fundamental train frequency, or other sources of repetitive impacts, for example, a cracked gear tooth).

Figure 16.11A and B shows the acceleration spectra from 0 to 25 kHz of a good bearing and a faulty bearing. Note that the spectrum is noticeably higher on the good bearing than on the faulty one, which confirms that comparative measurements should not be made between different measurement points or different machines. Absolute vibration levels do not provide a satisfactory indication of the condition of a machine; only changes

FIG. 16.10. Illustration of the application of envelope detection: (A) Unfiltered signal. (B) Frequency analysis of signal shown in (A). (C) Filtered signal. (D) Frequency analysis of signal shown in (C). (E) Envelope of filtered signal shown in (C). (F) Frequency analysis, in the low-frequency range, of the signal shown in (E), showing the impact frequency and harmonics.

FIG. 16.11. (A) Acceleration spectrum of a good bearing, from 0 to 25 kHz. (B) Acceleration spectrum of a faulty bearing from 0 to 25 kHz. (C) Envelope spectrum in 100-Hz range of the signal in (A) zoomed around 5,408 Hz with an 800-Hz frequency span. (D) Envelope spectrum in 100-Hz range of the signal in B zoomed around 5,408 Hz with an 800-Hz frequency span.

in level are relevant. Any simple method of bearing fault detection such as shock-pulse measurement, spike energy, kurtosis, or crest factor, was difficult to use on this specific machine, as a forced lubrication system gave repetitive pulses at a frequency of 5.4 Hz, independent of the rotating speed, which dominated the entire vibration signal. Figure 16.11C and D shows the analysis of the envelopes on the good and the faulty bearings obtained after zooming around 5,400 Hz with an 800-Hz frequency span. The only noticeable pattern on the good bearing comes from the forced lubrication system. In contrast, the result of the envelope analysis on the faulty bearings shows a complex pattern, and frequency information is necessary to confirm whether or not there is a ball-bearing fault. The following frequencies appear: 5.4 Hz (the repetition rate of the forced lubrication system on the actual bearing, and harmonics); 6.4 Hz (the repetition range of the forced lubrication system on adjacent bearings, and harmonics); and 15.43 Hz (the ball-passing frequency for an outer race defect, and its harmonics).

APPLICATION OF CEPSTRUM ANALYSIS

The use of cepstrum analysis (explained in more detail in Chap. 13) is particularly advantageous for detecting periodicities in the power spectrum (e.g., harmonics and sideband patterns) since it measures the frequency spacing between components very precisely.[8,9] Figure 16.12 shows the spectrum and the corresponding cepstrum analysis of a measurement made on an auxiliary gearbox driving a generator on a gas-turbine-driven oil pump. As a fault on one of the bearings develops, the first rahmonic appears and then increases at a quefrency equal to the reciprocal of the spacing in the frequency spectrum which corresponds to an outer race defect in one of the bearings.

Another advantage of cepstrum analysis is that one component in the cepstrum represents the global power content of a whole family of harmonics or sidebands. Thus it indicates the severity of a defect which does not manifest itself by a single component only in the spectrum but by a series of equally spaced components.

Figure 16.13 shows the increase in level for two different frequency components in the spectra from Fig. 16.12. Figure 16.13A represents the evolution of the harmonic component at 7.64 kHz (showing a clear ascending slope) and also the evolution of the harmonic component at 5.62 kHz (showing a rather horizontal slope). Figure 16.13B represents the changes on the first rahmonic in the cepstrum and shows the effective evolution of the

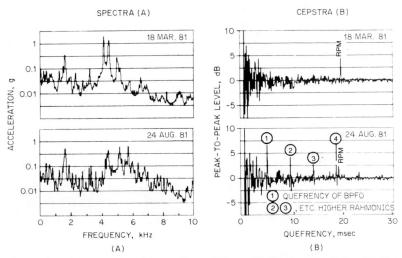

FIG. 16.12. Analysis of vibration of an auxiliary gearbox. (A) Spectrum analysis. (B) The corresponding cepstral analyses. The analyses are shown before and after the development of a bearing fault.[8]

FIG. 16.13. (A) Evolution of faults using harmonic at 7.64 kHz and at 5.62 kHz as parameter showing different trends and high spreading (case from Fig. 16.12). (B) Trend curve using the cepstrum component as the parameter, showing a smooth evolution.[8]

fault corresponding to a steep increase followed by stabilization as is typical for a spall in a rolling element bearing.

Envelope detection and cepstrum analysis make it possible to precisely determine the frequencies involved and are thus useful for condition monitoring of industrial machines. They permit a reliable diagnosis of defects. Also, the cepstrum appears to be an invaluable parameter for following the development of faults (for example, faults which develop in rolling element bearings, and damage to turbine blades) which shows up as a family of harmonics or sidebands.

TREND ANALYSIS

Trend analysis consists of making graphs of a condition-related parameter vs. time (date or running hours) to determine when the parameter is likely to exceed a given limit. The goal of a successful condition-monitoring program is to predict the date of an expected breakdown well in advance of its occurrence in order to shut down the machine in good time, order spare parts, and thereby minimize the shutdown time. Since vibration criteria indicate that equal changes on a logarithmic scale correspond to equal changes in severity, the trend should be taken in terms of decibel values. A linear trend on a logarithmic scale is found occasionally, but the actual trend may follow another course; for example, when the fault feeds back on the rate of deterioration (e.g., gear wear), the trend, when plotted on a logarithmic scale, may then be exponential. In some cases the fault changes suddenly in finite steps (for example, in the case of a spall caused by gradual subsurface fatigue), and it becomes very difficult to extrapolate to determine the date of the shutdown. To ensure accurate trend analysis, the following precautions should be taken:

1. Determination of a trend should be based on measurements of a parameter directly related to a specific type of fault, not based on measurements of over-all levels.

2. Diagnosis of faults should be done before attempting to interpret a trend curve (*a*) in order to select the appropriate parameter for the type of fault which is being monitored (it can be the level of an individual component, cepstrum component, broad-band level in a selected frequency range, etc.) and (*b*) in order to observe critically the results of the trend analysis to see if the linear interpolation is adequate or whether instead step or exponential changes can be expected.

3. Trend analysis should start from the last measurement before a significant change and not from the first measurement done.

USE OF COMPUTERS IN CONDITION-MONITORING PROGRAMS

Computers can be of great aid in a condition-monitoring program in handling, filing, and storing data and in performing tedious computations such as spectrum comparison and trend analysis. A condition-monitoring system which incorporates a computer includes:

1. A *recording device* for storing the analog time signals or frequency spectra. In a permanently installed monitoring system the analog time signal is directly connected to an analyzer.

2. An *analyzer* with both FFT narrow-band analysis and advanced diagnostic techniques (zoom, cepstrum) for diagnostics and constant-percentage–bandwidth analysis for fault detection. (If the latter is not included, the conversion from FFT to constant-percentage bandwidth can be done in the computer. Some analyzers can be used on both recording and analysis devices.)

3. A *computer and appropriate software* which will provide (*a*) management of the measurement program including route mapping, storage of reference spectra/cepstra, and new spectra/cepstra, (*b*) comparison of spectra and a printout of significant changes, and (*c*) trend analysis of any chosen parameter, including individual components, over-all level in a given frequency range, or cepstral components.

REFERENCES

1. Neale, M. et al.: "A Guide to the Condition Monitoring of Machinery," Department of Industry Committee for Terotechnology, Her Majesty's Stationery Office (London), 1979.
2. Mitchell, J. S.: "An Introduction to Machinery Analysis and Monitoring," Penwell Publishing Company, Tulsa, Okla., 1981.
3. Downham, E., and R. Woods: "The Rationale of Monitoring Vibration in Rotating Machinery in Continuously Operating Process Plant," *ASME Paper* No. 71-Vibr-96, 1971.
4. Howard, M. J.: "Induction Motor Magnetic Vibration," *Proc. Vibration Inst. Meet.*, Houston, April 19–21, 1983.
5. Randall, R. B.: "A New Method of Modeling Gear Faults," *ASME Paper* No. 81-Set-10, June 1981.
6. Sapy, G.: *Automatisme*, **xx**(10): Oct. 1975.
7. Barschdorff, D., W. Hensle, and B. Stühlen: *VGB Kraftwerkstechnik*, **59:** June 1979.
8. Bradshaw P., and R. B. Randall: "Early Detection and Diagnosis of Machine Faults on the Trans Alaska Pipeline," *ASME Conf., MSA Session*, Dearborn, Mich., Sept. 11–14, 1983.
9. Courrech, J.: "New Techniques for Fault Diagnostic in Rolling Element Bearings," *Proc. 40th Meet. of the Mechanical Failure Preventive Group*, National Bureau of Standards, Gaithersburg, MD, April 16–18, 1985.

17

STRAIN-GAGE INSTRUMENTATION

Earl J. Wilson
National Aeronautics and Space Administration

INTRODUCTION

The resistance strain gage may be employed in shock or vibration instrumentation in either of two ways. The strain gage may be the active element in a commercial or special-purpose transducer or pickup, or it may be bonded directly to a critical area on a vibrating member. Both of these applications are considered in this chapter, together with a discussion of strain-gage types and characteristics, cements and bonding techniques, circuitry for signal enhancement and temperature compensation, and related aspects of strain-gage technology.

The electrical resistance strain gage discussed in this chapter is basically a piece of very thin foil or fine wire which exhibits a change in resistance proportional to the mechanical strain imposed on it. In order to handle such a delicate filament, it is either mounted on or bonded to some type of carrier material and is known as the *bonded strain gage*.

The strain gage is used universally by stress analysts in the experimental determination of stresses. Since strain always accompanies vibration, the strain gage or the principle by which it works is broadly applicable in the field of shock and vibration measurement. Here it serves to determine not only the magnitude of the strains produced by the shock or vibration, but also the entire time-history of the event, no matter how great the frequency of the phenomenon.

BASIC STRAIN-GAGE THEORY AND PROPERTIES

The relationship between resistance change and strain in the foil or wire used in strain-gage construction can be expressed as

$$\frac{\Delta L}{L} = \frac{1}{K}\frac{\Delta R}{R}$$

or

$$K = \frac{\Delta R/R}{\Delta L/L}$$

(17.1)

where K is defined as the *gage factor* of the foil or wire, ΔR is the resistance change due to strain, R is the initial resistance, ΔL is the change in length, L is the original length of the wire or foil, and $\Delta L/L$ is the unit strain to which the wire or foil is subjected.

All materials do not exhibit this strain-sensitivity effect and different materials have different gage factors. Filament materials in common use in strain gages are Constantan (Ni 0.45, Cu 0.55), which has a gage factor of approximately +2.0, Iso-elastic (Ni 0.36, Cu 0.08, Fe 0.52, and Mo 0.005), which has a gage factor of about +3.5, and modified Karma (Ni 0.75, Cr 0.20, plus additions), which has a gage factor of +2.1.

STRAIN-GAGE CONSTRUCTION

Since the foil used in a strain gage must be very fine or thin to have a relatively high electrical resistance, it is difficult to handle. For example, the foil used in gages is about 0.1 mil in thickness. In the past, many strain gages have been constructed with fine-wire filaments, but this type of gage is now rarely used except in special or high-temperature applications.

Fig. 17.1. Typical construction of a foil strain gage.

In order to handle this foil, it must be provided with a "carrier medium," usually a piece of paper or plastic to which the filament is cemented. Lead wires are often provided on foil gages. The cement and paper or plastic sandwich perform another very important function in addition to providing ease of handling and simplicity of application. The cement provides so much lateral resistance to the foil that it can be shortened significantly without buckling; then compressive as well as tensile strains can be measured. Foil gages are formed in a grid as shown in Fig. 17.1.

TRANSVERSE SENSITIVITY

Because of its construction, a portion of the foil in each gage lies in the transverse direction and will respond to transverse strain. Therefore the gage factor K of a gage* is always slightly smaller than the gage factor of the material of which it is fabricated. One of the desirable features of foil-type gages is their low transverse sensitivity. In this case, the gage consists of a flat foil grid; a sufficiently large amount of the foil is

* In determining the *gage factor* of the gage it is assumed that the gage is mounted on a material having a Poisson's ratio of 0.285 and subjected to uniaxial stress in the direction of the gage axis.

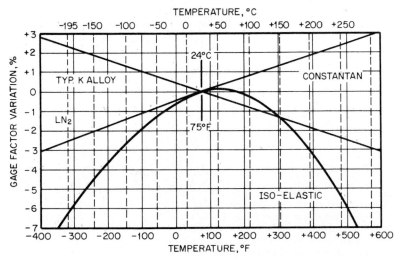

Fig. 17.2. Typical variation in the gage factor of strain-gage alloys as function of temperature.

left at the ends of each strand to reduce the transverse sensitivity of the gage to one-half the value for wire gages for some types and to essentially zero for others.

TEMPERATURE EFFECTS

The effects of temperature on the gage factor of several alloys are illustrated in Fig. 17.2. When a bonded strain gage is used in measurements, any change in resistance in the strain-gage measurement system is interpreted as resulting from a strain. If thermal expansion is not induced, then this change will result from a mechanical strain. However, if thermal expansion is induced, then there will be a change in resistance resulting from the mechanical strain; in addition, there will be a change in resistance resulting from the response of the strain gage to changes in temperature. The strain indication which results from such a temperature effect is known as an "apparent strain." Figure 17.3 shows typical apparent strain for three commonly used alloys. This effect is

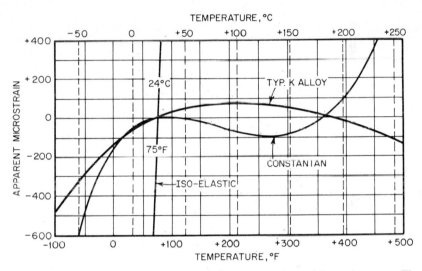

FIG. 17.3. Typical apparent strain for three alloys commonly used in strain gages. These data are based on an instrument gage factor of 2.00.

usually negligible in the measurement of dynamic strains, since the readout instrument associated with the strain gage usually does not respond to static or slow changes in its resistance. However, in the measurement of static strains, the effects of temperature represent the largest potential source of error and require some form of temperature compensation.[2]

STRAIN-GAGE CLASSIFICATIONS

Strain gages are classified in several ways. One classification cites the purpose for which the gage is to be used, that is, for static or dynamic strain measurement. Static gages are made up with Constantan foil, which has a minimum change of resistance with temperature. Dynamic strain gages are made up with Iso-elastic foil, which provides a greater gage factor than Constantan. The dynamic gages, while having a much greater resistance change for a given strain than the static gages, also are much more sensitive to changes in temperature. They are used only where the phenomenon to be measured is so short in time duration that no temperature change of any consequence can occur during the time of measurement. Gages also are available for the measurement of very large strains (up to 20 per cent) occurring in the plastic region of the material as distinguished from the more common gages which are used to measure elastic strains (up to 1 per cent).

STRAIN-GAGE SELECTION CONSIDERATIONS

In the case of shock measurements, a transient may be applied to the structure that is under investigation only once or it may be repetitive. Shock is of very short time duration, and the problem of temperature compensation is nonexistent because in most cases the temperature does not have time to change during the impact. For this reason a dynamic-type gage usually can be employed for the measurement of shock. This type of gage has the advantage of a higher gage factor than the static gage so that it will provide the greatest possible electrical signal for a given strain.

For vibration measurement the type of gage selected is dependent on the kind of information desired. If only the frequency of vibration and the magnitude of the cyclic stresses are desired, dynamic-type gages can be used since temperature changes will not affect the results obtained unless the temperature fluctuates at the same rate as the stress. If, however, a measurement of the static or slowly varying component of the stress is also to be determined (i.e., if the absolute values of the stresses are desired), a static-type gage must be employed. Since changes in temperature will affect the gage reading, temperature compensation must be incorporated to obtain true values of stress.

Gage selection is dependent on the space limitation and steepness of strain gradient in any region. The strain gage indicates the average strain over the length of the gage; in a region of steep strain gradient, this indicated value may be much less than the maximum strain. The shorter the gage used in such a region, the closer is the gage indication to the maximum strain (Fig. 17.4). However, two possible objectives must be considered quite carefully in selecting a gage for a particular installation: (1) the determination of the frequency of vibration, or comparison of relative amplitudes and frequencies with different conditions of excitation, and (2) the determination of the maximum stress pattern resulting from the vibration set up. In the first case there is considerable freedom with regard to the location of the gage on the structure, and therefore with the selection of the gage itself. In the second case severe restrictions exist in regard to the region of application of the gage and its possible dimensions. In general, very short gage-length gages are more difficult to apply properly. Therefore it is desirable to employ gage lengths of $\frac{1}{4}$ in. whenever possible. When the actual magnitude of the maximum stress resulting from shock or vibration is to be determined, a much more complicated system of gages must be employed. A single gage can be used in only the very limited case where a stress exists in one direction only, and that direction must be known. If stresses exist in several directions, or if the direction of a singly existing stress is unknown, a strain-gage rosette consisting of three or more gages must be employed.[3]

Fig. 17.4. Effect of gage length on indicated strain in the presence of a severe strain gradient. The shorter gage on the right indicates a higher strain. An infinitesimal gage length would be necessary to indicate the peak strain.

PHYSICAL ENVIRONMENT

The physical environment of the applied gage is an important factor which must be considered in gage selection and protective treatment. Temperature, pressure, humidity, oil, corrosive acid, abrasive action, and possible electromagnetic, neutron, and radiation fields are conditions which affect the choice of gage and its required protection. If high temperatures (up to 500°F or 260°C) are to be encountered, a Bakelite or other high-temperature-type gage must be selected. If even higher temperatures must be withstood, a ceramic-type gage should be employed. Gages of this sort are used at temperatures as high as 2000°F (1100°C). If the temperature never exceeds 200°F (95°C), however, any type of gage can be used. Most gages operate satisfactorily at very low temperatures.

ACCURACY CONSIDERATIONS

Gages must be selected with regard to the desired precision of the results. If only the frequency of the vibration or the duration of a shock wave is required, almost any gage, properly chosen for the temperature and humidity conditions to be encountered, gives quite satisfactory results. However, if the magnitude of the stresses produced is to be determined in addition, then considerable care must be exercised to select the proper gage to obtain the desired results. Not only must the gage be the proper one to portray the encountered strain faithfully, but precautions must be taken to install the gage correctly.

The testing "environment" can affect strain-gage accuracy in many ways. Magnetostrictive effects,[4] hydrostatic pressure,[5] nuclear radiation,[6] and high humidity are examples of conditions that may cause large strain-gage errors. Creep, drift, and fatigue life in the gages themselves may be important. In most normal environments these errors are either small or undetectable. Whenever unusual or harsh environments are encountered, it is wise to consult the strain-gage manufacturer to obtain recommendations for gage systems and estimates of expected accuracies.

BONDING TECHNIQUES

The proper functioning of a strain gage is completely dependent on the bond which holds it to the structure undergoing test. If the bond does not faithfully transmit the strain from the test piece to the wire or foil of the gage, the results obtained cannot be accurate. Failure to bond over even a minute area of the gage will result in incorrect strain indications. The greatest weakness in the entire technique of strain measurement by means of wire or foil gages is in the bonding of the gage to the test piece. Usually, the manufacturer of the strain gages will recommend cements which are compatible with their use and will provide instructions for their proper installation.

STRAIN-GAGE MEASUREMENTS

The resistance strain gage, because of its inherent linearity, very small mass, wide frequency response (from zero to more than 50,000 cps),[7] general versatility, and ease of installation in a variety of applications, is an ideal sensitive component for electrical transducers for use in shock and vibration instrumentation.[8] The Wheatstone bridge circuit, described in a subsequent section, can be used to extend the versatility of the strain gage to still broader applications by performing mathematical operations on the strain-gage output signals. The combination of these two devices can be used effectively for the measurement of acceleration, displacement, force, torque, pressure, and similar mechanical variables. Other useful attributes include the capacity for separation of forces and moments, vector resolution of forces and accelerations, and cancellation of undesired vector components.

The usual technique for employing a strain gage as a transducing element is to attach the gage to some form of mechanical member which is loaded or deformed in such a manner as to produce a signal in the strain gage proportional to the variable being measured. The mechanical member can be utilized in tension, compression, bending, torsion, or any combination of these. All strain-gage-actuated transducers can be considered as either force- or torque-measuring instruments. Any mechanical variable which can be predictably manifested as a force or a couple can be instrumented with strain gages.

There are a number of precautions which should be observed in the design and construction of custom-made strain-gage transducers.[9] First, the elastic member on which the strain gage is to be mounted should be characterized by very low mechanical hysteresis and should have a high ratio of proportional limit to modulus of elasticity (i.e., as large an elastic strain as possible). Although aluminum, bronze, and other metals are often employed for this purpose, steel is the most common material. An alloy steel such as SAE 4340, heat-treated to a hardness of RC 40, will ordinarily function very satisfactorily.

The physical form of the elastic member, and the location of the strain gages thereon, are not subject to specific recommendation, but vary with the special requirements of each individual instrumentation task. When no such requirements exist, a standard commercial transducer ordinarily should be used. In general, the shape of the member should be such as to (1) allow adequate space for mounting strain gages (preferably in regions of zero or near-zero strain gradient), (2) provide the desired natural frequency, (3) produce a strain in the gages which is great enough at low values of the measured variable to result in an output signal readily subject to accurate indication or recording, and not so great as to cause nonlinearities or abbreviated gage life at peak load values, (4) provide temperature compensation and/or signal augmentation (as described in a subsequent section) whenever feasible, and (5) allow for simplicity of machining, ease of gage attachment and wiring, and, if necessary, protection of the gages.

The strain gages should be cemented to the elastic member with the usual care and cleanliness necessary to all strain-gage applications, special attention being given to minimizing the bulk of the installation if the added mass is significant to the frequency response of the instrument. Other considerations vital to successful strain-gage-application technique are described elsewhere in this chapter.

DISPLACEMENT MEASUREMENT

Measurement of displacement with strain gages can be accomplished by exploiting the fact that the deflection of a beam or other loaded mechanical member is ordinarily proportional to the strain at every point in the member as long as all strains are within the elastic limit.

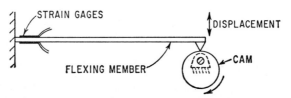

Fig. 17.5. Strain gages mounted on a cantilever beam for displacement measurement produce electrical signal proportional to cam motion.

For small displacements at low frequencies, a cantilever beam arranged as shown in Fig. 17.5 can be employed. The beam should be mounted with sufficient preload on the moving surface that continuous contact at the maximum operating frequency is assured. In the case of higher frequency applications the beam can be held in contact with the moving surface magnetically or by a fork or yoke arrangement, as illustrated in Fig. 17.6. It is necessary to make certain that the measuring beam will not affect the displacement to be instrumented, and that no natural mode of vibration of the beam itself will be excited.[10]

Fig. 17.6. Displacement transducer designed for continuous, positive contact with moving object.

The measurable displacement magnitude can be increased above that for the cantilever beam by employing other schemes such as the "clip gage" shown in Fig. 17.7. This gage is constructed by bonding strain gages to the upper and lower sides of a piece of channel-shaped spring steel, as shown in Fig. 17.7. The assembly is then clipped or otherwise mounted on the test specimen so that the legs deflect as the specimen is strained, thus straining the backbone of the clip gage to a greater or lesser extent. Any desired reduction in strain magnitude can be obtained in this manner by merely altering the proportions of the clip gage. Unfortunately, the maximum allowable frequency generally decreases as the displacement amplitude increases, since stiffness and natural frequency tend to change together. Displacement also can be measured through the use of the relative motion of a seismically mounted mass of much lower natural frequency than the applied frequency.

Fig. 17.7. Clip gage for instrumenting large displacements. Proportions of clip gage are designed to keep strain well within the proportional limit of the material.

VELOCITY

Velocities can be measured directly with strain-gage transducers only by producing a force such as viscous damping or hydro- or aerodynamic drag force which is uniquely related to velocity. Velocity indication also can be obtained with strain gages by differentiation of a displacement function or integration of an acceleration function. In either case, the transducer-design considerations correspond to those for force measurement described in the following section.

FORCE MEASUREMENT

The principle of force measurement with strain-gage-actuated transducers is very similar to that for displacement.[9] The procedure consists of placing a strain-gage-instrumented elastic member in series with the force to be measured. The strain in the transducer, and thus the output signal, is proportional to the force if all stresses are kept within the elastic limit. The proportionality constant between strain and force must be obtained by calibration if precise results are desired. Otherwise, tolerances on the gage factor of the strain gage, and uncertainty as to the elastic properties of the instrumented member, can produce errors of 5 per cent or greater—even for transducer configurations with readily calculable strain distributions.

Figure 17.8 illustrates a common form of force transducer, the cantilever beam. Strain gages are mounted on the top and bottom of the beam, producing double sensitivity (output) and virtually complete temperature compensation. While this type of trans-

FIG. 17.8. Cantilever force-measuring transducer, consisting of beam with load applied at free end. Gage strain is a linear function of the force if the proportional limit is not exceeded.

ducer is probably best suited to static or quasi-static measurements such as reaction forces, it also can be used very successfully for many shock and vibration problems as long as the natural frequency of the beam is higher than the frequency of the force being measured. The ring gage (Fig. 17.9) can be categorized with the cantilever beam, and is equally applicable to static or dynamic force measurement within the limitations imposed by its comparatively low natural frequency.

For most dynamic force-instrumentation problems a small compression or tension member (Fig. 17.10) is ordinarily employed. If the load is characterized by alternation between compression and tension, the transducer must be designed for a rigid, integral

FIG. 17.9. Ring gage for force measurement. This type of gage provides sensitive axial load measurement without undue loss of rigidity or ruggedness.

FIG. 17.10. Widely used commercial form of axial force transducer for large loads.

connection, with no backlash or clearance. This can be accomplished by employing threaded ends with lock nuts for joining the transducer to the remainder of the assembly. In many problems involving machine parts or other mechanical components it is possible to measure loads by applying strain gages to the machine member itself, necessitating calibration of the member to determine the relationship between force and strain.

PRESSURE

In hydraulic and aerodynamic devices, pressure fluctuations are often associated with vibration phenomena—either as cause or effect. Strain-gage transducers are widely used in such situations.[11]

Fig. 17.11. Piston-type pressure transducer with diaphragm seal for piston. Pressure load on piston head is sensed by strain gages on supporting column.

Fig. 17.12. Pressure pickup whose output is a function of diaphragm strain. As diaphragm deforms under pressure, strain is transmitted to gage to produce electrical signal.

Pressure pickups based on strain gages are commonly one of three principal types: piston, diaphragm, or tube. In the piston type the pressure acts against a freely movable flat surface (which may be either a piston or diaphragm), the motion of which is inhibited by an elastic member instrumented with strain gages to measure the force (Fig. 17.11).

Diaphragm-type pressure transducers, shown in Fig. 17.12, have the strain gages applied directly to the back surface of the diaphragm so that diaphragm strain is a measure of pressure.[12] The simplest form of pressure transducer to construct is the tube type, shown in Fig. 17.13. In this type, strain gages are applied to the outer surface of a tube which has the fluid pressure acting on its inner surface. It is sometimes necessary to thin the wall of the tube or to use a longitudinally crimped tube in order to increase the strain magnitude to a measurable level.

Fig. 17.13. Readily made pressure transducer consisting of length of tubing with strain gages attached. Dilation of the tube with pressure creates strain in gages.

As a convenient alternative, the bourdon tube in a conventional mechanical pressure gage can serve as the transducing element if strain gages are attached to it. The compressibility of the fluid contained in the tube must be considered for its effect on the frequency response of this type of unit. Pressure pickups should be calibrated statically, and preferably dynamically, prior to use.

ACCELERATION

At one time, wire or foil strain gages were used as the transducing elements in resistive accelerometers. Now silicon elements are usually used because of their higher sensitivity. See *Piezoresistive Accelerometers*, Chap. 12.

REFERENCES

1. Wu, C. T.: "Transverse Sensitivity of Bonded Strain Gages," *Experimental Mechanics, J. Soc. for Exptl. Stress Anal.*, November, 1962.
2. Hines, F. H., and L. J. Weymouth: Practical Aspects of Temperature Effects on Resistance Strain Gages, in M. Dean, III and R. D. Douglas (eds.), "Semiconductor and Conventional Strain Gages," Academic Press, Inc., New York, 1962.
3. Vigness, I.: "Magnetostrictive Effects in Wire Strain Gages," *Proc. Soc. Exptl. Stress Anal.*, **14**(2):139 (1957).
4. Milligan, R. V.: The Gross Hydrostatic-Pressure Effects as Related to Foil and Wire Strain Gages, "Experimental Mechanics," pp. 67ff, Society for Experimental Stress Analysis, Westport, Conn., February, 1967.
5. Vulliet, P. (ed.): *Proc. Western Regional Strain Gage Comm.*—1968 Spring Meeting, Marina Del Rey, Calif., Society for Experimental Stress Analysis, Westport, Conn.
6. Perry, C. C., and H. R. Lissner: "The Strain Gage Primer," pp. 117ff., McGraw-Hill Book Company, Inc., New York, 1955.
7. Bickle, L. W.: The Use of Strain Gages for the Measurement of Propagating Strain Waves, Proc. Tech. Comm. Strain Gages, Oct. 23, 1970, Society for Experimental Stress Analysis, Westport, Conn.
8. Norton, H. N.: "Handbook of Transducers for Electronic Measuring Systems," pp. 42ff., Prentice-Hall, Inc., Englewood Cliffs, N.J., 1969.
9. Motsinger, R. N.: Flexural Devices in Measurement Systems, in P. K. Stein (ed.), "Measurement Engineering," vol. 1 chap. 11, Stein Engineering Services, Phoenix, Ariz., 1964.
10. Cleveland, A. W.: *J. Soc. Auto. Eng.*, Vol. **59**:34ff., (1951).
11. Jasper, N. H.: *Proc. Soc. Exptl. Stress Anal.*, **8**(2):83 (1951).
12. Perry, C. C.: "Design Considerations for Diaphragm Pressure Transducers," *Tech. Note* TN-129, Micro-Measurements Division, Romulus, Mich., 1974.

18

CALIBRATION OF PICKUPS

John D. Ramboz
Consultant

M. R. Serbyn
National Bureau of Standards

Robert W. Lally
PCB Piezotronics, Inc.

INTRODUCTION

This chapter describes various methods of calibrating shock and vibration transducers, commonly called *vibration pickups*. The objective of calibrating a transducer is to determine its *sensitivity* or *calibration factor* as defined below.[1-4] This chapter is divided into three major parts which discuss absolute calibration methods, high-acceleration and shock calibration methods, and comparison methods of calibration. Finally, methods are described for the calibration of auxiliary circuits, including strain-gage bridge-type transducers, voltage amplifiers, charge amplifiers, and impedance converters.

PICKUP SENSITIVITY, CALIBRATION FACTOR, AND FREQUENCY RESPONSE

The *sensitivity* of a linear vibration pickup is the ratio of electrical output to mechanical input applied to a specified axis.[3,4] The sensitivity is specified at a discrete frequency. In general, the sensitivity includes both amplitude and phase information and is, therefore, a complex quantity. The sensitivity for all pickups varies with frequency. A typical response for a piezoelectric accelerometer, in both magnitude and phase, is shown in Fig. 18.1.

If the sensitivity is substantially independent of frequency over a specified range of frequencies, it is referred to as the *calibration factor* for that range.[3,4] There are many reasons why the sensitivity varies with frequency. Most important among these is the effect of the resonance of the seismic element, which causes the response to rise as the vibration period approaches the natural period of the pickup. Other factors which cause the sensitivity to vary with frequency are resonances in the pickup case and external electrical loading resulting from the limitations of auxiliary circuits. In some instances, damping fluid may develop springlike properties at high frequencies or the fluid may experience a change in temperature, thereby altering the damping rate. Joints which behave nearly elastically at low frequencies can develop appreciable damping at higher frequencies.

In the recommended frequency range of operation of a pickup, the sensitivity is usually nearly constant for a wide range of amplitudes. At amplitudes outside this range, the sensitivity may vary because of nonlinear performance of the mechanical or electrical elements, limitations set by mechanical stops, limited resolving abilities of the sensing element, electrical noise, or sticking where rubbing friction is present.

The *phase lag* for a vibration pickup is a measure of the time lag between the mechanical input and the electrical output. A distortion-free phase condition exists if the phase shift is 0 or 180°, or if it is proportional to the frequency. For example, as indicated in Chap. 12, accelerometers with damping about 0.65 of critical have a phase shift which is almost proportional to the frequency over their usable range. Pickups that are intended for measuring transient disturbances, vibrations containing several frequency components, and acoustic intensity should be checked to determine that the phase-angle distortion is acceptably small. Electronic filters and other signal conditioning equipment may add significant phase shift and thus produce phase-angle distortion of the input signal. Phase-angle distortion of the pickup may be compensated by associated electric circuits.

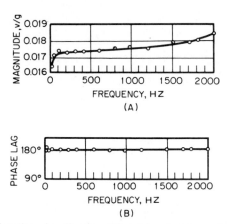

FIG. 18.1. Sensitivity function of a vibration pickup (piezoelectric accelerometer), calibrated by comparison with a reference standard. (*A*) Magnitude as a function of frequency. (*B*) Phase lag vs. frequency. (*After S. Levy and R. R. Bouche.*[5])

Displacements are usually expressed as single- or double-amplitude (peak-to-peak) values, while velocities are usually expressed peak, root-mean-square (rms), or average values. Acceleration and force generally are expressed as peak or rms values. The electrical output of the vibration pickup may be expressed as a peak, rms, or average value. The sensitivity magnitude or calibration factor is commonly stated in similarly expressed values, i.e., the numerator and denominator are both peak or both rms values. For example, a typical sensitivity for an accelerometer may be expressed in units of millivolts per unit acceleration, or simply mV/g. For some special applications it may be desirable to express the sensitivity in mixed values, such as rms millivolts per peak unit acceleration.

ABSOLUTE CALIBRATION METHODS

When the pickup sensitivity or calibration factor is obtained directly from a measurement of quantities which form the base of a system of units, the calibration is called *absolute*. The comparison method, by contrast, involves simultaneous measurement of the outputs of the pickup under test and a reference pickup of known and stable characteristics, with both pickups subjected to the same vibratory excitation. The comparison method is simple and relatively inexpensive as opposed to most of the absolute methods of calibration. Absolute methods are used to characterize reference pickups or when the minimum possible uncertainty is required. A number of accepted absolute methods exist. The most common are discussed in this section.

METHODS USING EARTH'S GRAVITATIONAL FIELD

CHATTER METHOD The chatter method utilizes the earth's gravitational field as a means of determining g. If a small mass (usually a ball bearing) is placed upon a vibrating surface which is perpendicular to the gravity vector and the mass is permitted to rest loosely on it, separation will occur whenever the acceleration exceeds $1g$. Separation will be maintained until the deceleration portion of the cycle. Separation and contact may be detected electrically or audibly.[6]

This method is useful only for vertical vibration at frequencies less than about 100 Hz and can indicate the $1g$ acceleration level to within about ± 1 to ± 5 per cent, depending on the mass, the vibrating surface, the frequency, and the means used to detect separation or contact.

TILTING-SUPPORT CALIBRATOR The earth's gravitational field provides a convenient means of applying small constant accelerations to a pickup. It is particularly useful in calibrating accelerometers whose frequency range extends to 0 Hz. A $2g$ change in acceleration may be obtained by first orienting the accelerometer with the positive direction of its sensing axis up, and then rotating the accelerometer through 180° so that the positive direction is down. (See the *inversion method* described in Chap. 15 in the section on *Field Calibration Techniques*.)

Increments of acceleration within the $1g$ range can be applied by means of a tilting-support calibrator with an accuracy better than $0.004g$ near the horizontal and less than $0.001g$ near the vertical.[3,7] The pickup to be calibrated is fastened as shown in Fig. 18.2 to a platform at the end of an arm. The arm may be set at any angle ϕ between 0 and 180° relative to the vertical. It is furnished with a pointer to indicate the angle ϕ. Care should be taken to level the base when the arm is set at $\phi = 0°$. Positioning of the arm angle ϕ to $\pm 0.2°$ or less is possible with an accurately divided circle. The component of sensed acceleration is given by

$$a = g \cos \phi \qquad (18.1)$$

FIG. 18.2. Tilting-support calibrator used to supply incremental static accelerations of $1g$ or less to accelerometers which have zero frequency response, such as strain-gage accelerometers. (*After ANSI standard S2.2-1959, R1981.*[3])

The calibration factor is obtained by plotting the response of the accelerometer as a function of the acceleration a, given by Eq. (18.1), for successive values of ϕ and then determining the slope of the straight line fitted through the points.

The pickup is also subjected to a component of acceleration at right angles to its sensitive direction, equal to

$$a_t = g \sin \phi \qquad (18.2)$$

Because of the transverse component, this method is not recommended for calibration of pickups for which the transverse sensitivity is significant. The extent to which the transverse component affects the calibration results may be checked by remounting the pickup at two additional positions on the mounting platform. The two new positions are selected by rotating the pickup 90 and 180° about its sensitive axis relative to the first position. The test is performed for each of the three different angular positions of the pickup on the pointer arm. Any measurable change in the results may indicate an undesirable influence of the transverse sensitivity. The maximum influence will occur when the pointer arm angle $\phi = 0°$.

PENDULUM CALIBRATOR   The pendulum calibrator is a simple device for generating a transient acceleration of as great as $10g$ with a duration on the order of a second.[3,8] It imparts to a pickup a transverse and an angular acceleration as well as the acceleration inward from its center of rotation. Therefore, it should be used with caution in calibrating pickups sensitive to these extraneous motions.

The pendulum calibrator in Fig. 18.3 consists of a beam and platform with its center-of-gravity pivoted so as to swing about a horizontal axis. The pickup to be calibrated is attached to the platform and is carefully aligned so that the pickup axis of sensitivity is along a radius of the beam. Electrical connections can be brought from the pickup along the beam bridging to the support frame. A well-constructed pendulum has negligible damping. If the pendulum is released to swing after being raised to an angle ϕ above the vertical, the increment a, in acceleration along the radius vector from the time of release to the bottom of the swing, is

QUICK
RELEASE
MECHANISM

ACCELEROMETER

CENTER OF GRAVITY

BEAM

$$a = \left[g + \frac{8\pi^2 l}{\tau^2} \right] (1 - \cos \phi) \quad (18.3)$$

where g = acceleration of gravity
l = distance from center-of-gravity of mass element to pivot
τ = period of pendulum for small amplitudes
ϕ = average of release angle and angle to which pendulum rises after passing bottom of its swing

The calibration factor normally is determined by plotting the peak-to-peak response

$$R = \frac{e_1 + e_2}{2} \quad (18.4)$$

Fig. 18.3. Pendulum calibrator for measuring the calibration factor. (*After H. Levy.*[8])

of the pickup as a function of the acceleration a as given by Eq. (18.3) for successive values of ϕ, and then finding the slope of the straight line fitted through the data. The error is ordinarily below ± 1 per cent if care is exercised.

A typical oscillogram record of an accelerometer response as a function of time is shown in Fig. 18.4.

ROTATING-TABLE (CENTRIFUGAL FIELD) METHOD. In the rotating-table method of calibration, the accelerometer is rotated at a uniform angular rate about a horizontal axis.[9,10] The experimental arrangement is shown in Fig. 18.5. The pickup is mounted so as to rotate about a horizontal axis; the sensitivity axis of the accelerometer rotates in a vertical plane. Thus it is possible to apply very low-frequency sinusoidal motion in the region where sinusoidal linear translation of large amplitude would otherwise be required to give useful accelerations. As a result, a sinusoidal acceleration, having a peak amplitude of $1g$ at the rotation frequency, is superposed on the centrifugal acceleration. By this method it is possible to obtain the response of the accelerometer under both static and dynamic conditions in the same test setup. The rotating-table method not only provides excitation at very low frequencies but also provides an accurate constant amplitude of acceleration. The basic consideration in applying this method is that the vibration pickup be sufficiently sensitive to have adequate output in the range of acceleration up to $1g$. The threshold sensitivity and the repeatability of the pickup under test must be such that a periodic excitation between $\pm 1g$ produces a response that is

FIG. 18.4. Typical oscillogram of an accelerometer response to the pendulum calibrator in Fig. 18.3. The release angle is ϕ_1, and the angle to which the pendulum rises after passing the bottom of its swing is ϕ_2. The average angle $(\phi_1 + \phi_2)/2$ corresponds to the average response $(e_1 + e_2)/2$. (*After H. Levy.*[8])

FIG. 18.5. Rotating-table calibrator used for both static and dynamic calibration, with a peak acceleration amplitude of $1g$. Table rotates in a vertical plane. (*After W. A. Wildhack and R. O. Smith.*[9])

representative of the instrument's behavior. Excitation less than $\pm 1g$ can be provided by inclining the axis of rotation. The useful frequency range over which this calibration technique applies is from about 0.5 to 45 Hz.

Bearing noise and dynamic unbalance limit the upper frequency range. Improved performance from that obtained from the calibrator shown in Fig. 18.5 can be obtained by incorporating air bearings and a magnetic-clutch drive;[10] the position of the pickup is adjusted by a fine lead screw and a series of adjustable counterbalance masses are arranged about the table's circumference.

The chief restrictions of the rotating-table method of dynamic calibration are (1) the rotation speed of a table is limited; (2) the acceleration level is limited to $1g$; and (3) the response to the centrifugal acceleration tends to become infinite at the natural frequency of the pickup. The output of the pickup must be measured with care. At frequencies below about 5 Hz, errors introduced by voltmeters tend to limit the measurement accuracy. Over all pickup calibration uncertainties of ± 1 per cent in the frequency range from 2 to 8 Hz are attainable, and ± 0.5 per cent in the frequency range from 8 to 45 Hz can be maintained.

GRAVIMETRIC FREE-FALL CALIBRATOR. This technique provides a quick and simple method of calibrating force and motion sensors.[11,12] The sensor to be calibrated is allowed to fall freely under the influence of gravity for an instant of time, and the peak signal representing an acceleration of $1g$, or representing a force on a falling test mass, is measured. Figure 18.6 shows a typical free-fall calibrator. It functions by transferring the motion caused by gravity and other forces into electrical signals. The structure includes an adjustable frame, an elastic suspension, and a cylindrical test mass. The method depends upon the behavior of the test mass and a monofilament nylon line when struck. Striking the suspension juction at the top end of the line causes the line to relax momentarily, thereby imparting a step-function gravitational force to the instrumented mass by allowing it to fall freely for a moment.

Either or both the force and acceleration signals are observed, and the peak response is measured with a peak-reading instrument, a digitizing oscilloscope, or analyzer. Figure 18.7A shows the arrangement for an accelerometer; Fig. 18.8A shows the arrangement for a force sensor. In either case, a quick momentary downward deflection is produced on the suspension at the junction. This action produces an output signal in the accelerometer which is proportional to the momentary free-fall $1g$ acceleration. Figure 18.7B shows a typical time response. The calibration factor for the accelerometer can be expressed as

$$S_a = \frac{E_a}{a} \qquad (18.5)$$

where S_a = accelerometer calibration factor
   E_a = peak value of voltage step
   a = acceleration, expressed as a fraction of acceleration of free-fall

If $a = 1g$, then Eq. (18.5) simply becomes $S_a = E_a$ in units of volts per g.

For the same excitation conditions as above, Fig. 18.8B shows a typical time response to a mass-loaded force sensor. The calibration factor for a force sensor can be expressed as

$$S_f = \frac{E_{fm}}{M_c a + M_e a} \tag{18.6}$$

where S_f = force-sensor calibration factor
   E_{fm} = peak value of force-sensor output when mass M_c is applied
   M_c = added calibration mass
   M_e = effective end mass of the sensor
   a = acceleration, expressed as a fraction of acceleration of free-fall

When $a = 1g$, Eq. (18.6) becomes $S_f = E_{fm}/(M_c + M_e)$. The value of the effective end mass is usually available from the manufacturer or it may be determined by using two or more masses and solving the resulting simultaneous equations. The amplitude linearity may be checked by using a series of masses. Figure 18.8C shows a typical output of the force sensor as a function of total mass.

This calibration method is satisfactory for many applications. However, the need for frequency- or phase-response information and the difficulties encountered in measuring low-level or fast-decaying signals often require more involved calibration procedures.

GRAVIMETRIC-ACCELERATION FORCE-RATIO METHOD.   Once a force sensor has been calibrated by the above free-fall method, it may be used to calibrate accelerometers, as shown in Fig. 18.9. The force sensor typically is mounted on an electrodynamic exciter. A known mass, M_c, and the accelerometer are mounted on the force sensor.

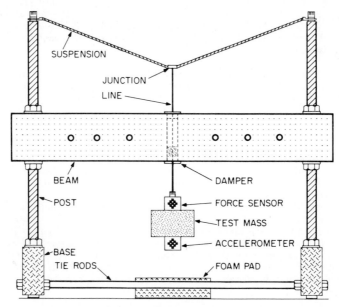

FIG. 18.6. Gravimetric calibrator. (*After D. Corelli and R. W. Lally.*[11])

FIG. 18.7. Free-fall calibration of accelerometers. (A) Arrangement of components. (B) Time-history of accelerometer response to momentary free-fall.

FIG. 18.8. Free-fall calibration of force sensors. (A) arrangement of components. (B) Time-history of force-sensor response to momentary free-fall. (C) Force-sensor output as a function of mass load.

FIG. 18.9. Measurements required by the gravimetric acceleration/force ratio method of calibrating accelerometers.

When excited, the accelerometer calibration factor can be expressed as

$$S_a = S_f(E_a/E_f)(M_c + M_e + M_a) \tag{18.7}$$

where S_a = calibration factor for accelerometer
S_f = calibration factor for force sensor
E_a = output voltage from accelerometer
E_f = output voltage from force sensor
M_c = calibration mass
M_e = effective end mass of force sensor
M_a = mass of accelerometer

For the most precise calibrations, the mass of the accelerometer mounting stud, the mass of the output connector, and the mass of a portion of the cable must be added to the accelerometer mass.

Where a calibrated accelerometer is available, a force sensor can be calibrated by the same method. Eq. (18.7) can be rearranged to yield the force-sensor calibration factor S_f.

Care must be exercised to avoid transverse motions. Also, it is essential to calibrate well below resonance frequencies. At higher frequenices, the calibration mass may no longer be considered as a rigid body and its elastic response may have to be taken into account. The accuracy of the gravimetric calibration results generally agrees within ½ per cent with laser-interferometer and reciprocity methods. A major uncertainty exists in the peak-signal determination since the value of the acceleration of gravity is not constant. However, this value generally varies less than 0.4 per cent around the world, and the local deviations in gravity are known and, thus, can be compensated for. This method is similar in principle to the impact-force shock calibration discussed in a later section.

METHODS USING THE CENTRIFUGE

SIMPLE CENTRIFUGE. A centrifuge provides a convenient means of applying constant acceleration to a pickup. Centrifuges can be obtained readily for acceleration levels up to $100g$ and have been made for use at much higher values. They are particularly useful in calibrating rectilinear accelerometers whose frequency range extends down to 0 Hz and whose sensitivity to rotation is negligible. Centrifuges are mounted so as to rotate about a vertical axis. Cable leads from the pickup, as well as power leads, usually are brought to the table of the centrifuge through specially selected low-noise slip rings and brushes.

To perform a calibration, the accelerometer is mounted on the centrifuge with its axis of sensitivity carefully aligned along a radius of the circle of rotation. If the centrifuge rotates with an angular velocity of ω rad/sec, the acceleration acting on the pickup is

$$a = \omega^2 r \tag{18.8}$$

where r is the distance from the center-of-gravity of the mass element of the pickup to the axis of rotation. If the exact location of the center-of-gravity of the mass in the pickup is not known, the pickup is mounted with its positive sensing axis first outward and then inward; then the average response is compared with the average acceleration acting on the pickup as computed from Eq. (18.8), where r is taken as the mean of the radii to a given point on the pickup case. The calibration factor is determined by plotting the output e of the pickup as a function of the acceleration a given by Eq. (18.8) for successive values of ω, and then determining the slope of the straight line fitted through the data.

DUAL CENTRIFUGE. The dual centrifuge is similar to the centrifuge described above, except that it has a second (satellite) table mounted on the main table.[13] The satellite table is driven such that it completes one revolution for each revolution of the main table. Each table rotates about a vertical axis. The pickup being calibrated is mounted on the satellite table with its sensitive axis parallel to a radius of the satellite table. The signal cable may be brought directly from the table near the center of the main table; slip rings are not necessary as with the regular centrifuge, although some care is required to secure the cable so that undue whipping action does not occur. The dynamic acceleration sensed by the accelerometer is given by

$$a = \omega^2 r \cos \omega t \qquad (18.9)$$

where a = acceleration
   r = distance from center of main table to center-of-gravity of accelerometer seismic element
   ω = angular velocity, rad/sec
   t = time

The dual centrifuge is capable of generating relatively large sinusoidal accelerations at low frequencies. For example, when the radius r = 25 cm, it is possible to attain an acceleration of $0.25g$ peak at 0.5 Hz and $125g$ peak at 11 Hz. Limitations in the use of the dual centrifuge are imposed by one's ability to balance the loads and to measure low-frequency voltages. The latter can be overcome by using modern instrumentation. Overall, the calibration factor of the accelerometer can be assessed with uncertainties on the order of ± 1 per cent.

DISPLACEMENT MEASUREMENT METHODS

The measurement of dynamic displacement is a common and valuable technique for the calibration of vibration pickups.[3] A variety of methods exist, and refined techniques have proven reliable. Both dynamic velocity and acceleration amplitudes can be accurately calculated by knowing the displacement time-history. Velocity is the first time derivative of displacement and acceleration is the second time derivative of displacement. These fundamental relationships are expressed in Chap. 1 [see Eqs. (1.1) through (1.4) for sinusoidal rectilinear motion].

The use of a factor ω or ω^2 to obtain velocity or acceleration parameters from a displacement measurement is valid only for sinusoidal motion. If harmonic distortion is present, errors will depend upon the order of the harmonic and the phase relationships. Experience has shown that if the total harmonic distortion in the acceleration waveform is less than a few per cent, calibration errors are negligible.

The pickup sensitivity is obtained by calculating the ratio of the appropriate output-to-input quantities.

VIBRATING WEDGE. One method of measuring relatively large displacement amplitudes is shown in Fig. 18.10.[3] A card is attached to the body whose vibratory motion is to be measured. This card contains two lines that intersect to form an angle as shown in the upper portion of the illustration. With the body at rest, the card is oriented so that a straight line pa is perpendicular to the expected motion. Line pb intersects line pa and forms a small angle with it. A scale is marked along line pa at each tenth of its length, and the length of line ab is set at 0.1 in. When vibrating, two angles apb and $a'p'b'$ are seen, corresponding to a peak-to-peak displacement, and an apparent intercept c appears. In the example, the intercept c falls 0.75 of the distance from p to a so that the peak-to-peak displacement amplitude pp' is read as 0.075 in. (The use of this method with the triangular

Fig. 18.10. Method of measuring displacements of relatively large amplitudes based on the vibrating-wedge technique. (*ANSI standard S2.2-1959, R1981.*[3])

area apb filled in to form a wedge results in a penumbra error, depending on the observer, and would require calibration.)

DIRECT-VIEWING MICROSCOPE. The displacement amplitude of the input vibration can be measured optically by observing a spot of high-intensity reflected light. A good source is an individual bright reflection from a field of emery cloth cemented to the edge of the calibrator table. Peak-to-peak displacement amplitudes of 10 mm can be measured with an accuracy of ±0.25 mm using conventional microscopes. With a high-quality microscope of high magnification and equipped with an optical micrometer, a peak-to-peak displacement on the order of 0.001 mm can be measured if care is taken to isolate the microscope from the vibration exciter. Stroboscopic illumination can be used to slow or stop the image. With care this method may be used at frequencies of several hundred hertz.

RONCHI RULING (OPTICAL GRATING). Electronic sensing of vibration displacement can be achieved by the use of two matched Ronchi rulings.[14] One ruling is rigidly mounted on the vibration exciter table, and the second ruling is attached onto a seismically stable and adjustable mount. It is important that the vibration exciter be fully decoupled from the fixed ruling. Figure 18.11 shows a block diagram of a typical system.

FIG. 18.11. Laboratory apparatus using Ronchi rulings to measure vibration displacement. (*After R. S. Koyanagi.*[14])

Either a laser or collimated white light may be used to illuminate the rulings. As the rulings are moved with respect to one another because of applied vibration, a first-order Moire fringe pattern causes light pulses to fall upon the photodetector. These pulses may be counted with an electronic counter. An output pulse will be generated for every displacement equal to the width of one line pair. (A line pair is one transparent line and one opaque line on a ruling.) The number of output pulses per vibration cycle is directly proportional to the vibration-displacement amplitude. For sinusoidal motion, the peak acceleration a can be calculated from

$$a = \frac{\pi^2 f^2 \nu}{n} \qquad (18.10)$$

where f = vibration frequency
ν = number of light pulses per vibration cycle
n = Ronchi ruling constant, number of line pairs per unit length

It is crucial that the rulings be precisely aligned. The line pairs on the stationary ruling must be parallel to the line pairs on the moving ruling. Additionally, the ruled surfaces must be parallel. Micrometer adjustments for rotation and translation of the stationary ruling are useful.

Ronchi rulings are commercially available with 200 line pairs per millimeter. By using such rulings, accelerometer calibration can be done for frequencies as high as several hundred hertz if the rating of the vibration exciter allows it. The calibration errors are chiefly due to displacement resolution, voltage measurement, and nonsinusoidal motion. Over-all uncertainties can generally be maintained at less than ±1 per cent.

FRINGE COUNTING INTERFEROMETER. An optical interferometer is a natural instrument for measuring vibration displacement. The Michelson and Fizeau interferometers are the most popular configurations. A modified Michelson interferometer is shown in Fig. 18.12.[14] A corner cube retroreflector is mounted on the vibration-exciter

FIG. 18.12. Typical laboratory setup for interferometric measurement of vibratory displacement by fringe counting. (*After R. S. Koyanagi.*[14])

table. A helium-neon laser is used as a source of illumination. The photodiode and its amplifier must have sufficient bandwidth (as high as 50 mHz) to accommodate the Doppler frequency shift due to the high vibration peak velocities. An electrical pulse is generated by the photodiode for each optical fringe passing it. The vibratory displacement amplitude is directly proportional to the number of fringes per vibration cycle. The peak acceleration can be calculated from

$$a = \frac{\lambda \nu \pi^2 f^2}{2}$$

(18.11)

where λ = wavelength of light
ν = number of fringes per vibration cycle
f = vibration frequency

Interferometric fringe counting is useful for vibration-displacement measurement in the lower frequency ranges, perhaps to several hundred hertz depending on the characteris-

FIG. 18.13. Simplified block diagram of a low-frequency vibration standard. (*After H. J. von Martens.*[16])

tics of the vibration exciter.[15] At the low end of the frequency spectrum, conventional procedures and commercially available equipment are not able to meet all the present requirements. Low signal-to-noise ratio, cross-axis components of motion, and zero-drifts are some of the problems usually encountered. In response to those restrictions an electrodynamic exciter for the frequency range 0.01 to 20 Hz has been developed.[16] It features a maximum displacement amplitude of 0.5 meter, a transverse sensitivity less than 0.01 per cent, and a maximum uncorrected distortion of 2 per cent. These characteristics have been achieved by means of a specially designed air bearing, an electro-optic control, and a suitable foundation. Figure 18.13 shows the main components of a computer-controlled low-frequency calibration system which employs this exciter. Its functions are (1) generation of sinusoidal vibrations, (2) measurement of rms and peak values of voltage and charge, (3) measurement of displacement magnitude and phase response, and (4) control of nonlinear distortion and zero correction for the moving element inside a tubelike magnet. Position of the moving element is measured by a fringe counting interferometer. Uncertainties in accelerometer calibrations using this system have been reduced to about 0.25 to 0.5 per cent, depending on frequency and vibration amplitude.

FRINGE DISAPPEARANCE INTERFEROMETER. The phenomenon of interference band disappearance in an optical interferometer can be used to establish a precisely known amplitude of motion. Figure 18.14 shows the principle of operation of the Michel-

Fig. 18.14. The principle of operation of a Michelson interferometer. (A) Optical system. (B) Observed interference pattern. (C) Variation of the light intensity along the X axis.

son interferometer employed in this technique.[17] One of the mirrors D, Fig. 18.14A, is attached to the mounting plate of the calibrator. Before exciting vibrations it is necessary to obtain an interference pattern similar to that shown in Fig. 18.14B.

When the mirror D vibrates sinusoidally with a frequency f and a peak displacement amplitude d, the time average of the light intensity I at position x, measured from a point midway between two dark bands, is given by

$$I = A + BJ_0\left(\frac{4\pi d}{\lambda}\right)\cos\left(\frac{2\pi x}{h}\right)$$

(18.12)

where J_0 = zero-order Bessel function of the first kind
A and B = constants of measuring system
h = distance between fringes, as shown in Fig. 18.14B and C

For certain values of the argument, the Bessel function of zero order is zero; then the fringe pattern disappears and a constant illumination intensity A is present. Table 18.1 shows the first six calculated values of displacement for which the fringe pattern disappears with a helium-neon laser source (λ = 632.8 nm). Electronic methods for more precisely establishing the fringe disappearance value of the vibratory displacement have been successfully used at the National Bureau of Standards[18,19] and elsewhere. The latter method has been fully automated using a desktop computer.

Table 18.1. Values of Peak Displacement d when $J_0(4\pi d/\lambda) = 0$,
with $\lambda = 632.8$ nm

	Peak displacement, d	
Root number	Nanometers	Microinches
1	121.10	4.7677
2	277.97	10.944
3	435.78	17.157
4	593.80	23.378
5	751.89	29.602
6	910.03	35.828

The use of piezoelectric exciters is common for high-frequency calibration of acceler-ometers.[20] They provide pistonlike motion and are structurally stiff at the lower fre-quencies where displacement noise is bothersome. When electrodynamic exciters are used with fringe disappearance methods, it is generally necessary to stiffen the armature suspensions to reduce the background displacement noise.[21]

SIGNAL-NULLING INTERFEROMETER. This method, although mathematically similar to fringe disappearance, relies on finding the nulls in the fundamental frequency component of the signal from a photodetector.[22,23] The instrumentation is, therefore, quite different, except for the interferometer. A typical arrangement is shown in Fig. 18.15. Laboratory environmental restrictions are much more severe for this method.

FIG. 18.15. Interferometric measurement displacements d as given by $J_1(4\pi d/\lambda) = 0$.

The interferometer apparatus should be well-isolated to ensure stability of the photo-detector signals. Air currents in the room may contribute to noise problems by physi-cally moving the interferometer components and by changing the refractive index of the air. An active method of stabilization has also been successfully tested.[24]

To make displacement amplitude measurements, a wave analyzer tuned to the fre-quency of vibration can be used to filter the photodetector signal. The filtered signal amplitude will pass through nulls as the vibration amplitude is increased, according to the following relationship:

$$I = 2BJ_1\left(\frac{4\pi d}{\lambda}\right) \tag{18.13}$$

where J_1 is the first-order Bessel function of the first kind and the other terms are as previously defined. The discrete amplitudes of vibration corresponding to the nulls of the Bessel function, when using a helium-neon laser, are shown in Table 18.2. The signal nulls may be established using a wave analyzer. The null amplitude will generally be 60 dB below the maximum signal level of the photodetector output.

Table 18.2. Values of Peak Displacement d when $J_1(4\pi d/\lambda) = 0$, with $\lambda = 632.8$ nm

Root number	Peak displacement, d	
	Nanometers	Microinches
1	192.96	7.596
2	353.29	13.909
3	512.27	20.168
4	670.94	26.415
5	829.41	32.654
6	987.77	38.888

The accelerometer output may be measured by an accurate voltmeter at the same time that the nulls are obtained. The sensitivity is then calculated by dividing the output voltage by the displacement. Because the filtered output of the photodetector is a replica of the vibrational displacement, a phase calibration of the pickup can also be obtained with this arrangement.[19,25]

FOLDED-BEAM INTERFEROMETER. The folded-beam interferometer is very similar to the Michelson interferometers described above. Its purpose is to increase the

Fig. 18.16. A typical arrangement for the folded-beam interferometer. (*After Robinson, Serbyn, and Payne*[22].)

resolution of the displacement measurement. This method is adaptable to either the fringe disappearance[22] or the fringe counting interferometric methods already discussed.[23] A folded-beam configuration is shown in Fig. 18.16. The laser beam is directed onto a mirror which is mounted on the moving element of the vibration exciter. The angle of incidence is selected to be slightly less than normal. The beam is then reflected to a small, flat, adjustable mirror, which returns the beam to the exciter mirror. This multiple reflection process is continued the desired number of times, after which the light passes through a beam splitter to combine with the reference beam. The small movable mirror has rotation and tilt adjustments to allow for beam steering. The number of folds can be varied by adjusting the distance between the steerable mirror and the vibration exciter, and by changing the angle of the adjustable mirror. Using from one to five folds, it is generally possible to maintain the acceleration amplitude within a required range. The measured vibration displacement is multiplied by the number of beam folds. Table 18.3 gives the peak acceleration expressed as multiples of g for various numbers of folds.[21]

The angle between the mirror mounted on the vibration exciter and the incident laser beam must be kept near 90° in order to minimize the error in vibration amplitude measurements. The error will be proportional to $(1 - \cos \theta)$, where θ is the angle between the incident beam and the normal. If the two beam splitters are 10 cm apart, the exciter should be located at least 120 cm from the beam splitter for the resulting error to be less than 0.1 per cent. Also, it will be necessary for the entire apparatus to be mounted on a rigid surface and be adequately isolated from floor vibrations.

Table 18.3. Peak Accelerations Expressed as Multiples of g, Measurable with the Folded-Beam Interferometer of Fig. 18.16, with $\lambda = 632.8$ nm

Freq., Hz	1 Fold	2 Folds	3 Folds	4 Folds	5 Folds	6 Folds
3,000	6.988					
3,500	9.511					
4,000	12.423	6.212				
4,500	15.722	7.861				
5,000	19.410	9.705	6.470			
5,500	23.486	11.743	7.829	5.872		
6,000	27.951	13.976	9.317	6.988	5.590	
6,500	32.803	...	10.934	8.201	6.561	
7,000	38.044	...	12.681	9.511	7.609	
7,500	43.673	...	14.558	10.918	8.735	
8,000	49.690	...	...	12.423	9.938	
8,500	56.095	...	...	14.024	11.219	9.349
9,000	62.889	...	...	...	12.578	10.482
9,500	70.071	...	...	...	14.014	11.679
10,000	77.641	...	...	...	15.528	12.940

RECIPROCITY CALIBRATION

The reciprocity calibration method is an absolute means for calibrating vibration exciters having a velocity coil or reference accelerometer. This method relates the pickup sensitivity to measurements of voltage ratio, resistance, frequency, and mass. For this method to be applicable, it is necessary that the exciter system be linear, i.e., that displacement, velocity, acceleration, and driver coil current each increase linearly with force and driver coil voltage. It is used chiefly with electrodynamic exciters,[26] but has also been used with piezoelectric vibration exciters.[27]

Reciprocity calibration procedures can generally be applied only under controlled laboratory conditions. Many precautions must be taken, and the process is usually time-consuming. Several variations of the basic approach have been developed at national standards laboratories;[5,28-30] the method described here is used at the National Bureau of Standards (U.S.).[3,5,22,26] The method consists of two laboratory experiments:

1. The measurement of the transfer admittance between the exciter's driver coil and the attached velocity coil or accelerometer.

2. The measurement of the voltage ratio of the open-circuit velocity coil or accelerometer and the driving coil while the exciter is driven by a second external exciter.

The use of a piezoelectric accelerometer is assumed here. The electrical connections for the transfer admittance and voltage ratio measurements are shown in Fig. 18.17.[26]

The relationship defining the transfer admittance is

$$\mathbf{Y} = \frac{\mathbf{I}}{\mathbf{E}_{12}} \tag{18.14}$$

where $\mathbf{Y}$ = transfer admittance
$\mathbf{E}_{12}$ = voltage generated in standard accelerometer and amplifier
$\mathbf{I}$ = current in driver coil

and the bold letters denote phasor (complex) quantities. The current is determined by measuring the voltage drop across a standard resistor. The phase, ψ_Y, of $\mathbf{Y}$ is measured with a phase meter having an uncertainty of $+0.1°$ or better. Transfer admittance measurements are made with a series of masses attached, one at a time, to the table of the exciter. Also, a zero-load transfer admittance measurement is made before and after attaching each mass. This zero-load measurement is denoted by $\mathbf{Y}_0$. Using the measured values of $\mathbf{Y}$ and $\mathbf{Y}_0$, graphs of the real and imaginary values of the ratio

$$\mathbf{T}_n = \frac{M_n}{\mathbf{Y} - \mathbf{Y}_0} \tag{18.15}$$

are plotted versus M_n for each frequency, where M_n is the value of the mass attached to the table. The zero intercepts, J_i and J_r, of the resulting nominally straight lines and their slopes, Q_i and Q_r, are computed by a weighted least-squares method.[26] The values of $\mathbf{Y}_0$ used in the calculations are obtained by averaging the values of the $\mathbf{Y}_0$ measurements before and after each measurement of $\mathbf{Y}$ using different masses. These computed values are used in determining the sensitivity of the standard.

TRANSFER-ADMITTANCE MEASUREMENT

VOLTAGE-RATIO MEASUREMENT

FIG. 18.17 Transfer-admittance and voltage-ratio-measurement circuit connections for the reciprocity calibration method in the Levy-Bouche realization.[26]

The ratio of two voltages, measured while the exciter is driven with an external exciter, is given by

$$\mathbf{R} = \frac{\mathbf{E}_{14}}{\mathbf{E}_{15}} \tag{18.16}$$

where $\mathbf{E}_{14}$ = voltage generated in standard accelerometer and amplifier, and $\mathbf{E}_{15}$ = open-circuit voltage in driving coil.

After $\mathbf{R}$, J_r, J_i, Q_r, and Q_i have been determined for a number of frequencies, f, the sensitivity of the exciter is calculated from the following relationship:

$$\mathbf{S} = \left[\frac{\mathbf{RJ}}{j2\pi f} \right]^{1/2} \left[1 + \frac{\mathbf{MQ}}{\mathbf{J}} \right] \tag{18.17}$$

where j = unit imaginary vector

$$\mathbf{J} = J_r + jJ_i$$
$$\mathbf{Q} = Q_r + jQ_i$$
$$\mathbf{M} = \frac{\mathbf{J}(\mathbf{Y} - \mathbf{Y}_0)}{1 - \mathbf{Q}(\mathbf{Y} - \mathbf{Y}_0)}$$

The sensitivity of the exciter is, therefore, determined from the measured quantities $\mathbf{Q}$, $\mathbf{J}$, $\mathbf{T}$, and f and from the masses M_n which are attached to the exciter table. The sensitivity as computed from Eq. (18.17) has the units of volts per meters per seconds squared if the values of the measurements are in the SI system. If the masses M_n are not in kilograms, appropriate conversion factors must be applied to the quantities $\mathbf{J}$, $\mathbf{Q}$, and $\mathbf{M}$. A commonly used engineering formula,[2] with the mass expressed in pounds and the sensitivity in millivolts per g, is

$$S = 2635 \left[\frac{\mathbf{RJ}}{jf} \right]^{1/2} \tag{18.18}$$

which also assumes that $MQ/J \ll 1$, a condition usually satisfied in practice. The use of a computer greatly facilitates the application of the reciprocity calibration process.

Reference 31 discusses the error analysis for a reciprocity calibration of an air-bearing exciter with a ceramic moving element. Assuming the errors to be uncorrelated, the estimate of uncertainty is ± 0.85 per cent. The frequency range where the reciprocity method has been found to be most accurate is from 100 to 1,000 Hz. References 32 to 40 contain additional information on the reciprocity method.

HIGH-ACCELERATION SHOCK AND CALIBRATION METHODS

Some applications in shock or vibration measurement require that high amplitudes be determined accurately. To ensure that the pickups used in such applications meet certain performance criteria, it is necessary to be able to calibrate at these high amplitudes. Several methods are available, depending on the application. Accelerations in excess of several hundred g are possible. The calibration methods discussed in this portion are dynamic methods and are divided into sinusoidal methods and impact (shock) methods.

SINUSOIDAL-EXCITATION METHODS. The use of a resonant bar to apply sinusoidal accelerations from calibration purposes has several advantages,[41–43] among them, an inherently constant frequency, very large amplitudes of acceleration (as high as $4,000g$), and low waveform distortion. A disadvantage of this type of calibrator is that only the resonance frequencies are available for use.

The bar can be supported at its nodal points and the pickup to be calibrated can be mounted at its mid-length location.[41,42] The bar can be energized by a small electromagnet or self-excited. Acceleration amplitudes of several thousand g can thus be obtained at frequencies ranging from several hundred to several thousand hertz. The bar also may be calibrated by clamping it at its midpoint and mounting the pickup at one end.[43] The displacement at the point of attachment of the pickup can be measured optically since displacements encountered are adequately large. The reasponse R and the peak-to-peak displacement $2d$ arc read for a series of values of driving force. The values of R are then plotted against $4\pi^2 f^2 d$, and the slope of the

FIG. 18.18. Resonant-bar calibrator with the pickup mounted at one end and a counterbalancing weight at the other. (*After E. I. Feder and A. M. Gillen.*[43])

line through the points is taken as the pickup sensitivity factor for the resonance frequency f of the bar. To calibrate a pickup at other frequencies, the bar must ordinarily be changed and the procedure repeated.

The resonant-bar calibrator shown in Fig. 18.18 is limited in amplitude primarily by the fatigue resistance of the bar.[43] Using aluminum bars, levels up to $500g$ have been attained without special design criteria. For levels up to $4,000g$ a bar material such as tempered vanadium steels machined to have a mounting boss at its mid-length is more suitable. The resonant bar is mounted at its mid-length on a conventional electrodynamic exciter. The accelerometer being calibrated is mounted at one end of the bar and an equivalent balance weight is mounted at the opposite end in the same relative position. (It should be noted that a pickup mounted at the end of a resonant bar is subjected to a rocking motion in addition to the desired translation. High-frequency pickups generally are unaffected by this extraneous rocking motion.)

Axial resonances of long rods and tubes have been used to generate motion for accurate calibration of vibration pickups over a frequency range from about 1 to 20 kHz and at acceleration levels up to $12,000g$.[44–46] The use of axially driven rods has an advantage over the beams discussed above in that no bending or lateral motion is present.[46] This minimizes errors from the pickup response to such unwanted modes and also from the direct measurement of the displacement having nonrectilinear motion.

The vibration amplitude can be measured by an optical method, discussed previously. As with many measurement methods, laboratory skills and techniques must be developed in order to achieve satisfactory results. When performing sinusoidal calibrations at very high levels, damage to the pickup or cable assembly may occur. Experience has shown that fatigue damage can be sustained by accelerometers at sinusoidal accelerations considerably less than the shock performance rating.[46] Furthermore, it is common that under some circumstances severe cable heating may occur—enough sometimes to melt the cable insulation.

SHOCK-EXCITATION METHODS. There are several methods by which a sudden velocity change may be applied to pickups designed for high-frequency acceleration measurement. Any method which generates a reproducible velocity change and time duration can be used to obtain the calibration factor.[a] Impactive techniques can be employed to obtain calibrations over an amplitude range from a few to over $100,000g$. An accurate determination of shock performance of an accelerometer depends not only upon frequency response and mechanical and electrical characteristics of the transducer, but also upon the characteristics of the instrumentation and recording equipment. It is often best to perform system calibrations to determine the linearity of the vibration pickup as well as the linearity of the recording instrumentation in the range of intended use. Several of the following methods make use of the fact that the velocity change during a transient pulse is equal to the time integral of acceleration:

$$v = \int_{t1}^{t2} a \, dt \qquad (18.19)$$

where the initial or final velocity is taken as reference zero and the integration is performed to or from the time at which the velocity is constant.[47]

In this section, several methods for applying known velocity changes v to a pickup are presented. The voltage output e of a pickup of calibration factor S_1 for an acceleration a is

$$e = \frac{S_1 a}{g} \qquad (18.20)$$

where S_1 is the calibration factor in volts per g (g being the acceleration of gravity). Combining the preceding two equations,

$$S_1 = \frac{g \int_{t1}^{t2} e \, dt}{v} \qquad (18.21)$$

Methods are presented in this section for measuring the integral in this equation. Having the value of both the integral and v, the calibration factor S_1 can be computed. The linear range of a pickup is determined by noting the magnitude of the velocity change v at which the calibration factor S_1 begins to deviate from a constant value. The minimum pulse duration is similarly found by shortening the pulse duration and noting when S_1 changes appreciably from previous values. Typical auxiliary circuit considerations are described in this section, and typical calibration methods are given.

Ballistic Pendulum Calibrator. The ballistic pendulum calibrator provides a means for applying a sudden velocity change to a vibration pickup. The calibrator consists of two masses which are suspended by wires or metal ribbons. These ribbons restrict the motion of the masses to a common vertical plane.[41,47,48] This arrangement, shown in Fig. 18.19, maintains horizontal alignment of the principal axes of the masses in the direction parallel to the direction of motion at impact. The velocity attained by the anvil mass as the result of the sudden impact is determined.

FIG. 18.19. Components arrangement of the ballistic pendulum with photodetector and light grating to determine the anvil-velocity change during impact. (*After R. W. Conrad and I. Vigness.*[47])

The accelerometer to be calibrated is mounted to an adapter which attaches to the forward face of the anvil. The hammer is raised to a predetermined height and held in the release position by a solenoid-actuated clamp. Since the anvil is at rest prior to impact, recording the measurement of the change in velocity of the anvil and transient waveform on a calibrated time base are required. One method of measurement of the velocity change is performed by focusing a light beam through a grating attached to the anvil, as shown in Fig. 18.19. The slots modulate the light beam intensity, thus varying the photodetector output which is recorded with the pickup output. Since the distance between grating lines is known, the velocity of the anvil is calculated directly, assuming that the velocity is essentially constant over the distance between successive grating lines. Other schemes may be used to measure velocity change, such as a sliding contactor or any suitable linear displacement device which can be timed as it crosses a sensitive region whose dimensions are known. The velocity of the anvil in each case is determined directly; the time relation between initiation of the velocity and the pulse at the output of the pickup is obtained by recording both signals on the same time base. The most frequently used method infers the anvil velocity from its vertical rise by measuring the maximum horizontal displacement and making use of the geometry of the pendulum system.

The duration of the pulse, which is the time during which the hammer and anvil are in contact, can be varied within close limits.[47] In Fig. 18.19 the hammer nosepiece is a disc with a raised spherical surface. It develops a contact time of 0.55 millisecond. For larger periods, ranging up to 1 millisecond, the stiffness of the nosepiece is decreased by bolting a hollow ring between it and the hammer. Pulses longer than 1 millisecond may be obtained by placing various compliant materials, such as lead, between the contacting surfaces.

Drop-Test Calibrator. Another impulsive device is the drop tester shown in Fig. 18.20. The pickup is attached to the hammer using a suitable adapter plate. An impact is produced as the guided hammer falls under the influence of gravity and strikes the fixed anvil. To determine the velocity change, measurement is made of the time required for a contactor to pass over a known region just prior to and after impact. The pickup output and the contactor indicator are recorded simultaneously in conjuction with a calibrated time base. The velocity change also may be determined by measuring the height h_1 of hammer drop before rebound and the height h_2 of hammer rise after rebound. The total velocity is calculated from the following relationship:

$$v = (2gh_1)^{1/2} + (2gh_2)^{1/2} \quad (18.22)$$

A total velocity change of 40 ft/sec (12.2 m/sec) is typical of the values which have been achieved by this type of machine.

Drop-Ball Shock Calibrator. Figure 18.21 shows a drop-ball shock calibrator.[49,50] The accelerometer to be calibrated is mounted on an anvil which is held in position by a magnet assembly. A large steel ball is dropped from the top of the calibrator impacting the anvil. The anvil (and mounted accelerometer) are accelerated in a short free-flight path. A cushion catches the anvil and accelerometer. Shortly after impact, the anvil passes through an optical timing gate of a known distance. From this, the velocity after impact can be calculated.

Acceleration amplitudes and pulse durations can be varied by selecting the mass of

FIG. 18.20. Components of a conventional drop tester used to apply a sudden velocity change to a vibration pickup. (*After R. W. Conrad and I. Vigness.*[47])

the anvil, mass of the impacting ball, and resilient pads on top of the anvil where the ball strikes. Common accelerations and durations are $100g$ at 3 milliseconds, $500g$ at 1 millisecond, $1,000g$ at 1 millisecond, $5,000g$ at 0.2 millisecond, and $10,000g$ at 0.1 millisecond.[49] With experience and care, shock calibrations can be performed with an uncertainty of about ±5 to ±10 per cent.

INTEGRATION OF ACCELEROMETER OUTPUT. Change-of-velocity methods for calibrating an accelerometer at higher accelerations than obtainable by methods previously discussed have been developed using specially modified ballistic pendulums, air guns, inclined troughs, and other devices.[48] Regardless of the device employed to generate the mechanical acceleration or the method used to determine the change of velocity, it is necessary to compare the measured velocity and the velocity derived from the integral of the acceleration waveform, as described by Eq. (18.19). An example of a typical acceleration-vs.-time waveform describing the characteristic response of an accelerometer is shown in Fig. 18.22. Electronic digitizers can be used to capture the waveform and produce a recording. Care must be exercised in selecting the time at which the acceleration waveform is considered complete, and its integral should be compared with the velocity.

The following linear relationship holds between acceleration a and voltage output e:

$$S' = \frac{e}{a/g} \tag{18.23}$$

where S' is the accelerometer calibration factor expressed in volts per g. Letting

$$\int_{t_1}^{t_2} e\,dt = A$$

where A is the area under the acceleration-vs.-time curve, the calibration factor for the test accelerometer expressed in terms of volts per g is

$$S' = \frac{Ag}{v} \tag{18.24}$$

If the output closely resembles a half-sine pulse, the area is equal to approximately $2h(t_2 - t_1)/\pi$, where h is the height of the pulse and $t_2 - t_1$ is its width.

The simplest integrating circuit is an R-C (resistance-capacitance) low-pass filter, but an operational amplifier of the type shown in Fig. 18.23 is much preferred. The output voltage from the circuit is inversely proportional to the time constant of the feedback network and directly proportional to the integral of the input voltage. The value of the input resistor should be high enough so that loading of the voltage source is not signifi-

STEEL BALL

BALL RELEASE GUIDE

GUIDE TUBE

ANVIL-HOLDING AND BALL-CATCHING ASSEMBLY

PHOTODIODE SWITCH ASSEMBLY HOUSING

LAMP ASSEMBLY HOUSING

CENTER-PLATE

TRIGGER OUTPUT PULSE TERMINALS AND POWER SWITCH

ACCELEROMETER UNDER TEST

ACCESS DOOR

MOUNTING AND PLUMB-ADJUSTING BASE

FIG. 18.21. Diagram of a drop-ball shock calibrator. The accelerometer being calibrated is mounted on an anvil which is held in place by small magnets. (*After R. R. Bouche.*[49])

cant; generally a value greater than 50 kilohms is sufficient. The value of the feedback capacitor is selected to be approximately equal to the period of the pulse divided by the value of R_1. The purpose of the shunting resistor R_s is to prevent small amplifier offset voltages from being accumulated on the capacitor and driving the amplifier into saturation. The value of R_s should be at least 10 times that of R_1.

FIG. 18.22. Typical acceleration-time record of a shock pulse resulting from a conventional impact calibrator. The dashed curve closely resembles a half-sine pulse. (*After R. W. Conrad and I. Vigness.*[47])

Several restrictions are imposed by this method of calibration and generally are related to the low-frequency response of the electrical integrator and the natural period of the accelerometer. In the case of piezoelectric pickups, the distortion in the response at low frequencies is limited by the time constant derived from the total capacitance of the accelerometer and cables and the input impedance of the signal conditioner. The response at high frequencies is limited by the natural period of the accelerometer and the nonlinearity of the integration network. The easiest and most accurate means to obtain the area under the acceleration-vs.-time waveform is to employ an electronic digitizer or digitizing oscilloscope, and then to process the results with the aid of a small computer or microprocessor-based system. In this way it is possible to perform an accurate integration and to apply scaling and correction terms. The accelerometer output and the velocity signal can both be digitized, and the processor can usually be programmed to perform the integration and entire calculation leading to the solution of Eq. (18.24) which gives the accelerometer calibration factor.

FIG. 18.23. A typical operation-amplifier integrator.

Auxiliary Circuit Effects. Voltage or charge amplifiers, filters, recorders, and analyzers ordinarily used as part of the measuring systems can significantly affect and distort the desired signals. Distortion in these auxiliary circuits can be more important in high-level high-frequency applications of shock measurements than in low-level applications. Therefore, the accurate determination of the performance of the accelerometer by the change-of-velocity method depends on the absence of frequency and phase distortion and upon linearity for the range of excitations expected for shock measurements. A low-pass filter may often be used to minimize the high-frequency ringing in the response. Where this is done, it is important that the linear range not be exceeded in portions of the circuit preceding the filter and that the upper cutoff frequency of the filter be well above the frequency range over which the calibration or measurement is desired. Many filters can appreciably contribute phase distortion which can alter the calibration results significantly. For minimum effect, the filter should have a constant phase shift of 0 to 180°, or a phase shift that varies linearly with frequency (constant time delay). In a finite sense, neither of these conditions can be completely satisfied, and a compromise must be accepted. One must evaluate the effects of this on the calibration process and determine if it is significant in comparison to the calibration uncertainties being sought.

IMPACT-FORCE SHOCK CALIBRATOR. The impact-force shock calibrator has a free-falling carriage and a quartz load cell.[50,51] The accelerometer to be calibrated is mounted onto the top of the carriage, as shown in Fig. 18.24. The carriage is suspended

Fig. 18.24. Impact-force calibrator with auxiliary instruments. (*After W. P. Kistler.*[51])

about ½ to 1 meter above the load cell and allowed to fall freely onto the cell.[51] The carriage's path is guided by a plastic tube. Cushion pads are attached at the top of the load cell to lengthen the impulse duration and to shape the pulse. Approximate haversines are generated by this calibrator. The outputs of the accelerometer and load cell are fed to two nominally identical charge amplifiers. The outputs of the two charge amplifiers connect to some means of measuring the transient outputs; peak-holding meters or a storage-type oscilloscope can be used.

During impact, the voltage produced at the output of the acceleration charge amplifier $e_a(t)$ is

$$e_a(t) = a(t)S_aH_a \tag{18.25}$$

where $a(t)$ = acceleration
$\quad S_a$ = calibration factor for accelerometer
$\quad H_a$ = charge amplifier gain

The output of the load cell charge amplifier $e_f(t)$ is

$$e_f(t) = F(t)S_fH_f \tag{18.26}$$

where $F(t)$ = force
$\quad S_f$ = calibration factor for load cell
$\quad H_f$ = charge amplifier gain

By using the relationship $F(t) = ma(t)$, where m is the falling mass, and combining Eqs. (18.25) and (18.26), one obtains

$$\frac{e_a(t)}{e_f(t)} = \frac{a(t)S_aH_a}{ma(t)S_fH_f} \tag{18.27}$$

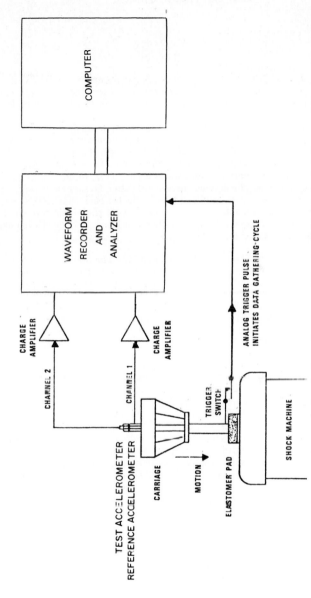

Fig. 18.25 Block diagram of a system for shock accelerometer calibration using Fourier-transform techniques. Sensitivity factor and phase are calculated as functions of frequency. (*After J. D. Ramboz and C. Federman.*[54])

and hence

$$S_a = \frac{e_a(t)}{e_f(t)} \frac{H_f}{H_a} \frac{m}{g} S_f \tag{18.28}$$

Customarily, in Eq. (18.28) the amplifier gains H_f and H_a are in units of volts per unit charge, mass m is in kilograms, g is the acceleration of gravity in meters per second squared, and the load-cell calibration factor S_f is in units of charge per unit force, pico-coulombs per newton; then S_a is in picocoulombs per g.

When calculating the mass, it is necessary to know the mass of the carriage, accelerometer, mounting stud, cable connector, and a short portion of the accelerometer cable. Experience has shown that for small coaxial cables, a length of about 2 to 4 cm is correct. When setting the value of charge gain H_f for the load cell charge amplifier, the terms mS_f/g of Eq. (18.28) can be computed and made a part of H_f to normalize the load cell channel output. By doing so, the accelerometer calibration factor can be read directly as $1/H_a$ on the accelerometer charge amplifier when adjusted such that $e_a(t) = e_f(t)$. This is done by successive trial drops and by adjusting H_a until the readings of the two peak meters are the same. Calibrations by this method can be accomplished with uncertainties generally between ±2 to ±5 per cent.

FOURIER-TRANSFORM SHOCK CALIBRATION. The previously discussed shock calibration methods yield the calibration factor for the accelerometer being tested; that is, they yield a single-value approximation of the magnitude of the sensitivity function. For many applications, a knowledge of the calibration factor may be sufficient. However, for shock standards or other critical applications, it is desirable to know the sensitivity in both magnitude and phase as a function of frequency. The shock calibration method described in this section employs Fourier techniques to transform time-domain data into frequency-domain data. By doing so, the frequency-dependent sensitivity function can be determined in both magnitude and phase.[51-55]

The laboratory equipment consists of a mechanical shock-generating machine, two accelerometers, a data-transfer system, and a small computer for data storage and processing. Figure 18.25 shows a block diagram of one such system.[54,55] The calibration of the test accelerometer is in terms of a reference standard. It is convenient to employ a piggyback accelerometer standard which permits the test accelerometer to be mounted directly on its top in a back-to-back configuration. The signals from each accelerometer pass through charge amplifiers into selectable-gain voltage amplifiers and into an analog-to-digital converter (ADC). Digital data may be stored in a buffer and then transferred to a larger computer where Fourier transforms can be done quickly and economically. A peak-holding meter, transient recorder, and oscilloscope are used to measure approximate acceleration peak values and to examine the time-domain waveshapes. A digital 10-bit word is generally sufficient for most work, and conversion rates of at least 100 kHz are desirable.

The computer performs a Fourier analysis by use of a fast Fourier transform (FFT).[54] Ratios of the test accelerometer output to the standard accelerometer output in the frequency domain yield the test accelerometer sensitivity in terms of the standard. Phase-

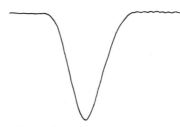

Fɪɢ. 18.26. A typical half-sine shock pulse generated by a pneumatic shock machine. Deceleration amplitude is 900g and pulse duration is 1 millisecond. (*After J. D. Ramboz and C. Federman.*[54])

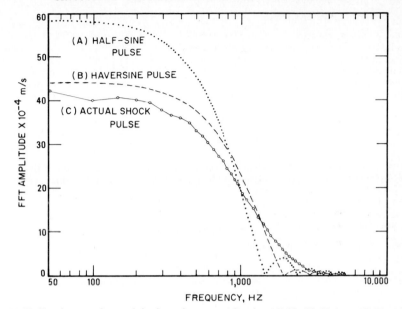

Fɪɢ. 18.27. Fourier transforms of shock accelerometer signals (*A*) Ideal half-sine. (*B*) Ideal haversine. (*C*) An actual shock pulse of 900*g* peak and a duration of 1 millisecond. The ideal pulses have the same peak acceleration and time durations as the actual pulse, which is shown in Fig. 18.26. (*Courtesy of the National Bureau of Standards.*)

calibration data is obtained by taking the difference between the test and standard phase spectra.

Figure 18.26 shows a typical half-sine shock pulse. Its Fourier transform is shown in Fig. 18.27. The range of usable frequencies is limited by the pulse shape and duration, sampling rate, and Fourier analysis capability. Generally, this type of calibration can yield complete calibration results over a frequency range from 0 to 10 kHz or more. Uncertainties are less than ±5 per cent.

SINUSOIDAL VIBRATION EXCITERS USED FOR CALIBRATION

MECHANICAL EXCITERS

Rectilinear motion can be produced by mechanical exciter systems of the type described under *Direct-drive Mechanical Vibration Machine* in Chap. 25. At low frequencies, a mechanical drive consisting of an electric motor which turns a cam or drives a linkage can be used to produce nearly sinusoidal motion. Rotating eccentric weights also can be used to produce sinusoidal motion of a spring-supported table. In both of these machines, the amplitude of the table motion is nearly independent of frequency. Their usable frequency range is from a few hertz to about 100 Hz. Although the displacement is essentially sinusoidal, appreciable distortion of acceleration from a pure sinusoid usually is present as a result of bearing tolerances and the geometry of linkage systems. Although mechanical exciters are inexpensive relative to other types, they have restricted application and are seldom used for high-quality calibrations.

For the calibration of vibration-measuring instruments at discrete frequencies below 5 Hz, a linear oscillator can be employed. Reference 56 describes a calibrator consisting of a spring-supported table which is guided vertically by air bearings. Its advantages are a clean waveform, resulting from free vibration, and large rectilinear displacements with little damping, made possible by the use of air bearings.

ELECTRODYNAMIC EXCITERS

The rectilinear electrodynamic exciter (see Chap. 25), which produces a constant force (acceleration), is widely used in calibration equipment. Its amplitude is displacement limited at the low-frequency end of the operating range, and current limited at the high-frequency end. Over a frequency range from about 1 to 10,000 Hz, electrodynamic calibrators are available which provide relatively undistorted sinusoidal motion.[2,14,22,31,57] Ordinarily, two or more exciters are needed to cover this broad frequency range since it is necessary to avoid resonance frequencies of the exciter. To accommodate the requirements of the seismic-frequency range (below 20 Hz) special electrodynamic exciters featuring long strokes and lightweight armatures have been designed. A commercially available unit is described in Ref. 58; Ref. 16 describes a state-of-the-art system developed by a national standards laboratory.

The construction of a typical electrodynamic vibration-pickup calibrator is illustrated in Fig. 18.28. The pickup to be calibrated is attached to the calibrator table. The arma-

FIG. 18.28. Typical electrodynamic vibration exciter used for calibration of small pickups. The major components are (1) ceramic moving element, (2) air bearing, (3) vibration absorbers, (4) permanent magnet, (5) vibration pickup, and (6) air inlet. (*After T. Dimoff.*[57])

ture and the table are as rigid as practical to minimize table bending and to increase the first axial resonance to as high a frequency as possible. Flexures or air bearings usually are used to constrain undesirable lateral motion. Frequently, calibrators of this type are mounted on trunnions to allow use in either a vertical or horizontal direction. Magnetic shielding or bucking magnetic coils can be used to reduce the stray magnetic field at the mounting table to nearly zero. This is important when calibrating pickups which are sensitive to magnetic fields, e.g., electrodynamic velocity pickups. To achieve isolation from building vibration, a calibrator may be mounted on a heavy rigid mass which in turn is supported by soft springs.

Calibration exciters often have a built-in accelerometer or velocity coil intended as a reference standard. Such internal standards are calibrated by methods described earlier in this chapter.

PIEZOELECTRIC EXCITERS

At frequencies greater than 1,000 Hz, piezoelectric exciters can be used to generate reliable uniaxial motion, i.e., exciters which have minimal transverse and rocking motions and a relatively flat frequency response over a wide range and which provide sufficient amplitude to make satisfactory calibrations.[20] A high-quality piezoelectric exciter is shown in Fig. 18.29. Discs of piezoelectric material are arranged such that alternate electrodes are connected electrically in parallel, but phased mechanically so that the displacement accumulates in series. When an alternating voltage is applied to the exciter, the piezoelectric material contracts and expands in direct proportion to the applied voltage.

The table material on which a pickup being calibrated is mounted must be stiff, dense, and rigid. The mechanical impedance of this surface must be greater than that of the

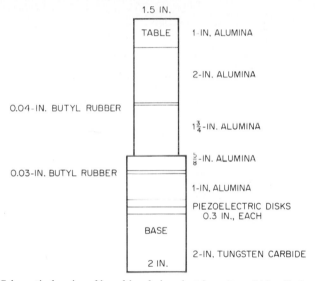

FIG. 18.29. Schematic drawing of broad-band piezoelectric exciter. (*After E. Jones, et. al.*[20])

pickup to be calibrated; otherwise calibration errors will result because of relative motion between the pickup and the surface location where displacement is measured. The base of the exciter must also possess high flexural rigidity and mechanical impedance so that the behavior of the exciter does not vary unduly with different supporting structures.

The displacement of the table top is commonly measured with the use of an optical interferometer described previously in this chapter. The piezoelectric exciter can also be employed to drive two pickups so that a comparison calibration can be made, assuming that one of the two pickups is a standard of known sensitivity.

CALIBRATION BY THE COMPARISON METHOD

A rapid and convenient method of obtaining the sensitivity factor of a vibration pickup is by direct comparison of the pickup voltage with that of a second pickup used as a standard and cailbrated for frequency and phase response by one of the methods described in this chapter.[3,4,22]

Pickups usually are mounted back to back as shown in Fig. 18.30. The most important consideration is to ensure that each pickup experiences the same motion. If both pickups are rectilinear and are placed on the calibrator table, the angular rotation of the table should be small to avoid any difference in excitation between the two pickup locations. The error due to rotation may be reduced by carefully locating the pickups firmly on opposite faces with the center-of-gravity of the pickups located at the center of the table. Relative differences in pickup excitation may be observed by reversing the pickup locations and observing if the voltage ratio is the same for both positions.

FIG. 18.30. Comparison method of calibration: Pickup 2 is calibrated against pickup 1 as the reference standard. The two pickups may be excited by any of the means described in this chapter. (*After ANSI standard S2.2 2-1959, R1981.*[3])

Calibration by the comparison method is limited to the range of frequencies and amplitudes for which the standard pickup has been previously calibrated. If both pickups are linear, the sensitivity S_2 of a vibration pickup can be calculated in both magnitude and phase from

$$S_2 = \frac{E_2}{E_1} S_1 \qquad (18.29)$$

The uncertainty in the pickup sensitivity obtained by the comparison method depends on the uncertainty in the measured voltage ratio and on the accuracy to which the sensitivity of the reference transducer is known. These factors are influenced by transverse motion of the exciter table, transverse sensitivity of the test transducer, and the load dependence of the reference transducer.[59]

Minicomputer and desktop computer controllers and the use of electronic voltage ratiometers can be utilized for comparison calibrations.[60,61] Correction factors and statistical procedures can easily be incorporated in the calibration algorithm. Figure 18.31

FIG. 18.31. System configuration for an automated comparison calibration system. (*Courtesy of the National Bureau of Standards.*[60])

shows how a comparison calibration system has been automated at the National Bureau of Standards.

SINUSOIDAL FREQUENCY-DWELL METHOD

A simple and convenient way to perform comparison calibrations is to fix the frequency at a desired value, adjust the vibration amplitude, and then make the necessary ratio measurements. The circuit shown in Fig. 18.31 is a simple and direct method of comparison calibration. The pickups are assumed to experience identical motion. Their outputs are processed through amplifiers (either voltage or charge amplifiers). The amplifier in the test channel may have a variable gain which has been calibrated. An attenuator is sometimes used in the reference channel (assuming this voltage output is greater than that of the test channel).

A single voltmeter can be employed to indicate the output of both channels.[62] A significant advantage results from the use of the same voltmeter to measure the voltages e_t and e_r. According to Eq. (18.29), the errors in the voltmeter tend to cancel out unless the voltages e_t and e_r are widely different in magnitude.

SINUSOIDAL SWEPT-FREQUENCY METHOD

A graphical plot of the test accelerometer sensitivity, or its deviation from a reference sensitivity, vs. frequency can provide a quick and convenient method of calibration.[62,63] A circuit like the one shown in Fig. 18.32 provides good resolution and a stable comparison

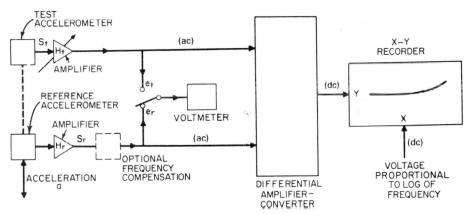

FIG. 18.32. An accelerometer-calibration plotting system using a differential amplifier, ac-to-dc converter, and an X-Y recorder. (*After K. Unholtz.*[62])

system. The system configuration is similar to that of the preceding method; however, a differential amplifier is used to measure the voltage differences between the two channels. The two ac voltages are made equal at some reference condition by adjusting amplifier gains. Recorder gain (ordinate) is adjusted so that a given difference equals a percentage of the test accelerometer calibration factor.

RANDOM-EXCITATION–TRANSFER-FUNCTION METHOD

The use of random vibration excitation and transfer-function analysis techniques can provide quick and accurate comparison calibrations.[64,65] Two pickups are mounted back to back on a suitable vibration exciter. Figure 18.33 shows a typical layout with the Fourier analyzer as a key piece of equipment. The two accelerometer channels are usually fed into the analyzer through a pair of low-pass (antialiasing) filters. The random signal to drive the exciter is generated under control of the analyzer.

This method provides a nearly continuous calibration over a desired frequency spectrum, with the resulting sensitivity factors having both amplitude and phase information. Since purely sinusoidal motion is not a requirement as in the other calibration methods, this lessens the requirements for the power amplifier and exciter to maintain low values of harmonic distortion. A very useful measure of process quality is obtained by computing the input-output coherence function, which requires knowledge of the input and output power spectra, the cross power spectrum, and the transfer function.

The sensitivity, regarded as a transfer function, can be described by

$$\mathbf{H}(f) = \frac{\mathbf{G}_{yx}}{G_{xx}} \tag{18.30}$$

where $\mathbf{G}_{yx}$ denotes the ensemble average of the cross power spectrum, and G_{xx} represents the ensemble average of the input power spectrum.

The question of whether the system output is totally caused by the system input is important. Noise or nonlinear effects can create large errors at various frequencies. The coherence function gives an indication of such errors. It is defined as

$$\gamma^2 = \frac{|\mathbf{G}_{yx}|^2}{G_{xx}G_{yy}} \qquad 0 < \gamma < 1 \tag{18.31}$$

where G_{yy} is the ensemble average of the output power spectrum. When $\gamma = 1$ at any frequency, the system transfer function has perfect causality at that frequency. The output is caused totally by the input. A low value of coherence function at given frequencies indicates that the transfer function has inaccuracies at those frequencies. When $\gamma = 0$, the output is caused totally by sources other than the input.

FIG. 18.33. Simplified block of a wide-band random-signal accelerometer calibration system using Fourier analysis. (*After K. Ramsey.*[64])

CALIBRATION OF AUXILIARY CIRCUITS

The calibration of auxiliary circuits for vibration pickups is as important as the calibration of the pickups themselves. Very often nonlinearities and apparent anomalies in the response can be traced to signal conditioners, cables, or recording equipment. A complete system checkout is, therefore, important in order to provide assurance that the vibration-measuring system will function properly under test conditions. In those instances where the pickup output is a voltage or charge, the auxiliary circuits can be calibrated by applying appropriate voltages or charges at selected frequencies and recording the response, thus establishing the scale factor for the instrumentation system.

STRAIN-GAGE BRIDGE-TYPE PICKUPS

Vibration pickups employing strain gages as sensitive elements may be of the wire- or foil-gage type, or of the semiconductor piezoresistive type. In all cases, the output is in the form of an electrical resistance change as a function of input vibration (usually acceleration).

A calculated value of resistance is introduced during the calibration such that the resulting indication is equal to that which would result from direct mechanical excitation. Since the resistance change appears as a step function, it is important that the auxiliary circuits have a flat response to 0 Hz. The circuit schematic of a typical system is shown in Fig. 18.34.[66–69] The calculated resistance corresponding to an equivalent acceleration is

$$R_c = \frac{10^6 r}{2NS} \tag{18.32}$$

where R_c = calibrating resistor value in ohms
$\quad r$ = resistance of each of the four bridge arms in ohms
$\quad S$ = calibration factor in microvolts open-circuit output per volt input applied across the bridge per unit acceleration determined by physical calibration
$\quad N$ = mechanical acceleration input simulated by the calibrating resistor

The accuracy of the calibrating resistor method depends upon the stability of the calibrating resistor R_c, the bridge-arm resistance r, and the calibration factor S.

Fig. 18.34. Schematic diagram for the resistance-change calibration of a Wheatstone-bridge unbonded-strain-gage accelerometer. (*After B. B. Helfand and J. Burns.*[66])

VOLTAGE AMPLIFIERS

The calibration of voltage amplifiers (solid-state emitter followers or operational amplifiers) used as signal conditioners can often be performed with sufficient accuracy with two voltmeters. One meter is used to measure the input voltage to the amplifier. This is typically in the millivolt range; therefore, care must be exercised so that hum, stray pickup, and ground loops do not affect the measurement. It may be necessary to use a precision voltage divider ahead of the amplifier, to measure the higher voltage at the input to the divider, and to apply the output of the divider to the amplifier input. Most voltage amplifiers have extremely high input impedance so that loading of the divider is minimized. A second voltmeter is used to measure the amplifier output voltage. Difficulties usually are encountered at very low frequencies where the voltmeter response is poor and calibration is inaccurate. For amplifiers having a gain of approximately unity, use of only one voltmeter is recommended as the errors in the voltage measurement will essentially divide out. When subsequently using the calibrated amplifier with piezoelectric accelerometers, corrections will have to be applied to the system for accelerometer, cable, and amplifier input capacitances. Equation (18.33) shows their relationship. Other information regarding amplifiers and cables is found in Chap. 12.

$$\text{System gain} = \text{amplifier gain} \times \frac{C_p}{C_p + C_c + C_i} \qquad (18.33)$$

where C_p = pickup capacitance
C_c = cable capacitance
C_i = input capacitance of amplifier

CHARGE AMPLIFIERS

To calibrate a charge amplifier, it is necessary to make a charge generator from a voltage generator and stable three-terminal standard capacitor. Figure 18.35 shows a typical calibration circuit. Generator voltage e_g is applied to the amplifier input through a three-terminal capacitor having a capacitance C_s. The charge amplifier gain can then be calculated from Eq. (18.34). If the generator voltage is measured in volts, the output voltage in millivolts, and the capacitance in picofarads, then Eq. (18.34) is expressed in millivolts per picocoulomb.

$$\text{Charge amplifier gain} = \frac{e_o}{e_g C_s} \qquad (18.34)$$

Fɪɢ. 18.35. Circuit connections for the calibration of a charge amplifier. Capacitor C_s simulates the piezoelectric pickup.

Typical values for the three-terminal capacitor are nominally 100 or 1000 picofarads. Some charge amplifiers have a built-in calibration capacitor so that a voltage generator can be connected directly to that test input and the gain determined. The value of the internal capacitance must be known and is usually given by the manufacturer of the amplifier.

IMPEDANCE CONVERTERS

Integrated circuit amplifiers which act as impedance converters are sometimes built into the base of the transducer.[70] This essentially eliminates the adverse and troublesome problems which can be associated with shunting cable capacitance and leakage resistance and with cable noise. A bias voltage must be supplied to the transducer having a built-in amplifier. Such supplies are either available from, or specified by, the manufacturer of the pickup. Unless special provisions are made in the transducer design, the amplifier cannot be calibrated separately. The transducer and amplifier must be treated as a system. The usable range of such transducers may be limited by temperature and shock.

REFERENCES

1. American National Standards Institute: "Selection of Calibration and Tests for Electrical Transducers used for Measuring Shock and Vibration," ANSI S2.11-1969 (R1978), New York.
2. Bouche, R. R.: "Calibration of Shock and Vibration Measuring Transducers," Shock and Vibration Monograph SVM-11, The Shock and Vibration Information Center, Washington, DC, 1979.
3. American National Standards Institute: "Methods for the Calibration of Shock and Vibration Pickups," ANSI S2.2-1959 (R1981), New York.
4. International Organization for Standardization: "Methods of Calibration of Vibration and Shock Pickups," ISO/DIS 5347, 1985. (Available from the American National Standards Institute, New York.)
5. Levy, S., and R. R. Bouche: *J. Res. Natl. Bur. Std.*, **57**:227 (1956).
6. Kissinger, C. W.: *Proc. 10th Annu. Meet. Instr. Soc. Amer.*, Paper 54-40-1, 1955.
7. Lederer, P. S., and J. S. Hilten: *Natl. Bur. Std. (U.S.) Tech. Note* 269, 1966.

8. Levy, H.: "Description of Pendulum Accelerometer Calibrator," *Naval Air Mat. Cen. (Philadelphia) Rept.* AML. NAM 2425, part V, Aug. 25, 1945.
9. Wildhack, W. A., and R. O. Smith: *Proc. 9th Annu. Meet. Instr. Soc. Amer.*, Paper 54-40-3, 1954.
10. Hilten, J. S.: *Natl. Bur. Std. (U.S.) Tech. Note* 517, March 1970.
11. Corelli, D., and R. W. Lally: "Gravimetric Calibration," *Third Int. Modal Analysis Conf.*, 1985.
12. Dally, J. S., and W. F. Riley: "Instrumentation for Engineering Measurements," John Wiley & Sons, Inc., New York, 1984.
13. Smith, R. D., E. A. Willis, and J. S. Hilten: *Natl. Bur. Std. (U.S.) J. Research*, Part C, 66D, 1962.
14. Koyanagi, R. S.: *Exp. Mech.* **15:**443 (1975).
15. Logue, S. H.: "A Laser Interferometer and its Applications to Length, Displacement, and Angle Measurement," *Proc. 14th Annu. Meet. Inst. Environ. Sci.*, 1968, p. 465.
16. von Martens, H. J.: "Representation of Low-Frequency Rectilinear Vibrations for High-Accuracy Calibration of Measuring Instruments for Vibration," *Proc. 2nd Symp. of IMEKO Tech. Comm. on Metrology-TC8*, Budapest, 1983.
17. Schmidt, V. A., S. Edelman, E. R. Smith, and E. Jones: *J. Acoust. Soc. Amer.*, **33:**748 (1961).
18. Schmidt, V. A., S. Edelman, and E. T. Pierce: *J. Acoust. Soc. Amer.*, **34:**455 (1962).
19. Payne, B. F., and M. R. Serbyn: "An Automated System for the Absolute Measurement of Pickup Sensitivity," *Proc. Natl. Conf. Stand. Lab.*, 1983 Workshop and Symposium, II, 11.1-11.22, Boulder, CO, July 18–21, 1983.
20. Jones, E., W. B. Yelon, and S. Edelman: *J. Acoust. Soc. Amer.*, **45:**1556 (1969).
21. Payne, B. F., R. S. Koyanagi, C. Federman, and E. Jones: *Instr. Soc. Amer. Preprint* ASI-74255, 1975.
22. Robinson, D. C., M. R. Serbyn, and B. F. Payne: *Natl. Bur. Std. (U.S.) Tech. Note* 1232, 1987.
23. Hohmann, P. *Akustica* **26:**122 (1972).
24. Serbyn, M. R., and W. B. Penzes: *Instr. Soc. Amer. Trans.*, **21:**55 (1982).
25. Serbyn, M. R., and F. A. Andrews: *J. Acoust. Soc. Amer.*, **46:**2 (1969).
26. Payne, B. F.: *Shock and Vibration Bull.*, **36,** pt. 6, (1967).
27. Ge., Lifeng: "The Reciprocity Method for Piezoelectric Calibrators," *Proc. ASME Des. Eng. Conf. Mech. Vibration and Noise*, Cincinnati, Ohio, Sept. 10–13, 1985.
28. Ge., Lifeng: *Kexue Tongbao Acad. Sci. China*, **29:**735 (1984).
29. Fromentin, J., and M. Fourcade: *Bull. d'Informations Sci, et Tech.*, (201):21, 1975.
30. Thompson, S. P.: *J. Acoust. Soc. Amor.*, **20:**637 (1948).
31. Dimoff, T., and B. Payne: *NBS Report* 9670, 1968.
32. Harrison, M., A. O. Sykes, and P. G. Marcotte: *J. Acoust. Soc. Amer.*, **24:**384 (1952).
33. Trent, H. M.: *J. Appl. Mech.*, **15:**49 (1948).
34. London, A.: *Natl. Bur. Std. (U.S.) Tech. News Bull.* **32,** January 1948.
35. Camm, J. C.: *Natl. Bur. Std. (U.S.) Rept.* 2651, 1953.
36. Bouche, R. R., and L. C. Ensor: *Shock and Vibration Bull.* **34,** pt. 4, (1965).
37. Ramboz, J. D.: *Natl. Bur. Std. (U.S.) Rept.* NBSIR 74-481, 1974.
38. Sheeks, O. P.: *Instr. Soc. Amer. Reprint* 68-581, 1968.
39. Kuhl, R.: *Bruel Kjaer Tech. Rev.*, **1:**26 (1971).
40. Ten Wolde, T.: *Akustica* **28:**23 (1973)
41. Perls, T. A., C. W. Kissinger, and D. R. Paquette: *Bull. Amer. Phys. Soc.*, **30:**34 (1955).
42. Tyzzer, F. G., and H. C. Hardy: *J. Acoust. Soc. Amer.*, **22:**454 (1950).
43. Feder, E. I., and A. M. Gillen: *IRE Trans. Instr.*, **6:**1, 1957.
44. Brennan, J. N.: *J. Acoust. Soc. Amer.*, **25:**610 (1953).
45. Nisbet, J. S., J. N. Brennan, and H. I. Tarpley: *J. Acoust. Soc. Amer.*, **32:**71 (1960).
46. Jones, E., S. Edelman, K. S. Sizmore: *J. Acoust. Soc. Amer.*, **33:**1462 (1961).
47. Conrad, R. W., and I. Vigness: *Proc. 8th Annu. Meet. Instr. Soc. Amer.*, Paper 11-3, 1953.
48. Perls, T. A., and C. W. Kissinger: *Proc. 9th Annu. Meet. Instr. Soc. Amer.* Paper 40-2, 1954.
49. Bouche, R. R.: *Endevco Corp. Tech. Paper* TP 206, April 1961.
50. Kelly, R. W.: *Instr. Soc. Amer. Preprint* M18-2-MISTIND-67, 1967.
51. Kistler, W. P.: *Shock and Vibration Bull.* **35,** pt. 4, (1966).
52. Favor, J. D.: *Instr. Soc. Amer. Preprint* P13-1-PHYMMID-67, 1967.
53. Favor, J. D.: *Shock and Vibration Bull.*, **37,** pt. 17, (1968).
54. Ramboz, J. D., and C. Federman: *Natl. Bur. Std. (U.S.) Rept.* NBSIR 74-480, March 1974.
55. Walston, W. H., and C. Federman: *Instr. Soc. Amer. Preprint* ASI-75257, 1975.
56. O'Toole, K. M., and B. H. Meldrum: *J. Sci. Instr. (J. Phys. E)*, (2)**1:**672 (1968).
57. Dimoff, T.: *J. Acoust. Soc. Amer.*, **40:**671, 1966.
58. Instruction Manual, Model 113-AB Electro-Seis Shaker, Acoustic Power Systems, Inc., Carlsbad, CA.
59. Lauer, G.: *Fortschr. Akust.* (DAGA '84) 275, 1984.

60. Payne, B. F.: "Automation of Vibration Testing at the National Bureau of Standards," *Proc. 30th Annu. Tech. Meet. Inst. Environ. Sci.*, Orlando, FL, May 1–3, 1984.
61. Benoit, P. E.: "An Automated Vibration Calibration System," *Proc. Natl. Conf. of Stds. Labs.*, 1982 Workshop, p. 51.
62. Unholtz, K.: *Instr. Soc. Amer. Preprint* M18-3-MESTIND-67, 1967.
63. Chernoff, R.: *Instr. Soc. Amer. Preprint* M18-3-MESTIND-67, 1967.
64. Ramsey, K.: *Instr. Soc. Amer. Preprint* ASI75256, 1975.
65. Hartz, K.: *Proc. Natl. Conf. of Stds. Labs. Symp.*, 1984.
66. Helfand, B. B., and J. Burns: "Calibration of Resistance-Bridge Transducer Circuits Under Temperature Extremes," *Statham Lab. Instr. Note* 14.
67. Oliver, F. J.: "Practical Instrumentation Transducers," 1971.
68. Sion, N.: *Inst. Cont. Syst.* **41:**107 (1968).
69. Stein, P. K.: "Some Properties of Input Conditioning Circuits for Wheatstone-Bridge-Based Transducers," *Fifth Transducer Workshop*, Document 112-68, White Sands Missile Range, 1967, p. 243–256.
70. Lally, R. W.: *Instr. Soc. Amer. Preprint* P4-2-PHYMMID-67, Sept. 1967.

19

VIBRATION STANDARDS

Paul H. Maedel, Jr.
Westinghouse Electric Corporation

Ronald L. Eshleman
The Vibration Institute

INTRODUCTION

A good standard must represent a consensus of opinion among users, be simple to understand, be easy to use, and contain no ambiguities or loopholes. Any standard must contain vital information that leads to common measurement procedures and evaluation of data that are compared with agreed-upon criteria. Standards are intended (1) to set up criteria for rating or classifying the performance of equipment or material, (2) to provide a basis for comparison of the maintenance qualities of pieces of equipment of the same type, (3) to test equipment whose continuous operation is necessary for industrial or public safety, (4) to provide a basis for the selection of equipment or material, and (5) to set up procedures for the calibration of equipment. Some standards establish classifications for equipment which is being rated and indicate how measurements are to be made and how the data, so obtained, are to be analyzed; they may also indicate how the equipment is to operate during the test procedure.

This chapter is mainly concerned with standards related to machinery vibration and its classification. Other chapters in the Handbook treat other types of shock and vibration standards. For example, Chap. 18 describes standards related to transducer calibration; Chap. 24 discusses recommended practices for the seismic qualification of equipment used in nuclear power stations and in other facilities where public safety is a concern; Chap. 26 discusses ANSI and military standards for the design of shock-testing machines; Chap. 39 describes standards for the evaluation, testing, and use of balancing machines; and Chap. 41 refers to ASTM test methods and military specifications for package cushioning materials.

STANDARDS ORGANIZATIONS AND GROUPS

In the field of vibration, the two recognized international organizations are the International Standards Organization (ISO), which is technology-oriented, and the International Electrical Commission (IEC), which is product-oriented. The ISO works in cooperation with national organizations, such as the American National Standards Institute (ANSI)—a nongovernmental institution that coordinates the development of voluntary national standards in the U.S.A.; outside the U.S.A., national standards organizations usually are government institutions. Vibration standards activity in the ISO is guided by Technical Committee 108, Mechanical Vibration and Shock; vibration standards in

ANSI are guided by Technical Committee S2, Mechanical Vibration and Shock. The secretariat for both organizations is held by the Acoustical Society of America. Vibration standards activity in the IEC is guided by Technical Committee 50, and in the U.S.A. by the Institute of Environmental Sciences. IEC works with trade associations, such as the National Electrical Manufacturers Association. Other trade organizations that have adopted formal vibration standards include the American Petroleum Institute, the Hydraulic Institute, and the Compressed Air and Gas Association. Various technical societies have been instrumental in the development of codes and standards concerned with vibration; such documents have been established by consensus of consumers and manufacturers and their use is voluntary. In addition, the Department of Defense has developed standards and specifications to ensure the quality of equipment which it procures.

MACHINERY VIBRATION STANDARDS AND CRITERIA

TYPES OF MACHINERY

Machinery can be subdivided into four basic categories for purposes of vibration measurement and evaluation:

1. *Reciprocating machinery having both rotating and reciprocating components*, such as diesel engines and certain types of compressors and pumps. Vibration is usually measured on the main structure of the machine at low frequencies.

2. *Rotating machinery having rigid rotors*, such as certain types of electric motors, single-stage pumps, and slow-speed pumps. Vibration is usually measured on the main structure (such as on the bearing caps or pedestals) where the vibration levels are indicative of the excitation forces generated by the rotor because of unbalance, thermal bows, rubs, and other sources of excitation.

3. *Rotating machinery having flexible rotors*, such as large steam turbine generators, multistage pumps, and compressors. The machine may be set into different modes of vibration as it accelerates through one or more critical speeds to reach its service speed. On such a machine, the vibration amplitude measured on a structure member may not be indicative of the vibration of the rotor. For example, a flexible rotor may experience very large amplitude displacements resulting in failure of the machine even through the vibration amplitude measured on the bearing cap is very low. Therefore, it is essential to measure the vibration on the shaft directly.

4. *Rotating machinery having quasi-rigid rotors*, such as low-pressure steam turbines, axial-flow compressors, and fans. Such machinery contains a special class of flexible rotor where vibration amplitudes measured on the bearing cap are indicative of the shaft vibration.

CLASSIFICATION OF SEVERITY OF MACHINERY VIBRATION

In the classification of severity of machinery vibration, the motion variable that is used (vibration displacement, velocity, or acceleration) depends on the type of standard, the frequency range, and other factors. In classifying machinery vibration in the range from 10 to 1,000 Hz, vibration velocity often is used because it is relatively independent of frequency in this range of frequencies and thus yields a simple measure of severity of a new or operating machine.

For simple harmonic motion, either peak or rms values of the motion variable may be used; however, for machines whose motion is complex, the use of these two indices provides distinctly different results, mainly because the higher frequency harmonics are given different weights. For rotating machinery whose rotational speed is in the range of 600 to 12,000 rpm, the rms value of the velocity amplitudes corresponds most closely with vibration severity. Therefore, the International Standards Organization has a special measure, *vibration severity*, which is defined as the highest value of the broad-band, root-mean-square value of the velocity amplitude in the frequency range from 10 to 1,000 Hz as evaluated on the structure at prescribed points (generally triaxial arrays on the bearing caps or pedestals).

ROTATING MACHINERY HAVING RIGID ROTORS

ISO standard 2372, "Mechanical Vibration of Machines with Operating Speeds from 10 to 200 rps—Basis for Specifying Evaluation Standard"[1] applies to rotating machinery having rigid rotors and to those machines having flexible rotors in which bearing cap vibration is a measure of the shaft motion. Vibration severity (defined above) covers those frequencies between 30 per cent of the rotational frequency to at least three times the rotational frequency; it includes the most common causes of rotating machinery vibration resulting from (1) nonsynchronous excitation such as rubs and rotor whirl, (2) rotor unbalance, (3) electric field excitation and harmonics thereof, and (4) harmonics of synchronous rotor excitation.

Table 19.1 lists the allowable vibration severity and examples of the following classes of rotating machinery.

Class I. Individual components, integrally connected with the complete machine in its normal operating conditions (i.e., electric motors up to 15 kilowatts).

Class II. Medium-sized machines (i.e., 15- to 75-kilowatt electric motors and 300-kilowatt engines on special foundations).

Class III. Large prime movers mounted on heavy, rigid foundations.

Class IV. Large prime movers mounted on relatively soft, light-weight structures.

Table 19.1. Vibration Severity Criteria
(After ISO IS 2372.[1])

RMS velocity ranges of vibration severity		Vibration severity* for separate classes of machines			
mm/s	in./sec	Class I	Class II	Class III	Class IV
0.28	0.01	A			
0.45	0.02	A			
0.71	0.03	A	A		
1.12	0.04	B	A	A	
1.8	0.07	B	B	A	A
2.8	0.11	C	B	B	A
4.5	0.18	C	C	B	B
7.1	0.28	D	C	C	B
11.2	0.44	D	D	C	C
18	0.71	D	D	D	C
28	1.10	D	D	D	D
45	1.77	D	D	D	D

*The letters A, B, C, and D represent machine vibration quality grades, ranging from good (A) to unacceptable (D).

Vibration severity is divided into four ranges, labeled A (the smoothest) through D (the roughest). The particular range selected by the user is based upon a number of considerations: (1) type and size of machine, (2) type and service expected, (3) mounting system, (4) effect of machinery vibration on the surrounding environment (e.g., instruments, adjacent equipment, and personnel). Machinery vibration is generally considered to be of equal severity if it exhibits the same rms velocity amplitude in the frequency range from 10 to 1,000 Hz.

An ISO standard for classifying the vibration severity of large rotating machines in situ, operating at speeds from 600 to 1,200 rpm, is shown in Table 19.2. It applies to large prime movers, class III and class IV as defined above. The severity rating of a machine depends on the classification of the supports in the mounting system for the machine. The supports are said to be *soft* if the fundamental frequency of the machine on its support is lower than its main excitation frequency; the supports are said to be *hard* if the fundamental frequency of the machine on its supports is higher than its main excitation frequency.

International and national standards that specify the requirements for instrumenta-

Table 19.2. Quality Judgment of Vibration Severity
(*After ISO IS 3945.*[2])

RMS velocity vibration severity		Support classification	
mm/s	in./sec	Rigid supports	Flexible supports
0.46	0.018	Good	Good
0.71	0.028		
1.12	0.044		
1.8	0.071		
2.8	0.11	Satisfactory	
4.6	0.18		Satisfactory
7.1	0.28	Unsatisfactory	
11.2	0.44		Unsatisfactory
18.0	0.71		
28.0	1.10	Unacceptable	
46.0	1.80		
71.0	2.80		Unacceptable

tion for measuring vibration severity in rotating and reciprocating machines are given in Refs. 4 and 5.

ELECTRIC MOTORS

Various international standards establish classification systems for permissible vibration in electric motors, trade associations, industries, and similar organizations in various countries. These classification systems are not identical; some standards are based on peak-to-peak displacement amplitude measurements, whereas others are based either on rms or peak velocity amplitudes. Each standard usually specifies the method of mounting the equipment under test and provides maintenance information and a guide on instrumentation to be used in the test-established measurement procedures.

International standard ISO 2373[6] is a special adaptation of ISO 2372 for electric motors. This standard applies to three-phase ac motors and dc motors with shaft heights* between 80 and 400 mm. (It does not apply to motor convertors, single-phase machines, or three-phase machines operated on single-phase systems.) The criterion for vibration severity (the same as for ISO 2372) is given in terms of the rms value of the velocity amplitude in the frequency range from 10 to 1,000 Hz which is measured with equipment which meets the requirements of ISO 2954.[4,5] Measurements are made on the machine installed on a free suspension (i.e., suspended from or mounted on an elastic support such as springs or rubber). The motor is operated at rated voltage and nominal frequency (for ac motors) and at its nominal speed. (For machines with several speeds or variable speeds, the tests are carried out at the various operations speeds.) Unless otherwise specified, measurements of vibration severity should be carried out under no-load operation at the temperature reached by the motor after a sufficient period of no-load operation. Table 19.3 lists the recommended limits of vibration severity for various size motors.[6]

TRADE ASSOCIATION STANDARDS. A standard for integral horseposer ac and dc motors is summarized in Table 19.4, which lists the maximum permissible values of vibration-displacement amplitude.[7] A standard procedure for mounting the motor, for measuring its vibration, and for operating it is specified.[8] Similar information is summarized in Table 19.5 for large induction motors[9,10] and in Table 19.6 for squirrel-cage induction motors.[11]

*The shaft height is the vertical distance from the base of the motor to the centerline of the shaft.

Table 19.3. Recommended Limits of Vibration Severity for Electric Motors
(After ISO IS 2373.[6])

Quality grade	Speed, rpm	Velocity amplitude (maximum rms values) for the following shaft heights h, in mm*					
		80 < h < 132		132 < h < 225		225 < h < 400	
		mm/s	in./sec	mm/s	in./sec	mm/s	in./sec
N (normal)	600 to 3,600	1.8	0.071	2.8	0.110	4.5	0.177
R (reduced)	600 to 1,800	0.71	0.028	1.12	0.044	1.8	0.071
	> 1,800 to 3,600	1.12	0.044	1.8	0.071	2.8	0.110
S (special)	600 to 1,800	0.45	0.018	0.71	0.028	1.12	0.044
	> 1,800 to 3,600	0.71	0.028	1.12	0.044	1.8	0.071

* A single set of values, such as those applicable to the 132- to 225-mm shaft height, may be used if shown by experience to be required. The shaft height is the vertical distance from the base of the motor to the centerline of the shaft.

Table 19.4. Maximum Permissible Vibration for Integral Horsepower Electric Motors
(After NEMA MG1-12.05.[7])

Speed, rpm*	Displacement amplitude (peak-to-peak), in.
3,000–4,000	0.001
1,500–2,999	0.0015
1,000–1,499	0.002
999 and below	0.0025

* For ac motors, use the highest synchronous speed. For dc motors, use the highest rated speed. For series and universal motors, use the operating speed.

Table 19.5. Maximum Permissible Vibration for Large Induction Motors
(After MEMA MG1-20.52.[9])

Speed, rpm	Displacement amplitude (peak-to-peak), in.
3,000 and above	0.001
1,500–2,999	0.002
1,000–1,499	0.0025
999 and below	0.003

Table 19.6. Maximum Permissible Vibration for Form-wound Squirrel-cage Induction Motors
(After API STD 541.[11])

Synchronous speed, rpm	Displacement amplitude (peak-to-peak), in.	
	Motor on elastic mount	Motor on rigid mount
720–1,499	0.002	0.0025
1,500–3,000	0.0015	0.002
3,000 and above	0.001	0.001

PUMPS

In Figs. 19.1 and 19.2, the Hydraulic Institute specifies acceptable vibration limits for horizontal clear-liquid pumps and horizontal nonclog pumps, respectively.[12] These curves are applicable under the following conditions. Pumps must be operating in a non-cavitating and nonseparating condition. Suction piping is arranged so as to provide a straight uniform flow to the pump. Pumping is connected in such a way so as to avoid

Fig. 19.1. Acceptable field vibration limits for horizontal clear-liquid pumps, with readings filtered and measured on bearing housing. (*After Hydraulic Institute.*[12])

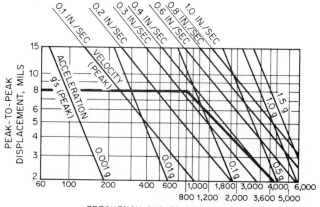

Fig. 19.2. Acceptable field vibration limits for horizontal nonclog pumps, with readings filtered and measured on bearing housing. (*After Hydraulic Institute.*[12])

undue strains on the pump. Shaft couplings are aligned to within the pump manufacturer's recommendations. These curves should be used as a general guide, keeping in mind that pumps which produce vibration amplitudes in excess of the values shown should be examined for defects or possible correction. Often, more important than the actual vibration amplitude itself, is the *change* in vibration amplitude over a period of time. Vibration in excess of the curve values may be acceptable if there is no increase over long periods of time and if there is no other indication of damage, such as an increase in bearing clearance or noise level.

RECIPROCATING PISTON COMPRESSORS

A German standard specifies applications for reciprocating piston compressors, having a power input or output greater than 100 kilowatts and speeds less than 3,000 rpm.[13] Typical examples of application are marine propulsion engines, marine auxiliary engines, engines operating in diesel-generator sets, and engines for diesel locomotives. The standard is limited to mechanical vibration in the frequency range from 2 to 300 Hz, measurable on the surface of the machine, at bearing or fastening points in one of three principal directions. In certain cases it is necessary to consider the displacement amplitude and acceleration amplitude in the evaluation as well as the velocity amplitude. Figure 19.3 shows the vibration severity limits as a function of vibration frequency for reciprocating machinery.

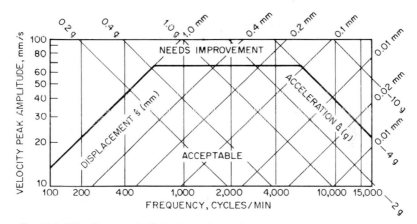

FIG. 19.3. Vibration severity limits for reciprocating machinery. (*After VDI 2063*,[13])

ROTATING MACHINERY HAVING FLEXIBLE ROTORS

Vibration criteria for rotating machinery having flexible rotors are given in ISO standard 7919/1, "Mechanical Vibration of Non-Reciprocating Machines—Measurements on Rotating Shafts and Evaluation," part I: "General Guidelines."[14] Guidelines are given for criteria which apply to shaft vibration measured close to the rotor bearings under normal operating conditions. According to this document, an over-all quality judgment of the vibratory state of the machine is made on the basis of both relative and absolute values of the displacement amplitude of the shaft. If only relative measurements are made, then the over-all quality rating of its vibratory state is made on the basis of both the relative vibration amplitudes as defined above and similar measurements made on nonrotating parts (see Ref. 1).

STEAM TURBINE GENERATOR SETS

The shaft vibration of large steam turbine generator sets (measured at or close to the bearings) can be classified on the basis of two criteria given in ISO standard 7919.[15]

Criterion I: The reliable and safe running of a machine requires that the displacement amplitude level remain below certain limits. The four quality zones are defined in terms of these limits as shown in Fig. 19.4 for absolute values of shaft vibration, and in Fig. 19.5 for shaft vibration relative to bearings. In both cases, measurements are made at, or close to, the main load-carrying journal bearings at the rated speed, under steady operating conditions.

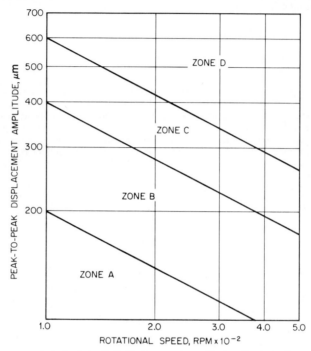

FIG. 19.4. Recommended absolute shaft displacement amplitude criteria for large turbine generator machinery vs. shaft rotational speed. Zone A—New machine, can be operated without restriction; Zone B—Acceptable for long-term operation; Zone C—Machine may be operated for a limited period at these values until a suitable opportunity arises for remedial action to be taken; Zone D—Vibration values which fall into this zone are considered to be of sufficient severity to cause damage. (*After ISO IS 7919, Part 2.*[15])

Criterion II: Changes in vibration amplitude may point to incipient damage or to some irregularity in operation, even though the limits referred to above are not exceeded. This criterion has limited application, since significant changes of varying magnitudes and rates occur in individual frequency components, but such changes may not necessarily result in a significant change in the over-all vibration amplitude of the shaft. For example, see Fig. 16.3. The propagation of a crack in a rotor can introduce a progressive change in the magnitude of the vibration components at multiples of rotational speed, but such changes may be small relative to the over-all vibration amplitude. Consequently, it may be difficult to identify the effects of the crack propagation by observing only the change in over-all vibration amplitude.

INDUSTRIAL TURBOMACHINERY

Standards which establish permissible vibration in turbomachinery (turbocompressors, steam-turbine generators, and electric drives) are useful for evaluating quality and establishing maintenance action. Most turbomachinery operates at a high speed, above its first or second critical speed. Hence measurements of shaft vibration provide a better indication of conditions of vibration of the machinery than do other measurements taken in other locations, e.g., at the bearing caps.

According to standards for single- and multistage steam turbines or special-purpose steam turbines, to avoid excessive vibration the critical speed (see Chap. 39) of a rigid-shaft turbine rotor should be at least 10 per cent higher than its maximum speed; the value of the first critical speed of a flexible-shaft turbine rotor should not exceed 60 per cent of the maximum continuous speed, nor should it be within 10 per cent of any operating speed.

Standards for acceptable limits of vibration exist for a restricted class of turbomachinery. Most of the standards are based on measured values of vibration displacements relative to the shaft of turbomachinery operating in situ. ISO 7919, part 3, provides guidelines for classifying turbomachinery based on measurements made close to the bearings under normal operating conditions.[16] As shown in Fig. 19.6, these guidelines are presented both in terms of values obtained during steady operation and in terms of changes from these steady values. In general, the vibratory condition of a machine is assessed by consideration of both the shaft vibration and the associated structural vibration (this standard complements ISO 2372). However, there are some changes in the vibratory state of the machine which, although significant, may not be detected by monitoring only the change of vibration-displacement amplitude. Such changes may only be identified by examining changes of individual frequency components.

FIG. 19.5. Recommended displacement amplitude criteria for shaft vibration measured relative to bearings for large turbine generator machinery vs. shaft rotational speed. Zone A—New machine, can be operated without restriction; Zone B—Acceptable for long-term operation; Zone C—Machine may be operated for a limited period at these vibration values until a suitable opportunity arises for remedial action to be taken; Zone D—Vibration values which fall into this zone are considered of sufficient severity to cause damage. (*After ISO IS 7919, Part 2.*[15])

In industrial turbomachinery, shaft displacement measurements usually are made relative to the bearings in the frequency range from 5 to 1,500 Hz. Because of the relatively high rotational speeds in such equipment, measuring methods using proximity transducers are generally preferable.

INDUSTRIAL GAS TURBINES

ISO standard 7919, part 4, applies to gas turbines having fluid film bearings, outputs greater than 3 megawatts, and shaft rotational speeds from 3,000 to 30,000 rpm.[17] The turbine configuration may be one of the following: (1) single-shaft constant-speed, (2) single-shaft variable-speed, and (3) separate shafts for hot gas generation and power delivery. This standard provides guidelines for classifying turbines based on measurements of shaft vibration made close to the bearings under normal operating conditions as shown in Fig. 19.7. These guidelines are presented in terms of both steady running vibrations and any changes which may occur in these steady values. In general, the vibratory condition of a machine is assessed by consideration of both the shaft vibration and the associated structural vibration (see Ref. 14). However, there are changes in the vibratory state of the machine which, although significant, may not be detected by monitoring only the change in displacement amplitude. Such changes may only be identified by examin-

FIG. 19.6. Recommended displacement amplitude criteria for shaft vibration measured relative to bearings for industrial turbomachinery vs. shaft rotational speed. Zone A—New machine, can be operated without restriction; Zone B—Acceptable for long-term operation; Zone C—Machine may be operated for a limited period at these vibration values until a suitable opportunity arises for remedial action to be taken; Zone D—Vibration values which fall into this zone are considered of sufficient severity to cause damage. (*After ISO IS 7919. Part 3.*[16])

ing changes of individual frequency components. The specified levels are expressed as the greater value of peak-to-peak displacement measured in two orthogonal directions as defined in the standard. In general, an over-all quality judgment of the vibratory state of the machine should be made on the basis of both shaft relative vibration levels and measurements made on nonrotating parts (see Ref. 1).

FIG. 19.7. Recommended displacement amplitude criteria for shaft vibration measured relative to bearings for industrial gas turbine machinery vs. shaft rotational speed. Zone A—New machine, can be operated without restriction; Zone B—Acceptable for long-term operation; Zone C— Machine may be operated for a limited period at these vibration values until a suitable opportunity arises for remedial action to be taken; Zone D—Vibration values which fall into this zone are considered of sufficient severity to cause damage. (*After ISO IS 7919, Part 4.*[17])

STEAM TURBINES FOR REFINERIES

Steam turbines used in refineries are horizontal, flexible rotor turbines used to drive equipment that usually is of relatively large horsepower, such as drives for centrifugal compressors, air blowers, reciprocating compressors, or generators. This category is not limited by steam conditions or turbine speed.[18]

AIR AND GAS CENTRIFUGAL COMPRESSORS

Vibration criteria established by the Compressed Air and Gas Institute for single and multistage in-line centrifugal and axial compressors in air and gas service are given in Fig. 19.8 for rotational speeds up to 15,000 rpm. The representative levels shown include both the vibration amplitudes and mechanical runout of the shaft. The *trip setting* is

FIG. 19.8. In-service vibration criteria for centrifugal compressors as a function of shaft speed. (*After Compressed Air and Gas Institute.*[20])

that displacement amplitude at which the compressor automatically shuts down. In case no automatic trip controls are provided, when the trip setting value is reached, the equipment should be shut down manually.

In general, a change in displacement amplitude of a compressor in operating service is of more significance than its absolute value. Hence any significant change in vibration, for the same operating conditions, should be immediately investigated and the cause of the change determined. The *change warning* value shown in the illustration is the increase in vibration over normal vibration level for the same operating conditions.

GEAR UNITS

An ISO Draft standard presents the methods for determining mechanical vibration of enclosed gear units which are individually housed.[21] This standard applies to a gear unit under test and operating within its design speed, load, temperature range, and lubrication for acceptance testing. Two types of vibration measurements are made on gear units which operate with oil-film journal bearings: (1) *shaft vibration*, and (2) *housing vibration*. Proximity-type transducers usually are used to measure the peak-to-peak value of the shaft's displacement relative to the housing, in the frequency range from 0 to 500 Hz. The transducers are located as close as is practical to a bearing; measurements should be

F‌IG. 19.9. Gear shaft vibration-displacement amplitude frequency, with parameters of shaft vibration displacement ratings (DR). (*After ISO Draft standard.*[21])

made in three orthogonal directions, one of which is parallel to the shaft axis. Only one axial transducer per shaft is necessary. Housing vibration is measured with a seismic transducer in the frequency range from 10 to 10,000 Hz. A classification system for gearboxes on the basis of shaft displacement and housing velocity measurements is given in Figs. 19.9 and 19.10.

VIBRATION STANDARDS FOR SHIPS

Vibration standards are used by the U.S. Navy to ensure structural integrity of ships and to minimize undesirable radiation of noise. One such specification requires that the

F‌IG. 19.10. Gear housing vibration-velocity amplitude frequency, with parameters of gear housing vibration velocity ratings (VR). (*After ISO Draft standard.*[21])

residual vibration of shipboard-mounted equipment have a magnitude less than that shown in Fig. 19.11;[19] these limits apply to systems having a vibration-mounting system with a natural frequency less than one-fourth that of the minimum rotational frequency of the unit. Vibration-displacement amplitudes are measured on the bearing housing in the direction of maximum amplitudes (except for turbines, where they are measured on the rotating shaft adjacent to the bearings). On constant-speed machines, measurements are made at the operating speed; on variable-speed machines, measurements are made at the maximum speed and at all critical speeds within the operating range. In the case of complex machinery such as reduction gearing and impellers, the peak vibration velocity must not exceed 0.15 in./sec; the displacement amplitude must not exceed the values given in Fig. 19.11 and must never exceed 1.25 mils.

Fig. 19.11. Maximum allowable machinery vibration on shipboard. (*After MIL-STD-167-1 Ships.*[19])

Hull vibration is excited by the propeller shaft and the propeller blade. Such vibration is greatest in the afterpart of the ship (except when one of the hull resonant frequencies coincides with a driving frequency). The displacement amplitudes often exceed 0.1 in. At higher speeds of the propeller shaft, the amplitudes of the high modes of vibration of the hull tend to be smaller. Vibration data indicate that the maximum displacement amplitude of the main structural members of the hull (measured at the antinodes and at the end of the ship) is smaller at the higher excited frequencies. When shipboard machinery and equipment are fastened to the hull, deck, and bulkhead stiffeners, the vibration amplitudes at the equipment supports are substantially the same as the main structural members of the hull. The amplitudes of vibration of the bow and stern are used as a measure of hull vibration. In some types of ships, the maximum displacement amplitude of hull vibration is less than 0.01 in. at propeller-blade frequencies. A displacement amplitude of less than 0.22 in. is considered satisfactory; if this value is exceeded, corrective measures are usually taken.

REFERENCES

1. International Standards Organization: "Mechanical Vibration of Machines with Operating Speeds from 10 to 200 rps—Basis for Specifying Evaluation Standards," ISO 2372-1974.
2. International Standards Organization: "The Measurement and Evaluation of Vibration Severity of Large Rotating Machines, in Situ, Operating at Speeds from 10 to 200 rps," ISO 3945-1980.
3. American National Standards Institute: "The Measurement and Evaluation of Vibration Severity of Large Rotating Machines, in Situ, Operating at Speeds from 10 to 200 rps," ANSI S2.41-1985.

4. International Standards Organization: "Mechanical Vibration of Rotating and Reciprocating Machinery—Requirements for Instruments for Measuring Vibration Severity," ISO 2954-1973.
5. American National Standards Institute: "Mechanical Vibration of Rotating and Reciprocating Machinery—Requirements for Instruments for Measuring Vibration Severity," ANSI S2.40-1984.
6. International Standards Organization: "Mechanical Vibration of Certain Rotating Electrical Machinery with Shaft Heights between 80 and 400 mm—Measurement and Evaluation of the Vibrating Severity," ISO 2373-1971.
7. National Electrical Manufacturers Association: "Dynamic Balance of Motor," MG1-12.05, June 1978.
8. National Electrical Manufacturers Association: "Method of Measuring the Motor Vibration," MG1-12.06, June 1978.
9. National Electrical Manufacturers Association: "Balance of Machines," MG1-20.52, October 1977.
10. National Electrical Manufacturers Association: "Method of Measuring the Motor Vibration," MG1-20.53, October 1977.
11. American Petroleum Institute: "Recommended Practice for Form-Wound Squirrel-Cage Induction Motors," API 541, December 1972.
12. Hydraulic Institute: "Acceptable Field Vibration Limits for Horizontal Pumps," 14th Ed., Centrifugal Pumps Applications.
13. Verlin Deutscher Ingenieure: "Measurement and Evaluation of Mechanical Vibrations of Reciprocating Machinery," VDI 2063, August 1985.
14. International Standards Orgnization: "Mechanical Vibration of Non-Reciprocating Machines—Measurements on Rotating Shafts and Evaluation," part 1: "General Guidelines," ISO 7919/1, 1986.
15. International Standards Organization: "Mechanical Vibration of Non-Reciprocating Machines," part 2: "Measurement and Evaluation of Shaft Vibration of Large Turbine-Generator Sets," ISO 7919/2, 1986.
16. International Standards Organization: "Mechanical Vibration of Non-Reciprocating Machines," part 3: "Measurement and Evaluation of Shaft Vibration of Industrial Turbo-Sets," ISO 7919/3, 1986.
17. International Standards Organization: "Mechanical Vibration of Non-Reciprocating Machines," part 4: "Measurement and Evaluation of Shaft Vibration of Industrial Gas Turbo-Sets," ISO 7919/4, 1986.
18. American Petroleum Institute: "Special-Purpose Steam Turbines for Refinery," API 612, 2nd ed., June 1979.
19. Military standard: "Mechanical Vibrations of Shipboard Equipment (Type I—Environmental and Type II—Internally Excited)," MIL-STD-167-1 (Ships), May 1, 1974.
20. Compressed Air and Gas Institute: "In-Service Vibration Standards for Centrifugal Compressors," 1963.
21. International Standards Organization Draft standard: "Determination of Mechanical Vibration of Gear Units during Acceptance Testing," ISO/TC-60/WG-9-N55, 1987.

20

INTRODUCTION TO DATA REDUCTION, TESTING, AND SPECIFICATIONS

Charles T. Morrow

Consultant

INTRODUCTION

This chapter serves as an introduction to chapters that follow on data reduction and testing. The mathematical treatment here is only descriptive because derivations are available in the later chapters. Furthermore, this chapter relates primarily to the development and production of mechanical, electrical, electronic, or hydraulically powered equipment, not structures. The development of a structure to withstand shock and vibration is dependent primarily on design, whereas development of equipment to withstand these environments is at the present time dependent primarily on test and redesign. There is as yet no requirement to minimize structural transmission to mounted equipment. Only the smallest missiles are subject in their entirety to shock or vibration test, although larger vehicles or portions of them are sometimes subject to test in acoustic noise. Shock and vibration engineers often participate directly in structural design. In respect to equipment, organizational constraints commonly limit them to a test responsibility.

GENERAL TYPES OF ENVIRONMENTS

The environments that must be considered in equipment design and testing include those listed in Table 20.1. Those printed in boldface are of special concern in this Handbook and of primary concern in this chapter. Except for sustained acceleration, these have the most complicated transmission patterns and are modified or intensified by mechanical resonance, and, therefore, are the ones appropriately described by spectra.

In practice, for economy of effort, most design and test approaches are based primarily on exposure to individual environments. Yet the environments of Table 20.1 (particularly the four printed in boldface) may occur simultaneously and are not entirely independent. For example, *noise* and *vibration* in any location tend to be proportional to each other; the same is true with *blast* and *shock*. At subsonic missile speeds in the atmosphere, rocket noise (propagating forward from the turbulent rocket exhaust) shakes the missile structure and therefore, in turn, shakes any equipment attached to the structure. Likewise, aerodynamic turbulence in contact with the airframe may produce vibration. A pyrotechnic blast, occurring, for example, when an explosive is used to initiate staging of a missile, excites transient vibration (shock) in any nearby structure and therefore, in turn, in any equipment attached to the structure.

ENVIRONMENTAL DISTINCTIONS

Practical distinctions between steady acceleration, vibration, and shock or between the corresponding acoustic environments in a surrounding fluid or gas cannot be made absolute, independent of the mechanical system subject to the environment. For example, an earthquake has the character of a shock for a tall building but not for typical aerospace equipment.

Any mechanical device which is more or less linear is characterized by one or more resonance frequencies and damping coefficients or by a corresponding set of transient (decaying) responses after any momentary excitation (shock). An excitation whose duration is comparable to, or less than, the response (or decay) times of these transients is called a *shock*. A dynamic excitation that is long by comparison with the response times is called a *vibration*. An acceleration that remains unchanged for a time interval which is long by comparison with the response times is called a *sustained acceleration* and can be simulated, if necessary, simply by using a centrifuge.

If a *periodic* excitation does not change its character significantly within the anticipated response times of equipment, it can be considered to be *steady-state* vibration. If the statistical measures of a *random* vibration do not change (or can be presumed not to change, from knowledge of the physical source of the excitation) within the response times and within the time necessary for a good statistical sample, the excitation may be considered to be *stationary* or *quasi-stationary* vibration.

Table 20.1. Various Types of Environments to Which Equipment May Be Exposed

Vibration	Shock
Acceleration (sustained)	**Acoustic noise and blast**
Atmospheric pressure	Temperature
Rain, hail, and snow	Humidity
Wind	Salt spray
Dust and sand	Fungus

Shock has one special characteristic: Transient (decaying) response in a device or equipment may persist long after the excitation has come to an end. In a device with a serious design defect, the largest response at a failure point may not occur until some time *after* the excitation has become negligible.

DESCRIPTIONS OF VIBRATION, SHOCK, AND ACOUSTIC ENVIRONMENTS

Descriptions of various types of environments including periodic, swept sine wave, random, and pseudorandom vibration are described in Chaps. 11, 13, and 21 through 23; these chapters also contain descriptions of shock environments. Some of these environmental descriptions used in data reduction are in part a matter of tradition and precedent and are dependent on conditions (such as computer availability) at the time precedent-setting decisions were made. Therefore, it is useful to consider some of these topics in historical perspective.

SHOCK SPECTRUM (RESPONSE SPECTRUM)

The *shock spectrum* (see Figs. 24.15 and 24.16), which is based on responses of a simple mechanical resonator (also called the *response spectrum*), was first applied by the U.S. Navy to shipboard shock. In contrast to the Fourier transform, it did not require a digital computer. It has remained the accepted foundation for Navy shock testing requirements. Further development of the approach took place in connection with U.S. Air Force missile programs. However, at the beginning, the Air Force chose time-domain descriptions for avionic shock testing requirements. Missile programs have undergone a vigorous movement away from time-domain to spectral descriptions. Only now, some 30 years after the introduction of the shock spectrum in USAF missile programs, is the use of the shock spectrum for avionics beginning to be officially accepted.

The shock spectrum has three basic forms with some additional variations. The peak acceleration response of the resonator, as a function of frequency, during the shock is known as the *primary* or *initial shock spectrum*. The peak acceleration response after shock termination (or decrease to negligible severity) is known as the *residual shock spectrum*. The peak acceleration response over all time (both during and after the shock) is known as the *maximax shock spectrum*. A nonzero damping for the resonator may be chosen for primary and maximax shock spectra but not for the residual shock spectrum unless the shock has a definite ending—otherwise the residual shock spectrum is not uniquely defined. The initial and maximax shock spectra may differ for positive vs. negative direction of acceleration. For the undamped residual shock spectrum, there is no such difference. The undamped residual shock spectrum has a simple relationship to the Fourier transform—as obtained by one method of computation, it permits a reconstruction based on the inverse Fourier transformation. The primary and maximax shock spectra have no such definite relationship to the Fourier transform.

SHIPBOARD SHOCK

The U.S. Navy specified, not a shock test environment directly, but a shock testing machine intended to simulate shipboard structure to some degree, and some prescribed adjustments. The machine contained a swinging hammer to impart a shock (see Chap. 26). However, the Navy permitted and even demanded some modifications of the machines to make the test shocks more realistic for particular situations. The realism of the shocks was verified by a spectrum from a reed gage (see Chap. 23) attached to the ship structure for actual shock data, and later to the shock testing machine. The reed gage is a collection of reeds tuned to different frequencies and made to mark their excursions, in response to a shock, on paper. It has no capability to distinguish residual shock spectra from maximax, but with the restricted frequency range of shipboard shock data reduction and the absence of any prescribed test pulses, this has not been a serious limitation in shipboard shock testing. Compared to the mounting platforms in the drop towers of Air Force programs, the Navy's shock testing machines permitted a relatively realistic simulation of dynamic loading of the deck by the equipment—an important consideration for massive equipment such as motor generator sets.

AVIONIC, MISSILE, AND SPACE SHOCK

Many years ago, the U.S. Air Force specified the shock environment directly, using standard acceleration pulses (such as the square wave and half-sine wave) which could be generated in a drop tower by controlled deceleration at the bottom of the drop. A terminal-peak sawtooth, generated originally by using a lead pellet as a decelerator, was introduced first in the Hughes Falcon missile program and in the USAF ICBM programs.[2] It had the advantage of having no nulls in its undamped residual shock spectrum (or in its Fourier transform). However, the time-domain descriptions implied time-domain data reduction. There was little room in airplane equipment racks for a reed gage, and no room in missiles. The reed gage was not developed for wide frequency range shock data reduction. Air Force contractors began using the actual shock durations and peak accelerations in determining the corresponding parameters for the specified pulse shapes. When the actual shocks consisted of relatively high-frequency transients, this tended to result in undertest at the frequencies of the transients and overtest at low frequencies.

In the middle 1950s, the shock spectrum was introduced into the USAF ICBM programs, from which it proceeded directly to space programs.[3] At the time of the first environmental specification, the only shock spectrum analyzer in existence was an experimental unit that had been constructed at the Hughes Aircraft Company for the Falcon missile program. Consequently, the shock specification requirement had to be written in such a way that it could be verified, at least in principle, by a reed gage. The terminal-peak sawtooth pulse was suggested in the specification as a way of achieving the spectrum, but the spectrum was the controlling requirement. The sawtooth shock, but not its spectrum, soon made its way to avionics, where it appeared first as an alternate requirement to the square wave and half-sine wave but is now the only pulse shape surviving in general specifications. The spectrum continued as the controlling requirement in ICBM and space programs. The commercial analog shock spectrum computers that soon became available were designed, except for a choice of damping, to do only what the reed gage could do—provide a maximax shock spectrum. As time went on, the damping preferred by specifications was increased from zero to make the spectrum more representative of the response of typical equipment resonators.

The resonators that are most likely to fail in use are not typical ones but lightly damped ones, particularly if they are at the end of a series of coupled resonators that are inadvertently tuned to approximately the same frequency. If it is necessary to make a choice, it is more important to provide adequate testing for such exceptional resonators than for typical ones that are more damped and have less response.

The shock spectrum assumed a dual role—shock description, for test purposes, and simulation of the equipment under development, for design applications. To this day, equipment simulation for design applications has seldom proceeded beyond the simple resonator and has therefore bypassed the interaction of one resonator with another.

Soon after the shock spectrum was introduced to the ICBM programs, a simple relationship was found between the *undamped residual shock spectrum* and the Fourier transform. The undamped residual shock spectrum has the practical advantage over the transform that it is expressed in terms of acceleration rather than in units of velocity and can, except at the lowest frequencies where displacement is important, be related more directly to endurance limits of the equipment.

PERSONAL DIGITAL COMPUTER APPLICATIONS

A shock or vibration program on a personal computer tends to run slower than on a dedicated computer such as a fast Fourier transform algorithm processor. However, while high computation speed is always a convenience, it is critical only for real-time spectral analysis of random vibration, for controlling a random vibration test. The personal computer provides the engineer with a convenient means for exploratory computations, free of any restrictions of past tradition, that may persist in instrumentation computers. With sufficient personal program development, the personal computer can even be used for shock data reduction.

A number of computer programs have been developed for the computation of shock spectra, usually written to do what the commercial instruments do. One algorithm which may be used with a personal computer permits computation of both the undamped residual shock spectrum and, with a sufficient choice of subroutines, the response of a mechanical system of arbitrary configuration.[5,6,7] The algorithm starts from the observation that any shock time-history can be approximated by a series of straight lines connecting points on the time-domain curve. Then, if the response of a mechanical system to discontinuities of acceleration and of acceleration slope is known, the response to any shock excitation can be computed.

PYROTECHNIC SHOCK ENVIRONMENT

In recent years, the term *pyrotechnic shock* has come to mean a high-acceleration (thousands of gs) high-frequency (thousands of hertz) shock observed on a structure near a pyrotechnic excitation source or possibly near a point of impact of an enemy shell.

Accelerometers for measurement of pyrotechnic shock should be small to avoid dynamic loading of the structure at high frequencies. Accelerometers may have a flat frequency response to 100,000 Hz or more when attached to a rigid mass, yet exhibit resonance peaks even below 2000 Hz when attached to an aerospace structure, especially when attached by means of a fixture to provide a prescribed orientation, because of the interaction of the mass of the accelerometer with the resilience of the structure and fixture. An accelerometer to be mounted on an aerospace structure for the measurement of shock (or vibration) above 2,000 Hz should be calibrated on a sample of the structure.

Any attachment fixture intended for use with the accelerometer should be present during the calibration.

The Fourier transform (or DFT) and undamped residual shock spectra are applicable to pyrotechnic shock as they are to ordinary shock. However, they encounter an overload problem at high frequencies and a signal-to-noise ratio problem at low frequencies. It may be necessary to integrate the accelerometer signal to a velocity signal, preferably in such a way as to avoid any noise and overload problems in the first stage of amplification, and to correct during data reduction. Any high-frequency accelerometer or accelerometer mounting resonance that cannot be excluded by a low-pass filter prior to amplification complicates the signal-to-noise ratio problem; any gain setting that avoids high-frequency overload leaves the low-frequency spectrum lost in electronic noise.

Any accelerometer signal offset produced by the shock, or any abrupt truncation of the high-frequency transients of the shock produced by data reduction (giving the effect of a net acceleration or velocity change), may result in an artificial boost in low-frequency spectra. Use of the maximax shock spectrum instead of the residual may compound the problem by permitting the initial spectrum to dominate the lower frequencies.

On first examination of the pyrotechnic shock problem, the high accelerations at high frequencies appear to be the major design challenge. However, the low-frequency part of the spectrum, even with relatively few gs, is often more severe than spectra of more ordinary shocks and may be more damaging than the high-frequency portion.

ACOUSTIC ENVIRONMENT

Most excitations of aerospace equipment, even though the primary mode of transmission to the equipment may be structural, result ultimately from acoustic sources—turbulence, rocket noise pyrotechnic blasts, etc. Fortunately, the *external* acoustic environment of an airplane, missile, or space vehicle does not exhibit resonance effects and does not require detailed frequency resolution in a spectrum. At sufficiently high frequencies (for wavelengths small compared to cavity dimensions) in the absence of acoustic absorption, the sound inside a closed cavity has the same level as that immediately outside. For closed cavities which are large compared to the wavelengths, with sound penetrating only by induced vibration in the aerospace shell rather than directly through openings of any significant size, small additions of acoustic absorption rapidly reduce the interior noise level. However, this is done primarily for personnel protection, seldom as yet for protection of airborne equipment.

Sound pressure is usually expressed on a logarithmic, rather than linear, scale (because the response of the ear is approximately logarithmic) and analyzed in octave or fractional octave bandwidths. For the estimation of structural response or equipment excitation, power spectral density of pressure can be measured directly or obtained by conversion. For small cavities or low frequencies, requiring better frequency resolution in spectra of internal noise, direct measurement of pressure power spectral density is preferable.

For acoustic blast, the Fourier transform or DFT is applicable to instantaneous acoustic pressure as it is to the instantaneous motions of mechanical shock. Alternately, it is possible to use a residual shock spectrum of pressure, detaching the concept from the mechanical resonator model and treating it as a purely mathematical operation like the DFT.

DATA ACQUISITION AND MEASUREMENT

Raw data are environmental data obtained directly from sensors (e.g., accelerometers, force gages, and strain gages), usually in the time domain without data reduction to deemphasize unimportant detail or to provide a more useful presentation. *Data measurement* is measurement of the environment, not necessarily part of an orderly plan, to obtain raw data or simply reduced data. *Data collection* is the collection of raw data and/or reduced data for some intended purpose. *Data acquisition* is the collection of environmental data by literature search, measurement, and/or data reduction according to some orderly plan and usually for specific purposes.

It might be assumed that all shock and vibration data that are important for specification writing, equipment design, or test can readily be made available to the responsible engineers. Such is not the case. At the beginning of a development program, it may be necessary to make decisions before the structure is available on which measurements can be made. There may be little choice but to examine any data that may be available from other similar programs, look at conditions specified in U.S. government general specifications, or use engineering judgment. For example, initial vibration estimates for the USAF ICBM program were extrapolations from the Corporal and Falcon missile programs, since no other reliable missile vibration data were available at the time. When measurements are possible, it may still be necessary to settle for a single measurement, in a single direction, for an entire piece of equipment or even an entire rack of equipment. Points at which one would like to measure may not be accessible for sensors. In missile programs, raw data on shock and vibration are transmitted to the ground by telemetry for recording and later data reduction. Since telemetry is also used for functional monitoring so as to permit diagnosis in the event of any malfunction, the channels for shock and vibration are limited in number.

DATA REDUCTION

Data reduction is the processing of data so as to present it in more usable or compact form and focus on the more important aspects for a particular application. This usually involves conversion of data from the time domain to the frequency domain—the generation of one or more spectra. Computation time is important, especially if it must be completed in real time, for controlling a vibration table. In the case of a mixture of shock and vibration, it is important, insofar as possible, to separate the two environments or reduce each in such a way that the other produces minimum error.

Particular aspects in the time domain, such as total velocity change or highest peak acceleration, are sometimes noted and recorded.

Data reduction is commonly an intermediate step toward the prescription of a design or test environment. These are related to, but not identical to, the actual environment. They have their own constraints, according to what is practical for use. A *design environment* is an environment prescribed for use in design, usually limited to simple criteria such as largest peak acceleration or largest value of a response spectrum. A *test environment* is an environment prescribed for use in environmental test, more realistic than design criteria but usually devoid of spectral detail, so as to simplify writing up specifications and setting up the test.

FAST FOURIER TRANSFORM (FFT) ALGORITHM

The most straightforward methods of digital computation of the discrete Fourier transform (DFT) are time-consuming. The DFT requires frequencies in a Fourier series and as many time instants n for data input as there are frequencies. Computation is therefore equivalent to evaluation of all elements of an $n \times n$ square matrix. However, if n is chosen as a power of 2, some redundancy in the matrix becomes evident—only n $\log_2 n$ independent computations need be carried out. The approach that utilizes this, known as the *fast Fourier transform (FFT) algorithm*, described in Chap. 13, is now the universal approach to DFT computation for shock description or as an intermediate step to power spectral density.[8]

WINDOWING

Windowing, or reducing the effective electronic gain so as to deemphasize portions of a sample during data reduction, is used to minimize certain errors in the computation of a DFT or of power spectral density for random vibration, or to decrease vibration error in the computation of a DFT for shock. It is usually accomplished digitally, by multiplying portions of the data by factors ranging from unity to zero. A comparison of various types is given in Table 13.2.

In general, windowing, except toward the end of the data, is insuitable for shock data reduction. Since a shock sample is usually selected to start just before the shock and last at least until the shock has become negligible, there can be no important discontinuities at the loop splice. Reducing the gain at the beginning might result in loss of important data. However, any vibration present with the shock sample can lead to significant error in DFT computation. It is beneficial to reduce the effective playback gain after the shock has terminated or decayed to negligibility, in effect substituting zeros for the data that the vibration would yield. On the other hand, if the undamped residual shock spectrum rather than the DFT is computed, stopping the computation at any instant can be equivalent to assuming that all subsequent accelerations are zero, so that windowing is needed only to avoid any discontinuity of acceleration or slope at data termination.

TESTING

An *environmental test* is a test of a device to a specified environment (or sometimes to the environment generated by a specified testing machine, as in the case of Navy shock testing) to determine whether this produces any deterioration of performance or any damage or malfunction. There are many kinds of environmental tests, distinguished not only by the kind of vibration or shock applied, but by the objectives. A *qualification test* is an environmental test, performed as an aid in determining whether a design is satisfactory for its intended purpose, in the anticipated environments, or if more development or production engineering should be carried out. A *development test* is a test similar in requirements to those of a qualification test, but it is performed in an exploratory manner earlier in the program as an aid to development. A *statistical reliability test* is a test performed a large number of times or for a long duration to obtain a statistically significant verification of an assigned reliability objective. A *production test* or quality control test is an environmental test, commonly but not necessarily less severe than a qualification test, applied as an aid in ensuring that a satisfactory quality or standard of work is maintained. A *screening test* is similar to a production test but longer in duration, with the intention of producing and correcting failures that would otherwise occur later while equipment was in use. If performed at all, it is performed, not on a sampling basis, but on the entire product prior to delivery and use. A *reliability growth test* is a test performed on a few items over an extended period of time, with corrective action including possible design changes after each failure. The failure rate is monitored so as to verify an improvement in reliability.

All environmental tests have the possibility of test item failure or may be intended to produce failures. These are of two extreme types. A *reliability failure* is the occurrence of any damage or malfunction or parameter out of tolerance that could make a system, in which the test item may be incorporated, impossible to use or useless in its mission. It is a catastrophic or potentially catastrophic failure. A *performance deterioration failure* is a parameter out of tolerance that would still leave the system usable, but perhaps less precise. It has a chance of being compensated for by other parameter errors in the same device or elsewhere in the system in which it is incorporated.

Performance deterioration failures have a more or less linear relationship to environmental severity and to resulting system performance deterioration, whereas reliability failures seldom have a linear relationship to either. An *accelerated test* is sometimes used with increased environmental severity to have an increased chance of detecting incipient reliability failures in limited time. If such an environment produces instead a performance deterioration failure, the parameter error should be scaled back as well as possible to an estimate for the actual environment (or the test should be rerun with more realistic severity) before a rejection is decided on.

All environmental tests are based on or imply an assumption about the test item and its failure mechanisms. The qualification test is based on an assumption that the test item may have design defects that are reproducible enough to be detected in a few tests of only a few test items. Statistical reliability testing implies in contrast that the failures are random in occurrence and have independent rates. The screening test is based on the assumption that in its absence the failure rate would have a high initial value, decreasing with the duration of use—the initial failures are made to occur during the test, thereby improving reliability during actual use. The reliability growth test is based on the assumption that many design defects will not have an effect until after a more extended period of use.

In addition, there are tests that use an environment but are not based on specifications and are not intended to produce failures. A *diagnostic test* is a test or analysis carried out as an aid in the diagnosis of some observed failure or suspected design deficiency. An *analytic test* is a test in which an environment is applied to a device to determine properties or parameters (e.g., resonance frequencies) of that device.

RANDOM VIBRATION TEST

A random vibration test (in which all vibration frequencies are present at once) is usually performed with an electrodynamic vibration testing machine (see Chap. 25) driven by electronic amplifiers with a power output capability of many kilowatts. The use of hydraulic shakers decreases the electronic power requirement and increases the excursion possible at low frequencies, but makes it more difficult to control the high-frequency end of the spectrum.

A random vibration system, shown in Fig. 25.11, includes an electronic random noise source, an equalizer (peak and notch filter) for the shaker armature and any mounting fixture, an equalizer for achieving a prescribed vibration spectrum, a multikilowatt electronic power amplifier, one or more accelerometers mounted on the shaker armature or mounting fixture to measure as directly as possible the excitation of the test item, shaker protective circuitry (to protect the armature and test item in the event of a large input signal to the system or a malfunction of the power amplifier), and a power spectral density computer (preferably, real-time) to verify and aid in the control of the spectrum of the excitation.

Since any power amplifier has limited power capability, when driven hard, it clips (to a slight degree) the tails of the distribution of the instantaneous values of the acceleration signal, so it becomes less accurately Gaussian. For this reason, specifications usually require that there be no clipping within three times the rms value of the output signal of the amplifier. However, moderate clipping has a negligible effect on the distribution of accelerations at a failure point, since the vibration (with one exception as noted below) tends to become more Gaussian as it is transmitted through the interior of the test item, unless it encounters a nonlinearity associated with a failure mechanism. However, clipping tends to scatter power about the spectrum and into any valleys created by notch filters to compensate for shaker resonances, so at shaker resonances the test item may be overtested.

Some digital noise generators of the pseudorandom type generate non-Gaussian narrow-band signals which are combined to produce a Gaussian wide-band signal. But transmission of the vibration through the interior of a test item tends to restore narrow-band characteristics at the failure points, in this case making the distributions less rather than more Gaussian. Tests using such noise generators may still be useful, but theoretical analyses of failure may be in error.

Automatic equalization saves much effort and time. Up to a point, rapid action of the equalizer provides closer control over the excitation spectrum without producing any undesirable effects. However, excessive rapidity can smooth out many of the normal variations of the peaks so as to make the vibration non-Gaussian.

The test system as discussed above is for a single axis of vibration only. In principle, it would be better to provide controlled vibration along two (or even three) axes simultaneously. However, attempts to do this almost always introduce resonances in the shaker assembly and make spectral control at high frequencies more difficult.

SWEPT SINE WAVE TEST

Until the middle 1950s, the swept sine wave test was performed by powering an electrodynamic shaker by a multi-speed-range variable-speed motor-generator set under manual control by the operator. More recently, random vibration test systems were adapted for this purpose by the addition of a signal source in the form of a beat-frequency oscillator that is swept steplessly under automatic control throughout the frequency range of interest. With provision for manual control at the will of the operator, this can also be used very conveniently for analytic tests—the measurement of resonance frequencies of an equipment.

The swept sine wave test has one significant advantage over the random vibration test: The timing of a failure during a sweep may permit it to be associated with a particular frequency—the frequency of a resonance that otherwise may not be readily accessible for measurement. However, compared to random vibration, it is a decelerated test and tends to take longer to produce failures because it excites each resonance only briefly during a sweep. When it is used in an attempt to simulate random vibration, precise comparison of the test with the actual environment is difficult.

When a swept sine wave is confined to the lowest frequencies (e.g., 2 to 50 Hz), its use can be justified more on the basis of beneficial influence on design than on realistic simulation, even though it may have originated in an attempt to simulate low-frequency vibration of ships or land vehicles. The simplest way for the designer to avoid trouble from the test is to stiffen the device or equipment until its resonance frequencies are all above the upper frequency limit of the test. If this is successful, the equipment will undoubtedly also survive and perform in the actual environment. Serious penalties in weight, cost, or development time are unlikely.

Wide-frequency-range swept sine wave tests usually have more or less constant acceleration at high frequencies, a constant displacement amplitude of ¼ or ½ in. at the lowest frequencies. It is often assumed that such displacements are realistic representations of actual low-frequency vibration. This is not the case, but the precise value in test is unimportant unless the design includes an isolator. Either displacement amplitude is sufficient to ensure that any low-frequency resonances other than those of a mechanical isolator will be moved to higher frequencies. Any such resonance will have an amplification of at least 3, perhaps 10 or more. If excited by ¼-in. displacement because it is not moved to higher frequencies, it will respond with ¾ in., perhaps 2½ in. or more, unless there is a collision. If excited by ½-in. displacement, it will respond with 1½ in., perhaps 5 in. or more. All these responses are greater than the part spacing in typical compact aerospace equipment design. Their discovery in test would be by a violent rattling that would immediately bring the operation to a stop for test item redesign.

DWELL TEST

The *dwell test* is a constant-frequency constant-amplitude sinusoid applied at a resonance frequency of the test item. It requires little special test apparatus. It is sometimes used as a supplement to a low-frequency swept sine wave, in the event that a resonance in the test frequency range is discovered. It is always a low-frequency test. It is seldom a realistic representation of the actual environment.

The dwell test is perhaps the only example of a deliberate attempt to influence design by a vibration test, as opposed to attempting to simulate the actual environment and investigate design adequacy. It is dependent almost entirely on precedent, not environmental data. It is often a crude requirement, when isolators or shipping container cushions are involved, with unnecessary penalties for the test item designer. It would be more subtle and beneficial if its frequency were tied to spectral peaks of the *excitation* or to frequencies of resonance of the *structure* to which the test item will be mounted while in use, as opposed to frequencies of the test item.

SHOCK TEST

The U.S. Navy shock testing machines, mentioned in the earlier section on the *Shock Spectrum (Response Spectrum)* are described in detail in Chap. 26.

The U.S. Air Force shock testing machines for applying standard acceleration pulses, usually to somewhat lighter test items than those intended for shipboard use, consist of a *drop tower* for a mounting platform carrying the test item, with a decelerating mechanism under the platform or at the base. Early drop towers used a spring under the platform (for a half-sine wave pulse) or a sand pit at the base (for a square wave). The terminal-peak sawtooth pulse was first achieved by having the platform squash a conical-topped pure lead pellet.[2]

Increasing use of the shock spectrum (i.e., the response spectrum) to describe shock severity has been accompanied by increasing use of random vibration test apparatus, with an electronic shock synthesizer, for shock testing.[9] This has the advantages of permitting more precise control over the spectrum and permitting the performance of a shock test in the same test apparatus as in a vibration test with no time required for separate setup. On the other hand, the drop tower has the advantage of being cheaper and less expensive than a random vibration system.

FIXTURE PROBLEMS IN TEST

A fixture is used between a test item and a shaker armature or a shock tower platform if the mounting hole locations do not correspond, or to permit applying the excitation along a different axis of the test item. Any practical fixture used to change the direction of excitation is likely to have resonances in the test frequency range. In principle, any consequent spectral peaks and valleys of an electronic shaker system in a vibration or shock test can be flattened out by electronic equalization, but damping of the fixture makes this easier. In contrast, no such equalization is possible in a shock test using a drop tower. Damping of fixtures has been much neglected in shock and vibration engineering, but it may be the only option for the reduction of the drop-tower fixture effect if resonances cannot be raised higher than the test frequency range.

EQUIVALENCE AND FAILURE MECHANISMS IN TEST

The concept of equivalence between periodic vibration and random vibration, in the sense of producing the same damage or malfunction, was introduced to encourage random vibration testing by showing that no general equivalence, independent of the test item, can exist. First, according to the argument, a reference periodic acceleration amplitude, resulting in the same simple damped resonator rms response as a random vibration of power spectral density $W(f)$ at the same frequency, was derived. This reference acceleration is dependent on the properties of the test item, which are seldom known in advance. To obtain a realistic equivalent acceleration, the reference acceleration must be multiplied by a factor dependent on the assumed mechanism of failure or malfunction—which is also a property of the test item, not known in advance. Thus, an equivalent acceleration can be based only on assumptions that the shock and vibration engineer chooses to make about the test item.

Despite this, testing or test evaluation in terms of equivalence is still carried out on occasion, when there is no other choice. The contractor may not have random vibration test apparatus for a qualification test or money for it in the budget. An equipment qualified by swept sine wave testing may be proposed for use in a new random vibration environment. Funding, schedule time, or lack of test apparatus may discourage retest with random vibration. Numerous equivalence formulas, based on various assumptions about the test items, are available in the literature for such occasions.[10]

The various possible malfunctions of an equipment in test are varied and difficult to list or predict in advance. Therefore, for theoretical purposes, fracture or collision of a part in test is assumed to be equivalent to failure, even though an actual failure may more often be known by the malfunction it creates. For shock, a single stress beyond the tensile strength is assumed to produce failure even though, with most practical materials, failure results from rapid fatigue by transient response persisting after the shock excitation. Such persistence, as opposed to magnitude of early response, is more a matter of the damping of the test item than of the character of the shock.

Part failure under vibration is more clearly a result of fatigue. The most commonly used foundation for theoretical analysis of fatigue accumulation is *Miner's hypothesis* (see Fig. 42.28). This assumes that when a material is subject to repetitive stress of varying peak value, partial damage accumulates linearly, independent of the order of the cycles, until the process meets a limit predictable from a laboratory fatigue test of the material. A more fundamentally practical approach would provide for an initial random distribution of surface defects. However, fatigue tests of materials are usually carried out on polished specimens to achieve a degree of repeatability. Even so, the resulting data exhibit a wide scatter that does not clearly favor one fatigue theory over another.

ACOUSTIC TESTS

At its best, acoustic testing is *system* testing, performed, not on individual parts or equipment, but on airframes or sections of airframes, with equipment mounted and operating. With the airframe excited by noise, much as it would be in flight, and the equipment interacting dynamically with the structure, much as they would in flight, it provides somewhat more realistic vibration environments for the equipment than is feasible in individual equipment tests. It also permits observing functional interactions of equipment and system malfunctions that are not present in individual equipment tests.

This kind of testing requires high-intensity noise generators, rather than ordinary loudspeakers, to achieve sound pressure levels comparable to those of flight. Compressed air is forced through an electronically controlled valve, which varies the flow to produce acoustic noise, and into a horn, which couples the sound more efficiently into the space containing the test item. Since random acoustic noise is almost always used, the electronic apparatus required is similar to that for random vibration testing, except that relatively little output power capability is necessary to operate the valve.

For small test items (e.g., small missiles), a progressive wave is applied, generated by a high-intensity noise generator near one end of the test item, and propagated through a tube extending beyond the other end.

For larger test items, a reverberant field is applied, using a large concrete-walled rectangular chamber. High-intensity noise generators with large low-frequency horns act as sound sources at lower corners. To reinforce the high-frequency noise, there may be an array of high-intensity noise generators with smaller, more rapidly flaring, high-frequency horns, within the chamber and pointed at the test item. Since a reverberant field is the combined result of excitation of many acoustic resonances within the chamber, it does tend to exhibit some narrow band statistical variation in sound pressure level with frequency and position.

When shock and vibration engineers first became aware of the acoustic environment, some papers appeared purporting to show that parts such as vacuum tubes should be tested in acoustic noise as well as vibration—the two environments produced different failures. In the exploratory comparison tests, however, the mounting conditions were seldom kept the same, since the two environments were thought to be independent. More fundamentally, in the absence of data on the actual part environments, either type of test is necessarily a crude simulation and using both cannot result in a clear refinement of part testing.

There has been a persistent tendency to treat the acoustic and mechanical environments as fundamentally independent. For example, it is common for specifications to require acoustic as well as mechanical testing of equipment if the actual sound pressure to be simulated is above a prescribed level. This is a simple criterion to apply, but otherwise it has little merit. A better criterion would be the measured relative sensitivity to noise and vibration, with the equipment mounted on a shaker and excited by a nearby loudspeaker.

ANALYTIC TEST

Analytic tests of equipment are carried out, usually using periodic vibration, at the option of the engineer (if his general program is approved), and not in accordance with any specification, to determine equipment properties as an aid to further development. In other words, they are measurements of resonance frequency (and damping) either directly on the device or, to gain schedule time, on an advance model or mock-up with similar dynamic properties. For such measurements it is important to have simple procedures that can be carried out with little cost and minimal instrumentation—to be supplemented on occasion by more sophisticated approaches, such as modal analysis (Chap. 21), as appropriate. It is important to have simple and direct approaches to design decision-making based on the data—otherwise the value of taking the data will be questioned.

Essentially no shock or vibration engineering was applied during World War II to the new electronic equipment that played so crucial a role. Good engineering practice was very nearly sufficient for the kinds of equipment then in use. Therefore, the appropriate objective for shock and vibration engineering, even with missile airborne equipment, is not to be exhaustively scientific or to be comprehensive in all measurements and decision-making processes, but to provide the most improvement in reliability for the effort, funding, and time expended, with minimal penalties in weight and in production cost.

Resonance frequencies and damping of equipment should not be measured by use of the signal from a single sensor—there must always be a comparison with the signal from another sensor closer to the source of excitation, usually close to a mounting point.[11] If the latter sensor is an accelerometer, the resonance frequencies measured will be those of the equipment as mounted to an infinite rigid mass. (If, on the other hand, this sensor is a force gage, or strain gage, fastened to a coupling member whose dynamic property is a simple stiffness, the resonance frequencies will be those of the equipment as mounted to an infinitely light and flexible source. For example, to measure the resonance frequencies of a missile as in flight, a force gage should be used at the excitation point in the laboratory.) Use of the excitation sensor at an appropriately chosen point eliminates any shaker resonances from the measurements.

For greatest sensitivity in the measurement of resonances by periodic excitation, both magnitude and phase comparisons of excitation and response should be utilized,[11] as, for example, by observing Lissajous figures on an oscilloscope screen as the excitation frequency is slowly varied.

It is best if the response sensor can be hand-held, moved from one point to another according to the interest of the operator, and oriented at will to measure response in various directions. Interior points of the equipment or mock-up can be reached by drilling several small holes in the equipment case, making a vibration probe by attaching a stiff pointed wire to an accelerometer, and inserting the wire through the middle of a hole. Such a sensor should be calibrated with the wire attached.[12] It is not necessary to perform an absolute calibration. It is merely necessary to verify the absence of probe resonances in the frequency range in which the probe is being used.

Should two or more resonances along a transmission path within a piece of equipment be found to be tuned to approximately the same frequency, large endpoint responses to both vibration and shock are likely, especially if (as is usually the case) the progression along the path is from heavy to lighter masses. Failure or malfunction is very likely, but the design defect can be corrected by detuning—by changing a stiffness or mass or adding a strut, usually without any serious weight penalty. Therefore, in equipment resonance measurements, special emphasis should be placed on looking for frequencies that coincide. Any frequency coincidence found should result in a violent motion of the final mass during the shock or vibration test. Analytic testing can be carried out in a simple way at minimum cost and can lead to simple redesign decisions. More subtle recommendations can be considered as the designer and the shock and vibration engineer gain experience in working together. At the present time, redesign after test is usually limited to stiffening a failed part, which may, however, tune the part more closely to the transmission path and increase its response.

SPECIFICATIONS AND THEIR ROLE

Specifications are documents, issued by the customer or contractor or both, that cover particular criteria for acceptability of what a contractor develops or produces under contract, or criteria for tests used to decrease the failure rate. Ultimately, the customer is usually a branch of the government such as the Navy, the Air Force, or the National Aeronautics and Space Administration. In development or production for commercial sale, the incentive of a free market is dominant and specifications of the types discussed in this chapter play at most a minor role. They are necessary as a protection to the customer when such a market does not exist—in other words when there is only one primary customer, who must contract for development and production. Maximizing the benefits of specifications and minimizing their cost and schedule penalties remain a continuing challenge.

Most contracts for development or production of a device contain a specification of the functional properties the device must have, as well as its weight and size and any power requirements, and a schedule and funding allotment. If an environmental specification is not initially part of the contract, it is soon added or becomes a supplementary document binding on the contractor.

A contract is usually a result of a "Request for Proposal" (RFP) or a "Request for Quotation" (RFQ) prepared by the customer which contains preliminary specifications, proposals prepared in response by one or more prospective contractors, and a negotiation between the customer and the winning contractor. For the development phase, there are usually several contractors competing for a contract; for production, there may at first be only one. The customer may change the specification requirements after choosing the winning contractor, but no contractor is privileged to change any requirement while preparing a proposal. Otherwise that contractor is judged unresponsive to the RFP and eliminated from the competition. In the proposal the contractor must agree to perform in accordance with all requirements, environmental and otherwise. If there is concern about the suitability of a requirement, the most the contractor can do safely is to propose an alternate approach for possible discussion later during contract negotiation or after being awarded the contract.

Some detailed design considerations, but *not design procedures*, are sometimes covered by specifications. For example, some materials, or the use of castings for certain parts, may be prohibited. But specifying a design procedure tends to stifle engineering competence and leaves the customer in the awkward position of taking the responsibility for the contractor's engineering. Likewise, the way in which the contractor organizes for development or production is seldom a subject for specification.

An *environmental specification* is a specification that describes the environmental conditions required for environmental test, without becoming deeply involved in details of testing.

A *test specification* is a comprehensive document prepared by the contractors. It may cover preparation of the test item, identification of all test equipment and instrumentation, description of any test fixtures, instructions for mounting any accelerometers or other sensors, step-by-step procedures for operating the test item (if operation is required) and the test apparatus, procedures for taking data on test item function and on applied environment, and performance acceptability criteria.

Several contracting branches of the U.S. government issue general environmental specifications for particular classes of equipment, often with choices of test environment according to conditions of use—e.g., the type of vehicle or airplane in which it will be installed or whether the test item is to be stored under a wing. New issues of such general specifications are negotiated in effect by a large committee with a customer official as chairperson with veto power. The engineers that will be most directly affected by the technical society with a major interest in the outcome are privileged to form their own committee to process and transmit recommendations.

At the time a development contract is awarded, there is a final negotiation to determine a specific environmental specification for the particular items to be developed. The engineers with the most direct responsibility for this may be the same ones that do the testing, but the equipment design engineers affected by the result have their say. Approval is required by both contractor and customer technical managers. Appropriate sections of any applicable general specifications may be referred to or quoted verbatim.

Missile programs and space programs that grew out of them are concerned with nonrecoverable hardware and critical reliability challenges. In respect to shock and vibration, the engineers have more freedom than in other programs to select their own test conditions, based on whatever data they have or whatever estimates they can make of the actual environment. The innovations in specifications and testing tend to take place first in these programs. Once an environmental specification for a particular program has been negotiated and approved, and has in effect become part of the contract, test specifications are negotiated. Design engineers have more say in the monitoring of any equipment malfunctions under environmental test. Test engineers have more say in the test methods. Approval by contractor program management is mandatory. As the tests take place, the results are monitored by customer representatives.

Shock and vibration engineers seldom participate directly in the *design* of the equipment to be tested.[11] Structural engineering has, in the natural course of time, incorporated some shock and vibration engineers into its design process. But shock and vibration engineering of equipment was created by environmental specifications. These deal only with testing. A development or production plant seldom has more than one environmental laboratory, and there is some feeling that test engineers should be impartial about the equipment they test—separated from the design engineers. Shock and vibration engineers are usually separated from equipment design engineers by at least one organizational barrier. Most shock and vibration papers that deal with design or appear to have design application are too complicated for negotiation across such barriers.

An environmental specification, by virtue of its special relationship to contracts, is in part a *legal document.* This fact, along with the difficulty of performing a scientific evaluation of test methods and the peculiar organizational constraints on the shock and vibration engineer, explains in part the major role that precedent plays in environmental test requirements.

More important, an environmental specification is a *technical management document* controlling indirectly how personnel are assigned, organized, and utilized.[11] Though such factors are not mentioned explicitly in a specification, they should not be neglected in evaluating its merit. The best specification is the one that gets the best performance from the contractor without unnecessary restrictions. Most specifications attempt to prescribe a simulation of the actual environment in which an equipment will be used or transported, but ease of test and whether test conditions have an important relationship to acceptable design are also important considerations.

Shock and vibration testing started with the unpretentious objective of obtaining a few design improvements, but there has been a strong tendency to regard environmental specifications as instruments for obtaining absolute proof of design adequacy. This disregards the inherent limitations of data acquisition and testing, which make it necessary to allow for some waivers and deviations to be negotiated between contractor and customer. It also disregards the fact that test conditions are established at the beginning of each program, when data on the actual environment are least reliable, and are difficult to change thereafter. Insistence on perfection in testing can lead to endless complications without proportionate reward in design improvement. The customer must have definite criteria for acceptance of developed or produced equipment, but these criteria cannot be made absolutely risk-free. It would be better if each specification contained a disclaimer at the beginning, stating that compliance with the requirements, though mandatory, will be interpreted by the customer only as *partial evidence* of design adequacy, along with any shock and vibration design efforts carried out and reported on by the customer.

STATISTICAL RELIABILITY TESTING

Specifications for statistical reliability testing (see the section on *Testing*), with numbers of test items selected according to statistical rules, represent the most extreme attempt at irrefutable proof of design adequacy to date. While they were not approved as originally proposed, they illustrate the complications that are inevitable in a rigid policy of verification.

In the middle of the 1950s, it was noted that (for failures with statistically independent failure rates) the reliability of a military system in use is the product of the reliabilities of its subsystems, components, or parts. This came to be known as the *product law.* Its use in assignment of subsystem and component reliability objectives, and statistical design of tests to verify that these objectives are met, appeared to offer a needed new approach to missile reliability.

To illustrate the product law, if a system with 10 components is to have a reliability of 0.9 (90 per cent, or one failure in tests of 10 units off the production line), the component reliabilities must be on the order of 0.99 (99 per cent, or one failure in tests of 100 components off the production line)—much closer to perfection. But making and testing 100 components gives only an even chance of observing one failure. More nearly 1,000 items and 1,000 tests are required, according to the assumptions underlying the product law, to obtain a statistically significant estimate of the component reliability. Specifications for qualification test often require only one test of one item.

Reliability objectives assigned on the basis of the product law cost little and are beneficial reminders of the risks of complexity in design. They were incorporated immediately into missile programs, even though environmentally induced failures are not statistically independent. Use of the objectives has since spread to other types of programs.

Reliability verification according to the product law at assembly levels much below that of a complete missile required many more components than were scheduled for assembly into missiles or for use as spares, for delivery to the customer. In practice, qualification tests, with more moderate demands on production for test, provided more information on design adequacy and design defects than appeared possible from statistical considerations. Statistical reliability testing was not approved for missile subsystems or components. More recently, it has found practical application with computer chips which have more random failures and are made in large quantities at low cost.

In the meantime, statistical reliability testing, with some modification of approach, found some acceptance for avionic equipment. Brief tests of many test items were replaced by a long test of one or two items. Interest shifted from measuring reliability to measuring the *mean time between failures* (*MTBF*). At first, the prescribed shock and vibration conditions were too mild to contribute to failure in test,[13] but this has been changed in the latest specification. Specifications for statistical reliability testing have become important to shock and vibration engineers but remain separate from specifications for qualification testing.

The original statistical approach to missile reliability focused on the mathematics of random failures to the exclusion of physical mechanisms of environmentally induced failure and on statistical error in test almost to the exclusion of systematic error. But for an infallible verification of design adequacy, systematic error must be no greater than random error. Accordingly, for infallible verification at levels of assembly below that of major airborne equipment for the missile, knowledge of the actual shock and vibration environments and the capability to simulate them must improve rapidly with the level of disassembly. In practice, the opposite is true. The environments are measured or *estimated* at the interfaces between the airframe and the equipment. Transmission to subassemblies is seldom investigated, and usually little is known about the interior dynamics of the equipment.

For equipment having systematic as well as random failure modes, establishment of a more realistic statistical law than the product law, like establishment of a completely defensible equivalence between periodic and random vibration, is dependent on extensive knowledge of the test item, which is better used to eliminate failure modes than to estimate better combinatorial rules.

Beyond some point, reliability verification competes with reliability attainment for funds, workers, and time.[11]

SCREENING SPECIFICATIONS

The philosophy of screening (see the section on *Testing*), like that of statistical reliability testing, assumes some randomness in failure modes, but goes further and assumes a time dependence—a failure rate decreasing initially with time, that is verifiable to some degree by test results. It amounts almost to a reversal of specification and management policy. It is applied not with the hope of avoiding failure but with the intent of producing failures[14–18] whether or not this causes the designers discomfort. In the absence of failures, the test conditions must be made more severe or the testing plan must be discontinued as useless. Screening is applied, not to verify reliability, but to improve the reliability of equipment to be delivered to the customer for use, by producing and correcting the initial failures and malfunctions in the laboratory rather than in the field. Since screening tests run counter to the expectations of people familiar primarily with other kinds of environmental testing, any specification for screening must make the assumptions and policy quite clear or they will drift away from their intended purpose.

Shock and vibration conditions for screening usually have the same severity as in qualification tests, but not necessarily. They could be less severe—or even more severe, depending on the failure rate produced. The desired failure rate, not realism, is the primary criterion for choice of severity.

SPECTRAL ENVELOPING

Shock and vibration spectra for specifications cannot contain all the detail of the spectra of actual environments, partly because reliable data are seldom available at the beginning of a development program and partly because spectral detail complicates the specifications and test procedures. Usually, an attempt is made to prescribe an upper envelope of the actual spectra, filling in the valleys.

As a development or production progresses, two risks emerge. There is a risk to the customer of undertest at a few frequencies. One way of dealing with this without renegotiating approved test conditions or unduly complicating the test procedures is to prescribe dwell tests *at frequencies of peak excitation* of the use environment, not at resonance frequencies of the test item. There is a risk to the contractor of overtest at a few frequencies. The best way of dealing with this is by waivers and deviations requested by the contractor for possible approval by the customer's representatives. Redesign because of overtest in turn costs the customer in dollars and schedule.

SUBASSEMBLY TEST CONDITIONS

Customer general specifications, and specifications agreed on by the contractor and customer at the beginning of the development, are directly applicable to particular equipment, not to their subassemblies. However, it is often considered desirable to have shock and vibration tests for subassemblies or even parts before they are incorporated into equipment, especially when they are developed and supplied by another vendor. Commonly, test conditions for equipment qualification tests are transcribed verbatim to tests for subassemblies. While this provides some ensurance that a subassembly will survive and operate when the complete equipment is subject to test, it completely ignores the resonant character of vibration and shock transmission. Whatever guidance it gives to subassembly designers tends to be misleading. Sometimes the test severity for subassemblies is increased over that used for the complete equipment, for better ensurance of reliability after the subassemblies are incorporated into the completed equipment. Such measures may be all that can be done routinely at the beginning of a development program, but by themselves they are inefficient.

The most important basis for improved design in such situations will be an exchange of information between equipment and subassembly designers about resonance frequencies. As the first few resonance frequencies of the equipment become known, it will be beneficial to supplement the initial subassembly specifications by dwell tests at these frequencies. A negotiation should then be carried out to decide whether equipment resonances or subassembly resonances should be changed to effect any necessary detuning. A similar strategy may sometimes also be beneficial for complete equipment in relation to the structure to which it will be mounted while in use.

SPECTRAL TOLERANCES

In the negotiation of environmental conditions for specifications, the customer has the greater responsibility for adequacy of test conditions, whereas the contractor has the greater responsibility to avoid overtest. In the first environmental specification for the USAF ICBM programs, the prescribed test spectra were minimum test conditions.[3] It was left to the contractors and in particular to the test engineers to negotiate within their companies any maximum vibration and shock environments for the test and to renegotiate within their companies in the event that these conditions proved to be so low as to result in too much constraint on the adjustment of the test apparatus or so high as to result in excessive overtest.

The more usual practice is to prescribe in the environmental specification a nominal spectrum with upper and lower tolerances. The upper tolerances define the maximum test conditions. It is convenient for the test engineer to know the upper tolerances from the beginning, without negotiation. However, if a test item is accidentally tested to an environment beyond the upper tolerance, but passes test, a management problem arises. Although the test item design has not been shown to be inadequate and retest to a less severe environment accomplishes nothing, the customer representative, adhering to the letter of the contract, will order a retest.[11]

REFERENCES

1. Morrow, C. T., and R. B. Muchmore: *J. Appl. Mech.*, **22:**366 (1955).
2. Morrow, C. T., and H. I. Sargeant: *J. Acoust. Soc. Amer.*, **28:**959 (1956).
3. Morrow, C. T.: *Shock and Vibration Bull.*, **45,** *part 5, 191 (June 1975).*
4. Southworth, H., Jr.: *Shock and Vibration Bull.*, **29,** part 4, 408 (1951).
5. Morrow, C. T.: *Shock and Vibration Bull.*, **52,** part 4, 37 (1982).
6. Morrow, C. T.: *Shock and Vibration Bull.*, **53,** part 1, 97 (1983).
7. Morrow, C. T.: *Shock and Vibration Bull.*, **54,** part 2, 131 (1984).
8. Cooley, J. W., and J. W. Tukey: *Math. Comput.*, **19:**297 (April 1965).
9. Favour, J. D., J. M. LeBrun, and J. P. Young: *Shock and Vibration Bull.*, **40,** part 2 (1982).
10. Fackler, W. C.: "Equivalence Techniques for Vibration Testing," Shock and Vibration Monograph 9, The Shock and Vibration Information Center, 1972.
11. Morrow, C. T.: "The Environmental Specification As a Technical Management Tool," The Shock and Vibration Information Center, 1981.
12. Morrow, C. T.: *J. Acoust. Soc. Amer.*, **20:**826 (1948)
13. Swet, B. H.: *Shock and Vibration Bull.*, **45,** part 2, 29 (1975).
14. Department of the Navy: NAVMAT P-9492, "Navy Manufacturing Screening Program," 1979.
15. Tustin, W.: *Sound and Vibration*, **6:** (1984).
16. Hobbs, G. K.: *Sound and Vibration*, **16:** (1984).
17. Hobbs, G. K.: *Sound and Vibration*, **5:** (1985).

21

EXPERIMENTAL MODAL ANALYSIS

Randall J. Allemang
University of Cincinnati

David L. Brown
University of Cincinnati

INTRODUCTION

Experimental modal analysis is the process of experimentally determining the modal parameters (frequencies, damping factors, and modal vectors) of a linear, time-invariant system. The modal parameters may be determined by analytical means, such as finite element analysis. A common reason for using experimental modal analysis is to verify the results of the analytical approach. Often, an analytical model does not exist and the modal parameters determined experimentally serve as the model for future evaluations, such as structural modifications. Predominantly, experimental modal analysis is used to explain a dynamics problem (vibration or acoustic) whose solution is not obvious from intuition, analytical models, or previous experience.

Experimental modal analysis began in the 1940s with work oriented toward measuring the modal parameters of aircraft so that the problem of flutter could be accurately predicted. At that time, transducers that measured dynamic force were primitive and the analog nature of the approach yielded a time-consuming process that was not practical for most situations. With the advent of digital minicomputers and the fast Fourier transform (FFT) in the 1960s, the modern era of experimental modal analysis was initiated. Today, experimental modal analysis represents an interdisciplinary field that makes use of aspects of signal processing, mechanics, vibrations, acoustics, control theory, and applied mathematics.

EXPERIMENTAL MODAL ANALYSIS OVERVIEW

The process of determining modal parameters from experimental data involves several phases. The success of the experimental modal analysis process depends upon (1) an understanding of the basis for each phase and (2) the establishment of specific goals for the test situation. Every phase of the process is affected by the goals which are established, particularly with regard to the reduction of errors associated with that phase. One possible delineation of these phases is:
1. Theoretical modal analysis
2. Experimental modal analysis
3. Modal data acquisition
4. Modal parameter estimation
5. Modal data presentation

Modal analysis theory refers to that portion of classical vibration theory that explains the existence of natural frequencies, damping factors, and modal shapes for linear systems. This theory includes both lumped-parameter, or discrete, models as well as continuous models. This theory also includes real (normal) modes as well as complex modes of vibrations as possible solutions for the modal parameters. Experimental methods of modal analysis involve the relationship between measured quantities and classical vibration theory, often represented as matrix differential equations. All modern methods of modal analysis are derived from the matrix differential equations but yield a final mathematical form in terms of measured data. These measured data can be raw input and

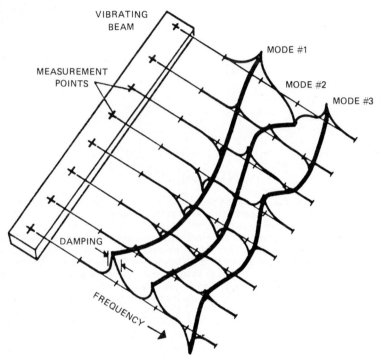

FIG. 21.1. Experimental modal analysis example. Since the modal frequency and modal damping, associated with each modal vector, do not change as a function of the frequency-response function location, the modal vectors can be estimated from the imaginary part of the frequency-response function evaluated at the modal frequency.

output data (in the time domain or frequency domain) or some form of processed data, such as impulse-response or frequency-response functions. Most methods involve processed data such as frequency-response functions in the estimation of modal parameters. The experimental modal analysis method establishes the form of the data that must be acquired.

The *modal data acquisition phase* involves the acquisition of data required to serve as input to the modal parameter estimation phase. Therefore, much care must be taken to ensure that the data match the requirements of the theory as well as the requirements of the numerical algorithm involved in the modal parameter estimation. The theoretical requirements involve concerns such as system linearity and time invariance of system parameters. The numerical algorithms are particularly concerned with the bias errors in the data as well as overall dynamic range considerations.

The *modal parameter estimation phase* is concerned with the practical problem of estimating the modal parameters, which are based upon a choice of mathematical model as justified by the experimental modal analysis method, from the measured data. Problems which occur at this phase most often arise from violations of assumptions used to justify previous phases. Serious theoretical problems such as nonlinearity cause serious problems in the estimation of modal parameters and may be reason to invalidate completely the experimental modal analysis approach. Serious practical problems, such as bias errors resulting from the digital signal processing, cause similar problems but are a function of data acquisition techniques which can be altered to minimize such errors. The modal parameter estimation phase is that point in the experimental modal analysis process where the errors of all previous work accumulate.

Modal data presentation represents the final phase of the process; it provides a physical interpretation of the modal parameters. It may simply be the numerical tabulation of the frequency, damping, and modal vectors, along with the associated geometry of the measured degrees-of-freedom. More often, modal data presentation is concerned with the plotting and animation of such information. This involves the additional information required to construct a three-dimensional representation of the test object. Primarily, this requires a display sequence which involves the order in which the degrees-of-freedom will be connected. The plotting and animation of the modal vectors may involve hidden line calculations which require surface definition as well.

Figure 21.1 illustrates all phases of the process. In this example, the first few modes of vibration of a continuous beam are evaluated. According to modal analysis theory, this is a linear system, so the modal vectors of this system should be real modes. The experimental modal analysis method used here is based upon the relationships of the frequency-response function to the matrix differential equations of motion. The imaginary part of the frequency-response function for each measured response degree-of-freedom and a common input degree-of-freedom is superimposed perpendicular to the beam. The modal data acquisition in this example involves the estimation of frequency-response functions for each degree-of-freedom shown. The frequency-response functions are complex-valued functions—only the imaginary portion of each function is shown. One method of modal parameter estimation suggests that for systems with light damping and widely spaced modes, the imaginary part of the frequency-response function, at the damped natural frequency, may be used as an estimate of the modal coefficient for that response degree-of-freedom. The damped natural frequency can be identified as the frequency of the positive and negative peaks in the imaginary part of the frequency-response functions. The damping can be estimated from the sharpness of the peaks. In this way, the modal parameters are estimated. Modal data presentation for this example is shown as the lines connecting the peaks. While animation is possible, a reasonable interpretation of the modal vector can be gained simply by plotting.

MODAL ANALYSIS THEORY

The application of modal analysis theory involves a thorough understanding of how the structural parameters of mass, damping, and stiffness relate to the impulse-response function (time domain), the frequency-response function (Fourier domain), and the transfer function (Laplace domain) for single and multiple degree-of-freedom systems.

SINGLE DEGREE-OF-FREEDOM SYSTEMS

A single degree-of-freedom system as presented and evaluated in the time, frequency (Fourier), and Laplace domains serves as the basis for many of the models that are used in modal parameter estimation. This single degree-of-freedom approach is trivial for the modal analysis case, but it is important because the multiple degree-of-freedom case can be viewed as simply a linear superposition of single degree-of-freedom systems.

The general mathematical representation of a single degree-of-freedom system is given by

$$m\ddot{x} + c\dot{x} + kx = f(t) \tag{21.1}$$

where m = mass constant
     c = damping constant
     k = spring constant
     $f(t)$ = force applied to mass m

and is represented schematically in Fig. 2.12. By setting $f(t)$ = 0, the homogeneous form of Eq. (21.1) can be solved [see Eqs. (2.13), (2.15), and (2.17)]. The general solution of Eq. (21.1) is

$$x(t) = A \exp{(p_1 t)} + B \exp{(p_2 t)} \tag{21.2}$$

where A and B are constants and the values of p_1 and p_2, for an underdamped system, are given by

$$p_r = \sigma_r + j\omega_r \tag{21.3}$$

where σ_r is the damping factor in radians per second and ω_r is the damped natural frequency in radians per second. For most real structures, unless active damping systems are present, the fraction of critical damping is rarely greater than 10 per cent. For such underdamped systems, the two roots, p_1 and p_2, are always complex conjugates and are referred to as the *poles of the system*. Also, the two coefficients, A and B, are complex conjugates of one another.

TIME DOMAIN: IMPULSE-RESPONSE FUNCTION. The impulse-response function of the single degree-of-freedom system can be determined from Eq. (21.2) assuming that the initial conditions are zero and that the system excitation, $f(t)$, is a unit impulse. The response of the system, $x(t)$, to such a unit impulse is known as the *impulse-response function, $h(t)$*, of the system. Therefore,

$$h(t) = A \exp{(p_1 t)} + A^* \exp{(p_1^* t)} \tag{21.4}$$

$$h(t) = \exp{(\sigma_1 t)}[A \exp{(+j\omega_1 t)} + A^* \exp{(-j\omega_1 t)}] \tag{21.5}$$

where p_1 = complex pole of system = $\sigma_1 + j\omega_1$
     σ_1 = damping factor
     ω_1 = damped natural frequency
     p_1^* = complex conjugate of pole p
     A = complex residue
     A^* = complex conjugate of residue A

Thus, the residue A controls the amplitude of the impulse response, the real part of the pole is the decay rate, and the imaginary part of the pole is the frequency of oscillation.

FREQUENCY DOMAIN: FREQUENCY-RESPONSE FUNCTION. Equation (21.1) is the time-domain representation of the system in Fig. 2.12. An equivalent equation of motion may be determined for the Fourier or frequency domain. This representation has the advantage of converting a differential equation to an algebraic equation. This is accomplished by taking the Fourier transform of Eq. (21.1). Thus, Eq. (21.1) becomes

$$[-m\omega^2 + jc\omega + k]X(\omega) = F(\omega) \tag{21.6}$$

Let $B(\omega) = -m\omega^2 + jc\omega + k$. Then Eq. (21.6) becomes

$$B(\omega)X(\omega) = F(\omega) \tag{21.7}$$

Eq. (21.7) is an equivalent representation of Eq. (21.1) in the Fourier domain. According to Eq. (21.14), the system response $X(\omega)$ is directly related to the system forcing function $F(\omega)$ through the quantity $B(\omega)$. If the system forcing function $F(\omega)$ and its response $X(\omega)$ are known, $B(\omega)$ can be calculated by

$$B(\omega) = \frac{F(\omega)}{X(\omega)} \tag{21.8}$$

If the input $F(\omega)$ is known,

$$X(\omega) = \frac{F(\omega)}{B(\omega)} \tag{21.9}$$

By defining $H(\omega) = 1/B(\omega)$, Eq. (21.9) becomes

$$X(\omega) = H(\omega)F(\omega) \tag{21.10}$$

The quantity $H(\omega)$ is known as the *frequency-response function* of the system. A frequency-response function relates the Fourier transform of the system input to the Fourier transform of the system response. From Eq. (21.10), the frequency-response function is defined as

$$H(\omega) = \frac{X(\omega)}{F(\omega)} \tag{21.11}$$

From Eq. (21.6), the frequency-response function can be written as

$$H(\omega) = \frac{1/m}{-\omega^2 + j(c/m)\omega + (k/m)} \tag{21.12}$$

The denominator of Eq. (21.12) is the *characteristic equation* of the system. The characteristic values of this complex equation usually are complex even though the equation is a function of a real-valued independent variable ω. The characteristic values of this equation are known as the *complex roots* of the characteristic equation or the *complex poles* of the system.

The frequency-response function $H(\omega)$ can now be rewritten as a function of the complex poles.

$$H(\omega) = \frac{1/m}{(j\omega - p_1)(j\omega \quad p_1{}^*)} \tag{21.13}$$

where $p_1 = \sigma_1 + j\omega_1$ is the complex pole and $p_1{}^* = \sigma_1 - j\omega_1$. Since the frequency-response function is a complex-valued function of a real-valued independent variable ω, the frequency-response function as shown in Fig. 21.2, is represented by a pair of curves.

LAPLACE DOMAIN: TRANSFER FUNCTION. Just as in the previous case for the frequency domain, the equivalent information can be presented in the Laplace domain by way of the Laplace transform [see Eq. (8.9a)]. The only significant difference in the development concerns the fact that the Fourier transform is defined from negative infinity to positive infinity while the Laplace transform is defined from zero to positive infinity with initial conditions. The Laplace representation, also, has the advantage of converting a differential equation to an algebraic equation. The development begins by taking the Laplace transform of Eq. (21.1).

$$[ms^2 + cs + k]X(s) = F(s) + (ms + c)X(0) + mX'(0) \tag{21.14}$$

Note that $X(0)$ and $X'(0)$ are the initial displacement and velocity at time $t = 0$.
If the initial conditions are zero, Eq. (21.14) becomes

$$(ms^2 + cs + k)X(s) = F(s) \tag{21.15}$$

Let $B(s) = ms^2 + cs + k$. Then, Eq. (21.15) becomes

$$B(s)X(s) = F(s) \tag{21.16}$$

Therefore, the transfer function can be defined in the same way that the frequency-response function was defined previously.

$$X(s) = H(s)F(s) \tag{21.17}$$

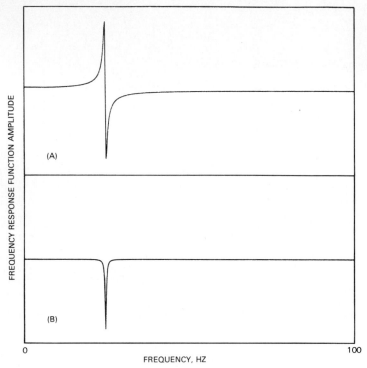

Fɪɢ. 21.2. Frequency-domain representation of a single degree-of-freedom frequency-response function. *(A)* Real part. *(B)* Imaginary part.

The quantity $H(s)$ is defined as the *transfer function* of the system which relates the Laplace transform of the system input to the Laplace transform of the system response. From Eq. (21.17), the transfer function can be defined as

$$H(s) = \frac{X(s)}{F(s)} \tag{21.18}$$

From Eq. (21.15), the transfer function can be written as

$$H(s) = \frac{1/m}{s^2 + (c/m)s + (k/m)} \tag{21.19}$$

This equation is valid under the assumption that the initial conditions are zero. The denominator term is called the *characteristic equation* of the system.

The transfer function, $H(s)$, can be rewritten as

$$H(s) = \frac{1/m}{(s + p_1)(s - p_1{}^*)} \tag{21.20}$$

Since the transfer function is a complex-valued function of a complex independent variable s, the transfer function is represented, as shown in Fig. 21.3, as a pair of surfaces. The variable s in Eq. (21.20) is a complex variable; i.e., it has a real part and an imaginary part. Therefore, it can be viewed as a function of two variables which represent a surface.

The definitions of undamped natural frequency, damped natural frequency, damping

(A)

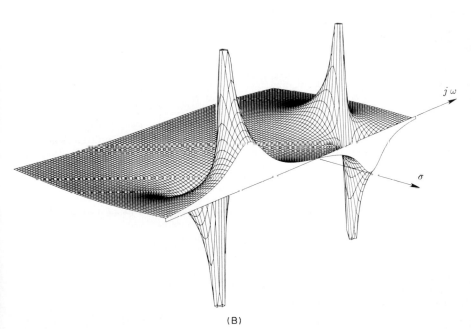

(B)

FIG. 21.3. Laplace-domain representation of a single degree-of-freedom transfer function. *(A)* Real part. *(B)* Imaginary part.

factor, fraction of critical damping, and residue are all relative to the information represented by Fig. 21.3. The projection of the pole locations onto the plane of zero amplitude yields the damping coefficient, damped natural frequency, undamped natural frequency, and the fraction of critical damping.

Equation (21.20) can be expressed in terms of partial fractions.

$$H(s) = \frac{1/m}{(s - p_1)(s - p_1{}^*)} = \frac{A}{(s - p_1)} + \frac{A^*}{(s - p_1{}^*)} \tag{21.21}$$

The residues of the transfer function are defined as being the constants A and A^*. The importance of the residues is that the residues are proportional to the modal coefficients of the system. The residues of the transfer function are also directly related to the amplitude of the impulse-response function. In general, the residue A can be a complex quantity. As shown for a single degree-of-freedom system, A is purely imaginary.

The Laplace transform formulation is the general case of the Fourier transform development if the initial conditions are zero. The frequency-response function is the part of the transfer function evaluated along the $s = j\omega$ axis.

From an experimental point of view, the complete transfer function cannot be estimated from the measured time-domain data. The frequency-response function is actually estimated from the measured time-domain data by means of the fast Fourier transform.

MULTIPLE DEGREE-OF-FREEDOM SYSTEMS

Modal analysis concepts are applied when a continuous, nonhomogeneous structure is described as a lumped-mass multiple degree-of-freedom system. Then, the modal frequencies, the modal damping, and the modal vectors, or relative patterns of motion, can be found by means of an estimate of the mass, damping, and stiffness matrices or from the measurement of the associated frequency-response functions. The two degree-of-freedom system shown in Fig. 21.4, the most basic example of a multiple degree-of-freedom system, is useful in discussing modal analysis concepts since a theoretical solution can be formulated in terms of the mass, stiffness, and damping matrices or in terms of the frequency-response functions.

The equations of motion for the system in Fig. 21.4, using matrix notation, are

$$\begin{bmatrix} m_1 & 0 \\ 0 & m_2 \end{bmatrix} \begin{Bmatrix} \dot{x}_1 \\ \ddot{x}_2 \end{Bmatrix} + \begin{bmatrix} c_1 + c_2 & -c_2 \\ -c_2 & c_2 + c_3 \end{bmatrix} \begin{Bmatrix} x_1 \\ \dot{x}_2 \end{Bmatrix} + \begin{bmatrix} k_1 + k_2 & -k_2 \\ -k_2 & k_2 + k_3 \end{bmatrix} \begin{Bmatrix} x_1 \\ x_2 \end{Bmatrix} = \begin{Bmatrix} F_1 \\ F_2 \end{Bmatrix}$$

$$\tag{21.22}$$

The development of the frequency-response function solution for the multiple degree-of-freedom case is described in a later section on the *Frequency-Response Function Method.* This development relates the mass, damping, and stiffness matrices to a transfer-function model involving multiple degrees-of-freedom. Since the transfer-function model can be considered as the general case of the frequency-response function, this demonstrates that the same modal information can be determined whether the mass, damping, and stiffness matrices are known or whether the associated frequency-response functions are measured. Just as in the analytical case where the ultimate solution can be described in terms of one degree-of-freedom systems, the frequency-response functions between any input and response degree-of-freedom can be represented as a linear superposition of the single degree-of-freedom models derived previously.

As a result of the linear superposition concept,

FIG. 21.4. Rigid-body representation of a multiple degree-of-freedom system. For this two degree-of-freedom system, m represents a lumped mass, c represents a lumped damper, k represents a lumped stiffness, x represents the displacement of the mass, and f represents the externally applied force. Subscript notation is used to indicate that the various parameters are unique and independent.

the equations for the impulse-response function, the frequency-response function, and the transfer function for the multiple degree-of-freedom system are defined as follows:

Impulse-response function:

$$h_{ik}(t) = \sum_{r=1}^{n} A_{ikr} \exp(p_r t) + A_{ikr}^* \exp(p_r^* t) \qquad (21.23)$$

Frequency-response function:

$$H_{ik}(\omega) = \sum_{r=1}^{n} \frac{A_{ikr}}{j\omega - p_r} + \frac{A_{ikr}^*}{j\omega - p_r^*} \qquad (21.24)$$

Transfer function:

$$H_{ik}(s) = \sum_{r=1}^{n} \frac{A_{ikr}}{s - p_r} + \frac{A_{ikr}^*}{s - p_r^*} \qquad (21.25)$$

where t = time variable
  ω = angular frequency variable
  s = Laplace variable
  i = measured degree-of-freedom (response)
  k = measured degree-of-freedom (input)
  r = modal vector number
  A_{ikr} = residue
  $A_{ikr}^* = Q_r U_{ir} U_{kr}$
  Q_r = modal scaling factor
  U_{ir} = modal coefficient
  p_r = system pole
  n = number of modal frequencies

The residue, A_{ikr}, in Eqs. (21.24) through (21.26) is the product of the modal deformations at the input k and response i degrees-of-freedom and a modal scaling factor for mode r. Therefore, the product of these three terms is unique, but each of the three individual terms is not unique.

DAMPING MECHANISMS

To evaluate multiple degree-of-freedom systems, the effect of damping on the complex frequencies and modal vectors must be considered. Types of damping include (1) structural damping, (2) viscous damping, and (3) coulomb damping. It is generally difficult to ascertain which type of damping is present in a particular structure. Most structures exhibit damping characteristics that result from a combination of several damping mechanisms located at several physical locations within the system. (Damping is considered in detail in Chap. 37.)

In order to avoid the difficulty of numerous mathematical representations for the different situations described above, modal analysis methods normally assume that the system damping matrix can be approximated as an equivalent viscous damping matrix. This assumption yields a single mathematical model that can be used to represent damping. Under this assumption, *proportional damping* is the case where the equivalent damping matrix is equal to a linear combination of the system mass and stiffness matrices. *Nonproportional damping* is the case where this linear combination cannot be found. For proportional damping, the coordinate transformation that diagonalizes the system mass and stiffness matrices also diagonalizes the system damping matrix. Therefore, when a system with proportional damping exists, that system of coupled equations of motion can be transformed to a system of equations that represent an uncoupled system of single degree-of-freedom systems that are easily solved.

EXPERIMENTAL MODAL ANALYSIS METHODS

Four general categories of experimental modal analysis methods are:
1. Forced normal mode excitation
2. Frequency-response function
3. Damped complex exponential response
4. Mathematical input-output model

In the past, the modal characteristics of mechanical systems have been estimated by techniques that fall into either the first or second category. The experimental modal analysis methods that fall into the last two categories are composite approaches that utilize elaborate parameter estimation algorithms based upon structural models.[1]

In evaluating experimental modal analysis methods, differences in the theoretical approach are of prime concern. Since most such methods involve a similar theoretical basis, the only significant areas of difference concern the concept of real vs. complex modal vectors and the explicit measurement of the input. The concept of a complex mode appears to be the most general consideration since it contains a real mode as a special case. Likewise, some experimental modal analysis methods do not require the measurement of the input. While this can be advantageous at times where the implicit nature of the input is known or assumed, it is prudent to measure the input when possible.

Beyond the direct theoretical differences, there are several key evaluation considerations which may or may not be a direct function of the theory. The availability of confidence factors, the potential for implementation, stability and precision of the solution algorithm, sensitivity to random and/or bias errors in the measured data, and the need for operator expertise may control the ability to estimate valid modal parameters. Through the knowledge of these aspects, with respect to other experimental modal analysis methods, the use of each of the methods may be enhanced because of this transfer of technology between the methods.

SYSTEM ASSUMPTIONS

In performing an experimental modal analysis, three basic assumptions are made concerning any structure:

1. *The structure is assumed to be linear.* This means that the response of the structure to any combination of forces, simultaneously applied, is the sum of the individual responses to each of the forces acting alone. This is a very good assumption for a wide variety of structures. When a structure is linear, its behavior can be characterized by a controlled excitation experiment in which the forces applied to the structure have a form convenient for measurement and parameter estimation rather than being similar to the forces that are actually applied to the structure in its normal environment. However, for many important kinds of structures, the assumption of linearity is not valid.

2. *The structure is assumed to be time invariant.* This means that the parameters that are to be determined are constants. In general, a system which is not time invariant has components whose mass, stiffness, or damping depend on factors that are not measured or are not included in the model. For example, if some components are temperature dependent, the temperature of the component is viewed as a time-varying signal, and, hence, the component has time-varying characteristics. Therefore, the modal parameters that would be determined by any measurement and estimation process would depend on the time (temperature may vary with time) at which the measurements were made. If the structure that is tested changes with time, then measurements made at the end of the test period determine a different set of modal parameters than measurements made at the beginning of the test period. Thus, the measurements made at the two different times will be inconsistent, violating the assumption of time invariance.

3. *The structure is assumed to be observable.* This means that the input-output measurements that are made contain enough information to generate an adequate behavioral model of the structure. Structures and machines which have loose components, or which have degrees-of-freedom of motion that are not measured, are not completely observable. Consider a description of the motion of a partially filled tank of liquid when complicated sloshing of the fluid occurs. Sometimes a sufficient number of measurements can be made

so that the system is observable under the form chosen for the model, and sometimes no realistic number of measurements will suffice until the model is changed. This assumption is particularly relevant to the fact that the data normally describe an *incomplete model* of the structure. This occurs in at least two different ways. First, the data are normally limited to a minimum and maximum frequency as well as to a limited frequency resolution. Secondly, no information is available relative to local rotations because of a lack of transducers available in this area.

Many other assumptions can be made with respect to the development of the theory concerning a modal analysis method. For example, Maxwell's reciprocity is normally assumed but generally not enforced by the modal parameter estimation algorithms. The modal analysis method may not allow the determination of repeated roots based upon a limited ability to process data or limited numerical algorithms. *Repeated roots* refer to the situation where one complex root or eigenvalue occurs more than once in the characteristic equation. Each root with the same value has an independent modal vector or eigenvector. This situation can only be detected by the use of multiple inputs or references. Detection is critical in developing a complete modal model so that arbitrary input-output information can be synthesized. While theoretical repeated roots are of debatable significance, the acquisition of data is normally restricted to a fixed frequency resolution which gives the possibility of a number of eigenvalues between the observed frequencies. This situation is a very real problem and is referred to as a *pseudorepeated root*.

EXCITATION ASSUMPTIONS

The primary assumption concerning the excitation of a linear structure is that the excitation is observable. Whenever the excitation is measured, this assumption implies that the measured characteristic properly describes the actual input characteristics. For the case of multiple inputs, the different inputs must often be uncorrelated for the computational procedures to yield a solution. In most cases this only means that the multiple inputs must not be perfectly correlated at any frequency. As long as the excitation is measured, the validity of these limited assumptions can be evaluated.

A number of techniques can be used to estimate modal characteristics from response measurements with no measurement of the excitation. If this approach is used, the excitation assumptions are imposing. If the excitation is not measured, accurate estimates of generalized parameters such as modal mass, stiffness, or damping cannot be generated. Even under the assumption that the estimation of these parameters is not required, all of these techniques have one further restriction: An assumption has to be made concerning the characteristics of the excitation of the system. Usually, it is assumed that the autospectrum of the excitation signal is sufficiently smooth over the frequency interval of interest.

In particular, the following assumptions about the excitation signal can be used:

1. *The excitation is impulsive.* The autospectrum of a short pulse (a pulse having a time duration much smaller than the period of the greatest frequency of interest) is nearly uniform, or constant in amplitude, and largely independent of the shape of the pulse.

2. *The excitation is white noise.* White noise has an autospectrum that is uniform over the bandwidth of the signal.

3. *The excitation signal is a step.* A step signal has an autospectrum that decreases in amplitude in proportion to the reciprocal of frequency. The step signal can be viewed as the integral of an impulsive signal.

4. *The excitation is a Wiener-Levy signal.* This signal can be treated as the integral of a white noise signal and has an autospectrum similar to that of the step signal.

5. *There is no excitation.* This is called the free-response or free-decay situation. The structure is excited to a condition of nonzero displacement or nonzero velocity or both. Then the excitation is removed, and the response is measured during free decay. This kind of response can be modeled as the response of the structure to an excitation signal that is a linear combination of impulsive and step signals.

When the excitation autospectrum is uniform, the autospectrum of the response signal is proportional to the square of the modulus of the frequency-response function. In this case, the poles of the response spectrum are the poles of the frequency response, which are the parameters of the system resonances. If the autospectrum is not uniform, then the excitation spectrum can be modeled as an analytic function, to a precision comparable to typical experimental error in the measurement of spectra. In this model, the excitation spectrum has poles that account for the nonuniformity of the spectrum amplitude. Therefore, the response signal can be modeled by a spectrum that contains zeros at the zeros of the excitation and the zeros of the frequency response, and contains poles at the poles of the excitation and at the poles of the frequency response. It is important that the force spectrum have no poles or zeros which coincide with poles of the frequency response.

For transient inputs, such as an impact or step relaxation, the assumption of smooth excitation spectra is generally true, but for operating inputs or inputs generated by an excited system, care must be taken to ensure that the input force spectrum is smooth. This is especially true for tests performed using a hydraulic or an electromechanical exciter, because the system being analyzed may "load" the exciter system (the structure's impedance is so low that the desired force level cannot be achieved within the constraint of small motion), causing a nonuniformity in the input force spectrum.

To determine the characteristics of the system from the response, it is necessary that the response have the same poles as the frequency response or that the analysis process corrects for the zeros and poles of the excitation. If the force input spectrum has a zero in the frequency range of interest, the pole location measured from the response spectrum will not match that of the frequency response. There is considerable interest in determining modal parameters from measured response data taken on operating systems, for example, turbulent flow over an airfoil. In such a case, poles that are system resonances should not be confused with those that exist in the output spectrum because of inputs.

In general, the poles of the response include those of the frequency response and of the input spectrum. Therefore, if the force is not measured, it is not possible, without some prior knowledge about the input, to determine if the poles of the response are truly system characteristics. If no poles or zeros exist in the force spectrum in the frequency range of interest, then any poles in the response in this range must be a result of the system characteristics. In order to avoid this potential problem, when the excitation can be measured, it is prudent to do so.

FORCED NORMAL MODE EXCITATION METHOD

The forced normal mode excitation method of experimental modal analysis is the oldest approach to the estimation of dynamic structural parameters, and the one first to use the application of multiple inputs in the estimation of modal parameters.[2] In this method (employed extensively in the aerospace industry for ground vibration testing of aircraft structures), a number of exciters are tuned to balance exactly the dissipative forces in a structure at certain frequencies. When force balance is achieved, the differential equations of motion describing the structure can be reduced to the undamped homogeneous differential equations of motion at that particular frequency. The theoretical basis for this testing may be explained in terms of the characteristic phase lag theory.[3] Further development of the practical application of this theory was enhanced by the concept of having an effective number of degrees-of-freedom.[4] This concept explains that the required number of exciters is a function of the effective number of degrees-of-freedom, not the total number of degrees-of-freedom. The effective number of degrees-of-freedom is a function of modal density and damping. Finally, the determinant of the real part of the frequency-response matrix may be utilized to locate damped natural frequencies and determine the effective number of degrees-of-freedom.[5] Since 1958 most work has been concerned with improvements in the implementation of the method, primarily the force appropriation.[5-7]

The forced normal mode excitation method works well in the presence of proportional damping but theoretically does not include the concept of complex modes of vibration. (*Proportional damping* is said to occur when the damping matrix, resulting from whatever damping mechanism is present, is proportional to the mass matrix, the stiffness matrix, or to some linear combination of the two.) Because of this theoretical limitation, the practical application of the 90° phase lag criteria is normally applied only to within plus or minus 10°.

FREQUENCY-RESPONSE FUNCTION METHOD

The frequency-response function method of experimental modal analysis is the most commonly used approach to the estimation of modal parameters. This method originated as a testing technique as a result of the use of frequency-response functions in the forced normal mode excitation method to determine natural frequencies and effective number of degrees-of-freedom.[8-10] With the advent of the computer and minicomputer, the frequency-response function method became a separate, viable technique.[11-34]

In this method, frequency-response functions are measured using excitation at single or multiple points. This excitation may be narrow band or broad band as well as being random or deterministic.[15,16,18-20] The frequency-response functions are used as input to a modal parameter estimation process to determine the modal parameters. Often, a curve-fitting procedure is used to evaluate the parameter estimation procedure. A weighted orthogonality check may be made of the modal vectors to evaluate the relationship of the estimates to some mathematical model. In order to ensure that all modal vectors have been found experimentally, a number of excitation points must be utilized, either one at a time or simultaneously. This reduces the possibility of trying to excite the system at a node of one of the modal vectors.

Frequency-response function technology involves obtaining and evaluating multiple independent modal vector estimates from multiple rows or columns of the frequency-response function matrix. A matrix inversion algorithm computes frequency-response functions in the presence of multiple uncorrelated random force inputs.[18,20,22-24] Statistical techniques involving single and multiple regression concepts determine the consistency of modal vectors estimated individually from multiple references.[18-21] Multiple reference modal parameter estimation algorithms estimate the best least-squares modal vector collectively from the multiple references.[18,22]

THEORY

The frequency-response function method begins with the matrix differential equation of motion description of an n degree-of-freedom system.

$$[m]\{\ddot{x}\} + [c]\{\dot{x}\} + [k]\{x\} = \{f\} \tag{21.26}$$

In this statement of the system equations of motion, only viscous damping appears although other damping mechanisms may be present. Only the viscous component of these other damping mechanisms actually contribute to the dissipation of energy.[13,32] Therefore C can be defined as the equivalent viscous damping matrix for the system. Taking the Laplace transform of Eq. (21.26), assuming that all initial conditions are zero, gives

$$[[m]s^2 + [c]s + [k]]\{X(s)\} = \{F(s)\} \tag{21.27}$$

In order to simplify Eq. (21.27) the concept of a *system impedance matrix*, $B(s)$, is defined.

$$[B(s)] = [[m]s^2 + [c]s + [k]] \tag{21.28}$$

Therefore,

$$[B(s)]\{X(s)\} = \{F(s)\} \tag{21.29}$$

The transfer function matrix $H(s)$ can be defined as

$$\{X(s)\} = [H(s)]\{F(s)\} \tag{21.30}$$

Therefore,

$$[H(s)] = [B(s)]^{-1} \tag{21.31}$$

The number of measurement points m cannot be greater than the number of elements of the excitation vector if the transfer function is to be equal to the left-handed inverse of the system matrix. The number of elements of the excitation vector can always be supplemented by zeros to equal m.

The inverse of the system impedance matrix can be given by

$$[H(s)] = [B(s)]^{-1} = \frac{[D(s)]}{\det[B(s)]} \tag{21.32}$$

where $[D(s)]$ is the adjoint matrix of $[B(s)]$ (transpose of the cofactor matrix of $[B(s)]$), and $\det[B(s)]$ is the determinant of $[B(s)]$

Since both the adjoint matrix of $[B(s)]$ and the determinant of $[B(s)]$ are polynomials in s, the elements of $[H(s)]$ are rational fractions in s.

Therefore, it is possible to represent any element of the transfer function matrix $[H(s)]$ in a partial fraction form.

If it is assumed that the poles of the system are of unit multiplicity and that the system is inherently underdamped (the poles occur in complex conjugate pairs), the transfer function matrix can be expressed as

$$[H(s)] = \sum_{r=1}^{n} \frac{[A_r]}{s - p_r} + \frac{[A_r{}^*]}{s - p_r{}^*} \tag{21.33}$$

where p_r = system pole = $\sigma_r + j\omega_r$
$\qquad j = \sqrt{-1}$
$\qquad \sigma_r$ = damping factor
$\qquad \omega_r$ = damped natural frequency
$\qquad [A_r]$ = residue matrix for mode r
$\qquad *$ = complex conjugate

The undamped natural frequency, Ω_r, is given by

$$\Omega_r = \sqrt{\sigma_r{}^2 + \omega_r{}^2} \tag{21.34}$$

The fraction of critical damping, ζ_r, is given by

$$\zeta_r = -\frac{\sigma_r}{\Omega_r} \tag{21.35}$$

The modal vectors can be described in terms of the adjoint matrix of $B(s)$. Using Eq. (21.32) with the following identity, the basis for the modal vectors can be developed.

$$[B(s)][B(s)]^{-1} = [I] \tag{21.36}$$

Substituting Eq. (21.32) into Eq. (21.36) yields

$$[B(s)][D(s)] = \det[B(s)][I] \tag{21.37}$$

Evaluating Eq. (21.37) at any one of the system roots, $s = p_r$, causes the determinant of $[B(s)]$ to go to zero.

$$[B(p_r)][D(p_r)] = [0] \tag{21.38}$$

Equation (21.38) can be rewritten using only the kth column of the adjoint matrix $[D(p_r)]$.

$$[B(p_r)]\{D(p_r)\} = \{0\} \tag{21.39}$$

Equation (21.39) looks exactly like the homogeneous solution of Eq. (21.27) when evaluated at any one of the system poles. Therefore, the kth column of the adjoint matrix

$[D(p_r)]$ is an estimate to within an arbitrary constant of the modal vector for that pole. Since the kth column was chosen arbitrarily, any column of the adjoint matrix is an estimate of the modal vector of that system pole to within a constant. Finally, from Eq. (21.32), it follows that any column of the transfer-function matrix, evaluated at a system pole, is also an estimate of the modal vector for that system pole to within a constant.

If the transfer-function matrix is limited to values of s having a zero real part, this special case results in the frequency-response matrix. Therefore, if the elements of the frequency-response matrix can be measured, each column contains information which can be used to estimate modal vectors. Since the frequency-response matrix is considered to be symmetric because of the Maxwell-Betti relations, each row also contains the information needed to estimate modal vectors.

In order to obtain estimates of the modal vectors, the frequency-response functions are used as input in a parameter-estimation scheme based on Eq. (21.33) or on the equivalent time-domain formulation given in Eq. (21.23), involving the impulse-response function.[11-18,27,35,36]

$$[H(\omega)] = \sum_{r=1}^{n} \frac{[A_r]}{j\omega - p_r} + \frac{[A_r{}^*]}{j\omega - p_r{}^*} \tag{21.40}$$

where ω = angular frequency variable
r = modal vector number
$[A_r]$ = residue matrix
p_r = system pole
n = number of modal frequencies

Therefore, Eq. (21.40) is the general matrix form that is used by modal parameter estimation algorithms as a mathematical model for the frequency-response function matrix. Often Eq. (21.40) is altered by an assumption of real modes, a specific damping mechanism, or known system poles. Under such assumptions, the evaluation of Eq. (21.40) for the remaining modal parameters becomes much simpler.

Multiple reference modal parameter estimation algorithms applied to the frequency-response function method involve the time- or frequency-domain models equivalent to the matrix forms of Eq. (21.23) or (21.24). Equation (21.40) is an example of the matrix form of Eq. (21.24). These multiple reference modal parameter estimation algorithms utilize as much of the redundant information within multiple rows and columns of the impulse- or frequency-response function matrix as is possible. The application of the polyreference time-domain method[35,36] to impulse-response functions is a primary example.

DAMPED COMPLEX EXPONENTIAL METHOD

The damped complex exponential response methods of experimental modal analysis are normally formulated to utilize data corresponding to the free decay of a system generated by the release of an initial condition, but they apply quite generally to impulse-response function data as well. Since impulse-response function data are scaled to include the forcing condition, use of this method yields properly scaled modal parameters that can be used to calculate generalized mass and stiffness. This would not be possible if free decay responses were used. The formulation of the impulse-response function generally involves the computation of the frequency-response function by use of a fast Fourier transform, potentially imposing bias errors, which may degrade the estimation of the modal parameters. Three implementations of the damped complex exponential method, of demonstrated capability, are the Ibrahim time-domain, polyreference time-domain, and the eigensystem realization algorithm approaches discussed below. The first two, as well as similar formulations,[26,27] have been explained in a unified damped complex exponential theory based on the polyreference time-domain theoretical approach.[51]

IBRAHIM TIME-DOMAIN (ITD) METHOD. One practical implementation of the damped complex exponential method is the Ibrahim time-domain (ITD) method,[38-42] developed to extract the modal parameters from the damped complex exponential response information. Digital free-decay response data are measured at various points on the structure. If response data from all the selected measurement positions cannot be obtained simultaneously because of equipment restrictions, a common position is retained between measurement groups. A recurrence matrix is created from the free-decay data; the poles of the system are computed from the eigenvalues of this matrix. The eigenvalues of the recurrence matrix are exponential functions of the poles of the system. The eigenvectors of the recurrence matrix are response residues, from which the mode shapes are determined.

This method computes the poles and residues based upon a specific initial vibration condition of the structure. A number of different initial conditions can be established, analogous to the practice of using several exciter positions in ordinary single input modal surveys, until all the important modes have been excited. All the modes cannot be established from one exciter position, and, likewise, all the modes cannot be determined from one initial condition.

Although this technique is based upon free-decay data, it can also be used with operating inputs if the free decay is computed from the operating inputs by using "random-decrement" averaging[40] or from measured autocorrelation and crosscorrelation functions. This can only be done if there are no poles or zeros in the input spectrum in the frequency range of interest.

An additional development with respect to this technique is the concept of a modal confidence factor.[38] The *modal confidence factor* is the complex scalar that describes the measured magnitude of the free-decay response of a single modal vector at a specific time, averaged degree-of-freedom by degree-of-freedom, compared to an estimate of that magnitude computed from the magnitude measured before or after that time. The modal confidence factor is based upon the modal deflection at a particular degree-of-freedom being related to the modal deflection at that same degree-of-freedom at any time earlier or later in the free-decay response. Therefore, if modal vectors are estimated from exponential decay data and two separate estimates are calculated from sets of data taken some fixed time Δt apart, the relationship between the measured second estimate of the modal vector and the calculated estimate of the modal vector based upon the measured first estimate of the modal vector is defined as the modal confidence factor. The purpose of the modal confidence factor is to provide an indicator for determining whether an estimated modal vector is real or computational.

POLYREFERENCE TIME-DOMAIN METHODS. Most of the comments relative to the ITD method also apply to the polyreference time-domain method.[35,36] Of particular importance is the increased requirement for computational power and the desirability of acquiring all response data simultaneously to reduce time-invariance problems.

The polyreference time-domain method utilizes all measured damped complex exponential information, from all references or initial conditions, simultaneously in the estimation of modal frequencies; it provides an estimation of a single modal coefficient for each measurement degree-of-freedom in the presence of multiple initial conditions or multiple inputs. Since the polyreference time-domain method involves multiple references, one significant advantage of the technique is the ability to detect poles of the system that are identical or nearly identical. Poles of the system that are identical are referred to as *repeated roots;* poles of the system that are nearly identical are referred to as *pseudorepeated roots.* The formulation of the algorithm such that constraints are included to account for redundant information is an advantage but requires that the total data set be acquired so as to match this assumption. Thus, the data acquisition best matches the analysis procedure when all of the data are acquired simultaneously.

Evaluation of the polyreference time-domain method based upon comparisons between experimentally measured and synthesized frequency-response functions is quite impressive when compared to other modal parameter estimation approaches.

EIGENSYSTEM REALIZATION ALGORITHM (ERA) METHOD. The eigensystem realization algorithm (ERA) method is a technique that is basically an extended version of the Ho-Kalman system realization algorithm.[43] It is similar to the other damped complex exponential methods in that it involves solutions of a matrix eigenvalue problem. The ERA method is most like the polyreference time-domain method in that multiple initial conditions can be involved in the matrix formulation. Therefore, repeated roots or eigenvalues can be identified with this method as in the polyreference time-domain approach. Other significant attributes of this approach include the extensive use of accuracy indicators to assess effects of noise and nonlinearities as well as rank information provided by singular value decomposition techniques.[43,45]

MATHEMATICAL INPUT-OUTPUT MODEL METHOD

The experimental modal analysis methods that are included within the category of mathematical input-output model methods are those that generally involve input and response data independently without the need for creating auto- and cross-moment functions. There is no other restriction with regard to time- or frequency-domain models, effective number of degrees-of-freedom, etc. On this basis, two methods are in use that can be described in this fashion. First, the autoregressive moving average (ARMA) method is a time-domain formulation that utilizes a pole-zero model as the basis for the description of the system characteristics. While this model is appropriate, it cannot easily be constrained to account for known system information. Additionally, although the current application of this technique does not involve multiple inputs, the theoretical background for the multiple input case is well developed.

The other method involves a reduced structural matrix model as the basis for the description of the system characteristics. This model involves the reduced mass, stiffness, and damping matrices with regard to the measured degrees-of-freedom. This model easily accounts for constraints such as known elements in the mass, stiffness, and damping matrices or known characteristics of the distribution within the matrices such as symmetric or banded characteristics. This method also incorporates the multiple input case routinely as known terms in the forcing vector of the matrix differential equation that serves as the mathematical model.

AUTOREGRESSIVE MOVING AVERAGE METHOD. One method of estimating the modal characteristics from input-output data is the autoregressive moving average (ARMA) method. This procedure has been applied to the determination of structural parameters.[46-50] With this technique the response data are assumed to be caused by a white random noise input to the structure. The technique computes the best statistical model of the system in terms of its poles (from the autoregressive part) and zeros (from the moving average part), as well as statistical confidence factors on the parameters. It is primarily used to estimate the characteristics of buildings being excited by wind forces. The data used in the computational process are the autocorrelation functions of the responses measured at various points on the structure. Since in the general case the inputs are not measured, the modal vectors are determined by referencing each response function to a single response to provide relative magnitude and phase information.

REDUCED STRUCTURAL MATRIX METHOD. Most of the methods for estimating reduced structural (mass, stiffness, damping) matrices from experimental data are based upon an indirect approach utilizing the estimated modal parameters to synthesize the reduced matrices.[51-53] There are several misconceptions concerning this technique. First, regardless of the approach used, the solution for the reduced matrices is not unique; there are many combinations of matrix relationships that can be generated from the given set of estimated modal parameters. Second, the reduced frequency range of the modal parameter estimates means that the matrices will be weighted to represent an incomplete model.[54] Third, the limitation of the precision of the modal parameter estimates as a result of commonly accepted experimental error tends to desensitize the process of estimating the reduced matrices. Finally, the problem of invalid modal parameter estimates obviously results in invalid estimates of reduced matrices.

MODAL DATA ACQUISITION

Acquisition of data, used in the formulation of a modal model, involves many important technical concerns. (See *Digital Vibration Data Analysis*, Chap. 27) The primary concern is the digital signal processing or the converting of analog signals into a corresponding sequence of digital values that accurately describe the time-varying characteristics of the inputs to and responses from a system. Once the data are available in digital form, the most common approach is to transform the data from the time domain to the frequency domain by use of a fast Fourier transform algorithm.

DIGITAL SIGNAL PROCESSING

This process of analog-to-digital conversion involves two separate concepts, each of which are related to the dynamic performance of a digital signal processing analyzer. *Sampling* is the part of the process related to the timing between individual digital pieces of the time history. *Quantization* is the part of the process related to describing an analog amplitude as a digital value. Primarily, sampling considerations alone affect the frequency accuracy while both sampling and quantization considerations affect magnitude and phase accuracy.[55-57] The two constraining relationships that govern the sampling process are known as Shannon's sampling theorem and Rayleigh's criteria. The selection of the sampling parameters by way of these constraints is summarized in Eqs. (27.5) through (27.9).

FAST FOURIER TRANSFORM. (Also see *FFT Analyzers*, Chap. 13) The fast Fourier transform algorithm is the basis for the formulation of any frequency-domain function in modern data acquisition systems. In terms of an integral Fourier transform, the function must exist for all time in a continuous sense in order to be evaluated. For the realistic measurement situation, data are available in a discrete sense over a limited time period. The fast Fourier transform, therefore, is based upon a set of assumptions concerning this discrete sequence of values. The assumptions can be reduced to one of two situations that must be met by every signal processed by the fast Fourier transform algorithm:

1. The signal must be a totally observed transient with respect to the time period of observation.
2. The signal must be composed only of harmonics of the time period of observation.

If one of these two assumptions is not met by any discrete history processed by the fast Fourier transform algorithm, then the resulting spectrum will contain bias errors accordingly.

ERRORS. The accurate measurement of frequency-response functions depends significantly upon the errors involved with the digital signal processing. In order to take full advantage of experimental data in the evaluation of experimental procedures and in verification of theoretical approaches, the errors in measurement, generally designated noise, must be reduced to acceptable levels. Various types of errors are discussed in other chapters of this Handbook.

TRANSDUCER CONSIDERATIONS

The transducer considerations are often the most overlooked aspect of the experimental modal analysis process. Transducer characteristics, described in Chap. 12, are primarily concerned with the magnitude and frequency limitations placed on a transducer because of its design. Measurement techniques are described in Chap. 15, and the calibration of transducers is treated in detail in Chap. 18.

It is also necessary to evaluate the transducer mounting system to ascertain whether it has compromised any of the transducer's specifications. This normally involves the possibility of relative motion between the structure under test and the transducer. Very often, the mounting systems which are convenient to use and allow ease of alignment with orthogonal reference axes are subject to mounting resonances which result in substantial relative motion between the transducer and the structure under test in the frequency range of interest. Therefore, the mounting system which should be used depends heavily upon the frequency range of interest and upon the test conditions. Test conditions include factors such as temperature, roving or fixed transducers, and surface irregularity. Mounting systems and measurement techniques are considered in Chap. 15.

FREQUENCY-RESPONSE FUNCTION ESTIMATION

The *frequency-response function* is the complex ratio of output response to input excitation as a function of frequency for a single input and a single output. It is directly releated to the concept of a transfer function. The *transfer function* is similarly defined except that it is a function of the complex Laplace variable *s*. If the real part of the Laplace variable is restricted to zero, the transfer function is equal to the frequency-response function. Frequency-response functions usually are used to describe the input-output force-displacement relationships of a system. Most often, the system is assumed to be linear and time invariant, although this is not necessary. Where assumptions of linearity and time invariance are not valid, the measurements of frequency-response functions are also dependent upon the independent variables of time and input. In this way, a conditional frequency-response function is measured as a function of other independent variables in addition to frequency.

The computation of the theoretical frequency-response function depends upon the transformation of data from the time to the frequency domain. The Fourier transform is used for this computation. The integral Fourier transform definition requires time histories from negative infinity to positive infinity. Since this is not possible experimentally, the computation is performed digitally using a fast Fourier transform algorithm which is based upon only a limited time history. In this way the theoretical advantages of the Fourier transform can be implemented in a digital computation scheme.

SINGLE INPUT THEORY. The frequency-response function can be computed in several ways through the use of the fast Fourier transform algorithm. In each case, the desired frequency-response function of the system is a function of frequency and position and all noise inputs are assumed to be zero. In all cases, only an estimate of the frequency-response function can be determined. Hereafter, no additional notation is used to denote that the frequency-response functions are estimates. The most common approach to the estimation of frequency response is to use a least-squares technique. This is one of the standard techniques for estimating parameters of noisy signals and, in this case, is based upon the auto-power and cross-power spectra. There is much interest concerning the formulation of the frequency-response function, in a least-squares sense, based upon a minimization of the noise on the input, the noise on the output, or the noise in a vector sense.[22] While justification can be made for the use of each approach based on the location and form of the noise, all methods yield the same answer as the noise is eliminated. The two common formulations of frequency response are based upon power spectra in which the phase information is preserved in the cross-power spectrum term.

In practice, the estimated values of the cross-power and auto-power spectra depend upon the form of the input signal (deterministic, random, etc.). Statistically, therefore, the power spectrum estimates are not unique unless the input is stationary. Even so, with nonstationary or deterministic inputs, the ratio of power spectra, used in the different formulations for the frequency-response function, is unique and, therefore, valid within the limitations of the frequency content of the input and linearity of the model.

One important consideration of the two common formulations for frequency-response function estimation is the behavior of each formulation in the presence of a bias error, such as leakage. In both cases, the estimate differs from the expected value particularly in the region of a resonance or antiresonance. The availability of cross-power and auto-power spectra allows the determination of another function, called *ordinary coherence*, which represents a level of confidence in the frequency-response function estimate. When the coherence is zero, the output is caused entirely by sources other than the measured input. In general, the coherence can be a measure of the degree of noise contamination in a measurement. Thus, with more averaging, the estimate of coherence may contain less variance, therefore giving a better estimate of the noise energy in a measured signal. This is not the case if the low coherence is due to bias errors such as nonlinearities, multiple inputs, or leakage. In all such cases, the estimated coherence function approaches, in the limit, the expected value of coherence at each frequency, dependent upon the type of noise present in the structure and measurement system. With more averaging, the estimate value of coherence does not increase; it always approaches the expected value from the upper side.

The coherence function indicates the degree of causality in a frequency-response function. If the coherence is equal to unity at any specific frequency, the system is said to have *perfect causality* at that frequency. Then the measured response power is caused totally by the measured input power (or by sources which are coherent with the measured input power). A coherence value less than unity at any frequency indicates that the measured response power has a value greater than can be explained from the measured input as a result of extraneous noise also contributing to the output power. Low coherence values do not necessarily imply poor estimates of the frequency-response function, but simply mean that more averaging is required for a reliable result.

Consider two special cases of low coherence. The first occurs when there is a leakage bias error in one or both of the input and output measurements. This causes the coherence in the area of the peaks of the frequency response to be less than unity. This error can be reduced by the use of techniques discussed in a later section on *Error Reduction Methods*. The second occurs when there is a significant propagation delay time between the input and output as may be the case with the acoustic measurements. If a propagation delay of length t is compared to a sample function of length T, a low estimate of coherence will be a function of the ratio t/T.[57] This propagation delay causes a bias error in the frequency response and should be removed prior to computation if possible.

MULTIPLE INPUT THEORY. Multiple input estimation of frequency-response functions is desirable for several reasons. The principal advantage is the increase in the accuracy of estimates of the frequency-response functions. During single input excitation of a system, there may be large differences in the amplitudes of vibratory motion at various locations because of the dissipation of the excitation power within the structure — especially when the structure is heavily damped. Small nonlinearities in the structure consequently cause errors in the measurement of the response. With multiple input excitation, the vibratory amplitudes across the structure typically are more uniform, with a consequent decrease in the effect of nonlinearities.

A second reason for improved accuracy is the increase in consistency of the frequency-response functions. When a number of exciter systems are used, the elements from columns of the frequency-response function matrix corresponding to those exciter locations are determined simultaneously. With the single input method, each column is determined independently, and it is possible for small errors of measurement (due to nonlinearities and time-dependent system characteristics) to cause a change in resonance frequencies, damping, or mode shapes among the measurements in the several columns. This is particularly important for the multiple reference modal parameter estimation algorithms, such as the polyreference[35,36] and the reduced matrix estimation methods[37,58,59] that use frequency-response functions for multiple columns simultaneously.

Another advantage of the simultaneous measurement of a number of columns of the frequency-response function matrix is the ability to use a linear combination of frequency-response functions in the same row of the matrix in order to enhance specific modes of the system.[18,22] This technique is analogous to the forced normal mode excitation experimental modal analysis in which a structure is excited by a forcing vector which is proportional to the modal vector of interest. For this analysis, the coefficients of a preliminary experimental modal analysis are used to weight the frequency-response functions, so that the sum emphasizes the modal vector that is sought. The revised set of conditioned frequency-response functions is analyzed to improve the accuracy of the modal vector.

An additional advantage of the multiple input excitation is a reduction of the test time. In general, when frequency-response functions are determined by multiple input estimation, they are obtained for all input locations in approximately the same time as is required for acquiring a set of frequency-response functions for one of the input locations, using a single input estimation method. This application of multiple input-output theory represents a very special case where everything about the inputs is known or can be controlled. The number of inputs, the location of the inputs, and the characteristics of the inputs are controlled by the test procedure. For the general case, none of these characteristics may be known.

The formulation of the least-squares method for determining the frequency-response functions in the presence of multiple inputs depends upon the minimization of the noise on the input, the noise on the output, or the noise in a vector sense. The solution of this minimization process yields a matrix equation similar in form to the single input case.[18,22] The primary concern in the solution of this matrix equation involves taking the inverse of a matrix containing the input auto-power and cross-power information. This inverse will exist and yield a unique solution for frequency-response functions if the inputs are not correlated. Evaluation of the correlation between inputs in the multiple input case is usually based upon the coherence functions computed between the inputs.

For the multiple input case, the concept of coherence requires expansion. *Ordinary coherence* is the squared correlation coefficient describing the linear relationship between any two single spectra. This is exactly the coherence function that is defined for single input measurement. Care must be taken in the interpretation of ordinary coherence when more than one input is present. The ordinary coherence of an output with respect to an input can be much less than unity even though the linear relationship between inputs and outputs is valid.

Partial coherence is defined as the ordinary coherence between two conditioned outputs, between two conditioned inputs, or between a conditioned input and a conditioned output. The output and input are conditioned by removing contributions from other input(s). The removal of the effects of the other input(s) is formulated on a linear least-squares basis. The order of removal of the inputs during "conditioning" has a definite effect upon the partial coherence if some of the inputs are mutually correlated. There is a partial coherence function for every input/input, output/output, and input/output combination for all permutations of condition. Partial coherence, especially between inputs, is useful for evaluating whether the multiple inputs are correlated. The determination of whether the multiple inputs are mutually related is critical to the matrix solution for the frequency-response functions. Therefore, except for the dual input case, the use of ordinary coherence between inputs does not allow the true degree of correlation between the inputs to be detected. Additional verification of the lack of correlation is determined from the partial coherences between all inputs.[21,22,55−57,60]

Multiple coherence is the correlation coefficient describing the linear relationship between an output and all known inputs. There is a multiple coherence function for every output. Multiple coherence can be used to evaluate the importance of unknown contributions to each output. These unknown contributions can be measurement noise, nonlinearities, or unknown inputs. Particularly, as in the evaluation of ordinary coherence in the single input case, a low value of multiple coherence near a resonance will often mean that the "leakage" error is present in the frequency-response function. Unlike the ordinary coherence function in the single input case, a low value of multiple coherence is not expected at antiresonances. The antiresonances for a given response degree-of-freedom and different inputs do not occur at the same frequencies.

There are a number of situations where the input power spectrum matrix may be singular at specific frequencies or frequency intervals:

1. One of the input auto-power spectra may be zero in amplitude over some frequency interval. When this occurs the input power spectrum matrix is singular over that frequency interval.

2. Two or more of the input signals may be fully coherent over some frequency interval. This can be evaluated and documented by computing the ordinary and conditioned partial coherences among the inputs.

3. Numerical problems, which cause the computation of the inverse to be inexact, may be present. This can happen when an auto-power spectrum is near zero in amplitude, when the cross-power spectra have large dynamic range with respect to the precision of the computer, or when the matrix is ill-conditioned because of nearly redundant input signals.

Of the variety of situations that can cause difficulties in the computation of the frequency-response functions, the highest potential for trouble is the case of coherent inputs. If two of the inputs are fully coherent, then there are no unique frequency-response functions associated with those inputs. This can be measured by computing the ordinary and conditioned partial coherence functions among all inputs. Although the signals used as inputs to the exciter systems must be uncorrelated random inputs, the response of the structure at resonance, combined with the inability to completely isolate the exciter systems from this response, will result in the ordinary or conditioned partial coherence functions with values other than zero, particularly at the system poles. As long as the coherence functions are not unity at these frequencies, a unique solution may be found for the frequency-response functions.

The multiple input estimation of frequency-response functions requires the solution of a set of simultaneous linear equations. The auto-power and cross-power spectra involved in the calculation of the multiple input case for the estimation of frequency-response functions should be computed from analog time data that have been digitized *simultaneously*. If data are not processed in this manner, many more averages will be required to reduce the variance on each individual auto-power and cross-power spectrum and the efficiency of the multiple input approach to the estimation of frequency-response functions will not be as attractive.

ERROR REDUCTION METHODS.   Several factors contribute to the quality of measured frequency-response function estimates (aside from measurement errors). For example, frequency-response and coherence functions have been defined as parameters of a linear system. Nonlinearities generally shift energy from one frequency to many new frequencies, in a way which may be difficult to recognize. The result will be a distortion in the estimates of the system parameters, which may not be apparent unless the excitation is changed. One way to reduce the effect of nonlinearities is to randomize these contributions by choosing a randomly different input signal for each of the n averages. Subsequent averaging reduces these contributions in the same manner that random noise is reduced. Another example involves control of the system input. The system should be excited at all frequencies for which data are required. It is important to ensure that the input power spectra does not have "holes" where there is little excitation energy. Otherwise, coherence will be very low and the variance on the frequency-response function will be large.

If the system is linear, the excitation is appropriate, and measurement errors are avoided, noise still will be present in the measurement process. Noise is a general designation describing the difference between the true value and the estimated value. A more exact designation is to view noise as the total error comprised of two terms, variance and bias. Each of these classifications is merely a convenient grouping of many individual errors which cause a specific kind of inaccuracy in the function estimate. The variance portion of the error essentially is Gaussian distributed and can be reduced by any form of synchronization in the measurement or analysis process. The bias or distortion portion of the error causes the expected value of the estimated function to be different from the true value. Normally, bias errors are removed if possible, but, if the form and the source of a specific bias error is known, many techniques may be used to reduce the magnitude of the specific bias error.

Four different approaches can be used to reduce the error involved in frequency-response function measurements in fast Fourier transform analyzers. *Signal averaging* can significantly reduce errors of both variance and bias and is probably the most general technique in the reduction of errors in frequency-response function measurement. *Selective excitation* is often used to verify nonlinearities or randomize characteristics. In this way, bias errors due to system sources can be reduced or controlled. The increase of *frequency resolution* through the use of large fast Fourier transform block sizes or through the use of a frequency-shifted fast Fourier transform can improve the frequency-response function estimate primarily by reduction of the leakage bias error due to the use of a longer time sample. The frequency-shifted fast Fourier transform by itself is a linear process and does not involve any specific error reduction characteristics compared to a baseband fast Fourier transform. Finally, the use of *weighting functions (windows)* is widespread. (See Table 13.13 for a comparison of various types of windows.)

Excitation.   Inputs which can be used to excite a system in order to determine frequency-response functions belong to one of two classifications: (1) *random signal*, or (2) *deterministic signal* (see Chap. 11).

The choice of input used to excite a system in order to determine frequency-response functions depends upon the characteristics of the system, upon the characteristics of the parameter estimation, and upon the expected utilization of the data. The characterization of the system is primarily concerned with the linearity of the system. As long as the system is linear, all input forms should give the same expected value. All real systems are nonlinear to some degree. Deterministic input signals result in frequency-response functions that are dependent upon the signal level and type. A set of frequency-response functions for different signal levels can be used to document the nonlinear characteristics of the system. Random input signals, in the presence of nonlinearities, result in a frequency-response function that represents the best linear representation of the nonlinear characteristics for a given level of random signal input. For small nonlinearities, use of a random input does not differ greatly from the use of a deterministic input.

The characterization of the parameter estimation is primarily concerned with the type of mathematical model being used to represent the frequency-response function. Generally, the model is a linear summation based upon the modal parameters of the system. Unless the mathematical representation of all nonlinearities is known, the parameter estimation process cannot properly weight the frequency-response function data to include nonlinear effects. For this reason, random input signals are prevalently used to obtain the best linear estimate of the frequency-response function when a parameter estimation process using a linear model is to be utilized.

The expected utilization of the data is concerned with the degree of detailed information required by any postprocessing task. For experimental modal analysis, this can range from implicit modal vectors needed for troubleshooting to explicit modal vectors used in an orthogonality check. As more detail is required, input signals, both random and deterministic, are needed to match the system characteristics and parameter estimation characteristics more closely. In all possible uses of frequency-response function data, the conflicting requirements of the need for accuracy, equipment availability, testing time, and testing cost normally reduce the possible choices of input signal.

For those excitation signals that require the use of an electromechanical exciter, Fig. 21.5 shows a typical test configuration; Fig. 21.6 shows a typical test configuration when

FIG. 21.5. Typical test configuration involving shaker excitation.

an impact form of excitation is to be used. The advantages and disadvantages of each excitation signal are summarized in Table 21.1. The following excitation signals are most commonly used:

Slow Swept Sine Wave. (Also see *Response to Swept Sine Wave*, Chap. 24) The slow swept sine wave signal is a periodic deterministic signal with a frequency that is an integer multiple of the fast Fourier transform frequency increment. Sufficient time is allowed in the measurement procedure for any transient response to the changes in frequency to decay so that the resultant input and response histories will be periodic with respect to the sample period. Therefore, the total time needed to compute an entire frequency-response function will be a function of the number of frequency increments required and the system damping.

Periodic Chirp. The periodic chirp is a fast swept sine wave signal that is a periodic deterministic signal and is formulated by sweeping a sine wave signal up or down within a frequency band of interest during a single sample period. Normally, the fast swept

FIG. 21.6. Typical test configuration involving impact excitation.

sine wave signal is made up of only integer multiples of the fast Fourier transform frequency increment. This signal is repeated without change so that the input and output histories are periodic with respect to the sample period.

Impact (Impulse) Excitation. The impact signal is a transient deterministic signal which is formed by applying an input pulse to a system lasting only a very small part of the sample period. The width, height, and shape of this pulse will determine the usable spectrum of the impact. The width of the pulse determines the frequency spectrum; the height and shape of the pulse controls the level of the spectrum. Impact signals are applied with some form of an instrumented hammer. While the concept is straightforward, effective utilization of an impact signal is very involved.[28]

Step Relaxation Excitation. The step relaxation signal is a transient deterministic signal which is formed by releasing a previously applied static input. The sample period begins at the instant that the release occurs. This signal is normally generated by the application of a static force through a cable. The cable is then cut or released through a shear pin arrangement.[30]

Pure Random Excitation. The pure random signal is an ergodic, stationary random signal which has a Gaussian probability distribution. In general, the frequency content of the signal contains all frequencies (not just integer multiples of the fast Fourier transform frequency increment) but may be filtered to include only information in a frequency band of interest. The measured input spectrum of the pure random signal is altered by any impedance mismatch between the system and the exciter.

Pseudorandom Excitation. The pseudorandom signal is an ergodic, stationary random signal consisting only of integer multiples of the fast Fourier transform frequency increment. The frequency spectrum of this signal has a constant amplitude with random phase. If sufficient time is allowed in the measurement procedure for any transient response to the initiation of the signal to decay, the resultant input and response histories are periodic with respect to the sample period. The number of averages used in the measurement procedure is only a function of the reduction of the variance error. In a noise-free environment, only one average may be necessary.

Periodic Random Excitation. The periodic random signal is an ergodic, stationary

Table 21.1. Summary of Characteristics of Excitation Signals Used for Frequency-Response Function Estimates

	Type of Excitation							
	Steady-state sine	Pure random	Pseudo-random	Random	Fast sine	Impact	Burst sine	Burst random
Minimize leakage	No	No	Yes	Yes	Yes	Yes	Yes	Yes
Signal-to-noise ratio	Very high	Fair	Fair	Fair	High	Low	High	Fair
RMS-to-peak ratio	Very high	Fair	Fair	Fair	High	Low	High	Fair
Test measurement time	Very long	Good	Very short	Fair	Fair	Very good	Very good	Very good
Controlled frequency content	Yes	Yes*	Yes*	Yes*	Yes*	No	Yes*	Yes*
Controlled amplitude content	Yes	No	Yes*	No	Yes*	No	Yes*	No
Removes distortion	No	Yes	No	Yes	No	No	No	Yes
Characterize nonlinearity	Yes	No	No	No	Yes	No	Yes	No

* Requires special hardware.

random signal consisting only of integer multiples of the fast Fourier transform frequency increment. The frequency spectrum of this signal has random amplitude and random phase distribution. Since a single history does not contain information at all frequencies, a number of histories must be involved in the measurement process. For each average, an input history is created with random amplitude and random phase. The system is excited with this input in a repetitive cycle until the transient response to the change in excitation signal decays. The input and response histories should then be periodic with respect to the sample period and are recorded as one average in the total process. With each new average, a new history, uncorrelated with previous input signals, is generated so that the resulting measurement will be completely randomized.

Random Transient (Burst Random) Excitation. A random transient signal is neither a completely transient deterministic signal nor a completely ergodic, stationary random signal but contains properties of both signal types. The frequency spectrum of this signal has random amplitude and random phase distribution and contains energy throughout the frequency spectrum. The difference between this signal and the periodic random signal is that the random transient history is truncated to zero after some percentage of the sample period (normally 50 to 80 per cent). The measurement procedure duplicates the periodic random procedure, but without the need to wait for the transient response to decay. The point at which the input history is truncated is chosen so that the response history decays to zero within the sample period. Even for lightly damped systems, the response history decays to zero very quickly because of the damping provided by the exciter system trying to maintain the input at zero. This damping, provided by the exciter system, is often overlooked in the analysis of the characteristics of this signal type. Since this measured input, although not part of the generated signal, includes the variation of the input during the decay of the response history, the input and response histories are totally observable within the sample period and the system damping is unaffected.

Increased Frequency Resolution. An increase in the frequency resolution of a frequency-response function affects measurement errors in several ways. Finer frequency resolution allows more exact determination of the damped natural frequency of each modal vector. The increased frequency resolution means that the level of a broadband signal is reduced. The most important benefit of increased frequency resolution is a reduction of the leakage error. Since the distortion of the frequency-response function due to leakage is a function of frequency spacing, not frequency, an increase in frequency resolution reduces the true bandwidth of the leakage error centered at each damped natural frequency. In order to increase the frequency resolution, the total time per history must be increased in direct proportion. This longer data acquisition time increases the variance error problem when transient signals are utilized for input and emphasizes any nonstationary problem with the data. The increase of frequency resolution often requires multiple acquisition and/or processing of the histories in order to obtain an equivalent frequency range. This increases the data storage and documentation overhead as well as extending the total test time.

The two methods of increasing the frequency resolution of a frequency-response function are (1) increasing the number of spectral lines in a baseband measurement and (2) reducing the bandwidth of the measurement while holding the number of spectral lines constant.

The frequency-shifted fast Fourier transform process for computing the frequency-response function has additional characteristics pertinent to the reduction of errors. Primarily, more accurate information can be obtained on weak spectral components if the bandwidth is chosen to avoid strong spectral components. The out-of-band rejection of the frequency-shifted fast Fourier transform is better than most analog filters that could be used in a measurement procedure attempting to achieve the same results. Additionally, the precision of the resulting frequency-response function improves because of the processor gain inherent in the frequency-shifted fast Fourier transform calculation procedure.

Weighting Functions. The use of weighting functions (data windows) is the most common method of reducing the leakage error in the frequency-response function.[55–57] (See Table 13.2 for a comparison of properties of varous types of windows.) Weighting functions may be applied to all classifications of signal averaging. The most common case is a weighting function equal to the inverse of the number of averages used in the estimate. When this weighting is used, the individual power spectra can be weighted at the end of the signal averaging or as an ongoing procedure referred to as *stable averaging*. This type of weighting introduces no further distortion in the frequency-response function estimate but also does not compensate for the leakage error.

MODAL PARAMETER ESTIMATION

Modal parameter estimation is the estimation of frequency, damping, and modal coefficients from the measured data which may be in (1) relatively raw form in terms of force and response data in the time or frequency domain or (2) in a processed form such as frequency-response or impulse-response functions. Most modal parameter estimation is based upon the measured data being the frequency-response function or the equivalent impulse-response function, typically found by inverse-Fourier-transforming the frequency-response function. Regardless of the form of the measured data, the modal parameter estimation techniques are divided into the two categories of (1) single degree-of-freedom (SDOF) approximations and (2) multiple degree-of-freedom (MDOF) approximations. Since the single degree-of-freedom equations are simply special cases of the multiple degree-of-freedom equations, all theoretical discussions are made only in terms of the multiple degree-of-freedom case. The various modal parameter estimation methods referred to in this section are shown in Table 21.2.

MATHEMATICAL MODELS

The most general model that can be used is one in which the elements of the mass, damping, and stiffness matrices are estimated, based upon measured forces and responses. Thus, the model that is used is a matrix differential equation such as
For the time domain:

$$[M]\{\ddot{x}\} + [C]\{\dot{x}\} + [K]\{x\} = \{f\} \tag{21.41}$$

For the frequency domain:

$$-[M]\{X\}\omega^2 + j[C]\{X\}\omega + [K]\{X\} = \{F\} \tag{21.42}$$

If Eq. (21.41) or (21.42) is used as the model for parameter estimation, the elements of the unknown matrices must first be estimated from the known force and response data measured in the time or frequency domain. Once the matrices have been estimated, the modal parameters can be found by the solution of the classic eigenvalue-eigenvector problem. Because of truncation of the data in terms of frequency content, limited numbers of degrees-of-freedom, and measurement errors, the matrices found by Eq. (21.41) or (21.42) are, in general, not directly comparable to matrices determined from a finite element approach. Instead, the matrices that are estimated simply yield valid input-output relationships and valid modal parameters. This is because there is an infinite number of sets of mass, damping, and stiffness matrices that yield the same modal parameters over a reduced frequency range limited to the dynamic range of the measurements. For this reason, Eq. (21.41) or (21.42) is often premultiplied by the inverse of the mass matrix so that the elements of the two matrices A and B are estimated

$$[I]\{\ddot{x}\} + [A]\{\dot{x} + [B]\{x\} = [M]^{-1}\{f\} \tag{21.43}$$

$$[I]\{X\}\omega^2 + j[A]\{X\}\omega + [B]\{X\} = [M]^{-1}\{F\} \tag{21.44}$$

Existing modal parameter estimation methods used in commercial modal analysis systems employ a model based upon measured frequency-response or impulse-response functions. While the exact model used as the basis for modal parameter estimation varies, almost all models used in conjunction with frequency-response function data can be described by a general model in the Laplace domain or by an equivalent general model in the time domain. The general model in the Laplace domain (transfer function) includes the frequency-domain model (frequency-response function) as a subset. A derivation has been given for the general formula of the transfer function of a multiple degree-of-freedom system with viscous or hysteretic damping.[11,12,32] For general viscous damping, the transfer function for a multiple degree-of-freedom mechanical system is

$$H_{ik}(\omega) = \sum_{r=1}^{n} \frac{A_{ikr}}{j\omega - p_r} + \frac{A_{ikr}^{*}}{j\omega - p_r^{*}} \tag{21.45}$$

where ω = angular frequency variable
i = measured degree-of-freedom (response)
k = measured degree-of-freedom (input)
r = modal vector number
A_{ikr} = residue
p_r = system pole
n = number of modal frequencies

Table 21.2. Summary of Characteristics of Modal Parameter Estimation Methods

	Modal Parameter Estimation Characteristics						
	Time or frequency domain	Single or multiple degrees-of-freedom	Global modal frequencies and damping factors	Repeated modal frequencies and damping factors	Global modal vectors	Global modal participation factors	Residuals
Quadrature amplitude	Freq.	SDOF	No	No	No	No	No
Kennedy-Pancu circle fit	Freq.	SDOF	No	No	No	No	Yes
SDOF polynomial	Freq.	SDOF	No	No	No	No	No
Nonlinear frequency domain	Freq.	MDOF	No	No	No	No	Yes
Complex exponential	Time	MDOF	No	No	No	No	No
Least-squares complex exponential (LSCE)	Time	MDOF	Yes	No	No	No	No
Ibrahim time domain (ITD)	Time	MDOF	Yes	No	Yes	No	No
Eigensystem realization algorithm (ERA)	Time	MDOF	Yes	Yes	Yes	Yes	Yes
Orthogonal polynomial	Freq.	MDOF	No	No	No	No	Yes
Polyreference time domain	Time	MDOF	Yes	Yes	No	Yes	No
Polyreference frequency domain	Freq.	MDOF	Yes	Yes	No	Yes	Yes
Direct parameter identification	Time	MDOF	Yes	Yes	Yes	Yes	No
Direct parameter identification	Freq.	MDOF	Yes	Yes	Yes	Yes	Yes

Other forms of this model are commonly used where certain assumptions or mathematical relationships are utilized. For example, an equivalent model can be found when the common denominator of Eq. (21.45) is formed yielding a polynomial numerator and polynomial denominator of maximum order $2n$. The denominator polynomial is then a function of the system poles. Often, an assumption is made concerning the modal vectors being normal (real) rather than complex. This reduces the number of unknowns that must be estimated by n.

Several common modal parameter estimation algorithms are based upon a time-domain model. In these cases, the general model represents the Laplace transform equivalent of the transfer function, the impulse-response function. Therefore, a mathematical expression for the impulse-response function can be obtained by a Laplace transform of Eq. (21.45)

$$h_{ik}(t) = \sum_{r=1}^{n} A_{ikr} \exp\left(p_r t\right) + A_{ikr}^* \exp\left(p_r^* t\right) \tag{21.46}$$

In general, the numerical considerations often determine which mathematical model will be most effective in the estimation of modal parameters. Time-domain models tend to provide the best results when a large frequency range or large numbers of modes exist in the data. Frequency-domain models tend to provide the best results when the frequency range of interest is limited and when the number of modes is small. While these are general considerations, the actual numerical implementation determines the ability of the algorithm to estimate modal parameters accurately and efficiently.

RESIDUALS

Continuous systems have an infinite number of degrees-of-freedom, but, in general, only a finite number of modes can be used to describe the dynamic behavior of a system. The theoretical number of degrees-of-freedom can be reduced by using a finite frequency range (f_a, f_b). Therefore, for example, the frequency response can be broken up into three partial sums, each covering the modal contribution corresponding to modes located in the frequency ranges $(0, f_a)$, (f_a, f_b), and (f_b, ∞) as shown in Fig. 21.7.

In the frequency range of interest, the modal parameters can be estimated to be consistent with Eq. (21.45). In the lower and higher frequency ranges, residual terms can be included to handle modes in these ranges. In this case, Eq. (21.45) becomes

$$H_{ik}(\omega) = P_{ik}(\omega) + \sum_{r=1}^{n} \frac{A_{ikr}}{j\omega - p_r} + \frac{A_{ikr}^*}{j\omega - p_r^*} + Q_{ik} \tag{21.47}$$

where ω = angular frequency variable
i = measured degree-or-freedom (response)
k = measured degree-of-freedom (input)
r = modal vector number
A_{ikr} = residue
p_r = system pole
n = number of modal frequencies
$P_{ik}(\omega)$ = lower residual
Q_{ik} = upper residual

In many cases the lower residual is called the *inertia restraint,* or *residual inertia,* and the upper residual is called the *residual flexibility.*[11] In this common formulation of residuals, both terms are real-valued quantities. The lower residual is a term reflecting the inertia or mass of the lower modes and is an inverse function of the frequency squared. The upper residual is a term reflecting the flexibility of the upper modes and is constant with frequency. Therefore, the form of the residual is based upon a physical concept of how the system poles below and above the frequency range of interest affect the data in the range of interest. As the system poles below and above the range of interest are located in the proximity of the boundaries, these effects are not the simple real-valued quantities noted in Eq. (21.47). In these cases, residual modes may be included in the model to partially account for these effects. When this is done, the modal parameters that are associated with these residual poles have no physical significance, but may be required in order to compensate for strong dynamic influences from outside the frequency range of interest. Using the same argument, the lower and upper residuals can take on any mathematical form that is convenient as long as the lack of physical significance is understood. Power functions of frequency (zero, first, and second order) are commonly used within such a limitation. In general, the use of residuals is confined to frequency-response function models. This is primarily due to the difficulty of formulating a reasonable mathematical model and solution procedure in the time domain for the general case that includes residuals.

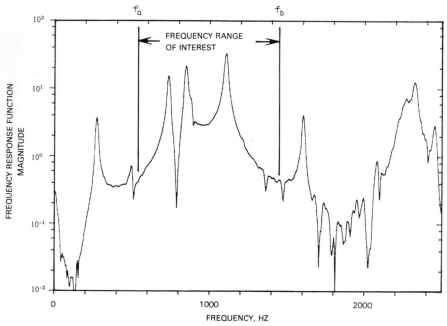

Fig. 21.7. Frequency-response function indicating the use of a partial frequency range in the modal parameter estimation. Note that f_a is the lowest frequency of interest and f_b is the highest frequency of interest.

ORDER OF THE MODEL

The *order* of the model equals the number of unknowns that must be estimated in the model. In the modal parameter estimation case, this refers to the frequency, damping, and complex modal coefficient for each mode of vibration at every measurement degree-of-freedom plus any residual terms that must be estimated. Therefore, the order of the model is directly dependent on the number of modal frequencies, n, that are to be estimated. For example, for a system with n modes of vibration, assuming that no residuals were required, $4n$ unknowns must be estimated. For cases involving real measured data, the order of the model is an extremely important decision since the estimates of modal parameters are affected by the choice of the order of this model. The difficulty in such cases arises from the inability to be certain that the correct order of the model has been chosen. If the number of modes of vibration is more or less than n, modes of vibration will be found that do not exist physically or ones that actually do occur will be missed. In addition, the values of the frequency, damping, and complex modal coefficient for the physically realizable modes of vibration will be affected by an inappropriate choice of the number of modes of vibration.

The number of modes of vibration is normally chosen between one and an upper limit, dependent on the memory limitations of the computational hardware. The true number of system poles is a function of the frequency range of the measurements used to estimate the modal parameters. By observing the number of peaks in the frequency-response function, the minimum number of system poles can be estimated. This estimate is normally low, based upon poles occurring at nearly the same frequency (pseudorepeated roots), limits on dynamic range, and poorly excited modes. For these reasons, the estimate of the correct order of the model is often in error. When the order of the model is other than optimum, the estimate of the modal parameters will be in error.

Many of the parameter estimation techniques that are used assume that only one mode exists in a limited range of interest and all of the other modes appear as residual terms. For this case Eq. (21.47) can be rewritten as

$$H_{ik}(\omega) = P_{ik}(\omega) + \frac{A_{ikr}}{j\omega - p_r^*} + \frac{A_{ikr}^*}{j\omega - p_r^*} + Q_{ik} \tag{21.48}$$

SOLUTION PROCEDURE

Equations (21.45) and (21.46) represent a nonlinear model in terms of the unknown modal parameters. This can be noted from the unknowns in the numerator and denominator of Eq. (21.45) and the unknowns as the argument of the transcendental functions of Eq. (21.46). The nonlinear aspect of the model must be treated in one of two ways:

1. By the use of an iterative solution procedure to solve the nonlinear estimation problem, allowing all modal parameters to vary according to a constraint relationship until an error criterion reaches an acceptably low value

2. By separating the nonlinear estimation problem into two linear estimation problems.

For the case of structural dynamics, the common technique is to estimate $2n$ frequencies and damping values in a first stage and then to estimate the $2n$ modal coefficients plus any residuals in a second stage.

In the iterative technique in the solution of the nonlinear estimation approach, a set of starting values must be chosen to initiate the sequence. The number and value of these starting values affect the final result. Poor initial estimates can lead to problems of convergence, as a result of which close operator supervision usually is required for a successful use of this technique.

An alternative method is to reformulate the nonlinear problem into a number of linear stages so that each stage is stable. The actual data that are used in the estimate of the modal parameters also affects the results. Based on the choice of the order of the model, n, there are $4n$ modal parameters to be estimated. If residuals, in one form or another, are also included, the number of modal parameters to be estimated will be slightly higher. The common method of solving for these unknown modal parameters is to find an equation involving known information for every unknown that will be found. In this case, the frequency-response function or impulse-response function that has been measured provides the known information and Eq. (21.45) or (21.46) can be repeated for different frequencies or time values in order to obtain the sufficient number of equations. These equations, for the linear case, can then be solved simultaneously for the unknown modal parameters. As an illustration of this relationship, consider a common modal parameter situation in which the number of modes in the frequency range of interest is between 1 and 30. Assuming the highest ordered model means that slightly more than 120 modal parameters must be estimated. From a single frequency-response measurement, 1,024 known values of the function will be available (512 complex values at successive values of frequency). Many more equations, based on the known values of frequency response, can be formed than is needed to find the unknown modal parameters. An obvious solution is to choose enough equations to solve for the modal parameters. The problem arises in determining what part of the known information is to be involved in the solution. As different portions of the known data (data near a resonance compared to an antiresonance, for example) are used in the solution, the estimates of modal parameters vary. As the quality of the data is marginal, this variance can be quite large. When the modal parameters that are estimated appear to be nonphysical, this is often the reason.

To solve this problem, all or a large portion of the data can be used if an averaging type of solution procedure is used. One procedure that is used to formulate the problem so as to minimize the squared error between the data and the estimated model is the least-squares error method. This is the most commonly used technique in the area of modal parameter estimation. If there are many more known pieces of information than unknowns that must be estimated, many more equations can be formed than are needed to solve for the unknowns. The least-squares error method allows for all of these redundant data to be used to estimate the modal parameters in a computationally efficient manner. The least-squares error method usually can be derived directly from the linear equations using a normal equations or pseudoinverse approach. In general, this procedure does not increase the memory or computational requirements of the computational hardware significantly. Any solution procedure that can be used is only estimating the "best" solution based upon the choice of the model, the order of the model, and the known measured data used in the model.

GLOBAL MODAL PARAMETERS

Theoretically, modal parameters are considered to be unique based upon the assumption that the system is linear and time invariant. Therefore, the modal frequencies can be determined from any measurement and the modal vectors can be determined from any reference condition. If multiple measurements or reference conditions are utilized, the possibility of several, slightly different, answers for each modal parameter exists. The concept of *global modal paramters*, as it applies to modal parameter estimation, means that there is only one answer for each modal parameter and that the modal parameter estimation solution procedure enforces this constraint. Every frequency-response or impulse-response function measurement theoretically contains the information that is represented by the characteristic equation, the modal frequencies, and damping. If individual measurements are treated in the solution procedure independent of one another, there is no guarantee that damping and a single set of modal frequencies will be generated. In a like manner, if more than one reference is measured in the data set, redundant estimates of the modal vectors can occur unless the solution procedure utilizes all references in the estimation process simultaneously. Most modal parameter estimation algorithms estimate the modal frequencies and damping in a global sense, but few estimate the modal vectors in a global sense.

MODAL PARTICIPATION FACTORS

A *modal participation factor* is a complex-valued scale factor that is the ratio of the modal coefficient at one reference degree-of-freedom to the modal coefficient at another reference degree-of-freedom. Modal participation factors can be computed any time that multiple reference data are measured, and such factors are used in multiple reference modal parameter estimation algorithms. Modal participation factors relate how well each modal vector is excited from each of the reference locations. This information is often used in a weighted least-squares error solution procedure to estimate the modal vectors in the presence of multiple references. Theoretically, these modal participation factors should be in proportion to the modal coefficients of the reference degrees-of-freedom for each modal vector. The use of modal participation factors in a solution procedure enforces the constraint concerned with Maxwell's reciprocity between the reference degrees-of-freedom. Most multiple reference modal parameter estimation methods estimate modal participation factors as part of the first stage estimation of global modal frequencies and damping.

REFERENCES

1. Allemang, R. J.: *Modal Testing and Model Refinement*, ASME AMD, **59**:1 (1983).
2. Lewis, R. C., and D. L. Wrisley: *Journal of Aeronautical Sciences*, **17**,(11):705 (1950).
3. De Veubeke, B. F.: *AGARD Rept.* 39, 1956.
4. Trail-Nash, R. W.: *J. Aeronaut. Sci.*, **25**(12):775 (1958).
5. Asher, G. W.: *Dynamics of Aeroelasticity*, Proc. Inst. Aeronaut. Sci., 1958, p. 69.
6. Ibanez, P.: *SAE Paper* No. 760873, 1976.
7. Morosow, G., and R. S. Ayre: *Shock and Vibration Bull.*, **48**(1):39 (1978).
8. Stahle, C. V., and W. R. Forlifer: *Flight Flutter Testing Symposium*, NASA-SP-385, 1958, p. 83.
9. Pendered, J. W., and R. E. D. Bishop: *J. Mech. Eng. Sci.*, **5**,(4):379 (1963).
10. Bishop, R. E. D., and G. M. L. Gladwell: *Phil. Trans., Royal Society of London*, **225**(A-1055):241 (1963).
11. Klosterman, A.: *Ph.D. Dissertation*, Univ. of Cincinnati, Mech. Eng. Dept., 1971.
12. Richardson, M., and R. Potter: *Instrument Society of America* ISA-ASI-74-250, 1974, p. 239.
13. Potter, R. W.: *Shock and Vibration Digest*, **7**(11):8 (1975).
14. Wada, B. K.: *SAE Paper* No. 751066, 1975.
15. Ramsey, K.: *Sound and Vibration*, **9**(11): (1975).
16. Ramsey, K.: *Sound and Vibration*, **10**(4): (1976).
17. Stahle, C. V.: *J. Environ. Sci.*, **21**(1):24 (1978).
18. Allemang, R. J.: *Ph.D. Dissertation*, Univ. of Cincinnati, Mech. Eng. Dept., 1980.
19. Brown, D. L., G. D. Carbon, and K. Ramsey: *SAE Paper* No. 770029, 1977.
20. Allemang, R. J., R. W. Rost, and D. L. Brown: *ASME Paper* No. 83-DET-73, 1983.
21. Allemang, R. J., and D. L. Brown: *Analysis Conf.*, 1982, p. 110.
22. Rost, R. W.: *Ph.D. Dissertation*, Univ. of Cincinnati, Mech. Eng. Dept., 1985.
23. Allemang, R. J., D. L. Brown, and R. W. Rost: *SAE Paper* No. 820193.
24. Allemang, R. J., D. L. Brown, and R. R. Rost: *U.S. Air Force Report* No. AFATL TR 84-15, 1984.
25. Prony, R.: *J. de l École Polytechnique (Paris)*, **1**(2): Floreal et Prairial, An. III, 24 (1795).
26. Spitznogle, F. R., et al.: *Texas Instruments, Inc. Report* No. U1-829401-5, *O.N.R. Contract* No. N00014-69-C-0315, 1971.
27. Brown, D. L., R. D. Zimmerman, R. J. Allemang, and M. Mergeay: *SAE Paper* No. 790221, *SAE Transactions*, **88**:828 (1979).
28. Halvorsen, W. G., and D. L. Brown: *Sound and Vibration*, 11(11):8 (1977).
29. Imes, R. S., W. P. Jennings, and N. L. Olsen: *AIAA Paper* No. 78-0505, 1978.
30. Stenger, G.: *M.S. Thesis*, Univ. of Cincinnati, Dept. Mech. Eng., 1979.
31. Elliott, K. B., and L. D. Mitchell: *Modal Testing and Model Refinement*, ASME AMD, **59**:63 (1983).
32. Richardson, M., and R. Potter: *Proc. 46th Shock Vibration Symp.* 1975.
33. Kennedy, C. C., and C. D. P. Pancu: *J. Aeronaut. Sci.*, **14**(11):603 (1947).
34. Coppolino, R. N.: *SAE Paper* No. 811046, 1981.
35. Vold, H., et al: *SAE Paper* No. 820194, 1982.
36. Vold, H., and T. Rocklin: *Proc. Intern. Modal Analysis Conf.*, 1982, p. 542.
37. Leuridan, J.: *Ph.D. Dissertation*, Univ. of Cincinnati, Dept. Mech. Eng., 1984.
38. Ibrahim, S. R.: *Shock and Vibration Bull.*, **18**(1):65 (1970).
39. Ibrahim, S. R., and E. C. Mikulcik: *Shock and Vibration Bull.*, **47**(4):183 (1977).
40. Ibrahim, S. R.: *AIAA Paper* No. 77-368, 1977.
41. Pappa, R. S., and S. R. Ibrahim: *Shock and Vibration Bull.*, **51**(3):43 (1981).
42. Pappa, R. S.: *AIAA Paper* No. 82-0768, 1982.
43. Pappa, R. S., and J. N. Juang: *AIAA Paper* No. 84-1070-CP, 1984.
44. Pappa, R. S. and J. N. Juang: *NASA-JPL Workshop, Identification and Control of Flexible Space Structures*, (Pasadena, CA.), June, 1984.
45. Juang, J. N., and R. S. Pappa: *AAS Paper* No. AAS-85-422, August, 1985.
46. Gersch, W.: *J. Sound Vibration*, **34**(1):63 (1974).
47. Gersch, W., and S. Luo: *J. Acoust. Soc. Amer.*, **51**(1):402 (1972).
48. Gersch, W., and D. R. Sharpe: *IEEE Trans. Automatic Control*, vol. AC-18, p. 367, August 1973.
49. Gersch, W., and D. A. Fouth: *IEEE Trans. Automatic Control*, vol. AC-19, p. 898, December 1974.
50. Pandit, S. M., and H. Suzuki: *ASME Paper* No. 79-DET-7, September, 1979.
51. Ross, R. G., Jr.: *SAE Paper* No. 710787, 1971.
52. Thoren, A. R.: *AIAA Paper* No. 72-346, 1972.
53. Potter, R. W., and M. Richardson: *Instrument Society of America* ISA-74-630, 1974.

54. Berman, A., and W. G. Flannelly: *AIAA J.*, **9**(8):1481 (1971).
55. Bendat, J. S., and A. G. Piersol: "Random Data: Analysis and Measurement Procedures," John Wiley and Sons, Inc., New York, 1971.
56. Otnes, R. K., and L. Enochson: "Digital Time Series Analysis," John Wiley and Sons, Inc., New York, 1972.
57. Bendat, J. S., and A. G. Piersol: "Engineering Applications of Correlation and Spectral Analysis," John Wiley and Sons, Inc., New York, 1980.
58. Leuridan, J., D. Brown, and R. Allemang: *AIAA Paper* No. 82-0767, *Proc. 23rd. Structures, Structural Dynamics and Materials Conf.*, Part 2, 1982, p. 548.
59. Leuridan, J., and J. Kundrat: *Proc. Intern. Modal Analysis Conf.*, 1982, p. 192.
60. Bendat, J. S.: *J. Sound Vibration*, **44**(3):311 (1976).

22

CONCEPTS IN VIBRATION DATA ANALYSIS

Allen J. Curtis
Hughes Aircraft Company

INTRODUCTION

This chapter discusses the mathematical concepts involved in the analysis of vibration data which are essentially steady-state or slowly time-varying processes. It provides the mathematical basis for data reduction procedures by which the original time-history of vibration is transformed to other forms required for particular applications. Classes of data discussed in this chapter include sinusoidal, periodic, complex, and random (shock and transient vibration are discussed in Chap. 23).

The basic objectives of vibration data analysis, discussed in Chap. 21, may be achieved by various means. No one method of analysis can be considered *the* correct one, since both the type of data and the use to be made of the results influence the appropriateness of a particular type of analysis. This chapter is restricted to a description of the basic principles and techniques of data analysis which will enable the reader to make a logical choice of the type of analysis and data presentation most suited to the particular problem at hand.

The techniques of data analysis are illustrated in this chapter by functional block diagrams and are necessarily described in terms associated with analog processing of the signals. With the advent of minicomputers and implementation of the fast Fourier transform algorithm, it became practical to perform these analyses digitally. Chapter 27 describes the use of digital computers for performing the types of vibration data analyses described in principle in this chapter.

The complete definition of a vibration condition requires the description of magnitude,* and its variation with both frequency and time. Various types of vibration are described mathematically, and the respective magnitude parameters defined explicitly or statistically. The determination of the statistical properties of random vibration is discussed, including the probability density and probability distribution functions. The determination of the frequency or spectral characteristics of the vibration magnitude at a given time is discussed, followed by a discussion of means of determining the time variation of the magnitude of the vibration. Finally, autocorrelation and crosscorrelation are discussed as means to determine both the statistical and spectral characteristics of vibration.

---

* The word "magnitude" is used in this chapter to indicate the severity of vibration without relating it specifically to one of the various mathematical expressions used to describe vibration quantitatively.

QUALITATIVE DESCRIPTION OF VIBRATION

This section considers various means which are commonly used in vibration analysis to describe time-varying functions such as displacement, velocity, acceleration, or force, and describes the principal characteristics of these functions.

Figure 22.1 illustrates three typical vibration waveforms or time-histories. The solid curve of Fig. 22.1A is the sum of the two sinusoidal vibrations shown by the dotted lines. Figure 22.1B represents broad-band random vibration, while Fig. 22.1C illustrates the waveform of a narrow-band random vibration. The latter has the appearance of a sine wave whose amplitude, represented by the dotted lines, varies in an unpredictable man-

FIG. 22.1. Typical vibration time-histories: (A) Sum of two sinusoids; (B) broad-band random vibration; (C) narrow-band random vibration.

FIG. 22.2. Results of data analysis of time-histories of Fig. 22.1: (A)· Sum of two sinusoids; (B) broad-band random vibration; (C) narrow-band random vibration.

ner. If the envelope of the waveform varies sinusoidally, the waveform is the sum of two sinusoids whose frequency difference is small compared to the sum, resulting in beats.

Figure 22.2 illustrates typical forms for describing, after suitable data analysis, the time-histories shown in Fig. 22.1. In Fig. 22.2A, the two sine waves are represented by vertical lines at their respective frequencies; the height of each line is proportional to the amplitude of the respective sine wave. This plot is known as a *line spectrum* or *discrete frequency spectrum*. The information also can be presented as a tabulation of frequencies and associated amplitudes. For complete definition of the waveform of Fig. 22.1A, it is necessary to specify the phase angle between the two sine waves at a convenient time origin. However, phase information seldom is required in vibration data analysis and the plot of Fig. 22.2A is the extent of the information usually required.

Figure 22.2*B* is a plot of the power spectral density* vs. frequency of broad-band random vibration showing the frequency content to be spread over a wide frequency range. Figure 22.2*C* illustrates the power spectral density of narrow-band random vibration showing the concentration of the frequency content in a narrow band of frequencies.

In general, three coordinates are necessary to describe a vibration time-history: (1) the magnitude of the vibration; e.g., the amplitude of sine waves or the power spectral density; (2) the variation of magnitude with frequency, i.e., spectral characteristics; and (3) the variation of the magnitude of the vibration at a particular frequency as a function of time.

The magnitude of a vibration may be represented in terms of its spectral characteristics and its variation with time by the surface shown in Fig. 22.3. The height of a point

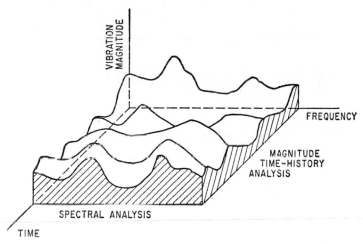

FIG. 22.3. Vibration magnitude surface: the height of the surface at any time and frequency represents the vibration magnitude at that time and frequency. Units of vibration magnitude depend on type of vibration, e.g., random, complex, or periodic.

on the surface raised over the time-frequency plane defines the magnitude of the vibration at a particular time and frequency. For example, the surface could represent the vibration encountered during the flight of an aircraft from take-off to landing. As the flight conditions change, both the vibration magnitude and the frequency characteristics of the vibration may change.

At a specific instant of time, say t_0, the vibration is described by its variation of magnitude with frequency in the plane normal to the time axis passing through t_0, i.e., in the plane labeled "Spectral Analysis" in Fig. 22.3 and shown in the two-dimensional plot of Fig. 22.4.

The variation of the magnitude of vibration at a particular frequency, say f_0, as a function of time is shown by passing a plane normal to the frequency axis through f_0, i.e., in the plane labeled "Magnitude–Time-history Analysis" in Fig. 22.3 and shown in the two-dimensional plot of Fig. 22.5.

In the case of truly steady-state vibration, i.e., vibration independent of time as during some laboratory vibration tests, the surface of Fig. 22.3 can be represented by the magnitude-frequency plot of Fig. 22.4. Similarly, a vibration consisting of a single sine

---

* Power spectral density is defined under *Spectral Analysis* [see Eq. (22.17) in this chapter; also see Eq. (11.16)]. It is a generic term used regardless of the physical quantity represented by the time-history. However, it is preferable to indicate the physical quantity involved. For example, the terms "mean-square acceleration density" or "acceleration spectral density" are used when the time-history of acceleration is to be described.

Fig. 22.4. Typical plot of spectral analysis.   Fig. 22.5. Typical plot of magnitude time-history analysis.

wave at a constant frequency whose amplitude varies slowly with time may be represented by a single amplitude-time plot such as Fig. 22.5.

BASIC TYPES OF TIME-VARYING FUNCTIONS

This section considers certain relations used to describe specific time-varying functions in vibration analysis and describes their principal characteristics. Types of functions discussed are sinusoidal, periodic,[1-3] phase-coherent,[4,5] complex, and random.[6-8, 29, 32] In the following sections, $\xi(t)$ is the function which represents the time variation of any physical parameter such as displacement, force, or acceleration.

SINUSOIDAL FUNCTIONS.   The simplest time-varying quantity of concern in vibration analysis is a sinusoidal function having constant amplitude ξ_0 and frequency f:

$$\xi(t) = \xi_0 \sin (\omega t + \theta) \tag{22.1}$$

where $\omega = 2\pi f$, and θ is the phase angle with respect to the time origin. The objective of data analysis on this waveform is the determination of ξ_0, ω, and, less frequently, θ.

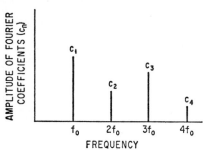

Fig. 22.6. Line spectrum to represent Fourier series of periodic function. Fundamental frequency $f_0 = 1/\tau$, where τ is period of function [see Eq. (22.8)].

PERIODIC FUNCTIONS.   If a function has a waveform which repeats itself exactly in consecutive time periods τ:

$$\xi(t) = \xi(t \pm n\tau) \tag{22.2}$$

where n is any integer; then it is said to be a *periodic function*[1-3] having a fundamental frequency $f = 1/\tau$. For example, a wing panel adjacent to a propeller may vibrate periodically at a fundamental frequency equal to the cyclic rate at which propeller blades pass the panel and also at higher frequencies equal to multiples of the blade passage frequency. A periodic function may be represented by the line spectra shown in Fig. 22.6.

PHASE-COHERENT FUNCTIONS.   A phase-coherent function[4,5] is defined as a sine wave whose amplitude and frequency vary smoothly with time. It is assumed that the changes in amplitude and frequency are significant in magnitude only over a time duration equal to a number of cycles of the waveform. Mathematically, it is expressed by

$$\xi(t) = \xi_0(t) \sin [\omega(t)t + \theta] \tag{22.3}$$

where ξ_0 and ω are functions of time and θ is the phase angle of the waveform at the time origin. A function of this type is useful in a description of vibration measured in a system whose natural frequencies vary with time, such as a rocket engine during the burning of the propellant. Equation (22.3) defines the input excitation to the test object during a sweep frequency vibration test.

COMPLEX FUNCTIONS. A time-history which is composed of a sum (or in the limit, integral) of sinusoidal functions which is not necessarily periodic is defined as a complex function. Such functions may be represented graphically by line spectra as shown in Fig. 22.6 except that the lines are not equally spaced along the frequency axis; when the complex function consists of an integral or continuous distribution of sinusoidal functions, it may be represented by a Fourier spectrum as discussed in Chap. 23.

TRANSIENT FUNCTION—SHOCK. A transient function is a time-varying function which is nonzero only over a given restricted time interval. Transients are discussed in Chap. 23.

RANDOM FUNCTIONS. A random time-function[6–9] (see Chap. 11) consists of a continuous distribution of sine waves at all frequencies, the amplitudes and phase angles of which vary in an unpredictable manner as a function of time. This unpredictability distinguishes random functions from "deterministic" or coherent functions (e.g., sinusoidal, periodic, and complex functions) in the following respect. Knowledge of the instantaneous value of a random function at a given time does not provide any information on the value which it will have at some specific later time. It is known only that there is a certain probability that the value will lie within a certain range of values, i.e., it can be described only in statistical terms. In contrast, if the value of a deterministic function is known at any time, the value at any other time can be determined.

DEFINITION OF MEAN, AVERAGE, AND MEAN-SQUARE VALUES.
Mean Value. The mean value of a function $\xi(t)$ is defined by

$$\bar{\xi} = \frac{1}{T} \int_0^T \xi(t)\, dt \tag{22.4}$$

where T is the "averaging" time.
Average Value. The term "average value" of a function $\xi(t)$ is employed to denote the time average of the magnitude of a function with zero mean value.

$$\overline{|\xi|} = \frac{1}{T} \int_0^T [\xi(t)]\, dt \tag{22.5}$$

Mean-square Value. The mean-square value of a function $\xi(t)$ is the time average of the square of the function:

$$\overline{[\xi]^2} = \frac{1}{T} \int_0^T [\xi(t)]^2\, dt \tag{22.6}$$

The rms value is the square root of the mean-square value.
For example, the mean, average, and mean-square values of a sine wave [Eq. (22.1)] averaged over an integer number of periods, are

$$\bar{\xi} = 0$$

$$\overline{|\xi|} = \frac{2}{\pi}\, \xi_0 \tag{22.7}$$

$$\overline{[\xi]^2} = \tfrac{1}{2}\xi_0^2$$

ANALYSIS OF PERIODIC FUNCTIONS

FOURIER SERIES

Periodic functions encountered in vibration analysis may be represented by a Fourier series[1–3] which consists of a sum of sine waves whose frequencies are all multiples of the fundamental frequency. The amplitudes of the sine waves are known as the Fourier coefficients c_n. An additional constant term equal to the mean value of the waveform during the period τ must also be included. The function is expressed mathematically as

$$\xi(t) \;=\; \sum_{n=1}^{\infty} a_n \sin n\omega t \;+\; \sum_{n=0}^{\infty} b_n \cos n\omega t \;=\; \sum_{n=0}^{\infty} c_n \sin \,(n\omega t + \theta_n) \tag{22.8}$$

where

$$c_n = +[a_n{}^2 + b_n{}^2]^{1\!/2} \qquad \theta_n = \sin^{-1} \frac{b_n}{c_n} = \cos^{-1} \frac{a_n}{c_n}$$

$$a_n = \frac{2}{\tau} \int_{-\tau/2}^{\tau/2} \xi(t) \sin n\omega t \; dt$$

$$b_n = \frac{2}{\tau} \int_{-\tau/2}^{\tau/2} \xi(t) \cos n\omega t \; dy$$

$$b_0 = \frac{1}{\tau} \int_{-\tau/2}^{\tau/2} \xi(t) \; dt = \bar{\xi}$$

If the function $\xi(t)$ is an *even function*, i.e., $\xi(t)$ equals $\xi(-t)$, the series consists of cosine terms only; for an *odd function*, $\xi(t)$ equals $-\xi(-t)$ and the series consists of sine terms only.

OBJECTIVE AND RESULTS OF ANALYSIS

The objective of the analysis of periodic functions is to determine the Fourier coefficients c_n and the phase angles θ_n of Eq. (22.8). By describing a function as a Fourier series, it is implied that the waveform during the period τ is repeated exactly in all successive periods between negative and positive infinite times and that all frequencies associated with the waveform are determined by τ, as shown by Eq. (22.8). No measured vibration data completely fulfill these requirements for description by a Fourier series. When conducting a Fourier series analysis, it is common practice to select what appears to be a typical section of the time-history and to treat it as if it were the time-history during one period of a periodic function. The consequences of this selection are (1) the time period chosen fixes the frequencies of the resulting line spectra, whether these are the actual frequencies present in the function or not, and (2) the signal is assumed, a priori, to consist of a sum of pure sinusoids.

The results of data analysis involving periodic vibration may be summarized by plotting the discrete frequencies and the corresponding amplitudes. For example, Fig. 22.2A applies to two superimposed sinusoidal functions of arbitrary frequencies f_1, f_2; Fig. 22.6 is an analogous plot of the Fourier coefficients of a periodic quantity. This form of data presentation may be abbreviated by plotting only points representing the upper ends of the spectrum lines in Figs. 22.2A and 22.2B. Representation of the data by such points is particularly advantageous for summarizing the results of analysis of the vibration of a number of data samples. For example, the vibration at a number of different flight conditions or on a number of different aircraft may be shown on a single plot, so that each point represents one measurement of a sinusoidal function at the given amplitude and frequency. If the number of points becomes excessive, different-sized symbols are used to represent given numbers, say 1, 10, 50, or 100 readings at the particular amplitude and frequency. The density of points in various frequency regions or at various amplitudes indicates, at least qualitatively, the frequency of occurrence of those values.

DETERMINATION OF FOURIER COEFFICIENTS

The Fourier coefficients of a periodic function are defined by Eq. (22.8) and may be determined by several alternate means. If an oscillographic recording of the function $\xi(t)$ during one complete period is obtained and the instantaneous value of $\xi(t)$ at N equally spaced time intervals is measured, the Fourier coefficients c_n and phase angles θ_n from $n = 0$ to $n = (N - 1)/2$ may be calculated by tabular methods.[1, 3, 10, 11] A detailed description of these tabular methods is given with respect to Fig. 38.12. Similar computations also may be carried out by digital computers.

A machine known as a Coradi* analyzer may be used to determine the first five Fourier coefficients from an oscillographic record, provided the time scale is expanded to yield 40 cm of record for one period of the function. The machine solves Eq. (22.8) by the use of planimeter-type mechanical integrators while the operator traces the time-history with a stylus. A similar mechanical analyzer known as the Harvey Harmonic Analyzer[11] is capable of determining the first fourteen coefficients. However, because of their complexity, such machines are rarely used. Instead, electrical analysis techniques are employed.

When the time-history is available as an analog voltage, filters may be used to determine the Fourier components as described later in this chapter under *Spectral Analysis*.

STATISTICAL ANALYSIS

Statistical analysis[13–20, 29–32] of a vibration time-history is carried out to determine the characteristics, e.g., sinusoidal and/or random, of the time-history so that the appropriate mathematical function for description of the vibration magnitude may be chosen. This section describes the statistical properties of random and sinusoidal functions, and discusses concepts of data analysis for determining the significant statistical parameters of the time-histories.

STATISTICAL PARAMETERS OF RANDOM FUNCTIONS

Statistical parameters used frequently in data analysis in addition to the mean and mean-square values defined by Eqs. (22.4) and (22.6) are the probability density function and probability distribution function.

PROBABILITY DENSITY FUNCTION. The probability density function $p(\xi/\sigma)$, illustrated in Fig. 22.7, defines the probability (or fraction of time, on the average) that

Fig. 22.7. Normalized probability density functions: (A) Gaussian or normal distribution [Eq. (22.10)]; (B) distribution of instantaneous values of a sine wave [Eq. (22.11)]. The curve (A) marked $\times 10^{-2}$ indicates hundredfold expansion of ordinate scale.

* The Henrici analyzer is a very similar device.

the magnitude of the quantity $\xi(t)$ will lie between two values. It is customary to normalize the curve by plotting the magnitude divided by the rms value σ as the abscissa. Then the probability that the magnitude lies between ξ/σ and $(\xi + d\xi)/\sigma$ is equal to $p(\xi/\sigma)d(\xi/\sigma)$, i.e., the shaded area shown in Fig. 22.7. Since it is certain, with probability 1.0, that the function $\xi(t)$ lies between plus and minus infinity, the area under the entire curve is unity.

PROBABILITY DISTRIBUTION FUNCTION. The probability distribution function (or *cumulative distribution function*) $P(\xi/\sigma \geq)$* defines the probability that the magnitude of ξ/σ will exceed a certain value. However, from the definition of the probability density function, $P(\xi/\sigma \geq)$ is equal to the area under the plot of $p(\xi/\sigma)$ as a function of ξ/σ between ξ/σ and infinity:

$$P(\xi/\sigma \geq) = \int_{\xi/\sigma}^{\infty} p\left(\frac{\xi'}{\sigma}\right) d\left(\frac{\xi'}{\sigma}\right) \tag{22.9}$$

where, for this equation and Eq. (22.12) only, ξ' is used as a dummy variable of integration for ξ.

The value of the distribution function $P(\xi/\sigma \geq)$ at the minimum possible value of ξ is $P(\xi_{min}/\sigma) = 1.0$ since ξ is never less than ξ_{min}; for the maximum possible value of ξ, $P(\xi_{max}/\sigma)$ is zero since ξ never exceeds ξ_{max}. Further, $P(\xi/\sigma \geq)$ must decrease monotonically between ξ_{min} and ξ_{max}, as illustrated in Fig. 22.11. Since the probability of exceeding a value is one minus the probability of not exceeding that value, the ordinate of Fig. 22.11 may be changed to $P(\xi/\sigma \leq)$ if the scale is changed to vary from 1 to 0 instead of 0 to 1.

COMPARISON OF PARAMETERS FOR SINUSOIDAL AND

RANDOM FUNCTIONS

Although a sinusoidal function is deterministic, the probability density and probability distribution of a sine wave may be determined for comparison with those of a random function. The comparison may be made on the basis of (1) the instantaneous values $\xi(t)$ of the function and (2) the peak values or maxima $\xi_p(t)$ of the function.

DISTRIBUTION OF INSTANTANEOUS VALUES OF $\xi(t)$**.** A comparison of the identifying characteristics of the probability densities and distributions for the instantaneous values of a sinusoidal function and the particular case of Gaussian random vibration is shown in Figs. 22.7, 22.8, and 22.9. The term "Gaussian" is used to describe a random function whose *instantaneous* value is defined by the Gaussian or normal probability density function [see Eq. (11.10)] given by

$$p(\xi/\sigma) = \frac{1}{\sqrt{2\pi}} e^{-(\xi^2/2\sigma^2)} \tag{22.10}$$

where σ is the rms value. Equation (22.10) is shown by curve A of Fig. 22.7. The prob-

FIG. 22.8. Normalized probability density functions: (A) Rayleigh distribution for peaks of narrow-band Gaussian vibration [Eq. (22.14)]; (B) distribution for peaks of sine-wave delta function at $\xi_p/\sigma = \sqrt{2}$. The curve (A) marked $\times 10^{-2}$ indicates hundredfold expansion of ordinate scale.

---

* The probability that ξ/σ is "greater than" is written $P(\xi/\sigma \geq)$; conversely, the probability that ξ/σ is "less than" is written $P(\xi/\sigma \leq)$. $P(\xi/\sigma \geq) = 1 - P(\xi/\sigma \leq)$.

FIG. 22.9. Probability distribution functions: (*A*) Instantaneous values of broad-band and narrow-band random vibration—Gaussian distribution [Eq. (22.12)]; (*B*) peaks of broad-band random vibration [Eq. (22.12)]; (*C*) peaks of narrow-band random vibration [Eq. (22.15)]; (*D*) instantaneous values of a sine wave [Eq. (22.13)]; (*E*) peak values of a sine wave. Multiply ordinate scale by factors marked adjacent to curves for large values of ξ/σ.

ability density function of the *instantaneous* value of a sinusoid is shown by curve *B* of Fig. 22.7 and is defined by

$$p(\xi/\sigma) \;=\; \frac{1}{\pi \sqrt{2 - (\xi/\sigma)^2}} \tag{22.11}$$

The probability density functions of Eqs. (22.10) and (22.11) are symmetrical about a mean assumed to be zero; then the probability that ξ exceeds a given absolute value (or magnitude) $|\xi|$ is twice the probability that it exceeds the same absolute value in either the positive or negative sense [see Eq. (22.9) or Fig. 22.7]. Therefore, it is convenient to plot the probability distribution function in terms of the absolute value of ξ, i.e., $P(|\xi|/\sigma\geq)$, as shown in Fig. 22.9.

The probability distribution functions for the Gaussian and sinusoidal functions are obtained by integration of Eqs. (22.10) and (22.11):

Gaussian:

$$P(|\xi|/\sigma>) \;=\; \frac{2}{\sqrt{2\pi}} \int_{\xi/\sigma}^{\infty} e^{-(\xi'^2/2\sigma^2)} d\left(\frac{\xi'}{\sigma}\right) \tag{22.12}$$

Sinusoid:

$$P(|\xi|/\sigma\geq) \;=\; \frac{2}{\pi} \cos^{-1} \frac{\xi}{\sigma\sqrt{2}} \tag{22.13}$$

The relations of Eqs. (22.12) and (22.13) are plotted as curves *A* and *B* in Fig. 22.9.

DISTRIBUTION OF PEAK VALUES (MAXIMA) OF $\xi(t)$. When the peak values or maxima of a function are considered, the two statistical functions differ from those found for the instantaneous values. For a sine wave, all maxima are of equal magnitude and the probability density function $p(\xi_p/\sigma)$ becomes a Dirac delta function[2] (see Chap. 23) as shown by curve *B* of Fig. 22.8. For broad-band Gaussian vibration, i.e., vibration with nonzero spectral density over a frequency bandwidth which is not small compared to the average or center frequency of the bandwidth, the distribution of peak values is normal, as shown by curves *A* of Fig. 22.7 and *B* of Fig. 22.9. However, for narrow-band Gaussian noise, i.e., noise with negligible spectral density except in a frequency bandwidth which is small compared to the center frequency, the distribution of peak values becomes the *Rayleigh distribution*.   The probability density and distribution functions for the Rayleigh distribution are defined by:

Probability density:

$$p(|\xi_p|/\sigma) = \frac{\xi_p}{\sigma}\, e^{-(\xi_p^2/2\sigma^2)} \tag{22.14}$$

Probability distribution:

$$P(|\xi_p|/\sigma) = e^{-\xi_p^2/2\sigma^2} \tag{22.15}$$

The relations given by Eqs. (22.14) and (22.15) are shown graphically by curve A of Fig. 22.8 and curve C of Fig. 22.9, respectively.

Figures 22.7 to 22.9 are plotted on the basis that the mean value of the function $\xi(t)$ is zero.   In the more general case of nonzero mean value, the abscissae of these figures are $(\xi - \bar{\xi})/\sigma$, where $\bar{\xi}$ is the mean value of the function $\xi(t)$.   However, in vibration data analysis, it is common practice to constrain the mean value to zero by use of a high-pass or bandpass filter prior to statistical analysis.   Thus these figures show the form usually encountered.

Figures 22.7 to 22.9 show that the probability density and probability distribution curves for sinusoidal and random functions differ considerably, thus providing identifying characteristics for each type of function.   When the time-history $\xi(t)$ is a combination of sinusoidal and random functions, the shape of the probability density and distribution curves depends on the relative magnitudes of each type of function.

RELATIONSHIP BETWEEN AVERAGE AND RMS VALUES.   The average value may be determined from the probability density function since $\overline{|\xi|}$ is represented by the distance from the vertical axis to the center-of-gravity of the area under the probability density function, for either positive or negative values (assuming a symmetrical density function):

$$\overline{|\xi|} = 2\sigma \int_0^\infty p\left(\frac{\xi}{\sigma}\right) \frac{\xi}{\sigma}\, d\left(\frac{\xi}{\sigma}\right) \tag{22.16}$$

where $\overline{|\xi|}/\sigma$ is indicated in Fig. 22.10 for the probability density functions for the positive values of (A) a Gaussian distribution and (B) a sinusoid.   For a sine wave, the average

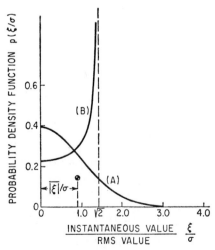

FIG. 22.10. Average value $\overline{|\xi|}$ of function with zero mean value: (A) Gaussian distribution $\overline{|\xi|}/\sigma = 0.798$; (B) sine wave $\overline{|\xi|}/\sigma = 0.900$.

value $\overline{|\xi|} = 0.900\sigma$; this is 0.636 times the amplitude. The average value of a Gaussian function $\overline{|\xi|} = 0.798\sigma$; it cannot be defined in terms of a peak value since the peak values vary from zero to infinity. However, physical processes usually are only approximately Gaussian in that the peak values are limited to some value between three and ten times the rms value.

PROBABILITY GRAPH PAPER

The characteristics of the probability density and distribution functions shown in Figs. 22.7 to 22.9 are emphasized by plotting the functions on probability paper.

GAUSSIAN DISTRIBUTION. Gaussian probability paper* is graph paper having a set of coordinates such that when the Gaussian distribution function [Eq. (22.12)] is plotted on it, a straight line is obtained, as shown in Fig. 22.11. For example, if the

FIG. 22.11. Probability paper for Gaussian distribution: (A) Gaussian probability distribution function gives a straight line where mean value $\overline{\xi}$ is at $P\,(\xi \geq) = 0.50$ and rms value σ is at 0.159 or 0.841; (B) probability distribution of a sine wave gives a nonlinear plot.

solid straight line in this figure represents the measured distribution function of the instantaneous value of $\xi(t)$, then $\xi(t)$ is random having a Gaussian distribution. In contrast, a sinusoidal function, for example, has a distribution function of the form shown by the dotted line of Fig. 22.11. Therefore, this type of paper is helpful in vibration analysis problems to determine the extent to which a time function has a normal distribution by observing the linearity of the plotted distribution function. To be useful in determining the randomness of $\xi(t)$, the probability distribution must be plotted when $\xi(t)$ is the *output of a relatively narrow-band filter* and the distribution function of the *instantaneous* values has been measured. A close approximation to the Gaussian dis-

* Available commercially, for example: Probability Scale Graph Paper No. 359-23, Keuffel & Esser Co., New York.

tribution can be synthesized by as few as seven appropriately chosen sinusoids,[21] and any complex function with broad frequency content will approximate a Gaussian distribution in accordance with the central-limit theorem.[9] Thus the probability distribution function of a broad-band signal is an insensitive measure of randomness.

The symmetry of the probability density function about the mean value also may be determined readily by use of this paper. If the shape of the curve is unchanged when it is rotated 180° about its intersection with the ordinate $P(\xi \geq) = 50$ per cent, the density function is symmetrical about the mean and the *mean value* $\bar{\xi}$ is the value of $\xi(t)$ at $P(\xi \geq) = 50$ per cent. The mean value of zero is plotted in Fig. 22.11. If the distribution function is linear as well as symmetrical when plotted on Gaussian probability paper, the *rms value* σ may be read from the curve at the 15.9 or 84.1 per cent points. The 2σ points at 2.3, or 97.7 per cent, and the 3σ points at 0.13, or 99.87 per cent, also may be used in Fig. 22.11. In Fig. 22.11, the rms value plotted is 2.0.

RAYLEIGH DISTRIBUTION. Rayleigh probability paper* is graph paper having a set of coordinates such that when the Rayleigh distribution function [Eq. (22.15)] is plotted on it, a straight line is obtained. This is shown in Fig. 22.12. The solid curve shows the distribution function of the *peak values* ξ_p of a narrow-band random function which has an rms value of 2.0. The analogous distribution function of the peak values of a sine wave is a horizontal line at the amplitude of the sine wave, as shown by the dotted curve of Fig. 22.12 for rms value of 2.0. Thus, the paper shown in Fig. 22.12 is useful in determining the randomness or coherence of $\xi(t)$ when $\xi(t)$ is the output of a relatively narrow-band filter and the distribution function of the *peak* values has been measured.

When data are presented in the form shown in Fig. 22.12, one curve is required for each narrow band of the signal which is analyzed. Since this may lead to a large number of plots, a technique of data presentation which includes all frequency bands and, at least qualitatively, a picture of the distribution of maxima is illustrated in Fig. 22.13. The abscissa is the center frequency of each band, and the ordinate is the magnitude of the maxima. Lines are then drawn between the values at each center frequency which represent the magnitude at or below which the indicated percentages of the peaks occur. For sinusoidal functions, all lines are coincident since all peaks are of equal magnitude.

DETERMINATION OF DISTRIBUTION FUNCTIONS

In practice, the characteristics of a signal $\xi(t)$ at a specific time and at a specific frequency cannot be determined, since only the instantaneous value of $\xi(t)$ is known at a specific time. Instead, the characteristics of the signal in a given restricted frequency bandwidth during a short period of time are determined. Thus the first step in a statistical analysis is to select a time interval short enough so that the nature and magnitude of $\xi(t)$ may be assumed constant, but long enough to give a statistically significant result.[6] Often, it is convenient to store the data sample in the form of a recording on a continuous loop of magnetic tape so that it may be played back repeatedly. Then a time sample of the signal may be analyzed to determine the probability distribution of either the in-

---

* Rayleigh probability paper can be constructed as follows:
The probability distribution function for the Rayleigh distribution, Eq. (22.15), can be written:

$$P(|\xi_p|/\sigma \leq) = 1 - e^{-\xi_{p^2}/2\sigma^2} \qquad \xi_p/\sigma = \left[2 \log_e \left(\frac{1}{1 - P} \right) \right]^{\frac{1}{2}}$$

Let $x(P)$ be the distance from the origin of a point on the abscissa corresponding to the ordinate ξ_p/σ. For the above equation to be a straight line

$$x(P) = a\frac{\xi_p}{\sigma} + b = a\left[2 \log_e \left(\frac{1}{1 - P} \right) \right]^{\frac{1}{2}} + b$$

The values of a and b are determined by the choice of origin and scale. If $x(0)$ is zero, then b also is zero and a is determined by the distance between $x(0)$ and $x(0.9999)$.

PROBABILITY DISTRIBUTION $P(|\xi_p|\geq)\times 100$

FIG. 22.12. Probability paper on which Rayleigh distribution function gives a straight line: (A) Rayleigh distribution for peaks of narrow-band random vibration with $\sigma = 2.0$; (B) distribution function of peaks of a sinusoid — $\sigma = 2.0$.

stantaneous or the peak values of the signal. The concepts involved in this determination are considered separately in the next two sections dealing with instantaneous and peak values, respectively. The instantaneous values of the *filtered* signal are indicated by $\xi'(t)$ while the peak values of the *filtered* signal are indicated by $\xi_p'(t)$.

INSTANTANEOUS VALUES. The probability distribution of instantaneous values is obtained for a set of discrete values of $\xi'(t)$ rather than as the continuous function that is illustrated in Fig. 22.9. The probability distribution function at a value ξ_1' is the probability $P(\xi' \geq)$ that $\xi'(t)$ exceeds ξ_1'.* Referring to Fig. 22.14, this is equivalent to summing all the Δt's during which $\xi'(t)$ exceeds ξ_1' or level L_1 and dividing by the total elapsed time. If this is done for a number of selected levels, $L_1, L_2, \ldots, L_n$, the probability distribution function at these levels is obtained. Figure 22.15A illustrates the block diagram of a system to make these measurements. The filtered signal $\xi'(t)$ is fed to an array of "discriminators" which determine whether a signal is greater or less than a preselected value. The levels of the discriminators are set at L_0, L_1, L_2, etc., usually in equal increments of voltage. The detail or resolution obtained is governed by the number of discriminator levels selected for measurement. The detail and accuracy yielded by ten levels for each polarity are sufficient for many engineering purposes. Since the resolution obtained improves with the number of counters registering, a variable gain

---

* The filtered signal $\xi'(t)$ may be passed first through a mechanism for changing the polarity (or sign) of the signal so that the positive and negative values of $\xi(t)$ can be analyzed separately, thus halving the number of discriminators and counters required for a given resolution. The time-history must be played back for twice as long but this usually will cause a rather small increase in the total analysis time.

Fig. 22.13. Distribution of peak accelerations as a function of frequency. Each curve represents the value below which the indicated percentage of peak values occurs in the response of the filter to $\xi(t)$, for the indicated center frequency of the filter.

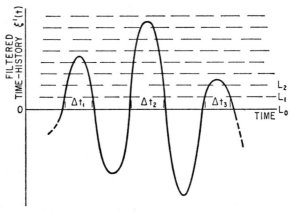

Fig. 22.14. Measurements required for statistical analysis: (A) To determine distribution of instantaneous values, sum of time increments Δt_i during which $\xi'(t)$ exceeds L_0, L_1, L_2 ... is measured; (B) to determine distribution of peak values, number of times that $\xi'(t)$ exceeds L_0, L_1, L_2 ... is measured.

control is needed to adjust the signal level, by a known amount, until an adequate number of counters are registering.

When the signal $\xi'(t)$ exceeds the level of a particular discriminator, a clock is started. It is stopped when $\xi'(t)$ next falls below that level and the counter counts the elapsed time between these events. At the end of the data sample, the readings of the counters are recorded by the read-out mechanism, a digital printer, for example. If each reading is divided by the total elapsed time of the data sample, the values obtained are the values

FIG. 22.15. Block diagrams showing steps in statistical analysis: (A) Instantaneous value analysis—automatic; (B) peak value analysis—automatic; (C) peak value analysis—manual. For each analysis, the center frequency of the filter remains constant.

of the probability distribution function at the values of the levels L_0, L_1, L_2, etc. If the values at adjacent levels are subtracted, the average probability density function between these two levels is obtained. The polarity of the signal is now changed and the process is repeated. However, using positive polarity, the distribution function $P(\xi' \geq)$ is obtained while the negative polarity yields $P(\xi' \geq)$. Since $P(\xi' \leq) = 1 - P(\xi' \geq)$, the two sets of values may be combined to give the complete distribution function. The character of the signal is then evaluated by the use of the probability paper illustrated in Fig. 22.11.

In addition to determining the characteristics of the signal, this analysis can be used to determine the mean square value of $\xi'(t)$, either by use of the probability paper if the signal is Gaussian or by computing the moment of inertia of the probability density function about the mean value [see Eq. (11.11)].

PEAK VALUES. The probability distribution of peak values or maxima $\xi'_p(t)$ of the filtered signal $\xi'(t)$ is obtained for a set of discrete values of $\xi'_p(t)$ rather than as the continuous function illustrated in Fig. 22.9. The probability distribution function at a value ξ'_{p1} is the probability $P(\xi'_p \geq)$ that a peak or maximum of $\xi'(t)$ exceeds ξ'_{p1}. Referring to Fig. 22.14, this is equivalent to counting the *number* of times that $\xi'(t)$ exceeds ξ'_{p1} (i.e., level L_1) and dividing by the total number of times that $\xi'(t)$ exceeds L_0. If this count is made for a number of levels, L_1, L_2, ..., the probability distribution function at these values is obtained.

Figure 22.15B illustrates the block diagram of a system for making such measurements. The filtered signal $\xi'(t)$ passes through a bank of level discriminators and causes the associated counters to register one count each time the level of the particular discriminator is exceeded. (Sometimes an array of discriminator-counter combinations is called a "pulse-height analyzer.") At the end of the data sample, the readings of the counters are recorded. Division of each count by the zero level count yields the probability distribution function $P(\xi'_p \geq)$ at the values equivalent to the levels L_1, L_2, ... If the values at adjacent levels are subtracted, the average probability density function between these two levels is obtained. The character of the signal then is evaluated by the use of the probability paper illustrated in Fig. 22.12.

For some end uses of the data, such as fatigue analysis, the probability distribution of the total or unfiltered signal $\xi(t)$ rather than the probability distribution of the filtered signal $\xi'(t)$ is desired. Then the distribution of the peaks of $\xi(t)$ is obtained by shunting out the filter in the block diagram of Fig. 22.15B as indicated by the dotted lines.

The systems illustrated by the block diagrams of Fig. 22.15A and B are relatively complex, and unless a large amount of analysis of this type is to be carried out, the cost of such systems may not be justified. A much simpler method is illustrated in Fig. 22.15C.

In this case a direct-writing recorder is employed and the number of peaks which exceed various levels are counted manually (see Fig. 22.14). In principle, the distribution of instantaneous values may be measured in this way, but practical difficulties make this method difficult to employ.

SPECTRAL ANALYSIS

The objective of spectral analysis[6, 14, 20, 22, 23, 24, 29–32] is to determine the variation of vibration magnitude with frequency. The narrowness of the bandwidth of the filter employed in the analysis determines the frequency resolution of the analysis and, therefore, the ability to detect the "fine-grain" variation of magnitude with frequency. The magnitude obtained at a particular frequency will be the average magnitude over the short time interval of data analyzed, during which the nature and magnitude of the vibration may be considered constant (see Figs. 22.3 and 22.4).

In preparing for spectral analysis, it is helpful if the characteristics of the time-history $\xi(t)$ are known from statistical analysis (or are known from previous experience) so that the most appropriate units to be employed for describing the vibration magnitude can be selected. In practice, the spectral analysis often is carried out first and the necessity for statistical analysis is judged from the appearance of the resulting spectrum.

DEFINITION OF POWER SPECTRAL DENSITY *

Power spectral density[6–9, 29] is defined as the limiting value of the mean-square response $\overline{[\xi']^2}$ of an ideal bandpass filter† to $\xi(t)$, divided by the bandwidth B of the filter, as the bandwidth of the filter approaches zero.

FIG. 22.16. Typical power spectral density plot of broad-band random function $\xi(t)$. Mean-square value of frequency content of $\xi(t)$ between f_a and f_b is equal to shaded area [Eq. (22.18)]. For "white noise" or flat spectrum, $W(f)$ is a constant.

An alternative definition[7] is as follows. If the function $\xi(t)$ is passed through an ideal low-pass filter‡ with cutoff frequency f_c, the mean-square response of the filter $\overline{[\xi']^2}$ will increase or decrease as f_c is increased or decreased, i.e., more or less of the function will be passed by the filter (assuming f_c is varied in a frequency range where the power spectral density is nonzero). The power spectral density $W(f)$ is the rate of change of $\overline{[\xi']^2}$ with respect to f_c, i.e.,

$$W(f_c) = \frac{d}{df_c} \{ \overline{[\xi']^2} \} \qquad (22.17)$$

Figure 22.16 illustrates a plot of power spectral density as a function of frequency obtained, for example, from spectral analysis of a random function. The mean-square value, or variance, of the frequency content of $\xi(t)$ between the frequencies f_a and f_b is equal to the shaded area of Fig. 22.16:

$$\sigma^2(f_a \le f \le f_b) = \int_{f_a}^{f_b} W(f) \, df \qquad (22.18)$$

* Power spectral density is defined by Eq. (22.17) and also by Eq. (11.16). It is a generic term used regardless of the physical quantity represented by the time-history. However, it is preferable to indicate the physical quantity involved. For example, the term "mean-square acceleration density" or "acceleration spectral density" is used when the time-history of acceleration is to be described.

† An ideal bandpass filter has a transmission characteristic which is rectangular in shape so that all frequency components within the filter bandwidth are passed with unity gain and zero phase distortion, while frequency components outside the bandwidth are completely removed. If the transmission characteristic is H instead of unity in the bandwidth B, the spectral density is obtained by dividing the mean-square response by $B \times H^2$ instead of B.

‡ An ideal low-pass filter is an ideal bandpass filter having a lower cutoff frequency of zero.

The rms value is σ. The mean-square value of the complete function $\overline{[\xi]^2}$ is given by Eq. (22.18) when f_a and f_b are zero and infinity, respectively; it is equal to the area under the entire spectrum. In the case of *white noise*, for which the spectral density is independent of frequency, i.e., $W(f) = W$, Eq. (22.18) simplifies to

$$\sigma^2 = W(f_2 - f_1) \tag{22.19}$$

where f_1 and f_2 are the limiting frequencies of the noise.

RELATIONSHIP OF POWER SPECTRAL DENSITY TO LINE SPECTRUM. The mathematical relationship between power spectral density $W(f)$ and the Fourier coefficients c_n is

$$W(f) = 2\pi W(\omega) = 2\pi \lim_{B \to 0} \frac{\overline{[\xi']^2}}{B} = 2\pi \frac{d}{d\omega} \overline{[(\xi')^2]}$$

$$= \frac{1}{2} - \sum_{n=1}^{\infty} c_n^2 \, \delta(f - f_n) \qquad [b_0 = 0] \tag{22.20}$$

where $\overline{[\xi']^2}$ is the mean-square response of the ideal filter with bandwidth B and center frequency $f = \omega/2\pi$; $\delta(f - f_n)$ is the Dirac delta function[2] (see Chap. 23).

The power spectral density of a sinusoidal function is a line spectrum since such a function has effectively zero bandwidth. This may be seen from Eq. (22.16) since $\overline{[\xi']^2}$ will change instantaneously as the cutoff frequency f_c increases through the frequency of the sine wave giving, theoretically, an infinite value for $W(f)$.

DETERMINATION OF POWER SPECTRAL DENSITY

One technique of obtaining a spectral analysis of a signal is to record it on a continuous loop of magnetic tape. As described in Chap. 11, the loop is played back repeatedly. Each time, an analysis is made at a slightly different frequency using, for example, a tunable filter. To obtain the complete spectrum, the process is repeated at enough frequencies to give a plot of magnitude vs. frequency. These steps are illustrated in the first two blocks of Figs. 22.17 and 22.18, which are block diagrams depicting examples of

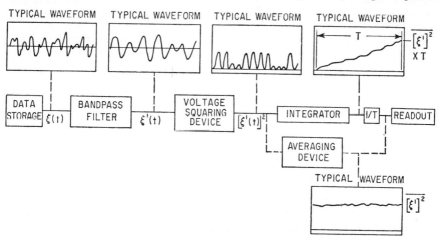

FIG. 22.17. Block diagram showing spectral analysis by the mean-square method. Dotted lines indicate alternative use of averaging device in place of integrator. Insets show typical waveforms at successive stages of analysis. Signal $\xi(t)$ is filtered to give $\xi'(t)$; squared to give $[\xi'(t)]^2$; integrated and divided by time of integration T to give $\overline{[\xi']^2}$, the mean-square value. Division by effective bandwidth of filter yields power spectral density of random function.

FIG. 22.18. Block diagram showing spectral analysis by averaging method. Insets show typical waveforms at successive stages of analysis. Signal $\xi(t)$ is filtered to give $\xi'(t)$; rectified (half or full wave) to give $|\xi'(t)|$; and averaged to yield $\overline{|\xi'|}$, the average value of $\xi'(t)$.

techniques employed in spectral analysis. Typical waveforms for one value of the center frequency of the filter are shown in the insets above the blocks. Techniques also are available for spectral analysis in real time.[12]

When a sample of a time-history of finite duration is employed to compute the power spectral density of a random function, it is assumed that (1) the function is ergodic, i.e., that averaging one time-history with respect to time yields the same result as averaging over an ensemble of time-histories at a given instant of time, and (2) that the function is *stationary*, i.e., that the power spectral density is independent of the sample of the time-history chosen. Further, the averaging time or sample duration must be long enough to yield a statistically significant value. Thus, the mean-square value obtained should not vary appreciably with a change in averaging time. The time over which a vibration record may be considered a stationary process and the need for a sufficiently long averaging time often are conflicting requirements.

MEASUREMENT OF POWER SPECTRAL DENSITY BY THE USE OF FILTERS

If a random function is applied to the input of a filter, the spectral density of the output is obtained by multiplying the spectral density of the input by the square of the transmission characteristic curve* at every frequency. First, consider the response of a filter to *white noise* (constant spectral density W). The output of an ideal bandpass filter is constant spectral density within the bandwidth B, and zero elsewhere. From Eq. (22.18), the mean-square output will be $W \times B$. The output of a practical filter (i.e., a filter having practically attainable characteristics in contrast to an ideal filter) has a spectral density which is directly proportional to the square of the transmission characteristic; the mean-square output is W multiplied by the area under the curve of the squared transmission characteristic. This area is defined as the effective bandwidth of the filter. However, due to the nonlinearities inherent in many filters, the effective bandwidth of a filter should be determined by measurement of the mean-square response of the filter to known white noise.

Now consider a random signal whose spectral density varies with frequency, as shown in Fig. 22.19B. It is desired to measure the spectral density by the use of filters, i.e., to perform a spectral analysis. In the case of an ideal filter, the mean-square output divided by B yields the mean (or average) value of the spectral density within the filter bandwidth [see Eq. (22.18) and Fig. 22.16]. When the signal is applied to a practical filter and the mean-square output is divided by the effective bandwidth, an approximation to the average spectral density within the filter bandwidth is obtained.

EFFECTS OF FILTER CHARACTERISTICS. The suitability of practical filters for the measurement of power spectral density depends on the rate of variation of the

---

* The ratio of the response amplitude to the input amplitude when the input is a sinusoidal function is the *transmission characteristic* or *frequency-response characteristic* of the filter. If this ratio is independent of the input amplitude, the filter is linear. The ratio usually is normalized so that the value at the center frequency is unity.

spectral density with frequency of the signal being analyzed, the shape of the transmission characteristic of the filter, and the time constant of the filter.

Figure 22.19*F* shows the spectral densities which would be measured using filters with the transmission characteristics shown in Fig. 22.19*C* and 22.19*D*, if the input power spectral density is actually as shown in Fig. 22.19*B*. The smoothing effect on the measured spectral density of the filter with the wider bandwidth is evident, particularly where the magnitude of the power spectral density is changing rapidly with frequency. Sharp peaks or valleys in the power spectral density are masked when the effective filter bandwidth approaches the bandwidth of the peak or valley. The power spectral density measured by the use of filters is exact only when an ideal filter is used, in the limit as the bandwidth approaches zero. Furthermore the time constant of the filter may affect the measurements by time-averaging the output signal.

MEASUREMENT OF LINE SPECTRA BY THE USE OF FILTERS

In the measurement of sinusoidal functions by the use of filters,[25, 26] the output amplitude is measured and plotted in place of the power spectral density used with random functions. The accuracy with which such measurements can be made depends upon the bandwidth of the filter. For example, Fig. 22.19*E* shows in solid lines the measured spec-

Fig. 22.19. Effects of filter bandwidth characteristics on measured spectra.

trum when the line spectrum of Fig. 22.19*A* is the input to the filters having the characteristics shown in Fig. 22.19*C*; the dotted lines in Fig. 22.19*E* show the spectrum obtained with the filter of Fig. 22.19*D*. The error introduced by the filter with the wider bandwidth is particularly evident when the sine waves are closely spaced in frequency. Assume now that the input $\xi(t)$ is a complex function comprised of several sinusoidal functions at unrelated frequencies. The output of the filter is the mean-square value of one or the sum of many sinusoids, depending on the spacing of the lines of the input spectrum, the breadth of the filter bandwidth, and the shape of the filter transmission characteristic. However, the output is often considered to be the mean-square value of a single sinusoidal function at the center frequency of the filter.

LINE SPECTRUM SUPERIMPOSED ON RANDOM VIBRATION SPECTRUM

The input signal $\xi(t)$ may be a "mixture" of a random function plus a single sinusoidal function within the bandwidth of the filter, as shown in Fig. 22.20. The mean-square output of the filter $\overline{[\xi']^2}$ is the sum of the mean-square values of the sinusoid and the frequency content of the random function within the filter bandwidth. Depending on the end use of the data and the relative magnitudes of the sinusoidal and random functions, it *may* be satisfactory to take the spectral density computed from this mean-square value $\overline{[\xi']^2}$ as a description of the vibration magnitude. Conversely, it may be preferable to consider $\overline{[\xi']^2}$ to be the mean-square value of a sinusoid at the filter center frequency. However, power spectral density should be used only to describe a random function. Periodic and complex functions should be represented by Fourier spectra, either line or continuous.

FIG. 22.20. Example of spectral analysis when signal contains narrow-band peak of random vibration or line spectrum superimposed on random vibration spectrum as shown by dotted line. Because of limitations of analysis equipment, line spectrum may give same appearance as narrow-band random vibration.

It is difficult to separate the random and coherent parts of a vibration time-history when such a combination occurs. For example, if the power spectral density of an acceleration time-history has been computed using a system such as illustrated in Fig. 22.17, and the spectrum has a peak in a narrow frequency band as shown in Fig. 22.20, it is necessary to know whether this peak represents a truly random function of increased intensity in this frequency band or a sinusoidal function superimposed on the smoother spectral density plot indicated by the dotted line in Fig. 22.20. Fundamentally, a sinusoidal function is represented by a line spectrum but may give the appearance of narrow-band random vibration because of the limitations of the spectral analysis techniques.[25] The probability density or probability distribution at the frequency of the peak in the spectrum may be used to determine the characteristics of the signal in this frequency region.

MEASUREMENT OF FILTER OUTPUT

The magnitude of the filtered signal is obtained in terms of (1) the mean-square value or (2) the average value.*

MEAN-SQUARE VALUE. Assume that statistical analysis has shown $\xi'(t)$ to be a random function. Then the mean-square value $\overline{[\xi']^2}$ must be obtained. Figure 22.17 shows one method of accomplishing this. The function $\xi'(t)$ is squared, integrated over the period of the data sample, and divided by the time of integration T to yield $\overline{[\xi']^2}$.

* See definition of average value in Eq. (22.5).

Typical waveforms during this process are shown. Alternatively, the integrator may be replaced by an averaging device, as shown by the dotted lines. The terminal value of the integrator output, divided by T, or the output of the averaging device is read out; division by the effective bandwidth of the filter yields the average power spectral density within the filter bandwidth.

AVERAGE VALUE.* Figure 22.18 is a block diagram showing an example of a system in which the filter output $\xi'(t)$ is half-wave (or full-wave) rectified and averaged. Thus the quantity measured by the read-out is $\overline{|\xi'|}$, the average value of $\xi'(t)$. If $\xi'(t)$ is known to be a random function, the relationships discussed under *Statistical Analysis* or empirical measurement of the response to a known spectral density can be used to relate the average value to the mean-square value, and thus to obtain the power spectral density.

EFFECT OF SIGNAL CHARACTERISTICS. When the magnitude of the filter output is obtained in terms of the mean-square value, as shown in Fig. 22.17, the read-out may be directly converted to the appropriate magnitude quantity, for example, power spectral density, the amplitude of a sine wave, etc. If the average value is obtained as shown in Fig. 22.18, the conversion of the read-out to an appropriate magnitude quantity is not direct, and may not even be possible. For example, if the filter output is the sum of two sinusoidal functions, the average value is a function of the phase angle between the sine waves, while the mean-square value is independent of phase angle. When the filter output is a mixture of a random function and a sinusoidal function, a similar indeterminacy exists in the average value. Thus, detailed knowledge of the characteristics of the function is required to interpret correctly the magnitude of $\xi(t)$ from its average value.†

MAGNITUDE TIME-HISTORY ANALYSIS

In magnitude time-history analysis, the variation of the magnitude‡ of the vibration as a function of time is examined. The result may be plotted as shown in Fig. 22.5 where the magnitude is computed as the average over a few seconds while the vibration data record has a duration of perhaps many minutes. In some cases, it may be desired to determine the time variation of the magnitude of the complete signal, e.g., the variation of the mean-square value $\overline{[\xi]^2}$ or, alternatively, the time-history of the magnitude of the vibration in a restricted frequency band. For example, suppose that spectral analysis of a number of time samples of the record shows that the vibration magnitude is concentrated in several narrow frequency bands, and that these bands are the only important components of the vibration. Then a magnitude time-history analysis would be carried out for these frequencies. However, if spectral analysis of a number of time samples revealed that the shape of the spectrum was approximately constant even though the magnitude varied, then a time-history of the magnitude of the complete signal would be appropriate. The amount of analysis carried out is determined by the completeness with which the vibration magnitude surface of Fig. 22.3 must be defined.

DETERMINATION OF MAGNITUDE TIME-HISTORY

Figure 22.21 shows block diagrams for carrying out a magnitude time-history analysis. The data storage contains the complete vibration record, rather than the short-duration samples used in spectral and statistical analysis. The signal $\xi(t)$ from the record is

---

* See definition of average value in Eq. (22.5).

† Similar effects must be taken into account in the use of voltmeters which, for example, respond to the average value of the input signal and have a scale which is graduated in terms of rms value.

‡ The word "magnitude" is used in this chapter to indicate the severity of vibration without specifically relating it to one of the various mathematical expressions used to describe vibration quantitatively.

Fig. 22.21. Block diagrams showing magnitude time-history analysis. To determine time-history of magnitude, "time record" blocks are used; magnitude distribution function is determined from discriminators and counters. Typical waveforms of magnitude time-history are shown by insets. For analysis of variation of magnitude of complete signal, filter is shunted as shown by dotted line. Filter center frequency is maintained constant during playback of entire vibration record. (Refer to *Statistical Analysis* and *Spectral Analysis* regarding counting and averaging techniques.) Diagrams show (*A*) mean-square value by successive integration; (*B*) mean-square value by averaging; (*C*) average value.

passed through a filter whose center frequency is held constant for the entire record. The mean-square value $\overline{[\xi']^2}$ or the average value $\overline{|\xi'|}$ of the output of the filter $\xi'(t)$ is obtained by one of the three alternative systems outlined in Fig. 22.21: (A) the mean-square value may be obtained by integration of $[\xi'(t)]^2$ during successive time intervals T; (B) the mean-square value may be obtained by averaging of $[\xi'(t)]^2$; or (C) the average value $\overline{|\xi'|}$ may be obtained. The results are recorded at the time record stations in Fig. 22.21. The filter may be shunted as shown by dotted lines to analyze the time variation of $\xi(t)$ rather than that of the filtered signal $\xi'(t)$.

When a running mean-square or running average value is obtained, as indicated in Fig. 22.21B and C, the value obtained at any instant is affected by the values at times previous to the time of observation, depending on the time constant of the averaging device. This tends to smooth out the time variation of the magnitude and is avoided when integration over consecutive periods is used because each integrand then is independent. Typical time records of the value of $\overline{[\xi']^2}$ or $\overline{|\xi'|}$ are shown in the insets adjoining the systems A, B, and C in Fig. 22.21. Such a time record permits correlation of the magnitude of the vibration with physical parameters. If the time-histories of the vibration magnitude are obtained for a number of filter bands and laid side by side, the vibration magnitude surface of Fig. 22.3 is synthesized.

MAGNITUDE DISTRIBUTION FUNCTION. In addition to the magnitude time-history, the total time that the magnitude is equal to or greater than a particular magnitude or set of magnitudes, i.e., the magnitude distribution function, may be determined, as indicated in Fig. 22.21. If $\overline{[\xi']^2}$ is obtained by integration [system (A)], the mean-square value is fed into a bank of discriminator-counter combinations which sense the number of times that the integrand exceeds various levels. If $\overline{[\xi']^2}$ is obtained by averaging [system (B)] or if $\overline{|\xi'|}$ is obtained [system (C)], the time that these quantities exceed various levels is measured by the bank of discriminator-clock-counter combinations. The function of discriminators and related equipment is discussed with reference to Figs. 22.18 and 22.19.

Figure 22.22 illustrates a manner of plotting the readings from the counters. The abscissa is normalized to indicate the percentage of the total time that the level is exceeded, while the ordinate scale corresponds to the magnitude of the level which is exceeded; e.g., magnitude L_1 corresponding to discriminator 1 is exceeded t_1 per cent of the time. The vibration magnitude distribution function illustrated in Fig. 22.22 is analogous to the probability distribution function shown in Fig. 22.9. The filter frequency and bandwidth must be specified unless the analysis represents the total (i.e., the unfiltered) signal.

FIG. 22.22. Distribution function of vibration magnitude for a filter having a center frequency f_0.

The time of occurrence of the various vibration magnitudes is lost and cannot be correlated with physical conditions. If the magnitude distribution functions are obtained for a number of frequencies and placed side by side, the vibration magnitude distribution surface is formed. This describes the total vibration signal in terms of magnitude, frequency, and duration in the most concise form possible.

CORRELATION ANALYSIS

CORRELATION FUNCTIONS AND THEIR USE

A correlation function[6, 7, 8, 9, 13, 29-32] defines the correlation between two parameters as a function of the times at which the parameters are observed. If the two parameters are the same except for the time of observation, i.e., $\xi(t)$ and $\xi(t + \tau)$, the function is known as an *autocorrelation function*. For example, the autocorrelation function of a single acceleration measurement may be used to determine both the nature of $\xi(t)$ and its power spectral density. If the two parameters $\xi_1(t)$ and $\xi_2(t + \tau)$ are physically distinct, the function is known as a *crosscorrelation function*. For example, the crosscorrelation between acceleration measurements at two different points of a structure may be determined for the purpose of studying the propagation of vibration through the structure. Crosscorrelation functions are not restricted to correlation of parameters with the same physical units; for example, one might determine the crosscorrelation between the applied force and the acceleration response to that force.

Correlation analysis of vibration data has many uses. For example, crosscorrelation data may be used to identify the normal modes of a vibrating structure; the autocorrelation function may be used to determine the power spectral density, and thus may be considered to be an alternative approach to spectral analysis. In the latter case, this technique has the advantage of yielding both the spectral density and a means of determining the characteristics of the signal, e.g., sinusoidal vs. random, etc., from the form of the autocorrelation function. Correlation techniques also are useful as a powerful method of detecting and analyzing a weak signal in the presence of noise.

Correlation analysis is concerned with stationary processes; in vibration analysis, it usually is carried out on a short interval of record during which physical parameters are assumed unchanged, as in spectral analysis.

AUTOCORRELATION FUNCTION

From Eq. (11.53), the autocorrelation function $R_\xi(\tau)$ of a time-varying parameter $\xi(t)$ is given by

$$R_\xi(\tau) = R_\xi(-\tau) = \lim_{T \to \infty} \frac{1}{2T} \int_{-T}^{T} \xi(t)\xi(t - \tau)\, dt = \overline{\xi(t)\xi(t - \tau)} \qquad (22.21)$$

Physically, $R_\xi(\tau)$ may be considered as the time average of the product of the instantaneous values of the function measured at two instants separated by a time interval τ. Setting $\tau = 0$ in Eq. (22.21) yields the mean-square value

$$R_\xi(0) = \overline{[\xi]^2} \qquad (22.22)$$

Since a parameter cannot be more closely related to another parameter than it is to itself, i.e., be better correlated, it follows that the maximum value of $R_\xi(\tau)$ occurs when $\tau = 0$.

The relationship between the autocorrelation function $R_\xi(\tau)$ and the power spectral density $W(f)$ is

$$W(f) = 4 \int_0^\infty R_\xi(\tau) \cos \omega\tau \, d\tau \qquad (22.23)$$

From Eq. (11.59), the inverse transformation is given by

$$R_\xi(\tau) = \frac{1}{2\pi} \int_0^\infty W(f) \cos \omega\tau \, d\omega \qquad (22.24)$$

Equation (22.23) is employed in the determination of the power spectral density of a random function from the autocorrelation function.

AUTOCORRELATION FUNCTION OF A SINE WAVE. Assume that $\xi(t)$ is a sine wave[13] as defined in Eq. (22.1). Then $R_\xi(\tau)$ is given by

$$R_\xi(\tau) = \frac{\xi_0^2}{2} \cos \omega\tau \qquad (22.25)$$

The relation given by Eq. (22.25) is shown by curve A in Fig. 22.23. This curve shows that the autocorrelation function of a sine wave is periodic with a frequency equal to the frequency of the sine wave. The physical interpretation is that a sine wave is unchanged by a translation of the time origin by an integer number of periods $2\pi/\omega$. The maxima of the autocorrelation function occur at $\tau = 2\pi n/\omega$ and are all equal to $R_\xi(0)$, i.e., the correlation at time delays τ equal to an integer number of periods $2\pi/\omega$ is the same as it is for zero time delay.

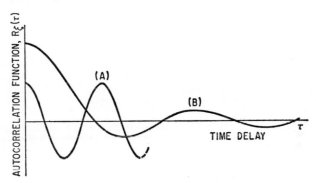

FIG. 22.23. Typical autocorrelation functions: (A) Sine wave [Eq. (22.25)]; (B) band-limited white noise [Eq. (22.26)].

AUTOCORRELATION FUNCTION OF WHITE NOISE. Suppose $\xi(t)$ represents band-limited white noise which has constant spectral density W up to a frequency f_c and zero for higher frequencies. According to Eq. (22.24), the autocorrelation function[13] is

$$R_\xi(\tau) = Wf_c \frac{\sin 2\pi f_c \tau}{2\pi f_c \tau}$$

$$R_\xi(0) = Wf_c = \overline{[\xi]^2} \tag{22.26}$$

Equations (22.26) define the decaying sinusoidal function shown by curve B in Fig. 22.23. As the cutoff frequency f_c approaches infinity, $R_\xi(\tau)$ approaches a delta function at the origin.

The autocorrelation function of the random process decreases rapidly as τ increases, and as the cutoff frequency f_c increases. Thus there is little correlation between the instantaneous values of $\xi(t)$ at different times. In the limit, as f_c approaches infinity, there is no correlation at all; i.e., the value of $\xi(t)$ is unpredictable from knowledge of its previous value $\xi(t - \tau)$. The autocorrelation function for a random function which has variable spectral density will have the same properties as curve B of Fig. 22.23, i.e., decreasing correlation for increasing τ and increasing bandwidth. Thus, the form of the autocorrelation function can be used to determine the character of $\xi(t)$, i.e., whether it is a random, complex, or periodic time function.

DETERMINATION OF AUTOCORRELATION FUNCTION. Figure 22.24 illustrates one method of obtaining a correlation analysis by analog means.[14, 20, 27, 28] (The arrangement is similar for either autocorrelation or crosscorrelation analyses.) The stored signal $\xi(t)$ is fed into a multiplier directly and through a variable time-delay τ. The time average of the multiplicand $\overline{\xi(t)\xi(t - \tau)}$ for various time-delays is obtained either by integration or averaging as shown by the alternative blocks in Fig. 22.24. The plot of the output vs. the time-delay is the autocorrelation function.

Another method of obtaining a correlation function from an analog record is to convert the analog record to digital form. Then a standard high-speed digital computer can be programmed to calculate the autocorrelation function and the power spectral density.

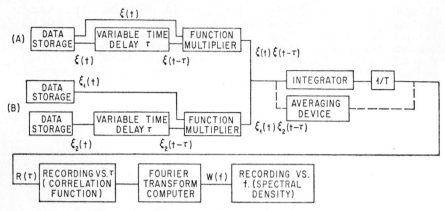

FIG. 22.24. Block diagram showing correlation analysis by analog means: (A) autocorrelation analysis; (B) crosscorrelation analysis. Dotted line shows alternative method of averaging multiplicand (see Fig. 22.17).

CROSSCORRELATION FUNCTION

The crosscorrelation function of two distinct time-varying parameters $\xi_1(t)$ and $\xi_2(t)$ is given by

$$R_{\xi_1\xi_2}(\tau) = \lim_{T \to \infty} \frac{1}{2T} \int_{-T}^{T} \xi_1(t)\xi_2(t - \tau)\,dt = \overline{\xi_1(t)\xi_2(t - \tau)} \qquad (22.27)$$

The magnitude of $R_{\xi_1\xi_2}(\tau)$ is a measure of the correlation of ξ_1 and ξ_2 when observed at times separated by a time τ.

Example 22.1. Assume ξ_1 is directly proportional to ξ_2 except for a time lag τ_1:

$$\xi_1(t) = A\,\xi_2(t - \tau_1) \qquad (22.28)$$

Then, by Eq. (22.27),

$$R_{\xi_1\xi_2}(\tau_1) = A\overline{[\xi_2]^2} = AR_{\xi_2}(0) \qquad (22.29)$$

Example 22.2. Assume ξ_1 and ξ_2 are sinusoidal functions at the same frequency:

$$\xi_1(t) = A\,\xi_0 \sin \omega(t - \tau_1) \qquad (22.30)$$

$$\xi_2(t) = \xi_0 \sin \omega t$$

Then the crosscorrelation function is

$$R_{\xi_1\xi_2}(\tau) = \frac{A[\xi_0]^2}{2} \cos \omega(\tau - \tau_1) \qquad (22.31)$$

This is the same form as the autocorrelation function for a sine wave given in Eq. (22.25) except for the time shift τ_1 (see Fig. 22.23). Thus, for example, the increase or decrease in the vibration displacement amplitude between points 1 and 2 may be determined from the cross-correlation function given by Eq. (22.31).

DETERMINATION OF CROSSCORRELATION FUNCTION. One method of obtaining a crosscorrelation function by analog means is illustrated at B in Fig. 22.24. One function $\xi_1(t)$ is fed directly to the multiplier and the second function $\xi_2(t)$ is fed through the variable time-delay to the multiplier. The time average of the multiplicand $\overline{\xi_1(t)\xi_2(t - \tau)}$ plotted vs. τ is the crosscorrelation function.

Crosscorrelation analysis also may be carried out by the use of standard high-speed digital computers which have been programmed for this purpose. Crosscorrelation analysis by digital means is almost identical to autocorrelation analysis since the only difference is that two distinct functions, $\xi_1(t)$ and $\xi_2(t)$, are employed instead of the one function $\xi(t)$ used in autocorrelation analysis.

REFERENCES

1. von Kármán, T., and M. A. Biot: "Mathematical Methods in Engineering," McGraw-Hill Book Company, Inc., New York, 1940.
2. Pipes, L. A.: "Applied Mathematics for Engineers and Physicists," 2d ed., McGraw-Hill Book Company, Inc., New York, 1958.
3. Campbell, G. A., and R. M. Foster: "Fourier Series for Practical Applications," D. Van Nostrand Company, Inc., Princeton, N.J., 1940.   Previously published as No. B-584 of the Bell System Series of Monographs, 1931.
4. Bradford, R. S.: "Phase-coherent Vibration in Missiles," *External Publ.* 349, Jet Propulsion Laboratory, Pasadena, Calif., October, 1956.
5. Bradford, R. S.: *Shock and Vibration Bull.* 23, p. 315.   Office of Secretary of Defense, Research and Development, Washington, D.C., February, 1957.
6. Crandall, S. H.: "Random Vibration," Technology Press of MIT, Cambridge, Mass., 1958.
7. Bendat, J. S.: "Principles and Applications of Random Noise Theory," John Wiley & Sons, Inc., New York, 1958.
8. Davenport, W. B., Jr., and W. L. Root: "An Introduction to the Theory of Random Signals and Noise," McGraw-Hill Book Company, Inc., New York, 1958.
9. Miller, K. S.: "Engineering Mathematics," Holt, Winston & Rinehart, Inc., New York, 1957.
10. Scarborough, J. B.: "Numerical Mathematical Analysis," 2d ed., The Johns Hopkins Press, Baltimore, 1950.
11. Manley, R. G.: "Waveform Analysis," John Wiley & Sons, Inc., New York, 1945.
12. Bickel, H. J.: "Spectrum Analysis with Delay-line Filters," IRE Wescon Convention Record (Part 8), 1959.
13. Knudtzon, N. H.: "Experimental Study of Statistical Characteristics of Filtered Random Noise," *MIT Research Lab. Electronics, Tech. Rept.* 115, July, 1949.
14. Symposium on Applications of Autocorrelation Analysis to Physical Problems, Woods Hole, Mass., June 13–14, 1949, published by U.S. Office of Naval Research, Washington, D.C., ATI No. 86556.
15. Whiteley, T. B.: "A Method for Amplitude Analysis of Vibration Data," *Tech. Memo.* 65–100, U.S. Naval Ordnance Lab., Corona, Calif., January, 1956.
16. Fine, A., T. B. Whiteley, D. Bell, and M. Buus: *Shock and Vibration Bull.* 23, p. 184. Office of Secretary of Defense, Research and Development, Washington, D.C., June, 1956.
17. White, H. E.: "An Analog Probability Density Analyzer," *MIT Research Lab. Electronics, Tech. Rept.* 326, April, 1957.
18. Meyer, D. D.: *Shock and Vibration Bull.* 23, p. 229.   Office of Secretary of Defense, Research and Development, Washington, D.C., June, 1956.
19. Baldwin, F. L.: *Shock and Vibration Bull.* 25, part I, p. 151.   Office of Secretary of Defense, Research and Development, Washington, D.C., December, 1957.
20. Carlson, E. R., C. C. Conger, J. C. Laurence, E. H. Meyn, and R. A. Yocke: *Proc. IRE*, 44:956 (1959).
21. Baruch, J. J.: *Shock and Vibration Bull.* 25, part II, p. 25.   Office of Secretary of Defense, Research and Development, Washington, D.C., December, 1957.
22. Stallard, R. L.: *Shock and Vibration Bull.* 24, p. 70.   Office of Secretary of Defense, Research and Development, Washington, D.C., February, 1957.
23. Stallard, R. L.: "The Validity of the Single Degree of Freedom System Analog for Vibration Analysis," *Johns Hopkins Univ. Appl. Phys. Lab. Rept.* APL/JHU CF-2518, May, 1956.
24. Gruen Applied Science Laboratories, Inc.: "Methods of Analyzing Shock and Vibration," *Rept.* 10002-F, July, 1957.
25. Granick, N.: "Status Report on Random Vibration Simulation," *WADC Tech. Note* 58–274, *ASTIA Document* No. AD 203125, March, 1959.
26. Haase, K. H., and F. Vilrig: "Errors in Spectrum Analysis by a Set of Narrowband Selecting Filters," *Air Force Cambridge Research Center Tech. Rept.* AFCRC—TR—56–121, *ASTIA Document* No. AD 110197, November, 1956.
27. Zabusky, N.: "The Mechanical Correlation Computer," *MIT Servomechanisms Lab., Eng. Rept.* 32, October, 1951.   *ASTIA Document* No. 45.
28. Howes, W. L., E. E. Callahan, W. D. Coles, and H. R. Mull: "Near Noise Field of a Jet-engine Exhaust," *NACA Rept.* 1338, 1957.   (Supersedes *NACA TN*'s 3763 and 3764.)

29. Bendat, Julius S., and Allan G. Piersol: "Random Data: Analysis and Measurement Procedures," Wiley-Interscience, New York, 1971.
30. Enochson, Loren D., and Robert K. Otnes: "Programming and Analysis for Digital Time Series Data," *Shock and Vibration Mon.* 3, Shock and Vibration Information Center, Washington, D. C., 1968.
31. Otnes, Robert K., and Loren Enochson: "Digital Time Series Analysis," Wiley-Interscience, New York, 1972.
32. Methods for Analysis and Presentation of Shock and Vibration Data, American National Standard, ANSI S2.10–1971.

23

CONCEPTS IN SHOCK DATA ANALYSIS

Sheldon Rubin

The Aerospace Corporation

INTRODUCTION

SCOPE

This chapter discusses the interpretation of shock measurements and the reduction of data to a form adapted to further engineering use. Methods of data reduction also are discussed. A shock measurement is a trace giving the time-history of a shock parameter over the duration of the shock. The shock parameter may define motion (such as acceleration, velocity, or displacement) or loading (such as force, pressure, stress, or torque). It is assumed that any corrections that should be applied to eliminate distortions resulting from the instrumentation have been made. The trace may be a pulse or transient vibration. The interpretation of periodic and random vibration measurements is discussed in Chap. 22.

Examples of sources of shock to which this discussion applies are aircraft landing, braking and gust loading; missile launching and staging; transportation of fragile equipment; accidental collision of vehicles; gunfire; explosions; and high-speed fluid entry.

ENGINEERING USES OF SHOCK MEASUREMENTS

Often, a shock measurement in the form of a time-history of a motion or loading parameter is not useful directly for engineering purposes. Reduction to a different form is then necessary, the type of data reduction employed depending upon the ultimate use of the data.

COMPARISON OF MEASURED RESULTS WITH THEORETICAL PREDICTION. The correlation of experimentally determined and theoretically predicted results by comparison of records of time-histories is difficult. Generally, it is impractical in theoretical analyses to give consideration to all the effects which may influence the experimentally obtained results. For example, the measured shock often includes the vibrational response of the structure to which the shock-measuring device is attached. Such vibration obscures the determination of the shock input for which an applicable theory is being tested; thus, data reduction is useful in minimizing or eliminating the irrelevancies of the measured data to permit ready comparison of theory with corresponding aspects of the experiment. It often is impossible to make such comparisons on the basis of original time-histories.

CALCULATION OF STRUCTURAL RESPONSE. In the design of equipment to withstand shock, the required strength of the equipment is indicated by its response to the shock. The response may be measured in terms of the deflection of a member of the equipment relative to another member or by the magnitude of the dynamic loads imposed upon the equipment. The structural response can be calculated from the time-history by known means; however, certain techniques of data reduction result in descriptions of the shock that are related directly to structural response.

As a design procedure it is convenient to represent the equipment by an appropriate model that is better adapted to analysis.* A typical model is shown in Fig. 23.1; it consists of a secondary structure supported by a primary structure. Depending upon the ultimate objective of the design work, certain characteristics of the response of the model must be known:

1. If design of the secondary structure is to be effected, it is necessary to know the time-history of the motion of the primary structure. Such motion constitutes the excitation for the secondary structure.

2. In the design of the primary structure, it is necessary to know the deflection of such structure as a result of the shock, either the time-history or the maximum value.

By selection of suitable data reduction methods, response information useful in the design of the equipment is obtained from the original time-history.

FIG. 23.1. Commonly used structural model consisting of a primary and a secondary structure. Each structure is represented as a lumped-parameter single degree-of-freedom system with the secondary mass m much smaller than the primary mass M so that the response of the primary mass is unaffected by the response of the secondary mass. The response of the primary mass to an input shock motion is the input shock motion to the secondary structure.

LABORATORY SIMULATION OF MEASURED SHOCK. Because of the difficulty of using analytical methods in the design of equipment to withstand shock, it is common practice to prove the design of equipments by laboratory tests that simulate the anticipated actual shock conditions. Unless the shock can be defined by one of a few simple functions, it is not feasible to reproduce in the laboratory the complete time-history of the actual shock experienced in service. Instead, the objective is to synthesize a shock having the characteristics and severity considered significant in causing damage to equipment. Then, the data reduction method is selected so that it extracts from the original time-history the parameters that are useful in specifying an appropriate laboratory shock test. The considerations involved in simulation are discussed in Chap. 24. Shock testing machines are discussed in Chap. 26.

EXAMPLES OF SHOCK MOTIONS

Five examples of shock motions are illustrated in Fig. 23.2 to show typical characteristics and to aid in the comparison of the various techniques of data reduction. The acceleration impulse and the acceleration step are the classical limiting cases of shock motions. The half-sine pulse of acceleration, the decaying sinusoidal acceleration, and the complex oscillatory type motion typify shock motions encountered frequently in practice.

* See Chap. 42 for a more complete discussion of models.

In selecting data reduction methods to be used in a particular circumstance, the applicable physical conditions must be considered. The original record, usually a time-history, may indicate any of several physical parameters; e.g., acceleration, force, velocity, or pressure. Data reduction methods discussed in subsequent sections of this chapter are applicable to a time-history of any parameter. For purposes of illustration in the following examples, the primary time-history is that of acceleration; time-histories of velocity and displacement are derived therefrom by integration. These examples are included to show characteristic features of typical shock motions and to demonstrate data reduction methods.

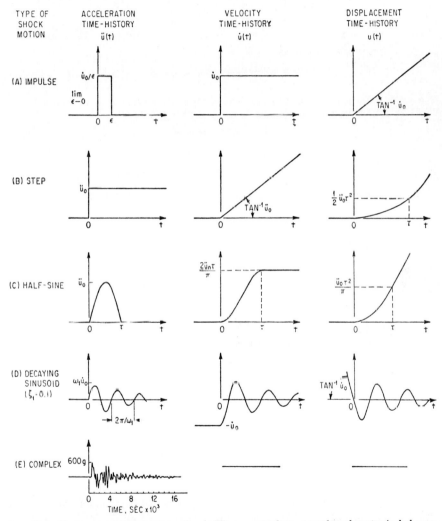

FIG. 23.2. Five examples of shock motions. These examples are used to show typical characteristics of shock motions and to aid in the comparison of various techniques of data reduction. The acceleration time-history is considered to be the primary description; and corresponding velocity and displacement time-histories are shown. [Mathematical descriptions appear in Eqs. (23.1) to (23.14).]

ACCELERATION IMPULSE OR STEP VELOCITY

The *delta function* $\delta(t)$ is defined mathematically as a function consisting of an infinite ordinate (acceleration) occurring in a vanishingly small interval of abscissa (time) at time $t = 0$ such that the area under the curve is unity. An acceleration time-history of this form is shown diagrammatically in Fig. 23.2A. If the velocity and displacement are zero at time $t = 0$, the corresponding velocity time-history is the velocity step and the corresponding displacement time-history is a line of constant slope, as shown in the figure. The mathematical expressions describing these time histories are

$$\ddot{u}(t) = \dot{u}_0 \delta(t) \tag{23.1}$$

where $\delta(t) = 0$ when $t \neq 0$, $\delta(t) = \infty$ when $t = 0$, and $\int_{-\infty}^{\infty} \delta(t)\, dt = 1$. The acceleration can be expressed alternatively as

$$\ddot{u}(t) = \lim_{\epsilon \to 0} \dot{u}_0/\epsilon \qquad [0 < t < \epsilon] \tag{23.2}$$

where $\ddot{u}(t) = 0$ when $t < 0$ and $t > \epsilon$. The corresponding expressions for velocity and displacement for the initial conditions $u = \dot{u} = 0$ when $t < 0$ are

$$\dot{u}(t) = \dot{u}_0 \qquad [t > 0] \tag{23.3}$$

$$u(t) = \dot{u}_0 t \qquad [t > 0] \tag{23.4}$$

ACCELERATION STEP

The *unit step function* $\mathbf{1}(t)$ is defined mathematically as a function which has a value of zero at time less than zero $(t < 0)$ and a value of unity at time greater than zero $(t > 0)$. The mathematical expressions describing the acceleration step are

$$\ddot{u}(t) = \ddot{u}_0 \mathbf{1}(t) \tag{23.5}$$

where $\mathbf{1}(t) = 1$ for $t > 0$ and $\mathbf{1}(t) = 0$ for $t < 0$. An acceleration time-history of the unit step function is shown in Fig. 23.2B; the corresponding velocity and displacement time-histories are also shown for the initial conditions $u = \dot{u} = 0$ when $t = 0$.

$$\dot{u}(t) = \ddot{u}_0 t \qquad [t > 0] \tag{23.6}$$

$$u(t) = \tfrac{1}{2}\ddot{u}_0 t^2 \qquad [t > 0] \tag{23.7}$$

The unit step function is the time integral of the delta function:

$$\mathbf{1}(t) = \int_{-\infty}^{t} \delta(t)\, dt \qquad [t > 0] \tag{23.8}$$

HALF-SINE ACCELERATION

A half-sine pulse of acceleration of duration τ is shown in Fig. 23.2C; the corresponding velocity and displacement time-histories also are shown, for the initial conditions $u = \dot{u} = 0$ when $t = 0$. The applicable mathematical expressions are

$$\ddot{u}(t) = \ddot{u}_0 \sin\left(\frac{\pi t}{\tau}\right) \qquad [0 < t < \tau] \tag{23.9}$$

$$\ddot{u}(t) = 0 \qquad \text{when } t < 0 \qquad \text{and } t > \tau$$

$$\dot{u}(t) = \frac{\ddot{u}_0 \tau}{\pi}\left(1 - \cos\frac{\pi t}{\tau}\right) \qquad [0 < t < \tau]$$

$$\dot{u}(t) = \frac{2\ddot{u}_0 \tau}{\pi} \qquad [t > \tau] \tag{23.10}$$

$$u(t) = \frac{\ddot{u}_0 \tau^2}{\pi^2}\left(\frac{\pi t}{\tau} - \sin\frac{\pi t}{\tau}\right) \qquad [0 < t < \tau]$$

(23.11)

$$u(t) = \frac{\ddot{u}_0 \tau^2}{\pi}\left(\frac{2t}{\tau} - 1\right) \qquad [t > \tau]$$

This example is typical of a class of shock motions in the form of acceleration pulses not having infinite slopes.

DECAYING SINUSOIDAL ACCELERATION

A decaying sinusoidal trace of acceleration is shown in Fig. 23.2D; the corresponding time-histories of velocity and displacement also are shown for the initial conditions $\dot{u} = -\dot{u}_0$ and $u = 0$ when $t = 0$. The applicable mathematical expressions are

$$\ddot{u}(t) = \frac{\dot{u}_0 \omega_1}{\sqrt{1 - \zeta_1^2}} e^{-\zeta_1 \omega_1 t} \sin\left(\sqrt{1 - \zeta_1^2}\,\omega_1 t + \sin^{-1}(2\zeta_1\sqrt{1 - \zeta_1^2})\right) \qquad [t > 0] \quad (23.12)$$

where ω_1 is the frequency of the vibration and ζ_1 is the fraction of critical damping corresponding to the decrement of the decay. Corresponding expressions for velocity and displacement are

$$\dot{u}(t) = -\frac{\dot{u}_0}{\sqrt{1 - \zeta_1^2}} e^{-\zeta_1 \omega_1 t} \cos\left(\sqrt{1 - \zeta_1^2}\,\omega_1 t + \sin^{-1}\zeta_1\right) \qquad [t > 0] \quad (23.13)$$

where $\dot{u}(t) = -\dot{u}_0$ when $t < 0$.

$$u(t) = -\frac{\dot{u}_0}{\omega_1\sqrt{1 - \zeta_1^2}} e^{-\zeta_1 \omega_1 t} \sin\left(\sqrt{1 - \zeta_1^2}\,\omega_1 t\right) \qquad [t > 0] \quad (23.14)$$

where $u(t) = -\dot{u}_0 t$ when $t < 0$.

COMPLEX SHOCK MOTION

The trace shown in Fig. 23.2E is an acceleration time-history representing typical field data. It cannot be defined by an analytic function; consequently, the corresponding velocity and displacement time-histories can be obtained only by numerical, graphical, or analog integration of the acceleration time-history.

CONCEPTS OF DATA REDUCTION

Consideration of the engineering uses of shock measurements indicates two basically different methods for describing a shock: (1) a description of the shock in terms of its inherent properties, in the time domain or in the frequency domain; and (2) a description of the shock in terms of the effect on structures when the shock acts as the excitation. The latter is designated reduction to the response domain. The following sections discuss concepts of data reduction to the frequency and response domains.

Whenever practical, the original time-history should be retained even though the information included therein is reduced to another form. The purpose of data reduction is to make the data more useful for some particular application. The reduced data usually have a more limited range of applicability than the original time-history. These limitations must be borne in mind if the data are to be applied intelligently.

DATA REDUCTION TO THE FREQUENCY DOMAIN

Any nonperiodic function can be represented as the superposition of sinusoidal components, each with its characteristic amplitude and phase.[1] This superposition is the Fourier spectrum, a plot of the amplitude and phase of the sinusoidal components into which the function can be decomposed. It is analogous to the Fourier components of a periodic function (Chap. 22). The Fourier components of a periodic function occur at discrete frequencies and the composite function is obtained by superposition of components. By contrast, the Fourier spectrum for a nonperiodic function is a continuous function of frequency and the composite function is achieved by integration. Applicable mathematical properties are given in the Appendix; the following sections discuss the application of the Fourier spectrum to describe the shock motions illustrated in Fig. 23.2.

ACCELERATION IMPULSE. Using the definition of the acceleration pulse given by Eq. (23.2) and substituting this for $f(t)$ in Eq. (23.57) of the Appendix,

$$\mathbf{F}(\omega) = \lim_{\epsilon \to 0} \int_0^\epsilon \frac{\ddot{u}_0}{\epsilon} e^{-j\omega t}\, dt \qquad (23.15)$$

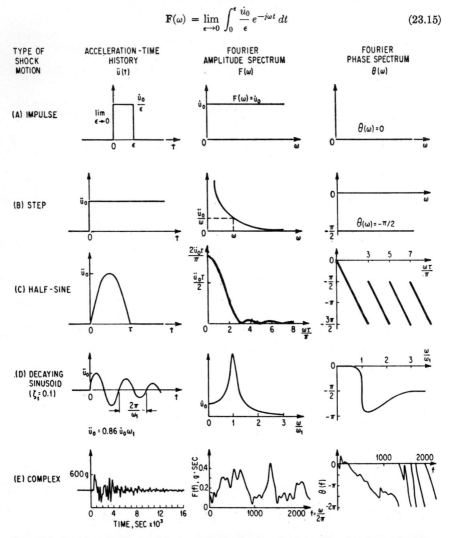

Fig. 23.3. Fourier amplitude and phase spectra for the shock motions in Fig. 23.2. These spectra represent the amplitude and phase of the continuous distribution of frequency components into which the shock motions can be decomposed. [Mathematical descriptions appear in Eqs. (23.15) to (23.27).]

Carrying out the integration,

$$\mathbf{F}(\omega) = \lim_{\epsilon \to 0} \frac{\dot{u}_0(1 - e^{-j\omega\epsilon})}{j\omega\epsilon} = \dot{u}_0 \quad (23.16)$$

The corresponding amplitude and phase spectra, from Eqs. (23.63) and (23.64) of the Appendix, are

$$F(\omega) = \dot{u}_0; \qquad \theta(\omega) = 0 \quad (23.17)$$

These spectra are shown in Fig. 23.3A. The magnitude of the Fourier amplitude spectrum is a constant, independent of frequency, equal to the area under the acceleration-time curve.

The physical significance of the spectra in Fig. 23.3A is shown in Fig. 23.4 where the rectangular acceleration pulse of magnitude $\dot{u}_0/\epsilon$ and duration $t = \epsilon$ is shown as approximated by superposed sinusoidal components for several different upper limits of frequency for the components. With the frequency limit $\omega_l = 4/\epsilon$, the pulse has a noticeably rounded contour formed by the superposition of all components whose frequencies are less than ω_l. These components tend to add in the time interval $0 < t < \epsilon$ and, though existing for all time from $-\infty$ to $+\infty$, cancel each other outside this interval so that $\ddot{u}$ approaches zero. When $\omega_l = 16/\epsilon$, the pulse is more nearly rectangular and $\ddot{u}$ approaches zero more rapidly for time $t < 0$ and $t > \epsilon$. When $\omega_l = \infty$, the superposition of sinusoidal components gives $\ddot{u} = \dot{u}_0/\epsilon$ for the time interval of the pulse, and $\ddot{u} = \dot{u}_0/2\epsilon$ at $t = 0$ and $t = \epsilon$. The components cancel completely for all other times. As $\epsilon \to 0$ and $\omega_l \to \infty$, the infinitely large number of superimposed frequency components gives $\ddot{u} = \infty$ at $t = 0$. The same general result is obtained when the Fourier components of other forms of $\ddot{u}(t)$ are superimposed.

FIG. 23.4. Time-histories which result from the superposition of the Fourier components of a rectangular pulse for several different upper limits of frequency ω_l of the components. The upper time-history ($\omega_l = 4/\epsilon$) has a noticeably rounded contour; the middle time-history ($\omega_l = 16/\epsilon$) is more nearly rectangular; and the lower time-history ($\omega_l = \infty$) is the complete Fourier representation of a rectangular pulse.

ACCELERATION STEP. The Fourier spectrum of the acceleration step does not exist in the strict sense (see Appendix) since the integrand of Eq. (23.56) does not tend to zero as $\omega \to \infty$. Using a convergence factor, the Fourier transform is found by substituting $\ddot{u}(t)$ for $f(t)$ in Eq. (23.73) of the Appendix:

$$\mathbf{F}(\omega - ja) = \int_0^\infty \ddot{u}_0 e^{-j(\omega - ja)t}\, dt = \frac{\ddot{u}_0}{j(\omega - ja)} \quad (23.18)$$

Taking the limit as $a \to 0$,

$$\mathbf{F}(\omega) = \frac{\ddot{u}_0}{j\omega} \quad (23.19)$$

The amplitude and phase spectra, from Eqs. (23.63) and (23.64), are

$$F(\omega) = \frac{\ddot{u}_0}{\omega}; \qquad \theta(\omega) = -\frac{\pi}{2} \quad (23.20)$$

These spectra are shown in Fig. 23.3B; the amplitude spectrum decreases as frequency increases, whereas the phase is a constant independent of frequency. Note that the spectrum of Eq. (23.19) is $1/j\omega$ times the spectrum for the impulse, Eq. (23.16), in accordance with Eq. (23.65) of the Appendix.

HALF-SINE ACCELERATION. Substitution of the half-sine acceleration time-history, Eq. (23.9), into Eq. (23.57) gives

$$\mathbf{F}(\omega) = \int_0^\tau \ddot{u}_0 \sin \frac{\pi t}{\tau} e^{-i\omega t} \, dt \tag{23.21}$$

Performing the indicated integration gives

$$\mathbf{F}(\omega) = \frac{\ddot{u}_0 \tau/\pi}{1 - (\omega\tau/\pi)^2} (1 + e^{-i\omega\tau}) \qquad [\omega \neq \pi/\tau]$$

$$\mathbf{F}(\omega) = -\frac{j\ddot{u}_0\tau}{2} \qquad [\omega = \pi/\tau] \tag{23.22}$$

Applying Eqs. (23.63) and (23.64) to find expressions for the spectra of amplitude and phase,

$$F(\omega) = \frac{2\ddot{u}_0\tau}{\pi} \left| \frac{\cos(\omega\tau/2)}{1 - (\omega\tau/\pi)^2} \right| \qquad [\omega \neq \pi/\tau]$$

$$F(\omega) = \frac{\ddot{u}_0\tau}{2} \qquad [\omega = \pi/\tau] \tag{23.23}$$

$$\theta(\omega) = -\frac{\omega\tau}{2} + n\pi \tag{23.24}$$

where n is the smallest integer that prevents $|\theta(\omega)|$ from exceeding $3\pi/2$. The Fourier spectra of the half-sine pulse of acceleration are plotted in Fig. 23.3C.

DECAYING SINUSOIDAL ACCELERATION. The application of Eq. (23.57) to the decaying sinusoidal acceleration defined by Eq. (23.12) gives the following expression for the Fourier spectrum:

$$\mathbf{F}(\omega) = \ddot{u}_0 \frac{1 + j2\zeta_1\omega/\omega_1}{(1 - \omega^2/\omega_1^2) + j2\zeta_1\omega/\omega_1} \tag{23.25}$$

This can be converted to a spectrum of absolute values by applying Eq. (23.63):

$$F(\omega) = \ddot{u}_0 \sqrt{\frac{1 + (2\zeta_1\omega/\omega_1)^2}{(1 - \omega^2/\omega_1^2)^2 + (2\zeta_1\omega/\omega_1)^2}} \tag{23.26}$$

A spectrum of phase angle is obtained from Eq. (23.64):

$$\theta(\omega) = -\tan^{-1} \frac{2\zeta_1(\omega/\omega_1)^3}{(1 - \omega^2/\omega_1^2) + (2\zeta_1\omega/\omega_1)^2} \tag{23.27}$$

These spectra are shown in Fig. 23.3D for a value of $\zeta = 0.1$. The peak in the amplitude spectrum near the frequency ω_1 indicates a strong concentration of Fourier components near the frequency of occurrence of the oscillations in the shock motion.

COMPLEX SHOCK. The complex shock motion shown in Fig. 23.3E is the result of actual measurements; hence, its functional form is unknown. Its Fourier spectrum must be computed numerically. The Fourier spectrum shown in Fig. 23.3E was evaluated digitally using 100 time increments of 0.00015 sec duration. The peaks in the amplitude spectrum indicate concentrations of sinusoidal components near the frequencies of various oscillations in the shock motion. The portion of the phase spectrum at the high frequencies creates an appearance of discontinuity. If the phase angle were not returned to zero each time it passes through −360°, as a convenience in plotting, the curve would be continuous.

APPLICATION OF THE FOURIER SPECTRUM. The Fourier spectrum description of a shock is useful in linear analysis when the properties of a structure on which the shock acts are defined as a function of frequency. Such properties are designated by the general term *transfer function;* in the shock and vibration technology, commonly used transfer functions are mechanical impedance, mobility, and transmissibility.

When a shock acts on a structure, the structure responds in a manner that is essentially oscillatory. The frequencies that appear predominantly in the response are (1) the preponderant frequencies of the shock and (2) the natural frequencies of the structure. The Fourier spectrum of the response $\mathbf{R}(\omega)$ is the product of the Fourier spectrum of the shock $\mathbf{F}(\omega)$ and an appropriate transfer function for the structure. This is indicated by Eq. (23.74) of the Appendix. For example, if $\mathbf{F}(\omega)$ and $\mathbf{R}(\omega)$ are Fourier spectra of acceleration, the transfer function is the transmissibility of the structure, i.e., the ratio of acceleration at the responding station to the acceleration at the driving station, as a function of frequency. However, if $\mathbf{R}(\omega)$ is a Fourier spectrum of velocity and $\mathbf{F}(\omega)$ is a Fourier spectrum of force, the transfer function is mobility as a function of frequency.

The Fourier spectrum also finds application in evaluating the effect of a load upon a shock source. A source of shock generally consists of a means of shock excitation and a resilient structure through which the excitation is transmitted to a load. Consequently, the character of the shock delivered by the resilient structure of the shock source is influenced by the nature of the load being driven. The characteristics of the source and load may be defined in terms of mechanical impedance or mobility (see Chap. 10). If the shock motion at the source output is measured with no load and expressed in terms of its Fourier spectrum, the effect of the load upon this shock motion can be determined by Eq. (23.79) of the Appendix. The resultant motion with the load attached is described by its Fourier spectrum.

The transfer function of a structure may be determined by applying a transient force to the structure and noting the response. This is analogous to the more commonly used method of applying a sinusoidally varying force whose frequency can be varied over a wide range and noting the sinusoidally varying motion at the frequency of the force application. In some circumstances, it may be more convenient to apply a transient. From the measured time-histories of the force and the response, the corresponding Fourier spectra can be calculated. The transfer function is the quotient of the Fourier spectrum of the force divided by the Fourier spectrum of the response, as indicated by Eq. (23.82) of the Appendix.

DATA REDUCTION TO THE RESPONSE DOMAIN

A structure or physical system has a characteristic response to a particular shock applied as an excitation to the structure. The magnitudes of the response peaks can be used to define certain effects of the shock by considering systematically the properties of the system and relating the peak responses to such properties. This is in contrast to the Fourier spectrum description of a shock in the following respects:

1. Whereas the Fourier spectrum defines the shock in terms of the amplitudes and phase relations of its frequency components, the response spectrum describes only the effect of the shock upon a structure in terms of peak responses. This effect is of considerable significance in the design of equipments (Chap. 42), and in the specification of laboratory tests (Chap. 24).

2. The time-history of a shock cannot be determined from the knowledge of the peak responses of a system excited by the shock; i.e., the calculation of peak responses is an irreversible operation. This contrasts with the Fourier spectrum where the Fourier spectrum can be determined from the time-history, and vice versa.

By limiting consideration to the response of a linear, viscously damped single degree-of-freedom structure with lumped parameters (hereafter referred to as a simple structure and illustrated in Fig. 23.5), there are only two structural parameters upon which the response depends: (1) the undamped natural frequency and (2) the fraction of critical

FIG. 23.5. Representation of a simple structure used to accomplish the data reduction of a shock motion to the response domain. The differential equation of motion is given by Eq. (23.31).

damping. With only two parameters involved, it is feasible to obtain from the shock measurement a systematic presentation of the peak responses of many simple structures. This process is termed *data reduction to the response domain*. This type of reduced data applies directly to a system that responds in a single degree-of-freedom; it is useful to some extent by normal-mode superposition to evaluate the response of a linear system that responds in more than one degree-of-freedom. The conditions of a particular application determine the magnitude of errors resulting from superposition.[2-4]

SHOCK SPECTRUM.* The response of a system to a shock can be expressed as the time-history of a parameter that describes the motion of the system. For a simple system, the magnitudes of the response peaks can be summarized as a function of the natural frequency or natural period of the responding system, at various values of the fraction of critical damping. This type of presentation is termed a shock spectrum. In the shock spectrum, or more specifically the two-dimensional shock spectrum, only the maximum value of the response found in a single time-history is plotted. The three-dimensional shock spectrum takes the form of a surface and shows the distribution of response peaks throughout the time-history. The two-dimensional spectrum is more common and is discussed in considerable detail in the immediately following section; for convenience, it is referred to hereafter simply as the *shock spectrum*. The three-dimensional spectrum is discussed in less detail in a later section.

Parameters for the Shock Spectrum. The peak response of the simple structure may be defined, as a function of natural frequency, in terms of any one of several parameters that describe its motion. The parameters often are related to each other by the characteristics of the structure. However, inasmuch as one of the advantages of the shock spectrum method of data reduction and presentation is convenience of application to physical situations, it is advantageous to give careful consideration in advance to the particular parameter that is best adapted to the attainment of particular objectives. Referring to the simple structure shown in Fig. 23.5, the following significant parameters may be determined directly from measurements on the structure:

1. Absolute displacement $x(t)$ of mass m. This indicates the displacement of the responding structure with reference to an inertial reference plane, i.e., coordinate axes fixed in space.

2. Relative displacement $\delta(t)$ of mass m. This indicates the displacement of the responding structure relative to its support, a quantity useful for evaluating the distortions and strains within the responding structure.

3. Absolute velocity $\dot{x}(t)$ of mass m. This quantity is useful for determining the kinetic energy of the structure.

4. Relative velocity $\dot{\delta}(t)$ of mass m. This quantity is useful for determining the stresses generated within the responding structure due to viscous damping and the maximum energy dissipated by the responding structure.

* This spectrum sometimes is designated the *response spectrum*.

5. Absolute acceleration $\ddot{x}(t)$ of mass m. This quantity is useful for determining the stresses generated within the responding structure due to the combined elastic and damping reactions of the structure.

The *equivalent static acceleration* is that steadily applied acceleration, expressed as a multiple of the acceleration of gravity, which distorts the structure to the maximum distortion resulting from the action of the shock.[5] For the simple structure of Fig. 23.5, the relative displacement response δ indicates the distortion under the shock condition. The corresponding distortion under static conditions, in a $1g$ gravitational field, is

$$\delta_{st} = \frac{mg}{k} = \frac{g}{\omega_n{}^2} \tag{23.28}$$

By analogy, the maximum distortion under the shock condition is

$$\delta_{\max} = \frac{A_{eq}g}{\omega_n{}^2} \tag{23.29}$$

where A_{eq} is the equivalent static acceleration in units of gravitational acceleration. From Eq. (23.29),

$$A_{eq} = \frac{\delta_{\max}\omega_n{}^2}{g} \tag{23.30}$$

The maximum relative displacement $\delta_{\max}$ and the equivalent static acceleration A_{eq} are directly proportional.

If the shock is a loading parameter, such as force, pressure, or torque, as a function of time, the corresponding equivalent static parameter is an equivalent static force, pressure, or torque, respectively. Since the supporting structure is assumed to be motionless when a shock loading acts, the relative response motions and absolute response motions become identical.

The differential equation of motion for the system shown in Fig. 23.5 is

$$-x(t) + 2\zeta\omega_n\dot{\delta}(t) + \omega_n{}^2\delta(t) = 0 \tag{23.31}$$

where ω_n is the undamped natural frequency and ζ is the fraction of critical damping. When $\zeta = 0$, $\ddot{x}_{\max} = A_{eq}g$; this follows directly from the relation of Eq. (23.29). When $\zeta \neq 0$, the acceleration $\ddot{x}$ experienced by the mass m results from forces transmitted by the spring k and the damper c. Thus, in a damped system, the maximum acceleration of mass m is not exactly equal to the equivalent static acceleration. However, in most mechanical structures, the damping is relatively small; therefore, the equivalent static acceleration and the maximum absolute acceleration often are interchangeable with negligible error.

Referring to the model in Fig. 23.1, suppose the equivalent static acceleration A_{eq} and the maximum absolute acceleration $\ddot{x}_{\max}$ are known for the primary structure. Then A_{eq} is useful directly for calculating the maximum relative displacement response of the primary structure. When the natural frequency of the secondary structure is much higher than the natural frequency of the primary structure, the maximum acceleration $\ddot{x}_{\max}$ of M is useful for calculating the maximum relative displacement of m with respect to M. The secondary structure then responds in a "static manner" to the acceleration of the mass M; i.e., the maximum acceleration of m is approximately equal to that of M. Consequently, both A_{eq} and $\ddot{x}_{\max}$ can be used for design purposes to calculate equivalent static loads on structures of equipment (see Chap. 42).

If the damping in the responding structure is large ($\zeta > 0.2$), the values of A_{eq} and $\ddot{x}_{\max}$ are significantly different. Because the maximum distortion of primary structures often is the type of information required and the equivalent static acceleration is an expression of this response in terms of an equivalent static loading, the following discussion is limited to shock spectra in terms of A_{eq}.

The response of a simple structure with small damping to oscillatory-type shock excitation often is substantially sinusoidal at the natural frequency of the structure, i.e., the envelope of the oscillatory response varies in a relatively slow manner as depicted in Fig. 23.6. The maximum relative displacement δ_{max}, the maximum relative velocity $\dot{\delta}_{max}$, and the maximum absolute acceleration $\ddot{x}_{max}$ are related approximately as follows:

$$\dot{\delta}_{max} = \omega_n \delta_{max}; \qquad \ddot{x}_{max} = \omega_n \dot{\delta}_{max}; \qquad \ddot{x}_{max} = \omega_n^2 \delta_{max} \qquad (23.32)$$

where the sign may be neglected since the positive and negative maxima are approximately equal. When applicable, these relations may be used to convert from a spectrum expressed in one parameter to a spectrum expressed in another parameter.

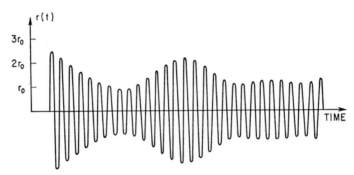

FIG. 23.6. Example of an oscillatory response time-history $r(t)$ for which the envelope of the response varies in a relatively slow manner. This type of response often results when an oscillatory-type shock motion acts on a simple structure with small damping. When such a response occurs, the response parameters of the simple structure are related in accordance with Eq. (23.32).

For idealized shock motions which often are approximated in practice, it is desirable to use a dimensionless ratio for the ordinate of the shock spectrum. Some of the more common dimensionless ratios are

$$\frac{g A_{eq}}{\ddot{u}_{max}} = \frac{\omega_n^2 \delta_{max}}{\ddot{u}_{max}}; \qquad \frac{\ddot{x}_{max}}{\ddot{u}_{max}}; \qquad \frac{\omega_n \delta_{max}}{\Delta \dot{u}}; \qquad \frac{\dot{\delta}_{max}}{\Delta \dot{u}}; \qquad \frac{\delta_{max}}{u_{max}}$$

where $\ddot{u}_{max}$ and u_{max} are the maximum acceleration and displacement, respectively, of the shock motion and $\Delta \dot{u}$ is the velocity change of the shock motion (equal to the area under the acceleration time-history). Sometimes these ratios are referred to as *shock amplification factors*.

Calculation of Shock Spectrum. The relative displacement response of a simple structure (Fig. 23.5) resulting from a shock defined by the acceleration $\ddot{u}(t)$ of the support is given by the Duhamel integral[6]

$$\delta(t) = \frac{1}{\omega_d} \int_0^t \ddot{u}(t_v) e^{-\zeta \omega_n (t - t_v)} \sin \omega_d(t - t_v) \, dt_v \qquad (23.33)$$

where $\omega_n = (k/m)^{1/2}$ is the undamped natural frequency, $\zeta = c/2m\omega_n$ is the fraction of critical damping and $\omega_d = \omega_n(1 - \zeta^2)^{1/2}$ is the damped natural frequency. The excitation $\ddot{u}(t_v)$ is defined as a function of the variable of integration t_v and the response $\delta(t)$ is a function of time t. The relative displacement δ and relative velocity $\dot{\delta}$ are considered to be zero when $t = 0$. The equivalent static acceleration, defined by Eq. (23.30), as a function of ω_n and ζ is

$$A_{eq}(\omega_n, \zeta) = \frac{\omega_n^2}{g} \delta_{max}(\omega_n, \zeta) \qquad (23.34)$$

If a shock loading such as the input force $F(t)$ rather than an input motion acts on the simple structure, the response is

$$\delta(t) = \frac{1}{m\omega_d} \int_0^t F(t_v) e^{-\zeta\omega_n(t-t_v)} \sin \omega_d(t - t_v) \, dt_v \tag{23.35}$$

and an equivalent static force is given by

$$F_{eq}(\omega_n,\zeta) = k\delta_{max}(\omega_n,\zeta) = m\omega_n^2\delta_{max}(\omega_n,\zeta) \tag{23.36}$$

The equivalent static force is related to equivalent static acceleration by

$$F_{eq}(\omega_n,\zeta) = mA_{eq}(\omega_n,\zeta) \tag{23.37}$$

It is often of interest to determine the maximum relative displacement of the simple structure in Fig. 23.5 in both a positive and a negative direction. If $\ddot{u}(t)$ is positive as shown, positive values of $\ddot{x}(t)$ represent upward acceleration of the mass m. Initially, the spring is compressed and the positive direction of $\delta(t)$ is taken to be positive as shown. Conversely, negative values of $\delta(t)$ represent extension of spring k from its original position. It is possible that the ultimate use of the reduced data would require that both extension and compression of spring k be determined. Correspondingly, a positive and a negative sign may be associated with an equivalent static acceleration A_{eq} of the support, so that A_{eq}^+ is an upward acceleration producing a positive deflection δ and A_{eq}^- is a downward acceleration producing a negative deflection δ.

For some purposes it is desirable to distinguish between the maximum response which occurs during the time in which the measured shock acts and the maximum response which occurs during the free vibration existing after the shock has terminated. The shock spectrum based on the former is called a *primary shock spectrum* and that based on the latter is called a *residual shock spectrum*. For instance, the response $\delta(t)$ to the half-sine pulse in Fig. 23.2C occurring during the period $(t < \tau)$ is the primary response and the response $\delta(t)$ occurring during the period $(t > \tau)$ is the residual response. Reference is made to primary and residual shock spectra in the next section on examples of the shock spectrum and in the section on the relationship between the shock spectrum and the Fourier spectrum.

Examples of Shock Spectra.* In this section the shock spectra are presented for the five acceleration time-histories in Fig. 23.2. These spectra, shown in Fig. 23.7, are expressed in terms of equivalent static acceleration for the undamped responding structure, for $\zeta = 0.1, 0.5$, and other selected fractions of critical damping. Both the maximum positive and the maximum negative responses are indicated. In addition, a number of relative displacement response time-histories $\delta(t)$ are plotted to show the nature of the responses.

Acceleration Impulse. The application of Eq. (23.33) to the acceleration impulse shown in Fig. 23.2A and defined by Eq. (23.1) yields

$$\delta(t) = \frac{\dot{u}_0}{\omega_d} e^{-\zeta\omega_n t} \sin \omega_d t \qquad [\zeta < 1] \tag{23.38}$$

This response is plotted in Fig. 23.7A for $\zeta = 0, 0.1$, and 0.5. The response peaks are reached at the times $t = (\cos^{-1} \zeta)/\omega_d$, $\cos^{-1} \zeta$ increasing by π for each succeeding peak. The values of response at the peaks are

$$\delta_{max}^{(i)}(\omega_n,\zeta) = \frac{\dot{u}_0}{\omega_n} \exp\left(-\frac{\zeta}{\sqrt{1 - \zeta^2}} [\cos^{-1} \zeta + (i - 1)\pi]\right) \quad [0 < \cos^{-1} \zeta \le \pi/2] \tag{23.39}$$

where i is the number of the peak ($i = 1$ for the first positive peak, $i = 2$ for the first negative peak, etc.).

---

* A large number of shock spectra, based on various response parameters, are given in Chap. 8.

RESPONSE SPECTRUM A_{eq}

RELATIVE DISPLACEMENT RESPONSE $\delta(t)$

FIG. 23.7. Time-histories of response to shock motions defined in Fig. 23.2 and corresponding shock spectra. Time-histories for several fractions of critical damping in the responding system are included, and the spectra are for corresponding fractions of critical damping. The spectra represent maximum values of response, as indicated by similar symbols. When the shock motions shown in Fig. 23.2 are superimposed on the time-histories of response, the shock motions are indicated by lines with hollow circles. Natural frequencies corresponding to the shock spectra are indicated where they have a significant relation to the frequency of the shock motion.

The largest positive response occurs at the first peak, i.e., when $i = 1$, and is shown by the solid dots in Fig. 23.7A. Hence, the equivalent static acceleration in the positive direction is obtained by substitution of Eq. (23.39) into Eq. (23.34) with $i = 1$:

$$A_{eq}^{+}(\omega_n,\zeta) = \frac{\omega_n \ddot{u}_0}{g} \exp\left(-\frac{\zeta}{\sqrt{1 - \zeta^2}} \cos^{-1} \zeta\right) \tag{23.40}$$

The equivalent static acceleration in the negative direction is calculated from the maximum relative deflection at second peak, i.e., when $i = 2$, and is shown by the hollow dots in Fig. 23.7A:

$$A_{eq}^{-}(\omega_n,\zeta) = \frac{\omega_n \ddot{u}_0}{g} \exp\left(-\frac{\zeta}{\sqrt{1 - \zeta^2}} (\cos^{-1} \zeta + \pi)\right) \tag{23.41}$$

The resulting shock spectrum is shown in Fig. 23.7A with curves for $\zeta = 0, 0.1, 0.5$, and 1.0. At any value of damping, a shock spectrum is a straight line passing through the origin. The peak distortion of the structure δ_{max} is inversely proportional to frequency. Thus, the relative displacement of the mass increases as the natural frequency decreases, whereas the equivalent static acceleration has an opposite trend.

Acceleration Step. The response of a simple structure to the acceleration step in Fig. 23.2B is found by substituting from Eq. (23.5) in Eq. (23.33) and integrating:

$$\delta(t) = \frac{\ddot{u}_0}{\omega_n^2} \left[1 - \frac{e^{-\zeta\omega_n t}}{\sqrt{1 - \zeta^2}} \cos(\omega_d t - \sin^{-1} \zeta)\right] \qquad [\zeta < 1] \tag{23.42}$$

The responses $\delta(t)$ are shown in Fig. 23.7B for $\zeta = 0, 0.1$, and 0.5. The response overshoots the value $\ddot{u}_0/\omega_n^2$ and then oscillates about this value as a mean with diminishing amplitude as energy is dissipated by damping. An overshoot to $2\ddot{u}_0/\omega_n^2$ occurs for zero damping. A response $\delta = \ddot{u}_0/\omega_n^2$ would result from a steady application of the acceleration $\ddot{u}_0$.

The response maxima and minima occur at the times $t = i\pi/\omega_d$, $i = 0$ providing the first minimum and $i = 1$ the first maximum. The maximum values of the relative displacement response are

$$\delta_{max}(\omega_n,\zeta) = \frac{\ddot{u}_0}{\omega_n^2} \left[1 + \exp\left(-\frac{\zeta i\pi}{\sqrt{1 - \zeta^2}}\right)\right] \qquad [i \text{ odd}] \tag{23.43}$$

The largest positive response occurs at the first maximum; i.e., where $i = 1$, and is shown by the solid symbols in Fig. 23.7B. The equivalent static acceleration in the positive direction is obtained by substitution of Eq. (23.43) into Eq. (23.34) with $i = 1$:

$$A_{qe}^{+}(\omega_n,\zeta) = \frac{\ddot{u}_0}{g} \left[1 + \exp\left(-\frac{\zeta\pi}{\sqrt{1 - \zeta^2}}\right)\right] \tag{23.44a}$$

The greatest negative response is zero; it occurs at $t = 0$, independent of the value of damping, as shown by open symbols in Fig. 23.7B. Thus, the equivalent static acceleration in the negative direction is

$$A_{eq}^{-}(\omega_n,\zeta) = 0 \tag{23.44b}$$

Since the equivalent static acceleration is independent of natural frequency, the shock spectrum curves shown in Fig. 23.7B are horizontal lines. The symbols shown on the shock spectra correspond to the responses shown.

The equivalent static acceleration for an undamped simple structure is twice the value of the acceleration step $\ddot{u}_0/g$. As the damping increases, the overshoot in response decreases; there is no overshoot when the structure is critically damped.

Half-sine Acceleration. The expressions for the response of the damped simple structure to the half-sine acceleration of Eq. (23.9) are too involved to have general usefulness.

For an undamped system, the response $\delta(t)$ is

$$\delta(t) = \frac{\ddot{u}_0}{\omega_n^2}\left(\frac{(\omega_n\tau/\pi)}{1-(\omega_n\tau/\pi)^2}\right)[\sin\omega_n t - (\omega_n\tau/\pi)\sin(\pi t/\tau)] \qquad [0 < t \leq \tau]$$

$$\delta(t) = \frac{\ddot{u}_0}{\omega_n^2}\left(\frac{(\omega_n\tau/\pi)}{1-(\omega_n\tau/\pi)^2}\right)2\cos\left(\frac{\omega_n\tau}{2}\right)\sin\left[\omega_n\left(t-\frac{\tau}{2}\right)\right] \qquad [t > \tau]$$

$$(23.45)$$

For zero damping the residual response is sinusoidal with constant amplitude. The first maximum in the response of a simple structure with natural frequency less than π/τ occurs during the residual response; i.e., after $t = \tau$. As a result, the magnitude of each succeeding response peak is the same as that of the first maximum. Thus the positive and negative shock spectrum curves are equal for $\omega_n \leq \pi/\tau$. The dot-dash curve in Fig. 23.7C is an example of the response at a natural frequency of $2\pi/3\tau$. The peak positive response is indicated by a solid circle; the peak negative response by an open circle. The positive and negative shock spectrum values derived from this response are shown on the undamped ($\zeta = 0$) shock spectrum curves at the right-hand side of Fig. 23.7C, using the same symbols.

At natural frequencies below $\pi/2\tau$, the shock spectra for an undamped system are very nearly linear with a slope $\pm 2\ddot{u}_0\tau/\pi g$. In this low-frequency region the response is essentially impulsive; i.e., the maximum response is approximately the same as that due to an ideal acceleration impulse (Fig. 23.7A) having a velocity change $\dot{u}_0$ equal to the area under the half-sine acceleration time-history.

The response at the natural frequency $3\pi/\tau$ is the dotted curve in Fig. 23.7C. The displacement and velocity response are both zero at the end of the pulse and hence no residual response occurs. The solid and open triangles indicate the peak positive and negative response, the latter being zero. The corresponding points appear on the undamped shock spectrum curves. As shown by the negative undamped shock spectrum curve, the residual spectrum goes to zero for all odd multiples of π/τ above $3\pi/\tau$.

As the natural frequency increases above $3\pi/\tau$, the response attains the character of relatively low amplitude oscillations occurring with the half-sine pulse shape as a mean. An example of this type of response is shown by the solid curve for $\omega_n = 8\pi/\tau$. The largest positive response is slightly higher than $\ddot{u}_0/\omega_n^2$ and the residual response occurs at a relatively low level. The solid and open square symbols indicate the largest positive and negative response.

As the natural frequency becomes extremely high, the response follows the half-sine shape very closely. In the limit, the natural frequency becomes infinite and the response approaches the half-sine wave shown in Fig. 23.7C. For natural frequencies greater than $5\pi/\tau$, the response tends to follow the input and the largest input response is within 20 per cent of the response due to a static application of the peak input acceleration. This portion of the shock spectrum is sometimes referred to as the "static region" (see *Limiting Values of Shock Spectrum* below).

The equivalent static acceleration without damping for the positive direction is

$$A_{\text{eq}}^{+}(\omega_n,0) = \frac{\ddot{u}_0}{g}\left(\frac{2(\omega_n\tau/\pi)}{1-(\omega_n\tau/\pi)^2}\right)\cos\left(\frac{\omega_n\tau}{2}\right) \qquad \left[\omega_n \leq \frac{\pi}{\tau}\right]$$

$$A_{\text{eq}}^{+}(\omega_n,0) = \frac{\ddot{u}_0}{g}\left(\frac{(\omega_n\tau/\pi)}{(\omega_n\tau/\pi)-1}\right)\sin\left(\frac{2i\pi}{(\omega_n\tau/\pi)+1}\right) \qquad \left[\omega_n > \frac{\pi}{\tau}\right]$$

$$(23.46)$$

where i is the positive integer which maximizes the value of the sine term while the argument remains less than π. In the negative direction the peak response always occurs during the residual response; thus, it is given by the absolute value of the first of the expressions in Eq. (23.46):

$$A_{\text{eq}}^{-}(\omega_n,0) = \frac{\ddot{u}_0}{g}\left(\frac{2(\omega_n\tau/\pi)}{1-(\omega_n\tau/\pi)^2}\right)\cos\left(\frac{\omega_n\tau}{2}\right) \qquad (23.47)$$

Shock spectra for damped systems can be found by use of an electrical analog or digital computer. Spectra for $\zeta = 0.1$ and 0.5 are shown in Fig. 23.7C.

The response of a damped structure whose natural frequency is less than $\pi/2\tau$ is essentially impulsive; i.e., the shock spectra in this frequency region are substantially identical to the spectra for the acceleration impulse in Fig. 23.7A. Except near the zeros in the negative spectrum for an undamped system, damping reduces the peak response. For the positive spectra, the effect is small in the static region since the response tends to follow the input for all values of damping. The greatest effect of damping is seen in the negative spectra because it affects the decay of response oscillations at the natural frequency of the structure.

Decaying Sinusoidal Acceleration. Although analytical expressions for the response of a simple structure to the decaying sinusoidal acceleration shown in Fig. 23.2D are available, calculation of spectra is impractical without use of a computer. Figure 23.7D shows spectra for several values of damping in the decaying sinusoidal acceleration. In the low-frequency region ($\omega_n < 0.2\omega_1$), the response is essentially impulsive. The area under the acceleration time-history of the decaying sinusoid is $\dot{u}_0$; hence, the response of a very low-frequency structure is similar to the response to an acceleration impulse of magnitude $\dot{u}_0$.

When the natural frequency of the responding system approximates the frequency ω_1 of the oscillations in the decaying sinusoid, a resonant type of build-up tends to occur in the response oscillations. The region in the neighborhood of $\omega_1 = \omega_n$ may be termed a quasi-resonant region of the shock spectrum. Responses for $\zeta = 0$, 0.1, and 0.5 and $\omega_n = \omega_1$ are shown in Fig. 23.7D. In the absence of damping in the responding system, the rate of build-up diminishes with time and the amplitude of the response oscillations levels off as the input acceleration decays to very small values. Small damping in the responding system; e.g., $\zeta = 0.1$, reduces the initial rate of build-up and causes the response to decay to zero after a maximum is reached. When damping is as large as $\zeta = 0.5$, no build-up occurs.

Complex Shock. The shock spectra for the complex shock of Fig. 23.2E are shown in Fig. 23.7E, as obtained with a direct-analog-type shock spectrum analyzer.[7] Time-histories of the response of a system with a natural frequency of 1,250 cps also are shown. The ordinate of the spectrum plot is equivalent static acceleration and the abscissa is the natural frequency in cycles per second. Three pronounced peaks appear in the spectra for zero damping, at approximately 1,250 cps, 1,900 cps, and 2,350 cps. Such peaks indicate a concentration of frequency content in the shock, similar to the spectra for the decaying sinusoid in Fig. 23.7D. Other peaks in the shock spectra for an undamped system indicate less significant oscillatory behavior in the shock. The two lower frequencies at which the pronounced peaks occur correlate with the peaks in the Fourier spectrum of the same shock, as shown in Fig. 23.3E. The highest frequency at which a pronounced peak occurs is above the range for which the Fourier spectrum was calculated.

Because of frequency-response limitations of the spectrum analyzer, the shock spectra do not extend below 200 cps. Since the duration of the complex shock of Fig. 23.2E is about 0.016 sec, an impulsive-type response occurs only for natural frequencies well below 200 cps. As a result, no impulsive region appears in the shock spectra. There is no static region of the spectra shown because calculations were not extended to a sufficiently high frequency.

In general, the equivalent static acceleration A_{eq} is reduced by additional damping in the responding structure system except in the region of valleys in the shock spectra where damping may increase the magnitude of the spectrum. Positive and negative spectra tend to be approximately equal in magnitude at any value of damping; thus, the spectra for a complex oscillatory type of shock may be based on peak response independent of sign to a good approximation.

Limiting Values of Shock Spectrum. The response data provided by the shock spectrum sometimes can be abstracted to simplified parameters that are useful for certain purposes. In general, this cannot be done without definite information on the ultimate use of the reduced data, particularly the natural frequencies of the structures upon which the shock acts. Two important cases are discussed in the following sections.

Impulse or Velocity Change. The duration of a shock sometimes is much smaller than the natural period of a structure upon which it acts. Then the entire response of the structure is essentially a function of the area under the time-history of the shock, described in terms of acceleration or a loading parameter such as force, pressure, or torque. Consequently, the shock has an effect which is equivalent to that produced by an impulse of infinitesimally short duration; i.e., an ideal impulse.

The shock spectrum of an ideal impulse is shown in Fig. 23.7A. All equivalent static acceleration curves are straight lines passing through the origin. The portion of the spectrum exhibiting such straight-line characteristics is termed the *impulsive region*.

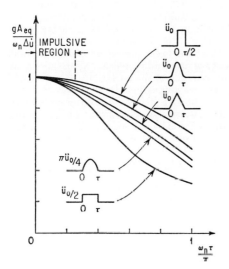

Fig. 23.8. Portions adjacent to the origin of the positive spectra of an undamped system for several single pulses of acceleration as shown. The ordinate is equivalent static acceleration A_{eq} normalized with respect to the peak impulsive response in g's, $\omega_n \Delta \dot{u}/g$, where $\Delta \dot{u}$ is the area under the corresponding acceleration time-history. An impulsive region is shown for which the deviation from impulsive response, $gA_{eq}/\omega_n \Delta \dot{u} = 1$, is less than 10 per cent.

Fig. 23.9. Portions of the positive shock spectra of an undamped system with high natural frequencies for several single pulses of acceleration. The ordinate is equivalent static acceleration A_{eq} normalized with respect to the peak acceleration of the pulse, $\ddot{u}_0/g$. A static region is shown for which the deviation from static response, $gA_{eq}/\ddot{u}_0 = 1$, is less than 20 per cent.

The shock spectrum of the half sine acceleration pulse has an impulsive region when ω_n is less than approximately $\pi/2\tau$, as shown in Fig. 23.7C. If the area under a time-history of acceleration or shock loading is not zero or infinite, an impulsive region exists in the shock spectrum. The extent of the region on the natural frequency axis depends on the shape and duration of the shock.

The portions adjacent to the origin of the positive shock spectra of an undamped system for several single pulses of acceleration are shown in Fig. 23.8. To illustrate the impulsive nature, each spectrum is normalized with respect to the peak impulsive response $\omega_n \Delta \dot{u}/g$, where $\Delta \dot{u}$ is the area under the corresponding acceleration time-history. Hence, the spectra indicate an impulsive response where the ordinate is approximately 1. The response to a single pulse of acceleration is impulsive within a tolerance of 10 per cent if $\omega_n < 0.25\pi/\tau$; i.e., $f_n < 0.4\tau^{-1}$, where f_n is the natural frequency of the responding structure in cps and τ is the pulse duration in seconds. This result also applies when the responding system is damped. Thus, it is possible to reduce the description of a shock pulse to a designated velocity change when the natural frequency of the responding structure is less than a specified value.   The magnitude of the velocity change is the area under the acceleration pulse:

$$\Delta \dot{u} = \int_0^\tau \ddot{u}(t) \, dt \qquad (23.48)$$

Peak Acceleration or Loading.   The natural frequency of a structure responding to a shock sometimes is sufficiently high that the response oscillations of the structure at its natural frequency have a relatively small amplitude.   Examples of such responses are shown in Fig. 23.7C for $\omega_n = 8\pi/\tau$ and $\zeta = 0, 0.1, 0.5$.   As a result, the maximum response of the structure is approximately equal to the maximum acceleration of the shock and is termed *equivalent static response.*   The magnitude of the spectra in such a static region is determined principally by the peak value of the shock acceleration or loading. Portions of the positive spectra of an undamped system in the region of high natural frequencies are shown in Fig. 23.9 for a number of acceleration pulses.   Each spectrum is normalized with respect to the maximum acceleration of the pulse.   If the ordinate is approximately 1, the shock spectrum curves behave approximately in a static manner.

The limit of the static region in terms of the natural frequency of the structure is more a function of the slope of the acceleration time-history than the duration of the pulse. Hence, the horizontal axis of the shock spectra in Fig. 23.9 is given in terms of the ratio of the rise time τ_r to the maximum value of the pulse.   As shown in Fig. 23.9, the peak response to a single pulse of acceleration is approximately equal to the maximum acceleration of the pulse, within a tolerance of 20 per cent, if $\omega_n > 2.5\pi/\tau_r$; i.e., $f_n > 1.25\tau_r^{-1}$, where f_n is the natural frequency of the responding structure in cps and τ_r is the rise time to the peak value in seconds.   The tolerance of 20 per cent applies to an undamped system; for a damped system, the tolerance is lower, as indicated in Fig. 23.7C.

The concept of the static region also can be applied to complex shocks.   Suppose the shock is oscillatory, as shown in Fig. 23.2E.   If the response to such a shock is to be nearly static, the response to each of the succession of pulses that make up the shock must be nearly static.   This is most significant for the pulses of large magnitude because they determine the ordinate of the spectrum in the static region. Therefore, the shock spectrum for a complex shock in the static region is based upon the pulses of greatest magnitude and shortest rise time.

THREE-DIMENSIONAL SHOCK SPECTRUM.[8]

In general, the response of a structure to a shock is oscillatory and continues for an appreciable number of oscillations.   At each oscillation, the response has an interim maximum value that differs, in general, from the preceding or following maximum value.   For example, a typical time-history of response of a simple system of given natural frequency is shown in Fig. 23.6; the characteristics of the response may be summarized by the block diagram of Fig. 23.10. The abscissa of Fig. 23.10 is the peak response at the respective cycles of the oscillation, and the ordinate is the number of cycles at which the peak response exceeds the indicated value.   Thus, the time-history of Fig. 23.6 has 29 cycles of oscil-

FIG. 23.10. Bar chart for the response of a system to a shock excitation. The chart represents a plane normal to the natural frequency axis in Fig. 23.11 and is obtained by counting the number of response maxima above various discrete increments of maximum response.   In concept, the width of the increments approaches zero and a line faired through the ends of the bars is a smooth curve.

lation exceeds $0.6r_0$, but only 2 cycles at which the peak response exceeds $2.0r_0$.

In accordance with the concept of the shock spectrum, the natural frequency of the responding system is modified by discrete increments and the response determined at each increment.   This leads to a number of time-histories of response corresponding to

Fig. 23.6, one for each natural frequency, and a similar number of block diagrams corresponding to Fig. 23.10. This group of block diagrams can be assembled to form a surface that shows pictorially the characteristics of the shock in terms of the response of a simple system. The axes of the surface are peak response, natural frequency of the responding system, and number of response cycles exceeding a given peak value. The block diagram of Fig. 23.10 is arranged on this set of axes at A, as shown in Fig. 23.11, at the appropriate position along the natural frequency axis. Other corresponding block diagrams are shown at B. The three-dimensional shock spectrum is the surface faired through the ends of the bars; the intercept of this surface with the planes of the block diagrams is indicated at C and with the maximum response–natural frequency plane at D. Surfaces are obtainable for both positive and negative values of the response, and a separate surface is obtained for each fraction of critical damping in the responding system.

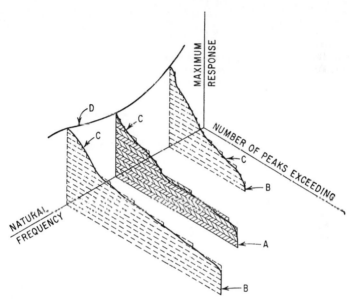

Fig. 23.11. Example of a three-dimensional shock spectrum, which is a plot of the maximum value of individual response oscillations as a function of the number of occurrences of these maxima and the natural frequency of the responding system. Bar charts are shown properly disposed with respect to the coordinate planes.

The two-dimensional shock spectrum is a special case of the three-dimensional surface. The former is a plot of the maximum response as a function of the natural frequency of the responding system; hence, it is a projection on the plane of the response and natural frequency axes of the maximum height of the surface. However, the height of the surface never exceeds that at one response cycle. Thus, the two-dimensional shock spectrum is the intercept of the surface with a plane normal to the "number of peaks exceeding" axis at the origin.

The response surface is a useful concept and illustrates a physical condition; however, it is not well adapted to quantitative analysis because the distances from the surface to the coordinate planes cannot be determined readily. A group of bar charts, each corresponding to Fig. 23.10, is more useful for quantitative purposes. The differences in lengths of the bars are discrete increments; this corresponds to the data reduction method in which the axis of response magnitudes is divided into discrete increments for purposes of counting the number of peaks exceeding each magnitude. In concept, the width of the increment may be considered to approach zero and the line faired through the ends of the bars represents the smooth intercept with the surface.

RELATIONSHIP BETWEEN SHOCK SPECTRUM AND FOURIER SPECTRUM.

Although the shock spectrum and the Fourier spectrum are fundamentally different, there is a partial correlation between them. A direct relationship exists between a running Fourier spectrum, to be defined subsequently, and the response of an undamped simple structure. A consequence is a simple relationship between the Fourier spectrum of absolute values and the peak residual response of an undamped simple structure.

For the case of zero damping, Eq. (23.33) provides the relative displacement response

$$\delta(\omega_n, t) = \frac{1}{\omega_n} \int_0^t \ddot{u}(t_v) \sin \omega_n(t - t_v) \, dt_v \tag{23.49}$$

A form better suited to our needs here is

$$\delta(\omega_n, t) = \frac{1}{\omega_n} \mathcal{J} \left[e^{j\omega_n t} \int_0^t \ddot{u}(t_v) \, e^{-j\omega_n t_v} \, dt_v \right] \tag{23.50}$$

The integral above is seen to be the Fourier spectrum of the portion of $\ddot{u}(t)$ which lies in the time interval from zero to t, evaluated at the natural frequency ω_n. Such a time-dependent spectrum can be termed a "running Fourier spectrum" and denoted by $\mathbf{F}(\omega, t)$:

$$\mathbf{F}(\omega, t) = \int_0^t \ddot{u}(t_v) e^{-j\omega t_v} \, dt_v \tag{23.51}$$

It is assumed that the excitation vanishes for $t < 0$. The integral in Eq. (23.50) can be replaced by $\mathbf{F}(\omega_n, t)$; and after taking the imaginary part

$$\delta(\omega_n, t) = \frac{1}{\omega_n} F(\omega_n, t) \sin [\omega_n t + \theta(\omega_n, t)] \tag{23.52}$$

where $F(\omega_n, t)$ and $\theta(\omega_n, t)$ are the magnitude and phase of the running Fourier spectrum, corresponding to the definitions in Eqs. (23.63) and (23.64). Equation (23.52) provides the previously mentioned direct relationship between undamped structural response and the running Fourier spectrum.

When the running time t exceeds τ, the duration of $\ddot{u}(t)$, the running Fourier spectrum becomes the usual spectrum as given by Eq. (23.57), with τ used in place of the infinite upper limit of the integral. Consequently, Eq. (23.52) yields the sinusoidal residual relative displacement for $t > \tau$:

$$\delta_r(\omega_n, t) = \frac{1}{\omega_n} F(\omega_n) \sin [\omega_n t + \theta(\omega_n)] \tag{23.53}$$

The amplitude of this residual deflection and the corresponding equivalent static acceleration are

$$(\delta_r)_{\max} = \frac{1}{\omega_n} F(\omega_n)$$

$$(A_{eq})_r = \frac{\omega_n^2 (\delta_r)_{\max}}{g} = \frac{\omega_n}{g} F(\omega_n) \tag{23.54}$$

This result is clearly evident for the Fourier spectrum and undamped shock spectrum of the acceleration impulse. The Fourier spectrum is the horizontal line (independent of frequency) shown in Fig. 23.3A and the shock spectrum is the inclined straight line (increasing linearly with frequency) shown in Fig. 23.7A. Since the impulse exists only at $t = 0$, the entire response is residual. The undamped shock spectra in the impulsive region of the half-sine pulse and the decaying sinusoidal acceleration, Fig. 23.7C and D, respectively, also are related to the Fourier spectra of these shocks, Fig. 23.3C and D, in a similar manner. This results from the fact that the maximum response occurs in the residual motion for systems with small natural frequencies. Another example is the entire negative shock spectrum with no damping for the half-sine pulse in Fig. 23.7C whose values are ω_n/g times the values of the Fourier spectrum in Fig. 23.3C.

METHODS OF DATA REDUCTION

Even though preceding sections of this chapter include several analytic functions as examples of typical shocks, data reduction in general is applied to measurements of shock that are not definable by analytic functions. The following sections outline data reduction methods that are adapted for use with any general type of function. Standard forms for presenting the analysis results are given in Ref. 9.

FOURIER SPECTRUM

The Fourier spectrum can be evaluated by either analog or digital techniques.[4] Block diagrams of two basic approaches to analog computation are shown in Fig. 23.12. Such

(A)

(B)

Fig. 23.12. Block diagrams for analog computation of Fourier spectra. (A) Direct implementation of Eqs. (23.60) and (23.61) and (B) bandpass filtering to determine spectrum magnitude.[4]

analyses can be performed in either a *parallel* or a *swept* fashion. In a parallel analysis, an array of oscillators or filters is implemented to determine all the spectral values of interest simultaneously. In a swept analysis, the computation is performed for one frequency at a time; the frequency is stepped or continuously varied slowly to determine the spectral values of interest in a serial fashion. For purposes of the data analysis the shock time-history can be made artificially periodic by repeating the shock at intervals of time τ. The Fourier spectrum then becomes a Fourier series involving discrete components at integer multiples of the fundamental frequency $1/\tau$.* The magnitude of the true continuous Fourier spectrum at the frequency of such a discrete component is

$$F\left(\frac{2\pi n}{\tau}\right) = \left(\frac{\tau}{2}\right) c_n \qquad (23.55)$$

where c_n is the amplitude of the nth discrete component at the frequency $\omega = 2\pi n/\tau$, as defined by Eq. (22.8).

* See Chap. 22 for the defining relationships of a Fourier series and a discussion of data-reduction techniques.

Digital evaluation of the Fourier spectrum is most efficiently accomplished by means of the fast Fourier transform (FFT).   Discussion of this approach appears in Chap. 27. Also see *FFT Analyzer*, Chap. 13.

SHOCK SPECTRUM

Both analog and digital computation techniques can be employed for evaluation of the shock spectrum.[4]   Block diagrams for analog computation are shown in Fig. 23.13. As with the Fourier spectrum, analyses can be performed in either a *parallel* or a *stepped* manner.   In a parallel analysis, the shock-spectrum values are obtained from simul-

Fig. 23.13. Block diagrams for analog computation of shock spectra using the simple system shown in Fig. 23.5.   (*A*) Active electronic analog, (*B*) passive electrical analog, (*C*) peak detection (positive, negative, or absolute) with timing control as required for primary, residual, or overall analysis.

taneous determinations of the responses of simple structures corresponding to all natural frequencies of interest.   A stepped analysis involves the determination of shock-spectrum results for one natural frequency and then a repeat of the determination for each successive value of the natural frequency.

Digital evaluation of the shock spectrum involves the use of a digital computer to determine the response of simple structures, followed by the detection of peak response. The response time-histories can be evaluated by the following techniques: (1) direct numerical or recursive integration of the Duhamel integral, Eq. (23.33), or (2) convolution or recursive filtering procedures.   These techniques are discussed in detail in Ref. 4.

In principle, both analog and digital techniques can be modified to determine the three-dimensional shock spectrum.   The modification involves the counting of the number of response maxima above various discrete increments of maximum response to obtain the results depicted in Fig. 23.11.

Reed Gage.   The shock spectrum may be measured directly by a mechanical instrument that responds to shock in a manner analogous to the data reduction techniques used to obtain shock spectra from time-histories.   The instrument includes a number of flexible mechanical systems that are considered to respond as single degree-of-freedom systems; each system has a different natural frequency and means are provided to indicate the maximum deflection of each system as a result of the shock. The instrument often is referred to as a *reed gage* because the flexible mechanical systems are small cantilever beams carrying end masses; these have the appearance of reeds.[10] The reed gage is described in Chap. 13.

The response parameter indicated by the reed gage is maximum deflection of the reeds relative to the base of the instrument; generally, this deflection is converted to equivalent static acceleration by applying the relation of Eq. (23.30). The reed gage offers a convenience in the indication of a useful quantity immediately and in the elimination of auxiliary electronic equipment. Also, it has important limitations: (1) the information is limited to the determination of a shock spectrum; (2) the deflection of a reed is inversely proportional to its natural frequency squared, thereby requiring high equivalent static accelerations to achieve readable records at high natural frequencies; (3) the means to indicate maximum deflection of the reeds (styli inscribing on a target surface) tend to introduce an undefined degree of damping; and (4) size and weight limitations on the reed gage for a particular application often limit the number of reeds which can be used and the lowest natural frequency for a reed. In spite of these limitations, the instrument sees continued use and has provided significant shock spectra where more elaborate instruments have failed.

APPENDIX 23.1

FOURIER INTEGRAL

Let the time-history of any motion or loading parameter (time-history of a shock) be designated $f(t)$. The corresponding Fourier spectrum or frequency spectrum $\mathbf{F}(\omega)$ is[*]

$$\mathbf{F}(\omega) = \int_{-\infty}^{\infty} f(t)e^{-j\omega t}\, dt \qquad (23.56)$$

The Fourier spectrum $\mathbf{F}(\omega)$ is a complex number used to describe the amplitude and phase of a sinusoid at the frequency ω. The complex amplitude of the Fourier component at the frequency ω is $\mathbf{F}(\omega)\, d\omega$, the area under the Fourier spectrum curve in the interval $d\omega$. The quantity $\mathbf{F}(\omega)e^{j\omega t}\, d\omega$ is the sinusoidal variation at the frequency ω; the integral sums these sinusoids. If it is assumed that the shock starts at $t = 0$, Eq. (23.56) becomes

$$\mathbf{F}(\omega) = \int_{0}^{\infty} f(t)e^{-j\omega t}\, dt \qquad (23.57)$$

If the duration of the time-history $f(t)$ has a finite limit, this limit is used in place of ∞ in Eq. (23.57). An alternate notation for $\mathbf{F}(\omega)$ is $\mathbf{F}[f(t)]$, which denotes the "Fourier transform of $f(t)$."

When $\omega = 0$, Eq. (23.56) becomes

$$\mathbf{F}(0) = \int_{-\infty}^{\infty} f(t)\, dt \qquad (23.58)$$

Thus, the zero frequency component of the Fourier spectrum of a shock is equal to the area under the time-history of the shock.

Since $\mathbf{F}(\omega)$ represents a distribution of amplitudes expressed in complex form, it may be written in terms of the real and imaginary parts:

$$\mathbf{F}(\omega) = \Re[\mathbf{F}(\omega)] + j\mathfrak{I}[\mathbf{F}(\omega)] \qquad (23.59)$$

where

$$\Re[\mathbf{F}(\omega)] = \int_{0}^{\infty} f(t)\cos \omega t\, dt \qquad (23.60)$$

$$\mathfrak{I}[\mathbf{F}(\omega)] = -\int_{0}^{\infty} f(t)\sin \omega t\, dt \qquad (23.61)$$

---

[*] Mathematical limitations on the function which permit this type of analysis do exist but these do not hinder engineering applications. See Refs. 11 to 13 for discussions of these limitations.

Alternatively, the Fourier spectrum may be defined in terms of its absolute value and phase angle:

$$\mathbf{F}(\omega) = F(\omega)e^{j\theta(\omega)} \qquad (23.62)$$

where $F(\omega)$ is the absolute value of $\mathbf{F}(\omega)$ and $\theta(\omega)$ is the phase angle. The absolute value and phase angle are related to the real and imaginary parts by

$$F(\omega) = \sqrt{\mathcal{R}^2[\mathbf{F}(\omega)] + \mathcal{I}^2[\mathbf{F}(\omega)]} \qquad (23.63)$$

$$\theta(\omega) = \tan^{-1}\left\{\frac{\mathcal{I}[\mathbf{F}(\omega)]}{\mathcal{R}[\mathbf{F}(\omega)]}\right\} \qquad (23.64)$$

Figure 23.14 shows the vector $\mathbf{F}(\omega)$, its magnitude $F(\omega)$ and phase angle $\theta(\omega)$, and its real and imaginary parts.

Equation (23.56) may be used with any shock parameter, such as acceleration, velocity, displacement, force, or pressure. The area under the curve of Fourier spectrum magnitude versus frequency has the same units as the time-history of the shock parameter.

The Fourier spectrum of the time-derivative of $f(t)$ is the product of $j\omega$ and the Fourier spectrum of $f(t)$:

$$\mathbf{F}\left[\frac{df(t)}{dt}\right] = j\omega\mathbf{F}[f(t)] \qquad (23.65)$$

Thus if $\ddot{u}(t)$, $\dot{u}(t)$, $u(t)$ are the acceleration, velocity, and displacement time-histories, respectively, of a given shock motion, their Fourier spectra are related as follows:

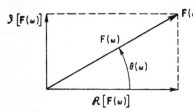

FIG. 23.14. Representation of the complex Fourier component at the frequency ω, its magnitude $F(\omega)$ and phase angle $\theta(\omega)$, and its real and imaginary parts, $\mathcal{R}[\mathbf{F}(\omega)]$ and $\mathcal{I}[\mathbf{F}(\omega)]$.

$$\mathbf{F}[\ddot{u}(t)] = j\omega\mathbf{F}[\dot{u}(t)] = -\omega^2\mathbf{F}[u(t)] \qquad (23.66)$$

$$\mathbf{F}[\dot{u}(t)] = j\omega\mathbf{F}[u(t)] \qquad (23.67)$$

The time-history of a function can be found by integrating its Fourier spectrum components:

$$f(t) = \frac{1}{2\pi}\int_{-\infty}^{\infty} \mathbf{F}(\omega)e^{j\omega t}\, d\omega \qquad (23.68)$$

Equations (23.68) and (23.56) are known as the inverse and direct Fourier transforms, respectively. The use of the factor $1/2\pi$ in Eq. (23.68) represents one formulation of the pair of Eqs. (23.56) and (23.68). In other formulations, the $1/2\pi$ appears in Eq. (23.56) and not in Eq. (23.68), or a $1/\sqrt{2\pi}$ appears in each. Caution should be exercised, when consulting treatises and tables of the Fourier transform, in determining which formulation of the transform is used. A useful compilation of direct and inverse Fourier transform pairs is given in Ref. 14.

The Fourier spectrum of a constant is a delta function (infinitely sharp spike) at zero frequency; hence, the addition of a constant to a function only alters the Fourier spectrum of that function at $\omega = 0$. Consequently, the relations for the spectra of acceleration, velocity, and displacement, Eqs. (23.66) and (23.67), are valid for all $\omega \neq 0$. This limitation is of little significance for engineering applications because the addition of a constant to a displacement time-history only changes the origin of the coordinate system. Similarly, the addition of a constant to a velocity time-history introduces another frame of reference which does not accelerate.

Equation (23.68) requires integration over negative frequencies. For engineering purposes, negative frequencies lack physical significance and a more desirable form of Eq. (23.68) is

$$f(t) = \frac{2}{\pi}\int_{0}^{\infty} \mathcal{R}[\mathbf{F}(\omega)] \cos \omega t \, d\omega \qquad [t > 0]$$

$$f(t) = 0 \qquad\qquad\qquad\qquad [t < 0]$$

$$(23.69)$$

Equation (23.69) is derived from (23.68) by including the restriction that $f(t) = 0$ for $t < 0$. For this to be true, $\int_0^\infty \Re[\mathbf{F}(\omega)]\cos\omega t\, d\omega$ must equal $-\int_0^\infty \Im[\mathbf{F}(\omega)]\sin\omega t\, d\omega$.

COMPARISON OF FOURIER AND LAPLACE TRANSFORMS*

The Laplace transform of $f(t)$ is defined, as a function of the complex variable p, by the integral

$$\mathbf{L}[f(t)] = \int_0^\infty f(t)e^{-pt}\, dt \tag{23.70}$$

where $p = a + j\omega$ for $a > 0$ and $\mathbf{L}[f(t)]$ denotes the "Laplace transform of $f(t)$." Equation (23.70) sometimes is written $\mathbf{L}[f(t)] = p\int_0^\infty f(t)e^{-pt}\, dt$. Caution should be exercised in determining the formulation of the transform. A comparison of Eqs. (23.70) and (23.57) reveals that the Laplace transform of a function approaches the Fourier transform of that function as $a \to 0$ or $p \to j\omega$, if the function is zero for all negative time.[12] The time-history is retrieved by the use of the inverse Laplace transform, involving integration in the complex plane:

$$f(t) = \frac{1}{2\pi j}\int_{a-j\infty}^{a+j\infty} \mathbf{L}[f(t)]e^{pt}\, dp \tag{23.71}$$

where a is chosen so that $f(t) = 0$ for $t < 0$. If $\mathbf{L}[f(t)]$ approaches $\mathbf{F}(\omega)$ as $p \to j\omega$ for *all* values of the frequency ω, then Eq. (23.71) provides a result identical to that of Eq. (23.68).

For problems associated with physically achievable shocks, the Fourier and Laplace transforms give identical results. The Fourier transform has certain mathematical disadvantages which the Laplace transform overcomes, and more extensive tables of the Laplace transform are available.[14] A detailed discussion of both transform methods and their limitations is found in Chap. 1 of Ref. 13.

A disadvantage of the Fourier transform method is that the integral defining the transform sometimes does not converge unless an extra convergence factor is provided. For example, the acceleration step in Fig. 23.2B has no Fourier transform in the strict sense since the integrand of Eq. (23.56) is $e^{-j\omega t}$ which does not have a well-defined limit as $t \to \infty$. The Fourier transform can be made useful in such instances by introducing a convergence factor e^{-at} and modifying the function $f(t)$ as follows:

$$f(t,a) = e^{-at}f(t) \tag{23.72}$$

The Fourier spectrum is

$$\mathbf{F}(\omega - ja) = \int_{-\infty}^\infty f(t)e^{-j(\omega-ja)t}\, dt \tag{23.73}$$

If the limit exists as $a \to 0$ of $\mathbf{F}(\omega - ja)$ $\mathbf{F}(\omega)$ can be called the Fourier transform of $f(t)$. The Laplace transform incorporates an appropriate convergence factor in its definition, thus guaranteeing the existence of the transform for practically all functions of possible interest in engineering applications.

Another disadvantage of the Fourier transform is the difficulty in obtaining the inverse transform; i.e., the application of Eq. (23.68). The Laplace transform achieves the inversion by integration in the complex p plane so that all the background of complex-function theory is available.

On the basis of these factors it may appear that the Laplace transform is the broader and more basic concept, and that the Fourier transform represents a special case of the Laplace transform. However, for purposes of physical interpretation of actual measured data, the Fourier spectrum has a greater intuitive appeal as a logical extension of the Fourier series concepts for periodic functions.

* The Laplace transform is discussed in detail in Chap. 8.

APPLICATIONS OF FOURIER INTEGRAL

DETERMINATION OF STRUCTURAL RESPONSE

If a shock loading parameter whose Fourier spectrum $\mathbf{F}(\omega)$ is defined by Eq. (23.56) acts upon a structure, the frequency content of the structural response is found from

$$\mathbf{R}(\omega) = \mathbf{H}(\omega)\mathbf{F}(\omega) \tag{23.74}$$

where $\mathbf{R}(\omega)$ is the Fourier spectrum of the response and $\mathbf{H}(\omega)$ is the applicable transfer function. The transfer function is the steady-state response per unit sinusoidal input as a function of the frequency ω. If $\mathbf{R}(\omega)$ represents a velocity response and $\mathbf{F}(\omega)$ represents a force input, $\mathbf{H}(\omega)$ is given the special designation of mobility $\mathfrak{M}(\omega)$; the reciprocal of $\mathbf{H}(\omega)$ is the mechanical impedance $\mathbf{Z}(\omega)$. [*Mobility* is sometimes called *mechanical admittance*.] Mechanical impedance and mobility are discussed in detail in Chap. 10. Equation (23.74) is applicable when the velocity and displacement of the structure are zero at $t = 0$. If the initial conditions are not zero, additional terms are required in Eq. (23.74). These terms can be obtained from the governing differential equation. For example, if the differential equation has the form

$$a\ddot{r} + b\dot{r} + cr = f(t) \tag{23.75}$$

the corresponding equation in the frequency domain is found by applying the Fourier transform to both the response $r(t)$ and the excitation $f(t)$[11, 13]

$$a[-\omega^2\mathbf{R}(\omega) - j\omega r(0+) - \dot{r}(0+)] + b[j\omega\mathbf{R}(\omega) - r(0+)] + c[\mathbf{R}(\omega)] = \mathbf{F}(\omega)$$

Solving for $\mathbf{R}(\omega)$,

$$\mathbf{R}(\omega) = \frac{\mathbf{F}(\omega) + a\dot{r}(0+) + [b + j\omega a]r(0+)}{-\omega^2 a + j\omega b + c}$$

The quantities $r(0+)$ and $\dot{r}(0+)$ are the values of the response and its derivative at $t = 0+$, the plus sign allowing for any discontinuity at $t = 0$. If $r(0+)$ and $\dot{r}(0+)$ are zero, the transfer function $\mathbf{H}(\omega)$ is found from Eq. (23.74):

$$\mathbf{H}(\omega) = \frac{1}{-\omega^2 a + j\omega b + c} \tag{23.76}$$

The response in the time domain may be found by the use of Eq. (23.69):

$$r(t) = \frac{2}{\pi} \int_0^\infty \Re[\mathbf{R}(\omega)] \cos \omega t \, d\omega \qquad [t > 0]$$

Substituting for $\mathbf{R}(\omega)$ from Eq. (23.74),

$$r(t) = \frac{2}{\pi} \int_0^\infty \Re[\mathbf{H}(\omega)\mathbf{F}(\omega)] \cos \omega t \, d\omega \qquad [t > 0] \tag{23.77}$$

Example 23.1. Consider the response of the single degree-of-freedom structure in Fig. 23.5 to an input acceleration shock $\ddot{u}(t)$. The transfer function for the relative displacement response per unit input sinusoidal acceleration is

$$\mathbf{H}(\omega) = \frac{\bar{\delta}e^{j\omega t}}{e\ddot{\mathbf{U}}^{j\omega t}} = \frac{1}{(\omega_n{}^2 - \omega^2) + j(2\zeta\omega\omega_n)}$$

where $\bar{\delta}$ and $\ddot{\mathbf{U}}$ are, respectively, complex amplitudes of relative displacement response and input sinusoidal acceleration at the frequency ω. From Eq. (23.77), the time-history of the response is

$$\delta(t) = \frac{2}{\pi} \int_0^\infty \Re \frac{\mathbf{F}(\omega)}{(\omega_n{}^2 - \omega^2) + j(2\zeta\omega\omega_n)} \cos \omega t \, d\omega \tag{23.78}$$

where $\mathbf{F}(\omega)$ is the Fourier spectrum of $\ddot{u}(t)$.

CALCULATION OF THE EFFECT OF STRUCTURAL LOAD CHANGE
ON A SHOCK SOURCE

An important application of the Fourier spectrum representation of a shock concerns the transmission of a shock from one structure to another structure. The combination of the structure from which the shock emanates and the process by which the shock is generated may be termed a mechanical shock source.[15] The other structure can be considered a load on this source; any change in the character of the load modifies the shock output of the source.

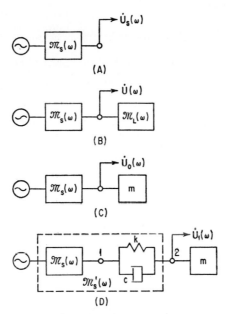

FIG. 23.15. Schematic representation of a mechanical source composed of a generator and an internal mobility.[16] The source is shown driving (A) no load, (B) a load with input mobility $\mathfrak{M}_L(\omega)$, (C) a mass load, and (D) a resiliently mounted mass.

A schematic representation of a simple mechanical source that is free at its output side is shown in Fig. 23.15A. The mobility of the source determined at its output junction is designated $\mathfrak{M}_s(\omega)$. The Fourier spectrum of the shock velocity delivered by the unloaded or free source is $\dot{U}_s(\omega)$.

A source driving a load having an input mobility $\mathfrak{M}_L(\omega)$ is shown in Fig. 23.15B. The output of the source in this condition is

$$\dot{U}(\omega) = \frac{1}{1 + [\mathfrak{M}_s(\omega)/\mathfrak{M}_L(\omega)]} \dot{U}_s(\omega) \qquad (23.79)$$

as $\mathfrak{M}_s/\mathfrak{M}_L$ approaches zero; i.e., when the impedance is large compared to the load impedance, the actual output shock approaches the free output of the source.

Example 23.2. Consider the effect of interposing a resilient mounting between a rigid mass and a source that drives the mass.[16] Figure 23.15C represents the direct mounting of a mass m to the source; the mobility of the mass m is $1/j\omega m$ (Chap. 10). Thus, the shock input to the mass is found by substituting in Eq. (23.79):

$$\dot{\mathbf{U}}_0(\omega) = \frac{1}{1 + j\omega m \mathfrak{M}_s(\omega)} \dot{\mathbf{U}}_s(\omega) \tag{23.80}$$

Figure 23.15D shows the resiliently mounted mass, the resilient mount being represented by a stiffness k and a viscous damping coefficient c. Since the shock input to the mass is of interest, the output junction of the source can be considered to be moved from point 1 to point 2; i.e., the resilient mount becomes part of the source. The modified source mobility is

$$\mathfrak{M}_s{}'(\omega) = \mathfrak{M}_s(\omega) + \frac{1}{c + (k/j\omega)}$$

The output of the free source $\dot{\mathbf{U}}_s(\omega)$ remains unchanged, and the load mobility remains $1/j\omega m$. The shock input to the mass is

$$\dot{\mathbf{U}}_1(\omega) = \frac{1}{1 + j\omega m \mathfrak{M}_s{}'(\omega)} \dot{\mathbf{U}}_s(\omega) \tag{23.81}$$

MEASUREMENT OF STRUCTURAL TRANSFER FUNCTION

The most straightforward technique for the measurement of a structural transfer function involves the application of a sinusoidal input motion and the measurement of the resulting sinusoidal output motion. The ratio of output to input as a function of frequency is, by definition, the transfer function. It may be desirable to determine the transfer function by applying an input shock and recording the output shock.

Given a structure which is initially at rest, a known shock input $f(t)$ is applied, producing a shock response $r(t)$. The direct application of Eq. (23.74) gives

$$\mathbf{H}(\omega) = \frac{\mathbf{F}(\omega)}{\mathbf{R}(\omega)} \tag{23.82}$$

where $\mathbf{F}(\omega)$ and $\mathbf{R}(\omega)$ are the Fourier spectra of $f(t)$ and $r(t)$, respectively. Accurate values of $\mathbf{H}(\omega)$ are obtained only in those frequency ranges where $f(t)$ has a significant frequency content. If the absolute value of $\mathbf{F}(\omega)$ is zero at certain frequencies, the value of $\mathbf{H}(\omega)$ is indeterminate at those frequencies.

REFERENCES

1. von Kármán, T., and M. A. Biot: "Mathematical Methods in Engineering," p. 388, McGraw-Hill Book Company, Inc., New York, 1940.
2. Rubin, S.: *J. Appl. Mechanics*, **25**:501 (1958).
3. Fung, Y. C., and M. V. Barton: *J. Appl. Mechanics*, **25**:365 (1958).
4. Kelly, R. D., and G. Richman: "Principles and Techniques of Shock Data Analysis," Shock and Vibration Information Center, Washington, D.C., 1969.
5. Walsh, J. P., and R. E. Blake: *Proc. Soc. Exptl. Stress Anal.*, **6**(2):150 (1948).
6. Timoshenko, S., and D. H. Young: "Advanced Dynamics," p. 49, McGraw-Hill Book Company, Inc., New York, 1948.
7. Caughey, T. K., and D. E. Hudson: *Proc. Soc. Exptl. Stress Anal.*, **13**(1):199 (1956).
8. Lunney, E. J., and C. E. Crede, *WADC Tech. Rept.* 57-75, 1958
9. American National Standards Institute: Methods for Analysis and Presentation of Shock and Vibration Data, ANSI S2.10-1971.
10. Rubin, S.: *Proc. Soc. Exptl. Stress Anal.*, **16**(2):97 (1956).
11. Pipes, L. A.: "Applied Mathematics for Engineers and Physicists," 2d ed., McGraw-Hill Book Company, Inc., New York, 1958.
12. Campbell, G. A., and R. M. Foster: "Fourier Series for Practical Applications," D. Van Nostrand Company, Inc., Princeton, N.J., 1940. Previously published as No. B-584 of the Bell System Series of Monographs, 1931.
13. Weber, E.: "Linear Transient Analysis," vol. II, chap. 1, John Wiley & Sons, Inc., New York, 1956.
14. Erdelyi, A.: "Tables of Integral Transforms," vol. I, McGraw-Hill Book Company, Inc., New York, 1954.
15. Molloy, C. T.: *J. Acoust. Soc. Amer.*, **29**:842 (1957).
16. Rubin, S.: *SAE Trans.*, **68**:318 (1960).

24

PART I: VIBRATION OF STRUCTURES INDUCED BY GROUND MOTION

W. J. Hall
University of Illinois

INTRODUCTION

SOURCE OF GROUND MOTION

Ground motion may arise from any number of sources such as earthquake excitation[1,2] (described in detail in this chapter), or high explosive[3] or nuclear device detonations.[4] In such cases, the source excitation can lead to major vibration of the primary structure or facility and its many parts, as well as to transient and permanent translation and rotation of the ground on which the facility is constructed. Detonations may result in drag and side-on overpressures, ballistic ejecta, and thermal and radiation effects.

Other sources of ground excitation, although usually not as strong, can be equally troublesome. For example, the location of a precision machine shop near a railroad or highway or of delicate laboratory apparatus in a plant area containing heavy drop forging machinery or unbalanced rotating machinery are typical of situations in which ground-transmitted vibrations may pose serious problems. Another different class of vibrational problems arises from excitation of the primary structure by other sources, e.g., wind blowing on a bridge, or people walking or dancing on a floor in a building. Vibration of the primary structure in turn can affect secondary elements such as mounted equipment and people located on a floor (in the case of buildings) and vehicles or equipment (in the case of bridges). A brief summary of such people-structure interaction is given in Ref 5.

The variables involved in problems of this type are exceedingly numerous and, with the exception of earthquakes, few specific well-defined measurements are generally available to serve as a guide in estimating the ground motions that might be used as computational guidelines in particular cases. A number of acceleration-vs.-time curves for typical ground motions arising from the operation of machines and vehicles are shown in Fig. 24.1. Another report arising from a rock quarry blast is shown in Fig. 24.2. Although the records differ somewhat in their characteristics, all can be compared directly with similar measurements of earthquakes, and response computations generally are handled in the same manner.

References 1 to 11 are typical of those containing data of the types described above; in some cases, these references contain suggested approaches for analyzing the data. In most cases, to analyze and evaluate such information one needs to (1) develop an understanding of the source and nature of the vibration, (2) ascertain the physical characteristics of the structure or element, (3) develop an approach for modeling and analysis, (4) carry out the analysis, (5) study the response (with parameter variations if needed), (6) evaluate the behavior of service and function limit states, and (7) develop, in light of the results of the analysis, possible courses of corrective action, if required. Merely changing

FIG. 24.1. Ground-acceleration-vs.-time curves for typical machine and vehicle excitations. *(A)* Vertical acceleration measured on a concrete floor on sandy loam soil at a point 6 ft from the base of a drop hammer. *(B)* Horizontal acceleration 50 ft from drop hammer. The weight of the drop hammerhead was approximately 15,000 lb, and the hammer was mounted on three layers of 12-in. by 12-in. oak timbers on a large concrete base. *(C)* Vertical acceleration 6 ft from a railroad track on the well-maintained right-of-way of a major railroad during passing of luxury-type passenger cars at a speed of approximately 20 mph. The accelerometer was bolted to a 2-in. by 2-in. by 2½-in. steel block which was firmly anchored to the ground. *(D)* Horizontal acceleration of the ground at 46 ft from the above railroad track, with a triple diesel-electric power unit passing at a speed of approximately 20 mph. *(E)* Horizontal acceleration of the ground 6 ft from the edge of a relatively smooth highway, with a large tractor and trailer unit passing on the outside lane at approximately 35 mph with a full load of gravel.[6]

the mass, stiffness, or damping of the structural system may or may not lead to acceptable corrective action in the sense of a reduction in deflections or stresses; careful investigation of the various alternatives is required to change the response to an acceptable limit.

FIG. 24.2. Typical quarry blast data. *(A)* Time-history of velocity taken by a velocity transducer and recorder. *(B)* Corresponding response spectrum computed from the record in *(A)* using Duhamel's integral.[3]

RESPONSE OF SIMPLE STRUCTURES TO GROUND MOTIONS

Four structures of varying size and complexity are shown in Fig. 24.3: (*A*) a simple, relatively compact machine anchored to a foundation, (*B*) a 15-story building, (*C*) a 40-story building, and (*D*) an elevated water tank. The dynamic response of each of the structures shown in Fig. 24.3 can be approximated by representing each as a simple mechanical oscillator consisting of a single mass supported by a spring and a damper as shown in Fig. 24.4. The relation between the undamped angular frequency of vibration $\omega = 2\pi f$, the natural frequency f, and the period T is defined in terms of the spring constant k and the mass m:

$$\omega^2 = \frac{k}{m} \tag{24.1}$$

$$f = \frac{1}{T} = \frac{\omega}{2\pi} = \frac{1}{2\pi}\sqrt{\frac{k}{m}} \tag{24.2}$$

In general, the effect of the damper is to produce damping of free vibrations or to reduce the amplitude of forced vibrations. The damping force is assumed to be equal to a damping coefficient c times the velocity $\dot{u}$ of the mass relative to the ground. The value of c at which the motion loses its vibratory character in free vibration is called the *critical damping coefficient;* for example, $c_c = 2m\omega$. The amount of damping is most conveniently considered in terms of the fraction of ζ critical damping [see Eq. (2.12)],

$$\zeta = \frac{c}{c_c} = \frac{c}{2m\omega} \tag{24.3}$$

For most practical structures ζ is relatively small, in the range of 0.05 to 0.2 (i.e., 0.5 to 20 per cent), and does not appreciably affect the natural period or frequency of vibration (see Ref. 1*b*).

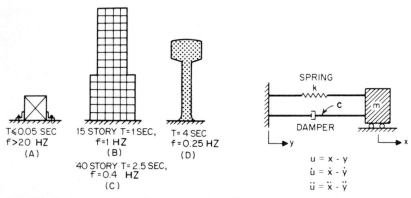

$T \leqslant 0.05$ SEC
$f > 20$ HZ
(A)

15 STORY T=1 SEC,
f=1 HZ
(B)

40 STORY T=2.5 SEC,
f=0.4 HZ
(C)

T= 4 SEC
f=0.25 HZ
(D)

$u = x - y$
$\dot{u} = \dot{x} - \dot{y}$
$\ddot{u} = \ddot{x} - \ddot{y}$

FIG. 24.3. Structures subjected to earthquake ground motion. *(A)* A machine anchored to a foundation. *(B)* A 15-story building. *(C)* A 40-story building. *(D)* An elevated water tank.

FIG. 24.4. System definition; the dynamic response of each of the structures shown in Fig. 24.3 can be approximated by this simple mechanical oscillator.

EARTHQUAKE GROUND MOTION

Strong-motion earthquake acceleration records with respect to time have been obtained for a number of earthquakes. Ground motions from other sources of disturbance, such as quarry blasting and nuclear blasting, also are available and show many of the same characteristics. As an example of the application of such time-history records, the recorded accelerogram for the El Centro, California, earthquake of May 18, 1940, in the north-south component of horizontal motion is shown in Fig. 24.5. On the same figure are shown the integration of the ground acceleration a to give the variation of ground velocity v with time, and the integration of velocity to give the variation of ground displacement d with time. These integrations normally require baseline corrections of various sorts, and the magnitude of the maximum displacement may vary depending on how the corrections are made. The maximum velocity is relatively insensitive to the corrections, however. For this earthquake, with the integrations shown in Fig. 24.5, the maximum ground acceleration is 0.32g, the maximum ground velocity is 13.7 in./sec (35 cm/sec), and the maximum ground displacement is 8.3 in. (21 cm). These three maximum values are of particular interest because they help to define the response motions of the various structures considered in Fig. 24.3 most accurately if all three maxima are taken into account.

RESPONSE SPECTRA

ELASTIC SYSTEMS

The response of the simple oscillator shown in Fig. 24.4 to any type of ground motion can be readily computed as a function of time. The maximum values of the response are of particular interest. These maxima can be stated in terms of the maximum strain in the spring $u_m = D$, the maximum spring force, the maximum acceleration of the mass (which is related to the maximum spring force directly when there is no damping), or a quantity, having the dimensions of velocity, which gives a measure of the maximum energy absorbed in the spring. This quantity, designated the pseudo velocity V, is defined in such a way that the energy absorption in the spring is $\frac{1}{2}mV^2$. The relations among

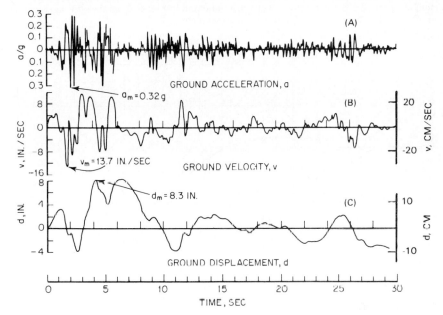

FIG. 24.5. El Centro, California, earthquake of May 18, 1940, north-south component. *(A)* Record of the ground acceleration. *(B)* Variation of ground velocity v with time, obtained by integration of *(A)*. *(C)* Variation of ground displacement with time, obtained by integration of *(B)*.

the maximum relative displacement of the spring D, the pseudo velocity V, and the pseudo acceleration A, which is a measure of the force in the spring, are

$$V = \omega D \tag{24.4}$$

$$A = \omega V = \omega^2 D \tag{24.5}$$

The pseudo velocity V is nearly equal to the maximum relative velocity for systems with moderate or high frequencies but may differ considerably from the maximum relative velocity for very low frequency systems. The pseudo acceleration A is exactly equal to the maximum acceleration for systems with no damping and is not greatly different from the maximum acceleration for systems with moderate amounts of damping, over the whole range of frequencies from very low to very high values.

Typical plots of the response of the system as a function of period or frequency are called *response spectra* (also called *shock spectra*). Plots for acceleration and for relative displacement, for a system with a moderate amount of damping and subjected to an input similar to that of Fig. 24.5, can be made. This arithmetic plot of maximum response is simple and convenient to use. Various techniques of plotting spectra may be found in the references cited at the end of this chapter, especially in Refs. 1c, 1d, and 11 to 24; also of interest are the techniques described in Ref. 6.

A somewhat more useful plot, which indicates at one and the same time the values for D, V, and A, is shown in Fig. 24.6. This plot has the virtue that it also indicates more clearly the extreme or limits of the various parameters defining the response. All parameters are plotted on a logarithmic scale. Since the frequency is the reciprocal of the period, the logarithmic scale for the period would have exactly the same spacing of the points, or in effect the scale for the period would be turned end for end. The pseudo velocity is plotted on a vertical scale. Then on diagonal scales along an axis that extends upward from right to left are plotted values of the displacement, and along an axis that extends upward from left to right the pseudo acceleration is plotted, in such a way that any one point defines for a given frequency the displacement D, the pseudo velocity V, and the pseudo acceleration A. Points are indicated in Fig. 24.6 for the several structures of Fig. 24.3 plotted at their fundamental frequencies.

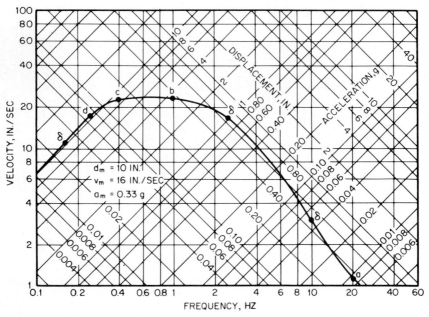

FIG. 24.6. Smooth response spectrum for typical earthquake.

A wide variety of motions have been considered in Refs. 12 to 14, ranging from simple pulses of displacement, velocity, or acceleration of the ground, through more complex motions such as those arising from nuclear-blast detonations, and for a variety of earthquakes as taken from available strong-motion records. Response spectra for the El Centro earthquake are shown in Fig. 24.7. The spectrum for small amounts of damping is much more jagged than indicated by Fig. 24.6, but for the higher amounts of damping the response curves are relatively smooth. The scales are chosen in this instance to represent the amplifications of the response relative to the ground-motion values of displacement, velocity, or acceleration.

The spectra shown in Fig. 24.7 are typical of response spectra for nearly all types of ground motion. On the extreme left, corresponding to very low frequency systems, the response for all degrees of damping approaches an asymptote corresponding to the value of the maximum ground displacement. A low-frequency system corresponds to one having a very heavy mass and a very light spring. When the ground moves relatively rapidly, the mass does not have time to move, and therefore the maximum strain in the spring is precisely equal to the maximum displacement of the ground. For a very high frequency system, the spring is relatively stiff and the mass very light. Therefore, when the ground moves, the stiff spring forces the mass to move in the same way the ground moves, and the mass therefore must have the same acceleration as the ground at every instant.

Hence, the force in the spring is that required to move the mass with the same acceleration as the ground, and the maximum acceleration of the mass is precisely equal to the maximum acceleration of the ground. This is shown by the fact that all the lines on the extreme right-hand side of the figure asymptotically approach the maximum ground-acceleration line.

For intermediate-frequency systems, there is an amplification of motion. In general, the amplification factor for displacement is less than that for velocity, which in turn is less than that for acceleration. Peak amplification factors for the undamped system (ζ = 0) in Fig. 24.7 are on the order of about 3.5 for displacement, 4.2 for velocity, and 9.5 for acceleration.

FIG. 24.7. Response spectra for elastic systems subjected to the El Centro earthquake for various values of fraction of critical damping ζ.

The results of similar calculations for other ground motions are quite consistent with those in Fig. 24.7, even for simple motions. The general nature of the response spectrum shown in Fig. 24.8 consists of a central region of amplified response and two limiting regions of response in which for low-frequency systems the response displacement is equal to the maximum ground displacement and for high-frequency systems the response acceleration is equal to the maximum ground acceleration. Values of the amplification factor reasonable for use in design are presented in the next sections.

DESIGN RESPONSE SPECTRA

A response spectrum developed to give design coefficients is called a *design response spectrum* or a *design spectrum*. As an example of its use in seismic design, for any given site, estimates are made of the maximum ground acceleration, maximum ground velocity, and maximum ground displacement. The lines representing these values can be drawn on the tripartite logarithmic chart of which Fig. 24.9 is an example. The heavy lines showing the ground-motion maxima in Fig. 24.9 are drawn for a maximum ground acceleration a of $1.0g$, a velocity v of 48 in./sec (122 cm/sec), and a displacement d of 36 in. (91.5 cm). These data represent motions more intense than those generally considered for any postulated design earthquake hazard. They are, however, approximately in correct proportion for a number of areas of the world, where earthquakes occur either on firm ground, soft rock, or competent sediments of various kinds. For relatively soft sediments, the velocities and displacements might require increases above the values corresponding to the given acceleration as scaled from Fig. 24.9, and for competent rock, the velocity and displacement values would be expected to be somewhat less. More detail can be found in Refs. 1c and d. It is not likely that maximum ground velocities in excess of 4 to 5 ft/sec (1.2 to 1.5 m/sec) are obtainable under any circumstances.

On the basis of studies of horizontal and vertical directions of excitation for various values of damping (Refs. 1c, 15, and 16), representative amplification factors for the 50th

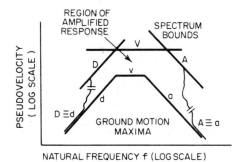

FIG. 24.8. Typical tripartite logarithmic plot of response-spectrum bounds compared with maximum ground motion.

FIG. 24.9. Basic design spectrum normalized to 1.0g for a value of damping equal to 2 per cent of critical, 84.1th percentile level. The spectrum bound values are obtained by multiplying the appropriate ground-motion maxima by the corresponding amplification value of Table 24.1.

and 84.1th percentile levels of horizontal response are presented in Table 24.1. The 84.1th percentile means that one could expect 84.1 per cent of the values to fall at or below that particular amplification. With these amplification factors and noting points B and A to fall at about 8 and 33 Hz, the spectra may be constructed as shown in Fig. 24.9 by multiplying the ground maxima values of acceleration, velocity, and displacement by the appropriate amplification factors. Further information on, and other approaches to, construction of design spectra may be found in Refs. 1c and d.

Table 24.1. Values of Spectrum Amplification Factors[16]

Percentile	Damping, per cent of critical damping	Amplification factor		
		D	V	A
50th	0.5	2.01	2.59	3.68
	2.0	1.63	2.03	2.74
	5.0	1.39	1.65	2.12
	10.0	1.20	1.37	1.64
84.1th	0.5	3.04	3.84	5.10
	2.0	2.42	2.92	3.66
	5.0	2.01	2.30	2.71
	10.0	1.69	1.84	1.99

RESPONSE SPECTRA FOR INELASTIC SYSTEMS

It is convenient to consider an elastoplastic resistance-displacement relation because one can draw response spectra for such a relation in generally the same way as the spectra were drawn for elastic conditions. A simple resistance-displacement relationship for a spring is shown by the light line in Fig. 24.10A, where the yield point is indicated, with a curved relationship showing a rise to a maximum resistance and then a decay to a point of maximum useful limit or failure at a displacement u_m; an equivalent elastoplastic resistance curve is shown by the heavy line. A similar elastoplastic resistance function, more indicative of seismic response, is shown in Fig. 24.10B. The ductility factor μ is defined as the ratio between the maximum permissible or useful displacement to the yield displacement for the effective curve in both cases.

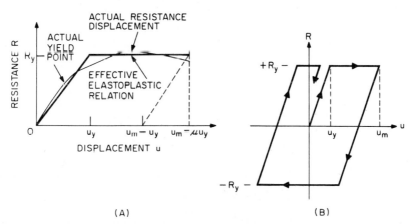

(A)   (B)

F_{IG.} 24.10. *(A)* Monotonic resistance-displacement relationships for a spring, shown by the light line; an equivalent elastoplastic resistance curve, shown by the heavy line. *(B)* A similar elastoplastic resistance function, more indicative of seismic response.

The ductility factors for various types of construction depend on the use of the building, the hazard involved in its failure, the material used, the framing or layout of the structure, and above all on the method of construction and the details of fabrication of joints and connections. A discussion of these topics is given in Refs. 1c, 15, and 16. Figure 24.11 shows acceleration spectra for elastoplastic systems having 2 per cent of critical damping that were subjected to the El Centro, 1940, earthquake. Here the symbol D_y represents the elastic component of the response displacement, but it is not the total displacement. Hence, the curves also give the elastic component of maximum displacement as well as the maximum acceleration A, but they do not give the proper value of maximum pseudo velocity. This is designated by the use of the V' for the pseudo velocity drawn in the figure. The figure is drawn for ductility factors ranging from 1 to 10.

FIG. 24.11. Deformation spectra for elastoplastic systems with 2 per cent of critical damping that were subjected to the El Centro earthquake.

A response spectrum for total displacement also can be drawn for the same conditions as for Fig. 24.11. It is obtained by multiplying each curve's ordinates by the value of ductility factor μ shown on that curve.

The following considerations are useful in using the design spectrum to approximate inelastic behavior. In the amplified displacement region of the spectra, the left-hand side, and in the amplified velocity region, at the top, the spectrum remains unchanged for total displacement and is divided by the ductility factor to obtain yield displacement or acceleration. The upper right-hand portion sloping down at 45°, or the amplified acceleration region of the spectrum, is relocated for an elastoplastic resistance curve, or for any other resistance curve for actual structural materials, by choosing it at a level which corresponds to the same energy absorption for the elastoplastic curve as for an elastic curve for the same period of vibration. The extreme right-hand portion of the spectrum, where the response is governed by the maximum ground acceleration, remains at the same acceleration level as for the elastic case, and, therefore, at a corresponding increased total displacement level. The frequencies at the corners are kept at the same values as in the elastic spectrum. The acceleration transition region of the response spectrum is now drawn also as a straight-line transition from the newly located amplified acceleration line and the ground-acceleration line, using the same frequency points of intersection as in the elastic response spectrum. In all cases the inelastic maximum acceleration spectrum and the inelastic maximum displacement spectrum differ by the factor μ at the same frequencies. The design spectrum so obtained is shown in Fig. 24.12.

The solid line $DVAA_0$ shows the elastic response spectrum. The heavy circles at the intersections of the various branches show the frequencies which remain constant in the construction of the inelastic design spectrum. The dashed line $D'V'A'A_0$ shows the inelastic acceleration, and the line $DVA''A_0''$ shows the inelastic displacement. These two differ by a <u>constant</u> factor μ for the construction shown, except that A and A' differ by the factor $\sqrt{2\mu - 1}$, since this is the factor that corresponds to constant energy for an elastoplastic resistance.

The modified spectrum to account for inelastic action is an approximation at best and should be used generally only for relatively small ductility values, for example, 5 or less. Additional information on the development of elastic and inelastic design response spectra may be found in Refs. 1c, 1d, and 12 to 27.

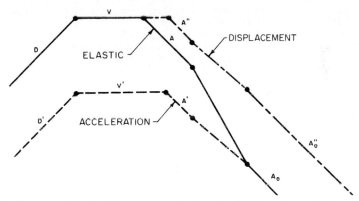

FIG. 24.12. The normal elastic design spectrum is given by $DVAA_0$. The modified spectrum (see text for rules for construction) representing approximately the acceleration or elastic yield displacement for a nonlinear system with ductility μ is given by $D'V'A''A_0$. The total or maximum displacement for the nonlinear system is given approximately by $DVA''A_0''$ and is obtained by multiplying the modified spectrum by the value μ.

MULTIPLE DEGREE-OF-FREEDOM SYSTEMS

USE OF RESPONSE SPECTRA

A multiple degree-of-freedom system has as many modes of vibration as the number of degrees-of-freedom. For example, for the shear beam shown in Fig. 24.13A the fundamental mode of lateral oscillation is shown in (B), the second mode in (C), and the third mode in (D). The number of modes in this case is 5. In a system that has independent (uncoupled) modes (this condition is usually satisfied for buildings) each mode responds to the base motion as an independent single degree-of-freedom system. Thus, the modal responses are nearly independent functions of time. However, the maxima do not necessarily occur at the same time.

For multiple degree-of-freedom systems, the concept of the response spectrum can also be used in most cases, although the use of the inelastic response spectrum is

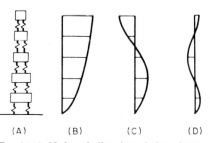

FIG. 24.13. Modes of vibration of shear beam. The first three (1, 2, 3) relative mode shapes are shown by (B), (C), and (D), respectively, for lateral vibration.

only approximately valid as a design procedure.[17,28,29] For a system with a number of masses at nodes in a flexible framework, the equation of motion can be written in matrix form as

$$M\ddot{u} + C\dot{u} + Ku = -M(\ddot{y})\,\{1\} \tag{24.6}$$

in which the last symbol on the right represents a unit column vector. The mass matrix M is usually diagonal, but in all cases both M and the stiffness matrix K are symmetrical. When the damping matrix C satisfies certain conditions, the simplest of which is when it is a linear combination of M and K, then the system has normal modes of vibration, with modal displacement vectors u_n.

FUNDAMENTAL MODE*

Procedures are available for the computation of the periods of vibration of undamped multiple degree-of-freedom systems (see Chap. 2). Damping of less than 20 per cent of critical affects the computed periods by less than 2 per cent. Hence, nominal damping does not affect the period appreciably.

Consider a system with a number of masses lumped at particular points, letting m_n represent the nth mass of the system, and assume that the system is vibrating in the jth mode. If the system is vibrating in a steady-state condition, without damping, the displacement u_{nj} of the nth mass can be written in the form

$$u_{nj} \sin \omega_j t \tag{24.7}$$

The acceleration experienced by the mass during its oscillatory motion is given by the second derivative with respect to time:

$$-\omega_j^2 u_{nj} \sin \omega_j t \tag{24.8}$$

The negative value of this acceleration, multiplied by the mass m_n, is considered to be a reversed effective force or inertial force applied at the point n. The inertial forces $\overline{Q}_{nj} \sin \omega_j t$ are considered to be applied to the structure at each mass point, where the coefficient of the sine term in the inertial-force expression has the form

$$\overline{Q}_{nj} = m_n \omega_j^2 u_{nj} \tag{24.9}$$

Since the inertial forces take account of the mass effects, the displacements of the structure due to the forces $\overline{Q}_{nj}$ must be precisely equal to the quantities u_{nj}. Consequently, in order to find the square of the angular frequency for the jth mode ω_j^2, it is necessary to find a set of displacements u_{nj} of such magnitudes that forces corresponding to each displacement multiplied by the local mass m_n, and by the square of the angular frequency for the jth mode ω_j^2, give rise to the displacements u_{nj}. Any procedure that establishes this condition will give both the modal frequencies and the modal deflection shapes. Multiplying the magnitudes of the modal deflections by a constant does not change the situation, since all the forces, and consequently all the deflections consistent with these forces, will be multiplied by the same constant. However, it is not possible without other knowledge of the situation to write down directly a correct set of displacements for the jth mode. Therefore, the calculations must make it possible to arrive at these deflections as a result of a systematic method of computation. The most useful procedures, at least for the determination of the fundamental mode, are Rayleigh's method, or modifications thereof, or methods based on a procedure of successive approximations developed originally by Stodola. A description of the successive-approximations procedure follows:

1. Assume a set of deflections at each mass point of magnitude u_{na}. Compute for these deflections an inertial force $\overline{Q}_{na}$ given by

$$\overline{Q}_{na} = m_n u_{na} \omega^2 \tag{24.10}$$

where the quantity ω^2 is an unknown angular frequency which may be carried in the calculations as an unknown.

---

* The following presentation is based on material from Ref. 29 by permission of the Portland Cement Association.

2. Apply these forces to the system and compute the corresponding deflections, designated by the symbol u_{nb},

$$u_{nb} = \overline{u}_{nb}\omega^2 \tag{24.11}$$

3. Make u_{nb} and u_{na} as nearly equal as possible. To do this, ω may be varied. The value of ω that gives the best fit is a good approximation of the angular frequency for the mode that corresponds to the deflection u_{nb}, which in general will be an approximation of the fundamental mode. In general, u_{nb} will be a better approximation for the fundamental mode shape than was u_{na}.

4. Repeat the calculations using u_{nb} as the starting point, thereby leading to a new derived deflection that will be an even better approximation.

In most cases, even with a very poor first assumption for the fundamental mode deflection, the process will converge with negligible errors to ω^2 in at most two or three cycles, and one can obtain a good approximation in only one cycle. However, the mode shape will not be so accurately determined unless the calculation is repeated several times.

If the quantity shown in Eq. (24.12) is made a minimum (in effect minimizing the square of the error between the derived deflection and the assumed deflection) the best value of ω^2 consistent with the assumed deflection curve can be determined by

$$\sum_n m_n(u_{nb} - u_{na})^2 = \text{minimum} \tag{24.12}$$

Substituting Eq. (24.11) into Eq. (24.12) and equating to zero, the derivative with respect to ω^2 gives

$$\omega^2 = \frac{\sum m_u u_{na}\overline{u}_{nb}}{\sum m_n \overline{u}_{nb}{}^2} \tag{24.13}$$

The value of ω^2 given by Eq. (24.13) exceeds, generally only slightly, the true value for the fundamental mode.

Rayleigh's method is probably the most widely used engineering procedure for computing the period of the fundamental mode. Without modification, however, it does not generally give accurate values of the mode shape. Rayleigh's method for calculating the fundamental frequency of a building frame can be related to the procedure described above by equating u_{na} to unity throughout the structure. In the case of a vertical or horizontal beamlike structure, the derived displacements $\overline{u}_{nb}$ will be proportional to the deflections of the structure because of forces equal to the weight of the structure. Since Rayleigh's procedure using $u_{na} = 1$ gives a quite accurate determination of the fundamental frequency, Eq. (24.13) yields a highly accurate value if any more reasonable deflection shape is assumed for the first mode.

The effect of foundation rotation, column shortening, or other contributions to deflection can be readily taken into account in both the successive-approximations procedure and the Rayleigh method.

HIGHER MODES

Several methods are available for computing the natural frequencies of modes higher than the fundamental mode for a multiple degree-of-freedom system. Two such procedures are described in Ref. 29. However, the high-speed digital computer can determine all the modes and frequencies of even highly complex systems in only a few minutes by use of standard programs, and detailed hand-calculation methods are no longer of interest except for approximating the fundamental frequency or period.

COMBINATIONS OF MODAL RESPONSES

When the modes and natural frequencies of the system are obtained, the modal responses are determined for each mode considering the participation factors, defined as follows for earthquake and general ground motion:

$$c_n = \frac{u_n{}^T M\{1\}}{u_n{}^T M u_n} \tag{24.14}$$

If the particular quantity desired—say the stress at a particular point, the relative displacement between two reference points, or any other effect—is designated by ω, then the modal values ω_n are determined for each mode and combined by use of the relations

$$\alpha_{\max} \leq \sum_n |c_n \alpha_n D_n| \tag{24.15}$$

$$\alpha_{\text{prob}} \approx \sqrt{\sum_n (c_n \alpha_n D_n)^2} \tag{24.16}$$

$$V_n = \omega_n D_n \qquad\qquad D_n = \frac{V_n}{\omega_n} \tag{24.17}$$

$$A_n = \omega_n^2 D_n = \omega_n V_n \qquad D_n = \frac{A_n}{\omega_n^2} \tag{24.18}$$

For an inelastic response, the quantities to be used are D_n', V_n', or A_n' from calculations such as those leading to Fig. 24.12. Equation (24.15) gives an upper bound to the value of α, and Eq. (24.16) gives the most probable value.

DESIGN

GENERAL CONSIDERATIONS

In the design of a building to resist earthquake motions, the designer works within certain constraints, such as the architectural configuration of the building, the foundation conditions, the nature and extent of the hazard should failure or collapse occur, the possibility of an earthquake, the possible intensity of earthquakes in the region, the cost or available capital for construction, and similar factors. There must be some basis for the selection of the strength and the proportions of the building and of the various members in it. The required strength depends on factors such as the intensity of earthquake motions to be expected, the flexibility of the structure, and the ductility or reserve strength of the structure before damage occurs. Because of the interrelations among the flexibility and strength of a structure and the forces generated in it by earthquake motions, the dynamic design procedure must take these various factors into account. The ideal to be achieved is one involving flexibility and energy-absorbing capacity which will permit the earthquake displacements to take place without generating unduly large forces. To achieve this end, careful design (with attention to continuity, redundancy, connections, strength, and ductility), control of the construction procedures, and appropriate inspection practices are necessary. The attainment of the ductility required to resist earthquake motions must be emphasized. Another important topic in design is that of soil-structure interaction. As the phenomenon becomes better understood, explicit provisions are increasingly showing up in the codes; for major critical structures, such as nuclear power plants, such effects are routinely being considered in design (Refs. 30 and 31).

Also, earthquake engineering research has led to many other design improvements, for example, provision for highway bridges[32] and gas- and liquid-fuel pipelines.[33]

EFFECTS OF DESIGN ON BEHAVIOR*

A structure designed for very much larger horizontal forces than are ordinarily prescribed will have a shorter period of vibration because of its greater stiffness. The shorter period results in higher spectral accelerations, so that the stiffer structure may attract more horizontal force. Thus, a structure designed for too large a force will not necessarily be safer than a similar structure based on smaller forces. On the other hand,

---

* This section is based on material from Ref. 29, by permission, with update modification.

a design based on too small a force makes the structure more flexible and will increase the relative deflections of the floors.

In general, yielding occurs first in the story that is weakest compared with the magnitudes of the shearing forces to be transmitted. In many cases this will be near the base of the structure. If the system is essentially elastoplastic, the forces transmitted through the yielded story cannot exceed the yield shear for that story. Thus, the shears, accelerations, and relative deflections of the portion of the structure above the yielded floor are reduced compared with those for an elastic structure subjected to the same base motion. Consequently, if a structure is designed for a base shear which is less than the maximum value computed for an elastic system, the lowest story will yield and the shears in the upper stories will be reduced. This means that, with proper provision for energy absorption in the lower stories, a structure will, in general, have adequate strength, provided the design shearing forces for the upper stories are consistent with the design base shear. The Uniform Building Code (UBC) recommendations are intended to provide such a consistent set of shears.

A significant inelastic deformation in a structure inhibits the higher modes of oscillation. Therefore, the major deformation is in the mode in which the inelastic deformation predominates, which is usually the fundamental mode. The period of vibration is effectively increased, and in many respects the structure responds almost as a single degree-of-freedom system corresponding to its entire mass supported by the story which becomes inelastic. Therefore, the base shear can be computed for the modified structure, with its fundamental period defining the modified spectrum on which the design should be based. The fundamental period of the modified structure generally will not be materially different from that of the original elastic structure in the case of framed structures. In the case of shear-wall structures it will be longer.

It is partly because of these facts that it is usual in design recommendations to use the frequency of the fundamental mode, without taking direct account of the higher modes. However, it is desirable to consider a shearing-force distribution which accounts for higher-mode excitations of the portion above the plastic region. This is implied in the UBC and SEAOC (Structural Engineers Association of California) recommendations by the provision for lateral force coefficients which vary with height. The distribution over the height corresponding to an acceleration varying uniformly from zero at the base to a maximum at the top takes into account the fact that local accelerations at higher levels in the structure are greater than those at lower levels, because of the larger motions at the higher elevations, and accounts quite well for the moments and shears in the structure. New provisions now under consideration will permit modal analysis, in addition to equivalent lateral load procedures, to be used under some conditions.[34,35] And, in some cases, direct integration procedures using time-history inputs are appropriate for special situations.

DESIGN LATERAL FORCES

Although the complete response of multiple degree-of-freedom systems subjected to earthquake motions can be calculated, it should not be inferred that it is generally necessary to make such calculations as a routine matter in the design of multistory buildings. There are a great many uncertainties about the input motions and about the structural characteristics that can affect the computations. Moreover, it is not generally necessary or desirable to design tall structures to remain completely elastic under severe earthquake motions, and considerations of inelastic behavior lead to further discrepancies between the results of routine methods of calculation and the actual response of structures.

The Uniform Building Code recommendations for earthquake lateral forces are, in general, consistent with the forces and displacements determined by more elaborate procedures. A structure designed according to these recommendations will remain elastic, or nearly so, under moderate earthquakes of frequent occurrence, but it must be able to yield locally without serious consequences if it is to resist an El Centro-type earthquake. Thus, design for the required ductility is an important consideration.

The ductility of the material itself is not a direct indication of the ductility of the structure. Laboratory and field tests and data from operational use of nuclear weapons indicate that structures of practical configurations having frames of ductile materials, or a combination of ductile materials, exhibit ductility factors μ ranging from a minimum of 3 to a maximum of 8. A minimum ductility factor of about 4 to 6 is a reasonable criterion for ordinary structures designed to UBC earthquake requirements.

As a result of the numerous earthquakes that have occurred throughout the world and of the resulting loss of life and property, seismic design codes have undergone major revisions. These revisions are based in part on the results of research and design study and in part on observational experience.

Beginning with the 1974 SEAOC code[36] and slightly later the UBC[37] provisions, the base shear is determined by

$$V_s = ZKCSIW \tag{24.19}$$

where Z = numerical coefficient related to expected severity of earthquakes in various regions of U.S.A.

K = numerical coefficient depending on type or arrangement of resisting elements

C = numerical coefficient dependent on period of structure

S = numerical coefficient reflecting site-structure resonance conditions (intended to reflect foundation conditions)

I = occupancy importance factor

W = total dead load plus applicable portions of other loads as specified

In general the seismic coefficients have been increased in comparison to earlier values and the approaches being adopted attempt to take more factors into consideration in arriving at the design base shear. The newer values are sometimes 1.5 to 2 times the values used previously.

Even more modern provisions [developed as a result of the Applied Technology Council (ATC) effort with later evaluation by the Building Seismic Safety Council (BSSC)] reflecting current structural dynamics practice are now being examined for code adoption.[34, 35] These provisions are drafted to make the analysis approach more rational, explicitly in handling of provisions, and to bring into practice the latest techniques, including use of design spectra, ductility provisions, limit stress, and building form.

Other material of importance in seismic design is given in Refs. 29 to 39.

SEISMIC FORCES FOR OVERTURNING MOMENT AND SHEAR

DISTRIBUTION

In general when modal analysis techniques are not used, in a complex structure or in one having several degrees-of-freedom, it is necessary to have a method of defining the seismic design forces at each mass point of the structure in order to be able to compute the shears and moments to be used for design throughout the structure. The method described in the SEAOC, UBC, or ATC provisions is preferable for this purpose. It is essentially the following:

1. Compute the total base shear corresponding to the seismic coefficient for the structure multiplied by the total weight.

2. Assign a proportion of the total base shear to the top of the structure, or use a special force distribution (ATC).

3. Assume a linear variation of acceleration in the structure from zero at the base to a maximum at the top.

4. Multiply the accleration assumed in step 1 by the mass at each elevation to find an inertial force acting at each level.

5. Adjust the assumed value of acceleration at the top of the structure in step 3 so that the total distributed lateral forces add up to the total base shear computed in step 1.

6. Use the resulting seismic forces, assigned to the various masses at each elevation, to compute shears and moments throughout the structure.

7. Make provisions to accommodate the tensions and compressions in the columns and walls of the structure due to the computed overturning moment at each elevation and at the base.

8. Notice that the overturning moment at the base causes a tilting of the base consistent with the foundation compliance and also may cause a partial uplift at one edge of the base. In the foundation design, consider the increased foundation compression due to such tilting.

DAMPING

The damping in structural elements and components and in supports and foundations of the structure is a function of the intensity of motion and of the stress or strain levels introduced within the structural component or structure, and is highly dependent on the makeup of the structure and the energy absorption mechanisms within it. For further details see Refs. 1 and 15 to 17.

GRAVITY LOADS

The effect of gravity loads, when the structures deform laterally by a considerable amount, can be of importance. In accordance with the general recommendations of most extant codes, the effects of gravity loads are to be added directly to the primary and earthquake effects. In general, in computing the effect of gravity loads, one must take into account the actual deflection of the structure, not the deflection corresponding to reduced seismic coefficients.

VERTICAL AND HORIZONTAL EXCITATION

Usually the stresses or strains at a particular point are affected primarily by the earthquake motions in only one direction; the second direction produces little if any influence. However, this is not always the case and is certainly not so for a simple square building supported on four columns where the stress in a corner column is in general affected equally by the earthquakes in the two horizontal directions and may be affected also by the vertical earthquake forces. Since the ground moves in all three directions in an earthquake, and even tilts and rotates, consideration of the combined effects of all these motions must be included in the design. When the response in the various directions may be considered to be uncoupled, consideration can be given separately to the various components of base motion, and individual response spectra can be determined for each component of direction or of transient base displacement. Calculations have been made for the elastic response spectra in all directions for a number of earthquakes. Studies indicate that the vertical response spectrum is about two-thirds the horizontal response spectrum, and it is recommended that a ratio of 2:3 for vertical response compared with horizontal response be used in design.

For parts of structures or components that are affected by motions in various directions in general, the response may be computed by either one of two methods. The first method involves computing the response for each of the directions independently and then taking the square root of the sums of the squares of the resulting stresses in the particular direction at a particular point as a combined response. Alternatively, one can use the second method of taking the seismic forces corresponding to 100 per cent of the motion in one direction combined with 40 per cent of the motions in the other two orthogonal directions, adding the absolute values of the effects of these to obtain the maximum resultant forces in a member or at a point in a particular direction, and computing the stresses corresponding to the combined effects. In general, this alternative method is slightly conservative.

A related matter that merits attention in design is the provision for relative motion of parts or elements having supports at different locations.

UNSYMMETRICAL STRUCTURES IN TORSION

In design, consideration should be given to the effects of torsion on unsymmetrical structures and even on symmetrical structures where torsions may arise from off-center loads and accidentally because of various reasons, including lack of homogeneity of structures or the presence of the wave motions developed in earthquakes. Most modern codes provide values of computed and accidental eccentricity to use in design, but in the event that analyses indicate values greater than those recommended by the code, the analytical values should be used in design.

SIMULATION TESTING

Simulation testing to create various vibration environments has been employed for years in connection with the development of equipment that must withstand vibration. Over the years such testing of small components has been accomplished on shake tables and involves many different types of input functions. As a result of improved development of electromechanical rams, large shake tables have been developed which can simulate the excitation that may be experienced in a building, structural component, or items of equipment, from various types of ground motions, including earthquake motions, nuclear ground motions, nuclear blast motions induced in the ground or in a structure, and traffic vibrations. Some of these devices are able to provide simultaneous motion in three orthogonal directions.

The matter of simulation testing became of great importance with regard to earthquake excitation because of the development of nuclear power plants and the necessity for components in these plants to remain operational for purposes of safe shutdown and containment and also because of the observed loss of lifeline items in recent earthquakes as, for example, communication and control equipment, utilities, and fire-fighting systems. It is common to require computation of floor response spectra[30] and to provide for equipment qualification. This latter subject, of great importance, is the subject of Part II of this chapter.

REFERENCES

1. Earthquake Engineering Research Institute Monograph Series, Berkeley, Calif. (1979–83).
   (*a*) Hudson, D. E.: "Reading and Interpreting Strong Motion Accelerograms."
   (*b*) Chopra, A. K.: "Dynamics of Structures—A Primer."
   (*c*) Newmark, N. M., and W. J. Hall: "Earthquake Spectra and Design."
   (*d*) Housner, G. W., and P. C. Jennings: "Earthquake Design Criteria."
   (*e*) Seed, H. B., and I. M. Idriss: "Ground Motions and Soil Liquefaction During Earthquakes."
   (*f*) Berg, G. V.: "Seismic Design Codes and Procedures."
   (*g*) Algermission, S. T.: "An Introduction to the Seismicity of the United States."
2. Bolt, B. A.: "Earthquake—A Primer," W. H. Freeman and Co., San Francisco, Calif., 1978.
3. Dowding, C. H.: "Blast Vibration Monitoring and Control," Prentice-Hall, Inc., Englewood Cliffs, N.J., 1985.
4. Glasstone, S., and P. J. Dolan: "The Effects of Nuclear Weapons," 3d ed., U.S. Dept. of Defense and U.S. Dept. of Energy, 1977.
5. Chang, F.-K.: "Psychophysiological Aspects of Man-Structure Interaction," in "Planning and Design of Tall Buildings," vol. 1*a*: "Tall Building Systems and Concepts," American Society of Civil Engineers, New York, N.Y., 1972.
6. Hudson, D. E.: "Vibration of Structures Induced by Seismic Waves," in C. M. Harris and C. E. Crede (eds.), "Shock and Vibration Handbook," Chap. 50, vol. III, McGraw-Hill Book Company, New York, 1961.
7. Ellingwood, B., and A. Tallin: *J. Struct. Eng.*, **110**:2 (1984). [See also article by same authors in same publication, **110**:10 (1984).]
8. Allen, D., and J. H. Rainer: *Can. J. Civil Eng.*, **3**:2 (1976).
9. Richart, F. E., Jr., J. R. Hall, Jr., and R. D. Woods: "Vibration of Soils and Foundation," Prentice-Hall, Inc., Englewood Cliffs, N.J., 1970.
10. Ward, H. S.: *J. Struct. Eng.*, **110**:10 (1984).
11. Abdel-Ghaffer, A. M., and R. H. Scanlan: *J. Eng. Mech.*, **111**:4, 1985.
12. Newmark, N. M., and A. S. Veletsos: "Design Procedures for Shock Isolation Systems of Underground Protective Structures," in "Report For Air Force Weapons Laboratory," vol. III, General American Transportation Corporation, RTD TDR 63-3096, June 1964.
13. Veletsos, A. S., and N. M. Newmark: *Proc. 2d World Conf. Earthquake Eng.*, Tokyo, **II**:895–912 (1960).
14. Veletsos, A. S., N. M. Newmark, and C. V. Chelapati: *Proc. 3d World Congr. Earthquake Eng.*, New Zealand, **2**:II-663 (1965).
15. Newmark, N. M., and W. J. Hall: "Development of Criteria for Seismic Review and Selected Nuclear Power Plant," *U.S. Nuclear Regulatory Commission Report* NUREG-CR-0098, 1978.

16. Hall, W. J.: *Nuclear Eng. Des.*, **69**:3, (1982).
17. Newmark, N. M., and E. Rosenblueth: "Fundamentals of Earthquake Engineering," Prentice-Hall, Inc., Englewood Cliffs, N.J., 1971.
18. Newmark, N. M., J. A. Blume, and K. K. Kapur: *J Power Div. Am. Soc. Civil Engrs.*, **99**(PO2):287 (November 1973).
19. Newmark, N. M., and W. J. Hall: *Natl. Bur. Std. Bldg. Sci.* (46)**1**:209 (February 1973).
20. Newmark, N. M., and W. J. Hall: *Proc. 5th World Conf. Earthquake Eng.*, International Association for Earthquake Engineering, Rome, **2**:2266 (1974).
21. Wiegel, R. L. (ed.): "Earthquake Engineering," Prentice-Hall, Inc., Englewood Cliffs, N.J., 1970.
22. Newmark, N. M., and W. J. Hall: *Proc. 4th World Conf. Earthquake Eng.*, Santiago, Chile, **II**:B4-37 (1969).
23. Newmark, N. M.: *Nucl. Eng. Des.*, **20**(2):303 (July 1972).
24. Riddell, R., and N. M. Newmark: *Proc. 7th World Conf. Earthquake Engieering*, vol. 4 (1980). (See also *Univ. of Ill. Civil Eng. Struct. Res. Report* No. 468, 1979.)
25. Nau, J. M., and W. J. Hall: *J. Struct. Eng*, **110**:7 (1984).
26. Zahrah, T. F., and W. J. Hall: *J. Struct. Eng.*, **110**:8 (1984).
27. *Proceedings of the 1st through 8th World Conferences on Earthquake Engineering*, International Association for Earthquake Engineering, Tokyo, Japan (1956, 1960, 1965, 1969, 1974, 1977, 1980, 1984).
28. Newmark, N. M.: "Current Trends in the Seismic Analysis and Design of High Rise Structures," in R. L. Wiegel (ed.), "Earthquake Engineering," Prentice-Hall, Inc., Englewood Cliffs, N.J., 1970, chap. 16, p. 403.
29. Blume, J. A., N. M. Newmark, and L. Corning: "Design of Multistory Reinforced Concrete Buildings for Earthquake Motions," Portland Cement Association, Chicago, 1961.
30. Stevenson, J. D., W. J. Hall, et al.: "Structural Analysis and Design of Nuclear Plant Facilities," American Society of Civil Engineers, *Manuals and Reports on Engineering Practice* No. 58, 1980.
31. Idriss, I. M., et al.: "Analysis for Soil-Structure Interaction Effects for Nuclear Power Plants," American Society of Civil Engineers, 1980.
32. "ATC-6 Seismic Design Guidelines for Highway Bridges," *Applied Technology Council Report* ATC-6, 1981.
33. *Technical Council on Lifeline Earthquake Engineering*, ASCE: "Guidelines for the Seismic Design of Oil and Gas Pipeline Systems," 1984.
34. "Tentative Provisions for the Development of Seismic Regulations for Buildings," *Applied Technology Council Publ.* ATC 3-06, *Nat. Sci. Fnd. Publ.* 78-8, 1978.
35. Building Seismic Safety Council: "NEHRP Recommended Provisions for the Development of Seismic Regulations for New Buildings," Wash., D.C., 1984.
36. "Recommended Lateral Force Requirements and Commentary," Seismology Committee, Structural Engineers Association of California, 1985.
37. "Uniform Building Code—1985 Edition," *International Conference of Building Officials*, Whittier, Calif., 1985.
38. Newmark, N. M., and W. J. Hall: "Dynamic Behavior of Reinforced and Prestressed Concrete Buildings Under Horizontal Forces and the Design of Joints (Including Wind, Earthquake, Blast Effects)," Preliminary Publication, 8th Congress, International Association Bridge and Structural Engineering, New York, 1968, pp. 585-613.
39. "Building Code Requirements for Reinforced Concrete," ACI 318-83, 1983, Appendix A, pp. 318-88 to 313-95.

24

PART II: SEISMIC QUALIFICATION OF EQUIPMENT

George D. Shipway
Wyle Laboratories

Daniel D. Kana
Southwest Research Institute

INTRODUCTION

Seismic qualification is the process of demonstrating, either analytically or experimentally, that equipment can perform its design function when subjected to a postulated seismic event (i.e., an earthquake). This process is particularly applicable to equipment that must perform a nuclear safety function, i.e., equipment whose operation is essential for the continued safety of a nuclear power plant. Depending on the particular equipment and the system it is operating in, the safety function may be a passive structural requirement (such as survival at a maximum dynamic stress level) or it may be the performance of an active event (such as the operation of a mechanical valve or an electrical relay), or a combination thereof. The safety function must be performed during and after the postulated seismic event. The dynamic characteristics of this event are a part of the specified nuclear plant design criteria.

It is a U.S. legal requirement that seismic qualification be performed for safety-related equipment to be used in nuclear power plants. This design requirement arises from the 1975 Code of Federal Regulations.[1] The section, "Design Bases for Protection Against Natural Phenomena," includes this quotation:

Structures, systems and components important to safety shall be designed to withstand the effects of natural phenomena such as earthquakes, tornadoes, hurricanes, floods, tsunami, and seiches without loss of capability to perform their safety functions. The design basis for these structures, systems, and components shall reflect:

(1) appropriate consideration of the most severe natural phenomena that have been historically reported for the site and the surrounding area, with sufficient margin for the limited accuracy, quantity, and period of time in which the historical data have been accumulated,

(2) appropriate combination of the effects of normal and accident conditions with the effects of the natural phenomena, and

(3) the importance of the safety functions to be performed.

Details for seismic qualification procedures whereby this regulation can be satisfied are given in national standards compiled by various committees of the Institute for Electrical & Electronic Engineers (IEEE) and the American Society of Mechanical Engineers (ASME), and in regulations set forth by the U.S. Nuclear Regulatory Commission (NRC), and are given in Refs. 2 to 4. These procedures apply only for equipment used in nuclear power plants. However, their use is recommended for equipment in nonnuclear installations that are important for public safety, or where the loss of the equipment could result in a large economic loss.

RESPONSE SPECTRUM USE IN SEISMIC QUALIFICATION

Earthquake motion measured on the ground or at a level where equipment is mounted within a building is a transient random motion as described in Part I of this chapter. It is most often characterized by means of a response spectrum (also referred to as a shock spectrum; see Chap. 23). A *response spectrum* is a plot of the maximum responses of an array of single degree-of-freedom oscillators as a function of oscillator frequency, in response to an applied transient base motion. It is an indirect method of defining vibratory motion. The response spectrum may be used (1) to describe the motion that equipment is expected to experience at its mounting during a postulated seismic event, (2) to analyze the movement of the test table to show that its motion is in compliance with the requirements for simulating the postulated seismic event, and (3) to describe the forcing function for computer analysis of structural models of equipment.

A response spectrum may be determined by applying the acceleration of a given waveform to an array of single degree-of-freedom oscillators as illustrated in Fig. 24.14. Figure 24.14A shows an array of n mechanical oscillators having resonance frequencies from f_1 to f_n. When a transient base motion such as that shown in Fig. 24.14B is applied to this array, each oscillator will respond differently. For example, the motion of the oscil-

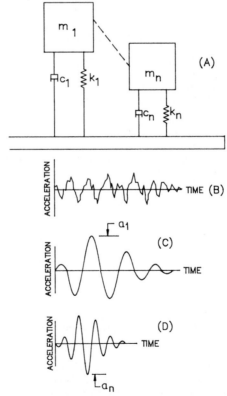

Fig. 24.14. *(A)* An array of single degree-of-freedom oscillators. *(B)* A transient which is applied to the array. *(C)* The motion of the first oscillator, which has a resonance at f_1; the peak absolute value is a_1. *(D)* The motion of the nth oscillator, which has a resonance of f_n; the peak absolute value is a_n.

lator whose resonance frequency is f_1 is shown in Fig. 24.14C and has a peak absolute value of acceleration a_1; the motion of the oscillator whose resonance frequency is f_n is shown in Fig. 24.14D and has a peak absolute value of acceleration a_n. The response spectrum is then obtained by plotting the peak absolute value of response of each oscillator as a function of its resonance frequency, as illustrated in Fig. 24.15.

In normal practice, the acceleration of the given waveform is analyzed by a computer which has been programmed to model the array of oscillators mathematically and perform the necessary computations to produce the response spectrum. Typically the resonance frequencies of the simulated oscillators are spaced at one-third- or one-sixth-octave intervals over the frequency range of interest (e.g., from 1 to 100 Hz, or from 1 Hz to an upper frequency limit which is higher than any frequency in the waveform being

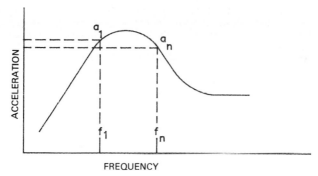

FIG. 24.15. The response spectrum for the array of oscillators shown in Fig. 24.14A to the motion shown in Fig. 24.14B. The first oscillator has a peak absolute value of a_1 and the nth oscillator has a peak absolute value of a_n.

analyzed). The value of damping assigned to the array of oscillators is typically in the range of 1 to 5 per cent of critical damping (i.e., $\zeta = 0.01$ to 0.05). This same waveform may be anlayzed for different values of damping to provide a family of response spectra, any one of which may be used to characterize the analyzed waveform.

The frequency at which the amplified portion of the spectrum is asymptotic to the lower constant portion of the spectrum is called the *cutoff frequency*. It indicates the upper limit of the frequency content of the waveform being analyzed. Oscillators tuned to higher frequencies respond rigidly without amplification of motion. This is called the *rigid region*. The acceleration response indicated by a constant value of the spectrum above the amplification cutoff frequency (see Fig. 24.16) is called *zero-period acceleration* (ZPA). A family of response spectra for a given waveform all converge to indicate the same zero-period acceleration regardless of the damping used in the analysis.

A response spectrum does not indicate the duration of the excitation waveform, nor does it tell at what point during the time period the peak response of each oscillator occurs. Although an analysis of a given waveform at a particular damping value yields a unique response spectrum, a response spectrum does not represent a unique waveform. In theory, a given response spectrum can be obtained from the analysis of an infinite number of different

FIG. 24.16. The zero-period acceleration of the response spectrum.

waveforms. The area under the spectrum is an approximate indication of the severity of the environment defined by the spectrum.

DEVELOPING REQUIRED RESPONSE SPECTRA FOR USE IN SEISMIC QUALIFICATION

The response spectra specified as a basis for seismic qualification of equipment are called the *required response spectra* (RRS). Such spectra are developed from analyses of historical records of earthquake ground motions, from soil-structure interaction, and from a structural analysis of the building which houses the equipment. The required response spectra are computed for various locations and elevations in the building structure and become the basis for definition of the motions for which the equipment must be qualified. The intent is to describe the test input requirements, not the anticipated response of a specific piece of equipment. In general, the response spectra specified for low elevations in a building are broad-band in nature and include frequencies up to 33 Hz. The response spectra for higher elevations are narrower in bandwidth because the building structure filters the motion; usually the cutoff frequency is in the range from 8 to 15 Hz.

Required response spectra are provided for two levels of earthquake severity for each power plant. One, the *safe-shutdown earthquake* (SSE), is the design basis for the plant and is the earthquake level at which all safety systems must function as required to be able to shut down the plant in a safe manner. The second, the *operational-basis earthquake* (OBE), is the largest earthquake which could reasonably be expected to affect the plant site during the operating life of the plant and is the earthquake level at which all safety systems necessary for continued operation are designed to remain fully functional. As an input to equipment being qualified for installation in the buildings, the OBE is usually 60 to 75 per cent of the SSE.

Commonly, seismic qualification regulations require the broadening of the frequency bandwidth of required response spectra in the area of any peaks by ±10 per cent to account for uncertainties of computed structural frequencies. The response spectra, so obtained, normally are specific to a particular section of a particular plant and are provided as requirements for seismic qualification of the equipment to be installed in that section. For generic, or more widely applicable, qualification of equipment, manufacturers and qualification engineers frequently produce composite spectra which envelop required spectra for various locations and applications in an effort to reduce the number of separate qualification tests required. Such enveloping must be done with caution since it may result in an environment which is significantly more severe than the equipment may be expected to experience in any specific application.

Response spectra generally are provided for the vertical direction and for the horizontal directions. To facilitate their use in both testing and analysis, spectra are frequently presented as a family of curves, each curve having a different value of damping.

MEASUREMENT OF DAMPING IN EQUIPMENT

Damping in the structure of a piece of equipment may vary with different modes of vibration of the structure (and hence frequency) and with response amplitude. In seismic qualification testing, damping usually is expressed as a per cent of critical damping. Damping in equipment often is difficult to measure with reasonable accuracy in any but the simplest of equipment. Common methods used for measuring damping in multiple degree-of-freedom systems are the half-power bandwidth and the logarithmic decrement, described in Chap. 37. Both methods provide reasonable results for damping below about 10 per cent of critical damping and for equipment of moderate complexity. The accuracy of such measurements depends on the ability to separate closely spaced modes of vibration of the equipment and to measure damping as a function of excitation level. Chapter 21 considers more sophisticated techniques which may be warranted if accurate results are important.

SEISMIC QUALIFICATION BY SIMULATION

Seismic qualification by simulation includes (1) observing the structural response and functional performance of equipment while it is subjected to an experimentally simulated seismic motion or (2) computing (by means of an analytical model) the structural response, stresses, and perhaps deformations, which result from an analytically simulated seismic motion. In either case, the motion should represent the safe-shutdown earthquake. Before this simulation takes place, seismic qualification requires environmental and vibratory aging of the equipment, as appropriate, so that it has a simulated age equal to its required life (usually defined as 40 years).

ENVIRONMENTAL AGING

The aging requirements for equipment depend on (1) the postulated service environmental conditions, (2) the materials in the equipment, (3) the system in which it is to operate, (4) the plant in which it is to be installed, and (5) the function of the equipment. In general, any aging mechanism (such as temperature or radiation), which has the potential for degrading the performance of the equipment under seismic conditions, must be accounted for prior to the seismic simulation. Some seismic qualification tests require accelerated aging since long service lives must be accounted for under more practical test durations.[4]

MECHANICAL AGING

Mechanical aging consists of subjecting equipment to any expected dynamic environment that may degrade its performance as the result of mechanical fatigue. Mechanical aging requirements (as appropriate to the equipment application) include: operational cycling, in-plant vibration, hydrodynamic vibration, and operational-basis earthquakes. Such vibration excitation, although generally low in amplitude, may occur a sufficient number of times to result in a significant contribution to mechanical fatigue. Operational cycling loads must be considered if the intended cyclical life is sufficient to induce mechanical fatigue which could degrade the operational capability of the equipment. Normal and transient in-plant vibration must be considered for equipment which may be located in an area where such vibration may be significant, e.g., in an area subject to pipe vibration.

HYDRODYNAMIC-INDUCED VIBRATION

Hydrodynamic-induced vibration (see Chap. 29) occurs in boiling-water reactor plants. Such transient motion is associated with the operation of the steam-suppression water pool, including the normal discharge of a safety relief valve (SRV), and a postulated loss-of-coolant accident (LOCA). The vented steam causes water surges, resulting in motion within the suppression pool which is transmitted to equipment mounted at various locations on the adjacent plant structures. Vibration related to safety relief valves occurs in the range from 20 to 100 Hz; although low in amplitude, such vibration may occur frequently. Such motion is commonly described by a response spectrum and may be applied as part of the vibratory aging requirement. Higher-amplitude vibration may originate from a loss-of-coolant accident wherein large amounts of steam are discharged into the suppression pool. Such vibration in the range from 20 to 40 Hz may be described by a response spectrum and applied as part of the SSE testing.

OPERATIONAL-BASIS EARTHQUAKES

Equipment is usually required to show no significant degradation of capability due to the aging effects of five operational-basis earthquakes. There may also be different equipment functional requirements during an operational-basis earthquake vs. a safe-shutdown earthquake.

OPERABILITY

Seismic qualification includes a demonstration of the capability of a piece of equipment to perform its safety function both during and after the earthquake simulation. The safety function is specific for each piece of equipment in each particular application. The safety function may, or may not, be the same during and after the earthquake. For example, an electrical component may be required to perform an active function both during and after the earthquake or it may have a passive function during the earthquake, merely needing to survive so that it can perform an active function after the earthquake. Another component may only be required to maintain structural integrity both during and after the earthquake. Thus the detailed functional requirements, and acceptance criteria, must be carefully considered and specified in a seismic qualification plan. The demonstration of complex functional performance often requires the simulation of more than one earthquake.

SEISMIC MOTION

Most safety-related equipment is mounted directly on a primary building structure—for example, on a concrete floor or wall. Depending on the location in the building, a varying amount of filtering by the building structure may alter the character of the earthquake-induced motion. The motion at the equipment mounting will be random in character, but its frequency content may be restricted to narrow bands near the building resonance frequencies. No matter where the mounting location, the *expected random motion* is defined by means of a response spectrum of appropriate frequency range and acceleration amplification resulting from the anticipated motion at that particular location.

Some safety-related components are mounted on substructures, such as pipes, which act as effective mechanical filters in transmitting the earthquake ground motion to the equipment. The resulting motion at the substructure will be predominantly narrow-band, or approximately sinusoidal at the resonance frequencies of the substructure, with a high amplification at one or more of the resonance frequencies. For this situation the test motion often is specified by a *required input motion* (RIM) of specified amplitude over a given frequency range. This motion may be simulated with an excitation composed of a sine-beat waveform consisting of a sine wave applied at a constant-center frequency, with the amplitude modulated at a lower frequency. The sine-beat motions are usually applied at one-third–octave frequency intervals over the range from 2 to 32 Hz.[3]

OPERATING LOADS

In general, seismic qualification requires the application of all normal operating loads (i.e., secondary operational influences) during the simulation of an earthquake. Examples of operating loads are voltage, current, pressure, temperature, and gravity. The application of only those normal operating loads which can be expected to interact with the seismic excitation can usually be justified. For example, suppose that in a particular electric device, the magnitude of the current has an influence on the device's ability to withstand seismic excitation, but the voltage does not. Then justification can be provided for applying the normal current but reducing the voltage during the seismic simulation.

SEISMIC QUALIFICATION TESTING PRACTICE

Seismic qualification of complex equipment which must demonstrate more than simple structural integrity is almost always done experimentally. Accelerated aging tests must be performed prior to the seismic simulation for all applicable aging phenomena, except for those that can be shown by analysis not to degrade the equipment. The effects of environmental factors (such as temperature) are usually applied in an accelerated test program; for example, higher test temperatures may be traded for shorter test durations.

Most qualification testing utilizes seismic excitation which is multifrequency in waveform, applied biaxially, i.e., both in a horizontal and vertical direction. Operating loads, as required for the particular equipment, are applied, and the equipment is subjected to simulated operational-basis earthquakes, followed by a simulated safe-shutdown earthquake.

In seismic qualification testing, the equipment to be tested is mounted on a vibration testing machine (also called a *shake table*) specially designed to produce simulated earthquake motions. The equipment is instrumented, tested, and monitored. There are two predominant types of seismic qualification testing: *proof testing* and *fragility testing*. In proof testing, it must be demonstrated that the equipment will continue to function when subjected to some specified condition of seismic excitation. For a specific installation, the equipment is subjected to seismic acceleration characteristic of the installation. This type of testing is costly because it may require multiple tests on similar equipment, i.e., one test for each installation. This has led to *generic proof testing*, in which composite test spectra are selected to encompass a variety of applications. The cost savings of fewer tests must be balanced against the potential higher cost of design and fabrication because the equipment will be subjected to more severe test motions than would be required for a single specified seismic excitation. In the second (lesser used) type of seismic qualification testing, called *fragility testing*, the maximum seismic loading that the equipment can withstand without malfunction is determined. Fragility testing is significantly more sophisticated and costly than proof testing, so it is used infrequently.

Systems, subsystems, or components may be subjected to a simulated seismic test in a particular qualification program, depending on the program's objectives. When the test item is a component (e.g., a device to be mounted inside an electrical cabinet), the test mounting must provide (1) simulation of the stiffness of the mounting interfaces, (2) provision for real or simulated functional interfaces, such as electric connections, to permit demonstration of proper performance, and (3) correct input spectra or required input motion.

MOUNTING

The vibratory environment inside the equipment subject to test may be affected significantly by the rigidity of its mounting. Therefore, in seismic qualification testing, equipment must be mounted on the test table of the testing machine in a manner that simulates the intended service mounting as closely as possible. This includes the use of supports and/or mass loading to account for the effect of interfaces such as wall supports, pipes, conduits, cables, and bus ducts.

MONITORING

To monitor vibratory motion and performance parameters, instrumentation must be installed on the test table and on and in the equipment subject to test. One accelerometer for each axis of input motion is used (1) for monitoring the test table motion, (2) for control purposes, and (3) for demonstrating compliance with the test requirements. Typically, several accelerometers are mounted on the equipment to monitor the response at various locations. Data showing the response of the equipment are particularly useful for equipment design purposes and for providing information on test amplitudes which may be required at a later time. The electrical outputs of the accelerometers are recorded for subsequent analysis.

Instrumentation must be used to monitor those functions and parameters which are required to demonstrate proper performance of the equipment before, during, and after the seismic simulation. The instrumentation required is specific to the equipment being qualified and is determined by the equipment operation and acceptance criteria. Typical functions or parameters which are monitored include: electrical continuity, change of state, current level, signal output, pressure, valve leakage, strain, and deformation.

EXPLORATORY TESTING

Exploratory testing is performed to determine the significant resonance frequencies in equipment. Such testing is seldom a qualification requirement, but it is usually included in the test program because it yields data useful in reviewing test results and in potential subsequent modifications. It is most often performed as a uniaxial, slowly swept sine wave over the range of approximately 1 to 40 Hz (higher if hydrodynamic loads are applicable) at an acceleration amplitude of $0.2g$ peak on the same test table as that used for the seismic testing. For the case of heavy and rigid equipment, flexibility of the test table can affect the test results. In such cases it may be desirable to mount the equipment on a rigid base and use one of the other methods, e.g., modal testing for determining resonance frequencies, mode shapes, and damping, as described in Chap. 21.

VIBRATORY AGING

Equipment which is located where it will experience a continuous, or frequent, low-level vibratory environment may require a fatigue test prior to the seismic qualification testing. The fatigue test is usually run at a somewhat higher vibration amplitude to shorten the test time. Pipe-mounted components typically are subjected to a swept sine wave from 5 Hz to 100 Hz to 5 Hz at a displacement of 0.025 in (0.63 mm) double amplitude up to 24 Hz, and an acceleration amplitude $0.75g$ peak above 24 Hz, for 90 minutes per axis.

Equipment which will be exposed to hydrodynamic excitation must be subjected to the simulated safety-relief-valve-related motions. In one typical simulation procedure, low-level random excitation is applied for 90 minutes along each orthogonal axis. The motions are specified by a response spectrum which is a part of the design criteria for the plant.

SEISMIC TESTING SEQUENCE

The general requirement in seismic qualification testing is that the equipment under test must be subjected to five simulated operational-basis earthquakes (OBE) followed by one safe-shutdown earthquake (SSE). Simulated normal operating conditions are applied (as required) during the operational-basis and safe-shutdown earthquakes. Any required aging tests (for example, involving temperature and radiation) must be done prior to the operational-basis earthquake. For those cases where there may be hydrodynamic excitation as a result of loss of coolant, the spectra for excitation are combined (by the square root of the sum of the squares method) with the operational-basis and safe-shutdown earthquake spectra; the resulting spectrum is used in synthesizing the test waveforms. In testing, an operational-basis earthquake is followed by the safe-shutdown earthquake, normally applied in one orientation; then it is repeated in the second (and third if required) orientation.

TEST WAVEFORMS

As indicated earlier, seismic excitation is comprised of motion containing many frequency components. Hence, in testing, motion comprised of multiple frequencies is preferred (except as indicated below under *Equipment Mounted on Substructures*) in order to yield a realistic simulation of the postulated motions and to provide the potential for exciting multimodal response in the equipment. Test waveforms are synthesized by mixing waveforms of several frequencies to obtain a multifrequency waveform Narrow-band frequency components are combined to form a waveform which has the necessary amplitude and frequency content to meet the required response spectrum. Each frequency component must exhibit stationarity, i.e., each must exhibit no significant periods of zero excitation. A test duration of 30 seconds is typical.

Figure 24.17 illustrates the steps in the process of producing a test table simulation of the motion postulated at the mounting location of the equipment. Figure 24.17A shows a hypothetical array of mechanical oscillators having resonance frequencies f_1 to f_n mounted on the building floor. A required response spectrum, shown in Fig. 24.17B, is obtained from the response of the oscillators to the postulated floor motion. This is the spectrum which must be simulated by the motion of the test table. An analysis of the motion of the test table [called the *test response spectrum* (TRS)] (see Fig. 24.17C) must envelop the required response spectrum, including the zero-period acceleration, but without excessive conservatism in either amplitude or frequency content.

A common difficulty encountered in synthesizing a waveform (the analysis of which will envelop the required response spectrum) is the occurrence of an excessively high peak acceleration of the synthesized motion and a resulting high zero-period acceleration of the test response spectrum. This is particularly common with the generic test method because the enveloping of several response spectra to produce a composite requires response spectrum results in a fictitious spectrum which cannot be synthesized with ordinary waveforms.

FIG. 24.17. Steps in the process of producing a test table simulation of the motion postulated at the mounting location of equipment under test. *(A)* A hypothetical array of mechanical oscillators having resonance frequencies from f_1 to f_n is mounted at the intended location of the equipment. *(B)* The response spectrum at the intended location from the response of the oscillators to the postulated building motion. *(C)* The test response spectrum obtained from an analysis of the test table motion. This curve should duplicate that shown in *(B)*.

For selected tests where the required response spectrum requires a motion having a very narrow frequency bandwidth, single-frequency waveforms may be used as an adequate simulation of the postulated motion. The sine-beat waveform is most commonly used for seismic qualification testing where single frequencies are applicable. A test employing such a waveform usually consists of the application of approximately five beats, with five to ten oscillations per beat, and includes a pause between the beats. The test response spectrum of the sine-beat waveform must envelop the required response spectrum including the zero-period acceleration. There are other, less commonly used, waveforms which are acceptable in certain circumstances.[2]

Response spectrum analysis of a test motion is usually performed at one-sixth-octave frequency intervals over the frequency range from 1 to 100 Hz, or from 1 to 200 Hz when hydrodynamic excitation is included. The damping must be the same, or greater than, the damping of the required response spectrum. When a family of curves for the required response spectrum is specified, the curve with a value of 5 per cent of critical damping is usually selected in testing. It is not necessary to correlate the damping used in the response spectrum analysis with the damping in the equipment.

MEASURING ZERO-PERIOD ACCELERATION

Determining the peak value of the test acceleration from the response spectrum often is difficult because of high-frequency motion which, when analyzed, yields a test response spectrum which may not drop to the zero-period acceleration (see Fig. 24.16). This is sometimes mistakenly interpreted as a high zero-period acceleration and, in turn, as a high peak acceleration of the waveform. This is particularly true of response measurements inside the cabinet of a piece of equipment, where rattling in the cabinet produces high frequencies. To make a more accurate determination of the peak acceleration in the seismic frequency range (i.e., at frequencies below the cutoff frequency), the waveform can be filtered prior to performing a second, or supplementary, response spectrum analysis. An easier method of determining the peak acceleration is to obtain a supplementary response spectrum analysis at a value of 10 per cent of critical damping or higher. Either method will indicate a zero-period acceleration which is much closer to the actual peak acceleration of the waveform. The peak acceleration also may be measured directly from a recording of the test waveform.

EQUIPMENT MOUNTED ON SUBSTRUCTURES

For equipment mounted on substructures such as pipes, non-steady-state sinusoidal excitation is postulated, as a result of mechanical filtering of the seismic motion by the substructure. *Required input motion* (RIM) is the technique used to define the required test motions for this application. For this test method, the *input motion* is defined as a sinusoidal acceleration whose amplitude is specified as a function of frequency. Required input motion testing to simulate operational-basis–earthquake aging is usually applied as two sequential swept sine waves from 2 Hz up to 35 Hz and then back down to 2 Hz at a rate of 1 octave/min at two-thirds of the safe-shutdown earthquake acceleration. The test motion for the safe-shutdown earthquake is usually applied as a continuous string of sine-beat waveforms (without a time delay between the beats) with 10 to 15 oscillations/beat, and phased such that any superposition of response motion is additive. The test motions are applied one frequency at a time, at one-third–octave intervals, over the range from 2 to 32 Hz, for a duration of 15 seconds at each frequency. Test velocity amplitudes of 35 in./sec (0.9 m/sec) from 2 to 8 Hz and acceleration amplitudes of 4.5g from 8 to 32 Hz are typical. The series of tests are applied uniaxially in each of the three principal axes in sequence. The demonstration of equipment operability must be performed at each frequency, and must be initiated during a nonzero portion of the waveform.

MULTIAXIS SEISMIC TESTING

Multiaxial seismic testing can be performed by simultaneously exciting two (biaxial) or three (triaxial) preselected orthogonal axes of the equipment. Biaxial testing is by far the most common test method. A horizontal axis is simultaneously excited with the vertical axis using independently synthesized and controlled excitation signals. The testing is repeated in a second orientation with the second horizontal axis combined with the vertical axis. When dependent motions are used (resulting in rectilinear motion), the test series is performed four times, i.e., in each direction of each horizontal axis.

Qualification standards permit testing along a single axis when it can be applied in a conservative manner, e.g., when the equipment can be shown to exhibit very low cross-axis response or when the postulated input motion can be shown to be essentially unidirectional. Accelerometers in the equipment to measure cross-axis response motions may be necessary to justify less than triaxial testing.

Triaxial testing with independent motions in each orthogonal axis is permitted by the standards and has been employed on a limited basis. There is no consensus on the merits of triaxial vs. biaxial testing.

SEISMIC TESTING MACHINES

Because of the high forces and large displacements typically required for seismic simulation, seismic testing machines are usually hydraulically driven. Typical machines provide displacements of 8 to 12 in. (200 to 300 mm) double amplitude, velocity amplitudes of from 35 to 50 in./sec (0.9 to 1.27 m/sec), and acceleration amplitudes of from 4 to 5g.

The large displacements, combined with test loads of 6,000 to 10,000 lbs (2,727 to 4,545 kg), limit the practical upper test frequency to approximately 100 Hz. Most testing machines operate best in the range from 1 to 70 Hz.

Biaxial hydraulic testing machines are of two basic designs. One, a true biaxial machine, may be driven by two (or more) vertical hydraulic cylinders and valves (actuators) and one (or more) hydraulic actuator for the horizontal axis. The actuators are controlled to provide independent translation for the two axes. A variety of techniques are used to control the other degrees-of-freedom of the test table. Another design, a psuedobiaxial machine, operates uniaxially along a diagonal direction—typically 45°, providing rectilinear motion, with dependent component motions in the horizontal and vertical axes.

True triaxial machines use hydraulic actuators in various designs to provide independent control of translational motions, and torque tubes for restraint of rotational motions.

A major design objective of most testing machines, particularly those of the multiaxis type, is to provide high driving stiffness of the hydraulic actuation system. This is desirable to control rotational degrees-of-freedom and to increase the frequency range for acceptable control and operation.

CALIBRATION

An iterative technique is used in the calibration of seismic testing machines. Such techniques vary in detail, but virtually all of them analyze the table motion resulting from the trial signal; then the signal is corrected; finally the motion resulting from the corrected signal is analyzed. This procedure is repeated until an acceptable motion of the table is achieved. For this reason calibration usually is performed on the bare table; then the equipment to be tested is mounted on the table after calibration. Since hydraulic seismic testing machines are relatively unaffected by the addition of the equipment to be tested, the first test motion is usually close to, but not exactly the same as, the calibration motion. To provide a margin for change and to account for the usual fluctuation in the test response spectrum, machines commonly are calibrated at values 10 to 20 per cent above the required response spectrum. Operational-basis earthquake tests provide a convenient way of checking calibration prior to the full-level safe-shutdown earthquake.

SEISMIC QUALIFICATION BY ANALYSIS

Seismic qualification of equipment by analysis generally includes (1) a representation of the equipment structure by an analytical model, (2) a prediction of the structure's displacement and stress responses when it is subjected to a prescribed motion excitation at the equipment mounting, and (3) an analytical verification which shows that the responses do not exceed prescribed allowable values. Detailed procedures typically used to accomplish this analysis are similar to those described in Part I of this chapter for the prediction of earthquake response in buildings and other large structures. However, an essential difference lies in the degree of refinement necessary to analytically synthesize a given equipment structure compared with that for large buildings. These details depend on the materials and mechanical construction used for a specific piece of equipment (see Refs. 2, 5, and 6).

To utilize the analytical approach, the satisfactory performance of the safety function must be demonstrable by analytical means, e.g., be able to show that structural stresses and deformations, which are within acceptable limits, can assure performance of the safety function. When environmental aging is required, the choice of the analytical approach depends on the feasibility of analyzing and exempting aging effects. A fatigue analysis is usually performed to show that stresses due to vibratory aging are low enough to avoid significant degradation of the equipment's integrity.

Generally, an analytical approach to equipment qualification is used only where the test approach cannot readily be employed (e.g., where the equipment is too large). Often, a combined test and analysis approach may also be used. This might include analytical modeling of a large system to predict response motions at various locations of subcomponents and the subsequent use of these motions to prescribe tests for the individual subcomponents. Such a combined approach is especially useful for those cases where the functional operation of the subcomponents cannot readily be described analytically.

Another type of combined test and analysis approach may be used to qualify equipment that already exists and is functioning in an operating plant. The equipment mounted in situ is first subjected to a resonance search procedure so that natural modes, frequencies, stiffnesses, masses, and damping can be assigned for a postulated analytical model.[7] This model is then subjected to a prescribed seismic motion, as described above, and the predicted displacements and stresses are compared with allowable values for the qualification.

REFERENCES

1. The 1975 Code of Federal Regulations, part 50, entitled "Licensing of Production and Utilization Facilities," Appendix A.
2. "Recommended Practices for Seismic Qualification of Class 1E Equipment for Nuclear Power Generating Stations," IEEE 344.
3. "IEEE Standard for Qualification of Safety Related Valve Actuators," IEEE 382.
4. "IEEE Standard for Qualifying Class 1E Equipment for Nuclear Power Generating Stations," IEEE 323.
5. deSilva, C. W.: *Dynamic Testing and Seismic Qualification Practice,* D. C. Heath & Co., Lexington, Mass., 1983.
6. Kana, D. D., et al.: "A Research Program for Seismic Qualification of Nuclear Plant Electrical and Mechanical Equipment," NUREG/CR-3892, vol. 1, U.S. Nuclear Regulatory Commission, Washington, D.C., 1984.
7. Bendat, J. S., and A. J. Piersol: "Engineering Applications of Correlation and Spectral Analysis," John Wiley & Sons, New York, N.Y., 1980.

25

VIBRATION TESTING MACHINES

Karl Unholtz

Unholtz-Dickie Corporation

INTRODUCTION

This chapter describes some of the more common types of vibration testing machines*
which are used for developmental, simulation, production, or exploratory vibration tests
for the purpose of studying the effects of vibration or of evaluating physical properties of
materials or structures. (*Fixed-frequency* vibrators, driven directly by the power line
and used for sorting, sifting, conveying, compacting, hammering, processing, etc., are
not included in this chapter.) A summary of the prominent features of each machine is
given. These features should be kept in mind when selecting a vibration testing machine for a specific test.

A vibration testing machine (sometimes called a *shake table* or *shaker* and referred to
here as a *vibration machine*) is distinguished from a vibration exciter in that it is complete
with a mounting table which includes provisions for bolting the test article directly to it.
A *vibration exciter*, also called a *vibration generator,* may be part of a vibration machine
or it may be a device suitable for transmitting a vibratory force to a structure. A
constant-displacement vibration machine attempts to maintain constant-displacement
amplitude while the frequency is varied. Similarly, a *constant-acceleration* vibration
machine attempts to maintain a constant-acceleration amplitude as the frequency is
changed.

The *load* of a vibration machine includes the item under test and the supporting structures that are not normally a part of the vibration machine. In the case of equipment
mounted on a vibration table, the load is the material supported by the table. In the
case of objects separately supported, the load includes the test item and all fixtures par-
taking of the vibration. The load is frequently expressed as the weight of the material.
The *test load* refers specifically to the item under test exclusive of supporting fixtures.
A *dead-weight load* is a rigid load with rigid attachments. For nonrigid loads the reaction of the load on the vibration machine is a function of frequency. The vector force
exerted by the load, per unit of acceleration amplitude expressed in units of gravity of
the driven point at any given frequency, is the *effective load* for that frequency. The
term *load capacity,* which is descriptive of the performance of reaction and direct-drive
types of mechanical vibration machines, is the maximum dead-weight load that can be
vibrated at the maximum acceleration rating of the vibration machine. The *load couple*
for a dead-weight load is equal to the product of the force exerted on the load and the
distance of the center-of-mass from the line-of-action of the force, or from some arbitrarily selected location (such as a table surface). The static and dynamic load couples
are generally different for nonrigid loads.

---

* Patents are owned by various manufacturers of vibration machines which cover many
aspects of design discussed here. Such discussion does not imply freedom from patent
restrictions.

The term *force capacity*, which is descriptive of the performance of electrodynamic shakers, is defined as the maximum rated force generated by the machine. This force is usually specified, for continuous rating, as the maximum vector amplitude of a sinusoid that can be generated throughout a usable frequency range. A corresponding maximum rated acceleration, in units of gravity, can be calculated as the quotient of the force capacity divided by the total weight of the coil table assembly and the attached dead-weight loads. The *effective force* exerted by the load is equal to the effective load multiplied by the (dimensionless) ratio $\mathfrak{g}$, which represents the number of units of gravity acceleration of the driven point.

DIRECT-DRIVE MECHANICAL VIBRATION MACHINES

The direct-drive vibration machine consists of a rotating eccentric or cam driving a positive linkage connection which forces a displacement between the base and table of the machine. Except for the bearing clearances and strain in the load-carrying members,

Fig. 25.1. Elementary direct-drive mechanical vibration machines: (*A*) Eccentric and connecting link. (*B*) Scotch yoke. (*C*) Cam and follower.

the machine tends to develop a displacement between the base and the table which is independent of the forces exerted by the load against the table. If the base is held in a fixed position, the table tends to generate a vibratory displacement of constant amplitude, independent of the operating rpm. Figure 25.1 shows the direct-drive mechanical machine in its simplest forms. This type of machine is sometimes referred to as a *brute force machine* since it will develop any force necessary to produce the table motion corresponding to the crank or cam offset, short of breaking the load-carrying members or stalling the driving shaft.

Fig. 25.2. Displacement adjusting method for a direct-drive vibration machine:
(*A*) $a = b$. Displacement $= (b - a) = 0$.
(*B*) $a \neq b$. Displacement $= 2c$ (peak-peak).

The direct-drive machines usually are rated in terms of pounds of test item that can be vibrated through the maximum displacement and frequency operating limits of the machine. (On some of the larger machines, the rating is based on a maximum value of the product of load weight plus table weight in pounds and the maximum acceleration, for a specified frequency range.) The operating limits are determined by allowable bearing loads based on the assumption that the test item behaves as a rigid mass and the bearing loads can be calculated as the product of mass and acceleration.

Figure 25.2 shows a method of providing stepless adjustability of crank offset between zero and full displacement. The boundary surfaces between shaft eccentric 1 and eccentric bushing 2 and between 2 and the connecting rod are circular cylinders. For a fixed-displacement setting, the eccentric 1 and the eccentric bushing 2 remain fixed relative to each other but rotate about the rotation center of the driving shaft. The displacement adjustment is made by rotating eccentric bushing 2 relative to shaft eccentric

Fig. 25.3. A sectional view of an inclined cylinder-type adjustable amplitude eccentric. (*Courtesy L.A.B. Corp.*)

1. In some machines it is necessary to stop the machine, unlock the cylindrical elements, reposition them according to a calibrated displacement scale of the desired displacement, and relock the position before resuming operation. In other machines, this adjustment can be made during machine operation.

Another basic mechanism for producing direct-drive vibration is an eccentric consisting of a cylinder inclined at a fixed angle with the main drive shaft on which it is mounted, as shown in Fig. 25.3. A bushing, which rotates with the shaft, and a connecting-rod bearing on the rotating bushing are assembled around the inclined cylinder. The bushing and the inclined cylinder are each pinned to the shaft, thereby being forced to turn with the shaft. The cylinder has a slot in the axial direction allowing axial motion of the cylinder while the axial position of the bushing and connecting rod is fixed. Thrust bearings (at T) position the inclined cylinder axially according to the setting of the amplitude control, thereby providing a stepless adjustment which can be made while the machine is operating.

The simplest direct-drive mechanical vibration machine is driven by a constant-speed motor in conjunction with a belt-driven speed changer and a frequency-indicating tachometer. Table displacement is set during shutoff and is assumed to hold during operation. An auxilary motor driving a cam may be included to provide frequency cycling between adjustable limits. More elaborate systems employ a direct-coupled variable-speed motor with electronic speed control, as well as amplitude adjustment from a control station. Machines have been developed which provide rectilinear, circular, and three-dimensional table movements—the latter giving complete, independent adjustment of magnitude and phase in the three directions.

PROMINENT FEATURES

1. Low operating frequencies and large displacements can be provided conveniently.
2. Theoretically, the machine maintains constant displacement regardless of the mechanical impedance of the table-mounted test item within force and frequency limits of the machine. However, in practice, the departure from this theoretical ideal is considerable, due to the elastic deformation of the load-carrying members with change in

output force. The output force changes in proportion to the square of the operating frequency and in proportion to the increased displacement resulting therefrom. Because the load-carrying members cannot be made infinitely stiff, the machines do not hold constant displacement with increasing frequency with a bare table. This characteristic is further emphasized with heavy table mass loads. Accordingly, some of the larger capacity machines which operate up to 60 cps include automatic adjustment of the crank offset as a function of operating frequency in order to hold displacement more nearly constant throughout the full operating range of frequency.

3. The machine must be designed to provide a stiff connection between the ground or floor support and the table. If accelerations greater than $1g$ are contemplated, the vibratory forces generated between the table and ground will be greater than the weight of the test item. Hence, all mass loads within the rating of the machine can be directly attached to the table without recourse to external supports.

4. The allowable range of operating frequencies is small in order to remain within bearing load ratings. Therefore, the direct-drive mechanical vibration machine can be designed to have all mechanical resonances removed from the operating frequency range. In addition, relatively heavy tables can be used in comparison to the weight of the test item. Consequently, misplacing the center-of-gravity of the test item relative to the table center for vibration normal to the table surface and the generation of moments by the test item (due to internal resonances) usually has less influence on the table motions for this type of machine than would other types which are designed for wide operational frequency bands.

5. Simultaneous rectilinear motion normal to the table surface and parallel to the table surface in two principal directions is practicable to achieve. It may be obtained with complete independent control of magnitude and phase in each of the three directions.

6. Displacement of the table is generated directly by a positive drive rather than by a generated force acting on the mechanical impedance of the table and load. Consequently, impact loads in the bearings, due to the necessary presence of some bearing clearance, result in the generation of relatively high impact forces which are rich in harmonics. Accordingly, although the waveform of displacement might be tolerated as such, the waveform of acceleration is normally sufficiently distorted to preclude recognition of the fundamental driven frequency, when displayed on a time base.

REACTION-TYPE MECHANICAL VIBRATION MACHINE

A vibration machine using a rotating shaft carrying a mass whose center-of-mass is displaced from the center-of-rotation of the shaft for the generation of vibration, is called a *reaction-type vibration machine*. The product of the mass and the distance of its center from the axis of rotation is referred to as the *mass unbalance*, the *rotating unbalance*, or simply the *unbalance*. The force resulting from the rotation of this unbalance is referred to as the *unbalance force*.

The *reaction-type vibration machine* consists of at least one[1] rotating-mass unbalance directly attached to the vibrating table. The table and rotating unbalance are suspended from a base or frame by soft springs which isolate most of the vibration forces from the supporting base and floor. The rotating unbalance generates an oscillating force which drives the table. The unbalance consists of a weight on an arm which is relatively long by comparison to the desired table displacement. The unbalance force is transmitted through bearings directly to the table mass, causing a vibratory motion without reaction of the force against the base. A vibration machine employing this principle is referred to as a reaction machine since the reaction to the unbalance force is supplied by the table itself rather than through a connection to the floor or ground.

CIRCULAR-MOTION MACHINE

The reaction-type machine, in its simplest form, uses a single rotating-mass unbalance which produces a force directed along the line connecting the center-of-rotation and the center-of-mass of the displaced mass. Referred to stationary coordinates, this force appears normal to the axis of rotation of the driven shaft, rotating about this axis at the rotational speed of the shaft. The transmission of this force to the vibration-machine table causes the table to execute a circular motion in a plane normal to the axis of the rotating shaft.

Figure 25.4 shows, schematically, a machine employing a single unbalance producing circular motion in the plane of the vibration-table surface. The unbalance is driven at various rotational speeds, causing the table and test item to execute circular motion at various frequencies. The counterbalance weight is adjusted to equal the test item mass moment calculated from d, the plane of the unbalance force, thereby keeping the com-

FIG. 25.4. Circular-motion reaction-type mechanical vibration machine.

bined center-of-gravity coincident with the generated force. Keeping the generated force acting through the combined center-of-gravity of the spring-mounted assembly eliminates vibratory moments which, in turn, would generate unwanted rotary motions in addition to the motion parallel to the test mounting surface. The vibration isolator supports the vibrating parts with minimum transmission of the vibration to the supporting floor.

Reference to the section on dynamic considerations for the reaction machine which appears later in this chapter shows that for a fixed amount of unbalance and for the case of the table and test item acting as a rigid mass, the displacement of motion tends to remain constant if there are no resonances in or near the operating frequency range. If constant table acceleration is desired, the unbalance force must remain constant, requiring the amount of unbalance to change with shaft speed.

FIG. 25.5. Rectilinear-motion reaction-type mechanical vibration exciter using a reciprocating mass.

RECTILINEAR-MOTION MACHINE

Rectilinear motion rather than circular motion can be generated by means of a reciprocating mass, as shown in Fig. 25.5. The shaft has an eccentric pin which engages the reciprocating mass which bears on guides in the frame. Rotation of the shaft causes reciprocating motion of the mass. The inertia forces caused by the reciprocating mass are transmitted to the frame in the direction shown, through the bearings supporting the shaft. This mechanism is referred to as a *Scotch yoke*. The frame may be attached to a vibration table or directly to a structure in order to excite the resonances in the structure. The crankpin can be counterbalanced, thereby eliminating most of the unwanted excitations at 90° to the motion of the reciprocating mass.

Rectilinear motions also can be produced with a single rotating unbalance by constraining the table to move in one direction. Two examples are shown in Fig. 25.6. Some fatigue testing machines use these configurations with springs *a* and *b* in Fig. 25.6*A* and spring *a* in Fig. 25.6*B* selected to have the proper stiffness in the direction of vibration so that the table and associated mass are resonant at a fixed operating frequency. Under these conditions, the force transmitted by the table to the item under test is approximately constant and independent of any changes in the mechanical mobility of the

FIG. 25.6. Rectilinear-motion reaction-type vibration machine: (*A*) Flexure constraint. (*B*) Radius arm with $d \ll r$.

load—provided the length of the eccentric arm is large compared to the table displacement. If the resonant frequency is below the operating frequencies and the rotating unbalance remains constant, then the displacement of the table tends to remain constant through the operating frequency range for rigid dead-mass table loads.

TWO ROTATING UNBALANCES. The most common rectilinear reaction-type vibration machine consists of two rotating unbalances, turning in opposite directions and phased so that the unbalance forces add in the desired direction and cancel in other

FIG. 25.7. Rectilinear-motion reaction-type mechanical vibration machine using two rotating unbalances: (*A*) Vibration perpendicular to table surface. (*B*) Vibration parallel to table surface.

directions. Figure 25.7 shows schematically how rectilinear motion perpendicular and parallel to the vibration table is generated. The effective generated force from the two rotating unbalances is midway between the two axes of rotation and is normal to a line connecting the two. In the case of motion perpendicular to the surface of the table, simply locating the center-of-gravity of the test item over the center of the table gives a proper load orientation. Tables are designed so that the resultant force always passes through this point. This results in collinearity of generated forces and inertia forces, thereby avoiding the generation of moments which would otherwise rock the table. In the case of motion parallel to the table surface, no simple orientation of the test item will achieve collinearity of the generated force and inertia force of the table and test item. Various methods are used to make the generated force pass through the combined center-of-gravity of the table and test item.

THREE ROTATING UNBALANCES. If a machine is desired which can be adjusted to give vibratory motion either normal to the plane of the table or parallel to the plane of the table, a minimum of three rotating unbalances is required. Inspection of Fig. 25.8 shows how rotating the two smaller mass unbalances relative to the single larger

(A)   (B)

FIG. 25.8. Adjustment of direction of generated force in a reaction-type mechanical vibration exciter: (A) Vertical force. (B) Horizontal force.

unbalance results in the addition of forces in any desired direction, with cancellation of forces and force couples at 90° to this direction. Although parallel shafts are usually used as illustrated, occasionally the three unbalances may be mounted on collinear shafts, the two smaller unbalances being placed on either side of the single larger unbalance to conserve space and to eliminate the bending moments and shear forces imposed on the structure connecting the individual shafts.

PROMINENT FEATURES

1. The forces generated by the rotating unbalances are transmitted directly to the table without dependence upon a reactionary force against a heavy base or rigid ground connection.

2. Because the length of the arm which supports the unbalance mass can be large, relative to reasonable bearing clearances and the generation of a force which does not reverse its direction relative to the rotating unbalance arm, the generated waveform of motion imparted to the vibration machine table is superior to that attainable in the direct-drive type of vibration machine.

3. The generated vibratory force can be made to pass through the combined center-of-gravity of the table and test item in both the normal and parallel directions relative to the table surface, thereby minimizing vibratory moments giving rise to table rocking modes.

4. The attainable rpm and load ratings on bearings currently limit performance to a frequency of approximately 60 cps and a generated force of 300,000 lb, respectively, although in special cases frequencies up to 120 cps and higher can be obtained for smaller machines.

ELECTRODYNAMIC VIBRATION MACHINE

GENERAL DESCRIPTION

A complete electrodynamic vibration test system is comprised of an electrodynamic vibration machine, electrical power equipment which drives the vibration machine, and electrical controls and vibration monitoring equipment.

The electrodynamic vibration machine derives its name from the method of force generation. The force which causes motion of the table is produced electrodynamically by the interaction between a current flow in the armature coil and the intense magnetic d-c field which passes through the coil, as illustrated in Fig. 25.9. The table is structurally attached to a force-generating coil which is concentrically located (with radial clearances) in the annular air gap of the d-c magnet circuit. The assembly of the armature coil and the table is usually referred to as the "driver coil-table" or "armature." The magnetic circuit is made from soft iron which also forms the *body* of the vibration machine. The body is magnetically energized, usually by two field coils as shown in Fig. 25.9C, generating a radially directed field in the air gap, which is perpendicular to the direction of current flow in the armature coil. Alternatively, in small shakers, the magnetic field is generated by permanent magnets. The generated force in the armature coil is in the direction of the axis of the coil, perpendicular to the table surface. The direction of the force is also perpendicular to the armature-current direction and to the air-gap field direction.

The table and armature coil assembly is supported by elastic means from the machine body, permitting rectilinear motion of the table perpendicular to its surface, corresponding in direction to the axis of the armature coil. Motion of the table in all other directions is resisted by stiff restraints. Table motion results when an a-c current passes through the armature coil. The body of the machine is usually supported by a base with a trunnion shaft centerline passing horizontally through the center of gravity of the body assembly, permitting the body to be rotated about its center, thereby giving a vertical or horizontal orientation to the machine table. The base usually includes an elastic support of the body, providing vibration isolation between the body and the supporting floor.

Where a very small magnetic field is required at the vibration machine table due to the effect of the magnetic field on the item under test, *degaussing* may be provided. Magnetic fields of 5 to 30 gauss several inches above the table are normal for modern machines with double-ended, center air-gap magnet designs, Fig. 25.9C, without degaussing accessories; in contrast, with degaussing accessories, magnetic fields of 2 to 5 gauss can be achieved.

Because of copper and iron losses in the electrodynamic unit, provision must be made to carry off the dissipated heat. Cooling by convection air currents, compressed air, or a motor-driven blower is used and, in some cases, a recirculating fluid is used in conjunction with a heat exchanger. Fluid cooling is particularly useful under extremes of hot or cold environments or altitude conditions where little air pressure is available.

MAGNET CIRCUIT CONFIGURATIONS

Three magnet circuit configurations which are used in the electrodynamic machines are shown schematically in Fig. 25.9. In Fig. 25.9A, the table and driver coil are located at opposite ends of the magnet circuit. The advantage of this configuration is that the location of the annular air gap, the region of high magnetic leakage flux, is spaced from the table and the body itself acts as a magnetic shield, resulting in lower magnetic flux density at the table. The disadvantage lies in the loss of rigidity in the connecting structure between the driver coil and the table because of its length. This configuration is usually cooled by convection air currents or by forced air from a motor-driven blower.

In Fig. 25.9B, the table is connected directly to the driver coil. This eliminates the length of structure passing through the magnet structure, thereby increasing the rigidity of the driver coil-table assembly and allowing higher operating frequencies. The leakage magnetic field in the vicinity of the table is high in this configuration. It is therefore difficult, if not impossible, to reduce the leakage to acceptable levels without adding extra length to the driver coil assembly, elevating the table above the air gap. The configuration in Fig. 25.9C has a complete magnet circuit above and below the annular air gap, thereby reducing external leakage magnetic field to a minimum. This configuration also increases the total magnetic flux in the air gap by a factor of almost 2 for the same diameter driver coil, giving greater force generation and a more symmetrical magnetic flux density along the axis of the coil. Hence more uniform force generation results when the driver coil is moved axially throughout its total stroke. All high-efficiency and high-performance electrodynamic vibration machines use the configuration shown in Fig. 25.9C.

Configurations B and C may use air cooling throughout, or an air-cooled driver coil and liquid-cooled field coil(s) or total liquid cooling.

FIG. 25.9. Three main magnet circuit configurations.

The main magnetic circuit uses d-c field coils for generating the high-intensity magnetic flux in the annular gap in all of the larger sizes and most of the smaller units. Permanent magnet excitation is used in small portable units and in some general-purpose units up to about 500 lb generated force.

FREQUENCY-RESPONSE CONSIDERATIONS

Testing procedures which call for sinusoidal motion of a vibration-machine table can be performed even though the frequency-response curve of the electrodynamic vibration machine is far from flat. For a test at a fixed frequency, the driving voltage e_d is adjusted until the table motion is equal in amplitude to that required by the test specifications. If the procedure calls for cycling the frequency between two frequency limits while keeping a constant displacement or acceleration, an automatic regulating device or servo control adjusts the driver-coil voltage as required to maintain the desired vibration machine table motion independent of the frequency of operation. This regulating device provides a correction at any frequency of operation within the testing frequency limits but it can correct for only one operating frequency at any instant of time. The closer the frequency response is to the desired variation in acceleration with frequency, the smaller the corrections in driver-coil voltage will be from the automatic regulator—thereby improving the attainable accuracy of the control.

Testing procedures which call for a random vibration source, covering a broad frequency band, require that the frequency-response characteristic of the system be adjusted. The frequency response can be adjusted to be constant throughout the band of frequencies of the test with attenuation above and below this band. In this case, a "white noise" generator at the input of the system produces an acceleration spectral density which is constant throughout the frequency range of the test. Alternatively, the frequency-response characteristic can be adjusted so as to approximate the variation in acceleration spectral density found under actual service conditions. In this case, a white noise generator at the input of the system produces a variation in acceleration spectral density at the table as a function of frequency duplicating the service conditions.

CONTROL SYSTEMS

SINUSOIDAL VIBRATION TESTING

A typical analog control system for sinusoidal vibration testing is shown in Fig. 25.10 (also see Fig. 27.2).   Voltage is supplied by a low-distortion sine-wave generator to an amplitude servo and programmer which adjusts the magnitude of this voltage; the adjustment is controlled by a voltage which is generated by the accelerometers.   The sine-wave signal, so modified, is fed to a low-distortion power amplifier which drives the vibration machine.

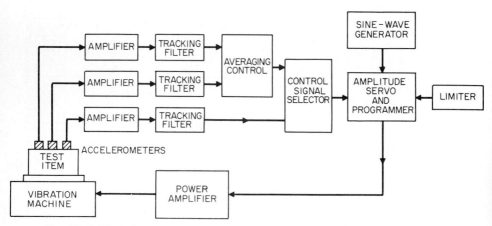

FIG. 25.10. Block diagram of a typical control system for sinusoidal vibration testing.

Sinusoidal motion of the vibration machine may induce harmonics, rattles, and other nonlinear response in the test item.   The tracking filters (narrow bandpass filters whose center frequency always corresponds exactly to the frequency of the sine-wave generator) connected to the outputs of the accelerometers reject any signals other than those corresponding to the driving frequency of the oscillator.   The function of the averaging control is to provide servo control on the arithmetic average of the individual accelerometer magnitudes.   The control signal selector limits the test to preset values.   The averager and/or the selector provide protection against excessive amplitude overtesting or undertesting.

RANDOM VIBRATION TESTING

A typical control system for random vibration testing is shown in Fig. 25.11.   The voltage, supplied by a random noise generator, is fed to an equalizer which shapes the spectrum of the random noise.   The equalizer is usually comprised of a number of narrow bandpass filters, each with an adjustable gain that divides the total frequency range of interest into contiguous narrowband segments.[10]   Digital techniques may be employed to achieve similar results (see Fig. 27.3).   The output of the equalizer provides the desired signal input to the power amplifier[2, 3] that drives the vibration machine.

Accelerometers are mounted on the item under test.   The output of these accelerometers is amplified and fed into an averaging control whose input switches between the signals from the various accelerometers.   Precautions must be observed in the use of such averaging controls.[5]   The mean-square acceleration density, provided by the spectrum analyzer, is displayed by any of several convenient techniques—for example, by means of an oscilloscopic display.

Fig. 25.11. Block diagram of a typical control system for random vibration testing.

SYSTEM RATINGS

The electrodynamic vibration machine system is rated: (1) in terms of the *peak value* of the sinusoidal generated force for *sinusoidal* vibration testing, and (2) in terms of the *rms* and *instantaneous* values of the maximum force generated under *random* vibration testing. In order to determine the acceleration rating of the system with a test load on the vibration table, the weight of the test load, assumed to be effective at all frequencies, must be known and used in the following expressions:

$$\mathfrak{g} = \frac{F}{W_L + W_T}$$

$$\mathfrak{g}_{\text{rms}} = \frac{F_{\text{rms}}}{W_L + W_T}$$

where $\mathfrak{g} = a/g$, a number expressing the ratio of the peak sinusoidal acceleration to the acceleration due to gravity (i.e., the peak sinusoidal acceleration in g's)

$\mathfrak{g}_{\text{rms}} = a_{\text{rms}}/g$, a number expressing the ratio of the rms value of random acceleration to the acceleration due to gravity

W_L = weight of load, lb

W_T = equivalent weight of table driver-coil assembly and associated moving parts, lb

F = rated peak value of sinusoidal generated force, lb

F_{rms} = rated rms value of random generated force, lb

The *force rating* of an electrodynamic vibration machine is the value of force which can be used to calculate attainable accelerations for any rigid-mass table load equal to (or greater than) the driver coil weight. It is not necessarily the force generated by the driver coil. These two forces are identical only if the operating frequencies are sufficiently below the axial resonant frequency of the armature assembly, where it acts as a rigid body. As the axial resonant frequency is approached, a mechanical magnification of the force generated electrically by the driver coil results. The design of the driving power supply takes into account the possible reduction in driver-coil current at frequencies approaching the armature axial resonant frequency, since full current in this range cannot be used without exceeding the rated value of transmitted force at the table —possibly causing structural damage.

In those cases where the test load dissipates energy mechanically, the system performance should be analyzed for each specific load since normal ratings are based on a dead-mass, nondissipative type of load. This consideration is particularly significant in resonant-type fatigue tests at high stress levels.[4]

PROMINENT FEATURES

1. A wide range of operating frequencies is possible, with a properly selected electric power source, from 0 to above 30,000 cps. Small, special-purpose machines have been made with the first axial resonant mode above 26,000 cps, giving inherently a resonance-free, flat response to 10,000 cps.

2. Frequency and amplitude are easily controlled by adjusting the power-supply frequency and voltage.

3. Pure sinusoidal table motion can be generated at all frequencies and amplitudes. Inherently, the table acceleration is the result of a generated force proportional to the driving current. If the electric power supply generates pure sinusoidal voltages and currents, the waveform of the acceleration of the table will be sinusoidal, and background noise will not be present. Operation with table acceleration waveform distortion of less than 10 per cent through a displacement range of 10,000-to-1 is common, even in the largest machines. Velocity and displacement waveforms obtained by the single and double integration of acceleration, respectively, will have even less distortion.

4. Random vibration, as well as sinusoidal vibration, or a combination of both, can be generated by supplying an appropriate input voltage.

5. A unit occupying a small volume, and powered from a remote source, can be used to generate small vibratory forces. A properly designed unit adds little mass at the point of attachment and can have high mobility without mechanical damping.

6. Leakage magnetic flux is present around the main magnet circuit. This leakage flux can be minimized by proper design and the use of degaussing coil techniques.

SPECIFICATIONS

DESIGN FACTORS

Force Output. The maximum vector-force output for sinusoidal excitation shall be given for continuous duty and may additionally be given for intermittent duty. When nonsinusoidal motions are involved, the force may additionally be given in terms of an rms value together with a maximum instantaneous value. The latter value is especially significant when random type of excitation is required.

In some cases of wide-frequency-band operation of the electrodynamic vibration machine, the upper frequencies are sufficiently near to the axial mechanical resonance of the coil-table assembly to provide some amplification of the generated force. Most system designs account for this magnification, when present, by reducing the capacity of the electrical driving power accordingly.

The peak values of the input electrical signal, for random excitation, may extend to indefinitely large values. In order that the armature coil voltage and generated force may be limited to reasonable values the peak values of excitation are clipped so that no maxima shall exceed a given multiple of the rms value. The magnitude of the maximum clipped output shall be specified preferably as a multiple of the rms value. If adjustments are possible, the range of magnitudes shall be given.

Weight of Vibrating Assembly. The weight of the vibration coil-table assembly shall be given. It shall include all parts which move with the table and an appropriate percentage of the weight of those parts connecting the moving and stationary parts giving an effective over-all weight.

Vibration Direction. The directions of vibration shall be specified with respect to the surface of the vibration table and with respect to the horizontal or vertical direction. Provisions for changing the direction of vibration shall be stated.

Unsupported Load. The maximum allowable weight of a load not requiring external supports shall be given for horizontal and vertical orientations of the vibration table. This load in no way relates to dyanmic performance, but is a design limitation, the basis of which may be stated by the manufacturer.

Static Moments and Torques. Static moments and torques may be applied to the coil-table assembly of a vibration machine by the tightening of bolts and by the overhang of the center-of-gravity of an unsupported load during horizontal vibration. The maximum permissible values of these moments and torques shall be specified. These loads in no way relate to the dynamic performance but are design limitations the basis for which may be stated by the manufacturer.

Total Excursion Limit. The maximum table motion between mechanical stops shall be given together with the maximum vibrational excursion permissible with no load and with maximum load supportable by the table.

Acceleration Limit. The maximum allowable table acceleration shall be given. (These large maxima may be involved in the drive of resonant systems.)

Stiffness of Coil-Table Assembly Suspension System

Axial Stiffness. The stiffness of the suspension system for axial deflections of the coil-table assembly shall be given in terms of pounds per inch of deflection. The natural frequency of the unloaded vibrating assembly may also be given. Provisions, if any, to adjust the table position to compensate for position changes caused by different loads shall be described.

Suspension Resonances. Resonances of the suspension system should be described together with means for their adjustment where applicable.

Axial Coil-Table Resonance. The resonant frequency of the lowest axial mode of vibration of the coil-table assembly shall be given for no load and for an added dead-weight load equal to one and to three times the coil-table assembly weight. If this resonant frequency is not obvious from measurements of the table amplitude vs. frequency, it may be taken to be approximately equal to the lowest frequency, above the rigid-body resonance of the table-coil assembly on its suspension system, at which the phase difference between the armature coil current and the acceleration of the center of the table is 90°.

Impedance Characteristics. When an exciter or vibration machine is considered independent of its power supply, information concerning the electrical impedance characteristics of the machine shall be given in sufficient detail to permit matching of the power-supply output to the vibration-machine input. It is suggested that consideration be given to providing schematic circuit diagrams (electrical and mechanical, or equivalent electrical) together with corresponding equations which contain the principal features of the machine.

Environmental Extremes. When it is anticipated that the vibration machine will be used under conditions of abnormal pressure and temperature, the following information shall be supplied as may be applicable: maximum simulated altitude (or minimum pressure) under which full performance ratings can be applied; maximum simulated altitude under which reduced performance ratings can be applied; maximum ambient temperature for rated output; low-temperature limitations; humidity limitations.

PERFORMANCE. The performance relates in part to the combined operation of the vibration generator and its power supply.

Amplitude-Frequency Relations. Data on sinusoidal operation shall be given as a series of curves for several table loads, including zero load, and for a load at least three times the weight of the coil-table assembly. Maximum loads corresponding to 20g and 10g table acceleration under full rated force output would be preferred. These curves should give amplitudes of table displacement, velocity, or acceleration, whichever is limiting, throughout the complete range of operating frequencies corresponding to maximum continuous ratings of the system. Additionally, the maximum rated force should be given. If this force is frequency-dependent, it should be presented as a curve with the ordinate representing the force and the abscissa the frequency.

If the system is for broad-band use, necessarily employing an electronic power amplifier, the exciting voltage signal applied to the input of the system shall be held constant and the output acceleration shall be plotted as a function of frequency with and without peak-notch filters or other compensating devices for the loads and accelerations indicated above. If the vibrator is used only for sinusoidal vibrations, and employs servo amplitude control, the curves should be obtained under automatic frequency sweeping conditions with the control system included. If cycling the exciter through its complete frequency range involves switching operations, such as may be required for matching impedances or for changing power sources, these discontinuities should be noted on the amplitude-frequency curves.

Waveform. Total rms distortion of the acceleration waveform at the center of the vibration table, or at the center on top of the added test weight, shall be furnished to show at least the frequencies of worst waveform under the test conditions specified under the above paragraph. The pickup type, and frequency range, shall be given together with the frequency range of associated equipment. It is desirable to have the over-all frequency range at least ten times the frequency of the fundamental being recorded. Tabular data on harmonic analysis may alternatively or additionally be given.

Magnetic Fields. The maximum values of constant and alternating magnetic fields, due to the vibration exciter, in the region over the surface of the vibration table should be indicated. If degaussing coils are furnished, these values should be given with and without the use of the degaussing coils.

Frequency Range. The over-all frequency range, and a division of frequency ranges for different alternators that supply the exciting current, if applicable, shall be given. A group of frequency ranges shall also be given for electronic power supplies if they require changes of their output impedance for the different ranges.

Automatic Frequency Control. The following factors shall be stated or provided: The range, or ranges, of the frequency bands that can be automatically cycled at full rated output or at reduced output if applicable; the minimum and maximum time for each frequency cycle; a curve of frequency vs. time for a maximum and a minimum cycle period and for the principal frequency ranges; the dwell time at the points of reversal of the direction of frequency change.

Frequency Drift. The probable drift of a set frequency shall be stated, together with factors that contribute to the drift. This shall apply for nonresonant loads.

Signal Generator. A vibration pickup, if built into the vibration machine, shall have calibrations furnished over a specified frequency and amplitude range.

INSTALLATION REQUIREMENTS. Recommendations shall be given as to suitable methods for installing the vibration machine and auxiliary equipment. Electrical and other miscellaneous requirements shall be stated.

HYDRAULIC VIBRATION MACHINE

The *hydraulic vibration machine* is a device which transforms power in the form of a high-pressure flow of fluid from a pump to a reciprocating motion of the table of the vibration machine. A schematic diagram of a typical machine is shown in Fig. 25.12. In this example, a two-stage electrohydraulic valve is used to deliver high-pressure fluid, first to one side of the piston in the actuator and then to the other side, forcing the actuator to move with a reciprocating motion. This valve consists of a pilot stage and power stage, the former being driven with a reciprocating motion by the electrodynamic driver. At the time the actuator moves under the force of high-pressure fluid on one side of the piston, the fluid on the other side of the piston is forced back through the valve at reduced pressure, and returned to the pump.

The electrohydraulic valve is usually mounted directly on the side of the actuator cylinder, forming a close-coupled assembly of massive steel parts. The proximity of the valve and cylinder is desirable in order to reduce the volume and length of the connecting fluid paths between the several spools and the actuator, thereby minimizing the effects of the compliance of the fluid and the friction to its flow. (Many types of electrohydraulic valves exist, all of which fail to meet the requirement of sufficient flow at high frequencies to give vibration machine performance equivalent to existing electrodynamic machine performance at 2,000 cps.)

P = PRESSURE
R = RETURN

ELECTRO
DYNAMIC
DRIVER

PILOT
STAGE

POWER
STAGE

TRANSDUCER

TRANSDUCER

TEST
ITEM

ACTUATOR

TABLE

FIG. 25.12. Schematic diagram of a typical hydraulic vibration machine.

OPERATING PRINCIPLE

In Fig. 25.12, the *pilot* and *power spools* of a hydraulic vibration machine are shown in the "middle" or "balanced" position, blocking both the pump high-pressure flow P and the return low-pressure flow R. Correspondingly, the piston of the actuator must be stationary since there can be no fluid flow either to or from the actuator cylinder. If the pilot spool is displaced to the right of center by a force from the electrodynamic driver, then high-pressure fluid P will flow through the passage from the pilot spool to the left end of the power spool, causing it to move to the right also. This movement forces the trapped fluid from the right-hand end of the power spool through the connecting passage, back to the pilot stage, and then through the opening caused by the displacement of the pilot spool to the right, to the chamber R connected to the return to the pump. Correspondingly, if the pilot spool moves to the left, the flow to and from the power spool is reversed, causing it to move to the left. For a given displacement of the pilot spool, a flow results which causes a corresponding velocity of the power spool. A displacement of the power spool to the right allows the flow of high-pressure fluid P from the pump to the left side of the piston in the actuator, causing it to move to the right and forcing the trapped fluid on the right of the piston to be expelled through the connecting passage to the power spool and out past the right-hand restrictions to the return fluid chamber R. The transducers shown on the power spool and the actuator shaft are of the differential transformer type* and are used in the feedback circuit to improve system operation and provide electrical control of the average (i.e., stationary) position of the actuator shaft relative to the actuator cylinder.

A block diagram of the complete hydraulic vibration machine system is shown in Fig. 25.13. The pump, in conjunction with accumulators in the pressure and return lines at the hydraulic valve, should be capable of variable flow while maintaining a fixed pressure. Most systems to date have required an operating pump pressure of 3,000 lb/in.^{2} The upper limit of efficiency of the hydraulic valve is approximately 60 per cent, the losses being dissipated in the form of heat. Mechanical loads are seldom capable of dis-

* See Chap. 15 for a discussion of such transducers.

FIG. 25.13. Block diagram—hydraulic vibration machine system.

sipating appreciable power; most of the power in the pump discharge is converted to a temperature rise in the fluid. Therefore a heat exchanger limiting the fluid temperature must be included as part of the system.

PROMINENT FEATURES

1. Large generated forces or large strokes can be provided relatively easily. Large forces and large velocities of motion, made possible with a large stroke, determine the power capacity of the system. For example, one hydraulic vibration machine has a peak output power of 450,000 lb-in./sec (519,000 kg-cm/sec) or approximately 34 hp with a single electrohydraulic valve. This power can be increased by the installation of several valves on a single actuator. Appreciable increases in valve flow can be realized by sacrificing high-frequency performance. Hence, the hydraulic vibration machine excels at low frequencies where large force, stroke, and power capacity are required.

2. The hydraulic machine is small in weight, relative to the forces attainable; therefore, a rigid connection to firm ground or a large massive base is necessary to anchor the machine in place and to attenuate the vibration transmitted to the surrounding area.

3. The main power source is hydraulic, which is essentially d-c in character from available pumps. The electrical driving power for controlling the valve is small. Therefore, the operating frequency range can be extended down to zero cps.

4. The magnetic leakage flux in the region of the table is insignificant by comparison with the electrodynamic-type vibration machine.

5. The machine, with little modification, is suitable for use in high- and low-temperature, humidity, and altitude environments.

6. The machine is inherently nonlinear with amplitude in terms of electrical input and output flow or velocity.

7. The waveform of table motion in general is not as good as that of the electrodynamic type, because of distortion introduced by the valve itself and friction in the actuator which is most noticeable at low drive levels.

8. The hydraulic machine actuator consists of a cylinder, piston, and shaft which are simple in construction and not easily damaged. Forces are generated hydraulically—no primary loads are carried through bearings.

9. Sufficient force capacity generally is available to support all rated test loads hydraulically in the vertical position without resorting to external supports.

10. Hydraulic power requires larger hose sizes than an electrical cable of equal capacity. Hydraulic fluid cleanliness, seepage, and leakage are problems not encountered in electrical power systems.

APPLICATION AND SELECTION OF VIBRATION MACHINES AND EXCITERS

Some of the most common types of vibration tests are listed in this section and the desirable characteristics of the test equipment are indicated. The type of vibration exciter or machine best fitted for a given test is suggested.

THE GENERATION OF PURE SINUSOIDAL RECTILINEAR MOTION FOR THE CALIBRATION OF VIBRATION PICKUPS

The generation of vibratory motion of suitable quality for calibration work, and the determination of the amplitude, as well as the frequency, of the generated vibration, is difficult. The deviations from ideal in the form of harmonics, unwanted lateral and rotational motions superposed on the desired rectilinear motion, lack of infinite stiffness in the vibration table, etc., which can be tolerated in the general vibration test, may be intolerable in calibration work. Therefore, a vibration machine designed specifically for calibration work may have limited usefulness for general vibration testing because it necessarily must be designed to accommodate small test loads to minimize design compromises. (See Chap. 18 for calibration techniques.)

DESIRED CALIBRATOR CHARACTERISTICS. A vibration machine for calibration work should provide the following:

1. Distortion-free sinusoidal motion.
2. True rectilinear motion normal to the vibration-table surface without the presence of any other motion.
3. A rigid table for all design loads at all operating frequencies.
4. Stepless variation of frequency and amplitude of motion within specified limits, easily adjustable during operation.
5. A table that remains at ambient temperature and does not provide either a source or sink for heat regardless of the ambient temperature.
6. A mounting area at the table free from electromagnetic disturbances.
7. Horizontal as well as vertical motion, if required.
8. Features allowing the use of absolute calibration methods when calibrating the calibrator.
9. Data which can be used to define the vibration imposed on the pickup.

If a range of calibrating frequencies from 2 to 10,000 cps is desired, the electrodynamic-type machine more nearly satisfies all the requirements of the ideal calibrator. For high-frequency (1,000 to 20,000 cps) and high-acceleration calibrations using steady sinusoidal vibration, the piezoelectric type has advantages over other types. The piezoelectric type is limited to very small displacements, making low-frequency calibrations difficult, and displacement must be measured by interferometric methods or external displacement-measuring pickups, such as the capacitance or variable-reluctance types. Mechanical direct-drive calibrators are no longer used because they have an upper-frequency limit of less than 100 cps, and because of the presence of background mechanical bearing noise and the difficulty of adjusting speed and amplitude.

DETERMINING THE NATURAL MODES OF A STRUCTURE

Vibration exciters are used to determine experimentally the natural modes of a structure. The frequency, the deflection form including nodal lines, and the damping factor associated with the natural mode are usually determined for all modes of vibration of interest. Usually, the dynamic characteristics of a structure are determined by operating at vibration levels which are small compared to damaging levels. If, in the exceptional case, nonlinear characteristics are suspected and are to be evaluated, then testing levels approaching operating levels may be desirable.

Structures are vibrated by introducing a vibratory force by a vibration exciter at one or more points on the structure while the structure is supported in an appropriate manner resembling the service configuration. An excellent example of natural mode testing is found in the "ground vibration testing" of winged aircraft where the results of the tests are used to check the calculated natural modes of the aircraft wings and tail as well as the complete aircraft.

In some cases, several natural modes may be sufficiently close in frequency so that they will be excited simultaneously from a single input-force location. Since the motions from the several modes add as vector quantities throughout the structure to give the actual structural response at any point, recognition of the individual natural modes becomes difficult. A method using multiple exciters appropriately distributed throughout the structure has been developed[6] and can be used to excite individual natural modes in a complex structure. An iteration process is used to arrive at the correct magnitude and phase of output force at each exciter to overcome locally the dissipation forces associated with the mode in question. In this manner, the excitation of modes closely adjacent in frequency is minimized.

DESIRED VIBRATION EXCITER CHARACTERISTICS. The desirable characteristics of a vibration exciter for determining the natural modes of vibration of a structure are:

1. Insignificant dynamic changes in the vibration characteristics of the structure should result from the addition of the exciter to the structure. For this condition to be met, the mass, stiffness, and damping introduced by the exciter must be small compared to the effective values of the structure at the point of attachment.

2. Force and frequency should be easily controlled by the operator at a position removed from the exciter. If the structure is large, a portable control station for frequency and amplitude is sometimes helpful. With a set position of the force control, it is desirable that the force remain constant, independent of frequency and mechanical impedance of the structure.

3. Stepless variation in force and frequency should be available to the operator. The operating frequency should correspond to the set frequency, independent of the mechanical structure characteristics.

4. It should be possible to interrupt the generation of force abruptly. Furthermore, the exciter should introduce negligible damping into the vibrating structure.

5. For the testing of complex structures, a number of exciters may be required—each with the force individually controllable in magnitude and phase characteristics.

6. The exciter should be portable and easily supported adjacent to the structure under test. Heavy and stiff connections to drive the exciter should be avoided where possible.

7. The exciter should generate good waveform in order to avoid the excitation of higher-frequency modes occurring at harmonics of the driven frequency.

8. Sufficient displacement capacity should be provided in the exciter to allow for acceptable vibratory displacements at the point of attachment, as well as inadvertent positional shifts between the structure and the supports of the stationary part of the exciter.

FATIGUE TESTING MACHINES

Many types of commercial equipment are available that are designed specifically for fatigue testing. These machines usually use (1) mechanical direct-drive, rotating unbalances, or (2) a hydraulic means of displacement or force generation. However, general-purpose vibration testing machines described earlier in this chapter may be employed.   In some test setups, force amplification is obtained by mechanical resonance.

The operating frequency of the testing machine should be as high as possible in order to shorten the time of the test, but the frequency is limited by the temperature rise of the test item if the endurance limits are dependent upon temperature.

A typical fatigue testing machine includes some or all of the following general features:

1. It should be capable of maintaining stress or strain in the test item within stated limits for the duration of the test.

2. Steady loads, applied simultaneously with vibratory loads, should be available when required by the test procedure.

3. Provisions should be available for maintaining elevated or reduced temperatures in the test item.

4. The exciting unit should operate without inadvertent transients or other malfunctions which could be transmitted to the test item, thereby complicating the analysis of the test results.

APPLICATION OF VIBRATION MACHINES AS FATIGUE TESTING MACHINES.   In selecting a vibration machine to perform a fatigue test, it is best to select a type that inherently produces the output desired.   For example, if a fixed displacement is desired, it is advantageous to select a direct-drive-type machine.   If a constant force is desired, it is advantageous to select a machine that incorporates a rotating unbalance at a fixed frequency since this type of machine generates a constant force against a point of high mechanical impedance.   If the mechanical impedance is not high, then a machine of the type shown in Fig. 25.6 can provide a constant force.

THE USE OF MECHANICAL RESONANCE.   In fatigue testing, vibration machines are used which inherently give a constant output; the machines are operated at frequencies removed from mechanical resonances; and they require only occasional manual adjustments for maintaining a constant output.   By making use of mechanical resonance for force amplification, with a given capacity of vibration exciter, a large increase in performance can be realized in some tests.   This increase is possible if the test fixture is carefully designed, and if a continuous control on the output of the exciter is provided.   In addition, the damping of the test item must be small to allow amplification of force by resonance.   Additional advantages from a resonant-system test include:

1. The generation of the desired motion, to the exclusion of all other motions and with good waveform.

2. The generation of bending stresses, direct tension-compression stresses, or torsion stresses with a general-purpose vibration exciter.

3. The reduction of damaging impact loads transmitted to the exciter which originate from free play or chatter in the test item, by allowing the use of a relatively large mass between the exciter and the test item.

4. The concentration of high stress in local areas of choice in order to reproduce service failures under laboratory testing procedures.   This is possible in many instances by selecting the proper fixture and natural mode of vibration.

For example, in Fig. 25.14, a resonant system for producing bending moments is shown. The massive legs provide a large mass moment of inertia at each end of the test item. The bending stiffness of the test item and the mass properties of the legs determine the resonant frequency.   Some axial force, in addition to the moments, is placed on the test item.   The axial force can be eliminated by extending the length of the legs equally on both sides of the test item.   A spring between the extremities of the legs can be used for a preload moment.   The exciter and test system are flexibly supported from the floor to eliminate the transmission of unwanted disturbances to the surrounding area and to eliminate unpredictable power losses through the supports.   A single vibration exciter is connected to one leg—although two may be used, one connected to each leg, if more power is required than is available from a single unit.   If two exciters are used, they are normally driven from a common power source to ensure synchronism.   A connection is used between the exciter and leg which gives axial stiffness and yet is soft to bending and

FIG. 25.14. A resonant fatigue test: bending stress.

shearing deflections, which is necessary to allow the leg to assume its pendulous motion while the exciter moves with rectilinear motion. Necked-down connectors on either end of a stiff member are satisfactory for this purpose. Moving the exciter toward (or away from) the test item along the leg increases (or decreases) the mechanical impedance, respectively, at the point of attachment, allowing maximum values of both force and velocity to be generated in the exciter for maximum power transfer to the test item. With good design, the desired mode of operation can be made high in frequency—consistent with dissipation and power considerations; the resonant frequency can be selected so that it is not near the natural frequencies of all other modes of the system and only a single mode of vibration is excited.

Figure 25.15 is similar to Fig. 25.14 except tension-compression forces are generated in the test item. In this case, the plate at the top of the legs serves as an ideal hinge.

FIG. 25.15. A resonant fatigue test: tension-compression stress.

It should be stiff in its own plane, but soft to bending.   Figure 25.16 shows a suggested configuration for torsion motion.   The mechanical impedance seen by the exciter in this case can be adjusted by changing the radial location of the point of exciter attachment and the ratio of the top and bottom mass moments of inertia.   Reducing the top inertia, with respect to the bottom inertia, reduces the amplitude of rotational displacement at the bottom for a given rotational strain in the test item—thereby increasing the mechanical impedance at that end.

Figure 25.17 illustrates how bending moments in the plane of the paper can be maximized at several locations along the length of the crankshaft.   In Fig. 25.17A, with the exciter connected in the center of the crankshaft, the natural mode shown gives a zero bending moment at the ends and a maximum moment at the center.   In Fig. 25.17B, using large mass moments of inertia at both ends, and driving horizontally at one end,

FIG. 25.16. A resonant fatigue test: torsion stress.

gives a nearly constant moment throughout the length of the shaft if the end masses are large compared to the distributed mass of the shaft.   In Fig. 25.17C, driving at one of the crankthrow bearings and operating at the mode shown gives a zero moment at each end and at the center, with the maximum moments occurring at the two positions between the three zero moment locations.   By adding large end-mass moments of inertia and exciting the mode in Fig. 25.17D, the maximum bending moment occurs at the two ends, with a zero moment in the center.   Tests of this kind permit failures to be duplicated in the laboratory which have been experienced in the field.   They also allow one to determine the effect of oil holes, fillets, section changes, materials, etc., on the endurance limit of the complete shaft.

In fatigue tests of the type described, it is usually necessary to measure stress or strain at an appropriate point on the test piece, and to use the voltage generated by the measuring device to control the operation of the system, so as to maintain, automatically, the set operating level.   Control can be performed manually by an operator.   In this case the operator must maintain a constant meter reading of stress or strain in the test article—this can be tedious.

In the examples shown, a large amplification of force can be achieved.   For example, in Fig. 25.15, amplifications of the vibration exciter force of 50 to 250 times in the test item are achieved readily.   However, if the item is tested to failure, within a relatively few cycles of stress near or beyond the yield point, a large part of the energy expended in deforming the test item is not returned to the system in the form of kinetic energy upon removal of the load.   In this case, large power may be dissipated, and the amplification of mechanical force attainable at resonance is reduced accordingly.

HEAVY END MOMENT
OF INERTIA WHEN
REQUIRED

VIBRATION
EXCITER

(A)

←— INPUT FORCE

(B)

(C)

(D)

Fig. 25.17. A resonant fatigue test: selective location of maximum bending moment.

Types of Exciters Employed in Resonant Fatigue Testing Machines. Vibration exciters which are applied to resonant fatigue testing include the following types: rotating unbalance, electrodynamic, hydraulic, electromagnetic, and pneumatic. In general, for this application, force-generating machines are employed rather than displacement-generating machines. The direct-drive mechanical-type exciter can be included if a spring is added between the output of the exciter and the input to the structure. By comparison with other types, the application of this exciter is awkward and difficult. The piezoelectric type is excluded because of its small displacements and small power capacity. The electromagnetic and pneumatic types offer the advantage of not requiring physical contact with the resonant member. The rotating-unbalance and hydraulic types can be applied to resonant fatigue tests although their characteristics of constant output force (with fixed frequency) and large output force potential, respectively, particularly suit them for nonresonant-type fatigue tests. The electrodynamic vibration exciter offers some advantages for resonant fatigue tests in the form of a wide frequency range of operation and ease of control.

RESONANT-LOAD EFFECTS ON THE VIBRATION EXCITER. In applying vibration exciters to resonant fatigue testing machines the following considerations are important:

1. With an approximately constant input force applied to a resonant structure, an increase in amplitude occurs at resonance, as a result of which the power absorbed by the structure is increased.

2. If the power originates from a variable-speed motor (e.g., from a rotating-unbalance or motor-alternator type of supply for the electrodynamic exciter), it may be impossible to adjust the operating frequency exactly to the resonant frequency. This is because of

the rapidly increasing torque placed on the motor with increasing frequency below resonance, and the rapidly decreasing torque with increasing frequency above resonance. Unless sufficient speed regulation exists, the motor may tend to avoid the speed corresponding to a structure resonance (1) by slowing down below resonance or (2) (with a slight upward adjustment in frequency setting) by increasing its speed and passing through the resonant frequency—stabilizing at a speed in excess of resonance. The use of an electronic amplifier and frequency source will avoid this problem since the mechanical load impedance is not reflected back to the oscillator which controls the frequency of operation.

3. If a measurement of the amplitude of motion at the structure is employed to control the amplitude of output of the exciter, note that a system (described in item 2 above) which is stable with respect to frequency and operable at resonance with fixed or manual amplitude control, may become unstable with respect to frequency if automatic amplitude control is added.

4. If the operating frequency is adjusted for maximum amplitude at resonance, assuming a force amplification of at least several times, any variations in the properties of the test item or vibration exciter frequency will cause the amplitude to diminish. It is desirable to maintain constant amplitude conditions for a fatigue test, but if inadvertent variations occur, decreases in amplitude can be corrected far more easily than increases in amplitude.

5. The electrodynamic vibration exciter will tend to adjust its driving force as required to maintain a constant amplitude of motion at a given operating frequency with a fixed position of the amplitude control, (a) if an electrodynamic vibration exciter is used below electrical resonance, (b) if the part of the impedance measured at the electrical terminals of the driver coil, due to the motion of the driver coil, is large compared to the impedance of the driver-coil electrical inductance and resistance, and (c) if the internal impedance of the power supply is small. By making use of this characteristic and by adjusting for the peak amplitude at the resonant frequency, as described in item 4, many different types of specimens can be fatigue tested in a resonant system without automatic controls and with only occasional minor adjustments to the operating controls.

6. Many vibration exciters are rated in terms of rigid-mass loads. In resonant tests, the exciter may easily attain accelerations in excess of those corresponding to full rated force, even with no mechanical load. These accelerations should be maintained below the manufacturer's stated maximum. In the case of Fig. 25.15, for example, for a given strain across the test article the acceleration at the exciter can be varied by moving the exciter along the length of the leg. Operation at resonance can absorb more power from the exciter than that anticipated by the manufacturer, who usually designs the system only (a) for mass load conditions, (b) for overloading of the driving motor, in the case of the rotating unbalance, or (c) for overloading of the motor-alternator, in the case of the electrodynamic exciter.[4] In the case of an electronic amplifier, sufficient voltage may not be provided to generate maximum velocity and force simultaneously in the electrodynamic exciter. For example, consider one commercially available electrodynamic vibration machine which is driven by an electronic amplifier power source. The system is rated at 1,200 lb peak sinusoidal generated force for any mass load on the table, the amplifier capacity is 2,900 watts, and the maximum table displacement is 1 in. peak-to-peak. Suppose that this vibration machine were attached to a resonant load and its acceleration were limited to 100g for structural strength reasons. If the mechanical impedance of the resonant load has a value such that a 1,200 lb peak sinusoidal force produces a 1-in. peak-to-peak displacement at approximately 44 cps (the frequency of maximum velocity at 100g acceleration), the mechanical power absorbed by the mechanical impedance would be 9,400 watts, or more than three times the power normally provided by the power source. This excludes the power loss in the electrodynamic exciter itself, which is significant.

7. If the driver-coil electrical impedance of an electrodynamic exciter is determined mainly by its mechanical motion, the driver-coil circuit must be protected against large currents and forces generated by the exciter should the mechanical connection between it and the structure fail, or should changes in the structure destroy the resonant operating condition as a result of incipient failure.

FIXED-FREQUENCY VIBRATION TEST

Some test specifications for equipment require that the test item be vibrated for extended periods at a frequency corresponding to the frequency of the most serious resonance in the equipment or, if no resonances are found, at some arbitrary nonresonant frequency. The amplitude of motion of the vibration table is specified as a function of frequency for the test. When searching for resonant frequencies, the frequency is slowly changed by adjusting the frequency manually. This search may be conducted at reduced amplitudes compared to those specified for the endurance test.

It may be possible to use a machine which has insufficient capacity when used to vibrate the loads directly. First search for resonances at reduced vibration levels which this machine will provide; then obtain the required output vibration force at the fixed-frequency endurance run by force amplification from a mechanical resonant system. A free-free uniform beam employed as a resonant system is simple to construct and has some advantage compared to many other types of resonant systems. As an example of such a system, see Fig. 25.18. Some adjustment in frequency can be obtained by adding weights at the extremities of the beam, as shown schematically in Fig. 25.19.

FIG. 25.18. A free-free beam resonant system.   FIG. 25.19. A schematic diagram of a free-free beam with concentrated weights at the center and at the ends.

The natural frequency of the lowest mode of a free-free beam system of Fig. 25.19 is, approximately,

$$f_n \approx \frac{\pi}{2} \left[\frac{EI(m_c + 2m_e + \mu l)}{l^3[2m_e(2m_c + \mu l) + m_c\mu l(3 - 8/\pi) + (\mu l)^2(1 - 8/\pi^2)]} \right]^{1/2} \tag{25.1}$$

where f_n = natural frequency, cps
    E = Young's modulus of beam material, lb/in.²
    I = area moment of inertia of beam cross section, in.⁴
    m_c = mass at center of beam, including that of test article and vibration exciter element attached to beam, lb sec²/in.
    m_e = mass at each end of beam, lb sec²/in.
    l = total separation of end masses m_e if m_e is large compared to μl over the same spanwise distance; otherwise, the length of the beam, in.
    μ = beam mass per inch of length, lb sec²/in.

The nodal lines pass through the beam parallel to the end and midway between the upper and lower surfaces at a distance from each end approximately equal to

$$d_n \approx \frac{l}{\pi} \sin^{-1} \left(\frac{[2\mu l/\pi] + m_c}{\mu l + 2m_e + m_c} \right) \tag{25.2}$$

where the angle is expressed in radians. Equations (25.1) and (25.2) are based on an assumed sinusoidal deflection form of the beam, and only translational motions are considered.

If the concentrated masses m_c and m_e are very small compared to the mass of the beam, the beam can be treated as a uniform beam without added loads. In this case, if the material of the beam is steel and it has a length l and a thickness t, both expressed in inches, the natural frequency is approximately $209,000 \, t/l^2$ cps; the nodal line is located approximately $0.23l$ in. from each end. The frequency is not dependent upon the width of the beam since both the mass and stiffness are directly proportional to the width. The test item is assumed to be a rigid mass connected to the center of the beam. The beam should be designed for vibratory stresses well below the endurance limit of its material at the maximum desired deflection.

Electromagnetic, hydraulic, or rotating-unbalanced exciters may be employed in this system. However, the electrodynamic vibration exciter is particularly well suited for this type of system because of its wide range of operating frequencies and ease of control. The system is awkward in that mechanical changes must be made to change the resonant frequency, but it has the advantage of imposing almost pure sinusoidal, rectilinear motion on the test item at its mounting points. This motion may be imparted to the test item even though its center-of-gravity is displaced horizontally from the center of its mounting, since the beam can be made quite massive and of large dimension by comparison to a typical vibration machine. A resonant-beam system[7] has been constructed which is adjustable in frequency from 5 to 500 cps, with mass loads ranging up to 200 lb, and which will impart $20g$ acceleration to the mass load with the maximum input force not exceeding 200 lb. This corresponds to a force amplification of 20 times throughout the frequency range. If the test item adds sufficient damping to the resonant beam mode or if it acts as a tuned vibration absorber at the resonant mode frequency, little or no amplification of vibration exciter force may be provided by a resonant beam system.

ENVIRONMENTAL VIBRATION TESTING

Specifications usually require that rectilinear motion be applied in each of three mutually perpendicular directions relative to the test item. With the exception of some of the mechanical types of vibration machines, all others generate motion normal to the vibration table surface, making it necessary for the test item to be reoriented relative to the table in order to produce vibratory motion in each of the three directions. This requires a versatile attachment between the table and test item or possibly three different attaching fixtures. If the direction of gravity is important to the test item and the vibration machine generates motion in one direction only, the complete machine must be rotated from vertical to horizontal, or vice versa, to meet the test conditions.

In order to obtain horizontal motion in the plane of a horizontal table, stiff tables, supported by elastic restraints which permit motion only in the horizontal direction, are used, driven usually by an electrodynamic machine rotated to give horizontal motion. For example, see Fig. 25.20A. Vibration in the plane of the table is more difficult to achieve than vibration normal to the plane of the table because the test article necessarily has dimensions of height separating its center-of-gravity vertically from the table surface. Accordingly, reactionary moments as well as forces are transmitted to the table. Tables for test items in excess of 50 lb weight are difficult to construct for use above 500 cps with acceptable performance; the large weight added to the vibration machine reduces the attainable vibration levels, reduces the frequency of the inevitable flexing and in-plane resonant modes of the horizontal table, and emphasizes the requirement of attachment to the vibration machine with sufficient rigidity. (Laminated Sitka-spruce-wood tables, having a depth approaching the horizontal dimension, and

Fig. 25.20. An auxiliary horizontal vibration table with vibration in the plane of the table: (A) Elastic support. (B) Fluid film support.

having metal inserts at attaching points, are used by one environmental testing laboratory.) Good performance is claimed by several users[8, 9] of a horizontal plate driven by an electrodynamic vibration machine. The plate is separated from a flat opposing surface of a massive block by an oil film, as shown in Fig. 25.20B. The thin oil film provides little shearing restraint but great stiffness normal to the surface, the stiffness being distributed uniformly over the complete horizontal area. Accordingly, a relatively light moving plate can be vibrated which has the properties of the massive rigid block in the direction normal to its plane.

In general, in the low-frequency category the mechanical direct-drive and hydraulic vibration machines are suitable for the lowest frequencies and largest displacements. If displacements are less than 1 in. and frequencies above 5 cps, the mechanical reaction-type machine may be considered. All three of these machines can be fitted with large massive tables, simplifying the attachment of the test article to the vibration table. The electrodynamic vibration machine can be used for displacements below 1 in. and frequencies above 2 to 5 cps. However, large, heavy test items in this case may require external flexible supports to avoid forcing the table against its stops. In the intermediate-frequency range, the electrodynamic and hydraulic vibration machines may be used. In the wide-band range, the electrodynamic machine is most widely used.

REFERENCES

1. Bernhard, R. K.: *Proc. ASTM,* **49**:1016 (1949).
2. Usher, T.: *AIEE Paper* No 58:1237, 1958.
3. Oleson, M. W., and R. E. Blake: *Publication Board Rept.* 132741, Washington, D.C. 1957.
4. Reen, G. K.: *MB Vibration Notebook,* **2**(1):3 (1956). MB Manufacturing Co., New Haven, Connecticut (1956).
5. Curtis, A. J.: *Shock and Vibration Bulletin No.* 42, *part III,* p. 1, U.S. Naval Research Laboratory, Washington, D.C., January, 1972.
6. Lewis, R. C., and D. L. Wrisley: *J. Aeronaut. Sci.,* **17**:705 (1950).
7. Granick, N., D. C. Kennard, and K. Unholtz: *Shock and Vibration Bulletin 17,* p. 84, U.S. Naval Research Laboratory, Washington, D.C., March, 1951.
8. Hansen, W. O.: 25th *Shock and Vibration Bulletin, part II,* p. 93, U.S. Naval Research Laboratory, Washington, D.C., December, 1957.
9. Klein, F. P.: *Rept.* GM-TN-0165-00141, Space Technology Laboratories, Inc., Los Angeles, May, 1958.
10. Klapper, J., and C. M. Harris: *J. Audio Eng. Soc.,* **8**:177 (1960).

26

SHOCK TESTING MACHINES

Edward W. Clements
Naval Research Laboratory

Jerry R. Sullivan
Naval Ship Engineering Center

Irwin Vigness
Naval Research Laboratory

INTRODUCTION

Equipment must be built sufficiently rugged to operate satisfactorily in the shock and vibration environments to which it will be exposed and to survive transportation to the site of ultimate use. To ensure that the equipment is sufficiently rugged and to determine what its mechanical faults are, it is subjected to controlled mechanical shocks on shock testing machines. The severity and nature of the applied shocks are usually intended to simulate environments expected in later use or to be similar to important components of those environments. However, a principal characteristic of shocks encountered in the field is their variety. These field shocks cannot be defined exactly. Therefore shock simulation can never exactly duplicate shock conditions that occur in the field.

There is no general requirement that a shock testing machine reproduce field conditions. All that is required is that the shock testing machine provide a shock test such that equipment which survives is acceptable under service conditions. Assurance that this condition exists requires a comparison of shock test results and field experience extending over long periods of time. This comparison is not possible for newly developed items. It is generally accepted that shocks that occur in field environments should be measured, and that shock machines should simulate the important characteristics of shocks that occur in field environments or have a damage potential which by analysis is shown to be similar to that of a composite field shock environment against which protection is required.

A *shock testing machine* (frequently called a *shock machine*) is a mechanical device that applies a mechanical shock to an equipment under test. The nature of the shock is determined from an analysis of the field environment.* Tests by means of shock machines usually are preferable to tests under actual field conditions for four principal reasons:

1. The nature of the shock is under good control, and the shock can be repeated with reasonable exactness. This permits a comparative evaluation of equipment under test and allows exact performance specifications to be written.

---

* A mechanical shock occurs when the position of an item is significantly changed in a relatively short time in a nonperiodic manner. It is characterized by suddenness and develops significant inertial forces in the item.

2. The intensity and nature of shock motions can be produced which represent an average condition for which protection is practical, whereas a field test may involve only a specific condition that is contained in this average.

3. The shock machine can be housed at a convenient location with suitable facilities available for monitoring the test.

4. The shock machine is relatively inexpensive to operate, so it is practical to perform a great number of developmental tests on components and subassemblies in a manner not otherwise practical.

SHOCK-MACHINE CHARACTERISTICS

DAMAGE POTENTIAL AND SHOCK SPECTRA

The damage potential of a shock motion is dependent upon the nature of an equipment subjected to shock, as well as upon the nature and intensity of the shock motion. To describe the damage potential, a description of what the shock does to an equipment must be given—a description of the shock motion is not sufficient. To obtain a comparative measure of the damage potential of a shock motion, it is customary to determine the effect of the motion on simple mechanical systems. This is done by determining the maximum responses of a series of single degree-of-freedom systems to the shock motion and considering the magnitude of the response of each of these systems as indicative of the damage potential of the shock motion to these standard systems. The responses are plotted as a function of these natural frequencies. A curve representing these responses is called a *shock spectrum*, or *response spectrum* (see Chap. 23). Its magnitude at any given frequency is a quantitative measure of the damage potential of a particular shock motion to a single degree-of-freedom system of that natural frequency. This concept of the shock spectrum originally was applied only to undamped single degree-of-freedom systems, but the concept has been extended to include systems in which any specified amount of damping exists.

The response of a simple system can be expressed in terms of the relative displacement, velocity, or acceleration of the system. It is customary to define velocity and acceleration responses as $2\pi f$ and $(2\pi f)^2$ times the maximum displacement response, where f is frequency expressed in hertz. The corresponding response curves are called *displacement, velocity*, or *acceleration shock spectra*. A more detailed discussion of shock spectra is given in Chap. 23.

MODIFICATION OF CHARACTERISTICS BY REACTIONS OF TEST ITEM

The shock motion produced by a shock machine may depend upon the mass and frequency characteristics of the item under test. However, if the effective weight of the item is small compared with the weight of the moving parts of the shock machine, its influence is relatively unimportant. Generally, however, the reaction of the test item on the shock machine is appreciable and it is not possible to specify the test in terms of the shock motions unless large tolerances are permissible. The test item acts like a dynamic vibration absorber (see Chap. 6) for fixed-base* natural frequencies of the test item (antiresonant frequencies of the shock-machine test-item system). If the item is relatively heavy, this causes the shock spectra of the exciting shock to have minima at these frequencies; it also causes its mounting foundation to have these minima during shock excitation at field installations. Shock tests and design factors are sometimes established on the basis of an envelope of the maximum values of shock spectra. However, maximum stresses in the test item will most probably occur at the antiresonant frequencies where the shock spectrum exhibits minimum values. To require that the item withstand the upper limit of spectra at these frequencies may result in overtesting and overdesign. Considerable judgment is therefore required both in the specification of shock tests and

---

* Fixed-base natural frequencies of a flexibly mounted item are those which would be calculated if the base on which the item is mounted were rigid and of infinite mass. They correspond to antiresonant frequencies of the system.

in the establishment of theoretical design factors on the basis of field measurements. See Chap. 42 for a more complete discussion of this subject.

DOMINANT FREQUENCIES OF SHOCK MACHINES

The shock motion produced by a shock machine may exhibit frequencies that are characteristic of the machine. The frequencies may be affected by the equipment under test. The probability that these particular frequencies will occur in the field is no greater than the probability of other frequencies in the general range of interest. A shock test, therefore, discriminates against equipment having elements whose natural frequencies coincide with frequencies introduced by the shock machine. This may cause failures to occur in relatively good equipment whereas other equipment, having different natural frequencies, may pass the test even though of poorer quality. Because of these factors, there is an increasing tendency to design shock machines to be as rigid as possible, so that their natural frequencies are above the range of frequencies that might be strongly excited in the equipment under test. The shock motion is then designed to be the simplest shape pulse that will give a desired shock motion or spectrum.

CALIBRATION

A *shock-machine calibration* is a determination of the shock motions, or spectra, generated by the machine under standard specified conditions of load, mounting arrangements, methods of measurement, and machine operation. The purpose of the calibration is not to present a complete study of the characteristics of the machine, but rather to present a sufficient measure of its performance to ensure the user that the machine is in a satisfactory condition. Measurements should therefore be made under a limited number of significant conditions that can be accurately specified and easily duplicated. Calibrations are usually performed with deadweight loads rigidly attached to the shock machine.

The statement of calibration results must include information relative to all factors that may affect the nature of the motion. These include the magnitude, dimensions, and type of load; the location and method of mounting of the load; factors related to the operation of the shock machine; the locations and mounting arrangements of pickups; and the frequency range over which the measurements extend.

SPECIFYING A SHOCK TEST

Two methods of specification are employed in defining a shock test: (1) a specification of the shock motions (or spectra) to which the item under test is subjected, and (2) a specification of the shock machine, the method of mounting the test item, and the procedure for operating the machine.[1]

The first method of specification can be used only when the shock motion can be defined in a reasonably simple manner, and when the application of forces is not so sudden as to excite structural vibration of significant amplitude in the shock machine. If equipment under test is relatively heavy, and if its normal modes of vibration are excited with significant amplitude, the shock motions are affected by the load; then the specified shock motions should be regarded as nominal. If comparable results are to be obtained for tests of different machines of the same type, the method of mounting and operational procedures must be the same.

The second method of specification for a shock test assumes that it is impractical to specify a shock motion because of its complexity; instead, the specification states that the shock test shall be performed in a given manner on a particular machine. The second method permits a machine to be developed and specified as a standard shock testing machine. Those who are responsible for the specification then should ensure that the shock machine generates appropriate shock motions. This method avoids a difficulty that arises in the first method when measurements show that the shock motions differ from those specified. These differences are to be expected if load reactions are appreciable and complex.

A shock testing machine must be capable of reproducing shock motions with good precision for purposes of comparative evaluation of equipment and for the determination as to whether a manufacturer has met contractual obligations. Moreover, different machines of the same type must be able to provide shocks of equivalent damage potential to the same types of equipment under test. Precision in machine performance, therefore, is required on the basis of contractual obligations and for the comparative evaluation of equipments even though it is not justified on the basis of knowledge of field conditions.

Sometimes equipment under test may consistently fail to meet specification requirements on one shock machine, but may be acceptable when tested on a different shock machine of the same type. The reason for this is that small changes of natural frequencies and of internal damping, of either the equipment or the shock machine, may cause large changes in the likelihood of failure of the item. Results of this kind do not necessarily mean that a test has been performed on a faulty machine; normal variations of natural frequencies and internal damping from machine to machine make such changes possible. However, standard calibrations of shock machines should be made from time to time to ensure that significant changes in the machines have not occurred.

SHOCK TESTING MACHINES[2]

CHARACTERISTIC TYPES OF SHOCKS

The shock machines described below are grouped according to types of shocks they produce. When a machine can be classified under several headings, it is placed in the one for which it is primarily intended. The types of shock, shown in Fig. 26.1, are classified as: (A), velocity shock; (B) through (D), simple shock pulses; (E), single complex shock; and (F), multiple shock. Shocks not included in this classification are grouped under a miscellaneous category.

These types of shock are defined as follows: A *velocity shock* consists of a sudden change of velocity. If the change of velocity takes place in a time that is short compared with principal natural periods of vibration of the item subjected to the shock, it is considered to be a sudden change of velocity. For a velocity shock, the magnitude of the acceleration involved in the velocity change is unimportant as the item is isolated from the maximum acceleration by the flexibility of its parts. The magnitude of the velocity change defines the intensity of the shock. A *simple shock pulse* is said to exist if the shape of the shock motion-time, or force-time, curve is known and can be expressed in practical mathematical form. Usual shapes include the half-sine wave acceleration pulse, the rectangular force pulse, and other simple-shaped motions or force functions. A *single complex shock* is any transient motion, or force function, caused by a single primary excitation, that cannot be expressed practically in mathematical terms. A *multiple shock* consists of a

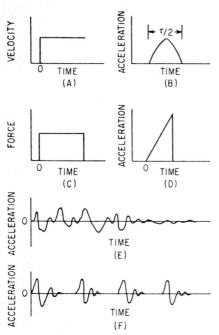

FIG. 26.1. Characteristic types of shocks. (A) Velocity shock, or step velocity change. (B) Simple half-sine acceleration shock pulse. (C) Rectangular force pulse. (D) Sawtooth acceleration pulse. (E) Single complex shock. (F) Multiple shock.

number of individual shocks repeated over a relatively long time interval. Shocks which do not fall into the above categories are labeled "miscellaneous."

VELOCITY SHOCK

DROP TESTS. A *drop tester* may be either of two types. It may consist of a hammer assembly that drops upon a test item in order that the resilience of the item (usually a cushioning material) can be determined; or it may be a device by means of which an item under test (a package or a packaged item) can be dropped in a prescribed manner onto a suitable surface.

DROP TESTS FOR SHIPPING CONTAINERS. Drop testers, in contrast to most shock machines, can be made in a variety of ways without appreciably altering the results of tests performed on them. The American Society for Testing Materials has stated, in part, that the shipping container drop tester shall be any suitable apparatus that conforms to the following requirements:[3]

1. Permits accurate prepositioning of the container to ensure a free-fall and impact at the exact places and in the direction desired.
2. Permits accurate and convenient control of the height of drop.
3. Utilizes lifting that will not damage the containers.
4. Permits a free unobstructed fall.
5. Provides for variations in height of drops within limits of anticipated requirements.
6. Provides a solid surface of concrete, stone, or steel of sufficient mass to absorb all shock without appreciable deflection.

DROP TESTS FOR AIR-DROP DELIVERY. The military services sometimes require that supplies be dropped from airplanes. Parachutes are used only if required for equipment survival and if maximum tolerable impact velocities are permitted. A simulation of these conditions requires impact velocities up to several hundred feet per second. In order to obtain such velocities within a relatively small distance, the item must be accelerated by a force in addition to that caused by gravity (this type of tester is considered in a later section), or very large drop heights must be employed.

A drop tester is shown in Fig. 26.2*A* where a drop height up to 275 ft (83.8 m) can be obtained.[4] The installation illustrated is used primarily to test package cushioning materials. Weights, guided by cables, are dropped onto the cushioning material, and measurements of the force-deflection curve are made by means of the arrangement shown in Fig. 26.2*B*. This installation also permits an inclined drop so that a horizontal component of velocity is obtainable.

INCLINED PLANE TEST. Some of the shocks received by packages because of careless handling, or encountered in railroad trains during switching, are simulated by the inclined plane test. One such test has been standardized as an *incline impact test for shipping containers*.[5] This is also referred to as a *Conbur test*. The testing machine employed for this test, shown in Fig. 26.3, consists of a rolling carriage and an inclined track which is terminated by a barrier at its lower end. The item under test is placed on the carriage so that it will strike the barrier when the carriage is rolled down the incline. This testing machine has been standardized so that different laboratories can obtain comparable results and so that the test may be cited in specifications. The standard provides detailed instructions relating to the construction, calibration, and use of this testing machine.

MACHINES FOR PRODUCING A SIMPLE SHOCK PULSE

Although shocks encountered in the field are usually complex in nature (for example, see Fig. 26.1*E*), it is frequently advantageous to simulate a field shock by a shock of mathematically simple form. This permits designers to calculate equipment response more easily and allows tests to be performed that can check these calculations. This technique is additionally justifiable if the pulses are shaped so as to provide shock spectra similar to those obtained for a suitable average of a given type of field condition. Machines are therefore built to provide these simple shock motions. However, note that the motions provided by actual machines are only ideally simple. The ideal outputs may be given as nominal values; the actual outputs can only be determined by measurement.

DROP TABLES. A great variety of drop testers are used to obtain acceleration pulses having magnitudes ranging from $80,000g$ down to a few g.[2] The machines each include a carriage (or table) on which the item under test is mounted; the carriage can be hoisted up to some required height and dropped onto an anvil. Guides are provided to keep the carriage properly oriented.

FIG. 26.2. (*A*) 275-ft (83.8-m) drop-test facility. (*B*) Force plate for force-displacement measurements. (*After Matlock, Ripperger, Turnbow, and Thompson.*[4])

When large velocity changes are required, the carriage may be accelerated downward by a means other than gravity. Frequently, parts of the carriage, associated with its lifting and guiding mechanism, are flexibly mounted to the rigid part of the carriage structure that receives the impact. This is to isolate the main carriage structure from its flexible appendages so as to retain the simple pulse structure of the stopping acceleration.

FIG. 26.3. Conbur tester; apparatus for incline-plane impact test. The carriage, which is 42 in. (107 cm) long and 36 in. (91 cm) wide, has a maximum run of 116 in. (3 m) and a maximum vertical drop of 20 in. (51 cm). The test item is placed on the carriage so that it strikes a 55- by 58½-in. (140- by 149-cm) barrier, consisting of 1⅝-in. (4.1-cm) planking when the carriage is rolled down the incline. *(L.A.B. Corp.)*

FIG. 26.4. Residual and over-all shock spectra of the half-sine acceleration pulse shown in the inset. The over-all spectrum is the usual shock spectrum, i.e., the maximum response regardless of when it occurs. S_a, S_v, and S_d are, respectively, acceleration, velocity, and displacement shock spectra expressed in units of in./sec², in./sec, and in. G is the acceleration expressed in units of gravity g. T is the pulse duration; f is frequency. If the pulse length is 0.006 seconds and the amplitude is $200g$ for a frequency of 100 Hz ($Tf = 0.6$), the over-all shock spectral values of S_a, S_v, and S_d are $340g$, 200 in./sec, and 0.32 in., respectively. *(Naval Res. Lab. Rept.)*

A plastic material, placed upon the point of impact of the anvil, can be used to control the acceleration vs. time characteristic and the shape of the acceleration pulse. These plastic pellets are usually made of lead, but a variety of materials and shapes generated by the impact of a drop on the pellet have been used. When rubberlike materials are used, the acceleration pulse can be made to approach the shape of a half-sine curve, Fig. 26.1B, which has the shock spectrum of Fig. 26.4.

A sawtooth acceleration pulse, Fig. 26.1D, has an acceleration shock spectrum that is relatively smooth with no extreme values which would make it discriminate for or against an item under test because of excessive test severity at certain frequencies.[6] Theoretically and experimentally obtained shock spectra for such a pulse are shown in Fig. 26.5A and B. Figure 26.5C illustrates the time-history of acceleration of a point on a beam when subjected to the sawtooth pulse shown on the same time scale. This type of pulse can be obtained relatively easily by dropping a rigid carriage onto a cylindrical lead pellet with a conical top. This form of excitation provides a satisfactory test for many types of shock environments where it is required that the shock spectrum rise to a maximum value within the first 100 Hz and remain constant thereafter. The amplitude and duration of the pulse may be modified by changing the dimensions of the pellet, so as to provide an appropriate response level and rise time for the spectral curves.

A typical machine of this type is shown in Fig. 26.6. The carriage consists of a rigid platform on which the specimen is mounted and which embodies the impacting surface on its lower side. The platform is usually cast of aluminum and designed to have no natural frequencies below about 1,000 Hz. The impacting surfaces are of hardened steel.

MISCELLANEOUS DROP TABLES. A great variety of drop-table machines are in existence.[2] One such machine, a 100-ft portable drop tester, is shown in Fig. 26.7A. While the impact velocity of the carriage corresponds to that attained by a free-fall of 100 ft (30.5 m), the height of the machine is only about 11 ft (3.4 m). The carriage is accelerated downwards by rubber shock cords and is arrested by a sand-filled rubber bag. Figure

Fig. 26.5. (A) Theoretical acceleration shock spectra for the sawtooth acceleration pulse shown in the inset. The responses during and after the pulse interval are shown separately. *(After Morrow and Sargeant.*[6]*)* (B) Experimental shock-response spectrum for a 100g, 6-millisecond, sawtooth shock pulse. (C) Response of a beam of 400-Hz natural frequency to the inset sawtooth shock pulse. *(After Jensen.*[7]*)*

26.7*B* indicates the complete acceleration-time curve. Bags of different shapes and stiffnesses permit the shape of the stopping pulse to be varied as indicated.

AIR GUNS. Air guns frequently are used to impart large accelerations to pistons on which items under test can be attached. The piston is mechanically retained in position near the breech end of the gun while air pressure is built up within the breech. A quick-release mechanism suddenly releases the piston, and the air pressure projects the piston down the gun barrel. The muzzle end of the gun is closed so that the piston is stopped by compressing the air in the muzzle end. Air bleeder holes may be placed in the gun barrel to absorb energy and to prevent an excessive number of oscillations of the piston between its two ends.

TEST
OBJECT

CARRIAGE

IMPACTING
SURFACES

FIG. 26.6. Drop table arrangement capable of being used with plastic pellets on the impacting surfaces so as to produce a pulse of desired shape. *(Barry Wright Corporation.)*

A variety of such guns, used by the Naval Surface Weapons Center, can provide acceleration pulses as shown in Fig. 26.8*A* and *B*. The peak accelerations may extend from a maximum of about 1,000*g* for the large-diameter (21 in, 53 cm) guns up to 200,000*g* for small-diameter (2 in, 5 cm) guns. The pulse length varies correspondingly from about 50 to 3 milliseconds. The maximum piston velocity varies from about 400 to 750 ft/sec (122 to 229 m/sec). The maximum velocities are not dependent upon piston diameter.

High-acceleration gas guns have been developed for testing electronic devices. The items under test are attached to the piston. The gun consists of a barrel (cylinder) that is closed at the muzzle end, but which has large openings to the atmosphere a short distance from the muzzle end. The piston is held in place while a relatively low pressure gas (usually air or nitrogen) is applied at the breech end of the gun. The piston is then released, whereby it is accelerated over a relatively long distance until it reaches the position along the length of the cylinder that is open to the atmosphere. This initial acceleration is of relatively small magnitude. After the piston has passed these openings, it is stopped by the compression of gas in the short closed end of the cylinder. This results in a reverse acceleration of relatively large magnitude. (Sometimes an inert gas, such as nitrogen, is used in the closed end to prevent explosions which might be caused by oil particles igniting under the high temperatures incident to the compression.) Thus, in contrast to the previously described devices, the major acceleration pulse is delivered during stopping rather than starting. An advantage of this latter technique is that the difficult problem of constructing a quick-release mechanism for the piston, which will work satisfactorily under the large forces exerted by the piston, is greatly simplified.

Fig. 26.7. (*A*) 100-ft (30-m) drop tester. (*B*) Acceleration-time curve. *(After U.S. Naval Ordnance Laboratory Report.)*

Fig. 26.8. Typical acceleration-time curves for (*A*) 5-in. (13-cm) air gun; (*B*) 21-in. (53-cm) air gun. Peak accelerations extend to 5,000*g*, with pulse lengths between 3 and 6 milliseconds for the 5-in. gun. The accelerations for the 21-in. gun extend to about 1,000*g*. *(After U.S. Naval Ordnance Laboratory Report.)*

CONSIDERATIONS RELATIVE TO SIMPLE SHOCK-PULSE MACHINES

In the above description of the output of shock machines designed to deliver simple shock pulses of adjustable shapes, it is assumed that the load imposed on the machine by the item under test has little effect upon the shock motions. This is true only when the effective weight of the load is negligibly small compared with that of the shock-machine mounting platform. If the effective weight of the load is independent of frequency, i.e., if it behaves as a rigid body, it is simple to compensate for the effect of the load by adjusting machine parameters. However, when the load is flexible and the reactions of excited vibrations are appreciable, the motions of the shock-machine platform are complex. Specifications involving the use of these types of machines should require that the mounting platform have no significant natural frequencies below a specified frequency. The weight of this platform together with that of all rigidly attached elements, exclusive of the test load, also should be specified. Pulse shapes then may be specified for motions of this platform or for the platform together with given deadweight loads. These may be specified as nominal values for test loads, but it is neither practical nor desirable to require that the pulse shape be maintained in simple form for complex loads of considerable mass.

SHOCK MACHINES FOR SINGLE COMPLEX SHOCK

Because of the infinite variety of shock motions possible under field conditions, it is not practical or desirable to construct a shock machine to reproduce a particular shock that may be encountered in the field. However, it is sometimes desirable to simulate some average of a given type of shock motion. To accomplish this may require that the shock machine deliver a complex motion. A shock of this type cannot be specified easily in terms of the shock motions, as the motions are very complex and dependent on the nature and mounting of the load. It is customary, therefore, to specify a test in terms of a shock machine, the conditions for its operation, and a method of mounting the item under test.

NAVY HIGH-IMPACT SHOCK MACHINE FOR LIGHTWEIGHT EQUIPMENT.[1,8,10,11] The Navy high-impact shock machine for lightweight equipment is designed to subject equipment to shocks of the nature and intensity that might occur on board a ship that is exposed to severe, but sublethal, noncontact underwater explosions. Such severe shocks produce motions that extend throughout the ship. This machine is used for equipment secured or mounted to bulkheads or decks and is shown in Fig. 26.9. The anvil A is struck on the back side by the pendulum hammer C, or the anvil is rotated 90° on a vertical axis and struck on the end by the pendulum hammer. The drop hammer B can be made to strike the top of the anvil, thus providing principal shock motions in the third orthogonal direction. Acceleration, velocity, and displacement measured at the center of an unloaded mounting plate (4A plate) are plotted in Fig. 26.10A. The weights of equipment tested on this machine are limited to about 350 lb (159 kg).

It can be seen that acceleration magnitudes extend to about 1,000g, velocities to 20 ft/sec (6.1 m/sec), and displacements to 1.7 in. (4.3 cm). Shock spectra of shock motions generated by this machine at two different locations on the mounting plate are shown in Fig. 26.10B. The spectrum for motion at the edge of the plate is typical of that for a relatively rigid structure—a straight line having a slope proportional to the velocity change resulting from the impact. The spectrum for the motion at the center of the plate illustrates the amplification of the spectrum level at a natural frequency of the plate (about 100 Hz) and some attenuation at higher frequencies.

NAVY HIGH-IMPACT SHOCK MACHINE FOR MEDIUM-WEIGHT EQUIPMENT.[1,10,11] This machine is used to test equipment that, together with its supporting structures, weighs up to 7,400 lb (3,357 kg). The machine, shown in Fig. 26.11, consists principally of a 3,000-lb (1,361-kg) hammer and a 4,500-lb (2,041-kg) anvil. Loads are not attached directly to the rigid anvil structure but rather to a group of steel channel beams which are supported at their ends by steel members, which in turn are attached to the anvil table. The number of channels employed is dependent upon the weight of the load and is such as to cause the natural frequency of the load on these channels to be about 60 Hz. The hammer can be dropped from a maximum effective height of 5.5 ft (1.68 m). It swings around on its axle so as to strike the anvil on the bottom, giving it an upward velocity. The anvil is permitted to travel a distance of up to 3 in. (7.6 cm) before being stopped by a ring of retaining bolts. The machine is secured to a large block of concrete which is mounted on springs to isolate the surrounding area from shock motions.

FIG. 26.9. Navy high-impact shock machine for lightweight equipment. (A) Anvil plate. The anvil plate can also be oriented in the plane of the paper for an "end" hammer blow. (B) Hammer for vertical blow. (C) Hammer for horizontal blow. (D) Restoring springs for vertical blow. (E) Rail support for end blow. An upper support angle, not shown, is also required. (F) Positioning springs for end blow.

The general nature of the shock is similar to that of the lightweight machine. Curves illustrating some shock motions generated by the machine are shown in Fig. 26.12. Figure 26.12A indicates the complex nature of the shock motions generated by this machine. Figure 26.12B shows that much of the high-amplitude high-frequency components of the shock motions is not transmitted to the load. Figure 26.12C and D illustrate, respectively, the velocity attained by the load and a shock spectrum of the motions near the base of a load. A complete study of the performance of this machine is given in Ref. 11.

NAVY FLOATING SHOCK PLATFORM (FSP) AND LARGE FLOATING SHOCK PLATFORM (LFSP).   The FSP and LFSP are also shock machines of the Navy high-impact class. They are rectangular barges fitted with semicylindrical canopies within

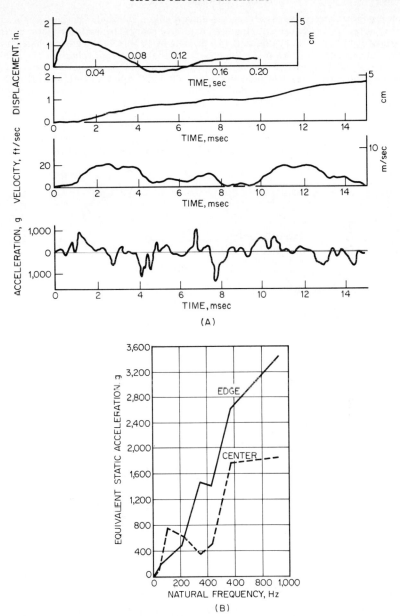

FIG. 26.10. (*A*) Shock motions of the center of an unloaded 4A mounting plate of the shock machine for lightweight equipment for a 5-ft back blow. All curves represent the same motion. (*After Oliver.[9]*) (*B*) Shock spectra for lightweight high-impact shock machine. Dotted line, center of 4A plate. Solid line, edge of 4A plate over channel. Five-foot back blow. (*After Walsh and Blake.[12]*)

which test items are installed as they are aboard ship. The shock motions comprising the test series are generated by detonating explosive charges beneath the water surface at various distances. The FSP is 28 ft (8.5 m) long by 16 ft (4.9 m) wide and has a maximum load capacity of 60,000 lb (27,216 kg). Its available internal volume is about 26 ft (7.9 m) by 14 ft (4.3 m) by 15 ft (4.6 m) high to the center of the canopy. The charges for the successive shots of the test sequence are all 60 lb (27 kg) at a depth of 24 ft (7.3 m) below the surface. The charge standoff, the horizontal distance from the near side of the FSP, is shortened for each shot to a final value of 20 ft (6.1 m).

The LFSP is 50 ft (15.2 m) long by 30 ft (9.1 m) wide with a maximum load capacity of 400,000 lb (181,440 kg) and an internal volume of about 48 ft (14.6 m) by 28 ft (8.5 m) by 34 ft (10.4 m) high to the center of the canopy. The charge size is 300 lb (136.1 kg), and the charge depth is 20 ft (6.1 m); the standoff is decreased for each shot to a final value

FIG. 26.11. High-impact shock machine for medium-weight equipment.

of 50 ft (15.2 m). Figures 26.13 and 26.14 show typical velocity-time curves and the variation of peak velocity and shock spectrum value with standoff for the FSP and LFSP, respectively. The FSP and LFSP are described in detail in Refs. 11 and 13.

SHOCK MACHINES FOR MULTIPLE IMPACT SHOCK

Many environments, particularly those involving transportation, subject equipment to a relatively large number of shocks. These are of lesser severity than the shocks of major intensity that have been considered above, but their cumulative effect can be just as damaging. It has been observed that components of an equipment that are damaged as a result of a large number of shocks of relatively low intensity are usually different from those that are damaged as a result of a few shocks of relatively high intensity. The damage effects of a large number of shocks of low intensity cannot generally be produced by a small number of shocks of high intensity. Separate tests are therefore required so that the multiple number of low-intensity shocks are properly emulated.

PACKAGE TESTER.[2] Packages that are not securely fastened may bounce while being transported by a vehicle. To simulate this mistreatment, a table is caused to vibrate in a vertical plane at a low frequency and at an acceleration amplitude slightly greater than 1g. The package then loses contact with the table at each cycle, subjecting the package to repeated impacts as it strikes the table. One of the commercially available package testers has been designed for various load ranges and for operation at 1.25g acceleration amplitude (0.5 in, 1.27 cm displacement amplitude) at about 5 Hz. The table is of steel construction with a wood top which is bordered by a fence in order to prevent the package from falling off.

REVOLVING-DRUM TUMBLING TEST.[2,14] A steel drum, having a cross-sectional shape of a regular hexagon and having circular end flanges so that the drum can be

FIG. 26.12. Medium-weight shock machine. (*A*) A typical acceleration curve (400-lb, 182-kg load, 3-ft hammer drop, 5,000-Hz low-pass filter). (*B*) Maximum accelerations of the anvil and of a deadweight load of 1,115 lb (506 kg) mounted on the anvil according to specifications; 1,000- and 300-Hz low-pass filters were used. (*C*) Maximum velocities attained by different deadweight loads. (*D*) A typical shock spectrum delivered by the shock machine for a 1-ft hammer drop and 500-lb (227-kg) load.

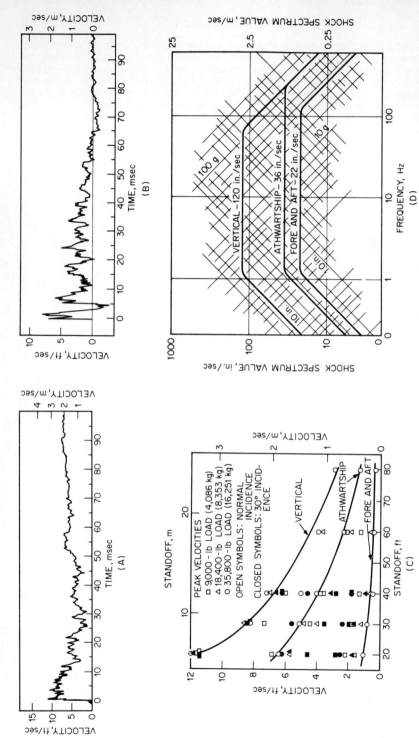

Fig. 26.13. (A) Typical vertical and (B) athwartship velocity-time curves from the FSP for an s shot at a 20 ft (6 m) standoff and 20 ft (6 m) depth. (C) Variation of peak velocity with standoff for 20 ft (6 m) charge depth. (D) Design shock spectra for the FSP. The lower cutoff frequency is 1.15 Hz for all directions. The upper cutoff frequencies are: vertical—67 Hz, athwartship—220 Hz, fore-and-aft—125 Hz.[11]

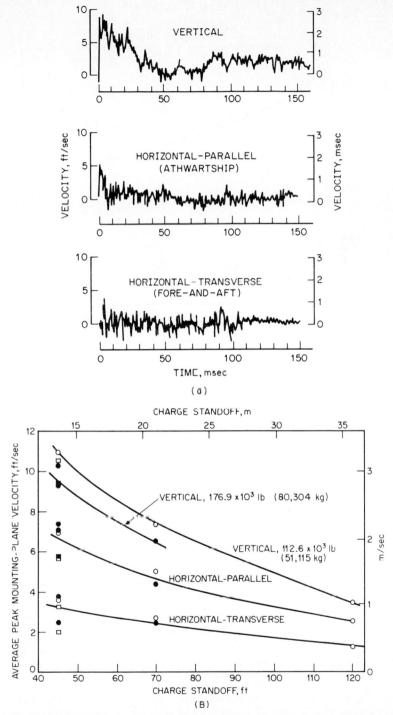

FIG. 26.14. (A) Typical velocity-time curves from the LFSP for a shot at 45 ft (14 m) and 20 ft (6 m) depth. (B) Variation of peak velocity with standoff for 20 ft (6 m) charge depth. (C) Variation of V_0, the vertical shock spectrum value, with standoff. (D) Design shock spectra for the LFSP.[13]

26–17

FIG. 26.14. (*Continued*)

rotated about its axis, is shown in Fig. 26.15. Baffles are located on the inside faces of the drum so that a container under test is carried around with the drum as it rotates, until a position is attained such that the container rolls over the baffle onto the area

FIG. 26.15. Revolving steel drum for testing of shipping containers.

below. The construction and operation of these drums have been standardized. Drums having diameters of 7 or 14 ft (2.1 or 4.3 m) are used; the former is rotated at 1⅞ rpm and the latter at 1 rpm. Tests generally are continued until a failure develops in the container. The relative qualities of containers are determined by their relative survival times.

BUMP TEST MACHINE. A machine has been designed in England which subjects equipment to repeated bumps at the rate of several bumps per second. Although the test conditions achieved by these machines are not necessarily intended to simulate the field conditions, experience has shown that items surviving the test conditions are satisfactory for transportation environments.

These machines are designed for loads up to 51 lb (23.1 kg) and up to 250 lb (113.4 kg). Each machine includes a table on which equipment is mounted. The table is raised by a cam and allowed to fall freely onto a rubber anvil; the drop height of the table and the hardness of the anvil are selected to produce a specified maximum acceleration and waveform. The drop height is adjustable between about ¼ in. (0.64 cm) and about 1¼ in. (3.18 cm); it is normally adjusted so that the maximum acceleration is about $40g$ (determined by acceleration measurements). The waveform of the acceleration pulse is approximately that of a half-period sine wave having a pulse duration on the order of 5 to 6 milliseconds.

A similar set of machines is described by the International Electrotechnical Commission.[15] The principal difference is that the impacting members for these latter machines are made of wood rather than rubber. The machines are calibrated by means of a Brinell-type test. The acceleration causes an iron cylinder, weighing 1 kg, to press a steel ball (diameter 3.97 mm) onto a 3-mm thick aluminum plate. The diameter of the indentation is determined. The ball is then statically pressed onto another part of the plate so as to give the same diameter impression. The nominal acceleration, in g, is equal to the static load in kilograms that gives the same diameter impression as the bump. The specified accelerations are given in terms of these nominal values. It is, perhaps, unfortunate that the measure of shock intensity is given in terms of acceleration; it would seem that the specification could just as well be in terms of impression diameter and thus prevent confusion with accelerations which might be measured by more accurate means.

ELECTRODYNAMIC BUMP TESTER. An electrodynamic vibration testing machine of the type described in Chap. 25 can provide multiple low-level shocks. Such a machine can be excited periodically by a condenser discharge or by electric current pulses of desired shapes. Maximum accelerations up to about $100g$ are obtainable for test items that are light compared with the armature of the machine.

MISCELLANEOUS SHOCK MACHINES

QUICK-STARTING CENTRIFUGE OR ROTARY ACCELERATOR.[2] A quick-starting centrifuge is used to quickly attain and to maintain an acceleration for a long period of time. The accelerator consists of a rotating arm which is suddenly set into motion by an air-operated piston assembly. The test object is mounted on a table attached to the outer end of the arm. The table swings on a pivot so that the resultant direction of acceleration is always along a fixed table axis. Initially the resultant acceleration is caused largely by angular acceleration of the arm so that this axis is in a circumferential direction. As the centrifuge attains its full speed, the acceleration is caused primarily by centrifugal forces so that this table axis assumes a radial direction. These machines are built in several sizes. They require between 5 and 60 milliseconds to reach the maximum value of acceleration. For small test items (8 lb, 3.6 kg), a maximum acceleration of $450g$ is attainable; for heavy test items (100 lb, 45.4 kg), the maximum value is $40g$.

WOX-7.[16] The WOX-7 is a device for shock testing weapons such as torpedoes or missiles. The test item is gripped in a frame structure and struck radially by a long cylindrical ram. The ram, which pivots about an axis above and parallel to that of the test item, is raised to some angle away from its rest position, released, and allowed to fall against a suitable impact pad. The drop angle and characteristics of the impact pad are adjusted to provide the desired motion time-history or shock spectrum for the shock experienced by the test item. A beam linkage arrangement provides the capability for axial shock also. The machine can accommodate test items up to 22 ft (6.7 m) long, weighing up to 5,000 lb (2,268 kg). The maximum peak velocity attainable is 18 ft/sec (5.5 m/sec) radial and 10 ft/sec (3.0 m/sec) axial.

OBSOLETE SHOCK MACHINES

Other shock machines which may be encountered are the JAN machine (or spring drop table), the medium-impact shock machine (or sand drop table), and the Hyge shock exciter. These machines were widely used in prior years but have now been superseded by other test machines, principally the various drop testers. They are described in Refs. 17, 18, and 19, respectively.

REFERENCES

1. Kennard, D. D., and I. Vigness: "Shock and Vibration Instrumentation," an ASME Publication, June 1956, p. 127.
2. Shock and Vibration Information Center, U.S. Naval Research Laboratory: "Index of Environmental Test Equipment in Government Establishments," November 1967.
3. ASTM standards, 1955, part 7, p. 1087.
4. Matlock, H., and E. A. Ripperger, J. Turnbow, and J. N. Thompson: *Shock and Vibration Bull.*, **25**, part 2, p. 144, December, 1957, Office of Secretary of Defense Research and Engineering.
5. ASTM standards, 1955, part 7, p. 1101.
6. Morrow, C. T., and I. Sargeant: *J. Acoust. Soc. Amer.*, **28**:959 (1956).
7. Jensen, G. A.: *Shock and Vibration Bull.* **25**, part 2, 137 (December 1957). Office of Secretary of Defense Research and Engineering, Washington 25, D.C.
8. The American National Standards Institute: "Specification for the Design, Construction and Operation of Class HI (High Impact) Shock-Testing Machines for Lightweight Equipment," S2.15-1972.
9. Oliver, R.: *Shock and Vibration Bull.* 3, *U.S. Naval Research Lab. Rept.* S-3106, May 1947, p. 10.
10. Military specification. "Shock Tests, HI (High-Impact); Shipboard Machinery, Equipment, and Systems, Requirements for," MIL-S-901C (Navy), January 15, 1963.
11. Clements, E. W.: "Shipboard Shock and Navy Devices for its Simulation," *U.S. Naval Res. Lab. Rept.* 7396, July 1972.
12. Walsh, J. P., and R. E. Blake: *Proc. Soc. Exp. Stress Anal.*, **6**(2):150 (1948).
13. Clements, E. W.: "Characteristics of the Navy Large Floating Shock Platform," *Naval Res. Lab. Rept.* 7761, July 15, 1974.
14. ASTM standards: "Standard Method of Test for Shipping Containers in Revolving Hexagonal Drum," 1955, part 7, p. 1907.
15. "Basic Climatic and Mechanical Robustness Testing Procedure for Components," Central Office of the International Electrotechnical Commission, Geneva, Switzerland, 1954.
16. Walsh, John J., Jr., and William W. Wassmann: "Environment Simulation," *Naval Surf. Weap. Ctr. Rept.* NSWC MP 81-55, 1981, p. 17.
17. "Shock-Testing Mechanism for Electrical Indicating Instruments," *ASA Publ.* C39.3-1948.
18. "American Standard Specification for Design, Construction, and Operation of Variable Duration, Medium-impact Shock-testing Machine for Lightweight Equipment," *ASA Publ.* S2.1-1961.
19. The Hyge Shock Tester, *Bull,* 4-70, Consolidated Electrodynamics Corp., Rochester, N.Y., February 1957.

27

APPLICATIONS OF DIGITAL COMPUTERS

Allen J. Curtis

Hughes Aircraft Company

INTRODUCTION

This chapter identifies and describes in general terms several basic categories of applications of digital computers to shock and vibration problems. Such categories are readily identifiable as (1) the use of computers to carry out numerical analyses of dynamic systems which are too complex and/or time-consuming to perform by hand calculation, and (2) the use of computers to perform other than analytical tasks which were performed previously by analog equipment or perhaps were not practical to perform. These categories tend to distinguish between analytical and experimental tasks. The major emphasis of this chapter is directed toward experimental applications.

FOR WHAT PROBLEMS SHOULD DIGITAL COMPUTATION BE USED?

The decision to employ a digital computer in the solution of a shock or vibration problem should be made with considerable care. Before a program of digital computation is decided on, the following questions should be answered.

1. Is existing software available to perform the required task?

2. If not, to what extent must the task or the software be modified in order to perform the task?

3. If no applicable software exists, what is the extent of the software development needed?

4. What are the detailed assumptions inherent in the software program (e.g., linearity, proportional damping, etc.)?

5. Is the software able to compute and output the information required (e.g., absolute vs. relative motion, phase relationships, etc.)?

6. What are the detailed input and output limitations of the program (e.g., types of input excitations, graphic outputs, etc.)?

7. How much computer time will be used to perform the task?

After satisfactory answers are obtained to such questions, the user must realize that the results obtained from the computer output can be no better than the input to the computer. For example, the quality of the natural frequencies and mode shapes obtained from a structural analysis computer program does not depend on the computer program but on the degree to which the mathematical model employed is representative of the mass, stiffness, and damping properties of the physical structure. Likewise, a spectral analysis of a signal with poor signal-to-noise ratio will provide an accurate spectrum of the signal *plus noise*, not of the signal.

ANALYTICAL APPLICATIONS

The development of large-scale computers having a very short cycle time (i.e., the time required to perform a single operation such as adding two numbers) and a very large memory now permits detailed analyses of structural responses to shock and vibration excitation which are many orders of magnitude more complex than was previously practical. A number of software programs have been developed to perform these analyses. In this chapter, they are categorized as *general-purpose programs* and *special-purpose programs*. References 1 and 2 contain extensive anthologies of both general-purpose and special-purpose analytical programs.

GENERAL-PURPOSE PROGRAMS

Programs may be classed as "general purpose" if they are applicable to a wide range of structures and permit the user to select a number of options, such as damping (viscous or structural), various types of excitation (sinusoidal or random vibration, transients), etc. Almost all general-purpose programs are limited to the analysis of linear systems.

FINITE-ELEMENT PROGRAMS. The most numerous programs are classed as "finite-element" or "lumped-parameter" programs.

In lumped-parameter programs, the structure to be analyzed is represented in a model as a number of point masses (or inertias) connected by massless, springlike elements. The points at which these elements are connected, and at which a mass may or may not be located, are the "nodes" of the system. Each node may have up to six degrees-of-freedom at the option of the analyst. The size of the model is determined by the sum of the degrees-of-freedom at the nodes and by the *dynamic* degrees-of-freedom (which is the number of degrees-of-freedom for which the mass or inertia is nonzero). The number of natural frequencies and modes which may be computed is equal to the number of dynamic degrees-of-freedom. However, the number of frequencies and modes which reliably represent the physical structure is generally only a fraction of the number which can be computed. Each program is limited in capacity to some combination of dynamic and zero mass degrees-of-freedom. The springlike elements are chosen to represent the stiffness of the physical structure between the selected nodes and generally may be represented by springs, beams, or plates of specified shapes. The material properties, geometric properties, and boundary conditions for each element are selected by the analyst.

In the more general finite-element programs, the springlike elements are not necessarily massless but may have distributed mass properties. In addition, lumped masses may be used at any of the nodes of the system.

The equations of motion of the finite-element model can be expressed in matrix form and solved by the methods described in Chap. 28. Regardless of the computational algorithms employed, the program computes the set of natural frequencies and orthogonal mode shapes of the finite-element system. These modes and frequencies are stored for future use in computing the response of the system to a specified excitation. For the latter computations, a damping factor must be specified. Depending on the program, this damping factor may have to be equal for all modes, or it may have a selected value for each mode.

The most sophisticated finite-element program, known as NASTRAN, was developed by the National Aeronautics and Space Administration (NASA).[3, 4, 5, 6] NASTRAN can be employed on most large-scale computer facilities. Many similar programs with fewer options and limited to fewer degrees-of-freedom are also available on commercial computing facilities. In addition, many large firms have developed their own programs.

Component Mode Synthesis.[7, 8, 9] The method of modeling described above leads to the prescribing of models with a very large number of degrees-of-freedom compared with the number of modes and frequencies actually of interest. Not only is this expensive, but it rapidly exceeds the capacity of many programs. To overcome these problems, component mode synthesis techniques have been developed. Instead of the modeling of an entire physical system, several models are developed, each representing a distinct identifiable region of the total structure and within the capacity of the computer program. The modes and frequencies of interest in each of these models are computed independently. A model of the entire structure is then obtained by joining these several models, using the component mode synthesis technique. This model retains the essential features of each substructure model and thus of the entire structure while using a greatly reduced number of degrees-of-freedom to do so.

Reduction of Model Complexity. Companion techniques, developed to reduce the cost of analysis and to permit the joining of several substructure models, are known as "reduction" methods, of which the Guyan reduction method[10, 11] is an example. Again, the objective is to reduce the mass and stiffness matrices to the minimum size consistent with retaining the modes and frequencies of interest together with other dynamic characteristics such as base impedance. It should be noted, however, that the Guyan reduction method yields a mass matrix which is nondiagonal (see Chap. 28) and which may be unacceptable for some computer programs. It is also of interest that the rigid-body mass properties, i.e., total masses and inertias of the structure, are not identifiable in the reduced mass matrix.

DISTRIBUTED (CONTINUOUS) SYSTEM PROGRAMS.[1, 2, 4] While still classified as general-purpose programs, a number of more specialized programs treating the analysis of distributed or continuous structural systems have been developed. These programs deal with beams, plates, shells, rings, etc., that is, standard structural members. They are general purpose in the sense that each program can be applied for a broad, selectable range of physical properties and dimensions of the particular structural shape. Not all programs will employ the same theory of elasticity; the user must thus examine the theoretical basis on which the program was developed. For example, the user must determine if the program includes such effects as rotary inertia or shear deformation.

SPECIAL-PURPOSE PROGRAMS[1, 2, 4]

The need for a special-purpose program may arise in several ways. First, for an engineering activity engaged in the design, on a repetitive basis, of what amounts analytically to the same structure, it may be economical to develop an analysis program which efficiently analyzes that particular structure. The analysis of vibration isolator systems or automobile suspension systems may be an example. Similarly, parametric studies of a particular structure, either to gain understanding or to optimize the design, may require a sufficient number of computer runs to justify the software development. A second type of special-purpose program includes programs which in some way perform an unusual type of analysis—for example, that of nonlinear systems.

Access to existing special purpose programs is generally more restricted than is access to general-purpose programs, in part because of their frequent proprietary nature and in part because of the investment in software required to develop them.

PREPROCESSORS AND POSTPROCESSORS

Experience with the general purpose analysis programs described above indicates two major shortcomings: (1) A large amount of expensive computer time is required to debug the models, i.e., to get them to run. (2) The large mass of tabulations and/or plots of the results of the analysis are very difficult to evaluate. To alleviate these problems, programs have been written, called *preprocessors* and *postprocessors*, which use sophisticated interactive graphics in combination with algorithms. Such programs greatly simplify the construction of the models, check the models, and condense the results of the analysis. These highly efficient programs often can be run very economically offline, independent of the larger computer required to exercise the model. Many organizations have developed their own preprocessors tailored to their product lines; commercially available processor software packages also are available.

EXPERIMENTAL APPLICATIONS

The classification "experimental applications" covers uses of computers which involve, in some way, the processing of shock and vibration information originally obtained during test or field operation of physical equipment. When this handbook was first published in 1961, experimental applications of digital computers were essentially non-existent. Two developments were crucial to the enabling of many of the applications described in later sections: (1) the recognition of the computational efficiency of the fast Fourier transform [FFT] algorithm,[12] and (2) the development of hardware FFT processors coupled with the development of minicomputers.

These developments now permit use of digital computers for such tasks as vibration data analysis, shock data analysis, and shock, vibration, and modal testing—all described in later sections. The information resulting from such applications is in digital form, which permits more sophisticated engineering evaluation of the information through further efficient digital processing, e.g., regression analysis, averaging, etc.

FUNDAMENTAL MATHEMATICAL RELATIONSHIPS[13, 14, 15]

A number of fundamental mathematical relationships between functions in the time and/or frequency domains are common to most experimental applications of digital computers to shock and vibration problems. These are summarized below. For clarity, the relationships are generally presented in terms of *continuous* functions though the reader must bear in mind that digital processing dictates that the functions can be computed only at a *discrete number* of frequencies or instants of time.

FOURIER TRANSFORM. The basic relationships used to transform information from the time domain to the frequency domain, or vice versa, are the Fourier transform pair (see Chap. 23).

$$S_x(f) = \int_{-\infty}^{\infty} x(t)e^{-i2\pi ft}\, dt \tag{27.1}$$

$$x(t) = \int_{-\infty}^{\infty} S_x(f)e^{i2\pi ft}\, df \tag{27.2}$$

where $x(t)$ is the time-history of a function x, and $S_x(f)$ is the Fourier transform of x, a complex quantity.

DISCRETE FOURIER TRANSFORM. In practical digital applications (also see *FFT Analyzers*, Chap. 13), $x(t)$ is known only for a finite time interval or record length T, at a total of N equally spaced time intervals Δt apart. The discrete Fourier transform (DFT) pair, equivalent to Eqs. (27.1) and (27.2), become

$$S_x(m\,\Delta f) = \Delta t \sum_{i=0}^{N-1} x(n\,\Delta t)e^{-i2\pi m\,\Delta f\, n\,\Delta t} \qquad m = 0, \ldots, \left(\frac{N}{2}\right) \tag{27.3}$$

$$x(n\,\Delta t) = \Delta f \sum_{m=0}^{N/2} S_x(m\,\Delta f)e^{i2\pi m\,\Delta f\, n\,\Delta t} \qquad n = 0, \ldots, (N-1) \tag{27.4}$$

The DFT Eq. (27.3) is correct only if $x(t)$ is periodic with period T. The set of S_x are the coefficients of the Fourier series expansion of $x(t)$ [see Eq. (22.8)]. If $x(t)$ is not periodic, the DFT treats $x(t)$ as if it were so and is thus only an estimate of the true Fourier transform of $x(t)$.

As a consequence of Shannon's theorem,[16] the DFT of $x(t)$ is defined only at $N/2$ frequencies at equally spaced intervals Δf apart, up to a maximum frequency F_{max}. The following fundamental relationships apply:

Block size:

$$N = \frac{T}{\Delta t} \qquad (27.5)$$

Frequency range:

$$F_{max} = 1/2\Delta t = \Delta f \left(\frac{N}{2}\right) \qquad (27.6)$$

Sampling rate:

$$\frac{1}{\Delta t} = 2 F_{max} \qquad (27.7)$$

Frequency resolution:

$$\Delta f = \frac{1}{T} \qquad (27.8)$$

Record length:

$$T = \frac{1}{\Delta f} = N \Delta t \qquad (27.9)$$

In performing a DFT analysis, assuming no restrictions on block size N, the parameters which may be selected are limited. The frequency range or the sampling rate, but not both, may be selected at will. Similarly, the frequency resolution or the record length may be chosen. If the block size is fixed, then only one of these four parameters may be selected. (See Fig. 13.16.)

The DFT, and thus any other quantities derived therefrom, is susceptible to certain errors, the most important of which is known as "aliasing."[13, 14, 15] Aliasing error arises when the signal $x(t)$ contains frequency components or "energy" above F_{max} of the DFT. Owing to the sampling or digitizing process, this energy will appear to be within the DFT frequency range. Therefore, to avoid aliasing error, either F_{max} must be large enough to include all significant frequency components of $x(t)$ or, for a given F_{max}, the components above F_{max} must be removed by analog filtering before digitization.

SPECTRAL DENSITY—FOURIER TRANSFORM. The concept of *power spectral density*, also called *auto spectral density*, is defined in analog terms in Chaps. 11 and 22. This quantity describes the frequency or spectral properties of a single time-history. A companion quantity, the *cross spectral density* describes the joint spectral properties of two time-histories. Both quantities are related to the Fourier transforms of the time histories through the following relationships:

Auto spectral density:

$$G_{xx}(f) = S_x(f)S_x{}^*(f) = |S_x(f)|^2 \qquad (27.10)$$

where $G_{xx}(f)$ is the one-sided auto spectral density and $S_x{}^*(f)$ is the complex conjugate of $S_x(f)$.

Cross spectral density:

$$G_{xy}(f) = S_x(f)S_y{}^*(f) \qquad (27.11)$$

$$G_{yx}(f) = S_y(f)S_x{}^*(f) \qquad (27.12)$$

where $G_{xy}(f)$ and $G_{yx}(f)$ are one-sided cross spectral densities and are complex quantities, and $S_x(f)$ and $S_y(f)$ are the Fourier transforms of $x(t)$ and $y(t)$, respectively.

When the Fourier transforms of $x(t)$ and $y(t)$ are obtained from digital data, i.e., DFT's, the spectral density functions are defined only at the frequencies for which the DFT is calculated. Further properties of the spectral density functions so obtained are described under *Vibration Data Analysis*.

SPECTRAL DENSITY—CORRELATION FUNCTIONS. The spectral density functions, which describe signal characteristics in the frequency domain, and the correlation functions, which describe signal characteristics in the time domain (see Chap. 22), are related through the Fourier transform as follows:

Auto:

$$G_{xx}(f) = 2 \int_{-\infty}^{\infty} R_{xx}(\tau) e^{-i2\pi f\tau} \, d\tau \qquad 0 \le f < \infty \tag{27.13}$$

$$R_{xx}(\tau) = \tfrac{1}{2} \int_{-\infty}^{\infty} G_{xx}(f) e^{i2\pi f\tau} \, df \qquad -\infty < \tau < \infty \tag{27.14}$$

Cross:

$$G_{xy}(f) = 2 \int_{-\infty}^{\infty} R_{xy}(\tau) e^{-i2\pi f\tau} \, d\tau \qquad 0 \le f < \infty \tag{27.15}$$

$$R_{xy}(\tau) = \tfrac{1}{2} \int_{-\infty}^{\infty} G_{xy}(f) e^{i2\pi f\tau} \, df \qquad -\infty < \tau < \infty \tag{27.16}$$

where $R_{xx}(\tau)$ is the auto correlation function at $x(t)$, and $R_{xy}(\tau)$ is the cross correlation function at $x(t)$ and $y(t)$.

FREQUENCY-RESPONSE FUNCTIONS. Frequency-response functions, synonymously known as *transfer functions*, describe in the frequency domain the relationship between the input and output or response of a linear system. This is shown schematically in Fig. 27.1 for a single input–single output system. (See Ref. 13 for multiple input-output systems.) The transfer function $H(f)$ describes the magnitude and phase of the response per unit sinusoidal input as a function of the input frequency. It is thus a complex quantity. Transfer functions can also be determined from the Fourier

FIG. 27.1. Schematic representation of transfer function $H(f)$.

transforms of the input and response time-histories and from the spectral densities of the input and response signals when the input is a random process. The governing relationships are

Fourier transforms:

$$S_y(f) = S_x(f)H(f) \qquad \text{or} \qquad H(f) = \frac{S_y(f)}{S_x(f)} \tag{27.17}$$

Auto spectral densities:

$$G_{yy}(f) = G_{xx}(f)|H(f)|^2 \qquad \text{or} \qquad |H(f)|^2 = \frac{G_{yy}(f)}{G_{xx}(f)} \tag{27.18}$$

Cross spectral densities:

$$G_{yx}(f) = G_{xx}(f)H_1(f) \qquad \text{or} \qquad H_1(f) = \frac{G_{yx}(f)}{G_{xx}(f)} \tag{27.19A}$$

$$G_{xy}(f)H_2(f) = G_{yy}(f) \qquad \text{or} \qquad H_2(f) = \frac{G_{yy}(f)}{G_{xy}(f)} \tag{27.19B}$$

The relationships above assume no extraneous signal or noise is present in either $x(t)$ or $y(t)$, in which case, $H_1(f)$ and $H_2(f)$ are equal. (See *Coherence Function* and *Modal Testing*, below.)

Only the transfer function magnitude can be determined from auto spectral densities; that is, phase information is unobtainable. This follows because the auto spectral densities themselves are void of phase information. In using cross spectral density, in which phase information is retained, both magnitude and phase of the transfer function is obtained. Depending on the relative characteristics of $x(t)$ and $y(t)$ and their "noisiness," it may be desirable to calculate $H(f)$ rather than the customary Eq. (27.19A).[45, 46]

COHERENCE FUNCTION. The coherence function, $\gamma_{xy}{}^2$, is a measure of the quality of the input, response and cross spectral densities, and the causality of input to response for a system as shown in Fig. 27.1.

$$\gamma_{xy}{}^2(f) = \frac{|G_{xy}(f)|^2}{G_{xx}(f)G_{yy}(f)} \qquad (27.20)$$

From Eqs. (27.18) and (27.19),

$$\gamma_{xy}{}^2(f) = \frac{|G_{xx}(f)H(f)|^2}{G_{xx}(f)|H(f)|^2G_{xx}(f)} = 1 \qquad (27.21)$$

assuming $x(t)$ and $y(t)$ are noise-free.

It can be shown[13] that, in practical applications,

$$0 \leq \gamma_{xy}{}^2 \leq 1.0$$

Values of $\gamma_{xy}{}^2$ less than unity indicate that the response is not attributable to the input, due, for example, to extraneous noise or nonlinearity of the system. In the frequency domain, the coherence function is analogous to the correlation coefficient in the time domain.

DIGITAL VIBRATION DATA ANALYSIS

The basic principles of vibration data analysis are described in Chap. 22. These principles are illustrated in terms of analog methods of processing data. The principles of analysis are not changed by a switch to digital methods and are not repeated. Only the basic digital methods are described. Various types of analyzers are described in Chap. 13.

Types of data analysis which may be performed entirely by digital computation parallel those performed by analog methods. Three basic types are spectral analysis, correlation analysis, and statistical or probability analysis. Common to all digital vibration data analysis is the implicit assumption that the data are sampled from a stationary or steady-state process; or that at least the data are varying slowly enough so the data sample may be considered stationary during the sample duration.

ANALOG-TO-DIGITAL CONVERSION AND DATA PREPARATION. The first step in any digital method is the analog-to-digital conversion (ADC) of the time-history or time-histories. This operation is generally built into self-contained digital analysis systems. However, when the digital processing is performed on a general-purpose scientific computer, this operation must usually be carried out at another facility.

A prime advantage of digital analysis methods is the requirement to play back the analog time-history of the signal no more than once for ADC, or alternatively, to obviate the necessity for recording the time-history if on-line analysis is desired. However, analog recording in parallel may be prudent for backup data purposes. Thus, for example, the problems associated with preparing a satisfactory tape-loop for repeated playback of the data sample to be analyzed by tunable analog filters are avoided.

For each of the digital data samples which constitute the digital record to be processed, the ADC process consists of two steps: (1) the analog signal is sampled by a sample-and-hold device, and (2) the analog voltage is converted to a binary digital number. Each sample is subject to two errors, which appear as digital noise on the record. First, the sample-and-hold device samples over a finite though small time interval. If the analog signal can vary significantly during this time interval, the "held" voltage will be inaccurate. Second, further error may be generated if the digital word length, i.e., number of bits, is insufficient. Since there is always an uncertainty equal to the least significant bit, this may represent a significant error, particularly for samples taken near a zero crossing.

When possible, it is desirable to preprocess the digital record prior to analysis to remove any wild points and any trends.[15] "Wild points" are defined as any individual digital samples (or possibly several successive samples) which are either unreasonably large or essentially zero for some generally unknown reason, such as analog tape spikes or dropouts or ADC momentary failure. "Trends" are defined as slowly varying changes in the data, such as a DC bias or a slow change in the running mean value of the signal. Trends can often be eliminated by the use of high-pass filtering of the signal prior to ADC. Alternatively, the digital record can be preprocessed to remove the calculated trend from each digital sample and a new record created.

The data must be preprocessed and digitized in a manner which avoids frequency aliasing. Aliasing errors may be avoided by using a sampling rate which is at least twice that of the highest significant frequency component in the signal. However, since this may be difficult before analysis, which has the objective of determining the frequency components up to some desired maximum frequency of interest, it is generally preferable to avoid aliasing errors by low-pass filtering prior to ADC. (See *Aliasing*, Chap. 13.)

SPECTRAL ANALYSIS. Spectral analysis (see Chap. 22) is performed to determine the frequency characteristics of a time-history. For deterministic signals, a line spectrum is desired; for random processes, a spectral density function is desired.

Deterministic Signal. Determination of the line spectrum of a deterministic signal $[x(t)]$ by digital computation requires only computation of the Fourier transform of the signal $S(x)$. The computed transform may be presented either in terms of magnitude $|S(x)|$ and phase or in terms of the real and imaginary parts for each computation frequency, or both.

Equations (27.5) through (27.9) define the relationships for selection of the analysis parameters. It should be noted that the Fourier transform treats the data as if it were periodic with period T. Thus, for a deterministic signal, the frequencies at which $S(x)$ is computed may not and probably will not coincide exactly with the actual frequency components of the signal. It is important, therefore, to select a frequency resolution, Δf, which is sufficiently small to minimize errors in both frequency and amplitude owing to this misalignment of actual and computed frequencies.

Random Signal. Determination of the spectral density function, either auto or cross, of a random signal may be performed in either of two alternative methods. The older and traditional method, known as the Blackman-Tukey method, arrives at the spectral density by first computing the correlation function [Eq. (22.21) or (22.27)], followed by Fourier transformation of the correlation function to obtain the spectral density [Eq. (27.13) or (27.15)]. The newer method (using the FFT algorithm), known as the Cooley-Tukey method,[12] obtains the spectral density directly from the Fourier transform of the time-history [Eqs. (27.10), (27.11), and (27.12)]. For applications requiring real-time, or more exactly, almost real-time, analysis the latter method is mandatory.

Equations (27.5) through (27.9) define the relationships for selection of the analysis parameters, using the FFT method. Indirectly, these equations also define the analysis parameters for correlation analysis when using the Blackman-Tukey method. Assuming the correlation functions are computed at incremental delay times, $\Delta\tau$ equal to the sampling interval, Δt, up to a maximum delay, m $\Delta\tau$, where m is the maximum lag number, then

$$\Delta f = \frac{1}{m\,\Delta\tau} \tag{27.22}$$

It should be noted that for spectral density analysis, these methods determine the average spectral density within each bandwidth, Δf wide, centered at the $N/2$ computation frequencies.

The maximum lag number m should be small compared with the total number of samples N. The effective record length for spectral analysis is thus $m \, \Delta t$ rather than the $N \, \Delta t$ needed to compute the correlation function.

Statistical Error. Spectral density functions computed by either of the methods described above are only estimates of the true spectral densities of the original analog time-histories. These estimates are subject to statistical error very similar to the statistical errors in analog processing. The normalized standard error ϵ of the spectral density estimate computed from a single FFT is 1.0 and is governed by the chi-squared distribution for two degrees-of-freedom. To obtain spectral density estimates with a more reasonable statistical error, further averaging is necessary. This may be accomplished by averaging over frequency and/or averaging over segments or records. In the first case, the spectral estimates in l adjacent frequency intervals are averaged to obtain an estimate of the average spectral density in a frequency bandwidth, $l \, \Delta f$. The standard error of the resultant spectral density then becomes $\epsilon = 1/\sqrt{l}$. Thus, statistical accuracy is gained at the expense of frequency resolution. In the second case, the spectral estimates obtained from a series of records, each $N \, \Delta t = T$ in duration, are averaged. This process is known as "ensemble averaging." The standard error for an ensemble average of q records is $1/\sqrt{q}$. Therefore, statistical accuracy is gained without loss of frequency resolution, for a constant block size N, by repeated spectral analysis. If both frequency and ensemble averaging are employed, the standard error becomes $1/\sqrt{ql}$. The error is chi-squared distributed with $2l$, $2q$, and $2ql$ degrees-of-freedom, respectively, for the three types of averaging.

Leakage. Spectral analysis by digital methods is subject to error known as "leakage."[13, 14, 15] Leakage arises from the truncation which occurs when either the FFT is computed over a finite frequency range, 0 to F_{max}, and the data is not periodic in the sample length (see Fig. 13.17) or the correlation function is computed over a finite maximum time delay, $m \, \Delta t$. This truncation causes undesirable side lobes in the effective filter shape of the transmission characteristics of a filter. These side lobes can be minimized by the use of one of several types of tapering functions instead of truncating in the digital processing. Commonly used functions are known as Parzen[13, 15] and Hamming[13, 15] windows. Properties of various types of windows are compared in Table 13.2.

TRANSFER FUNCTION MEASUREMENTS

Measurement of the complex transfer or frequency-response functions described by Eq. (27.19A), together with the auto spectral density functions [Eq. (27.10)] and the coherence function [Eq. (27.20)], can be carried out conveniently with a two-channel FFT analyzer, available commercially, described in Chap. 13. A wide range of analysis parameters are available. In performing transfer function measurements, the following considerations are important:

1. Aliasing errors should be avoided by filtering and/or by the appropriate selection of the digitizing frequency.

2. If the input and output signals are correlated, the statistical error inherent in the spectral density analysis is negligible in calculation of $H(f)$, i.e., shorter data sample durations can be tolerated.

3. Any relative time shifts between $x(t)$ and $y(t)$ due to head stack spacing of tape recorders, etc., will bias the effects of the results of phase measurements but will not change the magnitude of $H(f)$.

4. Auto spectral analysis can be performed from recorded signals played back in either the forward or reverse directions: if reverse playback is used, the complex conjugate of the transfer function is obtained.

5. While larger Δf improves auto spectral density estimates at the cost of spectral

smoothing, larger Δf degrades both the frequency resolution and the coherence function measurement.

Many of the FFT analyzers can also perform (or can be linked to desktop computers to perform) modal analyses from the array of transfer functions.

STATISTICAL ANALYSIS.   Chapter 22 described, in analog terms, the basic methods of determining the probability distribution functions of both instantaneous and peak values of a signal.   Performance of these types of analyses using digital computer techniques is no different in principle but is much more straightforward to implement. References 13, 14, and 15 contain detailed descriptions of the required digital operations.

An advantage gained by use of digital techniques is the ability to perform goodness-of-fit tests on the resultant probability density functions.   These tests determine (to some selected confidence level, and based on the number of digital samples) whether the distribution should or should not be considered Gaussian.

CORRELATION ANALYSIS.   Auto- and cross-correlation functions may be determined either from Fourier-transforming the auto- or cross-spectral density functions [Eqs. (27.14) and (27.16)] or directly from the time-histories.   The latter is generally the case during computation of PSD by the Blackman-Tukey method.   The principles of the latter method for digital processing are the same as described in Chap. 22 for analog processing.   Detailed digital procedures are described in Refs. 13, 14, and 15, and are clearly very simple, consisting merely of summing the products of pairs of samples.   The choice of record length, digitizing rate, and number of lags determines the statistical error of the correlation function estimate.   In addition, these choices define the frequency resolution, frequency range, and statistical error of spectra density, later computed from the correlation functions.

DIGITAL CONTROL SYSTEMS FOR VIBRATION TESTING

The vibratory motions prescribed for the majority of vibration tests consist of either sinusoidal, with fixed or variable frequency, or random motions.   The system employed to control the vibration level during the test utilizes the output signal from an accelerometer mounted at an appropriate location on the vibration exciter as the feedback signal to a servo system.   The servo system adjusts the driving signal to the power amplifier to maintain the desired test level.   Figure 27.2 shows this closed-loop system schematically.   Vibration testing machines are described in Chap. 25.

With the advent of FFT processors and minicomputers, it became possible to perform spectral analysis of random processes rapidly enough to permit the use of digital control systems for random vibration testing.   An extension of software also enables the system to control sinusoidal testing.

RANDOM VIBRATION TESTS.   Figure 27.3 shows a functional block diagram of a digitally controlled random vibration system.   Starting with the random number generator, which is the digital equivalent of an analog random noise generator, a complex

FIG. 27.2. Functional block diagram—closed-loop vibration system.

FIG. 27.3. Functional block diagram of random vibration digital control system.

Fourier spectrum is generated. The inverse Fourier transform converts this spectrum to a digital time-history which is used, after digital-to-analog conversion, to drive the power amplifier-electrodynamic shaker. The control accelerometer feedback signal is digitized and the auto spectral density computed by an FFT processor. After sufficient ensemble averaging, the auto-spectrum is compared with the specified test spectrum and a correction or modification of the driving spectrum is generated, thus closing the loop.

Two types of drive signals are employed for random vibration testing. (1) A purely random signal uses a completely new random time-history for each succeeding time-frame of data, obtained from the random number generator. In this case, the response time of the servo system is dependent on the time to ensemble-average the feedback spectrum sufficiently to obtain the required statistical accuracy. Typically, this time is 1 to 2 seconds, depending primarily on the chosen frequency resolution. (2) A pseudo-random signal (zero-variance signal) may be employed in order to achieve shorter servo response times, particularly during initial startup of the test. This type of signal consists of repeated use of the same time-frame of data, sometimes alternately played backward and forward, and is thus periodic. However, the time-frame itself displays the desired random characteristics. Since the time-frames are repetitive, the need to ensemble-average is obviated and the driving signal can be corrected every other time-frame.

SINUSOIDAL VIBRATION TESTS.[17, 18]   To perform sinusoidal vibration tests, the random noise generator in Fig. 27.3 is replaced by a digital sine-wave synthesizer which can be programmed. In addition, the feedback signal is analyzed to determine the amplitude of the sinusoidal signal rather than the spectral density of the signal for comparison to the reference (i.e., desired) amplitude.

DIGITAL CONTROL SYSTEMS FOR SHOCK TESTING

(The analysis of shock data is described in Chap. 23.) The shock-test machines described in Chap. 26 may be looked upon as devices for storing kinetic energy which can be suddenly released in order to create the desired shock or transient. For a number of reasons, it is desirable to perform shock or transient testing using electrodynamic or electrohydraulic vibration systems to induce the desired transient motion at the mounting points of the test item. The ability to employ this method is dependent on such parameters as the stroke or maximum allowable motion of the vibration exciter, the transient waveform, the mass of the test item, etc. Given that the required test is within the performance capability of an available vibration system, the ability to obtain and control the desired motion has been greatly expanded by the use of digital control equipment.

Shock-test requirements are specified in one of two alternative methods. The first, and most direct, method specifies a certain acceleration waveform, such as a half sine wave of specified duration and maximum acceleration. Most requirements of this type suffer from the following deficiency or indefiniteness. Most specified acceleration waveforms do not have a zero mean value and thus indirectly specify a certain velocity change. Since any laboratory test must start and eventually stop at rest, i.e., zero velocity change, additional accelerations, positive or negative, are applied before and/or after the specified waveform in order to give zero velocity change. These additional accelerations are minimized in amplitude by maximizing their duration as limited by the shock-test machine. The second method of specifying shock-test requirements employs shock spectra (see Chap. 23). The requirements must specify the frequency range, damping factor, type of spectrum, and either minimum or nominal values with an allowable tolerance on spectrum values.[19]

WAVEFORM CONTROL TESTS.[20]   Shock tests, defined in terms of waveform, are performed as follows:

1. With the test item installed on the vibration exciter, the complex transfer function between the drive signal input and the control accelerometer is measured. A short burst of random vibration or a suitable transient excitation may be used.

2. The complex Fourier spectrum of the desired waveform is calculated.

3. The Fourier spectrum of the desired waveform is divided by the complex transfer function of the system to obtain the complex Fourier spectrum of the required drive signal.

4. The inverse Fourier transform of this spectrum is computed to obtain the required drive signal time-history.

The cited reference describes in detail the application of digital control systems employing FFT processors and minicomputers to perform the steps above. If the specified waveform does not have a zero mean value, i.e., nonzero velocity change, the Fourier spectrum in step 2 must be calculated for a desired waveform which includes any additional accelerations required to give a zero velocity change or the vibration exciter may well become a launching pad. Within certain limitations with respect to duration and frequency content [see Eqs. (27.5) through (27.9)], an in-service transient acceleration waveform may be digitized for use as the control waveform in a laboratory test.

SHOCK-SPECTRUM CONTROL TESTS. When the shock test is specified in terms of a shock spectrum, no explicit requirements are placed on the acceleration waveform. Thus, a number of different waveforms may all meet the requirements. Waveforms such as a burst of random noise, a chirp (i.e., a very rapidly swept sinusoidal sweep), and combinations of exponentially growing and decaying sine waves (wavelets) have been used. To control shock tests in terms of a required shock spectrum the following strategy is employed.

1. Select the type of acceleration time-history to be employed, i.e., random burst, etc.

2. Calculate the amplitude and frequency characteristics of this waveform which will give the desired shock spectrum.

3. Perform an exploratory test, using the four steps to calculate the drive signal described above for waveform control test.

4. Calculate the shock spectrum of the resulting acceleration waveform measured by the control accelerometer.

5. Repeat steps 2, 3, and 4 as necessary until the desired shock spectrum is obtained.

Most digital control systems use wavelets to synthesize a waveform which exhibits the required shock spectrum. After this waveform is determined by means of the controller, the controller then reverts to its function of controlling the waveform for the actual test. Its adequacy (or inadequacy) is then demonstrated after the fact.

MODAL TESTING*

Modal tests, also known as *ground vibration tests* in the aircraft industry, are tests conducted to determine experimentally the natural frequencies, mode shapes, and associated damping factors of a structure. For many years, these tests have been conducted by exciting the system with sinusoidal excitation at a number of points of the structure. The driving frequency, the relative amplitudes, and phases of the excitations are adjusted to obtain the "purest" possible excitation of a single mode. The responses at locations throughout the structure then define the mode shape for that frequency. Damping factors can be measured by measurement of the bandwidth of the mode.

Mode shapes may also be measured by the use of broadband vibration or transient excitations applied at a single point of the structure for any one measurement. The responses to these excitations are measured at all points of interest. From these data, a set of transfer functions between the excitation and each response measurement location may be calculated for each excitation employed. The frequencies, mode shapes, and damping factors are then obtained from these measured transfer functions.

Either of the approaches given above involves the acquisition and analysis of a vast amount of data. Digital processing of these data on line, i.e., during the test, has greatly improved the efficiency and accuracy of this type of test, as well as the post-test data processing of the data to determine the desired information.

Digital computers have been applied in two distinct ways. First, for sinusoidal excitation, computers have been employed as an aid in obtaining the desired purity of the modal

* Also see Chap. 21.

excitation as well as in acquiring and processing data. Second, and more essentially, for broadband excitation (either random vibration or transient) FFT processing of the data is used to determine transfer functions essentially on line. In addition, the quality of the transfer function can be assessed immediately by computation and display of the coherence function [Eq. (27.20)]. If an unsatisfactory coherence function is obtained, the excitation can be repeated immediately until satisfactory data quality is achieved. However, care must be exercised in computing the coherence function. The coherence function obtained from raw spectral density estimates using the FFT method, i.e., from a single frame of data, will always be unity.[13] Reliable frequency-response-function and coherence-function estimates may be obtained either from smoothed spectral estimates obtained by FFT processes (Cooley-Tukey) or from spectral estimates obtained from correlation functions (Blackman-Tukey).

REFERENCES

1. Pilkey, W., K. Saczalski, and H. Schaeffer (eds.): "Structural Mechanics Computer Programs, Surveys, Assessments and Availability," University Press of Virginia, Charlottesville, Virginia, 1974.
2. Pilkey, W., and B. Pilkey: "Shock Vibration Computer Programs—Reviews and Summaries," SVM-10, Shock and Vibration Information Center, Washington, D.C., 1975.
3. NASA SP-222 (01): "The NASTRAN Users Manual," June, 1972 (and later updated editions).
4. NASA SP-221 (01): "The NASTRAN Theoretical Manual," June, 1972 (and later updated editions).
5. NASA SP-223: "The NASTRAN Programmers Manual," June, 1972 (and later updated editions).
6. MacNeal, R. H., and C. W. McCormick: "The NASTRAN Computer Program for Structural Analysis," *SAE Paper* 690612, October, 1969.
7. MacNeal, R. H.: *Computers Struct.*, **1**:581–601 (1971).
8. Hurty, W. C., J. D. Collins, and G. C. Hart: *Computers Struct.*, **1**:535–563 (1971).
9. Rubin, S.: "An Improved Component-Mode Representation," *AIAA Paper* 74-386, April, 1974.
10. Guyan, R. J.: *AIAA*, **3**(2):380 (February, 1965).
11. Kuhar, E. J., and C. V. Stahle: "A Dynamic Transformation Method for Modal Synthesis," *AIAA Paper* 73-396, March, 1973.
12. Cooley, J. W., and J. W. Tukey: *Math. Computing*, **19**:297 (April, 1965).
13. Bendat, Julius S., and Allan G. Piersol: "Random Data: Analysis and Measurement Procedures," John Wiley & Sons, Inc., New York, 1971.
14. Enochson, Loren D., and Robert K. Otnes: "Programming and Analysis for Digital Time Series Data," SVM-3, Shock and Vibration Information Center, Washington, D.C., 1968.
15. Otnes, Robert K., and Loren Enochson: "Digital Time Series Analysis," John Wiley & Sons, Inc., New York, 1972.
16. Shannon, C. E.: *Proc. IRE*, **37**:10 (1949).
17. Curtis, A. J., and J. G. Herrera: "Random-Vibration Test Level Control Using Input and Test Item Response Spectra," *Shock and Vibration Bull.* 37, pt. 3, p. 47, Shock and Vibration Information Center, Washington, D.C., January, 1968.
18. Blackman, R. B., and Tukey, J. W.: "The Measurement of Power Spectra," Dover Publications, Inc., New York, 1958.
19. Parzen, E.: "Stochastic Processes," Holden-Day, Inc., Publisher, San Francisco, 1962.
20. Curtis, A. J., N. G. Tinling, and H. T. Abstein, Jr.: "Selection and Performance of Vibration Tests," SVM-8, Shock and Vibration Information Center, Washington, D.C., 1971.
21. Chapman, C. P., J. Shipley, and C. L. Heizman: "A Digitally Controlled Vibration or Acoustics Testing System," *IES 1969 Proc.*, pts. I, II, and III, pp. 387–409.
22. Ratz, A. G., "Random Vibration Test Systems Using Digital Equalizers," *IES 1970 Proc.*, p. 75.
23. Nelson, D. B.: "Some Considerations in Design, Specification and Evaluation of Digital Control System for Random Vibration Testing," *ASME Paper* 71-Vibr-30, September, 1971.
24. Chapman, C. P.: "Computer-Controlled Vibration Test Systems: Criteria for Selection, Installation and Maintenance," *SAE Paper* 720819, October, 1972.
25. Heizman, C. L., and E. A. Sloane: "Evolution of Digital Vibration System," *SAE Paper* 720820, October, 1972.

26. Tebbs, J. D., and N. F. Hunter, Jr.: "Digital Control of Random Vibration Tests Using a Sigma V Computer," *IES 1974 Proc.*, pp. 36–43.
27. Tebbs, J. D., and D. O. Smallwood: "Extension of Control Techniques for Digital Control of Random Vibration Tests," *Shock and Vibration Bull.* 45, 1975.
28. Reiner, R. L., "A Digitally-Controlled Sweep Oscillator," *IES 1970 Proc.*, pp. 52–57.
29. Bosso, F. C.: "Sine-Test Electrodynamic Vibration Systems Using Digital Control," *IES 1970 Proc.*, pp. 61–69.
30. Norin, R. S.: "A Multi-Channel, Multi-Strategy Sinusoidal Vibration Control System," *IES 1974 Proc.*, pp. 222–228.
31. American National Standards Institute: "Methods for Analysis and Presentation of Shock and Vibration Data," ANSI S2.10-1971.
32. Favour, J. D., and J. M. LeBrun: "Final Report, Feasibility and Conceptual Design Study—Vibration Generator Transient Waveform Control System," NAS5-15171, NASA, 1969.
33. Favour, J. D., J. M. LeBrun, and J. P. Young: "Transient Waveform Control of Electromagnetic Vibration Test Equipment," *Shock and Vibration Bull.* 40, pt. 2, pp. 157–171, December, 1969.
34. Barthmaier, J. P.: "Shock Testing Under Minicomputer Control," *IES 1974 Proc.*, pp. 207–215.
35. Hunter, N. F., Jr.: "Transient Waveform Reproduction on Hydraulic Actuators Using a Nonlinear Gain Estimation Technique," *IES 1974 Proc.*, pp. 202–206.
36. Kao, G. C., K. Y. Chang, and W. W. Holbrook: "Digital Control Technique for Seismic Simulation," *Shock and Vibration Bull.* 43, pt. I, pp. 109–117, June, 1973.
37. Smallwood, D. O., and A. F. Wilte: "The Use of Shaker Optimized Periodic Transients in Matching Field Shock Spectra," *Shock and Vibration Bull.* 43, pt. 1, pp. 139–150, June, 1973.
38. Yang, R. C., and H. R. Saffell: "Development of a Waveform Synthesis Technique—A Supplement to Response Spectrum as a Definition of Shock Environment," *Shock and Vibration Bull.* 42, pt. 2, pp. 45–53, January, 1972.
39. Upton, P.: *Sound and Vibration,* **18:**18 (1984).
40. Rocklin, G. T., Crowley, J. and Vold, H.: "A Comparison of H_1, H_2, H_v Frequency Response Functions," *Proc. 3rd Intern. Modal Analysis Conf.*, Jan. 1985.

28

MATRIX METHODS OF ANALYSIS

Stephen H. Crandall
Massachusetts Institute of Technology

Robert B. McCalley, Jr.
General Electric Company

The mathematical language which is most convenient for analyzing multiple degree-of-freedom vibratory systems is that of *matrices*. Matrix notation simplifies the preliminary analytical study and in situations where particular numerical answers are required, matrices provide a standardized format for organizing the data and the computations. Computations with matrices can be carried out by hand or by digital computers. The availability of large high-speed digital computers makes the solution of many complex problems in vibration analysis a matter of routine.

This chapter describes how matrices are used in vibration analysis. It begins with definitions and rules for operating with matrices. The formulation of vibration problems in matrix notation then is treated. This is followed by general matrix solutions of several important types of vibration problems including free and forced vibrations of both undamped and damped linear multiple degree-of-freedom systems. The computational problems involved in handling matrices then are discussed.

MATRICES

Matrices [1] are mathematical entities which facilitate the handling of simultaneous equations. They are applied to the differential equations of a vibratory system as follows:

A single degree-of-freedom system of the type in Fig. 28.1 has the differential equation

$$m\ddot{x} + c\dot{x} + kx = F$$

where m is the mass, c is the damping coefficient, k is the stiffness, F is the applied force,

FIG. 28.1. Single degree-of-freedom system.   FIG. 28.2. Three degree-of-freedom system.

x is the displacement coordinate, and dots denote time derivatives. In Fig. 28.2 a similar three degree-of-freedom system is shown. The equations of motion may be obtained by applying Newton's second law to each mass in turn:

$$m\ddot{x}_1 \qquad + c\dot{x}_1 \qquad\qquad + 5kx_1 - 2kx_2 \qquad = F_1$$
$$2m\ddot{x}_2 \qquad + 2c\dot{x}_2 - 2c\dot{x}_3 - 2kx_1 + 3kx_2 - kx_3 = F_2 \qquad (28.1)$$
$$3m\ddot{x}_3 \qquad - 2c\dot{x}_2 + 2c\dot{x}_3 \qquad - kx_2 + kx_3 = F_3$$

The accelerations, velocities, displacements, and forces may be organized into columns, denoted by single boldface symbols:

$$\ddot{\mathbf{x}} = \begin{bmatrix} \ddot{x}_1 \\ \ddot{x}_2 \\ \ddot{x}_3 \end{bmatrix} \qquad \dot{\mathbf{x}} = \begin{bmatrix} \dot{x}_1 \\ \dot{x}_2 \\ \dot{x}_3 \end{bmatrix} \qquad \mathbf{x} = \begin{bmatrix} x_1 \\ x_2 \\ x_3 \end{bmatrix} \qquad \mathbf{f} = \begin{bmatrix} F_1 \\ F_2 \\ F_3 \end{bmatrix} \qquad (28.2)$$

The inertia, damping, and stiffness coefficients may be organized into square arrays:

$$\mathbf{M} = \begin{bmatrix} m & 0 & 0 \\ 0 & 2m & 0 \\ 0 & 0 & 3m \end{bmatrix} \qquad \mathbf{C} = \begin{bmatrix} c & 0 & 0 \\ 0 & 2c & -2c \\ 0 & -2c & 2c \end{bmatrix} \qquad \mathbf{K} = \begin{bmatrix} 5k & -2k & 0 \\ -2k & 3k & -k \\ 0 & -k & k \end{bmatrix} \qquad (28.3)$$

By using these symbols, it is shown below that it is possible to represent the three equations of Eq. (28.1) by the following single equation:

$$\mathbf{M}\ddot{\mathbf{x}} + \mathbf{C}\dot{\mathbf{x}} + \mathbf{K}\mathbf{x} = \mathbf{f} \qquad (28.4)$$

Note that this has the same form as the differential equation for the single degree-of-freedom system of Fig. 28.1. The notation of Eq. (28.4) has the advantage that in systems of many degrees of freedom it clearly states the physical principle that at every coordinate the external force is the sum of the inertia, damping, and stiffness forces. Equation (28.4) is an abbreviation for Eq. (28.1). It is necessary to develop the rules of operation with symbols such as those in Eqs. (28.2) and (28.3) to ensure that no ambiguity is involved. The algebra of *matrices* is devised to facilitate manipulations of simultaneous equations such as Eq. (28.1). Matrix algebra does not in any way simplify individual operations such as multiplication or addition of numbers, but it is an organizational tool which permits one to keep track of a complicated sequence of operations in an optimum manner.

DEFINITIONS

A *matrix* is an array of elements arranged systematically in rows and columns. For example, a rectangular matrix **A**, of elements a_{jk}, which has m rows and n columns is

$$\mathbf{A} = [a_{jk}] = \begin{bmatrix} a_{11} & a_{12} & \dots & a_{1n} \\ a_{21} & a_{22} & \dots & a_{2n} \\ \dots & \dots & \dots & \dots \\ a_{m1} & a_{m2} & \dots & a_{mn} \end{bmatrix}$$

The elements a_{jk} are usually numbers or functions but, in principle, they may be any well-defined quantities. The first subscript j on the element refers to the row number while the second subscript k refers to the column number. The array is denoted by the single symbol **A** which can be used as such during operational manipulations in which it is not necessary to specify continually all the elements a_{jk}. When a numerical calculation is finally required, it is necessary to refer back to the explicit specifications of the elements a_{jk}.

A rectangular matrix with m rows and n columns is said to be of order (m,n). A matrix of order (n,n) is a *square matrix* and is said to be simply a square matrix of order n. A matrix of order $(n,1)$ is a *column matrix* and is said to be simply a column matrix of order n. A column matrix is sometimes referred to as a *column vector*. Similarly, a matrix of order $(1,n)$ is a *row matrix* or a *row vector*. Boldface *capital* letters are used here to represent square matrices and *lower-case* boldface letters to represent column matrices or vectors. For example, the matrices in Eq. (28.2) are column matrices of order three and the matrices in Eq. (28.3) are square matrices of order three.

Some special types of matrices are:

1. A *diagonal matrix* is a square matrix $\mathbf{A}$ whose elements a_{jk} are zero when $j \neq k$. The only nonzero elements are those on the *main diagonal* where $j = k$. In order to emphasize that a matrix is diagonal, it is often written with small ticks in the direction of the main diagonal:

$$\mathbf{A} = \left\lceil a_{jj} \right\rfloor$$

2. A *unit matrix* or *identity matrix* is a diagonal matrix whose main diagonal elements are each equal to unity. The symbol $\mathbf{I}$ is used to denote a unit matrix. Examples are

$$\begin{bmatrix} 1 & 0 \\ 0 & 1 \end{bmatrix} \qquad \begin{bmatrix} 1 & 0 & 0 \\ 0 & 1 & 0 \\ 0 & 0 & 1 \end{bmatrix}$$

3. A *null matrix* or *zero matrix* has all its elements equal to zero and is simply written as zero.

4. The *transpose* $\mathbf{A}^T$ of a matrix $\mathbf{A}$ is a matrix having the same elements but with rows and columns interchanged. Thus, if the original matrix is

$$\mathbf{A} = [a_{jk}]$$

the transpose matrix is

$$\mathbf{A}^T = [a_{jk}]^T = [a_{kj}]$$

For example:

$$\mathbf{A} = \begin{bmatrix} 3 & 2 \\ -1 & 4 \end{bmatrix} \qquad \mathbf{A}^T = \begin{bmatrix} 3 & -1 \\ 2 & 4 \end{bmatrix}$$

The transpose of a square matrix may be visualized as the matrix obtained by rotating the given matrix about its main diagonal as an axis.

The transpose of a column matrix is a row matrix. For example,

$$\mathbf{x} = \begin{bmatrix} 3 \\ 4 \\ -2 \end{bmatrix} \qquad \mathbf{x}^T = [3 \quad 4 \quad -2]$$

Throughout this chapter a row matrix is referred to as the transpose of the corresponding column matrix.

5. A *symmetric matrix* is a square matrix whose off-diagonal elements are symmetric with respect to the main diagonal. A square matrix $\mathbf{A}$ is symmetric if, for all j and k,

$$a_{jk} = a_{kj}$$

A symmetric matrix is equal to its transpose. For example, all three of the matrices in Eq. (28.3) are symmetric. In addition the matrix $\mathbf{M}$ is a diagonal matrix.

MATRIX OPERATIONS

EQUALITY OF MATRICES. Two matrices of the same order are equal if their corresponding elements are equal. Thus two matrices $\mathbf{A}$ and $\mathbf{B}$ are equal if, for every j and k,

$$a_{jk} = b_{jk}$$

MATRIX ADDITION AND SUBTRACTION. Addition or subtraction of matrices of the same order is performed by adding or subtracting corresponding elements. Thus, $\mathbf{A} + \mathbf{B} = \mathbf{C}$, if for every j and k,

$$a_{jk} + b_{jk} = c_{jk}$$

For example, if

$$A = \begin{bmatrix} 3 & 2 \\ -1 & 4 \end{bmatrix} \qquad B = \begin{bmatrix} -1 & 2 \\ 5 & 6 \end{bmatrix}$$

then

$$A + B = \begin{bmatrix} 2 & 4 \\ 4 & 10 \end{bmatrix} \qquad A - B = \begin{bmatrix} 4 & 0 \\ -6 & -2 \end{bmatrix}$$

MULTIPLICATION OF A MATRIX BY A SCALAR. Multiplication of a matrix by a scalar c multiplies each element of the matrix by c. Thus

$$cA = c[a_{jk}] = [ca_{jk}]$$

In particular, the negative of a matrix has the sign of every element changed.

MATRIX MULTIPLICATION. If A is a matrix of order (m,n) and B is a matrix of order (n,p), then their *matrix product* $AB = C$ is defined to be a matrix C of order (m,p) where, for every j and k,

$$c_{jk} = \sum_{r=1}^{n} a_{jr} b_{rk} \tag{28.5}$$

The product of two matrices can be obtained only if they are *conformable*, i.e., if the number of columns in A is equal to the number of rows in B. The symbolic equation

$$(m,n) \times (n,p) = (m,p)$$

indicates the orders of the matrices involved in a matrix product. Matrix products are not commutative, i.e., in general,

$$AB \neq BA$$

The matrix products which appear in this chapter are of the following types:
    Square matrix × square matrix = square matrix
    Square matrix × column vector = column vector
    Row vector × square matrix = row vector
    Row vector × column vector = scalar
    Column vector × row vector = square matrix
In all cases, the matrices must be conformable. Numerical examples are given below.

$$AB = \begin{bmatrix} 3 & 2 \\ -1 & 4 \end{bmatrix} \begin{bmatrix} -1 & 2 \\ 5 & 6 \end{bmatrix} = \begin{bmatrix} -(3 \times 1) + (2 \times 5) & (3 \times 2) + (2 \times 6) \\ (1 \times 1) + (4 \times 5) & -(1 \times 2) + (4 \times 6) \end{bmatrix}$$

$$= \begin{bmatrix} 7 & 18 \\ 21 & 22 \end{bmatrix}$$

$$Ax = \begin{bmatrix} 3 & 2 \\ -1 & 4 \end{bmatrix} \begin{bmatrix} 5 \\ 3 \end{bmatrix} = \begin{bmatrix} (3 \times 5) + (2 \times 3) \\ -(1 \times 5) + (4 \times 3) \end{bmatrix} = \begin{bmatrix} 21 \\ 7 \end{bmatrix}$$

$$y^T A = [-2 \quad 1] \begin{bmatrix} 3 & 2 \\ -1 & 4 \end{bmatrix} = [-(2 \times 3) - (1 \times 1) \quad -(2 \times 2) + (1 \times 4)] = [-7 \quad 0]$$

$$y^T x = [-2 \quad 1] \begin{bmatrix} 5 \\ 3 \end{bmatrix} = (-10 + 3) = -7$$

$$xy^T = \begin{bmatrix} 5 \\ 3 \end{bmatrix} [-2 \quad 1] = \begin{bmatrix} -(5 \times 2) & (5 \times 1) \\ -(3 \times 2) & (3 \times 1) \end{bmatrix} = \begin{bmatrix} -10 & 5 \\ -6 & 3 \end{bmatrix}$$

The last product always results in a matrix with proportional rows and columns.
    The operation of matrix multiplication is particularly suited for representing systems of simultaneous linear equations in a compact form in which the coefficients are gathered

into square matrices and the unknowns are placed in column matrices. For example, it is the operation of matrix multiplication which gives unambiguous meaning to the matrix abbreviation in Eq. (28.4) for the three simultaneous differential equations of Eq. (28.1). The two sides of Eq. (28.4) are column matrices of order three whose corresponding elements must be equal. On the right, these elements are simply the external forces at the three masses. On the left, Eq. (28.4) states that the resulting column is the sum of three column matrices, each of which results from the matrix multiplication of a square matrix of coefficients defined in Eq. (28.3) into a column matrix defined in Eq. (28.2). The rules of matrix operation just given ensure that Eq. (28.4) is exactly equivalent to Eq. (28.1).

Premultiplication or postmultiplication of a square matrix by the identity matrix leaves the original matrix unchanged; i.e.,

$$IA = AI = A$$

Two symmetrical matrices multiplied together are generally not symmetric. The product of a matrix and its transpose is symmetric.

Continued matrix products such as **ABC** are defined, provided the number of columns in each matrix is the same as the number of rows in the matrix immediately preceding it. From the definition of matrix products, it follows that the *associative law* holds for continued products:

$$(AB)C = A(BC)$$

A square matrix **A** multiplied by itself yields a square matrix which is called the *square of the matrix* **A** and is denoted by A^2. If A^2 is in turn multiplied by **A**, the resulting matrix is $A^3 = A(A^2) = A^2(A)$. Extension of this process gives meaning to A^m for any positive integer *power m*. The extension to fractional and negative powers is described in connection with Eq. (28.44). Powers of symmetric matrices are themselves symmetric.

The rule for *transposition* of matrix products is

$$(AB)^T = B^T A^T$$

INVERSE OR RECIPROCAL MATRIX. If, for a given square matrix **A**, a square matrix A^{-1} can be found such that

$$A^{-1}A = AA^{-1} = I \qquad (28.6)$$

then A^{-1} is called the *inverse* or *reciprocal* of **A**. Not every square matrix **A** possesses an inverse. If the determinant constructed from the elements of a square matrix is zero, the matrix is said to be *singular* and there is no inverse. Every nonsingular matrix possesses a unique inverse. The inverse of a symmetric matrix is symmetric. The rule for the *inverse of a matrix product* is

$$(AB)^{-1} = (B^{-1})(A^{-1})$$

The solution to the set of simultaneous equations

$$Ax = c$$

where **x** is the unknown vector and **c** is a known input vector can be indicated with the aid of the inverse of **A**. The formal solution for **x** proceeds as follows:

$$A^{-1}Ax = A^{-1}c$$

$$Ix = x = A^{-1}c$$

When the inverse A^{-1} is known, the solution vector **x** is obtained by a simple matrix multiplication of A^{-1} into the input vector **c**.

The problem of calculating the inverse of given matrix **A** is one of the central computational problems associated with matrices. Routine procedures for both hand and machine computation exist.

QUADRATIC FORMS

A general quadratic form Q of order n may be written as

$$Q = \sum_{j=1}^{n} \sum_{k=1}^{n} a_{jk} x_j x_k$$

where the a_{jk} are constants and the x_j are the n variables. The form is quadratic since it is of the second degree in the variables. The laws of matrix multiplication permit Q to be written as

$$Q = [x_1 x_2 \ldots x_n] \begin{bmatrix} a_{11} & a_{12} & \ldots & a_{1n} \\ a_{21} & a_{22} & \ldots & a_{2n} \\ \ldots & \ldots & \ldots & \ldots \\ a_{n1} & a_{n2} & \ldots & a_{nn} \end{bmatrix} \begin{bmatrix} x_1 \\ x_2 \\ \ldots \\ x_n \end{bmatrix}$$

which is

$$Q = \mathbf{x}^T \mathbf{A} \mathbf{x}$$

Any quadratic form can be expressed in terms of a symmetric matrix. If the given matrix $\mathbf{A}$ is not symmetric, it can be replaced by the symmetric matrix

$$\mathbf{B} = \tfrac{1}{2}(\mathbf{A} + \mathbf{A}^T)$$

without changing the value of the form.

As an example of a quadratic form, the *potential energy* V for the system of Fig. 28.2 is given by

$$2V = 3kx_1{}^2 + 2k(x_2 - x_1)^2 + k(x_3 - x_2)^2$$

$$= 5kx_1x_1 - 2kx_1x_2$$

$$- 2kx_2x_1 + 3kx_2x_2 - kx_2x_3$$

$$- kx_3x_2 + kx_3x_3$$

Using the displacement vector $\mathbf{x}$ defined in Eq. (28.2) and the stiffness matrix $\mathbf{K}$ in Eq. (28.3), the potential energy may be written as

$$V = \tfrac{1}{2}\mathbf{x}^T \mathbf{K} \mathbf{x}$$

Similarly, the *kinetic energy* T is given by

$$2T = m\dot{x}_1{}^2 + 2m\dot{x}_2{}^2 + 3m\dot{x}_3{}^2$$

In terms of the inertia matrix $\mathbf{M}$ and the velocity vector $\dot{\mathbf{x}}$ defined in Eqs. (28.3) and (28.2) the kinetic energy may be written as

$$T = \tfrac{1}{2}\dot{\mathbf{x}}^T \mathbf{M} \dot{\mathbf{x}}$$

The *dissipation function* D for the system is given by

$$2D = c\dot{x}_1{}^2 + 2c(\dot{x}_3 - \dot{x}_2)^2$$

$$= c\dot{x}_1\dot{x}_1$$

$$+ 2c\dot{x}_2\dot{x}_2 - 2c\dot{x}_2\dot{x}_3$$

$$- 2c\dot{x}_3\dot{x}_2 + 2c\dot{x}_3\dot{x}_3$$

In terms of the velocity vector $\dot{\mathbf{x}}$ and the damping matrix $\mathbf{C}$ defined in Eqs. (28.2) and (28.3) the dissipation function may be written as

$$D = \tfrac{1}{2}\dot{\mathbf{x}}^T \mathbf{C} \dot{\mathbf{x}}$$

The dissipation function gives half the rate at which energy is being dissipated in the system.

While quadratic forms assume positive and negative values in general, the three physical forms just defined are intrinsically *positive* for a vibrating system with linear springs, constant masses, and viscous damping; i.e., they can never be negative for a real motion of the system. Kinetic energy is zero only when the system is at rest. The same thing is not necessarily true for potential energy or the dissipation function.

Depending upon the arrangement of springs and dashpots in the system, there may exist motions which do not involve any potential energy or dissipation. For example, in vibratory systems where rigid body motions are possible (crankshaft torsional systems, free-free beams, etc.) no elastic energy is involved in the rigid body motions. Also, in Fig. 28.2, if x_1 is zero while x_2 and x_3 have the same motion, there is no energy dissipated and the dissipation function is zero. To distinguish between these two possibilities, a quadratic form is called *positive definite* if it is never negative and if the only time it vanishes is when all the variables are zero. Kinetic energy is always positive definite, while potential energy and the dissipation function are positive but not necessarily positive definite. It depends upon the particular configuration of a given system whether or not the potential energy and the dissipation function are positive definite or only positive. The terms positive and positive definite are applied also to the matrices from which the quadratic forms are derived. For example, of the three matrices defined in Eq. (28.3), the matrices $\mathbf{M}$ and $\mathbf{K}$ are positive definite but $\mathbf{C}$ is only positive. It can be shown that a matrix which is positive but not positive definite is *singular*.

DIFFERENTIATION OF QUADRATIC FORMS. In forming Lagrange's equations of motion for a vibrating system,[*] it is necessary to take derivatives of the potential energy V, the kinetic energy T, and the dissipation function D. When these quadratic forms are represented in matrix notation, it is convenient to have matrix formulas for differentiation. In this paragraph rules are given for differentiating the slightly more general *bilinear form*

$$F = \mathbf{x}^T \mathbf{A} \mathbf{y}$$

where $\mathbf{x}^T$ is a row vector of n variables x_j, $\mathbf{A}$ is a square matrix of constant coefficients, and $\mathbf{y}$ is a column matrix of n variables y_j. In a quadratic form the x_j are identical with the y_j.

For generality it is assumed that the x_j and the y_j are functions of n other variables u_j. In the formulas below the notation $\mathbf{X}_u$ is used to represent the following square matrix:

$$\mathbf{X}_u = \begin{bmatrix} \dfrac{\partial x_1}{\partial u_1} & \dfrac{\partial x_2}{\partial u_1} & \cdots & \dfrac{\partial x_n}{\partial u_1} \\[2mm] \dfrac{\partial x_1}{\partial u_2} & \dfrac{\partial x_2}{\partial u_2} & \cdots & \dfrac{\partial x_n}{\partial u_2} \\[2mm] \cdots & \cdots & \cdots & \cdots \\[2mm] \dfrac{\partial x_1}{\partial u_n} & \dfrac{\partial x_2}{\partial u_n} & \cdots & \dfrac{\partial x_n}{\partial u_n} \end{bmatrix}$$

Now letting $\partial/\partial\mathbf{u}$ stand for the column vector whose elements are the partial differential operators with respect to the u_j, the general differentiation formula is

$$\frac{\partial F}{\partial\mathbf{u}} = \begin{bmatrix} \dfrac{\partial F}{\partial u_1} \\[2mm] \dfrac{\partial F}{\partial u_2} \\[2mm] \cdots \\[2mm] \dfrac{\partial F}{\partial u_n} \end{bmatrix} = \mathbf{X}_u \mathbf{A} \mathbf{y} + \mathbf{Y}_u \mathbf{A}^T \mathbf{x}$$

[*] See Chap. 2 for a detailed discussion of Lagrange's equations.

For a quadratic form $Q = \mathbf{x}^T \mathbf{A} \mathbf{x}$ the above formula reduces to

$$\frac{\partial Q}{\partial \mathbf{u}} = \mathbf{X}_u (\mathbf{A} + \mathbf{A}^T) \mathbf{x}$$

Thus whether $\mathbf{A}$ is symmetric or not this kind of differentiation produces a *symmetrical* matrix of coefficients $(\mathbf{A} + \mathbf{A}^T)$. It is this fact which ensures that vibration equations in the form obtained from Lagrange's equations always have symmetrical matrices of coefficients. If $\mathbf{A}$ is symmetrical to begin with, the previous formula becomes

$$\frac{\partial Q}{\partial \mathbf{u}} = 2\mathbf{X}_u \mathbf{A} \mathbf{x}$$

Finally, in the important special case where the x_j are identical with the u_j the matrix $\mathbf{X}_x$ reduces to the identity matrix yielding

$$\frac{\partial Q}{\partial \mathbf{x}} = 2\mathbf{A}\mathbf{x} \tag{28.7}$$

which is employed in the following section in developing Lagrange's equations.

FORMULATION OF VIBRATION PROBLEMS IN MATRIX FORM

This section begins with a generalized derivation of vibration equations in matrix form making use of Lagrange's equations.[2] This is followed by several particular examples illustrating the theory and indicating important areas of application. For simplicity of exposition, the examples have only 2 to 6 degrees-of-freedom. In practice, systems with 30 degrees-of-freedom are common and systems with over 100 degrees-of-freedom have been successfully treated.

Consider a holonomic linear mechanical system with n degrees-of-freedom which vibrates about a stable equilibrium configuration. Let the motion of the system be described by n *generalized displacements* $x_j(t)$ which vanish in the equilibrium position. The potential energy V can then be expressed in terms of these displacements as a quadratic form. The kinetic energy T and the dissipation function D can be expressed as quadratic forms in the generalized velocities $\dot{x}_j(t)$.

The equations of motion are obtained by applying Lagrange's equations

$$\frac{d}{dt}\left(\frac{\partial T}{\partial \dot{x}_j}\right) + \frac{\partial D}{\partial \dot{x}_j} + \frac{\partial V}{\partial x_j} = f_j(t) \qquad [j = 1, 2, \ldots, n]$$

The *generalized external force* $f_j(t)$ for each coordinate may be an active force in the usual sense or a force generated by prescribed motion of the coordinates.

If each term in the foregoing equation is taken as the jth element of a column matrix, all n equations can be considered simultaneously and written in matrix form as follows:

$$\frac{d}{dt}\left(\frac{\partial T}{\partial \dot{\mathbf{x}}}\right) + \frac{\partial D}{\partial \dot{\mathbf{x}}} + \frac{\partial V}{\partial \mathbf{x}} = \mathbf{f}$$

The quadratic forms can be expressed in matrix notation as

$$T = \tfrac{1}{2}(\dot{\mathbf{x}}^T \mathbf{M} \dot{\mathbf{x}})$$

$$D = \tfrac{1}{2}(\dot{\mathbf{x}}^T \mathbf{C} \dot{\mathbf{x}})$$

$$V = \tfrac{1}{2}(\mathbf{x}^T \mathbf{K} \mathbf{x})$$

where the *inertia matrix* $\mathbf{M}$, the *damping matrix* $\mathbf{C}$, and the *stiffness matrix* $\mathbf{K}$ may be taken as symmetric square matrices of order n. Then the differentiation rule (28.7) yields

$$\frac{d}{dt}(\mathbf{M\dot{x}}) + \mathbf{C\dot{x}} + \mathbf{Kx} = \mathbf{f}$$

or simply

$$\mathbf{M\ddot{x}} + \mathbf{C\dot{x}} + \mathbf{Kx} = \mathbf{f} \qquad (28.8)$$

as the equations of motion in matrix form for a general linear vibratory system with n degrees-of-freedom. This is a generalization of Eq. (28.4) for the three degree-of-freedom system of Fig. 28.2. Equation (28.8) applies to all linear constant parameter vibratory systems. The specifications of any particular system are contained in the *coefficient matrices* $\mathbf{M}$, $\mathbf{C}$, and $\mathbf{K}$. The type of excitation is described by the column matrix $\mathbf{f}$. The individual terms in the coefficient matrices have the following significance:

   m_{jk} is the momentum component at j due to a unit velocity at k,
   c_{jk} is the damping force at j due to a unit velocity at k,
   k_{jk} is the elastic force at j due to a unit displacement at k.

In certain applications it is more convenient to deal with $\mathbf{K}^{-1}$, the inverse of the stiffness matrix, than with $\mathbf{K}$. The elements α_{jk} of $\mathbf{K}^{-1}$ are called the *influence coefficients* and $\mathbf{K}^{-1}$ itself is sometimes called the *flexibility matrix*. The influence coefficients have the following significance:

   α_{jk} is the elastic displacement at j due to unit force statically applied at k.

When $\mathbf{K}$ is singular its inverse does not exist. This occurs in systems where rigid body motions are possible. A unit force cannot be statically applied to such a system. The result of applying a force is acceleration.

The general solution to Eq. (28.8) contains $2n$ constants of integration which are usually fixed by the n displacements $x_j(t_0)$ and the n velocities $\dot{x}_j(t_0)$ at some initial time t_0. When the excitation matrix $\mathbf{f}$ is zero, Eq. (28.8) is said to describe the *free vibration* of the system. When $\mathbf{f}$ is nonzero, Eq. (28.8) describes a *forced vibration*. When the time behavior of $\mathbf{f}$ is periodic and steady, it is sometimes convenient to divide the solution into a *steady-state response* plus a *transient response* which decays with time. The steady-state response is independent of the initial conditions.

COUPLING OF THE EQUATIONS

The off-diagonal terms in the coefficient matrices are known as *coupling terms*. In general, the equations have inertia, damping, and stiffness coupling; however, it is often possible to obtain equations that have no coupling terms in one or more of the three

FIG. 28.3. Coordinates (x_1, x_2) with uncoupled inertia matrix.

FIG. 28.4. Coordinates (x_1, x_2) with uncoupled stiffness matrix. The equilibrium length of the spring k_2 is L_2.

matrices. If the coupling terms vanish in all three matrices; i.e., if all three square matrices are diagonal matrices, the system of Eq. (28.8) becomes a set of independent uncoupled differential equations for the n generalized displacements $x_j(t)$. Each displacement motion is a single degree-of-freedom vibration independent of the motion of the other displacements.

The coupling in a system depends on the choice of coordinates used to describe the motion. For example, Figs. 28.3 and 28.4 show the same physical system with two different choices for the displacement coordinates.

The coefficient matrices corresponding to the coordinates shown in Fig. 28.3 are

$$\mathbf{M} = \begin{bmatrix} m_1 & 0 \\ 0 & m_2 \end{bmatrix} \qquad \mathbf{K} = \begin{bmatrix} k_1 + k_2 & -k_2 \\ -k_2 & k_2 \end{bmatrix}$$

Here the inertia matrix is uncoupled because the coordinates chosen are the absolute displacements of the masses. The elastic force in the spring k_2 is generated by the relative displacement of the two coordinates which accounts for the coupling terms in the stiffness matrix.

The coefficient matrices corresponding to the alternative coordinates shown in Fig. 28.4 are

$$\mathbf{M} = \begin{bmatrix} m_1 + m_2 & m_2 \\ m_2 & m_2 \end{bmatrix} \qquad \mathbf{K} = \begin{bmatrix} k_1 & 0 \\ 0 & k_2 \end{bmatrix}$$

Here the coordinates chosen relate directly to the extensions of the springs so that the stiffness matrix is uncoupled. The absolute displacement of m_2 is, however, the sum of the coordinates which accounts for the coupling terms in the inertia matrix.

FIG. 28.5. Two degree-of-freedom vibratory system. The equilibrium length of the spring k_1 is L_1 and the equilibrium length of the spring k_2 is L_2.

A fundamental procedure for solving vibration problems in undamped systems may be viewed as the search for a set of coordinates which simultaneously uncouples both the stiffness and inertia matrices. This is always possible. In systems with damping (i.e., with all three coefficient matrices) there exist coordinates which uncouple two of these, but in general it is not possible to uncouple all three matrices simultaneously.

The system of Fig. 28.2 provides an example of a three degree-of-freedom system with damping. The coefficient matrices are given in Eq. (28.3). The inertia matrix is uncoupled, but the damping and stiffness matrices are coupled.

Another example of a system with damping is furnished by the two degree-of-freedom system shown in Fig. 28.5. The excitation here is furnished by acceleration $\ddot{x}_0(t)$ of the base. This system is used as the basis for most of the numerical examples which follow. With the coordinates chosen as indicated in the figure, all three coefficient matrices have coupling terms. The equations of motion can be placed in the standard form of Eq. (28.8) where the coefficient matrices and the excitation column are as follows:

$$\mathbf{M} = \begin{bmatrix} m_1 + m_2 & m_2 \\ m_2 & m_2 \end{bmatrix} \qquad \mathbf{C} = \begin{bmatrix} c_1 + c_3 & c_3 \\ c_3 & c_2 + c_3 \end{bmatrix}$$

$$\mathbf{K} = \begin{bmatrix} k_1 + k_3 & k_3 \\ k_3 & k_2 + k_3 \end{bmatrix} \qquad \mathbf{f} = -\ddot{x}_0 \begin{bmatrix} m_1 + m_2 \\ m_2 \end{bmatrix} \qquad (28.9)$$

VIBRATION ISOLATION SYSTEMS

In Fig. 28.6 the motions in the xy plane of the symmetrical rigid body supported by four symmetrically located equal isolator springs are considered. Each of the four springs has spring constants k_x and k_y in the x and y directions, respectively. The mass of the body is m and the moment of inertia about the z axis is I_z. The free vibration equations of motion are derived elsewhere [3] in algebraic form. In matrix format they are

$$\begin{bmatrix} m & 0 & 0 \\ 0 & m & 0 \\ 0 & 0 & I_z \end{bmatrix} \begin{bmatrix} \ddot{x} \\ \ddot{y} \\ \ddot{\theta} \end{bmatrix} + \begin{bmatrix} 4k_x & 0 & 4k_x a \\ 0 & 4k_y & 0 \\ 4k_x a & 0 & 4(k_x a^2 + k_y b^2) \end{bmatrix} \begin{bmatrix} x \\ y \\ \theta \end{bmatrix} = 0$$

Note that the y equation is uncoupled from the other two but that the horizontal and rocking motions are coupled. This coupling could be removed by redesigning the spring attachments so that $a = 0$.

As a further extension of the suspension problem, consider a completely general rigid body, elastically supported at n points P_i by springs which have three mutually perpendicular stiffnesses k_{xi}, k_{yi}, k_{zi} for $i = 1, \ldots, n$, as shown in Fig. 28.7. The xyz axes are fixed in the body parallel to these directions through an arbitrary origin O. The inertia properties of the body are given by the coordinates $(\bar{x}, \bar{y}, \bar{z})$ of the center-of-mass

Fig. 28.6. Rigid body on isolator springs.

G, the mass m, and the moments I_x, I_y, I_z and products I_{xy}, I_{yz}, I_{xz} of inertia with respect to the xyz axes. The equations of motion for small free vibration can be placed in the form

$$\mathbf{M\ddot{x} + Kx = 0}$$

by using the following sixth-order displacement vector

$$\mathbf{x} = \begin{bmatrix} x_0 \\ y_0 \\ z_0 \\ \theta_x \\ \theta_y \\ \theta_z \end{bmatrix}.$$

where the first three elements describe the displacement of O and the second three elements describe the angular displacement of the $Oxyz$ frame. The matrices $\mathbf{K}$ and $\mathbf{M}$ are sixth-order symmetric square matrices.

$$\mathbf{K} = \begin{bmatrix}
\sum_{i=1}^{n} k_{xi} & 0 & 0 & 0 & \sum_{i=1}^{n} k_{xi}z_i & -\sum_{i=1}^{n} k_{xi}y_i \\
0 & \sum_{i=1}^{n} k_{yi} & 0 & -\sum_{i=1}^{n} k_{yi}z_i & 0 & \sum_{i=1}^{n} k_{yi}x_i \\
0 & 0 & \sum_{i=1}^{n} k_{zi} & \sum_{i=1}^{n} k_{zi}y_i & -\sum_{i=1}^{n} k_{zi}x_i & 0 \\
0 & -\sum_{i=1}^{n} k_{yi}z_i & \sum_{i=1}^{n} k_{zi}y_i & \sum_{i=1}^{n} (k_{yi}z_i^2 + k_{zi}y_i^2) & -\sum_{i=1}^{n} k_{zi}x_iy_i & -\sum_{i=1}^{n} k_{yi}x_iz_i \\
\sum_{i=1}^{n} k_{xi}z_i & 0 & -\sum_{i=1}^{n} k_{zi}x_i & -\sum_{i=1}^{n} k_{zi}x_iy_i & \sum_{i=1}^{n} (k_{zi}x_i^2 + k_{xi}z_i^2) & -\sum_{i=1}^{n} k_{xi}y_iz_i \\
-\sum_{i=1}^{n} k_{xi}y_i & \sum_{i=1}^{n} k_{yi}x_i & 0 & -\sum_{i=1}^{n} k_{yi}z_ix_i & -\sum_{i=1}^{n} k_{xi}y_iz_i & \sum_{i=1}^{n} (k_{xi}y_i^2 + k_{yi}x_i^2)
\end{bmatrix}$$

$$\mathbf{M} = \begin{bmatrix}
m & 0 & 0 & 0 & m\bar{z} & -m\bar{y} \\
0 & m & 0 & -m\bar{z} & 0 & m\bar{x} \\
0 & 0 & m & m\bar{y} & -m\bar{x} & 0 \\
0 & -m\bar{z} & m\bar{y} & I_x & -I_{xy} & -I_{xz} \\
m\bar{z} & 0 & -m\bar{x} & -I_{xy} & I_y & -I_{yz} \\
-m\bar{y} & m\bar{x} & 0 & -I_{xz} & -I_{yz} & I_z
\end{bmatrix}$$

Much of the elastic coupling is eliminated when the springs are located symmetrically. The inertia couplings are reduced if the origin 0 is taken through the center-of-mass G. If in addition the axes are principal axes of inertia, then there is no inertia coupling.

FIG. 28.7. General rigid body showing typical elastic suspension at point P_i.

TORSIONAL VIBRATION

Figure 28.8 shows a crankshaft of an engine; a mathematically idealized model is portrayed in Fig. 28.9. The torsional spring constants k_j represent the equivalent torsional stiffness of the complex crankshaft structure when supported in its bearings. The mass moments of inertia J_j represent the equivalent inertia of all the rotating and re-

FIG. 28.8. Crankshaft with flywheel.

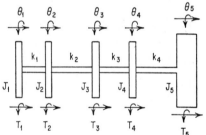

FIG. 28.9. Idealized equivalent torsional system for crankshaft.

ciprocating parts associated with the jth cylinder. The angles of rotation are θ_j and the applied torques are T_j. The equations of motion may be written in matrix notation as follows:

$$
\begin{bmatrix} J_1 & 0 & 0 & 0 & 0 \\ 0 & J_2 & 0 & 0 & 0 \\ 0 & 0 & J_3 & 0 & 0 \\ 0 & 0 & 0 & J_4 & 0 \\ 0 & 0 & 0 & 0 & J_5 \end{bmatrix} \begin{bmatrix} \ddot{\theta}_1 \\ \ddot{\theta}_2 \\ \ddot{\theta}_3 \\ \ddot{\theta}_4 \\ \ddot{\theta}_5 \end{bmatrix} + \begin{bmatrix} k_1 & -k_1 & 0 & 0 & 0 \\ -k_1 & k_1+k_2 & -k_2 & 0 & 0 \\ 0 & -k_2 & k_2+k_3 & -k_3 & 0 \\ 0 & 0 & -k_3 & k_3+k_4 & -k_4 \\ 0 & 0 & 0 & -k_4 & k_4 \end{bmatrix} \begin{bmatrix} \theta_1 \\ \theta_2 \\ \theta_3 \\ \theta_4 \\ \theta_5 \end{bmatrix} = \begin{bmatrix} T_1 \\ T_2 \\ T_3 \\ T_4 \\ T_5 \end{bmatrix}
$$

As the whole system can rotate as a rigid body without elastic deformation of the crankshaft, the potential energy is positive rather than positive definite and the stiffness matrix is singular.

THE MATRIX EIGENVALUE PROBLEM

In the following sections the solutions to both free and forced vibration problems are given in terms of solutions to a specialized algebraic problem known as the matrix eigenvalue problem. In the present section a general theoretical discussion of the matrix eigenvalue problem is given. Later, in the section on numerical methods, practical procedures for solving the eigenvalue problem are described.

The free-vibration equation for an undamped system,

$$\mathbf{M\ddot{x}} + \mathbf{Kx} = 0 \tag{28.10}$$

follows from Eq. (28.8) when the excitation $\mathbf{f}$ and the damping $\mathbf{C}$ vanish. If a solution for $\mathbf{x}$ is assumed in the form

$$\mathbf{x} = \Re \{\mathbf{v}e^{j\omega t}\}$$

where $\mathbf{v}$ is a column vector of unknown amplitudes, ω is an unknown frequency, j is the square root of -1, and $\Re \{\ \}$ signifies "the real part of," it is found on substituting in Eq. (28.10) that it is necessary for $\mathbf{v}$ and ω to satisfy the following algebraic equation:

$$\mathbf{Kv} = \omega^2 \mathbf{Mv} \tag{28.11}$$

This algebraic problem is called the *matrix eigenvalue problem*. Where necessary it is called the *real* eigenvalue problem to distinguish it from the *complex* eigenvalue problem described in the section on vibration of systems with damping. Equation (28.11) defines the central problem of this chapter. The coefficient matrices $\mathbf{K}$ and $\mathbf{M}$ are known; the problem is to determine the scalar ω^2 and vector $\mathbf{v}$. This is not a simple problem, nor does it have a simple easily understood solution; however, the fact that the matrices $\mathbf{K}$ and $\mathbf{M}$ are always symmetric in vibration problems and that $\mathbf{M}$ is always positive definite does guarantee that the nature of the solution is a good deal simpler than it otherwise might be.

To indicate the formal solution to Eq. (28.11), it is rewritten as

$$(\mathbf{K} - \omega^2\mathbf{M})\mathbf{v} = 0 \tag{28.12}$$

which can be interpreted as a set of n homogeneous algebraic equations for the n elements v_j. This set always has the trivial solution

$$\mathbf{v} = 0$$

It also has nontrivial solutions if the determinant of the matrix multiplying the vector v is zero, i.e., if

$$\det (\mathbf{K} - \omega^2\mathbf{M}) = 0 \tag{28.13}$$

When the determinant is expanded, a polynomial of order n in ω^2 is obtained. Equation (28.13) is known as the *characteristic equation* or *frequency equation*. The restrictions that $\mathbf{M}$ and $\mathbf{K}$ be symmetric and that $\mathbf{M}$ be positive definite are sufficient to ensure[4] that there are n real roots for ω^2. If $\mathbf{K}$ is singular, at least one root is zero. If $\mathbf{K}$ is positive definite, all roots are positive. The n roots determine the n *natural frequencies* $\omega_r(r = 1, \ldots, n)$ of free vibration. These roots of the characteristic equation are also known as *normal values, characteristic values, proper values, latent roots,* or *eigenvalues*. When a natural frequency ω_r is known, it is possible to return to Eq. (28.12) and solve for the corresponding vector $\mathbf{v}_r$ to within a multiplicative constant. The eigenvalue problem does not fix the absolute amplitude of the vectors $\mathbf{v}$, only the relative amplitudes of the n coordinates. In the sections which follow, the absolute amplitudes are determined in both free and forced vibration problems. There are n independent vectors $\mathbf{v}_r$ corresponding to the n natural frequencies which are known as *natural modes*. These vectors are also known as *normal modes, characteristic vectors, proper vectors, latent vectors,* or *eigenvectors*.

MODAL AND SPECTRAL MATRICES

The complete solution to the eigenvalue problem of Eq. (28.11) consists of n eigenvalues and n corresponding eigenvectors. These can be assembled compactly into matrices. Let the eigenvector $\mathbf{v}_r$ corresponding to the eigenvalue ω_r^2 have elements v_{jr} (the first subscript indicates which row, the second subscript indicates which eigenvector). The n eigenvectors then can be displayed in a single square matrix $\mathbf{V}$, each column of which is an eigenvector

$$\mathbf{V} = [v_{jk}] = \begin{bmatrix} v_{11} & v_{12} & \dots & v_{1n} \\ v_{21} & v_{22} & \dots & v_{2n} \\ \dots & \dots & \dots & \dots \\ v_{n1} & v_{n2} & \dots & v_{nn} \end{bmatrix}$$

The matrix $\mathbf{V}$ is called the *modal matrix* for the eigenvalue problem, Eq. (28.11).

The n eigenvalues ω_r^2 can be assembled into a diagonal matrix Ω^2 which is known as the *spectral matrix* of the eigenvalue problem, Eq. (28.11)

$$\Omega^2 = \begin{bmatrix} \ddots & & \\ & \omega_r^2 & \\ & & \ddots \end{bmatrix} = \begin{bmatrix} \omega_1^2 & 0 & \dots & 0 \\ 0 & \omega_2^2 & \dots & 0 \\ \dots & \dots & \dots & \dots \\ 0 & 0 & \dots & \omega_n^2 \end{bmatrix}$$

Each eigenvector and corresponding eigenvalue satisfy a relation of the following form:

$$\mathbf{K}\mathbf{v}_r = \mathbf{M}\mathbf{v}_r\omega_r^2$$

By using the modal and spectral matrices it is possible to assemble all of these relations into a single matrix equation

$$\mathbf{K}\mathbf{V} = \mathbf{M}\mathbf{V}\Omega^2 \tag{28.14}$$

Equation (28.14) provides a compact display of the complete solution to the eigenvalue problem Eq. (28.11).

PROPERTIES OF THE SOLUTION

The eigenvectors corresponding to different eigenvalues can be shown[5] to satisfy the following *orthogonality relations*. When $\omega_r^2 \neq \omega_s^2$,

$$\mathbf{v}_r^T\mathbf{K}\mathbf{v}_s = 0 \qquad \mathbf{v}_r^T\mathbf{M}\mathbf{v}_s = 0 \tag{28.15}$$

In case the characteristic equation has a p-fold multiple root for ω^2, then there is a p-fold infinity of corresponding eigenvectors. In this case, however, it is always possible[4] to choose p of these vectors which mutually satisfy Eq. (28.15) and to express any other eigenvector corresponding to the multiple root as a linear combination of the p vectors selected. If these p vectors are included with the eigenvectors corresponding to the other eigenvalues, a set of n vectors is obtained which satisfies the orthogonality relations of Eq. (28.15) for any $r \neq s$.

The orthogonality of the eigenvectors with respect to $\mathbf{K}$ and $\mathbf{M}$ implies that the following square matrices are *diagonal*.

$$\mathbf{V}^T\mathbf{K}\mathbf{V} = \begin{bmatrix} \ddots & & \\ & \mathbf{v}_r^T\mathbf{K}\mathbf{v}_r & \\ & & \ddots \end{bmatrix}$$

$$\mathbf{V}^T\mathbf{M}\mathbf{V} = \begin{bmatrix} \ddots & & \\ & \mathbf{v}_r^T\mathbf{M}\mathbf{v}_r & \\ & & \ddots \end{bmatrix} \tag{28.16}$$

The elements $\mathbf{v}_r^T\mathbf{K}\mathbf{v}_r$ along the main diagonal of $\mathbf{V}^T\mathbf{K}\mathbf{V}$ are called the *modal stiffnesses* k_r, and the elements $\mathbf{v}_r^T\mathbf{M}\mathbf{v}_r$ along the main diagonal of $\mathbf{V}^T\mathbf{M}\mathbf{V}$ are called the *modal masses* m_r. Since $\mathbf{M}$ is positive definite, all modal masses are guaranteed to be positive. When

$\mathbf{K}$ is singular, at least one of the modal stiffnesses will be zero. Each eigenvalue $\omega_r{}^2$ is the quotient of the corresponding modal stiffness divided by the corresponding modal mass; i.e.,

$$\omega_r{}^2 = \frac{k_r}{m_r}$$

In numerical work it is sometimes convenient to normalize each eigenvector so that its largest element is *unity*. In other applications it is common to normalize the eigenvectors so that the modal masses m_r all have the *same* value m where m is some convenient value such as the total mass of the system. In this case,

$$\mathbf{V}^T\mathbf{M}\mathbf{V} = m\mathbf{I} \tag{28.17}$$

and it is possible to express the inverse of the modal matrix $\mathbf{V}$ simply as

$$\mathbf{V}^{-1} = \frac{1}{m}\,\mathbf{V}^T\mathbf{M}$$

An interpretation of the modal matrix $\mathbf{V}$ can be given by showing that it defines a set of generalized coordinates for which both the inertia and stiffness matrices are uncoupled. Let $\mathbf{y}(t)$ be a column of displacements related to the original displacements $\mathbf{x}(t)$ by the following simultaneous equations:

$$\mathbf{y} = \mathbf{V}^{-1}\mathbf{x} \qquad \text{or} \qquad \mathbf{x} = \mathbf{V}\mathbf{y}$$

The potential and kinetic energies then take the forms

$$V = \tfrac{1}{2}\mathbf{x}^T\mathbf{K}\mathbf{x} = \tfrac{1}{2}\mathbf{y}^T(\mathbf{V}^T\mathbf{K}\mathbf{V})\mathbf{y}$$

$$T = \tfrac{1}{2}\dot{\mathbf{x}}^T\mathbf{M}\dot{\mathbf{x}} = \tfrac{1}{2}\dot{\mathbf{y}}^T(\mathbf{V}^T\mathbf{M}\mathbf{V})\dot{\mathbf{y}}$$

where, according to Eq. (28.16), the square matrices in parentheses on the right are *diagonal*; i.e., in the y_i coordinate system there is neither stiffness nor inertia coupling.

An alternative method for obtaining the same interpretation is to start from the eigenvalue problem of Eq. (28.11). Consider the structure of the related eigenvalue problem for $\mathbf{w}$ where again $\mathbf{w}$ is obtained from $\mathbf{v}$ by the transformation involving the modal matrix $\mathbf{V}$.

$$\mathbf{w} = \mathbf{V}^{-1}\mathbf{v} \qquad \text{or} \qquad \mathbf{v} = \mathbf{V}\mathbf{w}$$

Substituting in Eq. (28.11), premultiplying by $\mathbf{V}^T$ and using Eq. (28.14),

$$\mathbf{K}\mathbf{v} = \omega^2\mathbf{M}\mathbf{v}$$

$$\mathbf{K}\mathbf{V}\mathbf{w} = \omega^2\mathbf{M}\mathbf{V}\mathbf{w}$$

$$\mathbf{V}^T\mathbf{K}\mathbf{V}\mathbf{w} = \omega^2\mathbf{V}^T\mathbf{M}\mathbf{V}\mathbf{w}$$

$$(\mathbf{V}^T\mathbf{M}\mathbf{V})\Omega^2\mathbf{w} = \omega^2(\mathbf{V}^T\mathbf{M}\mathbf{V})\mathbf{w}$$

Now, since $\mathbf{V}^T\mathbf{M}\mathbf{V}$ is a diagonal matrix of positive elements it is permissible to cancel it from both sides, which leaves a simple diagonalized eigenvalue problem for $\mathbf{w}$:

$$\Omega^2\mathbf{w} = \omega^2\mathbf{w}$$

A modal matrix for $\mathbf{w}$ is the identity matrix $\mathbf{I}$ and the eigenvalues for $\mathbf{w}$ are the same as those for $\mathbf{v}$.

EIGENVECTOR EXPANSIONS

Any set of n independent vectors can be used as a basis for representing any other vector of order n. In the following sections, the eigenvectors of the eigenvalue problem of Eq. (28.11) are used as such a basis. An eigenvector expansion of an arbitrary vector $\mathbf{y}$

has the form

$$y = \sum_{r=1}^{n} v_r a_r \tag{28.18}$$

where the a_r are scalar *mode multipliers*. When y and the v_r are known, it is possible to evaluate the a_r by premultiplying both sides by $v_s{}^T M$. Because of the orthogonality relations of Eq. (28.15), all the terms on the right vanish except the one for which $r = s$. Inserting the value of the mode multiplier so obtained, the expansion can be rewritten as

$$y = \sum_{r=1}^{n} v_r \frac{v_r{}^T M y}{v_r{}^T M v_r} \tag{28.19}$$

or alternatively as

$$y = \sum_{r=1}^{n} \frac{v_r v_r{}^T M}{v_r{}^T M v_r} y \tag{28.20}$$

The form of Eq. (28.19) emphasizes the decomposition into eigenvectors since the fraction on the right is just a scalar. The form of Eq. (28.20) is convenient when a large number of vectors y are to be decomposed, since the fractions on the right, which are now square matrices, must be computed only once. The form of Eq. (28.20) becomes more economical of computation time when more than n vectors y have to be expanded. A useful check on the calculation of the matrices on the right of Eq. (28.20) is provided by the identity

$$\sum_{r=1}^{n} \frac{v_r v_r{}^T M}{v_r{}^T M v_r} = I \tag{28.21}$$

which follows from Eq. (28.20) because y is completely arbitrary.

An alternative expansion which is useful for expanding the excitation vector f is

$$f = \sum_{r=1}^{n} \omega_r{}^2 M v_r a_r = \sum_{r=1}^{n} M v_r \frac{v_r{}^T f}{v_r{}^T M v_r} \tag{28.22}$$

This may be viewed as an expansion of the excitation in terms of the *inertia force* amplitudes of the natural modes. The mode multiplier a_r has been evaluated by premultiplying by $v_r{}^T$. A form analogous to Eq. (28.20) and an identity corresponding to Eq. (28.21) can easily be written.

The above properties of the matrix eigenvalue problem of Eq. (28.11) are employed in the following section to obtain free and forced vibration solutions for systems without damping. Further properties of the eigenvalue problem including perturbation formulas and Rayleigh's quotient are described in the section on *Numerical Methods*.

VIBRATIONS OF SYSTEMS WITHOUT DAMPING

In this section the damping matrix C is neglected in Eq. (28.8), leaving the general formulation in the form

$$M\ddot{x} + Kx = f \tag{28.23}$$

Solutions are outlined for the following three cases: free vibration ($f = 0$), steady-state forced sinusoidal vibration ($f = \Re\{de^{i\omega t}\}$, where d is a column vector of driving-force amplitudes), and the response to general excitation (f an arbitrary function of time). The first two cases are contained in the third, but for the sake of clarity each is described separately.

FREE VIBRATION WITH SPECIFIED INITIAL CONDITIONS

It is desired to find the solution $x(t)$ of Eq. (28.23) when $f = 0$ which satisfies the initial conditions

$$x = x(0) \qquad \dot{x} = \dot{x}(0) \tag{28.24}$$

at $t = 0$ where $\mathbf{x}(0)$ and $\dot{\mathbf{x}}(0)$ are columns of prescribed initial displacements and velocities. The differential equation to be solved is identical with Eq. (28.10) which led to the matrix eigenvalue problem in the preceding section. Assuming that the solution of the eigenvalue problem is available, the general solution of the differential equation is given by an arbitrary superposition of the natural modes

$$\mathbf{x} = \sum_{r=1}^{n} \mathbf{v}_r(a_r \cos \omega_r t + b_r \sin \omega_r t)$$

where the $\mathbf{v}_r$ are the eigenvectors or natural modes, the ω_r are the natural frequencies, and the a_r and b_r are $2n$ constants of integration. The corresponding velocity is

$$\dot{\mathbf{x}} = \sum_{r=1}^{n} \mathbf{v}_r \omega_r(-a_r \sin \omega_r t + b_r \cos \omega_r t)$$

Setting $t = 0$ in these expressions and substituting in the initial conditions of Eq. (28.24) provides $2n$ simultaneous equations for determination of the constants of integration.

$$\sum_{r=1}^{n} \mathbf{v}_r a_r = \mathbf{x}(0) \qquad \sum_{r=1}^{n} \mathbf{v}_r \omega_r b_r = \dot{\mathbf{x}}(0)$$

These equations may be interpreted as eigenvector expansions of the initial displacement and velocity. The constants of integration can be evaluated by the same technique used to obtain the mode multipliers in Eq. (28.19). Using the form of Eq. (28.20), the solution of the free vibration problem then becomes

$$\mathbf{x}(t) = \sum_{r=1}^{n} \frac{\mathbf{v}_r \mathbf{v}_r{}^T \mathbf{M}}{\mathbf{v}_r{}^T \mathbf{M} \mathbf{v}_r} \left\{ \mathbf{x}(0) \cos \omega_r t + \frac{1}{\omega_r} \dot{\mathbf{x}}(0) \sin \omega_r t \right\} \qquad (28.25)$$

EXAMPLE 28.1. Consider the system of Fig. 28.5 with the following particular values:

$$m_1 = 1 \text{ lb-sec}^2/\text{in.} \qquad m_2 = 2 \text{ lb-sec}^2/\text{in.} \qquad c_1 = c_2 = c_3 = 0$$

$$k_1 = 3 \text{ lb/in.} \qquad k_2 = 0.5 \text{ lb/in.} \qquad k_3 = 1 \text{ lb/in.}$$

Substitution of these values into Eq. (28.9) yields the following coefficient matrices

$$\mathbf{M} = \begin{bmatrix} 3 & 2 \\ 2 & 2 \end{bmatrix} \qquad \mathbf{C} = 0 \qquad \mathbf{K} = \begin{bmatrix} 4 & 1 \\ 1 & 1.5 \end{bmatrix}$$

Both $\mathbf{M}$ and $\mathbf{K}$ are positive definite.

By use of numerical methods given later in this chapter, the solution of the eigenvalue problem can be found to be

$$\mathbf{V} = \begin{bmatrix} 0.2179 & -0.9179 \\ 1.0000 & 1.0000 \end{bmatrix} \qquad \Omega^2 = \begin{bmatrix} 0.7053 & 0 \\ 0 & 3.5447 \end{bmatrix}$$

$$\omega_1 = 0.8398 \text{ rad/sec} \qquad \omega_2 = 1.8827 \text{ rad/sec}$$

$$(28.26)$$

One of the terms needed for substitution in Eq. (28.25) is calculated below:

$$\frac{\mathbf{v}_1 \mathbf{v}_1{}^T \mathbf{M}}{\mathbf{v}_1{}^T \mathbf{M} \mathbf{v}_1} = \frac{\begin{bmatrix} 0.2179 \\ 1.0000 \end{bmatrix} [0.2179 \quad 1.0000] \begin{bmatrix} 3 & 2 \\ 2 & 2 \end{bmatrix}}{[0.2179 \quad 1.0000] \begin{bmatrix} 3 & 2 \\ 2 & 2 \end{bmatrix} \begin{bmatrix} 0.2179 \\ 1.0000 \end{bmatrix}} = \frac{\begin{bmatrix} 0.5782 & 0.5307 \\ 2.6537 & 2.4358 \end{bmatrix}}{3.0140}$$

$$= \begin{bmatrix} 0.1918 & 0.1761 \\ 0.8805 & 0.8082 \end{bmatrix}$$

Three similar matrices must be computed according to Eq. (28.25) to provide the complete solution which follows:

$$
\begin{bmatrix} x_1(t) \\ x_2(t) \end{bmatrix} = \begin{bmatrix} 0.1918 & 0.1761 \\ 0.8805 & 0.8082 \end{bmatrix} \begin{bmatrix} x_1(0) \\ x_2(0) \end{bmatrix} \cos 0.8398t
$$

$$
+ \begin{bmatrix} 0.2284 & 0.2097 \\ 1.0484 & 0.9623 \end{bmatrix} \begin{bmatrix} \dot{x}_1(0) \\ \dot{x}_2(0) \end{bmatrix} \sin 0.8398t
$$

$$
+ \begin{bmatrix} 0.8082 & -0.1761 \\ -0.8805 & 0.1918 \end{bmatrix} \begin{bmatrix} x_1(0) \\ x_2(0) \end{bmatrix} \cos 1.8827t
$$

$$
+ \begin{bmatrix} 0.4292 & -0.0935 \\ -0.4676 & 0.1019 \end{bmatrix} \begin{bmatrix} \dot{x}_1(0) \\ \dot{x}_2(0) \end{bmatrix} \sin 1.8827t
$$

The identity of Eq. (28.21) permits a partial check on the calculations. According to Eq. (28.21) the sum of the two square matrices multiplying the initial displacement vector $x(0)$ should be the unit matrix I. Note that this is actually the case.

STEADY-STATE FORCED SINUSOIDAL VIBRATION

It is desired to find the steady-state solution to Eq. (28.23) for single-frequency sinusoidal excitation f of the form

$$ f = \Re \{de^{j\omega t}\} $$

where d is a column vector of driving force amplitudes (these may be complex to permit differences in phase for the various components). The solution obtained is a useful approximation for lightly damped systems providing the forcing frequency ω is not too close to a natural frequency ω_r. For resonance and near-resonance conditions it is necessary to include the damping as indicated in the section which follows the present discussion.

The steady-state solution desired is assumed to have the form

$$ x = \Re \{ae^{j\omega t}\} $$

where a is an unknown column vector of response amplitudes. When f and x are inserted in Eq. (28.23), the following set of simultaneous equations for the elements of a is obtained

$$ (K - \omega^2 M)a = d \tag{28.27} $$

If ω is not a natural frequency, the square matrix $K - \omega^2 M$ is nonsingular and may be inverted to yield

$$ a = (K - \omega^2 M)^{-1}d $$

as a complete solution for the response amplitudes in terms of the driving-force amplitudes. This solution is useful if several force amplitude distributions are to be studied while the excitation frequency ω is held constant. The process requires repeated inversions if a range of frequencies is to be studied.

An alternate procedure which permits a more thorough study of the effect of frequency variation is available if the natural modes and frequencies are known. The driving-force vector d is represented by the eigenvector expansion of Eq. (28.22) and the response vector a is represented by the eigenvector expansion of Eq. (28.18):

$$ d = \sum_{r=1}^{n} \frac{M v_r v_r^T}{v_r^T M v_r} d \qquad a = \sum_{r=1}^{n} v_r c_r $$

where the c_r are unknown coefficients. Substituting these into Eq. (28.27), and making use of the fundamental eigenvalue relation of Eq. (28.11) leads to

$$ \sum_{r=1}^{n} (\omega_r^2 - \omega^2) M v_r c_r = \sum_{r=1}^{n} \frac{M v_r v_r^T}{v_r^T M v_r} d $$

This equation can be uncoupled by premultiplying both sides by $v_r{}^T$ and using the orthogonality condition of Eq. (28.15) to obtain

$$(\omega_r{}^2 - \omega^2)v_r{}^T M v_r c_r = v_r{}^T d$$

$$c_r = \frac{1}{\omega_r{}^2 - \omega^2} \frac{v_r{}^T d}{v_r{}^T M v_r}$$

The final solution is then assembled by inserting the c_r back into a and a back into x.

$$x = \Re \left\{ \sum_{r=1}^{n} \frac{e^{j\omega t}}{\omega_r{}^2 - \omega^2} \frac{v_r v_r{}^T}{v_r{}^T M v_r} d \right\} \tag{28.28}$$

This form clearly indicates the effect of frequency on the response.

EXAMPLE 28.2. Equation (28.28) is illustrated below for the system of Fig. 28.5. The eigenvectors and eigenvalues are given in Eq. (28.26). The square matrix for $r = 1$ is evaluated below.

$$\frac{v_1 v_1{}^T}{v_1{}^T M v_1} = \frac{\begin{bmatrix} 0.2179 \\ 1.0000 \end{bmatrix} [0.2179 \quad 1.0000]}{[0.2179 \quad 1.0000] \begin{bmatrix} 3 & 2 \\ 2 & 2 \end{bmatrix} \begin{bmatrix} 0.2179 \\ 1.0000 \end{bmatrix}} = \frac{\begin{bmatrix} 0.0475 & 0.2179 \\ 0.2179 & 1.0000 \end{bmatrix}}{3.0140}$$

$$= \begin{bmatrix} 0.0158 & 0.0723 \\ 0.0723 & 0.3318 \end{bmatrix}$$

Inserting this, and a similar matrix for $r = 2$, in Eq. (28.28) leads to the following representation for the steady state forced vibration:

$$\begin{bmatrix} x_1 \\ x_2 \end{bmatrix} = \Re \left\{ \frac{e^{j\omega t}}{0.7053 - \omega^2} \begin{bmatrix} 0.0158 & 0.0723 \\ 0.0723 & 0.3318 \end{bmatrix} \begin{bmatrix} d_1 \\ d_2 \end{bmatrix} \right.$$

$$+ \left. \frac{e^{j\omega t}}{3.5447 - \omega^2} \begin{bmatrix} 0.9842 & -1.0723 \\ -1.0723 & 1.1682 \end{bmatrix} \begin{bmatrix} d_1 \\ d_2 \end{bmatrix} \right\}$$

where d_1 and d_2 are the excitation amplitudes.

RESPONSE TO GENERAL EXCITATION

It is now desired to obtain the solution to Eq. (28.23) for the general case in which the excitation $f(t)$ is an arbitrary vector function of time and for which initial displacements $x(0)$ and velocities $\dot{x}(0)$ are prescribed. This kind of problem can be integrated directly as it stands by a step-by-step procedure[6] using a digital computer, or the system can be modelled on a large differential analyzer and the required excitations supplied by function generators. If the natural modes and frequencies of the system are available, it is again possible to split the problem up into n single degree-of-freedom response problems and to indicate a formal solution.

Following a procedure similar to that just used for steady-state forced sinusoidal vibrations, an eigenvector expansion of the solution is assumed

$$x(t) = \sum_{r=1}^{n} v_r c_r(t)$$

where the c_r are unknown functions of time and the known excitation $f(t)$ is expanded according to Eq. (28.22). Inserting these into Eq. (28.23) yields

$$\sum_{r=1}^{n} (M v_r \ddot{c}_r + K v_r c_r) = \sum_{r=1}^{n} \frac{M v_r v_r{}^T}{v_r{}^T M v_r} f(t)$$

Using Eq. (28.11) to eliminate $\mathbf{K}$ and premultiplying by $\mathbf{v}_r{}^T$ to uncouple the equation,

$$\ddot{c}_r + \omega_r{}^2 c_r{}^2 = \frac{\mathbf{v}_r{}^T \mathbf{f}(t)}{\mathbf{v}_r{}^T \mathbf{M} \mathbf{v}_r} \qquad (28.29)$$

is obtained as a single second-order differential equation for the time behavior of the rth mode multiplier. The initial conditions for c_r can be obtained by making eigenvector expansions of $\mathbf{x}(0)$ and $\dot{\mathbf{x}}(0)$ as was done previously for the free vibration case. Formal solutions to Eq. (28.29) can be obtained by a number of methods including Laplace transforms and variation of parameters. When these mode multipliers are substituted back to obtain $\mathbf{x}$ the general solution has the following appearance:

$$\mathbf{x}(t) = \sum_{r=1}^{n} \frac{\mathbf{v}_r \mathbf{v}_r{}^T \mathbf{M}}{\mathbf{v}_r{}^T \mathbf{M} \mathbf{v}_r} \left\{ \mathbf{x}(0) \cos \omega_r t + \frac{1}{\omega_r} \dot{\mathbf{x}}(0) \sin \omega_r t \right\}$$

$$+ \sum_{r=1}^{n} \frac{\mathbf{v}_r \mathbf{v}_r{}^T}{\omega_r \mathbf{v}_r{}^T \mathbf{M} \mathbf{v}_r} \int_0^t \mathbf{f}(t') \sin \{\omega_r(t - t')\} \, dt' \qquad (28.30)$$

This solution, while very general, is useful only if the integrals involving the excitation can be evaluated. If the elements $f_i(t)$ of $\mathbf{f}(t)$ are simple (e.g., step functions, ramps, single sine pulses, etc.), closed form solutions are available.[7] When the $f_i(t)$ are more complicated, it may be possible to evaluate the integrals numerically using a generalized Simpson's rule for trigonometric integrals.[8] It also may be possible to solve Eq. (28.29) on a small differential analyzer using a single function generator.

EXAMPLE 28.3. Consider the system of Fig. 28.5. Determine the response for a general excitation and also particularize this for the case of a velocity shock input. Since the treatment of initial conditions is given above, the initial displacements $\mathbf{x}(0)$ and velocities $\dot{\mathbf{x}}(0)$ are taken to be zero.

The natural modes and natural frequencies are given in Eq. (28.26). The square matrix for $r = 1$ required in Eq. (28.30) is calculated below.

$$\frac{\mathbf{v}_1 \mathbf{v}_1{}^T}{\omega_1 \mathbf{v}_1{}^T \mathbf{M} \mathbf{v}_1} = \frac{\begin{bmatrix} 0.2179 \\ 1.0000 \end{bmatrix} [0.2179 \quad 1.000]}{0.8398[0.2179 \quad 1.0000] \begin{bmatrix} 3 & 2 \\ 2 & 2 \end{bmatrix} \begin{bmatrix} 0.2179 \\ 1.0000 \end{bmatrix}}$$

$$= \begin{bmatrix} 0.0188 & 0.0861 \\ 0.0861 & 0.3951 \end{bmatrix}$$

A similar calculation is required for $r = 2$. Substituting in Eq. (28.30) yields the following formal solution for the response, starting from rest at $t = 0$, due to arbitrary excitation components $f_1(t)$ and $f_2(t)$:

$$\begin{bmatrix} x_1(t) \\ x_2(t) \end{bmatrix} = \begin{bmatrix} 0.0188 & 0.0861 \\ 0.0861 & 0.3951 \end{bmatrix} \int_0^t \begin{bmatrix} f_1(t') \\ f_2(t') \end{bmatrix} \sin \{0.8398(t - t')\} \, dt'$$

$$+ \begin{bmatrix} 0.5228 & -0.5695 \\ -0.5695 & 0.6205 \end{bmatrix} \int_0^t \begin{bmatrix} f_1(t') \\ f_2(t') \end{bmatrix} \sin \{1.8827(t - t')\} \, dt'$$

In Fig. 28.5, consider that the excitation is due to a prescribed base acceleration $\ddot{x}_0(t)$; then the excitation vector $\mathbf{f}$ is, according to Eq. (28.9),

$$\mathbf{f} = -\ddot{x}_0(t) \begin{bmatrix} m_1 + m_2 \\ m_2 \end{bmatrix} = -\ddot{x}_0(t) \begin{bmatrix} 3 \\ 2 \end{bmatrix}$$

With this substituted in the solution above and the matrix multiplication completed, the response is

$$\begin{bmatrix} x_1(t) \\ x_2(t) \end{bmatrix} = - \begin{bmatrix} 0.2284 \\ 1.0484 \end{bmatrix} \int_0^t \ddot{x}_0(t) \sin \{0.8398(t - t')\} \, dt'$$

$$- \begin{bmatrix} 0.4292 \\ -0.4676 \end{bmatrix} \int_0^t \ddot{x}_0(t) \sin \{1.8827(t - t')\} \, dt'$$

If the base sustains a *velocity shock* of magnitude V_0, so that $\ddot{x}_0$ vanishes at all t except at $t = 0$ where it is infinite in such a way that its integral is V_0, the solution can be further particularized as follows:

$$\begin{bmatrix} x_1(t) \\ x_2(t) \end{bmatrix} = - \begin{bmatrix} 0.2284 \\ 1.0484 \end{bmatrix} V_0 \sin (0.8398t) - \begin{bmatrix} 0.4292 \\ -0.4676 \end{bmatrix} V_0 \sin (1.8827t)$$

VIBRATION OF SYSTEMS WITH DAMPING

In this section solutions to the complete governing equation, Eq. (28.8), are discussed. The results of the preceding section for systems without damping are adequate for many purposes. There are, however, important problems in which it is necessary to include the effect of damping, e.g., problems concerned with resonance, random vibration,[9] etc. Information concerning damping in vibratory systems is surveyed in Ref. 10.

COMPLEX EIGENVALUE PROBLEM

When there is no excitation, Eq. (28.8) becomes

$$\mathbf{M\ddot{x}} + \mathbf{C\dot{x}} + \mathbf{Kx} = 0$$

which describes the free vibration of the system. As in the undamped case there are $2n$ independent solutions which can be superposed to meet $2n$ initial conditions. Assuming a solution in the form

$$\mathbf{x} = \mathbf{u}e^{pt}$$

leads to the following algebraic problem:

$$(p^2\mathbf{M} + p\mathbf{C} + \mathbf{K})\mathbf{u} = 0 \tag{28.31}$$

for the determination of the vector $\mathbf{u}$ and the scalar p. This is a *complex eigenvalue problem* because the *eigenvalue* p and the elements of the *eigenvector* $\mathbf{u}$ are, in general, complex numbers. Let a superscript C denote the complex conjugate of a quantity. If $\mathbf{u}$ and p satisfy Eq. (28.31), then so also do $\mathbf{u}^C$ and p^C. There are $2n$ roots in the negative half-plane which occur in pairs of complex conjugates or as real negative numbers. When the damping is absent, all roots lie on the imaginary axis; for small damping the roots lie near the imaginary axis. The corresponding $2n$ eigenvectors satisfy[11] the following *orthogonality* relations:

$$(p_r + p_s)\mathbf{u}_r{}^T\mathbf{M}\mathbf{u}_s + \mathbf{u}_r{}^T\mathbf{C}\mathbf{u}_s = 0$$

$$\mathbf{u}_r{}^T\mathbf{K}\mathbf{u}_s - p_r p_s \mathbf{u}_r{}^T\mathbf{M}\mathbf{u}_s = 0 \tag{28.32}$$

whenever $p_r \neq p_s$; they can be made to hold for repeated roots by suitable choice of the eigenvectors associated with a multiple root. Some indication of the relative computational effort required for systems with damping as compared to systems without damping can be had by comparing these orthogonality relations with Eq. (28.15), keeping in mind that the individual operations in Eq. (28.32) involve addition and multiplication of complex numbers. It is sometimes convenient to display a complex eigenvalue p_r in the form

$$p_r = \omega_r(-\zeta_r + j\sqrt{1 - \zeta_r^2}) \tag{28.33}$$

which is commonly employed in treating single degree-of-freedom systems. The parameter ω_r, called the *undamped natural frequency*, and the parameter ζ_r, called the *ratio of critical damping* (related to *quality factor* Q_r by the equation $Q_r = 1/2\zeta_r$), can be determined from the following quotients involving the eigenvector $\mathbf{u}_r$ and its conjugate $\mathbf{u}_r{}^C$:

$$\omega_r{}^2 = \frac{\mathbf{u}_r{}^T\mathbf{K}\mathbf{u}_r{}^C}{\mathbf{u}_r{}^T\mathbf{M}\mathbf{u}_r{}^C} \qquad 2\zeta_r\omega_r = \frac{\mathbf{u}_r{}^T\mathbf{C}\mathbf{u}_r{}^C}{\mathbf{u}_r{}^T\mathbf{M}\mathbf{u}_r{}^C} \tag{28.34}$$

FORMAL SOLUTIONS

If the solution to the eigenvalue problem of Eq. (28.31) is available, it is possible to exhibit a general solution to the governing equation

$$\mathbf{M}\ddot{\mathbf{x}} + \mathbf{C}\dot{\mathbf{x}} + \mathbf{K}\mathbf{x} = \mathbf{f} \tag{28.8}$$

for arbitrary excitation $\mathbf{f}(t)$ which meets prescribed initial conditions for $\mathbf{x}(0)$ and $\dot{\mathbf{x}}(0)$ at $t = 0$. The solutions given below apply to the case where the $2n$ eigenvalues occur as n pairs of complex conjugates (which is usually the case when the damping is light). This does, however, restrict the treatment to systems with nonsingular stiffness matrices $\mathbf{K}$ because if $\omega_r{}^2 = 0$ is an undamped eigenvalue the corresponding eigenvalues in the presence of damping are real. All quantities in the solutions below are *real*. These forms have been obtained by breaking down complex solutions[12] into real and imaginary parts and recombining. Introducing the notation

$$p_r = -\alpha_r + j\beta_r \qquad \mathbf{u}_r = \mathbf{v}_r + j\mathbf{w}_r$$

for the real and imaginary parts of eigenvalues and eigenvectors, note that

$$\alpha_r = \zeta_r\omega_r \qquad \beta_r = \omega_r\sqrt{1 - \zeta_r{}^2}$$

The general solution to Eq. (28.8) is then

$$\mathbf{x}(t) = \sum_{r=1}^{n} \frac{2}{a_r{}^2 + b_r{}^2} \{\mathbf{G}_r\mathbf{M}\dot{\mathbf{x}}(0) + (-\alpha_r\mathbf{G}_r\mathbf{M} + \beta_r\mathbf{H}_r\mathbf{M} + \mathbf{G}_r\mathbf{C})\mathbf{x}(0)\}e^{-\alpha_r t}\cos\beta_r t$$

$$+ \sum_{r=1}^{n} \frac{2}{a_r{}^2 + b_r{}^2} \{\mathbf{H}_r\mathbf{M}\dot{\mathbf{x}}(0) + (-\beta_r\mathbf{G}_r\mathbf{M} - \alpha_r\mathbf{H}_r\mathbf{M} + \mathbf{H}_r\mathbf{C})\mathbf{x}(0)\}e^{-\alpha_r t}\sin\beta_r t$$

$$+ \sum_{r=1}^{n} \frac{2}{a_r{}^2 + b_r{}^2} \mathbf{G}_r \int_0^t \mathbf{f}(t')e^{-\alpha_r(t-t')}\cos\beta_r(t - t')\,dt'$$

$$+ \sum_{r=1}^{n} \frac{2}{a_r{}^2 + b_r{}^2} \mathbf{H}_r \int_0^t \mathbf{f}(t')e^{-\alpha_r(t-t')}\sin\beta_r(t - t')\,dt' \tag{28.35}$$

where

$$a_r = -2\alpha_r(\mathbf{v}_r{}^T\mathbf{M}\mathbf{v}_r - \mathbf{w}_r{}^T\mathbf{M}\mathbf{w}_r) - 4\beta_r\mathbf{v}_r{}^T\mathbf{M}\mathbf{w}_r + \mathbf{v}_r{}^T\mathbf{C}\mathbf{v}_r - \mathbf{w}_r{}^T\mathbf{C}\mathbf{w}_r$$

$$b_r = 2\beta_r(\mathbf{v}_r{}^T\mathbf{M}\mathbf{v}_r - \mathbf{w}_r{}^T\mathbf{M}\mathbf{w}_r) - 4\alpha_r\mathbf{v}_r{}^T\mathbf{M}\mathbf{w}_r + 2\mathbf{v}_r{}^T\mathbf{C}\mathbf{w}_r$$

$$\mathbf{A}_r = \mathbf{v}_r\mathbf{v}_r{}^T - \mathbf{w}_r\mathbf{w}_r{}^T \qquad\qquad \mathbf{B}_r = \mathbf{v}_r\mathbf{w}_r{}^T + \mathbf{w}_r\mathbf{v}_r{}^T$$

$$\mathbf{G}_r = a_r\mathbf{A}_r + b_r\mathbf{B}_r \qquad\qquad \mathbf{H}_r = b_r\mathbf{A}_r - a_r\mathbf{B}_r$$

The solution of Eq. (28.35) should be compared with the corresponding solution of Eq. (28.30) for systems without damping. When the damping matrix $\mathbf{C} = 0$, Eq. (28.35) reduces to Eq. (28.30).

For the important special case of steady-state forced sinusoidal excitation of the form

$$\mathbf{f} = \mathcal{R}\{\mathbf{d}e^{j\omega t}\}$$

where $\mathbf{d}$ is a column of driving-force amplitudes, the steady-state portion of the response can be written as follows, using the above notation:

$$\mathbf{x}(t) = \Re\left\{ \sum_{r=1}^{n} \frac{2e^{j\omega t}}{a_r^2 + b_r^2} \frac{\alpha_r \mathbf{G}_r + \beta_r \mathbf{H}_r + j\omega \mathbf{G}_r}{\omega_r^2 - \omega^2 + j2\zeta_r\omega_r\omega} \mathbf{d}\right\} \qquad (28.36)$$

This result reduces to Eq. (28.28) when the damping matrix $\mathbf{C}$ is set equal to zero.

EXAMPLE 28.4.   Consider the system of Fig. 28.5 with the same mass and stiffness values as used heretofore but now including damping coefficients

$$c_1 = 0.10 \text{ lb-sec/in.} \qquad c_2 = 0.02 \text{ lb-sec/in.} \qquad c_3 = 0.04 \text{ lb-sec/in.}$$

The coefficient matrices of Eq. (28.9) thus have the following numerical values:

$$\mathbf{M} = \begin{bmatrix} 3 & 2 \\ 2 & 2 \end{bmatrix} \quad \mathbf{C} = \begin{bmatrix} 0.14 & 0.04 \\ 0.04 & 0.06 \end{bmatrix} \quad \mathbf{K} = \begin{bmatrix} 4 & 1 \\ 1 & 1.5 \end{bmatrix}$$

The complex eigenvalues p_r and eigenvectors $\mathbf{u}_r$ are given below, correct to four decimal places.

$$p_r = -\alpha_r + j\beta_r \qquad \mathbf{u}_r = \mathbf{v}_r + j\mathbf{w}_r$$

$$2\alpha_1 = 0.0279 \qquad \alpha_1 = \zeta_1\omega_1 = 0.0139 \qquad \zeta_1 = 0.0166$$

$$\beta_1 = 0.8397 \qquad\qquad \omega_1 = 0.8398 \qquad \omega_1^2 = 0.7053$$

$$2\alpha_2 = 0.1221 \qquad \alpha_2 = \zeta_2\omega_2 = 0.0611 \qquad \zeta_2 = 0.0324$$

$$\beta_2 = 1.8818 \qquad\qquad \omega_2 = 1.8828 \qquad \omega_2^2 = 3.5449$$

$$\mathbf{V} = \begin{bmatrix} 0.2179 & -0.9179 \\ 1.0000 & 1.0000 \end{bmatrix} \quad \mathbf{W} = \begin{bmatrix} 0.0016 & 0.0010 \\ 0 & 0 \end{bmatrix}$$

Note that this is a lightly damped system.   The damping ratios in the two modes are 1.66 per cent and 3.24 per cent respectively.   Alternatively, the quality factors are $Q_1 = 30.1$ and $Q_2 = 15.4$.   Using the above values to calculate a_r, b_r, A_r, B_r, G_r, and H_r for $r = 1$ and $r = 2$, on substitution into Eq. (28.35) the following general solution is found for vibration due to excitation $\mathbf{f}(t)$ which satisfies prescribed initial conditions for $\mathbf{x}(0)$ and $\dot{\mathbf{x}}(0)$ at $t = 0$:

$$\begin{bmatrix} x_1 \\ x_2 \end{bmatrix} = \left\{ \begin{bmatrix} 0.0014 & 0.0012 \\ -0.0012 & -0.0014 \end{bmatrix} \begin{bmatrix} \dot{x}_1(0) \\ \dot{x}_2(0) \end{bmatrix} \right.$$

$$+ \begin{bmatrix} 0.1919 & 0.1761 \\ 0.8805 & 0.8081 \end{bmatrix} \begin{bmatrix} x_1(0) \\ x_2(0) \end{bmatrix} \right\} e^{-0.0140t} \cos 0.8397t$$

$$+ \left\{ \begin{bmatrix} -0.0014 & -0.0012 \\ 0.0011 & 0.0014 \end{bmatrix} \begin{bmatrix} \dot{x}_1(0) \\ \dot{x}_2(0) \end{bmatrix} \right.$$

$$+ \begin{bmatrix} 0.8081 & -0.1761 \\ -0.8805 & 0.1919 \end{bmatrix} \begin{bmatrix} x_1(0) \\ x_2(0) \end{bmatrix} \right\} e^{-0.0611t} \cos 1.8818t$$

$$+ \left\{ \begin{bmatrix} 0.2285 & 0.2097 \\ 1.0485 & 0.9624 \end{bmatrix} \begin{bmatrix} \dot{x}_1(0) \\ \dot{x}_2(0) \end{bmatrix} \right.$$

$$+ \begin{bmatrix} 0.0017 & 0.0020 \\ 0.0141 & 0.0149 \end{bmatrix} \begin{bmatrix} x_1(0) \\ x_2(0) \end{bmatrix} \right\} e^{-0.0140t} \sin 0.8397t$$

$$+ \left\{ \begin{bmatrix} 0.4294 & -0.0936 \\ -0.4679 & 0.1020 \end{bmatrix} \begin{bmatrix} \dot{x}_1(0) \\ \dot{x}_2(0) \end{bmatrix} \right.$$

$$+ \begin{bmatrix} 0.0269 & -0.0053 \\ -0.0283 & 0.0056 \end{bmatrix} \begin{bmatrix} x_1(0) \\ x_2(0) \end{bmatrix} \right\} e^{-0.0611t} \sin 1.8818t$$

$$+ \begin{bmatrix} 0.0002 & 0.0004 \\ 0.0004 & -0.0011 \end{bmatrix} \int_0^t [f(t')e^{-0.0140(t-t')} \cos 0.8397(t - t')] \, dt'$$

$$+ \begin{bmatrix} -0.0002 & -0.0004 \\ -0.0004 & 0.0011 \end{bmatrix} \int_0^t [f(t')e^{-0.0611(t-t')} \cos 1.8818(t - t')] \, dt'$$

$$+ \begin{bmatrix} 0.0188 & 0.0861 \\ 0.0861 & 0.3953 \end{bmatrix} \int_0^t [f(t')e^{-0.0140(t-t')} \sin 0.8397(t - t')] \, dt'$$

$$+ \begin{bmatrix} 0.5230 & -0.5698 \\ -0.5698 & 0.6208 \end{bmatrix} \int_0^t [f(t')e^{-0.0611(t-t')} \sin 1.8818(t - t')] \, dt'$$

Inserting these same values into Eq. (28.36), the following expression is obtained for the steady-state forced sinusoidal vibration of the same system:

$$\begin{bmatrix} x_1 \\ x_2 \end{bmatrix} = \Re \left\{ \frac{e^{j\omega t} \left\{ \begin{bmatrix} 0.0158 & 0.0723 \\ 0.0723 & 0.3318 \end{bmatrix} + j\omega \begin{bmatrix} 0.0002 & 0.0004 \\ 0.0004 & -0.0011 \end{bmatrix} \right\}}{0.7053 - \omega^2 + 0.0279j\omega} \begin{bmatrix} d_1 \\ d_2 \end{bmatrix} \right.$$

$$\left. + \frac{e^{j\omega t} \left\{ \begin{bmatrix} 0.9842 & -1.0724 \\ -1.0724 & 1.1683 \end{bmatrix} + j\omega \begin{bmatrix} -0.0002 & -0.0004 \\ -0.0004 & 0.0011 \end{bmatrix} \right\}}{3.5449 - \omega^2 + 0.1221j\omega} \begin{bmatrix} d_1 \\ d_2 \end{bmatrix} \right\}$$

APPROXIMATE SOLUTIONS

For a lightly damped system the exact solutions of Eq. (28.35) and Eq. (28.36) can be abbreviated considerably by making approximations based on the smallness of the damping. A systematic method of doing this is to consider the system without damping as a base upon which an infinitesimal amount of damping is superposed as a perturbation. Using the calculus, it is possible to obtain the infinitesimal perturbation of the base solution which results. While only strictly valid for infinitesimal perturbations, this procedure provides useful approximations for small amounts of damping. In the section on numerical methods which follows, this process is carried out in detail to obtain complex eigenvectors and eigenvalues. It is found there that the effect of a small amount of damping on a natural vibration is (1) to introduce small exponential decay without altering the frequency and (2) to split the original real eigenvector into a pair of complex vectors each having the same real part as the undamped mode but having small conjugate imaginary parts. Thus in the perturbation solution the imaginary part of the eigenvalue p_r is simply $j\omega_r$, where ω_r is the corresponding undamped natural frequency and the real part of the eigenvector $\mathbf{u}_r$ is simply the corresponding undamped natural mode $\mathbf{v}_r$. The real part of the eigenvalue, i.e., $\alpha_r = \zeta_r \omega_r$, and the imaginary part of the eigenvector, i.e., $j\mathbf{w}_r$, are small quantities (strictly speaking, they are first-order infinitesimals).

This perturbation approximation can be continued into Eqs. (28.35) and (28.36) by simply neglecting all squares and products of the small quantities α_r, ζ_r, $\mathbf{w}_r$, and $\mathbf{C}$. When this is done it is found that the formulas of Eqs. (28.35) and (28.36) may still be used if the parameters therein are obtained from the simplified expressions below.

$$\left. \begin{aligned} \alpha_r &= \zeta_r \omega_r & \beta_r &= \omega_r \\ a_r &= -4\omega_r \mathbf{v}_r^T \mathbf{M} \mathbf{w}_r & b_r &= 2\omega_r \mathbf{v}_r^T \mathbf{M} \mathbf{v}_r \\ & a_r^2 + b_r^2 = 4\omega_r^2 (\mathbf{v}_r^T \mathbf{M} \mathbf{v}_r)^2 \\ \mathbf{A}_r &= \mathbf{v}_r \mathbf{v}_r^T & \mathbf{B}_r &= \mathbf{v}_r \mathbf{w}_r^T + \mathbf{w}_r \mathbf{v}_r^T \\ \mathbf{G}_r &= -4\omega_r (\mathbf{v}_r^T \mathbf{M} \mathbf{w}_r) \mathbf{v}_r \mathbf{v}_r^T + 2\omega_r (\mathbf{v}_r^T \mathbf{M} \mathbf{v}_r)(\mathbf{v}_r \mathbf{w}_r^T + \mathbf{w}_r \mathbf{v}_r^T) \\ \mathbf{H}_r &= 2\omega_r (\mathbf{v}_r^T \mathbf{M} \mathbf{v}_r) \mathbf{v}_r \mathbf{v}_r^T \end{aligned} \right\} \qquad (28.37)$$

For example, the steady-state forced sinusoidal solution of Eq. (28.36) takes the following explicit form in the perturbation approximation:

$$\mathbf{x}(t) = \Re \left\{ \sum_{r=1}^{n} \frac{e^{j\omega t}}{\mathbf{v}_r{}^T\mathbf{M}\mathbf{v}_r} \frac{\mathbf{v}_r\mathbf{v}_r{}^T + \frac{j\omega}{\omega_r}\left[\mathbf{v}_r\mathbf{w}_r{}^T - \left(2\frac{\mathbf{v}_r{}^T\mathbf{M}\mathbf{w}_r}{\mathbf{v}_r{}^T\mathbf{M}\mathbf{v}_r}\right)\mathbf{v}_r\mathbf{v}_r{}^T + \mathbf{w}_r\mathbf{v}_r{}^T\right]}{\omega_r{}^2 - \omega^2 + j2\zeta_r\omega_r\omega} \, \mathbf{d} \right\} \quad (28.38)$$

When this approximation is applied to the numerical example previously given it is found that the results are almost identical. There are a few entries which differ in the fourth decimal place. The above formula simplifies somewhat if $\mathbf{w}_r$ is chosen to be orthogonal to $\mathbf{v}_r$ with respect to $\mathbf{M}$ since this causes the coefficient of $\mathbf{v}_r\mathbf{v}_r{}^T$ to vanish. In the perturbation procedure described in the section on numerical methods, the eigenvectors possess this orthogonality property before they are normalized.

A cruder approximation, which is often used, is based on accepting the complex eigenvalue $p_r = -\alpha_r + j\omega_r$ but completely neglecting the imaginary part $j\mathbf{w}_r$ of the eigenvector $\mathbf{u}_r = \mathbf{v}_r + j\mathbf{w}_r$. It is thus assumed that the undamped mode $\mathbf{v}_r$ still applies for the system with damping. The approximate parameter values of Eq. (28.37) are further simplified by this assumption; e.g., $a_r = 0$, $\mathbf{B}_r = \mathbf{G}_r = 0$. The steady forced sinusoidal response of Eq. (28.38) reduces to

$$\mathbf{x}(t) = \Re \left\{ \sum_{r=1}^{n} \frac{e^{j\omega t}}{\omega_r{}^2 - \omega^2 + j2\zeta_r\omega_r\omega} \frac{\mathbf{v}_r\mathbf{v}_r{}^T}{\mathbf{v}_r{}^T\mathbf{M}\mathbf{v}_r} \, \mathbf{d} \right\} \quad (28.39)$$

This approximation should be compared with the undamped solution of Eq. (28.28), as well as with the exact solution of Eq. (28.36) and the perturbation approximation of Eq. (28.38). An idea of the error committed in the cruder approximation is provided by comparing the numerical solutions to the examples used to illustrate Eq. (28.28) and Eq. (28.36). Although large errors are possible at high frequencies, the discrepancies when $\omega = \omega_2$ (the highest natural frequency) are only of the order of 1 per cent.

NUMERICAL METHODS

THE REAL EIGENVALUE PROBLEM

Given the symmetric matrices $\mathbf{K}$ and $\mathbf{M}$ with $\mathbf{M}$ positive definite, the problem is to determine the eigenvectors $\mathbf{v}_r$ and the eigenvalues $\omega_r{}^2$ which satisfy

$$\mathbf{K}\mathbf{v} = \omega^2\mathbf{M}\mathbf{v} \quad (28.11)$$

Earlier in this chapter the basic properties of the solution are presented. Here a description is given of some additional properties of computational interest: Rayleigh's quotient, perturbation formulas, and Jacobi's method.

RAYLEIGH'S QUOTIENT.   If Eq. (28.11) is premultiplied by $\mathbf{v}^T$ the following scalar equation is obtained:

$$\mathbf{v}^T\mathbf{K}\mathbf{v} = \omega^2\mathbf{v}^T\mathbf{M}\mathbf{v}$$

The positive definiteness of $\mathbf{M}$ guarantees that $\mathbf{v}^T\mathbf{M}\mathbf{v}$ is nonzero so that it is permissible to solve for ω^2.

$$\omega^2 = \frac{\mathbf{v}^T\mathbf{K}\mathbf{v}}{\mathbf{v}^T\mathbf{M}\mathbf{v}} \quad (28.40)$$

This quotient is called "Rayleigh's quotient." It also may be derived by equating time averages of potential and kinetic energy under the assumption that the vibratory system is executing simple harmonic motion at frequency ω with amplitude ratios given by $\mathbf{v}$ or by equating the maximum value of kinetic energy to the maximum value of potential energy under the same assumption. Rayleigh's quotient has the following interesting properties.

1. When $\mathbf{v}$ is an eigenvector $\mathbf{v}_r$ of Eq. (28.11), then Rayleigh's quotient is equal to the corresponding eigenvalue $\omega_r{}^2$.

2. If $\mathbf{v}$ is an approximation to $\mathbf{v}_r$ with an error which is a *first-order* infinitesimal, then Rayleigh's quotient is an approximation to $\omega_r{}^2$ with an error which is a *second-order* infinitesimal; i.e., Rayleigh's quotient is *stationary* in the neighborhoods of the true eigenvectors.

3. As $\mathbf{v}$ varies through all of n-dimensional vector space, Rayleigh's quotient remains bounded between the smallest and largest eigenvalues.

A common engineering application of Rayleigh's quotient involves simply evaluating Eq. (28.40) for a trial vector $\mathbf{v}$ which is selected on the basis of physical insight. When eigenvectors are obtained by approximate methods, Rayleigh's quotient provides a means of improving the accuracy in the corresponding eigenvalue. If the elements of an approximate eigenvector whose largest element is unity are correct to k decimal places, then Rayleigh's quotient can be expected to be correct to about $2k$ significant decimal places. Several interesting numerical procedures[13,14] for solving eigenvalue problems make use of the stationary property of Rayleigh's quotient.

PERTURBATION FORMULAS. The perturbation formulas which follow provide the basis for estimating the changes in the eigenvalues and the eigenvectors which result from *small* changes in the stiffness and inertia parameters of a system. The formulas are strictly accurate only for infinitesimal changes but are useful approximations for *small* changes. They may be used by the designer to estimate the effects of a proposed change in a vibratory system and may also be used to analyze the effects of minor errors in the measurement of the system properties. Iterative procedures for the solution of eigenvalue problems[15-17] can be based on these formulas. They are employed here to obtain approximations to the complex eigenvalues and eigenvectors of a lightly damped vibratory system in terms of the corresponding solutions for the same system without damping.

Suppose that the modal matrix $\mathbf{V}$ and the spectral matrix Ω^2 for the eigenvalue problem

$$\mathbf{KV} = \mathbf{MV}\Omega^2 \qquad (28.14)$$

are known. Consider the perturbed eigenvalue problem

$$\mathbf{K_*V_*} = \mathbf{M_*V_*}\Omega_*{}^2$$

where

$$\mathbf{K_*} = \mathbf{K} + d\mathbf{K} \qquad \mathbf{M_*} = \mathbf{M} + d\mathbf{M}$$

$$\mathbf{V_*} = \mathbf{V} + d\mathbf{V} \qquad \Omega_*{}^2 = \Omega^2 + d\Omega^2$$

The perturbation formula for the elements $d\omega_r{}^2$ of the diagonal matrix $d\Omega^2$ is

$$d\omega_r{}^2 = \frac{\mathbf{v}_r{}^T \, d\mathbf{K} \, \mathbf{v}_r - \omega_r{}^2 \mathbf{v}_r{}^T \, d\mathbf{M} \, \mathbf{v}_r}{\mathbf{v}_r{}^T \mathbf{M} \mathbf{v}_r} \qquad (28.41)$$

Thus in order to determine the change in a single eigenvalue due to changes in $\mathbf{M}$ and $\mathbf{K}$, it is only necessary to know the corresponding unperturbed eigenvalue and eigenvector. To determine the change in a single eigenvector, however, it is necessary to know *all* the unperturbed eigenvalues and eigenvectors. The following algorithm may be used to evaluate the perturbations of both the modal matrix and the spectral matrix. Calculate

$$\mathbf{F} = \mathbf{V}^T d\mathbf{K} \, \mathbf{V} - \mathbf{V}^T \, d\mathbf{M} \, \mathbf{V}\Omega^2$$

and

$$\mathbf{L} = \mathbf{V}^T\mathbf{M}\mathbf{V}$$

The matrix $\mathbf{L}$ is a diagonal matrix of positive elements and hence is easily inverted. Continue calculating

$$\mathbf{G} = \mathbf{L}^{-1}\mathbf{F} = [g_{jk}] \qquad \text{and} \qquad \mathbf{H} = [h_{jk}]$$

where

$$h_{jk} = \begin{cases} 0 & \text{if } \omega_j{}^2 = \omega_k{}^2 \\ \dfrac{g_{jk}}{\omega_k{}^2 - \omega_j{}^2} & \text{if } \omega_j{}^2 \neq \omega_k{}^2 \end{cases}$$

Then, finally, the perturbations of the modal matrix and the spectral matrix are given by

$$dV = VH \qquad d\Omega^2 = \left[\diagdown g_{ii} \diagdown \right] \qquad (28.42)$$

These formulas are derived by taking the total differential of Eq. (28.14), premultiplying each term by V^T, and using a relation derived by taking the transpose of Eq. (28.14). Since eigenvectors are undetermined to the extent of a scale factor, the vector changes in the eigenvectors have a degree of indeterminateness. The particular selection in the above algorithm makes the change in each eigenvector orthogonal with respect to M to the corresponding unperturbed eigenvector, i.e.,

$$v_i{}^T M \, dv_i = 0$$

JACOBI METHOD. The most satisfactory general-purpose method for obtaining the complete solution to the real eigenvalue problem of Eq. (28.14) appears to be based on *diagonalization by successive rotation*.[18-20] This procedure involves a simple repetitive algorithm which is relatively easy to program. No special provisions for equal or nearly equal eigenvalues are required. The procedure has proved to be remarkably immune to serious round-off errors. The computational time is only proportional to n^3 although the proportionality constant is so large (roughly about 60) that the method is quite unsuited to hand computation even for small n.

The basic computational operation in the method is the resolution of a single symmetric matrix A into its modal matrix U and its spectral matrix Λ according to the relation

$$A = U\Lambda U^T \qquad (28.43)$$

where Λ is a diagonal matrix of the eigenvalues λ_r, and the columns of U are the eigenvectors u_r for the eigenvalue problem

$$Au = \lambda u \qquad \text{or} \qquad AU = U\Lambda$$

The modal matrix U in Eq. (28.43) has the additional property that its columns are normalized so that $u_r{}^T u_r = 1$, i.e., $U^T = U^{-1}$.

Before describing how the resolution of Eq. (28.43) is arrived at by successive rotations, we indicate how the solution to the real eigenvalue problem of Eq. (28.14) with *two* symmetrical matrices K and M can be obtained by two successive applications of Eq. (28.43).

It is first observed that real integer powers of A can be resolved in the form

$$A^m = U\Lambda^m U^T \qquad (28.44)$$

where Λ^m is a diagonal matrix of elements $\lambda_r{}^m$; i.e., the eigenvectors of A^m are identical with the eigenvectors of A and the eigenvalues of A^m are simply the mth powers of the corresponding eigenvalues of A. The resolution of Eq. (28.44) then can be used to *define* fractional powers of A. Negative powers can likewise be defined, provided none of the λ_r equal zero, i.e., provided A is nonsingular.

Returning to the eigenvalue problem of Eq. (28.14) it is supposed that the resolution of Eq. (28.43) can be applied to the inertia matrix M and in particular that the matrix $M^{-1/2}$ can be constructed according to Eq. (28.44). The problem of Eq. (28.14) now is reduced to an eigenvalue problem for a single symmetric matrix $A = M^{-1/2}KM^{1/2}$ with modal matrix $W = M^{1/2}V$ as follows:

$$KV = MV\Omega^2$$

$$KM^{-1/2}M^{1/2}V = M^{1/2}M^{1/2}V\Omega^2$$

$$(M^{-1/2}KM^{-1/2})(M^{1/2}V) = (M^{1/2}V)\Omega^2$$

$$AW = W\Omega^2$$

The solution of this last eigenvalue problem provides the spectral matrix Ω^2 of the original problem. To obtain the modal matrix V of the original problem, it is necessary to perform the matrix multiplication

$$V = M^{-\frac{1}{2}}W$$

which follows from inverting the definition of W.

It remains to indicate how the resolution of Eq. (28.43) is obtained by successive rotations. The procedure is an iterative process in which the matrix A is gradually transformed into its spectral matrix Λ by a sequence of two-dimensional rotation-of-axes transformations T_i. Starting with $A_0 = A$, the successive iterates are constructed as follows:

$$A_1 = T_1{}^T A_0 T_1$$

$$\cdot \quad \cdot \quad \cdot \quad \cdot \quad \cdot \quad \cdot \quad \cdot$$

$$A_i = T_i{}^T A_{i-1} T_i$$

Each transformation matrix T_i has the following form:

$$T_i = \begin{bmatrix} 1 & 0 & \cdots & 0 & 0 & 0 & 0 & \cdots & 0 \\ 0 & 1 & \cdots & 0 & 0 & 0 & 0 & \cdots & 0 \\ \cdot & \cdot & \cdot & \cdot & \cdot & \cdot & \cdot & \cdot & \cdot \\ 0 & 0 & \cdots & c & 0 & 0 & -s & \cdots & 0 \\ 0 & 0 & \cdots & 0 & 1 & 0 & 0 & \cdots & 0 \\ 0 & 0 & \cdots & 0 & 0 & 1 & 0 & \cdots & 0 \\ 0 & 0 & \cdots & s & 0 & 0 & c & \cdots & 0 \\ \cdot & \cdot & \cdot & \cdot & \cdot & \cdot & \cdot & \cdot & \cdot \\ 0 & 0 & \cdots & 0 & 0 & 0 & 0 & \cdots & 1 \end{bmatrix} \begin{matrix} \\ \\ \\ p \\ \\ \\ q \\ \\ \end{matrix}$$

with p and q column labels.

where

$$c = \cos \theta \qquad s = \sin \theta$$

The matrix T_i represents a rotation through the angle θ in the (p,q) plane. By suitably selecting θ, the off-diagonal elements in the (p,q) and (q,p) positions of A_i can be made to vanish. Thus in each step a single pair of off-diagonal elements is reduced to zero at the expense of altering all the elements in the 2 rows and 2 columns involved. The T_i may be chosen so that at each step the largest off-diagonal pair is annihilated, although satisfactory results have been obtained by systematically treating the off-diagonal elements in turn without regard to their magnitude. The process converges[21] in the limit with A_i approaching the spectral matrix Λ and the continued product of the T_i approaching the modal matrix U:

$$\lim_{i \to \infty} A_i = \Lambda \qquad \lim_{i \to \infty} (T_1 T_2 \ldots T_i) = U$$

For most matrices A, convergence to about eight decimal places is obtained after[18, 20, 22] six to ten complete sweeps through the off-diagonal elements. The individual T_i possess the property that $T_i{}^{-1} = T_i{}^T$ and so does their continued product. Thus the transpose of U is also its inverse. The computation can be programmed so that the major storage requirements are $(n^2 + n)/2$ locations for the elements of A_i and n^2 locations for the elements of the continued product of the T_i.

Several noniterative procedures for solving eigenvalue problems have been programmed for digital computers. One of the most attractive[18, 23] involves using the above rotation transformation matrices T_i, not to diagonalize A, but to put it in a *tri-diagonal* form in which the main diagonal together with the two adjacent diagonals contain the only nonzero elements. This reduction can be performed exactly in a fixed number of steps without any iteration. The tridiagonal form is especially convenient for obtaining the roots of the characteristic equation by inserting successive trial eigenvalues and applying the method of false position to the value of the determinant of the tridiagonal matrix. The torsional-vibration system of Fig. 28.9 possesses a tridiagonal matrix $A = M^{-1}K$. The most popular hand computation method for such a system is exactly this process of selecting trial frequencies and evaluating the remainder in a Holzer table.[24]

THE COMPLEX EIGENVALUE PROBLEM

Given the symmetric inertia, damping, and stiffness matrices $\mathbf{M}$, $\mathbf{C}$, and $\mathbf{K}$, the problem is to determine the complex eigenvalues $p_r = -\alpha_r + j\beta_r$ and the complex eigenvectors $\mathbf{u}_r = \mathbf{v}_r + j\mathbf{w}_r$ which satisfy Eq. (28.31).

$$(p^2\mathbf{M} + p\mathbf{C} + \mathbf{K})\mathbf{u} = 0 \qquad (28.31)$$

This problem is considerably more difficult than the real eigenvalue problem of Eq. (28.11), and much less attention has been given to efficient computational procedures. See, however, Refs. 25 and 26. Here only a *perturbation method* is given which furnishes a useful approximation solution for a lightly damped system when the solution to the corresponding undamped system is available.

When $\mathbf{C} = 0$ in Eq. (28.31), the complex eigenvalue problem reduces to the real eigenvalue problem of Eq. (28.11) with $p^2 = -\omega^2$. Suppose that the real eigenvalues ω_r^2 and real eigenvectors $\mathbf{v}_r$ are known. The perturbation formulas of Eqs. (28.41) and (28.42) can be used to estimate the perturbation of the rth mode due to the addition of small damping $\mathbf{C}$ by considering the damping to be a perturbation of the stiffness matrix of the form

$$d\mathbf{K} = j\omega_r\mathbf{C}$$

In this way it is found that the perturbed solution corresponding to the rth mode consists of a pair of complex conjugate eigenvalues

$$p_r = -\alpha_r + j\omega_r \qquad p_r^C = -\alpha_r - j\omega_r$$

and a pair of complex conjugate eigenvectors

$$\mathbf{u}_r = \mathbf{v}_r + j\mathbf{w}_r \qquad \mathbf{u}_r^C = \mathbf{v}_r - j\mathbf{w}_r$$

where ω_r and $\mathbf{v}_r$ are taken directly from the undamped system, and α_r and $\mathbf{w}_r$ are small perturbations which are given below. The real part of the eigenvalue which describes the rate of decay of the corresponding free motion is given by the following quotient:

$$2\alpha_r = 2\zeta_r\omega_r = \frac{\mathbf{v}_r^T\mathbf{C}\mathbf{v}_r}{\mathbf{v}_r^T\mathbf{M}\mathbf{v}_r} \qquad (28.45)$$

The decay rate α_r and damping ratio ζ_r for a particular r depend only on the rth mode undamped solution. The imaginary part of the eigenvector $j\mathbf{w}_r$ which describes the perturbations in phase is more difficult to obtain. All the undamped eigenvalues and eigenvectors must be known. Let $\mathbf{W}$ be a square matrix whose columns are the $\mathbf{w}_r$. The following algorithm may be used to evaluate $\mathbf{W}$ when the undamped modal matrix $\mathbf{V}$ is known. Calculate

$$\mathbf{F} = \mathbf{V}^T\mathbf{C}\mathbf{V}$$

and

$$\mathbf{L} = \mathbf{V}^T\mathbf{M}\mathbf{V}$$

The matrix $\mathbf{L}$ is a diagonal matrix of positive elements and hence is easily inverted. Continue calculating

$$\mathbf{G} = \mathbf{L}^{-1}\mathbf{F} = \lfloor g_{jk} \rfloor \qquad \text{and} \qquad \mathbf{H} = [h_{jk}]$$

where

$$h_{jk} = \begin{cases} 0 & \text{if } \omega_j^2 = \omega_k^2 \\ \dfrac{g_{jk}\omega_k}{\omega_k^2 - \omega_j^2} & \text{if } \omega_j^2 \neq \omega_k^2 \end{cases}$$

Then, finally, the eigenvector perturbations are given by

$$\mathbf{W} = \mathbf{V}\mathbf{H} \qquad (28.46)$$

The main diagonal elements g_{rr} of the matrix $\mathbf{G}$ are just the quotients of Eq. (28.45) and hence equal $2\alpha_r$.

REFERENCES

1. Hildebrand, F. B.: "Methods of Applied Mathematics," Chap. 1, Prentice-Hall, Inc., Englewood Cliffs, N.J., 1965.
2. Timoshenko, S., and D. H. Young: "Advanced Dynamics," Chap. 3, McGraw-Hill Book Company, New York, 1948.
3. Crede, C. E.: "Vibration and Shock Isolation," p. 45, John Wiley & Sons, Inc., New York, 1951.
4. Ref. 1, p. 76.
5. Ref. 1, p. 75.
6. Crandall, S. H.: "Engineering Analysis, A Survey of Numerical Procedures," p. 191, McGraw-Hill Book Company, New York, 1956.
7. See Chap. 8.
8. Tranter, C. J.: "Integral Transforms in Mathematical Physics," 2d ed., p. 67, Methuen & Co., Ltd., London, and John Wiley & Sons, Inc., New York, 1956.
9. Crandall, S. H.: "Random Vibration," Technology Press, M.I.T., Cambridge, Mass., 1958.
10. Ruzicka, J. E. (ed.): "Structural Damping," ASME, New York, 1959.
11. McCalley, R. B.: "Discussion of Ref. 12," *J. Appl. Mechanics*, **26**:306 (1959).
12. Foss, K. A.: *J. Appl. Mechanics*, **25**:361 (1958).
13. Crandall, S. H.: *Proc. Roy. Soc. (London)*: **A207**:416 (1951).
14. Crandall, S. H., and W. G. Strang: *J. Appl. Mechanics*, **24**:228 (1957).
15. Jahn, H. A.: *Quart. J. Mech. and Appl. Math.*, **1**:131 (1948).
16. Collar, A. R.: *Quart. J. Mech. and Appl. Math.*, **1**:145 (1948).
17. Mahalingham, S.: *J. Appl. Mechanics*, **25**:618 (1958).
18. White, P. A.: *J. Soc. Ind. Appl. Math.*, **6**:393 (1958).
19. Ref. 6, p. 118.
20. Gregory, R. T.: "Computing Eigenvalues and Eigenvectors of a Symmetric Matrix on the Illiac," *Math. Tables and Other Aids to Comp.*, **7**:215 (1953).
21. Henrici, P.: *J. Soc. Ind. Appl. Math.*, **6**:144 (1958).
22. Pope, D. A., and C. B. Tompkins: *J. Assoc. Comput. Mach.*, **4**:459 (1957).
23. Paige, L. J., and O. Taussky: "Simultaneous Linear Equations and the Determination of Eigenvalues," National Bureau of Standards, Applied Mathematics Series, No. 29, 1953.
24. See Chap. 38.
25. Wayland, H.: *Quart. Appl. Math.*, **2**:277 (1945).
26. Tarnove, I.: *J. Soc. Ind. Appl. Math.*, **6**:163 (1958).

ADDITIONAL REFERENCES

Bathe, Klaus-Jurgen, and E. L. Wilson: "Numerical Methods in Finite Element Analysis," Prentice Hall, Inc., Englewood Cliffs, N.J., 1976.
Blevins, R. D.: "Flow-Induced Vibrations," Van Nostrand Reinhold Co., New York, 1977.
Blevins, R. D.: "Formulas for Natural Frequency and Mode Shape," Van Nostrand Reinhold Co., New York, 1979.
Donea, J., ed.: "Advanced Structural Dynamics," Applied Science Publishers, London, 1979.
Jennings, A.: "Matrix Computation for Engineers and Scientists," John Wiley and Sons, New York, 1977.
Marguerre, K., and K. Woelfel: "Mechanics of Vibration," Sijthoff & Noordhoff, Alphen aan den Rijn, Netherlands, 1979.
Pilkey, W. D., and P. Y. Chang: "Modern Formulas for Statics and Dynamics," McGraw-Hill Book Co., New York, 1978.
Thomson, W. T.: "Theory of Vibrations with Applications," Prentice-Hall, Inc., Englewood Cliffs, N.J., 1981.
Vierck, R. K.: "Vibration Analysis," 2d ed., Harper & Row, New York, 1979.

29

PART I: VIBRATION OF STRUCTURES INDUCED BY FLUID FLOW

R. D. Blevins

Consultant

INTRODUCTION

Fluid around a structure can significantly alter the structure's vibrational characteristics. The presence of a quiescent fluid decreases the natural frequencies and increases the damping of the structure. A dense fluid couples the vibration of elastic structures which are adjacent to each other. Fluid flow can induce vibration. A turbulent fluid flow exerts random pressures on a structure, and these random pressures induce a random response. The structure can resonate with periodic components of the wake. If a structure is sufficiently flexible, the structural deformation under the fluid loading will in turn change the fluid force. The response can be unstable with very large structural vibrations—once the fluid velocity exceeds a critical threshold value.

Vibration induced by fluid flow can be classified by the nature of the fluid-structure interaction as shown in Fig. 29.1. Effects which are largely independent of viscosity include added mass and inertial coupling. Unsteady pressure on the surface of a structure, due to either variations in the free stream flow or turbulent fluctuations, induces a forced vibration response. Strong fluid-structure interaction phenomena result when the fluid force on a structure induces a significant response which in turn alters the fluid force. These phenomena are discussed in this section.

ADDED MASS AND INERTIAL COUPLING

If a body accelerates, decelerates, or vibrates in a fluid, then fluid is entrained by the body. This entrainment of fluid, called the *added mass* or *virtual mass effect*, occurs both in viscous and in inviscid, i.e., ideal, fluids. It is of practical importance when the fluid density is comparable to the density of the structure because then the added mass becomes a significant fraction of the total mass in dynamic motion.

Consider the rigid body shown in Fig. 29.2 which lies in a reservoir of incompressible inviscid irrotational fluid. The surface S defines the surface of the body. The body moves with velocity $U(t)$. From ideal flow theory, it can be shown that there exists a velocity potential $\Phi(x, y, z, t)$ which is a function of the special coordinates and time, such that the velocity vector is the gradient of a potential function.

$$\mathbf{V} = \nabla\Phi \tag{29.1}$$

$\mathbf{V}(x, y, z, t)$ is the fluid velocity vector. The potential function Φ satisfies Laplace's equation:[1,2]

$$\nabla^2\Phi = 0 \tag{29.2}$$

INERTIAL COUPLING EFFECTS	UNSTEADY FLOW INDUCED VIBRATION	FLOW-STRUCTURE COUPLED VIBRATION
1. Added mass 2. Inertial coupling 3. Instability due to parallel flow	1. Turbulence induced vibration 2. Ocean wave induced vibration 3. Sonic fatigue	1. Vortex induced vibration 2. Galloping and flutter 3. Fluid elastic instability

Fɪɢ. 29.1. A classification of flow-induced vibration.

The boundary condition is that on the surface of the body; the normal component of velocity must equal the velocity of the body:

$$\frac{\partial \Phi}{\partial \mathbf{n}} = \mathbf{V} \cdot \mathbf{n} \quad \text{on the surface } S$$

where $\mathbf{n}$ is the unit outward normal vector. The pressure in the fluid is given by the Bernoulli equation

$$p = -\rho \frac{\partial \Phi}{\partial t} - \frac{1}{2} \rho V^2$$

where ρ is the fluid density and V is the magnitude of $\mathbf{V}$. The force exerted by the fluid on the body is the integral of the fluid pressure over the surface.

$$\mathbf{F} = \int_S p\mathbf{n} \, dS$$

If the fluid is of infinite extent, then the solution of these equations is considerably simplified. The fluid force is[1]

$$\mathbf{F} = -\rho \frac{\partial}{\partial t} \int_S \Phi \mathbf{n} \, dS \tag{29.3}$$

and flow potential can be expressed as $\Phi = U(t)\phi(x', y', z')$, where x', y', and z' are coordinates that are fixed to the body and U is the flow velocity relative to the body. Substituting this potential in Eq. (29.3) yields the following force:

$$F = -m \frac{\partial U}{\partial t} \tag{29.4}$$

where the added mass m is

$$m = \rho \int_S \phi \frac{\partial \phi}{\partial n} \, dS \tag{29.5}$$

The added mass force Eq. (29.3) is zero for U and Φ independent of time, i.e., for steady translation. This is the D'Alembert paradox for an ideal inviscid fluid flow; the fluid force is not zero for steady translation in a viscous fluid.

As an example of added mass calculation, the potential for flow over a cylinder of radius a is

$$\phi = U \frac{r + a^2}{r} \cos \theta$$

where r = radial coordinate
   θ = angular coordinate
   U = flow velocity

The added mass per unit length is found from Eq. (29.5). The result is

$$m = \rho \pi a^2$$

where a is the cylinder radius. This added fluid mass is equal to the mass of fluid displaced by the cylinder.

In general, there will be an added mass tensor to represent the added mass for acceleration in each of the three coordinate directions:

$$m_{ij} = \rho \int_S \phi_j \frac{\partial \phi_i}{\partial n}\, dS$$

and an added mass tensor for rotation about the three coordinate axes. ϕ_i is the potential associated with flow in the i direction. Note that the added mass tensor is symmetric, i.e., $m_{ij} = m_{ji}$, but if the body is not symmetric, there is coupling between motions in the various coordinate directions.[1] For example, if a body is not symmetric about the X axis, acceleration in the X direction generally induces added mass force in the Y direction and a moment as well.

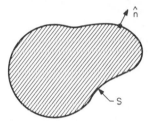

Since the added mass acts in phase with acceleration [Eq. (29.3)], the net effect of added mass is to increase the effective mass of the body and to decrease the natural frequencies. In general, added mass is only important to mechanical structures in dense fluids such as water. In gases, such as air, the added mass is ordinarily negligible except for very lightweight structures. Figure 29.3 gives added mass for various sections and bodies in large unrestricted reservoirs. Additional tables of added mass are given in Refs. 3 and 4.

Fig. 29.2. Fluid-filled region. Fluid density ρ.

If two structures are in close proximity, then the added mass will be a function of the spacing between the structures and inertial coupling will be introduced between the bodies. For example, consider a cylindrical rod centered in a fluid-filled annulus bounded by a cylindrical cavity shown in Fig. 29.4. The radius of the rod is a and the radius of the outer cylinder is b. The fluid forces exerted on the rod and outer cylinder because of their relative acceleration are[5]

$$F_1 = -m\ddot{x}_1 + (M_1 + m)\ddot{x}_2 \tag{29.6}$$
$$F_2 = (m + M_1)\ddot{x}_1 - (m + M_1 + M_2)\ddot{x}_2$$

where x_1, x_2 = displacement of inner rod and outer cylinder
   F_1, F_2 = force on inner rod and outer cylinder
   $m = \rho \pi a^2 (b^2 + a^2)/(b^2 - a^2)$, added mass of inner rod
   $M_1 = \rho \pi a^2$
   $M_2 = \rho \pi b^2$

These forces include not only added mass but also inertial coupling between the motion of the two structures. [These equations also apply for a sphere contained within a spherical cavity but here $m = (M_1/2)(b^3 - 2a^3)/(b^3 - a^3)$, $M_1 = \frac{4}{3}\rho\pi a^3$, and $M_2 = \frac{4}{3}\rho\pi b^3$.] Coupling is introduced between the cylinder and the rod through the fluid annulus. The coupling increases with the density of the fluid and decreases with increasing gap. If the cylinder and the rod are elastic, motion of either structure tends to set both structures into motion.

For example, consider an array of heat exchanger tubes contained within a shell. Water fills the shell and surrounds the tubes. If the tubes are widely spaced (more than about two diameters between centers), then the tubes are largely uncoupled and the effect of added mass is simply to reduce the tube natural frequencies by the addition of fluid equal to the displaced volume of the tubes. However, if the tubes are closely spaced, then

FIG. 29.3. Added mass for lateral acceleration.[3] The acceleration is left to right. b is the span for two-dimensional sections.

motion of one tube sets adjacent tubes and the shell into motion. Fluid-coupled modes of vibration will result in the tubes and the shell moving in fixed modal patterns as shown in Fig. 29.5. In Refs. 6 and 7, analysis is given for inertial coupling of a cylinder contained eccentrically within a cylindrical cavity, rows of cylinders, and arrays of cylinders.

FIG. 29.4. A rod in a fluid-filled annulus.

Added mass and inertial coupling occur in elastic and rigid bodies, but the added complexity of elasticity and the three-dimensional motions make a closed-form solution impossible for most elastic bodies. In the case of quasi-two-dimensional structures (such as long span tubes or rods), the axial variation in the motion occurs relatively slowly over the span, and two-dimensional results for sections are applicable. Concentric cylindrical shells coupled by a fluid annulus are important in the design of nuclear reactor containment vessels. Approximate solutions are required for both the vessels and the fluid. Reviews of the analysis of fluid coupled concentric vessels are given in Refs. 8 and 9.

Finite element numerical solutions, developed for an irrotational fluid, have been incorporated in the NASTRAN and other computer programs to permit solution for added mass and inertial coupling. These programs solve the fluid and structural problems and then couple the results through interaction forces.[10]

57.21 Hz   57.28 Hz   57.28 Hz

FIG. 29.5. Coupled modes of vibration of a bank of tubes in a dense fluid.[6]

VIBRATION OF OFFSHORE STRUCTURES

Ocean waves induce vibration of offshore structures, such as marine pipelines, oil terminals, tanks, and ships, by placing oscillatory pressure on the surface of the structure. These forces are often well-represented by the inviscid flow solution for many large structures such as ships and oil storage tanks. For most smaller offshore structures, viscous effects influence the fluid force and the fluid forces are determined experimentally.

Consider an ocean wave approaching the vertical cylindrical structure as shown in Fig. 29.6. The wave is propagating in the X direction. Using small-amplitude (linear) invis-

FIG. 29.6. A circular cylindrical structure exposed to ocean waves.

cid wave theory, the wave is characterized by the wave height h (vertical distance between trough and crest), its angular frequency ω, and the associated wavelength λ (horizontal distances between crests), and d is the depth of the water. The wave potential Φ satisfies Laplace's equation (Eq. 29.2) and a free-surface boundary condition (Ref. 11). The associated horizontal component of wave velocity varies with depth $-z$ from the free surface and oscillates at frequency ω.

$$U(t, z) = \frac{h\omega}{2} \frac{\cosh\left[2\pi(z + d)/\lambda\right]}{\sinh\left(2\pi d/\lambda\right)} \cos\left(\frac{2\pi x}{\lambda} - \omega t\right)$$ (29.7)

This component of wave velocity induces substantial fluid forces on structures, such as pilings and pipelines, which are oriented perpendicular to the direction of wave propagation.

The forces which the wave exerts on the cylinder in the direction of wave propagation (i.e., in line with U) can be considered the sum of three components: (1) a buoyancy force associated with the pressure gradient in the laterally accelerating fluid [Eq. (29.7)], (2) an added mass force associated with fluid entrained during relative acceleration between the fluid and the cylinder [Eq. (29.4)], (3) a force due to fluid dynamic drag associated with the relative velocity between the wave and the cylinder. The first two force components can be determined from inviscid fluid analysis as discussed previously. The drag component of force, however, is associated with fluid viscosity.

Thus, the in-line fluid force per unit length of cylinder due to an unsteady flow is expressed as the sum of the three fluid force components:

$$F = \rho A \dot{U} + C_I \rho A (\dot{U} - \ddot{x}) + \tfrac{1}{2}\rho \mid U - \dot{x} \mid (U - \dot{x}) D C_D \qquad (29.8)$$

where x = lateral position of structure in direction of wave propagation
    A = cross-sectional area = $\tfrac{1}{4}\pi D^2$ of cylinder having diameter D
    C_I = added mass coefficient, which has theoretical value of 1.0 for circular cylinder
    C_D = drag coefficient

This is the generalized form of the Morison equation, widely used to compute the wave forces on slender cylindrical ocean structures such as pipelines and piers.

If $\dot{x}$ and $\ddot{x}$ are set equal to zero in Eq. (29.8), the incline force per unit length on a stationary cylinder in an oscillating flow is obtained:

$$F(\dot{x} = \ddot{x} = 0) = C_m \rho \dot{A} U + \tfrac{1}{2}\rho \mid U \mid U D C_D \qquad (29.9)$$

Because of the absolute sign in the term $\mid U \mid U$, the force contains not only components at the wave frequency but also components associated with the drag at harmonics of the wave frequency. The resultant time-history of inline force due to a harmonically oscillating flow has an irregular form that repeats once every wave period.

If the flow oscillates with zero mean flow, $U = U_0 \cos \omega t$ as in Eq. (29.7), then the maximum fluid force per unit length on a stationary cylinder is

$$F_{\max} = \begin{cases} \rho A C_m \omega U_0 & \text{if } \dfrac{U_0}{\omega D} < \dfrac{C_m A}{C_D D^2} \\[2ex] \dfrac{1}{2}\rho U_0^2 D C_D + \dfrac{(\rho A C_m U_0 \omega)^2}{2\rho U_0^2 D C_D} & \text{if } \dfrac{U_0}{\omega D} > \dfrac{C_m A}{C_D D^2} \end{cases} \qquad (29.10)$$

If the cylinder is large (such as for a storage tank) with diameter D greater than the ocean wave height h and if the wavelength of the ocean wave is comparable to the diameter, then U_0 is small compared to ωD and the maximum force is given by the first alternative in Eq. (29.10). The drag force is negligible compared to the inertial forces for large cylinders. As a result, the ocean wave forces on large cylinders can be calculated using inviscid, i.e., potential flow, methods which are discussed in Refs. 11 and 12.

For the Reynolds number ranges typical of most offshore structures, measurements show that the inertial coefficient $C_m = 1 + C_I$ for cylindrical structures generally falls in the range between 1.5 and 2.0. $C_m = 1.8$ is a typical value. C_m decreases for very large diameter cylinders owing to the tendency of waves to diffract about large cylinders (Refs. 13 and 14). Similarly, measurements show that the drag coefficient falls between 0.6 and 1.0 for circular cylinders; $C_D = 0.8$ is a typical value.

Wave forces on elastic ocean structures induce structural motion. Since the wave force is nonlinear [Eq. (29.8)] and involves structural motion, no exact solution exists. One approach is to integrate the equations of motion directly by applying Eq. (29.8) at each spanwise point on a structure and then numerically integrate the time-history of deflection using a predictor-corrector or recursive relationship to account for the nonlinear term. A simpler approach is to assume that the structural deformation does not influence the fluid force and apply Eq. (29.9) as a static load. This static approximation is valid as long as the fundamental natural frequency of the structure is well above the wave frequency *and* the first three or four harmonics of the wave frequency. However, many marine structures are not sufficiently stiff to satisfy this condition.

One generally valid simplification for dynamic analysis of relatively flexible structures is to consider that the wave velocity is much less than the structural velocity so that $\mid U - \dot{x} \mid \simeq \mid U \mid$. With this approximation, application of Eq. (29.8) to a single degree-of-freedom model for a structure gives the following linear equation of motion:

$$(m + \rho A C_I)\ddot{x} + (2\zeta\omega_N + \tfrac{1}{2}\rho \mid U \mid D C_D)\dot{x} + kx = \rho A C_m \dot{U} + \tfrac{1}{2}\rho \mid U \mid U D C_D \qquad (29.11)$$

where m = structural mass per unit length
    k = stiffness
    ζ = structural damping

This equation is solved by expanding both $x(t)$ and $U(t)$ in a Fourier series and matching the coefficients.

The fluid forces contribute added mass and fluid damping to the left-hand side as well as forcing terms to the right-hand side. This equation may be simplified further by retaining only the first (constant) term in the series expansion for $|U|$ in the fluid damping term so that the equation becomes a classical forced oscillator with constant coefficient.[12]

Flexible structures will resonate with the wave if the structural natural period equals the wave period or a harmonic of the wave period. Since the wave frequencies of importance are ordinarily less than 0.2 Hz (wave period generally greater than one cycle per 5 sec), such a resonance occurs only for exceptionally flexible structures such as deep-water oil production risers and offshore terminals. The amplitude of structural response at resonance is a balance between the wave force and the structural stiffness times the damping. Since the wave force diminishes with increased structural motion [Eq. (29.8)], the resultant displacements are necessarily self-limiting. In other words, the response which would be predicted by applying Eq. (29.9) dynamically is overly pessimistic because the wave force contributes mass and damping to the structure as well as excitation as can be seen in Eq. (29.11).

The above discussion considers only fluid forces which act in line with the direction of wave propagation. These in-line forces produce an in-line response. However, substantial transverse vibrations also occur for ocean flows around circular cylinders. These vibrations are associated with periodic vortex shedding, which is discussed below. The models discussed in the following section for steady flow are applicable to vortex shedding in oscillatory flows provided that the period of shedding, based on the maximum oscillatory velocity, exceeds the wave period so that it is possible to fit one or more shedding cycles into the wave cycle.[13,14]

VORTEX-INDUCED VIBRATION

Many structures of practical importance such as buildings, pipelines, and cables are not streamlined but rather have abrupt contours that can cause a fluid flow over the structure to separate from the aft contours of the structure. Such structures are called *bluff bodies*. For a bluff body in uniform cross flow, the wake behind the body is not regular but contains distinct vortices of the pattern shown in Fig. 29.7 at a Reynolds number Re $= VD/\nu$ greater than about 50, where D is the width perpendicular to the flow and ν is the kinematic viscosity. The vortices are shed alternately from each side of the body in a regular manner and give rise to an alternating force on the body. Experimental studies have shown that the frequency, in hertz, of the alternating lift force is expressed as[16, 17]

$$f_s = \frac{SV}{D} \tag{29.12}$$

The dimensionless constant S called the *Strouhal number* generally falls in the range 0.25 $\geq S \geq 0.14$ for circular cylinders, square cylinders, and most bluff sections. The value of S increases slightly as the Reynolds number increases; a value of $S = 0.2$ is typical for circular cylinders.

The oscillating lift force imposed on a single circular cylinder of length L and diameter D, in a uniform cross flow of velocity V, due to vortex shedding is given by

$$F = \tfrac{1}{2}\rho V^2 C_L DLJ \sin (2\pi f_s t) \tag{29.13}$$

where the lift coefficient C_L is a function of Reynolds number and cylinder motion. The experimental measurements of C_L show considerable scatter with typical values ranging from 0.1 to 1.0. The scatter is in part due to the fact that the alternating vortex forces are not generally correlated on the entire cylinder length L. The spanwise correlation length l_c of vortex shedding over a stationary circular cylinder[16] is approximately three to seven diameters for $10^3 < \text{Re} < 2 \times 10^5$. In order to account for the effect of the spanwise correlation on the net force on the cylinder of length L, a factor J called the

Re < 5   REGIME OF UNSEPARATED FLOW

5 TO 15 ≤ Re < 40   A FIXED PAIR OF FOPPL VORTICES IN WAKE

40 ≤ Re < 90 AND 90 ≤ Re < 150 TWO REGIMES IN WHICH VORTEX STREET IS LAMINAR

150 ≤ Re < 300   TRANSITION RANGE TO TURBU-LENCE IN VORTEX

300 ≤ Re ≳ 3 X 10^5 VORTEX STREET IS FULLY TURBULENT

3 X 10^5 ≳ Re < 3.5 X 10^6

LAMINAR BOUNDARY LAYER HAS UNDERGONE TURBULENT TRANSITION AND WAKE IS NARROWER AND DISORGANIZED

3.5 X 10^6 ≤ Re

REESTABLISHMENT OF TURBU-LENT VORTEX STREET

Fig. 29.7. Regimes of fluid flow across circular cylinders.[15]

joint acceptance has been introduced on the right-hand side of Eq. (29.13). Two limiting cases exist for the joint acceptance.

$$J = \begin{cases} \left(\dfrac{l_c}{L}\right)^{1/2} & \text{if } l_c \ll L \\ 1 & \text{if fully correlated} \end{cases}$$

Thus, if a cylinder is much longer than three to seven diameters, the lack of spanwise correlation reduces the net vortex lift force [Eq. (29.13)] on the cylinder.

Motion of the cylinder in sympathy with vortex shedding greatly increases the correlation; it also affects the lift coefficient. Vortex shedding induces forces in the direction of fluid flow (i.e., the direction of drag). The drag force oscillation occurs at twice the vortex shedding frequency for single cylinders. However, the oscillating drag force is typically an order of magnitude smaller than the oscillating lift force.

If the vortex shedding frequency f_s is much different from the natural frequency f_n of a flexible cylinder and $C_L = 1$ and $J = 1$, then Eq. (29.13) gives a valid, conservative estimate of the vortex-induced lift force. Normally this off-resonance response is small compared with the resonance vibrations that occur as the shedding frequency approaches the natural frequency.

If the flow velocity is raised or lowered so that the shedding frequency approaches a natural frequency of the cylinder, the shedding frequency can synchronize with the natural frequency of a flexible cylinder. This condition is called *lock-in* or *synchronization*. Once lock-in has occurred, the shedding tends to persist at the structural natural frequency—even as the flow velocity is varied. In general, the larger the structural response to vortex shedding, the larger the range of lock-in.

Both the amplitude of the structural response and the velocity range over which lock-in persists are functions of the dimensionless reduced damping parameter δ_r:

$$\delta_r = \frac{2m(2\pi\zeta)}{\rho D^2}$$

where m = mass per unit length of cylinder, including added mass
ζ = damping factor for vibration in mode of interest, ordinarily measured in still fluid
ρ = fluid density
D = cylinder diameter

The lower δ_r, the greater the amplitude of the structural response and the greater the range of flow velocities over which lock-in occurs.[18] For lightly damped structures in dense fluids (such as marine pipelines), δ_r is on the order of 1 and lock-in can persist over a 20 per cent variation in velocity above and below that which produces resonance. Figure 29.8 shows the lock-in between an elastically mounted cylinder in an airflow and vortex shedding for two levels of damping. Synchronization between the structural motion and vortex shedding also increases the strength of the shed vortices, the correlation of the vortex shedding along the cylinder span, and both the steady and oscillating components of the drag on the cylinder.[12,17]

Within the synchronization band, substantial resonance vibration often occurs. Peak-to-peak vibration amplitudes of up to three diameters have been observed in water flows over cables and tubing. The vibrations are predominantly transverse to the flow and are self-limiting.[12] Lesser amplitude vibrations have also been observed in the drag direction at twice the vortex shedding frequency and at subharmonic frequencies of the vortex shedding frequency, i.e., at one-fourth, one-third, or one-half of the flow velocity required for synchronization,[20] $f_s = f_n$.

If a uniform elastic cylinder is subjected to a crossflow uniformly over its span, then the oscillating vortex-induced lift force is given by Eq. (29.13). At lock-in, the vortex shedding frequency equals the natural frequency of the nth vibration mode $f_s = f_n$, and the amplitude of the cylinder response is

$$\frac{y_n^*}{D} = \frac{C_L J \phi_n^*}{4\pi S^2 \delta_r} \qquad (29.14)$$

where y_n^* is the maximum value of the vibration mode over the span of the cylinder and ϕ_n^* is the maximum of the mode shape which has been normalized so that $\int_L \phi_n^2 \, dx = 1$. This equation is conservative if $C_L = J = 1$. However, Eq. (29.14) gives overly conservative predictions with $C_L = J = 1$ owing to the tendency of the actual lift coefficient to decrease at amplitudes in excess of 0.5 diameters and the lack of perfect spanwise correlation at lower amplitudes. Semiempirical correlations are given in Refs. 12 and 21. One of these correlations is[12]

$$\frac{y_n^*}{D} = \frac{0.07\gamma}{(\delta_r + 1.9)S^2} \left(0.3 + \frac{0.72}{(\delta_r + 1.9)S} \right)^{1/2} \qquad (29.15)$$

The mode shape ϕ has been normalized as before, and γ falls between 1.0 and 1.4. For a translating rigid rod ($\phi = 1$), $\gamma = 1$, for a cable or pipeline with a sinusoidal mode shape, $\gamma = 1.15$, and for a cantilever mode shape, $\gamma = 1.4$.

FIG. 29.8. Response of a spring-supported cylinder to vortex-induced vibration.[19]

Equation (29.15) correctly predicts the self-limiting behavior of the resonance vibrations. Setting damping to zero, $\delta_r = 0$, it follows that $y_n^*/D \simeq 1.5$, which is a typical vibration level for lightly damped marine cables in a current. See Fig. 29.9. Large amplitude vibrations also are associated with increased steady drag on the structure. Drag coefficients of up to 3.5 have been measured on resonantly vibrating marine cables as opposed to the typical value of 1.0 for a stationary cylinder.[22]

A number of fairings, strakes, and ribbons have been attached to the exterior of circular cylindrical structures to reduce vortex-induced vibrations as shown in Fig. 29.10. These devices act by disrupting the near wake and disturbing the correlation between the vortex shedding and vibration. They do, however, increase the steady drag from that which is measured on a stationary structure. Reviews of vortex suppression devices are given in Refs. 23 and 24.

FLUID ELASTIC INSTABILITY

Fluid flow across an array of elastic tubes can induce a dynamic instability, resulting in very large amplitude tube vibrations once the critical cross-flow velocity is exceeded. This is a relatively common occurrence in tube and shell heat exchangers. Once the critical cross-flow velocity is exceeded, vibration amplitude increases very rapidly with cross-flow velocity V, usually as V^n where $n = 4$ or more, compared with an exponent in the range $1.5 < n < 2.5$ below the instability threshold. This can be seen in Fig. 29.11 which shows the response of an array of metallic tubes to water flow. The initial hump is attrib-

FIG. 29.9. Maximum amplitude of vortex-induced vibration as a function of damping.[12]

uted to vortex shedding. The cross-flow velocity is defined as velocity perpendicular to the tube axis at the minimum gap between tubes. Once the critical velocity is exceeded, the very large amplitude vibrations usually lead to failures of the heat exchanger tubes.

Often the large amplitude vibrations vary in time; the amplitudes grow and fall about a mean value in pseudorandom fashion. Generally the tubes do not move independently but instead move in somewhat synchronized orbits with neighboring tubes. This orbital behavior has been observed in tests in both air and water with orbits ranging from near circles to nearly straight lines. See Fig. 29.12.

As the tubes whirl in orbital motion, they extract energy from the fluid (Refs. 12, 26, and 27). Below the onset of instability, energy which is extracted is less than the energy

FIG. 29.10. Methods of reducing vortex-induced vibration.

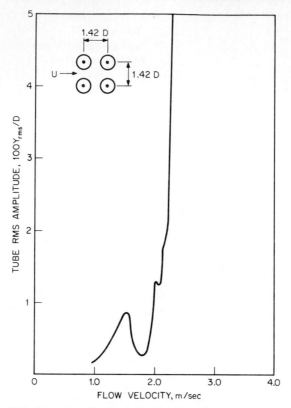

FIG. 29.11. Typical amplitude of vibration of a tube array in cross flow.[25]

which is expended in damping. Above the critical velocity, the energy extracted from the flow by the tube motion exceeds the energy expended in damping, so the vibrations increase in amplitude. Restricting the motion or introducing frequency differences between one or more tubes often increases the critical velocity for onset of instability. Such increases in critical velocity are generally no greater than about 40 per cent unless additional support is given to all tubes exposed to high velocity flow. Often the onset of instability is more gradual in a bank of tubes having tube-to-tube frequency differences than in a bank with identical tubes. Only a relatively small percentage of the tube will become unstable at one time. Flexible long-span tubes near areas of high flow velocity (such as at inlets) are most susceptible to the instability.

FIG. 29.12. Tube vibration patterns for fluid elastic instability.[26]

At cross-flow velocities beyond those which produce an onset of instability, damaging vibrations are encountered. The tube vibration amplitudes are limited by clashing with other tubes, by impacting with the tube supports, and by yielding of the tubes. Sustained operation in the unstable vibration regime ordinarily results in tube failure due to wear or propagation of cracks in the tubes. Fluid elastic instability is second only to corrosion as a cause of heat exchanger failure.

A displacement model for the fluid elastic forces is given in Ref. 12 which correctly predicts the observed onset of instability for most cases in air and gases. Results are less satisfactory in water or when the motion of some of the tubes is restricted. More complex models take into account velocity-induced forces as well as the displacement-induced forces.[27,28] These theories give somewhat better agreement with data over limited ranges, but none are entirely suitable for a design tool.

The most viable, practical procedure for predicting the onset of instability of closely spaced arrays of tubes to cross flow is to use the theoretical form given by the displacement mechanism but with parameters obtained by filling experimental data. The onset of instability is predicted as

$$\frac{V_{\text{crit}}}{f_n D} = C \left[\frac{m_t (2\pi\zeta)}{\rho D^2} \right]^a \qquad (29.16)$$

where V_{crit} = uniform cross flow averaged over minimum gap between tubes (If the velocity is nonuniform, then either the maximum can be used or a modal weighted average can be employed.)

f_n = fundamental natural frequency of tubing (Ordinarily the fundamental mode is most susceptible to instability.)

ζ = damping factor of fundamental mode (Typically ζ falls in the range between 0.01 and 0.03 for tubes with some intermediate supports. For rolled-in or welded-in tubes with no intermediate supports, ζ can be as low as 0.001.)

m_t = mass per unit length of tube including added mass and internal mass of fluid

ρ = fluid density

Fitting Eq. (29.16) to the available 174 data points for onset of instability[29] shown in Fig. 29.13 leads to the mean and lower-bound coefficients for the parameter C and the exponent a given in Table 29.1. The coefficient corresponding to the mean fit to the experimental data is C_{mean}; $C_{90\%}$ is the lower bound fit to the data such that 90% of the data are above the curve.

Most of the data used in this correlation come from tube arrays with center-to-center

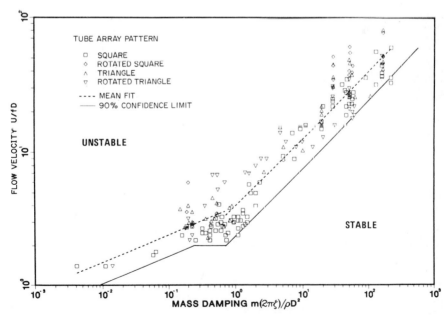

FIG. 29.13. Velocity for onset of instability of tube arrays in cross flow as a function of the damping parameter.[29]

spacing of between 1.25 and 2.0 diameters and with various array geometries. There is insufficient statistical evidence to determine if certain patterns are more or less susceptible to instability than others. Instability has been observed for both straight and curved tubes, tube rows, and tube arrays in a wide variety of tube patterns.

The most common means of increasing the resistance of an array of tubes to instability is to add intermediate supports to increase the natural frequency of the tubes. Details of the tube support (particularly the gap between the tube and the support) influence the resultant vibration. In general, smaller gaps tend to result in lower tube-support impact velocities and hence in lower tube wear.[30]

Table 29.1. Coefficients in Eq. (29.16) for Onset of Instability of Tube Arrays[29]

	$m_t(2\pi\zeta)/\rho D^2 < 0.7$	$m_t(2\pi\zeta)/\rho D^2 > 0.7$
C_{mean}	3.9	4.0
$C_{90\%}$	2.7	2.4
a	0.21	0.5
RMS error in fitted data for V_{crit},%	24.5	32.5

INTERNAL FLOW IN PIPES

Internal flow through a pipe decreases the natural frequency of the pipe. Sufficiently high internal velocity will induce buckling in a pipe supported at both ends since the momentum of fluid turning through a small angle of pipe deflection is greater than the stiffness of the pipe. If the pipe is restrained at only one end, the pipe will become unstable at high velocities like an unrestrained garden hose.

The equation of motion for a straight pipe conveying steady fluid flow is[31]

$$EI\frac{\partial^4 Y}{\partial x^4} + \rho A v^2 \frac{\partial^2 Y}{\partial x^2} + 2\rho A v \frac{\partial^2 Y}{\partial x \partial t} + M\frac{\partial^2 Y}{\partial t^2} = 0 \qquad (29.17)$$

where E and I are the modulus and moment of inertia of the pipe which conveys fluid of density ρ through the internal area A of the pipe at a steady velocity v; $Y(x, t)$ is the lateral deflection of the pipe which has total mass per unit length M. The first and last terms in Eq. (29.17) are the usual stiffness and mass terms. The middle terms are associated with fluid forces imposed on the pipe by the internal fluid as the pipe deflects slightly from its equilibrium position.

Although Eq. (29.17) is a linear partial differential equation with constant coefficients, its solution is difficult owing to the mixed derivative term (third term from the left). One technique used to solve the equation is to expand the solution in terms of the mode shapes of vibration which are obtained for zero flow, $v = 0$.

$$Y(x, t) = \Sigma_i \, a_i y_i(x) \sin \omega t \qquad (29.18)$$

where $y_i(x)$ are the mode shapes for zero flow that satisfy Eq. (29.17) and the boundary conditions on the ends of the pipe span. Equation (29.18) is substituted into Eq. (29.23), and the derivatives of $y_i(x)$ are expressed in terms of the orthogonal set $y_i(x)$

$$y_i'(x) = \Sigma_i b_i y_i(x)$$

Like terms in the series are equated.

For a uniform pipe with pinned ends, the result can be approximately expressed as a decrease in natural frequency due to flow.[12]

$$\frac{f}{f_1} = \left[1 - \left(\frac{v}{v_c}\right)^2\right]^{1/2} \qquad (29.19)$$

where f = fundamental natural frequency
f_1 = fundamental natural frequency in absence of flow
v_c = critical flow velocity

The critical flow velocity can be expressed as

$$v_c = \frac{\pi}{L}\left(\frac{EI}{\rho A}\right)^{1/2} \tag{29.20}$$

where L is the span of the pipe. As the flow velocity approaches v_c, the fundamental natural frequency f_1 decreases to zero. The pipe span spontaneously buckles at $v = v_c$.

The buckling velocity is a function of the boundary conditions on the ends of the pipe, and there can be vibration; these solutions for various boundary conditions are generally scaled by the velocity v_c [Eq. (29.20)]. In general, only exceptionally thin-walled flexible tubes with very high velocity flows, such as rocket motor feed lines and penstocks, are prone to vibration induced by internal flow. External parallel flow can also induce an analogous instability. (See the review given in Ref. 32).

REFERENCES

1. Newman, J. N.: "Marine Hydrodynamics," The MIT Press, Cambridge, Mass., 1977.
2. Lamb, H.: "Hydrodynamics," Dover Publications, New York, 1945. Reprint of the 6th ed., 1932.
3. Blevins, R. D.: "Formulas for Natural Frequency and Mode Shape," Kreiger, Malabar, Florida, 1984. Reprint of 1979 edition.
4. Wendel, K.: "Hydrodynamic Masses and Hydrodynamic Moments of Inertia," *The David W. Taylor Model Basin, Translation* 260, July 1956. Translation of "Hydrodynamische Massem und Hydrodynamische Massentragheitsmoments," *Jahrb. d. STG*, **44**, 1950.
5. Fritz, R. J.: *J. Eng. Industry*, **94**:167 (1972).
6. Chen, S-S: *J. Eng. Industry*, **97**:1212 (1975).
7. Chen, S-S: *Nucl. Eng. Des.*, **35**:399 (1975).
8. Brown, S. J.: *J. Pressure Vessel Tech.*, **104**:2 (1982).
9. Au-Yang, M. K.: *J. Vibration, Acoustics*, **108**:339 (1986).
10. Zienkiewicw, O. C.: "The Finite Element Method," 3d ed., McGraw-Hill Book Co., New York, 1977.
11. Ippen, A. T. (ed.): "Estuary and Coastline Hydrodynamics," McGraw-Hill Book Co., New York, 1966.
12. Blevins, R. D.: "Flow-Induced Vibration," Kreiger, Malibar, Fla., 1984.
13. Sarpkaya, T., and M. Isaacson: "Mechanics of Wave Forces on Offshore Structures," Van Nostrand Reinhold, New York, 1981.
14. Shaw, T. L. (ed): "Mechanics of Wave Induced Forces on Cylinders," Pitnam Advanced Publishing Program, San Francisco, 1979.
15. Lienard, J. H.: "Synopsis of Lift, Drag and Vortex Frequency Data for Rigid Circular Cylinder," Washington State University, College of Engineering, Research Division Bulletin 300, 1966.
16. Roshko, A.: "On the Development of Turbulent Wakes from Vortex Streets," *National Advisory Committee for Aeronautics Report* NACA TN-2913, 1953
17. Sarpkaya, T.: *J. Appl. Mech.*, **46**, 241 (1979).
18. Scruton, C.: "On the Wind Excited Oscillations of Stacks, Towers and Masts," *National Physical Laboratory Symposium on Wind Effects on Buildings and Structures, Paper* 16, 790, 1963.
19. Feng, C. C.: "The Measurement of Vortex-Induced Effects in Flow Past Stationary and Oscillating Circular and D-Section Cylinder," M.A.Sc. thesis, University of British Columbia, 1968.
20. Durgin, W. W., P. A. March, and P. J. Lefebvre: *J. Fluids Eng.*, **102**:183 (1980).
21. Griffin, O. M.: "Vibrations and Flow-Induced Forces Caused by Vortex Shedding," *Symposium of Flow Induced Vibrations*, **1**:1 ASME, 1984.
22. Vandiver, J. K.: "Drag Coefficients of Long Flexible Cylinders," *1983 Offshore Technology Conference, Paper* 4490, 1983, p. 405.
23. Zdravkovich, M. M.: *J. Wind Eng., Industrial Aerodynamics*, **7**:145 (1981).
24. Wong, H. Y., and A. Kokkalis: *J. Wind Eng. Industrial Aerodynamics*, **10**:21 (1982).
25. Chen, S-S, J. A. Jendrzejczyk, and W. H. Lin: "Experiments on Fluid Elastic Instability in a Tube Bank Subject to Liquid Cross Flow," *Argonne National Laboratory Report* ANL-CT-44, July 1978.
26. Connors, H. J.: "Fluid Elastic Vibration of Tube Arrays Excited by Cross Flow," Paper presented at the Symposium on Flow Induced Vibration in Heat Exchangers, ASME Winter Annual Meeting, December 1970.
27. Paidoussis, M. P.: *Nucl. Sci. Eng.*, **74**:31 (1983).

28. American Society of Mechanical Engineers, Symposium on Flow Induced Vibrations, *1* through *6*, ASME, N.Y., 1984.
29. Blevins, R. D.: *J. Sound Vibration*, **97:**641 (1984).
30. Blevins, R. D.: *J. Eng. Materials Tech.*, **107:**61 (1985).
31. Housner, G. W.: *J. Appl. Mech.* **19:**205 (1952).
32. Paidoussis, M. P., and P. Besancon: *J. Sound Vibration*, **76:**361 (1981).

29

PART II: VIBRATION OF STRUCTURES INDUCED BY WIND

A. G. Davenport and M. Novak
University of Western Ontario

INTRODUCTION

Vibration of significant magnitude may be induced by wind in a wide variety of structures including buildings, television and cooling towers, chimneys, bridges, transmission lines, and radio telescopes. No structure exposed to wind seems entirely immune from such excitation. The material presented here describes several mechanisms causing these oscillations and suggests a few simpler approaches that may be taken in design to reduce vibration of structures induced by wind. There is an extensive literature[1-5] giving a more detailed treatment of the subject matter.

FORMS OF AERODYNAMIC EXCITATION

The types of structure referred to above are generally unstreamlined in shape. Such shapes are termed "bluff bodies" in contrast to streamlined "aeronautical" shapes. The distinguishing feature is that when the air flows around such a bluff body, a significant wake forms downstream, as illustrated in Fig. 29.14. The wake is separated from the outside flow region by a shear layer. With a sharp-edged body (such as a building or structural number) as in Fig. 29.14, this shear layer emanates from the corner. With oval bodies such as the cylinder in Fig. 29.14, the shear layer commences at a so-called

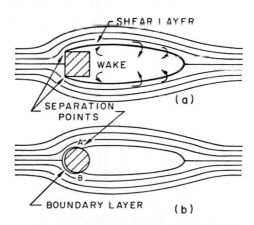

Fig. 29.14. Wake formation past bluff bodies: (a) sharp-edged body, (b) circular cylinder.

boundary layer on the upstream surface at points A and B (the separation points) and becomes a free shear layer. The exact position of these separation points depends on a wide variety of factors, such as the roughness of the cylinder, the turbulence in the flow, and the Reynolds number $R = VD/\nu$, where V = flow velocity, D = diameter of the body, and ν = kinematic viscosity.

The flow illustrated in Fig. 29.14 represents the time-average picture which would be obtained by averaging the movements of the fluid particles over a time interval that is long compared with the "transit time" D/V. The instantaneous picture of the flow may be quite different, as indicated in Fig. 29.15, for two reasons.

FIG. 29.15. Vortex street past circular cylinder ($R = 56$). (*After Kovasznay, Proc. Roy. Soc. London, 198, 1949.*)

First, if the flow is the wind, it is under almost all practical circumstances strongly turbulent; the oncoming flow will be varying continuously in direction and speed in an irregular manner. These fluctuating motions will range over a wide range of frequencies and scales (i.e., eddy sizes).

Second, the wake also will take on a fluctuating character. Here, however, the size of the dominant eddies (vortices) will be of a similar size to the body. The vortices tend to start off their career by curling up at the separation point and then are carried off downstream. Sometimes these eddies are fairly regular in character and are shed alternately from either side; if made visible by smoke or other means, they can be seen to form a more or less regular stepping-stone pattern until they are broken up by the turbulence or dissipate themselves. In a strongly turbulent flow, the regularity is disrupted.

The flow characteristics of the oncoming flow and the wake are the direct causes of the forces on the bodies responsible for their oscillation. The forms of the resulting oscillation are as follows.

1. *Turbulence-induced oscillations.* Certain types of oscillation of structures can be attributed almost exclusively to turbulence in the oncoming flow. In the wind these may be described as "gust-induced oscillations" (or turbulence-induced, oscillations). The gusts may cause longitudinal, transverse, or torsional oscillations of the structure, which increase with wind velocity (Fig. 29.16).

FIG. 29.16. Main types of wind-induced oscillations: (*a*) vibration due to turbulence, (*b*) vibration due to vortex shedding, and (*c*) aerodynamic instability.

2. Wake-induced oscillations. In other instances, the fluctuations in the wake may be the predominant agency. Since these fluctuations are generally characterized by alternating flow, first around one side of the body then around the other, the most significant pressure fluctuations act on the sides of the body in the wake behind the separation point (the so called after body); they act mainly laterally or torsionally and to a much lesser extent longitudinally. The resultant motion is known as "vortex-induced oscillation." Oscillation in the direction perpendicular to that of the wind is the most important type. It often features a pronounced resonance peak (Fig. 29.16b).

While these distinctions between gust-induced and wake-induced forces are helpful, they often strongly interact; the presence of free-stream turbulence, for example, may significantly modify the wake.

3. Buffeting by the wake of an upstream structure. A further type of excitation is that induced by the wake of an upstream structure (Fig. 29.17). Such an arrangement

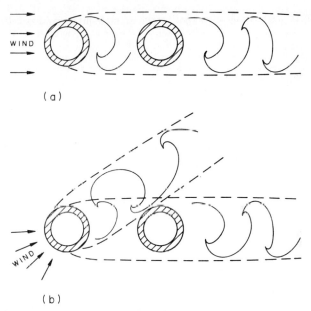

(a)

(b)

Fig. 29.17. Buffeting by the wake of an upstream structure.

of structures produces several effects. The turbulent wake containing strong vortices shed from the upstream structure can buffet the downstream structure. In addition, if the oncoming wind is very turbulent, it can cause the wake of the upstream structure to veer, subjecting the downstream structure successively to the free flow and the wake flow. This frequently occurs with chimneys in line, as well as with tall buildings.

4. Galloping and flutter mechanisms. The final mechanism for excitation is associated with the movements of the structure itself. As the structure moves relative to the flow in response to the forces acting, it changes the flow regime surrounding it. In so doing, the pressures change, and these changes are coupled with the motion. A pressure change coupled to the velocity (either linearly or nonlinearly) may be termed an "aerodynamic damping" term. It may be either positive or negative. If positive, it adds to the mechanical damping and leads to higher effective damping and a reduced tendency to vibrate; if negative, it can lead to instability and large amplitudes of movement. This type of excitation occurs with a wide variety of rectangular building shapes as well as bridge cross sections and common structural shapes such as angles and I sections.

In other instances, the coupling may be with either the displacement or acceleration, in which case they are described as either aerodynamic stiffness or mass terms, the effect of which is to modify the mass or stiffness terms in the equations of motion. Such modification can lead to changes in the apparent frequency of the structure. If the aerodynamic stiffness is negative, it can lead to a reduction in the effective stiffness of the structure and eventually to a form of instability known as "divergence." All types of instability feature a sudden start at a critical wind velocity and a rapid increase of violent displacements with wind velocity (Fig. 29.16c).

These various forms of excitation are briefly discussed in this chapter. Because all types of oscillations are influenced strongly by the properties of the wind, some basic wind characteristics are described first.

BASIC WIND CHARACTERISTICS

Wind is caused by differences in atmospheric pressure. At great altitudes, the air motion is independent of the roughness of the ground surface and is called the "geostrophic," or "gradient" wind. Its velocity is reached at a height called "gradient height," which lies between about 1,000 and 2,000 ft.

Below the gradient height, the flow is affected by surface friction, by the action of which the flow is retarded and turbulence is generated. In this region, known as the "planetary boundary layer," the three components of wind velocity resemble the traces shown in Fig. 29.18.

FIG. 29.18. Record of horizontal component of wind speed at three heights on 500 ft mast in open terrain. (*Courtesy of E. L. Deacon.*)

The longitudinal component consists of a mean plus an irregular turbulent fluctuation; the lateral and vertical components consist of similar fluctuations. These turbulent motions can be characterized in a number of different ways.

The longitudinal motion at height z can be expressed as

$$V_z(t) = \bar{V}_z + v(t) \tag{29.21}$$

where $\bar{V}_z$ = mean wind velocity (the bar denotes time average) and $v(t)$ = fluctuating component.

MEAN WIND VELOCITY. The mean wind velocity $\bar{V}_z$ varies with height z as represented by the mean wind velocity profile (Fig. 29.19). The profiles observed in the field can be matched by a logarithmic law, for which there are theoretical grounds, or by an empirical power law

$$\frac{\bar{V}_z}{\bar{V}_G} = \left[\frac{z}{z_G} \right]^{\alpha} \tag{29.22}$$

where $\bar{V}_G$ = gradient wind velocity, z_G = gradient height, and α = an exponent <1. Gradient height z_G and exponent α depend on the surface roughness, which can be characterized by the surface drag coefficient κ (here referenced to the wind speed at 10 meters).

A few typical values of these parameters are given in Fig. 29.19. The mean wind

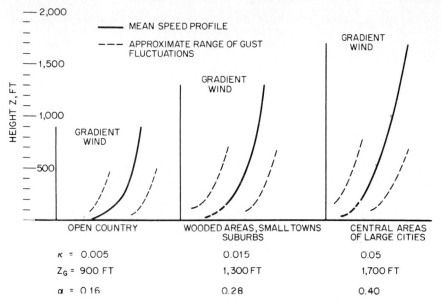

FIG. 29.19. Vertical profiles of mean wind velocity for three typical terrains.

profiles shown are characteristic of level terrain. They can significantly change, particularly in the lower region, when the air flow meets an abrupt change in surface roughness or terrain contour. A sudden increase in roughness reduces the wind speed near the ground while a hill accelerates the flow over its crest.

The mean wind profiles are useful when predicting the wind speed at a particular site. The gradient wind speed is estimated using data registered by the nearest meteorological stations at their standard height, which is usually 33 ft (10 meters).

The mean wind velocity generally depends on the period over which the wind speed is averaged. Periods from 10 to 60 min appear adequate for engineering considerations and usually yield reasonably steady mean values. The same duration is suitable to define the fluctuating wind component.

FLUCTUATING COMPONENTS OF THE WIND. The fluctuating components of the wind change with height less than the mean wind and are random both in time and space. The random nature of the wind requires the application of statistical concepts.*

The basic statistical characteristics of the velocity fluctuations are the intensity of turbulence, the power spectral density (power spectrum), the correlation between velocities at different points, and the probability distribution.

The intensity of turbulence is defined as $\sigma_v/\bar{V}_z$, where $\sigma_v = \sqrt{\overline{v^2(t)}}$ is the root-mean-square (rms) fluctuation in the longitudinal direction. The intensity of the lateral and vertical fluctuations can be described similarly. For wind, the intensity of turbulence is between 5 and 25 per cent. The magnitude σ_v also defines the probability distribution of the fluctuations which may be assumed to be Gaussian (normal).

The energy of turbulent fluctuations (gustiness) is distributed over a range of frequencies. This distribution of energy with frequency f can be described by the spectrum of turbulence (power spectral density) $S_v(f)$. The relationship between the spectrum and the variance is

$$\int_0^\infty S_v(f)\ df = \sigma_v^2$$

* Statistical approaches to random processes are described in Chap. 11 and Chap. 21.

which leads to another form of the spectrum known as the "logarithmic spectrum" $fS_v(f)/\sigma_v{}^2$. This form of the spectrum is dimensionless and preserves the relative contributions to the variance at different frequencies represented on a logarithmic scale; and its integral is

$$\int_0^\infty \frac{fS_v(f)}{\sigma_v{}^2}\, d\ln f = 1$$

The two forms of spectra are sketched in Fig. 29.20. A generalization of wind spectra for different wind velocities is possible if the frequency scale is so modified that

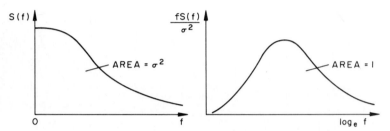

FIG. 29.20. Two different ways of presenting power spectral densities.

it too is dimensionless. The ratio $f/\bar{V}$ is the so-called "inverse wavelength" related to the "size" of atmospheric eddies. This may be expressed as a ratio to a representative length scale L, such as the wavelength of the eddies at the peak of the spectrum. The dimensionless frequency or inverse wavelength may now be written

$$\bar{f} = fL/\bar{V}$$

Under certain circumstances this relationship is also known as the Strouhal number or the reduced frequency.

It is generally found that while the length scale L in the oncoming flow corresponds to that of the turbulence itself (this in the natural wind is of the order of thousands of

FIG. 29.21. Universal spectrum of horizontal gustiness in strong winds and example of spectrum of fluctuations in wake.

feet), in the wake the governing length scale is of the same order as the diameter of the body D. This is illustrated in Fig. 29.21.

The spectrum of horizontal gustiness in strong winds is largely independent of height above the ground, is proportional to both the surface drag coefficient κ and the square of the mean velocity at the standard height of 10 meters, $\bar{V}_{10}$, and can be represented, with some approximations, as[6, 7]

$$S_v(f) = 4\kappa \bar{V}_{10}^2 \frac{L/\bar{V}_{10}}{(2 + \bar{f}^2)^{5\!/\!6}} \qquad (29.23)$$

in which f = frequency, cps, $\bar{f} = fL/\bar{V}_{10}$ where L = scale length $\approx$4,000 ft, and κ is given in Fig. 29.19. This spectrum is shown in Fig. 29.21.

The variance of the velocity fluctuations is

$$\sigma_v{}^2 = \int_0^\infty S_v(f)\,df = 6.68\kappa \bar{V}_{10}^2 \qquad (29.24)$$

It can be seen from Eqs. (29.23) and (29.24) that large velocity fluctuations can be expected in rough terrain where coefficient κ is large.

The spatial correlation of wind speeds at two different stations is described by the coherence function (coherence),

$$\gamma_{12}{}^2(f) = \frac{|S_{12}(f)|^2}{S_1(f)S_2(f)} \le 1 \qquad (29.25)$$

where $S_{12}(f)$ = cross spectrum (generally complex) between stations 1 and 2; $S_1(f)$ and $S_2(f)$ are power spectra of the two stations. The coherence function depends primarily on the parameter $\Delta z f/\bar{V}$, where Δz = separation and $\bar{V} = 1/2(\bar{V}_1 + \bar{V}_2)$ is the average wind speed. A suitable approximate function is

$$\sqrt{\text{Coherence}} = e^{-c(\Delta z f/\bar{V})}$$

where c is a constant having a value of approximately 7 for vertical separation and approximately 15 for horizontal separation. Coherence decreases with both separation and frequency. A more detailed discussion of wind characteristics is given in Refs. 1 and 7.

EXCITATION DUE TO TURBULENCE

When a structure is exposed to the effects of wind, the fluctuating wind velocity translates into fluctuating pressures, which in turn produce time-variable response (deflection) of the structure. This response is random and represents the basic type of wind-induced oscillations. The theoretical prediction of this oscillation is rather complex but can be reduced to a simple procedure suitable for design purposes. The discussion of the oscillation is therefore presented in two parts. In the first part, the basic theoretical steps are outlined. In the second part, the design procedure known as the gust-factor approach is given in more detail.

FUNDAMENTALS OF RESPONSE PREDICTION

If the area A of the structure exposed to wind is small relative to the significant turbulent eddies, the so-called quasi-steady theory for turbulence can be used to estimate aerodynamic forces. In the drag direction, the drag force

$$D(t) = \frac{1}{2}\rho C_D A V^2(t)$$
$$= \frac{1}{2}\rho C_D A \bar{V}^2 \left[1 + 2\frac{v(t)}{\bar{V}} + \frac{v^2(t)}{\bar{V}^2} \right]$$

where ρ = air density (normally equal to 0.0024 slugs/cu ft), and C_D = drag coefficient. If $v(t) \ll \bar{V}$, the squared term is ignored. The spectra of the fluctuating drag and velocity are then related as

$$\frac{S_D(f)}{\bar{D}^2} = 4\frac{S_v(f)}{\bar{V}^2} \qquad (29.26)$$

where the mean drag (static component of the drag) is

$$\bar{D} = \frac{1}{2}\rho C_D A \bar{V}^2 \tag{29.27}$$

and $S_v(f)$ is given by Eq. (29.23).

With large bodies the wavelength is comparable to the size of the body itself (that is, $f\sqrt{A}/\bar{V} \approx 1$) and it is necessary to modify the drag spectrum by the so-called "aerodynamic admittance function" $|X_{\text{aero}}(f)|^2$. This function[6] describes the modifying influence of any changes in effective drag coefficient, as well as the decrease in correlation of the eddies as the wavelength of the eddies approaches the diameter of the body. Thus, the modified drag spectrum is

$$\frac{S_D(f)}{\bar{D}^2} = 4|X_{\text{aero}}(f)|^2 \frac{S_v(f)}{\bar{V}^2}$$

If these forces act on an elastic spring-mass-damper system, the response of this system u will have a spectrum

$$\frac{S_u(f)}{\bar{u}^2} = |X_{\text{aero}}|^2 |X_{\text{mech}}|^2 \frac{4S_v(f)}{\bar{V}^2}$$

where static deflection $\bar{u} = \bar{D}/k$, k = stiffness constant, and the mechanical admittance function

$$|X_{\text{mech}}|^2 = \frac{1}{[1 - (f/f_0)^2]^2 + 4\zeta^2(f^2/f_0^2)}$$

where ζ = critical damping ratio, and f_0 = natural frequency of the system.

The transition from the spectrum of the wind-velocity fluctuations to the spectrum

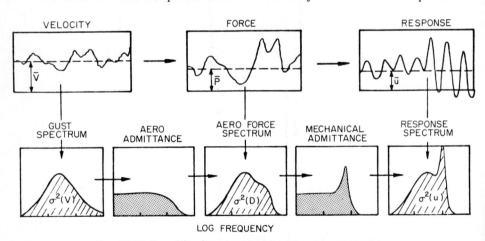

FIG. 29.22. Transition from gust spectrum to response spectrum.

of the response is shown diagrammatically in Fig. 29.22. The variance of the response σ_u^2 is obtained from the spectrum of the response,

$$\sigma_u^2 = \int_0^\infty S_u(f)\,df \tag{29.28}$$

The relationships above describe the mean and the variance of the response. For engineering purposes, it is also useful to define extreme values. It is often satisfactory to assume that the process in question is Gaussian with probability density function given by

$$p(u) = \frac{1}{\sqrt{2\pi}\,\sigma_u} e^{-(u-\bar{u})^2/2\sigma_u{}^2}$$

This distribution is fully described by the mean and the variance. Maximum values of the response during time T can be written as

$$u_{\max} = \bar{u} + g\sigma_u \qquad (29.29)$$

where g = peak factor. The average largest value of the peak factor in a period T can be estimated from[6]

$$g = \sqrt{2 \ln \nu T} + \frac{0.5772}{\sqrt{2 \ln \nu T}} \qquad (29.30)$$

where ν is an effective cycling rate of the process, generally close to the natural frequency. The relationship of the distribution of the largest peak value to the distribution of all values is shown in Fig. 29.23. As can be seen, when the period T or the

FIG. 29.23. Relationship of distribution of largest peak value to distribution of all values (for a stationary random process).

natural frequency increases, the expected peak displacement also increases. The factor g usually ranges between 3 and 5.

Further extension of the concept includes the cross correlation of the wind loads at different stations (e.g., heights), the shape of the vibration mode, and the nonuniformity of the mean flow. These factors can be included into the solution formulated in terms of modal analysis.*

With a prismatic structure, the displacement may be expressed in the form

$$u(z,t) = \sum_{j=1}^{\infty} q_j(t)\phi_j(z) \qquad (29.31)$$

where $q_j(t)$ = the generalized coordinate of the jth mode, and $\phi_j(z)$ = the jth mode of natural vibrations to an arbitrary scale.

With damping small and natural frequencies well separated, the cross correlation of the generalized coordinate can be neglected and the mean-square displacement (the variance) is

$$\overline{u^2(z,t)} = \sum_{j=1}^{\infty} \overline{q_j{}^2}\,\phi_j{}^2(z) \qquad (29.32)$$

* For detailed discussion of modal analysis, see Chaps. 2 and 7.

The variance of the generalized coordinate $\overline{q_j{}^2}$ is determined by the power spectrum of the generalized force Q_j. When the lateral dimension of the structure is small, only cross correlation in direction z need be considered. Then the power spectrum of the generalized force is

$$S_{Q_j}(f) = \int_0^H \int_0^H S_{12}(z_1, z_2, f) \phi_j(z_1) \phi_j(z_2) \, dz_1 \, dz_2 \tag{29.33}$$

where $S_{12}(z_1, z_2, f)$ = cross spectrum of the wind loads at heights z_1 and z_2, and H = height of the structure. With respect to Eq. (29.26), the cross spectrum of the wind loads can be expressed in terms of the power spectrum of the wind speed [Eq. (29.23)] and the coherence function, Eq. (29.25).

The variance of q_j is

$$
\begin{aligned}
\overline{q_j{}^2} &= \int_0^\infty \frac{1}{(2\pi f_j)^4 M_j{}^2} \frac{1}{[1 - (f/f_j)^2]^2 + 4\zeta^2 (f/f_j)^2} S_{Q_j}(f) \\
&\approx \frac{1}{64\pi^3 \zeta f_j{}^3 M_j{}^2} S_{Q_j}(f_j) + \frac{1}{(2\pi f_j)^4 M_j{}^2} \int_0^{f_j} S_{Q_j}(f) \, df
\end{aligned}
\tag{29.34}
$$

where f_j = jth natural frequency and generalized mass

$$M_j = \int_0^H m(z) \phi_j{}^2(z) \, dz \tag{29.35}$$

where $m(z)$ = mass of the structure per unit length. The approximate integration* of Eq. (29.34) yields the response composed of two parts, the resonance effect (the first term) and the background turbulence effect (the second term) (Fig. 29.24). The vari-

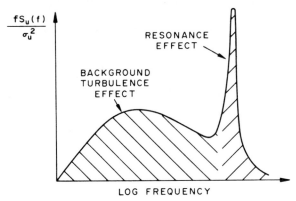

Fig. 29.24. Spectrum of structural response with indication of resonance effect and background turbulence effect.

ance of the displacement follows from Eq. (29.32), and its standard deviation (rms dynamic displacement) is $\sigma_u(z) = \sqrt{\overline{u^2(z,t)}}$. The peak response is established from Eq. (29.29) by means of the peak factor g [Eq. (29.30)] as in one degree-of-freedom. The mean deflection $\bar{u}(z)$ is the static deflection due to the mean wind $\bar{V}_z$.

Other analyses of slender structures are also available.[9, 10, 11] In applications to buildings and free-standing towers, the analysis can usually be limited to the first modal component in Eq. (29.32).

Application to buildings and structures with significant lateral dimension requires the incorporation of the horizontal cross correlation as well. A complete solution established by means of simplifying assumptions and numerical integrations is given below.

* The integral is similar to that appearing in Eq. (29.28) and can be evaluated accurately using the theory of residua. An example is given in Ref. 8.

GUST FACTOR APPROACH

The gust-factor approach is a design procedure derived on the basis of the theory above by means of a few simplifying assumptions. The approach given here is a modified version of the method described in Ref. 12 and adopted in Ref. 13. It considers only the response in the first vibration mode which is assumed to be linear. These assumptions are particularly suitable for buildings. The method yields all the data needed in design: the maximum response, the equivalent static wind load that would produce the same maximum response, and the maximum acceleration needed for the evaluation of the physiological effects of strong winds (human comfort).

The gust factor G is defined as the ratio of the expected peak displacement (load) in a period T to the mean displacement (load) $\bar{u}$. Hence, maximum expected response

$$u_{max} = G\bar{u} = \left(1 + g\frac{\sigma_u}{\bar{u}}\right)\bar{u} \qquad (29.36)$$

The gust factor is given as

$$G = 1 + g\sqrt{\frac{K}{C_e}\left(B + \frac{sF}{\zeta}\right)} \qquad (29.37)$$

where ζ = damping ratio and K = factor related to the surface roughness; this factor is equal to

0.08 for open terrain (zone A)
0.10 for suburban, urban, or wooded terrain (zone B)
0.14 for concentrations of tall buildings (zone C)

All the other parameters appearing in Eq. (29.37) can be obtained from Fig. 29.25. C_e = exposure factor based on the mean wind speed profile (coefficient α) and thus on surface roughness. For the three zones, the exposure factor is obtained from Fig. 29.25a for the height of the building H. C_e relates to wind pressure rather than speed. Hence, the mean wind speed at the top of the building is given by

$$\bar{V}_H = \bar{V}_{10}\sqrt{C_e}$$

where $\bar{V}_{10}$ = reference wind speed at the standard height of 10 meters. $\bar{V}_{10}$ can be obtained from meteorological stations. Velocity $\bar{V}_H$ is needed for determination of parameters s and F. Factors B, s, F, and g are given in Fig. 29.25c to f as a function of parameters indicated; D = width of the frontal area, and f_0 = the first natural frequency of the structure in cycles per second. The average fluctuation rate ν, on which the peak factor g depends, is evaluated from formula

$$\nu = f_0\sqrt{\frac{sF/\zeta}{B + sF/\zeta}} \qquad (29.38)$$

The peak factor g is plotted in Fig. 29.25f, assuming a period of observation $T = 3,600$ sec; it can also be calculated from Eq. (29.30).

The parameters given also yield the design wind pressure p, which produces displacement u_{max} if applied as a static load. This design pressure

$$p = qC_eGC_p \qquad (29.39)$$

where $q = \frac{1}{2}\rho\bar{V}_{10}^2$ is the reference mean-velocity pressure, and C_e = exposure factor. In this case, C_e varies continuously with the elevation according to Fig. 29.25a for pressures acting on the windward face of the structure; for the leeward face, C_e is constant and evaluated at one-half the height of the building. The quantity C_p = average pressure coefficient, which depends on the shape of the structure and the flow pattern around it. For a typical building with a flat roof and a height greater than twice the width, the coefficients are given for the windward and leeward faces in Fig. 29.25b together with the pressure distribution.

Fig. 29.25. Components of gust factor.

The peak acceleration A of a structure due to gusting wind is given by

$$A = u_{max} \frac{4\pi^2 f_0^2 g}{G} \sqrt{\frac{KsF}{C_e \zeta}}$$

where u_{max} = maximum deflection under the design pressure p. The other parameters are equal to those used in Eq. (29.37). When the acceleration exceeds about 1 per cent of gravity, the motion is usually perceptible. However, there are large differences in the perceptibility of motions having very low frequencies.[14, 15]

Similar approaches are given in Refs. 16 to 18.

EFFECT OF GUSTS ON CLADDING AND WINDOWS

Wind gusts produce local pressure on cladding and window panels of a building. Because the natural frequency of such a panel is very high compared with the frequency components of the wind-speed fluctuations, the panel displacement is essentially static. Its design may be based on the static displacement resulting from maximum expected pressure, which is the algebraic sum of the height and time-dependent exterior pressure (or suction) and the constant interior pressure (or suction). If the fluctuating component of the pressure $p(t)$ is considered to be a stationary random process, the exterior expected maximum pressure is

$$p_{\max} = \bar{p} \left(1 + g \frac{\sigma_p}{\bar{p}} \right) = \bar{p} G \qquad (29.40)$$

where $\bar{p} = \frac{1}{2}\rho C_p \bar{V}^2$ = mean pressure
$\qquad C_p$ = local pressure coefficient
$\qquad \sigma_p$ = standard deviation of the fluctuating pressure component
$\qquad g$ = peak factor given by Eq. (29.30)
$\qquad G$ = gust factor

To account for the sensitivity of glass to both static and dynamic fatigue, it has been suggested[19, 20] that g or G in Eq. (29.40) be multiplied by a wind-on-glass effect factor.

Factors g, $\sigma_p/\bar{p}$, and C_p are most reliably determined from wind-tunnel experiments. They strongly depend on location of the panel, wind direction, turbulence intensity and the local flow pattern determined by the shape of the building and its immediate environment. In full-scale experiments, values of g in excess of 10 have been observed in highly intermittent flow. Largest local pressure coefficients C_p (actually suctions) appear with skew wind at the leading edge of the building where a typical value is $C_p = -1.5$. In that part of the building exposed to free flow, a gust factor $G \approx 2.5$ is a reasonable estimate.[13, 21]

The interior pressure is not very high, but its magnitude and sign depend on openings and leakage.

Damage to windows may result from local wind pressure, but it also depends on material properties of glass and its fatigue. The fatigue limit of glass is only about 20 per cent of the instantaneous strength.[20]

VIBRATION DUE TO VORTEX SHEDDING

Vortex shedding represents the second most important mechanism for wind-induced oscillations. Unlike the gusts, vortex shedding produces forces which originate in the wake behind the structure, act mainly in the across-wind direction, and are, in general, rather regular. The resultant oscillation is resonant in character (Fig. 29.16b), is often almost periodic, and usually appears in the direction perpendicular to that of the wind. Lightly damped structures such as chimneys and towers are particularly susceptible to vortex shedding. Many failures attributed to vortex shedding have been reported.

When a bluff body is exposed to wind, vortices shed from the sides of the body creating a pattern in its wake often called the "Karman vortex street" (Fig. 29.15). The frequency of the shedding, nearly constant in many cases, depends on the shape and size of the body, the velocity of the flow, and to a lesser degree on the surface roughness and the turbulence of the flow. If the cross section of the body is noncircular, it also depends on the wind direction. The dominant frequency of vortex shedding, f_s, is given by

$$f_s = S \frac{\bar{V}}{D} \qquad \text{cps} \qquad (29.41)$$

where S = dimensionless constant called the Strouhal number, $\bar{V}$ = mean wind velocity, and D = width of the frontal area. The second dimensionless parameter is the Reynolds number $R = \bar{V}D/\nu$, where ν = kinematic viscosity. For air under normal conditions, $\nu = 1.6 \times 10^{-4}$ ft²/sec.

For a body having a rectangular or square cross section, the Strouhal number is almost independent of the Reynolds number.

For a body having a circular cross section, the Strouhal number varies with the regime of the flow as characterized by the Reynolds number. There are three major regions: the subcritical region for $R \gtrsim 3 \times 10^5$, the supercritical region for $3 \times 10^5 \gtrsim R \gtrsim 3 \times 10^6$, and the transcritical region for $R \gtrsim 3 \times 10^6$. Approximate values of the Strouhal number for typical cross sections are given in Table 29.2. The numbers

Table 29.2. Aerodynamic Data for Prediction of Vortex-induced Oscillations in Turbulent Flow

Cross section	Strouhal number S	Rms lift coefficient σ_L	Bandwidth B	Correlation length L (diameters)
Circular: region				
Subcritical	0.2	0.5	0.1	2.5
Supercritical	Not marked	0.14	Not marked	1.0
Transcritical	0.25	0.25	0.3	1.5
Square:				
Wind normal to face	0.11	0.6	0.2	3

given in this table are based on Refs. 1, 22, 23, and 24 and other measurements and may be used for turbulent shear flow.

PREDICTION OF VORTEX-INDUCED OSCILLATION

Although the mechanism of vortex shedding and the character of the lift forces have been the subject of a great number of studies,[25] the available information does not permit an accurate prediction of these oscillations.

The motion is most often viewed as forced oscillation due to the lift force, which, per unit length, may be written as

$$F_L = \frac{1}{2}\rho D\bar{V}^2 C_L(t) \qquad (29.42)$$

where $C_L(t)$ is a lift coefficient fluctuating in a harmonic or random way. (Some authors[26, 27] consider vortex shedding to be self-excitation, which does not seem necessary, however, for relatively small motions.) Hence, the solution of the response depends on the time-history assumed for $C_L(t)$.

HARMONIC EXCITATION OF PRISMATIC CYLINDERS BY VORTICES

Harmonic excitation represents a traditional model for vortex excitation, but it is really justified only for very low Reynolds numbers ($\gtrsim 300$) or possibly for large vibration where the motion starts controlling both the wake and the lift forces in the form of the "locking-in" phenomenon. Strongest oscillations arise at that wind velocity for which the frequency of vortex shedding f_s is equal to one of the natural frequencies of the structure f_j. This resonant wind velocity is, from Eq. (29.41),

$$V_c = \frac{1}{S}f_j D \qquad (29.43)$$

With free-standing towers and stacks, resonance in the first two modes is met most often; resonance with higher modes has been observed as well with guyed towers (Fig. 29.26). At the resonant wind velocity, the lift force is given by Eq. (29.42) in which $C_L(t) =$

FIG. 29.26. Vortex-induced oscillations in different modes measured on 1,000 ft guyed tower.[28]

$C_L \sin 2\pi f_j t$, and C_L = amplitude of lift coefficient. Assuming a uniform wind profile and a constant diameter D, the resonant amplitude of mode j at the critical wind velocity V_c is, from Eq. (29.31),

$$u_j(z) = \frac{\rho C_L}{16\pi^2 S^2} \frac{D^4}{\zeta M_j} \phi_j(z) \int_0^H \phi_j(z)\, dz \qquad (29.44)$$

where M_j is given by Eq. (29.35) and ζ = structural damping ratio. The formula can be further simplified if it is assumed that the lift force is distributed along the structure in proportion to the mode $\phi_j(z)$. (This assumption reflects the loss of spanwise correlation of the forces.) Then, with constant mass per unit length $m(z) = m$, the resonant amplitude at the height where the modal displacement is maximum:

$$u_j = \frac{\rho C_L}{16\pi^2 S^2} \frac{D^3}{\zeta m} \qquad (29.45)$$

For the first mode of a free-standing structure, this occurs at the tip. In higher modes, this amplitude appears at the height where local resonance takes place. For circular cylinders, a design value of the lift coefficient C_L is about $\sqrt{2}\,\sigma_L$. This simple formula can be used for the first estimate of the amplitudes that are likely to represent the upper bound. It is also indicative of the role of the diameter, mass, and damping of the structure. Approximate values of σ_L are given in Table 29.1.

RANDOM EXCITATION OF PRISMATIC CYLINDERS BY VORTICES

Even when vortex shedding appears very regular, the lift force and thus $C_L(t)$ are not purely harmonic but random. The power spectrum of the lift force per unit length is from Eq. (29.42).

$$S_L(f) = \left(\frac{1}{2}\rho D \bar{V}^2 \sigma_L\right)^2 S_L'(f) \tag{29.46}$$

where $\sigma_L = \sqrt{\overline{C_L{}^2(t)}}$ is the standard deviation of the lift coefficient and $S_L'(f)$ = normalized power spectrum of $C_L(t)$ for which

$$\int_0^\infty S_L'(f)\, df = 1 \tag{29.47}$$

With circular cylinders, the lift force is narrow-band random in the subcritical and transcritical[22, 23] ranges where the energy is distributed about the dominant frequency

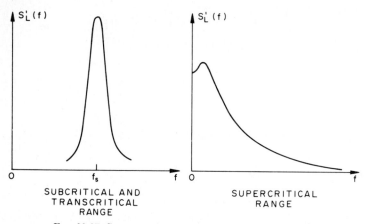

FIG. 29 27. Spectra of lift coefficient for circular cylinder.

f_s, given by Eq. (29.41) (Fig. 29.27a). Such spectra can be described by a Gaussian-type curve,

$$S_L'(f) = \frac{1}{\sqrt{\pi}\, Bf_s} \exp\left[-\left(\frac{1 - f/f_s}{B}\right)^2\right] \tag{29.48}$$

A few design values of bandwidth B are given in Table 29.2. In the supercritical range, the power spectrum is broad (Fig. 29.27b) and can be expressed as[29]

$$S_L'(f) = 4.8 \frac{1 + 682.2(fD/\bar{V})^2}{[1 + 227.4(fD/\bar{V})^2]^2} \frac{D}{\bar{V}} \tag{29.49}$$

Because the vortices are three-dimensional, a realistic treatment also requires the inclusion of the spanwise cross correlation of the lift forces. This can be done in terms of the "correlation length" L given in number of diameters.

Approximate values of L are given in Table 29.3. The correlation length decreases with turbulence[30] and shear and increases with aspect ratio $2H/D$ and the amplitude of the motion (Fig. 29.28).

Using the correlation length, the spectral density of the lift force, Eqs. (29.48) and (29.49), and a few further approximations, the vibration can be evaluated from Eqs.

(29.32) to (29.34). The root-mean-square (rms) displacement at height z in mode j is approximately[1]

$$\sqrt{\overline{u_j^2(z,t)}} = \frac{\pi^{1/4}\sigma_L\rho D^4\phi_j(z/H)}{\sqrt{B\zeta}\,(4\pi S)^2 M_j} C$$

where

$$C^2 = \frac{(H/D)^2}{1 + (H/2LD)} \int_0^1 \left(\frac{z}{H}\right)^{3\alpha} \phi_j^2\left(\frac{z}{H}\right) d\left(\frac{z}{H}\right)$$

where α = wind profile exponent (Fig. 29.19), and parameters $S, \sigma_L, B,$ and L are given in Table 29.1. The mode $\phi_j(z/H)$ is dimensionless, and consequently M_j is in slugs in this

FIG. 29.28. Variation of correlation length of vortex shedding with amplitude of motion and turbulence ($2a$ = double amplitude, turbulence intensity 10 per cent).

case. The peak response is $g\,\sqrt{\overline{u_j^2(z,t)}}$, where the peak factor g is given by Eq. (29.30). If it is larger than about 2 per cent of diameter D, locking-in may develop and the analysis should be repeated assuming harmonic excitation or at least random excitation with a significantly increased correlation length, as Fig. 29.28 indicates.

RANDOM EXCITATION OF TAPERED CYLINDERS BY VORTICES

Tapered cylinders, such as stacks, also vibrate due to vortex shedding; but less is known about the mechanism of excitation. It appears that the lift forces are narrow-band random with a rather small correlation length L and with the dominant frequency f_s given by Eq. (29.41). As the diameter is variable, local resonance between f_s and the natural frequency f_j takes place at different heights z_r. As the wind speed increases, the resonance first appears at the tip and shifts downward. The critical wind speed for each height follows from Eq. (29.43) with $D = D(z_r)$. The rms displacements at height H due to local resonance at height z_r can be obtained from an approximate formula,[32]

$$\sqrt{u_j{}^2(H,t)} = \sqrt{\frac{L}{2\pi^3\zeta\Psi}\,\frac{\sigma_L\rho D^4(z_r)\phi_j(z_r)}{8S^2 M_j}}\,\phi_j(H)$$

where

$$\Psi = \frac{dD(z_r)}{dz} + \frac{\alpha D(z_r)}{z_r}$$

or with a constant taper

$$\Psi = \frac{t}{H} + \frac{\alpha D(z_r)}{z_r}$$

where $t = D(0) - D(H)$ and α = the wind-profile exponent.

The other parameters can be taken from Table 29.2. The values listed for the trans-critical region may be adequate, inasmuch as most tapered stacks are large. The peak displacement is again obtained by means of the peak factor given by Eq. (29.30).

Maximum response of chimneys in the first mode usually results from local resonance at about $\tfrac{3}{4}$ H. The height of maximum excitation follows from condition $d[D^4(z)\phi_j(z)]/dz = 0$.

SUPPRESSION OF VORTEX-INDUCED VIBRATIONS

Vortex shedding may induce severe vibration of a cylindrical structure such as a chimney, free-standing tower, guyed mast, bridge columns, etc. Very strong oscillations have been observed[28, 31] in all-welded structures where the damping ratio is extremely low, sometimes less than 0.005.[8, 28] Welded structures are particularly prone to fatigue failure, as the endurance limit may be only a fraction of the strength if heavy notches, flaws, attachments, or other adverse details are present. In other cases, the motion is intolerable because of its physiological effects or swaying of antennas. For these reasons, suppression of vibration is often desirable.

In some cases, vibration can be reduced by increasing the structural damping. This can be accomplished by additional dampers attached to an independent support[28] or to a special mass suspended from the structure and suitably tuned or by hanging chains.[33] Columns of a few bridges were filled with gravel, sand, or plastic balls partly filled with oil. The increase in mass may be favorable but can reduce the original structural damping.

Another successful method of vibration control is to break down the wake pattern by providing the surface by helical "strakes" or "spoilers."[28, 31, 34] A suitable height of the spoilers is about $0.1D$ or more with a pitch of about $5D$. A significant drawback of the spoilers is that they considerably increase the drag, sometimes as much as 100 per cent or even more.[31, 35]

WAKE BUFFETING

If one structure is located in the wake of another, vortices shed from the upstream structure may cause oscillation of the downstream structure[36, 37] (Fig. 29.17). If the two structures differ greatly in size or shape, this excitation is usually not significant. Strong vibration of the downstream structure may arise when two or more structures are identical and less than about 10 diameters apart. Then the structure in the wake is efficiently excited by well-tuned wake buffeting and its own vortex shedding. Such excitation has been observed with stacks and bridges and to a certain degree with hyperbolic cooling towers.[36]

GALLOPING OSCILLATIONS

Vibrations due to turbulence and vortices discussed above are induced by aerodynamic forces which are, to a high degree, independent of the motion and act even on stationary bodies. Quite a different kind of oscillation is induced by the aerodynamic forces generated by the motion itself. Such forces may result from oscillatory changes in pressure distribution brought about by the continuous change in the angle under which the wind strikes the structure ("angle of attack"). This kind of oscillation often has a tendency to diverge; it is called, summarily, "aerodynamic instability," "flutter," or "self-excited oscillation." Sudden start and violent amplitudes are typical of such phenomena (Fig. 29.16c).

The mechanism of this oscillation is, in general, complex. The aerodynamic forces may be a function of the displacements (translation and rotation), vibration velocity, or both, and they may interact with turbulence and vortex shedding. The basic type of the self-excited oscillations is the lateral (across-wind) oscillation induced by aerodynamic forces which are related to vibration velocity alone. Such oscillation is referred to as "galloping." Typical features of galloping oscillation are motion in the direction perpendicular to that of the wind, sudden onset, large steady amplitudes increasing with wind velocity, and a frequency equal to the natural frequency. Galloping oscillation occurs in transmission lines and in a variety of structures having square, rectangular, or other sharp-edged cross sections.

The origin of galloping oscillation depends on the relation between lift and drag. If a body moves with a velocity $\dot{u}$ in a flow having velocity $\bar{V}$ perpendicular to its direction

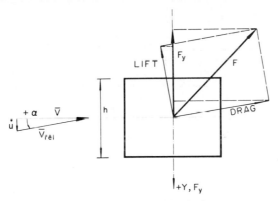

Fɪɢ. 29.29. Cross section in flow.

(Fig. 29.29), the aerodynamic force acting on the body is produced by relative wind velocity $\bar{V}_{\text{rel}}$. The angle of attack of relative wind is

$$\alpha = \arctan \frac{u}{\bar{V}} \tag{29.50}$$

The drag and lift components D and L of the aerodynamic force F are

$$D = C_D \frac{1}{2} \rho h l \, \bar{V}_{\text{rel}}^2$$

$$L = C_L \frac{1}{2} \rho h l \, \bar{V}_{\text{rel}}^2$$

where C_D and C_L are drag and lift coefficients at angle α (Fig. 29.30), h = depth of the cross section, and l = length of the body.

The component of force F into the direction of axis Y, therefore, is

$$F_y = -(C_D \sin \alpha + C_L \cos \alpha) \frac{1}{2} \rho h l \, \bar{V}^2 \sec^2 \alpha = C_{F_y} \frac{1}{2} \rho h l \, \bar{V}^2 \tag{29.51}$$

where

$$C_{F_y} = -(C_L + C_D \tan \alpha) \sec \alpha \tag{29.52}$$

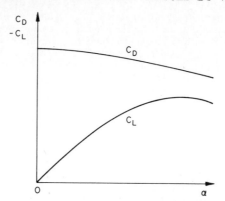

FIG. 29.30. Lift and drag as function of angle of attack.

The lateral force excites the vibration if the first derivative of C_{F_y} at $\alpha = 0$ is >0, hence

$$A_1 = \frac{dC_{F_y}}{d\alpha}\bigg|_{\alpha=0} = -\left(\frac{dC_L}{d\alpha} + C_D\right) > 0 \tag{29.53}$$

This condition for aerodynamic instability is known as "Den Hartog's criterion."[38] Substitution of Eq. (29.50) into Eq. (29.52) indicates that the aerodynamic forces depend on vibration velocity and thus actually represent the aerodynamic damping. This damping is negative if $A_1 > 0$. Because the system also has structural damping ζ, which is positive, the vibration will start only if the total available damping becomes less than 0. This condition yields the onset (minimum) wind velocity for galloping from the equilibrium (or zero displacement) position,

$$\bar{V}_0 = \zeta \frac{2\pi f_i h}{n A_1} \tag{29.54}$$

where f_i = natural frequency, $n = \rho h^2/(4m)$ = mass parameter, and m = mass of the body per unit length. Some values of coefficient A_1 are given in Table 29.3.

Table 29.3. Coefficients A_1 for Determination of Galloping Onset Wind Velocity (Infinite Prisms)

	Cross section (Side ratio)					
	Unstable in smooth flow			Stable in smooth flow		
	Square	Rect.	Rect.	Rect.	Rect.	D-section
$V \rightarrow$ Flow	$\dfrac{1}{1}$	$\dfrac{2}{3}$	$\dfrac{1}{2}$	$\dfrac{3}{2}$	$\dfrac{2}{1}$	2 * 1
Smooth	2.7	1.91	2.8	0	−0.03	−0.1
Turbulent ≈10 per cent intensity	2.6	1.83	−2.0	0.74	0.17	0

* Varies with Reynolds number.

Galloping oscillations starting from zero initial displacement can occur only when the cross section has $A_1 > 0$. Cross sections having $A_1 \leq 0$ are generally considered stable even though galloping may sometimes arise if triggered by a large initial amplitude.[41]

The response and the onset velocity are often very sensitive to turbulence. Some cross sections such as a flat rectangle or a D section are stable in smooth flow but can become unstable in turbulent flow.[41, 42] With other cross sections, turbulence may stabilize a shape that is unstable in smooth flow (see Table 29.2).

From Eqs. (29.51) and (29.52) the nonlinear, negative aerodynamic damping can be calculated[43] for inclusion in the treatment of the across-wind response due to atmospheric turbulence.

The prediction of oscillations for wind velocities greater than V_0 depends on the shape of the C_{F_y} coefficient and requires the application of nonlinear theory.[39–42] A few typical cases are shown in Fig. 29.31. The cases are typical of a square cross sec-

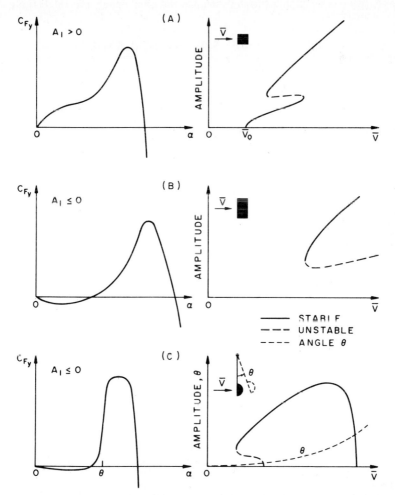

FIG. 29.31. Typical lateral force coefficients C_{F_y} and corresponding galloping oscillations: (a) vibration from equilibrium position, (b) vibration triggered by initial amplitudes, and (c) vibration with variable angle of attack.

tion, a flat rectangular section, and a D section whose angle of attack is allowed to change due to drag.   Similar response can be expected with other cross sections.

Torsion can also participate in galloping oscillations and play an important part in the vibration.   This is the case with angle cross sections[44] and bundled conductors.[45] The quasi-steady theory of pure torsional galloping can be found in Ref. 46.   A solution of coupled galloping is presented in Ref. 47.

Galloping often appears in overhead conductors which also vibrate due to vortex shedding.   Vortex shedding produces resonant vibration in a high-vibration mode. Galloping usually involves the fundamental mode and is known to occur when the conductor is ice-coated or free of ice.   The vibration often leads to fatigue failures, and various techniques are therefore used to reduce the amplitude.   This can be achieved by means of resonant dampers[48] consisting of auxiliary masses suspended on short lengths of cable which dissipate energy through the bending or aerodynamic dampers[49] consisting of perforated shrouds.   Vibrations of bundled conductors can be eliminated by twisting the bundle[45] and thereby changing the aerodynamic characteristics in the spanwise direction.

VIBRATION OF SPECIAL STRUCTURES

The basic types of vibration discussed above are common in many structures.   However, there are some special structures which would require individual treatment.   A few examples are cited below.

Guyed towers experience complicated vibration patterns because of the nonlinearity of the guys, the three-dimensional character of the response, the interaction between the guys and the tower, and other factors.[28, 50–52]

Hyperbolic cooling towers can suffer from some of the effects of wake buffeting[36] and are susceptible to turbulence.[53]   Information on vibration of a number of special structures can be found in Refs. 2 to 5.

REFERENCES

1. Davenport, A. G., et al.: "New Approaches to Design against Wind Action," to be published by the Faculty of Engineering Science, University of Western Ontario.
2. *Proc. IUTAM-IAHR Symp. Karlsruhe*, 1972.
3. *Proc. Conf. National Physical Laboratory, Teddinton, Middlesex*, 1963.
4. *Proc. Intern. Res. Seminar*, Ottawa, 1967.
5. *Proc. 3d Intern. Conf.*, Tokyo, 1971.
6. Davenport, A. G.: *Inst. Civil Eng. Paper* No. 6480, 449–472 (August, 1961).
7. Harris, R. I.: Seminar of Construction Industry Research and Information Association, paper 3, Institution of Civil Engineers, 1970.
8. Novak, M.: *Acta Tech. Czechoslovak Acad. Sci.*, **4**:375–404 (1967).
9. Vickery, B. J.: *J. Struct. Div. Am. Soc. Civil Engrs.*, **98**, 21–36 (January, 1972).
10. Davenport, A. G.: *Proc. Inst. Civil Engrs.*, **23**:449–472 (1962).
11. Etkin, B.: Meeting on Ground Wind Load Problems in Relation to Launch Vehicles, pp. 21.1–15, Langley Research Center, NASA, June, 1966.
12. Davenport, A. G.: *J. Struct. Div., Am. Soc. Civil Engrs.*, **93**, 11–34 (June, 1967).
13. "Canadian Structural Design Manual 1970," Suppl. 4, National Research Council of Canada, 1970.
14. Chen, P. W., and L. E. Robertson: *J. Struct. Div. Am. Soc. Civil Engrs.*, **98**, 1681–1695 (August, 1972).
15. Hansen, R. T., J. W. Reed, and E. H. Vanmarcke: *J. Struct. Div. Am. Soc. Civil Engrs.*, **99**, 1589–1605 (July, 1973).
16. Vellozzi, Y., and E. Cohen: *J. Struct. Div. Am. Soc. Civil Engrs.*, **94**, 1295–1313 (June, 1968).
17. Vickery, B.: U.S. Dept. of Commerce, *Nat. Bur. Std. Bldg. Sci.*, Ser. **30**:93–104.
18. Simiu, E.: *J. Struct. Div. Am. Soc. Civil Engrs.*, **100**, 1897–1910 (September, 1974).
19. Allen, D. E., and W. A. Dalgliesh: Preliminary Publication of IABSE Symposium on Resistance and Ultimate Deformability of Structures, pp. 279–285, Lisbon, 1973.
20. Dalgliesh, W. A.: *Proc. U.S.-Japan Res. Seminar Wind Effects Structures*, Kyoto, Japan, September, 1974.
21. Dalgliesh, W. A.: *J. Struct. Div. Am. Soc. Civil Engrs.*, **97**, 2173–2187 (September, 1971).

22. Roshko, A.: *J. Fluid Mech.*, **10**:345–356 (1961).
23. Cincotta, J. J., G. W. Jones, and R. W. Walker: Meeting on Ground Wind Load Problems in Relation to Launch Vehicles, pp. 20.1–35, Langley Research Center, NASA, 1966.
24. Novak, M.: Ref. 5, pp. 799–809.
25. Morkovin, M. V.: *Proc. Symp. Fully Separated Flows*, pp. 102–118, ASME, 1964.
26. Nakamura, Y.: *Rept. Res. Inst. Appl. Mech., Kyushu University*, **17**(59):217–234 (1969).
27. Hartlen, R. T., and I. G. Currie: *J. Eng. Mech. Div. Am. Soc. Civil Engrs.*, **70**, 577–591 (October, 1970).
28. Novak, M.: *Proc. IASS Symp. Tower-Shaped Steel r.c. Structures*, Bratislava, 1966.
29. Fung, Y. C.: *J. Aerospace Sci.*, **27**(11):801–814 (November, 1960).
30. Surry, D.: *J. Fluid Mech.*, **52**(3):543–563 (1972).
31. Wotton, L. R., and C. Scruton: Construction Industry Research and Information Association Seminar, Paper 5, June, 1970.
32. Vickery, B. J., and A. W. Clark: *J. Struct. Div. Am. Soc. Civil Engrs.*, **98**, 1–20 (January, 1972).
33. Reed, W. H.: Ref. 4, Paper 36, pp. 283–321.
34. Scruton, C.: National Physical Laboratory Note 1012, April, 1963.
35. Novak, M.: Ref. 4, pp. 429–457.
36. Scruton, C.: Ref. 4, pp. 115–161.
37. Cooper, K. R., and Wardlaw, R. L.: Ref. 5, pp. 647–655.
38. Den Hartog: *Trans., AIEE*, **51**:1074 (1932).
39. Parkinson, G. V., and J. D. Smith: *Quart. J. Mech. Appl. Math.*, **17**(2):225–239 (1964).
40. Novak, M.: *J. Eng. Mech. Div. Am. Soc. Civil Engrs.*, **95**, 115–142 (February, 1969).
41. Novak, M.: *J. Eng. Mech. Div. Am. Soc. Civil Engrs.*, **98**, 27–46 (February, 1972).
42. Novak, M., and Tanaka, H.: *J. Eng. Mech. Div. Am. Soc. Civil Engrs.*, **100**, 27–47 (February, 1974).
43. Novak, M., and Davenport, A. G.: *J. Eng. Mech. Div. Am. Soc. Civil Engrs.*, **96**, 17–39 (February, 1970).
44. Wardlaw, R. L.: Ref. 4, pp. 739–772.
45. Wardlaw, R. L., K. R. Cooper, R. G. Ko, and J. A. Watts: *Trans. IEEE*, 1975.
46. Modi, V. J., and J. E. Slater: Ref. 2, pp. 355–372.
47. Blevins, R. D., and W. D. Iwan: *J. Appl. Mech.*, **41**, no. 4, 1974.
48. "Overhead Conductor Vibration," Alcoa Aluminum Overhead Conductor Engineering Data, no. 4, 1974.
49. Hunt, J. C. R., and D. J. W. Richards: *Proc. Inst. Elec. Engrs.*, **116**(11):1869–1874 (November, 1969).
50. Davenport, A. G., and G. N. Steels: *J. Struct. Div. Am. Soc. Civil Engrs.*, **91**, 43–70 (April, 1965).
51. Davenport, A. G.: Engineering Institute of Canada, **3**:119–141 (1959).
52. McCaffrey, R. J., and Hartmann, A. J.: *J. Struct. Div. Am. Soc. Civil Engrs.*, **98**, 1309–1323 (June, 1972).
53. Hashish, M. G., and Abu-Sitta, S. H., *J. Struct. Div. Am. Soc. Civil Engrs.*, **100**, 1037–1051 (May, 1974).

30

THEORY OF VIBRATION ISOLATION

Charles E. Crede
California Institute of Technology

Jerome E. Ruzicka
Barry Wright Corporation

INTRODUCTION

Vibration isolation concerns means to bring about a reduction in a vibratory effect. A vibration isolator in its most elementary form may be considered as a resilient member connecting the equipment and foundation. The function of an isolator is to reduce the magnitude of motion transmitted from a vibrating foundation to the equipment, or to reduce the magnitude of force transmitted from the equipment to its foundation.

CONCEPT OF VIBRATION ISOLATION

The concept of vibration isolation is illustrated by consideration of the single degree-of-freedom system illustrated in Fig. 30.1. This system consists of a rigid body representing an equipment connected to a foundation by an isolator having resilience and energy-

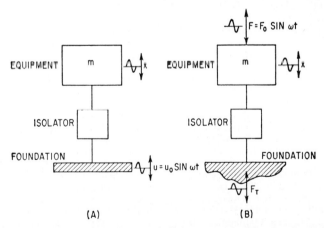

(A)   (B)

FIG. 30.1. Schematic diagrams of vibration isolation systems: (A) vibration isolation where motion u is imposed at the foundation and motion x is transmitted to the equipment; (B) vibration isolation where force F is applied by the equipment and force F_T is transmitted to the foundation.

dissipating means; it is unidirectional in that the body is constrained to move only in vertical translation. The performance of the isolator may be evaluated by the following characteristics of the response of the equipment-isolator system of Fig. 30.1 to steady-state sinusoidal vibration:

Absolute Transmissibility. Transmissibility is a measure of the reduction of transmitted force or motion afforded by an isolator. If the source of vibration is an oscillating motion of the foundation (motion excitation), transmissibility is the ratio of the vibration amplitude of the equipment to the vibration amplitude of the foundation. If the source of vibration is an oscillating force originating within the equipment (force excitation), transmissibility is the ratio of the force amplitude transmitted to the foundation to the amplitude of the exciting force.

Relative Transmissibility. Relative transmissibility is the ratio of the relative deflection amplitude of the isolator to the displacement amplitude imposed at the foundation. A vibration isolator effects a reduction in vibration by permitting deflection of the isolator. The relative deflection is a measure of the clearance required in the isolator. This characteristic is significant only in an isolator used to reduce the vibration transmitted from a vibrating foundation.

Motion Response. Motion response is the ratio of the displacement amplitude of the equipment to the quotient obtained by dividing the excitation force amplitude by the static stiffness of the isolator. If the equipment is acted on by an exciting force, the resultant motion of the equipment determines the space requirements for the isolator, i.e., the isolator must have a clearance at least as great as the equipment motion.

FORM OF ISOLATOR

The essential features of an isolator are resilient load-supporting means and energy-dissipating means. In certain types of isolators, the functions of the load-supporting means and the energy-dissipating means may be performed by a single element, e.g., natural or synthetic rubber. In other types of isolators, the resilient load-carrying means may lack sufficient energy-dissipating characteristics, e.g., metal springs; then separate and distinct energy-dissipating means (dampers) are provided. For purposes of analysis, it is assumed that the springs and dampers are separate elements. In general, the springs are assumed to be linear and massless. The effects of nonlinearity and mass of the load-supporting means upon vibration isolation are considered in later sections of this chapter.

Various types of dampers are shown in combination with ideal springs in the following idealized models of isolators illustrated in Table 30.1. Practical aspects of isolator design are considered in Chap. 32.

Rigidly Connected Viscous Damper. A viscous damper c is connected rigidly between the equipment and its foundation as shown in Table 30.1A. The damper has the characteristic property of transmitting a force F_c that is directly proportional to the relative velocity $\dot{\delta}$ across the damper, $F_c = c\dot{\delta}$. This damper sometimes is referred to as a *linear damper*.

Rigidly Connected Coulomb Damper. An isolation system with a rigidly connected Coulomb damper is indicated schematically in Table 30.1B. The force F_f exerted by the damper on the mass of the system is constant, independent of position or velocity, but always in a direction that opposes the relative velocity across the damper. In a physical sense, Coulomb damping is approximately attainable from the relative motion of two members arranged to slide one upon the other with a constant force holding them together.

Elastically Connected Viscous Damper. The elastically connected viscous damper is shown in Table 30.1C. The viscous damper c is in series with a spring of stiffness k_1; the load-carrying spring k is related to the damper spring k_1 by the parameter $N = k_1/k$. This type of damper system sometimes is referred to as a *viscous relaxation system*.

Elastically Connected Coulomb Damper. The elastically connected Coulomb damper is shown in Table 30.1D. The friction element can transmit only that force which is developed in the damper spring k_1. When the damper slips, the friction force F_f is independent of the velocity across the damper, but always is in a direction that opposes it.

Table 30.1. Types of Idealized Vibration Isolators

(A) RIGIDLY CONNECTED VISCOUS DAMPER	(B) RIGIDLY CONNECTED COULOMB DAMPER	(C) ELASTICALLY CONNECTED VISCOUS DAMPER	(D) ELASTICALLY CONNECTED COULOMB DAMPER
EXCITATION			
$u = u_0 \sin\omega t$ $F = F_0 \sin\omega t$	$u = u_0 \sin\omega t$ * OR $\ddot{u} = \ddot{u}_0 \sin\omega t$ * $F = F_0 \sin\omega t$	$u = u_0 \sin\omega t$ $F = F_0 \sin\omega t$	$u = u_0 \sin\omega t$ * OR $\ddot{u} = \ddot{u}_0 \sin\omega t$ * $F = F_0 \sin\omega t$
RESPONSE			
$x = x_0 \sin(\omega t + \theta^+)$ OR $\delta = \delta_0 \sin(\omega t + \theta^+)$ WHERE $\delta = x - u$ $F_T = (F_T)_0 \sin(\omega t + \theta^+)$	$x = x_0 \sin(\omega t + \theta^+)$ OR $\delta = \delta_0 \sin(\omega t + \theta^+)$ WHERE $\delta = x - u$ $F_T = (F_T)_0 \sin(\omega t + \theta^+)$	$x = x_0 \sin(\omega t + \theta^+)$ OR $\delta = \delta_0 \sin(\omega t + \theta^+)$ WHERE $\delta = x - u$ $F_T = (F_T)_0 \sin(\omega t + \theta^+)$	$x = x_0 \sin(\omega t + \theta^+)$ OR $\delta = \delta_0 \sin(\omega t + \theta^+)$ WHERE $\delta = x - u$ $F_T = (F_T)_0 \sin(\omega t + \theta^+)$
FREQUENCY PARAMETERS			
$\omega_0 = \sqrt{k/m} \ (c = 0)$	$\omega_0 = \sqrt{k/m} \ (F_f = 0)$	$\omega_0 = \sqrt{k/m} \ (c = 0)$ $\omega_\infty = \sqrt{(N+1)\frac{k}{m}} \ (c = \infty)$	$\omega_0 = \sqrt{k/m} \ (F_f = 0)$ $\omega_\infty = \sqrt{(N+1)\frac{k}{m}} \ (F_f = \infty)$
DAMPING PARAMETERS			
$c_c = 2\sqrt{km}$ $\zeta = c/c_c$	$\eta = \frac{F_f}{ku_0}$ $\xi = \frac{F_f}{m\ddot{u}_0}$ $\xi_F = \frac{F_f}{F_0}$	$c_c = 2\sqrt{km}$ $\zeta = c/c_c$	$\eta = \frac{F_f}{ku_0}$ $\xi = \frac{F_f}{m\ddot{u}_0}$ $\xi_F = \frac{F_f}{F_0}$

* PHYSICALLY, THESE FORMS OF EXCITATION ARE IDENTICAL. THEY ARE EXPRESSED IN TWO DIFFERENT MATHEMATICAL FORMS FOR CONVENIENCE IN DEFINING THE DAMPING PARAMETER FOR COULOMB DAMPING.

† IN VIBRATION ISOLATION, ONLY THE MAGNITUDE OF THE RESPONSE IS OF INTEREST; THUS, THE PHASE ANGLE USUALLY IS NEGLECTED.

Table 30.2. Transmissibility and Motion Response

Where the equation is shown graphically, the applicable figure is

Type of damper	Absolute transmissibility
Rigidly connected viscous damper.......	(a) $$T_A = \frac{x_0}{u_0} = \frac{F_T}{F_0} = \sqrt{\frac{1 + \left(2\zeta\frac{\omega}{\omega_0}\right)^2}{\left(1 - \frac{\omega^2}{\omega_0^2}\right)^2 + \left(2\zeta\frac{\omega}{\omega_0}\right)^2}}$$ Fig. 30.2
Rigidly connected Coulomb damper (see Note 1)......................	(d) $$(T_A)_D = \frac{x_0}{u_0} = \sqrt{\frac{1 + \left(\frac{4}{\pi}\eta\right)^2 (1 - 2\omega_0^2/\omega^2)}{\left(1 - \frac{\omega^2}{\omega_0^2}\right)^2}}$$ Fig. 30.5 (see Note 2)
Elastically connected viscous damper.....	(g) $$T_A = \frac{x_0}{u_0} = \frac{F_T}{F_0} = \sqrt{\frac{1 + 4\left(\frac{N+1}{N}\right)^2 \zeta^2 \frac{\omega^2}{\omega_0^2}}{\left(1 - \frac{\omega^2}{\omega_0^2}\right)^2 + \frac{4}{N^2}\zeta^2\frac{\omega^2}{\omega_0^2}\left(N + 1 - \frac{\omega^2}{\omega_0^2}\right)^2}}$$ Fig. 30.10 (see Note 3)
Elastically connected Coulomb damper (see Note 1)......................	(j) $$(T_A)_D = \frac{x_0}{u_0} = \sqrt{\frac{1 + \left(\frac{4}{\pi}\eta\right)^2 \left[\left(\frac{N+2}{N}\right) - 2\left(\frac{N+1}{N}\right)\left(\frac{\omega_0}{\omega}\right)^2\right]}{\left(1 - \frac{\omega^2}{\omega_0^2}\right)^2}}$$ Fig. 30.13 (see Notes 4 and 5)

NOTE 1: These equations apply only when there is relative motion across the damper.
NOTE 2: This equation applies only when excitation is defined in terms of displacement amplitude.
NOTE 3: These curves apply only for optimum damping [see Eq. (30.15)]; curves for other values of damping are given in Ref. 4.

INFLUENCE OF DAMPING IN VIBRATION ISOLATION

The nature and degree of vibration isolation afforded by an isolator is influenced markedly by the characteristics of the damper. This aspect of vibration isolation is evaluated in this section in terms of the single degree-of-freedom concept; i.e., the equipment and the foundation are assumed rigid and the isolator is assumed massless. The performance is defined in terms of absolute transmissibility, relative transmissibility, and motion response for isolators with each of the four types of dampers illustrated in Table 30.1. A system with a rigidly connected viscous damper is discussed in detail in Chap. 2, and important results are reproduced here for completeness; isolators with other types of dampers are discussed in detail in this chapter.

The characteristics of the dampers and the performance of the isolators are defined in terms of the parameters shown on the schematic diagrams in Table 30.1. Absolute transmissibility, relative transmissibility, and motion response are defined analytically in Table 30.2 and graphically in the figures referenced in Table 30.2. For the rigidly connected viscous and Coulomb-damped isolators, the graphs generally are explicit and complete. For isolators with elastically connected dampers, typical results are included and references are given to more complete compilations of dynamic characteristics.

for Isolation Systems Defined in Table 30.1

indicated below the equation. See Table 30.1 for definition of terms.

Relative transmissibility	Motion response
(b) $$T_R = \frac{\delta_0}{u_0} = \sqrt{\dfrac{\left(\dfrac{\omega}{\omega_0}\right)^4}{\left(1 - \dfrac{\omega^2}{\omega_0^2}\right)^2 + \left(2\zeta \dfrac{\omega}{\omega_0}\right)^2}}$$ Fig. 30.3	(c) $$\frac{x_0}{F_0/k} = \sqrt{\dfrac{1}{\left(1 - \dfrac{\omega^2}{\omega_0^2}\right)^2 + \left(2\zeta \dfrac{\omega}{\omega_0}\right)^2}}$$ Fig. 30.14
(e) $$(T_R)_D = \frac{\delta_0}{u_0} = \sqrt{\dfrac{\left(\dfrac{\omega}{\omega_0}\right)^4 - \left(\dfrac{4}{\pi}\eta\right)^2}{\left(1 - \dfrac{\omega^2}{\omega_0^2}\right)^2}}$$ Fig. 30.6 (see Note 2)	(f) $$\frac{x_0}{F_0/k} = \sqrt{\dfrac{1 - \left(\dfrac{4}{\pi}\xi\right)^2}{\left(1 - \dfrac{\omega^2}{\omega_0^2}\right)^2}}$$ (See Note 2)
(h) $$T_R = \frac{\delta_0}{u_0} = \sqrt{\dfrac{\dfrac{\omega^2}{\omega_0^2} + \dfrac{4}{N^2}\zeta^2 \dfrac{\omega^6}{\omega_0^6}}{\left(1 - \dfrac{\omega^2}{\omega_0^2}\right)^2 + \dfrac{4}{N^2}\zeta^2 \dfrac{\omega^2}{\omega_0^2}\left(N + 1 - \dfrac{\omega^2}{\omega_0^2}\right)^2}}$$ Fig. 30.11 (see Note 3)	(i) $$\frac{x_0}{F_0/k} = \sqrt{\dfrac{1 + \dfrac{4}{N^2}\zeta^2 \dfrac{\omega^2}{\omega_0^2}}{\left(1 - \dfrac{\omega^2}{\omega_0^2}\right)^2 + \dfrac{4}{N^2}\zeta^2 \dfrac{\omega^2}{\omega_0^2}\left(N + 1 - \dfrac{\omega^2}{\omega_0^2}\right)^2}}$$ Fig. 30.9 (see Note 4)
(k) $$(T_R)_D = \frac{\delta_0}{u_0} = \sqrt{\dfrac{\left(\dfrac{\omega}{\omega_0}\right)^4 + \left(\dfrac{4}{\pi}\eta\right)^2\left[\dfrac{2}{N}\dfrac{\omega^2}{\omega_0^2} - \left(\dfrac{N+2}{N}\right)\right]}{\left(1 - \dfrac{\omega^2}{\omega_0^2}\right)^2}}$$ Fig. 30.14 (see Notes 4 and 5)	

NOTE 4: These curves apply only for $N = 3$.
NOTE 5: This equation applies only when excitation is defined in terms of displacement amplitude; for excitation defined in terms of force or acceleration, see Eq. (30.18).

RIGIDLY CONNECTED VISCOUS DAMPER

Absolute and relative transmissibility curves are shown graphically in Figs. 30.2 and 30.3, respectively.* As the damping increases, the transmissibility at resonance decreases and the absolute transmissibility at the higher values of the forcing frequency ω increases; i.e., reduction of vibration is not as great. For an undamped isolator, the absolute transmissibility at higher values of the forcing frequency varies inversely as the square of the forcing frequency. When the isolator embodies significant viscous damping, the absolute transmissibility curve becomes asymptotic at high values of forcing frequency to a line whose slope is inversely proportional to the first power of the forcing frequency.

The maximum value of absolute transmissibility associated with the resonant condition is a function solely of the damping in the system, taken with reference to critical damping. For a lightly damped system, i.e., for $\zeta < 0.1$, the maximum absolute transmissibility [see Eq. (2.51)] of the system is [1]

$$T_{\max} = \frac{1}{2\zeta} \tag{30.1}$$

where $\zeta = c/c_c$ is the fraction of critical damping defined in Table 30.1.

* For linear systems, the absolute transmissibility $T_A = x_0/u_0$ in the motion-excited system equals F_T/F_0 in the force-excited system. The relative transmissibility $T_R = \delta_0/u_0$ applies only to the motion-excited system.

FIG. 30.2. Absolute transmissibility for the rigidly connected, viscous-damped isolation system shown at A in Table 30.1 as a function of the frequency ratio ω/ω_0 and the fraction of critical damping ζ. The absolute transmissibility is the ratio (x_0/u_0) for foundation motion excitation (Fig. 30.1A) and the ratio (F_T/F_0) for equipment force excitation (Fig. 30.1B).

FIG. 30.3. Relative transmissibility for the rigidly connected, viscous-damped isolation system shown at A in Table 30.1 as a function of the frequency ratio ω/ω_0 and the fraction of critical damping ζ. The relative transmissibility describes the motion between the equipment and the foundation (i.e., the deflection of the isolator).

FIG. 30.4. Motion response for the rigidly connected viscous-damped isolation system shown at A in Table 30.1 as a function of the frequency ratio ω/ω_0 and the fraction of critical damping ζ. The curves give the resulting motion of the equipment x in terms of the excitation force F and the static stiffness of the isolator k.

30–6

The motion response is shown graphically in Fig. 30.4. A high degree of damping limits the vibration amplitude of the equipment at all frequencies, compared to an undamped system. The single degree-of-freedom system with viscous damping is discussed more fully in Chap. 2.

RIGIDLY CONNECTED COULOMB DAMPER

The differential equation of motion for the system with Coulomb damping shown in Table 30.1B is

$$m\ddot{x} + k(x - u) \pm F_f = F_0 \sin \omega t \tag{30.2}$$

The discontinuity in the damping force that occurs as the sign of the velocity changes at each half cycle requires a step-by-step solution of Eq. (30.2).[2] An approximate solution based on the equivalence of energy dissipation involves equating the energy dissipation per cycle for viscous-damped and Coulomb-damped systems: [3]

$$\pi c \omega \delta_0{}^2 = 4F_f \delta_0 \tag{30.3}$$

where the left side refers to the viscous-damped system and the right side to the Coulomb-damped system; δ_0 is the amplitude of relative displacement across the damper. Solving Eq. (30.3) for c,

$$c_{eq} = \frac{4F_f}{\pi \omega \delta_0} = j\left(\frac{4F_f}{\pi \dot{\delta}_0}\right) \tag{30.4}$$

where c_{eq} is the *equivalent viscous damping coefficient* for a Coulomb-damped system having equivalent energy dissipation. Since $\dot{\delta}_0 = j\omega\delta_0$ is the relative velocity, the equivalent linearized dry friction damping force can be considered sinusoidal with an amplitude $j(4F_f/\pi)$. Since $c_c = 2k/\omega_0$ [see Eq. (2.12)],

$$\zeta_{eq} = \frac{c_{eq}}{c_c} = \frac{2\omega_0 F_f}{\pi \omega k \delta_0} \tag{30.5}$$

where ζ_{eq} may be defined as the *equivalent fraction of critical damping*. Substituting δ_0 from the relative transmissibility expression [(b) in Table 30.2] in Eq. (30.5) and solving for ζ_{eq}^2,

$$\zeta_{eq}{}^2 = \frac{\left(\dfrac{2}{\pi}\eta\right)^2 \left(1 - \dfrac{\omega^2}{\omega_0{}^2}\right)^2}{\dfrac{\omega^2}{\omega_0{}^2}\left[\dfrac{\omega^4}{\omega_0{}^4} - \left(\dfrac{4}{\pi}\eta\right)^2\right]} \tag{30.6}$$

where η is the Coulomb damping parameter for displacement excitation defined in Table 30.1.

The equivalent fraction of critical damping given by Eq. (30.6) is a function of the displacement amplitude u_0 of the excitation since the Coulomb damping parameter η depends on u_0. When the excitation is defined in terms of the acceleration amplitude $\ddot{u}_0$, the fraction of critical damping must be defined in corresponding terms. Thus, it is convenient to employ separate analyses for displacement transmissibility and acceleration transmissibility for an isolator with Coulomb damping.

DISPLACEMENT TRANSMISSIBILITY. The absolute displacement transmissibility of an isolation system having a rigidly connected Coulomb damper is obtained by substituting ζ_{eq} from Eq. (30.6) for ζ in the absolute transmissibility expression for viscous damping, (a) in Table 30.2. The absolute displacement transmissibility is shown graphically in Fig. 30.5, and the relative displacement transmissibility is shown in Fig. 30.6. The absolute displacement transmissibility has a value of unity when the forcing frequency is low and/or the Coulomb friction force is high. For these conditions, the friction damper is locked in, i.e., it functions as a rigid connection, and there is no relative motion across the isolator. The frequency at which the damper breaks loose, i.e., permits relative motion across the isolator, can be obtained from the relative displacement transmissibility expression, (e) in Table 30.2. The relative displacement is imaginary when $\omega^2/\omega_0^2 \leq (4/\pi)\eta$. Thus, the "break-loose" frequency ratio is *

$$\left(\frac{\omega}{\omega_0}\right)_L = \sqrt{\frac{4}{\pi}\,\eta} \tag{30.7}$$

FIG. 30.5. Absolute displacement transmissibility for the rigidly connected, Coulomb-damped isolation system shown at B in Table 30.1 as a function of the frequency ratio ω/ω_0 and the displacement Coulomb-damping parameter η.

FIG. 30.6. Relative displacement transmissibility for the rigidly connected, Coulomb-damped isolation system shown at B in Table 30.1 as a function of the frequency ratio ω/ω_0 and the displacement Coulomb-damping parameter η.

The displacement transmissibility can become infinite at resonance, even though the system is damped, if the Coulomb damping force is less than a critical minimum value. The denominator of the absolute and relative transmissibility expressions becomes zero for a frequency ratio ω/ω_0 of unity. If the "break-loose" frequency is lower than the undamped natural frequency, the amplification of vibration becomes infinite at resonance. This occurs because the energy dissipated by the friction damping force increases linearly with the displacement amplitude, and the energy introduced into the system by the excitation source also increases linearly with the displacement amplitude. Thus, the energy dissipated at resonance is either greater or less than the input energy for *all* amplitudes of vibration. The minimum dry-friction force which prevents vibration of infinite magnitude at resonance is

$$(F_f)_{\min} = \frac{\pi k u_0}{4} = 0.79\,k u_0 \tag{30.8}$$

where k and u_0 are defined in Table 30.1.

* This equation is based upon energy considerations and is approximate. Actually, the friction damper breaks loose when the inertia force of the mass equals the friction force, $m u_0 \omega^2 = F_f$. This gives the exact solution $(\omega/\omega_0)_L = \sqrt{\eta}$. A numerical factor of $4/\pi$ relates the Coulomb damping parameters in the exact and approximate solutions for the system.

As shown in Fig. 30.5, an increase in η decreases the absolute displacement transmissibility at resonance and increases the resonant frequency. All curves intersect at the point $(T_A)_D = 1$, $\omega/\omega_0 = \sqrt{2}$. With optimum damping force, there is no motion across the damper for $\omega/\omega_0 \leq \sqrt{2}$; for higher frequencies the displacement transmissibility is less than unity. The friction force that produces this "resonance-free" condition is

$$(F_f)_{op} = \frac{\pi k u_0}{2} = 1.57 \, k u_0 \qquad (30.9)$$

For high forcing frequencies, the absolute displacement transmissibility varies inversely as the square of the forcing frequency, even though the friction damper dissipates energy. For relatively high damping ($\eta > 2$), the absolute displacement transmissibility, for frequencies greater than the "break-loose" frequency, is approximately $4\eta\omega_0{}^2/\pi\omega^2$.

ACCELERATION TRANSMISSIBILITY. The absolute displacement transmissibility $(T_A)_D$ shown in Fig. 30.5 is the ratio of response of the isolator to the excitation, where each is expressed as a displacement amplitude in simple harmonic motion. The damping parameter η is defined with reference to the displacement amplitude u_0 of the excitation. Inasmuch as all motion is simple harmonic, the transmissibility $(T_A)_D$ also applies to acceleration transmissibility when the damping parameter is defined properly.

When the excitation is defined in terms of the acceleration amplitude $\ddot{u}_0$ of the excitation,

$$\eta_{\ddot{u}_0} = \frac{F_f \omega^2}{k \ddot{u}_0} \qquad (30.10)$$

where ω = forcing frequency, rad/sec
$\ddot{u}_0$ = acceleration amplitude of excitation, in./sec²
k = isolator stiffness, lb/in.
F_f = Coulomb friction force, lb

For relatively high forcing frequencies, the acceleration transmissibility approaches a constant value $(4/\pi)\xi$, where ξ is the Coulomb damping parameter for acceleration excitation defined in Table 30.1. The acceleration transmissibility of a rigidly connected Coulomb damper system becomes asymptotic to a constant value because the Coulomb damper transmits the same friction force regardless of the amplitude of the vibration.

ELASTICALLY CONNECTED VISCOUS DAMPER

The general characteristics of the elastically connected viscous damper shown at C in Table 30.1 may best be understood by successively assigning values to the viscous damper coefficient c while keeping the stiffness ratio N constant. For zero damping, the mass is supported by the isolator of stiffness k. The transmissibility curve has the characteristics typical of a transmissibility curve for an undamped system having

RATIO $\dfrac{\text{FORCING FREQUENCY}}{\text{UNDAMPED NATURAL FREQUENCY}} \dfrac{\omega}{\omega_0}$

FIG. 30.7. Comparison of absolute transmissibility for rigidly and elastically connected, viscous damped isolation systems shown at A and C, respectively, in Table 30.1, as a function of the frequency ratio ω/ω_0. The solid curves refer to the elastically connected damper, and the parameter N is the ratio of the damper spring stiffness to the stiffness of the principal support spring. The fraction of critical damping $\zeta = c/c_c$ is 0.2 in both systems. The transmissibility at high frequencies decreases at a rate of 6 db per octave for the rigidly connected damper and 12 db per octave for the elastically connected damper.

the natural frequency

$$\omega_0 = \sqrt{\frac{k}{m}} \tag{30.11}$$

When c is infinitely great, the transmissibility curve is that of an undamped system having the natural frequency

$$\omega_\infty = \sqrt{\frac{k + k_1}{m}} = \sqrt{N + 1}\, \omega_0 \tag{30.12}$$

where $k_1 = Nk$. For intermediate values of damping, the transmissibility falls within the limits established for zero and infinitely great damping. The value of damping which produces the minimum transmissibility at resonance is called *optimum damping*.

All curves approach the transmissibility curve for infinite damping as the forcing frequency increases. Thus, the absolute transmissibility at high forcing frequencies is inversely proportional to the square of the forcing frequency. General expressions for absolute and relative transmissibility are given in Table 30.2.

A comparison of absolute transmissibility curves for the elastically connected viscous damper and the rigidly connected viscous damper is shown in Fig. 30.7. A constant viscous damping coefficient of $0.2c_c$ is maintained while the value of the stiffness ratio N is varied from zero to infinity. The transmissibilities at resonance are comparable, even for relatively small values of N, but a substantial gain is achieved in the isolation characteristics at high forcing frequencies by elastically connecting the damper.

TRANSMISSIBILITY AT RESONANCE. The maximum transmissibility (at resonance) is a function of the damping ratio ζ and the stiffness ratio N, as shown in Fig. 30.8.[4] The maximum transmissibility is nearly independent of N for small values of ζ. However, for $\zeta > 0.1$, the coefficient N is significant in determining the maximum transmissibility. The lowest value of the maximum absolute transmissibility curves corresponds to the conditions of optimum damping.

FIG. 30.8. Maximum absolute transmissibility for the elastically connected, viscous-damped isolation system shown at C in Table 30.1 as a function of the fraction of critical damping ζ and the stiffness of the connecting spring. The parameter N is the ratio of the damper spring stiffness to the stiffness of the principal support spring.

MOTION RESPONSE. A typical motion response curve is shown in Fig. 30.9 for the stiffness ratio $N = 3$. For small damping, the response is similar to the response of an isolation system with rigidly connected viscous damper. For intermediate values of damping, the curves tend to be flat over a wide frequency range before rapidly decreasing in value at the higher frequencies. For large damping, the resonance occurs near the natural frequency of the system with infinitely great damping. All response curves approach a high-frequency asymptote for which the attenuation varies inversely as the square of the excitation frequency.

OPTIMUM TRANSMISSIBILITY. For a system with optimum damping, maximum transmissibility coincides with the intersections of the transmissibility curves for zero and infinite damping. The frequency ratios $(\omega/\omega_0)_{op}$ at which this occurs are different for absolute and relative transmissibility:

Absolute transmissibility:

$$\left(\frac{\omega}{\omega_0}\right)^{(A)}_{op} = \sqrt{\frac{2(N+1)}{N+2}}$$

(30.13)

Relative transmissibility:

$$\left(\frac{\omega}{\omega_0}\right)^{(R)}_{op} = \sqrt{\frac{N+2}{2}}$$

The optimum transmissibility at resonance, for both absolute and relative motion, is

$$T_{op} = 1 + \frac{2}{N}$$

(30.14)

The optimum transmissibility as determined from Eq. (30.14) corresponds to the minimum points of the curves of Fig. 30.8.

The damping which produces the optimum transmissibility is obtained by differentiating the general expressions for transmissibility [(g) and (h) in Table 30.2] with respect to the frequency ratio, setting the result equal to zero, and combining it with Eq. (30.13),

FIG. 30.9. Motion response for the elastically connected, viscous-damped isolation system shown at C in Table 30.1 as a function of the frequency ratio ω/ω_0 and the fraction of critical damping ζ. For this example, the stiffness of the damper connecting spring is three times as great as the stiffness of the principal support spring ($N = 3$). The curves give the resulting motion of the equipment in terms of the excitation force F and the static stiffness of the isolator k.

Absolute transmissibility: $(\zeta_{op})_A = \dfrac{N}{4(N+1)} \sqrt{2(N+2)}$

(30.15)

Relative transmissibility: $(\zeta_{op})_R = \dfrac{N}{\sqrt{2(N+1)(N+2)}}$

Values of optimum damping determined from the first of these relations correspond to the minimum points of the curves of Fig. 30.8. By substituting the optimum damping ratios from Eqs. (30.15) into the general expressions for transmissibility given in Table 30.2, the optimum absolute and relative transmissibility equations are obtained, as shown graphically by Figs. 30.10 and 30.11, respectively.[5] For low values of the stiffness ratio N, the transmissibility at resonance is large but excellent isolation is obtained at high frequencies. Conversely, for high values of N, the transmissibility at resonance is lowered, but the isolation efficiency also is decreased.

FIG. 30.10. Absolute transmissibility with optimum damping in elastically connected, viscous-damped isolation system shown at C in Table 30.1 as a function of the frequency ratio ω/ω_o and the fraction of critical damping ζ. These curves apply to elastically connected, viscous-damped systems having optimum damping for absolute motion. The transmissibility $(T_A)_{op}$ is $(x_o/u_o)_{op}$ for the motion-excited system and $(F_T/F_o)_{op}$ for the force-excited system.

FIG. 30.11. Relative transmissibility with optimum damping in the elastically connected, viscous-damped isolation system shown at C in Table 30.1 as a function of the frequency ratio ω/ω_o and the fraction of critical damping ζ. These curves apply to elastically connected, viscous-damped systems having optimum damping for relative motion. The relative transmissibility $(T_R)_{op}$ is $(\delta_o/u_o)_{op}$ for the motion-excited system.

ELASTICALLY CONNECTED COULOMB DAMPER

Force-deflection curves for the isolators incorporating elastically connected Coulomb dampers, as shown at D in Table 30.1, are illustrated in Fig. 30.12. Upon application of the load, the isolator deflects; but since insufficient force has been developed in the spring k_1, the damper does not slide and the motion of the mass is opposed by a spring of stiffness $(N + 1)k$. The load is now increased until a force is developed in spring k_1 which equals the constant friction force F_f; then the damper begins to slide. When the load is increased further, the damper slides and reduces the effective spring stiffness to k. If the applied load is reduced after reaching its maximum value, the damper no longer displaces because the force developed in the spring k_1 is diminished. Upon completion of the load cycle, the damper will have been in motion for part of the cycle and at rest for the remaining part to form the hysteresis loops shown in Fig. 30.12.

Because of the complexity of the applicable equations, the equivalent energy method is used to obtain the transmissibility and motion response functions. Applying frequency, damping, and transmissibility expressions for the elastically connected viscous damped system to the elastically connected Coulomb damped system, the transmissibility expressions tabulated in Table 30.2 for the latter are obtained.[6]

If the coefficient of the damping term in each of the transmissibility expressions vanishes, the transmissibility is independent of damping. By solving for the frequency ratio ω/ω_0 in the coefficients that are thus set equal to zero, the frequency ratios obtained define the frequencies of optimum transmissibility. These frequency ratios are given by Eqs. (30.13) for the elastically connected viscous damped system and apply equally well to the elastically connected Coulomb damped system because the method of equivalent viscous damping is employed in the analysis. Similarly, Eq. (30.14) applies for optimum transmissibility at resonance.

The general characteristics of the system with an elastically connected Coulomb damper may be demonstrated by successively assigning values to the damping force

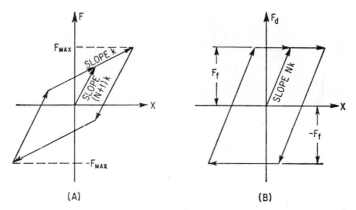

FIG. 30.12. Force deflection characteristics of the elastically connected, Coulomb-damped isolation system shown at D in Table 30.1. The force-deflection diagram for a cyclic deflection of the complete isolator is shown at A and the corresponding diagram for the assembly of Coulomb damper and spring $k_1 = Nk$ is shown at B.

while keeping the stiffness ratio N constant. For zero and infinite damping, the transmissibility curves are those for undamped systems and bound all solutions. Every transmissibility curve for $0 < F_f < \infty$ passes through the intersection of the two bounding transmissibility curves. For low damping (less than optimum), the damper "breaks loose" at a relatively low frequency, thereby allowing the transmissibility to increase to a maximum value and then pass through the intersection point of the bounding transmissibility curves. For optimum damping, the maximum absolute transmissibility has a value given by Eq. (30.14); it occurs at the frequency ratio $\left(\dfrac{\omega}{\omega_0}\right)_{op}^{(4)}$ defined by Eq. (30.13). For high damping, the damper remains "locked-in" over a wide frequency range because insufficient force is developed in the spring k_1 to induce slip in the damper. For frequencies greater than the "break-loose" frequency, there is sufficient force in spring k_1 to cause relative motion of the damper. For a further increase in frequency, the damper remains broken loose and the transmissibility is limited to a finite value. When there is insufficient force in spring k_1 to maintain motion across the damper, the damper locks-in and the transmissibility is that of a system with the infinite damping.

The "break-loose" and "lock-in" frequencies are determined by requiring the motion across the Coulomb damper to be zero. Then the "break-loose" and "lock-in" frequency ratios are

$$\left(\frac{\omega}{\omega_0}\right)_L = \sqrt{\frac{\left(\dfrac{4}{\pi}\eta\right)(N+1)}{\left(\dfrac{4}{\pi}\eta\right) \pm N}} \tag{30.16}$$

where η is the damping parameter defined in Table 30.1 with reference to the displacement amplitude u_0. The plus sign corresponds to the break-loose frequency while the minus sign corresponds to the lock-in frequency. Damping parameters for which the denominator of Eq. (30.16) becomes negative correspond to those conditions for which the damper never becomes locked-in again after it has broken loose. Thus, the damper eventually becomes locked-in only if $\eta > (\pi/4)N$.

DISPLACEMENT TRANSMISSIBILITY. The absolute displacement transmissibility curve for the stiffness ratio $N = 3$ is shown in Fig. 30.13 where $(T_A)_D = x_0/u_0$.

Fig. 30.13. Absolute displacement transmissibility for the elastically connected, Coulomb-damped isolation system illustrated at D in Table 30.1, for the damper spring stiffness defined by $N = 3$. The curves give the ratio of the absolute displacement amplitude of the equipment to the displacement amplitude imposed at the foundation, as a function of the frequency ratio ω/ω_o and the displacement Coulomb-damping parameter η.

Fig. 30.14. Relative displacement transmissibility for the elastically connected, Coulomb-damped isolation system illustrated at D in Table 30.1 for the damper spring stiffness defined by $N = 3$. The curves give the ratio of the relative displacement amplitude (maximum isolator deflection) to the displacement amplitude imposed at the foundation, as a function of the frequency ratio ω/ω_o and the displacement Coulomb-damping parameter η.

A small decrease in damping force F_f below the optimum value causes a large increase in the transmitted vibration near resonance. However, a small increase in damping force F_f above optimum causes only small changes in the maximum transmissibility. Thus, it is good design practice to have the damping parameter η equal to or greater than the optimum damping parameter η_{op}.

The relative transmissibility for $N = 3$ is shown in Fig. 30.14 where $(T_R)_D = \delta_0/u_0$. All curves pass through the intersection of the curves for zero and infinite damping. For optimum damping, the maximum relative transmissibility has a value given by Eq. (30.14); it occurs at the frequency ratio $\left(\dfrac{\omega}{\omega_0}\right)_{op}^{(R)}$ defined by Eq. (30.13).

ACCELERATION TRANSMISSIBILITY. The acceleration transmissibility can be obtained from the expression for displacement transmissibility by substitution of the effective displacement damping parameter in the expression for transmissibility of a system whose excitation is constant acceleration amplitude. If $\ddot{u}_0$ represents the acceleration amplitude of the excitation, the corresponding displacement amplitude is $u_0 = -\ddot{u}_0/\omega^2$. Using the definition of the acceleration Coulomb damping parameter ξ given in Table 30.1, the equivalent displacement Coulomb damping parameter is

$$\eta_{eq} = -\left(\frac{\omega}{\omega_0}\right)^2 \xi \qquad (30.17)$$

Substituting this relation in the absolute transmissibility expression given at j in Table 30.2, the following equation is obtained for the acceleration transmissibility:

$$(T_A)_A = \frac{\ddot{x}_0}{\ddot{u}_0} = \sqrt{\frac{1 + \left(\frac{4}{\pi}\xi\right)^2 \left(\frac{\omega^2}{\omega_0^2}\right) \left[\left(\frac{N+2}{N}\right)\left(\frac{\omega^2}{\omega_0^2}\right) - 2\left(\frac{N+1}{N}\right)\right]}{\left(1 - \frac{\omega^2}{\omega_0^2}\right)^2}} \qquad (30.18)$$

Equation (30.18) is valid only for the frequency range in which there is relative motion across the Coulomb damper. This range is defined by the "break-loose" and "lock-in" frequencies which are obtained by substituting Eq. (30.17) into Eq. (30.16):

$$\left(\frac{\omega}{\omega_0}\right)_L = \sqrt{\frac{\left(\frac{4}{\pi}\xi\right)(N+1) \pm N}{\frac{4}{\pi}\xi}} \qquad (30.19)$$

where Eqs. (30.16) and (30.19) give similar results, damping being defined in terms of displacement and acceleration excitation, respectively. For frequencies not included in the range between "break-loose" and "lock-in" frequencies, the acceleration transmissibility is that for an undamped system. Equation (30.18) indicates that infinite acceleration occurs at resonance unless the damper remains locked-in beyond a frequency ratio of unity. The coefficient of the damping term in Eq. (30.18) is identical to the corresponding coefficient in the expression for $(T_A)_D$ at j in Table 30.2. Thus, the frequency ratio at the optimum transmissibility is the same as that for displacement excitation.

An acceleration transmissibility curve for $N = 3$ is shown by Fig. 30.15. Relative motion at the damper occurs in a limited frequency range; thus, for relatively high frequencies, the acceleration transmissibility is similar to that for infinite damping.

OPTIMUM DAMPING PARAMETERS. The optimum Coulomb damping parameters are obtained by equating the optimum viscous damping ratio given by Eq. (30.15) to the equivalent viscous damping ratio for the elastically supported damper system, and replacing the frequency ratio by the

Fig. 30.15. Acceleration transmissibility for the elastically connected, Coulomb-damped isolation system illustrated at D in Table 30.1, for the damper spring stiffness defined by $N = 3$. The curves give the ratio of the acceleration amplitude of the equipment to the acceleration amplitude imposed at the foundation, as a function of the frequency ratio ω/ω_0 and the acceleration Coulomb damping parameter ξ.

frequency ratio given by Eq. (30.13). The optimum value of the damping parameter η in Table 30.1 is

$$\eta_{op} = \frac{\pi}{2} \sqrt{\frac{N+1}{N+2}} \tag{30.20}$$

To obtain the optimum value of the damping parameter ξ in Table 30.1, Eq. (30.17) is substituted in Eq. (30.20):

$$\xi_{op} = \frac{\pi}{4} \sqrt{\frac{N+2}{N+1}} \tag{30.21}$$

FORCE TRANSMISSIBILITY. The force transmissibility $(T_A)_F = F_T/F_0$ is identical to $(T_A)_A$ given by Eq. (30.18) if $\xi = \xi_F$, where ξ_F is defined as

$$\xi_F = \frac{F_f}{F_0} \tag{30.22}$$

Thus, the transmissibility curve shown in Fig. 30.15 also gives the force transmissibility for $N = 3$. By substituting Eq. (30.22) into Eq. (30.21), the transmitted force is optimized when the friction force F_f has the following value:

$$(F_f)_{op} = \frac{\pi F_0}{4} \sqrt{\frac{N+2}{N+1}} \tag{30.23}$$

To avoid infinite transmitted force at resonance, it is necessary that $F_f > (\pi/4)F_0$.

COMPARISON OF RIGIDLY CONNECTED AND ELASTICALLY CONNECTED COULOMB-DAMPED SYSTEMS. A principal limitation of the rigidly connected Coulomb-damped isolator is the nature of the transmissibility at high forcing frequencies. Because the isolator deflection is small, the force transmitted by the spring is negligible; then the force transmitted by the damper controls the motion experienced by the equipment. The acceleration transmissibility approaches the constant value $(4/\pi)\xi$, independent of frequency. The corresponding transmissibility for an isolator with an elastically connected Coulomb damper is $(N + 1)/(\omega/\omega_0)^2$. Thus, the transmissibility varies inversely as the square of the excitation frequency and reaches a relatively low value at large values of excitation frequency.

MULTIPLE DEGREE-OF-FREEDOM SYSTEMS

The single degree-of-freedom systems discussed previously are adequate for illustrating the fundamental principles of vibration isolation, but are an oversimplification in so far as many practical applications are concerned. The condition of unidirectional motion of an elastically mounted mass is not consistent with the requirements in many applications. In general, it is necessary to consider freedom of movement in all directions, as dictated by existing forces and motions and by the elastic constraints. Thus, in the general isolation problem, the equipment is considered as a rigid body supported by resilient supporting elements or isolators. This system is arranged so that the isolators effect the desired reduction in vibration. Various types of symmetry are encountered, depending upon the equipment and arrangement of isolators.

NATURAL FREQUENCIES—ONE PLANE OF SYMMETRY

A rigid body supported by resilient supports with one vertical plane of symmetry has three coupled natural modes of vibration and a natural frequency in each of these modes. A typical system of this type is illustrated in Fig. 30.16; it is assumed to be symmetrical with respect to a plane parallel with the plane of the paper and extending through the center-of-gravity of the supported body. Motion of the supported body in horizontal and vertical translational modes and in the rotational mode, all in the plane of the paper, are coupled. The equations of motion of a rigid body on resilient supports with six degrees-of-freedom are given by Eq. (3.31). By introducing certain types of symmetry and setting the excitation equal to zero, a cubic equation defining the free vibration of the system shown in Fig. 30.16 is derived, as given by Eqs. (3.36). This equation may be solved graphically for the natural frequencies of the system by use of Fig. 3.14.

SYSTEM WITH TWO PLANES OF SYMMETRY

A common arrangement of isolators is illustrated in Fig. 30.17; it consists of an equipment supported by four isolators located adjacent to the four lower corners. It is symmetrical with respect to two coordinate vertical planes through the center-of-gravity of the equipment, one of the planes being parallel with the plane of the paper. Because of this symmetry, vibration in the vertical

Fig. 30.16. Schematic diagram of a rigid equipment supported by an arbitrary arrangement of vibration isolators, symmetrical with respect to a plane through the center-of-gravity parallel with the paper.

translational mode is decoupled from vibration in the horizontal and rotational modes. The natural frequency in the vertical translational mode is $\omega_z = \sqrt{\Sigma k_z/m}$, where Σk_z is the sum of the vertical stiffnesses of the isolators.

Fig. 30.17. Schematic diagram in elevation of a rigid equipment supported upon four vibration isolators. The plane of the paper extends vertically through the center-of-gravity; the system is symmetrical with respect to this plane and with respect to a vertical plane through the center-of-gravity perpendicular to the paper. The moment of inertia of the equipment with respect to an axis through the center-of-gravity and normal to the paper is I_y. Excitation of the system is alternatively a vibratory force $F_x \sin \omega t$ applied to the equipment or a vibratory displacement $u = u_0 \sin \omega t$ of the foundation.

Consider excitation by a periodic force $F = F_x \sin \omega t$ applied in the direction of the X axis at a distance ϵ above the center-of-gravity and in one of the planes of symmetry. The differential equations of motion for the equipment in the coupled horizontal translational and rotational modes are obtained by substituting in Eq. (3.31) the conditions of symmetry defined by Eqs. (3.33), (3.34), (3.35), and (3.38). The resulting equations of motion are

$$m\ddot{x} = -4k_x x + 4k_x a\beta + F_x \sin \omega t$$

(30.24)

$$I_y\ddot{\beta} = 4k_x a x - 4k_x a^2 \beta - 4k_y b^2 \beta$$

$$- F_x \epsilon \sin \omega t$$

Making the common assumption that transients may be neglected in systems undergoing forced vibration, the translational and rotational displacements of the supported body are assumed to be harmonic at the excitation frequency. The differential equations of motion then are solved simultaneously to give the following expressions for the displacement amplitudes x_0 in horizontal translation and β_0 in rotation:

$$x_0 = \frac{F_x}{4k_z}\left(\frac{A_1}{D}\right) \qquad \beta_0 = \frac{F_x}{4\rho_y k_z}\left(\frac{A_2}{D}\right) \tag{30.25}$$

where

$$A_1 = \left(\frac{1}{\rho_y^2}\right)(\eta a_z^2 + a_x^2 - \eta\epsilon a_z) - \left(\frac{\omega}{\omega_z}\right)^2$$

$$A_2 = \frac{\epsilon}{\rho_y}\left(\frac{\omega}{\omega_z}\right)^2 + \frac{\eta}{\rho_y}(a_z - \epsilon) \tag{30.26}$$

$$D = \left(\frac{\omega}{\omega_z}\right)^4 - \left(\eta + \eta\frac{a_z^2}{\rho_y^2} + \frac{a_x^2}{\rho_y^2}\right)\left(\frac{\omega}{\omega_z}\right)^2 + \eta\left(\frac{a_x}{\rho_y}\right)^2$$

In the above equations, $\eta = k_x/k_z$ is the dimensionless ratio of horizontal stiffness to vertical stiffness of the isolators, $\rho_y = \sqrt{I_y/m}$ is the radius of gyration of the supported body about an axis through its center-of-gravity and perpendicular to the paper, $\omega_z = \sqrt{\Sigma k_z/m}$ is the undamped natural frequency in vertical translation, ω is the forcing frequency, a_z is the vertical distance from the effective height of spring (mid-height if symmetrical top to bottom) * to center-of-gravity of body m, and the other parameters are as indicated in Fig. 30.17.

Forced vibration of the system shown in Fig. 30.17 also may be excited by periodic motion of the support in the horizontal direction, as defined by $u = u_0 \sin \omega t$. The differential equations of motion for the supported body are

$$m\ddot{x} = 4k_x(u - x - a_z\beta)$$

$$I_y\ddot{\beta} = -4a_z k_x(u - x - a_z\beta) - 4k_z a_x^2\beta \tag{30.27}$$

Neglecting transients, the motion of the mounted body in horizontal translation and in rotation is assumed to be harmonic at the forcing frequency. Equations (30.27) may be solved simultaneously to obtain the following expressions for the displacement amplitudes x_0 in horizontal translation and β_0 in rotation:

$$x_0 = \frac{u_0 B_1}{D} \qquad \beta_0 = \frac{u_0 B_2}{\rho_y D} \tag{30.28}$$

where the parameters B_1 and B_2 are

$$B_1 = \eta\left(\frac{a_x^2}{\rho_y^2} - \frac{\omega^2}{\omega_z^2}\right) \qquad B_2 = \frac{\eta a_z}{\rho_y}\left(\frac{\omega}{\omega_z}\right)^2 \tag{30.29}$$

and D is given by Eq. (30.26).

NATURAL FREQUENCIES—TWO PLANES OF SYMMETRY. In forced vibration, the amplitude becomes a maximum when the forcing frequency is approximately equal to a natural frequency. In an undamped system, the amplitude becomes infinite at resonance. Thus, the natural frequency or frequencies of an undamped system may be determined by writing the expression for the displacement amplitude of the system in forced vibration and finding the excitation frequency at which this amplitude becomes infinite. The denominators of Eqs. (30.25) and (30.28) include the parameter D defined by Eq. (30.26). The natural frequencies of the system in coupled rotational and horizon-

---

* The distance a_z is taken to the mid-height of the spring to include in the equations of motion the moment applied to the body m by the fixed-end spring. If the spring is hinged to body m, the appropriate value for a_z is the distance from the X axis to the hinge axis.

tal translational modes may be determined by equating D to zero and solving for the forcing frequencies: [7]

$$\frac{\omega_{x\beta}}{\omega_z} \times \frac{\rho_y}{a_x} = \frac{1}{\sqrt{2}} \sqrt{\eta\left(\frac{\rho_y}{a_x}\right)^2\left(1 + \frac{a_z^2}{\rho_y^2}\right) + 1 \pm \sqrt{\left[\eta\left(\frac{\rho_y}{a_x}\right)^2\left(1 + \frac{a_z^2}{\rho_y^2}\right) + 1\right]^2 - 4\eta\left(\frac{\rho_y}{a_x}\right)^2}}$$

$$(30.30)$$

where $\omega_{x\beta}$ designates a natural frequency in a coupled rotational (β) and horizontal translational (x) mode, and ω_z designates the natural frequency in the decoupled vertical translational mode. The other parameters are defined in connection with Eq. (30.26). Two numerically different values of the dimensionless frequency ratio $\omega_{x\beta}/\omega_z$ are obtained from Eq. (30.30), corresponding to the two discrete coupled modes of vibration. Curves computed from Eq. (30.30) are given in Fig. 30.18.

The ratio of a natural frequency in a coupled mode to the natural frequency in the vertical translational mode is a function of three dimensionless ratios, two of the ratios relating the radius of gyration ρ_y to the dimensions a_z and a_x while the third is the ratio η of horizontal to vertical stiffnesses of the isolators. In applying the curves of Fig. 30.18, the applicable value of the abscissa ratio is first determined directly from the constants of the system. Two appropriate numerical values then are taken from the ordinate scale, as determined by the two curves for applicable values of a_z/ρ_y; the ratios of natural frequencies in coupled and vertical translational modes are determined by dividing these values by the dimensionless ratio ρ_y/a_x. The natural frequencies in coupled modes then are determined by multiplying the resulting ratios by the natural frequency in the decoupled vertical translational mode.

The two straight lines in Fig. 30.18 for $a_z/\rho_y = 0$ represent natural frequencies in decoupled modes of vibration. When $a_z = 0$, the elastic supports lie in a plane passing through the center-of-gravity of the equipment. The horizontal line at a value of unity on the ordinate scale represents the natural frequency in a rotational mode. The inclined straight line for the value $a_z/\rho_y = 0$ represents the natural frequency of the system in horizontal translation.

Calculation of the coupled natural frequencies of a rigid body on resilient supports from Eq. (30.30) is sufficiently laborious to encourage the use of graphical means. For general purposes, both coupled natural frequencies can be obtained from Fig. 30.18. For a given type of isolators, $\eta = k_x/k_z$ is a constant and Eq. (30.30) may be evaluated in a manner that makes it possible to select isolator positions to attain optimum natural frequencies.[8–10] This is discussed under *Space-Plots* in Chap. 3. The convenience of the approach is partially offset by the need for a separate plot for each

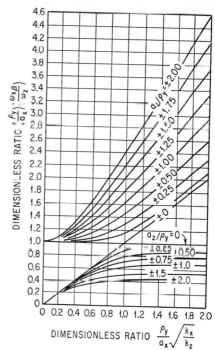

FIG. 30.18. Curves of natural frequencies $\omega_{x\beta}$ in coupled modes with reference to the natural frequency in the decoupled vertical translational mode ω_z, for the system shown schematically in Fig. 30.17. The isolator stiffnesses in the X and Z directions are indicated by k_x and k_z, respectively, and the radius of gyration with respect to the Y axis through the center-of-gravity is indicated by ρ_y.

value of the stiffness ratio k_x/k_z. Applicable curves are plotted for several values of k_x/k_z in Figs. 3.17 to 3.19.

The preceding analysis of the dynamics of a rigid body on resilient supports includes the assumption that the principal axes of inertia of the rigid body are respectively parallel with the principal elastic axes of the resilient supports. This makes it possible to neglect the products of inertia of the rigid body. The coupling introduced by the product of inertia is not strong unless the angle between the above-mentioned inertia and elastic axes is substantial. It is convenient to take the coordinate axes through the center-of-

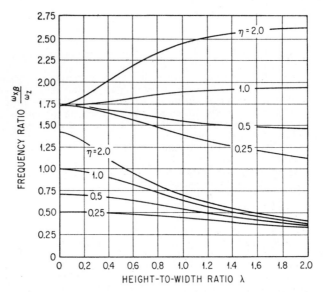

Fig. 30.19. Curves indicating the natural frequencies $\omega_{x\beta}$ in coupled rotational and horizontal translational modes with reference to the natural frequency ω_z in the decoupled vertical translational mode, for the system shown in Fig. 30.17. The ratio of horizontal to vertical stiffness of the isolators is η, and the height-to-width ratio for the equipment is λ. These curves are based upon the assumption that the mass of the equipment is uniformly distributed and that the isolators are attached precisely at the extreme lower corners thereof.

gravity of the supported body, parallel with the principal elastic axes of the isolators. If the moments of inertia with respect to these coordinate axes are used in Eqs. (30.24) to (30.30), the calculated natural frequencies usually are correct within a few per cent without including the effect of product of inertia. When it is desired to calculate the natural frequencies accurately or when the product of inertia coupling is strong, a calculation procedure is available that may be used for certain conventional arrangements using four isolators.[11]

The procedure for determining the natural frequencies in coupled modes summarized by the curves of Fig. 30.18 represents a rigorous analysis where the assumed symmetry exists. The procedure is somewhat indirect because the dimensionless ratio ρ_y/a_x appears in both ordinate and abscissa parameters, and because it is necessary to determine the radius of gyration of the equipment. The relations may be approximated in a more readily usable form if (1) the mounted equipment can be considered a cuboid having uniform mass distribution, (2) the four isolators are attached precisely at the four lower corners of the cuboid, and (3) the height of the isolators may be considered negligible. The ratio of the natural frequencies in the coupled rotational and horizontal translational modes to the natural frequency in the vertical translational mode then becomes a function of only the dimensions of the cuboid and the stiffnesses of the isolators in the several coordinate directions. Making these assumptions and substituting in Eq. (30.30),

$$\frac{\omega_{x\beta}}{\omega_z} = \frac{1}{\sqrt{2}} \sqrt{\frac{4\eta\lambda^2 + \eta + 3}{\lambda^2 + 1} \pm \sqrt{\left(\frac{4\eta\lambda^2 + \eta + 3}{\lambda^2 + 1}\right)^2 - \frac{12\eta}{\lambda^2 + 1}}} \tag{30.31}$$

where $\eta = k_x/k_z$ designates the ratio of horizontal to vertical stiffness of the isolators and $\lambda = 2a_z/2a_x$ indicates the ratio of height to width of mounted equipment. This relation is shown graphically in Fig. 30.19. The curves included in this figure are useful for calculating approximate values of natural frequencies and for indicating trends in natural frequencies resulting from changes in various parameters as follows:

1. Both of the coupled natural frequencies tend to become a minimum, for any ratio of height to width of the mounted equipment, when the ratio of horizontal to vertical stiffness k_x/k_z of the isolators is low. Conversely, when the ratio of horizontal to vertical stiffness is high, both coupled frequencies also tend to be high. Thus, when the isolators are located underneath the mounted body, a condition of low natural frequencies is obtained using isolators whose stiffness in a horizontal direction is less than the stiffness in a vertical direction. However, low horizontal stiffness may be undesirable in applications requiring maximum stability. A compromise between natural frequency and stability then may lead to optimum conditions.

Fig. 30.20. Plan view of the equipment shown schematically in Fig. 30.17, indicating the uncoupled rotational mode specified by the rotation angle γ.

2. As the ratio of height to width of the mounted equipment increases, the lower of the coupled natural frequencies decreases. The trend of the higher of the coupled natural frequencies depends on the stiffness ratio of the isolators. One of the coupled natural frequencies tends to become very high when the horizontal stiffness of the isolators is greater than the vertical stiffness, and when the height of the mounted equipment is approximately equal to or greater than the width. When the ratio of height to width of mounted equipment is greater than 0.5, the spread between the coupled natural frequencies increases as the ratio k_x/k_z of horizontal to vertical stiffness of the isolators increases.

NATURAL FREQUENCY—UNCOUPLED ROTATIONAL MODE. Figure 30.20 is a plan view of the body shown in elevation in Fig. 30.17. The distances from the isolators to the principal planes of inertia are designated by a_x and a_y. The horizontal stiffnesses of the isolators in the directions of the coordinate axes X and Y are indicated by k_x and k_y, respectively. When the excitation is the applied couple $M = M_0 \sin \omega t$, the differential equation of motion is

$$I_z \ddot{\gamma} = -4\gamma a_x^2 k_y - 4\gamma a_y^2 k_x + M_0 \sin \omega t \tag{30.32}$$

where I_z is the moment of inertia of the body with respect to the Z axis. Neglecting transient terms, the solution of Eq. (30.32) gives the displacement amplitude γ_0 in rotation:

$$\gamma_0 = \frac{M_0}{4(a_x^2 k_y + a_y^2 k_x) - I_z \omega^2} \tag{30.33}$$

where the natural frequency ω_γ in rotation about the Z axis is the value of ω that makes the denominator of Eq. (30.33) equal to zero:

$$\omega_\gamma = 2\sqrt{\frac{a_x^2 k_y + a_y^2 k_x}{I_z}} \tag{30.34}$$

VIBRATION ISOLATION IN COUPLED MODES

When the equipment and isolator system has several degrees-of-freedom and the isolators are located in such a manner that several natural modes of vibration are coupled, it becomes necessary in evaluating the isolators to consider the contribution of the several modes in determining the motion transmitted from the support to the mounted equipment, or the force transmitted from the equipment to the foundation. Methods for determining the transmissibility under these conditions are best illustrated by examples.

For example, consider the system shown schematically in Fig. 30.21 wherein a machine is supported by relatively long beams which are in turn supported at their opposite ends by vibration isolators. The isolators are assumed to be undamped, and the excitation is

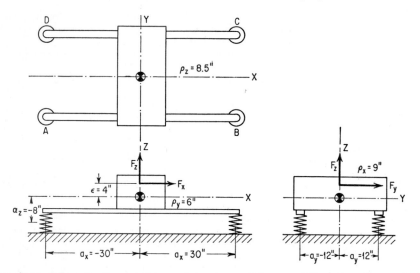

FIG. 30.21. Schematic diagram of an equipment mounted upon relatively long beams which are in turn attached at their opposite ends to vibration isolators. Excitation for the system is alternatively (1) the vibratory force $F_x = F_o \cos \omega t$, $F_z = F_o \sin \omega t$ in the XZ plane or (2) the vibratory force $F_y = F_o \cos \omega t$, $F_Z = F_o \sin \omega t$ in the YZ plane.

considered to be a force applied at a distance $\epsilon = 4$ in. above the center-of-gravity of the machine-and-beam assembly. Alternatively, the force is (1) $F_x = F_0 \cos \omega t$, $F_z = F_0 \sin \omega t$ in a plane normal to the Y axis or (2) $F_y = F_0 \cos \omega t$, $F_z = F_0 \sin \omega t$ in a plane normal to the X axis. This may represent an unbalanced weight rotating in a vertical plane. A force transmissibility at each of the four isolators is determined by calculating the deflection of each isolator, multiplying the deflection by the appropriate isolator stiffness to obtain transmitted force, and dividing it by $F_0/4$.

When the system is viewed in a vertical plane perpendicular to the Y axis, the transmissibility curves are as illustrated in Fig. 30.22. The solid line defines the transmissibility at each of isolators B and C in Fig. 30.21, and the dotted line defines the transmissibility at each of isolators A and D. Similar transmissibility curves for a plane perpendicular to the X axis are shown in Fig. 30.23 wherein the solid line indicates the transmissibility at each of isolators C and D, and the dotted line indicates the transmissibility at each of isolators A and B.

Note the comparison of the transmissibility curves of Figs. 30.22 and 30.23 with the diagram of the system in Fig. 30.21. Figure 30.23 shows the three resonance conditions which are characteristic of a coupled system of the type illustrated. The transmissibility remains equal to or greater than unity for all excitation frequencies lower than the highest resonant frequency in a coupled mode. At greater excitation frequencies, vibration isolation is attained, as indicated by values of force transmissibility smaller than unity.

The transmissibility curves in Fig. 30.22 show somewhat similar results. The long horizontal beams tend to spread the resonant frequencies by a substantial frequency in-

FIG. 30.22. Transmissibility curves for the system shown in Fig. 30.21 when the excitation is in a plane perpendicular to the Y axis. The solid line indicates the transmissibility at each of isolators B and C, whereas the dotted line indicates the transmissibility at each of isolators A and D.

crement, and merge the resonant frequency in the vertical translational mode with the resonant frequency in one of the coupled modes. A low transmissibility is again attained at excitation frequencies greater than the highest resonant frequency. Note that the transmissibility drops to a value slightly less than unity over a small frequency interval

FIG. 30.23. Transmissibility curves for the system illustrated in Fig. 30.21 when the excitation is in a plane perpendicular to the X axis. The solid line indicates the transmissibility at each of isolators C and D, whereas the dotted line indicates the transmissibility at each of isolators A and B.

FIG. 30.24. Schematic diagram of an equipment supported by vibration isolators. Excitation is a vibratory displacement $u = u_o \sin \omega t$ of the foundation.

between the predominant resonant frequencies. This is a force reduction resulting from the relatively long beams, and constitutes an acceptable condition if the magnitude of the excitation force in this direction is relatively small. Thus, the natural frequencies of the isolators could be somewhat higher with a consequent gain in stability; it is necessary, however, that the excitation frequency be substantially constant.

Consider the equipment illustrated in Fig. 30.24 when the excitation is horizontal vibration of the support. The effectiveness of the isolators in reducing the excitation vibration is evaluated by plotting the displacement amplitude of the horizontal vibration at points A and B with reference to the displacement amplitude of the support. Transmissibility curves for the system of Fig. 30.24 are shown in Fig. 30.25. The solid line in Fig. 30.25 refers to point A and the dotted line to point B. Note that there is no significant reduction of amplitude except when the forcing frequency exceeds the maximum resonant frequency of the system.

A general rule for the calculation of necessary isolator characteristics to achieve the results illustrated in Figs. 30.22, 30.23, and 30.25 is that the forcing frequency should be not less than 1.5 to 2 times the maximum natural frequency in any of six natural modes of vibration. In exceptional cases, such as illustrated in Fig. 30.22, the forcing frequency may be interposed between resonant frequencies if the forcing frequency is a constant.

Example 30.1. Consider the machine illustrated in Fig. 30.21. The force that is to be isolated is harmonic at the constant frequency of 8 cps; it is assumed to result from the rotation of an unbalanced member whose plane of rotation is alternatively (1) a plane perpendicular to the Y axis and (2) a plane perpendicular to the X axis. The distance between isolators is 60

FIG. 30.25. Displacement transmissibility curves for the system of Fig. 30.24. Transmissibility between the foundation and point A is shown by the solid line; transmissibility between the foundation and point B is shown by the dotted line.

in. in the direction of the X axis and 24 in. in the direction of the Y axis. The center of coordinates is taken at the center-of-gravity of the supported body, i.e., at the center-of-gravity of the machine-and-beams assembly. The total weight of the machine and supporting beam assembly is 100 lb, and its radii of gyration with respect to the three coordinate axes through the center-of-gravity are $\rho_x = 9$ in., $\rho_z = 8.5$ in., and $\rho_y = 6$ in. The isolators are of equal stiffnesses in the directions of the three coordinate axes:

$$\eta = \frac{k_x}{k_z} = \frac{k_y}{k_z} = 1$$

The following dimensionless ratios are established as the initial step in the solution:

$$a_z/\rho_y = -1.333 \qquad\qquad a_z/\rho_x = -0.889$$
$$a_x/\rho_y = \pm5.0 \qquad\qquad a_y/\rho_x = \pm1.333$$
$$(a_z/\rho_y)^2 = 1.78 \qquad\qquad (a_z/\rho_x)^2 = 0.790$$
$$(a_x/\rho_y)^2 = 25.0 \qquad\qquad (a_y/\rho_z)^2 = 1.78$$
$$\eta(\rho_y/a_x)^2 = 0.04 \qquad\qquad \eta(\rho_x/a_y)^2 = 0.561$$

The various natural frequencies are determined in terms of the vertical natural frequency ω_z. Referring to Fig. 30.18, the coupled natural frequencies for vibration in a plane perpendicular to the Y axis are determined as follows:
First calculate the parameter

$$\frac{\rho_y}{a_x}\sqrt{\frac{k_x}{k_z}} = 0.2$$

For $a_z/\rho_y = -1.333$, $(\omega_{x\beta}/\omega_z)(\rho_y/a_x) = 0.19; 1.03$. Note the signs of the dimensionless ratios a_z/ρ_y and a_x/ρ_y. According to Eq. (30.30), the natural frequencies are independent of the sign of a_z/ρ_y. With regard to the ratio a_x/ρ_y, the sign chosen should be the same as the sign of the radical on the right side of Eq. (30.30). The frequency ratio $(\omega_{x\beta}/\omega_z)$ then becomes positive. Dividing the above values for $(\omega_{x\beta}/\omega_z)(\rho_y/a_x)$ by $\rho_y/a_x = 0.2$, $\omega_{x\beta}/\omega_z = 0.96; 5.15$.
Vibration in a plane perpendicular to the X axis is treated in a similar manner. It is assumed that exciting forces are not applied concurrently in planes perpendicular to the X and Y axes; thus, vibration in these two planes is independent. Consequently, the example entails two independent but similar problems and similar equations apply for a plane perpendicular to the X axis:

$$\frac{\rho_x}{a_y}\sqrt{\frac{k_z}{k_y}} = 0.75$$

For $a_z/\rho_x = 0.889$, $(\omega_{y\alpha}/\omega_z)(\rho_x/a_y) = 0.57; 1.29$. Dividing by $\rho_x/a_y = 0.75$, $\omega_{y\alpha}/\omega_z = 0.76$; 1.72.
The natural frequency in rotation with respect to the Z axis is calculated from Eq. (30.34) as follows, taking into consideration that there are two pairs of springs and that $k_x = k_y = k_z$:

$$\omega_\gamma = \sqrt{\left(\frac{a_x{}^2 + a_y{}^2}{\rho_z{}^2}\right)\left(\frac{4k_z g}{W}\right)} = 3.8\omega_z$$

The six natural frequencies are as follows:
1. Translational along Z axis: ω_z
2. Coupled in plane perpendicular to Y axis: $0.96\omega_z$
3. Coupled in plane perpendicular to Y axis: $5.15\omega_z$
4. Coupled in plane perpendicular to X axis: $0.76\omega_z$
5. Coupled in plane perpendicular to X axis: $1.72\omega_z$
6. Rotational with respect to Y axis: $3.8\omega_z$
Considering vibration in a plane perpendicular to the Y axis, the two highest natural frequencies are the natural frequency ω_y in the translational mode along the Z axis and the natural frequency $5.15\omega_z$ in a coupled mode. In a similar manner, the two highest natural frequencies in a plane perpendicular to the X axis are the natural frequency ω_z in translation along the Z axis and the natural frequency $1.72\omega_z$ in a coupled mode. The natural frequency in rotation about the Z axis is $3.80\omega_z$. The widest frequency increment which is void of natural frequencies is between $1.72\omega_z$ and $3.80\omega_z$. This increment is used for the forcing frequency which is taken as $2.5\omega_z$. Inasmuch as the forcing frequency is established at 8 cps, the vertical natural frequency is 8 divided by 2.5, or 3.2 cps. The required vertical stiffnesses of the isolators are calculated from Eq. (30.11) to be 105 lb/in. for the entire machine, or 26.2 lb/in. for each of the four isolators.

INCLINED ISOLATORS

Advantages in vibration isolation sometimes result from inclining the principal elastic axes of the isolators with respect to the principal inertia axes of the equipment, as illustrated in Fig. 30.26. The coordinate axes X and Z are respectively parallel with the principal inertia axes of the mounted body, but the center of coordinates is taken at the elastic axis. The location of the elastic axis is determined by the elastic properties of the system. If a force is applied to the body along a line extending through the elastic axis, the body is displaced in translation without rotation; if a couple is applied to the body, the body is displaced in rotation without translation.

The principal elastic axes r, p of the isolators are parallel with the paper and inclined with respect to the coordinate axes, as indicated in Fig. 30.26. The stiffness of each isola-

FIG. 30.26. Schematic diagram of an equipment supported by isolators whose principal elastic axes are inclined to the principal inertia axes of the equipment.

tor in the direction of the respective principal axis is indicated by k_r, k_p. The principal elastic axis of an isolator is the axis along which a force must be applied to cause a deflection colinear with the applied force (see the section *Properties of a Biaxial Stiffness Isolator*).

Assume the excitation for the system shown in Fig. 30.26 to be a couple $M_0 \sin \omega t$ acting about an axis normal to the paper. The equations of motion for the body in the horizontal translational and rotational modes may be written by noting that the displacement of the center-of-gravity in the direction of the X axis is $x - \epsilon\beta$; thus, the corresponding acceleration is $\ddot{x} - \epsilon\ddot{\beta}$. A translational displacement x produces only an external force $-k_x x$, whereas a rotational displacement β produces only an external couple $-k_\beta\beta$. The equations of motion are

$$m(\ddot{x} - \epsilon\ddot{\beta}) = -k_x x$$

$$m\rho_e{}^2\ddot{\beta} - m\epsilon\ddot{x} = -k_\beta\beta + M_0 \sin \omega t$$

(30.35)

where ρ_e is the radius of gyration of the mounted body with respect to the elastic axis. The radius of gyration ρ_e is related to the radius of gyration ρ_y with respect to a line through the center-of-gravity by $\rho_e = \sqrt{\rho_y{}^2 + \epsilon^2}$, where ϵ is the distance between the elastic axis and a parallel line passing through the center-of-gravity. In the equations of motion, k_x and k_β represent the translational and rotational stiffness of the isolators in the x and β coordinate directions, respectively.

By assuming steady-state harmonic motion for the horizontal translation x and rotation β, the following displacement amplitudes are obtained by solving Eqs. (30.35):

$$x_0 = \frac{-M_0\epsilon\omega^2}{m[\rho_e{}^2(\omega^2 - \omega_\beta{}^2)(\omega^2 - \omega_x{}^2) - \epsilon^2\omega^4]}$$

(30.36)

$$\beta_0 = \frac{-M_0}{m\left[\rho_e{}^2(\omega^2 - \omega_\beta{}^2) - \dfrac{\epsilon^2\omega^4}{\omega^2 - \omega_x{}^2}\right]}$$

where $\omega_x = \sqrt{k_x/m}$ and $\omega_\beta = \sqrt{k_\beta/m\rho_e{}^2}$ are hypothetical natural frequencies defined for convenience. The natural frequencies $\omega_{x\beta}$ in the coupled x,β modes are determined by equating the denominator of Eqs. (30.36) to zero and solving for ω (now identical to $\omega_{x\beta}$):

$$\frac{\omega_{x\beta}}{\omega_x} = \sqrt{\frac{1 + \lambda_1{}^2 \pm \sqrt{(1 + \lambda_1{}^2)^2 - 4\lambda_1{}^2[1 - (\epsilon/\rho_e)^2]}}{2[1 - (\epsilon/\rho_e)^2]}} \qquad (30.37)$$

where λ_1 is a dimensionless ratio given by

$$\lambda_1 = \frac{(a_x/\rho_e)\sqrt{k_r/k_p}}{\cos^2\phi + (k_r/k_p)\sin^2\phi} \qquad (30.38)$$

The hypothetical natural frequency ω_x is

$$\omega_x = \sqrt{\frac{4k_p}{m}\left[\cos^2\phi + \frac{k_r}{k_p}\sin^2\phi\right]} \qquad (30.39)$$

The relation given by Eq. (30.37) is shown graphically by Fig. 30.27. The parameters needed to evaluate the natural frequencies by using this graph are calculated from the physical properties of the system and the relations of Eqs. (30.38) and (30.39). In addition, the distance ϵ between a parallel line passing through the center-of-gravity and the elastic axis must be known. The distance ϵ is determined by effecting a small horizontal displacement of the equipment in the X direction, and equating the resulting summation of elastic couples to zero:

$$\epsilon = a_z - \frac{a_x\,(1 - k_p/k_r)\cot\phi}{(k_p/k_r)\cot^2\phi + 1} \qquad (30.40)$$

where a_z is the distance between the parallel planes passing through the center-of-gravity of the body and the mid-height of the isolators, as shown in Fig. 30.26.

DECOUPLING OF MODES

The natural modes of vibration of a body supported by isolators may be decoupled one from another by proper orientation of the isolators. Each mode of vibration then exists independently of the others, and vibration in one mode does not excite vibration in other modes. The necessary conditions for decoupling may be stated as follows: The resultant of the forces applied to the mounted body by the isolators when the mounted body is displaced in translation must be a force directed through the center-of-gravity; or, the resultant of the couples applied to the mounted body by the isolators when the mounted body is displaced in rotation must be a couple about an axis through the center-of-gravity.

In general, the natural frequencies of a multiple degree-of-freedom system can be made equal only by decoupling the natural modes of vibration, i.e., by making $a_z = 0$ in Fig. 30.17. The natural frequencies in decoupled modes are indicated by the two straight lines in Fig. 30.18 marked $a_z/\rho_y = 0$. The natural frequencies in translation along the X axis and in rotation about the Y axis become

FIG. 30.27. Curves indicating the natural frequencies $\omega_{x\beta}$ in coupled modes with reference to the natural frequency in the decoupled (fictitious) horizontal translational mode ω_x for the system shown schematically in Fig. 30.26. The radius of gyration with respect to the elastic axis is indicated by ρ_e, and the distance between the center-of-gravity and the elastic center is ϵ. The dimensionless parameter λ_1 is defined by Eq. (30.38) and ω_x is defined by Eq. (30.39).

equal at the intersection of these lines; i.e., when $a_z/\rho_y = 0$, $k_x/k_z = 1$ and $\rho_y/a_x = 1$. The physical significance of these mathematical conditions is that the isolators be located in a plane passing through the center-of-gravity of the equipment, that the distance between isolators be twice the radius of gyration of the equipment, and that the stiffness of each isolator in the directions of the X and Z axes be equal.

When the isolators cannot be located in a plane which passes through the center-of-gravity of the equipment, decoupling can be achieved by inclining the isolators, as illustrated in Fig. 30.26. If the elastic axis of the system is made to pass through the center-of-gravity, the translational and rotational modes are decoupled because the inertia force of the mounted body is applied through the elastic center and introduces no

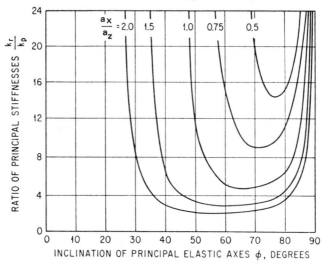

Fig. 30.28. Ratio of stiffnesses k_r/k_p along principal elastic axes required for decoupling the natural modes of vibration of the system illustrated in Fig. 30.26.

tendency for the body to rotate. The requirements for a decoupled system are established by setting $\epsilon = 0$ in Eq. (30.40) and solving for k_r/k_p:

$$\frac{k_r}{k_p} = \frac{(a_x/a_z) + \cot \phi}{(a_x/a_z) - \tan \phi} \tag{30.41}$$

The conditions for decoupling defined by Eq. (30.41) are shown graphically in Figs. 30.28 and 3.23. The decoupled natural frequencies are indicated by the straight lines $\epsilon/\rho_e = 0$ in Fig. 30.27. The horizontal line refers to the decoupled natural frequency ω_x in translation in the direction of the X axis, while the inclined line refers to the decoupled natural frequency ω_β in rotation about the Y axis.

PROPERTIES OF A BIAXIAL STIFFNESS ISOLATOR

A biaxial stiffness isolator is represented as an elastic element having a single plane of symmetry; all forces act in this plane and the resultant deflections are limited by symmetry or constraints to this plane. The characteristic elastic properties of the isolator may be defined alternatively by sets of influence coefficients as follows:

1. If the two coordinate axes in the plane of symmetry are selected arbitrarily, three stiffness parameters are required to define the properties of the isolator. These are the axial influence coefficients * along the two coordinate axes, and a characteristic coupling influence coefficient * between the coordinate axes.

* The influence coefficient κ is a function only of the isolator properties and not of the constraints imposed by the system in which the isolator is used. Both positive and negative values of the influence coefficient κ are permissible.

2. If the two coordinate axes in the plane of symmetry are selected to coincide with the principal elastic axes of the isolator, two influence coefficients are required to define the properties of the isolator. These are the principal influence coefficients. If the isolator is used in a system, a third parameter is required to define the orientation of the principal axes of the isolator with the coordinate axes of the system.

PROPERTIES OF ISOLATOR WITH RESPECT TO ARBITRARILY SELECTED AXES

A schematic representation of a linear biaxial stiffness element is shown in Fig. 30.29 where the X and Y axes are arbitrarily chosen to define a plane to which all forces and motions are restricted. In general, the deflection of an isolator resulting from an applied load is not in the same direction as the load, and a coupling influence coefficient is required to define the properties of the isolator in addition to the influence coefficients along the X and Y axes. The three characteristic stiffness coefficients that uniquely describe the load-deflection properties of a biaxial stiffness element are:

1. The influence coefficient of the element in the X coordinate direction is κ_x. It is the ratio of the component of the applied force in the X direction to the resulting deflection when the isolator is constrained to deflect in the X direction.

2. The influence coefficient of the element in the Y coordinate direction is κ_y. It is the ratio of the component of the applied force in the Y direction to the resulting deflection when the isolator is constrained to deflect in the Y direction.

3. The coupling influence coefficient is κ_{xy}. It represents the force required in the X direction to produce a unit displacement in the Y direction when the isolator is constrained to deflect only in the Y direction. (By Maxwell's reciprocity principle, the same force is required in the Y direction to produce a unit displacement in the X direction; i.e., $\kappa_{xy} = \kappa_{yx}$.)

Consider the isolator shown in Fig. 30.29 where the applied force F has components F_x and F_y; the resulting displacement has components δ_x and δ_y. From the above definitions of influence coefficients, the forces in the X and Y coordinate directions required to effect a displacement δ_x are

$$F_{xx} = \kappa_x \delta_x \qquad F_{yx} = \kappa_{yx} \delta_x \qquad (30.42)$$

The forces required to effect a displacement δ_y in the Y direction are

$$F_{xy} = \kappa_{xy} \delta_y \qquad F_{yy} = \kappa_y \delta_y \qquad (30.43)$$

The force components F_x and F_y required to produce the deflection having components δ_x, δ_y are the sums from Eqs. (30.42) and (30.43):

$$F_x = \kappa_x \delta_x + \kappa_{xy} \delta_y \qquad (30.44)$$

$$F_y = \kappa_{yx} \delta_x + \kappa_y \delta_y$$

If the three influence stiffness coefficients κ_x, κ_y and $\kappa_{xy} = \kappa_{yx}$ are known for a given stiffness element, the load-deflection properties are given by Eq. (30.44).

The deflections of the isolator in response to forces F_x, F_y are determined by solving Eqs. (30.44) simultaneously:

$$\delta_x = \frac{F_x \kappa_y - F_y \kappa_{xy}}{\kappa_x \kappa_y - \kappa_{xy}^2}$$

$$\delta_y = \frac{F_y \kappa_x - F_x \kappa_{xy}}{\kappa_x \kappa_y - \kappa_{xy}^2} \qquad (30.45)$$

FIG. 30.29. Schematic diagram of a linear biaxial stiffness element.

These expressions give the orthogonal components of the displacement δ for any load having the components F_x and F_y applied to a biaxial stiffness isolator. By substituting the relations of Eqs. (30.45) into Eq. (30.44), the following alternate forms of the force-deflection equations are obtained:

$$F_x = \left(\kappa_x - \frac{\kappa_{xy}^2}{\kappa_y} \right) \delta_x + \frac{\kappa_{xy}}{\kappa_y} F_y$$

$$F_y = \left(\kappa_y - \frac{\kappa_{xy}^2}{\kappa_x} \right) \delta_y + \frac{\kappa_{xy}}{\kappa_x} F_x$$

$$(30.46)$$

The specific force-deflection equations for a given situation are obtained from these general load-deflection expressions by applying the proper constraint conditions.

UNCONSTRAINED MOTION. The general force-deflection equations can be used to obtain the effective stiffness coefficients when the forces F_x and F_y shown in Fig. 30.29 are applied independently. The resulting deflection of the isolator is unconstrained motion, i.e., the isolator is free to deflect out of the line of force application. The force divided by that component of deflection along the line of action of the force is the effective stiffness k. When $F_y = 0$, the effective stiffness k_x resulting from the applied force F_x is obtained from Eq. (30.46):

$$k_x = \frac{F_x}{\delta_x} = \left(\kappa_x - \frac{\kappa_{xy}^2}{\kappa_y} \right)$$

$$(30.47)$$

When $F_x = 0$, the effective stiffness k_y in response to the applied force F_y is

$$k_y = \frac{F_y}{\delta_y} = \left(\kappa_y - \frac{\kappa_{xy}^2}{\kappa_x} \right)$$

$$(30.48)$$

For unconstrained motion, $k_x/k_y = \kappa_x/\kappa_y$; i.e., the ratio of the effective stiffnesses in two mutually perpendicular directions is equal to the ratio of the corresponding influence coefficients for the same directions.

CONSTRAINED MOTION. When the isolator is constrained either by the symmetry of a system or by structural constraints to deflect only along the line of the applied force, the effective stiffness is obtained directly by letting appropriate deflections be zero in Eq. (30.44):

$$k_x = \frac{F_x}{\delta_x} = \kappa_x \qquad k_y = \frac{F_y}{\delta_y} = \kappa_y$$

$$(30.49)$$

The force required to maintain constrained motion is found by letting appropriate deflections be zero in Eqs. (30.46). For example, the force that must be applied in the X direction to insure that the isolator deflects in the Y direction in response to a force F_y is

$$F_x = \frac{\kappa_{xy}}{\kappa_y} F_y$$

$$(30.50)$$

INFLUENCE COEFFICIENT TRANSFORMATION

Assume the influence coefficients κ_x, κ_y, and κ_{xy} are known in the X, Y coordinate system. It may be convenient to work with isolator influence coefficients in the X', Y' coordinate system as shown in Fig. 30.30. The X', Y' coordinate system is obtained by rotating the coordinate axes counterclockwise through an angle θ from the X, Y system. The influence coefficients with respect to the X', Y' axes are related to the influence coefficients with respect to the X, Y axes as follows:

$$\kappa_{x'} = \frac{\kappa_x + \kappa_y}{2} + \frac{\kappa_x - \kappa_y}{2} \cos 2\theta + \kappa_{xy} \sin 2\theta$$

$$\kappa_{x'y'} = \frac{\kappa_y - \kappa_x}{2} \sin 2\theta + \kappa_{xy} \cos 2\theta \tag{30.51}$$

$$\kappa_{y'} = \frac{\kappa_x + \kappa_y}{2} - \frac{\kappa_x - \kappa_y}{2} \cos 2\theta - \kappa_{xy} \sin 2\theta$$

The influence coefficient transformation of a biaxial stiffness isolator from one set of arbitrarily chosen coordinate axes to another arbitrarily chosen set of coordinate axes is described by the two-dimensional Mohr circle.[12] Since the influence coefficient is a

(A)   (B)

FIG. 30.30. Force and displacement transformation diagrams for a linear biaxial stiffness element.

tensor quantity, the following invariants of the influence coefficient tensor give additional relations between the influence coefficients in the X, Y and the X', Y' set of axes:

$$\kappa_x + \kappa_y = \kappa_{x'} + \kappa_{y'}$$

$$\kappa_x \kappa_y - \kappa_{xy}{}^2 = \kappa_{x'} \kappa_{y'} - \kappa_{x'y'}{}^2 \tag{30.52}$$

PRINCIPAL INFLUENCE COEFFICIENTS

The set of axes for which there exists no coupling influence coefficient are the principal axes of stiffness (*principal elastic axes*). These axes can be found by requiring $\kappa_{x'y'}$ to be zero in Eq. (30.51) and solving for the rotation angle corresponding to this condition. Letting θ' represent the angle of rotation for which $\kappa_{x'y'} = 0$:

$$\tan 2\theta' = \frac{2\kappa_{xy}}{\kappa_x - \kappa_y} \tag{30.53}$$

By substituting this value of the angle of rotation into the general influence coefficient expressions, Eqs. (30.51), the following relation is obtained for the principal influence coefficients:

$$\kappa_p, \kappa_q = \frac{\kappa_x + \kappa_y}{2} \pm \sqrt{\left(\frac{\kappa_x - \kappa_y}{2}\right)^2 + \kappa_{xy}{}^2} \tag{30.54}$$

where p and q represent the principal axes of stiffness. The principal influence coefficients are the maximum and minimum influence coefficients that exist for a linear biaxial stiffness isolator. In Eq. (30.54), the plus sign gives the maximum influence coefficient whereas the minus sign gives the minimum influence coefficient. Either κ_p or κ_q can be the maximum influence coefficient, depending on the degree of axis rotation and the relative values of κ_x, κ_y, and κ_{xy}.

INFLUENCE COEFFICIENT TRANSFORMATION FROM THE PRINCIPAL AXES

The influence coefficient transformation from the principal axes p, q is of practical interest. The influence coefficients in the XY frame of reference are determined from Eq. (30.51) by setting $\kappa_{x'y'} = \kappa_{pq} = 0$, $\kappa_{x'} = \kappa_p$, $\kappa_{y'} = \kappa_q$, and $\theta = \theta'$. The influence coefficients in the XY frame-of-reference may be expressed in terms of the principal

Fig. 30.31. Mohr-circle representation of the stiffness transformation from the principal axes of stiffness of a biaxial stiffness element. The p, q axes represent the principal stiffness axes and the X, Y axes are any arbitrary set of axes separated from the p, q axes by a rotation angle θ'.

influence coefficients as follows:

$$\kappa_x = \kappa_p \cos^2 \theta' + \kappa_q \sin^2 \theta' = \frac{\kappa_p + \kappa_q}{2} + \frac{\kappa_p - \kappa_q}{2} \cos 2\theta'$$

$$\kappa_{xy} = (\kappa_p - \kappa_q) \sin \theta' \cos \theta' = \frac{\kappa_p - \kappa_q}{2} \sin 2\theta' \qquad (30.55)$$

$$\kappa_y = \kappa_p \sin^2 \theta' + \kappa_q \cos^2 \theta' = \frac{\kappa_p + \kappa_q}{2} - \frac{\kappa_p - \kappa_q}{2} \cos 2\theta'$$

The transformation from the principal axes in the form of a two-dimensional Mohr's circle is shown by Fig. 30.31. This circle provides quick graphical determination of the three influence coefficients κ_x, κ_y, and κ_{xy} for any angle θ' between the P and X axes, where θ' is positive in the sense shown in the inset to Fig. 30.31.

Example 30.2. Consider the system shown schematically by Fig. 30.26. The transformation theory for the influence coefficient of a biaxial stiffness element may be applied to develop the effective stiffness coefficients for this system. The center of coordinates for the XZ axes is at the elastic center of the system. The principal elastic axes of the isolators p, r are oriented at an angle ϕ with the coordinate axes X, Z, respectively.* The position of the elastic center is determined by effecting a small horizontal displacement δ_x of the body, letting δ_z be zero and equating the summation of couples resulting from the isolator forces. The

---

* The properties of a biaxial stiffness element may be defined with respect to any pair of coordinate axes. In Fig. 30.26, the principal elastic axis q is parallel with the coordinate axis Y; then the analysis considers the principal elastic axes p, r which lie in the plane defined by the XZ coordinate axes.

forces F_x and F_z are determined from Eqs. (30.44):

$$F_x = \kappa_x \delta_x = k_x \delta_x \qquad F_z = \kappa_{zz} \delta_z = k_{zz} \delta_z$$

Each of the forces F_x acts at a distance $-a_{ze}$ from the elastic center; the force F_z at the right-hand isolator is positive and acts at a distance a_x from the elastic center whereas the force F_z at the left-hand isolator is negative and acts at a distance $-a_x$ from the elastic center. Taking a summation of the moments:

$$-2a_{ze}F_x + 2a_xF_z = 0$$

Substituting the above relations between the forces F_x, F_z and the influence coefficients κ_z, κ_{zz} into Eqs. (30.55), and noting that $\theta' = 90° - \phi$ (compare Figs. 30.30 and 30.26), the following result is obtained in terms of principal stiffnesses:

$$\frac{a_{ze}}{a_x} = \frac{F_z}{F_x} = \frac{\kappa_{zz}}{\kappa_x} = \frac{(k_r - k_p) \sin \phi \cos \phi}{k_r \sin^2 \phi + k_p \cos^2 \phi}$$

Substituting $\epsilon = a_z - a_{ze}$ in the preceding equation, the relation for ϵ given by Eq. (30.40) is obtained.

Since the equations of motion are written in a coordinate system passing through the elastic center, all displacements in this frame-of-reference are constrained. Therefore, the effective stiffness coefficients for a single isolator may be obtained from Eq. (30.55) as follows [see Eq. (30.49)]:

$$k_x = \kappa_x = k_r \sin^2 \phi + k_p \cos^2 \phi$$

$$k_z = \kappa_z = k_r \cos^2 \phi + k_p \sin^2 \phi$$

These effective stiffness coefficients define the hypothetical natural frequency ω_x given by Eq. (30.39) as well as the uncoupled vertical natural frequency ω_z. Since four isolators are used in the problem represented by Fig. 30.26, the translational stiffnesses given by the above expressions for k_x and k_z must be multiplied by 4 to obtain the total translational stiffness.

The effective rotational stiffness of a single isolator k_β can be obtained by determining the sum of the restoring moments for a constrained rotation β. When the body is rotated through an angle β, the displacements at the right isolator are $\delta_x = -a_{ze}\beta$ and $\delta_z = a_x\beta$, where a_{ze} is a negative distance since it is measured in the negative Z direction. The sum of the restoring moments is $(F_za_x - F_xa_{ze})$, where F_x and F_z are the forces acting on the right isolator in Fig. 30.26. The forces F_x and F_z may be written in terms of the influence coefficients and the displacements δ_x and δ_z by use of Eq. (30.44) to produce the following moment equation:

$$M_\beta = k_\beta\beta = \beta[k_x a_{ze}^2 - 2k_{xz}a_{ze}a_x + k_z a_x^2]$$

where the effective rotational stiffness k_β of a single isolator is

$$k_\beta = k_x a^2 - 2k_{xz}a_{ze}a_x + k_z a_x^2$$

The distance a_{ze} can be eliminated from the expression for rotational stiffness by substituting $a_{ze} = a_x F_z/F_x$ obtained from the summation of couples about the elastic center:

$$k_\beta = a_x^2 \left(\frac{k_x k_z - k_{zz}^2}{k_x} \right)$$

The numerator of this expression can be replaced by $k_r k_p$ [see Eq. (30.52)] where the r, p axes are the principal elastic axes of the isolator and $k_{rp} = 0$. Also, k_x can be replaced by its equivalent form given by Eq. (30.55). Making these substitutions, the effective rotational stiffness for one isolator in terms of the principal stiffness coefficients of the isolator becomes

$$k_\beta = \frac{a_x^2 k_p}{\sin^2 \phi + (k_p/k_r) \cos^2 \phi}$$

Since four isolators are used in the problem represented by Fig. 30.26, the rotational stiffness given by the above expression for k_β must be multiplied by 4 to obtain the total rotational stiffness of the system.

NONLINEAR VIBRATION ISOLATORS

In vibration isolation, the vibration amplitudes generally are small and linear vibration theory usually is applicable with sufficient accuracy.* However, the static effects of nonlinearity should be considered. Even though a nonlinear isolator may have approximately constant stiffness for small incremental deflections, the nonlinearity becomes important when large deflections of the isolator occur due to the effects of equipment weight and sustained acceleration. A vibration isolator often exhibits a stiffness that increases with applied force or deflection. Such a nonlinear stiffness is characteristic, for example, of rubber in compression or a conical spring.

In Eq. (30.11) for natural frequency, the stiffness k for a linear stiffness element is a constant. However, for a nonlinear isolator, the stiffness k is the slope of the force-deflection curve and Eq. (30.11) may be written

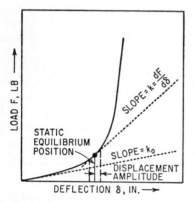

$$\omega_n = 2\pi f_n = \sqrt{\frac{g(dF/d\delta)}{W}} \qquad (30.56)$$

where W is the total weight supported by the isolator, g is the acceleration of gravity, and $dF/d\delta$ is the slope of the line tangent to the force-deflection curve at the static equilibrium position. Vibration is considered to be small variations in the position of the supported equipment above and below the static equilibrium position, as indicated in Fig. 30.32. Thus, the natural frequency is determined solely by the stiffness characteristics in the region of the isolator deflection.

Fig. 30.32. Typical force-deflection characteristic of a tangent hardening isolator.

NATURAL FREQUENCY

In determining the natural frequency of a nonlinear isolator, it is important to note whether or not all the load results from the dead weight of a massive body. The force F on the isolator may be greater than the weight W because of a belt pull or sustained acceleration of a missile. Then the load on the isolator is

$$F = n_g W \qquad (30.57)$$

where n_g is some multiple of the acceleration of gravity. For example, n_g may indicate the absolute value of the sustained acceleration of a missile measured in "number of g's."

CHARACTERISTIC OF TANGENT ISOLATOR. It is convenient to define the force-deflection characteristics of a nonlinear isolator having increasing stiffness (hardening characteristic) by a tangent function: [13]

$$F = \frac{2k_0 h_c}{\pi} \tan\left(\frac{\pi \delta}{2h_c}\right) \qquad (30.58)$$

where F is the total force applied to the isolator, k_0 is the stiffness of the isolator at zero deflection, δ is the deflection of the isolator, and h_c is the characteristic height of the isolator. The force-deflection characteristic defined by Eq. (30.58) is shown graphically in Fig. 30.33A. The characteristic height h_c represents a height or thickness characteristic of the isolator which may be adjusted empirically to obtain optimum agreement, over the deflection range of interest, between Eq. (30.58) and the actual force-deflection curve for the isolator.

The stiffness of the tangent isolator is obtained by differentiation of Eq. (30.58) with

* If the vibration amplitude is large, nonlinear vibration theory as discussed in Chap. 4 is applicable.

respect to δ:

$$k = \frac{dF}{d\delta} = k_0 \sec^2\left(\frac{\pi\delta}{2h_c}\right) = k_0\left[1 + \left(\frac{F\pi}{2k_0h_c}\right)^2\right] \tag{30.59}$$

The stiffness-deflection relation defined by Eq. (30.59) is shown graphically in Fig. 30.33B.

Replacing the load F by n_gW in Eq. (30.59) and substituting the resulting stiffness relation into Eq. (30.56):

$$f_n\sqrt{h_c} = 3.13 \sqrt{2.46n_g^2\left(\frac{W}{k_0h_c}\right) + \left(\frac{k_0h_c}{W}\right)} \tag{30.60}$$

The relation defined by Eq. (30.60) is shown graphically in Fig. 30.34. The ordinate is

FIG. 30.33. Elastic properties of a tangent isolator in terms of its characteristic height h_c and stiffness k_0 at zero deflection: (A) dimensionless force-deflection curve; (B) dimensionless stiffness-deflection curve.

FIG. 30.34. Natural frequency f_n of a tangent isolator system when a portion of the total load applied to the isolator is nonmassive. The weight carried by the isolator is W and the sustained acceleration parameter is n_g, a multiple of the gravitational acceleration. The characteristic height is h_c and the stiffness at zero deflection is k_0.

the natural frequency f_n (cps) times the square root of the characteristic height of the isolator (in.). The theoretical and experimental force-deflection curves for the isolator are matched to establish the numerical value of the characteristic height. For a given value of the acceleration parameter n_g, the natural frequency of the isolation system is determined by h_c and $W/k_0 h_c$.

The deflection of the isolator under a sustained acceleration loading is obtained by substituting Eq. (30.57) into the general force-deflection expression, Eq. (30.58), and solving for the dimensionless ratio δ/h_c:

$$\frac{\delta}{h_c} = \frac{2}{\pi} \tan^{-1}\left(\frac{\pi n_g}{2} \cdot \frac{W}{k_0 h_c}\right) = \frac{2}{\pi} \tan^{-1}\left[15.37\left(\frac{n_g}{h_c f_{n_0}{}^2}\right)\right] \qquad (30.61)$$

A reference natural frequency f_{n_0} is the natural frequency that occurs when the isolator is not deflected by the dead-weight load; i.e., $n_g = 0$. The nomograph of Fig. 30.35 gives the deflection ratio δ/h_c and the frequency ratio f_n/f_{n_0}.[14] The value of the parameter $15.37(n_g/h_c f_{n_0}{}^2)$ is transferred by a horizontal projection to the coordinate system for the curves. Values for the natural frequency ratio f_n/f_{n_0} are read from the lower abscissa scale and values for the deflection ratio δ/h_c are read from the upper abscissa scale.

Example 30.3. A rubber isolator having a characteristic height $h_c = 0.5$ in. (determined experimentally for the particular isolator design) has a natural frequency $f_n = 10$ cps for small deflections and a fraction of critical damping $\zeta = 0.2$. The equipment supported by the isolator is subjected to a sustained acceleration of $11g$. It is desired to determine the absolute transmissibility of the isolation system when the forcing frequency is 100 cps, and to determine the deflection of the isolator under the sustained acceleration.

Referring to the nomograph of Fig. 30.35, a straight line is drawn from a value of 10 on the f_{n_0} scale to 0.5 on the h_c scale. A second straight line is drawn from the intersection of the first line with the R scale through the value $n_g = 11$. The second line intersects the left side

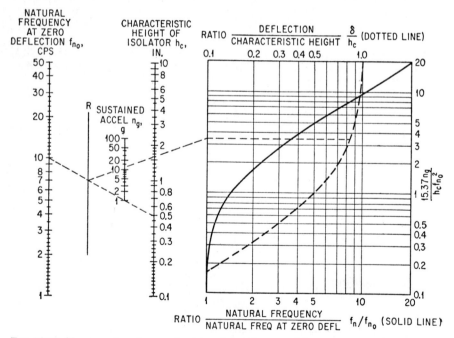

Fig. 30.35. Nomograph and curve for determining the natural frequency and deflection of an isolation system incorporating a tangent isolator when a portion of the total load applied to the isolator is nonmassive.

of the coordinate system and is extended horizontally so that it intersects the solid and dotted curves. The intersection points indicate that the natural frequency ratio $f_n/f_{n_0} = 3.5$ and the deflection ratio $\delta/h_c = 0.81$. The deflection of the isolator at equilibrium as a result of the sustained acceleration is $0.81h_c = 0.405$ in. The undamped natural frequency for the sustained acceleration of $11g$ is $f_n = 3.5 \times 10 = 35$ cps. The natural frequency also can be obtained from Fig. 30.34 by noting that $\dfrac{W}{k_0 h_c} = \dfrac{g/h_c}{(2\pi f_{n0})^2} = 0.196$ [see Eq. (30.60) when $n_g = 0$].

Then for $n_g = 11$, $f_n = 24.5/\sqrt{0.5} = 35$ cps.

From Fig. 30.2 the transmissibility for $\zeta = 0.2$, $f/f_n = 100/35 = 2.88$ is 0.22. In the absence of the sustained acceleration, the corresponding transmissibility would be 0.042 as obtained from Fig. 30.2 at $f/f_n = 100/10 = 10$. Thus, the transmissibility at 100 cps under a sustained acceleration of $11g$ is five times as great as that which would exist for a dead-weight loading of the isolator.

FIG. 30.36. Minimum natural frequency $f_{n(\min)}$ of a tangent isolator system as a function of (1) the characteristic height h_c of the isolator and (2) the sustained acceleration n_g expressed as a multiple of the gravitational acceleration.

FIG. 30.37. Weight loading W_0 required to cause a tangent isolator to have a minimum natural frequency $f_{n(\min)}$, as a function of the stiffness k_0 at zero deflection.

Minimum Natural Frequency. The weight W_0 for which a given tangent isolator has a minimum natural frequency is [15]

$$W_0 = \frac{2k_0 h_c}{\pi n_g} = \frac{k_0 g}{2\pi^2 (f_n)_{\min}^2} \qquad [f_n = \text{minimum}] \qquad (30.62)$$

where the minimum natural frequency $(f_n)_{\min}$ is defined by

$$(f_n)_{\min} = \frac{1}{2}\sqrt{\frac{n_g g}{\pi h}} \qquad (30.63)$$

The minimum natural frequency is shown graphically in Fig. 30.36 as a function of the characteristic height h_c and the sustained acceleration parameter n_g. The weight W_0 required to produce the minimum natural frequency $(f_n)_{\min}$ is shown graphically in Fig. 30.37 as a function of the initial stiffness k_0 and the minimum natural frequency $(f_n)_{\min}$. When the isolator is loaded to produce the minimum natural frequency, the isolator deflection is one-half the characteristic height ($\delta = h_c/2$) and the stiffness under load is twice the initial stiffness ($k = 2k_0$).

ISOLATION OF RANDOM VIBRATION

In random vibration, all frequencies exist concurrently, and the amplitude and phase relations are distributed in a random manner. A trace of random vibration is illustrated in Fig. 11.1A. The equipment-isolator assembly responds to the random vibration with the substantially single-frequency pattern shown in Fig. 11.1B. This response is similar to a sinusoidal motion with a continuously and irregularly varying envelope; it is described as narrow-band random vibration or a random sine wave.

FIG. 30.38. Required clearance expressed in inches rms for a damped isolator with viscous damping when subjected to random vibration defined by a flat spectrum of mean-square acceleration density $W_e(f)$. The natural frequency of the isolator system in cycles per second is f_n.

The characteristics of random vibration are defined by a frequency spectrum of power spectral density (see Chaps. 11 and 22). This is a generic term used to designate the mean-square value of some magnitude parameter passed by a filter, divided by the bandwidth of the filter, and plotted as a spectrum of frequency. The magnitude is commonly measured as acceleration in units of g; then the particular expression to use in place of power spectral density is mean-square acceleration density, commonly expressed in units of g^2/cps. When the spectrum of mean-square acceleration density is substantially flat in the frequency region extending on either side of the natural frequency of the isolator, the response of the isolator may be determined in terms of (1) the mean-square acceleration density of the isolated equipment and (2) the deflection of the isolator at successive cycles of vibration.

The mean-square acceleration densities of the foundation and the isolated equipment are related by the absolute transmissibility that applies to sinusoidal vibration:

$$W_r(f) = W_e(f)T_A{}^2 \tag{30.64}$$

where $W_r(f)$ and $W_e(f)$ are the mean-square acceleration densities of the equipment and the foundation, respectively, in units of g^2/cps and T_A is the absolute transmissibility for the vibration-isolation system [see Eq. (11.31)].

The severity of the vibration experienced by the isolated equipment may be expressed in terms of the rms value of acceleration at the foundation by integrating the mean-square acceleration density given by Eq. (30.64):

$$\ddot{x}_{\text{rms}} = \sqrt{\int W_e(f) T_A{}^2 \, df} \qquad (30.65)$$

where $\ddot{x}_{\text{rms}}$ is the rms acceleration of the equipment and the integration is carried out over the frequency interval for which $W_e(f)$ is defined.

The clearance in the isolator is obtained from Eq. (11.38) by substituting (1) $W_e(f)$ in units of g^2/cps for $2\pi W_e(\omega_n)/g^2$ where $W_e(\omega_n)$ is in units of $(\text{in./sec}^2)^2/(\text{rad/sec})$, (2) $\omega_n = 2\pi f_n$, and (3) $(T_A)_{\text{max}} = 1/2\zeta$. The rms relative deflection δ_{rms} for a rigidly connected viscous damped isolator is

$$\delta_{\text{rms}} = \sqrt{\frac{g^2}{32\pi^3} \frac{(T_A)_{\text{max}} W_e(f)}{f_n{}^3}} = 12.25 \sqrt{\frac{(T_A)_{\text{max}} W_e(f)}{f_n{}^3}} \quad \text{in.} \qquad (30.66)$$

where $W_e(f)$ is the mean-square acceleration density of the foundation measured in g^2/cps. Equation (30.66) applies only if the fraction of critical damping $\zeta = c/c_c$ for the isolation system is relatively small; i.e., $(T_A)_{\text{max}}$ is relatively large. The rms isolator clearance (relative motion between the equipment and the foundation) given by Eq. (30.66) is shown graphically by Fig. 30.38.

Example 30.4. Suppose the vibration of the foundation is defined by a flat spectrum of mean-square acceleration density of $0.2g^2/\text{cps}$ over a frequency band from 10 to 500 cps (wide relative to the width of the absolute transmissibility curve of the isolator in the region of resonance). The isolator has a natural frequency of 25 cps and damping defined by $(T_A)_{\text{max}} = 5$ ($\zeta = 0.1$). Entering Fig. 30.38 at $(T_A)_{\text{max}} W_e(f) = 5 \times 0.2 = 1.0$ on the abscissa scale, the rms isolator clearance as read from the ordinate is $\delta_{\text{rms}} = 0.093$ in.

The deflection of the isolator varies from cycle to cycle. If the vibration has truly normal (Gaussian) characteristics, as discussed in Chaps. 11 and 22, very large values of amplitude occur occasionally. Then bottoming of the isolator cannot be prevented while maintaining the isolator clearance reasonably small; rather, it can be made less frequent by increasing the isolator clearance. For example,* if the clearance of the isolator (in plus and minus directions along axis of vibration) is made $3\delta_{\text{rms}}$, it is probable that the isolator will bottom once in each 100 cycles of vibration; if the clearance is $4.3\delta_{\text{rms}}$, bottoming is probable once in 10^4 cycles; and if the clearance is $5.3\delta_{\text{rms}}$, bottoming is probable once in 10^6 cycles. It is common in certain testing procedures to limit the maximum amplitude to three times the rms value; then it is unlikely that bottoming will occur during tests if the clearance is made somewhat greater than $3\delta_{\text{rms}}$.

When the spectrum of mean-square acceleration density defining vibration of the support is not flat in the region of isolator natural frequency, the integration leading to the simple result of Eq. (30.66) cannot be carried out analytically. An equivalent result can be obtained by graphical integration of Eq. (11.32) where $A(j\omega)$ is a transfer function relating the acceleration of the support to the deflection of the isolator. When the mean-square acceleration density of the foundation $W_e(f)$ is not flat in the region of resonance, the rms isolator clearance is given approximately by

$$\delta_{\text{rms}} = 9.75 \sqrt{\frac{\int_{f_1}^{f_2} W_e(f) T_A{}^2 \dfrac{df_n}{f_n}}{f_n{}^3}} \quad \text{in.} \qquad (30.67)$$

where T_A† is the absolute transmissibility of a rigidly connected viscous-damped isolation

---

* These probabilities are calculated from Eq. (11.39). Also, see Fig. 11.10.

† Equation (30.67) is approximate since the absolute transmissibility function T_A is used in place of a transfer response function (determined from a rigorous analysis) that has a frequency dependency similar to T_A in the region of resonance. It is desirable to use the absolute transmissibility T_A in the relation defined by Eq. (30.67) since this property of an isolation system is most frequently available by experimental means.

system; f_n is the natural frequency of the isolation system, cps; and $W_e(f)$ is the mean-square acceleration density of the foundation, in units of g^2/cps between the frequency limits f_1 and f_2.

A procedure for evaluating Eq. (30.67) graphically is shown in Fig. 30.39. The square of the absolute transmissibility curve is plotted on the left side of the figure to a dimensionless frequency scale that is divided, in the example shown, into increments $\Delta f_n/f_n = 0.25$. Smaller increments may be used for more accurate results. In the example shown, the area of each block is $1 \times 0.25 = 0.25$, where the differential df_n/f_n is replaced by the finite incremental value $\Delta f_n/f_n$. The spectrum of mean-square acceleration density is drawn at the right side

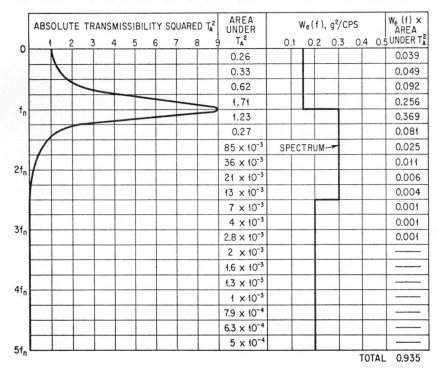

AREA UNDER T_A^2	$W_e(f)$ × AREA UNDER T_A^2
0.26	0.039
0.33	0.049
0.62	0.092
1.71	0.256
1.23	0.369
0.27	0.081
85×10^{-3}	0.025
36×10^{-3}	0.011
21×10^{-3}	0.006
13×10^{-3}	0.004
7×10^{-3}	0.001
4×10^{-3}	0.001
2.8×10^{-3}	0.001
2×10^{-3}	——
1.6×10^{-3}	——
1.3×10^{-3}	——
1×10^{-3}	——
7.9×10^{-4}	——
6.3×10^{-4}	——
5×10^{-4}	——

TOTAL 0.935

Fig. 30.39. Procedure for determining graphically the required isolator clearance when the excitation is random vibration. The example is for an isolation system with absolute transmissibility $T_A = 3$ ($\zeta = 0.17$) and natural frequency $f_n = 10$ cps; the mean-square acceleration density $W_e(f)$ varies with frequency according to the spectrum shown in the figure.

of the figure. For each frequency increment $\Delta f_n/f_n$, the product of $W_e(f)$ and the area under the curve of transmissibility squared is entered in the right-hand column. The sum of this column is $\int_0^\infty W_e(f) T_A^2 \, df_n/f_n$; δ_{rms} may be calculated from Eq. (30.67) using the appropriate value of the isolator natural frequency f_n. The necessary clearance and probability of bottoming can be determined as discussed above with reference to a flat spectrum of mean-square acceleration density.

VIBRATION ISOLATION OF NONRIGID BODY SYSTEMS

This section discusses the effect of nonrigidity of structures to which isolators are attached, in contrast to the rigid body theory discussed in the preceding sections of this chapter. Two aspects of vibration isolation are discussed in detail:

1. The equipment is attached by an isolator to a vibrating foundation which contains a power source. The vibration transferred to the equipment is determined.

2. A force-generating piece of equipment is connected by an isolator to a foundation which is at rest initially. The response of the equipment, the force transmitted to the foundation, and the resulting vibration of the foundation are determined.

The excitation in both cases is steady-state sinusoidal vibration. All forces and velocities are considered colinear, and all elements are considered linear. The equipment is represented by a model having one input terminal and one output terminal; the foundation is represented by a similar model.

MECHANICAL POWER SOURCES

A *mechanical power source* is a means to generate sinusoidally varying forces and motions, and an associated elastic system through which the forces and motions are transmitted to the output terminal. Such a source is shown schematically in Fig. 30.40. Practically, measurements can be made only at the power source output. The force applied by the source to any system attached to the source output is represented by F_s and v_s represents the resulting velocity at the source output; F_s and v_s are related by [16]

$$F_s = F_s^b - Z_s v_s = Z_s(v_s^f - v_s) \quad (30.68)$$

where F_s^b is the *blocked output force* * of the power source, v_s^f is the *free output velocity* * of the power source, and Z_s is the point

Fig. 30.40. Schematic diagram of a power source characterized by the point impedance Z_s at the power source output terminal.

mechanical impedance of the power source output. (See Chap. 10 for a detailed discussion of mechanical impedance.) The blocked output force of the source is the force transmitted to a body of infinitely great mechanical impedance attached at the source output. The free output velocity of the source is the velocity at the output with no load attached. The point mechanical impedance of the power source output Z_s is the ratio of a force applied at the source output to the resulting velocity at the output, with the source inactive. The force, velocity, and impedance are related by

$$Z_s = \frac{F_s^b}{v_s^f} \quad (30.69)$$

IMPEDANCE PROPERTIES OF MECHANICAL ELEMENTS

A mechanical element may be represented by a generalized linear mechanical system having an input terminal i and an output terminal j, as shown by Fig. 30.41. The dynamic properties of the element may be described conveniently by the mechanical impedances (see Chap. 10) associated with the element:

Z_i^{jf} = point impedance at terminal i with terminal j free ($F_j = 0$)
Z_i^{jb} = point impedance at terminal i with terminal j blocked ($v_j = 0$)
$Z_{ij}^f = Z_{ji}^f$ = free transfer impedance between terminal i and terminal j; i.e., ratio of force applied at either one of the terminals to resulting velocity at the other terminal when the latter is free
$Z_{ij}^b = Z_{ji}^b$ = blocked transfer impedance between terminal i and terminal j; i.e., ratio of force developed at either one of the terminals when blocked to the velocity applied at the other terminal

Other impedance terms can be defined by substitution of the proper superscripts and subscripts in the above definitions.

* These quantities are the force and velocity *amplitudes* (complex) of sinusoidal quantities but are referred to as "force" and "velocity" for convenience.

Definitions for velocity and force transmissibility are:

$(T_v)_{ij}$ = velocity transmissibility from terminal i to terminal j; i.e., ratio of resulting velocity at terminal j to imposed velocity at terminal i when no load is attached at terminal j

$(T_F)_{ij}$ = force transmissibility from terminal i to terminal j; i.e., ratio of resulting force at terminal j to imposed force at terminal i when a body of infinitely great mechanical impedance is attached at terminal j

Other velocity and force transmissibilities can be defined by substitution of the proper subscripts in the above definitions. Note that the order of the subscripts in the transmissibility terms describes the direction in which the vibration is being transmitted. This

FIG. 30.41. Schematic diagram of a generalized linear mechanical system having an input at terminal i and an output at terminal j. The velocities v_i and v_j are of the same sign when in the same direction.

order is not significant when writing transfer impedance terms since the transfer impedance between any two terminals in opposite directions is equal.

Mechanical impedance identities used in the following analyses are:

$$Z_i^{if} Z_j^{tb} = Z_i^{jb} Z_j^{tf} = Z_{ij}^{f} Z_{ij}^{b}$$

$$\frac{Z_{ij}^{b}}{Z_{ij}^{f}} + 1 = \frac{Z_j^{tb}}{Z_j^{tf}} = \frac{Z_i^{jb}}{Z_i^{if}}$$

(30.70)

VELOCITY ISOLATION

In velocity isolation, a nonrigid equipment is mounted on a vibrating foundation by a generalized vibration isolator as shown in Fig. 30.42. The foundation is considered to have the properties of the power source shown in Fig. 30.40, and its point impedance measured at point 1 is designated Z_F. Without the equipment attached, the output of the foundation is vibration with a velocity v_F^f, the free source velocity of the foundation (Fig. 30.42A). When the equipment-isolator combination is attached to the foundation (Fig. 30.42B), the velocity at the foundation output changes to v_1, and velocity v_2 is transmitted to the input terminal of the equipment. In general, the equipment and the foundation are nonrigid; the isolator is characterized by mass, stiffness, and damping. The velocity at any other point on the equipment is designated by v_3. Transmissibility is referred to as velocity transmissibility; this is equal to displacement or acceleration transmissibility because the system is linear and the vibration sinusoidal.

GENERAL MODIFIED VELOCITY TRANSMISSIBILITY. The force F_1 applied to the isolation system at terminal 1 in Fig. 30.42B is first determined by considering the foundation as the power source and using Eq. (30.68):

$$F_1 = Z_F(v_F^f - v_1)$$

(30.71)

The *modified velocity transmissibility* $(T_v)_{F3}$ is the ratio of the equipment velocity v_3 to the foundation free velocity v_F^f. The velocity ratio $(T_V)_{F3} = v_3/v_F^f$ is designated a "modified" transmissibility since it is not a ratio of two velocities which exist simultaneously but rather the ratio of a resultant velocity at the equipment to the velocity of the foundation without equipment attached. When the foundation in Fig. 30.42 is rigid, $v_F^f = v_1$; then the modified transmissibility becomes the ratio of two concurrent velocity amplitudes and is a true transmissibility expression. The expression for modified velocity transmissibility is [17]

$$(T_v)_{F3} = \frac{v_3}{v_F^f} = \cfrac{1}{\cfrac{Z_{23}^f}{Z_{12}^b}\left[1 + \cfrac{Z_1^{2b}}{Z_F}\right] + \cfrac{Z_{23}^f Z_{12}^f}{Z_2^{3f}}\left[\cfrac{1}{Z_1^{2f}} + \cfrac{1}{Z_F}\right]} \qquad (30.72)$$

where Z_{23}^f = free transfer impedance of equipment between terminals 2 and 3. (This may be determined by applying a known vibratory force at 2 and noting the resulting velocity at 3.)

Z_{12}^b = blocked transfer impedance of the isolator

Z_{12}^f = free transfer impedance of the isolator

Z_1^{2b} = point impedance of isolator at terminal 1 with a body of infinitely great impedance attached at terminal 2

Z_1^{2f} = point impedance of isolator at terminal 1 with no load at terminal 2

Z_F = point impedance of foundation measured at terminal 1. (In general, this determines the impedance of the foundation uniquely because any other terminal is inaccessible.)

Z_2^{3f} = point impedance of equipment at terminal 2 with no load at terminal 3

Only three of the four isolator impedances appearing in Eq. (30.72) must be measured, the fourth being determined by Eq. (30.70). If the actual velocity v_3 at terminal 3 of the equipment is desired, the free velocity v_F^f (see Fig. 30.42A) of the foundation must be measured. Thus, seven measurements are required to obtain the velocity v_3 at a point on the nonrigid equipment. In the following sections, the generalities of Eq. (30.72) are removed and corresponding expressions are derived for more restricted conditions.

MODIFIED VELOCITY TRANSMISSIBILITY FOR A MASSLESS ISOLATOR. If the isolator is massless, the point and transfer impedances Z_1^{2f}, Z_{12}^f of the isolator become zero; the point and transfer impedances Z_1^{2b}, Z_{12}^b become equal and are denoted by

Fig. 30.42. Schematic diagram illustrating velocity isolation. The nonrigid foundation is considered a power source and is characterized by the impedance Z_F; initially it is free as illustrated at A and vibrates with a velocity v_F^f. A generalized system comprised of a nonrigid equipment and a generalized isolator is then attached to the foundation; this changes the foundation velocity to v_1 and transmits velocities v_2 and v_3 to the equipment as illustrated at B. The velocities are of the same sign when in the same direction.

Z_I. Then Eq. (30.72) becomes

$$(T_v)_{F3} = \frac{v_3}{v_F^f} = \frac{1}{Z_{23}^f(1/Z_I + 1/Z_F + 1/Z_2^{3f})} \tag{30.73}$$

Equation (30.73) can be written more conveniently in terms of mechanical mobility (see Chap. 10):

$$(T_v)_{F3} = \frac{v_3}{v_F^f} = \frac{\mathfrak{M}_{23}^f}{\mathfrak{M}_I + \mathfrak{M}_F + \mathfrak{M}_2^{3f}} \tag{30.74}$$

where $\mathfrak{M}_{23}^f = 1/Z_{23}^f =$ transfer mobility of equipment between terminals 2 and 3 with no load at terminal 3

$\mathfrak{M}_I = 1/Z_I^{2b} = 1/Z_{12}^b =$ mobility of isolator when measured with a load of infinite impedance attached at either terminal

$\mathfrak{M}_F = 1/Z_F =$ mobility of foundation

$\mathfrak{M}_2^{3f} = 1/Z_2^{3f} =$ mobility of equipment at terminal 2 with no load attached at terminal 3

The impedances are defined explicitly with reference to Eq. (30.72). Four measurements of impedance or mobility are required to determine the modified velocity transmissibility when the isolation system is massless.

MODIFIED VELOCITY TRANSMISSIBILITY TO INPUT TERMINAL OF NON-RIGID EQUIPMENT. The modified velocity transmissibility between the foundation and terminal 2 of the equipment may be determined from Eq. (30.72) by requiring terminals 2 and 3 to coincide and using the impedance identities stated by Eq. (30.70):

$$(T_v)_{F2} = \frac{v_2}{v_F^f} = \frac{Z_{12}^b}{(Z_1^{2b}/Z_F)(Z_E + Z_2^{1f}) + Z_E + Z_2^{1b}} \tag{30.75}$$

where $Z_E = Z_2^{3f} = Z_{23}^f$ represents the input point impedance of the equipment at terminal 2 for this case. This equation differs from Eq. (30.77) only in that the isolator may have mass. By use of the impedance identities stated by Eq. (30.70), the four isolator impedances indicated in Eq. (30.75) may be obtained from three measurements. In addition, the point impedances of the equipment and the foundation must be known to determine the velocity ratio v_2/v_F^f. However, if the velocity v_2 is required, v_F^f also must be measured, making a total of six measurements required. If the isolator has mass but the foundation is infinitely rigid, Eq. (30.75) becomes

$$(T_v)_{F2} = \frac{v_2}{v_F^f} = \frac{Z_{12}^b}{Z_E + Z_{12}^b} \tag{30.76}$$

If the isolator is massless, the modified velocity transmissibility between the foundation and terminal 2 of the equipment is

$$(T_v)_{F2} = \frac{v_2}{v_F^f} = \frac{Z_I}{Z_I + Z_E(1 + Z_I/Z_F)} = \frac{\mathfrak{M}_E}{\mathfrak{M}_I + \mathfrak{M}_F + \mathfrak{M}_E} \tag{30.77}$$

where the point impedance and mobility of the equipment at terminal 2 are Z_E and $\mathfrak{M}_E$, respectively, and the other parameters are defined with reference to Eqs. (30.72) to (30.74).

FORCE ISOLATION

Force isolation occurs where a power-generating equipment having the properties of a power source is attached to an initially motionless foundation by means of an isolator, as indicated schematically in Fig. 30.43. The equipment vibrates with a velocity v_E^f when unattached. After the equipment is attached to the isolator, the velocity of the equipment changes and vibration is transmitted to the foundation. The force transmitted to the foundation, the resulting velocity of the foundation, and the resulting velocity of the equipment are determined in the following sections.

The equipment containing the power source is characterized by its blocked output force $F_E{}^b$ (i.e., by the force applied at the output terminal when this terminal is attached to a body with an infinitely great impedance) and its point impedance Z_E as indicated in Fig. 30.43A.

MODIFIED FORCE TRANSMISSIBILITY. The *modified force transmissibility* $(T_F)_{E1}$ is defined as the ratio of the force F_1 transmitted to the foundation to the output

(A) (B)

Fig. 30.43. Schematic diagram illustrating force isolation. The nonrigid equipment is considered a power source and is characterized by the impedance Z_E; initially it is attached to a rigid foundation and transmits a force $F_E{}^b$, as illustrated at A. The equipment is then attached by means of a generalized isolator to a nonrigid foundation; this changes the output force from the equipment to F_2 and transmits a force F_1 to the foundation. The velocities v_1, v_2 are of the same sign when in the same direction.

force $F_E{}^b$ of the equipment with point 2 blocked. The force ratio $(T_F)_{E1} = F_1/F_E{}^b$ is considered a "modified" transmissibility since it is not a true transfer force ratio of two forces that exist simultaneously, rather, it is the ratio of the force experienced by the foundation to the force generated by the equipment if rigidly constrained. Since the systems considered in this section are linear, the modified force transmissibility $(T_F)_{E1}$ is equal to the modified velocity transmissibility $(T_v)_{F2}$ given by Eq. (30.75). This relation applies for a power-generating nonrigid equipment, a generalized isolator, and a nonrigid foundation.

Massless Isolator. If the isolator is massless, all impedances (point and transfer impedances) of the isolation system with the output blocked are replaced by Z_I; the isolator impedances with the output free become zero. Then the modified force transmissibility $(T_F)_{E1}$ is equal to the modified velocity transmissibility $(T_v)_{F2}$ given by Eq. (30.77).

Rigid Foundation. If the isolator has mass, stiffness, and damping but the foundation is rigid, $Z_F = \infty$ and the modified force transmissibility $(T_F)_{E1}$ is equal to $(T_v)_{F2}$ given by Eq. (30.76).

VELOCITY RESPONSE OF FOUNDATION. In general, the nonrigid foundation experiences a velocity in response to force applied by the power-generating equipment,

The foundation velocity v_1 is

$$v_1 = \frac{F_1}{Z_F} \tag{30.78}$$

where Z_F is the impedance of the foundation and F_1 is the force applied by the isolator. Since $(T_F)_{E1} = F_1/F_E^b$, the foundation velocity v_1 is

$$v_1 = \frac{F_E^b}{Z_F}(T_F)_{E1} \tag{30.79}$$

By substitution of the proper modified force transmissibility function for the conditions of a problem, this relation gives the foundation velocity resulting from the vibration generated by the equipment. According to Eq. (30.69), the blocked output force F_E^b may be expressed as

$$F_E^b = Z_E v_E^f \tag{30.80}$$

where Z_E is the point impedance of the equipment measured at terminal 2. Substituting Eq. (30.80) into Eq. (30.79), the foundation velocity response v_1 can be written in dimensionless form as follows:

$$\frac{v_1}{v_E^f} = \frac{Z_E}{Z_F}(T_F)_{E1} \tag{30.81}$$

where v_E^f is the velocity of the equipment at terminal 2 when no load is attached. For example, velocity response of the foundation when the isolator is massless is obtained by substituting $(T_v)_{F2}$ from Eq. (30.77) for $(T_F)_{E1}$ in Eq. (30.79):

$$v_1 = \frac{F_E^b}{Z_E + Z_F(1 + Z_E/Z_I)} \tag{30.82}$$

Equation (30.82) can be written nondimensionally by replacing F_E^b by its equivalent form defined by Eq. (30.80):

$$\frac{v_1}{v_E^f} = \frac{Z_I}{Z_I + Z_F(1 + Z_I/Z_E)} \tag{30.83}$$

VELOCITY RESPONSE OF EQUIPMENT. The output velocity v_E^f of the equipment which exists before the attachment of the isolator is changed to the velocity v_2 when the attachment is made because the equipment is nonrigid:

$$\frac{v_2}{v_E^f} = 1 - (T_F)_{E1}\left(\frac{Z_2^{1b}}{Z_{12}^b} + \frac{Z_{12}^f}{Z_F}\right) \tag{30.84}$$

where Z_{12}^b = transfer impedance across isolator from terminal 1 to terminal 2, with terminal 2 blocked
Z_{12}^f = transfer impedance across isolator from terminal 1 to terminal 2, with terminal 2 free
Z_2^{1b} = point impedance of isolator at terminal 2 with terminal 1 blocked
Z_F = point impedance of the foundation at terminal 1

If the isolator is massless, $Z_{12}^b = Z_2^{1b} = Z_I$ and $Z_{12}^f = 0$; then Eq. (30.84) becomes

$$\frac{v_2}{v_E^f} = 1 - (T_F)_{E1} = \frac{Z_I + Z_F}{Z_I + Z_F(1 + Z_I/Z_E)} \tag{30.85}$$

where $(T_F)_{E1}$ for a massless isolator is given by Eq. (30.79).

ISOLATOR EFFECTIVENESS

The effectiveness of a vibration isolator is a measure of the reduction of vibration which it effects. In concept, effectiveness may be indicated in terms of vibratory velocity or vibratory force. Effectiveness in terms of velocity is the ratio of the velocity $v_2^{(U)}$ to the velocity $v_2^{(I)}$ (see Fig. 30.44) where $v_2^{(U)}$ is the (unisolated) velocity transmitted to an equipment attached directly to the foundation and $v_2^{(I)}$ is the corresponding (isolated) velocity when an isolator is interposed between equipment and foundation; effectiveness in terms of force is the ratio of the force $F_1^{(U)}$ to the force $F_1^{(I)}$, as indicated in Fig. 30.45.[18,19]

The velocity v_2 at the input terminal of the equipment (terminal 2) is given by Eq. (30.75) in terms of the free velocity of the foundation v_F^f and several characteristic im-

$$E = \frac{v_2^{(U)}}{v_2^{(I)}} = \text{ISOLATOR EFFECTIVENESS}$$

FIG. 30.44. Schematic diagram of the systems that define isolator effectiveness in velocity isolation as illustrated by Fig. 30.42. The isolator effectiveness is the ratio of the velocity $v_2^{(U)}$ transmitted to the unisolated equipment, as illustrated at B, to the velocity $V_2^{(I)}$ transmitted to the equipment through the isolator, as illustrated at C. The foundation without the equipment attached is shown at A.

pedances. This velocity is designated as $v_2^{(I)}$ when the isolator is effective; when the isolator is considered as a rigid, massless connection, $Z_{12}^b = Z_2^{1b} = Z_1^{2b} = \infty$, $Z_2^{1f} = 0$ and the corresponding velocity at the equipment is designated $v_2^{(U)}$. Then the expression for effectiveness is

$$E = \frac{v_2^{(U)}}{v_2^{(I)}} = \frac{Z_E + Z_2^{1b} + (Z_1^{2b}/Z_F)(Z_E + Z_2^{1f})}{Z_{12}^b(1 + Z_E/Z_F)} \tag{30.86}$$

where Z_E is the point impedance of the equipment at terminal 2, Z_1^{2b} is the point impedance of the isolator at terminal 1 with terminal 2 blocked, Z_2^{1f} is the point impedance of the isolator at terminal 2 with terminal 1 free, and the other parameters are defined in connection with Eq. (30.84). Inasmuch as the expression for force transmissibility is identical to Eq. (30.75) for velocity transmissibility, Eq. (30.86) for effectiveness applies to effectiveness of force transmissibility as follows:

$$E = \frac{F_1^{(U)}}{F_1^{(I)}} \tag{30.87}$$

where $F_1^{(U)}$ is the force transmitted to the foundation with the equipment attached directly thereto (Fig. 30.45*B*) and $F_1^{(I)}$ is the corresponding force with the isolator interposed therebetween (Fig. 30.45*C*).

EFFECTIVENESS OF MASSLESS ISOLATOR. If the isolator is massless, $Z_{12}^b = Z_2^{1b} = Z_1^{2b} = Z_I$ and $Z_2^{1f} = 0$; then the isolator effectiveness defined by Eq. (30.86) becomes

$$E = \frac{Z_E + Z_F(1 + Z_E/Z_I)}{Z_E + Z_F} = \frac{\mathfrak{M}_E + \mathfrak{M}_I + \mathfrak{M}_F}{\mathfrak{M}_E + \mathfrak{M}_F} \qquad (30.88)$$

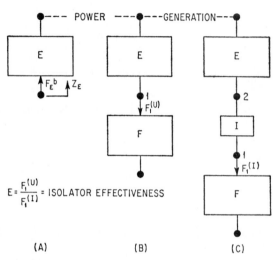

$$E = \frac{F_1^{(U)}}{F_1^{(I)}} = \text{ISOLATOR EFFECTIVENESS}$$

(A)   (B)   (C)

FIG. 30.45. Schematic diagram of the systems that define isolator effectiveness in force isolation as illustrated by Fig. 30.43. The isolator effectiveness is the ratio of the force $F_1^{(U)}$ transmitted to the unisolated foundation, as illustrated at *B*, to the force $F_1^{(I)}$ transmitted to the foundation through the isolator, as illustrated at *C*. The equipment is shown at *A* before being attached to the foundation.

NONRIGIDITY OF STRUCTURES

When the isolator is attached to a nonrigid structure of the equipment or foundation, the structure may vibrate with a relatively large amplitude if the excitation frequency coincides with a natural frequency of the structure. Thus, the isolator appears to be relatively ineffective at such frequencies and may afford little or no isolation. This ef-

FIG. 30.46. Schematic diagram of a free-free beam of length l supported by an isolator in the form of a massless linear spring. The beam is an example of a nonrigid equipment.

fect can be evaluated quantitatively if the mechanical impedance or mobility of the equipment and foundation are known at point of attachment of the isolator.

For example, consider a free-free beam of mass m_B attached to a vibrating support by a linear, massless spring of stiffness k, as shown in Fig. 30.46. There is no damping in the system. The expression for transmissibility v_2/v_1 given by Eq. (30.77) becomes applicable by setting the foundation impedance Z_F equal to infinity:

$$(T_v)_{12} = \frac{v_2}{v_1} = \frac{Z_I}{Z_E + Z_I} = \frac{1}{1 + Z_E/Z_I} \tag{30.89}$$

where Z_I is the mechanical impedance of the massless isolator with one end blocked and Z_E is the mechanical impedance at the center point of the free-free beam. Substituting the impedance of a spring $Z_I = -jk/\omega$ (see Table 10.2) and the impedance Z_E of a free-free beam at its center point, Eq. (30.89) becomes [20]

$$(T_v)_{12} = \frac{1}{1 - \left(\dfrac{f}{f_0}\right)^2 \left[1 - \displaystyle\sum_{n=1}^{\infty} \dfrac{[\phi_n(l/2)]^2(f/f_0)^2}{(f_n/f_0)^2 - (f/f_0)^2}\right]^{-1}} \tag{30.90}$$

where $\phi_n(l/2)$ = value of nth mode function at center point of free-free beam (see Chap. 7), dimensionless

f_n = natural frequency of nth mode of beam, cps

$f_0 = \dfrac{1}{2\pi} \sqrt{k/m_B}$ = natural frequency of beam (considered as a rigid body) on isolators, cps

$f = \omega/2\pi$ = forcing frequency, cps

The transmissibility as calculated from Eq. (30.90) and as determined experimentally is illustrated in Fig. 30.47 for three systems whose fundamental natural frequencies f_1 are two, five, and ten times the rigid body natural frequency f_0. The transmissibility in rigid body theory, Eq. (a) of Table 30.2 with $\zeta = 0$, is plotted in dash-dot lines for comparison with nonrigid body conditions.

As shown in Fig. 30.47, the rigid body theory is useful only for frequencies lower than the natural frequency of the beam in its fundamental mode. For example, when the natural frequency of the beam (considered as a rigid body) on the isolator is 0.2 times the fundamental natural frequency of the beam ($f_1/f_0 = 5$), as illustrated in Fig. 30.47B, the rigid body theory gives an accurate indication of transmissibility only for frequency ratios f/f_0 less than 2.5. This corresponds to a forcing frequency equal to 50 per cent of the lowest natural frequency of the beam. At higher frequencies, resonances of the beam become important and the rigid body theory is inapplicable. Generally similar results are obtained for other values of f_1/f_0, as indicated by Fig. 30.47A and C.

A result similar to that shown in Fig. 30.47 is obtained for force transmissibility when the foundation is nonrigid. The foundation responds with a large amplitude at its resonant frequencies; thus, the isolator appears to transmit large forces at such frequencies.

VIBRATION ISOLATION OF DAMPED STRUCTURES. The relatively large values of transmissibility in Fig. 30.47 result from vibration of the beam at its resonant frequencies; corresponding transmissibility curves are shown in Fig. 30.48 for a beam with significantly greater damping and with the same resonant frequencies.[21, 22] The transmissibility curves for the undamped beams and the rigid body are repeated from Fig. 30.47 for comparison. The damped beams consist of two laminates separated by a thin layer of viscoelastic damping material.

Even though the damping in the beam reduces the amplification at structural resonances, the over-all isolation at the higher forcing frequencies is not as efficient as the rigid body theory predicts. The mean value of transmissibility for all beams, regardless of the amount of damping, has an attenuation rate that is less than that predicted by rigid body theory. Therefore, structural resonances, whether highly damped or not, reduce the over-all efficiency of a vibration isolation system at high forcing frequencies.

Fig. 30.47. Acceleration transmissibility for the system shown in Fig. 30.46 for three degrees of equipment nonrigidity. The fundamental natural frequency f_1 of the beam is twice the natural frequency f_0 of the beam (considered as a rigid body supported by the isolator) at A. At B and C, the ratio f_1/f_0 is 5 and 10, respectively.

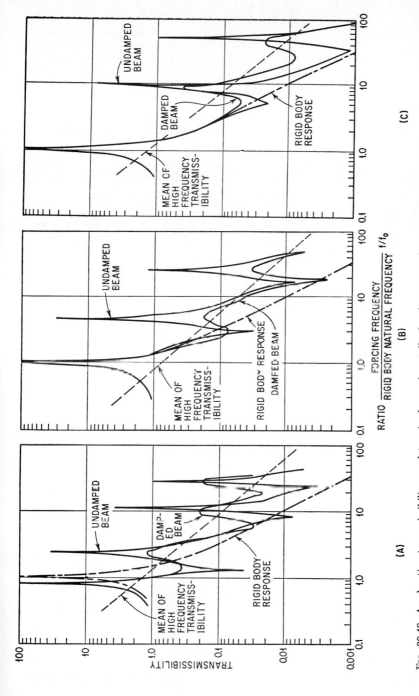

Fig. 30.48. Acceleration transmissibility as determined experimentally for the system shown in Fig. 30.46 for three degrees of equipment nonrigidity corresponding, respectively, to A, B, and C of Fig. 30.47. The damped beam is comprised of two steel members separated by a viscoelastic damping layer; the undamped beam is solid steel. The fundamental natural frequency f_1 of the beam is twice the natural frequency f_0 of the beam considered as a rigid body supported by the isolator, in the curves at A; corresponding ratios of f_1/f_0 are 5 and 10 at B and C, respectively.

WAVE EFFECTS IN ISOLATORS

When the forcing frequency becomes relatively high, standing waves tend to occur in an isolator and the classical theory of vibration isolation based upon a massless resilient element may not give acceptable results. The transmissibility may become relatively great at the standing-wave frequencies. It is difficult to determine by analytical means standing-wave frequencies of isolators which incorporate irregularly shaped metal or rubber springs. However, the principle can be demonstrated by a simplified model.

Consider a rigid equipment of mass m_E supported by a linear unidirectional isolator having a mass m_I as shown in Fig. 30.49. The transmissibility can be written in terms of displacement by using Eq. (30.76) and noting the equivalence of velocity and displacement transmissibility:

$$(T_v)_{12} = \frac{x_0}{u_0} = \frac{Z_{12}^b}{Z_E + Z_2^{1b}} \qquad (30.91)$$

where Z_{12}^b is the transfer impedance of the isolator with one end blocked, Z_2^{1b} is the point impedance of the isolator at station 2 with station 1 blocked, and Z_E is the point impedance of the equipment at terminal 2. Assuming the isolator to be a one-dimensional mechanical transmission line, the required impedances are [16, 23]

$$Z_{12}^b = \frac{Z_c}{\sinh \gamma l}$$

$$Z_2^{1b} = Z_c \left(\frac{\cosh \gamma l}{\sinh \gamma l} \right) = Z_c \coth \gamma l \qquad (30.92)$$

where the characteristic impedance Z_c is

$$Z_c = \frac{S\gamma}{\omega} [\omega\mu - j(c_0^2 \rho)] \qquad (30.93)$$

FIG. 30.49. Schematic diagram of a system used to demonstrate the effects on transmissibility of standing waves in the isolator. The rigid equipment having a mass m_E is supported by a unidirectional isolator having mass m_I.

and ρ is the mass density of the isolator material, S is the isolator cross section area undergoing vibration, μ is the viscosity of the resilient material of the isolator * and c_0 is the undamped phase velocity or classical velocity of sound for the resilient material of the isolator when the wavelength of vibration is much larger than the lateral dimension of the isolator. The complex propagation function γ is

$$\gamma = \alpha + j\beta \qquad (30.94)$$

The attenuation per unit length α (measure of damping in the material) and the phase function β may be determined by solving the general equations of motion that apply for the one-dimensional isolator.[24, 25] For low damping, α and β are defined by [26]

$$\alpha = \frac{\omega^2 \mu}{2\rho c_0^3} \qquad \beta = \frac{\omega}{c_0} \qquad (30.95)$$

If the impedance relations for the isolator given by Eq. (30.92) are substituted into the transmissibility expression given by Eq. (30.91), the following result is obtained:

$$(T_v)_{12} = \frac{1}{\cosh \gamma l + \dfrac{Z_E}{Z_c} \sinh \gamma l} \qquad (30.96)$$

* The coefficient of viscosity μ defines an effective viscous damping coefficient having a value $\mu S/l$ where S is the cross-section area and l is the length of the resilient material of the isolator.

Table 30.3. Characteristic Parameters Relating to Wave Effects in Undamped Isolators

PROPERTY	GENERAL	COMPRESSION	TORSION	SHEAR	HELICAL SPRING	FLUID	AIR
DIAGRAMS OF ISOLATION SYSTEMS							
CROSS-SECTION AREA	S	S_c	$\dfrac{\pi D^2}{4}$	S_s	$\dfrac{\pi^2 N D d^2}{4l}$	S_F	S_A
STATIC STIFFNESS	k_{st}	$\dfrac{S_c E}{l}$	$\dfrac{G I_p}{l}$	$\dfrac{S_s G}{l}$	$\dfrac{G d^4}{8ND^3}$	$\dfrac{S_{FK}}{l}$	$\dfrac{P_0 S_A^2 n}{V_0}$
MASS OF RESILIENT MATERIAL	m_I	$\rho S_c l$	$\dfrac{\pi \rho D^2 l}{4}$	$\rho S_s l$	$\dfrac{\pi^2 \rho N D d^2}{4}$	$\rho S_F l$	$\rho_0 V_0$
ELASTIC WAVE PROPAGATION VELOCITY c_0	$l\sqrt{\dfrac{k_{st}}{m_I}}$	$\sqrt{\dfrac{E}{\rho}}$	$\sqrt{\dfrac{G}{\rho}}$	$\sqrt{\dfrac{G}{\rho}}$	$\dfrac{ld}{\sqrt{2}\pi ND^2}\sqrt{\dfrac{G}{\rho}}$	$\sqrt{\dfrac{K}{\rho}}$	$\sqrt{\dfrac{P_0 n}{\rho_0}}$
CHARACTERISTIC IMPEDANCE Z_c	$\sqrt{k_{st} m_I}$	$S_c\sqrt{E\rho}$	$I_p\sqrt{G\rho}$	$S_s\sqrt{G\rho}$	$\dfrac{\sqrt{2}\pi d^3}{8D}\sqrt{G\rho}$	$S_F\sqrt{K\rho}$	$S_A\sqrt{P_0 \rho_0 n}$

DIMENSIONS OF THE RESILIENT ELEMENTS ARE INDICATED IN THE SKETCHES; RELEVANT PROPERTIES OF THE RESILIENT MATERIAL ARE:

ρ = MASS DENSITY, LB-SEC^2/IN.^{4}
V = VOLUME OF RESILIENT MATERIAL, IN.^{3}
n = RATIO OF SPECIFIC HEAT AT CONSTANT PRESSURE TO SPECIFIC HEAT AT CONSTANT VOLUME
P = PRESSURE, LB/IN.^{2}
E = YOUNG'S MODULUS, LB/IN.^{2}
G = MODULUS OF ELASTICITY IN SHEAR, LB/IN.^{2}
K = BULK MODULUS, LB/IN.^{2}
(SUBSCRIPT $_0$ INDICATES INITIAL CONDITIONS)

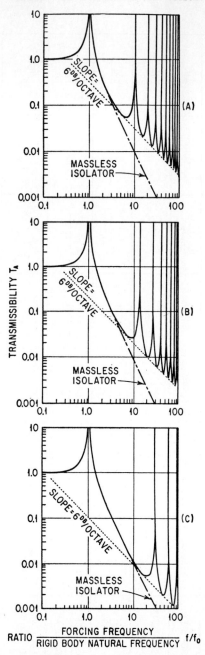

Fig. 30.50. Theoretical transmissibility curves for the system illustrated in Fig. 30.49. The ratio of the mass m_E of the equipment to the mass m_I of the isolator is 10 at A, 20 at B, and 100 at C. The isolator is undamped.

When the equipment is considered a rigid mass m_E, $Z_E = j\omega m_E$ and Eq. (30.96) becomes

$$(T_v)_{12} = \frac{1}{\cosh \gamma l + j\dfrac{\omega m_E}{Z_c} \sinh \gamma l} \tag{30.97}$$

The transmissibility expressions given by Eqs. (30.96) and (30.97) apply to unidirectional isolators for which the force-displacement equation is

$$F(x) = S(c_0^2 \rho + j\omega\mu)\frac{\partial \xi(x)}{\partial x} \tag{30.98}$$

where $F(x)$ is the force in the X coordinate direction, and $\xi(x)$ is the displacement of any cross section of the isolator from the equilibrium position in the X coordinate direction. Therefore, Eqs. (30.96) and (30.97) may be applied to isolators that can be classified as (1) resilient material strained in compression, torsion, or shear; (2) helical spring; and (3) fluid or air isolators. Expressions for c_0 and Z_c are given in Table 30.3 for the several classes of undamped isolators.[27]

For an undamped isolator, $\alpha = 0$ and $Z_c = \rho S c_0 = \sqrt{k_{st} m_I}$. Then Eq. (30.97) becomes

$$(T_v)_{12} = \frac{1}{\cos \dfrac{f}{f_0}\sqrt{\dfrac{m_I}{m_E}} - \dfrac{f}{f_0}\sqrt{\dfrac{m_E}{m_I}} \sin \dfrac{f}{f_0}\sqrt{\dfrac{m_I}{m_E}}} \tag{30.99}$$

The transmissibility defined by Eq. (30.99) is shown graphically in Fig. 30.50 as a function of the ratio of the forcing frequency f to the undamped natural frequency $f_0 = \dfrac{1}{2\pi}\sqrt{\dfrac{k_{st}}{m_E}}$ for three values of the ratio m_E/m_I.

The mass m_I is the mass of that portion of the isolator which contributes the resilience. Since no damping exists in the isolator, the transmissibility approaches infinity when the forcing frequency f approaches one of the standing-wave frequencies f_w. The valleys in the transmissibility curve between the standing-wave frequencies have minimum values that indicate a decrease in transmissibility at the rate of 6 db/octave. The lowest frequency at which standing-wave resonances occur increases as the mass of the equipment increases relative to the mass of the isolator.

The transmissibility curves shown in Fig. 30.51 are experimental results obtained from tests on isolators made from cylindrical samples of various rubber materials for a mass ratio $m_E/m_I = 20$.[28] A comparison of the experimental curves of Fig. 30.51 with the theoretical curves of Fig. 30.50B indicates the effects that damping has on the standing-wave resonances. Damping in the isolator limits the transmissibility at the standing-wave resonances to finite values. If sufficient damping exists in the isolator, the standing-wave

Fig. 30.51. Transmissibility curves for the system illustrated in Fig. 30.49 using various rubber compounds, as obtained experimentally to show the effect of isolator damping. The mass m_E of the equipment is 20 times as great as the mass m_I of the isolator. (*After A. O. Sykes.*[28])

effects of the isolator cause the isolation efficiency at high frequency to decrease without exhibiting any obvious resonant peaks.

The standing-wave frequencies do not occur exactly at the frequencies predicted by the analysis because the dynamic stiffness of a rubber isolator is greater than the static stiffness used to calculate the standing-wave frequencies. Thus, the experimental transmissibility curves shift to the right on the dimensionless frequency plot of Fig. 30.51. The different rubber compounds exhibit different dynamic-to-static stiffness ratios; thus, the curves shift by different increments along the horizontal axis.

The standing-wave frequency f_w is given approximately by the relation

$$\frac{f_w}{f_0} = n\pi \sqrt{\frac{m_E}{m_I}} \qquad [n = 1, 2, 3, \ldots] \qquad (30.100)$$

where $f_0 = \dfrac{1}{2\pi} \sqrt{\dfrac{k_{st}}{m_E}}$ is the undamped natural frequency, n represents the mode of the standing wave, and m_E, m_I are the mass of the equipment and isolator, respectively. The relation given by Eq. (30.100) is shown graphically by Fig. 30.52 where the ratio of the standing-wave frequency f_w to the undamped natural frequency f_0 is given as a function of the mass ratio m_E/m_I for the first ten standing-wave modes. Standing-wave frequencies determined from this graph agree with the resonances in the transmissibility curves shown in Fig. 30.50.

FIG. 30.52. Standing-wave frequencies for the system illustrated in Fig. 30.49. The standing-wave frequency f_w, which is referenced to the fundamental natural frequency f_0 of the isolator system, is given as a function of the mass ratio m_E/m_I and the mode number n of the standing-wave vibration.

REFERENCES

1. Crede, C. E.: "Vibration and Shock Isolation," John Wiley & Sons, Inc., New York, 1951.
2. Den Hartog, J. P.: *Trans. ASME*, APM-53-9, 1932.
3. Jacobsen, L. S.: *Trans. ASME*, APM-52-15, 1931.
4. Ruzicka, J. E., and R. D. Cavanaugh: *Machine Design*, Oct. 16, 1958, p. 114.
5. Ruzicka, J. E.: Unpublished work.
6. Ruzicka, J. E.: "Forced Vibrations in Systems with Elastically Supported Dampers," Master's Thesis, Massachusetts Institute of Technology, Cambridge, Mass., 1957.
7. Crede, C. E., and J. P. Walsh: *J. Appl. Mechanics*, 14:1A-7 (1947).
8. Lewis, R. C., and K. Unholtz: *Trans. ASME*, **69**:8 (1947).
9. Macduff, J. N.: *Prod. Eng.*, July, August, 1946.

10. de Gruben, K.: *VDI Zeitschrift*, **6**:41–42 (1942).
11. Crede, C. E.: *J. Appl. Mechanics*, **25**:541 (1958).
12. Timoshenko, S., and G. H. MacCullough: "Elements of Strength of Materials," 3d ed., p. 64, D. Van Nostrand Company, Inc., Princeton, N.J., 1949.
13. Mindlin, R. D.: *Bell System Tech. J.*, **24**(3–4):353 (1945).
14. Crede, C. E.: *Trans. ASME*, **76**(1):117 (1954).
15. Ruzicka, J. E.: *J. Eng. Industry (Trans. ASME)*, **83B**(1):53 (1961).
16. Molloy, C. T.: *J. Acoust. Soc. Amer.*, **29**:842 (1957).
17. Ruzicka, J. E.: Unpublished work.
18. Leedy, H. A.: *J. Acoust. Soc. Amer.*, **11**:341 (1940).
19. Sykes, A. O.: "Shock and Vibration Instrumentation," *ASME*, 1956, p. 1.
20. Ruzicka, J. E., and R. D. Cavanaugh: "Mechanical Impedance Methods for Mechanical Vibrations," *ASME*, 1958, p. 109.
21. Ruzicka, J. E.: Paper No. 100Y, SAE National Aeronautic Meeting, October, 1959.
22. Ruzicka, J. E.: *J. Eng. Industry (Trans. ASME)*, **83B**(4):403 (1961).
23. Sykes, A. O.: *Trans. SAE*, **66**:533 (1958).
24. Rayleigh, Lord: "The Theory of Sound," 2d ed., part II, p. 315, Dover Publications, New York, 1945.
25. Nolle, A. W.: *J. Acoust. Soc. Amer.*, **19**:194 (1947).
26. Harrison, M., A. O. Sykes, and M. Martin: "Wave Effects in Isolation Mounts," *David W. Taylor Model Basin Rept.* 766, 1952.
27. Ruzicka, J. E.: Unpublished work.
28. Sykes, A. O.: "A Study of Compression Noise Isolation Mounts Constructed from Cylindrical Samples of Various Natural and Synthetic Rubber Materials," *David W. Taylor Model Basin Rept.* 845, 1953.

31

THEORY OF SHOCK ISOLATION

R. E. Newton
Naval Postgraduate School

INTRODUCTION

This chapter presents an analytical treatment of the isolation of shock. Two classes of shock are considered: (1) shock characterized by motion of a support or foundation where a shock isolator reduces the severity of the shock experienced by equipment mounted on the support and (2) shock characterized by forces applied to or originating within a machine where a shock isolator reduces the severity of shock experienced by the support. In the simplified concept of shock isolation, the equipment and support are considered rigid bodies, and the effectiveness of the isolator is measured by the forces transmitted through the isolator (resulting in acceleration of equipment if assumed rigid) and by the deflection of the isolator. Linear isolators, both damped and undamped, together with isolators having special types of nonlinear elasticity are considered. When the equipment or floor is not rigid, the deflection of nonrigid members is significant in evaluating the effectiveness of isolators. Analyses of shock isolation are included which consider the response of nonrigid components of equipment and floor.

IDEALIZATION OF THE SYSTEM

In the application of shock isolators to actual equipments, the locations of the isolators are determined largely by practical mechanical considerations. In general, this results in types of nonsymmetry and coupled modes not well adapted to analysis by simple means. It is convenient in the design of shock isolators to idealize the system to a hypothetical one having symmetry and uncoupled modes of motion.

UNCOUPLED MOTIONS. The first step in idealizing the physical system is to separate the various translational and rotational modes, i.e., to *uncouple* the system. Consider the system of Fig. 31.1 consisting of a *homogeneous* block attached at the corners, by eight identical springs, to a movable rigid frame. The block and frame are constrained to move in the plane of the paper. With the system at rest, the frame is given a sudden vertical translation. Because of the symmetry of both mass and stiffness relative to a vertical plane perpendicular to the paper, the response motion of the block is pure vertical translation. Similarly, a sudden horizontal translation of the frame excites

Fig. 31.1. Schematic diagram of three degree-of-freedom mounting. Block and frame are constrained to move in plane of paper.

pure horizontal translation of the block. A sudden rotation about an axis through the geometric center of the block produces pure rotation of the block about this axis. This set of response behaviors is characteristic of an *uncoupled* system.

If the block of Fig. 31.1 is not homogeneous, the mass center (or center-of-gravity) may be at A or B instead of C. Consider the response to a sudden vertical translation of the frame if the mass center is at A. If the response were pure vertical translation of the block, the dynamic forces induced in the vertical springs would have a resultant acting vertically through C. However, the "inertia force" of the block must act through the mass center at A. Thus, the response cannot be pure vertical translation, but must also include rotation. Then the motions of vertical translation and rotation are said to be *coupled*. A sudden horizontal translation of the frame would still excite only a horizontal translation of the block because A is symmetrical with respect to the horizontal springs; thus this horizontal motion remains *uncoupled*. If the mass center were at B, i.e., in neither the vertical nor the horizontal plane of symmetry, then a sudden vertical translation of the frame would excite both vertical and horizontal translations of the block, together with rotation. In this case, all three motions are said to be *coupled*.

It is not essential that a system have any kind of geometric symmetry in order that its motions be uncoupled but rather that the resultant of the spring forces be either a force directed through the center-of-gravity of the block or a couple. If the motions are completely uncoupled, there are three mutually orthogonal directions such that translational motion of the base in any one of these directions excites only a translation of the body in the same direction. Similarly there are three orthogonal axes, concurrent at the mass center, having the property that a pure rotation of the base about any one of these axes will excite a pure rotation of the body about the same axis. The idealized systems considered in this chapter are assumed to have uncoupled rigid body motions.

ANALOGY BETWEEN TRANSLATION AND ROTATION. If the motions in translational and rotational modes are uncoupled, motion in the rotational mode may be inferred by analogy from motion in the translational mode, and vice versa. Consider the system of Fig. 31.1 and assume that the mass center is at C. For horizontal motion the differential equation of motion is *

$$m\ddot{\delta} + 4k\delta = -m\ddot{u} \qquad (31.1)$$

where δ = horizontal displacement of mass center of block relative to center-of-frame, in.
    m = mass of block, lb-sec^2/in.
    k = spring stiffness for each spring, lb/in.
    u = absolute horizontal displacement of center-of-frame, in.†
Equation (31.1) may be written

$$\ddot{\delta} + \omega_n{}^2\delta = -\ddot{u} \qquad (31.2)$$

where $\omega_n = \sqrt{4k/m}$, rad/sec, is the angular natural frequency in horizontal vibration. For rotation of the block the corresponding equation of motion is

$$I\ddot{\gamma}_r + 4k(a^2 + b^2)\gamma_r = -I\ddot{\Gamma} \qquad (31.3)$$

where I = mass moment of inertia of block about axis through C, perpendicular to plane of paper, lb-in.-sec^2
    a, b = distances of spring center lines from mass center (see Fig. 31.1), in.
    γ_r = rotation of block relative to frame in plane of paper, rad
    Γ = absolute rotation of frame in plane of figure, rad

Equation (31.3) may be written

$$\ddot{\gamma}_r + \omega_{n1}{}^2\gamma_r = -\ddot{\Gamma} \qquad (31.4)$$

where $\omega_{n1} = \sqrt{4k(a^2 + b^2)/I}$ is the angular natural frequency in rotation.

---

* It is assumed that forces in the four vertical springs have a negligible horizontal component at all times.

† In the equilibrium position the point C lies at the frame center.

Equations (31.2) and (31.4) are analogous; γ_r corresponds to δ, Γ corresponds to u, and ω_{n1} corresponds to ω_n. Because of this analogy, only the horizontal motion described by Eq. (31.2) is considered in subsequent sections; corresponding results for rotational motion may be determined by analogy.

CLASSIFICATION OF SHOCK ISOLATION PROBLEMS

It is convenient to divide shock isolation problems into two major classifications according to the physical conditions:

Class I. Mitigation of effects of foundation motion
Class II. Mitigation of effects of force generated by equipment

Isolators in the first class include such items as the draft gear on a railroad car, the shock strut of an aircraft landing gear, the mounts on air-borne electronic equipment, and the corrugated paper used to package light bulbs. The second class includes the recoil cylinders on gun mounts and the isolators on drop hammers, looms, and reciprocating presses.

(A)   (B)

Fig. 31.2. Idealized systems showing use of isolator with transmitted force $F(\dot{\delta},\delta)$ to protect equipment of mass m from effects of support motion u. In (A) the equipment is rigid and in (B) there is a flexible component having stiffness-damping characteristics $F_c(\dot{\delta}_c, \delta_c)$ and mass m_c.

The objectives in the two classes of problems are allied, but distinct. In Class I the objective is to limit the shock-induced stresses in critical components of the protected equipment. In Class II the purpose is to limit the forces transmitted to the support for the equipment in which the shock originates.

IDEALIZED SYSTEMS—CLASS I. The simplest approach to problems of Class I is through a study of single degree-of-freedom systems.[1] Consider the system of Fig. 31.2A. The basic elements are a mass and a spring-dashpot unit attached to the mass at one end. The block may be taken to represent the equipment,[*] and the spring-dashpot unit to represent the shock isolator. The displacement of the support is u. The equation of motion is

$$m\ddot{\delta} + F(\dot{\delta},\delta) = -m\ddot{u} \qquad (31.5)$$

where m = mass of block, lb.-sec^2/in.
δ = deflection of spring ($\delta = x - u$; see Fig. 31.2), in.
$F(\dot{\delta},\delta)$ = force exerted on mass by spring-dashpot unit (positive when tensile), lb
u = absolute displacement of left-hand end of spring-dashpot unit, in.

In the typical shock isolation problem, the system of Fig. 31.2A is initially at rest ($\dot{u} = \dot{\delta} = 0$) in an equilibrium position ($u = \delta = 0$). An external shock causes the support to move. The corresponding movement of the left end of the shock isolator is described in terms of the support acceleration $\ddot{u}$. Then Eq. (31.5) may be solved for the resulting extreme values of δ and $F(\dot{\delta},\delta)$, and these values may be compared with the permissible deflection and force transmission limits of the shock isolator. It also is necessary to determine whether the internal stresses developed in the equipment are excessive. If the equipment is sufficiently rigid that all parts have substantially equal accelerations, then the internal stresses are proportional to $\ddot{x}$ where $-m\ddot{x} = F(\dot{\delta},\delta)$.

[*] In this simplified and idealized analysis, the equipment is considered a rigid body. The effect of flexibility of elements comprising the equipment is considered in a later section.

A critical component of the equipment may be sufficiently flexible to have a substantially different acceleration than that determined by assuming the equipment rigid. If the total mass of such components is small in comparison with the equipment mass, the above analysis may be extended to cover this case. Equation (31.5) is first solved to determine not merely the extreme value of $F(\dot{\delta},\delta)$ but its time-history. Then the acceleration $\ddot{x}$ may be determined from the relation $\ddot{x} = -F(\dot{\delta},\delta)/m$. Now consider the system shown in Fig. 31.2B having a component of mass m_c and stiffness-damping characteristics $F_c(\dot{\delta}_c,\delta_c)$. The force $F_c(\dot{\delta}_c,\delta_c)$ transmitted to the mass m_c and the resulting acceleration $\ddot{x}_c = -F_c(\dot{\delta}_c,\delta_c)/m_c$ may be found by solving an equation that is analogous to Eq. (31.5) where $\ddot{x}$ is substituted for $\ddot{u}$, $\dot{\delta}_c$ for $\dot{\delta}$, and $F_c(\dot{\delta}_c,\delta_c)$ for $F(\dot{\delta},\delta)$.

(A)   (1)   (2)
(B)

FIG. 31.3. Idealized systems showing use of isolator with transmitted force $F(\dot{\delta},\delta)$ to reduce force transmitted to foundation when force F is applied to equipment of mass m. In (A) the foundation is rigid and in (B) it has mass m_F and stiffness damping characteristics $F_F(\dot{\delta}_F,\delta_F)$.

IDEALIZED SYSTEMS—CLASS II. Consider the system of Fig. 31.3A to represent the equipment (mass m) attached to its support by the shock isolator (spring-dashpot unit). The left end of the spring-dashpot unit is fixed to the supporting structure and there is a force F applied externally to the mass. The force F may be a real external force or it may be an "inertia force" generated by moving parts of the equipment. The equation of motion may be written

$$m\ddot{\delta} + F(\dot{\delta},\delta) = F \tag{31.6}$$

where F is the external force applied to the mass in pounds and the relative displacement δ of the ends of the spring-dashpot unit is equal to the absolute displacement x of the mass. Assuming the system to be initially in equilibrium ($\dot{\delta} = 0$, $\delta = 0$), Eq. (31.6) is solved for extreme values of δ and $F(\dot{\delta},\delta)$ since F is a known function of time. These are to be com-

SUB-SYSTEM 1   SUB-SYSTEM 2

FIG. 31.4. General two degree-of-freedom system. Figures 31.2B and 31.3B represent particular cases of the general system.

pared with the displacement and force limitations of the shock isolator. Often the supporting structure is sufficiently rigid that the maximum force in the isolator may be considered as a force applied statically to the support. Then the foregoing analysis is adequate for determining the stress in the support.

The load on the floor may be treated as dynamic instead of static by a simple analysis if the displacement and velocity of the support are negligible in comparison with those of the equipment. Consider the system of Fig. 31.3B where the supporting structure is represented as a mass m_F and a spring-dashpot unit in place of the rigid support shown in

Fig. 31.3A. The force acting on the supporting structure is a known function of time $F(\dot{\delta},\delta)$ as found from the previous solution of Eq. (31.6). To find the maximum force *within* the support structure requires a solution of an equation analogous to Eq. (31.6) where $\ddot{\delta}_F$ is substituted for $\ddot{\delta}$, m_F for m, $F_F(\dot{\delta}_F,\delta_F)$ for $F(\dot{\delta},\delta)$ and $F(\dot{\delta},\delta)$ for F. For engineering purposes it suffices to find the extreme values of δ_F and $F_F(\dot{\delta}_F,\delta_F)$. The first is needed to verify the assumption that support motion is negligible compared with equipment motion, and can be used to determine the maximum stress in the support. The second is the maximum force applied by the support structure to its base.

MATHEMATICAL EQUIVALENCE OF CLASS I AND CLASS II PROBLEMS. The similarity of shock isolation principles in Class I and Class II is indicated by the similar form of Eqs. (31.5) and (31.6). The right-hand side $(-m\ddot{u}$ or $F)$ is given as a function of time, and the extreme values of δ and $F(\dot{\delta},\delta)$ are desired. When the actual system is represented by two separate single degree-of-freedom systems as shown in Figs. 31.2B and 31.3B, the time-history of $F(\dot{\delta},\delta)$ is also required. Figure 31.4 may be considered a generalized form of the applicable system. In Class I, $F = 0$, $F_1(\dot{\delta}_1,\delta_1)$ represents the properties of the isolator and $m_2,F_2(\dot{\delta}_2,\delta_2)$ represents the component to be protected. In Class II, $u = 0$, $F_2(\dot{\delta}_2,\delta_2)$ represents the properties of the isolator, and $m_1,F_1(\dot{\delta}_1,\delta_1)$ represents the supporting structure.

The system of Fig. 31.4, with the spring-dashpot units nonlinear, requires the use of a digital or analog computer to investigate performance characteristics. Analytical methods are feasible if the system is linearized by assuming that each spring-dashpot unit has a force characteristic in the form

$$F(\dot{\delta},\delta) = c\dot{\delta} + k\delta \qquad (31.7)$$

where c = damping coefficient, lb-sec/in., and k = spring stiffness, lb/in. Even with this simplification, the number of parameters $(m_1,c_1,k_1,m_2,c_2,k_2)$ is so great that it is necessary to confine the analysis to a particular system. If the damping may be neglected [let $c = 0$ in Eq. (31.7)], then it is feasible to obtain equations in a form suitable for routine use.[2] Use of this idealization is described in the section on *Response of Equipment with a Flexible Component.*

A different form of idealization is indicated when the "equipment" is flexible; e.g., a large, relatively flexible aircraft subjected to landing shock. Then it is important to represent the aircraft as a system with several degrees of freedom. To find resulting stresses, it is necessary to superimpose the responses in the various modes of motion that are excited.

RESPONSE OF A RIGID BODY SYSTEM TO A VELOCITY STEP

PHYSICAL BASIS FOR VELOCITY STEP

The idealization of a shock motion as a simple change in velocity (velocity step) may form an adequate basis for designing a shock isolator and for evaluating its effectiveness. Consider the two types of acceleration $\ddot{u}$ vs. time t curves illustrated in Fig. 31.5A. The solid line represents a rectangular

Fig. 31.5. Acceleration-time curves (A) and velocity-time curves (B) and (C) for rectangular acceleration pulse (solid curves) and half-sine acceleration pulse (dashed curves). In (C) the time scale is compressed to one-tenth that used in (A) and (B).

pulse of acceleration and the dashed line represents a half-sine pulse of acceleration. Each pulse has a duration τ. In Fig. 31.5B, the corresponding velocity-time curves are shown. Each of these curves is defined completely by specifying the type of acceleration pulse (rectangular or half-sine), the duration τ, and the velocity change $\dot{u}_m$. The curves of Fig. 31.5B are repeated in Fig. 31.5C with the time scale shrunk to one-tenth. If τ is *sufficiently short*, the only significant remaining characteristic of the velocity step is the velocity change $\dot{u}_m$. The idealized velocity step, then, is taken to be a discontinuous change of $\dot{u}$ from zero to $\dot{u}_m$. A shock isolator characteristically has a low natural frequency (long period), and this idealization leads to good results even when the pulse duration τ is significantly long.

FIG. 31.6. Idealized system showing use of undamped isolator to protect equipment from effects of support motion u. The force transmitted by the isolator is $F_s(\delta)$.

GENERAL FORM OF ISOLATOR CHARACTERISTICS

The differential equation of motion for the undamped, single degree-of-freedom system shown in Fig. 31.6 is

$$m\ddot{\delta} + F_s(\delta) = -m\ddot{u} \qquad (31.8)$$

where m represents the mass of the equipment considered as a rigid body, u represents the motion of the support which characterizes the condition of shock, and $F_s(\delta)$ is the force developed by the isolator at an extension δ (positive when tensile). Equation (31.8) differs from Eq. (31.5) in that $F_s(\delta)$, which does not depend upon $\dot{\delta}$, replaces $F(\dot{\delta},\delta)$ because the isolator is undamped. The effect of a velocity step of magnitude $\dot{u}_m$ at $t = 0$ is considered by choosing the initial conditions: * At $t = 0$, $\delta = 0$ and $\dot{\delta} = \dot{u}_m$. A first integration of Eq. (31.8) yields

$$\dot{\delta}^2 = \dot{u}_m{}^2 - \frac{2}{m}\int_0^\delta F_s(\delta)\, d\delta \qquad (31.9)$$

At the extreme value of isolator deflection, $\delta = \delta_m$ and the velocity $\dot{\delta}$ of deflection is zero. Then from Eq. (31.9),

$$\int_0^{\delta_m} F_s(\delta)\, d\delta = \tfrac{1}{2}m\dot{u}_m{}^2 \qquad (31.10)$$

The right side of Eq. (31.10) represents the initial kinetic energy of the equipment relative to the support, and the integral on the left side represents the work done on the isolator. The latter quantity is equal to the elastic potential energy stored in the isolator, since there is no damping.

For the special case of a rigid body mounted on an undamped isolator, Eq. (31.10) suffices to determine all important results. In particular, the quantities of engineering significance are:

1. The maximum deflection of the isolator δ_m
2. The maximum isolator force, $F_m = F_s(\delta_m) = m\ddot{x}_m$ †
3. The corresponding velocity change $\dot{u}_m$

The interrelations of these three quantities are shown graphically in Fig. 31.7. The curve OAB represents the spring force $F_s(\delta)$ as a function of deflection δ. If point A corresponds to the extreme excursion, then its abscissa represents the maximum deflection δ_m.

FIG. 31.7. Typical force-deflection curve for undamped isolator. Shaded area represents kinetic energy of equipment following a velocity step $\dot{u}_m$ [Eq. (31.10)].

---

* These conditions correspond to a negative velocity step. This choice is made to avoid dealing with negative values of δ and $\dot{\delta}$. If $F_s(\delta)$ is not an odd function of δ, a positive velocity step requires a separate analysis.

† The maximum absolute acceleration $\ddot{x}_m$ of the equipment is related to the maximum force F_m by $F_m = m\ddot{x}_m$. It sometimes is convenient to express results in terms of $\ddot{x}_m$ instead of F_m.

The shaded area OAC is proportional to the potential energy stored by the isolator; according to Eq. (31.10), this is equal to the initial kinetic energy $m\dot{u}_m{}^2/2$. The maximum ordinate (at A) represents the maximum spring force F_m. [It is possible to have a spring force $F_s(\delta)$ which attains a maximum value at $\delta = \delta_f < \delta_m$. Then $F_m = F_s(\delta_f)$.]

The design requirements for the isolator usually include as a specification one or more of the following quantities:

1. Maximum allowable deflection δ_a
2. Maximum allowable transmitted force F_a
3. Maximum expected velocity step $\dot{u}_a$

It is important to observe that the limits 1 and 2 establish an upper limit $F_a\delta_a$ on the work done on the mass. It follows that $\dot{u}_a$ must satisfy the relation

$$F_a\delta_a \geq m\dot{u}_a{}^2/2$$

or the specifications are impossible to meet. The specifications may be expressed mathematically as follows:

$$\delta_m \leq \delta_a \qquad F_m \leq F_a \qquad \dot{u}_m \geq \dot{u}_a \qquad (31.11)$$

In many instances it is advantageous to eliminate explicit reference to the mass m. Then the allowable absolute acceleration $\ddot{x}_a$ of the mass is specified instead of the allowable force F_a where $F_a = m\ddot{x}_a$. With this substitution the second of Eqs. (31.11) is replaced by

$$\ddot{x}_m \leq \ddot{x}_a \qquad (31.12)$$

The acceleration $\ddot{x}$ is determined as a function of time by using $\dot{\delta}$ from Eq. (31.9) and finding the time t corresponding to a given value of δ:

$$t = \int_0^\delta \frac{d\delta}{\dot{\delta}} \qquad (31.13)$$

From Eq. (31.13) and the relation $\ddot{x} = F_s(\delta)/m$, the acceleration time-history is found.

The integrations required by Eqs. (31.9) and (31.13) sometimes are difficult to perform, and it is necessary to use numerical or graphical methods. Then a difficulty arises with the integral in Eq. (31.13): As δ approaches the extreme value δ_m, the velocity $\dot{\delta}$ in the denominator of the integrand approaches zero. The difficulty is circumvented by first using Eq. (31.13) to integrate up to some intermediate displacement δ_b less than δ_m; then the alternate form, Eq. (31.14), may be used in the region of $\delta = \delta_m$:

$$t = t_b + \int_{\dot{\delta}_b}^{\dot{\delta}} \frac{d\dot{\delta}}{\ddot{\delta}} \qquad (31.14)$$

where t_b is the time at which $\delta = \delta_b$, as determined from Eq. (31.13).

In the next three sections three different kinds of spring force-deflection characteristics $F_s(\delta)$ are considered. Equation (31.10) is applied to find the relation between $\dot{u}_m$ and δ_m. Curves relating $\dot{u}_m$, δ_m, and $\ddot{x}_m$ in a form useful for design or analysis are presented.

EXAMPLES OF PARTICULAR ISOLATOR CHARACTERISTICS

LINEAR SPRING. The force-deflection characteristic of a linear spring is

$$F_s(\delta) = k\delta \qquad (31.15)$$

where k = spring stiffness, lb/in. Using the notation

$$\omega_n = \sqrt{\frac{k}{m}} \qquad \text{rad/sec} \qquad (31.16)$$

the maximum acceleration is

$$\ddot{x}_m = \omega_n{}^2\delta_m \qquad (31.17)$$

SPRING FORCE, $F_s(\delta)$

k

1

0 d

DEFLECTION, δ

FIG. 31.8. Typical force-deflection curve for hardening spring (tangent elasticity) as given by Eq. (31.20). There is a vertical asymptote at the limiting deflection $\delta = d$.

From Eqs. (31.10) and (31.16), the relation between velocity change $\dot{u}_m$ and maximum deflection δ_m is

$$\dot{u}_m = \omega_n \delta_m \qquad (31.18)$$

Combining Eqs. (31.18) and (31.17),

$$\ddot{x}_m = \omega_n \dot{u}_m \qquad (31.19)$$

HARDENING SPRING (TANGENT ELASTICITY). The isolator spring may be nonlinear with a "hardening" characteristic; i.e., the slope of the curve representing spring force vs. deflection increases with increasing deflection. Rubber in compression has this behavior. A representative curve [1] having this characteristic is defined by

$$F_s(\delta) = \frac{2kd}{\pi} \tan \frac{\pi \delta}{2d} \qquad (31.20)$$

where the constant k is the *initial* slope of the curve (lb/in.) and a vertical asymptote is defined by $\delta = d$ (in.). Such a curve is shown graphically in Fig. 31.8. Using the notation of Eq. (31.16) and the relation $m\ddot{x}_m = F_s(\delta_m)$, Eq. (31.20) gives the following relation between maximum acceleration and maximum deflection:

$$\frac{\ddot{x}_m}{\omega_n{}^2 d} = \frac{2}{\pi} \tan \frac{\pi \delta_m}{2d} \qquad (31.21)$$

Note that ω_n, the angular natural frequency for a linear system, has the same meaning for small amplitude (small δ_m) motions of the nonlinear system. For large amplitudes the natural frequency depends on δ_m. Using Eq. (31.16), substituting for $F_s(\delta)$ from Eq. (31.20) in Eq. (31.10), and performing the indicated integration, the relation between velocity change and maximum displacement is

$$\frac{\dot{u}_m{}^2}{\omega_n{}^2 d^2} = \frac{8}{\pi^2} \log_e \left(\sec \frac{\pi \delta_m}{2d} \right) \qquad (31.22)$$

A graphical presentation relating the important variables $\dot{u}_m$, $\ddot{x}_m$, and δ_m is convenient for design and analysis. Such data are presented compactly as relations among the dimensionless parameters δ_m/d, $\dot{u}_m/\omega_n d$ and $\ddot{x}_m \delta_m/\dot{u}_m{}^2$. The physical significance of the ratio $\ddot{x}_m \delta_m/\dot{u}_m{}^2$ is interpreted by multiplying both numerator and denominator by m. Then the numerator represents the product of the maximum spring force $F_m (= m\ddot{x}_m)$ and the maximum spring deflection δ_m. This product is the maximum energy that *could* be stored in the spring. The denominator $m\dot{u}_m{}^2$ is *twice* the energy that *is* stored in the spring. The minimum *possible* value of the ratio $\ddot{x}_m \delta_m/\dot{u}_m{}^2$ is 1/2. Actual values of the ratio, always greater than 1/2, may be considered to be a measure of the departure from optimum capability.

In Fig. 31.9 the solid curve represents $\dot{u}_m/\omega_n d$ as a function of δ_m/d and the dashed curve shows the corresponding result for a linear spring [see Eq. (31.18)]. In Fig. 31.10 the solid curve shows $\ddot{x}_m \delta_m/\dot{u}_m{}^2$ vs. δ_m/d for an isolator with tangent elasticity. The dashed curve in Fig. 31.10 shows $\ddot{x}_m \delta_m/\dot{u}_m{}^2$ for a linear spring [see Eqs. (31.17) and (31.18)]; the ratio is constant at a value of unity because a linear spring is 50 per cent efficient in storage of energy, independent of the deflection.

SOFTENING SPRING (HYPERBOLIC TANGENT ELASTICITY). A nonlinear isolator also may have a "softening" characteristic; i.e., the slope of the curve represent-

FIG. 31.9. Dimensionless representation of relation between velocity step $\dot{u}_m$ and maximum isolator deflection δ_m for undamped isolators having tangent elasticity (solid curve) and linear elasticity (dashed curve). The natural angular frequency for small oscillations is ω_n and the limiting deflection is d (Fig. 31.8).

FIG. 31.10. Dimensionless representation of relation among velocity step $\dot{u}_m$, maximum transmitted acceleration $\ddot{x}_m$, and maximum isolator deflection δ_m for undamped isolators having tangent elasticity (solid curve) and linear elasticity (dashed curve). Ordinate is an inverse measure of energy-storage efficiency. The natural angular frequency for small oscillations is ω_n and the limiting deflection is d (Fig. 31.8).

ing force vs. deflection decreases with increasing deflection. The force-deflection characteristic for a typical "softening" isolator is [1]

$$F_s(\delta) = kd_1 \tanh \frac{\delta}{d_1} \tag{31.23}$$

where k is the initial slope of the curve. Figure 31.11 shows the form of this curve where the meaning of d_1 is evident from the figure. If $F_s(\delta)$ is replaced by $m\ddot{x}_m$, δ by δ_m, and k by $m\omega_n^2$, Eq. (31.23) becomes

$$\frac{\ddot{x}_m}{\omega_n^2 d_1} = \tanh \frac{\delta_m}{d_1} \tag{31.24}$$

where δ_m and $\ddot{x}_m$ are maximum values of deflection and acceleration, respectively, and ω_n may be interpreted as the angular natural frequency for small values of δ_m. To relate $\dot{u}_m$ to δ_m, substitute $F_s(\delta)$ from Eq. (31.23) in Eq. (31.10), let $\omega_n^2 = k/m$, and integrate:

$$\frac{\dot{u}_m^2}{\omega_n^2 d_1^2} = \log_e \left(\cosh^2 \frac{\delta_m}{d_1} \right) \tag{31.25}$$

A graphical presentation of the relation between $\dot{u}_m/\omega_n d_1$ and δ_m/d_1 is given by the solid curve of Fig. 31.12. The dashed curve shows the corresponding relation for a linear

FIG. 31.11. Typical force-deflection curve for softening spring (hyperbolic tangent elasticity) as given by Eq. (31.23). There is a horizontal asymptote at the limiting force kd_1.

spring. In Fig. 31.13 the solid curve represents $\ddot{x}_m \delta_m / \dot{u}_m^2$ as a function of δ_m/d_1. Note that, for large values of δ_m/d_1, the ordinate approaches the minimum value $1/2$ attainable with an isolator of optimum energy storage efficiency. The dashed curve shows the same relation for a linear spring.

LINEAR SPRING AND VISCOUS DAMPING. The addition of viscous damping can almost double the energy absorption capability of a linear shock isolator. Consider the system of Fig. 31.2A, with both spring and dashpot linear as defined by Eq. (31.7). Substituting $F(\delta,\dot\delta)$ from Eq. (31.7) in Eq. (31.5) gives the equation of motion. The initial conditions are $\dot\delta = \dot{u}_m$, $\delta = 0$, when $t = 0$; for $t > 0$, $\ddot{u} = 0$. Letting $c_c = 2m\omega_n$ and $\zeta = c/c_c$ [see Eq. (2.12)], the equation of motion becomes

$$\ddot\delta + 2\zeta\omega_n\dot\delta + \omega_n^2\delta = 0 \qquad (31.26)$$

Solutions of Eq. (31.26) for maximum deflection δ_m and maximum acceleration $\ddot{x}_m$ as

FIG. 31.12. Dimensionless representation of relation between velocity step $\dot{u}_m$ and maximum isolator deflection δ_m for undamped isolators having hyperbolic tangent elasticity (solid curve) and linear elasticity (dashed curve). The natural angular frequency for small oscillations is ω_n and the characteristic deflection d_1 is defined in Fig. 31.11.

FIG. 31.13. Dimensionless representation of energy-storage capabilities of undamped isolators having hyperbolic tangent elasticity (solid curve) and linear elasticity (dashed curve). The ordinate is an inverse measure of energy-storage efficiency. The characteristic deflection d_1 is defined in Fig. 31.11.

functions of ζ are available [1] in analytical form; the solutions are shown graphically in Figs. 31.14 and 31.15. In Fig. 31.14, the dimensionless ratio $\ddot{x}_m / \dot{u}_m\omega_n$ is plotted as a function of the fraction of critical damping ζ. Note that the presence of small damping reduces the maximum acceleration. As ζ is increased beyond 0.25, the maximum acceleration increases again. For $\zeta > 0.50$, the maximum acceleration occurs at $t = 0$ and exceeds that for no damping; it is accounted for solely by the damping force $c\dot\delta = c\dot{u}_m$.

In Fig. 31.15 the parameter $\ddot{x}_m\delta_m / \dot{u}_m^2$ is plotted as a function of ζ. (As pointed out with reference to Fig. 31.10, $\ddot{x}_m\delta_m / \dot{u}_m^2$ is an inverse measure of shock isolator effectiveness.) Figure 31.15 shows that the presence of damping improves the energy storage effectiveness of the isolator even beyond $\zeta = 0.50$. In the neighborhood $\zeta = 0.40$, the parameter $\ddot{x}_m\delta_m / \dot{u}_m^2$ attains a minimum value of 0.52—only slightly above the theoretical minimum of 0.50. This parameter has the value 1.00 for an undamped linear

FIG. 31.14. Dimensionless representation of maximum transmitted acceleration $\ddot{x}_m$ for an isolator having a linear spring and viscous damping.

FIG. 31.15. Dimensionless representation of energy absorption capability of an isolator having a linear spring and viscous damping. Ordinate is an inverse measure of energy absorption capability.

system, and even higher values for a hardening spring (see Fig. 31.10). On the other hand, $\ddot{x}_m \delta_m / \dot{u}_m{}^2$ may approach 0.50 when a softening spring is used.

True viscous damping of the type considered above is difficult to attain except in electrical or magnetic form. Fluid dampers which depend upon orifices or other constricted passages to throttle the flow are likely to produce damping forces that vary more nearly as the square of the velocity. Dry friction tends to provide damping forces which are virtually independent of velocity. The analysis of response to a velocity step in the presence of Coulomb friction is similar to that described in the section entitled *General Formulas—No Damping*. A general analytic method for finding the behavior of systems with nonlinear springs in the presence of velocity-squared damping is also available.[3]

Example 31.1. Equipment weighing 40 lb and sufficiently stiff to be considered rigid is to be protected from a shock consisting of a velocity step $\dot{u}_a = 70$ in./sec. The maximum allowable acceleration is $\ddot{x}_a = 21\,g$ (g is the acceleration of gravity) and available clearance limits the deflection to $\delta_a = 0.70$ in. Find isolator characteristics for: linear spring, hardening spring, softening spring, and linear spring with viscous damping.

Linear Spring. Taking the maximum velocity $\dot{u}_m$ equal to the expected velocity $\dot{u}_a$ and using Eqs. (31.18) and (31.11),

$$\delta_m = \frac{\dot{u}_m}{\omega_n} \le \delta_a; \quad \text{or} \quad \omega_n \ge \frac{70 \text{ in./sec}}{0.70 \text{ in.}} = 100 \text{ rad/sec}$$

From Eqs. (31.19) and (31.12), $\ddot{x}_m = \omega_n \dot{u}_m \le \ddot{x}_a$. Then

$$\omega_n \le \frac{\ddot{x}_a}{\dot{u}_m} = \frac{21 \times 386 \text{ in./sec}^2}{70 \text{ in./sec}} = 116 \text{ rad/sec}$$

Selecting a value in the middle of the permissible range gives $\omega_n = 108$ rad/sec [17.2 cps]. The corresponding maximum isolator deflection is $\delta_m = 0.65$ in. and the maximum acceleration of the equipment is $\ddot{x}_m = 7{,}580$ in./sec^2 = 19.6g. The isolator stiffness given by Eq. (31.16) is

$$k = m\omega_n{}^2 = \frac{40 \text{ lb}}{386 \text{ in./sec}^2} \times (108 \text{ rad/sec})^2 = 1{,}210 \text{ lb/in.}$$

If, as is usually the case, the isolation is provided by several individual isolators in parallel, then the above value of k represents the sum of the stiffnesses of the individual isolators.

Hardening Spring. The tangent elasticity represented by Eq. (31.20) is assumed. Since the linear spring meets the specifications with only a small margin of safety, it is inferred that the poorer energy storage capacity of the hardening spring shown by Fig. 31.10 will severely limit the permissible nonlinearity. Using the specified values as maxima,

$$\frac{\ddot{x}_m \delta_m}{\dot{u}_m{}^2} = \frac{\ddot{x}_a \delta_a}{\dot{u}_a{}^2} = \frac{(21 \times 386) \times 0.70}{(70)^2} = 1.16$$

From Fig. 31.10:

$$\frac{\delta_m}{d} = 0.54; \quad \text{thus } d = \frac{0.70}{0.54} = 1.30 \text{ in.}$$

From Fig. 31.9:

$$\frac{\dot{u}_m}{\omega_n d} = 0.58; \quad \text{thus } \omega_n = \frac{70}{1.30 \times 0.54} = 93 \text{ rad/sec } [14.8 \text{ cps}]$$

The initial spring stiffness k from Eq. (31.16) is

$$k = \frac{40}{386}(93)^2 = 896 \text{ lb/in.}$$

Because the selected linear spring provides a small margin of safety and the hardening spring provides none, superficial comparison suggests that the former is superior. Various other considerations, such as compactness and stiffness along other axes, may offset the apparent advantage of the linear spring. Moreover, a shock more severe than that specified could cause the linear spring to bottom abruptly and cause much greater acceleration of the equipment.

Softening Spring. The hyperbolic tangent elasticity represented by Eq. (31.23) is assumed. The softening spring has high energy-storage capacity as shown by Fig. 31.13. By working to sufficiently high values of δ_m/d_1, it is possible to utilize this storage capacity to afford considerable overload capability. Choose $\ddot{x}_m = 20g$ and $\delta_m/d_1 = 3$. From Fig. 31.13, $\ddot{x}_m \delta_m / \dot{u}_m{}^2$ = 0.645 at $\delta_m/d_1 = 3$. Then

$$\delta_m = 0.645 \frac{(70)^2}{20 \times 386} = 0.41 \text{ in.} \qquad d_1 = \frac{\delta_m}{3} = 0.137 \text{ in.}$$

From Fig. 31.12, $\dot{u}_m/\omega_n d_1 = 2.15$ at $\delta_m/d_1 = 3$.

Then
$$\omega_n = \frac{70}{2.15 \times 0.137} = 238 \text{ rad/sec } [37.9 \text{ cps}]$$

The initial spring stiffness k from Eq. (31.16) is

$$k = \frac{40}{386}(238)^2 = 5{,}870 \text{ lb/in.}$$

This initial stiffness is much greater than those found for the linear spring and hardening spring. Accordingly, for small shocks (small $\dot{u}_m$) the isolator with the softening spring will induce much higher acceleration of the equipment than will those with linear or hardening springs. This poorer performance for small shocks is unavoidable if the isolator with the softening spring is designed to take advantage of the large energy-storage capability under extreme shocks.

Linear Spring and Viscous Damping. The introduction of viscous damping in combination with a linear spring [Eq. (31.7)] affords the possibility of large energy dissipation capacity without deterioration of performance under small shocks. From Fig. 31.15, the best performance is obtained at the fraction of critical damping $\zeta = 0.40$ where $\ddot{x}_m \delta_m / \dot{u}_m^2 = 0.52$. If the maximum isolator deflection is chosen as $\delta_m = 0.47$ in. (67 per cent of δ_a), then

$$\ddot{x}_m = 0.52 \frac{\dot{u}_m^2}{\delta_m} = 5{,}450 \text{ in./sec}^2 = 14.1g$$

(This acceleration is 67 per cent of $\ddot{x}_a$.)   From Fig. 31.14:

$$\frac{\ddot{x}_m}{\dot{u}_m \omega_n} = 0.86 \text{ at } \zeta = 0.40$$

Then
$$\omega_n = \frac{5{,}450}{0.86 \times 70} = 90 \text{ rad/sec } [14.3 \text{ cps}]$$

The spring stiffness k from Eq. (31.16) is

$$k = \frac{40}{386}(90)^2 = 840 \text{ lb/in.}$$

The dashpot constant c is

$$c = 2\zeta m \omega_n = 2 \times 0.40 \times \frac{40}{386} \times 90 = 7.46 \text{ lb-sec/in.}$$

Example 31.2. The procedure for the analysis of a given isolator subjected to a specified velocity step differs somewhat from that used in design. For the isolators designed in Example 31.2, it is specified that each is subjected to a velocity step $\dot{u}_m = 84$ in./sec (20 per cent greater than in Example 31.1). Determine the maximum isolator deflection δ_m and maximum equipment acceleration $\ddot{x}_m$.

Linear Spring

From Eq (31.18):
$$\delta_m = \frac{\dot{u}_m}{\omega_n} = \frac{84}{108} = 0.78 \text{ in.}$$

From Eq. (31.17):

$$\ddot{x}_m = \omega_n^2 \delta_m = (108)^2 \times 0.78 = 9{,}110 \text{ in./sec}^2 = 23.6g$$

The influence of the 20 per cent increase of velocity for this (linear) system is to increase both δ_m and $\ddot{x}_m$ by 20 per cent over the values from Example 31.1.

Hardening Spring. From Example 31.1, $d = 1.30$ in. and $\omega_n = 93$ rad/sec. Then

$$\frac{\dot{u}_m}{\omega_n d} = \frac{84}{93 \times 1.30} = 0.69$$

From Fig. 31.9, the corresponding value of δ_m/d is 0.63. Thus, $\delta_m = 0.63 \times 1.30 = 0.82$ in. From Fig. 31.10, at $\delta_m/d = 0.63$, $\ddot{x}_m \delta_m / \dot{u}_m^2 = 1.26$. Therefore

$$\ddot{x}_m = \frac{1.26 \times (84)^2}{0.82} = 10{,}850 \text{ in./sec}^2 = 28.1g$$

Compared with the results of Example 31.1, δ_m and $\ddot{x}_m$ are 17 per cent and 34 per cent greater, respectively.

Softening Spring. From the given data,

$$\frac{\dot{u}_m}{\omega_n d_1} = \frac{84}{238 \times 0.137} = 2.58$$

From Fig. 31.12 the corresponding value of δ_m/d_1 is 4.0; thus, $\delta_m = 4.0 \times 0.137 = 0.55$ in. From Fig. 31.13, at $\delta_m/d_1 = 4.0$, $\ddot{x}_m\delta_m/\dot{u}_m{}^2 = 0.603$; thus

$$\ddot{x}_m = \frac{0.603 \times (84)^2}{0.55} = 7,730 \text{ in./sec}^2 = 20g$$

Comparing results with those of Example 31.1, δ_m is increased by 34 per cent and $\ddot{x}_m$ is unchanged.

Linear Spring—Viscous Damping. For $\zeta = 0.4$, Fig. 31.14 gives $\ddot{x}_m/\dot{u}_m\omega_n = 0.86$; thus, $\ddot{x}_m = 0.86 \times 84 \times 90 = 6,500$ in./sec^2 = 16.9g. From Fig. 31.15, at $\zeta = 0.4$:

$$\frac{\ddot{x}_m\delta_m}{\dot{u}_m{}^2} = 0.522; \text{ thus, } \delta_m = \frac{0.522 \times (84)^2}{6,500} = 0.57 \text{ in.}$$

Again comparing with the results of Example 31.1, both δ_m and $\ddot{x}_m$ are increased by 20 per cent. This is the result of the linearity of the system.

Of the four isolator designs determined in Example 31.1, only the softening spring and the linear spring with viscous damping can meet the original deflection and acceleration limits (0.70 in. and 21g) when the velocity step is increased in magnitude by 20 per cent. This limitation is illustrated by using the specified acceleration (21g) and deflection (0.70 in.) limits, together with the increased velocity (84 in./sec), and calculating

$$\frac{\ddot{x}_m\delta_m}{\dot{u}_m{}^2} = \frac{(21 \times 386) \times 0.70}{(84)^2} = 0.804$$

According to Fig. 31.10, neither a linear spring nor a hardening spring can be found to meet a required value of $\ddot{x}_m\delta_m/\dot{u}_m{}^2$ less than unity. On the other hand, Fig. 31.13 shows that hyperbolic tangent springs operating beyond $\delta_m/d_1 = 1.4$ may be found to meet or exceed the requirements. Similarly, Fig. 31.15 shows that viscous damped linear systems with ζ greater than 0.08 can be designed to meet or exceed the requirements.

RESPONSE OF RIGID BODY SYSTEM TO ACCELERATION PULSE

The response of a spring-mounted rigid body to various acceleration pulses provides useful information. For example, it establishes limitations upon the use of the velocity step in place of an acceleration pulse and is significant in determining the response of an equipment component when the equipment support is subjected to a velocity step. Additional useful information is afforded by comparing the responses to acceleration pulses of different shapes.

For positive pulses ($\ddot{u} > 0$) having a single maximum value and finite duration, three basic characteristics of the pulse are of importance: maximum acceleration $\ddot{u}_m$, duration τ, and velocity change $\dot{u}_c$. A typical pulse is shown in Fig. 31.16. The relation among acceleration, duration and velocity change is

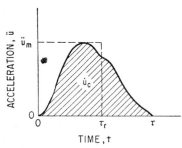

FIG. 31.16. Typical acceleration pulse with maximum acceleration $\ddot{u}_m$ and duration τ. The shaded area represents the resulting velocity change $\dot{u}_c$. The dashed rectangle represents an *equivalent rectangular pulse* of duration τ_r having the same velocity change $\dot{u}_c$.

$$\dot{u}_c = \int_0^\tau \ddot{u} \, dt \qquad (31.27)$$

where the value of the integral corresponds to the shaded area of the figure. The *equivalent rectangular pulse* is characterized by (a) the same maximum acceleration $\ddot{u}_m$ and (b) the same velocity change $\dot{u}_c$. In Fig. 31.16, the horizontal and vertical dashed lines outline the equivalent rectangular pulse corresponding to the shaded pulse. From condition (b) above and Eq. (31.27), the *effective duration* τ_r of the equivalent rectangular pulse is

$$\tau_r = \frac{1}{\ddot{u}_m} \int_0^\tau \ddot{u} \, dt \qquad (31.28)$$

where τ_r may be interpreted physically as the *average width* of the shaded pulse.

FIG. 31.17. Rectangular acceleration pulse.

RESPONSE TO A RECTANGULAR PULSE

The rectangular pulse shown in Fig. 31.17 has a maximum acceleration $\ddot{u}_m$ and duration τ; the velocity change is $\dot{u}_c = \ddot{u}_m \tau$. The response of an undamped, linear, single degree-of-freedom system (see Fig. 31.6) to this pulse is found from the differential equation obtained by substituting in Eq. (31.8) $F_s(\delta) = k\delta$ from Eq. (31.15) and $\omega_n^2 = k/m$ from Eq. (31.16):

$$\ddot{\delta} + \omega_n^2 \delta = -\ddot{u}_m \qquad [0 \leq t \leq \tau] \qquad (31.29)$$

$$\ddot{\delta} + \omega_n^2 \delta = 0 \qquad [t > \tau] \qquad (31.30)$$

Using the initial conditions $\dot{\delta} = 0$, $\delta = 0$ when $t = 0$, the solution of Eq. (31.29) is

$$\delta = \frac{\ddot{u}_m}{\omega_n^2} (\cos \omega_n t - 1) \qquad [0 \leq t \leq \tau] \qquad (31.31)$$

For the solution of Eq. (31.30), it is necessary to find as initial conditions the values of $\dot{\delta}$ and δ given by Eq. (31.31) for $t = \tau$. Using these values the solution of Eq. (31.30) is

$$\delta = \frac{\ddot{u}_m}{\omega_n^2} [(\cos \omega_n \tau - 1) \cos \omega_n(t - \tau) - \sin \omega_n \tau \sin \omega_n(t - \tau)] \qquad [t > \tau] \quad (31.32)$$

The motion defined by Eqs. (31.31) and (31.32) is shown graphically in Fig. 31.18 for $\tau = \pi/2\omega_n$, π/ω_n and $3\pi/2\omega_n$.

In the isolation of shock, the extreme absolute acceleration $\ddot{x}_m$ of the mass is important. Since $\ddot{x}_m = \omega_n^2 \delta_m$ [Eq. (31.17)], $\ddot{x}_m$ is found directly from the extreme value of δ. As indicated by Fig. 31.18, for values of τ greater than π/ω_n, the extreme (absolute) value of

FIG. 31.18. Response curves for an undamped linear system subjected to rectangular acceleration pulses of height $\ddot{u}_m$ and various durations τ. The angular natural frequency of the system is ω_n.

FIG. 31.19. Maximum acceleration spectrum for a linear system of angular natural frequency ω_n. Support motion is a rectangular acceleration pulse of height $\ddot{u}_m$.

δ encountered at $t = \pi/\omega_n$ is never exceeded. For values of τ less than π/ω_n, the extreme value occurs after the pulse has ended $(t > \tau)$ and is the amplitude of the motion represented by Eq. (31.32). This amplitude may be written

$$\delta_m = 2 \frac{\ddot{u}_m}{\omega_n{}^2} \sin \frac{\omega_n \tau}{2} \tag{31.33}$$

The extreme absolute values of the acceleration $\ddot{x}_m$ are plotted as a function of τ in Fig. 31.19. Note that the extreme value of acceleration is twice that of the acceleration of the rectangular pulse.

HALF-SINE PULSE

Consider the "half-sine" acceleration pulse (Fig. 31.20A) of amplitude $\ddot{u}_m$ and duration τ:

$$\ddot{u} = \ddot{u}_m \sin \frac{\pi t}{\tau} \qquad [0 \le t \le \tau]$$

$$\ddot{u} = 0 \qquad\qquad [t > \tau] \tag{31.34}$$

From Eq. (31.28), the effective duration is

$$\tau_r = \frac{2}{\pi} \tau \tag{31.35}$$

The response of a single degree-of-freedom system to the half-sine pulse of acceleration, corresponding to Eqs. (31.31) and (31.32) for the rectangular pulse, is defined by Eq. (8.32).

VERSED SINE PULSE

The versed sine pulse (Fig. 31.20B) is described by

$$\ddot{u} = \frac{\ddot{u}_m}{2} \left(1 - \cos \frac{2\pi t}{\tau} \right) = \ddot{u}_m \sin^2 \frac{\pi t}{\tau} \qquad [0 \le t \le \tau]$$

$$\ddot{u} = 0 \qquad\qquad\qquad [t > \tau] \tag{31.36}$$

 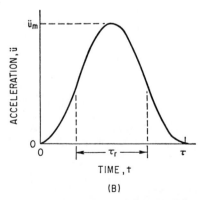

Fig. 31.20. Half-sine acceleration pulse (A) and versed sine acceleration pulse (B). The *effective duration* is τ_r [Eq. (31.28)].

The effective duration τ_r given by Eq. (31.28) is

$$\tau_r = (\tfrac{1}{2})\tau \tag{31.37}$$

The response of a single degree-of-freedom system to a versed sine pulse is defined by Eq. (8.33). The responses to a number of other types of pulse and step excitation also are defined in Chap. 8.

COMPARISON OF MAXIMUM ACCELERATIONS

VELOCITY STEP APPROXIMATION. A comparison of values of $\ddot{x}_m$ resulting from various acceleration pulses with that resulting from a velocity step is shown in Fig. 31.21.

FIG. 31.21. Dimensionless representation of maximum transmitted acceleration $\ddot{x}_m$ for the undamped linear system of Fig. 31.6. Support motion is a rectangular, half-sine, or versed sine acceleration pulse.

The maximum acceleration induced by a velocity step is $\omega_n \dot{u}_m$ [see Eq. (31.19)]. The abscissa $\omega_n \tau_r$ is a dimensionless measure of pulse duration. The effect of pulse shape is imperceptible for values of $\omega_n \tau_r \lesssim 0.6$. For pulses of duration $\omega_n \tau_r \lesssim 1.0$, the effect of pulse shape is small and the maximum possible error resulting from use of the velocity step approximation is of the order of 5 per cent.

EFFECTS OF PULSE SHAPE. The effects of pulse shape upon the maximum response acceleration $\ddot{x}_m$ for values of $\omega_n \tau_r > 1.0$ are shown in Fig. 31.22. The ordinate $\ddot{x}_m/\ddot{u}_m$ is the ratio of maximum acceleration induced in the responding system to maximum acceleration of the pulse. All three pulses produce the highest value of response acceleration when $\omega_n \tau_r \simeq \pi$. Physically, this corresponds to an effective duration τ_r of one-half of the natural period of the spring-mass system. For longer pulse durations the curves for half-sine and versed sine pulses are similar. For pulse durations beyond the range of Fig. 31.22 ($\omega_n \tau_r > 16$), the half-sine and versed sine curves approach the limiting ordinate $\ddot{x}_m/\ddot{u}_m = 1$. This corresponds physically to approximating a static loading of the spring-mass system. A limiting acceleration ratio $\ddot{x}_m/\ddot{u}_m = 2$ is encountered for all rectangular pulses of duration greater than the half-period of the spring-mass system. A more extensive study of responses to a variety of pulse shapes is included in Chap. 8.

Fig. 31.22. Shock transmissibility for the undamped linear system of Fig. 31.6 as a function of angular natural frequency ω_n and effective pulse duration τ_r. Support motion is a rectangular, half-sine, or versed sine acceleration pulse.

SHOCK SPECTRUM

The abscissa $\omega_n \tau_r$ in Fig. 31.22 may be treated as a measure of pulse duration (proportional to τ_r) for a given spring-mass system with ω_n fixed. Alternatively, the pulse duration may be considered fixed; then the curves show the effect of varying the natural frequency ω_n of the spring-mass system. Each of the curves of Fig. 31.22 shows the maximum acceleration induced by a given acceleration pulse upon spring-mass systems of various natural frequencies ω_n; thus, Fig. 31.22 may be used to determine the required natural frequency of the isolator if $\ddot{x}_m$ and $\ddot{u}_m$ are known, and the pulse shape is defined.

Each curve shown in Fig. 31.22 may be interpreted as a description of a pulse, in terms of the response induced in a system subjected to the pulse. The curve of maximum response as a function of the natural frequency of the responding system is called a shock spectrum or response spectrum. This concept is discussed more fully in Chap. 23. A pulse is a particular form of a shock motion; thus, each shock motion has a characteristic shock spectrum. A shock motion has a characteristic effective value of time duration τ_r which need not be defined specifically; instead, the spectra are made to apply explicitly to a given shock motion by using the natural frequency ω_n as a dimensional parameter on the abscissa. By taking the isolator-and-equipment assembly to be the responding system, the natural frequency of the isolator may be chosen to meet any specified maximum acceleration $\ddot{x}_m$ of the equipment supported by isolators. Spectra of maximum isolator deflection δ_m also may be drawn, and are useful in predicting the maximum isolator deflection when the natural frequency of the isolator is known.

When damping is added to the isolator, the analysis of the response becomes much more complex. In general, it is possible to determine the maximum value of the response acceleration $\ddot{x}_m$ only by calculating the time-history of response acceleration over the entire time interval suspected of including the maximum response. A digital computer has been used to find shock spectra for "half-sine" acceleration pulses with various

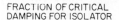

FRACTION OF CRITICAL
DAMPING FOR ISOLATOR

FIG. 31.23. Shock transmissibility for the system of Fig. 31.2A with linear spring and viscous damping. Support motion is a half-sine acceleration pulse of height $\ddot{u}_m$ and effective duration τ_r. Curves are for discrete values of the fraction of critical damping ζ in the isolator as indicated.

fractions of critical damping in the responding system, as shown in Fig. 31.23. Similar spectra could be obtained to indicate maximum values of isolator deflection. In selecting a shock isolator for a specified application, it may be necessary to use both maximum acceleration and maximum deflection spectra. This is illustrated in the following example.

Example 31.3. A piece of equipment weighing 200 lb is to be isolated from the effects of a vertical shock motion defined by the spectra of acceleration and deflection shown in Fig. 31.24. It is required that the maximum induced acceleration not exceed $7g$ (2,700 in./sec²). Clearances available limit the isolator deflection to 2.25 in. The curves in Fig. 31.24A represent maximum response acceleration $\ddot{x}_m$ as a function of the angular natural frequency ω_n of the equipment supported on the shock isolators. The isolator springs are assumed linear and viscously damped, and separate curves are shown for values of the damping ratio $\zeta = 0$, 0.1, 0.2, and 0.3. The curves in Fig. 31.24B represent the maximum isolator deflection δ_m as a function of ω_n for the same values of ζ.

Consider first the requirement that $\ddot{x}_m < 2{,}700$ in./sec². In Fig. 31.24A, the horizontal dashed line indicates this limiting acceleration. If the damping ratio $\zeta = 0.3$, then the angular natural frequency ω_n may not exceed 38.5 rad/sec on the criterion of maximum acceleration. The dashed horizontal line of Fig. 31.24B represents the deflection limit $\delta_m = 2.25$ in. For $\zeta = 0.3$, the minimum natural frequency is 30 rad/sec on the criterion of deflection. Considering both acceleration and deflection criteria, the angular natural frequency ω_n must lie between 30 rad/sec and 38.5 rad/sec. The spectra indicate that both criteria may be just met with $\zeta = 0.2$ if ω_n is 35 rad/sec. Smaller values of damping do not permit the satisfaction of both requirements.

Conservatively, a suitable choice of parameters is $\zeta = 0.3$, $\omega_n = 35$ rad/sec. This limits

$\ddot{x}_m$ to 2,500 in./sec² and δ_m to 2.0 in. The spring stiffness k is

$$k = \omega_n{}^2 m = (35)^2 \times \frac{230}{386} = 730 \text{ lb/in.}$$

If the equipment is to be supported by four like isolators, then the required stiffness of each isolator is $k/4 = 182.5$ lb/in.

(A)

(B)

Fig. 31.24. Shock spectra: (A) maximum acceleration and (B) maximum isolator deflection for Example 31.3.

RESPONSE OF EQUIPMENT WITH A FLEXIBLE COMPONENT

IMPACT WITH REBOUND

Consider the system of Fig. 31.4. The block of mass m_1 represents the equipment and m_2 with its associated spring-dashpot unit represents a critical component of the equipment. The left spring-dashpot unit represents the shock isolator. It is assumed here that $m_1 \gg m_2$ so that the motion of m_1 is not sensibly affected by m_2; larger values of m_2 are considered in a later section. Consider the entire system to be moving to the left at uniform velocity when the left-hand end of the isolator strikes a fixed support (not shown). The isolator will be compressed until the equipment is brought to rest. Following this the compressive force in the isolator will continue to accelerate the equipment toward the right until the isolator loses contact with the support and the rebound is complete. This type of shock is called *impact with rebound*. Practical examples include the shock experienced by a single railroad car striking a bumper at the end of a siding and that experienced by packaged equipment, shock-mounted inside a container of small mass, when the container is dropped upon a hard surface and then rebounds.

The procedure for finding the maximum acceleration $\ddot{x}_{2m}$ of the component, assuming the component stiffness to be linear and neglecting component damping, is:

1. Using the known striking velocity determine, from velocity step results (Figs. 31.9, 31.10, 31.12 to 31.15), the maximum deflection δ_{1m} of the isolator and the maximum acceleration $\ddot{x}_{1m}$ of the equipment.

2. From Eq. (31.28), find the effective duration τ_r for the acceleration time-history $\ddot{x}_1(t)$ of the equipment.

3. From the shock spectra corresponding to the acceleration pulse $\ddot{x}_1(t)$, find the maximum acceleration $\ddot{x}_{2m}$ of the component.

Details of the procedure using the isolators of Example 31.1 are considered in Example 31.4.

Example 31.4. Let the equipment of Example 31.1 weighing 40 lb have a flexible component weighing 0.2 lb. By vibration testing, this component is found to have an angular natural frequency $\omega_n = 260$ rad/sec and to possess negligible damping. For the isolators of Example 31.1, it is desired to determine the maximum acceleration $\ddot{x}_{2m}$ experienced by the mass m_2 of the component if the equipment, traveling at a velocity of 70 in./sec, is arrested by the free end of the isolator striking a fixed support. The four cases are considered separately. It is assumed that the component has a negligible effect on the motion of the equipment because $m_2 \ll m_1$.

Linear Spring. From the results of Example 31.1, it is known that $\omega_n = 108$ rad/sec and that the maximum acceleration of the equipment as found from Eq. (31.19) is

$$\ddot{x}_{1m} = 7{,}580 \text{ in./sec}^2 = 19.6g$$

This acceleration occurs at the instant when the isolator deflection has the extreme value $\delta_{1m} = 0.65$ in.* Subsequently the isolator spring continues to accelerate the equipment until the isolator force is zero and the rebound is complete. Since there is no damping, the rebound velocity equals the striking velocity (with opposite sign). The velocity change $\dot{x}_{1c}$ is twice the striking velocity and the effective duration τ_r [Eq. (31.28)] is

$$\tau_r = \frac{\dot{x}_{1c}}{\ddot{x}_{1m}} = \frac{2 \times 70}{7{,}580} = 0.0185 \text{ sec}$$

The acceleration time-history of the equipment is a half-sine pulse as represented in Fig. 31.20 (the ordinate is $\ddot{x}_1$ instead of $\ddot{u}$).

Since the equipment is the "support" for the component, the response of the latter may be found from results developed for the response of a rigid body whose support experiences a half-sine pulse of acceleration. The half-sine curve of Fig. 31.22 gives the desired information if the following interpretations are made: For $\ddot{x}_m/\ddot{u}_m$ read $\ddot{x}_{2m}/\ddot{x}_{1m}$; for $\omega_n \tau_r$ read $\omega_{n2}\tau_r$. Now $\omega_{n2}\tau_r = 260 \times 0.0185 = 4.80$. From Fig. 31.22, $\ddot{x}_{2m}/\ddot{x}_{1m} = 1.66$, and $\ddot{x}_{2m} = 1.66 \times 7{,}580 = 12{,}600$ in./sec$^2 = 32.6g$.

Hardening Spring. From Example 31.1, the maximum equipment acceleration is $\ddot{x}_{1m} = 21g = 8{,}100$ in./sec^2. Since the velocity change $\dot{x}_{1c}$ is twice the striking velocity, the effective duration τ_r [Eq. (31.28)] is

$$\tau_r = \frac{\dot{x}_{1c}}{\ddot{x}_{1m}} = \frac{2 \times 70}{8{,}100} = 0.0173 \text{ sec}$$

With a hardening isolator spring, the shape of the acceleration pulse $\ddot{x}_1(t)$ experienced by the equipment varies considerably as the maximum deflection δ_{1m} approaches the upper limit d. Up to $\delta_{1m}/d = 0.5$, the shape is closely approximated by a half-sine pulse. For $\delta_{1m}/d = 0.8$, a symmetric triangular pulse is a good approximation. For higher values of δ_{1m}/d, the pulse is very sharply peaked. The maximum response curve for a half-sine pulse is given in Fig. 31.22. The corresponding curve for a symmetric triangular pulse (Fig. 8.18b) is similar to that for the versed sine pulse, though lying generally below the latter. Inasmuch as the curve for the versed sine pulse is below that for the half-sine pulse, it is conservative to use the half-sine pulse for all values of δ_{1m}/d. Accordingly, $\omega_{n2}\tau_r = 260 \times 0.0173 = 4.50$. From the half-sine curve of Fig. 31.22, $\ddot{x}_{2m}/\ddot{x}_{1m} = 1.69$, and $\ddot{x}_{2m} = 1.69 \times 8{,}1000 = 13{,}700$ in./sec$^2 = 36.4g$.

Softening Spring. From Example 31.1, the maximum equipment acceleration $\ddot{x}_{1m}$ is

$$\ddot{x}_{1m} = 20g = 7{,}720 \text{ in./sec}^2$$

The effective duration τ_r [Eq. (31.28)] is

$$\tau_r = \frac{\dot{x}_{1c}}{\ddot{x}_{1m}} = \frac{2 \times 70}{7{,}720} = 0.0181 \text{ sec}$$

The shape of the acceleration pulse $\ddot{x}_1(t)$ for the equipment varies markedly as the departure from linearity increases (increasing values of δ_{1m}/d_1). The pulse shape is found by first performing the integration of Eq. (31.9) with $F_s(\delta)$ as given by Eq. (31.23). The result supplies

---

* If the equipment (Fig. 31.4) is moving toward the left when the isolator contacts the support, the extreme value of δ_{1m} is negative. It suffices to deal here with absolute values.

the integrand required for Eq. (31.13)   The integration of the latter equation is performed numerically.   Results [1] show that the pulse shape undergoes a rapid transition from the half-sine pulse at very small values of δ_{1m}/d_1 to shapes that are closely approximated by the trapezoidal pulse of Fig. 31.25.   Note that the pulse of Fig. 31.25 requires three parameters to fix it completely: the maximum acceleration $\ddot{x}_{1m}$; the effective duration τ_r; and the ratio τ_r/τ where τ is the actual duration and $\tau_r = \tau - \tau_1$.   From results of the numerical integra-

FIG. 31.25. Symmetric trapezoidal acceleration pulse.

FIG. 31.26. Dimensionless representation of effective duration τ_r of acceleration pulse experienced by equipment during impact with rebound.   Isolator has undamped hyperbolic tangent elasticity.   Maximum isolator deflection is δ_{1m} and characteristic deflection d_1 is defined in Fig. 31.11.

tions of Eq. (31.13), the curve of Fig. 31.26 is constructed to show τ_r/τ as a function of the deflection ratio δ_{1m}/d_1.

To find the maximum acceleration $\ddot{x}_{2m}$ of the component, the maximum response curves (shock spectra) of Fig. 31.27 are used.   These curves are constructed for symmetric trapezoidal pulses (Fig. 31.25).   The top curve ($\tau_r/\tau = 1.0$) corresponds to the limiting (rectangular) form.   The dashed curve ($\tau_r/\tau = 0.64$) represents response to a half-sine pulse.

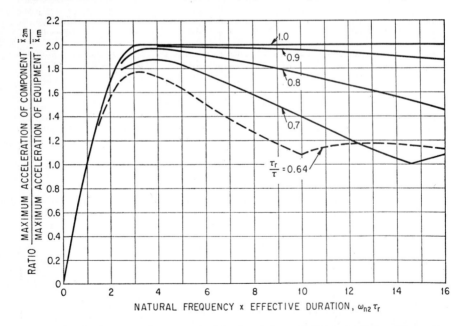

FIG. 31.27. Shock spectra for component having undamped linear elasticity with angular natural frequency ω_{n2}.   Equipment motion is symmetric trapezoidal acceleration pulse (Fig. 31.25).   Curves are for five discrete values of the ratio of effective pulse duration to actual duration τ_r/τ.

The value of δ_{1m}/d_1 corresponding to the maximum acceleration $\ddot{x}_{1m}$ of the equipment is (from Example 31.1) $\delta_{1m}/d_1 = 3$. From Fig. 31.26: $\tau_r/\tau = 0.88$. Now $\omega_n \tau_r = 260 \times 0.0181 = 4.7$. Using Fig. 31.27, linear interpolation between the curves for $\tau_r/\tau = 0.8$ and $\tau_r/\tau = 0.9$ gives $\ddot{x}_{2m}/\ddot{x}_{1m} = 1.98$ and $\ddot{x}_{2m} = 1.98 \times 7{,}720 = 15{,}300$ in./sec^2 = 39.6g.

Linear Spring and Viscous Damping. The presence of damping in the isolator adds several complications: (1) the rebound velocity is no longer equal to the striking velocity; (2) the acceleration pulse of the equipment is not symmetrical and returns to zero before the isolator deformation δ_{1m} returns to zero; and (3) the pulse shape varies greatly with damping ratio ζ_1. Shock spectra for acceleration pulse shapes corresponding to damping ratios of particular interest ($0.10 < \zeta_1 < 0.40$) are not available. However, for single acceleration pulses which do not change sign, it is conservative to assume that the maximum acceleration $\ddot{x}_{2m}$ of the component is twice the maximum acceleration $\ddot{x}_{1m}$ of the equipment. Using the results of Example 31.1, the maximum acceleration of the component is $\ddot{x}_{2m} = 2\ddot{x}_{1m} = 2 \times 5{,}450 = 10{,}900$ in./sec^2 = 28.2g.

IMPACT WITHOUT REBOUND

When impact of the isolator occurs without rebound, it must be recognized that the equipment-and-isolator system continues to oscillate until the initial kinetic energy is dissipated. Consider the system of Fig. 31.4; it consists of equipment m_1, shock isolator (left spring-dashpot unit), and flexible component (subsystem 2). The system is initially at rest. The left end of the shock isolator is attached to a support (not shown) which is given a velocity step of magnitude $\dot{u}_m$ at $t = 0$. The subsequent motion of the support is $u = \dot{u}_m t$. Determine the maximum force F_{1m} transmitted by the isolator, the maximum isolator deflection δ_{1m}, and the maximum acceleration $\ddot{x}_{2m}$ of the component.

Solutions are available only for linear systems, i.e., linear springs and viscous damping. Two such simplified analyses of this problem are included in the following sections: (1) The influence of damping is considered, but the component mass m_2 is assumed of negligible size relative to m_1 and (2) damping is neglected but the effect of the mass m_2 of the component upon the motion of the system is considered.

COMPONENT MASS NEGLIGIBLE. Assume that $m_1 \gg m_2$ so that the motion x_1 of the equipment may be determined by neglecting the effect of the component. Then the extreme value of the force F_{1m} transmitted by the isolator and the extreme deflection δ_{1m} of the isolator occur during the first quarter-cycle of the equipment motion; they may be found from Figs. 31.14 and 31.15 in the section on *Response of a Rigid Body System to a Velocity Step*. The subsequent motion of the equipment is an exponentially decaying sinusoidal oscillation or, if there is no damping in the isolator, a constant-amplitude oscillation. If the component also is undamped, an analytic determination of the component response is not difficult. The motion consists of harmonic oscillation at the frequency ω_{n1} of the equipment oscillation and a superposed oscillation at the frequency ω_{n2} of the component system. Since the oscillations are assumed to persist indefinitely in the absence of damping, the extreme acceleration of the component is the sum of the absolute values of the maximum accelerations associated with the oscillations at frequencies ω_{n1} and ω_{n2}. In the particular case of resonance ($\omega_{n1} = \omega_{n2}$), the vibration amplitude of the component increases indefinitely with time. Because actual systems always possess damping (usually a considerable amount in the isolator), solutions of this type tend to be unduly conservative for engineering applications.

The equation of motion for the viscous damped component is a special case of Eq. (31.5) with $F(\delta, \dot{\delta})$ as given by Eq. (31.7). If appropriate subscripts are supplied and customary substitutions are made, the equation is

$$\ddot{\delta}_2 + 2\zeta_2 \omega_{n2} \dot{\delta}_2 + \omega_{n2}{}^2 \delta_2 = -\ddot{x}_1 \tag{31.38}$$

Analytic solutions of Eq. (31.38) to find the acceleration $\ddot{x}_2 = \ddot{x}_1 + \ddot{\delta}_2$ of the component are too laborious to be practical. However, results have been obtained by analog computation [1,4] and are shown in Fig. 31.28. The ordinate is the ratio of the maximum acceleration $\ddot{x}_{2m}$ of the component to the maximum acceleration $\dot{u}_m \omega_{n2}$ [see Eq. (31.19)] that the component would experience if the shock isolator were rigid. The abscissa is the ratio of the undamped natural frequency ω_{n2} of the component to the undamped natural frequency ω_{n1} of the equipment on the isolator spring. Curves are given for several different values of the fraction of critical damping ζ_1 for the isolator. For all curves the fraction of critical damping for the component is $\zeta_2 = 0.01$. The effect of isolator damping in reducing the maximum acceleration $\ddot{x}_{1m}$ of the component is great in the neighborhood of $\omega_{n2}/\omega_{n1} = 1$. Above $\omega_{n2}/\omega_{n1} = 2$, small damping ($\zeta_1 \leq 0.1$) in the isolator has little effect and large damping may significantly increase the maximum acceleration of the component.

The ordinate in Fig. 31.28 represents the ratio of the maximum acceleration of the component to that which would be experienced with the isolator rigid (absent); thus, it may properly be called *shock transmissibility*. If shock transmissibility is less than unity, the isolator is beneficial (for the component considered). An isolator must have a natural frequency significantly less than that of the critical component in order to reduce the transmitted acceleration. If there are several critical components having different natural frequencies ω_{n2}, each must be considered separately and the natural frequency of the isolator must be significantly lower than the lowest natural frequency of a component.

TWO DEGREES-OF-FREEDOM—NO DAMPING. This section includes an analysis of the transient response of the two degree-of-freedom system shown in Fig. 31.4, neglecting the effects of damping but assuming the equipment mass m_1 and the component mass m_2 to be of the same order of magnitude. The equations of motion are

$$m_1 \ddot{\delta}_1 + k_1 \delta_1 = k_2 \delta_2 - m_1 \ddot{u}$$

$$m_2 \ddot{\delta}_2 + k_2 \delta_2 = -m_2 \ddot{\delta}_1 - m_2 \ddot{u}$$

(31.39)

FIG. 31.28. Shock transmissibility for a component of a viscously damped system with linear elasticity. The maximum acceleration of the component is $\ddot{x}_{2m}$ and the velocity step (support motion) is $\dot{u}_m$. The effect of the component on the equipment motion is neglected. (*After R. D. Mindlin.*[1])

FIG. 31.29. Shock transmissibility for component of system of Fig. 31.4 under impact at velocity $\dot{u}_m$ without rebound. Component and isolator have undamped linear elasticity. Ordinate is ratio of maximum acceleration of component $\ddot{x}_{2m}$ to maximum acceleration which would be experienced by component with a rigid isolator. Curves are for four discrete values of the mass ratio m_2/m_1 as indicated.

where k_1 = stiffness of isolator spring, lb/in., and k_2 = stiffness of component, lb/in. The system is initially in equilibrium; at time $t = 0$, the left end of the isolator spring is given a velocity step of magnitude $\dot{u}_m$. Initial conditions are: $\dot{\delta}_1 = \dot{u}_m$, $\dot{\delta}_2 = 0$, $\delta_1 = \delta_2 = 0$. Equations (31.39) may be solved simultaneously [2] for maximum values of the acceleration $\ddot{x}_{2m}$ of the component and maximum deflection δ_{1m} of the isolator:

$$\ddot{x}_{2m} = \frac{\dot{u}_m \omega_{n2}}{\left[\left(\frac{\omega_{n2}}{\omega_{n1}} - 1\right)^2 + \frac{m_2}{m_1}\left(\frac{\omega_{n2}}{\omega_{n1}}\right)^2\right]^{1/2}} \tag{31.40}$$

$$\delta_{1m} = \frac{\dot{u}_m}{\omega_{n1}} \frac{1 + \frac{\omega_{n2}}{\omega_{n1}}\left(1 + \frac{m_2}{m_1}\right)}{\left[\left(\frac{\omega_{n2}}{\omega_{n1}} + 1\right)^2 + \frac{m_2}{m_1}\left(\frac{\omega_{n2}}{\omega_{n1}}\right)^2\right]^{1/2}} \tag{31.41}$$

where $\ddot{x}_{2m}$ = maximum absolute acceleration of component mass, in./sec²; δ_{1m} = maximum deflection of isolator spring, in.; ω_{n1} * = angular natural frequency of isolator $(k_1/m_1)^{1/2}$, rad/sec; and ω_{n2} * = angular natural frequency of component $(k_2/m_2)^{1/2}$, rad/sec. Equation (31.40) is shown graphically in Fig. 31.29. The dimensionless ordi-

* The natural frequencies ω_{n1} and ω_{n2} are hypothetical in the sense that they do not consider the coupling between the subsystems.

nate is the ratio of maximum acceleration $\ddot{x}_{2m}$ of the component to the maximum acceleration $\dot{u}_m \omega_{n2}$ which the component would experience with no isolator present. The abscissa is the ratio of component natural frequency ω_{n2} to isolator natural frequency ω_{n1}. Separate curves are given for mass ratios $m_2/m_1 = 0.01, 0.1, 0.3,$ and 1.0. Equation (31.41) is shown graphically in Fig. 31.30. The ordinate is the ratio of the maximum isolator deflection δ_{1m} to the deflection $\dot{u}_m(1 + m_2/m_1)^{1/2}/\omega_{n1}$ which would occur if component stiffness k_2 were infinite. The abscissa is the ratio of natural frequencies ω_{n2}/ω_{n1}, and curves are given for values of $m_2/m_1 = 0.1$ and 1.0.

FIG. 31.30. Dimensionless representation of maximum isolator deflection in system of **Fig. 31.4** under impact at velocity $\dot{u}_m$ without rebound. Component and isolator have undamped linear elasticity. Ordinate is ratio of maximum isolator deflection δ_{1m} to the isolator deflection which would result if the component stiffness were infinite. Curves are for two discrete values of the mass ratio m_2/m_1 as indicated.

Figure 31.29 shows that the effect of the mass ratio m_2/m_1 upon the maximum component acceleration $\ddot{x}_{2m}$ is very great near resonance ($\omega_{n2}/\omega_{n1} \simeq 1$). As ω_{n2}/ω_{n1} increases above resonance, the effect of finite component mass steadily decreases. Figure 31.30 shows that except for small values of ω_{n2}/ω_{n1} the effect of finite component mass on the maximum isolator deflection δ_{1m} is slight. As ω_{n2}/ω_{n1} increases, the curves for all mass ratios asymptotically approach the ordinate 1.0.

The factor $(1 + m_2/m_1)^{1/2}$ in the ordinate parameter of Fig. 31.30 is introduced because the total equipment mass is $m_1 + m_2$. For the limiting case of rigid equipment (k_2 infinite), the natural frequency ω_n is given by

$$\omega_n{}^2 = \frac{k_1}{m_1 + m_2} \qquad \omega_n = \frac{\omega_{n1}}{(1 + m_2/m_1)^{1/2}}$$

Substituting this relation in Eq. (31.18) and solving for δ_{1m}:

$$\delta_{1m} = \dot{u}_m(1 + m_2/m_1)^{\frac{1}{2}}/\omega_{n1}$$

This is in agreement with the result given by Eq. (31.41) as ω_{n2}/ω_{n1} approaches infinity.

Example 31.5. Equipment weighing 152 lb has a flexible component weighing 3 lb. The angular natural frequency of the component is $\omega_{n2} = 130$ rad/sec. The equipment is mounted on a shock isolator with a linear spring $k_1 = 2,400$ lb/in. and having a fraction of critical damping $\zeta_1 = 0.10$. Find the maximum isolator deflection δ_{1m} and the maximum component acceleration $\ddot{x}_{2m}$ which result when the base experiences a velocity step $\dot{u}_m = 55$ in./sec.

Consider first a solution assuming that m_2 has a negligible effect on the equipment motion:

$$m_1 = \frac{152 \text{ lb}}{386 \text{ in./sec}^2} = 0.393 \text{ lb-sec}^2/\text{in.}$$

$$\omega_{n1} = \sqrt{\frac{k_1}{m_1}} = \sqrt{\frac{2,400}{0.393}} = 78.1 \text{ rad/sec [12.4 cps]}$$

Figure 31.14 gives $\ddot{x}_{1m}/\dot{u}_m\omega_{n1} = 0.88$ and Fig. 31.15 gives $\ddot{x}_{1m}\delta_{1m}/\dot{u}_m^2 = 0.76$ for $\zeta_1 = 0.1$. Then

$$\delta_{1m} = \frac{0.76}{0.88} \times \frac{\dot{u}_m}{\omega_{n1}} = \frac{0.76 \times 55}{0.88 \times 78.1} = 0.61 \text{ in.}$$

In finding $\ddot{x}_{2m}$ it is assumed that damping of the component has the typical value $\zeta_2 = 0.01$. Using $\omega_{n1}/\omega_{n2} = 130/78.1 = 1.67$, Fig. 31.28 gives $\ddot{x}_{2m}/\dot{u}_m\omega_{n2} = 1.15$; then $\ddot{x}_{2m} = 1.15 \times 55 \times 130 = 8,230$ in./sec$^2 = 21.3g$.

A second solution, taking into consideration the mass m_2 of the component, may be obtained if the damping is neglected. From Eq. (31.41),

$$\delta_{1m} = \frac{\dot{u}_m}{\omega_{n1}} \frac{1 + \dfrac{\omega_{n2}}{\omega_{n1}}\left(1 + \dfrac{m_2}{m_1}\right)}{\left[\left(\dfrac{\omega_{n2}}{\omega_{n1}} + 1\right)^2 + \dfrac{m_2}{m_1}\left(\dfrac{\omega_{n2}}{\omega_{n1}}\right)^2\right]^{\frac{1}{4}}}$$

$$= \frac{55}{78.1} \times \frac{1 + 1.67(1 + \frac{3}{152})}{[(2.67)^2 + \frac{3}{152}(1.67)^2]^{\frac{1}{4}}} = 0.71 \text{ in.}$$

From Eq. (31.40):

$$\ddot{x}_{2m} = \dot{u}_m\omega_{n2}\left[\left(\frac{\omega_{n2}}{\omega_{n1}} - 1\right)^2 + \frac{m_2}{m_1}\left(\frac{\omega_{n2}}{\omega_{n1}}\right)^2\right]^{-\frac{1}{4}}$$

$$= 55 \times 130[(0.67)^2 + \frac{3}{152}(1.67)^2]^{-\frac{1}{4}}$$

$$= 10,070 \text{ in./sec}^2 = 26.1g$$

This example is too complex for a practicable solution when damping and the mass effects are considered together. However, the two above solutions may be taken conservatively as limiting conditions; it is unlikely that the actual acceleration and deflection would exceed the maxima of the limiting conditions.

SUPPORT PROTECTION

This section considers conditions in which the shock originates within the equipment (e.g., guns and drop hammers). Attention is first given to determining the response of the support for such equipment in the absence of a shock isolator. The effect of a shock isolator introduced to protect the support from excessive loads is considered later.

EQUIPMENT RIGIDLY ATTACHED TO SUPPORT

If the equipment is rigidly attached to the support, the support and equipment may be idealized as a single degree-of-freedom system for purposes of a simplified analysis. Consider the system of Fig. 31.3B with the spring-dashpot unit 2 assumed to be rigid. The mass m represents the equipment, and the mass m_F represents, with spring and dashpot assembly (1), the support. The force F, applied externally to the equipment, is taken to be a known function of time. The equation of motion is

$$(m_F + m)\ddot{\delta} + F(\dot{\delta},\delta) = F$$

Considering only force-time relations $F(t)$ in the form of a single pulse, the analogous mathematical relations of Eqs. (31.5) and (31.6) are used by defining the impulse J applied by the force F as

$$J = \int_0^\tau F\,dt \qquad (31.42)$$

where τ is the duration of the pulse.

SHORT-DURATION IMPULSES. If τ is short compared with the half-period of free oscillation of the system, then the results derived in the section on *Response of a Rigid Body System to a Velocity Step* may be applied directly. An impulse J of negligible duration acting on the mass m produces a velocity change $\dot{u}_m$ given by

$$\dot{u}_m = \frac{J}{m} \qquad (31.43)$$

The subsequent relative motion of the system is identical with that resulting from a velocity step of magnitude $\dot{u}_m$.

If the damping capacity of the support is small, then velocity step results derived for linear springs, hardening springs, and softening springs are applicable. If the damping of the support may be represented as viscous and the stiffness as linear, then the linear-spring viscous damping results apply. In most installations it is sufficiently accurate to consider the support an undamped linear system.

A structure used to support an equipment generally has distributed mass and elasticity; thus the application of an impulse tends to excite the structure to vibrate not only in its fundamental mode but also in higher modes of vibration. The mass-spring-dashpot system shown in Fig. 31.3B to represent the structure would have equivalent mass and stiffness suitable to simulate only the fundamental mode of vibration. In many applications, such simulation is adequate because the displacements and strains are greater in the fundamental mode than in higher modes. The vibration of members having distributed mass is discussed in Chap. 7, and the formulation of models suitable for use in the analysis of systems subjected to shock is discussed in Chap. 42.

LONG-DURATION IMPULSES. If the duration τ of the applied impulse exceeds about one-third of the natural period of the equipment-support system, application of velocity step results may be unduly conservative. Then the results developed in the section on *Response of Rigid Body System to Acceleration Pulse* are applicable. The mathematical equivalence of Eqs. (31.5) and (31.6) is based on identifying $-m\ddot{u}$ in the former with F in the latter. Accordingly, if the shape of the force F vs. time curve is similar to the shape of the curve of acceleration $\ddot{u}$ vs. time, then the response of a system to an acceleration pulse may be used by analogy to find the response to a force pulse by making the following substitutions:

$$\ddot{u}_m = \frac{F_m{}^*}{m} \qquad \tau_r = \frac{J}{F_m}$$

where F_m is the maximum value of F, $\ddot{u}_m$ is the maximum value of $\ddot{u}$ and τ_r is the effective duration.

EQUIPMENT SHOCK ISOLATED

IDEALIZED SYSTEM. When a shock isolator is used to reduce the magnitude of the force transmitted to the support, the idealized system is as shown in Fig. 31.4. Sub-

---

* If the mathematical equivalence is literally applied, F_m/m is analogous to $-\ddot{u}_m$, not $\ddot{u}_m$. Since acceleration pulse results are given in terms of extreme *absolute* values, the sign is not important.

system 2 represents the equipment (mass m_2) mounted on the shock isolator (right-hand spring-dashpot unit). Subsystem 1 is an idealized representation of the support with effective mass m_1 and with stiffness and damping capacity represented by the left spring-dashpot unit. The free end of the latter unit is taken to be fixed ($u = 0$).

It is assumed that the system is initially in equilibrium ($\dot{\delta}_1 = \dot{\delta}_2 = 0$; $\delta_1 = \delta_2 = 0$) and that force F (positive in the $+X$ direction) applies an impulse J to m_2. Analysis is simplified by treating the duration τ of impulse J as negligible. This assumption, always conservative, usually is warranted if the natural frequency of the shock isolator is small relative to the natural frequency of the support.

SYSTEM SEPARABLE. In many applications the support motion $x_1 (= \delta_1)$ is sufficiently small compared with the equipment motion x_2 that the equipment acceleration $\ddot{x}_2$ is closely approximated by $\ddot{\delta}_2$ where $\ddot{x}_2 = \ddot{\delta}_2 + \ddot{x}_1$. Using this approximation, the analysis is resolved into two separate parts, each dealing with a single degree-of-freedom system.

If the system consists only of linear elements as defined by Eq. (31.7), the equation of motion of the equipment mounted on the shock isolator (subsystem 2 of Fig. 31.4) is

$$\ddot{\delta}_2 + 2\zeta_2\omega_{n2}\dot{\delta}_2 + \omega_{n2}{}^2\delta_2 = 0 \tag{31.44}$$

where $\omega_{n2}{}^2 = k_2/m_2$ and $\zeta_2 = c_2/2m_2\omega_{n2}$. The initial conditions are: $\delta_2 = 0$, $\dot{\delta}_2 = \dot{u}_m = J/m_2$ when $t = 0$. Because of the similarity of Eqs. (31.26) and (31.44), and the respective initial conditions, the maximum equipment acceleration $\ddot{x}_{2m}$ and the maximum isolator deflection δ_{2m} may be found from Figs. 31.14 and 31.15.

The differential equation for the motion of the support in Fig. 31.4 is

$$\ddot{\delta}_1 + 2\zeta_1\omega_{n1}\dot{\delta}_1 + \omega_{n1}{}^2\delta_1 = -\frac{m_2}{m_1}\ddot{x}_2 \tag{31.45}$$

where $\omega_{n1}{}^2 = k_1/m_1$ and $\zeta_1 = c_1/2m_1\omega_{n1}$. The initial conditions are $\dot{\delta}_1 = 0$, $\delta_1 = 0$.

The solution of Eq. (31.45) is formally identical with that of Eq. (31.38) because the equations differ only by the interchange of the numerical subscripts and the presence of the factor m_2/m_1 on the right-hand side of Eq. (31.45). The solutions of Eq. (31.45) as obtained by analog computer [1] are shown in Fig. 31.31. The ordinate is the ratio of the maximum force F_{1m} in the support to the quantity $J\omega_{n1}$. The latter quantity is the maximum force which would be developed in an undamped, linear, single degree-of-freedom support of mass m_1 and stiffness k_1 if the impulse J were applied directly to m_1. The abscissa in Fig. 31.31 is the ratio of the undamped support natural frequency ω_{n1} to the undamped isolator natural frequency ω_{n2}. Curves are drawn for various values of the fraction of critical damping ζ_2 for the isolator, assuming that the fraction of critical damping ζ_1 for the support is constant at $\zeta_1 = 0.01$.

Figure 31.31 appears to show that the presence of an isolator increases the maximum force F_{1m} transmitted by the support if the natural frequencies of isolator and support are nearly equal. This conclusion is misleading because the analysis assumes that the support deflection δ_1 is small compared with the isolator deflection δ_2—a condition which is not met in the neighborhood of unity frequency ratio. A more realistic analysis involves the two degree-of-freedom system discussed in the next section.

TWO DEGREE-OF-FREEDOM ANALYSIS. This section includes an analysis of the system of Fig. 31.4 considered as a coupled two degree-of-freedom system where both the support and isolator are linear and undamped $[F_1(\delta_1,\dot{\delta}_1) = k_1\delta_1, F_2(\delta_2,\dot{\delta}_2) = k_2\delta_2]$. This analysis makes it possible to consider the effect of deflection of the support on the motion of the equipment. Fixing the support base ($u = 0$), the equations of motion may be written

$$\ddot{\delta}_1 + \omega_{n1}{}^2\delta_1 = \frac{m_2}{m_1}\omega_{n2}{}^2\delta_2$$

$$\tag{31.46}$$

$$\ddot{\delta}_2 + \omega_{n2}{}^2\delta_2 = -\ddot{\delta}_1$$

FIG. 31.31. Dimensionless representation of maximum force in support F_{1m} resulting from action of impulse J on equipment. Isolator and support have linear elasticity and fraction of critical damping ζ_2 and ζ_1, respectively. Ordinate is the ratio of maximum force in support to that which would result from direct application of the impulse to the support with equipment absent. (*After R. D. Mindlin.[1]*)

Assuming that the impulse J has negligible duration, the initial conditions are: $\delta_1 = 0$, $\dot{\delta}_2 = J/m_2$, $\delta_1 = \delta_2 = 0$.

The solution of Eqs. (31.46) parallels that of Eqs. (31.39); the resulting expressions for the maximum isolator deflection δ_{2m} and force F_{1m} applied to the support are

$$\delta_{2m} = \frac{J}{m_2 \omega_{n2}} \left[1 + \frac{m_2/m_1}{(1 + \omega_{n1}/\omega_{n2})^2} \right]^{-\frac{1}{2}} \tag{31.47}$$

$$F_{1m} = J \omega_{n1} \left[\left(1 - \frac{\omega_{n1}}{\omega_{n2}} \right)^2 + \frac{m_2}{m_1} \right]^{-\frac{1}{2}} \tag{31.48}$$

The maximum deflection of the isolator given in Eq. (31.47) is shown graphically in Fig. 31.32. For small values of the ratio of support natural frequency to isolator natural frequency, the flexibility of the support may significantly reduce the maximum isolator deflection, especially if the mass of the support is small relative to the mass of the equipment. For large values of the frequency ratio, the effect of the mass ratio is small.

Maximum values of force in the support, given by Eq. (31.48), are shown in Fig. 31.33. The maximum deflection of the floor is the maximum force F_{1m} divided by the stiffness of the floor. The effect of mass ratio is profound for small values of the frequency ratio. The curves of Figs. 31.31 and 31.33 show corresponding results, the former including damping and the latter including the coupling effect between the two systems. The analysis which ignores the coupling effect may grossly overestimate the maximum force applied to the support at low values of the frequency ratio. At high values of the fre-

FIG. 31.32. Dimensionless representation of maximum isolator deflection δ_{2m} resulting from action of impulse J on equipment. Isolator and support have undamped linear elasticity. Ordinate is the ratio of maximum isolator deflection to that which would occur with a rigid support. Curves are for three discrete values of the mass ratio m_2/m_1 as indicated.

quency ratio, the two analyses yield like results if the fraction of critical damping in the isolator is less than about $\zeta_2 = 0.10$. The two methods are compared in Example 31.6.

Example 31.6. A forging machine weighs 7,000 lb exclusive of the 600-lb hammer. It is mounted at the center of a span formed by two 12-in., 50 lb/ft I beams having hinged ends and a span $l = 18$ ft. The hammer falls freely from a height of 60 in. before striking the work. Determine:

a. Maximum force F_{1m} in the beams and maximum deflection δ_{1m} of the beams if the machine is rigidly bolted to the beams.

b. The maximum force F_{1m} in the beams and the maximum deflection δ_{0m} of an isolator interposed between machine and beams.

Solution.

a. When the machine is bolted rigidly to the beams, the system may be considered to have only a single degree-of-freedom. The mass is that of the machine, plus the hammer, plus the effective mass of the beams. For the machine: $m_2 = (7,000 + 600)/386 = 19.2$ lb-sec²/in. The effective mass of the beams is taken as one-half of the actual mass:

$$m_1 = \frac{2(0.5)(18)(50)}{386} = 2.33 \text{ lb-sec}^2/\text{in.}$$

$$m = m_1 + m_2 = 21.5 \text{ lb-sec}^2/\text{in.}$$

The stiffness of the beams is

$$k = 2\frac{48EI}{l^3} = 2\frac{48 \times (30 \times 10^6) \times 302}{(18 \times 12)^3} = 123,000 \text{ lb/in.}$$

The natural frequency of the machine-and-beams system is

$$\omega_n = \sqrt{\frac{k}{m}} = \sqrt{\frac{123,000}{21.5}} = 75.6 \text{ rad/sec [12.0 cps]}$$

FIG. 31.33. Dimensionless representation of maximum force in support F_{1m} resulting from action of impulse J on equipment. Isolator and support have undamped linear elasticity. Ordinate is the ratio of maximum force in support to that which would result from direct application of the impulse to the support with equipment absent. Curves are for three discrete values of the mass ratio m_2/m_1.

If the impact between the hammer and work is inelastic and its duration is negligible, the resulting velocity $\dot{u}_m$ of the machine may be found from conservation of momentum. The impulse J is the product of weight of hammer and time of fall:

$$J = (600) \left(\frac{2 \times 60}{386}\right)^{1/2} = 335 \text{ lb-sec}$$

Then $\dot{u}_m = J/m = 335/21.5 = 15.6$ in./sec. If the damping of the beams is neglected, the maximum beam deflection is found from Eq. (31.18):

$$\delta_{1m} = \frac{\dot{u}_m}{\omega_n} = \frac{15.6}{75.6} = 0.21 \text{ in.}$$

The maximum force in the beams is the product of beam stiffness and maximum deflection:

$$F_{1m} = k\delta_{1m} = 25,300 \text{ lb}$$

b. An isolator having a stiffness $k_2 = 36,000$ lb/in. and a fraction of critical damping $\zeta_2 = 0.10$ is interposed between the machine and beams. The "uncoupled natural frequencies" defined in connection with Eqs. (31.40) and (31.41) are

$$\omega_{n2} = \sqrt{\frac{k_2}{m_2}} = \sqrt{\frac{36,000}{19.2}} = 43.3 \text{ rad/sec [6.9 cps]}$$

$$\omega_{n1} = \sqrt{\frac{k_1}{m_1}} = \sqrt{\frac{123,000}{2.33}} = 230 \text{ rad/sec [36.6 cps]}$$

Consider first that the system is separable. Figures 31.14 and 31.15 give, respectively: $\ddot{x}_{2m}/\dot{u}_m\omega_{n2} = 0.88$; $\ddot{x}_{2m}\delta_{2m}/\dot{u}_m{}^2 = 0.76$. Substituting $\dot{u}_m = J/m_2 = 17.4$ in./sec and solving for δ_{2m},

$$\delta_{2m} = \frac{0.76 \times 17.4}{0.88 \times 43.3} = 0.35 \text{ in.}$$

Entering Fig. 31.31 at $\omega_{n1}/\omega_{n2} = 5.3$, $F_{1m}/J\omega_{n1} = 0.23$. Then

$$F_{1m} = 17,700 \text{ lb}$$

Thus, the effect of the isolator is to reduce the maximum load in the beams from 25,300 lb to 17,700 lb. An isolator with less stiffness would permit a further reduction of this force at the expense of greater machine motion.

Consider now that the floor and machine-isolator systems are coupled, and use the two degree-of-freedom analysis which neglects damping. From Eq. (31.47):

$$\delta_{2m} = \frac{J}{m_2\omega_{n2}}\left[1 + \frac{m_2/m_1}{\left(1 + \dfrac{\omega_{n1}}{\omega_{n2}}\right)^2}\right]^{-\frac{1}{2}}$$

$$= \frac{335}{19.2 \times 43.3}\left[1 + \frac{19.2/2.33}{(1 + 5.3)^2}\right]^{-\frac{1}{2}} = 0.37 \text{ in.}$$

From Eq. (31.48):

$$F_{1m} = J\omega_{n1}\left[\left(1 - \frac{\omega_{n1}}{\omega_{n2}}\right)^2 + \frac{m_2}{m_1}\right]^{-\frac{1}{2}}$$

$$= 335 \times 230\left[(1 - 5.3)^2 + \frac{19.2}{2.33}\right]^{-\frac{1}{2}} = 14,900 \text{ lb}$$

Thus, the two results for the isolator deflection δ_{2m} differ only slightly, but the two degree-of-freedom analysis gives a maximum load in the beams about 16 per cent smaller than that obtained by assuming the systems to be separable.

REFERENCES

1. Mindlin, R. D.: *Bell System Tech. J.*, **24**:353 (1945).
2. McCalley, R. B., Jr.: "Velocity Shock Transmission in Two Degree Series Mechanical Systems," Atomic Energy Commission Contract W-31-109 Eng-52, Knolls Atomic Power Laboratory, Schenectady, March, 1956.
3. Klotter, K.: *J. Appl. Mechanics*, **22**:493 (1955).
4. Crede, C. E.: "Vibration and Shock Isolation," John Wiley & Sons, Inc., New York, 1951.

32

TYPES AND CHARACTERISTICS OF VIBRATION ISOLATORS

R. H. Racca
Barry Wright Corporation

INTRODUCTION

This chapter describes various types of vibration isolators as well as their characteristics. It treats the practical aspects of isolators, including elastomeric springs, plastic springs, and metal springs. Each type is described in terms of its typical configurations (such as shear and compression for elastomeric springs) and its application characteristics (such as load deflection and damping). Then detailed information is presented regarding commercial vibration isolators. This chapter supplements the theoretical treatment of vibration isolation and shock isolation discussed in Chaps. 30 and 31, respectively. It is followed by Chap. 33, which is devoted to air springs and servo-controlled isolation systems, and Chap. 34, which illustrates the application of isolators to different types of engineering problems and discusses the selection of isolators. Chapter 35 is entirely devoted to the properties of rubber and rubberlike materials (i.e., elastomers).

Many different materials are used as the resilient elements of isolators. The material used denotes the type of isolator. Each material has its peculiar advantages and disadvantages.

Elastomer (synthetic and natural rubber) isolators find wide application because they may be conveniently molded into many desired shapes and stiffnesses, have more internal damping than metal springs, usually require a minimum of space and weight, and can be bonded to metallic inserts adapted for simplified attachment to the isolated structures.

Plastic isolators have performance characteristics similar to those of the rubber-to-metal type of isolators used in equivalent configurations. Advantages include low cost and outstanding uniformity; disadvantages include a maximum operating temperature usually limited to 120°F (29°C).

Metal springs are commonly used where the static deflection required is large, where temperature or other environment conditions make elastomer unsuitable, and (in some circumstances) where a low-cost isolator is required. Where isolation is required at an even lower frequency, pneumatic isolators provide unusual advantages. They can be used in many of the same applications as metal springs but without certain disadvantages of the latter.

Other resilient materials, not discussed in this chapter, include felt, cork, sponge rubber, and some compound materials. (The characteristics of many cushioning materials used in packaging design are given in Chap. 41.) These materials are generally cut from large slabs and are used flat. They lack the ready adaptability of elastomeric parts which are molded into shape and are frequently adhered to metal inserts for easy application. Cork is often used to support relatively large concrete foundations. Felt is sometimes used under machinery; its application is convenient because it can be cemented in place and is generally unaffected by oils and other foreign substances. Glass fibers impregnated with a suitable binder may be used to form the resilient element of vibration isolators. Such isolators are relatively immune to environmental influences and are used extensively in military aircraft. Numerous other materials compounded in slab form are used for vibration isolation, particularly in conjunction with machinery.

CHARACTERISTICS OF ELASTOMERIC SPRINGS

THE FAMILY OF ELASTOMERS

The most commonly used type of isolator is the elastomeric isolator. Elastomers are natural rubbers or rubberlike materials. They are able to sustain large deformations and then return to their approximatley original state with virtually no damage. Elastomeric isolators are superior to other types of isolators in that for a given amount of elasticity, deflection capacity, energy storage, and dissipation they require less space and less weight; also, they may be molded into many different configurations of many different materials—generally at a lower cost than other types of isolators.

Elastomers have exceptional extensibility and deformability: They can be utilized at elongations of up to about 300 per cent, with ultimate elongations of some elastomers up to about 1,000 per cent. They may be stressed as much as or more than 1,000 to 1,500 lb/in.^{2} (0.145 to 0.218Pa) before their elastic limit is reached. Their great capacity for storing energy allows them to tolerate high stress. When the stress is released, there is virtually total recovery from the deformation. The inherent damping of elastomers is often useful in preventing an excessive vibration amplitude at resonance, an amplitude much lower than would occur if coil metal springs were used. Elastomeric materials provide excellent resistance to the transmission of noise through structures because of their relatively low specific acoustic impedance, which is an important consideration where acoustic requirements are a key factor.

Of the various elastomers, *natural rubber* probably embodies the most favorable combination of mechanical properties, such as minimum drift, maximum tensile strength, and maximum elongation at failure. Its usefulness is restricted by its limited resistance to deterioration under the influence of hydrocarbons, ozone, and high ambient temperatures. *Neoprene* and *Buna N (nitrile)* exhibit superior resistance to hydrocarbons and ozone, *Buna N* being particularly satisfactory for applications involving relatively high ambient temperatures. *Buna S* is a good general-purpose synthetic rubber for use in vibration isolators.

Silicone rubber is one of the costliest elastomers. Its properties are remarkably stable, and it provides effective isolation over a very wide temperature range: −65 to +350°F (−54 to 177°C). By comparison, Neoprene is limited in use to a range of about −40 to +200°F (−40 to 93°C). The upper temperature limit depends on the properties of the particular compound, the degree of deterioration which is permissible as a result of continued exposure at high temperatures, and the duration of exposure. For silicone a temperature substantially greater than 300°F (149°C) is permissible for several hours. The outstanding ability of silicone elastomers to withstand extremes of temperature is offset somewhat by inferior strength, tear resistance, and abrasion resistance.

Isolators fabricated of elastomers are complex in behavior because of the viscoelastic nature of elastomers in performance (somewhere between a solid and a liquid), because of their indefinite yield point, and because their physical properties vary with time, temperature, and environment. For example, rubber is a substantially incompressible material (it has a Poisson's ratio of approximately 0.5). Thus stiffness of a rubber spring when it is strained in compression depends, to a considerable extent, on the area of the surface available for lateral expansion. In contrast, the stiffness of a rubber spring in shear is substantially independent of the shape of the rubber member. As a rough rule of thumb, it may be assumed that the minimum likely compression stiffness of a given rubber spring is five times that of its shear stiffness. The maximum compression stiffness may be several times as great as the minimum value if lateral expansion of the rubber is constrained. See Chap. 35 for more detailed information concerning elastomers, as well as for definitions of many of the terms used in this chapter.

FAILURE FATIGUE AND PREMATURE FAILURE

Regardless of geometry, both elastomers and metals exhibit fatigue failure as a result of repeated cyclic loadings. Unlike metals, elastomers do not experience catastrophic-type fatigue failure. Instead the failure begins as tear at the point of highest cyclic shear strain, which is generally on the outer extremity (therefore, visible in many cases), and gradually propagates through the body of the elastomer. The result is a gradual reduction in stiffness that usually becomes apparent before there is total failure. The effect of strain on the fatigue life of rubber specimens, both in tension and compression, is illustrated is Fig. 35.11.

Fɪɢ. 32.1. Elastomeric isolators. (*A*) Bonded isolators. (*B*) Unbonded isolators. (*After R. Racca.*[7])

Most elastomeric isolators should not be subject to large static strains over long periods of time. An isolator with a large static deflection may give satisfactory performance temporarily, but the deflection tends to creep excessively over a long period. In general, elastomers should not be statically strained continuously more than 10 to 15 per cent in compression, nor more than 25 to 50 per cent in shear.

A factor contributing to the premature failure of an elastomeric isolator is the effect of the minimum strain on fatigue life as shown in Fig. 35.11. For elastomers which crystallize under high strains (such as Neoprene and natural rubber), fatigue life is greatly increased if the minimum cyclic stress is always either plus or minus and never passes through zero. Proper static precompression of the isolator within the limits specified above is often an effective way to prevent the minimum cyclic stress from passing through zero under dynamic conditions. Local stress concentrations, resulting in premature failure, often can be avoided by using fillets, radii, and generous overhangs of the elastomeric section. Sharp corners of metal inserts and support structures should be carefully rounded off wherever they contact the elastomer. Metal snubbing washers and/or support structures in contact with the elastomer should be large enough to prevent their edges from cutting into the elastomeric surfaces.

BONDED VERSUS UNBONDED ISOLATORS

Elastomeric isolators may be designed in both bonded and unbonded configurations. Figure 32.1 illustrates both types. In the bonded isolator, metal inserts are bonded to the elastomer on all load-carrying surfaces (see Fig. 32.1*A*). In the unbonded or semi-bonded isolator, the elastomeric load-bearing surface rests directly on the support structure (see Fig. 32.1*B*). Bonded parts usually cost more because of the special chemical preparation required to achieve a bond with a strength in excess of the elastomer itself. Bonded parts are generally preferred since they may be more highly stressed for a given deflection. With higher stress they provide higher spring constants and higher elastic energy-storage capacity.

Bonded isolators can be designed to provide proper load distribution in shear, compression, tension, or combination loading. A more uniform stress distribution in the elastomer is obtained by bonding inserts on all the load-bearing elastomer surfaces. The bonded inserts reduce unit stress by distributing the stress more uniformly throughout the volume of the elastomer. In contrast, unbonded parts usually fail to distribute the load uniformly, resulting in local areas of stress concentration in the elastomer body which shorten the life of the isolator.

A significant difference between bonded and unbonded elastomeric isolators relates to how elastomers behave under load. When an elastomer pad is compressed under load, its volume remains constant—only its shape is changed. The rubber bulges under load. When this ability to bulge is controlled, the load-deflection characteristics of the isolator are controlled. In a bonded isolator, the load-carrying surfaces have a fixed degree of bulge because the elastomer cannot move along the bond line, so it remains in a fixed position regardless of the load or environmental conditions.

In an unbonded isolator, this is not the case. The ability of the elastomer to bulge depends to a considerable degree on the maintenance of friction at the elastomer-support-structure interface. When all surfaces are clean and dry, the difference between the ability to bulge of a bonded and an unbonded isolator is negligible. But if oil or sand works its way into the elastomer-metal interface of the unbonded isolator, the ability of the elastomer to bulge is greatly increased; consequently its original load-deflection characteristics no longer exist. Then the isolator can exhibit load-deflection characteristics that are 50 per cent less than when the equipment was new; in many cases, this can cause the isolator to malfunction. Thus where consistent load-deflection characteristics are required for the life of the equipment, bonded isolators should be used. While the initial cost of unbonded isolators is lower, in many applications the cost of extra machining of the support structure and the reduced service life may well make unbonded isolators a poor selection.

TYPES OF LOADING

Elastomer isolators may be used with different types of loading: compression, shear, tension, or buckling—or with any combination of these.

COMPRESSION LOADING. The word *compression* is used to indicate a reduction in the dimension (thickness) of an elastomeric element in the line of the externally applied force. The stiffness characteristic of elastomers stressed in compression exhibits a nonlinear (hardening) which becomes especially pronounced for strains above 30 per cent. Compression loading, illustrated in Fig. 32.2A, is the most effective when used with simple unbonded isolators and where gradual snubbing (motion limiting) is required. Compression loading is frequently employed to provide a low initial stiffness for vibration isolation and a relatively high final stiffness to limit the dynamic deflections under shock excitations. Because of the nonlinear hardening characteristics of compression loading, it is the least effective type of loading for energy storage and, therefore, is not recommended where the attenuation of force or acceleration transmission is the primary concern. (The energy stored by any spring is the area under the load-deflection curve.)

SHEAR LOADING. Shear loading, illustrated in Fig. 32.2B, refers to the force applied to an elastomeric element so as to slide adjacent parts in opposite directions. An almost linear spring constant up to about 200 per cent shear strain is characteristic of elastomer stress in shear. This linear spring constant makes shear loading the preferred type for vibration isolators because it provides a constant frequency response for both small and large dynamic shear strains in a simple spring-mass system. Shear loading is also useful for shock isolators where attenuation force or acceleration transmission is important, because of its more efficient energy-storage capacity when compared to compression loading. However, care must be exercised to ensure that the expected dynamic loads do not result in shear strains that exceed the limits of the elastomer or that abrupt bottoming of the supported equipment does not occur. A modification of shear loading, sometimes listed as a separate type, is *torsional loading*, shown in Fig.

32.2*C*. It consists of winding up a sandwich of laminated sections to strain the elastomer in torsion. When the strain exceeds about 150 per cent, considerable axial thrust loads on connecting members are induced.

TENSION LOADING. Tension loading, illustrated in Fig. 32.2*D*, refers to an increase in the dimension (thickness) of an elastomeric element in the line of the externally

F $_{IG}$. 32.2. Load-deflection characteristics of typical elastomeric isolators. *(After R. Racca.[7])*

applied force. The stiffness characteristic of elastomers stressed in tension exhibit a nonlinear (softening) spring constant. For a given deflection, tension loading stores energy more efficiently than either shear or compression loading. Because of this, tension loading has been occasionally used for shock isolation systems. However, in general, tension loading is not recommended because of the resulting loads on the elastomer-to-metal bond which may cause premature failure.

BUCKLING LOADING. Buckling loading, illustrated in Fig. 32.2*E*, occurs when the externally applied load causes an elastomeric element to warp or bend in the direction of the applied load. Buckling stiffness characteristics may be used to derive the benefits of both softening stiffness characteristics (for the initial part of the load-deflection curve) and hardening characteristics (for the later part of the load-deflection curve). The buckling mode thus provides a high energy-storage capacity and is useful for shock isolators

Fɪɢ. 32.3. Maximum amplitude transmissibility for a viscous-damped single degree-of-freedom system when the forcing frequency changes at a constant rate in the region of the natural frequency of the system. *(After F. M. Lewis.[3])*

where force or acceleration transmission is important and where snubbing (motion limiting) is required under excessively high transient dynamic loads. This type of stiffness characteristic is exhibited by certain elastomeric cushioning foam materials and by specially designed elastomeric isolators. However, it is important to note that even simple compressive elements will buckle when the slenderness ratio (the unloaded length-width ratio) exceeds 1.6.

Combinations of types of loading described above are commonly used, which result in combined load-deflection characteristics. Consider, for example, a compression-type isolator which is installed at an angle instead of in the usual vertical position. Under these conditions, it acts as a compression-shear type of isolator when loaded in the vertical downward direction. When loaded in the vertical upward direction, it acts as a shear-tension combination type of isolator.

DAMPING

Damping, to some extent, is inherent in all resilient materials. The damping characteristics of elastomers vary widely. A tightly cured elastomer may (within its proper operating range) store and return energy with more than 95 per cent efficiency, while elastomers compounded for high damping have less than 30 per cent efficiency.

Damping increases with decreasing temperature because of the effects of crystallinity and viscosity in the elastomer. If the isolator remains at a low temperature for a prolonged period, the increase in damping may exceed 300 per cent. Damping quickly decreases with low-temperature flexure, because of the crystalline structure deterioration and the heat generated by the high damping.

Where the nature of the excitation is difficult to predict (for example, random vibration), it is desirable that the damping in the isolator be relatively great. Damping in an isolator is of the greatest significance at the resonance frequency. So it is desirable that

FIG. 32.4. Fraction of critical damping exhibited during vibration of a typical elastomeric material strained in shear, as a function of amplitude of dynamic shear strain for different frequencies. *(After G. W. Painter.[4])*

isolators embody substantial damping when they may operate at resonance, as is the case when the excitation is random over a broad frequency band or even momentary (as in the starting of a machine with an operating frequency greater than the natural frequency of the machine on its isolators). The relatively large amplitude commonly associated with resonance does not occur instantaneously, but rather requires a finite time to build up. If the forcing frequency is varied continuously as the machine starts or stops, the resonance condition may exist for such a short period of time that only a moderate amplitude builds. For a linear single degree-of-freedom system in which the forcing frequency varies uniformly through resonance, Fig. 32.3 shows the maximum amplitude transmissibility; the abscissa is the ratio of the rate of change of forcing frequency in hertz to the square of the natural frequency of the isolator expressed in hertz. Curves given for various values of the fraction of critical damping show that the rate of change of forcing frequency is of little importance for highly damped isolators but is of considerable importance for lightly damped isolators.

In general, damping in an elastomer increases as the frequency increases. The data of Figs. 32.4 and 32.5 can be used to predict transmissibility at resonance by estimating the frequency and the amplitude of dynamic shear strain; then the fraction of critical damping is obtained from the curves and used with Eq. (30.1) to calculate transmissibility at resonance.

USE OF AUXIALIARY DAMPERS. For some purposes, the damping in the load-carrying element may be adequate; for other purposes, the addition of a separate damper may be required. To determine whether auxiliary dampers are required, energy dissipation requirements should be compared with that available from internal damping of the main load-carrying element. Additional damping may be obtained by incorporating mechanical or other energy dissipation devices into a dynamic system. These may be either friction or fluid dampers built into isolator housings, or devices installed separately from the isolators. For example, most vehicular shock absorbers are fluid dampers used to augment a coil-spring and/or leaf-spring suspension system.

FIG. 32.5. Damping exhibited during vibration of Buna N synthetic rubber and natural rubber strained shear, as a function of frequency and amplitude of dynamic shear strain. *(After G. W. Painter.[4])*

STATIC AND DYNAMIC STIFFNESS

When the main load-carrying spring is made of rubber or of other similar organic materials, the natural frequency calculated using the stiffness determined from a static load-deflection test of the spring almost invariably is lower than the dynamic value. Thus the dynamic modulus appears greater than the static modulus. The ratio of moduli is approximately independent of the velocity of strain and has a numerical value generally between 1 and 3. This ratio increases significantly as the durometer increases, as illustrated in Fig. 32.6.

PLASTIC SPRINGS

Vibration and shock isolators are fabricated of resilient plastics which have performance characteristics similar to those of the rubber-to-metal type of isolators of equivalent configuration. The structural elements are manufactured from a rigid thermoplastic, and the resilient element is manfactured from a thermoplastic elastomer. These elements are compatible in the sense that they are capable of being bonded one to another by fusion. The most commonly used materials are polystyrene for the structural elements and butadiene styrene for the resilient elastomer.

The advantages of this type of spring are (1) low cost, (2) exceptional uniformity from piece to piece in dynamic performance and dimensional stability, and (3) very close tolerances which can be maintained. A disadvantage is that the operating termperature usually has a maximum value of about 120°F (49°C) and a minimum value of −40°F (−40°C). Within this operating range, plastic springs may be the appropriate choice when low cost and high volume are requirements. Some of the commercially available plastic isolators have snap-in attachments, which are especially advantageous in robotic assembly.

METAL SPRINGS

This section describes metal springs of the following types: helical springs (coil springs), ring springs, Belleville (conical or conical-disc) springs, involute (volute) springs, leaf and cantilever springs, and wire-mesh springs.

For all ferrous materials used in metal springs (except stainless steel), the values of the modulus of elasticity E and the modulus of rigidity G are 30×10^6 and 11.5×10^6 lb/in.^{2} (4370 and 1670 Pa), respectively, at room temperature. For Type 302 stainless steel, the corresponding values are 28×10^6 and 10×10^6 lb/in.^{2} (4070 and 1450 Pascals).

Physical properties, chemical compositions, and SAE numbers for steel bars commonly used for springs are given in Table 32.1 along with principal uses and special properties. Physical properties include tensile and torsional ultimate strengths and elastic limits, modulus of elasticity, and shear modulus. Similar data are given in Table 32.2 for flat spring steels and in Table 32.3 for nonferrous spring materials.

FIG. 32.6. Ratio of dynamic modulus of rubber to its static modulus, as a function of durometer.[5]

HELICAL SPRINGS (COIL SPRINGS)

Helical springs (also known as coil springs) are made of bar stock or wire coiled into a helical form as illustrated in Fig. 32.7. The load is applied along the axis of the helix; in a compression spring the helix is compressed; in a tension spring it is extended. The helical spring has a straight load-deflection curve as shown in Fig. 32.8. The spring is the simplest and most widely used energy-storage spring. Energy stored by the spring is represented by the area under the load-deflection curve (see Fig. 32.8).

Helical springs have the inherent advantages of low cost, compactness, and efficient use of material. They are widely used for shock and vibration control in machines.

Helical springs can be designed which have a low natural frequency when fully loaded. For example, such springs having a natural frequency as low as 2 Hz are relatively common. However, the static deflection of such a spring is about 2.4 in. (61 mm). For such a large static deflection, the spring must have adequate lateral stability or the mounted equipment will tip to one side. Therefore, all forces on the spring must be along the axis of the spring. For a given natural frequency, the degree of lateral stability depends on the ratio of coil diameter to working height. Lateral stability also may be achieved by the use of a housing around the spring which restricts its lateral motion. Helical springs provide little damping, which results in transmissibilities at resonance of 100 or higher. They effectively transmit high-frequency vibratory energy and, therefore, are poor isolators for structureborne noise paths unless used in combination with a resilient pad which provides the required high-frequency attenuation.

BASIC FORMULAS
Stress. The uncorrected shearing stress T in round-wire helical springs loaded axially by a force F in pounds (Fig. 32.8) is

$$T = \frac{8FD}{\pi d^3} \quad \text{lb/in.}^2 \tag{32.1}$$

Table 32.1. Properties and Compositions of Spring-steel Bars* (SAE† Committee[10] Associated Spring Corp.[11])

Name and composition	Chemical composition (major elements), per cent	Tensile properties		Torsional properties		Principal uses and special properties
		Ultimate strength, 10^3 lb/in.^{2}	Elastic limit, 10^3 lb/in.^{2}	Ultimate strength, 10^3 lb/in.^{2}	Elastic limit, 10^3 lb/in.^{2}	
Carbon steel, SAE 1085	C 0.80–0.93 Mn 0.70–1.00	175 to 230	130 to 175	115 to 150	80 to 105	Used for hot coiled and flat springs; has good hardenability
Carbon steel, SAE 1095	C 0.90–1.03 Mn 0.30–0.50	170 to 220	125 to 170	110 to 145	75 to 100	Used for heavy hot coiled and flat springs; does not fully harden in oil quench when diameter is greater than 0.375 in.
Alloy steel, SAE 4068	C 0.63–0.70 Mn 0.75–1.00 Si 0.20–0.35 Mo 0.20–0.30	200 to 270	175 to 240	145 to 200	105 to 145	Used for hot coiled springs, also for hot formed leaf springs
Chrome-vanadium alloy steel, SAE 6150	C 0.48–0.53 Cr 0.80–1.10 Mn 0.70–0.90 V 0.15 min Si 0.20–0.35	200 to 250	180 to 230	140 to 175	100 to 130	Resists heat better than carbon steels
Chrome-silicon alloy steel, SAE 9254	C 0.50–0.60 Mn 0.50–0.80 Si 1.20–1.60 Cr 0.50–0.80	250 to 325	220 to 300	160 to 200	130 to 160	Resists heat well to 450° F; used at high stresses
Silicon-manganese alloy steel, SAE 9260	C 0.55–0.65 Mn 0.70–1.00 Si 1.80–2.20	200 to 250	180 to 230	140 to 175	100 to 130	

* Sizes available 0.375 to 2.25 in. in diameter. Modulus of elasticity $E = 30 \times 10^6$ lb/in.^{2}; shear modulus $G = 11.5 \times 10^6$ lb/in.^{2}. (For hot-wound springs, use slightly lower values of E and G.) For larger sizes, SAE 8660 and 9262 may be used.
† Society of Automotive Engineers.

Table 32.2. Properties and Composition of Flat Cold-rolled Steel* (SAE† Committee[10])

| Name of material | Chemical composition (major elements), per cent | Tensile properties | | Principal uses and special properties |
		Ultimate strength, 10^3 lb./in.^{2}	Elastic limit, 10^3 lb./in.^{2}	
Flat spring steel, SAE 1050..........	C 0.48–0.55 Mn 0.60–0.90	160 to 270	120 to 210	Used for thin flat springs where severe forming is required
Flat spring steel, SAE 1060..........	C 0.54–0.66 Mn 0.60–0.90	160 to 270	120 to 210	Used for flat springs and stampings where higher carbon steel is not needed
Flat spring steel, SAE 1074..........	C 0.68–0.80 Mn 0.50–0.80	160 to 280	125 to 230	Very popular steel for miscellaneous flat springs Gives excellent life
Clock spring steel, SAE 1095..........	C 0.89–1.04 Mn 0.30–0.50	180 to 320	140 to 260	Supplied in heat-treated state Used for clock and motor springs and flat springs less than 0.06 in. thick

* Modulus of elasticity $E = 30 \times 10^6$ lb./in.^{2}
† Society of Automotive Engineers.

Table 32.3. Properties and Compositions of Nonferrous Spring Materials* (SAE Committee[10])

Name of material	Available wire sizes, in.	Typical chemical composition (major elem.), per cent	Tensile properties			Torsional properties			Principal uses and special properties
			Ultimate strength, 10^3 lb/in.^{2}	Elastic limit, 10^3 lb/in.^{2}	Young's modulus E, 10^6 lb/in.^{2}	Ultimate strength, 10^3 lb/in.^{2}	Elastic limit, 10^3 lb/in.^{2}	Shear modulus G, 10^6 lb/in.^{2}	
Spring brass	0.028 to 0.500	Cu 67.0, Zn 33.0	100 to 130	40 to 60	15	60 to 90	30 to 50	5.5	For low cost and low stresses / For electrical conductivity
Nickel silver ASTM B206-54 (wire) ASTM B122-55T (strip)	0.020 to 0.250	Cu 55.0, Zn 27.0, Ni 18.0	130 to 150	80 to 100	16	80 to 100	60 to 70	5.5	Superior to brass / Also used for its silver color
Phosphor bronze ASTM B159-54 (wire) ASTM B103-55 (strip)	0.028 to 0.500	Cu 90.0 min, Sn 8.0, P 0.3 max	100 to 150	60 to 110	15	70 to 110	50 to 85	6.5	For electrical conductivity / For higher stresses than brass
Silicon bronze	0.028 to 0.500	Cu 97.5, Si 2.5	100 to 150	60 to 110	15	70 to 110	50 to 85	6.5	Similar to phosphor bronze; less costly
Beryllium copper ASTM B197-52 (wire) ASTM B196-52 (rod) ASTM B194-55 (strip)	0.010 to 0.250	Cu 98.0, Be 2.0	160 to 200	110 to 150	19	100 to 130	70 to 100	7.0	High mechanical properties / Excellent electrical conductor; low hysteresis / Corrosion resistance like copper / Properties based on age-hardened conditions
Monel	0.028 to 0.500	Ni 65.0, Cu 35.0	130 to 170	90 to 125	26	80 to 110	60 to 80	9.5	Fair electrical conductor; slightly magnetic, moderate stresses to 400°F
"K" Monel	0.028 to 0.500	Ni 66.0, Cu 31.0, Al 3.0	160 to 200	110 to 150	26	100 to 125	70 to 90	9.5	High stresses to 450°F / Nonmagnetic to −150°F / Properties based on age-hardened conditions
Permanickel	0.028 to 0.500	Ni 97.0 min, Mn 0.4, Ti 0.4	180 to 230	130 to 170	30	110 to 150	80 to 110	11.0	High stresses to 550°F / Properties based on age-hardened condition
Inconel	0.028 to 0.500	Ni 76.0, Cr 16.0, Fe 8.0	160 to 200	110 to 150	31	100 to 125	70 to 90	11.0	High stresses to 700°F / Nonmagnetic to −40°F
Inconel X	0.028 to 0.500	Ni 70.0 min, Cr 16.0, Fe 8.0, Ti 2.5	160 to 200	110 to 150	31	100 to 125	70 to 90	11.0	Can be used up to 900°F / Properties based on age-hardened condition / Nonmagnetic to −280°F
Ni-Span C †		Ni 42.0, Ti 2.3, Cr 5.2, Fe balance	200	110	27.5			9.4	Zero temperature coefficient of modulus −50 to 150°F

* All materials may be obtained either cold-rolled of cold-drawn and all are corrosion resistant.
† Data on this alloy from Associated Spring Co.[4]

where d is the wire diamater in inches and D is the mean coil diameter in inches. The maximum shearing stress in the round wire is

$$T_{max} = K \frac{8FD}{\pi d^3} \quad \text{lb/in.}^2 \tag{32.2}$$

where K is the stress concentration factor which is a function of the spring index, $C = D/d$, and ranges from about 1.1 when $C = 14$ to about 2.0 when $C = 2$.

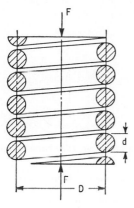

FIG. 32.7. Cross section of a helical spring of diameter D and wire diameter d, showing the direction of the applied force F.

FIG. 32.8. Load-deflection curve for a helical spring.

Static Deflection. The static deflection δ of a round-wire helical spring is

$$\delta = \frac{8FD^3 n}{Gd^4} \quad \text{in.} \tag{32.3}$$

where n is the number of active coils in the spring and G is the modulus of rigidity in pounds per square inch.

The axial stiffness k is

$$k = \frac{F}{\delta} = \frac{Gd^4}{8nD^3} \quad \text{lb/in.} \tag{32.4}$$

If the ends are squared and ground,[8] the number of active coils, n, used in Eq. (32.4) to calculate deflection may be taken equal to the total turns (tip to tip of bar) minus 1¾ turns.[9] For extension springs with full end loops, n may be taken as 1 plus the number of coils between points where the loops begin.

DYNAMICAL EFFECTS. A spring which is subject to rapid variation in load may experience additional dynamical effects which often increase the stress seriously. For example, particularly severe effects occur when a harmonic in the motion of the spring end has a frequency which coincides with one of the natural frequencies of the spring. Usually, the lowest natural frequency is of the most importance. For a spring compressed along the axis of the helix, the first mode of vibration (corresponding to the lowest natural frequency) has a node in the middle of the spring; maximum motion of the coils occurs at points one-fourth and three-fourths of the length from the end.

Natural Frequencies. The lowest natural frequency for a helical spring fully loaded in the axial direction is

$$f_n = \frac{2d}{\pi D^2 n} \sqrt{\frac{Gg}{32\gamma}} \quad \text{Hz} \tag{32.5}$$

(A) FULL LOOP   (B) HALF LOOP   (C) SMALL OFF-SET HOOK   (D) SMALL EYE   (E) LONG ROUND END HOOK

(F) V HOOK   (G) EXTENDED-EYE   (H) CONED .END TO HOLD LONG SWIVEL EYE   (I) CONED END WITH SWIVEL BOLT

FIG. 32.9. End designs for extension-type helical springs.

where d = wire diameter, in.
  D = coil diameter, in.
  G = modulus of rigidity, lb/in.^{2}
  g = acceleration of gravity = 386 in./sec^2 (980 cm/sec^2)
  γ = density of spring material, lb/in.^{3}
  n = number of active coils

For steel springs, $G = 11.5 \times 10^6$ lb/in.^{2} (1668 N/m^2) and $\gamma = 0.285$ lb/in.^{3} (7.9 g/cm^3). The natural frequency for a steel spring loaded axially is

$$f_n = \frac{14,000d}{D^2 n} \quad \text{Hz} \tag{32.6}$$

For a spring having one end free and the other fixed, Eqs. (32.5) and (32.6) apply where n is taken as twice the actual number of active turns in the spring.

For helical extension springs, particular consideration must be given to the spring ends. Tension springs may be wound with various types of end configurations, as illustrated in Fig. 32.9—some designs result in considerably higher stress concentration in the ends than other designs. The higher stress results in a considerable decrease in strength, particularly under fatigue loading. For example, lower stress concentration effects are present in the full loop shown in Fig. 32.9A. If the hook is at the side as in Fig. 32.9C, the moment arm of the load is much larger than that which would exist if a purely axial load were applied, thus resulting in a large increase in stress concentration. The coned-end designs shown in Fig. 32.9H and I have relatively low stress concentration in the end turns. For this reason threaded plugs which may be screwed to the ends of plain close-wound helical springs are used frequently.

HELICAL SPRINGS OF SQUARE OR RECTANGULAR BAR SECTION. Springs of square or rectangular bar section, such as that shown in Fig. 32.10, may be used to provide maximum energy storage within a limited space. This permits more material to be provided within a given outside diameter and length. Hence, greater energy storage is possible. However, it is difficult to obtain material in square or rectangular bars of quality that is equal to that of round bars.

The uncorrected stress T and the static deflection δ for a helical spring of square bar or wire section loaded axially by a force F in pounds are given by

$$T = \frac{2.4FD}{a^3} \quad \text{lb/in.}^2 \tag{32.7}$$

$$\delta = \frac{5.58FD^3 n}{Ga^4} \quad \text{in.} \tag{32.8}$$

where D = mean coil diameter, in.
  a = side of square section, in.
  G = modulus of rigidity, lb/in.^{2}
  n = numbers of active coils

For helical springs of square bar section, the corrected stress is found by multiplying the stress Eq. (32.7) by a factor slightly less than the stress concentration factor K for round wire, described above.[12]

RING SPRINGS

A ring spring, shown in Fig. 32.11A, absorbs the energy of motion in a few cycles—dissipating it as a result of friction between its sections. With a high load capacity for its size and weight, a ring spring absorbs linear energy with minimum recoil. Its linear load-deflection characteristic, shown in Fig. 32.11B, is similar to that of the helical spring; the isolator exhibits a maximum stress in the object being stopped which is equal to twice the average stress. Springs of this type often are used for loads of from 4,000 to 200,000 lb (1,814 to 90,720 kg), with deflections from 1 to 12 in. (25 to 305 mm).

BELLEVILLE SPRINGS

Belleville springs (also called *coned-disc springs*), illustrated in Fig. 32.12, absorb more energy in a given space than helical springs. Springs of this type are excellent for large loads and small deflections. They are available as assemblies, arranged in stacks. Their inherent damping characteristics are like those of leaf springs; oscillations quickly stop after impact. Coned discs of this type of spring have diametral cross sections and loading, as shown in Fig. 32.12. The shape of the load-deflection curve depends primarily on the ratio of the unloaded cone (or disc) height h to the thickness t. Some load-deflection curves are shown in Fig. 32.13 for different values of the ratio h/t, where the spring is supported so that it may deflect beyond the flattened position. For ratios of h/t approximately equal to 0.5, the curve approximates a straight line up to a deflection equal to half the thickness; for h/t equal to 1.5, the load is constant within a few per cent over a considerable range of deflection. Springs with ratios h/t approximating 1.5 are known

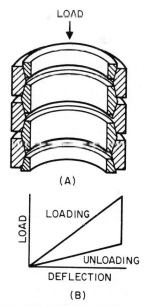

(A)

(B)

FIG. 32.10. Helical spring of diameter D, fabricated of square wire which is axially loaded by a force F. The wire becomes somewhat trapezoidal because of the coiling operation.

FIG. 32.11. Ring spring. (A) Cross section. (B) Load-deflection characteristic when it is loaded and when it is unloaded.

as *constant-load* or *stiffness springs*. Stacking Belleville springs in parallel (Fig. 32.14*A*) increases the load capacity; stacking the springs in series (Fig. 32.14*B*) increases the

F<small>IG</small>. 32.12. A Belleville spring comprised of coned discs of thickness *t* and height *h*, axially loaded by a force *F*.

deflection for a given load. The series arrangement should not be used for ratios *h/t* greater than 1.3 since instability and an irregular load-deflection characteristic may result.

Advantages of Belleville springs include (1) a small space requirement in the direction of the applied load, (2) the ability to carry lateral loads, (3) load-deflection characteristics which may be changed by adding or removing discs.

Disadvantages include nonuniformity of stress distribution, particularly for large ratios of outside to inside diameter. Design formulas for springs of this type are given in Refs. 14 and 15.

(A) STACKED IN PARALLEL

Fig. 32.13. The load-deflection characteristic for a Belleville spring having various ratios of *h/t*.

(B) STACKED IN SERIES

F<small>IG</small>. 32.14. Belleville spring. (*A*) Stacked in parallel. (*B*) Stacked in series.

F<small>IG</small>. 32.15. An involute spring. (*A*) Side view. (*B*) Cross section. (*C*) Load-deflection characteristic.

INVOLUTE SPRINGS

An involute spring shown in Fig. 32.15A and B can be used to better advantage than a helical spring when the energy to be absorbed is high and space is rather limited. Isolators of this type have a nonlinear load-deflection characteristic, as illustrated in Fig. 32.15C. They are usually much more complex in design than helical springs.[16]

LEAF AND CANTILEVER SPRINGS

Leaf springs are somewhat less efficient in terms of energy-storage capacity per pound of metal as compared with helical springs. However, leaf springs may be applied to function as structural members.[13] A typical semielliptic leaf spring is shown in Fig. 32.16. Design formulas for springs of this type are given in Ref. 13.

FIG. 32.16. Semielliptic leaf spring. FIG. 32.17. Simple cantilever spring.

SIMPLE CANTILEVER SPRING

Springs made from flat strip or bar stock may be formed into more or less complicated shapes. They often may be considered as simple cantilevers loaded by force F in pounds (Fig. 32.17). The deflection δ at the free end and the maximum bending stress σ are

$$\delta = \frac{4Fl^3}{Ebt^3} \quad \text{in.} \tag{32.9}$$

$$\sigma = \frac{6Fl}{bt^2} \quad \text{lb/in.}^2 \tag{32.10}$$

where l = length
b = width
and t = thickness (as shown in Fig. 32.17), in.
E = Young's modulus, lb/in.^{2}

Where the strip is wide compared to the thickness, as in most practical cases, slightly lower deflections than those calculated from Eq. (32.9) result.

This difference may reach values up to 10 per cent and occurs because lateral expansion or contraction of elements near the surface of the strip is prevented.

TRAPEZOIDAL-PROFILE CANTILEVER SPRING

Somewhat more efficient use of material is obtained by the use of the cantilever spring of trapezoidal profile, as shown in Fig. 32.18. The maximum bending stress is given by Eq. (32.10), where b is taken as the width at the built-in end. The deflection δ at the free end is obtained by multiplying the deflection of the simple cantilever [Eq. (32.9)] by a factor k_1, which depends on a/b and is given by the curve of Fig. 32.18 where a is the width at the loaded end.

WIRE-MESH SPRINGS

Knitted wire mesh acts as a resilient cushion with high damping characteristics and nonlinear spring constants. A circular knitting process is used to produce a mesh of multiple interlocking springlike loops. A wire-mesh spring, shown in Fig. 32.19, has a multidirectional orientation of the spring loops, i.e., each loop can move freely in three directions, providing a two-way stretch. Under tensile or compressive loads, each loop behaves like a small spring; when stress is removed, the loop immediately returns to its original shape. Shock loadings are limited only by the yield strength of the mesh material used. The mesh cushions, enclosed in springs, have characteristics similar to a spring and dashpot.

FIG. 32.18. Curve for finding factor K by which the deflection of the rectangular-profile cantilever should be multiplied to obtain the deflection of the trapezoidal-profile spring.

Commonly used wire-mesh materials include such metals as stainless steel, galvanized steel, Monel, Inconel, copper, aluminum, and nickel. Wire meshes of stainless steel can be used outside the range of -65 to $350°F$ (-53 to $177°C$) to which elastomers are restricted; furthermore, stainless steel is not affected by various environmental conditions that are destructive to elastomers. Wire-mesh springs can be fabricated in numerous configurations, with a broad range of natural frequency, damping, and radial-to-axial stiffness properties. Figure 32.20 shows a comparison between the load-deflection characteristics of a nonlinear knitted-wire-mesh isolator with those of a linear spring.

Wire-mesh isolators have a wide load tolerance coupled with overload capacity. The nonlinear load-deflection characteristics provide good performance, without excessive deflection, over a wide load range for loads as high as four times the static load rating.

FIG. 32.19. Wire-mesh spring, shown in section.

Stiffness for the nonlinear curve increases with load, resulting in increased stability and gradual absorption of overloads. Energy absorption increases in the amount indicated by the shaded area in Fig. 32.20. An isolation system has a natural frequency proportional to the ratio of stiffness to mass; therefore, if the stiffness increases in proportion to the increase in mass, the natural frequency remains constant. This condition is approached by the load-deflection characteristics of mesh springs. Figure 32.20 illustrates the following distinct advantages of such a nonlinear system: increased stability; resistance to bottoming out of the mounting system under transient overload conditions; increased shock protection; greater absorption of energy during the work cycle; and negligible drift rate.

Critical damping of 15 to 20 per cent at resonance is generally considered desirable for a wire-mesh spring (see Fig. 32.21). Environmental factors such as temperature, pressure, and humidity affect this value little, if at all. Damping varies with deflection: high damping at resonance and low damping at higher frequencies.

INTERACTION WITH SUPPORT STRUCTURE

An isolator must deflect if it is to isolate vibration; generally, the greater the deflection, the greater the isolation. The isolator functions by being soft enough to allow relative vibration amplitudes without transmitting excessive force to the support structure. It is often assumed in the selection of vibration isolators that the support structure is a rigid mass with infinite stiffness. This assumption is not true since if the foundation were infinitely stiff, it would not respond to a dynamic force and the isolator would not be needed. Since the foundation does respond to dynamic forces, its response must affect the components that are flexibly attached to it. In reality the support structure is a spring in series with the isolator, and springs in series carry the same force and deflect proportionally to their respective spring constants. Thus if the stiffness of the isolator is high compared to the stiffness of the foundation, the foundation will deflect more than

FIG. 32.20. The load-deflection characteristic of nonlinear wire-mesh spring of increasing stiffness compared with that for a linear spring.

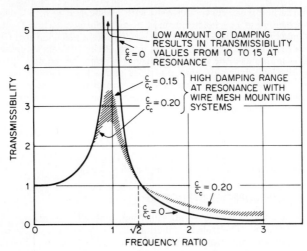

FIG. 32.21. Transmissibility vs. frequency for wire-mesh springs illustrating the effects of various amounts of damping. A fraction of critical damping, C/C_c, of 0.15 to 0.20 is usually desirable.

(A) (B)

FIG. 32.22A. Unbonded cup-type isolator. *Features:* low-profile compact design; all-attitude performance—can be used in tension, compression, or shear with equal performance; fail-safe design; snubbing in all directions; load ratings from 8 to 1,800 lb (3.6 to 816 kg); natural frequency down to 12 Hz; radial to axial dynamics stiffness ratio approximately equal to 1. *Typical applications:* refrigeration units, air conditioners, auxiliary motors, generators, fans, pumps, power plants, hydraulic and pneumatic equipment, installations where high transient (shock) loads are expected.

FIG. 32.22B. Semispherical elastomeric isolator. *Features:* bonded construction; nonlinear load-deflection characteristics; excellent energy-storage capacity; high deflection capability—good vibration and shock isolation; load ratings from 5 to 30 lb (2.3 to 13.6 kg); natural frequency down to 12 Hz; radial to axial dynamic stiffness equal to 1. *Typical applications:* electronic components, fans, portable generators, air conditioners, centrifuges, communication equipment, computers, applications where high energy storage is required.

the isolator and will actually nullify or limit the isolation needed from the isolator itself. To achieve maximum efficiency from the selected isolator, the spring constant of the support structure should be at least 10 times that of the spring constant of the isolator attached to it.[6] This will assure that at least 90 per cent of the total system spring constant is contributed by the isolators and only 10 per cent by the support structure.

CHARACTERISTICS OF COMMERCIAL ISOLATORS

Isolators are commercially available in many resilient materials, in countless shapes and sizes, and with widely diverse characteristics. In the U.S.A. alone there are more than 125 elastomeric isolator manufacturers, each offering a range of models in a variety of elastomeric compounds and natural rubber. The number is significantly higher if manufacturers of plastic, metal, pneumatic, and other-material isolators are included.

The properties of a given isolator are dependent not only on the material of which the isolator is fabricated but also on its configuration and the over-all construction with respect to the structural material used within the body of the isolator.

Many different specific types of isolators are available for application to original equipment manufacturer (OEM) machinery and equipment. Each type fills a different range of loads, sizes, stability, damping, and isolation efficiency requirements. Some of the more commonly used isolators types, illustrated in Fig. 32.22A to Z are given along with their performance characteristics (in terms of features and benefits and typical applications).

(C)   (D)

Fig. 32.22C. General-purpose compression isolator. *Features:* fail-safe design; snubbing in all directions; load ratings from 200 to 1,600 lb (91 to 720 kg); natural frequency down to 8 Hz; radial to axial stiffness ratio approximately equal to 0.3. *Typical applications:* internal-combustion engines, cabs, transmissions, pumps, motors, generators, compressors, installations where high transient (shock) loads are expected in the axial direction.

Fig. 32.22D. Bonded general-purpose isolator. *Features:* compact low-profile design; can be made fail-safe with special washer; load ratings from 40 to 1,800 lb (18 to 816 kg); natural frequency down to 7 Hz; radial to axial dynamic stiffness ratio equal to 1. *Typical applications:* internal-combustion engines, compressors, pumps, printing machinery, punch presses, fans, blowers, machine tools, rolling mills, installations where high transient (shock) loads are not expected, unless snubbing washer is included.

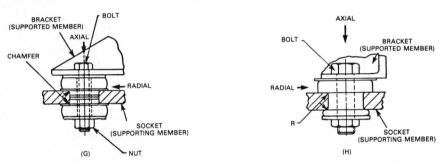

FIG. 32.22*E*. Fully bonded general-purpose isolator. *Features:* compact size; fail-safe design; snubbing in all directions; load ratings from 12.5 to 2,700 lb (5.7 to 1,225 kg); equal stiffness in axial and radial directions. *Typical applications:* internal-combustion engines, transmissions, cabs, air conditioners, pumps, radiators, generators, motors, compressors, in installations where high transient (shock) loads are expected.

FIG. 32.22*F*. Semibonded general-purpose isolator. *Features:* all-attitude capacity; fail-safe design; load ratings from 40 to 4,600 lb (18 to 2,090 kg); natural frequency down to 10 Hz; radial to axial dynamic stiffness ratio approximately equal to 1. *Typical applications:* internal-combustion engines, transmissions, cabs, radiators, air conditioners, pumps, motors, generators, compressors, in installations where high transient (shock) loads are expected.

FIG. 32.22*G*. Elastomeric sandwich isolator. *Features:* compact three-piece construction; fail-safe design; load ratings from 110 to 550 lb (50 to 250 kg); natural frequency down to 20 Hz in axial direction; axial to radial stiffness ratio from 4:1 to 7:1. Ideal isolator for focused systems. *Typical applications:* Internal-combustion engines, cabs, motors, generators, transmissions, pumps, installations where high transient (shock) loads are expected.

FIG. 32.22*H*. General-purpose semibonded isolator. *Features:* compact one-piece construction; fail-safe design; snubbing in all directions; load ratings from 75 to 2,100 lb (34 to 954 kg); natural frequency down to 10 Hz; radial to axial dynamic stiffness ratio approximately equal to 1. *Typical applications:* internal-combustion engines, transmissions, radiators, air conditioners, pumps, motors, generators, compressors, installations where high transient (shock) loads are expected.

Fig. 32.22*I*. Unbonded plate- and cup-type isolator. *Features:* compact, low-profile; fail-safe design—integral snubbing washers; snubbing in all directions; load ratings from 4 to 30 lb (2 to 14 kg); natural frequency down to 15 Hz; radial to axial dynamic stiffness ratio of 0.6. *Typical applications:* electronic components, motors, fans, pumps, communication equipment, air conditioners, refrigeration units, hydraulic and pneumatic equipment, installations where high transient (shock) loads are expected.

Fig. 32.22*J*. Elastomeric plate-form isolator. *Features:* compact construction; fail-safe design; all-attitude performance—can be used in tension, compression, or shear with equal performance; ease of installation—built-in snubbing washers; load ratings from 1 to 18 lb (0.5 to 8 kg); natural frequency down to 15 Hz; radial to axial dynamic stiffness ratio approximately equal to 1. *Typical applications:* pumps, fans, electronic and computer components, refrigeration units, installation where high transient (shock) loads are expected.

Fig. 32.22*K*. Bonded holder-type isolator. *Features:* standardized mounting dimensions; fail-safe design; low cost; load ratings from 2 to 310 lb (0.2 to 140 kg); natural frequency down to 7 Hz; radial to axial stiffness varies from 1:1 for low-frequency units to 2:1 for standard units. *Typical applications:* computers, printers, motors, electronic components, air conditioners, refrigeration units, portable generators, pumps, installation where high transient (shock) loads are not expected.

Fig. 32.22*L*. Cylindrical shear isolator. *Features:* compact design; low cost; ease of installation; load ratings from 9 to 140 lb (4 to 64 kg) in shear, 60 to 260 lb (27 to 118 kg) in compression; natural frequency down to 5 Hz in shear, 10 Hz in compression; shear to compression stiffness ratio of 0.13. *Typical applications:* internal-combustion engines, radiators, pumps, motors, generators, compressors, installation where high transient (shock) loads are not expected. Very good for focused systems.

FIG. 32.22*M*. General-purpose elastomeric isolator. *Features:* bonded construction; load ratings from 60 lb to 250 lb (27 to 113 kg); natural frequency down to 7 Hz; horizontal to vertical dynamic stiffness ratio of 0.4. *Typical applications:* electronic components, motors, fans, pumps, portable generators, air conditioners, centrifuges, compressors, vibratory feeders, woodworking and packaging machinery, used primarily for stationary applications where high transient (shock) loads are not expected.

Fig. 32.22*N*. General-purpose elastomeric compression isolator. *Features:* low-profile; compact design; bonded construction; load ratings from 700 to 4,400 lb (320 to 2,000 kg); natural frequency down to 8 Hz; horizontal to vertical dynamic stiffness ratio of 0.4. *Typical applications:* motors, fans, compressors, pumps, generators, metal and woodworking machinery, vibratory feeders, installations where high transient (shock) loads are not expected.

FIG. 32.22*O*. Elastomeric machinery isolator. *Features:* compact design, low-profile; leveling bolt attachment; special foundations or lagging machines to foundations not required; load ratings from 10 to 13,000 lb (4.5 to 5,900 kg); natural frequency down to 12 Hz. *Typical applications:* punch presses, milling and boring machines, injection molding machines, lathes, grinders, cold headers, drill presses, jig borers, printing machinery, installations that are stationary.

FIG. 32.22*P*. General-purpose elastomeric pneumatic isolator. *Features:* compact design; automatic leveling; zero static deflection under load; nearly constant natural frequency over rated load range; load ratings from 25 to 19,200 lb (11 to 8,700 kg); natural frequency down to 3 Hz; radial to axial dynamic stiffness ratio of unity. *Typical applications:* centrifuges, vibratory feeders, pumps, compressors, refrigeration units, air-conditioning equipment, fans, punch presses, cabs, precision optical equipment.

FIG. 32.22Q. Plate-form plastic isolator. *Features:* low cost, low profile; outstanding dimensional control; load ratings from 1.6 to 70 lb (0.7 to 32 kg); natural frequency down to 18 Hz; operating temperature −40 to +120°F (−40 to +50°C). *Typical applications:* data processing equipment, computers, printers, business machines, medial electronics and instrumentation, can be used wherever elastomeric isolators are used and where high temperature is not a problem.

FIG. 32.22R. Free-standing plastic isolator. *Features:* low cost, low-profile design; outstanding dimensional control; snap-in attachment for robotic assembly; load ratings from 0.6 to 6 lb (0.27 to 2.7 kg); natural frequency down to 20 Hz; nonlinear load-deflection characteristics—excellent shock attenuation; temperature range −40 to +120°F (−40 to +50°C). *Typical applications:* specifically designed for the computer and electronics industry, can be used wherever elastomer isolators are used and where high temperature is not a problem.

FIG. 32.22S. All-metal spring mesh-cup-type isolator. *Features:* fail-safe construction; load ratings from 0.55 to 75 lb (0.25 to 34 kg); natural frequency down to 8 Hz; amplitude-sensitive damping—transmissibility at resonance 4:1; radial to axial dynamic stiffness ratio approximately 0.3; wide temperature range −130 to +375°F (−90 to +190°C). *Typical applications:* electronic communication, electromechanical equipment that must operate under extreme environmental conditions.

FIG. 32.22T. All-metal spring mesh-plate-type isolator. *Features:* compact design; fail-safe construction; load ratings from 0.5 to 13 lb (0.2 to 6 kg); natural frequency down to 12 Hz; amplitude-sensitive damping—transmissibility at resonance 3.5:1; equal stiffness in axial and radial directions; wide temperature range −130 to +375°F (−90 to +190°C). *Typical applications:* electronic, communication, electromechanical equipment that must operate under extreme environmental conditions.

(U)

(V)

(W)

Fɪɢ. 32.22*U*. All-metal mesh isolator. *Features:* fail-safe design; high inherent damping; high overload capacity—can withstand shock loads as high as 100 g; load ratings from 70 to 1,600 lb (32 to 725 kg); natural frequency down to 14 Hz; wide temperature range −130 to +375°F (−90 to + 190°C). *Typical applications:* mobile engines, generators, pumps, compressors, heavy-duty applications where severe transient (shock) loads are encountered.

Fɪɢ. 32.22*V*. Wire-cable isolator. *Features:* compact construction, fail-safe design; amplitude-sensitive damping; transmissibility at resonance approximately 3:1 at high dynamic inputs; load ratings from 1 to 6,000 lb (0.5 to 2,720 kg); natural frequency down to 8 Hz; wide temperature range −100 to +500°F (−73 to +260°C); all-attitude shock protection. *Typical applications:* electronic and communications equipment, instruments, shipping containers, manufacturing machinery, installations where severe transient (shock) loads and environmental conditions are expected.

Fɪɢ. 32.22*W*. Air-damped metal-spring isolator. *Features:* natural frequency down to 7 Hz; amplitude-sensitive damping; high damping at resonance with maximum transmissibility of 3.5; good high-frequency isolation; load ratings from 0.03 to 35.0 lb (.14 to 16 kg); radial to axial dynamic stiffness ratio of 0.3; wide temperature range −65 to +300°F (−54 to + 149°C). *Typical applications:* delicate optical, processing, computer, and communication equipment.

Fig. 32.22X. Metal-spring isolators with friction damping. *Features:* compact design; high damping—transmissibility at resonance as low as 2:1; load ratings from 0.25 to 40 lb (0.1 to 18 kg); natural frequency down to 10 Hz; operating temperature range −85 to +250°F (−65 to 120°C). *Typical applications:* electronic and communication equipment that must operate in a severe broad-band vibration environment, installations where high transient (shock) loads are not expected.

FIG. 32.22Y. Metal-spring machinery isolator with friction damping. *Features:* adjustable friction damping; load ratings from 100 to 13,500 lb (45 to 6,125 kg); natural frequency down to 2.5 Hz; leveling attachment bolt. *Typical applications:* pumps, blowers, compressors, auxiliary power generators, high-speed punch presses, metal shears, drop hammers, installations that are stationary.

FIG. 32.22Z. Metal-spring machinery isolator. *Features:* low cost; load ratings from 10 to 12,400 lb (8 to 5,625 kg); natural frequency down to 2.5 Hz; leveling attachment bolt. *Typical applications:* pumps, fans, compressors, auxiliary power generator, installations that are stationary and operate at or near constant speed.

REFERENCES

1. Crede, C. E.: "Vibration and Shock Isolation," John Wiley & Sons, Inc., New York, 1951, p. 290.
2. Geiger, C. E., and J. P. Walsh: *J. Appl. Mech.*, **14:**A-7 (1947).
3. Lewis, F. M.: *Trans. ASME*, **54:**253 (1932).
4. Painter, G. W.: "Dynamic Properties of BTR Elastomer," *Paper* 83B, SAE National Meeting, Sept. 29. to Oct. 4, 1958.
5. Ref. 1, p. 221.
6. Racca, R.: "How to Select Power-Train Isolators for Good Performance and Long Service Life," *Paper* 821095, *SAE International Off-Highway Meeting and Expositon*, Sept. 13–16, 1982.
7. Racca, R.: "How to Select Isolators for a Long Service Life," *Design News*, Jan. 1982.
8. "Metals Handbook," 8th ed., American Society for Metals, Cleveland, Ohio, 1961, vol. 1, pp. 160–174.
9. Wahl, A. M.: "Mechanical Springs," 2d ed.; McGraw-Hill Book Company, Inc., New York, 1963.
10. "Design and Application of Helical and Spiral Springs," SAE J795a, Society of Automotive Engineers, Warrendale, Pa., 1973.
11. "Design Handbook," Associated Spring Corp., Bristol, Conn., 1970.
12. Liesecke, G; *Z. VDI*, **77:**425, 892 (1933).
13. "Manual on Design and Application of Leaf Springs," SAE J788a, Society of Automotive Engineers, Warrendale, Pa., 1970.
14. "Manual on Design and Application of Coned Disk Springs or Belleville Springs," SAE J798, Society of Automotive Engineers, Warrendale, Pa., 1971.
15. Almen, J. O., and A. Laszlo: *Trans. ASME*, **58:**303 (1936).
16. "Design and Manufacture of Volute Springs," SAE SP-11, Society of Automotive Engineers, Warrendale, Pa., 1945.

33

AIR SUSPENSION AND ACTIVE VIBRATION - ISOLATION SYSTEMS

Rodney W. Horning
Barry Wright Corporation

Dale W. Schubert
Barry Wright Corporation

INTRODUCTION

This chapter discusses (1) air springs for use in vibration-isolation systems and (2) active vibration-isolation systems.

Vibration isolation is achieved conventionally by supporting a body on a vibration isolator, as discussed in detail in Chap. 30. As a general rule, the natural frequency of the body-and-isolator system must be substantially lower than the excitation frequency. The deflection of the isolator under the deadweight load of the supported body is inversely proportional to the square of the natural frequency. Thus, the static deflection is large when the natural frequency is low. The combined requirement of a large static deflection and a large load-carrying capacity requires the storage of large quantities of potential energy in the isolator. The material used to construct mechanical springs has a limited energy-storage capacity per unit volume of material. Therefore, springs which must carry large loads and provide large static deflections tend to become excessively bulky.

A pneumatic spring is a spring which employs gas as its resilient element. Since the gas is usually air, such a spring usually is called an *air spring*. It does not require a large static deflection; *this is because the gas can be compressed to the pressure required to carry the load while maintaining the low stiffness necessary for vibration isolation.* The energy-storage capacity of air is far greater per unit weight than that of mechanical spring materials, such as steel and rubber. The advantage of air is somewhat less than would be indicated by a comparison of energy-storage capacity per pound of material because the air must be contained. However, if the load and static deflection are large, there usually is a large weight reduction resulting from the use of air springs. Because of the efficient potential energy storage of springs of this type, their use in a vibration-isolation system can result in a natural frequency for the system which is almost 10 times lower than for a system employing vibration isolators made from steel and rubber springs.

As an additional step toward providing improved vibration isolation, the isolator force acting on the suspended body can be servo-controlled to alter the passive vibration-isolation characteristics by controlling the motions of the suspended body at and below the passive natural frequency. Such servo-controlled isolation systems, referred to as *active vibration-isolation systems,* are described in a later section of this chapter.

AIR SPRINGS

An *air spring* consists of a sealed pressure vessel, with provision for filling and releasing a gas, and a flexible member to allow for motion. The spring is pressurized with a gas which supports the load. Air springs generally have lower resonance frequencies and smaller over-all length than mechanical springs having equivalent characteristics; they are employed where low-frequency vibration isolation is required. Air springs may require more maintenance than mechanical springs and are subject to damage by sharp and hot objects. The temperature limits are also restricted compared to those of mechanical springs.

There are numerous types of air springs. Figure 33.1 shows four of the most common types. The spring shown in Fig. 33.1*A* is available with one, two, and three convolutions. It has a very low minimum height and a stroke that is greater than its minimum height. The rolling lobe spring shown in Fig. 33.1*B* has a large stroke capability for applications which require large axial displacements. This isolator is frequently used in vehicle applications. The rolling diaphragm spring shown in Fig. 33.1*C* has a small stroke and is employed to isolate low-amplitude vibration. The spring shown in Fig. 33.1*D* has a low height and a small stroke capability. The thick elastomer sidewall can be used to cushion shock inputs. The isolators shown in Fig. 33.1*A* and *B* may have insufficient lateral stiffness for use without additional lateral restraint.

Since an air spring provides a simple means for resiliently supporting loads, it finds application as the load-carrying element in a vibration-isolation system. In the air spring shown in Fig. 33.2*A*, the load F supported by the spring is the product of the gage pressure P_g and the effective area S, $F = P_g S$. For a given area, the pressure may be adjusted to carry any load within the strength limitation of the cylinder walls. Since the cross section of many types of air springs may vary, it is not always easy to determine. For example, the spring shown in Fig. 33.1*A* has a maximum effective area at the minimum height of the spring and a smaller effective area at the maximum height. The spring illustrated in Fig. 33.1*B* is acted on by a piston which is contoured to vary the effective area. In vehicle applications this is often done to provide a low spring stiffness near the center of the stroke and a higher stiffness at both ends of the stroke in order to

Fɪɢ. 33.1. Four common types of air springs. (*A*) Air spring with convolutions. (*B*) A rolling lobe air spring. (*C*) Rolling diaphragm air spring. (*D*) Air spring having a diaphragm and an elastomeric sidewall.

limit the travel. The effective areas of the springs illustrated in Fig. 33.1C and D are usually constant throughout their stroke; the elastomeric diaphragm of the spring shown in Fig. 33.1D adds significantly to its stiffness.

STIFFNESS

The stiffness k of a spring of effective area S is the derivative of the force F with respect to the displacement δ:

$$k = \frac{dF}{d\delta} = S\frac{dP}{d\delta}$$

The stiffness of the air spring of Fig. 33.2A is derived from the gas laws governing the pressure and volume relationship. Assuming adiabatic compression, the equation defining the pressure-volume relationship is

$$PV^n = P_iV_i^n \qquad (33.1)$$

where P_i = absolute gas pressure at reference displacement
V_i = corresponding volume of contained gas
n = ratio of specific heats of gas, 1.4 for air

Fig. 33.2. Illustrations of a single-acting (A) and a double-acting (B) air spring consisting of a piston and a cylinder.

Written in terms of the force applied to the air spring, Eq. (33.1) becomes

$$\frac{F}{S}V^n = P_iV_i^n \qquad (33.2)$$

The relationship between the volume V and the deflection δ of the spring of Fig. 33.2A is

$$V = V_i - S\delta$$

Substituting this expression into Eq. (33.2) and taking the derivative $dF/d\delta$, the stiffness k as a function of displacement δ at any position is

$$k = \frac{dF}{d\delta} = \frac{nP_iS^2}{V_i}\left[\frac{1}{1-(S/V_i)\delta}\right]^{n+1}$$

Assuming that the area S is constant, if the change in volume is small relative to the initial volume V_i (i.e., if $S\delta \ll V_i$), then the expression for stiffness reduces to

$$k = \frac{nP_iS^2}{V_i} \qquad (33.3)$$

For cases where the effective load-supporting area of the air spring varies with the deflection δ, i.e., $S = f(\delta)$, the pressure P and area S both must be treated as variables. The resulting expression for stiffness is

$$k = \frac{dF}{d\delta} = S\frac{dP}{d\delta} + P\frac{dS}{d\delta}$$

Variation of the area S with deflection provides a means of minimizing the nonlinearity inherent in the load-deflection characteristics of a constant-area air spring. Controlled variation of the load-area relationship also provides means for obtaining any desired nonlinearity other than that normally available in the constant-area air spring. The height of the column of air, h, under the piston of area S in Fig. 33.2A is given by

$$h = \delta n + \frac{14.7}{P_g} \delta n \tag{33.4}$$

where P_g is the gage pressure, and δ is the static deflection of an equivalent spring.

The natural frequency f_n of an undamped air spring is expressed by

$$f_n = 3.13 \left(\frac{k}{W} \right)^{1/2} \tag{33.5}$$

where W is the weight in pounds.

For most air springs the stiffness of the spring and the volume V is furnished by the manufacturer as a function of the position and pressure. The stiffness of the spring varies with the pressure or load and is not a constant as with many types of mechanical springs. For most air springs the stiffness increases linearly as a function of the absolute internal pressure (gage pressure + atmospheric pressure).* The mass supported by the spring is proportional to the internal gage pressure. The natural frequency of an air spring increases or decreases much less as a function of load then a mechanical spring does. The change in the natural frequency Δf of an air spring is given by

$$\Delta f = f_n \left\{ 1 - \left[\frac{P_1/(P_1 + 14.7)}{P_2/(P_2 + 14.7)} \right]^{1/2} \right\} \tag{33.6}$$

where P_1 = original gage pressure
P_2 = new gage pressure
f_n = original natural frequency

Therefore, large changes in load at high pressure have little effect on the natural frequency while large changes in load at low pressures have a much greater effect.

ADDED RESERVOIR VOLUME. The axial stiffness of an air spring is inversely proportional to the volume as shown by Eq. (33.3). To increase the volume of an air spring, an air reservoir may be connected to the air spring via a large passageway. The increased volume then reduces the natural frequency as the square root of the increased volume. If the air volume is too large, an air spring, which has an effective area S that decreases as the spring is compressed, can become unstable.

If $P_1 S_1 = P_2 S_2$, the natural frequency is 0 Hz; however, if $P_1 S_1 > P_2 S_2$, the spring constant is negative and the spring is unstable, where P_1 and S_1 are the initial pressure and area, and P_2 and S_2 are the pressure and area when the spring is compressed. The diaphragm or flexible sidewall of most air springs contributes to the axial spring stiffness. This effect is most pronounced at very small displacements. It is usually very difficult to obtain natural frequencies of much less than 1 Hz even with a very large reservoir, because of the stiffness of the spring's diaphragm or sidewall.

TRANSVERSE (LATERAL) FEATURES. The transverse (i.e., lateral) stiffness of air springs illustrated in Fig. 33.1A and B varies from very small to moderate; the natural frequencies for such springs varies from 0 to 3 Hz. The spring illustrated in Fig. 33.1C has a higher transverse stiffness, with natural frequencies ranging from 2 to 8 Hz. The spring illustrated in Fig. 33.1D has a moderate transverse stiffness; the natural frequency varies in the range from 3 to 5 Hz. If an installation requires the selection of an air spring having insufficient transverse stiffness, additional springs in the transverse direction are often employed for stability, as shown in Fig. 33.3.

RATIO OF SPECIFIC HEATS. At frequencies above 3 Hz, the compression of gases used in air springs tends to be adiabatic and the ratio of specific heats n for both air and

---

* Atmospheric pressure at sea level equals 14.7 lb/in.^{2} in the English system of units, and it equals 1 Pa in the International System of Units.

nitrogen has a value of 1.4. At frequencies below approximately 3 Hz, the compression tends to be isothermal and the ratio of specific heats n has a value of 1.0, unless the spring is thermally insulated. For thermally insulated springs, the transition from adiabatic to isothermal occurs at a frequency of less than 3 Hz. Gases other than air which are compatible with the air spring materials can also be used. For example, sulphur hexafluoride (SF_6) has a value of n equal to 1.09—a value that reduces the axial spring stiffness by 22 per cent; it also has a considerably lower permeation (leakage through the air spring material) rate than air which may reduce the frequency of recharging (repressurizing) for a closed (passive) air spring.

Fig. 33.3. An air spring used to support a load and provide vibration isolation in the vertical direction. In addition, air springs are provided on the sides to increase the transverse stiffness.

SELECTING AN AIR SPRING

In selecting air springs, the following information should be known:
1. Function of the device to be isolated.
2. Total weight to be isolated (including the device and any necessary frame or foundation that must be isolated).
3. Configuration of the device and the location of the center-of-gravity.
4. Number and location of the mounting points.
5. Weight to be supported at each mounting point.
6. Space available for the air spring.
7. Displacement required (stroke).
8. Definition of the disturbances including axis, frequency, amplitude, waveform, and origin.
9. Transmissibility required.
10. Air or gas pressure available for charging the air springs.

STEPS IN SELECTING AN AIR SPRING

1. Determine the number of air springs required.
2. Determine the load capacity and effective area required for each air spring, based on the available air or gas pressure.
3. Determine the area and height available for each air spring.
4. Determine the vertical and lateral natural frequencies and spring constants required to meet the isolation requirements.
5. Determine the air spring stroke required.
6. Select air springs meeting all of the preceding criteria. It is generally advantageous from the standpoint of both size and isolation to select the lowest-load-capacity air spring that will support the load and meet the other criteria.

Example 33.1. Isolate an air compressor and motor mounted on a rigid base. The reduction in the vibration forces transmitted to the floor must be a minimum of 80 per cent. The total weight of the compressor, motor, and base is 10,000 lb (4,545 kg), and the center-of-gravity is at the center of the frame. The disturbing vibration force (from the compressor) is 1,000 lb (454 kg) at 600 rpm in the vertical axis. All other forces are much lower. The base can be supported at the four corners, and there is unlimited space available for the isolators. The available air pressure is 80 lb/in.^{2} (5.44 Pa).

Step 1. Four isolators are required.
Step 2. The total weight to be isolated is 10,000 lb (4,545 kg) and the center-of-gravity is at the

center. Therefore, the load on each isolator is 2,500 lb (1,136 kg). The effective area required is 2,500 lb/80 lb/in.^{2} = 31 in.^{2} (200 cm^2).

Step 3.   The space available for the isolators is unlimited. Therefore, any size isolator can be selected.

Step 4.   The vibration-isolation system must provide a minimum of 80 per cent isolation at 600 rpm (10 Hz) in the vertical axis. Using transmissibility curves, an air spring with a natural frequency of 3 Hz in the vertical direction will provide 80 per cent isolation (i.e., a transmissibility of 0.2) at 10 Hz. The spring constant k for each isolator can be determined from Eq. (33.5):

$$k = \left(\frac{f_n}{3.13}\right)^2 W \qquad k = \left(\frac{3}{3.13}\right)^2 \times 2,500 = 2,448 \text{ lb/in. (44 kg/mm)}$$

Step 5.   The motion of the isolated motor, compressor, and base will be 0.02 in. (0.5 mm) double amplitude. The air spring must have a minimum stroke of 0.02 in. (0.5 mm).

Step 6.   Select an air spring that meets all the above criteria. Since the required stroke is small, a single-convolution air spring, Fig. 33.1*A*, or the air springs shown in Fig. 33.1*C* or *D* would be suitable. The air springs must have an area slightly greater than 31 in.^{2} (200 cm^2) and a stiffness constant of 2,448 lb/in. (44 kg/mm) or less.

DOUBLE-ACTING SPRING

The stiffness expression given by Eq. (33.3) is for a single-acting spring, illustrated in Fig. 33.2*A*. The stiffness of the double-acting air spring illustrated in Fig. 33.2*B* is given by the expression

$$k = \frac{nP_1S_1^2}{V_1} + \frac{nP_2S_2^2}{V_2} \tag{33.7}$$

If the force on the air spring is zero when the piston is in its central position, $P_1 = P_2 = P_i$ and $V_1 = V_2 = V_i$. Then the stiffness is

$$k = \frac{2nP_iS^2}{V_i}$$

This equation assumes $S_1 = S_2 = S_i$; this is approximately correct when the piston rod is small compared to the piston area.

The advantage of the double-acting spring of Fig. 33.2*B* over the single-acting spring of Fig. 33.2*A* is its ability to provide a restoring force in either the positive or negative direction. The single-acting spring is used in applications where the force acting on the spring is unidirectional. Two examples of such applications are automobile suspensions and vibration isolators for industrial machinery. An example of a double-acting air spring application is found in vibration-isolation systems for equipment used on rocket-propelled missiles which experience large sustained accelerations.

FIG. 33.4. A self-damped double-acting air spring in which the damping is achieved by placing a capillary flow resistance between the cylinder of the air spring and the connecting surge tanks.

DAMPED AIR SPRING

The preceding discussion of a closed air spring is limited to the stiffness of the spring. Air springs have some inherent damping that is developed by damping in the flexible diaphragm or sidewall, friction, damping of the gas, and nonlinearity. The damping varies with the vibration amplitude; however, it generally is between 1 and 5 per cent of critical damping. If additional damping is required, an air spring that is self-damped, as illustrated in Fig. 33.4, may be designed. It is shown in Ref. 1 that the stiffness of a spring of this type is given by

$$k = \frac{F}{\delta} = \frac{2nP_iS^2}{V_c}\left[1 + \frac{\dfrac{1}{V_t/V_c}}{1 + (V_t/nC_rP_i)D}\right] \qquad (33.8)$$

EFFECT OF FREQUENCY. As the piston of the device illustrated in Fig. 33.4 is moved so that δ is positive, the gas in the B side of the cylinder is compressed and flows through the connecting pipe to the region of lower pressure in the surge tank. If the piston is left in its new position, the pressure in the tank eventually becomes equal to the pressure in the cylinder and the flow of air stops. If the piston now is moved back to its original position, the expansion of the air in the cylinder results in a lower pressure and the air flows in the reverse direction until the pressures in the tank and cylinder again become equal. Similar conditions apply to the A side of the piston. Thus, the pressure developed when the piston is held displaced from its neutral position is a function of the combined volumes of the cylinder and tank. This then is also true when the piston moves with a low-frequency vibration. Then the expression for stiffness in Eq. (33.8) becomes

$$k_0 = \frac{2nP_iS^2}{V_t + V_c} \qquad \text{lb/in.} \qquad (33.9)$$

As the vibration frequency increases, the quantity of air which can move from one volume to the other decreases because there is less time for flow to take place. Eventually, for relatively high frequencies, essentially no flow can take place and the surge tank becomes effectively cut off from the cylinder volume. At high frequencies, the stiffness of the system is

$$k_\infty = \frac{2nP_iS^2}{V_c} \qquad \text{lb/in.} \qquad (33.10)$$

At low frequencies, the compression process tends to become isothermal rather than adiabatic. The ratio of specific heats n which has a value of 1.4 for adiabatic conditions, has a value of 1.0 for isothermal conditions. For frequencies below approximately 3.0 Hz, isothermal conditions exist unless special pains have been taken to insulate the system thermally. For thermally insulated systems, the transition to an isothermal process occurs at frequencies below 3.0 Hz.

ACTIVE VIBRATION-ISOLATION SYSTEMS

An isolator that must operate over a relatively large load range and that has the requirement for a low natural frequency, as in the spring system of an automobile, presents design problems which cannot be overcome by use of either the mechanical spring or the air spring alone. Because of the low stiffness of the spring, a small change in the weight of the supported body results in a large change in the static deflection of the spring. In many instances, this cannot be tolerated. For example, an automobile with a spring system having an undamped natural frequency of 0.75 Hz requires springs having a static deflection of 17.5 in. (0.445 m) [see Eq. (2.10)]. A 30 per cent change in the weight supported by the automobile suspension results in a 5.25-in. (0.133-m) change in its normal height. This change in position is too large to allow proper operation of the springs over the required load range, i.e., approximately equal clearance travel in the up and down directions. A similar problem arises in a rocket-propelled missile where a substantially constant acceleration is sustained for an appreciably long time. This deflects an isolator in the same manner as would an increase in weight of the supported equipment. The clearance around such equipment must be large to provide for unobstructed displacement of the equipment under the influence of large steady acceleration experienced by the missile. Space on such missiles is extremely limited.

The large deflection experienced by a conventional spring element can be reduced by use of a servomechanism which compensates for load changes, thereby maintaining the supported body at a given position with respect to its vehicle. Figure 33.5 illustrates such a device schematically. As the force F is applied slowly, the spring of stiffness k stretches and the supported body of mass m moves downward. The relative motion between the body and the reference plane of the support causes the relative displacement sensing device to respond, requiring power to be supplied to the servo actuators which move the frame upward until the body returns to its original position with respect to the support plane. This position of equilibrium exists until another change in the force occurs. The element which produces the change in position of the frame is a servomechanism arranged to maintain the relative displacement δ at zero in the presence of the constant force F.

The isolation system illustrated in Fig. 33.5 always seeks as its equilibrium position a location at a distance h above the reference plane of the support, independent of the origin and magnitude of a steady force applied to the supported body. There is no change in the position of the body in response to a very slowly applied load, e.g., whether the isolator undamped natural frequency is 1 or 10 Hz. Thus certain disadvantages attendant on the use of isolators of low natural frequency are removed. When the load is applied suddenly, however, the relative displacement regulation servomechanism may be unable to respond fast enough to compensate for the tendency of the supported body to change position relative to the support; then the isolator can experience a significant deflection.

FIG. 33.5. Schematic diagram of an active vibration-isolation system which maintains the supported body m a fixed distance h from the reference plane of the support—irrespective of the steady force F applied to the supported body.

Active isolation systems are employed when the damped natural frequency of the isolation system must be relatively low—with the added requirement that the supported body be maintained at a relatively constant distance from the base to which it is attached. The position of the body, i.e., the deflection of the isolation system, can be controlled by means of a servomechanism. Likewise, low-frequency vibration isolation along with the isolation system resonance, i.e., resonance frequency and peak transmissibility, can be controlled by means of another servomechanism. Both these servomechanisms are made up of three basic components: (1) a sensing element which generates a signal proportional to a measurable isolation system motion, (2) computational elements to proportionally scale and integrate or differentiate the sensing element response, and (3) a motor, or actuator, element to generate a force applied to the supported body which is proportional to the signal from the computational element. The force output of the motor, or actuator, then supplements the spring forces to help support the supported body and provide forces which modify the isolator passive response to lessen or eliminate the effects of the system resonance. The servomechanisms formed are negative feedback regulators, a form of automatic control system. The term *negative feedback* is used since the force motor is arranged so that the feedback force applied to the supported body opposes its movement. An isolation system incorporating one or more servomechanisms is called an *active isolation system.*

The computational element for the elimination of the isolator static deflection is that of an integrator and scaling term called a *controller gain.* This combination of sensing, computation, and actuation provides what is known as *integral control*, since the feedback force is proportional to the time integral of the sensor response. The computational element for the control of the system resonance and low-frequency vibration isolation require only a scaling term. This combination of control elements is called *proportional control*, since the feedback force is proportional to the sensor response. The feedback elements added to a conventional isolation system must have an over-all characteristic such that the output force is proportional to the sensed function times the control function of the computational element. The control function describes the operation of the computational element, which can be a simple constant as in proportional control, an integration function as in integral control, or an equation describing the action of one or more electric circuits. This corresponds to a spring which provides an output force proportional to the deflection of the spring, a viscous damper which provides a force proportional to the rate of deflection of the damper, or an electric circuit which produces a force signal proportional to the dynamics of a spring and viscous damper, in series, undergoing a motion proportional to the sensor response.

The sensing and actuation devices which provide integral control of the isolator relative displacement may take many forms. For example, the sensing element which measures the position of the supported body relative to the reference plane of the support may be a differential transformer which produces an electrical signal proportional to its extension relative to a neutral position. The sensing element is attached at one end to the supported body and at the other end to the isolator support structure in a manner so that the sensor is in its neutral position when the supported body is at its desired operating height. The electrical signal is integrated and amplified in the computational element providing electric power to operate an electric motor actuation device. The differential transformer-integrator-motor system produces a force proportional to the integral of the signal from the differential transformer. The operation of this servomechanism can be visualized in the following manner:

1. A force F of constant magnitude F_0 is applied to the supported body causing a relative deflection of the isolator spring element.

2. The sensing element, in this case a differential transformer, applies an electrical signal that is proportional to the isolator relative displacement to the integration and scaling functions in the computational element.

3. The response of the computational integration function generates an electrical signal that continues to increase in magnitude so long as the relative displacement δ is not zero.

4. The signal from the computational element is applied to the motor element which generates a force in a direction that decreases the isolator deflection; the motor force follows the computational element signal and continues to increase in magnitude so long as the relative deflection δ is not zero.

5. At some point in time the force from the motor output will exactly equal the constant force F_0, requiring a relative displacement of zero.

6. The output from the differential transformer is zero; thus the output from the computational element integration function no longer increases but is maintained at a constant magnitude required for the motor element to generate a force exactly equal to the constant force, F_0, applied to the supported body.

The isolation system remains in this equilibrium condition until the force applied to the supported body changes and causes a nonzero signal to be generated by the sensing element, and the process starts all over again. Alternatively, a proportionally scaled signal from the differential transformer may be used to operate an electromechanical servo valve—the flow response of the servo valve being proportional to its excitation signal. The servo valve fluid flow output is directed into the chamber of an air spring to produce the desired force applied to the supported body. The control function remains integral in nature since the actuator's internal pressure responds to the volume output from the servo valve, which is the integral of its flow output. Thus, in this case, no electrical integration of the sensor signal is needed. It is also possible to operate a mechanical servo valve through a direct mechanical coupling whereby the motion of the suspended body with respect to its support is used directly to provide the required servo valve actuation. The possible combination of elements and control devices is almost limitless. The choice of a suitable combination of sensor, computation element, and actuator is dictated by the type of power available, the supported body size, the weight, and the type of application, e.g., spacecraft, aircraft, automotive, or industrial.

The second servomechanism, providing active control of the system resonance and low-frequency vibration isolation, consists of a velocity sensor, proportional computational element, and a motor actuation device that may also take on many forms. The velocity

FIG. 33.6. Schematic diagram of an active vibration-isolation system which acts like a passive vibration-isolator mass and spring element with a viscous damping element connected between the supported body and motionless fixed space. The active damping servomechanism can eliminate the isolation system resonance thereby providing vibration isolation starting at zero frequency.

of the supported body may be sensed by integrating the response of an accelerometer or through the use of an electromechanical sensor measuring velocity directly. Figure 33.6 illustrates the elements of this servomechanism; the servo amplifier contains the system electronic devices which form the computational element as well as the power elements required to operate the force actuator. The motor element is contained partly in the servo amplifier and partly in the force actuator. This shows that the three basic elements of a servomechanism are not always self-contained devices, but are made up from the combined operation of system hardware components. The force actuation device usually consists of an electromagnetic force actuator consisting of a strong magnet and coil similar to the type utilized in high-quality loudspeakers. This arrangement generally provides sufficient force to operate active isolation systems used to attenuate naturally occurring ground and building structure vibration. Where a larger force is required, as is the case where the system is subjected to larger-magnitude vibratory excitations or to vibratory forces applied directly to the supported body, the force motor may consist of driving elements similar to electrodynamic vibration exciters (see Chap. 25). The electronic amplifiers driving the force motor must have a frequency response extending down to zero frequency, so as not to introduce timing errors into the control signal that can significantly alter the response of the servomechanism. The velocity sensor-amplifier-motor system comprising this servomechanism applies a force to the supported body that is proportional to the body's velocity and thus acts in the same manner as a viscous damper connected to the supported body at one end and to motionless fixed space at the other end. This produces a form of damping within the active vibration-isolation system which cannot be synthesized using passive damping elements alone. The action of this velocity-controlled servomechanism is referred to as *active damping*, and the active damping scaling term G_2, relating the supported body velocity to the force applied to the mass m, when divided by the critical damping term for the passive spring and mass elements $2\sqrt{km}$ is commonly referred to as the *active fraction of critical damping* G_2/c_c.

An active isolation system usually is described by a cubic or higher-order differential equation; because of the complexity of these equations, it is difficult to visualize the effect of changes in the system constants on the performance of the isolation system. This is particularly true when the actual nonideal response characteristics of the system sensing, computational, and motor elements are included in the system differential equation of motion and when additional computational elements, called *compensation circuits*, are added. The compensation circuits are used to alter the system frequency response, i.e., resonance frequency and peak transmissibility. It is not uncommon in working with active vibration-isolation systems of the type presented here to have differential equations as high as the twelfth order or more. The field of automatic control system synthesis has devised methods to deal with differential equations of such high orders from both a theoretical analysis and an actual system hardware point of view.

Because integral feedback of displacement requires that energy be fed into the isolation system, it is possible to make the active system dynamically unstable by improper proportioning of its constants. An active vibration-isolation system that is dynamically unstable will undergo continuously increasing mechanical oscillations which, when not limited by available power, will increase until the system is destroyed. Therefore, one of the factors in achieving a satisfactory active isolation system is the determination of the margin of dynamic stability of the entire system. Here too, the field of automatic control systems has devised methods to establish the system margin of dynamic stability. The margin of dynamic stability is a measure of the degree of change in system constants that is required for the active isolation system to become unstable.

In the case of a conventional passive vibration-isolation system, it is possible to determine many of the performance characteristics from the constants appearing in the differential equation. For example, the transmissibility T of the conventional system at the condition of resonance is approximately

$$T_r \simeq \frac{\sqrt{km}}{c} = \frac{1}{2(c/c_c)} \qquad c/c_c < 0.2 \tag{33.11}$$

Similarly, the resonance frequency ω_r is approximately equal to the undamped natural frequency

$$\omega_r \simeq \sqrt{\frac{k}{m}} \qquad c/c_c < 0.2 \tag{33.12}$$

At high frequencies ($\omega \gg \omega_n$) the transmissibility of the conventional system approaches the asymptotic value

$$T_i = \frac{c/c_c}{\omega/\omega_n} \qquad \omega \gg \omega_n \tag{33.13}$$

The transmissibility curve of a conventional isolator, given by Fig. 2.17, may be estimated from Eqs. (33.11) to (33.13) without plotting the transmissibility equation point by point. Somewhat similar relationships can be obtained for an active system if its equation of motion is not higher than the second order. A convenient way to obtain rules of thumb for the design of an active vibration-isolation system is to compare the characteristic properties of a conventional vibration-isolation system with those of the same isolation system having active elements which provide integral relative displacement force feedback and proportional velocity force feedback added in parallel with a spring isolation element. The velocity feedback gain G_2 generally has a larger effect on the system response than the relative displacement gain term G_1. The feedback gain terms relate the sensed system motion term to the force applied to the supported body; therefore, the units of the velocity feedback gain term G_2 are the same as a viscous damper, or force per unit velocity; the gain term G_1 for the integral relative displacement feedback has no passive counterpart and has units of force per unit displacement times time. The active damping term dominates the system differential equation affecting the system response both above and below the undamped natural frequency while the effect of the relative displacement feedback on system performance is confined mainly to the frequency region below the undamped natural frequency. Setting the integral relative displacement gain term G_1 to zero gives an approximation for transmissibility of the active isolation system:

$$T = \sqrt{\frac{1}{[1 - (\omega/\omega_n)^2]^2 + [2(G_2/c_c)(\omega/\omega_n)]^2}} \tag{33.14}$$

Using the above equation, the following response estimations can be formulated. The system transmissibility T at a frequency equal to the undamped natural frequency ω_n, formed by the passive spring and mass elements k and m, is

$$T_n = \frac{1}{2G_2/c_c} \qquad \omega = \omega_n \tag{33.15}$$

The resonance frequency is less than the system undamped natural frequency, and with an active fraction of critical damping term of one or larger, there is no system resonance, i.e., at all frequencies the system transmissibility is less than one. In the case where the relative displacement feedback gain is not zero, the mechanics of the system must always form a resonance condition. At excitation frequencies well above the system undamped natural frequency, the transmissibility of the active isolation system approaches the asymptotic value

$$T_i = (\omega/\omega_n)^2 \qquad \omega \gg \omega_n \tag{33.16}$$

In the above response estimation relationship function, the system transmissibility at the undamped natural frequency is less than unity when the velocity feedback gain exceeds a value giving an active fraction of critical damping of 0.5; i.e., $G_2/c_c = 0.5$. With an active fraction of critical damping of unity, the system transmissibility at the undamped natural frequency is 0.5. Active vibration-isolation systems of this type typically exhibit velocity feedback gain magnitudes yielding an active fraction of critical damping ranging from a low of about 0.5 to a high of about 5. The incorporation of the integral relative displacement feedback servomechanism in conjunction with the velocity feedback servomechanism and the passive system elements forms a system described by a third-order differential equation. A resonance condition occurs well below the undamped natural frequency when the active fraction of critical damping is 0.5 or more. The simplified response estimations of transmissibility are valid for frequencies at and above the system undamped natural frequency in instances where the active fraction of critical damping is 0.5 or greater. As the active fraction of critical damping is decreased, the resonance frequency approaches the undamped natural frequency with an increasing peak transmissibility and an eventual dynamically unstable system.

The resonance frequency and peak transmissibility, in an ideal active isolation system, are a function of the passive system constants and the two feedback gain terms. In a nonideal active isolation system, there are many other factors that influence the system resonance characteristics, such as the low-frequency response of the velocity sensor or a more complex passive system formed from many mass and spring elements. The resonance characteristics of the active isolation system are manipulated by use of compensation functions formed using electric networks in the computation element of the velocity servomechanism. The function of the compensation networks utilized is to alter the nature of the velocity feedback signal applied to the motor element, in a manner that provides for a dynamically stable system, and to raise or lower the resonance frequency, peak transmissibility, and transmissibility frequency response above the resonance frequency. The use of system compensation circuitry is extensive in the field of automatic control system synthesis as well as with active vibration-isolation systems, which are a type of automatic control system. The result of system compensation yields active vibration-isolation systems with response characteristics similar but not limited to the response of the ideal system. The analysis of the transient and frequency-response characteristics of an active vibration-isolation system having ideal elements shows many of the advantages of actual active vibration-isolation systems when compared to the response of passive system elements alone.

The element, in an active vibration-isolation system, that provides integral control of relative displacement strives to maintain the supported body at a constant distance from the support base to which it is attached. When a step function of force is applied to the supported body, the response of the system gives a measure of the effectiveness of the element in performing the desired function. A comparison of the transient response of the active isolation system, i.e., one having integral relative displacement and absolute velocity force feedback, with that of the conventional passive isolation system described in Chap. 30 illustrates the advantage obtained from integral relative displacement feedback.

TRANSIENT RESPONSE

Referring to Fig. 2.12, the equation of motion for the mass of the passive isolation system is

$$m\ddot{x} + c\dot{x} + kx = F(t) \tag{33.17}$$

where the force $F(t)$ is a step function of force having a magnitude $F = F_0$ when $t > 0$ and $F = 0$ when $t < 0$. Writing the Laplace transform of Eq. (33.17)

$$\mathcal{L}[x(t)] = X(s) = \frac{F_0}{ms} \frac{1}{s^2 + (c/m)s + k/m} \tag{33.18}$$

where $X(s)$ designates the Laplace transform of x, a function of time.* Letting $c/m = 2(c/c_c)\omega_n$ and $k/m = \omega_n^2$, Eq. (33.18) may be written as

$$X(s) = \frac{F_0}{ms} \frac{1}{s^2 + 2(c/c_c)\omega_n s + \omega_n^2} \tag{33.19}$$

The time solution of Eq. (33.19) is a damped sinusoid offset by the deflection of the spring caused by the constant force F_0. A typical time solution is shown by curve A of Fig. 33.7. The deflection of the isolator can be calculated by applying the final value theorem of Laplace transformations. This theorem states that if the Laplace transform of $x(t)$ is $X(s)$ and if the limit $x(t)$ as $t \to$ infinity exists, then

$$\lim_{s \to 0} sX(s) = \lim_{t \to \infty} x(t) \tag{33.20}$$

Applying the final value theorem using the Laplace transform of the passive isolator responding to the step function of force, Eq. (33.19), shows that the final deflection of the isolator is

$$\lim_{s \to 0} sX(s) = \lim_{t \to \infty} x(t) = \frac{F_0}{m\omega_n^2} \tag{33.21}$$

From Eq. (33.21), the mass takes a new position of static equilibrium at a distance $F_0/(m\omega_n^2)$ from the original position as $t \to$ infinity. The final deflection term may be eliminated from Eq. (33.21) by adding an integral relative displacement control servomechanism. This added element produces a force proportional to the integral of displacement x with respect to time. The system damping element is replaced by an active damping control servomechanism. Active damping in this case acts in the same manner as the passive damping element used for Eq. (33.17) since x is the only system motion. The differential equation of motion for the supported body of the active isolation system is

$$m\ddot{x} + G_2\dot{x} + kx + G_1 \int x\, dt = F(t) \tag{33.22}$$

---

* Laplace transforms are discussed in detail in Chap. 8.

The Laplace transform of the active isolation system differential equation is

$$\mathcal{L}[x(t)] = X(s) = \frac{F_0}{ms} \frac{1}{s^2 + (G_2/m)s + k/m + G_1/(ms)} \tag{33.23}$$

Placing the above equation in a form similar to Eq. (33.19) gives

$$X(s) = \frac{F_0}{m} \frac{1}{s^3 + 2(G_2/c_c)s^2 + \omega_n^2 s + [G_1/(m\omega_n^3)]\omega_n^3} \tag{33.24}$$

The term G_2/c_c represents the active fraction of critical damping. The term containing the active relative displacement feedback gain $G_1/(m\omega_n^3)$ is called the *dimensionless rela-*

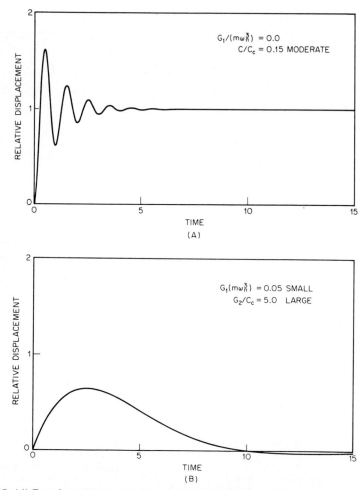

FIG. 33.7. (*A*) Transient response of a passive vibration-isolation system to a step-in force. (*B*), (*C*), and (*D*) show the transient response of an active vibration-isolation system to the same force step for different values of integral relative displacement and proportional velocity gains. The response is changed by changes in the feedback gain magnitude. In (*D*) the system is unstable as a result of the improper selection of the servomechanism constants; as a result the oscillations become increasingly large.

tive displacement feedback gain. The use of the dimensionless gain terms, active fraction of critical damping and dimensionless relative displacement feedback gain, allows the response characteristics of the active isolation system to be represented in a generalized manner where the numerical values of the passive system elements are not required.

Applying the final value theorem to the transient response of the active isolation system represented by Eq. (33.24) gives the deflection of the supported body in its final equilibrium position.

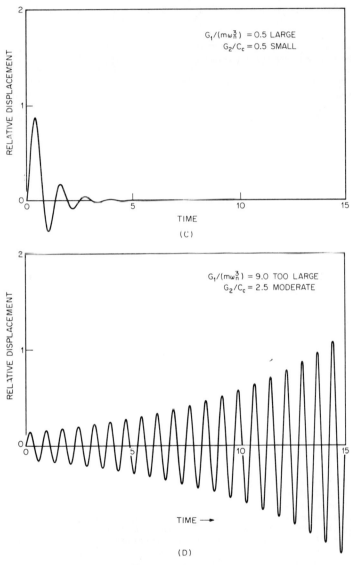

Fig. 33.7. (*Continued*)

$$\lim_{s \to 0} sX(s) = \lim_{t \to \infty} x(t) = 0 \qquad\qquad (33.25)$$

The final equilibrium position for the supported body of the active isolation system is zero so long as the dimensionless relative displacement feedback gain is not zero. The final position of the supported body is zero even with a very small dimensionless relative displacement feedback gain because of the integration operation provided by the relative displacement servomechanism. The magnitude of the two servomechanism gain terms affect the motion of the supported body during the transient. Figure 33.7A shows the transient response of a passive vibration-isolation system to a step in force which is applied to the supported body. In Fig. 33.7B, C, and D the transient response of an active system subjected to the same step force is shown for various values of the dimensionless feedback gain. The two servomechanisms in the active isolation system interact, but their effect can be generalized:

1. Increasing the magnitude of the dimensionless relative displacement gain increases the rate at which the system relative displacement approaches the final equilibrium position.

2. Increasing the active fraction of critical damping decreases the peak magnitude of the system relative displacement during the transient event and lowers the damped natural frequency.

The degree of oscillation exhibited by the active isolation system is a function of the magnitude and relative magnitude of the dimensionless gains of the two servomechanisms. In general, small magnitudes of the dimensionless relative displacement gain and large magnitudes of the active fraction of critical damping lead to little system oscillation as depicted by the curve of Fig. 33.7B. Likewise, large magnitudes of the dimensionless relative displacement gain and small magnitudes of the active fraction of critical damping tend to increase the amount of oscillation. The dimensionless relative displacement gain in relation to the active fraction of critical damping can be increased too much and will then produce a condition of instability, as shown by the curve of Fig. 33.7D. The conditions resulting in system instability are presented in the last section of this chapter.

The relative displacement response of this ideal active isolation system to constant acceleration of the isolator support, such as produced by gravity or by the sustained acceleration of a missile, cannot be represented by applying a constant force to the supported body as is frequently done with passive vibration-isolation systems. The reason for this is that active vibration-isolation systems which utilize absolute motion feedback, as in active damping of the type presented in this chapter, respond differently to forces applied to the supported body than to a constant acceleration of the support. In the case of a constant force applied to the supported body, presented above, the velocity servomechanism output force approaches zero as the transient motions of the system die out. In the case of a constant acceleration of the support, the velocity of the supported body continually increases in a manner similar to the increase in velocity of the support. The output of the velocity servomechanism increases constantly with time since the output force is proportional to the velocity of the supported body. This leads to a system which cannot work because the velocity servomechanism will rapidly reach its maximum force output at which time all active damping is lost. Active isolation in this situation is reobtained by placing an electric filter in the active damping servomechanism computational element. The filter used forms a control function which produces a zero output for a ramp input. The use of such a filter is part of the compensation process often required with automatic control systems; this process is presented in more detail in the next section.

Many active isolation systems of the ideal type presented in this chapter are used to isolate angular vibration, on which gravity has no effect. The active isolation of angular vibration uses the same system equations presented above except that the motions are angular, the mass is a moment of inertia, and the passive spring element applies a torque to the supported body that is proportional to the relative rotational displacement between the supported body and the support. The integral relative displacement servomechanism operates by measuring the rotation of the supported body relative to the support and applies a torque to the supported body that is proportional to the time integral of the sensed rotation. The relative angular displacement may be sensed using a rotational differential transformer or a linear potentiometer.

The active damping servomechanism operates by sensing the absolute rotational velocity of the supported body using a rate gyroscope which has an output response proportional to its rotational velocity: The active damping torque applied to the supported body is proportional to the output of the rate gyroscope. Many times the passive spring element is replaced by a servomechanism where the integral relative displacement control function in the computational element of the servomechanism is modified to produce an output proportional to the sum of the relative displacement and its first integral. Such a servomechanism has proportional plus integral control.

STEADY-STATE RESPONSE

A comparison of the steady-state response of the active and passive vibration-isolation systems illustrates some of the advantages and disadvantages associated with a servo-controlled vibration-isolation system. In Fig. 33.6, assume that $F(t) = 0$ and that the vibration excitation is caused by the motion $u(t)$ of the support base. Then the equation of motion for the supported body of the active vibration-isolation system having both the active damping servomechanism as shown by Fig. 33.6 and the integral relative displacement control servomechanism as shown by Fig. 33.5 is

$$m\ddot{x} + G_2\dot{x} + kx + G_1 \int x\, dt = ku + G_1 \int u\, dt \qquad (33.26)$$

The response of this isolation system, when the vibration excitation $u(t)$ is sinusoidal in nature and steady with respect to time, may be expressed in terms of transmissibility

$$T = \sqrt{\frac{[G_1/(m\omega_n^3)]^2 + (\omega/\omega_n)^2}{(\omega/\omega_n - \omega^3/\omega_n^3)^2 + [G_1/(m\omega_n^3) - 2(G_2/c_c)(\omega^2/\omega_n^2)]^2}} \qquad (33.27)$$

Figure 33.8 is a plot of Eq. (33.27) for four values of the relative displacement dimensionless gain term and six values of the velocity dimensionless gain terms, $G_1/(m\omega_n^3)$ and G_2/c_c, respectively. The corresponding expression for the transmissibility for the conventional passive vibration-isolation system, illustrated in Fig. 2.12, is given by Eq. (2.41); this equation differs from that for an active system, i.e., Eq. (33.27), because of the nature of the force feedback terms acting upon the supported body. At high frequencies well above the vibration-isolation system undamped natural frequency ω_n the active and passive system transmissibility equations differ because of the presence of a damping term in the numerator of the passive system equation. At these higher frequencies, the passive system transmissibility has the characteristic that as $\omega \to$ infinity, $T \to 2(c/c_c)(\omega_n/\omega)$. The active system, however, tends to act as an undamped vibration-isolation system wherein the transmissibility at high frequencies has the characteristic that as $\omega \to$ infinity, $T \to \omega_n^2/\omega^2$. Thus the active vibration-isolation system provides a lower transmissibility at frequencies above the system natural frequency, especially for large values of the active and passive damping terms.

At excitation frequencies close to the system natural frequency, both the active and passive vibration-isolation systems exhibit a resonance condition when the system damping terms are small. The peak value of the system transmissibility at the system resonance frequency is controllable by the addition of damping. In the passive vibration-isolation system, as the fraction of critical damping is increased, the peak transmissibility is lowered, reaching a value of unity for an infinite value of the fraction of critical damping. Although the passive system damping controls the peak transmissibility, high values of damping greatly degrade the system's main function of isolating vibration; in fact very large magnitudes of the system damping term yield little to no vibration isolation, since the damper tends to become a rigid link between the isolation system vibrating base and the supported body. The effect of damping on the active vibration-isolation system is similar to that of the passive vibration-isolation system when the active fraction of critical damping is small. However, as the active system damping is increased,

FIG. 33.8. Steady-state frequency response for an active vibration-isolation system having an ideal active damping servomechanism. The transmissibility is plotted against the frequency ratio ω/ω_n. In (A) there is no integral relative displacement control servomechanism, i.e., $G_1/(m\omega n^3)$ $= 0$; in (B), (C), and (D) such a control mechanism has be added and this ratio has values of 0.1, 0.2, and 0.5, respectively. For each of these illustrations a set of curves is shown for the following values of the ratio G_2/C_c: 0.2, 0.5, 1, 2, 5, and 10. Changes in the servomechanism feedback constants affect the response characteristics through their dynamic interactions which alter the frequency response at low excitation frequencies.

an increasingly more rigid link is placed between the supported body and motionless space; thus increasing the active fraction of critical damping always decreases the system transmissibility at frequencies above the natural frequency. With a relative displacement gain G_1 of zero, the active system resonance will disappear when the active fraction of critical damping exceeds unity, as is shown by the curve of Fig. 33.8A. With an active fraction of critical damping of unity the peak transmissibility is also unity and occurs at

FIG. 33.8. (*Continued*)

zero frequency, and for all other frequencies the system transmissibility is less than one having the approximate magnitude of $1/[2(G_2/c_c)(\omega/\omega_n)]$ at frequencies from zero to about twice the system natural frequency and ω_n^2/ω^2 at higher frequencies.

The addition of the relative displacement integral control has little influence on transmissibility at high frequencies and thus has no important effect on the ability of the complete system to isolate vibration. However, the effect at lower frequencies is significant as is shown in Fig. 33.8B, C, and D. As the dimensionless gain $G_1/(m\omega_n^3)$ of the displacement control loop is increased, the transmissibility of the system in the region of resonance increases. If the dimensionless displacement gain term equals twice the active fraction of critical damping, the active isolation system becomes dynamically unstable. Under these conditions, if the supported body receives the slightest disturbance, a system oscillation will develop and continue indefinitely, as would be the case with a passive system without damping. Increasing the relative displacement gain term above this critical value results in a condition where the system's automatic control functions continually add energy to the supported body and passive spring element in the form of ever-increasing oscillations, which continue to increase in amplitude until motor saturation or destruction of the system occurs.

STABILITY OF ACTIVE SYSTEMS

Operation of a dynamically unstable active vibration-isolation system exhibits one or more of the following characteristics:

1. The active isolation system acts like an undamped passive vibration-isolation system.

2. The system exhibits oscillations that increase with time and can become very large in magnitude.

3. The system moves to one of its excursion stroke limits and stays there.

The ensurance of a dynamically stable active isolation system is important at both the design and hardware stages of development and can become a complex design task. Much of the field of automatic control system analysis and synthesis deals with establishing the limits of feedback gains beyond which the system becomes unstable. A brief review of stability analysis is given in Ref. 1.

REFERENCES

1. Harris, C. M., and C. E. Crede (ed.): "Shock and Vibration Handbook," 2d ed., McGraw-Hill Book Company, Inc., New York, 1976, chap. 33.

34

APPLICATION OF ISOLATORS

Harry L. Hain
Lord Corporation

John J. Heintzel
Lord Corporation

Charles J. Leingang
Lord Corporation

INTRODUCTION

Chapter 32 describes the types and characteristics of isolators (also called *mounts*) used to control shock and vibration. This chapter shows (1) how to select the most appropriate isolator type and characteristic and (2) how to select the correct size of isolator. Such selection procedures apply to the usual engineering problem; for problems that are more complex, guidance is provided. Examples of the application of isolators which are described include electric motors; gears, bearings, and cams; reciprocating machines; pumps; piping; electronic equipment; aerospace equipment; road vehicles; shipboard equipment; and shipping containers for fragile items. Consideration also is given in this chapter to isolated foundations, inertia blocks, and flexible conduits. Air-spring isolators are described in Chap. 33.

The basic objectives in using isolators are (1) to protect the supporting structure and adjacent systems from vibration and shock disturbances originating in certain equipment and (2) to protect sensitive equipment from shock and vibration emanating from the structure on which the equipment is mounted. When such disturbances are present, properly applied isolators can allow equipment and systems to function as intended and can lengthen equipment life. Isolation of shock and vibration can also provide a more comfortable environment for equipment users, building occupants, and passengers in vehicles.

FACTORS CONSIDERED IN ISOLATOR SELECTION

Stiffness and damping are the basic properties of an isolator which determine its use in a system designed to provide vibration isolation and/or shock isolation. These properties usually are found in isolator supplier literature. However, there are other important factors which must be considered in the selection of an isolator:
1. Source of dynamic disturbance
2. Type of dynamic disturbance
3. Direction of dynamic disturbance (to determine axes of isolator loading)

4. Allowable response of a system to dynamic disturbance
5. Space and location available for the isolator
6. Weight and center-of-gravity of supported equipment (to determine load per isolator)
7. Space available for equipment motion
8. Ambient environment
9. Available isolator materials
10. Desired service life
11. Fail-safe installation
12. Cost

SOURCE OF DYNAMIC DISTURBANCE

The source of a dynamic disturbance (shock or vibration) influences the selection of an isolator in several ways. For example, a decision can be made whether to isolate the source of the disturbance or to isolate the item being disturbed. This decision affects which isolator is to be used. Consider the operation of a heavy punch press which has an adverse effect on a nearby electronic instrument. Isolation of the punch press would reduce this effect but would require fairly large isolators which might have to be resistant to grease or oil. In contrast, isolation of the instrument would also provide the required protection, but the required isolators would be smaller and (since grease or oil would not be a consideration) could be fabricated of a preferred elastomer.

A knowledge of the source of the vibration can aid in defining the problem to be solved. Within a given industry there may be published material describing problems similar to the one under consideration. Such material may describe possible solutions plus equipment fragility and/or dynamic characteristics of the equipment.

TYPE OF DYNAMIC DISTURBANCE

Two types of dynamic disturbance can require isolation: (1) vibration and (2) shock. As discussed in Chaps. 30 and 31, vibration may be periodic, random, or some combination thereof; shock excitation can have various pulse shapes and durations. A knowledge of the dynamic disturbance is very important in the choice of an isolator.

If only a vibration disturbance is present, a small isolator normally is suitable; vibration amplitudes usually are small relative to shock amplitudes. If a shock disturbance is the primary problem, then a larger isolator with more internal space for motion is required.

In selecting an isolator, it is necessary to ensure (1) that there is enough deflection capability in the isolator to accommodate the maximum expected motions from the dynamic environment; (2) that the load-carrying capacity of the isolator will not be exceeded (the maximum loads due to vibration and/or shock should be calculated and checked against the rated maximum dynamic load capacity of the isolator); and (3) that there will be no problem as a result of overheating of the isolator or fatigue deterioration due to long-term high-amplitude loading.

DIRECTION OF DYNAMIC DISTURBANCE

A factor that must be considered in the selection of an isolator is that of the directions (axes) of the dynamic disturbance. If the vibration or shock input occurs only in one direction, usually a simple isolator can be selected; its characteristics need be specified along only one axis. In contrast, if the vibration or shock is expected to occur along more than one axis, then the selected isolator must provide isolation (and its characteristics must be specified) along all the critical axes. For example, consider an industrial machine which produces troublesome vibration in the vertical direction and which must be isolated from its supporting structure. In this case, a standard plate-type isolator (see Fig. 32.22K) may be used. This isolator is stiffer in the horizontal direction than in the vertical direction which is the axis of the primary disturbance; the horizontal stiffness does not significantly affect the motion of the isolator in the vertical direction. Such horizontal stiffness adds to the lateral stability of the installation.

ALLOWABLE RESPONSE OF A SYSTEM TO DYNAMIC DISTURBANCE

The *allowable response* of a system is defined as the maximum allowable transmitted shock or vibration and the maximum displacements due to such disturbances. The allowable response of a system can be expressed in any of the following ways:

1. Maximum acceleration loading due to a shock input.
2. Specific system natural frequency and maximum transmissibility at that frequency.
3. Maximum acceleration, velocity, or displacement allowable over a broad frequency range.
4. The allowable level of vibration at some critical frequency or frequencies.
5. Maximum displacement due to shock loading.

The maximum acceleration which a piece of equipment can withstand without damage or malfunction is often called *fragility*. The definition of some allowable response is necessary for an appropriate isolator selection. If fragility data are not available for the specific equipment or installation at hand, then examples of similar situations should be used as a starting point.

The support structure characteristics can also affect the selection of isolators. For example, a beam or plate supporting a piece of equipment has resonances which could cause amplification of vibration levels; these resonance frequencies *must* be avoided in relation to isolated system natural frequencies. In addition, suppose an isolator were chosen only for its load-carrying capability, with no regard for the fragility of a piece of equipment in a specific frequency range. Then, the natural frequency of the system might be incorrectly placed such that a resonance within the equipment might be excited by the isolation system.

SPACE AND LOCATIONS AVAILABLE FOR ISOLATORS

Vibration and shock isolation should be considered as early as possible in the design of a system, and an estimate of isolator size should be made based on isolator literature. (See examples under *Isolator Selection.*) The size of the isolator depends on the nature and magnitude of the expected dynamic disturbances and the load to be carried. Typical literature describes the capabilities of isolators based on such factors.

The locations of isolators are very important to the dynamics of the equipment mounted on them. For example, a center-of-gravity installation, as shown in Fig. 34.1, allows the mounted equipment to move only in straight translational modes (i.e., a force at the center-of-gravity does not cause rotation of the equipment). This minimizes the motion of the corners of the equipment and allows the most efficient installation from the standpoint of space requirements.

If the isolators cannot be located so as to provide a center-of-gravity installation, then the system analysis is more difficult and more space must be allowed around the equipment to accommodate rocking motion (i.e., rotational modes) of the system. Finally, the isolators must be double-checked to ensure that they are capable of withstanding the additional loads and motions from the nontranslational movement of the equipment. This is particularly true when the center-of-gravity is a significant distance above or below the plane in which the isolators are located. *Rule of thumb: The distance between the isolator plane and the center-of-gravity should be equal to or less than one-third of the minimum spacing between isolators.* (For example, in Fig. 30.21 the overhang a_z should be no more than one-third of $2a_y$.) This helps to minimize rocking of the equipment and the resultant high stress in the isolators.

WEIGHT AND CENTER-OF-GRAVITY OF SUPPORTED EQUIPMENT

The weight and location of the center-of-gravity of the supported equipment should be determined. The location of the center-of-gravity is necessary for calculating the load supported on each mount. It is best to keep the equipment at least statically balanced (essentially equal deflections on all isolators). The preferred approach is to use the same isolator at all points, choosing isolator locations such that static loads (and thus deflections) are equalized. If this is not practical, isolators of different load ratings may be required at different support points on the equipment.

The size of the equipment and the mass distribution are important in dynamic analyses of the isolated system.

Fig. 34.1. Examples of center-of-gravity installations of vibration isolators. (*A*) Symmetrical spacing where a = half the lateral distance between isolators and b = half the fore-and-aft distance between isolators (plan view). (*B*) Center-of-gravity horizontal support (elevation). (*C*) Center-of-gravity diagonal support (elevation). (*D*) Center-of-gravity vertical support (elevation). (*Lord Corporation.*)

SPACE AVAILABLE FOR EQUIPMENT MOTION

The choice of an isolator may depend on the space available (commonly called *sway space*) around a piece of equipment. The spring constant of the isolator should be chosen carefully so that motion is kept within defined space limits. The motion which must be considered is the sum of (1) the static deflection due to the weight supported by the isolator, (2) the deflection caused by the dynamic environment, and (3) the deflection due to any steady-state acceleration (such as in a maneuvering aircraft).

If there is a problem of excessive motion of the supported mass on the isolator, then a *snubber* (i.e., a device which limits the motion) can be used. A snubber may be an elastomeric compression element designed into an isolator, such as the area below the small channel in Fig. 32.22*M*. It may also be a compression pad installed separately in the system at the point of excessive motion. The spring constant of such a snubber must be carefully selected to avoid transmission of high-impact loads into the supported equipment.

AMBIENT ENVIRONMENT

The environment in which an isolator is to be used affects its selection in two ways:

1. Some environmental conditions may degrade the physical integrity of the isolator and make it nonfunctional.

2. Some environmental conditions may change the operating characteristics of an isolator, without causing permanent damage.

This may alter the characteristics of the isolation system of the supported equipment; for example, frequency responses could change significantly with changes in the ambient temperature. Thus, it is important to determine the operating environment of the isolation system and to select isolators that will function with desired characteristics in this environment.

AVAILABLE ISOLATOR MATERIALS

Vibration and shock isolators are available in a wide variety of materials and configurations to fit many different situations, as described in Chap. 32. The type of isolator is chosen for the load and dynamic conditions under which it must operate. The material from which the isolator is made depends to a great extent on the ambient environment of an application and somewhat on the dynamic properties required. Chapters 32 and 35 provide guidance for the choice of isolators and isolator materials.

The all-metal and metal-spring, friction-damped isolators are used primarily where operating temperatures are too high for elastomeric isolators. They can be used in a variety of applications. However, the characteristics of these isolators can cause problems in high-load situations, and the variability in performance due to assembly tolerance build-up negates their use in highly sensitive applications.

By far, the majority of isolators in use today are elastomeric. The development of a vast array of elastomeric compounds has made it possible to use this type of isolator in almost any environment. Within a given type of elastomer it is a simple matter to vary the stiffness (modulus, durometer) of the compound; this gives much flexibility in adapting an isolator to an application without changing the isolator's geometry.

A summary of some of the more common elastomer families which can be used in an isolator is given in Table 35.3. The more exotic materials, such as silicones and fluoroelastomers, are more costly than the others, but their unique properties are needed for some isolator installations. Natural rubber is the most widely used elastomer in isolators. Chapter 35, *Rubber Springs*, gives a more detailed coverage of the use of elastomers in isolators.

Since the selection of material for an isolator depends so much on the environment in which the mount will be used, it is very important to learn as much as possible about the operating and storage environments.

DESIRED SERVICE LIFE

The expected, or desired, length of service for an isolator can affect the type and size of the vibration isolator which is selected. For example, an isolator which must operate for 2,000 hours under a given set of conditions typically is larger than one which must operate for only 500 hours under the same conditions.

In general, empirical data are used to estimate the operating life of an isolator. Accurate descriptions of the dynamic disturbances and ambient operating environment expected are needed to make an estimate of isolator life. A knowledge of the specific material and design factors in an isolator are necessary to make an estimate of fatigue life. Such information is best provided by the original manufacturer/designer of the isolator.

FAIL-SAFE INSTALLATION

A *fail-safe isolator installation* is one in which there is a positive metal-to-metal inter-lock should the resilient spring in the isolator fail. Thus in the event of failure, the equip-ment is supported in place until the failed isolator can be replaced. In some cases, the isolator itself contains this fail-safe feature; if it does not, this feature may be provided by the mounting system.

SHOCK AND VIBRATION ISOLATOR SPECIFICATIONS

Often, shock and vibration isolators are overspecified; this can cause needless compli-cation and increased cost. *Overspecification* is the practice of arbitrarily increasing shock or vibration load values to be safe (to make certain that the isolators have been chosen with a high margin of safety at the maximum load capability). In the case of vibration isolators, this could cause a vastly different performance in the test laboratory than in actual service. The best isolator specification is one which defines the critical properties of the isolation system and the specific environment in which the system will operate. Extraneous requirements cause needless complications. For example, if the vibration level is an acceleration of $\pm 1g$, it is not advisable to specify $\pm 2g$ to be safe. Likewise, it is inadvisable to rigidly apply an entire specification to an isolator installa-tion when only a small part of the specification is applicable.

Typically, specifications to which vibration and shock isolators are designed will include requirements regarding (1) vibration amplitudes, (2) shock amplitudes, (3) load to be carried, (4) required protection for equipment, (5) temperatures to be encountered (environmental factors, in general), and (6) steady acceleration loads superimposed on dynamic loading.

A good example of such a specification is U.S. document MIL-STD-810. It provides most of the above information for various possible installations of equipment.

PARALLEL AND SERIES COMBINATIONS OF ISOLATORS

When a number of isolators are used in a system, they are usually combined either in *parallel* or in *series*, or in some combination thereof.

ISOLATORS IN PARALLEL

Most commonly isolators are arranged in parallel. Figure 34.2 depicts two isolators schematically as springs in parallel. Here the static spring constant k for the system, determined at the center-of-gravity, is given by

$$k = \frac{(a + b)^2}{a^2/k_2 + b^2/k_1} \tag{34.1}$$

where k_1 = spring constant of isolator 1
k_2 = spring constant of isolator 2
a = distance between isolator 1 and center-of-gravity
b = distance between isolator 2 and center-of-gravity

If the spacing of isolators is equal ($a = b$) and the same isolator is used at both loca-tions ($k_1 = k_2$), then according to Eq. (34.1), $k = 2k_1$.

The total weight W (acting at the center-of-gravity) is divided between the two isola-tors as follows:

$$W_1 = W \frac{b}{a + b} \tag{34.2}$$

$$W_2 = W \frac{a}{a + b} \tag{34.3}$$

The total weight W always equals $W_1 + W_2$. If $a = b$, the static load on each isolator is one-half the total, so that

$$W_1 = W_2 = \tfrac{1}{2}W \tag{34.4}$$

The static deflection δ_{st} of each isolator is found by dividing the static load on the isolator by the particular spring constant (e.g., for k_1, $\delta_{st} = W_1/k_1$).

Similarly, the system static deflection (at the center-of-gravity) is found by dividing the total weight by the system spring constant, or

$$\delta_{st} = \frac{W}{k} \tag{34.5}$$

When isolator spacing is symmetrical in relation to the center-of-gravity and the same isolator is used at all support points, then the static deflection is the same at each isolator.

FIG. 34.2. Schematic diagram of two springs in parallel. Spring loads are added to get total weight. Plates are assumed to be massless.

FIG. 34.3. Schematic diagram of two springs in series. Spring deflections are added to get total deflection, but each spring carries total load.

The problems in Examples 34.1 through 34.5 which follow are based on the use of isolators in parallel.

ISOLATORS IN SERIES

When two isolators are combined in series, as shown in Fig. 34.3, the system spring constant is given by

$$k = \frac{1}{1/k_1 + 1/k_2} = \frac{k_1 k_2}{k_1 + k_2} \tag{34.6}$$

Each of the isolators in series supports the total weight, so that

$$W_1 = W_2 = W \tag{34.7}$$

The static deflection of each isolator is found by dividing the total weight W by the particular spring constant; the system deflection is calculated by Eq. (34.5) and is the sum of the deflections of the individual isolators.

ISOLATOR SELECTION

Simple shock and vibration problems involving translational motion can be classified into two categories:

1. *Force excitation* (see Figs. 2.10 and 30.43). A shock or vibration disturbance originates in the supported equipment. The isolators which support the equipment reduce the transmission of force to the supporting structure, thus protecting the structure or foundation, for example, in isolating a vehicle chassis from the vibration of an internal-

Table 34.1. Equations for the Isolation of Vibration

If you want ↓ \ You need →	Static stiffness, k_{st}	Dynamic stiffness,* k	Ratio of forcing frequency to natural frequency $\frac{f}{f_n}$ or $\frac{\omega}{\omega_n}$	Fraction of critical damping,† ζ	Transmissibility, T (See Fig. 2.17)	Static load, weight, W, or mass, m	Force applied (magnitude), F_o	Use these equations for linear springs
Static deflection, δ_{st}	•					•		$\delta_{st} = \dfrac{W}{k_{st}} = \dfrac{mg}{k_{st}}$ [See Eq. (2.10)]
Dynamic displacement, $|x|$		•	•	•			•	$|x| = \dfrac{F_o k}{\sqrt{[1-(f/f_n)^2]^2 + (2\zeta f/f_n)^2}}$ [See Table 30.2c]
Transmitted force, F_T					•		•	$F_T = TF_o$ [See Table 30.2a]
Natural frequency, f_n		•				•		$f_n = \dfrac{1}{2\pi}\sqrt{\dfrac{k}{m}} = \dfrac{1}{2\pi}\sqrt{\dfrac{kg}{W}}$ [See Eq. (2.8)]
Transmissibility at resonance for viscous damped isolators				•				$T_{max} = \dfrac{1}{2\zeta}$ [See Eq. (30.1)] for small values of ζ; otherwise use Fig. 30.8.

* The dynamic stiffness for elastomers, using natural rubber as an example, varies from about 1.1 times the static value (for soft stock) to about 3 times the static value (for midrange stock). The ratio of dynamic to static stiffness depends upon type of elastomer, modulus, frequency, dynamic strain level, and ambient temperature. The isolator manufacturer can provide specific data on performance under vibratory conditions.

† The fraction of critical damping for widely used elastomers varies from about 0.05 (for natural rubber) to about 0.20 (for highly damped stocks). See Table 35.4.

combustion engine or in reducing the transmission of machine vibration to adjacent structures.

2. *Motion (velocity) excitation* (see Figs. 2.11 and 30.42). Shock and vibration are transmitted through a supporting structure to equipment which is caused to move. The transmitted motion and force are reduced by mounting the equipment on isolators, for example, to protect equipment from vibrating floors or to protect buildings and contents from seismic (ground) disturbances.

The following examples present specific applications within the above categories. They show how isolators may be selected for some simple shock and vibration problems, but the steps used are basic and can be extended to many other situations. In the solution of these problems, the following simplifying assumptions are made:

1. The effect of damping is negligible, a valid assumption for many isolator applications.

2. All modes of vibration are uncoupled (see Figs. 3.15, 30.17, and 34.1).

3. The static and dynamic spring constants of the isolators are equal, valid for low-modulus elastomers with little damping.

A number of basic equations used in isolator selection are included in Table 34.1.[1]

EXAMPLE 34.1. ISOLATION OF VIBRATORY FORCE

Problem. An electric motor and pump assembly, rigidly mounted on a common base, transmits vibration to other components of a hydraulic system. The weight of the assembly and base is 140 lb (63 kg). Four isolators are to be located at the corners of the rectangular base. The center-of-gravity is centrally located in the horizontal plane near the base. The lowest vibratory forcing frequency is 1,800 rpm and is a result of rotational unbalance. There also are higher frequencies due to magnetic forces and pump forces. The excitation is in both the horizontal and vertical directions.

Objective. To reduce the amount of vibration transmitted to the supporting structure and thus to other system components. A vibration-isolation efficiency of 70 to 90 per cent is usually possible to attain; here a value of 80 per cent is selected.

Solution

1. First find the transmissibility T which corresponds to the required vibration isolation I of 80 per cent ($I = 0.8$) from

$$I + T = 1 \qquad (34.8)$$

so that $T = 0.2$.

2. Then determine the value of the lowest forcing frequency in hertz (cycles per second) by dividing 1,800 rpm by 60 sec/min, yielding a value of $f = 30$ Hz. The lowest forcing frequency is used because this is the worst condition. If high isolation is attained at this frequency, isolation will be even better at higher frequencies.

3. Next, find the natural frequency f_n that the isolated system needs to provide a transmissibility T for a forcing frequency of f Hz. The following equation, a variation of Eq. (2.27), assumes zero damping; Eq. (2.41), which includes damping, may be used in special cases.

$$T = \frac{1}{(f/f_n)^2 - 1} \qquad (34.9)$$

Using $T = 0.2$ and $f = 30$ Hz yields $f_n = 12.2$ Hz.

4. Now determine the static deflection δ_{st} required to provide a natural frequency f_n from Eq. (2.10).

$$f_n = \frac{1}{2\pi} \sqrt{\frac{g}{\delta_{st}}} \qquad (34.10)$$

This yields a value of $\delta_{st} = 0.066$ in. (1.67 mm) for $f_n = 12.2$ Hz.

The same results may be obtained by using the isolation efficiency chart, Fig. 34.4. To use this chart, find the point at which the horizontal line for $f = 30$ Hz intersects the

Fig. 34.4. Isolation efficiency chart. Vibration isolation occurs only when the ratio of the forcing frequency to the natural frequency is greater than $\sqrt{2}$. Enter the chart at the forcing frequency. Project a horizontal line to the desired vibration-reduction percentage curve. From that intersection project a vertical line to the required static deflection of the isolator system. (*Lord Corporation.*)

diagonal line for 80 per cent isolation (reduction in vibration). From the intersection point, project a vertical line to read the values of $\delta_{st} = 0.066$ in. (1.67 mm) and $f_n = 12.2$ Hz.

5. Then calculate the total spring constant k required to provide a natural frequency f_n from Eq. (2.8)

$$f_n = \frac{1}{2\pi} \sqrt{\frac{kg}{W}} \qquad (34.11)$$

yielding a value of $k = 2,120$ lb/in. (371.3 N/mm).

6. Next, determine the spring constant of each of the n isolators (acting in parallel) which are used to support the mass m from

$$k_i = \frac{k}{n} \qquad (34.12)$$

Here there is one isolator in each corner, so that $n = 4$, yielding a value of $k_i = 530$ lb/in. (92.8 N/mm).

7. Now calculate the static weight (load on each isolator). Since the center-of-gravity of the supported mass is centrally located in the horizontal plane, the weight is divided equally among all isolators. Determine the weight supported by each isolator from

$$W_i = \frac{W}{n} \qquad (34.13)$$

Here $W = 140$ lb (63 kg) and $n = 4$, so that $W_i = 35$ lb (15.8 kg).

8. Choose an isolator having a spring constant of 530 lb/in. (92.8 N/mm) and the capacity to support a static weight of at least 35 lb (15.8 kg). Try to select an isolator that is available from stock; use a catalog published by an isolator manufacturer or distributor. In this case, a standard plate-form isolator in a holder is recommended because it is readily available, relatively inexpensive, and easy to install. It provides a low stiffness (and therefore high isolation) in the vertical direction. The stiffness in the horizontal direction is somewhat higher but will still provide adequate isolation. An isolator of this type is shown in Fig. 32.22K. It should be installed so that the axis of the center bolt is vertical. A steel snubbing washer is used at each end of the center tube (1) to provide a fail-safe isolator and (2) to limit vertical deflection caused by shock overload.

EXAMPLE 34.2. ISOLATION OF VIBRATORY MOTION

Problem. An electronic device weighing 24 lb (10.8 kg) is mounted on a machine. The machine vibrates with a displacement amplitude of 0.02 to 0.03 in. (0.51 to 0.76 mm) in the vertical and horizontal directions in a frequency range between 20 and 40 Hz. The device is to be mounted on four isolators (one in each corner of the base). The center-of-gravity of the isolated system is centrally located in the horizontal plane.

Objective. To isolate the electronic device from motion of the machine by supporting the device on isolators. A vibration isolation of 85 per cent is required.

Solution. Although this system is motion-excited, the steps necessary to provide vibration isolation are the same as in the force-excited case of Example 34.1. The same equations are used here.

1. First find the transmissibility T corresponding to 85 per cent isolation from Eq. (34.8). For $I = 0.85$, the equation yields $T = 0.15$.

2. Determine the natural frequency f_n required to provide a transmissibility $T = 0.15$ from Eq. (34.9). The value for the forcing frequency should be $f = 20$ Hz because this is the worst expected condition (highest transmissibility). For $T = 0.15$ and $f = 20$ Hz, Eq. (34.9) yields $f_n = 7.2$ Hz.

3. Now calculate the static deflection δ_{st} needed to provide a natural frequency $f_n = 7.2$ Hz from Eq. (34.10). This yields $\delta_{st} = 0.19$ in. (4.8 mm).

4. Next, find the total spring constant k required to provide a natural frequency of 7.2 Hz. This can be calculated by Eq. (34.11) or from

$$k = \frac{W}{\delta_{\text{st}}}$$ (34.14)

For $W = 24$ lb (10.8 kg) and $\delta_{\text{st}} = 0.19$ in. (4.8 mm), $k = 126$ lb/in. (22.1 N/mm).

5. Determine the spring constant k_i of each of the four isolators (acting in parallel) from Eq. (34.12). Since the center-of-gravity is centrally located in relation to the isolators, $k_i = k/4$ or approximately 32 lb/in. (5.6 N/mm).

6. Select an isolator having a spring constant of 32 lb/in. (5.6 N/mm); in a catalog look for a part that can support a static load of at least 6 lb (2.7 kg). Here, for the reasons stated in Example 34.1, a plate-form isolator is the proper choice. But in this case, choose a low-frequency isolator, similar to Fig. 32.22K but having a special elastomer contour to provide a lower spring constant.

EXAMPLE 34.3. SHOCK PROTECTION

Problem. A business machine is to be isolated so that it will not experience damage during normal shipping. The unit can withstand 25g of shock without damage. The suspended weight of 125 lb (56.2 kg) is to be equally distributed on four isolators. The disturbances expected are those from normal transportation handling, with no damage allowed after a 30-in. (762-mm) flat bottom drop. The peak vibration disturbances are normally in the range of 2 to 7 Hz.

Objective. To limit acceleration on the machine to 25g using the 30-in (762-mm) drop test as a simulation of the worst expected shock conditions. A natural frequency between 7 and 10 Hz is desired to avoid the peak vibration frequency range and still provide good shock protection.

Solution.

1. First, solve for the dynamic deflection δ_d (displacement) of the machine required to limit acceleration to $\ddot{x}_F$ (expressed in gs) when the item is dropped from a height h using

$$\delta_d = \frac{2h}{\ddot{x}_F}$$ (34.15)

where $\ddot{x}_F =$ the fragility factor [maximum allowable value of n_g in Eq. (30.57)]; also see Example 41.1. Here $\ddot{x}_F = 25$ and $h = 30$ in. (762 mm) so that $\delta_d = 2.4$ in. (61 mm).

2. Then determine the required *dynamic* natural frequency f_n to result in a dynamic deflection δ_d from a variation of Eq. (2.8)

$$f_n = \frac{1}{2\pi} \sqrt{\frac{k \, \ddot{x}_F g}{W}} = \frac{1}{2\pi} \sqrt{\frac{\ddot{x}_F g}{\delta_d}}$$ (34.16)

Using $\ddot{x}_F = 25$ and $W = 125$ lb (56.2 kg) yields $f_n = 10$ Hz, within the target range for natural frequency. The graph in Fig. 41.4 yields the same f_n and δ_d when using $h = 30$ in. (762 mm) and a maximum acceleration of 25g.

3. Calculate the system *dynamic* spring constant k required to provide a dynamic deflection δ_d from

$$k = \frac{\text{force}}{\text{deflection}} = \frac{\ddot{x}_F W}{\delta_d}$$ (34.17)

yielding $k = 1,302$ lb/in. (228 N/mm) for the system. Note: For force $F = n_g W$, see Eq. (30.57).

4. Next, determine the *static* spring constant of the system. Assume the use of natural rubber here; then the static spring constant k_{st} is approximately equal to the dynamic spring constant k or 1,302 lb/in. (228 N/mm). Note: For shock conditions, $k \simeq 1.3 k_{\text{st}}$ for medium-damped elastomers and $k \simeq 1.5 k_{\text{st}}$ for high-damped elastomers.

5. Calculate the static spring constant k_i of the n isolators (acting in parallel) from Eq. (34.12)

$$k_i = \frac{k}{n}$$

Here $n = 4$, yielding a value of $k_i \simeq 325$ lb/in. (56.9 N/mm).

6. Since the total weight is distributed equally on each isolator, calculate the static load on each from Eq. (34.13).

$$W_i = \frac{W}{n}$$

This yields $W_i \simeq 31$ lb (14 kg).

7. Sandwich-type isolators are generally used to protect fragile items during shipment. The construction is typically two flat plates with threaded fasteners installed, bonded on either side of an elastomeric pad. Determine the minimum thickness of the elastomer (between the plates) needed to keep dynamic strain at an acceptable level. Use the following rule of thumb for natural rubber:

$$t_{min} = \frac{\delta_d}{1.5} \tag{34.18}$$

For $\delta_d = 2.4$ in. (61 mm), minimum elastomer thickness $t_{min} = 1.6$ in. (40.6 mm).

8. Now choose a sandwich isolator for this application. One type is shown in Fig. 32.22L. Sandwich configuration permits sufficient deflection in two directions (shear) to absorb high shock loads. The installed cost is low. Sandwich isolators are readily available in a wide range of sizes, spring constants, and elastomers. In a catalog, select a part that has the capacity to support a static shear load of 31 lb (14 kg), has a minimum elastomer thickness of 1.6 in. (40.6 mm), and has a shear spring constant of 325 lb/in. (56.9 N/mm).

9. In designing or choosing the container, certain criteria must be considered. The four isolators should be installed equidistant from the center-of-gravity in the horizontal plane, oriented to act in shear in the vertical and fore-and-aft directions. The isolators should be attached on one end to a cradle which carries the machine and on the other end to the shipping container. There must be enough space allowed between the mounted unit and the container to prevent bottoming (contact) at impact. A good rule is to allow a clearance space equal to $1.4\delta_d$. In this example the clearance space is a minimum of 3.4 in. (86.4 mm).

EXAMPLE 34.4. SHOCK AND VIBRATION ISOLATION COMBINED

Problem. A portable engine-driven air compressor, with a total weight of 2,500 lb (1,126 kg), is noisy in operation. An isolation system is required to isolate engine disturbances and to protect the unit from over-the-road shock excitation. The engine and compressor are mounted on a common base which is to be supported by four isolators. The weight is not equally distributed. At the engine end the static load per isolator is 750 lb (338 kg); at the compressor end the static load per isolator is 500 lb (225 kg). The lowest frequency of the disturbance is at engine speed. The idling speed is 1,400 rpm, and the operating speed is 1,800 rpm. The unit is expected to be subjected to shock loads due to vehicle frame twisting when transported over rough roads.

Objective. To control force excitation vibration and provide secondary shock isolation. A compromise is required; the isolation system must have a stiffness that is low enough to isolate engine idling disturbance but high enough to limit shock motion. A system having a natural frequency of 12 to 20 Hz in the vertical direction is usually adequate. (Note: The tires and basic vehicle suspension will provide the primary shock protection.)

Solution.

1. First assume that the natural frequency of the system in the vertical direction is 12 Hz.

2. Next, convert the engine speeds to hertz (cycles per second) for use in the calculations. Divide the rpm values by 60 sec/min, yielding force frequencies f of 23.3 Hz at idling speed and 30 Hz at operating speed.

3. Then calculate the transmissibility T for f_n = 12 Hz from Eq. (34.9).

At idling speed, using f = 23.3 Hz yields T = 0.36 (36 per cent). At operating speed, using f = 30 Hz yields T = 0.19 (19 per cent).

Use of Eq. (34.8) gives a vibration isolation T of 0.64 (64 per cent) at idling speed and 0.81 (81 per cent) under normal operation. For both conditions, performance with a natural frequency of 12 Hz is satisfactory.

4. Now determine the required static deflection δ_{st} to provide a natural frequency f_n from Eq. (34.10). For f_n = 12 Hz, this yields δ_{st} = 0.068 in. (1.73 mm).

5. Select a center-bonded (center-bolt) isolator for both ends of the unit. This type of part (Fig. 32.22H) is simple and rugged and gives protection against shock loads expected here. It should be installed so that the axis of the bolt is vertical and the static weight rests on the wide disk portion. After installation, this isolator provides cushioning against upward (rebound) shock loads as well as against downward loads. The isolation system is fail-safe and inexpensive. Each of the two isolators at the engine end should have a static load-carrying capacity of at least 750 lb (338 kg). Each of the two isolators at the compressor end should be able to support at least a 500-lb (225-kg) static load. For all isolators the static deflection should be close to 0.068 in. (1.73 mm) to give the desired natural frequency of 12 Hz.

EXAMPLE 34.5. NOISE CONTROL

Problem. (Also see Example 30.4.) A mechanical impact printer unit in an input-output console is the source of structureborne noise in the frequency range from 50 to 1,000 Hz. The noise is generated by impact forces from the printing operation. The unit weighs 200 lb (90 kg). The base of the unit is designed to accommodate four isolators in the horizontal plane. The center-of gravity is centrally located, close to the base.

Objective. For maximum noise reduction the best possible isolation system is desired, depending on the space available for isolators and equipment motion. Try for a minimum isolation of 98 per cent (at the lowest forcing frequency).

Solution.

1. Determine the system static deflection needed to provide a vibration isolation of 98 per cent (I = 0.98) at a forcing frequency f = 50 Hz. The isolation efficiency chart, Fig. 34.4, shows that a static deflection δ_{st} of 0.2 in. (5.1 mm) is required. Since the four isolators act in parallel, the static deflection of each isolator is the same as the system deflecton.

2. Then calculate the static load on each isolator from Eq. (34.13). For W = 200 lb (90 kg) and n = 4, W_i = 50 lb (22.5 kg).

3. Now choose an isolator capable of supporting a minimum static load of 50 lb (22.5 kg) while providing a minimum static deflection of 0.2 in (5.1 mm). To fit the available space, try a small channel-type isolator (Fig. 32.22M). This type of isolator has a high load capacity and a low vertical spring constant (in shear). It is also easy to install and gives built-in protection against metal-to-metal bottoming (contact) under shock load.

4. A manufacturer's catalog shows an isolator with a rated load of 50 lb (22.5 kg) and a static deflection of 0.25 in. (6.3 mm) at that load.

5. For the isolator shown in the catalog, determine the natural frequency from Eq. (34.10). Using δ_{st} = 0.25 in. (6.3 mm), the equation gives f_n = 6.3 Hz.

6. Then calculate the transmissiblity from Eq. (34.9). For f = 50 Hz and f_n = 6.3 Hz, the resulting T = 0.016 (i.e., 1.6 per cent). Using this value of T in Eq. (34.8) yields an isolation better than 98 per cent, thus exceeding the performance goal.

CHECKING THE ISOLATOR INSTALLATION

There are usually two basic causes for unsatisfactory performance of an isolation system: (*a*) the isolator has been selected improperly or some important system parameter has been overlooked, (b) the isolator has been installed improperly. The following criteria can help obviate problems that can otherwise cause poor performance.

1. The isolator should not be overloaded.

a. If a coil spring has been used, adjacent coils should not be in contact or close to being in contact. There should be adequate space between coils at normal static load, with no possibility of bottoming at the maximum load.

b. If an elastomeric compression-type isolator has been used, there should be no evidence of excessive bulge. As a guideline, the compression strain (ratio of deflection to original rubber thickness) at static load should not exceed 0.15. Overloading can shorten isolator life and affect performance. Because an elastomeric element loaded in compression has a nonlinear spring constant, overloading causes the effective dynamic spring constant to be higher than intended. This raises the system natural frequency and reduces the efficiency of isolation.

c. If an elastomeric shear-type isolator is used, the static shear strain ratio usually should not exceed 0.30. The suspended unit should not bottom out under shock conditions. If so, more clearance is needed or compression snubbers should be employed to reduce the maximum shock load on the unit.

2. The isolators and isolated equipment should be able to move freely under vibration and shock excitation.

a. No part of the isolation system should be short-circuited by a direct connection rather than a resilient support.

b. Flexible conduits should be installed, where required, on vibrating machines. Conduits and piping should not short-circuit or restrict the operation of the isolation system.

c. The vibrating equipment should not contact adjacent equipment or a structural member. Space should be provided to avoid contact.

d. There should be no metal-to-metal contact of unattached moving components within an isolator. Use of an elastomeric element may be beneficial in such a case if adequate clearance cannot be provided.

e. If an elastomeric pad has been installed beneath a machine, the resilient pad should not be short-circuited by hard-bolting the machine to its foundation. This problem can be remedied by designing a resilient connection at the attachment points.

f. If travel limit rods are used, the stop nuts should have been released to allow the spring to function as intended.

g. If the isolator has an adjustment bolt, the bolt should not touch any solid object which prevents the isolator from moving freely.

h. On a construction site, all debris, such as concrete rubble or pieces of wood, should be removed from under steel frames or inertia blocks supporting vibrating equipment. Such obstructions could seriously hinder the isolation system.

3. The isolators should be positioned properly. The specification should be followed exactly in regard to isolator orientation.

a. The load on the isolator should be along the axis designed to carry the load. The isolator should not be distorted. Unless the isolator has built-in misalignment capability, installation misalignment can affect performance and shorten isolator life.

b. The isolator should not be installed upside down if the design is such that this is important. A pipe hanger is an example; some of these are assemblies, designed to be installed with a certain end up. If this type of hanger is installed upside down, its performance can be affected.

4. If an elastomeric mount is used, clearance should be provided so that there is no solid object cutting the elastomer. There should be no evidence of bond separation between the elastomer and metal parts in the isolator.

Cuts and tears in the elastomer surface can propagate during operation and destroy the spring element. If there are bonded surfaces in the isolator, a bond separation also can cause problems; growth in the separation can affect the performance of the isolator and ultimately cause failure.

5. The static deflection of all isolators should be approximately the same. There should be no evidence of improper weight distribution. Excessive tilt of the mounted equipment may affect its performance. For economic reasons and simplicity in installation, it is desirable to use the same isolator at all points in the system. In such a case,

it is not usually a problem if the various isolators have slightly unequal static deflections. However, if one or more isolators exhibit excessive deflection (see number 1 above), then corrective measures are required. If the spacing between isolators has been determined improperly, a correction of the spacing to equalize the load may be all that is required. If this is impractical, an isolator having a higher spring constant can be used at points supporting a higher static load. This will tend to equalize deflection.

ISOLATED FOUNDATIONS AND INERTIA BLOCKS

In certain applications, it is not desirable or feasible to mount a machine directly on vibration isolators. Instead, the machine is attached to a relatively heavy and rigid block (usually made of concrete) which is then supported by suitable isolators; the use of such a massive block (called an *inertia block*) is desirable for the following reasons:

1. If a machine (such as a diesel engine, forging hammer, or vibration testing machine) generates relatively large forces during its operation, the over-all movement of the machine on its isolation system tends to become excessive unless its effective mass is substantially increased. This increase in effective mass can be achieved by rigidly attaching the machine to an inertia block and then mounting the inertia block on isolators.

2. In applications in which the frequency of excitation is low, the natural frequency of the system must be very low to provide low transmissibility and therefore good vibration isolation. A problem often arises with a machine intended to be mounted only at its base, because a low-stiffness base-mounted system tends to be unstable and to allow excessive motion. Effective isolation may be difficult to achieve because of coupled natural frequencies. A mounting arrangement with isolators angled upward may be used to move the system's elastic center closer to the center-of-gravity (see Fig. 30.26). This will reduce the effect of coupling of frequencies, improve the vibration isolation, and reduce motion. It may be more feasible to decouple the system by mounting the machine rigidly to an inertia block (to lower the center-of-gravity of the assembly) and to suspend the inertia block on isolators located in the same horizontal plane as the center-of-gravity.

3. Many types of equipment provide relatively nonmassive mounting structures for the attachment of isolators. The low mechanical impedance of such structures makes it impossible to attain a high degree of vibration isolation. The mechanical impedance of the supporting structure can be increased by rigidly attaching the equipment on an inertia block and then mounting the inertia block on isolators.

4. Some types of equipment do not operate properly unless supported by a rigid structure. This applies to certain types of machine tools that are not inherently rigid and therefore need a rigid support to maintain the prescribed accuracy. In other types of machinery (such as printing presses) consisting of articulated components, a rigid support may be needed to maintain the proper alignment of working parts.

An inertia block adapted to meet the requirements outlined above may be installed either above the floor level or within a pit below floor level. Isolators employed to support the block may be made of rubber, steel springs, or other suitable resilient material. The required size of the inertia block depends on the reason for its use and on the type and size of equipment. The purpose of the inertia block may be to provide rigidity; the size is determined by the momentum available for transfer to the block and by the permissible motion of the machine. The desired natural frequency for the isolation system usually is established by the operating characteristics of the mounted equipment and the isolation required as illustrated in the examples given above.

Figure 34.5 shows an inertia block supported above floor level. Commercially available isolators can be used effectively with an inertia block; particular advantages result from the use of isolators having a height adjustment feature to facilitate leveling of the block. The inertia block shown in this illustration may be poured without a form, using the floor to close the lower side of the frame and covering the floor with plastic sheeting before pouring the concrete. Upon hardening of the concrete, the inertia block is lifted from the floor by turning the lifting screws.

In the arrangement shown in Fig. 34.6, the coil-spring isolators are placed upon pedestals located near the edges of the inertia block. A vertical rod extends downward from the strongback over the several springs and is attached at its lower end to a structural member embedded in the inertia block. This pendulum-type suspension ensures stability of the system and makes possible a relatively low natural frequency in the horizontal mode of vibration.

Figure 34.7 shows how resilient pads can be used in constructing an inertia block below floor level. A concrete pit of the required size is lined with the resilient material. Then this material is covered with plastic sheeting, and the concrete is poured to form an inertia block. The desired natural fre-
quency can be otained by using resilient pads of the appropriate thickness and area.

To obtain a low natural frequency for the isolated system, a large static deflection is required. Then, steel-spring vibra-
tion isolators are used as illustrated in Fig. 34.8. The concrete inertia block is poured

FIG. 34.5. Inertia block supported on isolators. The foundation may be constructed by covering the floor with plastic sheeting, setting the frame upon the sheeting, and pouring concrete to fill the frame. After the concrete sets, the foundation is raised off the floor and made operative by turning the lifting screws. (*The Vibration Eliminator Company.*)

FIG. 34.6. Foundation above floor level sup-
ported by coil springs located on pedestals arranged around the periphery of the foundation. The support rods afford pendulum action and make possible relatively low natural frequencies in horizontal modes. (*The Korfund Company, Inc.*)

around a structural-steel frame having protrusions which provide a means of attachment for the isolators. If the inertia block is relatively small, commercially available isolators can be used.

If the isolators are located substantially below the center-of-gravity of the inertia block, some of the natural frequencies of the system are relatively low because of coupling of rotational and translational modes. (For example, see Fig. 30.18.) A tendency toward

FIG. 34.7. Independent foundation flush with general floor level. Stiffness of resilient pad is deter-
mined by type of material and its area and thickness; space between pads is filled with inactive fill, i.e., with material particles of negligible stiffness. The foundation may be poured directly on plastic sheeting placed over the pad. Alternate arrangements are shown on opposite ends. At left, the pad serves as a form for the concrete; where less horizontal stiffness is required, removable forms may be used to provide construction shown at right. (*The Vibration Eliminator Company.*)

instability thus is introduced, an effect which becomes more important if the machine generates large horizontal forces during normal operation. This limitation can be minimized by installing the isolators in positions closer to the upper surface of the inertia block, supported upon abutments extending inward from the walls of the pit. A more refined version of this concept is the T-shaped inertia block illustrated in Fig. 34.9. With such a design it is possible to place the iso-

lators in the same horizontal plane as the center-of-gravity and to equalize, approximately, the natural frequencies in all six modes of vibration. This requires coil springs having equal horizontal and vertical spring constants. If the distances between the springs when viewed in elevation from either of two sides are made equal to two times the respective radii of

Fig. 34.8. Independent foundation adapted to be located in a pit; the upper surface of the foundation is flush with or below the floor level. The foundation is supported by isolators having coil springs as the principal resilient load-carrying means. (*The Korfund Company, Inc.*)

Fig. 34.9. Independent foundation supported by coil springs and arranged to equalize substantially the natural frequencies in all six natural modes of vibration.[2]

gyration, the vertical and rotational natural frequencies are equalized (see Fig. 30.18). The natural frequency in rotation about the vertical axis cannot be controlled independently, but usually is approximately equal to the natural frequencies of the other modes.

When the natural frequency of an inertia block on coil springs is relatively low, the possibility of instability is always present because the only horizontal constraint is that provided by the horizontal spring constant of the load-carrying springs. The stability of a spring in this sense is discussed in Chap. 32. If there are doubts concerning stability, additional springs should be added to provide horizontal stability.

FLEXIBLE CONDUIT AND TUBING

The advantages provided by the use of vibration isolators may be nullified if equipment mounted on isolators is attached to a structure by means of rigid piping or conduit. Instead, flexible conduit or tubing should be used (1) to permit the mounted equipment to move, (2) to prevent the transmission of vibration to the structure, and (3) to prevent failure of the conduit or a fitting due to metal fatigue resulting from vibration. To obtain effective isolation, relative motion between the mounted equipment and its supporting structure must be permitted. Rigid conduit attached to this equipment can reduce the effectiveness of the isolators by restricting the necessary relative motion in the system.

Rubber hose often is used instead of rigid tubing. However, if the temperature is too great, if the pressure is too high, or if chemically active fluids are transmitted, metal hose or tubing must be used.

The use of flexible conduit or tubing prevents the transmission of vibration through the conduit or tubing but not the transmission of vibration through the fluid which it may contain. Thus flexible conduit can be very effective in air systems but is less effective in fluid systems.

In the installation of flexible conduit or tubing, the following design principles are important:

1. Attach the conduit to the mounted machine near the point of minimum vibratory movement in order to minimize the transmission of vibration through the conduit; the opposite end of the conduit should be attached to the most massive structure available in order to reduce the influence of transmitted vibration on this structure. If necessary, this principle can be effected by running a length of rigid conduit from the point of minimum movement to the location where the flexible portion can best be installed. The rigid part of the connection is then carried by the machine without affecting isolation.

2. Relatively large loops of flexible hose should be employed in the conduit where one end experiences a large displacement relative to the other end or where a very low conduit stiffness is required. Usually the vibration amplitude is small and the flexible portion of the conduit can be installed straight but with sufficient play to avoid vibration transmission.

3. If a loop is used, the flexible conduit should be installed so that the vibratory motion is essentially distributed over the full length of the loop. If the movement occurs only adjacent to the ends of the conduit, excessive wear resulting from stress concentration at the end connections may shorten the life of the conduit.

Fig. 34.10. Cutaway view of flexible seamless tubing of (*A*) annular and (*B*) spiral types.

A common type of flexible metal conduit is the corrugated tubing illustrated in Fig. 34.10. It is available with the corrugations having an annular configuration as shown in Fig. 34.10*A* or a helical configuration as shown in Fig. 34.10*B*. It is made from seamless, welded, or soldered tubing in diameters from ¼ to 4 in. (6.4 to 101.6 mm), depending upon the materials and styles offered by various manufacturers. Welded tubing with diameters as large as 20 in. (508 mm) is also available in corrugated form. Corrugated conduit is commonly made from bronze, steel, and stainless steel, but it is also available in a variety of alloys. The seamless or welded tubing is not susceptible to leakage; therefore, it is satisfactory for containing volatile substances and gases under pressure.

Flexible conduit as obtained from the manufacturer is normally equipped with integral fittings at the ends. Lengths of tubing are available in standard sizes or to custom dimensions determined by the application.* Corrugated tubing for pressure lines is normally enclosed within a flexible braid to prevent elongation and ultimate bursting of the tubing, the braid being attached to the end fittings. It is acceptable for use without a braid only for such applications as transmitting air or gas under low pressure.

FLEXIBLE COUPLINGS[3]

Flexible couplings help to reduce shock and vibration, to eliminate noise problems, to minimize destructive forces transmitted to other components, and to provide improved operator comfort.

---

* Manufacturers of flexible conduit and tubing include Flexonics, Inc., Bartlett, Ill. 60103; and Titeflex Corporation, Springfield, Mass. 01109.

Engines (particularly of the internal-combustion type) drive other equipment such as transmissions, hydraulic pumps, and compressors. The driven equipment, attached directly to the engine, may be sensitive to torsional vibration and require protection to prevent damage to gears and other components. Isolation can be provided by means of a torsionally flexible coupling that is capable of transmitting the driving torque, while providing isolation of torsional vibration coming from the engine. Torsionally flexible couplings are used mostly with reciprocating engines.

If the driven equipment is not attached directly to an engine flywheel, the shaft connection cannot be made torsionally stiff enough to place the first-mode critical speed above the operating speed range. When critical speed occurs within the operating speed range, the driveline components (such as transmission parts) vibrate at resonance; this causes early failure of these components. In addition, there may be angular and parallel misalignment of the drive and the shaft to which it is connected. In this case, a flexible coupling must not only have torsional isolation capability, but must also accommodate the angular and parallel misalignment.

Mechanisms other than resilient couplings may be used to reduce torsional vibration, but they are generally not satisfactory. For example, stiffening the drive shaft increases critical speed, but stiffening is often not practical because it results in an increase in cost and weight. Increasing the inertia of the flywheel reduces the critical speed, but this solution is limited by available space, cost, and weight. A coupling with a low torsional stiffness in series with the driveline usually is the best solution. This shifts the natural frequencies downward, away from the excitation frequencies of the operating equipment. A flexible coupling will not last long if the driveline system operates at or near resonance. A torsional analysis (see Chap. 38) should be performed on each new application to ensure that critical speeds are not in the operating range. Such an analysis will provide the basis for proper stiffness selection for good performance. The engine mass, flexible coupling mass, driveline mass, and torsional stiffness are the minimum parameters required for a torsional analysis. Such a torsional analysis considers the mass-spring systems for the rotating elements in calculating the natural frequency and compares the natural frequency to the torsional excitation. (See Chap. 38 for a discussion of mass-spring systems.)

The choice of an elastomer for a flexible coupling depends on the environmental conditions. A majority of applications use natural rubber when ambient operating temperatures do not exceed 180°F (82°C) and oil is not present. If oil is present, Neoprene may be used. At higher temperatures, various synthetic elastomers such as silicone may be used; if oil is present, fluorosilicone is more suitable.

INDUSTRIAL MACHINERY

ELECTRIC MOTORS

The torque delivered by an electric motor is the result of the action of the magnetic field of the stator upon current-carrying conductors of the armature. The nominally steady torque has superimposed fluctuations at a frequency that is a function of the motor speed and the number of poles on the stator. Vibration results from two sources:

1. Fluctuation in the torque, characterized as a rotational reaction upon the stator and thus upon the motor supports.

2. Unbalance of the rotating parts, characterized as a rotating radially directed force vector.

Isolation of vibration due to torque fluctuations is achieved by mounting the stator on an isolation system that provides a low stiffness in rotation about the shaft center. The system's torsional natural frequency should be substantially lower than the frequency of the torque fluctuation. One way in which this can be done is illustrated in Fig. 34.11. An isolator, consisting of an elastomeric ring, encircles the hub of the stator and is attached at its periphery to the motor support. The isolator in shear acts torsionally; the modulus of elasticity of elastomers in shear is relatively low. With the location of the elastomer ring close to the axis of rotation, the stiffness of the system is low in the rotational mode and a low torsional natural frequency can be achieved. In contrast, the stiffness of the isolator is comparatively high in a direction perpendicular to the shaft. The high radial spring constant results from the relatively high modulus of elasticity of elastomers in compression. Because of the annular shape of the isolator, equal reaction is provided for radial forces in all directions. The isolator provides the required high reaction to forces applied by belts and gears, yet does not sacrifice the low torsional stiffness needed to effectively isolate the torque fluctuation.

Vibration caused by unbalance of the rotating parts occurs at the rotational frequency of the motor. Because the unbalance introduces a radial force, the mounting system must have a somewhat low radial stiffness to provide good vibration isolation. This low stiffness may affect the operation of belt and gear drives. However, the rotating parts of electric motors usually are well-balanced; the expected vibratory disturbances due to unbalance are low in magnitude and are not a significant factor in the selection of an isolator.

GEARS, BEARINGS, AND CAMS

Vibration resulting from the operation of gears, bearings, and cams usually has its source in friction between sliding or rolling members, in impacts between loose-fitting parts, and/or in turbulent action of lubricating fluids.

Where gears are the major source of noise, the predominant frequency is the tooth frequency (i.e., the product of the rotational frequency of the gear and the number of teeth on its periphery); higher harmonics of the tooth frequency may be present, but subharmonics are seldom encountered. The vibraton caused by bearings and fluid turbulence tends to lack discrete frequencies. Adequate isolation of vibration originating in gears and bearings generally can be attained with an isolation system having a natural frequency under 25 Hz. The forces are small and cause no appreciable over-all motion of the mounted machine. Care should be taken to avoid resonance with any predominant frequency, such as the rotational frequency or tooth frequency.

The operation of cams may involve either continuous sliding or rolling contact, or momentary contact in the nature of an impact. In the former, vibration resulting from cam action is very similar to that resulting from the operation of gears and bearings Impact forces are different, and the required treatment depends upon their magnitude. For example, in a loom the impact force exerted by the cam may be sufficiently great to displace the entire machine. Then the isolators must be capable of large deflections. In other instances, the impacts associated with cam action are relatively small. For this latter condition, isolators designed in accordance with the principles of isolating gear and bearing vibration usually provide satisfactory results.

FANS

A squirrel-cage fan, driven by an electric motor, commonly is used to distribute air throughout a heating and air-conditioning system. The rotor of the fan is a drum having a relatively large number of longitudinally extending blades arranged around the periphery. Air enters the fan by flowing into the drum and is expelled radially between the blades into the duct system. There are two predominant sources of vibration: (1) rotor unbalance and (2) aerodynamic forces resulting from passage of the blades past the duct opening. The frequency of the former is the rotational speed of the rotor (rotor frequency); the frequency of the latter (blade frequency) is the product of rotor speed and number of blades on the rotor. Usually, the blade frequency is used as a basis for isolator selection. However, in a large central air-conditioning system which employs a large

FIG. 34.11. Electric motor equipped with vibration isolators in the form of elastomeric rings surrounding the hubs of the motor end bells. (*General Electric Company.*[4])

fan operating against high static pressure, the rotor frequency must be used in isolator selection when the forces resulting from the rotor unbalance are significant.

Where the isolators are selected on the basis of the blade frequency, the forcing frequency is relatively high; there is no problem in obtaining a low natural frequency compared to the forcing frequency. For example, if a squirrel-cage fan has 50 blades and the rotor operates at a speed of 900 rpm (15 Hz), the forcing frequency (i.e., the blade frequency) is 750 Hz; an isolation system having a natural frequency as high as 75 Hz provides good isolation of vibration. In contrast, if isolators are selected on the basis of the rotor frequency, a system natural frequency lower than 10 Hz may be required to attain adequate isolation. Such fans should be supported on concrete floors or roof panels at least 6 in. (15 cm) thick. For supports of less rigidity, it is advisable to mount the fan upon a concrete inertia block which is in turn supported by isolators (see the section on *Isolated Foundations and Inertia Blocks*). The mass of the inertia block should be at least equal to the mass of the fan-and-motor assembly.

Fig. 34.12. Squirrel-cage fan with electric motor mounted upon integral base fabricated from structural steel and supported by vibration isolators. (*The Vibration Eliminator Company.*)

Smaller fans are constructed of relatively light sheet metal, and the motors are furnished separately. A conventional installation practice is to mount the fan and motor on an integral base fabricated from structural steel. The base is then mounted on vibration isolators. The rigidity of the base maintains the necessary alignment between motor and fan and adds rigidity to the light sheet-metal structure of the fan. If the mounted structure is not sufficiently rigid, the isolators will not function effectively and the intended isolation may not be achieved. A typical fan and motor mounted upon an integral base are shown in Fig. 34.12.

PUMPS

Pumps may be classified as *centrifugal* or *reciprocating*. The former usually has a multivane rotor that operates at a relatively high speed. There may be (1) vibration at the rotational frequency as a result of mass unbalance of the rotor, (2) vibration at the vane frequency, as the moving vanes pass close to the fixed vanes, and (3) random vibration resulting from forces created by turbulent flow of liquid within the pump.

Centrifugal pumps are relatively easy to isolate because their operating speed is sufficiently high to allow the use of isolators of relatively high stiffness and because the oscillating forces are so small that the vibration amplitude of the body of the pump is not appreciable.

In reciprocating pumps, the predominant sources of vibration are unbalanced rotating and reciprocating parts, and torque impulses associated with movement of the fluid. Reciprocating pumps characteristically are low-speed machines; the number of cylinders often is small. The forcing frequency is low so that isolators having low stiffness are required to obtain effective isolation. Thus the reciprocating pump is subject to a large static displacement when it is mounted on vibration isolators. If this is unacceptable, the motion of the pump may be reduced by mounting it on an inertia block such as that shown in Fig. 34.7. Where the motion of the pump is large, flexible hose or conduit should be used instead of rigid pipe in all connections to the pump (see the section on *Flexible Conduit and Tubing.*)

PIPING

Piping often is an objectionable source of vibration and noise in buildings unless effective isolation is provided.[5,6] A common method of vibration control is to suspend the piping from the floor slab above using resilient hangers or to support it from below using resilient floor mounts.

Piping connected to vibration-isolated equipment should be installed so that it does not place strain on the isolators supporting the equipment and induce misalignment in them. The equipment must be able to move freely on its isolators if the system is to be effective.

Vibration isolators for heavy piping usually are of the steel-spring type. A steel spring often is used in conjunction with a Neoprene pad acting in series with it. The function of the Neoprene pad is to impede the transmission of vibratory energy at high frequencies which may be transmitted along the coils of the spring. The coil spring rests on a Neoprene spring seat. Metal-to-metal contact between the spring, its housing, and the rod (which supports the load) is prevented by a Neoprene collar through which the rod passes. Elimination of contact between metal surfaces helps to minimize noise and vibration transmission.

A vibration-isolator hanger can be used to suspend a single pipe; a trapeze support can be used to suspend several pipes.

Where pipes, ducts, or conduits penetrate a building partition, they should not be in solid contact with the partition but should be isolated from it. The space between the pipe and the partition it penetrates [usually about ½ in. (12.7 mm)] should be packed with a material such as fiberglass. A nonhardening, resilient sealant should be applied on both sides of the wall to reduce the transmission of sound from one side of the partition to the other through the packed material.

ELECTRONIC EQUIPMENT

Some types of industrial electronic equipment must be protected from severe shock and vibration. In general, moderately damped elastomeric isolators which support static loads and provide low natural frequencies in all directions are satisfactory for this purpose; one such isolator is of the grommet type. This is a small unbonded isolator that snaps into a hole in a plate, providing an installation similar to the one in Fig. 32.22K.

AEROSPACE EQUIPMENT

DYNAMIC ENVIRONMENT

Installations of aerospace equipment are subject to shock and vibration excitation from a number of sources. Vibrations may be periodic (as is the case in piston engine or turboprop-driven aircraft), or they may be random (as is the case in military aircraft). The levels of random vibration can be broad-band and exceptionally high in level—especially in high-speed, highly maneuverable aircraft used at low altitudes. In such cases, many electronic devices cannot function without vibration protection. In addition to engine-induced or aerodynamically induced vibration, the firing of munition (for example, a multibarrel cannon firing at a high rate) from an aircraft can be a significant source of vibration. Shock environments also must be considered in aerospace installations. A typical shock source is that due to the landing of an aircraft. Of particular severity is the arrested landing of a carrier-based aircraft. Since such a shock is transmitted to the equipment through the aircraft landing gear, it is sinusoidal (half-sine) in nature and may be analyzed as such. Another major source of shock excitation is that due to the launching of missiles from an aircraft. This type of shock is normally triangular in shape.

The predominant specification in use for the isolation of electronic equipment in U.S. military aircraft is MIL-STD-810.[7] This specification details all of the various vibration and shock tests which are typically required. It divides them by type of vehicle or aircraft and by location of the equipment in the aircraft. A similar specification for commercial aircraft is DO-160 (available from the Radio Technical Commission for Aeronautics, Washington, D.C.).

TYPES OF SUPPORTED EQUIPMENT

The selection of isolators for a specific application depends, in part, on the function of the equipment being supported. For example, isolators supporting an aircraft engine must be capable of reacting against high-level excitation resulting from maneuver and torque as well as from the inherent vibratory disturbances. In addition, these isolators must be fail-safe, i.e., they must not fail to hold the engine in place even if the resilient element is severely damaged.

Communications equipment (such as radios) may require isolation from shock and vibration but does not require any special considerations for limiting angular motions.

Some examples of critical equipment in aircraft are guidance/navigation systems and radars. These instruments are sensitive to position, alignment, and angular motions. Thus, besides providing protection from the dynamic environment, an isolator for this type of equipment must be analyzed for its ability to maintain the orientation of the supported equipment. The isolator, typically elastomeric, must be selected to have low set and creep and high returnability after loading. Thus the type of material in the isolator must be carefully considered along with the physical construction of the isolator. Silicone is often used because of environmental considerations. The isolator should be of fully bonded construction, not semibonded or unbonded.

The position of the supported element depends on the characteristics of the individual isolators, the geometry of the installation, and the relative stiffness of isolators with respect to one another. To maintain high rotational stability, the isolators should be separated geometrically as far as possible. In critical applications, the dynamic stiffnesses of the isolators to be used (as a set) to support a piece of equipment can be matched closely. Even though apparently identical isolators are used at all locations in an installation, very critical applications may require very close matching of the dynamic properties of the isolators. This helps to avoid introduction of dynamic unbalance due to isolator mismatch.

AMBIENT ENVIRONMENT

The ambient environment for aerospace applications of vibration isolators [from -65 to $250°F$ (-54 to $120°C$)] can be an important consideration in the selection of the material of which an isolator is fabricated; typically, elastomeric isolators are used. Of the available choices, silicone elastomers usually are best from the standpoint of temperature. Their properties vary least with temperature, and they have the best resistance to environmental damage. The properties of metal isolators are not affected by temperature changes in the above range; therefore, such isolators can be used where dynamic performance and equipment orientation are not overly critical. They are mainly used where temperatures range very high, above several hundred degrees Fahrenheit, which may be the case in installations on or near hot sections of aircraft engines.

Humidity and pressure have no effect on the operation of typical isolators. One area of potential damage is that resulting from exposure to a fluid. If the fluid is potentially damaging, an isolator fabricated of a special material such as fluorosilicone may be selected or the elastomer may be treated with a fluid-resistant coating.

A variety of isolators are used for aerospace applications. For example, Fig. 32.22*I* shows one such isolator. An isolator of this type is chosen because of its predictability of performance for critical installations.

SYSTEM DYNAMICS

Generally, isolators are chosen to have dynamic spring constants which provide system natural frequencies in the 20- to 30-Hz range. This gives excellent isolation of high-frequency disturbances while still affording reasonable stability of the system. However, system natural frequencies may properly be positioned in many other ranges, depending on the dynamic disturbance, the type of equipment being isolated, and the required degree of protection.

ROAD-VEHICLE SUSPENSIONS

Vibration that exists in a road vehicle may result from excitation by the power source, excitation resulting from movement of the vehicle over irregularities in the surface of the road, or from auxiliary equipment in the vehicle. The engine, usually an internal-combustion engine which operates at a relatively high frequency, is usually not an important source of vibration within the vehicle because the engine itself is supported by isolators. However, in exceptional cases (such as large trucks and army vehicles) vibration from the engine may be significant—depending on how well the engine suspension is designed.

In the conventional road vehicle, road irregularities may excite each of four elastic systems separately (see Fig. 42.4):

1. The body of the vehicle supported by the primary suspension.
2. The unsprung weight (wheels, axles, and drive mechanism) supported between the primary suspension and the tires.
3. The engine and its transmission supported by isolators attached to the mainframe of the vehicle.
4. The body structure which can experience structural vibration at its natural frequencies.

The natural frequencies of these systems are primary factors in determining design requirements for conventional isolators (for air-supported and servo-controlled suspension systems, see Chap. 33).

1. The natural frequency of the vehicle body supported by the primary suspension is usually between 1 and 4 Hz, the higher value generally applying to military vehicles. It is not practical to use isolators which provide a lower natural frequency than the natural frequency of the primary suspension. Therefore, it is unlikely that conventional isolators supporting the other systems can provide significant protection to equipment with respect to the over-all motion of the vehicle body.

2. The natural frequency of the unsprung weight (sometimes called the *wheel hop frequency*) usually is between 10 and 20 Hz, the higher value generally applying to military vehicles. Isolators for the suspension of the systems other than the primary suspension can be provided with natural frequencies in the range from 5 to 7 Hz (although additional clearances may be needed, depending on the design). Also, the larger size of these isolators may create space problems. If isolators of lower stiffness cannot be used, then avoiding resonance is the next best alternative. In this latter case, the natural frequencies should be higher than the 20-Hz unsprung natural frequency. A 25-Hz value may be a suitable objective.

3. The natural frequencies of the engine supported by isolators attached to the mainframe are usually higher than that of the body of the vehicle supported by the primary suspension and are selected to avoid resonances of the body structure. Many modern automobiles are small and have a low over-all vehicle weight. Their engines are small but the ratio of body-frame weight compared to the engine weight is larger than that of larger automobiles, thus requiring engine suspensions which are less stiff. Thus the vehicle is more susceptible to being excited by irregularities in the surface of the road, creating disturbances that can be felt by the driver. Extra damping in the engine suspension is required to alleviate this problem.

4. The natural frequencies of the structural members of the vehicle body usually are higher than the natural frequencies associated with the primary suspension system. A principal function of isolators is to alleviate the effects of the resonance conditions that would develop when the natural frequency of an important member of the equipment coincides with a natural frequency of a structural member supporting the equipment.

SHIPBOARD APPLICATIONS

Isolators are used in shipboard applications (1) to protect sensitive equipment from shock and vibration and (2) to prevent structureborne noise (resulting from vibration) from being transmitted from the ship to the water.

Vibration levels within a ship are typically quite low in amplitude and do not present a major threat to the integrity of shipboard-mounted equipment. However, such vibration may be the source of ship noise which is radiated into the water, making the ship detectable by enemy passive sonar. The use of isolators for noise reduction may be important even though vibration protection per se is not normally a concern. Noise reduction is enhanced by the use of isolators.

The primary task of shipboard isolators is shock protection. The shock disturbances specified for shipboard equipment are those anticipated from enemy action and can be in the range from 300g to 400g. Most equipment cannot withstand such accelerations without some protection against shock.

Since shipboard vibration is low in frequency and a high degree of shock protection is required, isolators having low spring constants are selected to provide low system natural frequencies. Most shipboard isolator systems provide natural frequencies in the range between 5 and 15 Hz. With these extremely low natural frequencies, the magnitude of the displacement of the suspended equipment can be quite high; the isolators selected must be able to accommodate these displacements, sometimes as high as 3 to 4 in. (76 to 102 mm).

The ambient environments in most shipboard applications do not pose a problem in isolator selection. Natural rubber or Neoprene elastomers are commonly used in these installations.

Typical dynamic disturbances and ambient environments for shipboard applications are given in U.S. military documents such as MIL-M-17185, MIL-STD-167, and MIL-S-901. The application and design of isolators for U.S. Navy shipboard use are covered in a number of documents, e.g., MIL-HDBK-149, NAVSEA 0900-LP-089-5010, and NAVSHIPS 0940-000-3010.

SHIPPING CONTAINERS

When sensitive equipment is shipped by common carrier from one location to another, it may be subject to potentially damaging shock and vibration, the magnitude of which depends on the particular mode of shipment. Therefore, it is important to know the intended manner of shipping. Specifications such as MIL-STD-810 and FED-STD-101 offer some guidance as to dynamic disturbance from various types of commercial carriers. High-amplitude, low-frequency vibration (in the range from 2 to 7 Hz) may be present having a magnitude of acceleration as large as $\pm 1g$ to $\pm 1.5g$. In addition, shock disturbances may arise from a variety of sources such as dropping of the shipping container.

Shipping containers usually are not exposed to severe temperatures; a range from -40 to 180°F (-40 to 82°C) is typical. Natural rubber isolators are used in a number of shipping-container applications, although natural rubber does stiffen significantly at lower temperatures. A number of synthetic elastomers are better suited to low-temperature environments; they also provide higher damping to limit shock motions. Since shipping containers may be exposed to outdoors, resistance of isolator materials to weathering (sunlight, ozone, etc.) should be considered in container system design.

Figure 32.22L shows one example of a sandwich-type isolator which can be used to protect equipment during shipping. Such isolators are chosen for simplicity of installation, low cost, and predictability of performance. They should usually be selected to provide natural frequencies in the range between 7 and 10 Hz.[8] Depending on the degree of protection required, a large displacement may be needed, and space must be allotted in the container for this motion to occur. For example, as much as 4 in. (102 mm) of sway space (clearance) may be required for some critical (low fragility) applications.

A comprehensive treatment of packaging design is given in Chap. 41.

REFERENCES

1. "Designer's Guide," *Publication* No. PB-8000c, Lord Corporation, Erie, Pa., 1986, p. 8.
2. Crede, C. E., and J. P. Walsh: *J. Appl. Mech.*, **14**:A-7 (1947).
3. Barickman, J. R.: "Flexible Elastomeric Couplings for Torsional Vibration Isolation," *SAE Earthmoving Industry Conference, Paper* No. 840792, April 1984.
4. Crede, C. E.: "Vibration and Shock Isolation," John Wiley & Sons, Inc., New York, 1951, p. 290.
5. Harris, C. M.: "Handbook of Noise Control," 2nd ed., McGraw-Hill Book Co., New York, 1979.
6. Jones, R. S.: "Noise and Vibration Control in Buildings," McGraw-Hill Book Co., New York, 1984.
7. "Avionics Isolator Application Guide," *Publication* No. TF-6104b, Lord Corporation, Erie, Pa., 1985.
8. "Shipping Container Mounts Application Guide," *Publication* No. TF-6701, Lord Corporation, Erie, Pa., 1985.

35

PROPERTIES OF RUBBER

William A. Frye
Inland Division
General Motors Corporation

INTRODUCTION

Rubber may be defined as an elastomeric material which can be or already is modified to a state exhibiting small plastic flow, high elongation, and high speed of retraction. This definition includes natural rubber, which is made from the milky juice of the *Hevea brasiliensis* tree, and a large family of synthetic rubbers made of polymers and copolymers of various organic compounds. The adjectives *elastomeric* and *rubberlike* are terms often used to describe rubber or similar materials. The term *polymer* is used to designate rubber in the "raw" or uncompounded state. The designation and composition of some

Table 35.1. Designation and Composition of Common Polymers

Polymer designation*	Common name	Chemical composition
NR	Natural rubber	Natural polyisoprene
IR	Isoprene	Synthetic polyisoprene
SBR	SBR	Copolymer of butadiene and styrene
BR	Polybutadiene	Polybutadiene
IIR	Butyl	Copolymer of isobutylene and isoprene
CR	Neoprene	Polychloroprene
EPDM	EPDM	Terpolymer of ethylene, propylene, and a diene
PO	Propylene Oxide	Copolymer of allyl glycidyl ether and propylene oxide
ECO, CO	Epichlorohydrin	Epichlorohydrin polymer and copolymer
CIIR	Chlorobutyl	Chloro-isobutene-isoprene
BIIR	Bromobutyl	Bromo-isobutene-isoprene

* American Society for Testing Materials Method D1418

of the common polymers are shown in Table 35.1. The term *vulcanizate* indicates a vulcanized rubber compound.

This chapter describes the properties of rubber which are of importance in vibration- and shock-isolation applications, how these properties are influenced by environmental conditions, and presents basic information for the design of rubber vibration and shock isolators, i.e., rubber springs.

Rubber has a low modulus of elasticity, and is capable of sustaining deformation of as much as 1,000 per cent. After such deformation, it quickly and forcibly retracts to essentially its original dimensions. It is resilient, and yet exhibits internal damping. Rubber can be processed into a variety of shapes and can be adhered to metal inserts or mounting plates. It can be compounded to have widely varying properties. The load-deflection curve of a rubber isolator can be altered by changing its shape. Rubber will not corrode and normally requires no lubrication. Because of these properties, rubber is uniquely suited for application to isolation problems.

In spite of continued advances, rubber technology remains more of an art than a science, for a number of reasons: (1) Rubber itself is difficult to analyze and describe accurately because it is made up of very large and complicated organic molecules. (2) Commercial rubber compounds are made up of a great number of ingredients, many of which are also organic compounds. (3) Variations in processing methods cause variations in the properties of a rubber part. (4) Rubber manufacturing companies tend to keep compounding and processing information confidential.

The essential steps in producing a molded rubber part are (1) mastication of the polymer and compounding ingredients to produce an uncured compound, (2) cooling of this compound to prevent vulcanization while it is transported and temporarily stored at the molding location, (3) vulcanization or cure in the mold under heat and pressure, and (4) cooling and trimming.

PROPERTIES OF RUBBER

The physical properties of a vulcanized rubber specimen are dependent on the stiffness of the rubber and the size and shape of the test specimen. They are also dependent on the specimen temperature and rate of straining the specimen to an extent far greater than most materials. Therefore, it is convenient to consider the static properties of rubber and then the effect of environment and strain rate on these properties.

STATIC PHYSICAL PROPERTIES OF RUBBER

A typical tensile stress-strain curve for rubber is shown in Fig. 35.1A; similar curves for rubber in compression and in shear are given in Fig. 35.1B and C, respectively. Note

STRAIN
(A)

STRAIN
(B)

STRAIN
(C)

FIG. 35.1. Stress-strain curves for rubber in (A) tension, (B) compression, and (C) shear.

that in these curves Hooke's law of elasticity is not followed. This fact makes it difficult to specify the modulus of elasticity of the material, the property which is probably the most important for isolator design purposes. In addition, the compression stress-strain curve is greatly influenced by the *shape factor* of the specimen used, i.e., whether the specimen is relatively tall with a small cross section, or relatively short with a large cross section.* For these reasons, it is common to specify the durometer hardness of rubber

---

* Rubber is essentially an incompressible substance that deflects by changing shape rather than changing volume. Poisson's ratio is approximately 1/2. Values of Poisson's ratio to five significant figures are given in Ref. 25 for five compounds of natural and synthetic rubber. Average values for each compound are: natural rubber, 0.49935; SBR, 0.49894; butyl, 0.49691; Neoprene, 0.49950; and Nitrile, 0.49712.

compounds.   The *durometer* is a pocket-type instrument which, when applied by hand, forces a small indentor into the rubber.

The resistance to indentation is indicated by a pointer and a graduated scale.[1]   The higher the reading, the harder the rubber.   Rubber hardness is not important of itself; it is merely an approximate and convenient measurement which is related to modulus of elasticity and which is independent of a specimen shape factor.   Because of the visco-elastic nature of rubber, a durometer reading reaches a maximum as soon as the instrument is completely applied to the specimen and then falls several points during the next 5 to 15 sec.   The reading is also sensitive to indentor side loads caused by the manner in which the instrument is applied.   These factors make it difficult to obtain good agreement between readings made in various laboratories.

OTHER BASIC PROPERTIES.   Basic rubber properties which are often specified, in addition to *hardness,* are *tensile strength, ultimate elongation,* and *compression set.* Tensile strength and ultimate elongation are determined by pulling to destruction small dumbbell-shaped specimens which have been cut from laboratory cure sheets.[1]   During this test, the tensile stress at a specified elongation may be determined.   This value is called *modulus* and is specified as (for example) "modulus at 200 per cent" to avoid confusion with the general term "modulus of elasticity."   Tensile test values for samples taken from finished articles may be somewhat different because of variance in processing and cures and because of the directional properties (anisotropy) of the material.   Rubber seldom is stressed to a value approaching its ultimate tensile strength or elongation when used in an isolator.   (Bond strength is usually the limiting factor.)   Hence these properties are important only as they relate to fatigue strength, tear resistance, and similar properties.   Rubber compounds commonly used for vibration isolators have tensile strengths ranging from 500 to 3,500 lb/in.^{2} (35 to 250 kg/cm^2).   The ultimate elongation of such vulcanizates is from 150 to 600 per cent.

Compression set is determined by compressing a 1.129-in. (28.4-mm) diameter specimen which is 0.50 in. (12.7 mm) thick to a preset deflection and exposing it at an elevated temperature.[1]   After exposure, the specimen is allowed to recover for one-half hour and measured.   *Per cent compression set* is defined as the decrease in thickness divided by the original deflection.   Typical rubber compounds used for vibration isolators have compression set values of from 10 to 50 per cent.   The exposure may consist of 22 hr at 158°F (70°C), or 70 hr at 212°F (100°C), depending on the intended use of the isolator.[1]

A "static" load-deflection curve for a rubber vibration isolator is shown in Fig. 35.2.   The area between the loading and unloading curves represents hysteresis or damping.   Both the stiffness and the hysteresis of the isolator depend on the specimen temperature and the rate of straining.   It is also necessary to make three or more loading cycles in order to obtain a stable load-deflection curve.   The flexing causes a slight softening of the material which is semipermanent in nature.

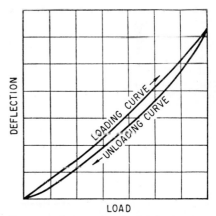

Fig. 35.2. Compression load-deflection curve for a rubber isolator.   The data for this curve were obtained on the third loading cycle.

This change is evidently caused by internal rearrangements of the carbon black structure which reverts to its original state after standing for several hours. It is common to specify that load-deflection readings be made at 73°F (23°C) and 0.125 in./min (0.318 cm/min) strain rate, flexing the specimen to the test load, or above, three times at those same conditions.

ENVIRONMENTAL FACTORS

AGING. The chemical reaction of vulcanization often continues at a very slow rate after a rubber part has been completed and placed in service. Therefore the hardness (stiffness) of rubber increases with time. At room temperature, the increase is quite rapid for the first few hours, and after a few days it drops to a very low value.

If the hardness of a typical vibration isolator were measured 3 days after vulcanization and again after a 10-year room-temperature exposure, a hardness increase of 5 durometer units might be found.

Indoor or shelf aging causes a small reduction in the tensile properties of rubber because the oxygen of the air attacks the molecular structure of the rubber. This effect is accelerated and magnified for test purposes by exposing tensile test specimens in the shape of dumbbells in a circulating air oven.[1] Exposures of 70 hr at 158°F (70°C), or at 212°F (100°C), are used—depending on the intended service. Measurements are made on aged and unaged samples, and the per cent reduction in tensile strength and in ultimate elongation as a result of the exposure are computed. Hardness increase is also measured in this test. Typical isolator rubber compounds may have a tensile strength reduction of as much as 25 per cent, ultimate elongation reduction of as high as 40 per cent, and a hardness increase of as much as 20 durometer units. Compounds of natural rubber and SBR normally used in vibration isolators are not suitable for continued use at service temperatures above 175°F (79°C). Chloroprene rubber compounds should not be used above 220°F (104°C). Propylene oxide and EPDM should not be used above 250°F (121°C). Continued exposure to high temperatures furthers the cure, promotes oxidation, and causes plasticizers and other ingredients to volatilize. Natural rubber is subject to *reversion*, in which the rubber becomes soft and tacky. Some of the other polymers undergo additional molecular "cross-linking" which causes a pronounced increase in hardness.

When rubber is exposed to direct weathering, it is exposed to oxygen, ultraviolet light, and to the leaching action of rain. The factors combine to promote oxidation. Sunlight entering the atmosphere generates ozone, which attacks the surface of the stressed rubber.[2] Ozone is especially deleterious to natural rubber, SBR (styrene-butadiene rubber), BR (butadiene rubber), and nitrile rubber (NBR) because it attacks the rubber molecule at the double bonds (the chemically unsaturated area) and produces surface cracks. These cracks can grow and cause complete failure of a rubber part if the part is continuously stressed. Other polymers have more inherent ozone resistance because they have fewer unsaturated areas in their molecular chains. The ozone resistance of the unsaturated polymers can be markedly improved by protective compounding ingredients. Ozone cracking will not occur if the rubber is unstressed. Although some reaction takes place between unstressed rubber and ozone, subsequent stretching will not reveal any cracks. There is a critical elongation at which ozone cracking is most severe. This elongation varies somewhat with the composition of the compound and the polymer used.[3] Threshold strains exist below which ozone damage is very small.[4] These strains are 7 to 9 per cent for natural rubber, SBR, and nitrile rubber compounds, 18 per cent for chloroprene rubber (CR), and 26 per cent for butyl rubber (IIR). Most polymers are not subject to ozone attack at temperatures below 0°F (−18°C).[5] Ozone is one constituent of "smog" and can occur in concentrations which are quite deleterious to rubber.

Ozone resistance usually is determined by exposing stressed specimens in a test chamber at 100°F (37.8°C) in an atmosphere of 50 parts of ozone per hundred million parts of air by volume.[1] Extruded specimens and actual parts such as weatherstrips are stressed by looping them around a 2-in.-diameter mandrel. Thinner samples are stressed by stretching them on a frame by an amount equal to 20 per cent of their lengths. The specimens are mounted, allowed to rest for 24 hr, and then exposed. They are examined frequently (perhaps every 2 hr) until the first cracking is noted and then are examined each day. The test is usually stopped after 3 days' exposure. At each examination, the specimens are rated for number and size of cracks. Vibration isolators normally are not subject to direct weathering, but sometimes operate near electrical equipment which may generate ozone by corona discharge. The correlation between ozone tests and actual weather exposure tests is poor. The actual exposure test normally is used for final evaluation and for acceptance tests, and the ozone test is used for development purposes.

IMMERSION IN FLUIDS. The properties of rubber are altered by exposure to fluids; the fluids present in vibration-isolation applications are usually petroleum oils or solvents. When immersed in these petroleum oils, natural rubber, SBR, isoprene, butadiene, butyl rubber, and silicone rubber soften, swell, and suffer a reduction in tensile strength and ultimate elongation. In a light petroleum oil, natural rubber might swell as much as 200 per cent and retain less than 20 per cent of its original tensile strength and ultimate elongation. Natural, SBR, and butyl rubbers usually are not tested for oil or solvent resistance. Two reference fuels and two reference oils have been established for immersion tests of other polymers. These tests are made by exposing tensile dumbbells and volume-swell specimens in the test fluid for 70 hr at 212°F (100°C) or 302°F (150°C), depending on the intended service. These samples, and unexposed duplicate samples, are then tested and per cent reduction in tensile strength and ultimate elongation are computed. Per cent volume swell and hardness change in durometer units are also found.[1] A typical chloroprene rubber compound, when exposed in a lubricating oil for 70 hr at 212°F (100°C), might suffer a 40 per cent reduction in tensile strength and a 30 per cent reduction in ultimate elongation. The volume might increase 15 per cent and the hardness might decrease 10 durometer units. Rubber compounds used for vibration isolators normally are not exposed to oil or solvents. It is usually more economical to locate the isolator in an area not exposed to these fluids, or to shield it, than to use an oil-resisting polymer.

LOW-TEMPERATURE PROPERTIES. When exposed to low temperatures, rubber becomes harder, stiffer, and less resilient. These changes are caused by changes in the mobility of the rubber molecule: (1) Simple temperature effects bring about an increase in stiffness and a decrease in resilience. They are caused by decreased thermal activity of the rubber molecule and are complete as soon as the specimen reaches thermal equilibrium. (2) Second order transition of the rubber molecule also causes an increase in stiffness and a decrease in resilience; however, the change occurs in a short time at a definite temperature, which depends on both the polymer and the compounding ingredients. This transition results from a change in the type of motion of certain segments of the rubber molecule, and is accompanied by a change in the coefficient of thermal expansion and in the specific heat of the rubber. Normal rubber compounds are rendered unserviceable by simple temperature effects before the second-order transition effects become evident. (3) Crystallization causes hardening at temperatures above those necessary for second-order transition. It occurs over a long period of time, but is accelerated by stretching the specimen. It is accompanied by a decrease in volume and evolution of heat. Crystallization occurs most rapidly at an "optimum crystallization temperature" which is specific to the base polymer. The approximate optimum crystallization temperature for natural rubber is −13°F (−25°C); for chloroprene rubber it is 14°F (−10°C); and for SBR it is −49°F (45°C). Butyl rubber does not crystallize unless stretched, and nitrile rubbers do not crystallize at all. Natural rubber, SBR, and silicone rubber have good low-temperature flexibility.

Figure 35.3 shows the effect of low temperature on the torsional modulus of elasticity of some tread-type rubber compounds. The modulus of each compound at various temperatures is shown relative to its own modulus at 77°F (25°C). Low-temperature stiffening is reduced by plasticizer addition to an extent which depends greatly on the amount and type of plasticizers used in the particular rubber compound. High plasticizer content generally has a detrimental effect on rubber-to-metal adhesion.

As a result of hysteresis, a rubber compound produces internal heat as it is dynamically strained, and its stiffness decreases. As the temperature of a rubber compound is lowered, hysteresis increases rapidly until the second-order transition temperature is reached. There is a simultaneous increase in stiffness. In a system which applies a constant dynamic force to a rubber isolator, there will be a "barrier" temperature below which the stiffness effect overcomes the hysteresis effect and the isolator will *not* warm up

Fɪɢ. 35.3. Increase in torsional modulus of elasticity of various elastomers as a function of temperature. (*After Gehman.*[22])

as a result of hysteresis.[6] This barrier temperature depends greatly on the composition of the rubber compound, the shape of the isolator, the frequency and amplitude of vibration, and the heat losses to the surroundings. Barrier temperatures for typical compounds of several polymers are shown in Table 35.2.

Standard low-temperature tests for rubber include various methods of determining Young's modulus, hardness increase, and brittle point.[1] The method used to obtain the data in Fig. 35.3 utilizes a specimen 1.5 in. (38.2 mm) long, 0.125 in. (3.2 mm) wide, and 0.085 in. (2.2 mm) thick. The specimen is placed in an instrument containing one of three standard torsion spring wires. The twist of the specimen is determined at 73.4°F (23°C) and after 5 min at each test temperature. The temperatures used vary with the rubber being tested and may be as low as −94°F (−70°C) for natural rubber. The temperatures at which the relative torsional modulus of elasticity is 2, 5, 10, and 100 are determined from a plot of the data. In the brittleness method, samples 0.070 in. (1.8 mm) thick and 0.10 in. (2.5 mm) wide are immersed in a low-temperature bath and then impacted by a striker traveling at a velocity of 6.5 ft/sec (198 cm/sec). The test is repeated with new specimens at various temperatures until the temperature at which 50 per cent of the samples fail is determined.

Table 35.2. Rubber Mounting Barrier Temperature[6]

Polymer	Temperature	
	°F	°C
NR..........	−42	−41
SBR.........	−24	−31
CR..........	−18	−28
IIR..........	+2	−17
NBR.........	+16	−9

DRIFT. Drift is the characteristic of rubber by which it continues to deform after the initial deformation caused by an applied stress. *Relative drift* is defined as

$$d = \frac{\delta_t - \delta}{\delta} \tag{35.1}$$

where d is per cent relative drift at time t, δ is deflection at 1 min, and δ_t is deflection at time t. Drift is also called *creep* and *strain relaxation*. It is related to compression set and stress relaxation—all are manifestations of the flow properties of rubber. The theory of these effects has been developed by many investigators.[7-9] It is thought that the short-time effects are connected with the stability of the secondary molecular double bonds of the rubber compound, but that the long-time effects are associated with oxidation which causes scission of the molecular chain and chain cross-linking. Drift characteristics of several natural rubber compounds loaded in shear are shown in Fig. 35.4.

FIG. 35.4. Static drift of three natural rubber compounds in shear relative to deflection at one minute. The test specimens were subjected to a load of 60 lb/in.² (4.2 kg/cm²). (*After Morron.*[23])

Drift is quite dependent on the composition of the rubber compound and on the state of cure. Since the long-term effects are chemical in nature, temperature plays an important role. This effect is shown in Fig. 35.5.

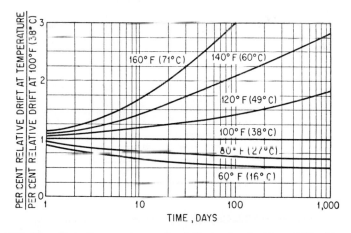

FIG. 35.5. The effect of temperature on the drift of natural rubber. (*After Morron.*[23])

Drift in compression is less than drift in shear, which in turn is less than drift in tension. Recommended maximum stress values normally are used to avoid excessive drift in an isolator.

Because of the long time required for actual drift tests, the compression set test is used to evaluate compounds.

POLYMER PROPERTIES. A qualitative comparison of the properties of various polymers is shown in Table 35.3. The composition of a rubber compound has a pro-

Table 35.3. Relative Properties

Poymer designation	Common name	Shore A hardness range	Max. tensile strength		Compression set	Tear resistance	Resilience		Heat resistance	Outdoor aging resistance
			lb/in.²	kg/cm²			Room temp.	High temp.		
NR......	Natural	30–100	4,000	280	Good	Good	High	High	Fair	Fair
SBR.....	SBR	40–100	3,000	210	Good	Fair	Fairly high	Fairly high	Fair	Fair
CR......	Neoprene	40–95	3,000	210	Poor (GN) Good (W)	Good	Fairly high	Fairly high	Good	Excellent
IIR......	Butyl	40–75	2,000	140	Fair	Good	Low	Fairly high	Good	Good
EPDM...	EPDM	45–100	2,000	140	Fair	Fair	Fairly high	Fairly high	Excellent	Excellent
NBR.....	Nitrile	20–100	2,500	176	Good	Fair	Medium	Medium	Good to excellent	Poor
PO......	Propylene Oxide	45–80	2,000	140	Fair	Fair	Fairly high	Fairly high	Good to excellent	Excellent
--	Thiokol	20–80	1,300	91	Poor	Good	Medium	Fairly high	Fair	Excellent
Si.......	Silicone	20–90	1,000	70	Excellent	Poor	Fairly high	Medium	Excellent	Excellent
CSM.....	Hypalon	45–95	2,800	197	Fair	Fair to good	Fairly high	High	Excellent	Excellent
ACM.....	Polyacrylate	40–90	1,800	127	Good	Fair to poor	Low	High	Excellent	Excellent
FPM.....	Fluororubber	60–90	3,000	210	Excellent	Fair to poor	Medium	Medium	Excellent	Excellent
ECO	Epichlorohydrin	35–80	2500	176	Fair to good	Good	Medium to high	Medium to high	Excellent	Excellent

* The relative ease of obtaining good adhesion to metal without employing costly metal treatments and elements.

nounced effect on its properties, and the comparisons shown are for each polymer as "normally" compounded.

The selection of the polymer to be used for any purpose is always a compromise. If high resilience, abrasion resistance, and good low-temperature flexibility are required, then natural rubber or isoprene is the best choice. If extreme temperature ranges are encountered, silicone rubber may be selected. If excellent weather resistance is of primary importance, EPDM or Hypalon might be selected. Butyl exhibits low permeability and low resilience at normal and low temperatures.

The cost of raw polymer does not represent the cost of a rubber compound in a finished part, because some rubber compounds require greater processing time and effort. Furthermore, the longer cures which are necessary for some polymers increase cost, and special ingredients necessary to modify the polymer properties to suit the application may be expensive.

DYNAMIC PROPERTIES OF RUBBER

MODULUS AND DAMPING. Rubber is not perfectly elastic—it exhibits internal damping and its stiffness tends to increase as the frequency of loading is increased. The action of rubber can be represented by an idealized mathematical model to which the measured performance can be compared. The idealized behavior most nearly approximating that of rubberlike materials is known as *linear viscoelastic behavior*. The mechanical model used for mathematical derivation is shown in Fig. 35.6. (Also see Chap. 29 for a discussion of electrical models.) If this model is subjected to a sinusoidal force

of Various Polymers

Low temperature flexibility	Specific gravity	Abrasion resistance	Adhesion strength*	Flex life	Solvent resistance		Air permeability	Moisture resistance	Flame resistance
					Hydrocarbons	Oxygenated			
Excellent	0.93	Excellent	Excellent	Excellent	Poor	Good	Fairly high	Fair to good	Poor
Good	0.94	Excellent	Excellent	Good	Poor	Good	Fairly high	Good to excellent	Poor
Fair	1.23	Good	Good	Good	Good	Poor	Moderate	Fair to good	Excellent
Fair	0.92	Fair to good	Fair	Fair to good	Poor	Good	Very low	Excellent	Poor
Excellent	0.90	Fair to good	Fair	Fair	Poor	Good	Low	Excellent	Poor
Fair	1.00	Good	Good	Fair	Excellent	Poor	Low	Excellent	Poor
Excellent	1.01	Fair to good	Fair	Fair	Good	Poor	Moderate	Excellent	Poor
Good	1.34	Fair	Poor	....	Excellent	Good	Low	Fair	Poor
Excellent	0.95	Poor	Poor	Good	Fair	Good	Fairly high	Excellent	Fair
Fair	1.10	Excellent	Poor	Good	Good	Fair	Moderate	Excellent	Good
Poor	1.09	Good	Poor	Good	Excellent	Poor	Moderate	Fair	Poor
Fair	1.85	Good	Fair	Good	Excellent	Fair	Moderate	Excellent	Good
Excellent	1.27	Fair to good	Fair	Good	Excellent	Good	Low	Good	Good

$F = F_0 \sin \omega t$, the response is given by[10]

$$x = \frac{F_0}{k_1 + k_2} \left[\sin \omega t + \frac{k_2 \sin \omega t}{k_1(1 + \omega^2 \psi^2)} - \frac{k_2 \omega \psi \cos \omega t}{k_1(1 + \omega^2 \psi^2)} \right] \tag{35.2}$$

where k_1 is the stiffness of spring 1, k_2 is the stiffness of spring 2, c is the viscous damping coefficient, and

$$\psi = c \left(\frac{1}{k_1} + \frac{1}{k_2} \right) \tag{35.3}$$

The terms in brackets in Eq. (35.2) represent components of the model response. The first term represents the ordinary elastic response which is in-phase with the force and is independent of frequency; the second term is also in-phase with the force, but is frequency-dependent and represents the elastic component of viscoelastic response; the third term is frequency-dependent and is 90° out-of-phase with the force; it is responsible for the energy losses.

A number of methods have been devised for the measurement of the dynamic properties of vulcanizates and of isolators.[25] Such methods usually assume the simpler mechanical model shown in Fig. 35.7. The results are usually expressed as the "complex modulus" (i.e., the modulus of the entire model) and the damping "coefficient." They may also be expressed as the "storage mod-

FIG. 35.6. Mechanical model for viscoelastic behavior of rubberlike materials with spring constants k_1 and k_2 and damping constant c.

Fig. 35.7. Simplified mechanical model used for dynamic tests of isolators.

ulus" [the first two terms in Eq. (35.2)] and the "loss modulus" [the third term in (Eq. 35.2)]. Damping is also expressed as the loss tangent:

$$\tan \phi = \frac{\text{loss modulus}}{\text{storage modulus}} \qquad (35.4)$$

EFFECT OF TEMPERATURE AND FREQUENCY

The dynamic properties of an elastomer are influenced by a number of factors. The general effects of temperature and frequency are shown in Fig. 35.8. (The storage modulus and loss tangent are shown on considerably different scales for clarity; the loss tangent is only a fraction of the storage modulus.) The exact shapes of these curves depend on the polymer and the specific composition used.

Figure 35.9 shows the response of typical isolator compounds made with several common polymers, as measured with the resonant beam apparatus.[25] These results, with the exception of IIR, represent a portion of the curve in Fig. 35.8A to the right of T_g.

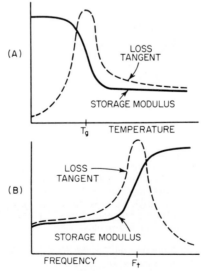

Fig. 35.8. Generalized response of vulcanizates to (A) temperature change at low frequency and (B) frequency change at normal temperature. T_g is the glass transition temperature ($-99°F$ or $-72.8°C$ for natural rubber). F_t is the transition frequency (10^6 to 10^8 cps).

Fig. 35.9. The effect of temperature on (*A*) the dynamic spring rate and (*B*) the damping coefficient of typical isolator compounds using several polymers.

EFFECT OF AMPLITUDE

Most compounds for vibration isolators contain carbon black to give the compound the necessary strength. The dynamic properties of vulcanizates loaded with carbon black are quite dependent on the strain applied as those properties are measured. This effect is shown in Fig. 35.10. The high modulus at low strains is related to carbon black

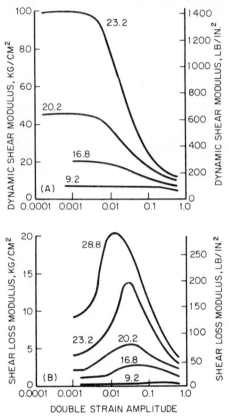

Fig. 35.10. The effect of amplitude of deformation on (A) the dynamic modulus and (B) the loss modulus of IIR vulcanizates. The numbers on the curves represent the percentage of carbon black by volume. (*After Payne.*[24])

structure and to the Van der Waals forces between the carbon black particles. For example, a specimen containing 23.2 per cent black has a modulus at 0.001 strain of 100 kg/cm². If that specimen is strained to 0.8 double amplitude, it has a modulus of 12 kg/cm². If it is then immediately tested at 0.001 strain, the modulus is still 12 kg/cm². If permitted to rest for over 24 hours, the carbon black structure will reform, and the modulus at 0.001 strain returns to nearly 100 kg/cm². These specific data are for carbon black in butyl rubber, but similar effects have been found with other reinforcing agents or fillers in other polymers.[24]

CRITICAL DAMPING

Representative values of the fraction of critical damping are given in Table 35.4. The range given for each polymer is necessary because of differences in properties brought about by hardness and by different compounding ingredients, and does not include the effect of temperature.

Table 35.4. Representative Values of Fraction of Critical Damping at Room Temperature

Polymer	Fraction of critical damping c/c_c
SBR.....................	0.05–0.15
Natural rubber............	0.01–0.08
Chloroprene rubber........	0.03–0.08
Butyl rubber..............	0.05–0.50

FATIGUE. A rubber specimen subjected to a tensile stress approaching its tensile strength will continue to elongate with time and eventually will rupture. This process is known as *static fatigue;* it is the end result of the creep process. Dynamic fatigue occurs in a specimen subjected to an alternating stress centered about zero. In most vibration

FIG. 35.11. The effect of strain on fatigue life of rubber specimens tested in tension and compression. (*After Caldwell, Merrill, Sloman, and Yost.*[12])

isolators, the fatigue process is some combination of static and dynamic fatigue. The results of an investigation of the dynamic fatigue characteristics of 50 durometer isolators strained in tension and compression are shown in Fig. 35.11. Dynamic fatigue life is plotted as a function of the per cent minimum strain, for fixed values of dynamic strain. The latter two parameters are computed in the following way. The per cent minimum strain is

$$\frac{l_{min} - l_0}{l_0} (100) \qquad \text{per cent} \qquad (35.5)$$

and the per cent dynamic strain is

$$\frac{l_{max} - l_{min}}{l_0} (100) \qquad \text{per cent} \qquad (35.6)$$

where l_0 is the unstrained length of the specimen, l_{min} is the minimum strained length, and l_{max} is the maximum strained length. Figure 35.11 shows that (1) for small strains there is a pronounced minimum in the fatigue life when the sample is returned to zero at the end of the stroke; (2) the point of maximum fatigue life shifts toward lower minimum strain as the dynamic strain is increased; and (3) fatigue life decreases as dynamic strain is increased. This latter effect may be caused partially by the greater heat generated. Specimens tested in shear also have a minimum life when returned to zero strain at one end of the stroke.[12] A summary of the fatigue life of shear specimens as a function of dynamic strain is given in Table 35.5.

Table 35.5. Fatigue Life in Cycles of Shear Specimens as a Function of Dynamic Strain for Various Lateral Strains[12]

Dynamic strain, per cent	Lateral strain		
	Zero	12.5 per cent compression	25 per cent tension
-25 to $+25$........	7×10^6	20×10^6	12×10^6
0 to 50.............	1×10^6	2×10^6	2×10^6
75 to 125...........	15×10^6	2×10^6	40×10^6

Stress concentrations reduce the fatigue life of rubber isolators. Sudden changes in rubber section, sharp-edged inserts, protruding boltheads, and weld flash should be avoided. In many cases, fatigue failures will develop inside the rubber at some distance from the sharp edge of a metal insert.

DYNAMIC AGING. Dynamic stresses accelerate most aging effects, because dynamic strain exposes new areas of the molecular structure to attack. The cracks which form in a rubber flex test specimen are caused by a combination of dynamic fatigue and oxidation or ozone attack. One mechanism of protection against ozone is the inclusion of a wax in the rubber compound. The solubility of the wax is low and it *blooms* (i.e., migrates to the surface) and forms a protective coating. Such a compound has poor ozone resistance under dynamic conditions because the wax film is broken. A rubber compound containing wax usually is not used for bonded vibration isolators, since wax has a detrimental effect on rubber-to-metal adhesion. Flexible coatings are sometimes used when rubber isolators are exposed to abnormal ozone concentrations. Another method of surface protection is surface chlorination, in which chlorine is added at the molecular double bonds. This treatment also greatly reduces the coefficient of friction of the rubber on smooth, dry hard surfaces.

CLASSIFICATIONS AND SPECIFICATIONS

The ASTM Standards on Rubber Products[1] includes many test methods, specifications, recommended practices, and definitions covering the entire field of rubber products. The following are of particular interest in shock and vibration control.

ELASTOMERIC MATERIALS FOR AUTOMOTIVE APPLICATIONS

This classification system is identified as ASTM Designation D-2000 or SAE Standard J200. Although the word "automotive" is included in the title, this standard is widely used in many fields to specify the particular type of compound desired for a specific application. It enables the user to select a *type* of elastomer based on heat resistance and a *class* of elastomer based on oil resistance. Grade numbers allow specification of basic properties such as hardness and tensile strength, and suffix letters are added to signify special requirements such as compression set and adhesion strength. A tabular format is provided in order to guide the user to compounds which are commercially feasible and to prevent specification of combinations of properties which are impossible to obtain.

There are similar systems for specifying latex foam rubber (ASTM D-1055) and sponge rubber (ASTM D-1056).

FORCED VIBRATION TESTING OF VULCANIZATES

A recommended practice for forced vibration testing is given in ASTM D-2231. General background information specific to the design of isolators for automobiles is given in the proceedings of a symposium on vibration and noise in motor vehicles.[26]

DESIGN OF RUBBER ISOLATORS

METHODS OF LOADING RUBBER

COMPRESSION LOADING. Rubber is practically incompressible, having a bulk modulus of approximately 345,000 lb/in.^{2} (24,300 kg/cm^2) (Poisson's ratio approximately 1/2). Therefore when rubber is loaded in compression, the sides bulge out and the stress-strain curve increases in slope as the deflection is increased, as shown in Fig. 35.1B. The curves shown are for normal condition of no slippage at the loaded rubber face. If the loaded surface of the rubber is lubricated thoroughly, the stress-strain curve for small strains is fairly linear and the *compression modulus* is

$$E = \frac{F}{S_1}\left(\frac{h}{h - h_1}\right) \tag{35.7}$$

where E is the elastic modulus in compression in pounds per square inch, F is the compressive force in pounds; h is the unstrained height in inches; h_1 is the height in inches under a load of F in pounds, and S_1 is the area at that load in square inches. If S is the original area, then $Sh = S_1 h_1$ (for an incompressible material), and the expression may be written as

$$E = \frac{F}{S}\left(\frac{h_1}{h - h_1}\right) \tag{35.8}$$

This relation is valid only for a specimen that is well lubricated and which is allowed to expand freely in the unloaded areas. Although rubber mountings usually are not used in such a manner, this scheme establishes a relation between hardness and modulus in compression which is independent of shape factor. This relationship is valuable as a design aid.

The shape of the load-deflection curve of a dry rubber specimen depends on the relative

FIG. 35.12. Loads for 10 per cent deflection for rubber specimens having various hardness values and shape factors. (*After Kimmich.*[14])

dimensions of the specimen and on the hardness of the rubber compound. This is illustrated by the data of Fig. 35.12. The shape factor (along the abscissa) is defined as

$$\frac{\text{Loaded area}}{\text{Free area}}$$

For a rectangular specimen loaded vertically, the shape factor is the area of the loaded cross section divided by the area of the four sides.

FIG. 35.13. Nomograph for determining relation of force and deflection for bonded rectangular rubber pads loaded in compression. A straight line is drawn intersecting the l/w and w/t scales at appropriate values. This determines a value of A, from which a second straight line is drawn to the E scale. To determine the deflection resulting from any unit force, a straight line is drawn from the force scale to the deflection scale, intersecting the B scale at the same point as the line from A to E. (*Courtesy of United States Rubber Company.*[18])

The nomograph shown in Fig. 35.13 applies to the design of bonded isolators loaded in compression. It covers the range of sizes and shapes normally used for mountings, and makes use of the compression modulus given by Eq. (35.8). The moduli shown in this figure were determined on the loading cycle with the samples at 70°F (21°C) after several loading and unloading cycles. Calculations made using this nomograph agree well with measured data.[14, 15, 16]

Rectangular Isolators. In Fig. 35.13, the A scale represents the ratio w/t for a square-shaped isolator. The E scale relates hardness and modulus; the B scale is merely a transfer device; and other quantities are defined in the legend. The nomograph is essentially an equation with six unknowns, l, w, t, F, E, and per cent compression. Given any five unknowns, the other can be found.

Example 35.1. Assume that a rubber mounting is 2 in. by 2 in. by 0.8 in. thick. Also assume that it is made of 50-durometer stock and that it is to have a compressive deflection of 10 per cent. What load will the mounting support?

$$\frac{l}{w} = 1 \qquad \text{and} \qquad \frac{w}{t} = 2.5$$

Connecting l/w and w/t yields a value on the A scale of 2.5. Connecting this value with the 50-durometer (375 lb/in.^{2}) value on the E scale gives a B-scale value of approximately 3.2. Connecting this value with the 10 per cent value yields a stress value of 80 lb/in.^{2} Since the mounting area S equals 4 in.^{2}, $F = 320$ lb.

Example 35.2. A machine weighing 12,000 lb is to be supported with four symmetrically located compression loaded isolators. Since the application involves infrequent vibration, a loading of 200 lb/in.^{2} and a deflection of 20 per cent are permissible (large static deflections contribute to high compression set; 20 per cent strain is the maximum recommended value). It is desired to use 60-durometer stock. What are the dimensions of square-shaped isolators to meet these requirements?

Connecting 200 lb/in.^{2} and 20 per cent yields a B-scale value of approximately 3.6. Connecting this value with 60 durometer (545 lb/in.^{2}) gives an A-scale value of approximately 2.

$$S = \frac{3,000}{200} = 15 \text{ in.}^2$$

Since $l = w$, $w^2 = 15$, and $w = 3.87$ in. From the nomograph for $l/w = 1$, $w/t = 2$, and $t = w/2 = 1.94$ in.

If a rectangular instead of square sandwich is desired, values of l/w and w/t can be obtained by cut-and-try alignments on those scales. This change would not affect the performance of the isolator, because the static deflection and pound per square inch loading would remain constant.

Cylindrical Isolators. For approximate calculations, using Fig. 35.13, a cylindrical isolator may be considered as a square isolator of the same thickness whose sides are equal to the cylinder diameter. Empirical determinations show that the A-scale value is actually 0.93 times the value for a square isolator because of the reduction of rubber area. When an approximation is made, the actual unit loading should be used for the F/S scale.

Triangular Isolators. A triangular isolator may be considered to be approximately equivalent to the largest cylindrical isolator of the same thickness which can be inscribed within it. For an equilateral triangle, the diameter of such a cylinder is

$$D = 0.577 \times \text{length of one side}$$

For any triangle, the diameter of the inscribed circle is

$$D = \frac{2}{\Sigma} [(\Sigma - x)(\Sigma - y)(\Sigma - z)]^{\frac{1}{2}}$$

where x, y, and z are the sides and

$$\Sigma = \tfrac{1}{2}(x + y + z)$$

Design Stress. The nomograph in Fig. 35.13 covers a wide range of possible designs. Designs which fall within its scope are not necessarily satisfactory for creep and flex life. Often there are unknown factors in a particular application such as actual operating temperature, magnitude of occasional overloads, amount of exposure to solvents, actual long-term creep rate of the compound used, etc. It is common practice to use a rule-of-thumb maximum design stress to cover these unknowns. A value of 90 lb/in.^{2} (6.3 kg/cm^2) is reasonable for an average compression isolator. This value can be modified to allow for operating conditions and expected life; for example, a value of 60 lb/in.^{2} (4.2 kg/cm^2) is used for some automotive engine mounts, and values as high as 400 lb/in.^{2} (28.1 kg/cm^2) are used.

The drift or the compression set of an isolator can cause misalignment of connecting shafts, pipes, etc. In addition, the reduction in thickness causes an increase in isolator stiffness which is not desired. Compression-loaded rubber isolators are used widely because the drift in compression is less than that in tension or shear and because the rubber-to-metal bond is not subjected to tensile loads.

TENSION LOADING. Rubber vibration isolators seldom are designed to operate primarily in tension because of the lower allowable bond stresses, increased importance

Fig. 35.14. Stress-strain curves for rubber of various hardnesses in tension. (*After U.S. Rubber Co.*[18])

of tear resistance, and lower creep limit.   The relationships between tensile stress, tensile strain, and Shore "A" hardness for a typical compound are shown in Fig. 35.14.   The curves cover only small strains.   From these curves an approximate tension modulus can be computed for design purposes.

SHEAR LOADING. Rubber is often used in shear because large deflections in one direction can be obtained with good stability in the other directions and because the characteristics of a shear mounting can be predicted with greater ease.   A typical shear

Fig. 35.15. A typical shear isolator, showing applied force F, length l, width w, and thickness t of rubber and deflection δ of isolator.

Fig. 35.16. Shear stress-strain curves for rubber of various hardnesses. (*After Crede.*[11])

mounting is shown in Fig. 35.15.   Shear stress-strain curves for typical rubber compounds are shown in Fig. 35.16.   The curves represent values taken on the loading cycle at 70°F (21°C) after several loading cycles.   These curves are approximately linear for small strains.   Because of this linearity and the fact that the stress-strain curves are not dependent on the shape factor, durometer hardness can be related to the shear modulus computed from the slope at the origin. This relation is shown in Fig. 35.17. With this relation the load-deflection characteristics of the shear mountings can be formulated as:

$$\frac{F}{S} = G\frac{\delta}{t}$$

where F = applied force, lb; S = area of rubber in shear $(2 \times l \times w)$, in.^{2}; G = modulus of elasticity in shear, lb/in.^{2}; δ = deflection, in.; and t = rubber thickness, in.

FIG. 35.17. Shear modulus of rubber for small deflections as a function of durometer hardness. (*After Crede.*[11])

Example 35.3. A double-shear sandwich, as in Fig. 35.15, supports a load of 500 lb. Each l and w are 3 in. and t is 0.75 in. The rubber is 40 durometer. Find the static deflection of the isolator.

The area is $S = 2 \times 3 \times 3 = 18$ in.^{2} Since two sides support the load, the shear stress is

$$\frac{F}{S} = \frac{500}{18} = 27.8 \text{ lb/in.}^2$$

The resulting shear strain may be found from Fig. 35.16 as 0.4. Thus the deflection is

$$\delta = 0.4 \times 0.75 = 0.3 \text{ in.}$$

Example 35.4. It is desired to support a 400-lb instrument with four single shear isolators, each of which is to have a static deflection of 20 per cent. If 50-durometer rubber is used, what size isolators are required?

The force per isolator $F = 100$ lb, and G is found from Fig. 35.17 as 88 lb/in.^{2} The strain $\delta/t = 0.2$ is given.

Thus

$$\frac{100}{S} = 88(0.2)$$

$$S = 5.68 \text{ in.}^2$$

Assume $l = w$; then $w = (5.68)^{1/2} = 2.38$ in. To determine the thickness of the sandwich, it is necessary to choose a static deflection commensurate with good stability, say $\frac{1}{4}$ in.

$$\frac{0.25}{t} = 0.2 \qquad t = 1.25 \text{ in.}$$

The bond stress is $100/5.68 = 17.6$ lb/in.^{2}, which is satisfactory.

Design Guess. As in the case of compression loading, it is usually necessary to use a recommended design stress figure. A design bond shear stress of 30 lb/in.^{2} (2.1 kg/cm^2) will be adequate for most situations. Some automotive engine mounts are designed at 20 lb/in.^{2} (1.4 kg/cm^2) and values as high as 100 lb/in.^{2} (7.0 kg/cm^2) are used.

If bond stress is *not* the limiting factor (i.e., if bond stress is quite low), a maximum shear strain of 50 per cent is recommended as a design figure.[11]

REFERENCES

1. American Society for Testing Materials, Standards on Rubber Products.
2. Crabtree, J., and A. R. Kemp: *Ind. Eng. Chem.*, **38**:278 (1946).
3. Newton, R. G. J.: *Rubber Research*, **14**:27 (1945).
4. Edwards, D. C., and E. B. Storey: *Trans. Inst. Rubber Ind.*, **31**:45 (1955).
5. Cuthbertson, G. R., and D. D. Dunnom: *Ind. Eng. Chem.*, **44**:834 (1952).
6. Morron, J. D., R. C. Knapp, E. F. Linhorst, and P. Viohl: *India Rubber World*, **110**:521 (1944).
7. Tobolsky, A. V., I. B. Prettyman, and J. H. Dillon: *J. Appl. Phys.*, **15**:380 (1944).
8. Peterson, L. E., R. L. Anthony, and E. Guth: *Ind. Eng. Chem.*, **34**:1349 (1942).

9. Beatty, J. R., and A. E. Juve: *India Rubber World*, **121**:537 (1950).
10. Gehman, S. D.: *Rubber Chem. and Technol.*, **30**:1202 (1957).
11. Crede, C. E.: "Vibration and Shock Isolation," John Wiley & Sons, Inc., New York, 1951.
12. Cadwell, S. M., R. A. Merrill, C. M. Sloman, and F. L. Yost: *Ind. Eng. Chem.*, **12**:19 (1940).
13. SAE (Society of Automotive Engineers) Handbook, 1959.
14. Kimmich, E. G.: *Rubber Chem. and Technol.*, **14**:407 (1941).
15. Smith, J. F. D.: *J. Appl. Mechanics*, A-13, March, 1938.
16. Chambers, G. G.: "The Development of a Rubber Engineering Design Handbook," General Motors Institute, 1957.
17. Smith, J. F. D.: *J. Appl. Mechanics*, **6**:159 (1939).
18. "Some Physical Properties of Rubber," United States Rubber Company, 1941.
19. Winspear, G. C. (ed.): "Vanderbilt Rubber Handbook," R. T. Vanderbilt Co., 1958.
20. Glossary of Terms Relating to Rubber and Rubber-like Materials, *ASTM Publ.* 184, 1956.
21. Whitby, G. S. (ed.): "Synthetic Rubber," John Wiley & Sons, Inc., New York, 1954.
22. Gehman, S. D., D. E. Woodford, and C. S. Wilkinson: *Ind. Eng. Chem.*, **39**:1108 (1947).
23. Morron, J. D.: *ASME Paper* 46-SA-18, presented June, 1946.
24. Payne, A. R., and R. E. Whittaker: *Rubber Chem. and Technol.*, **44**:440 (1971).
25. Hillberry, B. M.: "The Measurement of the Dynamic Properties of Elastomers and Elastomeric Mounts," Society of Automotive Engineers, January, 1973.
26. *Symp. Vibration and Noise in Motor Vehicles*, Automotive Division, Institute of Mechanical Engineers, London, 1972.

36

MATERIAL DAMPING AND SLIP DAMPING

L. E. Goodman
University of Minnesota

INTRODUCTION

The term *damping* as used in this chapter refers to the energy-dissipation properties of a material or system under cyclic stress but excludes energy-transfer devices such as dynamic absorbers. With this understanding of the meaning of the word, energy must be dissipated within the vibrating system. In most cases a conversion of mechanical energy to heat occurs. For convenience, damping is classified here as (1) material damping and (2) system damping. Material properties and the principles underlying measurement and prediction of damping magnitude are discussed in this chapter. For application to specific engineering problems, see Chap. 37.

MATERIAL DAMPING

Without a source of external energy, no real mechanical system maintains an undiminished amplitude of vibration. *Material damping* is a name for the complex physical effects that convert kinetic and strain energy in a vibrating mechanical system consisting of a volume of macrocontinuous (solid) matter into heat. Studies of material damping are employed in solid-state physics as guides to the internal structure of solids. The damping capacity of materials is also a significant property in the design of structures and mechanical devices; for example, in problems involving mechanical resonance and fatigue, shaft whirl, instrument hysteresis, and heating under cyclic stress. Three types of material that have been studied in detail are:

1. Viscoelastic materials. The idealized linear behavior generally assumed for this class of materials is amenable to the laws of superposition and other conventional rheological treatments including model analog analysis. In most cases linear (Newtonian) viscosity is considered to be the principal form of energy dissipation. Many polymeric materials as well as some other types of materials may be treated under this heading.

2. Structural metals and nonmetals. The linear dissipation functions generally assumed for the analysis of viscoelastic materials are not, as a rule, appropriate for structural materials. Significant nonlinearity characterizes structural materials, particularly at high levels of stress. A further complication arises from the fact that the stress and temperature histories may affect the material damping properties markedly; therefore, the concept of a stable material assumed in viscoelastic treatments may not be realistic for structural materials.

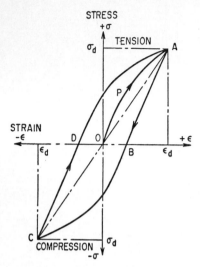

FIG. 36.1. Typical stress-strain (or load-deflection) hysteresis loop for a material under cyclic stress.

3. Surface coatings. The application of coatings to flat and curved surfaces to enhance energy dissipation by increasing the losses associated with fluid flow is a common device in acoustic noise control. These coatings also take advantage of material and interface damping through their bond with a structural material. They are treated in detail in Chap. 37.

Material damping of macrocontinuous media may be associated with such mechanisms as plastic slip or flow, magnetomechanical effects, dislocation movements, and inhomogeneous strain in fibrous materials. Under cyclic stress or strain these mechanisms lead to the formation of a stress-strain hysteresis loop of the type shown in Fig. 36.1. Since a variety of inelastic and anelastic mechanisms can be operative during cyclic stress, the unloading branch AB of the stress-strain curve falls below the initial loading branch OPA. Curves OPA and AB coincide only for a perfectly elastic material; such a material is never encountered in actual practice, even at very low stresses. The damping energy dissipated per unit volume during one stress cycle (between stress limits $\pm\sigma_d$ or strain limits $\pm\epsilon_d$) is equal to the area within the hysteresis loop $ABCDA$.

SLIP DAMPING

In contrast to material damping which occurs within a volume of solid material, slip damping arises from boundary shear effects at mating surfaces, or joints between distinguishable parts. Energy dissipation during cyclic shear strain at an interface may occur as a result of dry sliding (Coulomb friction), lubricated sliding (viscous forces), or cyclic strain in a separating adhesive (damping in viscoelastic layer between mating surfaces).

SIGNIFICANCE OF MECHANICAL DAMPING AS AN ENGINEERING PROPERTY

Large damping in a structural material may be either desirable or undesirable, depending on the engineering application at hand. For example, damping is a desirable property to the designer concerned with limiting the peak stresses and extending the fatigue life of structural elements and machine parts subjected to near-resonant cyclic forces or to suddenly applied forces. It is a desirable property if noise reduction is of importance. On the other hand, damping is undesirable if internal heating is to be avoided. It also can be a source of dynamic instability of rotating shafts and of error in sensitive instruments.

Resonant vibrations of large amplitude are encountered in a variety of modern devices, frequently causing rough and noisy operation and, in extreme cases, leading to seriously high repeated stresses. Various types of damping may be employed to minimize these resonant vibration amplitudes. Although special damping devices of the types described in Chap. 6 may be used to transfer energy from the system, there are many situations in which auxiliary dampers are not practical. Then accurate estimation of material and slip damping becomes important.

When an engineering structure is subjected to a harmonic exciting force $F_g \sin \omega t$ an induced force $F_d \sin(\omega t - \varphi)$ appears at the support. The ratio of the amplitudes, F_d/F_g, is a function of the exciting frequency ω. It is known as the vibration amplification factor. At resonance, when $\varphi = 90°$, this ratio becomes the resonance amplification factor[2] A_r:

$$A_r = \frac{F_d}{F_g} \qquad (36.1)$$

This condition is pictured schematically in Fig. 36.2 for low, intermediate, and high damping (curves 3, 2, 1, respectively).

The magnitude of the resonance amplification factor varies over a wide range in engineering practice.[3] In laboratory tests, values as large as 1,000 have been observed. In actual engineering parts under high stress, a range of 500 to 10 is reasonably inclusive. These limits are exemplified by an airplane propeller, cyclically stressed in the fatigue range, which displayed a resonance amplification factor of 490, and a double leaf spring with optimum interface slip damping[5] which was observed to have a resonance amplification factor of 10. Because of the wide range of possible values of A_r, each case must be considered individually.

FIG. 36.2. Effect of material and slip damping on vibration amplification. Curve (1) illustrates case of small material and slip damping; (2) one damping is large while other is small; (3) both material and slip damping are large.

METHODS FOR MEASURING DAMPING PROPERTIES

STRESS-STRAIN (OR LOAD-DEFLECTION) HYSTERESIS LOOP

The hysteresis loop illustrated in Fig. 36.1 provides a direct and easily interpreted measure of damping energy. To determine damping at low stress levels requires instruments of extreme sensitivity. For example, the width (DB in Fig. 36.1) of the loop of chrome steel at an alternating direct-stress level of 103 MPa* is less than 2×10^{-6}. High-sensitivity and high-speed transducers and recording devices are required to attain sufficient accuracy for measurement of such strains. For metals in general extremely long gage lengths are required to measure damping in direct stress by the hysteresis loop method if the peak stress is less than about 60 per cent of the fatigue limit. Under torsional stress, however, greater sensitivity is possible and the hysteresis loop method is applicable to low stress work.

PROCEDURES EMPLOYING A VIBRATING SPECIMEN

The following methods of measuring damping utilize a vibrating system in which the deflected member, usually acting as a spring, serves as the specimen under test. For example, one end of the specimen may be fixed and the other end attached to a mass which is caused to vibrate; alternatively, a freely supported beam or a tuning fork may be used as the specimen vibrating system.[4] In any arrangement the damping is computed from the observed vibratory characteristics of the system.

In one class of these procedures the rate of decay of free damped vibration is measured. Typical vibration decay curves are shown in Fig. 36.3. The measure of damping

---

* 1 MPa = 10^6 N/m^2 = 146.5 lb/in^2 (103 MPa = 15,000 lb/in^2).

usually used, the *logarithmic decrement,* is the natural logarithm of the ratio of any two successive amplitudes [See Eq. (37.6)]:

$$\Delta = \ln \frac{x_n}{x_{n+1}} \simeq \frac{\Delta x}{x_n} \qquad (36.2)$$

The relation between logarithmic decrement and other units used to measure damping is given in Eq. (36.16). Vibration decay tests can be performed under a variety of stress and temperature conditions, and may utilize many different procedures for releasing the specimen and recording the vibration decay. It is essential to minimize loss of energy either to the specimen supports or in acoustic radiation.

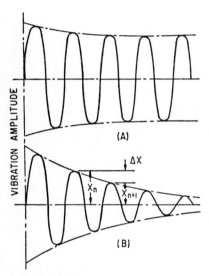

A second class of vibrating specimen procedures makes use of the fact, illustrated in Fig. 36.2, that higher damping is associated with a broader peak in the frequency response or resonance curve. If the exciting force is held constant and the exciting frequency varied, measurement of the steady-state amplitude of motion (or stress) yields a curve similar to those shown in Fig. 36.2. The damping is then determined by measuring the width of the curve at an amplification factor equal to $0.707A_r$. If a horizontal line drawn at this ordinate intercepts the resonance curve at frequency ratios f_1/f_n and f_2/f_n,

$$\Delta = \pi \left(\frac{f_2}{f_n} - \frac{f_1}{f_n} \right) \qquad (36.3)$$

FIG. 36.3. Typical vibration decay curves: *(A)* low decay rate, small damping, and *(B)* high decay rate, large damping.

The quantity $(f_2 - f_1)$ is the *bandwidth at the half-power point.* This procedure has the advantage of requiring only steady-state tests.

As in the case of the free-decay procedure, only the relative amplitude of the response need be measured. However, the procedure does impose a particular stress history. If the system behavior should be markedly nonlinear, the shape of the resonance curve will not be that assumed in the derivation of Eq. (36.3).[5]

If a system is operated exactly at resonance, the resonance amplification factor A_r is the ratio of the (induced) force F_d to the exciting force F_g [see Eq. (36.1)]. In direct application of this equation, F_g is usually made controllable and F_d computed from strain or displacement measurements. The principle has been applied to the measurement of damping in a large structure[6] and in simple test specimens. It can take account of high stress magnitude and of stress history as controlled variables. The natural frequency of vibration of a specimen can be altered so that damping as a function of frequency may be studied, but it is usually difficult to make such measurements over a wide frequency range. This technique requires accurately calibrated apparatus since measurements are absolute and not relative.

LATERAL DEFLECTION (OR LATERAL FORCE) OF

ROTATING CANTILEVER METHOD

The principle of the lateral deflection method is illustrated in Fig. 36.4. If test specimen S is loaded by arm-weight combination A—W, the target T deflects vertically downward from position 1 to position 2. If the arm-specimen combination is rotated by

END VIEW OF TARGET T

FIG. 36.4. Principle of rotating cantilever beam method for measuring damping.

spindle B, as in a rotating cantilever-beam fatigue test, target T moves from position 2 to position 3 for clockwise rotation. If the direction of rotation is counterclockwise, the target moves from position 3 horizontally to position 4. The horizontal traversal H is a direct measure of the total damping absorbed by the rotating system.[7]

A modification of the lateral deflection method is the *lateral force method*. The end of the rotating beam is confined and the lateral confining force is measured instead of the lateral deflection H. This modification is particularly useful for measurements of low modulus materials, such as plastic and viscoelastic materials.[8]

The advantages of the rotating cantilever beam method[9] are (1) the test variables, stress magnitude, stress history and frequency, may be easily and independently controlled so that this method is satisfactory for intermediate and high stress levels, and (2) it yields not only data on damping but also fatigue and elasticity properties.

The disadvantages of this method are (1) the tests are rather time-consuming, (2) accuracy is often questionable at low stress levels (below about 20 per cent of the fatigue limit) due to the small value of the horizontal traversal H, and (3) the method can be used under rotating-bending conditions only.

HIGH-FREQUENCY PULSE TECHNIQUES

A sequence of elastic pulses generated by a transducer such as a quartz crystal cemented to the front face of a specimen is reflected at the rear face and received again at the transducer. The frequencies are in the megacycle range. The velocity of such waves provides a measure of the elastic constants of the specimen; their decay rates provide a measure of the material damping.[10] This technique has been widely employed in the study of the viscoelastic properties of polymers and the elastic properties of crystals. So far as measurement of damping is concerned, it is open to the objection that the attenuation may be due to scattering by imperfections rather than to internal friction.

FUNDAMENTAL RELATIONSHIPS

Two general types of units[11] are used to specify the damping properties of structural materials: (1) the energy dissipated per cycle in a structural element or test specimen and (2) the ratio of this energy to a reference strain energy or elastic energy. Absolute damping energy units are:

D_0 = *total damping energy* dissipated by entire specimen or structural element per cycle of vibration, N.m/cycle

D_a = *average damping energy*, determined by dividing total damping energy D_0 by volume V_0 of specimen or structural element which is dissipating energy, N.m/m³/cycle

D = *specific damping energy*, work dissipated per unit volume and per cycle at a point in the specimen, N.m/m³/cycle

Of these absolute damping energy units, the total energy D_0 usually is of greatest interest to the engineer. The average damping energy D_a depends upon the shape of the specimen or structural element and upon the nature of the stress distribution in it, even though the specimens are made of the same material and have been subjected to the same stress distribution at the same temperature and frequency. Thus, quoted values of the average damping energy in the technical literature should be viewed with some reserve.

The specific damping energy D is the most fundamental of the three absolute units of damping since it depends only on the material in question and not on the shape, stress distribution, or volume of the vibrating element. However, most of the methods discussed previously for measuring damping properties yield data on total damping energy D_0 rather than on specific damping energy D. Therefore, the development of the relationships between these quantities assumes importance.

RELATIONSHIP BETWEEN D_0, D_a, AND D

If the specific damping energy is integrated throughout the stressed volume,

$$D_0 = \int_0^{V_0} D \, dV \tag{36.4}$$

This is a triple integral; $dV = dx \, dy \, dz$ and D is regarded as a function of the space coordinates x, y, z. If there is only one nonzero stress component the specific damping energy D may be considered a function of the stress level σ. Then

$$D_9 = \int_0^{\sigma_d} D \frac{dV}{d\sigma} \, d\sigma \tag{36.5}$$

In this integration V is the volume of material whose stress level is less than σ. The integration is a single integral and σ_d is the peak stress. The integrands may be put in dimensionless form by introducing D_d, the specific damping energy associated with the peak stress level reached anywhere in the specimen during the vibration (i.e., the value of D corresponding to $\sigma = \sigma_d$). Then

$$D_0 = D_d V_0 \alpha \tag{36.6}$$

where

$$\alpha = \int_0^1 \left(\frac{D}{D_d} \right) \frac{d(V/V_0)}{d(\sigma/\sigma_d)} \, d\left(\frac{\sigma}{\sigma_d} \right) \tag{36.7}$$

The average damping energy is

$$D_a = \frac{D_0}{V_0} = D_d \alpha \tag{36.8}$$

The relationship between the damping energies D_0, D_a, and D depends upon the dimensionless damping energy integral α. The integrand of α may be separated into two parts: (1) a damping function D/D_d which is a property of the material and (2) a volume-stress function $d(V/V_0)/d(\sigma/\sigma_d)$ which depends on the shape of the part and the stress distribution.

RELATIONSHIP BETWEEN SPECIFIC DAMPING ENERGY AND STRESS LEVEL

Before the damping function D/D_d can be determined, the specific damping energy D must be related to the stress level σ. Data of this type for typical engineering materials are given in Figs. 36.10 and 36.11. These results illustrate the fact that the damping-stress relationship for all materials cannot be expressed by one simple function. For a large number of structural materials in the low-intermediate stress region (up to 70 per cent of σ_e the fatigue strength at 2×10^7 cycles), the following relationship is reasonably satisfactory:

$$D = J \left(\frac{\sigma}{\sigma_e} \right)^n \tag{36.9}$$

Values of the constants J and n are given in Table 36.5 and Fig. 36.10. In general, $n = 2.0$ to 3.0 in the low-intermediate stress region but may be much larger at high stress levels. Where Eq. (36.9) is not applicable, as in the high stress regions of Figs. 36.10 and 36.11 or in the case of the 403 steel alloy of Fig. 36.9, analytical expressions are impractical and a graphical approach is more suitable for computation of α.

VOLUME-STRESS FUNCTION

The volume-stress function (V/V_0) may be visualized by referring to the dimensionless volume-stress curves shown in Fig. 36.5. The variety of specimen types included in

FIG. 36.5. Volume-stress functions for various types of parts. (See Table 36.1 for additional details on parts.)

this figure [tension-compression member (1) to turbine blade (9)] is representative of those encountered in practice. These curves give the fraction of the total volume which is stressed below a certain fraction of the peak stress. In a torsion member, for example, 30 per cent of the material is at a stress lower than 53 per cent of the peak stress. The volume-stress curves for a part having a reasonably uniform stress, i.e., having most of its volume stressed near the maximum stress, are in the region of this diagram labeled H. By contrast, curves for parts having a large stress gradient (such as a notched beam in

Table 36.1. Expressions and Values for α and β/α for Various Stress Distribution and Damping Functions

	Type of specimen and loading as designated in Fig. 36.1	Volume-stress function V/V_0	Dimensionless damping energy integral α for various damping functions					Dimensionless strain energy integral β	β/α if $n = 8$
			General case $D = f(\sigma)$	For any value of n	For special case $D = J(\sigma/\sigma_e)^n$				
					$n = 2.4$	$n = 8$	$n = 1$		
1	Tension-compression member		1	1	1	1	1	1.0	1
2	Cylindrical torsion member or rotating beam	$\left(\dfrac{\sigma}{\sigma_d}\right)^2$	$\left[1+\dfrac{\sigma_d}{2D_0}\dfrac{dD_0}{d\sigma_d}\right]^{-1}$	$\dfrac{2}{n+2}$	0.45	0.20		0.5	2.5
3	Rectangular beam under uniform bending	$\dfrac{\sigma}{\sigma_d}$	$\left[1+\dfrac{\sigma_d}{D_0}\dfrac{dD_0}{d\sigma_d}\right]^{-1}$	$\dfrac{1}{n+1}$	0.29	0.11		0.33	3.0
4	Cylindrical beam under uniform bending	$\dfrac{2}{\pi}\left[\dfrac{\sigma}{\sigma_d}\sqrt{1-\left(\dfrac{\sigma}{\sigma_d}\right)^2}+\sin^{-1}\left(\dfrac{\sigma}{\sigma_d}\right)\right]$		$\dfrac{1}{\sqrt{\pi}}\dfrac{2}{n+2}\dfrac{\Gamma\left(\dfrac{n+1}{2}\right)}{\Gamma\left(\dfrac{n+2}{2}\right)}$	0.21	0.055		0.24	4.5
5	Diamond beam under uniform bending	$2\dfrac{\sigma}{\sigma_d}-\left(\dfrac{\sigma}{\sigma_d}\right)^2$	$\left[1+\dfrac{2\sigma_d}{D_0}\dfrac{dD_0}{d\sigma_d}+\dfrac{\sigma_d^2}{2D_0}\dfrac{d^2D_0}{d\sigma_d^2}\right]^{-1}$	$\dfrac{2}{n^2+3n+2}$	0.13	0.022		0.17	7.7
6	Rectangular beam having bending moment shown $M_z=\dfrac{x}{L}M_0$	$\dfrac{\sigma}{\sigma_d}\left[1-\log_e\dfrac{\sigma}{\sigma_d}\right]$	$\left[1+\dfrac{3\sigma_d}{D_0}\dfrac{dD_0}{d\sigma_d}+\dfrac{\sigma_d^2}{D_0}\dfrac{d^2D_0}{d\sigma_d^2}\right]^{-1}$	$\dfrac{1}{(n+1)^2}$	0.088	0.012		0.11	9.1
7	$M_z=\left(\dfrac{x}{L}\right)^2 M_0$	$2\sqrt{\dfrac{\sigma}{\sigma_d}}-\dfrac{\sigma}{\sigma_d}$	$\left[1+\dfrac{5\sigma_d}{D_0}\dfrac{dD_0}{d\sigma_d}+\dfrac{2\sigma_d^2}{D_0}\dfrac{d^2D_0}{d\sigma_d^2}\right]^{-1}$	$\dfrac{1}{2n^2+3n+1}$	0.051	0.0065		0.067	10.0
8	Tuning fork in bending	$K\dfrac{\sigma}{\sigma_d}\left[1-\log_e\dfrac{\sigma}{\sigma_d}\right]$	$K\left[1+\dfrac{3\sigma_d}{D_0}\dfrac{dD_0}{d\sigma_d}+\dfrac{\sigma_d^2}{D_0}\dfrac{d^2D_0}{d\sigma_d^2}\right]^{-1}$	$\dfrac{K}{(n+1)^2}$ For $K = 0.8 \rightarrow$	0.091	0.0099		0.089	9.0

Note: $\beta/\alpha = 1$ for all cases if $n = 2$.

which very little volume is at the maximum stress and practically all of the volume is at a very low stress) are in the G region.

In order to illustrate representative values of α for several cases of engineering interest, the results of selected analytical and graphical computations[3] are summarized in Table 36.1 and in Fig. 36.6. In Fig. 36.6 the effect of the damping exponent n on the value

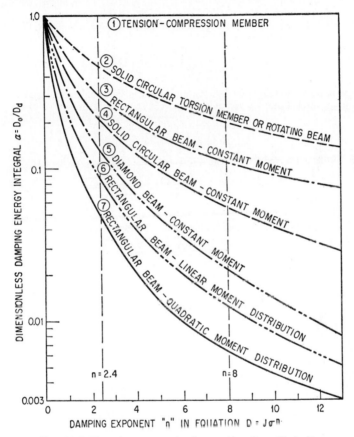

Fig. 36.6. Damping exponent n in equation $D = J(\sigma/\sigma_e)^n$.

of α for different types of representative specimens is illustrated. Note the wide range of α encountered for $n = 2.4$ (representative of many materials at low and intermediate stress) and for $n = 8$ (representative of materials at high stress, as shown in the next section).

RATIO OF DAMPING ENERGY TO STRAIN ENERGY

Owing to the complexity of the sources of material damping, the use of relative damping energy units does not produce all the advantages which might otherwise be associated with a nondimensional quantity. One motivation for the use of such units, however, is their direct relation to several conventional damping tests. The *logarithmic decrement* Δ is defined by Eq. (36.2). Other energy ratio units are tabulated and defined below. In this chapter, the energy ratio unit termed *loss factor* is used as the reference unit.

In defining the various energy ratio units, it is important to distinguish between loss factor η_s of a specimen or part (having variable stress distribution) and the loss factor η for a material (having uniform stress distribution). By definition the loss factor of a specimen (identified by subscript s) is

$$\eta_s = \frac{D_0}{2\pi W_0} \qquad (36.10)$$

where the total damping D_0 in the specimen is given by Eq. (36.6). The total strain energy in the part is of the form

$$W_0 = \int_0^{V_0} \frac{1}{2} \left(\frac{\sigma^2}{E}\right) dV = \frac{1}{2} \left(\frac{\sigma_d^2}{E}\right) V_0 \beta \qquad (36.11)$$

where E denotes a modulus of elasticity and β is a dimensionless integral whose value depends upon the volume-stress function and the stress distribution:

$$\beta = \int_0^1 \left(\frac{\sigma}{\sigma_d}\right)^2 \frac{d(V/V_0)}{d(\sigma/\sigma_d)} \, d\left(\frac{\sigma}{\sigma_d}\right) \qquad (36.12)$$

On substituting Eq. (36.6) and Eq. (36.11) in Eq. (36.10), it follows that

$$\eta_s = \frac{E \, D_d}{\pi \, \sigma_d^2} \frac{\alpha}{\beta} \qquad (36.13)$$

If the specimen has a uniform stress distribution, $\alpha = \beta = 1$ and the specimen loss factor η_s becomes the material loss factor η; in general, however,

$$\eta = \frac{ED_d}{\pi\sigma_d^2} = \eta_s \frac{\beta}{\alpha} \qquad (36.14)$$

Other energy ratio (or relative energy) damping units in common use are defined below:

For specimens with variable stress distribution:

$$\eta_s = (\tan \phi)_s = \frac{\Delta_s}{\pi} = \frac{\psi_s}{\pi} = \left(\frac{\delta\omega}{\omega_n}\right)_s = \frac{1}{(A_r)_s} = \frac{1}{Q_s} = \frac{ED_d}{\pi\sigma_d^2} \left(\frac{\alpha}{\beta}\right) \qquad (36.15)$$

For materials or specimens with uniform stress distribution:

$$\eta = \tan \phi = \frac{\Delta}{\pi} = \frac{\psi}{\pi} = \frac{\delta\omega}{\omega_n} = \frac{1}{A_r} = \frac{1}{Q} = Q^{-1} = \frac{ED_d}{\pi\sigma_d^2} \qquad (36.16)$$

where η = loss factor of material = dissipation factor (high loss factor signifies high damping)

$\tan \phi$ = loss angle, where ϕ is phase angle by which strain lags stress in sinusoidal loading

$\psi = \pi\eta$ = specific damping capacity

$\delta\omega/\omega_n$ = (bandwidth at half-power point)/(natural frequency) [see Eq. (36.3)]

A_r = resonance amplification factor [see Eq. (36.1)]

$Q = 1/\eta$ = measure of the sharpness of a resonance peak and amplification produced by resonance

The material properties are related to the specimen properties as follows:

$$\psi = \psi_s \frac{\beta}{\alpha} \qquad \Delta = \Delta_s \frac{\beta}{\alpha} \qquad A_r = (A_r)_s \frac{\alpha}{\beta} \qquad (36.17)$$

Thus, the various energy ratio units, as conventionally expressed for specimens, depend not only on the basic material properties D and E but also on β/α. The ratio β/α depends on the form of the damping-stress function and the stress distribution in the specimen. As in the case of average damping energy, D_a, the loss factor or the logarithmic decrement for specimens made from exactly the same material and exposed to the same stress range, frequency, temperature, and other test variables may vary significantly if the shape and stress distribution of the specimen are varied. Since data expressed as logarithmic decrement and similar energy ratio units reported in technical literature have been obtained on a variety of specimen types and stress distributions, any comparison of such data must be considered carefully. The ratio β/α may vary for specimens of exactly the same shape if made from materials having different damping-stress functions.[3]   For different specimens made of exactly the same materials, the variation in β/α also may be large, as shown in Fig. 36.7. For example, for a material and

Fig. 36.7. Effect of damping exponent n on ratio β/α for $D = J\sigma^n$.   Curves are (1) tension-compression member; (2) solid circular torsion member or rotating beam; (3) rectangular beam–constant moment; (4) solid circular beam–constant moment; (5) diamond beam–constant moment; (6) rectangular beam–linear moment distribution; and (7) rectangular beam–quadratic moment distribution.

stress region for which damping exponent $n = 2.4$ (characteristic of metals at low and intermediate stress), the value of β/α shown in Table 36.1 varies from 1 for a tension-compression member to 1.6 for a rectangular beam with quadratic moment distribution. If $n = 8$ (characteristic of materials at high stress), the variation is from 1 to 10, and larger for beams with higher stress gradient.

It is possible, for a variety of types of beams, to separate the ratio β/α into two factors:[12] (1) a cross-sectional shape factor K_c which quantitatively expresses the effect of stress distribution on a cross section, and (2) a longitudinal stress distribution factor K_s which expresses the effect of stress distribution along the length of the beam.   Then

$$\frac{\beta}{\alpha} = K_s K_c \qquad (36.18)$$

If material damping can be expressed as an exponential function of stress, as in Eq. (36.9), some significant generalizations can be made regarding the pronounced effect of the damping exponent n on each of these factors. Some of the results are shown in Fig. 36.8 for beams of constant cross-section. These curves indicate that high values of K_s and K_c are associated with a high damping exponent n, other factors being equal; K_c is high when very little material is near peak stress. For example, compare the diamond cross-section shape with the I beam, or compare the uniform stress beam with the cantilever.

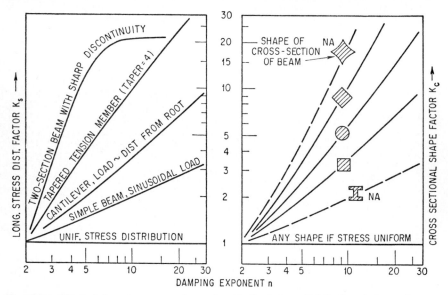

FIG. 36.8. Effect of damping exponent n on longitudinal stress distribution factor and cross-sectional shape factor of selected examples.

In much of the literature of damping, the existence of factors α and β (or K_s and K_c) is not recognized; the unstated assumption is that $\alpha = \beta = 1$. As discussed above, this assumption is true only for specimens under homogeneous stress.

Relative damping units such as logarithmic decrement depend on the ratio of two energies, the damping energy and the strain energy. Since strain energy increases with the square of the stress for reasonably linear materials, the logarithmic decrement remains constant with stress level and is independent of specimen shape and stress distribution only for materials whose damping energy also increases as the square of the stress [$n = 2$ in Eq. (36.9)]. For most materials at working stresses, n varies between 2 and 3 (see Fig. 36.10) but for some (Fig. 36.9) it is highly variable. In the high stress region, n lies in the range 8.0–20.0 (Fig. 36.10). In view of the broad range of materials and stresses encountered in design, the case $n = 2$ must be regarded as exceptional. Thus, logarithmic decrement is a variable rather than a "material constant." Its magnitude generally decreases significantly with stress amplitude. When referring to specimens such as beams in which all stresses between zero and some maximum stress occur simultaneously, the logarithmic decrement is an ambiguous average value associated with some mean stress. Published data require careful analysis before suitable comparisons can be made.

VISCOELASTIC MATERIALS

Some materials respond to load in a way that shows a pronounced influence of the rate of loading. Generally the strain is larger if the stress varies slowly than it is if the stress reaches its peak value swiftly. Among materials that exhibit this viscoelastic behavior are high polymers and metals at elevated temperatures as well as many glasses, rubbers, and plastics. As might be expected, these materials usually also exhibit creep, an increasing deformation under constant applied load.

When a sinusoidal exciting force is applied to a viscoelastic solid, the strain is observed to lag behind the stress. The phase angle between them, denoted φ, is the *loss angle*. The stress may be separated into two components, one in phase with the strain and one leading it by a quarter cycle. The magnitudes of these components depend upon the material and upon the exciting frequency, ω. For a specimen subject to homogeneous shear ($\alpha = \beta = 1$),

$$\gamma = \gamma_0 \sin \omega t \tag{36.19}$$

$$\sigma = \gamma_0[G'(\omega) \sin \omega t + G''(\omega) \cos \omega t] \tag{36.20}$$

This is a linear viscoelastic stress-strain law. The theory of linear viscoelasticity is the most thoroughly developed of viscoelastic theories. In Eq. (36.20) $G'(\omega)$ is known as the "storage modulus in shear" and $G''(\omega)$ is the "loss modulus in shear" (the symbols G_1 and G_2 are also widely used in the literature). The stiffness of the material depends on G' and the damping capacity on G''. In terms of these quantities the loss angle $\varphi = \tan^{-1}$ (G''/G'). The *complex*, or *resultant*, modulus in shear is $G^* = G' + iG''$. In questions of stress analysis, complex moduli have the advantage that the form of Hooke's law is the same as in the elastic case except that the elastic constants are replaced by the corresponding complex moduli. Then a correspondence principle often makes it possible to adapt an existing elastic solution to the viscoelastic case. For details of viscoelastic stress analysis see Refs. 42 and 43.

The moduli of linear viscoelasticity are readily related to the specific damping energy D introduced previously. For a specimen in homogeneous shear of peak magnitude γ_0, the energy dissipated per cycle and per unit volume is

$$D = \int_0^{2\pi/\omega} \sigma \left(\frac{d\gamma}{dt}\right) dt \tag{36.21}$$

In view of Eqs. (36.19) and (36.20) this becomes

$$D = \int_0^{2\pi/\omega} \gamma_0{}^2\omega(G' \sin \omega t + G'' \cos \omega t) \cos \omega t \, dt$$

$$= \pi\gamma_0{}^2 G''(\omega) \tag{36.22}$$

It is apparent from Eq. (36.22) that linear viscoelastic materials take the coefficient $\omega = 2$ in Eq. (36.9). These materials differ from metals, however, by having damping capacities that are strongly frequency- and temperature-sensitive.[10]

DAMPING PROPERTIES OF MATERIALS

The specific damping energy D dissipated in a material exposed to cyclic stress is affected by many factors. Some of the more important are:

1. Condition of the material
   a. In virgin state:
       Chemical composition; constitution (or structure) due to thermal and mechanical treatment; inhomogeneity effects
   b. During and after exposure to pretreatment, test, or service condition:
       Effect of stress and temperature histories on aging, precipitation, and other metallurgical solid-state transformations
2. State of internal stress
   a. Initially, due to surface-finishing operations (shot peening, rolling, case hardening)

    b. Changes caused by stress and temperature histories during test or service

3. Stress imposed by test or service conditions

    a. Type of stress (tension, compression, bending, shear, torsion)

    b. State of stress (uniaxial, biaxial, or triaxial)

    c. Stress-magnitude parameters, including mean stress and alternating components; loading spectrum if stress amplitude is not constant

    d. Characteristics of stress variations including frequency and waveform

    e. Environmental conditions:

        Temperature (magnitude and variation) and the surrounding medium and its (corrosive, erosive, and chemical) effects

Factors tabulated above, such as stress magnitude, history, and frequency, may be significant at one stress level or test condition and unimportant at another. The deformation mechanism that is operative governs sensitivity to the various factors tabulated.

Many types of inelastic mechanisms and hysteretic phenomena have been identified,[10, 14, 15] as shown in Table 36.2. The various damping phenomena and mechanisms may be classified under two main headings: *dynamic hysteresis* and *static hysteresis*.

Materials which display dynamic hysteresis (sometimes identified as viscoelastic, rheological, and rate-dependent hysteresis) have stress-strain laws which are describable by a differential equation containing stress, strain, and time derivatives of stress or strain. This differential equation need not be linear, though, to avoid mathematical complexity, much of existing theory is based on the linear viscoelastic law described in the previous section. One important type of dynamic hysteresis, a special case identified as *anelasticity* or *internal friction*, produces no permanent set after a long time. This means that if the load is suddenly removed at point *B* in Fig. 36.1, after cycle *OAB*, strain *OB* will gradually reduce to zero as the specimen recovers (or creeps negatively) from point *B* to point *O*.

A distinguishing characteristic of anelasticity and the more general case of viscoelastic damping is its dependence on time-derivative terms. The hysteresis loops tend to be elliptical in shape rather than pointed as in Fig. 36.1. Furthermore, the loop area is definitely related to the dynamic or cyclic nature of the loading and the area of the loop is dependent on frequency. In fact, the stress-strain curve for an ideally viscoelastic material becomes a single-valued curve (no hysteretic loop) if the cyclic stress is applied slowly enough to allow the material to be in complete equilibrium at all times (oscillation period very much longer than relaxation times). No hysteretic damping is produced by these mechanisms if the material is subjected to essentially static loading. Stated differently, the static hysteresis is zero.

Static hysteresis, by contrast, involves stress-strain laws which are insensitive to time, strain, or stress rate. The equilibrium value of strain is attained almost instantly for each value of stress and prior stress history (direction of loading, amplitudes, etc.), independent of loading rate. Hysteresis loops are pointed, as shown in Fig. 36.1, and if the stress is reduced to zero (point *B*) after cycle *OAB*, then *OB* remains as a permanent set or residual deformation. The two principal mechanisms which lead to static hysteresis are magnetostriction and plastic strain.

Table 36.2 also shows the simplest representative mechanical models for each of the behaviors classified. In these models *k* is a spring having linear elasticity (linear and single-valued stress-strain curve), *C* is a linear dashpot which produces a resisting force proportional to velocity, and *D* is a Coulomb friction unit which produces a constant force whenever slip occurs within the unit, the direction of the force being opposite to the direction of relative motion. More sophisticated models have been found to predict reliably the behavior of some materials, particularly polymeric materials.

Any one of the mechanisms to be discussed may dominate, depending on the stress level. For convenience, *low stress* is defined here as a (tension-compression) stress less than 1 per cent of the fatigue limit; *intermediate* stress levels are those between 1 per cent and 50 per cent of the fatigue limit of the material; and *high* stress levels are those exceeding 50 per cent of the fatigue limit.

Table 36.2. Classification of Types of Hysteretic Damping of Materials

	Types of material damping		
	Dynamic hysteresis		Static hysteresis
Name used here............	Dynamic hysteresis		Static hysteresis
Other names..............	Viscoelastic, rheological, and rate-dependent hysteresis		Plastic, plastic flow, plastic strain, and rate-independent hysteresis
Nature of stress-strain laws	Essentially linear. Differential equation involving stress, strain, and their time derivatives		Essentially nonlinear, but excludes time derivatives of stress or strain
Special cases and description	*Anelasticity.* Special because no permanent set after sufficient time. Called "internal friction"		
Simplest representative mechanical model	VOIGT UNIT (C, k) ±F ANELASTICITY; MAXWELL UNIT (k, C) ±F		(k, D) ±F / Plastic strain
Frequency dependence........	Critically at relaxation peaks		No, unless other mechanisms present
Primary mechanisms	Solute atoms, grain boundaries. Micro- and macro-thermal and eddy currents. Molecular curling and uncurling in polymers.	Magnetoelasticity	Plastic strain
Value of n in $D = JS^n$	2	3—up to coercive force	2–3 up to σ_L / 2 to > 30 above σ_L
Variation of η with stress	No change, since $n - 2 = 0$	Proportional to σ since $n - 2 = 1$	Small increase up to σ_L / Large increase above σ_L
Typical values for η.....	Anelasticity: <0.001 to 0.01 / Viscoelasticity: <0.1 to >1.5	0.01 to 0.08	0.001 to 0.05 up to σ_L / 0.001 to >0.1 above σ_L
Stress range of engineering importance	Anelasticity—low stress. Viscoelasticity—all stresses	Low and medium. Sometimes high	Medium and high stress
Effect of fatigue cycles	No effect	No effect	No effect up to σ_L / Large changes above σ_L
Effect of temperature	Critical effects near relaxation peaks	Damping disappears at Curie Temp.	Mixed. Depends on type of comparison
Effect of static preload		Large reduction for small coercive force	Either little effect or increase

DYNAMIC HYSTERESIS OF VISCOELASTIC MATERIALS

The linearity limits of a variety of plastics and rubbers are summarized in Table 36.3. While the stress limits are of the same order of magnitude for plastics and rubbers, the strain limits are much smaller for the former class of materials. Within these limits the dynamic storage and loss moduli of linear viscoelasticity may be used.

Table 36.3. Linearity Limits for a Variety of Plastics and Rubber

Material	Stress limit in creep MPa	Strain limit in relaxation
Polymethylmethacrylate............	10	
Polystyrene......................	5	
Plasticized polyvinyl chloride.......	1	0.1–1.0 %
Polythene........................	12	
Phenolic resins...................	10	
Polyisobutylene..................		50 %
Natural rubber...................	1–10	100 %
GR-S............................		100 %

Note: 1 MPa $= 10^6$ N/m^2 $= 146.5$ lb/in.^{2}

One distinguishing characteristic of the dynamic behavior of viscoelastic materials is a strong dependence on temperature and frequency.[16] At high frequencies (or low temperature) the storage modulus is large, the loss modulus is small, and the behavior resembles that of a stiff ideal material. This is known as the "glassy" region in which the "molecular curling and uncurling" cannot occur rapidly enough to follow the stress. Thus the material behaves essentially "elastically." At low frequencies (or high temperature) the storage modulus and the loss modulus are both small. This is the "rubbery" region in which the molecular curling and uncurling follow the stress in phase, resulting in an equilibrium condition not conducive to energy dissipation. At intermediate frequencies and temperatures there is a "transition" region in which the loss modulus is largest. In this region the molecular curling and uncurling[16] is out of phase with the cyclic stress and the resulting lag in the cyclic strain provides a mechanism for dissipating damping energy. The loss factor also shows a peak in this region although at a somewhat lower frequency than the peak in G''. Since the damping energy is proportional to G'', the specific damping curve also has its maximum in the transition region. Most engineering problems involving vibration are associated with the transition and glassy regions. In Table 36.4, values of G' and G'' are given for a variety of rubbers and plastics. In many cases the references from which these values are quoted contain additional useful information.

Metals at low stress exhibit certain properties that constitute dynamic hysteresis effects. Peaks are observed in curves of loss factors vs. frequency of excitation. For example, under conditions that maximize the internal friction associated with grain boundary effects polycrystalline aluminum will display a loss factor peak as high as $\eta = 0.09$. But for most metals, the peak values are less than 0.01. Although the rheological properties of metals at low stress can be described in terms of anelastic properties (rheology without permanent set), a more general approach which includes provisions for permanent set is required to specify the rheological properties of metals at high stress. This approach is best described in terms of static hysteresis.

Table 36.4. Typical Moduli of Viscoelastic Materials

(Two values are given: the upper value is G'; the bottom value is G''.
Moduli units are megapascals, MPa)

Material	Ref.	Frequency, cps				Temperature, °C
		10	100	1,000	4,000	
Polyisobutylene	33		0.512	1.31	2.36	−60–100
			0.410	1.76	4.50	
M 169A Butyl gum	35		0.480	1.40	2.70	21–65
			0.502	1.32	2.88	
Du Pont fluoro rubber, (Viton A)	33		2.00	4.54	7.93	0–100
			1.60	8.41	27.0	
Silicon rubber gum	35		0.05	0.08		21–65
			0.02	0.04		
Natural rubber	36		0.33	0.50		25
			0.02	0.02		
3M tape No. 466 (adhesive)	33		0.81	2.52	15.3	25
			0.95	4.59	13.0	
3M tape No. 435 (sound damping tape)	33		0.28	0.55	0.87	−40–60
			0.16	0.37	0.63	
Natural rubber (tread stock)	37	3.91	4.91			−30–75
		0.68	0.97			
Thiokol M-5	37	7.86	8.34			−30–75
		3.91	10.27			
Natural gum (tread stock)	37	0.73				−30–75
		0.07				
Filled silicone rubber	35		2.00	2.50	3.41	21–65
			0.26	0.44	0.58	
Polyvinyl chloride acetate	35		1.26	3.20	6.60	21–65
			1.44	2.32	5.78	
X7 Polymerized tung oil with polyoxane liquid	35			17.0	39.0	21–65
				9.45	20.8	
Du Pont X7775 pyralin	38	4.50	12.0	45.0		15 100
		2.51	9.45	28.3		
Polyvinyl butyral	38	30.0	200.0	600.0		−45–100
		3.1	12.5	37.6		
Polyvinyl chloride with dimethyl thianthrene	39		0.35	0.65		
			0.21	0.97		

Note: 1 MPa = 10⁶ N/m² = 10⁻³ kN/mm² = 146.5 lb/in.²

STATIC HYSTERESIS

The metals used in engineering practice exhibit little internal damping at low stress levels. At intermediate and high stress levels, however, magnetostriction and plastic strain can introduce appreciable damping. The former effect is considered first.

Ferromagnetic metals have significantly higher damping at intermediate stress levels than do nonferromagnetic metals. This is because of the rotation of the magnetic domain vectors produced by the alternating stress field.[16, 17] If the specimen is magnetized to saturation, most of the damping disappears, indicating that it was due primarily to magnetoelastic hysteresis. Figure 36.9 shows the loss factor of three metals, each heat-treated for maximum damping. The damping of 403 steel (ferromagnetic material with 12% Cr and 5% Ni) is much higher than that of 310 steel (nonferromagnetic with 25% Cr and 20% Ni). Most structural metals at low and intermediate stress exhibit loss factors in the general range of 310 steel until the hysteresis produced by plastic strain becomes significant. The alloy Nivco 10[17] (approximately 72% Co and

23% Ni), developed to take maximum advantage of magnetoelastic hysteresis, displays significantly larger damping than other metals.

The damping energy dissipated by magnetoelastic hysteresis increases as the third power of stress up to a stress level governed by the magnetomechanical coercive force; thus, the loss factor should increase linearly with stress. Nivco 10 follows this relationship for the entire range of stress shown in Fig. 36.9. Beyond an alternating stress governed by the magnetomechanical coercive force, i.e., beyond approximately 34.5

Fig. 36.9. Comparison of internal friction and damping values for different inelastic mechanisms.

MPa (5,000 lb/in²) for the 403 steel, the damping energy dissipated becomes constant. Since the elastic energy W_0 continues to increase as the square of the alternating stress, the value of loss factor (ratio of the two energies) decreases with the inverse square of stress. The curve for 403 steel in Fig. 36.9 at stresses between 62 MPa (9,000 lb/in²) and 103 MPa (15,000 lb/in²) demonstrates this behavior.

Magnetoelastic damping is independent of excitation frequency, at least in the frequency range that is of engineering interest. Magnetoelastic damping decreases only slightly with increasing temperature until the Curie temperature is reached, when it decreases rapidly to zero. Static stress superposed on alternating stress reduces magnetoelastic damping.[17, 18]

It is not entirely clear at this time what mechanisms are encompassed by the terms plastic strain, localized plastic deformation, crystal plasticity, and plastic flow in a range

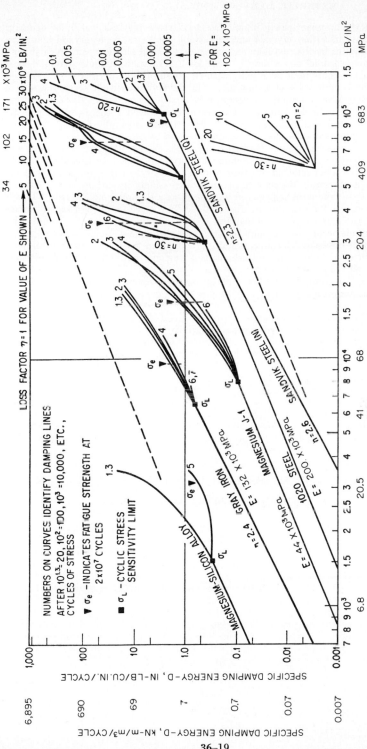

Fig. 36.10. Specific damping energy of various materials as a function of amplitude of reversed stress and number of fatigue cycles. Number of cycles is 10 to power indicated on curve. For example, a curve marked 3 is for 10^3 or 1,000 cycles. *Note*: 6.895 kN.m/m³ = 1 in.lb/in.³ and 1 MPa = 10^3 N/m². = 10^3 N/m² = 10^{-3} kN/mm² = 146.5 lb/in.²

36–19

of stress within the apparent elastic limit. On the microscopic scale, the inhomogeneity of stress distribution within crystals and the stress concentration at crystal boundary intersections produce local stress high enough to cause local plastic strain, even though the average (macroscopic) stress may be very low. The number and volume of local sites so affected probably increase rapidly with stress amplitude, particularly at stresses approaching the fatigue limit of a material. On the submicroscopic scale, the role of dislocations, their kind, number, dispersion, and lattice anchorage in the deformation process still remains to be determined. The processes involved in these various inelastic behaviors may be included under the general term "plastic strain."

At small and intermediate stress, the damping caused by plastic strain is small, probably of the same order as some of the internal friction peaks discussed previously and much smaller than magnetoelastic damping in many materials. In this stress region, damping generally is not affected by stress or strain history. However, as the stress is increased, the plastic strain mechanism becomes increasingly important and at stresses approaching the fatigue limit it begins to dominate as a damping mechanism. This is shown by the curves for titanium in Fig. 36.9.[18] In the region of high stress, microstructural changes and metallurgical instability appear to be initiated and promoted by cyclic stress. This occurs even though the stress amplitude may lie below the apparent elastic limit (that observed by conventional methods) and the fatigue limit of the material. This means that damping in the high stress region is a function not only of stress amplitude but also of stress history.

In Fig. 36.9, for example, the lower of the two curves for titanium indicates the damping of the virgin specimen and the upper curve gives the damping after 10,000 stress cycles.

The general position as regards stress history is given in Fig. 36.10. Below a certain peak stress, σ_L, known as the "cyclic stress sensitivity limit," the curve of damping vs. stress is a straight line on a log-log plot and displays no stress-history effect. The limit stress σ_L usually falls somewhat below the fatigue strength of the material. Above σ_L, stress-history effects appear; the curve labeled 1.3 indicates the damping energy after $10^{1.3} = 20$ cycles and the curve labeled 6 after 10^6 or 1 million cycles. To facilitate comparisons between the reference damping units, loss factor η and D under uniform stress $(\alpha/\beta = 1)$, the loss factor also is plotted in Fig. 36.10. Since the relationship between D and η depends on the value of Young's modulus of elasticity E, a family of lines for the range of $E = 34 \times 10^3$ to 205.0×10^3 MPa (5×10^6 to 30×10^6 lb/in²) is shown for $\eta = 1$. The lines for the other values of η correspond to a value of $E = 102.0 \times 10^3$ MPa (15×10^6 lb/in²).

COMPARISON OF VARIOUS MATERIAL DAMPING MECHANISMS AND REPRESENTATIVE DATA FOR ENGINEERING MATERIALS

The general qualitative characteristics of the various types of damping are summarized in Table 36.2 by comparing the effects of different testing variables. The data tabulated indicate that, in general, anelastic mechanisms do not contribute significantly to total damping at intermediate and high stresses; in these regions magnetoelastic and plastic strain mechanisms probably are the most important from an engineering viewpoint.

Damping vs. stress ratio data have been determined for a variety of common structural materials at various temperatures.[2] Some of these data are listed in Table 36.5 (all tests at 0.33 cps). For a large variety of structural materials (not particularly selected for large magnetoelastic or plastic strain damping), the data are found to lie within a fairly well-established band shown in Fig. 36.11. The approximate geometric-mean curve is shown. Up to the fatigue limit, that is up to $\sigma_d = \sigma_e$, the specific damping energy D is given with sufficient accuracy by the expression

$$D = J \left(\frac{\sigma}{\sigma_e} \right)^{2.4} \tag{36.23}$$

where $\qquad\qquad\qquad J = 6.8 \times 10^{-3}$

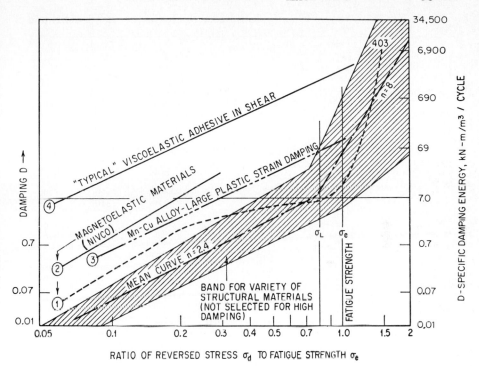

FIG. 36.11. Range of damping properties for a variety of structural materials. The shaded band defines the damping for most structural materials. 1 kN.m/m³ = 0.146 in.lb/in.³

if D is expressed in SI units of MN.m/m³/cycle and

$$J = 1.0$$

if D is expressed in units of in.lb/in.³/cycle.

The approximate bandwidth about this geometric mean curve for the various structural materials included in the band is as follows: from $\frac{1}{3}$ to 3 times the mean value at a stress ratio of 0.2 or less; from $\frac{1}{5}$ to 5 times at a ratio of 0.6; from $\frac{1}{10}$ to 10 times at a ratio of 1.0.

Also shown in Fig. 36.11 for comparison purposes are data for four materials having especially high damping. Materials 1 and 2 are the magnetoelastic alloys Nivco 10 and 403. Nivco 10 retains its high damping up to the stresses shown (data not available at higher stresses). However, the 403 alloy reaches its magnetoelastic peak at a stress ratio of approximately 0.2 and increases less rapidly beyond this point; when plastic strain damping becomes dominant (at stress ratio of approximately 0.8), damping increases very rapidly. By contrast, material 3, a manganese-copper alloy with large plastic strain damping, retains its high damping up to and beyond its fatigue strength.[20] Material 4 is a "typical" viscoelastic adhesive (G'' = 0.95 MPa = 138 lb/in²), assuming that the permissible cyclic shear strain is unity (experiments show that a shear strain of unity does not cause deterioration in this adhesive even after millions of cycles).[21] The magnetoelastic material has a damping thirty times as large as the average structural material in the stress range shown in Fig. 36.11, and the viscoelastic damping is over ten times as large as the magnetoelastic damping.

The range of D observed for common structural materials stressed at their fatigue limit is 0.003 to 0.7 MN.m/m³/cycle with a mean value of 0.05 (0.5 to 100 in. lb/in.³/cycle

Table 36.5. Static, Hysteretic, Elastic, and

Material*	Static properties			Fatigue behavior		
	Modulus of elasticity E, MPa 10^{-4}	Yield stress (0.2 % offset), MPa	Tensile strength, MPa	Fatigue strength, σ_e, MPa	Cyclic stress sensitivity limit σ_L, MPa	Stress ratio σ_L/σ_e
N-155 (superalloy)........	20.	410.	810.	360.	220.	0.62
Lapelloy (superalloy)........	22.	764.	880.	490.	490.	1.00
Lapelloy (480°C).....	17.5			270.	310.	1.14
RC 130B (titanium).........	11.5	950.	1,040.	590.	650.	1.10
RC 130B (320°C)....	9.9			430.	340.	0.81
Sandvik (O & T) steel..............	19.9	1,210.	1,400.	630.	680.	1.09
SAE 1020 steel.......	20.1	320.	490.	240.	200.	0.85
Gray iron............	13.2		140.	65.	44.	0.69
24S-T4 aluminum....	7.2	330.	500.	180.	160.	0.88
J-1 magnesium.......	4.4	230.	310.	120.	55.	0.47
Manganese-copper alloy.............		410.	610.	130.	120.	0.95

Note: 1 MPa = 10^6 N/m² = 146 lb/in.²
(Includes test temperature if above room temperature.)

with a mean value of 7).   For materials stressed at a rate of 60 cps under uniform stress distribution (tension-compression), 16.4 cm³ (1 in³) of a typical material will safely absorb and dissipate 48 watts (0.064 hp).   Some high damping materials can absorb almost 746 watts (1 hp) in the safe-stress range, assuming no significant frequency or stress-history effects.

SLIP DAMPING

INTRODUCTION

In some cases the hysteretic damping in a structural material is sufficient to keep resonant vibration stresses within reasonable limits.   However, in many engineering designs, material damping is too small and structural damping must be considered.   A structural damping mechanism which offers excellent potential for large energy dissipation is that associated with the interface shear at a structural joint.

The initial studies[22, 23, 24] on interface shear damping considered the case of Coulomb or dry friction.   Under optimum pressure and geometry conditions, very large energy dissipation is possible at a joint interface.   However, the application of the general concepts of optimum Coulomb interface damping to engineering structures introduces two new problems.   First, if the configuration is optimum for maximum Coulomb damping, the resulting slip can lead to serious corrosion due to chafing; this may be worse than the high resonance amplification associated with small damping.   Second, for many types of design configurations, the interface pressure or other design parameters must be carefully optimized initially and then accurately maintained during service; otherwise, a small shift from optimum conditions may lead to a pronounced reduction in total damping of the configuration.[22]   Since it usually is difficult to maintain optimum pressure, particularly under fretting conditions, other types of interface treatment have been developed.   One approach is to lubricate the interface surfaces.   However, the maintenance of a lubricated surface often is difficult, particularly under the large normal pres-

Fatigue Properties of a Variety of Metals

$D = J\left(\dfrac{\sigma}{\sigma_e}\right)^n,$ $\sigma \leq \sigma_L$		D $\dfrac{\sigma}{\sigma_L} = 1$	D $\dfrac{\sigma}{\sigma_e} = 0.6$	D at $\sigma/\sigma_e = 1$		D at $\sigma/\sigma_e = 1.2$	
J	$n,$ dimensionless			After $10^{1.3}$ cycles	After 10^6 cycles	After $10^{1.3}$ cycles	Maximum number of cycles
8.8	2.5	2.7	2.7	310.	170.	1,230.	1,500.
30.*	2.4*	10.9	4.0	11.	11.	55.	170.
24.	2.2	34.	8.2	26.	26.	41.	48.
14.	2.0	14.	4.4	12.	12.	18.	24.
17.	1.9	12.	6.1	13.	34.	30.	170.
16.	2.3	19.	5.5	16.	16.	31.	200.
4.3	2.0	3.1	1.6	4.5	140.	34.	680.
12.	2.4	4.5	3.4	14.	8.2	22.	16.
3.9	2.0	3.0	1.4	6.8	4.1	6.	15.
3.1	2.0	0.7	0.9	7.5	3.4	24.	7.
96.	2.8	82.	22.	89.	89.	170.	140.

Note: 1 kN.m/m³/cycle = 0.146 in.lb/in³/cycle.
* Up to $\sigma = 96$ MPa (14,000 lb/in²); at $\sigma = 204$ MPa (30,000 lb/in²) $n = 1.5$.

sure and shear sometimes necessary for high damping. Therefore, a more satisfactory form of interface treatment is an adhesive separator placed between mating surfaces at an interface. The function of the separating adhesive layer is to distort in shear and thus to dissipate energy with no significant Coulomb friction or sliding and therefore no fretting corrosion. The design of such layers is discussed in Chap. 37.

DAMPING BY SLIDING

The nature of interface shear damping can be explained by considering the behavior of two machine parts or structural elements which have been clamped together. The clamping force, whether it is the result of externally applied loads, of accelerations present in high-speed rotating machinery, or of a press fit, produces an interface common to the two parts. If an additional exciting force F_g is now gradually imposed, the two parts at first react as a single elastic body. There is shear on the interface but not enough to produce relative slip at any point. As F_g increases in magnitude, the resulting shearing traction at some places on the interface exceeds the limiting value permitted by the friction characteristics of the two mating surfaces. In these regions microscopic slip of adjacent points on opposite sides of the interface occurs. As a result, mechanical energy is converted into heat. If the mechanical energy is energy of free or forced vibration, damping occurs. The slipped region is local and does not, in general, extend over the entire interface. If it does extend over the entire interface, gross slip is said to occur. This usually is prevented by the geometry of the system.

The force-displacement relationship for systems with interface shear damping is shown in Fig. 36.12. Since there are many displacements which can be measured, the displacement which corresponds to the exciting force which acts on the system is taken as a basis. Then the product of displacement and exciting force, integrated over a complete cycle, is the work done by the exciting force and absorbed by the structural element. As shown in Fig. 36.13, there is an initial linear phase OP during which behavior is entirely elastic.   This is followed, in general, by a nonlinear transition phase PB during which slip progresses across the contact area.   The phase PB is nonlinear, not because of any plastic behavior, but simply because the specimen is changing in stiffness as slip progresses.   After the nonlinear phase PB, there may be a second linear phase BC during which slip is present over the entire interface. The existence of such a phase requires some geometric constraint which prevents gross motion even after slip has progressed over the entire contact area.   If no such constraint is provided, F_g cannot be allowed to exceed the value corresponding to point B. If it should exceed this gross value, slip would occur.

FIG. 36.12.   Force-displacement   hysteresis loop under Coulomb friction.

FIG. 36.13.   Friction force-slip relationship under Coulomb friction.

If the clamping force itself does not produce any shear on the interface and if the exciting force does not affect the clamping pressure, the force-displacement curve is symmetrical about the origin O.   These conditions are at least approximately fulfilled in many cases.   If they are not fulfilled, the exciting force in one direction initiates slip at a different magnitude of load than the exciting force in the opposite direction. This is the case pictured in Fig. 36.12.   With negative exciting force, slip is initiated at P' which corresponds to a force of considerably smaller magnitude than point P. However, the force-displacement curve is always symmetrical about the mid-point of PP' (intersection of dotted lines in Fig. 36.12).

The force-displacement curve has been followed from point O to point C.   If now a reduction in the exciting force occurs, the curve proceeds from C in a direction parallel to its initial elastic phase.   Eventually, as unloading proceeds, slip is initiated again. Its sense is now opposite to that which was produced by positive force.   The curve continues to point B', where slip is complete, and then along a linear stretch to C', where the exciting force has its largest negative value.   As the force reverses, the curve becomes again linear and parallel to OP.   Slip eventually occurs again and covers the interface at B.   The hysteresis loop is closed at C.

The energy dissipated in local slip can be found by computing the area enclosed by the force-displacement hysteresis loop.   It usually is simpler, however, to determine the energy loss at a typical location on the interface by analysis, and then to integrate over the area of the interface.   In this mode of procedure, interest centers on the frictional force per unit area $\mu\sigma$ and the relative displacement Δs of initially adjacent points on opposite sides of the interface.

The so-called "slip-curve" illustrating the relationship between $\mu\sigma$ and Δs is shown in Fig. 36.13.   Before the exciting force is applied, conditions are represented by point O'' which corresponds to point O in Fig. 36.12.   The initial elastic phase during which there is no slip is represented by $O''P''$ (note that the normal pressure σ may change during this phase).   The phase during which slip occurs only over part of the interface is represented by the curved line $P''B''$; it corresponds to PB in Fig. 36.12.   After slip has progressed over the entire interface, the normal force *vs.* relative-displacement relation is linear. This phase is represented by the curve $B''C''$ in Fig. 36.13 and by BC in Fig. 36.12. When the exciting force has reached its maximum value, a second nonslip phase $C''D''$ ensues. This is followed by slip along the curve $D''E''F''$ until the exciting force reaches its maximum negative value. As the exciting force completes its period, there is a non-slip phase $F''G''$ followed by slip along $G''C''$. The lengths $C''D''$ and $F''G''$ are equal and the curves $G''C''$ and $D''F''$ are congruent (F'' corresponds to C'' and D'' corresponds to G'').

If the point in question is at an element of area $dx\,dz$ of the xz interface, the energy dissipated in slip is proportional to the area enclosed by the slip curve. Because of the congruence of the curved portions of the diagram and the parallelism of the linear portions, this area can be expressed in terms of the total slip and the pressures at two instants during the loading cycle. Integrating over the entire interface,

$$D_0 = -\mu \iint [\sigma(E'') + \sigma(Q'')]\,\Delta s_{\text{tot}}\,dx\,dy \qquad (36.24)$$

In this expression, the parameters $\sigma(E'')$ and $\sigma(Q'')$ and the total slip Δs_{tot} are functions of x and z. They are the normal stresses at points E'' and Q'' of Fig. 36.13, located midway between the vertical lines $G''F''$ and $C''D''$. Since the pressures σ are always compressive (negative) and the total slip is always taken as a positive quantity, the negative sign is required to ensure a positive energy dissipation. Equation (36.24) is of little engineering value in itself because the stresses are functions of F_g as well as of x and z. In many of the problems which are of design interest, however, the shear on the interface is produced primarily by the exciting force and not by the initial clamping pressure. Conversely, the clamping pressure is not greatly affected by the addition of the time-varying exciting force. Under these circumstances, the slip curve of Fig. 36.13, like the force-displacement curve of Fig. 36.12, is symmetric about the point O''. Points P'' and Q'' then coincide, and the mean ordinate of the slip curve is that corresponding to point O''. Then Eq. (36.24) reduces to

$$D_0 = -4\mu \iint \sigma(O'')\,\Delta s_{\max}\,dx\,dz \qquad (36.25)$$

where $\sigma(O'')$ is the clamping stress corresponding to zero exciting force. It may be determined by any of the well-known methods of stress analysis. In most cases $\sigma(O'')$ can be determined without any reference to the existence of an interface. The term $\Delta s_{\max}$ represents the arc length of the maximum relative displacement, the so-called "scratch path." It is a function of the maximum value of F_g as well as of position on the interface. It may be inferred from Eq. (36.25) that energy dissipation due to interface shear is small both at very low clamping pressures and at very high ones. In the former case, $\sigma(O'') = 0$; in the latter case, $\Delta s_{\max} = 0$. It follows that for any distribution of clamping pressure there is an optimum intensity of clamping force at which the energy dissipation due to interface shear is a maximum. The maintenance of this optimum pressure is essential to the utilization of this form of damping. From the shape of the force-displacement curve $OPBC$ shown in Fig. 36.12, it is evident that systems in which interface shear damping plays a significant role behave like softening springs. This means that instability and jump phenomena may occur at frequencies below the nominal resonant frequency.[5]

In the case of plane stress, the thickness of the material is t and Eq. (36.25) becomes

$$D_0 = -4\mu t \int \sigma(O'')\,\Delta s_{\max}\,dx \qquad (36.26)$$

The slip can be related to stress through Hooke's law:

$$\Delta s = E^{-1} \int (\Delta\,\sigma_x)\,dx \qquad (36.27)$$

This indicates that any discontinuity in displacement is associated with a discontinuity in the component of stress parallel to the interface. These displacement discontinuities due to slip are members of a class of generalized dislocations whose existence has been demonstrated theoretically.[25] If Eq. (36.27) is substituted in Eq. (36.26), the energy dissipation can be expressed in terms of stress alone:

$$D_0 = -4\mu E^{-1} t \int_0^l \sigma(O'') \left[\int_0^x (\Delta \sigma_x)_{\max} dx' \right] dx \qquad (36.28)$$

The computation of energy dissipation per cycle D_0 is the first step in the prediction of the dynamic amplification factor to be expected in service. For interface shear damping, an elementary theory permits the dynamic amplification factor to be estimated even though the system behavior is nonlinear. The technique employs an averaging method. Denoting the displacement corresponding to the exciting force by the symbol v,

$$v = v_d \cos \omega t \qquad \text{and} \qquad F_g = F_m \cos (\omega t + \varphi) \qquad (36.29)$$

where v_d is the peak dynamic displacement, F_m is the peak exciting force, and φ is the loss angle. One relationship between these quantities is obtained by making the average value of the virtual work vanish during each half-cycle of the steady-state forced vibration:

$$\int_0^{\pi/\omega} [mv + kv - F_g] \cos \omega t \, dt = 0 \qquad (36.30)$$

In this integration, the stiffness k changes as slip progresses across the interface. If the hysteresis loop of Fig. 36.12 is replaced by a parallelogram, only two phases, elastic and fully slipped, need be considered. Denoting the stiffness (i.e., the ratio of exciting force to displacement) in the unslipped condition by the symbol k_e and the reduced stiffness in the fully slipped condition by the symbol k_s, the phase angle φ and the dynamic amplification factor A may be related by Eq. (36.30) to the duration of the elastic phase t':

$$\left(1 - \frac{k_s}{k_e}\right)(\omega t' + \sin \omega t') = \pi \left(\frac{m\omega^2 k_s}{k_e} + \frac{\cos \varphi}{A}\right) \qquad (36.31)$$

where A is the conventional dynamic amplification factor, i.e., $A = v_d k_e / F_m$. The duration of the elastic phase is given by the first of Eqs. (36.29) with $v = v_d - 2v_s$, where v_s is the displacement at which slip first occurs. Then eliminating t' from Eq. (36.31):

$$\frac{\cos \varphi}{A} = \frac{1}{\pi}\left(1 - \frac{k_s}{k_e}\right)\left[2\frac{v_s k_e}{AF_m}\sqrt{1 + \frac{v_s k_e}{AF_m}} + \cos^{-1}\left(1 - 2\frac{v_s k_e}{AF_m}\right)\right] - \frac{m\omega^2 k_s}{k_e} \qquad (36.32)$$

Equation (36.32) gives the relation between phase lag φ and amplification factor A. A second relationship between these quantities is found from the consideration that the energy dissipated during each half cycle of forced motion must be $D_0/2$:

$$\int_0^{\pi/\omega} F_g \frac{dv}{dt} dt = \frac{1}{2}D_0 \qquad \text{or} \qquad \sin \varphi = \frac{D_0 k_e}{\pi F_m^2 A} \qquad (36.33)$$

Equations (36.32) and (36.33) serve to determine the dynamic amplification factor A, after D_0 has been estimated. Conversely, they serve to estimate the amount of energy which must be dissipated per cycle to produce a given reduction in the amplification factor by interface shear. A detailed analysis of response to a parallelogram hysteresis loop has been made.[40] Hysteresis loops other than parallelograms also have been studied.[26] At resonance, $\varphi = 90°$ and

$$A = A_r = \frac{D_0 k_e}{\pi F_m^2} \qquad (36.34)$$

In general, the energy dissipation does not increase as rapidly as the square of the peak exciting force; consequently, the resonance amplification factor decreases as the exciting force increases. As a result, structures in which interface shear predominates tend to be self-limiting in their response to external excitation.

The foregoing discussion is based on the premise that changes in the exciting force do not materially affect the size of the contact area. There is an important class of problems for which this assumption is not valid, namely, those in which even the smallest exciting force produces some slip. An example of this type of joint is the press-fit bushing on a cylindrical shaft. If the ends of the shaft are subjected to a cyclic torque, part of this torque is transmitted to the bushing. Each part of the compound torque tube carries a moment proportional to its stiffness. Transmission of torque from the shaft to the bushing is effected by slip over the interface. The length of the slipped region grows in proportion to the applied torque. There is no initial elastic region such as OP or $O''P''$ in Figs. 36.12 and 36.13. If the peak value of the exciting torque is not too large, the fully slipped region BC or $B'C'$ in Fig. 36.12 never occurs. In these cases, Eqs. (36.31) to (36.34) are not applicable because there are no assignable constant values of k_s and k_e. A variety of simple cases of this type which occur in design practice have been analyzed.[27] They include the cylindrical shaft and bushing in tension and torsion, and the flexure of a beam with cover plate.

Another important case in which the smallest exciting force may produce slip arises in the contact of rounded solids. If these are pressed together by normal forces along the line joining their centers, a small contact region is formed. Subsequent application of a cyclic tangential force produces slip over a portion of the contact region even if the peak tangential force is not great enough to effect gross slip or sliding. This situation has been analyzed by Mindlin and by Deresiewicz.[45] Their conclusions have been verified experimentally by others.

REFERENCES

1. Phillips, E. M., and R. W. Weymouth: *SAE Paper* 438, January, 1955.
2. Lazan, B. J.: "Fatigue," chap. II, American Society for Metals, 1954.
3. Podnieks, E., and B. J. Lazan: *Wright Air Development Center Tech. Rept.* 55-284, August, 1955. Cochardt, A. W.: *J. Appl. Mechanics*, **21**:257 (1954).
4. Lazan, B. J.: *Trans. ASME*, **65**:87 (1943).
5. Goodman, L. E., and J. H. Klumpp: *Wright Air Development Center Tech. Rept.* 56-291, September, 1956.
6. Von Heydekampf, G. S.: *Proc. ASTM*, **31** (pt. II): 157 (1931).
7. Lazan, B. J.: *Trans. ASM*, **12**:499 (1950).
8. Maxwell, B.: *ASTM Bull.* 215, July, 1956, p. 76.
9. Lazan, B. J.: *Gen. Elec. Corp. Lab. Rept.* R 55GL129, January, 1955.
10. Nowick, A. S.: "Progress in Metal Physics," vol. 4, chap. I, p. 29, Interscience Publishers, Inc., New York, 1953.
11. Podnieks, E. R., and B. J. Lazan: *Wright Air Development Center Tech. Rept.* 56-44, 1956.
12. Lazan, B. J.: *J. Appl. Mechanics*, **20**:201 (1953).
13. Staverman, A. J., and F. Schwarzl: "Linear Deformation Behaviour of High Polymers," chap. I of "Theorie und molekulare Deutung technologischer Eigenschaften von hochpolymeren Werkstoffen," p. 71, Springer-Verlag, Berlin, 1956.
14. Zener, C.: "Elasticity and Anelasticity," University of Chicago Press, Chicago, 1948.
15. Wert, C.: "The Metallurgical Use of Anelasticity" in "Modern Research Techniques in Physical Metallurgy," American Society for Metals, Cleveland, Ohio, 1953.
16. Alfrey, T., Jr.: "Mechanical Behavior of High Polymers," Interscience Publishers, Inc., New York, 1948.
17. Cochardt, A.: *Scientific Paper* 8-0161-P7, Westinghouse Research Laboratories, West Pittsburgh, Pa., Nov. 6, 1956.
18. Person, N., and B. J. Lazan: *Proc. ASTM*, **56**:1399 (1956).
19. Demer, L. J., and B. J. Lazan: *Proc. ASTM*, **53**:839 (1953).
20. Torvik, P.: Appendix 72fg, *Status Rept.* 58-4 by B. J. Lazan, University of Minnesota, Wright Air Development Center, Dayton, Ohio, Contract AF-33(616)-2803, Dec. 31, 1958.
21. Whittier, J. S., and B. J. Lazan: Appendix B, *Prog. Rept.* 57-6, Wright Air Development Center, Dayton, Ohio, Contract AF-33(616)-2803, December, 1957.
22. Goodman, L. E., and J. H. Klumpp: *J. Appl. Mechanics*, **23**:421 (1956).
23. Lazan, B. J., and L. E. Goodman: "Shock and Vibration Instrumentation," p. 55, ASME, New York, 1956.
24. Pian, T. H. H., and F. C. Hallowell: *Proc. First U.S. Natl. Congr. Appl. Mechanics*, June, 1951, p. 97.

25. Bogdanoff, J. L.: *J. Appl. Phys.*, **21**:1258 (1950).
26. Rang, E.: *Wright Air Development Center Tech. Rept.* 59-121, February, 1959.
27. Goodman, L. E.: "A Review of Progress in Analysis of Interfacial Slip Damping," in "Structural Damping," American Society of Mechanical Engineers, New York, December, 1959.
28. Mentel, T. J.: *Wright Air Development Center Tech. Rept.* 58-547, December, 1958 (ASTIA Doc. 206,667).
29. Kerwin, E. M., Jr.: *J. Acoust. Soc. Amer.*, **31**:952 (1959).
30. Whittier, J. S.: *Wright Air Development Center Tech. Rept.* 58-568, May, 1959 (ASTIA Doc. 214,381).
31. Lazan, B. J.: *Proc. SESA*, **15**(1):1 (1957).
32. Kaelble, D. H.: private communication to the author.
33. Dalquist, C. A.: private communication to the author.
34. Fitzgerald, E., L. Grandine, and J. Ferry: *J. Appl. Phys.*, **24**:650 (1953).
35. Hopkins, I. L.: *Trans. ASME*, **73**:195 (1951).
36. Zapas, L. J., S. L. Shufler, and T. W. Dewitt: *J. Polymer Sci.*, **18**:245 (1955).
37. Phillipoff, W.: *J. Appl. Phys.*, **24**:685 (1953).
38. Rorden, H., and A. Grieco: *J. Appl. Phys.*, **22**:842 (1951).
39. Fitzgerald, E., and J. D. Ferry: *J. Colloid. Sci.*, **8**:1 (1953).
40. Caughey, T. K.: *J. Appl. Mechanics*, **27**:640 (1960).
41. Lazan, B. J.: "Damping of Materials and Members in Structural Mechanics," Pergamon Press, New York, 1968.
42. Flugge, W.: "Viscoelasticity," Blaisdell Publishing Company, a division of Ginn and Company, Waltham, Mass., 1967.
43. Lee, E. H.: Viscoelasticity, in W. Flugge (ed.), "Handbook of Engineering Mechanics," McGraw-Hill Book Company, New York, 1962.
44. Mindlin, R. D., W. F. Stubner, and H. L. Cooper: *Proc. Soc. Exp. Stress Anal.*, **5**(2), 69–87, 1948.
45. Deresiewicz, H.: Bodies in Contact with Applications to Granular Media, in G. Herrmann (ed.) "R. D. Mindlin and Applied Mechanics," Pergamon Press, New York, 1974.

37

APPLICATION OF DAMPING TREATMENTS

David I. G. Jones

Materials Laboratory

Wright-Patterson Air Force Base

INTRODUCTION TO THE ROLE OF DAMPING MATERIALS

The damping of an element of a structural system is a measure of the rate of energy dissipation which takes place during cyclic deformation. In general, the greater the energy dissipation, the less the likelihood of high vibration amplitudes or of high noise radiation, other things being equal. *Damping treatments* are configurations of mechanical or material elements designed to dissipate sufficient vibrational energy to control vibrations or noise. Proper design of damping treatments requires the selection of appropriate damping materials, location(s) of the treatment, and choice of configurations which assure the transfer of deformations from the structure to the damping elements.[1-6] These aspects of damping treatments are discussed in this chapter, along with relevant background information including:

1. Test techniques for measuring properties of damping materials.[7-11]
2. Methods for characterizing and displaying behavior of damping materials.[12-14]
3. Data on damping properties of commercial materials.[15-17]
4. Analytical means for predicting the effect of damping treatments on structural response.[18,19]

MECHANISMS AND SOURCES OF DAMPING

INTERNAL MECHANISMS OF DAMPING

There are many mechanisms whereby vibrational energy is dissipated in the form of heat within the volume of a material element as it is deformed.[20] Each such mechanism is associated with internal atomic or molecular reconstructions of the microstructure or with thermal effects. Only one or two mechanisms may be dominant for specific materials (metals, alloys, intermetallic compounds, etc.) under specific conditions, i.e., frequency and temperature ranges, and it is necessary to determine the precise mechanisms involved and the specific behavior on a phenomenological, experimental basis for each material specimen. Most structural metals and alloys have relatively little damping under most conditions, as demonstrated by the ringing of sheets of such materials after being struck. Some alloy systems, however, have crystal structures specifically selected for their relatively high damping capability; this is often demonstrated by their relative deadness under impact excitation. The damping behavior of metallic alloys is generally nonlinear and increases as cyclic stress amplitudes increase. Such behavior is difficult to predict because of the need to integrate effects of damping increments which vary with the cyclic stress amplitude distribution throughout the volume of the structure as it vibrates in a particular mode of deformation at a particular frequency. The prediction processes are complicated even further by the possible presence of external sources of damping at joints and interfaces within the structure and at connections and supports. For this reason, it is usually not possible, and certainly not simple, to predict or control the initial levels of damping in complex built-up structures and machines. Most of the current techniques of increasing damping involve the application of polymeric or elastomeric materials which are capable (under certain conditions) of dissipating far larger amounts of energy per cycle than the natural damping of the structure or machine without added damping.

POLYMERIC AND ELASTOMERIC MATERIALS

A mechanism commonly known as *viscoelastic damping* is strongly displayed in many polymeric, elastomeric, and amorphous glassy materials. The damping arises from the relaxation and recovery of the molecular chains after deformation. A strong dependence exists between frequency and temperature effects in polymer behavior because of the direct relationship between temperature and molecular vibrations. A wide variety of commercial polymeric damping material compositions exist, most of which fit one of the main categories listed in Table 37.1.

Table 37.1. Typical Damping Material Types

Acrylic rubber	Natural rubber	Urethane
Butadiene rubber	Nitrile rubber (NBR)	Vinyl
Butyl rubber	Polyisoprene	Polysulfone
Chloroprene (e.g., Neoprene)	Polysulfide	Polyvinyl chloride (PVC)
Fluorocarbon	Silicone	Polymethyl methacrylate (Plexiglas)
Fluorosilicone	Styrene-butadiene (SBR)	Nylon

EXTERNAL MECHANISMS OF DAMPING

Structures and machines can be damped by mechanisms which are essentially external to the system or structure itself.[15] Such mechanisms, which can be very useful for vibration control in engineering practice (discussed in other chapters), include:

1. Acoustic radiation damping, whereby the vibrational response couples with the surrounding fluid medium, leading to sound radiation from the structure.
2. Fluid pumping, in which the vibration of structure surfaces forces the fluid medium within which the structure is immersed to pass cyclically through narrow paths or leaks between different zones of the system or between the system and the exterior, thereby dissipating energy.[21]
3. Coulomb friction damping, in which adjacent touching parts of the machine or structure slide cyclically relative to one another, on a macroscopic or a microscopic scale, dissipating energy.[22–25]
4. Impacts between imperfectly elastic parts of the system.

HOW DAMPING TREATMENTS WORK

Damping treatments are configurations of polymeric damping materials and stiff or massive elements whose function is to amplify, as far as possible, the cyclic deformation of the damping material element which results from vibration of the structure to which the treatment is attached. This maximizes the energy dissipated per cycle.

Constrained-Layer Damping Treatment. A stiff metallic outer constraining layer forces the inner polymeric damping layer to deform cyclically in shear as the substrate surface deforms flexurally.

Tuned Elastomeric Damper. A mass between the elastomeric damping material and the structure surface is selected to amplify the extensional deformation of the elastomeric element in relation to the amplitude of cyclic displacement at the point of attachment of the damper.

ECONOMIC VALUE OF DAMPING

Mass and stiffness may be used to control deformations in a system under expected service loads, over a wide range of frequencies of excitation, but particular frequencies (i.e., the resonance frequencies) exist for which such control often is not adequate, and it is usually in the region of resonance frequencies that damping is most useful. Damping, if properly applied in a system, can control excessive amplitudes of vibration at the resonance frequencies and hence work with the other parameters to achieve a satisfactory design. The response of a system to sinusoidal excitation, random excitation, or impulsive excitation involves the resonance frequencies in different but related ways. For sinusoidal excitation, the resonant response amplitudes reach large magnitudes which are in inverse proportion to the amount of damping in each mode of vibration. For random response, the modes are all involved simultaneously and the root-mean-square response depends essentially on the inverse square root of the damping. For impulsive excitation, the rate of decay of the response is directly proportional to the damping. In each case, it is advantageous to increase the damping as far as possible without compromising other aspects of the performance of the system.

Noise Control. Damping is very useful for the control of noise radiation from vibrating surfaces, or the control of noise transmission through a vibrating surface. The noise is not reduced by sound absorption, as in the case of an applied acoustical material, but by decreasing the amplitudes of the vibrating surface.[26] For example, in a diesel engine, many parts of the surface contribute to the over-all noise level, and the contribution of each part can be measured by the use of the acoustic intensity technique[26] or by blanketing off, in turn, all parts except that of interest. If many parts of an engine contribute more or less equally to the noise, significant amplitude reductions of only one or two parts (whether by damping or other means) leads to only very small reductions of the over-all noise, typically 1 or 2 dB.

Product Acceptance. Damping can often contribute to product acceptance, not only by reducing the incidence of excessive noise, vibration, or resonance-induced failure but also by changing the "feel" of the product. The use of mastic damping treatments in car doors is a case in point. While the treatment may achieve some noise reduction, it may be the subjective evaluation by the customer of the solidity of the door which carries the greater weight.

Active Vibration Control. Robotic devices are finding widespread application in industrial production, and passive damping can be expected to have an increasing role.[27] (See the section on *Active Vibration-Isolation Systems* in Chap. 33.)

Failure/Fatigue Control. Damping can be used to control excessive resonance vibrations which may cause high stresses, leading to premature failure.[15] It should be used in conjunction with other appropriate measures to achieve the most satisfactory approach. Damping is used for additional reduction of response when this cannot be achieved for all resonances, and only for those resonances of most concern. For random excitation, on the other hand, it is not possible to detune a system and design to keep random stresses within acceptable limits without ensuring that the damping in each mode at least exceeds a minimum specified value. This is the case for sonic fatigue of aircraft fuselage, wing, and control surface panels when they are excited by jet noise or boundary layer turbulence-induced excitation. In these cases, structural designs have evolved toward semi-empirical procedures, but damping levels are a controlling factor and must be increased if too low.[28–30]

Simplified Maintenance. A useful by-product from reduction of resonance-induced fatigue by increased damping, or by other means, can be the reduction of maintenance costs.

TYPES OF DAMPING TREATMENTS

The main categories of damping treatments include free-layer treatments, constrained-layer treatments, integral damping treatments, tuned dampers, and damping links.

Free-Layer Damping Treatments. The free-layer treatment, illustrated in Fig. 37.1, is a macroscopically homogeneous single layer (usually) of stiff lossy polymeric material applied in considerable thickness to the surface of the structure.[18]   As the structure vibrates, the outer layers of the treatment, in particular, are deformed in extension-compression and dissipate energy.   The treatment should be as stiff and lossy as possible. The point of diminishing returns, where further increases of stiffness or material damping capability are counterproductive, lies far beyond the capabilities of current materials.*

Constrained-Layer Treatments.   The constrained-layer treatment consists of a thin layer of soft polymeric material† backed by a layer of metal or stiff plastic.[19]   In this case, the damping material deforms in shear.   The treatment should be as thick as practically possible, up to the point where the constraining layer is as thick as the base structure, but the polymeric material thickness is usually kept relatively low.   The stiffness of the inner polymeric layer must be chosen so that the coupling to the structure is not too low or too high.   Figure 37.1 illustrates some simple layered damping concepts.   Damping

(A)

(B)

(C)

FIG. 37.1. Additive layered damping treatments.   (*A*) Free-layer treatment.   (*B*) Constrained-layer treatment.   (*C*) Multiple constrained-layer treatment.

tapes lie within the category of constrained damping treatments, but the adhesive and constraining layers are usually very thin and flexible, are available only in combined form, and are relatively easy to apply to a structure surface in either single- or multiple-layer pair configurations.   For special applications, particular variations of the layered damping concepts can be used to provide enhanced damping properties, usually at the cost of increased complexity and expense.[31–33]

* Manufacturers of free-layer materials include: H. L. Blachford Inc. (Aquaplas); Lord Kinematics, Erie, Pa. (LD-400, LDS-500); Soundcoat Inc, Brooklyn, N.Y. (GP-1, GP-2, EPOXY 10).

† Manufacturers of constrained-layer materials include: 3M Co., St. Paul, Minn. (Scotchdamp ISD-110, 112, 113, 830, 3M-428, 3M-467); Soundcoat Inc, Brooklyn, N.Y. (Dyad 601, 606, 609, Soundfoil D, N, M, LT); Antiphon Inc, Newcastle, Del. (Antiphon 13); EAR Corporation, Indianapolis, Ind. (ISODAMP C1002, C2003); Eurovib, Ecully, France (Nepurane PL100, PL2000); Speciality Products, Newark, Del. (Noise Damper TND); General Electric, Philadelphia, Pa. (GE SMRD 100F 90).

Integral Damping Treatments. Some damping treatments are applied or added not after a structure has been partly or fully assembled but during the manufacturing process itself. Some examples are illustrated in Fig. 37.2. They include laminated sheets* which are used for construction assembly, or for deep drawing of structural components in a manner similar to that for solid sheets, and also faying surface damping which is introduced into the joints during assembly of built-up, bolted, riveted, or spot-welded structures. The conditions at the bolt, rivet, or weld areas critically influence the behavior of the damping configurations and make analysis particularly difficult because of the limited control of conditions at these points. Finite element analysis may be one of the few techniques for such analysis.

Fig. 37.2. Some basic integral damping treatments. (*A*) Laminate. (*B*) Faying surface damping.

Tuned Dampers. The tuned damper (Fig. 37.3) is essentially a single degree-of-freedom mass-spring system having its resonance frequency close to the selected resonance frequency of the system to be damped, i.e., tuned.[34] As the structure vibrates, the damper elastomeric element† vibrates with much greater amplitude than the structure at the point of attachment and dissipates significant amounts of energy per cycle, thereby introducing large damping forces back to the structure which tend to reduce the amplitude. The system also adds another degree of freedom, so two peaks arise in place of the single original resonance. Proper tuning is required to ensure that the two new peaks are both lower in amplitude than the original single peak. The damper mass should be as large as practicable in order to maximize the damper effectivness, up to perhaps 5 or 10 per cent of the weight of the structure at most, and the damping capability of the resilient element should be as high as possible.

Damping Links. The damping link (Fig. 37.3) is another type of discrete treatment, joining two appropriately chosen parts of a structure.[35] Damping effectiveness depends on the existence of large relative motions between the ends of the link and on the existence of unequal stiffnesses or masses at each end. Figure 37.3 shows the conceptual form of a unique frequency-sensitive antinodal damping treatment.[3] The deformation of the structure when it is bent leads to shear deformation of the viscoelastic elements, as well as to extensional deformation, which is related to relative motion between the beam and the structure. These deformations of the viscoelastic material lead to energy dissipation by the damper.

* Manufacturers of laminates include: Antiphon, Newcastle, Del. (MPM, MPM HT-100); Specialty Products, Newark, Del. (Noiseless steel); Anatrol Corporation, Cincinnati, Ohio.

† Manufacturers of elastomeric materials include: US Rubber Company (Paracril-BJ, Paracril-D); Dow Corning, Midland, Mich. (silicones); General Electric, Philadelphia, Pa. (silicones).

Fig. 37.3. Some basic discrete damping treatments. (A) Tuned damper. (B) Damping link. (C) Frequency-sensitive antinodal device.[3]

RATING OF DAMPING EFFECTIVENESS

MEASURES OR CRITERIA OF DAMPING

There are many measures of the damping of a system.[4] Ideally, the various measures of damping should be consistent with each other, being small when the damping is low and large when the damping is high, and having a linear relationship with each other. This is not always the case, and care must be taken, when evaluating the effects of damping treatments, to ensure that the same measure is used for comparing behavior before and after the damping treatment is added. The measures discussed here include the loss factor η, the fraction of critical damping (damping ratio) ζ, the logarithmic decrement Δ, the resonance or quality factor Q, and the specific damping energy D. Table 37.2 summarizes the relationship between these parameters, in the ideal case of low damping in a single degree-of-freedom system. Some care must be taken in applying these measures for high damping and/or for multiple degree-of-freedom systems, and especially to avoid using different measures to compare treated and untreated systems.

Loss Factor. The loss factor η is a measure of damping which describes the relationship between the sinusoidal excitation of a system and the corresponding sinusoidal response. If the system is linear, the response to a sinusoidal excitation is also sinusoidal and a loss factor is easily defined, but great care must be taken for nonlinear systems because the response is not sinusoidal and a unique loss factor cannot be defined. Consider first an inertialess specimen of linear viscoelastic material excited by a force $F(t) = F_0 \cos \omega t$, as illustrated in Fig. 37.4. The response $x(t) = x_0 \cos (\omega t - \delta)$ is also harmonic at the frequency ω as for the excitation but with a phase lag δ. The relationship between $F(t)$ and $x(t)$ can be expressed as

$$F = kx + \frac{k\eta}{\omega} \frac{\partial x}{\partial t} \tag{37.1}$$

Table 37.2. Comparison of Damping Measures

Measure	Damping ratio	Loss factor	Log dec	Quality factor	Spec damping	Amp factor
Damping ratio	ζ	$\dfrac{\eta}{2}$	$\dfrac{\Delta}{\pi}$	$\dfrac{1}{2Q}$	$\dfrac{D}{4\pi U}$	$\dfrac{1}{2A}\,^{*}$
Loss factor	2ζ	η	$\dfrac{2\Delta}{\pi}$	$\dfrac{1}{Q}$	$\dfrac{D}{2\pi U}$	$\dfrac{1}{A}\,^{*}$
Log decrement	$\pi\zeta$	$2\pi\eta$	Δ	$\dfrac{\pi}{2Q}$	$\dfrac{D}{4U}$	$\dfrac{2\pi}{A}\,^{*}$
Quality factor	$\dfrac{1}{2\zeta}$	$\dfrac{1}{\eta}$	$\dfrac{\pi}{2\Delta}$	Q	$\dfrac{2\pi U}{D}$	A^{*}
Spec damping	$4\pi U\zeta$	$2\pi U\eta$	$4U\Delta$	$\dfrac{2\pi U}{Q}$	D	$\dfrac{2\pi U}{A}\,^{*}$
Amp factor	$\dfrac{1}{2\zeta}$	$\dfrac{1}{\eta}$	$\dfrac{\pi}{2\Delta}$	Q	$\dfrac{2\pi U}{D}$	A^{*}

* For SDOF system only.

where $k = F_0/x_0$ is a stiffness and $\eta - \tan\delta$ is referred to as the *loss factor*. The phase angle δ varies from $0°$ to $90°$ as the loss factor η varies from zero to infinity, so a one-to-one correspondence exists between η and δ. Equation (37.1) is a simple relationship between excitation and response which can be related to the stress-strain relationship because normal stress $\sigma - F/S$ and extensional strain $\varepsilon = x/l$. This is a generalized form of the classical Hooke's law which gives $F = kx$ for a perfectly elastic system. The loss factor, as a measure of damping, can be extended further to apply to a system possessing inertial as well as stiffness and damping characteristics. Consider, for example,

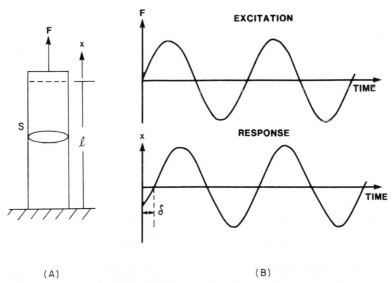

(A) (B)

FIG. 37.4. Linear viscoelastic behavior of a sample under sinusoidal loading, described in terms of response and excitation as functions of time. (*A*) Specimen. (*B*) Response and excitation.

the one degree-of-freedom linear viscoelastic system shown in Fig. 37.5. The equation of motion is obtained by balancing the stiffness and damping forces from Eq. (37.1) to the inertia force $m(d^2x/dt^2)$:

$$m\frac{d^2x}{dt^2} + kx + \frac{k\eta}{\omega}\frac{dx}{dt} = F_0 \cos \omega t \tag{37.2}$$

The steady-state harmonic response, after any start-up transients have died away, is illustrated in Fig. 2.22. If k and η depend on frequency as is the case for real materials, then the maximum amplitude at the resonance frequency $\omega_r = \sqrt{k/m}$ is equal to $F_0/k(\omega_r)\eta(\omega_r)$, while the static response, at $\omega = 0$, is equal to $F_0/k(0)\sqrt{1 + \eta^2(0)}$. The amplification factor A is approximately equal to $1/\eta(\omega_r)$, provided that $\eta^2(0) \ll 1$. Furthermore, the ratio $\Delta\omega/\omega_r$, where $\Delta\omega$ is the separation of the frequencies for which the response is $1/\sqrt{2}$ times the peak response, is known as the *half-power bandwidth* (see Fig. 2.22). It is also equal to η, provided that $\eta^2 \ll 1$. In summary, therefore,

$$\eta = \frac{1}{A} = \frac{\Delta\omega}{\omega_r} \tag{37.3}$$

This relationship between η and $1/A$ is applicable only for a single degree-of-freedom system and may not be directly applicable for more complex systems such as beams,

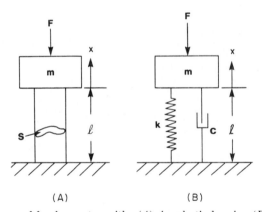

(A)   (B)

Fɪɢ. 37.5. Single degree-of-freedom system with: *(A)* viscoelastic damping; *(B)* viscous damping.

plates, or more complex structures. The measure $\Delta\omega/\omega_r$ is applicable for more complex systems, as well as single degree-of-freedom systems. For large values of η, on the order of 0.2 or greater, none of these measures of damping agree exactly, even for an ideal linear single degree-of-freedom system, but each measure is at least self-consistent. The stiffness and loss-factor parameters defined here do not specify any particular model of material behavior. For example, k and η could be constants as for hysteretic damping or they could be functions of frequency, temperature, specimen composition and shape, or amplitude as for a nonlinear material. The model with constant k and η is not too useful over a wide frequency range, and such behavior is impossible over an infinite frequency range, but these parameters can vary quite slowly with frequency for some particular material compositions. If k and η vary strongly with frequency, or amplitude, then the various definitions of the loss factor must be used with great care, since each measure gives different results.

Fraction of Critical Damping. The fraction of critical damping (damping ratio) is a measure of one very specific mechanism of damping, i.e., viscous damping which is proportional to velocity. If the damping forces acting on a single degree-of-freedom mass-spring system, illustrated in Fig. 37.5, satisfy this type of relationship, then the equation of motion for harmonic excitation is

$$m \frac{d^2\ddot{x}}{dt^2} + c\dot{x} + kx + = F_0 \cos \omega t \tag{37.4}$$

The response depends on m, k, and a parameter $c/2\sqrt{km}$ which involves c, k, and m and is known as the *fraction of critical damping* (damping ratio). This parameter, labeled ζ, controls the peak amplitude, the half-power bandwidth, and the resonance frequency ω_r:

$$x_{max} = \frac{F_0}{2k\zeta \sqrt{1 - \zeta^2}} \qquad x(0) = \frac{F_0}{k}$$

$$\omega_r = \sqrt{(k/m)(1 - \zeta^2)} \qquad \frac{\Delta\omega}{\omega_r} = 2\zeta \tag{37.5}$$

The plot of $x(\omega)$ versus frequency ω, for specific values of m and k is very similar to those for the viscoelastic damping, provided that $\eta \doteq 2\zeta$. The distinction between viscous and hysteretic damping (constant k, and η) is not at once apparent. Equations (37.2) and (37.4) convey the difference, since the damping coefficient in Eq. (37.2) decreases in proportion to $1/\omega$ as ω increases, while that in Eq. (37.4) is constant with frequency, at least for the hypothetical cases considered here. Figure 37.6 shows plots of response versus frequency based on the solutions of these equations of motion for each type of damping. Some differences arise at low frequency, but they are not very great except for very high values of damping. For high values of damping, neither η nor ζ are linearly related to the bandwidth ratio $\Delta\omega/\omega_r$. Figure 37.7 shows the variation of $\Delta\omega/\omega_r$ with η and 2ζ for values of η which are not small. Limits exist beyond which the ratio $\Delta\omega/\omega_r$ does not give a good estimate of η or ζ.

Logarithmic Decrement. When a damped system is struck by an impulsive load or is released from a displaced position relative to its equilibrium state, a decaying oscillation usually takes place as illustrated in Fig. 2.8. A measure of damping called *logarithmic*

(A)   (B)

Fɪɢ. 37.6. Comparison of viscous and hysteretic damping of a single degree-of-freedom system with: (A) low damping ($\eta = 0.1$, $\zeta = 0.05$); (B) high damping ($\eta = 1.0$, $\zeta = 0.5$).

Fɪɢ. 37.7. Variation of loss factor (η) and two times the fraction of critical damping (2ζ) of a single degree-of-freedom system with $\Delta\omega/\omega_r$.

decrement Δ is defined as the natural logarithm of the ratio of amplitudes of successive peaks [see Eq. (2.19)]:

$$\Delta = \ln \frac{x_1}{x_2} = \ln \frac{x_n}{x_{n+1}} \tag{37.6}$$

This definition is useful only if these ratios are equal for the various cycles, i.e., for specific types and amounts of damping. The measure is useful for viscous and hysteretic damping, within limits. For viscous damping, the solution of Eq. (37.4) for an impulsive excitation $F\,\delta(t)$ is obtained.[15]

$$x = \frac{F}{\sqrt{km(1 - \zeta^2)}} \, e^{-t\sqrt{k/m}} \sin t\sqrt{(k/m)(1 - \zeta^2)} \tag{37.7}$$

so that

$$\Delta = \frac{\pi\zeta}{\sqrt{1 - \zeta^2}} \tag{37.8}$$

for small ζ. If ζ approaches 1.0, the response becomes aperiodic and a logarithmic decrement cannot be defined or related to ζ. The loss factor in Eq. (37.1) also can be related to the transient response of a single degree-of-freedom mass-spring system, subject to an impulsive excitation. Consider the impulsive excitation $F(t)$ to be modeled as a spike of the form of a delta function at time $t = 0$. Then the equation of motion, in the form of Eq. (37.2), cannot be written directly, but if $F(t)$ and $x(t)$ are both described in terms of their corresponding Fourier transforms, then $\overline{F}(\omega) = \int_{-\infty}^{\infty} F(t) \exp(-j\omega t)\, dt = F$ and $\overline{x} = F/(k - m\omega^2 + jk\eta)$. The inverse Fourier transform gives

$$x(t) = \frac{1}{2\pi} \int_{-\infty}^{\infty} \frac{F e^{j\omega t}\, d\omega}{k - m\omega^2 + jk\eta} \tag{37.9}$$

This equation contains real and imaginary parts, but using the fact that $\exp(j\omega t) = \cos \omega t + j \sin \omega t$ and if $k(\omega) = k(-\omega)$ and $\eta(-\omega) = -\eta(\omega)$, then it may be shown that $x(t)$ is given by[15]

$$x(t) = \frac{F}{\pi} \int_0^\infty \frac{(k - m\omega^2) \cos \omega t + k\eta \sin \omega t}{(k - m\omega^2)^2 + (k\eta)^2} \, d\omega \qquad (37.10)$$

For k and η constant over all frequencies from zero to infinity, problems arise regarding $x(t)$ being finite for values of t less than zero, i.e., before the impulse is applied, and this is physically impossible. The problem is that k and η cannot be constants for real systems over any extremely wide frequency range, no matter how close to constant they may be over a limited frequency range. For small values of η, however, a useful and accurate solution is given by

$$x(t) = \frac{F}{\sqrt{km}} e^{-\frac{1}{2}\eta t \sqrt{k/m}} \sin t\sqrt{k/m} \qquad (37.11)$$

$$\Delta = \pi\eta/2 \qquad (37.12)$$

Comparing Eqs. (37.8) and (37.12) gives

$$\eta = 2\zeta \qquad (37.13)$$

These various equations relate η, Δ, and ζ for viscous and viscoelastic damping of single degree-of-freedom systems. The relationships usually agree well for low values of η and ζ ($\eta < 0.2$ or so), but for higher values the comparisons are not so precise. Figure 37.8, for example, shows comparisons between viscous, hysteretic, and viscoelastic damping for response to impulsive excitation.[15] For low damping, the three models agree well, but for high damping the differences are considerable.

Quality Factor. The quality factor Q is defined as $\omega_r/\Delta\omega$, so

$$Q = \frac{1}{\eta} \qquad (37.14)$$

For a single degree-of-freedom $Q = A$, but this is not the case for multiple degree-of-freedom systems.

Specific Damping Energy. Another useful measure of material damping is the amount of energy dissipated per unit volume per cycle, known as the *specific damping energy*.[20] For a damping material specimen subject to an applied external force $F(t) = F_0 \cos \omega t$ the specific damping energy D is equal to

$$D = \oint F \, dx = \int_0^{2\pi/\omega} F \frac{dx}{dt} \, dt \qquad (37.15)$$

For a viscoelastic material obeying Eq. (37.1)

$$D = F_0 x_0 k\eta \sqrt{1 + \eta^2} \qquad (37.16)$$

But $F_0 = kx_0 \sqrt{1 + \eta^2}$, also from Eq. (37.1), so

$$D = \pi x_0^2 k\eta \qquad (37.17)$$

The specific damping energy D increases as the square of the amplitude of vibration x_0 for linear viscoelastic materials, so it is clearly desirable to ensure that the damping material is strained as vigorously as possible in order to maximize D and hence the damping of the system. This has an important bearing on the choice of location within a vibrating system for application of a damping treatment. Furthermore, both k and η must be as large as possible to ensure maximum energy dissipation in the system, but this can be done only to the extent that further increases of k and η do not reduce x_0. While D is related to k and η for linear viscoelastic materials, this is not possible for nonlinear materials or for high cyclic strain levels where nonlinear behavior occurs; the value of D is then, of itself, often used as a measure of over-all damping performance.

$$\eta = 2\zeta = \frac{2\Delta}{\pi} = \frac{1}{Q} = \frac{D}{2\pi U} = \frac{1}{A} = \frac{\Delta\omega}{\omega_r} \tag{37.18}$$

COMPARISON OF DAMPING MEASURES. It is important when analyzing tests to determine the effects of damping treatments on dynamic response to be consistent in the use of these damping measures, and to recognize that they are not completely equivalent, especially over wide frequency ranges or for multimodal response.

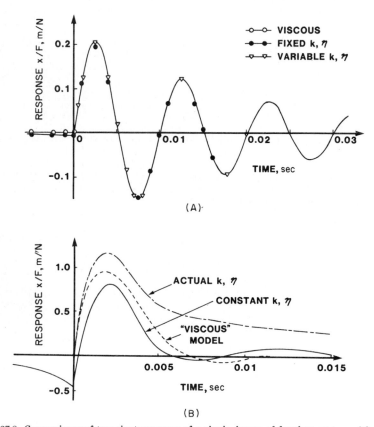

FIG. 37.8. Comparisons of transient response of a single degree-of-freedom system with viscous, hysteretic, and viscoelastic damping for (A) low damping, and (B) high damping.

Effects of Mass and Stiffness. Changing the mass or stiffness of a single degree-of-freedom mass-spring system without changing any other parameters leads to a change of resonance frequency, and when the frequency changes over a wide range, the differences of viscous and hysteretic damping become more apparent. For viscous damping, the fraction of critical damping $\zeta = c/2\sqrt{km}$ changes as k or m change, whereas for hysteretic damping ζ does not change, at least within a limited frequency range. Although viscous and hysteretic damping measures are related by the simple relationship $\eta = 2\zeta$ for a single mode at a particular frequency, they do not remain equivalent as the frequency changes, and significant differences in response may be observed.

TESTS FOR DETERMINING DAMPING MATERIAL PROPERTIES

TEST INSTRUMENTATION AND INTERPRETATION OF TEST DATA. Analog instrumentation has been used to apply sine-sweep excitation or random excitation to the test article. Now, digital instrumentation can be used to perform the same type of swept-sine-wave excitation by advancing stepwise rather than continuously through the frequency range selected. The main advantage of digital test systems is that transient excitation can be applied readily, the transient response can be recorded digitally in the time domain, and a Fourier transform can be performed using a fast Fourier transform (FFT) algorithm to place the response in the frequency domain for analysis of modal behavior such as resonance frequencies, mode shapes, and modal loss factors.[36]

Analog Sine-Sweep Tests. In the case of analog swept sine-wave testing of a structure an exciter is driven by the output of an analog sweep-frequency oscillator; the sweep rate is sufficiently low to avoid transient response during each sweep through a resonance peak. The response is measured by, for example, an accelerometer, and the signal is plotted on an X-Y plotter after passing through an A-to-D converter. The signals can be stored on magnetic tape for later playback to obtain meter readings of excitation and response signals at any frequency. Provided every instrument in the chain is properly calibrated and introduces minimal phase changes between incoming and outgoing signals, the resonance frequency and the modal loss factor can be determined readily.

Digital Sine Sweep Tests. The digital equivalent of swept sine wave excitation makes the exciter frequency dwell for a selectable period of time at a number of selected frequencies, with almost instantaneous change from one frequency value to the next. The choice of dwell time and frequency interval are now the most important parameters controlling the degree of dynamic equilibrium achieved during a test. The frequency points can be chosen close enough to be virtually indistinguishable from a continuous curve (as for analog swept sine wave testing), although test time can be saved by choosing the widest intervals commensurate with the goals of the test. Digital swept sine wave testing is usually microcomputer controlled. Digital swept sine wave testing is particularly appropriate for nonlinear systems, because analysis of the test results is usually easier for sinusoidal excitation than for other types, but it is also very useful for linear systems.

Impact Tests. For linear systems, impact testing is widely used. The structure is excited by impact at one or more points in succession by an instrumented hammer having a force gage at the impact point to measure the transient applied force. The response is measured at one or more points (often the same ones at which the force is applied) by an accelerometer or other transducer. The input force and the response in one or more channels are sampled digitally in the time domain and are stored digitally at preselected time intervals and for a selected period of time in a microprocessor. The signals are then transformed to the frequency domain by an FFT algorithm. Ideally, the response of the system to swept sine wave excitation having a fixed amplitude (described as the ratio of response to excitation, i.e., compliance or receptance) should be the same for analog swept sine wave, digital swept sine wave, or impact/FFT techniques. In practice, difficulties of ensuring that conditions are truly identical for each test, and varying limits of dynamic range for the various test instruments, result in discrepancies. For well-conducted tests, the differences should not be great.

Scale Models. Sometimes a structure or machine is not available for test. Finite element analysis can be used to predict the anticipated response behavior, but modeling is a difficult art and experimental confirmation of predictions is at the very least desirable. Scale models of the system can sometimes be built,[37] using metal or plastics such as Plexiglas or Lexan. To scale the response of the model to the full-scale system properly, consider a simple element of such a structure as illustrated in Fig. 37.9. The element illustrated can deform flexurally or longitudinally under the forces indicated. The element mass m is equal to ρwhl. The extensional stiffness k_e is equal to $2Ew(h/l)$ to a first approximation, and the flexural stiffness k_b is equal to $\frac{1}{4} Ew(h/l)^3$. Thus the fundamental longitudinal mode of vibration has a frequency $f_{le} = (\frac{1}{2}\pi) \sqrt{2k_e/m} = (\frac{1}{2}\pi) \sqrt{E/\rho l^2}$ and the fundamental bending mode has a frequency $f_{lb} = (\frac{1}{2}\pi) \sqrt{2k_b/m} = (\frac{1}{2}\pi) \sqrt{(E/\rho l^2)} (h/l)^2$. The frequency of model and full-scale structures depend only on $\sqrt{E/\rho l^2}$ for h/l constant. Geometric scaling is therefore a necessity. If the same exciting

FIG. 37.9. Modeling of a simple structure. (A) Beam element. (B) Discrete model.

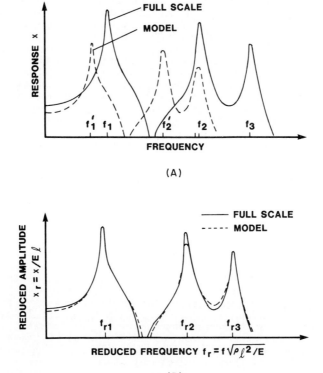

FIG. 37.10. Correlation of model and full-scale dynamic response behavior. (A) Model and full-scale test data. (B) Correlation of model and full-scale response data.

forces are applied to model and full-scale systems, the response amplitude will be proportional to Ew or EI, so x/EI will be the same for model and full-scale systems. Similarly, $f \sqrt{\rho l^2/E}$ will be the same for model and full-scale. These elementary considerations, applied with care, are useful to translate from model to full-scale response for geometrically scaled systems. Figure 37.10 illustrates the process of correlating model and full-scale data in an idealized situation.

VIBRATING BEAM TEST METHODS. The vibrating beam test methods are frequently used to measure the extensional or shear complex modulus properties of damping materials.[7,15,16,18] The dynamic response behavior of the beam, first in the undamped uncoated form and then with an added damping layer or added constrained configuration, is measured for several modes of vibration and over a range of temperatures. At each temperature, the measured damped resonance frequency f_n, the undamped resonance frequency f_{on}, and the loss factor η_n in the nth mode of vibration are measured and used in an appropriate set of equations to deduce the Young's modulus E, or the shear modulus G, and the loss factor η of the damping material at a number of discrete frequencies and temperatures.

Various configurations of cantilever beams are used to measure viscoelastic material damping properties in tension-compression or shear at low cyclic strain amplitudes. Figure 37.11 illustrates some of the configurations used. The damping layers are bonded to

FIG. 37.11. Cantilever beam damping material test configurations. (A) Nonsymmetric Oberst beam. (B) Symmetric modified Oberst beam. (C) Symmetric sandwich beam. (D) Symmetric constrained-layer beam.

the base beams by means of a stiff adhesive such as an epoxy. This bonding is very important and must be done well using an adhesive which is stiff in comparison to the damping layer and is very thin. The thickness ratio h_2/h_1 generally lies in the range 0.1 $\leq h_2/h_1 \leq 2.0$, and the length l is about 5 to 10 in. (12.7 to 25.4 cm). The base beam material is typically aluminum, steel, or a stiff epoxy or epoxy matrix composite material having low intrinsic damping. Great care must be taken to ensure that the temperature range of the tests is not excessive in relation to the behavior of the base beam, and in particular to allow for the effect of temperature on the base beam properties such as Young's modulus, the resonance frequencies, and the modal loss factor in the absence of the damping layer. The vibration test is conducted allowing the specimen to soak at each selected temperature for several minutes, often 10 or 15 minutes, to be sure of thermal equilibrium; then the beam is excited by means of a noncontacting transducer or by impact, and the resulting response in the frequency domain is measured, either through swept sine wave excitation or FFT analysis of the transient response signal in the time domain. At each temperature, several resonance frequencies and modal loss factors are measured over a wide range of frequencies. The test is then repeated after thermal equilibrium has been reached at the next selected temperature. The data obtained for the first mode is usually not used because of the low frequency involved and the high amplitudes and high modal damping of the base beam, as well as because of errors in the analysis when sandwich beams are used. Such vibrating beam tests are widely used for measuring viscoelastic material damping properties for shear and extensional deformation and form the basis for an ASTM standard.[7]

Free-Layer Beam Method. For the beam coated as a free layer on one side, as illustrated in Fig. 37.11A, the Young's modulus E_2 and the loss factor η_2 of the damping material are calculated from the equation for a free-layer treatment.[18] These equations must be used with care. For example, if the modal loss factor η_n in the nth mode is greater than 0.3 or so at any temperature, it is difficult to measure either f_n or η_n with sufficient accuracy. Also, if the ratio of stiffness of the coated to the uncoated beam is less than about 1.1, errors in the measured values of f_n and η_n are magnified in the calculated values of E_2 and η_2. This can happen at high temperatures, when the damping material layer is very soft relative to the base beam material. The onset of this type of problem can be delayed to higher temperatures by increasing h_2, although this can lead to excessive values of η_n at lower temperatures, or by using an epoxy-matrix composite beam. The resonance frequencies of the uncoated beam f_{on} must be measured as a function of temperature in separate tests. Another problem which is occasionally encountered in this type of unsymmetric specimen is excessive bending of the beam at high and low temperatures, resulting from unequal thermal expansion of the base beam and the viscoelastic material.

Symmetric Free-Layer Method. To alleviate the problem of excessive bending of the beam at high and low temperatures, as a result of unequal thermal expansion coefficients of the damping layer and the base beam, and to simplify the equations, the base beam can be coated symmetrically as illustrated in Fig. 37.11B.[15,38] The total amount of damping material is unchanged since each layer can be half the thickness used when coated on one side. The error magnification as the stiffness ratio approaches unity is high. As before, the initial damping of the base beam must be low, with modal loss factors of generally less than 0.01.

Sandwich Beam Technique. To measure material damping properties in shear, symmetric sandwich cantilever beams are used as illustrated in Fig. 37.11C. The equations usually used to determine the shear modulus G_2 and the loss factor η_2 from the measured modal loss factors and resonance frequencies neglect bending of the middle layer, so the sandwich beam test is usually used for measuring complex moduli of soft viscoelastic materials, for which the ratio E_2/E_1 is usually less than 0.01. Care must be taken to ensure good bonding between the two beams and the inner viscoelastic layer and at the spacers, and also that the resonance frequencies of each beam are identical. Errors greater than 1 per cent are unacceptable. Another variation on the sandwich beam approach is the symmetric constrained-layer configuration illustrated in Fig. 37.11D. The equations are simplified because the neutral axis now lies along the center of the composite beam and only a single base beam is required.[15]

GEIGER THICK-PLATE TEST METHOD. The Geiger thick-plate method is of importance because it is widely used to describe damping materials in the automotive industry.[3] It makes use of a large flat plate, suspended freely from four points selected to be at or near the nodal lines of the first free-free mode, to which is bonded the damping layer being evaluated. The rate of decay of vibration amplitude (expressed in decibels per second) is measured and serves as a measure of the effectiveness of the damping layer. Figure 37.12 illustrates a typical test setup. The system can be excited by an impulsive force, measured through a force gage, and applied by a hammer, or by an electromagnetic exciter or an electrically actuated impeller, or by sine-wave or random excitation. The response can be picked up by an electromagnetic transducer, in which case cross talk with the excitation transducer must be avoided by adequate separation or by the use of capacitive or electro-optical transducers or by a miniature accelerometer. The measured output can be displayed in many ways, including a decaying sinusoidal trace representing response to an impulsive excitation (a measure of the logarithmic decrement), or a frequency domain display in the region of the fundamental free-free mode (loss factor measure). The observed logarithmic decrement or loss factor value is a measure of the damping of the plate/damping material system and depends on the plate and treatment thicknesses. The free-layer treatment equations used for the vibrating beam tests may also be used with the Geiger plate test provided that the same conditions are satisfied. In particular, the treatment thickness must be sufficient to make the ratio of the stiffness of the coated plate to that of the uncoated plate greater than about 1.05. The size of the specimen and the use of only one mode makes this condition somewhat less restrictive than for the beam tests, for which the specimens are much smaller.

Fig. 37.12. Geiger plate test configuration.

SINGLE DEGREE-OF-FREEDOM RESONANCE AND IMPEDANCE TESTS.

Digital test instrumentation and data analysis techniques make it relatively easy to conduct vibration tests directly on relatively small samples of damping materials and to readily determine the damping properties. Typical test configurations are illustrated in Fig. 37.13.

Resonance Test. For a resonance type of test, the specimen is driven inertially by a large vibration table (see Chap. 25), usually by swept sine wave excitation.[38,39] The input and output accelerations are usually measured by accelerometers, and the response parameter of interest is the amplification $A = x/x_0$ as a function of frequency, where x is the amplitude of displacement of the mass and x_0 is that of the shaker table. At resonance, the maximum value of x/x_0 is observed along with the resonance frequency for each temperature. The modulus and loss factor of the specimen material are determined from

$$\eta = \frac{1}{\sqrt{A^2 - 1}} \tag{37.19}$$

$$E = \frac{m_e \omega_r^2 S}{l} \tag{37.20}$$

$$G = \frac{m_e \omega_r^2 S}{t} \tag{37.21}$$

for tension-compression or shear. For tension-compression, $S = wh$ where w is the width, h is the thickness, and l is the length. For shear, $S = 2wl$ and h is the thickness of one material layer. The effective mass m_e includes the added mass m and one-third of the mass of the damping material, to a first approximation. The ratio l/h or l/w, whichever is smaller, must be greater than 1.0 or shape effects must be taken into account. For shear, h/l must be smaller than about 0.2, for the same reason. For highly damped materials, for which x/x_0 does not exceed 1.0 by a significant amount, considerable error in measuring A and η will be encountered, but the method is very effective for values of η less than about 0.5. In this method, data are obtained at only one frequency; the mass m must be changed to obtain data at other frequencies. Care must be taken to avoid sagging or creep of the specimen at high temperatures and to ensure that thermal equilibrium has been achieved. A thermocouple placed within the volume of the specimen

(A)

(B)

Fɪɢ. 37.13. Single degree-of-freedom test configurations. (A) Resonance test, shaker sine-sweep excitation. (B) Impedance test.

material may be necessary, particularly for tests at high strain amplitudes where internal heating of the specimen by energy dissipation from damping may lead to wide differences between true specimen temperature and the temperature of the surroundings.

Impedance Test.[39,40] If the specimens are excited by a driver through a force gage, then the response measure used to characterize the system behavior is the compliance or receptance x/F, where F is the driving force measured by the force gage and x is the response at the same point, measured by an accelerometer, for example. If the mass m is large compared with the mass of the specimen, as illustrated in Fig. 37.13B, then one may add one-third the mass of the specimen to m to give the effective mass m_e of the equivalent single degree-of-freedom system, so that

$$\frac{x}{F} = \frac{1}{k(1 + j\eta) - m_e\omega^2} \tag{37.22}$$

If this is expressed instead in terms of the ratio F/x, the dynamic stiffness at the driving point, which is directly related to the driving-point impedance, then

$$\kappa = k - m_e\omega^2 + jk\eta \tag{37.23}$$

which shows that the direct dynamic stiffness is a linear function of ω^2 and the quadrature dynamic stiffness $\kappa_Q = k\eta$. Figure 37.14 illustrates the variation of κ_D and κ_Q with ω^2 for k constant and η varying with frequency. The stiffness constant k is obtained from the intercept on the vertical axis or from the zero crossing point $\omega_r{}^2$, with $k = m_e\omega_r{}^2$. The mass m_e can be estimated from the slope of the line of κ_D versus ω^2. It is not difficult to obtain good measurements of k and η by this type of test approach from about $0.2\omega_r$ to $3\omega_r$, so data can be obtained quite easily over about a decade of frequency instead of at only a single frequency as for the resonance method.

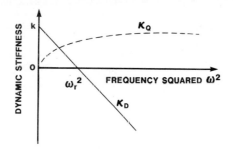

FIG. 37.14. Dynamic stiffness as a function of frequency squared for a single degree-of-freedom system with hysteretic damping.

COMMERCIAL TEST SYSTEMS FOR MEASURING DAMPING MATERIAL BEHAVIOR. Many commercial systems are available for measuring the complex modulus properties of viscoelastic damping materials. All are based on some kind of deformation mode of a sample of the material, measurement of the corresponding excitation forces and displacements, and analysis of the data to obtain the material properties.[8–10] Each system has advantages and disadvantages, but when due care is exercised, good results usually can be obtained with each system. Particular care should be taken to read, understand, and follow the manufacturer's instructions. For example, in some tests such as monitoring cure cycles of epoxies, the temperature sweep rate can be quite high in order to keep up with the reaction. This is acceptable if one is monitoring the progress of the cure cycle, but it may not be acceptable if one seeks to measure the damping properties at a state approximating thermal equilibrium. For thermal equilibrium to be maintained, temperature sweep rates well below 1°F (0.5°C) per minute are usually recommended, and even lower rates may be required for large specimens.

PROPERTIES OF DAMPING MATERIALS

CHARACTERIZATION AND MODELING

For engineering purposes, each damping material sample is a unique object exhibiting behavior under dynamic excitation which can be determined only from experiment. The behavior of viscoelastic damping materials under the action of general exciting loads can be very complex and difficult to understand or interpret, but the response to simple types of excitation may be used to form the basis of an analytical framework which in turn can be used to more easily understand the effects of more complicated loading. Examples of simple loading actions include:

1. Sudden application and maintenance of a static load, as illustrated in Fig. 37.15*A*. The response of the damping material sample as it seeks a new equilibrium position provides valuable information.

2. Sudden application of an impulsive load, as illustrated in Fig. 37.15*B*.

3. Application of a sinusoidally varying excitation, as illustrated in Fig. 37.15*C*.

The response of a damping material sample to these and other types of loading actions depends on a large number of factors, including the specimen geometry; the material composition, processing, and molecular structure; the temperature; the frequency or rate of change and magnitude of the excitation. For low-amplitude excursions of the sample the behavior is often linear, so the effects of several loadings applied simultaneously can be added algebraically. This is especially applicable for cyclic loading, because most such loadings can be expressed as linear summations, finite or continuous, of sinusoidal signals.

FIG. 37.15. Response of a damping material sample to various types of excitations. (*A*) Response to step load. (*B*) Response to impulsive load. (*C*) Response to harmonic excitation.

COMPLEX MODULUS. Employing the complex notation $x = x_0 e^{j\omega t}$ for the sinusoidal displacement of a damping material sample, Eq. (37.1) can be written in the form

$$F_0 = k(1 + j\eta)x_0 \qquad (37.24)$$

where $F = F_0 e^{j\omega t}$ is the exciting force and $k(1 + j\eta)$ is referred to as a *complex stiffness*. Using the fact that $k = ES/h$ for a cylindrical sample, the stress σ is equal to F/S, and the strain ε is equal to x/h, this becomes

$$\sigma = E(1 + j\eta)\varepsilon \qquad (37.25)$$

where $E(1 + j\eta)$ is known as a complex modulus with E being the Young's modulus and η the loss factor as before.

EFFECTS OF FREQUENCY AND TEMPERATURE ON COMPLEX MODULI. Using the complex modulus representation of viscoelastic material stress-strain behav-

(A)

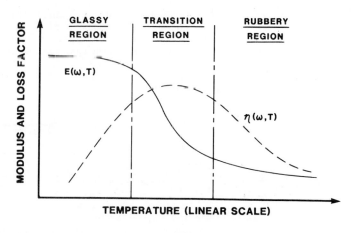

(B)

FIG. 37.16. Effects of frequency and temperature on viscoelastic material behavior. (*A*) Effects of frequency at constant temperature. (*B*) Effects of temperature at constant frequency.

ior, the modulus E and loss factor η are functions of frequency and temperature. While details of behavior vary from material to material, in general, E (or G) is a minimum at zero frequency and rises monotonically to a maximum at infinite frequency. As the frequency varies from zero to infinity, the loss factor varies from a low value through a rapid transition to a maximum and then to a low value at high temperature. These aspects of damping material behavior are illustrated in Fig. 37.16 for a material having a single transition.

Temperature-Frequency Equivalence (Reduced Frequency). Figure 37.17 illustrates the variation of Young's modulus E and loss factor η with frequency. These data demon-

FIG. 37.17. Idealized experimental complex modulus data plotted against frequency at various temperatures.

strate an important phenomenon: If the curves of E and η versus log frequency are shifted horizontally, they can be fitted together to form a continuous set of curves as illustrated in Fig. 37.18. This behavior is known as the *frequency-temperature correspondence principle* and is a consequence of the fundamental relationship between temperature and molecular vibrations, which in turn are closely related to the frequency of excitation of mechanical vibrations in the material sample. For engineering design purposes, this principle is an important tool in interpolating and correlating experimental data. The factor by which the frequency f (in hertz) must be multiplied in order to move it to the correct place in the reduced-frequency graph is referred to as the *shift factor* α_T and is a function of temperature only for each material composition. A specific temperature is chosen, arbitrarily or in accordance with standard procedures according to the needs of the user, to be the reference temperature and is usually designated as T_0. A graph of α_T versus temperature is illustrated in Fig. 37.19. The precise shape of the curve, and in particular the slope and curvature near $T = T_0$, depends on the particular material composition. The reference temperature T_0 is usually chosen to be within the transition temperature range of the material. The quantity $f\alpha_T$, having the dimensions of a frequency, is referred to as the *reduced frequency*.

FIG. 37.18. Hypothetical reduced-frequency plot of complex modulus data, in the form of a graph of modulus and loss factor versus reduced frequency.

A Nomogram for Data Display (Reduced Temperature Nomogram).[12] While the reduced-frequency plot of viscoelastic material damping properties is very useful for reduction of test data, the separate graphs of complex moduli versus reduced frequency and of the shift factor versus temperature lead to some inconvenience in the interpolation of data and sometimes lead to confusion in the use of published data because of the omission of the α_T versus temperature information. A major improvement in this respect has been achieved by a very simple process for combining all relevant data on a single graph, in the form of a display referred to, perhaps inaccurately, as a *reduced-temperature nomogram.* The process of creating the nomogram uses the reduced-frequency plot, but the upper horizontal scale is used to plot the temperature and the right vertical scale is used

FIG. 37.19. Hypothetical plot of shift factor versus temperature.

for frequency, as illustrated in Fig. 37.20. When T is equal to the reference temperature T_0 and the frequency f is equal to 1 Hz, $f\alpha_T = 1$, and this corresponds to point A. At the other extreme, when the frequency f is equal to 10,000 Hz and $T = T_0$, $f\alpha_T = 10^4$ and the point B on the horizontal line $f = 10^4$ represents the other end of the isotherm $T = T_0$, which is a straight line of unit slope. As another example, for $T = T_2$, $\alpha_T > 1$ and the point A' on the horizontal line $f = 1$ Hz represents the corresponding value of $f\alpha_T$. Similarly for all other temperatures. The result is a set of isotherms crossing the diagram diagonally. The intersection of the horizontal f line with the diagonal T line gives the proper value of $f\alpha_T$, and the vertical intercepts with E and η give the correct values of E and η at the selected frequency and temperature. Some practice is useful when producing such diagrams from experimental data, and some variation of plots will always occur because of the scatter in normal experimental data.

Example 37.1 (Generation of Nomogram for a Butyl Rubber Composition). Consider the data given in Table 37.3 for a typical Butyl rubber composition. The data,

FIG. 37.20. Hypothetical reduced-temperature nomogram plotting modulus and loss factor versus frequency, temperature, and reduced frequency.

Table 37.3. Measured Complex Modulus Properties for Butyl Rubber Sample

Beam	Temp, °F	Mode	f_{on}, Hz	f_n, Hz	η_n	$E_1 \times 10^6$, lb/in.^{2}	E_2, lb/in.^{2}	η_2
Aluminum	−25	2	128.8	107.1	.415	9.832	7.41 E4	0.966
		3	361.3	324.4	.570	9.869	1.03 E5	1.12
	+1	2	128.2	86.1	.200	9.741	1.40 E4	1.59
		3	359.5	253.7	.265	9.770	2.56 E4	1.27
		4	705.1	522.3	.353	9.788	3.83 E4	1.26
	+32	2	127.4	81.7	.036	9.620	4.11 E3	0.878
		3	357.4	231.3	.069	9.657	5.99 E3	1.18
		4	700.8	460.3	.089	9.669	9.08 E3	1.04
		5	1,158.4	769.4	.154	9.667	1.15 E4	1.45
	+60	2	126.5	80.6	.019	9.484	2.79 E3	0.666
		3	354.8	225.9	.030	9.517	2.66 E3	1.11
		4	695.8	445.0	.042	9.532	3.54 E3	1.17
		5	1,150.1	739.9	.054	9.529	4.71 E3	1.15
		6	1,713.6	1112.2	.067	9.480	6.46 E3	1.05
Epoxy matrix composite	+122	2	96.3	44.8	.032	2.417	1.23 E3	0.14
		3	270.1	126.5	.048	2.425	1.30 E3	0.20
		4	529.7	251.8	.069	2.429	1.47 E3	0.264
		5	875.6	413.9	.087	2.428	1.41 E3	0.344
	+250	2	92.6	43.2	.020	2.234	1.16 E3	0.086
		3	259.8	121.5	.018	2.243	1.19 E3	0.076
		4	509.4	239.8	.020	2.246	1.26 E3	0.081
		5	842.0	393.4	.022	2.246	1.18 E3	0.093
		6	1,254.6	580.5	.027	2.234	1.08 E3	0.123

measured by the vibrating beam technique at several temperatures and for several modes of vibration, are plotted in Fig. 37.21 for several temperatures as a function of frequency. The Young's modulus E and the loss factor η are plotted over two decades of frequency on a logarithmic scale. Data were available at temperatures ranging from −25 to +250°F (−32 to 121°C) and the frequency range was from about 50 to 1,000 Hz. The reference temperature T_0 was arbitrarily selected as about 40°F (4°C), and the data were shifted horizontally for each temperature so as to create the curves of E and η versus $f\alpha_T$ as shown in Fig. 37.22. The shift needed for several sets of points is indicated in this figure, and Fig. 37.23 summarizes the required values of α_T as a function of temperature. Note that the shifts are estimates based on a subjective evaluation of the smoothest graphs of E and η versus reduced frequency. The process can be automated, but the exact outcome will vary in apparent shape and scaling in a manner which depends on the skill and judgment of the operator, in the case of manual data reduction, or on the precise type of error minimization process used, in the case of automatic data reduction. Only in the case of completely error-free data can there be complete agreement between all data reduction exercises, and this is a rare event. Figure 37.24 summarizes the final nomogram for the Butyl rubber sample.

TYPES OF MATERIALS AVAILABLE. Some of the many types of polymeric damping materials which are available commercially usually are one of the following types:[15]

1. Mastic treatment materials
2. Cured elastomers and polymers
3. Adhesives
4. Damping tapes
5. Laminates

FIG. 37.21. Plots of modulus and loss factor versus frequency at various temperatures for Butyl rubber.

FIG. 37.22. Plot of Young's modulus E and loss factor η versus reduced frequency for Butyl rubber, showing amounts and directions of shifts at each temperature.

FIG. 37.23. Plot of shift factor versus temperature for Butyl rubber composition.

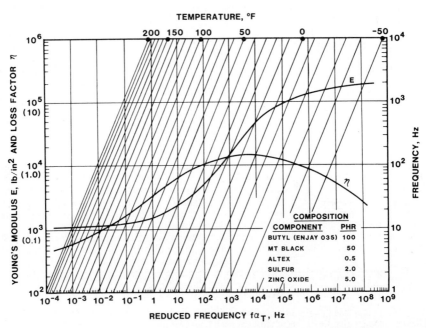

FIG. 37.24. Reduced temperature nomogram for Butyl rubber composition.

SELECTION AND DESIGN OF DAMPING TREATMENTS

SELECTION OF DAMPING TREATMENTS

AREA COVERAGE REQUIREMENTS. For layered damping treatments (free layer, constrained layer, damping tapes or laminates) one usually applies the treatment to all accessible areas of the structure for which damping is needed. Examples for which one might provide such coverage include (1) skin-stringer panels of airplane fuselage or wing structure, for which damping treatment designs are often needed to reduce noise transmission or resonance fatigue, (2) diesel-engine oil pan covers and side panels where the damping is needed to reduce noise radiation into the surroundings, (3) automotive body surfaces where noise transmission into the interior may need to be reduced. In each of these types of applications, the criteria which each damping treatment must meet include at a minimum (1) effectiveness, (2) low weight, (3) long-term durability and stability, and (4) low cost. Each of these criteria is difficult to meet. Effectiveness, in particular, depends not only on the area covered by the damping treatment, but also on the thickness and the material used for the treatment. Low weight and low cost are best achieved by careful design and experimental verification of the thickness, area coverage, and treatment location and type, along with careful integration of the treatment into the manufacturing process. Long-term durability and stability depend on a proper choice of materials to operate well for extended periods in the required environment, including effects of temperature extremes, maximum cyclic stresses experienced, effects of oil or other chemicals expected in the environment, effects of oxidation and radiation, including ultraviolet light and x-rays where relevant. All of these factors often make the selection of damping treatment and coverage requirements a very complex process. The process for deciding what material to use to achieve the damping characteristics desired, in the required modes of vibration, must be tempered by consideration of all the other requirements. This means that the most effective treatment may on occasion have to be rejected in favor of a less effective treatment in order to meet cost, weight, or durability criteria. The first and most important step is to design an effective treatment which actually achieves the vibration and noise reduction goals.

Antinodal Placement/Multiple Mode Requirements. Damping treatments which depend on local relative displacement or acceleration, such as tuned dampers or damping links, must be placed at or near antinodes (i.e., where the vibration amplitude is largest) in order to be effective. Thus a damping treatment may be effective for some modes but not for others which have nodes where the damping is placed; more than one damper or link may be needed to achieve the desired amplitude reductions in all modes of interest. Damping treatments which depend on local in-plane surface strains, such as free- or constrained-layer treatments, can be placed near antinodal areas of modes of vibration which display significant amounts of bending when vibrating. The requirements for several modes are rarely compatible, and uniform coverage may be required to damp a number of modes effectively.

AMOUNTS OF MATERIAL REQUIRED

Free-Layer Treatments. For treatments of this type, illustrated in Fig. 37.1, local panel weight increases up to 30 per cent may often be needed to increase the damping of the structure in several modes of vibration to an acceptable level. Greater weight increases usually lead to diminishing returns. This weight increase can be offset to some degree if the damping is added early in the design, by judicious weight reductions achieved by proper sizing of the structure to take advantage of the damping.

Constrained-Layer Treatments. Because of their greater efficiency, such treatments, illustrated in Fig. 37.1, may require less than a 20 per cent weight increase. Additional weight savings can be achieved if the damping is required in only a few very specific modes so that the damping treatment can be specially designed or placed at antinodes.

Built-in Damping. For laminates or joint damping, as illustrated in Fig. 37.2, the weight increase can be zero or even negative.

Discrete Tuned Dampers (Fig. 37.3A) or Damping Links (Fig. 37.3B). The weight increase needed to add significant damping in a single mode is usually smaller than for a layered treatment, perhaps 5 per cent or less.

EFFECTS OF INITIAL DAMPING. This is the damping which occurs in a structure before any damping treatments or devices are added. It arises from many sources, including joint damping, acoustic radiation damping, and internal damping. Estimates of the initial damping of a structure or machine have a very important bearing on the design of additive damping treatments. If, for example, a structure has few joints or interfaces as in the case of brazed, welded, or cast structures, then the initial modal loss factors can be 0.001 or less. In such a case, an increase of modal loss factor by an order of magnitude would imply damped structure loss factors of about 0.01, which may often be achieved quite readily. For built-up structures, such as bolted engine assemblies, the initial modal loss factors may be very high, on the order of 0.02 to 0.05, so that a treatment capable of adding an increment of 0.01 to the modal loss factor would be of no great value and increases of modal loss factors by 0.05 or 0.10 increments may be required but would be much more difficult to achieve.

EFFECTS OF TEMPERATURE AND FREQUENCY

Temperature is the most important parameter of all, so far as damping treatment design is concerned.[15] If the treatment is required to perform at a single temperature, there is often a wide choice of damping materials, unless the temperature is very high. Then the treatment design is optimized to make the best possible use of the given modulus and loss factor for the material selected. If, however, the treatment is expected to perform over a wide temperature range, a design which is optimum at the center of the range may not necessarily be optimum at the temperature extremes; if the temperature range is wide enough, it may be very difficult to achieve adequate levels of damping over the entire range. The selection of a damping treatment is a compromise in which many variables are selected to ensure an adequate level of performance over the full temperature range. For example, a damping material could be selected which has a loss factor less than the highest attainable in exchange for a lower rate of change of the modulus with temperature, or one might design the damping treatment using two or more viscoelastic materials, each having peak damping performance individually at different temperatures. These problems are more complex when the treatment is expected to survive, even if not to perform well, outside the selected temperature range.

The effect of *frequency* on complex modulus properties of viscoelastic materials is not as great as that of temperature unless the frequency range of concern is unusually wide. In the usual range of engineering interest, say from 50 to 5,000 Hz, the variation of properties is not very great, so the effects of frequency might appear to be of no great consequence. However, two other effects are associated with frequency. One is the wavelength of each particular mode of vibration of the structure, which varies in inverse proportion to the frequency and affects the performance of constrained-layer damping treatments through the influence of a shear parameter. The other is the tuning of damping links and tuned dampers. Such treatments usually are effective over only very limited frequency ranges, although for some types of structures such as skin-stringer panels used in aircraft construction, the frequency range of damper effectiveness may be as wide an an octave. The temperature and frequency ranges over which effective damping is required strongly influences the types of design and material selected for the application.

STRAIN EFFECTS. High-amplitude response of structures can lead to nonlinear behavior of both structure and dampers. The problem of separating damping effects from nonlinear effects is difficult and makes test results difficult to interpret. For testing purposes, it is usually best to conduct treatment damping performance verification tests at low-vibration amplitudes and to use the tests at higher amplitudes to verify durability under the expected vibrational environment. High-strain effects can also include internal heating of the viscoelastic materials as a result of the cyclic deformation, and the thermal flow of this dissipated energy may then be an important consideration in design. Figure 32.4 illustrates the effect of cyclic strain amplitude on the complex modulus properties of an elastomeric damping material.

DESIGN OF ADDITIVE DAMPING TREATMENTS

FREE-LAYER TREATMENTS. The mechanism of energy dissipation in a free-, or unconstrained-, layer treatment is the cyclic extensional deformation of the imaginary fibers of the damping layer during each cycle of flexural vibration of the base structure, as illustrated in Fig. 37.25.[15] The presence of the free layer changes the apparent flexural rigidity of the base structure in a manner which depends on the dimensions of the two layers involved and the elastic moduli of the two layers. The treatment depends for its effectiveness on the assumption, usually well-founded, that plane sections remain plane. The treatment fiber labeled yy in Fig. 37.25 is extended or compressed during each half of a cycle of flexural deformation of the base structure surface, in a manner which depends on the position of the fiber in the treatment and the radius of curvature of the element of length Δl, and can be calculated on the basis of purely geometric considerations. One fiber in particular does not change length during each cycle of deformation and is referred to as the *neutral axis*. For the uncoated plate or beam, the neutral axis is the center plane, but when the treatment is added, it moves in the direction of the treatment and its new position is calculated by the requirement that the net in-plane load across any section remain unchanged during deformation. The basic equations for predicting the modal loss factor η for the given damping layer loss factor η_2 and for predicting the direct flexural rigidity $(EI)_D$ as a function of the flexural rigidity $E_1 I_1$ of the base beam are well known.[15] Figure 37.26 describes the variation of η/η_2 and $(EI)_D/E_1 I_1$ with h_2/h_1 and E_2/E_1.

FIG. 37.25. Free-layer treatment. (*A*) Undeformed. (*B*) Deformed.

Limitations of Free-Layer Treatment Equations. The classical equations for free-layer treatment behavior are approximate. The main limitation is that the equations are applicable to beams or plates of uniform thickness and uniform stiff isotropic elastic characteristics with boundary conditions which do not dissipate or store energy during vibration. These boundary conditions include the classical pinned, free, and clamped conditions. Another limitation is that the deformation of the damping material layer is purely extensional with no in-plane shear, which would allow the "plane sections remain plane" criterion to be violated. This restriction is not very important unless the damping layer is very thick and very soft ($h_2/h_1 > 10$ and $E_2/E_1 > 0.001$). A third limitation is that the treatment must be uniformly applied to the full surface of the beam or plate, and especially that it be anchored well at the boundaries so that plane sections remain plane in the boundary areas where bending stresses can be very high and the effects of any cuts in the treatment can be very important. Other forms of the equations can be derived for partial coverage or for nonclassical boundary conditions.

Effect of Bonding Layer. Free-layer damping treatments are usually applied to the substrate surface through a thin adhesive or surface treatment coating. This adhesive layer should be very thin and stiff in comparison with the damping treatment layer in order to minimize shear strains in the adhesive layer which would alter the behavior of

(A)

(B)

Fig. 37.26. Graphs of (A) $\sqrt{(EI)_D/E_1I_1}$ vs. h_2/h_1, and (B) η/η_2 vs. h_2/h_1, for a free-layer treatment.

the damping treatment.[15,39] The effect of a stiff thin adhesive layer is minimal, but a thick stiff layer alters the treatment behavior significantly.

CONSTRAINED-LAYER DAMPING TREATMENTS

The mechanism of energy dissipation in a constrained-layer damping treatment is quite different from the free-layer treatment, since the constraining layer helps induce relatively large shear deformations in the viscoelastic layer during each cycle of flexural deformation of the base structure, as illustrated in Fig. 37.1.[15,19] The presence of the constraining viscoelastic layer-pair changes the apparent flexural rigidity of the base structure in a manner which depends on the dimensions of the three layers involved and the elastic moduli of the three layers, as for the free-layer treatment, but also in a manner which depends on the deformation pattern of the system, in contrast to the free-layer treatment. A useful set of equations which may be used to predict the flexural rigidity and modal damping of a beam or plate damped by a full-coverage constrained-layer treatment are given in Ref. 15. These equations give the direct (in-phase) component $(EI)_D$ of the flexural rigidity of the three-layer beam, and the quadrature (out-of-phase) component $(EI)_Q$ as a function of the various physical parameters of the system, including the thicknesses h_1, h_2, and h_3, the moduli $E_1 (1 + jn_1)$, $E_2 (1 + jn_2)$, $E_3 (1 + jn_3)$ and the shear modulus of the damping layer $G_2 (1 + jn_2)$.

Shear Parameter. The behavior of the damped system depends most strongly on the shear parameter

$$\mathbf{g} = \frac{G_2(\lambda/2)^2}{E_3 h_3 h_2 \pi^2} \tag{37.26}$$

which combines the effect of the damping layer modulus with the semiwavelength of the mode of vibration, the modulus of the constraining layer, and the thicknesses of the damping and constraining layers. The other two parameters are the thickness ratios h_2/h_1 and h_3/h_1. Figure 37.27 illustrates the typical variation of η_n/η_2 and $(EI)_D/E_1 I_1$ with

FIG. 37.27. Typical plots of (A) η/η_2, and (B) $(EI)_D/E_1 I_1$ versus shear parameter $g(h_2/h_1 = 0.10,\ \eta_2 = 0.1)$.

the shear parameter **g** for particular values of h_2/h_1 and h_3/h_1. These plots may be used for design of constrained layer treatments. Note that η_n will be small for both large and small values of **g**. For **g** approaching zero, G_2 or $\lambda/2$ may be very small or E_3, h_3, and h_2 may be very large. This could mean that while G_2 might appear at first sight to be sufficiently large, the dimensions h_2 and h_3 are nevertheless too large to achieve the needed value of **g**. This could happen for very large structures, especially for high-order modes. On the other hand, for **g** approaching infinity, G_2 or $\lambda/2$ may be large, or E_3, h_2, or h_3 may be very small.

Effects of Treatment Thickness. In general, increasing h_2 and h_3 will lead to increased damping of a beam or plate with a constrained-layer treatment, but the effect of the shear parameter will modify the specific values. It is seen from Fig. 37.27 that the influence of h_3/h_1 is stronger than that of h_2/h_1, and that as h_2/h_1 approaches zero, η_n/η_2 does not approach zero but a finite value. This behavior seems to occur in practice and accounts for the very thin damping layers, 0.002 in. (0.051 mm) or less, used in damping tapes. A practical limit of 0.001 in. (0.025 mm) is usually adopted to avoid handling problems.

Effect of Initial Damping. If the base beam is itself damped, with η_1 not equal to zero, then the damping from the constrained-layer treatment will be added to η_1 for small values of **g**. The general effect is readily visualized, but specific behavior depends on treatment dimensions and the value of the shear parameter.

Equivalent Free Layer. While the equations of Ref. 15 may be used to predict the effect of single or multiple layer-pair constrained damping treatments for a pinned-pinned beam, or for other classical boundary conditions using an equivalent semiwavelength, they cannot be used for stiffened plates or for partial coverage. For these cases, one may take advantage of the fact that a multiple layer-pair constrained damping treatment of the type illustrated in Fig. 37.1 may be analyzed, from the point of view of the structure to which it is applied, as if it were an equivalent free-layer treatment having equivalent Young's modulus E_e and equivalent loss factor η_e. The equivalent free-layer properties can be calculated by first calculating $(EI)_D/E_1 I_1$ and η_n/η_2 by the equations of Ref. 15 in terms of the shear parameter **g**, and then using the equations for a free-layer treatment in reverse to calculate E_e and η_e. These values depend on the shear parameter **g** and the thickness ratio h_3/h_2 but do not depend strongly on the number of layer pairs N. Predicted values of η_e/η_2 and E_e/E_3 may be plotted against **g**/N on the basis of plots of η_n/η_2 and f_n versus **g**. Such results are summarized in Fig. 37.28.[40,41] Note in particular that

FiG. 37.28. Effective modulus and loss factor versus $\mathbf{g}_n = \mathbf{g}/N$ for multiple constrained-layer treatment ($h_3/h_2 = 1.0$).

the results are plotted against g/N rather than g, and that this manner of plotting collapses the predicted data onto a single set of curves. The data fit is a good approximation for engineering purposes, especially for calculations that could not otherwise be performed except by finite element methods. A matter of some importance in regard to this calculation is the need to set η_3 equal to zero at each cycle of the calculation for values of N equal to 2, 3, 4, etc. Failure to do this means that the constraining layer has damping which is added to that resulting from the current shear layer and that η_e/η_2 rapidly rises above 1.0, which is physically impossible. This is an additional limitation of the constrained-layer treatment equations for calculation of multiple-layer treatments.

DISCRETE DAMPING TREATMENTS

In its simplest form, a tuned damping device is a single degree-of-freedom mass-spring-damper system attached to a vibrating structure at or near an antinodal point and tuned so that its individual natural frequency is close to that of the mode of the system for which damping is required. The mass, stiffness, and loss factor of the tuned damper are, at least in principle, controllable variables which may be chosen to achieve the best damper design for each given application.[15]

REFERENCES

1. Harris, C. M., and C. E. Crede: "Shock and Vibration Handbook," 1st ed., McGraw-Hill Book Company, Inc., New York, 1961.
2. Harris, C. M., and C. E. Crede: "Shock and Vibration Handbook," 2d ed., McGraw-Hill Book Company, Inc., New York, 1976.
3. Hamme, R. N.: Ref. 1, Chap. 37.
4. Plunkett, R.: Ref. 2, Chap. 37.
5. Crede, C. E.: Ref. 1, Chap. 32.
6. Richards, E. J., and D. J. Mead (eds.): "Noise and Acoustic Fatigue in Aeronautics," John Wiley & Sons, Inc., New York, 1968.
7. American Society for Testing and Materials: "Standard for Measuring Vibration Damping Properties of Materials," E 756-80, 1980.
8. "DuPont Dynamic Mechanical Analyzer (DMA)," DuPont Instruments, Wilmington, Del., 1986.
9. Rheometrics Dynamic Spectrometer (Model RDS-LA)," Rheometrics, Inc, Union, N.J., 1986.
10. "Viscoelasticimetre," Metravib, Ecully, France (in U.S., Ceeris International Inc., Washington, D.C.), 1986.
11. Pillot, C. G., J. F. Gerard, and B. Duperray: "A Critical Review of the Different Methods of Measurement of Dynamic Moduli and Loss Factor of Viscoelastic Materials," Paper presented at ASA 99th Annual Meeting, Atlanta, April 1980.
12. Jones, D. I. G.: *J. Aircraft*, **8**(8):644 (1981).
13. Rogers, L. C.: *Shock and Vibration Bull.*, **51**(1):55 (1981).
14. Bagley, R. L., and P. J. Torvik: *Shock and Vibration Bull.*, **49** (1979).
15. Nashif, A. D., et al.: "Vibration Damping," John Wiley & Sons, Inc., New York, 1985.
16. Ruddell, M., et al.: AFWAL-TR-82-4167, Air Force Wright Aeronautical Laboratories, 1982.
17. Hanson, W. J., and G. A. Hampel: *Sound and Vibration*, **18**:22, July 1984.
18. Oberst, H.: *Acustica*, **4**:181 (1952).
19. Ross, D., E. E. Ungar, and E. M. Kerwin, Jr.: "Damping of Plate Flexural Vibrations by Means of Viscoelastic Laminae," *Proc. Colloquim on Structural Damping*, ASME, New York, 1959, p. 49.
20. Lazan, B. J.: "Damping of Materials and Members in Structural Mechanics," Pergamon Press, New York, 1968.
21. Ungar, E. E.: *Shock, Vibration and Associated Environment Bulletin*, **33**(4):89 (1964).
22. Plunkett, R.: "Friction Damping," *ASME Publication* AMD **38**:65 (1981).
23. Griffin, J. H.: *J. Eng. Power*, **102**(2):329 (1980).
24. Srinivasan, A. V., and D. G. Cutts: *ASME Paper* 82-GT-162, 1982.
25. Jones, D. I. G., and A. Muszynska: *Shock and Vibration Bull.* **69**(2):87 (1979).
26. Baxa, D. E. (ed.): "Noise Control in Internal Combustion Engines," John Wiley & Sons, Inc., New York, 1982.
27. Kwakernaak, H., and R. Sivan: "Linear Optimal Control Systems," John Wiley & Sons, Inc., Interscience, New York, 1972.
28. Rudder, F. F., Jr., and H. E. Plumblee, Jr., *Flight Dynamics Lab Report* AFFDL-TR-74-112, WPAFB, Ohio, May 1975.

29. Clarkson, B. L.: *J. Roy. Aeronaut. Soc.*, **662:**603 (1962).
30. Jones, D. I. G., and W. J. Trapp: *J. Sound Vibration*, **17**(2):157 (1971).
31. Patel, B. M., G. E. Warnaka, and D. J. Mead: *Shock and Vibration Bull.*, **48:**39 (1978).
32. Miller, H. T.: *Shock and Vibration Bull.*, **38**(2):29, (1968).
33. Plunkett, R., and C. T. Lee: *J. Acoust. Soc. Am.*, **48:**150 (1970).
34. Snowdon, J. C.: "Vibration and Shock in Damped Mechanical Systems," John Wiley & Sons, Inc., New York, 1968.
35. Jones, D. I. G.: *Shock and Vibration Bull.*, **36**(4):9 (1967).
36. DeMichele, D. J. (ed.): *Proc. Intern. Model Analysis Confs. I to III*, Union College, NY, 1982–85.
37. Wilbur, J. B., and C. H. Norris: "Structural Model Analysis," in M. Hetenyi (ed.): "Handbook of Experimental Stress Analysis," John Wiley & Sons, Inc., New York, 1950, Chap. 15.
38. Cannon, C. M., A. D. Nashif, and D. I. G. Jones: *Shock and Vibration Bull.*, **38**(3):151 (1968).
39. Warnaka, G. E.: *ASME Paper* 67-VIBR-41, 1967.
40. Jones, D. I. G., A. D. Nashif, and M. L. Paris: *J. Sound Vibration*, **29**(4):423 (1973).
41. Knighton, D. L.: "The Design of Multiple Constrained Layer Damping Treatments," Paper presented at ASA Meeting, Cambridge, Mass., 1979.

38

TORSIONAL VIBRATION IN RECIPROCATING AND ROTATING MACHINES

Ronald L. Eshleman
Vibration Institute

Frank M. Lewis
Massachusetts Institute of Technology

INTRODUCTION

Torsional vibration is an oscillatory angular motion causing twisting in the shaft of a system; the oscillatory motion is superimposed on the steady rotational motion of a rotating/reciprocating machine. Even though the vibration cannot be detected without special measuring equipment, its amplitude can be destructive. For example, gear sets that alter speeds of power transmission systems transmit the vibration to the casing. Similarly, slider crank mechanisms in engines and compressors convert torques to radial forces that are discernable to human perception but are not measurable because of the insensitivity of test equipment and background noise. If gearboxes or reciprocating machines are part of a drive train, excess noise and vibration can indicate trouble. Standards and measurement methods dealing with acceptable magnitudes of radial vibration are provided in Chap. 19.

Motion is rarely a concern with torsional vibration unless it affects the function of a system. It is stresses that affect the structural integrity and life of components and thus determine the allowable magnitude of the torsional vibration. Torsional vibratory motions can produce stress reversals that cause metal fatigue. Components tolerate less reversed stress than steady stress. In addition, stress concentration factors associated with machine members decrease the effectiveness of load-bearing materials.

Figure 38.1 illustrates the twisting of a shaft of an electric motor-compressor system. The torsional mode shape associated with the first torsional natural frequency is shown in Fig. 38.2. A coupling in the power train allows for misalignment in the assembly. The mode shape shows that the stiffness of the coupling is much less than that of other shaft sections. This is indicated by the large slope (change in angular displacement) of the mode shape at the coupling. The coupling will be the predominant component in the motor-compressor system governing the torsional natural frequency associated with the mode.

Torsional vibration is usually a complex vibration having many different frequency components. For example, shock resulting from abrupt start-ups and unloading of gear teeth causes transient torsional vibration in some systems; start-up of synchronous electric motor systems may cause torsional resonance. Random torsional vibration caused by gear inaccuracies and ball bearing defects is relatively common in rotating machines.

CRITICAL SPEEDS AND FORCED RESPONSE

The crankshaft of a reciprocating engine or the rotors of a turbine or motor, and all moving parts driven by them, comprise a torsional elastic system. Such a system has several modes of free torsional oscillation. Each mode is characterized by a natural frequency and by a pattern of relative amplitudes of parts of the system when it is oscillating at its natural frequency. The harmonic components of the driving torque excite vibration of the system in its modes. If the frequency of any harmonic component of the torque is equal to (or close to) the frequency of any mode of vibration, a condition of resonance exists and the machine is said to be running at a *critical speed*. Operation of the system at such critical speeds can be very dangerous, resulting in fracture of the

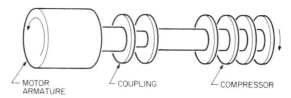

MOTOR
ARMATURE   COUPLING   COMPRESSOR

FIG. 38.1. Schematic drawing illustrating the twisting of the shaft of a motor-compressor system.

FIG. 38.2. Torsional mode shape for the motor-compressor system shown in Fig. 38.1.

shafting. Operation at other critical speeds can result in rapid wear of bearings, gears, and other components and in undesirable vibration of the machine.

The number of complete oscillations of the elastic system per unit revolution of the shaft is called the *order of a critical speed*. An order of a critical speed that corresponds to an harmonic component of the torque from the engine as a whole is called a *major order*. A critical speed also can be excited from that corresponds to the harmonic component of the torque curve of a single cylinder. The fundamental period of the torque from a single cylinder in a four-cycle engine is 720°; the critical speeds in such an engine can be of ½, 1, 1½, 2, 2½, etc., order. In a two-cycle engine only the critical speeds of 1, 2, 3, etc., order can exist. All critical speeds except those of the major orders are called *minor critical speeds;* this term does not necessarily mean that they are unimportant.

A dynamic analysis of an engine involves several steps. Natural frequencies of the modes likely to be important must be calculated. The calculation is usually limited to the lowest mode or the two lowest modes. In complicated arrangements, the calculation of additional modes may be required, depending on the frequency of the forces causing the vibration. Vibration amplitudes and stresses around the operating range and at the critical speeds must be calculated. A study of remedial measures is also necessary.

Example 38.1. A six-cylinder four-cycle engine with 120° crank spacings has three equally spaced firing impulses per revolution. What are the critical speeds?

The major orders of 3, 6, 9, 12, etc., are obtained from a phase diagram that considers the engine force balance at frequencies of 3, 6, 9, 12, etc., times rotational speed. The critical speeds occur at

$$\frac{60 f_n}{q} \quad \text{rpm} \tag{38.1}$$

where f_n is the natural frequency of one of the modes in Hz, and q is the order number of the critical speed. Although many critical speeds exist in the operating range of an engine, only a few are likely to be important.

MODELING

The torsional elastic system of a drive unit and its associated machinery is a complicated arrangement of mass and elastic distribution. The complete mechanical system can include the drive unit, couplings, gearboxes or other speed-changing devices, and one or more driven units. This complicated system is made amenable to mathematical treatment by representing it as a model—a simpler system that is substantially equivalent dynamically. The equivalent system usually consists of lumped masses which are connected by massless torsionally elastic springs as illustrated in Fig. 38.3. The masses are placed at each crank center and at the center planes of actual flywheels, rotors, propellers, cranks, gears, impellers, and armatures.[1,2,3,4,5]

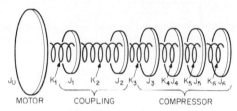

FIG. 38.3. A model of the motor-compressor system shown in Fig. 38.1, consisting of a series of masses connected by massless torsionally elastic springs (K = stiffness, lb·in./rad; J = polar moment of inertia, lb·in.·sec²).

The torsional calculation is made not for the drive unit alone but for the complete system, including all driven machinery. On an engine it is usually possible to consider such parts as camshafts, pumps, and blowers either as detached from the engine (if they are driven elastically) or as additional rigid masses at the point of attachment to the crankshaft (if the driver is relatively rigid). If there is doubt, these parts should be included in the torsional calculation as elastically connected masses and removed if the natural frequencies do not change after the parts are removed from the model.

CALCULATION OF POLAR MOMENTS OF INERTIA

Circular Disc or Cylinder Rotating about a Perpendicular Axis. The polar moment of inertia, essential in modeling torsional vibration, often is easy to calculate. The general form is $J = \int r^2 \, dm$, where r is the instantaneous radius, and dm is the differential mass. The formula for the polar moment of inertia of a circular disc or cylinder rotating about a perpendicular axis is

$$J = \frac{\pi d^4 l \gamma}{32g} \quad \text{lb-in.-sec}^2 \tag{38.2}$$

where J = polar moment of inertia, lb-in.-sec²
γ = material density, lb/in.³
d = diameter of disc or cylinder, in.
l = axial length of disc or cylinder, in.
g = acceleration due to gravity, 386.4 in./sec²

Table 38.1 contains formulas for some elements commonly used in torsional models. Also see Ref. 6.

Piston and Connecting Rod. The piston and connecting rod shown schematically in Fig. 38.4 introduce a variable-mass problem, the solution of which is complex. The exact

Table 38.1 Formulas for Torsional Stiffness

K = TORSIONAL STIFFNESS, LB - IN./RAD; G= SHEAR MODULUS, LB/IN²
ω = ROTATIONAL SPEED, RAD/SEC

SPRINGS IN SERIES	K_1 K_2	$K = \dfrac{1}{1/K_1 + 1/K_2}$
SPRINGS IN PARALLEL	K_1 K_2	$K = K_1 + K_2$
HOLLOW CIRCULAR SHAFT	d D ℓ	$K = \dfrac{\pi}{32}\dfrac{G(D^4 - d^4)}{\ell}$
TAPERED CIRCULAR SHAFT	d D ℓ	$K = \dfrac{3\pi}{32}\dfrac{d^4}{\ell(n + n^2 + n^3)}$ $n = \dfrac{d}{D}$
TWO GEARED SHAFTS (REFERRED TO SHAFT 1)	ω_2 K_2 D K_1 d ω_1	$K = \dfrac{K_1 K_2}{n^2 K_1 + K_2}$ $n = \dfrac{D}{d} = \dfrac{\omega_1}{\omega_2}$

FIG. 38.4. Schematic diagram of a crank and connecting rod.

solution[4] shows that the effect of the piston and connecting rod can be closely approximated by representing them as a concentrated rotor of polar inertia J defined by

$$J = \left[\frac{W_p}{2} + W_c\left(1 - \frac{h}{2}\right)\right]\frac{R^2}{g} \qquad \text{lb-in.-sec}^2 \qquad (38.3)$$

where W_p = weight of piston, piston pin, and cooling fluid, lb
W_c = weight of connecting rod, lb
h = fraction of rod length from crank pin to center-of-gravity
R = crank radius, in.

The polar inertias of the crank webs (see Fig. 38.5), the crankpin, and the journal sections are added to that given by Eq. (38.3). These polar inertias should be calculated with the best obtainable accuracy. The following procedure is recommended:

Let the crank web be intercepted by a series of concentric cylinders of radius R. The polar inertia of the crank web is defined by

$$J = \left(\frac{\gamma}{g} \int_{Rb}^{Rm} R^2 S \, dR \right) + J_b \qquad (38.4)$$

where γ = weight density of the crank web, lb/in.^{3}
R_m = maximum radius of the crank web, in.
R_b = radius of base cylinder (see Fig. 38.5), in.
J_b = polar inertia of portion of crank web within base cylinder, lb-in.-sec^2

The integral in the above expression for J is the area of the R^2S curve between the values of radii R_b and R_m.

FIG. 38.5. View of a crank web in a plane normal to the crankshaft axis.

For the crank web shown in Fig. 38.5 the area S is defined as $S = bR\theta/57.3$, where θ is measured in degrees. The polar inertia can then be expressed as

$$J = \left(\frac{\gamma}{57.3g} \int_{Rb}^{Rm} bR^3\theta \, dR \right) + J_b \qquad \text{lb-in.-sec}^2$$

The same procedure is used to calculate the polar inertia of propellers and other irregular parts. In a marine propeller of ogival sections, i.e., flat driving face, circular arc back, and elliptically developed outline (do not use for other shapes), the polar inertia (excluding hub) is given by

$$J = 0.0046 \frac{nD^3bt}{g} \qquad \text{lb-in.-sec}^2$$

where n = number of blades
D = diameter of propeller, in.
b = maximum blade width, in.
t = maximum blade thickness at one-half radius (axis to tip), in.

For propellers, pumps, and hydraulic couplings an addition must be made for the virtual inertia of the entrained fluid. For marine propellers this is ordinarily assumed at 26 per cent of the propeller inertia. Virtual inertias for pumps are not known accurately, but it can be assumed that half the casing is filled with rotating fluid.

EXPERIMENTAL DETERMINATION OF POLAR MOMENT OF INERTIA. For complex shaft elements such as couplings or small flywheels, it is often easier to determine the polar moment of inertia experimentally than to calculate it. In one experimental technique the element is suspended from three equally spaced vertical wires as shown in Fig. 38.6. The element whose polar moment of inertia is to be measured is hung on the cables and set into torsional motion. Then the period of vibration is measured. The experimentally determined period of torsional vibration, the weight of the element, the length of the suspending cables, and the radius of attachment of the cables are used to determine the polar moment of inertia from the following formula.

Fig. 38.6. Experimental determination of the polar moment of inertia. An element of weight W is suspended by three wires and the period of the torsional motion is determined.

$$J = \frac{Wr^2\tau^2}{(6.28)^2 l} \quad \text{lb-in.sec}^2 \tag{38.5}$$

where J = polar moment of inertia
τ = period of vibration, sec/cycle
W = weight of element, lb
l = length of cables, in.
r = radius of suspending cables, in.

CALCULATION OF SHAFT STIFFNESS. The stiffness of a circular shaft is the most common elastic element encountered in the modeling process. Table 38.1 shows some common formulas used to calculate torsional stiffness of a hollow circular shaft, a tapered circular shaft, and two geared shafts. The stiffness is referred to the rotational speed of shaft No. 1. The inertia of geared shafts is obtained in a similar manner.

CALCULATION OF CRANKSHAFT STIFFNESS. The crankshaft stiffness is the most uncertain element in a torsional vibration calculation. Shaft stiffness can be measured experimentally either by twisting a shaft with a known torque or from the observed values of the critical speeds in a running engine. Alternatively, it can be calculated from semiempirical formulas such as those given in Ref. 6. Those given by Eqs. (38.6), (38.7), and (38.8) are recommended. Refer to Figs. 38.4 and 38.5 for definitions of the dimensions; l_e is the length of a solid shaft of diameter D_s equal in torsional stiffness to the section of crankshaft between crank centers.

Wilson's formula.[5]

$$\frac{l_e}{D_s^{\,4}} = \frac{b + 0.4d_s}{D_s^{\,4} - d_s^{\,4}} + \frac{a + 0.4D_c}{D_c^{\,4} - d_c^{\,4}} + \frac{r - 0.2(D_s + D_c)}{hW^3} \tag{38.6}$$

Ziamanenko's formula.[6]

$$\frac{l_e}{D_s^{\,4}} = \frac{b + 0.6hD_s/b}{D_s^{\,4} - d_s^{\,4}} + \frac{0.8a + 0.2(W/r)D_s}{D_c^{\,4} - d_c^{\,4}} + \frac{r^{3/2}}{hW^3 D_c^{\,1/2}} \tag{38.7}$$

Constant's formula.[7]

$$\frac{l_e}{D_s^{\,4}} = \frac{1}{\alpha_1\alpha_2\alpha_3\alpha_4}\left(\frac{b}{D_s^{\,4} - d_s^{\,4}} + \frac{a}{D_c^{\,4} - d_c^{\,4}} + \frac{0.94}{hW^3}\right) \tag{38.8}$$

where α_1, α_2, α_3, α_4 are modifying factors, determined as follows:

$$\alpha_1 = 1 - \frac{0.0825}{\sqrt{\dfrac{W_s - d_s}{2W_s} + \dfrac{W_c - d_c}{2W_c} - 0.32}} \tag{38.9}$$

If the shaft is solid, assume $\alpha_1 = 0.9$. The factor α_2 is a web-thickness modification determined as follows: If $4h/l$ is greater than $\frac{2}{3}$, then $\alpha_2 = 1.666 - 4h/l$. If $4h/l < \frac{2}{3}$, assume $\alpha_2 = 1$. The factor α_3 is a modification for web chamfering determined as follows: If the webs are chamfered, estimate α_3 by comparison with the cuts on Fig. 38.7: Cut AB and $A'B'$, $\alpha_3 = 1.000$; cut CD alone, $\alpha_3 = 0.965$; cut CD and $C'D'$, $\alpha_3 = 0.930$; cut EF alone, $\alpha_3 = 0.950$; cut EF and $E'F'$, $\alpha_3 = 0.900$; if ends are square, $\alpha_3 = 1.010$. The factor α_4 is a modification for bearing support given by

$$\alpha_4 = \frac{Al^3 w}{D_c{}^4 - d_c{}^4} + B \tag{38.10}$$

For marine engine and large stationary engine shafts: $A = 0.0029$, $B = 0.91$
For auto and aircraft engine shafts: $A = 0.0100$, $B = 0.84$
If α_4 as given by Eq. (38.10) is less than 1.0, assume a value of 1.0.

The Constant's formula, Eq. (38.8), is recommended for shafts with large bores and heavy chamfers.

FIG. 38.7. Schematic diagram of one crank of a crankshaft.

The shafting of an engine system may contain elements such as changes of section, collars, shrunk and keyed armatures, etc., which require the exercise of judgment in the assessment of stiffness. For a change of section having a fillet radius equal to 10 per cent of the smaller diameter, the stiffness can be estimated by assuming that the smaller shaft is lengthened and the larger shaft is shortened by a length λ obtained from the curve of Fig. 38.8. This also may be applied to flanges where D is the bolt diameter. The stiffening effect of collars can be ignored.

The stiffness of shrunk and keyed parts is difficult to estimate as the stiffening effect depends to a large extent on the tightness of the shrunk fit and keying. The most reliable values of stiffness are obtained by neglecting the stiffening effect of an armature and assuming that the armature acts as a concentrated mass at the center of the shrunk or keyed fit. Some armature spiders and flywheels have considerable flexibility in their arms; the treatment of these is discussed in the next section *Geared and Branched Systems*. Properties of numerous types of torsionally elastic couplings are available.[6]

GEARED AND BRANCHED SYSTEMS. The natural frequencies of a system containing gears can be calculated by assuming a system in which the speed of the driver unit is n times the speed of the driven equipment. Multiply all the inertia and elastic constants on the driven side of the system by $1/n^2$, and calculate the system's natural frequencies as if no gears exist. In any calculations involving damping constants on the driven side, these constants also are multiplied by $1/n^2$. Torques and deflections thus obtained on the driven side of this substitute system, when multiplied by n and $1/n$, respectively, are equal to those in the actual geared system. Alternatively, the driven side can be used as the reference; multiply the driver constants by n^2.

Where two or more drivers are geared to a common load, hydraulic or electrical couplings may be placed between the driver and the gears. These serve as disconnected clutches; they also insulate the gears from any driver-produced vibration. This insulation is so perfect that the driver end of the system can be calculated as if terminating at the coupling gap. The damping effect of such couplings upon the vibration in the driver end of the system normally is quite small and should be disregarded in amplitude calculations.

The majority of applications without hydraulic or electrical couplings involve two identical drivers. For such systems the modes of vibration are of two types:

1. The *opposite-phase* modes in which the drivers vibrate against each other with a node at the gear. These are calculated for a single branch in the usual manner, terminating the calculation at the gear. The condition for a natural frequency is that $\beta = 0$ at the gear.

FIG. 38.8. Curve showing the decrease in stiffness resulting from a change in shaft diameter. The stiffness of the shaft combination is the same as if the shaft having diameter D_1 is lengthened by λ and the shaft having diameter D_2 is shortened by λ. (*F. Porter.*[8])

2. The *like-phase* modes in which the two drivers vibrate in the same direction against the driven machinery. To calculate these frequencies, the inertia and stiffness constants of the driver side of one branch are doubled; then the calculation is made as if there were only a single driver. The condition for a natural frequency is zero residual torque at the end.

If the two identical drivers rotating in the same direction are so phased that the same cranks are vertical simultaneously, all orders of the opposite-phase modes will be eliminated. The two drivers can be so phased as to eliminate certain of the like-phase modes. For example, if the No. 1 cranks in the two branches are placed at an angle of 45° with respect to each other, the fourth, twelfth, twentieth, etc., orders, but no others, will be eliminated. If the drivers are connected with clutches, these phasing possibilities cannot be utilized.

In the general case of nonidentical branches the calculation is made as follows: Reduce the system to a 1:1 gear ratio. Call the branches a and b. Make the sequence calculation for a branch, with initial amplitude $\beta = 1$, and for the b branch, with the initial amplitude the algebraic unknown x. At the junction equate the amplitudes and find x. With this numerical value of the amplitude x substituted, the torques in the two branches and the torque of the gear are added; then the sequence calculation is continued through the last mass.

The branch may consist of a single member elastically connected to the system. Examples of such a branch are a flywheel with appreciable flexibility in its spokes or an armature with flexibility in the spider. Let I be the moment of inertia of the flywheel rim and k the elastic constant of the connection. Then the flexibly mounted flywheel is equivalent to a rigid flywheel of moment of inertia

$$I' = \frac{I}{1 - I\omega^2/k} \qquad (38.11)$$

NATURAL FREQUENCY CALCULATIONS

If the model of a system can be reduced to two lumped masses at opposite ends of a massless shaft, the natural frequency is given by

$$f_n = \frac{1}{2\pi} \sqrt{\frac{(J_1 + J_2)k}{J_1 J_2}} \qquad \text{Hz} \qquad (38.12)$$

The mode shape is given by $\theta_2/\theta_1 = -J_1/J_2$.

FIG. 38.9. Schematic diagram of a shaft represented by three masses.

For the three-mass system shown in Fig. 38.9, the natural frequencies are

$$f_n = \frac{1}{2\pi} \sqrt{A \pm (A^2 - B)^{1/2}} \qquad \text{Hz} \qquad (38.13)$$

where $A = \dfrac{k_{12}(J_1 + J_2)}{2J_1 J_2} + \dfrac{k_{23}(J_1 + J_2)}{2J_1 J_2}$

$B = \dfrac{(J_1 + J_2 + J_3)k_{12}k_{23}}{J_1 J_2 J_3}$

In Eqs. (38.12) and (38.13) the ks are torsional stiffness constants expressed in lb-in./rad. The notation k_{12} indicates that the constant applies to the shaft between rotors 1 and 2. The polar inertia J has units of lb-in.-sec^2.

The above formulas and all the developments for multimass torsional systems that follow also apply to systems with longitudinal motion if the polar moments of inertia J are replaced by the masses $m = W/g$ and the torsional stiffnesses are replaced by longitudinal stiffnesses.

HOLZER METHOD. If there are more than three masses, the method* of Holzer can be used to calculate the natural frequencies. A frequency is assumed; then starting at one end of the system, a balance of torques and displacements is obtained step by step. The final external torque required to achieve balance is called the *residual torque.* If the residual torque is zero, the assumed frequency is one of the natural frequencies of the

* This method was first proposed in 1912 in graphical form[9] and later in a tabular form.[10] This sequence calculation has the advantages of simplicity, ready extension to forced vibration, and ease of programming on a digital computer. The tabulation given here is in a more consistent form.

system. The mode associated with this natural frequency is identified by the number of changes of sign of the amplitudes at the various rotors. A plot, Fig. 38.10A, of residual torque vs. frequency yields all the natural frequencies of the system.

In many internal-combustion engines, the polar inertias of the rotors representing the cranks and pistons are equal, as are the stiffness constants between these rotors. Advantage can be taken of this fact to reduce the labor of computation considerably. Let I be one of the equal polar moments of inertia and k be one of the equal spring constants. Divide all the polar moments of inertia by I, and divide all the spring constants by k. The new system is dynamically similar to the original one. It is called the *unity system*.

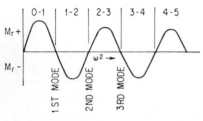

FIG. 38.10A. Typical curve of residual torque M_r as a function of the assumed frequency, ω. The numbers within the brackets at the top of the illustration represent the number of changes in the sign of β that are set forth opposite the rotor number in the sequence calculation.

Then if ω_n is the natural angular frequency of the actual system and ω_s that of the unity system,

$$\omega_n{}^2 = \omega_s{}^2 \frac{k}{I} \quad \text{rad/sec} \quad (38.14)$$

This "unity form" of the calculation is a labor-saving technique which is useful only when there are a number of equal masses and/or stiffnesses in the system; whether calculations are made using the unity form or the original constants is a matter of choice.

Before starting the sequence calculation it is necessary to know approximately the frequencies of the desired modes. Usually not more than the two lowest are required. For

FIG. 38.10B. Moments of inertia I and stiffness k for an eight-cylinder engine driving a generator. The constants for the actual system are shown in the upper scale and for the unity system in the lower scale.

an internal-combustion engine the following procedure is generally satisfactory. Add all the polar inertias within the engine part of the system and assume that 40 per cent of this is placed at the crank farthest from the flywheel or generator (No. 1 crank). Find the stiffness with reference to the flywheel, or the generator if no flywheel is fitted, to the No. 1 crank. If there is no flywheel, the system reduces to a two-rotor system, and Eq. (38.12) gives approximately the lowest natural frequency of the system. If there is a flywheel, the system reduces to one of three rotors; then Eq. (38.13) gives the approximate values of the two lowest natural frequencies of the system. For complicated systems involving numerous driven rotors no general rules can be given.

Example 38.2. Figure 38.10B (upper scale) shows the actual mass and elasticity distribution of an eight-cylinder, four-cycle engine. Four of the cranks are fitted with counterweights so that the crank inertias are not the same. Dividing the moments of inertia by 35.8 and the stiffness constants by 175.7×10^6, the constants for the unity system are obtained, as shown in Fig. 38.10B (lower scale). The sum of the inertia constants through rotor 9 is 7.152. The stiffness of the shaft to rotor 10 is

$$\frac{1}{7 + (1/1.32) + (1/6.06)} = 0.126$$

The two-rotor system is therefore represented by a two degree-of-freedom model. The approximate value of ω_s^2 is calculated from Eq. (38.12), where $I_1 = 0.40 \times 7.152 = 2.861$ and $I_2 = 42.3$:

$$\omega_s^2 = \frac{0.126 \times 45.16}{2.861 \times 42.3} = 0.047 \quad (\text{rad/sec})^2$$

The Holzer calculation is started with $\omega_s^2 = 0.047$. The calculation is arranged as shown in Table 38.2. The rotor numbers and their inertia constants I are placed on alternate lines. The β values opposite the rotor numbers define the first mode relative amplitude curve, and the torques between rotor numbers are the shaft torques between the corresponding rotors. The intermediate lines are designated line 1-2, line 2-3, etc. The

Table 38.2. Holzer Calculations for First Mode of System Shown in Fig. 38.10B
$$(\omega_s^2 = 0.047)$$

Rotor number	I	$I\omega_s^2$	Deflection β	Torque M, in.-lb	k
1	1.000	0.0470	1.0000	0.0470	
			0.0470	0.0470	1.00
2	0.722	0.0339	0.9530	0.0323	
			0.0793	0.0793	1.00
3	0.722	0.0339	0.8737	0.0296	
			0.1089	0.1089	1.00
4	1.000	0.0470	0.7648	0.0360	
			0.1449	0.1449	1.00
5	1.000	0.0470	0.6199	0.0291	
			0.1740	0.1740	1.00
6	0.722	0.0339	0.4459	0.0151	
			0.1891	0.1891	1.00
7	0.722	0.0339	0.2568	0.0076	
			0.1967	0.1967	1.00
8	1.000	0.0470	0.0601	0.0028	
			0.1511	0.1995	1.32
9	0.265	0.0125	−0.0910	−0.0011	
			0.0327	0.1984	6.06
10	42.500	1.9975	−0.1237	−0.2471	
			Residual torque $M_r =$	−0.0487	

stiffness constants are placed on the intermediate lines in the last column. For the trial $\omega_s^2 = 0.047$, the $I\omega_s^2$ column is computed next. An amplitude of 1.0 is taken for rotor 1. Then the inertia reaction of rotor 1 is $0.047 \times 1.0 = 0.047$; it is placed in the M column on line 1. The torque in the shaft section between rotors 1 and 2 is equal to this inertia reaction and is placed in the M column in line 1-2. This torque divided by the stiffness, 1.00 in shaft section 1-2, is the relative deflection of rotors 1-2 and is placed in the β column of line 1-2. Subtracting this relative deflection 0.0470 from the deflection of rotor 1 gives the deflection 0.9530 of rotor 2. This deflection multiplied by 0.0339 ($I\omega_s^2$ for rotor 2) gives the inertia reaction 0.0323 of rotor 2. The inertia reaction of rotor 2 added to the torque in section 1-2 gives the torque in section 2-3. Thus, Table 38.2 is carried through step by step. Note that in lines 8-9 and 9-10, k is no longer unity. The final figure in the torque column is called the *residual torque*. All additions and subtractions are made in the algebraic sense, i.e., on line 9 the deflection $-0.0910 = 0.0601 - 0.1511$.

Table 38.3. Holzer Calculations for First Mode of System Shown in Fig. 38.10B
$$(\omega_s^2 = 0.0457)$$

Rotor number	I	$I\omega_s^2$	Deflection β	Torque M, in.-lb	k
1	1.000	0.0457	1.0000	0.0457	
			0.0457	0.0457	1.00
2	0.722	0.0330	0.9543	0.0315	
			0.0772	0.0772	1.00
3	0.722	0.0330	0.8771	0.0290	
			0.1062	0.1062	1.00
4	1.000	0.0457	0.7709	0.0351	
			0.1413	0.1413	1.00
5	1.000	0.0457	0.6296	0.0288	
			0.1701	0.1701	1.00
6	0.722	0.0330	0.4595	0.0152	
			0.1853	0.1853	1.00
7	0.722	0.0330	0.2742	0.0090	
			0.1943	0.1943	1.00
8	1.000	0.0457	0.0799	0.0036	
			0.1499	0.1979	1.32
9	0.265	0.0121	-0.0700	-0.0008	
			0.0325	0.1971	6.06
10	42.500	1.9422	-0.1025	-0.1990	
			Residual torque $M_r =$	-0.0019	

The residual torque has the following physical significance: If a simple harmonic torque of amplitude M_r and angular frequency ω_s is applied to rotor 10 of the system, then all the rotors will oscillate with the amplitudes as given in the β column and the torques will be as given in the M column. If such a Holzer calculation is made over a sufficient range of values of ω_s to cover all the natural frequencies, M_r will follow the general shape of the curve shown in Fig. 38.10A. The number of crossings of zero amplitude, excluding the origin, is equal to the number of rotors minus one. The location of the trial values of ω_s^2 on this diagram can be determined from the sign of M_r and the number of changes of sign of the relative amplitudes β opposite the rotor numbers in the sequence table. In the above example, there is one change of sign of β and M_r is negative. Therefore the assumed value of ω_s^2 lies between the first and second modes.

To find the first mode, assume a new trial value of $\omega_s^2 = 0.045$. This leads to $M_r = +0.0253$. Interpolating for a zero M_r between $\omega_s^2 = 0.047$ and 0.045 yields 0.0457. Table 38.3 is constructed for this value and represents the first mode conditions with a frequency error of less than one one-thousandth, a much higher accuracy than is justified by the data on which it is based.

The second mode frequency corresponding to the condition in which there are two nodes in the engine shaft is approximated by

$$\omega_s^2 = \frac{22}{n^2}$$

where n is the number of cylinders. In the example of Fig. 38.10B

$$\omega_s^2 = \frac{22}{64} = 0.345$$

By the Holzer calculation, this is found to be somewhat low, and a value which will make M_r nearly zero is $\omega_s^2 = 0.360$, as indicated in Table 38.4. The natural frequencies of the actual system are given by Eq. (38.14).

$$\omega_n^2 = \frac{175.7 \times 10^6 \omega_s^2}{35.8} = 4,910,000\omega_s^2$$

Then the natural frequencies in the first and second modes are:

1st mode: $\omega_s^2 = 0.0457$; $\omega^2 = 224,000$; $f = 75.3$ Hz or 4,318 cpm

2nd mode: $\omega_s^2 = 0.3600$; $\omega^2 = 1,765,000$; $f = 212$ Hz or 12,720 cpm

The critical speeds are given by Eq. (38.1) and expressed as follows:

Order	4	4½	5	5½	6	6½	7	7½	8	12
1-noded (rpm)....	1,125	1,000	900	820	750	692	643	600	562	375
2-noded (rpm)....	...	...	...	...	...	...	...	...	1,670	1,115

Table 38.4 Sequence Calculations for Second Mode of System Shown in Fig. 38.10B
$(\omega_s^2 = 0.360)$

Rotor number	I	$I\omega_s^2$	Deflection β	Torque M, in.-lb	k
1	1	0.036	1.0000	0.36	
			0.3600	0.36	1.00
2	0.722	0.260	0.6400	0.166	
			0.5260	0.526	1.00
3	0.722	0.260	0.1140	0.0296	
			0.5556	0.5556	1.00
4	1	0.36	−0.4416	−0.1580	
			0.3976	0.3976	1.00
5	1	0.36	−0.8392	−0.3020	
			0.0956	0.0956	1.00
6	0.722	0.260	−0.9348	−0.2430	
			−0.1474	−0.1474	1.00
7	0.722	0.260	−0.7874	−0.2045	
			−0.3519	−0.3519	1.00
8	1	0.36	−0.4355	−0.1570	
			−0.3850	−0.5089	1.32
9	0.265	0.095	−0.0505	−0.0048	
			−0.0848	−0.5137	6.06
10	42.3	15.23	+0.0343	0.5230	
			Residual torque		
				$M_r = +0.0093$	

The maximum operating speed of the engine is 950 rpm. Not all of the critical speeds in the above table are necessarily important. The damping and dynamic torques determine the importance of the critical speed.

TRANSFER MATRIX METHOD. The transfer matrix method is an extended and generalized version of the Holzer method. Matrix algebra is used rather than a numerical table for the analysis of torsional vibration problems. The transfer matrix method is used to calculate the natural frequencies and critical speeds of other eigenvalue problems.

The transfer matrix and matrix iteration (Stodola) methods are numerical procedures. The fundamental difference between them lies in the assumed independent variable. In any eigenvalue problem, a unique mode shape of the system is associated with each natural frequency. The mode shape is the independent variable used in the matrix iteration method. A mode shape is assumed and improved by successive iterations until the desired accuracy is obtained; its associated natural frequency is then calculated.

A frequency is assumed in the transfer matrix method, and the mode shape of the system is calculated. If the mode shape fits the boundary conditions, the assumed frequency is a natural frequency and a critical speed is derived. Determining the correct natural frequencies amounts to a controlled trial-and-error process. Some of the essential boundary conditions (geometrical) and natural boundary conditions (force) are assumed,

FIG. 38.11. Typical torsional vibration model.

and the remaining boundary condition is plotted vs. frequency to obtain the natural frequency; the procedure is similar to the Holzer method. For example, if the torsional system shown in Fig. 38.11 were analyzed, the natural boundary conditions would be zero torque at both ends. The torque at station No. 1 is made zero, and the torsional vibration is set at unity. Then M_4 as a function of ω is plotted to find the natural frequencies. This plot is obtained by utilizing the system transfer functions or matrices. These quantities reflect the dynamic behavior of the system.

No accuracy is lost with the transfer matrix method because of coupling of mode shapes. Accuracy is lost with the matrix iteration method, however, because each frequency calculation is independent of the others. A minor disadvantage of the transfer matrix method is the large number of points that must be calculated to obtain an M_4 vs ω curve. This problem is overcome if a high-speed digital computer is used.

A typical station (No. 4) from a torsional model is shown in Fig. 38.11. This general station and the following transfer matrix equation are used in a way similar to the Holzer table to transfer the effects of a given frequency ω across the model (see the example of an engine driving a generator shown in Fig. 38.10B):

$$
\left| \begin{array}{c} \theta \\ M \end{array} \right|_n = \left[\begin{array}{cc} 1 & 1/k \\ -\omega^2 J & -(\omega^2 J/k) + 1 \end{array} \right]_n \left| \begin{array}{c} \theta \\ M \end{array} \right|_{n-1}
$$

where θ = torsional motion, rad.
M = torque, lb-in.
ω = assumed frequency, rad/sec
J = station inertia, lb-in.-sec²
k = station torsional stiffness, lb-in./rad

The stiffness and polar moment of inertia of each station are entered into the equation to determine the transfer effect of each element of the model. Thus, the calculation begins with station No. 1, which relates to the first spring and inertia in the model of Fig.

38.11. The equation gives the output torque, M_1, and output motion, θ_1, for given input values, usually 0 and 1, respectively. The equation is used on station No. 2 to obtain M_2 output and θ_2 output as a function of M_1 output and θ_1 output. This process is repeated to find the value of M and θ at the end of the model. This calculation is particularly suited for the digital computer.

FORCED VIBRATION RESPONSE

The torsional vibration amplitude of a modeled system is determined by the magnitude, points of application, and phase relations of the exciting torques produced by engine or compressor gas pressure and inertia, and by the magnitudes and points of application of the damping torques. Damping is attributable to a variety of sources, including pumping action in the engine bearings, hysteresis in the shafting and between fitted parts, and energy absorbed in the engine frame and foundation. In a few cases, notably marine propellers, damping of the propeller predominates. When an engine is fitted with a damper, the effects of damping dominate the torsional vibrations.

Techniques available for calculation of vibration amplitudes include the exact solution of differential equations, the energy balance method, the transfer matrix method, and modal analysis. The techniques are implemented on lumped parameter or finite-element models.

FIG. 38.12. Schematic diagram of a shaft with two rotors, showing positions of excitation and damping.

EXACT METHOD FOR TWO DEGREE-OF-FREEDOM SYSTEMS. The lowest mode of vibration of some systems, particularly marine installations, can be approximated with a two-mass system; an excitation is applied at one end and damping at the other.

Referring to Fig. 38.12, the torque equations for rotors I_1 and I_2 are

$$I_1 \omega^2 \theta_1 - k(\theta_1 - \theta_2) + M_e = 0$$
$$I_2 \omega^2 \theta_2 + k(\theta_1 - \theta_2) - jc\omega\theta_2 = 0 \tag{38.15}$$

The natural frequency is given by

$$\omega^2 = \frac{k(I_1 + I_2)}{I_1 I_2}$$

The shaft torque is $M_{12} = k(\theta_1 - \theta_2)$. If the above equations are solved, the amplitude of M_{12} at resonance is

$$|M_{12}| = k|\theta_1 - \theta_2| = M_e \frac{I_2}{I_1} \sqrt{1 + \frac{kI_2(I_1 + I_2)}{I_1 c^2}} \tag{38.16}$$

Since with usual damping the second term under the radical is large compared with unity, Eq. (38.16) reduces to

$$|M_{12}| \simeq \frac{M_e}{c} \frac{I_2}{I_1} \sqrt{(I_1 + I_2)\frac{I_2 k}{I_1}} \tag{38.17}$$

The torsional damping constant c of a marine propeller is a matter of some uncertainty. It is customary to use the "steady-state" value as derived in Ref. 3. This is approximated by

$$c = \frac{4M_{mean}}{\Omega} \quad \text{in.-lb/rad/sec}$$

where Ω = angular speed of shaft in radians per second. Considerations of oscillating airfoil theory indicate that this is too high and that a better value would be

$$c = \frac{2.3M_{mean}}{\Omega} \quad \text{in.-lb/rad/sec} \tag{38.18}$$

Equation (38.17) is applicable only when $I_1/I_2 > 1$. If used outside this range with other types of damping neglected, fictitiously large amplitudes will be obtained. Equation (38.17) gives the resonance amplitude, but the peak may not occur exactly at resonance. The complete amplitude curve is computed by the methods discussed in the following section.

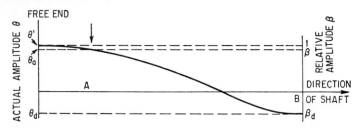

FIG. 38.13. Diagram of actual amplitude θ and relative amplitude β as a function of position along shaft. Excitaton is at A, and B is the position where damping is applied. The No. 1 cylinder is at the free end of the crankshaft.

ENERGY BALANCE METHOD. Both rational and empirical formulas for the resonance amplitudes of systems without dampers can be based on the energy balance at resonance. It is assumed that the system vibrates in a normal mode and that the displacement is in a 90° phase relationship to the exciting and damping torques. The energy input by the exciting torques is then equal to the energy output by the damping torques. Unless the damping is extremely large, this assumption gives a very close approximation to the amplitude at resonance.

Figure 38.13 shows a curve of relative amplitude in the first mode of vibration. Assume that a cylinder acts at A. Let the actual amplitude at A be θ_a and the amplitude relative to that of the No. 1 cylinder be β. The β values are taken from the column opposite each rotor number in the sequence calculation for the natural frequency. At a point such as B, where damping may be applied, let the actual amplitude be θ_d and the amplitude relative to the No. 1 cylinder be β_d.

The energy input to the system from the cylinder acting at A is

$$\pi M_e \theta_a \quad \text{in.-lb/cycle}$$

and the energy output to the damper is

$$\pi c \omega \theta_d^2 \quad \text{in.-lb/cycle}$$

where c^* is the damping constant action of the damper at B. Equating input to output,

$$M_e \theta_a = c \omega \theta_d^2 \tag{38.19a}$$

* The symbol c is used in this chapter to denote a torsional damping coefficient.

Let θ' be the amplitude at the No. 1 cylinder produced by the cylinder acting at A. Then $\theta_a/\theta' = \beta$ and $\theta_d/\theta' = \beta_d$. Substituting in Eq. (38.19a) gives

$$\theta' = \frac{M_e\beta}{c\omega\beta_d^2} \tag{38.19b}$$

If all the cylinders act, and if damping is applied at a variety of points, the total amplitude at the No. 1 cylinder is

$$\theta = \Sigma\theta' = \frac{M_e\Sigma\beta}{\omega\Sigma c\beta_d^2} \tag{38.20}$$

where $\Sigma\beta$ is taken over the cylinders and $\Sigma c\beta_d^2$ is taken over the points at which damping is applied. This formula can be applied directly when the magnitude and points of application of the damping torques are known. For the great majority of applications, where the damping is unknown, a number of empirical formulas have been proposed with coefficients based on engine tests. These formulas may give an amplitude varying 30 per cent or more from test results if applied to a variety of engines. Better agreement should not be expected, for even identical engines may have amplitudes differing as much as 2 to 1, depending on length of service, bearing fits, mounting, variation in the harmonic excitation because of different combustion rates, and other unknown factors.[11]

Good results have been obtained using the Lewis formula[2]

$$M_m = \Re M_e\Sigma\beta \tag{38.21}$$

The maximum torque at resonance in any part of the system is M_m; the exciting torque per cylinder is M_e. The vector sum over the cylinders of the relative amplitudes as taken from the sequence calculation for a natural frequency (Table 38.3 for example) is $\Sigma\beta$. It is determined as follows.

For a four-cycle engine construct a phase diagram of the firing sequence in which 720° corresponds to a complete cycle of a single cylinder, or two revolutions. The phase relationship for a critical of order number q is obtained by multiplying the angles in this diagram by $2q$, with the No. 1 crank held fixed. The β values assigned to each direction then are obtained from the values corresponding to each cylinder in the β column of the sequence calculation. Then $\Sigma\beta$ is the vector sum. The summation extends only to those rotors on which exciting torques act.

In a two-cycle engine the β phase relations are determined by multiplying the crank diagram by q, holding the No. 1 cylinder fixed.

Table 38.5 shows the $\Sigma\beta$ phase diagrams and $\Sigma\beta$ values for the one-noded mode of the example of Table 38.2, with a firing sequence 1, 6, 2, 5, 8, 3, 7, 4. The firing sequence is drawn first; then the angles of this diagram are multiplied by 2, 3, 4, etc., in succeeding diagrams. After multiplication by eight for the fourth order, the diagrams repeat. Diagrams which are equidistant in order number from the 2, 6, 10, etc., orders are mirror images of each other and have the same $\Sigma\beta$. The numerical values of $\Sigma\beta$ in Table 38.5 have been obtained by calculation, summing the vertical and horizontal components.

The empirical factor $\Re$ is determined by the measurement of amplitudes in running engines (Table 38.6).

The exciting torque per cylinder, M_e in Eq. (38.20), is composed of the sum of the torques produced by gas pressure, inertia force, gravity force, and friction force. The gravity and friction torques are of negligible importance; and the inertia torque is of importance only for first-, second-, and third-order harmonic components.

INERTIA TORQUE. A harmonic analysis of the inertia torque of a cylinder is closely approximated by[2,8]

$$M = \frac{W}{g}\,\Omega^2 r\left(\frac{\lambda}{4}\sin\theta - \frac{1}{2}\sin 2\theta - \frac{3}{4}\lambda\sin 3\theta - \frac{\lambda^2}{4}\sin 4\theta \cdots\right) \tag{38.22}$$

where $W = W_p + hW_c$ $\Bigg\}$ see Fig. 38.5 and Eq. (38.4)
$\quad\lambda = r/l$
$\quad\Omega$ = angular speed, rad/sec

It is usual to drop all terms above the third order.

GAS-PRESSURE TORQUE. A harmonic analysis of the turning effort curve yields the gas-pressure components of the exciting torque. The turning effort curve is obtained from the indicator card of the engine by the graphical construction shown in Fig. 38.14.

For a given crank angle θ, let the gas pressure on the piston be P. Erect a perpendicular to the line of action of the piston from the crank center, intersecting the line of the connecting rod. Let the intercept Oa on this perpendicular be y. Then the torque M for angle θ is given by

$$M = PSy \qquad (38.23)$$

where S is the piston area.

Table 38.5. Phase Diagrams and Deflections for the Mode Calculated in Table 38.3

PHASE DIAGRAMS	$\Sigma\beta$	ORDERS		cn	β
FIRING SEQUENCE 4 ↑ 6 7 —✳— 2 3 ⟋ ⟍ 5 8	0.778	1/2, 4½, 8½, – – <hr> MIRROR IMAGE FOR 3½, 5½, 7½, – –		1 2 3 4 5 6 7 8	1.0000 0.9543 0.8771 0.7709 0.6296 0.4595 0.2742 0.0799
x2 OR 6 1 8 4 ——— 3 5 ⎪ 6 2 7	0.169	1, 5, 9, — <hr> MIRROR IMAGE FOR 3, 7, 11			5.0455
x3 OR 5 1 3 ✳ 5 2 —✳— 7 4 ⎪ 6 8	1.549	1½, 5½, ·9½, + – <hr> MIRROR IMAGE FOR 2½, 6½, 10½, – –			
x4 1,2,7,8 ↕ 3,4,5,6	0.4287	2, 6, 10, – –			
x8 1,2,3,4,5,6,7,8 ↕ 	5.0455	4, 8, 12 MAJOR ORDERS			

Table 38.6. Empirical Factors for Engine Amplitude Calculations

Bore	Stroke	$\Re$
20 in. × 24 in. or larger		50–60
8 in. × 10 in.		40–50
4 in. × 6 in. or smaller		35

HARMONIC ANALYSIS. If an instrumental harmonic analyzer, such as the Coradi, is not available, the harmonic analysis can be made by any one of several rules based on selected ordinates. There are two types:

1. The Runge rules use the same set of ordinates for all harmonics. A 48-ordinate rule of this type is described in Refs. 1 and 12. It requires the scaling of 48 ordinates and numerous multiplications and additions, and is not recommended unless a digital computer is available.

2. The Fischer-Hinnen and Lewis rules use a different set of ordinates for each harmonic. They do not require the scaling of ordinates, and numerical work is reduced to a single multiplication. A single harmonic component, which in many cases is all that is needed, can be obtained in a few minutes independently of all the others.

FIG. 38.14. Schematic diagram of crank and connecting rod used in plotting torque curve.

FIG. 38.15. Harmonic analysis chart, fourth-order spacing. For the Fischer-Hinnen rule use only the ordinates marked x; the a_n or $b_n = (\frac{1}{2}q)\Sigma$ ordinates, where full lines indicate positive values and dashed lines negative values. For the three-ordinate Lewis rule, use all ordinates, full lines positive, dashed lines negative. Then a_q or $b_q = (\frac{2}{3}\pi q)\Sigma$ ordinates.

The Fischer-Hinnen rule is stated as follows: To determine the qth-order coefficients, divide the period into $4q$ intervals and erect ordinates at each division. Beginning at the left end designate them as y_0, y_1, y_2, etc., to y_{4q}. Then

$$a_q = \frac{1}{2q}[(y_1 + y_5 + y_9 + \cdots + y_{4q-3}) - (y_3 + y_7 + \cdots + y_{4q-1})] \quad \text{(sine terms)}$$

$$b_q = \frac{1}{2q}[(y_0 + y_4 + \cdots + y_{4q-4}) - (y_2 + y_6 + \cdots + y_{4q-2})] \quad \text{(cosine terms)}$$

This addition is accomplished most easily as follows: Adopt a length as standard for the period τ; 12 in. is convenient. Any curves to be analyzed should be plotted to this standard period. A series of charts is made on tracing cloth or transparent plastic, one for each order number. The ordinates are ruled on these charts, the plus ordinates in full lines and the minus ordinates in dashed lines. Such a chart, as ruled for the fourth-order coefficients, with the ordinates marked X, is shown in Fig. 38.15. It is laid over the curve to be analyzed so that the end points marked "sine" coincide with the ends of the curve period, and so that the base line of the chart is just below the lowest point of the curve, the exact position being immaterial. The positive ordinates are summed by marking them off in succession on one edge of a strip of paper; the negative ordinates are marked on the other edge. The harmonic coefficient is the difference of the sum of the positive and negative ordinates divided by the total number of ordinates $2q$. To obtain the cosine coefficients of the same order number, the chart is shifted until the points marked "cosine" coincide with the end points of the period; then the ordinates are summed as before.

In the analysis of the turning effort curve of a two-cycle engine, it is convenient to treat the compression portion by reversing it end to end and inverting. Sum the ordinates

CARD	M.I.P
1	0
2	25.3
3	49.0
4	73.1
5	92.6
6	117.4
7	140.9

FIG. 38.16. Standard diesel indicator diagrams for engine with $l/r = 4.25$. The mean indicated pressure is shown in the inset table. (*F. M. Lewis.*[2])

FIG. 38.17. Turning effort curves derived from indicator cards of Fig. 38.16 for a single-cylinder engine of 1-in.[2] piston area and 1-in. crank radius. (*F. M. Lewis.*[2])

from this curve, and add to the sum from the expansion portion. Thus, only the left half of the analysis chart is used. The divisor is $2q$.

For a four-cycle engine the fundamental period is two revolutions, or 720°, but from 180 to 540° the gas-pressure torque is negligible. Charts of the type shown in Fig. 38.15 having a 12″ period are used for integral order coefficients; and an additional set, constructed according to the same rule, is required for the fractional order coefficients. The compression portion of the turning effort curve from 540 to 360° is inverted and reversed end to end and laid in the 0 to 180° interval so that the 720° point coincides with the 0° point; then the summation from this portion is added to the summation from the expansion portion. The divisor for the four-cycle engine is now $4q$ instead of $2q$ as for the two-cycle engine.

For the low-order numbers the above rule makes the coefficients a function of a small number of ordinates and the accuracy may be low. The Lewis rules provide greater accuracy.[13] The three-ordinate rule is shown in Fig. 38.15. Lay in additional ordinates 0.532/4q to the right and left of the ordinates marked x and sum as before. Then

$$a_q \text{ or } b_q = \frac{2}{3\pi q} \times \Sigma \text{ ordinates}$$

Rules for five and seven ordinates per half of the interval $\tau/2q$ are also available[13] and have still higher accuracy. The rules are based on the assumption that within this interval the portion of the curve to be analyzed is a parabola of the second, fourth, or sixth order for the three-, five-, and seven-ordinate rules, respectively. In most cases, this is a very close approximation, but no rule based on the use of a limited number of ordinates can be more than an approximation.

THE COEFFICIENTS OF STANDARD INDICATOR CARDS. In some cases it may be possible to use prepared analyses obtained from standard sets of indicator cards. One such set of diesel indicator cards and the harmonic analysis of the corresponding turning effort curves are shown in Figs. 38.16 to 38.19. Except for the first, second, and third orders, the resultant harmonic coefficients are given by

$$h = \sqrt{a^2 + b^2}$$

Then the exciting torque per cylinder is

$$M_q = Srh$$

The coefficients are for four-cycle engines; for two-cycle engines multiply by 2; there are no fractional orders.

FIG. 38.18. Harmonic coefficients of sine and cosine terms for first, second, and third orders for indicator diagrams shown in Fig. 38.16. (*F. M. Lewis.*[2])

(A)

(B)

FIG. 38.19. Resultant harmonic coefficients (A) for lower orders of indicator diagrams shown in Fig. 38.16 and (B) for sixth and higher orders. (*F. M. Lewis.*[2])

For the first, second, and third orders, the sine inertia component of the turning effort curve is computed by Eq. (38.22) and added algebraically to the sine gas-pressure component. Then the resultant M_e is the square root of the sums of the squares of the sine and cosine components. Analyses for other standard diesel indicator cards are given in Ref. 14.

The indicator cards of modern high-speed diesel engines with supercharging may differ considerably from the standard cards of Fig. 38.16, and a full analysis may be necessary. Approximate values of the coefficients for such engines can be obtained by multiplying the coefficients of Figs. 38.18 and 38.19 by the ratio of the maximum combustion pressure

Table 38.7. Harmonic Components of Gas-pressure Torque for a Four-cycle Gasoline Engine

Order	½	1	1½	2	2½	3	3½	
Resultant co-efficient for h	34		31		17		10	
Sine coefficient for a		31		21		11.5		
Cosine coefficient for b		10		−4		4.5		

Order	4	4½	5	5½	6	6½	7	7½	
h	7.5	5.5	4	3.2	2.9	2.45	2	1.68	

Order	8	8½	9	9½	10	10½	11	11½	12
h	1.5	1.32	1.15	1.10	0.88	0.78	0.71	0.58	1.52

$$h, a, b = \frac{\text{mean indicated pressure}}{100} \times \text{coefficients above.}$$

$$M_e = Sr \times (h \text{ or } a \text{ or } b)$$

Table 38.8. Amplitude Estimates for First Mode of Diesel Engine Calculated in Table 38.3

Order	Rpm	$\Sigma\beta$	h	M_e, in.-lb	M_m, in.-lb	Stress	θ_1 rad
4	1,125	5.055	29.5	11,300	2,860,000	49,800	0.08257
4½	1,000	0.777	23.5	9,000	349,000	6,070	0.0105
5	900	0.164	18.5	7,080	58,000	1,005	
5½	820	1.54	15.0	5,730	441,000	7,700	0.0127
6	750	0.430	12.0	4,580	98,500	1,715	
6½	692	1.54	8.8	3,370	259,000	4,500	0.0075
7	643	0.164	6.5	2,490	20,400	355	
7½	600	0.777	5.5	2,110	81,900	656	
8	562	5.055	3.9	1,490	377,000	6,550	0.0108
12	375	5.055	1.4	537	136,000	2,380	0.0039
4	950	. . .	29.5	11,300	230,000	4,010	

to that shown by the cards of Fig. 38.16, with the additional multiplication by 2 for a two-cycle engine. The higher-order coefficients are extremely sensitive to the details of the fuel-injection and combustion process; unless good experimental indicator cards are available for harmonic analysis, it may not be possible to obtain accuracy in estimating exciting torques.

For a four-cycle gasoline engine the harmonic coefficients are more nearly proportional to the mean indicated pressure and are approximated by the values in Table 38.7.

Example 38.3. Table 38.8 gives the amplitude estimate for the first mode of the diesel engine, the frequency of which was calculated in Table 38.3. The engine is four-cycle, eight-cylinder with

9-in. bore by 12-in. stroke. The values of speed in rpm are as previously calculated, and $\Sigma\beta$ is taken from Table 38.6. The harmonic coefficients h are obtained from a harmonic analysis of an indicator card. Then the exciting torques are

$$M_e = \frac{\pi}{4} \times 9^2 \times 6 \times h = 383h \qquad \text{in.-lb}$$

For the maximum torques, M_m, a value of $\Re = 50$ was assumed in the equation

$$M_m = \Re M_e \Sigma\beta \qquad \text{in.-lb}$$

The crankpin diameter is 6½ in. so that the nominal stress is

$$\tau = \frac{16 M_m}{\pi \times (6.5)^3} = 0.0174\, M_m \qquad \text{lb/in.}^2$$

From Table 38.3 (unity system) the torque in shaft section 8–9 is 0.1979 for unity amplitude at the No. 1 cylinder. Therefore the vibration amplitude at the No. 1 cylinder in the actual installation is

$$\theta = \frac{M_m}{0.1979 \times 175.7 \times 10^6} = \frac{M_m}{34.7 \times 10^6} \qquad \text{rad}$$

AMPLITUDES AT NONRESONANT FREQUENCIES

A calculation of the nonresonant or "forced" vibration amplitude is required in some cases to define the range of the more severe critical speeds, particularly with geared drives; it also is required in the design of dampers. The calculation is readily made by an extension of the Holzer table sequence method illustrated in Table 38.2. In the sequence the initial amplitude is treated as an algebraic unknown θ. At each cylinder where an exciting torque acts, this torque is added. Assume first that there are no damping torques. Then the residual torque after the last rotor is of the form $a\theta + b$, where a and b are numerical constants resulting from the calculation. Since the residual torque is zero, $\theta = -b/a$.

The amplitude and torque at any point of the system are found by substituting this numerical value of θ at the appropriate point in the tabulation. At frequencies well removed from resonance, damping has but little effect and can be neglected. Damping can be added to the system by treating it as an exciting torque equal to the imaginary quantity $-jc\omega\theta$, where c is the damping constant and θ is the amplitude at the point of application. Relative damping between two inertias can be treated as a spring of a stiffness constant equal to the imaginary quantity of $+jc\omega$.

For the major critical speeds the exciting torques are all in phase and are real numbers. For the minor critical speeds the exciting torques are out-of-phase; they must be entered as complex numbers of amplitude and phase as determined from the phase diagram (discussed under *Resonant Amplitudes*) for the critical speed of the order under consideration. With damping and/or out-of-phase exciting torques introduced, a and b in the equation $a\theta + b = 0$ are complex numbers, and θ must be entered as a complex number in the tabulation in order to determine the angle and torque at any point. The angles and torques are then of the form $r + js$, where r and s are numerical constants and the amplitudes are equal to $\sqrt{r^2 + s^2}$.

While the calculation can be made with the moments of inertia, stiffness constants, and exciting torques of the actual system, the unity form of the calculation used in frequency calculations [see Eq. (38.14)] saves considerable labor. In the forced-vibration calculation the amplitude of the exciting torque also is assumed to be unity. It is advisable, particularly when the calculation is made for damper design, to start the sequence at the flywheel end of the system. The introduction of exciting and damping terms thus is deferred as long as possible.

The relationship between the actual and the unity system is given in Table 38.9. Here $I = J/g$ is the moment of inertia by which all the moments of inertia are divided and k

is the common stiffness constant by which all the stiffness constants are divided. A damping constant c in the actual system becomes

$$c_s = \frac{c}{\sqrt{Ik}}$$

in the unity system. Exciting torques are assumed to have unity amplitude. After the torque in any part of the unity system is found, the torque in the actual system is determined by multiplying by M_e, and deflections in the actual system are obtained by multiplying those in the unity system of M_e/k.

Table 38.9. Relation between Parameters in Actual and Unity Systems

	Actual	Unity
Inertia constant......	$I = J/g$	1
Spring constant.......	k	1
Damping constant....	c	$c_s = \dfrac{c}{\sqrt{Ik}}$
Torques.............	M_e	1
Deflections..........	$\dfrac{M_e\theta}{k}$	θ

Example 38.4. Determine the amplitude of the fourth-order vibration at the maximum speed of 950 rpm for the eight-cylinder, four-cycle engine driving a generator shown in Fig. 38.10B.

The frequency in the unity system corresponding to fourth-order vibration at 950 rpm in the actual system is given by

$$\omega_s{}^2 = \left(\frac{2\pi \times 950 \times 4}{60}\right)^2 \times \frac{35.8}{175.7 \times 10^6} = 0.0321$$

$$\omega_s = 0.179$$

The calculation is carried out in Table 38.10. Assuming no damping, the residual torque is

$$M_r = 0.3946\theta + 5.852 = 0$$

from which $\theta = -14.85$.

The maximum torque occurs in section 8–9; it is given by

$$M_{8,9} = 1.3726\theta = -20.4$$

In section 6–7 the torque is given by

$$M_{6,7} = 1.326\theta + 1.977 = -17.73$$

Fourth-order torque M_e in the actual engine is 11,300 in.-lb (Table 38.8); in the actual engine the fourth-order torque in section 8–9 is

$$M = 20.4 \times 11,300 = 231,000 \text{ in.-lb}$$

At the No. 1 cylinder the amplitude in the unity system is

$$-7.7401\theta - 24.748 = 90.3 \text{ rad}$$

In the actual system the corresponding amplitude is

$$\frac{90.3 \times 11,300}{175.7 \times 10^6} = 0.0058 \text{ rad}$$

The problem also can be solved by assuming that a concentrated damping of constant $c = 1,470$ in.-lb/rad acts at the No. 1 crank. In the unity system the corresponding damping constant is

$$c_3 = \frac{1,470}{\sqrt{35.8 \times 175.7 \times 10^6}} = 0.0185$$

At the No. 1 crank the torque $-jc_s\omega_s\theta$ is added; this torque is

$$M = -j \times 0.0185 \times 0.175 \times (-7.7401\theta - 24.748) = j(0.0256\theta + 0.0819)$$

The resulting residual torque is given by

$$M_r = 0.3946\theta + 5.852 + j(0.0256\theta + 0.0819) = 0$$

Solving for θ,

$$\theta = \frac{-5.852 - 0.0819j}{0.3946 + 0.0256j}$$

The amplitude of the deflection is

$$|\theta| = \sqrt{\frac{(5.852)^2 + (0.0819)^2}{(0.3946)^2 + (0.0256)^2}} = 14.85$$

which is the same as if damping were not present; thus, at this speed the effect of the external damping is completely negligible.

APPLICATION OF MODAL ANALYSIS TO ROTOR SYSTEMS

Classical modal analysis of vibrating systems (see Chap. 21) can be used to obtain the forced response of multistation rotor systems in torsional motion. The natural frequen-

Table 38.10. Sequence Calculations for Forced Vibration of Engine Generator Shown in Fig. 38.10*B* at Operating Speed of 950 rpm ($\omega_s{}^2 = 0.0321$)

Rotor number	I	$I\omega_s{}^2$	β Inertia $\theta \times$	β Exciting	M, in.-lb Inertia $\theta \times$	M, in.-lb Exciting	k
10	42.5	1.366	1.0000	...	1.366		
			0.226	...	1.366	...	6.06
9	0.265	0.0085	0.774	...	0.0066		
			1.042	...	1.3726	...	1.32
8	1	0.0321	−0.208	...	−0.0086		
			1.3640	1.000	1.3640	1.000	1.00
7	0.722	0.0231	−1.632	−1.000	−0.0377	−0.023	
			1.3263	1.977	1.3263	1.977	1.00
6	0.722	0.0231	−2.9583	−2.977	−0.0682	−0.069	
			1.2581	2.908	1.2581	2.908	1.00
5	1	0.0321	−4.2164	−5.885	−0.1350	−0.189	
			1.1231	3.719	1.1231	3.719	1.00
4	1	0.0321	−5.3395	−9.604	−0.1715	−0.308	
			0.9516	4.411	0.9516	4.411	1.00
3	0.722	0.0231	−6.2911	−14.015	−0.1452	−0.324	
			0.8064	5.087	0.8064	5.087	1.00
2	0.722	0.0231	−7.0975	−19.102	−0.1638	−0.441	
			0.6426	5.646	0.6426	5.646	1.00
1	1	0.0321	−7.7401	−24.748	−0.2480	−0.794	
					0.3946	5.852	

cies and mode shapes of the system are found using the transfer matrix method. The response of the rotor to periodic phenomena (not necessarily a harmonic or shaft frequency) is determined as a linear weighted combination of the mode shapes of the system. Heretofore with this technique, damping has been entered in modal form; the damping forces are a function of the various modal velocities. The formation of equivalent viscous damping constants that are some percentage of critical damping is required. The critical damping factor is formed from the system modal inertia.[23]

The modal analysis technique can be used for a torsional distributed mass model of engine systems using modal damping; nonsynchronous speed excitations are allowed. The shaft sections of the modeled rotor have distributed mass properties and lumped end masses (including rotary inertia). A transfer matrix analysis is performed to obtain a finite number of natural frequencies. The number required depends on the range of forcing frequencies used in the problem. The natural frequencies are substituted back into the transfer matrices to obtain the mode shapes. A function consisting of a weighted average of the mode shapes is formed and substituted into

$$\Theta(x, t) = \sum_{n=1}^{N} \alpha_n(x) f_n(t)$$

where Θ = torsional response
 α_n = normal modes
 f_n = periodic time-varying weighting factors

The function $f_n(t)$ is determined from the ordinary differential equations of motion and is a function of the forcing functions, rotor speed, modal damping constants, and mode shapes of the system.

DIRECT INTEGRATION

Direct integration of equations of motion of a system utilize first- or second-order differential equations. The method is fundamental for linear and nonlinear response problems.[24] Any digitally describable vibration or shock excitation can be carried out with this method.

Direct integration can be used on nonlinear models and arbitrary excitation, so it is one of the most general techniques available for response calculation. However, large computer storage is required, and large computer costs are usually incurred because small time- or space-step sizes are needed to maintain numerical stability. An adjustable step integration routine such as predictor-corrector helps to alleviate this problem. Such a numerical integration must be started with another routine such as Runge-Kutta.

Direct integration is particularly useful when nonlinear components such as elastomeric couplings are involved or when the excitation force varies in frequency and magnitude. Direct integration is used for analysis of synchronous motor start-ups in which the magnitude of the torque varies with rotor speed and the frequency is two times the slip frequency—starting at twice the line frequency and ending at zero when the rotor is locked on synchronous speed. Examples of this type of analysis are given in Refs. 24 and 25.

PERMISSIBLE AMPLITUDES

Failure caused by torsional vibration invariably initiates in fatigue cracks that start at points of stress concentration—e.g., at the ends of keyway slots, at fillets where there is a change of shaft size, and particularly at oil holes in a crankshaft. Failures can also start at corrosion pits, such as occur in marine shafting. At the shaft oil holes the cracks begin on lines at 45° to the shaft axis and grow in a spiral pattern until failure occurs. Theoretically the stress at the edges of the oil holes is four times the mean shear stress in the shaft, and failure may be expected if this concentrated stress exceeds the fatigue limit of the material. The problem of estimating the stress required to cause failure is further complicated by the presence of the steady stress from the mean driving torque and the variable bending stresses.

In practice the severity of a critical speed is judged by the maximum nominal torsional stress

$$\tau = \frac{16M_m}{\pi d^3}$$

where M_m is the torque amplitude from torsional vibration and d is the crankpin diameter. This calculated nominal stress is modified to include the effects of increased stress and is compared to the fatigue strength of the material.

U.S. MILITARY STANDARD

A military standard[22] issued by the U.S. Navy Department states that the limit of acceptable nominal torsional stress within the operating range is

$$\tau = \frac{\text{ultimate tensile strength}}{25} \quad \text{for steel}$$

$$\tau = \frac{\text{torsional fatigue limit}}{6} \quad \text{for cast iron}$$

If the full-scale shaft has been given a fatigue test, then

$$\tau = \frac{\text{torsional fatigue limit}}{2} \quad \text{for either material}$$

Such tests are rarely, if ever, possible.

For critical speeds below the operating range which are passed through in starting and stopping, the nominal torsional stress shall not exceed 1¾ times the above values.

Crankshaft steels which have ultimate tensile strengths between 75,000 and 115,000 lb/in.² usually have torsional stress limits of 3,000 to 4,600 lb/in.²

For gear drives the vibratory torque across the gears, at any operating speed, shall not be greater than 75 per cent of the driving torque at the same speed, or 25 per cent of full-load torque, whichever is smaller.

MEASURES OF CONTROL

The various methods which are available for avoiding a critical speed or reducing the amplitude of vibration at the critical speed may be classified as:

1. Shifting the values of critical speeds by changes in mass and elasticity
2. Vector cancellation methods
3. Change in mass distribution to utilize the inherent damping in the system
4. Addition of dampers of various types

SHIFTING OF CRITICAL SPEEDS

If the stiffness of all the shafting to a system is increased in the ratio a, then all the frequencies will increase in the ratio $\sqrt{a}$, provided that there is no corresponding increase in the inertia. It is rarely possible to increase the crankshaft diameters on modern engines; in order to reduce bearing pressures, bearing diameters usually are made as large as practical. If bearing diameters are increased, the increase in the critical speed will be much smaller than indicated by the $\sqrt{a}$ ratio because a considerable increase in the inertia will accompany the increase in diameter. Changes in the stiffness of a system made near a nodal point will have maximum effect. Changes in inertia near a loop will have maximum effect while those near a node will have little effect.

By the use of elastic couplings it may be possible to place certain critical speeds below the operating speed where they are passed through only in starting and stopping; this leaves a clear range above the critical speed. This procedure must be used with caution because some critical speeds, for example the fourth order in an eight-cylinder, four-cycle engine, are so violent that it may be dangerous to pass through them. If the acceleration through the critical speed is sufficiently high, some reduction in amplitude may be attained,[16] but with a practical rate the reduction may not be large. The rate of deceleration when stopping is equally important. In some cases mechanical clutches disconnect the driven machinery from the engine until the engine has attained a speed above dangerous critical speeds. Elastic couplings may take many forms including helical springs arranged tangentially, flat leaf springs arranged longitudinally or radially, various arrangements using rubber, or small-diameter shaft sections of high tensile steel.[20]

VECTOR CANCELLATION METHODS

CHOICE OF CRANK ARRANGEMENT AND FIRING ORDER. The amplitude at certain minor critical speeds sometimes can be reduced by a suitable choice of crank arrangement and firing order (i.e., firing sequence). These fix the value of the vector sum $\Sigma\beta$ in Eq. (38.21), $M_m = \Re M_e \Sigma\beta$. But considerations of balance, bearing pressures, and internal bending moments restrict this freedom of choice. Also, an arrangement which decreases the amplitude at one order of critical speed invariably increases the amplitude at others. In four-cycle engines with an even number of cylinders, the amplitude at the half-order critical speeds is fixed by the firing order because this determines the $\Sigma\beta$ value. Tables 38.11 to 38.18 list the balance and torsional-vibration characteristics for the crank arrangements and firing orders, for six to ten cylinders, having the most desirable prop-

Table 38.11. Balance and Torsional-vibration Characteristics for Six-cylinder, Four-cycle Engine Having 120° Crank Spacing

Cranks at 1, 6 0° 2, 5 120° 3, 4 240°	$\Sigma\beta$ of orders				C_1	C_2
	½, 2½, 3½, 5½, 6½, 8½,	1½ 4½ 7½	1, 2 4, 5 7, 8	3 6 9		
Firing sequence	...	...	...	12		
1, 5, 3, 6, 2, 4	0.577	1.5	0	3.5	0	0
1, 2, 4, 6, 5, 3	1.2	0.166	0	3.5	0	0

Values of 0 in the $\Sigma\beta$ column indicate small but not necessarily 0 values for actual β distributions.

Table 38.12. Balance and Torsional-vibration Characteristics for Six-cylinder, Four-cycle Engine Having 60° Crank Spacing

Firing sequence	$\Sigma\beta$ of orders				C_1	C_2
	1, 5, 7, 11	2, 4, 8, 10	3, 9	6, 12		
1, 5, 3, 4, 2, 6	0	0.577	1.5	3.5	0	$2\sqrt{3}$
1, 5, 3, 6, 2, 4	0.577	0	1.5	3.5	$2\sqrt{3}$	0

erties. All first- and second-order forces are balanced for these arrangements. The unbalanced couple coefficients C_1 and C_2 are defined as follows:

First-order unbalanced couple:

$$M_1 = \frac{C_1(W_p + hW_c)ra\Omega^2}{g} \quad \text{in.-lb} \tag{38.24}$$

Table 38.13. Balance and Torsional-vibration Characteristics for Eight-cylinder, Four-cycle Engine Having 90° Crank Spacing

FIRING ORDER	$\Sigma\beta$· OF ORDERS					C_1	C_2
	1/2, 7½, 8½, 3½, 4½	1½, 5½, 6½	2, 6, 10	13, 58, 7	4, 8		
1,6,2,5,8,3,7,4	0.745	1.44	0	0	4.5	0	0
1,6,2,4,8,3,7,5	0.686	1.48	0	0	4.5	0	0
1,3,2,5,8,6,7,4	1.48	0.686	0	0	4.5	0	0
6,3,5,7,8,6,4,2	1.74	0.176	0	0	4.5	0	0
1,7,4,3,8,2,5,6	0.176	1.74	0	0	4.5	0	0

```
CRANK ARRANGEMENT  #1,2,3            #4,5

                      1 8               1 8
                 45 ──┼── 36      27 ──┼── 36
                      2 7               4 5
```

Table 38.14. Balance and Torsional-vibration Characteristics for Eight-cylinder, Two-cycle Engine Having 45° Crank Spacing

Firing order	$\Sigma\beta$ of orders					C_1	C_2
	1, 7, 9	2, 6, 10	3, 5, 11	4, 12	8, 16		
1, 8, 2, 6, 4, 5, 3, 7	0.056	0	0.79	2.0	4.5	0.448	0
1, 7, 4, 3, 8, 2, 5, 6	0.175	0	1.61	0	4.5	1.405	0
1, 6, 5, 2, 7, 4, 3, 8	0.112	0	1.58	0.5	4.5	0.897	0

Table 38.15. Balance and Torsional-vibration Characteristics for Seven-cylinder, Two- or Four-cycle Engine Having 51.4° Crank Spacing

No.	Crank sequence	$\Sigma\beta$ of orders				C_1	C_2
		1, 6, 7 2½ 4½ 9½	2, 5, 9 1½ 5½ 8½	3, 4, 10 ½ 6½ 7½	7 3½		
1	1, 6, 3, 4, 5, 2, 7	0.04	0.14	1.4	4	0.27	1.01
2	1, 5, 2, 4, 6, 3, 7	0.56	0.04	1.2	4	3.90	0.25
3	1, 4, 7, 2, 3, 5, 6	0.01	1.30	0.6	4	0.08	9.10
4	1, 5, 4, 3, 6, 2, 7	0.18	0.21	1.4	4	1.30	1.50
5	1, 6, 4, 3, 5, 2, 7	0.04	0.33	1.4	4	0.30	2.30

where W_p is the weight of the piston, piston pin, and cooling fluid in pounds and W_c is the weight of the connecting rod in pounds [see Eq. (38.4)].

Second-order unbalanced couple:

$$M_2 = \frac{C_2(W_p + hW_c)r^2 a\Omega^2}{gl} \quad \text{in.-lb} \tag{38.25}$$

where Ω = angular speed of shaft, rad/sec
  a = cylinder spacing (assumed equal), in.
  l = connecting-rod length, in.

Other terms are as defined with respect to Eq. (38.3).

Table 38.16. Balance and Torsional-vibration Characteristics for Nine-cylinder, Two- or Four-cycle Engine Having 40° Crank Spacing

No.	Crank sequence	Order of critical $\left\{\begin{array}{l}1, 8, 10 \\ 3\frac{1}{2} \\ 5\frac{1}{2}\end{array}\right.$	$\begin{array}{c}2, 7, 11 \\ 2\frac{1}{2} \\ 6\frac{1}{2}\end{array}$	$\begin{array}{c}3, 6, 12 \\ 1\frac{1}{2} \\ 7\frac{1}{2}\end{array}$	$\begin{array}{c}4, 5, 13 \\ \frac{1}{2} \\ 8\frac{1}{2} \\ 9\frac{1}{2}\end{array}$	$\begin{array}{c}9 \\ 4\frac{1}{2} \\ 13\frac{1}{2}\end{array}$	C_1	C_2
				$\Sigma\beta$ of orders				
1	1, 9, 2, 7, 4, 5, 6, 3, 8	0.022	0.055	0.019	1.8	5.0	0.2	0.5
2	1, 9, 2, 5, 7, 6, 3, 4, 8	0.10	0.19	0.94	1.8	5.0	0.8	1.5
3	1, 6, 8, 2, 4, 7, 5, 3, 9	0.10	0.48	1.77	0.91	5.0	0.8	3.8
4	1, 6, 7, 2, 5, 8, 3, 4, 9	0.11	0.14	1.95	0.6	5.0	0.9	1.1
5	1, 9, 2, 7, 5, 3, 6, 4, 8	0.14	0.35	0.51	1.7	5.0	1.1	2.8
6	1, 8, 7, 3, 2, 9, 4, 5, 6						0.028	8.5

Table 38.17. Balance and Torsional-vibration Characteristics for Ten-cylinder, Four-cycle Engine Having 72° Crank Spacing

For all orders given, $C_1 = C_2 = 0$
$\Sigma\beta = 0$ for all whole numbered orders except 5, 10, etc.

No.	Crank sequence	Order of critical $\left\{\begin{array}{l}\frac{1}{2}, 4\frac{1}{2} \\ 5\frac{1}{2} \\ 9\frac{1}{2} \\ 10\frac{1}{2}\end{array}\right.$	$\begin{array}{c}1\frac{1}{2} \\ 3\frac{1}{2} \\ 6\frac{1}{2} \\ 8\frac{1}{2}\end{array}$	$\begin{array}{c}2\frac{1}{2} \\ 7\frac{1}{2} \\ 12\frac{1}{2}\end{array}$	5, 10, 15
		$\Sigma\beta$ of orders			
1	**1**, 7, 9, 3, 6, 10, 4, 2, 8, 5	0.56	1.95	0.1	5.5
2	1, 3, 4, 5, 9, 10, 8, 7, 6, 2	1.95	0.56	0.1	5.5
3	1, 8, 5, 7, 2, 10, 3, 6, 4, 9	0.09	1.00	2.5	5.5
4	1, 7, 3, 9, 5, 10, 4, 8, 2, 6	1.00	0.09	2.5	5.5

The values of $\Sigma\beta$ are calculated by assuming $\beta = 1$ for the cylinder most remote from the flywheel, assuming $\beta = 1/n$ for the cylinder adjacent to the flywheel (where n is the number of cylinders), and assuming a linear variation of β therebetween. In any actual installation $\Sigma\beta$ must be calculated by taking β from the relative amplitude curve; however, if the $\Sigma\beta$ as determined above is small, it also will be small for the actual β distribution. These arrangements assume equal crank angles and firing intervals. The reverse arrangements (mirror images) have the same properties.

V-TYPE ENGINES. In V-type engines, it may be possible to choose an angle of the V which will cancel certain criticals. Letting ϕ be the V angle between cylinder banks, and q the order number of the critical, the general formula is

$$q\phi = 180°, 540°, 1080°, \text{ etc.} \tag{38.26}$$

For example, in an eight-cylinder engine the eighth order is canceled at angles of 22½°, 67½°, 112½°, etc.

In four-cycle engines, ϕ is to be taken as the actual bank angle if the second-bank cylinders fire directly after the first and as $360° + \phi$ if the second-bank cylinders omit a revolution before firing. In the latter case the cancellation formula is

$$(\phi + 360°)q = n \times 180° \tag{38.27}$$

where $n = 1, 3, 5$, etc. For example, to cancel a 4.5-order critical the bank angle should be

$$\phi = \frac{180°}{4.5} = 40° \qquad \text{for direct firing}$$

or

$$\phi = \frac{11 \times 180°}{4.5} - 360° = 80° \qquad \text{for the 360° delay}$$

Table 38.18. Balance and Torsional-vibration Characteristics for Ten-cylinder, Two-cycle Engine Having 36° Crank Spacing

No.	Crank sequence	$\Sigma\beta$ of order						C_1	C_2
		1, 9, 11	2, 8, 12	3, 7, 13	4, 6	5	10		
1	1, 9, 3, 7, 5, 6, 4, 8, 2, 10	0.00	0.10	0.00	1.11	2.8	5.5	0.00	0.90
2	1, 6, 10, 2, 4, 9, 5, 3, 7, 8	0.03	0.16	1.97	0.48	0.1	5.5	0.28	1.56
3	1, 9, 5, 6, 2, 10, 4, 3, 8, 7	0.03	0.03	0.53	0.80	1.5	5.5	0.33	0.28
4	1, 9, 6, 3, 5, 10, 2, 4, 8, 7	0.04	0.19	1.86	0.12	1.1	5.5	0.45	1.90
5	1, 7, 6, 4, 8, 2, 9, 3, 5, 10	0.05	0.22	0.65	1.90	0.3	5.5	0.51	2.18
6	1, 5, 9, 4, 3, 8, 7, 2, 6, 10	0.06	0.04	1.95	0.50	0.3	5.5	0.57	0.45
7	1, 6, 9, 2, 7, 4, 8, 3, 5, 10	0.06	0.08	1.42	1.56	0.5	5.5	0.60	0.80
8	1, 9, 4, 5, 8, 3, 6, 7, 2, 10	0.06	0.06	0.16	1.80	1.3	5.5	0.62	0.62
9	1, 10, 3, 5, 8, 4, 7, 2, 9, 6	0.07	0.26	0.98	1.76	0.1	5.5	0.65	2.56
10	1, 9, 4, 6, 3, 10, 2, 7, 5, 8	0.09	0.00	1.00	0.00	2.5	5.5	0.90	0.00
11	1, 6, 10, 2, 5, 9, 3, 4, 7, 8	0.10	0.20	1.87	0.72	0.3	5.5	0.98	2.00
12	1, 9, 4, 5, 8, 3, 7, 6, 2, 10	0.10	0.08	0.10	1.87	1.1	5.5	1.00	0.77

CANCELLATION BY SHIFT OF THE NODE. If an engine can be arranged with approximately equal flywheel (or other rotors) at each end so that the node of a particular mode is at the center of the engine, $\Sigma\beta$ will cancel for the major orders of that mode. This procedure must be used with caution because the double flywheel arrangement may reduce the natural frequency in such a manner that low-order minor criticals of large amplitudes take the place of the canceled major criticals.

REDUCTION BY USE OF PROPELLER DAMPING IN MARINE INSTALLATIONS. From Eq. (28.17) it is evident that the torque amplitude in the shaft can be reduced below any desired level by making the flywheel moment of inertia I_1 of sufficient magnitude. The ratio of the propeller amplitude to the engine amplitude increases as the flywheel becomes larger; thus the effectiveness of the propeller as a damper is increased.

DAMPERS

Many arrangements of dampers can be employed (see Chap. 6). In each type there is a loose flywheel or inertia member which is coupled to the shaft by:

1. Coulomb friction (Lanchester damper)
2. Viscous fluid friction
3. Coulomb or viscous friction plus springs
4. Centrifugal force, equivalent to a spring having a constant proportional to the square of the speed (pendulum damper) (see Chap. 6)

Each of these types acts by generating torques in opposition to the exciting torques.

The Lanchester damper illustrated in Fig. 6.35 has been entirely superseded by designs in which fluid friction is utilized. In the Houdaille damper, Fig. 38.20, a flywheel is mounted in an oiltight case with small clearances; the case is filled with silicone fluid. The damping constant is

$$c = 2\pi\mu \left[\frac{r_2^3 b}{h^2} + \frac{1}{2} \frac{(r_2^4 - r_1^4)}{h_1} \right] \qquad \text{in.-lb-sec} \qquad (38.28)$$

where μ is the viscosity of the fluid and r_1, r_2, b, h_1, and h_2 are dimensions indicated in Fig. 38.20.

(A)   (B)

FIG. 38.20. Schematic diagram of dampers. *(A)* Houdaille type. *(B)* Paddle type.

FIG. 38.21. Schematic diagram of bonded rubber damper.

FIG. 38.22. Resonance curves for various conditions of auxiliary mass dampers: (1) damper free, $c = 0$; (2) damper locked, $c = \infty$; (3) auxiliary mass coupled to shaft by damping.

The paddle-type damper illustrated in Fig. 38.20 utilizes the engine lubricating oil supplied through the crankshaft. It has the damping constant

$$c = \frac{3\mu d^2 (r_2^2 - r_1^2)^2 n}{h^3 \left[\dfrac{d}{b_1} + \dfrac{d}{b_3} + \dfrac{4(r_2 - r_1)}{b_1 + b_2} \right]} \qquad \text{in.-lb-sec} \qquad (38.29)$$

where n is the number of paddles, μ is the viscosity of the fluid, and b_1, b_2, r_1, r_2, and d are dimensions indicated in Fig. 38.20. Other types of dampers are described in Ref. 5.

The effectiveness of these dampers may be increased somewhat by connecting the flywheel to the engine by a spring of proper stiffness, in addition to the fluid friction. In one form, Fig. 38.21, the connection is by rubber bonded between the flywheel and the shaft member. The rubber acts both as the spring and by hysteresis as the energy absorbing member. See Chaps. 35, 36, and 37 for discussions of damping in rubber. Dampers without and with springs are defined here as *untuned* and *tuned viscous dampers*, respectively.

In many cases the mode of vibration to be damped is essentially internal to the engine. Then the damper is located at the end of the engine remote from the flywheel. If the mode to be damped is essentially one between driven masses, other locations may be desirable or necessary.

DESIGN OF THE UNTUNED VISCOUS DAMPER, EXACT PROCEDURE

The first step in the design procedure is to make a tentative assumption of the polar moment of inertia of the floating inertia member. If the damper is attached to the forward end of the crankshaft with the primary purpose of damping vibration in the engine, the size should be from 5 to 25 per cent, depending on the severity of the critical to be damped, of the total inertia in the engine part of the system, excluding the flywheel.

Usually it is advantageous to minimize the torque in a particular shaft section. This may be done as follows: For a series of frequencies plot the resonance curve of this torque, first without the floating damper mass and then with the damper mass locked to the damper hub. Plot the curves with all ordinates positive. The nature of such a plot is shown in Fig. 38.22. The point of intersection is called the *fixed point*. The plot is shown as if there were only one resonant frequency. Usually only one is of interest, and the curves are plotted in its vicinity. If the plot were extended, there would be a series of fixed points.

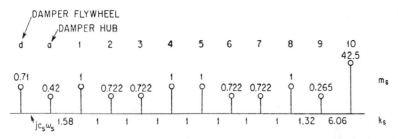

FIG. 38.23. Constants for system of Fig. 38.10*B* when damper is added at end of crankshaft.

If a damping constant is assigned to the damper and the new resonance curve plotted, it will be similar to curve 3 in Fig. 38.22 and will pass through the fixed point. If there is no other damping in the system except that in the damper, all of the resonance curves will pass through the fixed points, independent of the value assigned to the damping constant.[17] Therefore, the amplitude at the fixed point is the lowest that can be obtained for the assumed damper size. If this amplitude is too large, it will be necessary to increase the damper size; if the amplitude is unnecessarily small, the damper size can be decreased. When a satisfactory size of damper has been selected, it is necessary to find the damping constant which will put the resonance curve through the fixed point with a zero slope. Assume a value of ω^2 slightly lower than its value at the fixed point, and compute the amplitude at that value of ω^2 with the damping constant c entered as an algebraic unknown. Equating this amplitude to that at the fixed point, the unknown damping constant c can be calculated. Repeat the calculation with a value of ω^2 higher than the fixed point value by the same increment. The mean of the two values of c thus obtained will be as close to the optimum value as construction of the damper will permit. In constructing these resonance curves, it is not necessary to construct complete curves over a wide range of frequencies but only over a short interval in the vicinity of the fixed point. Considerable labor is saved by making the calculation in the unity system with conversion to the actual system as a final step.

Figure 38.23 shows the constants in the unity system for the example of Fig. 38.10*B*, when a damper is added at the end of the crankshaft. A floating rotor having 10 per cent of the total engine inertia is assumed; a is the fixed hub attached to the shaft. The "fixed point" is first located by calculating $M_{9,10}$ for $\omega_s^2 = 0.035$, 0.037, and 0.038. Table 38.19 gives the calculation for $\omega_s^2 = 0.037$.

In making the plot of the resonance curves, it is convenient to plot $1/M_{9,10}$ to avoid the resonant peak near $\omega_s^2 = 0.035$. Then $1/M_{9,10}$ is obtained from Table 38.19 as follows: Taking the case of the free damper, $\omega_s^2 = 0.037$. For zero residual torque: $0.1474\theta + 5.121 = 0$, or $\theta = -5.121/0.1474$.

The torque in the shaft between rotors 9 and 10 is $M_{9,10} = 1.57\theta$. The reciprocal is

$$\frac{1}{M_{9,10}} = \frac{0.1474}{5.121 \times 1.57} = 0.01832$$

Adding the damper d to the hub a, the corresponding value is

$$\frac{1}{M_{9,10}} = \frac{-0.0846}{4.391 \times 1.57} = -0.01225$$

Table 38.19. Sequence Calculations for System of Fig. 38.10B with Damper at End of Crankshaft ($\omega_s^2 = 0.037$)

Rotor number	I	$I\omega_s^2$	β		M, in.-lb		k
			Inertia $\theta \times$	Exciting	Inertia $\theta \times$	Exciting	
10	42.5	1.570	1.0000	...	1.570		
			0.259	...	1.570		6.06
9	0.265	0.0113	0.741	...	0.0083		
			1.191	...	1.5783		1.32
8	1	0.037	−0.450	...	−0.0166		
			1.5617	1.000	1.5617	1.000	1.00
7	0.722	0.0267	−2.0117	−1.000	−0.0538	−0.027	
			1.5079	1.973	1.5079	1.973	1.00
6	0.722	0.0267	−3.5196	−2.973	−0.0935	−0.080	
			1.4144	2.893	1.4144	2.893	1.00
5	1	0.037	−4.9340	−5.866	−0.1830	−0.217	
			1.2314	3.676	1.2314	3.676	1.00
4	1	0.037	−6.1654	−9.542	−0.2280	−0.352	
			1.0034	4.324	1.0034	4.324	1.00
3	0.722	0.0267	−7.1688	−13.866	−0.1910	−0.371	
			0.8124	4.953	0.8124	4.953	1.00
2	0.722	0.0267	−7.9812	−18.819	−0.2130	−0.513	
			0.5994	5.450	0.5994	5.450	1.00
1	1	0.037	−8.5806	−24.269	−0.3170	−0.898	
			0.1785	3.510	0.2824	5.552	11.58
a	0.42	0.0155	−8.7591	−27.779	−0.1350	−0.431	
					0.1474	5.121	
					0.2824	5.552	
$a+d$	1.13	0.0418	−8.7591	−27.779	−0.3670	−1.161	
					−0.0846	4.391	
			$-0.767j/c_s$	$-26.7j/c_s$	0.1474	5.121	$0.192c_s j$
d	0.71	0.0263	$\left\{\begin{array}{c}-8.7591 \\ +0.767j/c_s\end{array}\right\}$	$\left\{\begin{array}{c}-27.779 \\ +26.7j/c_s\end{array}\right\}$	$\left\{\begin{array}{c}-0.231 \\ +0.020j/c_s\end{array}\right\}$	$\left\{\begin{array}{c}-0.730 \\ +0.703j/c_s\end{array}\right\}$	
		$\omega_s = 0.192$			$\left\{\begin{array}{c}-0.084 \\ +0.020j/c_s\end{array}\right\}$	$\left\{\begin{array}{c}4.39 \\ +0.703j/c_s\end{array}\right\}$	

$$\theta = \frac{-4.39 - 0.703j/c_s}{-0.084 + 0.020j/c_s} \qquad\qquad |\theta|^2 = \frac{19.3 + 0.495/c_s^2}{0.0069 + 0.0004/c_s^2}$$

Table 38.20 gives $1/M_{9,10}$ for the other frequencies. The reciprocal $1/M_{9,10}$ of the torque is plotted in Fig. 38.24 as a function of ω_s^2; all ordinates are plotted with positive sign. The intersection of the curves for free and locked damper is at $\omega^2 = 0.0375$; $1/M_{9,10} = 0.016$ in the unity system. In terms of the actual engine, this indicates that with an optimum damper the vibratory torque will be $1/0.016 = 62.5$ times the exciting torque of one cylinder. For the eighth-order critical, the resonant amplitude calculation gives $M_e = 1,490$; then, in the engine

$$M_{9,10} = 93,200 \text{ in.-lb}$$

The calculation for optimum damping constant c_s is continued in Table 38.19. The damping constant is treated as an algebraic unknown so that the damper "spring" constant is $jc_s\omega_s$. Now let $M_{9,10}$ at $\omega_s^2 = 0.037$ equal 62.5, its value at the fixed point. Then $|\theta| = 62.5/1.570 = 39.8$.

Table 38.20. Summary of $\dfrac{1}{M_{9,10}}$ from Table 38.19 for Other Frequencies

ω_s^2	$\dfrac{1}{M_{9,10}}$	
	Free	Locked
0.035	−0.0261	0.000487
0.037	−0.01832	0.01225
0.038	−0.0137	0.0187

The damper "spring" constant $0.192c_s j$ is entered in the k column of Table 38.19 opposite the damper d. Then the relative amplitudes β in the "inertia" and "exciting" columns are the quotients M/k:

$$\frac{0.1471}{0.192c_s j} = \frac{0.767j}{c_s}$$

$$\frac{5.121}{0.192c_s j} = -\frac{26.7j}{c_s}$$

These values are entered in the β column and subtracted from the β values at rotor a, giving the β values at the damper rotor d. The amplitudes at this rotor multiplied by 0.0263 give the inertia torque reaction of damper rotor d. Adding this to the torque between a and d, the sum is equated to zero, and from this equation θ is solved for. Then the amplitude $|\theta|^2$ is obtained in the lower line of Table 38.19 by squaring and adding the real and imaginary terms of numerator and denominator. Substituting $\theta = 39.8$,

$$19.3 + \frac{0.495}{c_s^2} = (39.8)^2\left(0.0069 + \frac{0.0004}{c_s^2}\right)$$

Solving, $c_s^2 = 0.0172$, or $c_s = 0.131$. A similar calculation for $\omega_s^2 = 0.038$ leads to $c_s = 0.122$. A mean of these values is $c_s = 0.126$.

FIG. 38.24. Variation of torque in system of Fig. 38.23 when damper is either free or locked. The dots indicate computer points.

To check this value it is substituted in the $|\theta|^2$ equations of Table 38.19; then $M_{9,10}$ is obtained as follows:

$$M_{9,10} = 61.9 \quad \text{for} \quad \omega_s^2 = 0.037$$

$$M_{9,10} = 61.3 \quad \text{for} \quad \omega_s^2 = 0.038$$

Thus, the mean value of c_s places the fixed point very close to the maximum of the damped resonance curve.

It is of interest to find the fourth-order forced amplitude at 950 rpm with the damper in place. Table 38.21 shows the final items of Table 38.19 for the forced-vibration calculation with its continuation through the damper. The torque in section 9–10 is now 30.9 times the fourth-order exciting torque of one cylinder as against 20.3 without the damper (per Table 38.10); thus, the addition of the damper has increased the fourth-order stress by 50 per cent. This is because the addition of the damper inertia has lowered the frequency of the fourth-order critical, but the damper is of insufficient size to give any effective control over the fourth-order amplitude. It is extremely difficult to arrange a friction damper which will be effective for the lowest-order major criticals of four-cycle engines. For six- and eight-cylinder engines, the torque which must be developed in the damper is of the order of magnitude of the engine torque and requires very large dampers. In some cases a solution can be found with pendulum dampers.

In the actual engine the damper constant is

$$c = 0.131 \sqrt{35.8 \times 175.7 \times 10^6} = 10,400 \text{ in.-lb/rad/sec}$$

The polar inertia of the floating member is

$$J = 35.8 \times 0.71 \times 386 = 9,800 \text{ in.}^2\text{-lb}$$

A damper of the Houdaille type having the dimensions $r_2 = 8.25$ in., $r_1 = 4.125$ in., and $b = 5$ in. will have the polar inertia 9,800 in.^{2}-lb. Viscosities in the range of 20,000 to 60,000 centipoises are used.

To convert to inch-pound-second units, centipoises are multiplied by 1.45×10^{-7}. Assuming 40,000 centipoises fluid, $\mu = 0.0058$ reyn, and equal clearances $h_1 = h$, the expression for clearances obtained from Eq. (38.29) is

$$h = \frac{2\pi}{c} \mu \left[r_2^3 b + \frac{1}{2} (r_2^4 - r_1^4) \right] \tag{38.30}$$

which gives the damper clearance $h = 0.015$ in.

Table 38.21. Final Items of Table 38.19 Continued through the Damper
$(\omega_s^2 = 0.0321)$

Rotor	I	$I\omega_s^2$	β		M		k
			Inertia $\theta \times$	Exciting	Inertia $\theta \times$	Exciting	
1	1	0.0321	-7.7401	-24.748	-0.2480	-0.794	
			0.2490	3.710	0.3946	5.852	1.58
a	0.42	0.0135	-7.9891	-28.458	-0.1010	-0.384	
			$-13j$	$-242j$	0.2936	5.468	$0.0226j$
d	0.71	0.0228	-7.9891	-28.458	-0.162	-0.650	
			$+13j$	$+242j$	$0.296j$	$5.5j$	
				$RM \rightarrow$	0.1316	4.818	
					$0.296j$	$5.5j$	

$$|\theta| = \sqrt{\frac{(4.818)^2 + (5.5)^2}{(0.1316)^2 + (0.296)^2}} = 22.6$$

$|M_{9,10}|$ (unity system) $= 22.6 \times 1.366 = 30.9$

$M_{9,10}$ (actual system) $= 30.9 \times 11,300 = 349,000$ in.-lb

For a damper of the paddle type of Fig. 38.20, a diameter of 16.25 in. is taken together with the following dimensions: $r_2 = 7.16$ in., $r_1 = 3.58$ in., $d = 4.58$ in., $b_1 = b_2 = b_3 = 0.9$ in., and $n = 8$ pockets. The viscosity of SAE No. 30 oil at an operating temperature of 150°F is approximately 27 centipoises. Substituting in Eq. (38.30) and solving for h

$$h = 0.025 \text{ in. clearance}$$

To find the torque acting on the damper, refer to Table 38.20. Substituting $c_s = 0.126$ in the θ equation,

$$\theta = \frac{-4.39 - 5.58j}{-0.084 + 0.1586j}$$

But $M_{ad} = (0.1474\theta + 5.121)$ in.-lb reduces with the substitution of θ to

$$M_{ad} = \frac{1.079 - 0.013j}{0.084 - 0.1586j}$$

The absolute value is $|M_{ad}| = 6.01$, and in the actual system $|M_{ad}| = 6.01 \times 1,490 = 8,950$ in.-lb. Note that this is of the same order of magnitude as $M_e \Sigma \beta$. It is not necessarily the maximum torque in the damper, but it is close to the maximum. To find the maximum it is necessary to make a similar calculation for several adjacent frequencies.

The oil pressure in the pockets produced by this torque is

$$\frac{8,950}{8 \times 5.37 \times 3.58 \times 4.58} = 12.8 \text{ lb/in.}^2$$

Oil must be supplied at a pressure in excess of this.

To find the oil pressure for the fourth-order critical at 950 rpm, M_{ad} is obtained from Table 38.21:

$$M_{ad} = 0.2936\theta + 5.468$$

where

$$\theta = \frac{-4.818 - 5.5j}{0.1316 + 0.296j}$$

Substituting this expression for θ in the expression for M_{ad},

$$M_{ad} = \frac{0.71}{0.1316 + 0.296j}$$

The absolute value is $|M_{ad}| = 2.18$. In the actual engine, the torque is $M_{damper} = 2.18 \times 11,300 = 24,600$ in.-lb, giving a pressure of 35.3 lb/in.²

The above calculations are for a major critical with the excitations of the various cylinders all in-phase. In theory, for minor criticals where the excitations are out of phase, fixed points do not exist and this procedure does not apply. An approximation to optimum conditions for this case is obtained by assuming that the single exciting torque $M_e \Sigma \beta$ acts at the No. 1 cylinder. This approximation can be used for major criticals as well.

TWO-MASS APPROXIMATION

If the system is replaced by a two-mass system in the manner utilized to make a first estimate (see the section *Natural Frequency Calculations*) of the one-noded mode, the results are further approximated by the following formulas:

For such a two-mass plus damper system the amplitude at the fixed point is given by[17]

$$M_{12} = M_e \Sigma \beta \left(\frac{2I_2 + I_d}{I_d} \right) \tag{38.31}$$

where $M_e = Srh$ is the exciting torque per cylinder. The optimum damping is

$$c = \left[\frac{KI_2 I_d^2 (2I_1 + 2I_2 + I_d)}{I_1(I_2 + I_d)(2I_2 + I_d)} \right]^{1/2} \quad \text{in.-lb/rad/sec} \tag{38.32}$$

where I_1 = polar moment of inertia for flywheel or generator
I_2 = 40 per cent of engine polar moment of inertia taken up to flywheel
I_d = polar moment of inertia of damper floating element
k = stiffness from No. 1 crank to flywheel

Application to the engine discussed with reference to Table 38.21 gives

$$M_{9,10} = 74,300 \text{ in.-lb} \quad \text{and} \quad c = 9,700 \text{ in.-lb/rad/sec}$$

This compares with $M_{9,10} = 93,200$ and $c = 10,400$ as previously calculated.

TUNED VISCOUS DAMPERS

The procedure for the design of a tuned viscous damper is as follows:
1. Assume a polar inertia and a spring constant for the damper. As a first assumption, adjust the spring constant so that if f is the frequency of the mode to be suppressed and f_n is the natural frequency of the damper, assuming the hub as a fixed point,

$$\frac{f_n}{f} = 0.8$$

2. Plot the resonance curves of M for a particular section, first for the damper locked, then with zero damping but the damper spring in place. All ordinates are plotted positive. The curves have the general form of those shown in Fig. 38.25. They intersect in two fixed points through which all resonance curves pass, irrespective of the damping constant in the damper. If the fixed point a is higher than b, assume a lower constant for the damper spring and recalculate the M curve. If a is lower than b, do the reverse. Thus adjust the damper spring constant until a and b are of equal height. If this amplitude M is higher than desired, it is necessary to repeat the calculation with a larger damper.

With the spring and damper mass adjusted, a direct calculation (similar to that for the untuned damper) can be made to determine the damping constant c_r which will give the resonance curve the same ordinate at an intermediate frequency indicated by point c as at a and b. Figure 38.25B shows the resonance curve of an ideally adjusted damper.

Fig. 38.25. Curves of torque vs. square of frequency for auxiliary mass damper.

3. For a range of frequencies, using the inertia, spring, and damping constants as determined above, compute the amplitude of the damper mass relative to its hub by a forced-vibration calculation. In this calculation the damper spring constant becomes the complex number ($k + jc\omega$). The load for which the damper springs must be designed is k times the relative amplitude of the damper mass to its hub. The torque on the damper is approximately $M_e \Sigma \beta$.

PENDULUM DAMPERS

The principle of a pendulum damper is shown in Fig. 38.26A. (Also see Chap. 6.) The hole-pin construction usually used, which is equivalent to that of Fig. 38.26A, is shown in Fig. 38.26B. It is undesirable to have any friction in the damper. The damper produces an effect equivalent to a fixed flywheel, and the inertia of this flywheel is different for each order of vibration.

The design formulas for the pendulum damper are as follows:[3, 18] If the length L is made equal to

$$L = \frac{R}{1 + q_0{}^2} \qquad (38.33)$$

the damper is said to be tuned to order q_0. For excitation of q_0 cycles per revolution, it will act as an infinite flywheel, keeping the shaft at its point of attachment to uniform rotation insofar as q_0 order vibrations are concerned. But other orders of vibration may exist in the shaft.

FIG. 38.26. Pendulum-type damper. The arrangement is shown in principle at A, and the Chilton construction is shown schematically at B.

If the shaft at the point of attachment of the damper is vibrating with order q and amplitude θ, the maximum link angle ψ (see Fig. 38.26) is

$$\psi = \frac{\theta q^2 (1 + q_0{}^2)}{q_0{}^2 - q_2} \qquad \text{rad} \qquad (38.34)$$

The torque exerted by a single element of the damper is

$$M = \left| \frac{WR^2}{1 - q^2/q_0{}^2} + J \right| \frac{q^2 \Omega^2 \theta}{g} \qquad \text{in.-lb} \qquad (38.35)$$

where W is the weight of an element and J is the polar inertia of an element about its own center-of-gravity. The J term is equivalent to an addition to the damper hub. Dropping this term, the damper is equivalent to a flywheel of polar inertia

$$J_d = \frac{WR^2}{1 - q^2/q_0{}^2} \qquad \text{in.}^2\text{-lb} \qquad (38.36)$$

For $q < q_0$ this is a positive flywheel, for $q = q_0$ an infinite flywheel, and for $q > q_0$ a negative flywheel. Omitting the J term and eliminating θ between Eqs. (38.32) and (38.33),

$$\psi = \frac{M(1 + q_0{}^2)g}{q_0{}^2 WR^2 \Omega^2} \qquad \text{rad} \qquad (38.37)$$

APPLICATION TO A RADIAL AIRCRAFT ENGINE. Assume a nine-cylinder engine with the damper fitted to suppress the four and one-half order vibration. The damper would be tuned as an infinite flywheel with distance between hole centers

$$L = \frac{R}{1 + (4.5)^2}$$

In Eq. (38.35), M becomes the four and one-half order exciting torque. Then ψ is chosen as the permissible angle of pendulum swing, approximately 20 to 30°. The necessary polar inertia WR^2 of the damper is given by Eq. (38.35).

IN-LINE DIESEL ENGINE. As applied to a diesel engine, the above procedure is much more difficult. The exciting torques in diesel engines are nearly independent of speed. Hence from Eq. (38.37) it is evident that ψ will be inversely proportional to Ω^2. Thus for a variable-speed engine the damper size is fixed by the low-speed end of the range; if ψ is to kept in the 20 to 30° limit, the size may be excessive. This difficulty usually can be overcome by tuning the damper as a negative flywheel, thus acting to raise the undesired critical above the operating range while keeping ψ to a reasonable limit at low speed. The procedure is as follows:

Assuming a damper size and a q/q_0 ratio, a forced-vibration calculation is made starting at the flywheel end, for the maximum speed of the engine. In this calculation the damper is treated as a fixed flywheel of polar inertia $n \{[WR^2(1 - q^2/q_0^2)^{-1}] + J\}$ plus the inertia of the fixed carrier which supports the moving weights, where n is the number of weights. This calculation will yield θ, the amplitude at the damper hub, and the maximum torque in the engine shaft. Then ψ is given by Eq. (38.24). If either the shaft torque or the damper amplitude ψ is too large, it is necessary to increase the damper size and possibly adjust the q/q_0 ratio as well. A similar check for ψ is made at the low-speed end of the range with further adjustment of WR^2 and q/q_0 if necessary.

With a pendulum damper fitted, the equivalent inertia is different for each order of vibration so that each order has a different frequency. A damper tuned as a negative flywheel for one order becomes a positive flywheel for lower orders; thus, it reduces the frequencies of those orders, with possibly unfortunate results.

In in-line engines the application of a pendulum damper may be further complicated by the necessity of suppressing several orders of vibration, thus requiring several sets of damper weights. Alternatively, both a pendulum- and viscous-type damper may be fitted to an engine.

In general, the pendulum-type dampers are more expensive than the viscous types. Wear in the pins and their bushings changes the properties of the damper, thus requiring replacement of these parts at intervals.

The effect upon the fourth-order amplitude at 950 rpm of fitting a pendulum damper adjusted to act as a negative flywheel, for the generator-engine previously used, is calculated as follows: A damper is assumed which has four damper weights, each of 23.6 lb with their centers-of-gravity at a radius of 8.625 in. They will be tuned to $q_0 = 3.7$ order. The length of link is calculated from Eq. (38.31):

$$l = \frac{8.625}{1 + (3.7)^2} = 0.59 \text{ in.}$$

For fourth-order vibration the four weights are equivalent to a flywheel of polar inertia calculated from Eq. (38.36):

$$J_d = \frac{4 \times 23.6 \times (8.625)^2}{1 - (4/3.7)^2} = -41{,}250 \text{ in.}^2\text{-lb}$$

The hub inertia is assumed to be 3,000 in.²-lb; then the net inertia is $-38{,}250$ in.²-lb. In the unity system the moment of inertia is

$$I_d = \frac{-38{,}250}{35.8 \times 386} = -2.77$$

Table 38.22 gives the last lines of Table 38.10 with the calculation continued through the damper. The fourth-order torque in the engine shaft becomes 107,300 in.-lb, as compared with 230,000 in.-lb before the damper was fitted.

The angle at the damper in the unity system is

$$\theta_d = -7.9891\theta - 28.458 = 32.042 \text{ rad}$$

In the actual system this corresponds to

$$\theta_d = \frac{32.042 \times 11,300}{175.7 \times 10} = 0.00206 \text{ rad}$$

The link angle is calculated from Eq. (38.34):

$$\psi = \frac{0.00206 \times 16[1 + (3.7)^2]}{3.7^2 - 4^2} = 0.2 \text{ rad}$$

which is an acceptable amplitude.

A similar calculation is made at the low-speed end of the operating range. If the angle ψ is too large at the low-speed end, it is necessary to adjust WR^2 or q_0, or both.

Table 38.22. Continuation of Table 38.10 to Include Pendulum-type Damper ($\omega_s^2 = 0.0321$)

Rotor	I	$\omega_s^2 = 0.0321$ $\quad I\omega_s^2$	β Inertia $\theta \times$	β Exciting	M Inertia $\theta \times$	M Exciting	k
1	1.00	0.0321	-7.7401	-24.748	-0.2480	-7.94	
			0.2490	3.710	0.3946	5.852	1.58
d	-2.77	-0.0890	-7.9891	-28.458	+0.7110	2.540	
				$RM \rightarrow$	1.1056	8.392	

$$\theta = -8.392/1.1056 = -7.58$$

$$M_{9,10} \text{ (unity system)} = 1.366\theta = 10.38$$

$$M_{9,10} \text{ (actual system)} = 10.38 \times 11,300 = 107,300 \text{ in.-lb}$$

REFERENCES

1. Den Hartog, J.: "Mechanical Vibration," Dover Publications, Inc., New York, 1956.
2. Lewis, F. M.: *Trans. Soc. of Naval Arch. Marine Engrs.*, **33**:109 (1925).
3. Seward, H. L.: "Marine Engineering," vol. 2, chap. 2, The Society of Naval Architects and Marine Engineers, New York, 1944.
4. Biezeno, C. B., and Grammel, R.: "Engineering Dynamics" (translation), Blackie & Son, Ltd., Glasgow, 1946.
5. Wilson, W. K.: "Practical Solutions of Torsional Vibration Problems," John Wiley & Sons, Inc., New York, 1942.
6. Nestorides, E. J.: "A Handbook of Torsional Vibration," Cambridge University Press, 1958.
7. Constant, H.: *Brit. Aero. Res. Comm. R and M*, No. 1201, 1928.
8. Porter, F.: *Trans. ASME*, **50**:8 (1928).
9. Gümbel: *Inst. Naval Architects* (*England*), **54**:219 (1912).
10. Holzer: "Die Berechnung der Drehschwingungen," Springer-Verlag, Berlin, 1921.
11. Draminsky, P.: *Proc. Inst. Mech. Engrs.* (*London*), **159**:416 (1948).
12. Klock, N.: *Trans. ASME*, **62**:A-148 (1940).
13. Lewis, F. M.: *Trans. ASME* **57**:A-137 (1935).
14. Porter, F.: "Evaluation of Effects of Torsional Vibration," Society of Automotive Engineers, War Engineering Board; also *Trans. ASME*, **65**:A-33 (1943).

15. Dorey, S. F.: *Proc. Inst. Mech. Engrs.* (*London*), **159:**399 (1948).
16. Lewis, F. M.: *Trans. ASME*, **54:**253 (1932).
17. Lewis, F. M.: *Trans. ASME*, **78:**APM 377 (1955).
18. Zdanowich and T. S. Wilson: *Proc. Inst. Mech. Engs.* (*London*), **1943:**182 (1940).
19. Porter, F.: *Trans. ASME*, **75:**241 (1953).
20. Editorial Report on Flexible Couplings, *Engineers Digest*, **18:**271 (1957).
21. Roelig, H.: *ZVDI, Engineers Digest*, **87:**347 (1943); **1:**497 (1943).
22. U.S. Navy Department: "Military Standard Mechanical Vibrations of Mechanical Equipment." MIL-STD-167 (SHIPS).
23. Eshleman, R. L.: "Torsional Response of Internal Combustion Engines," *Trans. ASME*, **96**(2):441 (1974).
24. Anwar, I.: "Computerized Time Transient Torsional Analysis of Power Trains," *ASME Paper* No. 79-DET-74, 1979.
25. Sohre, J. S.: "Transient Torsional Criticals of Synchronous Motor-Driven, High-Speed Compressor Units," *ASME Paper* No. 66-FE-22, June 1965.

39

BALANCING OF ROTATING MACHINERY

Douglas G. Stadelbauer
Schenck Trebel Corporation

Douglas Muster
University of Houston

INTRODUCTION

The demanding requirements placed on modern rotating machines and equipment—for example, electric motors and generators, turbines, compressors, and blowers—have introduced a trend toward higher speeds and more stringent acceptable vibration levels. At lower speeds, the design of most rotors presents few problems which cannot be solved by relatively simple means, even for installations in vibration-sensitive environments. At higher speeds, which are sometimes in the range of tens of thousands of revolutions per minute, the design of rotors can be an engineering challenge which requires sophisticated solutions of interrelated problems in mechanical design, balancing procedures, bearing design, and the stability of the complete assembly. This has made balancing a first-order engineering problem from conceptual design through the final assembly and operation of modern machines.

This chapter describes some important aspects of balancing, such as the basic principles of the process by which an optimum state of balance is achieved in a rotor, balancing methods and machines, and definitions of balancing terms. The discussion is limited to those principles, methods, and procedures with which an engineer should be familiar in order to understand what is meant by "balancing."

In addition to unbalance, there are many possible sources of vibration in rotating machinery; some of them are related to or aggravated by unbalance so that, under appropriate conditions, they may be of paramount importance. However, this discussion is limited to the means by which the effect of once-per-revolution components of vibration (i.e., the effects due to mass unbalance) can be minimized.

DEFINITIONS[1]

Amount of Unbalance. The quantitative measure of unbalance in a rotor (referred to a plane) without referring to its angular position; obtained by taking the product of the unbalance mass and the distance of its center of gravity from the shaft axis.

Angle of Unbalance. Given a polar coordinate system fixed in a plane perpendicular to the shaft axis and rotating with the rotor, the polar angle at which an unbalance mass is located with reference to the given coordinate system.

Balance Quality Grade. For rigid rotors, the product of the specific unbalance and the maximum service angular speed of the rotor.

Balancing. A procedure by which the mass distribution of a rotor is checked and, if necessary, adjusted in order to ensure that the vibration of the journals and/or forces on the bearings at a frequency corresponding to operational speed are within specified limits.

Balancing Machine. A machine that provides a measure of the unbalance in a rotor which can be used for adjusting the mass distribution of that rotor mounted on it so that once-per-revolution vibratory motion of the journals or force on the bearings can be reduced if necessary.

Bearing Support. The part, or series of parts, that transmits the load from the bearing to the main body of the structure.

Center-of-Gravity (Mass Center). The point in a body through which passes the resultant of the weights of its component particles for all orientations of the body with respect to a uniform gravitational field.

Compensating (Null Force) Balancing Machine. A balancing machine with a builtin calibrated force system which counteracts the unbalanced forces in the rotor.

Correction Plane Interference (Cross Effect). The change of balancing-machine indication at one correction plane of a given rotor, which is observed for a certain change of unbalance in the other correction plane.

Correction Plane Interference Ratios. The interference ratios (I_{AB}, I_{BA}) of two correction planes A and B of a given rotor are defined by the following relationships:

$$I_{AB} = \frac{U_{AB}}{U_{BB}}$$

where U_{AB} and U_{BB} are the unbalances referring to planes A and B, respectively, caused by the addition of a specified amount of unbalance in plane B; and

$$I_{BA} = \frac{U_{BA}}{U_{AA}}$$

where U_{BA} and U_{AA} are the unbalances referring to planes B and A, respectively, caused by the addition of a specified amount of unbalance in plane A.

Critical Speed. A characteristic speed at which resonances of a system are excited. (The significant effect at critical speed may be motion of the journals or flexure of the rotor—depending on the relative magnitudes of the bearing stiffnesses.)

Couple Unbalance. That condition of unbalance for which the central principal axis intersects the shaft axis at the center of gravity.

Dynamic (Two-Plane) Balancing Machine. A centrifugal balancing machine that furnishes information for performing two-plane balancing.

Dynamic Unbalance. The condition in which the central principal axis is neither parallel to, nor intersects, the shaft axis.

Field Balancing Equipment. An assembly of measuring instruments for providing information for performing balancing operations on assembled machinery which is not mounted in a balancing machine.

Flexible Rotor. A rotor not satisfying the definition of a rigid rotor.

Flexural Critical Speed. A speed of a rotor at which there is maximum bending of the rotor and where flexure of the rotor is more significant than the motion of the bearings.

Flexural Principal Mode. For undamped rotor–bearing systems, that mode shape which the rotor takes up at one of the (rotor) flexural critical speeds.

High-speed Balancing (Relating to Flexible Rotors). A procedure of balancing at speeds where the rotor to be balanced cannot be considered rigid.

Initial Unbalance. Unbalance of any kind that exists in the rotor before balancing.

Journal Axis. The straight line joining the centroids of cross-sectional contours of the journal.

Low-speed Balancing (Relating to Flexible Rotors). A procedure of balancing at a speed where the rotor to be balanced can be considered rigid.

Minimum Achievable Residual Unbalance. The smallest value of residual unbalance that a balancing machine is capable of achieving.

Modal Balancing. A procedure for balancing flexible rotors in which balance corrections are made to reduce the amplitude of vibration in the separate significant principal flexural modes to within specified limits.

Multiplane Balancing. As applied to the balancing of flexible rotors, any balancing procedure that requires unbalance correction in more than two correction planes.

Perfectly Balanced Rotor. A rotor the mass distribution of which is such that it transmits no vibratory force or motion to its bearing as a result of centrifugal forces.

Permanent Calibration. That feature of a hard bearing balancing machine which permits it to be calibrated once, and then remain calibrated for any rotor within the capacity and speed range of the machine.

Plane Separation. Of a balancing machine, the operation of reducing the correction-plane interference ratio for a particular rotor.

Principal Inertia Axis. For each set of Cartesian coordinates at a given point, the values of the six moments of inertia of a body $I_{x_i x_j}$ $(i, j = 1, 2, 3)$ are in general unequal; for one such coordinate system the moments $I_{x_i x_j}$ $(i \neq j)$ vanish. The values of $I_{x_i x_j}$ $(i = j)$ for this particular coordinate system are called the principal moments of inertia and the corresponding coordinate directions are called the principal axes of inertia.

$$ I_{x_i x_j} = \int_m x_i x_j \, dm \qquad \text{if } i \neq j $$

$$ I_{x_i x_j} = \int_m (r^2 - x_i^2) \, dm \qquad \text{if } i = j $$

where $r^2 = x_1^2 + x_2^2 + x_3^2$ and $x_i x_j$ are Cartesian coordinates.

Residual Unbalance. Unbalance of any kind that remains after balancing.

Resonance Balancing Machine. A balancing machine having a balancing speed corresponding to the natural frequency of the suspension-and-rotor system.

Rigid Rotor. A rotor is considered rigid when its unbalance can be corrected in any two (arbitrarily selected) planes and, after the correction, its unbalance does not significantly change (relative to the shaft axis) at any speed up to maximum service speed and when running under conditions which approximate closely those of the final supporting system.

Rotor. A body, capable of rotation, generally with journals which are supported by bearings.

Single-plane (Static) Balancing Machine. A gravitational or centrifugal balancing machine that provides information for accomplishing single-plane balancing.

Static Unbalance. That condition of unbalance for which the central principal axis is displaced only parallel to the shaft axis.

Unbalance. That condition which exists in a rotor when vibratory force or motion is imparted to its bearings as a result of centrifugal forces.

BASIC PRINCIPLES OF BALANCING

Descriptions of the behavior of rigid or flexible rotors are given as introductory material in standard vibration texts, in the references listed at the end of the chapter, and in the few books devoted to balancing. A similar description is included here for the purpose of examining the principles which govern the behavior of rotors as their speed of rotation is varied.

PERFECT BALANCE

Consider a rigid body which is rotating at a uniform speed about one of its three principal inertial axes. Suppose that the forces which cause the rotation and support the body are neglected; then it will rotate about this axis without wobbling, i.e., the principal axis (which is fixed in the body) coincides with a line fixed in space (Fig. 39.1). Now construct circular, concentric journals around the axis at the points where the axis protrudes from the body, i.e., on the stub shafts whose axes coincide with the principal axis. Since the axis does not wobble, the newly constructed journals also will not wobble. Next, place the journals in bearings which are circular and concentric to the principal axis (Fig. 39.2). It is assumed that there is no dynamic action of the elasticity

FIG. 39.1. Rigid body rotating about principal axis.

FIG. 39.2. Balanced rigid rotor.

of the rotor and the lubricant in the bearings. A rigid rotor constructed and supported in this manner will not wobble; the bearings will exert no forces other than those necessary to support the weight of the rotor. In this assembly, the radial distances between the center-of-gravity of the rotor and the journal and bearing axes are zero. This rotor is said to be *perfectly balanced*.

RIGID-ROTOR BALANCING—STATIC UNBALANCE

Rigid-rotor balancing is important because it comprises a substantial and probably the greater portion of balancing work done in industry. By far the greatest number of rotors manufactured and installed in equipment can be classified as "rigid" by definition. All balancing machines are designed to perform rigid-rotor balancing.*

Consider the case in which the journals are concentric to a line fixed in the body other than the principal axis, as illustrated in Fig. 39.3. In practice, with even the closest manufacturing tolerances, the journals are never concentric with the principal axis of the rotor. If concentric rigid bearings are placed around the journals, thus forcing the rotor to turn about the journal axis, a variable force is sensed at each bearing.

FIG. 39.3. Unbalanced rigid rotor.

The center-of-gravity is located on the principal axis, and is not on the axis of rotation (journal and bearing axes). From this it follows that there is a net radial force acting on the rotor which is due to centrifugal acceleration. The magnitude of this force is given by

$$F = m\epsilon\omega^2 \tag{39.1}$$

where m is the mass of the rotor, ϵ the eccentricity or radial distance of the center-of-gravity from the axis of rotation, and ω is the rotational speed in radians per second. Since the rotor is assumed to be rigid and thus not capable of distortion, this force is balanced by two reaction forces. There is one force at each bearing. Their algebraic sum is equal in magnitude and opposite in sense. The relative magnitudes of the two forces depend, in part, upon the axial position of each bearing with respect to the center-of-gravity of the rotor. In simplified form, this illustrates the "balancing problem." One must choose a practical method of constructing a perfectly balanced rotor from this unbalanced rotor.

The center-of-gravity may be moved to the journal (bearing) axis (or as close to this axis as is practical) in one of two ways. The journals may be modified so that the jour-

* Field balancing equipment is specifically excluded from this category since it is designed for use with both rigid and flexible rotors.

nal axis and an axis through the center-of-gravity are moved to essential coincidence. From theoretical considerations, this is a valid method of minimizing unbalance caused by the displacement of the center-of-gravity from the journal axis, but for practical reasons it is difficult to accomplish. Instead, it is easier to achieve a radial shift of the center-of-gravity by adding mass to or subtracting it from the mass of the rotor; this change in mass takes place in the longitudinal plane which includes the journal axis and the center-of-gravity. From Eq. (39.1), it follows that there can be no net radial force acting on the rotor at any speed of rotation if

$$m'r = m\epsilon \tag{39.2}$$

where m' is the mass added to or subtracted from that of the rotor and r is the radial distance to m'. There may be a *couple*, but there is no net *force*. Correspondingly, there can be no *net bearing reaction*. Any residual reactions sensed at the bearings would be due solely to the couple acting on the rotor.

If this rotor-bearing assembly were supported on a scale having a sufficiently rapid response to sense the change in force at the speed of rotation of the rotor, no fluctuations in the magnitude of the force would be observed. The scale would register only the dead weight of the rotor-bearing assembly.

This process of *effecting essential coincidence between the center-of-gravity of the rotor and the axis journal is called "single-plane (static) balancing."* The latter name for the process is more descriptive of the end result than the procedure that is followed.

If a rotor which is supported on two bearings has been balanced statically, the rotor will not rotate under the influence of gravity alone. It can be rotated to any position and, if left there, will remain in that position. However, if the rotor has not been balanced statically, then from any position in which the rotor is initially placed it will tend to turn to that position in which the center-of-gravity is lowest.

As indicated below, single-plane balancing can be accomplished most simply (but not necessarily with great accuracy) by supporting the rotor on flat, horizontal ways and allowing the center-of-gravity to seek its lowest position. It also can be accomplished in a centrifugal balancing machine by sensing and correcting for the unbalance force characterized by Eq. (39.1).

RIGID-ROTOR BALANCING—DYNAMIC UNBALANCE

When a rotor is balanced statically, the journal axis and principal inertial axis may not coincide; single-plane balancing ensures that the axes have only one common point, namely, the center-of-gravity. Thus, perfect balance is not achieved. To obtain perfect balance, the principal axis must be rotated about the center-of-gravity in the longitudinal plane characterized by the journal axis and the principal axis. This rotation can be accomplished by modifying the journals (but, as before, this is impractical) or by adding masses to or subtracting them from the mass of the rotor in the longitudinal plane characterized by the journal axis and the principal inertial axis. Although adding or subtracting a single mass may cause rotation of the principal axis relative to the journal axis, it also disturbs the static balance already achieved. From this it can be deduced that a couple must be applied to the rotor in the longitudinal plane. This is usually accomplished by adding or subtracting two masses of equal magnitude, one on each side of the principal axis (so as not to disturb the static balance) and one in each of two radial planes (so as to produce the necessary rotatory effect). Theoretically, it is not important which two radial planes are selected since the same rotatory effect can be achieved with appropriate masses, irrespective of the axial location of the two planes. Practically, the choice of suitable planes can be vital. Usually, it is best to select planes which are separated axially by as great a distance as possible in order to minimize the magnitude of the masses required.

The above process of *bringing the principal inertial axis of the rotor into essential coincidence with the journal axis is called "two-plane (dynamic) balancing."* If a rotor is balanced in two planes, then, by definition, it is balanced statically; however, the converse is not true.

FLEXIBLE-ROTOR BALANCING[2-4]

If the bearing supports are rigid, then the forces exerted on the bearings are due entirely to centrifugal forces caused by the residual unbalance. Dynamic action of the elasticity of the rotor and the lubricant in the bearings has been ignored.

The portion of the over-all problem in which the dynamic action and interaction of rotor elasticity, bearing elasticity, and damping are considered is called flexible rotor or modal balancing.

CRITICAL SPEED. Consider a long, slender rotor, as shown in Fig. 39.4. It represents the idealized form of a typical flexible rotor, such as a paper machinery roll or

Fig. 39.4. Idealized flexible rotor.

turbogenerator rotor. Assuming further that all unbalances occurring along the rotor caused by machining tolerances, inhomogeneities of material, etc. are compensated by correction weights placed in the end faces of the rotor, and that the balancing is done at a low speed as if the rotor were a rigid body.

Assume there is no damping in the rotor or its bearing supports. Consider a thin slice of this rotor perpendicular to the journal axis (see Fig. 39.5A). This axis intersects the slice at its equilibrium center E when the rotor is not rotating, provided that deflection due to gravity forces is ignored. The center-of-gravity of the slice is displaced by δ from E due to an unbalance in the slice (caused by machining tolerances, inhomogeneity, etc., mentioned above) which was compensated by correction weights in the rotor's end planes. If the rotor starts to rotate about the journal axis with an angular speed ω, then the slice starts to rotate in its own plane at the same speed about an axis through E. Centrifugal force $me\omega^2$ is thus experienced by the slice. This force occurs in a direction perpendicular to the journal axis and may be accompanied by similarly caused forces at other cross sections along the rotor; such forces are likely to vary in magnitude and direction. They cause the rotor to bend, which in turn causes additional centrifugal forces and further bending of the rotor.

At every speed ω, equilibrium conditions require for one slice that the centrifugal and restoring forces be related by

$$m(\delta + x)\omega^2 = kx \qquad (39.3)$$

where x is the deflection of the shaft (the radial distance between the equilibrium center and the journal axis) and k is the shaft stiffness (Fig. 39.5B). In Fig. 39.5, the centrifugal and restoring forces are plotted for various speeds ($\omega_1 < \omega_2 < \omega_3 < \omega_4 < \omega_5$). The point of intersection of the lines representing the two forces denotes the equilibrium condition for the rotor at the given speeds. For this ideal example, as the speed increases, the point which denotes equilibrium will move outward until, at say ω_3, a speed is reached at which there is no resulting force and the lines are parallel. Since equilibrium is not possible at this speed, it is called the critical speed. *The critical speed ω_n of a rotating system corresponds to a resonant frequency of the system.*

At speeds greater than ω_3 ($\equiv \omega_n$), the lines representing the centrifugal and restoring forces again intersect. As ω increases, the slope of the line $m\omega^2(x + \delta)$ increases correspondingly until, for speeds which are large, the deflection x approaches the value of δ, i.e., the rotor tends to rotate about its center-of-gravity.

UNBALANCE DISTRIBUTION. Apart from any special and obvious design features, the axial distribution of unbalance in the slices previously examined along any rotor is likely to be random. The distribution may be significantly influenced by the presence of large local unbalances arising from shrink-fitted discs, couplings, etc. The rotor may also have a substantial amount of initial bend which may produce effects similar to those due to unbalance. The method of construction can influence signifi-

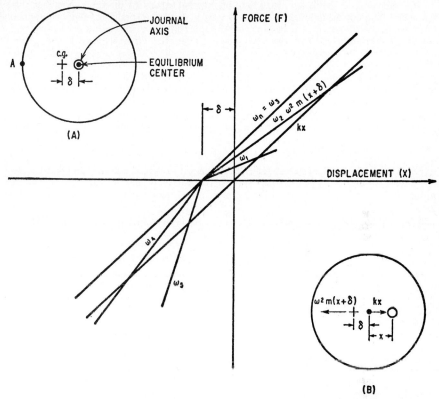

Fig. 39.5. Rotor behavior below, at, and above first critical speed.

cantly the magnitude and distribution of unbalance along a rotor. Rotors may be machined from a single forging or they may be constructed by fitting several components together. For example, jet-engine rotors are constructed by joining many shell and disc components, whereas alternator rotors are usually manufactured from a single piece of material, although they still have additional components fitted.

The unbalance distributions along two nominally identical rotors may be similar but rarely identical.

Contrary to the case of a rigid rotor, distribution of unbalance is significant in a flexible rotor because it determines the degree to which any bending or flexural mode of vibration is excited. The resulting modal shapes are reduced to acceptable levels by flexible-rotor balancing, also called "modal balancing."*

RESPONSE OF A FLEXIBLE ROTOR TO UNBALANCE. In common with all vibrating systems, rotor vibration is the sum of its modal components. For an *undamped* flexible rotor which rotates in flexible bearings, the flexural modes coincide with principal modes and are plane curves rotating about the axis of the bearing. For a *damped* flexible rotor, the flexural modes may be space (three-dimensional) curves rotating about the axis of the bearings. The damping forces also limit the flexural amplitudes at each critical speed. In many cases, however, the damped modes can be treated approximately as principal modes and hence regarded as rotating plane curves.

* All modal balancing is accomplished by multiplane corrections; however, multiplane balancing need not be modal balancing, since multiplane balancing refers only to unbalance correction in more than two planes.

The unbalance distribution along a rotor may be expressed in terms of modal components. The vibration in each mode is caused by the corresponding modal component of unbalance. Moreover, the response of the rotor in the vicinity of a critical speed is usually predominantly in the associated mode. The rotor modal response is a maximum at any rotor critical speed corresponding to that mode. Thus, when a rotor rotates at a speed near a critical speed, it is disposed to adopt a deflection shape corresponding to the mode associated with this critical speed. The degree to which large amplitudes of rotor deflection occur in these circumstances is determined by the modal component of unbalance and the amount of damping present in the rotor system.

If the modal component of unbalance is reduced by a number of discrete correction masses, then the corresponding modal component of vibration is similarly reduced. The reduction of the modal components of unbalance in this way forms the basis of the modal balancing technique.

FLEXIBLE-ROTOR MODE SHAPES. The stiffnesses of a rotor, its bearings, and the bearing supports affect critical speeds and therefore mode shapes in a complex manner. For example, Fig. 39.6 shows the effect of varying bearing and support stiff-

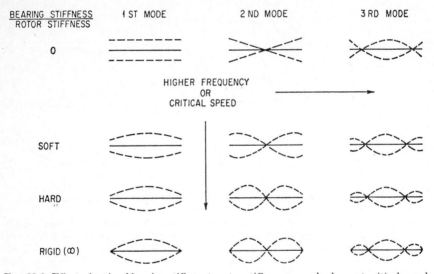

Fɪɢ. 39.6. Effect of ratio of bearing stiffness to rotor stiffness on mode shape at critical speeds.

ness relative to that of the rotor. The term "soft" or "hard" bearing is a relative one, since for different rotors the same bearing may appear to be either soft or hard. The schematic diagrams of the figure illustrate that the first critical speed of a rotor supported in a balancing machine having soft-spring-bearing supports occurs at a lower frequency and in an apparently different shape than that of the same rotor supported in a hard-bearing balancing machine where the bearing support stiffness approximates service conditions.

To evaluate whether a given rotor may require a flexible-rotor balancing procedure, the following rotor characteristics must be considered:

1. Rotor configuration and service speed.
2. Rotor design and manufacturing procedures. Rotors which are known to be flexible or unstable may still be capable of being balanced as rigid rotors.

ROTOR ELASTICITY TEST. This test is designed to determine if a rotor can be considered *rigid* for balancing purposes or if it must be treated as *flexible*. The test is carried out at service speed either under service conditions or in a high-speed, hard-bearing balancing machine whose support-bearing stiffness closely approximates that of the final supporting system. The rotor should first be balanced. A weight is then added in each end plane of the rotor near its journals; the two weights must be in the same angular position. During a subsequent test run, the vibration is measured at both bearings. Next, the rotor is stopped and the test weights are moved to the center of the rotor, or to a position where they are expected to cause the largest rotor distortion; in another run the vibration is again measured at the bearings. If the total of the first readings is designated x, and the total of the second readings y, then the ratio between $(y - x)/x$ should not exceed 0.2. Experience has shown that if this ratio is below 0.2, the rotor can be corrected satisfactorily at low speed by applying correction weights in two or three selected planes. Should the ratio exceed 0.2, the rotor usually must be checked at or near its service speed and corrected by a modal balancing technique.

HIGH-SPEED BALANCING MACHINES. Any technique of modal balancing requires a balancing machine having a variable balancing speed with a maximum speed at least equal to the maximum service speed of the flexible rotor. Such a machine must also have a drive-system power rating which takes into consideration not only acceleration of the rotor inertia but also windage losses and the energy required for a rotor to pass through a critical speed. For some rotors, windage is the major loss; such rotors may have to be run in vacuum chambers to reduce the fanlike action of the rotor and to prevent the rotor from becoming excessively hot. For high-speed balancing installations, appropriate controls and safety measures must be employed to protect the operator, the equipment, and the surrounding work areas.

FLEXIBLE-ROTOR BALANCING TECHNIQUES. Flexible rotor balancing consists essentially of a series of individual balancing operations performed at successively greater rotor speeds:

At a low speed, where the rotor is considered rigid. (Low speed balancing of flexible rotors usually is performed only in a balancing machine.)

At a speed whereby significant rotor deformation occurs in the mode of the first flexural critical speed. (This deformation may occur at speeds well below the critical speed.)

At a speed whereby significant rotor deformation occurs in the mode of the second flexural critical speed. (This applies only to rotors with a maximum service speed affected significantly by the mode shape of the second flexural critical speed.)

At a speed whereby significant rotor deformation occurs in the mode of the third critical speed, etc.

At the maximum service speed of the rotor.

The balancing of flexible rotors requires experience in determining the size of correction weights when the rotor runs in a flexible mode. The process is considerably more complex than standard low-speed balancing techniques used with rigid rotors. Primarily this is due to a shift of mass within the rotor (as the rotational speed changes), caused by shaft and/or body elasticity, asymmetric stiffness, thermal dissymmetry, incorrect centering of rotor mass and shifting of windings and associated components, magnetic forces, and fit tolerances and couplings.

Before starting the modal balancing procedure, the rotor temperature should be stabilized in the lower- or middle-speed range, until unbalance readings are repeatable. This preliminary warmup may take from a few minutes to several hours depending on the type of rotor, its dimensions, its mass, and its pretest condition.

Once the rotor is temperature-stabilized, the balancing process can begin. Several weight corrections in both end planes and along the rotor surface are required. Two axial planes, at right angles, are fixed in the rotor. Their line of intersection is the journal axis. Preferably, unbalance corrections should be made in these planes. This procedure permits all unbalance vectors (in a given correction plane) to be resolved into 90° components.

In the commonly practiced, comprehensive modal balancing technique, unbalance correction is performed in several discrete modes, each mode being associated with the speed

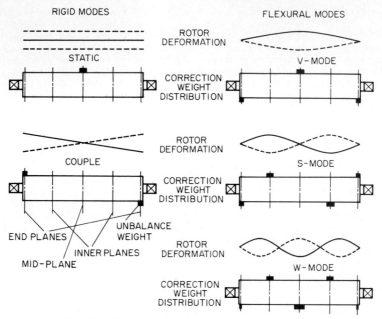

F<small>IG</small>. 39.7. Rotor mode shapes and correction weights.

range in which the rotor is deformed to the mode shape corresponding to a particular flexural critical speed. Figure 39.7 shows a rotor deformed in five of the mode shapes of Fig. 39.6; the location of the weights which provide the proper correction for these mode shapes is indicated.

First, the rotor is rotated at a speed less than one-half the rotor's first flexural critical speed and balanced using a rigid-rotor balancing technique. Balancing corrections are performed at the end planes in the X and Y axes to reduce the original amount of unbalance to three or four times the final balancing tolerance.

Correction for the First Flexural Mode (V Mode). The balancing speed is increased until rotor deformation occurs, accompanied by a rapid increase in unbalance indication at the same angular position for both end planes. Unbalance corrections for this mode are made in at least three planes. Due to the bending of the rotor, the unbalance indication is not directly proportional to the correction to be applied. A new relationship between unbalance indication and corresponding correction weight must be established by test with trial weights. Such weights are first added in the correction plane nearest the middle of the rotor, in perpendicular components using the previously established axial planes fixed in the rotor. For large turbo-generator rotors the resultant of these components should be in the range of 30 to 60 oz-in./ton of rotor weight. Two additional corrections are added in the end planes diametrically opposite to the center weights, again in component form, each equal to one-half the magnitude of the corresponding center weights. This process may have to be repeated a number of times, each run reducing the magnitude of the weight applications until the residual unbalance is approximately 1 to 3 oz-in./ton of turbo-generator rotor weight. Then the speed is increased slowly to the maximum service speed; at the same time, the unbalance indicator is monitored. If an excessive unbalance indication is observed as the rotor passes through its first critical speed, further unbalance corrections are required in the V mode until the maximum service speed can be reached without an excessive unbalance indication. If a second flexural critical speed is observed before the maximum service speed is reached, the additional balancing operation in the S mode must be performed, as indicated below.

Correction for the Second Flexural Mode (S Mode). The rotor speed is increased until significant rotor deformation due to the second flexural mode is observed. This is indicated by a rapid increase in unbalance indication measured in the end planes at angular positions opposite to each other. Unbalance corrections for this S mode are made in at least four planes, as indicated in Fig. 39.7. The weights placed in the end correction planes, again in components, must be diametrically opposed; on the idealized symmetrical rotor, each end-plane weight must be equal to one-half the correction weight placed in one of the inner planes. Of primary concern is that the S-mode weight set will not have any influence on the previously corrected mode shapes. The correction weight in each inner plane must be diametrically opposed to its nearest end-plane correction weight. The procedure to determine the relationship between unbalance indication and required correction weight is similar to that used in the V-mode procedure, described above. The S-mode balancing process must be repeated until an acceptable residual unbalance is achieved.

Corrections for the Third Flexural Mode (W Mode). The rotor speed is increased further until significant rotor deformation due to the third flexural mode is observed. Corrections are made in the rotor with a five-weight set (shown in Fig. 39.7) and in a manner similar to that used in correcting for the first and second flexural modes.

If the service-speed range requires it, higher modes (those associated with the nth critical speed, for example) may have to be corrected as well. For each of these higher modes, a set of $(n + 2)$ correction weights is required.

Final Balancing at Service Speed. Final balancing takes place with the rotor at its service speed. Correction should be made only in the end planes. The final balance tolerance for large turbo-generators, for example, will normally be in the order of 1 oz-in./ton of rotor weight. If the the rotor cannot be brought into proper balance tolerances, the S-mode, V-mode, and W-mode corrections may require slight adjustment.

To achieve repeatability of the correction effects, the duration of each run must remain constant and the same balancing speed for each mode must be accurately maintained. Depending on the size of the rotor, the number of modes that must be corrected, and the ease with which weights can be applied, the entire process may take anywhere from 3 to 30 hours.

The relative position of the unbalance correction planes shown in Fig. 39.7 applies to symmetrical rotors only. Rotors with axial asymmetry generally require unsymmetrically spaced correction planes. In case of assembled rotors which may "take a set" at or near service speed (e.g., shrunk-on turbine stages find their final position), only preliminary unbalance corrections are made at lower speeds to enable the rotor to be accelerated to service or overspeed, the latter being usually 20 per cent above maximum service speed. Since the "set" creates new unbalance, the normal balancing procedure is only commenced after the initial high speed run.

Computer programs are available which facilitate the selection of the most appropriate correction planes and the computation of correction weights by the influence coefficient method.[5,6] Other flexible rotor balancing techniques rely mostly on experience data available from previously manufactured rotors of the same type, or correct only for flexural modes if no low-speed balancing equipment is available.[7]

SOURCES OF UNBALANCE

Sources of unbalance in rotating machinery may be classified as resulting from

1. Dissymmetry
   (Core shifts in casting, rough surfaces on forgings, unsymmetrical configuration)
2. Nonhomogeneous material
   (Blowholes in cast rotors, inclusions in rolled or forged materials, slag inclusions or

variations in crystalline structure caused by variations in the density of the material)
3. Distortion at service speed
   (Blower blades in builtup design)
4. Eccentricity
   (Journals not concentric or circular, matching holes in builtup rotor not circular)
5. Misalignment of bearings
6. Shifting of parts due to plastic deformation of rotor part
   (Windings in rotor)
7. Hydraulic or aerodynamic unbalance
   (Cavitation or turbulence)
8. Thermal gradients
   (Steam-turbine rotors)

Often, balancing problems can be minimized by careful design in which unbalance is carefully controlled. When a part is to be balanced, large amounts of unbalance require large corrections. If such corrections are made by removal of material, additional cost is involved and part strength may be affected. If corrections are made by addition of material, cost is again a factor and space requirements for the added material may be a problem.

Manufacturing processes are the major source of unbalance. Unmachined portions of castings or forgings which cannot be made concentric and symmetrical with respect to the journal axis introduce substantial unbalance. Manufacturing tolerances and processes which permit any eccentricity or lack of squareness with respect to the journal axis are sources of unbalance. The tolerances, necessary for economical assembly of several elements of a rotor, permit radial displacement of parts of the assembly and thereby introduce unbalance.

Limitations imposed by design often introduce unbalance effects which cannot be corrected adequately by refinement in design. For example, electrical design limitations impose a requirement that one coil be at a greater radius than the others in a certain type of universal motor armature. It is impractical to design a compensating unbalance into the armature.

Fabricated parts, such as fans, often distort nonsymmetrically under service condi-

STATIC
UNBALANCE

COUPLE
UNBALANCE

(A)

(B)

Fig. 39.8. Effect of static and couple unbalance on free rotor motion.

tions.   Design and economic considerations prevent the adaptation of methods which might eliminate this distortion and thereby reduce the resulting unbalance.

Ideally, rotating parts always should be designed for inherent balance, whether a balancing operation is to be performed or not.   Where low service speeds are involved and the effects of a reasonable amount of unbalance can be tolerated, this practice may eliminate the need for balancing.   In parts which require unbalanced masses for functional reasons, these masses often can be counterbalanced by designing for symmetry about the journal axis.

MOTIONS OF UNBALANCED ROTORS

In Fig. 39.8 a rotor is shown spinning freely in space.   This corresponds to spinning above resonance in soft bearings.   In Fig. 39.8*A* only static unbalance is present and the center line of the shaft sweeps out a cylindrical surface.   Figure 39.8*B* illustrates the motion when only couple unbalance is present.   In this case, the center line of the rotor shaft sweeps out two cones which have their apexes at the center-of-gravity of the rotor.   The effect of combining these two types of unbalance when they occur in the same axial plane is to move the apex of the cones away from the center-of-gravity. In many cases, there will be no apex and the shaft will move in a more complex combination of the motions shown in Fig. 39.8.

OPERATING PRINCIPLES OF BALANCING MACHINES[8, 9]

This section describes the basic operating principles and general features of the various types of balancing machines which are available commercially.   With this type of information, it is possible to determine the basic type of machine required for a given application to stated specifications.

Every balancing machine must determine by some technique both the magnitude of a correction weight and its angular position in each of one, two, or more selected balancing planes.   For single-plane balancing this can be done statically, but for two- or multiplane balancing it can be done only while the rotor is spinning.   Finally, all machines must be able to resolve the unbalance readings, usually taken at the bearings, into equivalent corrections in each of the balancing planes.

On the basis of their method of operation, balancing machines and equipment can be grouped in two general categories:
1. Gravity balancing machines
2. Centrifugal balancing machines and field balancing equipment.

In the first category, advantage is taken of the fact that a body free to rotate always seeks that position in which its center of gravity is lowest.   Gravity balancing machines, also called *nonrotating balancing machines*, include horizontal ways, knife-edges or roller arrangements, and vertical pendulum types.   All are capable of only detecting and/or indicating static unbalance.

In the second category, the amplitude and phase of motions or reaction forces caused by once-per-revolution centrifugal forces resulting from unbalance are sensed, measured, and indicated by appropriate means.   Field balancing equipment provides sensing and measuring instrumentation only; the necessary measurements for balancing a rotor are taken while the rotor runs in its own bearings and under its own power.   However, on a centrifugal balancing machine, the rotor is supported by the machine and rotated around a horizontal or vertical axis by the machine's drive motor.   Balancing-machine instrumentation differs in specific features from field balancing equipment which simplify the balancing process.   A centrifugal balancing machine (also called a *rotating balancing machine*) is usually capable of measuring static unbalance (by means of a *single-plane rotating balancing machine*) or static *and* dynamic unbalance (by means of a *two-plane rotating balancing machine*).   Only a two-plane rotating balancing machine can detect couple unbalance or dynamic unbalance.

GRAVITY BALANCING MACHINES

First, consider the simplest type of balancing—usually called "static" balancing, since the rotor is not spinning. In Fig. 39.9A, a disc-type rotor on a shaft is shown resting on knife-edges. The mass added to the disc at its rim represents a known unbalance. In this illustration, in Fig. 39.8, and in the illustrations which follow, the rotor is assumed to be balanced without this added unbalance weight. In order for this balancing procedure to work effectively, the knife-edges must be level, parallel, hard, and straight.

In operation, the heavier side of the disc will seek the lowest level—thus indicating the angular position of the unbalance. Then, the magnitude of the unbalance usually is determined by an empirical process, adding mass in the form of wax or putty to the light side of the disc until it is in balance, i.e., until the disc does not stop at the same angular position.

In Fig. 39.9B a set of balanced rollers or wheels is used in place of the knife-edges. These have the advantage of permitting the rotor to turn without, at the same time, moving laterally.

In Fig. 39.9C, a setup for another type of static, or "nonrotating," balancing procedure is shown. Here the disc to be balanced is supported by a flexible cable, fastened to a point on the disc which coincides with the center of the shaft and is slightly above the normal plane containing the center-of-gravity. As shown in Fig. 39.9C, the heavy side will tend to seek a lower level than the light side, thereby indicating the angular position of the unbalance. The disc can be balanced by adding weight to the diametrically opposed side of the disc until it hangs level. In this case, the center-of-gravity is moved until it is directly under the flexible support cable.

In modified versions of this setup, the cable is replaced by a hardened ball-and-socket arrangement (used on many automobile wheel balancers) or by a spherical air bearing (used on some industrial and aerospace balancers).

Static balancing is satisfactory for rotors having relatively low service speeds and axial lengths which are small in comparison with the rotor diameter. A preliminary static unbalance correction may be required on rotors having a combined unbalance so large that it is impossible in a dynamic, soft-bearing balancing machine to bring the rotor up to its proper balancing speed without damaging the machine. If the rotor is first balanced statically by one of the methods just outlined, it is usually possible to decrease the combined unbalance to the point where the rotor may be brought up to balancing speed and the residual unbalance measured. Such preliminary static correction is not required on hard-bearing balancing machines.

CENTRIFUGAL BALANCING MACHINES

The following procedures may be used to balance the rotor shown in Fig. 39.8B. First, select the planes in which the correction weights are to be added; these planes should be as far apart as possible and the weights should be added as far out from the shaft as feasible to minimize the size of the weights. Next, by a balancing technique, determine the size of the required correction weight and its angular position for each correction plane. To implement these procedures two types of machines, *soft-bearing* and *hard-bearing* balancing machines, which are described below, are employed.

SOFT-BEARING BALANCING MACHINES. Soft-bearing balancing machines permit the idealized free rotor motion illustrated in Fig. 39.8B; but on most machines the motion is restricted to a horizontal plane (as shown in Fig. 39.10). Furthermore, the bearings (and the directly attached components) vibrate in unison with the rotor, thus adding to its mass. The restriction on the vertical motion does not affect the amplitude of vibration in the horizontal plane, but the added mass of the bearings does. The greater the combined rotor-and-bearing mass, the smaller will be the displacement of the bearings, and the smaller will be the output of the devices which sense the unbalance.

Consider the following example. Assume a balanced disc (see Fig. 39.11) having a

(A) HORIZONTAL WAYS   (B) ROLLER   (C) PENDULUM

FIG. 39.9. Static (single-plane) balancing devices.

FIG. 39.10. Motion of unbalanced rotor and bearings in flexible-bearing, centrifugal balancing machine.

FIG. 39.11. Displacement of center-of-gravity because of unbalance.

weight W of 1,000 gm, rotating freely in space. An unbalance weight w of 1 gm is then added to the disc at a radius of 10 mm. The unbalance causes the center-of-gravity of the disc to be displaced from the journal axis by

$$e = \frac{wr}{W + w} = 0.00999 \text{ mm}$$

Since the addition of the weight of the unbalance to the rotor causes only an insignificant difference, the approximation $e \approx wr/W$ is generally used. Then $e \approx 0.01$ mm.

If the same disc with the same unbalance is rotated on a single-plane balancing machine having a bearing and bearing housing weight W' of 1,000 gm, the displacement of center-of-gravity will be significantly reduced because the bearing and housing weight is added to the weight W of the disc. The center-of-gravity of the combined vibrating components will now be displaced by

$$e' = \frac{wr}{W + W'} \approx 0.005 \text{ mm}$$

The conversion of unbalance into displacement of center-of-gravity as shown in the example above also holds true for rotors of greater axial length which normally require correction in two planes. However, such rotors are prone to have unbalance other than static unbalance, causing an inclination of the principal inertia axis from the journal axis. In turn, this results in a displacement of the principal inertia axis from the journal axis in the bearing planes of the rotor, causing the balancing machine bearings to vibrate.

To find the bearing displacement or bearing vibration amplitude resulting from a given unbalance is more involved than finding the center-of-gravity displacement, because other factors come into play, as is illustrated by Fig. 39.12.

FIG. 39.12. Displacement of principal axis of inertia from shaft axis at bearing.

The weight and inertia of the balancing machine bearings and directly attached vibratory components are usually not known. In any case, they are usually small in relation to the weight and the inertia of the rotor and can generally be ignored. On this basis, the following formula may be used to find the approximate bearing displacement d:

$$d \approx \frac{wr}{W} + \frac{wrhs}{g(I_x - I_z)}$$

where d = displacement at bearing of principal inertia axis from journal axis
   r = distance from journal axis to unbalance weight
   h = distance from center-of-gravity to unbalance plane
   s = distance from center-of-gravity to bearing plane
   g = gravitational constant
   I_x = moment of inertia around transverse axis X
   I_z = moment of inertia around principal axis Z

From the above can be seen that the relationship between bearing motion and unbalance in a soft-bearing balancing machine is complex. Therefore, a direct indication of unbalance can be obtained only after calibrating the indicating elements to a given rotor by use of test weights which produce a known amount of unbalance.

HARD-BEARING BALANCING MACHINES. Hard-bearing balancing machines are essentially of the same construction as soft-bearing balancing machines except that their bearing supports are significantly stiffer in the horizontal direction. This results in a horizontal critical speed for the machine which is several orders of magnitude greater than that for a comparable soft-bearing balancing machine. The hard-bearing balancing machine is designed to operate at speeds well below its horizontal critical speed. In this speed range, the output from the sensing elements attached to the balancing-machine bearing supports is directly proportional to the centrifugal force resulting from unbalance in the rotor. The output is not influenced by bearing mass, rotor weight, or inertia, so that a permanent relation between unbalance and sensing element output can be established. Unlike soft-bearing balancing machines, the use of test weights is not required to calibrate the machine for a given rotor.

MEASUREMENT OF AMOUNT AND ANGLE OF UNBALANCE. Both soft- and hard-bearing balancing machines use various types of sensing elements *at the rotor-bearing supports* to convert mechanical vibration into an electrical signal. On commercially available balancing machines, these sensing elements employed are usually velocity-type pickups, although certain hard-bearing balancing machines use magnetostrictive or piezoelectric pickups.

Three basic methods are used to obtain a reference signal by which the phase angle of the amount-of-unbalance indication signal may be correlated with the rotor. On end-drive machines (where the rotor is driven via a universal joint driver or similarly flexible coupling shaft) a phase reference generator, directly coupled to the balancing machine drive spindle, is used. On belt-drive machines (where the rotor is driven by a belt over the rotor periphery) or on air-drive or self-drive machines, a stroboscopic lamp flashing once per rotor revolution or a photoelectric cell is employed to obtain the phase reference. Both the stroboscopic light and photoelectric cell require a reference mark on the rotor to obtain the angular position of unbalance in the rotor. The output from the phase-reference sensor and the pickups at the rotor bearing supports are processed in various ways by different manufacturers. Generally, the processed signals result in an indication representing the amount of unbalance and its angular position. In Fig. 39.13 block diagrams are shown for typical balancing instrumentation. In Fig. 39.13A an indicating system is shown which uses switching between correction planes (i.e., single-channel instrumentation). This is generally employed on balancing machines with stroboscopic-angle indication and belt drive. In Fig. 39.13B an indicating system is shown with two-channel instrumentation. Combined indica-

FIG. 39.13. Block diagrams of typical balancing-machine instrumentations. (A) Amount of unbalance indicated on analog meters, angle by strobe light. (B) Combined amount and angle indication on vectormeters.

tion of amount of unbalance and its angular position is provided on a vectormeter having an illuminated target projected on a screen. Displacement of the target from the central zero point provides a direct visual representation of the displacement of the principal inertia axis from the journal axis. Concentric circles on the screen indicate the amount of unbalance, and radial lines indicate its angular position.

INDICATED AND ACTUAL ANGLE OF UNBALANCE. An *unbalanced rotor* is a rotor in which the principal inertia axis does not coincide with the journal axis. When rotated in its bearings, an unbalanced rotor will cause periodic vibration of, and will exert a periodic force on, the rotor bearings and their supporting structure. If the structure is rigid, the force is larger than if the structure is flexible. In practice, supporting structures are neither rigid nor flexible but somewhere in between. The rotor-bearing support offers some restraint, forming a spring-mass system with damping having a single resonance frequency. When the rotor speed is below this frequency, the principal inertia axis of the rotor moves outward radially. This condition is illustrated in Fig. 39.14. If a soft pencil is held against the rotor, the so-called high spot is marked at the same angular position as that of the unbalance. When the rotor speed is increased, there is a small time lag between the instant at which the unbalance passes the pencil and the instant at which the rotor moves out enough to contact it. This is due to the damping in the system. The angle between these two points is called the "angle of lag." (See Fig. 39.14B.) As the rotor speed is increased further, resonance of the rotor and its supporting structure will occur, at this speed the angle of lag is 90°. As the rotor passes through resonance there are large vibration amplitudes and the angle of lag changes rapidly. As the speed is increased to more than twice the critical speed, the angle of lag approaches 180°. At speeds greater than approximately twice the critical speed, the rotor tends to rotate about its principal inertia axis; the angle of lag (for all practical purposes) is 180°.

The changes in the relative position of pencil mark and unbalance as shown in Fig. 39.14 of a statically unbalanced rotor occurs in the same manner on a rotor with dynamic unbalance. However, the center-of-gravity shown in the illustrations then represents the position of the principal inertia axis in the plane at which the pencil is applied

Fig. 39.14. A soft pencil is held against an unbalanced rotor. (*A*) A high spot is marked. (*B*) The angle of lag. Angle of lag between unbalance and high spot increases from 0° (*A*) to 180° (*D*) as rotor speed increases.

to the rotor shaft. Thus, the indicated angle of lag and displacement amplitude refer only to that particular plane and generally differ from any other plane in the rotor.

Angle of lag is shown as a function of rotational speed in Fig. 39.15: (*A*) for soft-bearing balancing machines whose balancing-speed ranges start at approximately twice the critical speed; and (*B*) for hard-bearing balancing machines. The effects of damping also are illustrated. Here the resonance frequency of the combined rotor-bearing support system is usually more than three times greater than the maximum balancing speed.

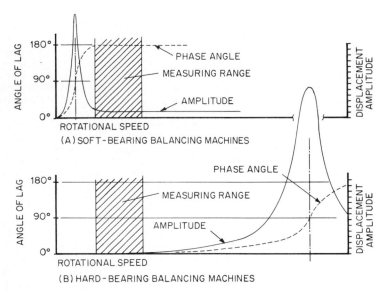

Fig. 39.15. Phase angle (angle of lag) and displacement amplitude vs. rotational speed in soft-bearing and hard-bearing balancing machines.

PLANE SEPARATION. Consider the rotor in Fig. 39.10 and assume that only the unbalance weight on the left is attached to the rotor. This weight causes not only the left bearing to vibrate but to a lesser degree the right. This influence is called "cross effect." If a second weight is attached in the right plane of the rotor as shown in Fig. 39.10, then the direct effect of the weight in the right plane combines with the cross effect of the weight in the left plane, resulting in a composite vibration of the right bearing. If the two unbalance weights are at the same angular position, the cross effect of one weight has the same angular position as the direct effect in the other rotor end plane; thus, their direct and cross effects are additive (Fig. 39.16*A*). If the two unbalance weights are 180° out of phase, their direct and cross effects are subtractive (Fig. 39.16*B*). In a hard-bearing balancing machine the additive or subtractive effect depends entirely on ratios between the axial positions of the correction planes and bearings. On a soft-bearing machine, this is not true, because the masses and inertias of the rotor and its bearings must be taken into account.

If the two unbalance weights on the rotor (Fig. 39.10) have an angular relationship other than 0 or 180°, then the cross effect in the right bearing has a different phase angle than the direct effect from the right weight. Addition or subtraction of these effects is vectorial. The net bearing vibration is equal to the resultant of the two vectors, as shown in Fig. 39.17. The phase angle indicated by the bearing vibration does not coincide with the angular position of either weight. This is the most common type of unbalance (dynamic unbalance of unknown amount and angular position).

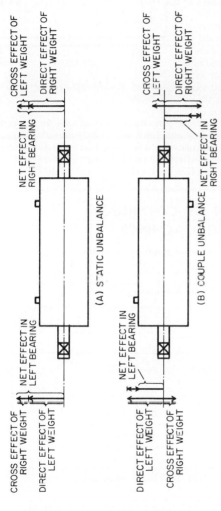

(A) STATIC UNBALANCE

CROSS EFFECT OF RIGHT WEIGHT
DIRECT EFFECT OF LEFT WEIGHT
NET EFFECT IN LEFT BEARING

NET EFFECT IN RIGHT BEARING
CROSS EFFECT OF LEFT WEIGHT
DIRECT EFFECT OF RIGHT WEIGHT

(B) COUPLE UNBALANCE

DIRECT EFFECT OF LEFT WEIGHT
CROSS EFFECT OF RIGHT WEIGHT
NET EFFECT IN LEFT BEARING

NET EFFECT IN RIGHT BEARING
CROSS EFFECT OF LEFT WEIGHT
DIRECT EFFECT OF RIGHT WEIGHT

FIG. 39.16. Influence of cross effects in rotors with static and couple unbalance.

DIRECT EFFECT OF LEFT WEIGHT
RESULTANT EFFECT IN BEARING
CROSS EFFECT OF RIGHT WEIGHT

RESULTANT EFFECT IN BEARING
CROSS EFFECT OF LEFT WEIGHT
DIRECT EFFECT OF RIGHT WEIGHT

FIG. 39.17. Influence of cross effects in rotors with dynamic unbalance. All vectors seen from right side of rotor.

This interaction of direct and cross effects could cause the balancing process to be a trial-and-error procedure. To avoid this, balancing machines incorporate a feature called "plane separation" which eliminates the influence of cross effect.

Cross effect may be eliminated by supporting the rotor in a cradle which rests on a knife-edge and spring arrangement, as shown in Fig. 39.18. Either the bearing-support members of the cradle or the pivot point are movable so that one unbalance correction plane always can be brought into the plane of the knife-edge. Any unbalance in this plane is prevented from causing the cradle to vibrate. Unbalance in one end plane of the rotor is measured and corrected. The rotor is turned end for end, so that the knife-edge is in the plane of the first correction. Any vibration of the cradle is now due solely to unbalance present in the plane that was first over the knife-edge. Corrections are applied to this plane until the cradle ceases to vibrate. The rotor is now in balance. If it is again turned end for end, there will be no vibration. Mechanical plane separation cradles restrict the rotor length, diameter, and location of correction planes; thus, electronic devices are used to accomplish the function of plane separation.

FIG. 39.18. Plane separation by mechanical means.

CLASSIFICATION OF CENTRIFUGAL BALANCING MACHINES

Centrifugal balancing machines may be categorized by the type of unbalance the machine is capable of indicating (static or dynamic), the attitude of the journal axis

Table 39.1. Classification of Balancing Machines

Principle employed	Unbalance indicated	Attitude of shaft axis	Type of machine	Available classes
Gravity (nonrotating)	Static (single-plane)	Vertical	Pendulum	Not classified
		Horizontal	Knife-edges	
			Roller sets	
Centrifugal (rotating)	Static (single-plane)	Vertical	Soft-bearing	
			Hard-bearing	
		Horizontal	Not commercially available	
	Dynamic (two-plane); also suitable for static (single-plane)	Vertical	Soft-bearing	II, III
			Hard-bearing	III, IV
		Horizontal	Soft-bearing	I, II, III
			Hard-bearing*	IV

* When suitably equipped, these machines may also be used for balancing flexible rotors.

of the workpiece (vertical or horizontal), and the type of rotor-bearing-support system employed (soft- or hard-bearing). The four classes included in Table 39.1 are described below.

Class I: Trial-and-error Balancing Machines. Machines in this class are of the soft-bearing type. They do not indicate unbalance directly in weight units (such as ounces or grams in the actual correction planes) but indicate only displacement and/or velocity of vibration at the bearings. The instrumentation does not indicate the amount of weight which must be added or removed in each of the correction planes. Balancing with this type of machine involves a lengthy trial-and-error procedure for each rotor, even if it is one of a presumably identical series. The unbalance indication cannot be calibrated for specified correction planes because these machines do not have the feature of plane separation. Field balancing equipment usually falls into this class.

FIG. 39.19. A permanently calibrated balancing machine, showing five rotor dimensions used in computing unbalance. (See Class IV.)

Class II: Calibratable Balancing Machines Requiring a Balanced Prototype. Machines in this class are of the soft-bearing type using instrumentation which permits plane separation and calibration for a given rotor type, if a balanced master or prototype rotor is available. However, the same trial-and-error procedure as for class I machines is required for the first of a series of presumably identical rotors.

Class III: Calibratable Balancing Machines Not Requiring a Balanced Prototype. Machines in this class are of the soft-bearing type using instrumentation which includes an integral electronic unbalance compensator. Any (unbalanced) rotor may be used in place of a balanced master rotor. In turn, plane separation and calibration can be achieved with the aid of precisely weighed calibration weights temporarily attached in each of two correction planes of the first of a series of rotors. This class includes soft-bearing machines with electrically driven shakers fitted to the vibratory part of their rotor supports, and machines with microprocessor instrumentation using influence coefficients.

Class IV: Permanently Calibrated Balancing Machines. Machines in this class are of the hard-bearing type. They are permanently calibrated by the manufacturer for all rotors falling within the weight and speed range of a given machine size. Unlike the machines in other classes, these machines indicate unbalance in the first run without individual rotor calibration. This is accomplished by the incorporation of an analog or digital computer into the instrumentation associated with the machine. The following five rotor dimensions (see Fig. 39.19) are fed into the computer: distance from left correction plane to left support; distance between correction planes; distance from right correction plane to right support; and r_1 and r_2 radii of the correction weights in the left and right planes, respectively. The instrumentation then indicates the magnitude and angular position of the required correction weight for each of the two selected planes.

The null-force balancing machine is in this class. Although no longer manufactured, it is still used. It balances at the same speed as the natural frequency or resonance of its suspension system (including the rotor).

BALANCING-MACHINE EVALUATION[10]

To evaluate the suitability of a balancing machine for a given application, it is first necessary to establish a precise description of the required machine capacity and performance. Such description often becomes the basis for a balancing-machine purchase specification. It should contain details on the range of workpiece weight, the diameter, length, journal diameter, and service speed, and whether the rotors are rigid or flexible, their application, available line voltage, etc. Such information enables the machine vendor to propose a suitable machine. Next the vendor's proposal must be evaluated not only on compliance with the purchase specification but also on the operation of his machine and its features. In describing the machine, the vendor should conform with the applicable standards. Once the machine is purchased and ready for shipment, compliance with the purchase specification and vendor proposal should be verified. Depending on circumstances, such verification is usually repeated after installation of the machine at the buyer's facility.

Precise testing procedures vary for different fields of application. Table 39.2 lists a number of standards for testing balancing machines used in the United States and Canada.

Table 39.2. Standards for Testing Balancing Machines

Application	Title	Issuer	Document no.
General industrial balancing machines	Balancing Machines–Description and Evaluation	International Standards Organization (ISO)	DIS 2953
Jet engine rotor balancing machines (for two-plane correction)	Balancing Equipment for Jet Engine Components, Compressor and Turbine, Rotating Type, for Measuring Unbalance in One or More Than One Transverse Planes	Society of Automotive Engineers, Inc. (SAE)	ARP 587A
Jet engine rotor balancing machines (for single-plane correction)	Balancing Equipment for Jet Engine Components, Compressor and Turbine, Rotating Type, for Measuring Unbalance in One Transverse Plane	Society of Automotive Engineers, Inc. (SAE)	ARP 588A
Gyroscope rotor balancing machines	Balancing Machine–Gyroscope Rotor	Defense General Supply Center, Richmond, Va.	FSN 6635–450–2208 NT
Field balancing equipment	Field Balancing Equipment–Description and Evaluation	International Standards Organization (ISO)	ISO 2371

UNBALANCE CORRECTION METHODS[11]

Corrections for rotor unbalance are made either by the addition of weight to the rotor or by the removal of material (and in some cases, by relocating the journal axis). The selected correction method should ensure that there is sufficient capacity to allow correction of the maximum unbalance which may occur. The ideal correction method permits reduction of the maximum initial unbalance to less than balance tolerance in a single correction. However, this is often difficult to achieve. The more common methods, described below, e.g., drilling, usually permit a reduction of 10:1 in unbalance if carried out carefully. The addition of weight may achieve a reduction as great as 20:1 or higher, provided the weight and its position are closely controlled. If the method selected for reduction of maximum initial unbalance cannot be expected to bring the rotor within the permissible residual unbalance in a single correction, a preliminary correction is made. Then a second correction method is selected to reduce the remaining unbalance to less than its permissible value.

UNBALANCE CORRECTION BY THE ADDITION OF WEIGHT TO THE ROTOR

1. *The addition of wire solder.* It is difficult to apply the solder so that its center-of-gravity is at the desired correction location. Variations in diameter of the solder wire introduce errors in correction.

2. *The addition of bolted or riveted washers.* This method is used only where moderate balance quality is required.

3. *The addition of cast iron, lead, or lead weights.* Such weights, in incremental sizes, are often used to correct large initial unbalance.

4. *The addition of welded weights.* Resistance welding provides a means of attaching large correction weights, although the total weight and center-of-gravity may be changed somewhat due to the weld.

UNBALANCE CORRECTION BY THE REMOVAL OF WEIGHT

1. *Drilling.* Material is removed from the rotor by a drill which penetrates the rotor to a measured depth, thereby removing the intended weight of material with a high degree of accuracy. A depth gage or limit switch can be provided on the drill spindle to ensure that the hole is drilled to the desired depth. This is probably the most effective method of unbalance correction.

2. *Milling, shaping, or fly cutting.* This method permits accurate removal of weight when the rotor surfaces, from which the depth of cut is measured, are machined surfaces and when means are provided for accurate measurement of cut with respect to those surfaces; used where relatively large corrections are required.

3. *Grinding.* In general, grinding is used as a trial-and-error method of correction. It is difficult to evaluate the actual weight of the material which is removed. This method is usually used only where the rotor design does not permit a more economical type of correction.

MASS CENTERING

A procedure known as "mass centering" is used to reduce unbalance effects in rotors. A rotor is mounted in a balanced cage or cradle which, in turn, is rotated in a balancing machine. The rotor is adjusted radially with respect to the cage, until the unbalance indication is zero; this provides a means for bringing the principal inertia axis of the rotor into essential coincidence with the journal axis of the balanced cage. Center drills (or other suitable tools guided along the axis of the cage) provide a means of establishing an axis in the rotor about which it is in balance. The beneficial effects of mass centering are reduced by any subsequent machining operations.

BALANCING OF ROTATING PARTS

MAINTENANCE AND PRODUCTION BALANCING MACHINES

Balancing machines of this type fall into three general categories: (1) universal balancing machines, (2) semiautomatic balancing machines, and (3) fully automatic balancing machines with automatic transfer of work. Each of these has been made in both the nonrotating and rotating types. The rotating type of balancer is available for rotors in which corrections for balance are required in either one or two planes.

Universal balancing machines are adaptable for balancing a considerable variety of sizes and types of rotors. These machines commonly have a capacity for balancing rotors whose weight varies as much as 100 to 1 from maximum to minimum. The elements of these machines are adapted easily to new sizes and types of rotors. The amount and location of unbalance are observed on suitable instrumentation by the machine operator as the machine performs its measuring functions. This category of machine is suitable for maintenance or job-shop balancing as well as for many small and medium lot-size production applications.

Semiautomatic balancing machines are of many types. They vary from an almost universal machine to an almost fully automatic machine. Machines in this category may perform automatically any one or all of the following functions in sequence or simultaneously: (1) retain the amount of unbalance indication for further reference, (2) retain the angular location of unbalance indication for further reference, (3) measure amount and position of unbalance, (4) couple the balancing-machine driver to the rotor, (5) initiate and stop rotation, (6) set the depth of a correction tool from the indication of amount of unbalance, (7) index the rotor to a desired position from indication of unbalance location, (8) apply correction of the proper magnitude at the indicated location, (9) inspect the residual unbalance after correction, and (10) uncouple the balancing-machine driver. Thus, the most complete semiautomatic balancing machine performs the complete balancing process and leaves only loading, unloading, and cycle initiation to the operator. Other semiautomatic balancing machines provide only means for retention of measurements to reduce operator fatigue and error. The equipment which is economically feasible on a semiautomatic balancing machine may be determined only from a study of the rotor to be balanced and the production requirements.

Fully automatic balancing machines with automatic transfer of the rotor are also available. These machines may be either single- or multiple-station machines. In either case, the parts to be balanced are brought to the balancing machine by conveyor, and balanced parts are taken away from the balancing machine by conveyor. All the steps of the balancing process and the required handling of the rotor are performed without an operator. These machines also may include means for inspecting the residual unbalance as well as monitoring means to ensure that the balance inspection operation is performed satisfactorily.

In single-station automatic balancing machines all functions of the balancing process (unbalance measurement, location, and correction) as well as inspection of the complete process are performed in a single checking at a single station. In a multiple-station machine, the individual steps of the balancing process may be done at individual stations. Automatic transfer is provided between stations at which the amount and location of unbalance are determined; then the correction for unbalance is applied; finally, the rotor is inspected for residual unbalance. Such machines generally have shorter cycle times than single-station machines.

FIELD BALANCING EQUIPMENT[12]

Many types of vibration indicators and measuring devices are available for field balancing operations. Although these devices are sometimes called "portable balancing machines," they never provide direct means for measuring the amount and location of the correction required to eliminate the vibration produced by the rotor at its supporting bearings. It is intended that these devices be used in the field to reduce or eliminate vibration produced by the rotating elements of a machine under service conditions. Basically, such a device consists of a combination of a transducer and meter which provides an indication proportional to the vibration magnitude. The vibration magnitude indicated may be displacement, velocity, or acceleration, depending on the type of transducer and readout system used. The transducer can be held by an operator, the hands of the operator providing a seismic mounting. A probe held against the vibrating machine is presumed to cause the transducer output to be proportional to the vibration of the machine. At frequencies below approximately 15 cps, it is almost impossible to hold the transducer sufficiently still to give stable readings. Frequently, the results obtained depend upon the technique of the operator; this can be shown by obtaining measurements of vibration magnitude on a machine with the transducer held with varying degrees of firmness.

Transducers having internal seismic mountings can be clamped firmly to the machine for measurement of vibration magnitude. This type of transducer should not be used where the frequency of the vibration being measured is less than three times the natural frequency of the transducer. The principles of vibration measurement are discussed more thoroughly in Chap. 12.

A transducer responds to all vibration to which it is subjected, within the useful frequency range of the transducer and associated instruments. The vibration detected on a machine may come through the floor from adjacent machines, may be caused by reciprocating forces or torques inherent in normal operation of the machine, or may be due to unbalances in different shafts or rotors in the machine. Location of the transducer on the axis of angular vibration of the machine can eliminate the effect of a reciprocating torque; however, the simple vibration indicator cannot discriminate between the other vibrations unless the magnitude at one frequency is considerably greater than the magnitude at other frequencies. For balancing it may be useful to have the magnitude indicated in displacement units.

Velocity and acceleration are functions of frequency as well as amplitude; therefore, they may be less convenient to use for reduction of unbalance effects where more than one frequency is present. Velocity- and acceleration-type transducers may be used if suitable integrating devices are introduced between the transducer and the meter. A suitable filter following the output of an electromechanical transducer may be introduced to attenuate frequencies other than the wanted frequency.

The approximate location of the unbalance may be determined by measuring the phase of the vibration. Phase of vibration may be measured by a stroboscopic lamp flashed each time the output of an electrical transducer changes polarity in a given direction. Phase also may be determined by use of a phase meter or by use of a wattmeter.

BALANCING OF ASSEMBLED MACHINES

The balancing of rotors assembled of two or more individually balanced parts and the balancing of rotors in complete machines are done frequently to obtain maximum reduction in vibration due to unbalance. In many cases the complete machine is run under service conditions during the balancing procedure.

The requirement for assembly balance often is made necessary by conditions dictated by machining operations and assembly procedures. For example, a balanced flywheel mounted on a balanced crankshaft may not produce a balanced assembly. When pistons and connecting rods are added to the above assembly, more unbalance is introduced. Such resultant unbalance effects can sometimes only be reduced by balancing the engine in assembly. The probable variation of unbalance in an assembly of balanced components is best determined by statistical methods.

Assemblies such as gyros, superchargers, and jet engines often run on antifriction bearings. The inner races of these bearings may not have concentric inside and outside surfaces. The eccentricity of the bearing races makes assembly balancing on the actual bearings desirable. In many cases such balancing is done with the stator supporting the antifriction bearings. This ensures that balance is achieved with the bearing race exactly in the position of final assembly.

PRACTICAL CONSIDERATIONS IN TOOLING A BALANCING MACHINE

SUPPORT OF THE ROTOR

The first consideration in tooling a balancing machine is the means for supporting the rotor. Various means are available, such as twin rollers, plain bearings, rolling element bearings (including slave bearings), gas bearings, nylon V blocks, etc. The most frequently used and easiest to adapt are twin rollers. A rotor should generally be supported at its journals to assure that balancing is carried out around the same axis on which it rotates in service.

Rotors which are normally supported at more than two journals may be balanced satisfactorily on only two journals provided that

1. All journal surfaces are concentric with respect to the axis determined by the two journals used for support in the balancing machine.

2. The rotor is rigid at the balancing speed when supported on only two bearings.

3. The rotor has equal stiffness in all radial planes when supported on only two journals.

If the other journal surfaces are not concentric with respect to the axis determined by the two supporting journals, the shaft should be straightened. If the rotor is not a rigid body or if it has unequal stiffness in different radial planes, the rotor should be supported in a (nonrotating) cradle at all journals during the balancing operation. This cradle should supply the stiffness usually supplied to the rotor by the machine in which it is used. The cradle should have minimum weight when used with a soft-bearing machine to permit maximum balancing sensitivity.

Rotors with stringent requirements for minimum residual unbalance and which run in antifriction bearings, should be balanced in the antifriction bearings which will ultimately support the rotor. Such rotors should be balanced either in (1) special machines where the antifriction bearings are aligned and the outer races held in half-shoe-type bearing supports, rigidly connected by tie bars, or (2) in standard machines having supports equipped with V-roller carriages.

Frequently, practical considerations make it necessary to remove antifriction bearings after balancing, to permit final assembly. If this cannot be avoided, the bearings should be match-marked to the rotor and returned to the location used while balancing. Anti-friction bearings with considerable radial play or bearings with a quality less than ABEC (Annular Bearing Engineers Committee) Standard grade 3 tend to cause erratic indications of the balancing machine. In some cases the outer race can be clamped tightly enough to remove excessive radial play. Only indifferent balancing can be done when rotors are supported on bearings of a grade lower than ABEC 3.

When maintenance requires antifriction bearings to be changed occasionally on a rotor, it is best to balance the rotor on the journal on which the inner race of the anti-friction bearings fits. The unbalance introduced by axis shift due to eccentricity of the inner race of the bearing then can be minimized by use of high-quality bearings to ensure minimum eccentricity.

BALANCING SPEED

The second consideration in tooling a balancing machine for a specific rotor is the balancing speed. For rigid rotors the balancing speed should be the lowest speed at which the balancing machine has the required sensitivity. Low speeds reduce the time for acceleration and deceleration of the rotor. If the rotor distorts nonsymmetrically at service speed, the balancing speed should be the same as the service speed. Rotors in which aerodynamic unbalance is present may require balancing under service conditions. Some machines show the effect of unbalance produced by varying electrical fields caused by changes in air gap and the like. Such disturbance can be reduced only by balancing (at service speed) if the disturbing frequency is identical to the service speed.

DRIVE FOR ROTOR

A final consideration in tooling a balancing machine for a specific rotor is the means for driving the rotor. For balancing rotors which do not have journals, the balancing machine may incorporate in its spindle the necessary journals; alternatively, an arbor may be used to provide the journal surfaces. An adapter must be provided to adapt the shaftless rotor to the balancing-machine spindle or arbor. This adapter should provide the following:

1. Rotor locating surfaces which are concentric and square with the spindle or arbor axis.

2. Locating surfaces which hold the rotor in the manner in which it is held in final assembly.

3. Locating surfaces which adjust the fit tolerance of the rotor to suit final assembly conditions.

4. A driver to ensure that a fixed angular relation is maintained between the rotor and the adapter.

5. Means for correcting unbalance in the adapter itself.

If the rotor to be tooled has its own journals, it may be driven through: (1) a universal joint or flexible coupling drive from one end of the rotor, (2) a pin-type drive from one end of the rotor, (3) a belt over the periphery of the rotor, or over a pulley attached to the rotor, or (4) air jets or other power means by which the rotor is normally driven in the final machine assembly.

The choice of universal joint or flexible coupling drive attached to one end of the rotor can affect the residual unbalance substantially. Careful attention must be given to the surfaces on the rotor to which the coupling is attached to ensure that the rotor journal axis and coupling are concentric (for example within 0.001 in. total indicator reading) when all fit tolerances and eccentricities have been considered. The weight of that part of the balancing machine drive which is supported by the rotor during the balancing operation, expressed in ounces (and, in this example multiplied by one-half of the total indicator reading or 0.0005 in.) must be less than the permissible residual unbalance in ounce-inches. Adjustable means must be provided in the coupling drive of the balancing machine to apply corrections for balancing the coupling. The adjustments may have to be effective in each of the correction planes of the rotor in an amount equal to at least twice the permissible residual unbalance. For convenience, the coupling should be movable to the rotor for attachment and away from the rotor for uncoupling.

The choice of a pin type of drive can seriously affect the residual unbalance. This drive generally takes the form of one or two driving pins in the balancing-machine face plate, and pins projecting from the adapter for the rotor toward the balancing-machine face plate. Slots or flexible members are used to transmit torque from the balancing-machine spindle through the pins. Misalignment of the rotor axis with respect to the spindle axis of the balancing machine introduces forces which have substantial effect on the indication of unbalance. The effect of misalignment on unbalance indication increases as the driving torque increases. Variations in the weight of successive rotors of the same type and wear of the balancing-machine members which support the rotor make it difficult to maintain alignment of the balancing-machine spindle axis and the rotor axis. For proper balancing, failure of the two axes to intersect by 0.005 in. at the point of the drive and failure of the two axes to be parallel within 0.005 in./ft should not introduce an error in amount of unbalance indication exceeding the permissible residual unbalance. The driver which attaches to the rotor must locate from surfaces of the rotor which are concentric with the journal axis because an accumulation of fit tolerances and eccentricities introduces an error in the result.

A belt drive can transmit only limited torque to the rotor. Driving belts must be extremely flexible and of uniform thickness. Driving pulleys attached to the rotor should be used only when it is impossible to transmit sufficient driving torque by running the belt over the rotor. Pulleys must be as light as possible, must be dynamically balanced, and should be mounted on surfaces of the rotor which are square and concentric with the journal axis. The belt drive should not cause disturbances in the unbalance indication exceeding one-quarter of the permissible residual unbalance. Rotors driven by belt should not drive components of the balancing machine by means of any mechanical connection.

The use of electrical means or air for driving rotors may influence the unbalance readout. To avoid or minimize such influence, great care should be taken to bring

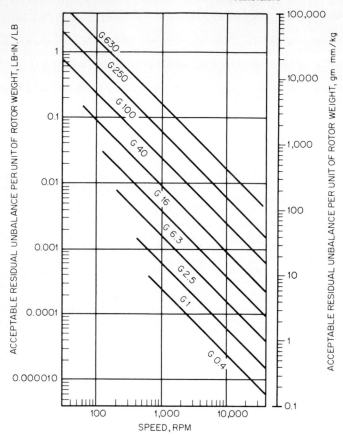

FIG. 39.20. Residual unbalance corresponding to various balancing quality grades, *G. Note:* 1 gm. mm/kg is equivalent to a displacement of the center-of-gravity of 0.001 mm = 39 μin.

in the power supply through very flexible leads, or have the airstream strike the rotor, at right angles to the direction of the vibration resulting from unbalance.

If the balancing machine incorporates filters tuned to a specific frequency only, it is essential that means be available to control the rotor speed to suit the filter setting.

BALANCE CRITERIA

Achieving better balance in rotors is primarily an art; therefore, it is often difficult for an engineer normally conversant with balancing methods and techniques to decide which particular balancing method to employ, the rotational speed for balancing to be used, and at what particular point in a production line the balancing procedure should be inserted. The appropriate choice of a balance criterion is likely to be an even greater problem.

A suitable criterion of the quality of balancing required would appear to be the running smoothness of the complete assembly; however, many other factors than unbalance contribute to uneven running of machines (for example, bearing dissymmetries, mis-

Table 39.3. Balance Quality Grades for Various Groups of Rigid Rotors[13]

Balance quality grade	Type of rotor
G4,000.........	Crankshaft drives of rigidly mounted slow marine diesel engines with uneven number of cylinders.
G1,600.........	Crankshaft drives of rigidly-mounted large two-cycle engines.
G630..........	Crankshaft drives of rigidly mounted large four-cycle engines; crankshaft drives of elastically mounted marine diesel engines.
G250..........	Crankshaft drives of rigidly mounted fast four-cylinder diesel engines.
G100..........	Crankshaft drives of fast diesel engines with six or more cylinders; complete engines (gasoline or diesel) for cars and trucks.
G40...........	Car wheels, wheel rims, wheel sets, drive shafts; crankshaft drives of elastically mounted fast four-cycle engines (gasoline or diesel) with six or more cylinders; crankshaft drives for engines of cars and trucks.
G16...........	Parts of agricultural machinery; individual components of engines (gasoline or diesel) for cars and trucks.
G6.3..........	Parts or process plant machines; marine main-turbine gears; centrifuge drums; fans; assembled aircraft gas-turbine rotors; fly wheels; pump impellers; machine-tool and general machinery parts; electrical armatures.
G2.5	Gas and steam turbines; rigid turbo-generator rotors; rotors; turbo-compressors; machine-tool drives; small electrical armatures; turbine-driven pumps.
G1............	Tape recorder and phonograph drives; grinding-machine drives.
G0.4..........	Spindles, disks, and armatures of precision grinders; gyroscopes.

Note: In general, for rigid rotors with two correction planes, one-half the recommended residual unbalance is to be taken for each plane; these values apply usually for any two arbitrarily chosen planes, but the state of unbalance may be improved upon at the bearings; for disc-shaped rotors, the full recommended value holds for one plane.

alignment, aerodynamic and hydrodynamic effects, etc.). In addition, there is no simple relation between rotor unbalance and vibration amplitude measured on the bearing housing. Many factors, such as proximity of resonant frequencies, fit, and machining errors, may influence over-all vibration level considerably. Therefore, a measurement of the vibration amplitude will not indicate directly the magnitude of unbalance or whether an improved state of unbalance will cause the machine to run smoother. For certain classes of machines, particularly electric motors and large turbines and generators, voluminous data have been collected which can be used as a guide for the establishment of criteria for such installations.

Table 39.3 and Fig. 39.20 show a classification system for various types of representative rotors, based on a document—ISO Standard 1940–1973, "Balance Quality of Rotating Rigid Bodies"—of the International Standards Organization. Balance quality grades are grouped according to numbers with the prefix G; the vertical scales in Fig. 39.20 indicate the maximum permissible residual unbalance per unit of rotor weight (at various maximum service speeds shown on the bottom scale) expressed in English and SI units. The residual unbalance is equivalent to a displacement of the center-of-gravity. The recommended balance quality grades are based on experience with various rotor types, sizes, and service speeds; they apply only to rotors which are rigid throughout their entire range of service speeds. Balance criteria for flexible rotors are discussed in ISO 5406[14] and listed in ISO 5343.[2]

REFERENCES

1. "Balancing—Vocabulary," ISO 1925-1981, and ANSI S2.7, 1982.
2. "Criteria for Evaluating Flexible Rotor Balance," ISO 5343-1983; ANSI S2.43-1984.

3. Federn, K.: "Auswuchttechnik," Springer Verlag, Berlin, W. Germany, 1977.
4. Federn, K.: *Werkstatt u. Betrieb*, **86**:243 (1953).
5. Giers, A.: "Rechnergestütztes Auswuchten elastischer Rotoren," Sonderdruck, June 1974.
6. Rieger, N. F.: Balancing High-Speed Rotors to Reduce Vibration Levels, *ASME Design Conf.*, Chicago, Ill., May 8, 1972.
7. Moore, L. S., and E. G. Dodd: *G.E.C.J.*, **31**(2) (1964).
8. Schneider, H.: "Balancing Technology," Carl Schenck AG, Darstadt, W. Germany, 1977.
9. Stadelbauer, D. G.: "Fundamentals of Balancing," 2d ed. Schenck Trebel Corp., Deer Park, N.Y., 1983.
10. "Balancing Machines—Description and Evaluation," ISO/DIS 2953-1984.
11. Senger, W. I.: *Machine Design*, **16**:131 (1944); **17**:127 (1945).
12. "Field Balancing Equipment—Description and Evaluation," ISO 2371-1974; ANSI S2.38-1982.
13. "Balancing Quality of Rotating Rigid Bodies," ISO 1940-1973; ANSI S2.19-1975.
14. "The Mechanical Balancing of Flexible Rotors," ISO 5406-1980; "Procedures for Balancing Flexible Rotors," ANSI S2.42-1982.

40

MACHINE-TOOL VIBRATION

J. G. Bollinger
University of Wisconsin, Madison

INTRODUCTION

The machining of metals and other materials is invariably accompanied by relative vibration between workpiece and tool. These vibrations are due to one or more of the following causes: (1) inhomogeneities in the workpiece material; (2) disturbances in the workpiece or tool drives, including the feed drive; (3) interrupted cutting; (4) vibration generated in other machines and transmitted through the foundation; (5) vibration generated by the cutting process itself (machine-tool chatter).

The tolerable level of relative vibration between tool and workpiece, i.e., the maximum amplitude and to some extent the frequency, depends on whether the machining process is of the roughing or finishing kind. In the roughing process the tolerable level is determined mainly by detrimental effects of the vibration on tool life (see *The Effect of Vibration on Tool Life*) and by the noise which is frequency generated. In finishing operations, surface finish and machining accuracy determine the tolerable level.

VIBRATION DUE TO INHOMOGENEITIES IN THE WORKPIECE

Hard spots in the material being worked impart small shocks to the tool and workpiece, as a result of which free vibrations are set up. If these transients are rapidly damped out, their effect is usually not serious; they simply form part of the general "background noise" encountered in making vibration measurements on machine tools. Cases in which transient disturbances do not decay but build up to vibrations of large amplitudes (as a result of dynamic instability) are of great practical importance, and are discussed later.

When machining under conditions resulting in discontinuous chip removal, the segmentation of chip elements results in a fluctuation of the cutting thrust. If the frequency of these fluctuations coincides with one of the natural frequencies of the structure, forced vibration of appreciable amplitude may be excited.[1] However, it is not clear whether the segmentation of the chip is a primary effect or whether it is produced by other vibration without which continuous chip flow would be encountered.

The breaking away of a built-up edge from the tool face also imparts impulses to the cutting tool which result in vibration. However, marks left by the built-up edge on the machined surface are far more pronounced than those caused by the ensuing vibration; it is probably for this reason that the built-up edge has not been studied from the vibration point of view. The built-up edge frequently accompanies certain types of vibration (chatter), and instances have been known when it disappeared as soon as the vibration was eliminated.

DISTURBANCES IN THE WORKPIECE AND TOOL DRIVE

Vibrations arise in drives that impart rotational motions (spindle and rotary table drives) and in drives that impart rectilinear motion (feed drives). Forced vibrations result from rotating unbalanced masses, gear and belt drives, bearing irregularities, and periodic but nonsinusoidal forces in drive motors. Free vibrations result from shocks caused by reciprocating unbalanced masses, a type of self-induced vibration frequently encountered with slide drives (stick-slip), and from spurious transient motions in the equipment.

VIBRATION CAUSED BY ROTATING UNBALANCED MEMBERS

When the speed of rotation of some unbalanced member falls near one of the natural frequencies of the machine-tool structure, forced vibration of large amplitude is induced which may affect both surface finish and tool life. This vibration can be eliminated by careful balancing, the procedure being basically similar to that described in Chap. 39.

Fig. 40.1. Grinding marks resulting from unbalance of grinding wheel or some other component. (A) Cylindrical grinding; (B) peripheral grinding. Marks which are unequally spaced or which have a varying slope are due to inhomogeneities in the wheel.

When designing a new machine, a great deal of trouble can be forestalled by placing rotating components in a position in which the detrimental effect of their unbalance is likely to be relatively small. Motors should not be placed on the top of slender columns, and the plane of their unbalance should preferably be parallel to the plane of cutting. In some cases, vibration resulting from rotating unbalanced members can be eliminated by applying the usual vibration-isolation techniques (Chaps. 30 and 32); for example, the coolant pump may be supported by vibration isolators.[2]

Among all machining processes, grinding is most sensitive to vibration because of the high surface finish resulting from the operation. In cylindrical grinding, marks resulting from unbalance of the grinding wheel or of some other component are readily recognizable. They appear in the form of equally spaced, continuous spirals with a constant slope, as shown in Fig. 40.1A. From these marks, the machine component responsible for their existence is found by considering that its speed in rpm must be equal to $\pi DN/a$ where D is the workpiece diameter in inches, a is the pitch of the marks in inches, and N is the workpiece speed in rpm. An analogous procedure also can be applied to peripheral surface grinding. Marks produced in one pass of the wheel are shown in Fig. 40.1B. The speed of the responsible component in rpm is equal to the number of marks (produced in one pass) which fall into a distance equal to that traveled by the workpiece (or wheel) in 1 min.

MARKS CAUSED BY INHOMOGENEITIES IN THE GRINDING WHEEL. Although grinding marks usually indicate the presence of a vibration, this vibration may not necessarily be the primary cause of the marks. Hard spots on the cutting surface of the wheel result in similar, though generally less pronounced, marks. Grinding wheels usually are not of equal hardness throughout.[3] A hard region on the wheel circumference becomes rapidly glazed in use and establishes itself as a high spot on the wheel (since it retains the grains for a longer period than the softer parts). However, these high spots eventually break down or shift to other parts of the wheel; in cylindrical grinding, this manifests itself as a sudden change of the slope of the spiral marks. Marks which appear to be due to an unbalanced member rotating at two or three times the speed of the wheel and which are nonuniformly spaced are always due to two or three hard spots.

Owing to the sensitivity of high precision grinding to inhomogeneities in the grinding wheel, a vibration test has been suggested with which the presence of cracks and hard spots can be detected.[4] This test is based on the fact that in any elastic system of cylindrical symmetry (for instance, a disc or a cup wheel) the presence of imperfections (inhomogeneities or dimensional deviations from perfect cylindrical symmetry) results in a splitting up of the natural frequencies of the various bending modes of vibration. Each bending mode of vibration (specified by the number of nodal circles and nodal diameters) can be excited at two distinct natural frequencies. The difference between these natural frequencies is, to some extent, a measure of the imperfections present. In a perfect system the natural frequencies coincide.[5]

THE EFFECT OF VIBRATION ON THE WHEEL PROPERTIES. If vibration exists between wheel and workpiece, normal forces are produced which react on the wheel and tend to alter the wheel shape and/or its cutting properties.[6, 7] In soft wheels the dominating influence of vibration appears to be inhomogeneous wheel wear, and in hard wheels inhomogeneous loading (i.e., packing of metal chips on and in crevasses between the grits). These effects result in an increased fluctuation of the normal force which produces further changes of the wheel properties. The over-all effect is that a vibration once initiated tends to grow.[8] When successive cuts or passes overlap, the inhomogeneous wear and loading of the wheel may cause a regenerative chatter effect which makes the cutting process dynamically unstable (see *Dynamic Stability*).

It is desirable to reduce vibration in so far as possible in the initial design. This necessitates careful balancing, well-adjusted bearings (see *Bearings*), a suitable belt (see *Belt Drives*), and a stiff machine construction (see *The Design of Vibration-free Machine Tools*).

GEAR DRIVES

All gear faults, eccentricities, pitch errors, profile errors, etc., produce nonuniform rotation which in some cases adversely affects surface finish, geometry, and possibly tool life. It is for this reason that in precision machines, where a high degree of surface finish is required, the workpiece or tool spindle usually is driven by belts. In this application, belts act like filters which suppress the high-frequency torsional oscillations.

Gear vibration[9, 10] is an important subject, particularly in precision machinery. Gear cutting and grinding, where precise kinematic relations must be maintained and surface finish is important, is an area in which gear-system-induced vibrations are detrimental. The dynamic behavior of gear trains has been extensively investigated by computer simulation techniques[11] to gain insight into how gear accuracy, stiffness variation, and tooth geometry relate to resulting vibration. Dynamic tooth load resulting from vibration in gear trains is also an important consideration in gear wear and noise generation.[12]

Tests with a lathe having a headstock gear with a deliberately large eccentricity (0.1 mm) showed that when the center distance was adjusted to maintain the gear in mesh without backlash, large amplitudes of vibration were set up when the frequency of tooth engagement coincided nearly with the natural frequency of the work.[13] The amplitude of the forced vibration resulting from gear eccentricity can be decreased by increasing backlash between the gears.

BELT DRIVES

Belt drives may give rise to both torsional and rectilinear forced vibrations. Any variation of the effective belt radius, i.e., the radius of the neutral axis of the belt around the pulley axis, produces a variation of the belt tension and the belt velocity. This causes a variation of the bearing load and of the rotational velocity of the pulley.

NORMAL GRINDING FORCE

(A)   (B)   (C)

Fig. 40.2. Direction of driving belt and its influence on performance. (A) Vibration is minimized when belt tension and normal grinding force point in the same direction. (B) Large amplitudes may arise when the normal grinding force is substantially equal to the belt tension. (C) Vibration due to centrifugal force is likely to be caused by an unbalance of wheel. (S. Doi.[8])

The effective pulley radius can vary as a result of (1) form faults of the pulley itself (eccentric rim or groove) or (2) form and/or structural faults in the belt (variation of belt profile or inhomogeneity of belt material).[14] Tests have shown that from the point of view of vibration, flat belts are preferable to V belts, partly because of the smaller disturbances generated by them and partly because the disturbing force is almost independent of the belt tension.[14]

Grinding is particularly sensitive to disturbances caused by belts. Seamless silk belts or a direct motor drive to the main spindle is recommended for high-precision machines.[8] Attention also should be paid to the direction of the belt tension with respect to the normal grinding force. Vibration is minimized when the belt tension and the normal grinding force point in the same direction, as shown in Fig. 40.2A. The clearance between bearing and spindle is thus eliminated. With the arrangement shown in Fig. 40.2B, large amplitudes of vibration may arise when the normal grinding force is substantially equal to the belt tension and/or the peripheral surface of the wheel is nonuniform. Tests indicate that with the arrangement shown in Fig. 40.2C, vibration due to the centrifugal force is likely to be caused by an unbalance of the wheel.[8]

The spindle pulley should preferably be placed between the spindle bearings (Fig. 40.3A) and not at the end of the spindle (Fig. 40.3B); the latter usually is done to achieve constructional simplicity. With the arrangement of Fig. 40.3B, the wheel displacement attributable to the normal

Fig. 40.3. Effect of relative position of grinding wheel, bearings, and driving pulley on grinding performance. (A) Driving pulley should be placed between bearings, as shown in (A). Arrangement shown in (B) is constructionally simpler but is more liable to cause trouble. (S. Doi.[8])

grinding force opposes that due to the belt tension; thus, conditions are basically similar to those of Fig. 40.2B.

BEARINGS

Dimensional inaccuracies of the components of ball or roller bearings and/or surface irregularities on the running surfaces (or the bearing housing) may give rise to vibration trouble in machines where high-quality surface finish is demanded. From the frequency of the vibration produced, it is sometimes possible to identify the component of the bearing responsible.[15, 16] For conventional bearings frequently used in machine tools, the outer race is stationary and the inner race rotates with n_i rpm; the cage velocity is of the order of $n_c \simeq 0.4n_i$, and that of the balls or rollers about $n_b \simeq 2.4n_i$. In some cases, a disturbing frequency of the order of $n_z = zn_c$ also can be detected where z is the number of rolling elements. This is the frequency with which successive rolling elements pass through the "loaded zone" of the bearing, which is determined by the direction of the load. These disturbing frequencies usually are not detectable with bearings having two rows of rolling elements, each unit of which lies halfway between units of the neighboring row.

From the point of view of vibration control, best results are achieved with bearings containing two rows, each having a large number of rolling elements and having a very small radial clearance. Where a large spindle stiffness is required, as in a lathe, the front spindle bearing may consist of two of such bearings, combined with a thrust bearing to take up the axial load acting on the spindle. Using such a bearing construction, a spindle supported only at its ends may have a greater stiffness and a higher natural frequency than a spindle supported in three places with single roller bearings.[17, 18]

The stiffness and damping of roller bearings increase with a decrease of the radial play. This is particularly pronounced when the radial play becomes negative, i.e., when the bearings are preloaded. Vibration is then reduced, and the surface finish and accuracy of the work improve. However, it is not clear whether preloading is practicable, because it is accompanied by an increase in heat generation and the likely decrease of bearing life.[19]

It is often claimed that sliding bearings have a greater damping capacity than roller bearings and are therefore superior with respect to vibration.[8] However, careful measurements show that the damping capacity of sliding bearings which have attained their steady working temperature is of the same order as that of roller bearings. As the temperature of the lubricating oil increases, the damping decreases. Consequently, vibration of undesirably large magnitude may arise only after the machine has been running for more than 30 min. The damping of sliding bearings is greater than that of roller bearings only if a highly viscous lubricating oil is used. Even under these conditions, the over-all damping effect is unlikely to be appreciable because the amplitude of vibration in the bearing is small.[16]

STICK-SLIP IN LINEAR DRIVES

The uniformity of feed motions is often disturbed by a phenomenon known as *stick-slip*.[20, 21] When motion of a tool support is initiated by engagement of a transmission, elements of the transmission suffer elastic deformation until the forces transmitted exceed the static frictional resistance of the tool support. Subsequently, the support commences to move, and the friction drops to its dynamic value. As a result of the drop of the friction force, the support receives a high acceleration and overshoots because of its inertia. At the end of the "jump," the transmission is wound up in the opposite sense; before any further motion can take place, this deformation must be unwound; this occurs during a period of standstill of the support. Subsequently, the phenomenon repeats itself. The physical sequence described falls into the category generally known as "relaxation oscillations"[23, 26] (Chap. 5).

The occurrence of stick-slip depends on the interaction of the following factors:[22] (1) the mass of the sliding body, (2) the drive stiffness, (3) the damping present in the drive,

(4) the sliding speed, (5) the surface roughness of the sliding surfaces, and (6) the lubricant used.[22] As a rule, it is encountered only at very low sliding speeds or when moving a mass of appreciable size a small distance from rest. Slide drives designed for stick-slip-free motion have small moving masses and a high drive stiffness. Excellent results also may be achieved by using cast iron and a suitable plastic material as mating surfaces.[24, 30] Pressurized slides are used with considerable success, particularly on servo-controlled machine tools. By keeping the oil film between the mating surfaces under a certain pressure, the possibility of mixed dry and viscous friction is eliminated and as a result stick-slip cannot arise. For many applications (measuring instruments, precision boring machines, etc.), slides in which sliding friction is replaced by rolling friction are used.[27–29]

VIBRATION DUE TO INTERRUPTED CUTTING

When machining interrupted surfaces, impulses of appreciable magnitude may be imparted to the tool which may lead to undesirable vibration. With single-pointed cutting tools, this vibration usually is not severe (although the tool may suffer) and is quickly damped out. In cylindrical grinding, when machining a workpiece which contains a slot or a keyway, visible marks frequently are observed near the "leaving edge" of the slot or keyway. These are due to a "bouncing" of the grinding wheel on the machined surface. They are eliminated or minimized by inserting a wooden plug or key into the recess.

Serious forced vibration owing to interrupted cutting is encountered when using cutting tools with multiple cutting edges, i.e., in the milling process.

VIBRATION TRANSMITTED FROM OTHER MACHINES

Vibration generated in machines (presses, machine tools, internal-combustion engines, compressors, etc.) is transmitted through the foundation to other machines and may set up forced vibration in these. In the workshop, the vibration of the floor contains a wide frequency spectrum;[31] it is almost inevitable that one of these frequencies should fall near a natural frequency of a particular machine tool. The amplitudes of this background vibration usually are small, and become troublesome only in the case of grinding machines and precision boring machines.

The isolation of vibration transmitted through the floor may be achieved by the customary methods of vibration isolation (Chaps. 30 and 32), i.e., the machine which generates the vibration or is to be protected from vibration is placed upon vibration isolators. In general, the first course is to be preferred, particularly when exciting forces result from presses or hammers.[32]

When applying vibration isolators to machine tools, some care must be exercised. The foundation constitutes the "end condition" of the machine-tool structure. Any alteration of the end condition affects the natural frequencies of the structure, the corresponding damping, equivalent stiffness, and the dynamic deflection curves (modes of vibration).[33–35]

In general, vibration isolators lower the natural frequencies, damping, and stiffness. The structure may thus become more susceptible to internal exciting forces, and its chatter behavior also may be affected in an undesirable way.[35] However, it is fortunate that only the lower modes of vibration are affected to an appreciable extent (primarily the rocking modes) and these do not play an important part in many machine tools (milling machines, grinding machines, etc.). In general, machine-tool structures which are very stiff by themselves (i.e., without being bolted down) can be placed on vibration isolators safely (milling machines, grinding machines, some lathes).[36, 37]

Isolators for medium loads usually contain rubber springs.[32, 38] Those suitable for very great loads (presses, hammers) are of all-metal construction, usually incorporating hydraulic or dry-friction dampers.[32, 39, 40] Sheets of cork, rubber, or plastic also have been used with good results as isolators.

The foundation of a machine tool has an important influence on the dynamic behavior of a machine structure due to excitations originating within the machine and those originating externally from the machine. This is particularly significant below 100 cps, as has been reported in studies directed at establishing important foundation parameters.[41]

MACHINE-TOOL CHATTER

INTRODUCTION

The cutting of metals is frequently accompanied by a violent vibration of workpiece and cutting tool known as machine-tool chatter. *Chatter* is a self-induced vibration which is caused not by external forces but which is induced and maintained by forces generated by the cutting process itself. It is highly detrimental to tool life[42–45] and surface finish,[44, 87] and is usually accompanied by considerable noise. Moreover, chatter also affects adversely the rate of production since, in many cases, its elimination can be achieved only by reducing the rate of metal removal.

Machine-tool chatter is characteristically erratic. Forced and free vibration effects in machine tools are usually detected in the development stage or during final inspection, and can be reduced or eliminated by known procedures. The inclination of a certain machine to chatter often remains unobserved in the plant of the machine-tool manufacturer; when finally encountered, its elimination from a particular machining process may be highly time-consuming and laborious.

A distinction can be drawn[46] between regenerative and nonregenerative chatter. The former occurs when there is an overlap in the process of performing successive cuts such that part of a previous cut surface is removed by a succeeding pass of the cutter. Under regenerative cutting, a displacement of the tool can result in a vibration of the tool relative to the workpiece resulting in a variation of the chip thickness. This in turn results in a variation in the cutting force during following revolutions. The regenerative chatter theory explains a wide variation of practical chatter situation in such operations as normal turning and milling.

Nonregenerative chatter is found in such operations as shaping, slotting, and screw-thread cutting. In this type of cutting, chatter has been explained through the principle of mode coupling.[8, 47–50] If a machine-tool vibration consists of two degrees-of-freedom, with the axes of major flexibilities normal and with a common mass, the dynamic motion can take an elliptical path. If the major axis of motion (axis with the greater compliance) lies within the angle formed by the total cutting force and the normal to the workpiece surface, energy can be transferred into the machine-tool structure, thus producing an effective negative damping. The width of cut for the threshold of stable operation is directly dependent upon the difference between the two principal stiffness values, and chatter tends to occur when the two principal stiffnesses are close in magnitude.

DYNAMIC STABILITY

Machine-tool chatter is essentially a problem of dynamic stability. A machine tool under vibration-free cutting conditions may be regarded as a dynamical system in steady-state motion. Systems of this kind may become dynamically unstable and break into oscillation around the steady motion. Instability is caused by an alteration of the cutting conditions produced by a disturbance of the cutting process (e.g., a hard spot in the material). As a result, a time-dependent thrust element dP is superimposed on the steady cutting thrust P. If this thrust element is such as to increase the original disturbance, oscillations will build up and the system is said to be unstable.

This chain of events is most easily investigated theoretically by considering that the incremental thrust element dP is a function not only of the original disturbance but also of the velocity of this disturbance. Forces which are dependent on the velocity of a displacement are damping forces; they are additive to or subtractive from the damping present in the system (e.g., structural damping, assumed to be viscous). When the damping due to dP is positive, the total damping (structural damping plus damping due to altered cutting conditions) also is positive and the system is stable. Any disturbance will then be damped out rapidly. However, the damping due to dP may be negative, in which case it will decrease the structural damping, which is always positive. If the negative damping due to dP predominates, the total damping is negative. Positive damping forces are energy absorbing. Negative damping forces feed energy into the system; when the total damping is negative, this energy is used for the maintenance of oscillations (chatter).

From the practical point of view, the fully developed chatter vibration (self-induced vibration) is of little interest. Production engineers are almost entirely concerned with conditions leading to chatter (dynamic instability). The build-up of chatter is very difficult to observe, and experimental work has to be carried out mainly under conditions which are only indirectly relevant to the problem being investigated. Experimental results obtained from fully developed chatter vibration may, in some instances, be not really relevant to the problem of dynamic stability.

The influence of the machine-tool structure on the dynamic stability of the cutting process is of great importance. This becomes clear by considering that with a structure (including tool and workpiece) of infinite stiffness, the cutting process could not be disturbed in the first place because hard spots, for example, would not be able to produce the deflections necessary to cause such a disturbance. Furthermore, it is clear that were the structural damping infinite, the total damping could not become negative and the cutting process would always be stable. This discussion indicates that an increase of structural stiffness and/or damping always has beneficial effects from the point of view of chatter.

In practically feasible machines, the interrelation between structural stiffness, damping, and dynamic stability is of considerable complexity. This is because machine-tool structures are systems with distributed mass, elasticity, and damping; their vibration is described by partial differential equations of intractable complexity. For this reason, the dynamic characteristics of the machine-tool structure are introduced into the analysis of dynamic stability by making certain simplifying assumptions. From this point of view, the existing theories of the dynamic stability of the cutting process can be divided into two groups. In the first group it is assumed that dynamic instability can arise only in one of the many modes of vibration of the structure. This assumption is legitimate when the modes of the structure are well separated and/or the geometry of the cutting tool does not permit the simultaneous excitation of several modes.

THE EFFECT OF VIBRATION ON TOOL LIFE

Inasmuch as the cutting speed and the chip cross section vary during vibration, it is to be expected that vibration affects tool life. The magnitude of this effect is unexpectedly large, even when impact loading of the tool is excluded. This has been shown by theoretical and experimental investigations of single-pointed cutting tools.[51, 43, 44, 52, 55]

Impact loading of the cutting edge may be very severe, and the elimination of vibration may multiply tool life 80 to 200 times.[45]

The life of face-mill blades may suffer considerably owing to torsional vibration executed by the cutter. A tool-life reduction of 30 per cent is not uncommon.[54] The torsional vibration need not necessarily be caused by dynamic instability of the cutting process but may be forced vibration, because of faulty gears or interrupted chip removal.

THE ELIMINATION OF VIBRATION

The vibration behavior of a machine tool can be improved by an increase of the equivalent static stiffness and the damping for the modes of vibration which result in a relative displacement between tool and workpiece. Furthermore, improvement can be achieved by appropriate choice of cutting conditions, tool design, and workpiece design. Stiffness and damping are important both for ordinary (free and forced) and self-induced (chatter) vibrations. Cutting conditions and tool design influence mainly self-induced vibration. In addition, the application of vibration dampers and absorbers has proven to be an effective technique for the solution of machine-vibration problems. These devices should be viewed as a functional part of a machine and not as an add-on to solve specific problems, although their past use has often taken that perspective.

THE ELIMINATION OF VIBRATION IN EXISTING MACHINES

Problems associated with free or forced vibration may be eliminated by getting rid of the source of the disturbance, isolating the source, absorbing the vibrational energy, or modifying the structure.

Cutting Conditions. In the elimination of chatter, cutting conditions are first altered. If this fails to achieve the desired result, an increase of stiffness between tool and workpiece and/or damping of disturbances impinging on the tool or improved tool design is resorted to.

Regenerative chatter usually is confined to definite speed ranges which are separated by chatter-free speeds. Other types of chatter are also usually confined to a certain speed range.[68] Thus, in some cases of regenerative chatter a small increase or decrease of speed may stabilize the cutting process. This is likely to be effective only when stiffness and/or damping are relatively large; otherwise the unstable speed ranges overlap. In most cases, very high cutting speeds are stable[48] and the consequent loss of tool life from speed may be compensated by the increase from elimination of vibration.

An increase of the feed rate is also beneficial in some types of machining (drilling, face milling, etc.). For the same cross-sectional area, narrow chips (high feed rate) are less likely to lead to chatter than wide chips (low feed rate) since the chip-thickness variation effect results in a relatively smaller variation of the cross-sectional area in the former (smaller dynamic cutting force).

Continuous variation of cutting conditions achieved through a programmed variation of spindle speed is a technique which has been investigated and under laboratory conditions appears to have some promise as a technique for chatter suppression.[97, 98]

Stiffness. Although the configuration of the structure is given in existing machines, the static and the equivalent static stiffness values may nevertheless vary within wide limits. High stiffness values are ensured by the use of steadies, by placing tool and workpiece in a position where the relative dynamic displacement between them is small (i.e., by placing them near the main column, etc.), by using rigid tools and clamps, by using jigs which rigidly clamp (and if necessary support) the workpiece, by clamping securely all parts of the machine which do not move with respect to each other, etc.[55]

In some cases chatter is eliminated by loosening the locks of slides. The conclusion often drawn from this is that flexible machines may be less inclined to chatter than stiff machines; i.e., rigidity sometimes is undesirable. When the rigidity of some machine element is reduced, greater damping may appear at the cutter; the increase of damping often outweighs the reduction in rigidity. Consequently, although a loss of rigidity is always undesirable, it may be tolerated when accompanied by a large increase of damping.

In some cases old machines are less likely to chatter than new machines of identical design. This may result from wear and tear of the slides which increases the damping and effects an improvement in performance. However, it would be wrong to conclude that lack of proper attention and maintenance is desirable. Proper attention to slides, bearings (minimum play), belts, etc., is necessary to satisfactory performance. It would be wrong also to conclude that a highly polluted workshop atmosphere is desirable because some new machines, even when not used but merely exposed to workshop dirt for a sufficiently long time, appear to improve in their chatter behavior. The explanation is that dirty slides increase the damping.

Damping. Damping should be increased without impairing the static stiffness and machining accuracy of the machine. This is achieved by the use of dampers. These are basically similar to those employed in other fields of vibration control (Chaps. 32 and 43). Dampers are effective only when placed in a position where vibration amplitudes are large.

The auxiliary mass absorber (Chap. 6) has been employed with considerable success on milling machines, radial drilling machines,[56] gear hobbing machines,[57] and grinding machines.   An interesting design variant of this type of absorber is shown in Fig. 40.4. In this design a plastic ring element is used which combines both the elastic and the damping element of the absorber.   The auxiliary mass may be attached to the top of a

(A)   (B)   (C)

Fig. 40.4. Auxiliary mass damper with combined elastic and damping element.   The combined element lies between two retainer rings, of which one (3) is attached with bolt 1 to the machine structure.   The other ring (2) takes the weight of the auxiliary mass.   (A) Arrangement when auxiliary mass is being supported.   (B) Arrangement when auxiliary mass is being suspended.   (C) Application of both types of arrangements to a hobbing machine.   (After F. Eisele and H. W. Lysen.[57])

column (Fig. 40.4C), as shown in Fig. 40.4A.   Alternatively, the auxiliary mass may be suspended on the underside of a table (Fig. 40.4C), using the design shown in Fig. 40.4B.   In either case, several plastic ring elements may support one large auxiliary mass, as shown in Fig. 40.4C.[57]

A variation of the *Lanchester damper* (Chaps. 6 and 38) is frequently used in boring bars to good advantage.[58]   This consists of an inertia weight fitted into a hole bored in the end of a quill, as shown in Fig. 40.5.   To ensure effective operation, a relatively small radial clearance of about 0.001 to 0.006 in. must be provided, depending upon the diameter of the inertia weight.   An axial clearance of about 0.006 to 0.010 in. is sufficient.   A smooth surface finish of both plug and hole is desirable.   The clearance values given refer to dry operation, using air as the damping medium.   Oil also can be used as a damping medium but it does not necessarily result in an improved performance. When applying oil, clearance gaps larger than those stated above have to be ensured,

Fig. 40.5. Lanchester damper for the suppression of boring-bar vibration. (After R. S. Hahn.[58])

depending on the viscosity of the oil.   Further means for the suppression of boring-bar vibration are discussed under *Boring Bars with Improved Resistance to Vibration*.

The impact damper shown in Fig. 40.6 also provides effective vibration control.[59] It consists of a bolt 1, a sleeve 2, a spring 3, and a cover 4.   It is set on the lathe tool post or on the tool, at a point where vibration of large amplitude arises normally.   In operation, the vibration of the tool results in an oscillatory motion of the inert mass 4 which, impinging on the bolt 1, absorbs energy. The important physical parameters of an impact damper are material, mass ratio, and stroke ratio (a function of the gap through which the free mass travels). Design charts have been prepared which facilitate the application of impact dampers to the solution of vibration problems found typically in machine tools.[60]

Dynamic absorbers can be designed to be self-optimizing. Thus, an absorber can be capable of self-adjustment of the spring rate to minimize vibration amplitude under changing excitation conditions.[61] The self-optimizing feature is achieved by placing vibration transducers on both the absorber mass and the main system. A control circuit measures the phase angle between the motions and activates a spring-modifying mechanism to maintain 90° phase difference between the two measured motions. It has been demonstrated that the 90° phase relationship guarantees minimum motion of the main vibrating mass.[61]

For the machining of long slender shafts, a hydraulically operated *support-vibration damper*[48, 63] can be used. This device consists of three cams which support the workpiece. The cams are hydraulically interconnected so that a floating support is ensured. Vibration energy is absorbed by viscous friction.

Tool Design. Sharp tools are more likely to chatter than slightly blunted tools. In the workshop, the cutting edge is often deliberately dulled by a slight honing. Consequently, a *vibro-damping bevel*[45] on the leading face of a lathe tool has been suggested. This bevel has a leading edge of −80° and a width of about 0.080 in. Tests show that the negative bevel does not in all cases eliminate vibration and that the life of the bevel is short.[64] Appreciably worn cutting edges cause violent chatter.[48]

Since narrow chips are less likely to lead to instability, a decrease of the approach angle of the cutting tool results in improved chatter behavior. With lathe tools, an increase of the rake angle results in improvement, but the influence of changes in the relief angle is relatively small.[48]

Fig. 40.6. Impact damper for lathe tools, consisting of bolt 1, sleeve 2, spring 3, and a cover 4. The oscillation of the inert mass 4, impinging on the bolt 1, absorbs vibratory energy. (*After D. I. Ryzhkov.*[59])

Theoretically, the tool shape should be such that any sudden increase of the feed rate meets the maximum possible resistance in the workpiece material. The penetration-rate coefficient Λ [see Eq. (40.1)] then is positive and large, and has an appreciable stabilizing influence. Such a stabilizing effect is produced by the chisel edge of a drill;[64, 65] it is not clear whether chisel-edge corrections, which decrease the static thrust, are also beneficial from the vibration point of view. Similarly, the wide face edge of certain face-mill blades prevents a "hacking-in" of these blades into the machined surface and appears to have a stabilizing influence.[46]

THE DESIGN OF VIBRATION-FREE MACHINE TOOLS. The following discussion is concerned mainly with problems of structural design, rather than with vibration sources.

Stiffness. In machine-tool practice, three different kinds of stiffness are considered.

1. *Static stiffness* k_s, defined as the ratio of the static force P_o, applied between tool and workpiece, to the resulting static deflection A_s between the points of application of force.

2. *Dynamic stiffness* k_d, defined as the ratio of the amplitude of a harmonic force P, applied between tool and workpiece, to the resulting resonance amplitude A_{max} between the points of application of force.

3. *Equivalent static stiffness* k, related to the dynamic stiffness by the expression $k = k_d Q$ where Q is the dynamic amplification factor.

A high static stiffness is necessary for machining accuracy. A high dynamic stiffness (i.e., large equivalent static stiffness and large damping) is necessary for satisfactory dynamic performance.

The stiffness of a structure is determined primarily by the stiffness of the most flexible component in the path of the force. Flexible members or "spots" show up in the static and dynamic deflection curves as sudden changes of their slope. In many cases the most flexible members are joints, i.e., bolted connections between relatively rigid elements such as column and bed, column and table, etc. Some points to be considered in the design of connections are illustrated in Fig. 40.7. To avoid bending of the flange in Fig. 40.7*A*, the

Fig. 40.7. Load transmission between column and bed. (*A*) Old design, relatively flexible owing to deformation of flange. (*B*) New design, bolt placed in a pocket (*A*) or flange stiffened with ribs on both sides of bolt (*B*). (*After H. Opitz.*[65])

bolts should be placed in pockets or between ribs, as shown in Fig. 40.7*B*. Increasing the flange thickness does not necessarily increase the stiffness of the connection since this requires longer bolts which are more flexible. There is an optimum flange thickness (bolt length), the value of which depends on the elastic deformation in the vicinity of the connection. Deformation of the bed is minimized by placing ribs under connecting bolts.[65, 66, 68]

The efficiency of bolted connections, and other static and dynamic structural problems, is conveniently investigated by model analysis techniques.[65, 67, 69, 73, 74] Figure 40.8 shows the results of successive stages of a model experiment in which the effect of the design of bolt connections on the bending rigidity (X and Y directions) and the torsional rigidity of a column were investigated. The relative rigidities are shown by the length of bars. In the design of Fig. 40.8*A*, the connection consists of 12 bolts ($\frac{5}{8}$ in. in diameter) arranged in pairs along both sides of the column. In the design of Fig. 40.8*B*, the number of bolts is reduced to 10 and arranged as shown. With the addition of ribs, shown in succeeding figures, the bending stiffness in the direction X was raised by 40 per cent, in the direction Y by 45 per cent, and the torsional stiffness by 53 per cent, compared to the original design.[65, 67]

Openings in columns should be as small as possible. Figure 40.9 shows the loss of static flexural stiffness k_{sx}, k_{sy}, and torsional stiffness $k_{s\theta}$, and the decrease of the flexural natural frequencies f_x, f_y, resulting from the introduction of a hole in a box-type col-

FIG. 40.8. Successive stages in the improvement of a flange connection.   (*H. Opitz.*[65])

FIG. 40.9. Influence of a hole in the wall of a box column on the static stiffness and natural frequency.   (*A*) Static stiffness; (*B*) natural frequency.   (*H. Opitz.*[65])

umn.[65, 67]   Smaller holes show relatively smaller decreases of stiffness and natural frequency than larger ones.   The torsional rigidity $k_{s\theta}$ of a box-type column is particularly sensitive to openings, as shown in Fig. 40.10.   Lids or doors used for covering these openings do not restore the stiffness.   The influence of covers depends on their thickness, mode of attachment, and design, as shown in Fig. 40.11.   However, covers may increase damping and thereby partly compensate for the detrimental effect of loss of stiffness.[65, 67]

Welded structural components not only are stiffer than cast-iron components but may also have a larger damping capacity.[70, 71]   This is because welds are never perfect;

Fig. 40.10. Torsional stiffness of box columns with different holes in walls. (*H. Opitz.*[65])

Fig. 40.11. Influence of cover plate and lid on static stiffness of box column. (*A*) Column without holes, (*B*) one hole uncovered, (*C*) hole covered with cover plate, and (*D*) hole covered with substantial lid, firmly attached. (*After H. Opitz.*[65])

consequently, rubbing takes place between joined members.[71] A considerable increase in damping can be achieved by using interrupted welds. This is of particular importance with welded ribs, which may be necessary not so much to increase rigidity but to prevent "drumming" (membrane vibration) of large unsupported areas.[65, 72, 73, 75]

The static and dynamic behavior of lathes is influenced significantly by the design of the spindle and its bearings. The static deflection of the spindle consists of two parts, X_1 and X_2, as shown in Fig. 40.12. The deflection X_1 corresponds to the deflection of a flexible beam on rigid supports, and X_2 corresponds to the deflection of a rigid beam on flexible supports which represent the flexibility of the bearings.   The deflection of the spindle amounts to 50 to 70 per cent of the total deflection, and the bearings 30 to 50 per cent of the total depending on the relation of spindle cross section to bearing stiffness and span.   The result is the same for roller or plain bearings of the same size.   The stiffness of needle roller bearings is about twice that of ordinary roller bearings.[65, 76, 113]

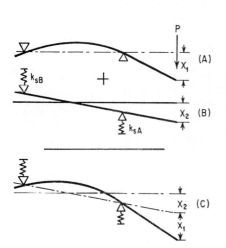

FIG. 40.12. Deflection of machine-tool spindle and bearings. A machine-tool spindle can be regarded as a beam on flexible supports. The total deflection under the force P consists of the sum of (A) the deflection X_1 of a flexible beam on rigid supports and (B) the deflection X_2 of a rigid beam on flexible supports. (*H. Opitz.*[65])

FIG. 40.13. Deflection of a beam on elastic supports as a function of the bearing distance.   Bearing stiffness k_A and k_B, spindle stiffness k_0. (*After H. Opitz.*[65])

The distance between the bearings has considerable influence on the effective stiffness of the spindle, as shown in Fig. 40.13.   The ordinate of the figure corresponds to the deflection in inches per pound and the abscissa represents the ratio of bearing distance b to cantilever length a.   The straight line refers to the deflection of the spindle, and the hyperbola refers to the deflection of the bearings.   The total deflection is obtained by the addition of the two curves; the minimum of the curve of total deflection corresponds to the optimum bearing distance.   For a short cantilever length a, the optimum value of b/a lies between 3 and 5; for a long cantilever length a, b/a is optimum at 2.

The optimum bearing span for minimum deflection may be computed by the use of nomograms which have been prepared for a variable cross-section spindle, with over-hanging end and supported by two bearings.[77]

It is often important to consider the dynamic behavior of a spindle before establishing an optimum bearing span.   Maximizing the stiffness of a spindle at one point does not establish any dynamic properties.   Care must be taken to investigate both bending and rocking modes of the spindle before accepting a final optimum span.   For example, a large overhand on the rear of a spindle could produce an undesirable low-frequency rocking mode of the spindle even if the "optimum span" as defined previously were satisfied.[18]

The influence of the ratio of bore diameter to outside diameter on the stiffness of a hollow spindle is shown in Fig. 40.14.   A 25 per cent decrease of stiffness occurs only at a diameter ratio of $d/D = 0.7$, where D is the outside diameter and d the bore diameter. This is important for the dynamic behavior of the spindle. A solid spindle has nearly the same stiffness, but a substantially greater mass. Consequently, the natural frequency of the solid spindle is considerably lower, which is undesirable.[65, 76, 18]

Fig. 40.14. Effect of bore diameter on stiffness of hollow spindle where k_1 = stiffness of solid spindle, k_2 = stiffness of hollow spindle, D = outer spindle diameter, d = bore diameter, J_2 = second moment of area of hollow spindle, and J_1 = second moment of area of solid spindle. The curve is defined by $k_2/k_1 = J_2/J_1 = 1 - (d/D)$.[4]   (*H. Opitz.*[65])

Fig. 40.15. Influence of various components on total damping of lathes. The major part of the damping is generated at the mating surfaces of the various components. (*K. Loewenfeld.*[78])

Damping. The over-all damping capacity of a structure is determined only to a small extent by the damping capacity of its individual components. The major part of the damping results from the interaction of joined components at slides or bolted joints. The interaction of the structure with the foundation also may produce considerable damping.[35] A qualitative picture of the influence of the various components of a lathe on the total damping is given in Fig. 40.15. The damping of the various modes of vibration differs appreciably; the values of the logarithmic decrement shown in the figure correspond to an average value for all the modes which play a significant part.[78]

The damping of various types of machine tool differs greatly, as indicated in Fig. 40.16. However, relatively little variation is found when comparing the damping of machines of the same type but different make.[78]

Fig. 40.16. Damping of machines and castings. Complete machines contain far greater damping than individual components. (*K. Loewenfeld.*[78])

Since damping in machine tools is due primarily to rubbing between mating surfaces, attempts have been made to increase this effect by attaching strips of sheet metal to the machine. Successive layers rub against each other when deformed and introduce damping. Sheets of plastic material may be placed between the strips. The design appears to be promising, particularly with welded structures where it suppresses "drumming" (membrane vibration) of unsupported walls.[79]

COMPUTER-AIDED DESIGN OF MACHINE TOOLS

As illustrated in the previous section, there is considerable data in the literature concerning the static and dynamic performance of machine-tool structures and elements as developed both analytically and experimentally. Experimental investigations have drawn upon both analytical computation and experimental measurement using actual machine elements and models of structures of machine elements. The desire to predict the static and dynamic behavior of a given specific machine-element design has led to the development of numerous computer techniques which employ both analog and digital computer systems.

USE OF ANALOG COMPUTER

The general-purpose analog computer is naturally suited for studying dynamic phenomena of machine elements. Fundamentally, any element or system which may be modeled mathematically by a differential equation or set of differential equations may be studied by electronic analog computation techniques. The problem of analyzing the dynamics of beams and structures may be handled in two ways. First, equations for the normal mode and lateral vibration may be set up, and a system of ordinary differential equations may be established whereby the independent variable is distance along the beam. The other approach is to establish a lump parameter model and replace the partial derivatives with finite differences. The latter technique offers several advantages for studying machine structures, in that variable cross sections, elastic supports, and the responses to forced excitations are readily handled.[80, 81]

An investigation of the static and dynamic behavior of the lathe spindle serves as an example of the application of the finite-difference techniques.[82] Once a model is functioning on the computer, it is possible to investigate a number of design concepts. For example, it is possible to optimize the bearing locations for maximum stiffness, explore the advantages of more than two supporting bearings, impose simulated cutting forces and investigate the chatter stability, and investigate vibration damping ideas.

USE OF DIGITAL COMPUTER

The utilization of the digital computer in analyzing the static and dynamic behavior of machine elements has become widespread in recent years. Lumped-parameter modeling for digital computation has been facilitated by the development of software programs for the stiffness matrix technique.[82] More recent utilization of finite-element methods has enhanced the computing capability and has extended the application of computers to investigating complex structural shapes and, in general, handling those problems which are not readily studied by beam methods.[83, 86]

REFERENCES

1. Landberg, P.: *Microtecnic*, **10**(5):219 (1956).
2. Saljé, E.: "Zweites Forschungs- und Konstruktionskolloquium Werkzeugmaschinen," p. 105, Vogel-Verlag, Coburg, 1955.
3. Späth, W.: *Werkstattstechnik und Maschinenbau*, **42**(2):59 (1952).
4. Rowe, R. G.: *Steel*, **126**(26):74, 84 (1950).
5. Tobias, S. A.: *Engineering*, **172**(4470):409 (1951).
6. Landberg, P.: *Microtecnic*, **11**(1):18 (1957).
7. Hahn, R. S.: *Trans. ASME*, Paper No. 58-A-97, 1958.
8. Doi, S.: *Trans. ASME*, **80**(1):133 (1958).
9. Strauch, H.: *Z. VDI*, **95**:159 (1953).
10. Harris, S. L.: *Proc. IME*, **172**(2):87 (1958).
11. Bollinger, J. G.: *Ind.-Anzeiger*, **46**:961 (June, 1963).
12. Bollinger, J. G., and M. Bosh: *Ind.-Anzeiger*, **86**(19):319 (March, 1964).
13. Doi, S.: *Mem. Fac. Eng., Nagoya Univ.*, **5**(2):179 (1953).
14. Lysen, H. W., and R. Schwaighofer: "Zweites Forschungs- und Konstruktionskolloquium Werkzeugmaschinen," pp. 121, 133, Vogel-Verlag, Coburg, 1955.
15. Schenk, O.: "Zweites Forschungs- und Konstruktionskolloquium Werkzeugmaschinen," p. 137, Vogel-Verlag, Coburg, 1955.
16. Lohmann, G.: "Zweites Forschungs- und Konstruktionskolloquium Werkzeugmaschinen," p. 139, Vogel-Verlag, Coburg, 1955.
17. Pittroff, H.: *Konstruktion*, **11**(3):106 (1959).
18. Bollinger, J. G., and G. Geiger: *Intern. J. Mach. Des. Res.*, **3**:193 (1964).
19. Piekenbrink, R.: "Drittes Forschungs- und Konstruktionskolloquium Werkzeugmaschinen und Betriebswissenschaft," vol. 2, p. D155, Vogel-Verlag, Coburg, 1958.
20. Bowden, F. P., and L. Leben: *Proc. Roy. Soc. (London)*, **169**:371 (1939).
21. Blok, H.: *SAE J.*, February, 1940.
22. Merchant, M. E.: "Characteristics of Typical Polar and Non-polar Lubricant Additives under Stick-slip Conditions," Meeting of the Society of Lubrication Engineers, Chicago, April, 1946.
23. Elyasberg, M. E.: *Stanki i Instrument*, **11**:1 (1951) and **12**:6 (1951); *Engineers' Digest*, **13**(9):298 (1952) and **13**(10):338 (1952).
24. Opitz, H., and E. Saljé: *Ind.-Anzeiger*, (63) (1952).
25. Beuerlein, P.: "Fortschrittliche Fertigung und moderne Werkzeugmaschinen," p. 117, Verlag W. Girardet, Essen, 1954.
26. Mikes, M.: *Czechoslovak Heavy Industry*, **1**(1):6 (1957).
27. Anon.: *Machinery (London)*, **89**(2292):912 (1956).
28. Galonska, D. A.: *Machinery (New York)*, **62**(8):159 (1956).
29. Havemeyer, H. R.: *Machine Design*, **29**(17):120 (1957).
30. Weiter, E. J., and A. O. Schmidt: *ASME Paper No. 57-SA-100*, 1957; *Engineers' Digest*, **18**(8):335 (1957).

31. Sadowy, M.: "Drittes Forschungs- und Konstruktionskolloquium Werkzeugmaschinen und Betriebswissenschaft," p. 127, Vogel-Verlag, Coburg, 1957.
32. Harris, C. M. (ed.): "Handbook of Noise Control," p. 13, McGraw-Hill Book Company, Inc., New York, 1957.
33. Sadowy, M.: "Drittes Forschungs- und Konstruktionskolloquium Werkzeugmaschinen und Betriebswissenschaft," p. 141, Vogel-Verlag, Coburg, 1957.
34. Tobias, S. A., and W. Fishwick: *Engineering*, **176**(4584):707 (1953).
35. Fishwick, W., and S. A. Tobias: *Engineering*, **185**(4808):568 (1958).
36. Eisele, F., and M. Sadowy: "Drittes Forschungs- und Konstruktionskolloquium Werkzeugmaschinen und Betriebswissenschaft," p. 167, Vogel-Verlag, Coburg, 1957.
37. Eisele, F., M. Sadowy, and E. Sun: "Drittes Forschungs- und Konstruktionskolloquium Werkzeugmaschinen und Betriebswissenschaft," pp. 163, 177, Vogel-Verlag, Coburg, 1957.
38. Anon: *Automobile Engr.*, **46**(4):150 (1956).
39. Nietsch, H. E.: *Tool Engr.*, **39**(1):105 (1957).
40. Hartz, H.: "Drittes Forschungs- und Konstruktionskolloquium Werkzeugmaschinen und Betriebswissenschaft," p. 153, Vogel-Verlag, Coburg, 1957.
41. Honshi, T.: *Ann. CIRP*, **22**:1 (1973).
42. Weilenmann, R.: *Werkstatt u. Betrieb.*, **87**(8):401 (1954).
43. Weilenmann, R.: *Werkstatt u. Betrieb.*, **89**(9):529 (1956).
44. Opitz, H., and E. Saljé: *Ind.-Anzeiger*, (45):690 (1954).
45. Klushin, M. I., and D. I. Ryzhkow: *Metalworking Production*, **100**:1153 (1956).
46. Tobias, S. A.: *Proc. IME*, **173**(18):474 (1959).
47. Saljé, E.: *Ind.-Anzeiger*, **36**:501 (1955).
48. Sokolowski, A. P.: "Präzision in der Metalbearbeitung," p. 345, VEB Verlag Technik, Berlin, 1955.
49. Hölken, W.: *Ind.-Anzeiger*, (1):1 (1958).
50. Lysen, H.: "Zweites Forschungs- und Konstruktionskolloquium Werkzeugmaschinen," p. 175, Vogel-Verlag, Coburg, 1955.
51. Tobias, S. A., and W. Fishwick: *Engineer*, **205**(5324):199 (1958) and **238**(5325):238 (1958).
52. Arnold, R. N.: *Proc. IME*, **154**:261 (1946).
53. Tourret, R.: "Performance of Metal-cutting Tools," p. 101, Butterworth & Co. (Publishers) Ltd., London, 1958.
54. Opitz, H.: *Microtecnic*, **12**(2):72 (1958).
55. Hinjink, J. A. W., and Vander Wolf, A. C. H.: *Ann. CIRP*, **22**, 1, 123 (1971).
56. Galloway, D. F.: *Proc. IME*, **170**(6):207 (1956).
57. Eisele, F., and H. W. Lysen: "Zweites Forschungs- und Konstruktionskolloquium Werkzeugmaschinen," p. 89, Vogel-Verlag, Coburg, 1955.
58. Hahn, R. S.: *Trans. ASME*, **73**(4):331 (1951).
59. Ryzhkov, D. I.: *Stanki i Instrument*, (3):23 (1953); *Engineers' Digest*, **14**(7):246 (1953).
60. Pinotti, P. C., and M. M. Sadek: *Proc. 11th Intern. MTDR Conf.*, September, 1970.
61. Bonesho, J. A., and J. G. Bollinger: *Proc. 7th Intern. MTDR Conf.*, pp. 229–241, Pergamon Press, New York, 1968. (See also *Machine Design*, **40**, 123, Feb. 29, 1968.)
62. Sadowy, M.: "Zweites Forschungs- und Konstruktionskolloquium Werkzeugmaschinen," p. 167, Vogel-Verlag, Coburg, 1955.
63. Podporkin, V. G.: *Stanki i Instrument*, (7):12 (1953).
64. Sharin, Yu. S., and V. B. Serebrovsky: *Stanki i Instrument*, (7):21 (1955).
65. Opitz, H.: "Conference on Technology of Engineering Manufacture," Paper 7, The Institution of Mechanical Engineers, London, 1958.
66. Piekenbrink, R.: "Untersuchungen an Elementen im Kraftfluss der Werkzeugmaschinen, Forschungsbericht," Verlag W. Girardet, Essen, 1957.
67. Bielefeld, J.: "Untersuchungen an Elementen im Kraftfluss der Werkzeugmaschinen, 7. Forschungsbericht," Verlag W. Girardet, Essen, 1957; *Ind.-Anzeiger*, (63):953 (1957).
68. Hölken, W.: *Ind.-Anzeiger*, (62):898 (1955).
69. Saljé, E.: *Ind.-Anzeiger*, (62):894 (1955).
70. Heiss, A.: *VDI-Forschungsheft 429*, **16**(429):7 (1949/1950).
71. Kienzle, O., and H. Kettner: *Werkstattstechnik u. Werksleiter*, **33**(9):229 (1939).
72. Eisele, F., and H. Drumm: "Drittes Forschungs- und Konstruktionskolloquium Werkzeugmaschinen und Betriebswissenschaft," vol. 2, p. D125, Vogel-Verlag, Coburg, 1957.
73. Loewenfeld, K.: "Zweites Forschungs- und Konstruktionskolloquium Werkzeugmaschinen," pp. 59, 65, 73, 83, Vogel-Verlag, Coburg, 1955.
74. Bollinger, J. G., and B. C. Cuppan: ASME Paper 65-MD-13, (1965).
75. Loewenfeld, K.: "Drittes Forschungs- und Konstruktionskolloquium Werkzeugmaschinen und Betriebswissenshaft" (Vorberichtsheft), pp. 49, 63, Vogel-Verlag, Coburg, 1957.
76. Honrath, K.: "Untersuchungen an Elementen im Kraftfluss der Werkzeugmaschinen, 7. Forschungsbericht," Verlag W. Girardet, Essen, 1957.

77. Terman, T., and J. G. Bollinger: *Machine Design*, **37**, 159, May 27, 1965.
78. Loewenfeld, K.: "Zweites Forschungs- und Konstruktionskolloquium Werkzeugmaschinen," p. 117, Vogel-Verlag, Coburg, 1955.
79. Loewenfeld, K.: "Drittes Forschungs- und Konstruktionskolloquium Werkzeugmaschinen und Betriebswissenshaft" (Vorberichtsheft), p. 105, Vogel-Verlag, Coburg, 1957.
80. Cuppan, B. C., and J. G. Bollinger: *Proc. 7th Intern. MTDR Conf.*, 191, 1967.
81. Bollinger, J. G.: *Proc. CIRP Intern. Conf. Mfg. Tech.*, 77, 1967.
82. Cuppan, B. C., and J. G. Bollinger: *Ann. CIRP*, **17**:243 (1969).
83. Opitz, H., and R. Noppen: *Ann. CIRP*, **22**(2):227 (1973).
84. Hinduja, S., and A. Cowley: *Proc. 12th Intern. MTDR Conf.*, 455, 1971.
85. Yoshimura, M., and T. Hoshi: *12th Intern. MTDR Conf.*, 439, 1971.
86. Hinduja, S., and A. Cowley: *Ann. CIRP*, **21**(1):113 (1972).
87. Takemura, T., T. Kitamura, and T. Hoshi: *Ann. CIRP*, **22**(1):121 (1974).
88. Inamura, T., and Sata, T.: *Ann. CIRP*, **22**(1):119 (1974).
89. Eisele, P. T., and R. F. Griffin: *Trans. ASME*, **75**(10):1211 (1953).

41

PACKAGING DESIGN

Masaji T. Hatae

Northrop Corporation

INTRODUCTION

Packaging is the technology of preparing an item for shipment in such a manner as to minimize the damage resulting from environmental hazards encountered during shipment. Such environmental hazards include moisture, temperature, dust, shock, vibration, etc. This chapter considers only the influence of shock and vibration.

The shock and vibration experienced by goods during shipment result from the vibration of a cargo-carrying vehicle, the shock resulting from the impacts of railroad cars, the shock caused by the handling (e.g., dropping) of packages, etc. Such goods are protected from damage by isolation. The theory and practice of vibration and shock isolation are discussed in Chaps. 30 to 33. This chapter discusses the application of the principles of isolation to the design of packages. In general, this consists of placing the packaged item within a container, and interposing resilient means between the item and the container to provide the necessary isolation. Such resilient means is known as *package cushioning*. The required degree of protection should be provided with minimum packaging and shipping costs through optimum combination of labor, materials, and package volume and weight. The means of protection may range from a simple corrugated carton to a more complex system such as a part "floated" in distributed cushioning material in a wooden box or by a spring suspension in a reusable metal drum.

In concept, the package cushioning is designed to protect an item of known strength from the known shock and vibration existing in the particular environment. Practically, package design cannot be pursued so rationally because the strength of equipment and the characteristics of the environment often are not known with the necessary accuracy. As a consequence, the strength of the equipment often must be inferred from exploratory tests or estimated on the basis of experience. Similarly, environmental conditions are simulated by simplified tests (e.g., drop tests) that represent environmental conditions of maximum expected severity. Then the design of the package cushioning proceeds in a straightforward manner using the methods and data set forth in this chapter.

FACTORS IN DESIGN OF PACKAGE CUSHIONING

This section discusses factors that are significant in the design of package cushioning.

FRAGILITY OF EQUIPMENT

The capability of an item to withstand shock and vibration is defined in terms of its "fragility." The term "fragility" is used as a quantitative index of the strength of the equipment when subjected to shock and vibration. The quantitative index is expressed as a maximum permissible acceleration $\ddot{x}_F$. In addition to the value of $\ddot{x}_F$, it is necessary to know the natural frequencies and damping characteristics of the equipment. With the foregoing information the package designer could proceed to select cushioning with the desired characteristics to protect the equipment during shipment. Unfortunately, it is difficult to determine this information for many types of equipment. As a consequence of this difficulty, several alternate procedures may be used to obtain information of use for package design.

ANALYTICAL METHODS. It is difficult to determine analytically (i.e., by theoretical analysis and calculation) the fragility of a complex device or structure. Sometimes it is possible to simplify or idealize the structure to a form that is more amenable to analysis, and the strengths of principal component structures may be calculated. This analytical basis for design is discussed in detail in Chap. 42.

LABORATORY TEST METHOD. Shock and vibration tests may be carried to failure in order to determine the fragility level of products. This method is particularly suited to products manufactured in large quantity where unit cost is small and tests-to-failure may be performed. Various types of laboratory shock test apparatus, such as drop test machines, hydraulic shock testing machines and pendulum impact machines, are available for testing a wide range of sizes and weights of equipment. These testing machines, which are described in Chap. 26, provide control of acceleration level, time duration of pulse, and shape of pulse. Laboratory vibration testing machines, described in Chap. 25, are available in various ranges of frequency and force output.

Shock tests-to-failure can provide data concerning the maximum allowable accelerations for particular input pulse shapes. These tests-to-failure define fragility levels for shock and vibration in terms of the characteristics of the testing machine. In general, such information is useful in the design of package cushioning if the testing machines are adjusted to have the same characteristics as the motion transmitted by the cushioning to the equipment or if the cushioning is made to have the characteristics of the testing machine.

Most products, particularly military equipment, are designed and tested according to requirements stipulated in contractual specifications; in general, such requirements relate to ultimate use rather than to packaging. The levels of shock and vibration to which the equipment is designed and tested most often represent the *minimum* acceptable resistance to damage; they do not represent the maximum level that the item can withstand without failure, nor are they generally related to the requirements for successful packaging. Strict adherence to these minimum values in package design often results in high costs of packaging and shipping.

ENGINEERING ESTIMATE. Field personnel experienced in the handling and operation of specific items (products) often can provide information on the ability of these items to withstand drops from various heights when protected by certain cushioning materials. These heights of drop can be converted into peak acceleration levels if the dynamic properties of the cushioning materials are known. Then the fragility is defined quantitatively for certain conditions. For example, assume that it has been determined that an item, blocked and braced in a corrugated carton, is able to withstand the shock that results from dropping the carton from a height of 12 in. Information on height of drop may be transformed to the parameters of rise time and peak acceleration by assuming a sinusoidal acceleration pulse. The applicable relations are given in Fig. 41.1.

Estimated rise times to maximum acceleration for various drop conditions are given in Table 41.1. Further refinements in the values may be made by actual instrumented drop tests using various types of containers and materials.

The fragility index of equipment as estimated from experience in dropping similar equipment applies only to the peak acceleration. The natural frequencies and damping

FIG. 41.1. Relation of drop height and rise time to maximum acceleration for a packaged item which experiences a half-sinusoidal pulse. The relation is defined mathematically as $\ddot{x}_1{}^2 = 0.0128ht_r{}^{-2}$, where $\ddot{x}_1$ = maximum acceleration, g; h = height of drop, in.; and t_r = rise time, sec.

present in the equipment are not indicated. In general, the natural frequencies of important components of the equipment are substantially greater than the natural frequency of the package cushioning; under these conditions, damage tends to be directly proportional to maximum acceleration and it is possible to use the maximum acceleration as an index of fragility (see Chap. 42).

Table 41.1. Approximate Sinusoidal Rise Time for Various Drop Conditions

Condition	Rise time, sec
(1) Unpackaged; metal container...	0.002
(2) Wooden box.................	0.004
(3) Carton.....................	0.000
(4) Approx. 1 in. latex hair........	0.008
(5) Approx. 3 in. latex hair........	0.015

ENVIRONMENTAL CONDITIONS

In general, the shock received during handling (for example, from being dropped) is greater than that experienced in being transported in a vehicle; such shocks usually determine the severity of a shock test.

The standard shock test for packages is the drop test in which the package is dropped from a predetermined height onto a rigid floor. The height of drop is determined by the type of handling the package may receive during shipment. For example, packages from 0 to 50 lb may be considered within the "one man throwing limit"; i.e., packages of such weight may be thrown easily onto piles or in other ways severely mishandled due to their light weight. Packages weighing between 50 and 100 lb may be considered within a "one man carrying limit"; such packages are somewhat heavy to be thrown but can be carried and dropped from a height as great as shoulder height. A "two man dropping limit" may apply to a weight range between 100 to 300 lb; the corresponding drop height for this mode of handling may be waist height. A further range may be from 300 to 1,000 lb; packages in this range would be handled with light cranes or lift trucks and may be subjected to shock from excessive lifting or lowering velocities. Finally, very heavy packages weighing over 1,000 lb would be handled by heavier equipment with correspondingly more skill; any drops would be from very small heights. Similarly, the size of the package classifies the type of handling into one man, two men, light equipment, or heavy equipment with corresponding drop heights. Thus, the drop heights for package shock tests are derived from the type of handling to which the package is most likely to be subjected during a shipping cycle; the type of handling is dependent on the size and weight of the package.

In addition to drop heights that vary with package size and weight, another factor in shock testing is the orientation of the package at impact. For example, small, light-weight packages are likely to be subjected to free fall drops onto sides, edges, and corners of the package. Larger, heavier packages handled by light or heavy equipment are likely to encounter drops of the type where one end rests on the floor and the other end is dropped (bottom rotational drop).

PROPERTIES OF CUSHIONING MATERIALS

The static and dynamic properties of cushioning materials that are commonly used in package design are included below in the section *Properties of Cushioning Materials*.

SHOCK ISOLATION

In the analytical treatment of shock isolation in packages, the idealized system illustrated in Fig. 41.2*A* is considered. It consists of a heavy rigid outer container, a packaged item assumed to be a rigid body, and a shock-isolation medium having characteristics which are considered below. The container is dropped from a characteristic height *h* upon a rigid floor, and the protection afforded to the packaged item by the isolation medium is indicated by the maximum acceleration experienced by the item. For

Fig. 41.2. Schematic arrangements of isolation media in shipping containers: *A* — axes of isolation medium perpendicular to faces of container; *B* — axes of isolation medium inclined to faces of container.

purposes of analysis, it is convenient to assume a medium having a linear stiffness k; then results are modified to include the effects of the nonlinearity in actual media.

The motion of the container as it falls freely from the height h is

$$u = \tfrac{1}{2} gt^2 \qquad (41.1)$$

The equation of motion for the packaged item is

$$m\ddot{x} = mg + k\delta \qquad (41.2)$$

where $\delta = u - x$ is the deflection of the isolation medium, m is the mass of the packaged item, and g is the acceleration of gravity. Substituting $u - x = \delta$ in Eq. (41.2) and

Fig. 41.3. Natural frequency of a package as a function of the weight of the packaged item and the stiffness of the isolation medium. These curves are defined by Eq. (41.5).

using the expression for u given by Eq. (41.1), the equation of motion of the packaged item is

$$\ddot{x} + \omega_n^2 x - y + \frac{\omega_n^2 gt^2}{2} \qquad (41.3)$$

The solution of this equation is

$$x = A \sin \omega_n t + B \cos \omega_n t + \frac{gt^2}{2} \qquad (41.4)$$

where $\omega_n = \sqrt{k/m}$ is the natural angular frequency in radians per second of the packaged item supported by the isolation medium. The natural frequency is expressed in units of cycles per second, as a function of the weight W of the packaged item:

$$f_n = \frac{1}{2\pi} \sqrt{\frac{kg}{W}} \qquad (41.5)$$

This relation is shown graphically in Fig. 41.3.

The coefficients in Eq. (41.4) are evaluated from the conditions at time $t = 0$. At this instant, the packaged item is in equilibrium between the gravitational force and the upward force applied by the isolation medium; i.e., there is no downward acceleration until the container has traveled a sufficient distance to reduce the magnitude of the force applied by the isolation medium. Then, at time $t = 0$, $\ddot{x} = 0$ and Eq. (41.4) becomes

$$x = \frac{g}{\omega_n^2} \cos \omega_n t + \frac{gt^2}{2} \qquad (41.6)$$

Thus, the packaged item falls at the same rate as the container, plus a superimposed oscillatory motion initiated at the moment of dropping. The velocity is found by differentiating Eq. (41.6):

$$\dot{x} = -\frac{g}{\omega_n} \sin \omega_n t + gt \qquad (41.7)$$

The velocity of the packaged item after the container has fallen from height h is determined by substituting $t = \sqrt{2h/g}$ in Eq. (41.7). The maximum possible velocity of the packaged item as the downward velocity of the container is arrested occurs when $\sin \omega_n \sqrt{2h/g} = -1$. The magnitude of the maximum velocity is

$$\dot{x}_{max} = \sqrt{2gh} + \frac{g}{\omega_n} \qquad (41.8)$$

The resulting response of the shock-isolation medium is determined by considering the container to rest upon the floor and the packaged item to have an initial downward velocity $\dot{\delta}_{max} = \dot{x}_{max}$ as given by Eq. (41.8). The equation of motion for the packaged item is

$$m\ddot{\delta} = -k\delta + mg \qquad (41.9)$$

The solution of this equation is

$$\delta = A \sin \omega_n t' + B \cos \omega_n t' + \frac{g}{\omega_n^2} \qquad (41.10)$$

where $t' = 0$ at the moment that the downward velocity of the container is arrested. Initial conditions are $\delta = 0$, $\dot{\delta} = \dot{x}_{max}$ from Eq. (41.8). Substituting these initial conditions, Eq. (41.10) becomes

$$\delta = \left(\frac{\sqrt{2gh}}{\omega_n} + \frac{g}{\omega_n^2} \right) \sin \omega_n t' + \frac{g}{\omega_n^2} (1 - \cos \omega_n t') \qquad (41.11)$$

In general, the term g/ω_n^2 is relatively small in packaging applications. Then the term $g/\omega_n^2 \cos \omega_n t'$ may be neglected and the maximum deflection δ_{max}, at time $t' = \pi/(2\omega_n)$, is

$$\delta_{max} = \left(\frac{\sqrt{2gh}}{\omega_n} + \frac{g}{\omega_n^2} \right) + \frac{g}{\omega_n^2} = \left(\frac{\sqrt{2gh}}{\omega_n} + \frac{2g}{\omega_n^2} \right) \quad \text{in.} \qquad (41.12)$$

The maximum acceleration experienced by the packaged unit is found by differentiating Eq. (41.11) * twice with respect to time, neglecting the term $g(\cos \omega_n t')/\omega_n^2$ and taking the maximum value at time $t' = \pi/(2\omega_n)$:

$$\ddot{\delta}_{max} = \omega_n \sqrt{2gh} + g \quad \text{in./sec}^2 \qquad (41.13)$$

The maximum acceleration $\ddot{\delta}_{max}$ may be expressed as a dimensionless multiple $(\ddot{\delta}_g)_{max}$ of

---

* In this portion of the analysis, the container is assumed to rest upon the floor; thus, $\ddot{\delta} = \ddot{x}$ is the acceleration of the packaged item.

the gravitational acceleration by dividing Eq. (41.13) by g:

$$(\ddot{\delta}_g)_{max} = \frac{\omega_n \sqrt{2gh}}{g} + 1 \qquad (41.14)$$

Solving Eq. (41.14) for the natural frequency ω_n and dividing the result by 2π to obtain units of cycles per second,

$$f_n = \frac{1}{2\pi} \sqrt{\frac{g[(\ddot{\delta}_g)_{max} - 1]^2}{2h}} \qquad \text{cps} \qquad (41.15)$$

The relation of Eq. (41.15) is shown graphically by the solid lines in Fig. 41.4.

The relation between maximum deflection δ_{max} and maximum acceleration $(\ddot{\delta}_g)_{max}$ is obtained by combining Eqs. (41.12) and (41.14):

$$(\ddot{\delta}_g)_{max} = \frac{2h + 3\dfrac{\sqrt{2gh}}{\omega_n} + \dfrac{2g}{\omega_n{}^2}}{\delta_{max}} \qquad (41.16)$$

For most packaging applications, the first term in the numerator of Eq. (41.16) is much larger than the other terms; thus, it is a reasonable approximation to neglect the latter and write Eq. (41.16) in the following form:

$$(\ddot{\delta}_g)_{max} \simeq \frac{2h}{\delta_{max}} \qquad (41.17)$$

This relation is shown by the dotted lines in Fig. 41.4. For a successful package design, it is necessary that $\ddot{x}_g$ be less than $\ddot{x}_F$.

NATURAL FREQUENCY, CPS

MAXIMUM ACCELERATION, g

Fig. 41.4. Natural frequency and maximum displacement of the packaged item in terms of its maximum acceleration and the drop height of the package. These curves are based on Eqs. (41.15) and (41.17).

Example 41.1. Consider an item weighing 25 lb that is to be packaged in such a manner that it does not receive an acceleration greater than $30g$ ($\ddot{x}_F = 30g$) when the package is dropped from a height of 36 in. Determine the spring characteristics and maximum displacement required to attain the desired maximum acceleration, and indicate the natural frequency of the packaged item on the isolation system.

Entering Fig. 41.4 at a value of $30g$ on the abscissa and 36 in. on the ordinate, the following values are obtained by interpolation between pairs of solid and dotted diagonal lines:

From the solid lines:

$$\text{Natural frequency} = 10.7 \text{ cps}$$

From the dotted lines:

$$\text{Maximum spring deflection} = 2.4 \text{ in.}$$

Then, entering Fig. 41.3 at a value of 25 lb on the abscissa scale, and reading on the ordinate scale at the diagonal line for 10.7 cps:

$$\text{Total spring stiffness} = 300 \text{ lb/in.}$$

INFLUENCE OF DAMPING ON SHOCK ISOLATION

The presence of damping in the isolation medium affects the response to shock. This subject is considered in detail in Chap. 31.

APPLICATIONS OF SHOCK-ISOLATION THEORY

TENSION-SPRING SUSPENSION. When the springs are arranged parallel to the direction of motion, as shown in Fig. 41.2*A*, the stiffness indicated by Fig. 41.3 is used directly in the design of the springs (see Chap. 34 for design data on metal springs, and Chap. 35 for data on rubber springs). The total stiffness is divided by the number of springs to obtain the stiffness per spring. However, if the springs extend between the interior corners of the container and the exterior corners of the packaged item, as indicated in Fig. 41.2*B*, several additional factors must be considered in determining the spring characteristics: (1) the axes of the springs are not parallel with the deflection, (2) the angle of the spring axes varies with deflection, and (3) if tension springs are used, only a portion of the total spring length is active. A detailed analysis of the tension spring package and a step-by-step procedure to determine the spring characteristics are given in Refs. 1 and 2.

BASE-MOUNT PACKAGE. For large and heavy equipment, it is common practice to mount the equipment on a base which is provided with skids to allow for fork lift entry during handling. The use of a skidded base on a large container dictates handling and transportation of the package with the base down at all times. Then isolation need be provided only for bottom rotational drops, and side and end impacts, as discussed previously under shock environment. An isolation system applicable to such a condition is illustrated in Fig. 41.5. An analysis of a base-mount package with linear isolators for both end impact and rotational drop is included in Chap. 3.

CUSHION DESIGN USING FLAT SHEETS. A common method of packaging embodies the use of an isolation medium in sheet form of standard thicknesses; the sheet is cut into pads that are

FIG. 41.5. Typical base-mounted package utilizing loading platform and skids. The sides and top of the container are mounted upon the base.

inserted between container and the packaged item on the several sides thereof. The theory of shock isolation with linear springs is applicable only for relatively small deflection of the isolation medium; for larger deflections, the stiffness becomes nonlinear and a different type of analysis must be used. In general, the stiffness cannot be defined analytically; * rather, the characteristics of the isolation medium are defined by stress-strain curves and by curves of maximum transmitted acceleration as a function of static loading stress for various heights of free fall of the container. These curves are obtained by physical tests of the material and are used to design the isolation medium, using the procedure described in the following paragraphs.

For a given isolation medium, the significant parameters in the design of a package are thickness of the medium, height of free fall of the package, unit loading on the medium, and maximum acceleration experienced by the packaged item. Unit loading is defined in terms of static conditions; it is expressed as a "static stress" representing the weight of the packaged item divided by the area of the medium in engagement with one side of the item. Isolation media are commercially available in a sequence of standard thicknesses; for example, thicknesses of 2, 4, and 6 in. are commonly available. Consequently, the design data are given for corresponding thicknesses. Similarly, the height of free fall is defined in discrete increments corresponding to accepted specification practices; the isolation at intermediate heights can be determined by interpolation. The shock isolation properties are defined by several families of curves. For example, Fig. 41.15 shows maximum acceleration of the packaged item as a function of static stress for several thicknesses of "Latex Hair, Firm" when the height of free fall is 12 in. (Corresponding data on the same material for other heights of free fall are given in Figs. 41.16 and 41.17.)

In the design of a package, data of the type given in Figs. 41.15 to 41.32 are applied by selecting the figure corresponding to the applicable height of free fall, entering the figure on the ordinate scale at a value of maximum acceleration corresponding to the fragility of the packaged item, and determining the required static loading from the curves representing various thicknesses of isolation medium. This determines the requirements of the design for purposes of shock isolation; other requirements are set forth below under *Vibration Isolation* and *Properties of Cushioning Materials*. The complete procedure for designing the isolation medium is discussed under *Design Technique—Nonlinear Cushioning*.

Comparison of Flat and Corner Drops. The selection of cushion dimensions is determined from the dynamic properties of cushioning materials. Such properties are determined by dropping an equipment mock-up flat onto cushion specimens. However, package tests require drops on corners and edges. The use of flat-drop data in cushion design has been justified by a series of tests summarized in Fig. 41.6.[3] These tests involved dropping a package consisting of a wooden block cushioned on all six sides and enclosed in an exterior container. Three accelerometers were mounted on the block with their directions of sensitivity along the three mutually perpendicular axes of symmetry of the block. The package was suspended above a rigid surface and its orientation was carefully measured. It was allowed to fall freely and the three accelerometer measurements were recorded. The first series of tests began with a flat drop. Subsequent drops from the same height were made with varying angles of orientation until a corner drop was reached. The angle of orientation was then further varied until the tests were concluded with a drop on an edge of the package. This procedure was repeated using 2- and 3-in. thick cushioning. The second series of tests began from a flat drop, continued to an edge, and concluded with a flat drop on the side 90° from the original impacting side. The acceleration measurements are plotted in Fig. 41.6 with a faired curve for each thickness. The accelerations are the vector resultants of the three accelerometer recordings.

VIBRATION ISOLATION. In general, the primary consideration in the design of a cushion system is the protection of the packaged item against shock. However, if the

* An analytical treatment of shock isolation wherein the characteristics of nonlinear isolators are defined analytically is included in Chap. 31.

item (or a component of the item) to be packaged has a resonant frequency falling within the range of vibration frequencies encountered during transportation, the vibration-isolation characteristics of the cushion system must be considered.

Fig. 41.6. Effect of container orientation upon maximum acceleration experienced by a packaged item during drop tests: A — sequence of orientations is flat to corner to edge; B — sequence of orientations is flat to edge to flat. (*L. W. Gammel and J. L. Gretz.*[3])

The theory of vibration isolation is discussed in detail in Chap. 30; equations for transmissibility with several types of damping are given in Table 30.2 and transmissibility with viscous damping is shown graphically in Fig. 2.17. It is highly desirable that the isolator have significant damping, so that the vibration amplitude does not become excessive if environmental conditions include vibration having a frequency equal to the natural frequency of the isolator. Furthermore, it is important that the natural frequency of the isolator be substantially less than the natural frequencies of vulnerable components of the equipment so that the excitation of the components is small.

Example 41.2. Consider the system shown schematically in Fig. 41.7 where m_1 represents the total mass of the packaged item, m_2 represents the mass of an element or component of the equipment (m_2 is assumed small relative to m_1), and $\ddot{u}$ is the acceleration of the vibration associated with the environmental condition. The environment is characterized by vibration having an acceleration amplitude of $1.3g$ at 6 cps (the natural frequency of the isolator) and an acceleration amplitude of $3g$ at 200 cps (the natural frequency of the element). The fragility $\ddot{x}_F$ of the critical element is $30g$ and its fraction of critical damping ζ_2 is 0.001. An undamped tension spring package with natural frequency f_1 has been designed. The fraction of critical damping ζ_1 for an undamped tension spring package has been experimentally determined to be 0.004.[4] Determine the response of the critical element at isolator resonance and element resonance.

Response of Element at Isolator Resonance. The transmissibility of the isolator at resonance is defined by $T_R = 1/(2\zeta)$ [see Eq. (30.1)]:

$$T_R = \frac{1}{2 \times 0.004} = 125$$

Then, the acceleration of m_1 at 6 cps is

$$\ddot{x}_1 = T_R \times \ddot{u} = 125 \times 1.3g = 163g$$

The acceleration $\ddot{x}_2$ of the element is approximately the same as that of the equipment; this is characteristic of elements of high natural frequency (relative to the natural frequency of the isolator).

Response of Element at Element Resonance. The transmissibility of the isolator at element resonance is $T = 0.0008$ [see Eq. (30.1)]. Then the acceleration of m_1 is

$$\ddot{x}_1 = T \times \ddot{u} = 0.0008 \times 3g = 0.0024g$$

The transmissibility of the element at resonance is $T_R = \dfrac{1}{2\zeta_2}$ [see Eq. (30.1)] = 500. Thus, the acceleration of the element is $\ddot{x}_2 = T_R \times \ddot{x}_1 = 500 \times 0.0024g = 1.2g$.

The results calculated above are tabulated in Table 41.2 and plotted in Fig. 41.8. Similar results are included for fractions of critical damping $\zeta_1 = 0.03$ and 0.10 characteristic of a damped spring suspension system. The addition of damping to the tension spring reduces the acceleration experienced by the component as a result of vibration at the resonant frequency of the spring but increases the acceleration resulting from vibration at the resonant frequency of the element.

FIG. 41.7. Model consisting of a damped two degree-of-freedom system to represent a packaged item and a critical element thereof

FIG. 41.8. Maximum acceleration amplitude $\ddot{x}_2$ of critical element m_2 in Fig. 41.7 for the conditions indicated in Example 41.2. The curves refer to different values of the fraction of critical damping ζ_1 for the package cushioning as follows: curve 1: $\zeta_1 = 0.004$; curve 2: $\zeta_1 = 0.03$; curve 3: $\zeta_1 = 0.10$.

Table 41.2. Element Acceleration for Tension Spring Suspension Package

(See Example 41.2)

Type of package	Fraction of critical damping	Response at package resonance				Response at element resonance			
		T_p	$\ddot{x}_1$, g	T_e	$\ddot{x}_2$, g	T_p	$\ddot{x}_1$, g	T_e	$\ddot{x}_2$, g
Undamped tension spring....	0.004	125	163	1	163	0.0008	0.0024	500	1.2
Damped tension spring....	0.03	16.7	21.7	1	21.7	0.002	0.006	500	3.0
Damped tension spring....	0.10	5.0	6.5	1	6.5	0.007	0.021	500	10.5

Damping of Tension Spring Package. One simple and economical means of providing damping is to stuff wool felt into the centers of the spring coils throughout the extended spring length. Fractions of critical damping in the range from 0.03 to 0.10 can be obtained by this method, indicating a range of maximum transmissibility from 16.7 to 5, respectively.

ANALYTICAL DETERMINATION OF NATURAL FREQUENCY. The relationships between transmissibility and frequency of vibration and between the natural frequency of the cushion system and the static stress on the cushion determine the degree of vibration isolation afforded by a cushion. Because of the many variables and unknowns involved in analyses of the materials commonly used as package cushioning, transmissibility data must be obtained empirically. For example, transmissibility curves for latex hair and polyester urethane foam are given in Figs. 41.37 and 41.38.

The natural frequency of a nonlinear isolator is

$$f_n = \frac{1}{2\pi} \sqrt{\frac{dF/d\delta}{W/g}} \tag{41.18}$$

where f_n = natural frequency, cps
    F = force, lb
    δ = deflection, in.
    W = weight supported by isolator, lb
    g = gravitational acceleration = 386 in./sec^2

The force F is equal to the weight W and $dF/d\delta$ is the slope of force-displacement curve at the force $F = W$. These may be expressed in terms of stress and strain as follows:

$$\sigma = \frac{F}{S} = \text{stress, lb/in.}^2$$

$$\epsilon = \frac{\delta}{t} = \text{strain, in./in.}$$

where S = cushion area, in.^{2}
    t = cushion thickness, in.,
Then Eq. (41.18) becomes

$$f_n = \frac{1}{2\pi} \sqrt{\frac{d\sigma/d\epsilon}{Wt/gS}} = 3.13 \sqrt{\frac{d\sigma/d\epsilon}{Wt/S}} \tag{41.19}$$

Applying the stress-strain relation $d\sigma/d\epsilon$ for a material, Eq. (41.19) gives the relationship between natural frequency f_n and static stress W/S for various cushion thicknesses t. Data giving this relationship for several package cushioning materials are presented in Figs. 41.39 through 41.41.

Equation (41.19) applies only for small vibration amplitudes because $d\sigma/d\epsilon$ does not remain constant throughout a vibration cycle of large amplitude; thus discrepancies between the actual and calculated natural frequencies may occur for large vibration amplitudes. Most materials also exhibit some difference between static and dynamic stiffness. Testing demonstrates fair correlation between actual and calculated values of natural frequency for small vibration displacement amplitudes; however, discrepancies may occur if large displacements are involved. Thus, the curves of static stress vs. strain should be used only to obtain the first approximation to the natural frequency of the isolator.

PROPERTIES OF CUSHIONING MATERIALS

This section considers cushioning materials such as latex hair, various plastic foams etc., which are supplied either in sheet form or in molded shapes; as package cushioning they are interposed between the container and the packaged item. Although such materials may exhibit linear force-deflection characteristics for small deflections, efficient package design involves large deflections and consequent nonlinearity of the cushioning materials.

MATERIAL CLASSIFICATION

All materials utilized to provide cushioning (i.e., shock and vibration isolation) in packaging may be classified as either of two general types—*elastic* (flexible or resilient) and *nonelastic* (crushable). Materials which, when deflected during impact, sustain no more than an arbitrarily selected amount of permanent deformation are placed in the first category; materials that sustain a greater amount are placed in the second category. Certain factors influence the choice of this arbitrarily selected level. One of the most important of these is that a sufficient thickness of material should be left after deformation to provide the required isolation on successive impacts. A permanent deformation of not more than 10 per cent after compression to a strain of 65 per cent is used as a criterion in the package design field.

Materials described in this section may be used to satisfy package design requirements other than those for cushioning, for example, to position an item within the package.

MATERIAL PROPERTIES

The fundamental property to be considered in application of a material in shock and vibration isolation is the manner in which it stores, absorbs, and dissipates energy. This characteristic is basic to the classification of materials set forth in the preceding paragraph.

Package cushioning may be designed analytically if an equation can be written for the force-deflection characteristics under dynamic conditions[1] (see Chap. 31). A quasi-analytical concept minimizes the ratio of transmitted force to stored energy.[5] The most directly useful method makes use of experimentally determined properties of the cushioning materials in a straightforward design procedure, as described in subsequent sections of this chapter.

PROPERTIES FOR ISOLATION. The properties of materials that are most useful in packaging design for shock isolation have been standardized as follows:

1. Compression stress vs. strain characteristics
2. Maximum acceleration vs. static stress characteristics, employing drop height and material thickness as parameters
3. Creep characteristics

Requirements covering these properties for elastic-type cushioning materials are set forth in a military specification on elastic-type cushioning materials.[6] Methods of testing for determination of these properties are detailed in Refs. 6 to 9.

FIG. 41.9. Compression stress vs. strain for latex hair.

Data of the types listed above are presented in this chapter for a number of specific materials commonly used for package cushioning. Static stress vs. strain characteristics are given in Figs. 41.9 to 41.14; these data are used to determine the degree of static

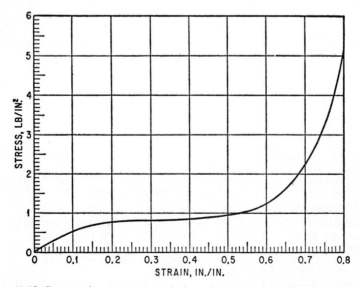

FIG. 41.10. Compression stress vs. strain for polyester urethane foam (2.2 lb/ft³).

compression of the cushion due to the weight of the item to be packaged. Maximum acceleration vs. static stress data are given in Figs. 41.15 to 41.32; the dynamic shock loading on the packaged item for various package drop heights can be determined from these data. Creep characteristics are given in Figs. 41.33 to 41.35; these data are used to determine the thickness loss of the cushion during storage. Because of wide differences in manufacturing processes and techniques associated with production of package cushioning materials, and the large number of manufacturers of each type of material, a great variation exists in the characteristics of commercially available materials. For this reason, the data presented here represent only an average for a particular material and condition.*

Fig. 41.11. Compression stress vs. strain for polyether urethane foam (2.0 lb/ft³).

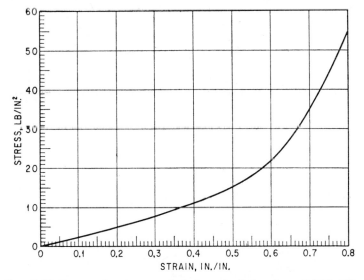

Fig. 41.12. Compression stress vs. strain for polyethylene foam (1.75 lb/ft³).

* The data used to plot Figs. 41.9 to 41.41 inclusive were made available through the courtesy of Rockwell International, Inc. and Ref. 10.

Fig. 41.13. Compression stress vs. strain for cellulose wadding.

Fig. 41.14. Compression stress vs. strain for "AirCap" SD 240.

Fig. 41.15. Maximum acceleration vs. static stress for latex hair, firm—12-in. drop.

FIG. 41.16. Maximum acceleration vs. static stress for latex hair, firm—24-in. drop.

FIG. 41.17. Maximum acceleration vs. static stress for latex hair, firm—30-in. drop.

FIG. 41.18. Maximum acceleration vs. static stress for polyester urethane (2.2 lb/ft³)—12-in. drop.

Fig. 41.19. Maximum acceleration vs. static stress for polyester urethane (2.2 lb/ft³)—24-in. drop.

Fig. 41.20. Maximum acceleration vs. static stress for polyester urethane (2.2 lb/ft³)—30-in. drop.

Fig. 41.21. Maximum acceleration vs. static stress for polyether urethane (2.0 lb/ft³)—12-in. drop.

Fig. 41.22. Maximum acceleration vs. static stress for polyether urethane (2.0 lb/ft³)—24-in. drop.

Fig. 41.23. Maximum acceleration vs. static stress for polyether urethane (2.0 lb/ft³)—30-in. drop.

FIG. 41.24. Maximum acceleration vs. static stress for polyethylene foam (2.0 lb/ft³), 12-in. drop.

FIG. 41.25. Maximum acceleration vs. static stress for polyethylene foam (2.0 lb/ft³), 24-in. drop.

FIG. 41.26. Maximum acceleration vs. static stress for polyethylene foam (2.0 lb/ft³), 30-in. drop.

FIG. 41.27. Maximum acceleration vs. static stress for cellulose wadding, 12-in. drop.

FIG. 41.28. Maximum acceleration vs. static stress for cellulose wadding, 24-in. drop.

FIG. 41.29. Maximum acceleration vs. static stress for cellulose wadding, 30-in. drop.

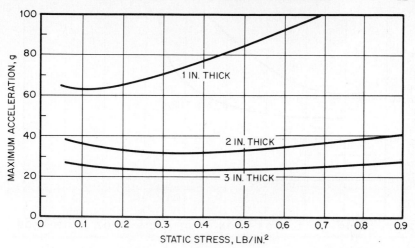

FIG. 41.30. Maximum acceleration vs. static stress for "AirCap" SD 240, 12-in. drop.

FIG. 41.31. Maximum acceleration vs. static stress for "AirCap" SD 240, 24-in. drop.

FIG. 41.32. Maximum acceleration vs. static stress for "AirCap" SD 240, 30-in. drop.

FIG. 41.33. Creep characteristics for latex hair, firm.

FIG. 41.34. Creep characteristics for polyester urethane (2.2 lb/ft³).

FIG. 41.35. Creep characteristics for polyether urethane (2.2 lb/ft³).

DESCRIPTION OF MATERIALS. This section includes a brief physical description of the various materials covered in this chapter, including their uses and characteristics.

Latex Hair. Curled animal hair (mostly from hogs but with some cattle tail and horse mane hair), bonded with natural latex (in molded shapes) or neoprene (in sheet form);

this material is most efficiently used to cushion items at static stresses up to approximately 0.3 lb/in.² and is available in various densities (normally referred to as soft, medium, and firm).

Urethane Foam. An open-cell plastic foam available in various densities, depending on the manufacturer; two forms of this material are available—one derived from a polyester and one from a polyether. They have somewhat different static stress-strain characteristics. Most commercially available materials are useful in approximately the same static stress range as latex hair, but some are available for use at a static stress as great as approximately 1.0 lb/in.²

Polyethylene Foam. A blown plastic foam, it is manufactured in various densities (approximately 2.0 to 9.0 lb/ft³). This is a stiff material but is very efficient in cushioning applications at static stresses from approximately 1.0 to 2.0 lb/in.²; in other applications the material can be cut very thin and used as a protective wrap.

Cellulose Wadding. A wood-fiber product most efficiently used as a protective wrap, but also used in some applications as cushioning where a static stress of less than 0.1 lb/in.² is required. For cushioning purposes, this material should not be used on extremely delicate items.

AirCap. A laminate of two layers of barrier-coated polyethylene film, one layer of which is embossed with rows of cells forming an air encapsulated cushioning material.

TEMPERATURE EFFECTS ON ISOLATION. The effects of temperature, both high and low, on the isolation characteristics of materials must be considered in the selection of the proper material for use as package cushioning. Most materials in common usage for package cushioning exhibit a marked increase in stiffness as the temperature decreases from 70°F (21.1°C), and a corresponding decrease in stiffness as the temperature is increased from 70°F. The variation in stiffness with temperature is considerably more pronounced in some materials than in others. For example, fibrous glass cushioning exhibits very little change in stiffness with temperature whereas the stiffness of polyester urethane foam increases rapidly as the temperature decreases. The change in the stiffness of latex-bound hair with variations in temperature may be considered approximately average for most cushioning materials. Because of wide variations in procedures and equipment used in testing to determine the effects of temperature on the isolation characteristics of cushioning materials, some published data of this type are not reliable. For this reason, only data on the effects of temperature on the stress vs. strain relationship for latex hair are presented here (see Fig. 41.36) as an indication of material performance.

Fig. 41.36. Compression stress vs. strain for latex hair, firm, at various temperatures.

DESIGN TECHNIQUE—NONLINEAR CUSHIONING

The following procedure can be used as a guide in the development of design techniques utilizing distributed cushioning material for shock and vibration isolation:

1. Define the shock environment in terms of drop height; define the vibration environment in terms of frequency and amplitude.

2. Establish the fragility $\ddot{x}_F$ of the packaged item (see Chap. 42) and the characteristics (natural frequency and damping) of the critical components of the item.

3. Provide protection for the packaged item by means such as blocking and bracing of protruding parts, distribution of load to avoid stress concentrations, and preservation methods. (Interior containers adapted to house the packaged item may serve as protection when the item is rigidly secured within, thus approximating a single degree-of-freedom system.)

4. From the weight and size of the packaged item (or interior container plus contents), determine the static stress for each face; then from the maximum acceleration vs. static stress curves (Figs. 41.15 to 41.32) select the optimum material and the required thickness.

5. Check the static stress vs. strain curve (Figs. 41.9 to 41.14) for the applicable material to determine the static deflection.

6. Check the creep properties (Figs. 41.33 to 41.35) and, when applicable, make allowance for any loss in thickness during storage. The required initial thickness is

$$t_{\text{req}} = \frac{t_s}{1 - C} \tag{41.20}$$

FIG. 41.37. Transmissibility curve for latex hair, firm.

FIG. 41.38. Transmissibility curve for polyester urethane (2.2 lb/ft³).

where t_{req} = initial thickness required to compensate for creep, in.
  t_s = thickness required for shock isolation, in.
  C = drift, the ratio of thickness loss from creep to original thickness
  7. To avoid localized buckling of the isolation medium, stability must be considered so that the cushion returns to its original position to provide protection for more than a single drop.   A rule-of-thumb stability relation requires that

$$\frac{t}{\sqrt{S}} \leq \frac{4}{3} \qquad (41.21)$$

where t = thickness of cushion pad, in.
  S = area of cushion pad, in.^{2}
  8. For vibration isolation, determine the natural frequency of the isolation system (Figs. 41.39 to 41.41) and, from the transmissibility curves of Figs. 41.37 and 41.38, calculate the vibration of the packaged item at the resonant frequencies of the isolation system and the critical component.

Fig. 41.39. Natural frequency vs. static stress for latex hair, firm.

Fig. 41.40. Natural frequency vs. static stress for polyester urethane (2.2 lb/ft^3).

F𝐢𝐠. 41.41. Natural frequency vs. static stress for polyethylene foam (1.75 lb/ft³).

9. Select a method of cushioning to provide the exact amount of material required by the analysis. Wrapping a part with the material or use of other methods which provide a larger effective cushion area than required may be detrimental because the static stress is decreased and the maximum acceleration may be increased.

10. Select the outer container dimensions to provide the proper thickness of material and sufficient clearances to obtain the required deflection.

Example 41.3. A packaged item is capable of withstanding a maximum acceleration of $30g$ without damage. A critical component in this item has a natural frequency of 250 cps and an equivalent fraction of critical damping of 0.001. A criterion for rough handling has been established as a drop from a height of 30 in. The anticipated transportation environment has the following vibration characteristics: Displacement amplitude of 0.5 in. from 2 to 5 cps, and acceleration amplitude of $3g$ from 5 to 300 cps. The storage period is 6 months. An interior container with proper blocking and bracing, and necessary preservation requirements, has the following dimensions: Length, 24 in.; width, 18 in.; and height, 12 in. The gross weight of the interior container plus contents is 40 lb.

1. From the dimensions of the interior container and its gross weight, the available cushioning area and static stress for each container face are determined:

Container face	Available area, in.²	Static stress, lb/in.²
Top and bottom........	24 × 18 = 432	0.0925
Sides.................	24 × 12 = 288	0.139
Ends.................	18 × 12 = 216	0.185

2. For this application, latex hair (firm) is to be used. From Fig. 41.17, for a 30-in. drop and $30g$ maximum acceleration, the thickness required is 3½ in. for top and bottom, 4 in. for the sides, and 5½ in. for the ends. Adjustment of the static stress on all faces to 0.0925 lb/in.² (total area on each face = 432 in.²) yields optimum design, resulting in a thickness of 3½ in. on each container face. The area on each face can be made equal by overlapping the cushioning materials in the manner illustrated in Fig. 41.42.

3. A static stress of 0.0925 lb/in.² produces a strain of approximately 0.025 (from Fig. 41.9), or a static deflection of 0.0875 in. This relatively small deflection does not significantly change the nominal dimensions of the package.

4. From Fig. 41.33, a static stress of 0.18 lb/in.² for a duration of 6 months results in an increase in strain of 0.167 in./in., i.e., in a thickness loss of approximately 16.7 per cent as

PACKAGED ITEM

DUNNAGE LOCATED AROUND PART TO PREVENT MOVEMENT IN ALL DIRECTIONS WITHIN INTERIOR CONTAINER

INTERIOR CONTAINER 24 IN. x 18 IN. x 12 IN. OUTSIDE DIMENSIONS

MOISTURE-VAPOR PROOF BARRIER

EXTERIOR CONTAINER 32 IN. x 26 IN. x 20 IN. INSIDE DIMENSIONS

PLYWOOD INSERT FOR INCREASING BEARING AREA

LATEX HAIR CUSHION PADS 24 IN. x 18 IN. x 4 IN. THICK CEMENTED TO EXTERIOR CONTAINER AS SHOWN

FIG. 41.42. Package cushioning design using distributed cushioning on six sides.

a result of creep. Assuming a linear relationship between stress and drift, the expected drift after 6 months under load at a static stress of 0.0925 lb/in.² is approximately 8 per cent. Then, from Eq. (41.20), the initial thickness required to compensate for drift is

$$t_{\text{req}} = \frac{3.5}{1 - 0.08} = 3.8 \text{ in.}$$

Rounding this result off to the nearest nominal value, cushioning material 4 in. thick should be used.

5. Checking the stability by substituting the thickness and area in Eq. (41.21),

$$\frac{t}{\sqrt{S}} = \frac{4}{\sqrt{432}} = 0.193$$

This is less than ⅓, and the design is not susceptible to localized buckling.

6. From Fig. 41.39, the natural frequency of the isolation system for a static stress of 0.0925 lb/in.² and 4-in. thickness is 30 cps.

7. At "package resonance," the impressed acceleration is $3g$ and (from Fig. 41.37) the interior container acceleration is $4 \times 3g = 12g$. The acceleration of the element also is $12g$. At resonance of the element, the acceleration of the interior container is $0.002 \times 3g = 0.006g$, and the acceleration of the element is $500 \times 0.006g = 3g$.

REFERENCES

1. Mindlin, R. D.: *Bell System Tech. J.*, **24**:353 (1945).
2. Franklin, P. E.: "Packaging for Shock and Vibration Protection with Tension Spring Suspensions," *Missile Division Rept.* AL-2608, Contract AF33(600)-28469, North American Aviation, Inc., 1957.
3. Gammell, L. W., and J. L. Gretz: "Report on Effect of Drop Test Orientation on Impact Accelerations," Physical Test Laboratory, Texfoam Division, B. F. Goodrich Sponge Products Division of B. F. Goodrich Company, Shelton, Conn., 1955.
4. Hatae, M. T.: "Shock and Vibration Tests of a Tension Spring Package," *Rept.* NA-54-1003, North American Aviation, Inc., 1954.
5. Janssen, R. R.: "A Method for the Proper Selection of a Package Cushion Material and Its Dimensions," *Rept.* NA-51-1004, North American Aviation, Inc., 1952.
6. "Military Specification—Cushioning Material, Elastic Type General," *USAF Specification* MIL-C-26861, 1959.
7. ASTM D 1372–64 Package Cushioning Materials, American Society for Testing Materials, 1964.
8. ASTM Method D 1596-64 Shock Absorbing Characteristics of Packaging Cushioning Materials, American Society for Testing Materials, 1964.
9. ASTM Method D 2221-68 Creep Properties of Package Cushioning Materials, American Society for Testing Materials, 1968.
10. "Military Standardization Handbook-Package Cushioning Design," *USAF Handbook* MIL-HDBK-304A, Sept. 25, 1974.

42

THEORY OF EQUIPMENT DESIGN

Edward G. Fischer
Westinghouse Electric Corporation

INTRODUCTION

The design of equipment to withstand dynamic loads is one of the more difficult and less developed aspects of shock and vibration control. The difficulty is inherent, in part, in the physical conditions because the nature and magnitude of such loads depend upon the dynamic characteristics of the equipment; thus, the loads cannot be determined until the equipment has been designed. Therefore, the designer must design for loads that are necessarily unknown. Such a problem is approached logically by first preparing an initial design based upon the best judgment of the designer. A subsequent analysis would show weaknesses and other undesirable features of the design, and point to desirable modifications. Thus, the design would be developed by cut-and-try methods.

Design by cut-and-try methods using only analytical procedures is not profitable in general because several important considerations are difficult to evaluate. For example, even though the dynamic loads could be determined by analysis, it is difficult to determine the resulting stresses if the equipment has an ordinary degree of complexity. Furthermore, it is difficult to evaluate the properties of materials under dynamic strain. Accordingly, it is preferred design practice to calculate strength requirements for only major and important structural members of the equipment. An initial prototype equipment is then constructed for test in which actual operating conditions preferably are duplicated. For military equipment, this is generally impractical and laboratory tests are used to simulate actual operating conditions. Laboratory testing and the simulation of actual conditions are discussed in Chap. 24. Neither designing nor testing to withstand dynamic loads is an exact science; hence, discrepancies frequently occur and must be resolved by experience and intuition. This chapter offers such guidance as the state of the art permits,

It is important to consider the design of the initial prototype from both the analytical and practical points of view. Much can be done at that stage of the design. Optimum stiffnesses can be determined for the interrelated structural members of the equipment with the objective of minimizing the stresses in such members. The forces experienced by components of the equipment also are minimized. The required strengths of these members then can be calculated. The succeeding sections of this chapter set forth procedures for determining strengths and stiffnesses of principal structural members, and outline applicable information on the selection and use of engineering materials in equipment that is required to withstand shock and vibration. A number of techniques, generally qualitative in nature, have been developed as a result of much practical experience in design. These are summarized in Chap. 43.

An important part of the design process is redesign to correct the deficiencies revealed by tests of the prototype. Such tests reveal not only the weaknesses of the equipment but also useful information, for example, on the natural frequencies of structural elements or the degree of damping present. With this information, new analyses can be carried out or original analyses modified using actual rather than assumed information.

Although methods for the design of equipment to withstand dynamic loads are not well developed in general, it is possible to apply known methods in particular cases. Some of these cases are discussed in other chapters. The design of internal-combustion engines and other types of rotating and reciprocating machinery is discussed in Chap. 38. Dynamic absorbers and auxiliary mass dampers are discussed in Chap. 6, and the use of damping materials is discussed in Chap. 37. An analysis of self-excited vibration is included in Chap. 5, and the isolation of vibration and shock is covered in Chaps. 30 to 33.

Although the connotation of the words "shock and vibration" is somewhat broader, this chapter is limited to the design of equipment that must withstand the motion of the support to which the equipment is attached. The motion may be of a continuing oscillatory nature, known broadly as vibration, or of a transient nature involving acceleration and/or displacements of large magnitude.

DYNAMIC LOAD

When an equipment is subjected to shock or vibration, the structural members of the equipment deflect or deform. The deflection is said to be caused by the *dynamic load;* i.e., the force required to cause structural members of the equipment to deflect in response to the motion imposed by the support. For practical purposes, the dynamic load may be considered as the static or dead-weight load multiplied by a *dynamic load factor.* The members must be designed to withstand, without failure, the stress imposed by the dynamic load. In general, the stress in the member subjected to a dynamic load is equal to that caused by a static load of the same magnitude. It does not necessarily follow, however, that a design stress which applies to static conditions also applies to dynamic conditions. For example, a structure designed to withstand only a static load may be satisfactory if the stress does not exceed a certain percentage of the ultimate stress. On the other hand, a structure may be unsatisfactory under a dynamic load inducing the same stress level because the structure may fail under repeated occurrences of the stress. In other words, if the dynamic load is repeated many times, a lower value of acceptable stress must be used to recognize the effect of fatigue of the material.

The magnitudes of the dynamic loads in an equipment subjected to shock and vibration usually are large relative to the static loads; thus, it is common practice in design to neglect static loads unless they are known to be significant relative to dynamic loads.

The physically observable effect of the application of a dynamic load is the deflection of a structural member. To require that the member have an acceptably low stress as a consequence of such deflection would introduce an uncommon concept in design because structures, in general, are designed to withstand designated *loads.* By considering the dynamic load to be equivalent to a static load that would cause the same deflection, the designer acquires a familiar working tool in that he can apply the principles of static loading. The process of designing for shock and vibration consists of (1) determining the magnitudes and directions of the dynamic loads and (2) designing the structure to withstand a hypothetical static load equivalent to the dynamic load, keeping in mind certain qualifying conditions related to the stress level and the distribution of stress.

METHODS OF ANALYSIS

USE OF MODELS

The first step in carrying out an analysis involving dynamic loading is to represent the structure by an *equivalent model* of rigid masses connected by massless springs and dampers. Theoretically, it is then possible to determine from any known excitation the motion of each mass and the consequent deflection of each spring. The force acting on each spring is known directly from its deflection and stiffness; thus, the springs can be designed with the strength required to withstand such forces. Therefore, it may be said that the equipment has adequate structural strength to withstand the excitation.

The type of model used for the analysis depends upon the accuracy required in the results and the effort that can be applied to the analysis. For example, consider the

FIG. 42.1. Rigid body supported by an elastic beam with hinged ends.

FIG. 42.2. Single degree-of-freedom system model of beam in Fig. 42.1, neglecting mass of beam.

problem of designing the simple beam supporting a rigid body of weight W as shown in Fig. 42.1. If the mass of the beam is small relative to that of the rigid body, a model that is adequate for most purposes is illustrated in Fig. 42.2 where the stiffness k and the damper c represent corresponding properties of the beam. However, if the mass of the beam is of the same order of magnitude as the mass of the supported body, parts of the beam tend to vibrate with respect to each other. This introduces dynamic loads that cannot be neglected if accurate results are desired. A model that includes the mass of the beam is shown in Fig. 42.3 where the sum of the incremental masses m' equals the total mass of the spring.

A model may be constructed with any degree of complexity required. For example, the model of an automobile in Fig. 42.4 could be used to determine the effect of shock absorbers on riding comfort or to calculate the maximum force applied to the seat springs. In this instance, the discrete masses of the model correspond to similarly discrete bodies of the actual automobile. By contrast, a model suitable for investigating the strength and stiffness of the guided missile of Fig. 42.5A under bending loads would appear as in Fig. 42.5B. The total mass of the missile, distributed more or less uniformly along its length, is represented by the sum of the several rigid masses of the model; the stiffness of the missile in bending is represented by the stiffnesses in bending of the members connecting the masses of the model. As an objective in setting up the model, the natural frequencies of model and actual missile should agree.

In general, complex models are useful only for investigating the capability of a particular equipment to withstand a particular excitation. However, the development of high-speed digital computer facilities has made it practical to investigate the dynamic response of a large number of possible combinations of masses, springs, and dampers in a complex model and eventually draw general conclusions. It must be cautioned that in order for the model to be authentic, the computer programming should first demonstrate that it can reproduce the results of simple static deflection and snap-back natural frequency tests of the original physical system. There is almost no limit to the degree of sophistication (i.e., yielding elements, random excitation, etc.) that can be achieved with computer-aided analyses, provided that the results make sense in terms of engineering experience.

A model may be used to study equipment design problems arising from any type of dynamic excitation. The general designation "shock and vibration" used to describe an environment is separable for purposes of analysis into (1) periodic vibration, (2)

FIG. 42.3. Model of beam in Fig. 42.1. The distributed mass effect of the beam can be included in the model where the sum of the incremental masses m' equals the total mass effect of the spring.

transient vibration or shock, and (3) random vibration. The response of mechanical systems (a "system" as discussed generally in other chapters is equivalent to a "model"

FIG. 42.4. Model of an automobile. The discrete masses of the model correspond to similarly discrete bodies of an actual automobile.

in this chapter) to these various classes of environmental conditions is discussed in several other chapters as follows:

Periodic vibration: Chaps. 2, 6, and 22.

Transient vibration or shock: Chaps. 8 and 23.

Random vibration: Chaps. 11 and 22.

The design of the model and the interpretation of the response are discussed in succeeding sections of this chapter.

FIG. 42.5. (A) Schematic diagram of a missile with the total mass distributed more or less uniformly along its length. (B) Model of the missile wherein the total mass is subdivided into a series of masses connected by members having the appropriate stiffnesses in bending.

TYPES OF MODELS. The usefulness of a model for initial design purposes tends to increase as its simplicity increases. A model that attempts to simulate an equipment in detail cannot be constructed until after an equipment is designed; thus, it becomes the means to check a design rather than to implement the design initially. On the other hand, a less complex model can simulate primary or important structures in a more general sense and can be evaluated in more general terms. For example, the model shown in Fig. 42.2 is defined completely by its natural frequency and fraction of critical damping; its response to any given excitation can be expressed concisely in terms of these parameters. By contrast, the model shown in Fig. 42.4 involves five masses and five springs; the relatively large number of possible permutations demonstrates the impracticality of expressing the results concisely or in a form suitable for ready reference. Techniques for the application of models in design are described below under *Structural Design Considerations* for various types of excitation.

Single Degree-of-freedom Model. The model shown in Fig. 42.2 is a spring-mass-damper system where the rigid mass $m = W/g$ is assumed to move along a straight line. The mass can be located at any time t by the single coordinate x; therefore the system is referred to as having a single degree-of-freedom. When the mass is displaced from its equilibrium position, it has only one mode of vibration; this occurs in free vibration at the single natural frequency of the system. The massless spring has a linear force-deflection relationship defined by the spring constant k. Similarly, the massless damper exhibits a linear force-velocity relationship denoted by the damping coefficient c. The damping in a typical mechanical system usually is so small that it has a negligible effect on the value of the natural frequency. Hence, the free vibration of the mass in the single degree-of-freedom model can be considered to be a simple harmonic motion. The undamped natural frequency from Eq. (2.8) is

$$\omega_n = \sqrt{\frac{kg}{W}} = \sqrt{\frac{g}{\delta_{st}}} \quad \text{rad/sec} \tag{42.1}$$

where δ_{st} is the static deflection of the spring caused by the weight W.

The analogy between the structure of Fig. 41.1 and the model of Fig. 42.2 is established as follows: The rigid body of weight W is supported at the mid-position of the elastic beam of length l and moment of inertia (of the cross-sectional area) I; the modulus of elasticity of the beam material is E. The single degree-of-freedom model of Fig. 42.2 is applicable only when the weight of the beam is small compared to that of the rigid body W. The static deflection directly under the body W is[1] $\delta_{st} = Wl^3/48EI$. The spring constant of the elastic beam is $k = W/\delta_{st}$ which, when substituted into the first expression of Eq. (42.1), gives the natural frequency for lateral vibration of the beam carrying the weight W:

$$\omega_n = \sqrt{\frac{48EIg}{Wl^3}} \quad \text{rad/sec} \tag{42.2}$$

The model of Fig. 42.2 represents the beam of Fig. 42.1 when their natural frequencies are equal.

The model shown in Fig. 42.2 can be used to represent many different types of structures, generally beams or plates. The natural frequencies of such structures can be calculated from the analyses in Chap. 7 or the tables of formulas in Chap. 1. If the structure carries a load that may be considered a rigid body whose mass is large relative to the mass of the structure, the single degree-of-freedom model of Fig. 42.2 is likely to be quite satisfactory.

If the load carried by the structure is relatively small compared with the mass of the structure, the model of Fig. 42.2 cannot be used in evaluating the actual equipment when the higher modes of vibration are of interest. The multiple degree-of-freedom model based upon the normal mode concept (see *Multiple Degree-of-Freedom Model* below) then becomes applicable. However, if only the fundamental mode of vibration is of interest, a portion of the weight of the beam may be lumped with the load and treated as a single degree-of-freedom system (see Table 7.2).

Two Degree-of-freedom Model. When the equipment includes two structures joined together in such a way that the vibration of one is dependent upon the vibration of the other, the single degree-of-freedom model of Fig. 42.2 may not be appropriate. The two degree-of-freedom model of Fig. 42.6A then may be used. This simulates, for example, the structure of Fig. 42.7 where a load of mass m_2 supported by a beam of stiffness k_2 is mounted upon a load of mass m_1 that in turn is supported by a beam of stiffness k_1. One type of analysis applies if the mass m_2 is small relative to the mass m_1 (uncoupled two degree-of-freedom model), and another type of analysis applies if the two masses are of the same order of size (coupled two degree-of-freedom model). These two types of analysis are discussed separately under *Design for Vibration Conditions* and *Design for Shock Conditions*.

In the model of Fig. 42.6A, two spring-mass-damper systems are connected in a series system where the rigid masses m_2 and m_1 are both assumed to move along the same vertical line. The individual masses can be located at any time t by the two coordinates x_2 and x_1; thus, the system is referred to as having two degrees-of-freedom. The massless springs have linear force-deflection relationships denoted by the spring constants k_2 and k_1. Similarly, the massless dampers exhibit linear force-velocity relationships denoted by the damping coefficients c_2 and c_1. The system consisting of k_1, m_1, c_1 is designated

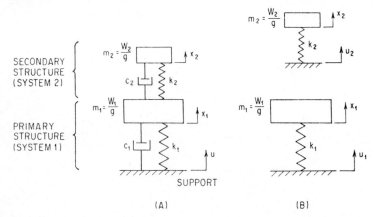

FIG. 42.6. (A) Two degree-of-freedom model consisting of a primary structure and a secondary structure, and (B) the uncoupled primary and secondary structures.

the *primary structure* because, generally, it represents a principal structure of the equipment or a major component thereof. Similarly, the system consisting of k_2, m_2, c_2 is designated the *secondary structure* because it represents a structure or component to which the excitation is transmitted by the primary structure. In some of the analyses that follow, the equations are simplified by omitting the terms for damping forces.

When the two masses m_2 and m_1 are arbitrarily displaced from their equilibrium positions and released, the system shown in Fig. 42.6A exhibits a complicated motion composed of two superimposed free vibrations at two different (*coupled*) natural frequencies. In the mode of vibration at the lower of these frequencies, the masses m_2

FIG. 42.7. Assembly of beams to which a two degree-of-freedom model is applicable.

and m_1 vibrate in phase; in the mode at the higher frequency of vibration, the two masses vibrate in opposition to each other. In general, the two natural frequencies can be determined as follows:[2] The differential equation of motion is derived for each mass in the system. Then, by means of the operational calculus, these equations are converted into two simultaneous algebraic equations. The determinant of the coefficients of x_2 and x_1 gives the characteristic equation which is solved for the unknown natural frequencies of the system. Sometimes it is more fruitful to discuss the analytical results in terms of the *uncoupled* natural frequencies defined according to Eq. (42.1) for the simple spring-mass models in Fig. 42.6B: $\omega_2 = \sqrt{k_2/m_2}$ and $\omega_1 = \sqrt{k_1/m_1}$. In the following analyses the uncoupled natural frequencies are used in place of the more complicated expressions for the actual natural frequencies.

Multiple Degree-of-freedom Model. Models with more than two degrees-of-freedom also may be used as a basis for design to withstand shock and vibration. Such models are used to represent equipment that may be idealized by several rigid masses and massless springs or, alternatively, by a continuous member whose vibration as an elastic body is significant. Response of such a model to excitation occurs in its normal modes; vibration in each of such modes occurs independently in the other normal modes and can be evaluated as the vibration of a single degree-of-freedom system. The fundamental concept of normal modes is discussed in Chap. 2. The vibration of an elastic body in terms of its normal modes is considered in Chap. 7.

DESIGN FOR VIBRATION CONDITIONS

The most important requirement in the design of equipment to withstand vibration is to avoid the large amplitude of vibration that accompanies a condition of resonance. Experience indicates that vibration troubles, whether leading to failure or improper operation of equipment or to personal discomfort, usually result from a condition of resonance. The excitation may be either that generated by normal operation of the machine or that coming from the environment in which the machine operates. The discussion in this chapter is limited to the design of equipment to withstand environmental conditions; a qualitative discussion of techniques to minimize the vibration generated by the machine is included in Chap. 43.

Techniques for minimizing the relatively large amplitude of vibration that occurs at the condition of resonance depend upon the nature of the excitation:

1. If the excitation is a continuing periodic vibration at a nonchanging frequency or frequencies, it is preferable that all natural frequencies of the equipment occur at other than the excitation frequencies.

2. If the excitation is (a) periodic vibration whose frequency varies between limits or (b) random vibration consisting of all frequencies within a defined band, it may not be feasible to design the equipment so that its natural frequencies do not coincide with excitation frequencies. Under these conditions, damping must be added to the vibrating structure.

SINGLE DEGREE-OF-FREEDOM MODEL. When the model of Fig. 42.2 is acted upon by a periodic excitation whose frequency coincides with the natural frequency of the model, the amplitude of vibration of the model is indicated quantitatively by Fig. 2.13. The relative motion or spring deflection is shown in Fig. 42.8. It usually is preferable to eliminate resonance by stiffening the spring k of the model; this increases the natural frequency and moves the operating condition to the left of the resonance peak in Fig. 42.8 where the spring deflection is small. In some instances, an increase in stiffness is not possible; a decrease in stiffness accomplishes the objective of avoiding resonance but may have the result not only of removing desirable rigidity from the equipment but also of making the spring deflection equal to or greater than the amplitude of vibration. The natural frequency of the model in Fig. 42.2 varies directly as the square root of the spring stiffness [see Eq. (42.1)]; consequently, a relatively large change in stiffness is required to effect an incremental change in natural frequency.

If the natural frequency of the structure cannot be modified or if the excitation covers a band of the frequency spectrum, it may be impossible to avoid a resonant condition. Under these circumstances, the desired low amplitude can be attained only by ensuring that the structures of the equipment are damped sufficiently. In general, metals commonly used in structures do not have sufficient hysteresis damping to maintain vibration amplitudes at an acceptable level if a resonant condition exists. Frequently, damping may be attained by use of bolted or riveted joints, or by use of damping materials in or on the structure. In extreme cases, discrete dampers must be added to achieve adequate control of vibration. These various techniques are discussed in Chap. 43 under *Damping of Structures*.

FIG. 42.8. Relative motion magnification factor for a single degree-of-freedom system with viscous damping. The forcing frequency is ω, the undamped natural frequency ω_n and the fraction of critical damping ζ.

UNCOUPLED TWO DEGREE-OF-FREEDOM MODEL. When the mass m_2 is small, it is unable to influence the motion of mass m_1. Then the vibratory motion of m_1 in system 1 of the model shown in Fig. 42.6A becomes the excitation for system 2, and the above discussion of the single degree-of-freedom model becomes applicable to evaluating the response of system 2. A condition of resonance of system 2 with its excitation should be avoided; i.e., the natural frequency of system 2 should be different from that of system 1. Preferably, the natural frequency of system 2 should be greater than the natural frequency of system 1; otherwise, the deflection of the spring k_2 may equal or exceed the vibratory displacement of the mass m_1. A system 2 with a natural frequency twice as great as that of system 1 experiences a vibration amplitude 1.35 times as great, as indicated in Fig. 2.13. This is a tolerable degree of amplification; it usually is achievable practically, and constitutes a good design objective. It has wide application to equipment that involves elastic systems supported on elastic systems, and should be adopted as a primary objective in the design of equipment to withstand vibration.

If the natural frequency criteria of the above paragraph can be met, damping requirements for system 2 may not be stringent. The excitation for system 2 is the vibration of system 1; the magnitude of such excitation is determined to a considerable extent by the damping in system 1. Thus, a basic principle of design where system 2 is of negligible mass is (a) to design system 2 to have a natural frequency at least twice as great as that of system 1 and (b) to introduce damping in system 1 to limit its amplitude of vibration if this is made necessary by the nature of the excitation. Methods for adding damping to structures are discussed in Chap. 43 under *Damping* in *Design of Structures*.

COUPLED TWO DEGREE-OF-FREEDOM MODEL. The analysis of an equipment represented by a two degree-of-freedom model as shown in Fig. 42.6A must consider coupling between the branches of the system when the two masses m_2, m_1 are of the same order of size. In other words, it cannot be considered that system 1 responds as a single degree-of-freedom system and imposes this response as the excitation for system 2.   The spring k_2 and damper c_2 apply significant forces to system 1; this is knows as a *coupling effect* and causes the entire assembly to respond as a single system.   The analysis of such a system is set forth in Chap. 6; it can be used to predict the response of the model to a known excitation.   The effect of changes in the physical properties of the model can be determined and related to the actual equipment.

DESIGN FOR SHOCK CONDITIONS

The design of equipment to withstand shock involves the formulation of a model to represent the equipment and the determination of the response of the model to the shock.   Theoretically, the design of equipment to withstand shock can proceed if the shock is defined as the time-history of the motion (displacement, velocity, or acceleration) of the support for the equipment.   As pointed out below (see *Single Degree-of-Freedom Model*), the shock spectrum is an adequate definition of the shock motion for purposes of equipment design.   Known methods can be applied to determine the maximum response of the model, and design requirements for the actual equipment can be inferred from this response.   Considerations in selecting a model are discussed above under *Types of Models*.

SINGLE DEGREE-OF-FREEDOM MODEL.   In the design of equipment to withstand shock, it is common practice to assume that failure occurs because the maximum allowable stress or acceleration is exceeded in an element of the equipment.   Stress of this magnitude is assumed to lead to failure should it occur once.*   The method of analysis involves a determination of the maximum stress in the model corresponding to the maximum deflection of spring k or maximum acceleration of mass m in Fig. 42.2.

FIG. 42.9. Typical shock spectra for an undamped system.   The solid line represents acceleration in the same direction as the principal excitation; the broken line represents acceleration in the opposite direction.   (*After J. E. Ruzicka.*[3])

The Shock Spectrum.   For a given shock motion, the maximum acceleration $\ddot{x}$ of mass m is a function of the natural frequency f_n of the model and its fraction of critical damping ζ.   This maximum acceleration $\ddot{x}$ can be plotted as spectra of the natural frequency f_n with ζ as the parameter of the family of spectra.   Typical spectra[3] are shown in Fig. 42.9.   Each shock motion has characteristic shock spectra; this is discussed in Chap.

* In laboratory testing, consideration sometimes is given to the cumulative effect of repeated occurrences of stress.   This concept also has application to the design of equipment; it is discussed in detail in Chap. 24.

23 where it is pointed out that the spectra constitute a convenient definition of a shock motion. An important aspect of such a definition is that it provides information directly useful for design purposes; i.e., a known value for the maximum acceleration of the mass of the single degree-of-freedom model makes possible the design of the spring. Methods for determining the shock spectrum and examples of spectra for typical excitations are given in Chaps. 8 and 23.

In using the spectra of Fig. 42.9 for design purposes, the chart is entered on the abscissa scale at the natural frequency of the model. The corresponding maximum acceleration of mass m, usually expressed as a multiple of the gravitational acceleration, is read from

FIG. 42.10. Spectra of Fig. 42.9 redrawn to indicate maximum relative displacement and maximum relative velocity as well as maximum acceleration.

the ordinate scale. Inasmuch as the acceleration of the model under static conditions may be considered as $1g$, the ordinate indicates the factor by which the stress under static conditions must be multiplied to give the stress under conditions of shock. This maximum acceleration is approximately equal to the *equivalent static acceleration,*[4] a concept* that permits the application of design methods for static loads to the design of equipment required to withstand dynamic loading. The spectrum shown by solid lines in Fig. 42.9 represents acceleration in the same direction as the pulse; the spectrum shown by dotted lines represents accelerations in the opposite direction, usually caused by free vibration of the model after the excitation has disappeared.

In general, the spectrum has lower maximum values of acceleration as damping is added to the model. Therefore, the use of the spectrum for the undamped model is conservative in design work; it is not excessively conservative for spectra of the type

---

* See Chap. 23 for the distinction between maximum acceleration and equivalent static acceleration. For practical design purposes, the two may be considered equal.

shown in Fig. 42.9 where the excitation is a single pulse of acceleration. If the excitation is oscillatory, however, there is a tendency for a model whose natural frequency is equal to the frequency of oscillation to respond in a resonant manner; then the discrepancy between the spectra for damped and undamped models may be relatively great. If the spectra for the undamped model tend toward values that are great relative to spectra for damped models, a representative value of damping should be chosen as the basis of design. (See *Approximate Methods*, below, for examples of spectra for damped systems and spectra plotted to dimensionless coordinates.) Otherwise, the undamped model is preferable because the spectra are more readily computed.

The spectra of Fig. 42.9 may be redrawn, as shown in Fig. 42.10, to indicate maximum relative displacement and maximum relative velocity as well as maximum acceleration. Relative displacement is the deflection of spring k in Fig. 42.2. This information is of importance, for example, in determining whether the deflection of a shaft would be great enough to permit gears to become momentarily disengaged and then reengaged in a different position.

Approximate Methods. Ideally, the concept of designing to withstand shock contemplates the availability of a time-history of the shock motion, and the computation of spectra as shown in Fig. 42.10. Time-histories and/or facilities for computation sometimes are not available. It becomes instructive, then, to examine typical spectra and to compare them with the corresponding time-histories to determine trends that are useful for design purposes. It may be convenient to employ dimensionless or normalized parameters to evaluate important characteristics of the response; this leads directly to useful design methods.

Acceleration Pulse. An idealized version of a typical shock motion is the half-sinusoidal acceleration pulse defined by

$$\ddot{u} = \ddot{u}_o \sin \omega_p t \qquad \left[0 \leq t \leq \frac{\pi}{\omega_p} \right]$$

$$\ddot{u} = 0 \qquad \left[\frac{\pi}{\omega_p} \leq t \right]$$

(42.3)

where $\ddot{u}$ is the acceleration of the support for the model at any time t, $\ddot{u}_o$ is the peak acceleration, and ω_p is the angular frequency associated with the acceleration pulse. The time duration of the pulse is $\tau = \pi/\omega_p$. With the system initially at rest, the time-histories of acceleration, as given by Eq. (42.3), and velocity $\dot{u}$ and displacement u, as obtained from integration of Eq. (42.3), are shown in Fig. 42.11.

The transient response of the model of Fig. 42.2 to the pulse defined by Eq. (42.3) is illustrated in Fig. 42.12[5] where the acceleration $\ddot{x}$ of the mass m is plotted as a function of time from initiation of pulse $\ddot{u}$. Acceleration $\ddot{x}$ is plotted as a multiple of the peak pulse acceleration $\ddot{u}_o$, and time t is plotted as a multiple of the pulse duration τ. The half-sinusoidal acceleration pulse defined by Eq. (42.3) is the shaded area bounded by a dotted line.

The time-history of the response acceleration $\ddot{x}$ is plotted in Fig. 42.12 for three different values of the ratio of pulse duration to natural period of model: $\tau = 1\frac{2}{3}(2\pi/\omega_n); (2\pi/\omega_n);$ and $\frac{1}{4}(2\pi/\omega_n),$

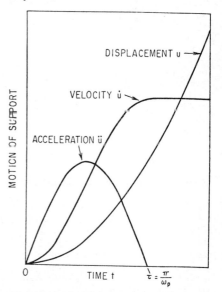

FIG. 42.11. Velocity and displacement as a function of time, corresponding to acceleration pulse defined by Eq. (42.3).

where ω_n is the natural frequency of the model [Eq. (42.1)]. The acceleration $\ddot{x}$ of the mass m can be equal to, greater than, or less than the peak acceleration of the support. In general, the acceleration of the support is not an indication of the forces induced in the model.

When the angular natural frequency of the model is appreciably larger than the frequency ω_p associated with the pulse, the response of the model approximates but does not exactly duplicate the acceleration pulse of the support. For example, when $\tau = 1\frac{9}{3}(2\pi/\omega_n)$ the response has superimposed small-amplitude vibration at the natural frequency ω_n of the model. The subsequent free vibration of the system represents the maximum *negative* response to the original positive (or upward) pulse. When $\tau = (2\pi/\omega_n)$ the maximum acceleration of the mass m is about 1.7 times the acceleration of

FIG. 42.12. Response of an undamped, single degree-of-freedom system to the half-sinusoidal pulse of acceleration shown shaded. Systems with different natural periods are included. (*J. M. Frankland.*[5])

the support. When $\tau = \frac{1}{4}(2\pi/\omega_n)$, the peak acceleration of the model occurs after the acceleration of the original pulse becomes zero and is about 0.9 times the peak acceleration of the pulse.

The shock response spectra for the half-sinusoidal acceleration pulse of Eq. (42.3) are shown in Fig. 42.13 for several values of the fraction of critical damping ζ [see Eq. (2.12)]. These spectra summarize the maximum values of the response for the time-histories of Fig. 42.12, for a wide range of frequency relations. The ordinate is the ratio of the peak acceleration $\ddot{x}_{max}$ of the response to the peak acceleration $\ddot{u}_{max}$ of the pulse where $\ddot{u}_{max}$ is used in place of $\ddot{u}_o$ to suggest that this discussion is applicable to other than the sinusoidal pulse defined by Eq. (42.3). The numerical value of the ratio $\ddot{x}_{max}/\ddot{u}_{max}$ can be interpreted as the shock transmissibility since it represents both spring and damper forces. For very high values of damping, this system acts as a rigid connection so that the acceleration ratio quickly becomes unity below a frequency ratio of 1. For the curve with zero damping, the acceleration ratio can be interpreted as the shock amplification [dynamic load factor $= \omega_n^2(x - u)_{max}/\ddot{u}_{max}$], since it then represents only the spring force, or relative motion. When system damping is included, Fig. 8.43 shows the shock amplification factors which are somewhat different from shock transmissibility values. For very high values of damping (not shown in Fig. 8.43), the shock amplification factors can decrease below unity as the relative motion $x - u$ approaches zero.[3]

When the frequency ratio ω_n/ω_p is large, the acceleration ratio approaches unity; i.e., the maximum acceleration experienced by the model is approximately equal to that of the pulse. In other words, when a structure is stiff and the support motion is applied slowly, the maximum acceleration of the support can be used to determine the dynamic load imposed upon structures of the equipment. This relation is independent of the

Fig. 42.13. Maximum dynamic response factor (shock transmissibility) for the half-sinusoidal pulse of acceleration. The fraction of critical damping in the system is indicated by ζ. (*After J. E. Ruzicka.*[3])

pulse shape; however, a pulse with superimposed high-frequency vibration cannot be interpreted in this manner because the superimposed vibration may induce a resonant response not contemplated by the preceding analysis. Furthermore, irregular pulses are sometimes made to appear smooth by limited high-frequency response of the instrumentation.

Starting Velocity. In Fig. 42.13 the equation of the tangent to the response spectrum for zero damping, at the origin, is

$$\frac{\ddot{x}_{max}}{\ddot{u}_{max}} = 2\left(\frac{\omega_n}{\omega_p}\right) \tag{42.4}$$

This straight line represents a good approximation to the spectrum for $\zeta = 0$ when $\omega_n/\omega_p \leq 0.50$. The velocity change $\dot{u}_o$ of the support is the integral of Eq. (42.3) with respect to time:

$$\dot{u}_o = \frac{2\ddot{u}_o}{\omega_p} \quad \text{in./sec} \tag{42.5}$$

Combining Eqs. (42.4) and (42.5) and letting $\ddot{u}_{\max} = \ddot{u}_o$,

$$\ddot{x}_{\max} = \omega_n \times \dot{u}_o \quad \text{in./sec}^2 \qquad \left[\frac{\omega_n}{\omega_p} \leq 0.50\right] \tag{42.6}$$

where ω_n is in units of radians per second and $\dot{u}_o$ is in units of inches per second. Thus, the maximum acceleration $\ddot{x}_{\max}$ of the mass m in Fig. 42.2 is proportional to the magnitude of the velocity change $\dot{u}_o$ of the support and to the natural frequency ω_n of the model, provided the acceleration pulse duration τ is short compared to the natural period $2\pi/\omega_n$ of the model. This relation is convenient when the duration of the shock motion is small compared with the natural periods of primary structures; it finds particular application in simplified aspects of design using the single degree-of-freedom model, as well as in shock isolation (Chap. 31) and packaging (Chap. 41).

The maximum deflection $\delta_{\max}$ of a single degree-of-freedom model of natural frequency ω_n, when ω_n is small relative to $2\pi/\tau$, is

$$\delta_{\max} = \frac{\dot{u}_o}{\omega_n} \quad \text{in.} \qquad \left[\frac{\omega_n}{\omega_p} \leq 0.50\right] \tag{42.7}$$

where $\dot{u}_o$ and ω_n have the units defined for Eq. (42.6).

In applying the relation of Eqs. (42.6) and (42.7), note that $\ddot{x}_{\max}$ and $\delta_{\max}$ are independent of the shape of the pulse and are determined solely by the area under the acceleration-time curve. If the excitation is irregular or vibratory, these relations are applicable if $\dot{u}_o$ is taken as the integral of the acceleration-time curve where $\omega_p = \pi/\tau$ and τ is the duration of the excitation.

UNCOUPLED TWO DEGREE-OF-FREEDOM MODEL. When the secondary structure of an equipment is small relative to the primary structure, the response of the model to a shock motion can be evaluated by first finding the response of the primary structure and using this response as the excitation for the secondary structure. For example, Fig. 42.12 gives several time-histories of the acceleration response of an undamped, single degree-of-freedom model to a half-period sinusoidal pulse of acceleration. The continuing periodic vibration of the primary structure appears to the secondary structure as steady-state vibration; thus, criteria established for design to withstand vibration constitute a good guide for design to withstand shock when the equipment can be represented by an uncoupled two degree-of-freedom model.

In accordance with the principle of design to withstand vibration, the secondary structure should have a natural frequency at least twice as great as that of the primary structure. The acceleration amplitude of the secondary structure then is 1.35 times as great as that of the primary structure (see *Design for Vibration Conditions—Uncoupled Two Degree-of-freedom Model*). The maximum acceleration of the primary structure is found by methods indicated under *Design for Shock Conditions—Single Degree-of-freedom Model*. The primary structure must be designed to withstand this acceleration; the secondary structure must be designed to withstand an acceleration 1.35 times as great.

If the natural frequencies of the primary and secondary structures are equal, the secondary structure may experience excessive vibration unless either the primary or secondary structure embodies significant damping. For example, Fig. 42.14 indicates the maximum acceleration of the secondary structure with reference to the maximum acceleration of the primary structure when the shock motion is a sudden velocity change $u = \dot{u}_o t$ of the support for the primary structure.[6] When there is significant damping in the primary and/or secondary structure, the maximum acceleration of the secondary structure is relatively small.

COUPLED TWO DEGREE-OF-FREEDOM MODEL. When the mass of the secondary structure is of the same order of magnitude as that of the primary structure, it is necessary to consider the response of the composite model of primary and secondary structures. The response of this model to any excitation can be determined by known

Fig. 42.14. Ratio of maximum acceleration of secondary structure to maximum acceleration of primary structure (see Fig. 42.6A) when support has motion $u = \dot{u}_o t$ and uncoupled natural frequencies of the structures are equal $\left(\text{i.e., } \omega_1 = \sqrt{\dfrac{k_1}{m_1}} = \omega_2 = \sqrt{\dfrac{k_2}{m_2}}\right)$. The mass m_2 of the secondary structure is negligible relative to the mass m_1 of the primary structure. (*C. E. Crede.*[6])

methods. For initial design work, it is convenient to generalize the excitation as a sudden motion change of the support, $u = \dot{u}_o t$. This is a good approximation when the velocity change takes place during a time interval that is small relative to the smallest natural period of the composite model.

The equations of motion for the system shown in Fig. 42.6A, when the excitation is a starting velocity $\dot{u}_o$ applied at the support, give results that can be expressed in terms of an equivalent starting velocity applied individually to each of the uncoupled spring-mass models in Fig. 42.6B; i.e., the maximum response of each model is equivalent to its maximum response as a part of the two degree-of-freedom model. The equivalent starting velocities are designated by $\dot{u}_2$ for the secondary structure and $\dot{u}_1$ for the primary structure. They are expressed in terms of the starting velocity $\dot{u}_o$ as:[7, 8]

$$\frac{\dot{u}_2}{\dot{u}_o} = \frac{1}{\sqrt{\left(1 - \dfrac{\omega_2}{\omega_1}\right)^2 + \dfrac{m_2}{m_1}\left(\dfrac{\omega_2}{\omega_1}\right)^2}} \tag{42.8}$$

$$\frac{\dot{u}_1}{\dot{u}_o} = \frac{1 + \left(1 + \dfrac{m_2}{m_1}\right)\dfrac{\omega_2}{\omega_1}}{\sqrt{\left(1 + \dfrac{\omega_2}{\omega_1}\right)^2 + \dfrac{m_2}{m_1}\left(\dfrac{\omega_2}{\omega_1}\right)^2}} \tag{42.9}$$

where ω_1 and ω_2 are defined under *Two Degree-of-freedom Model.*

Figure 42.15 shows a contour chart of $\dot{u}_2/\dot{u}_o$ plotted against the stiffness ratio k_2/k_1, the mass ratio m_2/m_1, and the frequency ratio ω_2/ω_1. Figure 42.16 shows a contour chart of a factor D plotted against the same stiffness, mass, and frequency ratios. The

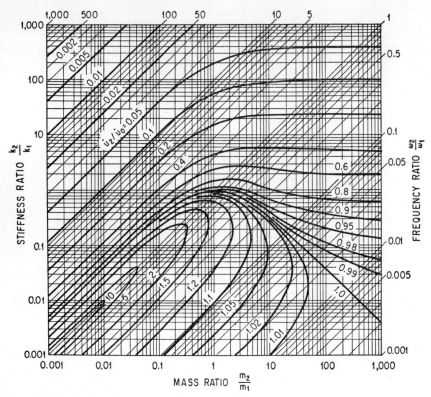

FIG. 42.15. Equivalent velocity change $\dot{u}_2$ for secondary structure relative to $\dot{u}_o$ where displacement of support for primary structure is $u = \dot{u}_o t$. The natural frequencies are $\omega_1 = \sqrt{\dfrac{k_1}{m_1}}$, $\omega_2 = \sqrt{\dfrac{k_2}{m_2}}$ where m_1, m_2, k_1, k_2 are defined in Fig. 42.6A. (R. B. McCalley, Jr.[3])

factor D is related to the velocities $\dot{u}_1$ and $\dot{u}_o$ by

$$\frac{\dot{u}_1}{\dot{u}_o} = D \sqrt{1 + \frac{m_2}{m_1}} \qquad (42.10)$$

The maximum relative motion $(x_1 - x_2)_{max}$; i.e., the maximum deflection $(\delta_2)_{max}$ of the spring k_2, is

$$(\delta_2)_{max} = \frac{\dot{u}_2}{\omega_2} = \left(\frac{\dot{u}_o}{\omega_2}\right)\left(\frac{\dot{u}_2}{\dot{u}_o}\right) \qquad \text{in.} \qquad (42.11)$$

where $\dot{u}_2/\dot{u}_o$ is obtained from Fig. 42.15. The contour chart of Fig. 42.15 is particularly useful where the characteristics of the secondary structure are fixed and a primary structure must be selected to reduce the shock experienced by the secondary structure. The equivalent static acceleration of the secondary structure is

$$(\ddot{x}_2)_{max} = \frac{k_2(\delta_2)_{max}}{m_2} \qquad \text{in./sec}^2 \qquad (42.12)$$

Fig. 42.16. Contour chart for factor D in Eq. (42.10). The physical conditions are described with reference to Fig. 42.15. (*R. B. McCalley, Jr.*[3])

The maximum relative motion $(u - x_1)_{max}$; i.e., the maximum deflection $(\delta_1)_{max}$ of the spring k_1, is

$$(\delta_1)_{max} = \frac{\dot{u}_1}{\omega_1} = \left(\frac{\dot{u}_o}{\omega_1}\right) D \sqrt{1 + \frac{m_2}{m_1}} \quad \text{in.} \quad (42.13)$$

where D is obtained from Fig. 42.16. The equivalent static acceleration of the primary structure is

$$(\ddot{x}_1)_{max} = \frac{k_1(\delta_1)_{max}}{m_1} \quad \text{in./sec}^2 \quad (42.14)$$

The response data given by the contour plots of Figs. 42.15 and 42.16, referring to the coupled two degree-of-freedom model of Fig. 42.6A, are shown in Fig. 42.17 as spectra of the frequency ratio ω_2/ω_1. Maximum deflection of the springs k_2, k_1 and the corresponding equivalent static acceleration for the respective structures are directly proportional, respectively, to the equivalent starting velocities $\dot{u}_2$, $\dot{u}_1$. These relations are defined by Eqs. (42.11) and (42.12); values for $\dot{u}_2$, $\dot{u}_1$ are given by the ordinates of Fig. 42.17.

Figure 42.17 illustrates the importance of maintaining appropriate relations between the natural frequencies ω_2 and ω_1:

1. If $\omega_2 = \omega_1$, the equivalent starting velocity of the secondary structure becomes great if its mass m_2 is small relative to the mass m_1 of the primary structure. This approaches the uncoupled model discussed in an earlier section.

Fig. 42.17. Equivalent starting velocity ratios for (A) secondary and (B) primary structures. The motion of the support is $u = \dot{u}_0 t$. (C. M. Friedrich.[7])

2. If $\omega_2 > 2\omega_1$, the equivalent starting velocity of the secondary structure tends toward relatively low values for any ratio of the masses m_2 and m_1. The equivalent starting velocity of the primary structure tends toward larger though moderate values as the frequency ratio ω_2/ω_1 increases.

Thus, a design criterion for coupled as well as uncoupled two degree-of-freedom models is a natural frequency of the secondary structure at least twice as great as that of the primary structure.

EXAMPLES OF TYPICAL REQUIREMENTS. Shock testing machines play an important role in the design of equipment to withstand shock. Their importance is derived, in part, from the difficulty of applying the foregoing design principles to all details of a complex equipment because the equipment cannot be represented by a simple model. A shock test confirms calculations that were made and takes the place of calculations that could not be made. Furthermore, even though analytical techniques can be applied, the conditions of shock often are not defined well enough to serve as a basis for design. An equipment then is designed to withstand the shock created by a shock testing machine. In general, the shock created by the testing machine does not duplicate that experienced in service; however, it does represent an explicit requirement for design purposes and presumably takes into account the variances from one shock motion to another in actual service.

Shock testing machines are discussed, and several common machines are illustrated, in Chap. 26. The specification of laboratory tests to simulate actual service conditions is considered in Chap. 24. Most shock testing machines are adapted to several alternate modes of operation, at least to the extent of modifying the severity of the shock. Compilations of data are available to define the characteristics of the shock motion of commonly used shock testing machines. As examples of design requirements, typical shock spectra are shown in Fig. 42.18 for several machines. These spectra are useful directly with single degree-of-freedom models as a design procedure for equipment required to withstand the shock produced by the respective machines.

EFFECT OF EQUIPMENT WEIGHT. The weight of the equipment sometimes is a factor of importance when selecting strength requirements for the equipment. Heavy equipment often receives shock of less severity than light equipment because (1) sources of shock may involve only finite energy and have less effect upon the heavy equipment or (2) the structures that transmit the forces have finite strength and thereby limit the amount of energy that reaches the equipment. Many types of shock testing machines

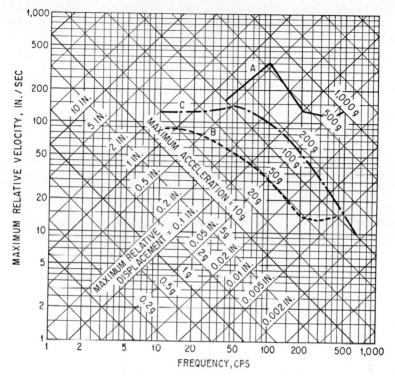

FIG. 42.18. Typical shock spectra for several shock testing machines as examples of design requirements: (*A*) MIL-S-901B, (*B*) MIL-S-4456 using sand, and (*C*) MIL-S-4456 using lead. The designation "MIL" denotes a military specification number.

recognize these limitations and deliver a shock motion whose severity tends to decrease with an increase in the weight of the equipment under test. For example, machines with swinging hammers tend to deliver a shock whose severity decreases as the equipment weight increases, even though the height of the hammer drop may be increased somewhat as the equipment weight increases.

Analytical methods for design of shock-resistant equipment employ one or two degree-of-freedom models that are tacitly assumed to be of negligible mass; i.e., the equipment offers no impedance to the shock motion. The effect of equipment weight in reducing the severity of shock is a factor of considerable significance; sometimes it is recognized empirically where design criteria are set forth. For example, the Bureau of Ships of the U.S. Navy specifies the strength requirements for bolts used to mount equipment on naval ships. The specification requires that the mounting bolts for light equipment be capable of withstanding a greater equivalent static acceleration than the bolts for mounting heavy equipment, as illustrated in Fig. 42.19. Strictly speaking, Fig. 42.19 applies only to mounting bolts and associated

FIG. 42.19. Typical empirical design curve indicating required strength of mounting bolts as a function of equipment weight. This applies to some classes of naval service.

mounting brackets; generally, the trend may be recognized and applied to the over-all design where there is reason to believe that the shock severity would be decreased by the weight of the equipment.

STRUCTURAL DESIGN CONSIDERATIONS

ENERGY STORAGE CAPACITY

The deflection of structural members that characterizes the response of equipment to shock and vibration involves the temporary storage of strain energy in such members. In a given member, strain energy increases as stress increases. The severity of shock that an equipment may withstand without damage is reached when the stress in the most highly stressed member reaches the maximum allowable value. Any material in the structure that does not at the same time reach its maximum allowable stress may be considered as contributing to a loss in efficiency of design. This material not only tends to increase the cost and weight of the equipment but, more important, may increase the weight carried by other structures with a consequent increase in strength requirements.

Efficiency is determined by the type of loading, the constraints on the structure, and the shape of the cross section. For example, the potential energy stored in a uniform bar in pure tension is the product of the average force $F/2$ and the deflection δ:

$$V = \frac{F\delta}{2} = \frac{(\sigma S)(l\epsilon)}{2} \qquad \text{in.-lb} \tag{42.15}$$

where σ = tensile stress, lb/in.^{2}
$\quad S$ = cross-sectional area of the bar, in.^{2}
$\quad l$ = length of the bar, in.
$\quad \epsilon$ = strain, in./in.

The substitutions made in Eq. (42.15) are $F = \sigma S$ lb and $\delta = l\epsilon$ in. By making the additional substitutions $V = Sl$ in.^{3} and $\epsilon = \sigma_m/E$ in./in., Eq. (42.15) can be expressed as

$$\frac{F\delta}{2} = \eta \frac{\sigma_m{}^2 V}{2E} \qquad \text{in.-lb} \tag{42.16}$$

where V = total volume of material being stressed, in.^{3}
$\quad E$ = Young's modulus, lb/in.^{2}
$\quad \sigma_m$ = maximum stress (uniform for bar in tension), lb/in.^{2}
$\quad \eta$ = efficiency (0 to 1)

The efficiency η is the fraction of the total strain energy, based on a uniform stress σ_m in all elements, that is stored by the structure. Solving Eq. (42.16) for η,

$$\eta = \frac{F\delta E}{\sigma_m{}^2 V} \tag{42.17}$$

When each element in the structure has the same stress, the material is being used at maximum efficiency and $\eta = 1$.

For a beam in bending, the efficiency usually is a small fraction. For a cantilever beam with a concentrated load F at the free end, the deflection δ at the free end[1] is $Fl^3/3EI$ and the stress σ_m in the outermost fiber at the built-in end of the beam is Flc/I. The moment of inertia (of the cross-sectional area S) is $I = S\rho^2$ in.^{4}, where ρ is the radius of gyration and c is the distance of the outer fiber from the neutral axis. Substituting in Eq. (42.17),

$$\eta = \frac{\rho^2}{3c^2} \tag{42.18}$$

Values of ρ and c are known for various cross sections.[9] For a cantilever beam with end load, the corresponding values of efficiency are $\eta = \frac{1}{12}$ for a circular section, $\frac{1}{9}$ for a rectangular section, $\frac{1}{6}$ for a thin-walled tube, and about $\frac{1}{3}$ for an I beam. These increasing values reflect the fact that each successive shape of cross section further concentrates the material in the outermost fibers where the bending stress reaches a maximum.

The efficiency η depends not only upon the distribution of stress over the cross section but also upon the distribution along the length of the beam. The relatively low efficiency of only $\frac{1}{3}$ for the cantilever I beam occurs because the bending moment in a cantilever beam varies from a maximum at the built-in end to zero at the free end. By contrast, when a pure moment instead of a concentrated load is applied at the free end of the cantilever beam, the bending moment is uniform along the length of the beam and the efficiency increases by a factor of 3 to become $\eta = \rho^2/c^2$. This is approximately unity for the I beam.

The foregoing examples are included in Table 42.1 along with the efficiencies of other common structural members for various cross sections.[7] Column 1 gives the general

Table 42.1. Efficiencies of Common Structural Members as Defined by Eq. (42.17)

(C. M. Friedrich.[7])

Type of structure	Type of cross section		
	(1) Uniform	(2) Rectangular	(3) Circular
Bar in tension..............	1	1	1
Cantilever beam with concentrated load at end.....	$\rho^2/3c^2$	$\frac{1}{6}$	$\frac{1}{12}$
Simply supported beam with concentrated load at any point..................	$\rho^2/3c^2$	$\frac{1}{6}$	$\frac{1}{12}$
Built-in beam with concentrated load at center......	$\rho^2/3c^2$	$\frac{1}{6}$	$\frac{1}{12}$
Simply supported beam with moment at center......	$\rho^2/3c^2$	$\frac{1}{6}$	$\frac{1}{12}$
Built-in beam with moment at center...............	$\rho^2/2c^2$	$\frac{1}{6}$	$\frac{1}{8}$
Cantilever beam with moment at end.............	ρ^2/c^2	$\frac{1}{3}$	$\frac{1}{4}$
Simply supported beam with uniform load.............	$8\rho^2/15c^2$	$\frac{8}{45}$	$\frac{2}{15}$
Beam with one built-in end, one simply-supported end, load at center...........	$7\rho^2/27c^2$	$\frac{7}{81}$	$\frac{7}{108}$

ρ = radius of gyration of the beam cross section,
c = distance of outermost fibers from the neutral axis of the beam.

expression for efficiency applicable to any uniform cross section, for the particular type of loading and constraints on the structure. Columns 2 and 3 give the efficiencies for rectangular and circular sections, respectively, and are obtained by substituting appropriate values for ρ and c in the expressions of column 1.

ELASTIC CHARACTERISTICS OF BEAMS AND PLATES

Analytical methods of design to withstand shock and vibration make extensive use of models to determine the response of an equipment or structure to the shock and vibration. Generally, the model is an elastic system of concentrated masses, massless springs, and massless dampers whose natural frequencies can be related to the natural frequencies of the actual structure. The forces and deflections associated with the response of the model then are transformed to stresses and strains in the actual structure. The section of this chapter on *Models* shows the relation between a model and a simple beam with center load. This section extends the treatment to other types of beams, and includes analyses directed toward the ready calculation of bending moments and reaction forces.

UNIFORM BEAM WITH DISTRIBUTED LOAD. The spring stiffness for lateral deflection of a uniform beam can be generalized as[10]

$$k_b = n_b \left(\frac{EI}{l^3}\right) = n_b \left(\frac{ES\rho^2}{l^3}\right) \qquad \text{lb/in.} \tag{42.19}$$

where E = Young's modulus, lb/in.^{2}
I = moment of inertia of cross section, in.^{4}
S = cross-sectional area of beam, in.^{2}
ρ = radius of gyration of cross section, in.
l = length of beam, in.

The numerical coefficient n_b depends upon the type of loading and the end constraints on the beam. For the beam shown in Fig. 42.1, the coefficient is $n_b = 48$.

The natural frequency of a uniform beam is[10]

$$\omega_n = \sqrt{\frac{EI}{\nu l^4}} = N_b \left(\frac{\rho}{l^2}\right) \sqrt{\frac{E}{\mu}} \qquad \text{rad/sec} \tag{42.20}$$

where ν = mass per unit length of beam, lb-sec^2/in.^{2}
μ = mass density of the material, lb-sec^2/in.^{4}

Hence, $\nu = \mu S$. The numerical coefficient N_b depends upon the end constraints of the beam. For the beam shown in Fig. 42.1 (without the rigid body W), $N_b = \pi^2$ for the lowest natural frequency. Values of N_b for higher natural frequencies, and for other constraints and types of loading, are given in Tables 7.3 and 7.5.

By considering a beam to be analogous to a single degree-of-freedom system having the stiffness k_b given by Eq. (42.19) and an effective mass m_e, Eq. (42.1) may be written as follows for the natural frequency of the beam:

$$\omega_n = \sqrt{\frac{k_b}{m_e}} \qquad \text{rad/sec} \tag{42.21}$$

Solving for m_e and substituting k_b and ω_n from Eqs. (42.19) and (42.20), respectively,

$$m_e = \left(\frac{n_b}{N_b^2}\right) m_b \qquad \text{lb-sec}^2/\text{in.} \tag{42.22}$$

where $m_b = \mu S l$ is the total mass of the beam.

UNIFORM BEAM WITH CONCENTRATED LOAD. The effects upon the natural frequency of the mass of both the rigid body W and the beam shown in Fig. 42.1 may be included by writing Eq. (42.21) as follows:

$$\omega_n' = \sqrt{\frac{k_b}{m_l + m_e}} \qquad \text{rad/sec} \tag{42.23}$$

where m_l is the mass of the load (rigid body W), m_e is given by Eq. (42.22), and k_b is given by Eq. (42.19). By using Eqs. (42.19) and (42.22), the natural frequency ω_n' can be expressed in terms of the dimensions and properties of the actual beam:

$$\omega_n' = \frac{\dfrac{\rho}{l^2} \sqrt{\dfrac{E}{\mu}}}{\xi_\delta} \qquad \text{rad/sec} \tag{42.24}$$

where

$$\xi_\delta = \sqrt{\frac{1}{N_b^2} + \frac{1}{n_b}\left(\frac{m_l}{m_b}\right)}$$

For the beam and load shown in Fig. 42.1, $\xi_\delta = \sqrt{0.0103 + 0.0208 \, m_l/m_b}$.

When the duration τ of a shock motion is small relative to the natural period $2\pi/\omega_n'$ of the beam, the shock motion can be assumed to consist of the starting velocity $\dot{u}_0$ applied at the beam supports in a direction transverse to the beam. Then, Eq. (42.7) leads to the following expression for the maximum beam deflection δ_m at the concentrated load m_l:

$$\delta_m = \frac{\dot{u}_o}{\omega_n'} = \xi_\delta \dot{u}_o \left(\frac{l^2}{\rho}\right) \sqrt{\frac{\mu}{E}} \quad \text{in.} \tag{42.25}$$

The maximum bending moment M_m and the maximum force F_m at the end supports are determined from the maximum deflection:[1]

$$M_m = \xi_M \dot{u}_o S\rho \sqrt{E\mu} \quad \text{lb-in.} \tag{42.26}$$

$$F_m = \xi_F \dot{u}_o \left(\frac{S\rho}{l}\right) \sqrt{E\mu} \quad \text{lb} \tag{42.27}$$

Table 42.2 gives values of the dimensionless factors ξ_δ, ξ_M, and ξ_F to be used in Eqs. (42.25) to (42.27) for calculating the maximum deflection δ_m, the maximum bending

Table 42.2. Dimensionless Factors * in Eqs. (42.25) to (42.27) for Calculating the Response of Beams to a Velocity Shock $u = \dot{u}_o t$

(*C. M. Friedrich.*[7])

Type of structure	Maximum deflection ξ_δ Eq. (42.25)	Bending moment ξ_M Eq. (42.26)	End reaction ξ_F Eq. (42.27)
Simply supported beam with mass at center (deflection and moment at center, reaction at simple support).	$\sqrt{0.0103 + 0.0208\, m_l/m_b}$	$12\xi_\delta$	$24\xi_\delta$
Built-in beam with mass at center (deflection at center, moment and reaction at built-in end).	$\sqrt{0.00199 + 0.00521\, m_l/m_b}$	$24\xi_\delta$	$96\xi_\delta$
Cantilever beam with mass at free end (deflection at free end, moment and reaction at built-in end).	$\sqrt{0.0806 + 0.333\, m_l/m_b}$	$3\xi_\delta$	$3\xi_\delta$
Beam with one built-in end, one simply supported end, with mass at center (deflection near center, moment at built-in end), reaction at:			
Built-in end	$\sqrt{0.00422 + 0.00932\, m_l/m_b}$	$20.1\xi_\delta$	$73.6\xi_\delta$
Simple support	$\sqrt{0.00422 + 0.00932\, m_l/m_b}$	$20.1\xi_\delta$	$33.5\xi_\delta$

* The dimensionless factors depend upon the type of loading and the end supports for the beam as well as the following:

m_l = mass of load   lb-sec^2/in.
m_b = total mass of beam   lb-sec^2/in.

moment M_m, and the maximum end support force F_m, respectively, for several types of loading and end support.[7] The natural frequency ω_n' also can be found from Eq. (42.25):

$$\omega_n' = \frac{\dot{u}_o}{\delta_m} = \frac{\rho}{\xi_\delta l^2} \sqrt{\frac{E}{\mu}} \quad \text{rad/sec} \tag{42.28}$$

CIRCULAR PLATE WITH CONCENTRATED CENTER LOAD. The elastic behavior of plates is analogous to that of beams. For example, the stiffness of a uniform circular plate with a built-in circumferential edge and supporting a load at the center is[10]

$$k_p = n_p \left(\frac{E t^3}{R^2} \right) \qquad \text{lb/in.} \tag{42.29}$$

where t is the thickness and R the radius of the plate, in inches. This expression is similar to Eq. (42.19). The constant n_p has a value of 4.60 when the plate is made of steel having a Poisson's ratio $\gamma = 0.3$. The expression for the natural frequency of the plate in its fundamental mode is[10]

$$\omega_n = N_p \left(\frac{t}{R^2} \right) \sqrt{\frac{E}{\mu}} \qquad \text{rad/sec} \tag{42.30}$$

This expression is similar to Eq. (42.20). The constant N_p has a value of 3.09 when the plate is made of steel having a Poisson's ratio $\gamma = 0.3$. Finally, the expression for the natural frequency ω_n', which accounts for the equivalent mass m_e of the vibrating plate as well as the concentrated load m_l, is

$$\omega_n' = \frac{\dfrac{t}{R^2} \sqrt{\dfrac{E}{\mu}}}{\xi_\delta'} \qquad \text{rad/sec} \tag{42.31}$$

where

$$\xi_\delta' = \sqrt{\frac{1}{N_p^2} + \left(\frac{\pi}{n_p} \right) \frac{m_l}{\mu t (\pi R^2)}}$$

Substituting the numerical values of N_p and n_p for a circular plate built in at the edge and carrying a concentrated load m_l at the center, $\xi_\delta' = \sqrt{0.1047 + 0.683 \, m_l/(\mu t \pi R^2)}$. For plates with other edge conditions and other types of loading, procedures have been worked out for computing the natural frequencies.[11]

MECHANICAL PROPERTIES OF ENGINEERING MATERIALS

The mechanical properties of a material are determined by standardized measurements of the reactions of the material to specified physical conditions. Under the dynamic conditions of shock and vibration, materials are likely to react differently than under conventional static testing. Differences in behavior are ascribable to the rapid application and short duration of the loading, and to repeated loading.

The important static properties are ultimate tensile strength, yield strength, modulus of elasticity, elongation at failure, and reduction of area. Hence, a standard tensile test can be used for comparison of materials subject to static loading. This is acceptable if each material remains within its elastic limit and dynamic effects do not become important. However, under dynamic loading the yield strength and ultimate strength depend upon the strain rate which in turn depends upon the geometry of the structure and the type of loading. As a consequence, dynamic properties are not standardized easily. They usually are expressed as ultimate strength and yield strength variation with strain rate, impact strength variation with temperature, fatigue behavior, and energy dissipation by internal damping. A limited amount of information in these categories is available.[12]

Undoubtedly there are additional dynamic properties of importance that remain to be formulated and evaluated. In particular, more work must be done in determining material strength to resist rapid repeated loading at a stress slightly less than the yield strength. For the case of transient loading, there is doubt concerning the amount that the yield strength of a material increases under dynamic loading. Usually, it is advisable in the design of equipment to withstand shock and vibration to assume no increase in yield strength for any material.

STATIC PROPERTIES

A standard tensile test supplies a stress-strain curve which can be used to define many of the mechanical properties of a material. Hooke's law applies at stresses below the proportional limit. The extension e is directly proportional to the load F and to the length of the test bar l, and inversely proportional to the cross-sectional area S. A proportionality constant, the initial slope of the stress-strain curve, is called the modulus of elasticity or Young's modulus E:

$$e = \frac{Fl}{SE} \quad \text{in.} \tag{42.32}$$

For example, with a low- or medium-carbon steel, the extension e of a test bar with gage length $l = 2$ in. is about 0.002 in. at the proportional limit. With a Young's modulus for steel of $E = 30 \times 10^6$ lb/in.^{2}, Eq. (42.32) gives a tensile stress (force per unit area) of $F/S = 30,000$ lb/in.^{2}

Fig. 42.20. Conventional stress-strain diagram for a metal with a yield point.

Most metals can be stressed slightly higher than the proportional limit without showing permanent deformation upon removal of the load. The maximum stress that can be attained without permanent deformation is called the elastic limit. Mild steel and a few rather uncommon metals exhibit a distinct yield point. This is defined as the stress at which permanent deformation begins suddenly and continues with no increase in stress. However, most materials continue to yield gradually after passing the elastic limit; there is no characteristic dip at strains somewhat greater than the yield point, as in the stress-strain curve of Fig. 42.20.

Various definitions of the yield condition have been proposed,[13] but they are not pertinent to the subsequent discussion of dynamic properties. In the United States of America the term *yield strength* has been adopted as standard.[14] The yield strength of a material is that stress which produces, on unloading, a permanent strain of 0.002 in./in. (0.2 per cent).

When the loading on a ductile test bar is increased beyond the yield strength, large permanent deformations occur and lead to actual fracture of the material. At the same time there is an appreciable reduction in cross-sectional area of the test bar at the section where fracture occurs. The shape of the stress-strain curve shown in Fig. 42.20 is the result of calculating an arbitrary nominal stress based on the original cross-sectional area. The ultimate strength is the nominal stress found by dividing the maximum load achieved in a tensile test by the original area of the test bar cross section.

If the tensile force on a test bar of ductile material (e.g., low-carbon steel) is continuously increased, the area of the bar at the cross section of eventual failure begins to

Table 42.3. Mechanical Properties of Typical Cast Irons

(A. Vallance and V. Doughtie.[16])

Material	Ultimate strength		Endurance limit in reversed bending, lb/in.^{2}, σ_e	Brinell hardness number	Modulus of elasticity		Elongation in 2 in., %
	Tension, lb/in.^{2}, σ_u †	Compression, lb/in.^{2}, σ_u			Tension and compression, lb/in.^{2}, E	Shear, lb/in.^{2}, G	
Gray, ordinary......................	18,000	80,000	9,000	100–150	10–12,000,000	4,000,000	0–1
Gray, good *	24,000	100,000	12,000	100–150	12,000,000	4,800,000	0–1
	16,000						
Gray, high grade....................	30,000	120,000	15,000	100–150	14,000,000	5,600,000	0–1
Malleable, S.A.E. 32510.............	50,000	120,000	25,000	100–145	23,000,000	9,200,000	10
Nickel alloys:							
Ni-0.75, C-3.40, Si-1.75, Mn-0.55 * ..	32,000	120,000	16,000	200	15,000,000	6,000,000	1–2
	24,000			175			
Ni-2.00, C-3.00, Si-1.10, Mn-0.80 * ..	40,000	155,000	20,000	220	20,000,000	8,000,000	1–2
	31,000			200			
Nickel-chromium alloys:							
Ni-0.75, Cr-0.30, C-3.40, Si-1.90, Mn-0.65......................	32,000	125,000	16,000	200	15,000,000	6,000,000	1–2
Ni-2.75, Cr-0.80, C-3.00, Si-1.25, Mn-0.60......................	45,000	160,000	22,000	300	20,000,000	8,000,000	1–2

* Upper figures refer to arbitration test bars. Lower figures refer to the center of 4-in. round specimens.
† *Flexure:* For cast irons in bending, the modulus of rupture may be taken as 1.75 σ_u (tension) for circular sections, 1.50 σ_u for rectangular sections and 1.25 σ_u for I and T sections.

Table 42.4. Mechanical Properties of Typical Carbon Steels

(A. Vallance and V. Doughtie.[16])

Material	Ultimate strength		Yield strength		Endurance limit in reversed bending lb/in.^{2}, σ_e	Brinell hardness number	Modulus of elasticity		Elongation in 2 in., %
	Tension, lb/in.^{2}, σ_u	Shear, lb/in.^{2}, σ_u	Tension and compression, lb/in.^{2}, σ_y	Shear, lb/in.^{2}, σ_y			Tension and compression, lb/in.^{2}, E	Shear, lb/in.^{2}, G	
Wrought iron...............	48,000	50,000	27,000	30,000	25,000	100	28,000,000	11,200,000	30–40
Cast steel:									
Soft.....................	60,000	42,000	27,000	16,000	26,000	110	30,000,000	12,000,000	22
Medium.................	70,000	49,000	31,500	19,000	30,000	120	30,000,000	12,000,000	18
Hard....................	80,000	56,000	36,000	21,000	34,000	130	30,000,000	12,000,000	15
SAE 1025:									
Annealed...............	67,000	41,000	34,000	20,000	29,000	120	30,000,000	12,000,000	26
Water-quenched *........	78,000	55,000	41,000	24,000	43,000	159	30,000,000	12,000,000	35
	90,000	63,000	58,000	34,000	50,000	183			27
SAE 1045:									
Annealed...............	85,000	60,000	45,000	26,000	42,000	140	30,000,000	12,000,000	20
Water-quenched *........	95,000	67,000	60,000	35,000	53,000	197	30,000,000	12,000,000	28
	120,000	84,000	90,000	52,000	67,000	248			15
Oil-quenched *..........	96,000	67,000	62,000	35,000	53,000	192	30,000,000	12,000,000	22
	115,000	80,000	80,000	45,000	65,000	235			16
SAE 1095:									
Annealed...............	110,000	75,000	55,000	33,000	52,000	200	30,000,000	12,000,000	20
Oil-quenched *..........	130,000	85,000	66,000	39,000	68,000	300	30,000,000	11,500,000	16
	188,000	120,000	130,000	75,000	100,000	380			10

* Upper figures: steel quenched and drawn to 1300°F. Lower figures: steel quenched and drawn to 800°F. Values for intermediate drawing temperatures may be approximated by direct interpolation.

decrease (neck down) as the maximum force is approached; then further extension takes place with decreasing force. This ability of the material to flow without immediate rupture is called ductility; it is defined as the per cent reduction of cross-sectional area measured at the section of fracture. Another measure of ductility is the per cent elongation of a 2-in. gage length; this depends upon the shape and size of the specimen (cross section and gage length).

Table 42.5. Mechanical Properties of Copper-Zinc Alloys (Brass)

(A. Vallance and V. Doughtie.[16])

Type of material	Ultimate strength Tension, lb/in.^{2}, σ_u	Yield strength Tension, lb/in.^{2}, σ_y	Endurance limit, lb/in.^{2}, σ_e	Brinell hardness number	Modulus of elasticity Tension and compression lb/in.^{2}, E	Elongation in 2 in., %
Commercial bronze (90 Cu, 10 Zn):						
Rolled, hard....................	65,000	63,000	18,000	107	15,000,000	18
Rolled, soft....................	35,000	11,000	12,000	52	15,000,000	56
Forged, cold...................	40,000–65,000	25,000–61,000	12,000–16,000	62–102	15,000,000	55–20
Red brass (85 Cu, 15 Zn):						
Rolled, hard....................	75,000	72,000	20,000	126	15,000,000	18
Rolled, soft....................	37,000	14,000	14,000	54	15,000,000	55
Forged, cold...................	42,000–62,000	22,000–54,000	14,000–18,000	63–120	15,000,000	47–20
Low brass (80 Cu, 20 Zn):						
Rolled, hard....................	75,000	59,000	22,000	130	15,000,000	18
Rolled, soft....................	44,000	12,000	15,000	56	15,000,000	65
Forged, cold...................	47,000–80,000	20,000–65,000		63–133	15,000,000	30–15
Spring brass (75 Cu, 25 Zn):						
Hard..........................	84,000	64,000	21,000	107 *	14,000,000	5
Soft..........................	45,000	17,000	17,000	57 *	18,000,000	58
Cartridge brass (70 Cu, 30 Zn):						
Rolled, hard....................	100,000	75,000	22,000	154	15,000,000	14
Rolled, soft....................	48,000	30,000	17,000	70	15,000,000	55
Deep-drawing brass (68 Cu, 32 Zn):						
Strip, hard...................	85,000	79,000	21,000	106 *	15,000,000	3
Strip, soft....................	45,000	11,000	17,000	13 *	15,000,000	55
Muntz metal (60 Cu, 40 Zn):						
Rolled, hard....................	80,000	66,000	25,000	151	15,000,000	20
Rolled, soft....................	52,000	22,000	21,000	82	15,000,000	48
Tobin bronze (60 Cu, 39.25 Zn, 0.75 Sn):						
Hard..........................	63,000	35,000	21,000	165	15,000,000	35
Soft..........................	56,000	22,000		90	15,000,000	45
Manganese bronze (58 Cu, 40 Zn):						
Hard..........................	75,000	45,000	20,000	110	15,000,000	20
Soft..........................	60,000	30,000	16,000	90	15,000,000	30

* Rockwell hardness F.

Other static properties of materials find application in the design of equipment to withstand shock and vibration. The modulus of rigidity G is the ratio of shear stress to shear strain; it may be determined from the torsional stiffness of a thin-walled tube of the material. The value of G for steel is 12×10^6 lb/in.^{2} The modulus of volume expansion (bulk modulus) κ is the inverse of the fractional change in volume which accompanies the application of unit hydrostatic pressure. The value of κ for steel is 25×10^6 lb/in.^{2} The inverse of the modulus of volume expansion is the compressibility. Poisson's ratio γ is the ratio of the lateral unit strain to the normal unit strain in the elastic range of the material. This ratio evaluates the deformation of a material that occurs perpendicular to the direction of application of load. The value of γ for steel is 0.3. From geometrical considerations it is known that the three moduli of elasticity E, G, and κ, and Poisson's ratio γ, are related to each other.[15] More complete data on materials and their properties as used in machine and equipment design are compiled in available references.[16, 17] Values of the static properties of typical engineering materials are given in Tables 42.3 to 42.5.

DYNAMIC PROPERTIES

When a load is applied to a structure suddenly, the dynamic properties of the material may be significant. Dynamic properties can be considered as modifications of static

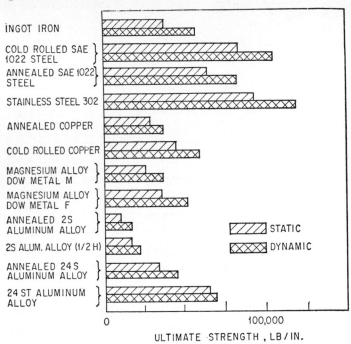

FIG. 42.21. Static and dynamic values of the ultimate strength of several metals where the dynamic strengths were obtained at impact velocities of 200 to 250 ft/sec. (*D. S. Clark and D. S. Wood.*[18])

properties. The most significant modifications are increases in ultimate strength and yield strength; other effects of some importance are an apparent lack of ductility at very high strain rates and delay in initiation of yielding under sudden application of load.

FIG. 42.22. Effect of strain rate on mechanical properties of mild steel. (*M. J. Manjoine.*[20])

STRAIN-RATE EFFECTS. The use of high-speed tensile testing machines shows that an increase in the rate of deformation raises both the yield strength and the ultimate strength of certain metals, as well as the entire stress level of the stress-strain curve. Figure 42.21 shows the static and dynamic values of the ultimate strengths of several metals where the dynamic strengths are determined at impact velocities of 200 to 250 ft/sec.[18] Impact tests of projectiles with striking velocities up to 2,750 ft/sec show as much as a 300 per cent increase in compressive strength for some types of steel.[19]

The influence of strain rate on the tensile properties of mild steel at room temperature is shown in Fig. 42.22. The marked difference between the yield stress and ultimate stress at low rates of strain disappears at high rates of strain.[20]   Figure 42.22 also shows that the ultimate stress remains practically unchanged for strain rates below 1 in./in./sec; in this limited range the stress-strain curve of most engineering metals is not raised appreciably.[21]   Mild steel is an exception in which the yield stress is influenced markedly by strain rate in the range from 0 to 1 in./in./sec.

TOUGHNESS AND DUCTILITY.   It is useful to evaluate the total energy needed to fracture a test bar under tension; this energy is a measure of the toughness of the material. The area under the typical stress-strain diagram shown in Fig. 42.20 gives an approximate measure of the fracture energy per unit volume of material.   However, the true fracture energy depends upon the true stress and true strain characteristics which take into account the nonuniform strain resulting from the reduction of area upon necking down of the test bar.   Calculated values of the fracture energy for various metals are given in Table 42.6.   Tough materials (e.g., wrought iron and low- or medium-carbon steel) exhibit high unit elongation and are considered to be ductile.   By contrast, cast iron exhibits practically no elongation and is considered to be brittle.   If only the elastic

Table 42.6. Fracture Energy or Toughness of Different Materials
(*J. M. Lessells.*[13])

Material	Condition	Yield strength, lb/in.^{2}, σ_y	Tensile strength lb/in.^{2}, σ_u	Unit elongation, in./in., ξ	Toughness or fracture energy, in.-lb/in.^{3}	Modulus of resilience, in.-lb/in.^{3}
Wrought iron.......	As received	24,000	47,000	0.50	17,700	7
Steel (0.13% C)....	As received	26,000	54,000	0.44	17,600	11
Steel (0.25% C)....	As received	44,000	76,000	0.36	21,000	24
Steel (0.53% C)....	Oil-quenched and drawn	86,000	134,000	0.11	12,000	100
Steel (1.2% C).....	Oil-quenched and drawn	130,000	180,000	0.09	10,800	280
Steel (spring).......	Oil-quenched and drawn	140,000	220,000	0.03	4,400	320
Cast iron..........	As received	...	20,000	0.005	70	1
Nickel cast iron.....	As received	20,000	50,000	0.10	3,500	9
Rolled bronze......	As received	40,000	65,000	0.20	10,500	60
Duralumin.........	Forged and heat-treated	30,000	52,000	0.25	10,200	17

strain energy up to the proportional limit is included, the resulting stored energy per unit volume is called the *modulus of resilience.*   Values for this property also are given in Table 42.6.

CRITICAL STRAIN VELOCITY.   When a large load is applied to a structure very suddenly, failure of the structure may occur with a relatively small over-all elongation. This has been interpreted as a brittle fracture, and it has been said that a material loses its ductility at high strain rates.   However, an examination of the failure shows normal ductility (necking down) in a region close to the application of load.   Large stresses are developed in this region by the inertia of the material remote from the application of the load, and failure occurs before the plastic stress waves are transmitted away from the point of load application.   This effect is important only where loads are applied very suddenly, as in a direct hit by a projectile on armor plate.   In general, equipment is mounted upon structures that are protected from direct hits; the resilience of such structures prevents a sufficiently rapid application of load for the above effect to be of significance in the design of equipment.

DELAY IN INITIATION OF YIELD.   Yielding of a structure made of ductile material does not occur immediately upon sudden application of load.   Rather, yielding occurs an instant of time later.   This delay in initiation of yield is a function of the material, the stress level, and the temperature.   Thus, a material may be stressed substantially above its yield strength for a short period of time without yielding.   For mild steel at room temperature, the delay time is of the order of 0.001 sec. For repeated applications of load, the material has a memory; i.e., the durations of load are additive to determine the time of yielding. An equipment subjected to shock or vibration experiences an oscillatory stress pattern wherein the higher stresses occur repeatedly. The durations of these stresses add quickly to a time greater than the delay time for common materials; thus, the effect is of little significance in the design of equipment to withstand shock and vibration.

FATIGUE BEHAVIOR. The ability of a material to withstand repeated applications of stress is termed its endurance strength or fatigue strength. Endurance tests commonly are made by subjecting a test specimen to a stress pattern in which the stress varies sinusoidally with time. The test specimen usually* is a cantilever or simply supported beam with a dead-weight load to induce the desired stress; the cyclic stress pattern is attained by rotating the beam about its longitudinal axis. Rotation of the beam is continued until rupture occurs; this gives a point on a curve of stress vs. number of cycles to failure. The test is repeated using identical specimens with different stress, and the results are plotted as logarithm of stress vs. logarithm of number of cycles to failure. Typical curves are shown in Fig. 42.23.[22]

FIG. 42.23. Typical endurance curves. (*H. F. Moore and J. B. Kommers.*[22])

The sharp break and horizontal continuation of the curves in Fig. 42.23 indicate the stress level at which the material can endure an infinite number of stress reversals without failure. Sometimes, the curve gradually turns horizontal at its left-hand end. In other words, the stress required to cause failure after only a few cycles may be independent of the number of cycles.[23] Representative tensile and fatigue properties of steel are shown in Table 42.7. Similarly, Table 42.8 shows the tensile and fatigue properties of some nonferrous metals and alloys.

NOTCHED-BAR IMPACT TESTS. The Charpy and Izod notched-bar impact tests measure the energy absorbed by a standard specimen when it is fractured by a pendulum.[13] They are of value in showing quickly the effects of temperature, heat-treatment, and composition on the ductility of different metals. Typical values of impact energy are given in Table 42.6.

Repeated-impact notched-bar tests were at one time considered important but this practice has been discontinued. The notch sensitivity of the material influences the results which then become difficult to interpret as quantitative values of resistance to impact loads.[24] Figure 42.24, showing the results of such tests, is of general interest because the curves start with the impact strength for a single blow and follow the characteristic shape of conventional vibration endurance curves.[25]

* Numerous special types of fatigue tests are made, including alternating tension and compression, alternating torsion, alternating tension or torsion superimposed upon a non-varying component of tension or torsion, alternating combined tension and torsion, and many others.

Table 42.7. Tensile and Fatigue Properties of Steels

(*J. M. Lessells.*[13])

Material	State	Yield strength, lb/in.^{2}, σ_y	Tensile strength, lb/in.^{2}, σ_u	Elongation, %	Reduction of area, %	Endurance limit, lb/in.^{2}, σ_e	Ratio σ_e/σ_u
0.02% C............	As received	19,000	42,400	48.3	76.2	26,000	0.61
Wrought iron........	As received	29,600	47,000	35.0	29.0	23,000	0.49
0.24% C............	As received	38,000	60,500	39.0	64.0	25,600	0.425
0.24% C............	Water-quenched and drawn	45,600	67,000	38.0	71.0	30,200	0.45
0.37% C............	Normalized	34,900	71,900	29.4	53.5	33,000	0.46
0.37% C............	Water-quenched and drawn	63,100	94,200	25.0	63.0	45,000	0.476
0.52% C............	Normalized	47,600	98,000	24.4	41.7	42,000	0.43
0.52% C............	Water-quenched and drawn	84,300	111,400	21.9	56.6	55,000	0.48
0.93% C............	Normalized	33,400	84,100	24.8	37.2	30,500	0.36
0.93% C............	Oil-quenched and drawn	67,600	115,000	23.0	39.6	56,000	0.487
1.2% C............	Normalized	60,700	116,900	7.9	11.6	50,000	0.43
1.2% C............	Oil-quenched and drawn	130,000	180,000	9.0	15.2	92,000	0.51
0.31% C, 3.35% Ni.	Normalized	53,500	104,000	23.0	45.0	49,500	0.47
0.31% C, 3.35% Ni.	Oil-quenched and drawn	130,000	154,000	17.0	49.0	63,500	0.41
0.24% C, 3.3% Ni, 0.87% Cr	Oil-quenched and drawn	128,000	138,000	18.2	61.8	68,000	0.49

Table 42.8. Tensile and Fatigue Properties of Nonferrous Metals

(*J. M. Lessells.*[13])

Material	State	Tensile strength, lb/in.^{2}, σ_u	Endurance limit or fatigue strength, lb/in.^{2}, σ_e	N_1,[*] millions of cycles	N_2,[†] millions of cycles	Ratio σ_e/σ_u
Aluminum..........	...	22,600	10,500	100	6	0.46
Duralumin..........	Rolled	51,000	14,000	400	>400	0.27
Duralumin..........	Annealed	25,200	10,000	200	>200	0.40
Duralumin..........	Tempered	51,300	12,000	400	4½	0.24
Magnesium..........	Extruded	32,500	8,000	200	2	0.25
Magnesium alloy (4% Al)...........	...	35,200	12,000	600	½	0.34
Magnesium alloy (4% Al, 0.25% Mn).	...	39,000	15,000	100	1	0.38
Magnesium alloy (6.5% Al)..........	...	41,200	13,000	600	½	0.31
Magnesium alloy (6.5% Al, 0.25% Mn).....	...	44,500	15,000	100	½	0.34
Magnesium alloy (10% Cu)..........	...	39,000	12,000	600	½	0.31
Electron metal......	...	36,600	17,000	200	30	0.47
Copper.............	Annealed	32,400	10,000	500	20	0.31
Copper.............	Cold-drawn	56,200	10,000	500	>500	0.18
Brass (60–40)........	Annealed	54,200	22,000	500	>500	0.44
Brass (60–40)........	Cold-drawn	97,000	26,000	500	50	0.27
Naval brass.........	...	68,400	22,000	300	10	0.32
Aluminum bronze (10% Al)..........	As cast	59,200	23,000	60	3	0.39
Aluminum bronze (10% Al)..........	Heat-treated	77,800	27,000	40	1	0.35
Bronze (5% Sn)......	Annealed	45,600	23,000	1000	10	0.50
Bronze (5% Sn)......	Cold-drawn	85,000	27,000	500	50	0.32
Manganese bronze....	As cast	70,000	17,000	150	20	0.24
Nickel..............	Annealed	70,000	28,000	100	50	0.40
Monel metal........	Hot-rolled	90,000	32,000	450	>450	0.36

[*] N_1 = cycles on which σ_e is based.
[†] N_2 = cycles at which σ-N curve becomes and remains horizontal.

FIG. 42.24. Impact endurance curves. (*D. J. McAdam, Jr.*[25])

THE SIGNIFICANCE OF MATERIAL PROPERTIES IN DESIGN

The method of designing equipment to withstand shock and vibration follows conventional methods of design in that it involves (1) determination of the loads, (2) calculation of the resultant stresses, and (3) selection of a suitable material. The loads and the nominal stresses are determined most conveniently by constructing a hypothetical model of the equipment as described in preceding sections of this chapter. This section discusses some of the important considerations in adapting the results of the model study to the selection of appropriate materials. Significant properties of materials are set forth in the above section under *Mechanical Properties of Engineering Materials*.

If the most severe environmental condition encountered by an equipment is a shock experienced only once or a few times, the most important consideration is to design the equipment to withstand a sudden application of a large load with its resultant stresses. Equipment that must endure sustained vibration of large magnitude must withstand repeated application of load of possibly smaller magnitude than that experienced during shock. The relative importance of these types of loading must be determined for each application; however, they are discussed in the following paragraphs under *Strength Requirements* and *Fatigue Behavior*.[23]

STRENGTH REQUIREMENTS

The most important consideration involving ultimate strength is the attainment of a proper balance between stress and ductility. A model, in so far as it can represent an equipment effectively, indicates the maximum strain if the equipment characteristics are linear. Where nonductile materials are used or where permanent deformation of a structure cannot be tolerated, the equipment must be designed so that the strain does not exceed the strain at the elastic limit. Usually, permanent deformation of some structural members of an equipment is permissible. Permanent deformation of other members may not be permissible because it would cause misalignment of parts whose proper operation depends upon accurate alignment. A concept that has application to the design of yielding members is discussed in the following paragraphs.

Experience indicates that the yield strength of a part may be exceeded at the maximum strain experienced during shock, provided the material yields under the shock loading. This concept may be applied only if it is possible to specify a maximum plastic deflection of the structure which will not lead to malfunctioning of the equipment. It is difficult to determine the degree of deformation that occurs at stresses greater than the yield strength because the structure becomes nonlinear and the linear model theory is only approximately applicable. However, semi-empirical procedures have been developed to specify permissible stresses.

An empirical design procedure that has application in the design of members subjected to bending contemplates that some predetermined percentage of the cross-sectional area of the member has a calculated stress greater than the yield strength of the material.[26] It is necessary that the material have a high ductility. In a typical application of this concept, 25 per cent of the cross-sectional area of the beam has a stress greater than the yield strength. For example, applying the concept to a beam in bending, the stress at maximum deflection is calculated; if the beam is linear without regard to the magnitude of the stress, the distribution of stress over a rectangular cross section is as shown in Fig. 42.25A. The beam is designed so that the stress is above the yield strength over 25 per cent of the cross-sectional area. Then the actual stress distribution, assuming a stress-strain curve of the type shown in Fig. 42.20, is shown in Fig. 42.25B. The permanent deformation of the beam is less than the maximum deflection because 75 per cent of the material, adjacent to the neutral axis, remains elastic and tends to restore the beam to its original shape.

FIG. 42.25. Stress distribution in a beam subject to bending with (A) calculated linear stress distribution so that the stress is above the yield strength over 25 per cent of the cross-sectional area, and (B) the actual stress distribution if the material has a stress-strain curve similar to that shown in Fig. 42.20.

Another empirical design procedure establishes a maximum allowable stress under shock loading that depends upon the function served by a particular member of the equipment. When a material has sufficient ductility to exhibit an elongation at failure of at least 10 per cent in 2 in., the allowable stress σ_a based on permissible yielding can be expressed:

$$\sigma_a = \sigma_y + C(\sigma_u - \sigma_y) \qquad \text{lb/in.}^2 \qquad (42.33)$$

where σ_y is the yield strength for a 0.2 per cent permanent extension, σ_u is the ultimate tensile stress, and the constant C depends upon the function served by a particular member. For bolts which are inaccessible for retightening, $C = 0$. Where dimensional stability is important, $C = 0.2$. If some yielding can be tolerated, as in piping, $C = 0.5$. For calculating the bearing stress between flat or mating surfaces or where permanent deformation can be tolerated, $C = 1.0$.

Although the yield strength and ultimate strength of mild steel show an increase as the rate of strain increases (Fig. 42.22), this effect is of very limited significance in the design of equipment to withstand shock and vibration. In general, a strain rate great enough to cause a significant increase in strength occurs only closely adjacent to a source of shock, as at the point of impact of a projectile with armor plate. Equipment seldom is subjected to shock of this character. In a typical installation, the structure interposed between the equipment and the source of shock is unable to transmit large forces suddenly enough to cause high strain rates at the equipment. Furthermore, the response of a structure to a shock is oscillatory; maximum strain rate occurs at zero strain, and vice versa. The data of Fig. 42.22 represent conditions where maximum stress and maximum rate of strain occur simultaneously; thus, they do not apply directly to the design of shock resistant equipment. The use of statically determined yield strength and ultimate strength for design purposes is a conservative but not overly conservative practice.

FATIGUE BEHAVIOR

Failure of equipment by repeated occurrence of stress usually is the result of a condition of resonance with consequent excessive stress. Several approaches for controlling the stress are suggested in Chap. 43. The effectiveness of such control determines the extent to which the fatigue properties of the material become significant. Most of the data defining the fatigue behavior of material are in the form of Fig. 42.23. In some applications, the stress on the ordinate of Fig. 42.23 may be applied directly to the calculated stress in a member of the equipment. Often, considerations that cannot be deduced directly from Fig. 42.23 become important: (1) the alternating stress may be superimposed upon a steady stress, (2) the stress may be greater than that calculated because of stress concentration, and (3) the stress amplitude may not be constant at each cycle of stress.

When a steady stress is combined with a variable stress, the resultant strength is determined from the Goodman or Gerber relations shown in Fig. 42.26[27] and from the applicable fatigue behavior of the type shown in Fig. 42.23. The abscissa in Fig. 42.26

FIG. 42.26. Graphical representation of Goodman and Gerber relations. (*H. J. Gough.*[27])

is the ratio of the steady stress existing in the part to the tensile strength of the material. The ordinate is the ratio of the allowable variable stress in the part to the endurance limit found from Fig. 42.23. The limiting cases shown are as follows: (1) For a steady stress equal to zero, the allowable variable stress equals the endurance limit; and (2) for a variable stress equal to zero, the allowable steady stress is the tensile or compressive strength. The straight line indicates that for a steady stress equal to half the tensile strength, the allowable variable stress is only half the endurance limit. The Gerber relation is in good agreement with test data. However, the Goodman relation gives a smaller allowable stress and may be preferred for more conservative design.

The fatigue strength of a machine member is greatly affected by its geometric shape. Figure 42.27 shows typical values of stress concentration factor at the change in diameter of a circular rod subjected to a bending moment M_B.[28] The stress concentration factor can be interpreted as the ratio of actual maximum stress in the fillet to the nominal stress calculated on the basis of the diameter d. The stress concentration increases with the sharpness of the radius r, and depends upon the ratio of diameters D/d.

To provide the data required for the design of equipment subjected to loads of different magnitudes, endurance testing must be done with *different* stress levels applied to a single specimen before failure is reached. Such data are not available to any appreciable degree, and it becomes necessary to adapt data of the type shown in Fig. 42.23 to circumstances of varying stress amplitude. Several different cumulative damage criteria have been suggested. One hypothesis[29] assumes that the percentage of fatigue damage is a linear function of the number of cycles. For example, referring to Fig. 42.28,[30] if a specimen is subjected to a stress σ_1 for n_1 cycles, σ_2 for n_2 cycles, etc., failure tends to

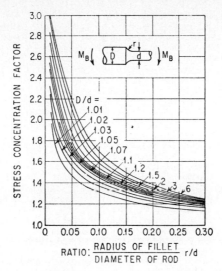

FIG. 42.27. Values of stress concentration factor K for fillets at change in diameter of a round rod in bending. (*R. E. Peterson.*[28])

FIG. 42.28. Typical stress distribution and stress-cycle curve illustrating the application of Miner's hypothesis for cumulative fatigue damage. (*N. M. Newmark.*[30])

occur when

$$\frac{n_1}{N_1} + \frac{n_2}{N_2} + \frac{n_3}{N_3} + \cdots = 1 \qquad (42.34)$$

where N_1, N_2, ... are cycles-to-failure at stresses σ_1, σ_2, ..., respectively. Limited experimental results indicate that the factor 1.0 in Eq. (42.34) varies between 0.3 and 3.0, depending on the order in which the stress cycles are applied and possibly on other influences. A somewhat more complicated hypothesis[31] assumes a nonlinear accumulation of damage.

REFERENCES

1. Roark, R. J.: "Formulas for Stress and Strain," 3d ed., p. 100, McGraw-Hill Book Company, Inc., New York, 1954.
2. Den Hartog, J. P.: "Mechanical Vibration," 3d ed., p. 105, McGraw-Hill Book Company, Inc., New York, 1947.
3. Ruzicka, J. E.: *Sound and Vibration*, **4**(9):10 September, 1970.
4. Walsh, J. P., and R. E. Blake: *Proc. SESA*, **6**(2):152 (1948).

5. Frankland, J. M.: *Proc. SESA*, **6**(2):11 (1948).
6. Crede, C. E., and M. C. Junger: "A Guide for Design of Shock Resistant Naval Equipment," U.S. Navy Department, Bureau of Ships, *NAVSHIPS* 230-660-3C, p. 17, 1949.
7. Friedrich, C. M.: "Shock Design Notes," unpublished report, Westinghouse Electric Corporation, Atomic Power Division, Pittsburgh, Pa., June, 1956.
8. McCalley, R. B., Jr.: *Supplement to Shock and Vibration Bull.* 23 (unclassified), Office of Secretary of Defense, June, 1956.
9. Ref. 1, p. 70.
10. Ref. 2, p. 457.
11. Stokey, W. F., and C. F. Zorowski: *J. Appl. Mechanics*, **E26**(2):210 (1959).
12. Rinehart, J. S., and J. Pearson: "Behavior of Metals Under Impulsive Loads," The Haddon Craftsmen, Inc., Scranton, Pa., 1954.
13. Lessells, J. M.: "Strength and Resistance of Metals," 1st ed., p. 7, John Wiley & Sons, Inc., New York, 1954.
14. ASTM Book of Standards, part 3, p. 117, 1958.
15. Sokolnikoff, I. S.: "Mathematical Theory of Elasticity," 2d ed., p. 65, McGraw-Hill Book Company, Inc., New York, 1956.
16. Vallance, A., and V. Doughtie: "Design of Machine Members," Chap. II, 3d ed., McGraw-Hill Book Company, Inc., New York, 1951.
17. Hoyt, S. L.: "Metal Data," 2d ed., Reinhold Publishing Corporation, New York, 1952.
18. Clark, D. S., and D. S. Wood: *Trans. ASM*, **42**:45 (1950).
19. Whiffin, A. C.: *Proc. Royal Society (London)*, **194A**:300 (1948).
20. Manjoine, M. J.: *J. Appl. Mechanics*, **66**:A215 (1944).
21. MacGregor, C. W., and J. C. Fisher: *J. Appl. Mechanics*, **67**:A217 (1945).
22. Moore, H. F., and J. B. Kommers: "The Fatigue of Metals," Chap. VI, McGraw-Hill Book Company, Inc., New York, 1927.
23. Pope, J. A.: "Metal Fatigue," Chap. II, Chapman & Hall, Ltd., London, 1959.
24. Gillett, H. W.: *Steel*, Nov. 22, 1943, and Feb. 14, 1944.
25. McAdam, D. J., Jr.: *Proc. ASTM*, **23**:56-105 (1923).
26. Graziana, D. J.: *Prod. Eng.*, **44**:44, Sept. 29, 1958.
27. Gough, H. J.: "The Fatigue of Metals," D. Van Nostrand Company, Inc., Princeton, N.J., p. 69, 1924.
28. Peterson, R. E.: "Stress Concentration Design Factors," p. 75, John Wiley & Sons, Inc., New York, 1953.
29. Miner, M. A.: *J. Appl. Mechanics*, **67**:A159 (1945).
30. Murray, W. M.: "Fatigue and Fracture of Metals," p. 197, John Wiley & Sons, Inc., New York, 1952. (See Paper 10 by N. M. Newmark.)
31. Corten, H. T., and T. J. Dolan: *Intern. Conf. Fatigue of Metals, IME and ASME*, September, 1956.

43

PRACTICE OF EQUIPMENT DESIGN

Edward G. Fischer
Westinghouse Electric Corporation

Harold M. Forkois
U.S. Naval Research Laboratory

INTRODUCTION

Equipment of a type required to withstand shock and vibration generally consists of a housing or chassis to provide structural strength and an array of functional components supported by the chassis. A suitable design is characterized by (1) properly selected or designed components, (2) chassis mounting to minimize damage from shock and vibration, and (3) a chassis capable not only of withstanding shock and vibration but also of providing a degree of protection to the components. The design of chassis and other structures is discussed in Chap. 42 from a quantitative viewpoint; this chapter discusses (1) chassis design in a general or qualitative sense, (2) the selection and mounting of components and (3) design to minimize vibration. Many of the examples refer explicitly to electronic equipment but the general principles have wider application.

DESIGN OF STRUCTURES

One of the more troublesome problems in the design of equipment is attainment of the proper balance between flexibility and rigidity of structures. As shown analytically in Chap. 42, the maximum acceleration experienced by a component is determined primarily, for a particular shock motion, by the natural frequency of the structure supporting the component. The acceleration decreases as the natural frequency decreases. Thus, components that are susceptible to damage from shock may be given a degree of protection by mounting them on relatively flexible structures; however, if the equipment is required to withstand vibration as well as shock, it is possible that the flexibility introduced to attenuate the shock may lead to failure as a result of vibration.

Failure or damage from vibration usually is the result of a resonant condition; i.e., a structure or component has a natural frequency that coincides with a frequency of the forcing vibration. In some applications, e.g., on naval ships, the maximum frequency of the forcing vibration is relatively low and it becomes feasible in most instances to design structures whose natural frequencies are greater than the highest forcing frequency. On the other hand, the vibration in aircraft and missiles is characterized by relatively high frequencies; as a consequence, it is not feasible to design structures whose natural frequencies are higher than the forcing frequencies. A condition of resonance often must be tolerated. In extreme conditions, the effect of the resonant condition can be alleviated by the provision of damping or energy dissipation (see the section on *Damping*). In general, this is not necessary for structural members if the natural frequencies are chosen wisely.

Because of the many different types of components and equipments in common use and the wide range of environmental conditions encountered, it is not possible to establish specific design objectives for natural frequencies of structures. In general, a structure should be designed to be stiff rather than flexible because (1) most components of modern design are more shock-resistant than commonly anticipated and do not require the protection afforded by structures of low natural frequency, and (2) failure of structures by repeated reversal of stress during vibration is a common occurrence of damage. The stress in a structure tends to decrease when the natural frequency increases, provided the severity of the excitation and the fraction of critical damping for the structure remain unchanged. Furthermore, some environmental conditions embody a lower severity at the higher vibration frequencies; thus, a resonance at a high frequency may become less damaging. However, if a component is known to be vulnerable to damage by shock or if this is demonstrated during a shock test, the desired protection often may be attained by introducing a structure of lower flexibility. Such a structure may be the chassis of the equipment, a bracket to support a component, or an isolator (see Chaps. 30 to 33).

In typical electronic equipment, the electronic components are mounted upon a chassis to form a unit that has a discrete function. The unit is made small enough to permit ready maintenance and servicing. In some instances, the function of the unit may be complete; then the unit is considered a complete equipment and is provided with independent mounting means. In other instances, several units are required to fulfill a function; such units are mounted in a single cabinet and interconnected electrically and/or mechanically to perform a function. Particular design problems associated with the housing for the equipment and involving ability to withstand shock and vibration are (1) design of the unit chassis, (2) design of the cabinet, and (3) mounting of the unit chassis in the cabinet.[1]

CHASSIS DESIGN

In general, unit chassis are either (1) formed from sheet metal or (2) cast from a light metal (usually aluminum or magnesium) and machined to form intimate supports for the components. The former is more common and less costly to manufacture, but tends to

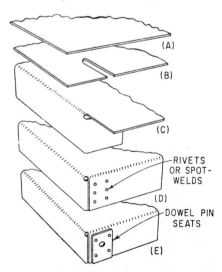

Fig. 43.1. Good chassis construction, showing front-panel mounting screws and hole-plate reinforcement for dowel pins in the rear. (*H. M. Forkois and K. E. Woodward.*[1])

Fig. 43.2. A method of forming chassis which produces stiff corners. (*H. M. Forkois and K. E. Woodward.*[1])

lack adequate rigidity. The casting has the desirable property of relatively great rigidity, but refinements in design and manufacture are necessary to keep the weight within reasonable limits. Because of its rigidity, the cast chassis is not only less likely to fail as a result of vibration but it also minimizes damage from broken leads and collision of components that occur when the chassis is unduly flexible. The flexibility of structure required for protection from shock can be provided in the support for the chassis.

The most common type of sheet-metal chassis is a shallow rectangular "box" formed from a single piece of sheet metal; the front panel is added and rigidly attached to the "box" by gussets, as shown in Fig. 43.1. The chassis usually is oriented with the flanges or sides of the "box" extending downward. The sequence of illustrations in Fig. 43.2 shows one method of forming chassis corners. All bending radii should be greater than the thickness of the sheet to reduce the effects of stress concentration. The overlapped

FIG. 43.3. View of underside of chassis showing stiffener added to support heavy component on upper side of chassis. (*H. M. Forkois and K. E. Woodward.*[1])

portions of the sheet can be joined by rivets or spot-welds (see *Methods of Construction*) to provide a stiff support for dowel pins or other attaching means. It is common practice to mount large components, such as transformers and electron tubes, on the upper side of the chassis; smaller components, such as resistors and capacitors together with substantially all wiring, are mounted on the lower side of the chassis.

Special care should be exercised to ensure maximum rigidity of the sheet-metal chassis. The overlapped flanges at the corners are areas of inherent rigidity and should be used to support the chassis. For example, the gussets for reinforcing the front panel and the plates to seat the dowel pins are attached at the corners as illustrated in Fig. 43.1. Where possible, heavy components such as transformers should be mounted adjacent the corners to attain a maximum ratio of chassis stiffness to component weight. If a heavy component cannot be mounted adjacent a corner, it should be mounted adjacent an edge; if it must be mounted near the center of the chassis, a stiffener should be provided on the lower side of the chassis, as shown in Fig. 43.3. This type of construction is necessary to obtain the required stiffness but is undesirable in that it limits the location of components and interferes with the wiring.

Where an equipment is comprised of a single chassis, it is common practice to provide an enclosure to protect the equipment during handling and during exposure to atmospheric contamination; such an enclosure is not intended as a support for the equipment and is not required to withstand the forces introduced by shock and vibration conditions. For example, in one type of application commonly used with vibration isolators in aircraft equipment, a supporting rack has forward-facing tapered pins adjacent the rear edge. These pins extend through clearance holes in the enclosure and engage dowel pin seats in the rear of the chassis, as illustrated in Fig. 43.1. Similarly, clamps at the forward edge of the rack engage the front panel of the chassis directly, and loads are not carried by the enclosure. Frequently, two or more chassis, each with its individual enclosure, are mounted independently upon one rack.

FIG. 43.4. Frame construction for cabinet showing clearance holes for drawer or chassis front-panel thumbscrews. The frame members are welded, riveted, or bolted together. (*H. M. Forkois and K. E. Woodward.*[1])

CABINET DESIGN

When several chassis are included in a single cabinet, it is common to design the cabinet to be both an enclosure and a structure capable of withstanding the forces introduced by shock and vibration conditions. The chassis usually must be readily removable from the cabinet for adjustment and maintenance. Suitable methods of mounting the chassis in the cabinet are discussed under *Chassis Mounting Methods.* If the objective of a rigid chassis is achieved, criteria to be applied in design of the cabinet are (1) optimum rigidity in view of shock and vibration conditions and (2) adequate structural strength. Suggestions for cabinet design are included in the following paragraphs.

A typical cabinet has horizontal dimensions corresponding to the width and depth of a chassis, and a height equal to the total heights of the chassis. The chassis are arranged one above the other, and are mounted independently to the cabinet. Except for very light equipment, the frame of the cabinet is assembled from structural steel or aluminum members welded, riveted, or bolted together, as illustrated in Fig. 43.4. Gussets are provided at the joints of structural members where they would not interfere with installation and removal of the chassis; they should be substantially the same thickness as the frame members. Generally, a cabinet constructed in this manner is notably lacking in rigidity, and all opportunities for stiffening should be employed. Usually, it is possible to cover the sides, back, top, and bottom with relatively light-gage sheet metal, securely attached to the structural members. Preferably, such sheets should be welded to the structural members to take full advantage of available rigidity, and they may be provided with welded or formed stiffeners, as illustrated in Figs. 43.5 and 43.6. It is difficult to achieve the desired rigidity in the front face of the cabinet because the requirement for removal of chassis leaves this face almost unstiffened. The several smaller front

FIG. 43.5. Welded channel stiffener to increase cabinet rigidity. (*H. M. Forkois and K. E. Woodward.*[1])

FIG. 43.6. Coined stiffener to increase cabinet rigidity. (*H. M. Forkois and K. E. Woodward.*[1])

panels of the individual chassis do not contribute as much stiffness as a single large sheet, and the screws and clamps commonly used to secure the chassis are not as effective as welding.   This lack of rigidity is particularly noticeable if the equipment is supported at the bottom and braced to a bulkhead by a support adjacent only the upper rear edge.

A cabinet with several chassis commonly weighs several hundred pounds; the re-

Fig. 43.7. Bottom view of cabinet construction showing channel stiffeners and shock isolator. (*H. M. Forkois and K. E. Woodward.*[1])

sultant loads at points of support may become very great under conditions of shock. Particular care in design is required to ensure that excessive deformation does not occur. For expedience of installation, it is preferable that the cabinet be mounted with a minimum of bolts.   Usually it is necessary to provide additional structural members to distribute the forces from the bolts over much of the base area of the cabinet.   Figure

Fig. 43.8. Structural stiffeners for bulkhead attachment near top of cabinet. (*H. M. Forkois and K. E. Woodward.*[1])

43.7 is a bottom view of a cabinet showing an arrangement of channel stiffeners for distributing the forces. Such stiffeners are of particular convenience when used with shock isolators because the isolators nest in the channels, but they can also be used with direct mounting. Figure 43.8 shows a type of stiffener that may be used near the top of a cabinet for attachment to a bulkhead.

METHODS OF CONSTRUCTION

In selecting a method of construction, it is important to consider the suitability of the structure for use under conditions of shock and vibration. The greatest cause of damage in a structure subjected to vibration is the large vibration amplitude that tends to occur during vibration at resonance of lightly damped structures. The interface friction between bolted members or riveted members introduces damping to a significant degree (see Chap. 36). Therefore, bolted and riveted structures may be better adapted than welded structures to withstand vibration, particularly if damping is important. However, a welded structure may be designed with less stress concentration and with better ability to maintain alignment of component parts; thus, it may be the most suitable method of construction even though larger amplitudes of vibration result at resonance. A qualitative indication of damping in various structures is given in the section on *Damping;* important considerations in various types of construction are discussed in the following sections.[2, 3]

WELDED JOINTS

Welded joints must be well designed, and effective quality control must be imposed upon the welding conditions. The most common defect is excessive stress concentration which leads to low fatigue strength and, consequently, to inferior capability of withstanding shock and vibration. Stress concentration can be minimized in design by reducing the number of welded lengths in intermittent welding. For example, individual welds in a series of intermittent welds should be at least $1\frac{1}{2}$ in. long with at least 4 in. between welds. Internal crevices can be eliminated only by careful quality control to ensure full-depth welds with good fusion at the bottom of the welds. Welds of adequate quality can be made by either the electric arc or gas flame process. Subsequent heat-treatment to relieve residual stress tends to increase the fatigue strength.

SPOT-WELDED JOINTS. Spot welding is quick, easy, and economical but should be used only with caution when the welded structure may be subjected to shock and vibration. Basic strength members supporting relatively heavy components should not rely upon spot welding. However, spot welds can be used successfully to fasten a metal skin or covering to the structural framework. Even though improvements in spot welding techniques have increased the strength and fatigue properties, spot welds tend to be inherently weak because a high stress concentration exists in the junction between the two bonded materials when a tension stress exists at the weld. Spot-welded joints are satisfactory only if frequent tests are conducted to show that proper welding conditions exist. Quality can deteriorate rapidly with a change from proved welding methods, and such deterioration is difficult to detect by observation. However, accepted quality-control methods are available and should be followed stringently for all spot welding.

RIVETED JOINTS. Riveting is an acceptable method of joining structural members when riveted joints are properly designed and constructed. Rivets should be driven hot to avoid excessive residual stress concentration at the formed head and to ensure that the riveted members are tightly in contact. Cold-driven rivets are not suitable for use in structures subjected to shock and vibration, particularly rivets that are set by a single stroke of a press as contrasted to a peening operation. Cold-driven rivets have a relatively high probability of failure in tension because of residual stress concentration, and tend to spread between the riveted members with consequent lack of tightness in the joint. Joints in which slip develops exhibit a relatively low fatigue strength.

BOLTED JOINTS. Except for the welded joints of principal structures, the bolted joint is the most common type of joint. A bolted joint is readily detachable for changes in construction, and may be effected or modified with only a drill press and wrenches as equipment. However, bolts tend to loosen and require means to maintain tightness (see section on *Locking Devices*). Furthermore, bolts are not effective in maintaining alignment of bolted connections because slippage may occur at the joint; this can be prevented by using dowel pins in conjunction with bolts or by precision fitting the bolts; i.e., fitting the bolts tightly in the holes of the bolted members. Other characteristics of bolted joints are discussed in the following paragraphs.

BOLTS AND NUTS

Bolts and nuts are made by various methods from different materials. Grades of steel bolts as standardized by the SAE and the ASTM are shown in Table 43.1. The ordinary type (18-8) stainless steel bolts have a low yield point (about 25,000 lb/in.²); thus, they tend to stretch and loosen under shock although they have a high ultimate strength. Where corrosion resistance as well as the ability to withstand shock is a consideration, bolts made from heat-treatable stainless-steel alloy having a high yield point should be

Table 43.1. Grades of Bolts (SAE and ASTM)
(*ASME Handbook.*[6])

SAE grade	ASTM designation	Tensile strength,* lb/in.²	Proof load or yield strength,* lb/in.²	Material
0	. . .	. . .	. . .	Low carbon
1	A307	55,000	. . .	Low carbon
2	. . .	69,000	55,000	Low carbon, stress-relieved
3	. . .	110,000	85,000	Medium carbon, cold-worked
5	. . .	120,000	85,000	Medium carbon, quenched and tempered
	A325	125,000	90,000	Medium carbon, quenched and tempered
6	. . .	140,000	110,000	Special medium carbon, quenched and tempered
7	. . .	130,000	105,000	Alloy steel, quenched and tempered
8	. . .	150,000	120,000	Alloy steel, quenched and tempered

* Properties for sizes ½ in. diameter and smaller. For complete data and for properties of larger bolts, see SAE Handbook and ASTM specifications.

Table 43.2. Comparison of Tensile Fatigue Strengths of Bolts with Rolled, Cut, and Milled Threads
(*H. Dinner and W. Felix.*[4])

Threading carried out by	Fatigue strength, lb/in.² (pulsating tension)	
	Unaged specimen	Specimen aged at 482°F
Rolling.............	28,450	28,450
Cutting.............	17,070	
Milling.............	17,070	12,800

FIG. 43.9. Shank of bolt undercut slightly smaller than root diameter of thread to obtain uniform elongation under shock load.

used. The fatigue strength of bolts under vibratory loading can be improved by initial cold working of the surfaces. Cold working may result from rolling of thread roots, shot peening of shanks, or rolling of the fillet between head and shank.[4,5] Benefits are derived from both the work hardening and the initial compression left in the surfaces. Table 43.2 gives data on the endurance limit of bolts in pulsating tension.[6]

The strength of a bolt when subjected to shock depends upon its capacity to absorb energy by elastic deflection. For optimum efficiency, all cross sections of the bolt should be designed to reach the yield stress simultaneously, where the stress is calculated by including stress concentrations. Figure 43.9 shows a bolt with the shank undercut slightly smaller than the root of the thread; therefore, the shank deflects equally throughout its length and stores energy without overstressing the material in the threads. This effect can be accentuated by increasing the length of the shank. The effect of a reduction in the shank diameter of bolts made from three classes of steel is shown in Table 43.3.[7] The composition and physical properties of the steels are given in the same table.

Stress raisers occur at abrupt changes in shape, such as shoulders, fillets, notches, grooves, and holes. They contribute to a lower endurance limit, and should be eliminated wherever possible. For example, Fig. 43.10A shows a stud with a smooth notch machined adjacent the last remaining thread. Maximum bending stress then occurs at the notch where the stress concentration is small, rather than at the root of a thread. The stud is closely fitted in the hole for a short length to reduce further the bending stress. Figure 43.10B shows a nut that is undercut so that the load is better distributed over several threads in the more flexible region adjacent the undercut. Plastic deformation of the first highly-loaded thread achieves somewhat the same result; this can be accomplished by a reverse taper or shorter pitch of the bolt threads relative to those of the nut. Subsequently, at lower bolt tensions, the load will be distributed more uniformly in the nut threads.[8,9]

Bolts usually are available in both coarse and fine thread series. Coarse thread bolts have deeper threads and need not be manufactured to as close a tolerance as fine thread bolts where care must be exercised to ensure a full thread depth. Fine thread bolts have a smaller helix angle; thus, they have less tendency to loosen when vibration is present.

It is good practice in assembling equipment intended to withstand shock and vibration to tighten a bolt until some yielding occurs in the bottom threads, thus ensuring maximum tightness. This reduces the stress change in a bolt when the loading is tension-directed and increases the resistance to fatigue.[10] Bolts may be preloaded by being

FIG. 43.10. Methods of relieving stress concentration at threads: (A) Smooth notch used to relieve bending stress at thread. (B) Undercut nut to attain more uniform stress distribution in bolt. (*R. A. MacGregor.*[8])

Table 43.3. Effect of Bolt Diameter and Material on Impact Fatigue Strength
(W. Staedel.[7])

Bolt design	Bolt body			Repeated impact fatigue limit					
	Diam., in.	Area, in.^{2}	Area, % root diam. area	Steel A		Steel B		Steel C	
				in.-lb	%	in.-lb	%	in.-lb	%
(15/32 design)	15/32	0.175	230	1.73	100	1.73	100		
(25/64 design)	25/64	0.122	160	2.60	150	2.26	130	~2.2	100
(5/64 R design)	5/16	0.070	100	4.34	250	2.78	160	~3.5	160
(1/8 R design)	9/32	0.060	78	5.21	300	2.95	170	5.21	240
(1/8 R design)	15/64	0.044	57	6.51	375	3.65	210		

	Composition, %					Ultimate, lb/in.^{2}	Elong. %	Red. in area, %	Bolt * ultimate, lb
	C	Mn	P	S	Si				
Steel A.........	0.15	0.65	0.095	0.186	0.04	78,200	12.8	56.0	9,390
Steel B.........	0.25	0.75	0.019	0.024	0.25	96,400	12.8	54.0	11,800
Steel C.........	0.33	0.70	0.039	0.027	0.18	85,300–99,600	16–10	...	11,800

* Static tensile strength of actual bolt (for the second of the designs illustrated above) having 5/32-in. length of free threads under nut.

initially heated, then lightly tightened in place, and finally allowed to cool. It is more common to apply a known torque to the bolt head or nut with a special wrench, and estimate the preloading from the following equation:

$$M = 0.2DF \qquad \text{lb-in.} \tag{43.1}$$

where M = torque applied to bolt, lb-in.

   D = diameter of bolt, in.

   F = preload tension in bolt, lb

Relaxation of bolt tension usually occurs to some extent in service, and a periodic check on preloading is advisable to ensure tight joints.

LOCKING DEVICES. Locking devices [11] include cotter pins, friction nuts, and lock washers. The castellated nut and cotter pin, as well as the locking wire inserted through holes in bolts and nuts, are devices to prevent positively the rotation of the nut relative to the bolt. Such devices must be removed each time the nut is tightened or loosened, and permit adjustment of the nut only at the discrete increments of the locking holes. In another type of locking device, additional thread friction is introduced by the design of the nut; it is a less positive means to retard loosening. Methods of introducing the friction are (1) the bolt is forced through an unthreaded, nonmetallic insert forming part of the nut; (2) the hole in the nut is elliptical and appreciable torque must be applied to turn it on the round bolt; and (3) the inner face of the nut is concave so that when tightened against a shoulder it distorts and binds the threads. In general, means to introduce additional friction are satisfactory when newly installed but tend to deteriorate with repeated use.

The split-ring lock washer is an initially flat washer that has been split and then sprung out of its plane slightly. It is flattened when the nut is tightened, and the resultant friction between washer and nut tends to prevent loosening of the nut. Its resilience enables it to maintain the friction even though the bolt deforms somewhat as a result of shock. Split-ring lock washers may become brittle and fracture; they then fall out of place and a loose joint results. Another type of lock washer embodies a sawtooth edge with an axial set to the teeth; it is adapted to wedge between the nut and the shoulder against which the nut is seated. The teeth are relatively rigid and cannot compensate for the additional space resulting from elongation of the bolt or indentation of the bolted members. This type of lock washer is not suitable for use in equipment subjected to severe shock.

CHASSIS MOUNTING METHODS

To give accessibility to the component parts of an equipment, it is common practice to mount the chassis in the cabinet by hinges or rollers. This imposes the general requirements: (1) the chassis should be movable readily to positions of accessibility, (2) the operators should be protected while the chassis are being so adjusted, and (3) when the chassis are closed, they should be attached securely to the cabinet and should add as much stiffness as possible thereto. Particular considerations that are important with different types of chassis are set forth in the following paragraphs.

DOOR-MOUNTED CONSTRUCTION. Equipment sometimes is designed with many of the parts mounted on a door or on hinged panels to permit better accessibility for maintenance. Two undesirable features of this type of design are (1) the design moves the center-of-gravity of the equipment forward in the cabinet, and (2) it is difficult to stiffen large door panels adequately when they are heavily weighted. Hinges and their associated structures must be designed to support the door adequately in the open position. Dowel pins for the opening side of the door should be provided to relieve the door fasteners from loads when closed. To ensure personnel safety, the opening and closing of heavy doors should be controlled by pivoted lead screws with suitable hand cranks for operation.

RETRACTABLE CHASSIS. It is preferable to mount components on a retractable chassis or directly on the cabinet structure, rather than on a door or hinged panel. There are principal requirements for a satisfactory retractable chassis: (1) the chassis must be constrained from bouncing and (2) the equipment loads must be transferred directly to the frame of the cabinet. (For large and heavy chassis, the additional use of front-panel dowel pins may be desirable. These pins slide into corresponding receiving-hole plates, or receptacles, attached to the chassis.) The front of the chassis is bolted to the frame by screw-type fasteners located around the periphery of the front panel. To eliminate alignment difficulties, the front-panel fasteners screw into "floating" nuts attached to the frame. It is important to place these fasteners at regular intervals around the entire periphery of the front panel to develop maximum stiffness, particularly in the front-to-back direction.

Chassis Slides. Accessibility to the chassis for maintenance and troubleshooting is improved by the use of ball-bearing slides with built-in tilting features. Slides are avail-

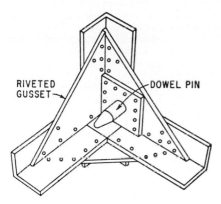

Fig. 43.11. Dowel pin arrangement for riveted or bolted gusset. (*H. M. Forkois and K. E. Woodward.*[1])

Fig. 43.12. Dowel pin arrangement for welded gusset. (*H. M. Forkois and K. E. Woodward.*[1])

able to permit tilting the chassis either up or down, and also to permit locking at a number of fixed angular positions. Since ball-bearing slides are not designed to carry large loads, it is important that the weight of the chassis in the locked-up position be supported entirely by the dowel pins and front-panel fasteners.

Dowel Pins. Dowel pins should be made of steel to avoid excessive abrasion. In general, they vary in diameter from $\frac{1}{4}$ to $\frac{1}{2}$ in., and should be no longer than 1 in. to minimize bending stresses. The pin receptacle should also be made of steel with a close fit to the pin. It usually is better practice to locate the pin on the frame of the cabinet and the pin receptacle on the chassis because the frame is better able to support the pin as a cantilever. Figure 43.11 illustrates a pin attached to a riveted or bolted gusset, and Fig. 43.12 shows a pin installed on a welded gusset. Maintaining the necessary alignment of close-fitting dowels (maximum clearance on diametral fit of 0.005 in.) is difficult, particularly where interchangeability of chassis is a requirement. Figure 43.13 shows a dowel pin clamped to the frame and held in place by serrated flange and washer. A maximum adjustment of $\frac{3}{16}$ in. in any direction is obtainable because of the oversize hole in the frame.

Quick-release Fasteners. Only a fraction of a turn is necessary to lock or release

Fig. 43.13. Adjustable dowel pin. A maximum adjustment of $\frac{3}{16}$ in. in any direction is obtainable as a result of the oversized hole in the frame. (*H. M. Forkois and K. E. Woodward.*[1])

Fig. 43.14. Reduced-shank front-panel thumbscrew which is threaded into a threaded hole in the chassis until the threaded portion passes through the hole; the reduced shank portion then rests in the threaded hole in a captive position which permits turning freely to engage a threaded hole in the frame. (*H. M. Forkois and K. E. Woodward.*[1])

a device in which a spring element is used to maintain tension in the "locked-up" position. Shock and vibration forces of low magnitude tend to overcome the spring force and cause looseness in the assembly. Under shock, the spring or the cam rider (the member which engages the spring and through which the loads are transmitted from the captive mass to the support) may deform or fracture. Hence, quick-release fasteners should not be used to carry large loads but rather may be used to mount such items as inspection plates or covers.

Front-panel Thumbscrews. Figure 43.14 illustrates a captive fastener used to secure the chassis to the frame of the cabinet. It is knurled and slotted to facilitate loosening and tightening. The thumbscrew is threaded into a threaded hole in the chassis until the threaded portion passes through the hole; the reduced shank portion then rests in the threaded hole in a captive position which permits turning freely to engage a threaded hole in the frame. Reduced shank diameters vary according to design requirements but should not be less than $\frac{3}{16}$ in. for lightweight or $\frac{5}{16}$ in. for mediumweight equipment. Corresponding thread sizes are $\frac{1}{4}$ in.-20 and $\frac{3}{8}$ in.-16.

DAMPING

A direct approach to the reduction of vibration resulting from resonance is to supplement the inherent hysteresis damping of materials by other means to dissipate energy. A structure fabricated from a number of component parts exhibits a marked degree of energy dissipation compared with a similar structure formed in a single piece. The method of fabrication influences the degree of energy dissipation. Bolted and riveted joints are most effective in dissipating energy because they permit a limited slippage at the interfaces between members in contact while maintaining pressure at the interface. Welded joints also exhibit considerable damping, less than bolted or riveted joints but substantially more than solid materials. This apparently is the result of some slippage between members in contact not restrained fully by the weld.

ENERGY DISSIPATION. The damping or energy dissipation resulting from slippage of parts in a fabricated structure is termed *slip damping*. The analytical basis for this type of damping is established in Chap. 36, and applications of the theory are discussed. The considerations involved in attaining optimum conditions also are discussed. Many structures encountered in the design of equipment are irregular and complex, and the theory given in Chap. 36 may be applied only with considerable difficulty. Various types of structures have been found to exhibit damping properties that are known within wide and approximate limits. Knowledge of such properties in a quantitative sense would assist in the design of equipment having desirable energy dissipation.

Frictional effects in composite structures are not well adapted to analysis except under ideal conditions. As guidance for the designer, it is convenient to define the damping in terms of the *equivalent viscous damping*.[12] The inverse of the fraction of critical damping ζ is proportional to the amplification at resonance [13] and makes it possible to estimate the severity of a resonance. Table 43.4 gives typical values of the fraction of critical damping ζ found in various types of structures. The damping varies widely from structure to structure of a given type; for example, in a given structure, it varies significantly with a change in tightness of bolts. Thus, Table 43.4 should be interpreted only as an indication of the relative magnitudes of damping in different types of structures, and of the variation in damping that may be expected from a change in design.

Damping can be added to members in flexure by utilizing the shear stress at the neutral plane to deform a layer of viscous material and attain a relatively high degree of energy dissipation. This sometimes is known as *sandwich construction*. The analytical

Table 43.4. Typical Values of Damping for Different Methods of Construction

Method of construction	Effective damping,* ζ
One-piece unit............	0.01
Welded assembly.........	0.02
Riveted assembly.........	0.04
Bolted assembly..........	0.05

* ζ = fraction of critical damping [see Eq. (2.12); also see Chap. 42].

basis for an understanding of this concept is included in Chap. 36. The concept as applied to the design of structural members is illustrated in Fig. 43.15 where a cantilever beam of rectangular cross section (inset A) is compared with a sandwich-type beam (inset B) having the same length and natural frequency.[14] The viscoelastic layer provides a partial constraint in shear between the separated portions of the beam; thus, the natural frequency of the sandwich-type beam is lower than that of a solid beam having the same over-all dimensions, but greater than that of one-half of the sandwich-type beam. The effective value of the fraction of critical damping $\zeta = c/c_c$ for the beam depends upon the viscosity of the viscoelastic layer and decreases as the load carried by the beam increases.

The sandwich-type construction has inherent limitations. For a given over-all dimension, a sandwich-type structure is less rigid than a structure using solid members.

Fig. 43.15. Comparison of structural damping in a conventional aluminum beam with that in a damped aluminum beam having a viscoelastic damping layer. The beam is a cantilever 11 in. long with cross section (A) for conventional beam and (B) for damped beam. Natural frequency is 18 cps. Transmissibility is ratio of amplitude at free end to amplitude at fixed end, during vibration at resonance. (*J. E. Ruzicka.*[14])

From a fabrication viewpoint, a sandwich-type structure is not well adapted to welding; appropriate bolted connections preferably should include means to relieve the viscoelastic layer of bolt forces. Finally, the viscosity of the viscoelastic layer tends to vary with temperature. Nevertheless, a large increase in damping is attainable by use of sandwich-type construction and may justify use of the concept for certain applications.

Members that deflect in flexure also may be damped by the application of vibration damping material that dissipates energy; this concept is discussed in detail in Chap. 37. Its effectiveness is a function of the energy dissipated by the undercoating relative to the elastic energy of the structure being damped. For this reason, it finds principal application to relatively light structures of large area, such as hoods and trunk covers of automobiles or enclosures for household appliances.

When vibration of an equipment or structure is characterized by a large amplitude at a definable point, a possible method of reducing the vibration level involves the attachment of a damper. Such a device dissipates energy as it extends and compresses. A damper can be attached between a vibrating structure and a relatively fixed structure, or between two structures vibrating in opposite directions. The design of dampers is discussed in Chap. 32.

AUXILIARY MASS DAMPERS. A tuned damper or dynamic vibration absorber may be attached directly to the vibrating structure; it consists of a simple mass-spring system that may be damped or undamped. Reduction of vibration is achieved because the mass vibrates out of phase with the vibration to be eliminated; sometimes this is called *sympathetic vibration*. The theory of the tuned damper and a number of applications are discussed in Chap. 6; applications to internal-combustion engines are discussed in Chap. 38.

An auxiliary mass damper without a spring or damping is designated an *acceleration damper*. It consists of small mass elements free to move or slide in a sealed container that is attached to the vibrating system. Incipient vibration is reduced by the multiple collisions and friction between the mass elements with consequent transfer of momentum and conversion of mechanical energy into heat. For example, when electrical contacts are closed, they tend to chatter because of the elastic impact; to minimize the chatter, a small container partially filled with small mass elements is attached to the movable contact. The abrupt motion of the contactor arm tosses the powder about inside the container. The result is a dull impact without rebound, because the relative motion and friction of the mass elements absorb the elastic energy involved in the stress waves set up by the impact.[15]

Analogous results are attainable with a single massive member contained within a box attached to a member whose vibration is to be reduced. Analyses of this damper [16, 17] indicate that its effectiveness is the result of momentum transfer induced by multiple impacts within the box; however, the analytical treatment is not sufficiently complete to permit design of a damper without supplementary test work.

An ideal material to use as the mass elements is tungsten powder. It has rough individual grains to create friction, is hard enough to minimize wear, and is sufficiently massive. A half-and-half mixture of 150–200 and 200–250 mesh powders is commercially available.*

Sand or steel shot may be used as the damping material where the quantity required would make tungsten powder prohibitively expensive. Some experimental results are available for the damping attainable from a flat, rectangular tray partially filled with steel shot of No 15 grit.[18] The tray was attached to a vibrating table and subjected to a vibratory motion in the lengthwise direction of the tray at a frequency of 13 cps. (The latter frequency happened to be of particular interest because it corresponded to the natural frequency of a radar antenna which was being set into vibration by gusts of wind.) The vibration table was put into motion by means of a rotating unbalance weight supported in bearings attached to the underside of the table. Various tray lengths and different weights of shot were tried; the width of the tray was fixed at 3 in.

---

* Westinghouse Electric Corporation, Lamp Division, Bloomfield, N.J., mixture specification 35-8-562.

Figure 43.16A shows that with a given rotating unbalance weight causing a harmonic forcing function $F = 1.7$ lb, the displacement amplitude could be reduced by adding more weight of damping material to the tray. At each of several different tray lengths an optimum weight of damping material was found beyond which the resulting displacement amplitude could no longer be reduced by the addition of damping material. The optimum weight of damping material was found to be larger as the length of the tray was increased. The results plotted in Fig. 43.16A indicate that energy was absorbed primarily because of sliding friction of the mass particles moving as a unit relative to the bottom and sides of the tray.

Figure 43.16B shows the results when the tests were repeated for the same weight of damping material $W = 0.91$ lb but located in trays of various lengths. The magnitude of the rotating unbalance weight was adjusted successively to establish various displacement amplitudes while maintaining the frequency at 13 cps; the equivalent viscous damping coefficient was calculated at each amplitude. For the shorter tray lengths, the damping coefficient increased rapidly as the displacement amplitude increased; however, it reached a peak value and then decreased gradually as the displacement amplitude was further increased. The exception was the 3-in. long tray, which was partitioned into three 1-in. long compartments, and gave the most effective damping action. The rising characteristic for the 3-in. tray shown in Fig. 43.16B indicates that the damping coefficient continues to increase with displacement amplitude and indicates a safe, stable design for the damper. By using partitions to reduce the lengthwise freedom for relative motion, the sliding friction is supplemented by the more effective turbulence of the damping material. The increased energy absorption indicated by the larger damping coefficient was caused principally by the multiple, inelastic collisions. However, depending upon circumstances, the net damping is the result of the combined sliding friction and inelastic collisions of the mass elements.

Fig. 43.16. Performance curves for steel shot damper attached to a table driven by a rotating unbalance at a frequency of 13 cps: (A) Variation of vibration amplitude as a function of weight of damping material for several lengths l of tray; driving force amplitude is 1.7 lb. (B) Damping coefficient as a function of vibration amplitude, calculated from driving force and vibration amplitude. Weight of damping material is 0.91 lb; the tray length $l = 3$ in. is divided into three 1-in. long compartments. ($G. O. Sankey$.[18])

MALFUNCTION OF EQUIPMENT

Failure of equipment as a result of shock and vibration may consist of (1) damage so severe that the ability of the equipment to perform its intended function is impaired permanently or (2) temporary disruption of normal operation in a manner permitting restoration of service by subsequent adjustment of the equipment or termination of the disturbance. For example, a common type of disruption involves excessive vibration of the elements within an electron tube; damage frequently does not occur but rather the electron tube generates spurious signals and is unable to perform its intended function. This difficulty is corrected by applying the principle set forth as a guide to the design of structures to withstand shock and vibration; i.e., a change in stiffness or damping of the electron tube elements is effected. However, another type of disruption may occur in a relay or circuit breaker where shock or vibration causes unintended and improper operation. Normal operation can be restored readily if the equipment is accessible to personnel for adjustment. Meanwhile, serious consequences may have developed from the disruption. The following paragraphs of this section discuss design principles that can be applied to minimize improper operation.

BALANCING OF MOVING MEMBERS

Where possible, pivoted or hinged members should be balanced statically to prevent maloperation. For example, Fig. 43.17 shows a circular rheostat with a single pivot arm

FIG. 43.17. Circular rheostat with a single pivot arm that is statically balanced. (*C. E. Crede and M. C. Junger.*[2])

that is statically balanced by the addition of a counterweight. Figure 43.18 shows a push-button station that is inherently unbalanced and vulnerable to maloperation as a result of accelerations along the line of motion of the push button. The pivoted arm and counterweight are added to prevent maloperation. The counterweight plus the assembly of push button with contacts are statically balanced about the fulcrum of the pivoted arm. Note that the statically balanced mechanisms shown in Figs. 43.17 and 43.18

FIG. 43.18. Statically balanced push button. The center-of-gravity of entire linkage is at fulcrum of counterweight arm. (*W. P. Welch.*[3])

would tend to malfunction if subjected to a shock motion having a strong rotational component.   In general, severe shock tends to involve translational motion primarily, and statically balanced mechanisms give adequate service.

In mechanisms involving more than one pivot, it is possible to make them unresponsive to rotational acceleration about any axis by arranging similar mechanisms to act in opposition or "back to back."   Typical designs having this feature are described in Ref. 2.

MANUAL AND AUTOMATIC LATCHES

Locking devices, such as manual or automatic latches, represent the most direct means of preventing maloperation.   For example, clamps can be added to hold a knife switch

FIG. 43.19. Manual latch with control lever and detent that is not statically balanced. (*C. E. Crede and M. C. Junger.*[1])

blade in either the open or the closed position.   Similarly, mechanisms can be used to block the tripping linkages of circuit breakers and contactors.   Such devices are undesirable expedients, because normal operation of the knife switch is impeded and the protective feature of the circuit breaker is nullified.   However, in some types of mechanisms, a manually operated latch can be used without impeding normal operation seriously. For example, Fig. 43.19 shows a control lever that is not statically balanced.   Means are provided to latch it to a selected notch in a lever guide.   A detent is pivoted to the lever and urged against the guide by a spring.   A sliding rod carried by the lever is depressed by the operator to disengage the detent and allow the lever to be moved to another notch.

Latches may be actuated automatically to prevent maloperation without interfering with normal operation.   Figure 43.20 shows a contactor locking device actuated by the electromagnet of the contactor.   The contacts are closed by energizing the solenoid and causing the armature to pivot against the force of the tension spring.   The magnetic flux from the solenoid rotates the latch against the force of the bias spring, thereby aligning the latch with the hole in the armature and allowing the armature to close in the normal manner.   When the solenoid is deenergized, the armature is blocked open by the latch.   This latch prevents accidental closing of the contacts as a result of acceleration toward the left; it does not prevent accidental opening due to acceleration toward the right.   However, accidental opening is less likely because the magnetic force tends to hold the contacts closed.

Figure 43.21 shows a contactor with a set of armature latches which are shock-actuated to block the contacts in either the open or closed position. The contacts are shown in Fig. 43.21*A* in the open position and are subsequently closed by vertical upward motion of the armature. A pair of armature-latches are mounted on pivots to the left of the

TENSION SPRING
ARMATURE PIVOT
ARMATURE
LATCH
HOLE IN ARMATURE
ELECTRO-MAGNET (SOLENOID)
BIAS SPRING
ELECTRICAL CONTACTS

Fɪɢ. 43.20. Magnetically actuated latch to prevent contact closure under shock when solenoid is deenergized. (*W. P. Welch.*[3])

crossbar. The pivot for the "out" latch shown in Fig. 43.21*B* is located to the right of the center-of-gravity; thus, a downward shock causes the latch to rotate in a clockwise direction against the force of the tension spring and engage the crossbar. This prevents upward motion of the armature and accidental closing of the contacts. Similarly, the "in" latch would rotate in a counterclockwise direction if the contactor is closed as shown in Fig. 43.21*C* and receives an upward shock. Thus, the "in" latch prevents accidental opening of the contactor. This type of latch is successful only if the latches respond more rapidly than the armature to all shock motions that are likely to occur. Since it is impossible to ensure this, the shock-actuated latch involves a degree of unreliability.

FIG. 43.21. (A) Contactor with shock-actuated antishock devices; (B) armature in position of "opened" contacts; (C) armature in position of "closed" contacts. (W. P. Welch.[3])

PRELOAD AND OVERCENTER

Electrical contacts can be prevented from chattering as a result of shock and vibration by interposing a spring that forces one contact against another in closed position. In the contact design shown in Fig. 43.22, the position of the contactor arm relative to the base is fixed when in closed position. This fixed position can maintain the compression spring in a deflected position with a consequent force being applied to the contacts.

FIG. 43.22. Spring preloading of movable contact to prevent separation (opening) due to shock and vibration.

A contactor may be made resistant to accidental opening and closing by the use of preloading springs to hold a crank mechanism in an overcentered position, as illustrated in Fig. 43.23. In either extreme position of the mechanism, the crank is rotated beyond its dead center and the crank pin engages a stop. The mechanism is held in its open position shown by Fig. 43.23A by means of springs A. It is held in its closed position shown by Fig. 43.23B by means of springs B. In the event of either an upward or

FIG. 43.23. Overcentered three-pole contactor in (A) open and (B) closed positions. (C. E. Crede and M. C. Junger.[2])

downward shock, the mechanism would respond only if the acceleration associated with the shock motion has the magnitude and direction necessary to move the crank over dead center against the opposing force of the preloading springs.

IRREVERSIBLE MECHANISMS

Shock resistance often can be obtained by employing an irreversible component or combination of elements. An irreversible component is one which responds only to a force or torque applied at the input side, remaining unresponsive to a force or torque applied at the output side. A common irreversible device is a screw and nut, in which the nut experiences translational motion as a result of a rotation of the screw. However, the screw cannot be made to rotate by applying an axial force to the nut unless the helix angle of the screw thread is made substantially larger than normally used. Similarly,

Fig. 43.24. Contactor actuated by worm and gear. (*C. E. Crede and M. C. Junger.*[2])   Fig. 43.25. Contactor actuated by spiral cam. (*C. E. Crede and M. C. Junger.*[2])

in a worm-and-gear combination, the gear can be rotated by rotating the worm, but usually not the reverse. The condition for reversibility is a function of the helix angle and coefficient of friction at the sliding surfaces.

The application of a worm and gear to a contactor is illustrated in Fig. 43.24. The arm which carries the movable contact is secured to the same shaft as the worm gear. Movement of the contact arm can be attained only by rotation of the worm gear shaft. Note that it is not necessary for the contact arm to be statically balanced to attain resistance to shock.

A contactor which employs a spiral cam is illustrated in Fig. 43.25. The movable contact is carried on one arm of a bell crank. The opposite arm of the bell crank carries a roller which rides in a spiral groove in the disc. The cam is driven with a rotary motion by a motor which controls the action of the contactor. It is not necessary that the bell crank be statically balanced.

DESIGN OF COMPONENTS

A record of damage sustained by equipment during laboratory tests is significant in emphasizing the considerations that are important in attaining resistance to damage from shock and vibration. Compilations of damage experiences are available.[1] These show that failure of principal structures and mounting brackets is the most common form of damage. During shock, structures may not have sufficient strength to withstand the forces that are applied; during vibration, resonant conditions may develop and the relatively undamped structures may fail from fatigue. Failure of electrical leads from fatigue is common, often because the leads are used improperly to support resistors and capacitors. Failure of electron tubes is common, but only because they are used in large quantities in electronic equipment; failure in terms of percentage of tubes used is not large. On the other hand, a relatively large percentage of incandescent lamps and cathode-ray tubes experience failure. In some instances, failure may be ascribed to the inherent properties of the components that failed; in other instances, failure is the result of improper installation. This section discusses design and selection of components to withstand shock and vibration, and the following section discusses the mounting of such components.

BRITTLE MATERIALS

Components which embody brittle material having important functional roles, such as glass, ceramics, and jewels, present particular problems in the protection of such material from fracture under shock. General design principles for the protection of brittle materials are difficult to formulate; however, several examples are given to show the circumstances of typical damage occurrence and corrective measures that have proved effective.

Fɪɢ. 43.26. Section through a flexibly mounted instrument bearing. (*R. D. Hickok.*[19])

Fɪɢ. 43.27. Edgewound resistors showing long, square-ended ceramic sections (*A*) and short round-ended sections (*B*). The latter has improved resistance to shock and vibration.

JEWEL BEARINGS. Damage to brittle materials is likely to occur because an applied force is excessive and/or because it is applied too suddenly. For example, the pivot bearings used for the pointers of instruments usually are made of hard, brittle material that tends to fracture if a large force is applied to it suddenly. Figure 43.26 shows a spring mounting for the bearing to reduce the severity and suddenness of load application, thereby minimizing the possibility of blunting the pivot or splitting the jewel. The shoulder on the shaft is provided to limit the axial travel so that the bearing cannot bottom forcibly in its retainer.[19]

GLASS AND CERAMICS. A brittle member may be damaged because it is constrained to experience a large deflection as a consequence of its attachment to a large

Fɪɢ. 43.28. Steps in development of correct strength and flexibility of the flange mounting for a glass window in a small instrument. (*C. E. Crede and M. C. Junger.*[2])

flexible member inherently capable of experiencing a greater deflection. For example, Fig. 43.27 shows an edge-wound resistor having a steel core capable of deflecting laterally when subjected to shock. The original long, square-ended ceramic sections shown in Fig. 43.27A tend to chip and split because they are incapable of participating in the lateral deflection of the core. Replacement of the original ceramic sections by shorter sections having rounded ends to permit knee action bending, as shown in Fig. 43.27B, increases the capability of the resistor to withstand shock.

Figure 43.28 shows the evolution in the design of a glass instrument window capable of withstanding severe shock. The glass window is sealed to a mounting flange, and must experience an equivalent distortion. In the design of Fig. 43.28A, the mounting flange was too flexible and its subsequent distortion under shock was greater than the

FIG. 43.29. Examples of shockproof designs and installations of incandescent lamps: (A) rubber insert between screw base and glass bulb; (B) rubber-covered socket and flexible cable; (C) flexible U-shaped flat spring mounting. (C. E. Crede and M. C. Junger;[2] W. P. Welch.[3])

glass could withstand. When the reinforcing ring shown in Fig. 43.28B was introduced to limit the distortion of the mounting flange, the glass failed because the forces transmitted by the rigid mounting arrangement to the glass were too great. A modified reinforcing ring, shown in Fig. 43.28C, reduced the magnitude of both the transmitted force and the mounting flange distortion, thereby minimizing damage to the glass window.

Figure 43.29 illustrates several methods of protecting incandescent lamps. The lamp may embody a rubber insert which separates the screw base from the bulb and filament, as shown in Fig. 43.29A. Flexible electrical conductors extend through the rubber insert and connect the filament to the screw base. Alternatively, the socket may be flexibly mounted to permit the use of a conventional lamp. For example, a rubber-covered socket may be suspended by a short length of flexible cable from an overhead structure, as shown in Fig. 43.29B. Excessive swinging may be prevented by using a guard ring attached to the rubber socket. Where excessive relative motion between the lamp and fixture must be avoided, the entire lighting fixture may be mounted on flexible U-shaped flat springs, as shown in Fig. 43.29C.

MECHANICAL MEMBERS OF INHERENT FLEXIBILITY

Equipment that includes long shafts is particularly susceptible to malfunction or other impairment of operation as a result of lateral deflection of a shaft. A shaft with a long distance between supporting bearings is inherently capable of relatively large lateral

TABS PUNCHED AND BENT FROM LIGHT PANEL

POINTER SHAFT

Fig. 43.30. Means to limit shaft deflection. (*C. E. Crede and M. C. Junger.*[2])

deflection without damage to the shaft itself; however, other damage may result from the deflection. For example, the shaft that carries the pointer of an electrical indicating instrument may deflect sufficiently to unseat the ends of the shaft from the jewel bearings. This type of damage may be minimized by the method illustrated in Fig. 43.30; it consists of tabs normally spaced from the shaft but arranged to prevent excessive deflection in event of severe shock. This concept may be applied to shafts having gears where a large lateral deflection of a shaft may cause temporary disengagement of gears, with the possibility of re-engagement in an incorrect relation.

WIRING AND CIRCUIT BOARDS

WIRES, CABLES, AND CONNECTORS

Wires, cables, and connectors are not in themselves susceptible to damage from shock and vibration. They can cause considerable difficulty, however, if improper methods are used in their application. Wires and cables have a high degree of flexibility as compared with other electronic component parts. This flexibility results from (1) the type of construction, (2) the kind of material, and (3) the high ratio of length to cross-sectional area. Because of this flexibility, the natural frequency of a span of wire or cable is low and may easily fall within the frequency range of existing vibration. The vibration of a wire or cable produces time-varying stresses which can lead to fatigue failure. The areas which usually are first to fail are the ends of the vibrating span, i.e., at the wire or cable terminations. Failures also may occur where the insulation rubs against an adjacent part, or where one conductor in a multiconductor cable rubs against another within the cable body. A cable passing through a hole in a partition should be protected by a grommet lining the hole.

Where several wires extend in the same direction, their stiffness may be increased by harnessing the wires with lacing cord to form an integral bundle. Harnessing also provides damping because of friction between the wires. The harnessed length should be supported adequately so that its weight will not load the wire termination points unduly. The relatively high degree of damping of stranded wire and cable is beneficial when they are subjected to flexing as a result of vibration[20] (see the discussion of *Cables* in Chap. 15).

MATERIALS AND CONSTRUCTION. Although all wire and cable are flexible when compared to other components, the degree of flexibility depends in part upon the material composition. The flexibility may vary widely with temperature, since many organic insulating materials stiffen considerably with decreased temperature and soften with increased temperature. Fluorinated hydrocarbons and silicone elastomers maintain flexibility at low temperatures better than most others.

LEAD LENGTH. The bending of a wire or cable may be reduced at its termination or support by shortening the length of lead between supports. This applies only in cases where the supports *do not* move relative to one another. The shortening of the span of wire or cable accomplishes two beneficial results: (1) it increases the natural frequency of the span and (2) it reduces the weight and inertial loads at each support. Where supports or terminals *do* move relative to one another, lead failures can be reduced considerably by allowing some slack in the lead. Lengthening a lead slightly (by about 20 per cent of its stretched length), as shown in Fig. 43.31, reduces the failure hazard considerably. The additional length, although lowering the natural frequency, prevents repeated straining of the lead. The use of stranded, insulated wire provides damping to limit the amplitude of vibration during resonant conditions.

METHODS OF ATTACHMENT. Wire or cable failures due to shock and vibration usually take place where the lead is interrupted, either at a termination point or at a support along the lead. Failures occur primarily because these locations are subjected to the greatest bending stress. When wire is terminated by soldering, the area most susceptible to shock and vibration lies between the end of the insulation and the rigidly "frozen" section at the terminal post shown in Fig. 43.32. This portion of the wire is subjected to severe flexing and undergoes fatigue very quickly. Figure 43.32 indicates how solder flows up the strands of the conductor by capillary action and concentrates all bending action into a very small area of wire. Clamping the wire to a structure can restrict the motion and reduce damage from flexing at the termination, as shown in Fig. 43.33.

Fig. 43.31. Examples of jumper leads: (*A*) Tightly stretched solid wire liable to failure if terminals experience relative motion. (*B*) Slightly bowed stranded wire with insulation; the extra length prevents excessive stress in the wire, and the damping provided by friction and insulation prevent large amplitudes of vibration. (*R. E. Barbiere and W. Hall.*[20])

Fig. 43.32. View of cable end illustrating flow of solder along strands to form a rigid mass of cable. Failure is most likely to occur at edge of solder-impregnated portion. (*R. E. Barbiere and W. Hall.*[20])

When wire is terminated by an end support lug, the lug should grip both the insulation and the conductor. When cable is terminated by a connector, the connector should be attached to the cable so that it grips the cable jacket to restrict flexing of individual wires within the cable body. Attaching a sleeve as shown in Fig. 43.34 prevents severe cable bending at a termination point. Bending of individual wires can be restricted by potting the cable end permanently into the connector. The connections are made in a normal manner; then the connector body is filled with a self-curing plastic material to encapsulate the end of the cable completely. The disadvantage to potting a cable is that potted components are difficult to repair.

PRINTED WIRING BOARDS. Printed wiring boards usually require a stiffening structure to prevent low resonant frequencies. The stiffening may take the form of a metal chassis, conformal coating, or complete potting in plastic. Small wiring boards may be rigid enough without additional stiffening. To apply conformal coating, the side of the board having the component parts is dipped into a clear plastic material that hardens to form a rigid assembly. Resistors and capacitors are mounted to the board by leads, and the transformers and other heavy parts are screwed or riveted directly to the board. Printed circuit boards can also be made rigid by complete potting in plastic.

FIG. 43.33. Examples of preferred (A) and inferior (B) methods of clamping cable ends. The relatively great length of unsupported cable is liable to experience excessive vibration. (R. E. Barbiere and W. Hall.[20])

FIG. 43.34. Restricting sleeve to prevent severe cable bending at a termination point. (R. E. Barbiere and W. Hall.[20])

DESIGN TO MINIMIZE VIBRATION

This section discusses means to reduce vibration which may occur during normal operating conditions of certain types of equipment. It is possible to reduce vibration by (1) eliminating the forcing function by counterbalancing (see Chap. 39), (2) isolating the vibration with isolators (see Chaps. 30 to 34), (3) increasing the damping of structural members (see Chap. 37), or (4) adding tuned dampers (see Chap. 6). Examples of the application of these techniques are given below. Further examples are given in Chap. 40 relative to the design of machine tools.

BEARINGS

Vibration of machinery may result from sleeve or journal bearings, and from antifriction bearings using balls or rollers. The conventional oil-lubricated, babbitted bearings often are sufficiently quiet and vibration-free. An occasional instability in a lightly loaded bearing, sometimes referred to as "oil-whip," usually can be cured by relocating oil grooves, reducing bearing clearances, or using an oil of different viscosity. Sleeve bearings made from Teflon, nylon, or similar materials may be used when loads are light and extra quiet operation is required.

Antifriction bearings must have small machining tolerances and superior surface finishing to be quiet. Inaccuracies in machining may lead to vibration. For example, the small induction motor shown in Fig. 43.35A was equipped with precision ball bearings and vibrated at a number of relatively low harmonic speeds as the de-energized motor coasted from its normal operating speed to a standstill. The vibration was a maximum in the axial direction, the vibration frequency was approximately the same value regardless of the motor speed, and an audible hum accompanied the motion. The axial mode of the system consisted of the motor frame vibrating in opposition to the rotor; the ball bearings at either end of the rotor provided nonlinear spring effect, as shown schematically in Fig. 43.35B. Usually, the outer race of one antifriction bearing was tight in the motor end bracket; the other bearing had a sliding fit but was "seated" and carried a small thrust load. The vibration was caused by minor discrepancies in the outer race of the press-fit ball bearing. It was not practical to require more precise machining operations. Instead, a relatively loose sliding fit was specified on both ball bearings so that the motor frame and rotor were no longer "spring-coupled" and the axial resonance could not materialize. A spring of low stiffness, such as a wavy washer or a rubber O ring, may be added to achieve a more definite assembly of the axial system.[22]

FIG 43.35. A-C motor with (A) tight and loose-fit antifriction bearings, and (B) equivalent axial vibration system. (*E. G. Fischer.*[22])

GEARS

Gears are a common source of vibration. They can be fabricated from nylon, rawhide, or plastic-impregnated layers of textile material when loads are light and quiet operation is desired. A compliant material sometimes may be used to separate the tooth-carrying rim from the web and hub. Usually, improper design and inaccurate manufacture are major causes of gear noise.[23]

Friction is an important source of noise, and can be reduced by adequate lubrication. For example, a dangerous situation involving cracked turbine discs was discovered in a geared-turbine drive for a low-speed induced-draft fan on a steam boiler. Investigation revealed high-frequency torsional vibration of the turbine shaft and corresponding harmonic tooth spacing patterns in the high-speed pinion. An abrasive type of fly ash had leaked into the oil lubrication system for the gear case. In the presence of this "grinding compound," the natural mode of torsional vibration of the turbine shaft and discs provided a feed-back mechanism for accentuating the original hobbing errors in the pinion. A filter should have been used to ensure a supply of clean oil for the gears.

VIBRATION ISOLATION

Periodic forces sometimes are unavoidable in machinery, and it may be necessary to isolate the resultant vibration from the foundation or from other parts of the equipment. In electrical machines the predominant vibration and noise are caused by periodic magnetic forces in the air gap between the rotor and stator. The excitation forces can be minimized by enlarging the air gap and by skewing the rotor and stator slots. The natural frequencies for the pertinent rotor and stator modes of vibration should be outside the range of known excitation frequencies.[24] Resilient supports may be used to isolate any remaining structure-borne vibration from the foundation.

REFERENCES

1. Forkois, H. M., and K. E. Woodward: "Design of Shock- and Vibration-resistant Electronic Equipment for Shipboard Use," U.S. Navy Department, Bureau of Ships, *NAVSHIPS* 900, 185A, 1957. Also, *U.S. Naval Research Lab. Rpt.* 4789, 1956.
2. Crede, C. E., and M. C. Junger: "A Guide for the Design of Shock Resistant Naval Equipment," U.S. Navy Department, Bureau of Ships, *NAVSHIPS* 250-660-30, 1949.
3. Welch, W. P.: "Mechanical Shock on Naval Vessels," U.S. Navy Department, Bureau of Ships, *NAVSHIPS* 250-660-26, 1946.
4. Dinner, H., and W. Felix: *Engineers Digest*, **3**(2):85 (1946).
5. Almen, J. O.: *Prod. Eng.*, **22**(4):153 (1951).
6. ASME Handbook: "Metals Engineering—Design" McGraw-Hill Book Company, Inc., New York, 1953.
7. Staedel, W.: *Mitt. Materialprüfungsanstalt Berlin-Lichterfelde West*, **4** (1933).

8. MacGregor, R. A., W. S. Burn, and F. Bacon: *Trans. North East Coast Inst. Engrs. & Shipbuilders*, **51** (1935).

9. Battelle Memorial Institute: "Prevention of Failure of Metals under Repeated Stress," Appendix 10, John Wiley & Sons, Inc., New York, 1941.

10. Almen, J. O.: *Trans. SAE*, **52**(4):151 (1944).   Also see *SAE Special Publication* 23.

11. Engineering Experiment Station, U.S. Navy Department, Annapolis, Md., Lock-Nut Test *Repts.* 7224, 7226, 9321, B-3984 and B-6180.

12. Jacobson, L. S.: *Trans. ASME*, APM-52-15, **52** (1930).

13. Den Hartog, J. P.: "Mechanical Vibrations," 4th ed, p. 87, McGraw-Hill Book Company, Inc., New York, 1956.

14. Ruzicka, J. E.: Paper No. 100Y, presented at *SAE National Aeronautic Meeting*, Oct. 5–9, 1959.

15. Johnson, W. V.: *Westinghouse Engineer*, May, 1942.

16. Grubin, C.: *Trans. ASME*, **78**:373 (1956).

17. Leiber, P., and D. P. Jensen: *Trans. ASME*, **67**:523 (1945).

18. Sankey, G. O.: "Some Experiments on a Particle or Shot Damper," unpublished memorandum, Westinghouse Research Laboratories, Pittsburgh, Pa., 1954.

19. Hickok, R. D., et al.: U.S. Patent 2,427,529, Sept. 16, 1947, for "Meter Pivot."

20. Barbiere, R. E., and W. Hall: "Electronic Designer's Shock and Vibration Guide for Airborne Applications," *WADC Tech. Rept.* 58-363, ASTIA No. AD 204095, December, 1958.

21. "Solid-State Power Circuits," RCA Designer's Handbook, Technical Series SP-52, RCA Solid State Division, Somerville, N.J., September, 1971.

22. Fischer, E. G.: *Conference Paper CP58-200, AIEE Winter Meeting*, New York, February 2–7, 1958.

23. Moeller, K. G. F.: "Gear Noise," chap. 23, in "Handbook of Noise Control," 1st ed., C. M. Harris (ed.), McGraw-Hill Book Company, Inc., New York, 1957.

24. Fehr, R. O., and D. F. Muster: "Electric Motor and Generator Noise," chap. 30, in "Handbook of Noise Control," 1st ed., C. M. Harris (ed.), McGraw-Hill Book Company, Inc., New York, 1957.

44

EFFECTS OF SHOCK AND VIBRATION ON MAN

Henning E. von Gierke
Aerospace Medical Research Laboratory, USAF Dayton, O.

David E. Goldman
The Medical College of Pennsylvania and Hospital, Philadelphia, PA

INTRODUCTION*

This chapter considers the following problems: (1) the determination of the structure and properties of the human body regarded as a mechanical as well as a biological system, (2) the effects of shock and vibration forces on this system, (3) the protection required by the system under various exposure conditions, the means by which this protection is to be achieved, and (4) tolerance criteria for shock and vibration exposure. Man, as a mechanical system, is extremely complex and his mechanical properties readily undergo change. There is very little reliable information on the magnitude of the forces required to produce mechanical damage to the human body. To avoid damage to humans while obtaining such data, it is necessary to use experimental animals for most studies on mechanical injury. However, the data so obtained must be subjected to careful scrutiny to determine the degree of their applicability to humans, who differ from animals not only in size but in anatomical and physiological structure as well. Occasionally it is possible to obtain useful information from situations involving accidental injuries to man, but while the damage often can be assessed, the forces producing the damage cannot, and only rarely are useful data obtained in this way. It is also very difficult to obtain reliable data on the effects of mechanical forces on the performance of various tasks and on subjective responses to these forces largely because of the wide variation in the human being in both physical and behavioral respects. Measurement of some of the mechanical properties of man is, however, often practicable since only small forces are needed for such work. The difficulty here is in the variability and lability of the system and in the inaccessibility of certain structures.

One section of this chapter is devoted to a discussion of methods and instrumentation used for mechanical shock and vibration studies on man and animals. Subsequent sections deal with the mechanical structure of the body, the effects of shock and vibration forces on man, the methods and procedures for protection against these forces, and safety criteria.

DEFINITIONS AND CHARACTERIZATION OF FORCES

CHARACTERIZATION OF FORCES. Forces may be transmitted to the body through a gas, liquid, or solid. They may be diffuse or concentrated over a small area.

* The contributions of J. T. Shaffer, Aerospace Medical Research Laboratory, USAF, are gratefully acknowledged.

They may vary from tangential to normal, and may be applied in more than one direction. The shape of a solid body impinging on the surface of the human is as important as the position or shape of the human body itself. All these factors must be taken into account in comparing injuries produced by vehicle crashes, explosions, blows, vibration, etc. Laboratory studies often permit fairly accurate control of force application, but field situations are apt to be extremely complex. Therefore it is often very difficult to predict what will happen in the field on the basis of laboratory studies. It is equally difficult to interpret field observations without the benefit of laboratory studies.

SHOCK. The term "shock" is used differently in biology and medicine than in engineering; therefore one must be careful not to confuse the various meanings given to the term. In this chapter the term *shock* is used in its engineering sense as defined in Chap. 1 of this Handbook. A *shock wave* is a discontinuous pressure change propagated through a medium at velocity greater than that of sound in the medium. In general, forces reaching peak values in less than a few tenths of a second and of not more than a few seconds' duration may be considered as shock forces in relation to the human system.

The term *impact* (i.e., a *blow*) refers to a force applied when the human body comes into sudden contact with a solid body and when the momentum transfer is considerable, as in rapid deceleration in a vehicle crash or when a rapidly moving solid body strikes a human body.

VIBRATION. Biological systems may be influenced by vibration at all frequencies if the amplitude is sufficiently great. This chapter is concerned primarily with the frequency range from about 1 cps to several hundred cycles per second. (Studies at higher frequencies are very useful for the analysis of tissue characteristics.)

METHODS AND INSTRUMENTATION

Most quantitative investigations of the effects of shock and vibration on humans are conducted in the laboratory in controlled, simulated environments. Meaningful results can be obtained from such tests only if measurement methods and instrumentation are adapted to the particular properties of the biological system under investigation to ensure noninterference of the measurement with the system's behavior. This behavior may be physical, physiological, and psychological although these parameters should be studied separately if possible. The complexity of a living organism makes such separation, even assuming independent parameters, only an approximation at best. In many cases if extreme care is not exercised in planning and conducting the experiment, uncontrolled interaction between these parameters can lead to completely erroneous results. For example, the dynamic elasticity of tissue of a certain area of the body may depend on the simultaneous vibration excitation of other parts of the body; or the elasticity may change with the duration of the measurement since the subject's physiological response varies; or the elasticity may be influenced by the subject's psychological reaction to the test or to the measurement equipment.

Control of, and compensation for, the nonuniformity of living systems is absolutely essential because of the variation in size, shape, sensitivity, and responsiveness of people, and because these factors for a single subject vary with time, experience, and motivation. The use of adequate statistical experimental design is necessary and almost always requires a large number of observations and carefully arranged controls.

PHYSICAL MEASUREMENTS

In determining the effects of shock and vibration on man, the mechanical force environment to which the human body is exposed must be clearly defined. Force and vibration amplitudes should be specified for the area of contact with the body. Vibration measurements of the body's response should be made whenever possible by noncontact methods. X-ray methods can be used successfully to measure the displacement of internal organs. Optical, cinematographic, and stroboscopic observation can give the displacement amplitudes of larger parts of the body. Small vibrations sometimes can be measured without contact by capacitive probes located at small distance from the (grounded) body surface (see *Variable Capacitance Pickups*, Chap. 14). If vibration pickups in contact with the body are used, they must be small and light enough so as not to introduce a distorting mechanical load. This usually places a weight limitation on the pickup of a few grams or less, depending on the frequency range of interest and the effective mass to which the pickup is attached. Figure 44.1 illustrates the effect of mass and size on the response of accelerometers attached to the skin overlying soft tissue.[1] The lack of rigidity of the human body as a supporting structure makes measurements

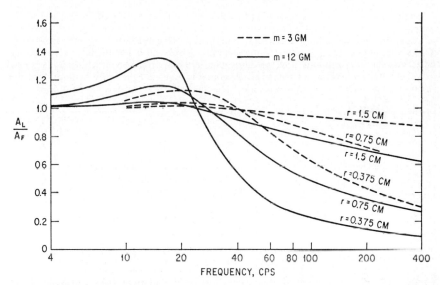

FIG. 44.1. Amplitude distortion due to accelerometers of different mass m and size which are attached to body surface over soft tissue of human subject exposed to vibration. The graph gives the ratio A_L/A_F of the response of the loaded to the unloaded surface for accelerometers having three different radii r. (*Values calculated from mechanical surface impedance data of Franke and von Gierke.*[1])

of acceleration usually preferable to those of velocity or displacement. The mechanical impedance of a sitting, standing, or supine subject is extremely useful for calculating the vibratory energy transmitted to the body by the vibrating structure.[2] The mechanical impedance of small areas of the body surface can be measured in many different ways (see Chap. 10), for example by vibrating pistons,[3] resonating rods, and acoustical impedance tubes.[4]

If the entire body is exposed to a pressure or blast wave in air or water, exact definition of the pressure environment is essential. The pressure distribution should be measured if possible.[5] If the environment deviates from free-field conditions it should be carefully specified because of its effect on peak pressure and pressure *vs.* time-history.[6]

BIOLOGICAL MEASUREMENTS

Instrumentation for the measurement of physiological properties such as blood pressure, respiration rate or depth, heart potentials, brain potentials, or galvanic skin response [7] must be carefully checked for freedom from artifacts when the subject, the instrument, or both are exposed to vibration, intense sound, or acceleration. For example, a bioelectronic harness,[8] such as is used in aviation medicine, is very useful for measurements of various physiological functions, but the proper functioning of instruments employed in such measurements must be verified for each environment.

If psychological experiments during exposure to the mechanical stimuli are planned, adherence to established procedures for such subjective tests is an absolute necessity to obtain valid results. The maintaining of a neutral situation with uniform motivation, subject instruction, and adequate statistical design of the experiment are some of the most important considerations. Care must be taken that the subject be not biased or influenced by environmental factors not purposely included in the test (e.g., the noise of a vibration table can act as a disturbing factor in a study of vibration effects).

SIMULATION OF MECHANICAL ENVIRONMENT

The desire to study the physical, physiological, and psychological responses of biological specimens in the laboratory, under well-controlled conditions, has led to the use of standard and specialized shock and vibration testing machines for experiments on man and animals. A summary of some machines employed in such tests is given in Fig. 44.2. An accurate simulation of the environmental conditions to which man is exposed frequently is not feasible for technical or economic reasons or may even be undesirable because of a need for more systematic investigation under somewhat simplified con-

TYPE OF MACHINE	APPLICATION OF FORCE	FORCE–TIME FUNCTION	FREQUENCY RANGE	MAXIMUM AMPLITUDE	REF NO.
SHAKE TABLE			MECHANICAL 0–50 CPS	UP TO 15g	10 2
			ELECTRODYNAMIC 15–1,000 CPS		
VERTICAL ACCELERATOR			0–10 CPS	±10 FT 3.7g PEAK	11
SHOCK MACHINE			DOWN TO T = 0.16 SEC $\tau = 2.10^{-3}$ SEC	PEAK AMPLITUDE 10^{-2} TO 10^{-1} CM	12
HORIZONTAL OR VERTICAL DECELERATOR OR ACCELERATOR (SLEDS ON TRACKS, CAR, DROP-TOWER)			RATE OF ACCELERATION UP TO 1,400 g/SEC	40g PEAK	9 13 14 15
BLAST TUBE FIELD EXPLOSION					11, 18 5 17
SIREN (AIRBORNE SOUND)			25–100,000 CPS	160–170 DB RE 0.0002 μBAR	
RESPIRATOR			0–15 CPS		19
VIBRATOR (SMALL PISTON)			0–10 MCPS		20 3 21
SHAKER ON CENTRIFUGE	ALTERNATING g FORCE IN ADDITION TO STATIC g FIELD		0–3 CPS		22
HEAD IMPACT MACHINE FOR DUMMY HEADS			DEPENDING ON STRUCTURE STRUCK	IMPACT VELOCITY 140 FT/SEC	23

Fig. 44.2. Summary of characteristics of shock and vibration machines used for human and animal experiments. Force-time functions are indicated schematically. Frequency range and maximum amplitudes refer to values used, not to capabilities of machines. References are to papers describing use of the machines for biological purposes.

ditions. Thus most investigations are limited to the study of a single degree-of-freedom at a time in which the human test specimen is vibrated only in one direction. Many fundamental studies are performed with sinusoidal forces. Usually mechanical and electrodynamic shake tables are employed for this purpose. Requirements for all shock and vibration machines used include: adequate safety precautions, safe and accurate control of the exposure, and sufficient load capacity for subject, seat, and instrumentation. Since the law of linear superposition is valid only in the linear physical domain, sinusoidal forces alone are not adequate for the study of nonlinear physical responses or physiological and psychological reactions to complex force functions. Therefore, some of the machines listed are designed uniquely for exposure of humans. One vertical accelerator, for example, employs a friction-drive mechanism to permit the simulation

FIG. 44.3. Ranges of time and acceleration obtainable with certain high-acceleration devices for tests on man. (*After Clark and Gray.*[24])

of large amplitude sinusoidal and random vibrations such as those encountered in buffeting during low-altitude high-speed flight or anticipated during the launch and reentry phases of spacecraft.[11] This device can be programmed with acceleration recordings obtained under actual flight conditions. Other machines for the study of human tolerance to ejection from high-speed aircraft (ejection seat) have upward or downward acceleration tracks with sliding seats projected by explosive charges. Horizontal tracks with rocket-propelled sleds, which can be stopped by special braking mechanisms, have been used to study the effects of linear decelerations similar to those occurring in automobile or aircraft crashes.[9] Studies of combinations of static acceleration and vibration have been carried out by mounting vibrators on centrifuges. Blast tubes, sirens, and body respirators are used to study the body's response to pressure distributions surrounding it. At low frequencies, the respirator is valuable in studying the response of the lung-thorax system. Small vibrating pistons, which are available for a wide frequency range, have been used in investigating the mechanical impedance of small surfaces, the transmission of vibration and the physiological reaction to localized excitation. The range of some high-acceleration devices is shown in Fig. 44.3.

SIMULATION OF HUMAN SUBJECTS

The establishment of limits of human tolerance to mechanical forces, and the explanation of injuries produced when these limits are exceeded, frequently requires experimentation at various degrees of potential hazard. To avoid unnecessary risks to humans, animals are used first for detailed physiological studies. As a result of these studies, levels may be determined which are, with reasonable probability, safe for human subjects. However, such comparative experiments have obvious limitations. The different structure, size, and weight of most animals shift their response curves to mechanical forces into other frequency ranges and to other levels than those observed on humans. These differences must be considered in addition to the general and partially known physiological differences between species. For example, the natural frequency of the thorax-abdomen system of a human subject is between 3 and 4 cps; for a mouse the same resonance occurs between 18 and 25 cps. Therefore maximum effect and maximum damage occur at different vibration frequencies and different shock-time patterns in a mouse than in a human. However, studies on small animals are well worth making if care is taken in the interpretation of the data and if scaling laws are established. Dogs, pigs, and primates are used extensively in such tests.

Many kinematic processes, physical loadings, and gross destructive anatomical effects can be studied on dummies which approximate a human being in size, form, mobility, total weight, and weight distribution in body segments.[25] Several such dummies are commercially available. In contrast to those used only for load purposes, dummies simulating basic static and dynamic properties of the human body are called "anthropometric" or "anthropomorphic" dummies. They are used extensively in aviation and automotive crash research. In other studies dummies are used in place of human subjects to evaluate protective seats and harnesses. In such dummies, an attempt is made to match the "resiliency" of human flesh by some kind of padding. However, they are crude simulations at best and their dynamic mechanical properties are, if at all, only reasonably matched in a very narrow low-frequency range. This and the passiveness of such dummies constitute important mechanical differences between them and living subjects.

Efforts have been made to simulate the mechanical properties of the human head in order to study the physical phenomena occurring in the brain during crash conditions.[23] Although these head forms only approximate the human head, they are very useful in the evaluation of the protective features of crash, safety, and antibuffet helmets. Plastic head forms, conforming to standard head measurements, are designed to fracture in the same energy range as that established for the human head. A cranial vault is provided to house instrumentation and a simulated brain mass with comparable weight and consistency (a mixture of glycerin, ethylene glycol, etc.). The static properties of the skin and scalp tissue are simulated with polyvinyl foam.

The static and dynamic breaking strength of bones, ligaments, and muscles and the forces producing fractures in rapid decelerations have been studied frequently on cadaver material.[26, 27, 28] Extreme caution must be exercised in applying elastic and strength properties obtained in this way to a situation involving the living organism. The differences observed between properties of wet, dry, and embalmed materials are considerable;[29] changes in these properties also result in changes in the force distribution of a composite structure. Thus a multitude of physiological, anatomical, and physical factors must be considered for each specific situation, in which the use of animals, dummies, or cadavers as substitutes for live human subjects is planned.

PHYSICAL CHARACTERISTICS OF THE BODY

ANATOMY

Structurally, the body consists of a hard, bony skeleton whose pieces are held together by tough, fibrous ligaments and which is embedded in a highly organized mass of connective tissue and muscle. The soft visceral organs are contained within the rib cage and the abdominal cavity.

An outline of the skeleton is shown in Fig. 44.4. The slightly bowed vertebral column is the central structural element. It consists of a number of individual vertebrae which have roughly cylindrical load-bearing elements separated by fibrocartilaginous pads. Near the lower end several vertebrae are fused together to form the sacrum which is a tightly fitting part of the pelvic girdle. The skull rests at the top of the column and is held in place by muscle and connective tissue as well as by ligaments. At the bottom of each side of the pelvic girdle is a roughly hemispherical hollow into which the head of the femur fits. Below the femur are the tibia and fibula, which in turn rest on the ankle and foot bones.

The intervertebral discs are dense fibrocartilaginous pads. The hip, knee, and ankle joints have cartilaginous layers on their articular surfaces as do also the joints of the upper

FIG. 44.4. Diagram of human skeleton. (*Goldman.*[34])

limbs. The foot has a tough connective tissue pad at the heel and an arrangement of bones which acts to distribute applied loads. All the joints are held together by ligaments which are flexible but relatively inextensible. These ligaments form a more or less crisscross lacing which permits movement of the joints in a suitable direction without stretching the ligaments themselves to an appreciable extent. The sacroiliac joint is held tightly and almost immovably. The rib cage and shoulder girdle are also dependent on muscle and connective tissue for their support.

In the "ideal" standing position, a plumb line through the center-of-gravity of the body passes through the lower lumbar and upper sacral vertebrae, slightly behind the hip joint sockets and a bit in front of the knee and ankle joints. Upward, the line passes in front of the thoracic curve of the vertebral column and through the support at the base of the skull. Vertical thrusts may be taken up by the compression of the joint pads and by bending of the vertebral column. There is often a slight forward turning moment at the pelvis, especially in older adults. Small deviations from postural symmetry may result in a markedly asymmetrical force distribution under the action of vertical thrusts.

The body musculature, supported from the skeleton by tendons and held together by a network of connective tissue fibers, forms a secondary supporting structure for the skeleton and joints. Fat and skin also contain connective tissue.

Fig. 44.5. Diagram of position of human viscera. (*Goldman.*[34])

In compression, soft tissues resemble water in their mechanical properties, but in shear they approximate stiff, nonlinear gels with internal losses.

The visceral organs (Fig. 44.5), contained within the thoracic cage and abdominal cavity, are soft tissue elements, separately encapsulated to slide freely over each other; they are supported individually by membranes and ligaments and collectively by the bone, muscle, and connective tissue surroundings. Their weights range from a fraction of an ounce to several pounds and most of the supporting membranes are quite flexible. The kidneys are embedded in fatty tissue and held by a connective tissue sheath in a depression in the posterior wall of the abdominal cavity. The stomach is supported by the esophagus; its displacement is restricted by the dome-shaped diaphragm, which is a large sheet of muscle separating the chest cavity from the abdominal cavity. The lungs, filled with tiny air sacs, are held in place by a combination of supports including a pressure differential. The heart is held in place by ligaments stretched longitudinally through the chest cavity and by large blood vessels. Considerable support also is provided by the diaphragm.

The brain and spinal cord have special protection. The brain is surrounded by liquid contained mostly in the subarachnoid space inside the skull. The spinal cord runs longitudinally through holes in the vertebrae lined with heavy ligaments and a membrane which forms a liquid-filled tube.

Fluid in the body consists of (1) 5 to 6 liters of blood in the heart, arteries, veins, and capillaries; (2) 100 to 150 cm^3 of cerebrospinal fluid surrounding the brain and spinal cord and also within the ventricular cavities of the brain; (3) the interstitial fluid found everywhere in the body as a bathing fluid for cells and tissues but nowhere in large reservoirs; and (4) liquid contained from time to time in the stomach, intestines, and bladder. Gas occurs in the sinuses of the skull, the oronasal cavity, the trachea, the lungs, and often in the stomach and intestines. The latter two organs often contain solid matter as well.

The cells of which the body is made have a wide variety of size and shape. Diameters range from about 0.001 in. to 0.01 in. Shapes may be spherical, disclike, columnar, flat or highly irregular. Many cells have filamentous processes projecting from them. The internal structure of cells is also very complex, containing salts, protein, carbohydrates, and many other substances. Approximately 60 to 80 per cent of the cell is water. Nuclei and other inclusion bodies are found. The rest of the cell is a viscous solution or gel with evidence of considerable submicroscopic structure.

Soft tissues consist of cells, held together by connective tissue and by intercellular links. Blood is a liquid containing nearly 50 per cent by volume of disc-like red cells, together with a few white cells. Soft tissues exhibit a wide variety of structures. Striped (voluntary) muscle consists of parallel bundles of long, thin cells which can be either relaxed or contracted. Control of contraction is provided by nerve fibers which act on small groups of muscle fibers. Thus a whole muscle can have graded degrees of contraction, even though the individual fibers have only two possible states. In this way the elastic stiffness of a muscle can vary widely. Smooth (involuntary) muscle occurs mostly in the walls of hollow organs such as the stomach, intestines, blood vessels, and other specialized organs. The heart consists of a specialized type of muscle fiber.

Nerve tissue is partly cellular (grey matter) or fiber (white matter). The latter contains considerable fatty material in the fiber sheaths.

The structure of bone is very complex. There is an outer layer of hard compact material underneath which is a layer of looser, spongelike bone so arranged as to produce a maximum of strength for commonly encountered stresses. The marrow of some bones contains blood-forming tissue.

The density of most soft tissue is between 1.0 and 1.2 with fatty tissue being somewhat lighter and bone somewhat heavier. Lung tissue is still lighter because of its air content.

PHYSICAL CONSTANTS AND MECHANICAL TRANSMISSION CHARACTERISTICS

USE OF THE PHYSICAL DATA. This section summarizes the passive mechanical responses of the human body and tissues exposed to vibration and impact. The data can be used to calculate quantitatively the transmission and dissipation of vibratory energy in human body tissue, to estimate vibration amplitudes and pressures at different locations of the body, and to predict the effectiveness of various protective measures. Table 44.1 lists some dynamic mechanical characteristics of the body and indicates some of the fields where these data find application. In cases where detailed quantitative investigations are lacking, the information may serve as a guide for the explanation of observed phenomena or for the prediction of results to be expected. Most physical characteristics of the human body presented in this section (except for the strength data) have been derived from the analysis of experimental data in which it is assumed that the body is a linear, passive mechanical system. This is an idealization which holds only for very small amplitudes. Therefore these data may not apply in analyses of mechanical injury to tissue. There is actually considerable nonlinearity of response well below amplitudes required for the production of damage. This is indicated, for example,

Table 44.1. Application of Mechanical Studies of Body

Dynamic mechanical quantity investigated	Field of application
Skull resonances and viscosity of brain tissue	Head injuries; bone-conduction hearing
Impedance of skull and mastoid.............	Matching and calibration of bone-conduction transducers; ear protection
Ultrasound transmission through skull and brain tissue	Brain tumor diagnosis; changes in central nervous system exposed to focused ultrasound
Sound transmission through skull and tissue	Bone-conduction hearing
Mechanical properties of outer, middle, and inner ear	Theory of hearing; correction of hearing deficiencies
Resonances of mouth, nasal, and pharyngeal cavities	Theory of speech generation; correction of speech deficiencies; oxygen mask design
Resonance of lower jaw....................	Bone-conduction hearing
Response of mouth-thorax system...........	Blast-wave injury; respirators
Propagation of pulse cardiac pressure.......	Circulatory physiology; hemodynamics
Heart sounds............................	Physiology of heart; diagnosis
Suspension of heart......................	Ballistocardiography; injury from severe vibrations and crash
Response of the thorax-abdominal mass system	Severe vibration and crash injury; crash protection
Impedance of subject sitting, standing, or lying on vibration platform	Isolation and protection against vibration and short-time accelerations; ballistocardiography
Impedance of body surface, surface wave velocity, sound velocity in tissue, absorption coefficient of body surface	Theory of energy transmission and attenuation in tissue; determination of tissue elasticity, viscosity and compressibility; determination of acoustic and vibratory energy entering the body; vibration isolation; design of vibration pickups; transfer of vibratory energy to inner organs and sensory receptors
Absorption of ultrasound in tissue...........	Theory of energy transmission on cellular basis; determination of doses for ultrasonic therapy

Table 44.2. Physical Properties of Human Tissue

	Tissue, soft	Bone, compact	
		Fresh	Embalmed, dry
Density, gm/cm^3.................	1–1.2	1.93–1.98	1.87
Young's modulus, dyne/cm^2.......	7.5×10^4	2.26×10^{11}	1.84×10^{11}
Volume compressibility,* dyne/cm^2.	2.6×10^{10}	...	1.3×10^{11}
Shear elasticity,* dyne/cm^2........	2.5×10^4	...	7.1×10^{10}
Shear viscosity,* dyne-sec/cm^2.....	1.5×10^2	...	
Sound velocity, cm/sec...........	$1.5–1.6 \times 10^5$	3.36×10^5	....
Acoustic impedance, dyne-sec/cm^3..	1.7×10^5	6×10^5	6×10^5
Tensile strength, dyne/cm^2.......	...	9.75×10^8	1.05×10^9
Shearing strength, dyne/cm^2, parallel	...	4.9×10^8	....
Shearing strength, dyne/cm^2, perpendicular....................	...	1.16×10^9	5.55×10^8

* Lamé elastic moduli.

by the data given in Fig. 44.6, which shows how the mechanical stiffness and resistance of soft tissue vary with static deflection. Bone behaves more or less like a normal solid; however, soft elastic tissues such as muscle, tendon, and connective tissue resemble elastomers with respect to their Young's modulus and S-shaped stress-strain relation. These properties have been studied in connection with the quasi-static pressure-volume relations of hollow organs such as arteries, the heart, and the urinary bladder,[30] assuming linear properties in studying dynamic responses. Then soft tissue can be described phenomenologically as a viscoelastic medium; plastic deformation need be considered

FIG. 44.6. Mechanical stiffness and resistance of soft tissue, per square centimeter, as a function of indentation (i.e., static deflection). The nonlinearity shows the effect of loading of body surface on surface impedance of soft tissues for two experimental human subjects A and B. (After Franke.[21])

FIG. 44.7. Simplified mechanical system representing the human body—standing on a vertically vibrating platform at low frequencies. (After Coermann et al.[33])

only if injury occurs. Physical properties of human body tissue are summarized in Table 44.2.

The combination of soft tissue and bone in the structure of the body together with the body's geometric dimensions results in a system which exhibits different types of response to vibratory energy depending on the frequency range: At low frequencies (below approximately 100 cps), the body can be described for most purposes as a lumped parameter system; resonances occur due to the interaction of tissue masses with purely elastic structures. At higher frequencies, through the audio-frequency range and up to about 100,000 cps, the body behaves more as a complex distributed parameter system—the type of wave propagation (shear waves, surface waves, or compressional waves) being strongly influenced by boundaries and geometrical configurations.

THE LOW-FREQUENCY RANGE. Simple mechanical systems, such as the one shown in Fig. 44.7 for a standing man, are usually sufficient to describe and understand the important features of the response of the human body to low-frequency vibra-

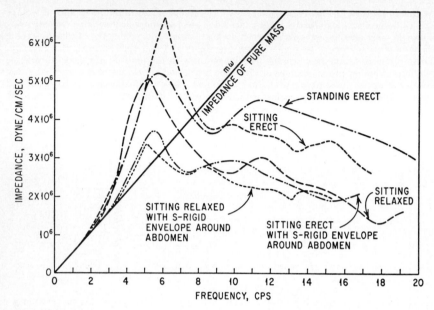

FIG. 44.8. Mechanical impedance of standing and sitting human subject vibrating in the direction of his longitudinal axis as a function of frequency. The effects of body posture and of a semirigid envelope around the abdomen are shown. The impedance of a mass m also is given. (*After Coermann* [32] *and Coermann et al.* [33])

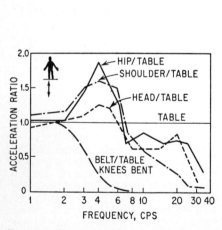

FIG. 44.9. Transmissibility of vertical vibration from table to various parts of the body of a standing human subject as a function of frequency. (*After Dieckmann;* [7] *data for transmission to belt, after Radke.* [35])

FIG. 44.10. Transmissibility of longitudinal vertical vibration from table to various parts of body of seated human subject as a function of frequency. (*After Dieckmann.* [7])

tions.[34, 36, 37] Nevertheless it is difficult to assign numerical values to the elements of the circuit, since they depend critically on the kind of excitation, the body type of the subject, and his position and muscle tone.

Subject Exposed to Vibrations in the Longitudinal Direction. The mechanical impedance of a man standing or sitting on a vertically vibrating platform is shown in Fig. 44.8. Below approximately 2 cps the body acts as a unit mass. For the sitting man, the first resonance is between 4 and 6 cps; for the standing man, resonance peaks occur at about 5 and 12 cps.[7, 32] The numerical value of the impedance together with its phase angle provides data for the calculation of the total energy transmitted to the subject.

The resonances at 4 to 6 cps and 10 to 14 cps are suggestive of mass-spring combinations of (1) the entire torso with the lower spine and pelvis, and (2) the upper torso with forward flexion movements of the upper vertebral column. The assumption that flexion of the upper vertebral column occurs is supported by observations of the transient response of the body to vertical impact loads and associated compression fractures. The greatest loads occur in the region of the eleventh thoracic to the second lumbar vertebra, which therefore can be assumed as the hinge area for flexion of the upper torso. Since the center-of-gravity of the upper torso is considerably forward of the spine, flexion movement will occur even if the force is applied parallel to the axis of the spine (also see Fig. 44.23). Changing the direction of the force so that it is applied at an angle with respect to the spine (for example, by tilting the torso forward) influences this effect considerably. Similarly the center-of-gravity of the head can be considerably in front of the neck joint which permits forward-backward motion. This situation results in forward-backward rotation of the head instead of pure vertical motion. Examples of relative amplitudes for different parts of the body when it is subject to vibration are shown in Fig. 44.9 for a standing subject, and in Fig. 44.10 for a sitting subject.[7] The curves show an amplification of motion in the region of resonance and a decrease at higher frequencies. The impedances and the transmissibility factors are changed considerably by individual differences in the body, the posture of the body, the type of support by a seat or back rest for a sitting subject, or by the state of the knee or ankle joints of a standing subject. The resonant frequencies remain relatively constant, whereas the transmissibility varies (for the condition of Fig. 44.10, values of transmissibility as high as 4 have been observed at 4 cps[35]). Above approximately 10 cps, vibration displacement amplitudes of the body are smaller than the amplitudes of the exciting table and they decrease continuously with increasing frequency. The attenuation of the vibrations transmitted from the table to the head is illustrated in Fig. 44.11. At 100 cps this attenuation is about 40 db. The attenuation along the body at 50 cps is shown in Fig. 44.12.

Between 20 and 30 cps the head exhibits a mechanical resonance, as indicated in Figs. 44.9 and 44.10. When subject to vibration in this range, the head displacement amplitude can exceed the shoulder amplitude by a factor of 3. This resonance is of importance in connection with the deterioration of visual acuity under the influence of vibration. Another frequency range of disturbances between 60 and 90 cps suggests an eyeball resonance.[36]

The mechanical impedance of the human body, lying on its back on a rigid surface and vibrating in the direction of its longitudinal axis, has been determined in connection with ballistocardiograph studies.[37] For tangential vibration, the total mass of the body behaves as a simple mass-spring system with the elasticity and resistance of the skin. For the average subject

FIG. 44.11. Attenuation of vertical and horizontal vibration for standing and sitting human subjects. (*Continuous lines after von Békésy.*[2] *Shaded area is range of values for 10 subjects after Coermann.*[36])

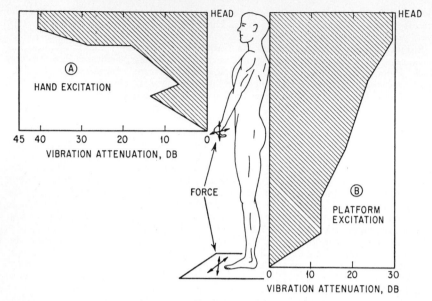

FIG. 44.12. Attenuation of vibration at 50 cps along human body. The attenuation is expressed in decibels below values at the point of excitation. For excitation of (A) hand, and (B) platform on which subject stands. (*After von Békésy.*[2])

FIG. 44.13. Typical response curves of the thorax-abdomen system of a human subject in the supine position exposed to longitudinal vibrations. The displacement of the abdominal wall (2 in. below umbilicus), the air volume oscillating through the mouth, and the variations in thorax circumference are shown per *g* longitudinal acceleration. (*From Coermann et al.*[33])

the resonant frequency is between 3 and 3.5 cps, and the Q of the system is about 3. If the subject's motion is restricted by clamping the body at the feet and at the shoulders between plates connected with the table, the resonant frequency is shifted to approximately 9 cps and the Q is about 2.5.

One of the most important sub-systems of the body, which is excited in the standing and sitting position as well as in the lying position, is the thorax-abdomen system.[33,38] The abdominal viscera have a high mobility due to the very low stiffness of the diaphragm and the air volume of the lungs and the chest wall behind it. Under the influence of both longitudinal and transverse vibration of the torso, the abdominal mass vibrates in and out of the thoracic cage. Vibrations take place in other than the (longitudinal) direction of excitation; during the phase of the cycle when the abdominal contents swing toward the hips, the abdominal wall is stretched outward and the abdomen appears larger in volume; at the same time, the downward deflection of the diaphragm causes a decrease of the chest circumference. At the other end of the cycle the abdominal wall is pressed inward, the diaphragm upward and the chest wall is expanded. This periodic displacement of the abdominal viscera has a sharp resonance between 3 and 3.5 cps (Fig. 44.13). The oscillations of the abdominal mass are coupled with the air oscillations of the mouth-chest system.[19, 33] Measurements of the impedance of the latter system at the mouth (by applying oscillating air pressure to the mouth) show that the abdominal wall and the anterior chest wall respond to this pressure. The magnitude of the impedance is minimum and the phase angle is zero between 7 and 8 cps. The abdominal wall has a maximum response between 5 and 8 cps, the anterior chest wall between 7 and 11 cps. Vibration of the abdominal system resulting from exposure of a sitting or standing subject is detected clearly as modulation of the air flow velocity through the mouth (Fig. 44.13). (Therefore at large amplitudes of vibration, speech can be modulated at the exposure frequency.) A lumped parameter model of the thorax-abdomen-airway system is used successfully to explain and predict these detailed physiological responses (Fig. 44.14). The same model can also be used, when appropriately excited, to describe the effects of blast, infrasound, and chest impact and to derive curves of equal injury potential, i.e., tolerance curves. Appropriate scaling laws have been applied to it to make it useful for the interpretation and extrapolation of experimental animal data (von Gierke, 1971).

Subject Exposed to Vibrations in the Transverse Direction. The physical response of the body to transverse vibration—i.e., horizontal in the normal upright position—is quite different from that for vertical vibration. Instead of thrust forces acting primarily along the line of action of the force of gravity on the human body, they act at right angles to this line. Therefore the distribution of the body masses along this line is of the utmost importance. There is a greater difference in response between sitting and standing positions for transverse vibration than for vertical vibration where the supporting structure of the skeleton and the spine are designed for vertical loading.

Data of transmission of vibration along the body are given in Fig. 44.45.[39] For a standing subject, the displacement amplitudes of vibration of the hip, shoulder, and head are about 20 to 30 per cent of the table amplitude at 1 cps and decrease with increasing frequency. Relative maxima of shoulder and head amplitudes occur at 2 and 3 cps, respectively. The sitting subject exhibits amplification of the hip (1.5 cps) and head (2 cps) amplitudes. All critical resonant frequencies are between 1 and 3 cps. The transverse vibration patterns of the body can be described as standing waves, i.e., as a rough approximation one can compare the body with a rod in which transverse flexural waves are excited. Therefore there are nodal points on the body which become closer to the feet as the frequency of excitation increases, since the phase shift between all body parts and the table increases continuously with increasing frequency. At the first resonant frequency (1.5 cps), the head of the standing subject has a 180° phase shift with respect to the table; between 2 and 3 cps this phase shift is 360°.[39]

There are longitudinal head motions excited by the transverse vibration in addition to the transverse head motions shown in Fig. 44.15 and discussed above. The head performs a nodding motion due to the anatomy of the upper vertebrae and the location of the head's center-of-gravity. Above 5 cps, the head motion for the sitting and stand-

FIG. 44.14. Five-degree-of-freedom body model. The model is used to calculate body deformation (thorax compression, pressure in the lungs, airflow into and out of the lungs, diaphragm and abdominal mass movement) as a function of external longitudinal forces (vibration or impact) and pressure loads (blast, infrasonic acoustic loads). It has also been used to calculate thorax dynamics under impacts to the chest wall, M_w.

FIG. 44.15. Transmission of transverse, horizontal vibration from table to various parts of sitting and standing human subject. (*After Dieckmann.*[39])

ing subjects is predominantly vertical (about 10 to 30 per cent of the horizontal table motion).

Vibrations Transmitted from the Hand. In connection with studies on the use of vibrating hand tools, measurements have been made of vibration transmission from the hand to the arm and body.[20, 40, 41] The mechanical impedance measured on a hand grip for a specific condition representative of hand tool use is presented in Fig. 44.16. The impedance has one maximum in the range below 5 cps, probably determined by the natural frequencies for transverse excitation of the human body between 1 and 3 cps. A second strong maximum appears between 30 and 40 cps; the effective mass of the hand [approximately 2.2 lb (1 kg)] is here in resonance with the elasticity of the soft parts on the inside of the hand. This elasticity is estimated to be 2.10^{-8} cm/dyne. With a practical hand tool which operates between 40 and 50 cps, the vibration amplitude decreases from the palm to the back of the hand by 35 to 65 per cent. Further losses occur between

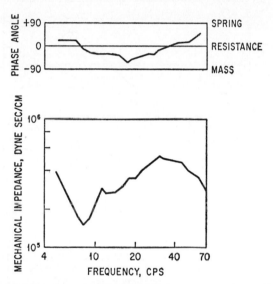

Fig. 44.16. Mechanical impedance and phase angle of arm measured at a vibrating hand grip. Elbow flexion 20 to 25°, static pressure on grip 22 lb. Measurements on one subject. (*After Dieckmann.*[40])

the hand and the elbow and the elbow and the shoulder. Figure 44.12, which shows this decrease of vibration amplitude from the hand to the head, indicates that the strongest attenuation occurs in the shoulder joint.

MIDDLE-FREQUENCY RANGE (WAVE PROPAGATION). Above about 100 cps, simple lumped parameter models become more and more unsatisfactory for describing vibration of tissue. At higher frequencies it is necessary to consider the tissue as a continuous medium for vibration propagation.

Skull Vibrations. The vibration pattern of the skull is approximately the same as the pattern of a spherical elastic shell.[42, 43] The nodal lines observed suggest that the fundamental resonant frequency is between 300 and 400 cps and that resonances for the higher modes are around 600 and 900 cps. The observed frequency ratio between the modes for the skull is approximately 1.7 while the theoretical ratio for a sphere is 1.5. From the observed resonances, the calculated value of the elasticity of skull bone (a value of Young's modulus = 1.4×10^{10} dynes/cm²) agrees reasonably well with static test results on dry skull preparations, but is somewhat lower than the static test data obtained on the femur (Table 44.2). Mechanical impedances of small areas on the skull over the mastoid area[44] have been measured for practical problems (Table 44.1).   The impedance of the skin lining in the auditory canal has been investigated and used in connection with studies on ear protectors.[45]

Vibration of the lower jaw with respect to the skull can be explained by a simple mass-spring system, which has a resonance, relative to the skull, between 100 and 200 cps.[46]

Impedance of Soft Human Tissue.   Mechanical impedance measurements of small areas (1 to 17 cm²) over soft human body tissue have been made with vibrating pistons between 10 and 20,000 cps.   At low frequencies this impedance is a large elastic reactance.   With increasing frequency the reactance decreases, becomes zero at a resonant frequency, and becomes a mass reactance with a further increase in frequency (Fig. 44.17).[3, 4, 47]   These data cannot be explained by a simple lumped parameter model, but

FIG. 44.17. Resistance and reactance of circular area, 2 cm in diameter, of soft tissue body surface as a function of frequency. Crosses and circles indicate measured values for reactance and resistance. Smooth curves calculated for 2-cm-diameter sphere vibrating in (A) viscoelastic medium with properties similar to soft tissue (parameters as in Table 44.2), (B) frictionless compressible fluid, and (C) incompressible viscous fluid. (*From von Gierke et al.*[47])

require a distributed parameter system including a viscoelastic medium—such as the tissue constitutes for this frequency range.[47, 48]   The high viscosity of the medium makes possible the use of simplified theoretical assumptions, such as a homogeneous isotropic infinite medium and a vibrating sphere instead of a circular piston.   The results of such a theory agree well with the measured characteristics.   As a consequence it is possible to assign absolute values to the shear viscosity and the shear elasticity of soft tissue (Table 44.2).   The theory together with the measurements show that, over the audio-frequency range, most of the vibratory energy is propagated through the tissue in the form of transverse shear waves—not in the form of longitudinal compression waves.   The velocity of the shear waves is about 20 m/sec at 200 cps and increases approximately with the square root of the frequency.   This may be compared with the constant sound velocity of about 1,500 m/sec for compressional waves.   Some energy is propagated along the body surface in the form of surface waves which have been observed optically.   Their velocity is of the same order as the velocity of shear waves.[47]

ULTRASONIC VIBRATION. Above several hundred thousand cycles per second, in the ultrasound range, most of the vibratory energy is propagated through tissue in the form of compressional waves; for these conditions, geometrical acoustics offer a good approximation for the description of their path. Since the tissue dimensions under consideration are almost always large compared to the wavelength (about 1.5 mm at 1 mcps) the mechanical impedance of the tissue is equal to the characteristic impedance, i.e., sound velocity times density. This value for soft tissue differs only slightly from the characteristic impedance of water.[49] The most important factor in this frequency range is the tissue viscosity, which brings about an increasing energy absorption with increasing frequency.[48] At very high frequencies this

Fig. 44.18. Approximate values of high-frequency sound attenuation in various tissues. (*After Goldman and Hueter* [49] *and Dussik et al.*[50])

viscosity also generates shear waves at the boundaries of the medium, at the boundary of the acoustic beams, and in the areas of wave transition to media with somewhat different constants (e.g., boundary muscle to fat tissue, or soft tissue to bone). These shear waves are attenuated so rapidly that they are of no importance for energy transport but are noticeable as increased local absorption, i.e., heating.

From 500 kcps to 10 mcps the attenuation coefficient describing the decrease of the sound intensity in a plane ultrasound wave is only in fair agreement with the value one would calculate from the tissue viscosity measured in the audio-frequency range (Table 44.2). The tissue deviates in this frequency range from the behavior of a medium with constant viscosity. In Fig. 44.18, attenuation coefficients measured in different types of tissue are summarized.[49, 50] On this graph a slope $\alpha/f^2 =$ constant (where f is the frequency) is indicative of classical viscous absorption with constant shear viscosity. A small slope, or a change in slope, indicates a change in viscosity with frequency (relaxation phenomenon). The graph gives only a few examples and typical functions from a large body of attenuation data available. The absorption of most soft tissues is in the range from 0.5 to 2 db/cm/mcps. In order of increasing absorption are: brain tissue, liver tissue, striated muscle, smooth muscle, kidney, skin, and tendon. Bone has the highest value of absorption—approximately 10 db/cm. The ultrasonic absorption coefficient depends very much on the structural features of the tissue and might well aid in obtaining a clearer quantitative picture of the mechanical structure of cells. Interesting in this respect are the acoustic anisotropies, i.e., cases where attenuation depends on the direction of propagation: fiber anisotropy has been found in the collinear fibers of striated muscle (Fig. 44.18) and layer anisotropy can occur in structures consisting of parallel layers of different tissue types such as in the abdominal wall.

MECHANICAL DATA FROM SHOCK FORCES

Very little numerical data on mechanical characteristics of the body are available from studies on shock or impact forces. Some evidence of resonances has been noted[51] but much of this is better obtained from vibration studies. The application of mechanical data which has been obtained from studies on vibration to exposure to shock and impact is discussed in the section on mechanical damage. Mechanical responses to shock and impact are, in general, extremely difficult to analyze numerically for basic body characteristics.

EFFECTS OF SHOCK AND VIBRATION

The motions and mechanical stresses resulting from the application of mechanical forces to the human body have several possible effects: (1) the motion may interfere directly with physical activity; (2) there may be mechanical damage or destruction; (3) there may be secondary effects (including subjective phenomena) operating through biological receptors and transfer mechanisms, which produce changes in the organism. Thermal and chemical effects are usually unimportant.

MECHANICAL INTERFERENCE. There are many ways in which forces can be applied to the body. The effects of these forces are quite varied. Certain types of displacement, velocity, or acceleration (if of sufficient magnitude) can be very disturbing to sensory and neuromuscular activities such as reading instruments, or making fine adjustments of controls or of the position of the body. Speech communication may be rendered difficult. Very little is known about how much of what kind of motion interferes with particular activities. When available, such information has meaning only in terms of tolerances permitted for the activities in question. For example, the disturbance of visual acuity arising from body vibration is not only frequency-dependent[36] (Fig. 44.19) but is roughly proportional to the amplitude of the vibration. This disturbance may be controlled either by changing the frequency, reducing the amplitude, or by decreasing the acuity required for the given task. Mechanical motion resulting from occasional shocks usually will not interfere directly with most tasks unless timed critically with respect to some operation or unless repeated at very short intervals. Such interference with task performance as does arise from shocks is more apt to be the result of biological responses or of actual damage.

MECHANICAL DAMAGE. Of the many kinds and degrees of mechanical damage to the human body arising from the application of mechanical forces, certain kinds have been singled out for attention and study because they are particularly common, dangerous, or disturbing in some special way. Among these, short of actual destruction, are bone fracture, lung damage, injury to the inner wall of the intestine, brain injury, cardiac damage, ear damage, tearing or crushing of soft tissues, and certain special types of chronic injury such as tendon or joint strains and the "white finger" syndrome of vibrating tool operators.

BIOLOGICAL RESPONSES. Mechanical stresses and motions may stimulate various receptor organs in the skin and elsewhere or may excite parts of the nervous system directly. The result may be reflex activity or modification of it. The stimuli

Fig. 44.19. Decrease of visual acuity of observers standing on vibrating platform. Each curve represents one subject. (*After Coermann.*[36])

initiate nervous system and hormonal activity which has a marked modifying action on many metabolic processes relating to food assimilation, muscular activity, reproductive activity, etc.[52] These changes are difficult to measure and correlate; they seem to differ in different species, at least in degree. Nevertheless, considerable indirect evidence exists for the reality of these response patterns and it usually is agreed that the general conclusions drawn from animal experimentation apply to man. Exposure to mechanical forces (when of great enough extent and duration) results, in part, in phenomena such as fatigue, changes in work capacity, ability to maintain attentiveness, etc. In response to acute stimulation, excitation of brain centers may produce emotional reactions such as fear or unpleasantness and lead to automatic or deliberate compensatory or protective behavior.

EFFECTS OF MECHANICAL VIBRATION

MECHANICAL DAMAGE. Damage is produced when the accelerative forces are high enough. However, experimental data can be obtained only from experiments on animals. Such data must be considered very carefully in applying the results to humans. Mice,[53] rats,[54] and cats[55, 56] have been killed by exposure to vibration. There is a definite frequency dependence of the lethal accelerations which coincides with the resonant displacement of the visceral organs, but which has been only partly established. Mice are killed at accelerations of 10 to 20g within a few minutes in the range 15 to 25 cps; above and below this frequency range, the survival time is longer. Rats and cats may be killed within 5 to 30 min at accelerations above about 10g. Post-mortem examination of these animals usually shows lung damage, often heart damage, and occasionally brain injury. The injuries to heart and lungs probably result from the beating of these organs against each other and against the rib cage. The brain injury, which is a superficial hemorrhage, may be due to relative motion of the brain within the skull, to mechanical action involving the blood vessels or sinuses directly, or to secondary mechanical effects. Tearing of intraabdominal membranes rarely occurs. Exposure for several minutes to peak acceleration of about 5g occasionally produces heart damage, as indicated by changes in the electrocardiogram.[56] An increase in body temperature is found on exposure to vibration. Since this occurs also in vibration experiments on dead animals it is probably mechanical in origin. Calculations of heat absorption based on body impedance data suggest that appreciable heat can be generated at large amplitudes. Exposure of monkeys to acceleration of 5g at a frequency of 10 and 20 cps for several hours seems to produce some damage to the vestibular system, but these findings require confirmation.[57] Observations on man[58] indicate that above about 3g, sharp pain in the chest may occur. Traces of blood occasionally have been found in the feces after exposure to acceleration of 6g at a frequency of 20 to 25 cps for about 15 min. This suggests mechanical damage to the intestine or rectum.

It is clear that several of the phenomena found in animals also must be possible in humans. Mechanical damage to heart and lungs, injury to the brain, tearing of membranes in the abdominal and chest cavities, as well as the above-mentioned intestinal injury, are all possible. Also it is likely that there is heating of the body when it is shaken. Acceleration-frequency curves for these effects have not been established. Because of the relatively greater visceral masses of the human, the minima of such curves, which would correspond to resonance ranges, must occur at relatively lower frequencies than in small animals. Subjective symptoms such as the occurrence of chest pain after exposure to acceleration of 3g (Table 44.3) may or may not be significant, although from the point of view of safety they must be taken seriously.

Chronic injuries may be produced by vibration exposure of long duration at levels which produce no apparent acute effects. In practice, such effects are usually found after exposure to repeated blows or to random jolts rather than to sinusoidal motion. When such blows are applied to the human body at relatively short intervals, the relation of the interval to tissue response times becomes very important. Exposure to such forces frequently occurs in connection with the riding of vehicles. Buffeting in aircraft or in high-speed small craft on the water, and shaking in heavy vehicles on rough surfaces, gives rise to irregular jolting motion. Acute injuries from exposure to these situations are rare, but complaints of discomfort and chronic minor injury are common. Truck and tractor drivers often have sacroiliac strain. Minor kidney injuries are occasionally suspected and, rarely, traces of blood may appear in the urine. The length of exposure and the details of the ways in which the body is supported play an important role.

Chronic injuries also are produced by localized vibration. A classical example of this is the pain and numbing of the fingers on exposure to cold which affects many people after several months of using such equipment as pneumatic hammers and drills or hand-held grinders or polishers. The heavier, slow-moving devices appear to produce more severe jolting. There is an extensive clinical literature on this subject[59] but little known of the mechanism of the injury or of the actual forces responsible. Many high-frequency components may be present.[41] The repeated insults to the tissues seem gradually to affect the capillaries and their nerve supply. Injuries resembling this have been produced in the feet of rats exposed to vibration at a frequency of 60 cps at an acceleration amplitude of 8 to 9g for 10 to 12 hr per day up to about 1,000 hr.[60]

PHYSIOLOGICAL RESPONSES. Rats exposed for 10 to 40 min a day for several months to vibration at a frequency of 12.5 cps at an acceleration amplitude of about 15g appear to show minor behavioral abnormalities.[61] The adrenal glands of rats show a rapid fall in ascorbic acid content on exposure to acceleration levels of a few tenths of a g unit at 5 to 10 cps,[62] and changes in the reproductive cycle and growth have been observed in rabbits exposed to vibration for several days.[64] Changes in respiration, heart activity and peripheral circulation have been observed in both man and animals as immediate and possibly transient responses to moderate vibration. Certain postural reflexes appear to be inhibited by vibratory motion.[64, 65]

SUBJECTIVE RESPONSES TO VIBRATION

The subjective responses of humans include perception, feelings of discomfort, apprehension, and pain. Since acceleration, frequency, mode of application, duration, and the situation of the subject are all involved, it is exceedingly difficult to find simple definitive ways of characterizing results. Early workers in this field[66-71] concerned themselves with whole-body exposure under conditions which were believed to be of practical interest. The results which they obtained were only very approximate since there appears to have been little control of subjective factors.

Generally there are three simple criteria for subjective responses to shock and vibration: the thresholds of perception, of unpleasantness, and of tolerance. The two latter are difficult to identify and reproduce, although agreement to within a factor of about 3 has been obtained. A compilation of these results based on exposures of about 5 to 20 min is given in Fig. 44.20.[72] For longer exposures, data are limited. Some information has been obtained on comfort and tolerance levels for aircraft pilots.[73] Very long exposure to vibration much above the level of perception seems to be irritating and fatiguing.

For short exposures, less than 5 min, a study has been carried out in the frequency range 1 to 15 cps.[74] Subjects were exposed to a specified acceleration amplitude until they could no longer tolerate it. They were then asked for their reactions and what their specific reason was for asking to be released. Table 44.3 shows a distribution of the major reasons for each of the frequencies used. Evidently no single criterion of tolerance was found although some reactions were more common than others. The estimated limits of tolerance for short exposures according to these criteria are given in Fig. 44.21.

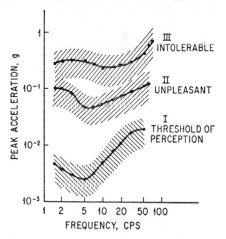

FIG. 44.20. Average peak accelerations at various frequencies at which subjects perceive vibration (I); find it unpleasant (II); or refuse to tolerate it further (III). Exposures of 5 to 20 min. Shaded areas are about one standard deviation on either side of mean. Data averaged from seven sources. (*Goldman.*[72])

EFFECTS OF MECHANICAL SHOCK

Mechanical shock includes several types of force application which have similar, though not identical, effects. Explosions, explosive compression or decompression, and impacts and blows from rapid changes in body velocity or from moving objects produce shock forces of importance. Major damage, short of complete tissue destruction, is usually to lungs, intestines, heart, or brain. Differences in injury patterns arise from differences in rates of loading, peak force, duration, localization of forces, etc.

BLAST AND SHOCK WAVES.[5, 17, 75, 110] The mechanical effects associated with rapid changes in environmental pressure are primarily localized to the vicinity of air-

Table 44.3. Criteria of Tolerance for Short Exposure to Vertical Vibration for Various Frequencies

(From Ziegenruecker and Magid; [74] *see Fig. 44.21)*

(Each cross indicates a decided comment from one of the ten human subjects used in this study as to his experiencing the symptom listed.)

Symptom → / Freq. ↓	Abdominal pain	Chest pain	Testicular pain	Head symptoms	Dyspnea	Anxiety	General discomfort
1 cps					xxxxx xxx		xxx
2 cps					xxxxx xxx		xxxx
3 cps	xx	xx			xxxxx	x	xxxxx
4 cps	xx	xx		xx	xxx	xx	xxxxx
5 cps		xxxx				x	xxxxx x
6 cps	xxx	xxxx		x			xxxx
7 cps	xx	xxxxx	x	x			x
8 cps	x	xxxx		x		xx	xxx
9 cps	xx	xxxx			x		xxxxx
10 cps	x	x	xxx	xx		x	
15 cps							xxxxx xxx

filled cavities in the body, i.e., the lungs and the air-containing gastrointestinal tract. Here, heavy masses of blood or tissue border on light masses of air. The local impedance mismatch can lead to a relative tissue displacement, which is destructive, by several different mechanisms. Starting with very slow differential pressure changes, of approximately 1 sec duration or longer, dynamic mechanical effects are unimportant; the static pressure is responsible for destructive mechanical stress or physiological response. Such pressure-time functions occur with the explosive decompression of pressurized aircraft cabins at high altitude and occur with the slow response of very well-sealed shelters to blast waves. If the pressure rise times or fall times are shortened (roughly to the order of tenths of seconds), the dynamic response of the different resonating systems of the body becomes important, in particular the thorax-abdomen system of Fig. 44.14. The dynamic load factor of the specific blast disturbance under consideration determines the response. Available data for single pulse, "instantaneously rising" pressures suggest the existence of a minimum peak pressure which corresponds to natural frequencies for dogs of between 10 and 25 cps;[17] for humans this frequency is lower. Sensitivity curves for blast exposure, i.e., curves of equal injury potential (maximum tolerable level) for various species are shown in Fig. 44.22. The theoretical curves are obtained by means of the thorax model of Fig. 44.14 after application of appropriate scaling laws to account for the different species sizes (von Gierke, 1971). For pressures with total durations of milliseconds or less and much shorter rise times (duration of wave, short compared to the natural period of the responding tissue), the effect and destruction seem

Fig. 44.21. Peak acceleration at various frequencies at which subjects refuse to tolerate further a short exposure (less than 5 min) to vertical vibration. The figures above the abscissa indicate the exposure time in seconds at the corresponding frequency. The shaded area has a width of one standard deviation on either side of the mean (10 subjects). (*Ziegenruecker and Magid.*[74])

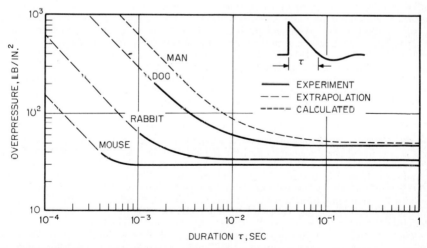

Fig. 44.22. Maximum tolerable blast overpressures for mouse, rabbit, dog, and man. The curve for man is calculated by means of a model of the type given in Fig. 44.14. Using the same model, dimensionally scaled to the animal sizes, results in curves matching closely the experimental data.

to depend primarily on the momentum of the shock wave. The mass m of an oscillatory system located in a wall or body surface, which is struck by a shock wave, is set into motion according to the relation $\int P_r \, dt = mv_0$, where P_r = reflected pressure at body surface, t = time, and v_0 = initial velocity. Experimental fatality curves on animals[5] generally show this dependence on momentum for short pressure phenomena (close to center of detonation) and the transition to a dependence on peak pressure for phenomena of long duration (far away from center). Fatal blast waves in air and water, for example, differ widely in peak pressure and duration (in air, 10 atm in excess of atmospheric pressure with a duration of 2.8 milliseconds, in water 135 atm in excess with a duration of 0.17 milliseconds) but their momenta are similar. In this most important range of short duration blasts, the mechanical effects are localized because of the short duration, i.e., the high-frequency content of the wave. The upper respiratory tract and bronchial tree, as well as the thorax and abdomen system, are too large and have resonant frequencies too low to be excited; there is no general compression or overexpansion of the thorax, which leads to pulmonary injury as in explosive decompression. The blast waves go directly through the thoracic wall, producing an impact or grazing blow. Inside the tissue, blast injury has three possible causes: (1) spalling effects, i.e., injuries caused by the tensile stresses arising from the reflection of the shock wave at the boundary between media with different propagation velocity [for example, subpleural pulmonary hemorrhages along the ribs]; (2) inertia effects which lead to different accelerations of adjacent tissues with various densities, when the shock wave passes simultaneously through these media; and (3) implosion of gas bubbles enclosed in a liquid. These phenomena are similar to the observations made with high-velocity missiles passing through water near air-containing tissues.[76, 77] The shock waves may produce not only pulmonary injuries but also hard, sharply circumscribed blows to the heart.

Of the injuries produced by exposure to high-explosive blast, lung hemorrhage is one of the most common. It may not of itself be fatal, since enough functional lung tissue may easily remain to permit marginal gas exchange. However, the rupture of the capillaries in the lung produces bleeding into the alveoli and tissue spaces, which can seriously hamper respiratory activity or produce various respiratory and cardiac reflexes. The heart rate is often very slow after a blast injury. Leakage of fluid through moderately injured, but not ruptured, capillaries may occur. There is also the possibility that air may enter the circulation to form bubbles or emboli and by reaching critical regions may impair fatally the heart or brain circulation, or produce secondary damage to other organs.[5] Fat emboli also may be formed and these, too, are capable of blocking vessels supplying vital parts. When gas pockets are present in the intestines, the shock may produce hemorrhage and in extreme cases rupture the intestinal wall itself.

The effects of underwater shock waves on man and animals are in general of the same kind as those produced by air blast. Differences which appear are those of magnitude and often depend on the mode of exposure of the body. A person in the water may, for example, be submerged from the waist down only; in this case, damage is practically confined to the lower half of the body so that intestinal, rather than lung, damage will occur.[78] Direct mechanical injury to the heart muscle and conducting mechanism is possible. Cerebral concussion resulting directly from exposure to shock waves is unusual. Neurological symptoms following exposure to blast, however, may include general depression of nervous activity sometimes to the point of abolition of certain reflexes. Psychological changes such as memory disturbances and abnormal emotional states are found sometimes. In extreme cases, there may be paralysis or muscular dysfunction. Unconsciousness and subsequent amnesia for events immediately preceding the injury result more commonly from blows to the head than from air blast. Recovery from minor concussion apparently may be complete, but repeated concussion may produce lasting damage.[79]

The ear is the part of the human body most sensitive to blast injury. Rupture of the tympanic membrane[17, 80] and injury to the conduction apparatus can occur singly or together with injury to the hair cells in the inner ear.[81] The two first-mentioned injuries may protect the inner ear through energy dissipation. The degree of injury depends on the frequency content of the blast pressure function. The fact that the ear's greatest mechanical sensitivity occurs at frequencies between 1,500 and 3,000 cps explains its vulnerability to short-duration blast waves. Peak pressures of only a few pounds per square inch can rupture the ear drum and still smaller pressures can damage the conducting mechanism and the inner ear. There are wide variations in individual susceptibility to these injuries.

IMPACTS, BLOWS, RAPID DECELERATION. This type of force is experienced in falls, in automobile or aircraft crashes, in parachute openings, in seat ejections for escape from high-speed military aircraft, and in many other situations. Interest in the body's responses to these forces centers on mechanical stress limits. The injuries which occur most often are bruises, tissue crushing, bone fracture, rupture of soft tissues and organs, and concussion. A bruise is a superficial area of slight tissue damage with rupture of the small blood vessels and accumulation of blood and fluid in and around the injured region. It is essentially a crushing injury produced by compression of the tissues, usually between the impinging solid and the underlying bone. It is extremely common and is readily repaired by the body itself. When the tissue is completely destroyed by crushing, the damage usually is irreparable. Bone fractures, like bruises, require that the forces be sustained long enough to produce appreciable displacements and properly concentrated stresses.

When soft tissues are displaced considerable distances by appropriate forces, so-called internal injuries, i.e., rupture of membranes or organ capsules, may take place. Such injuries are, in practice, more often produced by forces of relatively long duration and usually are dangerous.

Experiments have been carried out in which animals were embedded in plaster casts and then dropped about 20 ft onto a metal plate which had various degrees of cushioning. In this case the decelerative impact is well distributed. The degree of injury increases with the acceleration. Lung hemorrhage and laceration of liver and spleen capsules are most common. Damage to the diaphragm, the brain, and the bone structure occurs when the deceleration exceeds about $500g$ (velocity change 35 ft/sec).[82]

The obvious correlation between the response of the body system to continuous vibration and to spike and step-force functions may be used to guide and interpret experiments. The tissue areas stressed to maximum relative displacement at the various frequencies during steady-state excitation are naturally preferred target areas for injury under impact load if the force-time functions of the impacts have appreciable energy in these frequency bands, i.e., if the impact duration is of the same order of magnitude as the body's natural periods. If the impact exposure times are shorter, stress tolerance limits increase; if exposure times decrease to hundredths or thousandths of a second, the response becomes more and more limited and localized to the point of application of the force (blow). Elastic compression or injury will depend on the load distribution over the application area, i.e., the pressure, to which tissues are subjected. If tissue destruction or bone fracture occurs close to the area of application of the force these will absorb additional energy and protect deeper seated tissues by reducing the peak force and spreading it over a longer period of time. An example is the fracture of foot and ankle of men standing on the deck of warships when an explosion occurs beneath. The support may be thrown upward with great momentum; if the velocity reaches 5 to 10 ft/sec (which corresponds, under these conditions, to an acceleration of several hundred g) fractures occur.[83] However, the energy absorption by the fracture protects structures of the body which are higher up.

If the force functions contain extremely high frequencies, the compression effects spread from the area of force application throughout the body as compression waves. If these are of sufficient amplitude they may cause considerable tissue disruption. Such compression waves are observed from the impact of high-velocity missiles.

If the exposure to the accelerating forces lasts long enough so that (as in most applications of interest) the whole body is displaced, exact measurement of the force applied to the body and of the direction and contact areas of application becomes of extreme importance. In studies of seat ejection, for example, a knowledge of seat acceleration alone is not sufficient for estimating responses. One must know the forces in those structures or restraining harnesses through which acceleration forces are transmitted. The location of the center-of-gravity of the various body parts such as arms, head, and upper torso must be known over the time of force application so that the resulting body motion and deformation can be analyzed and controlled for protection purposes. In addition to the primary displacements of body parts and organs, there are secondary forces from decelerations if, due to the large amplitudes, the motions of parts of the body are stopped suddenly by hitting other body parts. Examples occur in linear deceleration where, depending on the restraint, the head may be thrown forward until it hits the chest or, if only a lap belt is used, the upper torso may jackknife and the chest may hit the knees.

FIG. 44.23. Prediction of spinal compression fracture under longitudinal acceleration loads, G_z. (A) Simple model for the study of body ballistics and injury prediction. Deflection of the spine δ is used to represent stress in the vertebral column. (B) Probability of spinal compression injury as a function of dynamic response index (DRI) [DRI $= \omega_n{}^2 \, \delta_{max}/g$, where $\omega_n = (K/M)^{1/2}$, $\delta_{max} =$ maximum deflection, and $g = 9.81$ m/sec²].

There is always the additional possibility that the body may strike nearby objects, thus initiating a new impact deceleration history.

Longitudinal Acceleration. The study of positive longitudinal (headward) acceleration of short duration is connected closely with the development of upward ejection seats for escape from aircraft. Since the necessary ejection velocity of approximately 60 ft/sec and the available distance for the catapult guide rails of about 3 ft are determined by the aircraft without much leeway, the minimum acceleration required (step function) is approximately 18.6g. Since the high jolt of the instantaneous acceleration increase is undesirable because of the high dynamic load factor in this direction for the frequency range of body resonances (compare with Fig. 44.10), slower build-up of the acceleration with higher final acceleration is preferable to prevent injury. Investigations show that the body's ballistic response can be predicted by means of analog computations making use of the frequency-response characteristics of the body.[84] The simplest analog used for the study of headward accelerations is the single degree-of-freedom mechanical resonator composed of the lumped-parameter elements of a spring, mass, and damper. A diagram of this model is shown in Fig. 44.23A. The model is used to simulate the maximum stress developed within the vertebral column (the first failure mode in this direction) for any given impact environment. Use of the model provides a basis for a probabilistic approach to injury prediction and permits the evaluation of simple

and complex acceleration waveforms in the same terms. The natural frequency, $\omega_n = \sqrt{K/M}$, for the model is 52.9 rad/sec and the damping ratio $c/2 \sqrt{KM}$ is 0.224. The dynamic response index (DRI) is defined as $\omega_n{}^2 \delta_{max}/g$. It has been determined that a DRI value of 18 corresponds to a probability of spinal injury of 55 per cent—a value frequently used for design purposes. This relation between DRI and probability of spinal injury (Fig. 44.23B) is based on extensive strength data on the vertebral column and on the evaluation of military injury statistics. (For references, see Symposium on Biodynamic Models and their Applications, 1970, and von Gierke, 1971.) Although more refined models of the human spine are being used[85] (Stech and Payne, 1969, von Gierke, 1971), the simple approach shown in Fig. 44.23 is the one most frequently applied for practical design problems.

Actual control or prevention of injury in this direction is critically dependent on optimal body positioning and restraint to minimize unwanted and forceful flexion of the spinal column.[86] The fracture tolerance limits are influenced by age, physical condition, clothing, weight, and many other factors which detract from the optimum. If the tolerance limits are exceeded, fractures of the lumbar and thoracic vertebrae occur first. While in and of itself this injury may not be classified as severe, small changes in orientation may be enough to involve the spinal cord, an injury which is extremely severe and may be life-threatening. Neck injuries from headward accelerations appear to occur at considerably higher levels.

In general, for vertical crash loads the same considerations apply as discussed for seat ejections, although no control over the build-up time of the acceleration is possible and more sudden onsets must be expected.

For negative (tailward) acceleration (downward ejection) no firm point for application of the accelerating force is accessible as for positive acceleration. If the force is applied as usual through harness and belt at shoulder and groin,[87] the mobility of the shoulder girdle together with the elasticity of the belts results in a lower resonant frequency than the one observed in upward ejection. To avoid overshooting with standard harnesses, the acceleration rise time must be at least 0.15 sec. This type of impact can excite the thorax-abdomen system (Fig. 44.14). The diaphragm is pushed upward by the abdominal viscera; as a result, air rushes out of the lungs (if the glottis is open) or high pressures develop in the air passages.[88] Tolerance limits for negative acceleration probably are set by the compression load on the thoracic vertebrae, which are exposed to the load of the portion of the body below the chest. This load on the vertebrae is higher than that for the positive acceleration case due to the greater weight; therefore a tolerance limit for acceleration has been set at $13g$. Shoulder accelerations of $13g$ have been tolerated by human subjects without injury, when the load was divided between hips and shoulders.

Transverse Accelerations. The forward-facing and backward-facing seated positions are most frequently exposed to high transverse components of crash loads. Human tolerance to these forces has been studied extensively by volunteer tests on linear decelerators,[9] in automobile crashes,[15] and by the analysis of the records of accidental falls.[89] The results indicate the importance of distributing the decelerative forces or impact over as wide an area as possible. The tolerable levels of acceleration amplitude of well over $50g$ ($100g$ and over for falling flat on the back with minor injuries, 35 to 40g for 0.05 sec voluntary tolerance seated with restraining harness) are probably limited by injury to the brain. An indication that the latter might be sensitive to, and based on specific dynamic responses, is the fact that the tolerance limit depends strongly on the rise time of the acceleration. With rise times around 0.1 sec (rate of change of acceleration $500g/\text{sec}$), no overshooting of head and chest accelerations is observed, whereas faster rise times of around 0.03 sec (1,000 to $1,400g/\text{sec}$) result in overshooting of chest accelerations of 30 per cent (acceleration front to back) and even up to 70 per cent (acceleration back to front). All these results depend critically on the harness for fixation and the back support used[9] (see *Protection Methods and Procedures*, below). These dynamic load factors indicate a natural frequency of the body system between 10 and 20 cps. Impact of the heart against the chest wall is another possible injury discussed and noted in some animal experiments.[85, 90]

The head and neck supporting structures seem to be relatively tough.[9, 15] Injury seems to occur only upon backward flexion and extension of the neck ("whiplash") when the body is accelerated from back to front without head support, as is common in rear-end automobile collisions.

Head Impact. Injuries to the head, beyond superficial bruises and lacerations, usually consist of concussion or fracture of the skull. The symptoms produced by head impact range from pain and dizziness through disorientation and depression of function to unconsciousness and loss of memory for events immediately preceding the injury.[79] Head injuries usually occur from heavy blows by solid objects to the head, rather than by accelerative forces applied to the body. Since approximately 75 per cent of airplane crash fatalities result from head injuries, and the latter are without doubt of similar importance in many other types of accident (athletics, etc.), the mechanisms leading to head injury have been the object of a large number of investigations.[23, 86, 91]

Neck Response. The limitations for forward and lateral bending of the neck are, for practical purposes, the anterior chest wall and the shoulders, respectively. Since the head is almost entirely held by the neck muscles, the absence of their supporting action gives any blow to the head or neck a "flying start." As a consequence, fractures or dislocations are more apt to result. Dislocations involving the first and second vertebral joints usually are less severe if the odontoid process is fractured (less damage to the spinal cord). If this is not the case, the spinal cord may be severed or crushed. The latter is the essential mechanism in the hanging of criminals. The energy release required for this is around 1,000 ft-lb (150 kg-m). As animal experiments indicate, damage done to the cervical cord at the first vertebra also may play a role in the generation of one type of concussion which had been attributed primarily to brain damage.[77]

Head Response. The elastic shell of the skull is filled with nerve tissue, blood, and cerebrospinal fluid, which have about the same density. The compressibility of the brain substance is very small (like water), and its shear modulus is very low. The viscosity of the brain tissue is around 20 dyne-sec/cm². The reaction of the head to a blow is a function of the velocity, duration, and area of impact, and the transfer of momentum. Near the point of application of the blow there will be an indentation of the skull. This results in shear strains in the brain in a superficial region close to the dent. Compression waves emanate from this area, which have normally small amplitudes since the brain is nearly incompressible. In addition to the forces on the brain resulting from skull deformation there are acceleration forces, which also would act on a completely undeformable skull. The centrifugal forces and linear accelerations producing compressional strains are negligible compared to the shear strains produced by the rotational accelerations. The distribution of the shear strains over the brain has been studied on models[92] and the motion of the brain surface has been observed in animals with sections of skull replaced by Lucite.[93] The maximum strains are concentrated at regions where the skull has a good grip on the brain owing to inwardly projecting ridges, especially at the wing of the sphenoid bone of the skull. Shear strains also must be present throughout the brain and in the brain stem. Many investigators consider these shear strains, resulting from rotational accelerations due to a blow to the unsupported head, as the principal event leading to concussion. Blows to the supported, fixed head are supposed to produce concussion by compression of the skull and elevation of cerebrospinal fluid pressure. Despite the general acceptance of rotational acceleration as one of the main causes of concussion, experimental data on this quantity are almost completely missing and concussion thresholds are discussed in terms of "available energy" (which is usually not the energy transferred to the head) and impact velocity.

In general it can be assumed that a high-velocity projectile (for example, a bullet of 10 gm with a speed of 1,000 ft/sec) with its high kinetic energy and low momentum produces plainly visible injury to scalp, skull, and brain along its path. The high-frequency content of the impact is apt to produce compression waves which in the case of very high energies may conceivably lead to cavitation with resulting disruption of tissue.[94] Skull fracture is not a prerequisite for these compression waves. However, if the head hits a wall or another object whose mass is large compared to the head's mass, the local, visible damage is small and the damage due to rotational acceleration may be large. Blows to certain points, especially on the midline, produce no rotation. Blows to the chin upward and sideward produce rotation relatively easily ("knock-out" in boxing). Therefore, it is almost impossible to define a concussion velocity or energy. Velocities listed in the literature for concussion from impact of large masses range from 15 to 50 ft/sec. At impact velocities of about 30 ft/sec, approximately 200 in.-lb of energy is absorbed in 0.002 sec, resulting in an acceleration of the head of 47g. Impact energies for compression concussion are probably approximately in the same range.

Skull Response and Skull Fracture.[95] Impact studies on cadaver skulls with strain gages show that a hammer blow to the bone itself lasts 2.5 to 5 × 10⁻⁴ sec. After the initial impact the bone oscillates for 2 to 4 × 10⁻³ sec with a frequency of approximately 700 cps, which agrees with the fundamental frequency found with continuous period excitation.[43] Scalp, skin, and subcutaneous tissue reduce the energy applied to the bone. If the response of the skull to a blow exceeds the elastic deformation limit, skull fracture occurs. Impact by a high-velocity, blunt-shaped object results in localized circumscribed fracture and depression. Low-velocity blunt blows, insufficient to cause depression, occur frequently in falls and crashes. Stresscoat studies of such situations reveal a general deformation pattern;[95] an inward bent area of the skull surrounds the contact area and is followed at a considerable distance by an outbending of the skull. Sometimes fairly symmetrical undulating patterns are observed, occasionally with an outbending of the contrecoup type, i.e., an outbending in the skull area opposite the point of contact. Since the bone, a brittle material, fails in tension, most linear skull fractures originate in the outer surface of the outward bent area surrounding the indentation. From this area the crack may spread toward the center of impact, which rebounds immediately after the blow and becomes an area of high tensile stress. Propagation in the direction opposite to the center of the impact takes place also. The size of the fracture line depends on the energy expended. Given enough energy two, three, or more cracks appear all radiating from the center of the blow. The skull has both weak and strong areas, each impact area showing well-defined regions for the occurrence of the fracture lines.

The total energy required for skull fracture varies from 400 to 900 in.-lb, with an average often assumed to be 600 in.-lb. This energy is equivalent to the condition that the head hits a hard, flat surface after a free fall from a 5-ft height. Skull fractures

FIG. 44.24. Head impact energy as affected by body attitude and barricade angle. (*Data from dummy drop tests after Dye.*[96])

occurring when a batter is accidentally hit by a ball (5 oz) of high velocity (100 ft/sec) indicate that about the same energy (580 in.-lb) is required. Additional energy 10 to 20 per cent beyond the single linear fracture demolishes the skull completely. Dry skull preparations required only approximately 25 in. lb for fracture. The reason for the large energy difference required in the two conditions is attributed mainly to the attenuating properties of the scalp.

For practical situations in automobile and aircraft crashes the form, elasticity, and plasticity of the object injuring the head is of extreme importance and determines its "head injury potential." For example, impact with a 90° sharp corner requires only a tenth of the energy for skull fracture (60 in.-lb) that impact with a hard, flat surface requires.[96] Head impact energies for various attitudes of the human body hitting a contact area at various angles are presented in Fig. 44.24. The conditions shown are representative of crash conditions involving unrestrained humans.

EFFECTS OF SHOCK AND VIBRATION ON TASK PERFORMANCE

Little is known about the effects of shock on task performance at accelerations which do not produce actual damage. Whole-body vibration affects task performance primarily through mechanical interference; performance capability can therefore be strongly influenced through vibration-insensitive human engineering of controls and displays. When mechanical interference is negligible, measurements of performance on very simple tasks have given equivocal or negative results.[36, 97] The extensive literature in this field has recently been reviewed by Shoenberger (1972) and Hornick (1973). Experiments relating to task performance and fatigue are extraordinarily difficult to carry out properly, and the results have limited practical application, i.e., are not easily applied from one situation to another, even if they appear similar. General guidance with respect to the effects of vibration environments on task performance and the effect of fatigue is given in Fig. 44.31. Recent very promising approaches in this area attack the problem by measuring the changes in the human-operator control parameters caused by the environment using the methodologies of human manual-control theory.

PROTECTION METHODS AND PROCEDURES

Protection of man against mechanical forces is accomplished in two ways: (1) isolation to reduce transmission of the forces to the man and (2) increase of man's mechanical resistance to the forces. Isolation against shock and vibration is achieved if the natural frequency of the system to be isolated is lower than the exciting frequency at least by a factor of 2. Both linear and nonlinear resistive elements are used for damping the transmission system; irreversible resistive elements or energy-absorbing devices can be used once to change the time and amplitude pattern of impulsive forces. Human tolerance to mechanical forces is strongly influenced by selecting the proper body position with respect to the direction of forces to be expected. Man's resistance to mechanical forces also can be increased by proper distribution of the forces so that relative displacement of parts of the body is avoided as much as possible. This may be achieved by supporting the body over as wide an area as possible, preferably loading bony regions and thus making use of the rigidity available in the skeleton. Reinforcement of the skeleton is an important feature of seats designed to protect against crash loads. The flexibility of the body is reduced by fixation to the rigid seat structure. The mobility of various parts of the body, e.g., the abdominal mass, can be reduced by properly designed belts and suits. The factor of training and indoctrination is essential for the best use of protective equipment, for aligning the body in the least dangerous positions during intense vibration or crash exposure, and possibly for improving operator performance during vibration exposure. The latter type of training may be helpful in anticipating and preventing man-machine resonance effects and in reducing anxiety which might otherwise occur.

PROTECTION AGAINST VIBRATIONS

The transmission of vibration from a vehicle or platform to a man is reduced by mounting him on a spring or similar isolation device, such as an elastic cushion.[7, 35] The degree of vibration isolation theoretically possible is limited, in the important resonance frequency range of the sitting man, by the fact that large static deflections of the man with the seat or into the seat cushion are undesirable. Large relative movements between operator and vehicle control interfere in many situations with man's performance. Therefore a compromise must be made. Cushions are used primarily for static comfort, but they are also effective in decreasing transmission of vibration above man's resonance range. They are ineffective in the resonance range and may even amplify the vibration in the subresonance range. In order to achieve effective isolation over the 2 to 5 cps range, the natural frequency of the man-cushion system should be reduced to 1 cps, i.e., the natural frequency should be small compared with the forcing frequency (see Chap. 30). This would require a cushion deflection of 10 in. (Fig. 44.25).

Examples of the mechanical impedance changes achieved for a human subject in the sitting position with various cushions are given in Fig. 44.26.[31] If a seat cushion without a back cushion is used (as is common in some tractor or vehicle arrangements), a condition known as "back scrub" (a backache) may result. Efforts of the operator to

FIG. 44.25. Static deflection of a mass on a spring (such as a man on a seat cushion) as a function of the natural frequency of this linear system. To obtain a natural frequency for the man-seat system which is lower than the resonant frequency of the man alone (5 cps) requires a large static deflection.

wedge himself between the controls and the back of the seat often tend to accentuate this uncomfortable condition. In ordinary passenger seats, the seat cushion does not alter the resonant frequency of the man-seat system significantly, so that no isolation is achieved in the frequency range below 5 cps. Sometimes some amplification is unavoidable. Nevertheless, the damping properties of the seat cushion are very important in attenuating the frequencies above resonance, which may be more important in automobile or aircraft transportation (see Fig. 44.33), for example. For severe low-frequency vibration and shock conditions, such as occur in tractors and other field equipment, suspension of the whole seat is superior to the simple seat cushion. Hydraulic shock absorbers, rubber torsion bars, coil springs, and leaf springs all have been successfully used for suspension seats. The differences between these springs appear to be small. However, a seat that is guided so that it can move only in a linear direction seems to be more comfortable than one in which the seat simply pivots around a center of rotation (Fig. 44.27).[7] The latter situation produces an uncomfortable and fatiguing pitching motion. Suspension seats can be built which are capable of preloading for the operator's weight so as to maintain the static position of the seat and the natural frequency of the system at the desired value. Suspension seats for use on tractors and on similar vehicles are available which reduce the resonant frequency of the man-seat system from approximately 4 to 1 cps[35] (measured in terms of the ratio of the acceleration

FIG. 44.26. Mechanical impedance of a sitting human subject with various cushions and without cushion (vertical, i.e., longitudinal vibrations). (*Dieckmann.*[31])

at the belt level of the man to the seat acceleration). The transmission ratio, which is between 2 and 2.5 at the resonant frequency for the man alone, is only 1.6 at 1 cps for the man on the suspension seat. It has not been possible to lower the resonant fre-

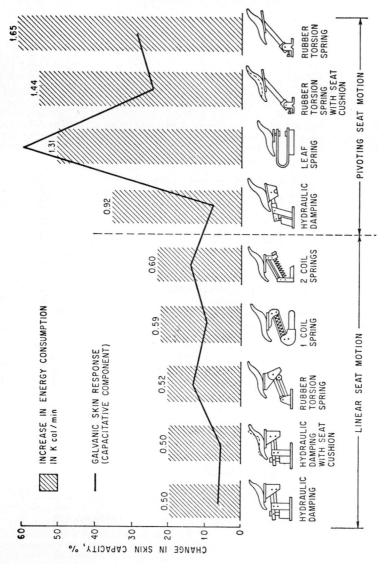

Fig. 44.27. Difference in vibration induced stress on a subject as a function of the seat design. Galvanic skin response and the increase in caloric energy consumption due to vibrations are plotted for the different seats indicated. The data represent averages over 10 tests of 15 min duration on five subjects with vibrations characteristic of tractor operation. The data prove the general superiority of seat designs which restrict seat vibrations to linear, compared to pivoting, motions. Subjective evaluations indicate a similar rank order as the physiological quantities in this graph. (*After Dieckmann.*[7])

quency below 2 to 3 cps, with a transmission ratio of 2.5 to 3, only with different types of foam and spring cushions. Therefore the use of elastic cushions alone results in an amplification of the vibration, as indicated by the impedance measurements of Fig. 44.26. These findings are confirmed not only by laboratory tests but also by field tests on tractors. (The superiority of man's legs to most seating devices, with respect to the transmission of vertical vibrations, is shown in Fig. 44.9.) Differences in positioning of the sitting subject also change the transmission as demonstrated by impedance measurements (Fig. 44.8).

For severe vibrations, close to or exceeding normal tolerance limits, such as those which may occur in military operations, special seats and restraints can be employed to provide maximum body support in all critical directions for the subject in the most advantageous position. In general, under these conditions, seat and restraint requirements are the same for vibration and rapidly applied accelerations. (A discussion of body restraints is given below.) Laboratory experiments[53] show that large protection can be achieved by the use of rigid or semirigid body enclosures. Immersion of the operator in a rigid, water-filled container with proper breathing provisions has been used in laboratory experiments to protect subjects against high, sustained static g loads.[98] This principle can be used to provide protection against high alternating loads.

The isolation of hand and arm against vibration from hand tools depends critically on the type, weight, and relative position of the tool. The model for the hand-arm system presented in Fig. 44.16 probably is adequate to estimate the effectiveness of isolation measures.

Since tolerance to continuous vibration depends critically on the exposure time, the control of the work hours in vibration environments or with vibrating tools is one of the most important protective measures. It can prevent cumulative permanent damage and reduce the possibility of accidents favored by vibration-aggravated fatigue.

PROTECTION AGAINST RAPIDLY APPLIED ACCELERATIONS (CRASH)

The study of automobile and aircraft crashes and of experiments with dummies and live subjects shows that complete body support and restraint of the extremities provide maximum protection against accelerating forces and give the best chance for survival.[9, 51] If the subject is restrained in the seat, he makes full use of the force moderation provided by the collapse of the vehicle structure, and he is protected against shifts in position which would injure him by bringing him in contact with interior surfaces of the cabin structure.[99, 100, 101] The decelerative load must be distributed over as wide a body area as possible to avoid force concentration with resulting bending moments and shearing effects. The load should be transmitted as directly as possible to the skeleton, preferably directly to the pelvic structure—not via the vertebral column.

Theoretically, a rigid envelope around the body will protect it to the maximum possible extent by preventing deformation. A body restrained to a rigid seat approximates such a condition; proper restraints against longitudinal acceleration shift part of the load of the shoulder girdle and arms from the spinal column to the back rest. Arm rests can remove the load of the arms from the shoulders.[51] Semirigid and elastic abdominal supports provide some protection against large abdominal displacements. The effectiveness of this principle has been shown by animal experiments[38] and by impedance measurements on human subjects (Fig. 44.8). Animals immersed in water, which distributes the load applied to the rigid container evenly over the body surface, or in rigid casts are able to survive acceleration loads many times their normal tolerance.

Many attempts have been made to incorporate energy-absorptive devices, either in a harness or in a seat with the intent to change the acceleration-time pattern by limiting peak accelerations. Parts of the seat or harness are designed with characteristics which become nonlinear at some given acceleration level. The benefits derived from such devices are usually small since little space for body or seat motion is available in airplanes or automobiles; furthermore, contact with interior cabin surfaces during the period of extension of the device is apt to result in more serious injury. For example (Fig. 44.28), consider an aircraft which is stopped in a crash from 100 mph in 5 ft; it is

FIG. 44.28. Crash deceleration and velocity of an aircraft as a function of time. The aircraft comes to a complete stop from 100 mph within 5 ft. An energy-absorbing device limiting the maximum acceleration on the passenger to 50g is assumed. It would require a displacement of 19 in.

subjected to a constant deceleration of 67g. An energy-absorptive device designed to elongate at 17g would require a displacement of 19 in. In traveling through this distance, the body or seat would be decelerated relative to the aircraft by 14.4g and would have a maximum velocity of 36.8 ft/sec relative to the aircraft structure. A head striking a solid surface with this velocity has many times the minimum energy required to fracture a skull. The available space for seat or passenger travel using the principle of energy absorption therefore must be considered carefully in the design. In the development of "catch-up" mechanisms for window washers and workers in similar situations, energy-absorptive elements in the form of undrawn nylon ropes may be applied. For seat belts and other crash restraint harnesses, extensible fabrics have been found to be extremely hazardous since their load characteristics cannot be sufficiently controlled.[102] Seats for jet airliners have been designed which have energy-absorptive mechanisms in the form of extendable rear-legs housing.[103] The maximum travel of the seats is 6 in.; their motion is designed to start between 9 and 12g horizontal load, depending on the floor strength. During motion, the legs pivot at the floor level—a feature considered to be beneficial if the floor wrinkles in the crash. Theoretically, such a seat can be exposed to a deceleration of 30g for 0.037 sec or 20g for 0.067 sec without transmitting a deceleration of more than 9g to the seat. However, the increase in exposure time must be considered as well as the reduction in peak acceleration. For very short exposure times where the body's tolerance probably is limited by the transferred momentum and not the peak acceleration, the benefits derived from reducing peak loads would disappear.

The high tolerance limits of the well-supported human body to decelerative forces (Figs. 44.34 to 44.41) suggest that in aircraft and other vehicles, seats, floors, and the whole inner structure surrounding crew and passengers should be designed to resist crash decelerations as near to 40g as weight or space limitations permit.[103] The structural members surrounding this inner compartment should be arranged so that their crushing reduces forces on the inner structure. Protruding and easily loosened objects should be avoided. To allow the best chance for survival, seats should also be stressed for dynamic loadings between 20 and 40g. Civil Air Regulations require a minimum static strength of seats of 9g.[104] A method for computing seat tolerance for typical survivable airplane crash decelerations is available for seats of conventional design.[105] It has been established that the passenger who is riding in a seat facing backward has a better chance to survive an abrupt crash deceleration since the impact forces are then more uniformly distributed over the body. Neck injury must be prevented by proper head support. Objections to riding backward on a railway or in a bus are minimized for air transportation because of the absence of disturbing motion of objects in the immediate field of view. Another consideration concerning the direction of passenger seats in aircraft stems from the fact that for a rearward-facing seat the center of passenger support is about 1 ft above the point where the seat belt would be attached for a forward-facing passenger. Consequently, the rearward-facing seat is subjected to a higher bending moment; in other words, for seats of the same weight the forward-facing seat will sustain higher crash forces without collapse. For the same seat weight, the rearward-facing seat will have approximately only half the design strength of the forward-facing seat and about one-third its natural frequency. A criterion for the selection of one type of seat over the other on the basis of allowable weight, seat tolerance to probable crash loads, and passenger injury has been proposed, but must await experiments for quantitative confirmation.[106]

Safety lap or seat belts are used to fix the occupants of aircraft or automobiles to their seats and to prevent their being hurled about within, or being ejected from, the car or aircraft. Their effectiveness has been proved by many laboratory tests and in actual crash accidents.[86] The belt load on the lower abdomen causes no severe intra-abdominal injury or injury to the lower spinal region—at least in most survivable crashes. A forward-facing passenger held by a seat belt flails about when suddenly decelerated; his hands, feet, and upper torso swing forward until his chest hits his knees or until the body is stopped in this motion by hitting other objects (back of seat in front, cabin wall, instrument panel, steering wheel, control stick, etc.). Since 15 to 18g longitudinal

(A)	(B)	(C)	(D)
SEAT BELT	SEAT BELT, #9 NYLON, 3" WIDE;	SEAT BELT, #9 NYLON, 3" WIDE;	SEAT BELT, #9 NYLON, 3" WIDE;
	SHOULDER STRAP, M–1G #0 NYLON, 1 3/4" WIDE	SHOULDER STRAP, #9 NYLON, 3" WIDE;	SHOULDER STRAP, #9 NYLON, 3" WIDE, 2 STRAPS;
		LEG STRAP, #9 NYLON, 3" WIDE, 2 STRAPS	LEG STRAP, #9 NYLON, 3" WIDE, 2 STRAPS;
			CHEST BELT, #9 NYLON 3" WIDE

FIG. 44.29. Protective harnesses for rapid accelerations or decelerations. The following devices were evaluated in sled deceleration tests:

   A. Seat belt for automobiles and commercial aviation.
   B. Standard military lap and shoulder strap.
   C. Like *B* but with thigh straps added to prevent headward rotation of the lap strap.
   D. Like *C* but with chest strap added. *(Stapp.[9])*

deceleration can result in three times higher acceleration of the chest hitting the knees, this load appears to be about the limit a human can tolerate with a seat belt alone. Approximately the same limit is obtained when the head-neck structure is considered.

Increased safety in automobile as well as airplane crashes can be obtained by distributing the impact load over larger areas of the body and fixing the body more rigidly to the seat. Shoulder straps, thigh straps, chest straps, and hand holds are additional body supports used in experiments. They are illustrated in Fig. 44.29. Table 44.4 shows the desirability of these additional restraints to increase possible survivability to acceleration loads of various direction.[51] In airplane crashes, vertical and horizontal loads must be anticipated. In automobile crashes, horizontal loads are most likely. The effectiveness of adequately engineered shoulder or chest straps in automobile crashes is illustrated in Fig. 44.30. Lap straps always should be as tight as comfort will permit to exclude available slack. During forward movement, about 60 per cent of the body mass is restrained by the belt, and therefore represents the belt load. In order to reduce slack and elongation under load, it is recommended that an 8,000-lb loop strength, 3-in. wide nylon belt replace the 3,000- to 4,000-lb, 3-in. lap belt currently in wide use.

Table 44.4. Human-body Restraint and Possible Increased Impact Survivability

(After Eiband[51])

Direction of acceleration imposed on seated occupants	Conventional restraint	Possible survivability increases available by additional body supports *
Spineward: Crew......................	Lap strap Shoulder straps	Forward facing: (a) Thigh straps (Assume crew members will be performing emergency duties with hands and feet at impact.)
Passengers.................	Lap strap	Forward facing: (a) Shoulder straps (b) Thigh straps (c) Nonfailing arm rests (d) Suitable hand holds (e) Emergency toe straps in floor
Sternumward: Passengers only............	Lap strap	Aft facing: (a) Nondeflecting seat back and (b) Integral, full-height head rest (c) Chest strap (axillary level) (d) Lateral head motion restricted by padded "winged back" (e) Leg and foot barriers (f) Arm rests and hand holds (prevent arm displacement beyond seat back)
Headward: Crew......................	Lap strap Shoulder straps	Forward facing: (a) Thigh straps (b) Chest strap (axillary level) (c) Full, integral head rest (Assume crew members will be performing emergency duties; extremity restraint useless.)
Passengers.................	Lap strap	Forward facing: (a) Shoulder straps (b) Thigh straps (c) Chest strap (axillary level) (d) Full, integral head rest (e) Nonfailing contoured arm rests (f) Suitable hand holds Aft facing: (a) Chest strap (axillary level) (b) Full, integral head rest (c) Nonfailing arm rests (d) Suitable hand holds
Tailward: Crew......................	Lap strap Shoulder straps	Forward facing: (a) Lap-belt tie-down strap (Assume crew members will be performing emergency duties; extremity restraint useless.)
Passengers.................	Lap strap	Forward facing: (a) Shoulder straps (b) Lap-belt tie-down strap (c) Hand holds (d) Emergency toe straps Aft facing: (a) Chest strap (axillary level) (b) Hand holds (c) Emergency toe straps
Berthed occupants..........	Lap strap	Feet forward: Full-support webbing net Athwart ships: Full-support webbing net

* Exposure to maximum tolerance limits (Figs. 44.34 to 44.41) requires straps exceeding conventional strap strength and width.

NO MOTORIST RESTRAINING DEVICE PASSENGER PROBABLE FATALITY	25 MPH	25 MPH	0 MPH	FRONT SEAT, PASSENGER SIDE, AS VIEWED FROM DRIVER'S SIDE WITH STEERING WHEEL AND DOOR REMOVED TO SHOW DUMMY MOTION.
LAP BELT PASSENGER PROBABLE FATALITY	25 MPH	25→0 MPH	0 MPH	FRONT SEAT, PASSENGER SIDE, AS VIEWED FROM DRIVER'S SIDE WITH STEERING WHEEL AND DOOR REMOVED TO SHOW DUMMY MOTION.
CHEST BELT DRIVER SURVIVED	25 MPH	25→0 MPH	0 MPH	FRONT PORTION OF CAR COLLAPSES UNDER HIGH DECELERATIVE FORCES BUT STEERING COLUMN REMAINS RELATIVELY INTACT. CAR CABIN AND ESPECIALLY DRIVER CONTINUE TO MOVE FORWARD WITH DRIVER STRIKING STEERING WHEEL.
SHOULDER BELT DRIVER SURVIVED	25 MPH	25→0 MPH	0 MPH	ACTION WAS SIMILAR TO CHEST BELT EXCEPT THAT HEAD DID NOT STRIKE STEERING WHEEL.
SHOULDER AND LAP BELT COMBINATION DRIVER SURVIVED	25 MPH	0 MPH	0 MPH	SKETCHES SUGGEST THAT BELT PERFORMANCE UNDER BARRIER IMPACT CONDITIONS RESTRAINED DUMMY FROM STRIKING ANY PART OF CAR INTERIOR. CAR TO CAR IMPACT USING THESE BELTS PROVIDED THE BASIS FOR THIS PARTICULAR PRESENTATION.

FIG. 44.30. Effect of varying safety-belt arrangements on driver and passenger for a 25-mph automobile collision with a fixed barrier. The sketches and evaluations are based on actual collision tests. (*UCLA-ITTE, Severy and Matthewson.*[99])

Double-thickness No. 9 undrawn nylon straps of 3 in. width are most satisfactory for all harnesses[60] with respect to strength, elongation, and supported surface area. If the upper torso is fixed to the back of the seat by any type of harness (shoulder harness, chest belt, etc.), the load on the seat is approximately the same for forward- and aft-facing seats. The difference between these seats with respect to crash tolerance as discussed above no longer exists. These body restraints for passenger and crew must be applied without creating excessive discomfort.[100, 107]

The dynamic properties of seat cushions are extremely important if an acceleration force is applied through the cushion to the body. In this case the steady-state response curve of the total man-seat system (Fig. 44.26) provides a clue to the possible dynamic load factors under impact. Overshooting should be avoided, at least for the most probable impact rise times. This problem has been studied in detail in connection with seat cushions used on upward ejection seats.[87] The ideal cushion is approached when its compression under static load spreads the load uniformly and comfortably over a wide area of the body and if almost full compression is reached under the normal weight. The impact acceleration then acts uniformly and almost directly on the body without intervening elastic elements. A slow-responding foam plastic, for example, with a thickness of 2 to 2.5 in. is very satisfactory in fulfilling these requirements.

PROTECTION AGAINST HEAD IMPACT

The impact-reducing properties of protective helmets are based on two principles: the distribution of the load over a large area of the skull and the interposition of energy-absorbing systems. The first principle is applied by using a hard shell, which is suspended by padding or support webbing at some distance from the head ($\frac{5}{8}$ to $\frac{3}{4}$ in.). High local impact forces are distributed by proper supports over the whole side of the skull to which the blow is applied. Thus, skull injury from relatively small objects and projectiles can be avoided. However, tests usually show that contact padding alone over the skull results in most instances in greater load concentration, whereas helmets with web suspension distribute pressures uniformly.[23] Since helmets with contact padding usually permit less slippage of the helmet, a combination of web or strap suspension with contact padding is desirable. The shell itself must be as stiff as is compatible with weight considerations; when the shell is struck by a blow, its deflection must not be large enough to permit it to come in contact with the head. For light industrial safety hats, molded bakelite reinforced with steel wire, laminated bakelite, or high-strength aluminum alloy are used; for football helmets, combination rubber and plastic compounds molded in a single shell or a vulcanized fiber shell have been used; for antibuffet helmets molded fiberglass flock, molded cotton fabric, or cloth laminates, all impregnated with plastic are some of the most commonly used materials.

Padding materials can incorporate energy-absorptive features. Whereas foam rubber and felt are too elastic to absorb a blow, foam plastics like polystyrene or Ensolite result in lower transmitted accelerations.

Most helmets constitute compromises among several objectives such as pressurization, communication, temperature conditioning, minimum bulk and weight, visibility, protection against falling objects, etc.; usually, impact protection is but one of many design considerations. The protective effect of helmets against concussion and skull fracture has been proved in animal experiments[108] and is apparent from accident statistics. However, it is difficult to specify the exact physical conditions for a helmet which can provide optimum impact protection.

PROTECTION AGAINST BLAST WAVES

Individual protection against air blast waves is extremely difficult since only very thick protective covers can reduce the transmission of the blast energy significantly. Furthermore, not only the thorax but the whole trunk would require protection. In animal experiments, sponge-rubber wrappings and jackets of other elastic material have resulted in some reduction of blast injuries.[5] Enclosure of the animal in a metal cylinder with the head exposed to the blast wave has provided the best protection—short of complete enclosure of the animal. Therefore it is generally assumed that shelters are the only practical means of protecting humans against blast. They may be of either the open or closed type; both change the pressure environment. Changes in pressure rise time introduced by the door or other restricted openings are physiologically most important.[17]

HUMAN TOLERANCE CRITERIA FOR SHOCK AND VIBRATION EXPOSURE

Only approximate limits for man's safety and performance under field conditions can be given, since the exact physical mode of action of the environment varies with man's unpredictable position and motion, and since biological variations with respect to physical, physiological, and psychological reactions make such limits statistical in nature. Individual cases may deviate considerably from the average. Therefore biological criteria must be used with caution; they are summarized here as rough guides for engineering purposes. Tolerance criteria, as well as examples of field environment, indicate only orders of magnitude—not rigid limits. For critical safety problems, a detailed study of the literature or expert consultation is indispensable.

VIBRATION EXPOSURE

Many methods have been developed to assess man's reaction to vibration in a quantitative manner, but most of these are based on a limited number, specific types, or a specific interpretation of experiments and contradict each other to a certain degree.

Whole-Body Exposure. In the absence of any clearly defined vibration-induced occupational disease and recalling the wide variability of the effects of vibration on human performance, it is difficult to select firm rating procedures considering all factors influencing such rating (magnitude, frequency, time function, intermittency, direction of vibration, and position of the body and its coupling to the vibrating surface).   In spite of these acknowledged difficulties, a general guide for the evaluation of human exposure to whole-body vibration has been developed and agreed upon as International Standard ISO 2631, 1974.   This guide gives positive quantitative guidance with respect to safety and acceptability of whole-body vibration for the 1 to 80 cps frequency band and is recommended for the evaluation of vibration environments in transportation vehicles (air, land, and water) and in industry (factories, agriculture).   This guide gives limits for the following items.   (1) The preservation of health or safety (called "exposure limits").   These limits should not be exceeded without special justification and awareness of a health risk.   (2) The preservation of working efficiency (called "fatigue, or decreased proficiency boundary").   It is obvious that this boundary cannot apply to all possible tasks; it applies to such typical tasks as flying an airplane or driving a tractor.   (3) The preservation of comfort ("reduced comfort boundary").   All three boundaries are a function of the average daily exposure time: all boundaries are high for short exposures such as a minute per day and are lowest for 24 hr, i.e., continuous exposure.   The "fatigue-decreased proficiency boundary" for longitudinal (head-to-foot or Z-axis) vibration is presented in Fig. 44.31A with daily exposure time as parameter.   In Fig. 44.31B the same boundaries are shown as a function of exposure time.   For transverse vibration (side-to-side or Y-axis and chest-to-back or X-axis) similar curves are shown in Fig. 44.31C and D.   The difference in the frequency-dependence between Fig. 44.31A and C is a result of the changed transmissibility for the different axis (see Figs. 44.9 and 44.15).   The boundaries are defined in terms of rms value for sinusoidal single frequency exposure or rms value in the third-octave bands for distributed vibra-

FIG. 44.31. Vibration exposure limits as a function of frequency and exposure time.   (*Figures A to D are based on ISO Standard 2631—1974.   Figures E and F are based on unpublished working papers of Subcommittee 4 "Effects of Shock and Vibration on Man" of ISO Technical Committee 108 "Mechanical Shock and Vibration."*)   (*A*) Longitudinal whole-body acceleration limits: fatigue—decreased proficiency boundary.

Fig. 44.31*B*. Longitudinal whole-body acceleration limits as a function of daily exposure time: fatigue-decreased proficiency boundary.

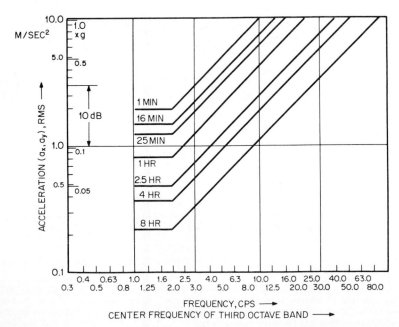

Fig. 44.31*C*. Transverse whole body acceleration limits: fatigue-decreased proficiency boundary.

FIG. 44.31D. Transverse whole-body acceleration limits as a function of daily exposure time: fatigue-decreased proficiency boundary.

tion.   For broad-band vibration, the rms value in each octave band should not exceed the limits shown; for vibration in more than one direction simultaneously, corresponding limits apply to each vectorial component in the three axes.   For evaluating time-varying exposure patterns over a day's period, the standard proposes a method for arriving at an "equivalent exposure with constant vibration magnitude."

The "exposure limits" for health and safety reasons are obtained by raising the boundaries of Fig. 44.31 A to D by a factor of 2 (6 dB higher).   To obtain the "reduced comfort boundaries," the values of the boundaries of Fig. 44.31 must be divided by a factor of 3.15 (10 dB lower).   In transportation vehicles the reduced comfort boundary is related to such functions as reading, writing, and eating.

The standard also gives useful guidance for measuring and reporting the vibration environment when its evaluation with respect to its effects on man is of primary concern.   Adherence to this standard will not only provide uniform evaluation of vibration environments but will also assist in collecting and evaluating more meaningful data and exposure histories than were previously available.

The ISO standard is only very tentative for vibrations having high crest factors (i.e., greater than 3).   Very few studies exist for such vibrations.

The results of one study on single-shock acceleration pulses (such as floor vibrations near drop forges or similar equipment) are presented in Fig. 44.32.[12]

Extension of the curves in Fig. 44.31A to D to frequencies below 1 cps is extremely difficult because it is not possible to use the same categories of human-exposure criteria:

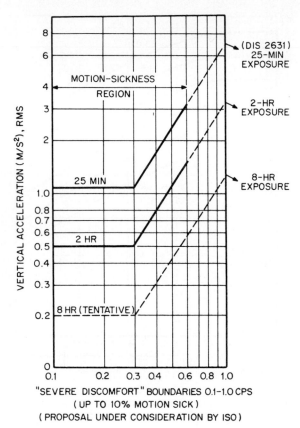

Fig. 44.31E. Severe discomfort boundaries for the frequency range of 0.1 to 1.0 cps (up to 10 per cent motion sickness).

motion sickness enters the picture, severly affecting comfort, work efficiency, and, under certain conditions, health and safety.   Motion sickness also depends on other environmental factors in addition to the physical motion.   It is very variable in its occurrence for the same subject and for different subjects, and it is, in general, very complex and hard to predict.   The best tentative generalized guidance which can be provided at the present time is given in Fig. 44.31E.

Fɪɢ. 44.31F. Vibration exposure boundaries for hand transmitted vibration.

Hand-Arm Vibrations. No generally accepted exposure limits for protecting against hand-transmitted vibration from tools such as pneumatic and electric hand tools used in the manufacturing, mining, or construction industries or chain saws used in forestry are available. Provisional exposure boundaries for routine daily exposures as a function of daily exposure time are provided in Fig. 44.31F. Adherence to the curves, which are based on the best data available, should reduce the risk of occupational vibration-induced disease to acceptable or negligible levels. The curves do not necessarily represent limits of safe exposure for all individuals and working conditions.

FIG. 44.32. Tolerance of human subjects in the standing or supine position to repetitive vertical impact pulses representative of impacts from pile drivers, heavy tools, heavy traffic, etc. Subjective reaction is plotted as a function of the maximum displacement of the initial pulse and its rise time. The numbers indicate the following reactions for the areas between the lines: I_a, threshold of perception; I_b, of easy perception; I_c, of strong perception, annoying; II_a, very unpleasant, potential danger for long exposures; II_b, extremely unpleasant, definitely dangerous. The decay process of the impact pulses was found to be of little practical significance. (*After Reiher and Meister.*[12])

(Biological effects of hand-transmitted vibration are very much influenced by the direction of the vibration transmitted to the hand, the method of working, proficiency, and the climatic conditions.)   The acceleration values of Fig. 44.31*F* are not to be exceeded by accelerations measured at the handtool interface for any direction of a rectangular coordinate system.

Typical Vibration Environments.   The range of vibration levels found in aircraft, trucks, and tractors is indicated in Fig. 44.33.   Individual points represent flight vibration data obtained in various types of military aircraft.[73]   The solid circles, squares, and triangles indicate vibrations at seat levels which were reported as excessive and undesirable in actual service; the open marks indicate conditions accepted as tolerable. The linearized dividing line between tolerable and excessive vibrations is the tolerance criterion, used as a long-time vibration tolerance criterion in military aviation before

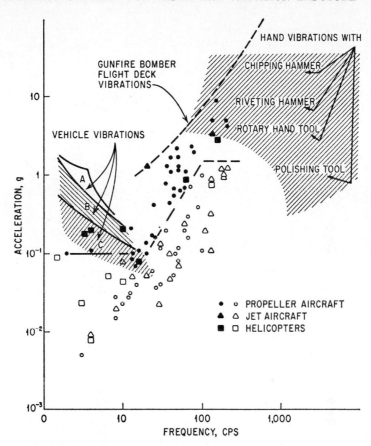

Fig. 44.33. Approximate vibration environments for automobiles, trucks, tractors, and aircraft. Peak values of acceleration are plotted as a function of frequency. Vibration levels observed on hands while operating various hand tools are also indicated. These levels are strongly dependent on tool design and type of operation. Most of the hand vibrations indicated are in the acceleration range at which chronic hand injuries may result. (*After Rathke,[25] Dieckmann,[7] Getline,[72] and Agate and Druett,[41] and others.*)

the ISO standard (Fig. 44.31) was available. The areas indicated for truck and tractor vibration[7, 35] are the range for the respective vibration maxima and do not represent spectra. Most vehicles have very pronounced natural frequencies excited according to ground conditions. Very generally, rubber-tired farm tractors, as well as trucks, have the maximum of their vertical accelerations in the 2 to 6 cps range, whereas large rubber-tired earth-moving equipment is in the 1.5 to 3.5 cps range; crawler-type tractors may have higher frequencies—of the order of 5 cps.

DECELERATION EXPOSURE, CRASH, AND IMPACT

The approximate maximum tolerance limits to rapid decelerations applied to a sitting human subject are summarized in Figs. 44.34 to 44.41.[51] The data are compared and summarized on the basis of trapezoidal pulses on the seat in all four acceleration directions with respect to a sagittal plane through the body axis. The limits as shown are

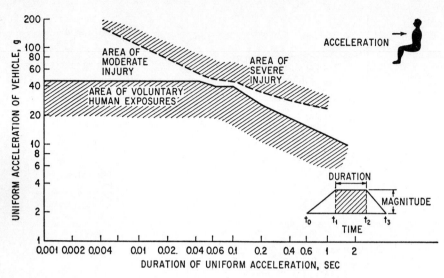

FIG. 44.34. Tolerance to spineward acceleration as a function of magnitude and duration of impulse. (*Eiband.*[51])

FIG. 44.35. Effect of rate of onset on spineward acceleration tolerance. (*Eiband.*[51])

FIG. 44.36. Tolerance to sternumward acceleration as a function of magnitude and duration of impulse. (*Eiband.*[51])

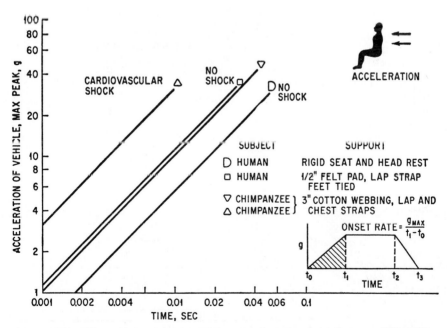

FIG. 44.37. Effect of rate of onset on sternumward acceleration tolerance. (*Eiband.*[51])

FIG. 44.38. Tolerance to headward acceleration as a function of magnitude and duration of impulse. (*Eiband.*[51])

FIG. 44.39. Effect of rate of onset on headward acceleration tolerance. (*Eiband.*[51])

FIG. 44.40. Tolerance to tailward acceleration as a function of magnitude and duration of impulse. (*Eiband.*[51])

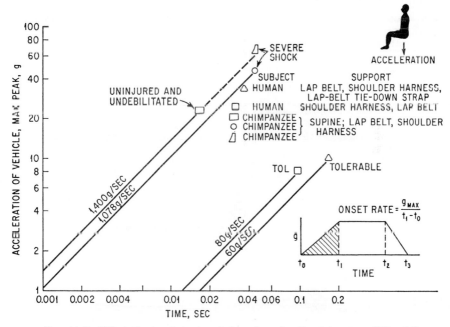

FIG. 44.41. Effect of rate of onset on tailward acceleration tolerance. (*Eiband.*[51])

based on experiments providing maximum body support (see Table 44.4), i.e., lap belt, shoulder harness, thigh and chest strap, and arm rests for the headward accelerations. The quantitative influence of the initial rate of change of acceleration is not too clearly established and not enough data are available for exact mathematical analysis of the influence of the total acceleration-time function.   Although the separation of this function into duration of (uniform) acceleration and onset rate is not completely satisfying, it constitutes the most complete analysis of the experimental evidence available.   Onset rates endured by various subjects therefore are summarized in separate graphs for the

FIG. 44.42. Longitudinal and normal crash loads on a pressurized transport aircraft hitting the ground at 35 mph under an impact angle of 15° and 29°. Acceleration levels in the aircraft are shown as a function of the distance from the point of contact (nose). (*After Preston and Pesman.*[109])

FIG. 44.43. Example of automobile head-on collision deceleration patterns as a function of time. The deceleration for the underbody under the seat (station 6) and the passenger's hip is plotted together with the load function of the seat belt. The data are for two cars engaged in experimental head-on collisions. (Impact speed 31 ft/sec. Kinetic energy of cars before impact approximately 45,000 ft-lb. Cars collapsed under impact approximately 1.7 ft.) (*Severy et al.*[100])

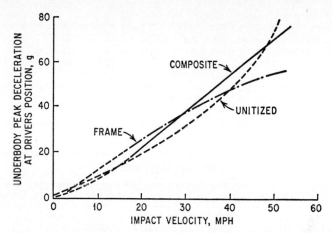

Fig. 44.44. Crash deceleration of the passenger compartment in head-on collisions of automobiles as a function of driving speed. The negligible difference between frame and unitized underbody construction also is shown. (*Severy et al.*[100])

different directions (Figs. 44.35, 44.37, 44.39, and 44.41). In applying these curves, caution must be exercised since they are based on well-designed body supports, minimum slack of the harnesses, heavy seat construction, young, healthy volunteer subjects, and subjects expecting the impact exposure. These curves constitute maxima in many respects, although further improvement in the protection methods certainly does not appear impossible. For example, the maximum limit for exposure to transverse front-to-back acceleration, as experienced in head-on automobile collisions, is indicated in Fig. 44.34 as between 40 and 50g for durations of less than 0.1 sec. For subjects without maximum upper torso restraint having only a lap belt or other types of abdominal restraints, this limit is estimated to be between 10 and 20g.

Approximate ranges for aircraft crash loads can be obtained from Fig. 44.42.[109] Horizontal crash loads, i.e., in the direction of the aircraft's longitudinal axis, increase with the crash angle to a maximum at 90°, whereas vertical loads reach their maximum approximately at 35°. The graph shows only one typical example; aircraft type, ground conditions, and point of initial crash contact have a strong influence in each individual case. For automobile head-on collisions, Fig. 44.43 shows typical deceleration patterns for the car structure under the seat and the passenger's hips; seat-belt loads are also indicated in the graph.[100] The two graphs in this figure are for two cars colliding with each other head-on. Figure 44.44 summarizes the results of many automobile crash

Fig. 44.45. Head-impact acceleration levels for the onset of cerebral concussion as a function of pulse duration. (*L. M. Patrick et al., Proc. Seventh Stapp Car Crash Conf., Ch. C. Thomas. Springfield, Ill., 1965.*)

experiments. The peak deceleration of the car body under the driver's compartment is plotted against the impact velocity. The difference in impact load between the frame and unitized underbody construction was negligible.

Examples of other types of short duration accelerations are given in Table 44.5.

With respect to head injury, the relation between trauma and mechanical insult is very complex and hard to compress into curves readily applicable for operational and/or design use. One curve frequently used for these purposes—and the one around which the performance specifications for most protective helmets (for example, ANSI Standard Z90.1—1971) are written—is presented in Fig. 44.45. It relates the onset of concussion to the magnitude and duration of the impact pulse to the head ("Wayne-State curve").

Table 44.5. Approximate Duration and Magnitude of Some Short-duration Acceleration Loads

Type of operation	Acceleration, g	Duration, sec
Elevators:		
Average in "fast service"......................	0.1–0.2	1–5
Comfort limit................................	0.3	
Emergency deceleration......................	2.5	
Public transit:		
Normal acceleration and deceleration...........	0.1–0.2	5
Emergency stop braking from 70 mph..........	0.4	2.5
Automobiles:		
Comfortable stop...........................	0.25	5–8
Very undesirable...........................	0.45	3–5
Maximum obtainable........................	0.7	3
Crash (potentially survivable)..................	20–100	<0.1
Aircraft:		
Ordinary take-off...........................	0.5	>10
Catapult take-off...........................	2.5–6	1.5
Crash landing (potentially survivable)..........	20–100	
Seat ejection...............................	10–15	0.25
Man:		
Parachute opening, 40,000 ft..................	33	0.2–0.5
6,000 ft..................	8.5	0.5
Parachute landing...........................	3–4	0.1–0.2
Fall into fireman's net.......................	20	0.1
Approximate survival limit with well-distributed forces (fall into deep snow bank).............	200	0.015–0.03
Head:		
Adult head falling from 6 ft onto hard surface....	250	0.007
Voluntarily tolerated impact with protective headgear.................................	18–23	0.02

REFERENCES

1. Franke, E. K., and H. E. von Gierke: Unpublished.
2. von Békésy, G.: *Akust. Z.*, **4**:360 (1939).
3. Franke, E. K.: *J. Appl. Physiol.*, **3**:582 (1951).
4. von Gierke, H. E.: *USAF Tech. Rept.* 6010, 1950.
5. German Aviation Medicine, World War II, vol. 2, Department of the Air Force, Government Printing Office, Washington, D.C., 1950.
6. Harris, C. M. (ed.): "Handbook of Noise Control," chaps. 16, 17, McGraw-Hill Book Company, Inc., New York, 1957.
7. Dieckmann, D.: *Intern. Z. angew. Physiol. einschl. Arbeitsphysiol.*, **16**:519 (1957).
8. Adams, O. S.: *USAF, WADC Tech. Rept.* 57-510, ASTIA Document AD 130983, 1958.
9. Stapp, J. P.: *USAF Tech. Rept.* 5915, pt. I, 1949, pt. II, 1951.
10. Forkois, H. M., and R. W. Conrad: *USNRL, Rept.* 4151, June, 1953.
11. von Gierke, H. E.: *Noise Control*, **5**:8 (1959).
12. Reiher, H., and F. J. Meister: *Forsch. Gebiete Ingenieurw.*, **3**:177 (1932).

13. Watts, D. T., et al.: *J. Av. Med.*, **18**:554 (1947).
14. Severy, D. M., and J. H. Matthewson: *Highway Research Board Bull.*, **91**:39 (1954).
15. Severy, D. M., and J. H. Matthewson: *Clin. Orthopaedics*, **8**:275 (1956).
16. Walker, R. Y., et al.: *USAMRL*, Ft. Knox, *Rept.* 321, November, 1957.
17. White, C. S., et al.: *AEC Civil Effects Test Group Rept.* WT-1179, Project 33.1, October, 1957.
18. Lee, R. N.: *U.S. Navy Mine Defense Lab.*, *Med. Research Rept. 2*, September, 1957.
19. Dubois, A. B., et al.: *J. Appl. Physiol.*, **8**:587 (1956).
20. von Békésy, G.: *Akust. Z.*, **4**:316 (1939).
21. Franke, E. K.: *USAF Tech. Rept.* 6469, April, 1951.
22. Hardy, J. D., and C. C. Clark: *Aero/Space Engineering*, **18**:48, June, 1959.
23. *Cornell Aeronaut. Lab. Rept.* 06-675-0-5, ONR project NM 118-389, April, 1951.
24. Clark, C. C., and F. Gray: *Proc. AGARD Aeromed. Panel*, Athens, 1959.
25. Hertzberg, H. T. E.: *USAF, WADC Tech. Rept.*, 56-30, May, 1958.
26. Evans, G. F., and M. Lebow: *J. Appl. Physiol.*, **3**:563 (1957).
27. Carothers, C. O., F. C. Smith, P. Calabresi: *USNMRI and NBS Rept.* NM 001 056.02.13, October, 1949.
28. Fasola, A. F., R. C. Baker, and F. A. Hitchcock: *USAF, WADC Tech. Rept.* 54-218, August, 1955.
29. Calabresi, P., and F. C. Smith: *USNMRI and NBS Memo Rept.* 51 2, February, 1951.
30. Stacy, R. W., et al.: "Essentials of Biological and Medical Physics," McGraw-Hill Book Company, Inc., New York, 1955.
31. Dieckmann, D.: *Intern. Z. angew. Physiol. einschl. Arbeitsphysiol.*, **17**:67 (1958).
32. Coermann, R. R.: *Human Factors*, **4**:227 (1962).
33. Coermann, R. R., et al.: *Aerospace Med.*, **31**:443 (1960).
34. Goldman, D. E.: In "Handbook of Noise Control," C. M. Harris (ed.), chap. 11, McGraw-Hill Book Company, Inc., New York, 1957.
35. Radke, A. O.: *Proc. ASME*, December, 1957.
36. Coermann, R. R.: *Jahrb. deut. Luftfahrtforschung*, **3**:111 (1938).
37. Von Wittern, W. W.: *Am. Heart J.*, **46**:705 (1953).
38. Roman, J., R. R. Coermann, and G. Ziegenruecker: *J. Av. Med.*, **29**:248 (1958).
39. Dieckmann, D.: *Intern. Z. angew. Physiol. einschl. Arbeitsphysiol.*, **17**:83 (1958).
40. Dieckmann, D.: *Intern. Z. angew. Physiol. einschl. Arbeitsphysiol.*, **17**:125 (1958).
41. Agate, J. N., and H. A. Druett: *Brit. J. Ind. Med.*, **4**:141 (1947).
42. von Békésy, G.: *J. Acoust. Soc. Amer.*, **20**:749 (1948).
43. Franke, E. K.: *USAF, WADC Tech. Rept.* 54-24, 1954.
44. Corliss, E. L., and W. Koidan: *J. Acoust. Soc. Amer.*, **27**:1164 (1955).
45. von Gierke, H. E.: *Noise Control*, **2**:37 (1956).
46. Franke, E. K., et al.: *J. Acoust. Soc. Amer.*, **27**:484 (1952).
47. von Gierke, H. E., et al.: *J. Appl. Physiol.*, **4**:886 (1952).
48. Oestreicher, H. L.: *J. Acoust. Soc. Amer.*, **23**:707 (1951).
49. Goldman, D. E., and T. F. Hueter: *J. Acoust. Soc. Amer.*, **28**:35 (1956).
50. Dussik, K. T., et al.: *Am. J. Phys. Med.*, **37**:107 (1958).
51. Eiband, M.: *NASA Memo* 5-10-59E.
52. Selye, H.: "Stress," Acta Inc., Montreal, 1950.
53. Roman, J.: *USAF, WADC Tech. Rept.* 58-107, April, 1958.
54. Schaefer, V. H., et al.: *USAMRL*, Ft. Knox, Ky., *Rept.* 390, June 20, 1959.
55. Fowler, R. C.: *Shock and Vibration Bull.* 22, *Suppl.*, Dept. of Defense, Washington, D.C., 1955.
56. Pape, R., et al.: *J. Appl. Physiol.*, **18**:1193 (1963).
57. Riopelle, A. J., M. Hines, and M. Lawrence: *USAMRL Rept.* 358, October, 1958.
58. White, D. C.: *USNADC*, Johnsville, Pa., *Rept.* NM 001 111 304.
59. Dart, E. E.: *Occup. Med.*, **1**:515 (1946).
60. Guillemin, V., and P. Wechsberg: *J. Av. Med.*, **24**:258 (1953).
61. Schaefer, V. H., et al.: *USAMRL*, Ft. Knox, Ky., *Rept.* 389, June 12, 1959.
62. Vollmer, E. P., and D. E. Goldman: Unpublished data.
63. Sueda, M.: *Mitt. med. Akad. Kioto*, **21**:1066 (1937).
64. Loeckle, W. E.: *Arbeitsphysiol.*, **13**:79 (1944).
65. Goldman, D. E.: *Am. J. Physiol.*, **155**:78 (1948).
66. Constant, H.: *J. Roy. Aeronaut. Soc.*, **36**:205 (1932).
67. Zand, S. J.: *SAE J.*, **31**:445 (1932).
68. Best, S. G.: *SAE J.*, **53**:648 (1945).
69. Reiher, H., and F. J. Meister: *Forsch. Gebiete Ingenieurw.*, **2**:381 (1931).
70. Jacklin, H. M., and G. J. Liddell: *Purdue Univ. Eng. Bull.*, *Research Series* 44, May, 1933.

71. von Békésy, G.: *Akust. Z.*, **4**:360 (1939).
72. Goldman, D. E.: *USNMRI Rept.* 1, NM 004 001, March, 1948.
73. Getline, G. L.: *Shock and Vibration Bull.* 22, *Suppl.*, Dept. of Defense, Washington, D.C., 1955.
74. Ziegenruecker, G., and E. B. Magid: *USAF, WADC Tech. Rept.* 59-18, 1959; also Magid, E. B., et al.: *Aerospace Med.*, **31**:915 (1960).
75. Clemedson, C. J.: *Physiol. Rev.*, **26**:336 (1956).
76. Harvey, E. N.: *Proc. Am. Phil. Soc.*, **92**:294 (1948).
77. Harvey, E. N., et al.: *Surgery*, **21**:218 (1947).
78. Greaves, F. C., et al.: *U.S. Naval Med. Bull.*, **41**:339 (1943).
79. Denny-Brown, D.: *Physiol. Rev.*, **25**:296 (1945).
80. Hoff, E. C., and L. J. Greenbaum: A Bibliographical Sourcebook of Compressed Air, Diving and Submarine Medicine, Dept. of the Navy, ONR and BuMed, vol. II, 1954.
81. Ruedi, L., and W. Furrer: "Das akustische Trauma," Karger, Basel, 1947.
82. Rushmer, R. F., E. L. Green, and H. D. Kingsley: *J. Av. Med.*, **17**:511 (1946).
83. Barr, J. S., R. H. Draeger, and W. W. Sager: *Mil. Surg.*, **91**:1 (1946).
84. Latham, F.: *Proc. Roy. Soc. (London)*, **B146**:121 (1957).
85. Hess, J. L., and C. F. Lombard: *J. Av. Med.*, **29**:66 (1958).
86. Pesman, G. J., and A. M. Eiband: *NACA Tech. Note* 3775, November, 1956.
87. Hecht, K. F., E. G. Sperry, F. J. Beaupre: *USAF, WADC Tech. Rept.* 53-443, November, 1953.
88. Shaw, R. S.: *J. Av. Med.*, **19**:39 (1948).
89. De Haven, H.: *War Med.*, **2**:586 (1942).
90. Joffe, M., and F. A. Hitchcock: *Fed. Proc.*, **8**:1 (1949).
91. Gross, G. A.: *USN BuAer Rept.* N1-R1, Bibliography on Head and Brain Injuries, January, 1955.
92. Holbourn, A. H. S.: *Lancet*, **2**:438 (1943).
93. Sheldon, C. H., R. H. Pudenz, and J. S. Restarski: *USNMRI Rept.* 2, Project X-182, January, 1946.
94. Ward, J. W., L. H. Montgomery, and S. L. Clark: *Science*, **107**:349 (1948).
95. Gurdjian, E. S., J. E. Webster, and H. R. Lissner: in "Medical Physics," Otto Glasser (ed.), Year Book Publishers, Inc., New York, 1950.
96. Dye, E. R.: *Clinical Orthopaedics*, **8**:305 (1956).
97. Mozell, M. M.: *USNADC Rept.* MA 5802, 1958.
98. Bondurant, S., et al.: *USAF, WADC Rept.* 58-156, 1958.
99. Severy, D. M., and J. H. Matthewson: *Proc. SAE*, January, 1956.   *Trans. SAE*, **65**:70 (1957).
100. Severy, D. M., J. H. Matthewson, and A. W. Seigel: *Proc. SAE*, National Passenger Car, Body and Materials Meeting, Detroit, March, 1958.
101. De Haven, H.: *Aeronaut. Eng. Rev.*, **5**:11 (1946).
102. Bierman, H. R.: *USNMRI, Rept.* 11, Project X-630, April, 1949.
103. Hawthorne, R.: *Space/Aeronautics*, October, 1958.
104. Civil Air Regulations ISO-C39.
105. Pinkel, I. I., and E. G. Rosenbert: *NACA Rept.* 1332.
106. Pinkel, I. I.: *Proc. ASME Aviation Conf.*, Los Angeles, March, 1959.
107. Matthewson, J. H., and D. M. Severy: *Highway Research Board Bull.* 73.
108. Denny-Brown, D., and W. R. Russell: *Brain*, **64**:93 (1941).
109. Preston, G. M., and G. J. Pesman: *Proc. Inst. Aeronaut. Sci.*, January, 1958.
110. Lee, R. H., F. I. Whitten, and F. W. Brown: *U.S. Naval Mine Defense Lab., Med. Research Rept.* 4, NM 640 123, June, 1959.
111. Fibikar, R. T.: *Prod. Eng.*, **27**:177, November (1956).

ADDITIONAL REFERENCES

TOLERANCE CRITERIA—SHOCK AND VIBRATION STANDARDS

"Evaluation of Human Exposure to Whole-Body Vibration," ISO 2631-1985; Part 1: General Requirements; Part 2: Evaluation of Human Exposure to Continuous Vibration and Shock Induced Vibration in Buildings (1 to 80 Hz); Part 3: Evaluation of Exposure to Whole-Body Z-Axis Vertical Vibration in the Frequency Range 0.1 to 0.63 Hz, Guide for the Evaluation of Human Exposure to Whole-Body Vibration, ANSI S3.18-1979, Standards Secretariat, Acoustical Society of America, New York.
"Guide to the Evaluation of the Response of Occupants of Fixed Structures, Especially Buildings and Off-Shore Structures, to Low Frequency Horizontal Motion (0.063 to 1 Hz)," ISO 6897, 1984.
"Hand-Arm (Segmental) Vibration, Threshold Limit Values for Chemical Substances in the Work Environment," adopted by Amer. Conf. of Governmental Industrial Hygienists for 1984–1985, Cincinnati, Ohio, 1984.

"Guidelines for the Measurement and the Assessment of Human Exposure to Hand-Transmitted Vibration," ISO/DIS 5349.2, 1984.

"Vibration and Shock-Mechanical Driving Point Impedance of the Human Body," ISO 5982, 1st ed., 1981-08-01.

Brinkley, J. W.: Acceleration Exposure Limits for Escape System Advanced Development, *SAFE J.* **15**(2):10 (1985).

"Guide to the Evaluation of Human Exposure to Vibration in Buildings," ANSI S3.29, 1983, Standards Secretariat, Acoustical Society of America, New York.

PHYSICAL RESPONSE OF THE HUMAN BODY—MODELING

Payne, P. R., and E. G. U. Band: "A Four Degree-of-Freedom Lumped Parameter Model of the Seated Human Body," Aerospace Med. Res. Lab., AMRL-TR-70-35, January, 1971.

von Gierke, H. E.: "Biodynamic Models and Their Applications," *J. Acoust. Soc. Amer.* **50**:1397 (1971).

Aerospace Med. Res. Lab., *Symp. Biodynamic Models and Their Applications*, AMRL-TR-71-29, December, 1971.

Orne, D., and Y. K. Liu: A Mathematical Model of Spinal Response to Impact, *J. Biomechanics*, **4**:49 (1971).

von Gierke, H. E. (ed.): "Models and Analogues for the Evaluation of Human Biodynamic Response, Performance and Protection," AGARD-CP-253, National Technical Information Service, Springfield," Va., 1978.

HUMAN RESPONSE TO VIBRATION

Harris, C. S., and R. W. Shoenberger: Effects of Frequency of Vibration on Human Performance, *J. Eng. Psychol.*, **5**(1):1 (1966).

"Ride and Vibration Data Manual," Society of Automotive Engineers, Information Report SAE J6a, 1965.

Shoenberger, R. W.: Human Response to Whole-Body Vibration, *Perceptual and Motor Skills*, **34**:127 (Monograph Suppl. 1-V34) (1972).

Hornick, R. J.: Vibration, in "Bioastronautics Data Book," 2d ed., NASA-3006, 1973.

Taylor, W. (ed.): "The Vibration Syndrome," Academic Press, London, 1974.

O'Hanlon, J. R. and M. E. McCauley: Motion Sickness Incidence as a Function of the Frequency and Acceleration of Vertical Sinusoid Motion, *Aerospace Med.* **45**(4):366 (1974).

"Vibration and Combined Stress in Advanced Systems," *AGARD Conf. Proc.* National Technical Information Service, Springfield, Va, 1975.

von Gierke, H. E., C. W. Nixon, and J. C. Guignard: Noise and Vibration, in "Foundations of Space Biology and Medicine," chap. 9, National Academy of Sciences, Washington, D.C., 1975.

Taylor, W. and Pelmear, P. L., (eds.): "Vibration White Finger in Industry," Academic Press, 1975.

Proceedings of the International Occupational Hand-Arm Vibration Conference (D. E. Wasserman, W. Taylor, eds.). U.S. Dept. of Health, Education and Welfare, (NIOSH) Publication No. 77-170, 1977.

Parsons, K. C., and M. J. Griffin: The Effect of the Position of the Axis of Rotation on the Discomfort Caused by Whole-Body Roll and Pitch Vibrations of Seated Persons. *J. Sound Vibration* **5**(1):127 (1978).

Griffin, M. J. and Witham, E. M.: Time Dependence of Whole-Body Vibration Discomfort, *J. Acoust. Soc. Am.* **68**(5)1522 (1980).

Guignard, J. C., and M. E. McCauley: Motion Sickness Incidence Induced by Complex Periodic Waveforms, *Aviat. Space Environ. Med.* **53**(6) 554 (1982).

Griffin, M. J., and C. H. Lewis: A Review of the Effects of Vibration on Visual Acuity and Continuous Manual Control, Part 1: Visual Acuity, Part II: Continuous Manual Control, *J. Sound Vibration* **56**(3):383 (1978).

Leatherwood, J. D., T. K. Dempsey, and S. A. Clevenson: A Design Tool for Estimating Passenger Ride Discomfort within Complex Ride Environments. *Human Factors* **22**:291 (1980).

Shoenberger, R. W.: Psychophysical Comparison of Vertical and Angular Vibrations, *Aviat. Space and Environ. Med.* **51**:759 (1980).

Griffin, M. J., et al.: Vibration and Comfort: I–IV, *ERGONOMICS*, **25**:7 (1982).

"Vibration Syndrome" Current Intelligence Bulletin 38, U.S. Dept. of Health and Human Services, DHHS (NIOSH) Publication No. 83-110, 1983.

HEAD IMPACT—PROTECTION

Caveness, W. F., and A. E. Walker (eds.): "Head Injury," J. B. Lippincott Company, Philadelphia, 1966.

"Specifications for Protective Headgear for Vehicular Users," American National Standards Institute, ANSI S90.1, 1971.

RESPONSE TO IMPACT OR BLAST

Snyder, R. G.: Impact, in "Bioastronautics Data Book," 2d ed., NASA SP-3006, 1973.

King, W. F. and A. J. Merts (eds.): "Human Impact Response," Plenum Press, New York, 1973.

Brinkley, J. W., and H. E. von Gierke: Impact Acceleration, in "Foundations of Space Biology and Medicine," vol. II, pt. 3, Chap. 3 National Academy of Sciences, Washington, D.C., 1975.

Hearon, B. F., J. A. Raddin, Jr., and J. W. Brinkley: "Evidence for the Utilization of Dynamic Preload in Impact Injury Prevention," AFAMRL-TR-82-6, Aerospace Med. Res. Lab., Wright-Patterson A.F.B., Ohio, 1982.

Haley, J. L., Jr. (ed): "Impact Injury Caused by Linear Acceleration: Mechanisms, Prevention and Cost," AGARD-CP-322, National Technical Information Services, Springfield, Va., 1982.

Hearon, B. F., and J. W. Brinkley: Psychomotor Performance after Forward-facing Impact, *Aviat. Space Environ. Med.* **56**(11):1043 (1985).

RESTRAINT—CRASH PROTECTION

Saczalski, K., et al. (eds.): "Aircraft Crash Worthiness," University of Virginia Press, 1975; U.S. Dept. of Transportation, 1975 Ride Quality Symposium, DOT-TSC-OSC-75-40, National Technical Information Services, Springfield, Va.

Desjardins, S. P., D. H. Laananen, and G. T. Singley III: "Aircraft Crash Survival Design Guide," USARTL-TR-79-22B, Vol. II, Applied Technology Laboratory, U.S. Army Research and Technology Laboratories, Fort Eustis, Va., 1980.

Aircraft Torso Restraint System, SAE Aerospace Standard AS-8043, Draft, revised May 30, 1985.

Chandler, R. F.: "Human Injury Criteria Relative to Civil Aircraft Seat and Restraint Systems," Society of Automotive Engineers, 1985.

Hearon, B. F., and J. W. Brinkley: Effect of Seat Cushions on Human Response to $+G_z$ Impact, *Aviat. Space Environ. Med.* **57**(2):113 (1986).

GENERAL—BIOMECHANICS

von Gierke, H. E.: Biodynamic Response of the Human Body, *Appl. Mech. Rev.* **17**:951 (1964).

von Gierke, H. E.: Transient Acceleration, Vibration and Noise Problems in Space Flight, in K. Schaefer (ed.) "Bioastronautics," MacMillan, New York, 1964.

Proc. 1st to 29th (1956 to 1985) Stapp Car Crash Conf., published each year by Society of Automotive Engineers, New York.

Fung, Y. C.: "Biomechanics-Mechanical Properties of Living Tissues," Springer-Verlag, New York, 1981.

INDEX

A-C generator, 6-22
Absolute transmissibility, 30-2
Absorbers:
  activated vibration, 6-34
  auxiliary mass (see Auxiliary mass absorbers)
  distributed mass (see Distributed mass absorbers)
  dynamic (see Dynamic absorbers)
  multimass, 6-19
  pendulum (see Pendulum absorber)
  shock, 1-24
  vibration, 6-1
Accelerated test, 20-7
Acceleration, 1-18, 2-11
  decaying sinusoidal, 23-5, 23-8, 28-18
  density, 22-3, 22-16
  equivalent static, 23-11, 42-10
  gravitational, value of, 3-3
  half-sine, 23-4, 23-8, 23-16
  longitudinal, of human body, 44-28
  measurement of, 12-4, 15-4
  peak, 23-20
  rapidly applied, protection of human beings from, 44-35
    (See also Crashes)
  resonance frequency, 2-15
  shock spectrum, 26-3
  spectral density, 22-3, 22-16
  sustained, 20-2
  transmissibility of, 30-9, 30-14, 30-50
  transverse, of human body, 44-29
  zero period, 24-2
  (See also High-acceleration calibration; High-acceleration devices)
Acceleration damper, 43-14
Acceleration impulse, 23-4, 23-6, 23-13
Acceleration limits in vibration machines, 25-18
Acceleration peaks, distribution of, 22-13
Acceleration pickups (see Accelerometer)
Acceleration pulse, 8-20, 42-11
  comparison of maximum acceleration, 31-17
  effects of pulse shape, 31-17
  response to, 31-14
  velocity-step approximation, 31-17
Acceleration step, 23-4, 23-7, 23-16
Accelerator, vertical, 44-5
Accelerometer:
  amplitude linearity, 12-9
  definition of, 12-1
  effects of acoustic excitation, 12-12
  environmental effects, 12-11, 15-2
  frequency response, 12-9, 15-7
  high-frequency response, 12-6, 12-10
  humidity effects on, 12-11

Accelerometer (Cont.):
  with internal preamplifier, 12-18, 13-2
  linearity, 12-10
  low-frequency response, 12-8, 12-10
  mounting of, 15-5
  phase shift, 12-11
  physical properties of, 12-12
  piezoelectric (see Piezoelectric accelerometer)
  piezoresistive (see Piezoresistive accelerometer)
  preamplifier, 12-18, 13-2
  response, 12-7
  seismic, 12-2
  sensitivity, 12-8
  servo, 14-7
  shear-type, 12-16
  temperature effects, 12-11
  transverse sensitivity, 12-9
Acceptable vibration limits (see Vibration severity)
Acoustic environment, 20-5
Acoustic excitation, 12-12
Acoustic tests, 20-11
Active vibration damping, 33-10
Active vibration isolation systems, 33-1, 33-7
  stability of, 33-20
  steady-state response, 33-17
  transient response, 33-13
Activated vibration absorber, 6-34
  antiroll fins, 6-35
Added mass in fluids, 29-1
Admittance, mechanical, 10-2
  (See also Mobility)
Aerodynamic data and vortex-induced oscillations, 29-30
Aerodynamic excitation, forms of, 29-17
Aerodynamic instability, 29-35
Aerospace equipment, 34-23
Aging:
  dynamic, 35-14
  environmental, 24-25
  mechanical, 24-25
  of rubber, 35-4
  vibratory, 24-28
Air compressors, 19-11
Air-drop delivery, 26-5
Air guns, 26-9
Air springs, 32-24, 33-2
  selection of, 33-5
  stiffness of, 33-3
Air suspension systems, 33-1
AirCap SD 240, 41-16, 41-23
Aircraft:
  crashes, 44-35
  ejection seat, 44-5, 44-28
  engines, 38-38, 38-40

Aircraft (*Cont.*):
  unmanned, 32-14
Aliasing, 13-19, 27-5, 27-8
Allowable response of a system, 34-3
Aluminum, 42-31
  alloy, 42-38
  bronze, 42-31
Ambient environment (*see* Environment)
Amplification, resonant (see *Q*)
Amplification factors, spectrum, 24-9
Amplification ratio, 12-17
Amplifiers, 10-32, 13-1
  charge, 13-2
  internal, 13-2
Amplitude, 1-18
  effect on rubber, 35-12
  resonance, 4-20
Amplitude demodulation, 13-47
Amplitude density, 11-1
Amplitude probability, 11-1
Amplitude ratio, 2-5
Analog, 1-18
  mechanical, human body, 44-11
Analog computer in design of machine
    tools, 40-17
Analog methods of transient analysis, 13-31
Analog sine-sweep test for damping, 37-13
Analog-to-digital conversion, 27-7
Analogy, 1-18
Analysis:
  cepstrum, 13-43
    power, 13-44
  of nonstationary signals, 13-36
  numerical methods of, 28-1, 28-25
    (*See also* Matrix)
  objectives of, 22-6
  of periodic functions, 22-5
  real-time, 13-28
  of stationary signals, 13-26
  statistical, 22-7, 27-1
  tracking, 13-29
  transient, 13-30
Analyzers, signal, 13-1, 13-11, 16-6
Anatomy, human, 44-20
  (*See also* Human body)
Anelasticity, 36-14
Angle:
  of attack, 29-35
  impedance, 10-2
  of lag, 39-17
  loss, 36-13
  phase (*see* Phase angle)
  of unbalance, 39-1, 39-17
Angular frequency, 1-18
  complex, 1-19
Angular mechanical impedance, 1-18
Angular natural frequency, 2-3
Animal studies, 44-5
Annealed copper, 42-28
Annulus, properties of, 3-46
Anthropometric dummies, 44-6
Anthropomorphic dummies, 44-6
Antialiasing, 13-41
  filter, 13-11
Antifriction bearings, 43-26
Antinode, 1-18
Antiresonance, 1-18
Antiroll tanks, 6-3, 6-23
Assembly, 11-2
Associative law, 28-5

Asymmetric shafting, 5-15
Asynchronous excitation, 4-9
Asynchronous quenching, 4-9
Attachment periods, wire and cable, 43-24
Audio frequency, 1-18
Auto spectral density (*see* Power spectral
    density)
Autocorrelation coefficient, 1-18
Autocorrelation function, 1-18, 11-12, 21-13,
    22-24
  determination of, 22-25
  of sine wave, 22-24
  of white noise, 22-24
Automatic balancing machines, 39-25
Automatic frequency control in vibration
    machines, 25-14
Automobiles, 34-25
  crashes, 44-29, 44-35
  model of, 42-4
  springs, 33-7
  use of elastomers in, 35-15
Autonomous systems, 4-14, 4-22, 4-25
Autoregressive moving average method, 21-
    17
Auxiliary circuits, calibration of, 18-32
Auxiliary mass absorbers, 40-10
  activated, 6-34
  applications of, 6-22
  with Coulomb damping, 6-15
  energy dissipation, 6-14
  optimum viscous damping, 6-14
Auxiliary mass damper, 1-18, 6-8, 32-7, 43-14
  antiroll tanks, 6-23
  applications, 6-22, 6-25
  balancing elements, 6-26
  Bibby coupling, 6-28
  in boring bar, 6-35
  chatter, 6-35
  distributed mass absorbers (*see* Distribut-
    ed mass absorbers)
  effects on primary system, 6-8
  fixed points, 6-9
  gear, 6-35
  Lanchester damper, 6-27
  locked frequency, 6-11
  mechanical impedance of, 6-2
  multimass absorbers, 6-19
  nonlinearity in spring, 6-17
  optimum damping, 6-10, 6-11, 6-14
  self-excited vibration, 6-35
  for steady table, 6-37
  tuning, 6-8, 6-10
  washing machines, 6-25
Average damping energy, 36-5
Average functions, 4-17
Average value, 11-4
  relation to rms value, 13-32, 22-10
Averaging, signal, 13-7
  synchronous, 13-42
Averaging ensemble, 27-9
Avionic shock, 20-3
Axial loads, effects on beams, 7-18
Axis:
  journal, 39-2
  of rotation, 3-30, 3-33

Background noise, 1-18
Balance:
  criteria for, 39-30

Balance (*Cont.*):
  perfect, 39-3
Balance characteristics, 38-28
Balance quality grades, 39-1, 39-31
Balancing:
  defined, 39-2
  flexible-rotor, 39-6
  high-speed, 39-2
  low-speed, 39-2
  modal, 39-3
  of moving masses, 43-16
  multiplane, 39-3
  principles of, 39-3
  rigid-rotor, 39-4, 39-5
  for rotating machinery, 6-26, 39-1
  single-plane, 39-3
  static, 39-13
  two-plane dynamic, 39-5
  (*See also* Unbalance)
Balancing equipment, field, 39-2, 39-26
Balancing machine(s):
  bearing support, 39-2
  center-of-gravity, 39-2
  centrifugal (*see* Centrifugal balancing
    machines)
  classification of, 39-22
  cross effect, 39-2, 39-22
  defined, 39-2
  dynamic, 39-2
  evaluation of, 39-23
  gravity, 39-13
  hard bearing, 39-17
  high-speed, 39-9
  maintenance and production, 39-25
  null force, 39-2
  operating principles of, 39-13
  resonance, 39-3
  single-plane, 39-3, 39-13
  soft bearing, 39-14
  standards for testing, 39-24
  universal, 39-25
  (*See also* Unbalance)
Balancing speed, 39-28
Balancing techniques, modal, 39-9
Ballistic pendulum calibrator, 18-20
Ballistocardiograph, 44-13
Bandpass filter, 1-18
  ideal, 22-16
Bandwidth, 2-15, 13-13
  effective, 1-18, 13-13
  nominal, 1-22
  selection of, 13-14
Barrier temperature for rubber, 35-5
Bars, longitudinal impact on, 9-8
Base-mounted package, 41-8
Beam:
  axially loaded, 7-18
  cantilever, 17-6, 17-8
  with concentrated loads, 1-14, 42-22
  continuous (*see* Continuous beams)
  damped elastically supported, 9-5
  distributed load, 42-21
  elastic characteristics, 42-2
  lateral, 7-12, 7-22, 7-25, 7-33
  method for measuring damping, 37-15
  modal analysis of, 21-2
  model of, 42-3
  multiple supported, 7-19
  natural frequencies of, 1-14, 1-15, 7-10, 7-
    20

Beam (*Cont.*):
  partially clamped, 7-22
  rotary motion and shearing forces, 7-16,
    7-35
  sections of, 1-14
  simple, 8-35
  strain and kinetic energies of, 7-3
  supported, 9-5
  transverse impact on, 9-3
  uniform (*see* Uniform beam)
  variable cross-section, 7-19
Beam theory, Timoshenko, 7-16
Bearing(s), 16-8, 16-15, 34-21, 43-26
  support, defined, 39-2
  vibration, 19-13, 40-5
Beats, 1-18, 11-10
  period of, 4-9
Belleville springs, 32-15
Belt drive vibration, 40-3
Belt friction, 4-4
Bending moments of loaded plates, 7-27
Bending stresses of plates, 7-26
Bernoulli's equation, 29-2
Bessel functions, 4-28
Bevel, vibro-damping, 40-11
Biaxial stiffness isolators, 30-28
Bibby coupling, 6-28
Bifilar pendulum absorber, 6-32
Bilinear form, 28-7
Bilinear restoring force, 4-24
Binomial theorem, 4-30
Biological measurements, 44-3
Biological response of human body, 44-20
Bistable vibration, 5-18
Black box systems, 10-33, 10-37
Blackman-Tukey method, 27-8
Blast tubes, 44-4
Blast waves:
  effect on human body, 44-23
  protection of human body from, 44-40
Blocked output force, 30-41
Blocked transfer impedance, 30-41
Blowers, air, mounting of, 43-32
Blows, effect on human body, 44-27
Bluff bodies, 29-7
Bode plot, 13-38
Body, human (*see* Human body)
Bogoliuboff's method, 4-15, 4-30
Bolts, 43-7
  fatigue, 43-7
  grade, 43-7
  joints, 43-7
Bonded isolators, 32-21
Bonded-layer damping, 37-29
Bonded rubber dampers, 38-31
Bonded strain gages, 17-1
Boring bar, 6-35
Branched system, 38-7
Brass, 42-27, 42-31, 42-37
Brittle fractures, 42-28
Brittle materials, 43-22
Broad-band random vibration, 1-19
Bronze, 42-27, 42-31
Brute force machine, 25-2
Buckling loading, 32-5
Buffeting:
  of structures, 29-19
  of wake, 29-34
Building design, 24-14
Buildings (*see* Structures)

Bulk modulus, 42-35
Bump test machine, 26-18
Buna N and buna S, 32-2
Burst random excitation, 21-25
Butt weld, 9-10

Cabinet design, 43-4
Cables:
  characteristic impedance of, 15-3
  noise, 15-15, 15-16
  in vibration measurements, 15-13
  and wires and connectors, 43-24
Calibration:
  of auxiliary circuits, 18-32
  of charge amplifiers, 18-33
  field, 15-11
  of impedance converters, 18-33
  mechanical impedance, 10-46
  overall measurement system, 15-11
  permanent, balancing machines, 39-3
  of shock machines, 26-3
  of strain gages, 18-32
  transducer, 15-11, 18-1
    (*See also* Transducer calibration methods)
  of voltage amplifiers, 18-33
  voltage substitution method, 15-13
Calibration factor, 18-1
Calibrators:
  shock, 18-19
  for transducers (*see* Transducer calibration methods)
  for vibration machines, 25-17
Cam, spiral, 43-21
Campbell diagram, 13-38
Cantilever springs, 32-17
Capacitance-type transducer, 14-8
Capacity:
  force, 25-2
  load, 25-1
Carbon steel, 42-26
Carton, 41-3
Cast iron, 42-26
Cast steel, 42-26
Causal signal, 13-45
Causality, error, 21-20
Cellulose wadding, 41-16, 41-21, 41-25
Cements for mounting transducers, 15-8
Center-of-gravity (mass center):
  defined, 39-2
  determination of, 3-4
  installation, 34-3
Center-of-gravity system, 3-24
  equivalent, 3-28
Center-of-mass, 3-3
Central impact of two spheres, 9-1
Central-limit theorem, 11-10
Centrifugal balancing machines, 39-14, 39-22
  classification of, 39-22
Centrifugal compressors, 16-8, 19-1
Centrifuge:
  quick-starting, 26-19
  use in transducer calibration, 18-8
Centroid, 3-4
Cepstrum analysis, 13-43, 16-16
  complex, 13-45
  power, 13-44

Ceramics, 43-22
  for transducers (*see* Piezoelectric accelerometer)
Characteristic equation, 21-6, 28-14
Characteristic impedance, 10-23, 15-3, 30-53
Characteristic vectors, 28-14
Charge preamplifier, 13-2
  calibration of, 18-34
Charge sensitivity, accelerometers, 12-17
Chassis design, 43-2, 43-10
Chatter, 5-16, 5-20, 40-7
  machine-tool, 40-7
  method of calibration, 18-3
Circle, properties of, 3-46
Circuit boards, 43-24
Circular frequency, 1-19
Circular-motion vibration machine, 25-4
Circular plate, 1-16, 7-30, 42-23
Circular rings, 7-36
Circular rods, 7-3, 7-5
Circular sector, properties of, 3-46
Classical method of solution, 7-2, 7-12, 7-36
Classical vibration analysis, 37-2
Clip gage, 17-7
Coefficient:
  critical damping, 2-4
  damping, 2-2
  equivalent damping, 37-2
  Fourier (*see* Fourier coefficients)
  influence, 28-9, 30-20, 30-28, 30-30, 30-31
  of restitution, 9-1, 9-4, 9-7
Coherence:
  multiple, 21-21
  ordinary, 21-19, 21-21
  partial, 21-21
Coherence functions, 22-5, 27-7
Coil springs, 4-2, 32-9
Coined stiffener, 43-4
Cold working, 43-8
Colliding bodies, 9-3
Column vector, 28-2
Comparison method of calibration, 15-11, 18-29
Compensating balancing machine, 39-2
Compensation circuits, 33-11
Complementary error function, 11-11
Complementary function, 8-3
Complex angular frequency, 1-19
Complex cepstrum, 13-45
Complex eigenvalue problem, 28-21, 28-29
Complex function, 1-19, 22-4
Complex modulus in shear, 36-13
Complex modulus damping:
  properties of, 37-24
  (*See also* Structural damping)
Complex-plane representation, 10-1
Complex shock, 23-3, 23-6, 23-8, 23-18
Complex vibration, 1-19
Compliance, 1-19, 10-4
Component, flexible, 31-20
Component design, 43-21
Compound pendulum, 2-26, 3-6
Compressibility, 42-25
Compression isolator, 32-21
Compression loading of rubber, 32-4, 35-15
Compression set of rubber, 35-3
Compression spring (*see* Helical compression springs)
Compression-type accelerometers, 12-15

Compressors, vibration standards for, 19-7
Computers:
  analog, in design of machine tools, 40-17
  digital (*see* Digital computers)
  use in condition monitoring, 16-18
Conbur tester, 26-7
Condition monitoring of machinery, 16-1,
  16-7
Conduit, flexible, 34-18
Cone, properties of, 3-52
Coned disc springs (*see* Belleville springs)
Connectors, wires, and cables, 43-24
Constant-acceleration vibration machine,
  25-1
Constant bandwidth, 13-14, 16-6
Constant-displacement vibration machine,
  25-1
Constant percentage bandwidth, 13-14, 16-6
Constantan foil, 17-3
Constrained-layer damping, 37-2, 37-4,
  37-27
Containers, shipping, 26-5, 34-26
Continued matrix products, 28-5
Continuous beams:
  multiple supports, 7-19
  natural frequencies, 7-21
Continuous system, 1-19
  approximation to, 28-13
  response of, 11-11
Control:
  of shock, 31-1
  of vibration, 30-1, 32-1
Control systems:
  random vibration testing, 25-10
  sinusoidal vibration testing, 25-10
  specifications for, 25-12
  system ratings, 25-11
  (*See also* Vibration machines)
Controller gain, 33-8
Conversion factors for simple harmonic mo-
  tion, 1-8
Convolution integral, 8-4
Coordinate:
  generalized, 2-19
  normal, 8-52
  principal, 8-52
Copper, 42-28, 42-31
Copper-zinc alloys (*see* Brass)
Coradi analyzer, 22-6
Corner drop, 41-9
Correction plane interference, 39-2, 39-21
Correlation analysis, 22-23
  with digital computers, 27-10
Correlation coefficient, 1-19
Correlation function, 1-19, 11-12, 27-6
  crosscorrelation function, 11-12, 22-23, 22-
    24, 22-26
  related to spectral density, 27-6
  (*See also* Autocorrelation function)
Cosines, direction, 3-6
Cotter pins, 43-10
Coulomb damper:
  elastically connected, 30-2, 30-12
  rigidly connected, 30-2, 30-7
Coulomb damping, 4-21, 8-50, 30-2, 31-12, 36-
  22
  in auxiliary mass absorber, 6-15
Coulomb friction (*see* Coulomb damping)
Couple unbalance, 39-2

Coupled modes, 1-19
  multiple degree-of-freedom system, 30-26
  natural frequencies, 3-15, 3-18, 3-21, 30-19
Coupled motion, 42-15
Coupled system, 31-2, 42-8
Coupling:
  Bibby, 6-28
  flexible, 34-19
  modal, 3-15
  torsionally elastic, 28-5
Coupling factor, 1-20
Coupling stiffness, 3-12
Coupling terms, 28-9
Crank spacing, 38-28
Crank web, 38-5
Crankshaft stiffness, 38-6
Crashes:
  automobile, 44-29, 44-35
  human exposure limits, 44-47
  protection for humans, 44-35
Creep characteristics, 41-13
  of cushioning materials, 41-15
  (*See also* Drift)
Crest factor, 13-6
Criteria:
  for balancing, 39-30
  for cumulative fatigue damage, 42-35
  for machinery vibration, 19-2
Critical damping, 1-19, 2-4
  equivalent fraction of, 30-7
  fraction of, 2-4
  of rubber, 35-12
Critical speed, 1-19, 5-19, 39-2, 39-6
  avoiding, 38-27
  defined, 39-2
  of flexible rotors, 39-2
  flexural, 39-2
  minor, 38-2
  order of, 38-2
  shifting the, 38-27
Cross-axis sensitivity (*see* Transverse sensitivity)
Cross effect, balancing-machine, 39-2, 39-22
Cross talk, 15-3
Crosscorrelation functions, 11-12, 13-47, 22-
  24, 22-26
Crushable materials, 41-13
Crystal pickups (*see* Piezoelectric accelerometer)
Cube, properties of, 3-51
Cubic equation:
  graphical solution, 3-16
  roots, 3-18
Cumulative distribution function, 22-8
Cumulative fatigue damage, 42-35
  concept of, 11-4
  criteria for, 42-35
Cup-type isolator, 32-23
Cushioning, 41-1
  nonlinear, 41-26
  package (*see* Package cushioning design)
Cushioning materials, 41-13
  creep characteristics of, 41-15
  described, 41-15
  static stress versus maximum accelera-
    tion, 41-15
  strain characteristics of, 41-14
  (*See also* Packaging materials)
Cushions, seat, 44-32, 44-39

Cycle, 1-19
Cylinders:
  inertia torque of, 38-17
  properties of, 3-52
  tapered, 29-33

D'Alembert's principle, 7-40
Damage fatigue (*see* Cumulative fatigue
  damage)
Damped complex exponential method, 21-15
Damped elastically supported beam, 9-5
Damped multiple degree-of-freedom sys-
  tems, 2-22
Damped natural frequency, 1-19, 2-4, 2-15
Damped structures, isolation of, 30-49
Damped vibration absorber, 6-1
Damper, 2-1, 2-2, 10-3
  acceleration, 43-14
  auxiliary mass (*see* Auxiliary mass damp-
    er)
  bonded rubber, 38-32
  Coulomb (*see* Coulomb damper)
  Houdaille-type, 38-32
  impact, 40-10
  Lanchester, 6-27, 38-32, 40-10
  mass, 6-1
    (*See also* Auxiliary mass damper)
  paddle-type, 38-32
  pendulum, 38-32
  viscous (*see* Viscous damper)
Damper-controlled system, 2-9
Damping, 1-19, 36-1, 43-12
  active, 33-10
  Coulomb (*see* Coulomb damping)
  criteria for, 37-6
  critical (*see* Critical damping)
  dynamic hysteresis, 36-14
  economic value of, 37-3
  effect of frequency on, 37-28
  energy (*see* Damping energy)
  and forced vibration, 7-43
  fundamental relationships, 36-5
  general, 4-20
  greater than critical, 2-4
  in isolators, 32-4
  lateral deflection method of measuring,
    36-4
  lateral force method of measuring, 36-5
  less than critical, 2-4
  linear, 4-19, 4-21, 4-35
  logarithmic decrement, 36-4
  of machine tools, 40-9, 40-16
  magnetoelastic, 36-18
  material, 36-1
  measurement of (*see* Damping measure-
    ments)
  measures of, 37-7
  mechanical, 36-2
  nonproportional, 21-9
  nth-power velocity, 4-21
  optimum, 6-11
    parameters for, 30-15
  of propellers, 38-31
  properties of materials, 36-13
  proportional, 21-9
  of rubber, 35-8
  shear parameter, 37-31
  by sliding, 36-23

Damping (*Cont.*):
  slip, 36-2, 36-22, 43-12
  sources of, 37-1
  static hysteresis, 36-14
  stress-strain loop for measuring, 36-3
  structural (*see* Structural damping)
  in structural elements, 24-17
  tests for determining (*see* Damping mea-
    surements)
  in torsional systems, 6-3
  uniform mass, 2-22
  velocity-squared, 4-11, 4-12, 31-12
  vibrating specimen used in measuring,
    36-3
  vibration damping tapes, 37-24
  of viscoelastic materials, 36-12
  viscous (*see* Viscous damping)
  volume-stress function, 36-7
  (*See also* Damper; Damping materials;
    Damping measurements; Damping
    treatments)
Damping coefficient, 2-2
Damping energy:
  average, 36-5
  ratio to strain energy, 36-9
  specific, 36-5, 37-11
    relation to stress level, 36-6
  total, 36-5
Damping layers:
  constrained, 37-2, 37-4, 37-27, 37-30
  free, 37-3, 37-29
Damping links, 37-5, 37-27
Damping materials:
  complex modulus of, 37-21
  effects of strain on, 37-28
  effects of temperature on, 37-28
  elastomeric, 37-2
  polymeric, 37-2
  properties of, 37-19, 37-28
  types of, 37-28
Damping matrix, 28-8
Damping measurements, 24-24, 36-3, 36-5,
    37-13
  free-layer method, 37-16
  Geiger thick-plate test, 37-16
  impact test, 37-13
  impedance test, 37-18
  loss factor, 37-17
  modulus, 37-17
  sandwich beam, 37-16
  scale model, 37-13
  sine-sweep tests, 37-13
  vibrating beam, 37-15
Damping mechanisms, 21-9, 37-1, 37-2
  linear, 36-20
  material, 36-20
  nonlinear, 36-20
Damping ratio (*see* Fraction of critical
    damping)
Damping treatments:
  area required, 37-27
  bonded-layer, 37-29
  constrained-layer, 37-2, 37-4, 37-27
  definition of, 37-1
  discrete, 37-6
  free-layer, 37-3, 37-24
  frequency sensitive, 37-5
  how they work, 37-2
  initial effects of, 37-28

Damping treatments (*Cont.*):
  integral, 37-5
  selection of, 37-27
  tuned dampers, 37-2
  types of, 37-3
Dashpot, 4-4
Data:
  collection of, 20-5
  raw, 20-5
Data acquisition, 20-5
Data preparation, A-to-D conversion, 27-7
Data reduction:
  to frequency domain, 23-5
  methods of, 23-23
  to response domain, 23-9
  shock, 23-1
  vibration, 22-1
Data windows (*see* Windows, data)
Dead-weight load, 25-1
Deadeners (*see* Damping materials)
Decaying sinusoidal acceleration, 23-5, 23-8, 23-18
Decaying sinusoidal shock, 23-3, 23-6
Deceleration effect on humans, 44-27, 44-47
Decibel, 1-19
Decoupled multiple degree-of-freedom system, 30-7
Decoupled natural frequencies, 3-26, 30-27
Decoupling of modes, 3-27, 3-28, 30-27
Decrement, logarithmic, 1-21, 2-5, 36-4, 36-9
Deflection:
  plate, 7-27
  static, 2-4
Deformation:
  from impact, 9-8
  permanent, 43-13
Degrees-of-freedom, 1-19, 2-16
  distributed systems, 7-1
  (*See also* Multiple degree-of-freedom system: Single degree-of-freedom system; Three degrees-of-freedom system; Two degrees-of-freedom system)
Delayed functions, 8-11
Delta function, 22-9, 23-4
Demodulation:
  amplitude, 13-47
  phase, 13-48
Demultiplication frequency, 4-8
Density, probability, 11-2, 22-7
Design:
  of equipment (*see* Equipment design)
  gust factor approach in, 29-13, 29-27
  lateral forces, 24-15
  limit, 9-10
  of machine tools, 40-17
  package (*see* Package cushioning design)
  response spectrum, 24-7
  spectrum, 24-11
  stress in rubber isolators, 35-17, 35-19
  of structures, 29-12, 43-1
Detectors, signal, 13-5, 13-7
  envelope, 13-8
  peak, 13-7
  rms, 13-6
Deterministic function, 1-19, 22-5
Deterministic signal, 27-8
Development test, 20-7
DFT (*see* Discrete Fourier transform)
Diagonal matrix, 28-3

Diagonalization by successive rotation, 28-27
Diesel engines, 38-40
Differential-transformer transducer, 14-6
Digital computers, 27-1
  analog-to-digital conversion, 27-7
  analytical application, 27-2
  applications of, 20-4
  coherence function, 27-7
  component mode synthesis in, 27-2
  correlation analysis with, 27-10
  in design of machine tools, 40-18
  deterministic signal, 27-8
  digital vibration data analysis, 27-7
  discrete Fourier transform (DFT) in, 20-6, 27-4
  distributed system programs, 27-3
  experimental applications, 27-3
  finite-element programs, 27-2
  Fourier transform, 27-4
  frequency-response functions, 27-6
  fundamental relationships, 27-4
  general-purpose programs, 27-2
  leakage error, 27-9
    (*See also* Leakage error)
  NASTRAN program, 27-2
  random signal, 27-8
  special programs, 27-3
  spectral analysis, 27-8
  spectral-density–correlation-function relationship, 27-5
  statistical analysis with, 27-10
  statistical error in, 27-9
Digital control systems:
  modal testing, 27-13
  random vibration tests, 27-10
  shock-spectrum control tests, 27-12
  for shock testing, 27-10
  sinusoidal vibration tests, 27-12
  for vibration testing, 27-10
  waveform control tests, 27-12
Digital filters, 13-11
  analysis with, 13-33
Digital recorders, 13-11
Digital sine wave sweep test, 37-13
Dirac delta function, 22-9
Direct-drive mechanical vibration machines, 18-27, 25-2
Direction cosines, 3-6
Disc springs (*see* Belleville springs)
Discrete damping treatments, 37-6
Discrete Fourier transform (DFT), 13-18, 20-6, 27-4
  compared with Fourier transform, 13-19
Discrete frequency spectrum, 22-3
Displacement, 1-19
  generalized, 28-8
  resonance frequency, 2-15
  response, 2-9
  shock spectra, 26-3
  transducer, 14-8
  transmissibility, 30-7, 30-14
Displacement maxima, 2-5
Displacement measurement, 14-2, 14-3, 15-4, 18-9
Dissipation function, 28-6
Distortion, 1-19
Distributed mass, 10-22

Distributed mass absorbers, 6-19
  damping, 6-20
  tuning, 6-20
Distributed mass dynamic absorbers, 6-19
Distributed system (continuous system), 7-1, 7-2
Distribution:
  Gaussian, 11-4, 22-7, 22-10
  probability (*see* Probability distribution)
  skewed, 11-5
Distribution function:
  cumulative, 22-8
  determination of, 22-13
  magnitude, 22-23
Dow metal F, 42-28
Drag, mechanical (*see* Hysteresis)
Drift:
  relative, 35-6
  of rubber, 35-6
Driving-point impedance, 1-20, 10-2, 10-7
Driving transform, 8-9
Drop-ball shock calibrator, 18-22
Drop conditions, 41-3
Drop tables, 26-6, 26-10
Drop test, 26-5
Drop towers, 44-5
Dry friction damping (*see* Coulomb damping)
Dry friction whip, 5-10
Dual centrifuge, 18-8
Ductility, 42-25, 42-28
Ductility factor, 24-10
Duffing's equation, 4-12
Duffing's method, 4-12
Duhamel's integral, 8-4
Dummies, anthropometric, 44-6
Duraluminum, 42-29, 42-31
Duration of shock pulse, 1-20
Durometer, 35-2
Dwell frequency, calibration method, 18-31
Dwell test, 20-9
Dynamic absorbers, 1-20, 6-1, 6-6
  a-c generator, 6-22
  amplitude of motion of, 6-7
  applications of, 6-22
  distributed mass, 6-19
  location, 6-7
  multimass, 6-19
  order of vibration, 6-28
  pendulum, 6-28
  sensitivity, 6-7
  tuning, 6-7
Dynamic aging of rubber, 35-14
Dynamic balancing, two-plane, 39-5
Dynamic balancing machine, 39-2
Dynamic hysteresis, 36-14
  of viscoelastic materials, 36-16
Dynamic load, 42-2
Dynamic load factor, 42-12
Dynamic mass, 12-25
Dynamic properties of rubber, 35-8
Dynamic range, 13-16
Dynamic shear strain in elastomers, 32-8
Dynamic stability in machine tools, 40-12
Dynamic stiffness, 12-25, 32-8, 37-19, 40-12
Dynamic unbalance, 39-2, 39-5
Dynamic vibration absorber, 1-20, 4-35

Ear damage, 44-26
Earthquake:
  ground motion, 24-4
  operational basis, 24-24
  safe-shutdown, 24-24
Earth's gravitation field, calibration using, 15-11, 18-5
Eddy-current transducers, 14-5
Effective bandwidth, 1-20, 13-13
Effective force, 25-2
Effective load, 25-1
Eigensystem realization algorithm method, 21-16
Eigenvalue problem, 28-14
  complex, 28-21, 28-29
  real, 28-25
Eigenvector, 28-14
Eigenvector expansions, 28-15
Ejection seat, 44-5, 44-27, 44-28
Elastic axes, principal, 3-11, 30-31
Elastic center of resilient element, 3-11
Elastic design spectrum, 24-11
Elastic materials, 41-13
  (*See also* Damping materials)
Elastic plate, 9-3, 42-41
Elastic properties:
  of beams, 42-21
  of metals, 36-22
Elastic waves, 9-3
Elastically connected Coulomb damper, 30-2, 30-12
Elastically connected viscous damper, 30-2, 30-9
Elasticity:
  hyperbolic tangent, 31-8
  modulus of, 42-24
  skin, 44-13
  tangent, 31-8
Elasticity test for flexible rotors, 39-8
Elastomeric damping materials, 37-2
Elastomers, 35-1, 32-2
  in automobiles, 35-15
  characteristics of, 32-2
  damping in, 32-7
  dynamic shear strain, 32-8
  failure in, 32-3
  (*See also* Rubber)
Electric motors, 34-20
  induction, 19-4
  spectrum analysis of, 16-12
  squirrel-cage, 19-4
  standards for, 19-4
Electrical wiring, 15-13, 43-26
Electrodynamic exciter, 18-28, 25-7
Electrodynamic transducers, 14-4
Electrodynamic vibration machines, 18-28, 25-7
Electronic equipment, 34-23
Electrostatic shields, 15-16
Electrostriction, 1-20
Ellipse, properties of, 3-54
Ellipsoid, properties of, 3-54
Endurance curves, 42-30
Energy:
  damping (*see* Damping energy)
  fracture, 42-28
  kinetic (*see* Kinetic energy)
  potential, 28-6

Energy (*Cont.*):
  strain (*see* Strain energy)
Energy balance method, 38-16
Energy dissipation, 6-14
Energy method, 9-10
Energy storage capacity, 42-19
Engine:
  amplitude calculations, 38-18
  diesel, 38-40
  gasoline, 22-10
  internal combustion, 6-26, 38-36
  radial aircraft, 38-40
  rocket, 11-1
  V-type, 38-36
Ensemble, 1-20, 11-3
  averaging, 27-8
Entrainment of frequency, 4-9
Envelope detection, 13-8, **16-14**
Enveloping, spectral, 20-16
Environment:
  acoustic, 20-5
  descriptions of, 20-2
  effect on rubber, 35-4
  effects on isolators, 34-24
  effects on transducers, 12-11, 15-2
  induced, 1-21
  mechanical simulation of, 44-4
  for packaging, 41-3
  test, 20-7
    (*See also* Environmental vibration testing)
Environmental aging, 24-25
Environmental conditions, simulation of, 44-4
Environmental specification, 20-13
Environmental vibration testing, 25-25
Equations:
  Bessel's, 4-28
  characteristic, 28-14
  cubic (*see* Cubic equation)
  Duffing's, 4-12, 4-29
  frequency, 2-18, 7-38, 28-14
  Hill's, 5-16
  Lagrangian (*see* Lagrange's equations)
  Rayleigh's, 4-6, 4-25
  subsidiary, 8-9
  Van der Pol's, 4-15, 4-17
Equipment:
  rigidly supported, 31-27
  shock isolated, 31-28
Equipment design, 42-1, 42-9, 42-19, 43-1
  metal properties, 42-32
  properties of materials, 42-24
Equivalent center-of-gravity system, 3-28
Equivalent fraction of critical damping, 30-7
Equivalent mass, 6-2, 6-4
Equivalent model, 42-2
Equivalent rectangular pulse, 31-14
  average width of, 31-14
  effective duration of, 31-14
Equivalent static acceleration, 23-11, 42-10
Equivalent static response, 23-20
Equivalent static stiffness, 42-12
Equivalent system, 1-20, 10-20
  Norton's, 10-25, 10-31
  pi, 10-25, 10-33
  T, 10-25, 10-33

Equivalent system (*Cont.*):
  Thévenin's, 10-24, 10-30
Equivalent viscous damping, 1-20, **30-7**, 43-12
Ergodic function, 22-17
Ergodic hypothesis, 11-3
Ergodic process, 1-20
  quasi-, 1-23
Error:
  aliasing, 27-5, 27-8
  leakage, 13-20, 21-21, 27-9
Error function, 11-11
Error reduction methods, 21-20, 21-22
Excitation, 1-20, 21-22
  acoustic, 12-12
  asynchronous, 4-9
  burst random, 21-25
  forms of, 8-3
  general, response to, 28-19
  impact, 21-24
  impulse, 8-20
  periodic random, 21-25
  pseudorandom, 21-25
  pulse types, 8-20
  pure random, 21-24
  random (*see* Random excitation)
  rectangular-step, 8-50
  rotating force, 3-34
  selective, 21-22
  single-pulse, 8-47
  sinusoidal, 2-6
  step-type function, 8-14
  swept sine wave, 21-24
  transient, 8-1
Excitation functions, 8-14
Excitation signals, 21-25
Exciters, vibration (*see* Vibration machines)
Experimental modal analysis, 24-1, 24-10
Exponential pulses, 8-25, 8-36, 8-41
Exponential window, 13-23

Fans, 32-4, 34-21
Fast Fourier transform (FFT), 20-6, 23-24
  analysis, 13-24
  analyzers, 13-18, 13-27, 13-4
Fatigue, 20-10
  behavior, 42-33
  in bolts, 43-33
  damage (*see* Cumulative fatigue damage)
  metal, 36-22, 42-29
  of rubber, 35-13
  static, 13-13
  steel, 42-30
  strength, 43-7, 43-9
Fatigue testing:
  machines for, 25-18, 25-22
  mechanical resonance in, 25-19
  vibration machines in, 25-19
Fault diagnosis in machinery, 16-4, 16-7
FFT (*see* Fast Fourier transform)
Fiber-optic sensor, 14-3
Field balancing equipment, 39-2, 39-26
Field calibration techniques, 15-11
Filters, 1-20
  antialiasing, 13-11
  bandpass, 1-18, 22-16
  bandwidth of, 13-13, 13-14

Filters (*Cont.*):
  digital, 13-12
  electrical, 13-11
  high-pass, 1-21
  low-pass, 1-21
  properties of, 13-14
  use of, 13-26, 13-30
Finite-element computer programs, 27-2
Firing order, 38-28
First probability distribution, 11-3
Fischer-Hinnen rule, 38-19
Fixed frequency vibrators, 25-1
Fixed points, auxiliary mass damper, 6-9
Fixed reference transducers, 12-2
Fixture problems, 20-10
Flat drop, 41-9
Flat plates (*see* Plate, flat)
Flattop window, 13-23
Flexible conduit and tubing, 34-18
Flexible couplings, 34-21
Flexible materials, 41-13
Flexible metal hose, 34-21
Flexible rotor(s), 38-1
  balancing, 39-6
    (*See also* Modal balancing)
  definition of, 39-12
  mode shapes, 39-8
Flexibility matrix, 28-9
Flexural critical speed, 39-2
Flight-deck vibration, 44-12
Floating shock platform, 26-12
Flow of liquids in pipes, 29-14
Flow-induced vibration in fluids, 29-1
Fluid bearing whip, 5-10
Fluid elastic instability, 29-10
Fluid flow in pipes, 29-14
Fluids, effect of, on rubber, 35-5
Flutter, 29-19
Flywheel, 6-31
FM (frequency modulation) tape recorders, 13-10
Foil:
  Constantan, 17-3
  Iso-elastic, 17-3
  strain-gage, 17-3
Folded-beam interferometer, 18-15
Force:
  effective, 25-2
  generalized, 2-20, 11-11, 28-8
  impulsive, 9-1
  lateral (*see* Lateral force)
  and mechanical impedance, 10-44
  and motion, measurement of, 10-44
  output, 25-12, 30-41
  restoring, 4-19, 4-24
  sinusoidal, 10-1
  strain-gage measurement of, 17-7
  unbalance, 25-4
Force capacity, 25-2
Force factor, 1-20
Force gages, 12-25
Force rating, 25-11
Force transmissibility, 2-6, 2-13, 30-16
Forced excitation, 34-7
Forced motion, 2-2, 2-6, 4-5, 4-11
  general equations, 2-19
  in multiple degree-of-freedom system, 2-19, 2-25
  (*See also* Forced vibration)

Forced normal mode excitation method, 21-12
Forced response, 38-2
Forced sinusoidal vibration, 28-18
Forced vibration, 1-20, 2-6, 4-11, 7-39, 28-9
  classical solution, 7-36
  damping and, 7-43
  foundation-induced shock, 3-38
  instabilities in, 5-18
  response of a modeled system, 38-15
  of rigid body on resilient supports, 3-29
  (*See also* Forced motion)
Foundation, 1-20
  isolated, 34-16
Foundation-induced velocity shock, 3-38, 30-45
Foundation-induced vibration, 3-29
Four-pole parameter, 10-34
Fourier analysis methods, 10-37
Fourier coefficients, 11-5, 22-5
  determination of, 22-6
Fourier integral, 23-25
  applications of, 23-28
Fourier series, 11-5, 22-25
Fourier spectrum (frequency spectrum), 23-25
  applications of, 23-9
  data reduction methods, 23-23
  discrete, 22-3
  example of, 23-6
  physical significance of, 23-7
  running, 23-22
  shock motions, 23-6
  and shock spectrum, 23-22
Fourier transform, 20-5, 27-4
  compared with DFT, 13-19
  discrete (DFT), 13-18, 20-6, 27-4
  fast (*see* Fast Fourier transform)
  and Laplace transforms, compared, 23-27
  related to spectral density, 27-5
  shock calibrator, 18-26
Fraction of critical damping, 1-20, 2-4, 12-3, 37-7, 37-8
  in elastomers, 32-7
Fracture, brittle, 42-28
Fracture energy, 42-28
Fragility, 34-3, 41-2
Frame-of-reference, inertial, 3-1
Free-fall calibration method, 15-11, 18-5
Free-layer damping, 37-3, 37-16, 37-27, 37-29
Free output velocity, 30-41
Free transfer impedance, 30-41
Free vibration, 1-20, 2-2, 2-17, 4-9, 28-9
  with damping, 2-4
  distributed systems, 7-1
  of multiple degree-of-freedom systems, 2-23
  with specified initial conditions, 28-16
Frequency, 1-20
  aliasing (*see* Aliasing)
  angular (*see* Angular frequency)
  circular (*see* Angular frequency)
  demultiplication, 4-8
  dwell 18-31
  effect on rubber, 35-10
  entrainment of, 4-9
  fundamental, 1-20
  natural (*see* Natural frequencies)
  pole-passing, 16-9

Frequency (*Cont.*):
  resonance (*see* Resonance frequency)
  spectrum (*see* Fourier spectrum)
  undamped natural, 28-22
  (*See also* High-frequency response)
Frequency domain, 21-4, 23-5
Frequency equation, 2-18, 7-38, 28-14
Frequency-response function, 21-4, 21-19
Frequency sampling, 13-21
Friction:
  belt, 4-4
  Coulomb (*see* Coulomb damping)
  internal (anelasticity), 36-14
  nut, 43-10
  (*See also* Hysteresis)
Fringe-counting interferometer, 18-11
Full-cycle pulses, 8-44
Function transform, 8-8
Function-transform pairs, 8-7
Fundamental frequency, 1-20
Fundamental mode of vibration, 1-21
  in multiple degree-of-freedom systems, 29-10

g, 1-21, 3-3
  (*See also* Center-of-gravity)
Gage factor, 17-2
Gages, strain (*see* Strain gages)
Galloping and flutter mechanism, 29-19
Gas compressors, 19-11
Gas-pressure torque, 38-18, 38-22
Gas turbines, 19-9
Gasoline engines, 38-22
Gaussian distribution, 11-4, 22-7, 22-10
Gears, 19-11, 34-21, 38-7, 43-27
  diagnosis of, 13-43, 16-5, 16-13, 16-16
  drive vibration, 40-3
Geiger thick-plate test, 37-16
Generalized coordinates, 2-19
Generalized displacements, 28-8
Generalized force, 2-20, 11-11, 28-8
Generalized mass, 2-20, 11-11
Generalized velocities, 28-8
Generators:
  a-c, 6-22
  steam-turbine, 19-7
  vibration (*see* Vibration machines)
Gerber relation, 42-34
Global modal parameters, 21-32
Glossary, 1-18
Goodman relation, 42-34
Graphic level recorder, 13-17
Graphical method:
  of integration, 4-21
  phase-plane, 8-5
Gravitational acceleration, value of, 3-3
Gravitational field, use in calibration, 18-5, 18-6
Gravity (*g*), 1-21, 3-3
  (*See also* Center-of-gravity)
Gravity balancing machines, 39-14
Gravity loads, 24-17
Ground loops, 15-17
Ground motion:
  caused by earthquakes, 24-4
  sources of, 24-1
  vibration in structures induced by, 24-1, 24-3
  (*See also* Response spectrum)

Ground vibration tests, 27-13
Grounding, electrical, 15-17
Guided missiles (*see* Missiles)
Gust factor in structural design, 29-13, 29-27
Guyan reduction method, 27-2
Gyration, radius of, 3-5
Gyroscopes, 6-24

Half-cycle sine pulses, 8-48
Half-power point, 2-15, 11-7
Half-sine acceleration, 23-4, 23-8, 23-16
Half-sine shock pulse, 23-3, 23-6, 26-4, 31-16
Hamming windows, 13-23, 27-9
Hand-held transducers, 15-10
Hanning windows, 13-23
Hard-bearing balancing machines, 39-17
Hard systems, 4-8
Hardening, 4-2, 4-6, 4-32, 4-34
Hardening spring, 4-6, 31-8, 31-12, 31-21
Harmonic, 1-21
Harmonic analysis, 38-19
Harmonic motion, 1-21
  simple (*see* Simple harmonic motion)
Harmonic resonance, 42-32
Harmonic response, 1-21
Harmonic vibration, 4-8
Heart damage, 44-22
Heat exchanger, 29-3
Helical compression springs, 1-12, 32-9
  (*See also* Springs)
Helmets, protective, 44-39
Hemisphere, properties of, 3-53
Hertz (frequency unit), 1-20
Hertz theory of impact, 9-2
High-acceleration calibration, 18-18
High-acceleration devices, 44-5
High-frequency response, transducers, 12-6
High-impact calibration, 18-24
High-impact shock machines, 26-11
Hilbert's transform, 13-45
Hill's equation, 5-16
Holder-type isolators, 32-23
Holzer method, 38-9
Hooke's law, 4-2, 36-25, 42-24
Hose, flexible metal, 34-21
Houdaille-type damper, 38-32
Hull vibration, 19-13
Human body:
  anatomy, 44-20
  biological responses, 44-20
  blast and shock wave effects, 44-23
  crash exposure limits, 44-47
  deceleration exposure limits, 44-47
  impact effect, 44-27
  impact tolerance limits, 44-47
  measurements on, 44-2
  mechanical analog, 44-11
  mechanical damage to, 44-20
  mechanical impedance of, 44-3, 44-12
  mechanical response of, 44-9
  mechanical shock effect, 44-23
  physical characteristics of, 44-6, 44-9
  physiological responses of, 44-22
  properties of, 44-1
  protection methods, 44-32, 44-35, 44-40
  shock and vibration machines for, 44-4
  simulation of, 44-5
  subjective responses to vibration, 44-23

Human body (*Cont.*):
  task performance, 44-31
  tolerance criteria, 44-40
  ultrasonic vibration effect, 44-19
  vibration environments, 44-46
  vibration exposure limits, 44-40
  vibration induced in, 3-29
Hydraulic vibration machines, 25-14
Hydrodynamically induced vibration, 24-25
Hyperbolic tangent elasticity, 31-8
Hysteresis, 2-13
  damping (*see* Structural damping)
  dynamic, 36-14, 36-16
  static, 36-14, 36-17
  (*See also* Friction)
Hysteresis loop in measuring damping, 36-3
Hysteresis loss, 2-15
Hysteretic properties of metals, 36-22
Hysteretic whirl, 5-6, 5-20

Ibrahim time-domain method, 21-15
Identity matrix, 28-3
Imbalance, unstable, 5-20
  (*See also* Unbalance)
Industrial machinery, 34-20
Impact, 1-21, 44-2
  on bars, 9-8
  central, 9-1
  effect on human body, 44-27
  endurance curves, 42-31
  fatigue strength, 43-9
  Hertz theory of, 9-2
  human tolerance limits, 44-47
  producing plastic deformation, 9-8
  protection of human head against, 44-39
  with rebound, 31-20
  without rebound, 31-23
  of solid sphere on plate, 9-3
  on structures, 9-1
  transverse, 9-4
Impact damper, 40-10
Impact excitation, 21-24
Impact force, 32-3
  calibration method, 18-24
Impact methods of transducer calibration, 18-1
Impact test:
  for damping, 37-13
  notched bar, 42-30
Impedance, 1-21
  angle, 10-2
  blocked transfer, 30-41
  cable, 15-3
  characteristic of, 10-23, 30-53
  driving-point, 1-20, 10-2, 10-7
  free transfer, 30-41
  of mass, 10-6
  mechanical (*see* Mechanical impedance)
  mechanical elements, 10-3, 30-41
  and mobility parameters, 10-33
  point, 30-41
  of a spring, 10-4
  transfer (*see* Transfer impedance)
Impedance converters, calibration of, 18-34
Impedance gage, 12-1
Impedance heads, 12-25, 12-26
Impulse, 1-21, 8-20, 31-28
  acceleration, 23-4, 23-6, 23-13

Impulse (*Cont.*):
  excitation, 8-20, 21-24
  long-duration, 31-28
  shock, 23-3, 23-6, 23-19
  short-duration, 31-28
Impulse-response function, 21-4
Impulse-type forces, 9-1
Impulsive force, 9-1
Impulsive region of shock spectrum, 23-19
Inclined plane test, 26-5
Inclined resilient supports, 3-25, 3-27, 3-31, 30-26
Indicator cards, 38-21
Induced environments, 1-21
Induction motors, 19-4
Inductive-type transducers, 14-4
Integration of signals, 13-3
  product of (*see* Product of inertia)
  residual, 21-29
Inertia axes, principal, 3-5, 39-3
Inertia blocks, 34-16
Inertia matrix, 28-8
Inertia restraint, 21-29
Inertia torque, 38-17
Inertial axes, principal, 39-3
Inertial coupling in fluids, 29-1
Inertial frame-of-reference, 3-1
Inertial properties of resiliently supported rigid body, 3-2
Influence coefficients, 28-9, 30-20, 30-28, 30-30
Ingot iron, 42-28
Initial conditions of a system, 8-14
Initial shock spectrum, 20-3
Initial unbalance, 39-2
Insertion loss, 1-21
Instability:
  aerodynamic, 29-26
  fluid elastic, 29-10, 29-12
  in forced vibration, 5-18
  lateral (*see* Lateral instability)
  parametric, 5-20
  (*See also* Whip; Whirl)
Instantaneous values, probability distribution of, 22-13
Instrumentation, 13-1
Integral:
  convolution, 8-4
  Duhamel's, 8-4
  Fourier, 23-25, 23-28
  particular, 8-3
  superposition, 8-4
Integral damping treatments, 37-5
Integration:
  graphical methods, 4-21
  phase-plane, 4-23
Integration networks, 13-3
Interferometers, 18-11, 18-13 to 18-15
  (*See also* Optical interferometric methods of calibration)
Internal combustion engines, 6-26, 38-36
Internal friction, 36-14
Inverse matrix, 28-5
Inverse transformation, 8-9
Inversion test method of calibration, 15-12
Involute springs, 32-17
Iron:
  cast, 42-26, 42-29
  ingot, 42-28

Iron (*Cont.*):
  wrought, 42-26, 42-29
Isochronous, definition of, 4-5
Isoclines, methods of, 4-22, 4-25
Iso-elastic foil, 17-3
Iso-elastic wire, 17-1
Isolated foundations, 34-16
Isolation, 1-21
  active (*see* Active vibration isolation systems)
  of damped structures, 30-49
  of nonrigid systems, 30-40
  shock (*see* Shock isolation)
  velocity, 30-42
  vibration (*see* Vibration isolation)
Isolator materials, 34-5
Isolators:
  air-spring, 33-1
  active (*see* Active vibration isolation systems)
  biaxial stiffness of, 30-28
  bonded, 32-3
  characteristics of, 32-2
  commercial, 32-21
  damping in, 32-4
  effectiveness of, 30-27
  effects of environment, 34-24
  efficiency of, 34-10
  elastomeric, 32-3
  installations of, checking, 34-14
  loading of, 32-4, 32-5
  machinery, 32-24
  massless, 30-43
  metal, 32-9
  minimum natural frequency, 30-37
  nonlinear (*see* Nonlinear isolator)
  nonmassive load, 30-35
  in parallel, 34-6
  plastic, 32-8, 32-25
  rubber (*see* Rubber isolators)
  selection of, 34-7
  semibonded, 32-22
  in series, 34-7
  service life, 34-5
  shear, 32-23
  shear strain in, 32-8
  shipboard, 34-26
  shipping container, 34-26
  shock protection, 34-6, 34-12
  specification of, 34-6
  tangent, 30-34
  types of, 32-23
  unbonded, 32-3, 32-20, 32-23
  undamped, 30-53
  wave effects in, 30-52
  wire mesh, 32-23

Jacobi methods, 28-27
Jerk, 1-21
Joint acceptance, 29-8
Joints, 43-6
Journal axis, 39-2
Jump phenomena, 4-7, 4-28, 4-33

Karma wire, 17-1
Kinetic energy:
  of beams, 7-3

Kinetic energy (*Cont.*):
  of plates, 7-3
  of rods, 7-3
Kirchhoff's laws, 10-24, 10-32
Kryloff's method, 4-17, 4-30

Lag, angle of, 39-17
Lagrange's equations, 2-25, 11-11, 28-8
  for compound pendulum, 2-26
  for U-tube, 2-27
Lagrange's method, 4-15
Lanchester damper, 4-27, 38-32, 40-10
Laplace domain, 21-5
Laplace transform, 8-6, 8-49
  and Fourier transform compared, 23-27
  in shock theory, 21-8
Laser-Doppler transducer system, 14-2
Latent vectors, 28-14
Lateral beam, 7-12, 7-22, 7-25, 7-33
Lateral force method of measuring damping, 36-5
Lateral forces:
  design, 29-13
  of earthquakes, 24-15
Lateral instability:
  due to asymmetric shafting, 5-15
  due to pulsating longitudinal loads, 5-16
  due to pulsating torque, 5-15
  of rotors, 5-15
Lateral sensitivity (*see* Transverse sensitivity)
Lateral vibration:
  of beams, 7-2, 7-12, 7-22, 7-25, 7-33
  of plates, 7-24
  of straight beams, 7-11
Latex hair, 41-16, 41-17, 41-24, 41-25
Leaf springs, 32-17
Leakage error, 13-20, 21-21, 27-9
Level recorder, 13-17
Lewis rule, 38-19
Liénard's method, 4-25
Limit cycles, 4-26
Limit design, 9-10
Line-drive preamplifier, 13-1
Line spectrum, 1-20, 22-3
  measurement of, 22-18
  relation to power spectral density, 22-17
  superimposition on random vibration, 22-20
Linear damping, 4-19, 4-21, 4-35
  equivalent, 37-2
Linear spring, 31-7, 31-12, 31-21
  with damping, 31-10, 31-13
Linear system, 1-21
  input-output relations, 11-7
  stepwise, 4-23
Linear viscoelastic stress-strain law, 36-13
Linearity limits for plastic and rubber, 36-16
Liquid flow (*see* Fluid flow in pipes)
Liquid pumps, 19-6
LMS (log-mean-square) detector, 13-6
Load:
  dead-weight, 25-1
  dynamic, 42-2, 42-12
  effective, 25-1
  pulsating longitudinal, 5-16

Load (*Cont.*):
  structural, shock source affected by
    changes in, 23-29
  in vibration machines, 25-1
Load capacity, 25-1, 32-2
Load couple, 25-1
Load-deflection characteristics:
  of elastomeric isolators, 32-5
  of metal springs, 32-13
  of wire mesh springs, 32-19
Load-deflection loop in measuring damping,
  36-3
Loading:
  buckling, 32-5
  compression, of rubber, 32-4, 35-15
  peak, 23-20
  random, 11-14
  shear, of rubber isolators, 32-4, 35-18
  tension, in rubber isolators, 32-5, 35-18
  torsional, 32-5
Lock-in, 29-9
Lock washers, 43-10
Log-mean-square detector, 13-6
Logarithmic decrement, 1-21- 2-5, 36-4, 36-9,
  37-7, 37-9
Logarithmic spiral, 4-23
Long-duration impulses, 31-28
Longitudinal impact on bars, 9-8
Longitudinal loads, instability due to, 5-16
Longitudinal vibration:
  circular rods, 7-5
  human body exposed to, 44-13
  rod with mass attached, 7-9
Longitudinal wave, 1-21
Loss angle, 36-13
Loss factor, 36-9, 37-7, 37-17
Loss modulus in shear, 42-27
Low brass, 42-27
Low-pass filter, 1-21
  ideal, 22-16
Lumped elements, 10-21
Lumped parameter systems, 2-1, 7-4

Machine(s):
  balancing (*see* Balancing machines)
  fatigue testing, 25-18
  maintenance of, 16-1
  reciprocating (*see* Reciprocating machin-
    ery)
  shock (*see* Shock machines; Testing ma-
    chines; Vibration machines)
Machine tool(s):
  boring bar, 6-35
  chatter, 40-7
  computer-aided design of, 40-17
  vibration-free design of, 40-11
Machinery:
  condition monitoring of, 16-1
  industrial, 34-20
  isolators, 32-24
  maintenance of, 16-1
  rotating (*see* Rotating machinery; Turbo-
    machinery)
Machinery vibration:
  classification of, 19-2
  criteria for, 19-2
  trouble-shooting chart, 16-10
Magnesium, 42-28, 42-31
Magnetic mounting of transducer, 15-9

Magnetic shields, 15-6
Magnetic tape recorders, 13-10
Magnetoelastic damping, 36-18
Magnetostriction, 1-21
Magnitude of vibration, 22-1
Magnitude distribution function, 22-23
Magnitude-time history, 22-3, 22-23
  determination of, 22-21
Manganese bronze, 42-27, 42-31
Marine propeller, 38-16
Mass, 2-1, 10-5
  center-of-mass, 3-3
  distributed, 10-22
  effect on natural frequencies, 6-7
  equivalent, 6-2, 6-4
  generalized, 2-20, 11-11
  impedance of, 10-6
  unbalance, 25-4
Mass center (*see* Center-of-gravity)
Mass centering, 39-25
Mass-controlled system, 2-9
Mass damper, 6-1
  auxiliary, (*see* Auxiliary mass damper)
Mass damping, uniform, 2-22
Mass-spring transducers, 12-2
Massless isolator, 30-43
Material(s):
  brittle, 43-22
  crushable, 41-13
  cushioning (*see* Cushioning materials)
  elastic, 41-13
  flexible, 1-13
  packaging, 41-13
  physical properties of, 42-24, 42-27
  properties in equipment design, 42-32
  resilient, 41-13
  static properties of, 42-24
  viscoelastic (*see* Viscoelastic materials)
Material damping, 36-1, 36-13, 36-20
Matrix, 28-2
  damping, 28-8
  diagonal, 28-3
  eigenvalue problem, 28-14, 28-21, 28-25,
    28-29
  eigenvector, 28-14, 28-15
  flexibility, 28-9
  identity, 28-3
  inertia, 28-8
  inverse, 28-5
  null, 28-3
  reciprocal, 28-5
  rectangular, 28-2
  singular, 28-5
  square, 28-2
  stiffness, 28-8
  symmetric, 28-3
  system impedance, 29-13
  transfer, 7-37
  transpose, 28-2
  unit, 28-3
  zero, 28-3
Matrix operations, 28-3
Matrix product, 28-4
Maximax response, 8-16, 8-47
Maximum shock spectrum, 20-3
Mean-square acceleration density, 22-3, 22-
  16
Mean-square value, 11-4, 11-6, 11-9, 22-5, 22-
  20
Mean time between failure (MTBF), 20-15

Mean value, 22-5
Mean wind velocity, 29-20
Measurement:
  of acceleration, 15-4, 16-2
  bandwidth selection, 16-6
  biological, 44-3
  of displacement, 14-3, 15-4, 16-2, 18-10
  instrumentation, 13-1
  location, selection of, 16-4
  mechanical impedance, 10-46, 12-25
  strain, 17-1
  system calibration, 15-11
  of velocity amplitude, 15-4, 16-2
Measurement planning, 15-1, 16-2
Measurement techniques, 15-1
Mechanical admittance, 10-2
  (*See also* Mobility)
Mechanical aging, 24-25
Mechanical damping, significance of, 36-2
Mechanical drag (*see* Hysteresis)
Mechanical elements:
  impedance, 10-3, 30-41
  mobility of, 10-16
  series, 10-8
Mechanical environment, simulation of, 44-4
Mechanical exciters (*see* Vibration machines)
Mechanical impedance, 6-2, 10-1, 10-2
  of auxiliary mass system, 6-2
  calculations of, 10-45
  data presentation, 10-38
  defined, 10-45
  of human body, 44-3, 44-12
  measurement of, 10-43, 10-46, 12-25
  (*See also* Impedance)
Mechanical power sources, 30-40
Mechanical resistance, 10-3
Mechanical shock, 1-22
  (*See also* Shock)
Mechanical vibration machines:
  direct-drive, 25-2
  reaction-type, 25-4
Metal:
  ductility, 42-48
  fatigue, 42-29
  nonferrous (*see* Nonferrous metals)
  properties of, 32-10 to 32-12, 36-22
  toughness of, 42-29
  yield, 42-29
Metal containers, 41-3
Metal helical springs, 32-9
Metal hose, flexible, 34-21
Metal springs, 32-9
Method of isoclines, 4-22, 4-25
Method of node forces, 10-28
Miner's hypothesis, 20-20
Minor critical speeds, 38-2
Missiles:
  model of, 42-4
  shock, 20-3
Mobility, 1-22, 10-1, 12-25
  driving-point, 10-2
  chart, 10-7
  and impedance parameters, 10-33
  of mechanical elements, 10-16
  (*See also* Mechanical impedance)
Modal analysis, 24-1, 27-12
  applied to rotor systems, 38-25
  experimental, 24-10

Model analysis (*Cont.*):
  single degree-of-freedom systems, 24-3
  test configurations, 21-23
  theory of, 21-3
  transducer considerations, 21-18
Modal balancing, 39-3
Modal confidence factor, 21-16
Modal coupling, 3-15
Modal data:
  acquisition of, 21-17
  presentation of, 21-3
Modal numbers, 1-22
Modal parameter:
  estimation, 21-26
  estimation phase, 21-2
  global, 21 32
Modal participation factor, 21-32
Modal responses, confirmation of, 24-13
Mode(s):
  coupled (*see* Coupled modes)
  decoupling of, 3-28, 30-27
  flexural principal, 39-2
  natural frequency, 2-20
  uncoupled, 1-25
  of vibration (*see* Fundamental mode of vibration; Normal mode of vibration), 1-22
Mode shapes of rotors, 39-8
Model:
  of automobile, 42-4
  of beam, 42-3
  equivalent, 42-2
  mathematical, 21-27
  of missile, 42-5
  multiple degree-of-freedom system, 42-6
  order of, 21-30
  single degree-of-freedom system, 42-4, 42-7
  of torsional elastic system, 38-3
  of torsional vibration, 38-14
  two degree-of-freedom system, 42-5, 42-7, 42-14
  types of, 42-4
  uses of, 42-2
Modulation, 1-22
Modulus:
  bulk, 42-25
  complex, in shear, 36-13
  dynamic, 10-2
  of elasticity, 42-24
  loss, in shear, 36-13
  of resilience, 42-38
  of rigidity, 42-25
  of rubber, 35-3, 35-8, 37-17
  storage, in shear, 36-13
  of viscoelastic materials, 36-17
  of volume expansion, 42-25
Moment(s), 11-4
  bending, of plates, 7-27
  of inertia, 3-5
    area, 1-14
    determination of, 3-6
    polar, 38-3, 38-6
    principal, 3-5
    transformation of axes, 3-5
  oscillating, 3-37
  static, and torque in vibration machines, 25-13
Monel metal, 42-31

Motion:
  coupled, 42-15
  and force, measurement of, 10-44
  forced (*see* Forced motion)
  of foundation, 2-14
  ground (*see* Ground motion)
  of response, 2-6, 30-2, 30-11
  rotational, 2-2
    of plate, 7-35
  seismic, 24-26
  shock (*see* Shock motion)
  of support, 2-14
    vibration resulting from, 7-42
  uncoupled, 31-21
Motion transmissibility, 2-6
Motor vehicles (*see* Automobiles; Vehicles)
Motors (*see* Electric motors; Induction motors)
Mount (*see* Isolators)
Mounting:
  of electrical components, 43-1
  transducer, 15-5
Movable-core differential-transformer transducer, 14-6
Multimass dynamic absorbers, 6-19
Multiple degree-of-freedom system, 1-22, 2-16, 21-8, 24-11, 30-16, 42-6
  coupled modes, 30-26
  damped, 2-22
  decoupled, 30-27
  forced motion of, 2-19, 2-25
  free motion of, 2-23
  fundamental mode of vibration in, 29-10
  natural frequencies, 3-16, 3-18, 3-21
  normal modes, 2-18
  one plane of symmetry, 30-16
  response, 11-11
  two planes of symmetry, 30-17
  uncoupled rotational mode, 30-21
  undamped, 2-16
  uniform mass damping, 2-23
  uniform viscous damping, 2-23
  (*See also* Three degree-of-freedom systems; Two degree-of-freedom systems)
Multiplane balancing, 39-3
Multiple impact shock machine, 26-14
Multiple input theory, 21-20
Multivibrator, 5-18
Muntz metal, 42-27

Narrow-band oscillator, 11-9
Narrow-band random vibration, 1-22, 11-12, 22-2
NASTRAN computer program, 27-2, 29-4
Natural frequencies, 1-22
  angular, 2-3
  beams, 1-13, 1-15, 7-10, 7-20
  compound pendulum, 1-17
  for coupled modes, 3-15, 3-18, 3-21, 30-19
  damped, 2-4, 2-15
  decoupled, 3-26, 30-27
  of distributed systems, 7-2, 7-4
  effect of mass on, 6-7
  of lumped-parameter systems, 7-4
  minimum, 30-37
  of a mode, 2-20

Natural frequencies (*Cont.*):
  of multiple degree-of-freedom systems, 3-16, 3-18, 3-21
  package, 41-5
  pendulum, 1-17
  plates, 1-16
  resiliently supported rigid body, 3-15, 30-19
  rotational, 3-19
  rotor shaft systems, 1-13
  shaft, 38-9
  single degree-of-freedom system, 2-3, 42-4
  spring systems, 1-12
  undamped, 1-25, 2-16, 28-22
  uniform beams, 7-14
Naval brass, 42-31
Navy shock testing machines, 26-11
Negative feedback, 33-8
Neoprene, 32-2
  (*See also* Elastomers)
Newkirk effect, 5-22
Newton's second law, 7-39
Nickel, 42-31
Nickel cast iron, 42-29
Nodal lines, distributed systems, 7-2
Nodal points, distributed systems, 7-2
Nodal surfaces, distributed systems, 7-2
Node, partial, 1-22
Node forces method, 10-28
Noise, 1-22
  background, 1-18
  cable, 15-15
  control of, 34-14, 37-3
  white, 1-26, 11-7, 22-16
    autocorrelation function of, 1-18
Noise suppression, 15-16
Nominal bandwidth, 1-22
Nominal stress, 42-25
Nonautonomous systems, 4-14, 4-17, 4-24
Nondestructive zoom, 13-23
Nonelastic materials, 41-13
Nonferrous metals, 42-31
Nonisochronous, definition of, 4-5
Nonlinear cushioning, 41-26
Nonlinear damping, 1-22, 5-18
Nonlinear isolator, 30-34, 31-8, 41-12
Nonlinear systems, single degree-of-freedom, 8-52
Nonlinear vibration, 4-1
  stability of, 4-28
Nonrigid body system, isolation of, 30-40
Nonrigidity of structures, 30-48
Nonsinusoidal functions, 10-37
Nonstationary signals, analysis of, 13-36
Normal coordinates, 8-52
Normal distribution (*see* Gaussian distribution)
Normal mode of vibration, 1-22
  concept in, 42-5
  distributed system, 7-2
  fluid-coupled, 29-4
  multiple degree-of-freedom system, 2-18
  orthogonality of, 7-5
  uniform beam, 7-14
Norton's equivalent system, 10-25, 10-31
  representation, 10-36
Notched-bar impact test, 42-30
Null force balancing machine, 39-2
Null matrix, 28-3

Numerical methods of analysis, 28-1, 28-25
(*See also* Matrix)

Oberst beam, 37-15
Ocean waves, 29-5
Odd function, 22-6
Offshore structures, 29-5
One-connection systems, 10-30
Operation transforms, 8-8
Operational-basis earthquake, 24-24, 24-25
Optical-electronic transducer systems, 14-2
Optical interferometric methods of calibration, 18-10
Optimum damping, 6-11
  parameters for, 30-15
Order:
  of critical speed, 38-2
  of a model, 21-30
  of vibration, 6-28
Ordinary coherence, 21-19
Orthogonal resilient supports, 3-15, 3-29, 3-33
Orthogonality, 2-18
  of normal modes, 7-5
Oscillating moment excitation, 3-37
Oscillation(s), 1-22
  flutter, 29-35
  galloping, 29-34
  relaxation, 4-9, 4-27, 40-5
  self-excited, 29-27
  structural, 29-18
  in turbulent flow, 29-30
  vortex-induced, 29-30
Oscillators:
  narrow-band, 11-9
  relaxation, 5-21
Output velocity, free, 30-41
Overall shock spectrum, 26-7
Ozone resistance of rubber, 35-4

Package:
  base-mounted, 41-8
  natural frequency of, 41-5
  shock tests for, 41-4
  tension-spring, 41-12
Package cushioning design, 41-8
  analytical methods, 41-2
  engineering estimates, 41-2
  laboratory test method, 41-2
Package tester, 26-3
Packaging, 41-1
  classification of, 41-13
Packaging materials, 41-13
Paddle-type damper, 38-32
Parabola, properties of, 3-47
Paraboloid, properties of, 3-54
Parallelogram, properties of, 3-45
Parameters:
  four-pole, 10-34
  impedance and mobility, 10-33
Parametric instability, 5-14
Partial fraction expansion, 8-13
Particular integral, 8-3
Parzen windows, 27-9
Passive-circuit transducer, 12-1
Peak acceleration, 23-20
  distribution of, 22-13
Peak loading, 23-20

Peak-to-peak value, 1-22
Peak values, 13-32, 22-15
  measurement of, 22-14
  probability distribution of, 22-14
Pendulum, 4-2, 4-5, 4-11
  calibration technique, 18-5
  compound, 2-26, 3-6
  torsional, 3-7
  trifilar, 3-8
Pendulum absorber, 6-28
  amplitude, 6-30
  bifilar, 6-32
  for linear vibration, 6-33
  nonlinearity, 6-31
  roller-type, 6-33
  tuning, 6-30
Pendulum damper, 38-39
Perception of vibration, 44-10
Perfect causality, 21-20
Performance:
  deterioration failure, 20-7
  of vibration machines, 25-13
Period, 2-3
  of beats, 4-9
Periodic chirp, 21-24
Periodic functions, 22-4
  analysis of, 22-5
Periodic quantity, 1-23
Periodic random, 21-25
Periodic vibration, 20-2
Permanent deformation, 41-13
Permissible stresses, 42-32
Perturbation formulas, 28-26
Phase angle, 2-4, 2-13
Phase-coherent functions, 22-4
Phase-coherent signals, 13-46
Phase demodulation, 13-48
Phase diagrams, 38-18
Phase-plane analysis, 4-27
Phase-plane-delta method, 8-49
Phase-plane graphical method, 8-5
Phase-plane integration, 4-23
Phase-plane methods, 4-24, 8-50
Phase-plane representation, 4-21
Phase-plane techniques, 4-27
Phase spectra, shock motions, 23-6
Phasor, 10-1
Physical pendulum calibrator, 18-4
Physiological responses of human body, 44-22
Pi equivalent systems, 10-25, 10-32
Picket-fence effect, 13-21
Pickup (*see* Piezoelectric accelerometer; Transducer)
Piezoelectric, definition of, 12-11
Piezoelectric accelerometer:
  amplification ratio, 12-7
  amplitude range, 12-19
  charge sensitivity, 12-7
  damping effects in, 12-7
  definition of, 12-1
  effects of shunt capacity on, 12-7
  effects of temperature on, 12-19
  electrical characteristics of, 12-7
  with internal preamplifier, 12-18, 13-2
  physical characteristics of, 12-17
  resonance frequency of, 12-7
  response curve, 12-14
  sensitivity, 12-7

Piezoelectric accelerometer (*Cont.*):
  shear-type, 12-16
  voltage sensitivity of, 12-7
Piezoelectric exciter, 18-29
Piezoresistive accelerometer:
  characteristics of, 12-22
  construction of, 12-21
  definition of, 12-1
Piping, 34-33
  fluid flow in, 29-14
Plane separation, 39-3
Plastic deformation from impact, 9-8
Plastic hinge, 9-8
Plastic isolators, 32-8, 32-25
Plastic waves, 9-9
Plastics, linearity limits for, 36-16
Plate:
  bending stresses, 7-26
  boundary conditions, 7-24, 7-30
  circular, 7-30, 42-23
  deflection of, 7-27
  elastic, 9-3, 42-21
  flat, characteristics of, 1-16
natural frequencies, 1-16
  rectangular, 7-3, 7-25, 7-30
    free vibration, 7-28
    natural vibration, 7-31
    skew, 7-30, 7-31
  Ritz's method, 7-29
  rotational motion, 7-35
  shearing forces, 7-35
  skew rectangular, 7-30
  square, 7-30
  trapezoidal, 7-30
  triangular, 7-30
  uniformly loaded: bending moments of, 7-27
    deflection of, 7-27
  vibration of, 7-24
Plate-form isolators, 32-23
Pneumatic isolation system (*see* Air suspension system)
Pneumatic spring (*see* Air springs)
Point impedance, 30-41
Poisson's ratio, 42-25
  for rubber, 35-2
Polar moment of inertia, 38-3, 38-6
Pole, 21-4
Pole passing frequency, 16-9
Polyester urethane, 41-17, 41-19, 41-24
Polyester urethane foam, 41-14, 41-15
Polyethylene foam, 41-15, 41-20, 41-25
Polymers, 35-1
  as damping materials, 37-2
  designation and composition of, 35-1
  properties of, 35-7
  (*See also* Rubber)
Polynomials of displacement, 4-10
Polyreference time-domain methods, 21-16
Post processors, 27-3
Potential energy, 28-6
Potentiometer-type transducer, 14-7
Power cepstrum, 13-44
Power sources, mechanical, 30-40
Power spectral density, 1-23, 11-5, 11-6, 22-16, 27-5
  determination of, 11-5, 22-17
  level, 1-23
  monatonic, 11-7

Power spectral density (*Cont.*):
  peaked, 11-7
  of random vibration, 22-3
  relation to line spectrum, 22-17
Preamplifiers, 13-1
  charge, 13-1
    calibration of, 18-34
  internal, 12-18, 13-2
  voltage, 13-1
    calibration of, 18-33
Preprocessors, 27-3
Pressure, strain gage measurement of, 17-9
Pressure transducers, 17-9
Primary shock spectrum, 20-3, 23-13
Primary structures, 42-6
Principal coordinates, 8-52
Principal elastic axes, 3-11, 30-31
Principal inertia axis, 39-3
Principal influence coefficients, 30-31
Principal moments of inertia, 3-5
Printed circuit boards, 43-25
Prism, properties of, 3-51
Probability amplitude, 11-1
Probability density, 11-2, 22-9
Probability density function, 11-3, 22-7
Probability distribution, 11-2, 22-9
  first, 11-3
  of instantaneous values, 22-13
  of peak values, 22-14
  second, 11-4
Probability distribution functions, 22-8
  Gaussian functions, 22-9
  Rayleigh distribution, 22-9
  sinusoids, 22-9
Probability graph paper, 22-10
Product of inertia, 3-5
  coupling, 30-20
  experimental determination of, 3-8
  transformation of axes, 3-5
Product law, 20-14
Production test, 20-7
Propagation constant, 10-23
Propeller(s):
  blade vibration, 19-13
  damping, 38-31
  marine, 38-16
  shaft vibration, 19-13
  whirl, 5-20
Proper vectors, 28-14
Proportional control, 33-8
Proportional damping, 21-12
Proportional limit, 42-24
Pseudorandom excitation, 21-25
Pseudorepeated roots, 21-16
Pseudovelocity, 24-5
Pulsating torque, 5-15
Pulse(s):
  acceleration (*see* Acceleration pulse)
  exponential, 8-25, 8-36, 8-31
  full-cycle, 8-44
  half-period sine, 26-7
  rectangular, 23-7
  response to, 31-15
  rise time, 1-23
  shock (*see* Shock pulse)
  sine (*see* Sine pulse)
  skewed versed-sine, 8-42
  symmetrical, 8-25
  trapezoidal, 31-22

Pulse(s) (*Cont.*):
  triangular, 8-31
  unsymmetrical, 8-29
Pulse-type excitation functions, 8-20
Pumps, 16-9, 19-6, 34-22
Pure random excitation, 21-24
Pyramid, properties of, 3-51
Pyrotechnic shock, 20-4

Q (quality factor), 1-23, 2-15, 37-7
Quadratic forms, 28-6
Qualification test, 20-7
Quantization, 21-18
Quasi-ergodic process, 1-23
Quasi-sinusoidal, 1-23
Quasi-stationary vibration, 20-2
Quasi-steady theory for turbulence, 29-23
Quenching, 4-9

Radial aircraft engine, 38-40
Radius of gyration, 3-5
Random-decrement averaging, 21-16
Random excitation, 11-9, 29-32, 29-33
  burst, 21-25
  method of calibration, 18-16
  periodic, 21-25
  pseudo-, 21-25
  pure, 21-25
  transient, 21-25
Random functions, 22-5
  compared with sinusoidal functions, 22-8
  statistical parameters, 22-7
Random input (excitation), 11-1, 11-9
Random loading, 11-14
Random output (response), 11-1
Random process, 11-3
Random signals:
  in computer technology, 27-8
  stationary, 13-26
Random sine wave, 11-10
  (*See also* Random vibration, narrow-
    band)
Random time function, 11-2
Random transient excitation, 21-25
Random vibration, 1-23, 11-1
  broad-band, 1-19
  example of, 22-2
  isolation of, 30-38
  narrow-band, 1-22, 11-12, 22-2
  power spectral density of, 22-3
  spectral analysis of, 11-6
  with superimposed line spectrum, 22-20
  testing, 20-8, 25-10, 27-10
Random walk problem, 11-10
Ratio of critical damping (*see* Fraction of
    critical damping)
Ratio of specific heats, 33-4
Rauscher's method, 4-13
Rayleigh distribution, 11-11, 22-8, 22-9, 22-
    12
  probability graph paper, 22-12
Rayleigh wave, 1-23
Rayleigh's method, 7-2
  for beams, 7-15
  for beams with masses attached, 7-22
  for plates, 7-29
  for rod with mass attached, 7-9

Rayleigh's method (*Cont.*):
  for variable-section beam, 7-19
Rayleigh's quotient, 28-25
Reaction-type vibration machine, 25-4
Real eigenvalue problem, 28-25
Real-time signal analysis, 13-28, 13-23, 13-33
Rebound:
  impact with, 31-20
  impact without, 31-23
Receptance, 10-2, 12-25
Reciprocal matrix, 20-5
Reciprocating machinery, 19-7
  torsional vibration in, 38-1
Reciprocity calibration, 18-16
Reciprocity theorem, 10-24
Recorders, magnetic tape, 13-10
  digital, 13-11
Rectangle, properties of, 3-45
Rectangular force pulse, 26-4
Rectangular matrix, 28-2
Rectangular plates, 7-3, 7-25, 7-30
  free vibration of, 7-28
  natural vibration of, 7-31
  skew, 7-30, 7-31
Rectangular pulse, 23-7
  equivalent (*see* Equivalent rectangular
    pulse)
  response to, 31-15
Rectangular shock motion, 23-7
Rectangular step excitation, 8-50
Rectilinear-motion vibration machine, 25-5
Red brass, 42-27
Reduced structural matrix method, 21-17
Reduced temperature nomogram, 37-23
Reed gage, 23-24
Reentry of spacecraft, 44-5
Relative drift, 35-6
Relative transmissibility, 30-2
Relaxation oscillations, 4-9, 4-27, 40-5
Relaxation oscillator, 5-18
Relaxation time, 1-24
Reliability failure, 20-7
Reliability growth test, 20-7
Reliability testing, 20-14
Repeated roots, 21-16
Request for proposal (RFP), 20-13
Request for quotation (RFQ), 20-13
Required input motion (RIM), 24-30
Required response spectrum (RRS), 24-24
Residual flexibility, 21-29
Residual imbalance, 39-3
Residual inertia, 21-29
Residual response amplitude, 8-16, 8-47
Residual shock spectrum, 20-3, 23-13, 26-7
Residual torque, 38-9
Residual unbalance, 39-3
Residuals, 21-29
Resilience, modulus of, 42-28
Resilient materials, 41-13
Resilient supports:
  idealized, 3-10
  inclined, 3-25, 3-27, 3-31, 30-26
  linear, 3-11
  orthogonal, 3-15, 3-29, 3-33
  properties of, 3-10
  radially inclined, 3-28
  (*See also* Rigid body, resiliently support-
    ed)
Resistance, mechanical, 10-3

Resistance strain gage, 17-5
Resonance, 2-15
  amplification (see *Q*)
  curve, 2-8
  harmonic, 4-32
  mechanical, in fatigue testing, 25-19
  superharmonic, 4-33
  transmissibility at, 30-10
  ultraharmonic, 4-33
Resonance amplitude, 4-20
Resonance frequency, 1-24, 2-15
  acceleration, 2-15
  displacement, 2-15
  velocity, 2-15
Response, 1-24
  allowable, 34-3
  continuous systems, 11-1
  curves, 2-8, 4-5
  forms of, 8-3
  harmonic, 1-21
  maxima, 8-16, 8-47
  multiple degree-of-freedom system, 11-11
  narrow-band oscillator, 11-9
  single degree-of-freedom system, 11-9
  structural, determining, 23-1, 23-28
  subharmonic, 1-25, 4-15, 4-33
  superharmonic, 1-25, 4-31
  transform, 8-9
  transient, 28-9, 33-13
  velocity, 30-45, 30-46
Response amplitude, residual, 8-16, 8-47
Response domain, data reduction to, 23-9
Response factors, 2-8, 2-11
Response spectrum, 8-14, 8-46, 24-4
  definition of, 24-22
  design, 24-7
  for earthquake, 24-6
  for elastic systems, 24-9
  required, 24-24
  test, 24-28
  use of, 24-11
  (*See also* Shock spectrum)
Restitution, coefficient of, 9-1, 9-4, 9-7
Restoring forces, 4-19
  bilinear, 4-24
Resultant modulus in shear, 36-13
Revolving-drum tumbling test, 26-13
Reynold's number, 29-7
Rigid body, resiliently supported, 3-1
  inertial properties of, 3-2
  modal coupling, 3-15
  natural frequencies, 3-15, 30-19
  with one plane of symmetry, 3-38
  response to acceleration pulse, 31-14
  response to velocity step, 31-5
  simplifying conditions, 3-14
  supported within rigid container, 3-38
  with three planes of symmetry, 3-24
  with two planes of symmetry, 3-19, 3-25,
    3-29, 3-31
  vibration of, 3-1
Rigid rotor(s):
  balancing grades for, 39-3
  definition of, 39-3
Rigidity, modulus of, 42-25
Rigidly connected Coulomb damper, 30-2,
  30-7
Rigidly connected viscous damper, 2-4, 3-2,
  30-5

Ring, circular, 7-36
Ring spring, 32-15
Rise time, 8-17
Ritz condition, 4-19
Ritz's method, 4-17, 4-19, 7-2
  for plates, 7-29
  for variable-section beam, 7-19
RMS (root-mean-square) value, 13-32, 22-5
  detectors, 13-6
  relation to average value, 22-10
Road vehicles, 34-25
  (*See also* Automobiles)
Rocket engine, 11-1
Rocket-propelled sleds, 44-5
Rocket track, 44-5
Rod:
  with attached disc, 7-7
  circular, 7-3, 7-5
  with mass attached, 7-9
  properties of, 3-50
Rolled bronze, 42-29
Roller-type pendulum absorber, 6-33
Ronchi ruling method, 18-10
Root-mean-square value (*see* RMS value)
Rotary accelerator, 26-19
Rotary motion and shearing force, beam, 7-
  16, 7-35
Rotating eccentric weight, 2-14
Rotating force excitation, 3-34
Rotating machinery, 5-2
  balancing of, 6-26, 39-1
  condition monitoring of, 16-1, 16-7
  fault detection in, 16-4
  maintenance of, 16-1
  self-excited vibrations, 5-21
  sources of unbalance in, 39-11
  spectra of, 16-7
  standards for, 19-3
  vibration trouble-shooting chart, 16-10
  (*See also* Balance; Rotors; Unbalance)
Rotating shafts, 5-7
Rotation, 2-2
  axis of, 3-30, 3-33
  successive, diagonalization by, 28-28
Rotational mechanical impedance (*see* Angular mechanical impedance)
Rotational motion, 2-2
  plate, 7-35
Rotational natural frequency, 3-19
Rotational stiffness, 3-12
Rotor(s):
  balancing, 39-4
  elasticity test, 39-8
  flexible (*see* Flexible rotors)
  lateral instability, 5-15
  rigid, 39-3
  whirl from fluid trapped in, 5-8, 5-20
  (*See also* Rotating machinery)
Rotor systems, 1-13
  modal analysis of, 38-25
Row vector, 28-3
Rubber:
  amplitude effect, 35-12
  barrier temperature for, 35-5
  butyl, properties of, 37-24
  classifications and specifications of, 35-14
  compression loading of, 35-14
  compression set of, 35-3
  critical damping, 35-12

Rubber (*Cont.*):
  drift of, 35-6
  dynamic aging of, 35-14
  dynamic properties of, 35-8
  effect of immersion in fluids, 35-5
  environmental factors, 35-4
  fatigue of, 35-13
  isolators (*see* Rubber isolators)
  loading methods, 35-15
  low-temperature properties, 35-5
  modulus of, 35-3, 37-24
  natural, 32-2
  ozone resistance of, 35-4
  Poisson's ratio, 35-2
  properties of, 35-2, 37-26
  silicone, 32-2
  static physical properties of, 35-2
  stress relaxation test for, 35-3
  temperature and frequency effect, 35-10
  tensile strength, 35-3
  ultimate elongation of, 35-3
  (*See also* Elastomers; Polymers; Vulcanizates)
Rubber isolators:
  compression loading, 32-4
  cylindrical, 35-17
  design of, 35-15
  design stress, 35-17, 35-19
  rectangular, 35-16
  shear loading, 32-23, 35-18
  tension loading, 32-5, 35-18
  torsional loading, 32-5
  triangular, 35-17
  (*See also* Isolators)
Rubberlike materials (*see* Elastomers)
Runge rules, 38-19
Running Fourier spectrum, 23-22

Safe-shutdown earthquake (SSE), 24-24
Sampling, 13-21, 21-18
Sampling frequency, effect of, 13-13, 13-40
Sandwich beam method, 37-16
Sandwich construction, 43-12
Scale model tests in damping, 37-13
Scaling, 13-29
Screening specifications, 20-15
Screening tests, 20-7
Seat belts, 44-37
Seat cushions, 44-32, 44-39
Second probability distribution, 11-4
Secondary structure, 42-6
Sector, circular, 3-46
Seismic differential-transformer transducer, 14-6
Seismic electrodynamic transducer, 14-4
Seismic forces, 24-16
Seismic motion, 24-1, 24-26
Seismic qualification, 24-21
  by analysis, 24-31
  exploratory testing, 24-27
  by simulation, 24-25
  testing practice, 24-26
  use of response spectrum in, 24-22
Seismic testing, 24-27
  machines for, 24-30
  multiaxis, 24-30
Seismic transducers, 12-2
Selective excitation, 21-22

Self-excited system, 4-4, 4-8
Self-excited vibration, 1-24, 4-4, 5-1, 5-21
  control of, 6-35
  identification of, 5-21
  rotating machinery, 5-20
Self-generating transducer, 12-1
Self-induced vibration (*see* Self-excited vibration)
Semicircle, properties of, 3-46
Semiparabola, properties of, 3-47
Sensitivity:
  charge, accelerometers, 12-17
  transverse (*see* Transverse sensitivity)
  voltage (*see* Voltage sensitivity of accelerometers)
Sensitivity factor, 18-1
Sensor, definition of, 12-1
Series-connected isolators, 34-7
Series-connected systems, 10-35
Series mechanical elements, 10-18
Servo accelerometer, 14-7
Servo-controlled isolation system (*see* Active vibration isolation systems)
Shaft:
  long, 43-23
  natural frequencies, 38-9
  rotating, 5-7, 16-8
  stiffness, 38-6
  in torsion, 1-13
Shafting, asymmetric, lateral instability due to, 5-15
Shake tables (shakers) (*see* Vibration machines)
Shannon's theorem, 27-4
Shear:
  complex modulus in, 36-13
  storage modulus in, 36-13
  strain in elastomers, 32-8
Shear deflection, 7-16
Shear isolators, 32-23
Shear loading of rubber isolators, 35-18
Shear parameter, effect on damping, 37-31
Shear-type accelerometers, 12-16
Shear wave, 1-24
Shearing force and rotary motion, beam, 7-16, 7-35
Shielding, cable, 15-16
Ship(s):
  isolators aboard, 34-26
  roll reduction, 6-23
  vibration in, 19-12
Shipping containers, 26-5, 34-26
Shipping cycle, 41-4
Shock:
  avionic, 20-3
  characteristic types of, 26-4
  complex (*see* Complex shock)
  control of, 31-1
  effects on human body, 44-1
  impulse, 23-3, 23-6, 23-19
  mechanical, 1-22
  missile, 20-3
  multiple, 26-4
  pyrotechnic, 20-4
  simple, 26-5
  space, 20-3
  step, 23-3, 23-6
  types of, 26-4
  velocity (*see* Velocity shock)

Shock absorber, definition of, 1-24
Shock calibrators, 18-19
Shock data reduction, 23-1
  frequency domain, 23-5
Shock isolation, 31-4, 34-6, 34-12, 41-4, 43-5
  approximation by uncoupled systems, 31-29
  of equipment, 31-28, 42-9
  idealized system, 31-28
  problem classification, 31-3
  theory of, 31-1
  translation and rotation, 31-2
Shock isolators (see Isolators)
Shock machines, 26-1
  air guns, 26-9
  bump tester, 26-18
  calibration of, 26-3
  characteristics of, 26-2
  Conbur tester, 26-7
  definition of, 26-1
  dominant frequencies of, 26-3
  drop tables, 26-6, 26-9
  floating shock platform, 26-12
  high-impact, 26-11, 26-14
  inclined plane impact tester, 26-7
  multiple-impact, 26-14
  Navy, 26-11, 26-12
  package tester, 26-15
  quick-start centrifuge, 26-19
  revolving-drum, 26-18
  rotary accelerator, 26-19
  types of, 26-4
  WOX-7, 26-19
Shock motion, 1-24
  complex, 23-3, 23-6
  decaying sinusoid, 23-3, 23-6
  Fourier spectra of, 23-6
  half-sine, 23-3, 23-6, 26-4
  impulse, 23-3, 23-6
  rectangular, 23-7
  saw-tooth, 26-4
  simple, 26-5
  single complex, 26-4
  step, 23-3, 23-6
  types of, 26-4
Shock mounts (see Isolators; Shock isolators)
Shock protection (see Shock isolation)
Shock pulse(s), 1-24
  duration of, 1-20
  half-sine, 23-3, 23-6, 31-16
Shock source, effect of structural load changes, 23-29
Shock spectrum, 20-2, 20-5, 23-10, 23-24, 31-38, 42-9, 42-18
  acceleration, 26-2
  calculation of, 23-12
  control tests for, 27-12
  data reduction methods, 23-24
  displacement, 26-2
  examples of, 23-13
  and Fourier spectrum, relationship between, 23-22
  impulsive region of, 23-19
  initial, 20-3
  limiting values of, 23-18
  maximum, 20-3
  overall, 26-7
  parameters for, 23-10

Shock spectrum (*Cont.*):
  primary, 20-3, 23-13
  reed gages in measuring, 23-24
  residual, 23-13, 26-3, 26-7
  static region of, 23-17
  three-dimensional, 23-20
  two-dimensional, 23-21
  undamped residual, 20-4
  velocity, 26-2
  (*See also* Response spectrum)
Shock testing, 20-9, 26-1, 41-4
  digital control systems for, 27-10
  machines for (*see* Shock machines)
  specifications in, 26-3
Shock testing machines (*see* Shock machines)
Shock waves, 44-23, 44-26
Short-duration impulses, 31-28
Signal(s), 1-24
  analysis, 13-1
    of nonstationary signals, 13-36
  averaging techniques, 13-7, 21-22
  detectors, 13-5
  enhancement, 13-42
  integration, 13-3
  quasi-periodic, 1-23
  stationary (*see* Stationary signals)
Signal analyzers, 13-1, 13-11
Signal generators, 13-3
Signal nulling interferometer, 18-14
Silicone rubber, 32-2
Simple beam, 8-53
Simple harmonic motion:
  conversion factors for, 1-8
  definition of, 1-24
Simple shock impulse, 26-4
Simulation:
  of human subjects, 44-5
  testing, 24-18
Sine pulse:
  half-cycle, 8-48
  seismic qualification by, 24-25
  versed, 31-16
Sine wave:
  autocorrelation function of, 22-24
  calibration using, 18-18, 18-27, 18-31
  random, 11-10
    (*See also* Random vibration, narrow-band)
  swept (*see* Swept sine wave)
Single degree-of-freedom system, 2-2, 28-1, 31-6
  acceleration, 2-11
  angular natural frequency, 2-3
  classical solution, 8-3
  damped, 8-48
  differential equation, 8-1
  forced vibration, 2-6
  initial conditions, 2-3
  maximum response, 8-16, 8-47
  model, 42-4, 42-7
  natural frequency of, 2-3, 42-4
  nonlinear, 8-52
  period of, 2-3
  phase angle, 2-13
  residual response, 8-16, 8-47
  response, 11-9
  response spectrum, 8-14, 8-46
  velocity, 2-11

Single input theory, 21-19
Single-plane balancing machine, 39-3
Single-pulse excitation, 8-5
Singular matrix, 28-5
Singular points, 4-22
Sinusoidal acceleration decaying, 23-5, 23-8, 23-18
Sinusoidal excitation, 2-6
Sinusoidal force, 10-1
Sinusoidal functions, 22-4
  compared to random functions, 22-8
Sinusoidal motion, definition of, 1-24
Sinusoidal velocity, 10-1
Sinusoidal vibration, steady-state forced, 28-18
Sinusoidal vibration testing, 25-10
Sinusoids, probability distribution function, 22-9
Skew rectangular plates, 7-30
Skewed distribution, 11-5
Skewed versed-sine pulse, 8-42
Sleds, rocket-propelled, 44-5
Sliding, damping by, 36-23
Slip damping, 36-2, 36-22, 43-12
Snubber, 1-24, 34-4
Soft-bearing balancing machine, 39-14
Soft system, 4-8
Softening, 4-2, 4-6, 4-12, 4-34
Softening springs, 4-6, 31-8, 31-12, 31-21
Space plots, 3-21
Space shock, 20-3
Spacecraft, reentry of, 44-5
  (See also Missiles)
Specific damping energy (see Damping energy, specific)
Specific heats, ratio of, 33-4
Specifications, 20-12
  for control systems, 25-12
  environmental, 20-13
  screening, 20-15
  test, 20-13
Spectral analysis, 22-3, 22-15
  in digital computer technology, 27-8
  random vibration, 11-6
Spectral density:
  cross, 27-5
  power (auto spectral density), 27-5
    (See also Power spectral density)
  related to correlation functions, 27-5
  related to Fourier transform, 27-5
Spectral enveloping, 20-16
Spectrum, 1-24
  amplification factors, 24-9
  line (see Line spectrum)
  response (see Response spectrum)
Speed, critical (see Critical speed)
Sphere(s):
  impact of, 9-1
  properties of, 3-53
Spot-welded joints, 43-6
Spring(s), 2-1, 10-4
  air, 33-1
  automobile, 33-7
  Belleville (see Belleville springs)
  brass, 42-27
  cantilever, 32-17
  coil, 4-2, 32-9
  combinations of, 1-12, 34-6
  coned disc (see Belleville springs)

Spring(s) (*Cont.*):
  constant, 2-1, 4-2, 32-5
  hardening, 4-6, 31-8, 31-12, 31-21
  helical (see Helical compression springs)
  impedance of, 10-4
  involute, 32-17
  leaf, 32-17
  linear, 31-7, 31-12, 31-21
    with damping, 31-10, 31-13
  metal, 32-9
  plastic, 32-8
  pneumatic (see Air springs)
  rate (see constant, *above*)
  ring, 32-14
  rubber, 32-2, 35-1
  softening, 4-6, 31-8, 31-12, 31-21
  stiffness, 2-1, 10-4
  wire mesh, 32-18
Spring brass, 32-12
Spring-controlled system, 2-9
Spring-mass system, 4-2
Spring-mounted table, 6-37
Square, properties of, 3-45
Square matrix, 28-2
Square plate, 7-30
Square-wave response, 8-44
Squirrel-cage motor, 19-2
Stability, 4-7
  of active servo systems, 33-20
  nonlinear vibration, 4-28
  of solutions, 4-23
Stability chart, 4-30
Stabilizers, gyro, 6-25
Standard deviation, 1-24, 11-5
Standard indicator cards, 38-21
Standards, vibration (see Vibration standards)
Standards organizations, 19-1
Standing wave, 1-24
Static acceleration, equivalent, 23-11, 42-10
Static deflection, 2-4
Static fatigue, 35-13
Static hysteresis, 36-14, 36-17
Static properties:
  of materials, 42-24
  of metals, 36-22
Static region of shock spectrum, 23-17
Static response, equivalent, 23-20
Static stiffness, 32-8, 40-12
  equivalent, 40-12
Static unbalance, 39-3, 39-4
Stationary function, 22-17
Stationary process, 11-3
Stationary signal(s), 1-24
  analysis of, 13-26
  calibration, 13-29
  deterministic, 13-26
  random, 13-26
  scaling, 13-29
Stationary vibration, 20-2
Statistical analysis, 22-7, 27-10
Statistical concepts in vibration, 11-1, 11-3
Statistical error using computers, 27-8
Statistical reliability test, 20-7, 20-14
Steady-state vibration, 20-2, 28-18
Steam turbines, 19-7
Steel, properties of, 32-10, 32-11, 42-26, 42-28, 42-30

Steel beams:
  uniform, 7-21
  variable section, 7-20
Step acceleration, 23-4, 23-7, 23-16
Step relaxation excitation, 21-24
Step shock, 23-3, 23-6
Stick-slip in linear drives, 4-5
Stick-slip rubs and chatter, 5-16, 5-20
Stiffness, 1-25, 2-1
  of air springs, 33-3
  coupling, 3-12
  crankshaft, 38-6
  dynamic, 32-8, 37-19, 40-12
  equivalent static, 40-12
  of helical springs, 32-9
  in isolators, 32-5
  in machine-tool vibration, 40-12
  rotational, 3-12
  shafts in torsion, 1-13
  of spring combinations, 1-12
  of springs, 2-1
  static, 32-8, 40-12
  torsional, 3-12, 38-4
Storage modulus in shear, 36-13
Strain, apparent, 17-3
Strain energy:
  beam, 7-3
  ratio to damping energy, 36-9
Strain gages:
  accuracy of, 17-5
  basic theory and properties, 17-1
  bonded, 17-1
  bonding techniques, 17-5
  calibration of, 18-32
  classifications of, 17-3
  construction of, 17-2
  displacement measurement, 17-6
  foil for, 17-2
  for force measurements, 17-7
  gage factor, 17-1, 17-2
  measurements, 15-4, 17-5
  physical environment of, 17-5
  pressure measurement, 17-9
  resistance, 17-5
  selection considerations, 17-4
  temperature effects, 17-3
  transverse sensitivity of, 17-2
  velocity measurements, 17-7
Strain rate, 42-27
Strain relaxation (see Drift)
Strength:
  tensile (see Tensile strength)
  ultimate, 42-25
  yield, 42-25
Strength requirements, 42-32
Stress(es):
  bending of plates, 7-26
  design, in rubber isolators, 35-17, 35-19
  nominal, 42-25
  permissible, 42-32
  volume-stress function, 36-7
Stress concentration factor, 42-34
Stress levels, 36-6, 36-14
Stress raisers, 43-8
Stress relaxation test for rubber, 35-3
Stress-strain loop in measuring damping,
    36-3
Stress-strain law, linear viscoelastic, 36-13
String, stretched, 4-3

Strouhal number, 29-7
Structural damping, 2-15, 24-17
  uniform, 2-24
Structural load change, effect on shock
    source, 23-29
Structural members, efficiencies of, 42-21
Structural motion induced by ground mo-
    tion, 24-1
Structural response, determining, 23-1, 23-
    28
Structural stiffeners, 43-5
Structural transfer function, 23-30
Structures:
  bluff bodies, 29-17
  buffeting of, 29-19
  damped isolation of, 30-49
  design of, 29-12, 43-1
  determining natural modes of, 25-18
  excitation of: due to air turbulence, 29-23
    due to fluid flow, 29-1
  flutter mechanisms, 29-19
  galloping mechanisms, 29-19
  gravity loads, 29-15
  gust effects on cladding and windows, 29-
    29
  impact effect, 9-1
  nonrigidity of, 30-48
  primary, 42-6
  secondary, 42-6
  vibration of: control by dampers, 6-34
    from ground motion, 24-3
    from vortex shedding, 29-29
    from wind, 29-17
Subharmonic, 4-25
Subharmonic response, 4-15, 4-33
Subharmonic vibrations, 4-8
Subsidiary equation, 8-9
Subsynchronous components, 16-8
Superharmonic resonance, 4-33
Superharmonic response, 4-31
Superharmonic vibrations, 4-8
Superposition, 10-29
Superposition integral, 8-4
Superposition principle, 8-14, 10-37
Superposition theorem, 10-24
Support:
  orthogonal, 3-15
  resilient (see Resilient support)
Support motion, vibration resulting from, 7-
    42
Support protection, 31-27
Support-vibration damper, 40-11
Suspension, tension-spring, 41-8
Suspension systems:
  active, 33-1
  air, 33-1
Sustained vibration, 20-2
Sway space, 34-4
Swept swine wave, 21-24
  test, 18-31, 20-8
Symbols, 1-5
Symmetric matrix, 28-3
Symmetric pulse, 8-25
Symmetry:
  three planes of, 3-24
  two planes of, 3-19
Sympathetic vibration, 43-14
Synchronization, 29-9
Synchronous averaging, 13-42

System calibration, 15-11
System impedance matrix, 21-13
System ratings for vibration machines, 25-11

T equivalent systems, 10-25, 10-33
Table, spring-mounted, 6-37
Tangent elasticity, 31-8
Tangent isolator, 30-34
Tanks, antiroll, 6-3, 6-23
Tape, vibration damping, 37-24
Tape recorders, 13-10
Taylor's expression, 4-30
Tensile strength:
 of nonferrous metals, 42-31
 of rubber, 35-3
 of steel, 42-30
Tension loading in rubber isolators, 32-5, 35-18
Tension specimen, 9-10
Tension-spring package, 41-12
Tension-spring suspension, 41-8
Terminology, 1-18
Test(s):
 accelerated, 20-7
 acoustic, 20-11
 analytical, 20-11
 drop, 26-5
 dwell, 20-9
 fixture problems, 20-10
 Geiger thick-plate, 37-16
 load, 25-1
 qualification, 24-2
 random, 20-8
 reliability, 20-14
 response spectrum, 24-28
 shock (*see* Shock testing)
 simulation, 24-18
 specifications, 20-13
 stress relaxation, 35-3
 swept sine wave, 20-8, 21-24
 vibration, fixed-frequency, 25-24
 (*See also* Testing)
Testing:
 digital control systems for, 27-9
 environmental, 25-25
 fatigue (*See* Fatigue testing)
 fixed-frequency, 25-24
 random vibration, 25-10, 27-10
 shock (*see* Shock testing)
 sinusoidal vibration, 25-10
 vibration (*see* Control systems; Vibration testing)
 (*See also* Environment, test; Environmental specification)
Testing machines:
 fatigue, 25-18
 shock, 26-1
 (*See also* Vibration machines)
Thermal inertia, 32-15
Thévenin's equivalent system, 10-24, 10-30
Three degrees-of-freedom system, 28-2
Three-dimensional shock spectrum, 23-20
Three planes of symmetry, 3-24
Three-scale method, 3-4
Tilting support calibrator, 18-3
Time domain, 21-4
Time function, random, assembly of, 11-2

Time-varying functions, 22-4
Time-window effect, 13-20
Timoshenko beam theory, 7-16
Tobin bronze, 42-27
Topological methods, 4-23
Torque:
 gas-pressure, 38-19
 pulsating, 5-18
 residual, 38-9
Torque curves, 38-19
Torsional elastic system, 38-3
Torsional impact on bars, 9-8
Torsional loading of isolators, 32-5
Torsional pendulum, 3-7
Torsional spring constant, 3-12
Torsional stiffness, 3-12, 38-4
Torsional stress, acceptable limits of, 38-27
Torsional vibration, 28-12
 auxiliary mass damper applied to, 6-26
 characteristics of, 38-28
 circular rods, 7-5, 7-7
 model of, 38-14
Total damping energy, 36-5
Toughness of metal, 42-28
Tracking analysis, 13-29
Trajectories, 4-21, 4-26
Transducer, 12-1
 acceleration-measuring, 12-4, 16-3
 capacitance-type, 14-8
 definition of, 12-1
 differential-transformer, 14-6
 displacement, 16-3
 eddy current, 14-5
 electrodynamic, 14-4
 fixed reference, 12-2
 frequency response, 12-1, 18-1
 hand-held, 15-10
 inductive, 14-4
 with internal preamplifier, 12-18, 13-2
 mass-spring, 12-2
 mounting, 15-5
 movable core, 14-6
 passive circuit, 12-1
 phase lag in, 18-2
 piezoelectric (*see* Piezoelectric accelerometer)
 piezoresistive (*see* Piezoresistive accelerometer)
 potentiometer-type, 14-7
 preamplifier, 13-1
 seismic, 12-2
 selection of, 15-4, 16-3
 self-generating, 12-1
 sensitivity factor, 18-1
 velocity, 14-1, 14-4, 16-3
Transducer calibration factor, 18-1
Transducer calibration methods:
 absolute, 18-2
 automated, 18-30
 with ballistic pendulum, 18-20
 centrifuge, 18-8
 chatter, 18-3
 comparison of, 15-11, 18-25
 displacement, 18-9
 drop-ball shock, 18-24
 earth's gravitational field, 15-11
 in the field, 15-11
 Fourier transform shock calibrator, 18-26
 free-fall, 15-11, 18-5

Transducer calibration methods (*Cont.*):
  frequency dwell, 18-31
  fringe, 18-11
  gravimetric, 18-5, 18-6
  impact, 18-19
  interferometer, 18-10 to 18-16
  inversion test, 15-12
  with microscope, 18-9
  optical grating, 18-10
  pendulum calibrator, 18-4
  random-excitation, 18-31
  reciprocity, 18-16
  Ronchi ruling, 18-10
  sinusoidal, 18-18, 18-31
  tilting support calibrator, 18-3
  transfer function, 18-31
  using vibration exciters, 18-27 to 18-29
  voltage substitution, 15-13
Transducer systems:
  laser-Doppler, 14-2
  optical-electronic, 14-2
Transfer function, 21-6, 23-9
  structural, 23-20
Transfer impedance, 10-2
  blocked, 30-41
  free, 30-41
Transfer matrix, 7-37, 38-14
Transform:
  driving, 8-9
  function, 8-7, 8-8
  Hilbert, 13-45
  Laplace (*see* Laplace transform)
  operation, 8-8
  response, 8-9
Transformation:
  inverse, 8-9
  method of calibration, 18-31
Transient analysis, 13-30, 13-37
Transient excitation, 8-1
Transient response, 28-9, 33-13
Transient vibration, 1-25
Transmissibility, 2-12, 2-14, 33-11
  absolute, 30-2
  of acceleration, 30-9, 30-14, 30-50
  in damped systems, 32-6
  displacement, 30-7, 30-14
  force, 2-6, 30-16, 30-45
  motion, 2-6
  optimum, 30-11
  relative, 30-2
  at resonance, 30-11
  sitting human, 44-12
  standing human, 44-12
  velocity, 30-42
Transmission, force, 2-13
Transmission loss, 1-25
Transpose matrix, 28-3
Transposition of matrix products, 28-5
Transverse impact, 9-4
Transverse sensitivity, 12-9
  of strain gages, 17-2
Transverse wave, 1-25
Trapezoid, properties of, 3-45
Trapezoidal plate, 7-30
Trapezoidal pulse, 31-22
Trend analysis, 16-17
Triangle, properties of, 3-44
Triangular plate, 7-30
Triangular pulse, 8-31

Triangular rubber isolators, 35-17
Triboelectric cable noise, 15-15
Trifilar pendulum, 3-8
Tubing, flexible, 34-18
Tumbling test, revolving drum, 26-13
Tuned vibration damper, 37-2, 37-27
  (*See also* Auxiliary mass damper; Dynamic vibration absorber)
Turbine:
  gas, 19-9
  generator, 19-7
  steam, 19-7
Turbocompressors, 19-8
Turbogenerator, 13-42
Turbomachinery, 5-13, 19-8
Turbulence:
  structural excitation due to, 29-23
  theory for, 29-23
  (*See also* Wind)
Turbulent flow, 29-30
Two-connection systems, 10-32
  vibration sources in, 10-36
Two degrees-of-freedom system:
  without damping, 31-24, 31-29
  model, 42-5, 42-7, 42-14
Two-dimensional shock spectrum, 23-21
Two-plane balancing, 39-5
Two planes of symmetry, 3-19

U tube, 2-27
Ultimate elongation of rubber, 35-3
Ultimate strength, 42-25
Ultraharmonic resonance, 4-33
Ultrasonic vibration, effect on humans, 44-19
Unbalance, 25-4
  amount of, 39-1
  angle of, 39-1, 39-17, 39-19
  correction of, 39-24
  couple, 39-2
  defined, 39-3
  dynamic, 39-2, 39-5
  initial, 39-2
  measurement of, 39-17
  residual, 39-3
  sources of, in rotating machinery, 39-11
  static, 39-3, 39-4
  (*See also* Balance)
Unbalance distribution, 39-6
Unbalance force, 25-4
Unbalanced rotor, 39-17
Unbonded isolators, 32-20, 32-23
Uncorrelated, 1-25
Uncoupled mode, 1-25
Uncoupled motions, 31-1
Uncoupled system, 31-2
Undamped isolators, 30-53
Undamped multiple degree-of-freedom system, 2-16
Undamped natural frequencies, 1-25, 28-22
Underwater shock waves, 44-26
Uniform beam, 1-15, 7-14, 42-22
  steel, 7-21
Uniform Building Code, 24-15
Uniform circular rods, 7-5
Uniform mass damping, 2-22
Uniform structural damping, 2-24

Uniform viscous damping, 2-23, 2-24
  bending moments, 7-27
  deflection, 7-27
Unit matrix, 28-3
Unit step function, 23-4
Unity system, 38-10
Unstable imbalance, 5-20
Unsymmetrical pulses, 8-29
Urethane foam, 41-25

V-type engines, 38-36
Value:
  average (see Average value)
  instantaneous, probability distribution of,
    22-13
  peak (see Peak values)
  peak-to-peak, 1-22
Van der Pol's equation, 4-8, 4-15, 4-17
Variance, 1-25, 11-5
Variation of parameters, 4-15
Vector:
  cancellation method, 38-28
  characteristics of, 28-14
  column, 28-2
  latent, 28-14
  proper, 28-14
  row, 28-3
Vehicles, 34-25
  (See also Automobiles)
Velocity, 1-25, 2-11
  foundation-induced shock, 3-38
  free-output, 30-41
  generalized, 28-8
  mean wind, 29-20
  measurements, 15-4
  resonance frequency, 2-15
  sinusoidal, 10-1
  starting, 42-13
  strain-gage measurements, 17-7
  transducer, 14-1, 14-4
Velocity, isolation, 30-42
Velocity response, 30-45, 30-46
Velocity shock, 3-38, 26-4, 26-5, 30-45
Velocity shock spectra, 26-3
Velocity squared damping, 4-11, 31-12
Velocity step, 31-5, 31-17
Velocity transmissibility, 30-42
Versed sine pulse, 31-16
Vertical accelerator, 44-5
Vibrating beam test for damping, 37-15
Vibrating wedge, 18-9
Vibration:
  acceleration (see Acceleration)
  basic theory of, 2-1, 3-29, 3-33
  bearing, 19-13
  belt drive, 40-3
  bistable, 5-18
  body-induced, 3-29
  complex, 1-19
  displacement measurement, 14-2, 16-2, 18-
    10
  dynamic absorbers, 4-35
  effects on human body (see Human body)
  flight deck, 44-12
  fluid flow-induced, 29-1
  forced (see Forced vibration)
  foundation-induced, 3-29
  free (see Free vibration)

Vibration (Cont.):
  fundamental mode (see Fundamental
    mode of vibration)
  harmonic, 4-8
  hull, 19-13
  isolators (see Isolators)
  lateral (see Lateral vibration)
  longitudinal (see Longitudinal vibration)
  machine-tool, 40-11
  machinery (see Machinery vibration)
  magnitude of, 22-1, 22-2
  measurements, 13-2, 15-1, 16-2
    (See also Measurement)
  nonlinear, 4-1, 4-28
  normal modes of (see Normal mode of vi-
    bration)
  of offshore structures, 29-7
  order of, 6-28
  perception of, 33-10
  period of, 2-3
  periodic, 20-2
  in pipes, 29-14
  of plate, 7-24
  propeller, 19-3
  propeller shaft, 19-13
  qualitative description of, 22-1
  quasi-stationary, 20-2
  random (see Random vibration)
  of resiliently supported body, 3-1
  resulting from motion of the supports, 7-
    42
  self-excited (see Self-excited vibration)
  ship, 19-13
  standards for (see Vibration standards)
  stationary, 20-2
  steady-state, 20-2, 28-18
  of structures (see Structures, vibration
    of)
  subharmonic, 4-8
  subjective responses of humans to, 44-23
  superharmonic, 4-8
  sympathetic, 43-14
  of systems: with damping, 28-21
    without damping, 28-16
  torsional (see Torsional vibration)
  transient, 1-25
  translational, 2-1
  ultrasonic, effects on humans, 44-19
  velocity (see Velocity)
  vortex-induced: in air, 29-34
    in liquids, 29-7, 29-11
Vibration absorbers, 4-35, 6-1, 6-34, 6-35
Vibration control, 30-1, 32-1, 34-1, 43-1
Vibration damping (see Damper; Damping)
Vibration data analysis, 27-7
Vibration energy dissipation, 6-14
Vibration exciters (see Vibration machines)
Vibration isolation, 28-10, 30-1, 32-1, 34-1,
    43-1
  active, 33-7
  cushioning materials for, 41-13
  damping in, 30-4
  design, 34-7
  of force, 30-44
  of nonrigid body systems, 30-40
  of random vibration, 30-48
  servo-controlled, 33-7
  theory of, 30-1
  (See also Isolators; Rubber isolators)

Vibration machines (shake tables), 25-1
  acceleration limit in, 25-13
  acceleration waveform in, 25-14
  amplitude-frequency relations in, 25-13
  application and selection of, 25-17
  automatic frequency control in, 25-14
  axial coil-table resonance in, 25-13
  calibrator characteristics of, 25-17
  circular-motion, 25-4
  constant acceleration, 25-1
  constant displacement, 25-1
  control systems (*see* Control systems)
  design factors, 25-12
  direct-drive mechanical, 25-2
  electrodynamic, 18-28, 25-7
  environmental extremes, 25-13
  in fatigue testing, 25-19
  force output, 25-14
  frequency drift in, 25-14
  hydraulic, 25-14
  impedance characteristics of, 25-13
  installation of, 25-14
  loading of, 25-1
  magnetic fields of, 25-14
  mechanical, 18-27, 25-2
  performance of, 25-13
  piezoelectric, 18-29
  reaction-type mechanical, 25-5
  rectilinear-motion, 25-5, 25-7
  signal generators for, 25-14
  static moment and torque in, 25-13
  stiffness of suspension system, 25-13
  unsupported load in, 25-13
  use of, in transducer calibration, 18-26
  weight of vibrating assembly, 25-12
Vibration-measuring instruments, 13-1, 14-1, 15-1
Vibration meter, 13-1, 13-9
Vibration mount (*see* Isolators)
Vibration pick-up (*see* Accelerometer; Transducer)
Vibration severity, 19-2
Vibration standards, 19-1
  bearing vibration, 19-12
  compressors, 19-7, 19-11
    air, 19-11
    centrifugal, 19-11
  electric motors, 19-4
  gas turbines, 19-9
  gear units, 19-11
  military, 38-27
  reciprocating machinery, 19-7
  rotating machinery, 19-3, 19-7
  shaft vibration, 19-11
  ships, 19-13
  turbine generators, 19-7
Vibration testing:
  digital control systems for, 27-11
  environmental, 25-25
  fixed frequency, 25-24
  forced, of vulcanizates, 35-15
  random, 21-23, 25-10, 27-11
  sinusoidal, 21-23, 25-10, 27-11
Vibratory aging, 24-28
Virtual work, theorem of, 7-40
Viscoelastic materials:
  damping of, 36-12
  dynamic hysteresis of, 36-16
  moduli of, 36-17

Viscoelastic materials (*Cont.*):
  properties of rubberlike materials, 35-8
  stress-strain law in, 36-13
Viscosity:
  absolute, 32-28
  apparent, 32-30
  dynamic, 32-28
  kinematic, 32-29
  temperature effects, 32-29
Viscous damper, 38-33
  elastically connected, 30-2, 30-9
  rigidly connected, 2-4, 30-2, 30-5
  tuned, 38-38
Viscous damping, 4-4, 8-48
  equivalent, 30-7, 43-12
  uniform, 2-23, 2-24
Viscous relation system, 30-2
Voltage preamplifier, 13-1
  calibration of, 18-33
Voltage sensitivity of accelerometers, 12-8, 12-17
Volume-stress function, 36-7
Vortex-induced oscillation, 29-30
Vortex-induced vibration:
  in air, 29-23, 29-34
  in liquids, 29-7, 29-11
Vortices, 29-30, 29-33, 29-34
Vulcanizates, 35-1, 35-15

Wakes, 29-17
Wave(s):
  elastic, 9-3
  longitudinal, 1-21
  ocean, 29-5
  plastic, 9-9
  Rayleigh, 1-22
  shear, 1-24
  shock (*see* Shock waves)
  sine (*see* Sine wave)
  standing, 1-24
  transverse, 1-25
Wave effects in isolators, 30-52
Wave interference, 1-25
Waveform control tests, 27-12
Wavelength, 1-26
Weight, 3-3
  effect on design, 42-19
  rotating eccentric, 2-14
Weight functions, 4-17
Weighting functions, 21-22, 21-26
  (*See also* Windows)
Weld, butt, 9-10
Welded joints, 43-6
Wheels, effect of vibration on properties of, 40-3
Whip, 5-20
  dry friction, 5-10
  fluid bearing, 5-10
Whipping, 5-2
Whirl, 5-20
  from fluid trapped in rotor, 5-8
  hysteretic, 5-6
  propeller, 5-13
  turbomachinery, 5-13
Whirling, 5-2
White noise, 1-28, 11-7, 22-16
  autocorrelation function of, 22-24
Wild points, 27-8

Wind:
  basic characteristics of, 29-20
  fluctuating components of, 29-21
  vibration of structures induced by, 29-17
  (*See also under* Aerodynamic; Gust factor
      in structural design; Turbulence)
Wind velocity, mean, 29-20
Windows, data, 13-21, 20-6, 21-22, 21-26
  exponential, 13-23
  flattop, 13-23
  Hamming, 13-23, 27-9
  Hanning, 13-23
  Kaiser-Bessel, 13-23
  Parzen, 27-9
  short rectangular, 13-23
  truncated Gaussian, 13-23

Wire-cable isolators, 32-26
Wire-mesh isolators, 32-18, 32-25
Wiring, electrical, 15-13, 43-24
Wrought iron, 42-26, 42-29

Yield point, 42-25
Yield strength, 42-25
Young's modulus, 42-24

Zero matrix, 28-3
Zero-period acceleration, 24-29
Zoom analysis, 13-23
Zoom FFT (fast Fourier transform), 13-23

About the editor

Cyril M. Harris is Charles Batchelor Professor of Electrical Engineering and Professor of Architecture at Columbia University in New York City. He holds a Ph.D. degree in physics from M.I.T. and has achieved international recognition in architectural acoustics and noise control. He is a member of both the National Academy of Sciences and the National Academy of Engineering and has been a recipient of the Gold Medal of the Acoustical Society of America, the Gold Medal of the Audio Engineering Society, the Franklin Medal, and the AIA Medal of the American Institute of Architects. Dr. Harris was the acoustical consultant for several prominent concert halls and auditoriums including the John F. Kennedy Center for the Performing Arts and the Metropolitan Opera House. He also is the editor of the *Handbook of Noise Control*, Second Edition, and *Dictionary of Architecture and Construction*, both published by McGraw-Hill.